International Thermal Spray Conference & Exposition (ITSC 2016)

Shanghai, China
10 - 12 May 2016

Volume 1 of 2

ISBN: 978-1-5108-4400-1

Printed from e-media with permission by:

Curran Associates, Inc.
57 Morehouse Lane
Red Hook, NY 12571

proceedings.com

Some format issues inherent in the e-media version may also appear in this print version.

Printed by Curran Associates, Inc. (2017)

For permission requests, please contact DVS Media GmbH
at the address below.

DVS Media GmbH
P. O. Box 10 19 65
40010 Dusseldorf
Germany

Phone: 49 (0)211 / 1591-0
Fax: 49 (0)211 / 1591-150

media@dvs-hg.de

Additional copies of this publication are available from:

Curran Associates, Inc.
57 Morehouse Lane
Red Hook, NY 12571 USA
Phone: 845-758-0400
Fax: 845-758-2633
Email: curran@proceedings.com
Web: www.proceedings.com

TABLE OF CONTENTS

VOLUME 1

LECTURES

COLD GAS SPRAYING I

NOVEL INDUSTRIAL APPLICATION

PROCESS DIAGNOSTICS, SENSORS & CONTROLS

CERAMIC COATINGS I

SUSPENSION SPRAYING I

WIND & OFFSHORE

NEW PROCESSES I

CERAMIC COATINGS II

YOUNG PROFESSIONALS SESSION

AUTOMOTIVE INDUSTRY

AVIATION INDUSTRY III

SUSPENSION SPRAYING III

PRE- & POST-TREATMENT

ARC SPRAYING & LASER SPRAYING

POWER GENERATION II

CHARACTERIZATION & TESTING METHODS

VOLUME 2

HVOF-/HVAF-SPRAYING

NEW PROCESSES II

POWER GENERATION III

CORROSION PROTECTION

EQUIPMENT / CONSUMABLES & POWDERS, WIRES, SUSPENSIONS

IRON-BASED COATINGS

POWER GENERATION IV

MEDICAL INDUSTRY

PLASMA SPRAYING

AMORPHOUS COATINGS

POSTERS

APPLICATIONS – AUTOMOTIVE INDUSTRY

APPLICATIONS – AVIATION INDUSTRY

APPLICATIONS – GENERAL PLANT ENGINEERING

APPLICATIONS – MEDICAL INDUSTRY

PROPERTIES – CORROSION PROTECTION

PROPERTIES – METAL COATINGS

PROPERTIES – POLYMER COATINGS

PROPERTIES – PRE- & POST-TREATMENT

PROPERTIES – PROPERTIES CHARACTERIZATION & TESTING METHODS

PROPERTIES – TRIBOLOGICAL COATINGS

PROPERTIES – WEAR PROTECTION

THERMAL SPRAY PROCESSES – ARC SPRAYING & LASER SPRAYING

THERMAL SPRAY PROCESSES – COLD GAS SPRAYING

THERMAL SPRAY PROCESSES – FLAME SPRAYING

THERMAL SPRAY PROCESSES – HVOF / HVAF SPRAYING

THERMAL SPRAY PROCESSES – MODELLING & SIMULATION

THERMAL SPRAY PROCESSES – NEW PROCESSES

THERMAL SPRAY PROCESSES – PLASMA SPRAYING

YOUNG PROFESSIONALS

International Thermal Spray

ITSC 2016

Conference & Exposition

Lectures and Posters
of the Conference in Shanghai/P.R. China
on May 10 – 12, 2016

Organized by

DVS – German Welding Society
ASM International – Thermal Spray Society (TSS)
IIW – International Institute of Welding

Supporting Societies:

Media Sponsors:

Bibliographic information published by the Deutsche Nationalbibliothek
The Deutsche Nationalbibliothek lists this publication in the Deutsche Nationalbibliografie; detailed bibliographic data are available in the Internet at http://dnb.dnb.de.

DVS-Berichte Volume 324

The conference abstracts are printed in form of manuscripts.

Print: Griebsch & Rochol Druck GmbH & Co. KG, Hamm

Preface

Shanghai is the largest Chinese city by population and the largest city proper by population in the world. It is one of the four direct-controlled municipalities of the People's Republic of China, with a population of more than 24 million as of 2013. It is a global financial center, and a transport hub with the world's busiest container port. Located in the Yangtze River Delta in East China, Shanghai sits on the south edge of the mouth of the Yangtze in the middle portion of the Chinese coast. The municipality borders the provinces of Jiangsu and Zhejiang to the north, south and west, and is bounded to the east by the East China Sea. Shanghai is a popular tourist destination renowned for its historical landmarks such as The Bund, City God Temple and Yu Garden as well as the extensive Lujiazui skyline and major museums including the Shanghai Museum and the China Art Museum. It has been described as the "showpiece" of the booming economy of mainland China.

In this extraordinary surrounding ITSC will take place in the Shanghai International Convention Center connected to the Oriental Riverside Hotel. It is located at the heart of Lujiazui, the dynamic financial district of Shanghai. The Oriental Pearl TV Tower is also next to the Convention Center and Hotel.

The Program Committee prepared the conference program very focussed on applications. Thermal spray technology has manifested its advances in lots of branches from medical industry over power generation (including conventional and alternative technologies) up to all kinds of mobility industries. The technical program of ITSC represents benefits and challenges of thermal spray solutions for all these trendsetting industries. Besides this, the technical program will also share the latest information about modern thermal spray processes, special coating characteristics and all kinds of consumables.

Another highlight of the technical program will be again the Session "Young Professionals". ITSC is still supporting young experts becoming part of the worldwide thermal spray family!

All practically interested experts will be addressed to get the latest information in the "Industrial Forum". The "Industrial Forum" will be open to all Expo only visitors and conference attendees.

The Three-Day Exposition is an integral part of ITSC 2016. If you are looking for new products and services, have questions or need answers, then ITSC 2016 is the right place for you.

ITSC exhibitors are experts who can offer attendees a wealth of problem-solving information and cost-saving opportunities. You will find information about equipment for thermal spraying, re- search and specialist institutes, applied research, and the latest innovations conveniently located in one big forum.

Come and enjoy countless fascinating experiences at ITSC 2016 in Shanghai!

Düsseldorf, May 2016

DVS – German Welding Society

Dr.-Ing. Roland Boecking
General Manager

Dipl.-Ing. Jens Jerzembeck
Head of Research and Technology

TIME SCHEDULE ITSC 2016 CONFERENCE AND EXPOSITION (Session Overview)

12:00 – 19:00 ITSC 2016 Exposition, Shanghai International Convention Center, Mandarin and Century Hall

Tuesday, May 10, 2016

Time	Auditorium	Yellow River Hall	Room 3E	Room 3C + 3D
09:00	**Opening:** ■ Welcome ■ Plenary Lectures			
10:30	Coffee Break, Shanghai International Convention Center, Foyers			
10:50	Cold Gas Spraying I	Novel Industrial Application	Process, Diagnostics, Sensors & Controls	Ceramic Coatings I
12:30	Lunch Break, Shanghai International Convention Center, Mandarin and Century Hall **Poster Session, Shanghai International Convention Center, Century Hall, Foyers**			
14:00	Suspension Spraying I	Wind & Offshore	New Processes I	Ceramic Coatings II
15:20	Coffee Break, Shanghai International Convention Center, Mandarin and Century Hall			
15:40	Session "Young Professionals"			
17:00	Award Ceremony: TSS Hall of Fame, ITSC Best Paper Awards and Oerlikon Metco Young Professionals Award			
17:40	End of ITSC 2016 Conference Program			
17:40	**Exhibitor Welcome Reception, Shanghai International Convention Center, Mandarin and Century Hall** **Poster Session, Shanghai International Convention Center, Century Hall, Foyers**			

09:00 – 17:00 ITSC 2016 Exposition, Shanghai International Convention Center, Mandarin and Century Hall

Wednesday, May 11, 2016

Time	Yellow River Hall	Room 3E	Room 3G	Room 3H + 3I + 3J	Room 3C + 3D
09:00	Automotive Industry	Cold Gas Spraying II	Wear Protection	Power Generation I	Industrial Forum
10:40	Coffee Break, Shanghai International Convention Center, Mandarin and Century Hall				
11:00	Aviation Industry I	Cold Gas Spraying III	Modeling & Simulation I	Power Generation – Fuel Cell I	
12:40	Lunch Break, Shanghai International Convention Center, Mandarin and Century Hall **Poster Session, Shanghai International Convention Center, Century Hall, Foyers**				
13:40	Aviation Industry II	Suspension Spraying II	Modeling & Simulation II	Power Generation – Fuel Cell II	Industrial Forum
15:20	Coffee Break, Shanghai International Convention Center, Mandarin and Century Hall				
15:40	Aviation Industry III	Suspension Spraying III	Pre & Post Treatment	Arc Spraying & Laser Spraying	
17:20	End of ITSC 2016 Conference Program				
18:30	**Networking Event, Huangpu Boat Trip – Presentation of JTST Award and TSS President's Award**				

09:00 – 14:00 ITSC 2016 Exposition, Shanghai International Convention Center, Mandarin and Century Hall

Thursday, May 12, 2016

Time	Yellow River Hall	Room 3E	Room 3G	Room 3C + 3D
09:00	Power Generation II	Characterization & Testing Methods	HVOF- / HVAF-Spraying	New Processes II
10:40	Coffee Break, Shanghai International Convention Center, Mandarin and Century Hall			
11:00	Power Generation III	Corrosion Protection	Equipment / Consumables & Powders, Wires, Suspensions	Iron-based Coatings
12:40	Lunch Break, Shanghai International Convention Center, Mandarin and Century Hall **Poster Session, Shanghai International Convention Center, Century Hall, Foyers**			
13:40	Power Generation IV	Medical Industry	Plasma Spraying	Amorphous Coatings
15:20	End of ITSC 2016 Conference Program			

Time	Friday, May 13, 2016
07:30	Industrial Tour SICCAS – Shanghai Institute of Ceramics, Chinese Academy of Sciences

Analytical Evaluation of Temperature Distribution within a Substrate under an Impinging Cold Spray Hot Air Jet

A. Mahdavi, A. McDonald, Edmonton, CA

The temperature distribution within the substrate under an impinging compressed hot air jet during cold-gas dynamic spraying was studied. This study was motivated by the significant impact of the gas-substrate heat exchange on the microstructure and quality of the coating in cold-gas dynamic spraying. In this regard, a mathematical model that was based on 2-D heat conduction was developed to determine the temperature distribution within the substrate. A low-pressure cold spray unit was utilized to cause the impingement of a compressible air jet on a flat substrate from a stationary nozzle. Fast-response thermocouples were used to determine the surface temperature of the substrate. Experimental data was used with the analytical solution of temperature distribution in order to estimate the heat transfer coefficient between the hot gas and the substrate. The results were compared to estimates of heat transfer coefficient that were obtained from numerical studies. It was found that the heat transfer coefficient was dependent on the distance from the stagnation point of the impinging air jet. It was also concluded that at higher air jet temperatures, the Nusselt number of the spreading air film in the vicinity of the stagnation point of the air jet could be affected by the external heat exchange of the air film with colder ambient air.

1 Introduction

Low-pressure cold spraying is a coating deposition method in which a supersonic gas flow is generated to transport un-melted metal or alloy powder particles to deposit with high impact forces upon a base substrate to form a highly adherent coating [1]. Due to the high speeds and impingement forces of the particles on the substrate, plastic deformation occurs, which leads to the formation of a dense adherent coating. The effect of the speed of the particles on the quality of the final coating has been studied extensively [2, 3]. Furthermore, in an experimental study by Limaye, *et al.* [4], it was shown that the heat transfer rate increased in all radial locations as a result of increasing the Mach number of the gas flow. In a parallel study, Irissou, *et al.* [5] showed that pre-heating the propellant gas jet in cold spray systems increased the velocities of the particles. Therefore, in light of the high temperature of the gas jet, a significant amount of thermal energy is expected to be exchanged between the impinging gas jet and the substrate.

The dominant mode of heat transfer between the impinging gas jet and the substrate is convection; therefore, the determination of the convective heat transfer coefficient is a key factor to estimate the heat transfer rate as well as the temperature distribution within the substrate. Several investigations have shed light on quantifying the heat transfer coefficient in various geometries and applications of impinging compressible air jets [6-8]. Rahimi, *et al.* [6] studied the effect of the fluid Mach number and pressure ratio, in addition to the Reynolds number, of the gas jet on the heat transfer coefficient of the impinging air jet. Zuckerman and Lior [7] discussed the phenomenon of jet impingement and proposed numerical modelling to study the heat transfer as a result of the jet impingement. Li, *et al.* [8], in a numerical and experimental study, investigated the temperature distribution within a solid body that was exposed to an air jet that was generated by a cold spray system. It was found that the maximum temperature was present at the stagnation point of the impinging air jet.

In a combined experimental-analytical study, Ryabinin, *et al.* [9] presented a numerical model to determine the Nusselt number of the impinging compressible gas jet that was produced by a cold spray unit. It was observed that the Nusselt number decreased as the distance from the stagnation point of the gas jet increased. The experimentally measured adiabatic wall temperature of the gas film over the substrate followed a similar trend in the study. As a complementary study, McDonald, *et al.* [10] used a 2-D finite difference method and the Nusselt numbers that were found by Ryabinin, *et al.* [9] to investigate the relationship between various factors in cold spraying such as stand-off distance, the substrate thickness, and the radius of the impinging gas jet on the temperature distribution of the substrate.

Although many investigators have conducted numerical and experimental studies to estimate the heat transfer coefficient between gas jets and substrates [6 - 9], analytical and mathematical modelling is lacking in this area.

In the present study, a mathematical model was employed to determine the temperature distribution of the substrate under an impinging air jet. Experiments were conducted and the results were utilized in a mathematical model to estimate the heat transfer coefficient of the impinging air jet that was generated by the cold spray system. The analytical results for the heat transfer coefficient were compared to the numerical results obtained from previous studies.

2 Experimental Method

A low-pressure cold spray unit (SST Series P, Centerline, Ltd., Windsor, ON, Canada) was utilized to generate the hot compressed air jet. The pressure and temperature range of the air were set at 635 kPa and 350 - 450°C, respectively at the cold spray console. A robot (HP20, Motoman, Yaskawa Electric Corp., Waukegan, IL, USA) was employed to direct the cold spray nozzle to the centre of the substrate in order to fix the stand-off distance at 10 mm. Stand-off distance is the distance between the nozzle exit and the substrate.

The substrates were aluminium (Al) and vitreous aluminosilicate fiber (ISAFORM 2300° L.O., Insulation Specialties of America, Inc., Wanatah, IN, USA). The geometry of the substrates was a disk with dimensions of 10 cm in radius and 7 mm in thickness. These substrates were mounted on a substrate holder. In the case of the Al substrate, in order to prevent heat exchange between the ambient air and the back surface of the Al substrate, the vitreous aluminosilicate fiber was used as the thermal insulation, as suggested in a previous study [9]. Figure 1 shows the schematic of experimental assembly of this present study. Fast-response J-type thermocouples were placed into the Al substrate, sufficiently close to the front surface, to measure the surface temperature during jet impingement and spreading.

The steady-state surface temperature distribution of a vitreous aluminosilicate fiber substrate was measured by an infrared camera (Model VIR50, FLIR Commercial Systems, Inc., Extech Instrument Division, Nashua, NH, USA). The surface of the vitreous aluminosilicate fiber substrate was painted in black with high-temperature paint so that the emissivity of the surface approached a value of 0.96, which approximately represents an ideal black-body.

3 Mathematical Model

The impingement of the hot air jet onto the substrate is expected to produce a sudden increase in temperature in the vicinity of the stagnation point of the jet. The dominant mode of heat transfer between the substrate and the impinging hot air gas is convection; therefore, estimates of the convective heat transfer coefficient (h) are needed in order to determine the temperature distribution within the substrate.

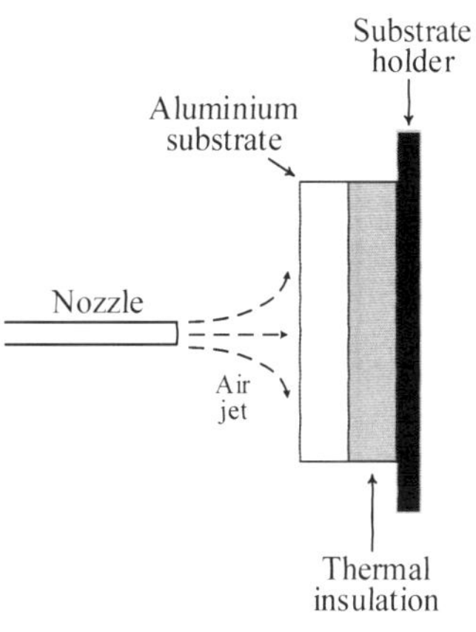

Fig. 1. Schematic of the experimental assembly for aluminium substrate under the impinging hot air jet.

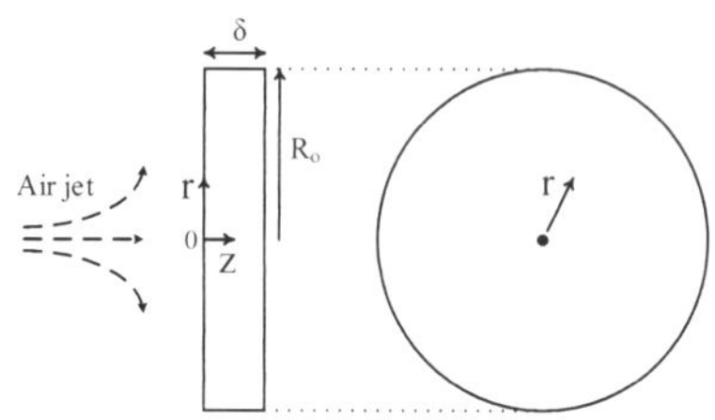

Fig. 2. Schematic of the geometry upon which the analytical model is based.

Generally, this is a three-dimensional heat conduction problem; therefore, a 3-D heat conduction model needs to be developed to determine the temperature distribution of the substrate. However, due to the axisymmetric nature of the impinging jet, the 3-D model can be reduced to a 2-D model [7]. Figure 2 shows the geometry upon which the model is based.

The governing equation for the steady-state 2-D heat conduction model is

$$\frac{1}{r}\frac{\partial}{\partial r}\left(r\frac{\partial T(r,z)}{\partial r}\right)+\frac{\partial^2 T(r,z)}{\partial z^2}=0\,, \tag{1}$$

and the boundary conditions for this problem are

$$-k\frac{\partial T(r,z=0)}{\partial z}=h(r)\left[T_{aw}(r)-T(r,z=0)\right], \tag{2}$$

$$-k\frac{\partial T(r,z=\delta)}{\partial z}=0\,, \tag{3}$$

$$T(r=0,z)=\text{Finite}\,, \tag{4}$$

$$T(r=R_o,z)=T_o\,, \tag{5}$$

where k is the thermal conductivity of the substrate. In Eq. (2), T_{aw} is the adiabatic wall temperature.

In Eq. (5), T_o is the surface temperature of the substrate, which is measured by mounting a J-type thermocouple at the location, R_o. Due to the negligible temperature variation across the cross section of the substrate, T_o represents the temperature across the entire thickness at R_o. This assumption is justified by considering the Biot number, Bi = $h\delta/k$, which is the ratio of the internal conduction and external convection resistances. The high thermal conductivity of Al (230 W/m-K) and the low thickness of the substrate may result in very small Biot numbers, which would validate the assumption.

Eq. (1) was solved by using the separation of variables method. The theory and procedure of the separation of variables method was described by Jiji [11]. The equation for determining the temperature distribution within the substrate is

$$T(r,z) = T_o + \sum_{m=1}^{\infty} \left[\frac{\int_{r=0}^{r=R_o} rF(r) J_0(\beta_m r) dr}{\int_{r=0}^{r=R_o} r J_0^2(\beta_m r) dr} \right] \times \tag{6}$$

$$\left[\frac{\exp(\beta_m z) + \exp(2\beta_m \delta) \exp(-\beta_m z)}{-k\beta_m \left(1 - \exp(2\beta_m \delta)\right)} \right] J_0(\beta_m r),$$

where J_0 is the Bessel Function of first kind. The eigenvalues, β_m are obtained from

$$J_0(\beta_m R_o) = 0. \tag{7}$$

In Eq. (6), $F(r)$ is

$$F(r) = h(r)\left[T_{aw}(r) - T(r, z = 0)\right], \tag{8}$$

where $h(r)$ is the radially varying heat transfer coefficient of the impinging air jet over the substrate surface.

4 Results and Discussion

4.1 Adiabatic Wall Temperature

Determination of the heat exchange between the impinging hot air jet and the substrate requires knowledge of the heat transfer coefficient of the air film that spreads over the substrate surface. In high velocity, compressible flows over surfaces, dissipative effects may be pronounced. Therefore, the adiabatic wall temperature (T_{aw}) of the gas will need to be estimated. The adiabatic wall temperature takes into account the effects of dissipation within the air film that flows over the substrate [12].

Based on the dependency of the adiabatic wall temperature on the flow regime and thermo-physical properties of the fluid, the analytical and computational determination of the adiabatic wall temperature required the use of complex mathematical modelling. The complexity was due to the determination of the temperature distribution across the thermal boundary layer of the flow over the substrate, upon which the adiabatic wall temperature depends. In boundary layer flow, the velocity profile of the fluid is coupled to the temperature distribution of the flowing fluid, and this coupling produces non-linear governing equations that are difficult to solve [11]. In order to reduce the complexity of the model and the solution, a vitreous aluminosilicate fiber was used as the substrate and the infrared camera was employed to measure the radial surface temperature of the substrate under the impinging hot air jet. Since the thermal conductivity of the vitreous aluminosilicate fiber is low, at 0.12 W/m-K, the surface temperature was approximately equal to the adiabatic wall temperature. A substrate with low thermal conductivity will restrict the conduction of heat into the material [9], thereby resulting in limited change in the surface temperature during the impingement and spreading of the air jet. Therefore, the surface temperature of the substrate would be equivalent to the adiabatic wall temperature of the air film.

Figure 3 shows the variation of the non-dimensional adiabatic wall temperature with the non-dimensional distance from the stagnation point of the impinging air jet onto the vitreous aluminosilicate fiber substrate. The results of this present study on the vitreous aluminosilicate fiber substrate have been compared with those on Lexan substrate that were presented in a previous study [9]. Similar to Ryabinin, *et al.* [9], the non-dimensional temperature and radial distance are

$$\theta_{aw} = \frac{T_{aw} - T_\infty}{T_g - T_\infty}, \tag{9}$$

$$r^* = \frac{r}{d_{nozzle}}, \tag{10}$$

where T_g and T_∞ are the gas temperature at the cold spray console and ambient temperature, respectively, and d_{nozzle} is the nozzle diameter, which was 6.3 mm in this study. The non-dimensional distance from the stagnation point of the impinging air jet was defined as r^*.

As shown in Fig. 3, in the vicinity of the stagnation point of the air jet, the measured non-dimensional adiabatic wall temperature differs from that presented by Ryabinin, *et al.* [9]. However, as the distance from the stagnation point increased, the results show good agreement with those obtained by Ryabinin, *et al.* [9]. The difference at the stagnation point is likely due to the temperatures of the air jet generated by the cold spray units. The temperature of the nitrogen gas jet that was used by Ryabinin, *et al.* [9] was fixed at 100°C, while the temperature of the air jet in the present study was set at 350°C. As the gas film temperature increased, the heat exchange between the hot gas film and the colder ambient air over the substrate increased. Given that a non-dimensional temperature, which is a ratio of temperature differences, was used in Fig. 3, and given the increased heat losses at the stagnation point in this study, the non-dimensional adiabatic temperature near the stagnation point was lower.

As the gas film spread over the substrate, its temperature decreased, the non-dimensional adiabatic wall temperature agrees with the one presented by Ryabinin et al. [9]. The minor deviation between the two measurements at distances further downstream from stagnation point was likely due to the difference between the thermal conductivity of the substrate materials. Since the thermal conductivity of the vitreous aluminosilicate fiber is lower than that of the Lexan, the measured adiabatic wall temperature of vitreous aluminosilicate fiber is expected to be slightly higher due to the restricted heat conduction into the material.

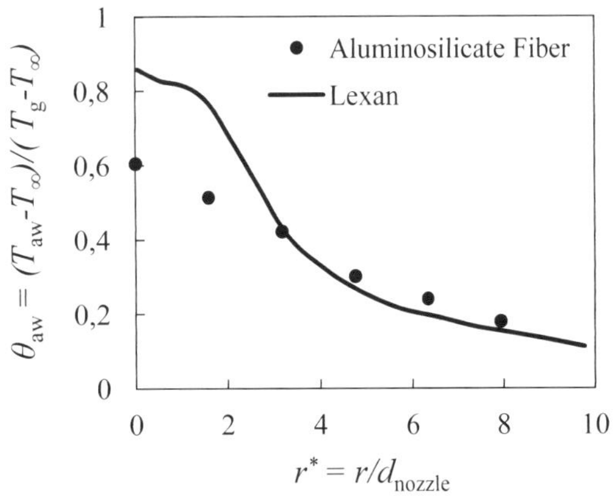

Fig. 3. Comparison of the adiabatic wall temperatures on vitreous aluminosilicate fiber and Lexan [9].

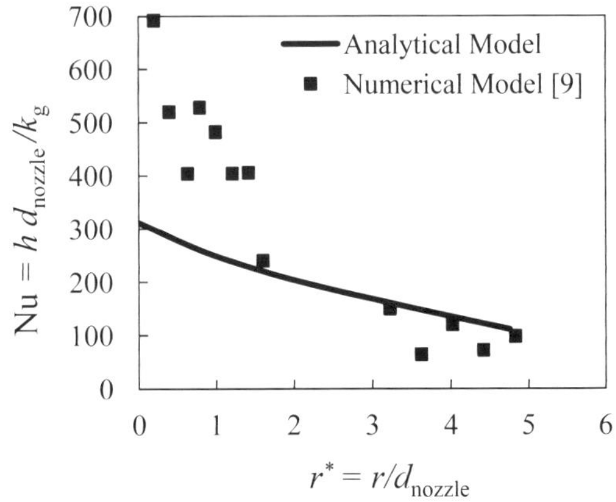

Fig. 4. Comparison of the Nusselt number with that of Ryabinin, *et al.* [9].

4.2 Heat Transfer Coefficient

The adiabatic wall temperature and the experimental data for the surface temperature of the Al substrate were substituted into Eq. (8), and a shooting method was applied to determine the spatially varying heat transfer coefficient and Nusselt number (Nu) of the air film over the substrate. Figure 4 shows the results for Nusselt number as a function of the non-dimensional radial distance from the stagnation point that was obtained from use of Eqs. (6) and (8). The results were compared with the numerical results that were obtained by Ryabinin, *et al.* [9]. It should be noted that the numerical results presented by Ryabinin, *et al.* [9] were obtained at a gas temperature of 100°C, while the gas temperature in the current study was 350°C.
Similar to Fig. 3, the results of the present study for Nusselt number differ from the numerical results that were obtained by Ryabinin, *et al.* [9] in the vicinity of the stagnation point of the impinging air jet. Due to the heat exchange between the hot air film and the colder ambient air, the heat transfer rate from the air film to the substrate decreased near the stagnation point. Since the heat transfer coefficient of the spreading air film is directly proportional to the heat transfer rate from the film to the substrate and the adiabatic wall temperature [7], it is expected that the heat transfer coefficient in the vicinity of the stagnation point would decrease. However, similar to the non-dimensional adiabatic wall temperature, as the radial distance from stagnation point increases, the analytical results come into good agreement with the numerical results of Ryabinin, *et al.* [9].
As shown in Fig. 4, the Nusselt number decreases as the radial distance from the stagnation point increases. It has been shown that for compressible flows, the Nusselt number is proportional to the square-root of the gas velocity within the air film [12]. Since the gas velocity decreases as the gas film spreads over the substrate, the Nusselt number decreases as the fluid flows away from the stagnation point. Several studies have placed emphasis on the fact that the Nusselt number decreases as the radial distance from stagnation point increases [8 - 10].

4.3 Validation of the Analytical Model

In order to validate the analytical results for Nusselt number presented in Fig. 4, the air was impinged at different temperatures upon the aluminium substrate. The cold spray nozzle was set at a stationary position, so that the stand-off distance (SOD) was fixed at 10 mm. The surface temperature was measured experimentally at steady-state condition. The substrate surface temperature was determined from Eq. (6) at different gas temperatures by using the Nusselt number estimates that were obtained in this study. The results for surface temperature were calculated and compared to the experimental data.
Figure 5 shows both the experimentally and analytically determined radial surface temperature distribution of the aluminium substrate under an impinging air jet at various temperatures. The analytical results show good agreement with the experimentally measured surface temperature.

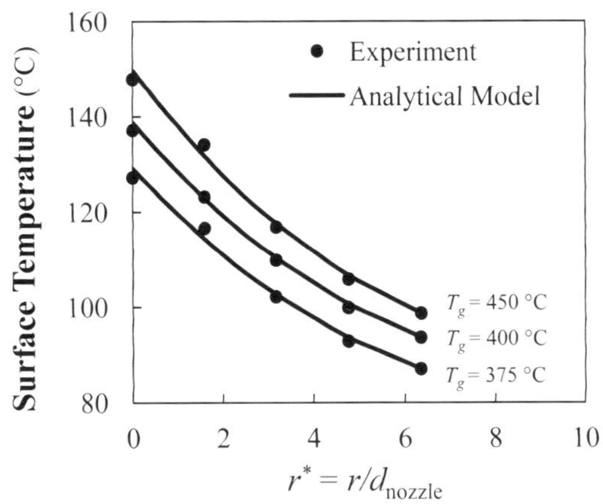

Fig. 5. Surface temperature of the aluminium substrate at three different gas temperatures as a function of the non-dimensional distance from the stagnation point of jet impingement.

The experimental and analytical results for surface temperature distribution were non-dimensionalized as

$$\theta_s = \frac{T_s - T_\infty}{T_g - T_\infty}. \tag{11}$$

Figure 6 shows the non-dimensional surface temperature as a function of the non-dimensional distance from the stagnation point of jet impingement. The figure shows that the results coalesced and there was good agreement between the analytical model and experimental results.

It was observed that the non-dimensional surface temperature did not change significantly as the distance from stagnation point of the impinging jet increased. This was due to the high thermal conductivity of the aluminium, which allowed for conduction of the heat through the substrate under the steady-state condition of the experiment and model. It is hypothesized that the agreement will not likely occur in substrates with lower thermal conductivities. This observation was made and reported in other studies [9, 10].

5 Conclusion

The analytical evaluation of the heat transfer coefficient of an impinging hot jet, generated by a cold spray unit, over various substrates was studied. A 2-D steady-state heat conduction equation was solved by using the separation of variables method and the temperature distribution within the substrate was determined. A low-pressure cold spray unit was employed to impinge the hot air jet upon the aluminium substrate. J-type thermocouples were used to measure the radial surface temperature distribution. Coupling the results of the experiment with the analytical solution for the temperature distribution, the heat transfer coefficient of the air film was analytically determined.

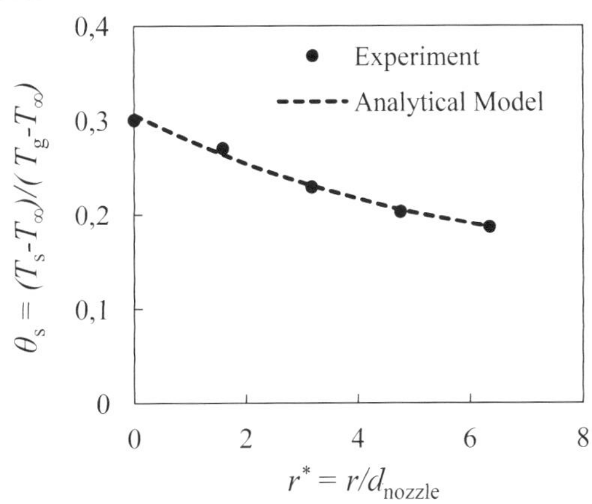

Fig. 6. Non-dimensional surface temperature of the aluminium substrate as a function of the non-dimensional distance from the stagnation point of jet impingement.

The results for the heat transfer coefficient and Nusselt number were in good agreement with the numerical results obtained in previous studies. However, it was observed that the Nusselt number that was estimated within the vicinity of the stagnation point of the air jet in the present study differed from that measured by previous investigators.

The Nusselt number that was determined from the analytical model of this study was used with various gas temperatures to determine the surface temperature distribution of the aluminium substrate. The results of the analytical model and the experiments were in good agreement.

6 Literature

[1] Kosarev, V.F., Klinkov, S.V., Alkhimov, A.P. and Papyrin, A.N.: On some aspects of gas dynamics of the cold spray process. J. Therm. Spray Technol., Vol. 12(2), (2003) pp. 265/81.

[2] Stoltenhoff, T., Kreye, H. and Richter, H.: An analysis of cold spray process and its coating. J. Therm. Spray Technol., Vol. 11, (2002) pp. 542/50.

[3] Schmidt, T., Gartner, F., Assadi, H. and Kreye, H.: Development of a generalized parameter window for cold spray deposition. Acta Mater, Vol. 54, (2006) pp. 729/42.

[4] Limaye, M.D., Vedula, R.P. and Prabhu, S.V.: Local heat transfer distribution on a flat plate impinged by a compressible round air jet. Int. J. Therm. Sci., Vol.49, (2010) pp. 2157/68.

[5] Irissou, E., Legoux J.G., Ryabinin. A., Jodoin, B. and Moreau. C.: Review on cold spray process and technology: part I – Intellectual Property. J. Therm. Spray Technol., Vol. 17, (2008) pp. 495/516.

[6] Rahimi, M., Owen, I. and Mistry, J.: Impingment heat transfer in an under-expanded axisymmetric air jet. Int. J. Heat Mass Transfer, Vol. 46, (2003) pp. 263/72.

[7] Zuckerman, N. and Lior, N.: Jet impingment heat transfer: physics, correlations, and numerical modelling. Adv. Heat Transfer, Vol. 39, (2006) pp. 565/631.

[8] Li, W., Yin, S., Guo, X., Liao, H., Wang, X. and Coddet, C.: An investigation on temperature distribution within the substrate and nozzle wall in cold spraying by numerical and experimental methods. J. Therm. Spray Technol., Vol. 21(1), (2012) pp. 41/48.

[9] Ryabinin, A.N., Irissou, E., McDonald, A. and Legoux, J.-G.: Simulation of gas-substrate heat exchange during cold-gas dynamic spraying. Int. J. Therm. Sci., Vol. 56, (2012) pp. 12/18.

[10] McDonald, A., Ryabinin, A.N., Irissou, E. and Legoux, J.-G.: Gas-substrate heat exchange during cold-gas dynamic spraying. J. Therm. Spray Technol., Vol. 22(2-3), (2013) pp. 391/97.

[11] Jiji, L.: Heat conduction, second edition., Begell House, New York, 2003.

[12] Shapiro, A.: The dynamics and thermodynamics of compressible fluid flow, Vol. II, The Ronald Press Company, New York, 1954.

Cold spray of agglomerated submicronic hydroxyapatite powders for biomedical applications

D.Moreau, L.Corté, F.Borit, V.Guipont, Evry / F

MINES ParisTech, PSL Research University, MAT - Centre des matériaux, Center for spray processing, CNRS UMR 7633, Evry, France

Hydroxyapatite (HA) ceramic coatings are used for orthopaedic implants to favour bone growth. The main conventional process is plasma spray of pure HA powder. A solid state deposition process like cold spray of HA powder might have a great potential to achieve fully crystalline HA coatings. Moreover, HA by cold spray could also be envisaged favourably in case of thermosensitive polymeric biomaterials, which would otherwise be degraded by plasma spray. In this work, two agglomerated hydroxyapatite powders, respectively with and without heat treatment after agglomeration, were cold sprayed for various spraying parameters. The cold spray process of pure HA coatings has been implemented through single track on two different biocompatible substrates: a metallic substrate (Ti-6Al-4V) and a polymeric substrate (poly (vinyl alcohol) (PVA)). The structure of the agglomerated HA powders and the corresponding features of the HA coatings were observed by SEM. For both types of substrates, it was shown that the agglomerated submicronic hydroxyapatite powders could lead to a homogenous layers of submicronic HA grains. In the case of particles without heat treatment, thick layers could be obtained due to the binding action of residual by-products of the chemical synthesis of HA particles. It was also shown that one HA layer remained adherent after immersion in water, thus enabling the fabrication of ceramic coated hydrogels.

1 Introduction

Hydroxyapatite (HA) ceramics are used to coat metallic dental and orthopaedic implants in order to achieve an osteo-active layer in direct contact with bony tissues [1-2]. The main conventional process is plasma spray of pure HA powder. However, the melting of HA during the plasma process drastically affects the phase composition and crystallinity of the HA coating. The spraying parameters have to be accurately controlled to maintain at least 50% of pure crystalline HA within the coating. Therefore, a solid state deposition process like cold spray of HA powder might have a great potential to achieve fully crystalline HA coatings. Moreover, other harmful thermal effects to the substrate from the plasma or from the hot particles could be avoided. Thus, HA by cold spray could also be envisaged favourably in case of thermosensitive polymeric biomaterials.

The literature about cold spray process applied to brittle and non-deformable materials like ceramics is not abundant. Ceramics like TiO_2, SiC WO_3 and HA were investigated with evidence of thick coating in case of TiO_2 [8-14]. However, some attempts to achieve HA coatings onto polymer substrates by cold spray have been reported [15-16]. In addition to the nature of the ceramic and its intrinsic mechanical properties, the role of powder morphology and grain size might play a key role to enable a solid state physical adhesion and consolidation to form a coating. In this context, ultrafine or nanostructured powders are of great interest. In this work, the cold spray process involving structured HA powders propelled at supersonic velocities was investigated in order to achieve an HA coating on both metallic and polymer materials of interest in biomedical industry.

Two agglomerated hydroxyapatite powders with submicronic primary grains, respectively with and without heat treatment after agglomeration were cold sprayed for various spraying parameters on metallic and polymeric substrates currently used in clinics. The main purpose was to investigate the role of the powder morphology and its degree of consolidation on the resulting impact behaviour and potential coating formation by the cold spray process.

The substrates studied here are made of metallic titanium alloy (Ti-6Al-4V) and thermoplastic polymer (poly(vinyl-alcohol) (PVA)), both of which are used as bulk materials for the reconstruction of hard and soft osteoarticular tissues. In the case of PVA, it is used in its hydrogel form composed of a cross-linked network of PVA macromolecules swollen with water. This kind of structure is very close to that of connective tissues. In particular, PVA hydrogels have been shown to form non-degradable hydrogels with a good history of biocompatibility [3]. Such PVA hydrogels are already used in clinic for cartilage replacement, and numerous researches are currently carried out to develop artificial meniscus, ligament, and intervertebral disc [4-7]. Coating with a bioactive ceramic coating is a promising way to strengthen the bone-implant interface of these PVA hydrogel implants. While the impacts and temperatures caused by cold-spray are certainly too severe for soft PVA hydrogels, they may be acceptable for dry PVA. Therefore, we propose to explore a two-step process to make PVA hydrogels coated by ceramics by firstly cold-spraying of HA on dry PVA and secondly swelling of the coated system in water.

2 Materials and methods

2.1 Powders and substrates

HA powders: Two types of hydroxyapatite powders were investigated. Commercial HA is synthetized with diamonium phosphate $(NH_4)_2HPO_4$ and calcium nitrate $Ca(NO_3)_2$ in distilled water with ammonium hydroxide NH_4OH to control the pH of the aqueous solution according to the following chemical reaction:

$$10\ Ca(NO_3)_2 + 6\ (NH_4)_2HPO_4 + 8\ NH_4OH \downarrow Ca_{10}(PO_4)_6(OH)_2 + 20\ NH_4NO_3 + 6\ H_2O$$

Then, filtration, atomization of the slurry and subsequent sieving led to a spherical agglomerated HA powder, namely the 'non-calcined HA' while further calcination and sintering at 1050 °C led to the corresponding 'calcined HA'. This latter HA powder is the commercial medical grade (25-60 µm) that could be used for plasma spray process but that was never used for cold spray. The purpose was to investigate the influence of the calcination/sintering consolidation of the starting HA powder on the resulting high velocity impact and potential HA coating formation. The particle size distribution (Malvern Mastersizer 3000) is summarized in **Table 1**. Both HA powders were with a similar range for the median size D_{50} with a narrow distribution range for the calcined HA while non-calcined HA was with a larger one.

Table 1: HA powders particles size (D in vol.%)

Particles	D_{10} (µm)	D_{50} (µm)	D_{90} (µm)
Calcined HA	27.6	41.8	61.4
Non-calcined HA	8.7	46.7	83.3

PVA substrates: Dry Poly (vinyl alcohol) (PVA) substrates were prepared by solvent-casting method. From PVA powder (Sigma-Aldrich), a 10wt% PVA aqueous solution was poured in Petri dishes and solvent was evaporated in vacuum-oven at 700 mbar and 80°C for 8h. Then it was cooled at room temperature for 12h. Translucid disks of dry PVA with thickness 540±170 µm were obtained. Dry PVA melting temperature is 225°C in air while PVA hydrogel after swelling completely dissolves in water after 1 hour upon heating above 85°C.

Ti-6Al-4V substrates: Ti-6Al-4V plates of 25×25×5mm^3 were prepared with and without sand blasting preparation using 300µm white corundum. The average roughness was Ra=2.5 µm and 0.85 µm respectively.

2.2 Cold spray experiments

A high pressure cold spray apparatus (Kinetiks K3000,Oerlikon Metco, Switzerland) was used with nitrogen and MOC nozzle. The powder feedrate was 7±1 g/min. PVA disk specimen was solidly hold in order to expose a 40 mm diameter zone to the supersonic flow of gas and particles. Single track with one pass was achieved with nozzle speed at 10 mm/s. These experiments were achieved only varying the stagnation gas temperature (T) and pressure (P) with also the stand-off distance (SoD).

3 Results and Discussion

3.1 Characteristics of the HA powders

Both powders were observed by SEM (Nova NanoSEM 450, FEI). SEM images with various magnifications on free or mounted and cross-sectioned powder are given in **Fig. 1** pour non-calcined HA and **Fig. 2** for calcined powder respectively.

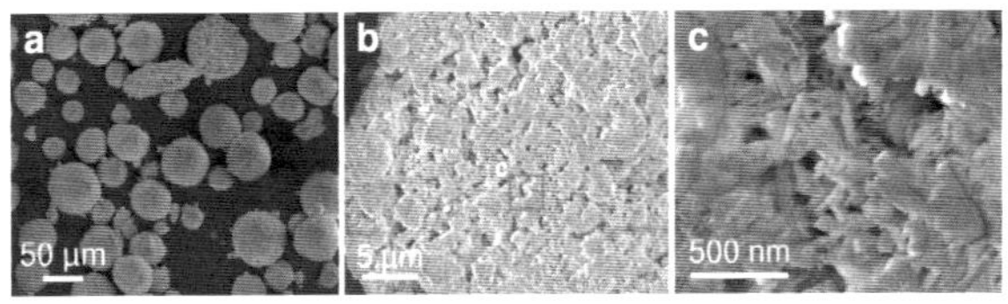

Fig. 1 SEM images non-calcined HA: a) free particles; b) details of crushed particle (SE Mode); c) zoom inside (b) with blocky and needle-like crystals.

Non-calcined HA (Fig. 1) exhibited a spherical shape with some porosity and inside grains with blocky or needle like primary particles. Calcined HA (Fig. 2) is also spherical with submicronic grains fused and sintered together in a highly porous microstructure. A porosity level of 57% was estimated by quantitative image analysis (imageJ) on a polished cross-section.

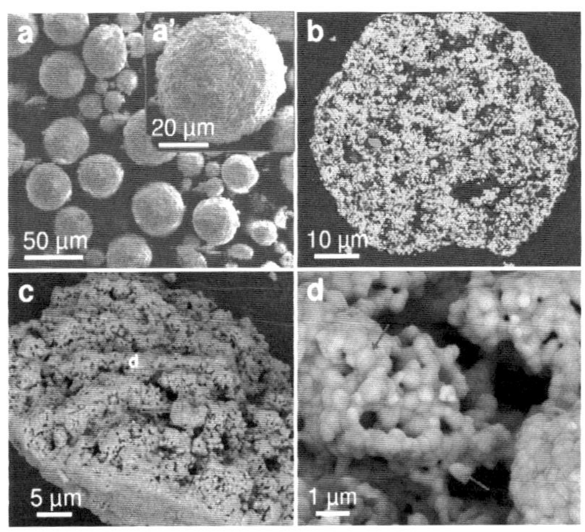

Fig. 2 SEM images of calcined HA: a-a') overview and details of free particles; b) Polished cross-section

(BSE mode); c) Crushed particle; d) Zoom inside (c) with submicronic fused/sintered particles.

The phase composition of both powders were analysed by X-ray diffraction (XRD) with CuKα radiation and Brucker D-500 diffractometer. Diffraction patterns ranging from 20 to 80° (step size 0.032°) are given in **Fig. 2**. Non-calcined HA was crystalline HA while calcined HA exhibited a few quantity of CaO.

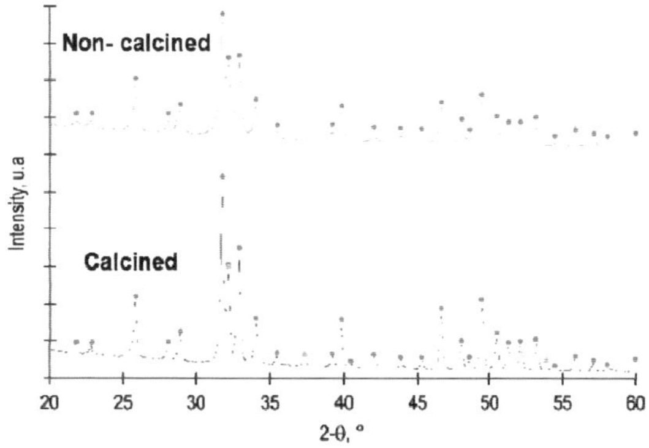

Fig. 3 XRD profiles of calcined and non-calcined HA powders, HA (#9-432) (●) and CaO (#37–1497) (▲).

FTIR analysis spectra (Bruker TENSOR 37 spectrometer) with 4 cm^{-1} resolution were carried out on HA pellets mixed with KBr. FTIR spectra of calcined and non-calcined HA particles are given in **Fig. 4**. Both powders exhibited typical 1100-900 cm^{-1} and 700-500 cm^{-1} bands for the phosphate groups v_3PO_4 and v_4PO_4 of apatite. In case of non calcined HA large absorption bands (3600 and 2800 cm^{-1}) are of absorbed water and bands centered at 1382 and 821 cm^{-1} correspond to NO^3 groups which are nitrate residues. Concerning the calcined powder, bands at 3647 and 3572 cm^{-1} correspond to the OH groups of calcium hydroxide ($Ca(OH)_2$). Another OH group is noticed at 630 cm^{-1}. Bands between 1550 and 1420 cm^{-1} and between 900 and 850 cm^{-1} are v_3CO_3 and v_2CO_3 groups of apatite carbonates.

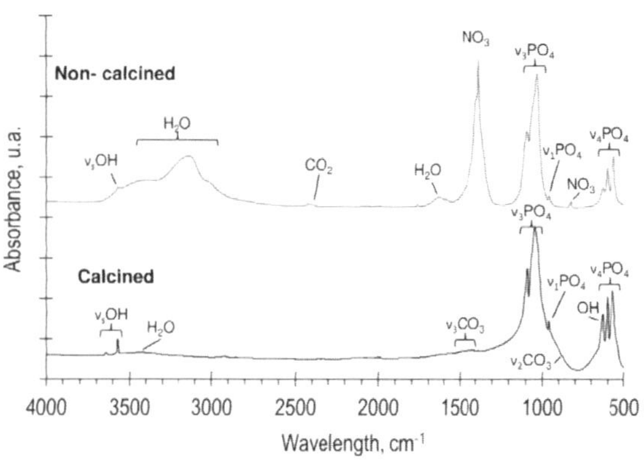

Fig. 4 FTIR spectra of calcined and non-calcined HA

3.1 Cold spray of calcined HA

Ti-6Al-4V substrates: In Fig.5 , Non Blasted (NB) and Sand Blasted (SB) Ti-6Al-4V substrates were investigated using both T=230 °C, P=1.0 MPa (A and C in Fig. 5), or T=200 °C , P=3.0 MPa (B and D in Fig. 5) with similar SoD=30 mm. From these observations, it evidenced the beneficial influence of roughness to achieve thin layer (~ 2-3 μm) of HA without spalled areas.

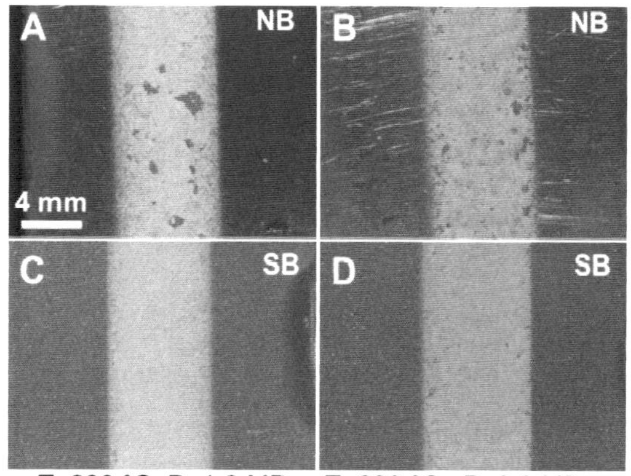

T=230 °C, P=1.0 MPa T=200 °C , P=3.0 MPa

Fig. 5 Top surface macroscopic views of calcined HA coatings on Non Blasted or Sand Blasted Ti-6Al-4V

Low temperature and pressure are enough to create an adhesive layer that consisted in a distribution of submicronic particles that were present in the starting porous and agglomerated powder (see Fig. 6.). The calcination/sintering process did not consolidate too much the primary HA particles together. Therefore they were easily separated and released during impact. This type of powder morphology might be used to taylor specific powder that can initiate original fragmentation and consolidation mechanisms that could be suitable for pure ceramics. Further consolidation between particles would be needed to get thicker coating. Such consolidation process are drastically influenced by the particle size which are submicronic and not nanometric in our case.

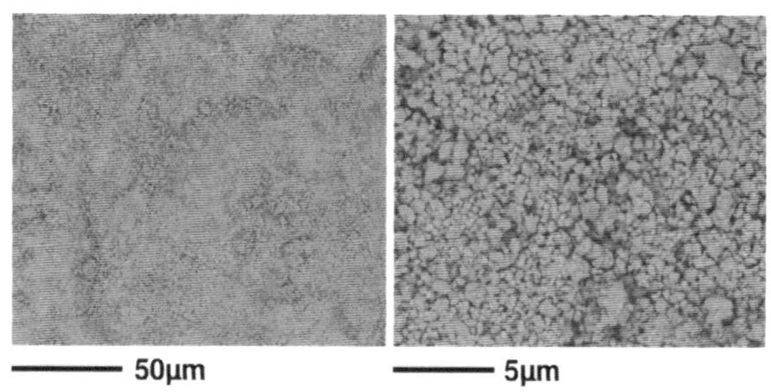

Fig. 6 Top surface SEM-BSE views of calcined HA coatings on SB Ti-6Al-4V (T=230 °C, P=1.0 MPa)

Dry PVA and PVA hydrogels substrates: Typical overviews after cold spray of calcined HA (T=200 °C,

P=1.0 MPa, SoD=30 mm) are given in Figures 7a and 7b in case of PVA hydrogel and dry PVA substrates respectively. It was obvious that swollen PVA substrate was fully damaged and completely flowed away by the gas stream and sprayed powder. Conversely, the dry PVA substrate resisted and a layer of HA was obtained without any macroscopic evidence of degradation. The white HA coating seemed homogeneous and adherent to the substrate.

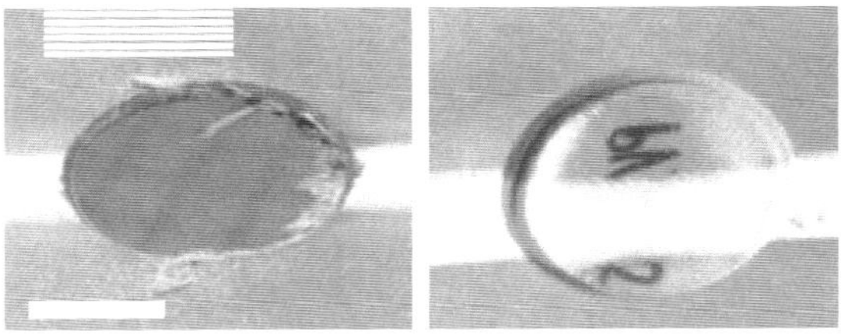

Fig. 7 Top surface macroscopic views of calcined HA deposited on a) hydrogel PVA; b) dry PVA

Other spraying parameters were further investigated only in case of dry PVA. These experiments are summarized in Fig. 8. The main limitation to achieve one HA layer is thermal degradation of the dry PVA substrate that turned brown. For the shorter stand-off distance, the temperature of the stagnation gas can be varied between 200 and 250°C for a pressure of 1 MPa without macroscopic evidence of substrate degradation. When the pressure is increased to the maximum of K3000, one HA coating can be still obtained for the lowest temperature of 200°C. Even if temperature higher than 250°C could be applied with longer stand-off distance, it did not drastically modify the damaging response of the PVA to the thermal flux except for SoD = 150 mm. For this latter, damaging effects were avoided and high energy parameters (T=300°C and P=2.0 MPa) could be applied. In the following, the PVA/HA interface and other HA coating's features for the lowest temperature (T=200°C) and pressure (P=1.0 MPa) at short spraying distance (SoD=30mm) were characterized and discussed.

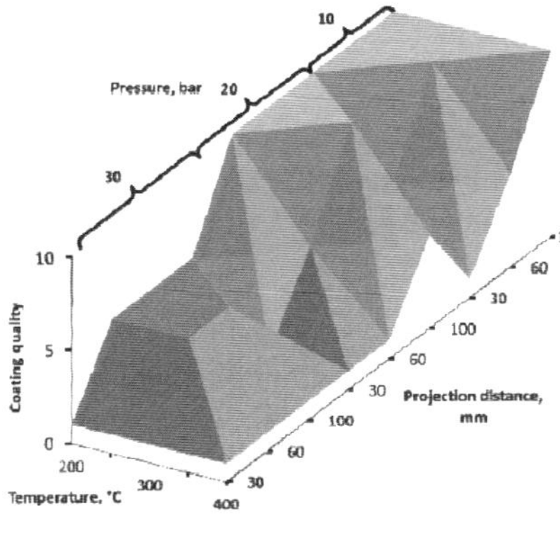

Fig. 8 Influence of main CS parameters on HA deposition and thermal degradation of dry PVA

From fractographic and top surface observations by SEM (Fig. 9) , it could be seen that an homogeneous HA layer could be obtained in the middle of the track. At the boundary, where fewer impacts occurred, single craters were obvious and evidenced the large deformation of the dry PVA substrate upon impacts. In case of biomedical applications, this roughening effect might be beneficial. Moreover, it can be observed that small HA particles were still attached in the centre of craters while PVA was ejected at the rim of the craters.

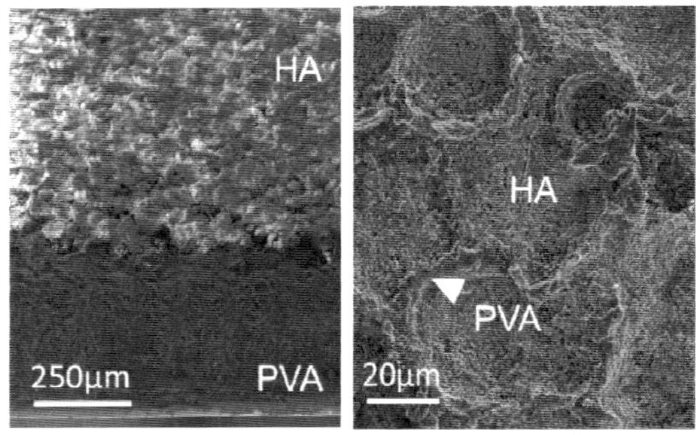

Fig. 9 SEM fractographic and top surface views in the centre (left) and at the boundary (right) of the HA track

The HA/PVA interface was observed by SEM at higher magnification (Fig. 10). The presence of submicronic primary HA particles that were released upon impact when calcined HA porous and agglomerated particles were broken. PVA was probably affected by heat during cold spray. Therefore HA submicronic particles seemed to be physically and strongly bonded to PVA.

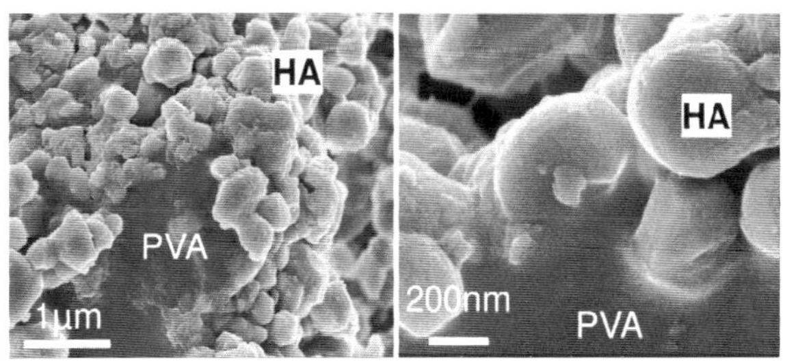

Fig. 10 SEM views of HA/PVA interface within a crater

The excellent adhesion of HA onto dry PVA was confirmed after swelling of the HA/PVA coating. The HA coating on dry PVA substrate was immersed in ultra-pure water for a week at 37 °C under magnetic stirring to reach a 50% equilibrium content of water in PVA. After this immersion period, it was observed that the PVA hydrogel obtained after swelling remained fully covered by the HA coating. This macroscopic observation was confirmed by SEM observations using BSE detector and low vacuum mode, which allow observations of hydrated PVA samples without sample preparation (Fig. 11). HA/PVA hydrogel was obtained and could be envisaged for biomedical application.

Fig. 11 HA/PVA hydrogel after 1 week immersion

3.2 Cold spray of non-calcined HA

Non-calcined HA powder was cold sprayed using similar parameters than for calcined HA powder (T=200°C, P=1.0 MPa, SoD =30 mm) and also by implementing a single track with one pass. A macroscopic view of the top and the side of a HA/Ti-6Al-4V specimen is given in Fig. 12. A very thick (~500µm) and homogenous HA coating was obtained on Ti-6Al-4V. Similar results were obtained on Dry PVA. The coating was very dense but with rather poor cohesion. From the macroscopic top surface view, it can be seen that two phases are coexisting within the HA layer.

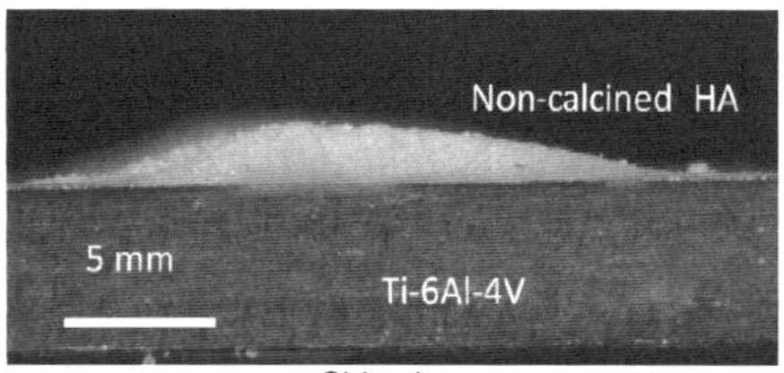

Side view

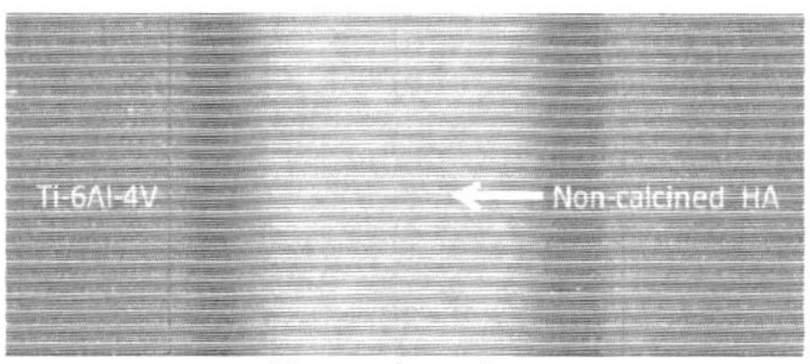

Top surface view

Fig. 12 Macroscopic views of non-calcined HA/Ti-6Al-4V specimen

It was confirmed by SEM observation (Fig. 13) where HA particles are present and surrounded by a glassy compound. This compound is probably made of hydrated nitrates species as by-products of the HA synthesis that were detected by FTIR in the starting non-calcined powder. Differential scanning calorimetry and thermogravimetric analyses confirmed the presence of a large amount (30% of mass loss) of a compound that degraded below 400°C. Such nitrates compounds are not degraded during cold spray and act as a binder within the coating and probably with the surface.

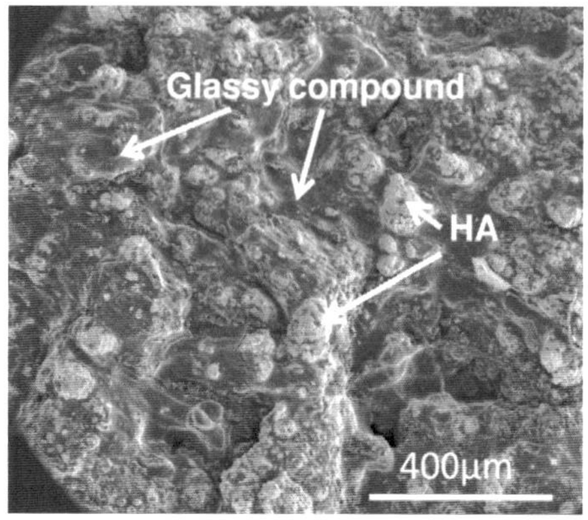

Fig. 13 SEM surface view non-calcined HA coating

4 Conclusion

In this work, the cold spray process was investigated in order to achieve an HA coating from two agglomerated and porous hydroxyapatite powders both with submicronic primary particles, respectively calcined and non-calcined. Two different types of substrates, metallic (Ti-6Al-4V) and polymeric (PVA) materials were also investigated because of their interests in implantology. It was found that both substrate materials exhibited a similar behaviour for both powders upon impact using low temperature and pressure.

The calcined HA powder broke during impact and released the submicronic primary particles that were previously agglomerated and probably poorly sintered in this case. The resulting film was made by a dense distribution of the submicronic particles. The roughness of the metallic substrate is beneficial to the adhesion. In case of PVA, deep craters that were formed by the HA particles increased the roughness during cold spray but no thermal degradation occurred macroscopically. In this case, a strong HA/PVA interface was obtained for each submicronic HA particle because of local heating of the PVA from the gas stream during impact. Thus, HA coated dry PVA could be swollen to form an original HA/PVA hydrogel coating without loss of HA when immersed for one week in water.

The non-calcined HA powder led to very thick (~500µm) that was very dense but with rather poor cohesion and adhesion. The HA deposit exhibited HA particles surrounded by a glassy compound which was probably the hydrated nitrates by-products remaining in a large amount in the non-calcined starting powder Such hydrated nitrates compounds are not degraded during cold spray and act as a binder within the coating and probably with the surface. Such experimental results could be used to discuss the design of powder involving rather low temperature binder that could be included in the starting powder to help the ceramic coating formation.

5 References

[1] Furlong and J. Osborn, J. Bone Joint Surg. Br., 1991, 73-B, pp. 741/45.
[2] J. Huracek and P. Spirig, Arch. Orthop. Trauma Surg., 1994, 113, 72–77.
[3] N. A. Peppas, J. Z. Hilt, A. Khademhosseini and R. Langer, Adv. Mater., 2006, 18, 1345–1360.
[4] F. V. Sciarretta, Eur. Rev. Med. Pharmacol. Sci., 2013, 17, 3031–3038.
[5] M. Kobayashi, Y.-S. Chang and M. Oka, Biomaterials, 2005, 26, 3243–3248.
[6] J. S. Bach, F. Detrez, M. Cherkaoui, S. Cantournet, D. N. Ku and L. Corté, J. Biomech., 2013, 46, 1463–1470.
[7] D. Darwis, P. Stasica, M. T. Razzak and J. M. Rosiak, Radiat. Phys. Chem., 2002, 63, 539–542.
[8] I. Burlacov, J. Jirkovský, L. Kavan, R. Ballhorn and R. B. Heimann, J. Photochem. Photobiol. Chem., 2007, 187, 285–292.
[9] J.-O. Kliemann, H. Gutzmann, F. Gärtner, H. Hübner, C. Borchers and T. Klassen, J. Therm. Spray Technol., 2011, 20, 292–298.
[10] G.-J. Yang, C.-J. Li, F. Han, W.-Y. Li and A. Ohmori, Appl. Surf. Sci., 2008, 254, 3979–3982.
[11] D. Seo, M. Sayar and K. Ogawa, Surf. Coat. Technol., 2012, 206, 2851–2858.
[12] Cold Spraying of TiO2 Photocatalyst Coating With Nitrogen Process Gas
[13] M. Yamada, H. Isago, H. Nakano, M. Fukumoto, J. Thermal Spray Technol., 2010, 19, pp. 1218/23
[14] H. Y. Lee, Y. H. Yu, Y. C. Lee, Y. P. Hong and K. H. Ko, J. Therm. Spray Technol., 2005, 14, 183–186.
[15] J. H. Lee, H. L. Jang, K. M. Lee, H.-R. Baek, K. Jin, K. S. Hong, J. H. Noh and H.-K. Lee, Acta Biomater., 2013, 9, 6177–6187.
[16] N. Sanpo, M. L. Tan, P. Cheang and K. A. Khor, J. Therm. Spray Technol., 2009, 18, 10–15.

6 Acknowledgments

The authors would like to thank M. Betbeder and F. Gaslain for their help on SEM and Jean-Dominique Bartout for granulometry measurements. The financial support from MINES ParisTech, the ESPCI ParisTech and the Institut Carnot M.I.N.E.S. (grant: HAP-Process 2012) is also acknowledged.

High Deposition Efficiency Additive Manufacturing of Titanium using Low Pressure Nitrogen Cold Spray

D. MacDonald, R. Fernández, F. Delloro, B. Jodoin / Ottawa, CA

Titanium parts are ideally suited for aerospace applications due to their unique combination of high specific strength and excellent corrosion resistance. However, the raw materials are expensive and challenging/costly to machine. Production of complex titanium aerospace parts through additive manufacturing looks promising but there are still many obstacles to overcome before reaching mass production.

The cold gas dynamic spraying process offers the potential for additive manufacturing of large titanium parts due to its reduced reactive environment, its simplicity to operate, and the high deposition rates it offers. A few challenges are to be addressed before the additive manufacturing potential of titanium by cold gas dynamic spraying can be reached. In particular, it is known that titanium is easy to deposit by cold gas dynamic spraying, but the deposits produced are usually porous when nitrogen is used as the carrier gas.

In this work, a method to manufacture low porosity titanium components at high deposition efficiencies is revealed. The components are produced using low pressure cold spray and nitrogen as the carrier gas. The microstructure and mechanical properties of additive manufactured titanium bulk parts are investigated.

1 INTRODUCTION

Cold Gas Dynamic Spraying (CGDS) is a solid state thermal spray process normally used for coatings and repairs. Over two decades of fundamental studies [1]–[3] have elucidated the theoretical principles of the process; which has led to the development of commercial equipment and commercial applications [4]–[6]. While no longer considered a novel spray process, particularly for repairs [7]–[9], it is still in its infancy as an additive manufacturing (AM) process [10]–[14].

The additive manufacturing of metals can be separated into two broad categories: Indirect methods, where a binder is used to bond the metal particles and post processing is used to remove the binder and consolidate the part, and direct methods, where the final part is created directly without a binder [15]. It is possible to fabricate intricate parts with good finishes with processes such as electron beam melting EBM [15]–[17]), however, this process is currently limited by a manufacturing rate of 6 grams/minute at the very high end for titanium (modified from [18], [19] using a density of titanium of 4.51g/cm^2 [20]). This rate of deposition leaves much room for improvement.

CGDS could be used as a direct process for AM, taking advantage of the large deposition rates possible with CGDS systems (up to 500 grams/minute [21]). This rate is an order of magnitude faster than the current fastest powder based AM technology (83 grams/minute [22]). Another advantage of the CGDS process is high adhesion to the substrate [23], which could be used to fabricate extensions on existing parts.

In the literature, helium or extremely high pressure and temperature nitrogen are required to create dense coatings [23], [24]. Schmidt et al. proposed that the critical velocity for titanium must be above 700 m/s for adiabatic sheer instability to occur [3]. This limits the possibility of using this process as fabrication tool due to high cost of equipment and operation. This work proposes using unique particle morphology to ensure high density deposits with low pressure nitrogen as carrier gas, making the process more economical.

2 EXPERIMENTAL

2.1 Feedstock Powder

The powder used for this study is commercially pure titanium powder from Cristal Metals (Cristal US Inc., Woodridge, IL, USA). This powder is produced using the Armstrong process, in which titanium chloride is reacted to form solid titanium [25]. This titanium holds a very unique, coral like morphology as seen in Fig. 1.

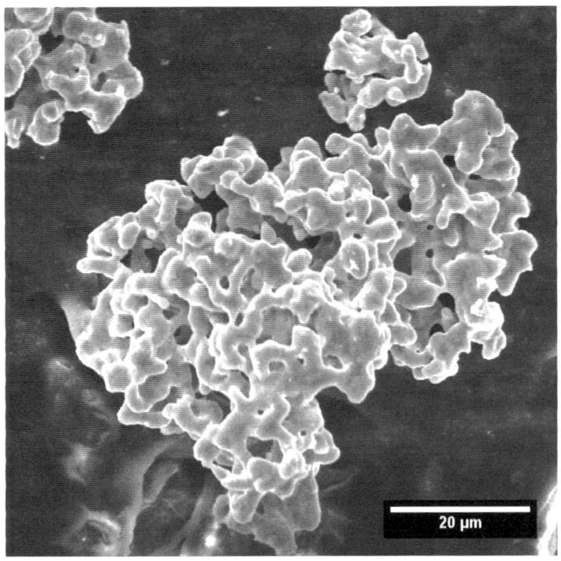

Fig. 1. SEM image of coral-shaped titanium powder.

This morphology is very different from other particles commonly using in the CGDS process, such as spherical gas atomized powder or mechanically crushed angular particles.

2.2 CGDS Process

The titanium powder was consolidated using the commercially available SST Series EP Cold Spray System (Centerline Ltd, Windsor, Canada). The system has a maximum operating gas temperature of 550°C and a maximum gas pressure of 3.8 MPa. A stainless steel De Laval nozzle with a 2 mm diameter throat, a 120 mm long diverging section, and an exit diameter of 6.6 mm was used. Powder was fed using a commercially available powder feeder from Praxair Surface Technologies (Model 1264, Praxair Surface Technologies, Concord, NH, USA).

In order to investigate and assess the effect of process gas pressure and temperature on the deposition behaviour of this powder, the parameters in Table 1 were implemented.

Table 1. CGDS Parameters during Testing

Parameter	Value
Gas Nature	Nitrogen
Gas Pressure	1.0 – 3.8 MPa
Gas Temperature	100 – 500°C
Powder Feeder Gas Flow Rate	1.7 m^3/h
Powder Feeder Gas Nature	Nitrogen
Standoff Distance	10 mm

2.3 Sample Characterization

As received and consolidated powders were investigated using a scanning electron microscope (EVO-MA10, Zeiss, UK) and an optical microscope (VHX-2000E, Keyence, Canada). Porosity levels were quantified through analysis of the optical images over the cross-section of the bulk parts produced.

Microhardness testing of the samples was conducted on polished samples using a Vickers microhardness tester (Struers Duramin-1) with a 300g load.

2.4 Particle Velocity Measurement

Particle velocity measurements were taken using the ColdSprayMeter (CSM) eVOLUTION (Tecnar Automation Ltd, St-Bruno, Canada). This system uses a continuous 2.4 W (λ=810 nm) laser with a dual spit photomask in order to perform in-flight diagnostic on individual particles. It provides precise velocity and diameter measurements of individual particles during flight.

2.5 Heat Treatment

In order to enhance the mechanical properties of the deposited parts, heat treatments were performed on selected samples. This was done under vacuum to prevent oxidation at the prescribed times and temperatures.

2.6 Mechanical Testing

Additional mechanical testing of selected samples was completed after heat treatment to assess the mechanical properties of the fabricated parts. These samples were tested using the standard test method for tension testing of metallic materials (ASTM E8) [26]. The dimensions of the CGDS fabricated tensile samples can be found in Fig. 2.

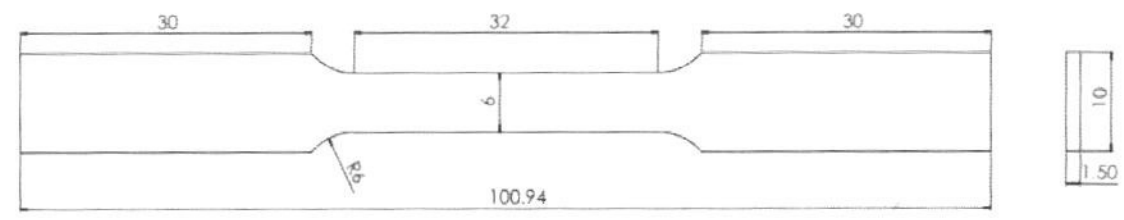

Fig. 2. Dimensions of tensile test specimens (in mm).

3 RESULTS AND DISCUSSION

3.1 Sample Characterization

Deposition efficiency was measured by weighing the samples before and after deposition and comparing the mass gain with the mass of powder fed. The powder feed rate was measured directly at the powder feeder exit in testing separate from the operation of the CGDS system. The deposition efficiency as a function of stagnation temperature for different pressures can be seen in Fig. 3(A). It is clear from this figure that deposition efficiency levels off at approximately 90%. Although high DE is expected with titanium, normally much higher velocities are required to approach 100% DE [24].

The density of the deposits as a function of spray process temperature for different pressures can be seen in Fig. 3(B). This figure shows that for all but the lowest parameters, the coatings have low porosity. An example of a deposit with low porosity can be seen in Fig. 4. It can be noted, that although the average of the entire cross section had a porosity of 3.3%, the majority of the porosity was contained in the edges of the deposition, which is a typical characteristic of CGDS titanium [27].

Particle velocities were taken 10 mm downstream from the nozzle exit without a substrate present. This velocity approximates the particle velocity (neglecting the effects of the bow shock) during impact. A mapping of the average particle velocities can be found in Fig. 3(C).

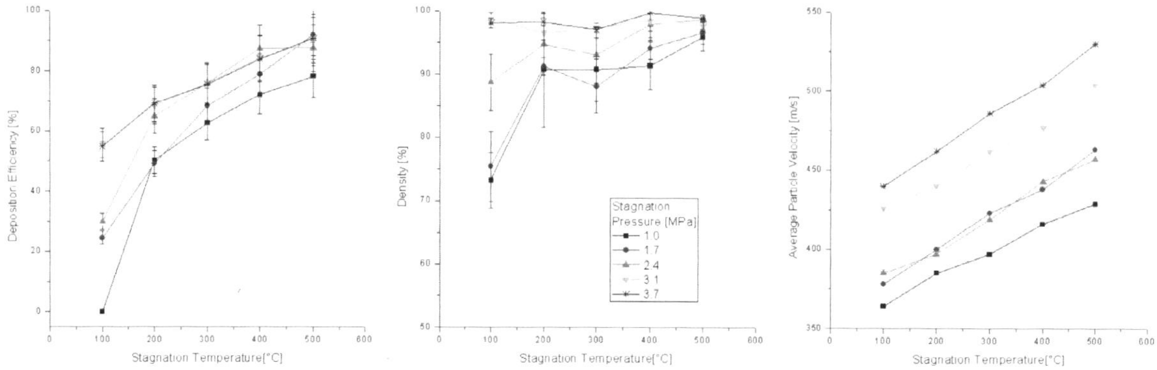

Fig. 3. Deposition efficiency (A), density (B), and average particle velocity (C) plotted against stagnation temperature for different stagnation pressures. NOTE: The average particle velocity represents the average of a normal distribution with SD = 57.36.

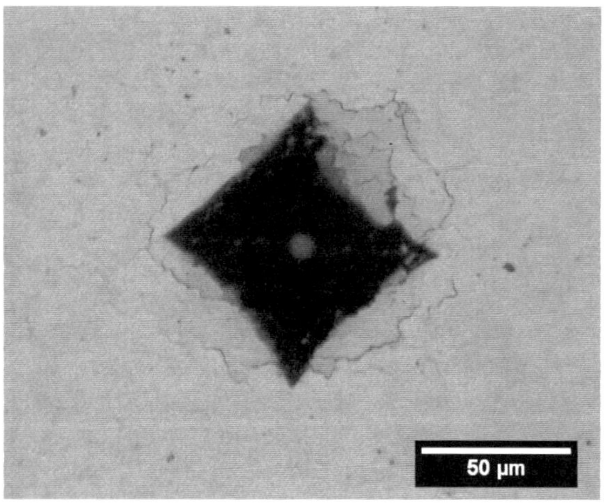

Fig. 4. Optical image showing a near fully dense deposition with cracking around Vicker's microhardness indentation (deposited with 1.7 MPa and 500°C).

These particle velocities were found to be well below the critical velocities shown in the literature for titanium, while still having a measured deposition efficiency up to 90%. This on its own shows that the values reported in the literature for titanium critical velocity may not be appropriate for non-traditional powder morphology such as the one used in this work. More so, the depositions also appear to be fully dense based on contrast optical imaging techniques. This result is unexpected since the particle velocities are well below the velocities required for dense coatings according to the literature. According to Wong et Al, a deposition of spherical particles at over 600m/s resulted in a coating of approximately 20% porosity and 46% DE [24]. In comparison, the highest velocity average reached in this study was 530m/s, which reached a DE of 90% and porosity of 1.1%. It is clear from this result that the powder morphology is a critical component of the CGDS process.

Microhardness testing of the bulk deposits was completed and ranged from 55.5 ±16.0 HV_{300g} to 87.2 ± 8.2 HV_{300g}. These values are substantially lower than expected values for bulk commercially pure titanium (145 HV [20]) and titanium produced through CGDS (>190HV [24]). These results are unforeseen, as in CGDS traditionally dense coatings have a high level of deformation and therefore a high level of cold working, resulting in higher hardness.

The image of the hardness test of as-sprayed parts shown in Fig. 4 provides some insight into the deposition behaviour of this powder. Cracks can be seen propagating from the indentation. These cracks are suspected to be following particle boundaries, demonstrating that there is very little cohesion between particles. This again is unexpected, since in CGDS the particle velocities required to get a dense coating usually results in strong particle to particle cohesion.

In this instance, the low cohesion and low hardness is likely a result of the low particle velocities required for deposition. It is hypothesized that during impact, the unique particle morphology allows the particles to anchor to one another while collapsing and flattening on themselves, resulting in a high DE, dense structure with very little cohesion between particles. The results of a simulation showing the impact behaviour of two particles with real geometry, properties, and particle velocities can be found in Fig. 5. The details of this simulation have been omitted from this proceeding for brevity. This figure acts as a visual aid showing how, as a result of powder morphology, the particle is able to collapse almost fully during impact. Fig. 5 not only shows the method for densification, but it sheds light on the how the particles would be able to interlock with each other without creating any substantial metallic bonds between the particles.

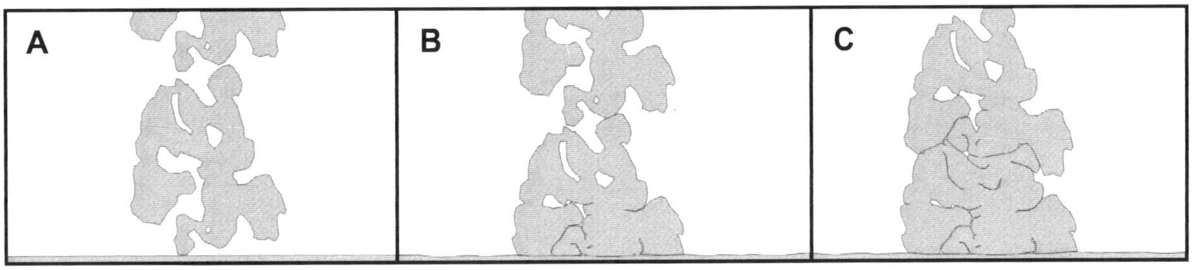

Fig. 5. Proposed densification method of coral-like powder showing a particle before impact (A), a particle during impact with substrate (B), and a particle after being impacted by a second particle (C).

3.3 Heat Treatment and Resulting Properties

The low hardness and poor cohesion of the fabricated components can be overcome by growing metallic bonds between particles through heat treatment. Three heat treatments based on existing literature were chosen for testing; 870°C for 36hours [28], 650°C for 8 hours [29], and 1000°C for 3 hours [30].

The CGDS spray parameters used to produce the components for this testing were 1.7 MPa and 500°C. These parameters were selected as the deposition efficiency and density did not appear to increase substantially beyond this point. SEM images of these samples after etching (solution of 1:1:20 of $HF:H_2O_2$:deionized water) can be seen in Fig. 6.

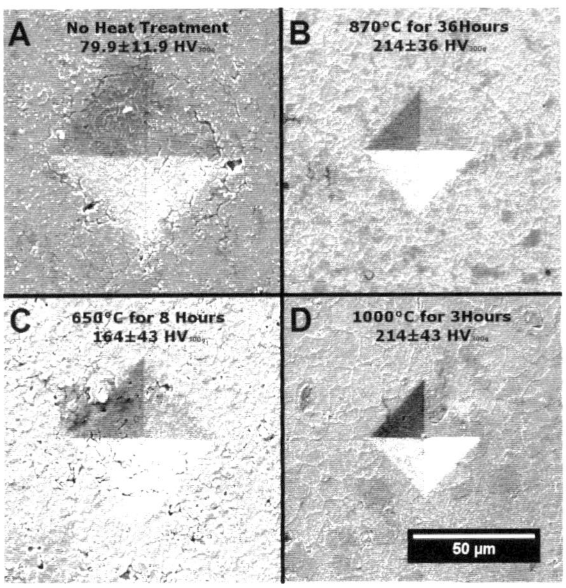

Fig. 6. SEM images of etched samples before heat treatment (A), after 870°C for 36 hours (B), 650°C for 8 hours (C), and 1000°C for 3 hours (D).

Fig. 6 also shows the microhardness values of the sample before and after the different heat treatments. In Fig. 6(A), cracking and delamination between particle boundaries can be seen around the edges of the microhardness indent. It can be observed in Fig. 6(B) and Fig. 6(D) that this cracking around the indent is no longer present. It is clear from this result that the cracks can no longer propagate through the particle boundaries, indicating that metallic bonding has occurred between the particles. For the heat treatment shown in Fig. 6(C), an increase in hardness is present indicating that the cohesion between the particles has increased, however, cracking at the particle boundaries is still visible indicating that the heat treatment process was not adequate to induce full metallurgical bonding between the particles.

It was found that for two of the heat treatments, the component microhardness was above that for bulk materials. This additional hardness is likely due to increased oxygen content present in the finished component. This small percentage of oxygen must have been present in the original powder as the CGDS process is done in an inert environment and the heat treatment occurs in a vacuum ($<10^{-7}$ torr).

Since the microhardness tests and density of the heat treated samples looked promising, especially the 1000°C for 3 hours, tensile samples (dimensions found in Fig. 2) were fabricated using CGDS. The samples were produced using the same parameters as the previous heat treatment testing (1.7 MPa and 500°C nitrogen). The samples were then separated from the substrate and heat treated at 1000°C for 3 hours before being machined to final dimensions. Stress relief annealing at 550°C for 2 hours [20] was done after machining.

The samples were tested in accordance with ASTM E8 [26] using a universal tensile testing machine. This resulted in an ultimate tensile strength (UTS) of 277± 88 MPa. The average value found appears to be above the minimum value listed for UTS in the literature of 240 MPa [20]. However, all three samples underwent brittle fracture will little plastic deformation.

The microstructure of the etched sample before and after the 1000°C for 3 hour heat treatment is shown in the high magnification SEM images (Fig. 7). The brittle failure and the large standard deviation that occurred during this test are unexpected. From Fig. 7(B), it is clear that the component recrystallized and that the particles were bonded during the heat treatment. Therefore, it would be expected that any residual stress or cold work would be eliminated, and the component should have failed in a ductile manner, similar to the bulk material.

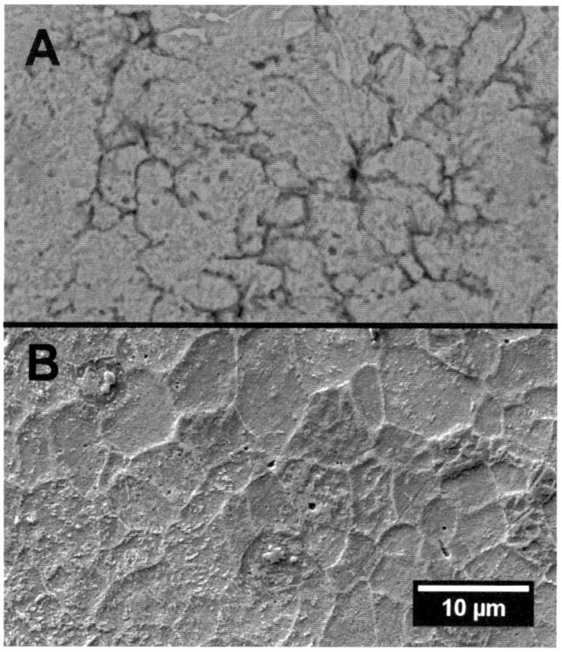

Fig. 7. High magnification SEM images of etched samples after deposition (A) and after heat treatment of 1000°C for 180 minutes (B).

However, post-fracture analysis showed that the fracture propagated from flaws created during final machining steps at the neck area of the sample. These flaws would have created stress concentrations during tensile testing, resulting crack propagation and sudden, brittle fracture. Therefore, the values for the UTS do not reflect the true strength of the material, but a minimum value for components without stress concentrations.

It should also be noted from Fig. 7(A), that the dark lines indicating a boundary between particles are remarkably similar to those proposed by the collapsing particles shown in Fig. 5.

4 CONCLUSION

Fully dense titanium parts were fabricated using CGDS as an AM process. This was accomplished using low pressure and temperature nitrogen as the process gas. The key to this success is attributed to the unique coral-shape powder morphology that allows the easy densification and deposition through CGDS. Although the properties of these components are not ideal in the as-manufactured state, a short heat treatment of 3 hours results in properties comparable to bulk material in terms of hardness and UTS.

CGDS as an AM tool does require additional post processing such as heat treatment and traditional machining; however this is common practice for all other metal additive manufacturing technologies [17]. In addition, since it is a direct method of AM and creates fully dense components, binder degassing and hot isostatic pressing are not required; which is common for many other AM processes.

This is a promising approach for CGDS as a method for AM of titanium parts, as previously only very high particle velocities could be used to create dense titanium components. These high particle velocities required very high process pressure and temperature when using nitrogen or required the use of helium as the process gas. With this powder, the operating costs could be substantially reduced and commercially available small handheld units could be integrated into an AM system.

5 ACKNOWLEDGMENTS

The authors of this paper would like to gratefully acknowledge the contributions of Damien Mangabhai and Cristal Metals.

6 LITERATURE

[1] J. R. Davis, *Handbook of Thermal Spray Technology*. ASM International, 2004.
[2] M. Grujicic, C. . Zhao, W. . DeRosset, and D. Helfritch, “Adiabatic shear instability based mechanism for particles/substrate bonding in the cold-gas dynamic-spray process,” *Mater. Des.*, vol. 25, no. 8, pp. 681–688, Dec. 2004.
[3] T. Schmidt, F. Gärtner, H. Assadi, and H. Kreye, “Development of a generalized parameter window for cold spray deposition,” *Acta Mater.*, vol. 54, no. 3, pp. 729–742, Feb. 2006.
[4] D. Wright, “Advancing Cold Spray Applications to Industry Markets,” in *AeroMat 2011*, 2011.
[5] D. Gasme, “Verfahren zum herstellen eines kochgeschirrs für einen induktionsherd (A method of manufacturing a cooking vessel for induction cooker),” European Patent 1 599 116 B12008.
[6] P. A. Kramer, “Spray processing of pourus medical devices,” US 7163715 B1.
[7] V. K. Champagne, “The repair of magnesium rotorcraft components by cold spray,” *J. Fail. Anal. Prev.*, vol. 8, no. 2008, pp. 164–175, 2008.
[8] D. MacDonald and B. Jodoin, “Study on the Restoration of Aluminum Aerospace IVD Coatings Using Cold Gas Dynamic Spray,” in *Thermal Spray 2013 - Innovative Coating*

Solutions for the Global Economy. Proceedings of the International Thermal Spray Conference, 2013, pp. 161–165.

[9] M. Yandouzi, S. Gaydos, D. Guo, R. Ghelichi, and B. Jodoin, "Aircraft Skin Restoration and Evaluation," *J. Therm. Spray Technol.*, vol. 23, no. December, pp. 1281–1290, 2014.

[10] Y. Cormier, P. Dupuis, B. Jodoin, and A. Corbeil, "Mechanical Properties of Cold Gas Dynamic-Sprayed Near-Net-Shaped Fin Arrays," *J. Therm. Spray Technol.*, vol. 24, no. February, pp. 476–488, 2014.

[11] A. Sova, S. Grigoriev, A. Okunkova, and I. Smurov, "Potential of cold gas dynamic spray as additive manufacturing technology," *Int. J. Adv. Manuf. Technol.*, vol. 69, no. 2013, pp. 2269–2278, 2013.

[12] P. Dupuis, Y. Cormier, A. Farjam, B. Jodoin, and A. Corbeil, "Performance evaluation of near-net pyramidal shaped fin arrays manufactured by cold spray," *Int. J. Heat Mass Transf.*, vol. 69, no. 2014, pp. 34–43, 2014.

[13] P. Heinrich, P. Richter, H. Holl, and E. Bahr, "Method for Producing a Pipe," US 2012/0273152 A12012.

[14] E. Calla, S. Pabla, and R. Goetze, "Turbine rotor fabrication using cold spraying," US 8261444 B22010.

[15] N. Guo and M. C. Leu, "Additive manufacturing : technology , applications and research needs," *Front. Mech. Eng.*, vol. 8, no. 3, pp. 215–243, 2013.

[16] W. E. Frazier, "Metal additive manufacturing: A review," *J. Mater. Eng. Perform.*, vol. 23, no. June, pp. 1917–1928, 2014.

[17] I. Gibson, D. Rosen, and B. Strucker, *Additive Manufacturing Technologies*. Springer, 2015.

[18] "Additive Manufacturing- Electron Beam Melting," 2012.

[19] J. Hiemenz, "Electron Beam Melting," *Adv. Mater. Process.*, no. March, pp. 45–46, 2007.

[20] M. Donachie, *Titanium: a technical guide*, Second. Materials Park, Ohio: ASM International, 2000.

[21] PLASMA GIKEN, "The Next Generaion in Cold Spray Technology: Introducing Cold Spray PCS-1000 and PCS-800."

[22] P. Calefati, "Borealis: Enlightening Next Generation of Material," in *Manufacturing processes for complex structures and geometries with efficient use of material*, 2014.

[23] D. Goldbaum, J. M. Shockley, R. R. Chromik, A. Rezaeian, S. Yue, J. Legoux, and E. Irissou, "The Effect of Deposition Conditions on Adhesion Strength of Ti and Ti6Al4V Cold Spray Splats," vol. 21, no. March, pp. 288–303, 2012.

[24] W. Wong, E. Irissou, A. N. Ryabinin, J. Legoux, and S. Yue, "Influence of Helium and Nitrogen Gases on the Properties of Cold Gas Dynamic Sprayed Pure Titanium Coatings," vol. 20, no. January, pp. 213–226, 2011.

[25] D. R. Armstrong, S. S. Borys, and R. P. Anderson, "Method of making metals and other elements from the halid vapor of the metal.," 59581061999.

[26] L. Alloy, S. Steel, M. Products, A. U. Tensile, and C. A. Force, "Standard Test Methods for Tension Testing of Metallic Materials 1," pp. 1–27, 2012.

[27] C.-J. Li and W.-Y. Li, "Deposition characteristics of titanium coating in cold spraying," *Surf. Coatings Technol.*, vol. 167, no. 2–3, pp. 278–283, Apr. 2003.

[28] T. Enjo, K. Ikeuchi, N. Akikawa, and T. Maruyama, "Effect of a-b Transformation on the Diffusion Welding of Titanium," *Trans. JWRI*, vol. 9, no. 1, pp. 61–67, 1980.

[29] K. Okazaki and H. Conrad, "Recrystallization and grain growth in titanium: I. characterization of the structure," *Metall. Trans.*, vol. 3, pp. 2411–2421, 1972.

[30] M. Pushkareva, J. Adrien, E. Maire, and A. Weck, "THREE-DIMENSIONAL INVESTIGATION OF VOID GROWTH LEADING TO FRACTURE IN COMMERCIALLY PURE TITANIUM," in *1st International Conference on 3D Materials Science*, 2012, pp. 61–66.

Anti-icing behaviour of thermally sprayed polymer coatings

Heli Koivuluoto, Christian Stenroos, Mikko Kylmälahti, Petri Vuoristo
Tampere University of Technology, Department of Materials Science, Tampere / FIN

Surface engineering shows an increasing potential to provide a sustainable approach to icing problems. Currently, several passive anti-ice properties adoptable to coatings are known, but further research is required to proceed for practical applications. This is due to the fact that icing reduces safety, operational tempo, productivity and reliability of logistics, industry and infrastructure. An icing wind tunnel and a centrifugal ice adhesion test equipment can be used to evaluate and develop anti-icing and icephobic coatings for a potential use in various arctic environments, e.g. in wind power generation, oil drilling, mining and in logistic industries. The present study deals with evaluation of icing properties of flame-sprayed polyethylene (PE) -based polymer coatings. In the laboratory-scale icing tests performed, thermally sprayed polymer coatings showed low ice adhesion compared with metals such as aluminium and stainless steel. The ice adhesion of the flame-sprayed PE coating was found to have approximately seven times lower ice adhesion values compared with metallic aluminium, indicating a very promising anti-icing behaviour.

1 Introduction

On-going climate change, opening of new logistic routes, energy and mineral resources as well as increasing tourism feed the growing activity in cold climate regions. One of the major challenges for operations in these areas is ice and snow accretion. Icing reduces safety, operational tempo, productivity and reliability of logistics, industry and infrastructure. Atmospheric icing on the surfaces is a very complex phenomenon and multiple factors affect the accretion process and behaviour of ice on the surface. [1] From a climatic point of view, several basic processes, water cycle, wind speed, the formation of the clouds, precipitation and presence of different phases (vapor, liquid, and solid) of atmospheric water influence greatly on the severity of icing, accretion rate and ice type. [2-4] Several factors affect the ice accretion and ice adhesion. For instance, surface topography, surface chemistry, wetting behaviour and ice characteristics, type, temperature and liquid water content, have been reported having an influence on ice adhesion, indicating the complexity of icing on the surfaces. [5-8]

Surface engineering can have a high potential as a sustainable approach in solving icing problems. Passive anti-ice coatings can hinder ice formation and icephobic surfaces can reduce the adhesion of ice to surfaces. At this moment, current commercial coatings with icephobic characteristics rely often on hydrophobicity, releasing of lubricants or melting point depressants and ablation. Research is additionally carried out on icephobic potential of super-hydrophobic surfaces [9], phase change materials [10], slippery liquid infused surfaces [11] and surface morphology [9,12]. All these anti-ice mechanisms show promising results in reducing ice accretion and adhesion. Nevertheless, so far these are functional only in specific icing conditions for a very limited length of time. Moreover, the wear resistance of these coatings is known to be poor and therefore, the current coatings are practical only in limited applications or the icephobic effect is insufficiently significant [13].

Generally, coatings have been presented to offer an environmentally friendly option, because no external energy is needed to remove ice and coatings can be tailored to be free of harmful chemicals. Different coating strategies have been presented, which can be roughly divided into polymeric coatings containing fluorine or silicone compounds and superhydrophobic polymer composite coatings. Their performance as icephobic coatings is based on the hydrophobicity, which is achieved by proper surface chemistry and roughness. In addition to the low ice adhesion surfaces, the future aim is to be able to prevent ice formation on the structures. [1]

At this moment, studies related to the icing behaviour of thermally sprayed polymer coatings are not available in literature. However, flame spraying is one of the common thermal spray method for spraying polymer coatings [14,15] and therefore, it is interesting to study the icing behaviour of these coatings. In a flame spraying of polymers, the material is fed in the form of powder to the flame formed by e.g., the mixture of acetylene-oxygen. In the flame, the powder melts and molten droplets are spread on the sprayed surface. Advantages of thermally sprayed polymer coatings are related to the low costs and high performance of the coatings [15,16].

Previously, flame-sprayed polyethylene (PE) coatings have been studied by Vuoristo et al. [17]. The research focused in the use of flame-sprayed PE coatings as natural gas pipe-line coatings. On the other hand, ultra-high molecular weight poly ethylene (UHMWPE) is known to have good protective properties and flame-sprayed UHMWPE coatings have also been studied elsewhere [18]. Flame-sprayed PE coatings have primary applications in corrosion protection of components and metal structures instead of paints and metallic coatings. Additionally, they can be used in other applications such as friction lowering, non-stick surfaces, decoration, etc. One benefit is also their applicability in difficult processing conditions. [14] Therefore, this research focuses on the icing behaviour of flame-sprayed PE-based coatings. Icephobic coatings have high potential to protect metal surfaces e.g., aluminum which is not ideal icephobic material, because it has a high surface energy value [19]. The aim of this study is to evaluate ice adhesion, wetting behaviour, surface roughness and erosion behaviour of flame-sprayed PE-based coatings and reference materials such as aluminium, stainless steel and PU-paint.

2 Experimental procedure

Polymer coatings were sprayed by using an acetylene-oxygen flame spray gun (Castodyn DS 8000). Powder was fed by powder feeder (Sulzer-Metco 4MP) using air as a carrier gas. Gas flows where set relatively low in order to prevent overheating and burning of the powder by the flame. Prior to spraying the samples where cleaned with alcohol. After spraying, a few re-melting passes where done without powder feeding in order to densify the structure. Commercially available and thermally sprayable polyethylene (PE) and ultra-high molecular weight polyethylene (UHMWPE) powders were used in the experiments. In addition, fine fluoropolymer perfluoroethylene-propylene (FEP) powder was mixed with PE powder (10, 25 and 50 wt.%). Substrate material was stainless steel. Reference surfaces were polished stainless steel, polished aluminium and polyurethane (PU) paint.

For the icing research, a novel icing wind tunnel was used in order to create the ice in outdoor conditions. The wind tunnel is placed in a cold climate room where the temperature can be varied from room temperature to – 40°C. After ice accretion, the ice adhesion was measured with a centrifugal ice adhesion test, placed also in cold climate room [1,20]. **Figure 1** presents the ice testing rig located in Tampere University of Technology.

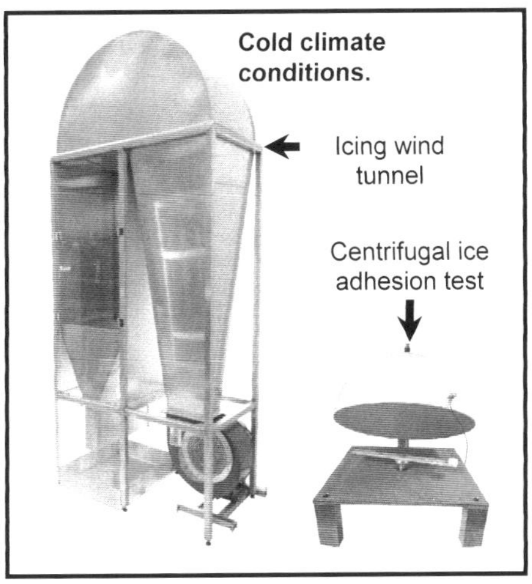

Fig. 1. Icing wind tunnel and centrifugal ice adhesion test at TUT [1,20].

The temperature was chosen as –10 °C and the air flow velocity was set to the maximum 25 m/s. The mixed ice was accreted with the 3.5 bar pressure corresponding to 31 µm volume median diameter of the water droplets. Different ice types e.g., rime, glaze and the mixture of those can be created in the icing wind tunnel, **Fig. 2**. More details of ice adhesion measurements can be found in the previous studies [1,20]. Ice adhesion was measured as an average of four measurements. An effect of surface quality on the ice adhesion was studied by measuring the ice adhesions of the coatings in as-sprayed and polished states.

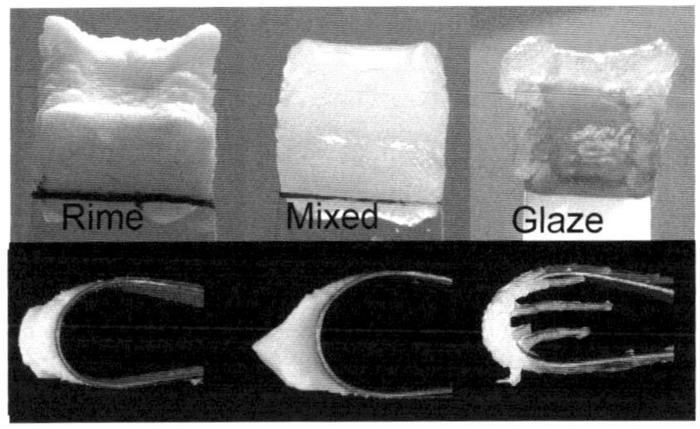

Fig. 2. Different ice types formed in the icing wind tunnel. Rime ice, glaze ice and mixture of them (mixed) on steel plates and aerodynamic profiles [1].

Contact angle (CA) measurements were carried out by using distilled water and KSV CAM200 equipment. The measurements were performed in the controlled environment, where relative humidity was 50 % and temperature 23°C. A droplet volume of 5 µl was used for static contact angle measurements. At least five droplets were used for determining the static contact angle.

Surface properties (surface profile and Sa-values) were analyzed by Alicona Infinite Focus G5 optical profilometer. Sa-values were measured from areas as large as possible in order to properly assess the surface roughness. Surface roughness, Ra-values, before and after erosion tests were measured with Mitutoyo SJ-301 surface roughness tester.

Dry particle erosion wear behavior of the coatings was studied with a centrifugal erosion tester (Erosion test 1). This test was performed for the polished samples. Samples were attached in the 90° sample holders. Rotation speed was set to 6000 rpm, which corresponds 80 m/s impact velocity of the abrasives. Samples were exposed total 5 kg of dry quartz sand (SiO_2) with a grain size of 0.1–0.6 mm. In order to calculate mass loss, samples were weighted before and after tests.

The micro-sandblasting test (Erosion test 2) was conducted to assess the wear behaviour of flame sprayed coatings. The distance between the abrasive gun and the sample was set to 30 mm. The exposed area of the sample was 25 mm in diameter. Sample surfaces were sand-blasted with Al_2O_3 abrasive (-340 +170 µm) under at a 2.5 bar pressure for 120 s. Mass losses and roughness changes were measured.

3 Results and discussion

3.1 Ice adhesion

Ice adhesions of the flame-sprayed PE-based coatings are presented in **Fig. 3**. Ice adhesion of flame-sprayed UHMWPE, PE and PE+25FEP coatings was measured in the as-sprayed state. Surface roughness plays a role in ice adhesion and therefore, surfaces were also polished in order to avoid /diminish the effect of surface roughness and focus on the effect of coating material itself. Flame-sprayed UHMWPE coating had higher ice adhesion with higher standard deviation compared with PE and PE+FEP coatings although it was smoother in the as-sprayed state (**Table 1**). The general finding is that ice adhesion increases as the surface roughness increases. [19,21,22] which is also seen here in the comparison between the as-sprayed and polished coating surfaces. Polished flame-sprayed PE-based coatings had similar ice adhesion values, around 54 kPa, indicating good icing behaviour. In this study, FEP increment in the powder did not affect much due to the low amount of FEP in the coating structure in all cases.

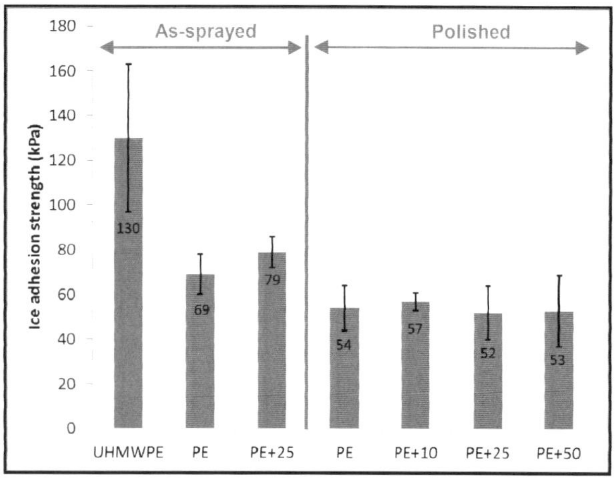

Fig. 3. Ice adhesion strengths (and standard deviations) of as-sprayed and polished flame-sprayed PE-based coatings.

Anti-icing behaviour of polished flame-sprayed PE-coatings was compared with the behaviour of metal sheets such as stainless steel and aluminium plates and reference PU-paint. Ice adhesions of these surfaces are presented in **Fig. 4**. Ice adhesion of flame-sprayed PE-coating was the lowest in this comparison. It was 5 times lower than the ice adhesion of stainless steel plate, 7 times lower than the ice adhesion of aluminium plate and 1.5 times lower than the ice adhesion of PU-paint. PU-paint had the second lowest ice adhesion in this study (3.3 times lower than stainless steel and 4.6 times lower than aluminium).

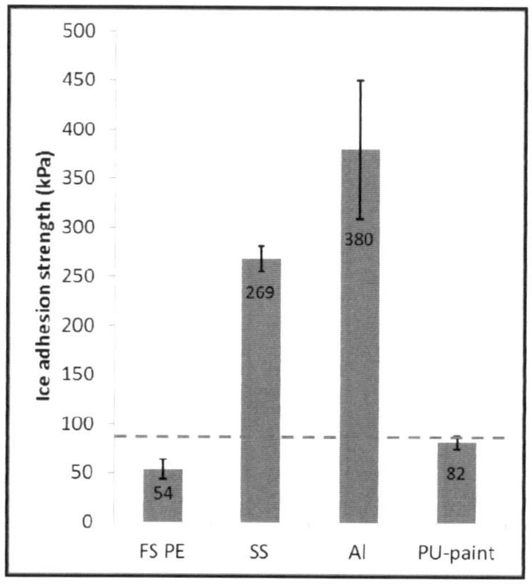

Fig. 4. Ice adhesion strengths (and standard deviations) of flame-sprayed PE (FS PE) coating, stainless steel (SS) bulk material, aluminium (Al) bulk material and PU-paint.

3.2 Surface properties

In addition to the surface roughness, wetting behaviour have been reported to have a relation with the icing behaviour of surfaces [1]. However, surface roughness has also great impact on the wetting characteristics. [22,23] so these factors are connected to each other. **Table 1** presents the contact angle, surface roughness and ice adhesion values of studied surfaces. Water contact angle values indicate the wetting behaviour. Hydrophilic surfaces possess contact angles below 90°. This means that the water will wet surface and penetrates into surface texture. When the water contact angle is over 90°, the surface is called hydrophobic. The wettability of reference metal surfaces, stainless steel and aluminium, is hydrophilic. These surfaces are widely used in outdoor applications so improvement of icephobicity is beneficial for these surfaces. Also the icing behaviour of aluminium is widely studied in literature [19,23-26]. The polished flame-sprayed PE-based coatings were hydrophobic whereas the as-sprayed coatings were slightly hydrophilic due to the surface roughness.

Table 1. Contact angle (CA), surface roughness (Sa) and ice adhesion of flame-sprayed PE-based coatings (as-sprayed and polished), reference metals and PU-paint.

Sample	Contact angle, CA (°)	Surface roughness, Sa (µm)	Ice adhesion (kPa)
FS UHMWPE (As-sprayed)	91	1.65	130
FS PE (As-sprayed)	79	2.05	69
FS PE+25 (As-sprayed)	85	5.44	79
FS PE (Polished)	97	0.641	54
FS PE+10 (Polished)	92	0.753	57
FS PE+25 (Polished)	91	1.13	52
FS PE+50 (Polished)	88	0.955	53
Stainless steel (SS, Polished)	61	0.230	269
Aluminum (Al, Polished)	66	0.255	380
PU-paint	79	2.24	82

It has been reported that different factors, such as surface roughness, wettability and surface energy, can have an influence on the ice adhesion strength (ice adhesion). For example, it has been proposed that the increasing water contact angle should decrease ice adhesion strength. The superhydrophopic surfaces have shown low ice adhesion values [27], but their performance cannot be guaranteed due to the frost formation on the surfaces and mechanical interlocking effect [23,25,26]. Therefore, a durability of the anti-icing surfaces is also a very important issue.

3.3 Wear behaviour

Flame-sprayed PE-based coatings have shown good anti-icing properties i.e. low ice adhesion values. In addition to that, the erosion behaviour of the coatings was evaluated. Two erosion tests were used: a dry particle erosion (traditional test for thermally sprayed coatings) and a micro-sandblasting test (modified for this study) in order to evaluate the coating performance and durability. **Table 2** presents the erosion test results. PU-paints have low erosion resistance and they worn out during both erosion tests. In the erosion test 1, PU-paint worn out after 1 kg test (total 5 kg) and therefore, the second erosion test (micro-sandblasting) was used. In the erosion test 2, erosive is smaller sized than in erosion test 1 but still PU-paint surface eroded totally and worn out.

Stainless steel surfaces eroded with a slight mass loss whereas flame-sprayed PE and PE+25FEP coatings slightly increased their mass in both erosion tests. Flame-sprayed coatings are softer and more elastic surfaces and therefore, sand particles impinged into the surfaces. The surfaces showed durability without erosion. However, the surfaces need to be optimized in order to avoid sand particle impingements in future. Also, Lima et al. [16] have reported the good corrosion behaviour and wear properties against abrasive wear for the flame-sprayed engineering polymer (PEI, PEEK, PA) coatings. The mass loss of stainless steel and the mass increment of flame-sprayed coatings were higher in the erosion test 2 due to the test setup (distance and pressure). Common for both erosion tests was that surface roughnesses (Ra-values) increased during the test. Roughness changes were higher in the erosion test 1 due to the larger erosive size. However, the roughness ratio (Ra_2/Ra_1) of stainless steel was higher than that of flame-sprayed coatings due to the different kind of wear behaviour of surfaces (softer vs. harder and/or elasticity vs. toughness).

Table 2. Dry particle erosion and micro-sandblasting erosion behaviour of flame-sprayed PE and PE+25FEP coatings and stainless steel (SS) and PU-paint references.

Erosion test 1: Dry particle erosion				
Sample	Ra_1 (before)	Mass loss (mg)	Ra_2 (after)	Ra_2/Ra_1
FS PE	0.34	-0.011	6.65	19
FS PE+25	0.67	-0.010	7.29	11
SS	0.04	0.031	3.36	88
PU-paint	1.00	-*	-	
Erosion test 2: Micro-sandblasting				
Sample	Ra_1 (before)	Mass loss (mg)	Ra_2 (after)	Ra_2/Ra_1
FS PE	0.34	-0.105	3.27	10
FS PE+25	0.67	-0.085	3.55	5
SS	0.04	0.160	1.23	32
PU-paint	1.00	-*	-	

*Worn out during the test

4 Conclusions

This study focused on the anti-icing behaviour of flame-sprayed PE-based polymer coatings. Icing can cause substantial problems for different industrial fields by decreasing the efficiency, safety and usability of operations. Therefore, new methods of preventing ice accretion on the surfaces or to diminish ice adhesion are welcome. Flame-sprayed PE and PE+25FEP coatings have shown the low ice adhesion values compared with stainless steel and aluminium metal surfaces indicating their potential to act as anti-icing surfaces. Polished flame-sprayed PE and PE+FEP coatings had lower ice adhesion values compared with corresponding as-sprayed coatings which indicate the effect of surface roughness together with wetting behaviour (polished had hydrophobic and as-sprayed hydrophilic wetting behaviour) on icing behaviour. Low ice adhesion reflects good anti-icing behaviour. Future research will be focused on improvement of coating performance in erosive conditions without losing anti-icing properties by avoiding erosive impingement.

5 Acknowledgements

The authors would like to thank M.Sc Henna Niemelä-Anttonen, of Tampere University of Technology, Department of Materials Science, for helping with the ice testing.

6 References

[1] Stenroos, C.: Properties of icephobic surfaces in different icing conditions, Master Thesis, Tampere University of Technology, Tampere, Finland, (2015), 108 p.

[2] Farzaneh, M.: Atmospheric icing of power networks. Springer, London, UK, (2008), 381 p.

[3] Kraj, A. and E. Bibeau: Phases of icing on wind turbine blades characterized by ice accumulation, Renewable Energy 35 (2010), 5, pp. 966/72.

[4] Fikke, S., J. Kristjánsson and B. Kringlebotn Nygaard: Modern meteorology and atmospheric icing, Atmospheric Icing of Power Networks, IWAIS XI, Montreal, Canada, (2005), pp. 1-29.

[5] Antonini, C.: Superhydrophobicity as a strategy against icing: Analysis of the water/surface dynamic interaction for icing mitigation. Università degli studi di Bergamo, (2011), 238 p.

[6] Safaee, A.: Nanostructured metal surfaces and their passivation for superhydrophoic and anti-icing applications, Université du Québec,(2008),182 p.

[7] Makkonen, L.: Ice Adhesion -Theory, Measurements and Countermeasures, Journal of Adhesion Science and Technology 26 (2012), pp. 413/45.

[8] Fortin, G. and J. Perron: Ice Adhesion Models to Predict Shear Stress at Shedding, Journal of Adhesion Science and Technology 26 (2012), 4-5, pp. 523/53.

[9] Cao, L., A. Jones, V. Sikka, J. Wu and D. Gao: Anti-Icing Superhydrophobic Coatings, Langmuir 25 (2009), 21, pp. 1244/48.

[10] Ryerson, C.C.: Assessment of Superstructure Ice Protection as Applied to Offshore Oil Operations Safety, ERDC/CRREL TR-09-4, (2009), p. 344.

[11] Kim, P., T.-S. Wong, J. Alvarenga, M.J. Kreder, W.E. Adorno-Martinez and J. Aizenberg: Liquid-Infused Nanostructured Surfaces with Extreme Anti-Ice and Anti-Frost Performance, ACS Nano 6 (2012), 8, pp. 6569/77.

[12] Jung, S., M. Dorrestijn, D. Raps, A. Das, C.M. Megaridis and D. Poulikakos: Are Superhydrophobic Surfaces Best for Icephobicity?, Langmuir 27 (2011), pp. 3059/66.

[13] Ryerson, C.C.: Ice protection of offshore platforms, Cold Regions Science and Technology 65 (2011), pp. 97-110.

[14] Winkler, R., F. Bűltmann, S. Hartmann and A. Jerz: Thermal Spraying of Polymers: Spraying Processes, Materials and New Trends, Thermal Spray 2003: Advancing the Science & Applying the Technology, C. Moreau, B. Marple (Eds.), ASM International, Ohio, USA, (2003), pp. 1635/38.

[15] Petrovicova, E. and L.S. Schadler: Thermal spraying of polymers, International Materials Reviews 47 (2002), 4, 22 pp. 169/90.

[16] Lima, C., N. de Souza and F. Camargo: Study of wear and corrosion performance of thermal sprayed engineering polymers, Surface & Coatings Technology 220 (2013), pp. 140/3.

[17] Vuoristo, P., E. Leivo, E. Turunen, M. Leino, P. Järvelä and T. Mäntylä: Evaluation of Thermally Sprayed and Other Polymeric Coatings for Use in Natural Gas Pipeline Components, Thermal Spray 2003: Advancing the Science & Applying the Technology, C. Moreau, B. Marple (Eds.), ASM International, Ohio, USA, (2003), pp. 1693/702.

[18] Gawne, D.T., Y. Bao and T. Zhang: Influence of Polymer Compostion on The Deposition of UHMWPE Coatings, Thermal Spray 2003: Advancing the Science & Applying the Technology, C. Moreau, B. Marple (Eds.), ASM International, Ohio, USA, (2003), pp. 1639/44.

[19] Yang, S., Q. Xia, L. Zhu, J. Xue, Q. Wang and Q. Chen: Research on the icephobic properties of fluoropolymer-based materials, Applied Surface Science 257 (2011), 1, pp. 4956/62.

[20] Koivuluoto, H., C. Stenroos, R. Ruohomaa, G. Bolelli, L. Lusvarghi and P. Vuoristo: Research on icing behavior and ice adhesion testing of icephobic surfaces, IWAIS XVI, June 28-July 3, (2015), Uppsala, Sweden, 6 p.

[21] Zou, M., S. Beckford, R. Wei, C. Ellis, G. Hatton and M. Miller: Effects of surface roughness and energy on ice adhesion strength, Applied Surface Science 257 (2011), 8, pp. 3786/92.

[22] Susoff, M., K. Siegmann, C. Pfaffenroth and M. Hirayama: Evaluation of icephobic coatings - Screening of different coatings and influence of roughness, Applied Surface Science 282 (2013), pp. 870/9.

[23] Bharathidasan, T., S. Kumar, M. Bobji, R. Chakradhar and B. Basu: Effect of wettability and surface roughness on ice-adhesion strength of hydrophilic, hydrophobic and superhydrophobic surfaces, Applied Surface Science 314 (2014), pp. 241/50.

[24] Menini, R. and M. Farzaneh: Elaboration of Al_2O_3/PTFE icephobic coatings for protecting

aluminum surfaces, Surface and Coatings Technology 203 (2009), 14, pp. 1941/46.
[25] S. Farhadi, M. Farzaneh, S. A. Kulinich, Anti-icing performance of superhydrophobic surfaces, Applied Surface Science 257 (2011), 14, pp. 6264/69.
[26] S. A. Kulinich, S. Farhadi, K. Nose, X. W. Du, Superhydrophobic surfaces: are they really ice-repellent?, Langmuir 27 (2011), 1, pp. 25-29.
[27] H. Dodiuk, S. Kenig, A. Dotan, Do Self-cleaning Surfaces Repel Ice ?, Journal of Adhesion Science and Technology 26 (2012), 4-5, pp. 701/14.

Investigation of Influencing Factors on the Transfer of Wire Arc Sprayed Zn Coatings for the Metallisation of Plastic Parts

K. Bobzin, M. Öte, M.A. Knoch, X. Liao, Surface Engineering Institute (IOT), Aachen / D
Ch. Hopmann, P. Ochotta, Institute of Plastics Processing (IKV), Aachen / D

In a current study, a new approach for the efficient production of metallised plastics parts which can be used for providing electric conductivity, electromagnetic shielding and decorative usage, the so called In-Mould-Metal-Spraying (IMMS), is being developed. A Zn coating is at first thermally sprayed onto mould cavity inserts (carrier body) using wire arc spraying and subsequently transferred onto a plastics carrier via injection moulding. Among others, a finely adjusted bonding strength between carrier body and coating is essential for a successful coating transfer. The influence of the carrier body and the surface pretreatment on the bonding strength at the coating/carrier body interface and subsequently on the transferability of the coatings was investigated. Carrier bodies made of three different steel types, pretreated by glass bead blasting and electrical discharge machining (EDM), were tested. The surface topography was characterised using confocal laser scanning microscopy. In the first series of experiments, the hardnesses and surface topographies of the carrier bodies were identified as possible influencing factors of coating transferability. Further experiments focusing on the bonding strength of the Zn coating to hardened and annealed carrier bodies were carried out to verify the influence of the carrier body hardness on the bonding strength.

1 Introduction

Parts made of a single type of material may not be able to fulfill all requirements under all the relevant circumstances due to the increasing complexity of the requirements from the operating conditions on them. Therefore, the use of so called "tailored design", which means the use of hybrid parts made of different types of materials, has been gaining more importance in practice. One of the major challenges related to this subject is the development of efficient production methods for manufacturing hybrid parts. Particularly for metal/plastics hybrid parts, nowadays there exist various concepts. By means of the well established In-Mould-Assembly (IMA) and Post-Mould-Assembly (PMA) the metallic part and the plastic part are joined during or after injection moulding [1-2]. Both methods require additional process steps for the manufacturing of metal parts and further handling steps, for example to place the metallic parts into the mould in case of IMA. The application of metallic coatings onto plastics parts with various coating technologies, such as physical vapour deposition (PVD), electroplating and thermal spraying, is the subject of research activities in literature. With the PVD technology, thin high-quality coatings can be applied on plastics parts [3]. However, the part size is restricted by the size of the vacuum chamber of the PVD equipment. In case of electroplating, on the other hand complex and elaborate procedures, such as activation or chemical pre-metallisation of the part surface, are required. Although the feasibility of a direct application of metallic coatings onto plastics parts by means of thermal spraying has been proven, warpage and polymer degradation due to the high-temperature particles and process gas might be challenging [4-6].

Considering the above mentioned characteristics of different metallization methods as well as their advantages and disadvantages, a new approach, the so called "In-Mould-Metal-Spraying" (IMMS), is being developed. The IMMS is based on the transfer of the thermally sprayed coatings from the mould insert (carrier body) to the plastics parts within the injection moulding process. The details of the process have been published [6-8]. One of the key factors, in order to achieve a successful coating transfer, is the adjustment of the coating bonding strength to the carrier body. It must be above a certain level to enable the coating buildup and the attachment required for handling operations. Furthermore, the coating must withstand the shear strain induced by the overflow of the viscous plastics melt during injection moulding. On the other hand, a high coating bonding strength to the mould insert would hinder a successful transfer of the coating to the plastics parts during demoulding.

It is well accepted that mechanical interlocking is the most important bonding mechanism for thermally sprayed coatings [9]. Therefore, a surface pretreatment prior to the coating process to roughen the substrate surface is usually necessary. In practice, grit blasting with a coarse blasting medium is often used for this purpose, as a surface topography with high roughness and severe undercuts can be fabricated by this method [9]. In case of IMMS, the roughness of the carrier body surface must however be kept within finely adjusted borders. Otherwise, either coating buildup or the transfer to the plastics part cannot be achieved. Therefore, one of the aims of this research work is finding a suitable surface pretreatment for a roughening of the carrier body.

2 Feasibility study

The focus of the feasibility study is finding a suitable surface pretreatment for the realization of the coating transfer. For this purpose, milled carbon steel 1.1730 (AISI SAE 1045) carrier bodies were used. The following methods were examined as surface pretreatment for the carrier body:

- Grit blasting with corundum F150
- Hand grinding with 180 grade abrasive paper
- Hand grinding with 400 grade abrasive paper

The corundum F150 has a particle size distribution of -106 +63 µm. By using this fine blasting medium, a low surface roughness can be achieved. The grinding with abrasive paper also allows the fabrication of a relatively smooth surface. The Ra values of the carrier bodies were measured by means of confocal laser scanning microscopy (CLSM) after the pretreatment. The grit blasted carrier body exhibits the highest Ra value of 4.1 µm, while the Ra values of the carrier bodies ground with 180 and 400 grade abrasive paper are 1.9 and 1.7 µm respectively. After the surface pre-treatment, Zn coatings were applied onto the carrier bodies by means of wire arc spraying using the process parameters shown in **Tab. 1**.

Tab. 1. Parameters for application of Zn coatings via wire arc spraying

Parameter	Value
Stand-off distance	150 mm
Current	150 A
Voltage	20 V
Air pressure	4 bar
Traverse velocity	500 mm/s

After coating deposition, the mould inserts were positioned in the mould cavity, and the coatings were back moulded with glass fibre reinforced polyamide 6 (PA6 GF30) from Lanxess Deutschland GmbH, Cologne, Germany. The coating on the grit blasted carrier body could not be completely transferred onto the plastic part. Coating residuals remained on the carrier body. The carrier body ground with 180 grade abrasive paper on the other hand enabled a complete coating transfer, see **Fig. 1**. Furthermore, on a carrier body which was ground by an abrasive paper with 400 grade grains, coating buildup could not be realized. This can be attributed to the smaller grain size of the abrasive particles of 400 grade paper than that of 180 grade abrasive paper, leading to a smoother surface topography of the carrier body resulting in a lower coating bond strength.

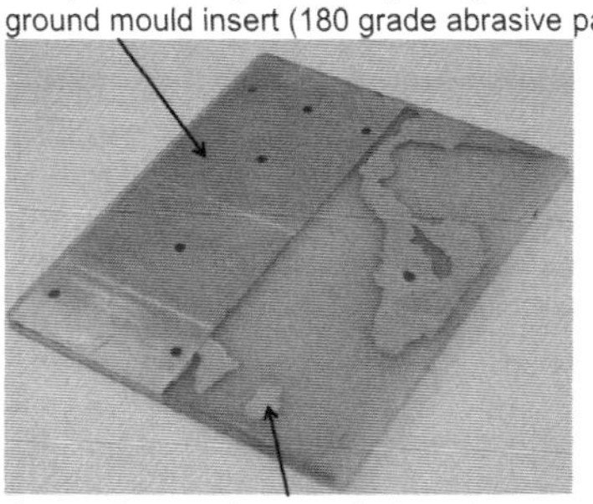

Fig. 1. Coating transfer using hand ground and grit blasted mould inserts

The results above have shown the significance of surface pretreatment for realising a sufficient coating transfer. Grit blasting with corundum generates surface topo-graphies with various undercuts as the angular blasting medium can penetrate deeply into the surface. This leads to a high coating adhesion. Although the trials with a ground carrier body were successful, the efficiency and reproducibility of grinding is pointed out to be limited. In further investigations, blasting with glass beads was employed as a pretreatment method. Due to the spherical shape of the glass beads, a lower penetration depth into the mould insert surface is expected, leading to lower roughness and reduced undercuts induced by the blasting. Glass beads GP90 with a particle size distribution of -150 +90 µm, which is similar to corundum F150, were used. Coatings applied onto a glass bead blasted carrier body could be completely transferred as it was the case with a carrier body ground with 180 grade abrasive paper.

3 Influence of the mould insert on the IMMS process

The wire arc sprayed Zn coating has shown good transferability in case of a carrier body made of carbon steel 1.1730, when suitable surface pre-treatment is used. The carbon steel 1.1730 is usually used for mould frames and add-on parts. In practice, high-strength steels are used for manufacturing mould inserts. Furthermore, the mould inserts may undergo different heat treatments, which lead to different properties, especially regarding the hardness. Therefore, the manufacturing method of mould inserts may influence their performance regarding coating transferability.

3.1 Coating bonding strength on hardened steel

In order to investigate the above mentioned effects, flat samples of tool steel 1.2842 (AISI O2) were used. The samples were annealed at their delivered condition and had a Vickers hardness of about 200 HV30. Six flat samples were used and categorized according to their hardness:

- Group 1: low hardness
- Group 2: moderate hardness
- Group 3: high hardness

The two samples of the group 1 remain at their delivered condition. The samples of group 2 and 3 were hardened to a hardness of about 300 HV30 and 790 HV30 respectively. One sample of each group was blasted with GP90, the other sample of each group was blasted using GP150 which has a particle size distribution of -250 +150 µm. A blasting pressure of 5 bar and nozzle stand-off distance of 100 mm were used during the blasting of all samples.

By means of CLSM, different sample surface topographies have been observed. In case of samples in group 1 and 2, the semi-spherical craters caused by the impact of glass beads can be recognized on the sample surface. The surface topographies of these samples are formed as a result of overlapping of various craters. For samples in group 3, the glass beads can barely penetrate the surface. The samples' surface remains mostly in its original state, only a few small craters can be found. The surface topographies of samples in different groups after glass bead blasting with GP150 are shown in **Fig. 2**.

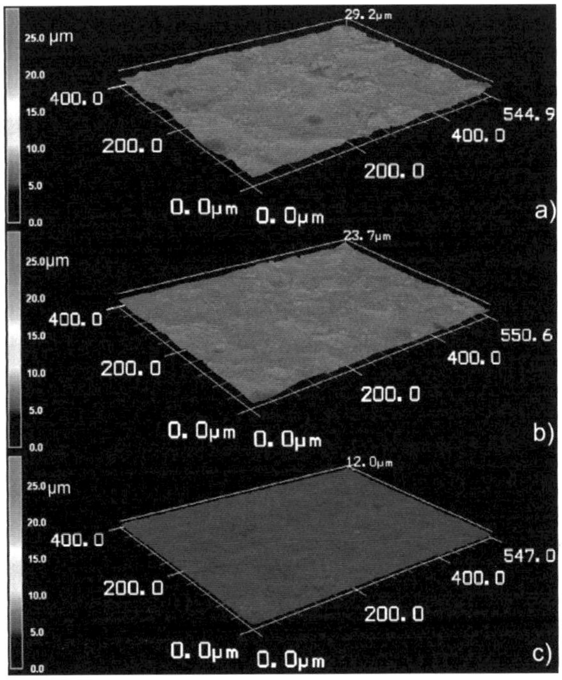

Fig. 2. Surface topographies of the samples in: a) group 1, b) group 2 and c) group 3, after glass beads blasting with GP150

The bonding strength of samples coated using the process parameters shown in **Tab. 1** were measured by means of PAThandy test. For this test method, a test dolly is firstly glued to the coating, and the coating around the sample is removed afterwards. The test dolly is then fixed in the test equipment with pressure cylinders. The cylinders press the sample with increasing force in the opposite direction so that the test dolly glued onto the coating is pulled away. The bonding strength is calculated by dividing the final force by the interface area between the test dolly and the coating. Although the removal of the surrounding coating can be realised for conventional coatings systems easily, this operation cannot be executed directly for coating systems which are intended to be transferred subsequently. Due to weak adhesion of the coating to the glass bead blasted samples, the test dolly detaches while removing the surrounding coating. Furthermore, the force needed to pull off a test dolly might be too low to be detected by the PAThandy equipment. In order to overcome this problem, a modification of the test method was implemented, see **Fig. 3**.

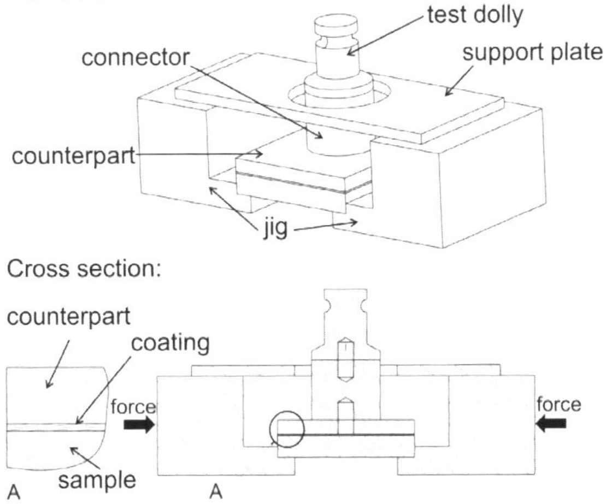

Fig. 3. Modification of the PAThandy measurement

A counterpart which has the same dimensions as the sample is glued to the sample, so that the removal of the surrounding coatings can be omitted. Moreover, the interface between the counterpart and the sample is larger than the area of the test dolly bottom surface. Therefore, the necessary pull-off force is increased in order to make sure that it can be measured by the PAThandy equipment. The test equipment can only fix test dollies with a certain geometry, therefore, the counterpart has to be connected to a standard test dolly through a connector. In the conventional test setup, the pressure cylinders of the test equipment press on the area of the sample, which is not covered by the test dolly. As the sample is completely covered by the counterpart in this modified setup, it must be fixed by two jigs. An additional support plate is used in order to apply the pull-off force to the counterpart.

With this measurement setup, the coating bonding strengths of the samples could be determined. The comparison of the surface roughness and coating bonding strength is shown in **Fig. 4**. As expected, roughness values decrease with increasing substrate hardness. Further-more, the use of coarser glass beads leads to higher Ra values. This effect can be attributed to the higher kinetic energy of the larger glass beads. On the samples in group 3, no coating buildup was possible. Except for the GP90 blasted sample in group 2, the coating bonding strengths correspond to the trend of Ra. The coating adhesion to the softer samples is in general higher than that to the harder ones. Moreover, higher coating bonding strength can be achieved in case of samples blasted with coarser glass beads. The results show that the surface pretreatment must be carefully adjusted with regard to the properties of the carrier body. Using a coarser blasting medium for a carrier body with high hardness seems to be promising in order to achieve a successful coating transfer. The increase of blasting pressure can also be considered for this purpose. Limited by the maximum power of the used blasting equipment, the latter hasn't been tested within the scope of this study.

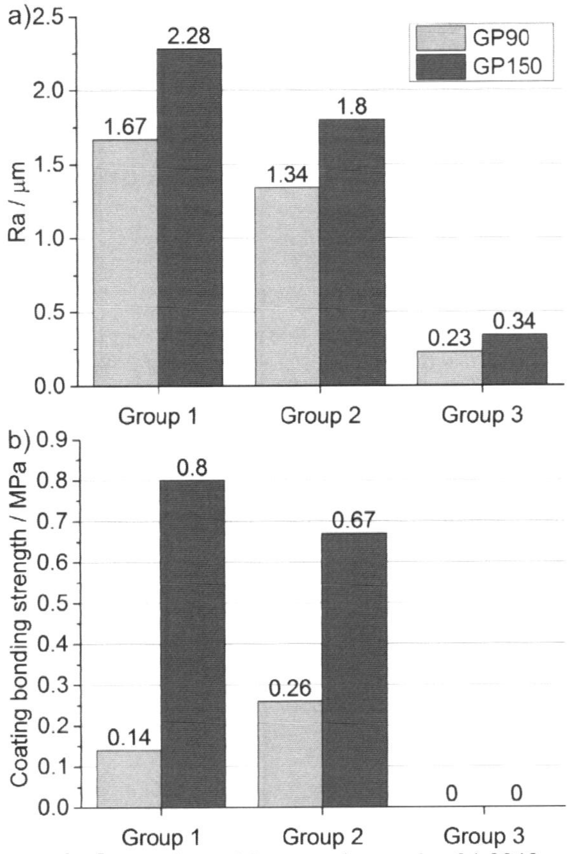

Fig. 4 Comparison of the samples made of 1.2842: a) Roughness; b) Coating bonding strength

3.2 Transfer experiment with mould insert manufactured by EDM

Mould inserts can not only be manufactured by milling but also by electrical discharge machining (EDM). The mould insert manufactured by EDM exhibits already a rough surface after manufacturing. This sample will hereafter be called EDM-sample. In order to investigate the possibility of realising a coating transfer with this kind of mould insert as carrier body for the coating, the following experiments were conducted. The tested carrier bodies are made of tool steel 1.2312 (AISI P20+S). A glass bead (GP150) blasted flat 1.2312 sample without EDM processing was used as a reference. The blasted sample and the EDM-sample exhibit hardness values of 298 HV30 and 340 HV30 respectively. These values are similar to the hardness of the samples from 1.2842 in the group 2 mentioned in section 3.1. The Ra value of the blasted 1.2312 sample is 2.02 µm, which is also similar to the corresponding sample in section 3.1. The measured Ra value of the EDM-sample is 4.50 µm, which means the EDM carrier body has a rougher surface than the glass bead blasted flat sample. After blasting the EDM-sample with GP150, the Ra value is reduced to 3.56 µm. However, a surface topography correspond-ing to overlapping impact craters of glass beads could not be observed, see **Fig. 5**.

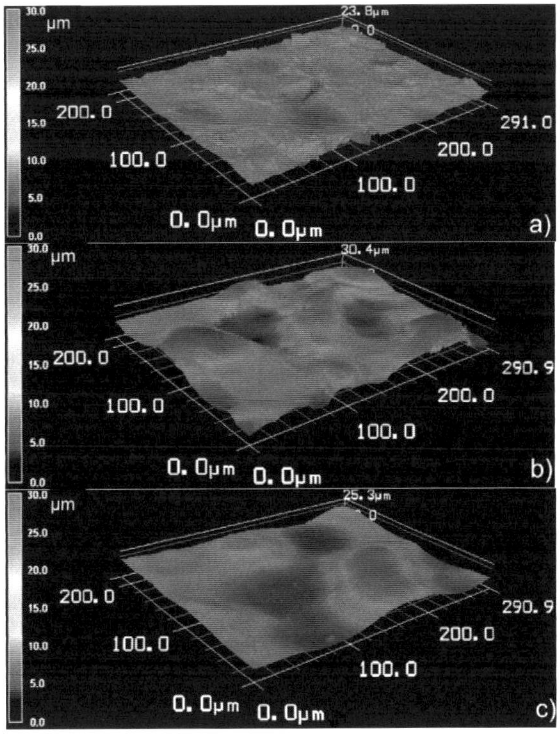

Fig. 5. Surface topographies of sample and mould inserts made of 1.2312: a) glass beads blasted flat sample; b) EDM-sample; c) glass beads blasted EDM-sample

A coating bonding strength of 0.67 MPa to the blasted sample could be measured using the modified PAThandy setup. No coating buildup was possible on the carrier bodies, although their Ra values are higher than that of the blasted sample. By means of hardness measurement with lower test force (HV0.1), a very hard surface layer with a hardness of 903 HV0.1 was found. This hard surface layer presumably prevents the penetration of the glass beads. Therefore, a surface topography similar to the blasted 1.2312 sample could not be achieved on the EDM carrier body. The manufacturing process has consequently an indirect influence on the coating bonding strength. Based on the fact that no coating buildup is possible on the rough surface generated by EDM, it is assumed that the high surface hardness might also have a direct influence on the coating bonding strength. In order to confirm this assumption, further investigations are conducted.

One requirement of the investigations is fabricating two groups of samples, whose main difference is their hardness. Samples made of carbon steel 1.1730 were firstly hardened and then grit blasted using same blasting parameters and corundum F16, which has a particle size distribution of -1,400 +1,000 µm. The first group of the samples, G1, including sample S1 and S2, is kept in this condition, while the second group of samples, G2, including sample S3 and S4, was tempered in vacuum. As a result, a similar surface topography can be achieved on all samples and it remains unchanged after tempering the samples. After coating deposition, the coating bonding strength was measured by means of PAThandy. During the removal of the coating around the test dollies, some of the substrate material was removed as well, in order to make sure that the test dollies are only joined to the coating directly underneath it. Due to the high hardness of the samples, a special cutting tool was used. Limited by the available cutting tool size, test dollies with a diameter of 8.16 mm were used. The samples' surface roughnesses at the area directly underneath the test dollies are more relevant to the coating bonding strength than the samples' average surface roughness. Therefore, before coating deposition, the samples' surface topographies were characterized using CLSM measurements at three measurement points per sample, which were also the measurement points for PAThandy test after coating deposition. In the small area underneath the test dollies, the required measurement length for Ra according to DIN EN ISO 4288 cannot be fulfilled. Therefore, the dimension of the measurement points for surface roughness was chosen as 3 x 3 mm. The arithmetic mean height of the measuring area, Sa, was chosen as the indicator for the surface roughness instead of Ra.

The comparison of roughnesses and hardnesses of the samples before and after the tempering process is shown in **Fig. 6**. The samples all exhibit hardness values between 600 and 700 HV30 after hardening. After tempering at 500 °C for 3 hours, the hardnesses of sample S3 and S4 are reduced to 382 HV30 and 371 HV30 respectively. After the tempering process, the roughnesses of S3 and S4 were measured once again at the same measuring points. No evident changes of the Sa values could be determined, which means the surface topography has not changed during the tempering process and the prerequisite for the further investigation is fulfilled.

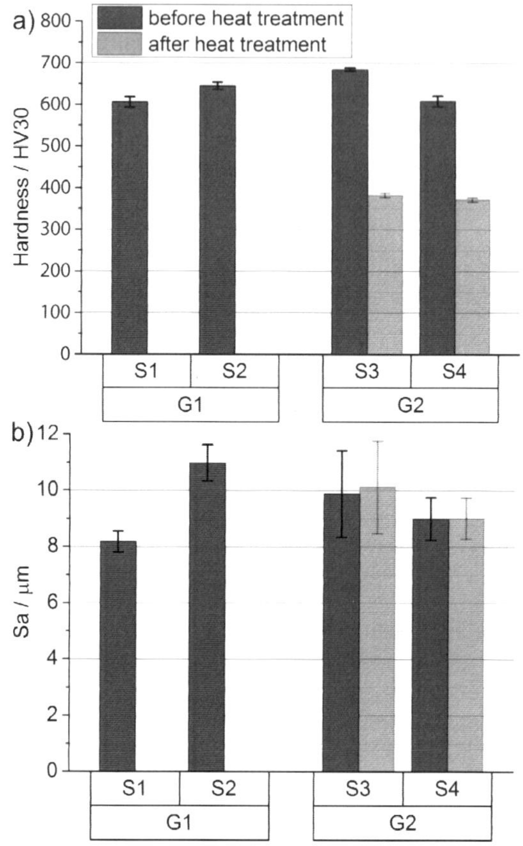

Fig. 6. Comparison of the samples before and after tempering in vacuum: a) Hardness; b) Sa

Zn coatings were applied onto all samples using the same process parameters. As mentioned above, coating bonding strength was characterised by means of PAThandy test. The average coating bonding strength of S3 and S4 is 4.7 MPa, which is about 24 % higher than the average bonding strength of S1 and S2. Furthermore, although sample S2 has the highest Sa value, the average coating bonding strength to S2 is still lower than to S3 and S4, see **Fig. 7**.

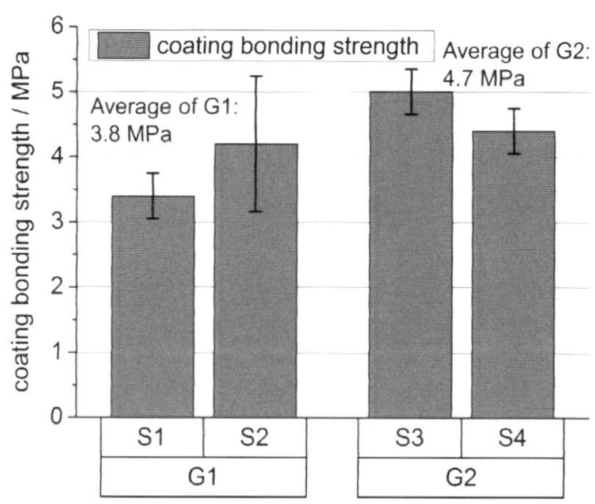

Fig. 7. Coating bonding strength on hardened and tempered samples

Based on the results above, it can be stated that the hardness seems to have a direct influence on the coating bonding strength. Further investigations are still needed to confirm this result. One assumption for the mechanism of this phenomenon is the different impact behaviour of molten particles on the hard and soft samples. The molten particles may undergo more severe splashing upon impact with a harder surface. Therefore, the coating material cannot penetrate into the roughness profile of the surface well, which leads to a weaker bonding between the coating and the sample. Due to the existence of the hard surface layer on the carrier body manufactured by means of EDM, this kind of carrier body may not suitable for the IMMS process. In order to confirm this assumption, further investigation, such as the splat-test, will be conducted in order to observe the particle morphology after impact on hard and soft samples.

4 Summary and outlook

The feasibility study shows that the transfer of thermally sprayed coating onto plastics parts during injection moulding is possible. The key factor for this is the fine adjustment of the coating bonding strength to the carrier body by choosing a suitable surface pretreatment method. The glass bead blasting could be characterised to be a suitable method.

Further investigations were carried out to investigate the influence of the carrier body on the coating transferability. Differently treated samples and carrier bodies made of different steel types were therefore tested regarding the bonding strength of the coatings. The hardness of the material significantly influences the results of the blasting process and thus the roughness value of the blasted surface. Therefore, the blasting para-meters should be adjusted regarding the hardness of the carrier body material. The investigations have also shown that the carrier body hardness might have a direct influence on the coating bonding strength. Further investigation is needed to understand the mechanism of this phenomenon.

In the current study, Zn coatings were chosen as the metallic coating for metallisation of plastics parts. The feasibility of applying other feedstock materials, e.g. Al and Cu, will be investigated in future work. Not only the carrier body material and hardness, but also further properties, e.g. geometry of the carrier body, can affect the coating transferability, which needs further investigation. Besides the carrier body, the process parameters of wire arc spraying, e.g. the current and the stand-off distance may also be influencing factors for the coating bonding strength and transferability. This aspect belongs also to the future research topics.

5 Acknowledgements

The authors would like to thank the German Research Foundation DFG for the support of the depicted research within the Cluster of Excellence "Integrative Production Technology for High-Wage Countries".

6 Literature

[1] Paul, H., Luke, M. and Henning, F.: Polymer Metal Hybrid Structures - Experimental Investigations of Deformation and Failure Behavior. Journal of Plastics Technology 10 (2014), Issue 4, pp. 118-141

[2] Endemann, U., Glaser, S. and Völker, M.: Kunststoff und Metall im festen Verbund. Kunststoffe 92 (2002), pp. 110-113

[3] Navinšek, B., Panjan, P. and Milošev, I.: PVD Coatings as An Environmental Clean Alternative to Electroplating and Electroless Processes. Surface and Coatings Technology 116-119 (1999), pp. 476-487

[4] Liu, A., Guo, M., Zhao, M. and Hu, M.: Arc Sprayed Erosion-Resistant Coating for Carbon Fiber Reinforced Polymer Matrix Composite Substrates. Surface and Coating Technology 200 (2006), Issue 9, pp. 3073-3077

[5] Robitaille, F., Yandouze, M., Hind, S. and Jodoin, B.: Metallic Coating of Aerospace Carbon/Epoxy Composites by the Pulsed Gas Dynamic Spraying Process. Surface and Coating Technology 203 (2009), Issue 19, pp. 2954-2960

[6] Bobzin, K., Öte, M., Linke, T.F., Liao, X., Hopmann, Ch. and Ochotta, P.: Metallisierung von Kunststoffoberflächen – ein neues Verfahren: „In-Mould-Metal-Spraying (IMMS)". Jahrbuch Oberflächentechnik 71 (2015), pp. 67-73.

[7] Bobzin, K., Öte, M., Linke, T.F., Liao, X., Hopmann, Ch. and Ochotta, P.: Application of Thermal Spraying for the Manufacturing of Metal/plastic components. Thermal Spray Bulletin 67 (2015), Issue 1, pp. 28-31

[8] Hopmann, Ch., Bobzin, K., Schoeldgen, R., Wunderle, J., Linke, T.F. and Ochotta, P.: IMKS and IMMS: Two Methods for the Production of Plastic Parts Featuring Metallic Areas. Journal of Polymer Engineering, (2014), online

[9] Crawmer, D.: Coating Structures, Properties, and Materials. In: Tucker RC (ed) ASM Handbook, Volume 5A, Thermal Spray Technology. ASM International (2013), Materials Park, Ohio, USA, pp. 24-60

Improved Reliability and Specific Design of Plasma Spray Processes

G. Mauer, K.-H. Rauwald, R. Mücke, R. Vaßen Jülich / D

Improving of process reliability is a prevalent issue in thermal spray technology. It is closely related to process design since both are intended to achieve specific quality characteristics by controlling the process. For this, implicit approaches are in demand to rapidly conclude on relevant coating properties in-situ instead of measuring them explicitly and ex-situ. Such approaches can be based on extrinsic as well as on intrinsic process variables having specific advantages and disadvantages, each.
In this work, the effects of these plasma spray process variables are systemized in process schemes. On this basis, different approaches to contribute to improved process reliability are described and assessed paying particular attention to in-flight particle diagnostics. Furthermore, process design by following such systematics reversely is discussed with respect to benefits and limitations. Finally, a new test applying spray bead analysis is introduced and first results are presented.

1 Introduction

Reliability basically means that a system operates as expected over a specified timeframe. For a thermal spray system, reliability corresponds to the probability of manufacturing a product successfully and consistently with respect to specific requisite service properties. As such, this term covers all aspects of repeatability and reproducibility [1]. It is a prevalent issue in thermal spray technology.
Reliability can be improved by monitoring specific process variables allowing conclusions on the coating service properties. In principle, corrective actions can be identified by applying such relationships inversely to control the process variability. And process parameters for new applications can be designed correspondingly.
In this work, the relationships between process parameters and coating properties are systemized and discussed with respect to their compatibility with approaches to improve process reliability and process design. A new supplementary test is introduced based on spray bead analysis.

2 Systematics of process variables

A basis for this work is a *process scheme* for plasma spraying systemizing process variables and relations which have an impact on the coating quality. Several examples can be found in the literature, e.g. in [2, 3, 4]. **Fig. 1** shows the first stage of the process scheme established in this work. It is continued to the right by the second stage scheme given in **Fig. 2**.
In principle, the relevant coating properties can be ascertained after finishing the process (ex-situ) *explicitly* on the sprayed part. However, to ensure specific quality characteristics already during the process, relevant variables have to be identified which can be monitored in-situ. This means that the coating properties are observed *implicitly*; i.e. it is assumed that the monitored variables are actually suggestive for the quality characteristics of the coating. They are predictive only if being significant, adequate, and sensitive enough. For the review of process parameters in this regard, several statistical methods are available to design and evaluate experimental plans [5].

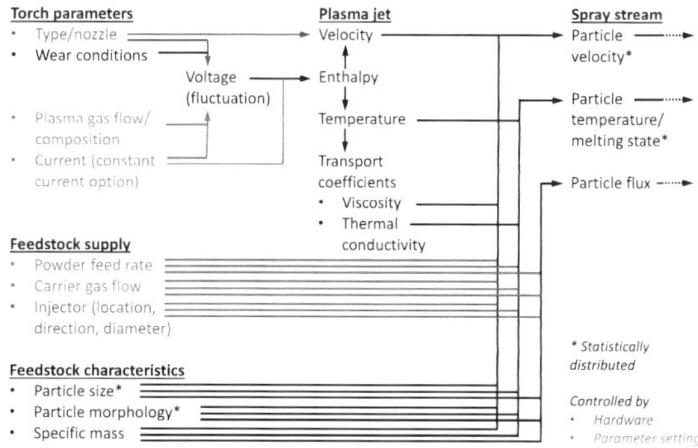

Fig. 1. First stage plasma spray process scheme.

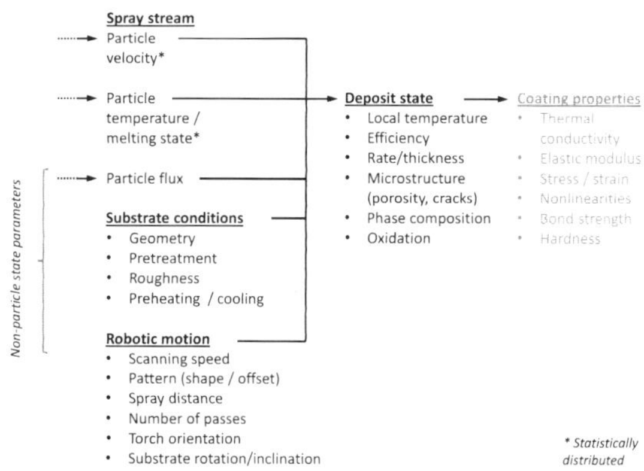

Fig. 2. Second stage plasma spray process scheme.

Heimann proposed a process scheme with a three-tiered hierarchy of parameters relating *extrinsic* ones and *intrinsic* ones [6]; first level: plasma spray system settings, second level: temperatures of plasma, particles, and substrate, third level: particle velocity, degree of melting, and degree of bonding. Although this structure differs partly from the scheme used in this work, it is apparent that intrinsic parameters are most influential as they are closely linked to the deposit state and bundle a variety of single effects.

3 Improving process reliability

Monitoring process parameters which are representative for coating quality is a key issue of any approach to improve process reliability. Depending on the category of the applied parameters, different approaches can be defined.

3.1 Implicit-extrinsic approaches

In this work, the torch parameters, the feedstock characteristics, the substrate conditions, and the robotic motion are designated as extrinsic operating parameters acting as input variables on the coating system, Fig. 1 and Fig. 2.

Monitoring spray parameters. Modern control units for plasma spray provide the possibility to log actual process data in real time and to detect deviations with respect to target tolerance bands. This refers in particular to extrinsic parameters (torch current, plasma & carrier gas flows, and powder feed rate). In Fig. 1, they are pooled in the group of torch parameters. It is obvious that they are detected distantly from the formation of the deposit state and do not comprise all impacting variables.

Feedstock characterization. Particle size and morphology are significant, statistically distributed feedstock properties. They have direct impact on the spray stream characteristics, Fig. 1, and may be subjected to variations. E.g. abnormalities in the particle size distribution can cause significant deviations in the deposition rate and coating porosity. In Forschungszentrum Jülich, whenever a new powder batch is started, it must undergo a standard characterization program, is designated individually and kept separately. The current batch in use must be documented for each spray run.

Monitoring substrate conditions. The most important substrate conditions impacting the deposition state are the geometry, pretreatment, roughness, preheating, and cooling, Fig. 2. Due to the variety of substrate geometries, manufacturing processes, and materials, general rules can hardly be established on how to ensure reproducible preconditions for coating. However, as substrate temperature has a significant impact on the deposit state, it should be monitored. In many cases, a pyrometer is applicable. Limited time and local resolution, possible hiding of the substrate by the torch, and unknown emissivities are issues to be considered in this context.

Verification of robotic motion. The parameters of robotic motion can be found in the process scheme in Fig. 2. While it is trivial to control the spray distance and the number of coating passes, this is rather demanding for the relative movement of torch and substrate. As the acceleration capacity of robots is always limited, the resulting restrictions for the scanning speed and path precision should be investigated at least initially to verify the system and learn about its limitations. **Fig. 3** gives two path-velocity diagrams for the horizontal direction of the meander-like motion of a six-axis robot at two different speed targets. They were obtained by two-fold integration of the torch accelerations measured by a calibrated 3D accelerometer (PCB Piezotronics Inc., Depew, NY, USA) which was mounted on the torch. The measurement range is ± 2 g at a frequency response of 0 to 250 Hz.

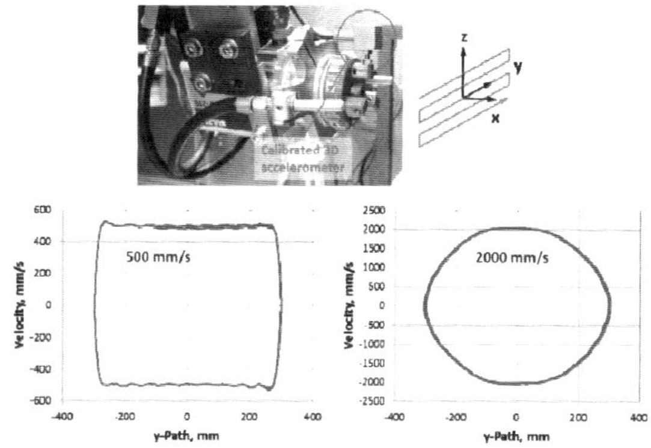

Fig. 3. Path-velocity diagrams for the horizontal (y-) direction of the a meander-like motion of a six-axis robot, meander width ±300 mm, speed targets 500 mm/s and 2000 mm/s.

It is obvious that there are large deviations from the target velocity at the beginning and end of the horizontal y-paths in case of the 2000 mm/s case. Large overruns are needed for acceleration and retardation to ensure a constant traveling velocity along the substrate. This must be known and taken into account when programming the robot.

3.2 Implicit-intrinsic approaches

Monitoring of extrinsic variables can be supplemented or replaced by the observance of intrinsic ones. They are associated with process sub-steps [1]. From Fig. 1 and Fig. 2 it is apparent that the development of the torch voltage, the plasma jet properties, and the spray stream characteristics impact the deposit state by increasing directness the more they are found downstream in the process scheme. The following approaches are based on such intrinsic parameters.

Voltage fluctuation analysis. The varying thickness and temperature of the cold-gas boundary layer between the arc and the anode nozzle wall are strongly correlated to the time-dependent voltage waveform [7]. Fig. 1 shows that these arc fluctuations are influenced by process settings (plasma gas flow and composition, current) as well as by wear of the torch electrodes [8] as this leads to a thinner boundary layer and a reduced motion of the arc root. As a consequence, the in-flight characteristics of the injected feedstock particles are affected [9]. Larger jet fluctuations can lead to a higher content of partially or unmelted particles and thus to increased porosities and lower deposition rates [10].
In principle, analyzing the torch voltage fluctuations could allow investigating the arc performance. In [11], correlations between torch power, arc root fluctuations and in-flight particle characteristics were identified.

However, these results were assessed to be still not sufficient for an accurate control of the spray process. Voltage frequency analyses in [12] revealed variances of the peaks between 5 kHz and 11 kHz when operation conditions (plasma gases, current) were changed or when the anode was eroded. Hence, such frequency analyses make only sense at constant torch parameters. However, they were suggested to give just slight response on increasing electrode wear [13]. Mainly final catastrophic erosion might be detectable reliably on this basis.

Power monitoring. In [11], it is shown that the time-dependent evolutions of the mean voltage exhibited a steady moderate decrease followed by a sharp drop and unstable course after fatal erosion had occurred [14]. As mentioned above, modern control units for plasma spray provide the possibility to log the development of actual process data. Thus, the torch power which is equivalent to the mean torch voltage at constant current operation can be monitored easily. **Fig. 4** gives the developments of the mean torch power and deposition rate during processing a batch of 38 aero-engine components at Forschungszentrum Jülich. Within this batch, the number of spray passes per part had to be increased from 32 to 34 to maintain the required coating thickness. The torch power dropped distinctly by 4%. Catastrophic electrode wear was still not visible. These results suggest that the torch power could be an apparently sensitive indicator to monitor the torch condition.

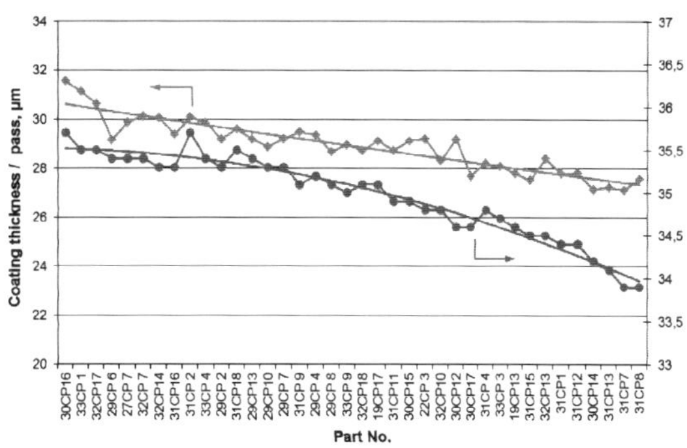

Fig. 4. Torch power and deposition rate per pass during processing a batch of 38 aero-engine components.

Fig. 5 gives torch power and deposition rate for another batch with 33 aero-engine components while depositing the first coating of a double layer system. **Fig. 6** shows it during deposition of the second layer. These two layers were deposited alternately. After spraying five parts, it was decided to replace the torch electrodes. Thereafter, the responses of the torch power and the deposition rate were ambiguous. While the torch power increased in both cases, the deposition rate gained with powder #1 but dropped spraying powder #2.

These examples show that power monitoring is of limited suitability for process monitoring. Furthermore it must be noted that these examples were obtained with a single cathode F4 torch. In contrast, modern three cathode torches show significantly smaller variations in torch power and a more stable operation at longer lifetimes.

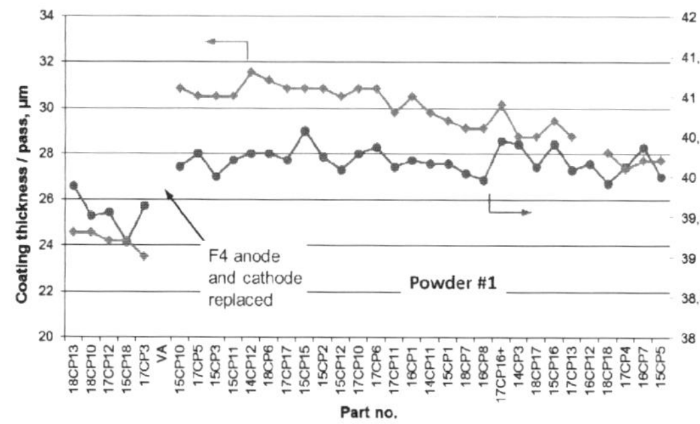

Fig. 5. Torch power and deposition rate per pass during processing a batch of 33 aero-engine components (1st coating layer).

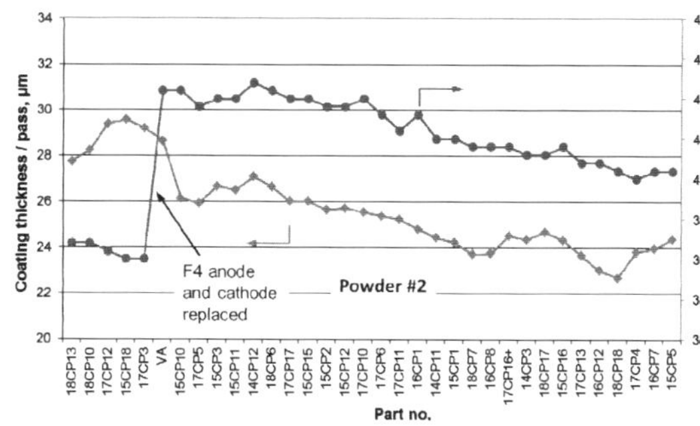

Fig. 6. Torch power and deposition rate per pass during processing a batch of 33 aero-engine components (2nd coating layer).

In-flight particle diagnostics and first order process maps. Fig. 2 shows that the spray stream variables of particle velocity and temperature have a direct impact on the deposition state. They are statistically distributed as particle size and morphology are. They can be measured in-flight by diagnostic systems [15, 16].

First order process maps link extrinsic parameters with intrinsic spray stream variables [17]. **Fig. 7** gives an example. The significance of the effects of specific parameters can be tested by statistical methods, e.g. by analysis of variances.

Some issues of first order process maps must be noted: The particle state does not fully describe the deposit formation since non-particle state parameters (feedstock flux, substrate conditions, torch motion) are not covered [18], see Fig. 1 and Fig. 2. Furthermore, in [19] it is reported that particle conditions contain only little information on process reproducibility as they tend to be not sufficiently sensitive. This relativizes some more general statements expressing that particle velocity and temperature explain the coating microstructure well [20] and can readily replace the influence of the spray gun parameters (current, plasma gas flow and composition, powder feed rate) [21].

In fact, the examples of measured particle characteristics given in **Fig. 8** and **Fig. 9** exhibit a distinct response on the replacement of the torch electrodes when spraying the first coating layers with powder #1 while no significant reaction is perceptible during spraying the second coating layers with powder #2. The spray runs and parts are the same as given in Fig. 5 and Fig. 6. These examples show that T-v-data can be meaningful to detect process deviation, but not necessarily in all cases.

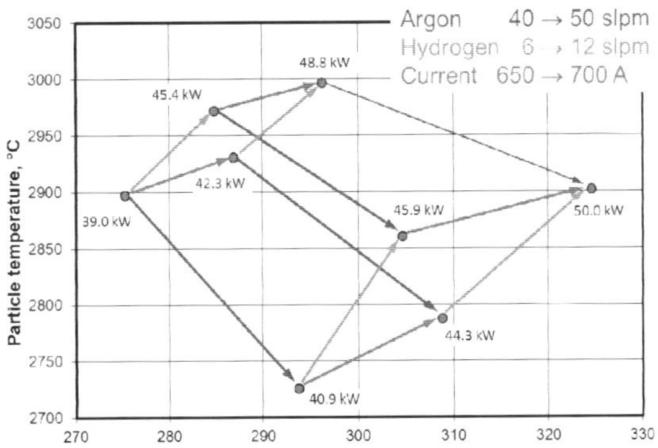

Fig. 7. Example of a first order process map for the application of a F4 torch.

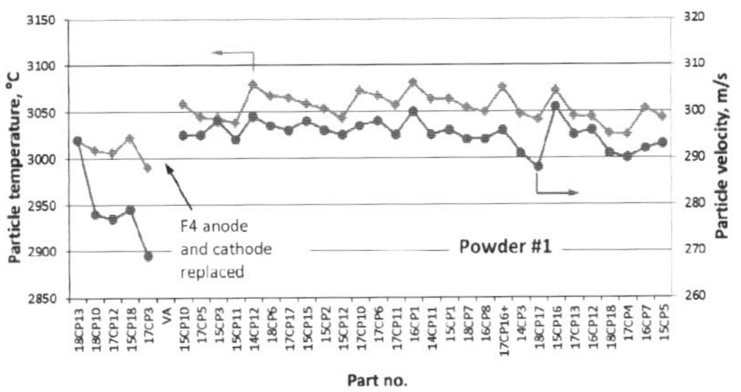

Fig. 8. Particle temperature and velocities measured by DPV-2000 during processing a batch of 33 aero-engine components (1st coating layer).

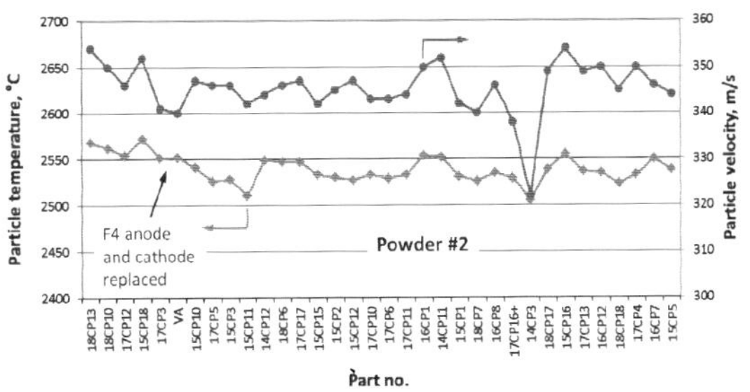

Fig. 9. Particle temperature and velocities measured by DPV-2000 during processing a batch of 33 aero-engine components (2nd coating layer).

Another issue is that the relations in first order process maps may not be invertible unique (i.e. they are not one-to-one corresponding in both directions). In [22] it is shown that similar particle states can be obtained by different F4 torch parameter sets. This ambiguity complicates the establishment of closed loop control systems based alone on particle diagnostics.

4 Process design using second order process maps

Process design is closely related to process control since also here process parameters must be adjusted to achieve particular coating properties. Second order process maps link intrinsic parameters and coating properties, e.g. particle characteristics with hardness or elastic modulus [23] (ex-situ measured). Similar to first order process maps it must be verified that these relationships are significant and sensitive. If so, coating properties can be predicted for specific process parameters. However, a small variability in particle state was found to correspond to a large variability in coating properties [19]. This complicates capturing all subtle changes in the process influencing the coating build-up. Second order process maps can also be applied reversely for process design. However, as for first order maps, it must be provided that they are invertible unique.

Fig. 10 gives an example of a combination of a first and second order process map as proposed in [20] with contour lines of constant porosity and deposition efficiency for spraying thermal barrier coatings by a Triplex II torch [24].

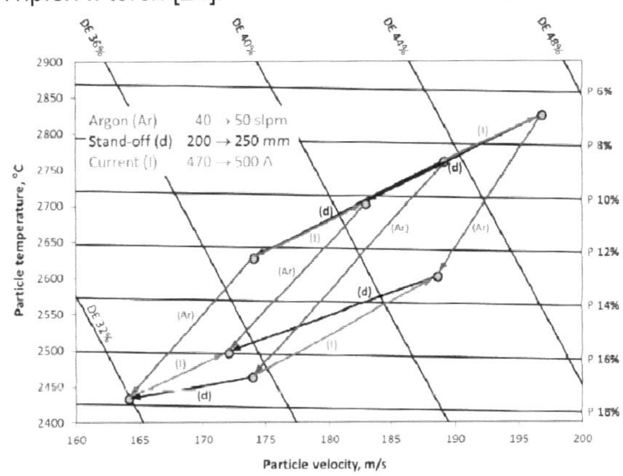

Fig. 10. Combination of a first and second order process map with contour lines of constant porosity (P) and deposition efficiency (DE) calculated by multiple linear regression [24].

Some researchers made attempts to generalize second order process maps by normalizing variables or applying dimensionless parameters like Reynolds numbers, melting and oxidation index [25, 26], or spray stream melting index [27], respectively. But obviously, such approaches do not solve the problem of the mentioned ambiguities.

5 A new approach of spray bead analysis

5.1 Background and motivation

Coating formation can be analyzed the more comprehensively, the more the observed variables have an immediate effect on the coating characteristics. It is only natural that sprayed test samples cover all influencing variables immediately without inferring coating

properties from any intermediate process variables based on process maps. Such a direct approach is the basic idea of in-situ curvature analysis [28, 29]. Here, a thin cantilever is coated and the development of its curvature is measured during deposition and cooling. From this data, residual stresses, Young's modulus [30], or non-linear elastic properties [31] can be determined.
Another variant of test samples are static spray spots or cones. They can be used e.g. to adjust the powder injection and to align multiple injectors relatively to the hot plasma core with the objective of optimum particle heating and an orientation-independent footprint of the plume [32]. Coating formation was investigated as well by observation of particle impacts during spraying of such cones [33]. Moreover, three-dimensional spray footprint profiles were used to derive flow distribution functions for coating simulations on free-formed workpiece surfaces [34].
At Forschungszentrum Jülich, plane specimens are sprayed with standard conditions at regular intervals to monitor process reliability. Thereby, the deposition rate (determined gravimetrically) and the porosity (measured by mercury intrusion porosimetry, MIP) proved to be sensitive indicators of process deviations.
Based on this experience, this approach should be improved; in particular, mercury intrusion porosimetry should be substituted since it requires freestanding coatings and shows only limited accuracy (according to the author's experience approx. ± 1.5% absolute volumetric porosity).

5.2 Experimental procedure

Beads were sprayed instead of static spots to allow starting the gun and stabilizing the process outside the sample. The beads are obtained by scanning the plasma torch on a line along plane 50 x 50 mm² stainless steel substrates. In contrast, static spots would imply the risk of overheating and partially melting or cracking. To avoid deformation of the specimen, substrates with 10 mm thickness were used. Three YSZ beads were sprayed in 25 passes at the same robot velocity as the nominal one. Thus, the applied conditions reflected typical operation. The bead masses were obtained by weighing the samples before and after spraying.

To determine the bead volume, a non-contact double-sided optical profilometer CT 350T (cyberTECHNOLOGIES GmbH, Eching-Dietersheim, Germany) with a 300 mm x-y-scanning stage (lateral resolution 50 nm) was applied in connection with a CHR 3000 chromatic white light interferometer (measurement range 3000 µm and resolution 0.10 µm). Data is acquired at a rate of 4 kHz.

5.3 Results and discussion

The volumetric bead porosities of the three samples were determined from the coating volumes and masses to 12.21%, 13.45%, and 11.37%, respectively corresponding to a mean of 12.34% and a standard deviation of 1.05%. Thus, the uncertainty of this first approach is slightly better than obtained by MIP.

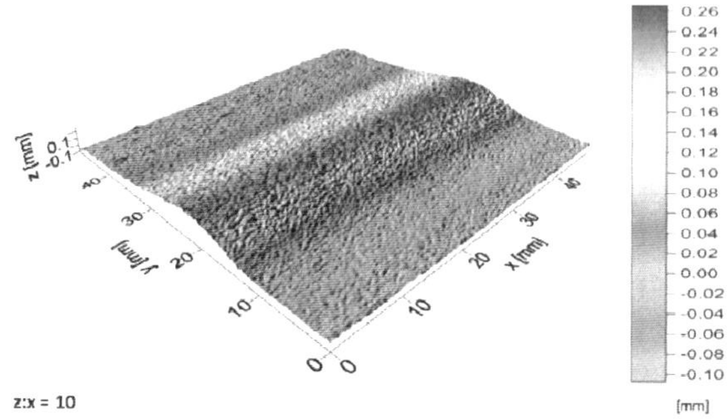

Fig. 11. Spray bead topography measured by a chromatic white light interferometer.

However, there is still a considerable potential for optimization. (1) Due to the limited accuracy of the scale, the coating mass was determined at an uncertainty of 0.01 g. In the investigated cases, this corresponds to a variation of 1.0% absolute porosity. Hence, a precision scale must be used. (2) The basic surface was averaged by selecting two uncoated polygons below and above the bead. Alternatively, this topography could be measured also by the profilometer before spraying to subtract it later from the bead profile. So, unevenness and roughness of the substrate could be considered. (3) Instead of the white light sensor, a DRS 500 laser triangulation sensor (measurement range 500 µm and resolution 0.125 µm) can be used because it penetrates the surface roughness less so that it is averaged.

6 Conclusions

- The observance of relevant process variables allows process monitoring and detection of deviations to a limited extend. Choosing extrinsic or intrinsic variables for this purpose, the limited process coverage and directness of impact on the deposit state must be considered. Furthermore, these parameters must be proven to be sufficiently sensitive in each spray case.
- Wear of the torch electrodes is an important reason for process deviations. It can be detected only by monitoring trends with respect to a known reference state. Since the formation of wear is dependent on the torch load collective, general thresholds for specific parameters are difficult to define.
- First order process maps are not generally applicable for process monitoring as particle characteristics can exhibit a lack of sensitiveness. The establishment of feedback control systems is aggravated by ambiguities.
- Coating properties may be predicted by means of second order process maps. But due to the lack of uniqueness, it is difficult to develop process settings reversely to achieve specific coating properties.
- Spraying standardized test samples is a promising

approach for process monitoring and detection of deviations since all relevant process variables are covered and coating characteristics are determined directly. The first approach to substitute MIP here yielded slightly improved uncertainties. Three options for further optimization were identified. Hereby, it is expected that the variability can be decreased to ± 0.3% absolute porosity.

Literature

[1] G. Dwivedi, T. Wentz, S. Sampath, T. Nakamura, Assessing Process and Coating Reliability Through Monitoring of Process and Design Relevant Coating Properties, J. Therm. Spray Technol. 19 (2010) Issue 4, p. 695-712

[2] M.Friis and C.Persson, Process Window for Plasma Spray Processes, in Thermal Spray 2001: New Surfaces for a New Millennium, C.C. Berndt, K.A. Khor, and E. Lugscheider, ed., ASM International, Materials Park, OH, 2001, pp. 1313-19.

[3] R.B. Heimann: Plasma Spray Coating - Principles and Applications. 2nd ed., Wiley-VCH, Weinheim, 2008

[4] P. Fauchais, M. Vardelle, A. Vardelle, Reliability of plasma-sprayed coatings: monitoring the plasma spray process and improving the quality of coatings, J. Phys. D: Appl. Phys. 46 (2013) 224016 (16 pp.)

[5] R.B. Heimann: Better Quality Control: Stochastic Approaches to Optimize Properties and Performance of Plasma-Sprayed Coatings. J. Therm. Spray Technol. 19 (2010), Issue 4, pp. 765-778

[6] R.B. Heimann, D. Lamy, and T. Sopkow, Optimization of Vacuum Plasma Arc Spray Parameters of 88WC12Co Alloy Coatings Using a Statistical Multifactorial Design Matrix, J. Can. Ceram. Soc. 59 (1990), Issue 3, p. 49.

[7] J.F. Coudert, M.P. Planche, P. Fauchais, Characterization of D.C. Plasma Torch Voltage Fluctuations, Plasma Chem. Plasma Process. 16 (1996), Issue 1 supplement, p. 211S-227S

[8] Z. Duan, J. Heberlein, Arc Instabilities in a Plasma Spray Torch, J. Thermal Spray Technol. 11 (2002) Issue 1, p. 44-51

[9] J.F. Bisson, B. Gauthier, C. Moreau, Effect of Plasma Fluctuations on In-Flight Particle Parameters, J. Thermal Spray Technol., 12 (2003) Issue 1, p. 38-43

[10] J.F. Bisson, C. Moreau, Effect of Plasma Fluctuations on In-Flight Particle Parameters: Part II, J. Thermal Spray Technol. 12 (2003) Issue 2, p. 258-264

[11] L. Leblanc, C. Moreau, The Long-Term Stability of Plasma Spraying, J. Thermal Spray Technol. 11 (2002) Issue 3, p. 380-386

[12] M. Vyšohlíd, J. Heberlein, Investigation of Arc Voltage Fluctuation in a Plasma Torch SG-100 operated with Ar/H_2, Thermal Spray 2004: Advances in Technology and Application, May 10-12, 2004 (Osaka, Japan), ASM International

[13] G. Mauer, J.-L. Marqués, R. Vaßen, D. Stöver, Detection of Wear in One-Cathode Plasma Torch Electrodes and its Impact on Velocity and Temperature of Injected Particles, J. Therm. Spray Technol. 16 (2007) Issue 5-6, p. 933-939

[14] H. Weckmann, S. Fiebig, A. Syed, J. Arnold, Investigation of Time-Dependent Instabilities of Plasma Spraying Process Using Online Diagnostic Systems, Thermal Spray 2006: Building on 100 Years of Success, B.R. Marple, M.M. Hyland, Y.-C. Lau, R.S. Lima, and J. Voyer, Ed., May 15-18, 2006 (Seattle, Washington, USA), ASM International

[15] G. Mauer, R. Vaßen, D. Stöver, Comparison and Applications of DPV-2000 and Accuraspray-g3 Diagnostic Systems, J. Therm. Spray Technol. 16 (2007) Issue 3, p. 414-424

[16] P. Fauchais and M. Vardelle, Sensors in Spray Processes, J. Therm. Spray Technol. 19 (2010), pp. 668-694

[17] A. Vaidya, V. Srinivasan, T. Streibl, M. Friis, W. Chi, S. Sampath, Process maps for plasma spraying of yttria-stabilized zirconia: An integrated approach to design, optimization and reliability, Mater. Sci. Eng., A 497 (2008), p. 239-253

[18] S. Sampath, V. Srinivasan, A. Valarezo, A. Vaidya, T. Streibl, Sensing, Control, and In Situ Measurement of Coating Properties: An Integrated Approach Toward Establishing Process-Property Correlations, J. Therm. Spray Technol. 18 (2009) Issue 2, p. 243-255

[19] V. Srinivasan, A. Vaidya, T. Streibl, M. Friis, S. Sampath, On the Reproducibility of Air Plasma Spray Process and Control of Particle State, J. Therm. Spray Technol. 15 (2006) Issue 4, p. 739-743

[20] M. Friis, C. Persson, Control of Thermal Spray Processes by Means of Process Maps and Process Windows, J. Therm. Spray Technol. 12 (2003) Issue 1, p. 44-52

[21] M. Friis, C. Persson, J. Wigren, Influence of particle in-flight characteristics on the microstructure of atmospheric plasma sprayed Yttria stabilized ZrO_2, Surf. Coat. Technol. 141 (2001), p. 115-127

[22] J. Colmenares-Angulo, K. Shinoda, T. Wentz, W. Zhang, Y. Tan, S. Sampath, On the Response of Different Particle State Sensors to Deliberate Process Variations, J. Therm. Spray Technol. 20 (2011) Issue 5, p. 1035-1048

[23] A. Valarezo, W.B. Choi, W. Chi, A. Gouldstone, S. Sampath, Process Control and Characterization of NiCr Coatings by HVOF-DJ2700 System: A Process Map Approach, J. Therm. Spray Technol. 19 (2010) Issue 5, p. 852-865

[24] G. Mauer, R. Vaßen, D. Stöver, Atmospheric plasma spraying of yttria-stabilized zirconia coatings with specific porosity, Surf. Coat. Technol. 204 (2009), p. 172-179

[25] W. Zhang, S. Sampath, A Universal Method for Representation of In-Flight Particle Characteristics in Thermal Spray Processes, J. Therm. Spray Technol. 18 (2009), p. 23-34

[26] H.-B. Xiong, L.-L. Zheng, L. Li, A. Vaidya, Melting and oxidation behavior of in-flight particles in plasma spray process, Int. J. Heat Mass Transfer 48 (2005), p. 5121-5133

[27] V. Srinivasan, S. Sampath, Estimation of Molten Content of the Spray Stream from Analysis of Experimental Particle Diagnostics, J. Therm. Spray Technol. 19 (2010) Issue 1-2, p. 476-483
[28] J. Matejicek, S. Sampath, In situ measurement of residual stresses and elastic moduli in thermal sprayed coatings. Part 1: apparatus and analysis, Acta Mater. 51 (2003), p. 863-872
[29] J. Matejicek, S. Sampath, D. Gilmore, R. Neiser, In situ measurement of residual stresses and elastic moduli in thermal sprayed coatings. Part 2: processing effects on properties of Mo coatings, Acta Mater. 51 (2003), p. 873-885
[30] G. Dwivedi, T. Nakamura, S. Sampath, Determination of Thermal Spray Coating Property with Curvature Measurements, J. Therm. Spray Technol. 22 (2013) Issue 8, p. 1337-1347
[31] T. Nakamura, Y. Liu, Determination of nonlinear properties of thermal sprayed ceramic coatings via inverse analysis, Int. J. Solids Struct. 44 (2007) p. 1990-2009
[32] S. Mihm, T. Duda, H. Gruner, G. Thomas, B. Dzur, Method and Process Development of Advanced Atmospheric Plasma Spraying for Thermal Barrier Coatings, J. Therm. Spray Technol. 21 (2012) Issue 3-4, p. 400-408
[33] K. Shinoda, H. Murakami, S. Kuroda, K. Takehara, S. Oki, In Situ Visualization of Impacting Phenomena of Plasma-Sprayed Zirconia: From Single Splat to Coating Formation, J. Therm. Spray Technol. 17 (2008) Issue 5-6, p. 623-630
[34] T. Wiederkehr, H. Müller, Acquisition and Optimization of Three-Dimensional Spray Footprint Profiles for Coating Simulations, J. Therm. Spray Technol. 22 (2013) Issue 6, p. 1044-1052

Particle in-flight velocity and dispersion measurements at increasing particle feed rates in Cold Spray

M. Meyer, S. Yin, *R. Lupoi, Trinity College Dublin, The University of Dublin, Department of Mechanical and Manufacturing Engineering, Parsons Building, Dublin 2, Ireland [*Contact email: lupoir@tcd.ie]

Cold Spray (CS) is attracting the interest of research and industry due to its rapid, solid-state particle deposition process and the respective advantages over conventional deposition technologies. The acceleration of the particles is critical to the efficiency of CS, and previous fluid dynamic investigations rarely consider the particle feed rate important. However, because higher particle loadings are typically used in the process, the effect of this cannot be assumed negligible. This study therefore investigates the particle velocities in the supersonic jet of an advanced CS system at low and high pressure levels and varying particle feed rates using Particle Image Velocimetry (PIV). The particle dispersion and velocity evolution along the jet axis were investigated for several feedstock materials. It was found that the average particle velocity noticeably decreases with increasing particulate loading in all cases. This effect is aggravated for lower working pressures, but mainly depends on the feedstock material, which implies more complex, volume-fraction related physics playing a role in this respect. Moreover, the velocity distribution and particle dispersion were also observed to be influenced by the feed rate, depending on the particle material. Increased particle feed rates hence affect the magnitude and distribution of impact velocity and consequently the efficiency of CS. In particular, numerical models neglecting this interconnection are required to be further improved, based on these experimental studies.

1 Introduction

Cold Spray (CS) is a coating manufacturing process, in which a feedstock powder is deposited onto a substrate by means of high velocity impacts and solid state deformation. As the process makes use of the high kinetic energy of the particles obtained from a supersonic process gas rather than their thermal energy, it allows for very low temperature levels compared to other coating technologies. It eliminates or minimizes the disadvantages of melting and hence it provides a possibility to coat oxidisation-sensitive materials and material combinations with different melting temperatures [1][2]. It is generally accepted that for deposition to occur a critical impact velocity must be crossed. This represents a threshold value that makes the particle acceleration a critical aspect of the technology [3][4]. Each material combination is characterized by a specific critical speed level. Both experimental and numerical studies on the two-phase nozzle flow were conducted in the past decades, yielding to the main parameters for gas and particulate acceleration. Primarily, the gas stagnation pressure and temperature [5][6], the gas species [7][8] and the powder injection conditions [9][10] are important. Another critical parameter is the particle material and size, as the larger and heavier particles are the less susceptible to the flow, having a much higher characteristic reaction time [11][12][13], whereas the particle shape is connected to the drag coefficient [14][15]. With the objective of analysing the gas and particle dynamics, different optical measurement techniques were employed in CS. Schlieren photography was frequently used to visualise the flow features. Only resolving the density gradients of the gas phase however impedes quantification of the velocity and particle motion. Therefore, non-intrusive velocity measurement techniques were used, firstly by Gilmore et al. [16] who recorded the particle velocity by a laser-two-focus (L2F) apparatus and, inter alia, could observe a particle feed rate link to exit velocity. Due to the low spatial resolution and precision of L2F, many researchers subsequently concentrated on doppler picture velocimetry (DVP), which is described in detail in [17] and is used for rather small particle densities, such that the particle loading effect was not of interest. Particle image velocimetry (PIV) or similar techniques, such as tracking techniques, could be used to record instantaneous particle velocity distributions throughout the field of measurement without any scanning procedure. In this concern, several researchers facilitated results based on a single-frame triple-pulse particle tracking technique [18][19][20]. In particular, Pardhasaradhi et al. [21] used this approach to measure the particle exit speed of different nozzles under varying conditions. It was found that the effect of the particle loading had a negligible counter-effect on the particle dynamics within the scope of measurement, attributed to a low mass fraction of the discrete phase. Nevertheless, depending on the particle feed rate and the relative gas consumption, the discrete phase loading can increasingly affect the acceleration process at higher feed rate levels. Results demonstrated by Samareh et al. [22] indicated that the gas flow structures change and the particle speed reduces as their mass fraction increases. Accordingly, a study by Lupoi [23] showed that a number of experimental observations with various nozzle geometries could not be explained by 1-way coupling CFD techniques. An improvement of the computations was shown to be achievable when increasing the detail of phase coupling for such experiments by Meyer & Lupoi [24]. As for higher particle feed rates, a work by Pattison et al. [25] showed an experimental optimisation involving a full 2D-velocity field deduction using PIV in its usual sense. Similarly, Zahiri et al. [26] demonstrated particulate plume characteristics in CS with PIV. Although a sufficiently high particle density is required to enable field measurements with PIV, these studies did not consult the links to velocity results. There is no doubt that the faster processing times can be achieved by higher the feed rates. In order to start understanding the detailed effect of the particle feed rate as a process parameter, this work aims to investigate the particle velocity in the jet of a Cold Spray system working under low pressure and high pressure conditions in dependence of the particle feed rate using PIV, in combination with a tracking algorithm for the identification of single particle vectors. Feedstock materials are varied and measured under different flow conditions.

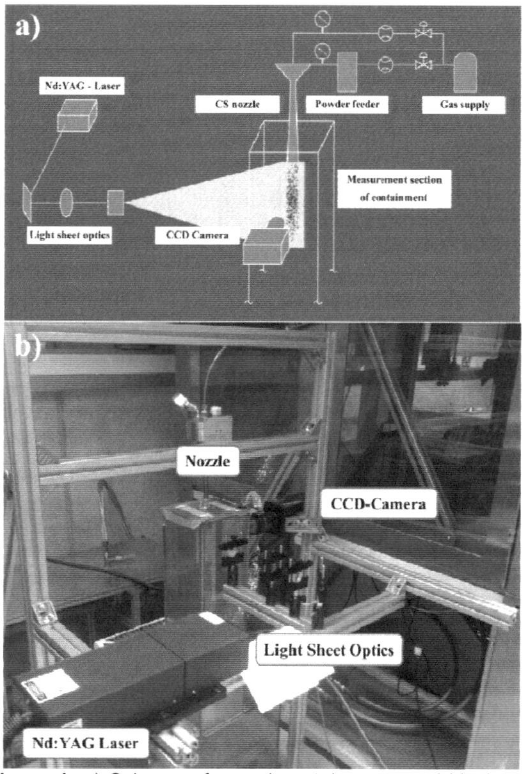

Figure 1. a) Scheme of experimental set-up and b) picture of the measurement system in the laboratory.

2 Methods

2.1 Cold Spray system and materials

Figure 1a) illustrates a schematic of the experimental arrangement, including PIV and the CS process. The nitrogen gas flow was provided by a supply, delivering the settings of 30bar and 15bar constant pressure at the nozzle inlet. The gas line was split into two lines, a main line, which was connected to the nozzle head, and the powder feeder line; both of which comprised flow meters and pressure gauges. The wheel type powder feeder included a load cell to measure the mass feed rate of powder. The setting of powder feed rate was increased according to the capabilities of the feeder for each material. Downstream, the gas–solid mixture then merged with the main flow in the nozzle head. The in-house nozzle-design of de-Laval converging/diverging shape had a full length of 210mm, a 2mm throat diameter and 6mm exit diameter. The velocity measurement area was oriented within the longitudinal direction and with a length of 50mm from the nozzle exit, such that the relevant region of the particle jet was covered. The nozzle sprayed into the measurement section of an enclosed area, which ensured undisturbed flow. Figure 1b) shows the CS-PIV setup in TCD.

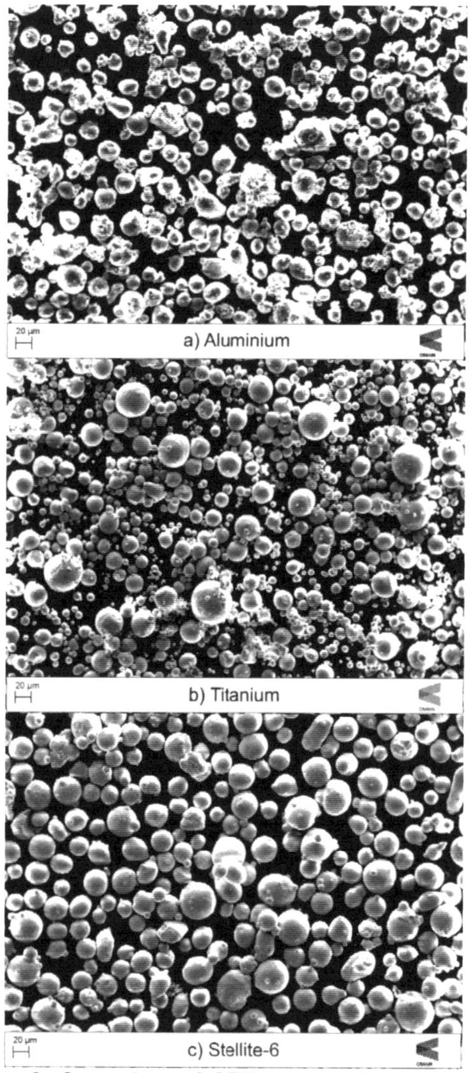

Figure 2. Comparison of SEM images of measured CS powders: a) Aluminium, b) Titanium, c) Stellite-6.

Figure 2 shows Scanning Electron Microscope (SEM) images of the three different materials which were used in this study. Commercially pure Aluminium and Titanium (CP2), were used as rather light-weight materials. In addition, Stellite–6, a cobalt–chrome alloy which is highly relevant for CS applications with well-known corrosion resistance characteristics, was compared. Table 1 summarises the material properties and testing conditions. It was not possible to obtain useful results for Titanium at 30bar pressure, frequently particles started glowing due to collisions with the nozzle wall, which interfered with the imaging techinique.

Feedstock material	Stellite-6 (ST)	Titanium (TI)	Aluminium (AL)
Density [kg/m-3]	8440	4500	2700
Pressure [bar-g]	30/15	15	30/15
Temperature [K]	290	290	290

Table 1. Poder materials and spray conditions

2.2 Measurement set-up

Particle Image Velocimetry was used to optically measure the particle velocity within the jet. They were illuminated with two subsequent laser pulses formed to a light sheet in the plane of measurement. A camera system captured two images of the scattered light respectively with varying pulse separation time depending on the conditions. These images were processed by a cross–correlation algorithm, deducing the particle displacement, which corresponds to velocity information by knowledge of the pulse separation. As the feedstock particles are measured directly in CS, the gas phase is not subject to measurement in this study. A Nd:YAG Laser with a wavelength of 532nm and 6ns pulse duration was used as the light source at 4Hz repetition rate. The light sheet thickness and width were approximately 1mm and 60mm respectively in order to obtain velocities of particles on the plan of symmetry. A monochromic camera with a resolution of 1280x1024 pixels and a 12bit dynamic range delivered the consecutive image pairs. The repetition rate was limited by the CCD readout time of the camera, such that the entire evolution of the particles through the flow field could not be tracked considering the high order of the present velocities. The pre–processing of images included a background subtraction and high–pass filtering to reduce stationary image features and low–frequency background variations. The processing involved a multi–pass cross–correlation with reducing interrogation window size and an additional tracking algorithm, which identified single particle vectors. The optical measurement error was estimated to be in the order of 1.3%.

3 Results and Discussion

3.1 Flow and loading conditions

As wheel speed of the powder feeder was increased, the mass feed rate of particles depends on the powder in use and the spray conditions. This results in varying ranges of feed rates for the different materials. For aluminium, having the lowest density, it was possible to increase the mass feed rate to just below 150g/min (corresponding to full wheel speed), whereas titanium could be fed at a maximum of 350g/min. For Stellite-6 even higher feed rates would have been possible, which would be beyond the scope of interest of this study. Figure 3 shows the connection between the particle feed rate and the corresponding relative mass loading Z as the ratio of the overall particle and gas mass flow rate in the nozzle. The gas flow conditions were measured to be constant for all measurements of each pressure setting, yielding 0.0136kg/s at 15bar and 0.0271kg/s at 30bar inlet pressure. Due to this, one can see linear the connection to the relative loading. The gas flow approximately doubles when increasing the pressure from 15 to 30 bar, which results in a steeper slope for the low pressure setting. One can hence expect any loading effect to be more drastic for such case – Z goes up to a maximum of 0.4, but even at lower values, the mass fraction is high enough for the phase interaction to become significant.

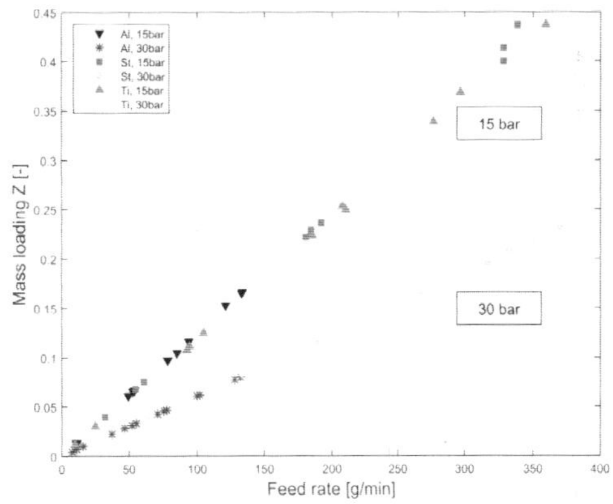

Figure 3. Verification of linear correlation between particle feed rate and raltive mass loading

3.2 Mean velocity change

In order to compare this phase-coupling effect for the chosen set of materials, one can investigate the particle exit velocity. As the experiments were conducted repeatedly and provided spatially resolved data, it is convenient to analyse the average change in particle velocity. This averaging was conducted over all measured particles, hence including changes with position, and over all image sets, as a temporal average. Figure 4 shows the respective results. The line segments are least-square linear fitting curves for the data points, clarifying the change in velocity. The error bars comprise 2σ-confidence intervals and the inherent uncertainty associated with the optical analysis.

With focus on the velocity values at very low feed rates, it can be seen that Stelllite-6 exhibits the lowest velocity, ca. 55m/s lower than Aluminium, which is due to the higher material density. Titanium, in turn, has a very similar average velocity, despite an intermediate density, because the amount of small particles is higher in its size distribution, as can be seen in Figure 2. As the pressure level is increased, this low-feed velocity increases by about 20% for both Stellite-6 and Aluminium, or by 75 and 90m/s respectively. Most importantly, one can clearly see a decrease in average velocity for all materials at given conditions. This trend appears to be approximately linear in the given range of feed rates and slightly less severe at higher pressure. The loss in particle velocity can be as high as 10% for Aluminium, and up to 16 and 20% for Stellite-6 and Titanium respectively, when taking into account the extended range of particle feed rates explored. It is very interesting to note that, depending on the material, the same amount of additional particle mass injected in the nozzle does not induce the same velocity loss. This implies that the slope for Stellite-6 is less steep than for Titanium, and the reaction of Aluminium is the most severe. According to the observation, the effect becomes more significant with decreasing density. Based on the conservation of momentum and energy, it can be argued that the velocity of the discrete phase must drop because its mass increased by a specific amount. However, for the different materials, the given mass feed rate corresponds to different numbers of particles and volumetric feed rates, and therefore to different volume fractions. For example, 100g/min feed rate corresponds to volume feed rates of 0.0118 and 0.037 litres/min for Stellite-6 and Aluminium respectively. The complex effects that are connected to a change in volume fraction seem to have an additional influence on the average particle dynamics. Such effects can be linked, for instance, to particle-particle collisions, change in turbulence and dissipation, as well as a dense regime in the nozzle throat region. In order to quantify these effects, it would be necessary to investigate the particle behaviour during acceleration, i.e. inside the nozzle.

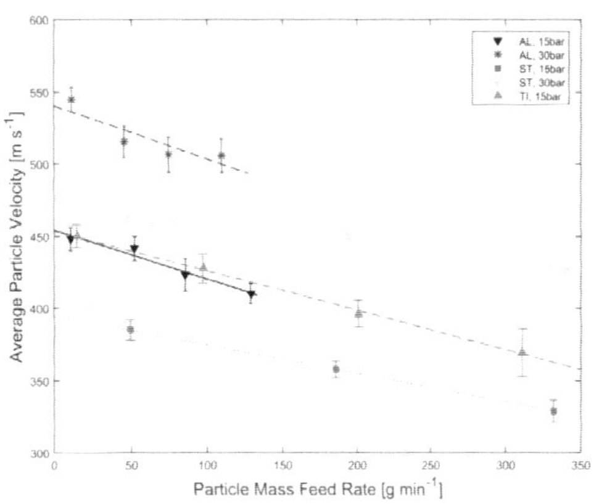

Figure 4. Comparison of material behaviour due to increasing particle feed rate using spatial and temporal average of particle velocity at the nozzle exit.

3.3 Profile evolution and distribution

Figure 5 shows a comparison of particle distributions depending on the loading for Aluminium and Stellite-6 at 15bar (a,b) and at 30bar (c,d) respectively. The positions of detected particles are projected in the plane of measurement, where the origin is on the centreline at the nozzle exit, and coloured by the their velocity magnitude. In this test, no substrate was present. At the nozzle exit, each plot shows a narrow stream of exiting particles that widens downstream until the plume approximately occupies the region from -5 to 5 mm. This downstream dispersion most likely results from the interaction of particles with turbulence, the shock waves forming in the gas jet and the shear layer at its edge. It is stronger for high loading cases, and Aluminium, providing the lighter particles, is more prone to this effect. The overall decrease in velocity due to growing feed rate can also be seen in this graphic, as the level of both the higher velocities in the jet core and the lower velocities at the boundaries decreases. The higher the loading of particles is, the stronger the downstream deceleration becomes. For both materials, one can record that the dispersion is weaker for higher pressure. Consequently, the more momentum the particles have, the less susceptible they are to unfavourable deviations and, at the lower pressure level, the gas jet is drastically over-expanded leading to stronger shock systems.

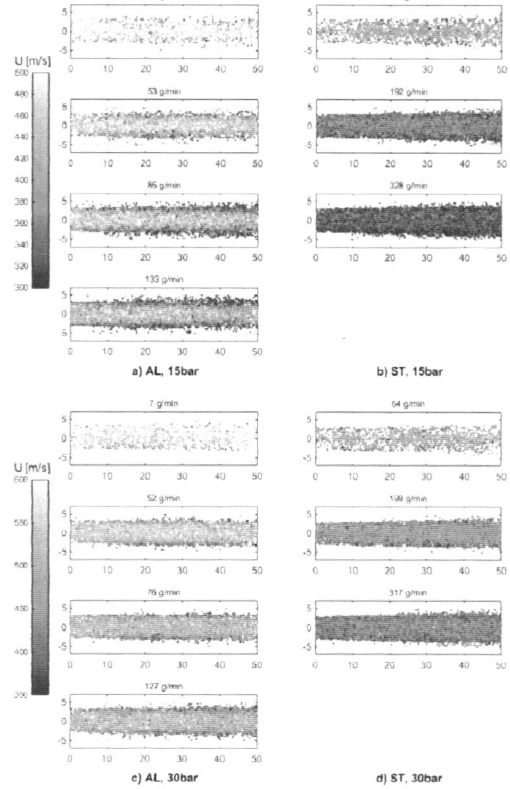

Figure 5. Scatter of single particle velocities in the x-y-measurement plane using the PIV (0 = nozzle exit). Comparison of Aluminium and Stellite-6 velocity distribution and particle dispersion for increasing pressure and mass loading.

It is interesting to investigate the particle velocity at locations where a substrate could be placed, when choosing larger possible stand-off distances, as it is an indicator of the expected trends for particle impact velocity. Figure 6 therefore displays particle velocity profiles in the radial direction of particles which are located between 40 and 50mm downstream of the nozzle exit, hence on the far right side of the plots in Figure 5. In this illustration, differences become clear and can be quantified more easily. In each subplot, the dots are data points for individual particle speeds, whereas the curves are second order fits to guide the eye for the comparison. When opposing the different materials, a weaker curvature of the profiles can be noticed when going from Aluminium through Titanium to Stellite-6. The maximum velocities of the profiles are located around the centreline and the radial velocity drop associated with the curvature easily reaches 200m/s for the light material, while for the alloy, it is only in the order of 50-70m/s. This observation shows that the light material is affected more than twice as much by the spreading of the gas jet. For higher pressure, the curvature is slightly higher than for smaller pressure due to the same reason, and one can register a shift of the maximum to higher velocities by about 23% for Aluminium and 21% for Stellite-6. This is approximately the amount by which the velocities reduce when finally comparing the low feed to high feed cases. Generally, the slowest particles in the dilute cases are as fast as the fastest in the dense flow case. This reduction in maximal velocity is stronger for Titanium and Stellite-6. In these cases, the velocity profile flattens as relatively fast particles tend to occupy the areas further away from the centreline. For Aluminium, the particles in these locations are rather slow. This is plausible as they are decelerated faster when entering regions of negative relative gas-particle velocity. In every case, the low feed results show particles mainly within the region of the nozzle exit diameter ($\pm r_{exit}$), whereas the high feed indicates a significantly stronger dispersion.

These observations are important considering effective deposition of different materials at increasing particle feed rates. The pressure level, the feed rate and the radial profiles affect the velocity of particles and hence the impact conditions. It is important to note, that due to the absence of a substrate, the additional effects of the bow shock system upon impact are not analysed. However, the analysed velocity reductions can lead to sub-critical impact velocities after passing through the shock and hence significantly reduce deposition efficiency.

4 Conclusions

In Cold Spray, the acceleration of the particles is critical to efficiency, and previous fluid dynamic investigations rarely consider the particle feed rate important. However, higher particle loadings can change the process, and induce noticeable effects on the particle dynamics. This study therefore experimentally investigated the particle velocities in the supersonic jet at lower and higher pressure level, analysing the effect of varying particle feed rates on Stellite-6, Aluminium and Titanium particles using PIV. The main novel findings are that the average particle velocity was reduced in every case when the particle loading was increased, in part as high as 20%. It was found that the materials behave slightly differently when changing this parameter. It is proposed, that complex processes which involve non-linear effects due to differing increase of volume fractions play a role in this respect. Moreover, the study found that the particle velocity decreases more strongly while traveling downstream at high loadings. Aluminium was found to be more prone to radial velocity decrease also for high feed rates, but all materials show intensified particle dispersion at large distances from the nozzle exit. Increased particle feed rates hence affect the magnitude and distribution of impact velocity and consequently the efficiency of Cold Spray. In particular, computational modelling approaches neglecting this interconnection are required to be further improved, based on experimental studies presented in this paper.

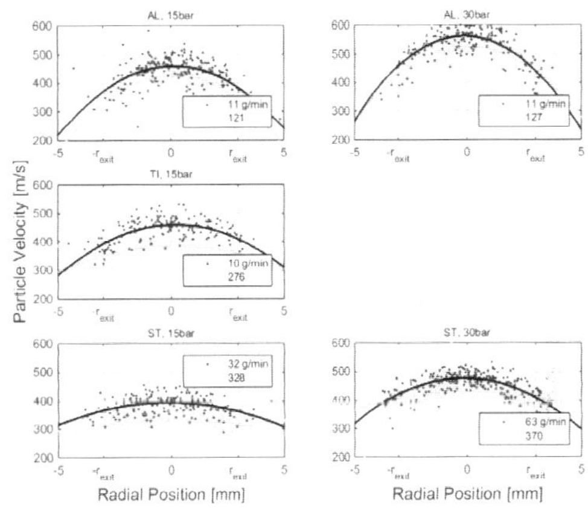

Figure 6. Radial velocity profiles at 40-50mm distance from nozzle exit by the PIV apparatus. Data points represent single particle velocities, lines indicate profile trend as 2nd order fits.

Acknowledgements

The authors wish to aknowledge FP7 Marie Curie Actions (Grant 333663) for the financial support and to the CRANN Advanced Microscopy Laboratory (AML) for the cooperation, as well as to Dr. Tim Persoons for the technical advice and support.

5 Literature

[1] A. Alkhimov, V. F. Kosarev, and A. Papyrin, "Method of Cold-Gas Dynamic Spraying," *Dokl. Acad. Nauk SSSR*, vol. 315, no. 5, pp. 1062–1065, 1990.

[2] A. Papyrin, "COLD SPRAY TECHNOLOGY," *Adv. Mater. Process.*, vol. 159, no. 9, pp. 49–51, 2001.

[3] H. Assadi, F. Gärtner, T. Stoltenhoff, and H.

Kreye, "Bonding mechanism in cold gas spraying," *Acta Mater.*, vol. 51, no. 15, pp. 4379–4394, 2003.
[4] M. Grujicic, C. L. Zhao, C. Tong, W. S. DeRosset, and D. Helfritch, "Analysis of the impact velocity of powder particles in the cold-gas dynamic-spray process," *Mater. Sci. Eng.*, vol. 368, no. 1–2, pp. 222–230, 2004.
[5] T. Han, Z. Zhao, B. a. Gillispie, and J. R. Smith, "Effects of Spray Conditions on Coating Formation by the Kinetic Spray Process," *J. Therm. Spray Technol.*, vol. 14, no. 3, pp. 373–383, 2005.
[6] E. H. Kwon, S. H. Cho, J. W. Han, C. H. Lee, and H. J. Kim, "Particle behavior in supersonic flow during the cold spray process," *Met. Mater. Int.*, vol. 11, no. 5, pp. 377–381, 2005.
[7] T. C. Jen, L. Li, W. Cui, Q. Chen, and X. Zhang, "Numerical investigations on cold gas dynamic spray process with nano- and microsize particles," *Int. J. Heat Mass Transf.*, vol. 48, no. 21–22, pp. 4384–4396, 2005.
[8] H. Katanoda, T. Matsuoka, and K. Matsuo, "Experimental study on shock wave structures in constant-area passage of cold spray nozzle," *J. Therm. Sci.*, vol. 16, no. 1, pp. 40–45, 2007.
[9] S. Yin, Q. Liu, H. Liao, and X.-F. Wang, "Effect of injection pressure on particle acceleration, dispersion and deposition in cold spray," *Comput. Mater. Sci.*, vol. 90, pp. 7–15, Jul. 2014.
[10] S. Yin, X. Suo, H. Liao, Z. Guo, and X. Wang, "Significant influence of carrier gas temperature during the cold spray process," *Surf. Eng.*, vol. 30, no. 6, pp. 443–451, 2014.
[11] F. Gärtner, T. Stoltenhoff, T. Schmidt, and H. Kreye, "The Cold Spray Process and Its Potential for Industrial Applications," *J. Therm. Spray Technol.*, vol. 15, no. 2, pp. 223–232, 2006.
[12] A. Sova, A. Okunkova, S. Grigoriev, and I. Smurov, "Velocity of the Particles Accelerated by a Cold Spray Micronozzle: Experimental Measurements and Numerical Simulation," *J. Therm. Spray Technol.*, vol. 22, no. 1, pp. 75–80, Nov. 2012.
[13] S. Li, B. Muddle, M. Jahedi, and J. Soria, "A Numerical Investigation of the Cold Spray Process Using Underexpanded and Overexpanded Jets," *J. Therm. Spray Technol.*, vol. 21, no. 1, pp. 108–120, Oct. 2011.
[14] X.-F. Wang, S. Yin, and B. P. Xu, "Effect of cold spray particle conditions and optimal standoff distance on impact velocity," *J. Dalian Univ. Technol.*, vol. 51, no. 4, pp. 498–504, 2011.
[15] W.-Y. Li and C.-J. Li, "Optimization of spray conditions in cold spraying based on numerical analysis of particle velocity," *Trans. Nonferrous Met. Soc. China (English Ed.*, vol. 14, no. 2, pp. 43–48, 2004.
[16] D. L. Gilmore, R. C. Dykhuizen, R. A. Neiser, T. J. Roemer, and M. F. Smith, "Particle Velocity and Deposition Efficiency in the Cold Spray Process," *J. Therm. Spray Technol.*, vol. 8, no. 4, pp. 576–582, 1999.
[17] H. Fukanuma, N. Ohno, B. Sun, and R. Huang, "In-flight particle velocity measurements with DPV-2000 in cold spray," *Surf. Coatings Technol.*, vol. 201, no. 5, pp. 1935–1941, 2006.
[18] B. Jodoin, F. Raletz, and M. Vardelle, "Cold spray modeling and validation using an optical diagnostic method," *Surf. Coatings Technol.*, vol. 200, no. 14–15, pp. 4424–4432, 2006.
[19] F. Raletz, M. Vardelle, and G. Ezo'o, "Critical particle velocity under cold spray conditions," *Surf. Coatings Technol.*, vol. 201, no. 5, pp. 1942–1947, 2006.
[20] X.-J. Ning, Q.-S. Wang, Z. Ma, and H.-J. Kim, "Numerical Study of In-flight Particle Parameters in Low-Pressure Cold Spray Process," *J. Therm. Spray Technol.*, vol. 19, no. 6, pp. 1211–1217, Sep. 2010.
[21] S. P. Pardhasaradhi, V. Venkatachalapathy, S. V Joshi, and S. Govindan, "Optical Diagnostics Study of Gas Particle Transport Phenomena in Cold Gas Dynamic Spraying and Comparison with Model Predictions," *J. Therm. Spray Technol.*, vol. 17, no. 4, pp. 551–563, Oct. 2008.
[22] B. Samareh, O. Stier, V. Lüthen, and A. Dolatabadi, "Assessment of CFD Modeling via Flow Visualization in Cold Spray Process," *J. Therm. Spray Technol.*, vol. 18, no. 5–6, pp. 934–943, Aug. 2009.
[23] R. Lupoi, "Current design and performance of cold spray nozzles: experimental and numerical observations on deposition efficiency and particle velocity," *Surf. Eng.*, vol. 30, no. 5, pp. 316–322, Jun. 2014.
[24] M. Meyer and R. Lupoi, "An analysis of the particulate flow in cold spray nozzles," *Mech. Sci.*, vol. 6, no. 2, pp. 127–136, 2015.
[25] J. A. Pattison, S. Celotto, a. Khan, and W. O'Neill, "Standoff distance and bow shock phenomena in the Cold Spray process," *Surf. Coatings Technol.*, vol. 202, no. 8, pp. 1443–1454, 2008.
[26] S. H. Zahiri, W. Yang, and M. Jahedi, "Characterization of Cold Spray Titanium Supersonic Jet," *J. Therm. Spray Technol.*, vol. 18, no. 1, pp. 110–117, Nov. 2008.

Gas and particle investigations during wire arc spraying process

M. Szulc*, J. Zierhut
Zierhut Messtechnik GmbH, Munich / D
*E-Mail: szulc@zierhut-messtechnik.de

A: Atzberger, J. Schaup, K. Hartz-Behrend, S. Zimmermann, J. Schein
Lab for Plasma Technology (LPT), Universitaet der Bundeswehr Muenchen, Munich / D

W. Krömmer, F. Lang
Linde Gases Division, Unterschleißheim / D

W. Mayr
Electrical Measurement Techniques, University of Applied Sciences Munich, Munich / D

Newly-developed wire arc spray technologies have improved the process stability and coating properties significantly. The gas injection, gas composition and interaction of gas flow and arc within a special wire arc spray process play a crucial role in separating the molten particles from the wire ends. Because of this fact, investigations with different gas compositions were made to characterize the dependencies between the gas flow, the gas velocities and resulting particle behavior, which determines the coating parameters. In order to investigate these topics a BFS research group was founded. The paper is organized as follows: In the first section an explanation of wire arc system and adjusted parameters is presented. In the second part the diagnostics are introduced and general setup and functionality of the used gas and particle investigation systems are shown. The third part discusses the results in more detail. An initial correlation of coating results with the individually detected particle parameters and measured gas parameters are evaluated for a first interpretation. The last part describes an outlook of future work and concludes the paper.

1 Introduction

The wire arc spraying process is increasingly used in different industrial branches due to its cost efficiency and high deposition rates. The popularity is understandable when compared to other coating processes, due to its easy handling, the lack of complicated hardware and the mobility of the system [1]. Hence, the application range stretches from anti-corrosion coatings for steel construction and off-shore industries to wear resistant coatings for moulding industry, metallization of plastic materials and automotive industry, where wire arc spraying is used for cylinder liner and piston coating.

Nonetheless, wire arc spraying has also some significant disadvantages. Due to its principle with two melting wire electrodes, the wires have to be made from conductive materials, may it be as solid or cored versions. From the molten wire ends, particles are detached and accelerated towards the coated target by the gas flow. The transient behaviour of arc attachment points, which are permanently exposed to the atomizing gas flow, cause a natural process fluctuation as presented in [2]. Furthermore, possible instabilities in the gas flow, as well as instabilities occurring in the wire feed system intensify the random arc movement. These fluctuations may lead to the formation of unmolten wire pieces (so called cold shuts) as shown in [3] and [4], different particle size distributions [5] and other defects in the coating or increased emission of nanoparticles. Especially the emission of nanoparticles are considered disadvantageous regarding workplace and environment pollution. Moreover, in a noticeable amount of applications dried, compressed air (short: DCA) is being used as atomizing gas what causes a stronger oxidation of coatings. As known from literature, the oxide content can be reduced by spraying with inert gases (e.g.. N_2). Lang showed in [6] that the amount of fine particles as well as the oxide content can be significantly reduced when inert gases are being used. It is also suggested, that a more uniform particle size distribution can be achieved. When the atomizing gas is varied parameters as density, viscosity, ionization energy and specific heat capacity change. These parameters have a substantial influence on the arc dynamics and hence on the spraying process in general [7]. Based on these assumptions one should be able to influence and improve the quality, i.e. hardness, oxide content, porosity, wear and corrosion resistance of the sprayed coatings by changing the atomizing gas mixture. To investigate these topics a BFS research group was founded. In the framework of this paper the influence of three different atomizing gases on the particle velocity, temperature and size distribution is examined and a first correlation with coating results is given.

2 Experimental Setup

2.1 Spraying system

For investigations within the BFS research group a special made wire arc spray gun, which allows to vary the position of the contact sleeves, the position of gas nozzle and the spray angle, has been developed. This spray gun has already been used by Eichler et al. in [4]

and has been used for particle investigations in this paper as well. For comparability reasons, the spraying angle and positions of aforementioned gun have been kept constant within this work. The spray gun has been coupled with a modified gas metal arc welding power supply, produced by EWM AG, Germany. The wire feed rate was set to 4 m/min and the arc voltage to 28.9 V. Resulting current values changed with the used atomizing gas and will be discussed later on.
The investigations have been carried out for three different atomizing gases: compressed air (DCA), pure nitrogen (N_2) and a nitrogen-hydrogen mixture of 4% by volume (N_2 + 4% H_2). Gas pressure has been kept constant at 5 bar. For particle measurements two materials have been used: an iron-based G3Si1 (0.08% C, 0.9% Si, 1.5% Mn, 0.015% P, 0.012% S) and a highly-popular in the industry 316L (17% Cr, 12.0% Ni, 2.5% Mo, 2.0% Mn, 1.0% Si, 0.08% C). The wire diameter of both materials was 1.6 mm. Three measurements per parameter set have been performed to determine process deviations.
Unlike the particle analysis, coating experiments had been done with an industrial wire arc spraying system SmartArc 350PP, produced by Oerlikon Metco, Switzerland. The wire feed rate remained constant, but the arc voltage has been adjusted to 28.0 V for DCA and 30.0 V for both nitrogen gases. The atomizing gas pressure has been kept constant at 2.4 bar. Due to these discrepancies, the coating analysis results cannot be directly compared with the particle diagnostic results, but can reflect trends to be expected in future investigations.

2.2 Diagnostic systems

To determine difference in the particle properties several diagnostic systems have been used. To observe droplet formation and detachment a high-speed-shadow-imaging system similar to the one used by Atzberger et al. [8] has been used. In this system the arc and plasma radiation is suppressed by means of a laser and shadows of the wire tips and detached droplets are projected onto the camera sensor. Due to the high frame rate of the camera (20000 fps) and some image processing algorithms it is possible to extract the size and velocity of the particles shortly after detachment. To do so, the images are binarized and a region-of-interest (ROI) is being set. Afterwards, the size of all particles within the defined ROI is determined and a mean particle size for each frame is calculated. Additionally, the particle velocity is established by cross-correlation as presented in [9]. Because of the chosen image section and the camera resolution the minimal particle diameter which can be detected is 64 µm. Hence the velocity has been calculated for particles with a diameter of 64 µm. For each parameter set as many as 8000 frames have been recorded. Parallel to that, arc voltage and current have been recorded using an oscilloscope.
Other particle diagnostics which have been used where the commercially available systems: SprayWatch produced by Oseir, Finland and DPV-2000 produced by Tecnar, Canada. The first one, SprayWatch, is an optical, camera based system which delivers particle velocity and temperature distribution as a direct result. The ROI of the system has a size of 21 x 17 mm^2. Similar to the high-velocity-shadow-imaging system, for each frame individual particles within a ROI of a size of 16 x 17 mm^2 are identified. Knowing the shutter time of the camera, as well as the scale, the position, direction and velocity can be calculated based on the length of the registered particle light traces. To achieve higher accuracy, since more particle light traces can be registered in one frame, a distribution function is formed. Furthermore, in the remaining ROI area of each frame two additional pictures with different spectral filters are enclosed to determine the particle temperature. Since the spectral images are taken simultaneously on a single chip and also the filter wavelengths are known (λ_1 = 700 nm and λ_2 = 850 nm) the particle temperature can be calculated using two-colour pyrometric rules. The grey level of each strip correlates to the intensity at a given wavelength and assuming black body radiation, the temperature for each pixel T_{px} can be derived from following equation as described in [10]:

$$T_{px} = \left(\frac{1}{\lambda_1} - \frac{1}{\lambda_2}\right) \frac{hc}{k \cdot \ln\left(\frac{I_1 \cdot \lambda_1^5}{I_2 \cdot \lambda_2^5} \cdot G\right)}$$

where λ_1 and λ_2 are the different wavelengths, I_1 and I_2 the detected intensities, h the Planck constant, c the speed of light, k the Boltzmann constant, and G a calibration factor. Based on the recorded data the particle size distribution can be calculated by the SprayWatch software.
To validate the obtained results a DPV-2000 diagnostic system has been used. Similarly to the previous one, it also allows to determine particle sizes, velocities and temperatures, but unlike SprayWatch or high-speed-shadow-imaging, the ROI of this system is relatively small with a volume of about 1 mm^3, hence data for individual particles can be acquired. The measurement is based on an optoelectronic sensor head with a two-slit mask. As a radiating particle flies through the ROI trigger signals from each slit are generated. Knowing the distance between the slits d_{slit} along with the magnification factor of the lens f_{magn}, the particle velocity v_p can be determined from the time difference Δt between the two trigger signals, as described by following equation:

$$v_p = \frac{d_{slit}}{\Delta t} f_{magn}$$

The particle temperature measurement is also based on the two-colour pyrometric rules. Here, a ratio of the radiation energies $E(\lambda_i)$ at two different wavelengths λ_i is being calculated assuming that the particles are grey-body emitters with equal emissivity at both spectral lines. Hence, the temperature T_p can be expressed by:

$$T_p = \left(\frac{1}{\lambda_1} - \frac{1}{\lambda_2}\right) \frac{c_2}{\left(\frac{E(\lambda_1)}{E(\lambda_2)} \cdot \left(\frac{\lambda_1}{\lambda_2}\right)^5\right)}$$

where c_2 = 1.438775·10$^{(-2)}$ mK. The radiation energy emitted at one wavelength is also used to determine the particle diameter d_p (under the assumption that the molten particles have a spherical or an almost spherical shape), which can be described by following equation:

$$d_p = \sqrt{\frac{E(\lambda_i)}{f_{diam}}}$$

To determine the particle diameter based on the radiation energy, the real emissivity of the particle has to be known. Usually a powder with known diameter distribution is measured in advance and the calibration factor f_{diam} determined. This approach may lead to uncertainties as the emissivity could be affected by temperature or the molten state of the particles as stated in [11]. Due to the relatively small ROI of the DPV-2000 system it is necessary to collect a sufficient number of datasets for a statistical assessment of the results.

Considering that both systems are based on optical sensors, the measurements will be affected by the arc radiation and thus it is not possible to measure particle parameters directly after detachment. Therefore the particle properties have been acquired at distances of 50, 100 and 150 mm, which includes the typical spraying distance range.

In addition to the particle diagnostics, coatings have been deposited with both wire materials for all three atomizing gases. The following coating analysis included the determination of hardness according to Vickers, oxide content and the deposition efficiency according to DIN EN ISO 17836 norm. The oxide content is a good indicator of corrosion resistance as well as hardness of coatings, depending on the used material. To determine the oxide content the coating has to be shed from the base substrate by using for ex. a three-point bending machine and coarsely milled subsequently. Then the milled granulate is being cleaned with acetone, as it can be contaminated with greases or other impurities during the shedding. Finally, the granulate is being analysed with a Leco TC-436DR nitrogen/oxygen detector. A defined amount of sample material is positioned in a graphite crucible and heated with a current of about 900 A up to 2400 °C, where the sample melts and evaporates. The gaseous oxides are being dispersed in a carrier gas, usually helium, and quantitatively analyzed by an infrared or thermal conductivity detector.

2.3 Validation of the particle diagnostic methods

The main part of the measurements within this work has been done with high-speed-shadow-imaging and the commercially available particle diagnostic system SprayWatch. Nonetheless, efforts have been made to compare both commercially available systems, the SprayWatch and the DPV-2000. The comparison data have been acquired just for one wire material, G3Si1, and one atomizing gas, DCA, at three distances: 50, 100 and 150 mm. As expected, the measured values differ between the systems, but it is apparent that similar trends can be observed, as presented in **Fig. 1**. It is also noteworthy, that the error bars for the DPV-2000 results are significantly larger. These discrepancies are probably due to the much higher accuracy of the latter. Nonetheless, to achieve accurate, reliable measurement results a significant amount of particles should be registered and analysed, which can only be done by extending the measurement time regarding the small ROI of that device. On the other hand, due to the bigger ROI area, the SprayWatch system analyses more particles in a shorter amount of time, and thus the measurement time can be reduced. It may lack the precision of the DPV-2000 system, but it seems to be sufficiently accurate to provide data for first interpretations, which is the aim of this work, and to show trends for further, more in-depth examinations for which the other system can be used. Therefore the SprayWatch system, along with the high-speed-shadow-imaging system, has been favoured for further investigations.

3 Results

As stated in the previous section various diagnostics have been used to determine the particle in-flight properties. Based on the high-speed-shadow-imaging recordings, average velocity and standard deviation within the first flight phase, i.e. shortly after material detachment took place, have been calculated for particles with a diameter of 64 µm and presented in **Tab. 1**. While recording, the arc current has also been measured and the mean value calculated. As can be seen, the mean current as well as average particle velocities vary significantly when the atomizing gas changes, especially between DCA and nitrogen based gases. Current changes can be explained as follows. It is known from literature that the arc voltage is influenced by the chosen gas mixture and, since the power source is voltage controlled, it tries to compensate for a changing voltage by adjusting the current value. This implies however, taking into account that the mean arc voltage calculated for each parameter set was constant at about 29.5 V, that the mean electrical power varies, since $\bar{P} = \bar{U} \cdot \bar{I}$.

Table 1: Mean arc current and average velocity of particles with a diameter of 64 µm determined by high-speed-shadow-imaging measurements. For DCA only one measurement for comparison reasons was carried out.

Atomizing gas	$\bar{I}$ $[A]$	$\sigma_{\bar{I}}$ $[A]$	$\bar{v}_{64\mu m}$ $[m/s]$	$\sigma_{\bar{v}}$ $[m/s]$
DCA	185.8	-	72.6	-
N_2	174.6	0.4	61.9	0.4
N_2 + 4% H_2	173.3	0.5	58.5	0.4

Furthermore, as also presented in **Tab. 1**, the gas velocities in the region of wire tips have to differ significantly, as the average velocities of particles with the same diameter of 64 µm vary for each atomizing gas, i.e. the particles have to experience different accelerations in that region due to different gas velocities.

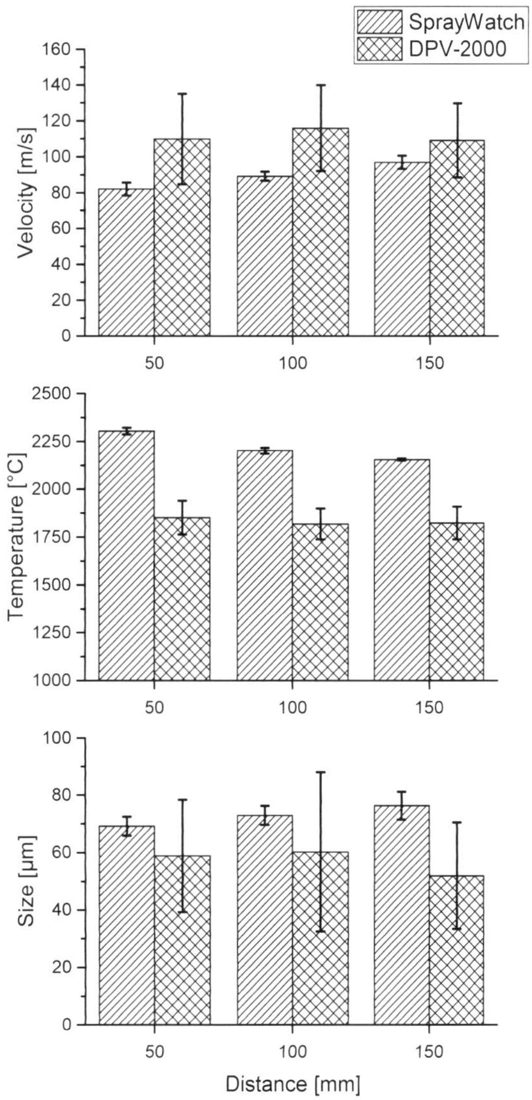

Figure 1: Comparison of the commercially available particle diagnostic systems. Measurements made for G3Si1 wire and compressed air as atomizing gas in different spraying distances.

Overall it can be stated that the mean electrical power and average particle velocity, and thus the atomizing gas velocity, for nitrogen-hydrogen gas mixture decrease when compared to pure nitrogen or DCA. Taken into account that the gas jet has to flow around the arc and thus is being heated by the arc, it can be assumed, that the decrease in electrical power results in a reduced heating of the gas. As a consequence, due to reduced heating, lower gas velocities in the wire tip region are attained and thus lower particle velocities can be measured. Based on this explanation, one may expect lower particle temperatures and particle velocities across the whole spraying distance, when nitrogen or a nitrogen-hydrogen atomizing gas mixture is used. These observations could be confirmed by the SprayWatch measurements for both materials, G3Si1 and 316L, made in a distance of 100 mm from the wire tips, **Fig 2**.

According to [12], due to different atomizing gas velocities, the smallest particle sizes from the cathodic wire, and also the anodic wire as stated in [7], can be expected when DCA is used for spraying. In the case when nitrogen is used, an increase in particle sizes in the wire tip region can be anticipated. Thus, the increase for nitrogen-hydrogen mixtures should be even more significant. To prove that, average particle sizes have been calculated from the high-speed-shadow-imaging recordings and presented in **Tab. 2**. The experimental results are in good agreement with the theoretical expectations. As shown in **Fig. 2**, similar trends can be observed at a 100 mm distance from the wire tips for both tested materials, although the average particle sizes are much smaller. This is to be expected, as the particles are being further atomized by the gas flow during their flight phase.

Table 2: Average particle sizes in the region of wire tips determined from high-speed-shadow-imaging measurements. For DCA only one measurement for comparison reasons was carried out.

Atomizing gas	d_{exp} [μm]	σ_d [μm]
DCA	206.0	-
N_2	216.7	4.4
N_2 + 4% H_2	226.4	4.1

Although a reduction in the mean electrical power, when nitrogen or nitrogen-hydrogen mixture were used instead of compressed air, could be observed, the coating experiments show a higher deposition efficiency for the G3Si1 wire material. Contrary, this is not the case for the 316L wire, as shown in **Fig. 3**. Further investigations, also with other materials, showed a correlation between the deposition efficiency and the melting temperature (TM) of the sprayed material. For materials with lower melting temperatures, as is the case with G3Si1 wire (TM 1450 °C), an increase in the deposition efficiency can be observed when nitrogen-hydrogen gas mixture is used. Materials with higher TM, here the 316L wire (TM 1750 °C), show an efficiency decrease, which could be explained as follows. When N_2 or N_2-H_2 is used as an atomizing gas, larger particle sizes in comparison to DCA are expected, as shown in **Fig. 2**. Larger particles cool down more slowly during the flight phase and therefore impinge at the coated surface with lower viscosity and surface tension, what results in a better grouting and hence better adhesion of the particles in relation to smaller ones. In all cases, the measured temperature values are higher than the melting temperature of the wire materials. In addition to that, much higher particle temperatures can be observed when DCA is used for both wire materials.

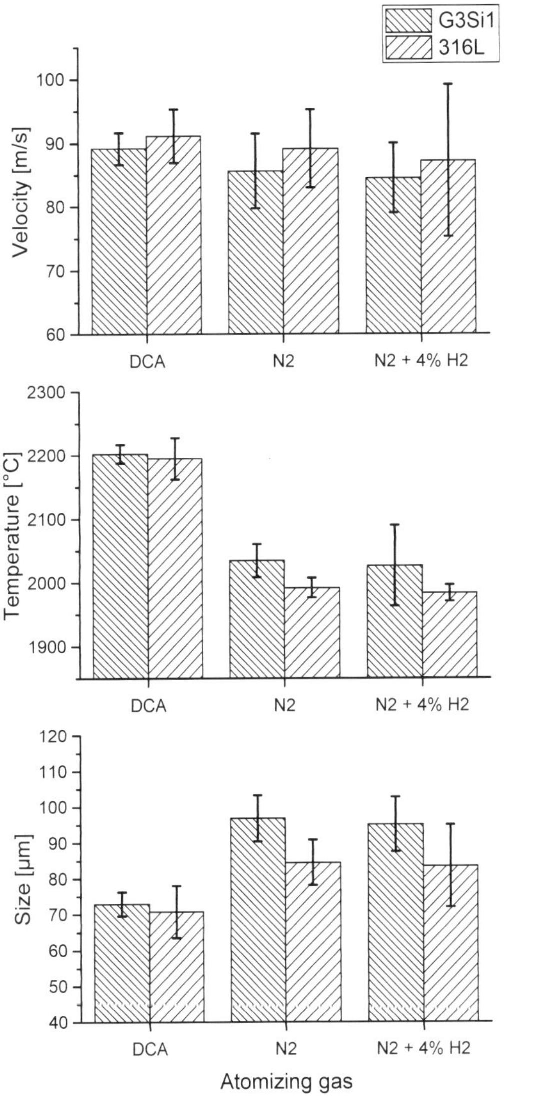

Figure 2: Velocity, temperature and particle size for different atomizing gases measured in a spray distance of 100 mm with SprayWatch system.

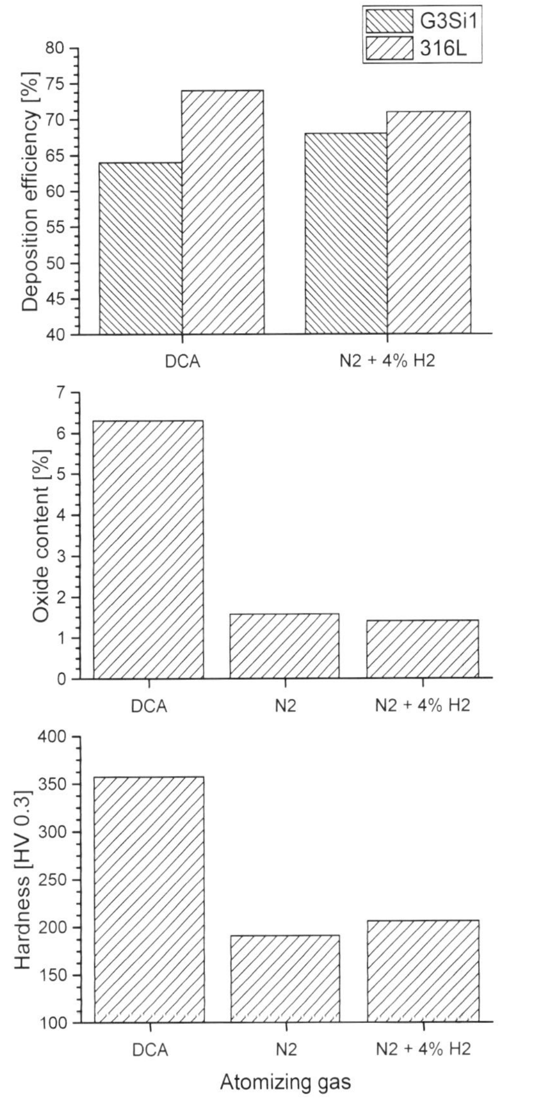

Figure 3: Deposition efficiency, oxide content and layer hardness for investigated wire materials when different atomizing gases are used.

For a higher temperature however, the temperature gradient between the molten particles and surrounding gas atmosphere has to be steeper, as the particle fly through the substantially cooler atmosphere. According to Newton's law of cooling, hotter particles are cooling faster than colder ones. So it can be assumed for materials with a higher melting temperature that the temperature gradient when sprayed with nitrogen or nitrogen-hydrogen mixture is not sufficient and so smaller particles can cool down during the flight to their TM and therefore adhere poorly, if at all, to the coated surface. This might explain the decreased deposition efficiency for materials with higher melting temperature, when these atomizing gas mixtures are used. To validate these assumptions, the temperature of individual particles should be determined, what can be done with the DPV-2000 system.

Besides the deposition efficiency, the oxide content of 316L layers has been determined. Oxides, as well as the carbon level, define such properties as conductivity, corrosion resistance or hardness of a coating depending on the alloy constituents of the sprayed wire material. Thus, the hardness according to Vickers has been measured in addition. As presented in **Fig. 3**, the oxide content is substantially reduced when DCA is substituted by N_2. The oxide reduction is even greater when a small amount of hydrogen (4% by vol.) is added to the atomizing gas, what corresponds well with the expectations. Since the layer hardness is material dependent, for materials with a low carbon content a reduction of oxides involves hardness decrease, what can be observed in the case of the 316L wire. Another reason for a higher layer hardness when using DCA for this specific material is the finer atomization of particles, as smaller particles have a larger specific surface area and thus a higher oxidation rate. For other wire materials in turn, nitrogen and nitrogen-hydrogen can bring qualitative benefits when a high layer hardness (for ex. 0.8% C steel) or conductivity (for ex. Cu) should be attained.

3 Conclusion and Outlook

As shown in this work, the selection of the atomizing gas has a significant effect on particle parameters such as size, velocity and temperature. Based on high-speed-shadow-imaging recordings, decreased particle velocities could be observed when the usually used for spraying compressed air has been substituted with nitrogen and further with a nitrogen-hydrogen gas mixture. This could be due to reduced gas velocities, as the arc heating has been diminished since the measured mean electrical power decreased for different gases. Furthermore, the analysis showed increased particle sizes. These results could be validated with SprayWatch measurements executed at a distance of 100 mm from the wire tips. Nonetheless, these effects should be further investigated experimentally, as the reduction in arc heating could be compensated with preheating of the atomizing gas. In addition to that, a first correlation with sprayed layers has been given. Here, a reduction in oxide content and layer hardness for the investigated wire material is observed, which is in good agreement with theoretical expectations. Interestingly, a decrease in deposition efficiency for this material (316L) could be seen. Contrary to that, an increased efficiency could be observed for another tested material (G3Si1). The investigations showed a correlation between the melting temperature of the sprayed wire material and resulting deposition efficiency. This is another interesting effect, which definitely needs further investigations and validation. A possible approach could be the temperature determination of individual particles in different positions, what could be done with the DPV-2000 diagnostic system. Moreover, a similar spraying setup should be used for both particle investigations and layer deposition.

4 Acknowledgment

The presented results were conducted in course of the project "Energy efficient production of thermally sprayed coatings through the use of a modified wire arc spraying process" funded by the Bayerische Forschungsstiftung (funding number AZ-1054-12).

5 Literature

[1] Z. Babiak and T. Wenz, „Grundlagen der thermischen Spritztechnik, Flamm- und Lichtbogenspritzen," in Moderne Beschichtungsverfahren, Weinheim, Wiley-VCH, 2005, pp. 131-149.

[2] M. von Wonderen, V. Sedov and I. MacMillan, „Newest light weight closed-loop arc spray gun," in Proceedings of the ITSC 2008, Maastricht, 2008.

[3] N. A. Hussary and J. V. R. Heberlein, „Effect of system parameters on metal breakup and particle formation in the wire arc spray process," Journal of Thermal Spray Technology, Vol. I, No. 16, pp. 140-152, 2007.

[4] S. Eichler, S. Kirner, S. Zimmermann, J. Schein, W. Mayr and J. Zierhut, „Detection and avoiding of unmelted wire pieces during wire arc spraying," in Proceedings of ITSC 2015, Long Beach, 2015.

[5] A. Pourmousa, J. Mostaghimi, A. Abedini and S. Chandra, „Particle size distribution in a wire-arc spraying system," Journal of Thermal Spray Technology, No. 14, pp. 502-510, 2005.

[6] F. T. Lang, Einflüsse des Zerstäubergases auf Umwelt und Beschichtung beim Lichtbogen-spritzen, Diploma thesis, Hochschule München, 2009.

[7] M. Kelkar and J. Heberlein, „Wire-arc spray modelling," in Plasma Chemistry and Plasma Processing, No. 22, pp. 1-25, 2002.

[8] A. Atzberger, S. Kirner, S. Zimmermann, G. Forster and J. Schein, „Investigation of a pulsed current wire arc spray process," in ITSC 2015 Proceedings, Long Beach, California, 2015.

[9] F. Bach, K. Möhwald, T. Rothardt, J. Prehm, L. Engl, K. Hartz and B. Drössler, „Particle Image Velocimetry in thermal spraying," Materials Science and Engineering: A, Vol. 383, pp. 146-152, 2004.

[10] H. Voggenreiter, J. Tewes, M. Hertter, T. Biermordt, M. Heinrich, S. Zimmermann, K. Landes and J. Schein, „Investigation and comparison of particle parameters determined with SprayWatch and LDA during the cold gas spraying process," in ITSC 2011 Proceedings, Hamburg, 2011.

[11] G. Mauer, R. Vaßen, S. Zimmermann, T. Biermordt, M. Heinrich, J.-L. Marques, K. Landes and J. Schein, „Investigation and comparison of in-flight particle velocity during the plasma-spray process as measured by Laser Doppler Anemometry and DPV-2000," Journal of Thermal Spray Technology Vol. 22, No. 4, 2013.

[12] J. Schaup, A. Atzberger, G. Huismann, K. Hartz-Behrend, S. Zimmermann, S. Kirner, M. Szulc, G. Forster and J. Schein, „Die Reise mit modernen Stromquellen und Brennersystemen hin zu optimiertem Lichtbogendrahtspritzen," in 10. Kolloquium Hochgeschwindigkeit-Flammspritzen, Erding, Germany, 2015.

Effects of Spray Kinetic Parameters on Coating Properties in Plasma Spray Process

Ting Liu, Yaping Ma, Lili Zheng, Hui Zhang, Guanzhong Zhang
Tsinghua University, Beijing, China

This study aims to develop a computational tool used for coating deposition simulation. Specifically, the interaction between spray kinetic parameters and coating properties (microstructure and thickness) is investigated. A three-dimensional coating build-up model is developed with actual in-flight particle locations and status measured in the experiments as model input. The spatial distribution of the splats is obtained from digital image processing and statistical analysis, while in-flight particles conditions before impact are measured by cross sectional scanning of DPV-2000 diagnostic system. The coating thickness uniformity under various scanning steps can be estimated by simulations. The developed model is applied to predict thickness, roughness and porosity of the coatings under various spray angles and moving speeds. It turns out that optimized scanning steps can be obtained for given spraying material, process and spray kinetic parameters. Appropriate step size can be determined in combination with other coating characteristics (thickness and residual stress). Results also indicate that strong influence of spray angle on the microstructure and thickness of the coatings. Reducing spray angle results in thinner coating with higher porosity. Torch scanning speed affects heat and mass transfer between plasma plume and substrate. A constant torch movement is preferred for homogeneous coating properties.

1 Introduction

Plasma spraying has been used for manufacturing a broad variety of functional coatings with high production and flexibility in operation. Sprayed feedstock materials (metals, alloys or ceramics et. al) can be heated into molten or semi-molten status when flying through plasma flame, and impacted on the substrate with certain momentum and energy. Microstructure and performance of the coatings are controlled by deformation, solidification and superposition of the impacting droplets, which are strongly affected by the kinetic parameters such as torch moving speed, angle and path on the substrate. Better quality of the coatings requires the optimal control of spray kinetic parameters. Common applied optimization methods are through trial and error approach, which is a time consuming and expensive process. It is therefore necessary to develop a numerical model to reveal the correlation between the kinetic parameters and coating characteristics.

It is widely accepted that the single deposit profile formed in thermal spray process can be described by a probability density function, whose parameters mainly depend on the material properties, spraying process parameters, spray angles and substrate conditions[1, 2]. Experimental results show that the deposit profile can be approximated by a Gaussian distribution[3, 4]. In some studies, the distribution pattern of particle sizes is considered to be the critical factor resulting in the eventual shape of the deposition, which is demonstrated to satisfy the log-Gaussian function[5-7]. Dyshlovenko et al. [6] performed a parametrical study on plasma spray process by numerical simulation. The particle diameters before impact were assumed to satisfy a log-normal distribution, while their temperatures and velocities were obtained through simulation. The calculated porosity of the coatings exhibited a certain deviation with the experimental results when the given distribution was different from real size distribution of arriving particles. Parizi et al.[8] developed a commercial software to model the coating formation. The impact conditions such as size, temperature and velocity of the particles were obtained directly from DPV-2000 measurements or certain distributions from prior experiments. Results demonstrated that more accurate coating properties can be obtained when the DPV-2000 data were directly applied into the software. Chen et al. [9] assumed that the deposition pattern was influenced by spatial distribution of individual droplets at cross section of impact on the substrate. The stochastic deposition of particle position was assigned according to the Gaussian distribution. Results indicated that such model can make a reasonable prediction of the change of coating porosity under various spray angles and scanning velocities. As a matter of fact, the coating deposition process is complicated which may be affected by a combination of various process parameters. It is difficult to realize accurate controlling of coating properties if the correlation between the status parameters and spatial distribution of inflight particles is ignored.

In this study, a coating build-up model is developed based on the stochastic analysis to setup several rules governing splat deformation, splat overlap and pore formation. A theoretical model is employed to estimate the actual spatial distribution of the particles at cross section before impact. In the meantime, the distribution of in-flight particle parameters at the corresponding cross section are measured by real-time diagnostic system of DPV-2000. Numerical simulations are performed to investigate the influences of spray kinetic parameters, such as torch scanning speed, spray angle and scanning step, on coating structure formation and coating properties.

2 Experimental approach

Experiments are carried out to measure the spatial and status distributions of arriving particles at normal and off-normal spray angles. ZB-80 plasma torch (Beijing Zhenbang Aerospace Precision Machinery Co, .Ltd, Beijing, China) is used in the experiments with operating conditions listed in Table 1. The spraying material is 8YSZ powder with particle size distribution from 50µm to 75µm.

Tab. 1. Plasma spray parameters for YSZ coating formation

Variables	Value
Input power (kW)	42
Current (A)	600
Primary gas flow rate, Ar (slm)	40
Secondary gas flow rate, H_2 (slm)	8.0
Carrier gas flow rate, Ar (slm)	5.0
Feeding rate (g/min)	2
Spray distance (mm)	90
Deposition efficiency (%)	40

Polished stainless steel substrate is fixed and preheated to capture the snapshot of the entire footprint of plasma plume at normal or off-normal spray angles. The splat map obtained is considered to reflect real spatial distribution of the particles before impact which can finally stick on the substrate to form the coatings. Figure 1 illustrates the schematic diagram of splat collection and deposition. A spraying process with experimental conditions is simulated. The coordinate system of the particle spatial distribution (x,y,z) will move along a predetermined path with a pre-defined speed and the original point in the coordinate system will be coincided with the centreline of plasma torch. The deposit morphology is characterized in a fixed coordinate system of (X,Y,Z), and its original point is coincided with the centre of substrate. The collected splat map is then photographed and analysed by a digital image processing to extract the coordinates of those splats in the (x,y,z) system. A full plume cross section scan is performed by DPV-2000, and the particle conditions are collected in cross section at 7×7 positions with the step size of 5mm. It is worth mention that the deposition efficiency for off-normal spraying is corrected according to total number of the particles at measuring points on the grid.

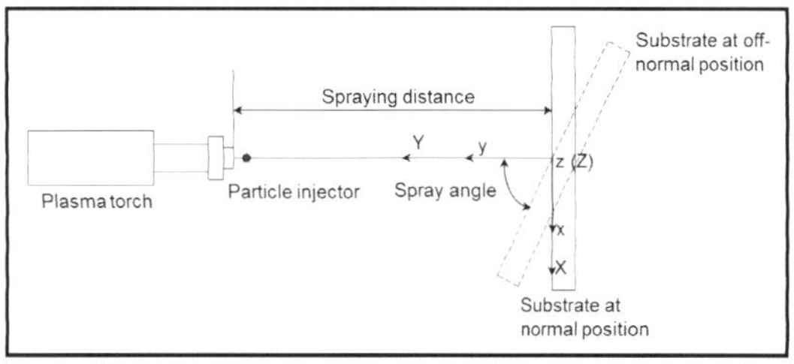

Fig. 1. Schematic view of splat collection and coating formation. The (x,y,z) coordinate system is used to characterize splats distribution; The (X,Y,Z) coordinate system is applied to coating deposition.

3 Modelling of coating build-up

The coating build-up process in plasma spraying is a multi-scale procedure with time scale ranging from microseconds to seconds. The time needed for droplet spreading and solidification is much shorter (about one-tenth) than the lag between two droplets hitting the same position. The coating deposition process can be divided into two independent parts based on the above evidence: splat formation and coating growth.

3.1 Splat formation model

When a molten or semi-molten droplet impacts on a substrate, it will flatten out and start solidification even before the flattening process is completed. The splat morphology and coating microstructure will be affected by droplet size, velocity, temperature and molten degree, as well as substrate condition. Besides, the splat formation becomes more complicated when substrate surface has highly complex topology. During the past decades, several theoretical models were developed to predict the flattening degree of droplet after impinging on a planar substrate, and the flattening ratio ξ of splat is defined as follows:

$$\xi = D/d \quad (1)$$

where D and d represent the diameter of the splat and arriving droplet, respectively.

Among those theoretical derivations, the most widely accepted approach is based on the macroscopic mechanical energy conservation:

$$\frac{d}{dt}(E_k + E_p + L_f) = 0 \quad (2)$$

where E_k, E_p and L_f are the kinetic energy, potential energy and work done due to friction, respectively. Madejski [10] proposed a velocity profile of a molten droplet after impact which satisfied the continuity and non-slip condition. In combination with viscous dissipation and surface tension, the flattening ratio of the molten droplet is a function of the Reynolds and Weber numbers as follows:

$$\frac{3\xi^2}{We} + \frac{1}{\mathrm{Re}}\left(\frac{\xi}{1.2941}\right)^5 = 1, We > 100, \mathrm{Re} > 100 \quad (3)$$

Delplanque and Rangel [11] adopted a new velocity profile for droplet flattening process and made correction to viscous dissipation. An improved model is proposed as follows:

$$\frac{3\xi^2}{We} + \frac{1}{\mathrm{Re}}\left(\frac{\xi}{1.1625}\right)^5 = 1, \mathrm{Re} > 140, We > 670 \quad (4)$$

In the case of plasma spraying, the term $3\xi^2/We$ is negligible according to McPherson[12]. The flattening ratio can therefore be simplified as follows:

$$\xi = 1.625\,\mathrm{Re}^{1/5} \quad (5)$$

In this study, Eq. 5 is adopted to calculate the flattening degree of the splats. Temperature, velocity and diameter of arriving particles are obtained through the measurements from DPV2000. In addition, the elongation effect of the splats due to off-normal spraying is neglected to simplify the modelling procedure.

3.2 Coating growth model

The spatial distribution of the splats collected on the substrate is considered to satisfy the Gaussian mixed distribution in this work. The probability distribution function $P(x,\mu,\Sigma)$ can be expressed as follows:

$$P(x,\mu,\Sigma)=\sum_{k=1}^{n} p_k N(x|\mu_k,\Sigma_k) \\ =\sum_{k=1}^{n} p_k \cdot \frac{1}{\sqrt{|\Sigma_k|\cdot(2\pi)^d}} \cdot e^{-\frac{1}{2}(x-\mu_k)\Sigma_k^{-1}(x-\mu_k)'} \quad (6)$$

where x is a two-dimensional vector with two components representing the coordinates of the splats in the planar of the substrate:

$$x=[x,z] \quad (7)$$

p_k is the probabilistic weighting of the k_{th} single peak Gaussian distribution, and the rest of the symbols with subscript "k" are parameters of this single peak distribution. For example, μ_k is a two-dimensional vector representing the statistical expectation of x in the k_{th} single peak distribution:

$$\mu_k=[\mu_{kx},\mu_{kz}] \quad (8)$$

Σ_k is a second order covariance matrix, which is defined as follows:

$$\Sigma_k=\begin{bmatrix} \sigma_{kx} & Conv_k(x,z) \\ Conv_k(x,z) & \sigma_{kz} \end{bmatrix} \quad (9)$$

where σ_{kx}, σ_{kz} and $Conv_k(x,z)$ are the variance of the coordinates x, z and covariance of those two coordinates. A series of random impact positions can be obtained through the probability density function of Eq. 6. The measured data from DPV-2000 are stored as database, and once the location of a particle is determined, its temperature, velocity and diameter are evaluated by searching and interpolating from the database.

The microstructure of thermal spray coatings is composed of splat-based elements formed through flattening and solidification of the droplets. The most common defects in thermal spray coatings can be generalized into three types: disk-like lamellar pores, globular pores and vertical micro cracks in both splats and coatings. The coating properties and performance are strongly influenced by the above imperfections.

A 3D stochastic model for coating deposition is developed based on random impact position, size, velocity and temperature of the particles obtained from the statistical analysis of the experiment data. In order to simplify the model for more efficient calculation, some assumptions are made as follows: i) particles are fully melted; ii) the splat is a regular disc-like shape without splashing; iii) the substrate is smooth and planar with constant temperature. The mechanism of coating formation is simplified as superposition of the splats one by one, and a set of rules are proposed to control the deposition process:

(1) The droplet follows exactly the shape of underlayer where the particles impact on.

(2) There will be a displacement of the centre of splat if the layer under the splat is not flat, and the displacement depends on the maximum height difference of the region under the splat.

(3) The void will not be filled if it is not on the top surface.

(4) If the gap on the surface is narrower than the height of the upper splat, it will not be filled and become a globular pore.

4 Results and discussion

4.1 Statistical analysis of Inflight particle status

The snapshot of splat distribution is collected on a polished substrate at the spraying distance of 90mm. The substrate is photographed (Fig. 2a) for statistical analysis, and image processing is adopted afterwards to extract the coordinates of individual splats on the map. Figure 2b shows the spatial distribution of the splats plotted from the data of image analysis, and it is obvious that most of the splats can be identified with their coordinates being recorded properly. Statistical analysis can then be performed to figure out the regularities of the splat distribution in Fig. 2a or 2b. A Gaussian mixture model is applied in this work to predict the spatial distribution of the particles. In order to test the feasibility of the above model, a bunch of random positions are generated according to this model and exhibited in Fig. 2c. Results indicate that the Gaussian mixture model is appropriate to estimate the impact location of the particles.

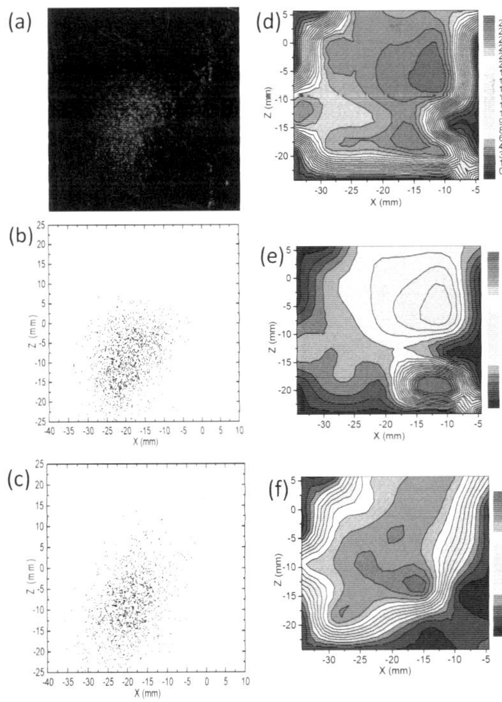

Fig. 2. Splat spatial distribution obtained through: (a) experiment; (b) image analysis; (c) statistical analysis,

and distribution of inflight particles: (d) temperature; (e) velocity; (f) diameter.

In-flight particle conditions are measured in the corresponding cross-section plane of plasma plume. The measuring area is 30mm by 30mm in size and DPV-2000 is applied to perform a cross-section scan at 7 by 7 grids with grid size of 5mm. Figure 2d-2f shows the measured temperature, velocity and diameter distributions of plasma plume. During coating deposition simulation, a random position of the particles is generated at first, and temperature, velocity and diameter of the particles are evaluated after interpolation with the experimental data.

4.2 Coating structures

Figure 3a and 3b shows the top and side views of the deposit after spraying along a single line parallel to the X axis at Z=2.5 mm. It is noticed that the coating deviates to an extent from the trajectory of the torch centreline in the Z direction, which is mainly resulted from the vertical deflection of plasma plume induced by the perpendicular injection mode of the powder. Figure. 3c exhibits the top view of a typical coating sprayed along a "z" type trajectory with scanning step of 10mm. It is clear that the thickness of the coating is inhomogeneous since the scanning step is too large in this case. Optimization of scanning trajectories is therefore needed to guide the industrial production with minimum time and money. Figure 3d shows the cross-section of the coating structure obtained by simulation, and the green parts represent the pores generated according to the rules presented early. The average coating thickness, surface roughness as well as porosity under different kinetic parameters (such as torch scanning speed and spray angle) can therefore be predicted by this model. In addition, the top and 3D view of the coating prepared under the optimized scanning step are shown in Fig. 3e and 3f, which demonstrate that in addition to the roughness on the coating surface induced by the inappropriate scanning steps (Fig. 3c), the random superposition of the splats during spraying is another important parameter of surface roughness.

4.3 Spraying trajectory optimization

The homogeneity of coating thickness will be largely influenced by the spraying trajectory. In the case of the most frequently-used "z" type path, the step size of the adjacent two paths is the most important parameter which needs to be optimized. Plasma spraying of YSZ coating is simulated with substrate size of 25mm by 25mm. The "z" type trajectory is assigned to the "plasma torch" with the step size ranging from 6mm to 9mm, and the scanning velocity is fixed to be 50mm/s for all cases. The top views of the sprayed coatings with different scanning steps are shown in Fig. 4a-4d. It is evident that for a given spray condition the coating morphology changes gradually from striped pattern to uniform pattern. Since

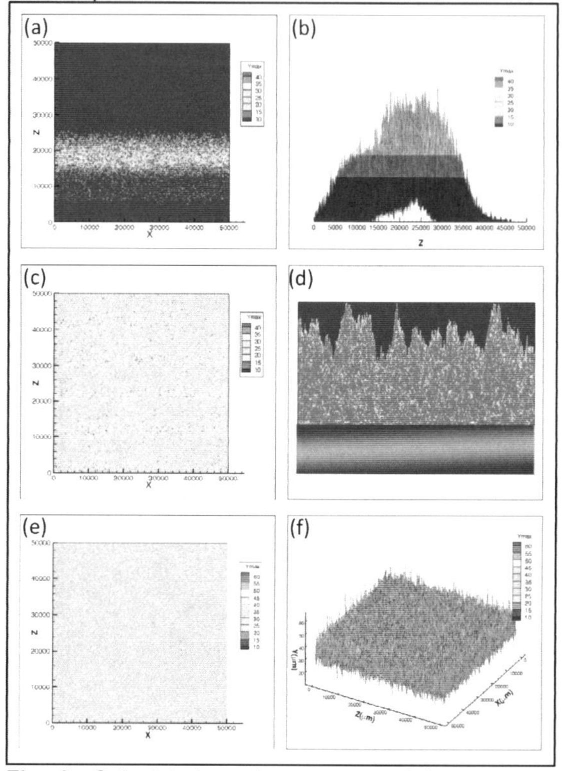

Fig. 3. Calculated coating structure: (a) top and (b) side views of deposit along a straight line, (c) top view and (d) cross section morphylogy of the deposit with step size of 10mm, (e) top and (f) 3D views of the deposit with optimized step size of 7mm.

the stochastic process will result in the oscillation of the coating thickness around its average. Statistical average is performed to investigate the inherent regularity of the deposition pattern under different scanning steps. The integral-averaged coating thickness distributions along the Z axis under different scanning steps are shown in Fig. 4e-4h. Results demonstrate that the coating thickness homogeneity can be firstly obtained at a statistical sense when the step size decreases to 7mm. Further decrease of the step size will result in a continuous increase of coating thickness, but the homogeneity of the coating thickness will be barely changed. On the other hand, more intensive heat transfer between coating and plasma flame will occur with the decrease of step size. Hence, the coating properties such as temperature and stress distributions will significantly change with the scanning path.

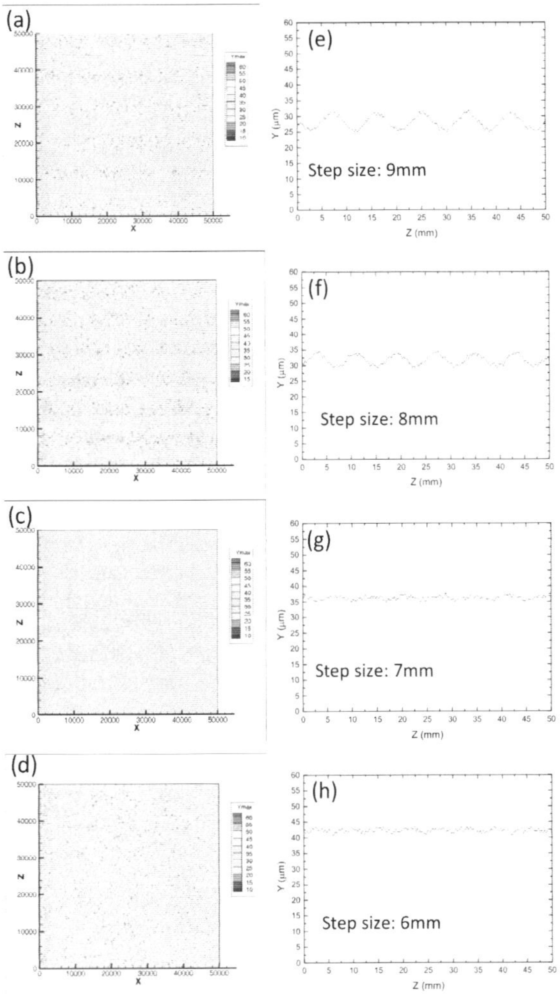

Fig. 4. Top view of the deposit with step size of (a) 9mm; (b) 8mm; (c) 7mm; (d) 6mm, and corresponding thickness distributions along the Z direction.

4.4 Influence of spray angle and scanning velocity

For plasma spraying on a complex surface, drastic change of spray angle or torch scanning velocity may occur in the region with large curvature, owing to the limitation of robot kinematics. It is therefore necessary to understand the influences of spray angle and scanning speed on coating microstructure. Splat maps at different spray angles are built up and inputted into the calculation program for coating prediction. The cross sectional profile of the deposit sprayed along a single line at spray angle of 50°, 70°and 90°are shown in Fig. 5. It is obvious that the deposition parameters, such as width, centre of mass and maximum height, are significantly influenced by the spray angle. The changes of width and centre line of deposit profile are mainly resulted in the shift of particle distribution in the spray core at different spray angles. But the maximum height of the deposit decreases when the spray angle changes from 90 ° to 50 ° since the deposition efficiency decreases with the decrease of spray angle. Trifa et al. [3] investigated the effect of spray angle on the deposit shape by experiments and similar tendency was reported in their experiments.

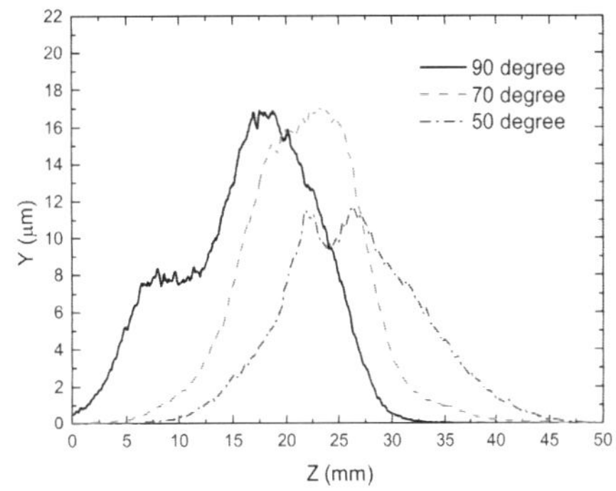

Fig. 5. Deposit profiles under different spray angles.

The coating structure properties including average thickness, surface roughness and porosity under different spray angles are calculated subsequently. Results, as shown in Fig. 6, illustrate that average coating thickness and roughness decreases to 25.5 μm and 3.6 μm when the spray angle decreases from 90° to 50°, but the coating porosity shows a reverse tendency with the spray angle. The coating thickness is largely controlled by the deposition efficiency during spray process, which decreases as the spray angle decreases. The surface roughness is induced not only by inappropriate scanning step but also by the random superposition of the splats during deposition. Meanwhile, the amplitude of random oscillation resulted from splat overlap may accumulate as the coating grows, which in turn causes a minor increase of roughness with the increase of spray angle. The mechanism of porosity formed in this simulation is due to the unfulfillment of the void between the splats, which are related to the distribution and deformation degree of the particles to a large extent. Berndt et al. [13] pointed out that off-normal spray would result in lower particle impact momentum, and this might lead to the increase of coating porosity.

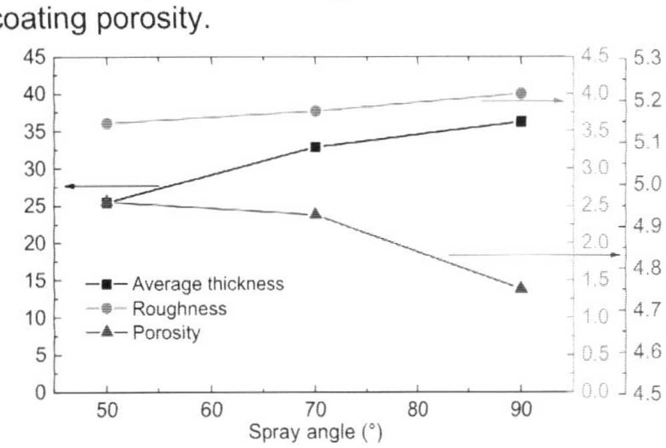

Fig. 6. Average coating thickness, roughness and porosity as a function of spray angles.

The coating thickness, surface roughness and porosity at different scanning velocities are also calculated, and results are exhibited in Fig. 7. Significant changes can be observed in coating thickness and roughness, which decrease from 36.2 μm and 3.99 μm to 1.81 μm and 0.90 μm, respectively when the torch scanning speed increases from 50 mm/s to 1000 mm/s. It is evident that total mass of the deposit will decrease as the scanning speed increases, which results in the decrease of the average coating thickness and surface roughness. Also, the increase of the scanning speed from 50 mm/s to 1000 mm/s is only associated with a small decline of 0.14% in the coating porosity.

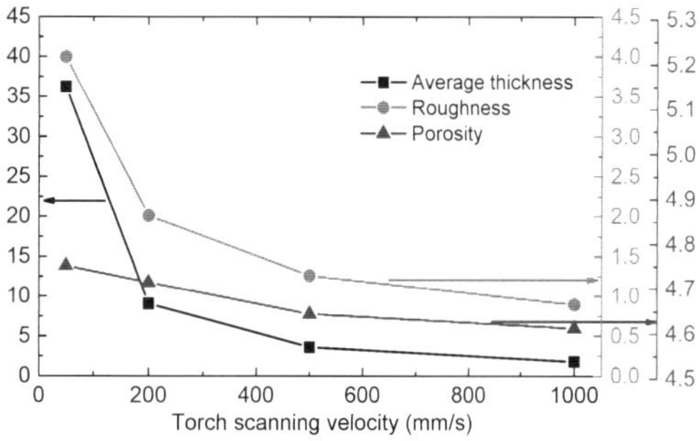

Fig. 7. Average coating thickness, roughness and porosity as a function of torch scanning speed.

5 Conclusions

A coating build-up model is developed with actual particles spatial and status distributions obtained as input data from the experiments. The model is able to simulate real spray process for parametric analysis, and it is applied to predict the coating structures as a function of spray kinetic parameters. For specific operating conditions, spray trajectories can be designed and optimized to increase the thickness uniformity of the coatings. At the meantime, the effects of spray angle and torch scanning speed on coating structure are investigated. Results indicate that the coating microstructure and thickness are strongly influenced by spray angle. Lower coating thickness and higher porosity can be obtained as spray angle decreases. The torch scanning speed is intensively influenced on heat and mass transfer between plasma plume and substrate, which will significantly influence the coating thickness. Experiments are conducted afterwards to validate the results from the simulations on predicting coating microstructure and thickness under different kinetic parameters.

6 Acknowledgments

This paper is supported by the Civil Aviation Science and Technology Project of China (Grant No. 20150215).

7 Nomenclature

d	in-flight droplet diameter, um
D	splat diameter, um
Re	Reynolds number, W m^{-1} K^{-1}
We	Weber number, m
x,y,z	coordinate system for particles distribution
X,Y,Z	coordinate system for coating deposition

Greek symbols

ξ	flattening ratio of splats

Subscripts

k	index

8 Literature

[1] Cai, Z., S. Deng, H. Liao, C. Zeng and G. Montavon: The Effect of Spray Distance and Scanning Step on the Coating Thickness Uniformity in Cold Spray Process. Journal of Thermal Spray Technology. 23 (2013), Issue 3, pp. 1/9.

[2] Fauchais, P., A. Vardelle, M. Vardelle and M. Fukumoto: Knowledge concerning splat formation: An invited review. Journal of Thermal Spray Technology. 13 (2004), Issue 3, pp. 337/60.

[3] Trifa, F.I., G. Montavon and C. Coddet: On the relationships between the geometric processing parameters of APS and the Al2O3–TiO2 deposit shapes. Surface & Coatings Technology. 195 (2005), Issue 1, pp. 54/69.

[4] Kout, A. and H. Müller: Parameter optimization for spray coating. Advances in Engineering Software. 40 (2009), Issue 10, pp. 1078/86.

[5] Dyshlovenko, S. and B. Pateyron: Numerical simulation of hydroxyapatite powder behaviour in plasma jet. Surface & Coatings Technology. 179 (2004), Issue 1, pp. 110/17.

[6] Dyshlovenko, S., L. Pawlowski, B. Pateyron, I. Smurov and J.H. Harding: Modelling of plasma particle interactions and coating growth for plasma spraying of hydroxyapatite. Surface & Coatings Technology. 200 (2006), Issue s 12–13, pp. 3757/69.

[7] Shi, D., M. Li and P.D. Christofides: Diamond Jet Hybrid HVOF Thermal Spray: Rule-Based Modeling of Coating Microstructure. Industrial & Engineering Chemistry Research. 43 (2004), Issue 14, pp. págs. 3653/65.

[8] Parizi, H.B., J. Mostaghimi, L. Pershin and H.S. Jazi: Analysis of the Microstructure of Thermal Spray Coatings: A Modeling Approach. Journal of Thermal Spray Technology. 19 (2010), Issue 4, pp. 736/44.

[9] Chen, Y., G. Wang and H. Zhang: Numerical simulation of coating growth and pore formation in rapid plasma spray tooling. Thin Solid Films. 390 (2001), Issue s 1–2, pp. 13/19.

[10] Madejski, J.: Droplets on impact with a solid surface. International Journal of Heat & Mass Transfer. 26 (1983), Issue 7, pp. 1095/98.

[11] Delplanque, J.P. and R.H. Rangel: An improved model for droplet solidification on a flat surface. Journal of Materials Science. 32 (1997), Issue 6, pp. 1519/30.

[12] Mcpherson, R.: Relationship Between The Mechanism of Formation, Microstructure and Properties of Plasma-Sprayed Coatings. Thin Solid Films. 83 (1981), Issue 3, pp. 297/310.
[13] Leigh, S.H. and C.C. Berndt: Evaluation of off-angle thermal spray. Surface & Coatings Technology. 89 (1997), Issue 3, pp. 213/24.

Synthesis and Thermophysical Properties of $Gd_2Zr_2O_7/SrZrO_3$ Composite as a Thermal Barrier Coating Material

W. Ma*, L.-L. Cai, H.-Y. Dong, Y. Bai, Y.-C. Yin, X.-H. Li
Inner Mongolia University of Technology, Hohhot 010051, P.R. China
**E-mail: w.ma@imut.edu.cn*

Abstract

The $(Gd_xSr_{1-x})ZrO_{3+0.5x}$ composite (x=0.3, G3S7; x=0.5, G5S5; x=0.7, G7S3) powders and bulk materials were in-situ synthesized by co-precipitation method and sintered by both pressureless sintering (PS) and spark plasma sintering (SPS), respectively. The phase stability of the composite powders and bulk materials were evaluated. The mechanical and thermophysical properties of the composite bulk materials were also investigated. The composite powders and bulk materials showed good phase stability from room temperature to 1400°C and after heat-treatment at 1450°C for a long period, respectively. The bulk G5S5 had the highest microhardness of 15.2±2.2 GPa compared with the bulk G7S3 (14.1±1.9 GPa) and G3S7 (12.5±1.6 GPa). The bulk G5S5 also had the highest Young's modulus of 266.5±13.9 GPa in comparison with the bulk G7S3 (261.5±13.8 GPa) and G3S7 (256.3±7.8 GPa). The bulk G3S7 has the highest fracture toughness of 1.96±0.11 $MPa\cdot m^{1/2}$, followed by the bulk G7S3 (1.63±0.07 $MPa\cdot m^{1/2}$) and G5S5 (1.60±0.09 $MPa\cdot m^{1/2}$). The CTEs of the bulk G3S7, G5S5 and G7S3 were $9.1\text{-}10.8\times10^{-6}$, $9.1\text{-}10.8\times10^{-6}$ and $9.3\text{-}10.8\times10^{-6}\ K^{-1}$ in a temperature range of 200-1350°C, respectively. The CTEs of the bulk G7S3 has no obvious abnormal change during measuring, indicating the phase transitions of $SrZrO_3$ were suppressed. The thermal conductivity of the bulk G7S3 was ~1.1 $W\cdot m^{-1}\cdot K^{-1}$ at 1000°C. Thus, the bulk G7S3 is considered as a promising thermal barrier coating material.

Introduction

Thermal barrier coatings (TBCs) are extensively applied to protect hot-section metallic components both in aero engines and stationary gas turbines, with significantly improved efficiencies and decreased emissions [1-3]. State-of-the-art TBCs are based on 7-8 wt.% Y_2O_3 partially stabilized ZrO_2 (8YSZ) due to its excellent thermophysical and mechanical properties [4,5]. However, 8YSZ cannot be long-term operated above ~1200°C because of phase transitions and accelerated sintering [6-8]. Therefore, much effort was devoted to develop new ceramic TBC candidates which can be operated at higher temperatures.

Among the new ceramic TBC candidate materials, $Gd_2Zr_2O_7$ and $SrZrO_3$ have attracted much attention. $Gd_2Zr_2O_7$ is considered as a promising TBC candidate material owing to its high melting point (>2300°C), phase stability up to the melting point, low thermal conductivity (~1.3 W/mK) [9,10], as well as superior CMAS resistance [11-13]. However, the relatively low CTE and poor fracture toughness of $Gd_2Zr_2O_7$ resulted in relatively short thermal cycling lifetime of its coatings.

$SrZrO_3$ has been proven as one of the most refractory oxides with melting point as high as 2650°C [14]. A fracture toughness of the monolithic $SrZrO_3$ was comparable to that of YSZ. The CTE of the $SrZrO_3$ coating is larger than the YSZ coating in the temperature range of 200 to 1200°C, it also exhibits superior sintering resistance [15]. However, $SrZrO_3$ undergoes three phase transitions from room temperature to high temperatures, and only the transformation from orthorhombic *Pnma* to pseudo-tetragonal *Imma* at 700°C involves a small volume change of ~0.14% [16].

Ceramic matrix composites (CMCs) design can be an effective method to overcome the intrinsic brittleness and lack of mechanical reliability of monolithic ceramic. CMCs may benefit other properties such as thermal expansion coefficient, thermal conductivity, as well as phase stability. The $La_2(Zr_{0.7}Ce_{0.3})_2O_7$ composite has good sintering resistance, low thermal conductivity compared to $La_2Zr_2O_7$ and $La_2Ce_2O_7$ [17]. The Al_2O_3/YSZ composite coatings exhibit better thermal insulation and superior thermal shock resistance than YSZ [18]. The $La_2Zr_2O_7$/YSZ nano-composite powder has much higher activation energy for grain growth of $La_2Zr_2O_7$ and YSZ than that of the single-phase counterparts [19], resulting in higher phase stability and sintering resistance.

In this contribution, the $(Gd_xSr_{1-x})ZrO_{3+0.5x}$ (x=0.3, 0.5, 0.7) composite was in-situ synthesized by co-precipitation and calcination method. The mechanical properties and thermophysical properties of the sintered $(Gd_xSr_{1-x})ZrO_{3+0.5x}$ composite were investigated.

Experimental Procedure

$Gd(NO_3)_3{\cdot}6H_2O$, $Zr(NO_3)_4{\cdot}5H_2O$ and $Sr(NO_3)_2$ (Tianjin Guangfu Fine Chemical Institute, AR, Tianjin, China) were used as starting materials. $Gd(NO_3)_3{\cdot}6H_2O$ and $Zr(NO_3)_4{\cdot}5H_2O$ with proper ratio were added into the distilled water to get a solution, followed by dropping the solution into dilute ammonia to get the suspension A, with continuously stirring and keeping the pH value as 10. The same procedure was repeated to get the suspension B using $Zr(NO_3)_4{\cdot}5H_2O$ and $Sr(NO_3)_2$, the only difference was that ammonium oxalate was used as a precipitant. The suspension A and B were mixed and stirred for 30 min, followed by washing to pH value as 7 using distilled water. The precipitates were dried at 120°C for 12 h, followed by heating at high temperatures in air to get $(Gd_xSr_{1-x})ZrO_{3+0.5x}$ (x=0.3, 0.5, 0.7) composite powders. For comparison, $Gd_2Zr_2O_7$ and $SrZrO_3$ were also prepared under the same preparation condition.

The bulk $(Gd_xSr_{1-x})ZrO_{3+0.5x}$ were cold pressed and sintered by two different sintering methods. One was pressureless sintering (PS) at 1600°C for 6 h, another was spark plasma sintering (SPS) (Model SPS-20T-10, Shanghai Chen Hua Electric Science and Technology Co. Ltd., China) at 1500°C for 5 min with heating rate of ~100°C/min. The pressureless sintered samples were used to characterize thermal expansion, thermal conductivity properties and microstructure, whereas the SPS sintered samples were used to measure the mechanical properties.

The CTEs and thermal diffusivities were recorded by a high-temperature dilatometer (Model DIL 402E, Netzsch, Selb, Germany) and a laser flash method (Model THETA, Netzsch, Selb, Germany), respectively. The density (ρ) of the sintered samples was measured according to the Archimedes' principle. The porosity of the bulk $(Gd_xSr_{1-x})ZrO_{3+0.5x}$ used for measuring thermal diffusivity was in a range of 8~15%. The thermal conductivity of the bulk samples was calculated with the Eq. (1) and then calibrated with Eq. (2):

$$\lambda = D_{th}(T)C_p(T)\rho(T) \quad \text{(Eq 1)}$$

$$\lambda_0 = \lambda / [1-(4\Phi/3)] \quad \text{(Eq 2)}$$

where λ, $D_{th}(T)$, $C_p(T)$, and $\rho(T)$, λ_0 and Φ are the thermal conductivity, the thermal diffusivity, the specific heat capacity, the measured density, the calibrated value of the thermal conductivity and the porosity, respectively. The specific heat capacities of the bulk $(Gd_xSr_{1-x})ZrO_{3+0.5x}$ were calculated using Neumann-Kopp rule [20].

For measurements of the mechanical properties, the sintered disk-shaped samples were mounted and polished to < 1 μm. Young's modulus and microhardness of the bulk $(Gd_xSr_{1-x})ZrO_{3+0.5x}$ were determined by an indentation instrument (Model Nano Indenter XP, Agilent Technologies, Inc., Santa Clara CA, USA) under a constant load of 100 mN. A fracture toughness of the bulk $(Gd_xSr_{1-x})ZrO_{3+0.5x}$ was determined using a conventional microhardness tester (Model HXD-1000TM/LCD, Shanghai Taiming Optical Instrument Co. Ltd., Shanghai, China) with applied loads of 1-20 Kg.

The phase analysis of the synthesized $(Gd_xSr_{1-x})ZrO_{3+0.5x}$ powder and the bulk $(Gd_xSr_{1-x})ZrO_{3+0.5x}$ was characterized by XRD (Model D/MAX 2200, Rigaku Co. Ltd., Japan). The microstructure and composition of the bulk $(Gd_xSr_{1-x})ZrO_{3+0.5x}$ were investigated by a field-emission transmission electron microscope (Model JEM-2100F, JEOL, Tokyo, Japan) with an energy dispersive X-ray spectrometer (EDX) (Model Inca, Oxford Instruments, Oxfordshire, U.K.).

Results and Discussion

Synthesis of the $(Gd_xSr_{1-x})ZrO_{3+0.5x}$ Ceramics and Their Phase Stability

The XRD patterns of the $(Gd_xSr_{1-x})ZrO_{3+0.5x}$ (x=0, 0.3, 0.5, 0.7, 1.0) powders calcined at 1400°C for 3 h are shown in Fig. 1. $SrZrO_3$ exhibits a perovskite structure. With an increase of Gd content in the $(Gd_xSr_{1-x})ZrO_{3+0.5x}$, the diffraction peaks of $SrZrO_3$ shift to the large 2θ-side (lattice parameter decreased) and those of $Gd_2Zr_2O_7$ to the small 2θ-side (lattice parameter increased), indicating that Gd^{3+} and Sr^{2+} are replaced by each other. Furthermore, it is clear that $Gd_2Zr_2O_7$ exhibits a disordered fluorite structure. That is unusual for $Gd_2Zr_2O_7$ after calcination at 1400°C, because $Gd_2Zr_2O_7$ undergoes a pyrochlore to fluorite (order-disorder) transition when heated above 1530-1550°C [21]. In this case, Gd^{3+} is substituted by Sr^{2+} with development of oxygen vacancy in $Gd_2Zr_2O_7$, resulting in the formation of the disordered fluorite structure.

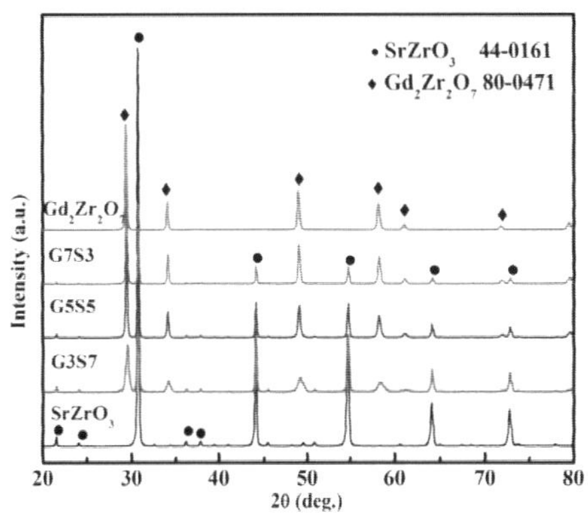

Figure 1: XRD patterns of the $(Gd_xSr_{1-x})ZrO_{3+0.5x}$ powders (x=0, 0.3, 0.5, 0.7, 1.0) calcined at 1400°C for 3 h.

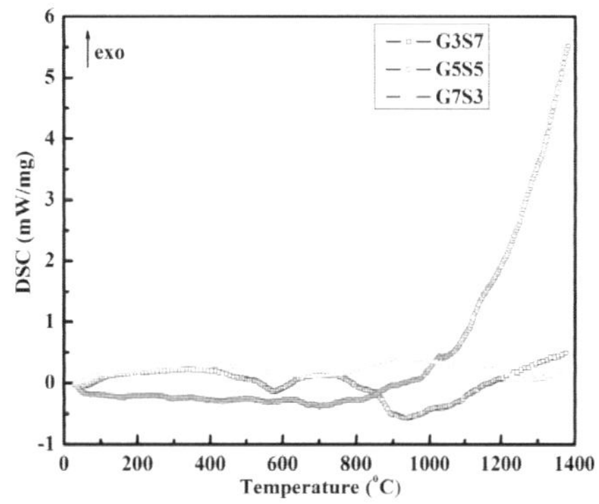

Figure 2: DSC curves of the G3S7, G5S5 and G7S3 powders synthesized at 1400°C for 3 h.

Figure 2 shows the DSC curves of the G3S7, G5S5 and G7S3 powders synthesized at 1400°C for 3 h. The DSC curve of the G3S7 powder has two obvious endothermal peaks at ~580 and ~930°C, which might correspond to the phase transitions of $SrZrO_3$ [22]. For both the G5S5 and G7S3 powders, especially the G7S3 powder, there is neither an endothermal peak nor an exothermal peak in the DSC curve, indicating superior phase stability.

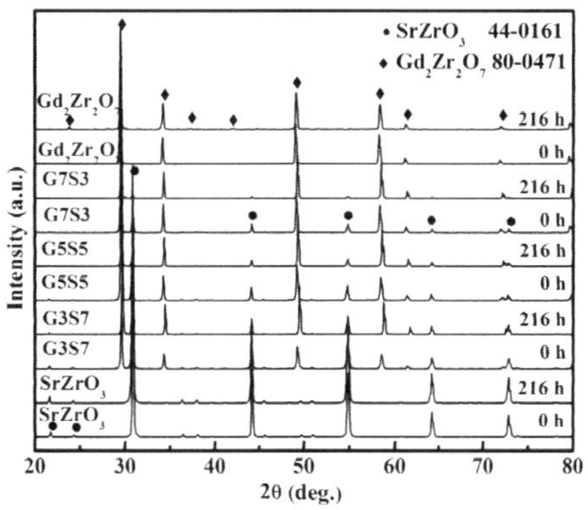

Figure 3: XRD patterns of the bulk $(Gd_xSr_{1-x})ZrO_{3+0.5x}$ (x=0, 0.3, 0.5, 0.7, 1.0) sintered at 1600°C for 6 h and after heat-treatment at 1450°C for 216 h.

The bulk $(Gd_xSr_{1-x})ZrO_{3+0.5x}$ were sintered at 1600°C for 6 h by pressureless sintering. The XRD patterns of the as-sintered bulk $(Gd_xSr_{1-x})ZrO_{3+0.5x}$ and the sintered bulk materials after heat-treatment at 1450°C for 216 h are shown in Fig. 3. The diffraction peaks of the as-sintered bulk $(Gd_xSr_{1-x})ZrO_{3+0.5x}$ become sharp compared to their powder counterparts, indicating well crystalized. After heat-treatment at 1450°C for 216 h, the diffraction peaks of the bulk $(Gd_xSr_{1-x})ZrO_{3+0.5x}$ (x=0, 0.3, 0.5, 0.7) do not change in comparison with the as-sintered ones, whereas the characteristic diffraction peaks of (311) and (511) at ~37° and ~44°, respectively, appear for $Gd_2Zr_2O_7$ with pyrochlore structure. It indicates that $Gd_2Zr_2O_7$ transformed from the disordered fluorite structure to the ordered pyrochlore structure after heat-treatment at 1450°C for 216 h, which was also reported by Subramanian [21].

Microstructure of the Bulk $(Gd_xSr_{1-x})ZrO_{3+0.5x}$ and Their Mechanical Properties

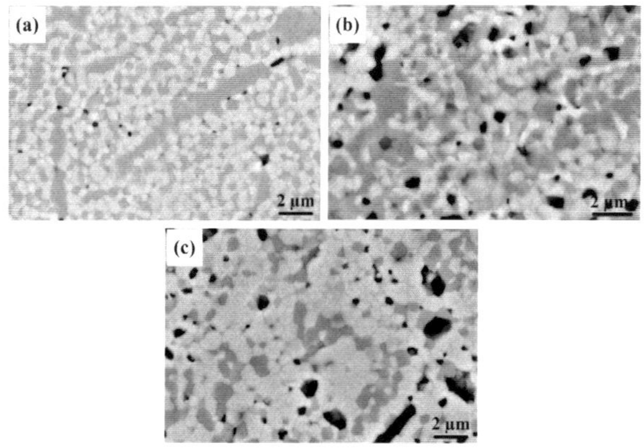

Figure 4: Microstructures of the bulk $(Gd_xSr_{1-x})ZrO_{3+0.5x}$: (a) x= 0.3, (b) x=0.5, (c) x=0.7.

The microstructures of the bulk $(Gd_xSr_{1-x})ZrO_{3+0.5x}$ (x=0.3, 0.5, 0.7) are shown in Fig. 4. The bulk G3S7, G5S5 and G7S3 are comprised of dark grey grains and light grey grains, which were confirmed as $SrZrO_3$ and $Gd_2Zr_2O_7$ by EDS analyses together with XRD results. The average grain size for both $SrZrO_3$ and $Gd_2Zr_2O_7$ are in a range of 1~2 µm. The $Gd_2Zr_2O_7$ grains are distributed in bulk G3S7 more homogenous compared with the bulk G7S3. And the $Gd_2Zr_2O_7$ grain size in the bulk $(Gd_xSr_{1-x})ZrO_{3+0.5x}$ becomes bigger with an increase of x value. This might be due to much higher volume fraction of $Gd_2Zr_2O_7$ than $SrZrO_3$ in the bulk G7S3, the growth of $Gd_2Zr_2O_7$ grains could not be suppressed effectively by $SrZrO_3$ grains.

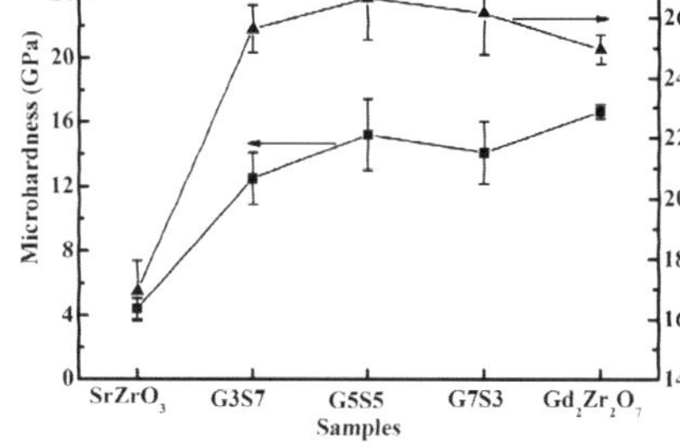

Figure 5: Microhardness and Young's modulus of the bulk $(Gd_xSr_{1-x})ZrO_{3+0.5x}$ (x=0, 0.3, 0.5, 0.7, 1.0).

The microhardness and Young's modulus of the bulk $(Gd_xSr_{1-x})ZrO_{3+0.5x}$ (x=0, 0.3, 0.5, 0.7, 1.0) are shown in Fig. 5. Both the bulk $SrZrO_3$ and $Gd_2Zr_2O_7$ have a minimum and a maximum microhardness of 4.4±0.6 GPa and 16.6±0.4 GPa, respectively. The bulk G5S5 has a higher microhardness of 15.2±2.2 GP, followed by the bulk G7S3 (14.1±1.9 GP) and G3S7 (12.5±1.6 GP). As for Young's modulus, the bulk G5S5 has the highest value of 266.5±13.9 GP, followed by the bulk G7S3 (261.5±13.8 GP), G3S5 (256.3±7.8 GP), $Gd_2Zr_2O_7$ (249.6±4.8 GP), as well as $SrZrO_3$ (169.5±10.1 GP). The microhardness and Young's modulus of the bulk G3S7, G5S5 and G7S3 are more comparable to that of the bulk $Gd_2Zr_2O_7$, the difference in microhardness and Young's modulus is possibly due to the different porosity in the bulk materials to some extent.

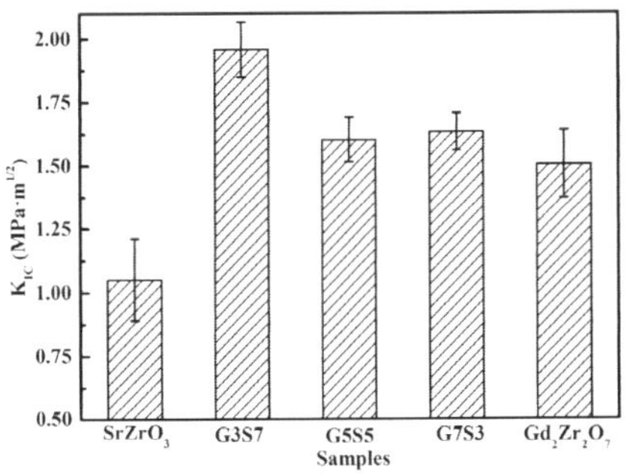

Figure. 6 Fracture toughness of the bulk $(Gd_xSr_{1-x})ZrO_{3+0.5x}$ (x=0, 0.3, 0.5, 0.7, 1.0).

The fracture toughness of the bulk $(Gd_xSr_{1-x})ZrO_{3+0.5x}$ is shown in Fig. 6. The bulk G3S7, G5S5 and G7S3 composites have higher fracture toughness than both the bulk $SrZrO_3$ (1.05±0.16 MPa·m$^{1/2}$) and $Gd_2Zr_2O_7$ (1.50±0.13 MPa·m$^{1/2}$). The bulk G3S7 has the highest fracture toughness of 1.96±0.11 MPa·m$^{1/2}$, followed by the bulk G7S3 (1.63±0.07 MPa·m$^{1/2}$) and G5S5 (1.60±0.09 MPa·m$^{1/2}$). The composites have more homogeneous microstructure with small grain size compared with single-phase $SrZrO_3$ and $Gd_2Zr_2O_7$, in which the grain growth for both $SrZrO_3$ and $Gd_2Zr_2O_7$ can be suppressed effectively. This gives a positive influence on increasing the fracture toughness of the bulk G3S7, G5S5 and G7S3 composites. Another reason might be due to the CTE mismatch between $Gd_2Zr_2O_7$ and $SrZrO_3$, which can produce micro cracks to some extent at the grain interfaces and suppress the crack growth further, resulting in higher fracture toughness.

Thermophysical Properties of the Bulk $(Gd_xSr_{1-x})ZrO_{3+0.5x}$

The coefficient of thermal expansion is an important thermal physical property for the bulk $(Gd_xSr_{1-x})ZrO_{3+0.5x}$. The results of the dilatometric measurement for the bulk $(Gd_xSr_{1-x})ZrO_{3+0.5x}$ are shown in Fig. 8. The CTE of the bulk $SrZrO_3$ is 8.7-10.6 × 10^{-6} K^{-1} (200-1100°C), whereas the CTEs of both the bulk G3S7 and G5S5 are 9.1-10.6 × 10^{-6} K^{-1} (200-1100°C). Among the investigated materials, The bulk G7S3 and $Gd_2Zr_2O_7$ exhibit the highest CTE of 9.3-10.7 × 10^{-6} K^{-1} (200-1100°C). The CTE of the bulk $SrZrO_3$ has two abnormal changes at ~750 and ~1130°C, which correspond to the phase transitions of orthorhombic↔pseudo-tetragonal and tetragonal↔cubic [16]. Both the phase transitions are suppressed gradually for G3S7 and G5S5, and fully suppressed for G7S3. This is in good agreement with the DSC measurement. The diffusion of Gd^{3+} into $SrZrO_3$ lattice during sintering might contribute to the phase stabilization of $SrZrO_3$ in the bulk $(Gd_xSr_{1-x})ZrO_{3+0.5x}$, and the generated compressive stress at the grain interface of $SrZrO_3$ and $Gd_2Zr_2O_7$ due to the CTE mismatch also might be a reason for the phase stabilization of $SrZrO_3$.

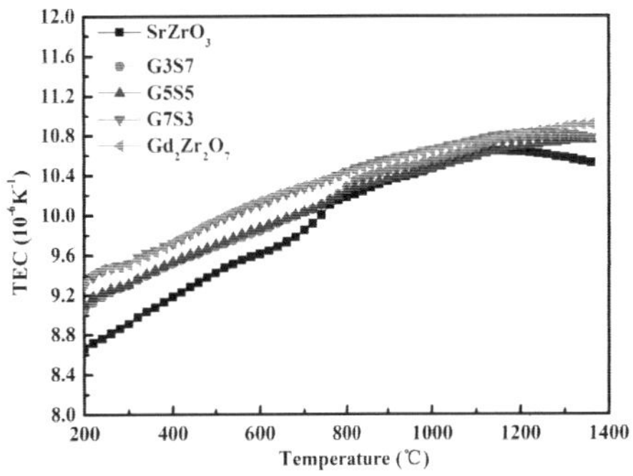

Figure 7: Coefficients of thermal expansion of the bulk $(Gd_xSr_{1-x})ZrO_{3+0.5x}$ (x=0, 0.3, 0.5, 0.7, 1.0).

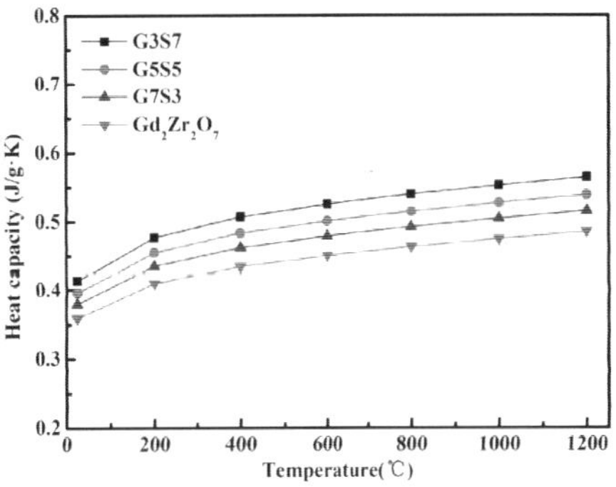

Figure 8: Heat capacities of the bulk $(Gd_xSr_{1-x})ZrO_{3+0.5x}$ (x=0.3, 0.5, 0.7, 1.0).

The heat capacities of the bulk $(Gd_xSr_{1-x})ZrO_{3+0.5x}$ are calculated by Neumann-Kopp rule and shown in Fig. 8. $Gd_2Zr_2O_7$ exhibits the lowest heat capacity among the investigated bulk materials, and the heat capacity increases with increasing the content of $SrZrO_3$ in the bulk $(Gd_xSr_{1-x})ZrO_{3+0.5x}$.

The variations in thermal diffusivity with temperature for the bulk $(Gd_xSr_{1-x})ZrO_{3+0.5x}$ are shown in Fig. 9. The thermal diffusivities of the bulk $(Gd_xSr_{1-x})ZrO_{3+0.5x}$ monotonically decrease with increase of temperature from 25 to 1200°C, which suggests a dominant phonon conduction behavior. The thermal diffusivities of the bulk G3S7 and G5S5 are identical, which are higher than the bulk G7S3, followed by the bulk $Gd_2Zr_2O_7$ with the lowest value.

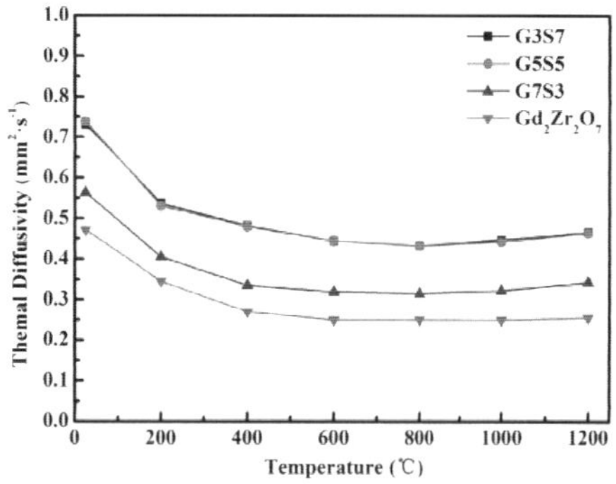

Figure. 9 Thermal diffusivities of the bulk ($Gd_xSr_{1-x})ZrO_{3+0.5x}$ (x=0.3, 0.5, 0.7, 1.0).

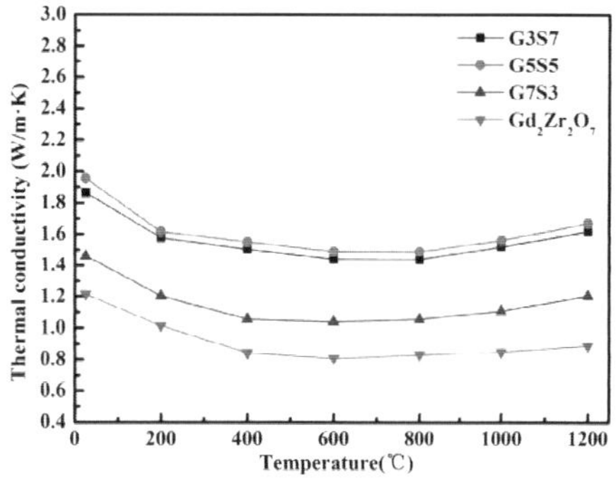

Figure. 10 Thermal conductivities of the bulk ($Gd_xSr_{1-x})ZrO_{3+0.5x}$ (x=0.3, 0.5, 0.7, 1.0).

The thermal conductivities of the bulk $(Gd_xSr_{1-x})ZrO_{3+0.5x}$ as a function of temperature are shown in Fig. 10, which has a similar changed tendency in comparison with the thermal diffusivities. The thermal conductivity of the bulk $Gd_2Zr_2O_7$ at 1000°C is 0.8 W/mK, which is about 73% of that of the bulk G7S3 (1.1 W/mK), as well as about half that of the bulk G5S5 (1.6 W/mK) and G3S7 (1.5 W/mK). Among the three composites, the thermal conductivity of the bulk G7S3 is much lower than that of $SrZrO_3$ (2.1 W/mK @1000°C) [15] and 8YSZ (2.1-2.2 W/mK @1000°C) [23], compared to the bulk G5S5 and G3S7.

Conclusions

The $(Gd_xSr_{1-x})ZrO_{3+0.5x}$ (x=0.3, 0.5, 0.7) composite powders were successfully synthesized by calcining the co-precipitated precursor at 1400°C, and the G7S3 powder has an excellent phase stability from room temperature to 1400°C. The bulk $(Gd_xSr_{1-x})ZrO_{3+0.5x}$ (x=0.3, 0.5, 0.7) has no phase change after heat-treatment at 1450°C for 216 h.

The bulk G3S7 has the highest fracture toughness of 1.96±0.11 MPa·m$^{1/2}$, followed by the bulk G7S3 (1.63±0.07 MPa·m$^{1/2}$) and G5S5 (1.60±0.09 MPa·m$^{1/2}$), which are higher than that of $SrZrO_3$ and $Gd_2Zr_2O_7$. The bulk G7S3 and $Gd_2Zr_2O_7$ exhibit the highest CTE of 9.3-10.7 × 10^{-6} K^{-1} (200-1100°C), which is higher than that of $SrZrO_3$. The obvious phase transitions at ~750°C and ~1130°C for $SrZrO_3$ has been suppressed in the bulk G7S3 effectively. The bulk G7S3 has the lowest thermal conductivity of 1.1 W/mK at 1000°C compared to the bulk G5S5 and G3S7, which is much lower than that of $SrZrO_3$ (2.1 W/mK @1000°C) and 8YSZ (2.1-2.2 W/mK @1000°C). Thus, the bulk G7S3 is considered as a promising thermal barrier coating material.

Acknowledgments

The authors gratefully acknowledge the financial support by the National Natural Science Foundation of China (51462026), the Inner Mongolia Natural Science Foundation (2014MS0509), and Shanghai technical platform for testing and characterization on inorganic materials (14DZ2292900).

References

[1] Miller R.A., "Current status of thermal barrier coatings - an overview," *Surf Coat Technol*, Vol. 30, No. 1 (1987), pp. 1-11.

[2] Padture N.P., Gell M., Jordan E.H., "Thermal barrier coatings for gas-turbine engine applications," *Science*, Vol. 296, No. 5566 (2002), pp. 280-284.

[3] Clarke D.R., Phillpot S.R., "Thermal barrier coating materials," *Mater. Today*, Vol. 8, No. 6 (2005), pp. 22-29.

[4] Nelson W.A., Orenstein R.M., "TBC experience in land-based gas turbines," *J Therm Spay Technol*, Vol. 6, No. 2 (1997), pp. 176-180.

[5] Nicholls J.R., Lawson K.J., Johnstone A., Rickerby D.S., "Method to reduce the thermal conductivity of EB-PVD TBCs," *Surf Coat Technol*, Vol. 151-152, (2002), pp. 383-391.

[6] Miller R.A., "Oxidation-based model for thermal barrier coating life," *J Am Ceram Soc*, Vol. 67, No. 8 (1984), pp. 517-521.

[7] Miller R.A., Smialek J.L., Garlick R.G., "Phase stability in plasma-sprayed partially stabilized zircoina-yttria," *In: Advances in Ceramics*, Cloumbus, OH, January 1981, pp. 241-253.

[8] Renteria A.F., Saruhan B., "Effect of ageing on microstructure changes in EB-PVD manufactured standard PYSZ top coat of thermal barrier coatings," *J Eur Ceram Soc*, Vol. 26, No. 12 (2006), pp. 2249-2255.

[9] Stöver D., Pracht G., Lehmann H., Dietrich M., Döring J., Vassen R., "New material concepts for the next generation of plasma-sprayed thermal barrier coatings," *J Therm Spay Technol*, Vol. 13, No. 1 (2004), pp. 76-83.

[10] Wu J., Wei X.Z., Padture N.P., Klemens P.G., Gell M., Garcia E., Miranzo P., Osendi M.I., "Low-thermal-conductivity rare-earth zirconates for potential thermal-barrier-coating applications," *J Am Ceram Soc*, Vol. 85, No. 12 (2002), pp. 3031-3035.

[11] Krämer S., Faulhaber S., Chambers M., Clarke D.R., Levi C.G., Hutchinson J.W., Evans A.G., "Mechanisms of cracking and delamination within thick thermal barrier

systems in aero-engines subject to calcium-magnesium-alumino-silicate (CMAS) penetration," *Mater Sci Eng A*, Vol. 490, No. 1-2 (2008), pp. 26-35.

[12] Drexlera J.M., Ortizb A.L., Padture N.P., "Composition effects of thermal barrier coating ceramics on their interaction with molten Ca-Mg-Al-silicate (CMAS) glass," *Acta Mater*, Vol. 60, No. 15 (2012), pp. 5437-5447.

[13] Drexler J.M., Chen X.H., Gledhill A.D., Shinoda K., Sampath S., Padture N.P., "Plasma sprayed gadolinium zirconate thermal barrier coatings that are resistant to damage by molten Ca-Mg-Al-silicate glass," *Surf Coat Technol*, Vol. 206, No. 19-20 (2012), pp. 3911-3916.

[14] Noguchi T., Okubo T., Yonemochi O., "Reactions in the System ZrO_2-SrO," *J Am Ceram Soc*, Vol. 52, No. 4 (1969), pp. 178-181.

[15] Ma W., Mack D.E., Vassen R., Stöver D., "Perovskite-type strontium zirconate as a new material for thermal barrier coatings," *J Am Ceram Soc*, Vol. 91, No. 8 (2008), pp. 2630-2635.

[16] Zhao Y.S., Weidner D.J., "Thermal expansion of $SrZrO_3$ and $BaZrO_3$ perovskites," *Phys Chem Miner*, Vol. 18, No. 5 (1991), pp. 294-301.

[17] Cao X.Q., Vassen R., Tietz F., Stöver D., "New double-ceramic-layer thermal barrier coatings based on zirconia–rare earth composite oxides," *J Eur Ceram Soc*, Vol. 26, No. 3 (2006), pp. 247-251.

[18] Tarasi F., Medraj M., Doaltabadi A., Lima R.S., Moreau C., "Thermal cycling of suspension plasma sprayed alumina-YSZ coatings containing amorphous phase," *J Am Ceram Soc*, vol. 95, No. 8 (2012), pp. 2641-2621.

[19] Jiang K., Liu S.B., Li C.C., "Nano-nano composite powders of lanthanum zirconate and yttria-stabilized zirconia by spray pyrolysis," *J Am Ceram Soc*, Vol. 96, No. 10 (2013), pp. 3296-3303.

[20] Leitner J., Chuchvalec P., Sedmidubsky D., Strejc A., Abrman P., "Estimation of heat capacities of solid mixed oxides," *Thermochim Acta*, Vol. 395, No. 1-2 (2002), pp. 27-46.

[21] Subramanian M.A., Aravamudan G., Subba Rao G.V., "Oxide pyrochlores - a review," *Prog Solid State Chem*, Vol. 15, No. 2 (1983), pp. 55-143.

[22] Howard C.J., Knight K.S., Kennedy B.J., Kisi E.H., "The structural phase transitions in strontium zirconate revisited," *J Phys Condens Matter*, Vol. 12, N0. 45 (2000), pp. L677-L683.

[23] Sodeoka S., Suzuki M., Inoue T., Ueno K., Oki S., "Thermal and mechanical properties of plasma sprayed ZrO_2-CeO_2-Y_2O_3 coatings," *Proceedings of the 9th National Thermal Spray Conference, Thermal Spray: Practical Solutions for Engineering Problems*, C.C. Berndt, Ed., ASM International, Cincinnati, OH, 1996, pp. 295-302.

Heat-shock properties in yttrium-oxide films synthesized from metal–ethylenediamine tetraacetic acid complex through flame-spray apparatus

D. Y. Enkin, Keiji Komatsu*, Keita Abe, Takashi Kosuta, Yutaka Ikeda, Atsushi Nakamura, Shigeo Ohshio, Hidetoshi Saitoh

We investigated the heat-shock properties of metal-oxide films synthesized from a metal‐ethylenediamine tetraacetic acid (EDTA) complex. Y_2O_3 films were synthesized on a stainless-steel (SUS) substrate from EDTA・Y・H through the combustion of H_2-O_2 gas. A cyclic heat-shock test was conducted on the fabricated Y_2O_3 films through exposure to the H_2-O_2 flame. The maximum temperatures on SUS were 450–728°C in the experiment. The existence of Y_2O_3 crystals was confirmed. Although the number of cracks, crack lengths, and cracks per unit area were increased by the heat shock, delaminations were not observed in the Y_2O_3 films. The results show that the prepared Y_2O_3 films have high thermal-shock resistance and are suitable for use as thermal barrier coatings.was 453°C owing to the high heat of evaporation of water.

1 Introduction

Nowadays, the level of desired properties in thermal barrier coatings (TBCs) for extreme environmental applications has increased [1][2]. Ceramic coating techniques, which are applied in aircraft technology, have been selected for coating on top of TBCs. Jet engines in aircraft often operate under extreme thermal-shock environments at a temperature of approximately 1400°C. Thermal shocks produce many issues; for instance, cracks are produced by thermal stress in ceramic films. Thermal stress occurs at grain boundaries with different diameters [3]. Such cracks lead to deformations and delaminations. Therefore, the accurate control of the microstructure of a ceramic film is required in TBC techniques. In the thermal-spraying deposition process, melted or semisolid in-flight particles collide with the substrate or are deposited in the ceramics layer and splatted. Subsequently, the splatted particles solidify. Consequently, pores are produced in the region between particles. Pores can be open or closed [4]. Pores penetrating the deposited layer are open pores. The relaxation of rapid thermal stress in a deposited film is accomplished by controlling the pores in the film's microstructure [5].

Recently, a new deposition technique using a metal–ethylenediamine tetraacetic acid (EDTA) complex was developed [6][7][8]. In this technique, metal–EDTA complex powders are introduced in a commercial flame-spraying apparatus. These complex powder are decomposed in the flame and forms metal-oxide particles, which are subsequently deposited on the substrate. Micron-order-thickness metal-oxide films with 6%–15% cross-sectional porosities were deposited on stainless-steel (SUS) substrates in several seconds[7]. This technique is considered inferior to established thermal-spraying techniques, such as plasma spraying. Dense metal-oxide films can be deposited using the plasma-spraying technique. However, the established thermal-spraying techniques require high costs and large equipment, in contrast to the meta–EDTA deposition technique. Metal-oxide films obtained through metal–EDTA deposition are expected to have high thermal shock because of their dense microstructure. Therefore, the metal–EDTA deposition holds potential as a TBC technique in extreme environments.

In this study, the heat-shock properties of yttrium-oxide (Y_2O_3) films synthesized using a metal–EDTA complex with flame-spray apparatus were investigated. Y_2O_3 was selected as it shows high plasma resistance [9]. The crystal structure and microstructure (surface and cross-sectional) were analyzed for the deposited films. A cyclic thermal-shock test [10][11] was conducted to investigate the thermal-shock properties of the deposited films. The deposited films were exposed to a H_2-O_2 flame, and the deposited film's microstructures were analyzed. Furthermore, the existence of cracks, deformation, and delamination in the deposited films was analyzed through the thermal-shock test.

2 Experimental

The EDTA complex, EDTA・Y・H (Chubu Chelest Co., Ltd.), was prepared for Y_2O_3 film syntheses[6]. A conventional flame-spray apparatus consisting of a feed unit (5MPE, Sulzer Metco) and spray gun (6P-II, Sulzer Metco) was used to perform reactive spraying. We prepared three types of carrier gases to compare the effect of these gas species on the properties of the deposited films. The EDTA・Y・H was then placed into a feed unit and transported through O_2 gas flow to the spray gun. The carrier gas-flow rate was 7.1 L/min. In this study, we selected hydrogen gas as the fuel. A mixture of H_2 and O_2 at flow rates of 32.6 and 43.0 L/min, respectively, was used as the flame gas. The EDTA・Y・H powder was mixed with the flame and reacted with oxygen after the thermal decomposition of EDTA. The reacted particles were then sprayed onto a stainless-steel substrate (SUS304, 30 × 50 × 1 mm) that was previously blasted by #60 alumina grit (99.7% purity, 212–250 μm particle size, Fuji Manufacturing Co., Ltd.), resulting in the deposition of a metal-oxide film. The stand-off distance between the spray gun and the substrate was 150 mm. The deposition duration was 4–10 s. The traverse rate of the gun was 50 mm/s. Subsequently, the spray nozzle was moved in a longitudinal direction, and deposition was performed at each area for approximately 6 s without pre-heating the substrate. For reference, the yield percentage (deposition efficiency) of this technique was measured to be approximately 60%.

Figure 1 shows the setup of the thermal-shock test for films deposited from EDTA・Y・H on the SUS substrate, in addition to a photograph of the experiment. The rotation apparatus, which has the capability to deposit twelve samples at a time, was fabricated in-house. Mortar was used in the apparatus for rolling. The rolling velocity of the apparatus was kept within the range of 45–90 rpm by controlling the applied voltage of a variable autotransformer. To investigate the effects of thermal shock, we used the rotation apparatus at only 45 rpm for control. To define the parameters for the thermal-shock test, the number of times the sample passed through the flame (number of scans) was counted. The numbers of scans used were 100, 300, and 500. Furthermore, we measured the substrate temperature in situ by using a K-type thermocouple. First, the thermocouple was combined with a data logger (EL-USB-TC-LCD, MK Scientific). Next, the thermocouple was inserted in a hole bored into the substrate and fixed to the substrate using a bolt to measure the in situ substrate temperature. Temperature profiles were obtained in each experiment by using analysis software. The maximum substrate temperatures during thermal-shock tests were measured.

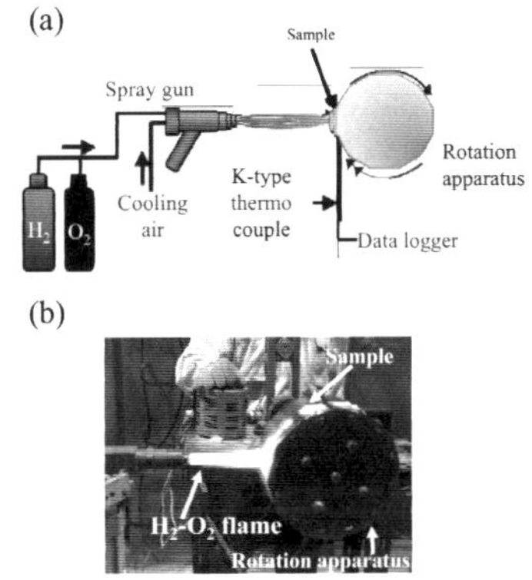

Fig. 1. Thermal-shock test for films deposited from EDTA・Y・H complex on SUS substrate. (a) Schematic diagram of the experiment. (b) Photograph of the experiment.

X-ray diffractometry (M03XHF22, MAC Science) with Cu Kα radiation was used to observe the crystal structure. The surface and cross-sectional morphologies of the films were observed using field-emission scanning electron microscopy (FE-SEM, JSM-6700F, JEOL). Secondary-electron and backscattered-electron (COMPO) images of the deposited films were acquired. The elements contained in the sample and their distributions were estimated using energy-dispersive X-ray (EDX) spectroscopy combined with FE-SEM. The porosity was measured by digitizing the obtained SEM images using commercial software (Image J).

The heat-shock properties of the deposited films were analyzed as follows. First, each obtained COMPO image was digitized using commercial software (Image J). Next, the number of cracks, crack lengths, and crack areas were measured from the digitized image. Furthermore, number of cracks per unit area was estimated using the following equation [12]:

Number of cracks per unit area (number/ μm^2)
= Number of cracks in the deposited film / Area of the deposited film excluding pore voids (μm^2)

3 Results and discussion

All deposited films on the SUS304 substrates appear white[7]. The existence of cracks or delaminations was determined according to the sample's appearance. We measured the thermal history during the thermal-shock test using the H_2–O_2 flame. The maximum temperatures on the SUS substrates were increased by increasing the scan time. The maximum temperatures were 450.5°C, 625.5°C, and 728.5°C at 100, 300, and 500 scans, respectively. The theoretical combustion-gas temperature of the H_2–O_2 flame was 2700°C [13], which is greater than the melting point of the Y_2O_3 crystal of 2425°C [14]. A difference in thermal expansion coefficients existed between the SUS substrate and the deposited material. Therefore, volume expansion stresses can lead to cracks in the coating upon cooling and may cause coating failure. Harsher conditions in the thermal-cycling test induce higher thermal stress in the entire sample. Consequently, a large amount of cracks was produced in the deposited films [15].

Figure 2 shows XRD profiles of the films deposited on SUS substrates. Crystal identification according to the International Centre for Diffraction Data (ICDD) card revealed that all deposited samples were composed of Y_2O_3 crystals with a cubic structure (ICCD card no. 00-041-1105). The crystal phase in the deposited films was not changed by the cyclic thermal-shock test.

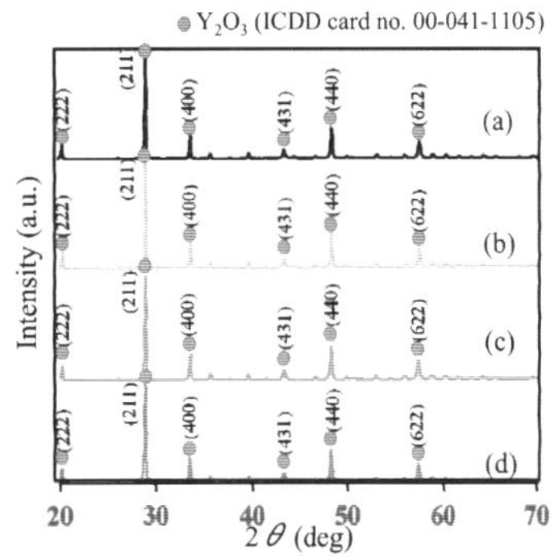

Fig. 2. XRD profiles at 2θ = 20° - 70° for films synthesized from EDTA・Y・H on SUS304 substrates by using flame-spray apparatus.

SEM images of the surfaces of the deposited films are shown in Fig. 3. The surface morphologies of the deposited films were uneven. Splatted particles and spherical particles approximately 1–10 μm in diameter were observed. Cracks or damages were not observed in the samples after the cyclic thermal test.

In previous reports, networks of cracks in deposited films have been expanded by increasing the number of thermal cycles. The cracks are thought to be produced by in-plane tensile stresses in surface layers [16][17][18].

Fig. 3. Surface SEM images of films synthesized on SUS304 substrates using the flame-spraying apparatus. (a) As-synthesized, (b) after heat-shock test (100 scans), (c) after heat-shock test (300 scans), (d) after heat-shock test (500 scans).

Y_2O_3 has high thermal stability. Many researchers have investigated the thermal conduction behaviors of Y_2O_3 for application as interlayers [19][20][21]. We observed changes in the cross-sectional microstructures by using the cyclic thermal test. Cross-sectional SEM and EDX images of the deposited films are shown in Figs. 4 and 5. Metal-oxide layers with dense microstructure were observed on the SUS304 substrate. The SUS substrate was also observed for any damages. Table 1 lists the average thicknesses and 2D porosities of the deposited films. The thicknesses of the deposited films were in the range of 8.8–13.5 μm. The 2D porosities were in the range of 5.1–7.2%. In addition, the elemental distributions of the Y_2O_3 films were observed using the EDX measurement. A uniform elemental distribution was observed in the Y_2O_3 films. In addition, any elemental differences caused by the cyclic thermal test were observed. A distinct elemental boundary between the Y and Fe atoms was confirmed. Thermal elemental diffusion was not observed after the cyclic thermal test. Microstructure behavior during thermal cycling of TBC coatings such as yttria-stabilized zirconia (YSZ) have been investigated [22][23][24]. In the present study, we attempted to observe the detailed microstructure of the deposited Y_2O_3 films through cyclic thermal tests.

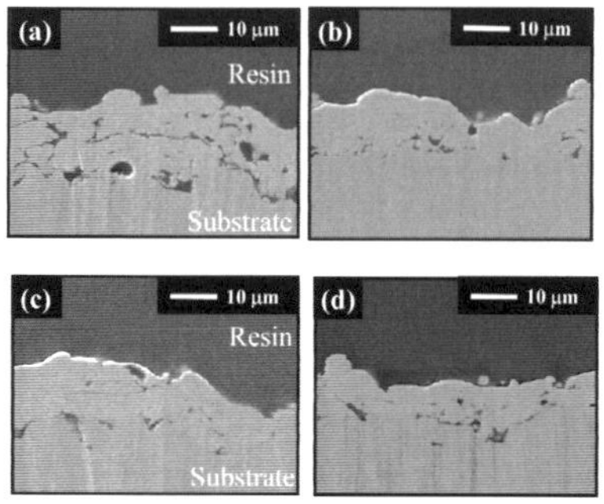

Fig. 4. Cross-sectional SEM images of films synthesized on SUS304 substrates using flame-spraying apparatus. (a) As-synthesized, (b) after heat-shock test (100 scans), (c) after heat-shock test (300 scans), (d) after heat-shock test (500 scans). (c) after heat-shock test (300 scans), (d) after heat-shock test (500 scans).

Table 1 Values of estimated film thickness and cross-sectional porosity of films synthesized on SUS304 substrates.

Sample	Average Thickness [μm]	Maximum thickness [μm]	Minimum thickness [μm]	2D porosity [%]
As-synthesized	13.5	17.2	11.4	7.2
After the 100 scans	10.0	15.8	5.3	6.2
After the 300 scans	10.2	14.8	2.6	6.5
After the 500 scans	8.8	13.7	5.3	5.1

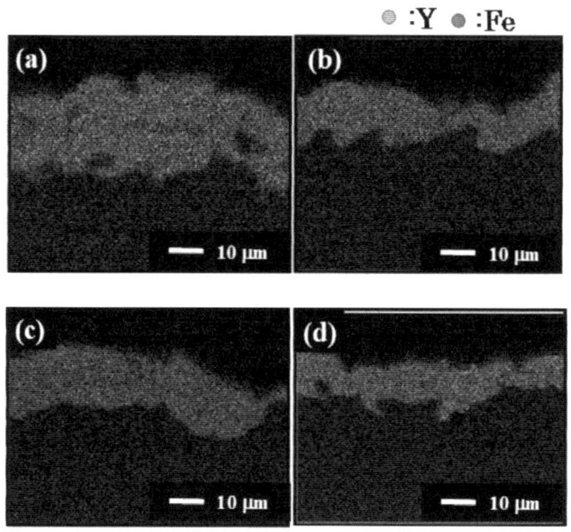

Fig. 5. Cross-sectional elemental EDX images of films synthesized on SUS304 substrates using flame-spraying apparatus. (a) As-synthesized, (b) after heat-shock test (100 scans), (c) after heat-shock test (300 scans), (d) after heat-shock test (500 scans)

Figure 6 shows cross-sectional COMPO images of the deposited Y_2O_3 films used for analyzing the detailed microstructure. The deposited Y_2O_3 films had various defects, such as micro-cracks, inter-splat voids, and open pores. Small structural cracks were observed in all Y_2O_3 films. For instance, the existence of many small structural cracks was confirmed from Fig. 6(c). Particularly, vertical segmentation cracks were observed in the films (Figure 7(a)). Furthermore, some coarse cracks horizontal to Y_2O_3 layers (Figure 7(b)) were formed at splat–splat boundaries and inside the layers. These increase the stress between the Y_2O_3 layer and SUS substrate during thermal cycles, which causes the initiation and propagation of cracks at the splat–splat boundaries, the Y_2O_3 inside, and the Y_2O_3 layer/SUS substrate interface. Although crack structures were observed, through-thickness cracks were not observed in the Y_2O_3 films synthesized with the EDTA・Y・H complex.

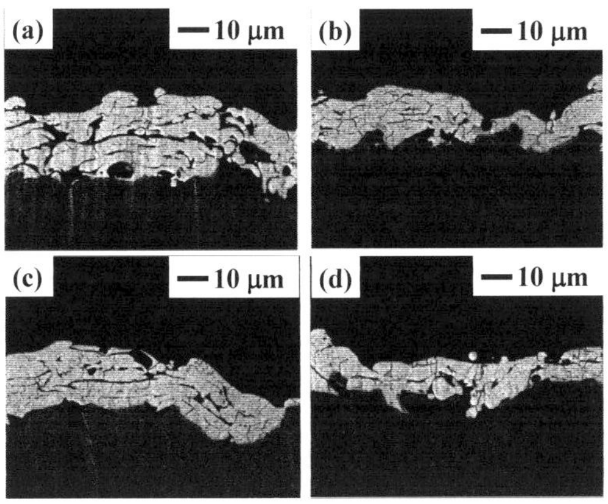

Fig. 6. Cross-sectional BSE (COMPO) images of films synthesized on SUS304 substrates using flame spraying apparatus. (a) As-synthesized, (b) after heat-shock test (100 scans), (c) after heat-shock test (300 scans), (d) after heat-shock test (500 scans).

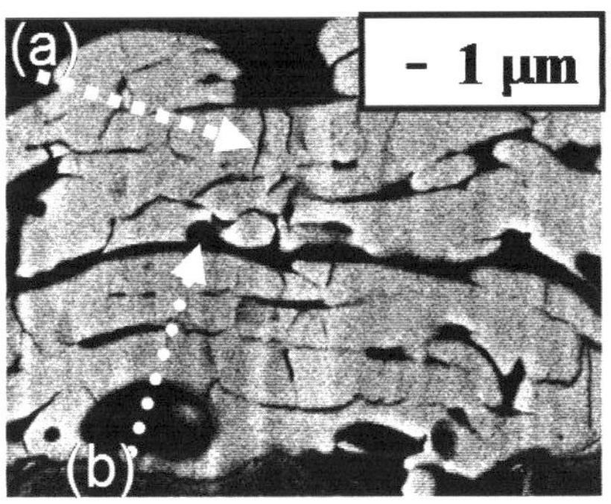

Fig. 7. Examples of cracks in the deposited film. (a) Vertical segmentation crack, (b) horizontal crack. The COMPO image of the as-synthesized sample was used.

We discuss the applicability of the Y_2O_3 films obtained from the EDTA・Y・H complex as TBCs. The Y_2O_3 films were formed by the decomposition of the EDTA・Y・H complex and oxidation. Subsequently, the Y_2O_3 particles were splatted on the SUS substrate, forming Y_2O_3 layers. Thermal-stress concentration occurs on the grain boundary of Y_2O_3 films during the deposition process. As proof of these phenomena, many exploded Y_2O_3 particles were observed in Fig. 3(d). The stresses produced many small cracks in the vertical direction, as shown in Fig. 7. The small cracks in the vertical direction occurred in all Y_2O_3 films. We also observed a decrease in sample thickness with increasing number of scans. This phenomenon is explained by delamination and deformation caused by the local thermal stress on the Y_2O_3 films. As mentioned above, the Y_2O_3 films formed from EDTA・Y・H have small cracks in the surface layers. Therefore, thermal stresses are focused on these small cracks. Stressed Y_2O_3 layers are eventually delaminated or deformed. The heat capacity of the Y_2O_3 films decreased with damage, and the temperature of the SUS substrate increased with damage. Finally, we attempt to analyze the Y_2O_3 films formed from the EDTA・Y・H complex quantitatively. Quantitative estimation was performed using cross-sectional COMPO images of the deposited samples. Table 2 lists values of the estimated crack numbers, sizes, and squares per unit area in the films synthesized on SUS substrates. These values increased after the thermal-shock test, and the increase was not commensurate with the number of scans. Although damage on the microstructure of the Y_2O_3 films was observed after the thermal-shock test, the degree of damage was low. The cracks did not progress or form a crack network. The damage causes visco-plastic deformations and delaminations in the deposited films. In a previous work, the stress at the tip of a crack located on the surface was higher than that at a crack located on the deposited layer, as revealed through simulations [25]. In addition, the presence of quenching stresses and thermal stresses during the cool-down that follows the deposition process was simulated using a finite element method (FEM) analysis [26]. In the present study, cracks produced in the deposition process progressed by thermal stress between the Y_2O_3 film and SUS substrate. Hence, the heat-shock properties of the Y_2O_3 films synthesized with the metal–EDTA complex using flame-spray apparatus are high. Therefore, the deposited films are expected to be candidates for next-generation TBCs.

Table 2 Values of estimated crack numbers, sizes, and squares per unit area in films synthesized on SUS substrates.

Sample	Numbers of crack	Crack's average length [μm]	Numbers of crack per unit area × 10^{-4} [Number/μm^2]
As-synthesized	25	1.6	1.27
After the 100 scans	33	1.7	2.77
After the 300 scans	29	1.6	1.86
After the 500 scans	26	1.5	1.84

4 Conclusion

Y_2O_3 films with high thermal-shock resistance were synthesized on SUS substrates with a metal- EDTA complex as the starting material. The EDTA・Y・H complex and commercial flame-spraying apparatus were used for the deposition of Y_2O_3 films. A cyclic thermal-shock test was conducted using rotation apparatus and H_2–O_2 combustion gas. Y_2O_3 films with 13.5-μm thickness and 7.2% porosity were synthesized. The maximum temperature during the thermal shock test was in the range of 450°C–728°C. Increases in the number of cracks, crack lengths, and number of cracks per unit area were confirmed in the thermal-shock test. The delamination of Y_2O_3 films was not observed.

Acknowledgment
This work was supported by JSPS KAKENHI Grant Number 26420697.

5 References

[1] A. Feuerstein, J. Knapp, T. Taylor, A. Ashary, A. Bolcavage, and N. Hitchman, Technical and Economical Aspects of Current Thermal Barrier Coating Systems for Gas Turbine Engines by Thermal Spray and EBPVD: A Review, J. Therm. Spray Tech., 17(2) (2008) 199–213.
[2] S. Matthews and B. James, Review of Thermal Spray Coating Applications in the Steel Industry: Part 1—Hardware in Steel Making to the Continuous Annealing Process, J. Therm. Spray Tech., 19(6) (2010) 1267–1276.
[3] A. Gilbert, K. Kokini, S. Sankarasubramanian, Thermal Fracture of Zirconia–Mullite Composite Thermal Barrier Coatings Under Thermal Shock: A Numerical Study, Surf. Coat. Tech., 203 (2008) 91–98.
[4] H. Guo, S. Kuroda, and H. Murakami, Microstructures and Properties of Plasma-Sprayed Segmented Thermal Barrier Coatings, J. Am. Ceram. Soc., 89(4) (2006) 1432–1439.
[5] A. Scrivani, G. Rizzi, U. Bardi, C. Giolli, M. Muniz Miranda, S. Ciattini, A. Fossati, and F. Borgioli, Thermal Fatigue Behavior of Thick and Porous Thermal Barrier Coatings Systems, J. Therm. Spray Tech., 16(5–6) (2007) 816–821.
[6] H. Akasaka, M. Ohto, Y. Hasebe, A. Nakamura, S. Ohshio, H. Saitoh, Yttria Coating Synthesized by Reactive Flame Spray Process Using Y-EDTA Complex, Surf. Coat. Tech., 205 (2011) 3877–3880.
[7] K. Komatsu, T. Sekiya, A. Toyama, Y. Hasebe, A. Nakamura, M. Noguchi, Y. Li, S. Ohshio, H. Akasaka, H. Muramatsu, H. Saitoh, Deposition of Metal Oxide Films from Metal–EDTA Complexes by Flame Spray Technique, J. Therm. Spray Tech., 23 (2014) 833–838.
[8] K. Komatsu, T. Sekiya, A. Toyama, A. Nakamura, I. Toda, S. Ohshio, H. Muramatsu, H. Saitoh, Synthesis of Y_2O_3 Films by Spray Coating with Milled EDTA•Y•H Complexes, Int. J. Chem. Nucl. Metallurg. Mater. Eng., 8(12) (2014) 1322–1326.
[9] J. Iwasawa, R. Nishimizu, M. Tokita, M. Kiyohara, K. Uematsu, Plasma-Resistant Dense Yttrium Oxide Film Prepared by Aerosol Deposition Process, J. Am. Ceram. Soc., 90(8) (2007) 2327–2332.
[10] A. Keyvani, M. Saremi, M. H. Sohi, Z. Valefi, A Comparison on Thermomechanical Properties of Plasma-Sprayed Conventional and Nanostructured YSZ TBC Coatings in Thermal Cycling, J. Alloys Compd. 541 (2012) 488–494.
[11] L. Sun, L.-L. Zhang, D. Zhao, Microstructure and Thermal Cycling Behavior of Nanostructured Yttria Partially Stabilized Zirconia (YSZ) Thermal Barrier Coatings, J. Rare Earth., 28, Spec. Issue (2010) 198–201.
[12] I. Ogata, K. Mizutani, K. Makino, Y. Kobayashi, Analysis of the Low Thermal Expansion Mechanism of Preferably Oriented Cordierite Ceramics, DENSO Tech. Review, 13 (2008) 112–118.
[13] M. Suga, Y. Kitamura, Kaenn-to-bakuhatu, Sankyo publishing Co., Ltd. p.64.
[14] K. Daijiten 3 (in Japanese), Kyoritsu Shuppan Co., Ltd., p. 895.
[15] G. Di Girolamo, F. Marra, C. Blasi, E. Serra, T. Valente, Microstructure, Mechanical Properties and Thermal Shock Resistance of Plasma Sprayed Nanostructured Zirconia Coatings, Ceram. Inter., 37 (2011) 2711–2717.
[16] H. Jamalin, R. Mozafarinia, R. S. Razavi, R. A. Pidani, Comparison of Thermal Shock Resistances of Plasma-Sprayed Nanostructured and Conventional Yttria Stabilized Zirconia Thermal Barrier Coatings, Ceram. Inter., 38 (2012) 6705–6712.
[17] Y. Bai, Z. H. Han, H. Q. Li, C. Xu, Y. L. Xu, C. H. Ding, J. F. Yang, Structure–Property Differences Between Supersonic and Conventional Atmospheric Plasma Sprayed Zirconia Thermal Barrier Coatings, Surf. Coat. Tech., 205 (2011) 3833–3839.
[18] A. Nusair Khan, J. Lu, Behavior of Air Plasma Sprayed Thermal Barrier Coatings, Subject to Intense Thermal Cycling, Surf. Coat. Tech., 166 (2003) 37–43.
[19] N. C. Shukla, H.-H. Liao, J. T. Abiade, F. Liu, P. K. Liaw, S. T. Huxtable, Thermal Conductivity and Interface Thermal Conductance of Amorphous and Crystalline $Zr_{47}Cu_{31}Al_{13}Ni_9$ Alloys with a Y_2O_3 Coating, Appl. Phys. Lett., 94 (2009) 081912.

[20] H.-S. Yang, J. W. Kim, G. H. Park, C. S. Kim, K. Kyhm, S. R. Kim, K. C. Kim, K. S. Hong, Interfacial Effect on Thermal Conductivity of Y_2O_3 Thin Films Deposited on Al_2O_3, Thermochim. Acta 455 (2007) 50–54 51.
[21] C. Durand, C. Valle´ e, V. Loup, O. Salicio, C. Dubourdieu, S. Blonkowski, M. Bonvalot, P. Holliger, O. Joubert, Metal–Insulator–Metal Capacitors Using Y_2O_3 Dielectric Grown by Pulsed-Injection Plasma Enhanced Metalorganic Chemical Vapor Deposition, J. Vac. Sci. Technol. A, 22 (2004) 655–660.
[22] H. B. Guo, R. Vaßen, D. Sto¨ver, Thermophysical Properties and Thermal Cycling Behavior of Plasma Sprayed Thick Thermal Barrier Coatings, Surf. Coat. Tech., 192 (2005) 48–56.
[23] S. Ahmaniemi, P. Vuoristo, T. Ma¨ntyla¨, C. Gualco, A. Bonadei, R. Di Maggio, Thermal Cycling Resistance of Modified Thick Thermal Barrier Coatings, Surf. Coat. Tech., 190 (2005) 378–387.
[24] Y. Bai, L. Zhao, J. J. Tang, S. Q. Ma, C. H. Ding, J. F. Yang, L. Yu, Z. H. Han, Influence of Original Powders on the Microstructure and Properties of Thermal Barrier Coatings Deposited by Supersonic Atmospheric Plasma Spraying, Part II: Properties, Ceram. Inter., 39 (2013) 4437–4448.
[25] L. Wang, Y. Wang, W. Q. Zhang, X. G. Sun, J. Q. He, Z. Y. Pan, C. H. Wang, Finite Element Simulation of Stress Distribution and Development in 8YSZ and Double-Ceramic-Layer $La_2Zr_2O_7$/8YSZ Thermal Barrier Coatings During Thermal Shock, Appl. Surf. Sci., 258 (2012) 3540–3551.
[26] P. Michlik, C. Berndt, Image-Based Extended Finite Element Modeling of Thermal Barrier Coatings, Surf. Coat. Tech., 201 (2006) 2369–2380.

Super-hydrophobic surface prepared by lanthanide oxide ceramics deposition through PS-PVD process

J. Li, C.X. Li, Q.Y. Chen / Xi'an / China

Super-hydrophobic surface has been received widespread attention in recent years. Both of the morphology and surface energy of the materials have a significant impact on super-hydrophobic property. A novel method to fabricate super-hydrophobic surface is introduced in the present paper. We try to construct rough structure using samaria-doped ceria by plasma spray-physical vapor deposition method followed by fluorination treatment. During the PS-PVD process, the samples were placed at the different deposition distance with a same straight line for shielding and depositing to realize the surface with different morphology. An island-liked structure consisted of nano particles like the hierarchical structure in lotus leaf were obtained by this process, which realize the super-hydrophobic characteristic of the coatings accompany with modification process. Meanwhile, the size of the nanostructure influenced the sliding behaviour of the coating surface. Furthermore, the resultant superhydrophobic surface could withstand repeated tearing by an adhesive tape, exhibiting an excellent stability under extreme conditions.

1 Introduction

Super-hydrophobicity is a kind of extremely common phenomenon in nature, such as lotus leaf and the wings of butterfly, which defined as a surface with contact angle larger than 150° and sliding angle less than 10° [1, 2]. Functional super-hydrophobic surface can get properties such as self-cleaning, corrosion resistance and anti-fogging. These applications are conducive to the promotion of resource conservation and environmental protection [3, 4]. The geometrical structure and the chemical composition are the two key factors which affect the surface wettability [5]. Generally, two conventional ways are used to produce super-hydrophobic surface including constructing rough structure on the intrinsic hydrophobic material and modifying a rough surface with low surface free energy [3].

At present, there are a variety of methods to prepare super-hydrophobic surface. However, whether the surface can be successfully applied in practice is the utmost concerned by the researchers. One of the most critical problem is how to improve the surface durability under the extreme working condition of the environment [6]. Recent studies demonstrate that the robust superhydrophobic surface is better choice to extend the practical lifespan of super-hydrophobic surfaces [7]. Ceramic as the wear-resisting materials have been widely applied in various industries and daily life which is a remarkable material for making super-hydrophobic surface [8]. In this study, samaria-doped ceria was used to prepare the surface with stability of super-hydrophobicity. A novel method plasma spray-physical vapor deposition was used to control the microstructure of coating surface with hierarchical structure to meet the requirements of structure for superhydrophobic property. According to previous research by our group, during the PS-PVD deposition process, the micro island-liked on coating surface can be easily formed in the form of vapor atoms or clusters [9]. Because of the micro-nano structures can affect the wettability of coating surface [10]. PS-PVD process will give the possibility to control the material grain size at the nanoscale through physical vapor deposition [9]. Meanwhile, The spical structure of the coating makes the surface topography with self-healing capability against the damaged outside.

2 Experimental

2.1 Materials

The commercially spherical agglomeration samaria-doped ceria (SDC) powder (S80736, TERIO Corporation, China) with size ranging from 5 to 50 μm was used as the feedstock material. The sintered alumina wafer with dimensions of Φ20×3 mm was employed as a substrate for coating deposition. Prior to spraying, the substrate was polished and decontaminated with alcohol and acetone. 1,1,2,2-Tetrahydroperfluorodecyltrimethoxysilane (FAS) (F-1061, SiKang, China) was employed to reduce the surface energy of as-sprayed coating.

2.2 Coating preparation method

The 80 kW plasma spraying system (MF-P 1000 APS/VPS, GTV, Germany) was used to deposit SDC coatings. The plasma torch with a gas shroud was fixed on a six-axis robot manipulator inside a 10 m^3 vacuum chamber with an automatic pumping system to assure that the work pressure was maintained at 100 Pa during spraying. The plasma torch was operated to generate an Ar-H_2 plasma jet. Ar gas was used as delivering powder gas. The powder was fed into a plasma jet using an internal powder port. The shroud was attached in the front of the plasma nozzle to restrain the expansion of plasma jet. **Tab. 1** shows the detailed spraying parameters. During the spraying, the axis of plasma jet was perpendicular to the alumina substrate surface and a shielding using graphite rod was set in front of the substrate. The samples were placed on the same straight line with different distance. Three different distance was chosen for the coating deposition, 300 mm, 400 mm and 500 mm, respectively. During the process, the plasma torch was not moved. The schematic diagram of this setting is presented in **Fig. 1**.

Tab. 1. The spraying process parameters

Parameters	Unit	Value
Chamber pressure	Pa	100
Current	A	750
Ar gas flow	SLPM	10
H_2 gas flow	SLPM	52
Power	kW	65
Spray distance	mm	300/400/500
Powder flow rate	g/min	6.5

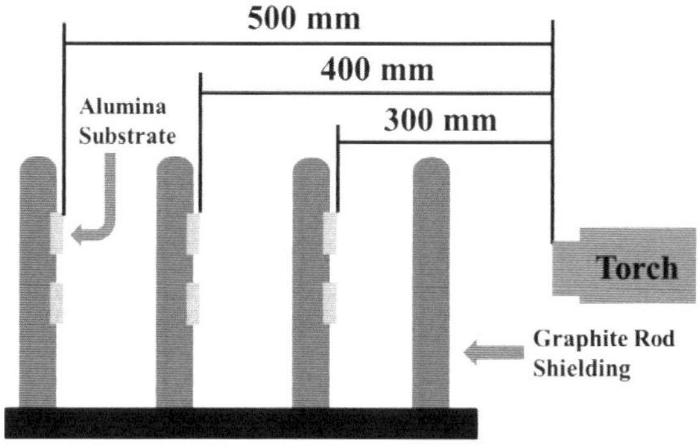

Fig. 1. Schematic diagram of the shielding position and different spray distance.

2.3 Surface modification

1 ml FAS and 99 ml isopropyl alcohol were mixed together to diluting FAS. Then 0.5 ml acetic acid was added to promote FAS hydration. After that, the solution was stirred 5 h using a magnetic stirrer to make FAS fully hydration. The clean and dried coating was immersed into the solution for 18 h and dried in the ambient atmosphere for 24 h. After solidification process, the surface was decontaminated with alcohol to remove the residual reagents by physical adsorption.

2.4 Characterization

The morphology of coating surface was characterized by field emission scanning electron microscopy (MIRA3 LMH, TESCAN, Czech Republic). The contact angle (CA) of water droplets on the coating surface was measured by dipping a distilled water drop whose volume was about 5 μl to the coating surface using a contact angle instrument (JC2000C4, Zhongchen, China). At least 5 droplets were used to measure each sample at different locations. The contact angle value was used to measure the hydrophobic performance of the coating. Furthermore, adhesive tape test was also employed to demonstrate the mechanical stability of the prepared super-hydrophobic surface. The surfaces were pressed with adhesive tapes applying and then detached. This process was carried out repeatedly 10 times, and the corresponding water contact angle and sliding angle after each test was recorded.

3 Results and Discussion

3.1 Microstructure of the SDC coatings

The typical surface morphologies of the SDC coating surfaces were shown in **Fig**. **2**(a-f). Images (a-c) correspond to the sample that was placed 300 mm away from the spray gun. Images (d-f) show the morphology of the sample that was placed 400 mm away from the spray gun. The microstructure of the sample placed furthest from the gun with 500 mm were shown in image (g-i). From the high magnification, all the coating surfaces were shown an island-like structure consisted of nano particles with angular shape. The hierarchical structure including micro-nano structure in SDC coating surfaces was particularly similar to the microstructure on lotus leaf. As a comparison, it can be seen that with the increasement of the shielding and sample position, the size of island-like structures decreased in size. Meanwhile, the size of nano structure also reduced gradually. The unit size of island-like structure decreased from about 5 μm to 1 μm with increasing the deposition distance from 300 mm to 500 mm. The size of nano particles was decreased from about 1 μm to less than 100 nm with the deposition distance decreased.

Fig. 2. The surface morphology of SDC coatings deposit on shadowed substrate with different deposition distance from the spray gun: (a-c) 300 mm; (d-f) 400 mm; (g-i) 500 mm respectively.

3.2 Surface wettability

After cooling down to room temperature naturally during the spraying process, the as-sprayed SDC coating surfaces produced a super-hydrophilic state. The water droplet could penetrate into the surface immediately just contacted with it. It is known that the surface wettability is affected by surface roughness as well as surface anergy. At atmospheric conditions, the surface of most given metal oxide are coverd with a thin layer, including an outmost layer of absorbed organic contamination consisted of C and O, an layer of chemisorbed water and a hydroxylated region, which has large impact on the wettability of the as-sprayed coating surface [11]. The hydroxylation layer has a strong affinity to form hyrogen bonds with water liquid leading the surface with a super-hyrophilic phenomenon [12]. Carbonly and carboxyl groups are the common organic contamination absorbed from atmospheric environment which are polar will serve a hyrophilic property [13]. Therefore, the metal oxide SDC presented an intrinsic characteristic of hydrophilicity. Meanwhile, Wenzle equation revealed that the intrinsic hydrophilic surface could get more hydrophilic with increase of the roughness [14]. The hierarchical structure gave a rise to a super-hydrophilic nature of as-sprayed coating surfaces.

After modified with a thin layer by FAS, the wettability of coating surfaces transformed from super-hydrophilic to super-hydrophobic property with contact angle as large as 155.5±0.3° and sliding angle less than 10°. **Fig. 3** shows the diagram of contact angle measurement of each coating surface. The contact angle values of three samples were very similar proved that the hierarchical structure on each surface could all trap a large fraction of air while the FAS was uesed to reduce the surface energy acquiring the surface to obtain large contact angle as described in Cassie-Baxter state. Meanwhile, it can be seen that the values of sliding angle gradually decreasd with the increase of deposition distance. According to the recent references, nanostructures are essential in farbricating super-hydrophobic surface with low sliding angle [3, 15]. The results in our study show a phenomenon that the sliding angle values were influenced by the size of the nanostructure which decreasd with the size of nanostructures decreased.

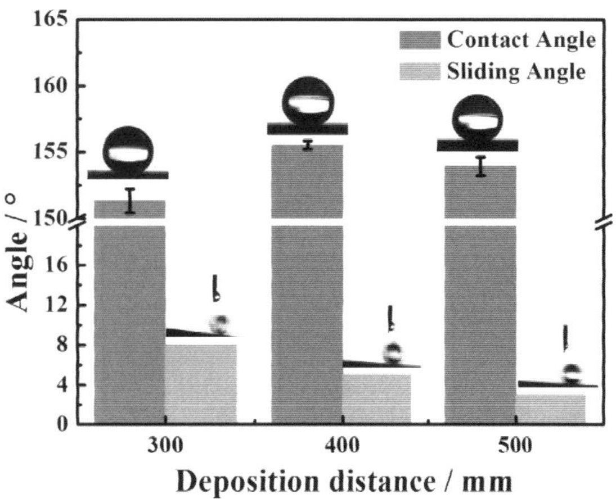

Fig. 3. The contact angle and sliding angle of SDC coating surfaces with different spray distance after chemical modification.

In Cassie-Baxter state [1], the equation shown in Eq. 1 describes that the contact angle for the heterogeneous interface composed of solid and air.

$$\cos\theta'=f\cos\theta+(1-f)\cos 180°=f\cos\theta+f-1 \qquad \textbf{Eq. 1}$$

For SDC coating surface, the composite interface consists of a solid–liquid contact fraction (f) and liquid–air contact fraction (1-f). θ' is the apparent contact angle, θ is the contact angle of smooth coating surface modified with FAS, which is a constant value. By the equation, the super-hydrophobicity can be enhanced by reducing the value of solid–liquid contact fraction (f). The extremely large portion of air cushion exposed to the water droplet will prevent the penetration of water droplets into the surface. Schematic of a water droplet on SDC coating surface was shown in **Fig. 4**(b). The micro-potrusion formed the main part of the air cushion between water droplet and coating surface, refer to Mark 1 in **Fig. 4**(a) and (b). As a result the of nanostructures with angular shape existing in the outermost layers of coating surface, when water droplet contacted with this structure, the formation of the secondary air cushion (Mark 2) leading to further reduce the solid–liquid contact fraction, which would increasing the contact angle, and enhancing the super-hydrophobicity. The actual measurement showed that the value of θ was about 107.5°. According to this equation, the solid–liquid contact fraction f of SDC coating surface was estimated to be about 0.13-0.18 basised on different morphologies of SDC surfaces. Herein, the area fraction of the air cushion trapped was 0.82-0.87.

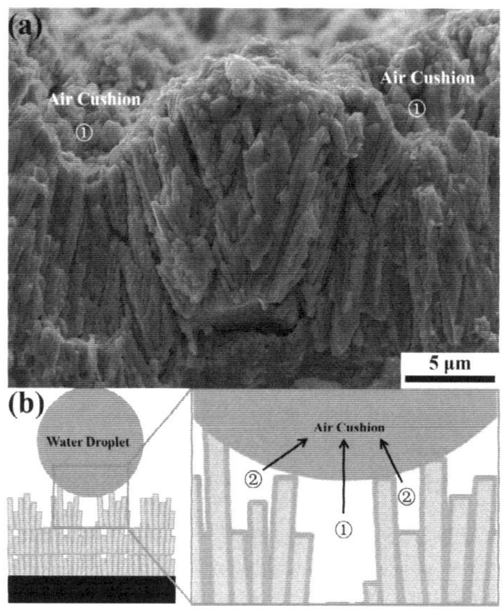

Fig. 4. The cross-sectional morphology of SDC coating deposited distance of 300 mm from spray gun (a). The effect of surface structure on super-hydrophobicity (b).

3.1 Surface stability

The mechanical stability of the SDC surface prepared by PS-PVD process was investigated by repeated tear tests with an adhesive tape. The surfaces were pressed with adhesive tapes applying and then detached. **Fig. 5** shows the contact angle measurements after each test. After being a 10-cycle tear test with adhesive tape, the water contact angle of the surface had only a slight decline and still higher than 150° while the sliding angle remained unchanged with 3°, which demonstrated the prepared SDC coating surface had an excellent stability under extreme conditions.

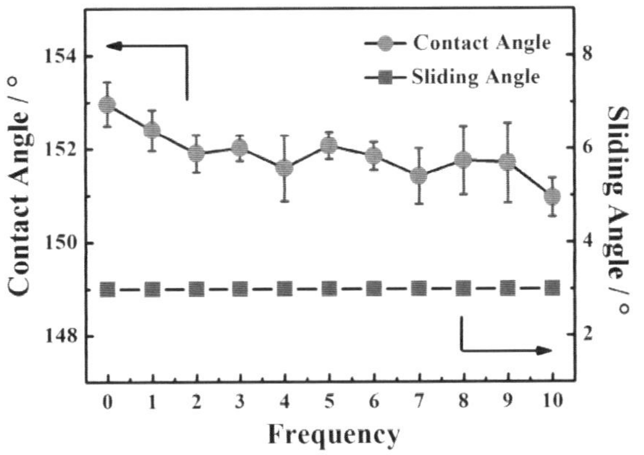

Fig. 5. Varition of water contact angle and sliding angle of SDC coating surface after each tear test.

From the cross-sectional morphology of SDC coating shown in **Fig. 4**(a), the island-like clusters composed of columnar structure stacked layer by layer to formed the deposited coating. Meanwhile, with the results of surface morphology, some pores and microcracks between each of the island-like structures were found inside the prepared coating. During FAS adsorption treatment, the mixed FAS solution could penetrate into the coating by capillary action. Dehydration-condensation reaction happened between the hydroxyl groups in FAS molecule and coating surface. Thus the FAS monolayer was successfully grafted on every corner inside the coating where the FAS solution could contacted with as shown in **Fig. 6**. After each tear test, the most superficial structure of coating surface was peeled off by adhesive tape. With the special stack structure of the coating, the new bare surface again had a hierarchical structure with chemical adsorption of low surface free energy as before. This superhydrophobic SDC coating surface with an excellent stability give a new possibility of applications in industry and daily life.

4 Conclusions

In conclusion, two steps method including PS-PVD process conbine with chemical modification were used to prepare super-hydrophobic SDC coating. The coating surfaces have the hierarchical structure with an island-like structure consisted of nano particles. The sliding behavior are influenced by the size of the nanostructure and the value declined with the size of nanostructures decreased. The prepared samples exhibit the contact angle as large as 155.5±0.3° and sliding angle as low as 3°. The solid–liquid contact fraction of the surface is estimated to be about 0.13-0.18. With the special stack structure, the surface was demonstrated an excellent stability by adhesive tape test, which make it possible in practical application.

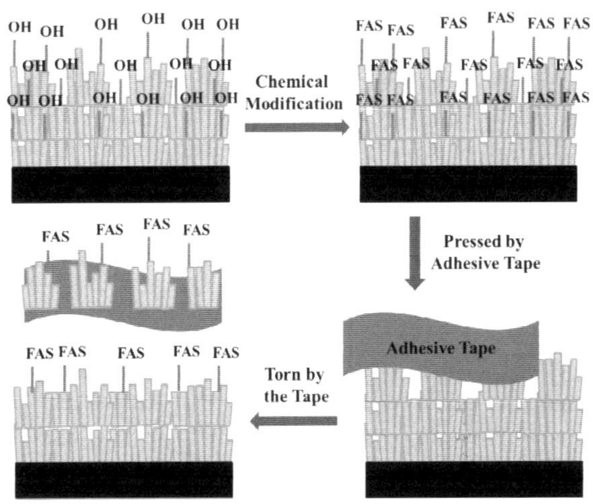

Fig. 6 The stability mechanism of super-hydrophobic surface during repeatedly tear test by adhesive tape.

5 References

[1] Bharat B., and Yongchae J.: Natural and biomimetic artificial surfaces for superhydrophobicity, self-cleaning, low adhesion, and drag reduction. Materials Science 56 (2011), pp. 1-108.

[2] Ogihara H., Xie J.: Simple method for preparing superhydrophobic paper: spray-deposited hydrophobic silica nanoparticle coatings exhibit high water-repellency and transparency. Langmuir 28 (2012) pp.4605-4608.

[3] Feng L. and Shuohong L.: Super-hydrophobic surfaces: from natural to artificial. Adv. Mater. 14 (2002), pp. 1867-1860.

[4] Akira N., and Kazuhito H.: Recent studies on super-hydrophobic films. monatshefte fuÈr Chemie 123 (2001), pp. 31-41.

[5] Jongmin L. and Gira Y.: Superhydrophobic Films of Electrospun Fibers with Multiple-Scale Surface Morphology. Langmuir 23 (2007), pp.7981-7989.

[6] Yang W., Xiaowei L.,: Fabrication of self-healing super-hydrophobic surfaces on aluminium alloy substrates. AIP Advance 5 (2015) , pp. 041314.

[7] Yang L. and Shanshan C.: All Spraying processes for the fabrication of robust, self-healing, superhydrophobic coatings. Adv. Mater. 26 (2014), pp.3344–3348.

[8] Gisele A. and Rajeev D.: Hydrophobicity of rare-earth oxide ceramics. Nature Material. 12 (2013), pp. 315-320

[9] Qingyu C. and Chengxin L.: Microstructure of YSZ coatings deposited by PS-PVD using 45 kW

shrouded plasma torch. aterials and Manufacturing Processes. 0 (2015), pp.1-9.
[10] Qingyu C. and Chengxin L.: Controlling grain size in columnar YSZ coating formation by droplet filtering assisted PSPVD processing. Rsc Advances. 5 (2015), 102126-102133.
[11] E. McCafferty and J. P. Wightman: Determination of the concentration of surface hydroxyl groups on metaloxide films by a quantitative XPS method. Surf. Interface Anal. 26 (1998), pp. 549-564.
[12] M.M. Gentleman and J.A. Ruud: Role of Hydroxyls in Oxide Wettability. Langmuir 26(2010), pp.1408–1411
[13] Jiangyou L. and Minlin Z.: Wettability conversion of ultrafast laser structured copper surface. J. Laser Appl. 27(2015), pp. S29107.
[14] Y.Y. Yan and N. Gao: Mimicking natural superhydrophobic surfaces and grasping the wetting process: areview on recent progress in preparing superhydrophobic surfaces. Adv. Colloid Interface Sci. 169 (2011), pp. 80-105.
[15] V. Khranovskyy and T. Ekblad: Surface morphology effects on the light-controlled wettability of ZnO Nanostructures. Applied Surface Science 258 (2012), pp. 8146–8152.

ZrB_2-based coatings by plasma spraying: Behavior under oxidizing and corrosive atmosphere at temperature exceeding 2000 °C

C. Barré, A. Quet, L. Bianchi, Monts/F, C. Zeppa, Le Haillan/F, C. Sanchez, Paris/F

Thermostructural composites are of great interest in aerospace field as structural components of hypersonic flights, re-entry vehicles or aeropropulsion systems. In order to overcome composite weakness in oxidizing environment at always increasing working temperatures, a solution consists in applying them protective coatings. Ultra-High Temperature Ceramics (UHTC) such as zirconium diboride (ZrB_2) combined with additives such as silicon carbide are potential candidates as coating materials.
In this study, ZrB_2-based coatings were prepared by plasma spraying and were evaluated under heat flux with a plasma jet at temperature greater than 2000 °C, in oxidative and corrosive atmosphere. The as-sprayed and oxidized specimens were characterized to understand the relationship between the starting microstructure of the coatings and their behavior at very high temperature.
Results allowed to distinguish the most promising composition and microstructure to protect composite materials. It is shown that the coating microstructure can significantly influence the kinetic of oxidation. Nano-scale additives in a homogeneously dispersed state are particularly advantageous. Moreover, a low amount of additives avoiding liquid phase formation in high quantity during oxidation tests is to be favored.

1 Introduction

Ultra-high temperature ceramics (UHTCs) are a family of refractory materials that are promising candidates for advanced aerospace systems [1]. UHTCs include borides, carbides, nitrides, and oxides of early transition metals. They show some of the highest available melting temperatures, greater than 3000 °C, excellent thermal conductivity (30-80 $W.m^{-1}.K^{-1}$) and low coefficient of thermal expansion (5-8.10^{-6} K^{-1}), a set of properties which gives them good thermal shock resistance. In comparison to carbides and nitrides, the diborides show lower melting point but higher thermal conductivity. Most of the studies, incidentally, have focused on borides such as zirconium diboride (ZrB_2) and hafnium diboride (HfB_2), mainly because of the high temperature stability of the ZrO_2 and HfO_2 scales resulting from the oxidation of these materials. Lighter and less expensive than HfB_2, ZrB_2 presents some advantages when considering aerospace applications.

Despite their high melting point, UHTCs are unsuited for many systems operating at high temperature, their poor oxidation resistance remaining a major bottleneck. Addition of silicon carbide improves the properties of ZrB_2 in oxidative environments [2, 3]. ZrB_2 containing around 20 vol% SiC was found to be the best tradeoff at temperature lower than 2000 °C [4]. Active oxidation of pure ZrB_2 occurs above 1100 °C due to volatilization of B_2O_3 [5], whereas the SiC-including ZrB_2 shows stable oxidation behavior up to 1600 °C [6]. For temperatures above 1100 °C, the addition of SiC promotes the formation of a borosilicate glassy layer which has a higher temperature capability than B_2O_3 alone. Beyond 1800 °C, the volatility of the glass layer components increases. Some solutions were investigated to improve the oxidation resistance of UHTCs [7]. Smaller SiC particle sizes lead to improved oxidation resistance [8]. The addition of transition metals such as Ta allows the formation of higher viscosity glasses and reduces oxidation rate up to 1600 °C, but the high quantity of liquid phases due to tantalum oxide is detrimental beyond 1900 °C [9]. Additives based on rare earth elements such as La were also investigated for temperatures up to 2400 °C [10].

As with other largely covalent bonded and refractory materials, processing of zirconium diboride is challenging and tend to require high temperature to densify them. Relatively untapped to manufacture these materials [13-16] with regard to powder metallurgy techniques such as hot isostatic sintering or spark plasma sintering [4, 6, 8-12], plasma spray process presents definitively numerous assets, foremost of which is its capacity to provide, in the form of coatings, a wide range of materials with high melting point on parts of varied shapes. As a downside, plasma-sprayed coatings are characterized by a high concentration of defects and are potentially less resistant to oxidation than denser materials.

Characterization means at temperatures higher than 2000 °C are relatively scarce. Customized high temperature testing solutions were used, including solar furnaces [17], oxyacetylene torch [18], arc jet [19].

This study deals with ZrB_2-based coatings prepared by inert plasma spraying, with SiC and Y_2O_3 as additives, and the investigation of their behavior under high heat flux and moist atmosphere by means of a plasma jet. Five types of coatings with different compositions and microstructures were performed. Comparison of the oxidation rate on an oxide scale thickness basis allows to identify the most oxidation-resistant sample.

2 Experimental

2.1 Plasma spray processing

Powder mixtures from commercially available ZrB_2, SiC and Y_2O_3 powders with, respectively, an average particle size of 32, 19 and 28 µm, were used as spray feedstock. The assessment of Y_2O_3 size and repartition was made by processing the SiC particles with $Y(OH)_3$ using the sol-gel technique.

A F4-VB type plasma torch from Oerlikon-Metco equipped a with 6 mm internal diameter nozzle was used. Coatings were carried out at atmospheric pressure under an inert atmosphere, provided by an argon flow, so as to limit the oxidation of boride particles. An Ar-He-H_2 mixture with a weight enthalpy of 39 kJ.g^{-1} was used as plasma gas. Spray distance was fixed at 120 mm. The coatings were deposited up to around 1000 µm on 50*50*10 mm^3 graphite and carbon-based composite substrates.

Five types of coatings were produced in order to determine the effect of SiC addition, Y_2O_3 rate and Y_2O_3 size and distribution: (i) pure ZrB_2, (ii) ZrB_2 including SiC particles, (iii) ZrB_2 with SiC particles and relatively high quantities of coarse Y_2O_3 splats, (iv) ZrB_2 with SiC particles and coarse Y_2O_3 splats in low amounts and, (v) ZrB_2 combined with SiC particles embedded in a colloidal sol of $Y(OH)_3$. The coatings (iii), (iv) and (v) are called, respectively, ZSY(H), ZSY(L), and ZSY(S).

2.2 Set-up for high temperature oxidation tests

The oxydation-resistant property of the coated samples was assessed out using a plasma jet in order to achieve temperatures exceeding 2000 °C.

The specimens were mounted in graphite holders. A two-color pyrometer was used to record sample surface temperature during heat-flux tests. The booth was maintained in an inert environment with argon, at atmospheric pressure. Moist environment was obtained by means of the addition of water into the plasma jet. Heat-flux conditions were provided by an O3C plasma gun from Oerlikon Metco. Corrosive and oxidative atmosphere was not characterized but it can be assumed that, upon vaporization of water, the plasma jet included dissociated species such as OH and, possibly some atomic and partially ionized species.

Runs began when the gun was placed above the samples, lasted for time lengths of 2 and 5 minutes and were ended by removing the plasma gun from the specimens. Tests were stopped earlier for samples that visually failed. All the runs were performed using the same freestream conditions. Sample load was considered as homogeneous, in terms of temperature and oxidizing species generated by the plasma (OH, O...), on a diameter of about 15 mm. Upon plasma flux exposure, the surface temperature rised at a heating rate of about 100 °C.s^{-1}, allowing to evaluate the thermal shock resistance, and reached a relative steady value in the range of 2250 to 2400 °C, depending on the specimens.

2.3 Microstructure and composition analysis

Open porosity of as-sprayed coatings was estimated by Archimedes method, using free-standing samples. The microstructural characteristics and compositions of the as-sprayed and post-test coatings were evaluated by scanning electron microscopy (SEM), electron probe microanalysis (EPMA) and X-ray diffraction (XRD). Semi-quantification of phases was performed by XRD analysis, using the RIR (Reference Intensity Ratio) method. The oxidation rate was determined from the difference between the starting thickness and the thickness of unreacted material that was measured in SEM cross-section micrographs.

3 Results

3.1 As-sprayed coatings

SEM cross-sections illustrating the microstructure of as-sprayed ZrB_2 and ZrB_2-SiC coatings are given in **Fig. 1 and 2**.

Fig. 1. SEM cross-section micrographs of as-sprayed (a) ZrB_2 and (b) ZrB_2-SiC coatings.

The coatings exhibit the typical lamellar structure of plasma-sprayed coatings. Open porosity is of the order of 15 % for the different coatings, **Table 1**. Pure ZrB_2 coatings tend to be denser than the coatings containing added phases, for which trapped SiC particles are likely to affect the stacking of ZrB_2 splats. Indeed, SiC does not have a melting temperature and decomposes around 2730°C. Therefore, its inclusion into the coating is allowed by ZrB_2 splats and can involve stacking defects. SEM image in backscattering mode allows to distinguish angular SiC particles into the diboride matrix, **Fig. 1b**.

Table 1. Characteristics of the as-sprayed coatings.

Coating	Open porosity (%)	SiC rate (vol%)	Y_2O_3 rate (vol%)
ZrB_2	12±2	-	-
ZrB_2-SiC	15±2	10±2	-
ZSY(H)	16±2	8±2	9±2
ZSY(L)	16±2	9±2	2±1
ZSY(S)	17±2	9±2	1±0,5

The distribution of the different elements in the ZSY(H) and ZSY(S) coatings is highlighted by EPMA analyses, **Fig 2**. Y_2O_3 splats and trapped SiC particles appear uniformly distributed through the coatings. The use of the sol-gel technique to process SiC particles with a sol of $Y(OH)_3$ allowed to introduce Y_2O_3 in low quantities and following a uniform and fine distribution, **Fig. 2b**.

Quantities of each compound were evaluated by XRD, **Table 1**. SiC rate can be considered as constant with a value of 10 vol%. ZSY(H) is characterized by a Y_2O_3 rate of 10 vol% whereas coatings ZSY(L) and ZSY(S) include a lower amount of Y_2O_3, about 2 vol%.

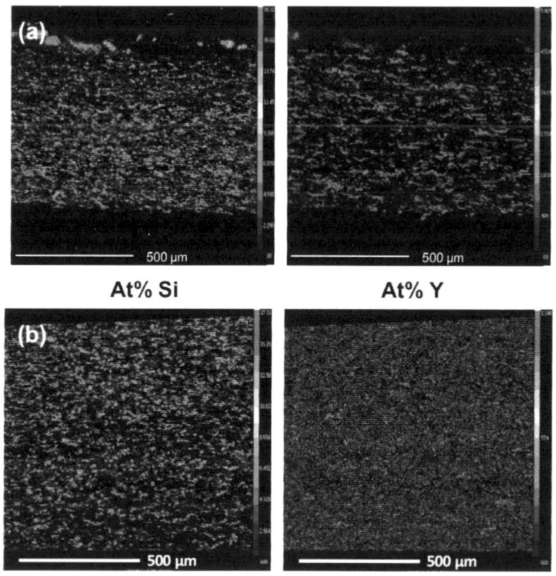

Fig. 2. EPMA cross-section analyses of as-sprayed (a) ZSY(H) and (b) ZSY(S) coatings.

3.2 Coatings after oxidation

Peak temperatures measured at the surface of the specimens during heat-flux tests as well as exposure time are given in **Table 2**.

Table 2. Peak temperature acquired during heat-flux tests, test duration and oxidation rate determined from post-test cross-sections.

Coating	Exposure time (min)	Peak temperature (°C)	Oxidation rate (µm.s^{-1})
ZrB_2	2	2350	-
ZrB_2-SiC	2	2230	4.2±0,3
ZrB_2-SiC	5	2280	2.2±0,2
ZSY(H)	2	2240	1.6±0,2
ZSY(H)	5	2350	2.6±0,2
ZSY(L)	0,5	2270	-
ZSY(S)	2	2220	3.3±0,3
ZSY(S)	5	2410	1.7±0,2

Macrographs in **Fig. 3** show the surface appearance of samples after the exposure to the plasma jet, in corrosive and oxidative conditions.

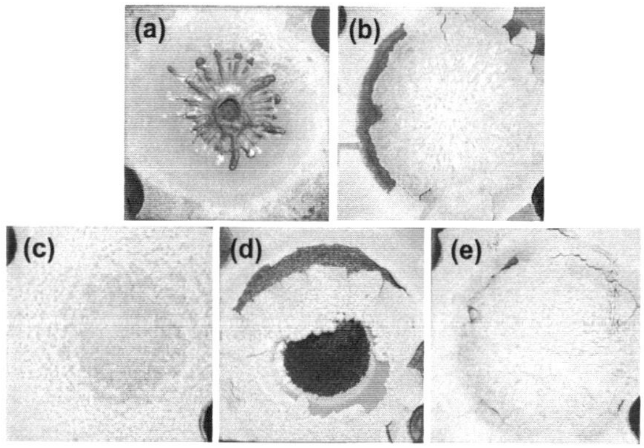

Fig. 3. Macrographs of coated samples after heat-flux plasma tests: (a) ZrB_2, (b) ZrB_2-SiC (5 min), (c) ZSY(H) (5 min), (d) ZSY(L) and (e) ZSY(S) (5 min).

ZrB_2 coating failed after 2 minutes upon the oxidative and corrosive plasma flux, **Table 2**. It exhibits a shape reminding a volcano after eruption with a crater in the middle of the sample and molten streams, **Fig. 3a**. The ZSY(L) coating containing coarse Y_2O_3 splats in low amounts underwent, after only 30 s, a severe spallation that exposed the substrate to the heat flux, **Fig. 3d**. The three other coatings provided oxidation protection during the time lengths of 2 and 5 minutes that implied a good adherence between the coating and the substrate and an efficient resistance to thermal shock upon the heating. During the cooling, the oxide scale cracked and spalled from the remaining unreacted boride coating, likely because of thermal-induced stresses and phase transformations. Extensive liquid phase formation, distinguishable in the form of elongated drops, is seen on the oxide scale of the ZSY(H) coating, **Fig. 3c**. ZrB_2-SiC and ZSY(S) coatings show identical surface morphology with some solidified liquid drops, in reduced amount compared to ZSY(H).

Regarding Y_2O_3-free coatings, XRD analyses show the surface oxidation product is monoclinic ZrO_2. For oxidized ZSY coatings, Y_2O_3 enabled the formation of a mixture of tetragonal and monoclinic ZrO_2.

SEM surface micrograph of the ZSY(H) sample after oxidation reveals drops forced outwardly and across the sample by the heat flux, **Fig. 4**. Orifices can be seen at the each starting point of drops.

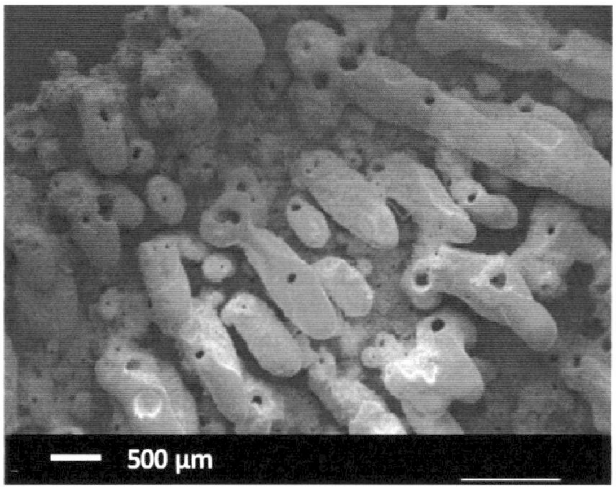

Fig. 4. SEM surface micrograph of 5-min oxidized ZSY(H) coating.

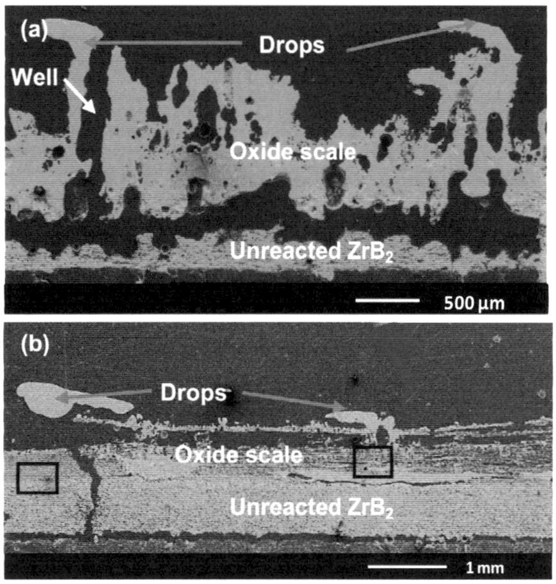

Fig. 5. SEM cross-section micrographs of (a) ZSY(H) and (b) ZSY(S) coatings after 5-min heat-flux plasma test.

Cross-sectional microstructures of post-test ZSY(H) and ZSY(S) samples are shown on **Fig. 5**. After exposure to plasma heat flux, coatings include an outer ZrO_2 layer and the unoxidized diboride. Drops are pointed out with arrows. The orifices seen in **Fig. 4** turned out to be tantamount to some wells through the oxide scale, **Fig. 5b**. **Fig. 6** shows ZrO_2 scale of ZSY(S) coating seen in **Fig. 5b**, with a higher magnification. Both porous and sintered areas are noticed.

Oxidation rates, evaluated from the thickness of unreacted diboride, indicate that the best performing coating, in this study, is ZSY(S) containing 10 vol% SiC and finely distributed 1 vol% Y_2O_3, **Table 2**.

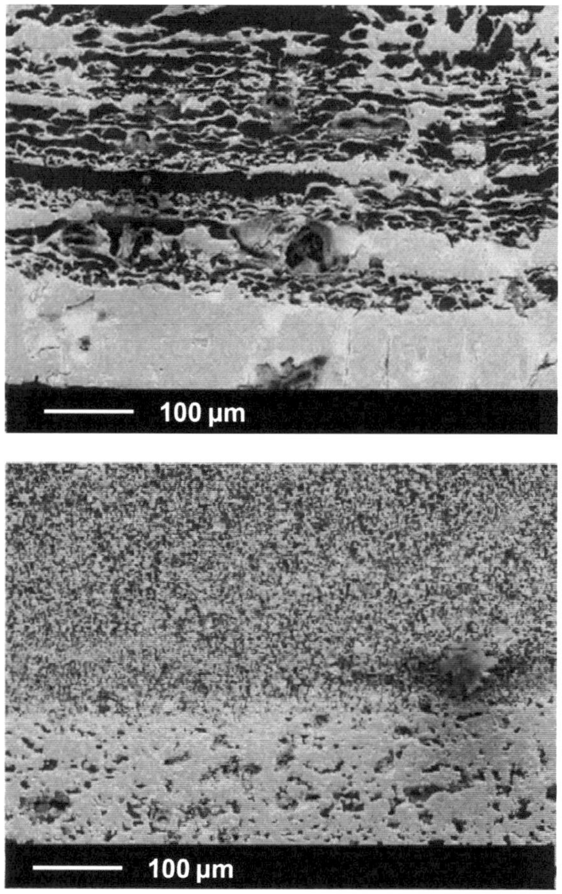

Fig. 5. SEM cross-section micrographs of ZSY(S) coating after 5-min heat-flux plasma test showing the zirconia scale.

4 Discussion

4.1 Effect of SiC addition

Addition of 10 vol% SiC into the plasma-sprayed ZrB_2 coating led to a completely different behavior of specimens during the heat-flux plasma tests. Whereas ZrB_2 coating failed after barely 2 minutes, 10 vol% SiC-including ZrB_2 coating ensured the protection of the substrate during at least 5 minutes and showed a residual non-oxidized ZrB_2 layer. Some assumptions can be advanced as possible explanations. Open porosity of ZrB_2 coating could not be able to inhibit the oxidant species from passing therethrough. Gases build up at the substrate/coating interface and the coating just bursts open. Further insight of liquid phase formation during the blow-up of the ZrB_2 coating could be gained by additional analyses. Regarding the coatings containing SiC particles, a transient protection can be assumed, allowing sintering of ZrO_2 produced from the oxidation of ZrB_2 and slowing down the progression of oxidant species into the coating. No protective glass scale was found at the top of the specimens containing SiC particles, but only a ZrO_2 layer that is consistent with the high volatility of SiO_2 and B_2O_3 phases at 2400 °C. However, the formation of transient glass phases can be assumed.

The key to the success of UHTCs seems to lie in the ability to generate dense materials. On the contrary, the specific morphology of plasma-sprayed coatings implies the presence of porosity. This caracteristic could be supposed to be a drawback to prevent oxygen permeation through the coating but it can turn into an asset when considering the volume changes of the materials when they oxidize and provided that a mechanism slows down the passage of oxidant species.

4.3 Effect of Y_2O_3 rate

Y_2O_3 was selected so as to improve the densification of the layer which formed on surface during the oxidation at 2000 °C. 2 vol% coarse Y_2O_3 splats were detrimental for the resistance of ZSY(L) coatings that failed by delamination in the early stage upon the plasma jet exposure. Increasing Y_2O_3 rate up to 10 vol% seemed to be advantageous to protect the substrate for 5 minutes. Nevertheless, it led to the formation of a high amount of viscous phases and the growth of many drops and orifices through the oxide scale. While these orifices allowed the release of gases and prevented from the coating spalling, they also provided easy ways for oxidant species that could explain the increase in oxidation rate noticed between 2-min and 5-min tests. The orifices related to the formation of liquid phases have a positive effect to release gases but it seems necessary to limit their number to prevent a fast access of oxidant species into the coating.

4.3 Effect of Y_2O_3 size and distribution

ZrB_2-SiC and ZSY(S) coatings showed a similar protective behavior upon the plasma jet exposure with a decrease in oxidation rate between 2-min and 5-min tests. A difference in oxidation rate was noticed, showing a slower oxidation kinetic for the coatings including SiC particles processed by $Y(OH)_3$ sol. The addition of nano-scale Y_2O_3 additives in a homogeneously dispersed state could improve sintering of ZrO_2 scale without leading to a delamination mechanism such as observed for ZSY(L) coatings including low amounts of coarse Y_2O_3 splats.

5 Conclusion

Several ZrB_2-based coatings were plasma-sprayed, including different rates of SiC particles (0 and 10 vol%) and Y_2O_3 splats (0, 2 and 10 vol%), and with different distribution of Y_2O_3 (coarse and fine). As this class of ceramics is proposed to be exposed to harsh environments, their behavior was studied under heat-flux at temperature exceeding 2000 °C under moist atmosphere. The facility used to perform high temperature tests was suitable to study oxidation and moist corrosion resistance at extreme temperature. Moreover, it allowed to test samples for thermal shock resistance. This makes eligible the plasma heat-flux test to drive the design of new and improved UHTC coatings in order to obtain desired performances.

This study showed that plasma-sprayed ZrB_2 coatings including SiC particles are likely to be relevant to protect composite materials in moist atmosphere at temperatures higher than 2000 °C. A low amount of additives, avoiding liquid phase formation in high quantity during oxidation tests, is to be favored. Moreover, nano-scale additives in a homogeneously dispersed state are particularly advantageous to slow down the kinetic of oxidation. At temperature higher than 2000 °C, for which top protective glass layer does not exist, the development of a dense oxide scale structure seems to be responsible for the good performance. The aim could be to target the optimal oxide microstructure by tuning the particle sizes and composition of the starting coating.

Our understanding of the complex oxide scales would be greatly enhanced by more detailed experimental characterizations and will lead, in further work, to the proposal of an oxidation mechanism.

6 Acknowledgments

The authors would like to thank gratefully M. Lahaye from PLACAMAT, Pessac, for EPMA analyses and V. Génissel for the plasma jet heat-flux tests.

6 Literature

[1] Opeka M.M., I.G. Talmy, and J.A. Zaykoski: Oxidation-based materials selection for 2000 °C + hypersonic aerosurfaces: theoretical considerations and historical experience. J. Mater. Sci. (2004), 39, pp. 5887/904.

[2] Clougherty, E., R. Pober, and L. Kaufman: Synthesis of oxidation resistant metal diboride composites. Transactions of the Metallurgical Society AIME 242 (1968), pp. 1077/82.

[3] Tripp W.C., H.H. Davis, and H.C. Graham: Effect of a SiC Addition on the Oxidation of ZrB_2. American Ceramic Society Bulletin 52, (1973), Issue 8, pp. 612/6.

[4] Han, W.B., P. Hu, X.H. Zhang, J.C. Han, and S.H. Meng: High-temperature oxidation at 1900 °C of ZrB2–*x*SiC ultra high temperature ceramic composites. J. Am. Ceram. Soc. 91 (2008), pp. 3328/34.

[5] Tripp, W.C. and H.C. Graham: Thermogravi metric study of the oxidation of ZrB_2 in the temperature range of 800 °C to 1500 °C: Journal of The Electrochemical Society 118 (1971), Issue 7, pp. 1195/99.

[6] Levine S.R., E.J. Opila, M.C. Halbig, J.D. Kiser, M. Singh, and J.A. Salem: Evaluation of ultra-high temperature ceramics for aeropropulsion use. J. Eur. Ceram. Soc. 22 (2002), pp. 2757/67.

[7] Eakins E., D.D. Jayaseelan, and W.E. Lee: Toward oxidation-resistant ZrB_2-SiC ultra high temperature ceramics. Metallurgical and Materials Transactions A 42A (2011), pp.878/87.

[8] Hwang S.S., A.L. Vasiliev, and N.P. Padture: Improved processing and oxidation-resistance of ZrB_2 ultra-high temperature ceramics containing SiC nanodispersoids. Materials Science and Engineering: A 464 (2007), Issues 1-2, pp. 216/24.

[9] Opila E., S. Levine, and J. Lorincz: Oxidation of ZrB_2- and HfB_2-based ultra-high temperature ceramics: Effect of Ta additions. Journal of Materials Science 39 (2004), Issue 19, pp. 5969/77.

[10] Zhang X.H., P. Hu, J.C. Han, L. Xu, and S.H. Meng: The addition of lanthanum hexaboride to zirconium diboride for improved oxidation resistance. Scripta Materialia 57 (2007), Issue 11, pp. 1036/39.

[11] Sciti D., L. Silvestroni, V. Medri, and F. Monteverde, (2014) Sintering and Densification Mechanisms of Ultra-High Temperature Ceramics, in Ultra-High Temperature Ceramics: Materials for Extreme Environment Applications (eds W. G. Fahrenholtz, E. J. Wuchina, W. E. Lee and Y. Zhou), John Wiley & Sons, Inc, Hoboken, NJ. doi: 10.1002/9781118700853.ch6.

[12] Bellosi A., F. Monteverde, and D. Sciti: Fast Densification of Ultra-High-Temperature Ceramics by Spark Plasma Sintering. International Journal of Applied Ceramic Technology 3 (2006), Issue 1, pp. 32/40.

[13] Tului, M., G. Marino, and T. Valente: Plasma spray deposition of ultra high temperature ceramics. Surface and Coatings Technology 201 (2006), Issue 5, pp. 2103/08.

[14] Yao X. H. Li, Y. Zhang, K. Li, Q. Fu, and H. Peng: Ablation Behavior of ZrB_2-Based Coating Prepared by Supersonic Plasma Spraying for SiC-Coated C/C Composites Under Oxyacetylene Torch. Journal of Thermal Spray Technology 22 (2013), Issue 4, pp 531/37.

[15] Barré C., A. Quet, L. Bianchi, C. Sanchez, and A. Sauveroche. Spray plasma processed ZrB_2-based coatings for oxidation protection. Thermal Spray 2014: Proceedings of the International Thermal Spray Conference (May 21–23, 2014, Barcelona, Spain), pp. 541/5.

[16] Niu Y., H. Wang, L. Huang, H. Li, X. Liu, X. Zheng, and C. Ding: Oxidation and Ablation Resistance of Low Pressure Plasma-Sprayed ZrB_2-Si Composite Coating. Journal of Thermal Spray Technology 23 (2014),Issue 3, pp. 470/6.

[17] Andreani A.-S., F. Rebillat, A. Poulon-Quintin: Oxidation mechanism of ZrB_2-SiC tested in a solar furnace above 2200°C. Advances in Science and Technology 65 (2010), pp. 124/29.

[18] Tan W., M. Adducci, and R. Trice: Evaluation of rare-earth modified ZrB_2–SiC ablation resistance using an oxyacetylene torch. Journal of the American Ceramic Society 97 (2014), Issue 8, pp. 2639/45.

[19] Monteverde F. and R. Savino: Stability of ultra high temperature ZrB_2–SiC ceramics under simulated atmospheric re-entry conditions. J. Eur. Ceram. Soc. 47 (2007), pp. 4797/805.

Microstructure and thermal property of atmospheric plasma sprayed $Yb_2Si_2O_7$ coating

Xin Zhong, Yaran Niu, Liping Huang, Xuebin Zheng, Chuanxian Ding
Key Laboratory of Inorganic Coating Materials, Shanghai Institute of Ceramics, Chinese Academy of Sciences, Shanghai 200050, China
Hong Li, Jinliang Sun
Research Center of Composite Materials, Shanghai University, Shanghai 200072, China

In the present work, $Yb_2Si_2O_7$ powder was synthesized by solid-state reaction using Yb_2O_3 and SiO_2 powders as starting materials. Atmospheric plasma spray was applied to fabricate $Yb_2Si_2O_7$ coating. The phase composition and microstructure of the coating were characterized. Some basic properties of the coating including density, open porosity and microhardness were detected. Its thermal stability was evaluated by thermogravimetry and differential thermal analysis (TG-DTA). The thermal diffusivity and thermal conductivity of the coating were measured. The results showed that the as-sprayed coating was mainly composed of crystalline $Yb_2Si_2O_7$ with a small amount of Yb_2SiO_5 and amorphous $Yb_2Si_2O_7$. The coating had a dense structure containing defects, such as pores, interfaces and microcracks. The TG-DTA results showed that there was almost no mass change from room temperature to 1200 °C, while a sharp exothermic peak appeared at around 1038 °C in DTA curve, which indicated that the amorphous phase crystallized. The thermal conductivity of the coating decreased with increase in temperature up to 600 °C and then followed by an increase at higher temperatures. The minimum value of the thermal conductivity of the $Yb_2Si_2O_7$ coating was about 0.68 W/(m·K).

1 Introduction

Silicon-based nonoxide ceramics, including SiC and Si_3N_4 and silicon-based ceramic matrix composites, are promising structural materials for application in gas turbine engines, attributing to their excellent high-temperature mechanical properties and good stability in dry air [1]. However, the corrosion resistance of these materials in combustion environments containing water vapor is relatively poor. The lack of environmental durability in combustion environments obstructs the applications of these materials [2]. To protect silicon-based ceramics and composites from corrosion by water vapor, environmental barrier coatings (EBCs) have been developed [3-4].

Plasma spray deposition method has been widely applied to fabricate a variety of surface protective coatings due to its high deposition rate, dense and uniform microstructure and capability of handling different substrates [5-6]. Atmospheric plasma spray (APS) process has been reported to fabricate EBCs. K.N. Lee et al. [7-8] prepared fully-crystalline mullite coating and BSAS coating on silicon carbide by plasma spray. And tri-layer $Yb_2SiO_5/Al_6Si_2O_{13}/Si$ EBCs have been deposited onto SiC substrates using APS approach [9-10].

Recent researches have demonstrated that rare-earth silicates, especially $Yb_2Si_2O_7$, are promising candidates of EBC materials due to their superior high-temperature stability, relatively low coefficient of thermal expansion (CTE), long durability in water vapor and excellent chemical compatibility with silicon-based ceramic materials [4, 11]. Some researchers reported the preparation methods of rare-earth silicates powders using hydrothermal solution, sol-gel synthesis and solid-state reaction [12-15]. Some efforts have been put to study the mechanical and thermal properties of bulk $Yb_2Si_2O_7$. Y.C. Zhou et al. [16] investigated the thermal and mechanical properties of β-$Yb_2Si_2O_7$ using a combination of first-principles calculations and experimental investigations and claimed that $Yb_2Si_2O_7$ is a damage tolerant ceramic. Z.L. Tian et al. [17] studied the damage tolerance and extensive plastic deformation of $Yb_2Si_2O_7$ from room to high temperatures. Their results indicate that $Yb_2Si_2O_7$ is a prospective EBC candidate with noticeable damage tolerance and extensive plastic deformation capability. The microstructure of coating is closely related to its fabrication process and is quite different from that of bulk materials, which would result in quite different properties. However, the fundamental investigations of the microstructure and thermal property of plasma-sprayed $Yb_2Si_2O_7$ coating are still lacking.

In this work, $Yb_2Si_2O_7$ powder was synthesized by solid-sate reaction method. And atmaspheric plasma spray (APS) was applied to fabricate $Yb_2Si_2O_7$ coating. The phase composition and microstructure of the powder and coating were characterized by x-ray diffraction (XRD), scanning electron microscopy (SEM) and x-ray energy-dispersive spectroscopy (EDS). The thermal stability was evaluated by thermogravimetry and differential thermal analysis (TG-DTA). Thermal conductivity of the coating was focused. The aim of this work is to study the microstructure and compositions of plasma-sprayed $Yb_2Si_2O_7$ coating and give some information about thermophysical property characteristics of this kind of coating.

2 Experimental procedures

2.1 Materials and sample preparation

The $Yb_2Si_2O_7$ powder was synthesized by solid-state reaction. SiO_2 (Slnopharm Chemical Reagent Co., Ltd.), and Yb_2O_3 (Shanghai Heli Rare Earth Material

Co., Ltd,) were used as starting materials. An ethanol-based suspension was prepared by mixing the Yb_2O_3 and SiO_2 powders in 1:2 molar ratios and then was ball-milled for 3 h with zirconia balls. After mixing, the suspension was completely dried in an oven and then calcined to obtain $Yb_2Si_2O_7$ according to the chemical reaction $2SiO_2 + Yb_2O_3 \rightarrow Yb_2Si_2O_7$, and then the as-synthesized powders were ground and granulated before spray process.

A plasma spray system (A-2000, Sulzer Metco AG, Switzerland) equipped with a F4-MB torch was applied to deposit the $Yb_2Si_2O_7$ coatings. Argon and hydrogen were used as plasma forming gases. To obtain free-standing samples used in this work, a coating of about 2 mm thickness were deposited onto a aluminium substrate and then mechanically peeled off for microstructure and property characterization.

2.2 Specimens characterization

The phase compositions of the fabricated powder and the coating were examined by XRD using a Rigaku D/Max2550 diffractometer with Cu-Kα (λ= 0.154056 nm) radiation. The particle size distribution was identified by laser diffraction on BT-9300Z system (Baite Instruments Ltd., China).The microstructure and chemical composition of the raw powder and as-deposited coating were characterized using SEM (S-4800, Hit-achi, Japan) with EDS (INCA SERIES, Oxford Instrument, UK). Open porosity and density of the coating were detected by the Archimedes method. Vickers hardness (Hv) was measured on the polished cross-section of the coating using an Instron Wilson-Wolpert Tukon 2100 B hardness tester under a load of 300 g with a dwell time of 15 s. The thermal stability of the as-deposited coating was examined using Differential Thermal Analysis (STA449C, Netzsch, Germany) operated in continuous heating mode from room temperature to 1400 °C at a rate of 10 °C /min under an argon atmosphere.
Thermal diffusivity (α) of the $Yb_2Si_2O_7$ coating was characterized using the laser flash method. The specimen with 11 mm in diameter and 0.8 mm in thickness was used. The surfaces of specimen were coated with a thin layer of graphite in order to prevent direct transmission of the laser through the specimen. The specific heat capacity (Cp), as a function of temperature, was calculated by the Neumann–Kopp rule with the heat capacity data of SiO_2 and Yb_2O_3 [18-19]. The specific heat capacity of $Yb_2Si_2O_7$ coating was measured using diamond DSC in the range from room temperature to 500 °C to fitting the calculated values. The thermal conductivity (κ) values were calculated by Eq. (1) with the heat capacity (Cp), thermal diffusivity (α) and density (ρ).

$$\kappa = Cp \cdot \alpha \cdot \rho \quad (1)$$

3 Results and discussion

3.1 Microstructure and composition characterizations

The morphology and particle size distribution of $Yb_2Si_2O_7$ powders are presented in Fig.1. It can be seen that the raw $Yb_2Si_2O_7$ particles are of irregular shape. The measured medium size (D50) of the powers measured is about 45 μm.

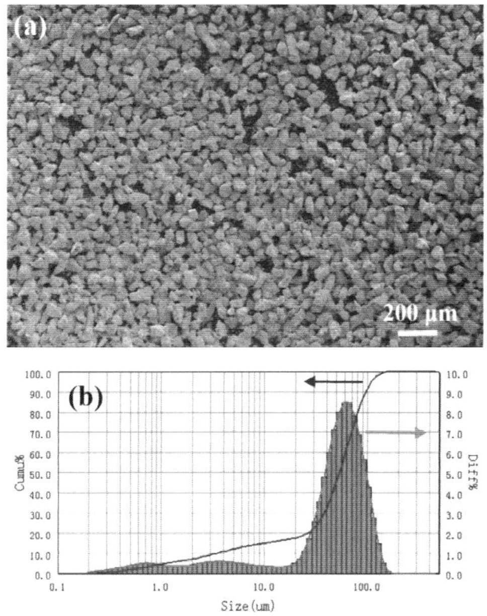

Fig. 1. Morphology (a) and particle size distribution (b) of $Yb_2Si_2O_7$ powder.

Figure 2 illustrates the XRD patterns of both the $Yb_2Si_2O_7$ powder and the as-sprayed coating. The phase composition of powders is $Yb_2Si_2O_7$ that crystallizes in monoclinic structure (JCPDS card No. 25-1345) [see Fig. 2(a)], indicating that solid-sate reaction method is suitable to prepare pure $Yb_2Si_2O_7$. Compared the XRD pattern of the as-sprayed coating with that of the $Yb_2Si_2O_7$ powder, it can be seen that some changes occurred after the spray process. Besides $Yb_2Si_2O_7$ phase, a bulged broad peak emerged, which indicates that certain amount of amorphous phase formed in the $Yb_2Si_2O_7$ coatings. The characteristic peak correponding to Yb_2SiO_5 phase with monoclinic structure was also identified.

The rapid cooling which freezes the melted $Yb_2Si_2O_7$ particles after the deposition process is believed to be closely related to the formation of the amorphous phase. In more detail, the $Yb_2Si_2O_7$ particles were mostly fully melted and became $Yb_2Si_2O_7$ melt when they passed through the plasma flame with an extremely high temperature. As a silicate, the anion radicals such as $[SiO_4]^{4-}$ were accumulated in the $Yb_2Si_2O_7$ melt. With the temperature decreased, more and more anion radicals assembled. The gather of anion radicals could produce a closely packed disordered structure of amorphous material and the liquid-solid interface energy increased, resulting in suppressed crystal nucleation. At the same time, the viscosity of components in the undercooled state increased rapidly because of the closely packed disordered structure, which restrained the long-range diffusion of component atoms, thus crystal growth was inhibited [20-21]. The emergence of Yb_2SiO_5 could be attributed to the dissosiation of $Yb_2Si_2O_7$. The chemical reaction $Yb_2Si_2O_7\,(l) \rightarrow Yb_2SiO_5\,(l) + SiO_2\,(g)$ probably occurred during the deposition process.

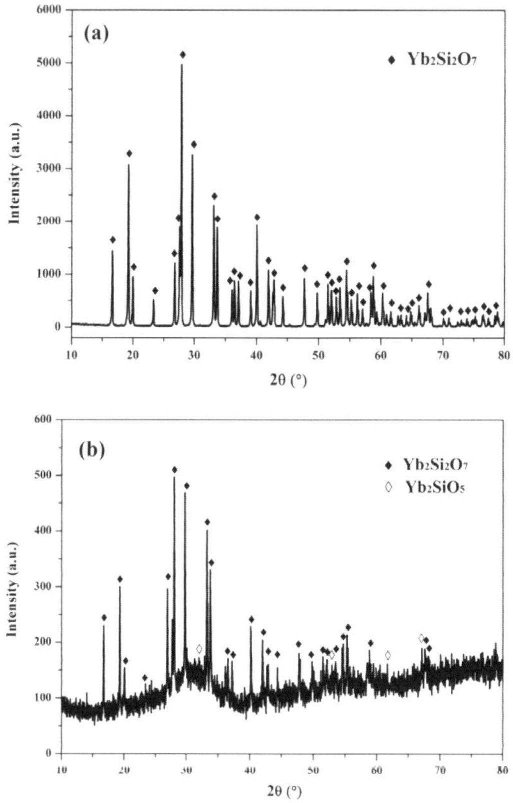

Fig. 2. XRD patterns of the $Yb_2Si_2O_7$ powder (a) and the as-sprayed coating (b).

The surface morphology of the $Yb_2Si_2O_7$ coating shows that the coating was mostly built up by well-flattened splats, as shown in Fig. 3(a). Partly melted region and some microcracks were also observed at the same time. The fractural morphology of the $Yb_2Si_2O_7$ coating is illustrated in Fig. 3(b), which gives more microstructure information. Differing from other plasma-sprayed ceramic coatings, the coating did not have an apparent lamellar structure with fine columnar grains. Defects such as interface between splats, microcracks and pores were observed in the inside of the coating. The presence of microcracks can be attributed to the the release of high stresses during cooling. The round pores inside the splats are mainly caused by the existence of some gases that had no time to escape due to the rapid overlapping of droplets.

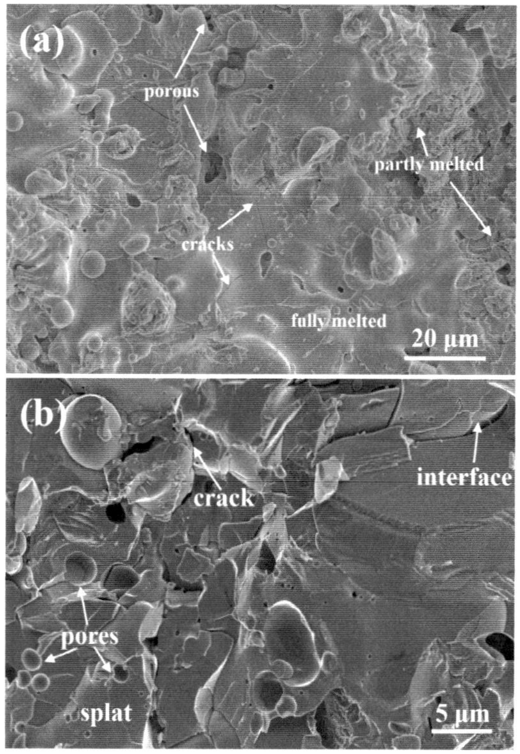

Fig. 3. SEM micrographs of the $Yb_2Si_2O_7$ coatings: (a) surface morphology, (b) fracture section morphology.

Figure 4(a) displays the SEI-SEM (secondary electron imaging of scanning electron microscopy) micrographs of the cross-section of the $Yb_2Si_2O_7$ coating. It can be seen that the coating was relatively dense. The open porosity measured by the Archimedes method was about 4.6%. Other than microcracks, typical two kinds of pores were found in the coating. One is coarse voids caused by the relatively poor bonding between splats, the other was fine pores arising from entrapped gases. The magnified BEI-SEM (back-scattered electron imaging of scanning electron microscopy) view of the polished cross section of the coating was examined, as shown in Fig. 4(b). A significant color variation among splats indicated compositional differences in the as-sprayed coating. The chemical compositions of these regions were determined by EDS. A summary of the composition data is given in Table 1. The results of spot 1, 2 and 3 revealed the appearance of $Yb_2Si_2O_7$. The atomic ratio of spot 4 was in agreement with Yb_2SiO_5 which indicated that dissosiation of $Yb_2Si_2O_7$ took place during deposition process, which was ingreement with the XRD result.

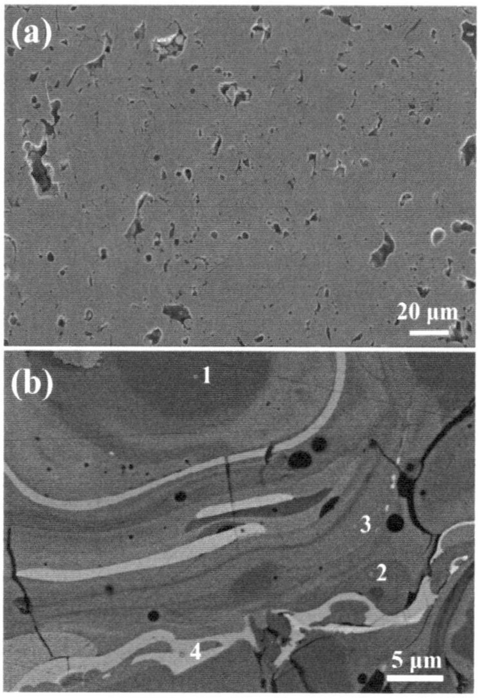

Fig. 4. Cross-sectional morphologies of $Yb_2Si_2O_7$ coating: SEI-SEM image (a); BEI-SEM image (b).

Table 1 Summary of EDS results of areas in Fig. 4(b)

Spectrum location	Yb (at %)	Si (at %)	O (at %)
Yb_2SiO_5	25.00	12.50	62.50
$Yb_2Si_2O_7$	18.18	18.18	63.64
Point 1	19.80	19.91	60.29
Point 2	19.83	19.06	61.11
Point 3	18.77	17.96	63.27
Point 4	28.21	10.94	60.86

3.2 Properties characterization

3.2.1 Thermal stability and microhardness

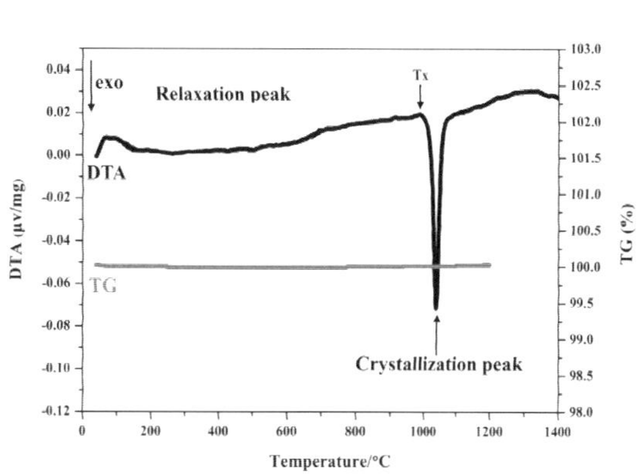

Fig. 5. TG-DTA curves of the $Yb_2Si_2O_7$ coating.

The typical TG - DTA curves of the $Yb_2Si_2O_7$ coating is shown in Fig. 5. It is found that there was almost no mass change during the heating process, while the DTA curve of the coating included a sharp exothermic peak at 1038 °C. The sharp peak is related to crystallization of $Yb_2Si_2O_7$ from amorphous phase into crystals, which is expected to spontaneously generate. Additionally, it is shown that the onset crystallization temperature (T_x) of the coating is about 1000 °C. Careful measurements of micro-hardness were made on cross sections of the as-sprayed coatings. The microhardness of the coatings was about 4.28 GPa which was only 58.6 % of that reported for the bulk material (7.30 GPa) [16]. As observed in Fig.3(b) and 4(a), there were many defects in the coating, such as pores and microcracks, which resulted in a low hardness. Certain amount of amorphous $Yb_2Si_2O_7$ phase in the coating also contributed to the low hardness.

3.2.2 Thermal conductivity

The value of the density of $Yb_2Si_2O_7$ coating was 5.67 $g{\cdot}cm^{-3}$. The heat capacity of the $Yb_2Si_2O_7$ coating is shown in Fig. 6, which increased as the temperature elevated. Figure 7 shows the thermal diffusivity and thermal conductivity of the coating in the temperature range of 200 °C-1200 °C. It can be seen that both curves decreased with the temperature increasing up to 600 °C and then increased after 600 °C. The decrease in thermal conductivity suggests a dominant phonon conduction behavior, similar to most ceramics [22]. And the thermal conductivity increases in the high-temperature region above 600 °C may be caused by two main reasons. On the one hand, the microstructure of the as-sprayed coating changed, such as microcrack healing and shrinkage of pores, due to sintering effects. On the other hand, the amorphous phase in the coating crystallized after reaching the crystallization temperature (about 1000 °C). The minimum value of the thermal conductivity (at 600 °C) of the as-sprayed coating is about 0.68 W/(m·K), which was much lower than that of the bulk material (2.1 W/(m·K) at 1273 K) [16].

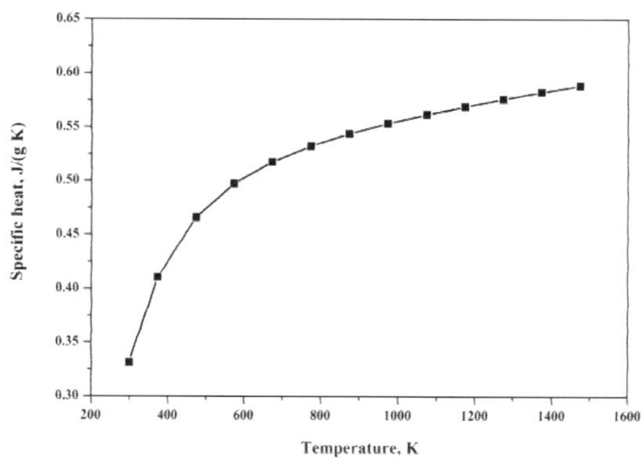

Fig. 6. Heat capacity of the $Yb_2Si_2O_7$ coating.

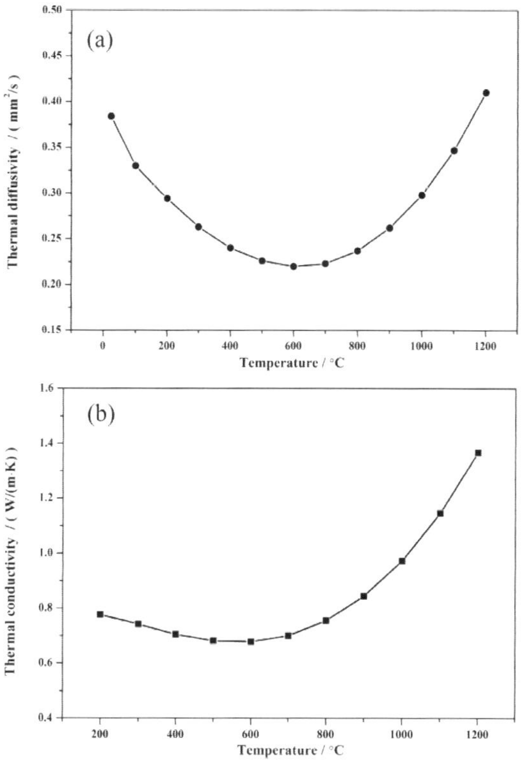

Fig. 7. Thermal diffusivity (a) and thermal conductivity (b) of the $Yb_2Si_2O_7$ coating.

4 Conclusions

The $Yb_2Si_2O_7$ coating was fabricated by atmospheric plasma spray using $Yb_2Si_2O_7$ powder synthesized by solid-state reaction. The thermal property of the coating was investigated. The microstructure observation exhibited that the coating was composed of dense structure with low porosity. Crystalline and amorphous $Yb_2Si_2O_7$ phases were detected and some Yb_2SiO_5 emerged which was resulted from the dissosiation of $Yb_2Si_2O_7$ during deposition process. The amorphous phase in the coating starts to crystallize at around 1000 °C. The micro-hardness of the as-sprayed coating is 4.28 GPa. The thermal diffusivity and thermal conductivity of the coating decrease with temperature lifting up to 600 °C and then followed by an increase with further temperature rise. The minimum value of the thermal conductivity of the as-sprayed coating was about 0.68 W/(m K) at 600 °C.

Acknowledgements

This work was supported by the National Natural Science Foundation (for Young Scholar) of China under grant 51102267.

References

[1] R. Raj: Fundamental research in structural ceramics for service near 2000 °C. Journal of the American Ceramic Society 76 (1993), Issue 9, pp. 2147/74.

[2] J. L. Smialek, R. C. Robinson, E. J. Opila, D. S. Fox, and N. S. Jacobson: SiC and Si_3N_4 recession due to SiO_2 scale volatility under combustor conditions. Advanced Composite Materials 8 (1999), Issue 1 pp. 33/45.

[3] K.N. Lee: Current status of environmental barrier coatings for Si-based ceramics. Surface and Coatings Technology 133-134 (2000), pp. 1/7.

[4] K.N. Lee, D.S. Fox, and N.P. Bansal: Rare earth silicate environmental barrier coatings for SiC/SiC composites and Si_3N_4 ceramics. Journal of the European Ceramic Society 25 (2005), pp. 1705/15.

[5] E. fender: Fundamental studies associated with the plasma spray process. Surface and Coatings Technology 34(1988), pp. 1/14.

[6] R. McPherson: The relationship between the mechanism of formation, microstructure and properties of plasma-sprayed coatings. Thin Solid Films 83 (1981), Issue 3, pp. 297/310.

[7] K.N. Lee, D.S. Fox, J.I. Eldridge, D. Zhu, R.C. Robinson, N.P. Bansal, and R.A. Miller: Upper temperature limit of environmental barrier coatings based on mullite and BSAS. Journal of the American Ceramic Society 86 (2003), Issue 8, pp. 1299/306.

[8] K.N. Lee, R.A. Miller, N.S. Jabocson: New generation of plasma-sprayed mullite coatings on silicon-carbide. Journal of the American Ceramic Society 78 (1995), Issue 3, pp. 705/10.

[9] B. T. Richards, and H. N. G. Wadley: Plasma spray deposition of tri-layer environmental barrier coatings. Journal of the European Ceramic Society 34(2014), Issue 12, pp. 3069/83.

[10] B. T. Richards, H. Zhao and H. N. G. Wadley: Structure, composition, and defect control during plasma spray deposition of ytterbium silicate coatings. Journal of Materials Science 50 (2015), pp. 7939/57.

[11] N. Maier, K.G. Nickel, and G. Rixecker: High temperature water vapor corrosion of rare earth disilicates (Y, Yb, $Lu)_2Si_2O_7$ in the presence of $Al(OH)_3$ impurities. Journal of the European Ceramic Society 27 (2007), pp. 2705/13.

[12] H.F. Chen, Y.F. Gao, Y. Liu, and H.J. Luo: Hydrothermal synthesis of ytterbium silicate nanoparticles. Inorganic Chemistry 49 (2010) pp. 1942/46.

[13] A.J. Fernández-Carrión, M.D. Alba, A. Escudero, and A.I. Becerro: Solid solubility of $Yb_2Si_2O_7$ in β-, γ- and δ-$Y_2Si_2O_7$. Journal of Solid State Chemistry 184 (2011) pp. 1882/89.

[14] C. Zhao, F. Wang, Y. J. Sun, and Y. C. Zhou: Synthesis and characterization of β-$Yb_2Si_2O_7$ powders. Ceramics International 39 (2013), Issue 5, pp. 5805/11.

[15] H.T. Sun, M. Fujii, N. Nitta, F. Shimaoka, M. Mizuhata, H. Yasuda, S. Deki, and S. Hayashi: Large-scale controllable synthesis and characterization of

ytterbium silicate nanostructures. Journal of the American Ceramic Society 91 (2008), Issue 5, pp. 4158/61.
[16] Y. C. Zhou, C. Zhao, F. Wang, Y. J. Sun, L. Y. Zheng, and X. H. Wang: Theoretical prediction and experimental investigation on the thermal and mechanical properties of bulk β-$Yb_2Si_2O_7$. Journal of the American Ceramic Society 96 (2013), Issue 12, pp. 3891/900.
[17] Z. L. Tian, L. Y. Zheng, J. M. Wang, and J. B. Yang: Damage tolerance and extensive plastic deformation of β-$Yb_2Si_2O_7$ from room to high temperatures. Journal of the American Ceramic Society 98(2015), Issue 9, pp. 2843/51.
[18] R. A. Swalin: Thermodynamics of solids. Wiley, New York, 1972, pp. 53/87.
[19] D. L. Ye: Inorganic material thermodynamics (2nd ed.). Metallurgical Industry Press, Beijing, 2002, pp. 923/62.
[20] D. Turnbull: Under what conditions can a glass be formed. Contemporary Physics 10 (1969), Issue 5, pp. 473/88.
[21] D.R. Uhlmann: Kinetic treatment of glass formation. Journal of Non-Crystalline Solids 7 (1972), Issue 4, pp. 337/48
[22] T. G. Xi: Thermophysical properties of inorganic materials. Shanghai Scientific and Technical Press, Shanghai, 1981.

Characterization of the Deposition Formation Dynamics of Suspension Plasma Spray Coatings using In-Situ Coating Property Measurements

Ramachandran Chidambaram Seshadri, Sanjay Sampath
Center for Thermal Spray Research (CTSR), Department of Materials Science and Engineering, The State University of New York @ Stony Brook, NY 11790. E mail: ram.seshadri@stonybrook.edu

Suspension plasma spraying (SPS) enables the production of finely grained nanometric or submicrometric coatings with unique mechanical, thermal, and structural properties. In SPS, a liquid carrier (ethanol/ water) is used to transport the submicrometric feedstock into the useful core centerline region of the plasma jet. This condition brings about a series of important changes in the deposition state and microstructure of the deposited coatings compared to traditional plasma sprayed coatings. Considering complex deposition dynamics of SPS deposition technique, there is an extensive need to better understand the relationship between plasma spray conditions and stress evolution in coating during deposition. In this investigation a submicron (300 nm) yttria stabilized zirconia (YSZ) particles suspended in ethanol was spray deposited using a single electrode cascaded arc plasma torch. The suspension contained 25 wt% solid content. The stresses generated (termed evolving stress) during the deposition of the coatings were calculated using the change in curvature of the substrate measured using an in-situ coating property (ICP) sensor. Multitudes of coating morphologies ranging between porous lamellar to heavily segmented columnar structures were observed under the considered deposition parameters. The mechanisms for formation of multitudes of SPS coating morphologies were proposed based on ICP results.

1.0 Introduction

The Suspension Plasma Spray (SPS) process was developed to overcome the particle entrainment issue associated with the injection of super fine (< 10µm) powders into the plasma jet. SPS uses ultrafine particles suspended in water/ ethanol, which are melted and then nearly accelerated to the plasma gas velocity and deposited on the substrate as ultra-thin "splats" **[1]**. Nanopowders in suspension have the potential to produce nanostructured ceramic coatings, which leads to improvements in certain desirable properties such as density, strength, fracture toughness and hardness **[2]**. SPS has been successfully employed to prepare coatings designated for SOFCs, TBCs, field emitters, photocatalytic applications, biomaterials, photovoltaics, as shown in examples of the literature **[3-8]**.

Most of the papers cited above are focused on examining the effects of the processing parameters on the properties of SPS coatings. More specifically, the relations between the characteristics of the stress evolution during the formation of different morphologies (dense columnar, cracked columnar, porous/ feathery columnar, dense vertically cracked, banded/ porous vertically cracked) of SPS coatings and the final coating properties have not been investigated. This research therefore strives to assess the in-situ stress evolution of SPS coatings sprayed under different plasma spray conditions which were found to possess distinctive microstructure and mechanical properties.

2.0 Experimental Work

The suspension used in this investigation was Oerlikon Metco 6608 (D50: 0.3 µm high purity HOSP 25 wt. % $8Y_2O_3$ ZrO_2 suspended in ethanol). The suspension was fed into the plasma using 5MPE-SFTM Suspension Powder Feed Unit (Oerlikon Metco, USA) and the operational aspects of the feeder is described elsewhere **[9]**. Due to the low arc root fluctuations and high throughput deposition capability the single electrode high efficiency cascaded arc plasma spray torch Sinplex Pro™ – 180 (Oerlikon Metco, USA) **[10, 11]** fitted with a 9 mm nozzle and Ø 0.8 mm ruby lined suspension injector was used to deposit coatings. The temperature and velocity of the suspension particle stream were measured using Accuraspray-g3™ (Tecnar Automation, Canada) ensemble particle sensor. Grit blasted strips having dimensions 228.6 X 25.4 X 2.3 mm^3 made out of AISI 304 stainless steel were used as substrates in this study. The SPS ceramic coatings were deposited on the bond coated strips. The parameters used to spray the bond coat and SPS ceramic coatings are displayed in **Tables 1 and 2** respectively.

The H_2 flow rate and the suspension feed rate were kept constant at 7 lpm and 14 gpm (in-terms of feed pressure-solid loading relationship) respectively. A six axis robot (Staubli RX90, France) programmed with ladder type spray pattern was used to spray the coatings. Further, the thickness of bond coat was 150 µm ± 15 µm and the thicknesses of the SPS coatings were maintained between 500 to 650 µm for all the experimental runs. The stress evolution of coatings while spraying over the stainless steel strips was measured using an In-situ Coating Property Sensor (ICP) **[12]**. The microstructural characteristics of the metallographically prepared SPS coatings were examined using Hitachi TM3000 (Angstrom Scientific Inc., USA) Scanning Electron Microscope (SEM).

Table 1: APS parameters used to deposit NiCrAlY bond coat

Powder	Ar (lpm)	H_2 (lpm)	I (Amperes)	V (Volts)	Carrier gas flow rate (lpm)	Powder feed rate (gpm)	Spray Distance(mm)	Raster Speed (mm/min)	Injection angle (Deg)
Amdry 421.1	45	8	330	92	3	30	100	1000	90

Table 2: Suspension spray parameters used to deposit the YSZ ceramic coatings with the resultant thickness/ pass and average substrate temperature measured during deposition

Coating ID	Ar Flow Rate (lpm)	Current (Amps)	Voltage (Volts)	Spray Distance (mm)	Raster Speed (mm/ min)	Injection Angle (Deg)	Avg. Substrate Temp. (°C)	Thickness/ Pass (µm)
C1	45	400	93	70	1000	90	570	11
C2	45	480	96	70	1000	90	650	13
C3	45	530	100	70	1000	90	720	16
C4	125	480	110	70	500	90	460	9
C5	75	480	98	70	500	90	590	14
C6	40	480	86	70	500	90	740	17
C7	40	520	99	60	1000	75	650	11
C8	40	520	99	60	1000	90	720	20
C9	40	520	99	60	1000	105	850	15
C10	35	420	91	50	500	75	654	33
C11	35	500	94	60	1000	90	730	18
C12	35	420	91	50	500	90	747	38

3.0 Results and Discussions

In this investigation the spray conditions were chosen to acquire different morphologies of SPS deposits by varying the plasma spray parameters namely the plasma input power, primary gas flow rate and injection angles. The conditions C1-C3 in **Table 2** were used to study the effects of plasma input power, condition sets C4-C6 and C7-C9 were used to study the effects of primary gas flow and injection angles on the coating formation. Conditions C10-C12 were chosen to get dense and cracked coating architectures.

3.2 Effects of SPS parameters on in-flight particle behavior and coating microstructure

The effects of input power on particle surface temperature and velocity responses are displayed in **Figure 1**. From the figure, the following points can be inferred. Surface temperatures of the particles under the three power levels are above the melting temperature of zirconia and the surface temperatures were found to be increasing with power levels with a slight increase in velocity. As the arc current increases, the net enthalpy in the plasma increases. This condition leads to a better in-flight particle melting and enables the particles to flatten better on the substrate surface **[13]**. It is observable that with increase in the plasma powder level, the SPS microstructure changes from puffy columnar microstructure to a dense columnar microstructure with cracks as seen in **Figure 1** (C1-C3). The particle temperature - velocity variations and the resultant microstructures with respect to the change in the primary gas flow rates are illustrated in **Figure 1** (C4-C6). If the primary gas flow rate is low, then it reduces the velocity of the plasma flame, which in turn increases the residential time of the particles in the plasma leading to better melting of the particles and vise versa for higher flow rates **[14]**. The particle velocity seems to increase concurrently with the primary gas flow rate with a simultaneous decrease in the particle surface temperature (C4-C6) as seen in **Figure 1**. From the micrographs shown in **Figure 1**, it is also observable in cross-sectional micrographs (C4-C6) that there is an increase in inter-columnar spacing and a reduction in the density of the deposited coating microstructure with the increase in the primary gas flow rates. The injection of the suspension into the highly viscous plasma plume is one of the most challenging aspect in SPS, as a non-ideal injection often prevents particles from entering the central area of the plasma plume.

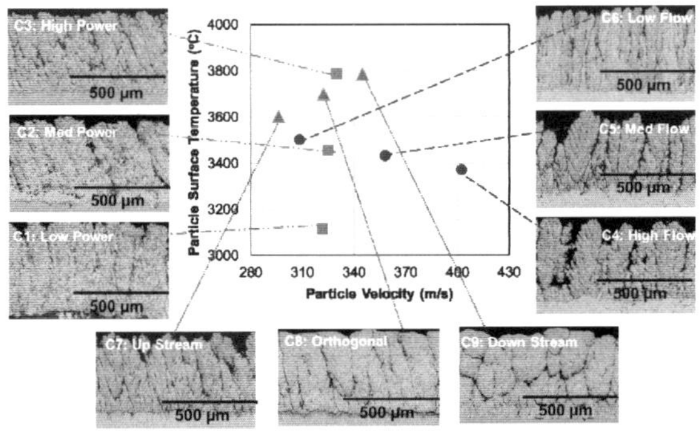

Figure 1. First order process map displaying the effect of plasma power, primary gas flow rate and angle of injection on SPS YSZ microstructure evolution

The basic three different injection configurations namely the upstream (75°), orthogonal (90°) and downstream (105°) injectors are considered in this investigation. The particle surface temperature and the velocities vary greatly with respect to the injection configurations, which could be visualized in **Figure 1** (C7-C9). The highest temperature and velocity was recorded for downstream injection (105°), which is followed by orthogonal injection (90°) and the lowest inflight temperature and velocity was recorded when upstream injection (75°) was adopted. It is put forth form **Figure 1** (C7-C9) that the microstructural variations are drastic with respect to injection configurations. Porous feathery columnar structure was observed in case of upstream (75°) injection **Figure 1** (C7), a dense columnar structure was observed in case of orthogonal (90°) injection (C8) and a defective columnar structure (C9) with a lots of inter-columnar voids and fanned columns are observed in case of downstream (105°) injection configuration.

3.3 In-situ stress evolution of SPS coatings

The time curvature plot displaying the curvature patterns obtained during the deposition of three distinct SPS coating microstructures is displayed in **Figure 2**. The plot represent the time-curvature measurements made during the deposition of coatings sprayed under conditions C3, C4 and C12. It is evident from **Figure 2** that the slope of curvature rise of ICP measurements during deposition suggests that lower and higher net-stresses were generated in coatings deposited under conditions C4 and C12 respectively and the slope for coating C3 falls in-between. In addition to the net curvature change it is also observed in **Figure 2** that the curvatures of coatings sprayed under C3 and C12 display a two stage curvature change. The first stage curvature change the during deposition gradually increases up to a certain number of coating passes i.e., 14th pass for C3 and 5th pass for C12 after this comes the second stage in which the curvature starts to decrease. The increase in curvature during the first stage might be related to the rapid stress buildup which when exceeds the strength of the coating causes the cracks to open up and propagate vertically during deposition itself depending on the cohesive strength of the coatings.

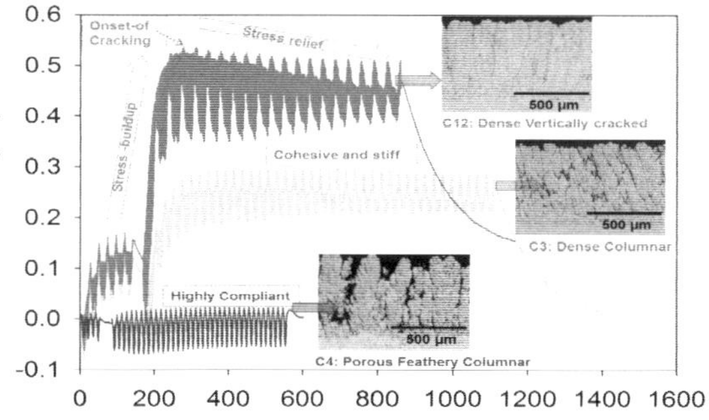

Figure 2. ICP Time Curvature plot displaying the curvature patterns obtained during the deposition of three distinct SPS coating microstructures

The crack opening and propagation would relieve the accumulated deposition stresses and the curvature starts to decrease accordingly in the second stage. The net increase and decrease of curvatures in the first and second stages during deposition depends upon the substrate temperature and deposition rate of the coatings. The peak stress value determined at end of the first stage of curvature change was considered for coatings which showed the double stage curvature change and those values are reported in this article. For the rest of the coatings, the conventional stress evaluation procedure was followed **[12]**. The stress modulus values derived from the ICP measurements are displayed in **Figure 3**. It is worth mentioning that the compliance of the coatings can be easily adjudged using the ICP curvature-microstructure relationship plot shown in **Figure 2**. The coating sprayed under condition C4 displays (**Figure 2**) a porous feathery columnar structure which is less cohesive and compliant and hence the curvature change value is low. A dense columnar coating microstructure was formed when sprayed with condition C3 (**Figure 2**) which is highly cohesive and stiff. A Dense Vertically Cracked (DVC) type coating with lots of inter pass bands forms when sprayed under condition C12 (**Figure 2**) and it is dense as well as compliant.

3.4 Stress-modulus-microstructure relationship among the coatings

The stress-modulus relationship among the coatings is displayed in **Figure 3**. This plot can be divided into four different zones depending upon the resulting microstructure evolution. The porous feathery columnar microstructure evolves in zone 1. In zone 2 defective and narrow columnar microstructures with and with-out cracks are formed. A competing phenomenon is observed in zone 3 in which dense columnar microstructure and broad columns with wide macro-fanning head/ stacking faults are formed. In zone 4 the column formation is heavily suppressed and DVC type of coatings are largely promoted. In order to better explain the stress modulus-microstructure relationship, the basic plasma parameter and the coating microstructure evolution has to be revisited.

3.4.1 Formation of porous feathery columnar microstructures and the resultant stress-modulus evolution

In this investigation the porous feathery columnar microstructures are formed when the suspensions are sprayed with conditions C1, C2, C4 and C5 (**Table 2**). These coatings fall within the zone 1 of modulus-microstructure plot shown in **Figure 3.** The mechanisms involved in the SPS comprise a series of successive stages **[15]**, the first of which is, the fragmentation of the liquid stream into fine liquid drops after its injection into the thermal plasma stream issuing out of the torch, this step is followed by vaporization of the solvent, release of the solid contents in the form of individual particles or agglomerates, possible disruption of part of agglomerates, melting of particles and agglomerates and deposition onto the substrate.

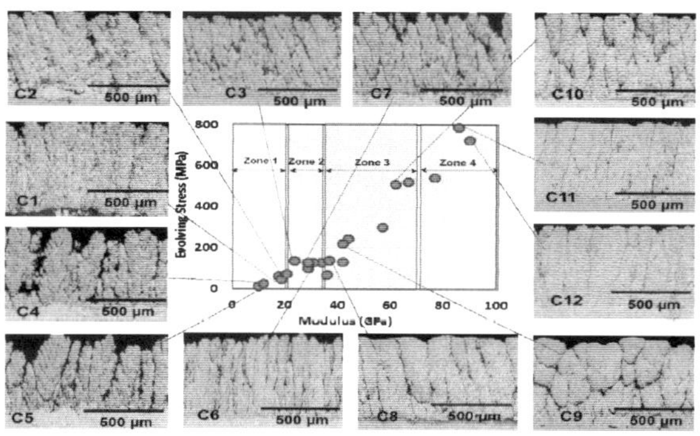

Figure 3. Stress – Modulus – Microstructure relationship among the coatings

The size of the solid particles, the possible presence of large agglomerates inside the original suspension, the size of the fragmented liquid drops and of the resulting agglomerates obtained after solvent vaporization, the cohesiveness of the latter are all important parameters determining the deposition behavior and the eventual coating microstructure **[16]**. The coatings C1 and C2 are sprayed under low powder levels. The resultant coatings are puffy columnar structure with lots of submicron pores. Under low power conditions, the atomizing effect of the plasma is reduced. It is, therefore, expected that the C1 and C2 conditions will generate larger suspension droplets and once the ethanol evaporates, larger agglomerated ceramic droplets are formed. The combination of low power and high volume of solvent for evaporation within the plasma further reduces the energy available for particle heating, thus larger particles arrive at the substrate without complete melting. This can be observed in the resultant microstructure that contains a large quantity of pores and spherical particles. Compared to condition C1 the coating sprayed with condition C2 is better melted and the coating seems to be denser due to higher input power which results in higher particle temperature and velocity (**Figure 1**). High particle melting and high substrate temperature during deposition aids lamellar compaction and higher thickness per pass, the proof for this is provided as data in **Table 2**. The conditions the suspension stream first encounters are the gases flowing in the periphery of the hot core of the plasma. The velocities of the primary/ plasma gas are high enough to fragment the suspension stream, vaporize the liquid phase of the resulting droplets, and propel the molten droplets towards the substrate. It should be noted that the higher the primary gas flow rate the higher will be the shear and drag forces acting on the fragmented droplets **[17]**. Under very high plasma gas flow rates the suspension stream will undergo catastrophic breakup. In this breakup regime, aerodynamic shear and drag forces repeatedly fragment the liquid until a critical droplet size is reached where the surface tension balances the drag forces **[18]**. The conditions C4 and C5 have very high plasma gas flow rates 125 lpm and 75 lpm respectively (**Table 2**). The temperature velocity plot (**Figure 1**) suggests that there is a concomitant increase in velocity and a corresponding decrease in the temperature. From **Figure 3** it is put forth that the coatings sprayed with conditions C4 and C5 have the lowest modulus and evolving stress values among the deposited coatings. Further with the increase in the primary gas flow rates the inter-columnar spacing is found to be increasing (**Figure 1**).

Since the fragmentation of suspension droplets are severe under high primary gas flow rates the diameters of droplets are smaller. This droplet size represents a multiple order of magnitude decrease in mass relative to the injected suspension stream that penetrated into the plasma flow. Consequently, the final fragments were assumed to lack the momentum necessary to move radially through the plasma flow. Since these smaller particle travel along with the plasma flow, they get deposited on the sides of the substrate asperities and starts to grow in lateral direction. The high flow rates reduces the residence time, furthermore the smaller particles cool faster and these form a porous feathery columnar microstructure. The high flow rates also blow away the poorly treated/ re-solidified or loosely bound particles from the spray bead edges thereby increasing the inter-columnar spacing. The thickness/ pass and substrate temperature have inverse proportional relationship with the primary gas flow rates (**Table 2**).

3.4.2 Formation of dense and defective columnar microstructures and the resultant stress-modulus evolution

The coatings sprayed under high power (C3), low primary gas flow rate (C6) and upstream injection (C7) were found to possess dense columnar microstructures these coatings and fall with in zone 2 of stress modulus relationship shown in **Figure 3**. Conditions C3 and C6 with high power level and low primary gas flow rate enables better melting of the particles and also transfers more heat to the substrate this scenario leads the splats to spread well and form to a stronger bonding with the substrate surface on solidification. Under high power and low gas flow rate conditions (C3 and C6) the plasma flame is likely to produce a larger fraction of molten particles and, generally, a higher deposition efficiency and thickness per pass. The experimental results shown in **Table 2** and **Figures 1** and **3** agree pretty well in describing these effects. The coatings sprayed under orthogonal (C8), downstream (C9) and condition C10 fall in zone 3 of the stress modulus relationship plot illustrated in **Figure 3.** The SPS process is very sensitive to changes in the process parameters. The method of injecting the suspension is one of the most crucial and influencing parameter. In this investigation the injection points/ angles were altered. From the temperature and velocity profiles shown in **Figure 1** (C7, C8, C9) it is quite obvious that the downstream (105°) injection imparts higher temperature and velocity to the injected particles followed by orthogonal (90°) and upstream injection (75°). This result is quite contrary to the general understanding that the residential time of the droplets in the plasma should be higher and the droplets should melt more effectively when injected upstream which is very near to the torch exit. The reasons for the aforementioned occurrence is because in this investigation the feed rate was kept constant and was not optimized for each plasma condition.

The feed rate and injection are not independently controllable in our feeding system as a matter of fact both are controlled by injection pressure which was maintained constant (**Table 2**). According to Delbos et.al., **[19]** the plasma flame can be divided into three different regions from the suspension injection point which are unmelted zone, spherical mass zone and liquid droplet zone. An optimally injected suspension stream will pass through all the aforementioned zones to form a coating deposit. The volume of these zones will change with respect to the point of injection and the plasma parameters. It is quite easy to understand that the injection point moves from near the gun exit up to a few centimeters from the gun exit with the increase in the angles of injection considered in this study. This on the one hand leads to a decreased dwell period in the plasma and on the other hand to an injection into a compartively low enthalpy plasma region. Both effects at least impair the coating structure. The suspension injected upstream (75°) has to act against the viscosity, shear and aerodynamic forces to enter the core/ center line of the plasma. Since the injection is not optimized the suspension droplets are not adequately treated in the plasma which led to lower temperature and velocity (**Figure 1**). This condition puts the coating in zone 2 microstructure category (**Figure 3**). It is surprising to note that the downstream (105°) injected droplets were recorded with high temperature and velocity values (**Figure 1**) even though the solution stream is injected farther away from the torch nozzle exit or into the colder regions of the plasma. The plausible explanation for the aforementioned phenomena is due to combustion of ethanol near the substrate. Once the solution stream is injected the combustion of the ethanol takes place which is highly exothermic and some investigators **[15]** have shown that the energy generated during ethanol combustion can improve the melting of droplets by contributing additional enthalpy to the plasma. The results shown in **Table 2** is a testimony to the fact that the moving the combustion zone location can in turn manipulate the thickness/ pass and heat transferred to the substrate during deposition. The heat transferred to the substrate was higher for downstream injection deposited coatings followed by orthogonal and upstream injection deposits (C7, C8 and C9 in **Table 2**). The fragmention of particles is suppressed when the suspension stream is injected downstream (C9) and this condition leads to the formation of large droplets/ agglomerated submicron particles ending up in the formation of fan headed heavily distorted wide columnar structures displayed in **Figures 1** and **3** (C9). The orthogonal injection (C8) might have directed the suspension stream into optimum injection point. The microstructure displays a dense wide columnar microstructure with a few stacking faults which is evidenced in **Figures 1** and **3**.

The stacking faults in coatings sprayed under conditions C7, C8 and C9 displayed in **Figures 1** and **3** might have occurred due to the deviation of submicron particles by the turning of the plasma plume near the substrate surface. Berghaus et.al., **[20]** proposed that if droplets generated by atomization in the plasma are smaller than a few microns in diameter, then their trajectory of deposition is influenced dramatically by the plasma flow close to the sample surface. This would result in an effective shallow impact angle for the particles. Van Every et.al, **[21]** further proposed that for such small droplets or particles, their deposition trajectory would result in the formation of columnar structures on surface asperities. The stress values (**Figure 3**) of the coatings deposited under C7 (upstream 75°) and C8 (orthogonal 90°) are comparable but the stress value of the downstream C9 (105°) deposited coating is little higher due to the formation of large droplets which gains more momentum in the plasma and compacts the underlying lamellae. There is also a slight change in the spray distance in the order of few centimeters with respect to change in injection configuration which could also affect the microstructure, stress and modulus values. The injection results presented in this article are still rudimentary in nature and an intense investigation using laser strobe shadow photography technique is being carried out in our center to study the effects of injection orientation and optimization of suspension stream to better explain the characteristics of the SPS deposits.

3.4.3 Formation of dense vertically cracked microstructures and the resultant stress-modulus evolution

Dense vertically cracked (DVC) type coatings are formed when the suspension is sprayed under the C11 and C12 conditions (**Table 2**). These coatings fall in zone 4 of stress modulus relationship plot shown in **Figure 3**. The stress modulus values of these coatings were the highest among the rest of the coatings deposited during the investigation. From the theory proposed by Van Every et.al, **[17]** it is inferred that the columnar characteristics decreased as the average YSZ droplet mass in the plasma increased. These characteristics were also found to decrease with increasing substrate temperature. This dependence was connected to an increase in the spreading of the impacting YSZ droplets with substrate/coating temperature, which reduced the coating surface asperity growth rate as seen in **Figure 3** (C11 and C12). There are also bands of porosity that spanned the coating cross-section, which could be visualized in **Figure 3** (C12). The C11 and C12 spray conditions are shown in **Table 2**. The C11 and C12 have low primary gas flow rates. But the difference is the current, spray distance and raster speeds employed. The porosity bands in the C11 coatings are narrower and the coating is dense and cracked, whereas the porosity bands in the C12 coatings are a little wider and the vertical cracks are also visible but horizontal braching cracks are absent (**Figure 3**). The net heat transferred to the substrate during deposition and the net thickness per pass is higher for C12 since it is deposited under a low raster speed (**Table 2**). The formation of DVC type coatings in suspension plasma-sprayed zirconia coatings has been previously reported **[22]**.

In the current investigation, the higher energy plasmas and shorter stand-off distances increased segmentation cracking. Under these conditions high velocity, highly melted large sized droplets impact the substrate and produce dense, well-adhered coatings. In addition, these conditions have high plasma-to-substrate heat fluxes, which create steep temperature gradients in the coatings and high surface temperatures. The formation of DVCs is promoted by the combined effects of high surface temperatures and a steep temperature gradient between the coating surface and the substrate. High surface temperatures improve inter-splat contact and better permit propagation of cracks through multiple splats **[18]**. During deposition, most coating material is deposited when surface temperatures and the temperature gradient are near their peak values. After the torch has passed, the temperature gradient disappears as the surface cools by radiative and convective heat loss to the surroundings and conductive heat dissipation to the underlying coating and substrate. During cooling, the coating develops tensile stresses as it cools much more than the substrate while its shrinkage is constrained by the thicker substrate. If the stress is large enough, it may be relieved by the formation of vertical cracks this phenomenon is very obvious in the curvature time plot shown in **Figure 2**. There is also another crack formation mechanism which might occur due to the partial sintering of the lamellae during deposition. It has been shown that non-negligible sintering of sub-micron YSZ particles can occur in very short periods of time (<10 ms) at 727 °C **[18]**. In the current investigation substrate temperatures as high as 730 °C and 747 °C (**Table 2**) were recorded during the spray deposition of C11 and C12 coatings and coating surface temperatures will be still higher which might promote vertical crack formation but this phenomenon still needs a thorough understanding and re-examination. The C11 and C12 microstructures displayed in **Figure 3** also reveals the presence of over spray/ inter pass bands. It is interesting that these bands are narrow and did not contribute the horizontal crack formation. These interpasss bands are formed by very fine particles that travel along the fringes of the plasma jet during flight. They are therefore not properly thermally treated and embedded in the microstructure as fine solid particles forming kind of nests **[22]**. Investigations are underway in our center to reduce/ eliminate these inter-pass bands by optimizing the injection, employing air knives oriented perpendicular/ at certain angles to the plasma plume in-order to blow-off the loose over spray particles and by optimizing the step/ overlap/ meander size which can help remelt/ sinter these submicron particles.

4.0 Summary and Conclusions

In this investigation the various morphologies of SPS coatings were deposited by changing the plasma spray parameters namely the input power, primary gas flow rate and angle of injection. The stress evolution during the formation of distinct SPS coatings were characterized using an In-situ coating property sensor. The stress modulus microstructure relationship indicates that there can be four different microstructural zones namely the porous feathery zone (low modulus and low stress), dense columnar zone, mixture of cracked columnar and defective columnar architecture zone and DVC zone (high modulus and high stress). From the ICP curvatures it was identified that the porous columnar structures are highly compliant, the dense columnar structures are stiff and the DVCs are dense and compliant. The dense columnar and DVC type microstructures displayed a twin stage curvature change which clearly indicated the stress build up, onset of cracking and the stress relief after axial crack propagation during coating deposition.

5.0 Aknowledgements

The authors would like to thank the Oerlikon Metco Team: Riston Rocchio-Heller, Jing Liu, Jose Colmenares-Angulo, Jonathan Gutleber, Ronald Molz, David Hawley and Montia Nestler for the help and support with the supply of Sinplex torch spares, Suspension feeder parts and Suspension precursor restocks.

6.0 References

[1] Gell, M., Jordan, E.H., Teicholz, M., Cetegen, B.M., Padture, N.P., Xie, L., Chen, D., Ma, X., and Roth, J: Journal of Thermal Spray Technology 17 (2008), Issue 1, pp. 124/35.

[2] Karthikeyan, J., Berndt, C.C., Reddy, S., Wang, J.Y., King, A.H., and Herman, H.: Journal of American Ceramic Society 81 (1998), Issue 1, pp. 121/28.

[3] Bouyer, E., Gitzhofer, F., and Boulos, M.: Thermal Spray: Practical Solutions for Engineering Problems, C.C. Berndt, Ed. (Materials Park, OH), ASM International, (1996), pp. 683/91.

[4] Monterrubio-Badillo, C., Ageorges, H., Chartier, T., Coudert, J.F., and Fauchais, P.: Surface and Coatings Technology 200 (2006), Issue 1, pp. 3743/59.

[5] Burlacov, I., Jirkowsky, J., Muller, M. and Heimann, R.B.: Surface and Coatings Technology 201 (2006), Issue 1, pp. 255/64.

[6] Tomaszek, R., Znamirovski, Z., Pawlowski, L. and Wojnakowski, A.: Surface and Coatings Technology 201 (2006), Issue 5, pp. 2099/102.

[7] Toma, F.L., Bertrand, G., Begin, S., Meunier, C., Barres, O., Klein, D. and Coddet, C.: Applied Catalysis B: Environmental 68 (2006), Issue 2, pp. 74/84.

[8] Kassner, H., Siegert, R., Hathiramani, D., Vassen, R., and Stoever, D.: Journal of Thermal Spray Technology 17 (2008), Issue 1, pp. 115/23.

[9] Elliot M. Cotler, Dianying Chen, and Ronald J. Molz, Journal of Thermal Spray Technology 20 (2011), Issue 4, pp. 967/73.

[10] Hawley, D., Molz, R. Chen, D.: Thermal Spray 2012: Proceedings from the International Thermal Spray Conference and Exposition, May 21–24, (2012), Houston, Texas, USA.

[11] Colmenares-Angulo, J., Sabouni, O., Ibanez, C.: Thermal Spray 2013: Proceedings from the International Thermal Spray Conference and Exposition, May 13–15, (2013), Busan, South Korea.

[12] Kentaro Shinoda, Jose Colmenares-Angulo, Alfredo Valarezo, and Sanjay Sampath: Journal of Thermal Spray Technology 21 (2012), Issue 6, pp. 1224/33.

[13] Streibl, T., Vaidya, A., Friis, M., Srinivasan, V., and Sampath, S.: Plasma Chemistry and Plasma Processing 26 (2006), Issue 1, pp. 73/102.

[14] Ramachandran, C.S., Balasubramanian, V., and Ananthapadmanabhan, P.V.: Journal of Thermal Spray Technology 20 (2011), Issue 3, pp. 590/607.

[15] Aurélien Joulia, Giovanni Bolelli, Enrico Gualtieri, Luca Lusvarghi, Sergio Valeri, Michel Vardelle, Sylvie Rossignol, Armelle Vardelle: Journal of the European Ceramic Society 34 (2014) Issue 1, pp. 3925/40.

[16] Pierre Fauchais and Ghislain Montavon: Journal of Thermal Spray Technology 19 (2010), Issue 2, pp. 226/39.

[17] Kent Van Every, Matthew, J.M. Krane, Rodney W. Trice, Hsin Wang, Wallace Porter, Matthew Besser, Daniel Sordelet, Jan Ilavsky, and Jonathan Almer: Journal of Thermal Spray Technology 20 (2011), Issue 4, pp. 817/28.

[18] Michael Marr and Olivera Kesler: Journal of Thermal Spray Technology 21 (2012), Issue 6, pp.1334/46.

[19] Delbos, C., Fazilleau, J., Rat, V., Coudert, J.F., Fauchais, P., Pateyron, B,: Plasma Chemistry and Plasma Processing 26 (2006), Issue 4, pp. 393/414.

[20] Oberste Berghaus, J., Bouaricha, S., Legoux, J.G., Moreau, C and Chraska T: Proceedings, of the International Thermal Spray Conference 2005 (ITSC 2005), May 2–4, 2005, Basel, Switzerland.

[21] Kent Van Every, Matthew John, M. Krane, Rodney W. Trice: Surface & Coatings Technology 206 (2012), Issue 2, 2464/73..

[22] Alexandre Guignard, Georg Mauer, Robert Vaßen, and Detlev Stover: Journal of Thermal Spray Technology 21 (2012), Issue 4, pp. 416/24.

Comparison of SFS, SPS and HVSFS for the Production of Photocatalytic Titania Coatings

B. Robinson[a], A. Tabecki[a], S. Paul[a], G. Shi[a], A. Mills[b], I. P. Parkin[c], J. A. Darr[c] and H. L. de Villiers Lovelock[a]
[a]TWI Granta Park, Great Abington, Cambridge, CB21 6AL, UK
[b]School of Chemistry and Chemical Engineering, Queen's University Belfast, Belfast, BT9 5AG, UK
[c]Department of Chemistry, University College London, 20 Gordon Street, London, WC1H 0AJ, UK

The photocatalytic capabilities of titanium dioxide are widely published. Reported applications of titania coatings include air purification, water purification and self-cleaning. Suspension spray has been highlighted as a possible route for the deposition of highly active nanostructured TiO_2 coatings. Published work has demonstrated the capabilities of suspension plasma spray (SPS) and high velocity suspension flame spray (HVSFS), however, little work exists for suspension flame spray (SFS). Herein, these three suspension spray processes are compared as regards their capability to produce photocatalytic TiO_2 coatings and their potential for industrial scale up. A range of coatings were produced for each process, manipulating coating parameters in order to vary phase composition, crystal size and other coating characteristics to effect activity. Coatings produced were found to vary significantly between the processes with SFS highlighted as the most effective technique with regards scale-up and coating photoactivity. SFS coatings were found to be up to 5 times more active than analogous coating produced by CVD.

1 Introduction

There are currently three suspension spray techniques reported in the literature which have been used to produce photocatalytic titania coatings; suspension plasma spray (SPS), suspension flame spray (SFS) and high velocity suspension flame spray (HVSFS).[1-3] Each of these techniques utilises different thermal spray torches, giving rise to different particle temperatures and velocities.[4] This has significant effects on coating microstructure and phase composition, which in turn affects coating photoactivity. There have been a number of papers which compare the photocatalytic properties of HVSFS and SPS TiO_2 coatings.[5] In all instances, SPS has been demonstrated as producing significantly more photoactive TiO_2 coatings than the HVSFS process. This is typically attributed to greater retention of the anatase phase in SPS coatings, due to the slower velocity of the jet, which allows resolidification of particles in flight (which crystallise as anatase phase).[6] However, there are only a limited number of examples reported where photoactive SFS titania coatings have been produced, and there are no examples where they have been compared directly with SPS and HVSFS.[2] Herein, the viability of the three above mentioned suspension spray techniques for producing photoactive titania coatings is assessed. Coatings were characterised by a number of methods including XRD, SEM, UV-Vis, and sacrificial photocatalytic hydrogen production test. The practical aspects of each of the techniques and which has the greatest potential for future industrial up-scale are also discussed.

2 Experimental

2.1 Spraying Processes

Suspension Flame Spraying (SFS) was performed using a Metco Flame 5P spray gun with external air atomised suspension injection, **Fig 1A**. The atomiser was constructed by modifying a commercial air brush. Air was used as the atomisation gas and acetylene as the fuel gas. The suspension was gravity-fed from a torch mounted reservoir at a rate of approximately 7 $mL\ min^{-1}$. Small oscillations in feed rate of approximate 1 $mL\ min^{-1}$ were observed. During spraying the flame spray torch was mounted on a 6-axis robot for manipulation.

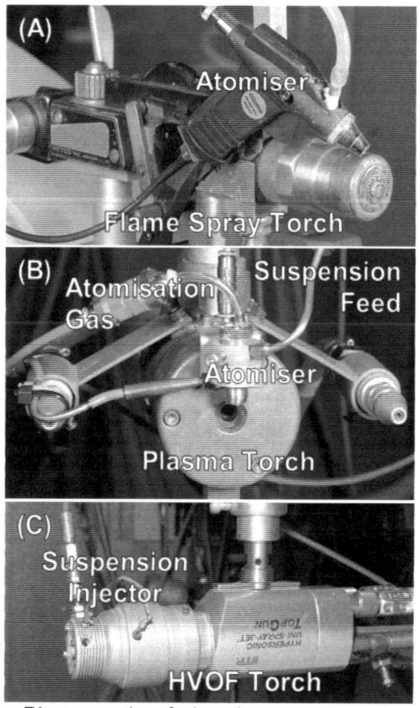

Fig. 1. Photograph of the three suspension spray systems, SFS (A), SPS (B) and HVSFS (C).

Suspension plasma spraying (SPS) was performed using a Praxair®SG-100 torch, **Fig. 1B**. Argon and argon/hydrogen mixtures were used as the plasma gases. The nanoparticle suspension was fed to the external atomiser at a flow rate of 5 mL min^{-1} using an ISCO 260D syringe pump. High velocity suspension spraying (HVSFS) was performed using a Top Gun® HVOF torch with propylene fuel, **Fig. 1C**. Suspension was fed using the same syringe pump through a 0.3 mm injector nozzle at a flow rate of 20 mL min^{-1}.

For all suspension spray trials, samples were water cooled to a temperature of approximately 55°C during spraying, using a custom built jig.

2.2 Materials

All suspension spray trials were performed using both 6 and 12 nm titania suspensions (6.2±2.2 nm and 11.8±1.8 nm primary crystallite size, agglomerate size of approximately 22 nm and 41 nm respectively) synthesised *via* continuous hydrothermal synthesis (CHFS).[7] All substrates used were 25 x 25 x 0.5 mm AISI 304 stainless steel plates, which had been etched for *ca.* 30 seconds to a finish of 202±12 nm Ra with a ferric acid solution.

2.3 Characterisation Methods

X-ray diffraction (XRD) was performed using a Bruker AXS D8 (GADDS) diffractometer, using a Cu Kα X-ray source. The Scherrer equation was used to estimate crystal size and phase composition was estimated by integrating and comparing the anatase (101) and rutile (110) peaks (Eq 1):

$$C_A = \frac{8I_A}{8I_A + 13I_R}$$

Equation 1. Equation for the determination of anatase phase composition percentage where I_A and I_R are the anatase (101) and rutile (110) relative peak intensities.[8]

Scanning electron microscopy (SEM) was performed using a LEO 1550 Field emission scanning electron microscope (FESEM). For cross-sectional SEM imaging, samples were mounted by vacuum impregnating them with epoxy resin containing conductive filler, before metallographic preparation. Micro-Raman spectroscopy was performed using a Renishaw inVia™ Raman microscope, measuring in the range 100 to 1000 cm^{-1}. UV-Vis spectroscopy was performed using a Lambda 950 UV/Vis/NIR spectrophotometer in reflectance mode in the range 200 to 1500 nm. Surface profilometry was performed using an Alicona Infinite Focus SL, using a 50x objective lens. Selection of λ_c filters was carried out in agreement with ISO 4288:1998.[9] Photocatalytic activity testing was performed using a sacrificial hydrogen generation test. Prior to testing, the backs of all samples were sputter coated with platinum. The photoactivity of the coating was measured by observing the rate of hydrogen production in a sacrificial solution of 0.1 M HCl in ethanol and water (1:1).[10-11] The solution was irradiated with two 8W 365 nm UV lamps at a distance of 8 cm. The gas produced was analysed in a Varian CP-3800 gas chromatograph every 15 minutes for a total of 120 minutes, to determine hydrogen production rate.

3 Results and Discussion

All of the coatings ranged in colour from light grey to dark blue. The change in colour, from the white of the initial feed material, was as a result of Ti^{3+} generation, **Fig. 2**. Coatings produced by SFS were the darkest, comparable to the SPS coatings produced in the Ar/H_2 plasma at the shortest stand-off distance, and did not vary in colour with stand-off distance. Coatings produced by HVSFS were the lightest in colour, indicating the lowest concentration of Ti^{3+}. Ti^{3+} gives rise to intermediate electronic states, which allow absorption of visible light (hence the colouration of the titania), which can result in enhanced photoactivity.[12] However, Ti^{3+} defects can also act as recombination centres, decreasing photoactivity.[13]

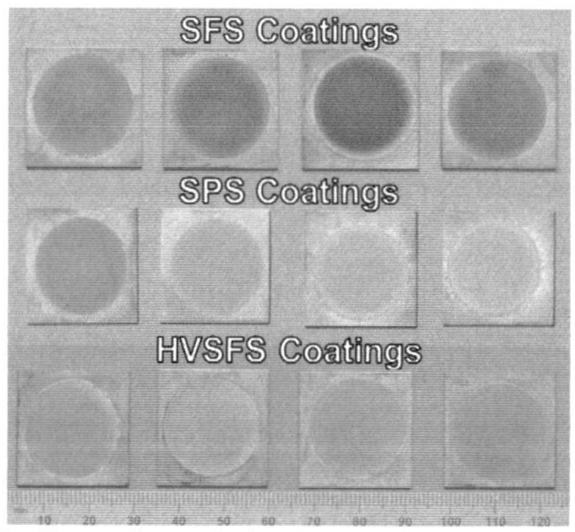

Fig. 2. Examples of titania coatings produced by SFS, SPS and HVSFS (from top to bottom).

A tape test, a method commonly used to determine the adhesive qualities of thin films, was performed on all coatings. SFS and SPS coatings were found to be adherent in all cases, while HVSFS coatings varried between conditions, with failures occuring in a number of instances. Adhesion testing based on ASTM C633-13 (Standard Test Method for Adhesion or Cohesion Strength of Thermal Spray Coatings) was not used due to the low thickness (<50 µm) of the coatings, which is typical for suspension spray coatings.[14]

Phase composition of the coatings was dependent on particle velocity and jet temperature, which affected the melting and resolidification pathways of the titania. In SFS, the low particle velocity provided a long dwell time in the flame allowing for significant melting. However, due to the short stand-off distance required (as a result of the low particle velocity), melting was still limited, showing increased phase transformation (and melting) with increased stand-off distance (75-95% anatase). This contrasted with the SPS and HVSFS coatings. SPS trials were in agreement with published work by Jaworski *et al.*, where increasing stand-off distance led to increased anatase content (90-100% anatase) due to homogeneous nucleation of anatase from molten titania inflight.[15] This was as a result of the high plasma temperature, causing rapid melting of the titania material and allowing changes in resolidification to occur by altering stand-off distance. In both SPS and SFS, phase composition was controlled by altering stand-off distance. Phase composition could not be manipulated with stand-off distance in HVSFS trials, in agreement with recent results reported by Toma *et al.*[16] This was due to the high particle velocity, allowing minimal time for the particles to melt. The high water content of the suspensions also reduced heating of the solids in the hottest area of the flame, and decreased the available timeframe for melting. This resulted in *ca.* 100% anatase for the HVSFS coatings (with trace rutile observed by micro-Raman spectroscopy); this is outside the often quoted optimal 80% anatase phase used in materials such as P25.[17] To observe the phase distribution of the coatings, micro-Raman spectroscopy was used. All coatings were found to be largely composed of anatase phase. Rutile phase was localised in small areas, *ca.* 1-5 µm in size, varying in concentration between the processes. These areas were observed as blue/black spherical structures. Structures such as these have been previously identified as deformed surface quenched agglomerates. Micro-Raman analysis also highlighted traces of rutile phase in coatings where only anatase had been detected using XRD (HVSFS and SPS/Ar coatings).

Under SEM observation, the most significant difference between coatings produced by the three approaches, was the quantity and appearance of deformed surface quenched titania, **Fig. 3**. The quantity of deformed material visible in the coatings decreased from SFS to SPS to HVSFS. This correlated with the phase composition of the coatings as this molten material appeared to be largely rutile (determined by micro-Raman). This change in deformed material concentration was due to the jet velocity and temperature as described above, with SFS having the slowest particle velocity and therefore, allowing significant time for particle heating/melting. The size of the quenched material also decreased from SFS to SPS to HVSFS, due to the increase in velocity/jet turbulence which disrupted agglomerate growth. This contributed to the lack of rutile phase in the HVSFS coatings. The small molten droplets could not retain heat for long and underwent homogenous nucleation of anatase phase inflight. Beyond the differences in quenched material, all coatings produced were dominated by the same fine granular zones similar to that of the feedstocks. This nanostructure was thought to be beneficial for photocatalytic activity by providing greater surface area for absorption.

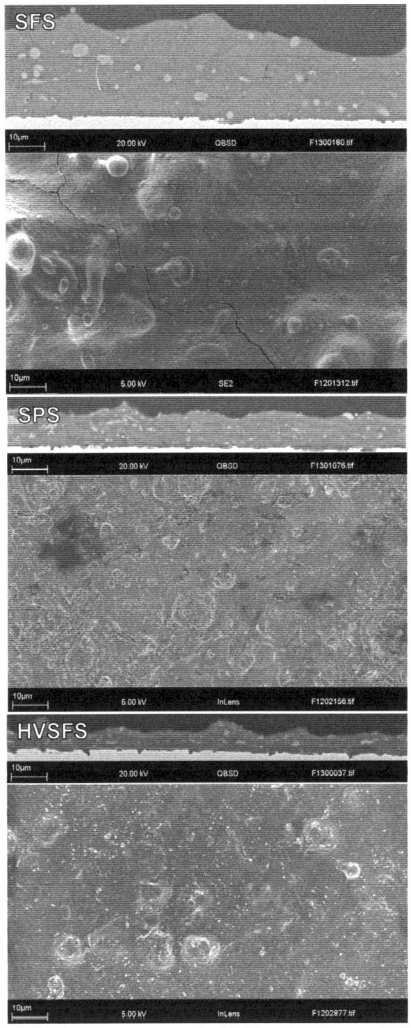

Fig. 3. SEM images of SFS, SPS and HVSFS coating surfaces showing the difference in deformed quenched titania along with example cross sections for each.

Surface roughness for HVSFS and SFS coatings, was found to be affected by coating mass, **Fig. 4**. This was attributed to a combination of the stacking of larger agglomerates and templating over the etched grooves of the substrate. No such process was observed for SPS coatings. Roughness has been found to be an important factor contributing to coating photoactivity, with larger Ra (average roughness) values resulting in greater photoactivities, in some cases. Ra was therefore considered when comparing these techniques. SPS produced the roughest coatings (Ra 2.4-12.3$\pm$0.4 µm) followed by SFS (Ra 1.7-5.6$\pm$0.3 µm) and then HVSFS (Ra 0.6-1.2$\pm$0.1 µm). This difference in coating roughness was attributed to the difference in coating mass, with HVSFS coatings having the lowest mass. However, the reason for the high roughness values for the SPS coatings is not known.

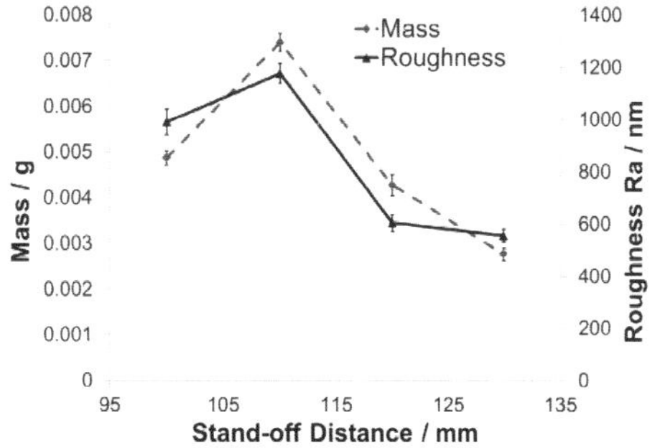

Fig. 4. The correlation between coatings mass and roughness, Ra, for 6 nm HVSFS coatings (B).

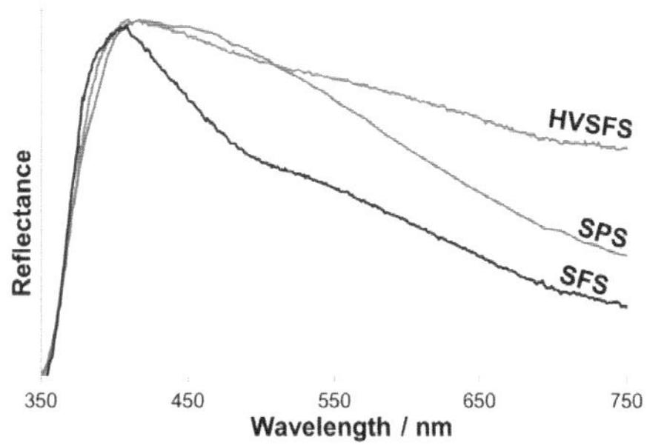

Fig. 5. UV-Vis spectra for SFS, SPS and HVSFS coatings.

Due to the differing temperature and velocity profiles of the three spray processes and their effect on phase composition, crystal size and Ti^{3+} concentration, UV-Vis spectroscopy was performed. SFS gave coatings with the darkest blue colouration, followed by SPS and HVSFS (least dark). This was observed *via* UV-Vis spectroscopy as a significant absorbance in the visible region (>450 nm) which decreased in intensity from SFS to SPS to HVSFS, **Fig 5**. This trend in concentration of Ti^{3+} was attributed to the relative velocities of the processes, with SFS providing the greatest inflight time for reduction reactions with combustion products (or hydrogen in the case of SPS) followed by SPS and HVSFS. Titania can also lose oxygen at temperatures of >500 °C, which is well within the temperature ranges reached during these processes.[18] As has been previously observed from SEM images, SFS coatings also underwent the greatest heating, with the greatest percentage of melted material, followed by SPS. This difference in particle heating was a result of the differing particle velocities of the three processes and also contributed to the differences in Ti^{3+} and visible light absorbencies of the coatings. No significant difference in band-gap was observed between the coatings produced by the three processes. All coatings had band-gaps in the range *ca.* 3.0 to 3.3 eV, typical for mixed phase titania (rutile 3.0 eV and anatase 3.2 eV).[19]

Photoactivity of SFS, SPS and HVSFS coatings was determined by a sacrificial hydrogen generation test. Samples were compared with compressed disks of P25, the highly photoactive titania material often used as a standard in photocatalytic testing, as reported by Elouali *et al.* [10] Samples were also compared to CVD titania coatings, previously reported by Hyett *et al.*, CVD being a competing coating method for the production of thin photoactive coatings.[11] These two sets of results could not be compared with each other due to the differing ways in which they were calculated in their respective publications. In both definitions of maximum hydrogen production rate ($\mu molhr^{-1}cm^{-2}$ and $\mu molhr^{-1}g^{-1}cm^{-2}$) the order of photoactivity was SFS>SPS>HVSFS, **Fig. 6**. This was attributed to a number of factors. Whilst there is still discussion as to the effect of phase composition on photoactivity and the optimal composition thereof (largely thought to be 80% anatase 20% rutile), a consensus is generally held that the presence of rutile significantly increases coating photoactivity.[20] This is ascribed to vectorial charge separation when the phases are in contact. It could be due to this that SFS coatings were more active than SPS and HVSFS coatings. The maximum hydrogen generation achieved by SFS was from a coating with *ca.* 80% anatase while the maximum achieved by HVSFS was for a coating with only trace rutile. The highest hydrogen generation attained by SPS was for a *ca.* 100% anatase coating, however, this was due to the recombination effects observed in SPS coatings containing rutile phase as a result of the presence of Ti^{3+} (cause by the use of Ar/H_2 plasma relative to Ar plasma).[1] SPS coating activity was found to be greater than that of HVSFS coatings, this was despite the fact that the coatings with the greatest activity in both instances had ca. 100% anatase phase composition with trace rutile. This difference in H_2 generation may have been as a result of the different surface roughness of coatings produced by SPS and HVSFS processes (6.1±0.1 µm and 1.0±0.1 µm, respectively), with the greater roughness of the SPS coatings meaning greater surface area for absorbance. From this it was clear that assigning the trend in photocatalytic activity to a single property of the coatings was not possible. Rather, coating activity has been observed to be affected by crystal size, phase composition, TiO_x concentration and surface roughness/area in these results alone, with more reported in the literature.[24] A challenge in suspension spray is therefore, to observe the effect of each of these aspects separately, with spray parameters having far reaching effects on coating properties. For example, to generate rutile in SPS coatings, for increased activity, hydrogen was added to the plasma. This also resulted in increased reduction of TiO_2, increasing recombination and ultimately decreasing overall activity. Similar observations of more complex coating parameters affecting coating activity have been reported previously.[21]

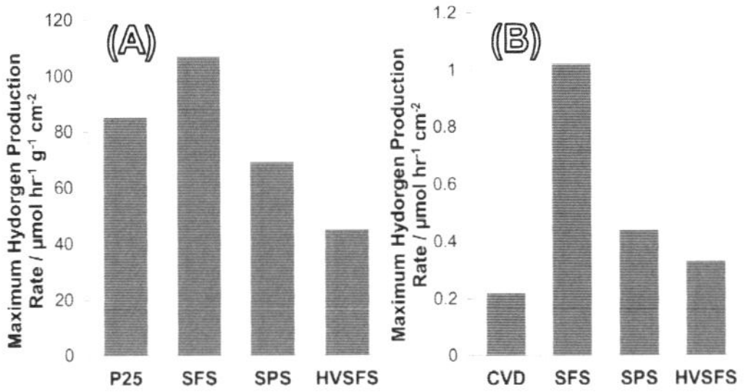

Fig. 6. Graph showing the maximum hydrogen production rates of SFS, SPS and HVSFS titania coatings relative to a sintered disks of P25 (A) and rutile CVD coatings (B).

When comparing SFS, SPS and HVSFS coatings with compressed P25, only SFS was observed to have a greater activity, **Fig. 6A**. Maximum hydrogen generation for the SFS process was observed to be 107 $\mu molhr^{-1}g^{-1}cm^{-2}$ relative to the 85 $\mu molhr^{-1}g^{-1}cm^{-2}$ for P25 discs, an increase of *ca.* 26%.[11] This demonstrates the high photoactivity of SFS coatings and suggests that SFS is a potential route to the production of photoactive coatings for applications such as water purification. When comparing suspension spray coatings produced herein with CVD analogues, SFS, SPS and HVSFS were all more active, producing up to 5 times more hydrogen in the case of SFS, **Fig. 6B**.[11] This demonstrated the potential of suspension spray techniques, particularly SFS, for the production of photoactive titania coatings relative to existing methods. It is also noted that suspension spray is a much more scalable technique than methods such as CVD and PVD, due to their use of expensive deposition or vacuum equipment. However, it should also be said that CVD coatings used for this comparison were composed of rutile phase and would therefore have decreased activity relative to an anatase alternative.

4 Scale-up, Cost and Technical Challenges

Suspension spray facilities are a combination of nano-suspension feeding equipment with conventional thermal spray torches (although some custom suspension spray torches are now commercially available). When comparing the scalability and expense of the suspension spray processes, the feeding equipment was ignored as this would not change significantly between torches in an external injection setup common for photocatalytic titania coatings and only the torch and controller were considered. Flame spray is the oldest, simplest and least expensive of the thermal spray techniques and costs significantly less than plasma spray or HVOF systems.[22] Flame spray also requires less operator training.

All suspension spray methods are typically carried out at a shorter stand-off distance than their thermal spray counter parts (e.g. HVSFS 100 mm, HVOF 200 mm).[23] As such, there is an increased thermal load on the substrate and jigging. Here, this issue was resolved by the use of a custom designed water cooled jig. However, this would be impractical in a mass production scenario. Other researchers have managed temperature through the use of proprietary gas cooling technology.[24] As such, it is clear that thermal loading is a challenge in suspension spraying. Of the three processes SFS, while having the shortest stand-off distance, transferred the least heat into the substrate/jigging due to the lower temperature and kinetic force of the jet relative to SPS and HVSFS respectively. SFS therefore presents fewer difficulties as regards thermal management.

SFS is the least common of the suspension spray techniques, with no commercial SFS equipment on offer. Suspension plasma spray particularly has gained significant attention for the generation of YSZ thermal barrier coatings.[25] As such, SFS is at a disadvantage as it may not have the same support and supply network as SPS and HVSFS going forwards.

The largest practical problem with suspension spray is blockages. In a production setting, blockages can be costly and time consuming.[26] However, during spray trials, no one individual process resulted in a greater level of observed blockages.

Considering all of the above issues, SFS appears to have a number of desirable characteristics compared with SPS and HVSFS, namely its relative simplicity, low expense and decreased substrate thermal load.

4 Conclusion

Coatings were deposited *via* SFS, SPS and HVSFS techniques and their general characteristics and photocatalytic capabilities evaluated. Suspension flame spray was able to produce phase controlled nanostructured coatings, as previously also reported for SPS.[1] SFS and SPS coatings were found to be adherent using a tape test, however failures were noted for HVSFS coatings. SFS coatings were shown to be up to 32% more photoactive in a hydrogen generation test than those produced by SPS. SFS is also a far more simple technique and suffered less from substrate overheating. However, far less is currently known about SFS titania coatings and associated technical challenges.

5 Acknowledgements

This research was partially funded by the UK's Technology Strategy Board (TSB) and the Engineering and Physical Sciences Research Council (EPSRC) under Project 100898 "Nanocrystalline Water Splitting Photodiodes II; Device Engineering, Integration & Scale-up". The associated ESPRC EngDoc sponsorship for B Robinson was co-funded by TWI Limited under UCL's Industrial Doctorate Centre.

6 References

[1] Robinson, B.W., *et al.*, Suspension plasma sprayed coatings using dilute hydrothermally produced titania feedstocks for photocatalytic applications. Journal of Materials Chemistry A (2015) 3, 24, pp. 12680/12689

[2] Liu, Y.X., *et al.*, Deposition, Characterization, and Enhanced Adherence of Escherichia coli Bacteria on Flame-Sprayed Photocatalytic Titania-Hydroxyapatite Coatings. Journal of Thermal Spray Technology (2013), 22, 6, pp. 1053/1062.

[3] Bolelli, G., *et al.*, Properties of High Velocity Suspension Flame Sprayed (HVSFS) TiO_2 coatings. Surface & Coatings Technology (2009), 203, pp. 1722/1732.

[4] Oksa, M., *et al.*, Optimization and Characterization of High Velocity Oxy-fuel Sprayed Coatings: Techniques, Materials, and Applications. Coatings (2011), 1, pp. 17/52.

[5] Toma, F.L., *et al.*, Microstructures and Functional Properties of Al_2O_3 and TiO_2 Suspension Sprayed Coating: An Overview. Thermal Spray 2009: Proceedings of the International Thermal Spray Conference, Marple, B. R., *et al.* (2009), pp. 168/173.

[6] Toma, F.L., *et al.*, Comparative study on the photocatalytic behaviour of titanium oxide thermal sprayed coatings from powders and suspensions. Surface & Coatings Technology (2009), 203, 15, pp. 2150/2156.

[7] Zhang, Z., *et al.*, Direct continuous hydrothermal synthesis of high surface area nanosized titania. Journal of Alloys and Compounds (2009), 476, 1, pp. 451/456.

[8] N. Berger-Keller, *et al.*, Microstructure of Plasma-Sprayed Titania Coatings Deposited from Spray-Dried Powder. Surface Coatings Technology (2007), 168, pp. 281/290

[9] BS EN ISO 4288:1998 Geometric product specification (GPS). Surface texture. Profile method: Rules and procedures for the assessment of surface texture. 1998, BSI.

[10] Elouali, S., *et al.*, Photocatalytic evolution of hydrogen and oxygen from ceramic wafers of commercial titanias. Journal of Photochemistry and Photobiology a-Chemistry. (2010), 216, 23, pp. 110/114.

[11] Hyett, G., *et al.*, Substrate-Dependant Ability of Titanium (IV) Oxide Photocatalytic Thin Films Prepared by Thermal CVD to Generate Hydrogen Gas from a Sacrificial Reaction. Chemical Vapor Deposition. (2010), 16, 10-12, pp. 301/304.

[12] Saputera, W.H., *et al.*, Ti^{3+}-containing titania: Synthesis tactics and photocatalytic performance. Catalysis Today. (2015), 246, pp. 60/66.

[13] Robinson, B.W., *et al.*, The Effect of Fuel Ratio on Photoactivity of Suspension Flame Sprayed Coatings. ITSC 2015, May 11-14, Long Beach, CA.

[14]ASTM C633-13, in Standard Test Method for Adhesion or Cohesion Strength of Thermal Spray Coatings. 2013.

[15] Jaworski, R., *et al.*, Influence of suspension plasma spraying process parameters on TiO_2 coatings microstructure. Journal of Thermal Spray Technology. (2008), 17, 1, pp. 73/81

[16] Toma, F.L., *et al.*, Parameters Influencing the Photocatalytic Activity of Suspension-Sprayed TiO_2 Coatings. Journal of Thermal Spray Technology, (2014), 23, 7, pp. 1037/1053.

[17] Wu, C.Y., *et al.*, Investigation on the synergetic effect between anatase and rutile nanoparticles in gas-phase photocatalytic oxidations. Catalysis Today. (2004), 93-5, pp. 863/869.

[18] Erne, M., *et al.*, Basic principles of reaching triboactive coatings by mixing of nanosized feedstock powders in the suspension plasma spraying process. Materialwissenschaft Und Werkstofftechnik. (2010). 41, 7, pp. 541/546.

[19] Luttrell, T., *et al.*, Why is anatase a better photocatalyst than rutile? - Model studies on epitaxial TiO_2 films. Scientific Reports. (2014), 4, pp. 8.

[20] Li, K., *et al.*, Enhanced Photoactivity of Fe plus N Codoped Anatase-Rutile TiO_2 Nanowire Film under Visible Light Irradiation. International Journal of Photoenergy. (2012), pp. 8

[21] Bannier, E., *et al.*, Microstructure and photocatalytic activity of suspension plasma sprayed TiO_2 coatings on steel and glass substrates. Surface & Coatings Technology. (2011), 206, 2-3, pp. 378/386.

[22] Champagne, V.K., The cold spray materials deposition process: Fundamentals and applications. 2007: Elsevier.

[23] Bolelli, G., *et al.*, Deposition of TiO_2 Coatings: Comparison between High Velocity Suspension Flame Spraying (HVSFS), Atmospheric Plasma Spraying and HVOF-spraying, Thermal Spray 2009: Expanding Thermal Spray Performance to New Markets and Applications, B.R. Marple, *et al.* (2009), ASM International: Las Vegas. pp. 207/212

[24] Joulia, A., *et al.*, Tailoring the Spray Conditions for Suspension Plasma Spraying. Journal of Thermal Spray Technology. (2015), 24, 1-2, pp. 24/29.

[25] Schlegel, N., *et al.*, Columnar Structured Thermal Barrier Coatings (TBCs) by Suspension Plasma Spraying (SPS), ITSC 2014. (2014), DSV: Barcelona, Spain.

[26] Toma, F.-L., *et al.*, Suspension Thermal Spraying – The Potential of a New Spray Technology. Thermal Spray Bulletin. (2010), 3, 1, pp. 24/29.

Influence of particle size and spray parameters on formation of dense yttria coatings by suspension plasma spraying

K. Fujimori, J. Kitamura, T. Wada, Tokyo / JPN, J. Liu, NY / USA, B. Callen, Alberta / CAN

Dense yttria coatings are expected in electronics industries because they have possibility to improve the properties of erosion resistance and decreasing particle size during dry etching processes compared with conventional atmospheric plasma spraying (APS). Suspension spraying is considered as one of the technologies to attain such coatings with affordable production cost and high deposition rate.
In this study, influence of particle size in ethanol based suspension and spray parameters, such as plasma gases, plasma power, nozzle size and spray distance have been investigated to form dense yttria coatings successfully using suspension plasma spraying (SPS) with cascaded arc torch. Particle size has much influence on formation of coatings, where sub-micron size is difficult to have hard and dense structure. Micron sized, 2 micron in this study, is found to be suitable in contrast. Coating structure has been changed with changing spray distance sensitively than APS, suggesting that careful and precise control will be required in parameter development.

1 Introduction

Thermal spray coatings provide some customized surface properties, such as wear resistance, corrosion resistance, thermal barrier and so on, with affordable production costs as well as relatively higher deposition rate mainly done in the air [1]. In general, coatings have micron sized pores, generated from the gap among flattened deposited powders, whose original size is typically a few ten microns [2]. The pores play a very important role for improving thermal barrier properties widely applied to aerospace and gas turbine applications [3]. In contrast, denser coatings are expected in the applications for wear and corrosion protection. High velocity oxy-fuel spraying (HVOF) has been successful to prepare very dense coatings of tungsten carbide and chromium carbide based materials mainly for wear protection. However, oxide ceramics are difficult to apply because of the limitation of flame temperature.

Suspension spraying is expected to become suitable process to attain very dense oxide ceramic coatings as fine particles of single micron or much smaller with solvent are possible to feed stably [4-6]. It can be considered that electronics industries are one of the possible applications for the suspension spraying. Currently, aluminum oxide (Al_2O_3) and yttrium oxide (yttria, Y_2O_3) coatings by atmospheric plasma spraying (APS) are widely applied to semiconductor and flat-panel-display (FPD) production equipment mainly used for dry etching process in the industries [7, 8]. Y_2O_3 coatings onto inside chamber wall are quite valuable to attain anti-plasma erosion resistance against halogen containing plasma, which is used to etch silicon wafer for device fabrication. This prolongs the mean time between maintenance and increases yield of the device by reduction of the particles dropped onto silicon wafer [7].

However, further improvement of the coatings or any alternative is continuously required because both reduction of generating the number of particles and their sizes are required from higher density integration of electronic devices. This suggests much denser coatings close to sintered bulk are desirable than current APS. In this paper, Y_2O_3 coating trial by suspension plasma spraying (SPS) have been conducted to make denser coatings as well as to study the influence of particle size in the suspension and several spray parameters. In situ measurement of inflight particles on SPS has been also examined to understand the coating formation mechanism.

2 Experimental

Ethanol based three suspensions used in this study are summarized in Table 1. Submicron sized yttria stabilized zirconia (8YSZ, ZrO_2-8wt%Y_2O_3) and fine yttria (FY) were used. Coarser yttria (CY) of 1.9 micron in Dv_{50} was also used to study the influence of particle size of the raw materials. Concentration of the solid content was 25wt% in each test type product.

Table 1. Ethanol based suspensions in this study

Name	Composition	Dv_{50} [μm]	Viscosity [cSt]	Solvent	Ratio [wt%]
YSZ	8YSZ	0.2	1.0	Ethanol	25
FY	Yttria (Y_2O_3)	0.4	0.9		
CY		1.9	0.65		

In plasma spraying, TriplexPro-210™ from Oerlikon Metco, Switzerland was applied because of its advantageous features of stable plasma both of partially and lower pulsation [9, 10]. They are considered to be beneficial for SPS in order to vaporize the solvent and melt the particles homogeneously [11]. As for suspension feeder, test type product of 5MPE-SF was used also from Oerlikon Metco [12]. All suspensions were easily redispersed by pot mill prior to putting into the container of the feeder. Suspension was injected into plasma jet vertically from the injector placed at downstream from the exit of the gun nozzle.

Velocity and temperature of inflight particles on SPS were measured using Accuraspray g-3 (Tecnar Automation, Canada). Substrates were aluminum alloy (A6061) for yttria, which was grit blasted prior to SPS.

Table 2 summarizes plasma spray parameters for YSZ. Plasma power, suspension feed rate and spray distance were fixed at 46 kW, 30 g/min and 50 mm, respectively. Spray parameters for yttria suspension are summarized in Table 3. Nos. 1 and 2 are the parameters where YSZ coatings are successfully formed. 6.5 mm nozzle, which is narrower than standard 9 mm, was also applied for yttria SPS trial. Both hydrogen and helium were fed as the secondary gas. Suspension feed rate of 30 g/min was same as the case of YSZ. Spray distances were ranging from 50~100 mm.

Table 2. Spray parameters for YSZ suspension

Name	Nozzle inner diameter [mm]	Ar / He flow rate [L/min]
Columnar	9	44 / 6
DVC		15 / 50

Table 3. Spray parameters for yttria suspension

No	Mater.	Nozzle inner diameter [mm]	Ar flow rate [L/min]	Sec. gas flow rate [L/min]	Power [kW]
1	FY, CY	9	40	H_2, 5	53
2	FY		15	**He, 50**	47
3	FY, CY		60	H_2, 10	61
4	FY		40	**He, 20**	48
5	FY	**6.5**	80	**He, 140**	70
6	CY	9	135	H_2, 5	68
7	CY	**6.5**	135	**He, 55**	69

Knife scratch test was conducted to check whether coating was successfully formed or not. In the case of unsuitable spray parameters, scratch was easily generated meaning that the deposited was just pressurized powder body (PPB), not coating. Microstructure and micro hardness were also investigated for the coatings.

3 Results and discussion

YSZ coatings were successfully prepared onto stainless steel substrate with CoNiCrAlY bondcoat. Typical microstructure of columnar and DVC (dense vertically cracked) can be seen in the previous report [13]. Measured velocity and temperature of the in-flight particles are summarized in Table 4. It should be noticed that measured distance of 90 mm is far from the actual spray distance of 50 mm. Since light emission by the plasma was very high, longer distance was necessary to measure using Accuraspray.

Both of temperature and velocity are relatively higher in DVC parameters at 90 mm suggesting that similar behaviour will be seen at 50 mm. It can be considered that lower viscous molten particles by high temperature can be flattened easily with much higher impact velocity onto the substrates that results in densification of the coating. Understanding of the tendency will support when applying to other oxide ceramic materials.

Table 4. Velocity and temperature of in-flight particles during YSZ-SPS

Name	Velocity [m/s]	Tempearture [°C]
Columnar	270	2400
DVC	305	2800

In spite of similar particle size as the YSZ, no hard yttria coatings were formed onto A6061 substrates from FY feedstock in all spray parameters from Nos. 1 to 5. All deposits were PPBs easily removed by knife scratch test. When using coarser particle of CY with Dv_{50} of 1.9 µm, deposits tend to be harder than those from FY at the same spray parameters of Nos. 1 and 3. However, deposits were still removed by knife scratch test. Deposit at No. 6 was still removed although it was much tougher very similar to hard coating. At No. 7, hard yttria coatings were successfully deposited at various spray distances (SD) of 50, 70 and 90 mm as shown in Fig. 1.

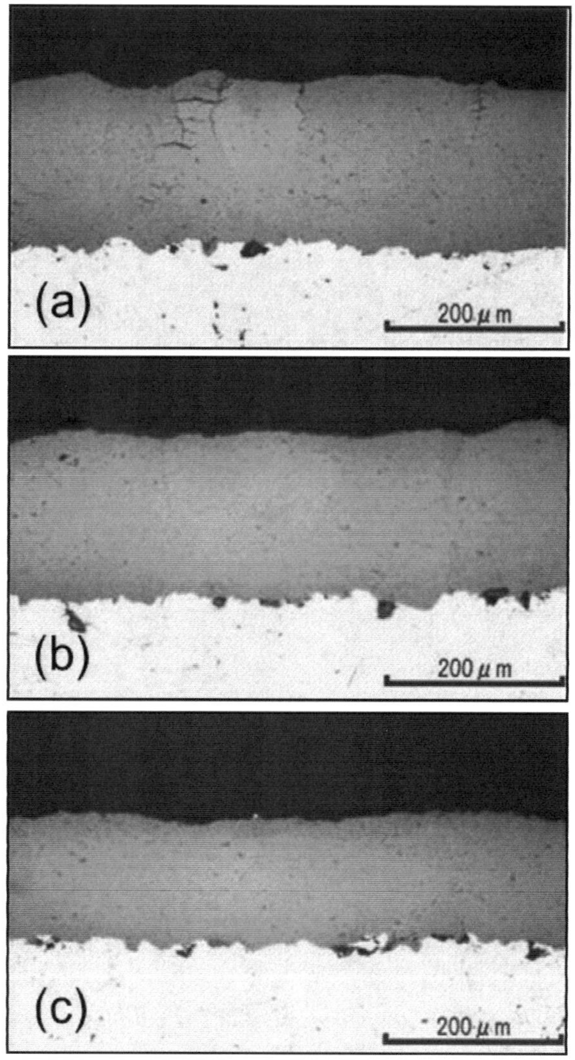

Fig. 1. Cross section of SPS-Yttria at No. 7 spray parameters with a different spray distance of (a) 50, (b) 70 and (c) 90 mm.

These results clearly show both of coarser particles and high-velocity plasma flame are necessary for yttria to form hard coatings compared with YSZ. Further discussions will be done later including the measurement of in-flight particles. As for microstructure, coating made at SD of 50 mm tends to have vertical cracks or nodules with relatively high density. They are decreased with increasing SD. Resultantly coating at SD of 90 mm has much homogeneous micro structure with lower density of nodules.

Deposition efficiency (DE), application rate (AR) described as μm/pass and micro Vickers hardness (HV) of the coatings are summarized in Table 5. DE and AR were decreased with increasing SD. On the other hand, harder coating is obtained at SD of 70 mm within this study. Inhomogeneous structure with many nodules at SD 50 mm might lower the micro hardness because nodules seem to have many structural defects. In spite of high homogeneity, coating at 90 mm shows lower hardness than others. It would come from lower coating density due to relatively weak impact and flattening of molten particles with lower velocity and/or lower temperature.

Table 5. DE, AR and HV of SPS-yttria coatings at No. 7 spray parameters

SD [mm]	DE [%]	AR [um/pass]	HV0.3 (Sigma)
50	67	4.8	519 (24)
70	53	4.0	540 (20)
90	39	2.3	509 (13)

Cross sectional images of secondary electron microscopy (SEM) for SPS-Yttria at No. 7 with SD of 90 mm are shown in Fig. 2, corresponding to Fig. 1(c). It has the most homogeneous microstructure in the study and is apparently very dense compared with the coatings by conventional APS. In magnified image shown in Fig. 2(b), very small pores whose size is a few micron or smaller are observed among the small sized splats. The microstructure, which can be seen at such a magnification, seems to be much inhomogeneous than bulk ceramics as one of the target. Bulk ceramics have quite homogeneous microstructure because they are typically fabricated by static sintering process. As for microhardness, SPS coatings of about 500HV have high hardness than APS of 400HV but are softer than the bulk Y_2O_3, whose hardness is about 700HV in general. Observed submicron sized pores can be easily considered to deteriorate the hardness of the SPS coatings. Therefore, this results show further improvement is essentially necessary to make better coatings similar to bulk.

Figure 3 shows velocity and temperature of the in-flight particles of coarser yttria at SD of 70 mm. No. 7, where coating has been formed, indicates apparently higher velocity of more than 1000 m/s than others. In contrast, particle temperature at No. 7 is the lowest, almost similar to No. 1 but still above the melting point of about 2410 °C.

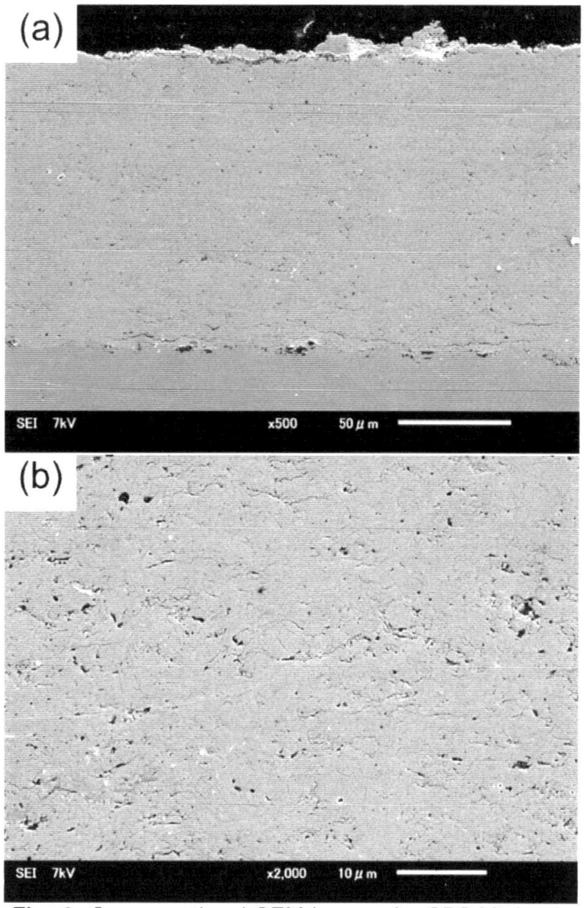

Fig. 2. Cross sectional SEM images for SPS-Yttria at No. 7 with SD of 90 mm corresponding to Fig. 1(c)

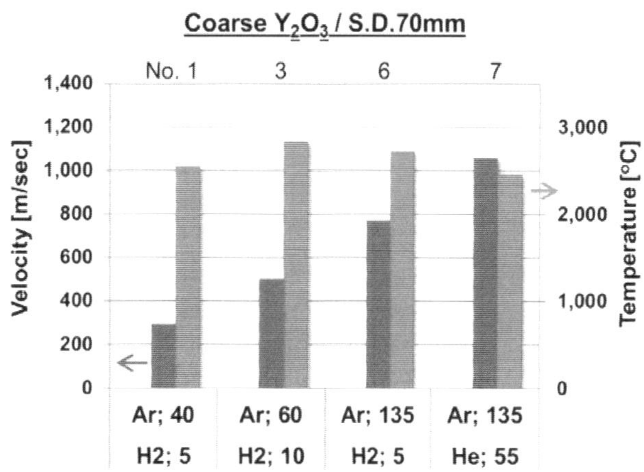

Fig. 3. Velocity and temperature of in-flight particles during SPS-coarse yttria (CY) at SD of 70 mm

Influence of SD (stand-off distance) on velocity and temperature of coarser yttria particles has been investigated. Figs. 4(a) and (b) show the result at No.

7 and 6. At No. 7, both of velocity and temperature show quite high about 1760 m/s and 2800 °C, respectively, at SD of 50 mm. Velocity lowers drastically with increasing SD to 1050 and 760 m/s at SD of 70 and 90 mm, respectively. Temperature lowered from 50 to 60 mm at first. After that, it kept around melting point temperature to SD of 100 mm. In spite of high homogeneity (a little nodules and cracks) in microstructure, relatively lower velocity of 760 m/s at 90 mm might result in lower coating hardness than higher velocity of 1050 m/s at 70 mm. Low coating hardness suggesting low coating density will be disadvantageous for high wear and corrosion resistance. This implies further improvement is necessary to attain both hard and homogeneous coating.

In No. 6, where hard PPB was formed, lower velocity and higher temperature were measured than No. 7 at the same SD. Table 6 summarizes coating thickness and weight of Nos. 6 and 7 parameters onto aluminum plates of 25x100 mm after 20 passes. This shows they are quite comparable although properties of the deposit are entirely different.

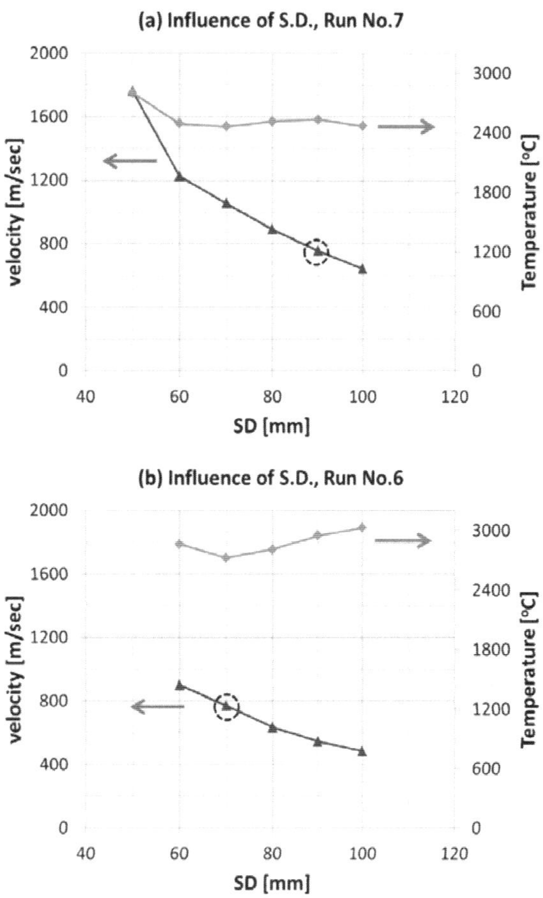

Fig. 4. Influence of SD (stand-off distance) on velocity and temperature of coarser yttria (CY) in-flight particles during SPS ((a) No. 7 and (b) No. 6).

Table 6. Coating thickness and weight of coarser yttria at Nos. 6 and 7 parameters

SD	Thickness [μm]		Weight [g]	
[mm]	No. 6	No. 7	No. 6	No. 7
50	126	123	1.04	0.98
70	94	94	0.72	0.71
90	44	61	0.49	0.51

When comparing with 90 mm in No. 7 and 70 mm in No. 6, highlighted by dot circles in Fig. 3, No. 6 shows much higher temperature with comparable velocity. It suggests that No. 6 at this SD seems to be suitable conditions for coating formation. However, the result was opposite, showing that further speculation of the mechanism should be considered.

In No. 7, higher plasma density in unit area should be attained possibly about double than No. 6 because of similar plasma input power and quite narrower nozzle about half in cross-section area. This might give much denser hot gas section at the center that has an ability to keep particles at high temperature still at impact onto the substrate. When impact, it can be easily considered that particle will be flattened and quenched simultaneously. If quenching occurs in shorter time than flattening, insufficient particle deformation will cause porous coating structure and the worst case will be not coating formation but pressurized powder body (PPB). In order to avoid rapid quenching, surrounding hot gas with much higher density with particles at impact seems to be advantageous qualitatively than that with lower density. Properties of the gases for the plasma might affect to keep molten state of the particles. Hydrogen and helium typically applied to secondary gas have thermal conductivity roughly 10 times higher than argon as primary gas. When comparing with No. 6 and 7, No. 7 has higher mixing ratio of secondary gas (helium) than No. 6 apparently. It can be assumed that this high temperature flame would have an ability to transfer the heat to particles in shorter time that contributes to keep particles at high temperature during the impact. As particle size is very small in SPS than conventional APS, such a phenomenon could be sometimes observed.

Fig. 5 shows measured velocity and temperature of in-flight particles of YSZ, FY and CY at all spray parameters with SD ranging from 50~100mm. In almost of all cases, temperature of FY particles was higher than melting point of cubic Y_2O_3. This suggests fine yttria will impact onto the substrate in molten state at the beginning. However, lower heat capacity due to small sized particle (small volume) might result in rapid quenching and formation of PPB. It seems No. 5 condition similar to No. 7 is not enough to keep high temperature for a long time to avoid PPB.

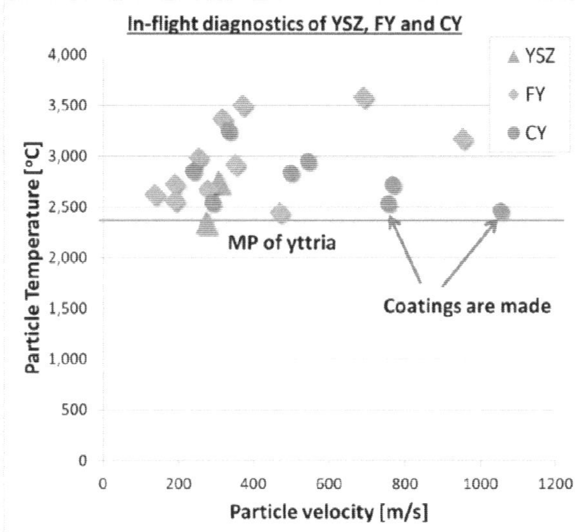

Fig. 5. Measured velocity and temperature of in-flight particles of YSZ, FY and CY at all spray parameters with SD ranging from 50~100mm

In spite of similar particle size as that of FY, coatings were successfully formed in YSZ. Measured velocity and temperature of YSZ in-flight particles were similar to those of FY and CY. This suggests difference of physical and thermal properties of material will play important roles to give the difference in coating formation or not. As for melting point, YSZ (cubic) has about 2690 °C, which is higher than that of cubic yttria about 2410 °C. This will be eliminated from the discussion because low melting point yttria is difficult to make coating. Most probable is thermal conductivity, where YSZ has very low of about 3 W/m/K in room temperature (RT). Yttria of about 15 W/m/K in RT is quite higher than YSZ. This high heat conduction will transfer the heat from hotter yttria particles to colder substrates rapidly. From the results, it can be said that quenching process of molten particles at impact to the substrate will be much important and should be considered carefully compared with conventional APS because of using very fine particles with lower heat capacity.

It is difficult to say it is typical in SPS because experimental conditions in this study are limited, ie, only ethanol solvent, only 25wt% of suspension concentration and limited spray parameters. Therefore further study will be necessary to clarify much more such as further investigation of formed coatings by X-ray diffraction, process study by collecting in-flight particles as well as splat observations, and coating trial for other ceramic materials of aluminum oxide and titanium oxide.

Table 7 summarized the coating trial of SPS yttria in this study. Using FY is unsuccessful to form hard coatings in all spray conditions. High velocity and relatively lower temperature of the particles are suitable for coating formation in CY. Lower velocity with high temperature was unsuitable relatively.

Table 7. Summary of the SPS yttria for coating formation

Velocity	**Low**	**Low**	**High**	**High**
Temp.	**Low**	**High**	**Low**	**High**
FY	No	No	No	No
CY	Not done	Medium	Good	N.A

5 Conclusions

In this study, influence of particle size in ethanol based suspension and spray parameters, such as plasma gases, plasma power, nozzle size and spray distance have been investigated to form dense yttria coatings successfully using suspension plasma spraying (SPS) with cascaded arc torch. Using sub-micron size Y_2O_3 suspension based on YSZ experiences was unsuccessful, just forming pressurized powder body. Coarse particles of about 2 micron in D_{50} with higher velocity were necessary to make dense and hard coatings in Y_2O_3 suggesting that material properties such as heat conductivity affect critically.
Coating structure has been changed with changing spray distance sensitively than APS, suggesting that careful and precise control will be required in parameter development. In-flight diagnostics suggests particles will impact at molten state onto the substrate implying that balance of quenching and flattening of the particles will affect coating formation or making pressurized powder body.

Referencesns

[1] Davis, J. R., Ed., Handbook of Thermal Spray Technology, ASM International (2004).
[2] Heimann, R.B., Plasma-Spray Coating, Principles and Applications, VCH Verlagsgesellschaft mbH, Germany (1996).
[3] Kulkarni, A.A., Sampath, S., Goland, A., Herman, H: Porosity-Thermal Conductivity Relationships in Plasma Sprayed Zirconia Coatings, Proc. International Thermal Spray Conf., (2000), pp. 1061/1066.
[4] Fauchais, P., Etchart-Salas, R., Rat, V., Coudert, J.F., Caron, N., Wittmann-Ténèze, K: Parameters Controlling Liquid Plasma Spraying: Solutions, Sols, or Suspensions, J. Thermal Spray Technol., Vol. 17(1), (2008), pp. 31/59.
[5] Fauchais, P., Rat, V., Delbos, C., Coudert, J.F., Chartier, T., Bianchi, L: Understanding of Suspension dc Plasma Spraying of Finely Structured Coating for SOFC, IEEE Trans. Plasma Sci., Vol. 33(2), (2005), pp. 920/930
[6] Fauchais, P., Montavon, G: Latest Developments in Suspension and Liquid Precursor Thermal Spraying, J. Thermal Spray Technol.,Vol. 19(1-2), (2009) pp. 226/239.
[7] Kobayashi, Y.: Current Status and Needs in the Future of Ceramics Used for Semiconductor

Production Equipment, Proc. 37th Seminar on High-Temperature Ceramics, K. Matsuhiro, Ed., July 19, 2005 (Osaka, Japan), Japan Ceramics Society, 2005, pp 1/7. (in Japanese)
[8] Fukumoto, M.: The Current Status of Thermal Spray in Asia, J. Therm. Spray Technol., 2008, 17(1), pp. 5/13.
[9] Bernecki, T. Varley, K. Rusch, W. Wlodarczyk, J. and Klein, J., Plasma Gun with Adjustable Cathode, US Patent 4,780,591, October 25, 1988.
[10] Hawley, D., Improved Quality, Reduced Cost and Cycle Times. Sulzer Technical Review, 2007(3), pp. 14/16.
[11] Sulzer Technical Review 2/2011, Liquid suspension spray – New challenges, new opportunities, (2011), pp. 26.
[12] Sulzer Metco Flyer, “5MPE-SF Material Feed System for Suspension Plasma Spray” (2013)
[13] Chen, D, Dambra, C, Sabouni, O.: Influence of Process Parameters and Feedstock Selection on Coating Microstructures with a Cascaded SinplexPro™ Plasma Torch, Proc. ITSC2014, electric file, (2014).

Improved corrosion resistance and electrical insulation behavior of plasma sprayed alumina coatings treated by inorganic sealant

Fang SHAO*, Huayu ZHAO, Kai YANG, Chenguang LIU, Shunyan TAO

The Key Laboratory of Inorganic Coating Materials, Shanghai Institute of Ceramics, Chinese Academy of Sciences, Shanghai 201899, PR China

In this study, alumina coatings have been deposited by atmospheric plasma spraying technique and sealed by inorganic sealant consisting of aluminum phosphate and small fraction of alumina nano particles. The sealing process was briefly described and the microstructure of the sealed coatings was examined. Potentiodynamic polarization, electrochemical impedance spectroscopy (EIS) and salt spray tests were used to evaluate anti-corrosive characteristics of the as-sprayed and sealed coatings, respectively. The corrosion potential of sealed coatings were much more positive than unsealed coatings. The corrosion current density of sealed coating was only about one sixth of that of unsealed coating. After sealing treatments, the impedance of the coating was also enhanced in both high and low frequencies range. After subjected to the salt fog test for 96 h, there were rusty corrosion products at the surface of the unsealed coatings while there was no visible corrosion on the sealed coatings. The results indicate that the sealing treatment is highly effective to inhibit corrosion. The electrical resistance was measured at room temperature under dry and humid conditions. It shows that the sealing treatment can enhance the anti-corrosion and insulating performance of alumina coatings by reducing the open porosity.

1 Introduction

Atmospheric plasma spraying (APS) is traditionally the most common spray process for the preparation of ceramic coatings due to their high melting points [1]. Especially, aluminum oxide (Al_2O_3) is widely used as protective material owing to its remarkable properties: a good anti-corrosion performance, an excellent dielectric behavior and good mechanical properties even at high temperatures [2]. Nevertheless, the thermal sprayed ceramic coatings have laminated orthotropic structures with penetrating, interconnected porosity [3]. This is a major problem for applications requiring anti-corrosion and/or electrical insulation properties since corrosive solutions and/or water vapor can percolate through the interconnected pore network, allowing corrosion of the undercoat and/or leading to a sharp decrease in insulation resistance (especially in high humid condition). Thus, post-treatments of the as-sprayed coatings are necessary to reduce the interconnected porosity.

Among several possible pre-treatments such as laser remelting [4], high temperature treatments [5], sealants permitting to densify the coatings appear in this context as a suitable technique. Because the high thermal inputs during the remelting or heat treating may induce microstructural changes of the substrate material, even lead to cracking of the coatings due to the large residual stress in some cases. Although organic sealants can provide some economic and technical benefits, their wear resistance and temperature resistance are poor [6]. In consideration of all thermal, chemical, and electrical requirements, inorganic sealants were studied. Due to the low solid yield and large shrinkage during heat treatment, the sealing efficiency of sol-gel treatments was very low [7]. Even worse, the high curing shrinkage may create local tensile fields, causing easier and larger particle detachments during the wear process [8]. Sodium silicate and other silicate were also used as inorganic sealants, however, ions such as Na^+, K^+ appear detrimental for the electrical insulation. In order to solve the above problems, aluminum phosphate was applied to fill the open and interconnected porosity by sealing. On the other hand, solid phases, such as fine powders or fibers were demonstrated to enhance the sealing efficiency and the mechanical properties of resin as advanced filler [9]. However, few research has been done on the aluminum phosphate with filler. And to our best knowledge, there were little systematic study (microstructure, corrosion resistance and electrical resistance) of the Al_2O_3 coatings sealed by aluminum phosphate with small fraction of Al_2O_3 nanoparticles.

Here, we prepared Al_2O_3 coatings by atmospheric plasma spraying (APS) process. In addition, aluminum phosphate with small fraction of Al_2O_3 nanoparticles were synthesized as sealant via a wet chemical method to further improve the coatings' quality. The influences of sealing treatment on the microstructures, the corrosion and dielectric behavior of Al_2O_3 coatings were investigated by scanning electron microscopy (SEM), electrochemical testing such as potentiodynamic polarization, electrochemical impedance spectroscopy (EIS), complemented by salt spray tests and electrical resistance.

2 Experimental procedure

2.1 Plasma Spraying

Stainless steel was used as the coating substrate. Prior to spraying, rectangular samples were sandblasted and then dipped into ethanol, cleaned with ultrasonic washer. The Multicoat™ atmospheric plasma-spraying system equipped with a F4-MB plasma gun (Sulzer Metco AG, Switzerland) was used to fabricate Al_2O_3 coatings. Commercial available fused and crushed Al_2O_3 powders were used as feedstock. The main plasma spraying parameters are exhibited in **Table 1**. The thickness of the as-sprayed coatings was controlled to be about 300 μm.

Table 1 Plasma-spraying operating parameters.

Parameters	value	Unit
Arc current	640-650	A
Primary plasma gas (Ar)	45-50	slpm
Secondary plasma gas (H_2)	7-9	slpm
Carrier gas (Ar)	3.0-4.0	slpm
Powder feed rate	40	g/min
Spray distance	100-120	mm

2.2 Sealing of the Coating

The as-sprayed Al_2O_3 coatings were sealed by aluminum phosphate having Al_2O_3 nanoparticles dispersed therein. Aluminum phosphate was prepared as follows [10]. Aluminum hydroxide ($Al(OH)_3$) was mixed with diluted orthophosphoric acid (~ 71% H_3PO_4) at about 75°C on a magnetic stirrer until it became clear. The $Al(OH)_3$:H_3PO_4 ratio was 1:4.2 by weight, which corresponded to the molar ratio P/Al of about 3. Then, 5 wt.% Al_2O_3 nanoparticles were added and well dispersed into the solution. The Al_2O_3 coatings were immersed into the sealants at 50°C for about 30 min with an ultrasonic bath and then standing at ambient temperature and atmospheric pressure for 12 h. Thereafter, the impregnated coatings were heat-treated with the following continuous procedure: 2 h at 100°C, followed by 2 h at 200°C, and 2 h at the final curing temperature of 250°C. The excess sealant was ground off by polishing, ensuring that the true surface of the coating was exposed for microscopic analysis, electrical and corrosion tests.

2.3 Characterization

The coating microstructure was characterized using a scanning electron microscope (SEM, TM3000, HITACHI, Japan) equipped with an energy-dispersive spectrometer (EDS). The accelerated voltage for EDS was 15 KV. The corrosion resistance of samples was evaluated by the electrochemical measurements and salt spray test. A three-electrode system including a working electrode, an auxiliary electrode and a reference electrode was used for electrochemical measurements. The working electrodes were made of the coating samples with an exposed area of 1.0 cm^2, the counter electrode is a platinum mesh and the reference electrode is a saturated calomel electrode (SCE) with a Luggin capillary positioned close to the working electrode surface in order to minimize ohmic potential drop. The experiment was conducted in unstirred 3.5 wt.% NaCl solution. Electrochemical systems Ametek 1287 & 1260 were used for the corrosion tests. The open circuit potential (E_{ocp}) was obtained after 30 min immersion to ensure it was in steady state. The polarization curves were carried out at a scan rate of 0.2 mV/s and a scan range of 0.2 V-0.4 V vs E_{ocp}. The corrosion potential and the corrosion current density were obtained through the linear analysis of Tafel approximation with CView software. Electrochemical impedance spectroscopy (EIS) curves of the coatings were measured at the open circuit potential. The frequency range was varied from 10^{-2} Hz to 10^{6} Hz. The salt spray test was conducted following the Chinese National Standard GB/T 10125-2012. During the salt spray test, the temperature, NaCl concentration, and pH were kept constant at (35±2)°C, (50±5) g/L and 6.5-7.2, respectively. The exposure duration lasted for 96 h. The DC insulating resistance of the as-sprayed samples was measured using an ultra-high resistance meter (Fluke 1555). Silver paint (1×1 cm^2) was used as a contact material and the test voltage was 500 V.

3 Results and discussion

3.1 Microstructure characterization

Fig. 1 exhibit the cross-sectional morphology of the coatings. Pores were clearly observed in the unsealed coating as shown in Fig. 1a and Fig. 1b. The topside of the sealed coating exhibits less porosity, and the sealant has penetrated into the coating to a depth of about 70 μm from the surface (Fig. 1c and Fig. 1d). Fig. 2 is the EDS patterns of the unsealed and sealed coatings, the detect regions were marked with white rectangle in Fig. 1b and Fig. 1d. The chemical compositions are listed in **Table 2**. There is the evident element of phosphorus in the cross section of sealed coatings. On the contrary, peaks corresponding to phosphorus were barely observed in the unsealed coatings. It strongly indicates that the sealant can be brought into the coatings through the structural defects (e.g. voids, cracks and gaps between the lamella).

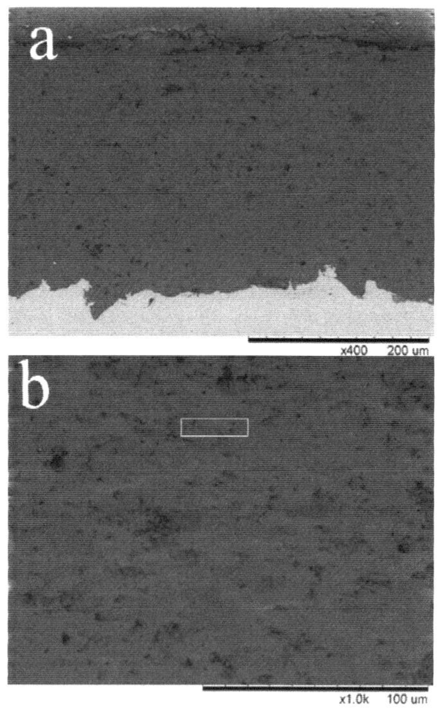

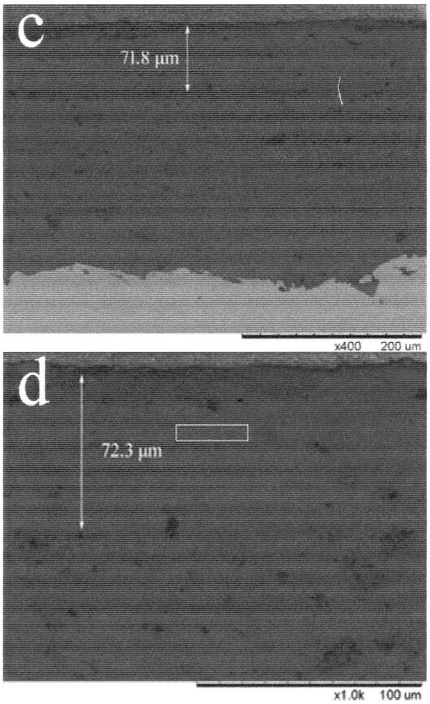

Fig. 1. The polished cross-section morphologies of the Al_2O_3 coatings (a) unsealed coating and (b) its high-magnification image; (c) sealed coating and (d) its high-magnification image.

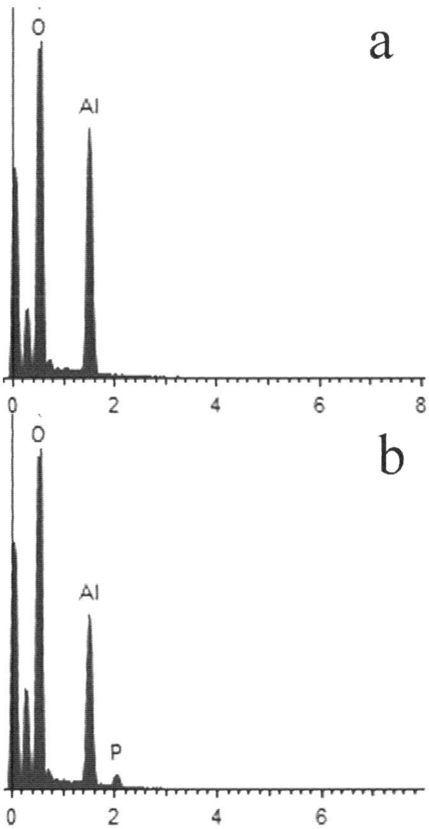

Fig. 2. EDS analysis of (a) unsealed coating and (b) sealed coating corresponding to the region marked with white rectangle in Fig. 1b and Fig. 1d.

Table 2 Chemical composition (atom%) of the unsealed and sealed Al_2O_3 coatings corresponding to **Fig. 2**.

Samples	O	Al	P
unsealed coatings	69.8	30.2	/
sealed coatings	75.2	21.7	3.1

3.2 Potentiodynamic polarization

The open-circuit potential of the unsealed and sealed Al_2O_3 coatings is -0.36 V and -0.14 V, respectively. Potentiodynamic polarization curves are shown in **Fig. 3**. **Table 3** presents the electrochemical values determined from the polarization curves. Higher values of the corrosion current density (I_{corr}) or a more negative value for the corrosion potential (E_{corr}) indicate that the coating-substrate system is less resistant to corrosion [11]. The corrosion potential of sealed coating was much more positive than the other two coatings. This meant that the sealing treatment suppressed the thermodynamic tendency of corrosion largely. The corrosion current density of sealed coating was only about one sixth of that of unsealed coating. The Polarization resistances (*Rp*) was calculated based on the Stern and Geary relationship. Consistent with the results of I_{corr} and E_{corr}, the values of *Rp* reveal that sealed coatings have better corrosion inhibition capability. To simplify the comparison of the coatings and to quantify the sealing properties, sealing efficiency (*Si*) was calculated from the polarization results. According to the results reported in other literatures, the open porosity (designated by P_o) can be calculated by the following equation:

$$P_o= \frac{I_c}{I_s}\times 100\% \qquad (1)$$

Where *Ic* and *Is* are the corrosion current density of coated and uncoated substrate in the same electrolyte, respectively[12]. *Si* is related to P_o as follows.

$$Si= [1-(\frac{P_o}{P_o^0})]\times 100\% \qquad (2)$$

P_o and P_o^0 are the open porosity of the sealed and unsealed coatings.

Thus, Eq. (3) can be derived to relate *Si* and I_{corr}.

$$Si= [1-(\frac{I_{corr}}{I_{corr}^0})]\times 100\% \qquad (3)$$

I_{corr} and I_{corr}^0 represent the corrosion current density of the sealed and unsealed coatings, respectively [13]. The calculated *Si* was 83.9%, which represented 83.9% open pores were sealed by modified aluminum phosphate sealant.

Table 3 Corrosion parameters obtained from potentiodynamic polarization measurements of the unsealed and sealed coatings in 3.5% NaCl electrolyte.

Samples	E_{corr} (mV)	I_{corr} (μA/cm^2)	Rp (KΩcm^2)	Si (%)
unsealed coatings	-328	3.1	74.8	/
sealed coatings	-170	0.5	323.3	83.9

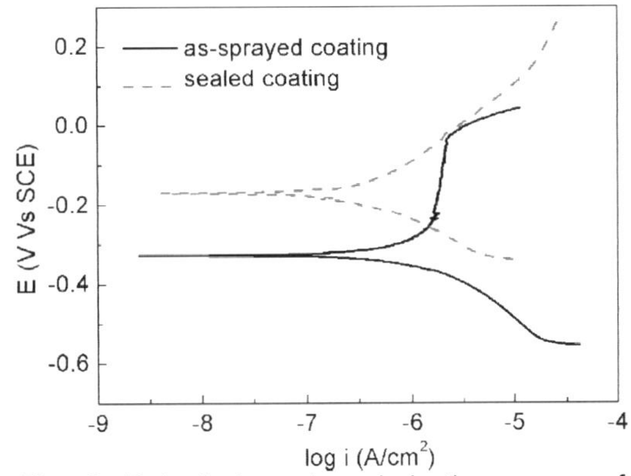

Fig. 3. Potentiodynamic polarization curves of the unsealed and sealed coatings in 3.5% NaCl electrolyte.

3.3 Electrochemical impedance spectroscopy

Impedance data have been used extensively in a semi-quantitative way for the detection of surface coatings in the field of aqueous corrosion science [14]. The EIS characteristics of Al_2O_3 coatings were examined at the open-circuit potential, as shown in **Fig. 4**. Compared with unsealed coatings, the radius of Nyquist plots increased obviously, indicating the decreased corrosion. In Bode plots (**Fig. 5**) the magnitude of the impedance |Z| and phase-angle are plotted respectively as a function of frequency (*f*). At high frequency range, the impedance spectra reflected the performance of the coating in corrosion solution, while at low-frequency, it is related to the corrosion at the substrate/solution interface [6]. After sealing treatments, the impedance were enhanced in both high frequency and low frequency range. This indicate that the corrosion resistance of the sealed coating are much greater than that of the unsealed coating.

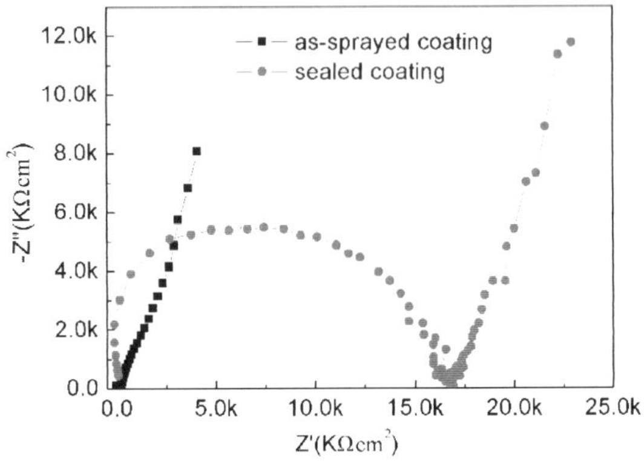

Fig. 4. Nyquist impedance plot of the as-sprayed and sealed coatings.

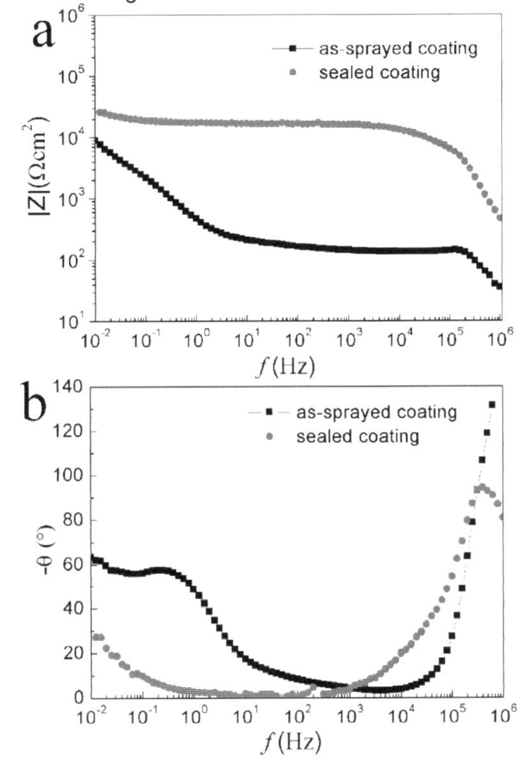

Fig. 5. Bode plots of the as-sprayed and sealed coatings.

The overall corrosion durability of the samples was studied using neutral salt spray (NSS) testing. The visual NSS results are presented in **Fig. 6**. After subjected to the salt fog for 96 h, there were rusty corrosion products on the surface of the unsealed coatings while there was no visible corrosion on the sealed coatings. The results indicate that the sealing treatment is highly effective to inhibit corrosion.

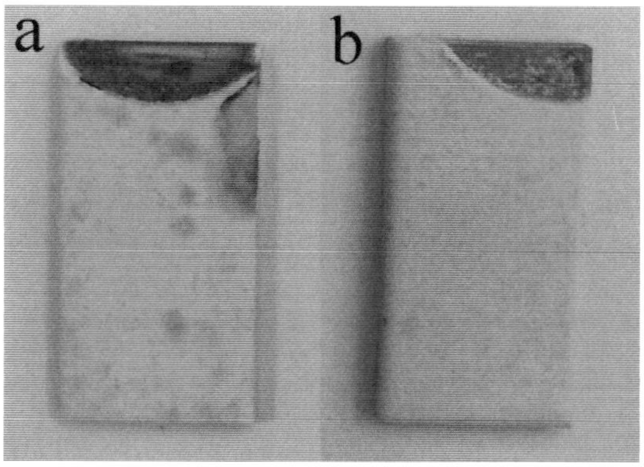

Fig. 6. Digital photos of the coatings after 96 h of the salt spray test: (a) without sealing treatment, (b) with sealing treatment.

3.4 Electrical resistance measurements

DC insulating resistances for coatings with and without sealing treatments are determined at room temperature in air at two different humidity levels (ambient humidity and humid condition) are shown in **Table 4**. The humid samples were obtained by immersing the coatings in tap water for about 2 min and then remove the superficial water with buckskin cloth. Silver paint (1×1 cm^2) was used as a contact material and the test voltage was 500 V. Compared with the values detected under ambient humidity, the DC electrical resistance reduced to about one tenth at humid condition for both unsealed and sealed coatings. The sensitive to humidity could be explained by the capillary condensation of water vapor in microstructural defects (cracks and open pores). However, the sealed coatings exhibited much better insulating properties even at humid condition.

Table 4 Average values of DC electrical resistance of unsealed and sealed coatings determined under different test conditions

Samples	R(MΩ) at RT, ambient humidity	R(MΩ) at RT, humid condition
unsealed coatings	78.9	1.4
sealed coatings	896.3	81.0

4 Conclusion

A comparative study on the characteristics (microstructure, corrosion resistance, electrical resistance, etc.) of unsealed and sealed plasma sprayed alumina coatings was performed. Sealing with modified aluminum phosphate (having Al_2O_3 nanoparticles dispersed therein) of a plasma-sprayed alumina coating decreased the porosity especially in the topside. The sealing treatments significantly improve the corrosion resistance as demonstrated by polarization, electrochemical impedance spectroscopy and neutral salt spray tests. The electrical insulating properties was also enhanced after sealing treatment, the sealed coatings exhibited much higher electrical resistance even at humid condition.

5 Acknowledgements

The study is jointly supported by National Nature Science Foundation of China (51302299) and Shanghai Nature Science Fund Project (13ZR1446000). Simultaneously, we are grateful to Prof. Shigang Xin and Prof. Yuzhi Zhang for providing the support for this work.

6 Literature

[1] W. Zorawski, A. Goral, O. Bokuvka, L. Litynska-Dobrzynska, K. Berent, Microstructure and tribological properties of nanostructured and conventional plasma sprayed alumina-titania coatings, Surf Coat Tech, 268 (2015) 190-197.
[2] G. Berard, P. Brun, J. Lacombe, G. Montavon, A. Denoirjean, G. Antou, Influence of a sealing treatment on the behavior of plasma-sprayed alumina coatings operating in extreme environments, J Therm Spray Techn, 17 (2008) 410-419.
[3] H.J. Kim, S. Odoul, C.H. Lee, Y.G. Kweon, The electrical insulation behavior and sealing effects of plasma-sprayed alumina-titania coatings, Surf Coat Tech, 140 (2001) 293-301.
[4] S. Ahmaniemi, P. Vuoristo, T. Mantyla, Improved sealing treatments for thick thermal barrier coatings, Surf Coat Tech, 151 (2002) 412-417.
[5] G.Z. Xie, X.Y. Lin, K.Y. Wang, X.Y. Mo, D.J. Zhang, P.H. Lin, Corrosion characteristics of plasma-sprayed Ni-coated WC coatings comparison with different post-treatment, Corros Sci, 49 (2007) 662-671.
[6] Z. Liu, D.R. Yan, Y.C. Dong, Y. Yang, Z.H. Chu, Z. Zhang, The effect of modified epoxy sealing on the electrochemical corrosion behaviour of reactive plasma-sprayed TiN coatings, Corros Sci, 75 (2013) 220-227.
[7] T. Troczynski, Q. Yang, G. John, Post-deposition treatment of zirconia thermal barrier coatings using sol-gel alumina, J Therm Spray Techn, 8 (1999) 229-234.
[8] J. Knuuttila, P. Sorsa, T. Mantyla, Sealing of thermal spray coatings by impregnation, J Therm Spray Techn, 8 (1999) 249-257.
[9] S.Q. Xia, L.G. Wang, E.Z. Wang, X.A. Yue, Study on the Wear Resistance of Al_2O_3 Particles Reinforced Epoxy Resin Composite Coating, Advances in Textile Engineering and Materials Iii, Pts 1 and 2, 821-822 (2013) 1148-1151.
[10] S. Ahmaniemi, J. Tuominen, P. Vuoristo, T. Mantyla, Sealing procedures for thick thermal barrier coatings, J Therm Spray Techn, 11 (2002) 320-332.
[11] Z. Liu, Z.H. Chu, Y.C. Dong, Y. Yang, X.G. Chen, X.J. Kong, D.R. Yan, The effect of metallic bonding layer on the corrosion behavior of plasma sprayed

Al_2O_3 ceramic coatings in simulated seawater, Vacuum, 101 (2014) 6-9.
[12] S. Liscano, L. Gil, M.H. Staia, Effect of sealing treatment on the corrosion resistance of thermal-sprayed ceramic coatings, Surf Coat Tech, 188 (2004) 135-139.
[13] Y.H. Yoo, D.P. Le, J.G. Kim, S.K. Kim, P. Van Vinh, Corrosion behavior of TiN, TiAlN, TiAlSiN thin films deposited on tool steel in the 3.5 wt.% NaCl solution, Thin Solid Films, 516 (2008) 3544-3548.
[14] M. Hasanzadeh, M. Shahidi, M. Kazemipour, Application of EIS and EN techniques to investigate the self-healing ability of coatings based on microcapsules filled with linseed oil and CeO_2 nanoparticles, Prog Org Coat, 80 (2015) 106-119.

Corrosion performance of damaged thermally sprayed aluminium in simulated deep sea environment

S Paul, Cambridge / UK

Thermally sprayed aluminium (TSA) has been used in offshore applications for decades. Their effectiveness in mitigating corrosion of steel structures in presence of seawater has been proven over the years. However, very little work is reported on the performance of TSA when damaged. Furthermore, data on the performance of damaged TSA in deep sea is virtually non-existent. The paper addresses this knowledge gap and reports the corrosion performance of damaged TSA in a simulated deep sea environment. Holiday amounting to ~3% of the sample area was drilled to expose the underlying carbon steel and the sample was exposed to synthetic seawater at 5°C for 30 days in an autoclave at 50MPa to simulate 5000m of water pressure. After testing, examination of the sample revealed the formation of calcareous deposit on the exposed steel surface with no visible steel corrosion product. Detailed microstructural characterisation of the calcareous deposit confirmed the formation of protective Mg-based layer with negligeable Ca-based material. Thus, one can conclude that TSA can protect carbon steel in deep sea environment even when damaged.

1 Introduction

Thermally Sprayed Aluminum (TSA) coatings provide long-term corrosion protection of offshore steel structures [1, 2]. Its low corrosion rate coupled with its ability to act as a evenly distributed anode makes it ideal for such applications [3-5]. This anodic nature of aluminium offers protection to carbon steel in seawater resulting in the formation of calcareous deposits on any exposed steel [6]. Such deposits have the favourable effect of reducing the cathode area and hence decreasing the corrosion rate [7].

Performance of TSA in protecting steel in marine environments has been widely investigated in long-term laboratory as well as on-site studies [2, 8-10]. It is known that TSA has the ability to provide a substantial level of cathodic protection to steel at ambient and elevated temperatures, even when damaged [5-7]. Its behaviour on steel in seawater with and without external cathodic protection has also been explored [5]. Additionally, work has been devoted to the study of the mechanism of calcareous deposit formation in seawater [4, 9-11]. The influence of parameters, such as temperature, pH and various ions present in seawater has been studied. Although, several papers have been published in the area, the research so far on calcareous deposits formation has focussed on ambient pressure exposure, typically at ambient temperatures.

However, the application of TSA is not limited to ambient conditions. For deep subsea pipelines where the temperature is low in most cases and the external pressure is usually high, the rate of formation of calcareous deposit may differ. This missing information is crucial as deep subsea pipelines may be in contact to the rough seabed where the erosion and wear is likely.

This paper addressed the above mentioned technology gap. Arc-sprayed aluminium-coated steel samples with controlled defect (holiday) were exposed to synthetic seawater at 5°C for 30 days in an autoclave at 50MPa to simulate 5000m of water pressure to study the protective performance of TSA.

2 Experimental

2.1 Materials

Carbon-Manganese steel was selected as the substrate material conforming to specification S355J2+N(2004). The coating material was 2.3mm diameter commercially 'pure' aluminium (99.5% Al) supplied by Metallisation Ltd. (UK). The nominal composition of the substrate and coating material is given in **Table 1**.

Table 1. Composition of steel substrate and Al wire (coating material).

Materials	Element (wt%)							
	C	Mn	Si	S	P	Fe	Cu	Al
Steel substrate	0.12	1.39	0.39	0.019	0.014	Bal.	-	-
Al wire	-	<0.01	0.07	-	-	0.21	<0.01	Bal

2.2 Coating production

The substrate was cut into 42×40×6mm flat panels and the edges were rounded off using a file prior to further preparation. Prior to coating, the substrates were degreased and then prepared by blasting with F60 and NK12 alumina to give a rough surface to which the molten and semi-molten spray particles can key onto.

Thermal spray coatings were prepared using a Metallisation 528 twin wire arc spray gun mounted on a 6-axis robot. Dry compressed air, supplied at 5.5bar, was used as the particle atomising and propellant gas. Coatings were produced on one surface using the spray parameters given in **Table 2** to attain a nominal coating thickness of 300μm.

Table 2. Spray parameters used for preparaing TSA coatings

Wire feed rate (g/min)	Spray distance (mm)	Increment step (mm)	Traverse speed (m/s)
98.7	95	15	0.5

2.3 Corrosion testing in an autoclave

TSA-coated specimens were taken and a 8 mm diameter slot drill was used to remove the coating to expose the steel substrate (~3% area of the coating surface). The side and edges of the specimens were covered with a curable polymer to protect the surfaces not intended to be exposed. The exposed TSA surface area was 15.5cm^2 (obtained by taking the holiday area into account). A glass-lined autoclave was used for the immersion test in 400mL ASTM D1141 synthetic seawater at 5°C [12]. The autoclave was pressurised to 50MPa by compressed air for 30 days.

2.4 Determination of corrosion rate

The corrosion rate (CR) was calculated by noting the amount of dissolved aluminium in seawater after teating. As no aluminium salt is used in the production of ASTM D1141 synthetic seawater all the Al^{3+} ions (y mg/L) present after testing would have come from the TSA coating. Since the volume of solution used is known, we can calculate the amount of aluminium lost from the surface of the coating in 30 days by taking the density of Al as 2.7g/cm^3 i.e. 0.4y/2700mg.

The amount of Al^{3+} in solution was converted to the thickness of coating lost during exposure (d mm) by noting the geometric area os the coating exposed (15.5cm^2). This was found to be

$d = 4y/(2700\times15.5)$ mm in 30 days, or

$d = 4y\times365/(2700\times15.5\times30)$ mm/year

2.5 Microstructural characterisation

After the autoclave exposure, changes to the coating appearance were recorded (i.e. the samples were photographed). The surfaces of the corroded specimens were examined by light optical microscopy and SEM (samples were gold coated prior to this), including element mapping obtained using energy dispersive X-ray spectroscopy (EDX). Further analysis of cross-sections was also carried out to understand the corrosion mechanism.

3 Results

3.1 Visual examination

The samples were photographed before and after exposure to synthetic seawater in an autoclave, **Fig. 1**.

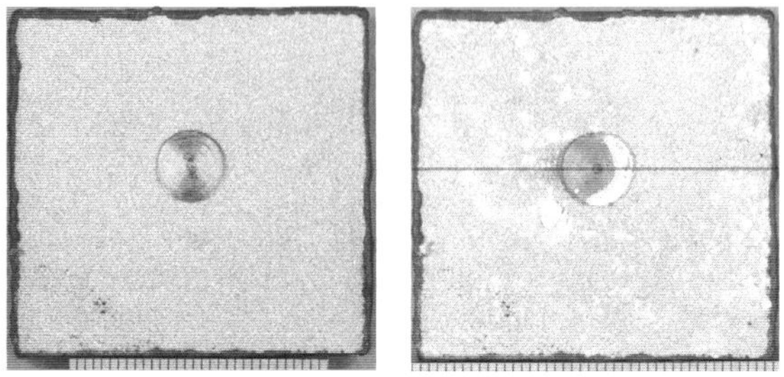

(a) (b)

Fig. 1. Photographs of samples before(a) and after (b) testing in synthetic seawater. The line in (b) shows the region along which the specimen was sectioned for further analysis.

The exposed steel (holiday) did not show any rust, **Fig. 1**. The presence of white corrosion product is evident on the surface of the TSA coating, **Fig 1b**. White deposit is also seen in the holiday region on the surface of exposed steel. The polymer used to protect the edges and the back surface performed well and no rust or corrosion product was seen in these regions.

3.2 Corrosion products on the surface of the coating

The white powdery deposit distributed randomly on the surface of the TSA coating after exposure was analysed using a SEM/EDX, **Fig 2**. The reason for this uneven distribution is not entirely clear. However, the nature and distribution of porosity in thermal spray coatings is often difficult to predict and this will be invariable linked to the corrosion behaviour.

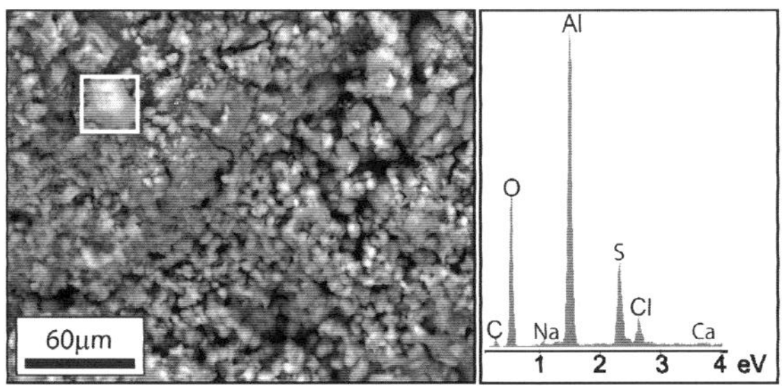

Fig. 2. SEM micrograph of the white deposit on the surface of the coating (a) and the associated EDX pattern for the highlighted region (b).

The SEM micrograph showed light and slightly dark regions in the deposit in the electron back scattered mode. The EDX spectra of the while spots showed Al and O (amongst other elements such as S, Cl and Na present in seawater), implying the presence of oxide and possibly hydrated oxides of alumina, **Fig. 2b**. The darker regions also had very similar EDX pattens (not shown). The reflectivity of the coating also changed i.e. metallic lusture offered by TSA when sprayed was lost after exposure due to the presence of the corrosion product layer on the TSA surface, as expected.

3.2 Microstructure of coating cross-section

Detailed microstructural examination of cross-sections of the holiday region revealed the formation of a thick scale covering part of the steel surface, **Fig. 3**. The holiday was deep enough to cut through the entire thickness of the coating. The thickness of deposits formed was upto 1.2mm in places. The deposit appeared to be porous in places with some dense regions. The growth of the deposit in some places were high which meant that part of the TSA was covered, **Fig. 3**. The contrast in the back-scattered electron (BSE) mode can be observed which indicates the presence of different compounds. The white regions (e.g. region 1 in **Fig. 3**) are expected to have different chemical composition from the slightly darker regions. Three regions were selected based on the contrast in the BSE image. These were marked as 1, 2 (**Fig. 3**), and 3 (**Fig. 4**).

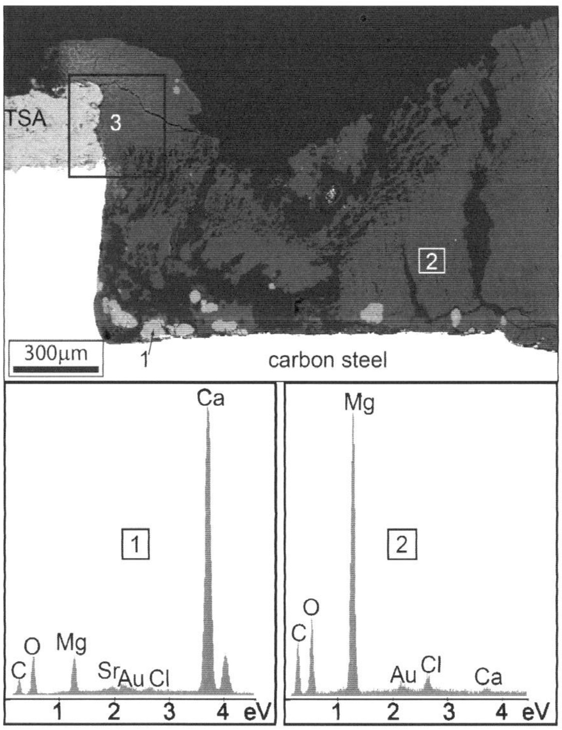

Fig. 3. TSA-coated specimen tested in boiling seawater (~101.5°C) showing SEM image of cross-section along the holiday and EDX traces of regions 1 and 2. Details of region 3 is given in Fig 4.

On closer examination of the deposit using EDX, it was observed that the darker coloured layer (region 2 in **Fig. 3**) predominantly contained Mg and O (indicating the presence of $Mg(OH)_2$). The other elements were either present due to the dried seawater (Ca, Cl) or a result of the sample preparation process (Au). The white areas (region 1 in **Fig. 3**) predominantly contained Ca, Mg, O and C. The C seen could possibly be due to the ingress of the carbonaceous mounting resin or due to the formation of a carbonate.

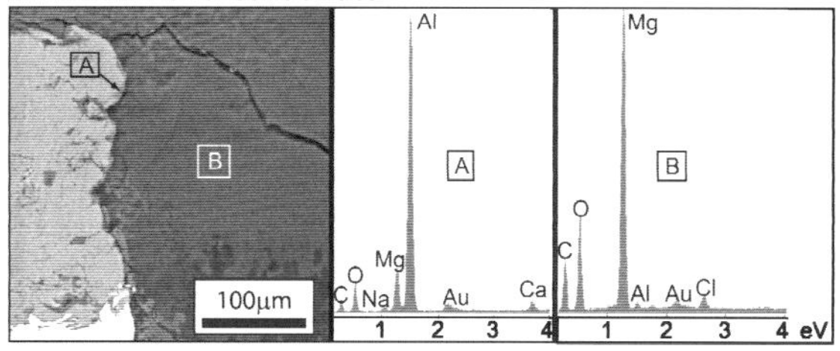

Fig. 4. Details of region 3 from Fig. 3 showing the EDX patterns of regions A and B.

The region close to the TSA layer when analysed in detail showed two different types of deposits/corrosion products, **Fig. 4**. Some cracks were also seen in the deposit close to the TSA layer. The layer adjascent to the TSA layer (A) had different composition to the layer further away from it (B). The EDX pattern of layer A showed the presence of Al, Mg, O, C, Na, Au and Ca. Of these, the presence of Al and O indicates the presence of alumina or hydrated alumina on TSA. However, the composition of the B layer was different and did not show much Al. It was predominantly a Mg-containing compound. This layer seems to have the same composition as the deposit formed away from the TSA layer.

3.4 Corrosion rate of TSA-coated steel

The amount of dissolved Al and Fe were found to be 12.93mg/L and 0.12mg/L, respectively. Taking the density of Fe and Al, and the surface areas exposed the dissolution rate was converted to a corrosion rate of 0.015mm/y for Al and 0.0003mm/y for carbon steel. These numbers are approximate values as the extrapolation of corrosion rate from short-term tests is prone to propagation of errors. Nonetheless, the negligeable loss of Fe from carbon steel in seawater elucidates the sacrificial protection offered by TSA in seawater.

4 Discussion

4.1 Corrosion of aluminium in seawater

Copper-free aluminium alloys are generally resistant to natural seawater. The corrosion mechanism of aluminium is considered to be the one of localized attack (i.e., pitting) as opposed to uniform or general corrosion [16]. Therefore, corrosion rate calculated from weight change can be misleading as this requires the assumption that the material is lost uniformly from the exposed surface. Rates of pitting usually range from 3 to 6 μm/year during the first year and from 0.8 to 1.5 μm/year averaged over a 10-year period [17]. These values are different from the ones presented here (15 μm/year) as the work reported here is on damaged TSA where general corrosion was assumed to calculate the corrosion rate. The duration of 30 days is not considered long enough to give the long-term performance of these coatings. The long-term tests usually give lower corrosion rates owing to the tendency of the older pits to become inactive. Another factor to consider is the pressure at which the tests were carried out. There are no reported corrosion rate values at 5km depth in seawater. Limited information on the corrosion behavior of aluminum alloys in deep seawater in very similar to that at the surface except that the effect of crevices is greater [17].

4.2 Mechanism of calcareous deposit formation

The morphology of the calcareous deposits formed on a cathodically polarised steel is temperature dependent [7]. At low temperatures (≤5°C) Mg-rich deposit is reported at ambient pressures, but at ambient temperatures and ambient pressures Ca-rich deposits are formed, presumably due to the difference in the temperature dependence of the solubility limits of $Mg(OH)_2$ and $CaCO_3$ [6, 13]. Solubility of $CaCO_3$ in water decreases while that of $Mg(OH)_2$ increases with increasing temperatures. For precipitation, supersaturation is essential and supersaturation of $Mg(OH)_2$ can arise near the steel surface as a consequence of the production of OH^-. The generation of OH- can also affect the solubility of Ca^{2+}, especially in the presence of dissolved CO_2.

The steel-seawater interface has a pH possibly above 9.2 (calculated at 18-20°C) which allows the precipitation of $Mg(OH)_2$. As the $Mg(OH)_2$ precipitates at the interface, the pH decreases away from the interface to a value where $Mg(OH)_2$ can no longer precipitate. Packter and Derry [14], and later Yang et al. [15] demonstrated that $Mg(OH)_2$ can co-precipitate with iron hydroxides. The pH at which this can occur was calculated to be ~2.3 [15], well below the pH at which precipitation of $CaCO_3$ can ensue. Hence, the deposit predominantly consists of $Mg(OH)_2$ close to steel with some specks of $CaCO_3$, sometimes embedded in the $Mg(OH)_2$ precipitate, **Fig. 5**.

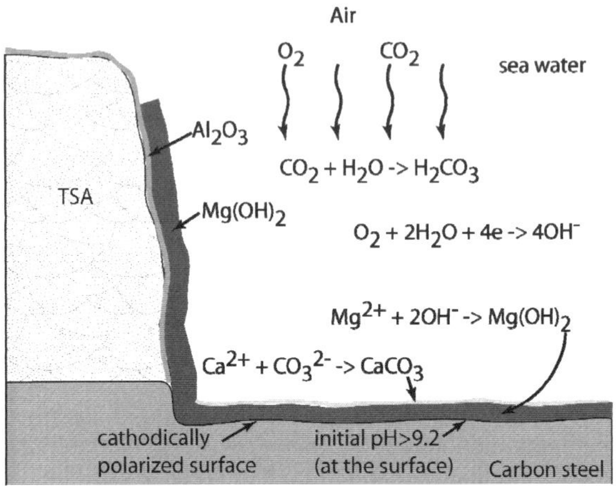

Fig. 5. Mechanism of calcareous deposit formation on steel surface polarised by TSA in seawater containing dissolved air at 5°C.

4.3 Corrosion performance of damaged TSA-coated steel in seawater

The galvanic series of materials in seawater gives a good indication regarding the anodic and cathodic behaviour of materials when in electrical contact. Auminium being more electronegative or active to carbon steel is expected to be anodic when sprayed on top by arc-spray. Aluminium being anodic to carbon steel in seawater protects it by being preferentially consumed, thus providing cathodic protection. Owing to this nature, aluminium is thermally sprayed onto carbon steel to protect offshore steel structures from the corrosive marine environment. The environment includes splash and tidal zones and submerged structures. Although, aluminium is anodic to steel, the primary function of the coating is to act as a barrier. This is the reason why it can protect steel in sreas where a continuous layer of electrolyte is not present such as in the splash and tidal zones. If the barrier is breached, the anodic nature of the coating becomes vital. Even in the absence of an external damage thermally sprayed coatings contain porosity due to the spraying process. The molten and semi-molten droplets of the consumable impinge on the substrate and form the coatings by overlapping pancake-shaoed splats. As the droplets impinge they form inter-splat pores and the cooling also causes shrinkage pores. The semi-molten and unmolten particles, and surface roughness also contribute to pores. Some of the internal pores are surface-connected and at times reach the substrate. The presence of surface-connected porosity (or through porosity) could result in the electrolyte (seawater) reaching the substrate, but since the substrate (steel) is cathodic to the coating (anode), corrosion of steel is not expected. Aluminium will dissolve while protecting the steel substrate from corrosion. However, there will be some corrosion of steel which is often negligeable. The rate of corrosion of steel will depend on, amongst other factors, the area of steel exposed, temperature, dissolved oxygen content, conductivity of the electrolyte etc.

The anode (aluminum in the case of TSA-coated steel) while protecting the steel polarises the exposed surface and this polarisation facilitates the cathodic reaction on the steel surface results in the production of hydroxyl ions (OH^-), which increases the interfacial pH. The generation of OH^- occurs either due to the reduction of dissolved oxygen or by the reduction of water (which occurs at slightly greater negative potential) [9, 11]. The generation of OH^- ions facilitates the precipitation of magnesium hydroxide from seawater. Moreover, the increased production of OH^- leads to changes in the stability of dissolved Ca^{2+} ions at the steel surface. The Ca^{2+} ions form the basis of calco-carbonic equilibrium which is governed by the dissolution and subsequent dissociation of CO_2 in seawater. The use of pressurized air in synthetic seawater to simulate deep sea environments can be questioned as the dissolved oxygen (DO) content in the test will be higher than that observed in reality. At ambient conditions the DO content in oxic surface seawater is 6-8mL/L. This can be significantly in subsea environments where the oxygen minimum zones can have DO below 0.1mL/L [18].

It is generally recognised that during subsea cathodic protection, the formation of calcareous deposit on exposed steel surface is favourable since it reduces the cathode area. These deposits on steel under cathodic protection comprises $CaCO_3$ and $Mg(OH)_2$ at ambient seawater temperatures [5-7]. The solubilities of the components of the calcareous deposit imply that predominantly $Mg(OH)_2$ will be deposited at 5°C in synthetic seawater kept at 500bar, very similar to observations at similar temperatures, but at ambient pressures. However, the formation of calcareous deposit and the subsequent reduction in the corrosion of exposed steel is in line with the observation of previous experimental work where an increase in exposure time resulted in a reduction in corrosion rate with time [6].

5 Conclusions

The following conclusions can be drawn from this work where damaged TSA was used to understand the mechanism of corrosion protection offered by TSA in a simulated deep sea environment at 5°C:

- The arc-sprayed Al coating protected the exposed steel (~3%) from seawater and no rust was seen on the holiday region.
- The calculated corrosion rate extrapolated from the 30 day test data was found to be 0.015mm/y for Al and 0.0003mm/y for carbon steel.
- The low corrosion rate and high integrity of TSA implies that it can provide long-term corrosion protection to carbon steel offshore structures in deep sea environments. The anodic nature of TSA also ensure the protection of steel even when TSA is damaged in certain regions.

6 Acknowledgements

The work was funded by TWI. The author would like to thank TWI staff, especially Ian Wallis, Andrew Tabecki and Sheila Stevens for their technical support.

7 References

[1] Paul, S.: Corrosion control for marine- and land-based infrastructure applications, ASM Handbook, Vol. 5A, pp. 248/52, ASM International, Ohio, 2013.

[2] Kuroda, S. et al.: An 18-year exposure test of thermal-sprayed Zn, Al, and Zn-Al coatings in marine environment. Corrosion 62 (2006), Issue 7, pp. 635/47.

[3] Paul, S. et al.: Corrosion testing of thermally sprayed aluminium. ITSC/2015, paper no. 38552 (Long Beach, CA: ASM International, 2015), pp 964/970.

[4] Paul, S. et al.: Thermally sprayed aluminum (TSA) coatings for extended design life of 22%cr duplex stainless steel in marine environments. Journal of Thermal Spray Technology 22 (2013), Issue 2-3, pp328/36.

[5] Gartland, P.O. and T.G. Eggen: Cathodic and anodic properties of thermally sprayed Al- and Zn-based coatings in seawater. Corrosion/90, paper no. 367 (Houston, TX: NACE, 1990), pp. 1/16.

[6] Paul, S. et al.: Electrochemical monitoring of thermally sprayed aluminium coatings on steel in a simulated marine immersion environment. Eurocorr/14, Paper No 7458, (Pisa, Italy: EFC, 2014), pp1/7.

[7] Paul, S: Corrosion performance of damaged thermally sprayed aluminium in synthetic seawater at different temperatures. Thermal Spray Bulletin 8 (2015), Issue 2, pp. 139/46.

[8] Fischer, K. et al: Performance history of thermally sprayed coatings in offshore service. Materials Performance 34 (1995), No. 4, pp. 27/35.

[9] Rousseau, C. et al.: Calcareous deposit formed under cathodic protection in the presence of natural marine sediments: A 12-month experiment. Corrosion Science 52 (2010), No. 6, pp. 2206/18.

[10] Kain, R.M., and E.A. Baker: Marine atmospheric corrosion museum report on the performance of thermal spray coatings on steel. ASTM Report STP 947 (Philadelphia, PA: ASTM, 1987), p. 211.

[11] Barchiche, C. et al.: Characterization of calcareous deposits by eletrochemical method role of sulphates, calcium concentration and temperature. Eletrochimica Acta 49 (2004), Issue 17-18, pp. 2833/39.

[12] ASTM D1141-98(2013), Standard Practice for the Preparation of Substitute Ocean Water, ASTM International, West Conshohocken, PA, 2013.

[13] Kunjapur, M.M., Hartt, W.H., and S.W. Smith: Influence of temperature and exposure time upon calcareous deposits, Corrosion 43 (1987), No. 11, pp. 674/79.
[14] Packter, A., and A. Derry: The coprecipitation of magnesium iron (iii) hydroxide powders from aqueous solution: precipitate compositions and coprecipitation and ageing mechanisms. Crystal Research and Technology 21 (1986), No. 11, pp. 1391/1400.
[15] Yang Y., Scantlebury, J.D., and E.V. Koroleva: A study of calcareous deposits on cathodically protected mild steel in artificial seawater. Metals 5 (2015), pp. 439/56.
[16] Lorking, K.F., and J.E.O. Mayne: The corrosion of aluminium. Journal of Applied Chemistry 11 (1961), No. 5, pp. 170/80.
[17] Ghali, E.: Aluminum and Aluminum Alloys, in Uhlig's Corrosion Handbook, 3rd Edition (Ed. R. W. Revie), pp. 715/45, John Wiley & Sons, Hoboken, NJ, USA, 2011.
[18] Levin, L.A.: Deep-ocean life where oxygen is scarce. American Scientist 90 (2002), pp. 436/44.

Production rate improvement in liquid fuel HVOF spraying by modified powder injection concepts

T. Schläfer, A. Wank, GTV Verschleiss-Schutz GmbH, Luckenbach / D

Coating production costs in large surface area coating jobs using liquid fuel HVOF spraying are dominated by powder, oxygen and machine costs. Powder costs are minimized by process optimization concerning deposition efficiency. Assuming high utilization oxygen and machine costs are minimized by maximizing deposition rate. Both deposition efficiency and deposition rate are system variables depending on choice of powder, torch and torch components as well as process parameters.
Among liquid fuel HVOF torches GTV K2 torch is already renowned for unsurpassed process stability and deposition rates in a large number of applications. Within the last decade there has been research on expansion nozzle and combustion chamber design that resulted in novel torch components that permit setting of spraying conditions close to those of cold gas spraying. Recent research focuses on powder injection design. In particular the increase of number of injection ports from 2 to 4 has been studied along with increasing powder feed rates to extreme values.
The capability of the novel powder injector block concerning expansion of deposition rate without a negative effect on deposition efficiency is analyzed for spraying of WC and Cr_3C_2 based cermets as well as Ni and Fe base powders. Influence of increased deposition rate on coating microstructure is analyzed by optical microscopy.

1 Introduction

The reduction of coating production costs for existing spray processes can be obtained by either improving the deposition efficiency (DE) or by increasing the spray material throughput, assuming that one of the factors does not affect the other in a negative way.
The DE directly influences the production costs related to the spray material. The material throughput strongly affects the productivity and thus all time-related production costs, such as costs for gases, etc, when spray gun parameter settings remain unchanged.

The combination of both, the deposition rate $\dot{m}_{powder,eff} = DE * \dot{m}_{powder}$ is one of the major factors affecting productivity and thus all spray time-related costs, such as costs for oxygen, other gases and kerosene or costs of labour.

Optimization of the DE and thus its effect on the productivity is strongly limited. With thermal spray processes there is often just little room for increasing DE and thus productivity. In contrast, there is a tremendous potential to improve productivity and time related costs by increasing the spray material throughput.

Liquid fuel HVOF spray guns already allow high powder feed rates. For example, AISI 316L powder is being processed in everyday industrial use at feed rates of up to 200 g/min. Here, the DE is approx. 65 %, leading to a deposition rate of 130 g/min. Cr_3C_2/NiCr 75/25 powders are being processed at feed rates of up to 130 g/min, a deposition rate of 65 g/min, respectively.

Increasing the powder feed rate is often limited by the spray gun (e.g. by the inner diameter of the powder ports) while, especially volumetric disk based powder feeders, such as the GTV PF feeders (GTV Verschleißschutz GmbH, Germany), allow feed rates of 200 g/min and more.

So, the aim is to enable the spray gun to be capable of processing strongly increased powder feed rates. For this purpose, a novel powder injector block, named "Quattro", was developed, which allows to utilize the high powder feed rates provided by the powder feeder at minimum effort. This component can be used with already existing K2 spray guns (GTV Verschleißschutz, Germany).

The novel powder injector block posesses 4 powder injector ports circumferentially offset by 90°, Figure 1. It can be operated with 4 separate powder feeders, though operation with just 2 feeders is also possible when the powder stream delivered by each feeder is split up into two, e.g. by installing Y-pieces into the powder feeding lines as depicted in Figure 1. Here, each of the two powder lines supplies powder to two powder ports facing each other.

If operation with two powder feeders is favoured, it has to be made sure that the powder feeders mass flow controller is capable of delivering twice the gas flow required per powder port, e.g. 9 slpm when spraying many cermet powders. It is obvious, that the powder feeder must be capable of delivering the desired feed rates, especially when spraying powders that exhibit a low apparent density such as typical agglomerated and sintered Cr_3C_2/NiCr powders.

Aim of this feasibility study is to investigate the influence of the powder feed rate at high levels on the deposition efficiency for several metal and cermet spray powders. In a second step, the influence on the coating quality is investigated. Within the frame of this work, the standard K2 spray gun and the K2 equipped with the novel Quattro powder injector block are employed.

Figure 1: Photograph of the K2 liquid fuel spray torch equipped with the 4 powder port injector block "Ouattro" and additional cooling nozzels. Each of the two powder hoses coming from the two powder feeders supplies two powder ports facing each other.

2 Experimental

All spray trials were carried out using an aspirating volumetric powder feeder type GTV PF MF 4/4 comprising four individual powder feed lines. With this feeder the carrier gas flow rate is controlled by mass flow controllers. Originally installed mass flow controllers allowed a gas flow rate of up to 15 slpm. With one powder feeding line, one mass flow controller respectively serving two powder ports of the gun, the carrier gas flow rate would have been too small to process cermet powders which require a flow rate of 9 slpm. Hence, bigger controllers, allowing flow rates of up to 30 slpm were installed.

Altogether 3 metallic and 3 cermet powder feedstocks were processed with the novel 4 port powder injector block as well as with the standard set-up for reference. Information about the powders is given in Table 1.

Table 1: Powders used within the frame of this work; PSD: particle size distribution

GTV.-Nr.	Description	PSD [µm]	Production process
80.20.1	NiCr 80 20	-53+20	gas atomised
81.43.1	FeNiCrMoCSi	-45+20	
80.46.1	AISI 316L	-53+20	
80.81.1	Cr_3C_2/NiCr 75 25	-45+15	agglomerated & sintered
80.76.1	WC/Co/Cr 86 10 4	-45+15	
80.75.1	$WCCr_3C_2$/Ni 93 7	-45+15	

Parameter settings and initial powder feed rates correspond to industrially established ones. When spraying with the novel Quattro injector block, the powder feed rates were successively increased to 400 g/min for the metallic powders, to 300 g/min and 500 g/min for cermet powders, respectively. Prior to all DE spray trials the precise powder feed rates were determined. The revolution speed of the powder feeder disks was adjusted so that the actual powder feed rate does not deviate more than 3 % from the nominal value. With all 6 powders investigations of the deposition efficiency in accordance to ISO 17836:2004 were carried out. Each measurement was repeated once. As test pieces plates made from mild steel were used. Prior to carrying out the DE tests, the surfaces were activated by grit blasting.

Spray torch parameters such as gas and kerosene flow rates as well as spray distance were not changed while increasing the powder flow rates or changing powder injector blocks. The carrier gas flow rate per powder port did remain unchanged, too. All parameters used during the DE investigations are given in Table 2.

Table 2: Spray parameters for DE investigations

Powder	Powder feed rate [g/min]	Kerosene flow rate [l/h]	Oxygen flow rate [slpm]	Carrier gas flow rate [slpm]	Nozzle type	Spray distance [mm]
80.20.1	100...400	26	900	7	150K	350
81.43.1	150...400	25	950	7	150K	350
80.46.1	200...400	25.5	925	8.5	150K	350
80.81.1	100...300	26	900	9	200/11	380
80.76.1	100...500	26	900	9	150K	350
80.75.1	100...500	26	900	9	200/12	450

For subsequent metallographic investigations, mild steel samples ø 40 mm x 10 mm were coated. Parameters concerning spray gun and component handling are given in Table 3.

Table 3: Additional parameters for producing samples for metallographic investigations

Powder	Track offset [mm]	Spray velocity [m/min]	Number of passes
81.43.1	8	200	36...12
80.81.1	5...8	90	17...11
80.76.1	5...8	72...265	30

With the two iron based materials GTV 80.46.1 (AISI 316L) and 81.43.1 (FeNiCrCSi) track offset and spray velocity were unchanged when the powder feed rates were increased. In order to receive uniform coating thicknesses of 400 µm (80.46.1) and 300 µm (81.43.1), the number passes was reduced with an increasing powder feed rate.

With the cermet powder 80.81.1 (Cr_3C_2NiCr 75/25) track offset and number of passes were varyied in order to receive a coating thickness of approx. 400 µm. 80.76.1 (WC/CoCr 86/10-4) was sprayed at a constant number of passes (30). In order to receive coatings with a thickness of approx. 300 µm and a constant single layer thickness of 10 µm, spray velocity and track offset were increased with increasing powder feed rate. Track offset was varyied between 5 mm and 8 mm. The required spray velocity v_S then was calculated in accordance to

$$v_S = \frac{DE \cdot \dot{m}_p}{s_t \cdot t_s \cdot \rho_c}$$

with:

DE: Deposition Efficiency
$\dot{m}_p$: Powder feed rate
s_t: Track offset
t_s: Single layer thickness
ρ_c: Coating density

From all coated coupons micrographs were produced and analyzed with regard to apparent coating porosity by digital image analysis (Clemex VISION Lite, Clemex Technologies, inc., Canada). Given values represent the average of 10 measurements. Coating thickness was determined using a light optical microscope and the imaging software DigiVID 8.05 (Mikrovid GmbH, Germany), 20 measurements were carried out for each sample. Vickers hardness of the coatings was determined utilizing a load force of 0.3 kg (Microhardnesstester FM-300, Future-Tech Corp., Japan), here 10 measurements were carried out for each sample. The average roughness Rz of all coatings was determined by means of the profile method.

3 Results

3.1 DE Investigations

Figure 2 depicts the summarized results of the DE investigations concerning the metallic feedstocks 80.20.1, 81.43.1 and 80.46.1 as a function of the type of the powder injector block and the nominal powder feed rate. Comparing the results achieved by using the two different powder injector blocks at low feed rates, a significant increase of the DE value stands out when the version comprising 4 powder ports is being used. For alle three powders an increase of 2 to 3 percentage points can be observed. 80.46.1 was processed with both torch configurations at an elevated powder feed rate of 200 g/min, here no difference of the DE values were observed.

When planning the experiments, a significant decrease of the DE was expected when increasing the powder feed rate. With 80.46.1 the DE did decrease from 68% to 65% when increasing the powder feed rate from 100 g/min to 400 g/min. With the iron based hard alloy powder 81.43.1 DE values between 64 % and 66 % were observed. Doubling the initial feed rate led to a minor decrease. At 400 g/min a higher value of 66% was observed. When spraying the NiCr powder 80.20.1, a different trend was observed; quatrupling the feed rate even exhibited an increase of the DE by 3 percentage points.

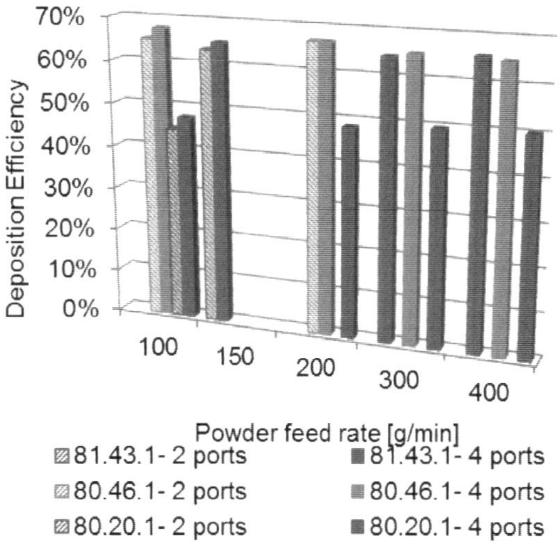

Figure 2: DE values vs type of injector block and powder feed rate for metallic powders 80.46.1, 81.43.1 and 80.20.1

With the cermet powders 80.76.1 and 80.81.1 similar effects as with the metallic powders were determined regarding DE and a change from the standard powder injector block to the novel Quattro block. At low powder feed rates of 100 g/min (80.75.1, 80.76.1), 130 g/min (80.81.1), respectively, the DE increases significantly. At an elevated feed rate of 200 g/min almost identical DE values were achieved for both types of powder injector blocks. For 80.75.1 an opposing effect was observed; the DE value decreases by 1 percentage point when processing the material with the Quattro block at a feed rate of 100g/min. At an elevated feed rate of 200 g/min the values achieved with the different components converge; with both, a DE of 52% was observed.

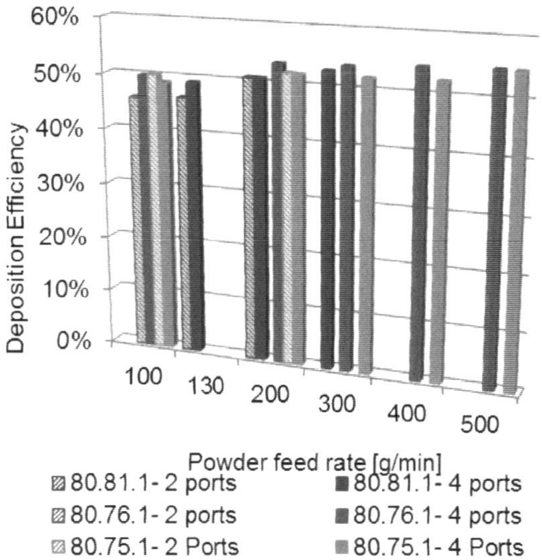

Figure 3: DE values vs type of injector block and powder feed rate for cermet powders 80.75.1, 80.76.1 and 80.81.1

Increasing the powder feed rate from 130 g/min to 300 g/min for 80.81.1 or from 100 to 500 g/min for 80.76.1 and 80.75.1 shows that the DE increases steadily when the novel injector block is used. The growth reaches a peak value of 6 percentage points when spraying 80.76.1 (absolute value of 55%).

3.2 Microhardness

The micro hardness HV0.3 of the cermet coatings 80.81.1 and 80.76.1 lay within the range typical for coatings sprayed with the liquid fuel HVOF torch K2. Comparing the coatings sprayed using the standard powder injector block and the novel one, only small differences are observed. The coatings sprayed using the novel powder injector block both exhibit slightly higher hardness values of 1,440 HV0.3 for 80.76.1 and 1,020 HV0.3 for 80.81.1. Increasing the powder feed rate up to 300 g/min for 80.81.1, the hardness values decrease marginally to 1,000 HV0.3. With 80.76.1 coatings produced with elevating powder feed rates a steady but small decrease of the coating hardness is determined. The lowest value of 1,260 HV0.3 is observed for the coating sprayed at a feed rate quintupled in comparison to the standard powder feed rate (Figure 4). Considering the relative standard deviation of the measurements which is between 4% and 6%, it can be stated, that when tripling the initial feed rate, no significant drop down of hardness is observed. The same statement can be made for the 80.81.1 coatings exhibiting a relative standard deviation between 6% and 8%.

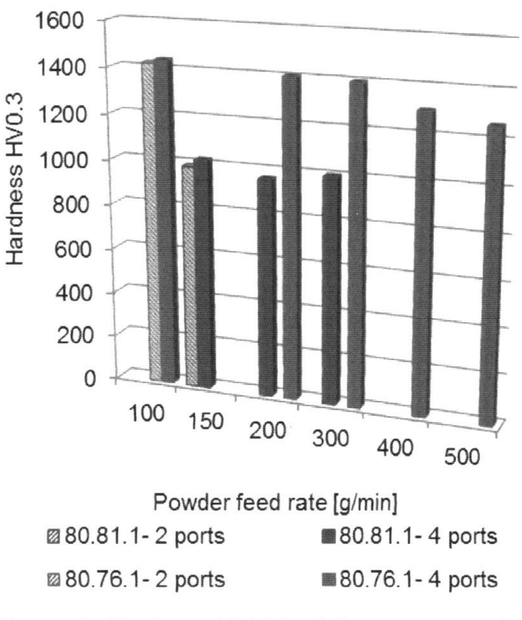

Figure 4: Hardness HV 0.3 of the cermet coatings 80.81.1 and 80.76.1 vs type of powder injector block and powder feed rate

With the coatings produced from 81.43.1, only small deviations of the hardness were determined when spraying with the different Injector blocks or when increasing the powder feed rate up to 400 g/min (± 15 HV0.3) as shown in Figure 5. The highest value of 630 HV0.3 was determined for the coating sprayed using the novel powder injector block at a powder feed rate of 150 g/min.

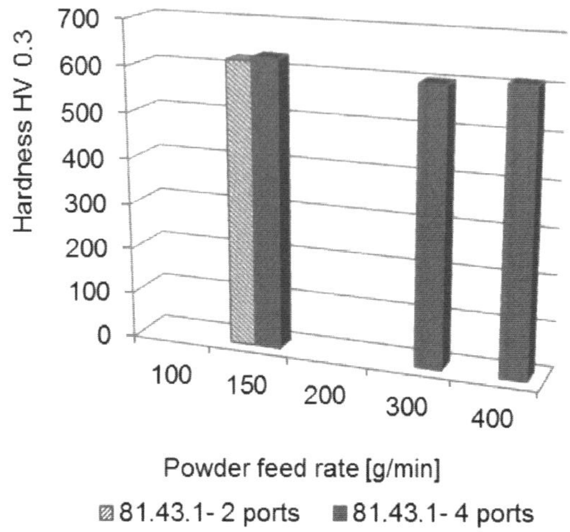

Figure 5: Hardness HV0.3 of the iron based 81.43.1 coatings vs type of powder injector block and powder feed rate

3.3 Apparent porosity

Figure 6 depicts the apparent porosity of the cermet coatings produced from 80.81.1 and 80.76.1. Looking at the apparent porosity of these coatings produced using the standard injector block and the novel one, a small increase is observed for the coatings produced with quadruple powder injection. For 80.81.1 the apparent porosity raises from 1.3% to 1.5%, for 80.76.1 from 0.8% to 1.0%. An increasing powder feed rate leads to a steady increase in porosity to 1.5% for 80.81.1 when sprayed at 300 g/min. With 80.76.1, a maximum of 2.4% is reached when quintupling the powder feed rate to 500 g/min.

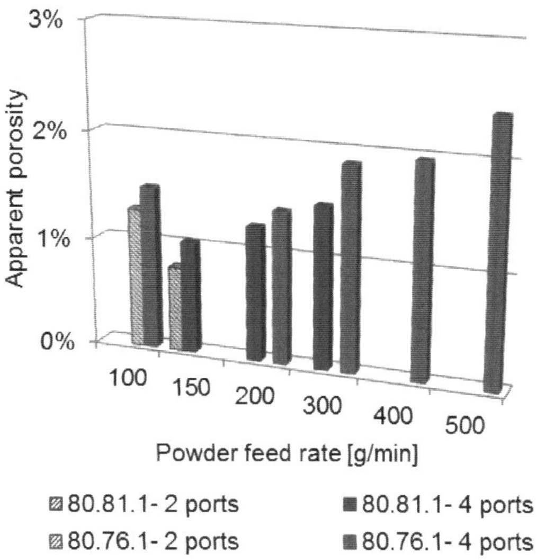

Figure 6: Apparent porosity of the cermet coatings produced from 80.81.1 and 80.76.1 powders vs type of powder injector block and powder feed rate

For the iron based 81.43.1 coatings similar trends are apparent (Figure 7). Here too, the effect is very small when changing to quadruple injection the total difference is less than 0.1 percentage points. With an increasing powder feed rate though, a significant growth of apparent porosity is observed, reaching a peak value of 4%.

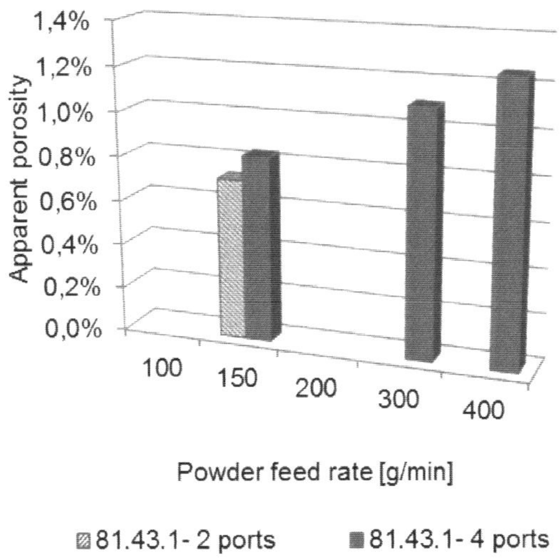

Figure 7: Apparent porosity of the iron based 81.43.1 coatings vs type of powder injector block and powder feed rate

Exemplarily, Figures 10 and 11 show micrographs of metallic and carbide based coatings. Figure 8 depicts 81.43.1 coatings sprayed at different powder feed rates, Figure 9 shows WC/CoCr (80.76.1) coatings.

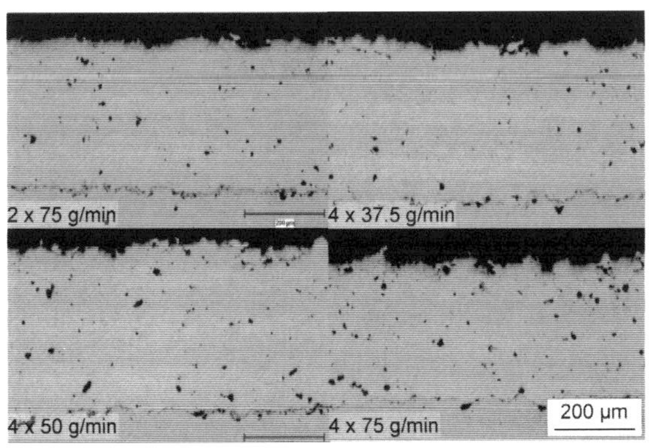

Figure 8: Micrographs of coatings produced from FeNiCrCSi powder 81.43.1 at different spray rates using the standard powder injector block and the novel Quattro block.

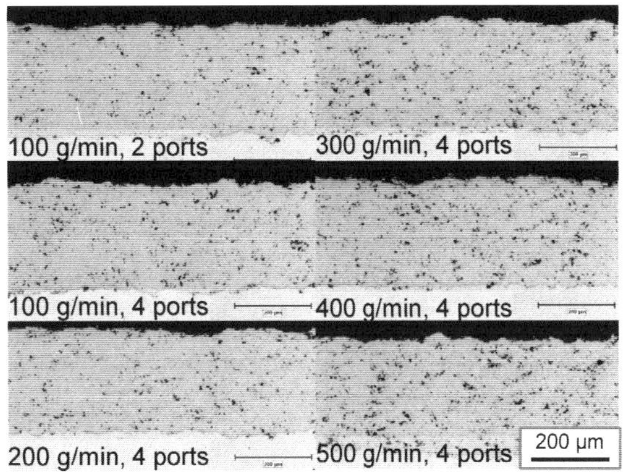

Figure 9: Micrographs of coatings produced from WC/CoCr powder 81.76.1 at different spray rates using the standard powder injector block and the novel Quattro block.

3.4. Roughness Rz

The results from roughness measurements for both cermet coatings as depicted in Figure 10, exhibit a small but neglible increase of roughness Rz of the 80.81.1 coatings when using the novel powder injector block. With 80.76.1 no significant increase is observed. Increasing the powder feed rates results in increasing roughness, which reaches a maximum of 27 µm when spraying at 500 g/min. Up to a feed rate of 300 g/min the increase appears neglible. Increasing the feed rate for 80.81.1 up to 300 g/min, a small but neglible increase is observed, too.

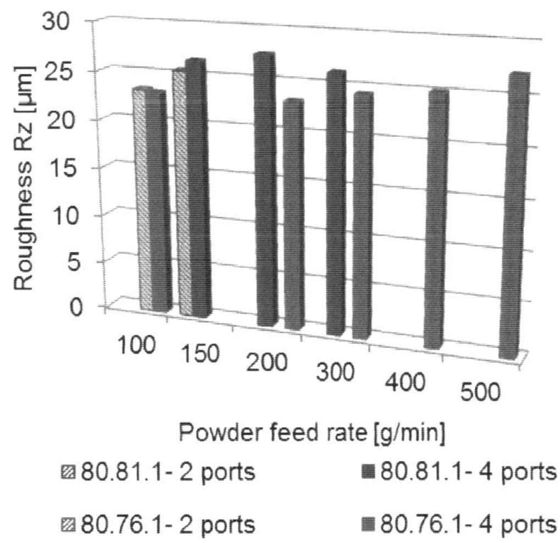

Figure 10: Average roughness Rz of the cermet coatings produced from 80.81.1 and 80.76.1 powders vs type of powder injector block and powder feed rate

Figure 11 shows the roughness values determined for the iron based coatings produced from 81.43.1 powder; changing from 2 to 4 powder ports leads to a marginally increased roughness, while increasing the powder feet rate up to 400 g/min has no negative effect on the roughness; the smallest value was determined for the coating sprayed with the novel injector block and a feed rate of 300 g/min.

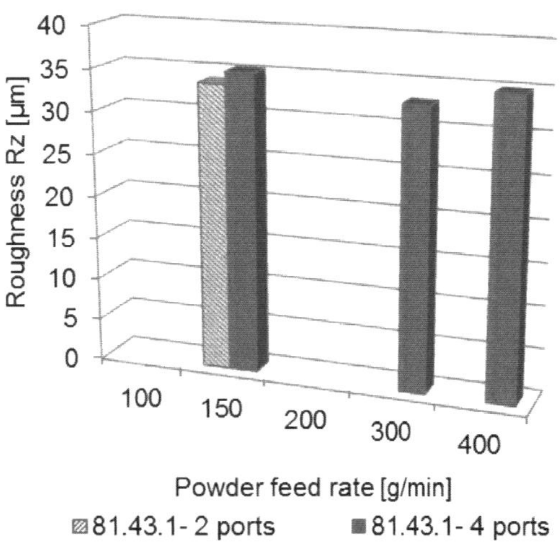

Figure 11: Average roughness Rz of the iron based 81.43.1 coatings vs type of powder injector block and powder feed rate

4 Summay and Conclusion

Within the frame of this study an increase in deposition efficiency for all three metal powders as well as for two out of three cermet powders (80.81.1 and 80.76.1) was proven when upgrading the K2 spray gun with the novel Quattro injector block. For the metal powders the increase is 2 – 3 percentage points. For abovementioned cermet powders an improvement of 4 – 5 percentage points was determined, when processing the powders at standard feed rates.

Furthermore, it was shown for 5 out of the 6 powders that the deposition efficiency rises with an increasing powder feed rate. Just for 316L (80.46.1) spray powder a drop down of the DE by 3 percentage points to 65 % could be observed when spraying at extreme powder feed rates of 300 and 400 g/min. The DE reached for these two parameter settings matches the value that was observed when using the standard K2 and adjusting a spray rate of 100 g/min.

Based on these results it can be stated that upgrading the K2 spray gun with the 4 port powder injector block offers significant potential to reduce powder related coating costs and, to a lesser degree, also time-related coating costs.

Concerning coating quality, only minor differences between caotings sprayed with the two powder injector blocks are are noticable. One should consider, that no spray parameter optimization for the novel Quattro injector block was carried out so far. All parameter settings that were used within the frame of this work originally were developed for the standard set-up. As expected, a drop down in coating quality appears when spraying at highly elevated spray rates.

In any case, if not the best possible coating quality is aimed for, extremely high spray rates feasible when upgrading the K2 spray gun with the novel Quattro powder injector block. Future work will address improving the coating quality at strongly elevated pray rates.

A hybrid thermal spray and sinter fusion process to apply wear and corrosion resistant coatings

Matthew Yao, Hakim Belhadjhamida, Kennametal Stellite, Inc./Canada
David Lee, Jack Zheng, Kennametal Inc./USA

Cobalt-based Stellite™ alloys such as Stellite 720 with nominal composition of Co-33Cr-18Mo-2.45C are known for their excellent wear and corrosion resistance. Applied as thin coatings, these alloys can be used for applications such as balls and seats used in the valve industry. Stellite 720 is difficult to apply by HVOF thermal spraying and to achieve fusion bonding by sintering. A hybrid HVOF thermal spray and sinter fusion process is developed to realize a thick coating with thickness greater than 0.025" (0.6 mm). In order to improve the spraying process and fusibility, the Stellite alloy composition is modified. Unlike the HVOF coating alone, the coatings formed by thermal spraying and sinter fusion are metallurgically bonded to the substrate and form a substantially continuous alloy coating on the surface of the component body. The microstructure of the hybrid processed coatings are examined and compared to HVOF coating microstructure of the same alloy. The wear and corrosion resistance of the hybrid processed coatings are evaluated. The coatings showed high hardness exceeding 60 HRC with excellent wear and corrosion resistance. Balls and seats are coated for valve applications.

1 Introduction

The commercial wear resistant Stellite alloys are derived from the Co-Cr-W-C family first investigated by Elwood Haynes in early 1900s [1]. Hardfacing alloys of the Co-Cr-W-C type exist in several modifications and it is generally known that the available range of commercial grades satisfies most of the industry requirements. However, the high carbon Co-Cr-W-C alloys may occasionally be inadequate especially where components of pumps, impellers, etc., must often withstand the simultaneous abrasive and corrosive action of media composed of a suspension of hard mineral particles in an aqueous solution. Such failures are encountered, for instance, in the superphosphate industry [2]. Early research [2] showed that adding Mo, singly or in combination with Ni and/or Cu, enhances greatly the corrosion and abrasion resistance of high carbon Co-Cr-W-C cast alloys. United States patent [3] has provided saw tip for attachment to saw blades for cutting wood. The saw tips formed from Co-Cr-Mo-C type Stellite alloys have demonstrated better wear and corrosion resistance compared with those made from conventional Co-Cr-W-C type Stellite alloys during wood cutting services. The Co-Cr-Mo-C type Stellite 700 series alloys developed at Kennametal Stellite, Inc. raise the standards for wear and corrosion resistant alloys [3-5]. These Stellite alloys have the unusual combination of excellent wear resistance and exceptional corrosion resistance in environments that are either reducing or complex.

Co-based Stellite alloys such as Stellite 720 with nominal composition of Co-33Cr-18Mo-2.45C are known for their excellent wear and corrosion resistance [3-5]. The Stellite 720 alloy can be used as a coating applied by slurry coating and can also be produced to be components by powder metallurgy process. A slurry coating process called Ultraflex [6], which is based on a powder metallurgy process where coatings are applied to substrates in a "green" state and then heated to fuse the coating into a dense, uniform, metallurgically bonded layer with almost no porosity, has been developed at Kennametal Inc. While the Ultraflex Stellite 720 thin coatings are readily available, the Stellite 720 is difficult to apply by HVOF thermal spraying and achieve fusion bonding by sintering. Trials have been conducted to apply Stellite 720 on Inconel 718 by HVOF process and showed that this coating does not have a high adherence to substrate. A basic concept in thermal spray is that there is a coating thickness limit for spraying many coating compositions on a given substrate. There are some tricks to be able to spray thicker coating, such as adding a WC/17Co layer as a "bond" coat for many Stellite alloys, or to spray the coating so as to create the coating in very high compressive state.

In order to improve the spraying process and fusibility and to manufacture coatings greather than 0.025" (0.6 mm), the Stellite 720 alloy composition is modified. Coatings formed by hybrid thermal spraying and sinter fusion are metallurgically bonded to the substrate and form a substantially continuous alloy coating on the surface of the component body and can be applied significantly thicker than previous versions of Stellite 720 alone.

2 Thick and thin coatings and claddings

2.1 Conforma cladding

To manufacture even thicker coatings/claddings for the difficult to apply alloys/materials, Kennamtal has a technology to offer a cloth process/product called Conforma Clad/Cladding [7]. The Conforma cladding thickness of greater than 0.020" (0.5 mm) can be readily achieved.

There are two general families of Confoma Clad to apply thick coatings/claddings. One contains tungsten carbide in a nickel-chrome matrix (**Fig. 1**). There are 3 primary grades WC200, WC210, WC219 which contain about 62%, 55% and 48% respectively tungsten carbide, uniformly distributed throughout the nickel-chrome matrix. Having the hardness in Rockwell C scale about 58-66, 54-60 and 50-56, respectively, the 3 tungsten carbide cladding grades provide exceptional wear and erosion protections. The carbide particles are finer than what is typically seen in hardfacing alloys. Flexible tungsten carbide cloth can conform to complex geometries (see the insert in **Fig. 1**). During the furnace cycle, the nickel-chrome alloy melts to fill in any porosity within the carbide layer and form a metallurgical bond to the underlying substrate.

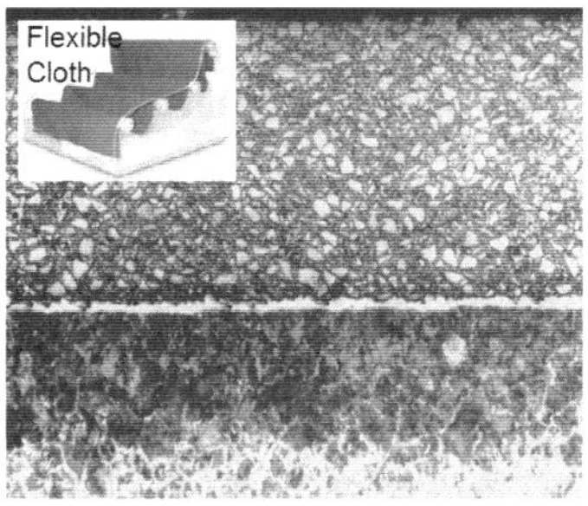

Fig. 1. A typical Conforma Clad WC grade applied to steel substrate.

The second family of Conforma Clad is alloy based cladding such as those based on Stellite and Tribaloy cobalt alloys (**Fig. 2**). For the cobalt alloys, instead of melting, the cloth is sintered during the furnace cycle. The cobalt alloys have less wear resistance than the carbide based coatings but are still more wear resistant than most substrate materials. The cobalt alloys have the advantage of having very good corrosion and high temperature oxidation properties. Consequently, they can be used in many areas where a good combination of wear, corrosion, and oxidation is needed.

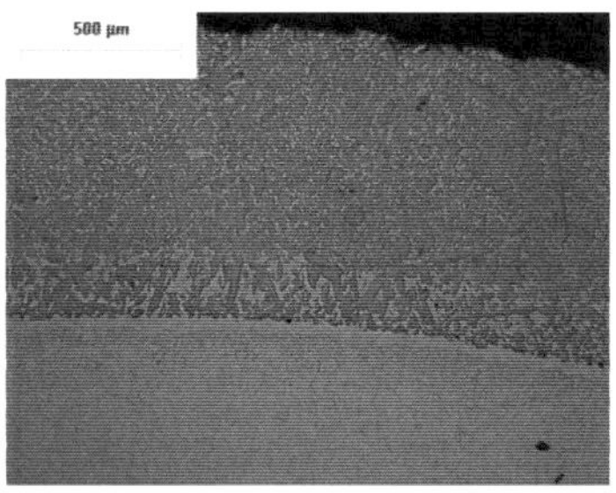

Fig. 2. A typical Conforma Clad cobalt alloy grade (modified Stellite 720) applied to steel substrate.

2.2 Ultraflex coating

In the Ultraflex™ process, carbide and/or alloy powders are mixed with a liquid medium to form a slurry or paste. This slurry can then be applied to parts either through mechanical means such as dipping, brushing, and troweling or by flow coat methods. Once the slurry is applied it is then dried to remove the excess moisture. Multiple coats can be applied to achieve the desired thickness of the final coating, typically in the 0.005" – 0.030" (0.12 - 0.75 mm) range. Once the coating has dried, the entire part is then heated in a vacuum furnace to fuse the coating. Fusing temperatures are typically 1000-1200°C (1830-2190°F) depending on the alloy. The high fusing temperatures will anneal most steel alloys. Subsequent heat treatment of the part can be performed to restore the mechanical properties of the underlying steel substrate if desired. A typical Stellite 720 alloy ultraflex thin coating is shown in **Fig. 3**. There is a true metallurgical bond with little cross dilution between the substrate and the coating. In addition to the bond between the substrate and coating, the bond between the Stellite alloy particles is also metallurgical. The dilution in the coating by the substrate is minimized.

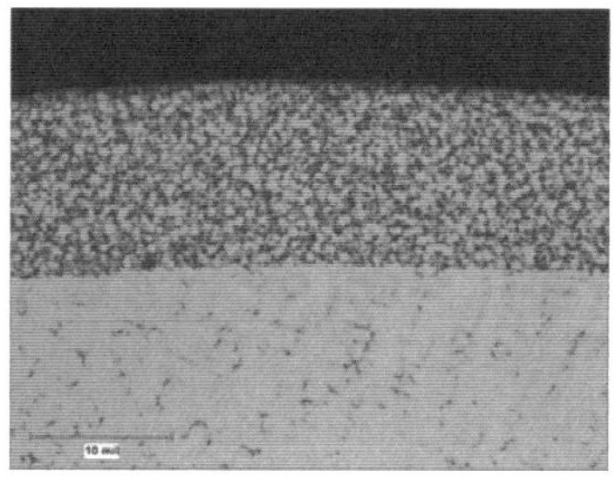

Fig. 3. Ultraflex Stellite 720 alloy thin coating with thickness of 0.012" (0.30 mm).

The 0.041" (1 mm) thick Ultraflex Stellite 720 alloy coating has been applied on a ball valve made of 316 stainless steel (**Fig. 4 & Fig. 5**). Cross-section shows adherent metallurgical bonding (**Fig. 5**). However, this thickness Stellite 720 Ultraflex thick coating showed surface divot/ fissure formation (**Fig. 4**).

Fig. 4. Thick Ultraflex Stellite 720 alloy coating with thickness of 0.041" (1 mm).

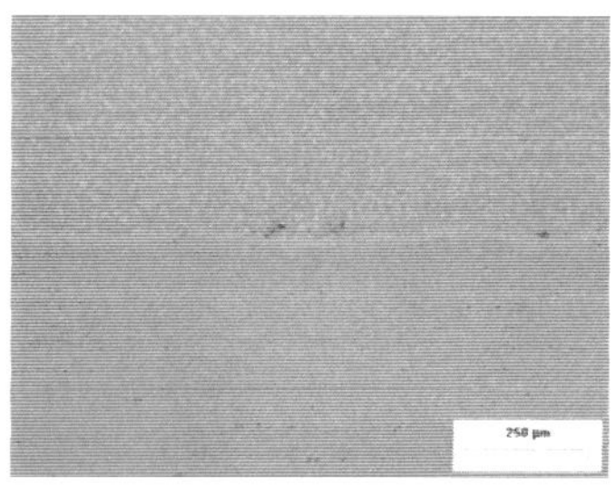

Fig. 5. Thick Ultraflex Stellite 720 alloy coating shows adherent metallurgical bonding.

2.3 HVOF thermal spray coating

Sound HVOF thermal spray coating of Stellite alloys such as Stellite 6 (JK576) can readily be applied. We sprayed Stellite 720 on 1 inch (25 mm) Inconel 718 cylinders with the same flame but with two different surface speeds and steps - 60 ft/min – step 0.10 inch – target thickness 0.040 inch (**Fig. 6**) and 300 ft/min – step 0.10 inch – target thickness 0.020 inch (**Fig. 7**). The powder feed rate was 21 g/min. Both coatings showed full delamination (see **Fig. 6** and **Fig. 7**).

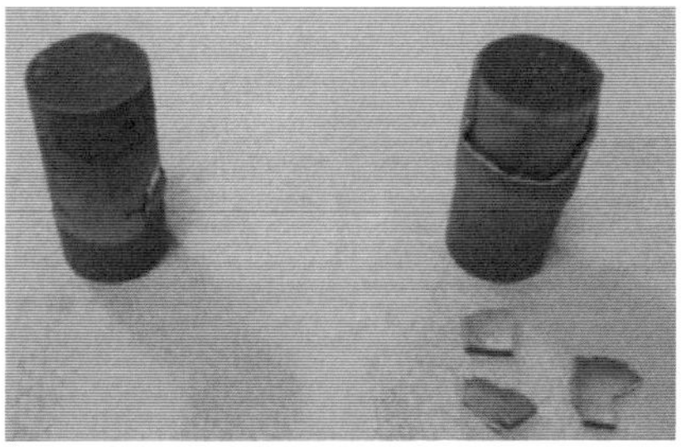

Fig. 6. Photograph of the two samples sprayed to a target thickness of 0.040" (1 mm).

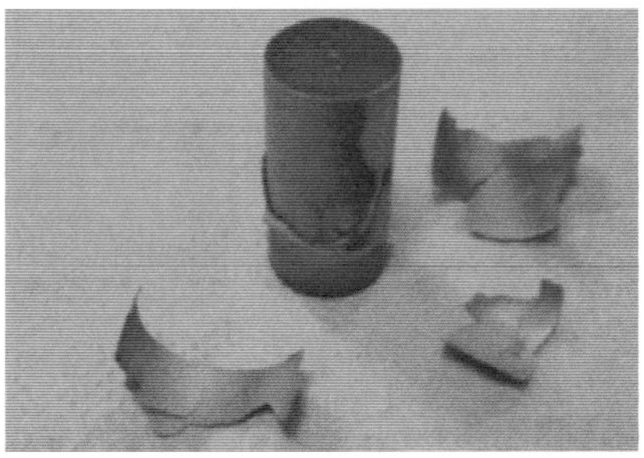

Fig. 7. Photograph of one sample sprayed to a target thickness of 0.020" (0.50 mm).

Stellite 720 was also sprayed onto 0.5" (12.7 mm) diameter AISI 1018 carbon steel shaft by HVOF thermal spray using the Stellite 6 (JK576) parameters. The attempt was another failure and showed coating cracks (**Fig. 8**). The experiments indicated that Stellite 720 alloy is not very friendly to the HVOF thermal spray process.

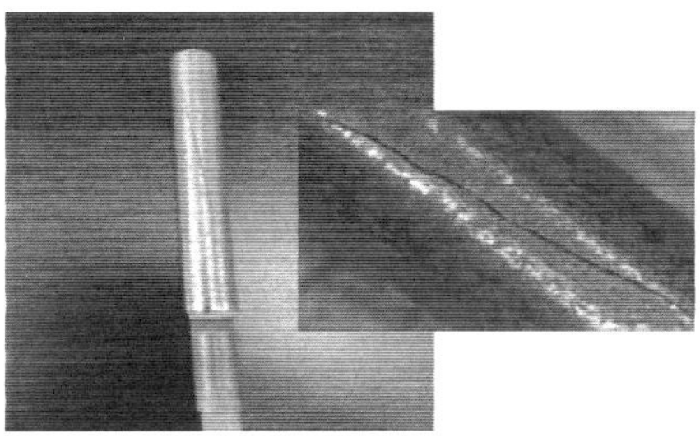

Fig. 8. As-Sprayed Stellite 720 on Inconel 718 shows cracks – as received, Sample ID 61014-4, Sprayed like Stellite 6 (JK576).

2.4 Hybrid thermal spray and sinter fusion coating

Valve industry often utilizes HVOF process to spray wear resistant coatings on balls and seats. Stellite 720 is known for its excellent wear resistance for the ball and seat applications. However, Stellite 720 is not quite friendly to HVOF process and currently only Ultraflex coating and cloth cladding methods are employed to form this alloy into a useful application. To use Stellite 720, Ultraflex slurry method was utilized but limited to thinner coating (<0.025" /0.64 mm). Conforma cloth cladding Stellite 720 on balls and seats also proved difficulties due to the part geometry. In order to coat thicker and process the part faster, thermal spray method is best suitable. The idea is to utilize spray and fuse method to obtain thicker coatings of difficult to form alloys such as Stellite 720 for balls and seats applications used in valve industry. The objective of this study is to develop processing parameters for spraying and fusing Stellite 720 on various substrates with coating thickness between 0.025" to 0.055" (0.64 mm to 1.40 mm). There are many uncertainties for the thick coating development. Stellite 720 is difficult to spray and fuse directly and the coating quality depends on the substrate material and geometry. This study will develop alloy composition, processing parameters and products for valve industry.
In order to improve the spraying process and fusibility, the Stellite alloy composition is modified. The spray and fusible alloy is obtained with the addition of nickel (Ni), boron (B) and silicon (Si). The modified Stellite alloy is called Alloy 720Mod. The addition of boron (B) and silicon (Si) in Stellite alloys lowers the melting points and increases the melting range. The solidus and liquidus temperatures and the melting range of the Stellite 720 and Alloy 720Mod alloys are shown in **Table 1**. The lower melting points and the wide melting range are advantageous to material's fusibility. The spraying ability and fusibility of the modified alloys increase with the amount of (B+Si)% additions. Some of the published properties for Alloy 720Mod, as an emerging material, are estimated and preliminary results such as the data are shown in **Fig. 9**. **Fig. 9** illustrates that the fusibility of Co-Cr-Mo-C alloys increases with the increase in (B+Si) contents. It is expected that the modified Alloy 720Mod will have better spraying ability and fusibility than Stellite 720 alloy.

Table 1. The solidus and liquidus temperatures and the melting range of the Stellite alloys

Alloy	Soildus, °C	Liquidus, °C	Range, °C
Stellite 720	1182	1228	46
Alloy 720Mod	1107	1165	58

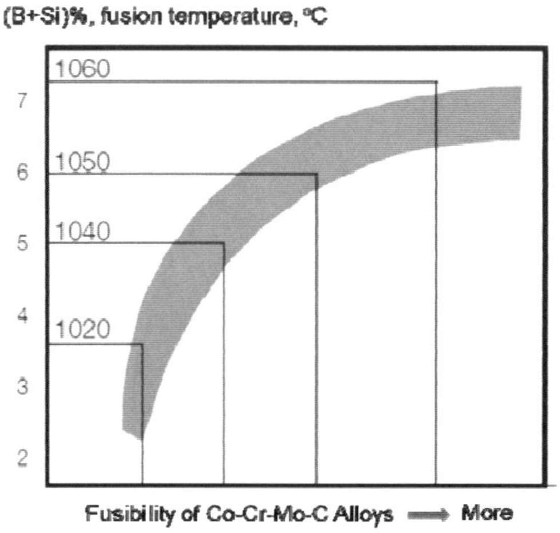

Fig. 9. The influence of B and Si contents on the fusibility of Co-Cr-Mo-C alloys.

Series of Alloy 720Mod samples were thermal sprayed to demonstrate our capability to spray and fuse high hardness coating on Inconel 718 material. A sprayed coupon and the fused coupon are demonstrated in **Fig. 10**. All samples sprayed with any parameters tested were free of cracks and delamination, indicating a large process window for spraying. Fusing was done at various temperatures, followed by furnace cooling to the aging temperature, and furnace cooling to under 500°F (260°C). The fusing temperature is chosen from 1900°F (1038°C) to 2100°F (1150°C). With this fusing temperature, the coating was of good quality, free of leakage (sagging), and was easy to polish which should give additional benefits for grinding and polishing. Typical microphotographs are shown in **Figs. 11** to **13**. One inch dameter Inconel 718 coupon was also coated by the hybrid thermal spray and sinter fusion process crack free and showed very smooth surface finish (**Fig. 14**). Inconel 718 balls are then coated (**Fig. 15**) which demonstrated a very smooth coated surface.

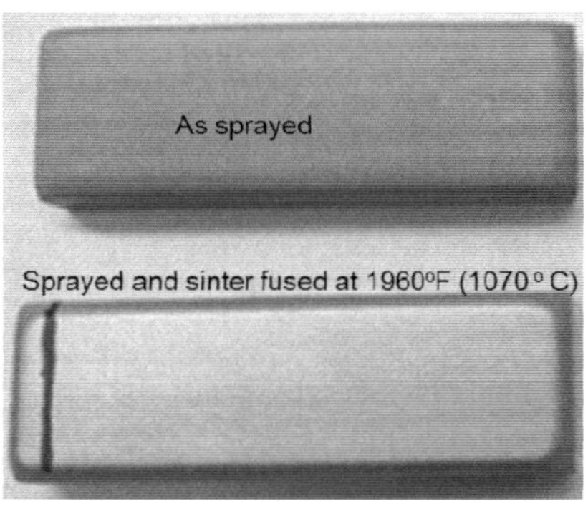

Fig. 10. Sprayed and the sprayed & fused Alloy 720Mod coupons.

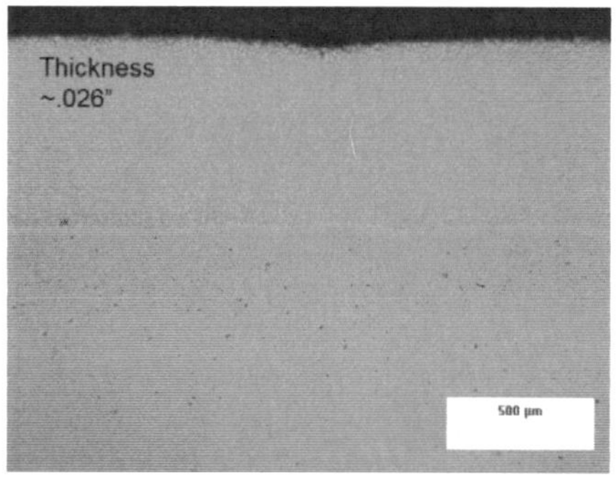

Fig. 11. Hybrid processed Alloy 720Mod coating on Inconel 718 (0.026" / 0.7 mm).

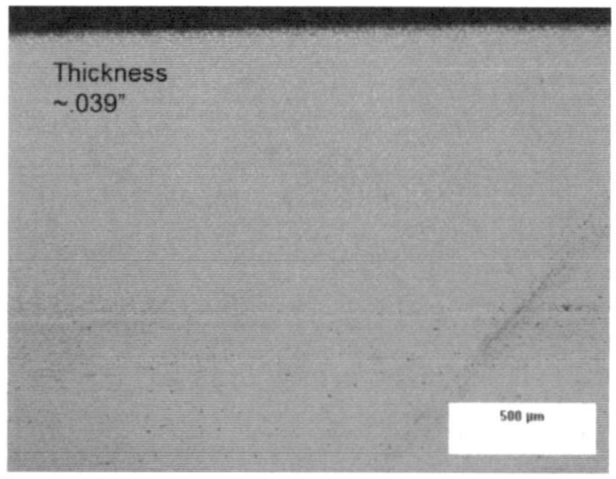

Fig. 12. Hybrid processed Alloy 720Mod coating on Inconel 718 (0.039" / 1.0 mm).

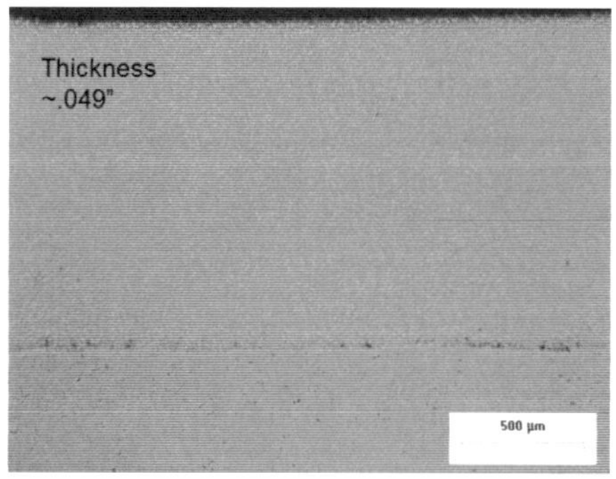

Fig. 13. Hybrid processed Alloy 720Mod coating on Inconel 718 (0.049" / 1.2 mm).

Fig. 14. Alloy 720Mod As-sprayed onto IN718 – Crack free.

Fig. 15. An Inconel 718 ball coated by hybrid thermal spray and sinter fusion process.

3 Discussion

Alloy 720Mod is one of the hardest Stellite alloys, harder than cast, Conforma Clad and Ultraflex versions of Stellite 720. **Fig. 16** and **Table 2** demonstrated the high hardness of Alloy 720Mod hybrid coating. The Vickers hardness in hybrid processed Alloy 720Mod coating is measured about 600 to 950 (**Fig. 16**). The hardness in Rockwell C scale on top of the hybrid processed Alloy 720Mod coating is measured to be about 62 as compared to 59 and 54 for the two high hardness commercial spray and fusion nickel-base hardfacing alloys Colomonoy 88 (Ni-15Cr-15.5W-0.6C-3B-4Si) and Deloro 75 (Ni-16Cr-2.5Mo-0.7C-3.5B-4.5Si), respectively.

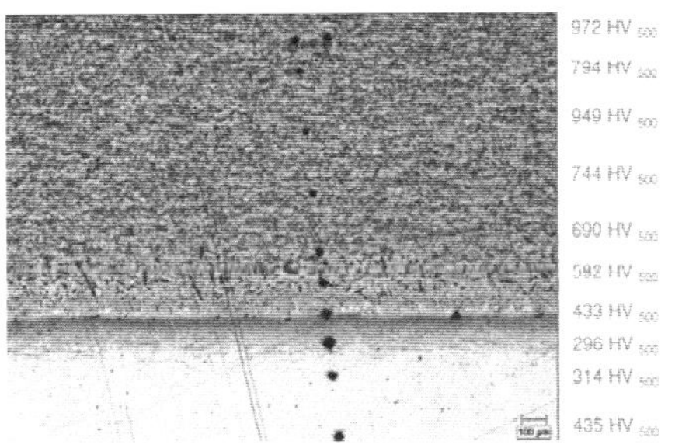

Fig. 16. The Vickers hardness of hybrid processed Alloy 720Mod coating.

Table 2. Hardness of Alloy 720Mod and commercial spray & fusion alloys Colomonoy 88 and Deloro 75

Hardness, HRC	#1	#2	#3	#4	#5	Average
Alloy 720Mod	62.1	59.5	62.9	62.8	62.1	61.9
Colmonoy 88	59.5	58.8	59.3	59.7	59.2	59.3
Deloro 75	54.1	54.5	54.6	54.5	54.4	54.4

Stellite 6 (Co-29Cr-4.5W-1.1C) is the most widely used cobalt base alloy. Consequently, it has recieved more academic research and in general it exhibits good all-around performance. It has excellent resistance to many forms of mechanical and chemical degradation over a wide temperature range, and retains a reasonable level of hardness up to above 800°C (1470°F). The abrasion wear resistance of the Stellite 720 and Alloy 720Mod was compared to that of Stellite 6 in **Fig. 17**. The Stellite 720 and Alloy 720Mod have 3 times and 4 times better abrasion wear resistance than Stellite 6. Thus the hybrid processed Alloy 720Mod can provide better wear protection than Stellite 6 alloy can.

W. Schwarz and H. Warlimont studied the ranges of amorphous alloy formation of Co-Cr-W-B-C and Co-Cr-Mo-B-C [8]. Results showed that the Co-based alloys with boron are superior to those with carbon only for abrasion wear protection. The boron modification in Alloy 720Mod could also provide the superior wear resistance.

As it is mentioned in section 2.1, the 3 Conforma Cladding grades WC200, WC210 and WC219 are among the most erosion resistant materials available in industry. With similar hardness, Alloy 720Mod (HRC62) has similar erosion wear resistance as WC219 (HRC50-56) (**Fig. 18**).

The effect of alloying elements of Cr, Mo and W on the crystallization temperature and hardness of Co-Cr-Mo-C and Co-Cr-Mo-W-C was examined by Inoue et al [9]. These amorphous alloys are limited to the carbon content between about 13 and 23 at%. The ranges of the alloying elements Cr, Mo and W were also determined. The hardness of these alloys increases with increasing alloying elements and the highest value attained is about 1450 DPH. It may be deduced that the high hardness of Alloy 720Mod may be partly attributed to the amorphous microstructure.

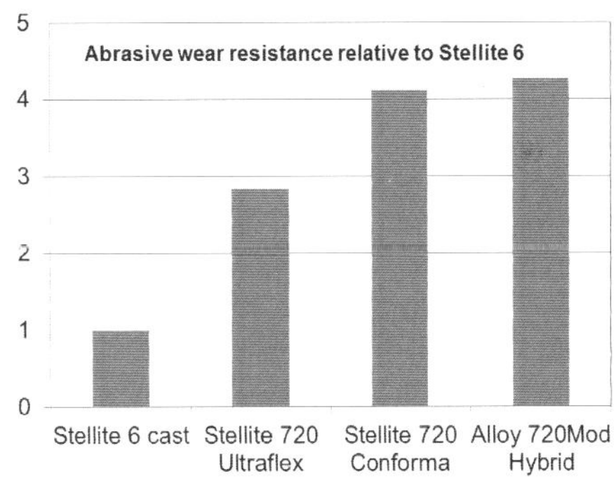

Fig. 17. Abrasion wear resistance of Stellite alloys

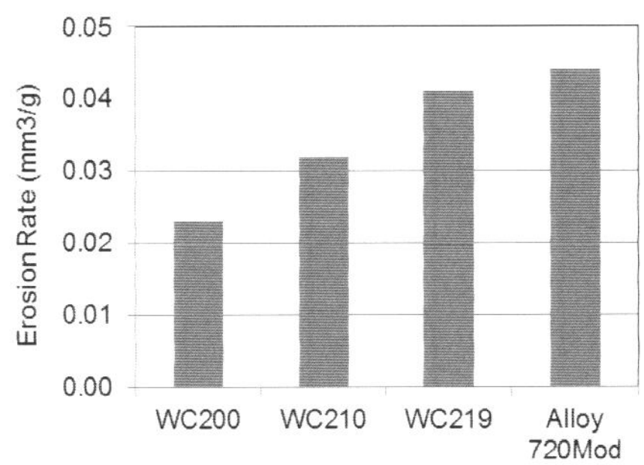

Fig. 18. Erosion wear resistance of Alloy 720Mod compared to the most wear resistant Conforma Claddings

Oliveira investigated Co-based Stellite SF20 (Co-19.5Cr-15W-13Ni-1.3C-3B-3Si) laser cladding of about 0.020" (0.5 mm) thick layer [10]. From the analysis of the various hardening mechanisms it can be concluded that the formation of the metal carbides $M_7(C,B)_3$ and $M_{23}(C,B)_6$, and to a smaller extent the presence of DO_{19} ordered Co_3W precipitates are the predominant factors determining the overall mechanical properties of laser coated Stellite SF20. Similarly, the metal carbides $M_7(C,B)_3$ and $M_{23}(C,B)_6$, and Mo-Si-rich intermetallic phases have strengthened the Alloy 720Mod coatings.
The corrosion resistance in aqueous hydrochloric acid and sulfuric acid solutions was determined for WC210 Conforma clad and Ultraflex Stellite 720 compared to Stellite 6 (**Fig. 19**). It is indicated that Co-Cr-Mo-C Stellite 720 alloy has the best corrosion resistance.

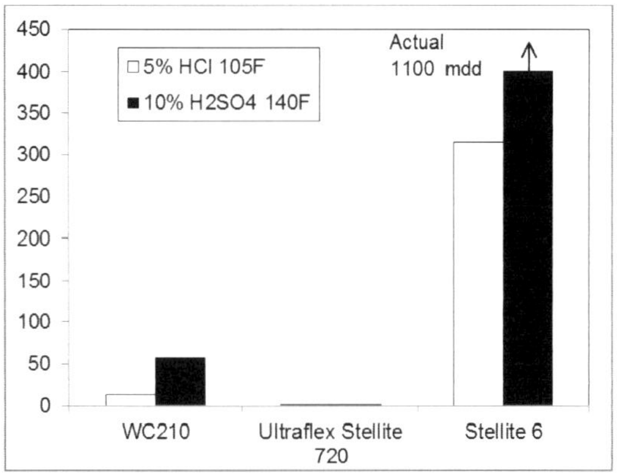

Fig. 19. Corrosion rate expressed in mdd in aqueous hydrochloric acid and sulfuric acid solutions

In summary, the good quality, crack free and smooth surface finish of Alloy 720Mod coating can be achieved by thermal spray and sinter fusion hybrid process. The optimium coating thickness is realized to be about 0.025" to 0.055" (0.64 mm to 1.40 mm) (**Fig. 20**).

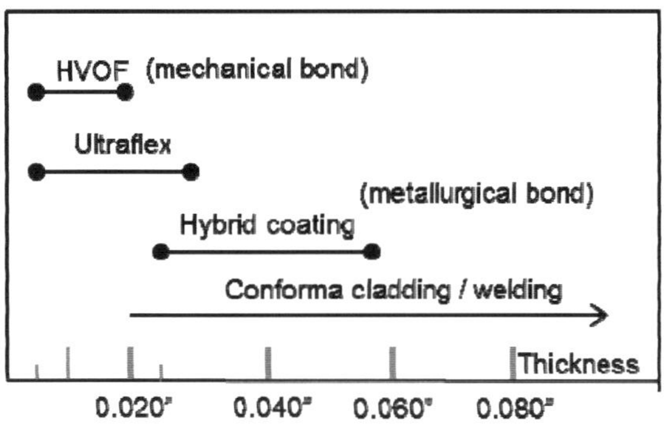

Fig. 20. The achievable Alloy 720Mod coating thickness by thermal spray and sinter fusion hybrid process.

4 Summary

This work introduces a boron and silicon modified Co-Cr-Mo-C Stellite alloy and a hybrid thermal spray and sinter fusion process that manufactures a coating with thickness greater than 0.025" and hardness greater than 60 HRC. The hybrid process significantly expands the manufacture window for wear resistant coatings. The Stellite coatings can be applied to many geometries that are not possible by traditional thermal spray and weld overlay methods. The hybrid process produces coatings that are dense, uniform and metallurgically bonded to the underlying substrate metals. Cobalt-based Stellite alloy coatings have good wear resistance combined with excellent corrosion and high temperature resistance for applications involving chemical attack and oxidation in addition to erosive wear. The hybrid coating process, the Ultraflex process, the HVOF thermal spray process, the Conforma cladding and the welading hardfacing form the wide range of coating and cladding solutions which Kennametal offers to its customers.

4 Literature

[1] Elwood Haynes: Metal alloy, US patent No. 873,745, Dec. 17, 1907.
[2] J.M. Drapier, A. Davin, A. Magnee, D. Coutsouradis, L. Habraken: Abrasion and corrosion resistant cobalt base alloys for hardfacing. Wear 33 (1975) pp. 271/82.
[3] J.B.C. Wu, D. Raghu, B. McKee: Saw blade tips and alloys therefor, US patent No. 6,479,014, Nov. 12, 2002.
[4] Stellite 700 Alloys – A new Stellite alloy series that resists wear and corrosion, Deloro Stellite Alloy Datasheet No. 2050.
[5] M.X. Yao, J.B.C. Wu, Y. Xie: Wear, corrosion and cracking resistance of some W- or Mo-containing Stellite hardfacing alloys. Materials Science and Engineering A407 (2005), pp. 234/44.
[6] D. Kimball, J. Faust and J. Dezelle: UltraflexTM coatings for protection of non-line of sight surfaces. ITSC 2015, May 11-14, 2015, Long Beach Convention & Entertainment center, Long Beach, CA, USA.
[7] E. J. Breton, J. M. Handzel and O. K. Tennant: Process for making wear-resistant coatings, US patent No. 6,649,682, Nov. 18, 2003.
[8] W. Schwarz and H. Warlimont: A new series of Co-based amorphous alloys and their applications as cladding materials. Materials Science and Engineering A266-228 (1997, pp. 1098/1101.
[9] U.O.B. de Oliveira: Chapter 4 An Engineering System: A Cobalt-Base Alloy in Laser Processing of Alloys: Processing, Microstructure and Structural Properties. MSC PhD thesis series 2007-01 ISSN 1570-1530.
[10] A. Inoue, T. Naohara and T. Masumoto: Crystallization temperature and hardness of new type cobalt-based amorphous alloys. Trans. JIM Vol20 (1979), pp. 329/33.

Coatings Deposited Using a Valve-less Detonation System

Rajan Bamola†, Thomas Ewell†, Paul Robinson†, and Robert Gansert§
† Surface Modification Systems, Inc., Santa Fe Springs, CA 90670 USA
§ Advanced Materials & Technology Services, Inc., Simi Valley, CA 93065 USA

Detonation (D-Gun) Spraying is one of the oldest yet least utilized thermal spray technologies in spite of these coatings being known to be harder and better adhered then those applied via plasma and even HVOF methods. Some of the reasons for the lower popularity of D-Gun spraying are the more cumbersome equipment, lower productivity and higher complexity of interaction between the equipment and powder. Much of this is due to the use of mechanical valving to feed in combustion gases and therefore necessitating the use of intermittent powder flows. In this study we report on a new valve-less detonation spraying device allowing the use of continuous powder feeding and capable of higher detonation frequency than through conventional valved D-Gun systems. Previous valve-less systems were unreliable in terms of firing and safety. Tungsten carbide cobalt chrome (86WC:10Co:4Cr) was used to investigate the process-coating interactions during detonation spraying of tungsten carbide.

1 Introduction

Detonation spraying (D-Gun) is a relatively old but little utilized technology considering its advantages of superior bond, density and hardness when compared against HVOF and plasma techniques, particularly when applying carbides. Other advantages of the D-Gun over HVOF are the lower operating cost and lower heat input into the substrate. Therefore, this technology has the potential of offering superior coatings at lower cost. The traditional D-Gun design consists of a cannon like water cooled tube with mechanical valves for gas metering and a spark plug close to the blind end [1]. A water cooled powder feed tube is usually placed slightly in front of the valves. Combustion mixtures typically of acetylene and oxygen are fed in the tube, ignited via the spark plug and the resulting detonation front carries the powder through the tube and towards the substrate. A nitrogen purge is utilized between firing to prevent backfiring. In contrast, a valve-less D-Gun system eliminates the mechanical valves but retains the spark plug and powder feed tube. Generally, gases are fed through a specially designed path where the gases mix and are fed into a combustion chamber [2]. This design prevents back flow into the gas train, therefore, nitrogen purging can be eliminated. One advantage of such a system is less dangerous propane can be used as the combustion gas rather than acetylene. Additionally, the firing rate and thus productivity can be increased as long as sufficient gas flow can be maintained to sustain ignition. Another key advantage over conventional D-Guns' is that powder can be fed continuously using standard high pressure powder feeders. An important differentiation between the two powder feed methods are that in the valve-less system air is used as the carrier gas not nitrogen. This becomes important in spraying of oxides and carbides. D-Gun systems can be viewed as a high velocity intermittent thermal spray technique. This paper reports on work conducted using a valve-less system.

Tungsten carbides are an industrially important and widely studied thermal spray coating material with abundant literature describing process-material-properties effects [3,4]. Tungsten carbide cobalt chrome, WC-CoCr (86:10:4), is widely used in aerospace and petroleum industry, therefore, this study concentrated on characterizing this composition using the valve-less detonation gun process.

All tungsten carbide-cobalt based thermal sprayed coatings are very sensitive to process conditions. In particular, oxidation can have a profound effect on the phase composition and thus the hardness and brittleness of the coating. Since the reaction

$$2C + O_2 = 2CO$$

is more thermodynamically favorable then oxidation of cobalt or tungsten [5], brittle phases such as W_2C and $Co_xW_yC_z$ (eta phase) are promoted. Furthermore, as the cobalt binder melts (MP = 1495ºC), WC can rapidly dissolve in it. The solubility of tungsten in cobalt is inversely related to the carbon content. Therefore, decarburization promotes the dissolution of tungsten in cobalt. Fernandes and Senos describe the phase transformations in cemented carbide systems in their review paper for the interested reader [6]. Decarburization and dissolution reactions lead to lower WC containing coatings resulting in reduced hardness and fracture toughness. Since valve-less D-Guns' have extended barrels (up to approximately 1500 mm) with oxygen rich environments, such reactions can be intensified. Decarburization and dissolution reactions are temperature dependent, therefore, the choice of fuel gas is of great importance. Acetylene and propane oxygen flame temperatures are 3480°C and 2526°C respectively, suggesting that valve-less systems may have an advantage in the spraying of tungsten carbides. In fact, patent US4902539 granted to Union Carbide Corporation attempts to solve this problem for the valved system by diluting acetylene with propylene to lower the combustion temperature while maintaining peak pressures [7]. Diluting with argon or nitrogen lowered the peak pressures faster than peak temperature thus lowering particle velocities while still maintaining high temperatures. This work reports the effects of oxygen, propane ratios, flow rates and combustion (shot) frequency on the hardness and microstructure of WC-CoCr coatings. Coating quality was determined on the basis of existing industrial specifications. AMS 2448 is usually the governing specification for HVOF tungsten carbides while Detonation Gun specifications are not widely found, MIL-HDBK-1886(AT) provides some guidelines as to the expected hardness and microstructure. AMS 2448 lists minimum hardness of 950 DPH while MIL-HDBK-1886 requirements are 1000-1150 DPH. Both standards require porosity of less than 1% [8,9]. In consideration of the above, in our case the acceptable microhardness was set at 1200 DPH_{300} with porosity less than 1% to demonstrate that industrial and military requirements can be met or exceeded using the current valve-less detonation system.

2 Experimental

All spray experiments were conducted using a robotically mounted system with a simple raster program to minimize process variations. Gas flows were controlled using mass flow meters. All aspects of the process sequence were controlled via a PLC. A cast and crushed powder sized from -44+15 microns was used. The powder characteristics are shown below, **Fig. 1**.

Fig. 1. WC-CoCr Cast and Crushed Powder

Coating characteristics were determined via optical and scanning electron microscopy. Porosity was determined using Paxit image analysis [10]. Hardness testing was conducted using a Phase II 900-392 Vickers Microhardness tester. The built in software is capable of statistical analysis. **Table 1** list the process conditions and sample designations for the experiments. The feed rate was kept constant. X-Ray Diffraction (XRD) analysis was conducted on the samples prepared in Table 1. The X-Ray Diffractometer was model, Lab XRD-6000, Shimadzu (Japan). The XRD diffraction patterns were analyzed with International Centre for Diffraction Data (ICDD), Release 2012.

Table 1 Experimental Conditions

Sample #	Distance (mm)	Oxygen/Propane (slm)	Oxygen/Propane ratio	Shot Frequency
1	300	72/12	6:1	2
2	250	48/12	4:1	2
3	250	64/16	4:1	4
4	250	60/18	3.3:1	2
5	250	63/18	3.5:1	2
6	250	70/20	3.5:1	2
7	250	77/22	3.5:1	2
8	200	82/23	3.5:1	8
9	250	82/23	3.5:1	8
10	300	82/23	3.5:1	8

3 Results

In the valve-less system the powder feed rate is constant and the shot frequency merely distributes the powder in larger or smaller increments per unit time depending on the respective frequency. In **Fig. 2** is a bar chart of microhardness (DPH_{300}) plotted as a function of test conditions. Samples 8-10 were generated at higher shot frequencies.

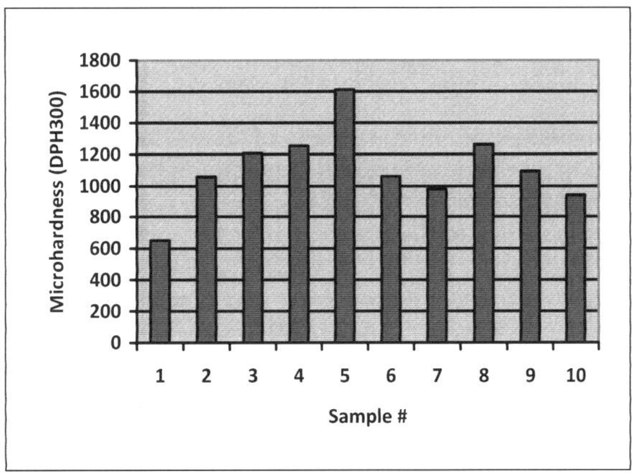

Fig. 2. Microhardness as a Function of Test Conditions

At the higher frequency, the 63/18 oxygen-to-propane flow rate could not sustain the flame adequately. Therefore, the flow rates were increased to 82/23 while maintaining the fuel-oxygen ratio. **Fig. 3** illustrates the relative deposition efficiency at stand-off distances from 200 to 300 mm. The deposition efficiency is observed with an increase in coating thickness with a change in stand-off distance. All tests were conducted at 8 shots/second.

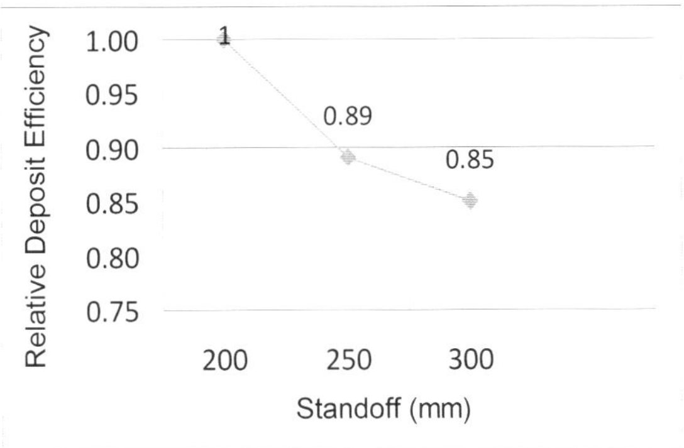

Fig. 3. Relative Deposit Efficiency as a function of Stand-off Distance

Additional samples were prepared to compare efficiencies at 4, 6 and 8 shots per second at the optimized ratio of 3.5:1 oxygen-to-propane (**Fig.4**). The flame stability could be maintained at a ratio of 63:18 oxygen-to-propane and for flow rates of 4 and 6 shots per second, but not at 8 shots/second. As mentioned above, at 8 shots/sec flame stability required flow rates of 82:23 slm oxygen-to-fuel for 8 shots per second.

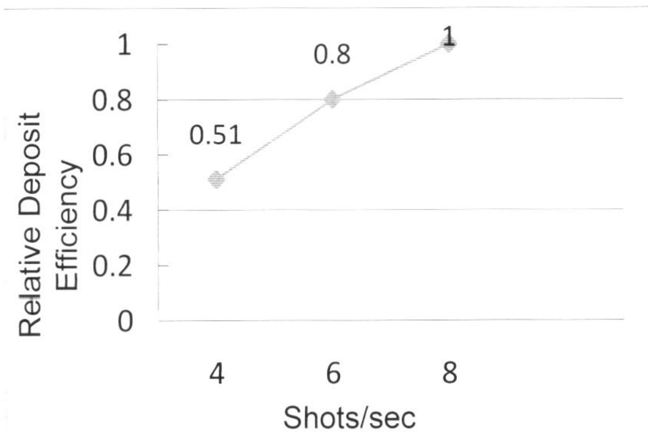

Fig. 4. Relative Deposit Efficiency as a function of Shot Frequency

While coating density is exceptionally high (e.g., +99%) in all cases, there exists microstructural differences (e.g., degree of decarburization) between the coatings as a function of the changes in detonation parameters. Scanning electron microscopy was conducted on all the test samples. **Figs. 5a** and **5b** show the back scattered images of the softest and hardest coatings, respectively.

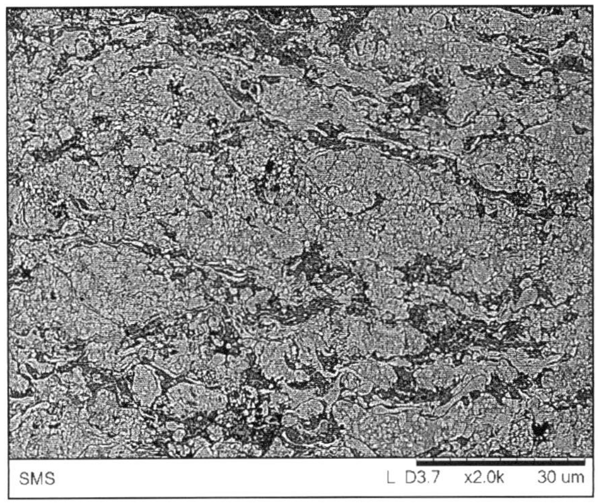

Fig. 5a. SEM Back Scatter Image, Test Sample #1, Cross-Section

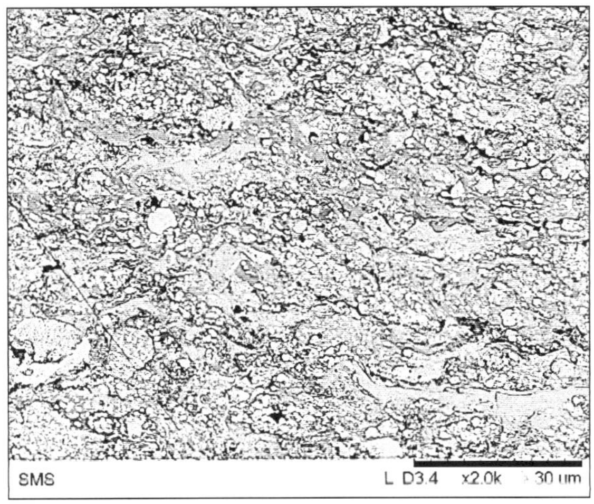

Fig. 5b. SEM Back Scatter Image, Test Sample #5, Cross-Section

X-Ray Diffraction (XRD) analysis was conducted on the samples prepared in Table 1, using a Shimadzu, model XRD-6000, diffractometer. The XRD diffraction patterns that were generated from these samples were analyzed with International Centre for Diffraction Data (ICDD). JCPDS were entered into the data base for the various tungsten carbide cobalt variants that could potentially be formed as a result of the detonation processes. **Fig. 6** provides the XRD patterns for detonation processes conducted at oxygen-to-fuel gas ratios 72:12, and 82:23. The various phases are identified in the two patterns.

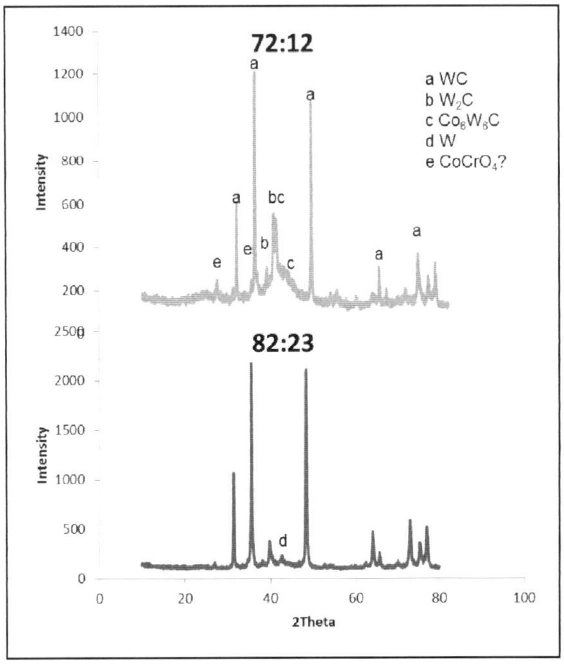

Fig. 6. X-Ray Diffraction Patterns, with Phases Identified for Detonation Process Conditions

4 Discussion

An investigation into valve-less detonation gun spraying of WC-CoCr was conducted. A variety of propane-oxygen flows and ratios were evaluated with the intent to form coatings with at least 1200 DPH which is industrially considered a higher hardness coating without being too brittle. MIL-HDBK-1886 "Tungsten Carbide Coating Detonation Process" lists required hardness as 1000-1150 DPH_{300} with porosity less than 1%. Reference 3 measured microhardness of WC-CoCr coatings applied using two HVOF systems DJ 2700 and JP 5000 and recorded hardness of 1231 DPH. In our case at high oxygen-propane ratios substantial oxidation occurs resulting in lower hardness. Interestingly, all coating microstructure showed less than 1% porosity regardless of gas flow and distance variation from 200-300 mm indicating that coating properties are determined by decarburization and dissolution reactions, except for samples 9 and 10 where porosity increases to 1.5% at 300 mm spray distance. A stoichiometric mix for combustion of propane requires 5:1 oxygen to propane ratio, therefore, at 6:1 ratio an excess of oxygen is created in a high temperature environment where decarburization and dissolution occur. Back scattered electron images of coating cross-sections sprayed at 6:1 oxygen-propane ratios also shows oxidation of the matrix (**Fig. 5a**) as well as the formation of $CoCrO_4$ (**Fig. 6**). This matrix oxidation is not observed in coatings sprayed under reducing conditions (**Fig. 5b**). The $CoCrO_4$ phase has been reported and formed at 600°C and 2000 bar in a reactor using an oxygen compressor [11]. At slightly reducing conditions of 4:1 oxygen-propane ratio the hardness increases with flow rate and shot frequency. At even more reducing parameters the hardness further increases, however, flame instability was observed at 3.3:1 ratios. A hardness of 1610 DPH was obtained at 3.5:1 ratio and at flow rates of 63:18 slm oxygen to fuel. Further increases in flow rates at this ratio decreased hardness. The most plausible explanation is that the temperature increases with flow rates thereby increasing dissolution resulting in the lower hardness observed. **Fig. 6** illustrates the effects of flow rates and oxygen-fuel ratio on the phase composition of coatings. Oxidizing combustion mixtures (72:12) show an increased amount of W_2C and eta phase Co_6W_6C than reducing mixtures (82:23). A broad mound can be seen in the case of the 72:12 flow rates in the region of 40-46 degrees. This can be attributed to the combination of peaks from eta and free tungsten phases. The eta phase could not be detected in the spectra for 82:23 flow rate, however, free tungsten could. An oxidation product $CoCrO_4$ could be indexed in both coatings with a reduced amount in 82:23 flow rate specimens. A literature search on XRD of sprayed WC-CoCr coatings has not revealed this phase before, however, the two major peaks match quite well. In the WC-Co system, decarburization leads to formation of W_2C and free tungsten. According to the ternary phase diagram the carbon loss leads to carbon content in a region where the eta phase is favored. Thermodynamically, the implication is that WC phase will now dissolve in cobalt to form the ternary eta phase.

The 3.5:1 ratio was considered to be the optimum based on microstructure and hardness at lower shot counts. In order to increase productivity shot rate was increased to 8 while increasing flow rates and maintaining 3.5:1 oxygen to propane ratio. **Figure 3** showed the hardness values as a function of temperature. Acceptable hardness of 1263 could be achieved at 200 mm. More importantly, the efficiency doubled from 4 shots at 63:18 flow rates to 8 shots/sec at 82:23 oxygen-propane flow rate (**Fig. 4**). It was noticed, microstructurally, at low shot frequency, that increased coating erosion could be seen from impingement of carbide particles. It is expected as the shot frequency is increased per flow rate reduction in particle temperatures and velocity will occur therefore less erosion of the coating will be observed. It was estimated from past experiments with DJ 2700 that that time to prepare similar samples thickness was improved by 30%. Since the samples are small not much can be inferred about industrial applications where workpieces are larger and heat effects are not as pronounced.

5 Conclusions

The use of propane as a fuel in a valve-less detonation system proved beneficial when applying tungsten carbide cobalt chrome coatings. Propane-oxygen flow rates, ratios and shot frequency could be changed to obtain coatings with hardness as high as 1600 DPH_{300}. A further advantage of this system is that dense coatings could be produced regardless of variations in process variations with the exception of distance. The most important parameters in controlling the amount of the WC phase were the oxygen-to-fuel ratio, flow rate and shot frequency.

6 References

[1] http://www.praxair-gmbh.de, PST Super D-Gun™.

[2] US 5542606 "Gas Detonation Spraying Apparatus" Kadyrov E. June 17, 1994.

[3] Schwetzke R., and Kreye H., Microstructure and Properties of Tungsten Carbide Coatings Sprayed with Various High Velocity Oxygen Fuel Spray Systems, Journal of Thermal Spray Technology, Vol 8(3) Sept. 1999, pp 433-439.

[4]http://www.asetsdefense.org/documents/Workshops/24thCadmiumPlatingMeeting/15.%20Nicoll%20hcat%20july%2021%201.pdf.

[5] Darken, L.S. and Curry R.W., Physical Chemistry of Metal, McGraw-Hill, N.Y. 1953.

[6] Fernandes C.M., and Senos A.M.R., "Cemented Carbides Phase Diagrams; A Review", Int. Journal of Refractory Metals and Hard Materials, vol 29 (4), 2011, pp 405-418.

[7] US 4902539 "Fuel-oxidant Mixture for Detonation Gun Flame Plating" Jackson, J.E., Feb 20, 1990.

[8] AMS 2448A "Application of Tungsten Carbide Coatings on Ultra High Strength Steels, High Velocity Oxygen/Fuel Process" 2008-06-04, SAE International.

[9] Mil-HDBK-1886(AT) Dept. of Defense Handbook "Tungsten Carbide-Cobalt, Detonation Process For Dept. of Defense", 27 March 1998.

[10] http://www.paxit.com/.

[11] Kume S., et.al "Generation of High Pressure Oxygen and Synthesis of $CoCrO_4$ with $CrVO_4$ Type Structure", Rev. Sci. Instrum. 42, 1856 (1971).

Columnar structured hydroxyapatite coating deposited by micro-plasma spray

Yiming.Wang*, Tingting.Fan, Xiaomei.Liu, Zheng.Zhou, Dingyong.He, Beijing / CHN

Bioactive hydroxyapatite (HA) coating is commonly applied for medical implant. The reliability and biocompatibility of which depends highly on its crytallinity and phase composition. In this work, micro-plasma spraying (MPS) process was conducted to produce coatings from spherical HA powder on Ti-6Al-4V. To understand the coating formation mechanism, surface temperature and velocity of flying HA particles were traced by online infrared monitoring equipment. An important feature for MPS is that the particle velocity is only 1/3 of that compared to conventional air plasma spraying (APS). Through X-ray diffraction (XRD) measurements, it is found out that MPS is feasible to fabricate HA coatings with high degree crystallinity and little amorphous/metastable phases are existed. Such HA coatings can exhibit remarkable stability of in HANK's solution. Moreover, strong (002) texture is fabricated by MPS under certain spray conditions. The corresponding microstructure refers to orderly-distributed columnar parallel to the coating deposition direction. In addition, the relationship among process parameter, texture density and stability in HANK's solution is investigated.

1 Introduction

Hydroxyapatite (HA) is the main mineral composition for human bone and teeth [1]. The artificial HA is long used in medical field due to its outstanding biocompatibility [2]. Whereas, because of its brittleness, HA is commonly applied as coating on metal substrate other than bulk structure material. The combination of metal substrate and HA coating is not only showing splendid mechanical properties, but also capable of inducing bone tissue growth after implantation [3]. Another advantage for HA coating is that it can prevent harmful metal ions from diffusion into body fluid, therefore enhance the safety of implant.

Among all the reported fabrication methods for HA coating, thermal spraying is the only method authorized by U.S. Food and Drug Administration (FDA) [4]. Owing to the acceptable mechanical properties of thermal sprayed HA coating as well as low cost. However, crucial problem for thermal sprayed HA coating still exists, which is mainly referring to the decomposition of HA, and amorphous (ACP) formation within coating. This problem is fatal because that either of them is instable in body fluid, and thus may cause unexpected coating failure ahead of service time. Therefore fabricating HA coating with seldom decomposition phases/ACP is one of our main goals to achieve. Moreover, the previous research shows that MPS is feasible to fabricated HA coating with high crystallinity [5]. Whereas the detail microstructure investigation of MPS HA coating is still rare. In addition, the stability of HA coating in simulated body fluid is important to predict the reliability of coating in service condition, and worth to be analysed.

2 Experimental Procedure

2.1 Micro-plasma spraying

MPS facility (WDP–1) used in this work is from our research group. Parameters for micro-plasma spraying are listed, **table 1**. The experiments were taken place in open air. It is apparent that different from conventional APS, the power of MPS is quite small and only 1.4 KW. Besides, the stand-off distance (SOD) refers to the distance from sample surface to plasma nozzle, and varies from 60 mm to 110 mm in this work.

Table 1. Parameters for micro-plasma spraying.

Micro-plasma spray parameters	
Voltage(V)	35
Current (A)	40
Working gas	Ar
Flow rate of plasma gas (L/min)	12
Deposition rate (mg/sec)	1.5
Stand-off distance (mm)	60-110

2.2 Characterisation methods

The phase composition of initial HA powder and MPS fabricated HA coatings were measured by XRD. The equipment used is D8 ADVANCE from BRUKER/AXS, Germany. The working voltage and current was set to be 40 KV and 20 mA, respectively. Scanning speed of 0.02°/s was used.

Microstructure investigations were carried out by cold field emission scanning electron microscopy. The equipment is SU8020, Japan. Working voltage was 10 KV and working distance was around 10 mm. Second electron detector was applied for image capturing.

Immersion test was conducted in HANK's solution, which was delivered from Sigma, USA. All the experiments proceeded with water bath container and a constant temperature of 37 °C was set. A vibration system was introduced into water bath container in oder to simulate the blood flow in human body.

3 Results and discussion

3.1 Temperature and Velocity of flying particle

Initial HA powders were purchased from medicoat, Switzerland. The geometry of HA particles are almost spherical. Particles with 45-63 µm in diameter are used in this work, **Fig. 1.**

Fig. 1. Hydroxyapatite powder

As mentioned above, HA coating fabricated by thermal spray suffers most from phase decomposition. According to [6] , the following chemical reactions may occur: At 800-900 °C, HA loses hydroxyl and Oxyhydroxyapatite (OHA) or Oxyapatite (OA) can be formed; At 1050-1400 °C, β - TCP and CaO can exist; At 1120-1470 °C: β - TCP turns into α – TCP; Above 1550 °C, HA melts and above 1730 °C, TCP melts. Therefore, knowing the temperature of flying particles during thermal spary is sort of guidance for understanding HA coating composition.

Under given experimental condition (table 1), the flying particles were traced at different SOD. Scattering data referring to particle velocity versus surface temperature is plotted, **Fig. 2**. In which the melting points of HA (T_m - *HA*, 1550 °C) and TCP (T_m - *TCP*, 1730 °C) are drawn as dash lines respectively.

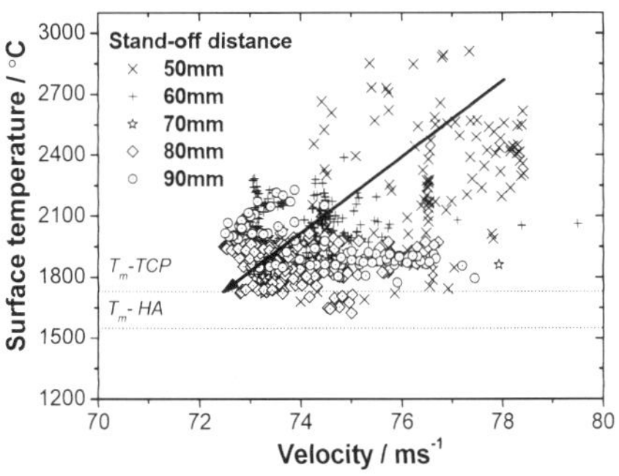

Fig. 2. Scattering plot for surface temperature versus velocity of flying HA particles

It is found out from Fig. 2 that, more than 92% of particles have surface temperatures higher than T_m - *TCP*. Implying most of HA particles are melted at least on the surface within experimental stand-off distance range. While specifically, when SOD is from 50 mm to 90 mm, the scattering data move from right-up region to left-bottom area. This trend along with SOD can be observed clearer in temperature and velocity versus SOD, **Fig. 3**. Which means that the surface temperature of particles decreases from 2400 °C (50 mm) to 1900 °C (90 mm), and the particle velocity is relatively stable at ~ 75 m/s all through. The measured temperature error at 50 mm SOD is ± 500 °C due to broad plasma width near nozzle. This may cause composition inhomogeneity within coating, therefore only SOD larger than 60 mm (include) is used in the following fabrication procedures.

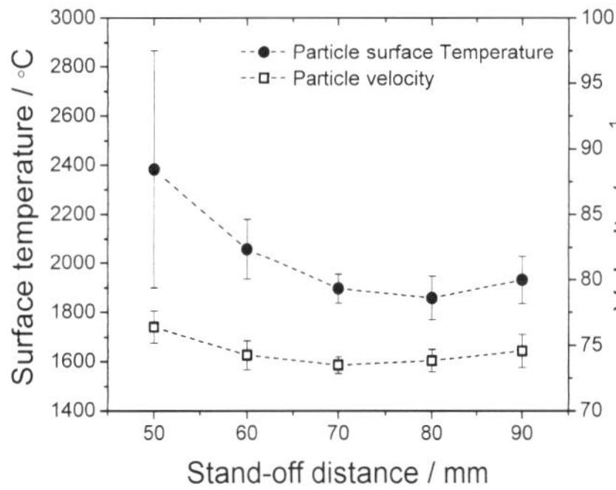

Fig. 3. Surface temperature and velocity of flying particles

3.2 Texture of HA coating

The hase composition patterns of MPS HA coatings are provided, **Fig. 4**. Initial HA powder is also shown for reference. It is seen that all the diffraction peaks from HA powder belong to HA itself, indicating high phase purity in as-received state. From the XRD patterns related to HA coatings with SOD of 60 mm to 110 mm. We observe that in general, the crystallinity is for HA coatings are higher than 95% when SOD is 60 mm and 90 mm. Which is consistence with the research from [5]. Aside with high cooling rate, vacancies are attempted to be formed when hydroxyl is absent in HA during heating [7]. Therefore ACP is usually formed in HA coating during conventional plasma spraying. During MPS procedure in our work, aqueous vapor is introduced around plasma nozzle, in order to prevent severe loss of OH^- and thus restrain the decomposition of HA. This method is also valid in other reports [8, 9]. Only slight amount of TCP is detected when SOD is larger than 90 mm, and no CaO, TTCP or apparent ACP is found under given experimental conditions for all samples. This indicates that the MPS fabricated HA coastings pocess rare unstable phases as for body fluid environment.

It is also noticed from Fig. 4 that no texture is existed in initial HA powder, the strongest diffraction crystal plane is (211). However after MPS deposition, an extreme strong texture is obtained for HA coatings with SOD of 60 mm. From which, the diffraction intensity from (002) is about 9 times of that from (211). This strong texture is larger than the reported result in [10]. In addition, our systematical research reveals that, increasing SOD leads to a decrease in the texture intensity in (002). When the SOD is of 110 mm, the texture is weak and diffraction intensity of (002) is fairly low, which is already comparable to the diffraction intensity of (211).

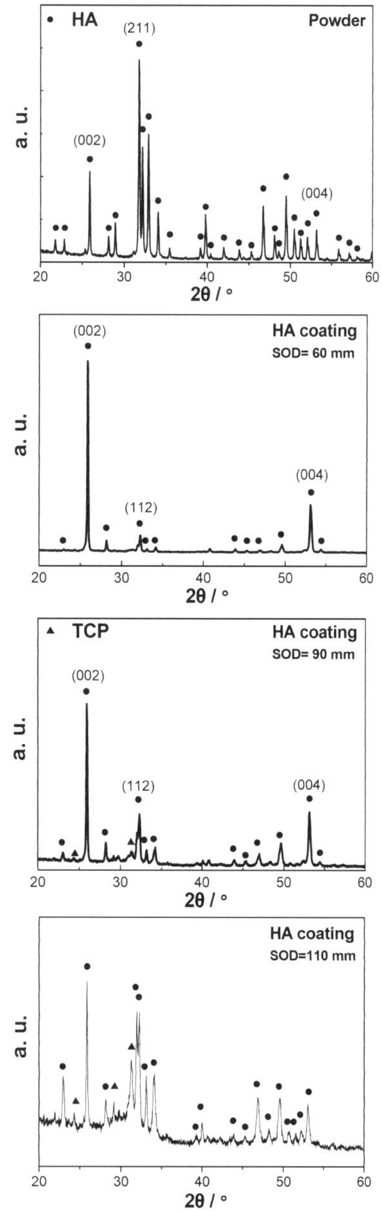

Fig. 4. XRD diagrams of HA powder, coatings with different stand-off distance.

The texture formation in HA coating is normally obtained by alternative fabrication method other than plasma spray, such as RF-magnetron sputter deposition [11]. Whereas coatings deposited by plasma spraying is usually end up with highly deformed splats, which is solidified before crystallization, due to the remarkable cooling rate compared to RF-magnetron sputter deposition. However, the coating thickness is quite limited when it is fabricated by RF-magnetron sputtering. The other method reported to obtain textured coating is related to laser remelting [12]. Assistant by which, the conventional plasma sprayed coating exhibit strong texture with coarse columnar structure parallel to deposition direction. While the results in our work show that MPS is a feasible method to deposite textured HA coating without post treatment, and the coating thickness is also sufficient.

Microstructure investigations of MPS fabricated HA coating reveal that columnar structure is formed during MPS in HA coatings. A typical microstructure is provided, **Fig. 5**. Two observation views are shown, including cross section (left) and from coating top (right), respectively. From which we can see that the columnar distributes in a perfect order. The length of each columnar layer is around 2 μm, which is similar to the splat thickness after deposition by MPS, according to our observation experiences. Thereafter it is assumed that the layer boundaries of columnar is originally splat boundaries. It is also seen that the columnar in Fig. 5 is about 200 nm in width, and this value varies at different area of coating, from 50 nm to 500 nm.

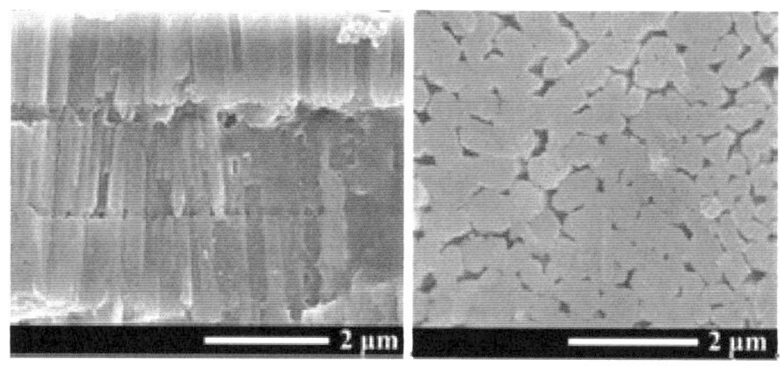

Fig. 5. Columnar microstructure in HA coating. Cross section (left) and from coating top (right).

In general, we also find out that the columnar region occupies at least 50 μm coating thickness from top surface, as for HA coating with SOD of 60 mm. In addition, the top view of columnar structure shows that they are not only distributed in specific cross section of HA coating but can through the coating area. These above observation results are consistence with the extreme strong texture we observe from XRD result (Fig. 4). Therefore, we can assume that the columnar grows along c-axis of HA crystal and form a (002) texture in HA coating during MPS.

During MPS process in this work, the velocity of flying particles is only 1/3 of that compared to conventional plasma spray. Therefore within the given SOD, the particles' duration in plasma jet is relatively long, and the heat transfer from surface to center of particle is more complete. Fully melt state is crucial for the columnar formation before solidification of splat after deposition. Furthermore, the heat input to substrate from plasma jet is higher when shorter SOD is used, and this is also constant during spraying procedure. Which acts as remelting of coating, and is key cause to slow down the crystallization process. Thus facilitates the columnar structure formation in coating. According to the equation in [13], beside small temperature gradient and post heat flux input, the splat cooling rate is also related to the splat thickness after deposition. In this work, the particle velocity is slow enough here by MPS procedure compared to APS. Therefore, instead of fabricating severely deformed thin splat (sometimes with splash), thick plats may form, and thus provide enough space for columnar growth, as is shown in Fig. 5 left.

3.3 Stability in HANK's solution

Specific positions near the sample center is traced and their microstructure changes after immersion tests are recorded. The difference microstructure changes behavior in HANK's solution for HA coating with SOD of 60 mm and 90 mm are provided, **Fig. 6**. Please note that different immersion days were applied for both samples.

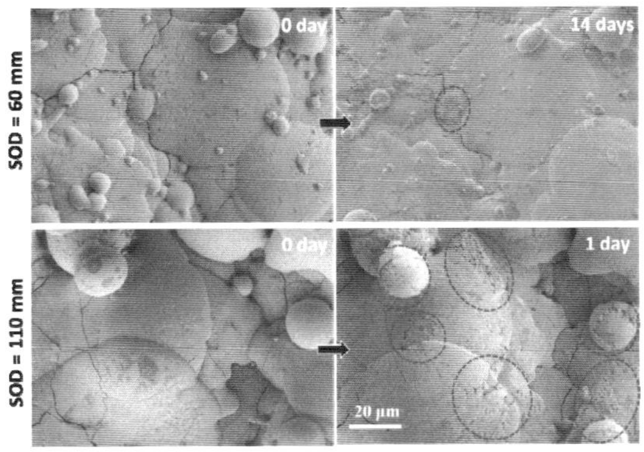

Fig. 6. HA coatings before and after immersion tests in HANK's solution.

We are able to notice from Fig. 6 that the stability for coating with SOD of 60 mm is remarkably high compared to its counterpart. After 14 days of immersion, the coating surface morphology of which rarely changed, except several small beads were washed out by fluid in HANK's solution. Moreover, the research shows that among all the crystalline, (002) has superior stability in body fluid compared to other crystal orientation [14]. (002) is also the favorite HA crystal orientation in human teeth [1].Therefore we optimistically assume that this strong (002) texture may show further advantages due to its bionic microstructure in the future study.

4 Conclusions

1) Through MPS process, high crystallinity and high phase purity HA coatings are successfully fabricated.
2) Orderly distributed columnar structure is formed in HA coating fabricated by MPS with SOD of 60 mm, corresponding to an extreme strong (002) texture. Which shows outstanding stability in HANK's solution even after 14 days of immersion.
3) The formation of columnar structure in HA coating requires following conditions: a. The flying particles must be fully melted before hitting to the substrate; b. The velocity of flying particles shall not be too fast in order to fabricate splat with enough thickness after deformation; c. Constant and sufficient heat flux input is supposed to be necessary during the solidification procedure.

Literature

[1] Park, J.: Bioceramics:Properties, Characterizations, and Applications, Springer Science+Business Media, LLC, New York, 2008.
[2] Albee, F.: Studies in bone growth: triple CaP as a stimulus to osteogenesis. J Ann Surg 71 (1920), pp. 32/36.
[3] Sun, L. Other: Material Fundamentals and Clinical Performance of Plasma-Sprayed Hydroxyapatite Coatings: A Review. Journal of Biomedical Materials Research Part B: Applied Biomaterials (maybe) (2002), pp. 570/592.
[4] http://www.fda.gov/
[5] Zhao, L. Other: Study on the influence of plasma spray processes and spray parameters on the structure and crystallinity of hydroxylapatite coatings. Materialwissenschaft und Werkstofftechnik 37 (2006), pp. 516/520.
[6] Levingstone, T.J.: Optimisation of Plasma Sprayed Hydroxyapatite Coatings, School of Mechanical and Manufacturing Engineering, , Dublin City University, Ireland, 2008.
[7] Gross, K.A. Other: Amorphous phase formation in plasma-sprayed hydroxyapatite coatings. Journal of biomedical materials research April (1998), pp. 407/414.
[8] Park, E. Other: Characterization of hydroxyapatite: Before and after plasma spraying. Journal of Materials Science: Materials in Medicine 13 (2002), pp. 211/218.
[9] Sridhar, T.M. Other: Sintering atmosphere and temperature effects on hydroxyapatite coated type 316L stainless steel. Corrosion Science 45 (2003), pp. 2337/2359.
[10] Garcia-Sanz, F.J. Other: Hydroxyapatite coatings: a comparative study between plasma-spray and pulsed laser deposition techniques. Journal of materials science: mateterials in medicine 8 (1997), pp. 861/865.
[11] Surmeneva, M.A. Other: The structure of an RF-magnetron sputter-deposited silicate-containing hydroxyapatite-based coating investigated by high-resolution techniques. Surface and Coatings Technology 218 (2013), pp. 39/46.
[12] Chen, C. Other: Influence of laser remelting on the microstructure and phases constitution of plasma sprayed hydroxyapatite coatings. Applied Surface Science 250 (2005), pp. 98/103.
[13] Wang, Y. Other: A transmission electron microscopy study of the microstructure and interface of zirconia-based thermal barrier coatings. Journal of Alloys and Compounds 619 (2015), pp. 820/825.

[14] Boyd, A.R. Other: Surface characterisation of the evolving nature of radio frequency (RF) magnetron sputter deposited calcium phosphate thin films after exposure to physiological solution. Surface and Coatings Technology 200 (2006), pp. 6002/6013.

Transition mechanism in flattening of plasma sprayed ceramic particles

M. Fukumoto, N. Maeda, B. Wei, D. Suzuki, Aichi / Japan

Transition in the flattening behavior of the thermally sprayed metallic materials has been recognized in our previous systematic investigations. A hypothesis, ultra-rapid cooled chill structure preferentially formed at the bottom part of the splat, may play an essential role for the generation of the disk splat, has been proposed. Universality of this hypothesis beyond material difference was verified experimentally by using several kinds of ceramic materials, having different thermal properties. To perform this, Al_2O_3, Y_2O_3 and YSZ particles were plasma sprayed onto AISI304 stainless steel substrate, and fraction of the disk splat with substrate temperature increase was investigated, followed by the precise observation on the cross-section microstructure of the splat. The results showed that unique amorphous and chill structure was observed in Al_2O_3 and Y_2O_3 splat at their bottom part, indicating that similar disk formation mechanism may act in these materials. On the other hand, only normal columnar structure was recognized in YSZ splat. As a result, the hypothesis was verified partially regardless of the materials difference.

1 Introduction

Main objective of our series of research on the single splat thermally sprayed is to establish the controllability of the thermal spray process. Since an individual particle thermally sprayed is a fundamental element for the coating formation, the flattening behavior of the sprayed particle onto the flat substrate surface has been precisely investigated. The results showed that a transitional phenomenon was recognized in the flattening behavior both with the substrate temperature and ambient pressure change. That is, the flattening of the splat changes transitionally from a distorted shape with splash to a disk-shaped splat at certain critical temperature and pressure values, respectively. A transition temperature, T_t and transition pressure, P_t were defined and introduced by the author [1], respectively, for those critical conditions. Correspondingly, it was found that the coating adhesion property changes transitionally on those critical temperature and pressure ranges, respectively. Based on the results, three dimensional transition curvatures by combining both changing curves was proposed for each sprayed material as a controlling principle for the practical usage in the thermal spray industries [2].
On the other hand, from an academic viewpoint, restraint mechanism for the splashing have to be explained based on the fluid dynamics viewpoint, namely, from the splashing parameter K. In the observations on the metallic materials, it was indicated that ultra-rapid solidified chill structure preferentially formed at the bottom part of the splat plays an essential role for the disk splat formation [3]. To verify the universality of this hypothesis beyond material difference between metal and ceramic, three kinds of ceramic powder particles with different thermal conductivity, Al_2O_3, Y_2O_3 and YSZ, were thermally sprayed onto the metallic substrate. The effect of ultra-rapid solidification at the bottom part of the splat on the flattening behavior of the sprayed particles was investigated in the present study.

2 Experimental Procedures

Commercially available metallic and ceramic powders with diameter of several tens micrometers were used as feedstock powders as shown in **Table 1**. AISI304 stainless steel plates with dimensions of 25mm × 25mm × 6mm finally polished their surfaces with 0.3μm Al_2O_3 buff were used as substrates. Sprayed particles were collected on the substrate surface using both atmospheric and low pressure plasma spray equipment. The splat shape was classified and the number of both disk splat and splash splat were counted on every trial. The fraction of the disk splat was defined as the ratio of the number of disk splat to the total splat. The cross section microstructure of the splat was precisely observed by means of a scanning ion microscope, SIM and transmission electron microscope, TEM after cutting the splat by focused ion beam microscope, FIB.

Table 1 Metallic and ceramic feedstock powders

	Cu	Al_2O_3	Y_2O_3	YSZ
Heat conductivity [$W\cdot m^{-1}\cdot K^{-1}$]	401	30.0	14.5	3.0
Melting point [K]	1,358	2,345	2,647	2,715
Specific heat [kJ/kg·K]	0.39	0.80	0.47	0.46
Specific gravity	8.9	3.9	5.0	6.0

3 Results and Discussion

3.1 Transition Mechanism in Metallic Materials

Generally, it is well known that the break-up phenomenon of the liquid film generated by the collision of the liquid particle onto the solid surface can be evaluated by the splashing parameter, *K*, in the fluid dynamics field [4]. *K* is defined as $We^{0.5}Re^{0.25}$, and it is based on the in-flight information of the liquid particle.

Here, We = $\rho dv^2/\gamma$, and Re = $\rho dv/\eta$; where ρ is density, d is diameter, v is velocity, γ is viscosity, and η is surface tension, respectively. K has a critical value, Kc = 57.7, and if the K value of the particle exceeds Kc, the liquid film shows the break-up after the collision onto the solid surface. To verify the K value of the regular thermally sprayed particles, measurement of the in-flight information was conducted by using the DPV-2000 system. By introducing the measured temperature into each physical constant (i.e., density, surface tension, and viscosity), the K value was calculated for a few kinds of sprayed metallic materials. As a results, it was found that K values were remarkably larger compared to Kc regardless of the sprayed material [2]. Namely, the regular-sprayed particle has enough driving force for the splashing. The results indicates that the essential point of the transition problem lies in not why and how the splashing occurs on the cold substrate surface, but why and how the disk splat appears on the hot substrate surface instead of the splash splat. Thus, we need to elicidate the mechanism on what factor acts as restraint to the splash formation.

To verify the transition mechanism in metallic materials, solidification microstructure of the splat was precisely observed. On the bottom surface of the splat, after peeled off from the substrate surface, obtained at heated substrate or reduced pressure conditions indicated that ring shaped dense zone with fine microstructure was observed near peripheral region. This fine microstructure near peripheral region indicates that rapid solidification might be occurred at this region. To confirm the solidification inside of the splat directly and to elucidate the effect of rapid solidification on the flattening, cross section microstructures of the splats were precisely observed by SIM. The results were summarized and shown in **Fig. 1**.

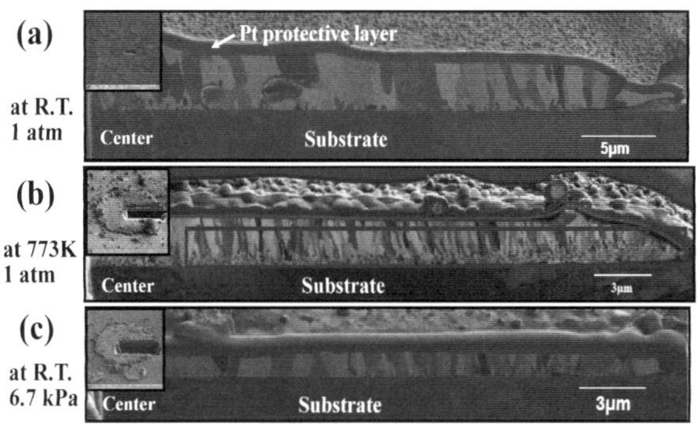

Figure 1. Cross section microstructures of the Cu splats obtained on the AISI304 stainless steel substrate at a) room temperature substrate under atmospheric pressure, b) 773K heated substrate under atmospheric pressure, c) room temperature substrate under 6.7 kPa pressure conditions.

In the splash-typed splat collected on the room temperature substrate under atmospheric pressure (**Fig. 1(a)**), coarser columnar grains perpendicular to the substrate surface are observed in whole part of the splat. On the other hand, in the disk-typed splat collected on the 773K heated substrate, finer columnar grains inside of the splat are observed in whole part of the splat (**Fig. 1(b)**). Finer structure indicates that faster solidification occurred in the disk splat given on the heated substrate. However, since it is well known that the columnar solidification structure is generally formed after the flowing of the liquid, it is clear that the solidification inside of the splat never affect the flattening of the splat. Hence, difference in the mean cooling rate, and solidification behavior inside of the splat cannot become a reason for the difference of the splat shape.

More precise observation was conducted on the splat, and a layer composed with equiaxial fine grain, so-called chill crystal structure which usually caused by ultra-rapid cooling, was finally found at the bottom part of the splat indicated as blue rectangular shown in **Fig. 1(b)** and **(c)**. The essential point is that chill crystal layer appeared on the heated substrate is much thicker than that on the unheated substrate. Moreover, the thickness of this ultra-rapid solidified layer becomes gradually thicker toward the periphery region of the splat compared with the central region. This indicates that ultra-rapid cooling and resultant ultra-rapid increase of the viscosity occurred near the peripheral region and this ultra-rapid material change acts as the restraint role for the splashing. This is exactly a new finding found in our experiments in the relating research field.

3.2 Transition Mechanism in Ceramic Materials

Universality of the new finding, namely ultra-rapid solidified chill structure in the splat has a restraint role for the splashing, obtained in the metallic materials was verified in the ceramic materials, since the similar transition behavior is recognized also in the ceramic splats. The change of the fraction of disk splat with

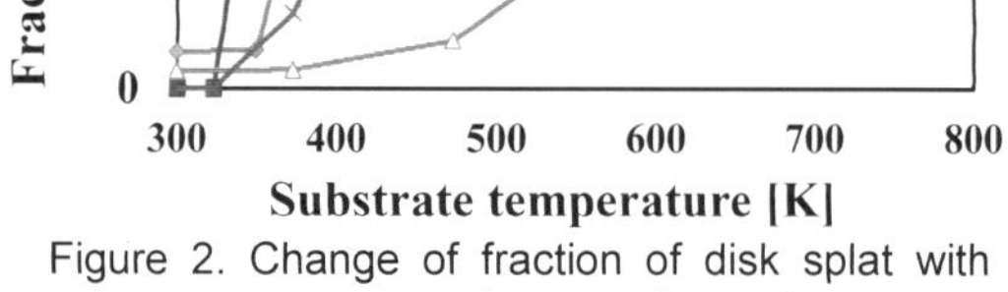

Figure 2. Change of fraction of disk splat with substrate temperature increase for each material.

substrate temperature increase for each sprayed material was summarized in **Fig. 2**, which includes the data on Cu particle. It was found from the figure that the ceramic material with higher thermal conductivity shows lower transition temperature. The result shows that the ceramic material having a higher heat conductivity tends to show a transitional behavior more easier, indicating that a certain solidification property of the material may affect the flattening. To verify this tendency more precisely, the cross-section microstructures of the disk splats for each ceramic material collected on the heated substrate surface was observed. The results were summarized and shown in **Fig. 3**.

In Al_2O_3 splat shown in **Fig. 3(a)**, having the highest heat conductivity, an amorphous layer, mostly half thickness of the splat, was observed at bottom part of the splat entirely whole position of the radius direction. In case of

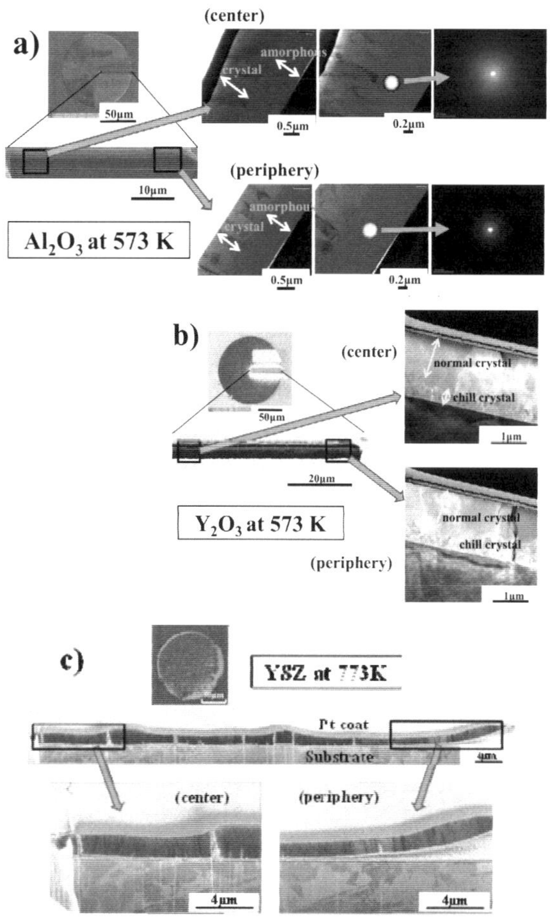

Figure 3. Cross section microstructures of ceramic splats obtained on heated substrate surface. a) Al_2O_3. b) Y_2O_3. c) YSZ.

Y_2O_3 shown in **Fig. 3(b)**, on the other hand, having a middle value of the heat conductivity, chill crystal layer at bottom part of the splat was observed entirely whole position of the radius direction. In these two materials, ultra-rapid solidification at bottom part of the splat may cause the disk splat, namely, the result supports the above hypothesis given in the metallic materials. However, in ZrO_2 splat shown in **Fig. 3(c)**, only the columnar structure similar to the metallic material was observed without any ultra-rapid solidified microstructure even in the disk splat. In this case, solidification cannot be used for the explanation in transition behavior. Some other mechanism, such as a rapid viscosity increasing inside of the splat due to the cooling, have to be adopted in the explanation for the appearance of the disk splat. More precise, systematic observation has to be done in the future study.

4 Summary

Ultra-rapid cooling may dominate the flattening of the thermal sprayed particles by arresting splashing, through the formation of chill crystal or rapid increase in viscosity of the metallic materials. Through the observation on the flattening behavior of the ceramic materials, the hypothesis namely the ultra-rapid solidification at bottom part of the splat may act as the restraint for the splashing, was verified both in Al_2O_3 and Y_2O_3 splats. However, since an exception was existed as a case of ZrO_2 splat, more precise, systematic observation has to be done in the future study.

5 References

[1] M. Fukumoto, S. Katoh, and I. Okane: Splat Behavior of Plasma Sprayed Particles on Flat Substrate Surface. Proceedings of the 14th International Thermal Spray Conference (1995), pp. 353/58.

[2] M. Fukumoto, M. Shiiba, H. Kaji, and T. Yasui: Three-dimensional transition map of flattening behavior in the thermal spray process. Pure Appl. Chem., **77** (2005), issue 2, pp. 429/42.

[3] M. Fukumoto, D. Suzuki and N. Maeda: Flattening mechanism in thermally sprayed ceramic particles. Proceedings of the 7th Asian Thermal Spray Conference, (2015), CD.

[4] C. Mundo, M. Sommerfeld, C. Tropea: Droplet-wall collisions: Experimental studies of the deposition and breakup process. Int. J. Multiphase Flow, **21** (1995), pp. 151/173.

Modification of thermal transport properties of Ar-H_2 working gases by calcium-phosphate precursors used for suspension plasma spraying of HA coatings

Rolando T. Candidato, Jr., Bernard Pateyron, Alain Denoirjean and Lech Pawlowski, Limoges, FR

The current study is mainly focused on the modification of thermal transport properties of plasma working gases and their interaction to calcium-phosphate solution used of hydroxyapatite (HA) synthesis via solution precursors plasma spraying (SPPS). Argon (45 slpm) and hydrogen (5 slpm) were the primary gas and secondary gas used. The concentration of solution was varied and it was injected at different feed rates. Thermal conductivity and modified ability of heating factor were calculated numerically and it was found out that these parameters increased with an addition of the solution. Moreover, these parameters were influenced by solution feed rate rather than its concentration. On the other hand, viscosity was not affected by the presence of solution regardless of the rate and concentration. The morphologies of HA coatings synthesized in the process were studied.

1 Introduction

The solution precursor plasma spraying (SPPS) is an emerging deposition technique which aims to develop nano-structured coatings. This technique uses liquid precursors injected into the plasma jet to form solid deposits onto a prepared substrate. Studies have been performed employing SPPS for producing materials intended for desired application such as in thermal barrier coatings [1], gas sensors [2] and biomaterials [3]. The resulting deposits show unique particle morphologies which highly depend on the thermal history of droplets undergoing chemical and physical transformations in plasma jet. Experimental examination on the thermo-physical changes of droplets is not an easy because of the high temperature of plasma jet and rapidity of reaction. Numerical analysis of droplet transformation during the spraying experiment is a way to understand the processes involved. Several numerical modeling works on the transformation of droplets in SPPS have been reported with emphasis on zirconium hydrochloride solution [4], lanthanum-strontium manganite solution [5] and cerium nitrate solution [6]. However, the analysis of thermodynamic properties is also important to be understood as it plays a significant role during the heat and momentum transfer between solution droplets and plasma working gases. During spraying, the evaporation of liquid from the solution precursor alters the composition of plasma working gases, thereby changing the thermal transport properties. The current work aims to study the effect of aqueous calcium-phosphate solution used for the synthesis of hydroxyapatite (HA) [7] on the thermal transport properties of the plasma working gases. Related works on the numerical analysis on the transport properties of HA based on powder plasma spraying (APS) [8] and suspension plasma spraying (SPS) [9] had been done previously. HA is particularly selected in this work since it is an important bioceramic material and that SPPS could be a way to deposit HA coatings with nanometric or sub-micron features useful for biomedical applications. The understanding of the interaction between solution precursors and plasma may be a step forward in controlling and optimizing the spray processes involved. The present paper analyzes two concentrations of solution precursors and two different liquid injection feed rates. The initial micrographs of the SPPS-HA coatings are also presented.

2 Experimental approach

2.1 Solution precursor preparation and spray parameters

The aqueous calcium-phosphate solution used is composed of calcium nitrate tetrahydrate ($Ca(NO)_3{\cdot}4H_2O$) and triethyl phosphite ($P(OEt)_3$). Two solution concentrations were used, namely: (i) high concentration (0.5M/0.3M); and (ii) low concentration (0.167M/0.1M). The aqueous solution mixture forms a complex ionic derivative composed of Ca^{2+}, NO_3^-, H^+ and the *Ca-O-P* bonds. The simplified chemical reaction taken from [10] may be expressed as:

$$P(OEt)_{3-x}(OH)_x + Ca^{2+} + (NO_3)^- \rightarrow (NO_3)^-(OH)\text{-}Ca\text{-}O\text{-}P(OH)(OEt)_{3-x} + H^+ + C_2H_5OH + H_2O \quad (1)$$

The plasma working gases were: Argon with 45 slpm flow rate and Hydrogen with 5 slpm. Solution injection feed rates were varied from low level at 38 mL/min to high level at 50 mL/min. The data of solutions used in the coating experiments are presented in Table 1.

Table 1: Variable solutions parameters used in spray experiments.

Sample Code	Feed rate (mL/min)	Concentration [M]
ClRl	38	0.167/0.1
ChRl	38	0.5/0.3
ClRh	50	0.167/0.1
ChRh	50	0.5/0.3

2.2 Calculation of thermal transport properties

The transport properties of the working gases were determined using *T&TWinner* software described elsewhere [11]. The computation was possible through evaluation of the molar fraction of the chemical composition of the working gases as well as calcium-phosphate solution used to synthesize HA. The different fluxes in the gas system are described by transport coefficients determined from the method of Chapman-Enskog [12], useful for the calculation of thermal conductivity and viscosity. Thermal conductivity controls the heat transfer while viscosity determines the momentum of the working gases. Supposing a particle having only single temperature in a gas having also one temperature, the heat transfer from plasma to particle is given by the following equation:

$$\pi d_p^2 h(T_g-T_p) = (1/6)\pi \rho_p c_p d_p^3 (dT_p/dt) \qquad (2)$$

where d_p-is diameter of particle, T_g-temperature of gas, T_p-temperature of particle, ρ_p – density of particle and c_p – its specific heat. The left hand side is the transferred energy from plasma which is related to the ability heating factor (AHF) parameter while the right hand side describes the energy absorbed by particle. AHF describes the ability of the working gas to melt the particles. After consideration of conditions described elsewhere [13], the final calculations were modified to only take the transport coefficient and is expressed as:

$$1/(T_g-300)(AHF\cdot v_g/L)^{1/2} = <\Lambda_g>/(<\eta_g>)^{1/2} \qquad (3)$$

Where v_g-is velocity of gas, L-length of high temperature zone of plasma jet, $<\Lambda_g>$-average thermal conductivity of gas and $<\eta_g>$-average dynamic viscosity of gas.

3 Results

3.1 Thermal transport properties

The coefficients of the thermal conductivity of the plasma working gases before and after introducing calcium-phosphate solution are displayed in Fig. 1. For the initial Ar-H_2 mixture, a peak from 2500 K to 5000 K range is due to the dissociation of H_2 to 2H. Injection of the calcium-phosphate solution resulted for about four (4) times increased in heat conductivity at this temperature range. A maximum peak from 10,000 K to 15, 000 K centered at 15,000 K is notable which might be attributed to the translation thermal conductivity of heavy species from solution as presented in Equation 1. Varying the concentrations of solution precursor seem not to change the heat conductivity. On the other hand, the injection feed rate has striking effect especially at 15,000 K where about 0.5 W/m·K increase was calculated when using high injection rate.

Fig. 2 shows the results of viscosity of the working gases before and after the introduction of the solution precursors with different injection feed rates and different concentrations of solution. The viscosity of the plasma working gases is not strongly affected by change in concentration and injection feed rate of solution. Near the 4000-5000 K range, a slight decrease of viscosity is seen. This is due to the vapors coming from the solvent and other chemical species produced from reaction of solution precursors .

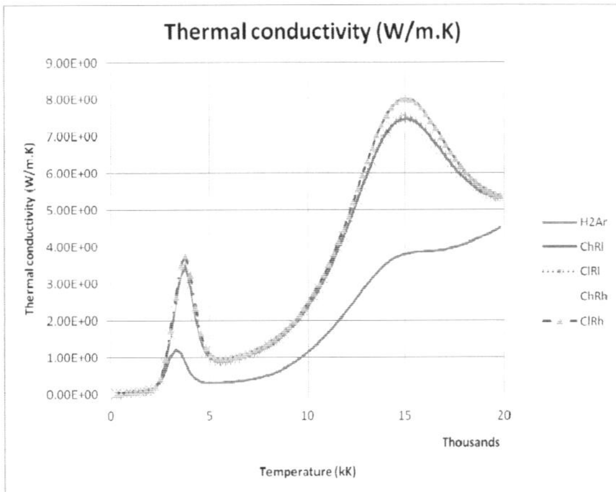

Fig. 1. Thermal conductivity vs. temperature of working gases (H2Ar) and after the injection of different solutions.

Finally, the modified ability of heating factor shown in Fig. 3 increases by more than three times when solution is introduced with respect to the inital Ar and H_2 mixture. The presence of ethanol and other by product of the reaction seem to account for such big difference. Solutions used at high injection feed rate has 10% higher AHF value than the AHF of solutions injected at lower rate.

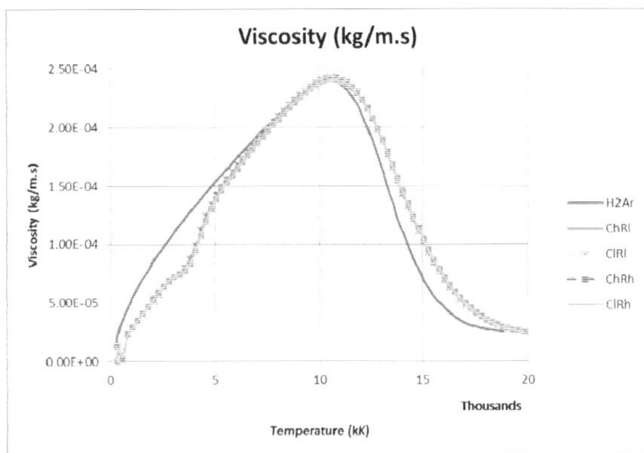

Fig. 2. Viscosity of working gases vs. temperature for viscosity of initial working gas Ar and H_2 (H2Ar) and the gases modified by addition of calcium-phosphate precursor.

3.2 Microstructures of coatings

The investigations of the microstructures of the coatings sprayed using varying concentrations of initial solution precursors and solution feed rates was done. Fig. 4 shows the SEM images for each considered spray condition. It was found that the morphology of coatings is not influenced by the solution injection rate but rather by its concentration . The coatings prepared from low concentration of solution precursors (Fig. 4a and 4c) have similar features. These coatings showed a compact and bulging structure (like a dome) on their surface. On the other hand, the coatings prepared from high concentration solution (Fig. 4b and 4d) are composed of agglomerated fine particles and solid spheres. Fully melted splats were not visible for all the coatings suggesting that the particle did not undergo sufficient heating. Regardless of the solution feed rate, the morphology of the coatings is solely dependent on the initial concentration of solution precursor.

4 Discussion

Solution precursor plasma spray, a relatively new spraying process has an advantage compared to the well established plasma spraying techniques as it avoids powder synthesis and, eventually, suspension preparation stages including milling and stabilization processes. The current work is focused on the analysis of the thermal transport properties of the initial plasma working gases and to their interaction with the calcium-phosphate solution. The thermal conductivity of plasma working gases which affects the transient heating of droplets was increased by a factor of four when the solution was introduced to the plasma jet. Solution feed rate, rather than solution concentration remarkably affected the heat conductivity with a strong peak of about 8.0 W/m·K at 15, 000 K.

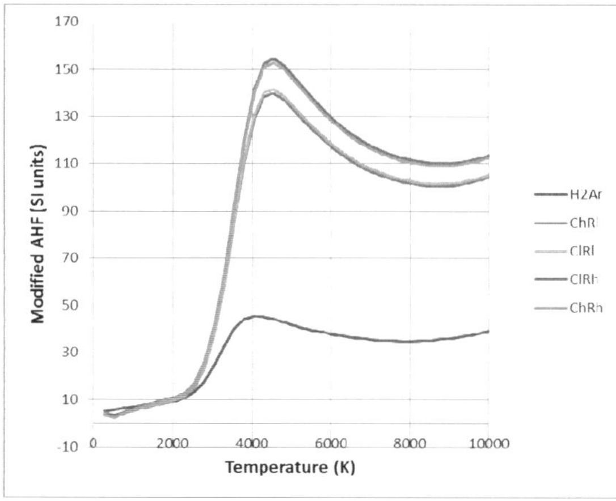

Fig. 3. Modified ability of heating factor against temperature of the plasma working gases before and after injection of solution.

This is associated to the evaporated heavy species especially when using 50 mL/min rate of calcium-phosphate solution. On the other hand, the concentration of solution and injection rate sparingly changed the viscosity of the original working gases. The presence of solution just slightly lowered the conductivity at the temperature of 4000 K. This result means that the momentum of working gases is not significantly affected with the presence of calcium-phosphate solution. In addition, the modified ability of heating factor was increased by more than three times when solution was introduced to the original plasma working gases. AHF is influenced more by the solution feed rate than the concentration of solution. It may be however deduced that the ability of heating of the plasma working gases is not that enough to fully melt the formed particles due to the absence of fully melted splat on the coatings presented in Fig. 4. This factor might be affected by the thermo-chemical reaction of solution inside the plasma jet essentially for large droplet sizes. It is noted also that the AHF value in this work is three times lower compared to the AHF value for suspension plasma spraying of HA under water-ethanol media [14]. Changing the flow rate of gases or solution feed rate might improved the value of modified AHF. The morphology of the coatings however showed that they were impacted more by the change in solution concentration rather than by the injection rate of solution. Numerical analysis on the phenomena involving solid particle formation should clarify these findings.

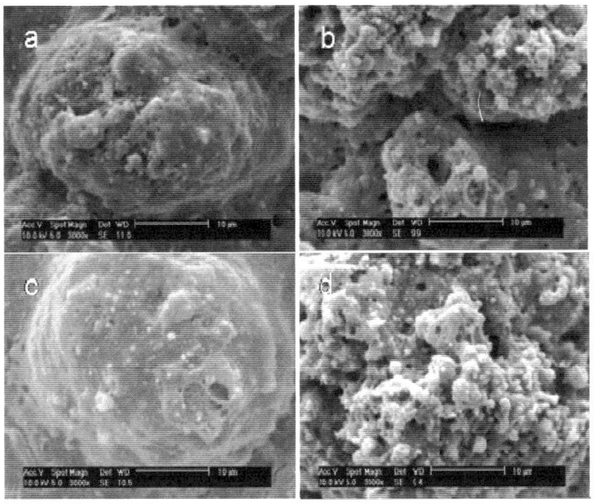

Fig. 4. SEM images of coatings prepared in different spray conditions: (a) ClRl, (b) ChRl, (c) ClRh and (d) ChRh showing the characteristics particle morphology.

5 Conclusion

The simulation on the interaction between calcium-phosphate solution and the Ar-H_2 mixture was done. Thermal transport properties of the plasma working gases were calculated before and after injection of the calcium-phosphate solution. Thermal conductivity greatly increased after injection of solution. It was found also to be influenced by the solution feed rate and not by the concentration of solution. The solution however has no significant effect to the momentum of working gases as the calculated viscosity was just slightly changed. Moreover, the modified ability of heating factor of Ar-H_2 plasma is at highest value when high solution feed rate was used regardless of the concentration of solution. But even so, the AHF value seems not enough to sufficiently heat the solution droplets. Further work should be devoted on exploring the thermo-physical phenomena involve during the process numerically and experimentally.

References

[1] Chen, D., Jordan, E. and Gell, M: Effect of solution concentration on splat formation and coating microstructure using the solution precursor plasma spray process. Surface & Coatings Technology 202 (2008) 2132/38.

[2] Chien, K. and Coyle, T.W: Rapid and continuous deposition of porous nanocrystalline SnO_2 coating with interpenetrating pores for gas sensor applications. Journal of Thermal Spray Technology 16 (2007) 886/92.

[3] Candidato, Jr. R.T., Sokołowski, P. Pawłowski, L. and Denoirjean, A: Preliminary study of hydroxyapatite coatings synthesis using solution precursor plasma spraying. Surface and Coatings Technology 277 (2015) 242/50.

[4] Hu, Y. and Shan, Y: Modeling of solution droplet evolution and particle morphologies in solution plasma spraying. Proceedings of the 20th International Symposium on Plasma Chemistry, Pennsylvania, USA, 2011.

[5] Shan, Y.G., Wang, Y.L. and Coyle, T: Analysis of deposits formation in plasma spraying wih liquid precursors. Applied Thermal Engineering 51 (2013) 690/97.

[6] Saha, A., Cetegen, B., Jordan, E., Özturk, A. and Basu, S: Thermo-physical processes in cerium nitrate precursor droplets injected into high temperature plasma. Surface and Coatings Technology 203 (2009) 2081/2091.

[7] J.D. Bronzino, The Biomedical Engineering Handbook, Second Edition, CRC Press LLC, USA and Springer-Verlag GmbH & Co. KG, Germany, 2000.

[8] Dyshlovenko, S., Pateyron, B., Pawłowski, L. and Murano, D: Numerical simulation of hydroxyapatite powder behaviour in plasma jet. Surface and Coatings Technology 179 (2004) 110/17.

[9] Pateyron, B., Pawłowski, L., Calve, N., Delluc, G. and Denoirjean, A: Modeling of phenomena occurring in plasma jet during suspension spraying of hydroxyapatite coatings. Surface and Coatings Technology 214 (2013) 86/90.

[10] Liu, D.M., Troczynski, T. and Tseng, W.T: Water based sol-gel synthesis of hydroxyapatite: process development, Biomaterials 22 (2001) 1721/30.

[11] Pateyron, B., Delluc, G. and Calvé, N: T&TWinner, the chemistry of non-line transport properties in interval 300K to 20000 K, Mécanique et industries 6 (2005) 651/54.

[12] Engelke, J.K. : Heat transfer to particles in plasma flame, Presented at the AIChE Meeting, Los Angeles, CA., USA, (1962).

[13] L. Pawłowski, The Science and Engineering of Thermal Spray Coatings, 2nd Edition, Wiley, Chichester, U.K., 2008.

[14] Pateyron, B., Calve, N. and Pawłowski, L: Influence of water and ethanol on transport properties of the jets used in suspension plasma spraying. Surface and Coatings Technology 220 (2013) 257/60.

Warm Spraying of high strength Ni-Al-bronze for cavitation protection

Sebastian Krebs, Hamburg / DE, Seiji Kuroda, Tsukuba / JP, Hiroshi Katanoda, Kagoshima / JP, Frank Gaertner, Hamburg / DE, Thomas Klassen, Hamburg / DE, Hiroshi Araki, Tsukuba / JP, Simon Frede, Hamburg / DE

Bronze materials such as Ni-Al Bronze show exceptional performances against corrosion, erosive wear and cavitation-erosion, due to their high fatigue strength and a high resistance against plastic deformation. These materials are therefore used for ship propellers, turbines, in pump systems, or for other applications with alternating stresses. Usually, the respective parts are cast. With the aim to use ressources more efficiently and to reduce costs, the aim of this study is to evaluate opportunities to apply bronze only as a coating to critical areas of the respective parts. First experiments with cold gas sprayed bronze coatings are promising and show good coating performance, allowing for maximum half the cavitation resistance as compared to the bulk material. Higher process gas temperatures and pressures would lead to higher impact velocities and temperatures, thus to more deformation upon impact of these high-strength materials. Consequently, coating quality could improve. However, process parameters in cold spraying are limited by nozzle clogging at temperatures above 700 °C. Warm Spraying, a nitrogen-cooled HVOF-process, provides similar particle impact velocities but reaches higher temperatures as compared to cold spraying, while still minimizing oxidation effects. Therefore, this study systematically evaluates the formation and performance of warm sprayed Ni-Al-Bronze coatings for different combustion pressures and nitrogen flow-rates. Additionally, substrate pre-heating was used to improve coating adhesion. The coatings show low porosities as well as high mechanical strengths, allowing for a cavitation resistance similar to that of the bulk material.

1 Introduction

By thermal and kinetic spraying, a wide range of coatings can be produced for different applications [1]. Protection or repair of parts in challenging environments are of particular interest. As one example, ship rudders suffer from corrosion, erosive wear and predominantly cavitation erosion. Failure or disfunctionalities can cause high expenses and restrict ship manoeuvrability. Cavitation erosion occurs if the local water pressure varies substantially. A decrease in pressure leads to evaporation and formation of vapour bubbles. As soon as these bubbles reach areas of higher local pressure, they collapse omnidirectional. But imploding on boundary surfaces, a directed water jet arises and hits the ships surface with up to 1 GPa [2]. This permanent loading and unloading leads to fatigue at the material surface and associated microstructural damage. In addition, cracks will form and result in material loss at exposed areas. To minimize this problem, materials with high resistance against plastic deformation are employed, i.e. materials that possess high hardness or yield strength and high fatigue strength [3,4]. Bronzes fulfil these requirements and are commonly used for ship propellers. Since bronzes are too expensive for producing ship rudders or other massive parts, different alternatives have been evaluated so far, i.e. steel cladding, synthetic protective coatings or fluid dynamic optimizations [5,6,7].

Thus, depositing a Ni-Al-Bronze as a dense and well adhering coating could serve as a protective layer for enhancing rudder life times or could be used as a cost efficient repair method. Typically, Cold Gas Spraying (CS) can provide pore-free coatings due to high kinetic impacts. However, the high material strength of the Ni-Al-Bronze requires high process temperatures for sufficient thermal softening and deformation of the feedstock material upon impact Furthermore, process gas temperatures above 650°C can lead to nozzle clogging. Thus, the available parameter range is limited and producing dense coatings with high cohesive strength proves as challenging. Warm Spraying (WS), also named Two-Stage HVOF or modified HVOF [8,9], is a spraying technology that closes the gap between CS and HVOF. Respective spray systems guarantee for high particle velocities, comparable to CS, but higher particle temperatures, and can deposit particles in solid, semi-solid or molten state [10,11]. In contrast to conventional HVOF equipment, a mixing chamber connects the combustion chamber and the nozzle throat (compare figure 1). This allows to feed an inert gas and to control the process temperatures within a

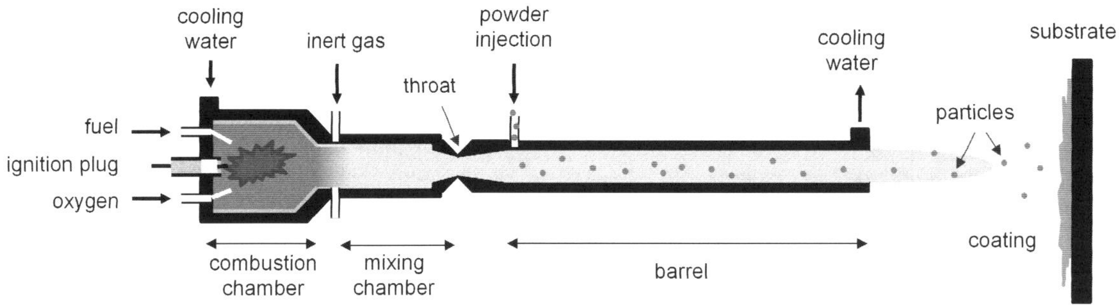

Figure 1: Schematic of the Warm Spray System, showing, as compared to HVOF, a mixing chamber between combustion chamber and nozzle throat to insert inert gas for better process temperature control.

wide range of temperatures, from 1000 K up to 2500 K [12]. As compared to HVOF spraying, WS decreases oxidation and leads to better coating cohesion [8].
Previous studies for CS revealed that for good coating cavitation resistance high material strength and good coating integrity is needed [13]. With respect to WS, this study focusses on good coating cohesion and investigates systematically the influences of the nitrogen flow-rate between 500 l/min and 1000 l/min, as well as the effect of the combustion pressure of Low-Pressure Warm Spray (LP-WS) at 1 MPa and High-Pressure Warm Spray (HP-WS) at 4 MPa. Additionally, LP-WS parameters were sprayed on pre-heated substrates to enhance coating adhesion.

2 Methods

2.1 Warm Spraying

Spray Parameter
For coating formation, a WS system (Plasma Giken Co., Japan) was used with a LP-WS gun modified from JP5000 and an original HP-WS gun with nitrogen as cooling gas and kerosene as fuel operating was used at stoichiometric conditions (λ = 1). An 8" barrel was employed and the spray distance was kept at 200 mm. The powder feed-rate was set to 62 g/min. All experiments were performed with a robot-traverse speed of 700 mm/s and a line spacing of 4 mm. In order to examine the influences on coating cohesion and adhesion, the nitrogen flow-rate was changed for LP-WS and the corresponding HP-WS parameter. Additionally, the substrate temperature was increased by resistance heaters to 300 °C for equivalent LP-WS parameters. To keep a stable combustion pressure during various nitrogen flow-rates, the oxygen and kerosene flow were adjusted, respectively (see table 1).

Table 1: Overview about the WS parameters.

gun system	combustion pressure [MPa]	T_{sub} [°C]	N_2 flow-rate [l/min]	kerosene [ml/min]	oxygen [l/min]
LP-WS	0,72	RT	500	347	714
	1		750	369	760
	1,05		1000	391	805
	0,72	300	500	347	714
	1		750	369	760
	1,05		1000	391	805
HP-WS	4	RT	N500	980	480
	4		N1000	780	380

Substrate and Powder Material
The coatings were sprayed onto ship-building steel GL-A (comparable to S235). The substrates (50x70x5 mm) were grid-blasted and cleaned with acetone within an ultrasonic bath. A thermocouple was fixed to the substrates for recording the temporal evolution of the temperature during the spray process. The mechanical properties of the substrate materials and of the spray material are summarized in table 2. As feed stock powder, an inert gas atomized spherical CuAl10Fe5Ni5

Table 2: Overview about the used substrate and spray material.

steel GL-A	Ni-Al-Bronze		
YS [MPa]	density	UTS	size
339	7,6 g/cm³	~740 MPa	d10 = 21 µm
UTS [MPa]	Tm	cp	d50 = 30 µm
424	1333 K	399 J/kgK	d90 = 50 µm

- powder from TLS Technik (Bitterfeld / DE) was used.

Simulations
For given spray parameter of the WS process, Katanoda et al. introduced a 1D-model to evaluate resulting impact conditions [9,10,11]. This model is based on following:

(a) thermodynamic and gas-dynamic calculations of the combustion and the mixing chamber
(b) quasi-one-dimensional calculations of the internal gas flow through the gun
(c) semi-empirical calculations of the jet flow
(d) calculation of particle velocity and temperature traveling in the gas flow [9].

Resulting simulation data supply information about the particle velocity and temperature and allows for estimating the ratio of molten fraction of particles.

2.2 Coating analyses

Optical Microscopy and porosity measurement
For analyzing coating microstructures and porosities, an optical microscope (OM) type DMRM with Axio-Vision® software from Leica (Germany) was used. Porosity was determined on the basis of optical micrographs according to ASTM-Standard E2109-1 by using the software KS300 from Zeiss (Germany),.

Hardness measurements
Vickers Hardness was measured on cross sections as HV0.3 using a load of 2,942 N according to the ASTM E384-10 standard, using an indentation tester type Miniload II from Leitz (Germany.

Cavitation testing
Cavitation resistance is tested in accordance with ASTM G32-10 by applying the indirect mode. Here the polished sample surface is kept at a distance of 0.5 mm to the sonotrode in distilled water at 22°C. The amplitude was set to 25 µm and the frequency to 20 kHz. The tests were performed with a test rig type BK101Z from KLN Ultraschall GmbH (Germany) using a sonotrode diameter of 15 mm. During 300 min exposure time, the sample mass loss and surface topography (by SEM) were subsequently recorded. As a comparable standard, the average erosion depth in

µm and maximum erosion rate (MER) in µm/h were determined.

Electrical conductivity

The coating electrical conductivity was measured on polished coating surfaces using an inductive methode. In accordance to ASTM E1004 standard, a frequency of 1250 kHz was applied by a Sigmascope® SMP10-HF and measuring sensor ES40HF from Helmut Fischer GmbH (Germany).

Adhesion testing

The bond strength specimen (Ø 25 x 25 mm) are glued to same geometry counterbodies using Ultrabond 100 glue. For the experiments, a tensile testing machine from Zwick/Roell (Ulm/Germany) was used in accordance with DIN EN 10002 part 1.

3 Results

3.1 Simulation Results

By the simulations, particle impact conditions of the warm sprayed bronze particles are revealed. Figure 2 shows the Window of Deposition as introduced by Assadi et. al for CS [14]. The figure shows that impact conditions for all spray parameters are well above the critical velocity for Ni-Al-Bronze. Applying high pressure

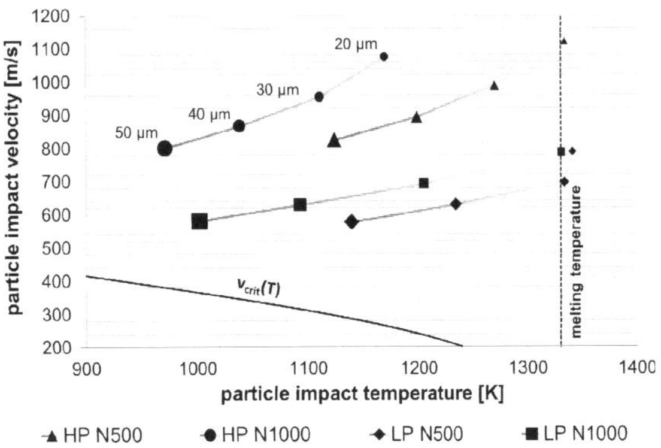

Figure 2: Window of Deposition of the highest (N1000) and lowest (N500) nitrogen flow-rate for LP-WS and HP-WS. showing the dependency of particle impact conditions for different particle sizes.

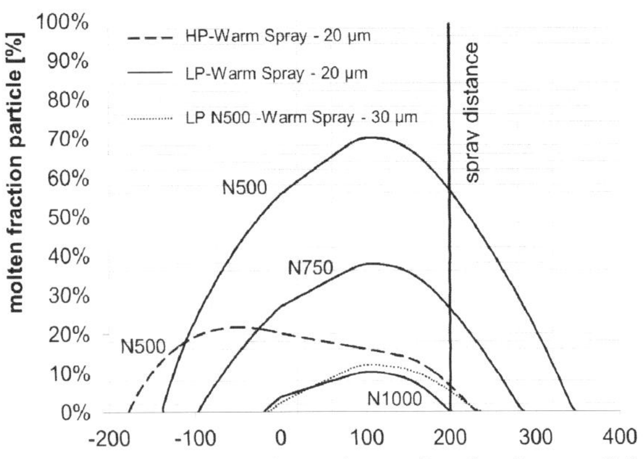

Figure 3: Overview about the molten fraction particle for different spray conditions.

can yield about 200 m/s higher particle impact velocities than achieved for lower pressure conditions. At the same time the particle impact temperature decreases about 100 K. The nitrogen flow-rate affects mainly the particle temperature whereas the particle velocity remains almost unchanged within a range of 10 m/s. Considering the particle sizes for each spray condition, larger particles remain cooler and are less accelerated than smaller particles. Smaller bronze particles between 20-30 µm reach the melting temperature and are partially molten, as shown in figure 3. At a spray distance of 200 mm from the nozzle exit (x = 0 mm), only parameter sets with low nitrogen rates (N500 & N750 LP-WS) indicated molten fractions.

3.2 Coating Microstructure

As stated in literature [13], the cavitation resistance is influenced by the coating microstructure. An overview of representative coating microstructures is given in figure 4. Figure 4a shows the microstructure of a LP-WS coating sprayed with hot spray conditions (N500). Here, the microstructure consists of large particles that indicate a low degree of deformation and smaller particles that are squeezed between the larger ones. A small fraction of pores or oxides is detected. The interface between coating and substrate shows larger non-bonded areas. In contrast, the coating sprayed with cooler conditions (N1000, figure 4b) does not show any non-bonded areas between substrate and coating, however, higher amounts of pores within the coating. Comparing coatings sprayed with hot conditions (N500) onto pre-heated and not pre-heated substrates (Fig 4a & c), the non-bonded area to the substrate decreases with substrate temperature, while slightly more pores arise in the coating. Spraying with a higher combustion pressure (Fig. 4d) and low nitrogen flow-rates, leads to an almost fully dense and well adhering coating. Particles are strongly deformed and particle-particle interfaces are hardy detectable.

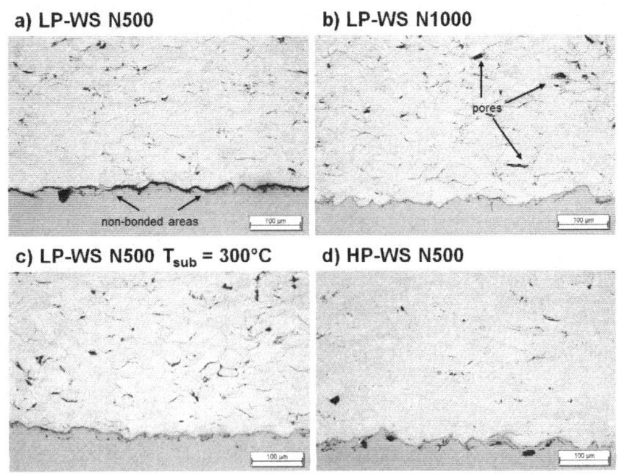

Figure 4: Coating microstructure of selected warm-sprayed Ni-Al-bronze coatings.

The correlation between coating hardness and porosity is given in Figure 5. For thermal spray coatings, it would be expected that increased porosity would result in lower hardness. The present results show an opposite trend and indicate that a decrease of nitrogen flow-rate reduces coating porosity and hardness. A further drop in hardness and porosity can be achieved using higher combustion pressures. Pre-heating the substrates leads to a stronger decrease in coating hardness at slightly higher porosity levels.

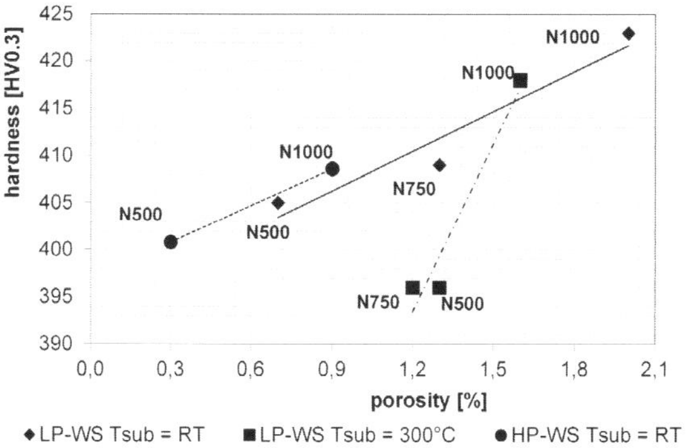

Figure 5: Relation between coating hardness and porosity of warm-sprayed specimen.

3.3 Coating Cohesion

The cohesive properties of the coatings were evaluated by cavitation tests and electrical conductivity analyses. Figure 6 gives an overview about the cavitation erosion depth during 300 min of testing. Generally, a fast initial

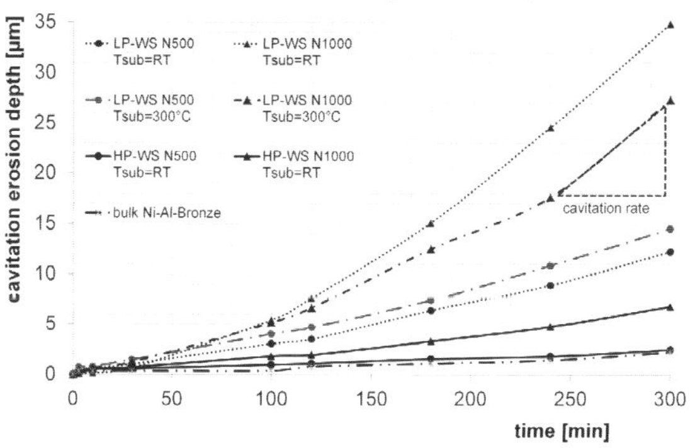

Figure 6: Cavitation erosion depth vs. exposure time.

loss of few loosely-bonded particles on the polished surface is observed. Further on, the erosion rate is comparably low until passing a certain incubation time of about 100 min. After this period, the characteristic mean cavitation rate is determined. The coatings show different erosion depths and rates depending on the spray parameter sets. The coatings sprayed with lower N_2 flow-rate and thus hotter spray conditions demonstrate lower erosion depths than those sprayed at cooler conditions. An increased substrate temperature affects mainly the coatings sprayed with 1000 l/min nitrogen flow and reduces their cavitation depth. Coatings sprayed with a lower nitrogen flow-rate show less sensitivity to the substrate temperature. Exploring the effect of a higher combustion pressure, the cavitation erosion depth is 5 - 6 times lower, approaching almost bulk-like properties.

Figure 7 shows the surface morphologies of the specimen after 300 min cavitation testing and reveals differ-

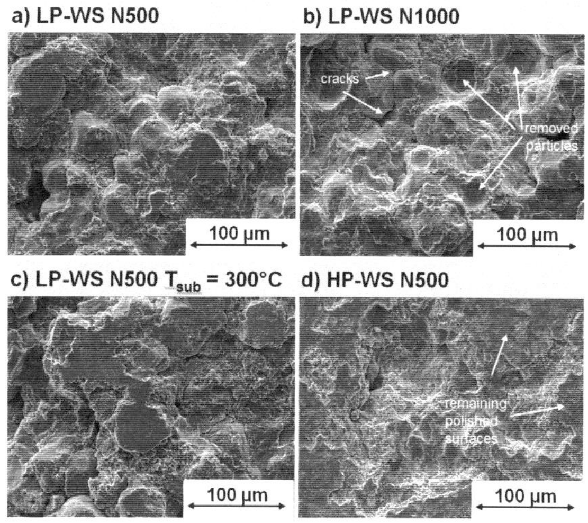

Figure 7: Surface morphologies of warm-sprayed coating surfaces after 300 min cavitation testing.

ences in wear behaviour and mass losses. Coatings sprayed with low nitrogen flow-rates (Fig. 7a & c) to some extent still show parts of the as polished surface and display adhering particles that are enclosed by strongly deformed and ruptured material fractions. These rough topographies seem to correspond to trans-particle or inter-splat failure. Differenences of wear mechanisms due to higher substrate temperatures cannot be detected. In contrast, coatings sprayed with higher nitrogen flow-rates (Fig.7b), suffer from more completely removed particles and fracture along particle-particle interfaces. Most of the fracture seems to occur across spray splats. For the hotter spray condition, in addition a significant amount of the originally polished surface is retained (Fig.7c). For the cases of using higher combustion pressures, only few particle shaped morphologies can be identified in the erosion pattern. (Fig.7d).

For cold spraying, the electrical conductivity could be used as a measure of coating intregrety and typically correlates with coating cohesion [13]. The general trend in figure 8 emphasizes that also for warm spraying good cavitation performance correlates with conductivity and thus good coating cohesion. For both chamber pressures, best results are obtained for

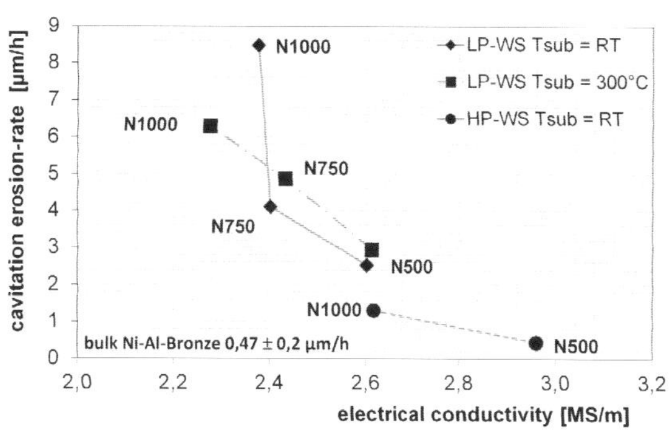

Figure 8: Relation between mean cavitation erosion rate and electrical conductivity.

coatings sprayed with low nitrogen flow-rates.

3.4 Coating Adhesion

The results for coating adhesion are summarized in figure 9. Two different trends for coatings sprayed with LP-WS are derived. Coatings sprayed onto substrates at room temperature show an increasing adhesion strength with increasing nitrogen flow-rate and thus cooler conditions. In contrast, the adhesion strength on pre-heated substrates show a reverse trend. Maximum adhesion strength could be achieved for the hottest spray (N500) conditions, exceeding the strength of the glue. Coatings sprayed with HP-WS shows better bond strength than those sprayed with LP-WS, with values

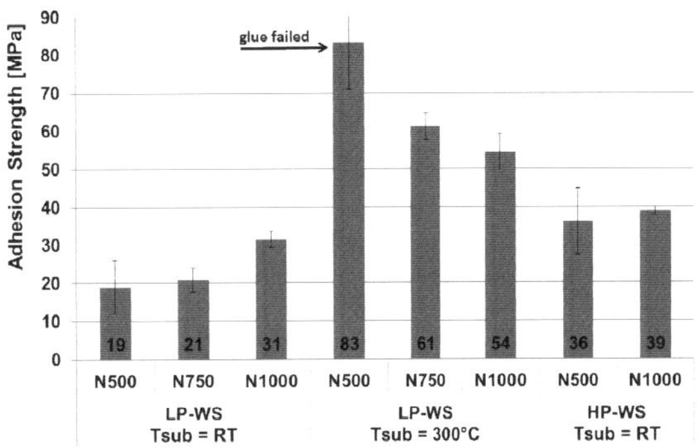

Figure 9: Results of the adhesion test.

between 36 and 39 MPa.

4 Discussion

4.1 Coating Cohesion

The results demonstrate that coating properties are mainly determined by combustion pressure and nitrogen flow-rates during spraying. The simulation data revealed that the nitrogen flow-rate basically affects the particle temperature, whereas the combustion pressure increases the particle velocity but decreases the particle temperature. The kinetic and the thermal contributions can be summed up to a energy balance, i.e. the overall energy of the impacting particle, as given in equation 1, here termed the particle impact energy, with m as particle mass, v as velocity, c_p as specific heat and T as temperature. For the specific heat a simplified expression is used without accounting for temperature dependencies.

$$E_{particle} = \frac{1}{2}mv^2 + mc_p\Delta T \qquad (1)$$

Correlations between the electrical conductivity and the cavitation-rate of the coating against particle impact energy for the mean particle size of 30 µm are displayed in figure 10. Clear trends allow to extrapolate or to predict coating performance on the basis of this impact energy. Higher impact energies, either thermal or kinetic, allow for better particle deformation upon impact and thus result in better coating cohesion, as shown by cavitation rates and confirmed by SEM investigations of the cavitated surfaces. At low impact energies, whole particles are removed and fracture

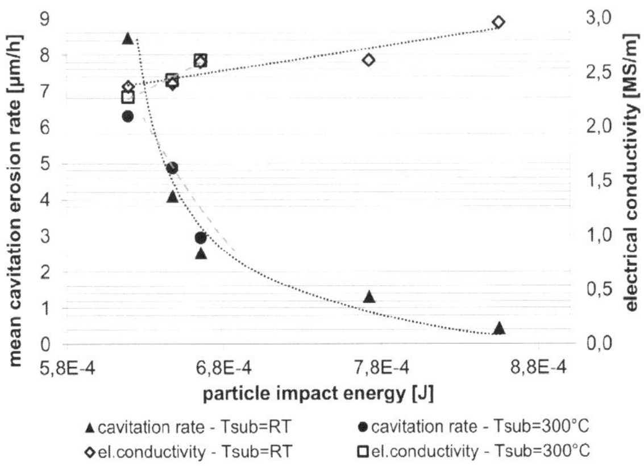

Figure 10: Correlation of coating properties like electrical conductivity and cavitation rate with particle impact energy, showing clear trends. The results are obtained for a particle diameter of 30 µm.

occurs at particle-particle interfaces. In contrast, coatings sprayed with higher impact energies are more cavitation resistant and show only minor loss by whole particle removement.

In addition, increasing impact energy leads to reduced porosity that can be attributed to more particle deformation at hotter or faster impact conditions. In the case of less cooling, the simulation results indicate that more particles are partly molten. These then could close pores or gaps within the coating. Associating molten particle fraction with high coating surface temperature, this could be an explanation for recrystallization effects causing the observed decrease of coating hardness, that may counteract any effect of particle work-hardening.

Regarding pre-heating of the substrates, the results indicate only minor influences on coating properties for spraying at high impact energies. Impacting particles then experience already a high degree of deformation and the slightly lower deceleration on thermally softened adhering spray layers does not play an important role. In contrast, coatings sprayed with cooler impact conditions and thus lower impact energy show slightly higher porosities on pre-heated substrates due to less deceleration and lower deformation upon impact. This can be confirmed by lower coating electrical conductivity and lower cavitation performance. Interestingly, the measurements of the substrate temperature during the spray process showed that after 3 to 6 passes, the substrate temperature of pre-heated and not pre-heated experiments reach a similar steady temperature regime between 300 °C and 340 °C. This observation allows for the conclusion that, in the present case, pre-heating mainly affects the first impacting particles on the substrate [15], but plays an important role for bond strength, especially for moderate or low spraying conditions.

4.2 Coating Adhesion

Regarding adhesion, the interplay between heat and deformation is rather complex. On the one hand, a higher surface temperature by substrate heating should result in easier deformation of the substrate and better bonding. On the other hand, if only one partner during the impact is soft, the other will deform less and adiabatic shear instabilities may not be as much confined to the interface, where bonding should occur. Furthermore, if both partners have a higher strength, the deceleration and deformation is confined to a small interface volume and in shorter time lead to pronounced formation of areas of shear instabilities.
This is in line with the results of this study. For LP-WS on cold substrates, higher impact energies do not lead to better coating adhesion. If spraying with cooler conditions (N1000), the particles have a higher strength and cause more deformation to the substrate. Thus, more ASI are formed at common interfaces to improve bonding.
In the same way, HP-WS increases the particle impact velocity and decreases slightly the particle impact temperature. This also results in a better formation of ASI and thus better adhesion than for LP-WS.
Coatings sprayed on pre-heated substrates show the highest adhesion strength, especially when spraying at hot conditions having a better fit between the substrate and material strength.
Summing up, best conditions for coating adhesion can be achieved when particles and substrate have a similar strength at adjusted temperatures.

5 Conclusion

The Warm Spray technology combines high powder temperatures and high velocities that result in high particle impact energies. The correlations show that highest particle impact energies lead to best coating performance. High impact energies improve bonding and the formation of shear instabilities. Furthermore molten and partly molten particles can act as a kind of bonding agent and improve coating cohesion by decreasing the number of non-bonded intrerfaces. Coating hardness plays a minor role in terms of cavitation resistance. However, best coatings showed almost bulk like properties and thus a high cavitation resistance. Such performance could be achieved by applying hottest spray conditions (N500) for HP-WS.
By substrate pre-heating, porosity increases and hardness decreases due to less deceleration upon impact. In contrast, coating adhesion increases significantly. Here a better match between strengths of thermally softened particles and substrate can be achieved.
In summary, this study proves that Warm Spraying is a promising alternative to current thermal and kinetic spraying of Ni-Al-Bronze coatings.

6 Acknowledgement

The authors would like to thank (in alphabetical order) T. Breckwoldt, T. Hiraoka, N. Németh, C. Schulze, M. Schulze, and U. Wagener, for their support in experiments and specimen characterization.

7 Literature

[1] Fauchais, P., Heberlein, J., Boulos, M., Thermal Spray Fundamentals – From Powder to Part, Springer US, 2014, p.14-72.
[2] Bourne, N.K., On the collapse of cavities, Shock Waves 11 (2002), p.447-455.
[3] Richmann, R.H., McNaughton, W.P.: Correlation of Cavitation Erosion behavior with Mechanical Properties of Metals, Wear 140, 1990, p. 63-82.
[4] Richmann, R.H., McNaughton, W.P.: A Metallurgical Approach to Improved Cavitation-Erosion Resistance, Journal of Material Engineering & Performances, 1997, p.633-641.
[5] Helma, S., Kavitationsbeobachtung in der Großausführung und am Modell, 14. SVA-Forum „Ruder in Theorie und Praxis", 2007, Potsdam/DE (German).
[6] Light, Kendrick H. Development of a cavitation erosion resistant advanced material system. Diss. University of Maine, 2005.
[7] Kumar, A., Boy, J., Zatorski, R., Stephenson, LD, Thermal Spray and Weld Repair Alloys for the Repair of Cavitation Damage in Turbines: Technical Note, Journal Thermal Spray Technology, 2005, 14(2), p 177 182.
[8] Kuroda, S., Watanabe, M., Kim, K.-H.Katanoda, H., Current Status and Future Prospects
of Warm Spray Technology, Journal of Thermal Spray Technology Vol. 20(4), June 2011, p.653-676.
[9] Katanoda, H., Morita, H., Komatsu, M., Kuroda, S., Experimental and Numerical Evaluation of the Performance of Supersonic Two-Stage High-Velocity Oxy-Fuel Thermal Spray (Warm Spray) Gun, Journal of Thermal Science Vol. 20(1), 2011, p.66-92.
[10] Watanabe, M., Brauns, C., Komatsu, M., Kuroda, S., Gaertner, F., Klassen, T., Katanoda, H., Effect of N2 flow-rate on microstructure and mechanical properties of metallic coatings by warm spray deposition, Surface & Coatings Technology 232, 2013, p.587-599.
[11] Kawakita, J., Fukushima, T., Kuroda, S., Katanoda, H., Matsuo, K., Fukanuma, H., Dense titanium coatings by modified HVOF spraying, Proc. Thermal Spray Conference 2005, Thermal Spray Connects, , May 2-4, 2005, Basel, Switzerland (Ed.E.Lugschneider), DVS-Verlag, Düsseldorf, Germany, 2005, p.705-710.
[12] Katanoda, H., Kiriaki, T., Tachibanaki, T., Kawakita, J., Kuroda, S., Fukuhara, M., Mathematical Modeling and Experimental Validation of the Warm Spray process, Journal of Thermal Spray Technology, Vol. 18(3), September 2009, p.401-410.
[13] Krebs, S., Gaertner, F., Klassen, T., Optimizing Cavitation Resistance - An Approach By Cold Spraying Of Several Bronze Materials -, Proc. Thermal Spray Conference, May 11-14, 2015, Long Beach, USA, 2015, ASM International, p.705-710.

[14] Assadi, H., Schmidt, T., Richter, H., Klieemann, J.-O., Binder, K., Gaertnerm F., Klassen, T., Kreye, H., On parameter selection in Cold Spraying, Journal of Thermal Spray Technology, 2009, 18(5-6), p.794-808.
[15] Villa, M., Haußler, F., Assadi, H., Gaertner, F., Klassen, T., Influence of Substrate on Cold Sprayed Titanium Coatings, Proc. Thermal Spray Conference, May 11-14, 2015, Long Beach, USA, 2015, ASM International, p.1047-1054.

In-flight Particle Diagnostics and Control for the Production of Thermally Sprayed Babbitt Coatings

[1]Nascimento, A. R. C., [1]Ettouil, F. B., [2]Savoire, S., [2]Schulz, R., [1]Moreau, C.
1- Concordia University, Department of Mechanical and Industrial Engineering
2- Institut de Recherche d'Hydro Québec

High-Sn bearing alloys have been used for over a century in different industries due to their unique tribological properties. While traditionally applied by casting, the thermal spray technique has gained its space in the bearing manufacturing market due to its high productivity and flexibility especially when it comes to the repair of components. Regardless of such facts, different spray techniques and process parameters are used for the production of Babbitt coatings and little is known of their effects on the bearing's performance. The present work reports the results of using the AccuraSpray and DPV-2000 diagnostic systems to characterize in-flight Babbitt particle characteristics during the production of HVOF coatings. Those characteristics were correlated with process deposition efficiency as well as the coatings general structure including porosity and intermetallic phase distribution. HVOF coatings of up to 370 μm thickness were produced. A significant refinement of intermetallic phase distribution was obtained, when compared with commercial flame-spray and arc-sprayed coatings.

1 Introduction

The term Babbitt or white metal refers to a group of tin (Sn) or lead (Pb) alloys. Babbitting therefore is the process by which those alloys are bonded to a stronger supporting base metal, usually steel, cast-iron or a bronze. They are widely used in hydrodynamic bearings due to their characteristics that include:

- ***Compatibility:*** because of its low hardness Babbitt operates as a sacrificial surface in the event of a contact with a more sensitive and expensive component. Moreover, it is known for its low tendency to gall to steels.
- ***Conformability:*** Its relative low stiffness allows Babbitt alloys to conform to bearing misalignments.
- ***Embedability:*** Babbitt bearings are able to retain dirt that eventually reaches the lubricating oil, avoiding further abrasion and oil heating.

Traditionally babbitting is accomplished by static or centrifugal casting. In both processes it is important to control the cooling of the assembly, ensuring that the solidification starts on the supporting metal/Babbitt interface and not from the surface of the Babbitt. This prevents poor bonding with the substrate and also segregation of chemical elements during solidification which would lead to inhomogeneous bearing properties. [1]

During thermal spraying of Babbitt alloys, the material is fed in the form of powder or wire and it is melted (or partially melted) by means of a flame (flame spray) or an arc (arc spray). The molten/semi-molten particles are then projected towards a target substrate by means of an accelerated gas. As a result, the solidification rates are much faster when compared to traditional casting techniques, which eliminates the problem of segregation of alloying elements such as Cu and Sb. Unlike cast coatings, thermally sprayed Babbitt coatings can be made thin which not only helps in saving material but also increases the coating load capacity and fatigue resistance [2].

Thermal spray coatings have inherent defects such as porosity and oxides that usually remain between the splats formed during molten particle deposition. On the one hand, a certain amount of porosity is believed to be beneficial in conditions like oil starvation, in which some oil droplets trapped in the pores come out of the Babbitt coating to compensate for the lack of oil between the sliding surfaces in a bearing [3]. On the other hand, if such porosity level is too high the coating mechanical properties will be severely reduced compromising the bearing's load capacity.

The use of Sn in higher velocity thermal spray techniques dates back to 2001 when Jeandin et. al. [4] used plasma spray to coat LiF windows. They successfully produced 500 µm thick coatings with significant closed porosity (4.1 – 5.1%). A more recent work by Legoux et. al. [5] shows that the cold spray deposition efficiency of pure Sn is very low (~5%) due to the high average particle velocity (500 m/s) which causes coating erosion instead of deposition.

Instead of very high temperatures or very high velocities, the present work reports attempts to produce Babbitt coatings in an intermediary condition exploring the heat generation of the Diamond Jet HVOF flame as well as avoiding excessive particle velocities by coupling to the DJ gun a radial injection system. The effects of the use of different particle size distributions and process parameters are investigated by in-flight particle diagnostic techniques and by evaluating deposition efficiencies and the final coating structures.

2 Methodology

The commercially available BabTec® 29240 powder was used. According to the manufacturer, the powder composition follows that of the ASTM B23 grade 2, which is 89Sn7.5Sb3.5Cu. The alloy's melting point is 241 °C and its density is 7.39 g/cm^3. The particle size distribution (PSD) was determined by laser diffraction in the Horiba® LEA-950 particle analyser. By using a

200 mesh screen, the powder was separated in two diameter groups: coarse powder (75-150 μm) and fine powder (20-75 μm).
To avoid excessive particle heating, oxidation and potential nozzle clogging, a radial injection system was developed and adjusted to the Diamond Jet (Oerlikon Metco) HVOF gun allowing the Babbitt powder to be injected perpendicularly to the nozzle's axis. The radial injector was adjusted to the DJ gun's nozzle at axial and radial distances of 2 cm.

A more traditional means to control particle melting and acceleration in HVOF is by changing the flame enthalpy i.e. changing the heat generated during the combustion of propylene and oxygen. In the present work, while maintaining a fixed fuel-to-oxygen ratio, three different flow rates were chosen yielding three varying flame temperatures. Table 1 shows the selected gas flow rates for the present study.

Table 1 - Gas flow rates used during HVOF spray of BatTec® powder.

Operation Conditions	***Flow Rate (SCFH)***		
	Air	**Oxygen**	**Propylene**
HE (High Enthalpy)	710	360	50
ME (Med Enthalpy)	550	290	40
LE (Low Enthalpy)	395	215	30

Nitrogen gas was used at 25 LPM to keep the HVOF gun chamber properly cooled. Argon was chosen as carrier gas for the Babbitt powder. After some tryouts, its flow rate was fixed at 8 LPM. The powder feed rate was fixed at 20 grams per minute. It is worth mentioning that the chosen process conditions significantly differ from those standardized by Oerlikon Metco. The general operation specifications for this equipment includes using propylene at 200 SCFH, air at 1000 SCFH and oxygen at 700 SCFH.
In order to assess the effects of different flame enthalpies as well as particle diameter on particle's in-flight temperature and velocity, the AccuraSpray-G3C (Tecnar Automation Ltd., St-Bruno, Canada) was used. During the first spray try-outs the optical intensity of the flame was measured along a sampling line positioned along the spray-axis by using AccuraSpray-G3C's charge-coupled device (CCD) camera. The low radiation intensity of the signals rendered temperature and velocity measurements insatisfactory. Therefore, the use of DPV-2000 (Tecnar Automation Ltd., St-Bruno, Canada) in laser scattering mode was necessary to properly measure particle velocity.
Deposition efficiency (DE) tests were performed at 200 mm stand-off distance and the gun's traverse speed was set to 1 m/s. DE tests for all of the three operation conditions (HE, ME, LE) were performed by translating the gun in front of carbon steel substrates (3 in x 3 in) in a single horizontal line (step = 0 mm) for a total of 100 times. The calculation of the DE was done by determining the ratio of the mass of material that was actually deposited on the substrate to the mass of powder that is sprayed while the spray gun was in front of the substrate. The effect of process parameters on the plume width and powder penetration as measured by the distance between the plume axis and the substrate mid-plane (or the gun's axis) was evaluated. **Figure 1** illustrates the parameters being investigated.

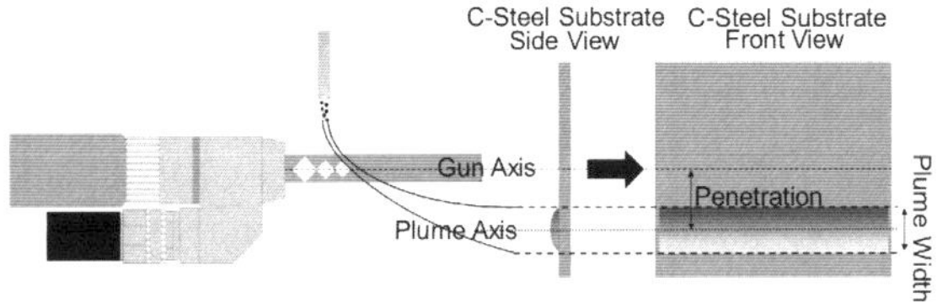

Fig. 1 Schematics of the experimental set-up to evaluate particle trajectories and deposition efficiencies.

After choosing the most favorable process conditions, Babbitt coatings were produced on Ni-Al coated steel coupons.
The analysis of the coating structures was performed by means of optical and electronic microscopy. Samples were molded in epoxy resin and then cut prior to grinding in SiC paper and polishing with Al_2O_3 (1 μm diameter) suspension. Chemical etching was applied to reveal intermetallic phase size distribution. A solution of 2% Nital was used as etchant.

3 Results

3.1 Particle In-Flight Diagnostics

Despite the use of radial injection and the low oxygen and fuel flow rates, some particles still achieved supersonic (>340 m/s) speeds and more importantly temperatures that exceeded tin oxide's (SnO and SnO_2) boiling and sublimation points (1527 K and 1850 K respectively). Regardless of the aforementioned values, the optical intensity measured by AccuraSpray's CCD camera was very low for all chosen process parameters and particle size groups studied. It is therefore very likely that a significant part of the collected optical signal was generated by the oxidation reaction of the Babbitt powder and not solely by the molten particle's thermal radiation. Moreover, the low optical intensity suggests that only a small fraction of the particles could generate enough radiation to be detected and therefore have its temperature and velocities measured.

Using DPV-2000 in laser scattering mode allowed a more reliable determination of particle velocities, which are summarized in **Table 2**. As expected, finer particles showed consistently higher velocities as compared to the coarse group especially for low and medium enthalpies. Another interesting remark is that the combination of lower flame enthalpy and external radial injection places the current process on the interface between HVOF and Cold Spray techniques. Such information has proven valuable for comprehending the deposition efficiency results.

Table 2 Results of Babbitt particle velocity measurements by using DPV-2000. (Error corresponds to standard deviation).

Particle Velocity (m/s)			
	Flame Enthalpy		
Size Group	***Low***	***Medium***	***High***
Coarse Particles	142 ± 62	214 ± 64	438 ± 76
Fine Particles	215 ± 42	323 ± 52	456 ± 67

3.2 Particle Trajectory and Deposition Efficiency

Figure 2 shows the narrowing of the plume width as the flame enthalpy increases. The values are averaged for both coarse and fine group of particles. Since no changes to the argon flow rate was applied, the increase in fuel and oxygen flow rates made it more difficult for the Babbitt particles to penetrate the flame. The same figure shows there is a good correlation between the particle's penetration and the plume width.

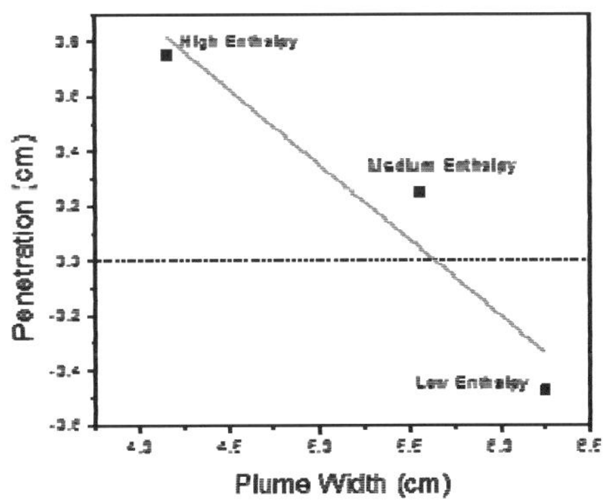

Fig. 2 Correlation between particle penetration and the plume width.

As shown in **table 3**, the deposition efficiency of coarse and fine particles increase as the flame enthalpy is reduced. The effect was more pronounced in the finer group of particles since those are more easily accelerated and heated during their flight. Schmidt et. al. [6] have calculated that the velocity range for proper cold spray deposition of Sn particles remains approximately within 150 m/s (critical threshold) to 375 m/s (erosion threshold). DPV-2000 measurements showed that at high flame enthalpy both particle size groups had average velocities above the erosion threshold but only the finer had a decrease in the deposition efficiency possibly due to the particle's higher temperature as indirectly assessed by AccuraSpray's CCD camera On the milder process condition range, the change in enthalpy from low to medium level causes 8.5% and 13.5% reduction in DE for the coarse and the fine group of particles respectively. In summary those results show that using coarser particles is more favorable when trying to deposit soft and low melting point materials.

Table 3 Deposition efficiency of BabTec® 29240 as a function of particle size and flame enthalpy.

Deposition Efficiency (%)			
	Flame Enthalpy		
Size Group	***Low***	***Medium***	***High***
Coarse Particles	46	42	42
Fine Particles	37	32	28

The values obtained for the deposition efficiency in this work are much superior to those previously reported by Legoux et. al. [5]. Tables 2 and 3 show that combining the control of flame enthalpy and particle size in HVOF process can provide industrially acceptable deposition rates for soft materials such as Sn-alloys.

3.3 Coating Structure

Figure 3 shows the cross section of the Babbitt coatings produced after 100 passes at low enthalpy. The use of the finer particles resulted in a 290 µm thick coating, while for the coarser group, the achieved thickness was 370 µm. This 27% increase in the coating's average thickness is, as previously shown, very similar to the increase in deposition efficiency (around 24%) when we use coarse particles instead of fine particles.

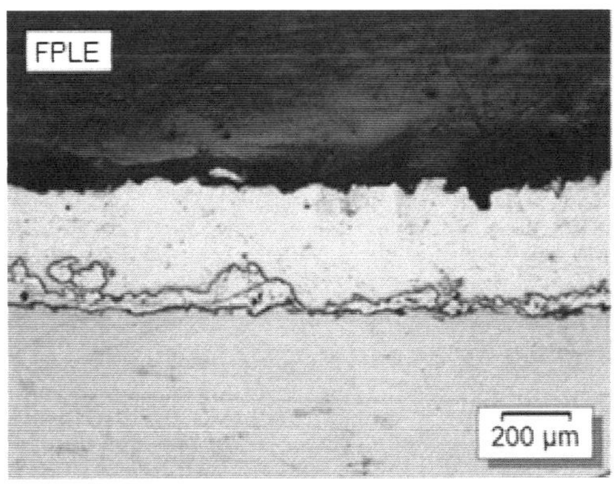

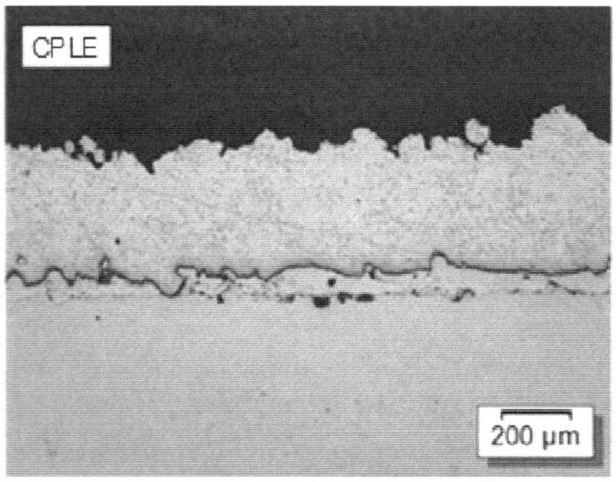

Fig. 3 Cross section of unetched Babbitt coatings. FPLE : fine particles low enthalpy. CPLE : coarse particles low enthalpy.

The coatings in **Figure 3** were chemically etched to reveal the size and distribution of intermetallic phases such as SnSb and Cu_6Sn_5. The results are displayed in **Figure 4**, and are compared with a cast, a flame spray and an arc-spray coating, all provided by Hydro-Quebec. It is shown how the HVOF coatings combine both low porosity common in cast materials with a fine precipitate (brighter phase) distribution, typical of thermal spray coatings in which the cooling rate is much higher.

4 Conclusions

During HVOF deposition of BabTec® 29240, as oxygen and fuel flow rates are reduced, so does the optical intensity measured by AccuraSpray's CCD camera, which is an indication of the reduction in particle temperature. Accordingly, DPV-2000 velocity diagnostics revealed particle velocity reduction caused by lower flame enthalpies. The diagnostics used in this work reveal that the selection process conditions allied with the use of radial injection rendered the particles characteristics that are closer to cold or warm spray processes. Moreover, it was shown that using particle size ranges within the typical 20-53 µm values is not the best strategy when attempting to produce coatings with low melting point materials. Particle sizes with diameters ranging from 75-150 µm outperformed those between 20-75 µm probably due to their lower velocity and less likeliness to cause erosion of the coating during deposition. For the case of BabTec® 29240 subsonic particle velocities are preferable.
It was shown that combining the use of radial injection and low oxygen and fuel flow rates could affect BatTec® 29240 deposition efficiency from 28% up to 46%, which is an industrially acceptable value.
Finally, in terms of coating structure, Babbitting by HVOF combines low porosity and fine intermetallic phases distribution, which cannot be found in other processes individually such as casting and other low velocity thermal spray techniques.

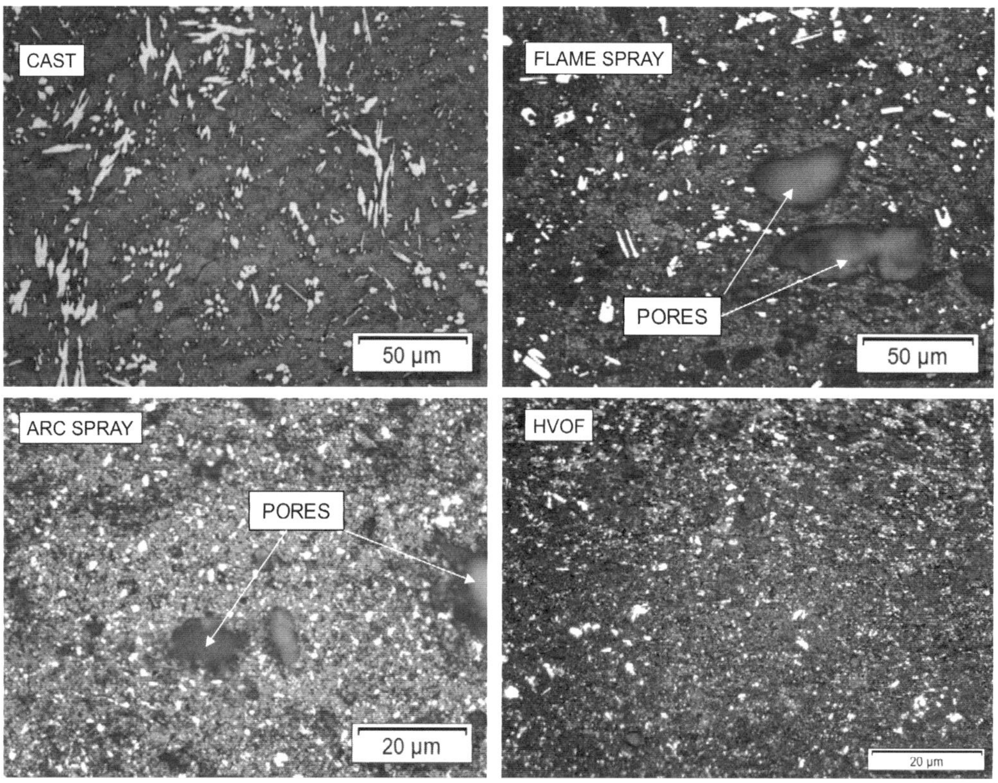

Fig. 4 Comparison of intermetallic phase distribution in cross section areas of Babbitt coatings produced by different processes.

5 References

[1] W. P. Bardet and D. J. Wengler, "Babbitting," in *Surface Engineering, Vol5, ASM Handbook*, 1994, pp. 372–377.

[2] E. S. Hedges, *Tin and its alloys*. London: Edward Arnold Publishers, 1960.

[3] F. Van Rodijnen, "Using Thermal Spray to Conserve Resources," *Welding Journal*, pp. 24–27, 2009.

[4] J. P. Beaulieu, F. Borit, V. Guipont, M. Jeandin, and C. Mabire, "Plasma Spraying of Tin on to a Monocrystalline Lithium Fluoride Window," in *Proceedings of the International Thermal Spray Conference*, 2001, pp. 79–85.

[5] J. Legoux, E. Irissou, and C. Moreau, "Effect of Substrate Temperature on the Formation Mechanism of Cold-Sprayed Aluminum, Zinc and Tin Coatings," *J. Therm. Spray Technol.*, vol. 16, no. 5, pp. 619–626, 2007.

[6] T. Schmidt, F. Gärtner, H. Assadi, and H. Kreye, "Development of a generalized parameter window for cold spray deposition," *Acta Mater.*, vol. 54, no. 3, pp. 729–742, Feb. 2006.

TGO formation and failure mode of TBC systems comprising PVD-AlO_x interlayers

I. Ali[1], D. Wett[2], T. Grund[1], T. Lampke[1]
[1]Institute of materials and surface engineering, Technische Universitaet Chemnitz, Chemnitz, Germany
[2]Institute of composite materials, Technische Universitaet Chemnitz, Chemnitz, Germany
Ibrahim.ali@mb.tu-chemnitz.de, Thomas.grund@mb.tu-chemnitz.de

Abstract

The physical characteristics and volume growth of TGOs in TBC systems lead to a premature TBC failure. The formation and growth of TGO layers were studied with respect to the modification of the bond coat by controlled addition of alloying elements, pre-oxidation, pre-diffusion oxidation and adding inter-layers at the bond coat/top coat interface. It is proved that enriching the BC/TC interface with α-Al_2O_3 is beneficial to an extended operational time by prolonging the steady-state growth stage of the TGO. The corresponding phase reactions in TBC systems with increased Al activity at the BC/TC interface, however, are under study. In this work the stage formation of TGO layers of TBC systems with PVD-AlO_x interlayers is described. The study uses thermal cyclic loading with dwell time at a maximum temperature of 1,150 °C. The crack formation in the ceramic top coat and the TGOs thickness are investigated by means of SEM/EDS after 6, 12, 24, 40 and 80 thermal cycles. The results plot the interface change and crack formation as a function of the thermal cycle number. The corresponding failure mode is discussed.

1 Introduction

Thermal barrier coatings (TBCs) are applied for the protection of metallic components during service in high-temperature environments [1 - 2]. TBC systems generally comprise [2 - 5] a super alloy substrate, an oxidation-resistant metallic bond coat (BC), an 0.25 to 0.35 mm thick layer of stabilized zirconia ceramic top coat (TC) and a layer of thermally grown oxides (TGO) which is formed in the interface between the metallic BC and the ceramic TC by artificial or operational oxidation. Today, standard TBC layers of 0.25 mm thickness reduce the surface temperatures of turbine components by up to 170 K compared to the surrounding gas temperature [6, 7]. This goes along with savings in cooling air (about 6 %) and fuel consumption (about 1.3 %) [8]. The oxidation-resistant metallic BC usually comprises a 0.1 - 0.15 mm thick M-CrAlX alloy layer where "M" generally represents a Ni-, Co- or NiCo-base, while "X" represents further alloying elements in small quantities, like Y, Hf, Si or Re [2, 9 - 14]. The main phases in Ni and NiCo based BC are γ-Ni and β-NiAl phases, representing the face-centered cubic phase with low Al content and the Al-rich body-centered cubic phase, respectively.

During service, the M-CrAlX BC material is subjected to oxidation and degradation. The degradation is characterized by the depletion of the Al-rich β-phase in the volume near the BC/TC interface. The depletion of Al is due to its use in forming an α-Al_2O_3-rich protective oxide layer which decreases the oxygen transport into the lower BC regions and to the substrate [2, 15-17]. Ongoing service and oxidation leads to a continuous degradation process of the BC material, until it loses the ability to form the protective TGO (α-Al_2O_3-rich) layer, detrimental oxides are increasingly formed and failure occurs. A BC or barrier layer that meets all requirements (TC adhesion, oxidation resistance, thermo-mechanical behavior, chemical protection) is under development [13]. A controlled TGO development or preferential formation of α-Al_2O_3 is one possibility to enhance operational times of TBC systems. High aluminum rich oxide activity within the BC/TC interface area would promote dense and hermetic α-Al_2O_3-rich TGO phases and thus would improve the TBC system performance. A systematic investigation of TBC systems that present high concentrations of AlOx in the BC/TC interface area is therefore promising. An approach to face this challenge is the application of thin interlayers. This could provide high concentrations of Al to quickly form alumina rich thin layers between BC and TC. Al ratios in artificially applied layers could furthermore purposefully provide "Al sources" and prevent uncontrolled Al depletion in regions beneath the TGO. Hence, the addition of thin PVD-interlayers of AlO_x into conventional TBC systems is applied to improve their durability. Therefore, investigation in this direction is presented.

2 Experimental work

The chosen sreference TBC system was air plasma sprayed (APS) and comprised an Inconel 600 (DIN 2.4816, UNS N06600) substrate, a CoNiCrAlY BC and a yttria-stabilized zirconia (YSZ) TC. The BC and TC were sprayed with an aimed thickness of 150 µm. The processed CoNiCrAlY powder (CoNi32Cr21Al8Y0.5, catalog number: 60.95.1, GTV GmbH, Luckenbach, DEU). The YSZ powder (YSZ 92/8, catalog number: 40.23.1, GTV GmbH, Luckenbach, DEU). The used APS system was an F6 plasma torch (GTV GmbH, Luckenbach, DEU). The applied APS parameters were standard parameters and were provided by the supplier. The interlayer were produced using DC magnetron sputtering physical vapor deposition (in the following referred to as PVD) in a vacuum chamber (working pressure 8×10^{-3} mbar) and using a pure-Al target (Al 99.999, EVOCHEM Advanced Materials GmbH, Offenbach, DEU). The sputtering time of the interlayer onto the BC is 20 min. The sputtering gas consisted of an Ar/O_2 (30 sccm/20 sccm) mixture. Thermal cycling tests were carried out up to 80 cycles. In each cycle, the specimens were exposed to air at 1,150 °C for 25 min, then cooled down to values between 100 and 150 °C using compressed air and heated up again to 1150 °C. Heating and cooling rate were + 400 K/min and - 400 K/min, respectively. For comparative reasons, the reference TBC system without interlayer was included into the thermal tests. The reference TBC system and the TBC system with AlO_x interlayer are abbreviated in the following as R-TBC and AlO_x-TBC respectively. After the thermal tests, all specimens were metallographically investigated, including scanning electron microscopy (SEM) and energy dispersive X-ray analysis (EDX). Image analysis (Fiji software) was used to measure the crack lengths and TGO thicknesses for samples after 6, 12, 24, 40 and 80 cycles. The TGO thicknesses at the interface were estimated based on the following formula [19]:

$$\delta eq = \sum \frac{\text{cross sectional area of TGO}}{\text{cross sectional length of TGO}}$$

SEM images with magnification of 100 and 400 were used for crack and TGO thickness measurements, respectively. The averaged values of 25 and 35 measurements, i.e. 25 and 35 images, have been considered for cracks and TGO thicknesses measurements, respectively.

3 Results

3.1. Microstructure

The microstructures of the as-sprayed R-TBC and AlO_x-TBC systems were presented in previous studies from the authors in Ref. [18] and are shown in **Fig.1**.

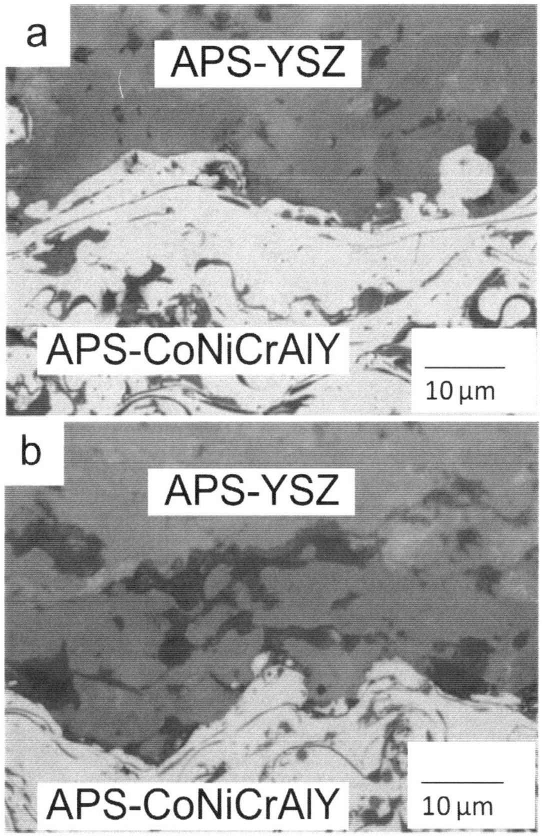

Fig 1: Optical microscope microstructure for a TBC system (a) without AlO_x interlayer (R-TBC), (b) with AlO_x interlayer (AlO_x-TBC) [18].

The as-sprayed microstructure reveal contact between the metallic BC and the ceramic TC as well as rough and uneven features at the BC/TC interface are also observed in both TBC systems. The microstrcucture after 40 thermal cycles is presented in Ref [18]. It was reported that after 40 thermal cycles, the AlO_x-TBC system as compared to R -TBC system shows limited cracks formation in the YSZ TC and limited TGO thickness at the BC/TC interface. In the present study, the microstructures of R-TBC and AlO_x-TBC systems after 80 thermal cycles are shown in **Fig. 2**. The AlO_x-TBC system as compared to R-TBC system shows limited TGO formation at the BC/TC interface. TGO formation in both TBC systems is marked in **Fig. 2**. Recent studies [10, 19] discussed the chemical characteristics and the effect of such oxide phases at the BC/TC interface.

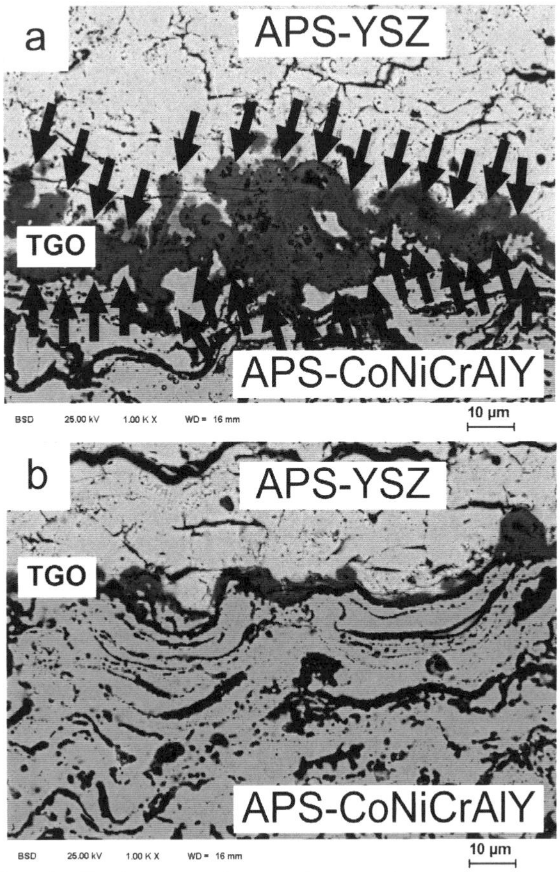

Fig 2: Backscattered scanning electron microscope image (BSD-SEM) microstructure of TBC systems after 80 thermal cycles (a) without AlO_x interlayer (R-TBC) (b) with AlO_x interlayer (AlO_x-TBC); TGO is marked in the BC/TC interface.

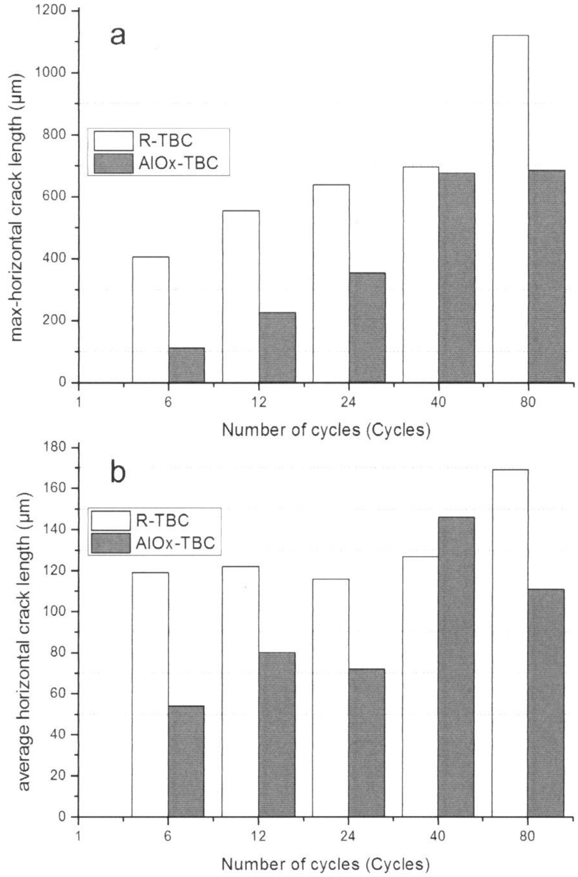

Fig 3: Effect of number of cycles on (a) average and (b) maximum horizontal crack length for R-TBC and AlO_x-TBC systems

3.2. Cracks Measurement

The effect of the number of thermal cycles on the average and maximum horizontal crack lengths in the R-TBC and AlO_x-TBC systems is shown in **Fig. 3**. As can be seen, increasing the number of cycles mostly increases the average and maximum crack length in both systems. The maximum horizontal crack length is a critical factor controlling the spallation of the YSZ TC layer. In comparison to the R-TBC, the AlO_x-TBC system shows a shorter maximum crack length. Accordingly, a lower spallation rate for the YSZ TC layer is expected for AlO_x-TBC systems.

3.4. TGO measurement

The effect of the number of thermal cycles on the TGO thickness in the BC/TC interface for R-TBC and AlO_x-TBC systems is shown in **Fig. 4**. As can be seen, increasing the number of cycles already increases the average, maximum and minimum TGO thickness in both systems. After 80 cycles the R-TBC system shows higher TGO thicknesses in comparison to the AlOx-TBC system. Since the final TGO thickness is a critical factor which controlling the spallation of the YSZ TC layer, as a result, the R-TBC system is expected to have higher failure rate (by spallation) and hence lower durability.

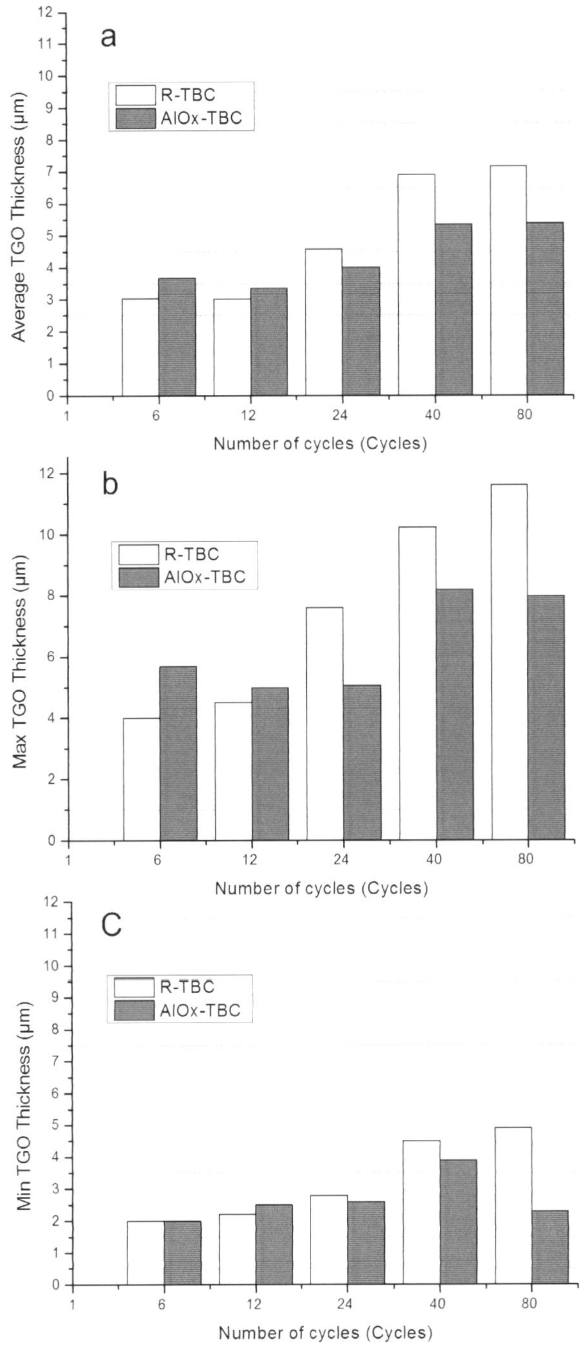

Fig 4: Effect of number of cycles on (a) average, (b) maximum and (c) minimum TGO thickness for R-TBC and AlOx-TBC systems

3.4. Discussion of loading condition and failure mode

R. Herzog et al. [2, 5] classified the load types and failure modes of TBC systems according different conditions. The applied loading condition in the present study was thermal-cyclic loading with dwell time at high temperature. The investigated microstructures revealed the expected mixed-failure mode showing crack paths across the TGO and the TC layer for both TBC system types. Basing on this, the slowed TGO growth and apparently stabilized TGO steady-state growth stage of the AlOx-TBC system are promising approaches to increase the service times of TBC applications by suchlike altered TBC systems.

4 Conclusion

An air plasma sprayed thermal barrier coating system showing a PVD-AlOx interlayer is addressed in the present study. The basing TBC system comprised a CoNiCrAlY bond coat and a YSZ top coat. For comparative reasons, an unaltered TBC system without PVD-AlOx interlayer - referred to as reference TBC system - was prepared and equally investigated. Repeated thermal cycling load with dwell time at high temperature was applied on both systems. Investigations were conducted after 6, 12, 24, 40 and 80 cycles. Crack lengths and TGO thicknesses were measured in the YSZ TC layer and at the BC/TC interface, respectively. The investigated microstructures revealed the expected mixed-failure mode in both TBC systems. In comparison to the reference TBC system, the altered system with PVD-AlOx interlayer showed a reduced crack formation in the ceramic TC as well as a slowed TGO formation and resulting TGO thickness at the BC/TC interface during and after the thermal treatment. The conducted approach hence is understood as a promising possibility to enhance the service time of TBC systems. While this first basic results are gained only for the specific APS sprayed TBC systems and used materials, a general effectiveness of artificial AlOx interlayers is assumed. Further investigations using a broader range of materials, spraying methods and parameters will be carried out.

5 References

[1] Zircotec coatings. Link, http://www.zircotec.com/page/exhaust-system-coatings/60, 19th Jan, 2013.

[2] Trunova, O. Beck, T. Herzog, R. Steinbrech, R. Singheiser, L. Damage mechanisms and lifetime behavior ofplasma sprayed thermal barrier coating systems for gas turbines-part 1: Experiments. Surf. Coat. Tech 202, (2008), pp. 5027/5032.

[3] Beck, T. Trunova, O. Herzog, R. Singheiser, L. TBCs for Gas Turbines under Thermomechanical Loadings: Failure Behaviour and Life Prediction. J. Energy and Power Engineering 7 (2013), pp. 647/653.

[4] Lima, C. Nin, J. Guilemany, J. Evaluation of residual stresses of thermal barrier coatings with HVOF thermally sprayed bond coats using the modified layer removal method (MLRM). Surf. Coat. Tech 200 (2006), pp. 5963/5972.

[5] Herzog, R, Trunova, O. Steinbrech, R. Singheiser, L. Failure modes and damage evolution of plasma sprayed thermal barrier coatings. First Japanese-German workshop on properties and performance

of thermal barrier coating systems and factors affecting it. (2006), Darmstadt, Presentation.
[6] Clarke, D. Phillpot, S. Thermal barrier coating materials. Materials today 8 (2005), issue 6, pp. 22/29.
[7] Lima, C. Guilemany, J. Adhesion improvement of thermal barrier coatings with HVOF thermally sprayed bond coats. Surf. Coat. Tech. 201 (2007), pp. 4694/4701.
[8] Miller, R. Levine, S. Stecura, S. Thermal barrier coatings for aircraft gas turbines. AIAA 18th aerospace and sciences meeting (1980), Pasadena, California.
[9] Limarga, A. Widjaja, S. Yip, T. Mechanical properties and oxidation resistance of plasma sprayed multilayered Al_2O_3/ZrO_2 thermal barrier coatings. Surf. Coat. Tech. 197 (2005), pp. 93/102.
[10] Chen, W. Wu, X. Marple, B. Patnaik, P. The growth and influence of thermally grown oxide in a thermal barrier coating. Surf. Coat. Tech. 201 (2006), pp. 1074/1079.
[11] Richer, P. Yandouzi, M. Beauvais, L. Jodoin, B. Oxidation behavior of CoNiCrAlY bond coats produced by plasma, HVOF and cold gas dynamic spraying. Surf. Coat. Tech. 204 (2010), pp. 3962/3974.
[12] Wang, Y. Sayre, G. Commercial thermal barrier coatings with a double-layer bond coat on turbine vanes and the process repeatability. Surf. Coat. Tech., 203 (2009), pp. 2186/2192.
[13] Darolia, R. Thermal barrier coatings technology: Critical review, progress update, remaining challenges and prospects. Int. Mater 58 (2013), issue 6, pp. 315/348.
[14] Munawar, A. Schulz, U. Cerri, G. Lau, H. Microstructure and cyclic lifetime of Gd and Dy containing EB-PVD TBCs deposited as single and double layer on various bond coats. Surf. Coat. Tech 245 (2014), pp. 92/101 (doi: 10. 1016/j.surfcoat 2014, 02.047).
[15] Yuan, F. Chen, Z. Huang, Z. Wang, Z. Zhu, S. Oxidation behavior of thermal barrier coatings with HVOF and detonation sprayed NiCrAlY bond coats. Corros Sci., 50 (2008), pp. 1608/1617.
[16] Kwon, J. Lee, J. Gil, J. Effect of bond coat nature and thickness on mechanical characteristic and contact damage of zirconia based thermal barrier coatings. Surf. Coat. Tech. 201 (2006), pp. 3483/3490.
[17] Busso, E. Qian, Z. Taylor, M. Evans, H. The influence of bond coat and top coat mechanical properties on stress development in thermal barrier coating systems. Acta Mater 57 (2009), pp.2349/2361.
[18] Ali, I. Wett, D. Grund, T. Nestler, D. Wielage, B. Lampke, T. Intermediate PVD layers as diffusion barriers in turbine coating systems. Proc ITSC 302 (2014), pp. 273/278, ISBN978-3-87155-574-9.
[19] Chen, W. Wu, X. Marple, B. Lima, R. Patnaik, P. Pre-oxidation and TGO growth behavior of an air plasma sprayed thermal barrier coating. Surf. Coat. Tech., 202 (2008), pp. 3787/3796.

3D tomography of plasma sprayed Cr-clad Ni on alumina

A.Y.X. Wang, P.R. Munroe, UNSW, Australia | A.T.T. Tran, M.M. Hyland, The University of Auckland, New Zealand

A mechanically clad Ni-Cr powder feedstock was thermally sprayed onto an alumina substrate held at room temperature by atmospheric plasma arc deposition at 2400°C. The resultant splats, with typical dimensions ~100 µm diameter and ~2-3 µm thickness, were analysed for features such as interfacial bonding, splat classification and, critically, Cr distribution. Using a slice-and-view sectioning technique in a dual-beam focussed ion beam – scanning electron microscope (FIB-SEM), a representative splat exhibiting both discrete Ni and Cr regions, identified by energy dispersive spectroscopy (EDS) analysis, was physically deconstructed then virtually reconstructed using specialized data visualisation software (Avizo Fire 8.1 and MeshMagic) to analyse individual layers within the splat. Although the powder feedstock featured discrete Ni particulates clad with clusters of Cr to form a porous coating, the solidified splat features distinct segregation of Ni and Cr layers, with little interaction between them. A model was formulated upon this analysis, showing that the distribution of the Cr cladding on the Ni particulate influences the amount and location of Cr around the solidified Ni splat.

1 Introduction

The discovery of depositing metallic nickel and chromium ions onto surfaces in the 1800s has led to significant advances in electroplating and other coating techniques. Both elements are essential ingredients in synthesizing stainless steel and are popular in high temperature and high performance applications (e.g. as nichrome wire in electric toasters and bond coats in jet engines) due to their abilities in minimising mismatches in thermal expansion coefficient and resistance to oxidation and corrosion.

Clad powders produce coatings with enhanced thermal and oxidation properties, which is especially useful in gas turbines of jet engines. Mechanically clad Cr on Ni is a readily available powder feedstock, where a higher percentage of Cr offers increased resistance to corrosion in coatings suitable for industrial coating applications. Typically, for post-spraying machining, a finer powder particulate size (e.g. 45 ± 5 µm) is used to produce coatings with finer splats [1]. However, this study was designed with the intent to analyse interactions between Ni and Cr splats. Therefore, a larger particle size distribution (106 ± 45 µm) was chosen to analyse solidified splat layers measuring several hundreds of nanometres in thickness.

Two-dimensional micrographs of splat cross-sections have been widely used to identify and characterise microstructural features, in particular the interfacial interactions between the splat and the substrate or between two splats [2-4]. However, due to the destructive nature of milling cross-sections and/or preparing transmission electron microscope (TEM) specimens, it is impossible to revisit a site to obtain further data regarding a specific splat's surroundings. Conclusions derived from analysing microstructural features are made through extrapolating 2D cross-sections. However, a near-exact replica of a desired feature can be obtained using 3D tomography to extract essential data on feature morphology and other interactions without sacrificing the immediate surroundings of the feature [5, 6].

In this study, a variety of characterisation techniques will be used, in conjunction with 3D tomography, to establish a relationship between the microstructure of the powder feedstock with observations made from both two-dimensional cross-sections and a three-dimensional reconstructed splat.

2 Experimental Methods

The as-received, as-sprayed substrate sample consisted of Cr-clad Ni splats on an untreated alumina substrate (average roughness (R_a) = 274 ± 12 nm, skew = -0.5 ± 0.1). The substrate was held at 25°C prior to spraying, and deposited with Ni-20vol%Cr powder feedstock (particle size distribution = 106 ± 45 µm) using a Sulzer Metco 9MB plasma torch. Spray parameters are listed in **Table 1**.

A number of microscopy methods were used to analyse the topographical and cross-sectional profiles of both splat and substrate. Methods included focussed ion beam (FIB) microscopy, scanning electron microscopy (SEM), back-scattered electron (BSE) imaging, transmission electron microscopy (TEM) and energy dispersive spectroscopy (EDS).

Observed features were translated into a three-dimensional reconstruction using interactive data visualisation software (Avizo Fire 8.1), to highlight interactions between the splat and the underlying substrate. A 3D modelling software (MeshMagic) was used to complement analysis.

The morphology of discrete particulates in the powder feedstock was initially analysed using two methods. Firstly, feedstock particulates were directly mounted onto carbon tape to reveal spatial and morphological information regarding the distribution of Cr deposits across the Ni particulates. Secondly, feedstock particulates were mounted in resin and mechanically polished to reveal cross-sectional information of individual grains and surrounding regions.

Table 1. Plasma spraying parameters of Ni and Ni-20vol%Cr deposits on alumina substrates.

Parameter	Value
Current (A)	900
Voltage (V)	40
Gas flow rate (nominal litres per minute, NLPM)	Argon: 50 Helium: 0.5
Powder feed rate (g/min)	6
Nozzle internal diameter (mm)	8
Spray distance (mm)	75

Table 2. SEM and FIB milling parameters used for serial sectioning of the splat.

Electron beam		Ion beam	
Voltage (kV)	15	**Voltage (kV)**	30
Current (pA)	500	**Current (pA)**	500
Signal	Secondary electrons	**Trench dimensions (µm)**	60 x 61 x 5
Working Distance (mm)	5	**No. of sections**	609
Pixels	2048 x 1536	**Section thickness (nm)**	~100
		Voxel ratio	1 : 1 : 2.6

As-received substrates were mounted using silver paste and coated with ~20 nm of gold, using an Emitech K550x Gold Sputter Coater. Both FEI xP200 FIB microscope and Carl Zeiss Auriga dual-beam SEM were used to characterize microstructural cross-sectional data of splats and to prepare specimens for TEM [4, 7]. TEM specimens were lifted out using ex-situ methods from the substrate using a Leica S6E stereo microscope and Kleindiek micromanipulator [4, 8] and examined in a FEI Tecnai TEM operating at 200 kV. EDS was performed on the TEM specimens to acquire elemental maps and line scans.

Serial sectioning of a splat region of interest was performed in a Carl Zeiss Auriga dual-beam microscope, after the splat had been identified using EDS, sputter coated with gold (**Fig. 1a**), protected with platinum (**Fig. 1b**) and trenches excavated to contain milled debris (**Fig. 1c**), using methods as described by Uchic *et al.* [5]. However, as both the splat and substrate surfaces were rough without a consistent feature to reference, an additional step was made during sample preparation, where a clearly defined line (labelled 'x') was milled across the length of a flat Pt strip (**Fig. 1c**), filled with a layer of sputtered gold for contrast and covered with an additional layer of Pt (**Fig. 1d**). This step resulted in a simple, yet effective, T-shaped feature, which acted as a tool for x and y alignment of each milled section (**Fig. 1f**) during the virtual reconstruction process. The region of interest for serial sectioning was then marked (**Fig. 1e**), and the automated process was started using the parameters listed in **Table 2**, with the aim of obtaining an average of 10 milled sections per feature for a representative splat [5].

A representative 3D model was reconstructed where individual phases of each section were identified using a variety of microscopy methods. EDS elemental maps were used to assist in calibrating grey levels of individual phases observed in FIB, SEM and TEM micrographs with each 3D section identified using Avizo Fire 8.1. The virtual model was then rendered to reveal a three-dimensional reconstruction of the splat and substrate, where features of interest can be explicitly examined in isolation [6].

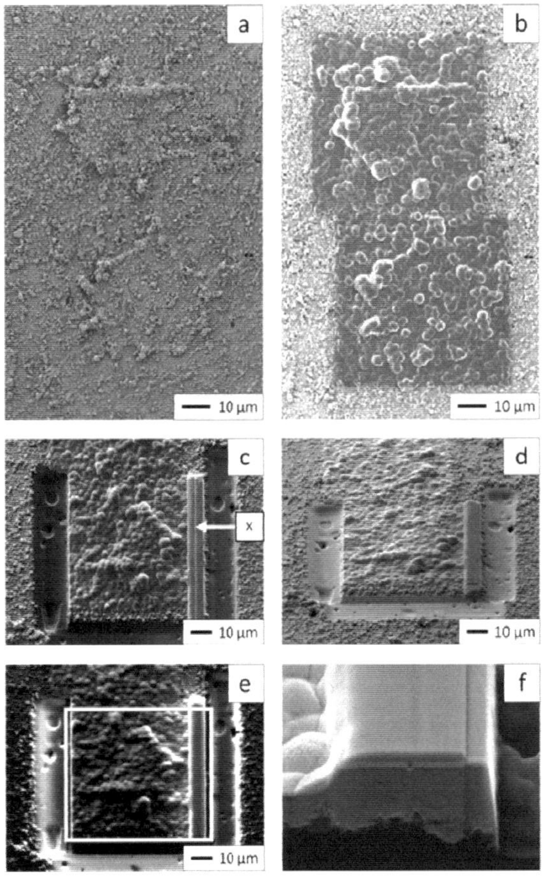

Fig. 1. SEM and FIB micrographs depicting the process taken to prepare the (a) splat of interest for serial sectioning after (b) Pt deposition, highlighting the addition of (c) milling a straight line at 'x' and (d) coated with gold used as an indicator after (e) serial sectioning the selected region for the (f) alignment of individual slides during data visualisation reconstruction.

3 Results and Discussion

It is relevant to note that the feedstock used in this study was manufactured using water-atomisation, resulting in an irregular powder particulate morphology, and sieved to obtain a large particle size distribution of 106 ± 45 µm. Generally, particles measuring 10 to 44 µm in diameter are the typical particle size range for plasma spray [1]. The irregular morphology of the powder feedstock inherently affects the spraying process, in terms of feedability, sprayability and vaporisation, which influences the coating's apparent density, porosity and amount of unmelted deposits [1]. Water-atomised powders also contain more impurities and oxygen content than gas-atomised powders, which may lead to increased porosity [9].

Cross-sectional micrographs of the powder feedstock, analysed using FIB (**Fig. 2a, b**), BSE (**Fig. 2c**) and EDS (**Fig. 2d**), exhibited large Ni particles (with lighter BSE contrast) surrounded by much smaller Cr particulates (with darker BSE contrast). Water atomisation generally produces highly irregular powder morphologies, however particulates observed at high magnification (**Fig. 2a**) appear to resemble irregular spheres. Also, no interaction or diffusion between Ni and Cr was observed (**Fig. 2c, d**). The observations are due to the water atomisation process, where Cr particulates were atomised and mechanically adhered onto the Ni particle. Surface grains of the Ni particle are plastically deformed during the process (**Fig. 2b**), which in turn altered the morphology of the Ni particle. Inter-particulate abrasion, possibly during powder transportation, caused the Cr particulates to detach from the Ni particle, resulting in a mixture of dissimilar-sized particulates.

The spraying temperature of 2400°C ensured that all Ni and Cr powder particulates were fully molten upon deposition (with homologous melting temperatures of 1455°C and 1907°C respectively) according to the Ni-Cr phase diagram [10], eliminating the possibility of a partially or unmolten particulate being deposited onto the substrate due to insufficient heating. However, it cannot be disregarded that the smaller molten deposits may have cooled and solidified in-flight between the spray nozzle and substrate. The smaller temperature difference between the spraying temperature and the melting temperature of Cr, in comparison to Ni, suggests that Cr is more likely to cool mid-flight, to be deposited as a solid particulate, provided the surface energy of the deposit is sufficient. Additionally, the volatility of both Ni and Cr at high spraying temperatures need to be considered. The exponentially large difference in vapour pressure between Ni (19.5 Pa) and Cr (774 Pa) at ~1726°C [11] provides a clear indication that Cr is more volatile and will evaporate at a faster rate than Ni at higher temperatures. Therefore, it is possible that Cr may have started to evaporate during the heating and deposition processes, thereby reducing the amount of Cr to impact the substrate surface and form splats. Whilst larger Cr feedstock particulates may have only partially evaporated, smaller Cr feedstock particulates may have completely evaporated, thus the distribution of Cr splats across the substrate surface may not accurately reflect the distribution of Cr within the powder feedstock. The same principle applies to Ni, but at a lesser magnitude due to its lower volatility.

Two models are proposed in **Fig. 3** to illustrate possible splat cross-sections dependant on whether Cr or Ni impacts the substrate first. It should be noted that there is variability in the size and shape of solidified splats depending on the dimensions of the original Ni and Cr particulates as well as the spatial distribution of Cr relative to the larger Ni grain. This further enforces the need for 3D tomography, to provide insight on a feature in its entirety instead of as a single cross-sectional image.

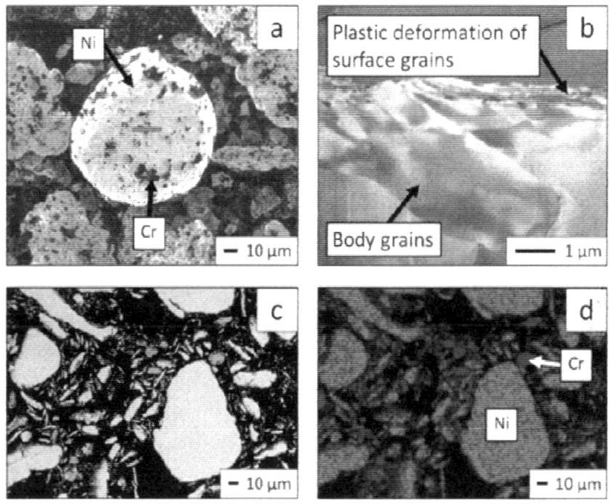

Fig. 2. FIB micrographs of Ni-20vol%Cr feedstock, (a) at high mag, (b) Ni particulate cross-section; (c) BSE image and (d) EDS elemental maps for Ni and Cr of mechanically polished feedstock.

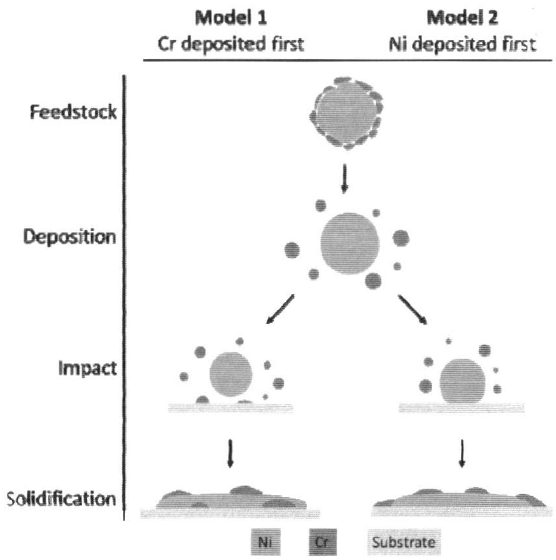

Fig. 3. Proposed models of the deposition process of Ni-20vol%Cr, where Cr is deposited first in Model 1 and Ni is deposited first in Model 2, resulting in different cross-sectional morphologies, which may alter the mechanical properties upon solidification.

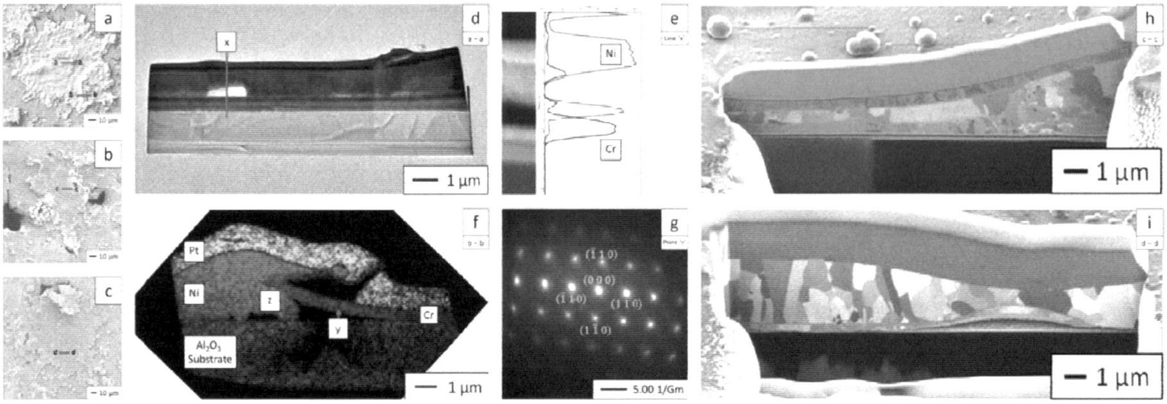

Fig. 4. Cross-sectional analysis of three Ni-20vol%Cr splats (a-c) on an alumina substrate at 25°C using FIB, SEM, TEM, EDS. (d) Bright field, (e) line scan of Line 'x' for Ni (red) and Cr (blue), (f) elemental mapping for Pt (green), Ni (red), Cr (blue) and substrate (purple), and (g) selected area diffraction of point 'y' of Cr in (f) was conducted on Splat (a), revealing a uniform layer of single-phase [001] zone axis BCC Cr crystals at the splat-substrate interface below Ni, with no interaction with Ni. (h) Cross-sectional FIB micrograph of Splat (b) showed a uniform layer of Cr above Ni with varying grain sizes; and (i) cross-sectional FIB micrograph of Splat (c) illustrated a lack of Cr on the splat, with distinct splat-substrate pores.

A large range of splats were examined. Cross-sectional views of sampled splats exhibited multiple interfacial layers, as well as pores within the splat, at the splat-substrate interface and at the Ni-Cr splat interface. Splat-substrate pores are often the result of splat material solidifying around a gas bubble either trapped within the deposit during deposition, or released from the substrate surface due to condensation of local adsorbates upon splat impact. However, the pore observed at point 'y' in Cr (**Fig. 4f**) is due to a splat periphery delaminating from the substrate surface. Here, the Cr splat periphery had thermally contracted from the residual heat during solidification, lifting off the substrate surface to expose an opening. A portion of molten Ni overlapping the Cr splat had flowed underneath the delaminated periphery at point 'z', but had fully solidified before the pore was completely filled, resulting in a splat-substrate pore.

Distinct Ni and Cr layers were identified, by grey-level contrast in FIB cross-sectional micrographs and EDS elemental mapping in both the SEM and TEM of three separate splats (**Fig. 4a-c**). **Fig. 4a** consists of multiple overlapping splats, resulting in a lamellae structure of Ni (lighter contrast) and Cr (darker contrast) in **Fig. 4d**. Line scans (**Fig. 4e**) at point 'x' of **Fig. 4d** showed no evident interdiffusion or melting between the two materials, whilst Ni and Cr EDS elemental maps (**Fig. 4f**) exhibited no evidence of Cr_2O_3. Selected area diffraction (SAD) on regions exhibiting a high concentration of Cr, such as point 'y' of **Fig. 4f**, produced a diffraction pattern shown in **Fig. 4g**. The pattern observed was consistent with a single-phase BCC Cr crystal structure with a lattice parameter of 2.885Å at the [001] zone axis, in agreement with works published by Loretto [12]. As the thermal conductivities of Ni and Cr are similar (91 $W{\cdot}m^{-1}{\cdot}K^{-1}$ and 94 $W{\cdot}m^{-1}{\cdot}K^{-1}$ at 25°C, respectively [13]), the higher atomic mass and lower volatility of Ni [11] serves as a greater driving force for Ni to cool and solidify at a faster rate. The sufficiently cooled Ni deposit then acts as a local heat sink during the deposition of subsequent Cr, to conduct interfacial heat as a function of length, cross-sectional area and temperature difference, according to **Eq. 1**.

Eq. 1. Rate of heat conduction (dQ/dt) as a function of material thermal conductivity (k), contact length (L), cross-sectional area (A) and temperature difference (dT).

$$\frac{dQ}{dt} = -k \cdot A \cdot \frac{dT}{L}$$

Thermal conductivity at the splat-substrate and splat-splat interfaces must be investigated to understand the mechanisms behind the solidification process. At the splat-substrate interface, Al_2O_3 has low thermal conductivity (~35 $W{\cdot}m^{-1}{\cdot}K^{-1}$ at 25°C [14]), therefore, in theory, heat transfer between the room temperature substrate and molten deposit is slow (**Eq. 1**), suppressing nucleation and promoting grain growth. At the Ni-Cr splat-splat interface, thermal conductivity between the two materials is similar, therefore heat transfer is accelerated, promoting nucleation and suppressing grain growth. However, due to the low volatility and higher atomic mass of Ni, and the effects of undercooling from the large temperature difference between the Al_2O_3 substrate at room temperature and the molten Ni deposit at ~2400°C, rapid quenching of Ni had occurred to form a thin ~250 nm layer of columnar Ni grains at the splat-substrate interface [15]. This observation is in alignment with the simulated deformation profiles of cooling plasma sprayed splats by Chae *et al.* [16]. As the deposit continues to cool, grain growth is promoted and larger discernible Ni grains are observed. Grains appear columnar in the Ni layers of both **Fig. 4h** and **Fig. 4i**, and the average grain size ranges from 0.1 µm at the splat-substrate interface to 1 μm closer to the splat surface.

Diffusion at the Ni-Cr splat-splat interface was not detected, as the Ni splat surface was solidified prior to Cr being deposited at 2400°C, which suggests the temperature of the Ni splat, when Cr impacted, was below its melting point of 1455°C. The similar thermal conductivity and small temperature difference of Ni and Cr promotes heat transfer from the molten Cr deposit to the solidified Ni splat (**Eq. 1**). Resultant Cr grains measured ~0.1 μm in diameter, as observed in **Fig. 4h**, similar to the Ni grains at the splat-substrate interface, elucidating that the nucleation rate of Ni at the point of impact with the cool substrate is comparable to the nucleation rate of Cr grains at the Ni-Cr interface, despite the temperature difference.

Measured dimensions of features in a 2D micrograph are only apparent as each grain is subjected to a different orientation and position upon milling. Electron backscatter diffraction (EBSD) can be used to identify and model grains in relation to one another. Similarly, 3D tomography was used to identify and analyse multiple aspects of a splat (e.g. splat material, size and shape), relative to its surrounding features (e.g. an overlapping splat).

Splats have been identified to exhibit features in agreement with Model 1 (Ni overlapping Cr, **Fig. 4f**), Model 2 (Cr overlapping Ni, **Fig. 4h**), and Ni splats without Cr present were detected (**Fig. 4i**). The Cr layer observed in **Fig. 4f** and **Fig. 4h** are of a uniform thickness (~500 nm) across the milled cross-section, with no interfacial interaction with either substrate or Ni layer. This lack of thickness variation in Cr and difference in grain size with Ni suggests that Cr was indeed deposited separately from the Ni splat. Additionally, good interfacial contact was observed at the splat-substrate and splat-splat interface in **Fig. 4d**, although delamination was observed at the Cr periphery in **Fig. 4f**. Splat-splat pores were observed in **Fig. 4d** and splat-substrate pores were observed in **Fig. 4i**. In both instances, pores were identified in the Ni region of the splat, suggesting poor bonding with Ni. Conversely, it was interesting to observe the precise contouring of Cr onto the Ni layer (**Fig. 4h**) within the short quenching timeframe, further illustrating superior bonding of Cr upon impact.

Three-dimensional virtual reconstruction of a splat revealed topological information of both the Ni and Cr splats, as well as splat-substrate and splat-splat interfacial interactions.

The plan-view of the virtual splat in **Fig. 5a** clearly shows Cr overlapping a Ni splat, which could not be distinguished from the splat in **Fig. 1a** prior to sectioning. The splashy Cr layer does not completely cover the Ni and appears rough when viewed from the side (**Fig. 5b**). When the layers are separated, a shallow imprint of the splat is observed on the substrate in **Fig. 5c**, suggestive of localised deformation of the substrate surface. However, it is unlikely that the alumina was indented during the spraying process due to its high hardness. Large voids are observed, but the presence of intermetallics, oxides or diffusion was not evident at the splat-substrate interface. The imprint is more likely a result of local thermal interactions on the substrate surface due to the hot Ni deposit.

One of the greatest advantages of combining 3D tomography with virtual data visualisation is the ability to calculate an isolated feature's volume with significantly greater reliability than a 2D micrograph. From measurements extracted from the segmented splat, it was calculated that the splat comprised of ~62vol% Ni and ~38vol% Cr, the substrate contained ~15vol% discernible pores and the sampled region consisted of ~13vol% porosity within and surrounding the splat.

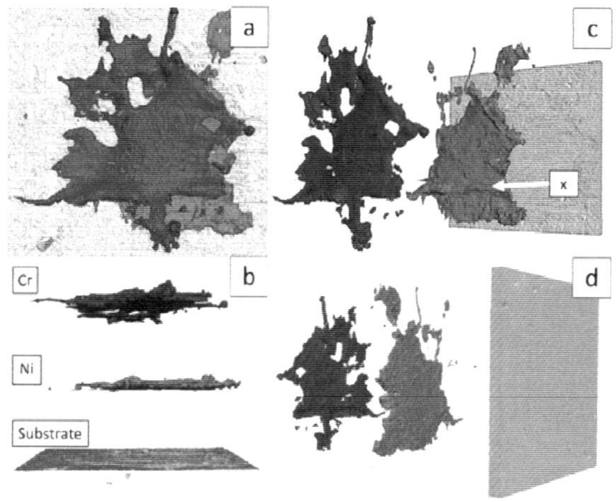

Fig. 5. Reconstructed Ni-20vol%Cr splat, using Avizo Fire 8.1; expanded using data visualisation software MeshMagic, showing (a) plan-view, (b) side-view and expanded (c) front and (d) underside of the splat, displaying components Cr (blue), Ni (red) and alumina substrate (yellow).

Superior spreading ability of Cr is observed in **Fig. 5**, as the outline of both an elevated feature (labelled 'x' on the Ni splat in **Fig. 5c**) and the Ni splat can be seen on the underside of the Cr splat in **Fig. 5d**, concurring with the observations shown in **Fig. 4h**. Furthermore, it can be observed that the morphology of the Cr regions in contact with Ni conformed well with the Ni splat, whilst regions of Cr away from the Ni splat body appeared splashy. Additionally, no observable chemical interaction was detected between Ni and Cr. It may be possible that a significant amount of Cr particulates, surrounding the Ni particle prior to heating, had unfurled and detached from the Ni in-flight allowing for Ni to impact, quench and solidify onto the substrate first before Cr covers the Ni splat. It is also possible that the splashy portion of the Cr splat is the result of excess Cr rebounding off the Ni splat upon impact, though neither the peripheral morphology of the splashy Cr portion or the grain size and shape of the Cr layer suggest that the splashy portion was a secondary deposit. Regardless, the effects of undercooling would be more apparent at the splat-substrate interface, due to the large temperature difference, which would ultimately limit the spreading extent of the initial deposit [15, 16]. The smaller temperature difference at the Ni-Cr interface, and their similar thermal conductivities, may have promoted spreading. Finally, it was observed in the FIB milled sections that the majority of Cr across the Ni splat was of uniform thickness, again comparable to observations in **Fig. 4d, f** and **h**, where no measurable interdiffusion was detected.

4 Conclusion

FIB, SEM, TEM and EDS analysis of Ni-20vol%Cr splats on an alumina substrate at 25°C revealed cross-sectional microstructures consistent with the proposed models. No inter-mixing or diffusion between the Ni and Cr layers was observed and chemically and structurally distinct splats were present.

When considering the rate of heat conduction and each material's respective thermal conductivity, it was deduced that the alumina substrate acted as a large heat sink to initially cool the deposit of first contact. Despite the low thermal conductivity of Al_2O_3, the large temperature difference between splat and substrate is a driving force for undercooling the deposit of first contact sufficiently, to achieve a solid splat surface, prior to subsequent deposits. This was deduced by studying changes in grain size in the Ni and Cr layers and noting the lack of interfacial diffusion. It is clear that the rate of heat conduction at the immediate Ni-substrate interface and Ni-Cr interface were comparable and that the Ni splat surface had solidified before Cr impacted, aiding in cooling subsequent Cr deposits.

3D tomography of a Ni-20vol%Cr splat on an alumina substrate at 25°C exhibited superior bonding of Cr on Ni and no splat-splat or splat-substrate interfacial interactions. Observations have been consistent in both 2D and 3D analysis conducted on this sample.

5 Acknowledgements

The author would like to thank Dr. Charlie Kong, Dr. Quadir Md Zakaria and Mr. Sean Lim of the Electron Microscope Unit in UNSW for their continued support and advice.

6 References

1. Davis, J.R., *Handbook of Thermal Spray Technology*. 2004: ASM International.
2. Brossard, S., Munroe, P.R., Tran, A., and Hyland, M.M., *Study of the Splat-Substrate Interface for a NiCr Coating Plasma Sprayed onto Polished Aluminum and Stainless Steel Substrates.* Journal of Thermal Spray Technology, 2010. **19**(1-2): p. 24-30.
3. Bolelli, G., Sabiruddin, K., Lusvarghi, L., Gualtieri, E., Valeri, S., and Bandyopadhyay, P.P., *FIB assisted study of plasma sprayed splat–substrate interfaces: NiAl–stainless steel and alumina–NiAl combinations.* Surface & Coatings Technology, 2010. **205**(2): p. 363-371.
4. Munroe, P.R., *The application of focused ion beam microscopy in the material sciences.* Materials Characterization, 2009. **60**(1): p. 2-13.
5. Uchic, M.D., Holzer, L., Inkson, B.J., Principe, E.L., and Munroe, P., *Three-Dimensional Microstructural Characterization Using Focused Ion Beam Tomography.* MRS Bulletin, 2007. **32**(5): p. 408-416.
6. McGrouther, D. and Munroe, P.R., *Imaging and Analysis of 3-D Structure Using a Dual Beam FIB.* Microscopy Research and Technique, 2007. **70**(3): p. 186-194.
7. Cairney, J.M., Munroe, P.R., and Hoffman, M., *The application of focused ion beam technology to the characterization of coatings.* Surface and Coatings Technology, 2005. **198**(1–3): p. 165-168.
8. Giannuzzi, L.A., Kempshall, B.W., Schwarz, S., Lomness, J., Prenitzer, B.I., and Stevie, F.A., *FIB Lift-Out Specimen Preparation Techniques*, in *Introduction to Focused Ion Beams - Instrumentation, Theory, Techniques and Practice*, L.A. Giannuzzi and F.A. Stevie, Editors. 2005, Springer Science + Business Media, Inc.: New York, USA. p. 201-228.
9. Dorfman, M.R., *Chapter 20 - Thermal spray coatings*, in *Handbook of Environmental Degradation of Materials*, M. Kutz, Editor. 2005, William Andrew Publishing: Norwich, NY. p. 405-422.
10. Reed, R.C., *The Superalloys: Fundamentals and Applications*. 2006: Cambridge University Press.
11. Haynes, W.M., *Vapor Pressure of the Metallic Elements — Data*, in *CRC Handbook of Chemistry and Physics, 96th Edition*. 2015, CRC Press: Boca Raton, Florida. p. 127.
12. Loretto, M.H., *Electron Beam Analysis of Materials*. 1993: Springer Netherlands.
13. Aylward, G.H. and Findlay, T.J.V., *SI Chemical Data*. 6th ed. 2008: Wiley.
14. Auerkari, P., *Mechanical and physical properties of engineering alumina ceramics.* 1996: Technical Research Centre of Finland Finland.
15. Mostaghimi, J. and Chandra, S., *Splat formation in plasma-spray coating process.* Pure Applied Chemistry, 2002. **74**(3): p. 441-445.
16. Chae, Y.K., Mostaghimi, J., and Yoshida, T., *Deformation and solidification process of a super-cooled droplet impacting on the substrate under plasma spraying conditions.*

Science and Technology of Advanced Materials, 2000. **1**(3): p. 147-156.

Pulsed Current Investigations and Effects of the Wire Arc Spraying Process

A. Atzberger[1], G. Huismann[2], M. Szulc[1], S. Kirner[1], S. Zimmermann[1], G. Forster[1] and J. Schein[1]
[1]Lab for Plasma Technology (LPT), University of the Federal Armed Forces Munich, Neubiberg, Germany
[2]Lab for Welding Technology, University of the Federal Armed Forces Hamburg, Hamburg, Germany
E-mail: alexander.atzberger@unibw.de

Modern power sources enable to take a new look at the long established core parameter of the wire arc spray process, the applied current. In standard DC mode, the movement of the arc between the wire tips is taking place randomly with a broad frequency range. Investigations regarding the behavior of the arc have been carried by the use of high speed image analysis. It was ascertained, that the position of the arc between the electrodes can be determined by taking a look at the voltage trace. In the next step, the fluctuations of the arc have been analyzed and a specific "natural fluctuation frequency" of the DC process has been determined.
Our investigations show that the arc fluctuations can be controlled by pulsing the current. When that specific pulsing frequency is applied, the arc fluctuation frequency correlates with the frequency of the pulsed current. By controlling the arc fluctuations differences in coating properties such as the oxide content occur. The investigations suggest that when using pulsed current the mechanical properties of the system used are decisive for the resulting coating and thus need to be adapted to the electrical parameters.

1 Introduction

Wire arc spraying is one of the oldest thermal spray processes, which is preferably used due to its low operating costs, easy handling and high deposition rates. In comparison to plasma spraying, the operating costs of a wire arc spraying system are only 10% per hour used [1]. With wire arc spraying being the most cost-efficient of all thermal spray processes, it is a technique commonly used in today's industries (e.g. automotive and aerospace industry). Thus, it has a broad range of applications, especially in terms of corrosion protection or the production of wear-resistant coatings [2].
Compared to other thermal spray processes the wire arc spraying has a rather simple setup. Two consumable wires are continuously fed towards each other by a wire feed unit. The conductive wires are guided by contact sleeves that are aligned to each other at a certain angle. A power source is connected to the contact sleeves and thus ignites an arc between the two wires. In order to transport the molten particles onto a substrate, a gas nozzle is positioned right behind the wires.
Once the arc between the electrodes is struck, it starts to move downstream because of the high velocity of the gas stream applied. By the movement of the arc along the wire tips, the wires begin to melt, the molten material is being blown out and eventually particles are detached due to the high gas velocity [3].
In terms of fluid flow the arc can be described as a medium with a different density, which makes it an obstacle that is being influenced by the gas flow of the atomizing gas. When the fluid and magnetic force (Lorentz force) on the arc is too strong, it will re-ignite in the smallest gap between the electrodes again [4]. These movements are called the fluctuation of the arc and they are taking place repetitively. However, these arc fluctuations are taking place randomly and could not be controlled so far [5]. There are different approaches to optimize the wire arc spraying process. Most investigations deal with the development of optimized nozzle geometries or a secondary atomizing gas flow. The goal is to achieve a uniform droplet size distribution and higher particle velocities in order to influence the quality of the coatings [6]. Besides that, the use of other process gases are investigated to directly influence the particle properties [7]. The remaining input parameter of the system is the power source.
Investigations of the plasma cutting process have shown that when a specific current pulse is applied the anodic arc attachment can be influenced – the voltage drops and the arc shortens respectively due to the changing attachment of the arc [8]. Having identified that experimental phenomenon for plasma cutting, the transfer to the wire arc spraying process has been carried out in the following investigations.

2 Process Conditions

In order to be able to adjust as many parameters as possible, an experimental wire arc spray system (EXP) was built at the LPT and used in the following experiments. An iron based wire material G3Si1 (composition: 0.08%C, 0.9%Si, 1.5%Mn, 0.015%P, 0.012%S) with a diameter of 1.6mm was used. Compressed air (CA) was used as atomizing gas with varying gas pressures. The experimental system is displayed in Fig. 1. As power supply serves the GMAW power source "Phoenix 451" by the company "EWM", which can be run in DC as well as in Pulse mode. The Phoenix power source is equipped with a second wire feeder and is responsible for the control of the wire feed rate (which is directly linked to the current). It is possible to adjust the two wire feeder rates separately. In further experiments an additional GTAW power source "Tetrix 300 puls" also by "EWM" was used. In that pulse combination, the Phoenix serves as DC source and the Tetrix adds current pulses onto the existing DC signal – thus a pulsed signal is being generated.

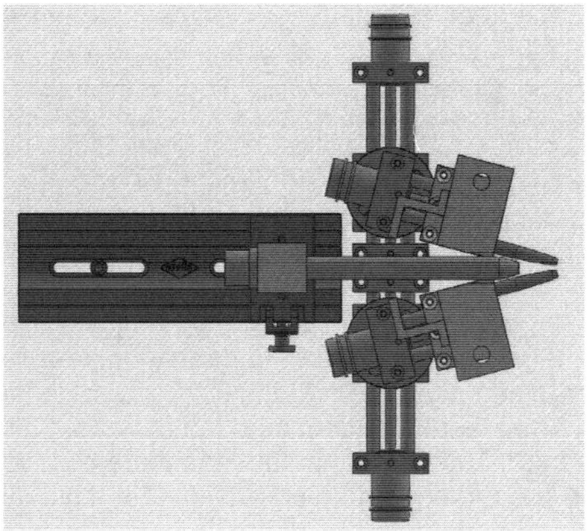

Figure 1: CAD model of the experimental wire arc spraying system.

Summing up, the GMAW and GTAW power sources serve as power supply with a wire feed unit (linked to the Phoenix PS), and the experimental wire arc spray system, as displayed in Fig. 1, is used. It should be mentioned that no wire straightening unit was used. Also, only one wire feed unit was used (per wire), not a push-pull system consisting of two wire feed units.

The adjustable parameters of the system are the gas pressure, the voltage and the wire feed rates of the Phoenix PS (and thus the current) as well as the pulse frequency (f_p), the base (I_b) and pulse current (I_p) and the pulse ratio (pulse %) of the Tetrix.

For all investigations in DC mode, the wire feed rate was set to 4 m/min, the voltage was set to 29 V and the average current to 190 A (Phoenix). When using the system in pulse mode, the settings of the Tetrix were I_b=1 A, I_p= 150 A, f_p=1.2 kHz and a pulse ratio of 20%. When modulating the pulsed current onto the DC signal, the Phoenix automatically reduced its DC current, so the resulting average current was 190 A in total, as set by default. The gas pressure was set to 200 - 400 kPa for all experiments and the distance between the experimental wire arc system and the substrate was 90 mm.

2.1 Experimental Setup and Analysis

Figure 2 shows the experimental setup of the High speed (HS)-imaging for the observation of the arc behavior as well as the particle generation. Besides that, HS-shadow imaging was used in order to supress the light of the plasma. By using this technique, the shadows of the wire tips and the molten droplets are made visible. For the HS-shadow imaging a combination of a laser (100 mW, 656 nm) in front of the wire arc system and an interference filter (656 nm) right before the HS camera is mounted. In order to illuminate a large area (and to see the secondary breakup of the particles), a lens is used to broaden the laser beam. When regular HS imaging is used, an incandescent light bulb is used to illuminate the background, so that the wires can be distinguished from the background.

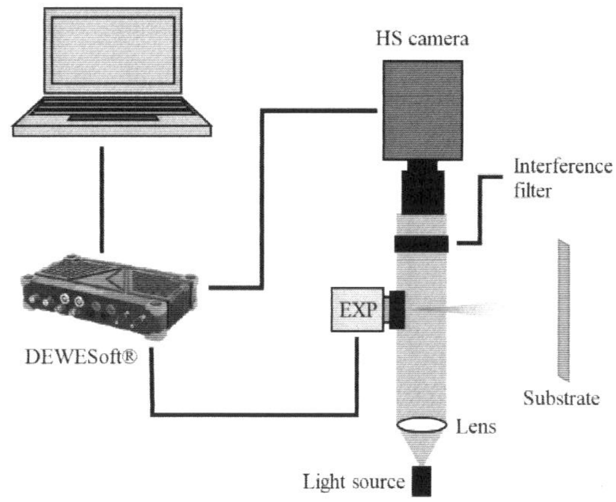

Figure 2: Experimental setup for the observation of the arc behaviour.

The DEWESoft® Sirius measurement unit collects voltage and current measurement data and displays it corresponding to the images acquired by the HS camera.

2.2 Results

With this setup, the aim was to obtain a deeper understanding of the process itself, the arc behavior and the effects of the voltage fluctuations. As mentioned above, the voltage fluctuations of the process have a significant influence on the particle properties and the resulting coatings.

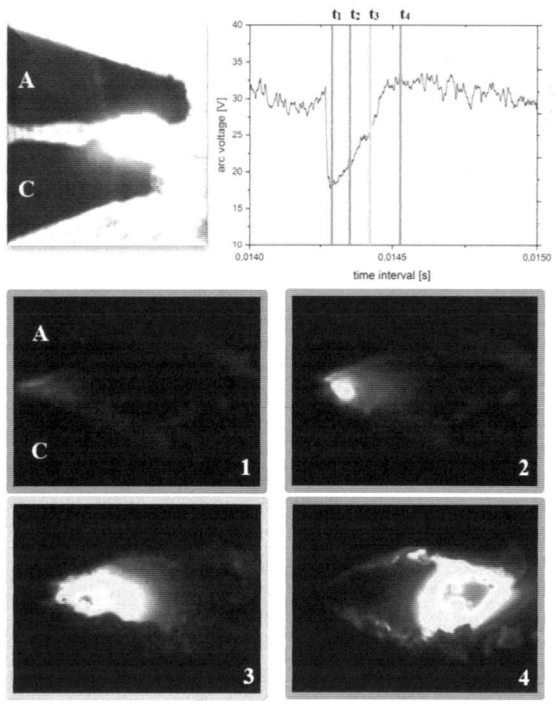

Figure 3: Visualization of the voltage fluctuation phases (1 - 4) by false color images with the respective voltage trace.

By having a HS camera available the frame rate was set to 70,000 fps to be able to investigate these fluctuations thoroughly. Therefore, images of the arc with corresponding voltage and current traces have been taken. Figure 3 displays the voltage trace of one fluctuation during the process. At certain time stamps (t_1 – t_4) pictures of the wire tips have been taken to evaluate the size of the arc.
Due to the the high intensity of the plasma, no light source was used. By displaying the images in false color pictures the intensity levels and thus the position of the arc can be made visible. From this evaluation one might assume that the length of the arc is directly linked to voltage of the process. In order to study this relationship scientifically, an in-house algorithm code has been written to display the length of the arc in every picture taken. The arc length algorithm calcutes the average length of the arc and thus determines a certain length of the arc in every single picture. First, the ROI in the set of pictures has to be set (Figure 3, image 1,
yellow rectangle).

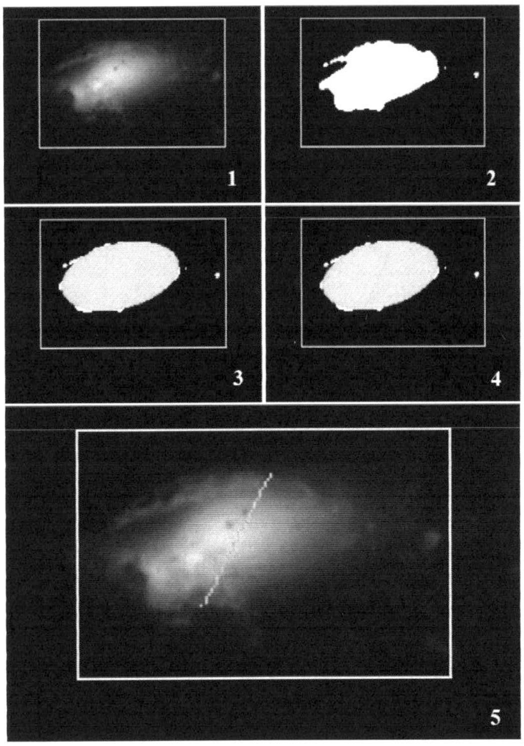

Figure 4: Steps in the arc length algorithm.

Next, the ROI is being binarized by an adjustable threshold (2). Then, an area-equivalent ellipse is being calculated with the values of the centroid, the eccentricity and the angular position (3, light blue). For every row of pixels in this ellipse, the median is being chosen. By doing so, the centerline of the ellipse is being displayed (4, green line). When comparing this calculated arc length to the orginal picture (5), an accordance to the actual arc length can be registered. Multiplying this arc length in pixel with the actual scale of the picture, an average arc length for every picture and a calculated arc length trace over time can be displayed.

The trace of the calculated lenght of the arc in comparison to the measured voltage of the process is displayed in Figure 5. The arc length displayed has values of zero for the times when it is re-igniting in the small gap between the electrodes. At these time stamps the dynamic range of the picture is underneath the threshold and thus no resulting arc length value can be determined. It is nevertheless valid due to the fact that the corresponding voltage at that time is at its lowest level as well. For illustration purposes the arc length calculated in Pixel was divided by a constant.

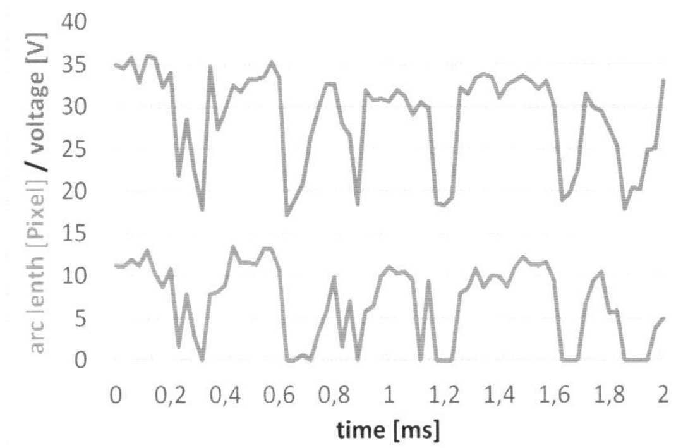

Figure 5: Calculated arc length (orange) correlating to the actual voltage trace (blue).

It can be seen that the length of the arc correlates to the voltage, which leads to the following relationship:

$$U_{arc} \sim E * l_{arc} \qquad (1)$$

This relationship has also been identified in Figure 3. Having illustrated this relationship with both false color image figures and an image processing algorithm, the voltage fluctuations over time were analysed next.
As Kirner et al. [] have shown, the periodicity of the voltage fluctuations rises with increasing gas pressure. Nevertheless, the fluctuations take place randomly and can not be controlled so far. Having analyzed regular DC measurements by a Fast Fourier Transformation, the frequency spectrum was in a very broad range (0.5 – 5 kHz). When taking these voltage fluctuations for granted, they are considered to be the "natural fluctuations" of the process further on.
For the experimental wire arc spray system, the region around 1.2 kHz showed a small peak compared to the remaining spectrum. Therefore, the experiments in pulsed mode have been conducted with that specific frequency.
In first experiments the built-in pulse mode of the Phoenix 451 GMAW power source was used. Experiments in a range between 0,5 – 2 kHz have been conducted. As it turned out, the current pulses at

a pulse frequency of 1,143 Hz showed an influence on the voltage trace, as displayed in Figure 6.

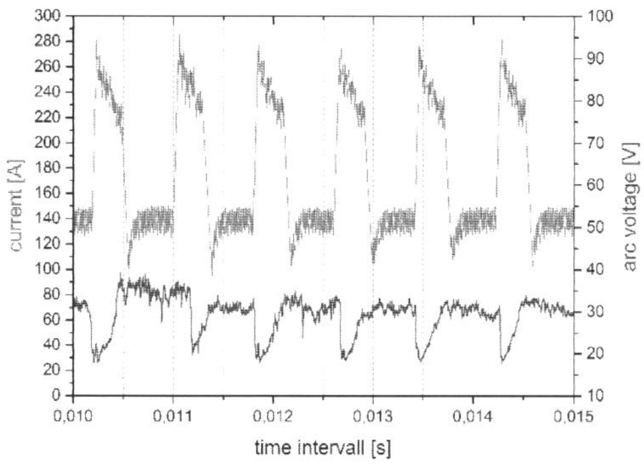

Figure 6: Current (red) and voltage (blue) trace of the EXP in pulse mode of the Phoenix 451 PS.

Frequencies above and underneath the 1,143 Hz did not show such a strong pulse-related fluctuation behavior. Since the pulse mode of the Phoenix 451 is not destined for high-frequent pulses it was not able to adjust the shape of the pulse any further. Thus the possibility of linking a GTAW power source to the GMAW source was tested. In that constellation, the Phoenix operates as a standard DC source, the regular parameters (JOB) were set and the Tetrix modulates the current on top of the DC signal. With the Tetrix being a GTAW power source it does not control the voltage (i.e. a constant arc length), but a constant current.

As it turns out, when pulsing the experimental system with the GMAW and GTAW combination, the resulting pulses show a constant "pulse shape". Further on, pulsing with the specific frequency of 1.2 kHz results in controlled voltage fluctuations. Figure 7 shows the current (red) and voltage (blue) trace of a pulsed wire arc spraying process.

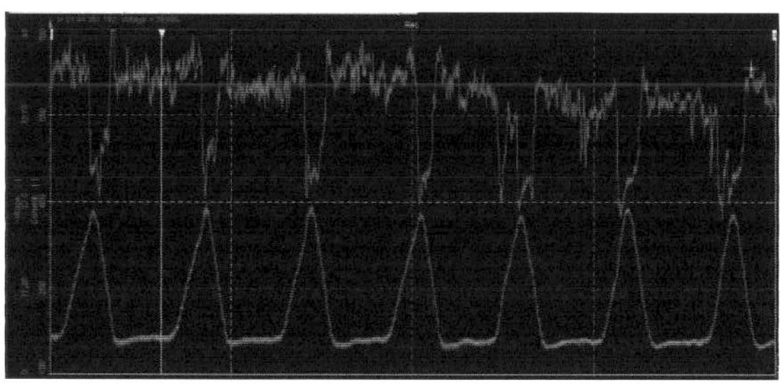

Figure 7: Actual current (red) and voltage (blue) trace of a pulsed wire arc spray process acquired with DEWESoft®.

It can be seen that the current pulses influence the voltage periodically and thus force the arc to re-ignite when a current pulse is applied. For every current pulse the voltage drops to a low level, i.e. a fluctuation takes place. When taking a look at the HS images, the arc does re-ignite in the small gap for every current pulse. That behavior underpins the assumption of a correlation between the voltage and the arc length as well.

By operating the experimental system with a pulsed current at a frequency of 1.2 kHz, it was possible to control the movement of the arc and thus the natural voltage fluctuations. These experiments have been carried out multiple times with the same result.

Being in control of the arc movement, the resulting coating has been investigated. At a gas pressure of 400 kPa for compressed air (CA) as atomizing gas, coatings for both the DC and pulsed process have been carried out, as displayed in Figure 8.

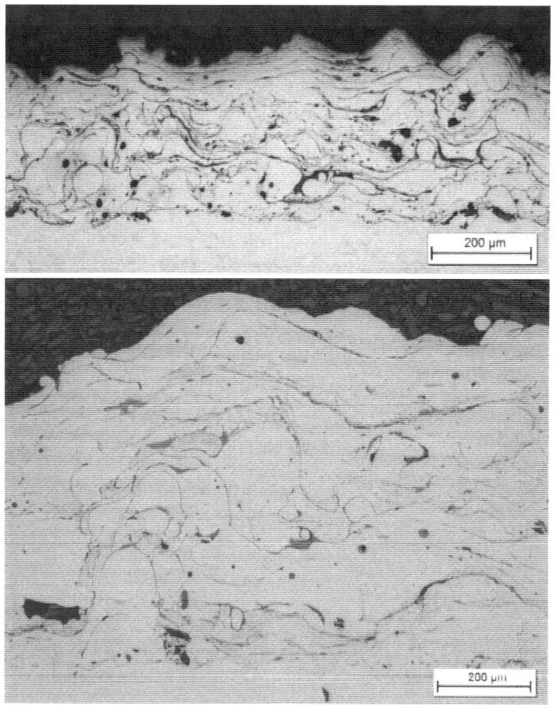

Figure 8: Microsections of coatings of G3Si1 and CA operated in DC (top) and pulsed mode (bottom) at 400 kPa.

The thickness of the coatings can not be considered since the DC sample was coated twice, the pulsed one five times. Apart from that, the pulsed coating shows a significant lower oxide content than the one operated in DC mode throughout the whole microsection. The typical lamellae (Fig. 8, above) of the DC process can rarely be found in the pulsed coating.

In order to demonstrate the influence of the gas pressure another sample with a gas pressure of 200 kPa has been coated. The differences of varying gas pressures for coatings in pulse mode can be seen in Figure 9.

The coating with a lower gas pressure (bottom) shows a larger size of the molten particles, larger oxide lamellae and larger pores compared to a higher gas pressure. With the particles being slower, the time for the particles to oxidize is larger and due to a slower gas stream the secondary break up of the particles is not as high, therefore the molten particles are larger.

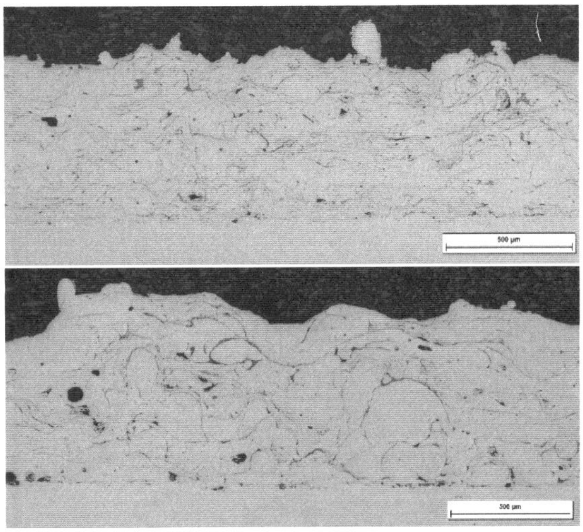

Figure 9: Microsection of coatings of G3Si1 and CA with a gas pressure of 400 kPa (top) and 200 kPa (bottom).

3 Conclusion

The aim of this paper was to investigate the influence of pulsed current and evaluate its effects of the wire arc spraying process. It was shown by both false color images and an in-house image processing algorithm, that the length of the arc is directly related to the corresponding voltage trace. By analyzing the frequency spectrum of the "natural" voltage fluctuations an adequate pulse frequency for the experimental wire arc spraying system was found. By applying current pulses with that specific frequency it is possible to control the movement in its periodicity. Simply put, the arc re-ignites for every current pulse applied.

It was also shown that by being able to control the voltage fluctuations the resulting coating shows a lower oxide content and less lamellae compared to a regular DC coating. Considering the investigated experimental system is a so-called "open" system, the differences in coating properties by adjusting one of the input parameters, the current, is tremendous.

However, it is necessary to mention that the reproducibility of these coating differences depends on multiple factors that are still subject to investigation. For one, the mechanical components (in this case the distance and the angle between the contact sleeves and the gas nozzle) have to be adjusted to its electrical counterpart. The transferibility to commercial spray torches still needs to be evaluated. Also, the exact influence of the current pulses on the particles is not yet finally clarified.

4 Outlook

Further investigations are necessary to evaluate the crucial factor of the pulsed current on the particles. To date it is not certain which aspect leads to a lower oxide content in the pulsed coating compared to the DC coating. Also, the influence of the current pulses on other material and gas combinations has to be evaluated. Lastly, it is necessary to develop a wire arc spray torch in which the mechanical and the electrical components are geared to each other.

5 Literature

[1] Jonke, D.P. : Private notice (2011).

[2] Pfender, E.: Thermal Plasma Technology: Where Do We Stand and Where Are We Going? Plasma Chemistry and Plasma Processes (1999), 19(1).

[3] Newberry, A.P.; Grant, P.S.: Large Arc Voltage Fluctuations and Droplet Formation in Electric Arc Wire Spraying. Powder metallurgy (2003), 46(3), pp. 229-235.

[4] Hussary, N.A.; Heberlein, J.V.R.: Atomization and Particle-Jet Interactions in the Wire-Arc Spraying Process. Journal of Thermal Spray Technology (2001), 10(4), pp. 604-610.

[5] Kirner, S.; Atzberger, A.; Zimmermann, S.; Schein, J.; Forster, G.: Investigation of a Pulsed Current Wire Arc Spray Process. Processdings of ITSC 2015 (2015), pp. 526-531.

[6] Wang, X.; Heberlein, J.V.R.; Pfender, E.; Gerberich, W.: Effect of Nozzle Configuration, Gas Pressure, and Gas Type on Coating Properties in Wire Arc Spray. Journal of Thermal Spray Technology (1999), 8(4), pp. 565-575.

[7] Schaup, J.; Atzberger, A.; Huismann, G.; Hartz-Behrend, K.; Zimmermann, S.; Kirner, S.; Szulc, M.; Forster, G.; Schein, J.: The journey with modern power sources und torch systems to optimized arc spraying. HVOF-Kolloquium (2015), pp. 35–44.

[8] Eichler, S.; Hussary, N.; Siewert, E.; Schein, J.: Investigation of the Anode Attachment Process in Plasma Arc Cutting. Journal of Physics: Conference Series (2014), 550(1), pp. 12007-12016.

Microstructure and Thermal conductivity of Fe-based amorphous coatings prepared by HVOF thermal spraying

L. Wang, Z. Zhou, H.H. Yao, Y.M. Wang, X. Wu, D.Y. He, Beijing/CN

Gas-atomized $Fe_{59}Cr_{12}Nb_5B_{20}Si_4$ amorphous powders were deposited to form amorphous coatings on the ASTM 4032 aluminum alloy substrate by HVOF thermal spraying under different conditions. The microstructural studies show that the coatings exhibit dense layered structure and relatively low porosity, less than 1.3 %. The coatings primarily consist of amorphous matrix with some precipitated nanocrystals, and the amorphous fractions of all the coatings are more than 40% by calculation from DSC results. Depending on the structure advantage, the coatings exhibit high microhardness, from 1230 HV to 1285 HV. Meanwhile, the bonding strength of as-deposited coatings increases with the rising flow of oxygen and kerosene, reached to 53.3 MPa. The ability of heat transport of as-deposited Fe-based coatings were investigated by calculation thermal conductivities from $k= a\rho C_p$. The results show that the thermal conductivity ascends slightly with the enlargement of spraying parameter in examined range, from 1.99 to 3.1 W/mK. However, thermal conductivities of all designed coatings are lower than metal-based materials. Furthermore, it is also notable that the microstructural characteristic and performance of the Fe-based coatings under different spraying conditions are similar, which manifests the $Fe_{59}Cr_{12}Nb_5B_{20}Si_4$ powder has a great broad adaptability for HVOF thermal spraying process. It is very important for practical engineering applications.

1 Introduction

Over the last years, the transportation industry has been facing increasing pressure to decrease vehicle weight in order to reduce fuel consumption and emission [1]. For instance, aluminum (Al) alloy replacing the traditional steel components is an efficient way for reduction weight increasingly. It is regarded that Al alloy has lower hardness and melt point compared to steel, which is adverse to extensive application. One promising method to solve this problem is to deposit protective coatings on the surface of Al alloy parts [2-3].

Thermal barrier coatings (TBCs), most commonly stabilized zirconia such as Yttria-Stabilized Zirconia (YSZ), are well-known for the good thermal insulating properties and high hardness. The use of TBCs can result in temperature reduction of as much as 200 °C at the metal surface, thereby improving the durability of the metal component as well as enhancing the fuel efficiency [3-4]. However, TBCs using the ceramic materials have their inherent brittle behavior, which leads to a poor toughness and unpredictable life-time or failure upon mechanical loading or thermal shock. Moreover, there is a mismatch in coefficients of thermal expansion (CTE) between ceramic TBCs and the metallic substrates to be protected. For example, yttria-stabilized zirconia displays a CTE of $10\times10^{-6}/^{o}C$ at room temperature while the CTE of Al is $24\times10^{-6}/^{o}C$[1].Therefore, the metal-based TBCs (MBTBCs) have been devised to overcome these problems. The nanostructured or amorphous metals are considered to emerge low thermal diffusivity and conductivity [5-6], which has attained an increasing interest. Shin [3] has reported that the thermal conductivity of an induction plasma-sprayed Fe-based amorphous coating is as low as 2 W/mK, which is close to tranditional YSZ coatings. Metallic glasses also perform unique combinations of mechanical, physical and chemical properties [7]; these materials are more attractive as protective coatings to withstand aggressive environments [7-10]. Up to now, amorphous metallic coatings have been produced by using most thermal spraying methods, such as High velocity oxygen fuel (HVOF) spraying [11], Plasma spraying [12], and Arc spraying [13]. Among of these Thermal spraying technologies, HVOF thermal spraying has received great attention because of the lower temperature, high particle flying speed, low porosity, high hardness, low oxide content, high adhension and compression stress[7,10-11].

Fe-based amorphous metallic glasses are considered to be extremely viable candidates as surface protective coatings owing to their superior corrosion and wear resistance, good magnetic and relatively low material cost [14-17]. However, researches regarding heat protection in Fe-based amorphous coatings are relatively few. In this article, an amorphous alloy system of $Fe_{59}Cr_{12}Nb_5B_{20}Si_4$ was selected to prepare amorphous coatings by HVOF spraying process and explore the potential as heat protective coating in transportation field. The microstructural characteristics of the amorphous coatings obtained under different spraying parameters were investigated, and the microhardness, bonding strength and thermal conductivity were also examined.

2 Experimental Procedure

2.1 Materials Selection and preparation

A $Fe_{59}Cr_{12}Nb_5B_{20}Si_4$ powder was prepared by high pressure nitrogen gas atomization method from mixtures of pure Cr (99.1 mass %), Fe (99.1 mass %), as well as pre-alloy Fe-Nb (65 mass %), Fe-B (19 mass %) and Fe-Si (75 mass %). The as-atomized powders with particle sizes in the range of 25-45 µm were used for spraying onto the ASTM 4032 aluminum alloy substrates(with dimensions of $150\times100\times10$ mm^3). The Fe-based coatings were fabricated by Tafa JP5000 HVOF spraying system (Praxair) under three different conditions with varied oxygen and fuel flow. The detailed spraying parameters for the HVOF process are presented in **Table 1.** Two kinds of specimens were deposited for different experiments: (1) the coatings were deposited on the top of surface of the ASTM 4032 aluminum for 400-500μm. These specimens were used to investigate the microstructure of the coatings (cutting with $15\times10\times10$mm^3) and wear behavior test. (2) The coatings were deposited on the substrates for nearly 1mm and then separated from the substrates. Circular discs with a dimension of Φ11mm×1mm^3 were for density and thermal diffusivity, respectively. Before spraying the substrates were degreased by acetone, dried, and then grit-blasted.

Table 1 Spraying parameters employed in the HVOF process

Parameter	H-1	H-2	H-3
Kerosene flow, GPH	5.6	6.2	6.8
Oxygen flow, SCFH	1800	2000	2200
Nitrogen flow, SCFH	26	26	26
Feed rate, r/min	5.0	5.0	5.0
Spraying distance, mm	380	380	380

2.2 Microstructure Characterization

The microstructure of the powders and as-deposited coatings was characterized by scanning electron microscopy (SEM, Quanta-200, FEI) and transmission electron microscopy (TEM, JEM-2100, JEOL). Thin coating slices for TEM were first peeled from substrates by mechanical method and ground to about 50 μm, then thinned by ion-beam milling technique until perforation. X-ray diffraction (XRD, D8 ADVANCE, BRUKER/AKS) analysis of the specimens was performed on an x-ray diffractometer (Cu Ka radiation) with 2θ diffraction angle ranging from 20°to 80°. Percentage of the porosity in the coatings was evaluated using image analysis on optical microscopy (OM, BX51M, OLYMPUS). Oxygen content of the powders and detached coatings were analyzed by a Oxygen/Nitrogen Determinator (G8 GALILEO, BRUKER). The thermal stability and amorphous fraction of the samples was examined by differential scanning calorimeter (DSC, STA 449F3, NETZSCH) in a continuous heating mode at a rate of 0.33 K/s from room temperature to 1000°C under argon atmosphere.

2.3 Coating properties

Vickers microhardness measurements were conducted on the cross section of the HVOF coatings with an applied load of 50 gf and indentation time of 10s, and calculated as an average of 15 measurements. The bonding strength was measured according to ASTM C633-01 standard using a universal mechanical test machine (QBD-100, Chembound, China). At least three tests were done for each sample and the averaged value was used as the indicator of the bonding strength.

The apparent density (ρ) of the sprayed coatings was measured by the Archimedes principle. Thermal diffusivity (α) measurement of the coatings was carried out at room temperature by using the laser flash thermal constant analyzer (LFA427, NETZSCH). Specific heat (C_p) of the coatings was also measured using the Differential scanning calorimetry (STA 449F3, NETZSCH) at room temperature. Then, the Thermal conductivity (k) was calculated from the relation $k= a\rho C_p$.

3 Results and Discussion

3.1 Coating Characteristic

The typical SEM image of the feedstock with diameters of 25-45 μm can be seen in **Fig.1.** It is clearly that the majority of the particles produced by gas atomization in the nitrogen atmosphere are spherical or near-spherical although some had small satellites attached. Most of powders exhibit smooth surface as the meaning of good fluidity, which is propitious to keep the stability and uniformity during spraying process.

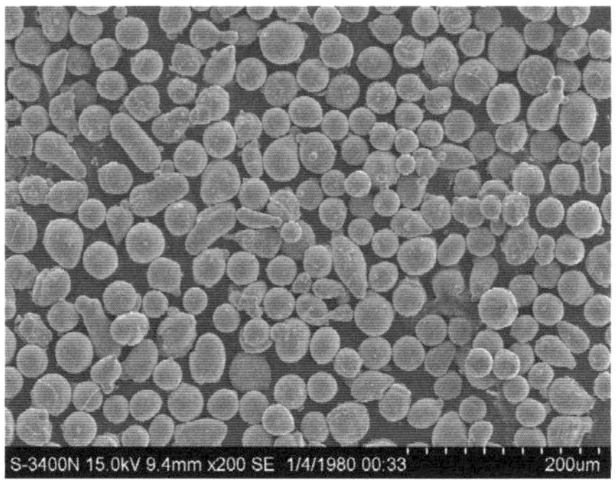

Fig.1. A typical SEM image of $Fe_{59}Cr_{12}Nb_5B_{20}Si_4$ alloy powders with 25-45 μm diameters

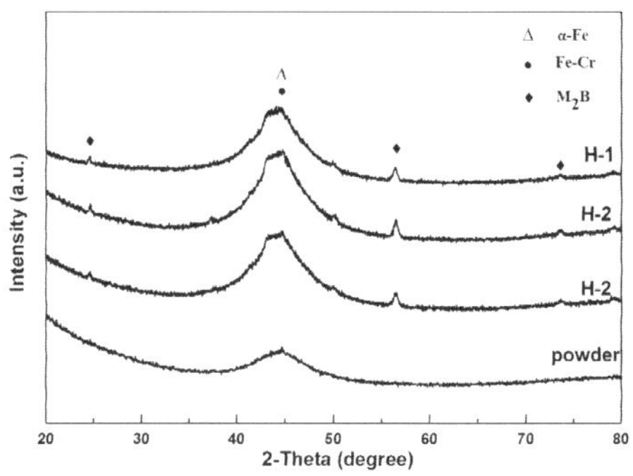

Fig.2. XRD patterns of the atomized powders and the as-deposited coatings H-1, H-2 and H-3.

The XRD patterns of the atomized powders and as-deposited coatings are shown in **Fig.2.** It is notable that a main broad halo peak appearing in all the samples indicates high amorphous phase content, which is primarily because of the high GFA of $Fe_{59}Cr_{12}Nb_5B_{20}Si_4$ alloy system. The reason of the high GFA for this alloy is according to the three empirical rules for the achievement of high GFA [18]. It is regarded that the alloy chosen here satisfies the three empirical rules for the stabilization of the super cooled liquid, leading to highly dense random packed atomic configurations, higher viscosity, and lower atomic diffusivity [18]. Moreover, the high cooling rate of HVOF process also contributes to the formation of amorphous phase in the coatings. Obvious, compared with the powder, few peaks of α-Fe, Fe-Cr and boride are also observed in as-deposited coatings.

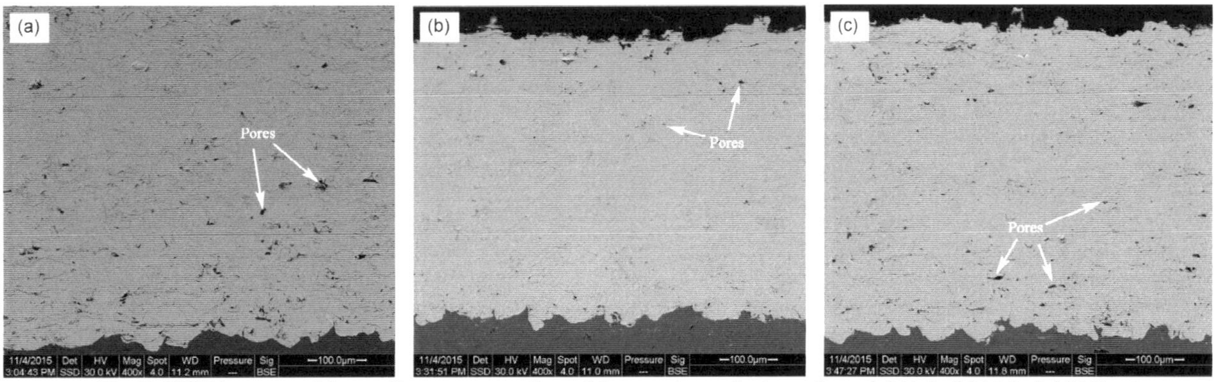

Fig.3. SEM images of the entire cross-section of the as-deposited coatings:(a) H-1, (b) H-2, (c) H-3.

Fig.3 reveals the typical region from cross-sections of coatings and substrates. Generally, the coatings have a dense layered structure as typical of HVOF thermally sprayed deposits. The coatings exhibit low oxide stringer that always generates in Fe-based and Ni-based amorphous coatings [10, 14], although some pores are visible as very dark regions from the images. To confirm this observation, the oxygen content of the powders and detached H-1, H-2, H-3 coatings were measured using O/N determinater, ascertained 0.017% and 0.14%, 0.16%, 0.18%, respectively, thus it could be concluded that slightly oxidation occurred during spraying process which is also attributes to the low temperature and high particles flying speed of HVOF. Moreover, the big pores located between flattened droplets are mainly caused by the loose packed layer structure or gas porosity phenomenon, while the small pores within the flattened particles originate from the shrinkage porosity [19]. It is also can be seen that the top layers seem more porous than the rest of coating. This microstructure characteristic may be significantly affected by the newly incoming particles with high velocity peening the previously deposited layers [10]. After analysis on optical microscopy, it is found that as the flow of oxygen and kerosene increase within the parameter rang examined, the porosity decreases from 1.3% down to 0.7%. But all the coatings express a low porosity less than 2%, which is the typical of HVOF thermal sprayed deposits. Furthermore, a few unmelted particles also exist, as can be seen from Fig. 3. The unmelted particles decreases with rising flow of oxygen and kerosene within the parameter rang examined.

TEM was undertaken to obtain more detail information on microstructure formation of H-2coating, shown in **Fig.4**. The selected area diffraction (SAD) pattern sited at the corner of micrograph characterize that the coating are primarily composed of amorphous phase, which is according with the results of XRD examination. Nevertheless, some nanocrystalline grains, as highly faulted particles obviously ascertained in the micrograph, were embedded in amorphous matrix with grain size in the range of 100-300 nm. It validates the appearance of boride peaks in XRD results, which is regarded that the generation of these nanocrystals are principally caused by heats accumulated inside the as-deposited coatings.

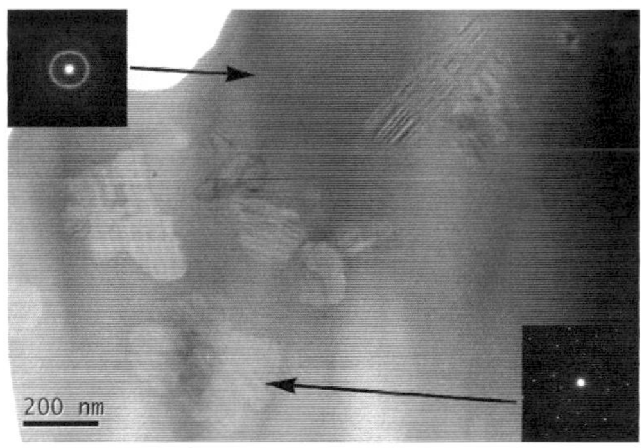

Fig.4. Bright-field TEM image and selected area electron diffraction (SAED) patterns of the as-deposited coating H-2.

To clarify the amorphous content of the as-deposited coatings quantitatively, DSC traces of the coatings and powders were examined as well as their crystallization enthalpy (*ΔH*), as shown in **Fig.5 and Table 2**. The onset crystallization temperature (*Tx*) of the coatings is higher than 630°C, which is similar to the amorphous powder. That is to say, the coatings with high thermal stability against crystallization can be used in practice reliably and widely. The volume fraction of amorphous phase of the coatings is calculated by comparing the heat of crystallization of the powder and coatings,

$$am.\% = \frac{\Delta H_{coating}}{\Delta H_{powder}}$$

The amorphous fraction (am.%) of the coatings are presented in **Table 2**. It reveals that the amorphous fractions in all the coatings are similar and higher than 40% with the different spraying conditions. This is also according to the preceding conclusions in XRD and TEM results that partial precipitates were precipitated in amorphous matrix. Moreover, the amorphous fraction of coatings is less than that of powder, owing to the relatively high spraying temperature and not sufficient enough cooling rate during the coating deposition process. Therefore, in this case, it could be much easier to obtain amorphous phase for gas atomization process than HVOF thermal spraying process. Nevertheless, as mentioned above, although the successive praying process has an adverse effect on the stability of amorphous phase, but the amorphous fraction of coatings prepared by different spraying parameters in a same level, which can be concluded that the amorphous fraction of HVOF sprayed $Fe_{59}Cr_{12}Nb_5B_{20}Si_4$ coatings is insensitive to parameters.

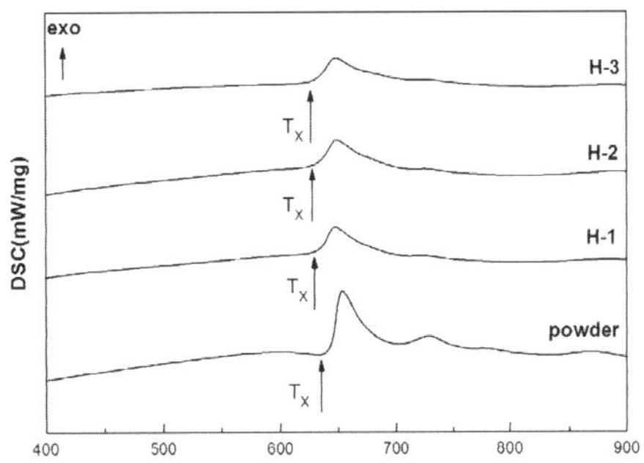

Fig.5. DSC traces of atomized powders, and the as-deposited coatings

Table 2 Crystallization temperature (T_X), crystallization enthalpy (ΔH), and amorphous fraction (am.%) of the powders, and coatings obtained by thermal analysis.

Specimen	Tx, °C	am.%
Powder	643.73	100
H-1	634.67	40.7
H-2	633.62	48.9
H-3	632.20	43.3

3.3 Hardness and Bonding strength

Fig.6 shows the hardness and bonding strength of coatings with different spraying conditions. The three coatings exhibit high microhardness, from 1100 to 1300HV. It manifest that the spraying process does not affect the hardness of coatings significantly. The high hardness for Fe-based coatings obtained now is mainly attributed to their relatively high amorphous content, which has been validated by microstructural analysis conducted above. Also, it is one of the most attractive characteristics of the BMGs, in which the amorphous coatings have retained. In general, the extremely dense random packed atomic configurations of the amorphous structure can effectively resist the plastic deformation arose by applied load, thus leading to the high hardness performance of the amorphous coatings, and metal atoms in the amorphous phase is disordered, does not exist grain boundaries, defects, segregation, etc, showing that the isotropic which also make the amorphous coatings have high microhardness. In addition, the dispersion strengthening of nanocrystals and the presence of the hard boride phase precipitated in amorphous matrix, also contributes to the high hardness of as-deposited coatings. Of course, the high particle flying speed of HVOF make the coating closely packed, low porous also contributes to this.

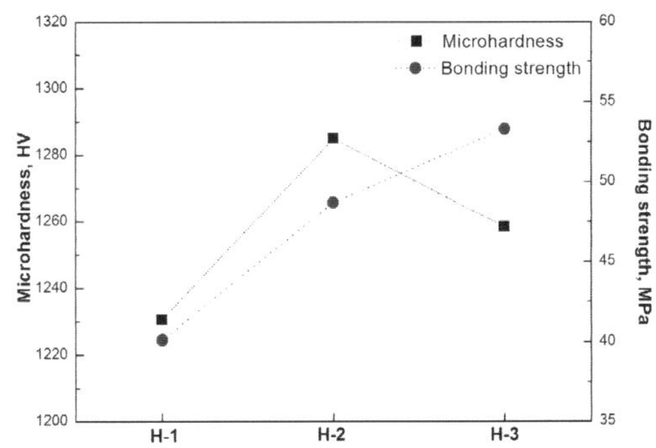

Fig. 6 Microhardness and bonding strength of HVOF coatings under different spraying condition

Meanwhile, the bonding strength of as-deposited coatings was also examined, as shown in **Fig. 6.** As a result, the bonding strengths of coatings exceed 40 MPa, which has been able to meet the requirements of vehicles engine. Besides, the bonding strength of coatings increases with the rising oxygen and fuel flow. Thereby, the H-3 coating has the highest value, reached to 53.3 MPa under the highest parameter. That is because, the flying particles with high speed can produce compressive stress on the surface of substrate, which is beneficial to coatings more compact mechanical bite with the substrate as well as a lower porosity..

3.4 Thermal Conductivity

The thermal conductivity directly indicates the ability of material to conduct heat [1]. The room temperature thermal conductivity (k) of the $Fe_{59}Cr_{12}Nb_5B_{20}Si_4$ coatings under different spraying conditions were calculated based on the measurement of densities (ρ), thermal capacities (Cp) and thermal diffusion coefficients (α) of coatings, presented in **Table 3.** The result shows that, the thermal conductivity of all the Fe-based coatings deposited with different spraying parameters is kept in a low level, less than 3.1 W/mK. Therein, *α, Cp* and *ρ* are rising with the increase of parameter, which leads to the thermal conductivity enlarging gradually. Therefore H-1 coating emerges the lowest thermal conductivity, as low as 1.99 W/mk, which is close to conventional YSZ TBCs. The reason could be that the amorphous structure and grain refinement are disadvantaged to transport heat of coatings [20-22].

Furthermore, with the spraying parameter increase, the thermal conductivity rise from 1.99 W/mK to 3.10 W/mK, but still at a low level compared with metal materials (304L stainless steel 15~16 W/mK), meanwhile, the bonding strength of coatings also rise from 40 MPa to 53.3 MPa, which express a practical significance for actual engineering.

Table 3 The densities, thermal capacity and thermal diffusion coefficient of coatings

Specimen	ρ (g/cm^3)	C_p (J/g·K) ×10^{-1}	α (cm^2/s) ×10^{-2}	*K* (W/mk)
H-1	6.856	3.886	0.750	1.99
H-2	6.879	5.030	0.771	2.66
H-3	7.029	5.151	0.857	3.10

4 Conclusions

Gas-atomized $Fe_{59}Cr_{12}Nb_5B_{20}Si_4$ amorphous powders were deposited to form amorphous coatings by HVOF thermal spraying under different conditions. The results of the present study can be summarized:

(1) The as-deposited coatings exhibit dense layered structure with the porosity below 1.3% and slight oxidation occurs during deposition. The coatings contain amorphous matrix with some precipitated nanocrystals, and the amorphous fractions of the coatings are more than 40%.

(2) Depending on the structure advantage, the coatings exhibit high microhardness, from 1230 HV to 1285 HV. The bonding strength of as-deposited coatings increases with the rising flow of oxygen and kerosene, reached to 53.3 MPa. Meanwhile the thermal conductivity ascends slightly with the enlargement of spraying parameter in examined range, from 1.99 to 3.1 W/mK, which are lower than metal-based materials significantly.

(3) The microstructural characteristic and performance of the Fe-based coatings under different spraying conditions are similar, including the microhardness, bonding strength and thermal conductivity. It manifests the $Fe_{59}Cr_{12}Nb_5B_{20}Si_4$ powder has a great broad adaptability for HVOF thermal spraying process, which has a great significance to the actual engineering application.

5 References

[1] Poirier, D., Lamarre, J.M. and Legoux, J.G.: Thermal cycling assessment of steel-based thermal barrier coatings for Al protection. Journal of Thermal Spray Technology 24(2014), Issue 1-2, pp. 175/84.

[2] Lugscheider, E., Dicks, R., Kowalsky, K., Cook, D., Nassenstein, K. andVerpoort, C.: A Materials System and Method of Its Application for the Wear Protection of Aluminium Engine Cylinder Bore Surfaces, Thermal Spray 2004: Advances in Technology and Applications, May 10-12, 2004, Osaka, Japan.

[3] Shin, D., Gitzhofer, F., Moreau, C.: Properties and induction plasma sprayed iron based nanostructured alloy coatings for metal based thermal barrier coatings. Journal of Thermal Spray Technology 16 (2005), Issue 1, pp. 118/27.

[4] Schlichting, K.W., Padture, N.P. and Klemens, P.G.: Thermal Conductivity of Dense and Porous Yttria Stabilized Zirconia. Journal of materials science 36(2001), Issue 12, pp. 3003/10.

[5] Shin, D., Gitzhofer, F. and Moreau, C.: Development of Metal based Thermal Barrier Coatings for Low Heat Rejection Diesel Engines, Proceedings of ITSC 2005, May 2-3, Basel, Swiss.

[6] Marr, M.A., Wallace, J.S., Pershin, L. and etc.: Preliminary testing of metal-based thermal barrier coating in a spark-ignition engine. Journal of Engineering for Gas Turbines and Power 132 (2010), Issue7, pp. 1/7.

[7] Zhou, Z., Wang, L., He, D.Y. and etc.: Microstructure and wear resistance of Fe-based amorphous metallic coatings prepared by HVOF thermal spraying. Journal of Thermal Spray Technology 19 (2010), Issue 6, S. 1287/93.

[8] Zhang, C., Liu, L., Chan, K.C. and etc.: Wear Behavior of HVOF-Sprayed Fe-Based Amorphous Coatings, Intermetallics 29(2012), pp. 80/85.

[9] Yang, Y., Zhang, C., Peng, Y. and etc.: Effects of Crystallization on the Corrosion Resistance of Fe-Based Amorphous Coatings, Corrosion Science 59(2012), pp. 10/19.

[10] Zhou, Z., Wang, L., He, D.Y. and etc.: Formation and corrosion behavior of Fe-based amorphous metallic coatings by HVOF thermal spraying, Surface and Coating Technology 204(2009), pp. 563/70.

[11] Bolelli, G., Bonferroni, B. and Laurila, J.: Micromechanical properties and sliding wear behavior of HVOF-sprayed Fe-based alloy coatings, Wear 276-277 (2012), pp. 29/47.

[12] Zhou, Z., Wang, L., He, D.Y. and etc.: Microstructure and Electrochemical Behavior of Fe-Based Amorphous Metallic Coatings Fabricated by Atmospheric Plasma Spraying. Journal of Thermal Spray Technology 20 (2011), Issue1-2, pp. 344/50.

[13] Lin, J.R., Wang, Z.H., Lin, P.H. and etc.: Microstructure and cavitation erosion behavior of FeNiCrBSiNbW coating Prepared by twin wires

arc spraying Process. Surface and Coatings Technology 240 (2014), Issue15, pp. 432/36.
[14] Branagan, D.J., Swank, W.D., Haggard, D.C. and Fincke, J.R.: Wear-Resistant Amorphous and Nanocomposite Steel Coatings, Metallurgical and Materials Transactions A 32(2001), Issue 10, pp. 2615/21.
[15] Otsubo, F. and Kishitake, K.: Corrosion Resistance of Fe-16%Cr-30%Mo-(C, B, P) Amorphous Coatings Sprayed by HVOF and PS Processes, Materials Transaction 46(2005), Issue 1, pp. 80/83.
[16] Guo, R.Q., Zhang, C., Chen, Q. and etc.: Study of Structure and Corrosion Resistance of Fe-Based Amorphous Coatings Prepared by HVAF and HVOF, Corrosion Science 53(2011), Issue 7, pp. 2351/56.
[17] Voyer, J.: Surface Wear Improvement of Al-Alloys by Amorphous Iron-Based Flame-Sprayed Coatings, Materials Science Forum 690(2011), pp. 405/08.
[18] Inoue, A. High strength bulk amorphous alloys with low critical cooling rates. Materials Transactions JIM 36 (1995), Issue 7, pp. 866/75.
[19] Sobolev, V. V. Guilemany, J. M: Investigation of Coating Porosity Formation During High Velocity Oxy-Fuel (HVOF) Spraying, Materials Letters 18 (1994), Issues 5-6, pp. 304/08.
[20] Wu, J., Guo, H.B., Zhou L.: Microstructure and Thermal Properties of Plasma Sprayed Thermal Barrier Coatings from Nanostructured YSZ. Journal of Thermal Spray Technology 19 (2010), Issue 6, pp. 1186/94.
[21] Shin, D., Gitzhofer, F., Moreau, C.: Thermal property evolution of metal based thermal barrier coatings with heat treatments. Journal of Material Science 42(2007), Issue 15, pp. 5915/23.
[22] Yang, H.-S, Eastman, J.A., Thompson, L.J. and Bai, G.R.: Grain-Size-Dependent Thermal Transport Properties in Nanocrystalline Yttria-Stabilized Zirconia. Materials Research Society Symposium Proceedings 703(2002), pp. 179/84.

Characterisation of Thermally Sprayed ZnAl15 Corrosion Protection Coatings for Offshore-Wind-Turbines

Bobzin, K.; Öte, M.; Knoch, M.A.

Due to strong and steady winds in offshore environments, offshore-wind-turbines (OWT) offer a great potential for satisfying ever growing power demands and reducing the carbon footprint. However, corrosion is one of the main failure causes of OWT tower segments. Therefore, corrosion-protection-systems allowing for a maintenance-free protection of OWTs over their lifetime are crucial. Especially considering that repair costs are approximately 75 - 100 times higher than the cost of initial coating application. The influence of using inert and active atomising-gases on process stability, morphology as well as corrosive resistance of the coatings is analysed. Furthermore, coatings applied with three different spray guns using 1.6, 2.5 and 3.2 mm wires are tested against a reference according to the industry standard in a long term salt spray test. Using the results of this test, the influence of the different atomizing gases on the corrosion behaviour of ZnAl15 coatings is regarded.

1 Introduction

Strong and steady winds in offshore environments offer a great potential for the effective use of renewable energies. Offshore-Wind-Turbines are necessary in order to harvest this source. Besides its advantages, offshore environments also pose several challenges to the offshore structures with regard to longevity and economic success. The high chloride content of seawater and maritime atmosphere, combined with the ample presence of water, promote the corrosion of steel structures. In order to prevent premature failure of the steel structures, duplex corrosion protection systems are commonly used. Such systems consist of a zinc-based metallization offering cathodic corrosion protection, a sealant and several layers of organic coatings offering passive corrosion protection to the whole system. The cathodic protection comes into effect if the organic coating is damaged. It has been shown that an undamaged, 100 µm thick ZnAl15 coating can protect the underlying steel for over a decade in the splash zone. However there are no systems available that allow a maintenance-free service over lifetime. Another challenge is the fact that offshore wind turbines are usually unmanned structures and thus every maintenance check requires sending specialists to these structures via boat or helicopter. Furthermore required repair works also require elaborate planning and fair weather conditions. Therefore each and any maintenance check and repair mission is highly expensive. For this reason, the development of corrosion protection systems allowing for a maintenance free operation over lifetime is part of current research activities [1–3].

2 Experimental

In order to investigate the influence of different atomizing gases on the coating's morphology, chemical composition, phase distribution and oxide content, twelve sample sets were coated on S355J2+N substrates and analysed. Compressed air (CA) was used as a reference; pure nitrogen (N_2) as an inert atomising gas and a nitrogen-hydrogen mix (N_2-H_2) was used as an inert atomising gas with an active component. Each gas was used to produce four sample sets with different combinations of stand-off

distance and atomising gas pressure, see **Tab. 1**. Voltage and current were adapted for each gas in order to achieve a stable process. A coating thickness of 100 µm was targeted.

Tab. 1: Parameters used for analysing the influence of different atomizing gases

Sample number	Atomizin g gas	Pressure [MPa]	Stand-off distance [mm]
1	CA	0.31	120
2	N_2	0.31	120
3	N_2- H_2	0.31	120
4	CA	0.41	120
5	N_2	0.41	120
6	N_2- H_2	0.41	120
7	CA	0.31	90
8	N_2	0.31	90
9	N_2- H_2	0.31	90
10	CA	0.41	90
11	N_2	0.41	90
12	N_2- H_2	0.41	90

For each sample set, two cross sections made of two samples which were produced with identical parameters, were prepared. These cross-sections were examined with regard to the overall coating morphology and special regard to the porosity. The porosity was calculated using an optical image analysis in which selected colours are manually assigned to the considered phases. Using the contrast between the substrate material, the coating and the pores, the internal porosity can be determined. For each cross section, a total of ten images were analysed, reducing the impact of possible process inhomogeneities resulting in a locally poor coating quality. These cross sections were furthermore investigated using scanning electron microscopy (SEM) and energy dispersive spectroscopy (EDS) in order to analyse the morphology and chemical composition. Special regard was paid to the presence of oxides and their percentage depending on the atomising gas. Further investigations were conducted using a microprobe.

Prior to the microprobe analysis, the samples were ground and polished in an anhydrous environment. After the polishing was finished, the samples were immediately placed in the vacuum chamber in order to exclude oxide formation due to preparation. Each sample was further investigated with regard to its phase composition. For this purpose, XRD scans were performed.

Furthermore free standing coatings were manufactured and analysed with regard to their chemical composition. In order to produce these samples, substrates made of Polypropylene (PP) were coated using 150 A, 20 V for a 2.5 mm diameter ZnAl15 wire and a stand-off distance of 100 mm. Since the chemical dissolution requires a certain amount of material for each considered element, a coating thickness of 500 µm was targeted instead of the usual 100 µm.

In addition to the study of the coating properties, emissions in the micron and submicron scale were measured. In order to identify the influence of the atomizing gases on emission behaviour, other process parameters such as stand-off distance, robot speed, current, pressure and total gas flow were kept constant. The voltage was adjusted in order to have stable processes for all gases. The robot mounted gun was repeatedly moved in constant meanders for 60 s during the measurement. Stationary flat samples were used as substrate material. The setup is illustrated in **Fig. 1**.

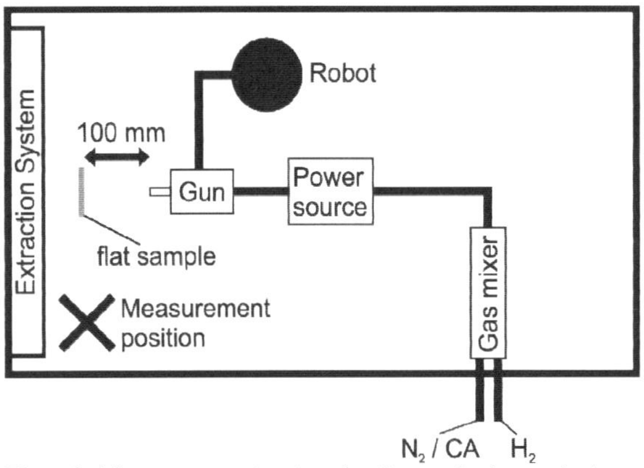

Fig. 1: Measurement setup for the emissions study

Subsequent to these analyses substrate plates were grit blasted to Sa 2.5 using steel shot. Afterwards the substrate material was coated using three different spraying systems and three different wire diameters with the three atomizing gases. For N_2 and N_2- H_2, the voltage had to be increased by ca. 5 – 10 % while keeping the current and wire feed speed constant for all considered gases. These samples, along with three ZnAl15 reference samples supplied by the Muehlhan AG, were tested in a neutral salt spray test over 2.500 hours. The reference samples were blasted and coated in accordance to the industrial standard. The test was done in accordance with DIN EN ISO 9227 [4]. During the salt spray test, the samples were periodically checked for superficial corrosion on a regular basis. After the test, the corrosion of each sample was assessed by analysing the cross section using light microscopy. During the preparation special care was paid to keeping the original surface structure and composition, e.g. not losing the corrosion product during the cutting process. For this purpose the samples were embedded using epoxy prior to cutting them into pieces for the metallographic preparation. The stabilization was supported by vacuum infiltration.

3 Results

Porosity shows different levels of sensitivity on stand-off distance and pressure variation depending on the atomizing gas used in the process. The lowest overall porosity was achieved with the nitrogen-hydrogen atomizing gas. This combination of process parameters and atomizing gas also had the lowest standard deviation of all sample sets. The process robustness however was superior for compressed air. Using nitrogen or nitrogen-hydrogen, a slight variation in the stand-off distance caused a significant increase of porosity, see **Fig. 2**. The XRD scans did not show any influence of the atomizing gas on the detected phases. Neither could any oxides be detected.

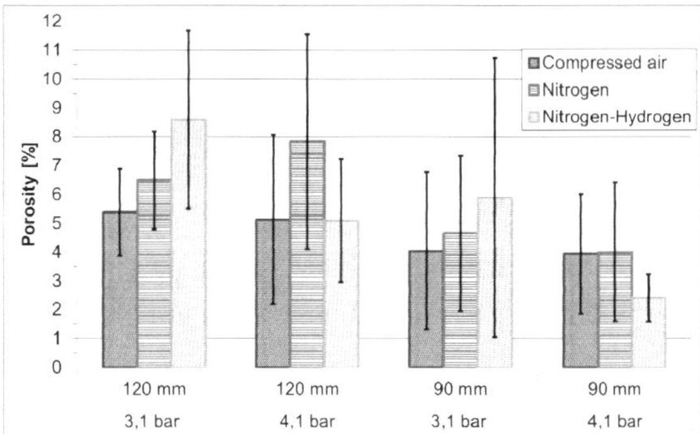

Fig. 2: Porosity at Variation of atomizing gas and process parameters

During the spraying process emission in the micron and submicron scale were measured. Just like the porosity, the emissions indicate an influence of the atomizing gas on the process. Within the considered scale, an increase of smaller particles could be found when switching from compressed air to nitrogen. By adding hydrogen to nitrogen, emission levels are reduced but still remained above those of the CA process. The overall measured emissions are shown in **Tab. 2**. It should be noted that the measurements for the micron scale emissions do contain all particles smaller than the stated value, e.g. all PM1 particles are also part of PM2.5.

Tab. 2: Emissions measured over 60 s

Size	Unit	CA	N_2	N_2-H_2
submicron	[10^7 pt/cm³]	2,61	2,42	2,37
PM1	[mg/m³]	1,75	2,28	2,04
PM2.5	[mg/m³]	1,77	2,30	2,05
PM4	[mg/m³]	1,94	2,41	2,17
PM10	[mg/m³]	2,34	2,78	2,44

Using SEM and EDS, the cross sections were further investigated. Regardless of the atomizing gas used during the coating process, oxides could be detected in every cross section. However, the amount of oxides could not be related to the process parameters. It should be noted that during the preparation of the cross sections, water was used up to a certain point. This water might have caused some oxidation and therefore the SEM results concerning oxygen levels in the coatings should be considered with care. Regardless of the preparation method, the zinc and aluminium levels can be measured by SEM with a high accuracy. Zinc and aluminium rich phases can be detected. The microstructure is very fine, indicating very high cooling rates. Furthermore the amount of re-solidified particles seems to depend on the combination of process parameters and atomizing gas. With compressed air, the least re-solidified particles were found whereas the nitrogen coating had the most, see **Fig. 3**.

Using the microprobe, no oxides could be detected in any sample. This indicates that there is no measureable oxidation during the spraying process. This result does not mean that there are no oxides in the coating; the resolution of the sensor was 10 nm. Therefore, any oxide lamellae smaller than 10 nm cannot be detected. However, it can be stated that the amount of oxygen in the coating is negligible.

The wet chemical analysis of the free standing coating shows that the zinc-aluminium ratio does not depend on the atomizing gas, see **Tab. 3** Furthermore small amounts of carbon were detected; most likely due to the method the samples were produced.

Tab. 3: Chemical composition of free standing coatings

	Zn [Mass-%]	Al [Mass-%]	C [Mass-%]
CA	Matrix	15,5	0,013
N_2	Matrix	15,8	0,015
N_2-H_2	Matrix	15,7	0,014

No sample failed during the salt spray test, with the criterion for failure being the appearance of macroscopic red rust. Blisters formed however under the zinc and aluminium corrosion products. A closer examination of these blisters indicated the formation of red corrosion products; most likely red rust. These red dots are very small and can only be detected under the microscope. In **Fig. 4** two very pronounced blisters are shown.

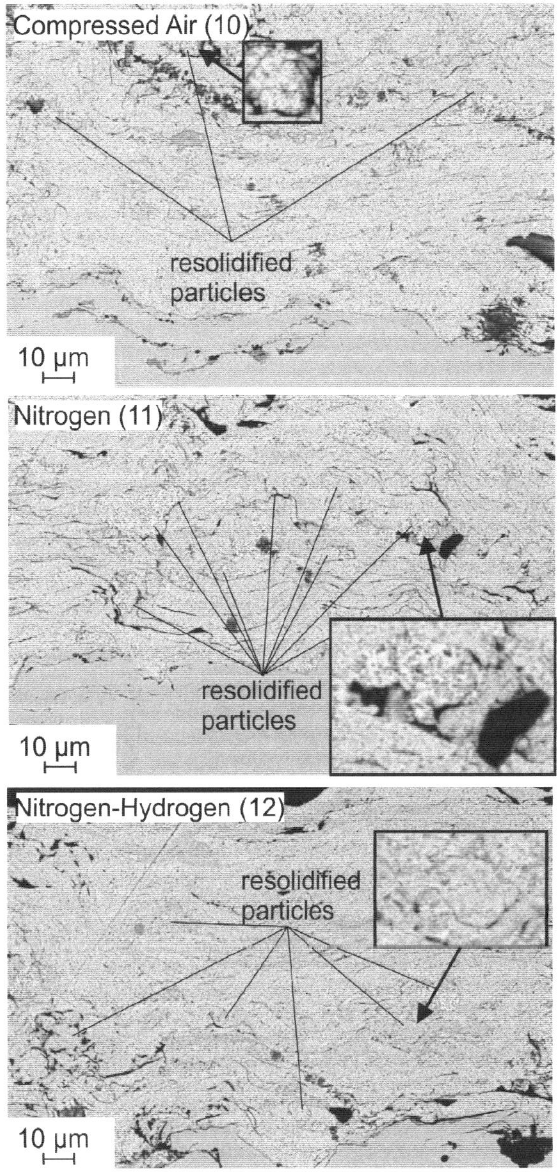

Fig. 3: SEM of ZnAl15 coatings on S355J2+N with atomizing gas variation

Therefore cross sections were prepared and analysed with a light microscope. The reference sample shows a higher level of oxidation of the coating than that of any other sample examined. Furthermore, the reference coating's thickness is very uneven and the coating has lots of cavities. Using another reference sample that was not exposed to the salt spray test shows that the irregular thickness and high porosity can be attributed to the blasting and coating parameters used for these samples.

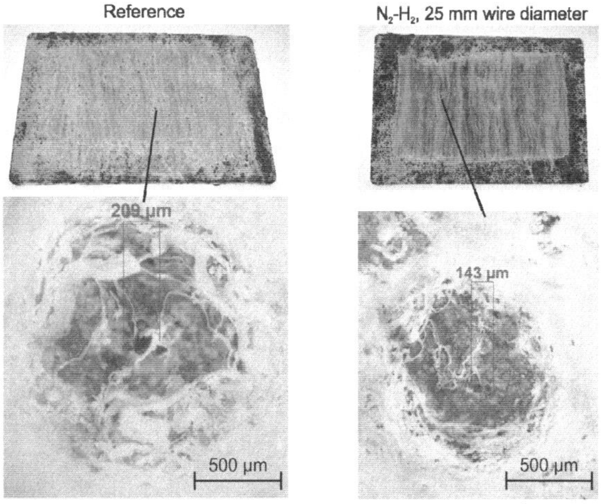

Fig. 4: Blisters on a sample and on the reference after 2.000 h salt-spray-test

Different levels of oxidation were found in the samples depending on the duration of the salt spray test. All samples exhibit a rather uniform coating thickness and level of porosity. No sample, including the industry reference, shows an indication that the substrate material was attacked. An influence of the atomising gas could not be detected

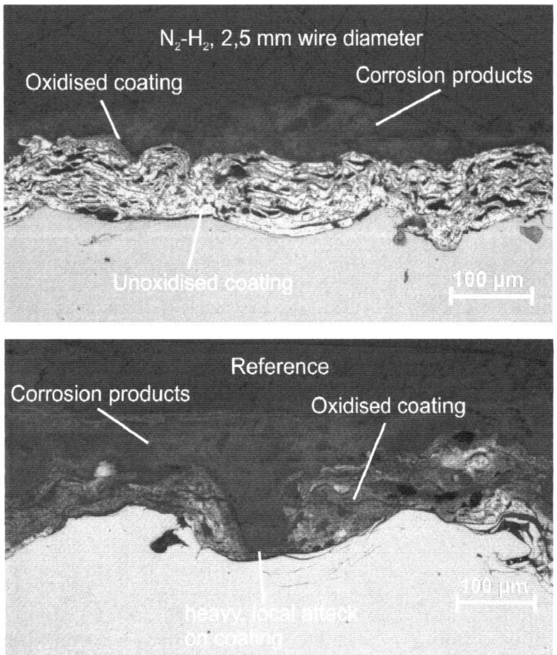

Fig. 5: Representative cross section through the samples shown in **Fig. 4** at 200x magnification

4 Discussion

The results show that switching the atomizing gas has an effect on the arc properties. The increased voltage necessary for maintaining a stable process indicates that the plasma properties within the arc are also affected. The resulting porosity, process robustness and detected emission are dependent on the atomizing gas as well. Since these parameters depend on the atomizing behaviour of the molten wire tip and the subsequent cooling during flight, an influence on the atomizing process can be assumed.

The chemical analysis of free standing coatings did not show a significant influence of the atomizing gas on the zinc-aluminium ratio. The detected carbon likely originates from PP particles left on the sample during the separation of the coating from the substrate. XRD scans on the sample 1 – 12 further indicate that the phase formation is not influenced by the atomizing gas.

SEM and EDS analysis indicate a dependency of the coatings' morphology on the atomizing gas. The higher occurrence of re-solidified particles for the considered inert gases supports this assumption. EDS analysis does not show any phases with an unexpected composition. The zinc-aluminium ratios correspond to the ZnAl phase diagram and are independent of the atomizing gas. Oxides were detected in every sample; a dependency on the atomizing gas could not be observed either. The detection limits of the Rutherford-Backscattering-Detector (RBSD) do not allow for a quantitative study using EDS. Furthermore, oxides may form during sample preparation and thus the measured oxygen content should be considered with care.

A microprobe analysis allows for quantitative measurements of the oxide content. Therefore, samples 10, 11 and 12 were examined using a microprobe. Great care was taken during the samples' preparation process to avoid the formation of oxides. Therefore these results give a detailed overview of the oxygen content within the as sprayed coating; the only limitation being the detection limit of 10 nm. Using this method no oxides could be detected in any coating. This indicates that independent of the atomizing gas, no significant oxidation takes places during the spraying process. A possible explanation for this phenomenon may be the formation of a very thin aluminium oxide layer on the interface between ZnAl15 melt and air. This oxide film naturally occurs on aluminium and forms the alumina passive layer. Such a layer hinders further oxidation by preventing oxygen from advancing beyond this layer. The ultimate thickness of natural alumina passive layers is assumed to be below the detection limit of 10 nm [5]. If such a layer is formed on the melt pool at the wire tip or on the surface of molten particles respectively; an influence on the atomizing process is also very likely. Consequently, an effect on the particle size would be expected. Different particle sizes, in turn, would have an effect on the cooling behaviour of particles and thus on the amount of re-solidified particles at a certain stand-off distance. Considering the different thermo-physical properties of the different atomizing gases as well as the different arc properties, an influence on the particle size and temperature is very likely. Newbery et al. observed similar a phenomenon for steel droplets [6]. The lack of any measureable oxides due to the formation of a very thin passive layer would explain the lack of any influence of the atomizing gases on the coatings chemical composition and its phases as well.

5 Conclusion

Different atomizing gases for arc-wire spraying were tested with regard to their influence on the coatings morphology, chemistry and corrosion resistance. ZnAl15 was selected as the coating material. It could be shown that the different gases have an influence on the atomizing process and thus on the particle size distribution as well as on the morphology and porosity. However, no influence was detected on the chemical composition of the coating. In fact no oxides could be found using a microprobe analysis with a spatial resolution of 10 nm. For nitrogen and nitrogen-hydrogen, a slight increase of the voltage is necessary in order to obtain a stable arc. However, the sensitivity of ZnAl15 to slight deviations of the ideal stand-off distance increases significantly if an oxygen-free atomizing gas is used instead of compressed air, see **Fig. 2**. Considering compressed air, the lack of measureable oxides within the coatings as well as the high process robustness may be caused by the formation of a thin alumina film on the outside of molten particles. Such a film would act as a passivation layer and thus prevent further chemical reactions of the melt with its surrounding atmosphere. The natural alumina layer's thickness is temperature dependant. For room temperature the ultimate thickness is roughly one order of magnitude smaller than the microprobe's detection limit [5]. The very small size as well as the high reaction speed of aluminium with oxygen, has so far prevented a proof that such oxides form during the spraying process of ZnAl15. On the other hand, there was no significant influence of the atomizing gas on the corrosion behaviour, whereas the different morphologies due to wire diameter and process parameters did have an influence on the corrosion behaviour.

In further studies several experimental ZnAl15 + X alloys will be tested. The X stands for Mg, Mn or Ti which will be added to the ZnAl15 in order to increase the corrosion resistance while keeping the cathodic protection of the underlying substrate material. Using a sealer in order to increase the corrosions resistance of thermally sprayed Zn-based coatings is a well-established method [1]. In order to take fully advantage of this effect, a sealer will be developed for the coatings of this study.

6 Acknowledgments

The work in this paper is performed in the context of the KOKON project. KOKON is funded by the German Federal Ministry of Economics and Technology (BMWi).

7 Literature

[1] DIN Deutsches Institut für Normung e.V., Thermisches Spritzen - Metallische und andere anorgangische Schichten - Zink, Aluminium und ihre Legierungen - Entwurf, Beuth Verlag GmbH, Berlin, 2014.

[2] M. Büteführ, Einfluss des Aluminiumgehaltes gespritzter Zinküberzüge auf den Korrosionsschutz von Stahl, Otto-Graf. Inst., Stuttgart, 2006.

[3] M. Büteführ, F. Prenger, Korrosionsverhalten von verschiedenen ZnAl Überzügen im maritimen Bereich: Ergebnisse von Langzeitauslagerungen, Dresden, 2015.

[4] DIN Deutsches Institut für Normung e.V., Korrosionsprüfungen in künstlichen Atmosphären – Salzsprühnebelprüfungen, Beuth Verlag GmbH, Berlin, 2012.

[5] M.S. Hunter, P. Fowle, Natural and Thermally Formed Oxide Films on Aluminum, Journal of the Electrochemichal Society 103 (1956), pp. 482–485.

[6] A.P. Newberry, P.S. Grant, R.A. Neiser, The velocity and temperature of steel droplets during electric arc spraying, Surface & Coatings Technology, 2005, pp. 91–101.

Spray parameter effect on HVAF sprayed (Fe,Cr)C-30FeNiCrSi hardmetal coatings

V. Matikainen, K. Khanlari, A. Milanti, H. Koivuluoto and P. Vuoristo
Department of Materials Science, Tampere University of Technology, Tampere / FIN

Thermal spray processes are known for their complexity and the numerous variables that affect the resulting coating properties. Systematic study of the main spray parameters is an often used method to determine the process operation window. The design of experiments approach was employed to map the selected main parameters of a high velocity air-fuel (HVAF) spray process. Commercial chromium carbide based material with NiFeCrSi matrix was sprayed with varying gas flow parameters and nozzle designs. The effect of these variables on particle in-flight parameters (temperature and velocity) was determined. Moreover, the effect of in-flight parameters on coating build-up and properties were analysed with a scanning electron microscope and x-ray diffraction methods. Also, abrasive wear tests were carried out to determine the wear resistance of the coatings. The study shows that sufficient heating is critical for the abrasion wear resistance of the studied material. The average particle velocity is significantly increased when using a nozzle with convergent-divergent geometry. The highest wear resistance was achieved with high fuel flow rates and with cylindrical or convergent-divergent nozzles.

1 Introduction

High velocity oxy-fuel (HVOF) flame spraying has long been the industrial standard process to produce hardmetal coatings. Good properties of the HVOF coatings are due to the high temperature supersonic gas stream which accelerates the particles to high velocities (500-800 m/s) and produces dense coatings. The development of the HVOF process has aimed for higher particle velocities, lower particle temperatures and denser coating structures. The main steps to achieve this have been 1) the introduction of convergent-divergent nozzles and 2) the liquid fuelled HVOF gun designs with powder injection located at the diverging section of the nozzle [1]. The modern HVOF spray equipment are robust but possess some restrictions regarding the overheating of fine particle size distributions and the spray rates.

High velocity air-fuel (HVAF) process has been increasingly studied as the alternative to the HVOF spraying. The HVAF process uses compressed air as the oxidizer to burn gaseous fuel (propane or propylene), which reduces significantly the maximum flame temperature compared to HVOF. This lower temperature provides sufficient heating to reach the melting temperature of the metallic matrix of the most common hardmetal materials while the supersonic gas stream provides very high particle velocities. The early HVAF guns suffered from low particle temperature and deposition efficiency, which was significantly improved by later developments introducing the new combustion chamber designs to stabilise the combustion of gaseous fuels [2].

The effect of process parameters of M3 HVAF gun on numerous responses have been studied by Lyphout et al. [3] and there exists several studies, which have compared different HVOF and HVAF processes [4,5]. However, the studies have mainly focused on WC-CoCr hardmetal materials and only few studies have focused on chromium carbide based materials [6,7]. Also, the previous studies of the HVAF process have not provided enough information about the actual process parameters and lack the information about the process parameter effects on particle in-flight parameters (velocity and temperature). This study focuses on three main process parameters to provide new information about the HVAF process and the produced particle in-flight parameters. Also, novel hardmetal material based on cheaper (Cr,Fe)C primary carbides is utilised to study the effect of the parameters on the coating microstructure and properties [8].

2 Experimental

2.1 Material and spray processing

Commercial agglomerated and sintered Amperit 575.074 (Fe,Cr)C-FeNiCrSi powder (H.C. Starck GmbH, Germany) with nominal particle size of -45+15 µm was sprayed on S235 low carbon steel and 316L stainless steel substrates (50x200x5 mm) with M3 HVAF spray gun (Uniquecoat Technologies LLC, USA). The schematic of the spray gun is presented in a previous study [9]. Propane was used as the fuel gas.

The HVAF process is generally pressure controlled and actual flows in slpm are not usually known. The process parameters are commonly expressed as inlet pressures of the process gases: *Air*, *Fuel1* and *Fuel2* in the case of the M3 HVAF gun. For further understanding of the process the *Fuel1* and *Fuel2* flows were measured by using rotameters and the resulting combustion chamber pressures were measured with a pressure gauge connected to the powder hose 20 cm upstream from the back of the HVAF gun.

Two-level full factorial experiment design with three factors was employed to study the parameter effects and interactions. The full factorial design includes all possible combinations of the selected parameters on two levels resulting in eight (2^3) different spray runs. "*Fuel1*", "*Fuel2*" and "*nozzle type*" were selected as the main process parameters as these were seen as the most useful parameters in practise to affect the heating of the feedstock powder. These parameters were changed on two levels according to **Table 1**. Other spray parameters and hardware were kept constant: 7.45 bar (108 psi) compressed air pressure setting, long combustion chamber, 300 mm spraying distance, 750 mm/s surface speed, 5 mm step, 100 g/min feedrate and 60 slpm carrier gas flow rate (N_2). Target coating thickness was 300 µm and compressed air cooling was used to maintain the substrate temperature below 200 °C.

Table 1. Fuel parameter settings used for the two nozzles (C: Cylindrical and CD: Convergent-divergent).

Sample	Nozzle	Fuel1 [bar (psi)]	Fuel2 [bar (psi)]
C1	4L0	7.24 (105)	7.24 (105)
C2	4L0	7.24 (105)	6.21 (90)
C3	4L0	6.84 (99)	7.24 (105)
C4	4L0	6.84 (99)	6.21 (90)
CD1	4L2	7.24 (105)	7.24 (105)
CD2	4L2	7.24 (105)	6.21 (90)
CD3	4L2	6.84 (99)	7.24 (105)
CD4	4L2	6.84 (99)	6.21 (90)

2.2 Diagnostics

The measurements of the spray stream characteristics and particle in-flight parameters were carried out with Spraywatch 4S online spray monitor (Oseir Ltd., Finland). The monitor is based on CCD camera, which constantly records the whole spray stream width, giving particle velocity and temperature distributions, spray stream width, direction and flux [10]. The particle temperature measurement is based on 2-color stripe-pyrometry. All measurements were carried out at the actual spray distance of 300 mm with 75 mm focus distance and 2 µs exposure time for the particle velocity image.

2.3 Coating characterisation and testing

The feedstock powder and the coating structures were studied with XL-30 scanning electron microscope (SEM, Philips, The Netherlands). Phase compositions of the feedstock powder and sprayed coatings were assessed with Empyrean x-ray diffractometer (XRD, PANalytical, The Netherlands) by using Co-Kα radiation (45kV and 40mA). Microhardness measurements were carried out with MMT-X7 vickers microindenter (Matsuzawa, Japan) by using 0.3 kg and 1.0 kg loads to calculate an average hardness value from 15 and ten indentations, respectively.

Abrasive wear resistance of the coatings was evaluated according to modified ASTM G65 standard, in which four coating samples were tested simultaneously with one 316L stainless steel reference sample. Samples were pressed against the rubber wheel with 23 N normal force for 60 min and 5900 m total wear distance. Quartz sand (0.1-0.6 mm) was used as the abrasive media with an approximate feedrate of 25 g/min per sample. The sample positions were circulated at 12 min intervals. Three samples per coating were tested to calculate an average wear rate.

3 Results and discussion

3.1 Feedstock material

The feedstock material consisted of coarse carbide particles embedded in metallic matrix and having the typical morphology of the agglomerated and sintered powder. The powder morphology and cross section images are presented in **Fig. 1**. The backscattered electron (BSE) images show the metallic matrix as white or light grey areas, while the carbides are darker grey particles. The outer surface of the particles is mainly covered by carbides with some small islands of metallic matrix in **Fig. 1a**. Majority of the particles showed very porous structure visible in **Fig. 1b**. The EDS analysis confirmed that the carbides consisted mainly of chromium with significant amount replaced by iron (14-17 wt.%). The presence of some nickel in the carbides could not be ruled out. The metallic matrix was nickel-based with iron, chromium and silicon alloying elements.

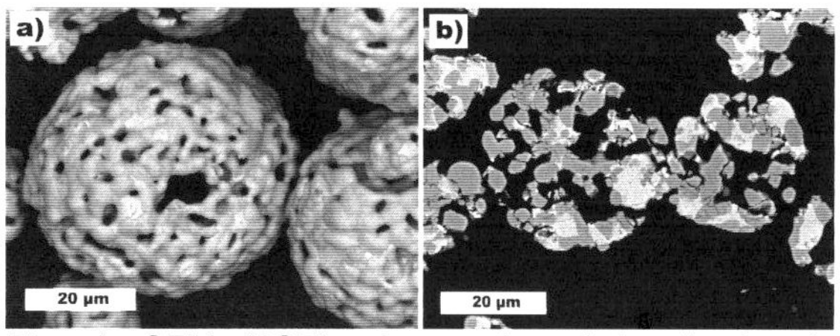

Fig. 1. SEM (BSE) micrographs of the feedstock powder a) morphology and b) cross section.

3.2 Process characteristics

The M3 HVAF gun comprises two nozzles: the primary nozzle and the secondary nozzle. Compressed air is used as an oxidizer to burn propane in the combustion chamber (*Fuel1*) and in the secondary nozzle (*Fuel2*). The air is also used to cool down the gun structures, combustion chamber and the secondary nozzle. The thermal efficiency of the process is very high as all the cooling air is projected out through the secondary nozzle where it is also participating in the combustion with *Fuel2,* acting as an afterburner. The typical 3rd generation gas and liquid fuelled HVOF processes use water cooling, to which 5-25 % of the generated thermal energy is lost [11]

The combustion chamber pressure is controlled by the compressed air inlet pressure and by both fuel inlet pressures: *Fuel1* and *Fuel2*. The increase of *Fuel1* inlet pressure increases the combustion chamber pressure, due to increased fuel flow, and decreases the *Fuel2* flow by increasing the back pressure for the *Fuel2* inlet line. Increase of the *Fuel2* inlet pressure, on the other hand, affects also the combustion chamber pressure by increasing the pressure downstream from primary nozzle. This increases the combustion chamber pressure and decreases slightly the *Fuel1* flow rate.

The combustion chamber pressures and propane fuel flow rates were measured for an extended parameter set that also contained the center points of each fuel parameter, i.e. additional points of 102 psi for *Fuel1* and 97.5 psi for *Fuel2*. The results are shown in **Fig. 2**, where coloured circular (cylindrical nozzle) and triangular (convergent-divergent nozzle) points belong to the parameter set presented in **Table 1**. It can be noticed that the combustion chamber pressures are significantly higher for the cylindrical nozzle due to more restricted gas flow though the long secondary nozzle compared to the expanding convergent-divergent nozzle. The higher chamber pressure decreases the pressure difference between the chamber and the *Fuel1* inlet regulator, and hence reduces the fuel gas flow into the chamber. In the case of the convergent-divergent nozzle, fuel gas flow rates were 10-25 % higher compared to the cylindrical nozzle when identical inlet pressure parameters were used.

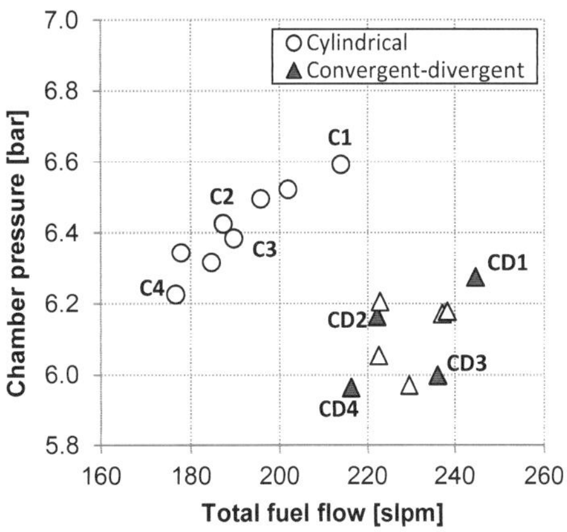

Fig. 2. Measured combustion chamber pressures and total fuel flow rates for the extended parameter set. Coloured points stand for the parameters used for the actual spray experiments.

3.3 Particle in-flight parameters

The particle temperature and velocity results from the diagnostic measurements are presented in **Fig. 3**. Both nozzle geometries seem to produce simultaneously increasing average particle temperatures and velocities when the fuel inlet pressure settings are increased. The measured particle temperatures are almost identical, reaching 1480 °C when the high flow rate parameters (105 psi / 105 psi) are used. The particle velocities, on the other hand, are systematically over 100 m/s higher for the convergent-divergent nozzle. This means that the average kinetic energy of the particles was increased by 25-34 % when convergent-divergent nozzle was used. The highest measured average velocities were 855 m/s and 970 m/s for the cylindrical and convergent-divergent nozzles, respectively. The measured velocities for both nozzles are higher than the particle velocities typically produced by the HVOF processes. For comparison, an average temperature and velocity of 1750 °C and 695 m/s were measured for the same material sprayed with DJ2700 (C_3H_8: 70 l/min, O_2: 240 l/min, air: 383 l/min, spraying distance: 230 mm, feedrate 40 g/min), respectively.

Main effect plots for mean particle temperature and velocity were produced based on the full factorial design and the parameters in **Table 1**. These plots are used to identify whether the selected factors, i.e. process parameters, have an effect on the selected responses, i.e. particle temperature and velocity. It is clear from **Fig. 4a** that the *Fuel1* setting has the most significant effect on the particle temperature followed by *Fuel2*. For the particle velocity, on the other hand, the nozzle design has the most significant effect followed by *Fuel1* setting, **Fig. 4b**.

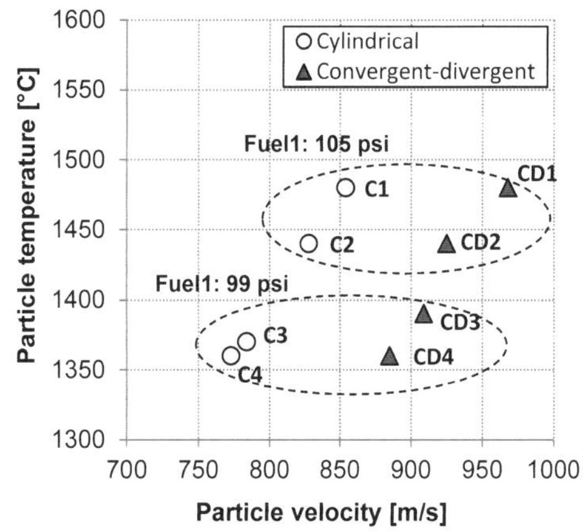

Fig. 3. Graph showing the measured average particle velocity and temperature values for different spray parameters.

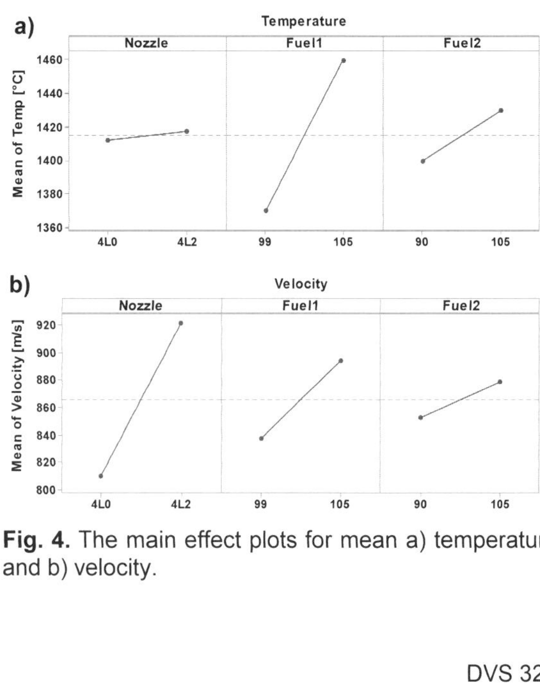

Fig. 4. The main effect plots for mean a) temperature and b) velocity.

3.4 Coating build-up

Based on the measured particle temperatures it could be expected that the deposition efficiency was comparable between the nozzles. However, the results in **Fig. 5** show that higher relative deposition efficiencies were achieved with the cylindrical nozzle design. While the measured particle temperatures are comparable between the two nozzles, it seems that the higher particle velocity slightly decreases the deposition efficiency. There are a number of reasons that could cause this effect. Firstly, the measured particle temperature is the surface temperature of the particles and does not represent the overall temperature or give information about the temperature history of the particle. Particle dwell time could be longer with the cylindrical nozzle due to the lower particle and gas stream velocity. Also, the temperature drop in the secondary nozzle could be lower compared to the convergent-divergent type nozzles. Secondly, it is possible that the higher velocity gas stream in the case of convergent-divergent nozzle deflects a fraction of the smaller/lighter particles when the gas stream meets the substrate surface and is directed parallel to the surface.

Besides deposition efficiency, the nozzle geometry was found to affect the spray stream width. The average spray stream width measured with the Spraywatch 4S spray monitor at the spray distance was decreased from 15.6 mm to 12.8 mm when the convergent-divergent nozzle was used instead of the cylindrical nozzle.

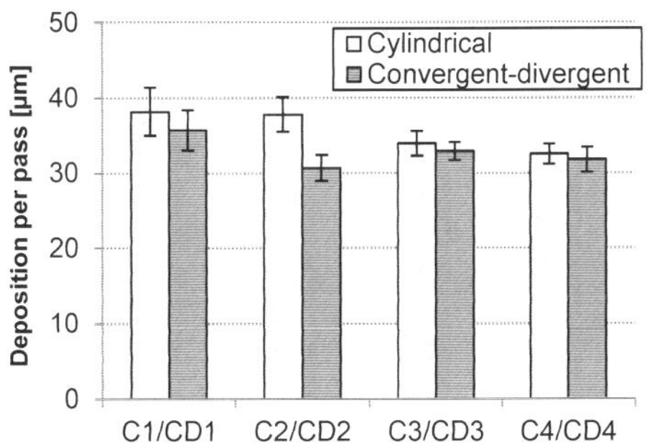

Fig. 5. Calculated deposition per pass for the two nozzles and four parameter sets.

3.5 Coating microstructure

The SEM study confirmed that the powder consisted of large carbide particles embedded in nickel-based metallic matrix. The XRD pattern of the powder (P) in **Fig. 6** confirms that the material consists of M_7C_3 carbides and metallic matrix alloy (Ni based). The patterns of the sprayed coatings (C1-C4: Cylindrical nozzle, CD1-CD4: Convergent-divergent nozzle) present the same structure as the feedstock powder. However, some changes can be detected when comparing the different parameter sets. The intensities of the nickel-based matrix peaks (at 52° and 60.2°) are decreasing with increasing particle temperature (C4→C1 and CD4→CD1). This is caused by the dissolution of the primary carbide particles during the spraying [12]. As a result, the carbon content of the matrix is locally increased creating carbon saturated areas that during rapid cooling could create amorphous or nanocrystalline structures. The dissolution of the primary carbides during spraying is demonstrated by the SEM (BSE) images of the C1 (**Fig. 7a**) and CD1 (**Fig. 7c**) coating microstructures. The metallic matrices of both coatings contain only few areas consisting of the brighter (original) metallic matrix (marked with white ovals). Instead, the structures are dominated by the different levels of grey indicating the dissolution of carbides.

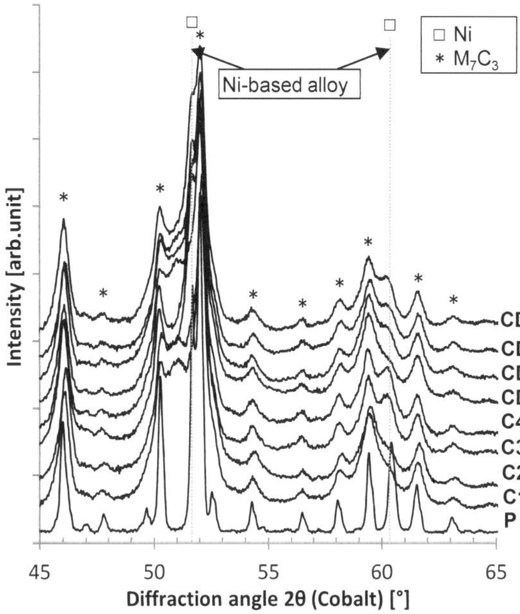

Fig. 6. The XRD patterns of the sprayed coatings and the powder (45-65°) measured with Co-tube. Dashed lines mark the peaks of the metallic Ni-based alloy.

The coatings C3 and C4 retain the most distinct metal matrix diffraction peaks of all the sprayed coatings. This can be observed also from the corresponding SEM images of the coating microstructures (**Fig. 7b** and **Fig. 7d**) that contain significantly larger amounts of the original metallic matrix. It appears that the largest fraction of the particles has not had enough time and heat to melt during the spray process and to cause carbide dissolution. These areas in the microstructures resemble the solid state coating build-up. Insufficient heating of the particles can hinder the particle deformation during the impact especially in the current case of hardmetal material with large primary carbide particles. This causes poorly bonded structures especially in locations where carbides have built-up in clusters, e.g. location pointed by an arrow in **Fig 7b**. Such weak areas provide low energy pathways for cracks to propagate and easy removal of coating material when exposed to wear. Also, corrosive electrolyte could easily penetrate through weakly bonded areas in the coating.

When comparing the microstructures of C4 to CD4 with similar particle temperature, it can be noticed that the higher measured particle velocity improves the bonding of the particles and decreases the amount of poorly bonded areas observed in coating C4.

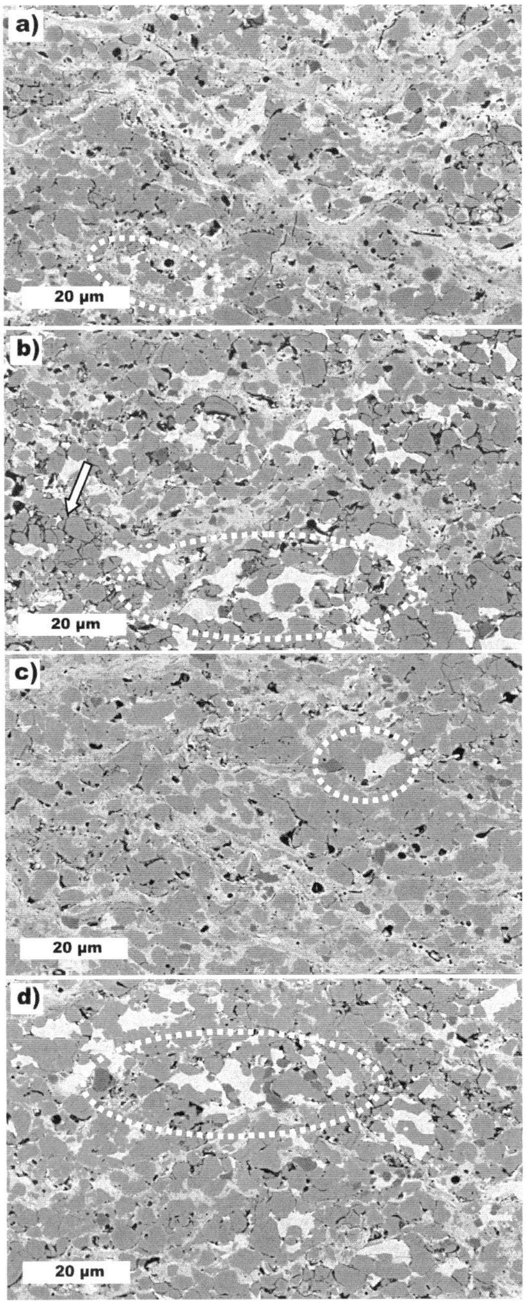

Fig. 7. SEM (BSE) micrographs of the as-sprayed coating microstructures: a) C1, b) C4, c) CD1 and d) CD4. The white ovals indicate metallic matrix without carbide dissolution and the white arrow poorly bonded carbide cluster.

3.6 Hardness and abrasive wear resistance

The measured Vickers microhardness values from the polished coating cross sections show little difference in **Fig. 8**. Only coatings C3 and C4 have clearly lower microhardnesses, while other coatings have an average hardness of 750-800 HV and show insignificant difference between the 0.3 kg and 1.0 kg loads. The low values of C3 and C4 are caused by the combination of low particle temperature and lowest measured particle velocity values.

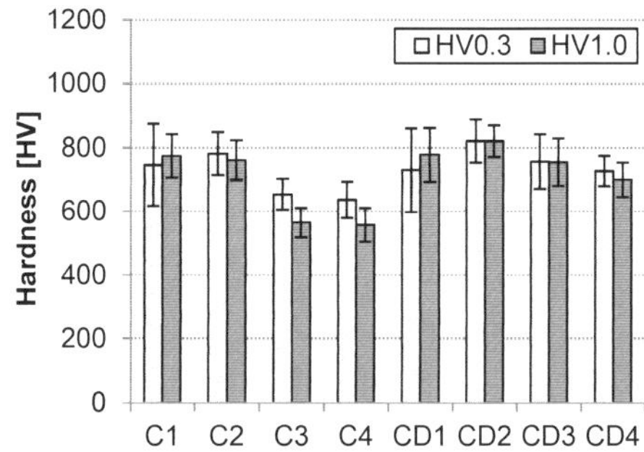

Fig. 8. Vickers hardness values of the coatings measured with 0.3 and 1.0 kg loads (C: Cylindrical, CD: Convergent-divergent).

The results from the abrasive wear tests are presented in **Fig. 9** as a function of the particle temperature. Lowest wear rates are produced with the convergent-divergent nozzle, while the coatings sprayed with the cylindrical nozzle are comparable. It appears that the particle temperature was effectively improving the wear resistance of the coatings, while high velocity had a positive effect. The velocity effect can be observed when comparing the coatings C3-C4 and CD3-CD4. At higher particle temperatures (105 psi *Fuel1*) the velocity has only marginal effect on the wear resistance. The results show that the C3 and C4 coatings with the lowest average microhardness values also experienced highest volume loss in the wear test. All coatings performed better than the 316L stainless steel reference sample with an average volume loss of 76.2 mm^3.

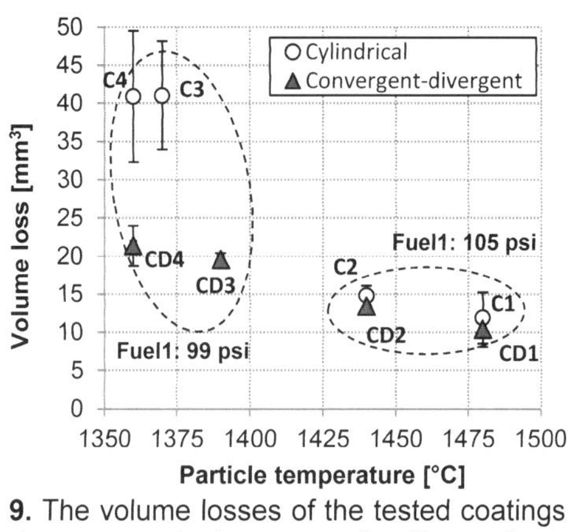

Fig. 9. The volume losses of the tested coatings after abrasive wear test as a function of particle temperature.

4 Conclusions

Commercial (Cr,Fe)-NiFeCrSi material was sprayed with a modern M3 HVAF spray gun. Cylindrical and convergent-divergent nozzle designs were used to spray coatings with four parameter sets. Following conclusions can be made from the results:

- The average particle velocity was increased approximately by 100 m/s when convergent-divergent nozzle was used instead of the conventional cylindrical nozzle with same process parameters. The nozzle geometry had the most significant effect on particle velocity.
- Comparable particle temperatures can be achieved with both nozzle shapes, while the fuel parameters have the most significant impact on the measured particle temperature.
- Highest deposition efficiencies were estimated for the cylindrical nozzle and high fuel flow parameters.
- Significant carbide dissolution was observed when high fuel flow parameters were used. This also resulted in good abrasive wear resistance.
- Volume loss of the 316L stainless steel after abrasive wear test was reduced from 76 mm^3 down to 10 mm^3 with the M_7C_3 based hardmetal coating.

5 Acknowledgements

The work has been done within FIMECC Ltd and its HYBRIDS program. We gratefully acknowledge the financial support from Tekes and the participating companies. The authors would also like to thank Mr. Mikko Kylmälahti of Tampere University of Technology for his help with the spray processing of the coatings.

6 Literature

[1] Wielage, B., A. Wank, H. Pokhmurska, T. Grund, C. Rupprecht, G. Reisel and E. Friesen: Development and trends in HVOF spraying technology. Surface & Coatings Technology 201 (2006), pp. 2032/37.

[2] Verstak A. and V. Baranovski: Activated combustion HVAF coatings for protection against wear and high temperature corrosion. Thermal Spray 2003: Proceedings of the International Thermal Spray Conference (2003), ASM International, pp. 535/41.

[3] Lyphout C., S. Björklund, M. Karlsson, M. Runte, G. Reisel, P. Boccaccio: Screening design of supersonic air fuel processing for hard metal coatings, Journal of Thermals Spray Technology 23 (2014), Issue 8, pp. 1323/32.

[4] Wang, Q., S. Zhang, Y. Cheng, J. Xiang, X. Zhao and G. Yang: Wear and corrosion performance of WC-10Co4Cr coatings deposited by different HVOF and HVAF spraying processes, Surface&Coatings Technology 218 (2013), pp. 127/36

[5] Bolelli, G., L.-M. Berger, T. Börner, H. Koivuluoto, L. Lusvarghi, C. Lyphout, N. Markocsan, V. Matikainen, P. Nylen, P. Sassatelli, R. Trache and P. Vuoristo: Tribology of HVOF- and HVAF-sprayed WC–10Co4Cr hardmetal coatings: A comparative assessment. Surface&Coatings Technology 265 (2015) pp. 125/44

[6] Matikainen, V., H. Koivuluoto and P. Vuoristo: Microstructural characteristics of different Cr_3C_2 coating compositions sprayed with HVOF and HVAF processes. Proceedings of SMT28 (2015), Valardocs, pp. 77/87.

[7] Hulka, I., V.-A. Şerban, K. Niemi, P. Vuoristo and J. Wolf: Wear Properties of CrC-37WC-18M Coatings Deposited by HVOF and HVAF Spraying Processes. Surface & Coatings Technology 210 (2012), pp. 15/20.

[8] Zimmermann, S., B. Gries and B. Brüning: New cermet powders for HVOF spraying with improved corrosion and oxidation resistance for offshore, mining and power generation applications. Thermal Spray Bulletin 2 (2011), pp. 94/100.

[9] Milanti, A., V. Matikainen, H. Koivuluoto, G. Bolelli, L. Lusvarghi and P. Vuoristo: Effect of spraying parameters on the microstructural and corrosion properties of HVAF-sprayed Fe-Cr-Ni-B-C coatings. Surface & Coatings Technology 277 (2015), pp. 81/90.

[10] Vuoristo, P., S. Ahmaniemi, S. Nuutinen, T. Mäntylä, E. Hämäläinen, N. Arola and J. Vattulainen: Optimisation and monitoring of spray parameters by a CCD camera based imaging thermal spray monitor. Thermal Spray 2001: New Surfaces for a New Millennium (2001), ASM International, pp. 727/36.

[11] Rusch, W.: Comparison of operating characteristics for gas and liquid fuel HVOF torches. Thermal Spray 2007: Global Coating Solutions (2007), ASM International, pp. 572/76.

[12] Zimmermann, S. and H. Kreye: Chromium carbide coatings produced with various HVOF spray systems. Thermal Spray: Practical Solutions for Engineering Problems (1996), pp. 147/52.

Fluid dynamics and modelling of plasma jet in PS-PVD process and Monte Carlo simulations of PS-PVD columns

Panpan Wang, Robert Mücke, Wenting He, Georg Mauer, Robert Vaßen,Jülich / D

Modelling of the supersonic compressible plasma flow has been developed to describe the thermodynamic and transport properties of the plasma spray physical vapor deposition process (PS-PVD) for typical processing parameters used for columns microstructure formation of thermal barrier coating (TBC). Commercial computational fluid dynamics software (ANSYS fluent 16.1) has been used for the simulations. The required properties of the plasma gas mixtures were obtained as a function of temperature and pressure from the thermodynamic calculations in chemical equilibrium (CEA program) with the effect of ionization. The carrier gas effects on the plasma fluid field was discussed. The boundary layer thickness was calculated near the sample wall. Two-dimensional Monte Carlo simulations have been conducted to provide insight on the evolution of PS-PVD columns. This model was implemented on a molecular scale that incorporates the effect of self-shadowing and vapor incidence angle.

1 Introduction

For the system sketched in **Fig.1.** the plasma stream leaves the torch in the form of a turbulent, compressible jet of hot ionized gas. The problem of practical importance is to calculate the velocity and temperature fields in the gas downstream of the torch for a given set of initial conditions at the exit. In general, the gas velocity and temperature at the exit from the torch are not known but they may be estimated from simple mass and energy conservation principles applied to the working gas of the torch since the torch power and the mass flow rate of gas are usually given. For simplicity it was assumed that the radial profiles of temperatures and velocity at the exit were 'top hat' distributions and that there were 55% energy losses from the torch.The velocity and temperature fields down streams of the torch exit were calculated from the appropriate equations of mass, momentum, and energy conservation which may be expressed in two-dimensional form by assuming that arc is symmetrical about the torch axis and that the flow is steady.Particular attention is directed toward complex flows where pressure gradients and streamline curvature effects are important. Boundary layer separation is the detachment of a boundary layer from the surface into a broader wake. Boundary layer separation occurs when the portion of the boundary layer closest to the wall or leading edge reverses in flow direction. The separation point is defined as the point between the forward and backward flow, where the shear stress is zero. The overall boundary layer initially thickens suddenly at the separation point and is then forced off the surface by the reversed flow at its bottom. The point where the dividing streamline attaches to the wall again is called the reattachment point. As the flow goes farther downstream it eventually achieves an equilibrium state and has no reverse flow.

Plasma spraying and electron beam physical vapor deposition are two processing methods used to deposit thermal barrier coatings. The former provides better thermal insulation but have lower mechanical durability due to internal pores [1-2]. These internal pores can affect the overall thermal conductivity in two ways. Firstly, pores that have a lower thermal conductivity than the matrix may lower the mean thermal conductivity and alter the heat conduction path. The mean diameter and orientations of pores also affect the thermal conductivity. Secondly, structural defects interrupt lattice continuity, so that phonons propagating around them are scattered, resulting in lower thermal conductivity in the region around the defects. Both effects are relevant in lowering the thermal conductivity.

In the present study, computer simulations have been performed in order to investigate the formation of the nanometer-sized pores and determine the physical deposition parameters influencing this phenomenon. The orientation of the pores and the total porosity are analyzed quantitatively. In our simulations, the formation of a nano-sized pore is a stochastic event originating from the random vertical and horizontal accumulation of the constituent particles. No assumptions are made about the formation of the macroscopic structure of the film other than it resulting from the cumulative deposition of individual particles, as macroscopic growth should occur as a natural consequence of the stochastic process on the microscopic level. The macroscopic structure was therefore also examined to confirm the utility of our simulation technique. In every case, however, columnar structures are observed only when the mobility of the deposited atoms is "limited". The self-shadowing effect will therefore increase as atomic mobility decreases and vapor incident angle increases.

2 Heat transfer in the fluid field and Monte Carlo Simulation

2.1 Geometry and initial coundary condtions

Because of the elliptic nature of the governing partial differential equations, boundary conditions are required along all domain boundaries for all dependent variables. The following boundary conditions are specified as shown in **Fig.1.** Environment pressure is constant, the value is 200 Pa.

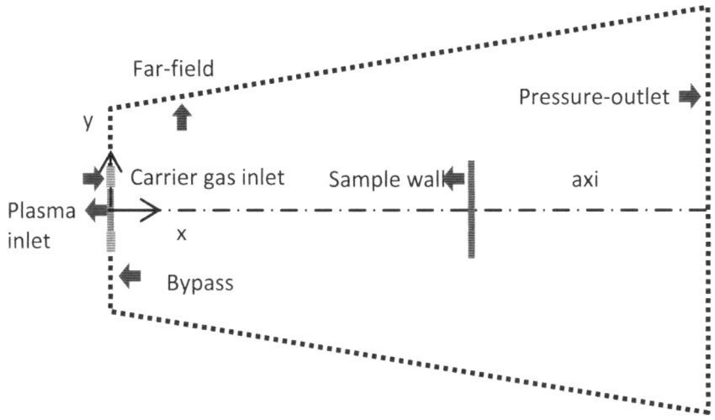

Fig.1. The schematic graph of the plasma flow field geometry and boundary conditions.

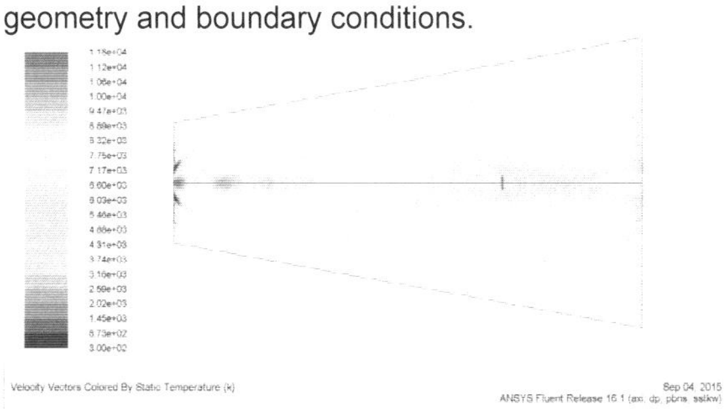

Fig.2. The contour of the plasma temperature flow field .

Table 1 Plasma spray parameters used in the modelling

Plasma gas	Ar 35 slpm He 60 slpm
Carrier gas	Ar 32 slpm
Net power	60 kW
Spray distance	1400mm

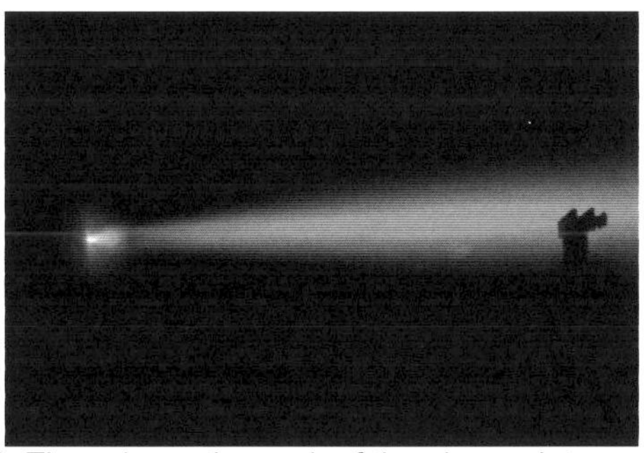

Fig.3. The schematic graph of the plasma jet.

Boundary I : At the nozzle exit (plasma inlet), Mass flow inlet, A elliptic nozzle exit pressure profile is assumed with the profile of nozzle exit

Boundary II : Far field, pressure outlet boundary conditions

Boundary III : Bypass, pressure outlet boundary conditions

Boundary IV : Axisymmetric axis, a zero flux of all quantities across a symmetry boundary. The normal gradients of all flow variables are therefore zero at the symmetry plane

Boundary V : Pressure outlet

Boundary VI : At the sample surface: no slip-wall

Boundary VII : Carrier gas inlet, when the carrier gas argon effect on plasma fluid is considered

2.2 2-D Heat Convection Steady-flow Equations[3]

Conservation of energy

For steady-state constant properties,

$$\underbrace{\frac{\partial(\rho C_P T u)}{\partial x}+\frac{\partial(\rho C_P T v)}{\partial y}}_{convection} = k\underbrace{\left(\frac{\partial^2 T}{\partial x^2}+\frac{\partial^2 T}{\partial y^2}\right)}_{heat\ diffusion} + \underbrace{\mu\emptyset}_{\substack{heat\ dissipation\\ source\ due\ to\ friction}}$$

$$\mu\emptyset = \mu\left\{\left(\frac{\partial u}{\partial y}+\frac{\partial v}{\partial x}\right)^2 + 2\left[\left(\frac{\partial u}{\partial x}\right)^2+\left(\frac{\partial v}{\partial y}\right)^2\right]\right\} \sim \mu\left(\frac{\partial u}{\partial y}\right)^2$$

The continuity equation (conservation of mass) for compressible boundary layers:

$$\frac{\partial(\rho u)}{\partial x}+\frac{\partial(\rho v)}{\partial y}=0$$

The momentum equations:

The x-direction momentum equation (radial direction)

$$u\frac{\partial u}{\partial x}+v\frac{\partial u}{\partial y}+\frac{1}{\rho}\frac{\partial p}{\partial x}=\gamma\frac{\partial^2 u}{\partial x^2}+\gamma\frac{\partial^2 u}{\partial y^2}$$

The y-direction momentum equation yields (axial direction)

$$u\frac{\partial v}{\partial x}+v\frac{\partial v}{\partial y}+\frac{1}{\rho}\frac{\partial p}{\partial y}=\gamma\frac{\partial^2 v}{\partial x^2}+\gamma\frac{\partial^2 v}{\partial y^2}$$

2.3 Fluid field discussions and results

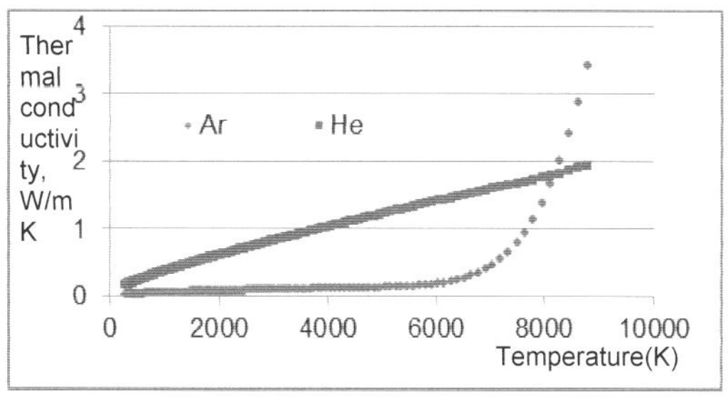

Fig.4. Thermal conductivity of argon and helium dependences on temperature at the constant pressure of 200 Pa.

As shown in **Fig.6,** Carrier gas a bit increase the temperature due to the thermal conductivity of helium is higher than argon when temperature lower than 8000K. Fourier's law, $\vec{q}=-k\nabla T,$ heat flow rate proportional to the negative temperature gradient and thermal conductivity.

Compared with **Fig.2.** and **Fig.3,** the plasma jet from modelling is comparable with the plasma jet from experiment.

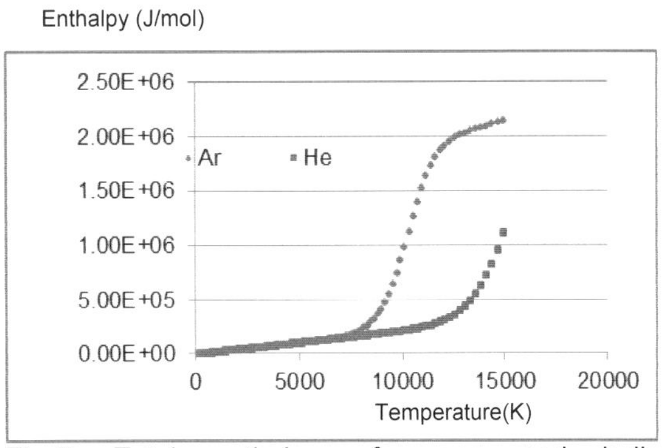

Fig.5. Total enthalpy of argon and helium dependences on temperature at the constant pressure of 200 Pa.

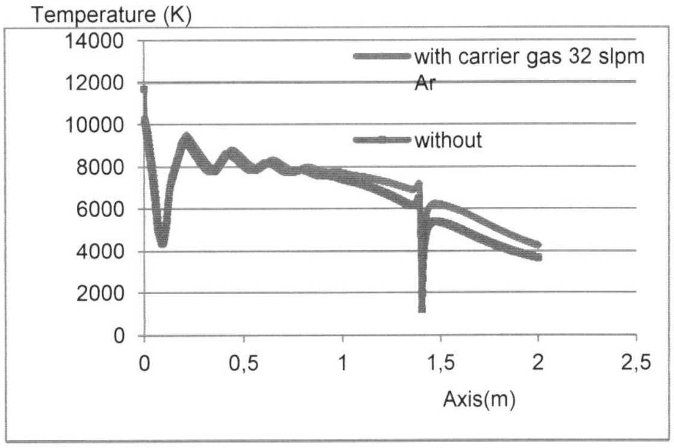

Fig.6. The effect of carrier gas on plasma flow temperature .

2.4 Flow near a solid boundary discussions and results

The Boundary-Layer Approximations, the outer region of the boundary layer [4] can be employed to simplify the boundary equations as follows:

1) Velocity in the x-direction(axial velocity) is greater than that in the y-direction(radial velocity), and velocity in the y-direction close to zero;

$$u \rightarrow \text{constant}\,; v \rightarrow 0$$

2) For the outer boundary condition , the pressure is constant;

$$P \rightarrow \text{constant As } y \rightarrow \propto;$$

3) Based on Bernoulli's theorem, $\frac{v^2}{2} + \frac{P}{\rho} = \text{constant}$, generally inner boundary layers, momentum equation is approximately stated as ;

$$\nabla p = \nabla^2 u$$

4) On the body surface is the no slip condition.

The thickness of the velocity boundary layer is normally defined as the distance from the solid body at which the viscous flow velocity is 99% of the freestream velocity. For The plasma spray process, due to the low pressure, the boundary layer thickness defined from the surface at which the pressure increase.

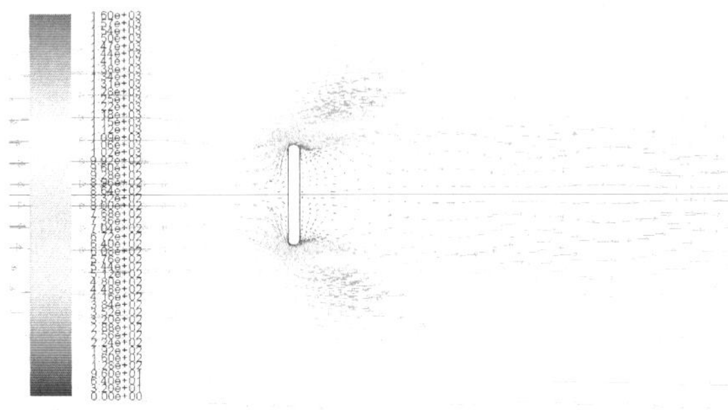

Fig.7. Vector of the velocity of the plasma flow field near the sample in two-dimensional domain.

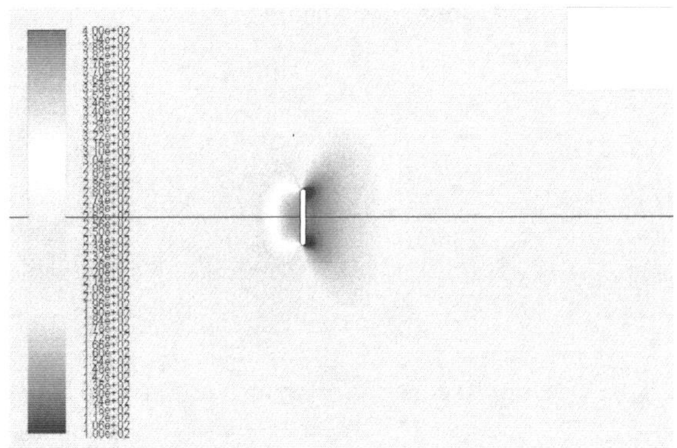

Fig.8. Contour of the plasma pressure near the sample in two-dimensional domain.

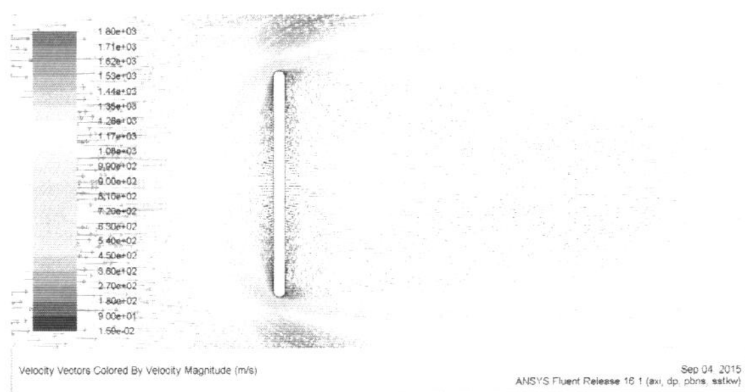

Fig.9. The velocity flow vectors colored by velocity distributes around the sample.

Pressure gradient is one of the factors that influence a flow immensely .The boundary layer thickness is nearly 15mm due to the pressure increase.

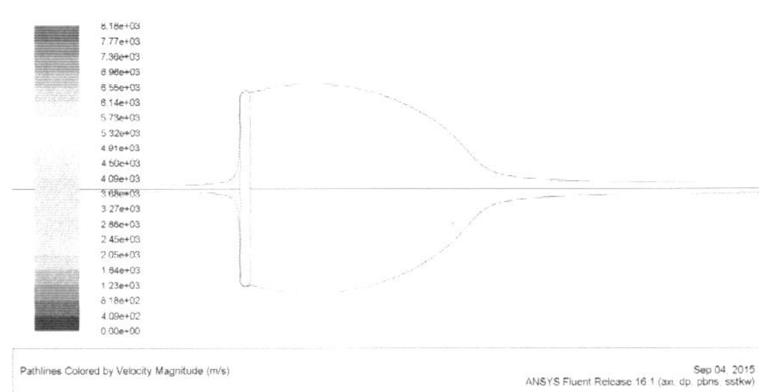

Fig.10. A single flow pathline colored by velocity magnitude distributions around the sample.

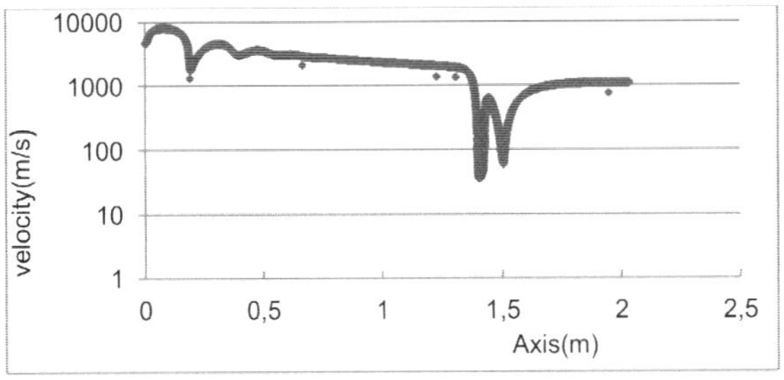

Fig.11. The velocity as a function of axis along the pathline shown in **Fig.10.**

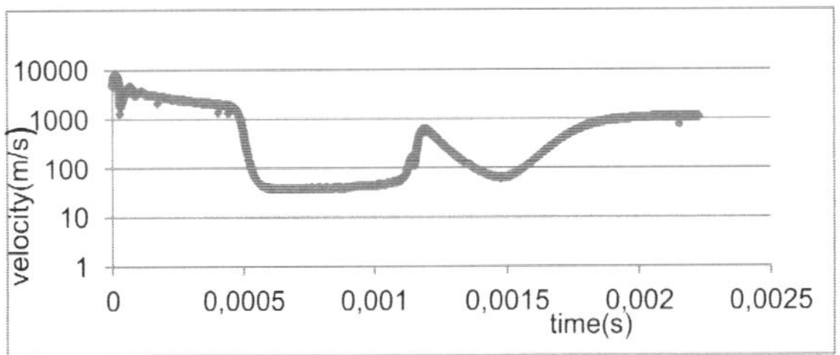

Fig.12. The flow velocity as a function of time along the pathline as shown in the **Fig.10.**

As shown in **Fig.9**, virtices are formed in the back of the sample. It would prove that there is a possibility to deposit the coating in the back of the sample, which is called non-line-of-sight (NLOS) effects. **Fig.10.** shows a single flow from the center of the nozzle outlet to go around the sample surface. **Fig.11.** and **Fig.12.** results are all according to **Fig.10.**

2.5 Monte Carlo simulation of physical vapor deposition

Simulation methodologies

Two-dimensional Monte Carlo simulations [5-6] have been conducted to provide insight on the evolution of columns. This model is implemented on a molecular scale that incorporates the effect of self-shadowing and vapor incidence angle. The results indicate that the pore configuration can be altered by changing the physical parameters of the deposition process. The vapor incidence angle α, is defined as the angle between the surface normal and the direction of vapor incidence. The mean orientation of the columns is such that they incline toward the angle of vapor incidence angle. If there are no obstacles in the trajectories of the particles, the deposited columns incline exactly in the direction from which the particles come. However, if the particles already deposited shadow the trajectories of the incoming vapor, the resulting deposition is not allowed.

Before the deposition begins, the atom and void occupy all substrate sites of the simulation cell (y = 0 if substrate is planar) one after another, whereas the rest of the lattice is empty. Each atom is dropped from a randomly chosen x position at a height y0, greatly exceeding that of the surface atoms. The atoms then follow linear trajectories (assuming no collision with gas atoms) in the direction of vapor incidence (varying between -90° and 90° according to a prescribed algorithm, as described below) until they touch the surface. Periodic boundary conditions are imposed in the x-direction while the depositing atoms are traveling.

For the square lattice three various algorithms to specify preferred direction were proposed. It allows the evaluated site to interact only with neighboring sites in horizontal and vertical directions. The neighbors in diagonal directions do not participate in the deposition.

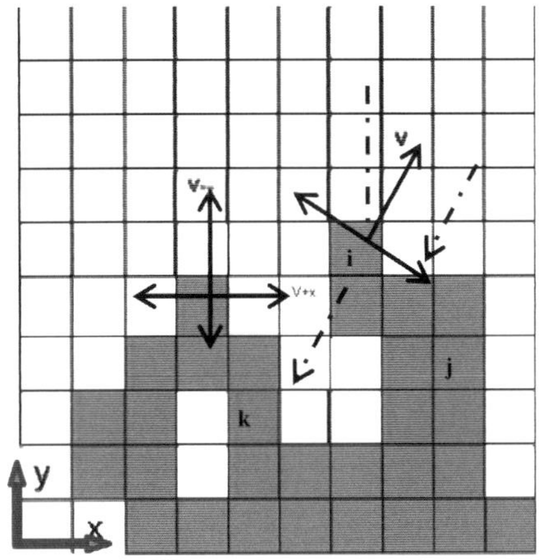

Fig.13. Schematics of the deposition process. Vertical and horizontal accumulation velocities, v_{+y}, v_{-y}, v_{+x}, and v_{-x}, are calculated form $v_{\|}$ and $v_{\perp}$. For deposition on particle j, accumulation in the +y and +x directions are allowed. However, for particle k, accumulation in the +x is allowed; +y direction is not allowed since depositied particle, i, blocks the incoming flux.

Assumptions

Two-dimensional model of physical vapor deposition coatings are based on the following approximations:

1) Surface diffusion was neglected;
2) particle coagulation and sintering (surface diffusion) were neglected;
3) vapor incident angle was considered;
4) the effect of self-shadowing on the columns growth was considered ;

Particles deposition process is simulated on a 500×500 square lattice with periodic boundary conditions. The probability of movement is dependent on the ratio of velocity. Monte Carlo algorithm was used to simulate the deposition of particles on a substrate in two dimensions:

1) Initializing: before the deposition begins, the atom and void occupy all substrate sites of the simulation cell (y = 0 if substrate is planar) one after another, whereas the rest of the lattice is empty;
2) The accumulation probabilities in vertical and horizontal direction(four directions);
3) Each square represents a single particle (one pixel in size)deposited from the vapor phase ;

4) The vapor incidence angle relative to the substrate normal ranges from -90 degree to 90 degree;
5) New molecules inherit all the properties of the parent particles;
6) Self-shadowing effect: the particles could not be deposited on sites where the incoming trajectory of the incident flux was blocked by already deposited particles;

Then the accumulation probabilities in the vertical and horizontal directions are given by, which is shown in **Fig.13.** above:

$$k = {}^{V_{\parallel}}/_{V_{\perp}} \quad (1)$$

$$P_{+x} = \frac{V_{+x}}{V_{//}} = \left(\sin(abs(\alpha)) + \frac{\cos(abs(\alpha))}{k}\right) * \cos(abs(\alpha)) * P_0 \quad (2)$$

$$P_{-x} = \frac{V_{-x}}{V_{//}} = \left(\frac{\cos^2(abs(\alpha))}{k}\right) * P_0 \quad (3)$$

$$P_{+y} = \frac{V_{+y}}{V_{//}} = \left(\cos(abs(\alpha)) + \frac{\sin(abs(\alpha))}{k}\right) * \cos(abs(\alpha)) * P_0 \quad (4)$$

$$P_{-y} = \frac{V_{-y}}{V_{//}} = \left(\frac{\sin(abs(\alpha)) * \cos(abs(\alpha))}{k}\right) * P_0 \quad (5)$$

P_0 Represents the saturation state. At any given time, one or more terms will be zero, depending on the number of exposed surfaces. The accumulation probabilities were set proportional to the velocities given in Eqs(1)-(5). One further limitation imposed on the crystallites' growth was that particles could not be deposited on sites where the incoming trajectory of the incident flux was blocked by already deposited within closed pores.

Oblique deposition

Inclination of the incident flux with respect to the substrate results in changes in the gross structure of the deposited film. The columnar crystallites are formed just as in the case of normal incidence. The surface of the film remains parallel to the substrate. However, the columnar growth direction does not correspond with the incident direction, which results in changes in the nano-pore formation mechanism. In this section, the effect of oblique incidence on the formation of intercolumnar and intracolumnar pores is discussed. **Fig.14-18** show columns structures formed using oblique incidence fluxes. Since the incident direction was inclined, the resulting columns are also inclined. It should be noted that the columns are not perfectly straight. Just above the substrate, the columns become gradually more inclined at a rate dependent on alpha. The columnar structure therefore becomes increasingly more noticeable with larger angles of vapor incidence.

At the same condition of the ratio of ${}^{V_{\parallel}}/_{V_{\perp}}$ and unsaturation ratio P_0, the larger incident angle, orientation of columns become more inclined. At the same condition of unsaturation ratio P_0 and incident angle, the larger values of the ratio ${}^{V_{\parallel}}/_{V_{\perp}}$, the width of columns is smaller.

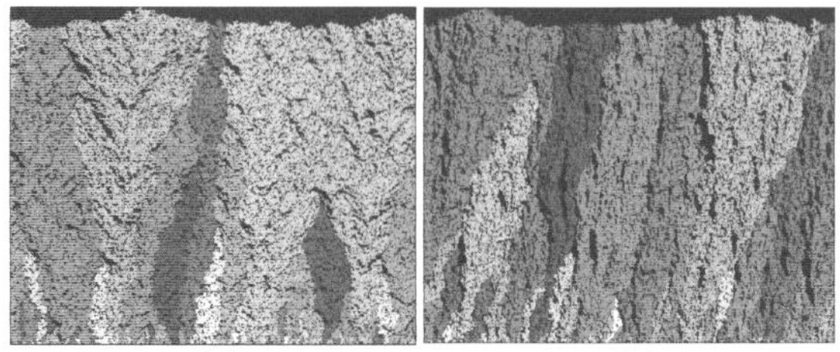
(a) (b)

Fig.14. Simulated structures deposited at oblique deposition angles ;(a)α=10, (b)α=30 , ${}^{V_{\parallel}}/_{V_{\perp}}$=1, P_0=0.01.

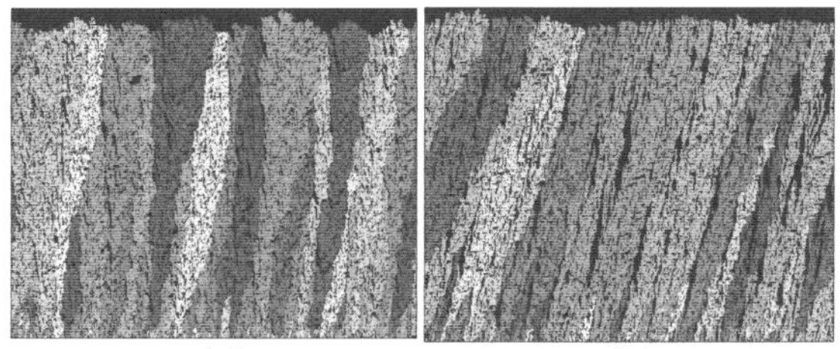
(a) (b)

Fig.15. Simulated structures deposited at oblique deposition angles ;(a)α=10, (b)α=30, ${}^{V_{\parallel}}/_{V_{\perp}}$=10, P_0=0.01.

When h is small, the width of the column is also small, indicating that horizontal growth was limited. At this stage, the difference in column heights is not large enough for the greater width of one column to block the growth of another. As h increases, the width of the column increases over a broad transition region until width of column is almost linear at high h.

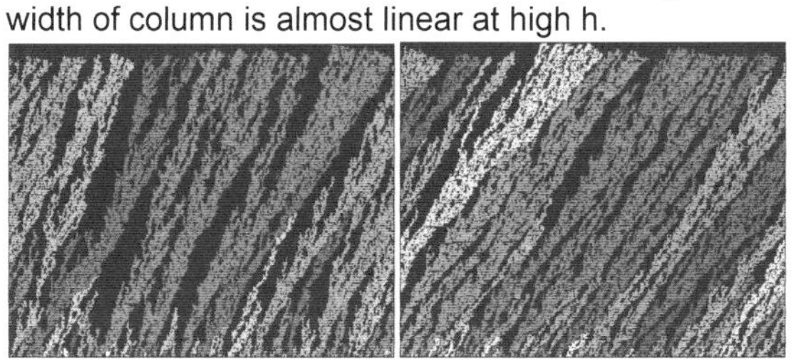
(a) (b)

Fig.16. Simulated structures deposited at oblique deposition angles ;(a) ${}^{V_{\parallel}}/_{V_{\perp}}$=1, (b) ${}^{V_{\parallel}}/_{V_{\perp}}$=10, α=60, P_0=0.01.

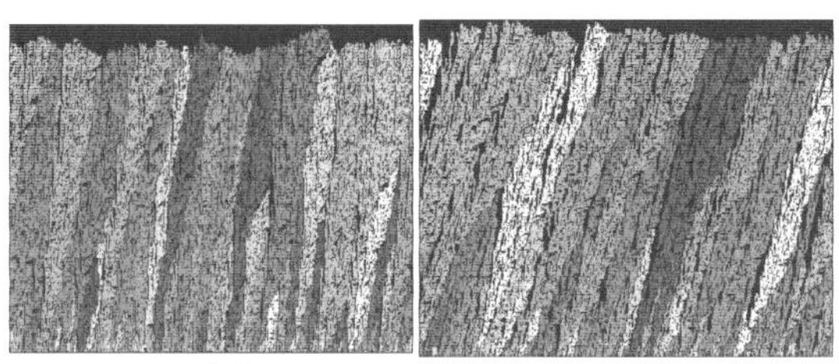
(a) (b)

Fig.17. Simulated structures deposited at oblique deposition angles ;(a)α=10, (b)α=30 ,${}^{V_{\parallel}}/_{V_{\perp}}$=100, P_0=0.01.

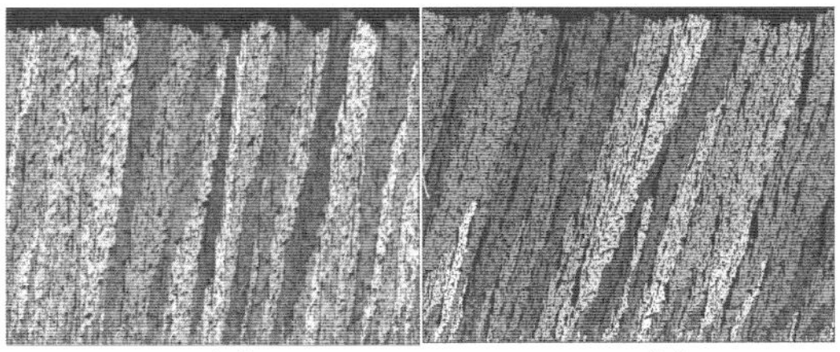

(a) (b)

Fig.18. Simulated structures deposited at oblique deposition angles; (a)α=10, (b)α=30, $V_{\parallel}/V_{\perp}$=1000, P_0=0.01.

Deposition normal to the substrate

In addition, it is expected that the deposition rate of the vapor particles affects the surface diffusion; absorbed atoms do not have time to diffuse if the deposition rate $V_{\parallel}$ is much faster than the diffusion speed $V_{\perp}$. In this study, surface diffusion was neglected. As a result, the nanometer-sized pores are well preserved. This assumption therefore represents a well-defined limit from which realistic phenomena can be easily extrapolated. Our simulations can be thought of as representing systems in which the deposition rate is vey high, such as is normally the case for PS-PVD.

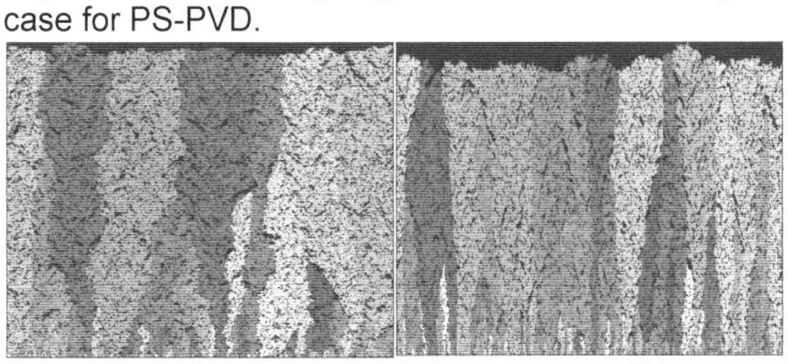

(a) (b)

Fig.19. Simulated structures deposited at normal incidence for (a) $V_{\parallel}/V_{\perp}$=1, (b) $V_{\parallel}/V_{\perp}$=10, P_0=0.01

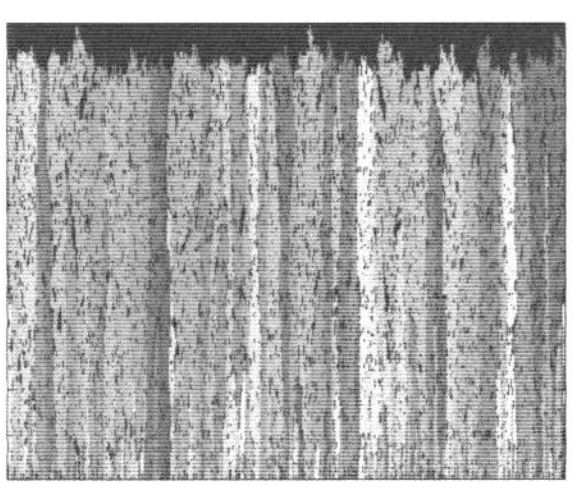

Fig.20. Simulated structures deposited at normal incidence for $V_{\parallel}/V_{\perp}$=100, P_0=0.01.

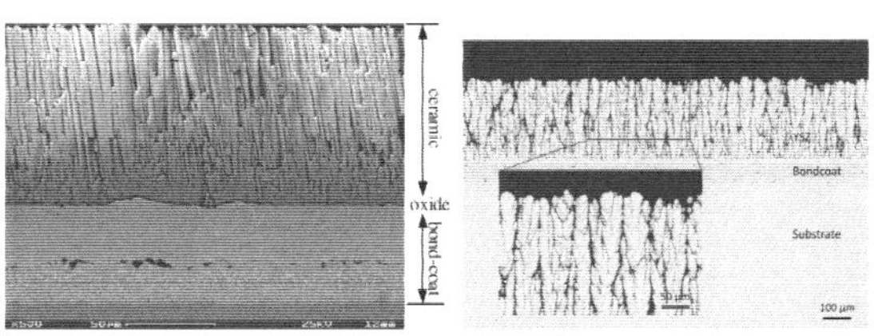

(a) (b)

Fig.21. SEM micrograph of EB-PVD system[7] and PS-PVD system[8].

The total area for 2D simulations of the intercolumnar pores did not change with the ratio $V_{\parallel}/V_{\perp}$, while the total area of intracolumnar pores decreased as the ratio increase.

The high degrees of mobility of sputtered atoms provide the mobility and promote the formation of smooth films. The large value of the surface diffusion coefficient relative to the bulk diffusion coefficient permits the formation of structures with sharp interfaces, because good films can form with plenty of surface mobility, yet with insufficient diffusion to allow mixing of layers. It was comparable with the results of PS-PVD system process, as shown in **Fig.21.**

3 Conlusions

The two dimensional plasma jet from modelling is comparable with the photograph of plasma jet. From modelling results, we can get the temperature and velocity plasma flow field. Near the sample wall, the boundary thickness was analyzed and calculated. It can also prove the possibility of non-line-of-sight.

For the Monte Carlo simulation, the columns growth were researched without considering the diffusion. the self-shadowing effect and vapor incidence angle were considered in the simulation.

Acknowledgment

I gratefully thank China Government Schlarship to sponsor me work at Juelich Forsungszentrum and Mr Metin Cakircali for the discussions of Fluent 16.1.

4 References

[1] Ö.Altun and Y.E.Böke.Effect of the microstructure of EB-PVD thermal barrier coatings on the thermal conductivity and the methods to reduce the thermal conductivity, archives of materials science and engineering, 40(2009), Issue 1, pp. 47-52.

[2] M.Caliez. Numerical simulation of EBPVD thermal barrier coatings spallation, Acta Materialia, 51(2003) Issue 4, pp. 1133/41.

[3] Je-Chin Han. Analytical Heat Transfer, pp. 45/53

[4] William S. Janna. Intoduction to Fluid mechanics, forth edition, pp. 630/31.

[5] M.Yoshiya. computer simulation of nano-pore formation in EB-PVD thermal barrier coatings, surface and coatings technology, 187(2004), pp. 399/407.

[6] S.Baek, V.Prabhu. Simulation model for an EB-PVD coating structure using level set method, Journal of Manufacturing Processes, 11(2009), Issue 1, pp. 1-7.

[7] M. Caliez, Numerical simulation of EBPVD thermal barrier coatings spallation Acta Materialia, 51(2003), Issue 4, pp. 1133/41.

[8] G. Mauer, Novel opportunities for thermal spray by PS-PVD, Surface and Coatings Technology, 268(2015), pp. 52/57.

Cold spray onto carbon fiber reinforced polymers for lightning strike protection

Hanqing Che, Stephen Yue, Phuong Vo, Montreal/CDN

Abstract: Carbon fibre reinforced polymers (CFRPs) are more and more used in a wide range of industries, but their low electrical conductivity has limited their application. Especially in the aerospace industry, the low conductivity of CFRP used in aircrafts would result in a lightning strike problem. To make the polymer composites electrically conductive, a conductive media must be either embedded into or coated onto the composites. During the past few years, metallization of CFRP has attracted increasing interest, and cold spray is one coating approach to achieve this. In this work, metallic powders were cold sprayed onto the CFRPs used in aircraft by using a low-pressure cold spray system. The coatings as well as the coating/substrate interfaces were characterized and the deposition mechanism onto the CFRP substrate was determined.

1 Introduction

Carbon fiber reinforced polymers are increasingly used in a wide range of industries due to their low density and high specific strength, but their low electrical conductivity has limited their application. To make the polymer composites electrically conductive, a conductive media must be either embedded into or coated onto the composites. For example, for the latest generations of aircrafts and helicopters which make extensive use of CFRPs, embedded metal meshes or ply-integrated interwoven wires are the usual materials of choice for lightning strike protection [1]. However, these embedded structures are not only difficult to manufacture, but also difficult to joint with other pieces, and to repair if any damage occurs. Alternatively, cold spray is one coating approach to achieve this. This process enables selective deposition only onto the places where high conductivity is required (e.g. lightning zone 1# in the aircraft). The cold sprayed coatings are also easy to repair if there is some damages, since this process has already been successfully applied as a mature repair method for damaged part [2].

As traditionally used to produce metallic coatings onto metallic substrates, cold spray directly onto polymer substrates has been rarely successful. It has been reported that erosion of the substrates is the key problem, especially for these reinforced with carbon/glass fibres [3-5]. Furthermore, for CFRPs with thermosetting matrix, which degrade at elevated temperature rather than softening, it is hardly possible to take advantages of local thermal softening of the substrates, thus even more difficult to obtain coatings [6]. Therefore, more investigations are indeed needed to achieve directly cold spraying onto CFRPs and understand the deposition mechanism.

In this work, copper and tin were chosen as the coating materials because of their excellent electrical and thermal conductivity, good mechanical properties, good cold sprayability and relatively low cost. Copper is more conductive, and has excellent cold sprayability; tin is less conductive as compared with copper, but was reported to be most successful when cold spraying onto polymeric substrates. Various different combinations of gas pressure and gas preheating temperature were used for the cold spray process with a low-pressure system. .

2 Experimental methodology

Commercial-purity copper and tin powders were chosen as the coating materials. The feedstock powders used in this work are listed in Table 1. The particle sizes of the feedstock powders were measured with a Horiba LA-920 Laser Scattering Particle Size Distribution Analyser. The scanning electron microscope images and cross-section optical micrographs of the feedstock powders are shown in Fig. 1. It can be seen from that the both powders are not perfectly spherical; tin contains a larger number of relatively small particles and has a smaller average particle size.

Table 1. Feedstock powders used in this work.

Powder	Morphology	Supplier	D_{avg}	Hardness
Cu	Relatively spherical	Plasma Giken	29 µm	55 HV
Sn	Relatively spherical	CenterLine, SST	17 µm	11 HV

The CFRPs used in this work were provided by Bombardier Aerospace (Montreal, Canada). For the cold spray campaigns, sheet sections of dimensions 7 x 7 cm were used as the substrates. Prior to the cold spray experiments, these sections were degreased with acetone, and no other surface preparation methods were adopted. Unlike the metal substrates, which are usually grit blasted before cold spray, the CFRP substrates would be eroded during grit blasting.

Cold spray experiments were carried out at the McGill-NRC cold spray facility at National Research Council Canada, Boucherville. A commercially available CenterLine SST system (Supersonic Spray Technologies, CenterLine Windsor Limited, Canada) was used to perform the cold spray. The CenterLine system enables the so-called "downstream injection" mode, during which particles are introduced into the main gas stream immediately after the throat, the narrowest orifice of the nozzle, so that the risk of throat clogging in the nozzle can be alleviated when using a gas temperature higher than the melting point of the metal. Nitrogen was used as the carrier gas, and other cold spray process parameters are listed in Table 2.

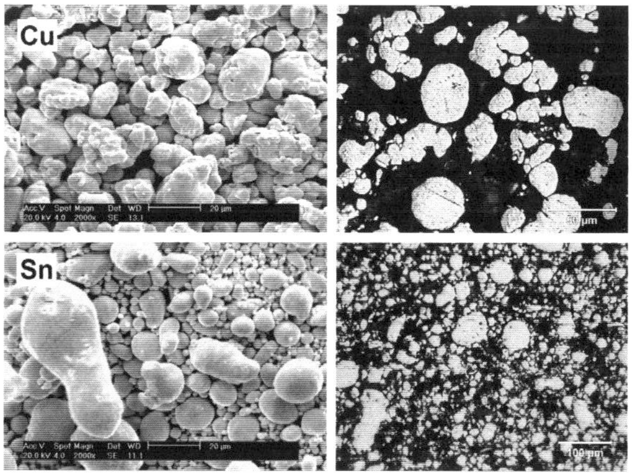

Fig.1. SEM images (left) and cross-sectional optical micrographs (right) of the feedstock powders.

Table 2. Process parameters for cold spray.

	Temp °C	Pressure MPa (psi)	Stand-off D mm	Gun travel speed mm/s	Feed rate g/min
Cu	425	0.34 to 1.38 (50 to 200)	18	25	~11
Sn	25 to 325	0.29 to 1.38 (42 to 200)	18	12.5 to 50	~13

3 Results

Figure 2 shows the deposition efficiency (DE) of spherical copper at various gas pressure conditions with a fixed gas temperature of 425°C. It can be seen that at all conditions performed, from 50 to 200 psi (0.34 to 1.38 MPa), the DE values are negative. This means that erosion of the substrate occurred and dominated despite the fact that the gas pressure was relatively low. It should also be noted that at 200 psi, the measured deposition efficiency was 7.24 per cent, but the actual erosion of the substrate was severe, more than 20 per cent of material, in thickness, was removed (the lightweight CFRP possesses a much smaller density, ~1.5g/cm^3, than metals, making the DE values look "small" in case the net weight change is caused by loss of CFRP). Due to the unsuccessful results of spraying copper, only tin was investigated thereafter in this work.

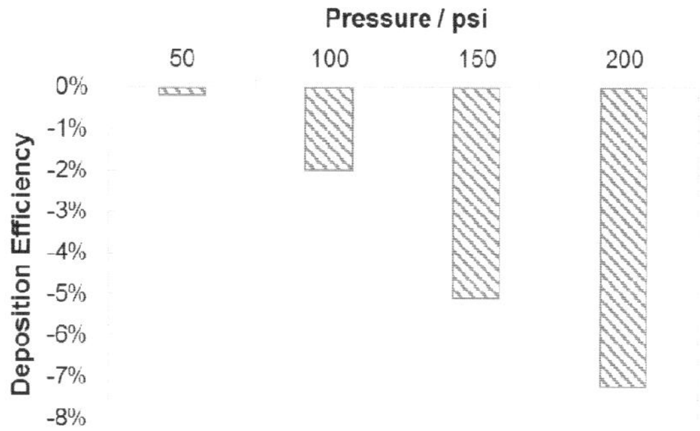

Fig. 2. Deposition efficiency of spherical copper at different gas pressure at 425°C.

The measured results of the DE of tin as function of gas preheating temperature and gas pressure are presented in Fig. 3. It can be seen that successful deposition of tin can be achieved with the low-pressure CenterLine system. When the gas temperature was fixed at 300°C, there was an abrupt increase in DE leading to a peak in deposition efficiency at a gas pressure of 60 psi (0.41 MPa), as shown in Fig. 3a. The deposition efficiency decreased to near zero when the gas pressure was increased to 150 psi (1.03 MPa), which is still low for conventional cold spray. On the other hand, when the gas pressure was fixed at 60 psi, there was a dramatic increase in DE when the gas temperature was increased from 280°C to 300°C, as shown in Fig. 3b. This abrupt change is possibly the indication of particle melting, since the gas preheating temperature, 300°C, is higher than the melting point of tin, 232°C. Whereas at relatively low temperature, DE was close to zero and even negative at 200°C due to slight erosion. The trend suggests that with further increase in gas preheating temperature, DE would continue to increase, so a trial spray at 325°C was performed, but the nozzle clogged despite the downstream injection mode was used.

The top morphology and cross-sectional microstructure of the tin coatings deposited under different gas pressures at 300°C are shown in Fig. 4 and Fig. 5, respectively. It can be seen clearly that good deposition and continuous coatings were achieved with tin when using the low-pressure system. At 42 psi (0.29 MPa, DE = 1%), the deposition of tin can be observed, but some uncoated area can also be found, indicating the deposition is non-continuous. At 60 psi (0.41 MPa), where the maximum DE (13%) was achieved, continuous coating was obtained and the top surface of the coating was relatively smooth. With increasing gas pressure, at 80 psi (0.55 MPa, DE = 7%), the coating was still continuous but some elliptical clusters emerge. These elliptical clusters have their longer axes in the vertical direction (aligned with gravity) and perpendicular to the gun travel direction. The reason of the formation of these clusters is unclear, but may be a combined effect of particle melting, particle bouncing and gravity. When the gas pressure was increased to 150 psi (1.03 MPa, DE = 0.6%), more and larger elliptical clusters and some grooves can be found. These grooves are essentially the carbon fibres coated with a thin layer of tin coating, which means erosion of the substrates nevertheless occurred despite the low-pressure.

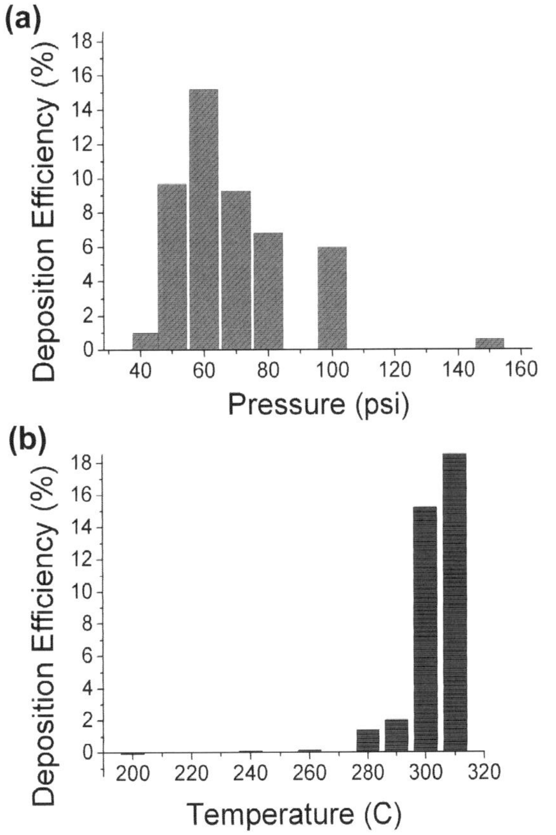

Fig. 3. Deposition efficiency of tin as a function of gas preheating temperature when sprayed at 300°C (a) and as a function of gas pressure when sprayed at 60 psi (b).

The cross-sectional micrographs in Fig. 5 are in agreement with the SEM top images. At 42 psi, although the deposition was non-continuous, the top surface of the CFRP remained smooth, without obvious signs of erosion. At 60 psi and 80 psi, the coatings were continuous, and the coating at 60 psi is approximately twice as thick as that at 80 psi, which agrees with the DE results. A close examination of the coating/substrate interfaces indicates that increasing gas pressure disturbed the smoothness of the interface, leading to waviness and irregularity at the interface. This might help to enhance mechanical interlock at the coating/substrate interface at low and medium gas pressure (e.g. 42 to 80 psi). However, further increase in gas pressure also led to erosion at the interface. At 150 psi, the deposition became non-continuous again, and signs of substrate erosion can be observed clearly: the surface epoxy was removed and the carbon fibres were exposed. The substrate erosion (removal) along with the tin deposition together resulted in a near-zero DE at this condition. It can be seen that in case that deposition and erosion occur simultaneously, the nominal DE is underestimated.

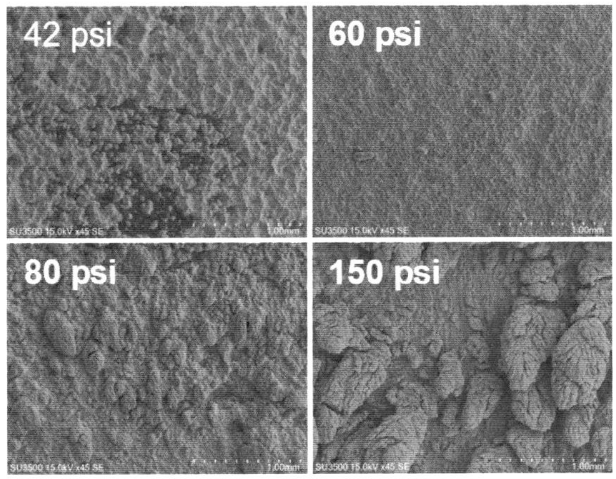

Fig. 4. SEM top images of the cold-sprayed tin coatings under different gas pressure at 300°C.

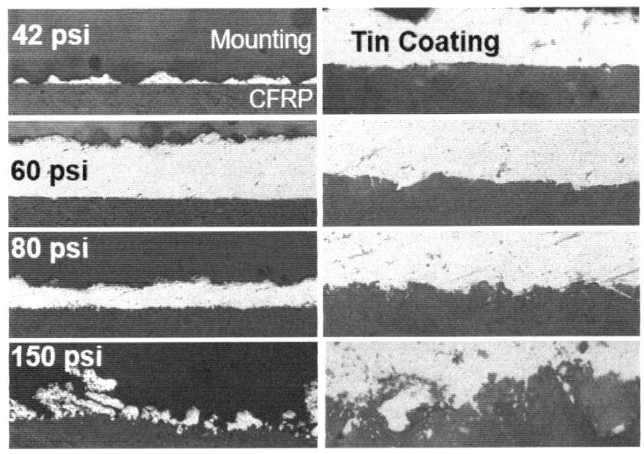

Fig. 5. Optical micrographs showing the cross-sections of cold-sprayed tin coatings under different gas pressure at 300°C at low magnification (left) and high magnification (right).

Figure. 6 presents the top surfaces of the samples after cold spray of tin at 60 psi at four different gas temperatures, 200°C, 240°C, 290°C and 310°C.

Figure 6a shows the sample after spray of tin at 200°C, at which the gas temperature was below the melting point of tin and DE was slightly negative (–0.1%). It can be seen clearly that no coating was deposited. Fragmentation of the surface epoxy, an early sign of erosion, was also observed, but no significant removal of the epoxy occurred due to insufficient kinetic energy. A trace amount of tin can be seen within the cracks in the surface epoxy. When the gas temperature was increased to 240°C, just above the melting point of tin (232°C), a number of small tin clusters were deposited on the CFRP, and slight surface erosion can be observed from the cross-section, as shown in Fig. 6b (DE = 0.1%). Again, it indicates that deposition of particles and erosion of the substrates can occur simultaneously, which cannot be differentiated from DE values, so this may lead to an underestimated DE value. However, such underestimation is usually negligible when the erosion is slight considering the relatively low density of the CFRP (~1.5 g/cm^3). At gas temperature of 290°C, some small areas of the CFRP were coated with a thin layer of tin, but the deposition was still non-continuous in long range (Fig. 6d, DE = 2%). At 310°C, continuous tin coating was obtained, as shown in Fig. 6d (DE = 20%).

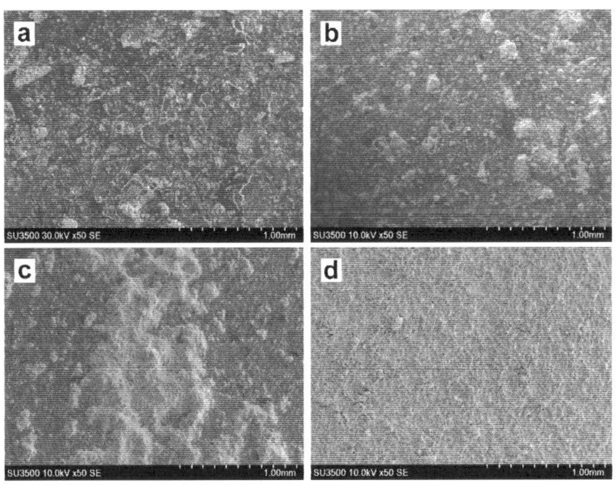

Fig. 6. SEM top images of the cold-sprayed tin coatings under different gas pressure at 60°C at (a) 200°C, (b) 240°C, (c) 290°C and (d) 310°C.

In particular, Fig. 7 presents a particle found at the top surface of the coating sprayed at 290°C. It can be seen the rim of this particle experienced noticeable change, as compared with the smooth surface of the starting powder. This may result from the melting of tin, considering the gas temperature is well above the melting point of tin. However, since the change was only found in the rim/outer surface of a particle while the particle as a whole basically remain spherical, it is reasonable to believe that only the outer surface of the tin particle melted while the core remained solid.

Fig. 7. SEM image showing a particle at the top surface of coating sprayed at 290°C.

4 Discussion

In this work, the unsuccessful trials with copper can be attributed to the failure in the development of the first layer. The copper particles cannot achieve mechanical interlocking with the CFRP substrates, but cause brittle fragmentation and material removal in the substrate surface. Even when spraying the relatively soft tin at 200°C, there were still difficulties in developing the first layer. When the gas temperature was increased to above 300°C, successful coatings were finally obtained by taking advantages of partial melting of tin particles. A "crack filling" mechanism is proposed to explain the success, as shown in Fig. 8. With this mechanism, when the tin particles enter the hot gas stream with a higher temperature than their melting point, the rims of these particles begin to melt. Due to the short residence time in the nozzle, the particles cannot wholly melt before they exit the nozzle, so the cores of each particles remain solid. When these partially molten particles bombard a brittle CFRP substrate, the impact generates a number of micro-cracks in the surface epoxy. Then the molten part of the tin particles get squeezed into these cracks and solidify, not only filling these cracks, preventing further erosion, but also achieving mechanical anchorage with the substrate and forming the first layer. Then the upcoming particles impact with this first layer and build up a coating, since the partially molten tin particles travelling at relatively low velocity cannot destroy the first layer whereas the velocity is still higher than the critical velocity for tin.

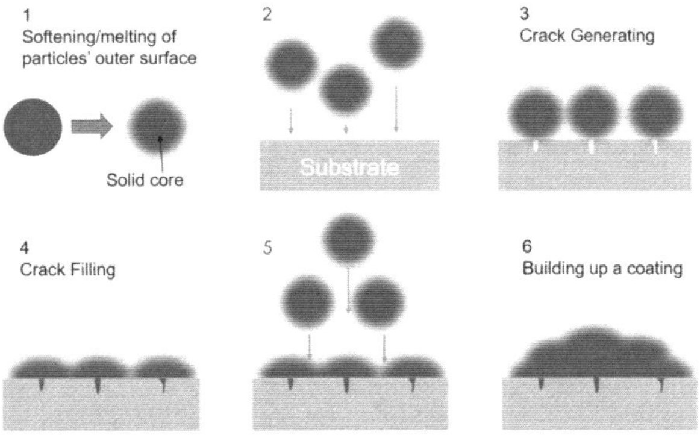

Fig. 8. Schematic of the crack filling mechanism.

5 Summary and conclusions

Continuous tin coatings were successfully obtained with the low-pressure cold spray system, whereas low-pressure cold spray of copper at various conditions still resulted in the erosion problem. Analysis of deposition efficiency indicates that medium gas pressure (60 psi) and high gas temperature (300°C and higher) are required for continuous deposition of tin onto the CFRP substrates. Examination of the coating morphology and microstructure reveals that tin particles partially melted when sprayed at gas temperature of 300°C and higher, and successful coatings were achieved by taking advantage of this partial melting. Accordingly, a "crack filling" mechanism is proposed to explain the deposition.

Acknowledgement

The authors wish to acknowledge the financial support of the Consortium for Research and Innovation in Aerospace in Quebec and the Natural Sciences and Engineering Research Council of Canada. The industrial partners, Bombardier Aerospace, Bell Helicopter, 3M Canada, and the collaborating universities, École Polytechnique de Montréal and Université du Québec à Montréal are gratefully acknowledged. National Research Council Canada, Boucherville is acknowledged for its contribution to the cold spray experiments.

References

[1] Black, S. Lightning strike protection strategies for composite aircraft. High-Performance Composites (2013), May.

[2] Irissou, E. Legoux, J.G. Ryabinin, A.N. Jodoin, B. Moreau, C. Review on cold spray process and technology: Part I-Intellectual property. Journal of Thermal Spray Technology. 17 (2008), pp. 495/516.

[3] Zhang, D. Shipway, P.H., McCartney, D.G. Cold gas dynamic spraying of aluminum: The role of substrate characteristics in deposit formation. Journal of Thermal Spray Technology 14 (2005), pp. 109/16.

[4] Lupoi, R. and O'Neill, W. Deposition of metallic coatings on polymer surfaces using cold spray. Surface and Coatings Technology 205 (2010), pp. 2167/73.

[5] Affi, J., Okazaki, H., Yamada, M., Fukumoto, M. Fabrication of aluminum coating onto CFRP substrate by cold spray. Materials Transactions 9 (2011), pp. 1759/63.

[6] Ganesan, A. YMFM. Cold spray coating deposition mechanism on the thermoplastic and thermosetting polymer substrates. Journal of Thermal Spray Technology (22) 2013, pp. 1275/82.

Application of cold spraying for contact reactive brazing of Mg alloy to steel

Hong Li [1], Hai-jie Nie [1], Wei-min Long [2], Su-juan Zhong [2], Zhuo Chen [1],
1 College of Materials Science and Engineering, Beijing University of Technology,Beijing /CN
2 State Key Laboratory of New Brazing Material and Technology, Zhengzhou Research Institute of Mechanical Engineering, Zhengzhou/CN

In the present study, application of cold gas spraying technology (CGS) for deposition of braze filler coatings is investigated. Pure copper coating were deposited on magnesium alloy AZ31B substrate by CGS. Then coated magnesium samples were joined to AlSi 1008 steel in vacuum furnace by contact reaction brazing (CRB) process. The microstructure of coated and brazed samples was investigated. The experimental results proved the feasibility of using cold spraying coatings as filler interlayer for precision brazing. Cold spraying performed well due to the low thermal impact on the copper powder and the magnesium substrates with low oxygen content. The bonding temperature influenced the dissolution of Cu coatings, as well as the eutectic reaction between Cu coating and magnesium substrate. The initial thickness of Cu coatings was 50μm while the brazed seam has a thickness of 300μm after eutectic reaction. Within a certain range, the shear strength of brazed joints showed a trend of increasing firstly and then decreasing as the heating temperature increased. Maximum shear strength 37MPa was gained at brazing temperature of 530°C and holding time of 60 min for brazed joint. Intermetallic phases and interfacial defects such as microcracks, unmelted coatings, overheated compounds were responsible for the low strength of the joints.

1 Introduction

The coatings produced by cold gas spraying (CGS) process can be widely applied not only as protective coatings but also functional coatings. It provides a promising filler coatings for welding applications such as fluxless brazing or TLP diffusion bonding processes. Compared with conventional thermal spraying processes under atmospheric conditions, it can produce compact coatings with higher density and lower oxygen content due to very low thermal load of spray particles during CGS process, resulting in high-quality brazed joints.

Wielage B. et al. [1] investigated the application of AlSi12, Zn-Al and Cu-Ni CGS coatings in ultrasonic brazing of Al, Mg and Ti light weight alloys. The joints showed comparable morphology like those produced with conventionally added fillers, i.e. wire, roll plating and foil. Thus it is demonstrated that CGS filler coatings are appropriate to be used in manufacturing heat exchanger segments. Bobzin K. et al. [2] reported the utilization of AlSi12 and AlSi10Cu4 CGS coatings in induction brazing of Mg-containing aluminium alloy 6063 and 5754. The resuls showed that the deposited coatings leads to a good metallurgical bond between the brazing alloy and the substrate. Anyhow, some oxide sacles prevented complete fusion of coatings during brazing, which responsible for the relatively low shear strength.

Although Magnesium alloys receiving great interest for automotive, electronics, transportation, aerospace industry because of its good mechanical and electrical properties and lightness, joining of magnesium alloys to other metals (such as Fe, Al...) is very difficult. The maximum solid solubility of Fe in Mg is 0.00041 at% according to the Mg-Fe binary alloy phase diagram[3]. Up to now, research on induction brazing of Mg in protective atmosphere using fluxes and fluxless brazing of Mg with ultrasonic-assisted have been conducted. However, most of the fluxes and filler metal have high melting temperature and low wettability for brazing of Mg to steel[4,5]. Tillmann W. et al. [6] prepared a 10μm-thick copper coatings both on Magnesium alloy AZ31B and AlSi1008 steel substrates by Arc-PVD. Diffusion brazing was performed and the joints achieved a maximum lap shear strength of 26MPa. Whereas compared with PVD coating in vacuum, CGS exhibits higher-efficiency, which is flexible and suitable for developing larger-sized components [7].

Contact reactive brazing (CRB) is an effective method for joining dissimilar materials which can form eutectic liquid phase[8]. It has no demand of brazing flux and can decrease heating temperature. The filler metals completely fills the joint and defects are avoided, which meet the requirement for precision brazing. Since CGS coatings of filler metals can retain almost their original compositions, it can serve as filler coatings for the formation of a eutectic reaction between substrates.

In the present work, we investigated the possibility of CRB brazing Mg alloy to steel in vacuum without any fluxes. Mg alloy were deposited with pure copper coatings as filler interlayer in advance. The influences of holding time on the dissolution of Cu coatings and the interface reaction were studied. In addition, the optimum parameters for CRB brazing were indentified for achieving maximum shear strength of brazed joints.

2 Materials and Experimental

Pure copper powder (average particle diameter≤25μm) were deposited onto specimen with 30mm×10mm×3mm dimensions for commercial magnesium alloy AZ31B sheet plate. The bond surfaces of AZ31B was prepared on SiC paper to 1000 grit and cleaned by alcohol. CGS system KM-CDS3.0 manufactured by Inovati was employed for deposition. Spraying parameters are optimized in order to increase deposition rate, efficency and coating density, see **Tab. 1**. Coatings with a thickness of roughly 50µm were sprayed.

Table 1. CGS process parameters.

He pressure KPa	He temperature °C	Gun trav. velocity $mm \cdot s^{-1}$	Powder feed rate $g \cdot min^{-1}$	Spray distance mm
620	15	160	5	12

For the brazing experiment, low carbon steel AlSi 1008 with 30mm×10mm×3mm dimensions were selected as substrate material. The AlSi1008 steel was grinded with SiC paper up to 600 grit and subsequently cleaned by an ultrasonic bath with alcohol. Coated AZ31B samples were assembled with prepared AlSi1008 steel in a fixture under a uniaxial pressure of 0.0375MPa, see **Fig.1**. CRB process was carried out in VDW30 vacuum furnace at pressures below 5×10^{-3}Pa. According to the Mg-Cu binary phase diagrams[9], a first eutectic reaction exists between magnesium and copper at a lower temperature of 485°C. The samples were heated to 485°C, 500°C, 515°C, 530°C and 545°C respectively and then held for 60min. Both CGS coatings and brazed joints were examined by means of stereo microscope, scanning electron microscope (SEM) and quantitative energy dispersive X-ray (EDS) analysis. Furthermore, the brazed seam was investigated by D8 ADVANCE X-ray diffractometer (XRD) in order to identify the interfacial phases accurately. The lap shear tensile strength of brazed samples was determined using an MTS-810 system at a speed of $0.5mm \cdot s^{-1}$. The hardness measurements of brazed joint were were also carried out using micro-Vickers hardness tester (HXD-100TM/LCD) under a load of 100N.

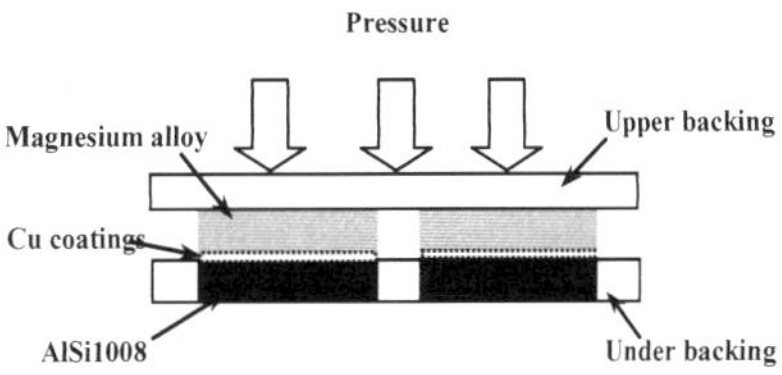

Fig. 1. Fixture for contact reaction brazing.

3 Results and Discussion

The cross-section of magnesium alloy AZ31B after CGS with copper coating is shown in Fig.2.

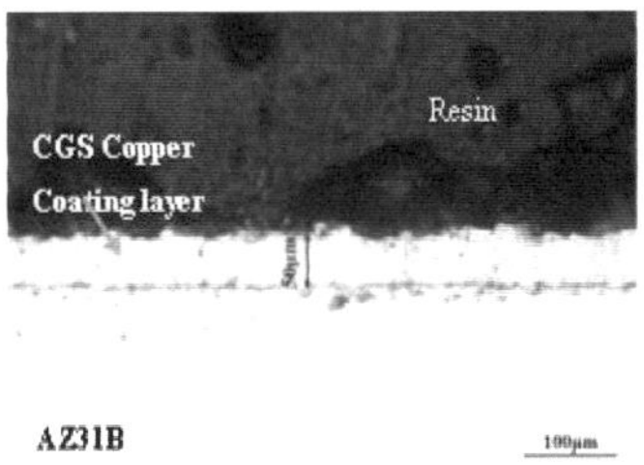

Fig. 2. Cross-section of interface of Cu coating/AZ31B.

The average thickness of copper coating deposited is about 50µm. While no oxide film were formed on AZ31B, which might prevent bonding. It can be attributed to the low-temperature process and less oxidation of CGS process. The coating formation is strongly controlled by the powders and the spray parameters.

Reliable brazing of deposited Mg substrate and AlSi 1008 steel was achieved by CRB technology in the present study. Fig.3 illustrates the influence of the bonding temperature on the shear strength of the AZ31B/Cu/1008 steel brazed joint at a dwell time of 60 min and a uniaxial pressure of 0.0375MPa. With the increase of heating temperature, the shear strength of brazed joints showes a trend of increasing firstly and then decreasing. When the bonding temperature is 485°C, the joint strength was found as only 13MPa. However, a further increase in the heating temperature to 530°C caused a significant increase in the bonding shear strength. Hence, the maximum strength of CRB joint under optimal conditions was 37MPa. At 545°C, the joint strength decreases dramatically while the strength values has a relatively large standard deviation.

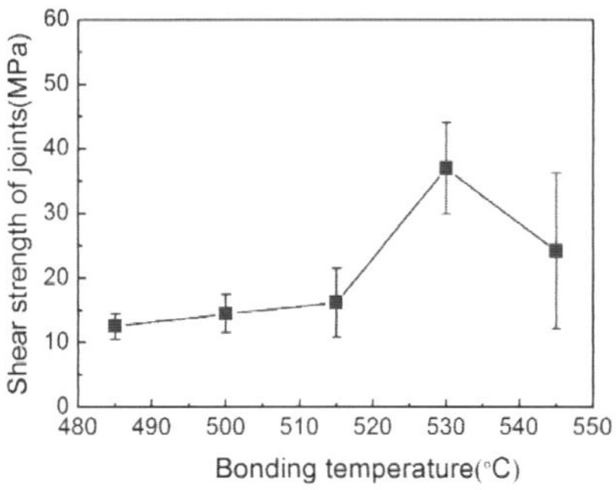

Fig. 3. Effect of bonding temperature on shear strength of AZ31B/Cu coating/1008 steel brazed joints.

Fig.4.(a)-(d) illustrates the microstructural evolution under different bonding temperature.

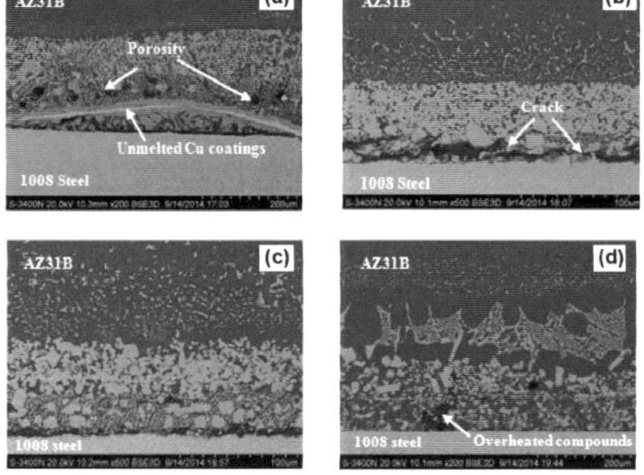

Fig. 4. Microstructures of AZ31B/Cu coating/1008 steel joints at different bonding temperature a)485°C, b)500°C, c)515°C and d)545°C.

In Fig.4(a), when the sample was heated up to 485°C (eutectic point) and held for 60 min, an unmelted, band-shaped copper coatings near interface areas can be obviously observed due to incomplete reaction between Mg and Cu under low temperature. In addition, a large amount of pores formed between the center zone of brazed seam and unmelted copper coatings. The interface of brazed joint was mainly composed of whitle α-Mg solid solution, Mg_2Cu eutectic phase and the ternary Mg-Cu-Al intermetallic phase, as well as unmelted copper coatings according to the EDS analysis. In Fig.4(b), when the temperature increases to 500°C, the joint was mainly consists of Mg-Cu-Al phase while no Mg_2Cu phase was found. Microcracks are found in the interface area between the brazed seam and steel substrate, which was considered to be caused by internal stress due to the excessive accumulation of Mg-Cu-Al phase. The propagation of cracks results in the unbonded area in steel side. These microcracks are expected to adversely affect the strength of the joint. In Fig.4(c), when the temperature increases to 515°C, Mg-Cu-Al phase scattered throughout the brazed seam with increasing grain size simultaneously. Indeed, a network of Mg_2Cu phase formed at the interfacial zone near steel side, as well as a few Mg-Cu-Al phase was embedded in Mg_2Cu phase matrix. It results in the increase of shear strength of brazed joint. Nevertheless, SEM images gave no indication of a reaction zone at the brazed seam/steel interface, thus the joint was actually mechanical bonded. In addition, when the brazing temperature increases to 545°C, it can be seen from Fig.4(d) that the microstructure of the brazed joint consists of a network of Mg_2Cu at the AZ31B/brazed seam interface, while the brazed seam/steel interface was composed mainly by α-Mg solid solution and Mg-Cu-Al phase, while no Mg_2Cu was found particularly. Simultaneously, due to the higher temperature, overheating of Mg compounds has occurred at the interfacial zone near steel side. Therefore, it is assumed that tha lack of Mg_2Cu phase and overheating phenomenon contribute to the deterioration of joint shear strength.

In order to further investigate the interfacial microstructure and reaction phases formed during brazing, the AZ31B/Cu coating/1008 steel joint brazed at 530°C for 60 min was observed carefully by SEM using backscattered electron mode. As shown in Fig.5, the joint can be classified into four characteristic zones (I, II, III and IV) according to their microstructural morphology and chemical composition.

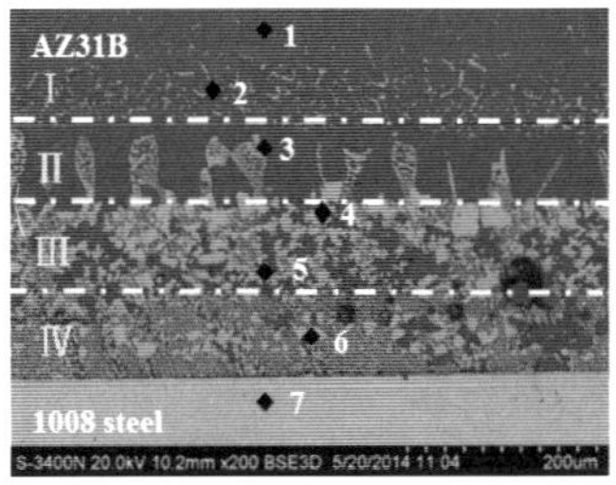

Fig. 5. Typical interfacial microstructure of AZ31B/Cu coating/1008 steel joint brazed at 530°C for 60 min.

It can be measured that the thickness of reaction zone is approximate 300μm after brazing process although the initial thickness of copper coating is only 50μm. The deposited copper coating clearly reacted with magnesium substrate. Meanwhile, Mg-Cu eutectic phase has good wettability on steel without defects in brazed seam under optimal conditions. According to EDS analysis, region I of the brazing seam region was mainly composed of granular or thin strip Mg-Cu-Al ternery phase and Mg_2Cu eutectic phase. Region II was composed of discontinuous reticulate Mg_2Cu phase. Furthermore, region III consisted of massive Mg-Cu-Al intermetallic phase and few Mg_2Cu phase. While region IV consisted of massive Mg-Cu-Al phase and continuous reticulate Mg_2Cu phase. In contrast, there was no clear interaction layer in between the brazed seam and steel.

In order to identify the interfacial phases in joint accurately, X-ray diffractometer was performed across the brazed seam and the XRD patterns is shown in Fig. 6. The analysis of XRD indicates that α-Mg solid solution, Mg_2Cu eutectic phase, Mg-Cu-Al ternery phase and Fe-rich phase were detected.

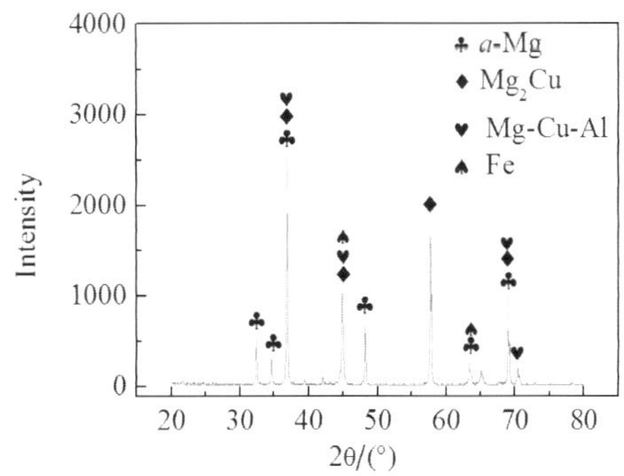

Fig. 6. XRD patterns of the cross-section around the brazed seam of AZ31B/Cu coating/1008 steel joint as shown in Fig. 5

Fig.7. shows the microhardness profile across the brazed AZ31B/Cu/1008 steel joint, as per hardness points 1-7 illustrated in Fig.5.

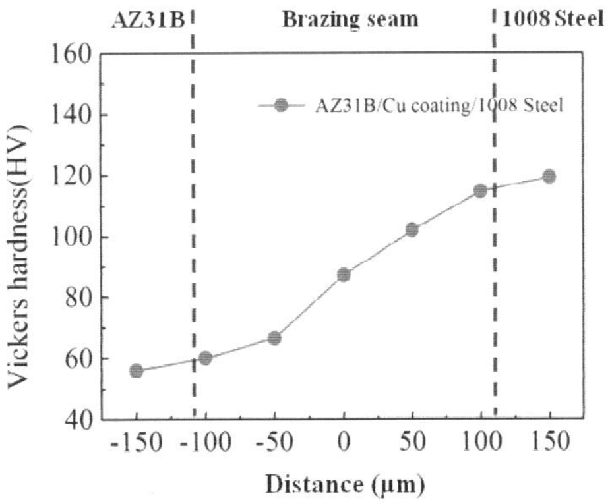

Fig. 7. Distribution of microhardness in the brazed seam AZ31B/Ni coating/1008 steel joint

Due to the presence of brittle intermetallic Mg-Cu-Al phase, the microhardness profile across the brazed seam increases gradually from left AZ31B side to right steel side. The average hardness of the brazed seam is approximaterly 86 Hv.

Fig.8. shows the comparision of shear strength of brazed Mg/steel joints by different researchers using different brazing method and different filler/coatings to be served as interlayer. Elthalabawy W.[10] et al. studied diffusion bonding of AZ31B/316L stainless steel using pure copper foil as interlayer. The maximum shear strength of brazed joint was 57MPa at brazing temperature 530°C for 60 min. However, among all research results, it can be found that the maximum strength of 96MPa was achieved for AZ31B/Q235 CRB joints by using Zn-*x*Al coated (hot-dip galvanizing) steel substrate[11]. Elrefaey A. [6] et al. performed diffusion brazing of AZ31B/AlSi 1008 by using 10μm-thick thin copper coatings deposited on Mg substrate by Arc-PVD when the sample was heated to 520°C and held for 15 minutes. The joint achieved average shear strength of 26MPa. Furthermore, Tachibana T.[12] et al. also investigated the weldability of diffusion bonded AZ31B/GI (a zinc coated steel) joints using zinc insert when heating at 342-412°C for a holding time from 0 to 45min, with a pressure of 5MPa. The maximum shear strength of joints reached 40MPa because Zn-Mg eutectic reaction efficiently remove the oxide films on the Mg alloy. Typically, our results shows the feasibility of brazing AZ31B/AlSi 1008 using a thick copper coatings (50μm) deposited on Mg substrate by cold gas spraying(CGS) serving as interlayer, which achieved a maximum strength of 37MPa under optimal conditions. In addition, compared with other methods, CGS shows the advantage of no oxidation and high quality thick coatings, which is more suitable for the formation of brazed seam, as well as high deposition rate on the bonding substrate with a large or complex surface.

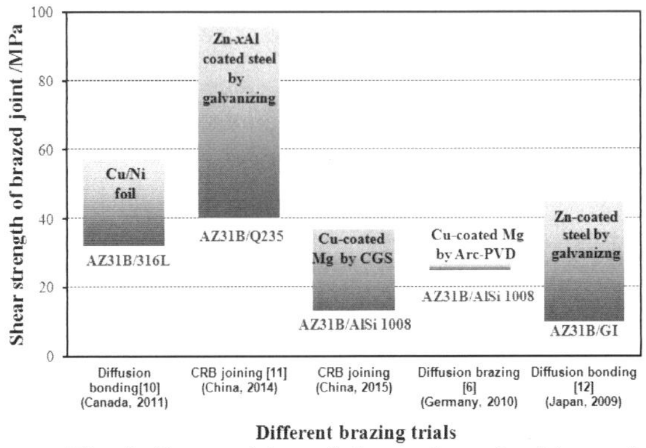

Fig. 8. Comparision of shear strength of brazed Mg/steel joints by different brazing trials

4 Conclusions

In the present study, pure copper powder was deposited onto Mg alloys AZ31B by cold spraying. A thick CGS coating layer has been employed successfully as filler interlayer for AZ31B/AlSi 1008 steel joints under vacuum by contact reactive brazing (CRB) process without any fluxes added. The shear strength of CRB joint achieved the maximum 37MPa at brazing temperature 530°C for 60 min. It was shown that the deposited copper coating clearly reacted with magnesium resulting in the formation of four characteristic zones (I, II, III and IV) in brazed seam. The interface phases of the brazed joint were mainly composed of Mg-Cu-Al ternary phase, eutectic phase Mg_2Cu and α-Mg solid solution. The microstructural evolution and the interfacial defects under temperature change were assessed. Under low temperature 485°C, an unmelted copper coatings near interface areas was obviously observed due to incomplete reaction between Mg and Cu. When the temperature increased to 500°C, the joint was mainly consists of Mg-Cu-Al phase while no Mg_2Cu phase was found. Microcracks were responsible for the unbonded area in steel side. When the temperature increased to 545°C, overheating of Mg compounds occurred at the interfacial zone near steel side, which contribute to the deterioration of joint shear strength. In summary, interfacial defects were responsible for the the low strength of the joints. Meanwhile, iIntermetallic phases caused the increase of microhardness across the brazed seam gradually from left AZ31B side to right steel side.

Acknowledgements

The authors thank State Key Laboratory of Advanced Brazing Filler Metals and Technology of Zhengzhou Research Institute of Mechanical Engineering(ZRIME) for financially supporting the research work within the scope of project SKLABFMT201303.

5 Literature

[1] Wielage, B., Podlesak H., Grund T., A. Wank: High resolution microstructural investigations of

interfaces between light weight alloy substrates and cold gas sprayed coatings. Proceedings of the ITSC 2005, Basel, CH
[2] Bobzin, K., Zhao.L.D., Ernst F.: Flux-free brazing of Mg-containing aluminium alloys by means of cold spraying, Front. Mech. Eng. China 2008, 3(4): 355–359
[3] Mao H.K., Bell P.M. Equations of state of MgO and Fe under static pressure conditions, J. Geophys. Res., 1979, 83: 4533-4536.
[4] Watanabe T., Komatsu S., Oohara K. Development of flux and filler metal for brazing Magnesium alloy AZ31B. Welding Journal, 2005, 84(3): 37-40
[5] Alexander E. S., Brazing Magnesium alloys and Magnesium matrix composites. Welding Journal, 2005, 84(10): 33-43.
[6] Elrefaey A.,Tillmann W. Application of novel brazing process for joining AZ31 magnesium alloy. Brazing, High Temperature Brazing and Diffusion Bonding, Germany, 2010, 361-365.
[7] Wielage, B., Grund T., Ahrens S. et al. CGS sprayed filler coatings for brazing of light weight alloys. Proceedings of the ITSC 2005, Basel, CH
[8] Lynch J.F., Feinstein L.,Huggins R.A. Brazing by the Diffusion-Controlled Formation of A Liquid Intermediate Phase. Welding Journal,1959,38(2): 85-89.
[9] The Materials Information Society, Binary Alloy Phase Diagram, 2nd edn, ASM International, Metals Park, OH, 1996, plus updates.
[10] Elthalabawy W., Khan T.. Liquid Phase Bonding of 316L Stainless Steel to AZ31 Magnesium Alloy. Materials Science and Technology, 2011, 27(1): 22-28.
[11] Xu Y.L. Study on the contact reaction brazing of dissimilar metals between Mg to steel.Ph.D thesis, Dalian Jiaotong University, 2014
[12] Tachibana T., Hojo S., Iwatani S. Effects of Zinc insert and Al content in Mg Alloy on the Bondability in Dissimilar Joints of Steel and Magnesium Alloys. Quarterly Journal of the Japan Welding Society, 2009, 27(2): 183s-186s.

Integration of a Thermal Spray Process into the GROB Production Line for Passenger Car Aluminum Engine Blocks of Combustion Engines into Mass Production

Bernhard GAND Eschweiler / D
Andreas WÖRFEL Mindelheim / D
Christoph MÖRZ Türkheim / D
Heribert SAULE Augsburg / D

Abstract - The passenger car manufacturers are using new technologies for improving the efficiency of combustion engines, reducing their emissions, fuel consumption and weight. A technology that can fulfil these requirements is the thermal spraying process of Aluminum cylinder bores. Therefore, GROB Werke have integrated their new **G**ROB **T**hermal **S**pray process (GTS) into their engine block production lines. A thermal spray production cell was developed and built by GROB Werke. In order to offer a turnkey solution to the customer GROB Werke are actively supporting automotive companies with their GROB-Knowhow in the thermal spray technology. With GROB as partner in development a thermally sprayed steel deposit layer can be developed and applied to the customer's needs for any passenger car engine architecture and cylinder bore geometry. Thecreation of highly specialized cylinder bore surfaces allows a topography design which heavily reduces wear and friction and is very resistive to chemical reactions induced by the fuel. A thermaly sprayed steel deposit layer produced with GTS will thereby help the passenger car manufacturers reaching their goals regarding efficiency, emissions, fuel consumption and weight.

1. Introduction

Terms like downsizing, increasing efficiency, weight reduction and improvement in performance are describing the actions taken by passenger car manufacturers at the moment aiming for the reduction of emissions and fuel consumption. Therefore the improvement of the overall performance of the engine is the central point in this process. Using new technologies the passenger car manufacturers are trying to increase the efficiency of combustion engines and reduce their weight in the same step. One of these technologies is dealing with the characteristics of the cylinder surfaces in combustion engines with an aluminum engine block. The solutions which are state of the art up to now will be replaced by a thin sprayed steel deposit layer.

Fig. 1. Integration of the GTS Process into the GROB Production Line of Aluminum Cylinder Block

The development of the thermal spray technology by the GROB Werke is the next step in extending the GROB production lines for engine blocks by a production cell for the thermal spraying with a GROB thermal spray process developed in the company, **Fig. 1**., **Fig. 2**. The strength of this super-thin steel deposit layer on an aluminum engine block cylinder bore surface lies within its extraordinary properties. This steel deposit will reduce the weight in comparison to conventional grey-cast cylinder liners and achieves – also in comparison to monolithic, hypereutectic aluminum engine block – an extremely low wear value, very low friction and a higher resistance to chemical reactions induced by aggressive fuels, too. Dropping the grey cast cylinder liners also reduces the residual stress after the solidification in the casting process of aluminum engine blocks. Moreover, this results in a higher thermal conductivity leading to a better heat dissipation and therefore increases the maximal static and dynamic load of the engine in operation. For the combustion engine powertrain of the future this technology is trend-setting and will thus be a fixed part in the GROB array of products.

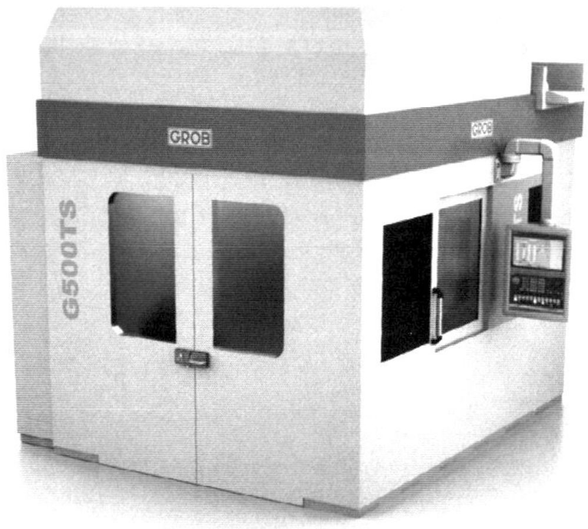

Fig. 2. GROB Thermal Spray Production Cell G500TS – Front Side

2. Preliminary Processes

Before the actual thermal spray process is applied several preliminary steps, described in the following, must be taken to successfully perform the spray procedure. Only an aluminum engine block casting in very high quality can be used for the thermal spray process.

The activation of the aluminum cylinder bore surface ensures that the steel deposit layer is able to adhere on this substrate. The cylinder bores are bored precisely in the right position in relation to the crankshaft. GROB offers a solution on the G-Module, where the activation is done by a GROB mechanical roughening profile. The quality of the activation is ensured by an optical inspection of the cutting edges of the tool inside the G-Module.

Fig. 3. Cylinder Surface activated with GROB mechanical Roughening Process

3. Mechanical Activation as Key

The adhesion between the substrate and the deposit to be sprayed is dominated by a mechanical interlocking between the substrate and the deposit. Using a scanning process the aluminum cylinder bore surface is inspected, searching for allowed porosities and different types of non-allowed damages that might appear. To scan the cylinder correctly by optical means the surface is cleaned before the inspection. The activation of the substrate surface is the crucial step before the thermal spray process. Not only is the mechanical adhesion influenced by this, but also by a subtle choice of the geometry of the activation profile the microstructure of the steel deposit layer can be influenced. This activation profile will be a vital factor to the properties of the final surface, **Fig. 3**.

4. The Thermal Spray Process

The large number of parameters in the GROB thermal spray process results in a high variety concerning the properties of the steel deposit. These parameters are strongly influencing the microstructure and microhardness of the deposit and therefore the mechanical and tribological behavior of the final surface, **Fig. 4**. By applying several diagnostic methods and quality checks, which are used by the GROB Werke to monitor the process and quality, a deposit layer with a reproducibly high layer quality and strictly defined state can be achieved. The quality management includes metallography, hardness tests, tests for defects and the live-surveillance systems to monitor the plasma and particle beam. [1], [2], [3]

Fig. 4. Cylinder Surface Sprayed with GROB Thermal Spray Process - GTS

5. Finalization of the Surface

The steel deposit layer in the cylinder is pre-machined precisely in a G-Module. The cylinder bore thereby reaches a state of high cylindricity and straightness. The intended topography of the super-thin steel layer is achieved by a honing process with up to 3 steps, **Fig. 5**. This creates the final diameter, surface structure and smoothness. The high, defined quality and structure of the surface is the key to the tribological behavior of the piston, piston ring and cylinder bore surface, and therefore the central aspect in achieving the improvements in comparison to the conventional state of the art technologies, e.g. in-casted cylinder liners.

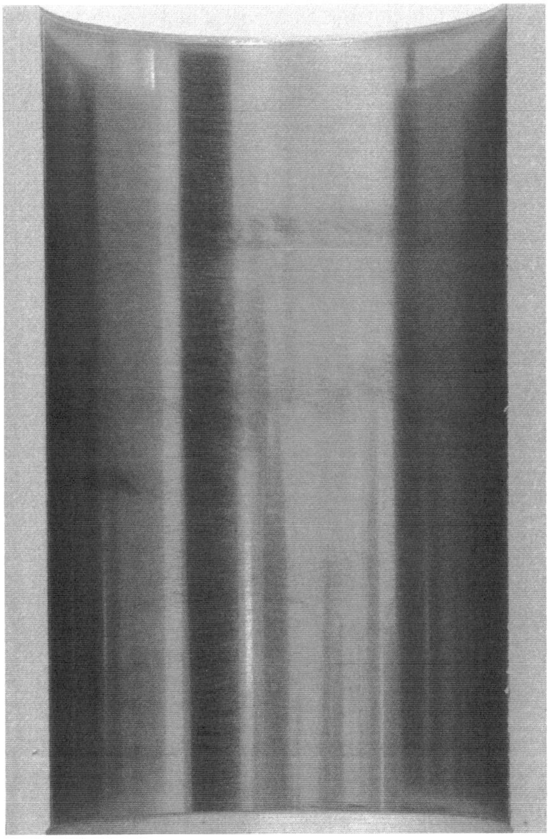

Fig. 5. Honed Cylinder Bore Surface Steel Deposit

6. GROB as Partner in Development

GROB develops and applies the whole process chain: Activation, cleaning, inspection, thermal spraying, process surveillance, realization of the final surface and quality management. GROB-Werke support their customers as partner in development in the individual application of the thermal spray process to the customer's engine architecture. The product palette reaches from a standard process to "Advances Surfaces", which are designed specifically to the customer's needs.

7. The GTS Process

The GTS (**G**ROB **T**hermal **S**pray) thermal spray process developed by GROB offers a high variability and reliability to the customer with low costs at the same time. It is a single-wire process where the wire is not rotating but the gun head rotates around the wire, **Fig. 6**. The process is designed to run constantly and stable without interruption, robustly forming a steel deposit of defined thickness, which eliminates the need for any layer thickness measurements.

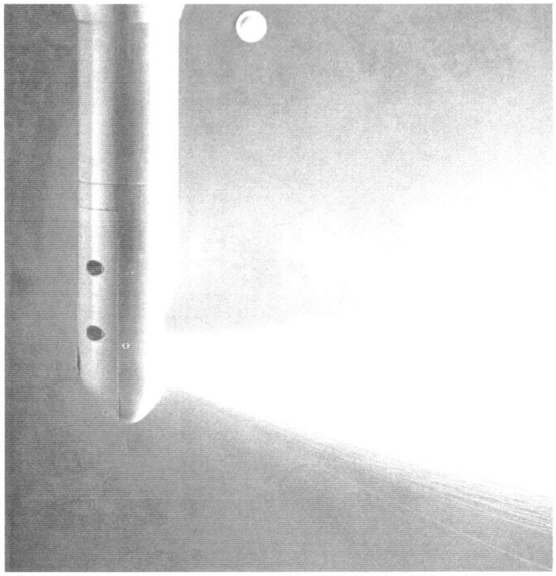

Fig. 6. Gun Head at Full Load running - GTS – GROB Thermal Spray Process

The dimensions of the thermal spray cell are based on the G-Module dimensions, which allows a simple integration in the production line with a linear gantry system, **Fig. 7**. The low investments and process costs and the very low cycle time compared to other, common thermal spray processes result in a high added value by the GTS-process, where the acquisition cost remain low.

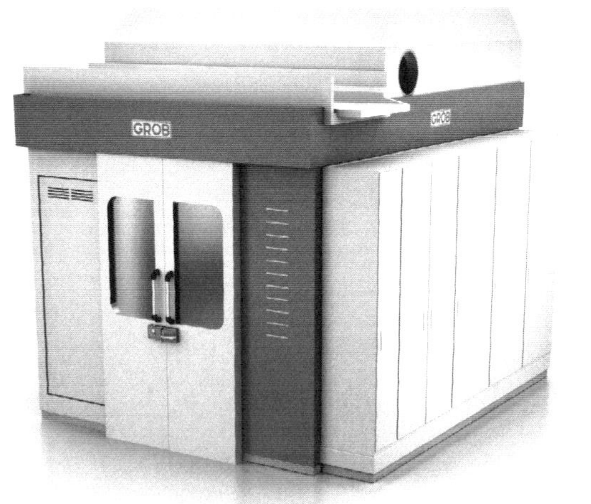

Fig. 7. GROB Thermal Spray Production Cell G500TS – Loading Side – Top or Back Side Loading possible

8. Future Developments

At the moment GROBWerke are working on the development of the final surface. With aggressive fuels entering the world market, e.g. ethanol, methanol or sulfur, steel layers which are resistant to corrosion offer another potential in increasing the life time of an engine. Not only is the microstructural composition of the deposit investigated but also the development of an advanced topography is focused. Regarding future "High Performance Tribological Systems" the surface properties will play a vital role in the development and creation of a low-friction, low-wear and sustainable combustion engine powertrain.

9. Literature

[1] GAND, Bernhard, High Performance Cylinder Bore Surfaces of Aluminum Crankcases. Thermal Spray Bulletin 1/2011

[2] GAND, Bernhard, Beschichtung von Zylinderlaufflächen in Aluminium-Kurbelgehäusen. MTZ Motortechnische Zeitschrift 02-2011

[3] GAND, Bernhard, SALENBAUCH, Hermann, AluThinFer–Zylinderlaufflächen – Verbesserung des Kraftstoffverbrauchs und des Verschleißverhaltens für Motoren mit niedriger und hoher thermo-mechanischer Beanspruchung. 5. VDI-Fachtagung Zylinderlaufbahn, Kolben, Pleuel – Baden-Baden 2010

Investigation of Amorphous Iron Based Thermal Barrier Coatings

K. Bobzin, M. Öte, T. F. Linke, T. Königstein, Aachen / D

Because of their favorable thermo-physical properties, good machinability and low material costs, coatings, which exhibit a highly amorphous microstructure, are currently in the focus of research. Especially their low thermal conductivities make them attractive as coatings for components in combustion engines. Considering the recrystallization temperature of the material, iron based coatings might serve as a new generation of thermal barrier coatings (TBCs) for low temperature systems, reducing thermal losses.
The objective of this research project is the development of highly amorphous, iron based coatings. For the manufacture of the coatings amorphous feedstock materials were developed and characterized regarding their morphologies, phase compositions and recrystallization temperatures. The coatings were deposited by air-plasma-spraying and high-velocity-oxygen-fuel-spraying with different parameter sets. Thereby the materials' ability of keeping an amorphous structure during spraying was investigated. In addition, particle velocities and temperatures were measured and compared to the recrystallization temperatures of the feedstock materials. Subsequently, the thermal diffusivity values were determined for selected parameter sets using free standing coatings, which confirm the potential of iron based TBCs.

1 Introduction

The demand for high performance and cost efficient materials increases steadily. Thereby innovative coating materials become more and more focus of research. One of the most important purposes of thermal spray coatings is thermal insulation in order to protect the bulk material from excessive heat or to reduce heat losses of energy conversion systems. In case of high temperature application ceramic materials such as Yittrium Stabilized Zirconia (YSZ) are established coating systems which have been investigated extensively. For example, research showed a relation between the thermal conductivity and the grain size of the material. By reducing the average grain size from 100 nm to about 10 nm a reduction of thermal conductivity by factor two was observed [1, 7]. Basically, this principle can be transferred to other materials, such as metals. In previous researches this phenomena has been investigated and confirmed using amorphous or nanocrystalline materials [2, 3]. Amorphous metals are called Metallic Glasses (MG). Due to its high wear and corrosion resistance, low thermal conductivity and economic advantages, iron based MGs are focus of many recent research activities.

The present work summarizes the results with regard to the manufacture of amorphous iron based metallic glass coatings and investigates the effect of different process technologies and parameter sets on the morphology of the coatings. Furthermore, the coatings were tested isothermally in a furnace. In order to proof the thermal insulation properties of MGs thermal diffusivities were measured by laserflash method.

2 Experiments

Based on the results from previous research, the chemical composition Fe72Si4B20Nb4 was selected and modified with regard to better glass forming ability and corrosion resistance [4]. Two modification to the above mentioned composition were designed by adding 5 at. % and 15 at. % of chromium and reducing the iron content, respectively. For the characterization of the feedstock material following methods and equipment were used

- Particle size distribution by optical means Morphology 2G (Malvern Instruments Ltd, Worcestershire, United Kingdom)
- Scanning electron microscopy (SEM) SEM LEO 1530 VP Gemini (Carl Zeiss AG, Oberkochen, Germany)
- Differential scanning calorimetry (DSC) SETSYS Evolution (SETARAM Instrumentation, Caluire, France)
- X-Ray diffraction (XRD) Seifert XRD 3003 (GE Sensing & Inspection Technologies GmbH, Hürth, Germany)
- Wet chemical analysis

In this study two spraying systems were utilized

- Air Plasma Spraying (APS) F4MB-XL (Oerlicon Metco, Winterthur, Switzerland)
- High Velocity Oxygen Fuel Spraying (HVOF) K2 (GTV-Verschleißschutz GmbH, Luckenbach, Germany)

All coatings were deposited on the aluminum substrates AlSi10. Prior to deposition, the substrates were grit blasted using F24 blasting medium and cleaned with ethanol. For the APS process the particle temperatures and velocities were recorded by the diagnostic system SprayWacht4s (Oseir Ltd., Tampere, Finland). After deposition, the coatings were characterized by following methods and equipment

- SEM/EDX
- XRD
- Laserfalsh FlashlineTM 4010 (Anter Corp., Pittsburgh, USA)

For the laserflash method free standing coatings were manufactured. Therefore the feedstock materials were deposited on smooth steel substrates and removed mechanically. After segmentation these samples were used for laserflash measurements. Furthermore, proper samples were selected for isothermal heat exposition tests in a furnace. Subsequently, the microstructures of heat treated and untreated samples were compared.

For the deposition APS and HVOF technology were used. Following process parameters are kept constant for all HVOF processes: kerosene fuel rate V_k = 20 l/min, carrier gas flow rate V_N = 8 l/min, gun velocity v_g = 1.0 m/s, nozzle length l_N = 145 mm and nozzle diameter D_N = 12 mm. Table 1 gives an overview of the parameter sets which were varied. In total five parameter sets were investigated for each feedstock material.

Table 1 Process matrix for HVOF

	Lambda (oxygen fuel rate [l/min])	Standoff distance [mm]
C01	1.15 (800)	250
C02	1.15 (800)	300
C03	1.15 (800)	350
C04	1.08 (750)	300
C05	1.30 (900)	300

Table 2 shows the test matrix which was used for the APS coatings. In comparison to the HVOF test matix, four factors were identified for variation. Due to the larger number of tests, only one powder was used to manufacture coatings considering all parameter sets. After the evaluation of polished cross sections and XRD-measurements, selected parameter sets were used to manufacture coatings with the other powders. For the APS processes the carrier gas flow rate V_{Ar} = 4 l/min, nozzle diameter D_N = 6 mm and gun velocity v_g = 1.2 m/s were kept constant.

Table 2 Process matrix for APS

	Ar flow rate [l/min]	H_2 flow rate [l/min]	Current [A]	Standoff distance [mm]
C06	37	10	550	120
C07	30	10	550	120
C08	37	8	550	120
C09	37	10	500	120
C10	44	10	550	120
C11	37	12	550	120
C12	37	10	600	120
C13	37	10	550	100

3 Results and Discussion

The results of wet chemical analysis show good agreement with the aimed composition

- Powder 1: Fe68.5B21.5Si4.8Nb5.3
- Powder 2: Fe61.7Cr6.1B22.0Si5.0Nb5.3
- Powder 3: Fe51.5Cr16.6B21.8Si4.9Nb5.2

With regard to the size distribution of the different feedstock materials no significant difference between powder 1, 2 and 3 was observed. Therefore only the 10th, 50th and 90th percentile values for powder 1 are presented in the following. For HVOF, it follows:

- 10th percentile: 20.80 µm
- 50th percentile: 31.75 µm
- 90th percentile: 44.15 µm

The APS powder particles are significantly larger:

- 10th percentile: 49.30 µm
- 50th percentile: 71.40 µm
- 90th percentile: 98.60 µm

SEM pictures of the particles show spherical shapes for all particles. Satellites were recognized for HVOF and APS powder. However, the effect seems to be more dominant in case of the coarser APS powder. Furthermore, powders 1 and 2 exhibit smooth surfaces. Particles of powder 3 exhibit a structured topography which is an indicator for a particular microstructure. This was confirmed by etched polished cross sections, see Fig. 1.

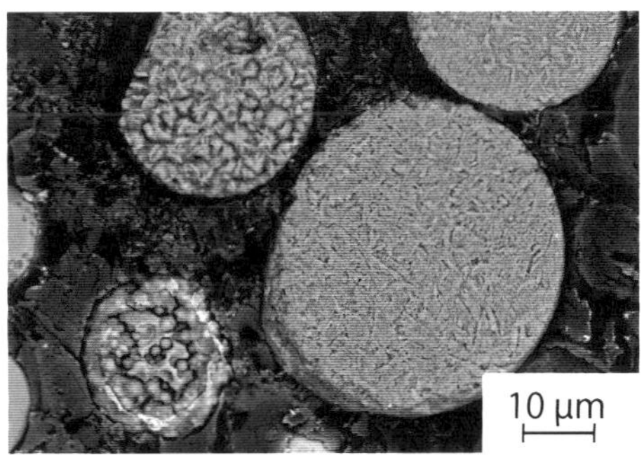

Fig. 1 SEM micrographs of etched powder particles of powder 3

The XRD-measurements of all feedstock materials show an amorphous or partially amorphous microstructure. In addition, DSC-measurements show distinctive exothermic peaks, indicating the recrystallization behavior of the material. No significant differences regarding the characteristic temperatures, such as glass transition temperatures (T_g), recrystallization temperatures (T_x), melting temperatures (T_m) and liquidus temperatures (T_l) was identified between HVOF and APS powders. For example, the temperatures measured for powder 1 correspond to: T_g = 540 °C, T_x = 580 °C, T_m = 1,125 °C and T_l = 1,155 °C. Increasing the chromium content to 15 % results in a temperature shift for all characteristic temperature of T = 60 °C to higher values. This has huge impact on the morphology of the deposited coatings due to inclusion of partially melted particles.

As mentioned above, no significant difference has been observed with regard to the T_g, T_x, T_m and T_l, between APS and HVOF powders. However, the exothermic energy peak describing the recrystallization behavior is smaller for the APS powders, see Fig. 2. This behavior can be attributed to a higher content of crystalline phases. Due to the lower surface to volume ratio of the APS powders compared to the HVOF powders, the particles cannot be cooled as rapidly as HVOF powders during the atomization process.

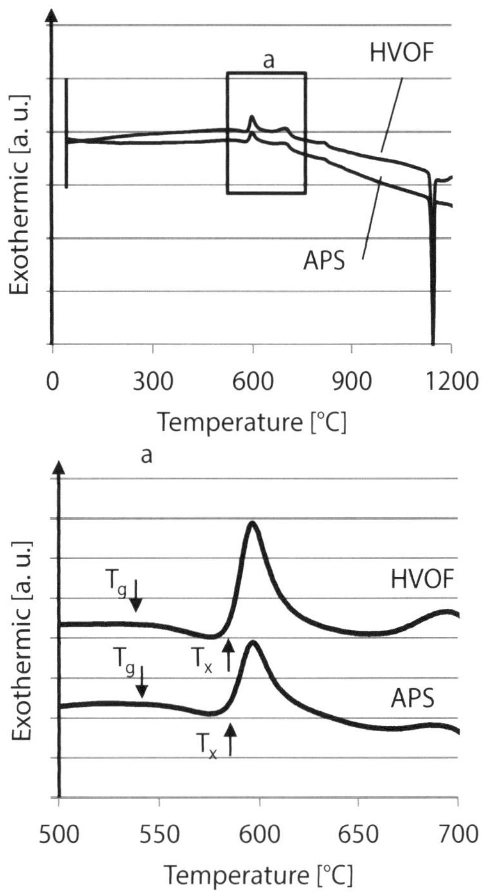

Fig. 2 DSC-measurements for APS and HVOF powder 1

For coatings deposited with HVOF, it can be seen that with higher chromium content the melting degree of particles is significantly reduced, see Fig. 3. As a consequence, the particles are more likely to bounce after impact and are lost in the overspray reducing the deposition efficiency. Furthermore, the continuous impact of unmelted particles results in a densification of the coating. Both, the lower deposition efficiency and the densification result in a reduced coating thickness.

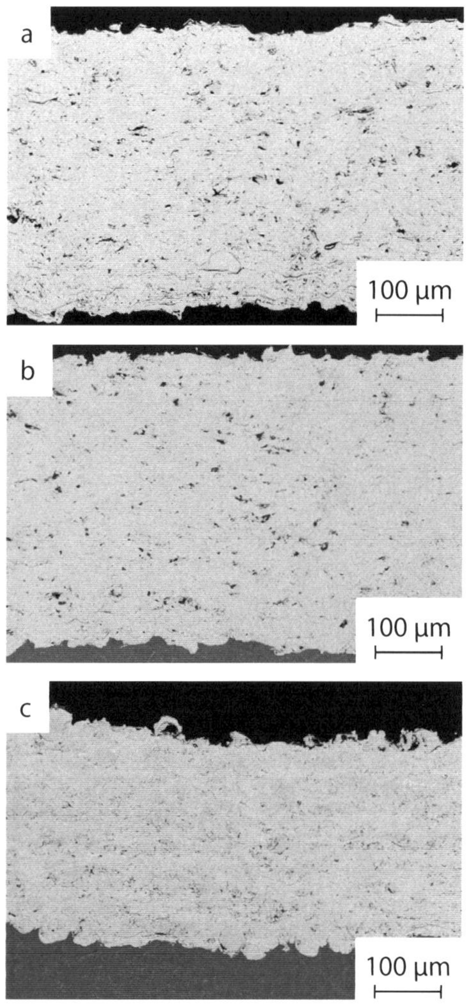

Fig. 3 Coatings deposited with parameter set C2 using powder 1 (a), powder 2 (b) and powder 3 (c)

The variation of standoff distance and lambda value has shown significant impact on the morphology, especially on the porosity and the interface areas between the lamella. From polished cross sections it can be seen, that larger standoff distance results in a higher porosity, especially for powder 2 and 3. In Fig. 4a and 4b the cross sections for powder 2 are presented exemplary. For larger standoff distance, the interface areas between the lamella are wider, presumably due to the lower particle velocity. The effect of different lambda values on the microstructure is not pronounced. It is assumed that rising lambda values result in higher particle velocities and reduced particle temperatures. The lower particle temperatures reduce the melting degree of the particles and consequently increase the porosity. This can be observed for powder 1. For powder 2 no significant difference in the microstructure was observed. Coatings deposited with powder 3 and higher lambda values exhibit a denser structure presumably due to further densification of the coatings, see Fig. 4c and 4d.

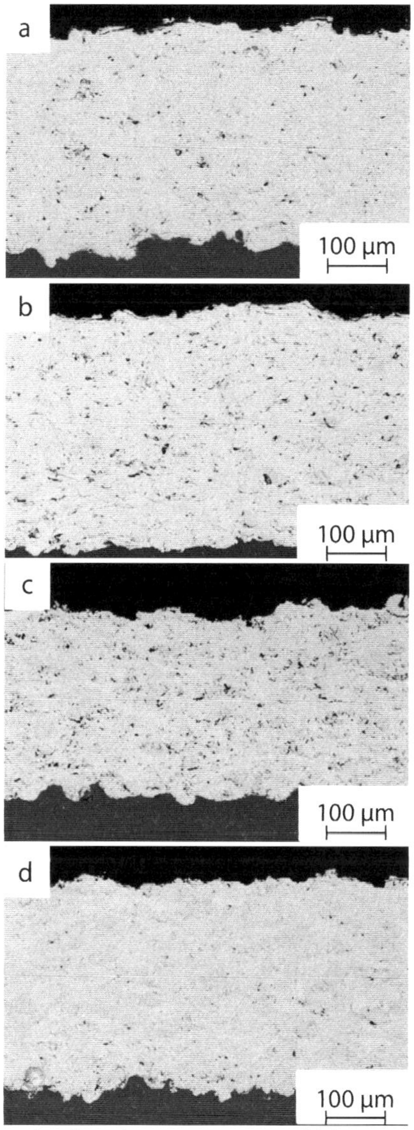

Fig. 4 Powder 2 and parameter sets C01 (a) and C03 (b) and powder 3 and parameter sets C04 (c) and C05 (d)

XRD-measurements of HVOF coatings do not show tremendous change in the crystalline structure. Only for powder 1 and a larger standoff distance (C03) two new peak values were identified at $2\theta = 25°$ and $2\theta = 45°$. The peak value at lower 2θ angle most likely corresponds to iron oxide. Polished cross sections confirmed this assumption.

In contrast to the HVOF coatings the APS coatings exhibit a more pronounced, inhomogeneous microstructure. The evaluation of the coating's cross sections, deposited with parameter sets listed in Table 2, show huge influences of hydrogen and the standoff distance on the morphology and the coatings thickness. Fig. 5 shows exemplarily the cross sections for different hydrogen flow rates. In micrograph 5a, it can be seen that the coating exhibits a higher porosity in comparison to that shown in micrograph 4b. It is assumed, that changing the hydrogen flowrate has large influence on the particle temperatures and velocities, resulting in different morphologies. In order to confirm this assumption, measurements of average particle temperatures ($T_{P,A}$) and velocities ($v_{P,A}$) were conducted. Changing the hydrogen flow rate from $V_{P,H}$ = 8 l/min to $V_{P,H}$ = 12 l/min results in a rise of particle temperature from $T_{P,A}$ = 1,760 °C to 2,190 °C. The particle velocities remain almost constant, $v_{p,A}$ = 125 m/s to 131 m/s. These results show, that with changing hydrogen flow rate the particle temperatures reacts more sensitively than the particle velocity. It is assumed, that the higher particle temperature is responsible for the higher degree of porosity.

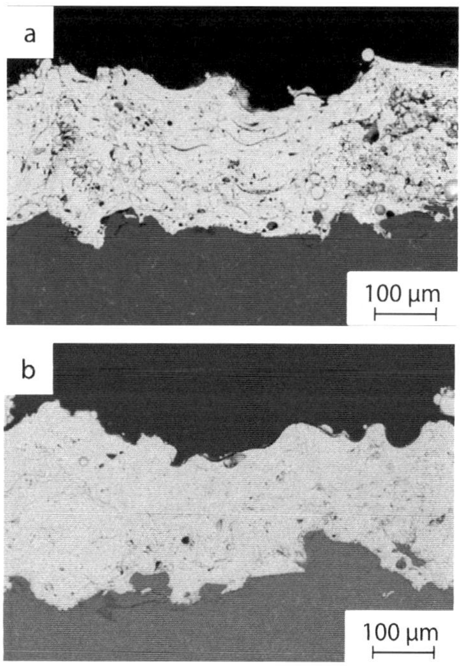

Fig. 5 Influence of hydrogen on the morphology of the coatings deposited with parameter sets C11 (a) und C08 (b)

With regard to particle temperatures and velocities the same tendency can be observed with the variation of the electric current. In contrast to that, a change of argon flow rate influence both, temperature and velocity of the particles. Measurements show a reduction of particle temperature from $T_{P,A}$ = 2,290 °C to 2,070 °C and a rise of velocity from $v_{P,A}$ = 118 m/s to 148 m/s. These observations are in consistency with the literature [8]. Besides the average values, the deviations of particle temperatures ($T_{P,D}$) and particle velocity ($v_{P,D}$) were measured. In all experiments a $T_{P,D}$ = 200 °C to 400 °C and a v_{pd} = 14 to 27 m/s was measured. These fluctuations are presumably responsible for the inhomogeneities in the morphologies. In addition and due to the higher process temperatures of APS compared to HVOF a higher melting degree of powder 3 can be observed, which was confirmed by flattend splats in polished cross sections.

After evaluation of the cross sections proper coatings were selected for heat exposition tests. Therefore the specimens were placed in an atmospheric oven at a temperature of 400 °C for 100 h. A comparison of the polished cross sections before and after heat exposition show a densification of the microstructure of powder 1, see Fig. 6. This can be attributed to sintering effects. Examinations of the interface and the top layer do not show thermally growing oxides.

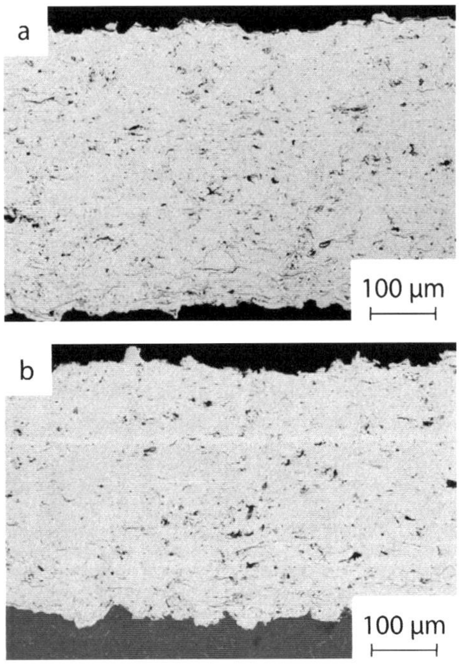

Fig. 6 Coatings deposited with powder 1 and parameter set C02 before (a) and after heat treatment

In order to investigate the insulation properties of the material, the thermal diffusivities were determined by laser flash method. Therefore free standing coatings were manufactured. The results show, that metallic glasses exhibit a 3 to 5 times lower thermal diffusivity values compared to S235JR, see Fig. 7, top diagram. Furthermore, all thermal diffusivities of coatings deposited with powder 2 and the parameter sets presented in Table 1, were investigated. It can be seen, that the different parameter sets and consequently the morphologies of the coatings have significant influence on the thermal diffusivity. It can be observed, that higher porosity and larger interface areas reduce the thermal diffusivity, see Fig. 7, bottom diagram.

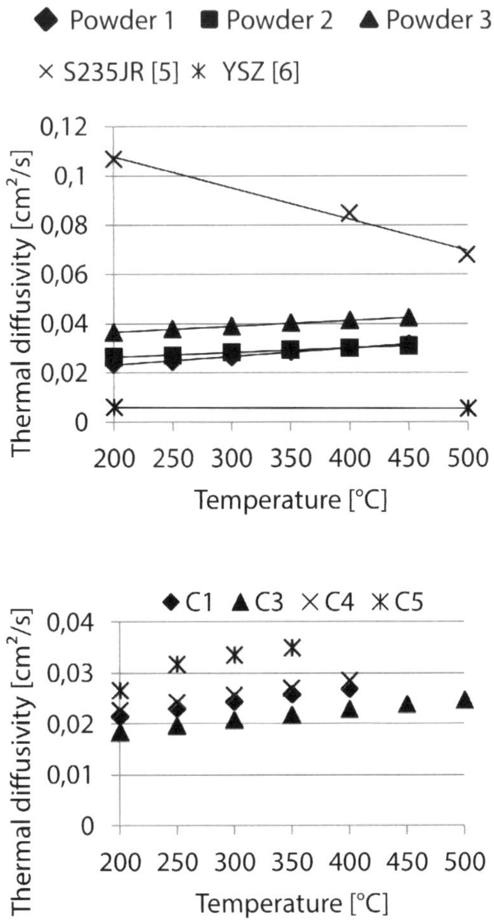

Fig. 7 Thermal diffusivities for HVOF coatings; top: different feedstock materials deposited with parameter set C02; bottom: powder 2 deposited with different parameter sets

4 Conclusion and future outlook

In this study amorphous metallic glass coatings were manufactured by HVOF and APS processes. The HVOF coatings exhibit a more homogenous microstructure than the APS coatings. Due to lower process temperatures of the HVOF process particles of powder 3 were not fully melted. Indeed a high degree of unmelted particles can be identified in the cross section. Using the APS process improves the melting degree of the particles. Selected samples were exposed to heat in a furnace. For powder 1 a densification due to sintering was observed. Investigations with regard to the thermal diffusivities show low values over a wide temperature range. Furthermore, it was shown that larger interface areas and higher degree of porosity affects the thermal diffusivity.

For future work, the application of the developed feedstock materials onto the inner geometries i. e. onto inner surface of cylinder liners by Plasma

Transferred Wire Arc (PTWA) technology will be investigated. For this purpose, the developed feedstock material needs to be produced as wire.

Acknowledgment

The authors would like to thank the Deutsche Forschungsgemeinschaft (DFG, German Research Foundation) for funding the project „Grundlegende Erforschung der Einsatzmöglichkeit von Fe-basierten TS-Schichten zur Wärmedämmung" (Basic research on the applicability of Fe-based TS-coating with the purpose of thermal insulation)

4 Literature

[1] H. Yang, J. A. Eastman, L. J. Thompson, G.R. Bai, Grain-Size-Dependent Thermal Transport Properties in Nanocrystalline Yttria-Stabilized Zirconia, Materials Research Society, 2002, 703, p 179-184

[2] D. Shin, F. Gitzhofer, C. Moreau, Properties of Induction Plasma Sprayed Iron Based Nanostructured Alloy Coatings for Metal Based Thermal Barrier Coatings, Journal of Thermal Spray Technology, 16 (1), p 118-127

[3] D. Shin, F. Gitzhofer, C. Moreau, Thermal property evolution of metal based thermal barrier coatings with heat treatments, Journal of Materials Science, 2007, 42, p 5915-5923

[4] K. Amiya, A. Urata, N. Nishiyama and A. Inoue, Fe-B-Si-Nb Bulk Metallic Glasses with High Strength above 4000 MPa and Distinct Plastic Elongation, Material Transactions, 2004, 45 (4), p 1214 – 1218

[5] M. Iordachescu, J. Ruiz-Hervias, D. Iordachescu, A. Valiente, L. Caballero, Thermal Influence of Welding Process on Strength Overmatching of Thin Dissimilar Sheets Joints, Iberian Conference on Fracture and Structural integrity, 2010

[6] L. Jina, P. Lib, H. Zhoua, W. Zhanga, G. Zhoua, C. Wanga, Improving thermal insulation of TC4 using YSZ-based coating and SiO2 aerogel, Progress in Natural Science: Materials International, 2014, 25, p. 141-146

[7] K. Bobzin, L. Zhao, M. Öte, T. Frederik Linke Deposition and characterization of thermal barrier coatings of ZrO2–4 mol.% Y2O3–1 mol.% Gd2O3–1 mol.% Yb2O3, Surface & Coatings Technology, 2015, 268, p. 205–208

[8] L. Zhao, K. Seemann, A, Fischer, E. Lugscheider, Study on atmospheric plasma spraying of Al2O3 using on-line particle monitoring, Surface and Coatings Technology, 2003, 168, p.186–190

Microstructure and Mechanical Properties of Cold Sprayed Ni-Al_2O_3 Coatings

W. Żórawski, Kielce /PL, A. Góral, Cracow / PL, M. Makrenek, Kielce / PL

This work concerns a study of the properties of a cold sprayed composite Ni-Al2O3 coatings. The Ni- Al_2O_3 powder was applied in the cold spraying process. The coatings were sprayed by means of Impact Innovations 5/8 system with three different contents of Al_2O_3 (20, 40 and 60 %wt.). The investigations revealed that the increase in ceramic content in the blend leads to an increase in Al_2O_3 amount in the cold sprayed Ni-Al_2O_3 coatings. The ceramic particles were mainly distributed between plastically deformed elongated nickel grains. It is also visible areas with a fragmentation of ceramic particles due to the impact of one or more incoming Al_2O_3 particles onto an embedded ceramic particle. All coatings were characterized by a high surface roughness, significantly higher than in the case of pure Ni coating. The coating Ni+40%wt Al_2O_3 showed the best retained balance between an amount of embedded Al_2O_3 particles and coating porosity. Mechanical properties (hardness and Young's modulus) were dependent on the different areas of composite coatings.

1 Introduction

Metal matrix composite coatings (MMC) belong to the group of the most promising materials that can find wide engineering application. They create the possibility of programming desired operating characteristics by optimizing their microstructure and properties, both by selection of the composition, as well as proper parameters of manufacturing process. Composite coatings containing embedded hard particles can achieve a wide spectrum of improved properties, resulting from a combination of features from a ductile matrix and the hard particles (eg. ceramic). This type of composition can provide ideal combination of toughness and hardness. New opportunities in this area creates the newest method, from among thermal spray techniques, that is a cold spraying. It has been widely investigated owing to possibility to achieve enhanced properties of coatings. This technique allows to eliminate the adverse effect of temperature on the particles of the coating materials and the substrate that occurs in conventional methods of thermal spraying, and therefore the coatings obtained are unattainable by other methods [1,2]. The coating is formed by deposition of highly deformed plastically at the moment of hitting the ground powder particles and mutual bonding occurring at high speed of deformation under adiabatic conditions and below the melting point of the metal [3,4]. Cold spraying is a high kinetic energy coating process, where fine powder particles are propelled by a supersonic gas jet towards a substrate [5-8]. The ability of a cold sprayed particle to undergo plastic deformation and adiabatic shear is a function of its mechanical (e.g. yield strength) and thermal (e.g. melting point) properties of the powder material as well as parameters such as the deposition temperature, particle velocity and particle size [9]. Low temperatures also aid in retaining the original powder chemistry and phases in the coating, with only changes due deformation and cold working.

The detailed characterization of microstructure and mechanical properties of cold sprayed Ni-Al_2O_3 coatings is subject of further researches. Especially changes of microstructures and properties of coatings depending on an alumina content are still little known.

2 Experimental

Composite coatings Ni-Al_2O_3 were cold sprayed from various mixtures of micron size powders: Ni (Metallisation P836) (-45 +11 μm) and Al_2O_3 (Sulzer Metco M 6051), (-31 +5,5 μm) by means of Impact Innovations 5/8 system. Aluminium alloy (Al 7075) characterized by high mechanical properties was used as a substrate. The coatings were prepared from the following powder mixtures (in brackets used abbreviation of coating destination are presented): 80 %wt. Ni + 20 %wt. Al_2O_3 (Ni20Al_2O_3), 60 %wt. Ni + 40 %wt. (Ni40Al_2O_3), 40 %wt. Ni + 60 %wt. Parameters of cold spray process of Ni-Al_2O_3 coatings were presented in Table 1. Morphology of powder feedstocks, as well microstructure of sprayed coatings were investigated both on surface and cross section of coatings-substrate systems using scanning electron microscopy (SEM; Jeol JSM-7100 F; E-SEM FEI XL 30). X-ray diffraction phase analysis was performed using Bruker D8 Discover (Co Kα radiation) and program Diffrac.Eva 3.0 with crystallographic database PDF-4+.

Table 1. Parameters of cold spraying of Ni-Al_2O_3 composite coatings.

Parameter	Value
Pressure, MPa	3,0
Temperature, °C	500
Spraying distance, mm	40
Powder output, feeder scale	3
Robot arm speed, m/s	500

3 Results and discussion

3.1. Characterization of Ni and Al_2O_3 powders

The grain sizes and morphology of their cross-section the two powders are shown in Fig. 1 and 2 respectively. Ni powder is produced by agglomeration and sintering of fine powder grains. Powder contain a very large amount of near spheroid-shaped grains, which causes that the powder grains are loose, compared to those with irregularly shaped grains. The angular shape of Al_2O_3 grains show that powder is obtained in the course of grinding. Cross-section image of Ni powder (Fig. 2a) showed that there were some small pores in the grains structure. On the other hand, cross-section of Al_2O_3 grains (Fig. 2b) do not reveal any porosity.

Fig. 1. Morphology of powders grains: a) Ni, b) Al_2O_3.

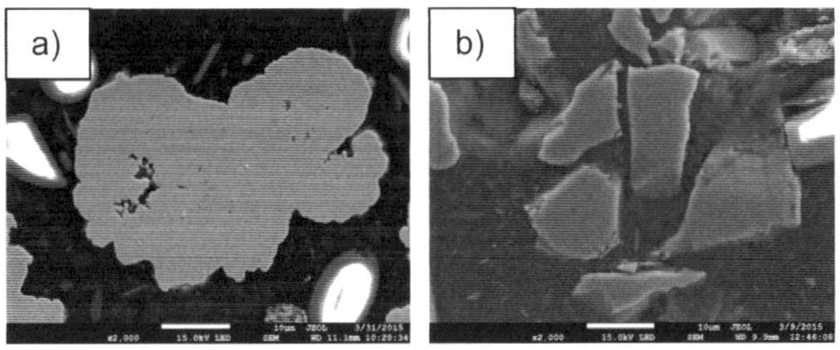

Fig. 2. Microstructure of cross section of powders: a) Ni, b) Al_2O_3.

3.2. Characterization of cold sprayed Ni-Al_2O_3 coatings

SEM examinations of the cold sprayed Ni-Al_2O_3 coatings showed that the addition of various amounts (20 - 60 %wt.) of micrometer sized Al_2O_3 powder has considerable effect on the obtained coating's microstructure (Fig. 3). As it was expected the increase in ceramic content in the blend lead to an increase in ceramic content in the coating, what also is confirmed by XRD analysis (Fig. 5). The microstructure of composite coatings comprised of adjoining plastically deformed nickel particles between which the ceramic particles were distributed. They are also visible areas with a fragmentation of ceramic particles due to the impact of one or more incoming ceramic particles onto other embedded ceramic particle. The most similar microstructure of coatings to that desirable seems to be obtained for the coating Ni40Al_2O_3. This coating showed the best retained balance between an amount of embedded Al_2O_3 particles and coating porosity. The presence of Al_2O_3 particles in nickel matrix is shown on the example of Ni60Al_2O_3 coating (Fig. 4). In this case the most ceramic phase was built in the coating microstructure, however it showed the biggest porosity. Moreover an efficiency of the deposition process in this case was significantly smaller than others, the coating had different thickness in different places visible in the cross section observations. It was also confirmed by a presence of peaks belonging to substrate (Al phase) on the diffraction pattern (Fig. 5).

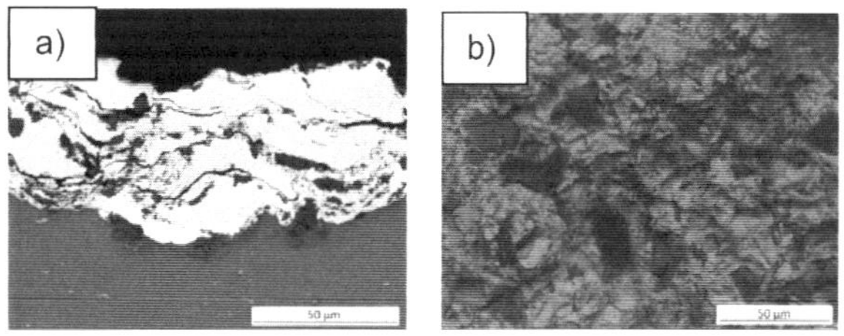

Fig. 3. Ni60Al_2O_3 coating: a) cross section, b) surface morphology.

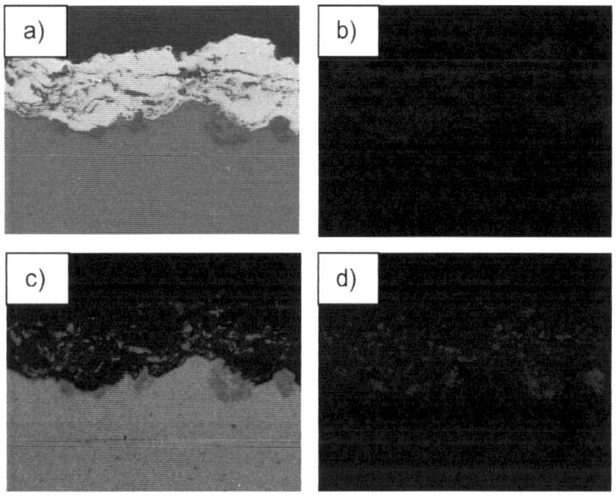

Fig. 4. The qualitative analysis of of Ni60Al_2O_3 cold sprayed coating: a) microstructure, b) Ni, c) Al, d) O_2.

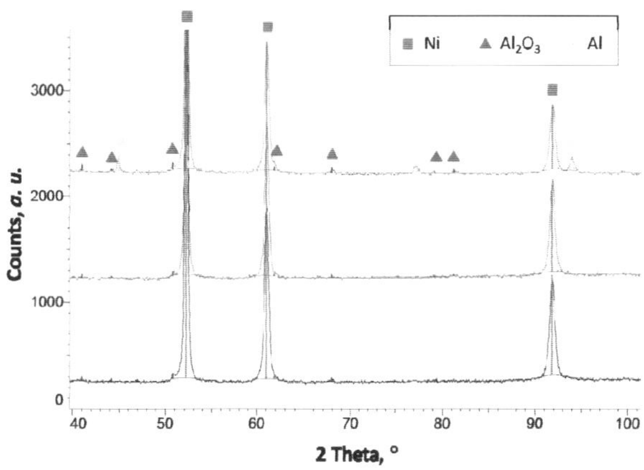

Fig. 5. XRD patterns of Ni-Al_2O_3 coatings cold sprayed from different mixtures of Ni and Al_2O_3 powders.

Based on the morphology analysis of the sprayed coatings it can be state all coatings were characterized by a high surface roughness, which is the result of both a broad particle size distribution of powders used, and the different effects of ceramic particles of various sizes on the sprayed surface. The parameters of surface geometry of cold sprayed coating are presented in Table 2. The highest arithmetic mean of the surface height Sa of Ni20Al_2O_3 sprayed coating was nearly two times higher than that showed for the Ni coating, S_a = 12.4 µm (Fig. 6). Moreover, the increase of Al_2O_3 particles in the blend caused further significant almost 50% growth of coating roughness (Fig. 10). A similar dependence was observed for the mean squared surface height, Sq. A negative value of asymmetry of the surface, Ssk indicated that both surfaces were flattened and the occurring peaks were rounded. Kurtosis Sku was responsible for steep irregularities and defects. A kurtosis of approximately 3 indicated that the distribution of the profile ordinate corresponded to a normal distribution the mean value. In the case of the maximum peak Sp, the maximum valley Sv and the maximum height Sz were nearly twice higher than in the case of cold sprayed Ni coating. It is due of the participation of Al_2O_3 grains which after the impact cause significant roughening of surface of Ni matrix.

Table 2. Surface geometry parameters according to ISO 25178.

Parameter	Surface of coating, µm		
	Ni20Al_2O_3	Ni40Al_2O_3	Ni60Al_2O_3
S_a	22.2	30.1	29.0
S_q	27.8	36.8	35.6
S_{sk}	-0.1	-0.24	0.04
S_{ku}	3.0	2.6	2.5
S_p	91.8	95.4	100.2
S_v	108.3	113.7	107.6
S_z	200.1	209.1	207.8

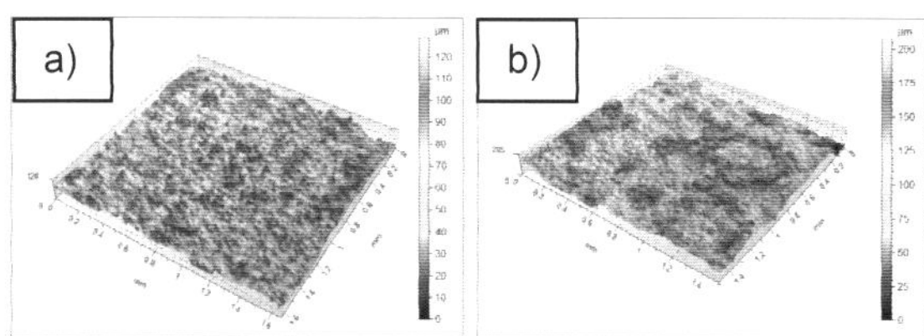

Fig. 6. Surface geometry of cold sprayed coating: a) Ni, b) Ni60Al_2O_3.

The Olivier and Pharr methodology was applied to measure hardness and Young's modulus of the investigated microstructures. Results were obtained on the base of analyses using a square grid of 49 nanoindentation tests on the sample of coatings The distributions of the nanomechanical properties were as then plotted as histogram distributions and maps of surface in which each hardness and Young's modulus result was showed on the charts with the same dimensions and locations as in the investigated areas. Histograms and probability distributions of the hardness and Young's modulus of the Ni40Al_2O_3 coating is shown in Fig. 7.

Nanoindentation tests showed a distribution of the mechanical properties (hardness and Young's modulus) which are related to the different areas (Ni matrix with Al_2O_3 particles) of composite coatings.

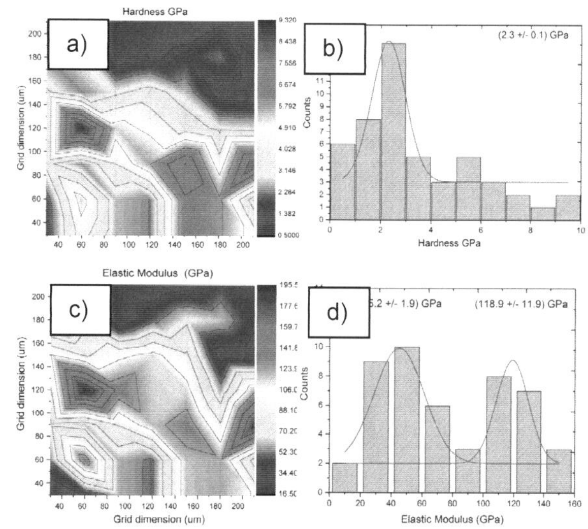

Fig. 7. Distribution of Ni40Al_2O_3 coating: a) hardness map, b) hardness histograms and probability, c) Young's modulus map, d) Young's modulus histograms and probability.

4 Conclusion

Increase of Al_2O_3 grains in the blend with Ni powder led to the increase of ceramic content in the composite coating. Efficiency of the deposition process decrease with growth of ceramic content in sprayed mixture of powders. The parameters of surface geometry of cold sprayed composite coating are over two times higher than that showed for the Ni coating. Mechanical properties (hardness and Young's modulus) were related to the different areas of composite coatings.

5 Literature

[1] W-Y Li, C. Zhang, H. Liao, J. Li, C. Coddet, Characterizations of cold-sprayed Nickel-Alumina composite coating with relatively large Nickel-coated Alumina powder, Surf Coat Tech 202 (2008) 4855-4860.

[2] A. Papyrin, Cold spray technology, Adv. Mater. Process 159 (2001) 49-51.

[3] R.C. Dykhuizen, M.F. Smith, D.L. Gilmore, R.A. Neiser, X. Jiang, S. Sampath, Impact of high velocity cold spray particles, J Therm Spray Technol 8 (1999) 559-564.

[4] J. Vlcek, L. Gimeno, H. Huber, E. Lugscheider, A systematic approach to material eligibility for the cold-spray process, J Thermal Spray Tech 14 (2005) 125-133.

[5] M. Grujicic, J.R. Saylor, D.E. Beasley, W.S. DeRosset, D. Helfritch, Computational analysis of the interfacial bonding between feed-powder particles and the substrate in the cold-gas dynamic-spray process Appl. Surf. Sci. 219 (2003) 211-227.

[6] R.S. Lima, J. Karthikeyan, C.M. Kay, J. Lindemann, C.C. Berndt, Microstructural characteristics of cold-sprayed nanostructured WC–Co coatings, Thin Solid Films 416 (2002) 129-135.

[7] T.H.V. Steenkiste, J.R. Smith, R.E. Teets, Aluminum coatings via kinetic spray with relatively large powder particles, Surf. Coat. Tech. 154 (2002) 237-252.

[8] T.H.V. Steenkiste, J.R. Smith, R.E. Teets, J.J. Moleski, D.W. Gorkiewicz, R.P. Tison, D.R. Marantz, K.A. Kowalsky, W.L. Riggs II, P.H.Zajchowski, B. Pilsner, R.C. McCune, K.J. Barnett, Kinetic spray coatings, Surf. Coat. Tech. 111 (1999) 62-71.

[9] T. Schmidt, F. Gärtner, H. Assadi, H. Kreye, Development of a generalized parameter window for cold spray deposition, Acta Mater 54 (3) (2006) 729-742.

Evaluation of High Temperature Oxidation Behavior of a Cold Spray Ni-20Cr Nano-Structured Coating

Manoj Kumar[1], Harpreet Singh[2*]
[1,2)] School of Mechanical, Materials & Energy Engineering, Indian Institute of Technology Ropar, Rupnagar, Punjab, India
* e-mail: harpreetsingh@iitrpr.ac.in, Tel.: +91- 9855709052

Narinder Singh[3]
[3)] Department of Chemistry, Indian Institute of Technology Ropar, Rupnagar, Punjab, India

S. V. Joshi[4], N. M. Chavan[5], S. Kumar[6]
[4,5,6)] International Advanced Research Centre for Powder Metallurgy & New Materials (ARCI), Hyderabad, India.

Abstract

In the current investigation, a nanostructured Ni-20Cr coating was formulated on SAE-213-T22 steel by cold-spraying. The Ni-20Cr coating powder was prepared by the ball milling route. The oxidation behavior of the coating was investigated under cyclic conditions at 900° C for 50 cycles. The kinetics of oxidation was established using weight change measurements for the uncoated and the coated boiler steel. The oxidized samples were characterized using X-ray Diffraction (XRD), Scanning Electron Microscopy/Energy Dispersive Spectroscopy (SEM/EDS), and X-ray mapping analyses. The nanostructured coating was found to have 2.2-times higher microhardness in comparison with the microstructured Ni-20Cr coating. The coating was successful in reducing the oxidation rate of the steel in terms of weight gain. The coating reduced the weight gain of the steel by 68%. The coating was found to have oxidation-protective oxides such as NiO, Cr_2O_3 in its oxide scale and was found to be spallation-free.

Introduction

India is a country which has abundance stocks of coal for next couple of decades, that iswhy it is used as the most important source of energy generation in power plants. However, the high-temperature degradation of the commonly used iron based boiler tubes, in a coal fired boilers,in India is a serious issue due to hot corrosion and erosion. This phenomenon of erosion-corrosion (E-C) continuously reduces the thickness of wall and finally puncture it which results in the leakage of the steam, leading to boiler downtime. This puncturing of the boiler tubes due to E-C is responsible for downtime of boilers generally by 50-75% of their total arrest time [1-2]. In the light of said facts the area of E-C protection of boiler tubes needs to be addressed on-priority [3-6]. Thermal spraying has achieved high reputation to develop E-C resistant coatings, due to its versatility and possibility for on-site applications in comparison with other alternatives of surface modification [7-12]. Thermal spray techniques generally involve the partially/fully melting of feed stock powders before deposition,however cold-spray process works well below the melting point of feedstock powders [13]. This avoidsin-flight oxidation of coating powders during deposition process.In cold spraying, a gas such as He, N_2 or air is used as a powder carrier gas. The powder particles are accelerated through a de Laval nozzle with the help of the carrier gas. Usually the carrier gas gets accelerated to supersonic velocities which further help to propel the powder particles at high velocities by momentum transfer principle. These particles subsequently impinge on the substrate surface and deform plastically and get deposited on the surface in the form of interlocked splats [14]. Furthermore, some portable CS systems have recently been developed for on-site applications, which is another attractive feature from engineering point of view. Since, CS is a solid state process; therefore it is considered as promising thermal spray process to develop nano-structured coatings with low oxidation content.

Nanocrystalline materials are classifiedbya physical dimension less than 100 nm [15-16]. The coatings developed from these nanostructured materials can offer several better properties such as hardness, ductility, corrosion and wear resistance in comparison with their conventional micron-sized counterparts [17-19]. It is learnt from the literature that the nano-structured coating materials are usually synthesized by mechanical milling of powders [20-24]. It is established that the nickel-based alloys possess several useful properties, such as wear, erosion and corrosion resistance, and good thermal conductivity. Moreover, nickel-based alloys, having high chromium content, are widely used corrosion resistant coating materials, since they offer good resistance to oxidation/corrosion [25-26]. Thermal spraying is one of the most suitable processes to deposit such alloy powders. Thermal sprayed coatings can provide suitable protection to the underlying substrate by selective oxidation of chromium to produce a protective oxide layer at high temperatures [27]. These coatings are known to form Cr_2O_3 protective oxide scales when exposed to high temperature. Cr_2O_3 is a thermodynamically stable phase upto very high temperature due to its very high melting point and forms a dense, continuous and adherent layer that inhibits interaction of oxygen with the underlying coating/substrate [28-29].

In view of the above facts, cold-spray process was used to deposit a pre-synthesized Ni-20Cr nanocrystalline powder on T22 boiler steel. In the present investigation, an attempt has been made to ascertain the high temperature oxidation behavior of this locally developed novel Ni-20Cr alloy powder at 900ºC for 50 cycles. The outcome of the present investigation shallbe useful to explore the use of this new coating material in boiler industry.

Experimental Details

The substrate material SAE 213-T22 (hereafter refered as T22) selected in the present study, is the most commonly used boiler tube material in Indian boiler industry. The chemical composition (wt%) of steel is Carbon 0.15, Manganese 0.3-0.6, Silicon 0.5, Sulphur 0.03, Phosphorous 0.03, Chromium 1.9-2.6, Molybdenum 0.87-1.13 and balance iron. The substrate material was procured from Cheema Boilers (Pvt.) Limited, Kurali (Punjab) India in the form of tubes. Specimens (20 mm × 15 mm × 5 mm in dimensions) were cut from T22 boiler steel and polished using standard metallurgical procedure. The steel samples were grit blasted with silicon carbide of grit 250 prior to the deposition of the coatings [30].

The nanocrystalline powder was synthesized by blending three types of powders Ni-powder-74 μm (N1), Ni nano-particles-67 nm (N2), Cr nano-particles-60 nm (N3) in composition (wt%) of 64%N1-16%N2-20%N3 using planetary ball mill. The detailed procedure has been reported elsewhere [31].The average particle size of Ni-20Cr alloy powder obtained after 20 hours of milling was 11 μm with a crystallite size of 10 nm [31].The cold spraying facility was extended by International Advanced Research Centre for Powder Metallurgy and New Materials (ARCI), Hyderabad, India. The details about the geometry of the nozzle and the parameters used in cold spray process have been reported elsewhere [30].

X-ray diffraction (XRD) analysis of the coated samples was done using an XRD machine (PANalytical,X'Pert-Pro, Netherland). Scanning electron microscopy/energy dispersive spectroscopy (SEM/EDS) (JEOL, JSM-6610LV, Japan) was done to study the surface and cross-sectional morphology/composition of the coating. Detailed procedure of characterization of as-sprayed coatings has already been reported elsewhere [32]. To investigate the cross-sectional micro-hardness of the investigated coating, a microhardness tester (Wilson, 402MVD, Illinois) was used at a load of 100 g with a dwell period of 10 s. Ten indentations (each) were performed at a distance of 50 μm, 100 μm, 150 μm, 200 μm and 250 μm from the coating-substrate interface in coating area along the lines parallel to substrate material. Average value was reported at the respective distances from the coating-substrate interface. Subsequently these average values of microhardness were plotted versus distance from the interface.

Cyclic isothermal air oxidation studies were carried out in a silicon carbide tube furnace (TGIS, Roorkee, India) fitted with a proportional-integral-derivative (PID) controller calibrated to ±1°C. Test specimens were exposed for 50 cycles each of which consisted of 1 h heating at 900°C followed by 20 min cooling at room temperature. The aim of the cyclic loading was to create accelerated conditions for the oxidation testing. The details of the oxidation study have been reported elsewhere [32]. The kinetics of the oxidation was established using this weight-change data. After the exposure for 50 cycles, the oxidized samples were characterized by XRD and SEM/Energy dispersive spectrometry (EDS) for the surface as well as cross-sectional analysis as per the procedure mentioned in the preceding section.

Results and Discussion

Microstructural Charactersiation

The XRD analysis of coating confirmed the formation of Ni as the primary phase, as shown in the inset of Fig. 1. This shows that oxidation of feedstock powder has not taken place during the coating process. The grain size of the coating was calculated using Scherrer's formula and was found to be 14 nm [33]. This confirms the formation of nano-structured coatings, which was further confirmed by TEM analysis (not shown here). A SEM micrograph of the as-sprayed steel with EDS analysis at various points is shown in Fig. 1. The coating is found to have a dense rock-like morphology with interlocked splats. It is perceptible that this type of microstructure is a result of appropriate coalescence of nano-sized powder particles. The coalescence is so strong that individual powder particles have lost their identity to get packed properly to give rocky appearance. The coating thickness was found to be 250 ± 10 μm measured from cross-sectional image as shown in Fig. 2.

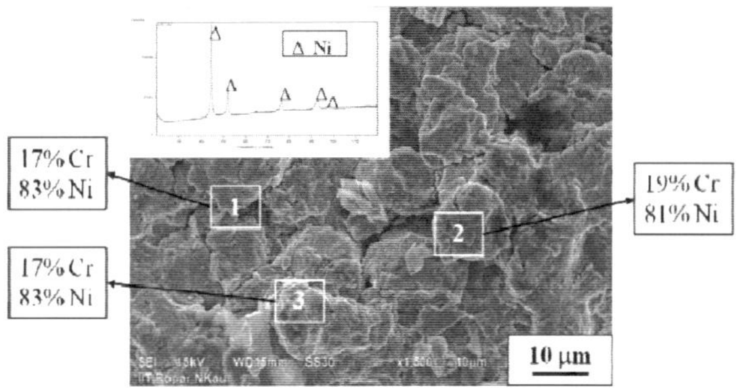

Figure1:SEM/EDS analysis of cold-sprayed nano-structured Ni-20Cr coating on T22 boiler steel along with XRD diffraction pattern (inset).

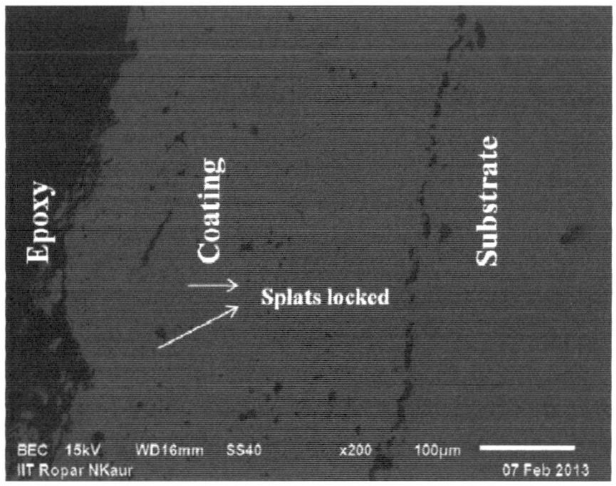

Figure 2: BSEI image of cross section of cold-sprayed Ni-20Cr coating on T22 boiler steel

Mechanical Charcterisation

Cross-sectionalmicrohardness vs distance from interface plot for the nanostructured Ni-20Cr coating is shown in Fig. 3. The measured average microhardness of investigated coating was found to be 539 Hv, while the average microhardness of steel is 210Hv. Further it is found that the microhardness of the investigated coating is double incomparison to micron-size Ni-20Cr cold-sprayed coating, on the same steel [34]. The improved value of microhardness of the nanostructured coating may be attributed to the refinement in grain size of the feedstock powder due to mechanical milling and further preservance of the grain size of the powder during deposition with cold spray process [19].

Cyclic Air-Oxidation Study

Weight gain data with respect to elapsed time for the uncoated and the coated T22 boiler steel is reported in Fig. 4. The total weight gain in case of uncoated T22 steel was found to be 193.67 mg/cm^2 while in case coated steel it was 63.15 mg/cm^2, which shows that the coating is successful to reduce the oxidation rate of the steel. The coating has reduced the weight gain by 67%, which is a substantial contribution.Further the plots between (weight change/area)^2 versus number of oxidation cycles (not shown here) were plotted, which revealed that bare and coated steel samples, by and large, showed conformance with the parabolic rate law, especially the coated steel, for which deviations were very meager. This indicated that the coating has shown a tendency to act as diffusion barrier to the oxidizing species. The parabolic rate constants for the bare and coated boiler steel, calculated based on 50 cycles' exposure data, are 2212x 10^{-10} g^2/cm^4/s and 192x 10^{-10} g^2/cm^4/s respectively. These values clearly indicate that the coating is successful in reducing the oxidation rate of the steel.

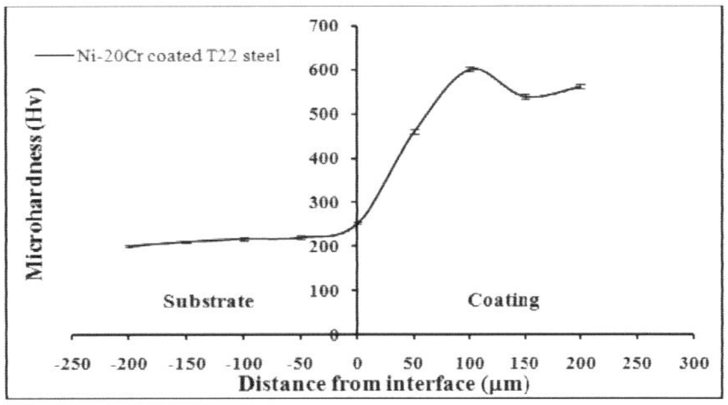

Figure3:Microhardness profiles of cold sprayed Ni-20Cr nanostructured coating on T22 steel.

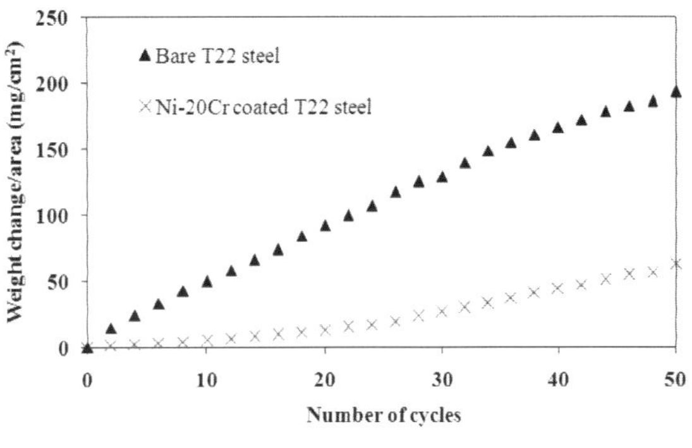

Figure 4: Weight change per unit area vs. number of cycles plot for uncoated and cold spray coated T22 boiler steel subjected to cyclic oxidation in air at 900°C for 50 cycles.

The XRD analysis of bare and coated samples subjected to cyclic isothermal oxidation in air at 900°C for 50 cycles is shown in Fig. 5. It is obvious from Fig. 5a,b that the oxide scales of the bare steel contain Fe_2O_3 as the main phase. On the other hand, the analysis of Ni-20Cr coated steels indicates the presence of NiO and Cr_2O_3 as the main phases (Fig. 5b), along with the nascent Ni in the oxide scale.

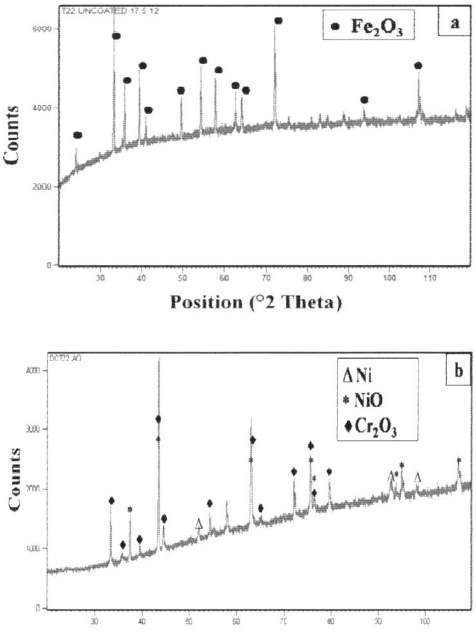

Figure 5: XRD profiles of bare and cold-spray coated T22 boiler steel subjected to cyclic air oxidation at 900°C for 50 cycles (a) bare T22 (b) Ni-20Cr coated T22 steel.

SEM/EDS analysis of the bare and the coated steel samples subjected to cyclic air oxidation at 900°C for 50 cycles is shown in Fig. 6. The SEM analysis of the bare T22 steel (Fig. 6a) indicates the presence of a granular scale [35], whereas the EDS analysis reveals that the scale is rich in Fe and O throughout the composition of the scale. This indicates the possibility of formation of Fe_2O_3 phase in the oxide scale, as has also been revealed by XRD analysis. A similar analysis of air-oxidized Ni-20Cr coated T22 steel shown in Fig. 6b respectively, reveals a scale with a typical pyramidal, crystalline structure, which is a typical NiO scale [36]. The EDS analysis indicates a significant concentration of Ni and O in the oxide scale along with significant amounts of Cr, which predicts the possibility of formation of oxides of Ni and/or Cr, as has been evaluated by XRD analysis also.

SEM micrograph and elemental variation across the cross-section of the bare and the coated T22 boiler steel subjected to cyclic oxidation in air at 900°C for 50 cycles are shown in Fig. 7. The oxide scale (Fig. 7a) looks to be loosely bounded to the uncoated steel by the end of exposure. The scale is found to have cracks both parallel, as well as, perpendicular to the surface of the sample. EDS analysis reveals the presence of P, Si, Fe, Cr and oxygen throughout the scale. The existence of significant amounts of oxygen throughout the scale thickness predicts the possibility of formation of oxides of these elements. The top portion of the scale (line 1) mainly contains Fe and O, which indicates the formation of oxides of Fe. The concentrations of Fe and O at point 2 are observed to be similar to that at point 1. Whereas the amount of Fe is lower at point 3, with an increase in the amounts of O and C.

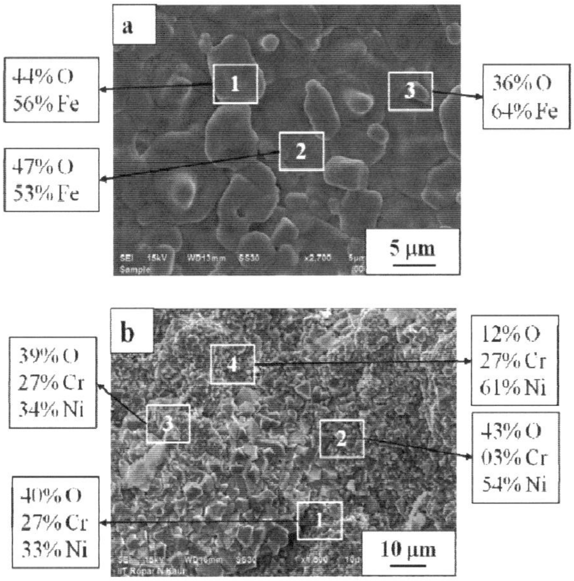

Figure 6:Surface morphology with EDS analysis for uncoated and cold-sprayed boiler steels subjected to cyclic oxidation in air at 900°C for 50 cycles (a) bare T22 steel (b) Ni-20Cr coated T22 steel.

Cross-sectional SEM analysis (Fig. 7b) of the Ni-20Cr coated steel revealed the presence of small cracks near the oxide-substrate interface through which the Fe might have diffused. The EDS analysis reveals the presence of significant amount of Ni, Cr and O at the top surface of scale (Line 1 to Line 2). At the interface (Line 3) of scale and substrate, presence of large amount of Fe and O is revealed, along with a small amount of Cr. The EDS measurements indicate that the dark colored channels in the oxide scale are most likely Fe_2O_3.

The bare T22 boiler steel sample has been found to have a loose oxide scale with the presence of some cracks along the interface, which may be due to formation of non-protective Fe_2O_3 phase, having weaker cohesion among its layers, however the scale on the coated steel was found to be intact. Furthermore, the Ni-20Cr coating has reduced the weight gain by 67%, therefore indicates that the coating can induce the oxidation resistance in the steel, which may be attributed to the presence of NiO and Cr_2O_3 phases in the oxide scale of the coatings, which are able to provide higher oxidation protection to the materials. The phenomenon of formation of NiO and Cr_2O_3 (Fig. 7b) due to selective oxidation of Ni-20Cr alloy was explained by Stott [29] according to which the fast growing NiO phase overgrows the Cr_2O_3 phase and a transient oxide scale layer of essential NiO develops, incorporating the other transient oxide phases. The Cr_2O_3 phase is said to be thermodynamical stable phase upto very high temperatures due to its very high melting point and forms a thick, constant and adherent layers that reduces contact of oxygen (O) with the underlying coating/substrate [29].

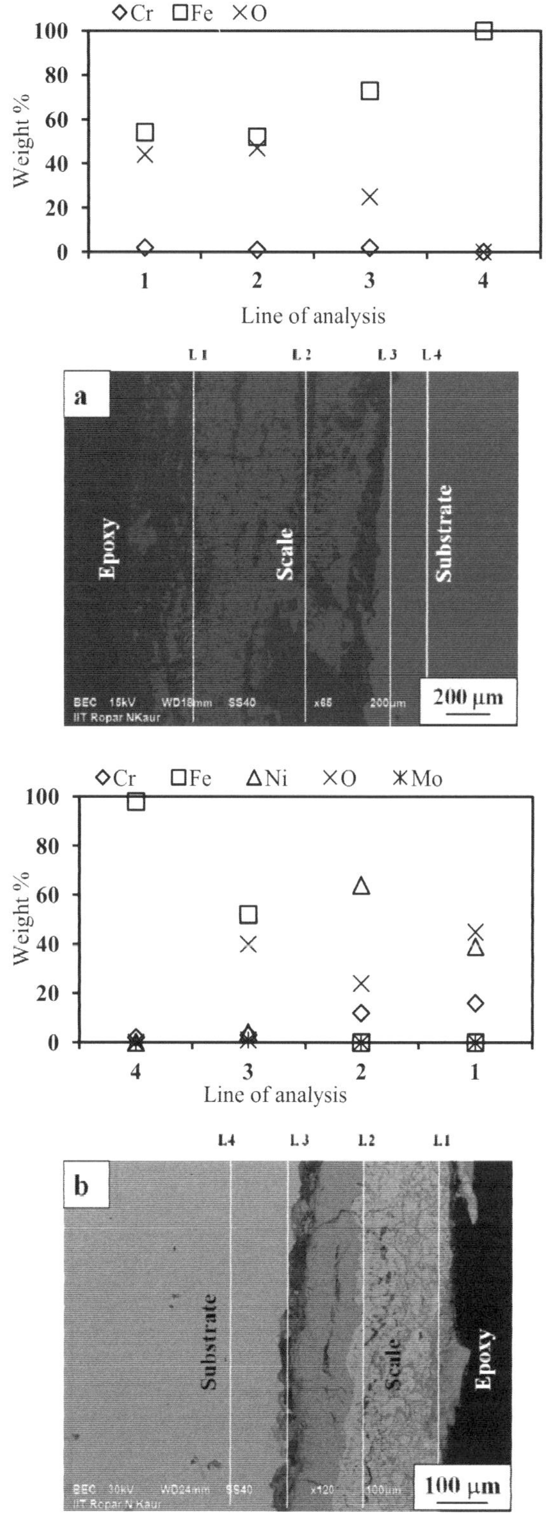

Figure 7: Oxide scale morphology and variation of elemental composition across the cross-section of (a) bare T22 boiler steel (b) Ni-20Cr coated steel subjected to cyclic oxidation in air at 900°C for 50 cycles.

Conclusions

1. Ni-20Cr nano-structured powder was successfully deposited on T22boiler steel by the cold-spray technique at the given process parameters. The thickness of the coating obtained was 250 ± 10 μm.
2. The coatings were found to have lamellar splat-like microstructure with intact coating-substrate interface. The investigated coating on T22 steel was found to have 539 Hv average microhardness, which is double than the micron-size Ni-20Cr cold-sprayed coating (247Hv).
3. The investigated Ni-20Cr coating was successful to reduce the weight gain of T22 steel by 67%. Thus offered better oxidation resistance, which might be attributed to the presence of protective NiO and Cr_2O_3 phases in their oxide scales.

Acknowledgement

Harpreet Singh et al. thankfully acknowledge the research grant from Department of Science and Technology, New Delhi, India (SR/S3/MERC/023/2010) for carrying out the R & D work on "Surface Engineering to Control Erosion-Corrosion of Steam Generating Plants by Nano-particle Coatings".

References

[1] Koch, G. H. *et al.*, Historic Congressional Study: Corrosion Costs and Preventive Strategies in the United States, Supplement to Material Performance (2002), pp. 1-11.

[2] Hidalgo, V. H., Varela, F. J. B., Rico, E. F., "Erosion Wear and Mechanical Properties of Plasma-Sprayed Nickel and Iron-Based Coatings Subjected to Service Conditions in Boilers," *Tribology International*, Vol. 30, No. 9 (1997), pp. 641-649.

[3] Stringer, J., Wright, I. G., "Current Limitations of High-Temperature Alloys in Practical Applications," *Oxidation of Metals*, Vol. 44, No. 1 (1995), pp. 265-308.

[4] Simms, N. J., Environmental Degradation of Boiler Components, In: Oakey JE, editor. Power Plant Life Management and Performance Improvement, Woodhead Publishing, (2011), pp. 145-179.

[5] Simms, N. J., Kilgallon, P. J., Oakey, J. E., "Fireside Issues in Advanced Power Generation Systems," *Energy Materials: Material Science and Engineerring for Energy Systems*, Vol. 2, No. 3 (2007), pp. 154-160.

[6] Simms, N. J., Kilgallon, P. J., Oakey, J. E., "Degradation of Heat Exchanger Materials under Biomass Co-firing Conditions," *Materials at High Temperature,* Vol. 24, No. 4 (2007), pp. 333-342.

[7] McCune, R. C. *et al.,* An exploration of the cold gas-dynamic spray method for several materials systmes, in *Advances in Thermal Spray Science and Technology*, Berndt, C. C. *et al.,* (Eds), ASM International, Materials Park, OH (1995), pp. 1-5.

[8] Young, M. A., "Health Hazards of Electroplating," *Journal of Occupational and Environmental Medicine*, Vol. **7** (1965), pp. 348-352.

[9] Choy, K. L., Chemical Vapor Deposition of Coatings, *Progress in Material Science*, Vol. 48 (2003), pp. 57-170.

[10] Tsai, W. T., Chen, H. P., Hsien,W. Y., "AReview of Uses, Environmental Hazards and Recovery/Recycle Technologies of Perfluorocarbons (PFCs) Emissions from the Semiconductor Manufacturing Processes," *Journal of Loss Prevention in Process Industries*, Vol. 15 (2002), pp. 65-75.

[11] Koropchak, J. A., Roychowdhury, S. B., "Evidence for Aerosol Ionic Redistribution within Aerosols Produced by Chrome Electroplating," *Environmental and Science Technology*, Vol. 24(1990), pp. 1861-1863.

[12] Pawlowski, L., The Science and Engineering of Thermal Spray Coatings, John Wiley and Sons, Ltd., (2008).

[13] Papyrin, A. *et al.,*Cold Spray Technology,Elsevier (2006).

[14] Lima, R. S., Karthikeyan, J., Kay, C. M., Lindemann, J., Berndt, C. C., "Microstructural Characteristics of Cold-Sprayed Nanostructured WC–Co Coatings," *Thin Solid Films*, Vol. 416, No. 1-2 (2002), pp. 129-135.

[15] Seigel, R. W., "Nanostructured Materials -Mind Over Matter,"*Nanostructured Material*, Vol. 3, No. 1-6 (1993), pp. 1-18.

[16] Cao, G., Nanostructures ,Nanomaterials - Synthesis. Properties and Applications, Imperial College Press (London 2004).

[17] Murty, B. S. *et al.,*Textbook of Nanoscience and Nanotechnology, Universities Press (India 2012).

[18] Marple, B. R., Voyer, J., Bisson J. F., Moreau, C., "Thermal Spraying of Nanostructured Cermet Coatings,"*Journal of Material Processing and Technology*, Vol.117, No.3 (2001) pp. 418-423.

[19] Witkin D. B., Lavernia, E. J., "Synthesis and Mechanical Behavior of Nanostructured Materials via Cryomilling, "*Progress in Materials Science*, Vol. 1 (2006) pp. 1-60.

[20] C. Suryanarayana, "Nanocrystalline Materials," *International Material Reviews*, Vol. 40, No. 2 (1995) pp. 41-64.

[21] Datta, M. K., Pabi, S. K., Murty, B. S., "Phase Fields of Nickel Silicides Obtained by Mechanical Alloying in the Nanocrystalline State," *Journal of Applied Physics*, Vol. 87, No. 12 (2000) pp. 8393-8400.

[22] Nayak, S. S., Pabi, S. K., Murty, B. S., "High Strength Nanocrystalline $L1_2$-Al_3 (Ti,Zr) Intermetallic Synthesized by Mechanical Alloying," *Intermetallics*, Vol. 15 (2007) pp. 26-33.

[23] Varalakshmi, S., Kamaraj, M., Murty, B. S., "Processing and Properties of Nanocrystalline CuNiCoZnAlTi High Entropy Alloys by Mechanical Alloying,"*Material Science and Engineering A*, Vol. 527 (2010) pp. 1027-1030.

[24] Pabi, S. K., Murty, B. S., "Synthesis of Nanocrystalline Alloys and Intermetallics by Mechanical Alloying,"*Bulletin of Material Science*, Vol. 19, No. 6 (1996) pp. 939-956.

[25] He, J., Julie, M., Schoenung, J. M., "Nanocrystalline Ni Coatings Strengthened with Ultrafine Particles"

Metallurgical and Materials Transaction A, Vol. 34 (2003) pp. 673-683.

[26] Zhao, S., Xie, X., Smith, G. D., Patel, S. J., "The Corrosion of INCONEL Alloy 740 in Simulated Environments for Pulverized Coal-fired Boiler" *Material Chemistry and Physics*, Vol. 90 (2005) pp. 275-281.

[27] Matthews, S. J., "Erosion-Corrosion of Cr_3C_2-NiCr High Velocity Thermal Spray Coatings", Ph.D. Thesis, The University of Auckland (2004).

[28] Wang, B., Gong, J., Wang, A. Y., Sun, C., Huang, R. F., Wen, L. S., "Oxidation Behaviour of NiCrAlY Coatings on Ni-Based Superalloy," *Surface and Coatings Technology,* Vol. 149 (2002) pp. 70-75.

[29] Stott, F. H., Principles of Growth and Adhesion of Oxide Scales, in: E. Lang (Ed.), The Role of Active Elements in the Oxidation Behavior of High Temperature Metals and Alloys, Lang, E. Ed., Elsevier Applied Science (London, 1998).

[30] Kumar, M., Singh, H., Singh, N., "Synthesis and Deposition of Ni-20Cr Powder Using Cold Spraying,"*Surface Engineering*, Vol.29, No. 6 (2013), pp. 419-426.

[31] Kumar, M., *et al.,* "Improving High Temperature Oxidation Resistance of a Boiler Steel by a Novel Ni-20Cr Nanostructured Coating" *Proc. of International Corrosion Prevention Symposium for Research Scholars (Corsym 2014)*, IIT Bombay, India, 2014.

[32] Kumar, M., Singh, H., Singh, N., "Production of Nano-Crystalline Ni-20Cr Coatings for High-temperature Applications,"*Journal of Thermal Spray Technology,* Vol. 23, No. 4 (2013) pp. 692-707.

[33] Klug H. P. *et al*, X-Ray Diffraction Procedures, Second ed., Wiley (NY, USA, 1974).

[34] Bala, N., Singh, H., Prakash, S., "Characterization and High-Temperature Oxidation Behavior of Cold-Sprayed Ni-20Cr and Ni-50Cr Coatings on Boiler Steels,"*Metallurgical and Materials Transaction A,* Vol. 42A (2011) pp. 3399-3416.

[35] Kumar., M., *et al.,* "High Temperature Oxidation Behavior of a Novel Ni-20Cr Alloy Powder Coating on T22 Steel," *Proc. of International Thermal Spray Conference (ITSC-2013),*Busan, Republic of Korea, 2013, pp.120-125.

[36] Khoi, N. N., Smeltzer, W. W., Embury,J. D., "Growth and Structure of Nickel Oxide on Nickel Crystal Faces," *Journal of Electrochemical Society*, (1975) pp. 1495-1503.

Deposition of Thick Copper Cold Spray Coatings on Steel for Used Nuclear Fuel Repository using Nitrogen and Pulsed Waterjet Surface Preparation

R. Fernández, D. MacDonald, A. Nastic, B. Jodoin, M. Vijay, A. Tieu/ Ottawa, CAN

Canada's nuclear reactors have produced nearly 2 million used fuel bundles since the 1960s and this number is expected to double over the next 40 years. For the long term containment of the used nuclear fuel, a multiple-barrier system has been proposed, which relies on the corrosion protection of the steel container using a layer of copper. Cold spray is being considered as an option to produce these copper coatings due to its high deposition rate and low oxidation during deposition. Previous investigations on the deposition of thick (4 mm) copper coatings using only nitrogen as process gas on carbon steel substrates have continually resulted in coating delamination. One approach to solve this delamination issue has been to first apply an initial copper layer to the steel substrate using helium as process gas to ensure coating adhesion. The current work aims at exploring the possibility of limiting the process gas to nitrogen by creating a substrate surface that will enhance mechanical anchoring through the use of the pulsed waterjet technique prior to the deposition of copper. The results of this work show that through the use of this surface preparation method, adhesion strength can be drastically increased and thick copper coatings with superior bond strength can be deposited using only nitrogen.

1 INTRODUCTION

Canada's nuclear reactors have produced almost 2 million used fuel bundles over the last decades and this number is expected to double over the next 40 years [1]. Only very recently has Canada decided the course of action to be taken to deal with the long term containment of the spent nuclear fuel. Assessment of long-term safe and secure management options has led to the development of an adaptive phased management approach and is consistent with the programs that have been developed in many other countries such as Sweden, the United Kingdom, Finland and France [4]–[7]. This approach includes centralized containment and isolation of used nuclear fuel in a deep underground repository, in a suitable rock formation. The deep geological repository is a multiple-barrier system designed to safely contain and isolate used nuclear fuel over the long term.

The purpose of this sealing system is to provide a physical and chemical environment that will limit the mobility of any radionuclides that may be released under potential conditions that deviate from the expected normal evolution of the repository [4], [5], [8], [9]. The used fuel container is one of the principal engineered barriers of this multi-barrier deep geological repository. The most important sought features of the used fuel container design are corrosion resistance, mechanical strength, capacity, and compatibility with surrounding sealing materials such as bentonite clay [4], [5]. The reference design of the used fuel container employs an outer corrosion-resistant material and an inner supporting material.

Similarly to existing used fuel container used in Europe, the inner vessel is made of steel [6], [7]. What differentiates the Canadian design is the thin corrosion-resistant outer shell that is proposed. As opposed to existing designs, this shell is thinner (4 mm thick) and consists of an integrated copper coating as opposed to a thick sleeved copper shell found in other designs [2], [4], [5].

A few options are being considered to produce the copper coating and one of them has been the use of cold spray. The process can achieve high deposition rate in combination with high deposition efficiency [10], [11] while preventing oxidation of copper during deposition. As such, cold spray offers an interesting engineering solution for the production of the expected large quantity of containers. Thorough studies have shown that the production of 4 mm thick copper coatings on steel is a challenging task. While the cold spray process generally allows building thick copper coatings, the adhesion strength becomes an issue when spraying on steel [12]. As such, even when using various copper powders (size range, manufacturers, oxide content, etc.) and exploring a wide range of spray parameters, coating delamination occurs when building up the copper coating [13], [14]. An innovative approach to solve this problem has been to first apply an initial copper layer to the steel substrate using helium as the process gas, followed by the build-up of the coating to the required thickness using nitrogen [14]. This two-step approach has resulted in extremely well-bonded thick copper coatings as the use of helium instead of nitrogen drastically increases the gas velocity and consequently the particle impact velocity, thus promoting more intense deformation of both the particles and the substrates. This results in an intricate bonding interface that promotes mechanical anchoring which translates into increased adhesion strength [14].

While this two-step approach provides a method to use cold spray to produce the required coatings with superior bond strength, the use of helium is not ideal as it is considerably more expensive than nitrogen and can be considered as a sensitive non-renewable resource. The current work aims at exploring the possibility of limiting the cold spray process gas to nitrogen by developing a substrate surface preparation that will promote mechanical anchoring during the cold spray of copper when using nitrogen to ensure proper coating adhesion. While grit blasting is extensively used as a surface preparation process in thermal spray processes [10], [11], [14]–[17], the surface topology/roughness level it can produce on steel parts is typically limited to roughness (R_a) below 10 µm. This roughness level might not be large enough to sufficiently enhance copper particle adhesion, and grit blasting presents the potential for adhesion strength to be reduced due to grit embedment. As such, it is proposed in this work to use the novel forced pulsed waterjet (FPWJ) technique to produce a surface with a larger R_a.

The forced pulsed waterjet (FPWJ) surface preparation technique relies on a high pressure and high frequency water jet to alter the surface of a substrate. This process generates an intricate surface topology and produces various roughness levels based on the process parameters used [17]–[19]. More details on FPWJ are given elsewhere [17]. Using this method, the FPWJ substrate preparation process will artificially induce intricate features on the substrate surface prior to spraying. It is expected that these features will promote mechanical anchoring of the particles, resulting in high levels of adhesion with the use of nitrogen.

2 EXPERIMENTAL PROCEDURES

2.1 Materials

The substrate material used in this study was low carbon steel A519 Grade 70. The substrates were machined into 25.4 mm diameter cylinders in accordance to the ASTM C-633 adhesion strength standard. 38mm x 38mm x 25.4mm flat substrates were machined for coating analysis. To ensure consistency between sprays, the substrates were thermally insulated during deposition using a high temperature polymer (polybenzimidazole) holder. The powder used in this investigation was a gas atomized copper with an average particle diameter of 8µm (CU-159, Praxair surface technologies).

2.2 Substrate Preparation

The substrate samples underwent a variety of surface preparations. Eight different substrate surface roughnesses were produced; two using conventional substrate preparation processes (polishing and grit blasting) and six using the FPWJ process. Samples were cleaned with acetone and dried with air immediately before cold spraying. The polished substrate was finished with a 3 µm particle fluid suspension. The grit basted substrate was prepared with an 850 µm grit (grit designation 20) using 400kPa air pressure at 45° angle. The parameters used for substrate prepared with FPWJ process can be seen in Table 1.

The six different roughness intensities generated using the FPWJ were achieved by holding the parameters identified in the top of Table 1 constant, while only varying the nozzle velocity between 160 mm/s and 600 mm/s. The velocity values were selected in order to obtain samples with evenly distributed roughness.

Table 1. Parameters for FPWJ Substrate Preparation

Parameters	Value
Stand Off	2
Water Pressure	69 MPa
Nozzle Diameter	1.4 mm
Pulse Frequency	40,000 Hz
Step Index	0.1 mm/pass
Sample	**FPWJ Nozzle Velocity (mm/s)**
FPWJ-1	600
FPWJ-2	500
FPWJ-3	420
FPWJ-4	350
FPWJ-5	290
FPWJ-6	160

2.3 Surfaces and Samples Characterisation

3D digital imaging by depth composition was performed with a digital microscope (VHX – 1000, Keyence Corporation, Osaka, Japan) on the substrates previous to the coating deposition. These images were then used to perform roughness measurements. Three 3D images where taken and analysed per sample and three samples were used per substrate preparation method. The coatings samples were cross-sectioned and analysed using an SEM (EVO MA-10, Carl Zeiss AG, Oberkochen, Germany).

2.4 Cold Spray Deposition

The cold spray system used to produce the coatings for this work is the commercially available EP Series SST Low Pressure Cold Spray System (Centerline (Windsor) Ltd., Windsor, Ontario, Canada). The system consists of a 15 kW heater with a maximum gas temperature of 500°C and a maximum gas pressure of 3.8 MPa. The De Laval nozzle used for this work has a throat diameter of 2 mm and a diverging section length and exit diameter of 120 mm and 6.6 mm respectively. The copper feedstock powder was fed using a commercially available powder feeder from Thermach Inc. (Model AT-1200HP). All substrates were coated using the spray parameters given in Table 2. After deposition, the samples successfully produced for adhesion strength measurements were machined to a nominal thickness of 600 µm.

Table 2. Cold Spray Parameters Used in this Study

Parameter	Value
Gas Temperature	500 °C
Gas Pressure	3.8 MPa
Gas Nature	Nitrogen
Traverse Speed	25 mm/s
Feed Rate	35 g/min
Standoff Distance	10 mm

2.5 Adhesion Strength Testing

The coatings adhesion strength was determined in compliance with the standard ASTM C-633. A thermally cured elastomeric adhesive (FM-1000) with a tensile strength of 76.9 MPa was used to glue the coating to a counter cylinder. The samples were placed in a vertical fixture and then cured in a 175°C oven for two hours to ensure the adhesive had properly cured. The testing was performed using a universal tensile testing machine. Three samples of each surface preparation were tested and the mean and standard deviation of each preparation were computed.

3 RESULTS AND DISCUSSION

3.1 Topology and Substrate Roughness

Topology images comparing the surfaces obtained prior to cold spraying are shown in Fig.1. Fehler! Verweisquelle konnte nicht gefunden werden.The figure exhibits the topology of a polished surface (Fig. 1(A)), a grit blasted surface (Fig. 1(B), and one FPWJ prepared surfaces (Fig.1C). Fig. 1(A), 1(B) and 1(C) were taken at the same magnification for comparison purposes and show a noticeable difference in topology between the surface preparation methods. FPWJ samples show intricate features that may act as anchoring points. The observable features are in the order of hundreds of microns. It should be noted that micron and sub-micron size features are also present in the FPWJ samples as can be seen in Fig. 1(D).

The roughness values obtained for each sample preparation method are found in Table 3. They are evenly spaced, ranging from less than a micron for the polished samples to values above 80 µm for the most aggressive FPWJ treatment.

Table 1: Roughness Values and Adhesion Strength for each Substrate Preparation

Sample	Roughness (µm)	Adhesion Strength (MPa)
Polished	0.7 ± 0.2	0.0 ± 0.0
Grit	5.2 ± 0.9	0.0 ± 0.0
FPWJ-1	16.4 ± 4.8	28.2 ± 7.2
FPWJ-2	32.0 ± 4.9	49.0 ± 5.8
FPWJ-3	44.9 ± 6.2	53.2 ± 4.3
FPWJ-4	58.0 ± 8.9	56.5 ± 2.5
FPWJ-5	63.3 ± 6.2	53.4 ± 5.3
FPWJ-6	83.0 ± 7.5	51.1 ± 6.3

3.2 Coating Characteristics and Adhesion Strength

No coatings were successfully adhered for the polished and grit blasted samples. This result is coherent with previous studies of this material couple where metallurgical bonding is expected to be low or

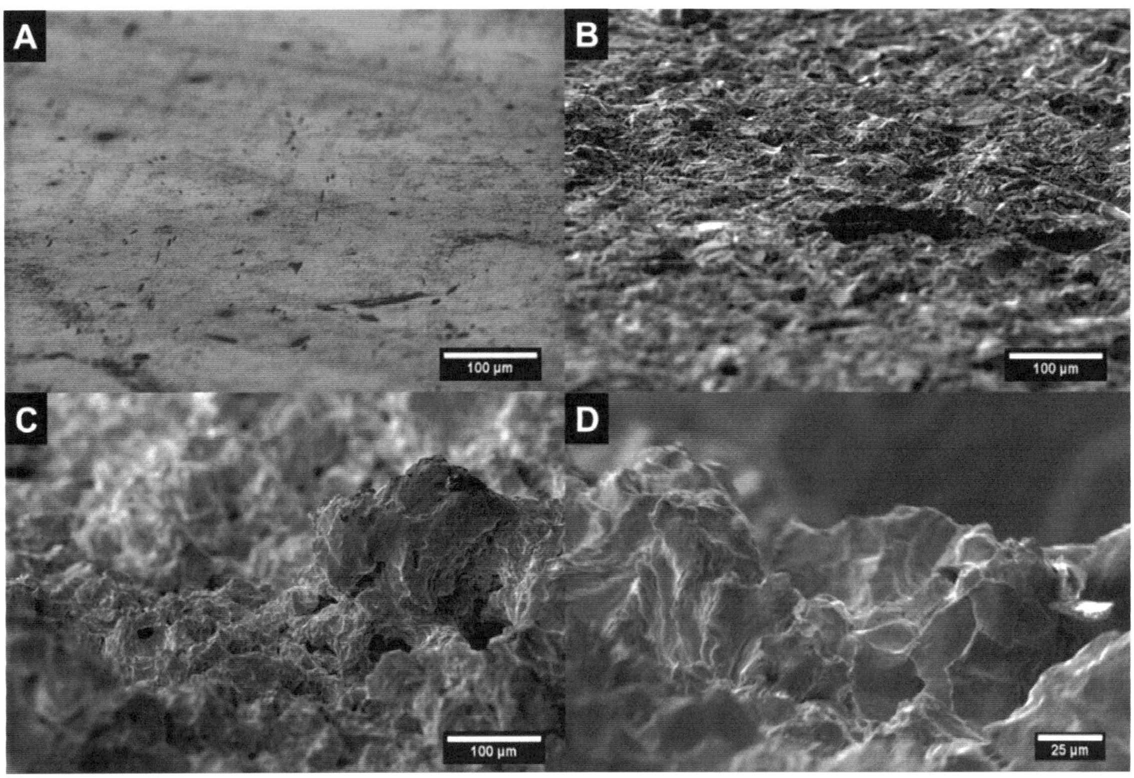

Fig. 1. Surface topology for different substrate preparations: (A) polished, (B) grit blasted, (C) FPWJ-2, and (D) FPWJ-2 details.

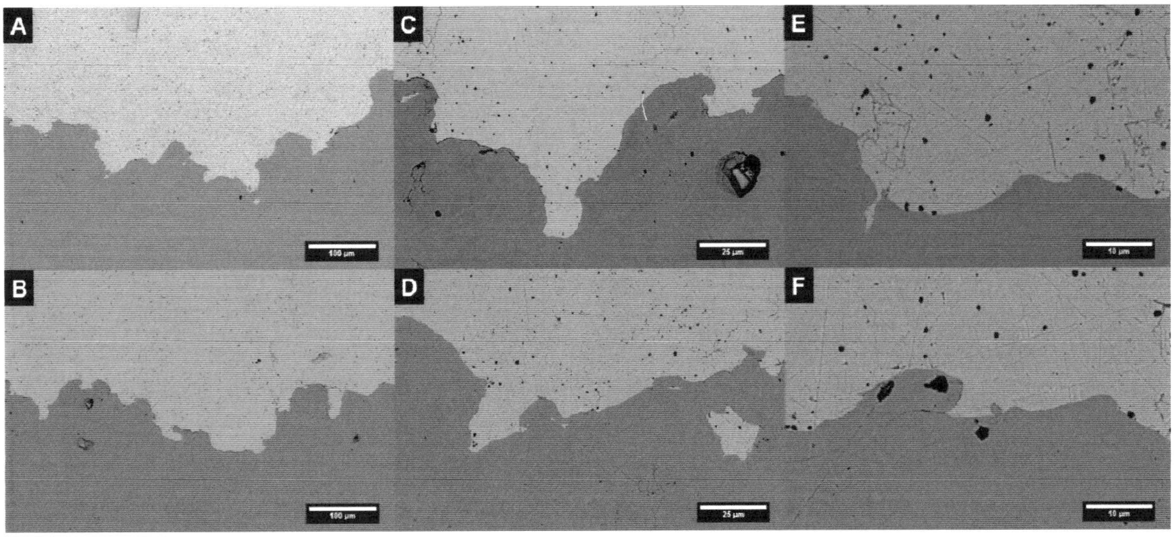

Fig. 2. Coating/substrate interfaces showing mechanical anchoring at different magnifications; (A) (C) (E) is FPWJ-2 and (B) (D) (F) is FPWJ-4.

null. During the deposition, thick coatings were built onto the substrate but they delaminated either during deposition when reaching a certain thickness or after deposition during cooling. The delaminated coatings showed a strong arc at the center suggesting the presence of compressive residual stresses. It is expected that the stress builds up with coating thickness, and it is thought that when a critical thickness is reached the stress at the interface overcomes the low adhesion strength of the coating, resulting in delamination.

For the deposition on FPWJ prepared substrates, all the coatings adhered to the substrates. This result alone is novel, as several researchers have focused on the deposition of copper on steel using nitrogen and delamination has always been an issue reported in each instances. This seems to be the first successful attempt to adhere a thick copper coating to steel using exclusively nitrogen and low spray parameters. The cross- sections of two of the coatings are shown in Fig. 2. It can be seen that the coatings are dense. The most noticeable characteristic of Fig. 2 is the degree of anchoring points between the coatings and the substrates. It is believed that it is through these anchoring points that the coatings were successfully built. It is clear that in these zones there is a strong interlocking of the materials. This is true at several magnifications, which gives an indication that this substrate preparation method produces macro and micro anchoring points. These points are expected to enhance the coating/substrate adhesion.

Adhesion testing in accordance with ASTM 633-C was performed for the samples; the results can be seen in Table 3 and in Fig. 3.

A general trend between roughness and adhesion strength can be identified. As the roughness increases it also increases the adhesion strength, reaching a plateau around 50 MPa at an R_a of 40 µm. Since the coatings sprayed on the polished and grit blasted substrates delaminated, their adhesion strength was considered zero. It is important to note that all samples failed in adhesion at the interface between the coatings and the substrates.

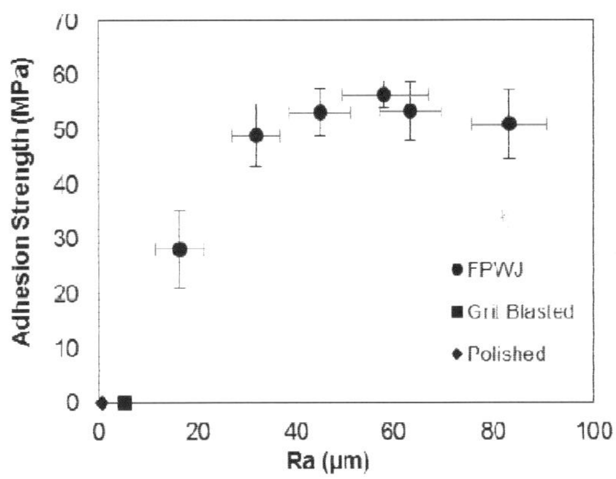

Fig. 3. Adhesion strength vs. substrate roughness.

It is believed that this material couple, copper and steel, are not very compatible in creating metallurgical bonding [20]–[23]. This is the reason that it hasn't been possible to obtain a coating using regular substrate preparation with cold spray and nitrogen. This low compatibility has been shown in other high strain processes [22], [23]. The result of this study suggests that the adhesion strength between copper and steel may rely solely on mechanical anchoring. The low roughness substrates lack the anchoring points and the adhesion strength to overcome the residual stresses that act to delaminate the coatings. However, at higher roughnesses mechanical anchoring becomes strong enough to sustain the residual stresses. The rougher the substrate, the more

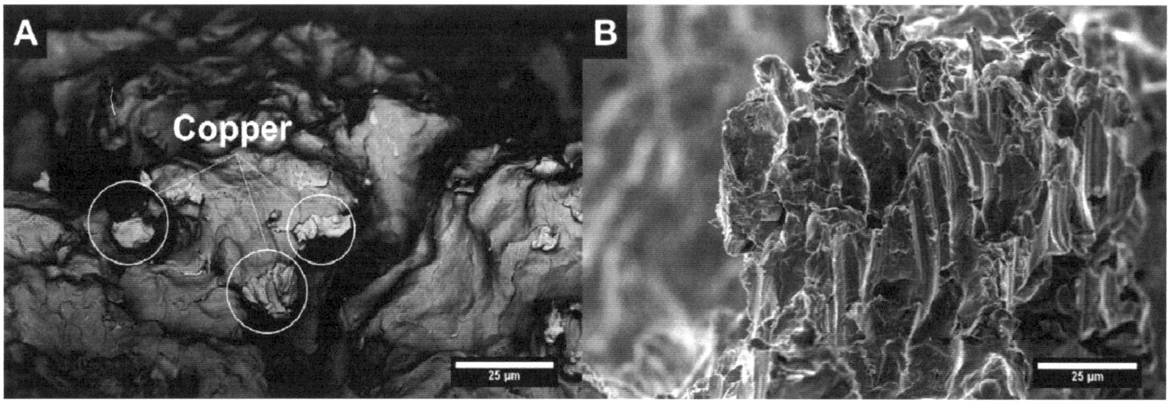

Fig. 4. Fracture surface topologies of FPWJ-2 of (A) substrate side, (B) coating side.

anchoring points are present to increase the mechanical adhesion.

This type of adhesion by anchoring points was verified upon inspection of the fractures surfaces after the adhesion tests, as shown in Fig 4. These figures show the fracture surfaces for the substrate and the coatings. In Fig. 4(A), the remnants of copper can be seen in the anchors of the steel. The copper side (Fig. 4(B)) shows distinct evidence of striping of the materials as it was pulled. These striation lines confirm that the bonding present at the interface is mechanical and that the bonding between particles is metallurgical.

Even though the results showed that the roughness promotes the mechanical bonding between coatings and substrates, the benefit of this anchoring point appears to have a limit, as is seen in the plateau of Fig.3.

The reason for this behaviour could be attributed to the method of roughness measurement. Ra may not be the most adequate method to characterize the surface in a FPWJ sample. Ra evaluates the roughness values based on the average height differences of peaks and valleys of the surface. Therefore, it is a vertical assessment of the surface. However, the number of anchoring points present in the surface is related with a planar dimension rather than a vertical one. As such, while two surfaces could have the same amount of anchoring points, one of them could be twice as deep giving a higher roughness but not necessarily more adhesion strength. It is possible that the process of FPWJ produces a certain number of anchoring points per area, and starting from the sample FPWJ-3 the substrate is already saturated with these anchoring points. This will result in similar adhesion of the coatings independently of the roughness measured on the surfaces.

This method of coating consolidation aligns with the one proposed by Maev et al. [13], [14] where helium was first used to create the adhesion between the copper and steel. Fig. 5 shows a superpossitoon of the interface obtained by Maev et al. with the one produces in this investigation. Although the topography is not exactly the same, the similarities between both interfaces is noticeable.

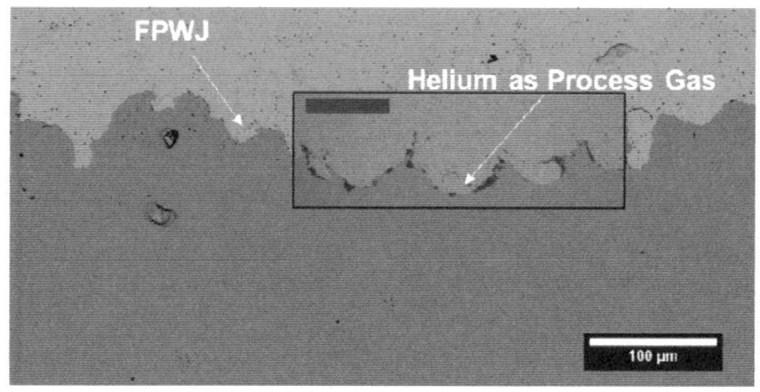

Fig. 5. Superposition of coating interfaces images prepared with FPWJ (main image) and with helium as process gas [14] (frame imaged).

4 CONCLUSIONS

This investigation focused on a technique to produce copper coatings on steel substrates using cold spray and solely nitrogen as the process gas. The study explored the use of FPWJ as a surface preparation method in order to overcome the delamination issues that occur with the use of cold spray and nitrogen with these materials. In order to accomplish this, samples were prepared by typical substrate preparation methods and FPWJ, resulting in a series of substrate roughness values. Samples were then sprayed to create dense and thick copper coatings and adhesion tests were performed. It was shown that while typical substrate preparation methods fail in producing an adhered coating, substrates prepared using FPWJ resulted in well adhered coatings independently of the substrate surface roughness obtained. It was presented that the increased adhesion strength is a result of anchoring features produced with FPWJ, these features interlocking with the incoming particles and allowing to form a strong mechanical bond.

5 REFERENCES

[1] M. Garamszeghy, “Nuclear Fuel Waste Projections in Canada - 2013 Update,” no. December, Nuclear Waste Management Organization, 2013.

[2] E. Irissou, D. P. P. Vo, P. Keech, and J. G. Legoux., “Cold sprayed corrosion protection coating for nuclear waste repository canister.,” in *Cold sprayed corrosion protection coating for nuclear waste repository canister.*, 2012.

[3] Nuclear Waste Management Organization, “Choosing a Way Forward,” *Backgrounder*, pp. 2–8, 2010.

[4] C. H. Boyle and S. A. Meguid, “Mechanical performance of integrally bonded copper coatings for the long term disposal of used nuclear fuel,” *Nucl. Eng. Des.*, vol. 293, pp. 403–412, Nov. 2015.

[5] P. G. Keech, P. Vo, S. Ramamurthy, J. Chen, R. Jacklin, and D. W. Shoesmith, “Design and development of copper coatings for long term storage of used nuclear fuel,” *Corros. Eng. Sci. Technol.*, vol. 49, no. 6, pp. 425–430, 2014.

[6] ANDRA, “No TitleEvaluation of the Feasibility of a Geological Repository in an Argillaceous Formation Meuse/Haute-Marne Site,” *Doss. 2005 Argile*, 2005.

[7] P. S. B. Hansson, J. Magnusson, “Underground Design Forsmark, Layout D2 – Layout and Construction Pla,” *SKB*, vol. Rapport R-, 2009.

[8] H.-J. Choi, M. Lee, and J. Y. Lee, “Application of a cold spray technique to the fabrication of a copper canister for the geological disposal of CANDU spent fuels,” *Nucl. Eng. Des.*, vol. 240, no. 10, pp. 2714–2720, Oct. 2010.

[9] S. K. Kim, M. S. Lee, H. J. Choi, J. W. Choi, and T.-W. Kwak, “Progress of a cost optimization for an HLW repository in Korea,” *Prog. Nucl. Energy*, vol. 51, no. 3, pp. 401–408, Apr. 2009.

[10] M. C. Spray, *Modern Cold Spray*. Cham: Springer International Publishing, 2015.

[11] A. N. Papyrin, *Cold Spray Technology*, vol. 1. Elsevier, 2007.

[12] R. G. Maev and E. Leshchinsky, “Low Pressure Gas Dynamic Spray : Shear Localization during Particle Shock Consolidation,” *Strain*, 2006.

[13] R. G. Maev, V. Leshchynsky, E. Strumban, D. Dzhurinskiy, J. Kocimski, and E. Maeva, “Structure and Mechanical Properties of Thick Copper Coating Made by Cold Spray,” *J. Therm. Spray Technol.*, Sep. 2015.

[14] E. . R.Gr. Maev, V.Leshchynsky, E.Strumban, D.Dzhurinskiy, “Influence of Grit Blasting on the Interface Roughness and Adhesion Strength of Cold Sprayed Copper Coatings,” in *International Thermal Spray Conference*, 2015, pp. 493–497.

[15] T. D. B. Jacobs, K. E. Ryan, P. L. Keating, D. S. Grierson, J. A. Lefever, K. T. Turner, J. A. Harrison, and R. W. Carpick, “The Effect of Atomic-Scale Roughness on the Adhesion of Nanoscale Asperities: A Combined Simulation and Experimental Investigation,” *Tribol. Lett.*, vol. 50, no. 1, pp. 81–93, Feb. 2013.

[16] A. F. Harris and A. Beevers, “The effects of grit-blasting on surface properties for adhesion,” *Int. J. Adhes. Adhes.*, vol. 19, no. 6, pp. 445–452, Dec. 1999.

[17] T. Samson, D. MacDonald, R. Fernández, and B. Jodoin, “Effect of Pulsed Waterjet Surface Preparation on the Adhesion Strength of Cold Gas Dynamic Sprayed Aluminum Coatings,” *J. Therm. Spray Technol.*, vol. 24, no. 6, pp. 984–993, Jul. 2015.

[18] M. M. Vijay, “US Patent - Apparatus and Method For Prepping A Surface Using a Coating Particle Entrained In A Pulsed Waterjet or Airjet,” 8,389,066 B2, 2013.

[19] M. M. Vijay, A. H. Tieu, W. Yan, and B. R. Daniels, “US Patent - Method And Apparatus For Prepping Surfaces With A High-Frequency Forced Pulsed Waterjet,” 8,550,873 B2, 2013.

[20] Z. Sun and J. C. Ion, “Laser welding of dissimilar metal combinations,” *J. Mater. Sci.*, vol. 30, no. 17, pp. 4205–4214, Sep. 1995.

[21] X. Yuan, K. Tang, Y. Deng, J. Luo, and G. Sheng, “Impulse pressuring diffusion bonding of a copper alloy to a stainless steel with/without a pure nickel interlayer,” *Mater. Des.*, vol. 52, pp. 359–366, Dec. 2013.

[22] G. R. Cowan and A. H. Holtzman, “Flow Configurations in Colliding Plates: Explosive Bonding,” *J. Appl. Phys.*, vol. 34, no. 4, p. 928, Jun. 1963.

[23] K. Raghukandan, “Analysis of the explosive cladding of cu–low carbon steel plates,” *J. Mater. Process. Technol.*, vol. 139, no. 1–3, pp. 573–577, Aug. 2003.

Optimization to fabrication method of Al-Ni composite for Shaped charge liner Applications

Seungtae Lee, Hyungkwon Park, Jaeick Kim, Changhee Lee*, See Jo Kim, and Seong Lee
Hanyang Univ., Haengdang 1-dong, Seongdong-gu, Seoul, Korea

Liners used in shaped charges (SC) must prepare strong penetration ability and explosive power. In preceding study, arrested reactive milling(ARM) and kinetic spray was utilized to enhance reactivity of reactive coating layer (i.e, Al-Ni) for SC liner. More investigation was requirded to detail fabrication conditions of ARM and kinetic spray. Microstructure, reactivity, deposition efficiency of these powders and coating layer were investigated to optimize the milling condition and heat treatment for powders. 2-1 coating layer had higher deposition efficiency than 2-3 coating layer. The SHS reaction behavior and reactivity of the ARMed Al-Ni layer had no considerable differences between 2-1 and 2-3 milling conditions. As a result, process conditions of 2-1 powder were more proper for SC liner.

1 Introduction

Shaped charge (SC) ammunition is a weapon that attacks enemy by penetrating a barrier and generating an explosion (Ref 1). The penetration ability and explosion performance of SC ammunition rely on the type of liner material. Here, penetration ability is a term that means how well the liner penetrates the barrier of target. Traditionally, Cu liner was used due to its high penetration ability. However, the explosive performance of Cu liner is relatively lower than other reactive materials, such as Al and Mg. In order to supplement the weak explosive power of Cu liner, it can be considered as a solution to produce reactive layer on the Cu liner

With regard to fabricating reactivity-enhancing coating layer (i.e., Al), the reactive materials must not react during the fabrication process. Kinetic spray was fabrication method to satisfy this reactivity issue. Kinetic spray process is a method to deposit feedstocks on target by accelerating particles to a supersonic velocity without any melting of the particles (Ref 2). Phase transformation and oxidation of the feedstock scarcely occur since the feedstock powder can be deposited under solid state (Ref 3, 4). A few studies have been conducted related to the fabrication of reactive layer on the Cu liner via kinetic spray process (Ref 5, 6). J. Won et al. (Ref 6) investigated the fabrication of an dense Al coating layer on Cu liner via kinetic spray process. Also, Byun et al. (Ref 7), Al- successfully fabricated Al-Ni composite layer on Cu liner using kinetic spray process, which is highly reactive self-propagating high-temperature synthesis (SHS) material (Ref 8). The reactivity was improved much more relative to the case of sole Al layer. However, considerable amount of Ni remained in Al-Ni composite layer even after the end of reaction between Al and Ni. Byun et al. (Ref 17), tried to resolve this problem by increasing the contact area between Al-Ni and decreasing the initiation temperature of SHS via arrested reactive milling. Arrested reactive milling (ARM) is one of the milling process, it is proper fabrication process for preservation of reactivity and heat valu. However, preceding study (Ref 17) did not investigate about detailed deposition efficiency and tendency of reactivity according to the various ARM conditions. Also, the problem that bonding between ARMed particles was degraded by remaining solid lubricant (Ref 18) should be solved.

In this study, an optimization of ARM condition was conducted to improve deposition efficiency and reactivity of ARMed Al-Ni powders in kinetic spray process. The reactivity, deposition efficiency, microstructure of ARMed powders and coating layer were investigated to secure the suitability for the SC liners.

2 Experimental Methods

Feedstock Characterization

Commercial Al (ALE16PB, Kojundo Co. Ltd) and Ni (INCO type 123, AOMETAL Co. Ltd.) with high purity (99.9%) were chosen as feedstock powder materials. **Figures 1(a-d)** show the particle shape of as-received Al and Ni powders. Al particle shows irregular rod shape and has the mean size of 22 μm. On the other hand, The shape of Ni particle was irregular and agglomerated form and it has the average size of around 12 μm. The Ni powder was chosen in the consideration of a similar trajectory of in-flight Al particles. The particle velocity of Al and Ni was numerically calculated using the following empirical equation (Ref 14).

$$V_p = \frac{1}{\frac{1}{M}\sqrt{\frac{M_{N_2}}{\gamma RT} + 0.85\sqrt{\frac{D}{x}\sqrt{\frac{\rho_p}{P_0}}}}}$$

Here V_p is the particle velocity, M is the Mach number of the nozzle, M_{N_2} is the molecular weight of N_2 gas, g is the specific heat of the gas, R is the gas constant, T is the gas temperature, D is the particle diameter, x is the axial position, q_p is the particle density, and P_0 is the stagnation pressure. Cu substrates (30 mm × 50mm× 3mm) were used and grit-blasting was used on Cu substrate prior to deposition.

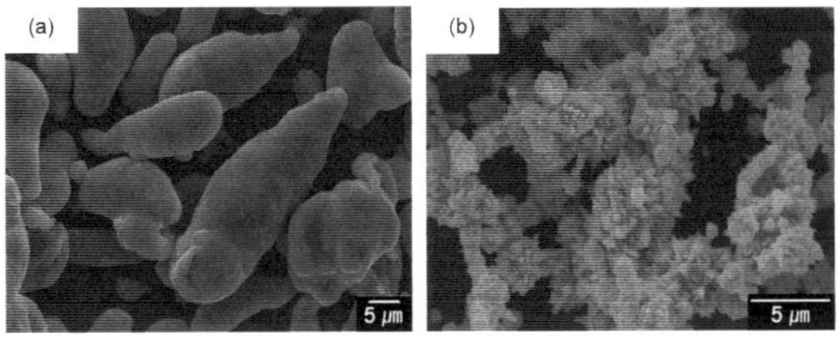

Fig. 1. (a) Aluminum powder morphology, (b) Nickel powder morphology

Arrested Reactive Milling

In order to prepare the Al-Ni composite powder for kinetic spraying, arrested reactive milling (ARM) method was performed using FRITSCH Planetary Mill Pulverisette 5. Al and Ni powders were mixed with the ratio of 1:1 to maximize reactivity and heat value (Ref 7). The powders were milled in stainless jars with SUS 316 balls (9.525Φ). Milling speed was fixed to 150 RPM, and ball and powder were input with the ratio of 5:1. Ar gas was input into the jar to avoid any reaction between Al and Ni. First, the milling was conducted for 1 and 2 hours (denoted as 1-0h and 2-0h). And then solid lubricant was additionally input to the 2-0h powder to decrease the particle size and milling was progressively carried out for 1 and 3 hours (denoted as 2-1h and 2-3h). Here, Stearic acid with 0.5 wt% was used as solid lubricant. Additionally, the case of solid lubricant input from the beginning stage was denoted as 0-1h.

Kinetic Spraying

To deposite ARMed Al-Ni composite powder on Cu substrate, kinetic spraying system (KINETIC 3000, CGT) was used with an anti-clogging polybenzimidazol nozzle. Process gas to accelerate the powder was used N_2 gas. Gas temperature and pressure was 400 oC, 27 bar with a sistance of 30 mm from the point of the nozzle to the substrate. The detail on kinetic spraying process conditions is summarized in **Table 1**.

Powder	Process gas	Gas temperature	Gas pressure	Stand-off distance	Transverse speed	Cycle
Blended 1:1 Al-Ni, ARMed Al-Ni	N_2	400 °C	27 bar	20 mm/s	30 mm	2

Table 1. Process parameter of the kinetic spray

Characterization of Powder and Coating

The microstructures of ARMed Al-Ni powders and Al-Ni coating layer were observed using optical microscope (OM, BX60M, OLYMPOS) and scanning electron microscope (SEM, JCM-5700, JEOL). to exam the chemical reaction between Al and Ni, X-ray diffractometer (XRD, New D8-Advance, Bruker-AXS). Particle size distribution of each powder was analysed via laser scattering particle size analyser (HELOS). The reactivity of the powders and coatings were estimated using differential scanning calorimetry (DSC, DSC404 F1, NETZSCH) under Air atmosphere with heating rates of 50 K/min.

3 Results and Discussion

Preparations of ARMed Al-Ni powder

Figures 2 and 3 show cross-sections and particle size distribution of Al-Ni powders. White and gray areas represents Ni and Al, respectively. As milling process was proceeded, Al and Ni powders were progressively agglomerated, as shown in **Figs. 2(a) and 2(b)**. Although two powders were mixed well to enhance reactivity, the composite particle was too large (d_{50} = 87.61 μm for 1-0h and 139.23 μm for 2-0h) to be used for kinetic spraying, as confirmed in **Fig. 3**. On the other hand, in the case of solid lubricant input from the beginning (0-1h), the powders of Al and Ni was simply fragmented to 13.3 μm without any mixing reaction between them, as shown in **Fig. 2(c) and Fig. 3**. Thus, it is known that stearic acid reduced the effect of cold welding between particles and thereby inhibits agglomeration and bonding (Ref 9, 13). However, in order to increase reactivity and heat value about the exothermic SHS reaction, The contact area between Al-Ni must be increased. Consequently, it is figured out that to form composite coating layer, it is necessary to implement agglomeration firstly and then fragmentation process with solid lubricant.

To manufacture the suitable particle size, the milling process was set to conduct without stearic acid 2 hour. After then, solid lubricant was additionally input to the mixed powder and milling was sequently performed for 1 and 3 hours. **Figures 2 (d) and (e)** indicate the cross-sectional OM micrograph of ARMed 2-1h and 2-3h, respectively. Powder morphology was irregularly spherical or rod shape. Al and Ni agglomerated and mixed each other effectively. In **Fig. 3**, it is known that the size distribution of 2-1h and 2-3h powders was significantly reduced from 2-0h as much as it could be deposited by kinetic spraying. In case of 2-3h powder, because it underwent more plastic deformation and fracture energy, it is thought that its morphology was longer and thinner than 2-1h powder.

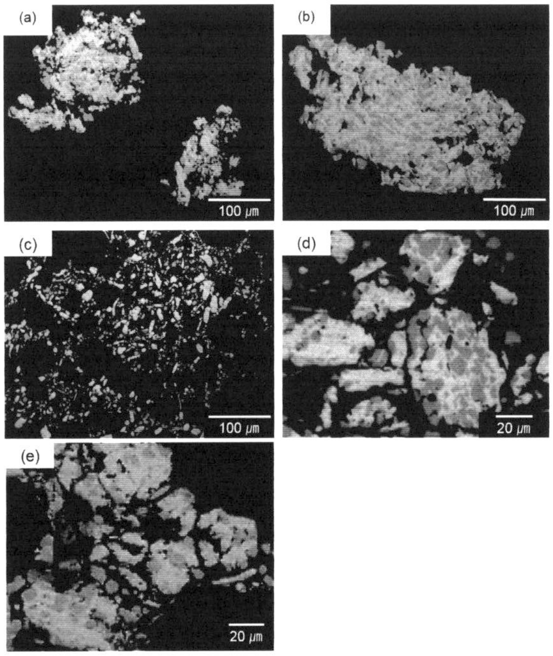

Fig. 2. Cross-sectioned OM image of ARMed Al-Ni powder: (a) 1-0h, (b) 2-0h, (c) 0-1h, (d) 2-1h, and (e) 2-3h

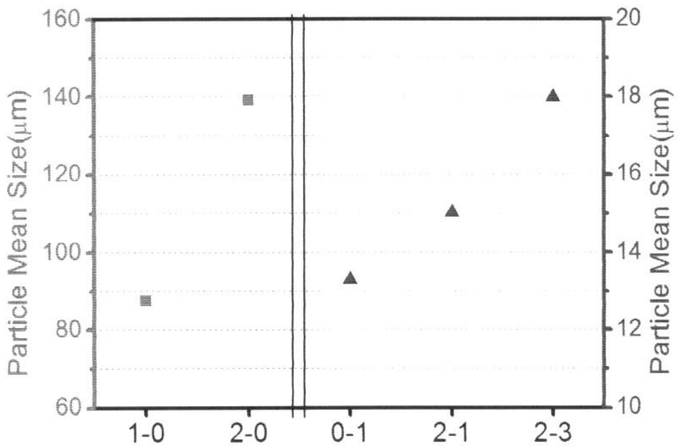

Fig. 3. Distribution graph of median value of powder size

<u>**Deposition by the Kinetic Spray Process**</u>

In case of as-prepared 2-1h and 2-3h, the powders were not successfully deposited by kinetic spraying. It was thought that the remained stearic acid in the powders disturbed the particle bonding during kinetic spray process. To confirm this hypothesis, it was attempted that only Al powder with 0.5 wt% stearic acid was tried to deposition. Consequently, the coating layer was almost not created similar to that of 2-1h and 2-3h powders. In this respect, 2-1h and 2-3h powders were heated to eliminat remained stearic acid at 573 K during 1 hour considering boliing point of stearic acid was 505K. In addition, to exam whether the Al-Ni intermetallic coupounds were formed or not during the heat treatment, the powders were analysed by XRD. **Figures 4(a)** indicate the XRD result of ARMed Al-Ni powder and it was compared with that of non-milled powder. In the graph, heated ARMed 2-1h, and 2-3h powders were marked as 2-1 and 2-3 ARMed HT, and non-heated 2-1h and 2-3h were marked as 2-1 and 2-3 ARMed. As shown, ARMed 2-1 and 2-3 HT powder were unchanged after heat treatmen and all of peaks were the same. Thus, it can be concluded that there was not any reaction during milling and post heat-treatment.

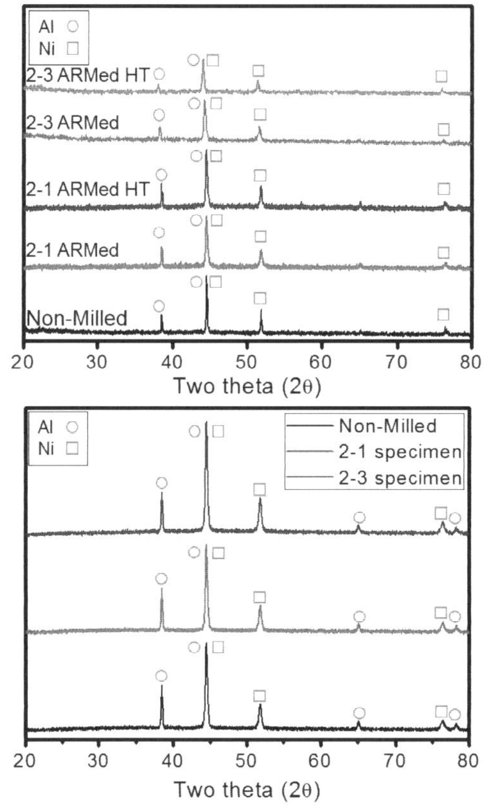

Fig. 4. XRD results of (a) ARMed Al-Ni powder and Non-Milled powder and (b) ARMed specimen and Non-Milled specimen

Finally, heated ARMed 2-1h and 2-3h powder was successfully deposited on Cu substrate by kinetic spray process. **Figure 5** shows cross-sectioned OM micrographs of kinetic sprayed simply blended, ARMed 2-1h and 2-3h coatings. In all specimens, porosity was hardly observed. That is, very dense coating layer was formed. However, the microstructures fabricated usnig ARMed were significantly different from simply blended powders, as confirmed in **Figs. 5(a), 5(c), and 5(e)**. Definitely, two phases of Al and Ni were dispersed more effectively with smaller areas in the case of ARMed powders, as shown in **Figs. 5(b), 5(d), and 5(f)**. Thus, it can be expected that the coating layer reacts with higher exothermic energy.

Figure 6 represents the thickness of each coating layer. The thickness of simply blended, ARMed 2-1h, and ARMed 2-3h were estimated to 438.2, 906.4, and 879.4 μm, respectively. It is certain that the deposition efficiency of ARMed composite powders was significantly higher than that of simply blended powder. Such increased deposition efficiency was attributed to the enhanced plastic deformability of ARMed composite powder, compared to the pure Ni powder. Originally, the Ni particle velocity under given process condition was ~650 m/s, which is lower than the critical velocity of Ni (~700 m/s) from preceding study (Ref 15). Thus, Ni particles are hard to be deposited successfully under used process condition. In contrast, the plastic deformability of ARMed composite particles was improved by bonded soft Al parts, so it is natural that the critical velocity of ARMed particles was reduced relative to the pure Ni. Such decrease of critical velocity might contribute to the improved deposition efficiency of ARMed composite powder.

To confirm the formation of intermetallic compound generated during kinetic spraying, the phases of three kinds of coating layers were analysed by XRD, as shown in **Fig. 4(b)**. Compared to peak positions of powder materials, new phase peak was not observed, and thus, it is concluded that the reaction of intermetallic compouds didn't occur.

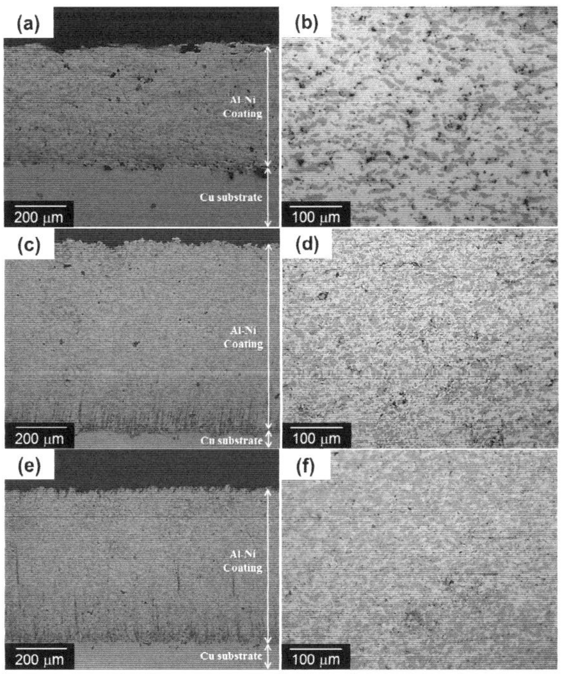

Fig. 5. Cross-sentioned OM micrographs: (a) simply blended, (c) 2-1h, and (e) 2-3h coating layer, and (b), (d),(f) higher magnification of (a),(c),(f).

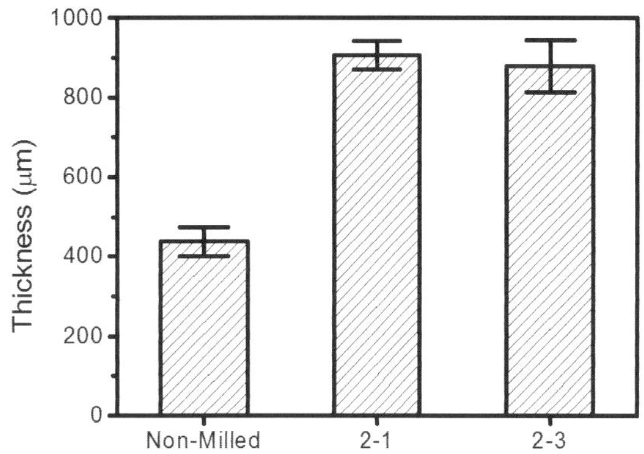

Fig. 6. Average thickness of 2-1, 2-3 specimens.

SHS reaction of powders and coating

Figure. 7(a) shows DSC thermogram of 2-1h and 2-3h powders. Exothermic reactions were assited by SHS reactions between Al and Ni phases at at 620 °C and it was resulted from the creation of intermetallic compounds (NiAl3, Ni2Al3, NiAl, and Ni3Al) (Ref 16). Since the SHS reactions are related to melting of Al, It is considered that mechanical activaton of the ARMed Al-Ni powders decreased SHS reaction to lower temperature. Exothermic heat values around peak between 2-1h and 2-3h powders were calculated to 32.5, 31.9 kJ/g. Consequently, Two powders had similar exothermic heat values.

On the other hand, the DSC results of 2-1h and 2-3h coating layers are shown in **Fig. 7(b)**. Two curves of 2-1h and 2-3h coatings are similar patterns, and thus, the specimens had similar exothermic heat values at peak. SHS reaction of coating peaks appeared at lower temperature (590 °C) than that of powders. It is considered that mechanical activation during kinetic spraying affect the SHS reaction temperature. Moreover, exothermic heat values of coating specimens were higher than that of powders. It is regarded that contact area between Al-Ni increased during kinetic spraying and this rise leaded to increase of SHS reactions.

4. Conclusions

For optimizing the reactivity and deposition quality of Al-Ni composite coating layer, various milling condition of ARM and fixed process parameter of kinetic spraying was proceeded. The major conclusions are summarized as follows;

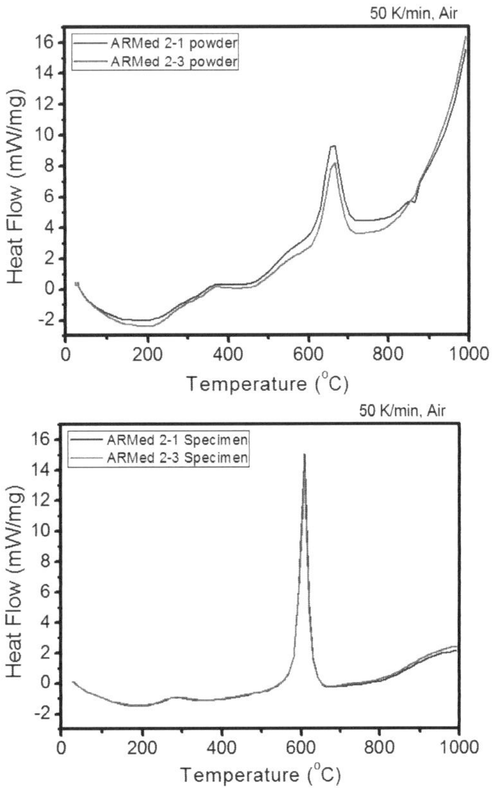

Fig. 7. DSC thermograms of (a) ARMed 2-1h and 2-3h powders and (b) ARMed 2-1h and 2-3h coatings.

(1) Although ARMed Al-Ni powders was heated up to 573K, significant amount of intermetallic compounds was not created.

(2) The deposition efficiency of 2-1 specimen that related to thickness of coating layer was higher than 2-3 specimen since 2-1 received lowder strain by ARM.

(3) Exothermic heat value of 2-1 powder had no meaningful difference compared to 2-3 powder. 2-1, 2-3 specimen had similar tendency in exothermic heat value.

As a results, It was demonstrated that 2-1 powder is more appropriate than 2-1 powder with regard to the reactivity and deposition efficiency. It is expected that optimizations with fabrication method can be utilized in application of SC liner.

5. Literature

[1] R.T. Ziemba, R.W. McLay, and J.A. Siewert, Shaped Charge Projectile System, US Patent 4,567,829, 4 Feb, 1986.

[2] M.F. Smith, J.E. Brockmann, R.C. Dykhuizen, D.L. Gilmore, R.A. Neiser, and T.J. Roemer, Cold Spray Direct Fabrication: High Rate Solid State Material Consolidation, MRS Symp. Proc., 542 (1999), pp 65/76.

[3] S. Yoon, H.J. Kim, and C. Lee, Deposition Behavior of Bulk Amorphous NiTiZrSiSn According to the Kinetic and Thermal Energy Levels in the Kinetic Spraying Process, Surf. Coat. Technol., 200 (2006), pp 6022/9.

[4] X.-T. Luo and C.-J. Li, Thermal Stability of Microstructure and Hardness of Cold-sprayed cBN/NiCrAl Nanocomposite Coating, J. Therm. Spray Technol., 21 (2012), Issue 3-4, pp 578/85.

[5] T. Langan, M.A. Riley, and W.M. Buchta, Reactive Shaped Charges Comprising Thermal Sprayed Reactive Components, US Patent 7,658,148 B2, 9 Feb, 2010.

[6] J. Won, G. Bae, K. Kang, C. Lee, S.J. Kim, K.-A. Lee, and S. Lee, Bonding, Reactivity and Mechanical Properties of the Kinetic Sprayed Deposition of Al for a Thermally Activated Reactive Cu Liner, J. Therm. Spray. Technol., 23 (2014), Issue 5, pp 818/26.

[7] G. Byun, J. Kim, C. Lee, S. Kim, and S. Lee, Property Evaluation of Kinetic Sprayed Al-Ni Composite Coatings, J. Weld. Join., 32 (2014), Issue 5, pp 72/9.

[8] J. Subrahmanyam and M. Vijayakumar, Review Self-propagating High-temperature Synthesis, J. Mater. Sci., 27 (1992), pp 6249/73.

[9] C. Suryanarayana, Mechanical Alloying and Milling, Prog. Mater.Sci., 46 (2001), pp 1/184.

[10] Jeremiah D.E. White, R.V. Reeves, S.F. Son, and A.S. Mukasyan, Thermal Explosion in Al-Ni System: Influence of Mechanical Activation, J. Phys. Chem. A, 113 (2009), pp 13541/7.

[11] S.G. Vadchenko, O.D. Boyarchenko, N.F. Shkodich, and A.S. Rogachev, Thermal Explosion in Various Ni-Al Systems: Effect of Mechanical Activation, Int. J. Self-Propag. High-Temp Synth., 22 (2013), Issue 1, pp 60/4.

[12] G. Byun, J. Kim, C. Lee, S. Kim, and S. Lee, Kinetic Spraying Deposition of Reactive-Enhanced Al-Ni Composite for Shaped Charge Liner Applications, J. Therm. Spray. Technol., 25 (2016), Issue 3, pp 484/93.

[13] M. Youseffi, N. Showaiter, and M.T. Martyn, Sintering and Mechanical Properties of Prealloyed 6061 Al Powder with and Without Common Lubricants and Sintering Aids, Powder Metall., 49 (2006), pp 86/95.

[14] J. Wu, H. Fang, S. Yoon, H.J. Kim, and C. Lee, Measurement of Particle Velocity and Characterization of Deposition in Aluminum Alloy Kinetic Spraying Process, Appl, Surf, Sci 252 (2005), pp. 1368/77.

[15] G. Bae, Y. Xiong, S. Kumar, K. Kang, C. Lee, General aspects of interface bonding in kinetic sprayed coatings, Acta Mater, 56 (2008) 4858/68.

[16] P. Swaminathan, M.D. Grapes, K. Woll, S.C. Barron, D.A. Lavan, and T.P. Weihs, Studying Exothermic Reactions in the Ni-Al System at Rapid Heating Rates using a Nanocalorimeter, J. Appl. Phys., 113 (2013).

Fractal Analysis of Wear Debris for Plasma Sprayed MoS_2/Ni-SiC-Y Solid Self-lubricant Coatings

Xiaojing Yuan, Bailin Zha, Li Jiang and Zhihang Yu
Xi'an Research institute of Hi-Tech, Shannxi, Province, China, 710025
E-mail: yxj2003@263.net

Abstract: A series of MoS_2/Ni-SiC-Y solid lubricant coatings were fabricated by plasma spraying using particle-reinforced composite powders. The yttrium powder improved the microstructure of the solid lubricant coatings increases the hardness. The fractal dimension declined with Y mass fraction at 50N. The fractal dimension of sliding wear debris was about 2.50-2.66 when the mass fraction of Y, was about 12 wt% in the MoS_2/Ni-SiC- Y solid lubricant coatings.

Keywords: frictional performance; plasma spray; solid lubricant coating; sliding wear

1 Introduction

Recently, self-lubricating coatings was focused by several researchers for its performance, which can be divided into three groups according to their operating temperature. MoS_2 solid lubricant coatings with graphite or their combination as additives, as well as urea formaldehyde resin (UFAR) as the binder, are used to keep the coefficient of friction nearly constant at room temperature. Conventional lubricant coatings include soft metal, transition metal chalcogenides (molybdenum and tungsten disulphides), graphite, and polymer lubricants. Some are effective including WS_2/MoS_2 multilayer films, Cr-doped Diamond-Like Carbon films and polyester coatings[1-4] . Self-lubricant coatings can be fabricated by assembling three different parts: the robust solid lubricant phase, the metallic solvent phase, and the solid phase which supports extreme load. The robust solid lubricant phase can affect frictional behavior in unfavorable conditions, particularly at elevated temperature, high vacuum and excessive load. The metallic solvent phase can fatigue and demonstrate wear-resistance with large diffusion coefficients on sliding surfaces for rapid formation of lubricating coatings. The solid phase can provide superior load carrying capacity for the coatings.

Additionally, several surface lubricant treatment technologies have been developed in tandem with solid lubricating techniques to produce self-lubricating coatings. These include thermal spraying, Physical Vapor Deposition (PVD), Chemical Vapor Deposition (CVD), Ion Beam Assisted Deposition (IBAD) and their combinations.[5-8] Production of thermal sprayed coatings containing liquid lubricants has not yet been achieved because of the difficulties associated with keeping a liquid in a condensed matrix during the spraying process. Synchronously, at extremely high loads, the equipment surface can fatigue and decrease the overall wear-resistance. This is especially true in weaponry equipment. The humidity and abnormal load can also modulate the friction at the surface while traditional lubricants wear away. All three of these parts for solid self-lubricating coatings, such as the robust solid lubricant phase, the metallic solvent phase, and the solid phase which supports extreme load, should be possessed synchronously.

So, in order to obtain homogeneous coatings under high load and environmental humidity, every phase of the MoS_2/Ni-SiC-Y self-lubricant coatings must be carefully controlled in the process of plasma sprayed coatings.

2 Experimental details and calculating methods

2.1 experiment data

Plasma spraying is a high-energy deposition technique that involves the ejection of molten powder, which is then deposited as a coating. During plasma spraying, the plasma torch provides extreme energy and temperature to deposit coatings minimizing the number of un-melted particles. A higher in-flight particle temperature during plasma spraying yields coatings with higher density and bonding strength, even if the material is under a definite melting point. The parameters of the plasma spraying technique are shown in Tab.1. To improve particle flight behavior during plasma spraying, the MoS_2/Ni-SiC-Y nanoparticles were agglomerated with a mechanical spraying granulating method listed in Tab.2. In the pretreatment process, the substrate of a 45# steel grade surface was sandblasted before powder deposition. The morphology of the powders and coatings was characterized by JSM-840, Tescan VEGALLXMU Scanning Electron Microscope. Furthermore, SEM morphology was utilized to investigate the cross-sectional microstructure of the coatings. The phase of the powders and coatings was studied with a D/max-3C X-ray spectrum.

Tab.1 Parameters of Plasma Spray Technique

	Parameters	Value
Grit-blasting	Material	Al_2O_3
	Particles size(μm)	880
	Gas pressure (MPa)	0.6
	Distance(mm)	100
APS parameters	Current (A)	500
	Voltage (V)	56.7
	Nitrogen (L/h, MPa)	40, 0.5
	Pressure of powder injection (L/h, MPa)	0.2, 0.3
	Stand-off distance(mm)	150
	Coatings thickness(mm)	about 0.3

Tab.2 components of composite self-lubricant powders

Sample No.	Description
P1	Deposited using 90wt.%(MoS_2/Ni)-10wt.%SiC composite powders
P2	Deposited using 84wt.%(MoS_2/Ni)-10wt.%SiC-4wt.%Y composite powders
P3	Deposited using 78wt.%(MoS_2/Ni)-10wt.%SiC-12wt.%Y composite powders
P4	Deposited using 64wt.%(MoS_2/Ni)-12wt.%SiC-24wt.%Y composite powders

According to the national standard of China, the friction and wear tests were carried out with the metal-metal (MM) wear experiment method. The upper steel rings (surface roughness Ra=0.8 μm) were kept still, while the lower hard steel discs and a MoS_2/Ni-SiC-Y electro-less plating (Ra=0.8 μm) was rotated under a load of (0-200N), a velocity of 200 $r\cdot min^{-1}$, and ambient temperature of 298K; the average diameter of the sliding wear track was 30 mm. The samples were treated with ethanol before and after each test. The frictional moment was measured with computer attached to the test rig. After sliding for a long distance, the wear loss of the discs was measured with an analytic balance, FA2104 electrical balance made by HANGPING Co., having an accuracy of 0.1 mg.

2.2 Fractal dimension calculation

The wear debris for MoS_2/Ni-SiC-Y lubricant coatings possesses the self- similar features, which can characterize the correlation between the wear surfaces and wear debris accumulation. Where, the logarithm of the wear debris accumulation number and of the debris morphology shows fractal characteristics. Since the fractal dimension of the wear debris accumulation well describe the change of the wear rate, it could be used as a criterion to distinguish the wear status.

If the value of the averaged area of the wear debris accumulation for the different area-scales L in this region, section S can satisfy the expression

$$S(X) \sim L^D \quad \text{(Eq.1)}$$

And the volume fraction of wear debris accumulation (f_{debris}) in the coatings can be calculated.

$$f_{debris} = C\varsigma^{3-D} \quad \text{(Eq.2)}$$

Here, C is constant which can be gained in with Eq.2. ς is the effective scale of wear debris accumulation from the coatings, which is $\varsigma = r_{\min} / r_{\max}$ in the computing region.

3 Results and Discussion

3.1 Microstructure of plasma sprayed MoS_2/Ni-SiC-Y lubricant coatings

The morphology of the agglomerated lubricant coatings is illustrated in Fig.1 (a-d). Coatings combine closely with substrate. Some particles were deformed in the plasma flow. The flattened particles spread out in the coatings and decreased the porosity. Fig.1 (e) shows the XRD pattern of plasma sprayed MoS_2/Ni-SiC-Y lubricant coatings. These coatings are mainly composed of Ni, MoS_2, SiC and Y. All of the compositions construct the structure of solid lubricant coatings. From the XRD pattern, some oxidation, such as SiO_2 and NiS, occurs for the particles flow injected into high temperature plasma flow in the building-up of coatings. This means that the classical structure of solid lubricant coatings can be deposited in the plasma spray process.

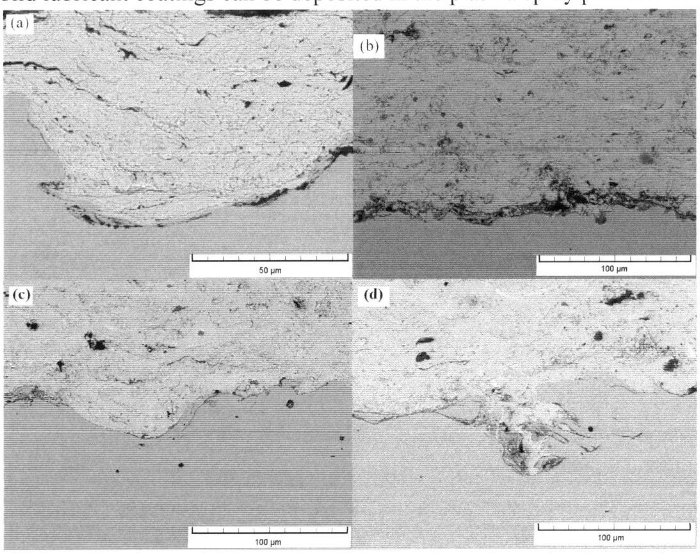

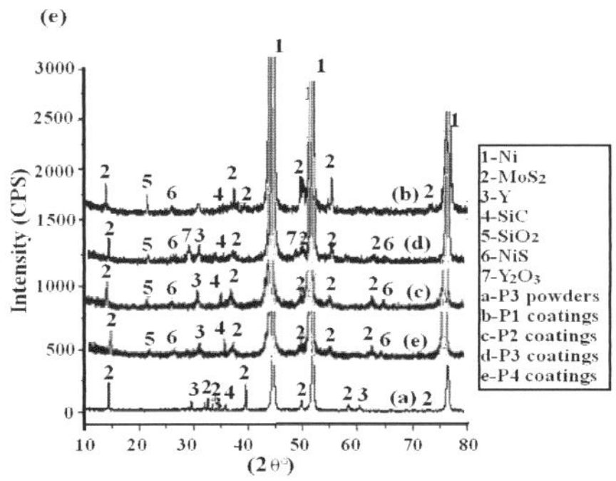

Fig.1 Morphology and XRD pattern of Plasma-sprayed MoS_2/Ni-SiC-Y lubricant coatings: (a-d) SEM of P1, P2, P3 and P4 coatings, respectively; (e) XRD of lubricant coatings

3.2 hardness of plasma-sprayed MoS_2/Ni-SiC-Y lubricant coatings

The hardness values of plasma-sprayed MoS_2/Ni-SiC-Y lubricant coatings are shown in Tab.3. The Rockwell hardness (P2, P3, and P4) was 95.3 HRB, 100.5 HRB and 98.1 HRB, respectively. The classical morphology of the hardness contour for plasma-sprayed MoS_2/Ni-SiC-Y (24 wt %) lubricant coatings is shown in Fig.2. The maximal strain occurred at the boundary of the coatings and the steel sphere during the extruding process between the steel sphere and MoS_2/Ni-SiC-Y (24 wt %) lubricant coatings Fig.2 (a). The crack in the coatings, shown in Fig.2 (b), can come from the focused and expanded stress in the process of the high load, which explains its free energy.

Tab.3 Hardness of Plasma sprayed MoS_2/Ni-SiC-Y lubricant coatings

Coatings No.	Maximal hardness(HRB)	Mean of hardness (HRB)
P2	95.1, 95.1, 95.7	95.3±0.4
P3	100.0, 100.9, 100.7	100.5±0.5
P4	98.4, 97.8, 98.2	98.1±0.3

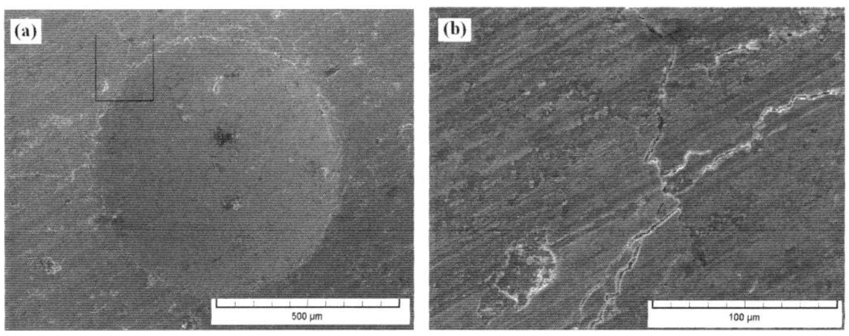

Fig.2 Destroyed surface of MoS_2/Ni-SiC-Y (24wt.%) lubricant coatings.

3.3 Wear property for MoS_2/Ni-SiC-Y lubricant coatings

During the dry sliding wear process, wear weight was another crucial factor for the performance of MoS_2/Ni-SiC-Y (12 wt.%) composite lubricant coatings. In this experiment, the FA2104 electrical balance, made by HANGPING Co., was utilized to measure wear weight loss. When the wear load was 50N, the wear weight of the MoS_2/Ni-SiC-Y composite lubricant coatings (P1, P2, P3 and P4) were 0.0196, 0.028, 0.002 and 0.0024, respectively (Tab.4). The optimal weight fraction of yttrium was 12 wt.% leading to the smallest wear weight loss. When the wear load was 100N, the wear weight loss of the MoS_2/Ni-SiC-Y composite lubricant coatings were 0.0272, 0.0381, 0.0324 and 0.0484, respectively. With increasing sliding wear load, the weight loss of coatings increased quickly. This means that the friction heat energy caused an increase in the friction surface temperature, which narrowed the bond strength of the flattened particles and structural strength of coatings.

Tab.4 wear weight of plasma sprayed MoS_2/Ni-SiC-Y composite lubricant coatings

Coatings	Wear load	Weight before wearing(g)	Weight after wearing(g)	Wearing weight (g)
P1	50N	10.1943	10.1747	0.0196
	100N	10.1918	10.1646	0.0272
P2	50N	10.2430	10.2402	0.00280
	100N	10.2464	10.2083	0.0381
P3	50N	10.8769	10.8749	0.00200
	100N	10.9718	10.9394	0.0324
P4	50N	10.7014	10.6990	0.00240
	100N	10.7144	10.6660	0.0484

3.4 wear debris for MoS_2/Ni-SiC-Y lubricant coatings

In order to explain the wear characteristics of the APS-deposited Ni-MoS_2-SiC-Y coatings, we analysed the wear debris of the coatings under the loads (50 and 100 N). The wear debris of the P3 coating is shown in Fig. 3. The large and polygonal debris particles were observed early in the wear test (for less than 1000 cycles, 50 N; see Fig.3 (a)). Smaller wear debris particles were observed in the middle stage (2000 cycles; see Fig.3 (b)), because the surface of the coating had become smoother. After 4000 cycles, the size of wear debris particles increased (see Fig.3 (c)). This indicates that the tangential strength of the coatings was higher during this stage. The debris particles of this coating for a load of 100 N were flat and spherical as well as agglomerated ellipsoids (see Fig.3 (d)-(f)). The images show debris produced for less than 1000 cycles (Fig.3 (d)), 4000 cycles (Fig.3 (e)), and more than 4000 cycles (Fig.3 (f)). These images suggest that the degradation of the Ni-MoS_2-SiC-Y coatings was mainly caused by plastic flow. At the same time, the presence of SiC, which can act as a stabiliser, allowed the tetragonal phase to be retained after the plasma-spraying process. This can improve the mechanical properties of the coatings at high temperature.

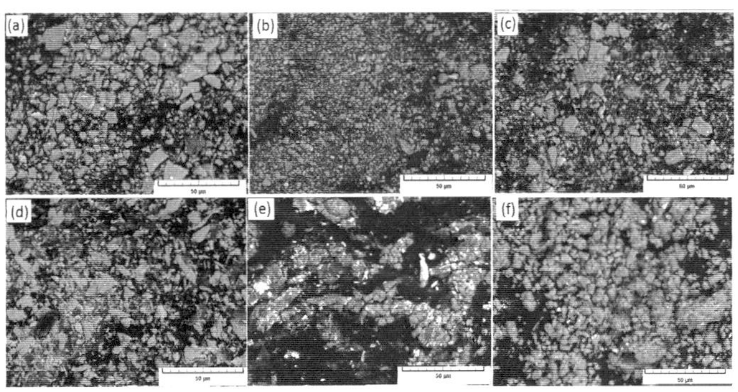

Fig.3 Wear debris of the APS-deposited MoS_2/Ni-SiC-Y(12wt%) coating: (a) 50 N, 1000 cycles; (b) 50 N, 2000 cycles; (c) 50 N, 4000 cycles; (d) 100 N, 1000 cycles; (e) 100 N, 2000 cycles; (f) 100 N, 4000 cycles

3.5 Fractal features of wear debris for MoS_2/Ni-SiC-Y lubricant coatings

The wear debris distribution can be analyzed as part of the fractal dimension, which was collected and statistically analyzed with microscope [9-11]. In the multi-scales micrograph for wear debris for MoS_2/Ni-SiC-Y lubricant coatings, the volume fraction of wear debris accumulation can be calculated with the fractal dimension. In order to simplify the computing procedures, the wear debris accumulation was assumed to be periodical, the debris elements were assumed to be a cube.

Shown in Fig.4, the fractal dimension of sliding wear debris for plasma-sprayed

MoS_2/Ni-SiC-Y composite lubricant coatings were calculated by box counting fractal dimension. The fractal dimension was made during month 2.50-2.66 for the 50N and 100N.

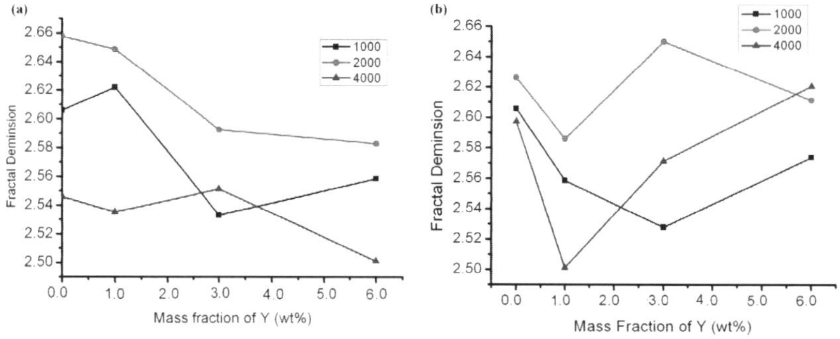

Fig.4 Fractal dimension of friction debris for the MoS_2/Ni-SiC-Y lubricant coatings (P3): (a) 50N; (b) 100N.

A larger fractal dimension value was found when the friction cycle was 2000 r in the 50N and 100N. The larger debris was accumulated due to the action of the surface energy, and thus the fractal dimension of debris became small. The fractal dimension particularly reduces by increasing the mass fraction of yttrium at 50N, Fig.4 (a). In contrast, when the friction load increased, the fractal dimension of the lubricant coatings was not influenced by the yttrium mass fraction, Fig.4 (b). In the first stages of sliding wear, the contacted area was compacted and contracted; high stress was transferred on the coating surface. There were a few prominent peaks and these reached yield strength and plastic deformation and even micro-cutting or peeling off. Here, the fractal dimension is inconsequential for the disperse wear debris. Then, when the pointy peaks were polished and cut, the contact surface was usually smoother, and the fractal dimension of the sliding wear debris increases because of the thin debris crushed the wear particle. However, the friction surface can become soft, and the wear process stable.

4 Conclusions

A series of MoS_2/Ni-SiC-Y solid lubricant coatings with a variety of particle reinforced composite powders was fabricated by a plasma spray technique. This research offers fabrication details and evaluation for self-lubricant coatings at elevated loads. Some results were shown as follows:

The additional yttrium particle can ameliorate the microstructure of MoS_2/Ni-SiC-Y solid lubricant coatings increasing the cohesive strength and hardness of the coatings. When the mass fraction of yttrium is about 12 wt%, the hardness increases to 100.5 HRB. The friction properties of the lubricant coatings under high load were studied. There is just one optimal mass fraction for wear weight at the smaller friction coefficients of 50N and 100N that value is 12 wt% Y.

Debris distribution can be presented by a fractal dimension. The fractal dimension decreases with increasing mass fraction of yttrium at 50N. When the friction load increases, the fractal dimension of the lubricant coatings is not influenced by the yttrium mass fraction. When the mass fraction of yttrium is 12 wt% the fractal dimension of sliding wear debris is about 2.50-2.66.

Acknowledgments

This research was partly supported by the National Natural Science Foundation of China (No.

51405497).

References

[1] Michael P C, Rabinowicz E, Iwasa Y. Friction and wear of polymeric materials at 293, 77 and 4.2 K. Cryogenics, 1991, 31(8): 695–704

[2] Watanabe S, Noshiro J, Miyake S. Friction properties of WS_2/MoS_2 multilayer films under vacuum environment. Surface & Coatings Technology, 2004, 188–189: 644–648

[3] Yu X, Wang C B, Liu Y, et al. Cr-doped DLC films in three midfrequency dual-magnetron power modes. Surface and Coatings Technology, 2006, 200(24): 6765–6769

[4] Zouari M, Kharrat M, Dammak M. Wear and friction analysis of polyester coatings with solid lubricant. Surface & Coatings Technology, 2010, 204(16–17): 2593–2599

[5] Kubart T, Polcar T, Kopecky L, et al. Temperature dependence of tribological properties of MoS_2 and $MoSe_2$ coatings. Surface & Coatings Technology, 2005, 193(1–3): 230–233

[6] Ma Guozheng, Xu Binshi, Wang Haidou, Wang Xiaohe, Li Guolu, Zhang Sen, Research on the microstructure and space tribology properties of electric-brush plated Ni/MoS_2–C composite coating, Surface & Coatings Technology 221 (2013) 142–149

[7] Gamulya G D, Kopeteva T A, Lebedeva I L, et al. Effect of low temperatures on the wear mechanism of solid lubricant coatings in vacuum. Wear, 1993, 160(2): 351–359

[8] D.B. Luo, V. Fridrici, Ph. Kapsa, Evaluating and predicting durability of bonded solid lubricant coatings under fretting conditions, Tribology International 44 (2011) 1577–1582

[9] LING F F, The possible role of fractal geometry in tribology, Tribological Transaction, 1989, 32: 497-505.

[10] HUANG Chuanhui, ZHU Hua, Relationship between Worn Surface Fractal Dimension and Wear Particle Accumulation Fractal Dimension, Tribology, 2003, Vol.23, No.2: 145-148(in China)

[11] K. Plumlee, C. J. Schwartz, Investigating UHMWPE wear mechanisms by decomposing wear debris distributions, wear, 271 (2011) 2208–2212

New hardfacing material with high impact wear resistance

Sven Bengtsson, Senad Dizdar, Barbara Maroli, Höganäs AB, Sweden and Luc Baguette, Höganäs Belgium SA Belgium

Many components in oil & gas, mining and agricultural applications are subjected to impact loading and wear in service. A common method to counteract these failure mechanisms is to apply a hard face, thus increasing service life. However, normally, materials with a high hardness exhibit a low impact resistance and vice versa so compromises are necessary. Typical deposition methods are metal inert gas (MIG) overlay welding, plasma transferred arc (PTA) -, and laser cladding. In the present work clad layers of a new iron based alloy, exhibiting an unusually high impact resistance combined with high hardness and good wear performance is presented. The material is compared to two commercially available materials, AISI D2 – a commonly used cold work tool steel and WC-NiCrSiB – a common choice for severe hard facing applications. The latter material is a metal matrix composite (MMC) formed by adding tungsten carbide particles to a self-fluxing nickel alloy.
The new material is a FeWCrCB tool steel, with fine carbides and borides in a martensitic matrix. The material solidifies in a narrow temperature range, limiting segregation during solidification. The first phase formed is an fcc phase that forms dendrites at most solidification rates followed by a eutectic structure consisting of austenite, carbides and borides. Following solidification the austenite is transformed to martensite.
The impact wear results show approximately 10x longer life compared to tungsten carbide – nickel alloy coating. The abrasive wear performance is close to that of the tungsten carbide nickel base material, but about three times better than AISI D2 tool steel.

1 Introduction

For applications were the wear and impact loading is severe, the number of viable alternatives is restricted. Typically, high hardness and impact toughness are mutually exclusive properties. Furthermore, if the wear and impact resistance is achived by a diffusion bonded coating, the bonding strength becomes critical. By using plasma transferred arc (PTA) cladding or laser cladding (LC) the clad is metallurgically bonden with the substrate. However, the materials used for wear protection are usually very hard, but also brittle and exhibiting low toughness. For coatings a typical material is a NiCrSiB base powder with WC particles creating a composite (MMC) which benefits from the hardness of the WC articles and bonding quality of the NiCrSiB matrix. However, in order to reach sufficient hardness and thus wear resistance, the amount of carbides must be quite high, which means that the ductility and impact wear resistance is suffering. Recently, Höganäs AB developed a new material targeting this application window. It is a Fe-Cr-W high speed steel that in the as-clad condition exhibit high hardness and high impact wear resistance. In the present work the the properties of coatings made by PTA and LC are compared to coating made with NiSiB/WC and regular tool steel AISI D2.

2 Experimental

The new powder is designated Rockit® 701 and it is a gas atomized Iron based powder with major alloying elements being Tungsten and Chromium. This powder is compared to a well known hardfacing powder comprising gas atomized NiCrSiB-based powder and WC powder [1]. The Ni-base powder used is a Chromium free NiSiB powder with a hardness of about 49 HRC. This powder is mixed with macro crystalline WC powder having a particles size in the range 53-150 µm. Here a mix with 50 wt.% WC powder was used. For convienince the mix is designated MMC (as in Metal Matrix Composite).

The PTA samples were coated in a Hettiger 200A PTA unit. The substrates were EN S235JR plates 125x40x20 mm in size. These substrates were not pre-heated before cladding. The laser cladding was performed by a Coherent 4000L direct diode laser with a spot size of 12x1 mm. The substrates were EN S235JR steel 100x70x10 mm in size and they were pre-heated to 100°C prior to cladding. For both methods the samples were allowed to cool to room temperature in a bed of insulating material ensuring a slow cooling rate.

The samples were first cleaned (CRC CRICK 110) and then coated with a highly penetrating red dye (CRC CRICK 120) which is pulled into surface defects or cracks, if any, through capillary forces. After 10 minutes excess dye was removed from the surface and a white developer (CRC CRICK 130) was applied. The developer draws the penetrant out of crevices, cracks or other hollow imperfections communicating with the surface.

The clads were sectioned transversal to the cladding direction, moulded in bakelite, ground and polished. The as-polished samples were examined for dilution of the cladding by the substrate material and for defects, such as pores, cracks, bonding defects, etc. The samples were then etched in 3.3 (100 ml 95 % Ethyl alcohol, 2 g Picric acid, 1 ml concentrated HCl).

Dilution of the coating material was estimated by measuring the chemistry of the coating using a portable XRF Gun (Thermo Scientific). By selecting an element that is abundant in the coating and scarse in the substrate the dilution is estimated by eq. 1.

$$\text{Dilution [\%]} = \frac{c_{Powder} - c_{Coating}}{c_{Powder} - c_{Substrate}} 100\% \qquad (1)$$

A Wolpert Universal hardness tester (HRC) was used for the hardness tests. The coatings were ground and five hardness indents were performed on the flat surface.

The liquidus and solidus temperatures were estimated using a Netzsch STA 449 Fe as a differential scanning calorimeter.

The micro hardness was measured using Matsuzawa MMT-7 micro hardness tester. A load of 25 g was used.

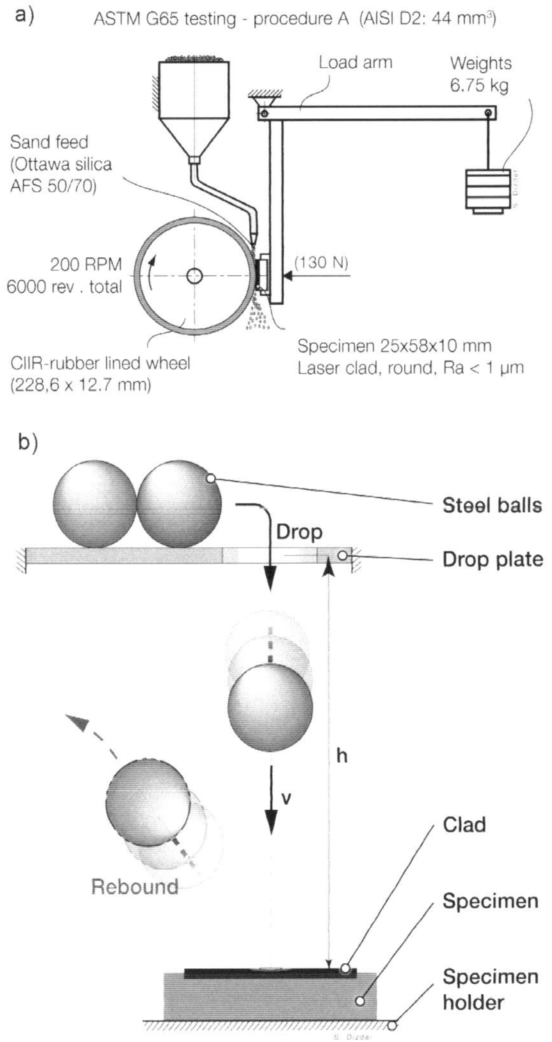

Fig. 1. a) Schematic of the abrasive wear testing according to ASTM G65 standard, procedure A.[2]. b) Schematic arrangement of the impact wear testing setup.

Low stress abrasive wear testing was performed according to ASTM G65 standard, procedure A [2], by using a commercial multiplex sand/wheel abrasion tribometer. Fig 1a shows a schematic of the testing. Five sample replicas per tests material were tested.

Impact wear was performed by using an in-house build tribometer, schematic of which is shown in Fig 1b. Standard ball bearing steel balls drop from pre-defined heights onto the cladded test specimen. The impacting ball does not rebound on the test specimen. A data point is total number of ball strikes for a pre-defined height i.e. impact energy until occurrence of a first circular crack around the impact dent.

3 Results

The apparent density is 4.2 g/cm3 and the Hall flow is 16 sec/50 g.

The PTA samples exhibited a dilution in the range 10-15% while the LC samples exhibited a dilution of 5-10%.

In Fig 2a a single track of PTA overlay weld is shown while a laser cladding is shown in Fig 2c. The corresponding liquid penetrant test results are shown in Fig 2b and 2d respectively. No defects can be seen except end of clad porosity as indicated by arrows in the figure.

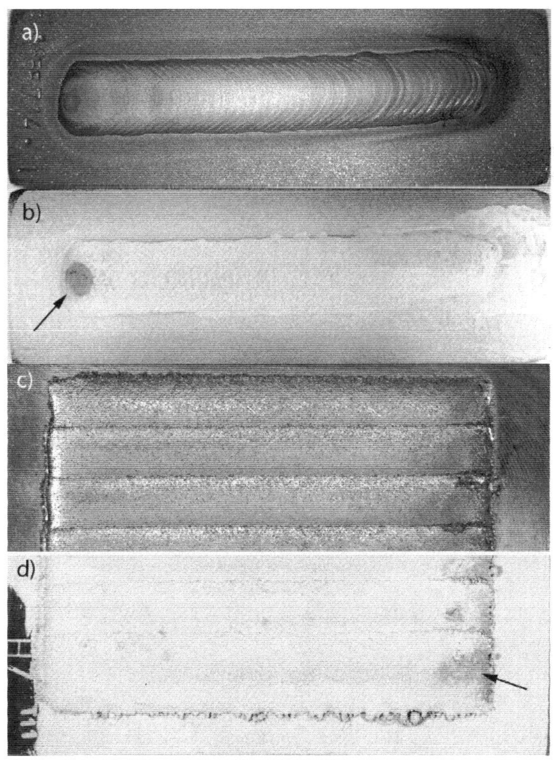

Fig. 2. a) Photograph of PTA overlay clad using Rockit 701 powder. b) Liquid penetrant test of a). c) Photograph of laser cladding using Rockit 701

powder. d) Liquid penetrant test of c). Arrows indicate end of clad defect.

The differential scanning calorimetry shows three major peaks; at 1140, 1204 and at 1320 °C, see Fig 3. The diagram shows that the alloy starts solidifying as the temperature drops past 1320 °C and that the solidification is complete at 1140 °C.

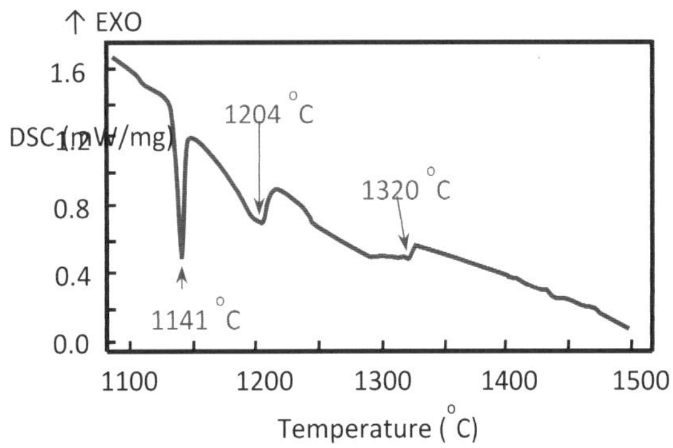

Fig. 3. Differential Scanning Calorimetry.

The coatings exhibited good bonding without porosity at the interface between substrate and coating for both PTA and LC as shown in Fig 4a and 4b respectively.

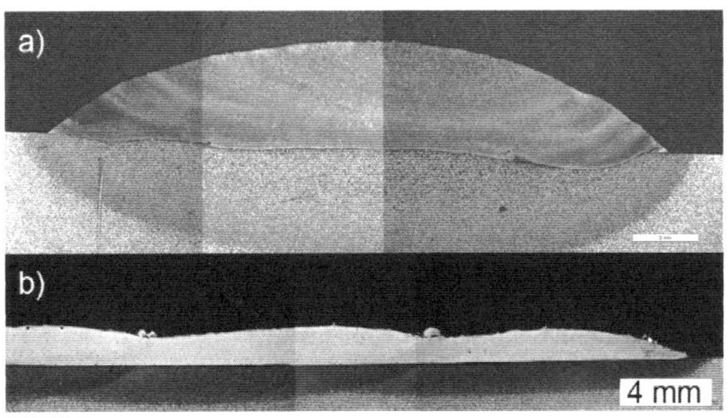

Fig. 4. a) Cross section of PTA clad layer in the etched condition. B)

The hardness of the as-clad material was 65 HRC for Rockit 701 and 66 HRC for the NiCr-WC composite material.

Typical microstructures in the PTA (a) and LC (b) conditions are found in Fig 5. The structure is very similar for the two clads, except for the size of features. Both dendrites and carbides and borides in the eutectic are larger in the PTA clad specimen. The structure consists of the primary phase that forms as dendrites or in some cases as cells and a eutectic structure. The primary phase transforms into martensite (M) upon cooling. In the PTA coating there is a small amount of lower bainite (LB). In the eutectic (E) both carbides and borides are found. Carbides and or borides (B/C) are found in the primary phase.

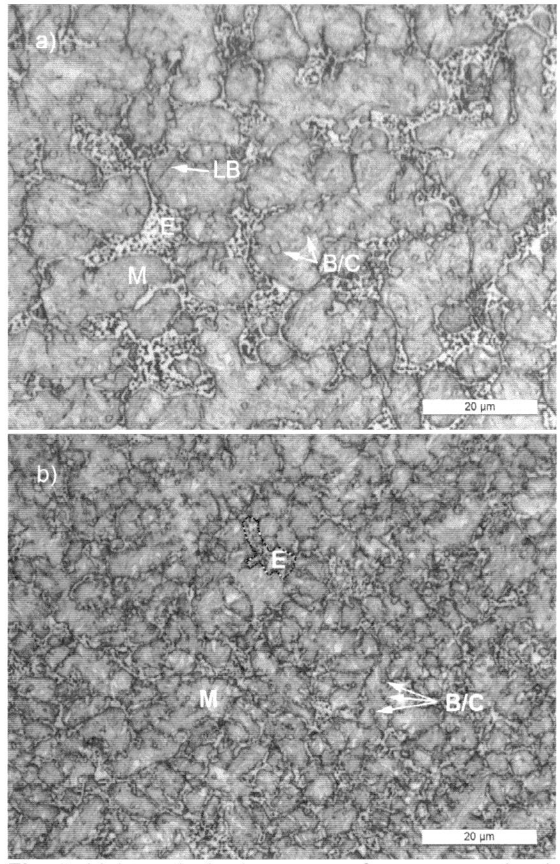

Fig. 5. Light optical microscopy of a cross section in the etched condition. a) PTA clad specimen. b) Laser clad specimen.

Fig 6a shows low stress abrasive wear achieved in ASTM G65 testing, 5 to 10 mm^3 for MMC clads, 15 mm^3 for Rockit 701 clads and 44 mm^3 for reference AISI D2 cold work tool steel. Wear number for MMC clads is specified as an interval with bottom and top limit. This because the wear depends on the clad microstructure and on the cladding outcome in particular. One and the same composition of the MMC powder mixes can, depending on cladding outcome, achieve unexpected variations in the clad's dilution, carbide dissolution and carbide sinking. For Rockit 701 on the other hand the carbides precipitate from the melt and their size depends on cooling rate. The mixing in the melt during cladding is strong which results in an even microstructure after solidification.

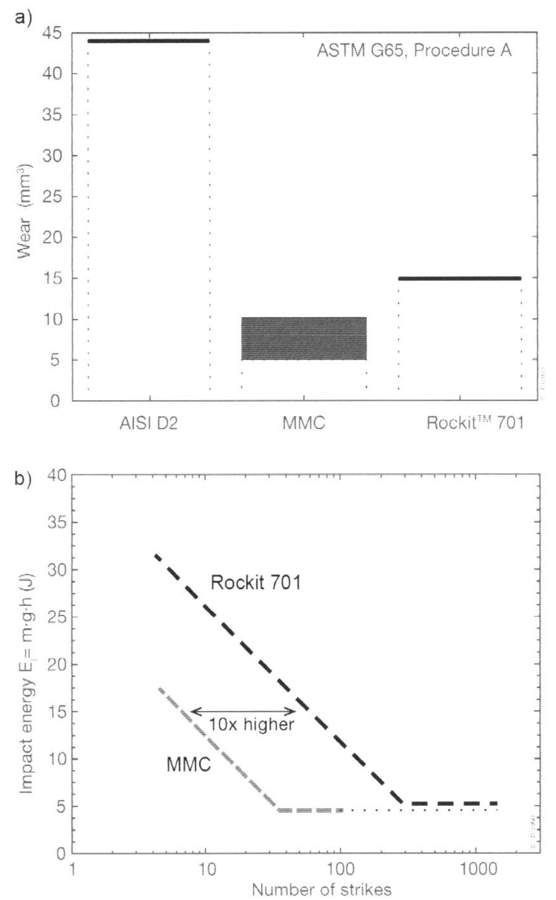

Fig 6. a) ASTM G65 wear b) impact energy as a function of number of strikes to achieve first crack in 50% of the tests.

Fig 6b shows impact energy as a function of number of strikes to achieve first crack in 50% of the tests. Rockit 701 curve shows a higher overload capacity and about ten times longer survival time. Thus, Rockit 701 is a preffered choice for combined impact- and abrasive wear exposed applications. When the impact wear exposition is not significantly present, then MMC clads are a preferred choice.

4 Discussion

The DSC measurements show that the poorly defined peak shapes at 1204 and 1320 C, in Fig 7, is consistent with solidification of a phase that allow change in chemical composition with change in temperature. However, as the temperature approaches 1140°C the reaction is quite strong at a well defined temperature, a behavior which is typical for solidification of a single element or a eutectic composition. The area covered by the peak is proportional to the product of heat of formation and the amount of phase formed. In the present case this means that the peaks at 1320 and 1204 °C indicates a quite small amount of material participating in the solidification. For the peak at 1140 °C, on the other hand, the shape indicates that a significant proportion of the material can be used.

Only a small fraction of primary phase is formed at high temperature. The melt maintains the fluidity to near solidus temperature. Result is good weldability. Furthermore, the matrix phase tranforms from austenite to martensite which causes a volume increase, which in turn causes a drop in residual (tensile) stresses, which in turn causes less risk for cracking during cooling. Martensite start/finish temperatures are high enough to allow pre-heating to temperatures up to 250°C.

In the abrasive wear test the MMC material outperforms Rockit 701 by 50%, while the wrought tool steel AISI D2 is significantly inferior. It can be noticed that the size of the hard phase is significantly different in the two clad cases, see Fig 7. The WC particles exhibit a size range of 50-150 µm, whereas the hard phases in the Rockit 701 is of the order 2-15 micron. It has been suggested that one factor in the good performance of the MMC is that the hard particles shadow the matrix after a running in period. The mean particle size of the abradent is of the order of 290 µm. In the case of Rockit 701 the abradent particle is significantly larger compared to the hard phases present.

In the impact wear test it is obvious that the material is required to deform plastically in order to avoid immediate crack formation. In Fig 8 the hit marks on a PTA clad MMC sample can be seen. The dark colour is in fact not a discolouration of the material, but an effect of the curvature of the plastically deformed region on the sample surface. The cracks appear around the hit marks, in this case after five hits with an energy of 21.4 J. In the present investigation it has not been possible to determine the root cause of the differences in performance between MMC and Rockit 701. In Fig 7 a lower energy limit is indicated. At the lowest testing energy of 5.2 J the MMC and Rockit 701 material were subjected to 600 and 1000 hits respectively without any cracks reaching the surface, despite clear indication of plastic deformation.

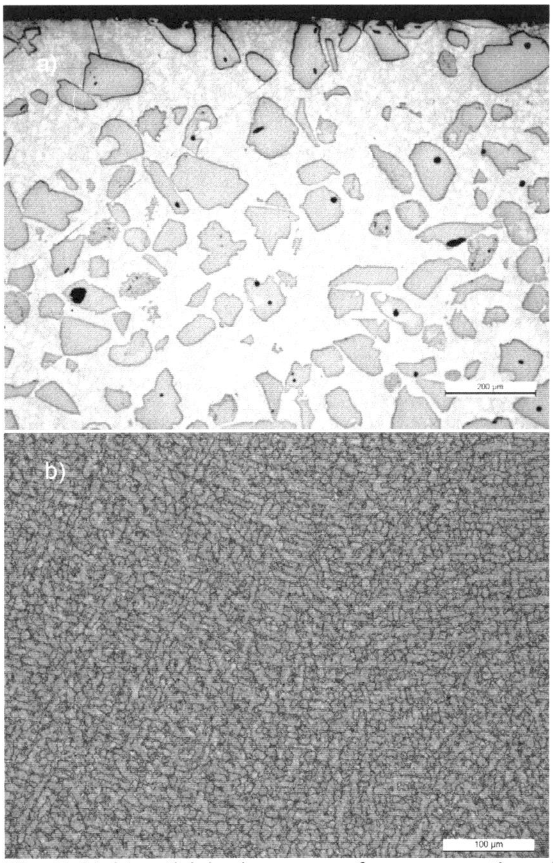

Fig. 7. Light opticial microscopy of a cross section of a) MMC showing large WC carbides in a matrix of NiSiB alloy and b) Rockit 701, both clads produced by PTA overlay welding. Observe the difference in magnification.

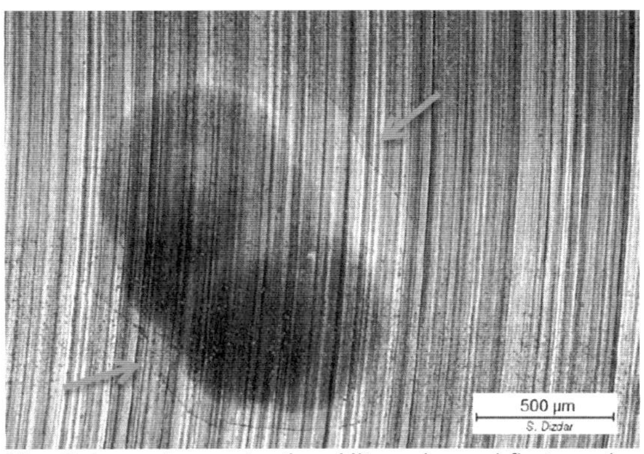

Fig. 8. Impact wear testing. Hit marks and first cracks on a MMC sample tested with five hits at an energy of 21.4 J.

5 Conclusions

A new Fe-based hard facing powder alloy grade has been presented. The powder was PTA- and laser cladded onto mild steel plates. The clad specimens were investigated for weldability, microstructure properties, abrasive- and impact wear. Following conclusions were met:

- Weldability is good assessed thorugh flaw-free clad seams and low tendency for cracking.
- Clad microstructure is very fine both for PTA and laser deposition. Hardness reaches 65 HRC.
- Impact wear performance of Rockit 701 clads exceeds that of MMC – it shows higher overload capacity and about 10 times longer survival time to 50% probability of a first crack occurrence.
- Abrasive wear performance according to ASTM G65 testing of Rockit 701 clads reaches 15 mm^3 compared to 10 mm^3 for the MMC clad.
- In overall, Rockit 701 is an Fe-based powder grade suitable for PTA- and laser cladding of structural components exposed to combined impact- and abrasive wear.

6 References

[1] Powder mix: 1559-00 / 4580 in "Thermal Surfacing: Powder Choice With Ease", Höganäs AB, 2016.

[2] ASTM G65 - Standard Test Method for Measuring Abrasion Using the Dry Sand/Rubber Wheel Apparatus.

7 Acknowledgements

The contribution of Mattias Weibull is gratefully appreciated, especially his endurance and patience in performing the welding and cladding in this work.

A new type high-performance thermal spray powder with WB reinforcement

Yifeng Xie[1], Lijie Xie[2], HaibinWang[2], Yue Yang[1], Chao Hou[2], Baogang Liu[1], Xiaoyan Song[2]*
[1] Luoyang Golden Egret Cemented Carbide Tool Co., Ltd
[2] College of Materials Science and Engineering, Key Laboratory of Advanced Functional Materials, Education Ministry of China, Beijing University of Technology, Beijing 100124, China
*Corresponding author, Email: xysong@bjut.edu.cn

Oral
Presenter: Baogang Liu

Abstract: The WB powder was firstly prepared by the Luoyang Golden Egret Cemented Carbide Tool Co., Ltd using a unique technology. A new type of thermal spray feedstock reinforced with WB powder was developed. The cermet coatings were fabricated by high velocity oxy-fuel (HVOF) thermal spraying and characterized on the phase constitution, microstructure and mechanical properties. The WB-reinforced coating was compared with the WC-based coating at the same preparation and testing conditions. It was found that the hardness and wear resistance of the WB-reinforced cermet coating are higher than those of the WC-based coating. By analyzing the failure behavior of the coatings, the effects of WB and the related mechanisms are proposed for the WB-reinforced cermet coating. The WB-containing powders and cermet coatings are promising in applications in the fields of hard materials and powder metallurgy.
Keywords: WB compound; thermal spray feedstock; cermet coating; wear resistance

1 Introduction

A series of W-B compounds, including W_2B, WB, WB_2, WB_3, WB_{12}, with a high melting point, hardness, electrical conductivity, thermal conductivity, wear resistance and chemical are widely used in many particular application fields, such as high temperature corrosion resistant electrode materials, casting mould, crucible and so on[1]. Owing to their complex structures and chemical bonds, the basic data and properties of these compounds have been explored. The experimental data show Vickers hardness of WB and WB_2 can reach 18.4GPa and 29.2GPa, respectively [2], even harder than WC, which was widely used in traditional WC-based cemented carbide. Figure 1 shows the gibbs free energy of some W-B compounds. It can be seen that the free energy of W-B series compounds is lower than that of WC, proving that the stability of W-B series compounds are higher than WC in theory.

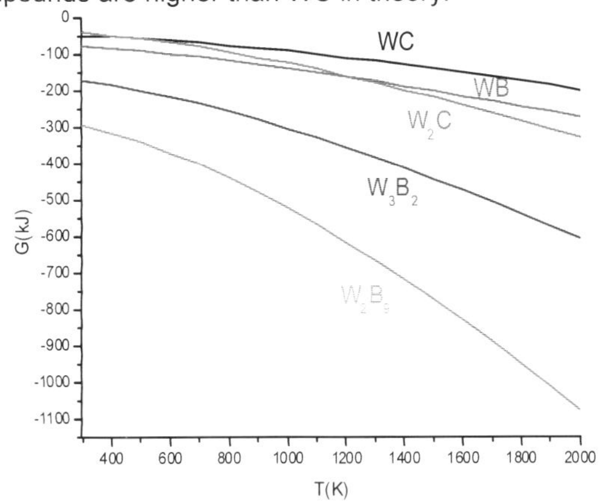

Fig. 1. Gibbs free energies of W-B compounds

Although W-B series compounds have more outstanding properties, higher hardness and high stability than WC, few previous literatures reported the influence of adding WB to cemented carbide coating. In the present study, the microstructure and mechanical properties WB powder and cemented coating added WB has been investigated.

2 Experimental

2.1 Materials

The materials in this study are produced by Luoyang Golden Egret Cemented Carbide Tool Co., Ltd, including WB powder, 60%WC-30%WB-10%Co composite powder and coating, 88%WC-12%Co and 86%WC-10%Co-4%Cr coating.

2.2 Characterization

The phases of the materials were analyzed by X-ray diffraction (XRD) using Cu K radiation in a Rigaku D/max-3c diffractometer.

The morphologies, microstructures and spectrum analysis of the powders and coatings were examined by Nova Nano SEM (With EDAX 9100 energy spectrum analyzer).

The sliding wear test was carried out using a CFT-1 reciprocating sliding tribometer. The silicon nitride ceramic balls with a radius of 5mm were used as the counter material. Before tests, all the coating samples were polished on the diamond resinpolishing discs with the diamond particle sizes of 20μm and 9μm and then were subjected to a cloth polishing using the diamond polishing paste with a particle size of 1.5μm. The sliding wear tests were conducted at the load of 80N, the sliding speed of 5m/min and the total time of 15min.

Microhardness tests were performed with a 300g load and a dwell time of 15s using a Vickers indenter on the polished cross-sections of the coatings.

3 Results and discussion

3.1 Characterization of WB powder and 60%WC-30%WB-10%Co composite powder

Figure 2 shows the XRD pattern of the WB powder produced by Luoyang Golden Egret Cemented Carbide Tool Co., Ltd. It indicates that the powder has WB as main phase and contain a little W_2B and W_2B_5 phases. The inhomogenous atom diffusion causes some region with rich W and some with rich B, and WB generates at first in the reaction between W and B, W_2B phase and W_2B_5 phase gradually appear as the holding time last. Figure 3 shows the microstructures and particle size distribution of WB powder. As observed from Figure 3, the shape of WB particle is irregular, and most of the particles are elliptical, which causes the low flowability. The average particle size is 3.1μm, ranging from 1μm to 6μm, which indicates the nonuniformity of the powder is serious.

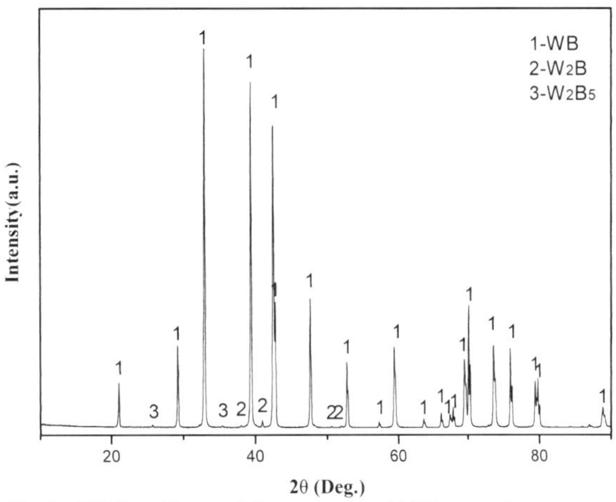

Fig.2. XRD pattern of the prepared WB powder

Fig.3. SEM micrograph and particle size distribution of WB powder

(a) (b)

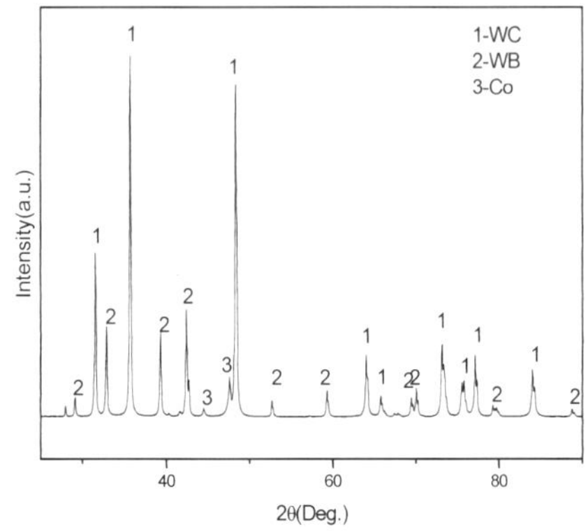

(c)

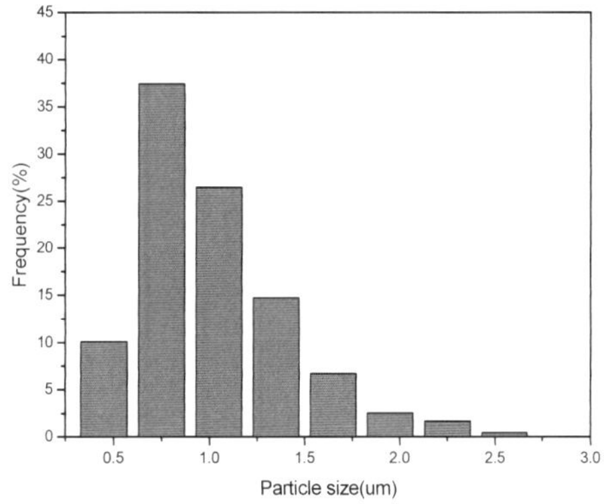

(d)

Fig. 4. SEM micrographs (a, b), XRD pattern (c) and particle size distribution (d) of the WC-WB-Co composite powder

The XRD pattern in Figure 4(c) demonstrates that the composite powder contains WC, WB and Co, and no other phase is detected. The SEM micrographs of 60%WC-30%WB-10%Co composite powder reveal that the particles aggregate seriously with irregular shape (Figure 4a, b). Figure 4(d) shows that the particle size ranges mainly from 0.6μm to 1.2μm with a mean particle size of 1.02μm.

3.2 Characterization of thermal spray feedstock reinforced with WB powder

The physical properties of the thermal spray feedstock are also analyzed. The distribution range, average apparent density and average liquidity of particle size are 15~45μm, 4.5g/cm^3 and 16s/50g, respectively.

3.3 Characterization of the of 60%WC-30%WB-10%Co coating

The cermet coatings were fabricated by high velocity oxy-fuel (HVOF) thermal spraying. Figure 5 shows the XRD patterns of 60%WC-30%WB-10%Co coating. It can be observed that besides the main phase of WC, the ternary compound WCoB and W_2CoB_2 are detected after thermal spraying, indicating that Co and WB have reacted as follow:

WB+Co=WCoB

2WB+Co=W_2CoB_2

The ternary compound WCoB and W_2CoB_2 have been reported in the literature. Zakhariev explored the influences of the boronizing pretreatment on the cemented carbide[3]. The results showed that the cemented carbide through boronizing had higher hardness and wear resistance. The microhardness of WC and WCoB measured in this study are 16.85GPa and 45GPa, respectively, belonging to the superhard materials.

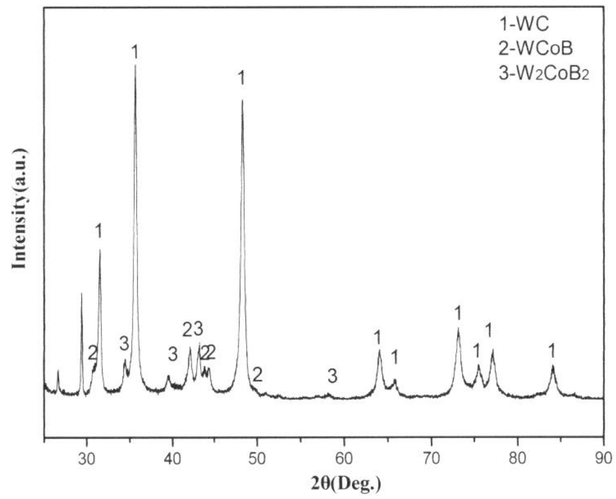

Fig. 5. XRD pattern of the 60%WC-30%WB-10%Co coating

Zhang et al investigated the microstructure and thickness of the boronizing layer in the cemented carbide, boronized by heating at the gas-solid state[4]. The results showed that the light gray phase surrounding the WC grains was W_2CoB_2, and some gray phase WCoB was also detected in the transition layer between the boronized layer and substrate. Zhou et al explored the influences of B addition on the WC -10%Co material[5]. The investigation showed that the phases of WCoB and CoB were detected surrounding the WC grains, just like the η phases in the cemented carbide, which was named as "pseudo η phases". Xu et al took WC, TiB_2 and Co powder as raw materials and prepared WCoB cermets materials by liquid-phase reaction sintering in vacuum furnace[6]. It was found that the cermets contain WCoB, W_2CoB_2, TiC, Co_2B and TiB_2 phases, and the wearing test showed that abrasive resistance was better than that of YG8 cemented carbide.

Figure 6 shows the microstructures of the two kinds of coating. As observed from Figure 6(a) and (b), besides the white phase of WC grains, some gray phases can be observed. The size distribution of these phases is uneven. Figure 6(b) clearly shows there are three kinds of grey phases, located in the 1#, 2# and 3# regions, respectively. The dark grey phase in 1# region should be Co phase, which is quite similar with the grey phase of 88%WC-12%Co coating. The result of EDS analysis of the 2# and 3# region are shown in table 1. Owing to the element of B is light element, which could not be detected by EDS analysis, the EDS results only show W and Co element. It is found that the molar ratio of W and Co in 2# and 3# regions are 56.37:43.63 and 70.13:29.87, which are close to 1:1 and 2:1, respectively. Combined with the color described in the previous literature (Reference), the grey phase and light grey phase in regions of 2# and 3# should be WCoB and W_2CoB_2, respectively.

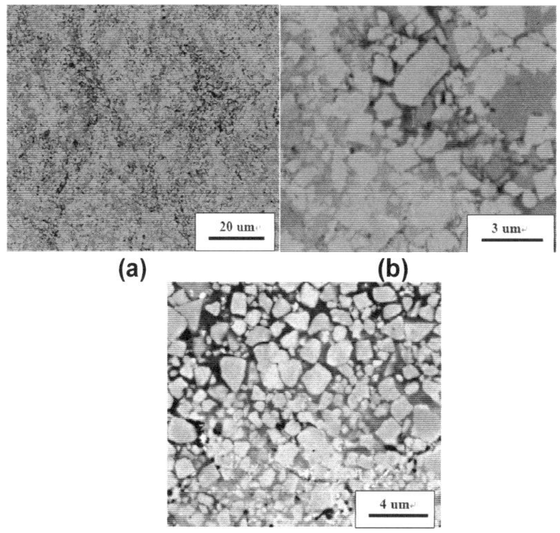

Fig.6. (a, b) SEM micrograph of 60%WC-30%WB-10%Co coating;(c) 88%WC-12%Co coating

Table 1 Spectrum analysis results of 2 and 3 region

Region	Atom fraction of W (%)	Atom fraction of Co (%)
2	56.37	43.63
3	70.13	29.87

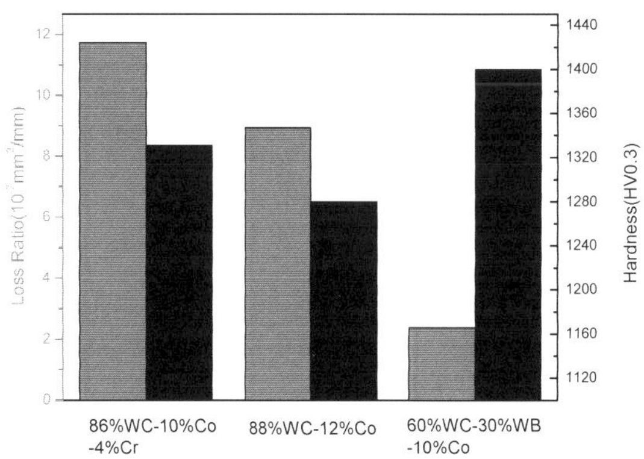

Fig. 7 Microhardness and loss ratio of the three kinds of coatings

The microhardness and wear resistance are tested as shown in Figure 7. The microhardness of 60%WC-30%WB-10%Co coating up to 1400$HV_{0.3}$, which is relatively higher than that of the conventional coatings of 88%WC-12%Co (1280 $HV_{0.3}$) and 86%WC-10%Co-4%Cr (1350 $HV_{0.3}$). However the loss ratios of these three kinds of coatings are quite different. It can be seen the loss ratio of 60%WC-30%WB-10%Co coating is $2.38\times10^{-7}mm^3/mm$, which is 1/4 and 1/5 of that of the 88%WC-12%Co ($8.93\times10^{-7}mm^3/mm$) and 86%WC-10%Co-4%Cr ($11.73\times10^{7}mm^3/mm$), demonstrating the wear resistance of the coating with WB reinforcement is better than that of the conventional coatings.

Figure 8 shows the microstructure of the wear surface of the coating. It can be seen from Figure 8(a) and (b) that there is no scratches on the wear surface of 60%WC-30%WB-10%Co coating. However a large number of scratches appear on the 88%WC-12%Co coating, and the deformation of the coating with WB addition is lighter than that of the conventional coating, which indicates coating with WB addition is much harder. However, a lot of pits can be observed on the wear surface of the 60%WC-30%WB-10%Co coating, as observed in Figure 8(a) and (c), due to the loss of Co as the binder in the reaction of Co and WB, decreasing the impact of bonding.

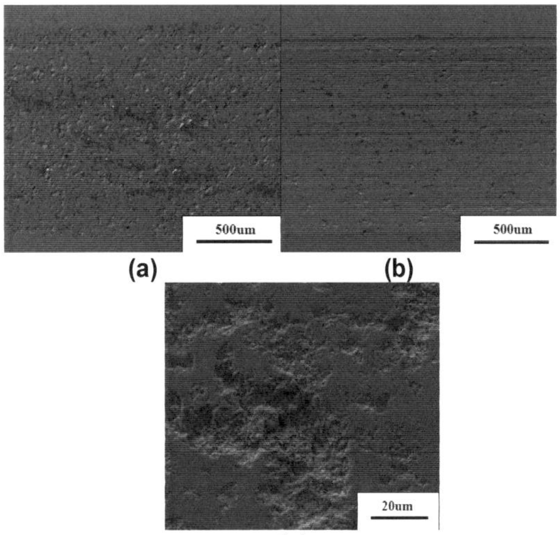

Fig.8. SEM micrographs of the wear surface: (a) 60%WC-30%WB-10%Co coating; (b) 88%WC-12%Co coating; (c) high magnification of 60%WC-30%WB-10%Co coating

4 Conclusion

Taking advantages of the high hardness and stability of W-B compounds, WB is added to the raw material to investigate its influences on the properties of the conventional WC-Co coatings. The following conclusions are drawn: Co reacts with WB during the thermal spraying, creating ternary grey phases of WCoB and W_2CoB_2;.as compared to the conventional coatings, the 60%WC-30%WB-10%Co coating has higher hardness and much better wear resistance.

5 Literature

[1] Liang Y C, Yuan X, Zhang W Q. Phys Rev B[J], 2011, 83(22): 1

[2] Dong W B. Study on tungsten borides synthesized at high pressure and high temperature [D], Changchun: Jilin University.

[3] Zakhariev Z, Marinov M. Superhard boride layer deposition on a carbide–cobalt hard alloy[J]. AlloyCompd, 1993,201:1-3.

[4] Zhang Z J, Lin G B, Qiu Z H, et al. Investigation on microstructure and thickness of boronizing layer in cemented carbide [J].Cemented Carbide, 2012, 29(2): 61-65.

[5] Zhou D L, Xie W, Huang W L. Study on the influence of B on mechanical properties of WC-10%Co alloy[J]. Cemented Carbide,2009, 26(1): 7-9.

[6] Xu M. Study on prepared of WCoB cermets by liquid phase reacting sintering and its properties[D].Wuhan: Wuhan University of Science and Technology.

Thermal Stability of Sub-micro Structured YSZ Coatings Deposited by APS and SPS

Minglong Zhai[a,b], Jiasheng Yang[a,b], Xinghua Zhong[a], Huayu Zhao[a], Chenguang Liu[a], Shunyan Tao[a,*]

[a] The Key Laboratory of Inorganic Coating Materials, Shanghai Institute of Ceramics, Chinese Academy of Sciences, Shanghai 201899, China

[b] University of Chinese Academy of Sciences, Beijing 100039, China

*Corresponding author: E-mail address: sytao@mail.sic.ac.cn (S. Tao).

The sub-micro structured YSZ coatings were deposited by atmospheric plasma spray (APS) and suspension plasma spay (SPS) using nano-structured feedstock, respectively. Free-standing YSZ coatings were isothermal treated in air from 1200 to 1600 ℃ for 24 h and at 1550 ℃ for 20 h to 100 h. The phase composition, microstructure, hardness and Young's modulus of the thermal aged coatings were examined. The research results showed that the as-sprayed coating was mainly composed of metastable tetragonal (t'-ZrO_2) phase. No evidence of any tetragonal-monoclinic phase transformation was detected in the coatings that is isothermally heat-treated for 24 h even though the heat treatment temperature is enhanced from 1200 to 1600 ℃, but the phase transformation occurred after 1550 ℃/40h heat treatment. During the process of the isothermal treatment, the grain size of the coatings has grown evidently. Experimental results show that APS coating has a lower grain growth activation energy. Two kinds of as-sprayed coatings have the similar hardness (*H*) and Young's modulus (*E*), the *H* and *E* values of the coatings improved significantly due to the isothermal treatment.

1 Introduction

Thermal barrier coatings (TBCs) are widely used to protect the substrate of high superalloy in the aero or land gas-turbine engine from high temperature. Common thermal barrier coatings are comprised with a superalloy substrate, bond-coat and ceramic layer. 6-8wt.% Y_2O_3 partially stability ZrO_2 (6-8YSZ) is the material that is widely used for depositing the ceramic layer by atmospheric plasma spraying (APS) or electron beam-physical vapor deposition (EB-PVD) [1-3]. For the plasma sprayed TBCs, the application process leads a unique microstructure with microcracks, porosity and roughly columnar microstructure of splats. This ceramic coating has a low thermal conductivity and a high coefficient of thermal expansion [4, 5]. For the TBCs prepared by EB-PVD, which have a fine columnar microstructure. This microstructure make the coating have a higher thermal shock resistance and higher thermal conductivity compared with the plasma sprayed coatings[6, 7]. In the past two decades, the TBCs with nanostructured zirconia ceramic coatings have been focused on. The nanostructured ceramic coatings fabricated by suspension plasma spray (SPS) or solution precursor plasma spray (SPPS) have attracted many researchers [8-10]. Because nanostructure area can be found in APS TBCs fabricated by nano-agglomerated powders, the potential excellent performance of the coating also draws a lot of study [11].

For the coatings, they experience harsher working environment during the gas turbine operation. The sintering of ceramic coating at high temperature is an important issue for the development of these advanced TBCs [12]. Sintering has a great influence on the phase composition, microstructure and physical properties. It has been shown that the hardness, elastic modulus and thermal conductivity of the as-sprayed TBCs dramatically changed upon annealing at high temperature, because of the micro-cracks and pores changing during the sintering process [4, 12, 13]. It is the metastable nontransformable tetragonal (t'-ZrO_2) phase that is a main phase in the as-sprayed coating. But, with the temperature and the grain size increasing during sintering process, the Y^{3+} will be easy segregation to the grain surface, which may lead to tetragonal (t-ZrO_2) to monoclinic (m-ZrO_2) phase transition. Meanwhile, the tetragonal to monoclinic phase transformation result in significant volume increase (approximately 3-5%), which may aggravate thermal stresses [14-16]. This change of the ceramic coatings may result in losing the protection to the substrate. So, it is necessary to study the phase and microstructure evolution of the ceramic coat with the variation of exposure temperature and dwell time.

The objective of the current work is to study the influence of thermal temperature and the dwell time on thermal stability of the sub-micro structured YSZ TBC. The phase composition, microstructure and mechanical properties of the coatings before and after thermal treated are evaluated.

2 Experimental Procedures

2.1 Coating Fabrication

7YSZ nano-agglomerated powder (Nanox S4007, Inframat Corporation, Farmington, CT) was used to deposit the ceramic coat with thickness about 700 μm by the atmospheric plasma spray system (F4MB-XL, Sulzer-Metco, Switzerland) on stainless steel substrates. The size of agglomerated powder is about 15-150 μm. The suspension of 7.5YSZ nano powders (Jiangsu Lida Hi-tech Special Materical Co., Ltd., China) was used to prepare another ceramic coat with thickness about 750 μm on the stainless steel substrate, which was adopted by the axial suspension plasma spraying system (SPS, Northwest Mettech Corp., North Vancouver, Canada). The percentage of the composition of solid phase in the suspension is about 20 vol.% and the particle diameter (D50) is about 0.06 μm. The spraying parameters of APS coating shows in the Table 1. The spraying parameters of the SPS coating can be seen in Ref. [5]. The free-standing specimens used for heat treat were separated from the stainless steel substrates. The thermal aging was performed in high temperature box furnace (HTK 20/17, thermConcept, Germany) at 1200, 1300, 1400, 1500 and 1600 ℃ for 24 h and at 1550 ℃ for 20, 40, 60, 80 and 100 h, respectively. The free standing samples were heated with the furnace temperature increase at about 10 ℃/min to the target temperature. When the holding time was over, the samples were cooled down to room temperature in the furnace.

Table 1.
spray parameters for nano-structured APS coating

Parameters	Value
Power(kW)	39-43
Ar flow-plasma (slpm)	30-36
H_2 flow-plasma (slpm)	7-9
Ar flow-carrier gas (slpm)	3.0-3.5
Powder feed rate (g/min)	30
Powder inject diameter(mm)	1.8
Spray distance (mm)	80-85

2.2 Coating Characterization

The phase composition of the as-sprayed coating and thermal treated samples were characterized by x-ray diffractometer (XRD, D/max 2550V, Rigaku Industrial Corporation, Japan). Due to the high sensitivity to the presence of monoclinic phase, the Raman spectroscopy was also used to detect the phase composition of the samples. Raman spectra (RS) were recorded using a Renishaw in Via Raman microscope system with a 532 nm laser excitation.
The field emission scanning electron microscope (FESEM, SU8220, HITACHI, Japan) and scanning electron microscope (SEM, TM3000, HITACHI, Japan) were used to observe the microstructure and grain size of the as-sprayed and thermal treated coating specimens. The image analysis was used to calculate the grain size.
The hardness (*H*) and Young's modulus (*E*) of the samples before and after thermal aging were measured by nanoindentation technique. The G-200 nanoindenter (Agilent Technologies, Oak, Ridge, USA) was used to investigate the value of *H* and *E*. There were ten indentations being measured at random locations on the polished cross sections, and the value at the depth about 500 nm of every indentation was selected to calculate the mean value as the needed hardness and Young's modulus. The micro hardness was tested by Vickers indenter (TUKON-2100B, Instron Co., USA) with 300g applied load.

3 Results and Discussion

3.1 Phase Stability

Phase composition of the as-sprayed coatings and samples after thermal aging at different temperatures for 24 h and at 1550 ℃ for different times are shown in the Fig. 1. It can be noted that there was only t'-ZrO_2 being found in the as-sprayed coating, meanwhile, as Zhao et al. have reported in Ref. 5, the as-sprayed SPS coating also exclusively consist of t'-ZrO_2. Due to the high quenching rate of molten particles, the t'-ZrO_2 may be formed by diffusionless transformation[17, 18]. At different temperatures for 24 h thermal aging, even though the temperature up to 1600 ℃, there was still no t-ZrO_2 → m-ZrO_2 transformation occurring(Fig. 1a). For the annealing at 1550 ℃ for different times, after 40 h, there was obvious monoclinic phase (m-ZrO_2) peak appearing (Fig. 1b). With the increase of annealing temperature or prolonged aging times, the diffusion of yttria lead to forming low-Y_2O_3 and high-Y_2O_3 tetragonal phase. During the process of cooling, the low-Y_2O_3 tetragonal may transform into m-ZrO_2 [18-20]. So, as the annealing time prolong at 1550 ℃, the t-ZrO_2 →m-ZrO_2 transformation occurred in the coating samples.
Raman spectra of the as-sprayed and thermal aged nano-structured APS coating samples is shown in Fig. 2. According to the reports, 176.8 cm^{-1} and 186.5 cm^{-1} are the characteristic peaks of monoclinic phase, 148.26 cm^{-1}, 257.8 cm^{-1}, 327.0 cm^{-1}, 470.7cm^{-1}, 611.6 cm^{-1} and 640 cm^{-1} are characteristic peaks of t-ZrO_2 [19, 21]. As the Fig. 2a showed, the peaks can be correspondence with the characteristic peaks of tetragonal in the reasonable scope of shifts. It can be confirmed that there was no t-ZrO_2 → m- ZrO_2 phase transition occurring at different temperatures for 24h annealing. The process of thermal treated at 1550℃ for 40 h, partial transformation to the m-ZrO_2 was found. The characteristic peak of monoclinic phase at 184.7 cm^{-1} and 195. 2 cm^{-1} appeared, and long-time

thermal aging at 1550 ℃ for 80 h and 100 h was particularly apparent.

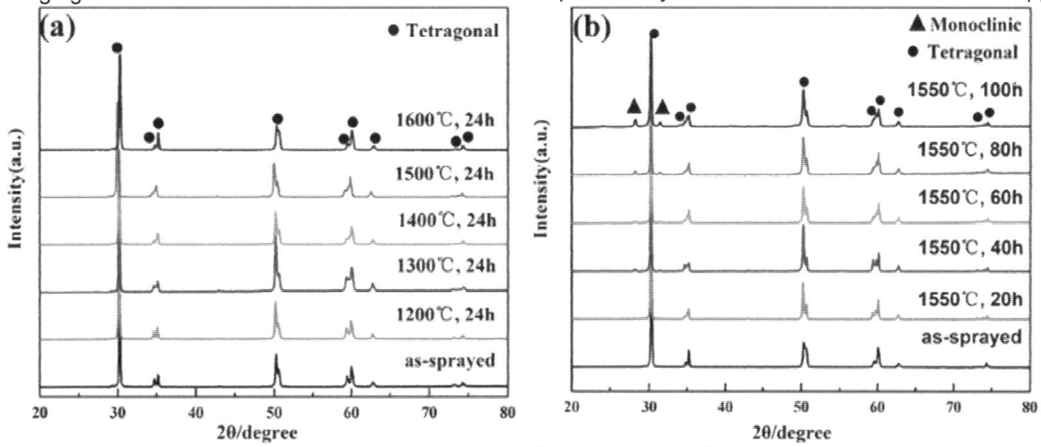

Fig. 1 Phase composition of the nano-structured APS coating before and after thermal treated: (a) samples thermal treated for 24 h at different temperatures, (b) samples thermal treated at 1550 ℃ for different times

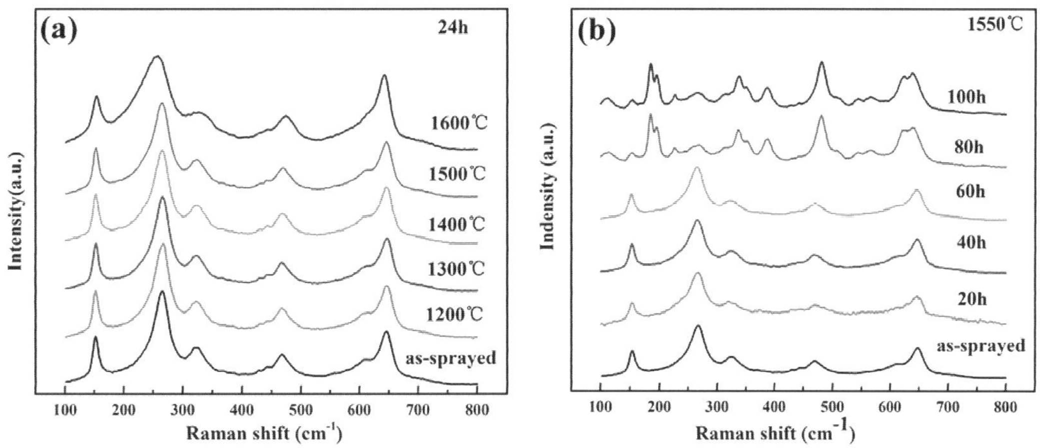

Fig. 2 Raman spectra of the samples before and after thermal treated: (a) samples thermal treated for 24 h at different temperatures, (b) samples thermal treated at 1550 ℃ for different times

The RS analysis result further verified the above results of XRD. The exposure time at high-temperature had more influence on the t-ZrO_2 → m-ZrO_2 phase transformation. As Zhao et al. has reported in Ref. [5], phase stability of SPS coating during thermal aging had a similar result with that of the nano-structured APS coating in this study. The microstructure of the SPS and APS coatings has no obvious effect on the phase stability at high temperature.

3.2 Microstructure of Coatings

Fig. 3 shows the microstructure of the as-sprayed nano-structured APS coating and SPS coating. As can be seen in Fig. 3a, the coating had the obvious segmentation cracks that were perpendicular to the coting surface, and some horizontally oriented branching cracks intersection with segmentation crack. The segmentation cracks contributed to increasing the thermal shock resistance, because these column gaps were similar with the columnar structure of EB-PVD coating [22, 23]. According to Fig. 3b, the nano-structured APS coatings was same with that of conventional APS coating. The apparent lamellar structure can be found and the lamellae consists of columnar grains. The microstructure of the SPS coating shows a great difference with the APS coating, which consisted of fine columnar grains and equiaxed grains (Fig. 3c). The nano-microporous can be found in the coatings, which is helpful to improve the insulation performance.

As shown in the Fig. 4 and Fig. 5, the grain growth occurred in the nano-structured APS coating during thermal aging. With the increase of temperature and prolonged time of annealing, the grain size had obvious increase. The value of the grain size based on image analysis is exhibited in the Fig. 6. As can be seen, the grain size (diameter of columnar crystal) as-sprayed nano-structured APS coating was about

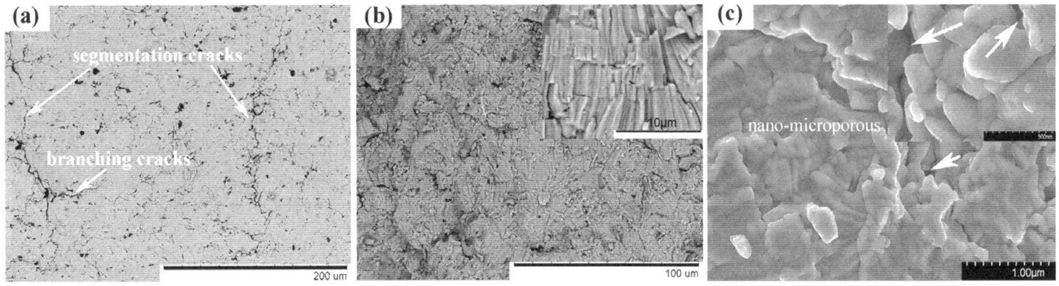

Fig. 3 Microstructure of the as-sprayed coating: (a) polished cross section of the APS coating, (b) fracture surface of the APS coating, (c) fracture surface of SPS coting

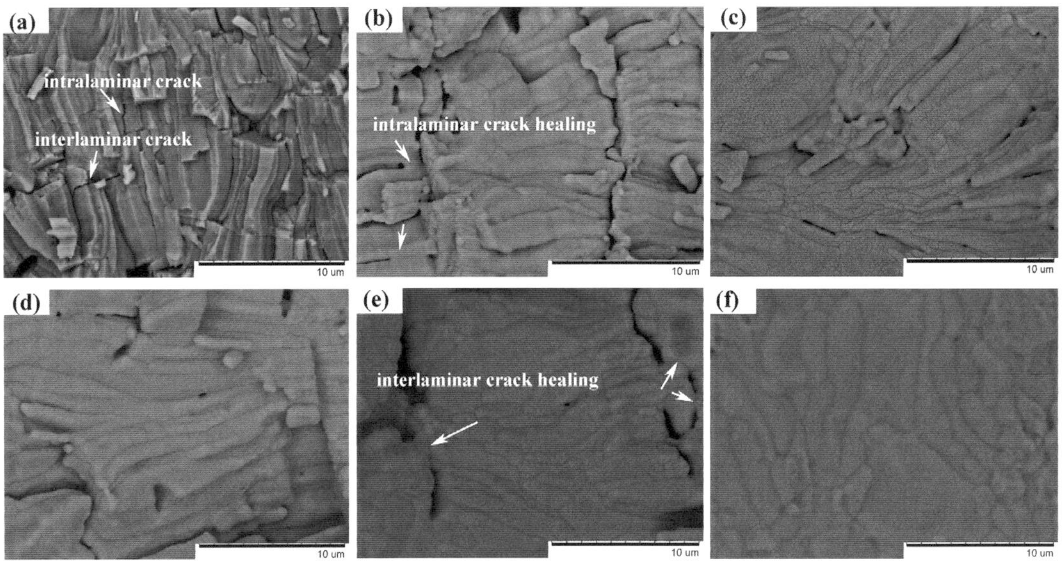

Fig. 4 Microstructure of the nano-structured APS coating before and after thermal treated at different temperatures for 24 h: (a) as-sprayed, (b) 1200 ℃, (c) 1300 ℃, (d) 1400 ℃, (e) 1500 ℃, (f) 1600 ℃

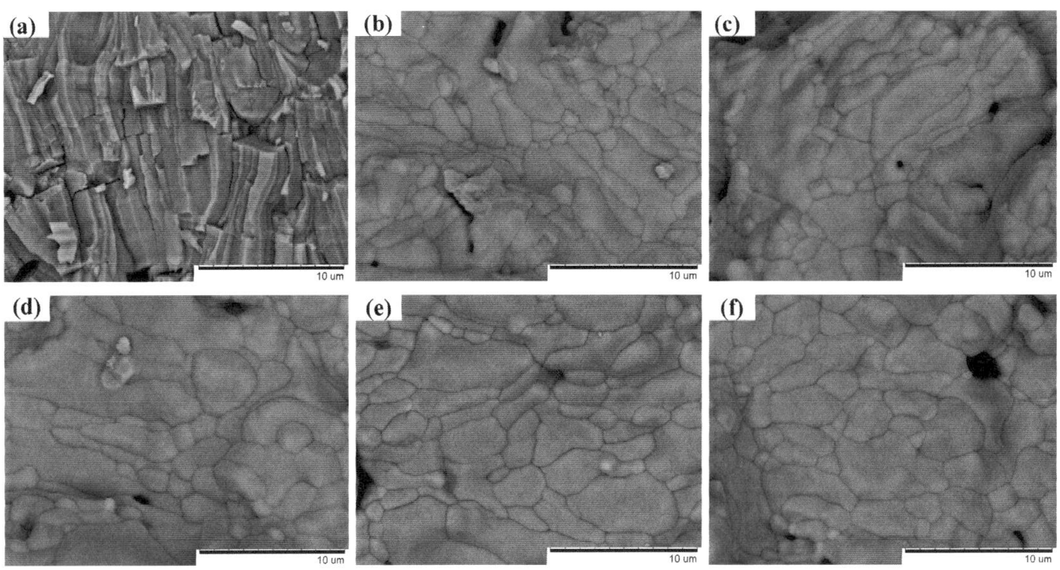

Fig. 5 Microstructure of the nano-structured APS coating before and after thermal aging at 1550 ℃ for different times: (a) as-sprayed, (b) 20 h, (c) 40 h, (d) 60 h, (e) 80 h, (f) 100 h

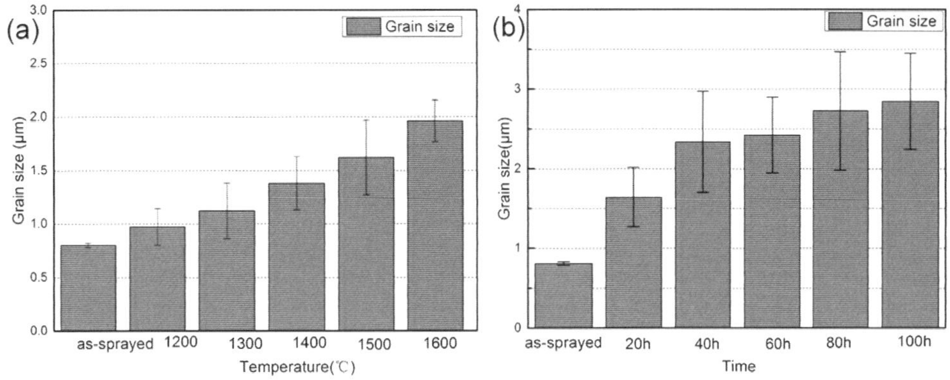

Fig. 6 Grain size of the nano-structured APS coating after thermal treated at different temperatures for 24 h (a) and at 1550 ℃ for different times (b)

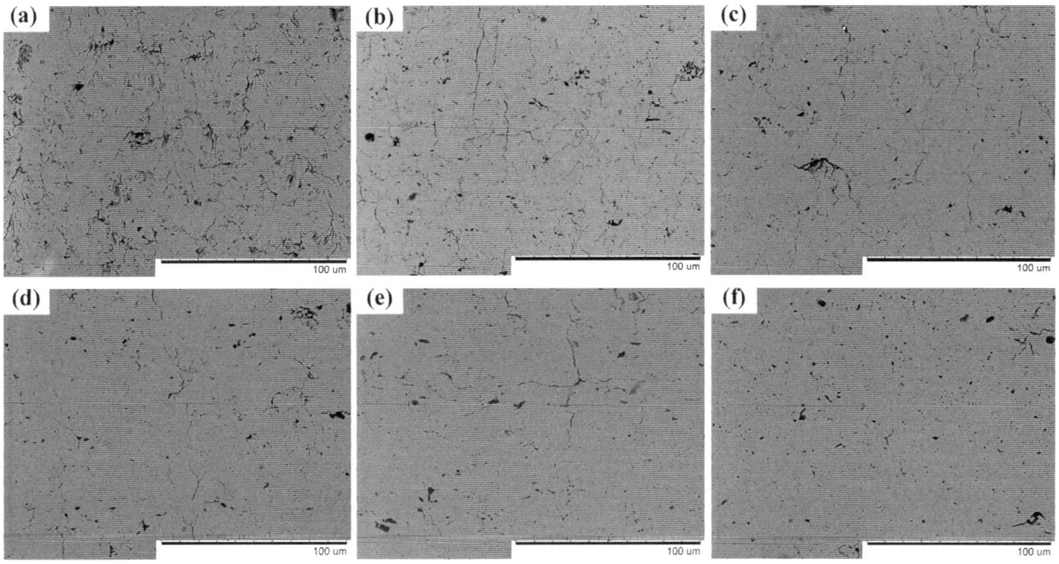

Fig. 7 Morphology of cross section of the nano-structured APS coating before and after themal aging at different temperatures for 24 h: (a) as-asprayed coating, (b) 1200 ℃, (c) 1300 ℃, (d) 1400 ℃ (e)1500 ℃ , (f) 1600 ℃

800nm, which was about 10 times larger than the grain size of SPS coating (80 nm) [5] and about 8 times larger than that of nano-structured APS coating (100nm) as Liang et al. researched [19]. The relatively larger grain size can be attributed to a well molten characteristic of the powder during spraying. After thermal aging at 1600 ℃ for 24h, the diameter of the columnar crystal increased to about 1.96 μm (Fig. 4f), which was about 2.5 times larger than that of the as-sprayed coating. During the annealing for 24 h, the cracks of intralaminar crack was healing (Fig. 4b), and then with the increase of the temperature the cracks of interlaminar crack was also healing (Fig. 4e). During thermal aging at 1550 ℃, with the time prolonging, columnar crystal transformed to equiaxed grains gradually. After 1550 ℃/100h thermal treated, the grain size was about 2.8 μm. It was about 3.5 times larger than that of the as-sprayed coatings. For the SPS coatings, the grain size had increased for 21 times and 26 times after thermal treated at 1600 ℃ for 24 h and at 1550 ℃ for 100 h, respectively[5]. Because the grain size was smaller in the SPS coating and grain surface energy was higher relatively, the grain growth was faster at high temperature. Fig. 7 and Fig. 8 are the polished cross section of APS coating before and after annealing. For the process of thermal aging at different temperatures for 24 h, with the increase of temperature, the microcracks and micropores have decreased, which confirmed the cracks and pores healing. After the process of annealing at 1550 ℃ for 20 h, the cracks and pores were healing and decrease quickly, with the time prolonged, the cracks and pores decreased slowly. But , after thermal treated at 1550 ℃ for 100 h, the micro-pores increased, it may arise from the tetragonal-monoclinic phase transformation. As shown in the Fig. 1b, m-ZrO_2 phase was found after annealing at 1550 ℃ for 100 h. The t- ZrO_2 transformation to m-ZrO_2 is with volume expansion about 3-5%, which may lead to the generation of pores.

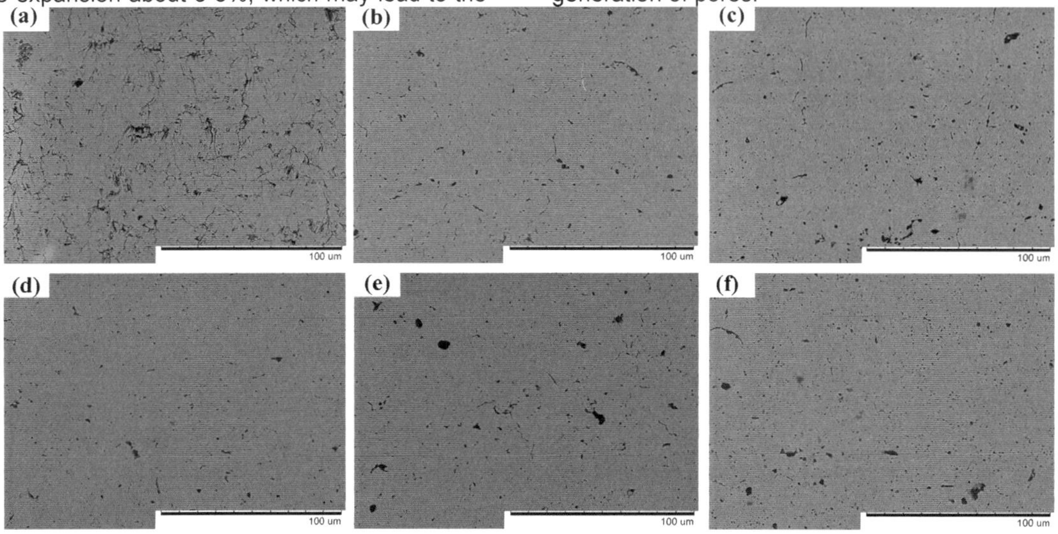

Fig. 8 Morphology of coss section of the nano-structured APS coating before and after thermal aging at 1550℃ for different times: (a) as-sprayed, (b) 20 h, (c) 40 h, (d) 60 h, (e) 80 h, (f) 100 h

3.3 Hardness and Young's modulus

As known that annealing has great influence on the microstructure of the coating, and the change of microstructure leads to the change of the mechanical properties. Fig. 9 shows the evaluation of hardness and Young's modulus of nano-structured APS coating during the process of thermal aging. For the as-sprayed APS coating, the values of the *H* and *E* were 10.8±1.9 GPa and 181.5±18.5 GPa, respectively. As was reported by Ghasemi, the values of *H* and *E* investigated by nanoindenter for the nano-structured coating varied from 5.531 to 16. 019 GPa and 85.233 to 202.94 GPa respectively [11]. The values of the as-sprayed nanostructured coatings of this study are found within this range. The values were also close to that of the SPS coating, the mean value of *H* and *E* were 10.6±0.7 and 174.5±5.9 GPa [5]. The relatively higher values of the *H* and *E* of the coating was attributed to good molten characteristics during spraying process. As shown in Fig. 9a, with the rises of annealing temperature, the values of the *H* and *E* had obvious increase compared with that of as-sprayed coting. After thermal aging at 1500 ℃ for 24 h, the *H* and *E* of the APS coating had the maximum mean values with 17.0±3.3 and 263.9±18.2 GPa, which increased approximately 60% and 51% compared with the as-sprayed coating(Fig. 9a). Fig. 9b showed the *H* and *E* of APS coating after annealing at 1550 ℃ for different times. After annealing at 1550 ℃ for 20h, the mean values of *H* and *E* increased to 11.8±0.8 and 217.0±7.8 GPa, but when the annealing time prolonging to 40 h, the *H* and *E* had a decrease, and then, with the annealing time prolonged to 80 h, the *H* and *E* increased to the maximum mean values with 13.3±1.2 and 227.7±15 GPa, which increased about 25% and 30%, respectively. After thermal aging for 100h, the *H* and *E* had a decrease again. This trend may be attributed to the t-ZrO_2 → m-ZrO_2 phase transformation after annealing at 1550 ℃ for 40h. For the SPS coating as, during annealing at 1550 ℃, the values of the H and E had a trend that increased firstly and then decrease with the prolonging of the annealing times[5]. For the conventional APS coating as A. Shinmi et al. reported, after annealing at 1400 ℃ for 100h , the values of the *H* and *E* had a decrease [24]. When the *H* and *E* decreased, t-ZrO_2 → m-ZrO_2 phase transformation occurred in these two coatings during annealing [5, 24]. As is discussed above, after thermal aging at 1550 ℃ for 40 h, the phase transformation also occurred at nano-structured APS coating. As has discussed above about the Fig. 8b, the micro-pores had a variation tendency that decreased firstly and then increase during the thermal aging at 1550 ℃ for different times. Which may also confirm the variation tendency of the *H* and *E* attributing to t- ZrO_2 → m-ZrO_2 phase transformation during thermal aging.

In order to further studying the hardness variation of the nano-structured APS coating and SPS coating during annealing, the Vickers hardness of the coatings were investigated. As shown in Fig. 10, the Vickers hardness with 300g ($HV_{0.3}$) load of APS coating and SPS coating were 919±21$HV_{0.3}$ and 950±50$HV_{0.3}$. A relatively high value of Vickers hardness for the SPS coating can be attributed to finer grains compared with APS coating. Because the nano-agglomerated powder had a good molten characteristic in this paper, the APS coating had a higher Vickers hardness than that of others research (about 730$HV_{0.3}$) [25]. During the process of thermal aging at different temperatures for

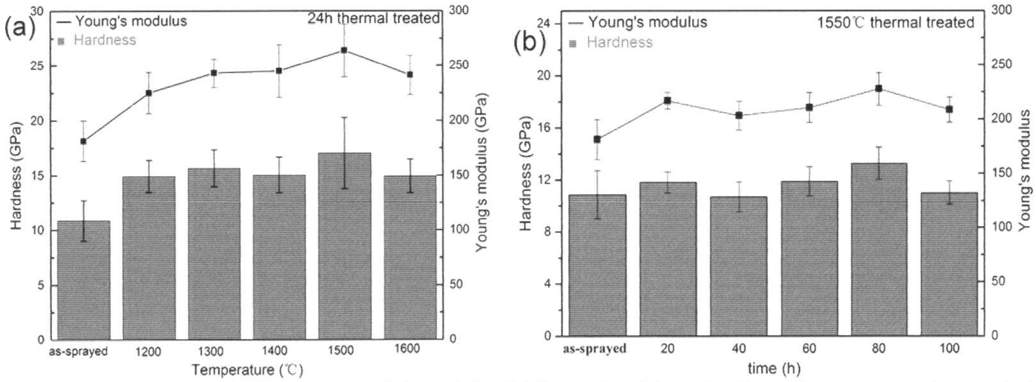

Fig. 9 Variation of hardness and Young's modulus of the APS coating fabricated by using nano-agglomerated powder before and after annealing: (a) annealing at different temperatures for 24 h, (b) annealing at 1550 ℃ for different times

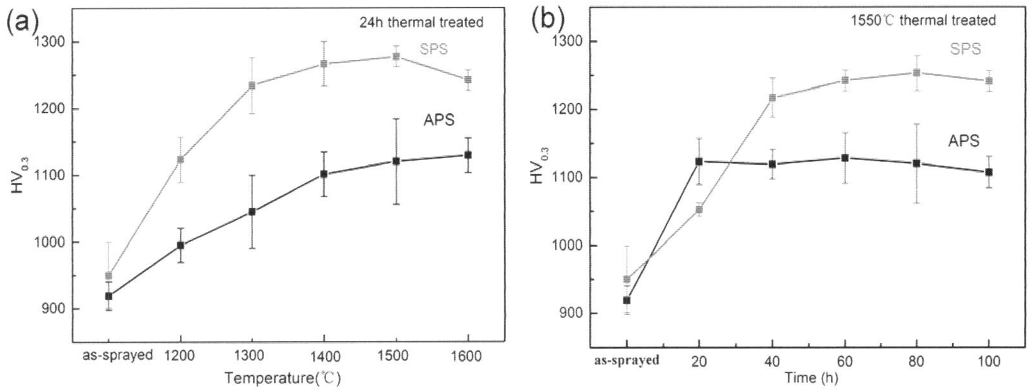

Fig. 10 Variation of Vickers hardness of APS coating fabricated by using nano-agglomerated powder and SPS coating before and after annealing: (a) annealing at different temperatures for 24 h, (b) annealing at 1550 ℃ for different times

24 h, with the increase of the temperature, the $HV_{0.3}$ of the APS coating and SPS coating gradually increased. After annealing at 1600 ℃ for 24 h, the coating had the maximum $HV_{0.3}$ values and the SPS coating had the maximum $HV_{0.3}$ after annealing at 1500 ℃ for 24 h. SPS coatings had a more obvious changes and a higher value of the $HV_{0.3}$ than that of the APS coating during annealing at different temperatures for 24 h (Fig.10a). During thermal aging at 1550 ℃ for different times, the $HV_{0.3}$ of two coatings have increased firstly and then decreased with the prolonging of the annealing times(Fig.10b). The $HV_{0.3}$ variation tendency of the coatings approximately agreed with the nanoindentation results above after thermal aging. Because the as-sprayed SPS coating consisted of nano grains, grain size was more likely to grow up at high temperature. The microstructure of SPS coating had a more obvious change, so, the mechanical properties had larger change compared with the APS coating during same thermal aging process in this study.

4 Conclusions

The nano-structured APS coatings fabricated by using nano-agglomerated powder were heat treated at 1200 -1600 ℃ for 24 h and at 1550 ℃ for 20-100 h, respectively. Annealing lead to the changes of phase composition and microstructures for the coating, these variations result in further changes of mechanical properties (Hardness and Young's modulus). The coating had a similar phase stability compared with SPS coating under the same thermal treated condition. There was no $t\text{-}ZrO_2 \rightarrow m\text{-}ZrO_2$ phase transformation during thermal aging from1200-1600 ℃ for 24h, and after thermal aging at 1550 ℃ for 40 h, the phase transformation occurred. The nano-agglomerated powder had a good molten characteristic during spraying, and a lamellar structure was formed. Because of crack healing and grain growth during annealing from 1200-1600 ℃ for 24 h, the hardness and Young's modulus significantly increased. For thermal aging at 1550 ℃, with the times prolonged, the hardness and Young's modulus increased at initial stage of annealing and then decrease. It may be caused by the $t\text{-}ZrO_2 \rightarrow m\text{-}ZrO_2$ phase transformation. The variation trend of hardness and Young's modulus of the APS coating and SPS coating was similar under the same thermal aging conditions. But the SPS coating had a greater degree of change.

5 References

[1] Vassen R.and A. Stuke, Other: Recent Developments in the Field of Thermal Barrier Coatings. J. Therm. Spray Technol. 18 (2009),Issue 2, pp. 181/86.

[2] Padture N. P.and M. Gell, Other: Materials science-Thermal barrier coatings for gas-turbine engine applications. Science 296 (2002),Issue 5566, pp. 280/84.

[3] Vassen R.and M. O. Jarligo, Other: Overview on advanced thermal barrier coatings. Surf. Coat. Technol. 205 (2010),Issue 4, pp. 938/42.

[4] Choi S. R.and D. M. Zhu, Other: Effect of sintering on mechanical properties of plasma-sprayed zirconia-based thermal barrier coatings. J. Am. Ceram. Soc. 88 (2005),Issue 10, pp. 2859/67.

[5] Zhao Y.and L. Wang, Other: Thermal Aging Behavior of Axial Suspension Plasma-Sprayed Yttria-Stabilized Zirconia (YSZ) Thermal Barrier Coatings. J. Therm. Spray Technol. (2014).

[6] Curry N.and K. VanEvery, Other: Thermal conductivity analysis and lifetime testing of suspension plasma-sprayed thermal barrier coatings. Coatings 4 (2014),Issue 3, pp. 630/50.

[7] Feuerstein A.and J. Knapp, Other: Technical and economical aspects of current thermal barrier coating systems for gas turbine engines by thermal spray and EB-PVD: A review. J. Therm. Spray Technol. 17 (2008),Issue 2, pp. 199/213.

[8] Fauchais P.and R. Etchart-Salas, Other: Parameters controlling liquid plasma spraying: Solutions, sols, or suspensions. J. Therm. Spray Technol. 17 (2008),Issue 1, pp. 31/59.

[9] Pawlowski L.: Suspension and solution thermal spray coatings. Surf. Coat. Technol. 203 (2009),Issue 19, pp. 2807/29.

[10] Fauchais P.and A. Joulia, Other: Suspension and solution plasma spraying. J. Phys. D-Appl. Phys. 46 (2013),Issue 22, pp. 14.

[11] Ghasemi R.and R. Shoja-Razavi, Other: Comparison of microstructure and mechanical properties of plasma-sprayed nanostructured and conventional yttria stabilized zirconia thermal bather coatings. Ceram. Int. 39 (2013),Issue 8, pp. 8805/13.

[12] Zhu D. M. and R. A. Miller: Sintering and creep behavior of plasma-sprayed zirconia- and hafnia-based thermal barrier coatings. Surf. Coat. Technol. 108 (1998),Issue 1-3, pp. 114/20.

[13] Moskal G., Microstructure and thermal diffusivity of micro- and nano- sized YSZ, in: T. Chandra, N. Wanderka, W. Reimers and M. Ionescu (Eds.) Thermec 2009, Pts 1-4, Trans Tech Publications Ltd, Stafa-Zurich, 2010, pp. 900/05.

[14] Ho C. J.,W. H. Tuan: Phase stability and microstructure evolution of yttria-stabilized zirconia during firing in a reducing atmosphere. Ceram. Int. 37 (2011),Issue 4, pp. 1401/07.

[15] Theunissen G.and A. J. A. Winnubst, Other: Surface and grain-boundary analysis of doped zirconia ceramics studied by aes and xps. J. Mater. Sci. 27 (1992),Issue 18, pp. 5057/66.

[16] Witz G.and V. Shklover, Other: Phase evolution in yttria-stabilized zirconia thermal barrier coatings studied by rietveld refinement of X-ray powder diffraction patterns. J. Am. Ceram. Soc. 90 (2007),Issue 9, pp. 2935/40.

[17] Cernuschi F.and P. G. Bison, Other: Thermophysical, mechanical and microstructural characterization of aged free-standing plasma-sprayed zirconia coatings. Acta Mater. 56 (2008),Issue 16, pp. 4477/88.

[18] Schulz U.: Phase transformation in EB-PVD yttria partially stabilized zirconia thermal barrier coatings during annealing. J. Am. Ceram. Soc. 83 (2000),Issue 4, pp. 904/10.

[19] Liang B.and C. X. Ding, Other: Phase composition and stability of nanostructured 4.7 wt.% yttria-stabilized zirconia coatings deposited by atmospheric plasma spraying. Surf. Coat. Technol. 200 (2006),Issue 14-15, pp. 4549/56.

[20] Muraleedharan K.and J. Subrahmanyam, Other: Identification of t' phase in ZrO_2-7.5 wt-percent y2o3 thermal-barrier coatings. J. Am. Ceram. Soc. 71 (1988),Issue 5, pp. C226/C27.

[21] Ishigame M.,T. Sakurai: Temperature-dependence of raman-spectra of ZrO_2. J. Am. Ceram. Soc. 60 (1977),Issue 7-8, pp. 367/69.

[22] Karger M.and R. Vassen, Other: Atmospheric plasma sprayed thermal barrier coatings with high segmentation crack densities: Spraying process, microstructure and thermal cycling behavior. Surf. Coat. Technol. 206 (2011),Issue 1, pp. 16/23.

[23] Zhao Y. X.and D. C. Li, Other: Thermal shock behaviors of YSZ thick thermal barrier coatings fabricated by suspension and atmospheric plasma spraying. Surf. Coat. Technol. 249 (2014), pp. 48/55.

[24] Shinmi A.and X. Zhao, Other: Mechanical and electrical characterisation of a phase transformation in thermal barrier coatings. Surf. Coat. Technol. 205 (2010),Issue 7, pp. 2678/86.

[25] Cernuschi F.and L. Lorenzoni, Other: Studies of the sintering kinetics of thick thermal barrier coatings by thermal diffusivity measurements. Journal of the European Ceramic Society 25 (2005),Issue 4, pp. 393/400.

Hot corrosion behavior of multi-layer suspension plasma sprayed $Gd_2Zr_2O_7$ /YSZ thermal barrier coatings

Krishna Praveen Jonnalagadda[1], Satyapal Mahade[2], Nicholas Curry[3], Xin-Hai Li[4], Nicolaie Markocsan[2], Per Nylén[2], Stefan Björklund[2], Ru Lin Peng[1]

[1] *Department of Management and Engineering, Linköping University, 581 83, Linköping, Sweden*
[2] *Department of Engineering Science, University West, Sweden*
[3] *Treibacher Industrie AG, Austria*
[4] *Siemens Industrial Turbomachinery AB, SE-61283 Finspång, Sweden*

Abstract: This study investigates the corrosion resistance of double layer $Gd_2Zr_2O_7$/YSZ, triple layer dense $Gd_2Zr_2O_7$/ $Gd_2Zr_2O_7$/YSZ and a reference single layer YSZ coating with a similar overall top coat thickness of 300-320 µm. All the coatings were manufactured by suspension plasma spraying (SPS), resulting in a columnar structure. Corrosion tests were conducted at 900 °C for 8 hours using vanadium pentoxide and sodium sulphate as corrosive salts at a concentration of 4 mg/cm^2. SEM investigations after the corrosion tests show that $Gd_2Zr_2O_7$ coatings exhibited lower corrosion resistance than the reference material, YSZ. Reaction between the corrosive salts and $Gd_2Zr_2O_7$ results in the formation of gadolinium vanadate ($GdVO_4$) along the top surface and between the columns. While the stresses due to phase transformation of zirconia can be relieved to some extent by realigning of the columns in the top coat, it is believed that $GdVO_4$ formation between the columns, along with low fracture toughness of $Gd_2Zr_2O_7$ had resulted in lower corrosion resistance. Furthermore, the presence of a relatively dense layer of $Gd_2Zr_2O_7$ on the top, as a preventive layer for salt infiltration, did not improve the corrosion resistance.

1 Introduction

Thermal barrier coatings (TBC) are widely employed in the field of gas turbines for protecting the components operating at high temperatures [1-5]. TBC system comprises of an intermediate metallic bond coat (of MCrAlY type where M is either Ni or Co or both, Y is a reactive element) and a ceramic top coat. TBC's are typically deposited using thermal spraying technique. Industry standard material for the ceramic coating currently is 7-8 Wt.% Yttria stabilized zirconia (YSZ). These coatings are functional up to 1200 °C, above which the non-transformable tetragonal prime *(t')* phase of zirconia starts to transform to yttria-rich cubic *(c)* and yttria-poor tetragonal *(t)* zirconia. Tetragonal zirconia is not stable at the operating temperatures of gas turbine and during cooling it transforms to monoclinic zirconia. This results in a volume change of 3-5% which is sufficient to induce cracking and therefore damage the coating [6]. In addition to having limitations of higher temperature operation, YSZ is also prone to corrosive attack form species such as V, Na and S which causes accelerated phase transformation of zirconia from t' to monoclinic [7-8].

With more demands now on increasing the turbine operating temperatures, there is a wide search for materials that could be a potential replacement for YSZ [9-10]. Among the new materials, gadolinium zirconate (GZ) appears promising with low thermal conductivity and high temperature phase stability compared to YSZ. One of the limitations with using gadolinium zirconate is that it is not thermodynamically compatible with alumina (typical oxide that grows between the bond coat and the top coat when exposed to high temperature). It forms a porous layer of $GdAlO_3$ thereby reducing the resistance of oxide layer [11]. In addition, the fracture toughness of gadolinium zirconate is lower compared to YSZ [12]. An approach to address this problem is to have multi-layer coatings with GZ forming the outer layer and YSZ as the base layer.

In addition to the new TBC chemistry, changing the deposition process can have significant effect on the overall lifetime of TBC's. Suspension plasma spraying (SPS), a recent advancement in the deposition techniques of TBC, has proven to improve the cyclic lifetime of coatings compared to conventional air plasma sprayed (APS) coatings [13-15]. SPS coatings have a columnar microstructure and exhibit better strain tolerance than APS YSZ. Thus a proper choice of material and deposition process may improve the overall lifetime of the coating.

The new materials and processes that are developed have to outperform the conventional YSZ if they are to be considered as a potential replacement. In this work the salt corrosion resistance, which is a major failure mode in industrial gas turbines, of a two layer gadolinium zirconate/YSZ and a three layer dense gadolinium zirconate/gadolinium zirconate/YSZ in the presence of vanadium pentoxide (V_2O_5) and sodium sulphate (Na_2SO_4) has been studied. A reference coating of 8 .Wt% YSZ has also been included for comparison.

2 Experimental Set-up

2.1 Materials

The TBC system studied consisted of the substrate,HastelloyX(Ni22Cr1.5Co0.5W9Mo18Fe1Si). On the top of the substrate, a bond coat was applied using the High Velocity Air Fuel system to obtain a nominal thickness of 220 µm. The bond coat composition was a cobalt based alloy (Co32Ni17Cr8Al0.5Y) Amdry 9951. SPS coatings were applied using the Axial III high power plasma torch and Nanofeed 350 suspension feeder. Three top coat types were produced for this study: a single layer YSZ as reference, a double layer gadolinium zirconate ($Gd_2Zr_2O_7$)/YSZ, and a triple layer dense $Gd_2Zr_2O_7$/$Gd_2Zr_2O_7$/YSZ. The triple layer system with relatively denser outer surface layer was produced for enhancing the erosion and corrosion resistance. All the top coat materials were deposited to achieve an overall coating thickness of roughly 300-320 µm.

2.2 Corrosion tests

Corrosion tests were conducted using vanadium pentoxide (V_2O_5) and sodium sulphate (Na_2SO_4) in the ratio of 55:45 Wt.%. The salt concentration used was 4 mg/cm^2 (based on unpublished earlier corrosion tests on APS coatings). The salts were mixed manually and spread over the surface of the sample that was then placed in the furnace and heated to 900 ^{o}C. After holding the samples for 8 hours, they were removed from the furnace and allowed to cool in air.

3. Results

3.1 As-sprayed microstructures

All the SPS coatings resulted in a columnar structure as can be observed form the as-sprayed microstructures in **Fig. 1a-1c** below. From the **Fig. 1a-1c**, a good interface (without any cracks or pores) between the layers in the top coat can be seen. Columnar boundaries or the gaps between the columns, formed during the deposition process, can be observed in the coatings. The third layer of the coating, though sprayed with SPS, does not have a columnar structure. This was achieved by changing the spray parameters during deposition. Having a more compact structure may slow down the infiltration rate compared to a columnar structure.

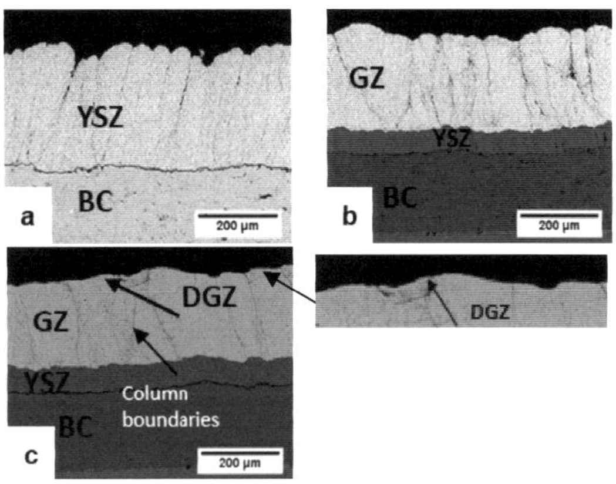

Fig. 1. As-sprayed microstructures of a) single layer YSZ, b) two layer $Gd_2Zr_2O_7$-(GZ)/YSZ, c) three-layer dense $Gd_2Zr_2O_7$(DGZ)/$Gd_2Zr_2O_7$/YSZ.

Porosity analysis in the top coat was done using ImageJ software (available in public domain) and the average porosity value in each coating system is approximately 13-15% (see **Fig. 2**).

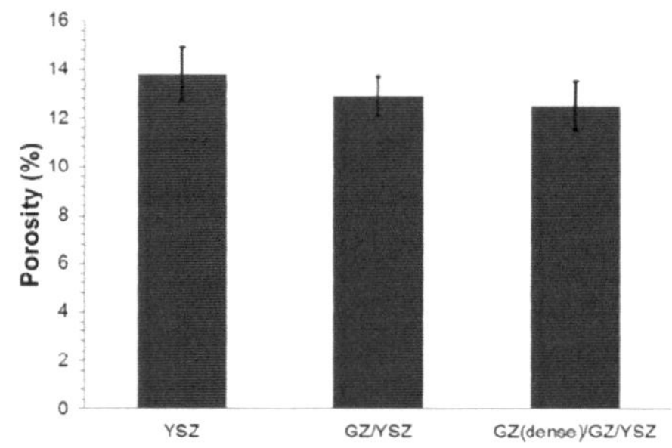

Fig. 2. Average porosity values in each coating system.

3.2 Top view of as corroded surfaces

Fig. 3a, 3b and **3c** show the top view of the as corroded surfaces in single, double and triple layer coating systems respectively. The dark areas shown in the figure represents the corroded yet unspalled regions. The bright areas indicate spalled regions. **Fig. 3d**, **3e** and **3f** show the high magnification SEM images from the top of the above coating systems. No spallation was observed in the single layer YSZ coating. For double and triple layer coating systems, partial spallation had occurred exposing the underlying surface. SEM images were taken in the region where there was no spallation to determine the extent of corrosive product formation on the top surface. YVO_4 rods can be observed all over the corroded surfaces in **Fig. 3d**. **Fig. 3e** and **3f** show the top view of the corroded surfaces in double and triple layer coatings. The rods observed in these figures are of gadolinium vanadate ($GdVO_4$). $GdVO_4$ rods are much thinner as can be seen from **Fig. 3e** and **3f**. The $GdVO_4$ rods are scattered all over the top surface in the two-layer coating while they appear to be clustered in the three-layer coating. Due to the low porosity in the top layer in three-layer coating system the corrosive salts may not penetrate much into the surface and hence $GdVO_4$ appear to be clustered.

XRD analysis (not shown in figure) confirms that the corrosive phases formed are YVO_4 and $GdVO_4$ on the as corroded samples. In as-sprayed samples, peaks of only tetragonal prime zirconia and cubic gadolinium zirconate were observed in YSZ and GZ respectively.

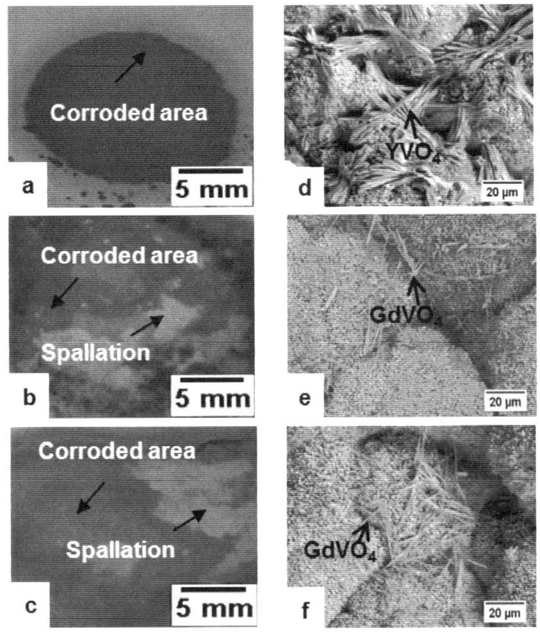

Fig. 3. Top view of the as corroded surfaces showing a) YSZ, b) Two layer GZ/YSZ with spalled areas, c) Three layer DGZ/GZ/YSZ with spalled areas, d)-f) correspond to a)-c), respectively, but in a higher magnification.

3.3 Cross-section analysis

Single layer YSZ coating

Fig. 4 shows the infiltration of the molten salts through the vertical columnar boundaries in the coating. These salts leach yttria from YSZ and subsequently cause phase transformation of ZrO_2. The stresses due to YVO_4 formation (see **Fig. 4**) and phase transformation of zirconia causes cracks in the coating as can be seen in the figure. However, the propagation of these cracks and the extent of damage is dependent on the material and as well as porosity in the coating. The horizontal cracks present in the coating did not result in the spallation of the affected regions within the testing time of the current experiment.

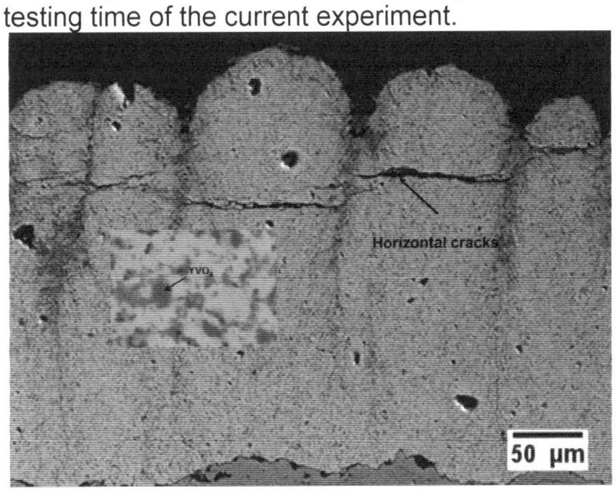

Fig. 4. horizontal cracks in the coating due to the stresses caused by phase transformation of zirconia.

Two-layer $Gd_2Zr_2O_7$ /YSZ coatings

For two-layer coating system the sample was sectioned to obtain information on how the spallation had occurred. **Fig. 5a** shows that spallation of the coating occurred in the second layer (gadolinium zirconate). The YSZ/GZ interface is still intact. From **Fig. 5b** and **5c**, a thin layer of $GdVO_4$ can be observed on the top surface and between the columns. $GdVO_4$ may be formed due to either of the chemical reactions shown by equation(1) or (2) where $NaVO_3$ is the product of reaction between vanadium oxide and sodium

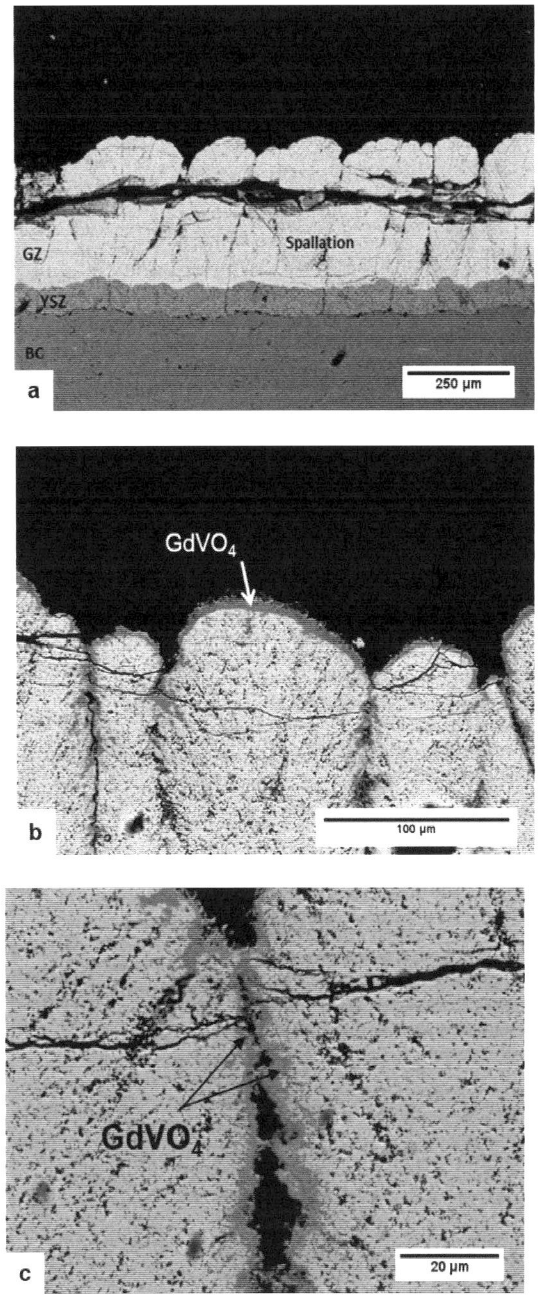

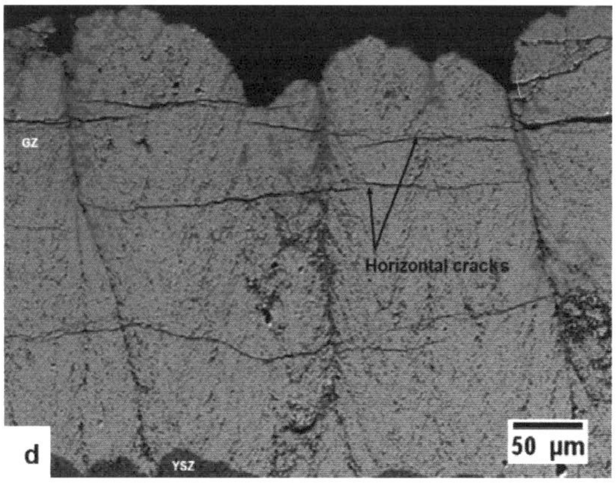

Fig. 5. Cross section of two layer coating system showing a) partial spallation in the second layer, b) continuous $GdVO_4$ formation on the top surface, c) $GdVO_4$ formation between the columns, d) horizontal cracks in the coating due to phase transformation of zirconia.

sulphate. Zirconia in c-$Gd_2Zr_2O_7$ gets transformed into m-ZrO_2 and phase transformation stresses cause cracks in the coating.

$Gd_2Zr_2O_7 + 2NaVO_3 \rightarrow 2GdVO_4 + m\text{-}ZrO_2 + 2Na_2O$ (1)

$Gd_2Zr_2O_7 + V_2O_5 \rightarrow 2GdVO_4 + m\text{-}ZrO_2$ (2)

Fig. 5d shows the cracks in the gadolinium zirconate layer. Comparing it with **Fig. 4**, the YSZ coating, the number of cracks is higher in gadolinium zirconate coating.

Triple layer dense $Gd_2Zr_2O_7$ / $Gd_2Zr_2O_7$ / YSZ coatings

Fig. 6a and **6b** show the cracks in both the second and third layers respectively. In this coating system, spallation had occurred in the second layer similar to a two layer system (not shown in the figure). The molten salts initially react with the compact layer of $Gd_2Zr_2O_7$ forming $GdVO_4$. The stresses here cannot be relieved as the layer is dense (low strain compliance) and causes cracks in the third layer and close to the interface between the third and the second layer. Moreover, due to the presence of column boundaries in the third layer (**Fig. 6a**), the salts infiltrate through them and cause damage from inside the coating (the second layer).

4 Discussion

From **Fig. 3d-3f** it can be observed that the volume fraction of $GdVO_4$, is significantly less than the volume fraction of YVO_4 in the reference coating. This is due to the fact that $GdVO_4$ is more stable thermodynamically than YVO_4 which will result in

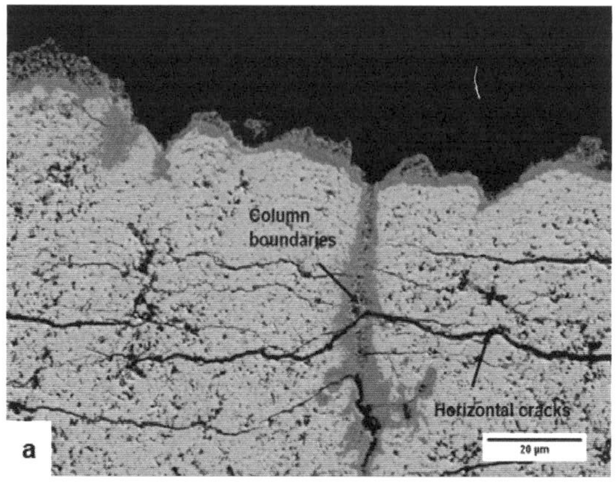

Fig. 6. Cross-section of corrosion tested three-layer $Gd_2Zr_2O_7$/YSZ coatings showing a)horizontal cracks due to phase transformation of zirconia, b)crack in the top layer of dense $Gd_2Zr_2O_7$.

lower growth rate of $GdVO_4$ [8].

In both the cases, the corrosive salts attack the stabilizers, yttria in the case of YSZ, and gadolinia in the case of gadolinium zirconate. As reported in the literature[8], basicity of yttria is higher than gadolinia which means lower reactivity for gadolinia as the corrosion reaction typically follows Lewis acid-base mechanism. Accordingly, the work by Habibi et.al [8] shows that APS coatings with gadolinium zirconate have better corrosion resistance than YSZ coatings in the presence of sodium sulphate and vanadium pentoxide. In the results presented here; higher number of cracks and spallation occurred in the gadolinium zirconate samples and not in the YSZ samples (though internal cracking was observed). This difference could be partly due to different microstructural features in the coating compared to the above referred research.

The columnar structure of the coating provides easy access to the corrosive salts in both YSZ and GZ. But due to the lower reactivity of gadolinia with the corrosive salts compared to yttria, the salts can a) penetrate much deeper, b) stay at the same location without infiltrating into the pores in the columns to the sides. This explains the formation of gadolinium vanadate ($GdVO_4$) between the columns and on the top surface. In the case of YSZ, as yttria is highly reactive with the corrosive salts, the salts have higher tendency to get infiltrated into the pores rather than staying between the columns. This again can explain why there was no corrosive product formed between the columns.

The corrosion induced damage depends on the growth stresses caused by the formation of the corrosive products and stresses caused due to phase transformation of zirconia. In the case of SPS coatings,
the stresses can be relieved to some extent by what can be called as column realignment. This is due to the minimal bonding of columns with their neighboring columns, the coatings can expand or contract when load is applied or with thermal stresses [16]. It is assumed that if anything were to prevent this column realignment (for instance the formation of $GdVO_4$ between the columns) then the stresses would be so high that it can lead to spallation. As a result, the slow reacting gadolinium zirconate is more susceptible to corrosive attack than YSZ. Damage due to corrosion can also be affected by porosity and fracture toughness. The porosity analysis on all the coating systems revealed that the values are quite close to each other. So the effect of porosity can be neglected. On the other hand, the fracture toughness values are lower in gadolinium zirconate compared to 7-8 YSZ [12]. This means the crack propagation is much faster in gadolinium zirconate compared to YSZ. Therefore, in this work, the lower corrosion resistance of gadolinium zirconate coatings is considered due to a combination of corrosive product formation between the columns and low fracture toughness.

A third, relatively dense, layer which was initially thought to improve the corrosion resistance did not show any improvement compared to the two layered coating. The reasons could be that the relatively dense layer still allowed the corrosive salts to infiltrate and it results in cracking in the third layer (see **Fig. 6b**). As the third layer is more compact, it has lower strain tolerance and due to that the stresses may not be relieved. However, the major damage in the three layer coating system is due to the salt infiltration through the column boundaries in the coating. These boundaries are difficult to avoid from a mechanical stand-point. It can be seen from **Fig. 6a** that the number of cracks in the second layer is higher than in the third layer. This means the damage is similar to a two layer coating system and the additional third layer could not improve the corrosion resistance.

It has to be noted that this type of corrosion induced damage can be rather extreme and an actual turbine component may or may not encounter these harsh conditions. Nevertheless, the results of these tests can be used for ranking the materials. It also shows that both the microstructural features as well as material properties need to be considered when designing coatings that will perform well against salt corrosion. A potential solution to the salt penetration problem with SPS coatings could be to use a more reactive material like YSZ as a top layer which would immobilize the salts and combat corrosion. As a continuation to this work, the microstructure of the top dense layer will be optimized and alternate materials instead of gadolinium zirconate as the third layer will be investigated.

5 Conclusion

This study provides information on corrosion resistance of multi-layer suspension plasma coatings. When exposed to a mixture of vanadium pentoxide and sodium sulphate at 900 °C, multi-layer gadolinium zirconate coatings exhibited lower corrosion resistance than the reference single layer YSZ coating. Reaction of the corrosive salts with gadolinium zirconate resulted in its decomposition to gadolinium vanadate (formed between the columns) and monoclinic Zirconia. Presence of $GdVO_4$ between the columns along with the low fracture toughness of gadolinium zirconate is thought to have reduce its corrosion resistance. Furthermore, a relatively dense third layer on the top did not result in improvement of corrosion resistance as the damage occurred more in the second layer due to the salt infiltration through the column boundaries.

6 Acknowledgement

Vinnova in Sweden is acknowledged for research funding.

7 Literature

[1] N.P. Padture, M. Gelll and E.H. Jordan, " Thermal barrier coatings for gas-turbine engine applications", Science(2002), Issue 5566, pp.280/84.

[2] B. Gleeson, " Thermal barrier coatings for aero-engine applications", Journal of Propulsion and Power(2006), Vol 22, pp. 375/83.

[3] DeMasi-Marcin and D.Gupta,"Protective coatings in the gas turbine engine", Surface and coatings technology (1994), Vol. 68-69, pp.1/9.

[4] M.J. Pomeroy, "Coatings for Gas Turbine Materials and Long Term Stability Issues", Materials & Design (2005), Vol. 26 No.3, pp. 223/31.

[5] G.W. Goward, "Progress in coatings for gas turbine foils", Surface and coating technology(1998), Vol. 108-109, pp. 73/79.

[6] S.Bose, "High temperature coatings", Butterworth-Heinemann (2007).

[7] Z.chen, S.Speakman, J.Howe, H.Wang, W. Porter and R. Trice, "Investigation of reactions between vanadium oxide and plasma-sprayed yttria-stabilized zirconia coatings", Journal of European Ceramic Society (2009) ,Vol 29, pp. 1403/1411.

[8] M.H.Habibi, L. Wang and S.M.Guo, "Evolution of hot corrosion resistance of YSZ, $Gd_2Zr_2O_7$, and $Gd_2Zr_2O_7$ + YSZ composite thermal barrier coatings in Na_2SO_4 + V_2O_5 at 1050 °C, Journal of European Ceramic Society (2012), Vol 32, pp. 1635/42.

[9] D.R.Clarke and S.R. Phillpot, "Thermal barrier coating materials", Materials Today(2005), Vol.8, Issue 6, pp. 22/29.

[10] G. Mauer, M.O. Jarligo, D.E. Mack and R. Vassen, " Plasma-Sprayed Thermal Barrier Coatings: New Materials, Processing Issues, and Solutions (2013), Journal of Thermal Spray Technology (2013), Vol.22, Issue 5, pp. 646/58.

[11] R.M. Leckie, s. Kraemer, M. Ruhle and C.G. Levi, "Thermochemical compatibility between alumina and ZrO_2 – $GdO_{3/2}$ thermal barrier coatings", Acta Materialia(2005), Vol. 53, Issue 11, pp. 3281/92.

[12] U.Bast and E.Schumann, "Development of novel oxide materials for TBC's", Ceramic Engineering Science Proceedings (2002), Vol.23, Issue 4, pp. 525/32.

[13] N.Curry, K. VanEvery, T. Snyder and N. Markocsan, "Thermal Conductivity Analysis and Lifetime Testing of Suspension Plasma-Sprayed Thermal Barrier", Coatings (2014), Vol.4, No.3, pp. 630/650.

[14] N. Curry, K. VanEvery, T. Synder, J. Susnjar and S. Bjorklund, "Performance Testing of SuspensionPlasma Sprayed Thermal Barrier Coating Produced with Varied Suspension parameters," Coatings(2015), Vol. 5, No.3, pp.338/356.

[15] S.Mahade, N.Curry, S.Bjorklund, N.Markocsan and P. Nylen, " Thermal conductivity and thermal cyclic fatigue of multi-layered $Gd_2Zr_2O_7$/YSZ thermal barrier coatings processed by suspension plasma spray," Surface & Coatings Technology (2015), Vol. 283, pp. 329/336.

[16] N. Curry, "Design of Thermal Barrier Coatings", PhD. Thesis, 2014.

Plasma Spraying of Suspensions with Hybrid Water-Stabilized Plasma Technology

R. Musalek, J. Medricky, J. Kotlan, T. Tesar, Z. Pala, F. Lukac, T. Chraska, Prague / CZ

Technology of water-stabilized plasma torch was recently significantly updated through introduction of a so-called hybrid concept which combines benefits of water stabilization and gas stabilization principles. High-enthalpy plasma provided by the WSP-H ("hybrid") torch may be used for thermal spraying of powders as well as liquid feedstocks with high feed rates. Results from three selected experiments with suspension plasma spraying with WSP-H technology are presented in this study. Shadowgraphy was used for optimization of the suspension injection and visualization of the liquid fragmentation in the plasma jet. Coating deposition on substrates attached to the rotating carousel with integrated temperature monitoring provided effective cooling and reproducibility of the deposition process. Deposition efficiency was evaluated according to EN ISO 17 836 standard. Possibility of deposition of coatings with various microstructures was demonstrated for three different ceramics (YSZ, YAG and Al_2O_3).

1 Introduction

Plasma spraying with liquid feedstock has recently opened new opportunities for thermal spray technology by spraying solutions and submicron-sized suspensions providing coatings with novel types of microstructures, e.g. columnar or dense structures [1–5]. On the other hand, new challenges also emerged, e.g. formulation of ready-to-spray suspensions, equipment development, proper injection and fragmentation of the liquid in the plasma jet, cooling of the substrates due to short stand-off distance, etc. [6–8].
Evaporation of liquid phase (solvent) from the injected liquid consumes significant portion of the plasma enthalpy. Spraying with high-enthalpy torches is therefore preferred in order to provide efficient deposition with considerable feed rates [9, 10].
In our previous study [11], feasibility of YSZ (yttria stabilized zirconia) suspension spraying was demonstrated for water-stabilized plasma torch developed at Institute of Plasma Physics CAS, v.v.i. (Czech Republic). In this study, three selected examples of coatings deposited with the latest generation of "hybrid" water-argon plasma torch WSP®-H 500 are presented. Deposition design was also updated, in particular in spraying onto samples mounted to a rotating carousel, which substantially helped to improve repeatability of the deposition and prevent excessive heating of the samples.
The most recent hybrid torch WSP®-H 500 used in this study combines principles of arc stabilization by gas flow and water vortex. First, argon plasma is generated in the tungsten cathode part (gas stabilization) and then enters water-stabilized section, where plasma enthalpy is substantially increased through water evaporation and steam ionization. Generated high-speed plasma can be used for cost-effective thermal spraying of powders and liquids with high throughput. More details on hybrid WSP-H technology are provided e.g. in [12–19].

2 Materials and Methods

Hybrid water-stabilized plasma torch WSP®-H 500 (ProjectSoft HK a.s., Czech Republic) was used for spraying (**Fig. 1**). Torch was operated at 500 A (~150 kW) and 15 slpm argon flow rate in the cathode section. Pneumatic feeding system was used for single point radial liquid injection.
Three ethanol-based suspensions were sprayed. Ready-to-spray suspensions of yttria-stabilized zirconia (YSZ, 8 wt. %) with d50 = 0.58 µm and yttrium aluminium garnet (YAG) with d50 = 1.46 µm were delivered from Treibacher Industrie AG (Austria). Concentrations of solids in the suspension were 25% and 40%, respectively. Al_2O_3 suspension with 10% solids load was prepared in-house from powder with nominal size 300 nm provided by Allied High Tech Products (USA).

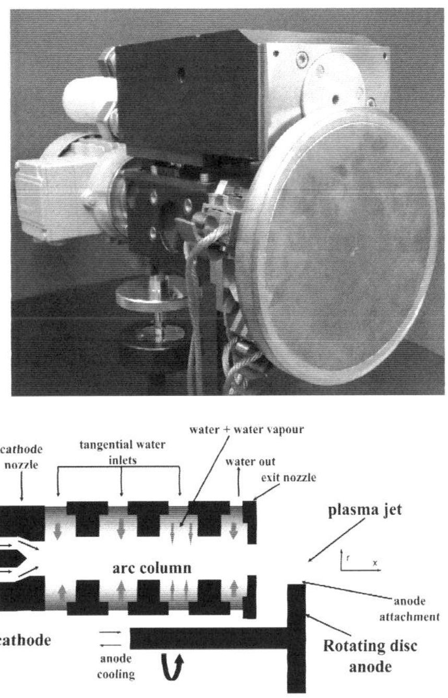

Fig. 1. Photo and scheme of hybrid WSP-H torch [19].

Injection optimization procedure was based on our previous study [20]. In order to provide optimum penetration of the liquid into the plasma jet core, uniform fragmentation and heat treatment of the deposited material, injection (feeding) angle was set to 25°. Spraying (stand-off) distance was set to 100 mm. Suspension was fed through calibrated sapphire nozzle with 0.35 mm internal diameter producing laminar stream. Feeding pressure was optimized for each suspension by incorporation of shadowgraphy system SprayCam (Control Vision Inc, USA), which also enabled observation of the liquid stream fragmentation in the plasma jet. Additional deposition parameters are listed in **Table 1**.

Substrates (20 x 30 x 2.5 mm) were prepared from AISI 304 stainless steel, grit-blasted by alumina grit, cleaned in aceton in ultrasonic bath and mounted for spraying to the rotating carousel (see **Fig. 2**). Relative speed between the plasma torch and the substrate surface was 650 mm/sec.

Fig. 2. Samples on carousel after deposition.

Samples were preheated prior to the deposition by a brief exposure to the plasma torch with deactivated suspension feeding. Carousel and mounted samples were continuously cooled from the side by compressed air in order to reduce the heat build-up. However, after each deposition cycle (3 x up&down robot strokes = 6 deposition passes), sample temperature was increased by several hundred degrees. The torch was therefore deflected from the substrates and brief cooling period followed. Deposition resumed after the substrate temperature was lowered to a predefined value. This procedure was repeated until preselected number of deposition cycles was achieved. In this way, highly reproducible and controlled deposition conditions were established. Example of sample temperature history as monitored by IR camera (front side) and thermocouple (attached to the back side of one of the substrates) is given in (**Fig. 3**).

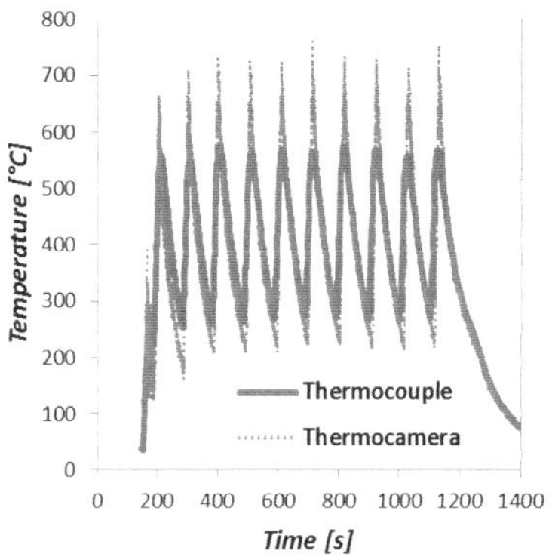

Fig. 3. Temperature record of Al_2O_3 deposition.

Cross-sections of the samples were prepared by standard materialographic procedure and observed by SEM microscope EVO MA 15 (Carl Zeiss, Germany). Coating thickness and approximate value of porosity was evaluated by image analysis of the cross-sections. XRD analysis was carried out with Bruker D8 Discover diffractometer (Germany).

3 Results

3.1 Deposition setup

Effective fragmentation of the injected laminar suspension stream in the plasma jet is illustrated in **Fig. 4**. Using shadowgraphy technique enabled optimization of the feeding pressure which was increased until the mean droplet trajectory was identical with the centerline of the plasma jet. After deposition, optimized feeding pressure was used for measurement of suspension feedrate which varied between 90 to 120 ml/min depending on the suspension type (see **Table 1**).

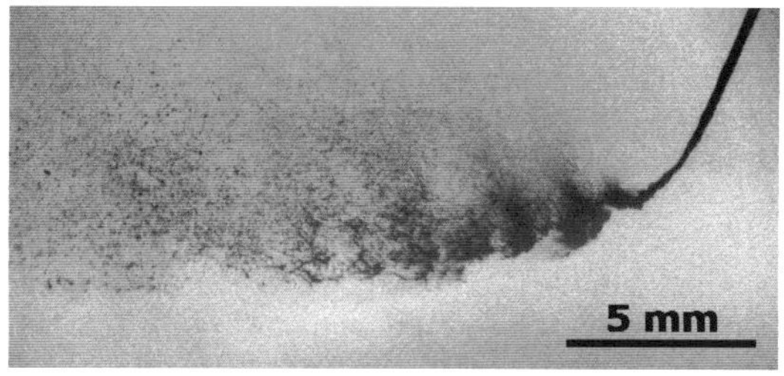

Fig. 4. Fragmentation of YAG suspension in the plasma jet. Feeding pressure 4 bars (92 ml/min).

Table 1: Deposition parameters.

Suspension	YSZ	YAG	Al_2O_3
Solid load [wt%]	25	40	10
Feeding distance [mm]	30	20	25
Feeding pressure [bar]	3.0	4.0	4.0
Feedrate [ml/min]	98	92	120
Preheating	yes	no	yes
Number of deposition cycles	15	3	10
Total number of deposition passes	90	18	60
Coating thickness [µm]	168 ± 12	99 ± 13	79 ± 6
Porosity (IA) [%]	25.0 ± 0.6	6.8 ± 1.0	5.5 ± 0.7

3.3 Deposits microstructure

For YSZ, formation of columnar microstructure was observed (**Fig. 5**). Between the columnar gaps, regions of highly sintered as well as of rather porous material with numerous detached spherical particles were observed. Interpass porosity boundaries were imperceptible. Overal porosity distinguishable by image analysis was about 25%. The XRD analysis confirmed that only desirable tetragonal phase was present within the deposit (**Fig. 6)** which confirms conclusions of our previous work obtained with the previous generation of WSP technology [11].

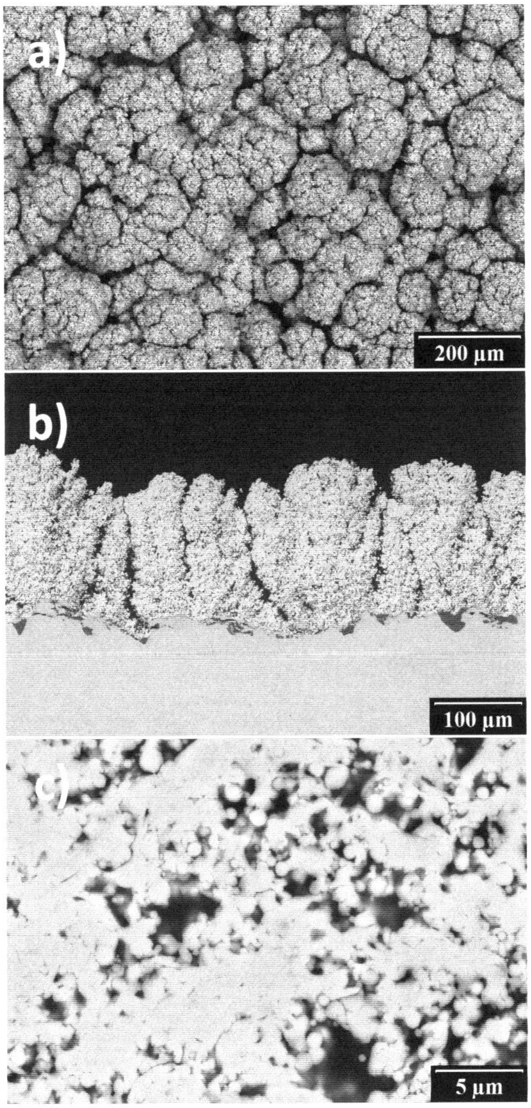

Fig. 5. Deposit of YSZ. a) Free-surface, b) cross-section and c) detail of microstructure.

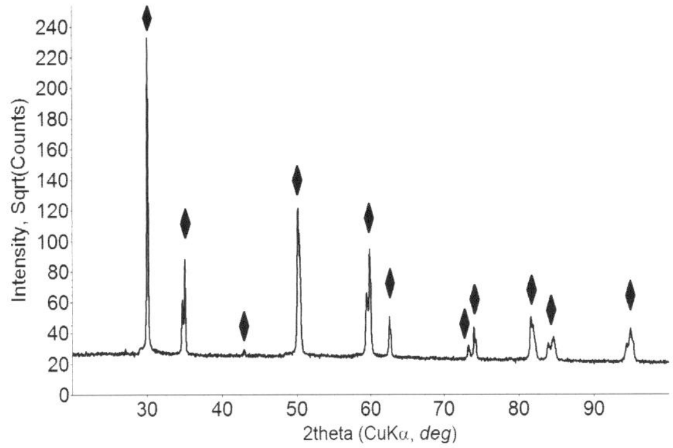

Fig. 6. XRD pattern of YSZ deposit. Diamonds indicate positions of peaks of tetragonal phase.

For YAG and Al_2O_3, deposition of a compact and dense coating uniformly covering the substrate was successfully achieved (**Fig. 7** and **8**). Splats were approximately 1 to 2 orders of magnitude smaller than those produced by conventional APS spraying with powder feedstock.

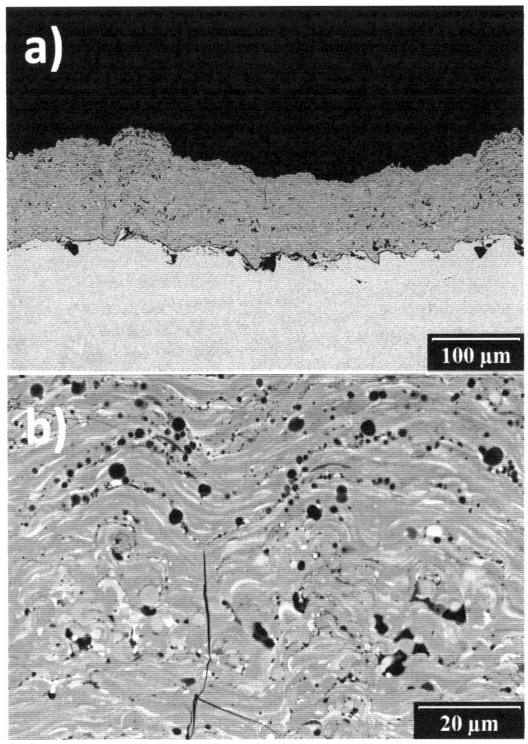

Fig. 7. Cross-section of YAG coating.

For both materials, porosity was in the range of 5 to 7 % as evaluated by image analysis of the cross-sections. However, nature of porosity was different; in the case of YAG, most of the porosity was due to the presence of spherical intrasplat pores whereas in the case of Al_2O_3, intersplat porosity was characteristic (compare **Fig. 7b** and **Fig. 8b**).

No preheating was used for YAG, as the deposits showed pronounced tendency to delamination during cooling phase when the preheating was used. However, localized short cracks were observed within the YAG microstructure even if no preheating was used (**Fig. 7b**). Formation of such cracks relieving internal stresses arising due to the CTE mismatch of the deposit and the substrate are quite typical for YAG coatings and in general illustrate excellent bonding between the deposited splats.

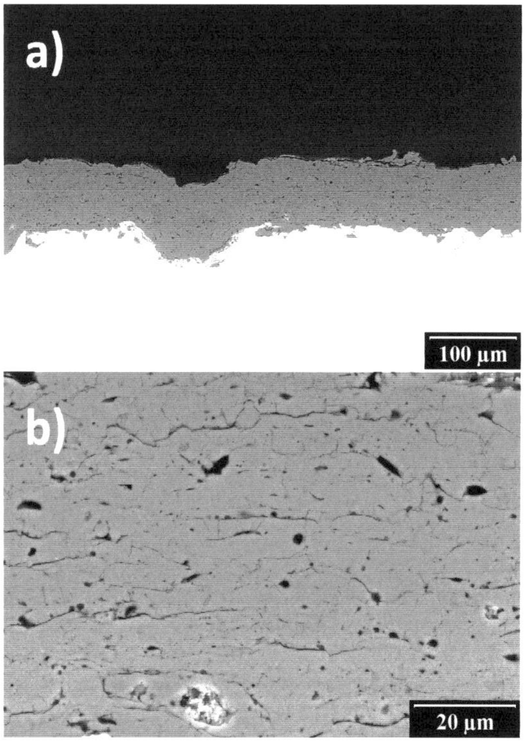

Fig. 8. Cross-section of Al_2O_3 coating.

3.1 Deposition efficiency (DE)

Due to the introduction of unified and repeatable deposition conditions, it was possible to evaluate relative deposition rate as a ratio of final coating thickness and total number of deposition passes (**Table 2**). Please note that the provided value is informatory and dependent on many variables, in particular mutual velocity and trajectory of the torch and the sprayed surface. However, its computation is for a fixed deposition setup straightforward and with knowledge of this value, it is possible to prepare coatings with predefined coating thickness. As expected, relative deposition rate is highly dependent on the solid load concentration being highest for the YAG suspension with the highest solid phase concentration.

Alternatively, deposition efficiency was estimated from the known total spraying time, solid load concentration, density of the solid phase and the solvent as a ratio of the experimental coating thickness (obtained from image analysis) and theoretical coating thickness computed under assumption that all of the solid phase was deposited.

In order to validate such DE estimate, deposition efficiency was evaluated for YSZ suspension according to EN ISO 17836 standard in the separate experiment. The spraying setup was the same as for previous spraying experiments, but this time carousel with mounted samples was replaced by a tube (168 mm diameter) also manufactured from AISI 304 stainless steel (**Fig. 9**).

Rotation speed of the tube was adjusted so that the mutual velocity of the torch and the sprayed surface was the same as for spraying with carousel. The tube was grit-blasted, cleaned in acetone and weighed before and after the deposition. Temperature of the sprayed surface was again monitored with the thermocamera in order to achieve comparable temperature history as for spraying with the carousel. From the known amount of spent suspension and weight of the deposit, deposition efficiency was evaluated. Experiment was repeated twice with the result DE = 54.3% and 52.4%.

From the comparison of estimated and measured DE values provided in **Table 2**, a very good agreement is apparent. It may be therefore assumed, that the DE estimates are legitimate also for YAG and Al_2O_3 deposition where a lack of suspension prevented to carry out DE evaluation according to EN ISO 17836.

It may be concluded that experimental deposition of YSZ was carried out with a deposition efficiency remarkably high for this type of coating with columnar microstructure and radial suspension injection. For both YAG and Al_2O_3 coatings, deposition efficiency was even higher in the range generally acceptable for conventional plasma spraying with powders. Moreover, further increase of DE value may be expected after optimization of the deposition process.

Table 2: Deposition efficiency.

Suspension	YSZ	YAG	Al_2O_3
Solid load [wt.%]	25	40	10
Deposition rate [µm/pass]	1.9 ± 0.1	5.5 ± 0.7	1.3 ± 0.1
DE estimate	54.6%	73.8%	64.1%
DE (EN ISO 17836)	53.4%*	-	-

* mean value from two measurements.

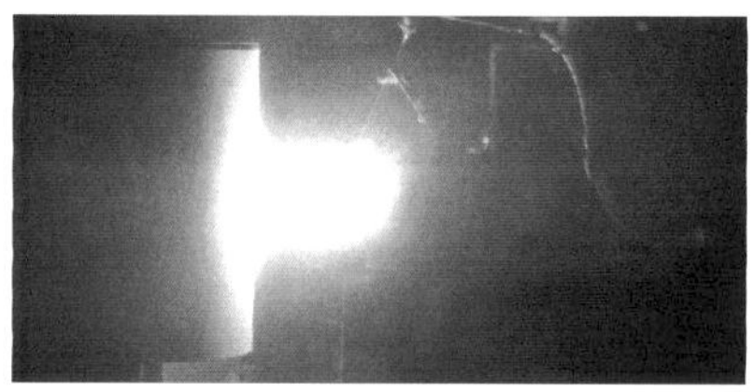

Fig. 9. Spraying on AISI 304 tube for DE evaluation.

4 Conclusions

Applicability of the latest generation of hybrid water-stabilized plasma technology for suspension plasma spraying was demonstrated. High enthalpy of the torch allowed for cost-effective suspension spraying with considerable feedrates (up to 120 ml/min) and industrially relevant deposition efficiency (up to 60 - 70%). Further increase in the feedrate may be expected as no unmolten material was observed within the deposits. This makes the WSP-H process potentially interesting for industrial suspension spraying of large-scale parts and thick coatings. Furthermore, capability of deposition of thin coatings from very fine powders (introduced into suspensions) and homogeneously covering the whole sprayed surface opens a another potential window for WSP-based technology which is sometimes considered to be best suited mainly for deposition of thick (even free-standing) deposits.
Integration of the latest generation of the hybrid WSP-H torch, shadowgraphy and suspension spraying on carousel with integrated online temperature monitoring enabled deposition of the coatings with highly reproducible conditions. Negligible interpass porosity within the deposits demonstrates succesful penetration of the injected liquid into the plasma jet core and stability of the deposition process.
Possibility of deposition of coatings with tailored microstructures was also demonstrated. A columnar microstructure of YSZ deposit with macroscopic porosity level of about 25% and purely tetragonal phase composition without detrimental Badelleyite seems promising for further development of this coating for example for sintering resistant and strain tolerant thermal barrier coatings. Contrary, possibility of deposition of dense deposits with low porosity was demonstrated for YAG and Al_2O_3.

5 Acknowledgements

Fruitful discussions with Nicholas Curry (Treibacher Industrie AG) as well as financial support provided by Czech Science Foundation through grant No. GA15-12145S (Physical aspects of plasma spray deposition from liquid feedstock) are gratefully acknowledged.

6 Literature

[1] Killinger, A., Gadow, R., Mauer, G., Guignard, A., Vaen, R., and Stöver, D.: Review of new developments in suspension and solution precursor thermal spray processes. Journal of Thermal Spray Technology 20 (2011), Issue 4, pp. 677–695.

[2] VanEvery, K., Krane, M. J. M., Trice, R. W., Wang, H., Porter, W., Besser, M., Sordelet, D., Ilavsky, J., and Almer, J.: Column formation in suspension plasma-sprayed coatings and resultant thermal properties. Journal of Thermal Spray Technology 20 (2011), Issue 4, pp. 817–828.

[3] Pawlowski, L.: Suspension and solution thermal spray coatings. Surface and Coatings Technology 203 (2009), Issue 19, pp. 2807–2829.

[4] Vassen, R., Kaner, H., Mauer, G., and Stöver, D.: Suspension plasma spraying: Process characteristics and applications. Journal of Thermal Spray Technology 19 (2010), Issue 1-2, pp. 219–225.

[5] Fauchais, P. and Montavon, G.: Latest developments in suspension and liquid precursor thermal spraying. Journal of Thermal Spray Technology 19 (2010), Issue 1-2, pp. 226–239.

[6] Fauchais, P., Vardelle, M., Goutier, S., and Vardelle, A.: Key Challenges and Opportunities in Suspension and Solution Plasma Spraying. Plasma Chemistry and Plasma Processing 35 (2014), Issue 3, pp. 511–525.

[7] Fauchais, P., Vardelle, M., Vardelle, A., and Goutier, S.: What Do We Know, What are the Current Limitations of Suspension Plasma Spraying? Journal of Thermal Spray Technology 24 (2015), Issue 7, pp. 1120–1129.

[8] Toma, F.-L., Potthoff, A., Berger, L.-M., and Leyens, C.: Demands, Potentials, and Economic Aspects of Thermal Spraying with Suspensions: A Critical Review. Journal of Thermal Spray Technology 24 (2015), Issue 7, pp. 1143–1152.

[9] Curry, N., VanEvery, K., Snyder, T., Susnjar, J., and Bjorklund, S.: Performance Testing of Suspension Plasma Sprayed Thermal Barrier Coatings Produced with Varied Suspension Parameters. Coatings 5 (2015), Issue 3, pp. 338–356.

[10] Ganvir, A., Curry, N., Björklund, S., Markocsan, N., and Nylén, P.: Characterization of Microstructure and Thermal Properties of YSZ Coatings Obtained by Axial Suspension Plasma Spraying (ASPS). Journal of Thermal Spray Technology 24 (2015), Issue 7, pp. 1195–1204.

[11] Mušálek, R., Bertolissi, G., Medřický, J., Kotlan, J., Pala, Z., and Curry, N.: Feasibility of suspension spraying of yttria-stabilized zirconia with water-stabilized plasma torch. Surface and Coatings Technology 268 (2015), pp. 58–62.

[12] Matějíček, J., Kavka, T., Bertolissi, G., Ctibor, P., Vilémová, M., Mušálek, R., and Nevrlá, B.: The role of spraying parameters and inert gas shrouding in hybrid water-argon plasma spraying of tungsten and copper for nuclear fusion applications. Journal of Thermal Spray Technology 22 (2013), Issue 5, pp. 744–755.

[13] Hrabovsky, M.: Thermal plasma generators with water stabilized arc. The Open Plasma Physisc Journal 2 (2009), Issue 1, pp. 99–104.

[14] Hlína, M. and Hrabovský, M.: Enthalpy Probe Diagnostics of Steam/Argon Plasma Jet. Plasma Physics and Technology 2 (2015),

Issue 2, pp. 142–145.
[15] Hrabovsky, M.: Generation of thermal plasmas in liquid-stabilized and hybrid dc-arc torches. Pure and Applied Chemistry 74 (2002), Issue 3, pp. 429–433.
[16] Kavka, T., Matějíček, J., Ctibor, P., Mašláni, A., and Hrabovský, M.: Plasma Spraying of Copper by Hybrid Water-Gas DC Arc Plasma Torch. Journal of Thermal Spray Technology 20 (2011), Issue 4, pp. 760–774.
[17] Kavka, T., Matějíček, J., Ctibor, P., and Hrabovský, M.: Spraying of Metallic Powders by Hybrid Gas/Water Torch and the Effects of Inert Gas Shrouding. Journal of Thermal Spray Technology 21 (2011), Issue 3-4, pp. 695–705.
[18] Kavka, T., Kopecky, V., Sember, V., Maslani, A., and Chumak, O.: Analysis of plasma jets generated in gas and gas-water torches. High Temperature Material Processes 11 (2007), Issue 1, pp. 59–70.
[19] Jeništa, J., Takana, H., Nishiyama, H., Bartlová, M., Aubrecht, V., Křenek, P., Hrabovský, M., Kavka, T., Sember, V., and Mašláni, A.: Integrated parametric study of a hybrid-stabilized argon–water arc under subsonic, transonic and supersonic plasma flow regimes. Journal of Physics D: Applied Physics 44 (2011), Issue 43, pp. 435204.
[20] Kotlan, J., Pala, Z., Musalek, R., and Ctibor, P.: On reactive suspension plasma spraying of calcium titanate. Ceramics International 42 (2016), Issue 3, pp. 4607–4615.

Thermal fatigue failure of thermal barrier coatings with a high-Cr MCrAlY bond coat

Kang Yuan[1], Krishna Praveen Jonnalagadda[2], Yueguang Yu[1], Ru Lin Peng[2], Xin-Hai Li[3], Xiaojuan Ji[1], Jie Shen[1]
[1]*Beijing General Research Institute of Mining and Metallurgy, Beijing 100160, China*
[2]*Department of Management and Engineering, Linkoeping University, SE-58183 Linkoeping, Sweden*
[3]*Siemens Industrial Turbomachinery AB, SE-61283 Finspong, Sweden*

Abstract. Thermal barrier coatings (TBCs) were air-plasma sprayed onto Hastelloy X substrates. The TBCs consisted of a high-Cr MCrAlY (M for Ni and Co) bond coat and a yttria-stabilized zirconia (YSZ) top coat. The TBC samples were thermally cycled between 100 °C and 1100 °C with 1 hour dwell time at 1100 °C. The thermal fatigue failure of the TBCs was investigated via microstructure analyses. The final fatigue failure of the TBCs was caused by the formation of interface-parallel cracks in the YSZ top coat. The formation of the cracks was found to be strongly related to the oxidation behaviour of the MCrAlY bond coat. The development of the oxide layers was therefore studied in detail. A thermo-kinetic model was also used to deepen the understanding on the elemental diffusion behavior in the materials.

1 Introduction

The lifetime of thermal barrier coatings (TBCs) is strongly related to the oxidation behaviour of MCrAlY (M for Ni/Co) bond coat (BC) and the fracture toughness of ytttia-stabilized zirconia (YSZ) top coat (TC) [1, 2]. During high temperature exposure, a thermally-grown oxide (TGO) layer is formed at the BC/TC interface. For a good oxidation resistance design, alumina is the preferred TGO since alumina scale is usually dense and pure and has good compatibility with both MCrAlY and YSZ coatings [3-5]. However, in a real service environment, corrosion could also occur when components are exposed to relatively lower temperatures. Then a high-Cr MCrAlY coating is needed especially in sulphur corrosion conditions [6, 7]. A problem for the high-Cr MCrAlY coatings, however, is the formation of MCr_2O_4 spinel oxides (M for Ni/Co) which weekens the interface strength and introduces more stresses in TC [1, 8]. To avoid the formation of the spinels but still form chromia layer, Yamano H. et al. applied chromate pretreatment on MCrAlY coatings to form a chromia-alumina bi-layer TGO structure [9]. Some efforts have also been put by other researchers to get oxidation-corrosion-resistant coatings e.g. "smart coatings" aiming at forming alumina scale at high temperatures and chromia scale at low temperatures [10].
In this study, TBCs with a high-Cr MCrAlY were tested under thermal cycling condition. It was found that chromia-alumina bilayer structure was formed in a specific microstructure, indicating that such an MCrAlY coating may potentially be used for both oxidation and corrosion applications. In this paper, the thermal cyclic fatigue behaviour of the TBCs associated with the oxidation behaviour of the MCrAlY coating was particularly investigated.

2 Experimentals

2.1 Materials and tests

MCrAlY BC (Ni-22Co-25Cr-7Al-0.5Y, wt.%) and (7-8%) YSZ TC were successively air-plasma sprayed (APS) on 25 mm-diameter Hastelloy X discs (Ni-1.5Co-18Fe-22Cr-0.6W-9Mo-1Si-0.1C, wt.%). The spray parameters are shown in **Table 1**; two BCs, namely *BC1* and *BC2*, were made using different spraying paramters.
Thermal cyclic fatigue (TCF) test was performed on the TBC samples in a cyclic furnace. The samples were cycled between 100 °C and 1100 °C with 1 hour dwell time at 1100 °C and 10 min cooling down to approximately 100 °C. The coating lifetime was defined as the time when 20% TC was spalled off from the sample. Scanning electron microscope (SEM) equipped with energy dispersive spectroscope (EDS) was used to analyze the microstructure and chemical composition in the materials.

Table 1. APS spraying parameters.

	Ar flow rate, L·min-1	H_2 flow rate, L·min-1	Plasma power /kW	Powder feed rate/ (g·min-1)	spray distance, mm
BC1	50	10	46	~35	140
BC2	50	10	46	~50	100
TC	38	13	35	~25	95

2.2 Simulation setting-up

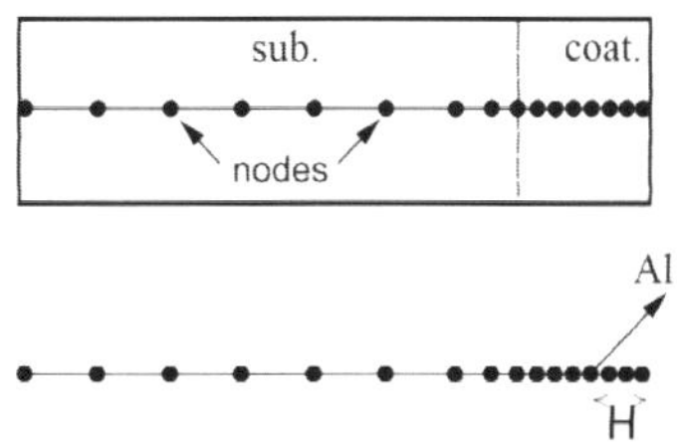

Fig. 1. Oxidation-diffusion model. Al oxidation was simulated by removing Al content at depth H = 10 μm under the MCrAlY interface.

An oxidation-diffusion model [11-13], as shown in **Fig. 1**, was used to simulate the elemental diffusion behaviour in the Hastelloy X - MCrAlY system. The surface oxidation of Al was considered and intersplat oxidation was ignored as well as the oxidation of other elements like Cr, Ni and Co. The oxidation of Al in the model was realized via removing Al content from the grids; the Al removal was set at distance = 10 µm based on the microstructure study that the alumina layer was formed about 10 µm (an estimated average value) below the BC/TC interface. Parabolic oxidation law was used as shown in Eq. 1.

$$h = (kp*t)^{1/2} \qquad (1)$$

In Eq. 1, h is the alumina thickness (µm) at thermal exposure time t (hour) and the oxidation constant k_p was 0.233 µm^2/h for 1100 °C [12]. Since the oxidation law was not necessarily accurate for a semi-quantitative calculation, the modelling results would be mainly used to deepen the understanding on the elemental diffusion behaviour rather than to predict the coating life.

3 Results and analyses

3.1 Failure of the TBCs

Fig. 2 gives the cross sections of the as-sprayed TBCs with *BC1* and *BC2*. Typical splat-on-splat microstructures were found. Intersplat oxidation occurred in BC during the APS spray process, and alumina was formed along the splats' boundaries. It was further observed that the intersplat oxide layers were thicker in *BC1* than that in *BC2*, which was due to the usage of different spray parameters (**Table 1**). Lower powder-feed rate and longer spray distance could introduce more heating and pre-oxidation on the MCrAlY powders in the plasma stream resulting in slightly higher intersplat oxidation in *BC1*. The two BCs had surface roughness about 8-10 µm (the roughness was obtained by image analyses over the cross sections of the materials with a cut-off length of 0.8 mm).

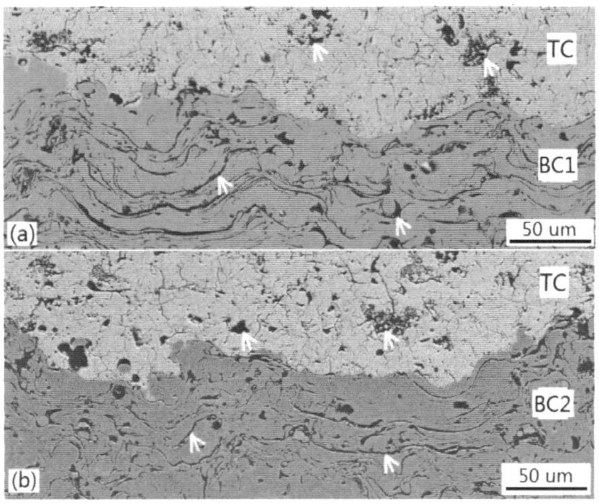

Fig. 2. Cross sections of as-sprayed TBC samples: (a) with *BC1*, (b) with *BC2*. The arrows denote pores and cracks in TCs and intersplat oxides in BCs.

Table 2. TCF lifetimes of TBCs (time to 20% TBC spallation).

Sample	TCF life
BC1+TC	650 cycles, failed
	> 600 cycles, unfailed at 600 cycles
BC2+TC	635 cycles, failed
	> 600 cycles, unfailed at 600 cycles

The TCF lifetimes of the TBCs are shown in **Table 2**. The TBCs failed at about 600-650 cycles, and no large difference of the lifetimes between *BC1+TC* and *BC2+TC* samples was shown. Microstructure has been carefully analysed in both the failed samples and in the interrupted test samples (which were removed from the TCF furnace before TBC spallation). No significant difference of the microstructures for TBC failure was observed between *BC1+TC* and *BC2+TC* samples. This indicates that the observed minor microstructure difference between *BC1* and *BC2* had little influence on the failure mechanisms of the TBC samples. In the following analysis, only the microstructure in *BC1+TC* samples will be shown.
Long horizontal cracks in TC parallel with the BC/TC interface were commonly observed in the TBC at 600 cycles (**Fig. 3**). Such horizontal cracks were also typically found in the failed samples, indicating that the TBC spallation was caused by such horizontal cracks.

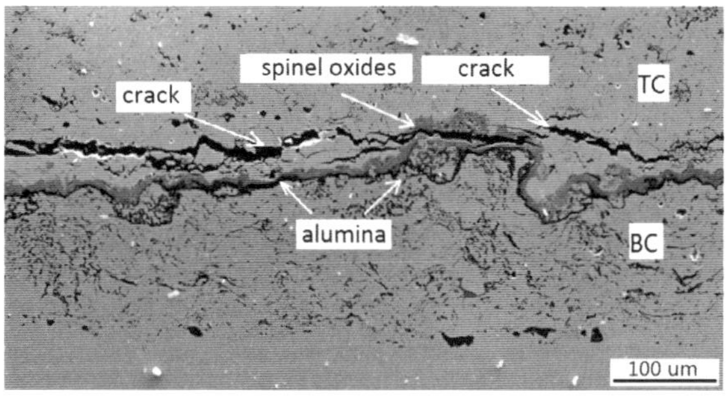

Fig. 3. A typical cross section of a TBC at 600 cycles (unfailed yet).

To analyse the formation and propagation of the horizontal cracks, the thermal cyclic oxidation behaviour of the MCrAlY coating at shorter cycles was studied. **Fig. 4** compares the typical microstructures of the oxides formed at the BC/TC interface at 200, 400 and 600 cycles. "Sandwich" oxide structure was formed at the interface, which was composed of an outer chromia layer, a Ni/Co rich metallic outer zone (OZ), and an internal alumina layer (see **Fig. 4a, b**). The outer oxide layer contained a small amount of $(Ni,Co)Cr_2O_4$ spinel oxides but chromia was dominant (see EDS mapping results in **Fig. 5**). In the Ni/Co rich OZ some chromia and alumina precipitates were found. At long cycles (e.g. 600 cycles in **Fig. 4c**), the OZ was significantly oxidized, and spinel oxides were greatly formed. After long exposure time, the chromia was missing in the outer oxide layer probably due to the following reaction (Eq.2):

$$Cr_2O_3 + (Ni,Co)_xO_y \rightarrow (Ni,Co)Cr_2O_4 \quad (2)$$

The overview of the oxide structure at different cycles is shown in **Fig. 6**. The "sandwich" oxide structure was not uniform at BC/TC interface (**Fig. 6a, b**). The thickness of the "sandwich" could reach 50 µm in some locations but may not be found in some other positions where a chromia-alumina bi-layer structure was formed instead. An average value of the OZ thickness was estimated as 10 µm according to SEM observation. The inhomogeneity of the "sandwich" oxide structure was probably due to the rough surface and intersplat oxidation. Spinels were significantly observed at 600 cycles, as shown in **Fig. 6c**. The local volumetric increase accompanying the formation of the spinel could induce extra tensile stresses in the neighbouring TC and promote the formation and propagation of horizontal cracks [1, 8], as shown in **Fig. 6c**.

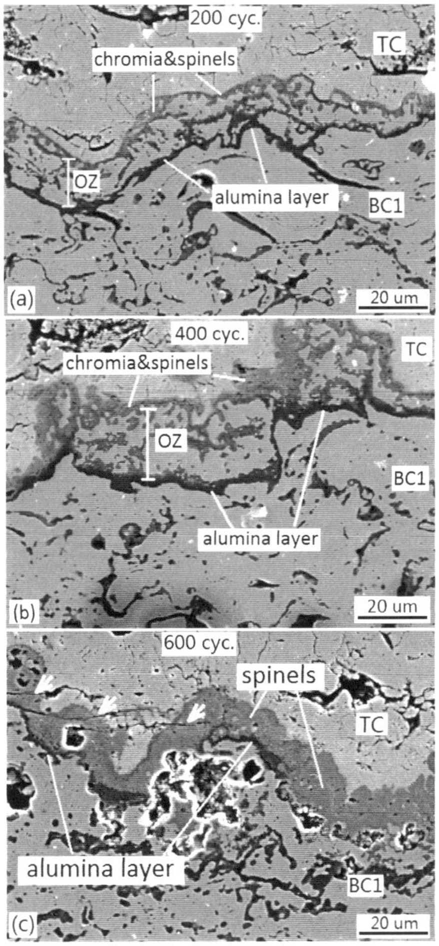

Fig. 4. Morphology of the oxidation near the *BC1* interface at (a) 200 cycles, (b) 400 cycles and (c) 600 cycles. The arrows in (c) denote the cracks due to the thermal fatigue.

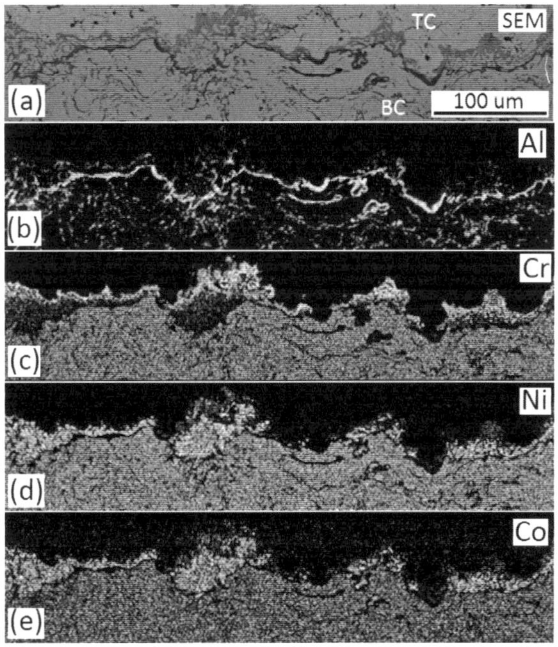

Fig. 5. EDS mapping results in the TBC at 200 cycles.

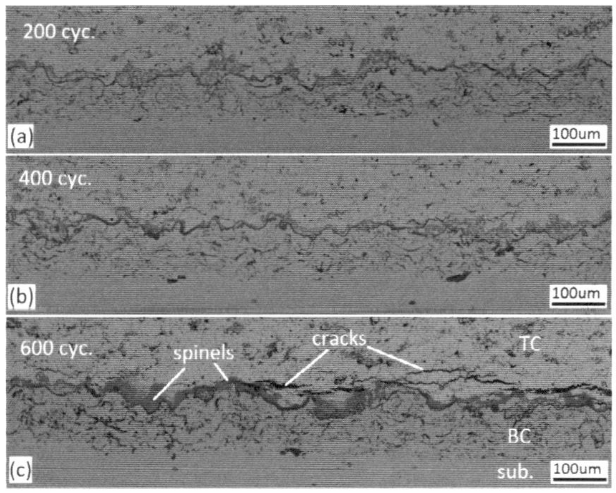

Fig. 6. Overview of the cross sections of the TBCs after (a) 200, (b) 400, and (c) 600 cycles.

Some researchers have shown that chromia and spinels were usually formed at high-Cr low-Al MCrAlY coatings at the transient stage of oxidation processes [9, 11, 14]. The outer chromia layer (as well as some spinels around the layer) in the "sandwich" structure could be formed in the transient stage of the TCF oxidation process. A similar "sandwich" structure was also observed in an oxidized HOVF-sprayed CoNiCrAlY coating (28% Cr, 7.5% Al) at 1100 ºC [11]. According to Wagner [15, 16], the transition from internal oxidation (forming continuous layer) to external oxidation (not forming continuous layer) occurs when the solute content (say Al here) is sufficient to form a critical volume fraction of precipitates at the reaction front. Before the formation of the internal continuous alumina layer, the inward diffusion of O competed with the outward supply of Al from the BC. Until to a certain distance in BC where O content became so low such that alumina precipitates can be kinetically grown to form a layer. After the continuous alumina layer was formed, the growth of the alumina became dominant in the following TCF process.

3.2 Elemental diffusion and thermo-kinetic simulation

Fig. 7a shows the development of Al content in OZ (Al_OZ) and the underlying BC (Al_BC) from experimental measurement and simulation (Al_sim); **Fig. 7b** schematically gives the OZ-alumina-BC microstructures in the MCrAlY to show where the Al content was measured. The experimental data showed that the Al content in the OZ (Al_OZ) decreased down to almost zero (< 0.1 wt.%) at quite early oxidation stage. The Al content under the alumina layer (Al_BC), however, was high enough to support the thickening of the alumina layer until up to about 600 cycles when the Al content became very low (< 0.5%) and the MCrAlY reached it's Al-depletion life.

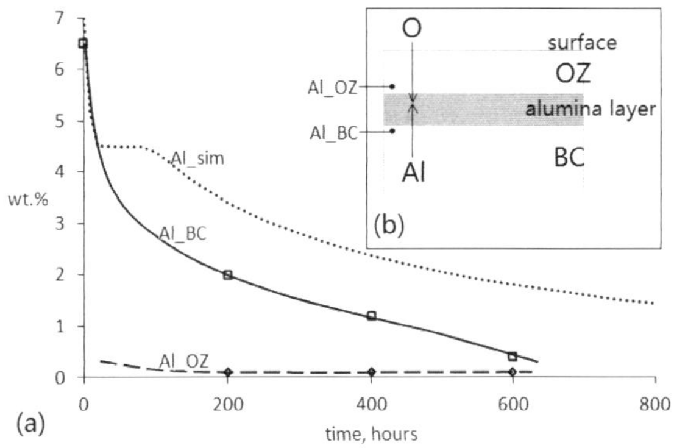

Fig. 7. (a) Development of the Al content by experiment and simulation. (b) Schematic image showing the OZ-alumina-BC structure. Al_sim (simulation result of the Al content under MCrAlY interface), Al_OZ (EDS-measured Al content above the alumina layer in the OZ), Al_BC (EDS-measured Al content under the alumina layer).

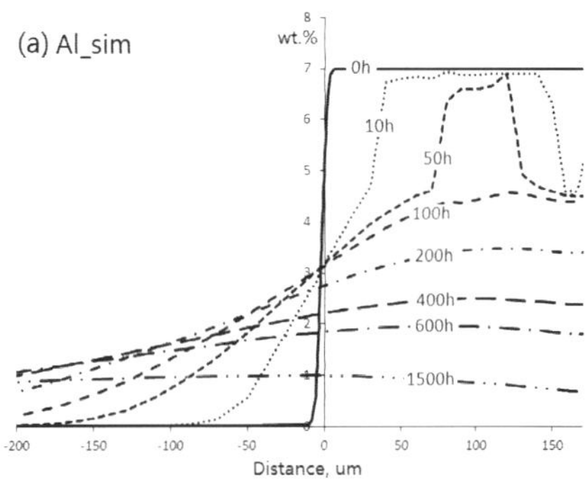

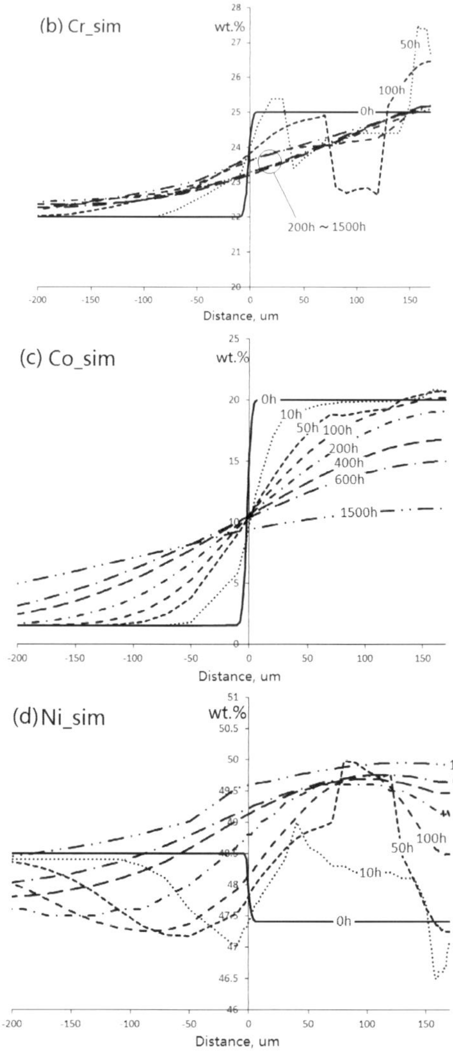

Fig. 8. Simulation results of the development of the elemental composition in the coating (0~170 µm) and the substrate (-1000~0 µm): (a) Al profiles, (b) Cr profiles, (c) Co profiles, (d) Ni profiles.

The simulation result in **Fig. 7a** under-estimated the oxidation rate because an estimated oxidation law of Al was used in the model and the oxidation of Cr and other elements as well as intersplat oxidation were not considered. Nevertheless, the simulation results were useful to understand the development trend of the elemental diffusion in the coating-substrate system. According to the simulation results in **Fig. 8**, the oxidation of Al (achieved by removing Al from the grids) succeeded to drive the decreasing of Al concentration in the BC. It can be also clearly seen that the changes of the coating composition were a co-effect by the oxidation and interdiffusion of elements via substrate-coating interface. Al, Cr and Co tended to diffuse into the substrate from the coating while Ni diffused in the opposite direction. The occurance of the Al diffusion from the coating into the substrate is usually adverse to the oxidation resistance at the BC interface since it caused an extra Al loss [17]. More work is needed for an accurate Al-depletion life prediction by measuring the oxidation rate of Al at BC/TC interface and intersplat boundaries [18].

4 Discussion

Fig. 9 schematically shows the microstructure development and the failure mechanism of the TBCs in the TCF process. In the early stage of the process, "sandwich" oxide structure is formed. After that, the outer chromia layer and Ni/Co outer zone (OZ) almost keep unchanged, and the dominant process becomes the thickening of the alumina layer in the "sandwich" structure ("II. in process"). Meanwhile, due to the mismatch of the thermal/mechanical properties between TC and BC (e.g. thermal expansion coefficient, E modulus), stresses can be generated at the BC/TC interface [8, 19]. The response against the stresses of the BC is usually plastic deformation and creep during cooling in the TCF process, but the ceramic TC , being brittle, almost always starts to crack [8, 19]. When the BC is depleted of Al, it loses ability to provide Al for continuous oxidation. Then, the oxidation of Ni/Co rich OZ occurs and spinels are largely formed at the coating interface. The formation of spinels causes local volumetric changes and additional tensile stresses in TC which promotes the formation of more cracks and crack propagation that eventually causes TBC spallation [1, 8] ("III. failure"). As the TBC failure is caused by the oxidation of MCrAlY BC and especially due to the formation of spinels at the end of coating's Al-depletion life, the failure agrees with a "chemical failure" mode [20, 21].

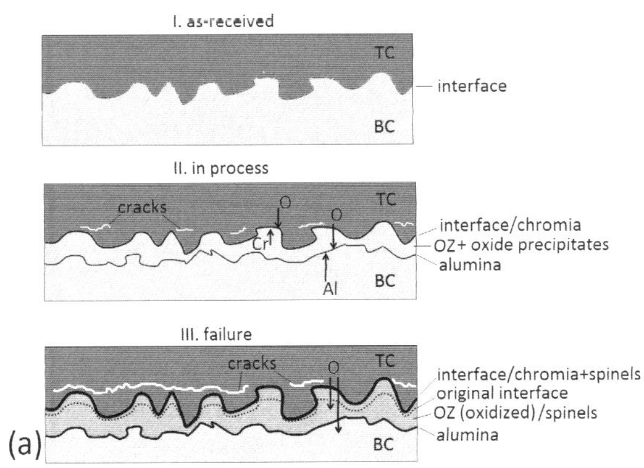

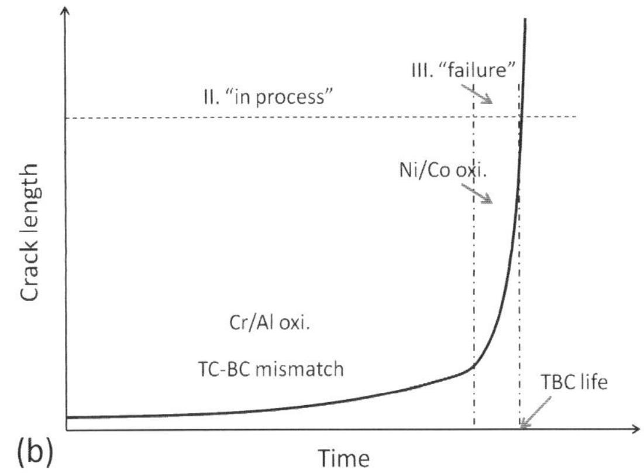

Fig. 9. (a) Schematic microstructure development showing the oxidation of MCrAlY BC and cracking in TC. (b) Schematic illustration of crack development in the TCF process.

Since the high-Cr MCrAlY used in this study can form chromia-alumina oxide structure, this coating may provide good performance in both oxidation and corrosion environments. However, since the MCrAlY's Al content is low (7%), the oxidation life of such coating is limited. To improve the Al-depletion life and also keep the chromia-alumina oxide structure, at least two approaches can be applied: 1) according to the results of oxidation-diffusion modelling that a large Al content in the MCrAlY inwards diffused into the none-Al subtrate, one can insteadly use a high-Al substrate (e.g. 3-6%) to slow down the Al inward diffusion; 2) or to spray a high-Al coating before spraying the high-Cr MCrAlY to form a composition-gradient coating; this approach is similar to Nicholls's "smart coating" concept [10].

5 Conclusions

In this paper, APS sprayed TBCs with a high-Cr MCrAlY bond coat (25% Cr, 7% Al) and YSZ top coat were tested under thermal cyclic fatigue (TCF) condition (1100 °C, 1 hour cycled). In the cyclic oxidation process, a "sandwich" oxide structure was formed at the MCrAlY/YSZ interface, containing an outer chromia layer, internal alumina layer and a Ni/Co rich metallic zone in between. Some cracks were formed in YSZ top coat near the MCrAlY interface during the TCF process but were in small size. Large horizontal cracks were formed and significantly propagated when the MCrAlY coating lost Al oxidation ability and spinels were largely formed. The formation of the large horizontal cracks was due to the formation of the spinels. Since such failure agreed with a "chemical failure" mode, the TBC life is expected to be further enhanced without losing the ability to form the "sandwich" oxide structure if a high-Al substrate is used or a high-Al coating is applied before spraying the high-Cr MCrAlY.

6 Achnowledgement

This research is financially supported by Beijing General Research Institute of Mining and Metallurgy (China), the Swedish governmental agency for Innovation Systems (Vinnova), and Siemens Industrial Turbomachinery AB (Sweden) in an international cooperation project for TBC development.

7 Literature

[1] Y. Li, C.-J. Li, Q. Zhang, G.-J. Yang, and C.-X. Li, "Influence of TGO composition on the thermal shock lifetime of thermal barrier coatings with cold-sprayed MCrAlY bond coat," *Journal of Thermal Spray Technology,* vol. 19, pp. 168-177, 2010.

[2] R. Vasen, M. O. Jarligo, T. Steinke, D. E. Mack, and D. Stover, "Overview on advanced thermal barrier coatings," *Surface and Coatings Technology,* vol. 205, pp. 938-942, 2010.

[3] M. Ahrens, R. Vasen, and D. Stover, "Stress distributions in plasma-sprayed thermal barrier coatings as a function of interface roughness and oxide scale thickness," *Surface and Coatings Technology,* vol. 161, pp. 26-35, 2002.

[4] A. M. Freborg, B. L. Ferguson, W. J. Brindley, and G. J. Petrus, "Modeling oxidation induced stresses in thermal barrier coatings," *Materials Science and Engineering: A,* vol. 245, pp. 182-190, 1998.

[5] M.-J. Pindera, J. Aboudi, and S. M. Arnold, "The effect of interface roughness and oxide film thickness on the inelastic response of thermal barrier coatings to thermal cycling," *Materials Science and Engineering: A,* vol. 284, pp. 158-175, 2000.

[6] S. Bose, *High Temperature Coatings*. Oxford: Butterworth-Heinemann, 2007.

[7] R. A. Rapp, "Hot corrosion of materials: a fluxing mechanism?," in *International Symposium on Oxidation and Corrosion of Advanced Materials and Related Phenomena*, UK, pp. 209-221.

[8] J. A. Haynes, M. K. Ferber, and W. D. Porter, "Thermal cycling behavior of plasma-sprayed thermal barrier coatings with various MCrAlX bond coats," *Journal of Thermal Spray Technology,* vol. 9, pp. 38-48, 2000.

[9] H. Yamano, K. Tani, Y. Harada, and T. Teratani, "Oxidation control with chromate pretreatment of MCrAlY unmelted particle and bond coat in thermal barrier systems," *Journal of Thermal Spray Technology,* vol. 17, pp. 275-283, Jun 2008.

[10] J. R. Nicholls, N. J. Simms, W. Y. Chan, and H. E. Evans, "Smart overlay coatings: concept and practice," *Surface and Coatings Technology,* vol. 149, pp. 236-244, 2002.

[11] K. Yuan, R. Eriksson, R. Lin Peng, X.-H. Li, S. Johansson, and Y.-D. Wang, "Modeling of microstructural evolution and lifetime prediction of MCrAlY coatings on nickel based superalloys during high temperature oxidation," *Surf. Coat. Technol.,* vol. 232, pp. 204-215, 2013.

[12] K. Yuan, R. Eriksson, R. Lin Peng, X.-H. Li, S. Johansson, and Y.-D. Wang, "MCrAlY coating design based on oxidation-diffusion modelling. Part I: Microstructural evolution," *Surface and Coatings Technology,* vol. 254, pp. 79-96, 2014.

[13] R. Eriksson, K. Yuan, X.-H. Li, and R. Lin Peng, "MCrAlY coating design based on oxidation¨Cdiffusion modelling. Part II: Lifing aspects," *Surf. Coat. Technol.,* vol. 253, pp. 27-37, 2014.

[14] Y. Li, C. J. Li, Q. A. Zhang, L. K. Xing, and G. J. Yang, "Effect of Chemical Compositions and Surface Morphologies of MCrAlY Coating on Its Isothermal Oxidation Behavior," *Journal of Thermal Spray Technology,* vol. 20, pp. 121-131, Jan 2011.

[15] C. Wagner, "Reaktionstypen bei der Oxydation von Legierungen," *Zeitschrift f¨¹r Elektrochemie, Berichte der Bunsengesellschaft f¨¹r physikalische Chemie,* vol. 63, pp. 772-782, 1959.

[16] N. Birks, G. H. Meier, and F. S. Pettit, *Introduction to the high-temperature oxidation of metals*. UK: Cambridge University Press, 2006.

[17] A. Rabiei and A. G. Evans, "Failure mechanisms associated with the thermally grown oxide in plasma-sprayed thermal barrier coatings," *Acta Materialia,* vol. 48, pp. 3963-3976, Sep 2000.

[18] R. Eriksson, K. Yuan, S. Johansson, and R. L. Peng, "Intersplat Oxidation of Atmospheric Plasma Sprayed MCrAlY Coatings," in *ASME Turbo Expo 2014: Turbine Technical Conference and Exposition*, pp. V006T22A017-V006T22A017.

[19] W. J. Brindley, "Thermal barrier coatings of the future," *Journal of Thermal Spray Technology,* vol. 6, pp. 3-4, Mar 1997.

[20] H. E. Evans and M. P. Taylor, "Oxidation of high-temperature coatings," *Proceedings of the Institution of Mechanical Engineers, Part G: Journal of Aerospace Engineering,* vol. 220, pp. 1-10, 2006.

[21] H. E. Evans, A. T. Donaldson, and T. C. Gilmour, "Mechanisms of breakaway oxidation and application to a chromia-forming steel," *Oxidation of Metals,* vol. 52, pp. 379-402, Dec 1999.

High Entropy Alloy Coatings for Application as Bond Coating for Thermal Barrier Coating Systems

Rabi Bhattacharya, Oleg N. Senkov, A.K. Rai, Dayton, OH / USA
Xinqing Ma, Peter Ruggiero, East Windsor, CT / USA

High-entropy alloys (HEAs) have potential to be used as high temperature materials and in coating material applications due to their combination of strength, ductility, thermal stability, corrosion and wear resistance. HEAs typically refer to alloys that are comprised of five or more elements, at or near equi-atomic compositions. Some of these alloys are solid solution alloys on simple underlying lattices such as face-centered cubic or body-centered cubic. In this paper, HEA coatings based on NiCoCrAlSi were deposited by high velocity oxy fuel (HVOF) and air plasma spray (APS) processes. The coatings were analyzed for phases, microstructure and composition using x-ray diffraction (XRD), scanning electron microscopy (SEM) and energy dispersive x-ray spectroscopy (EDS). The results showed that BCC/B2 phase is the major phase present in the as-produced HVOF coating that was diffusion heat treated at 1080°C. A small amount of an ordered L12 phase with a FCC crystal lattice was also found to be present. APS coatings of the same composition alloy showed a two-phase structure consisting of the L12 and BCC/B2 phases. The HEA coatings produced by HVOF were tested for oxidation resistance at 1050°C in air. The coating morphology and oxide scales were examined with the aim of developing a high quality bond coat for thermal barrier coating (TBC) systems.

1 Introduction

Thermal barrier coatings (TBC) for high temperature operation of propulsion and power generation systems have enabled substantial improvements in the performance and efficiency of gas turbine engines. Current TBC comprises a metallic bond coat for high temperature oxidation protection of the substrate and a top ceramic oxide material layer (COML) for thermal barrier [1, 2]. Bond coat serves two primary functions: 1) to improve the adherence of the TBC and 2) to protect the substrate from environmental attack if the COML does spall. This is achieved by making the bond coat composition such that it is a reservoir from which Al can diffuse out to form a protective α- Al_2O_3 thermally grown oxide (TGO) while maintaining cohesion with the COML without reacting with it. The TGO should remain elastic to the highest operating temperatures and not creep to prevent "rumpling" or cavitation on thermal cycling that can lead to the development of local separations at the TGO/TBC interface. At the same time, TBC has to operate at the highest temperature possible to minimize the amount of air used to cool the vanes and blades, without reacting with underlying substrate material, which is Ni-based superalloy for high temperature turbine engines. Overlay coatings of MCrAlY, where M is either Ni, Co or a mixture of these elements are generally used as bond coats. Co is better for hot corrosion and Ni is better for oxidation. Thus, a mixture of both is a compromise for protection from both forms of attack. Other elements are added to improve bond coat performance. Overlay coatings are generally applied by plasma spray. Frequently, a diffusion anneal treatment is carried out after coating to obtain diffusion between coating and substrate for good bonding and to homogenize as-sprayed microstructure.

The drive toward higher efficiency engines is demanding higher performance TBC coatings, both bond and top coats. High-entropy alloys (HEAs) have potential to be used as high temperature materials and in coating material applications due to their combination of strength, ductility, thermal stability, corrosion and wear resistance. HEAs typically refer to alloys that are comprised of five or more elements, at or near equi-atomic composition. Some of these alloys are solid solution alloys on simple underlying lattices such as face-centered cubic (FCC) or body-centered cubic (BCC). The basic principle behind HEAs is that solid-solution phases are stabilized by their significantly high entropy of mixing compared to intermetallic compounds, especially at high temperatures. This allows them to be synthesized and manipulated to result in possible compositions and combinations of properties required for a bond coat. Several HEAs have been shown to have very sluggish diffusion kinetics [3, 4] and improved oxidation resistance [5] in their bulk form. Not much work has been done on coating deposition of HEAs, in particular, for applications as bond coat in TBC system. Huang et al. have studied coatings of HEAs of compositions $AlSiTiCrFeCoNiMo_{0.5}$ and $AlSiTiCrFeNiMo_{0.5}$ deposited by thermal spray [6]. The as-sprayed coatings showed the same primary phase as the as-cast alloys except that the x-ray intensities of coatings were much lower than that from as-cast alloys. This was interpreted as due to rapid solidification induced saturation in solute concentration resulting in severe lattice distortion in the simple solution phase of BCC structure. It was found that hardness of the as-sprayed material was about one half of hardness in the as-cast state. The authors stated that this is reasonable since hard phases were inhibited by rapid solidification and lost their contributions in precipitation strengthening and composite strengthening. Aging at high temperatures up to 1100 °C resulted in an increase in hardness closer to the as-cast values. Oxidation studies revealed good oxidation resistance up to 1000 °C. The oxidation protection was due to the formation of chromium oxide. Recently, Yue et al. have fabricated an AlCoCrCuFeNi HEA coating using a two-step method involving plasma spray processing and laser re-melting for application on Mg alloy for corrosion resistance [7]. The as-sprayed coating was found to contain micro-porosity that disappeared after laser re-melting. The XRD results showed that both the as-sprayed and the laser re-melted coatings were composed of a BCC solid solution phase with a small amount of FCC phase, and no intermetallic compounds were found. Laser cladding was used by Ye et al. to form an AlxFeCoNiCuCr HEA coating to achieve high temperature hardness [8]. The coating demonstrated increasing hardness after holding at temperatures in the range 400-700 °C compared to as-produced state. This was explained by decomposition of a supersaturated solid solution and formation of intermetallic precipitates. Thus, it is clear that the previous work on HEA coatings did not address the requirements of bond coat for TBC system.

In this paper, we report the development of bond coats of TBC systems using novel high temperature HEA compositions. Selection of alloy composition was made from thermodynamic calculations. High velocity oxy fuel (HVOF) and atmospheric plasma spray (APS) processes were used to deposit the coatings.

2 Selection of HEA Coating Materials

The technical approach was to synthesize HEA alloys with compositions that have the following characteristics:

- High oxidation resistance: must produce thermally grown oxide (TGO), α-Al_2O_3 stabilized with α - Cr_2O_3. Thus, the alloy must contain sufficient amounts of Al and Cr.
- Strong bond with the substrate: Since the substrates are Ni-based alloys, the HEA alloy should contain Ni or elements forming extended solid solutions with Ni (Co, Cr, Cu, Fe, Mn, Pd, Pt).
- High fracture toughness: The alloy must contain sufficient amount of ductile solid solution phase.
- High hardness and wear resistance: The alloy must contain sufficient volume fraction of hard intermetallic phases, preferably as fine, homogeneously distributed precipitates. The goal is to produce an alloy with a fine mixture of two coherent phases, one of which is a disordered BCC crystal structure and another is an ordered B2 crystal structure. The mixture of these two phases will provide the necessary combination of mechanical properties (hardness, wear resistance and toughness). Age hardenable composition is one of the possibilities.
- Similar coefficient of thermal expansion (CTE) with substrate and top coat.
- High thermal stability: No first order phase transformations in the temperature range of use

In order to find a composition range where no phase transformation occurs, we conducted calculations of equilibrium phase diagram of $Alx(Co,Cr,Ni)_{88\text{-}x}Si_2$ system for 8%<x<35% and 26%<x<50%, as shown in **Fig. 1a and b**. The results show that there is a wide composition-temperature range, where no phase transformations occur in the $Alx(Co,Cr,Ni)_{88\text{-}x}Si_2$ system. Here, Al can vary from 31 at% to 42 at% and temperature from 500°C to 1300°C retaining the same phase content: B2+BCC+M_3Si. We believe that the alloys in this composition range should be a good choice for coatings, because oxidation, as well as heating and cooling cycles, would not cause any phase transformations. Thus, we have identified $Al_{30}Co_{22}Cr_{23}Ni_{23}Si_2$ (at%) alloy from the analysis as a promising alloy for bond coating.

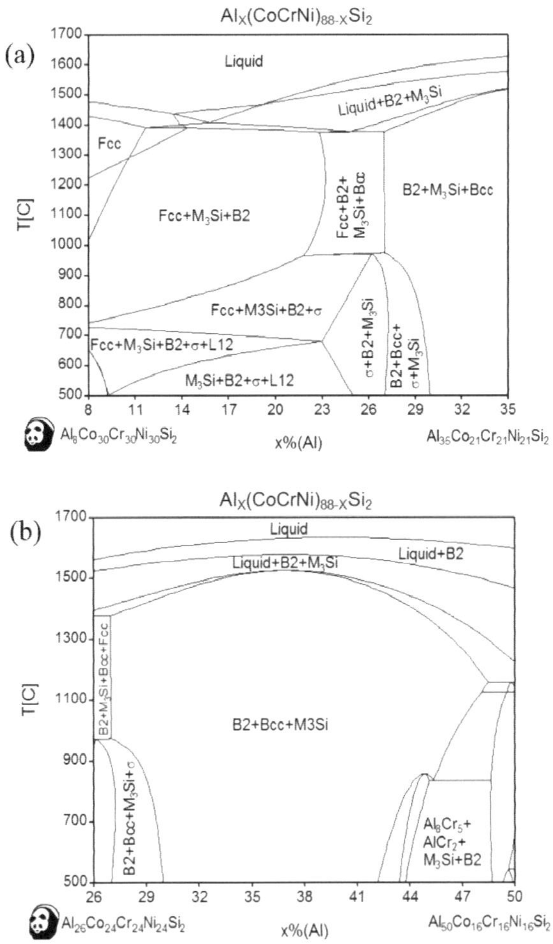

Fig. 1. Equilibrium phase diagram of (a) $Alx(Co,CrNi)_{88\text{-}x}Si_2$ alloy for 8%<x<35% and (b) $Alx(Co,CrNi)_{88\text{-}x}Si_2$ alloy for 26%<x<50%

3 Experimental

Powders of $Al_{30}Co_{22}Cr_{23}Ni_{23}Si_2$ (at%) of two sizes, -75/+45 µm and -45/+20 µm were obtained from F.J. Brodmann & Co.,L.L.C. The powder was made by mechanical alloying, according to the supplier. XRD analysis revealed that the powder samples consisted of a mixture of the pure elements (Fig. 2a). However, after sintering these powders at 1250°C for 6 hours in a high-purity argon, diffraction peaks from pure elements disappeared and an essentially single-phase BCC structure, with the lattice parameter a = 286.1 pm synthesized (Fig. 2b). A diffraction peak at at 2Θ=50.7° (d = 179.9 pm), which does not belong to the BCC phase indicates the presence of a minor secondary (likely silicide) phase in the produced HEA.

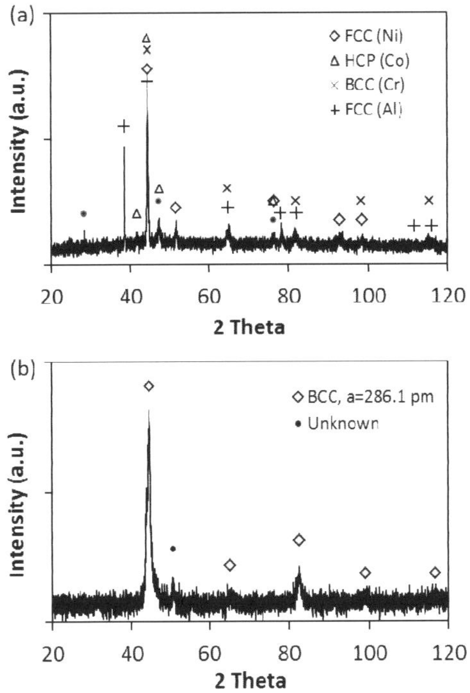

Fig. 2. X-ray diffraction patterns taken from (a) mechanically alloyed elemental powder and (b) a sample of the $Al_{30}Co_{22}Cr_{23}Ni_{23}Si_2$ (at%) HEA sintered from the powder.

Coatings were deposited by APS of $Al_{30}Co_{22}Cr_{23}Ni_{23}Si_2$ (at%) powders of size 75/+45 µm. HVOF Spray of $Al_{30}Co_{22}Cr_{23}Ni_{23}Si_2$ (at%) was used for powder size -45/+20 µm. Coatings were first heat treated to 1080°C for 4 hours in vacuum for diffusion bonding. All coatings were deposited on one side of 1" diameter polished discs of Haynes® 230 alloy. Coatings were analyzed for phase, microstructure and composition using x-ray diffraction (XRD) and scanning electron microscopy with energy dispersive x-ray analysis (SEM/EDAX).

4 Results and Discussion

Fig. 3a and Fig.3b show XRD patterns of APS and HVOF coatings, respectively, after diffusion treatment. The PS coating has a two-phase structure consisting of the FCC and BCC (or B2) phases. The volume fraction of FCC phase seems to be comparable with that of BCC. Small amount of Cr-doped alumina is also recognized. The major phase in the HVOF coating is BCC/B2. Small peaks from FCC phase and Al_2O_3 are also identified (**Fig. 3b**). The ordered B2 phase is recognized by the presence of a super-lattice (001) peak at 2Θ = 31.18°. High intensity of the fundamental diffraction peaks from BCC/B2 indicates that this phase is the major phase in the HVOF sample. Compared to the XRD spectra of the powder particles that showed the characteristics of an elemental mixture, the sprayed and bonded coating shows that it is an alloy with a primary BCC/B2 phase characteristic of HEA.

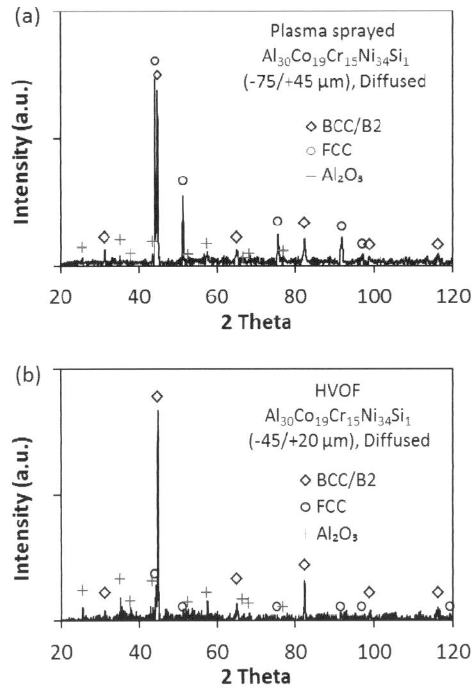

Fig. 3. X-ray diffraction patterns taken from $Al_{30}Co_{22}Cr_{23}Ni_{23}Si_2$ (at%) coatings: (a) APS, powder size -75/+45 µm and (b) HVOF spray, powder size -45/+20 µm

Since APS coatings did not show a predominantly solid solution alloy phase, these coatings were not investigated any further.

Secondary electron (SE) and backscatter electron (BSE) images of the HEA coating deposited on the Haynes 230 substrate using HVOF method are shown in **Figs. 4a** and **4b**, respectively. Note two characteristic layers of the coating. The bottom layer, which is in contact with the substrate, is denser than the top layer. Microstructure is representative of standard MCrAlY bond coat deposited by HVOF as reported in the literature [9]. Coating-substrate interface appears continuous and free from pores.

Oxidation tests were performed in a box furnace at 1050°C for 50, 100 and 200 hrs in atmospheric air. XRD patterns of HVOF sprayed coatings after oxidation showed minimal change from the as-deposited pattern. **Fig. 5** shows the XRD pattern of a sample oxidation tested at 1050°C for 100 hrs. Compared to XRD spectrum of as-deposited coating, **Fig. 3b**, very little change can be observed after oxidation. Primary BCC/B2 phase along with small peaks of FCC and Al_2O_3 are present after oxidation.

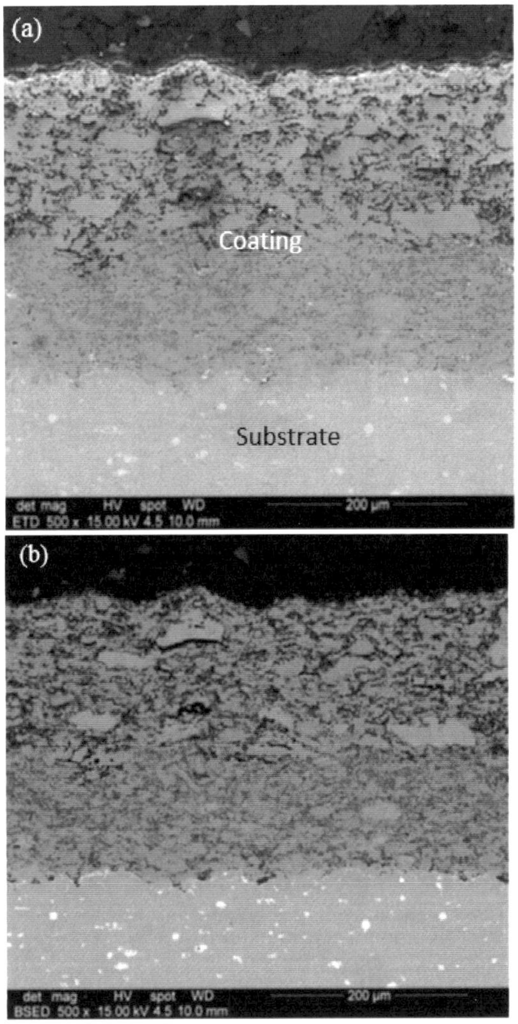

Fig. 4. Cross-section SEM images of HVOF $Al_{30}Co_{19}Cr_{15}Ni_{34}Si_1$ HEA coating: (a) secondary electron and (b) backscatter electron

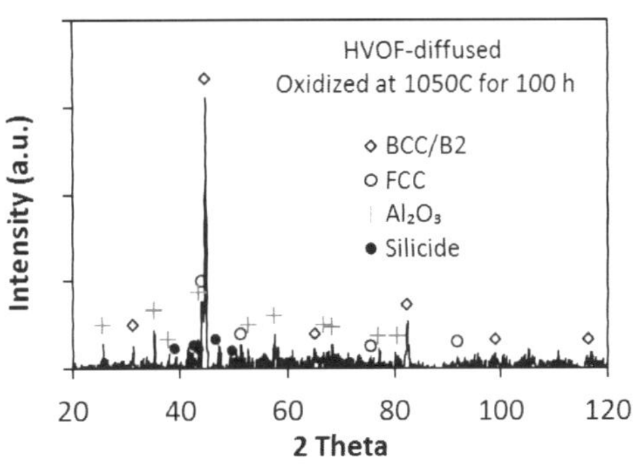

Fig. 5. XRD spectrum of HVOF $Al_{30}Co_{19}Cr_{15}Ni_{34}Si_1$ coating after oxidation in air at 1050°C for 100 hrs in air

Figs. 6 a,b shows SE and BSE images of HVOF coating after oxidation at 1050°C for 50 h in air. The coating consists of (i) non-oxidized particles of original material (they are gray color in **Fig. 6a**, (ii) oxide layer on the surfaces of the deposited particles (oxide is of dark color in **Fig. 6a,b** and (iii) pores (very bright color in **Fig. 6a**. Higher magnification SE image of the same is shown in **Fig. 7**. Three clearly distinguished constituents are identified as: non-oxidized particles (A), Al_2O_3 oxide (dark phase adjacent to the particles) (B), and pores (C) (shown as bright regions on the SE image). **Table 1** shows the average composition of the phases identified in **Fig. 7.**

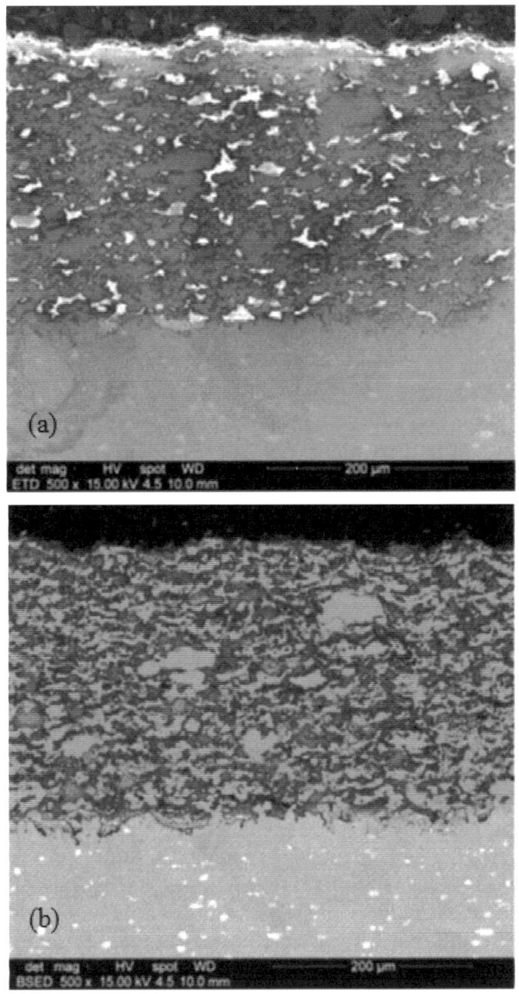

Fig. 6. HVOF coating made of a high entropy alloy (HEA) after oxidation at 1050°C for 50h. (a) Secondary electron and (b) backscatter electron SEM images.

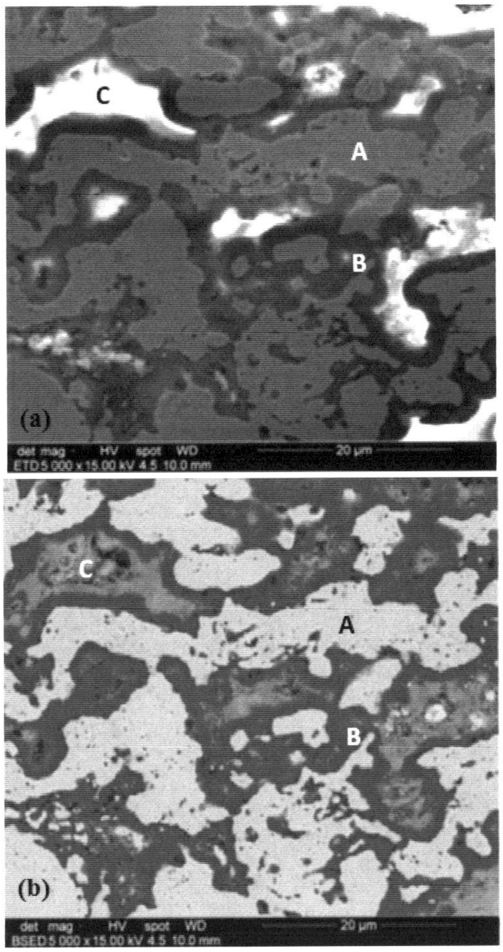

Fig. 7. HVOF coating made of a high entropy alloy (HEA) after oxidation at 1050°C for 50 h. (a) Secondary electron and (b) backscatter electron SEM images of the coating showing three distinguished constituents: Non-oxidized particles (A), Al_2O_3 oxide (dark phase adjacent to the particles) (B), and pores (C) (shown as bright regions on the SE image).

Table 1. Average chemical compositions (in at.%) of regions indicated in **Fig. 7** by letters A, B and C.

Element	O	Al	Co	Cr	Ni	Si
Particles (A)	2.01	1.07	42.43	17.21	35.25	2.03
Oxide (B)	50.34	44.96	1.25	2.34	0.88	0.23
Pores (C)	36.52	24.77	13.78	9.09	5.14	10.7

Fig. 8 shows the SEM images of the 1050°C for 200h oxidized sample. Thickness of the coating remained unchanged and no measurable oxide layer on the top of the surface was found. A darker layer between substrate and coating is likely an artifact caused by poor sample preparation (cleaning and drying after polishing). Gray particles in **Fig. 8b** are original non-oxidized particles. Dark regions between the particles are aluminium oxide doped with other alloying elements. The microstructures were analyzed in great detail at higher magnifications and are described elsewhere.

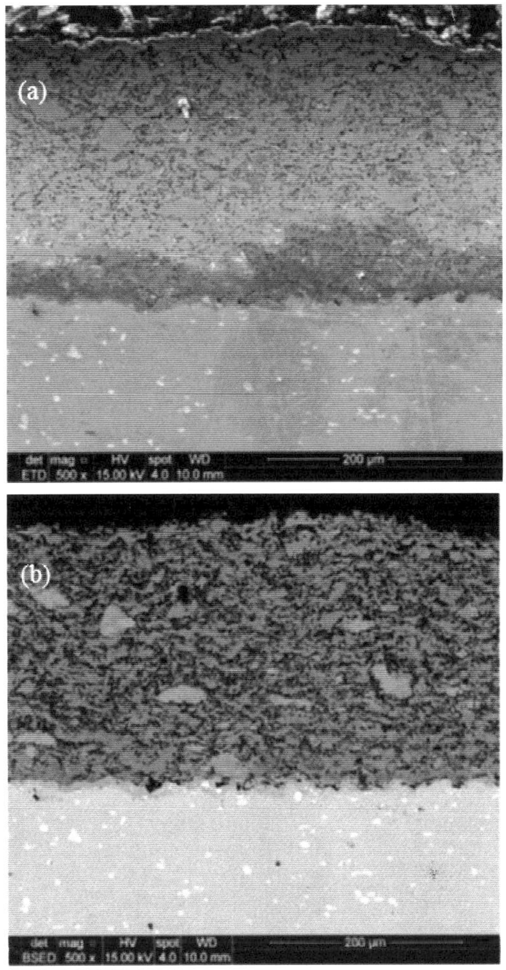

Fig. 8. (a) Secondary and (b) backscatter electron images of coating and substrate after annealing at 1050°C for 200h.

Selected HVOF HEA coated substrates were overlaid with atmospheric plasma spray (APS) YSZ top coats to a nominal thickness of 250 µm. These YSZ top coats are equivalent to that applied to land based turbine engine components at Curtiss Wright. The YSZ coatings were applied on one face that was coated with HEA bond coat. The other face is bare Hanes 230 alloy. Laboratory thermal cycle testing was conducted manually in a box furnace at 1150°C (±10°C) in air. A 60 min isothermal hold at 1150°C and forced air cooling to room temperature for 15 min after the sample was transferred to a cold stage was used as the cycle parameters. Specimens were visually inspected after each cycle and weight was measured after every 10 cycles. **Fig. 9a** shows the photos of the TBC sample before and after cyclic oxidation at 1150°C in air after 100 cycles. No spallation or damage of the TBC was observed. Weight gain chart in **Fig. 9b** shows no additional weight gain after the initial (transient oxidation stage) gain of about 0.2 mg/cm^2. This result can be directly compared with that of Haynes et al [10] where various MCrAlX bond coats with standard YSZ top coat on one side of Rene N5 substrate was evaluated in an exactly similar manner. Their reported weight gain after 100 cycles was higher than 1mg/cm^2. In conclusion, based on the results of oxidation weight gain and thermal cycling resistance, HEA bond coat has the potential to significantly increase the life of standard APS YSZ top coat compared with the current MCrAlY bond coat/YSZ TBC at about equal cost.

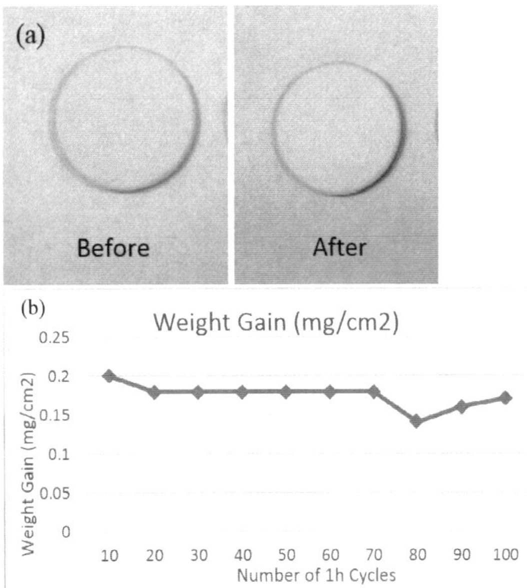

Fig. 9. Results of thermal cycle tests of HVOF HEA bond coat with YSZ top coat: (a) photo of the sample before and after 100 cycles at 1150°C, (b) weight gain. Note that only one face of the TBC specimen was coated.

5 Conclusions

- New High Entropy Alloy (HEA) composition was selected from thermodynamic calculations based on the requirements of a thermally stable solid solution alloy with predominantly BCC structure.
- Bulk alloy of selected material was synthesized and fully characterized to demonstrate HEA formation.
- Overlay HEA bond coats have been developed to provide durability to thermal barrier coating system (TBC) for gas turbine engine components. Results of this research clearly show promise of a superior oxidation resistant HEA bond coat as compared to standard MCrAlY bond coats.
- Air Plasma spray (APS) and High velocity oxy fuel (HVOF) coating deposition processes were utilized to synthesize the HEA coatings. APS coatings showed a mixture of FCC and BCC phases in nearly equal amount while the HVOF coating showed primarily BCC phase.
- HVOF Spray was done by using powder alloy composition of $Al_{30}Co_{22}Cr_{23}Ni_{23}Si_2$ (at%) produced by mechanical alloying. Coatings were first heat treated to 1080°C for 4 hours in vacuum for diffusion bonding. The HVOF coatings were predominantly BCC/B2 phase. These coatings were dense near the interface with the substrate but had porosity away from the interface. This has resulted in oxidation around the particle splats after exposure at 1050°C in air up to 200h thus densifying the coating. Besides the internal oxidation around the splat particles, no other change in microstructure or phase content was found after oxidation.
- Thermal cycling test of the HVOF bond coat with standard yttria-stabilized zirconia (YSZ) TBC as the top coat showed no spallation after 100 cycles at 1150°C in air. Minimal weight gain after first 10 cycles and then remaining at the same level clearly indicate the slow rate of growth of thermally grown oxide (TGO) layer.
- The results of this initial study emphasize the importance of optimizing the HEA composition, powder manufacture and coating deposition process to fully define the capability and limitations of HEAs for bond coating application.

Acknowledgements

This work was financially supported by Department of Energy's (DOE) Small Business Innovative Research Program, Grant No. DE-SC0013213. We thank Dr. Patcharin Burke, DOE Program Manager, for useful discussions.

6 Literature

[1.] Clarke, D.R., Oechsner, M. and Padture, N.P.: Thermal-barrier coatings for more efficient gas-turbine engines, MRS Bulletin, Vol. 37, October 2012, p. 891.
[2]. Walston ,W.S.: Coating and surface technologies for turbine airfoils, Superalloys, 2004, TMS 2004, p. 579.
[3.] Tsai, K.-Y., Tsai, M.-H. and Yeh, J.-W.: Sluggish diffusion in Co-Cr-Fe-Mn-Ni high-entropy alloys. Acta mater. 2013;61:4887-4897.
[4.] Murty, B.S., Yeh, J.W. and Ranganathan, S.: High Entropy Alloys, London, UK: Butterworth-Heinemann; 2014.
[5.] Zhang, Y., Zuo, T.T., Tang, Z., Gao, M.C., Dahmen, K.A., Liaw, P.K., et al.: Microstructures and properties of high-entropy alloys, Progress in Materials Science, 2014, 61:1-93.
[6.] Huang, P.K., Yeh, J.W., Shun, T.T. and Chen, S.K.: Advanced Engineering Materials, vol. 6, No. 1-2, p. 74 (2004).
[7.] Yue,T.M., Xie, H., Lin, X., Yang, H. and Meng, G.: Entropy, vol. 15, 2833 (2013).
[8.] Ye, X., Ma, M., Liu, W., Li, L., Zhong, M., Liu, Y. and Wu, Q.: Adv. in Mater. Sci. and Eng. 2011.
[9.] Choi, J.H. and Lee, D.B.: J. Ceramic Processing Research, 7 (2006) 136-139.

[10.] Haynes, J.A., Ferber, M.K. and Porter, W.D.: J. Thermal Spray Technology, 9 (2000) 38-48.

New Investigation on the High Temperature Capability Limits of APS YSZ Thermal Barrier Coatings at ~1500°C: Preliminary Laser-rig Testing Results

R. S. Lima and B. R. Marple, National Research Council of Canada (NRC), Boucherville, QC, Canada

The current effective high temperature capability limit of APS YSZ TBCs is ~1300°C. This is partially related to the formation of the so called non-transformable tetragonal *t'*-phase during the rapid quenching of the YSZ particles during spraying. The metastable *t'*-phase transforms slowly into the equilibrium tetragonal and cubic phases at temperatures ≥ 1300°C; which can lead to the formation of the monoclinic phase upon cooling to RT. These series of phase transformations will eventually lead to TBC failure due to cracking. An APS TBC was optimized to allow it to operate at ~1500°C temperature levels. The TBCs were sprayed onto CMSX-4 substrate pucks. By using a laser gradient rig, a temperature gradient was generated across the TBC/substrate system. The YSZ T-front temperature and substrate backside T-back temperature levels were ~1500°C and ~1000°C, respectively. Two types of cycles have been performed: (i) 5 min hot and 2 min cooling for 1000 cycles and (ii) 1 h hot and 2 min cooling for 10 cycles. In both cycle conditions, no TBC failure has been observed. Although preliminary, these results are promising regarding the potential to increase the effective high temperature capability limits of APS YSZ TBCs.

1 Introduction

1.1 Thermal Barrier Coatings (TBCs)

Thermal barrier coatings (TBCs) deposited via thermal spray processing are applied onto the static metallic super-alloy parts of the hot zones of jet turbine engines (e.g., combustion chambers) [1] or even on the exhaust nozzles of rocket engines [2]. TBCs have the potential to increase engine efficiency by allowing increased turbine operating temperatures [3]. On the other hand, TBCs can also maintain the current turbine combustion temperature levels, but at the same time reducing the air cooling requirements for the metallic parts. This reduction of the air cooling requirements could be translated into (i) an easier part manufacturing and/or (ii) more air to be introduced into the combustor to generate a leaner burn (i.e., excess air), which reduces the formation of nitrogen oxide (NOx) gases (environmental/health concerns) [4]. In all cases, TBCs reduce the metal part temperature and extend the lifespan of the component.

A thermally sprayed TBC system typically exhibits a bi-layered structure, i.e., a ceramic top coat (e.g., ZrO_2-8wt%Y_2O_3, a.k.a., YSZ) and a MCrAlY bond coat (BC) (M = Ni, Co, NiCo or CoNi) deposited onto a metallic super-alloy substrate. The thicknesses of turbine jet TBCs are typically 125-250 µm and 250-500 µm for the BC and YSZ top coat, respectively.

The main function of the MCrAlY BC is to (i) increase the high temperature oxidation/corrosion resistance of the underlying super-alloy by the slow growing of a thin thermally grown oxide (TGO) layer, (ii) increase the YSZ top coat adhesion to the component and (iii) reduce residual stress due to the mismatch of coefficient of thermal expansion (CTE) values between the ceramic top coat and metallic substrate.

The main function of the YSZ top coat is to provide a thermal insulating layer to avoid having the metallic components to reach their softening or melting temperatures. This is achieved due to the high melting point of the YSZ (~2700°C), as well as, its low thermal conductivity values (<2 W/mK).

1.2 Thermally Spraying TBCs

Air plasma spray (APS) processing is the standard technique employed to deposit the YSZ top coat. The MCrAlY BC can be deposited by different thermal spray systems, however, although not state-of-the-art, the APS technique is the work-horse system to spray BCs (mainly for internal diameters of combustion chambers and afterburners).

1.3 Turbine Entry/Inlet Temperatures

It is been reported by the manufacturer that the turbine entry/inlet temperature (i.e., the maximum flame temperature exiting the combustor before reaching the turbine blades) of the M88 Snecma engine of the Rafale fighter jet is 1577°C [5]. This temperature is higher than the melting points of aerospace metallic super-alloys (<1400°C). However, the metal parts and the TBC itself do not experience these high temperatures due to the air cooling that is employed to cool the un-coated back-side of the metallic substrate, as well as, the air film cooling that is injected through the cooling holes spread over the TBC-coated surface of the turbine part.

As a curiosity, APS YSZ TBCs have been tested in high heat flux (H_2/O_2) rocket engines by the National Aeronautics and Space Administration (NASA). The maximum gas combustion temperature reached was at the~1500°C level [6], i.e., similar to that reported for the M88 Snecma jet engine.

1.4 TBC Maximum Operating Temperatures

The high temperature (e.g., 1500°C) stable phases of YSZ are the tetragonal (*t*) and cubic (*c*) phases. Upon cooling to room temperature (RT) the *c*-phase can keep its structure, however, the *t*-phase undergoes the known stress-induced martensitic transformation to the monoclinic (*m*) phase of YSZ. This martensitic transformation is accompanied by a volume expansion of 3-5%. Under controlled conditions, it leads to the so-called transformation toughening of the partially stabilized zirconia (PSZ) [7]. However, on a large scale, it leads to material damage due to extensive micro-cracking [7]. It needs to be highlighted that this phase transformation is reversible and cyclic, i.e., *t*-phase transforms into *m* upon cooling to RT and *m*-phase transforms into *t* upon heating from RT.

The current effective high temperature capability limit of APS YSZ TBCs is related to the phase transformations that occur with the tetragonal (*t'*) phase of the YSZ at high temperature exposure. The so called non-transformable *t'*-phase is formed during the rapid quenching and solidification (~10^6 °C/s) of the YSZ molten particles during thermal spraying (as-sprayed coating). It is important to highlight the fact that the *t'*-phase is a single phase throughout the YSZ coating; and not a mixture of *t'* and other phases (unless semi-molten particles are embedded in the coating microstructure). The formation of the *t'*-phase is inherent to the thermal spray processing and any other technique that involves rapid YSZ solidification from the melt or vapour state.

It is important to point out that the t'-YSZ phase does not undergo the stress-induced martensitic transformation to the *m*-phase of YSZ when cooled from high temperature (e.g., 1500°C) to RT. Therefore, the presence of the *t'*-phase in TBCs is highly desirable due to the lack of phase transformation [8, 9].

However, the *t'*-phase is in fact, metastable, i.e., transforms slowly into the equilibrium tetragonal (*t*) and cubic (*c*) phases at temperatures ≥ ~1300°C [8, 9]. Upon cooling from high temperature to RT, the *t*-phase transforms into *m*-phase, as well as, the *c* one.

It needs to be stressed that the formation of the *m*-phase is detrimental to the mechanical integrity of the YSZ TBC, due to the volume changes (i.e., micro-cracking) that occur upon typical TBC thermal cycling, i.e., heating from RT to high temperature ($m \rightarrow t$) and cooling from high temperature to RT ($t \rightarrow m$) [8].

As the *t'*-phase begins its slow phase transformation to the *t* and *c* phases at ~1300°C, this temperature is typically considered as the maximum temperature limit for YSZ regarding TBC operations. As previously stated, the turbine entry/inlet temperature of the M88 jet engine is 1577°C [5]. This is one of the main reasons why there is the need for air film cooling on the surface of TBCs in turbine parts (in addition to back cooling), to avoid the TBC reaching temperatures higher than ~1300°C.

1.5 Bond Coat and TGO Effects

Although this paper is aiming at studying the maximum operation temperature limits of TBCs from the point of view of the YSZ top coat, the importance of the maximum operating temperature limits of the MCrAlY BC and the TGO formation, which are fundamental for TBC performance [10], cannot be forgotten. In fact, it has been reported that when a MCrAlY BC is isothermally exposed in air at temperature levels of 1250°C, the APS YSZ top coat spalls-off in 5 h upon cooling to RT [10]. Therefore, the need of obtaining a thermal gradient with a "steep slope", to avoid exposing the BC to temperature levels similar to those of the YSZ top coat, is paramount for this type of study. This is necessary to minimize the BC and TGO contribution to the YSZ top coat failure.

1.6 Objective

The objective of this present work is to obtain preliminary data to check the validity of the current 1300°C upper temperature limit of YSZ TBCs. To accomplish this objective, the thermal cycling performance of an optimized APS YSZ TBC, operating at temperatures of ~1500°C [5, 6] in a thermal gradient laser-rig, will be evaluated.

2 Experimental Procedure

2.1 Thermal Spraying

The substrate material employed in this study was the single-crystal CMSX-4 metallic super-alloy. The puck-shaped substrates (2.5 cm (1") diameter and 3.2 mm (1/8") thick) were degreased and grit-blasted on both sides prior to thermal spraying. The NiCoCrAlY+HfSi BC (Amdry 386-4, Oerlikon Metco, Westbury, NY, USA) and the YSZ top coat (Metco 204B-NS, Oerlikon Metco, Westbury, NY, USA) feedstock powders employed in this study were sprayed using an APS torch (Metco 3MB, Oerlikon Metco, Westbury, NY, USA). The YSZ TBC was optimized to yield low thermal conductivity values at high temperatures. The complete set of spray parameters and the optimization process is considered as intellectual property (IP) of the National Research Council of Canada (NRC) and cannot be disclosed. The BC and YSZ top coat thicknesses were ~175 µm and ~425 µm, respectively.

2.2 Microstructural Characterization

The TBC sample was vacuum impregnated in epoxy resin prior to cutting. The mounted sample was then cut to produce a cross-section, which was re-mounted in epoxy and polished according to standard metallographic procedures for TBCs. The cross-sectional microstructure was analyzed by scanning electron microscopy (SEM). The SEM pictures were taken in backscattered mode. X-ray diffraction (XRD), using CuKα radiation in Bragg-Brentano configuration, was employed to evaluate the phase compositions of the YSZ powder, as-sprayed and thermally-cycled TBCs. The XRD 2θ values ranged from 20° to 80° (scanning step size of 0.05° and step time of 2.5 s). The porosity of the TBC was measured on the cross-section via optical microscopy and image analysis at a magnification of 200X.

2.3 Thermal Conductivity

The thermal conductivity values of as-sprayed and annealed YSZ TBCs were calculated using the traditional thermal conductivity (TC) equation: K (W/mK) = α (m^2/s) x C_p (J/kgK) x d (kg/m^3); where K is TC, α is thermal diffusivity (TD), C_p is the specific heat of the YSZ and d is the coating density. The samples for TC measurements were free-standing YSZ coatings obtained by spraying the YSZ onto low carbon steel substrates detaching the coatings from the substrate via acid dissolution and, subsequently, drying them in a furnace at 150°C for 12 h. The free-standing coatings were heat treated in air at 1300°C (3 coatings) and 1500°C (3 coatings) for 10 h in a box furnace (heating rate 300°C/h and natural cooling to RT inside the furnace). The laser-flash technique (FlashLine 5000, Anter Corp., Pittsburgh, PA, USA) was employed to measure the TD values of the free-standing as-sprayed and annealed YSZ samples. For TD measurements, a ~1-2 µm thick graphite layer (Graphite Spray Paint CX-01, TA Instruments, New Castle, DE, USA) was sprayed on the YSZ samples to increase the absorption of the yttrium aluminum garnet (YAG) laser (λ = 1.1 µm) radiation energy and maximize the accuracy of the measurement of the infrared sensor. The TD measurements were performed at 1 atm (1 bar) with constant Ar gas flow of ~0.25 lpm. The TD values were measured at RT only for the as-sprayed coating and at different temperature intervals, from RT to 1300°C and from RT to 1500°C for the annealed ones. A total of 3 samples per YSZ coating were used for TD measurements.

The d values of the as-sprayed and annealed YSZ samples were measured via the Archimedes technique: d (kg/m^3) = $(m_{dry} / (m_{wet} - m_{immersed}))$ x 1000; where m is mass in grams. A total of 3 samples per coating were used to generate the density values. It has to be highlighted that the d values of the as-sprayed and annealed coatings were measured at RT and used to calculate the RT and high temperature TC values reported in this work.

The RT and annealed C_p values of the YSZ samples (at RT and high temperatures) were estimated based on the equation from the APS YSZ TBC database of Wang et al [11]:
C_p (J/kgK) = $(m_1 + m_2T + m_3T^{0.5} + m_4T^2 + m_5T^3)$ x 1000; where m_1 = 0.600045, m_2 = 4.296 x 10^{-5}, m_3 = 0.22845, m_4 = -24201, m_5 = 2.3668 x 10^{-6} and T is the temperature in Kelvin.

2.4 Thermal Gradient Laser Rig

The NRC thermal gradient laser rig is based on a 3 kW CO_2 laser, which produces a constant laser beam (10.6 µm wave-length) over the YSZ TBC top coat. The rig is set to create a laser spot size of about 2.5 cm (1") in diameter, in order to fit the dimensions of the substrate. A compressed air jet cools the back-side of the substrate. The rig is closed-loop computer-controlled; therefore, the temperature of the YSZ top coat (T-front) and the temperature of the un-coated back side of the substrate (T-back) are continuously recorded during the tests. This type of system has been previously developed at NASA by Zhu et al. [12]. The overall schematic of the rig is found in **Fig. 1**.

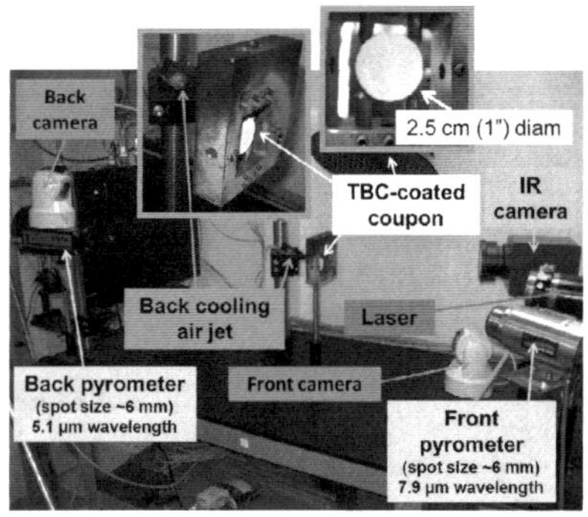

Fig. 1. Thermal gradient laser rig of the NRC.

The temperature of the YSZ top coat was monitored using a 1-color 7.9 µm wavelength pyrometer with a spot size of ~6 mm at the centre of the sample. The temperature of the un-coated back side was measured using a 1-color 5.1 µm wavelength pyrometer with a spot size of ~6 mm at the centre of the sample. An infra-red (IR) camera (7.5-14 µm spectral range) coupled with a CO_2 notch filter at 8 µm was also employed to measure the overall distribution of temperature on the surface of the YSZ top coat. The IR camera is used to guarantee a uniform laser beam and temperature over the surface of the TBC. The emissivity value for the YSZ was 0.96, as has been reported in the literature [13]. A CMSX-4 substrate was grit-blasted to remove surface contaminants and subsequently oxidized in air for 1 h at ~1000°C. After oxidation, a high heat enamel black spray paint (Tremclad, Rust-Oleum, Concord, ON, Canada) of know emissivity (0.96) was sprayed on half-side of the 2.5 cm (1") diameter substrate. The laser rig was used to heat the sample and the back-side pyrometer emissivity was set to 0.96 to measure the temperature of the half-coated side. The laser rig power was slowly increased until the back-side of the substrate reached a temperature of ~600°C, which is near the maximum operational temperature of the paint (650°C). The pyrometer measurement zone was then switched to the half un-coated side and the same laser power was applied on the sample. Then the emissivity of the pyrometer was manually changed until the half un-coated back-side exhibited a temperature of ~600°C. The emissivity value found was 0.8, which has also been reported in the literature for an oxidized CMSX-4, from 650°C to 1000°C [14].

2.5 Thermal Gradient Cycling

Two types of thermal gradient cycles were performed, short duration (5 min hot & 2 min cool) and long duration (1 h hot & 2 min cool). The back-cooling air jet was set at 330 lpm non-stop during the entire testing. The laser power was set constant at 600 W in cycles (5 min or 1 h). The combination of 600 W laser power over the YSZ and 330 lpm of air cooling on the substrate back-side was chosen because it generated the desired T-front and T-back values of ~1500°C [5, 6] and ~1000°C, respectively.

3 Results and Discussion

3.1 TBC Microstructure

The overall as-sprayed TBC microstructure is found in **Fig. 2**. No major cracking, debonding or delamination was observed. The YSZ top coat exhibited a lamellar micro-cracked morphology, with inter-splat and globular pores, which are desired for TBC applications. The BC exhibits a combination of splats, semi-molten particles and some oxides (dark greyish phases – atoms of lower atomic numbers appear darker in back-scattered images). The porosity values of the BC and YSZ top coat were 11±1% and 6±1%, respectively.

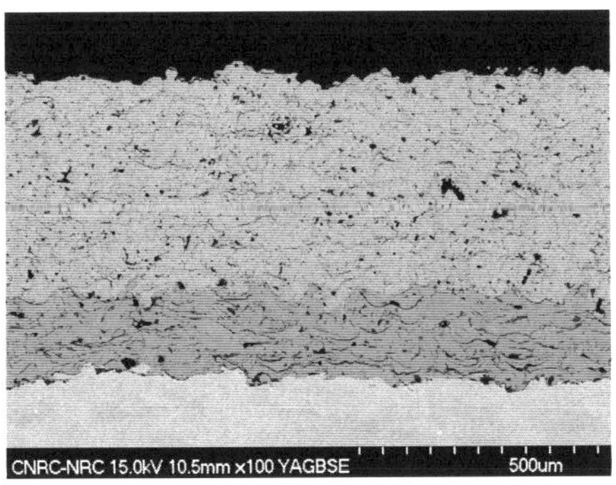

Fig. 2. Overall as-sprayed TBC microstructure.

3.2 Thermal Conductivity

The TC values for the as-sprayed and annealed YSZ TBCs can be found in **Fig. 3**. As expected, the TC values of the annealed coatings are higher than of the as-sprayed one at RT (~0.7 W/mK) due to coating sintering. For the YSZ annealed at 1300°C-10h, the TC value measured at 1300°C is ~1.0 W/mK; whereas, for the one annealed at 1500°C-10h, the TC value at 1500°C is ~1.2 W/mK.

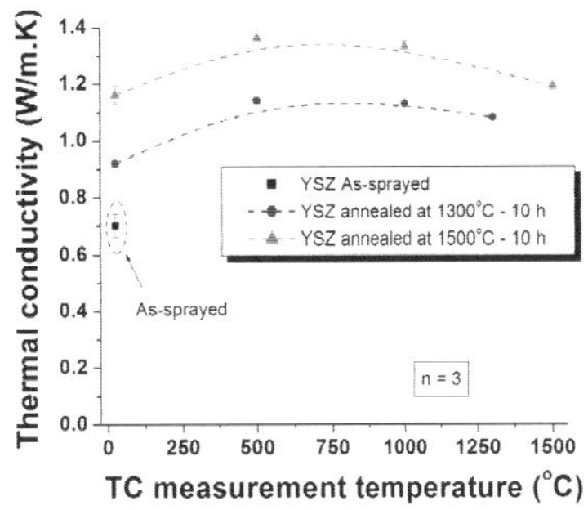

Fig. 3. TC values for as-sprayed and annealed YSZ.

Tan et al. [15], using the same type of TC measuring procedure, reported APS YSZ TC values of ~1.3 W/mK measured at 1200°C for a coating annealed at 1200°C-10h. Although this is not a large database, the TC values of the APS YSZ TBC developed for this study are lower than those reported in the literature, for similar TC measuring conditions. These low TC values also probably contributed to the "good performance" of these TBCs in thermal cycling, as shown in the next section. The low TC likely protected the BC, avoiding an un-stable TGO growth, which leads to TBC premature failure.

3.3 Thermal Cycling

Regarding the short-duration cycles (5 min hot & 2 min cool), for the first 60 cycles of 1000 cycles, the T-front (YSZ) and T-back (substrate backside) values were ~1500°C and ~975°C, respectively. For the last 60 cycles of 1000 cycles, the values were ~1475°C and ~875°C (**Fig. 4**).

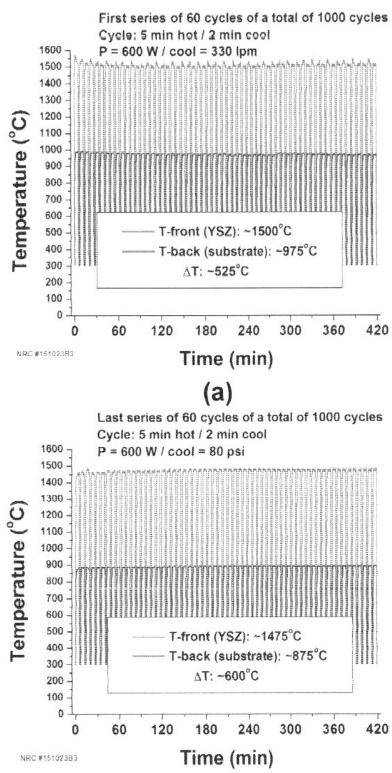

(b)

Fig. 4. First and last series of 60 cycles of a total of 1000 short-duration cycles.

After 1000 cycles, the TBC was intact, as observed by the images of the as-sprayed and 1000X-cycled TBCs, as well as, the YSZ IR camera image at the 1000th cycle (**Fig. 5**). If the TBC is spalling or about to spall, hot spots would appear in the IR camera picture. However, a well-distributed temperature is seen over the YSZ surface. It needs to be pointed out that for this TBC, the total time of high temperature exposure at ~1500°C was around 80 h (1000 hot cycles - 5 min).

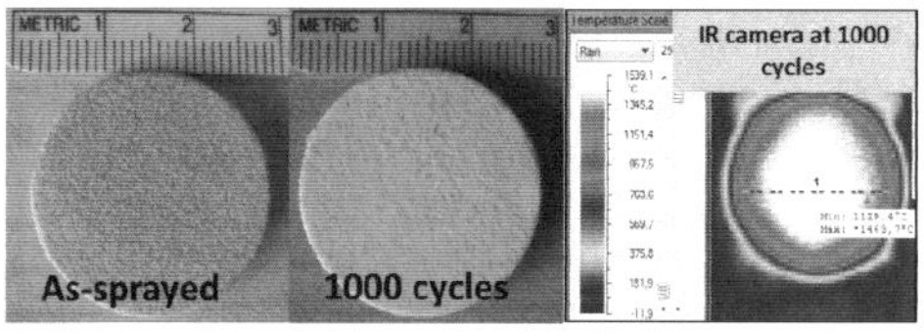

Fig. 5. Images of the as-sprayed and 1000X-cycled TBCs, as well as, YSZ IR camera picture.

Throughout the 1000 cycles, the T-front temperatures remained almost unchanged at ~1500°C; whereas, the T-back temperatures dropped from ~975°C to ~875°C after 1000 cycles (**Fig. 4**). This "counter-intuitive" event has been reported for laser-rig testing of TBCs [12]. Initially the T-front levels tend to drop due to YSZ sintering effects, i.e., an increase of TC values (**Fig. 3**) leading to an improved heat transfer to the substrate. However, after a continuous high temperature cycling, horizontal micro-cracks begin to propagate along the YSZ/BC interface due to TGO growth. These horizontal micro-gaps (i) reduce the heat propagation from the YSZ/BC interface to the metal substrate, (ii) thereby counter-acting the YSZ sintering effect (heat transfer is now reduced and "bounces back" to the YSZ, i.e., resulting in no major changes at T-front overall) but (iii) lowering the temperature of the back-side of the substrate (**Fig. 4**).

Regarding the long-duration cycles (1 h hot & 2 min cool), after 10 cycles, the T-front (YSZ) and T-back (substrate backside) values were ~1450°C and ~950°C, respectively (**Fig. 6**).

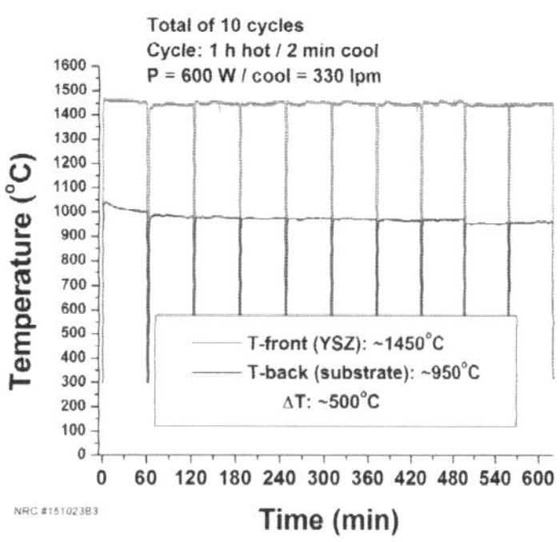

Fig. 6. One series of 10 long-duration cycles.

After 10 cycles, the TBC was intact, as observed by the images of the as-sprayed and 10X-cycled TBCs, as well as, the YSZ IR camera image at the 10th cycle (**Fig. 7**). It needs to be pointed out that for this TBC, the total time of high temperature exposure at ~1500°C was 10 h (1 h hot x 10 cycles).

It is important to highlight that for both types of cycles (i.e., short & long), the YSZ average heating rate from RT up to ~1500°C was ~60°C/s; whereas, the average YSZ cooling rate from ~1500°C to RT was ~13°C/s.

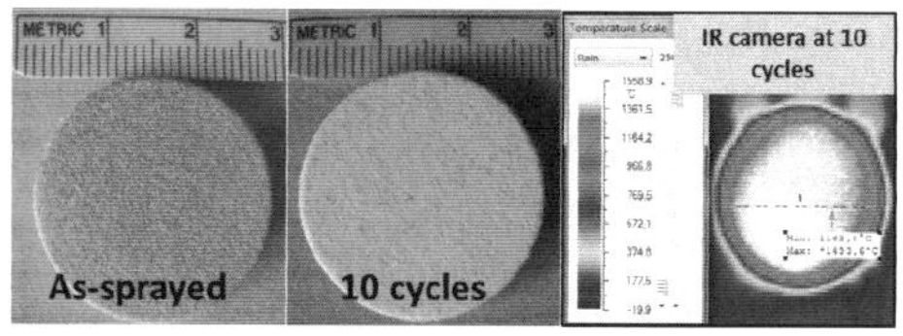

Fig. 7. Images of the as-sprayed and 10X-cycled TBCs, as well as, YSZ IR camera picture.

The high thickness of the YSZ TBC (400-500 µm range), in addition to the low YSZ TC value at 1500°C (**Fig. 3**), may have helped to avoid an un-stable TGO formation. Therefore, TBC spallation and even early spallation signs were not observed (**Figs. 5 and 7**).

3.4 XRD Evaluation

The XRD peaks were identified based on the powder diffraction files (PDFs) provided by the International Centre for Diffraction Data (ICDD) (Newtown Square, PA, USA). The presence of the *t'*-phase of YSZ is observed in the 2θ region from 72° to 76° of the XRD spectrum (**Fig. 8**) [9]. For the as-sprayed YSZ, it is possible to observe the two characteristic peaks of the *t'*-phase at ~73.4° (004) and ~74.2° (400) (PDF #17-0923), which is the desired phase for the YSZ for TBC applications, as reported by other authors [8, 9].

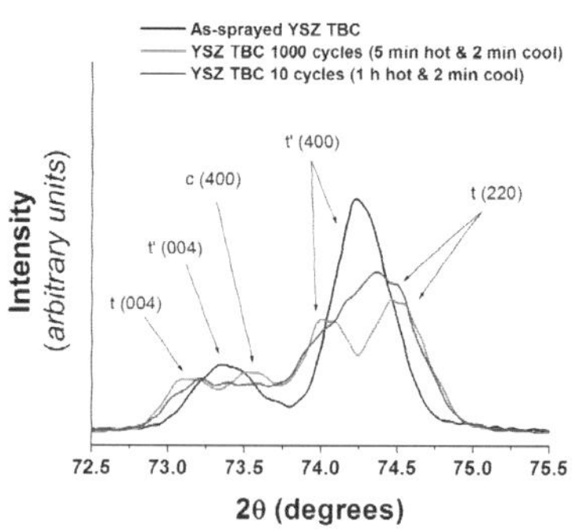

Fig. 8. XRD spectra of the TBCs from 72.5° to 75.5°.

For the two cycled YSZ TBCs, it is possible to observe the slow phase transformation of the *t'*-phase into the *t* and *c* phases. This is observed by the decreasing or decay of intensity of the *t'*-phase peaks and the arising of the *t* phase peaks at ~73.1° (004) and ~74.4° (220) (PDF #48-0224), as well as, the *c* phase at ~73.5° (400) (PDF #30-1468). It is important to point out that for both cycled coatings, the presence of the desired *t'* phase is still observed. This is probably linked to the lack of spallation failure of the TBCs in this study.

As previously stated, the formation of the *m* phase is detrimental to TBC performance. The *m* phase (PDF #37-1484), if present, is promptly observed in the 2θ region from 28° to 32° of the XRD spectrum (**Fig. 9**) [9]. Surpringly, the two highest intensity peaks (100% and 70%) of the *m* phase at ~28.2° (-111) and ~31.5° (111) are not observed for the as-sprayed and cycled TBCs. The absence of the *m* phase is probably another reason to explain the lack of spallation failure of the TBCs in this study. At ~30.3°, the XRD peak can represent the *c* (111) (PDF #30-1468), *t* (101) (PDF #48-0224) or *t'* (111) (PDF #17-0923) phases.

The absence of the *m* phase can be explained based on the work of VanValzah and Eaton [8]. APS YSZ TBCs were heat treated in air in a furnace. The volume percentage of the *m* phase in the as-sprayed coating was ~4%. After an isothermal exposure at 1500°C for 100 h, one TBC was air quenched to RT and another one was gently cooled inside the furnace down to RT. The *m* phase volume percentages for the air quenched and furnace cooled TBCs were ~12% and ~48%, respectively [8]. It was hypothesized that the high cooling rates (air quenched TBC) contracted the *t* phase, supressing the *t* → *m* transformation, i.e., supressing the volume expansion associated with this transformation [8]. As previously stated, for both types of cycles (i.e., short & long), the average YSZ cooling rate from ~1500°C to RT was ~13°C/s. This can be considered as a high cooling rate (when compared to furnace cooling), which may explain the absence of the *m* phase for the cycled coatings.

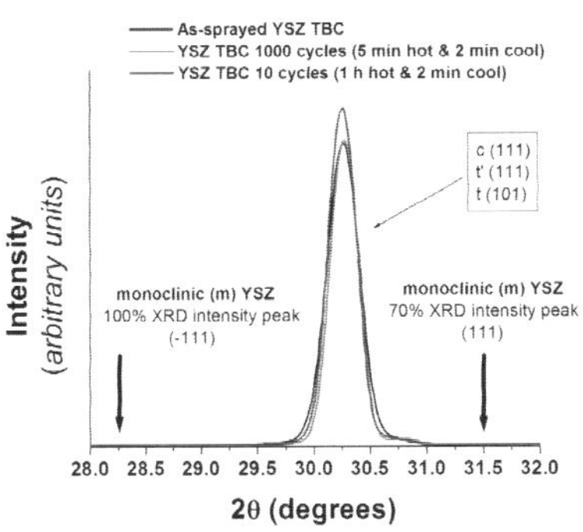

Fig. 9. XRD spectra of the TBCs from 28° to 32°.

4 Conclusions

The preliminary results have shown that optimized APS YSZ TBCs (i.e., low TC at high temperatures) may be able to operate above the 1300°C temperature limit without a catastrophic failure. TBC T-front and substrate backside (T-back) temperature levels of ~1500°C and ~1000°C, respectively, were achieved. No TBC spallation or even signs of spallation were observed after up to 1000 short cycles (5 min hot & 2 min cool) and 10 long cycles (1 h hot & 2 min cool). It is hypothesized that the (i) low thermal conductivity of the YSZ (~1.2 W/mK) at ~1500°C and high YSZ thickness (400-500 µm range) (which may have avoided an unstable TGO formation), (ii) the presence of the *t'* phase in the cycled TBCs and (iii) the rapid TBC cooling rates (~13°C/s), which led to (iv) the absence of the *m* phase in the cycled TBCs, probably contributed to this result. Further statistical evaluation (i.e., larger number of samples and longer testing time) is paramount to confirm these findings.

5 Literature

[1] D. J. Wortman, B. A. Nagaraj and E. C. Duderstadt, Thermal Barrier Coatings for Gas Turbine Use, *Materials Science and Engineering A*, 1989, **121**, p 433-440

[2] X. Viot, A. Souchier, G. Dantu and F. Lassoudiere, High Thrust Engine Demonstrations (AIAA 2009-5132), *45th AIAA/ASME/SAE/ASEE Joint Propulsion Conference & Exhibit*, Denver, Co, USA, 2009, p 1-8

[3] J. H. Perepezko, The Hotter the Engine the Better, Science, 2009, **326**, p 1068-1069

[4] www.faa.gov/about/office_org/headquarters_offices/apl/research/aircraft_technology/cleen/reports/media/TAPS_II_Public_Final_Report.pdf (January 12, 2016).

[5] www.snecma.com/military-engines/training-and-combat-aircraft/m88 (January 12, 2016).

[6] J. A. Nesbitt, Thermal Modeling of Various Thermal Barrier Coatings in a High Heat Flux Rocket Engine, *Surface & Coatings Technology*, 2000, **130**, p 141-151

[7] R. Stevens, Engineering Properties of Zirconia, *Engineered Materials Handbook – Vol 4: Ceramic and Glasses*, S. J. Schneider Jr., Ed., ASM International, Materials Park, OH, USA, 1991, p 775-786

[8] J. R. VanValzah and H. E. Eaton, Cooling Rate Effects on the Tetragonal to Monoclinic Phase Transformation in Aged Plasma-sprayed Yttria Partially Stabilized Zirconia, *Surface & Coatings Technology*, 1991, **46**, p 289-300

[9] J. R. Brandon and R. Taylor, Phase Stability of Zirconia-based Thermal Barrier Coatings Part I: Zirconia-yttria Alloys, *Surface & Coatings Technology*, 1991, **46**, p 75-90

[10] R. A. Miller and C. E. Lowell, Failure Mechanisms of Thermal Barrier Coatings Exposed to Elevated Temperatures, *Thin Solid Films*, 1982, **95**, p 265-273

[11] H. Wang, R. B. Dinwiddie and W. D. Porter, Development of a Thermal Transport Database for Air Plasma Sprayed ZrO_2-Y_2O_3 Thermal Barrier Coatings, *Journal of Thermal Spray Technology*, 2010, **19**(5), p 879-883

[12] D. Zhu, S. R. Choi and R. A. Miller, Development and Thermal Fatigue Testing of Ceramic Thermal Barrier Coatings, *Surface and Coatings Technology Technology*, 2004, **188-189**, p 146-152

[13] L. Moulla, M. P. Planche and C. Coddet, Effective Spectral Emissivity Measurement of Thermally Sprayed YSZ Thermal Barrier Coating at High Temperature using

a Wide-band Single Wavelength Pyrometer, Proceedings of the 2006 International Thermal Spray Conference, May 15-18, Seattle, WA, USA, 2006, 1-4
[14] S. Alaruri, L. Bianchini and A. Brewinton, Effective Spectral Emissivity Measurements of Superalloys and YSZ Thermal Barrier Coating at High Temperatures using a 1.6 μm Single Wavelength Pyrometer, *Optics and Lasers in Engineering*, 1998, **30**, p 77-91
[15] Y. Tan, J. P. Longtin, S. Sampath and H. Wang, Effect of Starting Microstructure on the Thermal Properties of As-sprayed and Thermally Exposed Plasma-sprayed YSZ Coatings, *Journal of the American Ceramic Society*, 2009, **92**(3), p 710-716

Thermal and mechanical properties of novel lanthanum zirconate based thermal barrier coatings

Xingye Guo[1], Zhe Lu[2], Yeon-Gil Jung[2] , Li Li[3] , James Knapp[3], Jing Zhang[1]

[1] Indiana University-Purdue University Indianapolis, IN 46202, USA
[2] Changwon National University, Changwon, Gyeongnam 641-773, Republic of Korea
[3] Praxair Surface Technologies Inc., Indianapolis, IN 46222, USA

Lanthanum zirconate ($La_2Zr_2O_7$) is one of the proposed next generation thermal barrier coating materials with a low thermal conductivity and a superior high temperature stability. In this study the $La_2Zr_2O_7$ and conventional 8wt% yttria stabilized zirconia (8YSZ) coatings were deposited using a plasma spray technique. The single and double-layered coating architectures were designed for evaluation. Coating's thermal and mechanical properties were characterized. Thermal shock tests were implemented using jet engine thermal shock (JETS) tests. Although confirmed with a lower thermal conductivity in this test, $La_2Zr_2O_7$ coatings present lower bond cap tensile strength and poor erosion resistance comparing to conventional 8YSZ coatings. The double-layered structure, with a $La_2Zr_2O_7$ coating on the top of a porous 8YSZ sublayer coating, has a better thermal shock performance than either single layer $La_2Zr_2O_7$ coatings or double-layered structure with a dense 8YSZ sublayer.

1 Introduction

Modern gas-turbine engines demand advanced thermal barrier coatings (TBCs) that can be employed at high operating temperatures above 1473 K (1200 °C). With the benefit of high operation temperature, the energy efficiency of the gas-turbine can be greatly increased. [1-5]. Conventional TBCs are 7 – 8 wt% yttria stabilized zirconia (8YSZ) coatings deposited by air plasma spraying (APS) or electron beam physical vapor deposition (EB-PVD). As the coating is deposited, the non-transformable tetragonal (T′) forms upon rapidly quenching. At elevated temperature above 1473 k (1200 °C) for sufficient amounts of time, both transformable tetragonal (T) and cubic (F) phases form. The transformable tetragonal (T) is capable of forming the monoclinic (M) phase upon cooling due to the partitioning of yttria at high temperatures, which causes 3-5% volume expansion that is detrimental to coating life[6]. Lanthanum zirconate ($La_2Zr_2O_7$, LZ) has been proposed as a promising advanced TBC material candidate [7-9]. LZ has no phase change from room temperature to its melting point, which is 2573 K (2300 °C). Comparing with YSZ, it has a lower thermal conductivity (1.5 W/m/K for bulk LZ and 2.1 – 2.2 W/m/K for bulk YSZ), lower sintering ability, and lower coefficient of thermal expansions (CTEs, 9.1 – 9.7×10^{-6} K^{-1} for LZ and 10.5 – 11.5×10^{-6} K^{-1} for YSZ) [8]. The high temperature phase stability of LZ has been studied using synchrotron X-ray diffraction (XRD) at Argonne National Laboratory. The results showed no phase transformation in the temperature range of 303 – 1673 K (30 – 1400 °C) [10-12]. Vassen *et al.* studied the thermal cycling behavior of TBCs with the single-layer (LZ or 8YSZ layer) and double-layer (LZ top layer on 8YSZ sublayer) coatings at the temperature range between 1473 K (1200 °C) and 1723 K (1450 °C). The results showed that the single-layer LZ coating had a rather poor thermal cycling performance at temperatures above 1573 K (1300°C), and the double-layer systems showed similar results to the single-layer 8YSZ coatings at temperatures below 1573 K [9]. Lee *et al.* investigated the thermal cycling behavior of 8YSZ TBC systems, indicating that the microstructural features are an important factor, and it is necessary to optimize the porosity in thick coatings [13].

The present study focuses on the comprehensive characterization of the thermal and mechanical properties of LZ based TBCs. In this investigation the thermal conductivity and CTEs of LZ coating are measured. The thermal shock and thermal cycling performance of the single and double layered LZ based TBCs are investigated using jet engine thermal shock (JETS) tests. The results are compared with conventional 8YSZ coatings.

2 Experimental details

2.1 Coating preparation

NiCrAlY powder, Ni-211 (Praxair Surface Coating Technologies, Indianapolis, IN), was selected as a bond coat feedstock. The bond coat thickness is 180 ~ 200 µm. AMPERIT 825.001 and AMPERIT 827.006 (HC Starck, Munich, Germany) were selected as top coat feestock powders to produce the dense and porous 8YSZ coatings. LAO-109-1 (Praxair Surface Coating Technologies, Indianapolis, IN) was the powder used to produce LZ coating. All coating processs were conducted at Praxair Surface Technologies, Inc. using a proprietary shrouded plasma technique. Two different types of double-layer TBC samples were deposited: LZ coating on porous 8YSZ layer and LZ coating on dense 8YSZ layer. Additionally, two types of single-layer TBC samples, single-layer LZ and single-layer porous 8YSZ, were produced for comparison purposes. All prepared TBC samples are listed in Table 1.

Table 1. Sample summary

Type of TBC	Top coats	Thickness
1	Single-layer $La_2Zr_2O_7$	424 μm
2	Single-layer porous 8YSZ	452 μm
3	Porous 8YSZ+ $La_2Zr_2O_7$	138 + 305 μm
4	Dense 8YSZ + $La_2Zr_2O_7$	140 + 292 μm

2.2 Testing method

The density measurement for LZ top coat was conducted using 432 μm thick free standing coatings produced by depositing onto aluminum substrates, which was then leached away using a diluted NaOH solution. The density was measured following the ASTM standard B328-94, which is based on the Archimedes' principle.
Thermal diffusivity was measured through a laser flash diffusivity system DLF 1200 (TA Instrument , Delaware). Thermal conductivity, k, was then calculated from thermal diffusivity α (measured), specific heat capacity C_p,(from literature [7]) and density ρ(measured): $k = \alpha \cdot C_p \cdot \rho$.
CTEs (sample size 5 mm in width, 25 mm in length and 0.43 mm in thickness) were measured by using a BAEHR dilatometer DIL 802 (Now belongs to TA instrument, Delaware) from room temperature to 1673 K (1400 °C).
Erosion resistance tests were performed following ASTM G76-13. Aluminum oxide sand with a particle size of 50 μm was used as erosive material. 600-g sands were used for each sample. The speed of sand flow stream is 6 g/s with an impingement angle of 20°. The erosion rate (μg/g) is used to evaluate the erosion resistance property in this experiment, which is calculated by dividing TBCs erosion weight (μg) by the abrasive flow weight (g).
Bond cap tensile strength tests were conducted to investigate the adhesion/cohesion bond strength following ASTM C633. FM 1000 adhesive film was used to glue the samples to the bond bars. The adhesion or cohesion bond strength can be calculate from dividing maximum load by the coating area.
JETS tests were conducted to investigate the thermal shock and thermal cycling performances. During JETS tests, TBC samples were heated to 1505 K (1232 °C) at the front surface for 20 sec, and then were cooled by compressed N_2 for 20 sec, followed by ambient cooling for 40 sec. The heating and cooling cycle was repeated until TBC samples failed. The failure criteria in JETS tests are typically more than 20 % spallation of the TBC surface [14]. Both the front and back side temperatures were measured during the test through two-color pyrometers to investigate the temperature gradient across sample thickness. The TBC samples were sectioned and polished according to the ASTM standard E1920-30 to examine their microstructures using both optical microscopy and scanning electron microscopy (SEM, JEOL Model JSM-5610, Japan).

3 Results and discussion

3.1 Coating microstructure

The microstructure of as coated TBCs are illustrated in Fig.1. The measured average density and porosity for LZ layer are 5.31g/cm^3 and 11.54%, respectively. Porosities in the dense and porous 8YSZ layers are 7.24 % and 15.95 %, respectively.

(a)

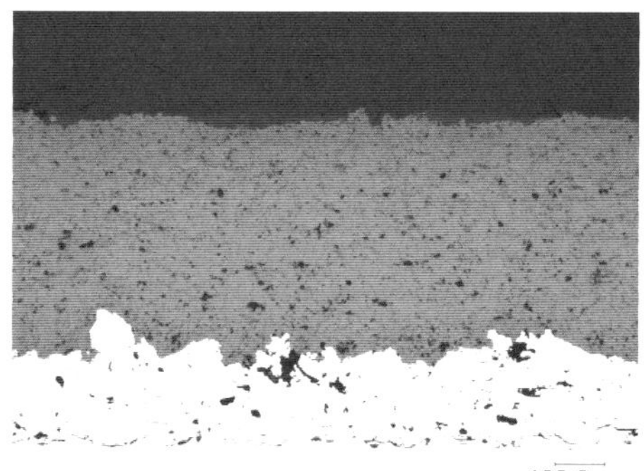

(b)

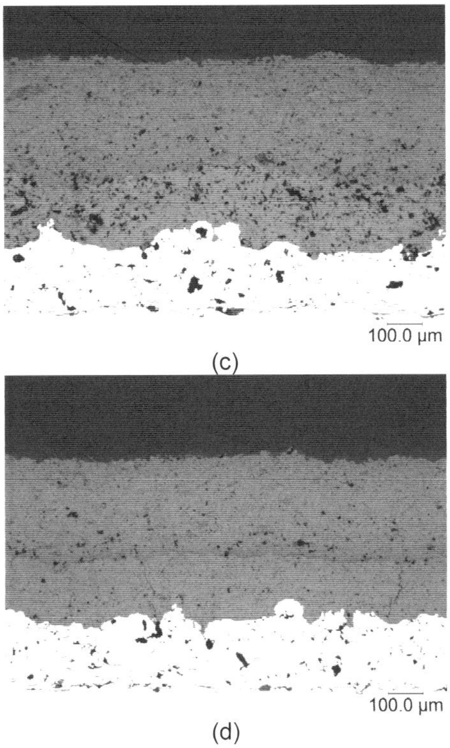

Fig. 1. Microstructures of as coated samples. (a) Single- layer porous 8YSZ; (b) Single-layer LZ; (c) ; LZ on Porous 8YSZ; (d) LZ on Dense 8YSZ.

3.2 Coating Thermal Properties

The thermal conductivities of the LZ and porous 8YSZ are plotted in Fig. 2, respectively. Three thermal conductivity measurements were conducted for the LZ at each temperature setting (579 K, 677 K, 579 K, 780 K, 974 K, and 1174 K). The measured average thermal conductivity of the LZ is about 0.59 – 0.68 W/m/K at a temperature range of 297 to 1172 K (24 – 899 °C), which is about 25% lower than that of porous 8YSZ (0.84 – 0.87 W/m/K) at same temperature range.

The CTE values measured for the LZ coating employed in this study are shown in Fig. 3. The CTE values of the LZ are about 9 – 10×10^{-6} K^{-1} from 400 K (127 °C) to 1600 K (1327 °C), which are close to the reported literature data [15-17]. The CTE values of the LZ are about 10% lower than those of 8YSZ, which is 10 – 11×10^{-6} K^{-1} in the same temperature range [7, 18]. Due to the difference of CTE between the 8YSZ and the LZ, a thermal mismatch stress could be generated at the interface between 8YSZ sublayer and LZ top layer during the thermal shock and thermal cycling tests.

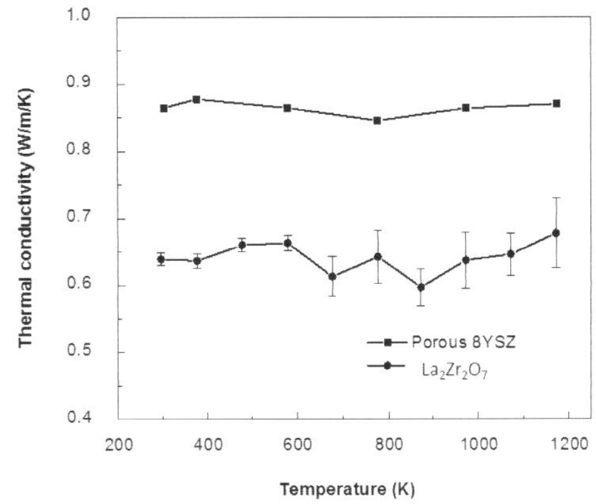

Fig. 2. Thermal conductivities of $La_2Zr_2O_7$ and 8YSZ as a function of temperature.

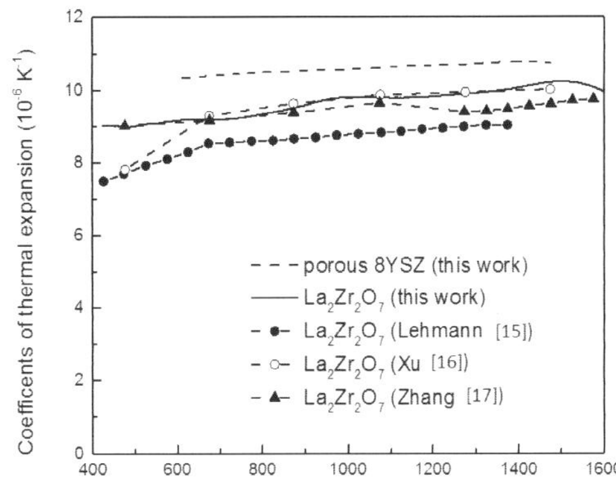

Fig. 3. Cofficient of thermal expansion values of $La_2Zr_2O_7$ and 8YSZ as a function of temperature.

3.3 Coating mechanical properties

Only the single layer samples were tested for bond cap tensile strength. As shown in Fig. 4, the 8YSZ coating has a higher bond cap strength than that of the $La_2Zr_2O_7$ coating. When the tensile stress is lower than the adhesive/cohesive bond strength, the TBCs will be intact. Local fracture could happen when tensile load in the coating layer equals the adhesive/cohesive bond strength. There are typically two expressions for the failure mechanism: (1) residual stress enhances defects inside the ceramic layer, which will lead to delamination (2) tensile stress enhances the tendency of cracking. Failure occurred when overload stress applied [19].

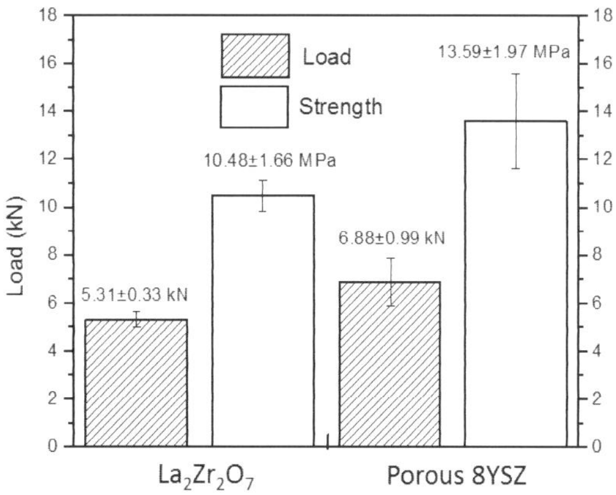

Fig. 4. Load and bond strength tensile test results

Fig. 5 shows the cross section SEM images of the TBCs system after the bond strength tensile tests. The spallation occurs predominantly near the interface between the top and bond coats. For the LZ, it apperenatly failed within the ceramic layer which indicates a lower LZ coating cohesive strength, which is probably assoicated with its lower fracture toughness than the 8YSZ (2.2 – 3.3 MPa·m$^{1/2}$ for 8YSZ and 1.1 ± 0.2 MPa·m$^{1/2}$ for LZ) [7, 20-21].

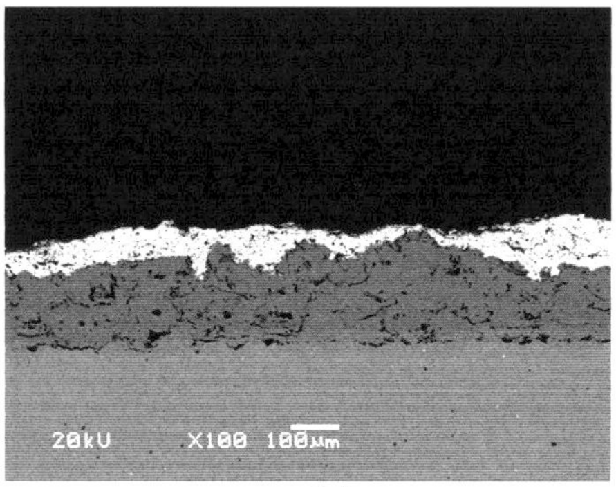

(a)

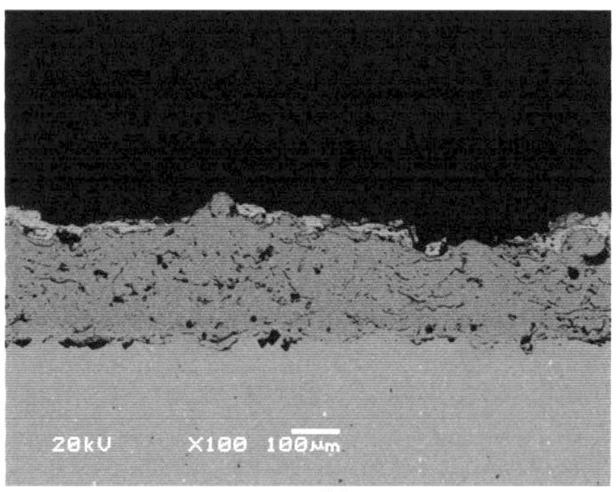

(b)

Fig. 5. SEM images of interface layer after bond strength tensile tests for single-layer coatings: (a) LZ and (b) 8YSZ.

Fig. 6 summarizes the erosion testing results for TBC systems indicated in Table 1. Single layer porous 8YSZ coating has the lowest erosion rates among the tested samples. The erosion rates of all LZ coating layers are much higher than the 8YSZ. However, the doulbe layer coating, LZ layer on dense 8YSZ, has relatively lower erosion rates than other samples with LZ layers. Fig.7 illustrates the cross-section image of eroded samples. Generally speaking the eroded area is quite large in LZ coating layer. As shown in Fig. 7(b), although the erosded area is relatively small compare to other samples with LZ layers, the delamination of 8YSZ from the substrate was observed. For the double layer coating system with the porous 8YSZ and LZ layers shown in Fig. 7(c), both layers are removed in the center, however porous 8YSZ sublaye has less eroded area. The interfaces between each layers seems intact from the photos. In Fig. 7(d), although large area of LZ layer has been removed , the dense 8YSZ layer is almost intact. The interface between dense 8YSZ and substrate is also intact.

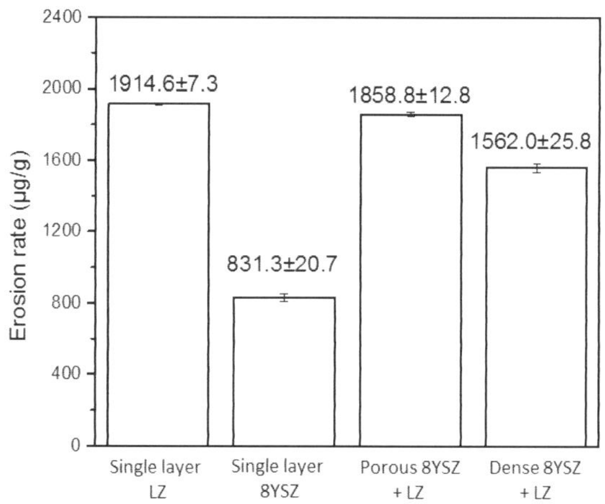

Fig. 6. Erosion test results comparison

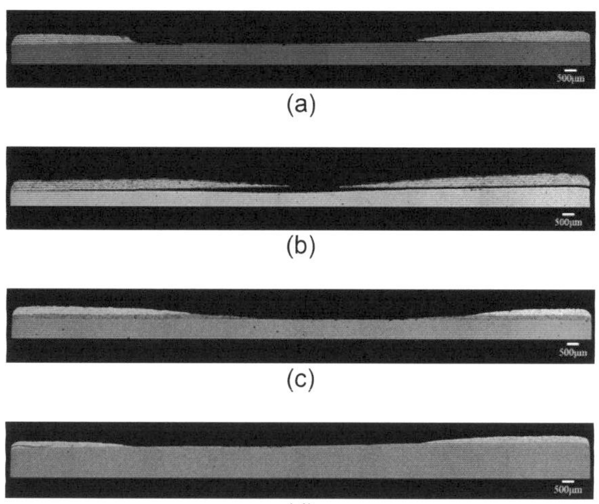

(a)

(b)

(c)

(d)

Fig. 7 Crosse-section images of tested erosion samples: (a) Single layer LZ; (b)Single layer 8YSZ; (c)Porous 8YSZ+LZ and (d) Dense 8YSZ+ LZ.

3.4 Coating thermal shock behaviors

The number of cycles-to-failure and final top coat photos after JETS testing are illustrated in Fig. 8. The three black marks on the edge of the samples were induced by the prones of sample holders. During JETS tests, the single-layer LZ coating completely spalled at 25 cycles(Fig 8a), and the double-layer coating with LZ on top of the dense 8YSZ was completely spalled in about 885 cycles(Fig8d). Delamination happened near the interface between the dense 8YSZ and the LZ. However, the double-layer coating with LZ on top of the porous 8YSZ was only partially failed at the edge after 2000 cycles and the cracked edge area was about 20% of the total area of the top coat. In the remaining area, the LZ top coat was still bonded with the porous 8YSZ layer(Fig 8c). The single-layer porous 8YSZ coating was intact after 2000 cycles(Fig8b).

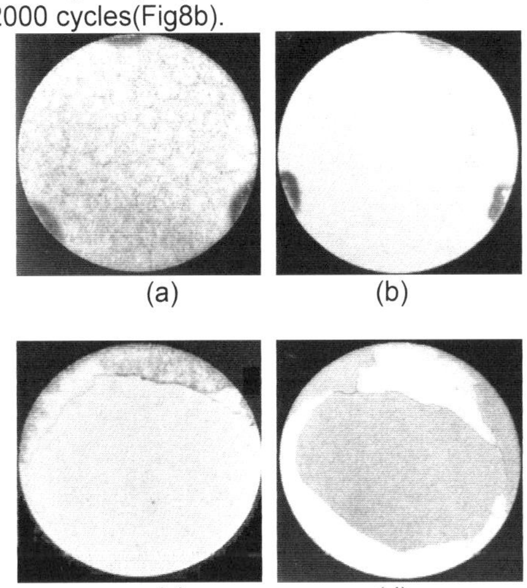

(a) (b) (c) (d)

Fig.8 Photos of the TBC samples after JETS tests: (a) Single layer LZ, 25 cycles; (b)Single layer 8YSZ, 2000 cycles; (c)Porous 8YSZ+LZ, 2000 cycles and (d) Dense 8YSZ+ LZ, 885 cycles

4 Conclusion

In this study lanthanum zirconate ($La_2Zr_2O_7$, LZ) coatings, both single-layer and double-layer, were fabricated, and their basic thermal and thermomechanical properties were characterized. The conclusions are summarized below.

1. The measured average thermal conductivity of LZ is 0.59 – 0.68 W/m/K in the temperature range of 297 to 1172 K (24 -899 °C), which is ~ 25% lower than that of porous 8YSZ in the same temperature range.
2. The CTE values of LZ coating are approximately 9 – 10×10^{-6} K^{-1} in the temperature range of 400 – 1600 K (127 - 1327 °C), which are about 10 % lower than those of 8YSZ at the same temperature range. The large CTE difference between the bond coat and the LZ leads to large thermal stresses, which is one of main reasons to cause the failure of the single-layer LZ coating during Thermal shock testing.
3. $La_2Zr_2O_7$ coatings present lower bond cap tensile strength and poor erosion resistance comparing to conventional 8YSZ coatings. The results suggest a poor foreign object damage resistance for LZ coatings.
4. JETS tests show that the double-layer coating with LZ on top of the porous 8YSZ has a better thermal shock and thermal cycling performances than those of the single-layer LZ coating and the double-layer coating with LZ on top of the dense 8YSZ sublayer. The results suggest that the porous 8YSZ could be used as a buffer layer for LZ based TBC system to improve the durability during service

5. Acknowledgements

This work was supported by the United State Department of Energy (Grant No. DE-FE0008868, program manager: Richard Dunst) and Indiana University-Purdue University Indianapolis Research Support Funds Grant (RSFG) and by a National Research Foundation of Korea (NRF) grant funded by the Korean Government (MEST) (2011-0030058) and by Changwon National University in 2015~2016.

6 Literature

[1] Clarke, D.R., M. Oechsner, and N.P. Padture, Thermal-barrier coatings for more efficient gas-turbine engines. MRS Bulletin, 2012. 37(10): p. 891-898.

[2] Evans, A.G., et al., Mechanisms controlling the durability of thermal barrier coatings. Progress in Materials Science, 2001. 46(5): p. 505-553.

[3] Clarke, D. and C. Levi, Materials design for the next generation thermal barrier coatings. Annual Review of Materials Research, 2003. 33(1): p. 383-417.

[4] Miller, R.A., Current status of thermal barrier coatings — An overview. Surface and Coatings Technology, 1987. 30(1): p. 1-11.

[5] Vaßen, R., et al., Overview on advanced thermal barrier coatings. Surface and Coatings Technology, 2010. 205(4): p. 938-942.

[6] Scott, H. G. "Phase Relationships in the Zirconia-Yttria System", J. of Material Science,Vol 10 (No.9), 1975, pp. 1527-1535

[7] Vassen, R., et al., Zirconates as New Materials for Thermal Barrier Coatings. Journal of the

American Ceramic Society, 2000. 83(8): p. 2023-2028.
[8] Cao, X.Q., R. Vassen, and D. Stoever, Ceramic materials for thermal barrier coatings. Journal of the European Ceramic Society, 2004. 24(1): p. 1-10.
[9] Vaßen, R., F. Traeger, and D. Stöver, New Thermal Barrier Coatings Based on Pyrochlore/YSZ Double-Layer Systems. International Journal of Applied Ceramic Technology, 2004. 1(4): p. 351-361
[10] Zhang, J. Novel Functional Graded Thermal Barrier Coatings in Coal-fired Power Plant Turbines, 2015 NETL Crossingcutting Research Review Meeting, Pittsburgh, PA, April 27-30, 2015. 2015; Available from: https://www.netl.doe.gov/File%20Library/Events/2015/crosscutting/Crosscutting_20150428_1300A_IUPUI.pdf.
[11]. Zhang, J., et al., Microstructural non-uniformity and mechanical property of air plasma-sprayed dense lanthanum zirconate thermal barrier coating. Materials Today: Proceedings, 2014. 1(1): p. 11-16.
[12] Zhang, J., et al. Quantitative Analysis of Pore Morphology in Lanthanum Zirconate Thermal Barrier Coating. in Materials Science & Technology 2014 (MS&T). 2014. 2061 - 2068.
[13] Lee, P.-H., et al., Thermal cycling behavior and interfacial stability in thick thermal barrier coatings. Surface and Coatings Technology, 2010. 205(5): p. 1250-1255.
[14] Bolcavage, A., et al., Thermal shock testing of thermal barrier coating/bondcoat systems. Journal of Materials Engineering and Performance, 2004. 13(4): p. 389-397
[15]. Lehmann, H., et al., Thermal Conductivity and Thermal Expansion Coefficients of the Lanthanum Rare-Earth-Element Zirconate System. Journal of the American Ceramic Society, 2003. 86(8): p. 1338-1344.
[16] Xu, C., et al., Theory of the thermal expansion of Si and diamond. Physical Review B, 1991. 43(6): p. 5024-5027.
[17] Zhang, J., et al., Thermal expansion and solubility limits of cerium-doped lanthanum zirconates. Journal of Alloys and Compounds, 2012. 525(0): p. 78-81.
[18] Hayashi, H., et al., Thermal expansion coefficient of yttria stabilized zirconia for various yttria contents. Solid State Ionics, 2005. 176(5–6): p. 613-619.
[19] M. Gell, E. Jordan, Bond strength and stress measurements in thermal barrier coatings, 1995.
[20] K. Jiang, S. Liu, G. Ma, L. Zhao, Ceramics International, 40 (2014) 13979-13985.
[21] G.K. Beshish, C.W. Florey, F.J. Worzala, W.J. Lenling, JTST, 2 (1993) 35-38.

Study of the Properties of Cold-Sprayed In718 Deposits

R. Huang and H. Fukanum, Plasma Giken Co. Ltd., /Japan

Inconel 718 alloy is precipitation-hardenable nickel-based superalloy which has decent corrosion resistance, high strength at ambient temperature, and excellent creep and fatigue strength at high temperature. Thus this alloy has been used for a variety of applications such as gas turbines, jet engines, steam generators and fission and fusion reactor structures. The deposition of pore-free nickel-based superalloy coatings with excellent machnical properties is of great interest for engines design and gas turbine applications, both in case of maintenance repair and overhaul operations as well for mechanical and chemical protection purposes in aeronautics and energy fields. In this investigation, Inconel 718 was deposited by cold spray process with nitrogen and helium gas. The particle velocities were maseured experimentally.The Inconel 718 particles splats and coating microstructures were observed. The mechanical properties of cold-sprayed Inconel 718 deposits including hardness, bonding strength and tensile strength were also investigated. The results showed that the deposits sprayed with helium gas had a little better mechanical properties than which deposited with nitrogen gas. Still the deposits prepared with nitrogen gas with post-heat treatment illustrated excellent mechanical properties simliar to that of Inconel 718 itself.

1 Introduction

Cold spray is an emerging spray coating technology that was first developed in the mid 1980's at the Institute of Theoretical and Applied Mechanics in former Soviet Union [1]. In Cold spraying process, spray particles are injected into a supersonic jet by compressed gas and accelerated to a high velocity (300-1200 m/s). The deposition of particles takes place through the intensive plastic deformation upon impact in solid state at a temperature well below the melting point of spray materials [2]. As a result, spray particles experience little oxidation or decomposition in cold spray process[3 and 4]. So far, cold spray was used to spray not only ductile materials such as copper [5 and 6], aluminium [7], nickel [8], nickel based alloys [9], zinc [10] but also metal matrix composites [11], cermets [12] and ceramic materials [13].

As a low-temperature coating processing technology, the most outstanding advantage of the cold-sprayed coatings is the little oxidation or degradation to the feedstock material during the process. The coating deposited by cold spray is extremely dense with high adhesive and cohesive strength[14-19]. Therefore, the cold-sprayed coatings have excellent electrical, thermal and mechanical properties. Moreover, cold spray process can produce deposits without thickness limitation with high deposition rate and high efficiency. Cold spray process characteristics and its deposits properties make this process perfectly suitable in spray forming or restoring used parts. Therefore, cold spray process is promising to deposit Inconel 718 coatings with excellent mechnical properties on engine and gas turbine parts in aeronautics and energy fields.

In this study,Inconel 718 was deposited by cold spray process with nitrogen and helium gas. The particle velocities were measured experimentally and the splats of Inconel 718 on stainless steel substrate were observed. The adhesive strength of Inconel 718 coatings on Inconel 718 substrate was measured. The microstructure, hardness, and tensile strength were also tested.

2 Experimental procedure

2.1 Feedstock Powder and Cold Spray Process

Inconel 718 powder (PG-AMP-1070) with commercial availability from Plasma Giken Co., Ltd. was used for cold spray process. The morphology of the powder is presented in **Fig.1**(a).The particles show perfect spherical shape. The powder size distribution was characterized by a laser diffraction particle size analyzer (Seishin Trading Co., Ltd. Kobe, Japan). The volume distribution of particle diameter is shown in Fig.1(b). It can be seen that the powder diameter ranges from 8 to 50 μm with average volume diameter about 22 μm.

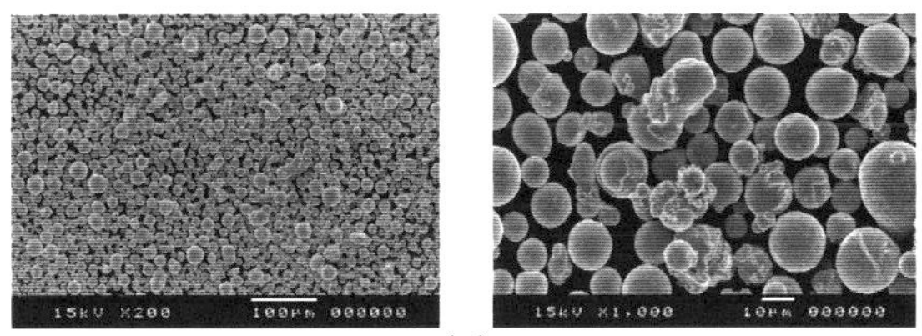

(a)

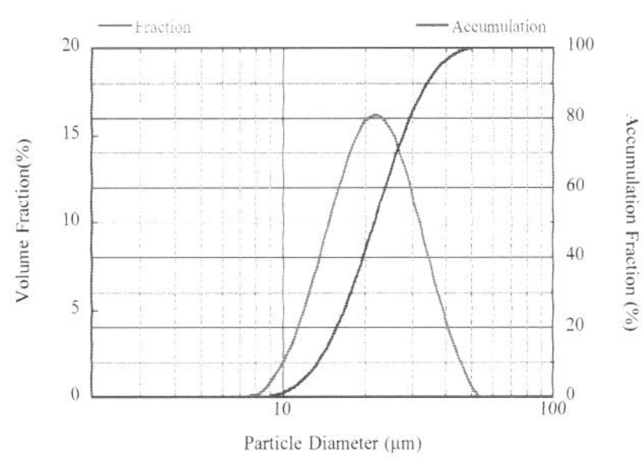

(b)

Fig.1. Morphology (a) and diameter distribution (b) of Inconel 718 powder.

DVS 324

In this study, a commercial high-pressure cold spray system PCS-1000 designed by PLASMA GIKEN CO., LTD. was used to prepare the Inconel 718 coatings. The convergence-divergence nozzle of cold spray gun is optimized one to get a higher particle velocity. it employs water cooling system for the cooling of nozzle wall which alleviates nozzle clogging and therefore guarantees high system reliability. Both helium and nitrogen gas were employed as working gas. The detailed spray conditions for cold spraying are shown in Table 1.

Table 1: The spray conditions

Gas Type	N_2	He
Working gas pressure (MPa)	3, 5, 7	3
Working gas temperature (^{o}C)	600, 800, 1000	1000
Spray distance (mm)	30	
Powder feed rate (g/min)	100	

2.2 The Measurement of In-flight Particle Velocity

The in-flight particle velocity under the above spraying conditions was measured at the centre line of the gas flow, using CSM Evolution system (Tecnar Automation Ltd., St-Bruno,Québec, Canada). The substrate was removed during the particle velocity measuring process. For cold spray process, because of the low temperature of particles, the radiation intensity emitted from the in-flight particles is too weak to be detected by the optical sensor. Therefore, a high-power diode laser was equipped in the CSM Evolution system to beam the in-flight particles. By detecting the monochromatic light scattered by particles, the velocity of particles can be measured by CSM Evolution system.In this study, the velocity measurement was taken at a point 30 mm away from the spraying gun exit.

2.3 Tests of Coatings Properties

In order to analyze the bonding, splats of Inconel 718 on stainless steel substrates were prepared. The morphologies of the splatted Inconel 718 particles were examined by an emission scanning electron microscope (JSM-5200LV, JEOL, Japan). Cross-sections of as-sprayed coatings were prepared vertically to their surfaces by a conventional mechanical polishing method with SiC paper and diamond suspensions. The mirror-polished cross-sections were examined with a digital microscope (VHX-900, Keyence, Japan). The micro-hardness of coatings was tested using micro-hardness tester FM-700 (Future-Tech. Crop) under the load of 300 g and the load-time of 15 seconds.
In conventional adhensive strength test for thermal spray coatings, a epoxy resin adhesive is commonly employed to glue the coated tensile bar with a bar no coated. The testing method is tremendously restricted by the strength of glue. The strength of epoxy resin adhesive is usually no more than 70 MPa, and consequently this testing method cannot be employed to measure coating with bond strength higher than that of the adhesive.In this study, a special adhensive strength test method was employed to measure the bonding strengt of cold-sprayed Inconel 718 coatings on Inconel 718 substrates. The bonding test method is shown in **Fig. 2**. Firstly, a sample with thick coating was prepared as shown in Fig. 2 (a),secondly the sample was machined into the shape as shown in Fig. 2 (b). Finally, the sample was pushed directly with a special jig as shown in Fig. 2 (c).The as-sprayed Inconel 718 coatings were heat treated at 990°C for 4 hours under argon atomosphere in order to improve the bonding at the coating/substrate interface. The adhesive strength of Inconel 718 coatings both before and after heat treatment was measured.

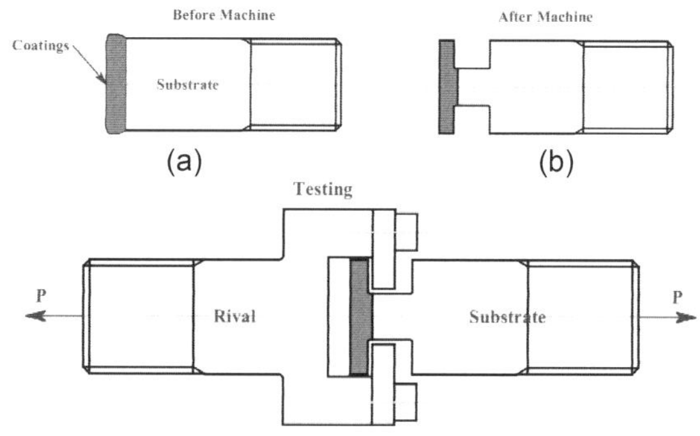

Fig.2. The special method to test the adhesive strength of coating.

Spray Inconel 718 coatings with thickness more than 5mm and then remove the coatings from their substrates machine them into tensile specimens according to the standard of ASTM E8. Heat treated these specimens at the same condition as did in adhesive strength test. Both specimens as-sprayed and specimens heat-treated were measured with the tensile testing equipment: AG-X (100KN), manufactured by Shimadzu Co., Ltd. Two cameras were equipped with the tensile machine in order to measure the extension of the specimen within the gage length during tensile test.

3 Results and Discussions

3.1 Particle Velocity

The particle velocity distribution measured at gas pressure of 5 MPa is shown in **Fig. 3**(a). It can seen that the particle velicities rang from 400 to 1200 m/s. The particle velocity increses as gas temperature gets higher. From the particle velocity distribution profiles, the average particle velocity can be obtained. The average particle velocities under different spray conditions are shown in Fig. 3.(b). It reveals that the particle velocity increases with the increase of not only gas temperature but also gas pressure. The average particle velocity of Inconel 718 particles reach about 830 m/s at the gas pressure of 7 MPa and gas temperature of 1000°C with nitrogen gas. It seems that the valid way to obtain a higher particle velocity is to develop a higher pressure and temperature cold spray system. It has been widely understood that particle velocity is a key factor to the properties of cold-sprayed coating. Delevoping a high pressure and high temperature cold spray system benefits to prepare the coating with more excellent properties.

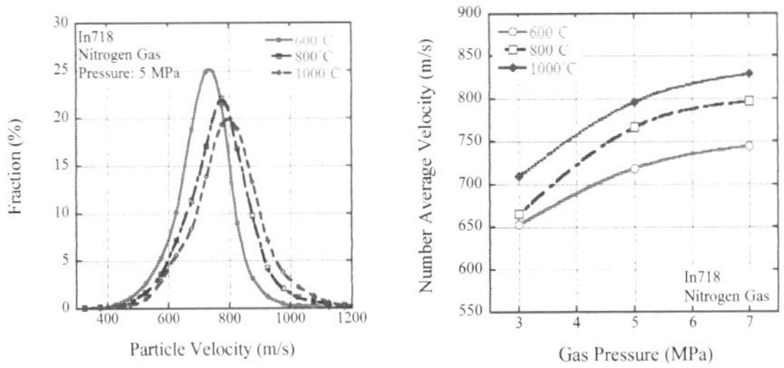

(a) Distributions of particle velocity (b) Average particle velocities

Fig. 3. The particle velocities measured

3.2 Splats and Cross-sections

Figure 4 shows the morphologies of the cold-sprayed Inconel 718 splats on stainless steel substrates under different spray conditions. Comparing (a),(b) and (d) in Fig. 4, more deformation to the Inconel 718 particles can be observed with the increase of working gas temperature, moreover the particle be embedded into substrate deeper. This phenomena can be explained from two sides. When the gas temperature increases, it makes the particle temperature higher which consequently results in easier particle deformation; meanwhile increased gas temperature generates higher particle velocity as shown in Fig. 3. which furtherly benefits to particle deformation during the particle impact process on substrate. Comparing (c) to (d) in Fig. 4, Inconel 718 particle embedded itself into substrate more deeply too as the working gas pressure increased. This is due to the particle velocity increases with the increase of gas pressure as shown in Fig. 3, which consequently causes severer deformation during the impact process.

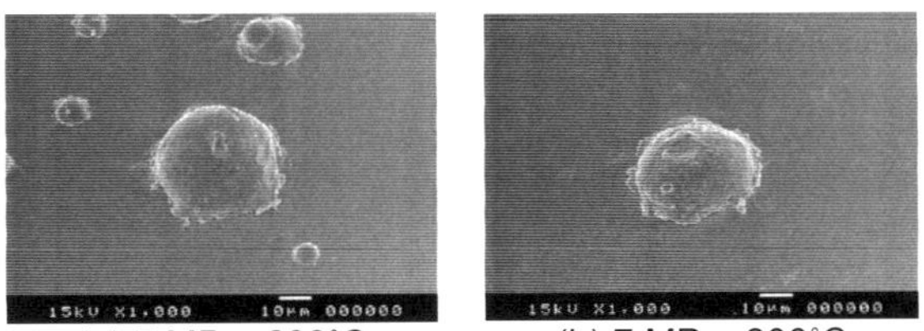

(a) 7 MPa, 600°C (b) 7 MPa, 800°C

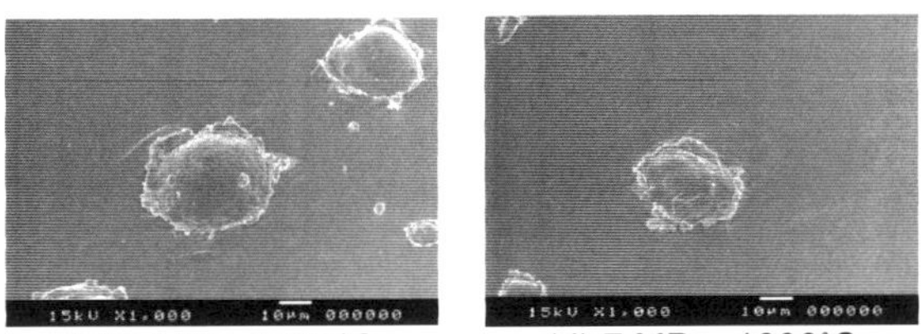

(c) 3 MPa, 1000°C (d) 7 MPa, 1000°C

Fig. 4. Inconel 718 splats on stainless steel substrate

The cross-sections of the cold-sprayed Inconel 718 coatings at different spray conditions are shown in Fig.5. Comparing (a), (b) and (c) in Fig. 5, it can be seen that the increase of gas pressure generated much denser coatings owing to the increase of particle velocity. Comparing (a) with (d) in Fig. 5, it is obvious that the coating sprayed with He gas is extremely much denser than that sprayed with N_2 gas. With He gas accelerating the particle , higher particle velocity can be obtained because it has much higer sonic speed than N_2 gas. From the coatings microstructure shown in Fig. 5, the porosity of coatings can be meansured with image analysis method. The measured porosity values are shown in Table 2. It reveals that the porosity of He gas sprayed coating is extremely low with almost no pores.

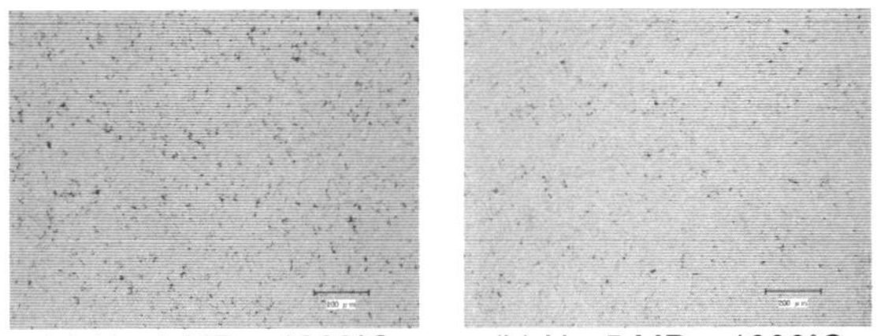

(a) N_2, 3 MPa, 1000°C (b) N_2, 5 MPa, 1000°C

(c) N_2, 7 MPa, 1000°C (d) He, 3 MPa, 1000°C

Fig. 5. Cross-sections of Inconel 718 coating

Table 2: The porosity of Inconel 718 coatings

Spray Conditions	Porosity (%)
N_2, 3 MPa, 1000°C	5.3
N_2, 5 MPa, 1000°C	2.3
N_2, 7 MPa, 1000°C	1.8
He, 3 MPa, 1000°C	0.2

3.3 Adhesive strength of coatings

The adhensive strength of cold-sprayed Inconel 718 coatings on Inconel 718 substrates is shown in Fig. 6. It can be seen that the adhesive strength experiences a modest increase as the working gas pressure rises when using nitrogen as working gas. This is caused by a higher particle velocity with the increase of gas pressure, and consequently better bonding with the substrate achieved as shown in Fig. 4. The adhensive strength significantly increased after coating heat treatment at 990°C for 4 hours in argon atmosphere, which is because diffusion between coating and substrate strengthened the coating/substrate bonding. After heat treatment, the adhesive strength of Inconel 718 coating on Inconel 718 substrate reached about 600 MPa even with nitrogen gas. Therefore, cold spray process has a high possibility applying to restoring or repairing defective Inconel 718 parts. With helium gas, higer particle velocity can be obtained than that with nitrogen gas. Therefore, significantly high adhesive strength more than 400 MPa was obtained even without heat treatment. The adhensive strength of Inconel 718 coating on Inconel 718 substrate with helium gas even increased to about 900 MPa after heat treatment owing to the diffusion between coating and substrate like the coatings sprayed with nitrogen gas.

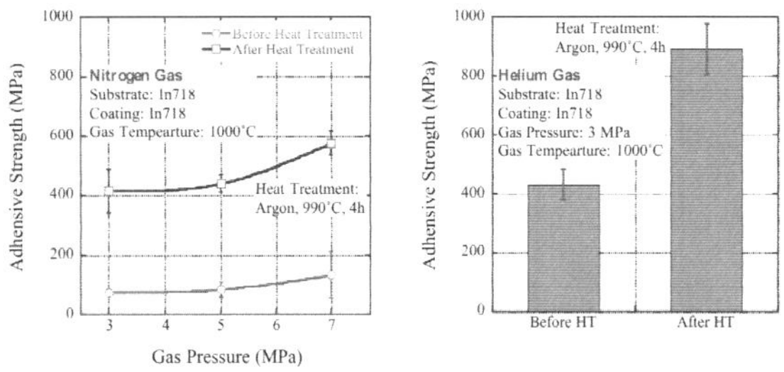

(a) Nitrogen gas (b) Helium gas

Fig. 6. Adhesive strength of Inconel 718 coatings on Inconel 718 substrate

3.4 Mechnical Properties

The measured mirco-hardness of cold-sprayed Inconel 718 coatings is shown in Fig. 7. With nitrogen gas, the hardness increases with the increase of working gas pressure which leads to denser coating as shown in Fig. 5. In common, after heat treatment, the coatings hardness decreases because the heat treatment can in some degree eliminates the work hardening occurring in the deposition process. But it can be seen that the coating deposited at 3 MPa with nitrogen gas hardly decreases after heat treatment. This coating itself is slightly porous as shown in Fig. 5 (a) with low hardness, after coating heat treatment, due to the work harding elimination,it becomes a bit softer, but the diffusion inside coating during heat treatment improves the particles bonding and also can restore some coating defects. These two factors counteracts and leads to coating hardness sprayed at 3 MPa changed little after heat treatment. With helium gas, the deposited coating has higher hardness of about 600 beceuse the extremely dense coating as show in Fig. 5 (d). After heat treatment, the hardness decreased to below 400 due to the elimination of work hardening. Comparing the hardness of heat-treated coatings sprayed by nitrogen and helium in Fig. 7 , the hardness of helium sprayed coating (about 400) is harder than that (about 350) sprayed with nitrogen gas. It is because denser coating can be achieved with helium gas than that with nitrogen gas as shown in Fig. 5. It seems that the coating porosity influences its micro-hadness even after heat treatment.

Figure 8 shows the tensile test results of cold-sprayed Inconel 718 coatings according to the standard of ASTM E8. The mechanical properties of cold-sprayed Inconel 718 coating from Fig. 8 is summerized in Table 3. The tensile strength of coating sprayed with helium gas is stronger than that sprayed with nitrogen gas. Lower elongation can be observed for the as-spray coatings with nitrogen gas as well as with helium gas. After heat treatment, both the tensile strength and elongation improved due to the diffusion happened during heat treatment process. The Young's Modulus of nitrogen gas sprayed Inconel 718 coating is lower than that of coatings sprayed by helium gas; even after heat-treatmeant it is still lower than that of coatings sprayed by helium gas due to the defects diffenence inside the coatings shown in Fig. 5. Heat treatment can improve the bonding between particles via diffusion, leading to slight improvment in coatings' Young's Modulus. However, the defects (especially the pores) inside the coating cannot be completely fixed by heat treatment process. Therefore, the Young's Modulus of nitrogen gas sprayed Inconel 718 coating is much lower than that of the referenced bulk material even after heat treatment. On the contrast, the Young's Modulus of helium gas sprayed Inconel 718 coating is very close to the value of the referenced bulk material owing to its fewer defects.

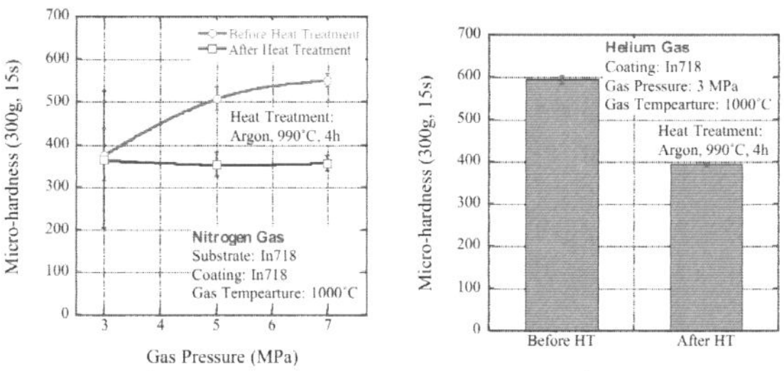

(a) N_2 (b) He

Fig. 7. Micro-hardness of Inconel 718 coatings

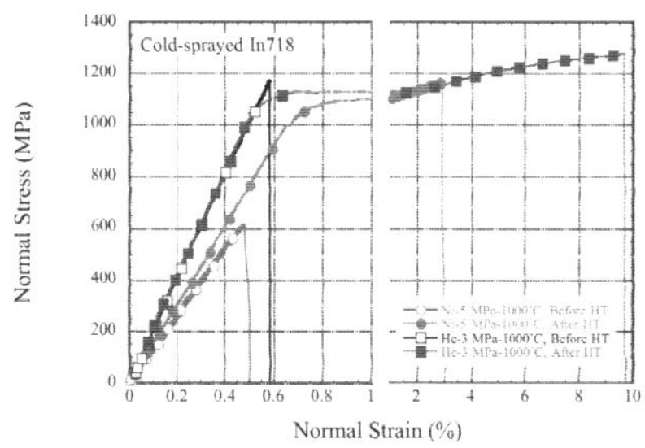

Fig.8. Tensile test results of cold-sprayed Inconel 718 coatings.

Table 3: Mechanical properties of In718 coating

Conditions	Tensile Strength (MPa)	Elogation (%)	Young's Modulus (GPa)
N_2, Before HT	611	0.48	130
N_2, After HT	1160	2.94	153
He, Before HT	1168	0.58	201
He, After HT	1272	9.64	206
Bulk(ref)	1240	12	211

The fracture surface was observed by SEM after tensile test in order to understand its fracture mechanism. Figure 9 shows the fracture surface of Inconel 718 coatings.When using nitrogen gas, the rupture of as-aprayed Inconel 718 coating without heat treatment happened at the particle boundary as shown in Fig. 9 (a). No dimple can be observed according to the fractography. It seems that de-cohesive rupture occurred to the as-sprayed Inconel 718 coating since its cohesion of particles is quite poor without heat treatment; plus the defects and pores inside the coating accelerated the failure during tensile test. This shows its strength is lower than that of the corresponding bulk material. After heat treatment, certain diffusion happened among the particle interfaces of the coating sprayed by nitrogen gas which significantly improved the cohesive strength of coating; some dimples can be observed at the fracture surface as shown in Fig. 9 (b). It seems that the fracture of the coating is micro-void coalescence rupture because some plastic deformation can be observed. However, the defects such as pores still existed even after heat treatment, so the failure occured at the early plastic deformation stage during the tensile test. When using helium gas, even though with the high tensile strength close to that of the referenced bulk material, the as-sprayed Inconel 718 coating still has some micro-defects inside especially at the deformed particle interface. Therefore, almost no dimple can be observed at the facture surface as shown in Fig. 9 (c). After heat treatment, The diffusion improved the particles cohesion and plenty of dimples can be observed shown in Fig. 9 (d) where we can see the coating fracture is micro-void coalescence rupture and its elongation is as good as that of the referenced bulk material.

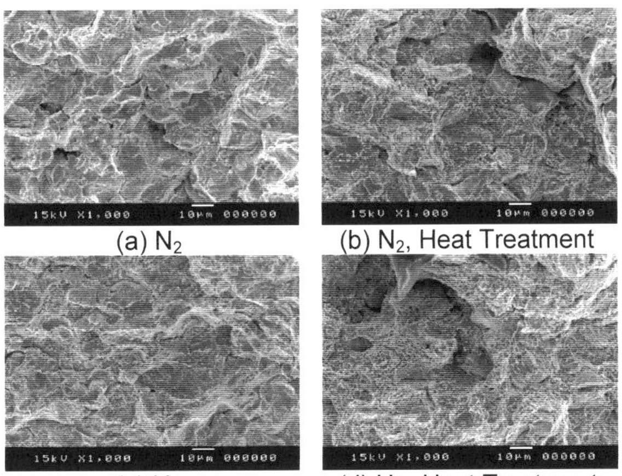

(a) N_2 (b) N_2, Heat Treatment

(c) He (d) He, Heat Treatment

Fig. 9. Morphologies of ruptured In718 coating after tensile test.

4 Conclusions

In this study, Inconel 718 coatings were prepared with cold spray process. In-flight particle velocity of Inconel 718 particles was measured experimently. The morphologies of Inconel 178 splats and microstructures of Inconel 178 coating were observed. And the mechanical properties of the cold-sprayed coatings were tested. The results reveals that the in-flight particle velocity increases with the increase of working gas temperature and/or gas pressure. Higher particle velocity can generate more deformation on the substrates and particles, consequently, particles are embedded into the substrates more deeply which can generate higher coating adhesive strength with substrate and denser coatings theirselves. When particles accelerated by helium gas at an extremely higher velocity, extremely dense Inconel 718 coating almost without pores can be achieved and the coating showed significantly strong adhesive strength more than 400 MPa on Inconel 718 substrate. Dense coating has excellent mechanical properties such as high hardness and strong tensile strength close to the value of the corresponding bulk. In general, heat treatment can eliminate the work hardening of cold-sprayed coatings which decreases its micro-hardness. The diffusion during heat treatment can amend some defects inside coatings and thus improve the tensile strength and elongation of cold-sprayed Inconel 718. After heat treatment, Inconel 718 coating even sprayed by nitrogen gas has a high bonding strength about 600 MPa with Inconel 718 substrate, and more than 1100 MPa tensile strength. Therefore, with great improvement in coating bonding strength, tensile strength and elongation brought by after heat treatment, It is a prospective way to repair or restore defective Inconel 718 parts with cold-sprayed Inconel 718 coating using nitrogen gas; when using helium gas, it proves much more feasible technically because the as-sprayed Inconel 718 coating exhibited an excellent properties both in bonding strength and tensile strength even without heat treatment.

5 References

[1]. P. Richer, B. Jodoin, K. Taylor, E. Sansoucy, M. Johnson, L. Ajdelsztajn and Davis, Effect of particle geometry and substrate preparation in cold spray, *Thermal Spray 2005: Internal Thermal Spray Conference*

[2]. W.-Y. Li and C.-J. Li, Effect of annealing treatment on the microstructure and properties of cold-sprayed Cu coating, *Thermal Spray 2005: Internal Thermal Spray Conference*

[3]. T. Stoltenhoff, H. Kreye, and H. J. Richter, An Analysis of the Cold Spray Process and Its Coatings, *J. Therm. Spray Technol.*, 2002, **11** (4), p 542-550.

[4]. R. C. Mcune, W. T. Donlon, E. L. Cartwright, Characterization of Copper and Steel Coatings Made by the Cold Gas-Dynamic Spray Method, *Thermal Spray: Practical Solutions for Engineering Problems*, C.C. Berndt, Ed., Oct 7-11, 1996 (Cincinnati, OH), ASM International, 1996, p 397-403.

[5]. E. Calla, D.G. McCartney, P.H. Shipway, Deposition of copper by cold gas dynamic spraying: an investigation of dependence of microstructure and properties of the deposits on the spraying conditions, *Thermal Spray 2004: Advances in Technology and Application*, May

10-12, 2004 (Osaka, Japan), ASM International, 2004, p 382-387
[6]. T. Xiong, Z. Bao, T. Li, Z. Li, Study on cold-sprayed copper coating's properties and optimizing parameters for the spraying process, *Thermal Spray 2005: Explore its surfacing potential*, May 2-4, 2005 (Basel, Switzerland), ASM International, 2005, p 178-184
[7]. R. Morgan, P. Fox et al., Analysis of cold gas dynamically sprayed aluminum deposits, *Mater. Lett.*, 2004,**58**, p 1317-1320.
[8]. M. K. Decker, R. A. Neiser, D. Gilmore, H.D. Tran, Microstructure and Properties of Cold Spray Nickel, *Thermal Spray 2001: New Surfaces for a New Millennium*, C.C. Berndt, K.A. Khor, and E.F. Lugscheider, Ed., May 28-30, 2001 (Singapore), ASM International, 2001, p 433-439.
[9]. F. Raletz, G. Ezo'o, M. Vardelle, M. Ducos, Characterization of cold-sprayed nickel-base coatings, *Thermal Spray 2004: Advances in Technology and Application*, May 10-12, 2004 (Osaka, Japan), ASM International, 2004, p 344-349
[10]. C-J. Li, W-Y Li, Impact Fusion Phenomenon during Cold Spraying of Zinc, *Thermal Spray 2004: Advances in Technology and Application*, May 10-12, 2004 (Osaka, Japan), ASM International, 2004, p 356-361
[11]. D. Morelli, A. Elmoursi, T. Vansteenksite, D. Gorkiewicz, B. Gillispie, Kinetic Spray of Aluminum Metal Matrix Composites for Thermal Management Applications, *Thermal Spray 2003: Advancing the Science and Applying the Technology*, B.R. Marple and C. Moreau, Ed., May 5-8, 2003 (Orlando, FL), ASM International, 2003, Vol. 1, p 85-90
[12]. J. Karthikeyan, C. M. Kay, J. Lindemann, Cold Sprayed nanostructured, WC-Co, *Thermal Spray 2001: New Surfaces for a New Millennium*, C.C. Berndt, K.A. Khor, and E.F. Lugscheider, Ed., May 28-30, 2001 (Singapore), ASM International, 2001, p 383-387.
[13]. C-J. Li, G-J. Yang, X-C. Huang, W-Y. Li, A. Ohmori, Formation of TiO2 photocatalyst Through cold spraying, *Thermal Spray 2004: Advances in Technology and Application*, May 10-12, 2004 (Osaka, Japan), ASM International, 2004, p 315-319
[14]. R. Huang and H. Fukanuma, Study of the Influence of Particle Velocity on Adhesive Strength of Cold Spray Deposits, J. Therm. Spray Technol., 2012, 21(3-4), p 541-549
[15]. F. Gärtner, T. Stoltenhoff, J. Voyer, H. Kreye, S. Riekehr and M. Koçak, Mechanical properties of cold-sprayed and thermally sprayed copper coatings, Surf. Coat. Technol., 200 (24), 2006, p 6770–6782
[16]. B. AL-Mangour, P. Vo, R. Mongrain, E. Irissou, and S. Yue, Effect of Heat Treatment on the Microstructure and Mechanical Properties of Stainless Steel 316L Coatings Produced by Cold Spray for Biomedical Applications, J. Therm. Spray Technol., 2014, 23(4), p 641-652
[17]. W. Wong, E. Irissou, P. Vo, M. Sone, F. Bernier, J.-G. Legoux, H. Fukanuma and S. Yue, Cold Spray Forming of Inconel 718, J. Therm. Spray Technol., 2013, 22(2-3), p 413-421
[18]. Renzhong Huang, Wenhua Ma and Hirotaka Fukanuma, Development of ultra-strong adhesive strength coatings using cold spray, Surf. Coat. Technol., 258 (2014) p 832–841
[19]. Renzhong Huang, Michiyoshi Sone, Wenhua Ma, Hirotaka Fukanuma, The effects of heat treatment on the mechanical properties of cold-sprayed coatings, Surf. Coat. Technol., 261 (2015) p 278–288

Advanced diamond-reinforced copper composite coatings via cold spray and material characterization

Shuo Yin, Barry Aldwell and Rocco Lupoi
Trinity College Dublin, The University of Dublin, Department of Mechanical and Manufacturing Engineering, Parsons Building, Dublin 2, Ireland

Diamond-reinforced copper matrix composites (DCMC) has great potentials for heat sinks applications due to their excellent thermal properties. Cold Spray, as a relatively emerging coating technique, is able to fabricate coatings or bulk materials at entire solid-state, significantly lowering the risk of oxidation, phase transformation and high thermal residual stress because of the low processing temperature. In this paper, thick DCMC coatings were fabricated via cold spray using copper-clad diamond powder or its mixture with pure copper powder. It was found that cold spray due to its low processing temperature was able to avoid the graphitization of diamond in the DCMC coatings. Using the pure clad diamond powder only, the diamond in the original feedstock was almost completely retained in the coating which contained more than 40 wt.% diamond. Such high mass fraction of diamond has never been achieved in previous cold spray works using pre-mixed powders. The mechanically mixed powder of clad diamond and copper also exhibited great retainbility for diamond; the diamond fraction in the coating was even larger than that in the original feedstock. Besides, the most important finding here is the additional copper powders acted as a buffer, effectively preventing the fracture of diamond in the coating.

1 Introduction

Copper has been widely applied as heat sinks material for a long time due to its high thermal conductivity. However, the rapid growth of the modern industry has brought an urgent demand of the heat sink materials with higher thermal performance. Diamond is known to possess much higher thermal conductivity and lower thermal expansion coefficient than copper, but it is really difficult to be machined due to the extremely high hardness, which significantly limits its direct application as heat sink material. Therefore, a kind of advanced material, metal matrix composite (MMC) consisting of copper and diamond, was developed to improve the thermal performance of pure copper heat sink. In many studies, such diamond-reinforced copper matrix composite (DCMC) exhibited excellent thermal performance, hence sparking significant interests from both science and industrial communities [1–11].

Cold Spray, as an emerging coating technique is capable to deposit metals, MMCs and even ceramics, thereby attracting great interests over decades [1]. In this process, feedstock materials in the form of micro-size powders accelerated by the supersonic driving gas through a de-Lavel nozzle, impact onto the substrate to form the coating. During the deposition process, the feedstock remains solid state without any melting; coating is formed through the metallurgical or mechanical bonding at the interface of adjacent particles and coating/substrate. Therefore, defects like oxidation, thermal residual stress and phase transformation which always appear in powder metallurgy, pressure infiltration and thermal spray can be considerably avoided in the cold sprayed coating [2]. Furthermore, with cold spray, coatings can be deposited on various similar or dissimilar substrates; also the thickness growth is almost unlimited for most metals and MMCs, which allows cold spray to act as a kind of additive manufacturing technique for producing bulk materials.

Very few attempts of diamond-reinforced metal matrix composites (DMMC) coatings via cold spray have been carried out in previous works for improving the coating wear-resistance capability [3,4], hardness and Young's module [5,6]. In these works, powders were normally mechanically mixed before spraying. By this means, the fraction of diamond in the coating significantly reduces compared with the original feedstock, which significant lowered the coating performance and resulted in a vast waste of the expensive diamond [3,4,7]. Ball milling as an alternative way is able to guarantee the minimum loss of the diamond during the deposition. But the graphitization and serious fracture of the diamond in the DMMC coating also resulted in poor-quality coating [5,6]. Besides, all the existing studies on DMMC via cold spray focused on the improvement of coating mechanical properties. In this paper, thick DCMC coating as a great potential candidate for heat sinks materials was fabricated via cold spray technique. Special copper-clad diamond powder or its mixture with pure copper powder were selected as the feedstock to fabricate the coating.

2 Experimental methodology

2.1 Coating fabrication and powders

MMC coatings were deposited on aluminium substrates using a home-made cold spray system (Trinity College, Ireland). The system consists of high pressure gas from bottles, a gas heater, a powder feeder, a CNC working platform for controlling the substrate movement, a de-Laval nozzle and a computer control system. In this work, nitrogen and helium were applied as the driving gas to deliver and accelerate the powders. Copper powders (-38+15 μm, Safina, Czech Republic), copper-claded diamond powders (-53+45 μm, Element Six, Ireland) and their mixture were used as the feedstock. Figure 1 shows the morphology and characterization of each powders observed by SEM (Carl Zeiss ULTRA, Germany). The copper-clad diamond powder is made of three different prats, namely, an inside diamond powder, a very thin interbedded nickel layer and an outside electroless copper cladding, which can be seen from Fig. 1c showing the powder cross-section prepared by the FIB (DB235, FEI Strata, USA). The reason for introducing nickel layer is because nickel provides much better bonding with diamond than copper. The thickness of the metal cladding layer is about 2-5 μm, leading to an approximate 50% mass fraction in a single grain. Detailed powders information and working conditions are provided in Table 1.

Table 1. Powders and working conditions

	Powders	Gas	P	T
P	Copper	N_2	3.0MPa	350°C
P1	Copper + Clad diamond (8:1)	He	2.0MPa	25°C
P2	Copper + Clad diamond (1:1)	He	2.0MPa	25°C
P3	Clad diamond	He	2.0MPa	25°C
P4	Clad diamond	He	2.5MPa	25°C

For convenience, coatings fabricated with different powders and working conditions were labelled as P to P5. The clad diamond fraction in the original feedstock increases from P1 to P3, while P3 and P4 are pure clad diamond without copper mixture. Note that the pure copper coating marked as 'P' was only used as a benchmark for the comparison with other DCMC coatings.

Fig. 1. Morphology and characterization of copper and copper-clad diamond powders used in this study. (a) copper, (b) copper-clad diamond and (c) cross-section of copper-clad diamond

2.2 Materials characterization

The as-sprayed coatings were examined by an X-Ray diffractometer (Siemens D500, Germany) with the Co (λ=1.789 Å) source at a current of 40 mA and voltage of 35kV to examine whether graphitization of the diamond occurs during the coating fabrication process. For analysing the coating microstructure, the as-sprayed coatings were post-processed to produce the polished and fractured cross-section. Samples for polish treatment were firstly cut, then grounded by SiC sand papers and finally polished by 0.05 μm Al_2O_3 solution. The fracture surface was obtained by bending the sample with pliers. The coating microstructures were then characterized by a SEM (Carl Zeiss ULTRA, Germany). Besides, the element analysis on the coating surface and cross-section was also performed with an EDS unit (Oxford Instruments INCA system, UK) equipping on the SEM system. In order to achieve an accurate measurement, five zones were randomly selected from the coating surface or polished cross-section and the measured data was then averaged.

2 Results and discussion

2.1 Coatings properties

Figure 2 shows the digital photos of as-sprayed DCMC coatings with different powders and working conditions. The coating thickness reached more than 5mm for each case, much higher than the CVD coating [8]. If necessary, this thickness can be further increased to form the bulk DCMC. Besides, it is obvious that the coating colour changed from red to grey as the clad diamond mass fraction increased gradually (from P to P3). To interpret this phenomenon, Fig. 3 provides the EDS element analysis on the coating surfaces. Nickel which was originally coated by copper cladding can be clearly detected on the substrate surface. This fact means that the clad diamond powder experienced large deformation during the deposition process; nickel cladding was fractured and then exposed to the outside environment. Nickel is known to present grey colour, thus coatings with clad diamond had different colour as compared to the pure copper coating. Furthermore, the nickel mass fraction on the substrate surface was found to increase gradually, while the copper mass fraction showed a revered trend. The increased nickel and decreased copper on the coating surface resulted in the coating colour to change from red to grey.

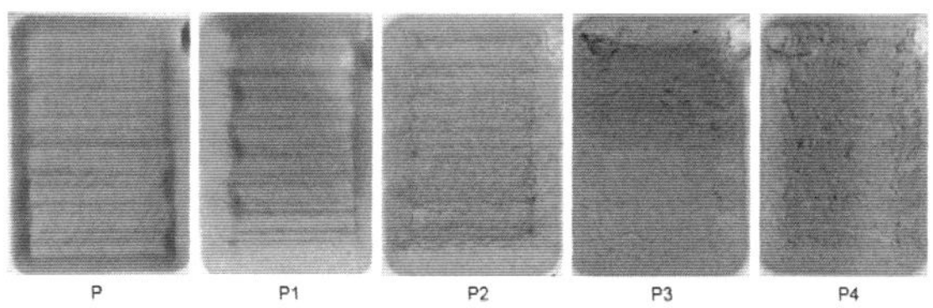

Fig. 2. Digital photos of the coatings fabricated with different powders and working conditions

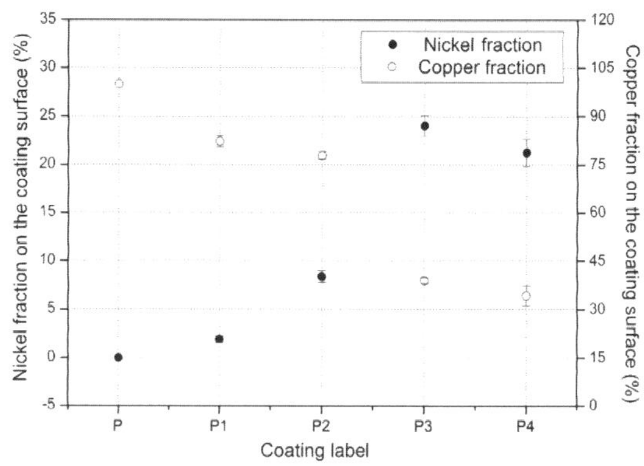

Fig. 3. EDS element analysis on the coating surface

Graphitization of the diamond during the fabrication of DCMC always occurs when using sintering or infiltration technique due to the high processing temperature [9,10,3,4]. However, the comparison of XRD spectra shown in Fig. 4 indicated that no graphitization occurred during the fabrication of cold sprayed DCMC coatings due to the relatively low processing temperature. This fact clearly showed the advantage of cold spray over other techniques in avoiding the diamond graphitization. Moreover, from Fig.4, it is also found that the peak of nickel and diamond phase exhibited significant difference, which suggested that the diamond mass fraction in different coatings must differ from each other.

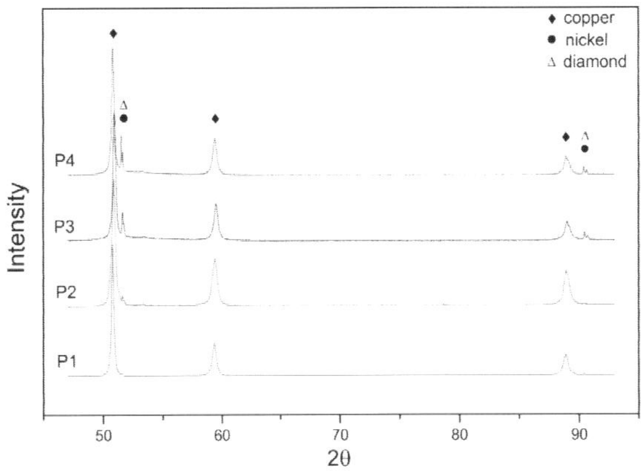

Fig. 4. XRD spectra of different coatings

Therefore, in order to quantitatively analyse the diamond mass fraction inside the coatings, Fig. 5 provides the diamond mass fraction measured at the coating cross-section by EDS. It is clearly seen that P3 and P4 coatings fabricated with pure clad diamond contained more than 40 wt.% of diamond. In comparison with the original feedstock, the diamond mass fraction in the coating was almost unchanged. Such high fraction has never been achieved in cold or thermal sprayed DCMC coatings.

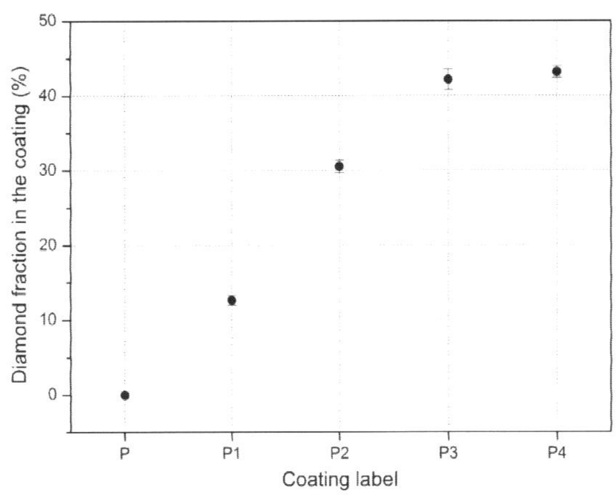

Fig. 5. EDS element analysis on the coating cross-section

In addition, for the coatings fabricated with mixed powders (copper + clad diamond), the diamond mass fraction reached 12.6 wt.% for P1 and 30.5wt.% for P2. The contents were even higher than those in the original feedstock, which implied that clad diamond was easier to deposit and had higher deposition efficiency than copper. In previous cold spray works where mechanically mixed metal-diamond powders were used as the feedstock [7,11,12], most of the diamond could not successful deposit and thus the diamond mass fraction in the coating was normally very low. The excellent performance of clad diamond in this work clearly indicated that it will be a promising feedstock and have great potential to fabricate cold sprayed DCMC coatings.

3.2 Coating microstructure

Diamond fraction and purity (non-graphitization) are critical to the coating performance, while coating microstructure is also very important and thus worthy to be further investigated. Figure 6 shows the cross-sectional SEM images of the DCMC coatings fabricated with different powders and working conditions. From Fig. 6a-d, it is clearly seen that the diamond phase was uniformly distributed in the coating for each sample. It was known that the uniformity of diamond in the DCMC has decisive effect on the final coating thermal properties. In this work, the cold sprayed DCMC coatings fabricated with the clad diamond powders exhibited excellent uniformity, providing an essential condition for the high thermal performance. Besides, an obvious difference in the diamond fraction between each sample can be found from Fig. 6a-d. P3 and P4 coatings contained more diamond than P2 coating and much more than P1 coating, which is consistent with the EDS element analysis shown in Fig. 5.

Furthermore, the magnified images of the cross-section is given in Fig. 6e-h to clarify the diamond fracture behaviour in the coating. It is interesting to find that the diamond in the coating fabricated with mixed powders (P1 and P2) was hardly fractured after deposition. Diamond in the P1 coating almost had no damage, while cracks and few small diamond pieces can be found in P2 coating. However, for P3 and P4 coatings fabricated with pure clad diamond powders, the diamond phase was significantly fractured into small pieces, particularly for the P4 coating fabricated with higher impact velocity. The reason for this difference is attributed to the additional copper powders. For P1 and P2, the soft copper powders could act as a buffer to the clad diamond powders. When diamond particles impacted on the soft copper, the kinetic energy will be absorbed through the plastic deformation of copper. The impacting stress was therefore insufficient to damage the diamond. Nevertheless, for P3 and P4 coatings, the outside metal layer was very thin, most of the impact energy was directly imposed on the inside diamond due to the absent of the buffer materials Hence, most of the diamond was fractured during the deposition. P4 coating fabricated at higher inlet pressure therefore had more fractured diamond pieces than P3 coating.

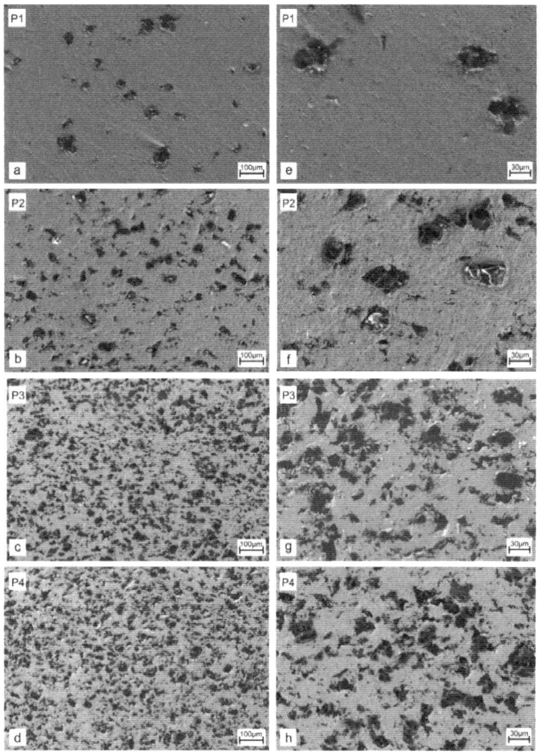

Fig. 6. Cross-sectional SEM images of MMC coatings fabricated with different powders and working conditions

To further clarify the coating microstructure and formation mechanism, Fig. 7 provides the SEM images of the coating fracture surface. As can be seen from Fig. 7a and b, the diamond in P1 coating was intact without any damage, while P2 coating had both an undamaged diamond and a cracked diamond, which is consistent with the cross-sectional view as show in Fig. 6e and 6f. In addition, the metal jet and metallic bonding features can be clearly observed in the P1 and P2 coatings. This fact means that the bonding in P1 and P2 is based on the conventional formation mechanism in cold spraying; coating was formed through the plastic deformation and adiabatic shear of copper powders and copper cladding. However, for P3 coating shown in Fig. 7c, a seriously cracked diamond and the surrounding small diamond pieces can be clearly observed. Particularly, in the P4 coating, diamond was completely fractured into many small pieces. For these two cases, differing from the P1 and P2 coatings, bonding only occurred between the copper claddings, and the coating was formed through fractured diamond uniformly dispersing into the metal phase [13]. Besides, another interesting phenomenon is most of the fracture happened at the metal-diamond interface. This phenomenon means that the metal-to-metal bonding strength in the coating was sufficiently high although the copper cladding is very thin.

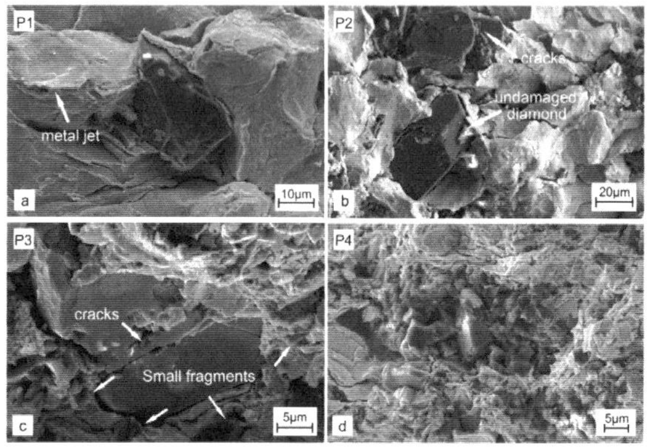

Fig. 7. SEM images of the fracture surface of MMC coatings fabricated with different powders and working conditions

Conclusions

In this paper, a range of diamond-reinforced copper matrix composites (DCMC) were fabricated via the cold spray technique using the copper-clad diamond powder or its mixture with pure copper powder. Copper-clad diamond is a novel powder consisting of an inside diamond powder, a very thin interbedded nickel layer and an outside electroless copper cladding. The experimental results indicated that all the as-sprayed DCMC coatings fabricated with different powders had thickness of more than 5 mm, much thicker than those produced by CVD. When using pure clad diamond powder as the feedstock, the coating was formed through the bonding between copper claddings and the fractured diamond uniformly dispersing into the metal phase. The diamond in the feestock was fully transfered in the coatings. . The diamond mass fraction reached more than 40 wt.%. Such level has not been achieved in previous cold spray works with conventional powders. As for the mechanically mixed powder of clad diamond and copper, the coating was formed through the plastic deformation and adiabatic shear of copper powder and copper cladding, which is similar with the conventional formation mechanism of cold spray coating. The coating also exhibited great retainbility for diamond; the diamond fraction in the coating was even larger than that in the original feedstock. Besides, because of the additional copper powders acting as a buffer, the fracture of diamond in the coating was effectively prevented. In summary, cold spray technique and clad diamond had great potential to produce the DCMC coatings.

Acknowledgments

The authors would like to thank the CRANN Advanced Microscopy Laboratory (AML) of Trinity College and Dr. Yingchun Xie (Université de technologie Belfort-Montbéliard, France) for the technical support, and Element Six for the provision of the diamond powders.

References

[1] A. Papyrin, Cold Spray Technology, Advanced Materials and Processes, 159 (2001), pp. 49/51.

[2] M. Grujicic, C. Zhao, W. DeRosset, D. Helfritch, Adiabatic shear instability based mechanism for particles/substrate bonding in the cold-gas dynamic-spray process, Materials and Design, 25 (2004), pp. 681/688

[3] L. Yang, B. Li, J. Yao, Z. Li, Effects of diamond size on the deposition characteristic and tribological behavior of diamond/Ni60 composite coating prepared by supersonic laser deposition, Diamond and Related Materials, 58 (2015), pp. 139/148.

[4] J. Yao, L. Yang, B. Li, Z. Li, Beneficial effects of laser irradiation on the deposition process of diamond/Ni60 composite coating with cold spray, Applied Surface Science, 330 (2015), pp. 300/308.

[5] D.J. Woo, F.C. Heer, L.N. Brewer, J.P. Hooper, S. Osswald, Synthesis of nanodiamond-reinforced aluminum metal matrix composites using cold-spray deposition, Carbon, 86 (2015), pp. 15/25.

[6] D.J. Woo, B. Sneed, F. Peerally, F.C. Heer, L.N. Brewer, J.P. Hooper, et al., Synthesis of nanodiamond-reinforced aluminum metal composite powders and coatings using high-energy ball milling and cold spray, Carbon, 63 (2013), pp. 404/415.

[7] H.J. Kim, D.H. Jung, J.H. Jang, C.H. Lee, Study on Metal/Diamond Binary Composite Coatings by Cold Spray, Materials science forum, 534 (2007), pp. 441/444.

[8] B.Q. Wan, X.Y. Sun, H.T. Ma, R.F. Feng, Y.S. Li, Q. Yang, Plasma enhanced chemical vapor deposition of diamond coatings on Cu–W and Cu–WC composites, Surface Coatings Technology, 284 (2015), pp. 133/138.

[9] A.M. Abyzov, M.J. Kruszewski, Ł. Ciupiński, M. Mazurkiewicz, A. Michalski, K.J. Kurzydłowski, Diamond–tungsten based coating–copper composites with high thermal conductivity produced by Pulse Plasma Sintering, Materials and Design, 76 (2015), pp. 97/109.

[10] W.Z. Shao, V. V. Ivanov, L. Zhen, Y.S. Cui, Y. Wang, A study on graphitization of diamond in copper-diamond composite materials, Materials Letters, 58 (2004), pp. 146/149.

[11] H. Kwon, S. Cho, A. Kawasaki, Diamond-Reinforced Metal Matrix Bulk Materials Fabricated by a Low-Pressure Cold-Spray Process, Materials Transactions, 56 (2015), pp. 108/112.

[12] S. Shin, Y. Xiong, Y. Ji, H.J. Kim, C. Lee, The influence of process parameters on deposition characteristics of a soft/hard composite coating in kinetic spray process, Applied Surface Science, 254 (2008), pp. 2269/2275.

[13] B. Aldwell, S. Yin, K.A. Mcdonnell, D. Trimble, T. Hussain, R. Lupoi, A novel method for metal – diamond composite coating deposition with cold spray and formation mechanism, Scripta Materialia, 115 (2016), 10/13.

Effect of Content and Type of Reinforcing Particles on the Hardness and Wear Rate of Low-Pressure Cold-Sprayed TiC and B_4C-Based Metal Matrix Composite Coatings

Y. T. R. Lee, H. Ashrafizadeh, G. Fisher, A. McDonald, Edmonton / Canada

A low-cost, low-pressure (less than 1 MPa) cold spray unit was used to deposit B_4C and TiC-based metal matrix composite coatings. B_4C and TiC powders were individually and mechanically blended at 50 wt.%, 75 wt.%, and 92 wt.% carbide content with Ni powder comprising the remainder of the mixture. Scanning electron microscopy, Vickers micro-hardness, and ASTM Standard G65 dry abrasion wear testing were conducted to determine the properties and performance of the coatings. The particle velocity of the reinforcing carbide particles upon impact was calculated by using a mathematical model that was based on the dynamics and thermodynamics of compressible fluid flow through a converging-diverging nozzle. The model showed that the kinetic energy of the TiC particles prior to impact was three times smaller than that of the B_4C particles. This led to a maximum carbide content in the B_4C-Ni coatings of 8 wt.% (23 vol.%) compared to that in TiC-Ni coatings of 18 wt.% (29 vol.%). This was due to reduced fracture and rebound of the TiC particles, which possessed lower kinetic energy on impact. The B_4C-Ni coatings had the highest hardness value of 421 kg/mm^2, even though the carbide content was as low as 8 wt.% (23 vol.%). This suggests that the high hardness of the B_4C reinforcing particles will result in high overall coating hardness with lower carbide content. Despite the fact that the hardness values are within the range of the cold-sprayed WC-12Co-Ni-based coatings, the wear resistance of the deposited coatings (wear rates of 435 x 10^{-6} mm^3/Nm for B_4C-Ni and 433 x 10^{-6} mm^3/Nm for TiC-Ni) are high. Further development is required to enable the use of B_4C-Ni and TiC-Ni coatings in high wear resistance applications.

1 Introduction

Corrosion and wear are the main causes of failure of and leakage from components and parts used in the Oil & Gas industry [1]. Although corrosion and wear cannot be completely eliminated, the negative outcomes can be controlled by the use of appropriate protective and preventative measures. Thermal spray deposition is one of the methods used for the fabrication of protective wear and corrosion resistant coatings on components used in the Oil & Gas industry [2]. Thermal spraying is an industrial process in which a heat source is used to melt and accelerate particles. The particles impact and spread onto a substrate to form a coating. WC-Co-based metal matrix composite (MMC) coatings have been used in the Oil & Gas industry due to their excellent resistance to sliding, abrasion, and erosion [3].

WC-Co-based MMC coatings can be deposited by thermal spray processes such as high-velocity oxy-fuel (HVOF) or plasma spraying [4, 5]. The high process temperatures of HVOF and plasma spraying can change the chemical composition of the sprayed material and decarburization may occur during deposition of carbides such as WC [4, 6, 7]. To that end, low temperature spraying techniques such as cold gas dynamic spraying are of interest.

Cold gas dynamic spraying ("cold spraying" hereafter) is a low temperature deposition process, where small metal particles (1 – 50 µm) are accelerated to high velocities (300 – 1200 m/s) prior to impact [8]. The coating is formed due to the plastic deformation of particles that impact the substrate with high velocities. Cold spraying is usually conducted at temperatures well below the melting point of the feedstock material and, therefore, the negative effects of high temperature, such as changes in chemical composition or degradation of the feedstock material are minimized or eliminated [9].

Melendez *et al.* [10] showed that a low-cost low-pressure cold spraying system can be used to create WC-12Co-Ni MMC coatings with wear rates comparable to that of HVOF sprayed WC-Co coatings. The wear resistance of MMC coatings is a function of their microstructure. This includes the percentage of reinforcing particles, the mean free path between the reinforcing particles, and the size and hardness of the reinforcing particles [11]. There are carbides with higher hardness than that of WC (2900 kg/mm^2) such as B_4C (3950 kg/mm^2) and similar hardness such as that of TiC (2980 kg/mm^2) [12]. The higher reinforcing particle hardness may provide further improvement in the hardness and wear resistance of cold-sprayed coatings beyond that which was observed in WC-12Co-based coatings. In addition, the density of these alternative carbides (B_4C is 2.52 g/cm^3 and TiC is 4.93 g/cm^3) are lower than that of WC (15.8 g/cm^3). Thus, they may be accelerated to higher velocities during the cold spray deposition process to improve deposition efficiency, coating overall hardness, and coating density.

The objectives of this study were to: a) evaluate the possibility of deposition of hard B_4C and TiC reinforcing particles with Ni as the matrix by using a low-pressure cold spray system, b) study the effect of the velocity and kinetic energy of the sprayed particles on the deposition efficiency, and c) study the microstructure, hardness, and wear resistance of the cold-sprayed B_4C-Ni and TiC-Ni MMC coatings.

2 Experimental Method

2.1 Powder and Substrate

Fused and crushed B_4C (Panadyne, Montgomeryville, PA, USA) and TiC (Pacific Particulate Materials Ltd., Port Moody, BC, Canada) powders were used as the reinforcing particles in the MMC's in this study. The average particle diameter function in an image analysis software (ImagePro, Media Cybernetics, Bethesda, MD, USA) was employed to measure the particle size distributions of the powders. The powder size distribution of B_4C and TiC was 39 ± 10 µm (n = 42) and 23 ± 7 µm (n = 28), respectively. The carbide powders were mechanically blended with Ni powder that was specially designed for cold spraying (N5001, CenterLine Ltd., Windsor, ON, Canada). The compositions of each mechanical blend are presented in Table 1. Figure 1 shows the morphology of the B_4C, TiC, and Ni powders. Low carbon steel was used as the substrate. The substrates were grit blasted with alumina grit #24 (Manus Abrasive Systems Inc., Edmonton, AB, Canada) to promote adhesion of the coating.

2.2 Cold Spray Deposition

A low pressure cold spray system (SST series P, CenterLine, Ltd., Windsor, ON, Canada) was used to deposit the powder blends. The cold spray nozzle was attached to a robot (HP-20, Motoman, Yaskawa Electric Corp., Waukegan, IL, USA) to allow for automated control of the deposition process. Compressed air was used as the working fluid in the cold spray system and the air pressure was set to 634 kPa (92 psig). The temperature of the compressed air was 550°C. The distance between the nozzle and the substrate, the stand-off distance, was set to 5 mm. The traverse velocity was set to 5 mm/s for all powder blends, except for the 92 wt.% B_4C + 8 wt.% Ni powder blend where a traverse speed of 2.5 mm/s was used.

2.3 Coating Characterization

The coatings were cut into representative cross sections. The cross sections were cold-mounted in epoxy (Allied High Tech Products, Inc., Rancho Dominguez, CA, USA). Each cold-mounted cross section was ground using 180, 240, 360, 400, 600, 800, and 1200 grit paper (LECO, Mississauga, ON, Canada) then polished using 3 µm and 1 µm diamond slurries (LECO, Mississauga, ON, Canada).

A scanning electron microscope (EVO MA 15, Zeiss, Cambridge, UK) equipped with energy dispersive X-ray (EDX) spectroscopy was used to analyze the cross-sections of the coatings. The imaging software was used to determine the carbide content in the coatings and the mean free path between the reinforcing carbide particles. A minimum of five images (n = 5) were used for analysis.

Mean free path is the average distance between reinforcing particles, and it was calculated by using [13]:

$$\lambda = \frac{1 - V_p}{N_L}, \qquad (1)$$

where, λ is the mean free path in microns (µm), V_p is the volume fraction of reinforcing particles, and N_L is the number of intercepts per unit length of line (µm^{-1}). A minimum of five random lines were drawn on each of the images used to determine the carbide content in the coating. The number of particle intercepts were then manually counted and applied to Eq. 1, along with the carbide volume percentage in the coating. An arithmetic mean of the mean free path was found from the mean free path of each line.

A micro-hardness Vickers indenter (MVK-H1, Mitutoyo, Aurora, IL, USA) was used to determine the hardness of the deposited coatings. The indentations were conducted using 300 gf with a dwell time of 10 s, in accordance with ASTM Standard E384 [14]. The spacing between the indentations was selected to be four times the length of the indentation diagonal, in accordance with ASTM Standard C1327 [15].

Table. 1. Powder blend compositions.

Powder blend	Powder (wt.%)		
	B_4C	TiC	Ni
50 wt.% B_4C + 50 wt.% Ni	50	-	50
75 wt.% B_4C + 25 wt.% Ni	75	-	25
92 wt.% B_4C + 8 wt.% Ni	92	-	8
50 wt.% TiC + 50 wt.% Ni	-	50	50
75 wt.% TiC + 25 wt.% Ni	-	75	25
92 wt.% TiC + 8 wt.% Ni	-	92	8

(a) (b) (c)

Fig. 1. Morphology of a) B_4C, b) TiC, and c) Ni powder particles.

2.4 Dry Abrasion Testing

Dry abrasion testing was conducted in accordance with ASTM Standard G65 [16]. The rotational speed of the wheel was set to 200 revolutions per minute (rpm) and a load of 130 N was applied to the specimen while an abrasive sand media (AFS50/70 silica, US Silica, Ottawa, IL, USA) was made to flow at a rate of 365 g/min between the specimen and the rotating wheel. The coatings were then tested according to Procedure E of the Standard at 1000 revolutions, which is equivalent to a distance of 718 m. The coupons were weighed before and after the abrasion test to measure their mass loss. The mass loss was converted to volume loss by dividing the mass loss by the density of the coating. The wear rate (W), in mm^3/Nm, was determined by dividing the volume loss (Δv) by the applied load (P) and the sliding distance (s) as shown in Eq. (2):

$$W = \frac{\Delta v}{Ps}. \qquad (2)$$

3 Mathematical Model for Determination of Velocity of the Sprayed Powder Particles

The velocity of the sprayed powder particles at the nozzle exit was determined by employing the mathematical model that was developed by Dykhuizen *et al.* [17]. The model was premised on the principles of dynamics and thermodynamics of compressible fluid flow through a converging-diverging nozzle. The geometry and dimensions of the nozzle are shown in Fig. 2. Table 2 details all the variables that were used in the model. The following assumptions were made in order to derive the model: (a) The gas flow was one dimensional and isentropic (adiabatic and frictionless); (b) The gas was treated as ideal; (c) The specific heats of the gas were constant; and (d) The carrier gas was supplied from a large chamber where its velocity was zero. Based on the geometry of the nozzle, the Mach number at each section of the diverging section of the nozzle was determined by [18]:

$$M=\left[k_1\frac{A}{A^*}+(1-k_1)\right]^{k_2}, \qquad (3)$$

where $k_1=218.0629-243.5764\gamma+71.7925\gamma^2$ and $k_2=-0.12245+0.28130\gamma$.

The Mach number was determined at a given cross-sectional area. Once the Mach number was known, the pressure, temperature, velocity, and density of the gas at any point was determined by using Eqs. (4) to (7) where [17, 18]:

$$P=P^*\left(\frac{\gamma+1}{2+(\gamma-1)M^2}\right)^{\frac{\gamma}{\gamma-1}}, \qquad (4)$$

$$T=\frac{T_0}{\left(1+\left(\frac{\gamma-1}{2}\right)M^2\right)}, \qquad (5)$$

$$V=M\sqrt{\gamma RT}\text{, and} \qquad (6)$$

$$\rho=\frac{\rho_0}{\left(\left(1+\frac{\gamma-1}{2}\right)M^2\right)^{\frac{1}{\gamma-1}}}. \qquad (7)$$

The validity of Eqs. (4) to (7) are based on the assumption that there are no shocks in the nozzle. This condition requires that the shock pressure, P_s, be higher than that of the ambient pressure. In other words, P_s, as determined by Eq. (8) should be larger than the ambient pressure [17, 18]. This condition was verified in this study and was based on the chosen spraying parameters and the nozzle geometry. Equation (8) for the shock pressure is:

$$P_s=P_e\left(\frac{2\gamma}{\gamma+1}M_e^2-\frac{\gamma-1}{\gamma+1}\right). \qquad (8)$$

The particle velocity was determined according to the second law of motion by assuming that the drag force was the only force that was applied on the accelerating particles. The second law states that:

$$m_p a_p=F_D=\frac{1}{2}C_D A_p \rho\left(V_g-V_p\right)^2, \qquad (9)$$

where $m_p=\frac{1}{6}\pi d_p^3\rho_p$ and $A_p=\frac{1}{4}\pi d_p^2$.

C_D, the drag coefficient, was determined by [19]:

$$C_D=\frac{24}{Re_p}\left[\frac{\left(1+0.15Re_p^{0.687}\right)\left(1+e^{-\left(0.427/M_p^{4.63}\right)-\left(3/Re_p^{0.88}\right)}\right)}{1+\left(M_p/Re_p\right)\left(3.82+1.28e^{-1.25M_p/Re_p}\right)}\right], \qquad (10)$$

where $Re_p=\left|V-V_p\right|\frac{\rho d_p}{\nu}$ and $M_p=\frac{\left|V-V_p\right|}{\sqrt{\gamma WRT}}$.

By using the chain rule to simplify Eq. (9) and using Euler's numerical integration method, the equation of particle velocity was determined as [20]:

$$V_{p,x}=\left(V_{p,x-1}^2+\frac{1.5\Delta x C_D \rho_A}{d_p\rho_p}\left(V-V_{p,x-1}\right)^2\right)^{0.5}. \qquad (11)$$

Equation (11) was solved at spatial intervals of Δx = 0.1 mm through the length of the nozzle.

4 Results and Discussion

4.1 Velocity of Powder Particles

The spraying parameters employed in this study, coupled with the measured size of the powder particles and their density were selected as the inputs for the mathematical model. Figure 3 shows a typical graph for the velocities of the air and the sprayed particles through the divergent section of the nozzle. The position at zero meters in the figure refers to the inlet of the nozzle (the origin as shown in Fig. 2). The particle velocities at the exit of the nozzle were similar, at 345 m/s and 376 m/s for the TiC and B_4C particles, respectively (see Fig. 3). However, due to differences in the mass and size of the particles, the kinetic energy ($k=\frac{1}{2}mV^2$) of the particles was not the same and it was determined to be 5.5 mJ and 1.9 mJ for the B_4C and TiC particles, respectively.

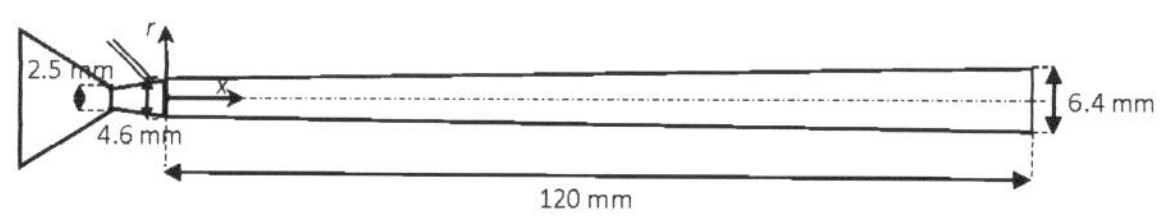

Fig. 2. Geometry of the converging-diverging nozzle of the cold spray system.

Table 2. Model variables.

General Symbols		Greek Symbols	
a	Acceleration (m/s^2)	ρ	Density (kg/m^3)
A	Cross section area (m^2)	γ	Ratio of specific heats
C_D	Drag coefficient		
d	Diameter (m)	ν	Kinematic viscosity (m^2/s)
F_D	Drag force (N)		
m	Mass (kg)	Subscripts	
M	Mach number	0	Stagnation condition
P	Pressure (kPa)		
P_s	Shock pressure (kPa)	e	Nozzle exit condition
R	Gas constant (J/mol.K)		
T	Temperature (K)	p	Particle condition

t	Time (s)	*Superscripts*	
V	Velocity (m/s)	*	Nozzle throat condition
W	Molecular weight (kg/mol)		
x	Axial distance (m)		

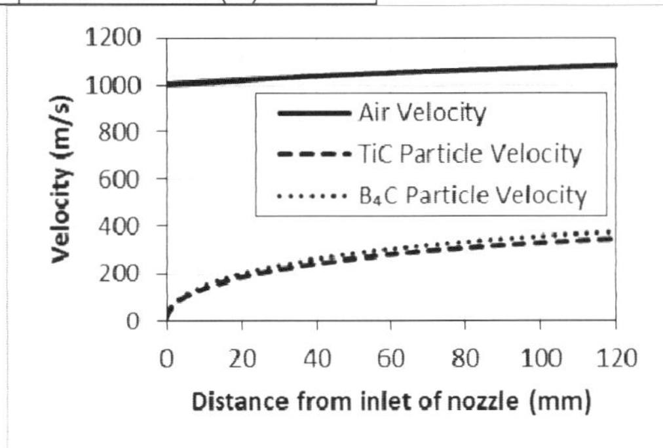

Fig. 3. Air and particle velocities through the diverging section of the nozzle (P_0 = 92 psig and T_0 = 550°C).

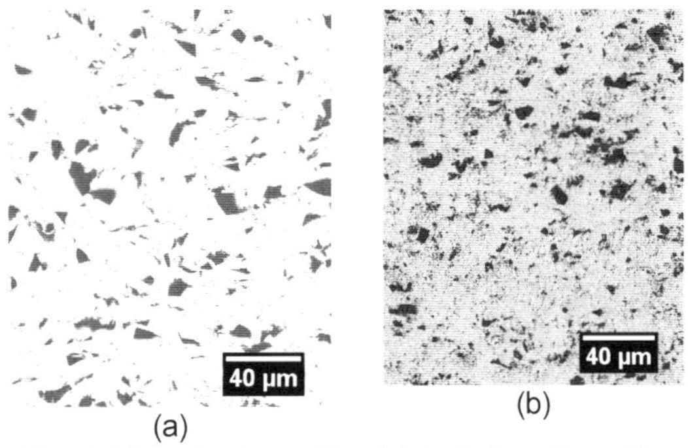

Fig. 4. Microstructure of the fabricated coating with feedstock composition of a) 75 wt.% B_4C + 25 wt.% Ni and b) 75 wt.% TiC + 25 wt.% Ni.

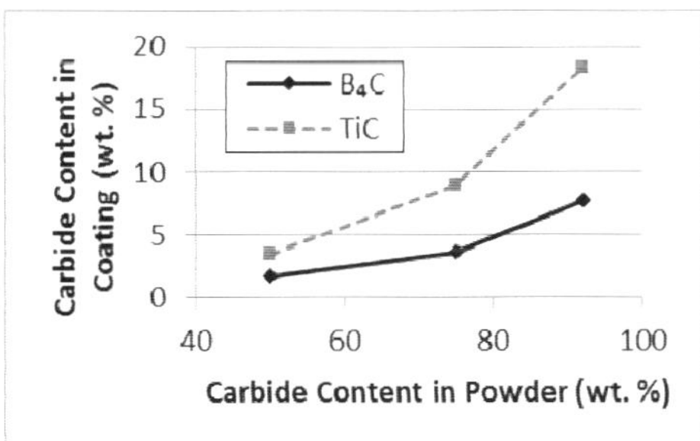

(a)

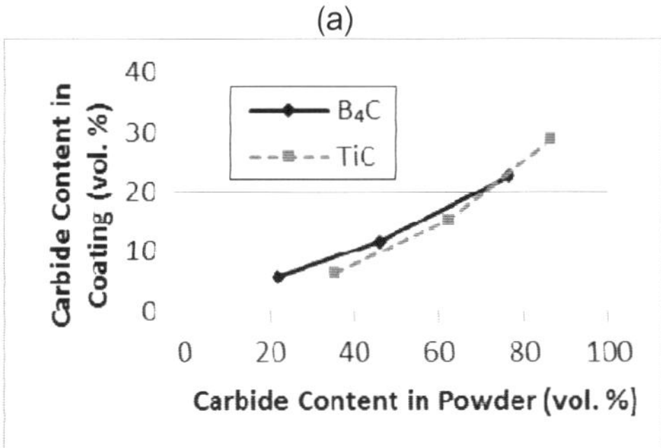

(b)

Fig. 5. Curves of carbide content in the coating versus carbide content in the mechanical powder blend in terms of a) weight percentage and b) volume percentage.

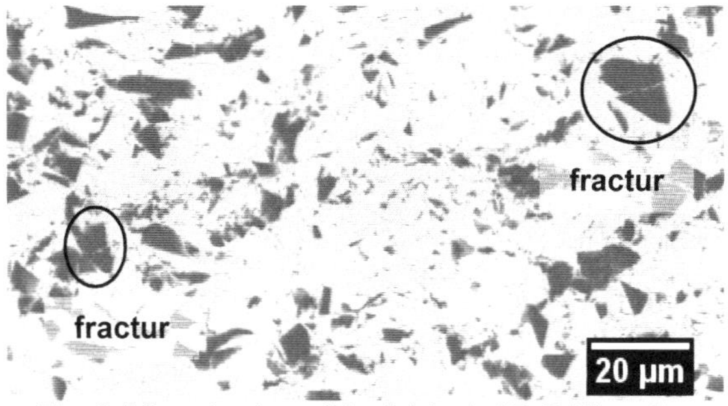

Fig. 6. Microstructure of the fabricated B_4C-Ni coating highlighting particle fracture.

4.2 Carbide Content in the Coatings

Scanning electron microscope (SEM) images were used to provide qualitative and quantitative information about the MMC coatings. Figure 4 shows representative microstructures of the coatings that were fabricated in this study. Given that the microstructures of the B_4C-Ni and TiC-Ni coatings were similar EDX analysis determined that the dark areas in the images were the carbide particles and that the grey areas were Ni.

Image analysis was used to determine the porosity and volume percentage of carbide particles in the coatings. In the TiC-Ni coatings, it was found that the porosity was less than 1.8 vol.% in all the samples. Measurement of porosity of the B_4C-Ni coatings was not possible due to the low contrast between the B_4C particles and the pores. Melendez *et al.* [9] also showed that with WC-12Co-Ni cold-sprayed coatings, the porosity was also below 2%. With these facts in mind, the porosity of the MMC coatings was assumed to be negligible.

Figure 5a shows the measured weight percentage of the carbide content in the coating versus the carbide content of the sprayed powder blend. The carbide content in the coating increased by increasing the carbide content in the mechanical powder blend. The deposition efficiency of TiC carbides was higher for all powder blends (see Fig. 5a) and a maximum carbide content of 8 wt.% (23 vol.%) and 18 wt.% (29 vol.%) were measured for B_4C and TiC, respectively. Figure 5b presents the data on carbide content in terms of volume percent to provide a direct comparison between the MMC coatings.

In low pressure cold spraying, the critical velocity of particles is not typically attained and the coating is formed by consolidation of the hard cermet particles into the previously deposited layers of softer Ni [21]. The fact that the B_4C particles, with lower deposition efficiency, had a kinetic energy three times higher than that of TiC upon deposition suggests that the higher kinetic energy of B_4C particles has led to greater fracture and rebound and less deposition of those hard particles. The inherent brittleness of the B_4C particles probably contributed to the amount of fracture that occurred. The fractured particles and fragments that were observed in the SEM micrographs

such as the one shown in Fig. 6 supports the argument that
the higher kinetic energy of the brittle B_4C particles contributed to the fracture of the particles.

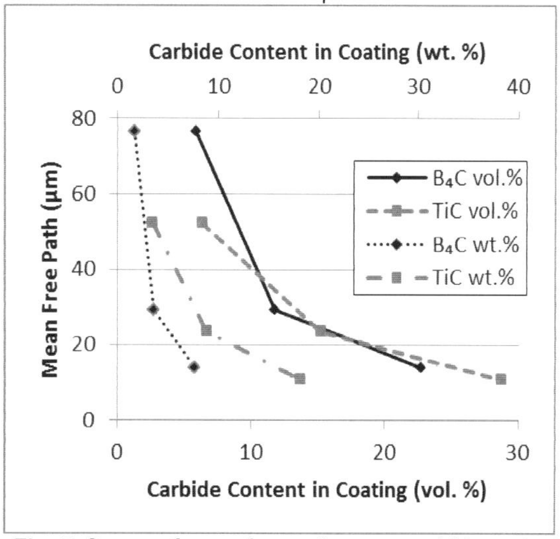

Fig. 7. Curves of mean free path versus carbide content in coating.

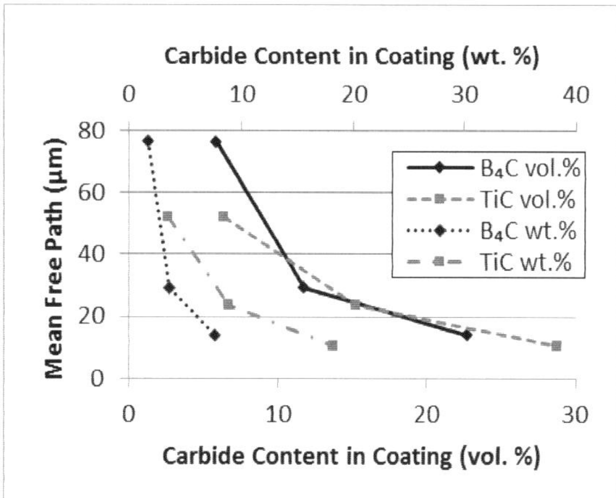

Fig. 8. Curves of hardness versus carbide content in the coatings.

Figure 7 shows the relationship between mean free path and the carbide content within the coating. A shorter mean free path was observed for coatings with higher carbide content. This was expected since the higher content of reinforcing particles in the coating results in a higher density of reinforcing particles, which reduces the space between those particles.

4.3 Vickers Micro-Hardness

Figure 8 shows a curve of the Vickers micro-hardness versus the carbide content in the coating. The deposited coatings became harder with the increase in reinforcing particle content in the coating due to the shorter mean free path between the carbide particles (see Fig. 9). Coatings with lower mean free path have greater load sharing between the reinforcing particles, which results in an improvement in the hardness [5, 22]. It can also be seen that in Fig. 8, for a similar volume percent of carbide in the coating, the B_4C-Ni MMC coating has a higher hardness due to the higher hardness of B_4C (3,950 kg/mm^2) when compared to TiC (2,980 kg/mm^2) [12].

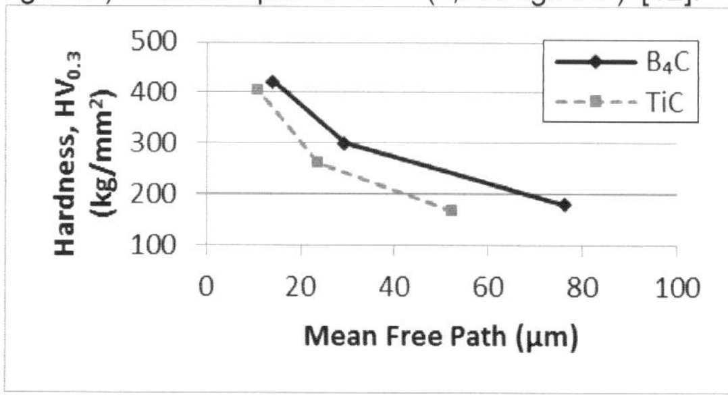

Fig. 9. Curves of hardness versus mean free path in the coatings.

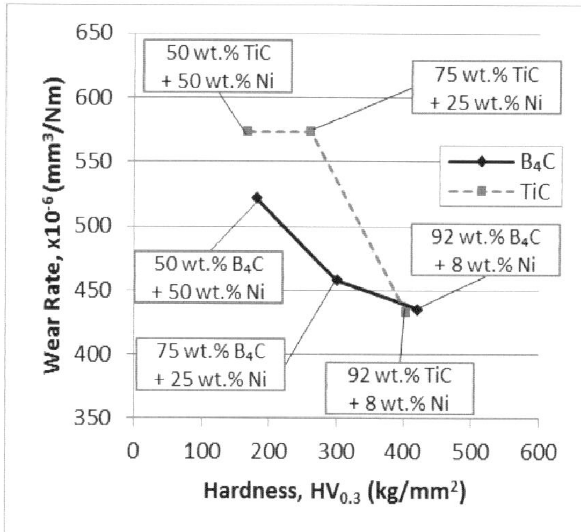

Fig. 10. Curve of wear rate versus hardness.

This observation may be explained by considering the rule of mixtures, where the maximum theoretical hardness of the composite is a volume-weighted average of the hardness values for the individual phases.

4.4 Relationship between Hardness and Wear Rate

Figure 10 shows curves of wear rate versus the hardness of the coatings. For the two types of coatings that were explored in this study, an improvement in wear resistance at higher hardness was observed. Minimum wear rates of 435 and 433 x 10^{-6} mm^3/Nm were achieved for B_4C-Ni and TiC-Ni coatings, respectively. For the majority of MMC compositions that were studied, the wear rates of the B_4C-Ni coatings were lower than those of the TiC-Ni coatings for similar hardness values, which suggests that the B_4C-Ni coatings had higher toughness. However, Melendez *et al.* [10] reported a wear rate of approximately 70 x 10^{-6} mm^3/Nm for cold-sprayed WC-12Co-Ni MMC coatings with hardness of around 400 kg/mm^2. Given that the wear rate of B_4C and TiC-Ni coatings are almost 6 times higher than that of WC-12Co-Ni coatings for a hardness of 400 kg/mm^2 is evidence suggestive of the low toughness of the coatings that were explored in this study. Additionally, the fact that the wear rate of the deposited coatings is higher than thewear rate range (4 to 20 x 10^{-6} mm^3/Nm) of cold sprayed and HVOF sprayed WC-Co-based coatings (one of the most commonly used coating materials in the Oil & Gas industry) [3, 11], suggests that further study is required with respect to cold-sprayed B_4C-Ni and TiC-Ni MMC coatings for application in the Oil & Gas industry for purposes of wear protection.

5 Conclusion

Metal matrix composite (MMC) coatings with hard reinforcing particles of B_4C and TiC with Ni as the binder were successfully deposited by a low-cost low-pressure cold spraying. The maximum carbide content of 8 wt.% (23 vol.%) and 18 wt.% (29 vol.%) was obtained for the B_4C-Ni and TiC-Ni coatings, respectively. The determination of the velocities of the sprayed particles and the kinetic energies before impact was accomplished by use of a mathematical model. It was found that higher kinetic energy of brittle B_4C particles led to more fracture and rebounding of the reinforcing particles and lower carbide content in the final coating. Due to the high hardness of the deposited carbide particles, hardness values as high as 421 kg/mm^2 for B_4C-Ni and 403 kg/mm^2 for TiC-Ni was achieved for reinforcing particle contents of 8 wt.% (23 vol.%) and 18 wt.% (29 vol.%), respectively. However, the coatings were found to have low toughness with wear rates that were almost 6 times higher than those of cold-sprayed WC-12Co-Ni MMC coatings for similar hardness values. The low wear resistance of the deposited coatings (wear rates of 435 x 10^{-6} mm^3/Nm for B_4C and 433 x 10^{-6} mm^3/Nm for TiC) suggests that these coatings will require further development to enable their application where high resistance to wear is desired.

6 References

[1] A. Al-Hamed, *et al.*, “Investigation of HVOF thermal sprayed nanostructured WC-12Co mixed with Inconel-625 coatings for oil/gas applications,” 2013, pp. 215–225.

[2] V. R. S. Sá Brito, *et al.*, “Corrosion resistance and characterization of metallic coatings deposited by thermal spray on carbon steel,” *Mater. Des.*, vol. 41, pp. 282–288, Oct. 2012.

[3] S. Al-Mutairi, *et al.*, “Microstructural characterization of HVOF/plasma thermal spray of micro/nano WC–12%Co powders,” *Surf. Coat. Technol.*, vol. 264, pp. 175–186, Feb. 2015.

[4] D. A. Stewart, *et al.*, “Abrasive wear behaviour of conventional and nanocomposite HVOF-sprayed WC–Co coatings,” *Wear*, vol. 225–229, pp. 789–798, Apr. 1999.

[5] H. Chen and I. . Hutchings, “Abrasive wear resistance of plasma-sprayed tungsten carbide–cobalt coatings,” *Surf. Coat. Technol.*, vol. 107, no. 2–3, pp. 106–114, Sep. 1998.

[6] J. M. Guilemany, *et al.*, “Characterization of the W2C phase formed during the high velocity oxygen fuel spraying of a WC + 12 pct Co powder,” *Metall. Mater. Trans. A*, vol. 30, no. 8, pp. 1913–1921, Aug. 1999.

[7] C. Verdon, *et al.*, “A study of high velocity oxy-fuel thermally sprayed tungsten carbide based coatings. Part 1: Microstructures,” *Mater. Sci. Eng. A*, vol. 246, no. 1–2, pp. 11–24, May 1998.

[8] A. Papyrin, *Cold spray technology*. Amsterdam; London: Elsevier, 2007.

[9] N. M. Melendez and A. G. McDonald, “Development of WC-based metal matrix composite coatings using low-pressure cold gas dynamic spraying,” *Surf. Coat. Technol.*, vol. 214, pp. 101–109, Jan. 2013.

[10] N. M. Melendez, *et al.*, “Effect of reinforcing particles on the wear rate of low-pressure cold-sprayed WC-based MMC coatings,” *Wear*, vol. 306, no. 1–2, pp. 185–195, Aug. 2013.

[11] K.-H. Zum Gahr, *Microstructure and wear of materials*. Amsterdam ; New York: Elsevier, 1987.

[12] A. Nino, *et al.*, “Indentation Size Effect for the Hardness of Refractory Carbides,” *Mater. Trans.*, vol. 51, no. 9, pp. 1621–1626, 2010.

[13] M. Kouzeli and A. Mortensen, “Size dependent strengthening in particle reinforced aluminium,” *Acta Mater.*, vol. 50, no. 1, pp. 39–51, Jan. 2002.

[14] ASTM E384-11e1, Standard Test Method for Knoop and Vickers Hardness of Materials , ASTM International, West Conshohocken, PA, 2011.

[15] ASTM C1327-15, Standard Test Method for Vickers Indentation Hardness of Advanced Ceramics, ASTM International, West Conshohocken, PA, 2015.

[16] ASTM G65-04(2010), Standard Test Method for Measuring Abrasion Using the Dry Sand/Rubber Wheel Apparatus, ASTM International, West Conshohocken, PA, 2010.

[17] R. C. Dykhuizen and M. F. Smith, “Gas Dynamic Principles of Cold Spray,” *J. Therm. Spray Technol.*, vol. 7, no. 2, pp. 205–212, Jun. 1998.

[18] M. Grujicic, *et al.*, “Analysis of the impact velocity of powder particles in the cold-gas dynamic-spray process,” *Mater. Sci. Eng. A*, vol. 368, no. 1–2, pp. 222–230, Mar. 2004.

[19] U. Prisco, “Size-dependent distributions of particle velocity and temperature at impact in the cold-gas dynamic-spray process,” *J. Mater. Process. Technol.*, vol. 216, pp. 302–314, Feb. 2015.

[20] H. Z. Li, *et al.*, “Analysis and modelling of particle velocities in micro-abrasive air jet,” *Int. J. Mach. Tools Manuf.*, vol. 49, no. 11, pp. 850–858, Sep. 2009.

[21] P. L. Fauchais, *et al.*, *Thermal Spray Fundamentals*. Boston, MA: Springer US, 2014.

[22] Z. F. Zhang, *et al.*, “Wear of ceramic particle-reinforced metal-matrix composites: Part II A model of adhesive wear,” *J. Mater. Sci.*, vol. 30, no. 8, pp. 1967–1971, 1995.

Influence of phosphorus and tin content on various characteristics of cold sprayed-copper coatings

Y.Yamauchi, S.Miyaji, M.Akabayashi, R.Tanaka, F.Yoshino, H.Hideshima, T.Hanamachi, S.Hirano, N.Aikawa,
NHK Spring Co., Ltd., Yokohama/Japan

Conventional copper coatings made by cold spray are known as a brittle material while having a high tensile strength. In addition, electrical conductivity of the cold sprayed copper coating is also inferior as compared with that of bulk material. These behaviors are due to high dislocation density and fine crystal grains of the cold sprayed coating. In order to solve these problems, conventionally, it has been attempted to recover the metallic structure by heat treatment. In this study, focusing on phosphorus and tin that significantly affect copper property, we have aimed at improving various characteristics of cold sprayed copper coating without heat treatment by using the optimization of impurity content in the production process of water atomized copper powder. As a result, we have achieved a dense cold sprayed coating with a copper powder which is suitably adjusted the content of phosphorus and tin. It is confirmed that the coating has not only high ductibility but equal electrical conductivity to bulk material without heat treatment.

1 Introduction

The cold spray is a new coating technology that is expected as a novel industrial process [1-6]. In the case of spraying copper powder to an aluminum substrate, it is known as a combination that can be obtained relatively high deposition efficiency [7-9] and bonding strength [9, 10]. The copper-aluminum bonding material is required to be used for conductive parts and heat dissipation parts such as heatsinks, electrodes, connectors for electric vehicles, heavy electric machineries, and several other applications because it has excellent electrical/thermal properties and lightweight.

However, conventional copper coatings made by cold spray are known as a brittle material despite having a high tensile strength [11, 12]. In addition, electric conductivity and thermal conductivity of the cold sprayed copper coating are also inferior as compared with those of bulk material [11, 13]. These behaviors are due to high dislocation density and fine crystal grains of the cold sprayed coating [14].

In order to solve these problems, conventionally, it has been attempted to recover the metallic structure by heat treatment. Heat treatment causes an increase in product cost due to a necessity of some equipments and an addition to manufacture process whereas characteristics can be improved [15-17].

In previous studies regarding copper bulk material, impurity concentration in the copper are significantly related to the conductivity and recrystallizaion temperature of the copper has been revealed. Especially, phosphorus content is generally known as an element which significantly reduces the conductivity [18]. It is also known that tin content is effective to changing the recrystallization temperature [19].

These phenomena have been known for a long time, but the research reports of the copper powder and coatings in the cold spray method are rarely found.

In this study, we have aimed at the realization of the various characteristics of cold sprayed copper coating those are either equalling or surpassing to copper bulk material by utilizing the heat input during cold spray process without heat treatment after coating.

In this paper, the influence of phosphorous and tin contents in water atomized copper powder on various characteristics of cold sprayed copper coating have been investigated.

2 Experimental procedure

2.1 Materials

A pure aluminium plate (50 × 50 × 3 mm) was used as substrate. To investigate the influence of phosphorus and tin, water atomized copper powder (Cu-A, Nippon Atomized Metal Powders, Japan) which is five different impurity concentrations as shown in **Table 1** with the same atomization conditions was prepared. Copper powder have the least phosphorus content is referred to as "P-L" and the most one is referred to as "P-H". Tin content are also referred to as "S-L" and "S-H" as well as phosphorus. Copper powder called "N" respectively has an intermediate content of phosphorous and tin.

The morphology is non-spherical, the average particle size is approximately 37 µm, and purity is more than 99.8%.

Table 1. Impurity concentration of the copper powders.

L/N	P [mass%]	Sn [mass%]	O [mass%]
N	0.012	0.041	0.070
S-L	0.012	0.009	0.050
S-H	0.016	0.198	0.050
P-L	0.002	0.061	0.180
P-H	0.028	0.063	0.050

2.2 Spray condition

A commercially available cold-spray system (PCS1000, Plasma Giken Co. Ltd., Japan) was employed to make copper coatings by using nitrogen gas. The cold spray parameters are listed in **Table 2**.

The specimens were deposited up to approximately 8 mm thick onto the substrate by using aforementioned powders.

Table 2. Spray condition

Process gas	N_2
Gas temperature [°C]	800
Gas pressure [MPa]	3
Powder feed rate [g/min.]	300
Spray distance [mm]	25
Pass	8
Traverse speed [mm/sec.]	200

2.3 Evaluations

About copper powder, the observation of cross-section microstructure was only conducted. About copper coatings, the evaluation of the microstructure, electrical conductivity, and the 4 point bending strength were carried out.

The microstructures of the copper coatings were examined by scanning electron microscope (SEM, Carl Zeiss, Sigma, Germany) after being filled with resin and optically polished. The porosity in the copper coatings was calculated by image processing from 500 times the backscattered electron images. In order to observe more details of microstructure, inverse pole figure (IPF) showing the crystal orientation, the average crystal grain size, and the Kernel average misorientation (KAM) which means local strain distribution were evaluated by the electron backscatter diffraction (EBSD, JOEL, JSM-7000F, Japan) after cross section polisher processing. The grain was defined as a crystal in more than 0.05 CI (Confidence Index) value and more than 5° crystal misorientation. Twin was calculated as $\Sigma 3$ {111} coherent twin boundary. The microstructures in the copper coating of 2 x 2 mm above 1 mm from substrate interface were observed.

Evaluation of electrical conductivity was carried out by using four-terminal method with 2 x 2 x 40 mm specimens that were prepared by electrical discharge wire processing from the horizontal direction of the copper coatings. The measurement distance was 35 mm and the rated current was 1 A. The measured resistance value was converted to IACS (International Annealed Copper Standard) that defined the conductivity of pure copper as 100%.

The 4-point bending test was conducted with the same shape specimens as those of conductivity measurement, 10mm inner-fulcrum distance, 30mm outer-fulcrum distance and 3mm/min displacement rate. The curvature radius of the rotary type support point was R2.5. All measurements were evaluated by the average value of three to five points.

3 Results and discussion

3.1 Microstructures

Fig.1 shows the cross section microstructure of the copper powder and coatings. In only “P-L” powder, a lot of fine pores were observed in the copper powder. Also in the coating, it was confirmed that “P-L” coating had some micro pores compared to the other coatings. Generally, phosphorous is intentionally added to the copper melt in order to an increase the fluidity and deoxidation during copper powder production by a water atomizing method. In the production process of the water atomized copper powder, when the phosphorous content is extremely low, oxygen density increased in the copper melt because there is no deoxidation effect. It is presumed that outgas generates and many pores remain in the powder when the copper melt that contains a large amount of dissolved oxygen becomes a powder by an atomization process. As a result, it is considered that the fine pores remained in the “P-L” coating based on the “P-L” powder containing many pores.

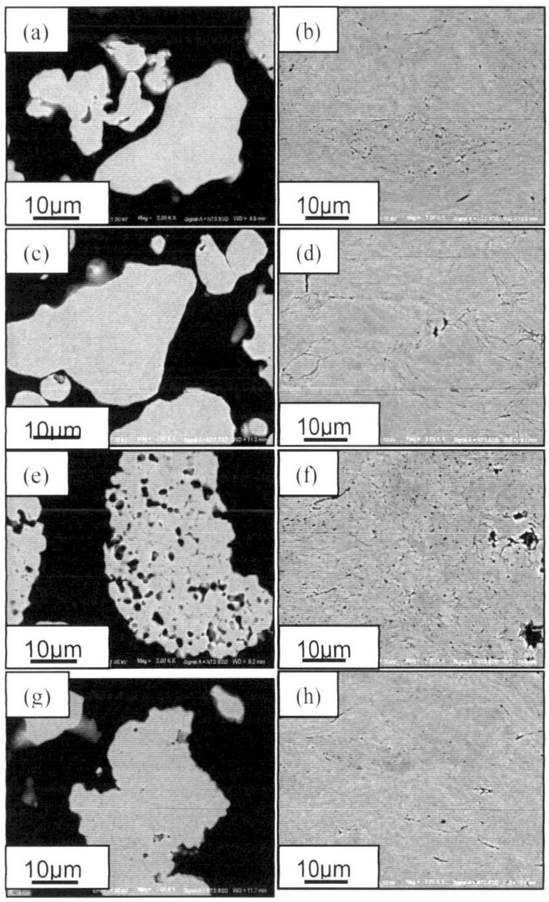

Fig.1. Cross section microstructures of (a) S-L powder, (b) S-L coating, (c) S-H powder, (d) S-H coating, (e) P-L powder, (f) P-L coating, (g) P-H powder, (h) P-H coating.

Subsequently, the coatings were also analyzed by EBSD, the results are shown in **Fig.2** and **Fig.3**. **Fig.2** shows an average of grain size, and **Fig.3** shows KAM and twin length.
It was observed that the grain coarsening and the twin length increased with decreasing the tin content but KAM which indicates local strain distribution decreased.
From these results, it is considered that the recrystallization in the copper coating occured with decreasing the tin content, in the case of having a high tin content, the efficient recovery did not occur due to a high temperatures of the recrystallization temperature. It is inferred that the observed twin seems an annealed twin, and occured the grain coarsening while the strain relieved by the recovery. It was assumed that the dynamic recrystallization occurs in the copper coating by the heat input during cold spray coating process since it was observed such a difference with a slight deference in the tin content despite a same spray condition and non heat treatment.

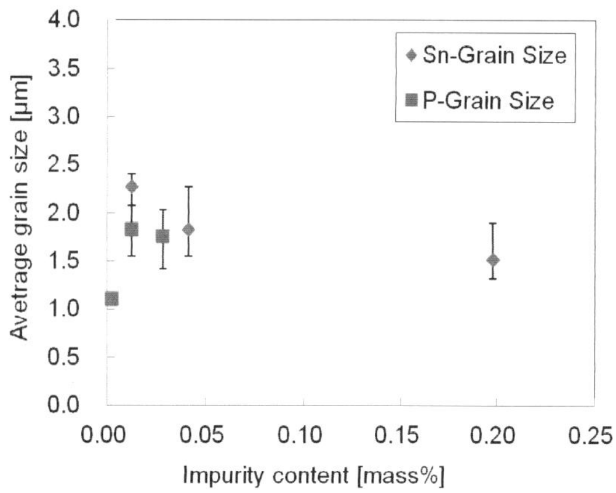

Fig.2. Dependence of average grain size of copper coating on impurity content.

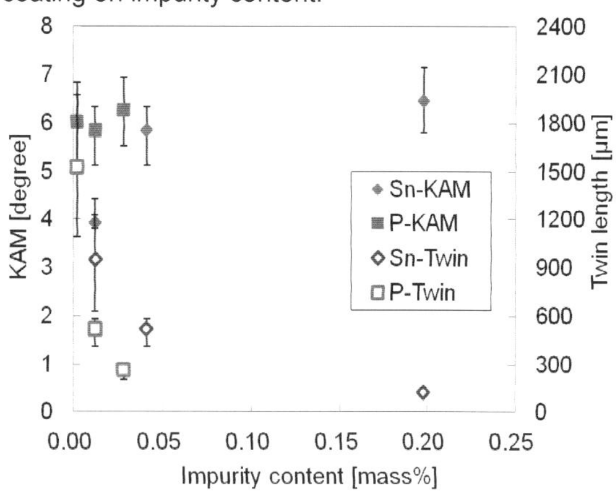

Fig.3. Dependence of KAM and Twin of copper coating on impurity content.

Focusing on the phosphorous content, it was not observed the grain coarsening and KAM value did not decrease though the annealing twin occurred most often in the "P-L" coating.
This phenomenon is supposed that did not lead to the grain growth due to the influence of some fine pores in the copper coating although obviously occurs the recovery in a grain. Moreover, it can be easily presumed that the growth between grains cross over the particle is even more difficult than that between grains in a particle.

3.2 Electrical properties

The result of measurement of the electrical conductivity is shown in **Fig.4**. The electrical conductivity was improved with the reduction of the tin content, it reached up to approximately 103 % that is comparable to bulk material. Although the same tendency was observed regarding phosphorous content, the electrical conductivity was slightly reduced in the lowest phosphorous content, "P-L" coating. It is considered two reasons in this phenomenon, the first one is assumed to be due to oxygen content. The oxygen in the copper is known as an impurity to decrease the conductivity [20]. The second one is based on the porosity in the copper coating.
Thus, influence of impurities of phosphorous and tin in cold spray copper coating on the electrical characteristics is approximately equal to bulk material. The resistance increases with the increase of the impurity content and the phosphorous influences more than tin. However, it's not necessarily appropriate to suggest that the lower phosphourous content is better because the pores remains in the coating.

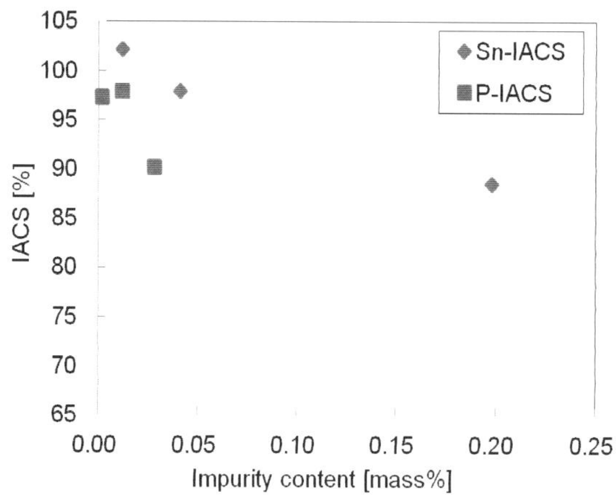

Fig.4. Dependence of electrical conductivity of copper coating on impurity content.

As a result, the conductivity is improved with decreasing the impurity concentration except for the influence of the porosity caused by phosphorous depletion in powder. We cannot conclude that this is the influence of the difference in the progress of dynamic recrystallization due to the low impurity concentration. It can be simply considered that good resistivity is achieved because of the low impurity concentration.

3.3 Mechanical properties

Fig.5. shows a four point bending strength and a displacement at breakage in copper coatings. It was observed to increase the bending strength and improved the ductility with decreasing tin content. In particular, "S-L" coating that has the lowest tin content is observed obvious ductility and reached displacement up to approximately 3.0 mm.

In contrast, it was confirmed that the bending strength decreased although the ductility improved with decreasing the phosphorous content. This difference between tin and phosphorous content behaviors can be explained with a coating microstructure described above so far.

In the case of having a high tin content, the recrystallizaion does not proceeed so much in the copper coating and it remains a lot of work-hardened organization that seems to generate during deposition. Moreover, it fractures at between the particles because the grain growth that enhances the adhesion between the particles does not proceed.

If the tin content is low, it is considered that the ductility occurs due to a recovary caused by the dynamic recrystallization in the copper coatings.

Especially, since the grain growth of the neighboring grains in between the patricles would improve the adehesion at between the particles and may reduce the vulnerable site where is an origin of the fracture, it is inferred that obtained a high strength. Based on these reasons, the brittle fracture occurs in the formar case, and the latter is considered to become a coating of high strength and ductility.

Regarding phosphorous conntet, it can be considered as well as tin content, however, there are many fine pores derived from the powder in the coating if the phosphorous contet is extremely low. Thus, the culcurated bending strength might have been assessed to be low due to decrease of the cross-section area , likewise, these pores would be an origin of the fracture.

In order to support these phenomena, fracture surfaces were also investigated after bending test. As seen in **Fig.6**, the dimples characterized by the ductile fracture were formed on the the fracture serface of the "S-L" coating that shows the ductile fracture.

Contrarily, the fracture surface of the "S-H" coating that little shows the ductility was a planer fracture surface so that it is expeced that the rupture occured in the particle interface.

Also in the fracture surfaces in phosphorous content, some dimples were partially observed , not so much as the "S-L" coating, and a correlation with the fracture shape and the amount of displacement are revealed.

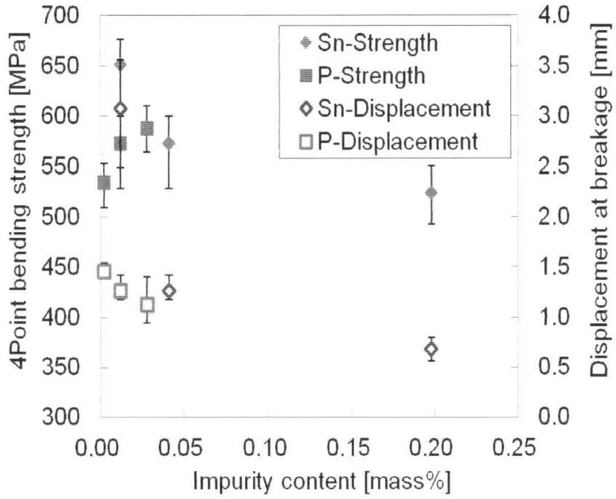

Fig.5. Dependence of bending strength and displacement of copper coating on impurity content.

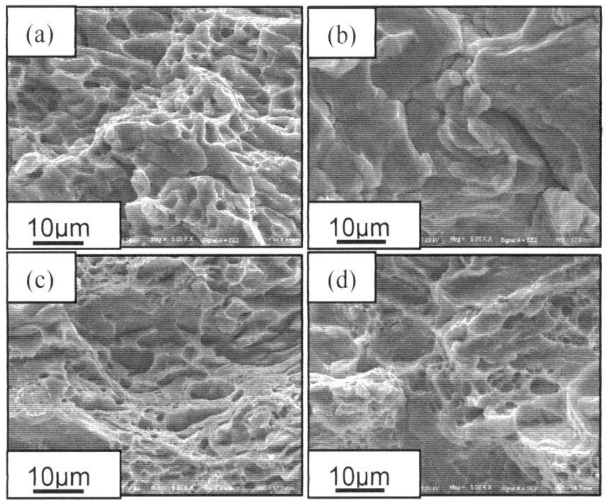

Fig.6. Fracture surface of (a) S-L coating, (b) S-H coating, (c) P-L coating, (d) P-H coating.

4 Conclusions

The influence of phosphorus and tin contents in cold sprayed copper coatings were investigated. According to the experimental results obtained in this study, the following conclusions can be drawn:

1. In the copper coatings, the lower phosphorous and tin contents, the more recrystallization by heat input during cold spray coating process tends to occur. However, it's not necessarily appropriate to suggest that the lower phosphourous content is better because the pores derived from the powder remain in the coating.
2. The conductivity of the copper coating is improved with decreasing of the phosphorous and tin contents except in the case of extremely low phosphorous content; it reached a maximum IACS 103 %.

3. With the reduction of the phosphorous and tin contents in the coating, the displacement tend to be improved and it is also confirmed that the coating has an obvious ductility. If the phosphorous content is extremely low, it is difficult to achieve the high strength due to the influence of pores.
4. In the case of using the water atomized copper powder, adjusting the phosphorous and tin contents, the electrical and mechanical properties nearly equal to those of the copper bulk material were achieved without heat treatment.

5 Literature

[1] A.P.Alkimov, A.N.Papyrin et al.: Gas Dynamic Spraying Method for Applying a Coating. USB-5302414, (1994), U.S.Patent.

[2] Kazuhiro Sakaki: Overview of cold spray technology and col sprayed light metals coalsings. Journal of Japan Institute of Light Metals, 56, (2006), Issue 7, pp. 376/385.

[3] Kazuhiro Sakaki: Promising Field in New Thermal Spray Technology "Cold Spray". Journal of Japan Thermal Spray Socieety, 47, (2010), Issue 3, pp. 113/119.

[4] Hirotaka Fukanuma: The development of High Temperature and Pressure Cold Spray Equipment. Journal of Japan Thermal Spray Socieety, 47, (2010), Issue 4, pp. 179/188.

[5] Makoto Watanabe, Seiji Kuroda, KeeHyun Kim: Bonding Mechanisms of Solid Particles in Cold Spray and Warm Spray Depositions. Journal of Japan Thermal Spray Socieety, 48, (2011), Issue 1, pp. 11/17.

[6] S. Marx, A. Paul, A. Kohler, and G. Huttl: Cold Spraying: Innovative Layers for New Applications. Journal of Thermal Spray Technology, 15, (2006), Issue 2, pp. 177/183.

[7] Victor Champagne, Dennis Helfritch,Phillip Leyman, Robert Lempicki and Scott Grendahl: The effects of gas and metal characteristics on sprayed metal coatings. Modelling Simul. Mater. Sci. Eng, 13 (2005), pp. 1/10.

[8] D.L. Gilmore, R.C. Dykhuizen, R.A. Neiser, T.J. Roemer, and M.F. Smith: Particle Velocity and Deposition Efficiency in the Cold Spray Process. Journal of Thermal Spray Technology, 8, (1999), Issue 4, pp. 576/582.

[9] T. Schmidt, F. Gaerter, H. Kreye: New Developments in Cold Spray Based on Higher Gas and Particle Temperature. Journal of Thermal Spray Technology, 15, (2006), Issue 4, pp. 488/494.

[10] Victor K. Champagne, Jr., Dennis Helfritch, Phillip Leyman, Scott Grendahl, and Brad Klotz: Interface Material Mixing Formed by the Deposition of Copper on Aluminum by Means of the Cold Spray Process. Journal of Thermal Spray Technology, 14, (2005), Issue 3, pp. 330/334.

[11] Heli Koivuluoto, Andrew Coleman, Keith Murray, Martin Kearns, and Petri Vuoristo: High Pressure Cold Sprayed (HPCS) and Low Pressure Cold Sprayed (LPCS) Coatings Prepared from OFHC Cu Feedstock: Overview from Powder Characteristics to Coating Properties. Journal of Thermal Spray Technology, 21, (2012), Issue 5, pp. 1065/1075.

[12] R.C. McCune, W.T. Donlon, O.O. Popoola, and E.L. Cartwright: Characterization of Copper Layers Produced by Cold Gas-Dynamic Spraying. Journal of Thermal Spray Technology, 9 (2000), Issue 1, pp. 73/82.

[13] Yoshiyasu Itoh, Syoko Suyama, Hirotaka Fukanuma: Thermal and Electrical Properties of Copper Coatings Produced by Cold Spray. Journal of the Society of Materials Science, Japan,59 (2010), Issue 2, pp. 143/148.

[14] Yuji Ichikawa, Yusuke Watanabe, Isamu Nonaka, Hideo Miura: Microstructure evaluation of cold-sprayed copper coatings by EBSD. Journal of Japan Thermal Spray Society, 50, (2013), Issue 4, pp. 170/175.

[15] Wen-Ya Li, Chang-Jiu Li, and Hanlin Liao: Effect of Annealing Treatment on the Microstructure and Properties of Cold-Sprayed Cu Coating. Journal of Thermal Spray Technology, 15, (2006), Issue 2, pp. 206/211.

[16] E. Calla, D.G. McCartney, and P.H. Shipway: Effect of Deposition Conditions on the Properties and Annealing Behavior of Cold-Sprayed Copper. Journal of Thermal Spray Technology, 15, (2006), Issue 2, pp. 255/262.

[17] P. Sudharshan Phani, D. Srinivasa Rao, S.V. Joshi, and G. Sundararajan: Effect of Process Parameters and Heat Treatments on Properties of Cold Sprayed Copper Coatings. Journal of Thermal Spray Technology, 16, (2007), Issue 3, pp. 425/434.

[18] Shigenori Hori, Hideo Tai and Hiroaki Katayama: Recrystallization Temperatures of Dilute of Copper Alloys. Japan Inst. Metals, 45, (1981), Issue 12, pp. 1223/1228.

[19] Norio Furushiro, Yoshimasa Takayama, and Shigenori Hori: Effect of Trace Elements on Dynamic Recrystallization Behaviour of Copper. The society of Materials Science, Japan, 360, (1983), Issue 32, pp. 44/48.

[20] Metal data book. The Japan institute of Metals and Materials, Japan, (1984), Issue 2, pp. 159/170.

Implementing the equation of state to account for the density variation at the inlet boundaries for high velocity oxy gas fuel process

M. Aldosari, K.Y. Benyounis, and J. Stokes
School of Mechanical and Manufacturing Engineering, Dublin City University, Dublin, Ireland

Abstract

The gas dynamics of high velocity oxy fuel (HVOF) can be described by the conservative continuum equations of mass continuity, Navier-Stokes, a form of κ-ε model for turbulence and the equation of state, to account for the density variation in the flow field. Regarding combustion, a reduced kinetic (eddy-dissipation model) of one-step global reaction has been validated against measurements in previous work in the literature and good agreement was found. However, in all of the previous studies the inlet condition of the gases into the domain was considered to be of constant mass flow inlet without taking into account the variation of the densities of the gases. In this study, the equation of state is implemented into ANSYS FLUENT code to correlate the variation of density to the temperature and pressure as this may reveal more about the flow behaviour in the HVOF process. It was concluded that the pressure and temperature of the gases prior to entering the combustion chamber has a pronounce effect on the combustion pressure and the compressibility behaviour of the gases before expanding to the ambient pressure. For the velocity, the influence of pressure and temperature of the gases in the range considered has a marginal effect on the velocity behaviour except at the exit of the nozzle and the subsequent compressibility behaviour. As for the temperature, no considerable effect was observed for the range considered in this study.

Keywords: equation of state, boundary condition, numerical simulation, HVOF

1 Introduction

Engineering components service life is strongly affected by degradation due to wear, corrosion, thermal stresses and other surface phenomena. The cost of degradation to materials used in engineering components is estimated at billions of dollars in the United States of America alone, typically 3-5% of its GDP [1]. Thermal spray coating techniques are considered as one of the most promising solutions to this problem [2, 3].

Amongst different thermal spray technologies developed so far, High velocity oxy fuel is considered the most widely used process due to its ability to produce coating with low porosity or more dense compared to other thermal spray technology.

The HVOF process is a combustion-based process in which a mixture of hydrocarbon fuel and oxidant is introduced into a combustion chamber to react and produce a high pressure products. The products of combustion are then allowed to expand through a converging-diverging nozzle (de Laval Nozzle) or converging-parallel sided barrel nozzle to produce a supersonic jet that carries the powders to the substrate to be coated.

The dense coating produced by HVOF process is attributed to its ability to supply the coating powders (particles) with relatively low to medium temperature and high velocity. The temperature and velocity of the particles upon impact on the substrate are considered the most important parameters regarding the coating microstructure and quality [4-8]. Numerous experimental work has been conducted to optimize the impact particle velocity and temperature for different application as in [9]. The particles velocity and temperature are primarily influenced by the continuum with which they exchange thermal and kinetic energy.
However, due to the complexity of the physics involved in the process, experimental approach might not fully reveal about the interaction of the physics involved in the process [10].

It is therefore the role of the computational modelling approach to fill this gap and to complement the experimental approach to reveal more about the interaction of the physics involved in the process.

Regarding the computational approach, extensive work has been done to analyse the gas dynamics of the process. The earliest computational work of Power et al. [11] and Smith et al. [12] analysed the axisymmetric internal and external flow separately using modern CFD techniques. Oberkampf and Talpallikar [13] studied the same nozzle of the work of Power et al. and to computationally model the process with full coupling between the internal and external flow by using CFD-ACE commercial software.

A series of research was conducted by Li et al. to formulate a mathematical model based feedback control system, and to analyse the combustible flow and particle size behaviour within HVOF process [10,14-15].

In the previous mentioned work, the fuel was either propylene or a liquid propane with low boiling point.

Kamnis et al. was the first to computationally investigate the combustion of liquid kerosene [16-18].

Nevertheless, in all of the previous work the mass flow rate at the inlet of the nozzle was considered to be constant even in the case of gas fuel such as propylene, and relating the variation of density to the temperature and pressure was not considered. In this study the variation of density of propylene fuel is related to temperature and pressure of the gas via the equation of state at the inlet of the nozzle to computationally study the effect pressure and temperature of the gas at the inlet of the nozzle on the gas dynamics flow of the HVOF process.

2 Model development

In this study a1050DJ gun manufactured by Oerlikon Metco, Switzerland, available in DCU facility was chosen as a case study. The gun operate on both propylene and propane. The gun has a combustion chamber followed by a converging part and then a parallel sided barrel. The gun has eight circular inlet for the fuel and oxidant mixture and one axial circular inlet for the carrier gas which is nitrogen.

Due to the simplicity and axisymmetric of the geometry of the gun and, a 2-D axisymmetric model was adopted in this study to simulate both the internal and external flow. A quadrilateral mesh was applied all over the domain, the mesh was clustered in the region of high gradient of the solution variables, particularly along the centreline of the free jet region and in the internal domain of the gun. The grid independency solution was checked and 29771 cells was determined to be the limit for grid independency. The solution domain is shown in **Fig. 1.**

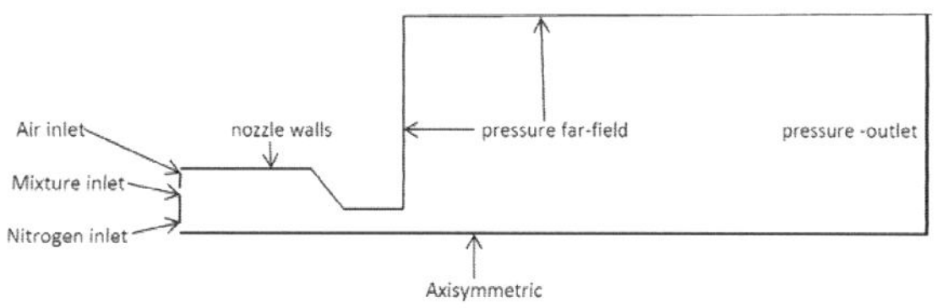

Fig. 1. Schematic diagram of the model boundary conditions applied to the DJ1050 Oerlikon Metco gun.

2.1 Mathematical Models

The commercial software of ANSYS FLUENT was used to simulate the process. The mathematical models of mass continuity, Navier-Stokes and energy equations together with the equation of state available in the code were implemented to solve for the continuum flow field. However, the flow in HVOF process is a high Reynold's number turblent compressible flow and a direct solution for these fundamental conservative equations is limited by the avaiblable computation capacity.

Applying Reynold's or Favre averaging, the aforementioned governing equations can be simplified by decomposing the solution variables into a mean (time or density averaged) and fluctuating components without resolving the small scale turbulent fluctuations.
Reynold's and density averaging for any solution variable takes the form:
Reynold's (or time) averaging, given by :

$$\varphi = \overline{\varphi} + \hat{\varphi} \; with \; \overline{\varphi} = \frac{1}{\Delta t}\int_{t_0}^{t_0+\Delta t} \varphi \; dt \; and \; \overline{\hat{\varphi}} = 0. \quad (1)$$

$$\varphi = \overline{\varphi} + \tilde{\varphi} \; with \; \tilde{\varphi} = \frac{\overline{\rho\varphi}}{\bar{\rho}} \; and \; \overline{\tilde{\varphi}} = 0. \quad (2)$$

Taking the time average of pressure and density, and density average of all the other quantities, once obtains the following continuity and momentum equations written in the cartesian tensor form [14]:

$$\frac{\partial \bar{\rho}}{\partial t} + \frac{\partial(\bar{\rho}\,\tilde{v}_j)}{\partial x_j} + = 0 \quad (3)$$

where ρ is the density, P is the static pressure, x is the coordinate, μ is the molecular viscosity, δ_{ij} is the Kronecker delta and $i, j \; and \; l$ are the indices $\overline{-\rho v''v''}$ is the Reynold's stress term representing the effect of turbulence, which can be related to the mean velocity gradients according to the Boussinesq hypothesis.

$$\frac{\partial(\bar{\rho}\,\tilde{v})}{\partial t} + \frac{\partial(\bar{\rho}\,\tilde{v}_i\,\tilde{v}_j)}{\partial x_j}$$

$$= -\frac{\partial \bar{p}}{\partial x_i} + \left[\frac{\partial}{\partial x_j}\mu\left(\frac{\partial \tilde{v}_i}{\partial x_j} + \frac{\partial \tilde{v}_j}{\partial x_i} - \frac{2}{3}\delta_{ij}\frac{\partial \tilde{v}_l}{\partial x_l}\right)\right]$$

$$+ \frac{\partial}{\partial x_j}\left(-\overline{\rho v''v''}\right), \qquad i = 1,2,3 \quad (4)$$

$$-\overline{\rho v''v''} = \mu_t\left(\frac{\partial \tilde{v}_i}{\partial x_j} + \frac{\partial \tilde{v}_j}{\partial x_i}\right)$$

$$-\frac{2}{3}\left(\bar{\rho}k + \mu_t\frac{\partial \tilde{v}_l}{\partial x_l}\right)\delta_{ij} \quad (5)$$

where μ_t is the turbulent viscosity and k is the turbulence kinetic energy.

The RNG $k - \epsilon$ turbulence model was used in the simulation to account for the turbulence fluctuations in the flow. The RNG $k - \epsilon$ model takes the form:

$$\frac{\partial(\bar{\rho}k)}{\partial t} + \frac{\partial(\bar{\rho}\tilde{v}_j k)}{\partial x_j} = \frac{\partial}{\partial x_j}\left[\alpha_k(\mu + \mu_t)\frac{\partial k}{\partial x_j}\right]$$

$$+G_k - \bar{\rho}\epsilon - Y_M \quad (6)$$

$$\frac{\partial(\bar{\rho}\epsilon)}{\partial t}+\frac{\partial(\bar{\rho}\tilde{v}_j\epsilon)}{\partial x_j}=\frac{\partial}{\partial x_j}\left[\alpha_\epsilon(\mu+\mu_t)\frac{\partial\epsilon}{\partial x_j}\right]$$

$$+C_{1\epsilon}\frac{\epsilon}{k}G_k-C_{2\epsilon}\bar{\rho}\frac{\epsilon^2}{k}-R_\epsilon \qquad (7)$$

Where ϵ is the turbulence dissipation rate, G_k is the generation of turbulence kinetic energy due to the mean velocity gradients, Y_M is the contribution of the fluctuating dilatation in compressible turbulence to the overall dissipation rate, R_ϵ is the additional term in ϵ equation. $C_{1\epsilon}$ = 1.42, $C_{2\epsilon}$= 1.68, and $\alpha_k$are the inverse effective Prandtle numbers for both the kinetic energy and the dissipation rate of turbulence respectively.

The finite rate eddy dissipation was used to solve for the combustion. The model was chosen to simplify the modelling of the process since under the assumption of the model, the Arrhenius kinetic rate of reactions can be ignored and the reaction rate is assumed to be limited by the turbulence mixing rate. Under this assumption the chemical species transport and reaction can be mathematically represented by following convection-diffusion equation:

$$\frac{\partial(\bar{\rho}Y_i)}{\partial t}+\frac{\partial(\bar{\rho}\,Y_i\,\tilde{v}_j)}{x_j}=-\frac{\partial(J_i)}{\partial x_j}+R_i$$

$$i=1,\ldots,N-1 \qquad (8)$$

Where J_i is the diffusion flux of species i calculated by Maxwell-Stefan equations, R_i is net rate of production of each species by the chemical reaction, and N is the number of the total number of species involved in the reaction. The energy equation takes the form:

$$\frac{\partial(\bar{\rho}\,H)}{\partial t}+\frac{\partial}{\partial x_j}[\tilde{v}(\bar{\rho}H+\bar{P})]$$

$$=\frac{\partial}{\partial x_j}\left[\alpha C_p(\mu+\mu_t)\frac{\partial T}{\partial x_j}+\tilde{v}(\mu+\mu_t)\left(\frac{\partial\tilde{v}}{\partial x_i}+\frac{\partial\tilde{v}}{\partial x_j}-\frac{2}{3}\delta_{ij}\frac{\partial\tilde{v}}{\partial x_l}\right)-\sum_{i=1}^{N}j_iH_i\right]+S_E \qquad (9)$$

Where T is the temperature, H is the total enthalpy defined by

$$H=\sum_i H_iY_i \qquad (10)$$

and S_E is the source term, which is the heat generated by the exothermic reaction. It is know that the products of combustion of hydrocarbon fuel at high temperatures will dissociate into a number of species with low molecular weight [7,13,16]. However, given the oxy/fuel ratio and combustion pressure and temperature, the equilibrium composition of the reaction can be determined by chemical equilibrium composition program developed by McBride and Gordon [19].The programme calculates the equilibrium composition based on minimization of Gibbs free energy principle.
In this paper propylene was considered as the fuel and oxygen the oxidant. The flow of air was considered constant and the products of the reaction were CO_2 and H_2O were assumed to decompose into nine species according to the following form [14] :

C_3H_6 + 4.5O_2 => n_1 CO + n_2 CO_2 + n_3 H + n_4 H_2 + n_5 H_2O + n_6 O + n_7 OH + n_8 O_2 (11)

It should be mentioned that if the dissociation of the products of the reaction was achieved, then the exothermic temperature of the reaction could be over-predicted.

2.2 Results and discussion

As pointed out, the purpose of this study is to investigate the effect of pressure and temperature of the processing gases at the inlet of the computational domain on their densities [equation 12] and the consequence effects on the flow field of HVOF process. In all the previous studies constant mass flow rate boundary condition was assigned as an inlet zone to the computational domain. The conversion was found to be calculated by simply multiplying the volume flow rate of the gases by their densities at the standard conditions of 1 bar and 273.15 K such as in the study of Martinez et al. [20] It is known that the density of the gas can be related to the pressure, temperature and volume via different equations of states and each equation of state gives different accuracy [21]. However, in most of engineering cases the equation of state in the form of:

$$p\,\dot{v}=\rho\,R\,T \qquad (12)$$

which was applied to take into account the effect of temperature and pressure on the mass flow rate of the processing gases in HVOF process.

First, the density of standard conditions (1 bar and 273.15 K) were assumed for the baseline parameters recommended by the manufacturer and then the density variation against pressure and temperature was investigated. The parameters range are shown in **Table 1.** The behaviour of gases for the baseline condition are shown in **Figure 2.** The combustion pressure computed by the simulation was 4.4 bar which was in a quite good agreement with the manufacturer measured pressure which is 4.2 bar. It can be noted from the graph that the pressure in the combustion chamber and the converging part of the nozzle is almost constant at 4.4 bar, and the pressure decreased as the gas entered the barrel part of the nozzle until it expands to a slightly over atmospheric pressure, which indicates that jet was of over-expanded kind.

Table.1:Parameters used in the model [22]

p bar	T Kelvin	$\dot{V}$ C_3H_6+O_2 SLPM	$\dot{V}$ N_2 SLPM	$\dot{m}$ C_3H_6+O_2 g/s	$\dot{m}$ N_2 g/s
1.0(BL)	273	329	325	18	8
1.5	283	329	325	25	11
2.0	293	329	325	33	14
2.5	302	329	325	40	17

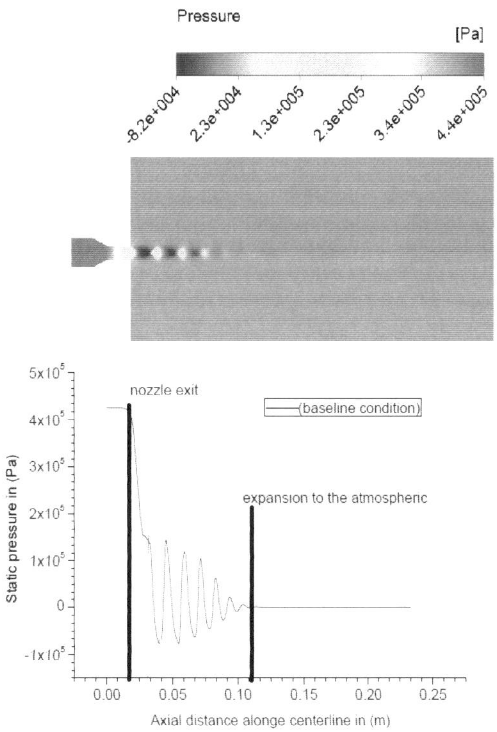

Fig.2. Contour plot and profile of static pressure of gases for the baseline condition

The velocity profile for the baseline condition is shown in **Fig.4.** In contrary to the gas behaviour in the converging-diverging nozzle (de-Laval nozzle), the gas did not have any supersonic behaviour inside the gun and the supersonic behaviour begins at the exit of the nozzle. The gas velocity reaches up to approximately 2000 m/s, but typically of around 1300 m/s which is in a good agreement with the velocity documented by the manufacturer [23,7]. The sonic condition occurs at the exit of the nozzle while in the de-Laval nozzle it normally occurs at the throat of the nozzle (corresponding to the minimum area in the nozzle) as shown in **Fig.5.** The velocity profile for the other cases along with the base line condition is shown in **Fig.6.** For the four cases, variation of the inlet mass flow rate, related to the variation of density with pressure and temperature, has an effect only at the compressibility behaviour at the exit of the nozzle. For the temperature, the simulation results shows that variation of pressure and temperature in the range considered has a marginal effect on the static temperature of the gases in the flow as in **Fig.7.**

3 Summary and conclusion

In this study, the calculation of mass flow rate as an inlet condition for a computational study was related to the pressure and temperature via a common form of the equation of state. This form of the equation was used by the manufacturer for converting the volume flow rate to the correspondent mass flow rate at the standard condition of 1bar and 273.15 K.

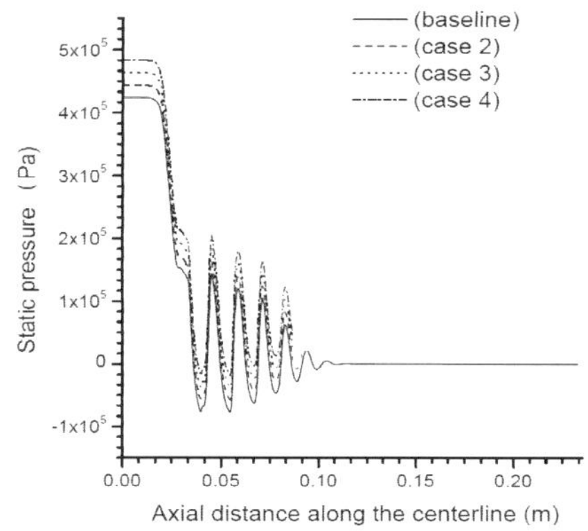

Fig.3 Static pressure profile for the four cases considered.

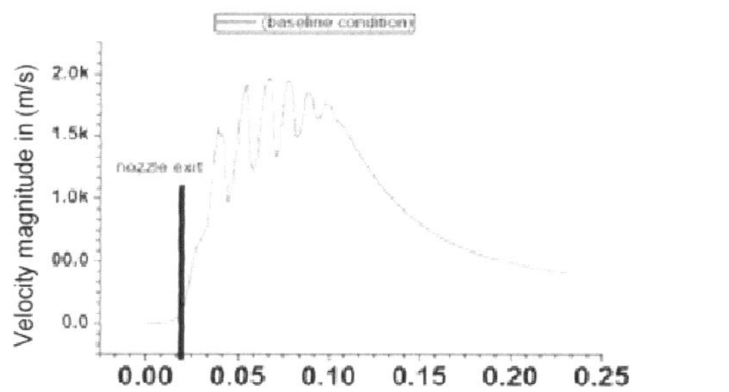

Fig.4. Velocity profile for the baseline conditio

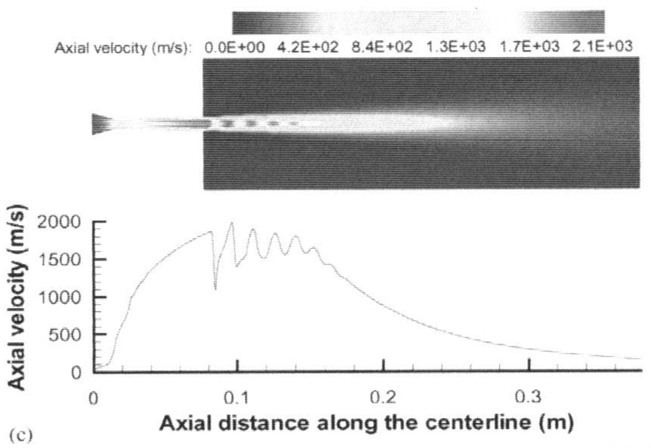

Fig.5. Velocity profile for the baseline condition ref [7]

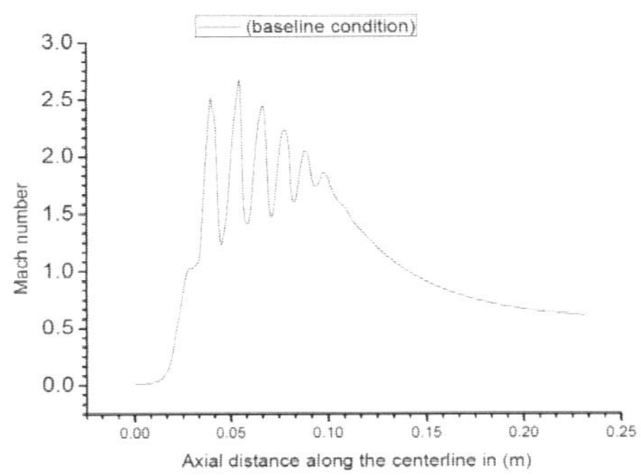

Fig.6. Mach number profile for the baseline condition

In all the previous computational work, mass flow rate inlet condition was assigned without taking into account the effect of pressure and temperature in the calculation of the mass flow rate and consequence effect on the flow of the gas. In this work, a range of temperature and pressure for the calculation of the mass flow rate was considered and the results could be summarized as follow:

The pressure and temperature of the gases prior to entering the combustion chamber has a pronounce effect on the combustion pressure and the compressibility behaviour of the gases before expanding to the ambient pressure.

The mass flow rate at the critical area (chocking area) can be related to the stagnation pressure through the following equation :

$$\dot{m}_g = \rho_t v_t A_t = \frac{P_0}{\sqrt{T_0}} A_t \left[\frac{\gamma \bar{M}_{pr}}{R} \left(\frac{2}{\gamma + 1} \right)^{(\gamma+1)/(\gamma-1)} \right]^{1/2} \quad (13)$$

and the stagnation pressure is dependent on the state of the gases before entering the combustion chamber. So, the state of the gases as they inlet the combustion chamber has to be investigated more deeply as that might assist in the design process of the nozzle

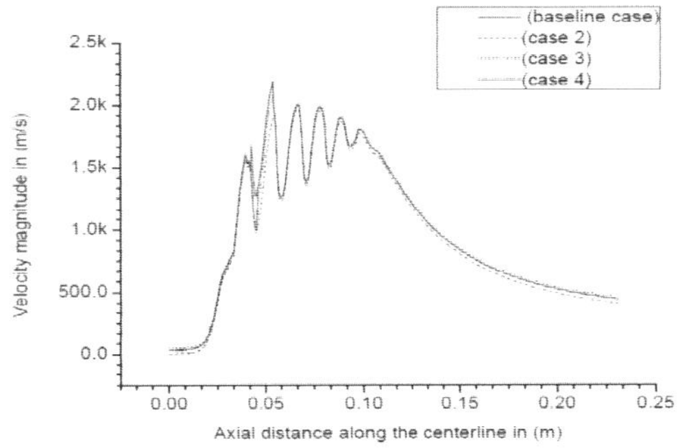

Fig.7. Velocity magnitude for the four cases considered

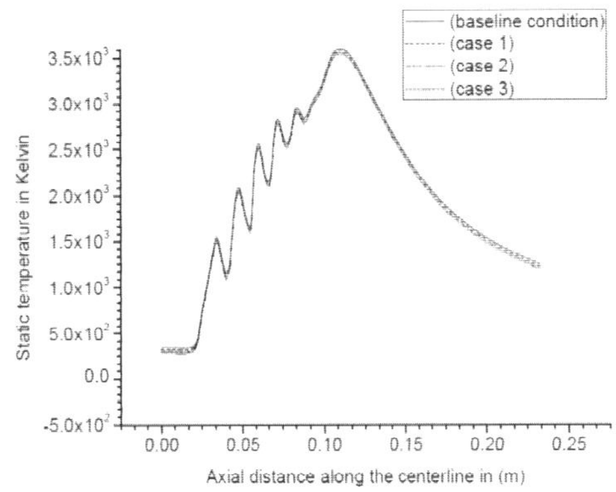

Fig.8. Static temperature for the four cases considered

For the velocity, the influence of pressure and temperature of the gases in the range considered has a marginal effect on the velocity behaviour except at the exit of the nozzle and the subsequent compressibility behaviour. And for the temperature, no considerable effect was observed for the range considered in this study.

references

[1] Cheng, D., Trapaga, G., McKelliget, J., 2003, "Mathematical Modelling of High Velocity Oxygen Fuel Thermal Spraying of Nanocrystalline Materials: An Overview," Modelling and Simulation in Materials Science and Engineering, 11(1) pp. R1

[2] Davis, J.R., 2004, "Handbook of thermal spray technology," ASM international, .

[3] Li, M., and Christofides, P. D., 2009, "Modeling and Control of High-Velocity Oxygen-Fuel (HVOF) Thermal Spray: A Tutorial Review," Journal of Thermal Spray Technology, 18(5-6) pp. 753-768.

[4] Pawlowski, L., 2008, "The science and engineering of thermal spray coatings," John Wiley & Sons, .

[5] Oksa, M., Turunen, E., Suhonen, T., 2011, "Optimization and Characterization of High Velocity Oxy-Fuel Sprayed Coatings: Techniques, Materials, and Applications," Coatings, 1(1) pp. 17-52.

[6] Li, M., and Christofides, P. D., 2002, "Feedback control of nanostructured coatings processing accounting for powder size distribution," American Control Conference, 2002. Proceedings of the 2002, Anonymous IEEE, 2, pp. 1019-1027.

[7] Li, M., and Christofides, P. D., 2005, "Multi-Scale Modeling and Analysis of an Industrial HVOF Thermal Spray Process," Chemical Engineering Science, 60(13) pp. 3649-3669.

[8] Gaona, M., Lima, R. S., and Marple, B. R., 2008, "Influence of Particle Temperature and Velocity on the Microstructure and Mechanical Behaviour of High Velocity Oxy-Fuel (HVOF)-Sprayed Nanostructured Titania Coatings," Journal of Materials Processing Technology, 198(1) pp. 426-435.

[9] Bozorgtabar, M., Salehi, M., Rahimipour, M., 2010, "Influence of High Velocity Oxy-Fuel Parameters on Properties of Nanostructured TiO2 Coatings," Bulletin of Materials Science, 33(6) pp. 671-675.

[10] Li, M., and Christofides, P. D., 2005, "Multi-Scale Modeling and Analysis of an Industrial HVOF Thermal Spray Process," Chemical Engineering Science, 60(13) pp. 3649-3669.

[11] Power G D, Barber T J and Chiappetta L M 1992 AIAA Paper No 92-3598

[12] Smith E B, Power G D, Barber T J and Chiappetta L M 1992 proc. Ineternational Thermal Spraying Conf. (Orlando, FL, 1992) p 805

[13] Oberkampf W L and Talpalikar M 1994 Thermal Spray Industrial Applications (Boston, MA: ASM International) p 381

[14] Li, M., Shi, D., and Christofides, P. D., 2004, "Diamond Jet Hybrid HVOF Thermal Spray: Gas-Phase and Particle Behavior Modeling and Feedback Control Design," Industrial & Engineering Chemistry Research, 43(14) pp. 3632-3652.

[15] Li, M., and Christofides, P. D., 2004, "Feedback Control of HVOF Thermal Spray Process Accounting for Powder Size Distribution," Journal of Thermal Spray Technology, 13(1) pp. 108-120.

[16] Kamnis, S., and Gu, S., 2006, "Numerical Modelling of Propane Combustion in a High Velocity Oxygen–fuel Thermal Spray Gun," Chemical Engineering and Processing: Process Intensification, 45(4) pp. 246-253.

[17] Gozali, E., Kamnis, S., and Gu, S., 2013, "Numerical Investigation of Combustion and Liquid Feedstock in High Velocity Suspension Flame Spraying Process," Surface and Coatings Technology, 228pp. 176-186.

[18] Kamnis, S., and Gu, S., 2006, "3-D Modelling of Kerosene-Fuelled HVOF Thermal Spray Gun," Chemical Engineering Science, 61(16) pp. 5427-5439.

[19]McBride, B. J., and Gordon, S., 1996, "Computer Program for Calculation of Complex Chemical Equilibrium Compositions and Applications: II. Users Manual and Program Description," NASA Reference Publication, **1311**pp. 84-85.

[20] B. Martinez, G. Mariaux, A. Vardelle, G. Barykin, and M. Parco., 2009, , " Numerical Investigation of a Hybrid HVOF-Plasma Spraying Process," Journal of Thermal Coating Technology, 18(5-6)pp. 909-920

[21] Moran, M.J., Shapiro, H.N., Boettner, D.D., 2010, "Fundamentals of engineering thermodynamics," John Wiley & Sons,.

[22] Thorpe, ML and Richter, HJ., 1992, " A pragmatic analysis and comparison of HVOF processes," Journal of Thermal Spray Technology, 1(2)pp.161-170.

[23] http://www.oerlikon.com/metco/en/products-services/coating-equipment/thermal-spray/spray-guns/spray-guns-hvof/hvof-diamond-jet/

Modelling of coating profile in cold spray and its application in offline programming software

C. CHEN, Y. XIE, S. DENG, C. Verdy, H. LIAO, IRTES – LERMPS, Université de Bourgogne Franche-Comté (UBFC), Université de Technologie de Belfort-Montbéliard(UTBM), Site de Sévenans, 90100 Belfort Cedex, France

Due to the various advantages of cold spray technology, its potential applications are additive manufacturing and dimensional recovery of damaged parts. In order to obtain a flat as-sprayed coating surface and reduce extra machining work, it is essential to optimize the robot trajectory and the operating parameters for a desired coating thickness distribution. In this study, a numerical model of single coating profile based on standard experimental results was established, which included the effects of spray angle, nozzle traverse speed as well as scan step. Afterwards, the numerical model was extended to the coating thickness distribution on the entire substrate surface by integrating the thickness value on each mesh-grid node. Meanwhile experimental studies were also made to validate and compare with the numerical model. At last, the functions of coating profile simulation and coating thickness prediction were added into the addin extensive software "Thermal Spray Toolkit" (TST) which is developed in the offline programming software RobotStudio™. It enables the simulation and prediction of coating profile in 2D and coating thickness distribution on substrate surface after the trajectory generation process, which is based on the real-time kinematics data obtained from robot process simulation results.

1 Introduction

Among various spray technologies, cold spray has drawn more and more attentions due to its low porosity, high adhesion strength and low particle oxidation. In this process, particles are accelerated to high velocity ranging from 300 to 1200 m/s by the heated and compressed driving gas through a converging-diverging nozzle, depositing onto the substrate or the layer already deposited in solid state with relatively low temperature [1]. Differing from traditional thermal spray processes where molten or semi-molten particles deposit at a low velocity, the low temperature and high velocity of cold sprayed particles upon impact can avoid the occurrence of particle oxidation as well as local thermal residual stresses [2]. Due to its features of high deposition efficiency, high adhesion strength, low oxidation and residual stress, cold spray is considered as an effective technology of additive manufacturing. The small heat transfers in cold spray process compared with other additive manufacturing technologies like selective laser melting (SLM) and direct metal deposition (DMD), can largely retain the microstructure and properties in mechanical and chemical of feedstock powders. Meanwhile, the controllable spray jet in cold spray enables coating deposition on free-form workpieces and fabrication of complex forms. Thus, the coating thickness predication based on operating parameters have become a crucial part in the process of cold spray additive manufacturing. An effective coating thickness modelling can be used to optimize the nozzle trajectory as well as the operating parameters including nozzle traverse speed, spray angle, off-stand distance, and scan step.

Djurić et. al [3] developed a metal spray deposition model to simulate the spatial mass flux distribution produced by the nozzle. The deposition efficiency was included by transferring the non-linear inverse problem to a boundary-value problem. Rayment et. al [4] investigated the distribution of temperature and temperature variance on the substrate by using the same model, which aimed at the path planning optimization as well as the elimination of the thermal residual stress and distortion of the sprayed steel shell. Duncan et. al [5] also used the numerical model developed by Djurić to optimize the path separation in spray coating, which used the sampling theory to transfer the problem into the spatial frequency domain. However, the studies mentioned above have not included the influence of angular spray, which has become common cases due to the compromise on workpiece geometry or spray strategy. Fasching et. al [6] achieved a coating thickness distribution with low standard deviation by optimizing robotic trajectory, which was realized by using a nozzle spray tilting model. Similarly, Leigh et. al [7] evaluated the effects of spray angle on coating profile by various coating properties like micro-hardness, tensile adhesion strength of the plasma sprayed coating

In this study, a numerical model of coating profile based on the standard Gaussian distribution was made and then added to the offline programming software. With the numerical model, it is able to simulate the coating thickness distribution with various spray parameters, such as spray angle, scan step and nozzle traverse speed. Particularly, based on a polar coordinate transformation, the coating profile of the angular spray cases was studied. The experimental validation of the numerical model was made with cold-sprayed aluminum coating on stainless steel substrate by varying operating parameters such as scan step between two paths and spray angle. Afterwards, the coating thickness model was integrated to the offline programming software RobotStudio ™ as a module in the addin extensive software "Thermal Spray Toolkit" (TST) [8, 9]. Combined with other functions in TST such as trajectory generation [10-12] on different kinds of substrate surface, the coating thickness simulation based on real-time kinematics data enables the optimization of operating parameters and robot trajectory.

2 Coating profile modelling

According to the central limit theorem, the averages of random variables can be considered normally distributed when the amount of variable is sufficiently large. Thus, in the case of thermal spray process, the feedstock jet distribution out of nozzle as well as the coating profile distribution on substrate can be approximated by the mathematical expression of Gaussian distribution, as shown in Fig. 1. As a result, for the coating deposited by thermal spray process, the coating thickness distribution can be expressed as the Gaussian approximation as shown in the equation below [13].

$$\phi = \zeta(\theta)\int \frac{A}{\sigma\sqrt{2\pi}} e^{-\left(\frac{(x-\mu_x)^2}{2\sigma^2}+\frac{(y-\mu_y)^2}{2\sigma^2}\right)} dxdy \qquad (1)$$

Where A is the amplitude factor in correlation with the feedstock flowrate, σ is the standard deviation of the coating profile, (μ_x, μ_y) is the center coordinate of the coating profile on substrate and $\zeta(\theta)$ is the deposition efficiency in function of the spray angle. The single coating profile can be obtained by integration on the substrate surface. Based on standard experimental data, it is able to obtain the values of each variable, which can be considered as a standard basic model. Based on this standard model, the coating thickness distribution under different operating parameters can be obtained. Currently, in this work, the effects of deposition efficiency have not been taken into account.

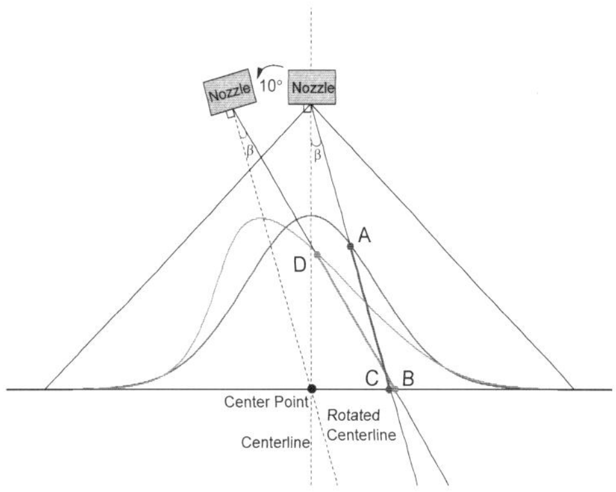

Fig. 1. Schematic of coating profile distribution for perpendicular (blue line) and angular (red line) spray cases. The nozzle is rotated 10° around the center origin point anticlockwise.

Generally, the spray angle is 90° in order to obtain a maximum deposition efficiency and coating quality. However, in the practical spray process, due to the limits of workpiece geometry and working condition, the angular spray usually appears as a compromise on the spray strategy. As a result, the coating profile simulation in off-angle spray cases were investigated. As shown in Fig. 1, for the perpendicular spray case the coating profile is conical and symmetric with the central line of nozzle.

The coating profile in off-angles cases can be deducted by transforming the perpendicular spray model in the Cartesian coordinate system to the polar coordinate system. The spray cone in polar coordinate system is then divided a series of rays with constant interval angle.

In the polar coordinate system, any point on coating surface in the perpendicular spray case can be described by two variables. One is the deflection angle θ between each ray and the central line and the other one is the spray distance at this angle from the impacting point on the substrate to the point at coating profile. It is able to describe the coating profile as a function of the deflection angle and the corresponding spray length. As a result, in the off-angle spray cases, the coating profile can be established by referring to the standard model in perpendicular case. In the case where the nozzle is rotated 10° anticlockwise in Fig. 1, the rotated coating profile consists of a series of points at each deflection angle, which are defined by the distance L from the impacting point on the substrate. And the distance L of off-angle spray case at each deflection angle can be referred from the perpendicular spray model. For example at the deflection angle of β as indicated in Fig. 1, the spray distance AC at perpendicular spray case has the same value as the one of BD at angular spray cases. The impacting point at each deflection angle is obtained by perspective projection through the nozzle exit point at the substrate surface. As a result, by applying the spray length at each deflection angle, it can obtain the corresponding points at the coating surface within the rotated spray cone area. The coating profile for angular spray cases is given in Fig. 1, where angular coating profile is marked as red, and perpendicular one is marked as blue.

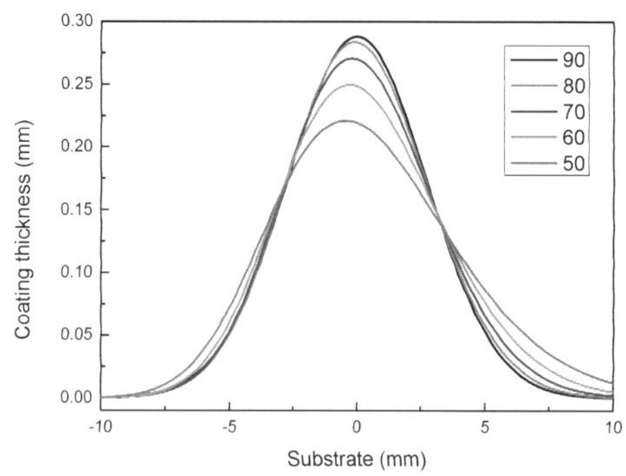

Fig. 2 Coating profile at different spray angles

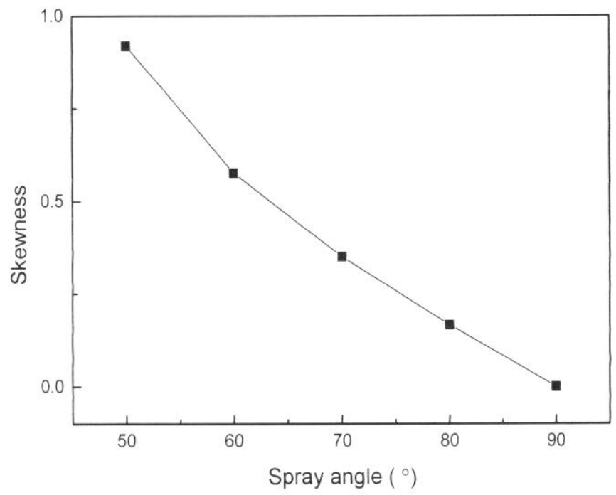

Fig. 3 The skewness of coating profile at different spray angle

As shown in Fig. 2, the coating profiles at different spray angles from 90° to 50° are given by the methodology above, where the nozzle is rotated anticlockwise, standoff distance is 30 mm and nozzle traverse speed is 150 mm/s. It is found that for off-angle spray cases, as spray angle decreasing from 90°, the mass distribution is gradually concentrated on the left side, and its asymmetry becomes more evident. Due to the decrease of spray angle, more particles are concentrated at one side, which causes the decrease of maximum coating profile height as it is observed in Fig. 2. In order to evaluate the coating profile variation with spray angle, the parameter of skewness, which is usually used to characterize the symmetry of the probability distribution of a set of random values, was applied and its variation is given in Fig. 3. For the spray angle of 90, the skewness of the coating thickness distribution is zero, which indicates the symmetric distribution. With the decrease of spray angle, the skewness increases, which indicates the asymmetric distribution as well as the fact that the coating profile is gradually concentrated on the same direction of nozzle rotation. The skewness variation with the spray angle has a good accordance with the coating profile variation in Fig. 2.

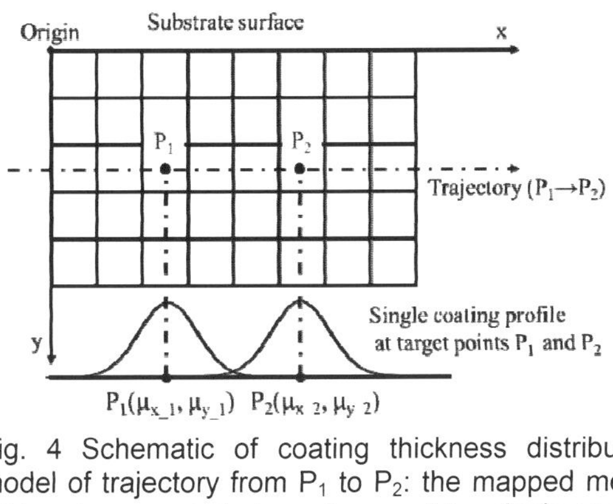

Fig. 4 Schematic of coating thickness distribution model of trajectory from P_1 to P_2: the mapped mesh-grid nodes on substrate surface and the single coating profiles at target points P_1 and P_2.

Based on the numerical model of single coating profile, it is able to build the model of coating thickness distribution for a robot trajectory on the substrate surface. As shown in Fig. 4, a simple trajectory consisting of target points P1 (μ_{x_1}, μ_{y_1}) and P2 (μ_{x_2}, μ_{y_2}) with constant time step is given as a demonstration. The substrate surface is treated as a mapped mesh-grid with constant grid size as shown in Fig. 4. The distance between two adjacent target points is defined as the advancement based on the nozzle speed at P_1 and the time step. As a result, for each target point P (μ_x, μ_y) on the trajectory, the thickness value at each mesh-grid node can be obtained by substituting the node coordinate (x, y) into the corresponding single coating profile model $\phi(\sigma, \mu_x, \mu_y)$. By integrating the node value at each time step on the entire substrate surface, the coating thickness distribution of the robot trajectory on the substrate can be obtained. Thus, it is able to describe the influence of nozzle traverse speed in the model of coating thickness distribution, meanwhile the effects of operating parameters like spray angle have already been included in the single coating profile model.

3 Experimental and numerical results

In order to study the effects of spray operating parameters on coating thickness and validate the proposed coating thickness modelling method, the experimental study realized by cold spray was made. Cold sprayed coating was produced by homemade cold spray system (LERMPS, UTBM, France), which was equipped with a de-Laval type converging-diverging nozzle. The nozzle was cooled by a circulating water system. The pure Al powder (ECKA Granules Metal Powders Ltd, Germany) with spherical morphology and the size range between10 and 45 µm was selected as the feed stock. The stainless steel substrate was grit-blasted before spray process. High pressure compressed nitrogen gas was applied as the propellant gas with a temperature of 450°C and pressure of 2.8 MPa. The standoff distance was 30 mm away from the substrate. The spray parameters such as scan step and spray angle were studied respectively. The scan steps from 2 mm to 6 mm and spray angles from 90° to 60° were investigated. Single scan was made for the cases of spray angle study and a round-trip scan was made for the scan step study. Afterwards, the coating thickness variation was measured by the profile meter (AltiSurf 500, Altimet, France) and compared with the numerical model results.

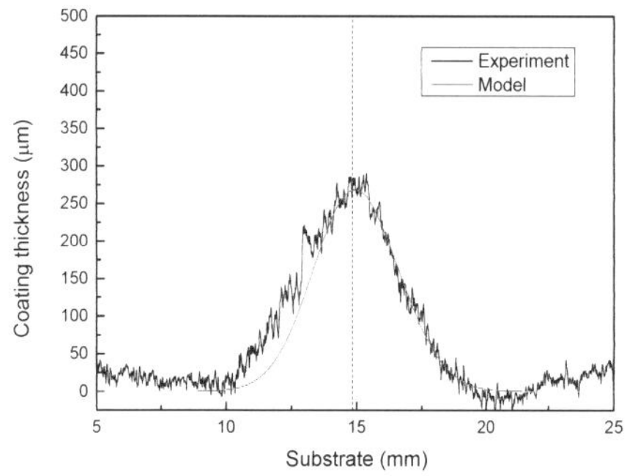

Fig. 5 Comparison of coating profile at spray angle of 80° between experimental and simulation results

The typical angular spray case of 80° was chosen to compare with the numerical model results. As shown in Fig. 5, the coating profiles at spray angle of 80° were compared between the experimental and simulation results. Compared with the nozzle centerline in perpendicular spray case at the position of 15 mm, dash line that indicates the maximum coating profile height point shows the asymmetry of coating profile and its concentration at the left side. Besides, a good consistency is also observed in Fig. 5 between the experimental data and numerical modeling, which supports the validation of the model.

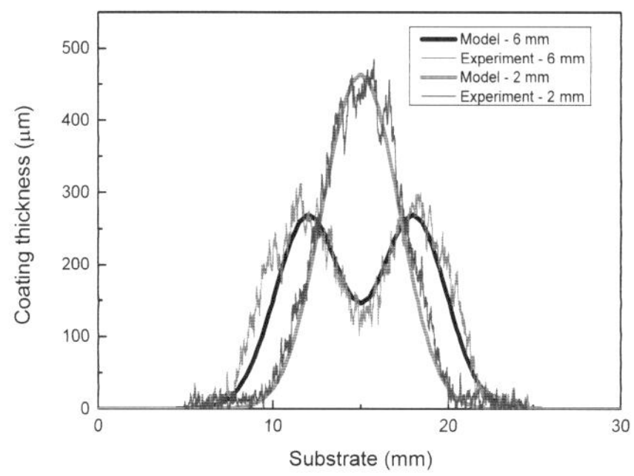

Fig. 6 Comparison of coating profile at different scan steps between experimental and simulation results

Besides, the effects of scan step on coating profiles were also investigated. The comparison between experimental and simulation results at scan steps of 2 mm and 6 mm are shown in Fig. 6. It is observed that in the scan step of 6 mm, two peaks appear at the coating profile, which indicates an undesired coating surface flatness. However, in the case of scan step of 2 mm, only one peak at the coating profile appears and no obvious deviation between two individual coating profiles is found, which means that a good coating surface flatness can be achieved. Combined with the work by Fasching [6], high deposition flatness and uniformity can be achieved by specifying the scan step as one standard deviation of the coating profile distribution. The standard deviation value of the coating profile by single individual nozzle path is 2.77 mm. Thus, it can also be concluded that similar result is obtained that the scan step of one standard deviation can ensure a desired flatness of coating surface.

4 Coating thickness simulation in RobotStudio™

As reported by Deng et. al [8, 9, 14], an addin extensive software called "Thermal Spray Toolkit" (TST) was developed in the offline programming software RobotStudio™. It aims to provide a complete solution for the application of robot in thermal spray process, which consists of 3 modules which cover the functions like trajectory generation and real-time robot trajectory monitoring. Among the modules in TST, PathKit provides the function of trajectory generation on different kinds of substrate surface. After the trajectory generation, the coating profile under different operating parameters can be simulated in the module ProfileKit, and the real-time robot trajectory can be monitored in the module MonitorKit. At the same time, efforts have been made to improve the functions in TST, such as the application of external axis [11], trajectory generation with mesh-grid method on curved surface [12]. In this study, in order to optimize the trajectory to by predicting the coating thickness, the module ProfileKit is improved by introducing the functions of off-angle coating profile simulation and real-time coating thickness simulation on substrate surface.

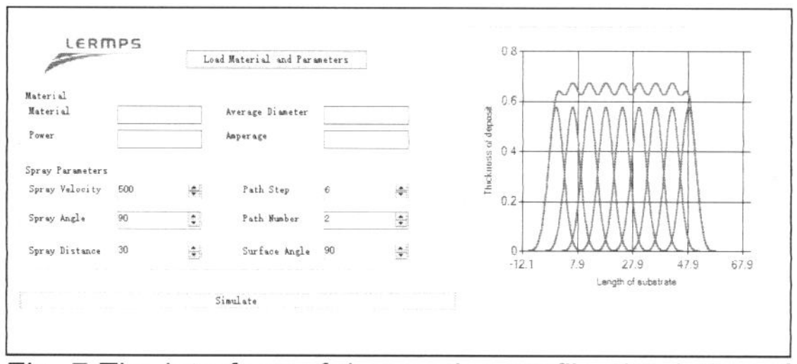

Fig. 7 The interface of the coating profile simulation of TST in RobotStudio™ and results demonstration

As shown in Fig. 7, the user-interface (UI) of coating profile simulation is developed in TST on, which is divided into the parameter input panel on the left and the result display panel on the right. By altering the operating parameters such as spray angle, nozzle traverse speed, scan step and clicking the 'simulate' button in the parameter input panel, each coating profile of the 9 paths with the same scan step and the final coating thickness distribution, which is the superposition of all 9 coating profiles, are shown in the panel on the right. The final coating profile is marked as red curve and the coating profile of each path is marked as blue curve. As the example is shown in Fig. 7, the obvious coating thickness fluctuation is observed in the case of 6 mm scan step, whereas the standard deviation of single coating profile is 2.7 mm. As the results in Fig. 8 shows, the fluctuation of coating surface when the scan step is 6 mm is within expectation. The influences by operating parameters like nozzle speed and spray angle can also be observed from the coating profile simulation by altering the corresponding parameters based on standard model obtained from experiments. With this function, the user is able to investigate the influence of each operating parameters on the final coating profile after optimize these parameters to get a uniform coating thickness distribution.

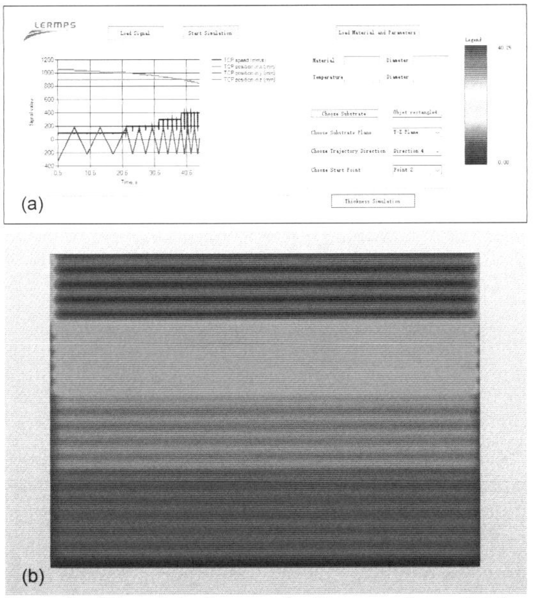

Fig. 9 The interface of the coating thickness distribution simulation of TST in RobotStudio™ (a) and the results of example trajectory on plane substrate surface (b)

As known to all, the stability of robot performance plays a crucial role in the spray process. For example, unstable and fluctuation of nozzle traverse speed influenced by robot performance can result in an unwanted coating surface flatness. Thus, the real-time coating thickness simulation of the robot trajectory on substrate surface based on kinematics data is introduced to TST, which aims to investigate the influence of robot kinematics and operating parameters on coating thickness distribution, and provide the evidences for the optimization of robot trajectory. As the ProfileKit shows in Fig. 9 (a), the kinematics data collection widgets appear on the left where the robot process simulation and kinematics data collection are controlled and displayed. Before the coating thickness simulation, the kinematics data such as TCP (Tool Center Point) speed and TCP position are recorded along the robot process simulation and presented as a diagram in the UI of ProfileKit, which are completed by the button 'Load Signal' and 'Start Simulation'. Afterwards, the coating thickness on the substrate surface is simulated by applying the kinematic data obtained in the process simulation to the coating thickness model. In this step, user has specified the substrate surface and starting point with the widgets on the right of UI. The coating thickness distribution result is then presented as a colored parametric surface on substrate specified by coordinates and corresponding thickness value, with the color specified by the value of thickness. And the color legend is given on the right side, which indicates the range of color variation as well as the range of thickness value.

A demonstration example is given in the Fig. 9 (b), where the coating thickness distribution on the substrate with size of 210 mm * 300 mm is simulated by the ProfileKit. The perpendicular spray robot trajectory consists of 22 round-trip paths with scan step of 10 mm. The nozzle traverse speed for every 5 round-trip paths gradually increases from 100 mm/s to 200 mm/s, 300 mm/s and 400 mm/s. Based on the coating profile simulation results as shown in Fig. 6, it can be expected that an obvious coating thickness fluctuation is formed and the coating thickness decreases for every 5 round-trip paths as the decrease of nozzle speed. As the result is shown in Fig. 9 (b), the obvious color differences are observed between two adjacent round-trip paths, which indicates the coating thickness fluctuation between two adjacent paths. Combined with the color legend in Fig. 9 (b), it is found that the coating thickness gradually decreases for every 5 paths as the decrease of nozzle speed. At last, it can be concluded that the coating thickness prediction function in ProfileKit of TST can be used for the coating thickness simulation of the robot trajectory on the substrate surface based on the standard experimental model.

5 Conclusion

In order to obtain a flat as-sprayed coating surface and reduce extra machining work, a coating profile simulation model is developed in this study by taking into account of the effects from various operating parameters in thermal spray process. It enables the optimization of robot trajectory and operating parameters based on the coating thickness predication. Meanwhile the coating thickness distribution model is added to the offline programming software RobotStudio. The main conclusion of this study is made as below.

i. A numerical model of single coating profile based on standard experimental results was established, which included the effects of spray angle, nozzle traverse speed as well as scan step. Meanwhile experimental studies were also made to validate and compare with the numerical model.

ii. Afterwards, the numerical model was extended to the coating thickness distribution on the entire substrate surface by integrating the thickness value on each mesh-grid node. The effects from nozzle traverse speed. Meanwhile the influence of parameters like spray angle and scan step are also included by the single coating profile model above.

iii. At last, the functions of coating profile simulation and coating thickness prediction were added into the addin extensive software "Thermal Spray Toolkit" (TST). It enables the simulation of single coating profile and coating thickness distribution on substrate surface based on the real-time kinematics data obtained from robot process simulation results.

6 Acknowledgement

The authors would like to acknowledge the support by China Scholarship Council (CSC) and Marie Curie FP7-IPACT-268696(EU).

7 Literature

[1] Li, W. and D. Zhang. Other: Modelling of impact behaviour of cold spray particles: review. Surface Engineering 30 (2014), Issue 5, pp. 299-308.

[2] Bala, N. and H. Singh. Other: Cold spray coating process for corrosion protection: a review. Surface Engineering 30 (2014), Issue 6, pp. 414-21.

[3] Djurić, Z. and P. Grant: An inverse problem in modelling liquid metal spraying. Applied Mathematical Modelling 27 (2003), Issue 5, pp. 379-96.

[4] Rayment, T. and P. Grant: Modeling the heat flow in spray formed steel shells for tooling applications. Metallurgical and Materials Transactions B 37 (2006), Issue 6, pp. 1037-47.

[5] Duncan, S. and P. Jones. Other: A frequency-domain approach to determining the path separation for spray coating. Automation Science and Engineering, IEEE Transactions on 2 (2005), Issue 3, pp. 233-39.

[6] Fasching, M. and F. Prinz. Other: Planning robotic trajectories for thermal spray shape deposition. Journal of Thermal Spray Technology 2 (1993), Issue 1, pp. 45-57.

[7] Leigh, S. and C. Berndt: Evaluation of off-angle thermal spray. Surface and Coatings Technology 89 (1997), Issue 3, pp. 213-24.

[8] Deng, S. and H. Liao. Other. Robotic trajectory autogeneration in thermal spraying. *International Thermal Spray Conference* 2005 (Seattle, Washington, USA). 2005. p 481 - 85

[9] Deng, S. and Z. Cai. Other: Application of robot offline programming in thermal spraying. Surface & Coatings Technology 206 (2012), Issue 19-20, pp. 3875-82.

[10] Fang, D. and S. Deng. Other. Automatic generation of robot trajectory for free-form surfaces in thermal spraying. *Thermal Spray 2011: Proceedings of the International Thermal Spray Conference*. 2011 (Hamburg, Germany). 2011. p 1110 - 14

[11] Deng, S. and D. Fang. Other: Application of External Axis in Robot-Assisted Thermal Spraying. Journal of Thermal Spray Technology 21 (2012), Issue 6, pp. 1203-15.

[12] Cai, Z. and H. Liang. Other: Computer-Aided Robot Trajectory Auto-generation Strategy in Thermal Spraying. Journal of Thermal Spray Technology 24 (2015), Issue 7, pp. 1235-45.

[13] Bolot, R. and S. Deng. Other: A Coupled Model Between Robot Trajectories and Thermal History of the Workpiece During Thermal Spray Operation. Journal of Thermal Spray Technology 23 (2014), Issue 3, pp. 296-303.

[14] Deng, S. and H. Li. Other. New Functions of Thermal Spray Toolkit A Software Developed for Off-line and Rapid Robot Programming. *International Thermal Spray Conference*. 2006 (Seattle, USA). 2006. p 1437 - 42

Thermally sprayed Al-Mn coatings for wet corrosion protection as alternative to cadmium

A. Foerg, G. Wolf, P.J. Masset, Sulzbach-Rosenberg / D
L. Libralesso, Liège / B

Cadmium coatings are still used in aerospace industry for parts in corrosive environments such as actuators or landing gears. Since cadmium is a highly toxic heavy metal, it is subject to strong restrictions, namely REACh. Alternatives include aluminum-zinc alloys which possess a similar electronegative potential and act as sacrificial anode on steel. The present work describes preliminary results for cadmium substitution by using an aluminum-manganese (Al-Mn) system with 25 at.% manganese deposited on 42CrMo4 steel by atmospheric plasma spraying (APS). The prepared alloy powder for the APS process was characterized by laser granulometry and SEM/EDX. A fraction with particle sizes ranging from 10 to 60 µm was used for thermal spray process. The APS coatings were characterized by optical image analysis, Vickers hardness and tested in Standard Salt Spray (SSP). The Al-Mn coatings exhibited significantly higher hardness in comparison to pure aluminum. Red iron corrosion appeared during SSP testing after 240 h but remained moderate even after 1000 h. The Al-Mn material itself showed only a low oxidation rate but worsened adhesion to the substrate if compared to as-coated state. Future work will mainly focus on the material composition to achieve more ductile coatings.

1 Introduction

Cadmium once was the most common coating material if it comes to wet corrosion protection. Because of its many mechanical and electrical advantages such as good lubricity, ductility, brazeability, and conductivity cadmium was used in many industrial applications such as aircraft, offshore components or outdoor locks and fasteners. [1] It is one of the few coating materials which offers sacrificial corrosion protection because of its electronegative potential according to steel. This behaviour still exists when the coating is scratched and the steel substrate is exposed to the atmosphere. [2]
Electroplating as coating process is very simple and available for a wide range of substrate shapes and sizes. However, cadmium is a highly toxic heavy metal which leads to severe health and security issues for workers and environment. It is not only critical in its deposition process like hard chromium plating but in skin contact itself. That means, each task like maintenance or cleaning of any coated piece causes significant risks. [3]
Consequently, cadmium is forbidden under many EU-rules and restricted under REACh (Registration, Evaluation, Authorisation and Restriction of Chemicals) [4]. The only application fields nowadays remain in aircraft and military industry since up to now, no alternative systems correspond to the safety regulations of this sector [5]. In other industries, cadmium is banned completely and common steel is substituted by corrosion resistant alloys so that no protective coating is needed.
Metals with similar electronegativity as cadmium are aluminum and zinc. As a consequence, research is focussed on their alloys to bring solutions up to an industrial level. Coatings are prepared via aqueous and non-aqueous electroplating, vapor deposition (PVD, CVD), vacuum processes, slurry applications, and thermal spraying (arc spray, flame spray, cold spray). Improved coating performance is also possible by embedding aluminum and zinc flakes in polymer or ceramic matrices. [6]
The aircraft industry introduced much effort in an effective electroplated aluminum-manganese (Al-Mn) system that could work with the cadmium plating installations. However, during this process visibly fumes are emitted, representing economical loss and increased expenditure for safety measures in form of a ventilation system, both resulting in higher costs. In addition, many countries restrict facilities which are emitting visibly fumes because of their potential for environmental damage. At the end, this led to the cease of the research program. [7]
Comparative projects dealed with the Al-Mn electrodeposition out of ionic liquids for the corrosion protection of NdFeB magnets. These coatings exhibited adequate features: good adhesion, high hardness and sacrificial corrosion behavior. Furthermore, no danger for workers and the environment was reported. [8] Current investigations deal with manganese contents ranging from 13 at.% to 25 at.% [9].
In order to further decrease health and security issues and costs the Al-Mn system was transferred into thermal spraying in the current approach. A powder with a distinct chemical composition was prepared as feedstock for a thermal spray coating which was then tested towards its corrosion performance.

2 Experimental procedure

2.1 Materials

Flat samples of 42CrMo4 steel (1.7225) were used as substrates. Al-Mn alloy powder containing 25 at.% manganese was gas atomized at Fraunhofer UMSICHT under nitrogen atmosphere using pure aluminium and manganese as starting material. The pure metals were molten in a graphite crucible and superheated to a temperature of nearly 200 °C above the melting point of the alloy prior to atomization. The as-atomized powder was then sieved in a suitable fraction for thermal spray process. Electroplated cadmium coatings were provided by Airbus Helicopters Deutschland GmbH and used as reference for corrosion tests.

2.2 Thermal Spraying

Prior to coating, the steel substrates were degreased and grit blasted with corundum F46 powder with 3 bar and an impact angle between 45 ° and 90 °. The Al-Mn coatings were deposited using a Sulzer Metco APS system with a F4 spray gun. The applied thickness of the coatings was roughly 100 µm.

2.3 Characterization

The particle size distributions of the as-atomized and sieved alloy powder were measured using a Sympatec HELOS Lasergranulometer with dry particle injection. The measurement range of this unit is between 0.5 µm and 350 µm.
Further investigations on the powder have been performed via scanning electron microscopy (SEM; Leo 1450VPSE, Zeiss, Germany) equipped with a X-Max 50 mm² silicon drift detector (Oxford instruments, United Kingdom) for energy dispersive X-ray spectroscopy (EDX).
For the image analysis of the coating cross sections, a digital microscope (Keyence, VHX 2000, Japan) was employed.
A Vickers diamond indenter (Micromet1, Micro Hardness tester, Buehler, USA) was used for the hardness evaluation on polished cross sections of the coated samples. A 50 g load was applied. The measurements were carried out according to ISO 4516 [11].

2.4 Corrosion Testing

The resistance against wet corrosion was investigated by Standard Salt Spray (SSP) testing according to ISO 9227 [10]. The edges of the 10 x 15 cm² coated steel specimens were protected by an appropriate tape. The surface to be characterized was placed vertically in the SSP chamber with a tilt angle of 20 °. The samples were kept in the SSP chamber for 1000 h. The samples were checked every 24 h in order to evaluate the corrosion progress.

3 Results

3.1 Powder

During atomization, the powder particles have been collected in a can at the bottom of the plant and, simultaneously, in a second can under the cyclone where the powder is separated from the atomizing gas. The cyclone fraction (71 wt.% of the collected powder) is considerably finer than the plant fraction, **Fig. 1**. For the APS process, the atomizer fraction was sifted and sieved to obtain a powder particle size fraction ranging from 10 to 60 µm. The powder particles majorly show spherical shape without agglomeration, **Fig. 2**.

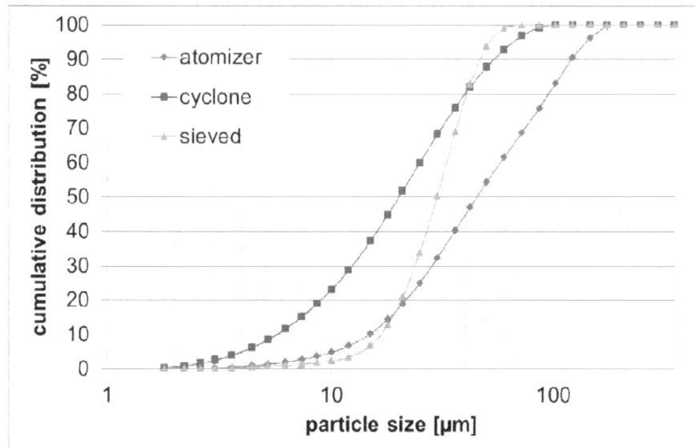

Fig. 1. Particle size distributions of alloy powder.

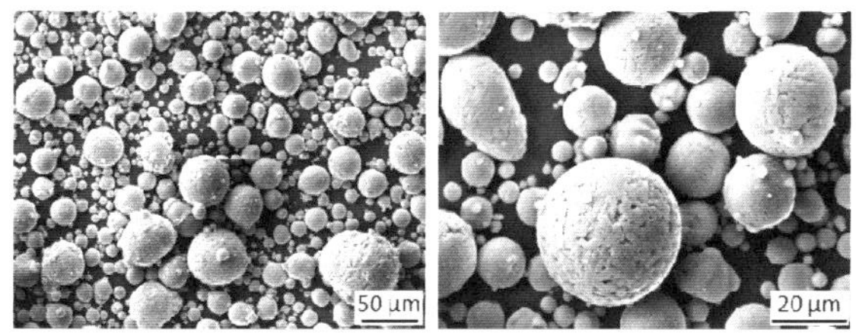

Fig. 2. SEM-pictures of the alloy powder at different magnifications.

The measured chemical composition of the powder corresponds well with the desired percentage of manganese in the alloy, **Table 1**.

Table 1. Chemical composition of the powder (at.%).

	measured ø	deviation σ
Al	73.94	0.48
Mn	26.06	0.48

3.2 Coating

The microstructure of the APS coating exhibits a quite large fraction of pores and oxide phases (alumina), **Fig. 3**. A non-uniform melting behaviour of the powder particles in the plasma and/or inhomogeneities in the spray process which disturb the layer formation may have caused these structural defects. Moreover, micro-cracks in the thermally sprayed coatings can be observed which may have its origin from the brittleness of the alloy. All these factors lower the cohesion within the coating and may induce additional stress that is relieved by the formation of cracks.
The chemical composition of the coatings analysed via EDX correspond to the measured percentages in the powder, **Table 1**.
The Vickers hardness of the APS Al-Mn coating is 501 ± 2 $HV_{0.05}$ (556 ± 55 $HV_{0.05}$ for bulk material) which corresponds to the hardness of electrodeposited Al-Mn coatings reported earlier [12].

In comparison to an aluminum thermal spray coating reference with 30 ± 2 $HV_{0.05}$ (39 ± 1 $HV_{0.05}$ for 99.9 % pure bulk aluminum), the addition of manganese results in a significant increase of the hardness, **Fig. 4**.

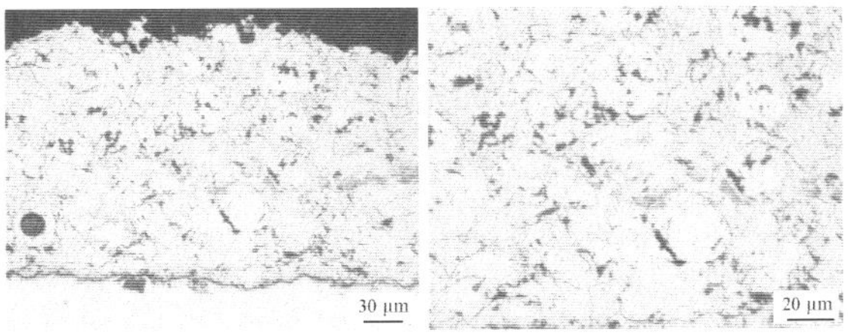

Fig. 3. Microstructure of Al-Mn coating at different magnifications.

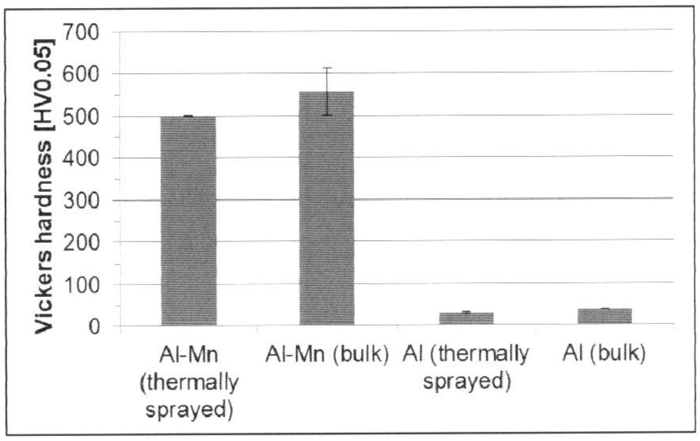

Fig. 4. Vickers hardness of different coating materials.

With regard to the application as cadmium substitute, the hardness is particularly important for wear behaviour. Cadmium itself possesses a very high ductility, also compared to aluminum. Currently, cadmium coatings are therefore chromated in order to raise their rather low resistance against mechanical loads [6]. Creating an alloy layer that combines corrosion and wear protective properties allows the substitution of two hazardous coating systems and by that, lowers the process costs.

3.3 Wet corrosion

After 24 h of SSP testing, a few, but large holes, similar to pitting corrosion were observed on the Al-Mn coating, **Fig. 5**. The blank substrate got visible at these locations. After 48 h dark streaks appeared on the specimen, starting around these holes and leading down from them. The number and size of these holes remained constant during the exposure time, only the streaks worsened. Red rust became visible after 240 h of testing. After 1000 h, locally single bulks with cracks could be observed on the surface of the coating.
The origin of these defects can be related to the structural defects described before as a consequence of the alloy brittleness and imperfections during the spraying process. These local defects allow the penetration of the salt solution through the micro-cracks. As a consequence, the underlying substrate corroded and provoked voluminous oxidation products at the interface that detached parts of the Al-Mn layer.

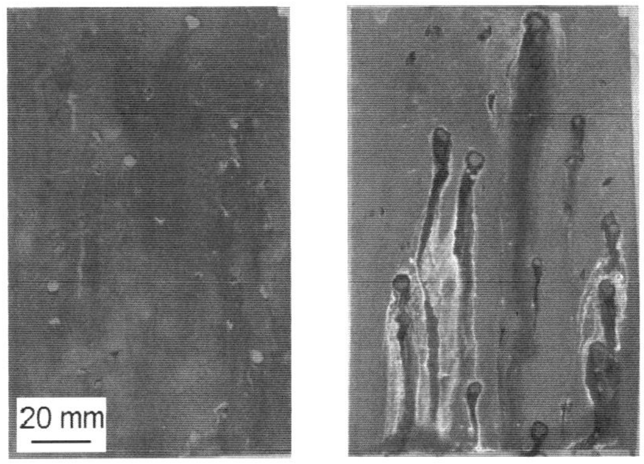

Fig. 5. Al-Mn APS coating after 24 h (left) and 240 h (right) SSP testing.

EDX analysis confirmed that the exposure to the salt spray led to oxidation of the coating surface with preferential reaction of aluminum to alumina. The presence of manganese or contaminations of iron explains the darkening of regions. According to the EDX analysis, the substrate surface that is visible through the pittings is covered with oxides of potassium, chromium and iron.
Cross sections before and after SSP testing were prepared in order to further examine the corrosion attack, **Fig. 6**. Originally, the coating thickness was not homogeneous and varied between 100 µm and approximately 150 µm. After SSP testing the thickness was still approximately 100 µm. However, its adhesion to the substrate was affected significantly as can be seen by the clear horizontal crack at the interface which might be explained due to the lower ductility of the Al-Mn alloy. Its surface is slightly oxidized.

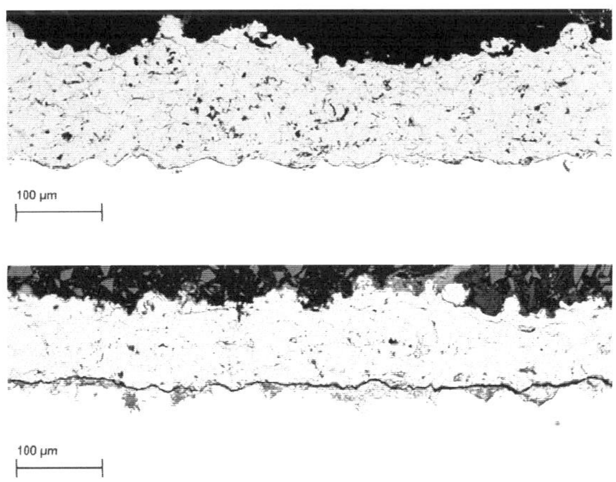

Fig. 6. Cross sections of the tested Al-Mn coatings before (upper) and after SSP testing (lower SEM photo).

4 Conclusion

Al-Mn powder with 25 at.% manganese has been produced by nitrogen gas atomization. The chemical analysis showed a good agreement with the theoretical Al-Mn ratio. The particle size range between 10 and 60 µm was used for coating manufacturing via thermal spraying. The main results obtained in this work are:

- The coating hardness was significantly increased by the addition of manganese which fulfills the criterium of enhanced wear resistance.
- Due to the brittle nature of the Al-Mn alloy the APS coatings exhibited a high portion of pores and micro-cracks.
- The coating itself showed only moderate corrosive attack under SSP. Red rust first appeared after 240 h and was located at coating defects.
- After 1000 h SSP testing the investigated coating locally detached from the substrate.
- The corrosion attack obviously starts at coating defects such as micro-cracks. The origin of these defects can be explained by the brittleness of the alloy material itself, a non-uniform melting behavior of the powder and/or inhomogeneities in the spraying process.

As a consequence, the Al-Mn system shows good properties in respect to the hardness increase and corrosion stability. However, the coating structure needs to be improved. The points identified for potential optimization are the ductility of the Al-Mn composition and the thermal spray process parameters.
A higher ductility might be achieved by lowering the manganese content whereas adapted spray parameters may prevent structural defects due to a better melting and a more homogeneous deposition of the powder particles. The optimization of these points represents a high potential for an adequate substitution of cadmium by a thermally sprayed Al Mn coating system.

5 Literature

[1] Brown, C.J., Smith, M.W., Zeng, L., Haylock, L., Gurrola, B. and Youngblood, D.: Evaluation of Alternatives to Electrodeposited Cadmium for Threaded Fasteners Applications. SAE Interantional (2009), pp. 1/13.

[2] Press release MVL (2007), available under http://www.mvl-lampertheim.de/documents/Metall_Veredlung_Lampertheim_PresseInfo_Cadmium.pdf, accessed on: 17.02.2016..

[3] Agüero, A., del Hoyo, J.C., García de Blas, J., García, M., Gutiérrez, M., Madueño, L. and Ulargui, S.: Aluminum slurry coatings to replace cadmium for aeronautic applications. Surface and Coatings Technology 213 (2012), pp. 229/238.

[4] http://www.reach-info.de/, accessed on: 01.12.2015.

[5] Fasuba, O.A., Yerokhin, A., Matthews, A. and Leyland, A.: Corrosion behaviour and galvanic coupling with steel of Al-based coating alternatives to electroplated cadmium. Materials Chemistry and Physics 141 (2013), Issue 1, pp. 128/137.

[6] Legg, K.: Choosing Cadmium Plate Alternative, at: AeroMat (2002), Orlando, available under http://www.rowantechnology.com, accessed on: 03.02.2016.

[7] Naval Air Systems Command, Materials Engineering Division, Final Report on Aluminum Manganese Molten Salt Plating WP 9903 (2006).

[8] Chen, J., Xu, B. and Ling, G.: Amorphous Al-Mn coating on NdFeB magnets: Electrodeposition form $AlCl_3$-EMIC-$MnCl_2$ ionic liquid and its corrosion behavior. Materials Chemistry and Physics 134 (2012), Issues 2-3, pp. 1067/1071.

[9] Bielewski, M.: Replacing cadmium and chromium, in: RTO-AG-AVT-140 – Corrosion Fatigue and Environmentally Assisted Cracking in Aging Military Vehicles, RTO/NATO, 2011, pp. 315/336.

[10] ISO 9227:2012, Corrosion tests in artificial atmospheres - Salt spray tests.

[11] ISO 4516:2002, Metallic and other inorganic coatings - Vickers and Knoop microhardness tests.

[7] J. Ding, B. Xu and G. Ling, Al-Mn coating electrodeposited from ionic liquid on NdFeB magnet with high hardness and corrosion resistance, Applied Surface Science 305 (2014), pp. 3097313.

Effect of chemical surface treatment of polymer composites on the formation of erosion-resistant coatings

Axelle Elrikh[1,2], Morgane Dolhen[1], Simon Goutier[1], Gordon Armstrong[2], Armelle Vardelle[1]
[1]Univ. Limoges, CNRS, ENSCI, SPCTS, UMR 7315, F-87000 Limoges, France
[2]Materials & Surface Science Institute, University of Limerick, Co. Limerick, Ireland
axelle.elrikh@unilim.fr

Carbon fiber reinforced composites (CFRP) exhibit light weight, high stiffness and strength along the direction of reinforcement. They also have good corrosion resistance and low thermal expansion coefficient, compared to metals. However, they are more sensitive to the impact of solid particles (e.g. sand particles) and droplets (e.g. rain droplets) which have a critical impact on their reliability in service. Various solutions are already used to prevent erosion of CFRPs (e.g. metallic caps, organic paints, tapes), but these do not demonstrate sufficient durability and noticeably increase the components weight. Solutions under development involve sol-gel and thermal sprayed coatings. However, the polymer matrix's low surface energy and low chemical reactivity proved to be an issue, when bonding the metallic or ceramic coating to the polymer surface.
Previous works carried out in this laboratory have shown that a way to improve coating deposition was to remove the superficial layer of polymer to favor the coating formation on the carbon fibers. This can be done by using a chemical etching that removes the superficial resin. This study aims at understanding the effects of this chemical etching on the composite surface and on the formation of plasma-sprayed ceramic coating on the composite.

1 Introduction

Carbon fiber reinforced polymers, also known as CFRPs, are today a material of choice to replace aluminum and titanium alloys in the aerospace industry, for parts such as stabilizers, engine body, inlet fans and valves, but also wings and fuselage. Indeed, thanks to their low density (1600kg/m^3) and improved fatigue resistance [1], CFRPs are now widely used in aircraft applications. Today, aircraft such as the Airbus A350 and Boeing 787 have a composite content higher than 50wt% [2-3] and a wet surface ratio higher than 90%. This comes along with a significant reduction in the aircraft's mass and the amount of fuel that needs to be carried. However, CFRPs show poor resistance to erosion damage caused by rain droplets or particles impacting on the aircraft during flight [4].

Current approaches to erosion protection involve sacrificial layers of resin added during composite manufacturing, epoxy tapes, and metallic caps. A more lasting solution would be to apply an erosion-resistant coating on the composite surface. Numerous studies have been conducted on applying such coatings using various surface technologies, such as electrodeposition [5,6], physical vapor deposition (PVD) [7,8], cold spray [9] and thermal spray [10–13]. Among these techniques, plasma spraying could be a suitable coating process as it is already used in the aircraft industry and makes it possible to deposit ceramic wear-resistant coatings, such as alumina.

However, previous works in this laboratory highlighted the difficulty to obtain coatings with an acceptable adhesion strength on CFRPs [14], because of the intrinsic properties of the polymer matrix, i.e. low surface energy and hydrophobicity [15]. To overcome this problem, coatings on polymer composites have been applied, using a composite lay-up modified to add a metallic tie-layer during composite manufacturing [16] though such modified composite would require certification to be used in the aerospace industry. Recent works have also shown it is possible to use a metal bond coat prior to ceramic coating deposition to improve the adhesion between the ceramic coating and polymer matrix [17-19]. However this approach makes the coating system deposition trickier; it also increases the number of interfaces and associated risks of failure. Another approach could be to use a specific surface preparation prior to coating deposition.

Most of the plasma spray deposition procedures involve the roughening of the surface by grit-blasting. Nevertheless, grit blasting breaks the carbon fibers and delaminates the composite plies, causing irreversible damage to the composite [8]. Another surface preparation method, recognized as a valuable tool to improve the adhesion properties of polymers, consists of adjusting the surface free energy by introducing functional groups (e.g.; hydroxyl, carboxylic acid groups) at the polymer surface. However, this surface functionalization is generally time limited, and the covalent bonds between the surface and functional groups could be broken if a substrate preheating stage is used prior to coating deposition and also during the plasma spray process by the heat input by the plasma jet and impacting particles. A final possible CFRP surface preparation is chemical etching that can selectively remove the excess of polymer at the surface [17–19] by dissolving or degrading it depending on the etchant used; it also increases the surface roughness without mechanically damaging the carbon fibers.

The main aim of this study is, therefore, to use chemical etching to remove the problematic epoxy layer from the CFRP surface, and make it suitable for plasma spraying of a ceramic coating. The effect of etching time on polymer top layer removal, surface roughness changes and chemistry was investigated.

2 Research Strategy

The excess of epoxy on the composite surface, preventing the adhesion of inorganic coatings to the CFRP surface, will be first removed by a chemical etchant. Therefore, a preliminary step consisted of testing a range of etchants on the composite used in this work. These etchants were chosen from commonly used etchants, and etchants whose successful use on composite substrates had been reported in literature. Alkaline solutions [17] and phenol based solution [19] did not effectively removed the epoxy surface excess, while sulfuric acid [18] was found to be effective when used as a concentrated solution. This etchant was thus chosen and the composite samples were etched with this acid solution for a range of times.

Epoxy removal was observed by optical and scanning electron microscopy. The CFRP's surface profiles and roughness were measured before and after etching. Also water contact angle measurements were conducted to investigate the possible changes in surface wettability brought about by etching.

Finally, an alumina coating was plasma sprayed on the etched composite. Alumina was chosen because of its low density and general good mechanical properties for erosion protection [20].

3 Materials and Methods

3.1 Composite material

The composite material consisted of a Cycom 890® epoxy matrix reinforced with 4 plies of unidirectional PAN-coated carbon fibers (CF), arranged in 0/90 layup, as shown in **Fig. 1**. The sheet, 2-mm thick, was manufactured using resin transfer molding. The composite service temperature is 200°C, which is the polymer matrix service temperature [21]. Plates of 10 x 15mm were cut using a diamond saw for etching tests and plasma spraying.

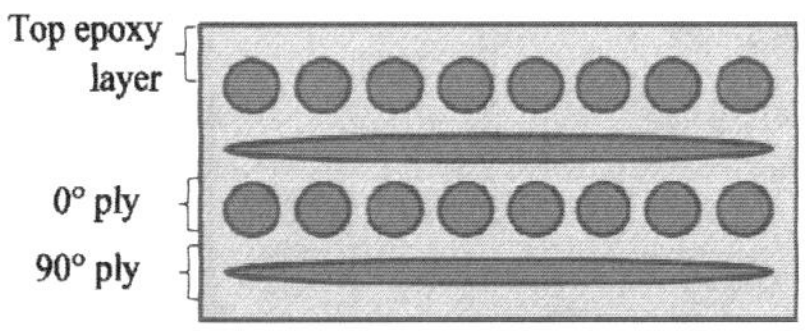

Fig. 1. Fiber orientation and crimping weave. The drawing shows the epoxy top layer.

3.1 Chemical etching

The chemical etchant was H_2SO_4, 95-98% (Sigma Aldrich). Before etching, the composite plates were successively cleaned in acetone, ethanol and distilled water ultrasonic baths. Etchings were conducted at ambient temperature, by dipping the composite surface in acid for the determined etching time. Once etched, samples were rinsed with distilled water and put in an ultrasonic bath to remove the epoxy residues and finally left to dry in air.

The etching times were set by first determining the time needed to damage the composite. The latter showed fiber and matrix delamination when etched with the sulfuric acid for 15min; therefore, shorter etching times (5 and 10 min, respectively) were selected to avoid irreversible damage to composite.

3.2 Air Plasma Spraying

To study the effect of plasma-sprayed particle parameters (velocity and temperature) on alumina coating deposition, two sets of plasma parameters were used. Set 1 led to a high-enthalpy plasma and fully melted particles while Set 2 led to a lower plasma enthalpy, but higher particle velocities. The spraying parameters are summarized in **Table 1**.

Before spraying, the composite samples were pre-heated at 190°C during 10min to allow the surface desorption of adsorbates and condensates. This temperature was then maintained during coating deposition. Particle speed and temperature were measured at the substrate standoff distance using the SprayWatch diagnostics system (Oseir®).

Table 1. Alumina coating spraying parameters

	Set 1	**Set 2**
Plasma gas flow rate (L/min)	Ar: 33 H_2: 10	Ar: 70 H_2: 10
Plasma specific enthalpy (MJ/kg)	12	9
Particle speed (m/s)	200	450
Particle temperature (°C)	2600	2100
Arc Current (A)	600	
Spraying distance (mm)	80	
Ar carrier gas flow rate (L/min)	30.5	
Powder feed rate (g/min)	24	
Substrate temperature (°C)	190	
Number of passes	30	
Sample rotation speed (mm/s)	300	
Sample translation speed (mm/s)	100	

3.3 Sample Characterization

3.3.1 Surface morphology

Etched composites were first observed using an optical microscope (Nikon Eclipse LV) to detect the change in the epoxy outer layer and carbon fibers. Multiple surface images were taken then binarized to distinguish between the exposed carbon fibers and remaining epoxy. The percentage of exposed carbon fibers on the etched composite surface was measured using the Matrox Calculator software v.4.1.

The etched samples were then observed using a scanning electron microscope (SEM) (Phillips XL30) with a voltage of 15kV and secondary electron detection, after Au-Pt surface metallization. A coat of Ag paint was put on the top of some samples to ease SEM observation.

3.3.2 Surface roughness

Surface profiles and roughness of etched samples were measured using a roughness sensor (Surtronic 25+) according to the ISO 3274 standard. Results were treated with the TalyMap Silver software v.4.1. For each sample, the surface profiles and roughness were measured five times with a sampling length of 8mm. Two parameters were investigated:
- R_a, the arithmetic average roughness,
- and R_z, the average distance between the highest peak and the deepest valley in five sampling lengths.

3.3.3 Composite wettability

The composite wettability was quantified by using the static sessile drop method. A water droplet with a volume of 10µL was deposited on the composite surface and pictures of the droplet were captured with a CCD camera whose resolution was 2560 x 1920 pixels associated with an objective lens whose focal length was 91mm. Contact angles were measured using a custom software.

4 Results and discussions

4.1 Surface morphology

Both the 5 and 10-min etched samples showed a significantly amount of exposed carbon fibers (CF) compared to the as-received CFRP. In the case of the 10-min etched composite, some fibers were delaminated from the surface and peeled off from the sample. Optical micrographs of the CFRP surface are shown in **Fig. 2** and the percentage of exposed fibers for each sample is given in **Table 2**.

As expected, the 10-min etched sample showed more exposed CF than the 5-min one. The peeling fibers suggested that etching damaged the carbon fiber/epoxy matrix interface. This interface allows the load transfer from the brittle epoxy to the mechanically-stronger fibers and is responsible for the good mechanical properties of CFRPs. Damaging this interface prior to plasma spraying will cause the fibers to delaminate during coating deposition, and the plasma-sprayed coating would not adhere to the substrate.

Etching the composite resulted in a slight increase in R_a for all etched samples while the increase in R_z was more significant (**Table 2**). The partial removal of epoxy, with regions of residual epoxy remaining on the surface, was responsible for the roughness peak increase. 5-min and 10-min etched samples did not show appreciable roughness differences as the etchant began to attack the carbon fiber/matrix interface. Although this interface was degraded, the surface profile of the CF was largely unaffected, and so the measured surface roughness changed little.

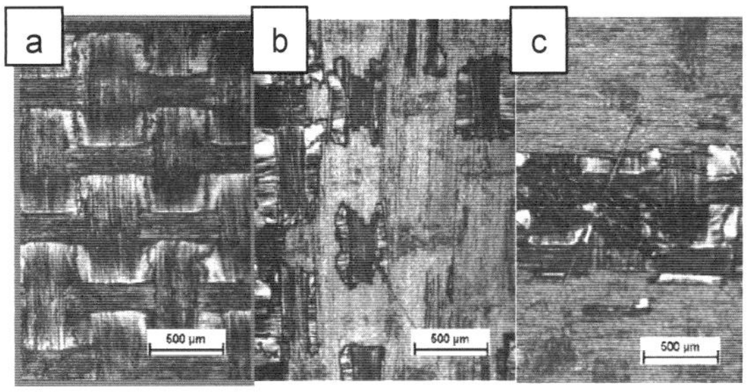

Fig. 2. Surface morphology of as-received composite (a); after 5-min etching (b); and 10-min etching (c).

Table 2. Percentage of exposed carbon fibers (CF) and surface roughness after etching according to etching time

Etching time	% of exposed CF	R_a (µm)	R_z (µm)
As-received	0	3.2	13.7
5 min	26	4.0	20.0
10 min	70	4.3	22.2

Fig. 3 shows a SEM picture of the carbon fiber/matrix interface. **Fig. 3.a** is a picture of the as-received composite surface. The epoxy surface layer showed a waviness caused by the underlying presence of carbon fibers. Microcracks were present at the epoxy surface. The 5-min etched sample (**Fig. 3.b)** showed sharp-edged zones of remaining epoxy on the top of the fibers, with fibers still embedded within the polymer matrix. 10-min etched samples presented cracks between matrix and fibers, and delamination of the first layer of carbon fibers (**Fig. 3.c**). Fewer epoxy-rich zones remained on the surface. Finally, on both samples, the surface of the individual carbon fibers surface did not appear to be mechanically damaged by the etching process.

As explained in section 4.1, the CF/matrix interface plays a major role on composite properties. The presence of delaminated fibers, observed in **Fig. 3.c**, showed that the CF/matrix interface in 10-min etched composites were too damaged such that these samples were no longer suitable for plasma spraying.

Hence, to avoid CF/matrix interface damage when etching the composite, a compromise must be found:
- The epoxy excess must be removed to expose the carbon fibers that will allow the formation of the plasma-sprayed coating,
- But the matrix/fibers interface must be preserved to maintain the composite mechanical properties.

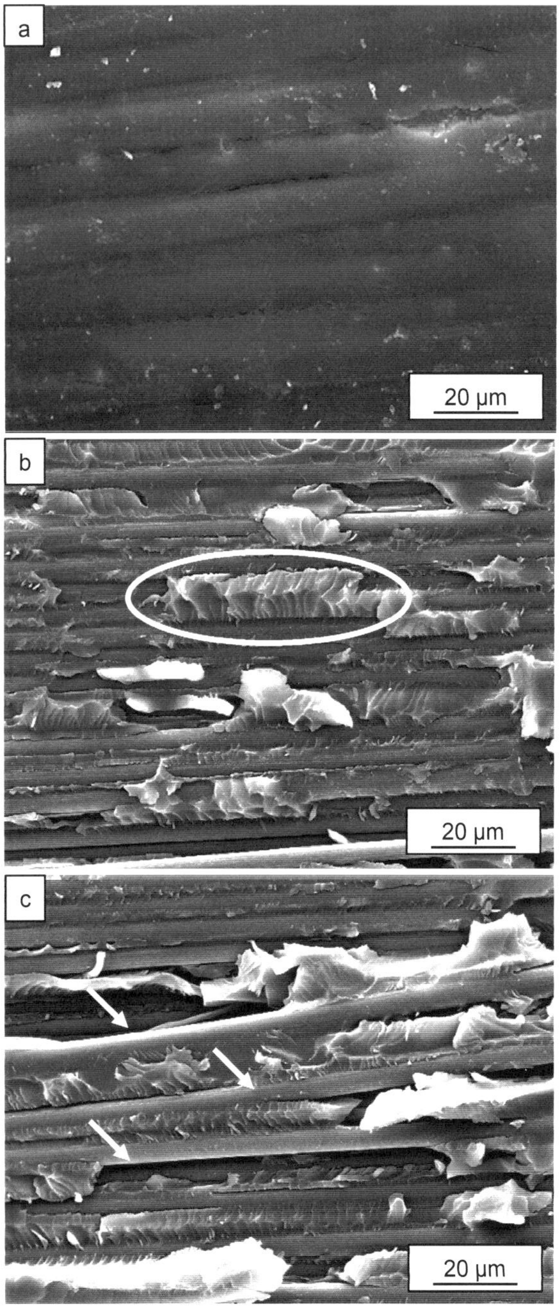

Fig. 3. SEM observation of the surface of the (a) as-received sample, (b) 5-min etched sample with epoxy rich zones (circled), (c) 10-min etched sample with delaminated fibers (highlighted).

4.2 Wettability

Wettability experiments were conducted on the as-received and etched samples. The measured contact angles are presented in **Table 3**.

The water droplet was dewetted on as-received composite that had a top-layer of epoxy (Fig. 1). As this epoxy was removed by the etchant, the water contact angle significantly decreased as the droplet was mostly standing on carbon fibers. The more carbon fibers were exposed, the more the water droplet wetted the CFRP substrate.

The improved wettability of etched surfaces should result in a better splat formation during air plasma spraying: instead of dewetting and scattering into multiple droplets, the molten particle is expected to flatten and cool in the form of a splat on the surface.

Table 3. Water droplet contact angles on as-received and etched composites

Sample	Contact angle (°)
As received	81.5
5-min etched	60.5
10-min etched	47.5

4.3 Air plasma sprayed alumina coating

Air plasma spraying was first conducted on the 5-min etched samples using the Set 1 plasma parameters, which are commonly used to spray ceramic materials. The long enough residence time of particles within the plasma jet combined to a high-enthalpy plasma jet allowed the particles to fully melt before impacting on the surface.

The composite surface before spraying is shown in **Fig. 4.a**. The remaining zones of epoxy top layer after spraying using Set 1 plasma parameters are visible in **Fig. 4.b**.

A very heterogeneous coating was achieved on this sample. Clusters of alumina formed on the carbon fibers and edges of the regions with residual epoxy. The sharp edges of the latter acted as mechanical anchors. The remaining epoxy did not allow the growth of a continuous coating, and only a few alumina clusters, anchoring on the surface roughness, were observed on the composite surface. The impacting droplets were dewetted by the epoxy matrix which, therefore, must be completely removed to promote homogenous coating formation. To do this, a second set of spray parameters was designed to obtain partially melted particles but with higher kinetic energy at impact (Set 2, **Table 1**). The idea was to use the impacting particles as high temperature grit blasting to remove the epoxy remaining on the etched surface.

Fig. 4.c shows the 5-min etched samples after spraying with the Set 2 plasma parameters. The impacting alumina particles successfully removed the remaining epoxy from the CFRP surface.

It is proposed that the partially melted particles obtained under Set 2 plasma parameters deformed at impact on the substrate, absorbing a part of the impact energy.

underlying CFRP. Although some breaking of carbon fibers was observed, the alumina particles achieved a "softer" grit blasting and caused less damage to the carbon fibers compared with the conventional grit blasting process that uses solid alumina particles. Thus, it was decided to conduct the alumina coating deposition in two stages: first using the Set 2 plasma parameters as a surface preparation step and second using the Set 1 plasma parameters to apply the alumina coating.

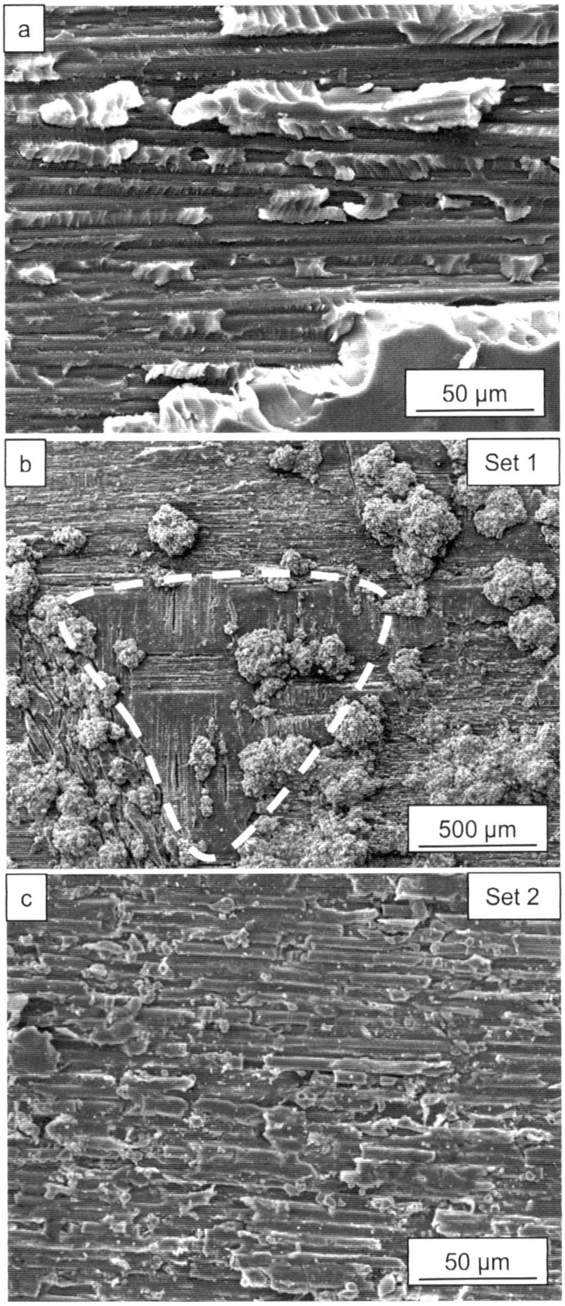

Fig. 4. (a) Surface of a 5-min etched sample before spraying. Residual epoxy may be observed on the underlying carbon fiber exposed during etching. (b) Surface of 5-min etched sample after air plasma spraying using Set 1 parameters. Alumina clusters did not adhere to the epoxy rich zones (highlighted) remaining after etching. (c) Surface of 5-min etched sample after high-temperature blasting using Set 2 parameters.

The heat input to the substrate by the impacting and cooling down particles also burned the epoxy remaining after etching at the composite surface. However, they did not cause internal damage to the

Fig. 5 shows the coating achieved with this 2-stage plasma spray procedure on the 5-min etched sample. A more homogeneous coating is observed, without the formation of clusters. A few carbon fibers, broken during the first stage conducted with the high-velocity partially melted particles, are embedded within the coating.

The combination of both sets of parameters, with a preliminary epoxy removal using chemical etching, allowed the formation of a continuous coating (about 50µm thick) on the composite surface. Optimizing the spraying parameters of both Set 1 and Set 2 plasma parameters should help to minimize the breaking of carbon fibers, and is the object of current work.

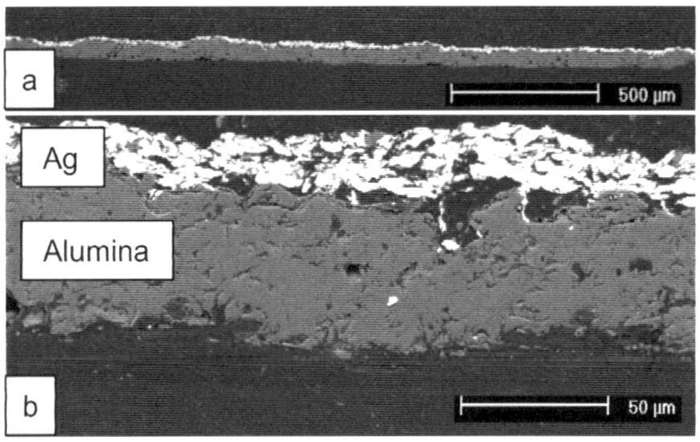

Fig. 5. Alumina coating achieved with the two-stage plasma spray procedure. (a) coating-substrate overview and (b) coating-substrate interface.

5 Conclusion

The aim of this study was to improve the surface properties of carbon fiber reinforced polymers (CFRP) to allow the deposition of a ceramic erosion-resistant coating using air plasma spraying (APS). To avoid the mechanical damage induced by grit blasting prior to plasma coating deposition, chemical etching was first used to remove the majority of the superficial epoxy, responsible for the poor adhesion of ceramic coatings.

Chemical etching proved to be an efficient way to remove the top layer of epoxy but too long etching times (10min) caused unacceptable composite damage, especially at the carbon fiber/matrix interface. A shorter etching time (5min) was thus used for sample surface preparation.

However, the remaining regions of epoxy on the etched surface prevented the formation of a homogeneous alumina coating. Clusters of alumina were observed on sharp epoxy edges and exposed carbon fibers. Therefore, a different set of plasma spray parameters was used; it yielded high velocity and partially melted particles at impact making it possible to remove the epoxy left on the surface after the chemical etching but it also resulted in the breaking of some carbon fibers. Finally a surface sample preparation combining chemical etching and plasma spraying with high-velocity and partially molten alumina particles was used and then permitted to achieve a continuous 50µm thick alumina coating.

The work-in-progress deals with the optimization of the plasma spray parameters to ensure the composite's mechanical integrity. It is also concerned with the investigation of the effect of the chemical etching on the upper surface chemistry of the composite and, in particular, how it affects its wettability.

6 Acknowledgements

The authors would like to acknowledge the support of the Ulysses program, funded jointly by the Irish Research Council and the French Ministry of Foreign Affairs, and of the French Embassy in Ireland.

7 Literature

[1] C. Soutis, Carbon fiber reinforced plastics in aircraft construction, Materials Science and Engineering: A, Issue 412 (2005), pp. 171–176.

[2] G. Hellard, Composites in Airbus, (2008).

[3] J. Hale, AERO - Boeing 787 from the Ground Up, (2006).

[4] K.V. Pool, C.K.H. Dharan, I. Finnie, Erosive wear of composite materials, Wear, Issue 107 (1986), pp. 1–12.

[5] P. Lammel, A.H. Whitehead, H. Simunkova, O. Rohr, B. Gollas, Droplet erosion performance of composite materials electroplated with a hard metal layer, Wear, Issue 271 (2011), pp. 1341–1348.

[6] P. Lammel, L.D. Rafailovic, M. Kolb, K. Pohl, A.H. Whitehead, G. Grundmeier, et al., Analysis of rain erosion resistance of electroplated nickel–tungsten alloy coatings, Surface and Coatings Technology, Issue 206 (2012), pp. 2545–2551.

[7] C. Maurer, U. Schulz, Solid particle erosion of thick PVD coatings on CFRP, Wear, Issue 317 (2014), pp. 246–253.

[8] A. Elrikh, S. Goutier, R. Bignon, G. Armstrong, A. Vardelle, Improving the adhesion of wear resistant coatings on aerospace polymer composites, International Thermal Spray Conference Proceedings, Long Beach (CA), USA, (2015).

[9] F. Alonso, I. Fagoaga, P. Oregui, Erosion protection of carbon—epoxy composites by plasma-sprayed coatings, Surface and Coatings Technology, Issue 49 (1991), pp. 482–488. [10] M. Ivosevic, R. Knight, S.R. Kalidindi, G.R. Palmese, J.K. Sutter, Adhesive/cohesive properties of thermally sprayed functionally graded coatings for polymer matrix composites, Journal of Thermal Spray Technology, Issue 14 (2005), pp. 45–51.

[11] A. Liu, M. Guo, M. Zhao, H. Ma, S. Hu, Arc sprayed erosion-resistant coating for carbon fiber reinforced polymer matrix composite substrates, Surface and Coatings Technology, Issue 200 (2006), pp. 3073–3077.

[12] S. Guanhong, H. Xiaodong, J. Jiuxing, S. Yue, Parametric study of Al and Al2O3 ceramic coatings deposited by air plasma spray onto polymer substrate, Applied Surface Science, Issue 257 (2011), pp. 7864–7870.

[13] A. Liu, M. Guo, J. Gao, M. Zhao, Influence of bond coat on shear adhesion strength of erosion and thermal resistant coating for carbon fiber reinforced thermosetting polyimide, Surface and Coatings Technology, Issue 201 (2006), pp. 2696–2700.

[14] A. Elrikh, S. Goutier, G. Armstrong, A. Vardelle, Droplets Impacts on Carbon Fiber Reinforced Polymers in Plasma Spray Conditions, RIPT, Limoges, France (2015).

[15] F. Awaja, M. Gilbert, G. Kelly, B. Fox, P.J. Pigram, Adhesion of polymers, Progress in Polymer Science, Issue 34 (2009), pp. 948–968.

[16] F. Robitaille, M. Yandouzi, S. Hind, B. Jodoin, Metallic coating of aerospace carbon/epoxy composites by the pulsed gas dynamic spraying process, Surface and Coatings Technology, Issue 203 (2009), pp. 2954–2960.

[17] X. Roizard, M. Wery, J. Kirmann, Effects of alkaline etching on the surface roughness of a fibre-reinforced epoxy composite, Composite Structures, Issue 56 (2002), pp. 223–228.

[18] V. Prysiazhnyi, M. Stupavská, J. Ráheľ, C. Kleber, M. Černák, L.D. Rafailović, A comparison of chemical and atmospheric plasma assisted copper plating on carbon fiber reinforced epoxy polymer surfaces, Surface and Coatings Technology, Issue 258 (2014), pp. 1082–1089.

[19] J.L. Graham, T. Yamada, Understanding the science behind how methylene chloride/phenolic chemical paint stippers remove coatings (2011).

[20] D.I. Pantelis, P. Psyllaki, N. Alexopoulos, Tribological behaviour of plasma-sprayed Al_2O_3 coatings under severe wear conditions, Wear, Issue 237 (2000), pp. 197–204.

[21] Cycom 890 RTM Technical Data Sheet, Cytec Solvey Group (2014).

Erosion Behavior of Gadolinium Zirconate/YSZ Multi-Layered Thermal Barrier Coatings Deposited by Suspension Plasma Spray

Satyapal Mahade[1*], Nicholas Curry[2], Stefan Björklund[1], Nicolaie Markocsan[1], Per Nylén[1], Robert Vaßen[3]

[1]Department of Engineering Science, University West, Sweden
[2]Treibacher Industrie AG, Austria
[3]Institute of Energy Research (IEF-1), Forschungszentrum Jülich, Germany

Abstract: Yttria stabilized zirconia (8YSZ) is the standard ceramic material for thermal barrier coating (TBC) applications. However, above 1200°C, it has limitations such as poor sintering resistance & susceptibility to CMAS (Calcium Magnesium Alumino Silicates) degradation. Gadolinium zirconate (GZ) is considered as one of the promising top coat candidates for TBC applications at high temperature (>1200 °C) due to its lower thermal conductivity, good sintering resistance and CMAS infiltration resistance. Single layer 8YSZ, double layer GZ/YSZ and triple layer GZdense/GZ/YSZ TBCs were deposited by suspension plasma spray (SPS). Microstructural analysis was carried out by SEM (scanning electron microscopy). Phase analysis of as sprayed TBCs was carried out using XRD (X ray diffraction). The as sprayed multi-layered TBCs were subjected to erosion test at room temperature and their erosion resistance was compared with single layer 8YSZ. It was observed that the erosion resistance of 8YSZ TBC was higher than GZ/YSZ multi-layered TBCs at room temperature. Among the multi-layered TBCs, triple layer TBC was slightly better than double layer in terms of erosion resistance.

1 Introduction

Thermal barrier coatings (TBC) are used in hot sections of gas turbine engine components for insulating the underlying metallic material from being exposed to higher temperatures. YSZ is the standard ceramic material used for TBC applications due to its excellent thermal barrier properties. Efficiency of a gas turbine can be increased by increasing the operating temperature [1]. However, above 1200 °C, YSZ undergoes significant sintering, undesirable phase transformation and CMAS (calcium magnesium alumino silicates) degradation [2-6]. New ceramic materials such as pyrochlores offer excellent thermal barrier properties [7]. Among the pyrochlores, gadolinium zirconate (GZ) is an interesting choice as it is resistant to CMAS attack [8]. However, GZ has a poor thermochemical compatibility with the thermally grown oxide as it has a tendency to react with alumina to form $GdAlO_3$ [9]. This results in the consumption of protective alumina layer and leads to lowering of thermal cyclic life of single layer GZ TBC [10]. Therefore, double layered gadolinium zirconate/YSZ TBCs are widely investigated [10-12].

In TBCs, better erosion resistance of the ceramic top coat is highly desirable as it has a direct influence on the TBC performance. Recently, Nicholls et al demonstrated that EB-PVD (Electron Beam Physical Vapor Deposition) deposited TBCs possess better erosion resistance compared to APS (Atmospheric Plasma Spray) due to their columnar microstructure [13-14]. However, EB-PVD is a comparatively expensive process with lower deposition rates than APS. Suspension plasma spray (SPS), which is a modification of APS process, makes it possible to deposit columnar coatings [11,15] coupled with higher deposition rates than EB-PVD process.

So far, very few attempts have been made to evaluate the erosion resistance of SPS deposited TBCs with columnar microstructure. In this study, single layer 8YSZ along with GZ/YSZ double layer TBC were deposited by SPS. Additionally, a triple layer TBC comprising relatively denser 30 µm thick GZ layer on top of GZ/YSZ TBC was deposited by SPS in order to impart a better erosion and CMAS attack resistance. These GZ/YSZ multi-layered TBCs were evaluated for their erosion resistance at room temperature and compared with 8YSZ single layer TBC.

2 Experimental Details

2.1 Specimen preparation

Stainless steel substrates of 50mm x 50mm x 3 mm dimensions were grit blasted using alumina particles of 220 µm grit size to create a roughness profile of 3 µm on R_a scale. AMDRY 386-2 of NiCrAlY composition was deposited on the grit blasted substrates using the HVAF (High velocity air fuel) process (M3 gun, UniqueCoat, Virginia USA) . For this study, three different suspensions were provided by Treibacher Industrie AG, Austria. First suspension was 8YSZ with a median particle size of 500 nm suspended in alcohol with a solid load of 25 wt%. Second suspension comprised of GZ suspended in alcohol with a median particle size of 500 nm and solid load of the suspension being 25 wt%. Third suspension was GZ with a median particle size D_{50} of 1 µm and solid load content in the suspension being 40 wt%.

Axial suspension plasma spray process was used to deposit the coatings using Axial III plasma gun and Nanofeed 350 suspension feeding system (Northwest Mettech Corp. Vancouver, Canada). Details of the spray process are described in previous work [11]. Two different spray parameter sets were chosen to deposit the TBCs. First parameter set was meant to create a columnar microstructure whereas the second parameter set was chosen to obtain a denser microstructure, as reported in Table. I.

Table. I Spray parameters used for depositing different layers

Parameter	Columnar layer	Dense layer
Solid load (wt %)	25	40
Particle size (nm)	500	1000
Power (kW)	122	89
Standoff distance (mm)	100	70
Enthalypy (kJ)	11.2	12.5

Three variations in the ceramic coatings were sprayed, as shown in Fig.1.

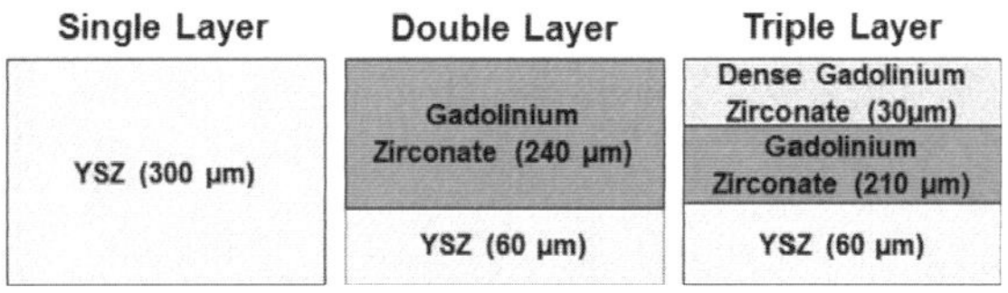

Fig. 1. Different TBC architectures deposited by SPS

2.2 Coating characterization

The cross section & top surface of as sprayed TBCs were analyzed by scanning electron microscopy (Hitachi TM 3000, Japan) in back scattered electron mode. Twenty five different SEM micrographs at low magnification (250 X) were used to measure the thickness of as sprayed TBCS. The phase composition of top surface of as sprayed TBCs was analyzed using Seifert TT 3003 X Ray Diffractometer. Porosity measurements were done using water intrusion method. Details of the procedure are discussed in our previous work [11]. Erosion test of as sprayed TBCs at room temperature was performed at Forschungszentrum Jülich, Germany (IEF-1) according to GE specification no. ES0TF121. Aluminum oxide particles of 240 # grit size were used as the eroding media. The test specimen were held at an angle of 20 ± 2 degrees to the axis of blast nozzle and the thickness loss at the end of 90 seconds was measured using a 0.250 inch diameter ball tip micrometer. After the test, time to spray the aluminum oxide powder (seconds) and depth of erosion (mils) was measured and the erosivity number (seconds/mils) of the as sprayed TBCs was obtained.

3 Results & Discussion

3.1 Microstructure analysis

The as sprayed TBC microstructures were analyzed by SEM. Single layer 8YSZ TBC had a columnar microstructure, as seen in Fig. 2(a). Thickness of YSZ layer was found to be approximately 300 μm. The top view surface morphology showed microstructure resembling a cauliflower, as seen in Fig. 2(b).

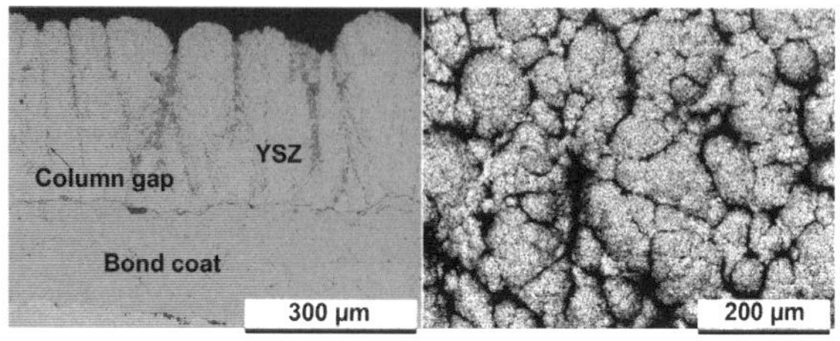

Fig. 2. SEM micrographs of single layer 8YSZ (a) cross section (b) top surface morphology

The double layer GZ/YSZ TBC also showed a columnar microstructure, as seen in Fig 3(a). Thickness of YSZ layer was found to be approximately 60 μm whereas the GZ layer had a thickness of approximately 240 μm. The total thickness of ceramic layer was approximately 300±10 μm. The top surface morphology of the double layer TBC showed a cauliflower look alike microstructure, as seen in Fig. 3(b).

Fig. 3. SEM micrographs of double layer GZ/YSZ TBC (a) cross section (b) top surface morphology

In the case of triple layer TBC sample, as shown in Fig. 4(a), a denser GZ layer of approximately 30 μm thickness could be seen on top of the relatively porous GZ layer of 210 μm thickness and YSZ as the underlying layer with an approximate thickness of 60 μm. The cross sectional SEM microstructure is columnar. A higher solid load content (40 wt %) in the suspension along with a higher median particle size (1 μm) leads to formation of a larger droplet while spraying results in the formation of relatively denser layer. The top surface morphology of triple layer TBC resembles cauliflower microstructure and looks relatively denser compared to the top surface morphology of double layer TBC, as seen in Fig.4 (b).

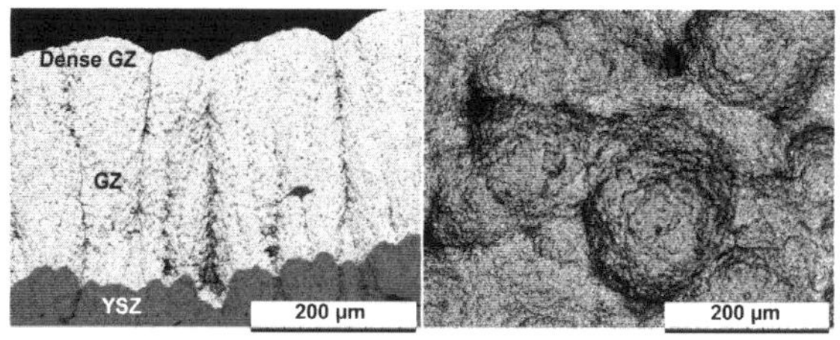

Fig.4 SEM micrographs of triple layer TBC (a) cross section (b) top surface morphology

3.2 Porosity measurement

Porosity content of the as sprayed TBC samples was estimated by water intrusion method and the mean values along with standard deviation are reported. All the pores in the coating which are accessible by water can be measured by this method (column gaps & open pores). It was observed that the porosity content of single layer YSZ TBC had the highest porosity content of approximately 21 % whereas the double layer TBC had an overall porosity content of approximately 18%. Triple layer TBC had the lowest porosity content of approximately 17% due to the presence of a relatively denser GZ layer on top, as shown in Fig. 5.

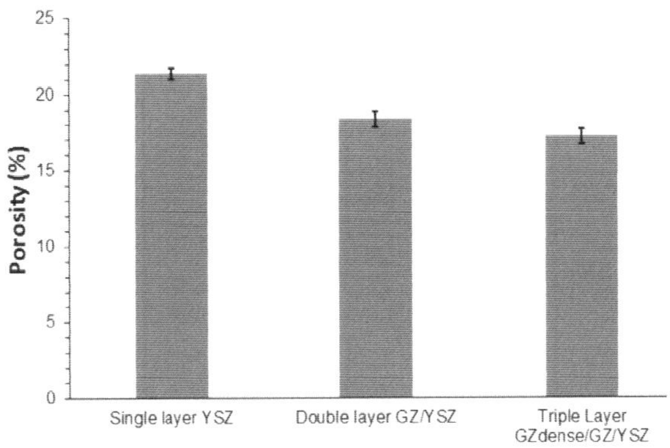

Fig.5 Porosity content of the as sprayed TBCs estimated by water intrusion method

3.3 Phase analysis

Phase analysis of the top surface as sprayed TBCs was carried out using XRD. The peaks obtained by XRD were identified using JCPDS standard, as shown in Fig. 6.

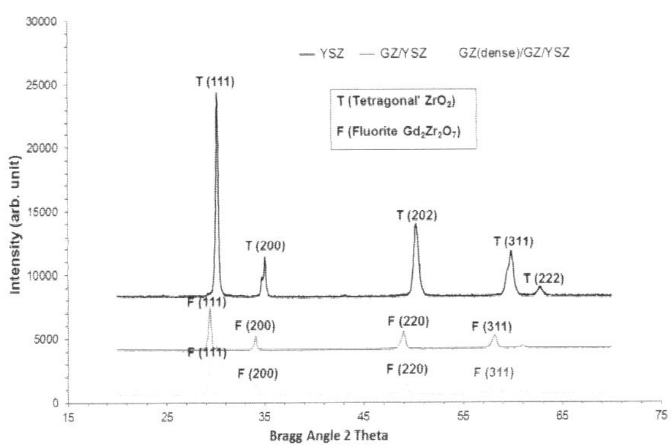

Fig.6 XRD analysis of the as sprayed TBCs

Single layer 8 YSZ showed tetragonal prime zirconia phase, which is an oversaturated phase of tetragonal zirconia with yttria. This metastable phase is highly desirable as it remains stable from room temperature to 1200 °C without undergoing phase transformation and hence enhances the thermal cyclic life time of TBC. In the case of double and triple layered TBCs, cubic defect fluorite phase was found. The defect fluorite phase transforms to cubic pyrochlore structure when exposed to higher temperature. Defect fluorite is a disordered variation of pyrochlore structure where the cations and oxygen vacancies occupy random positions in the cubic crystal structure. Change in suspension for depositing the relatively denser GZ layer which comprised of higher solid load content (40 wt %) and higher median particle size (1 µm) of GZ did not alter the stoichiometry.

3.4 Erosion test

Erosion test of the as sprayed TBCs was performed at room temperature and their mean erosivity number along with standard deviation is plotted in Fig. 7.

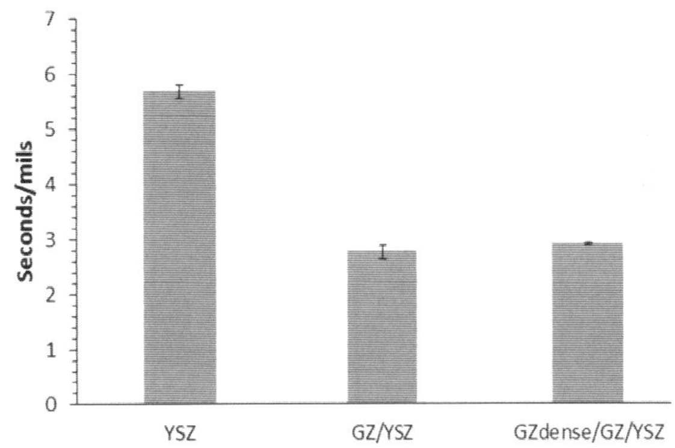

Fig. 7 Erosion resistance of the as sprayed TBCs

Erosivity number is the ratio of time of exposure (seconds) to loss in thickness (mils). Higher the erosivity number of a TBC, higher is its erosion resistance. Among the as sprayed TBCs, double layer GZ/YSZ had the lowest erosion resistance. Triple layer GZ dense/GZ/YSZ TBC had a slightly better erosion resistance than double layer TBC due to the presence of relatively denser GZ on top. It was demonstrated by Ramachandran et al in the case of APS (atmospheric plasma spray) deposited TBCs that the erosion resistance depends on porosity content of the coating. Higher the porosity content of a TBC, the lower its erosion resistance, for the same given material composition [16]. Single layer 8YSZ TBC showed a higher erosion resistance which was approximately two times better than the GZ/YSZ double layer TBC. 8YSZ TBC had a higher porosity content compared to GZ/YSZ multi-layered TBCs but still showed a better erosion resistance. A similar observation of poor erosion resistance of APS sprayed pyrochlore based TBCs compared to YSZ was reported by Ramachandran et al [16].

SEM analysis of cross section of eroded TBC samples was carried out. Single layer 8YSZ showed a lower coating thickness loss compared to multi-layered TBCs after the erosion test at room temperature, as seen in Fig. 8(a). Also, delamination within YSZ and close to the point of impact with eroding media was observed at a higher magnification, as shown in Fig. 8(b). Similar delamination crack within YSZ was reported for EB-PVD processed YSZ TBC after the erosion test [17]. In the case of GZ/YSZ double layer TBC, thickness loss of GZ layer was higher than single layer YSZ TBC. However, there were no visible TBC delaminations seen on the eroded GZ/YSZ TBC and the cross sectional SEM micrograph looked similar to the as sprayed microstructure but with a reduced thickness of GZ layer; as seen in Fig. 8(c).

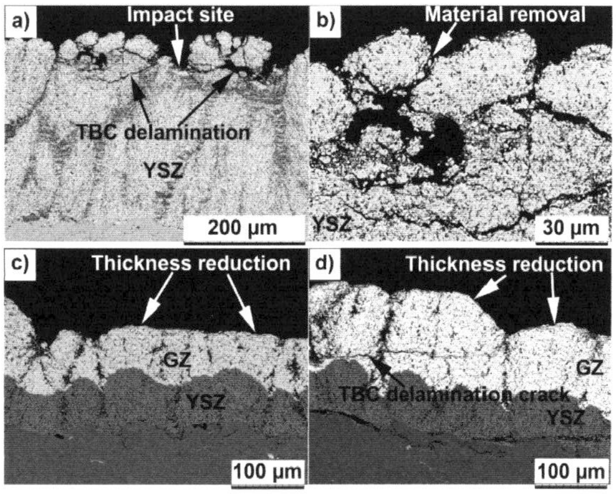

Fig. 8 SEM micrographs of eroded TBCs (a) single layer YSZ (b) YSZ at higher magnification (c) GZ/YSZ double layer (d) GZdense/GZ/YSZ triple layer

Possible explaination for the lower erosion resistance of GZ than YSZ could be due to the lower fracture toughness of GZ compared to YSZ [13].

In the case of eroded triple layer TBC, as seen in Fig. 8(d), thickness reduction was relatively lower compared to double layer TBC. Also, a horizontal delamination crack was seen in triple layer TBC near to GZ/YSZ interface. Reason for the slight improvement in erosion resistance of triple layer TBC compared to double layer TBC could be due to longer time taken by alumina particles to erode the denser top GZ layer.

4 Conclusion

In this study, current industry standard 8YSZ single layer TBC along with GZ/YSZ multi-layered TBCs were deposited by suspension plasma spray process and their erosion resistance at room temperature was compared. It was observed that:-

- Columnar microstructure in single layer 8YSZ and GZ/YSZ multi-layered TBCs could be created.

- Single layer 8YSZ had a higher erosion resistance than GZ/YSZ multi-layered TBCs despite of its higher porosity content among the as sprayed TBCs.

- A denser GZ top layer in the case of triple layer TBC helped in slightly improving the erosion resistance compared to GZ/YSZ double layer TBC.

Acknowledgements

The authors would like to acknowledge financial support from KK foundation (Dnr 20140130), Sweden. The authors would also like to thank Mr. Matin Tandler from Institute of Energy Research, Forschungszentrum Jülich, for carrying out the erosion test.

References

[1] D. Stöver and C. Funke, "Directions of the development of thermal barrier coatings in energy applications," J. Mater. Process. Technol., vol. 92–93, pp. 195–202, Aug. 1999.

[2] A. Cipitria, I. O. Golosnoy, and T. W. Clyne, "A sintering model for plasma-sprayed zirconia TBCs. Part I: Free-standing coatings," Acta Mater., vol. 57, no. 4, pp. 980–992, Feb. 2009.

[3] M. P. Borom, C. A. Johnson, and L. A. Peluso, "Role of environmental deposits and operating surface temperature in spallation of air plasma sprayed thermal barrier coatings," Surf. Coat. Technol., vol. 86–87, no. PART 1, pp. 116–126, 1996.

[4] C. Mercer, S. Faulhaber, A. G. Evans, and R. Darolia, "A delamination mechanism for thermal barrier coatings subject to calcium–magnesium–alumino-silicate (CMAS) infiltration," Acta Mater., vol. 53, no. 4, pp. 1029–1039, Feb. 2005.

[5] J. Wu, H. Guo, M. Abbas, and S. Gong, "Evaluation of plasma sprayed YSZ thermal barrier coatings with the CMAS deposits infiltration using impedance spectroscopy," Prog. Nat. Sci. Mater. Int., vol. 22, no. 1, pp. 40–47, Feb. 2012.

[6] L. Li, N. Hitchman, and J. Knapp, "Failure of Thermal Barrier Coatings Subjected to CMAS Attack," J. Therm. Spray Technol., vol. 19, no. 1–2, pp. 148–155, Jul. 2009.

[7] R. Vaßen, M. O. Jarligo, T. Steinke, D. E. Mack, and D. Stöver, "Overview on advanced thermal barrier coatings," Surf. Coat. Technol., vol. 205, no. 4, pp. 938–942, Nov. 2010.

[8] J. M. Drexler, A. L. Ortiz, and N. P. Padture, "Composition effects of thermal barrier coating ceramics on their interaction with molten Ca–Mg–Al–silicate (CMAS) glass," Acta Mater., vol. 60, no. 15, pp. 5437–5447, Sep. 2012.

[9] R. M. Leckie, S. Krämer, M. Rühle, and C. G. Levi, "Thermochemical compatibility between alumina and ZrO_2–$GdO_{3/2}$ thermal barrier coatings," Acta Mater., vol. 53, no. 11, pp. 3281–3292, Jun. 2005.

[10] E. Bakan, D. E. Mack, G. Mauer, and R. Vaßen, "Gadolinium Zirconate/YSZ Thermal Barrier Coatings: Plasma Spraying, Microstructure, and Thermal Cycling Behavior," J. Am. Ceram. Soc., vol. 97, no. 12, pp. 4045–4051, Dec. 2014.

[11] S. Mahade, N. Curry, S. Björklund, N. Markocsan, and P. Nylén, “Thermal conductivity and thermal cyclic fatigue of multilayered $Gd_2Zr_2O_7$/YSZ thermal barrier coatings processed by suspension plasma spray,” Surf. Coat. Technol., vol. 283, pp. 329–336, Dec. 2015.

[12] C. Jiang, E. H. Jordan, A. B. Harris, M. Gell, and J. Roth, “Double-Layer Gadolinium Zirconate/Yttria-Stabilized Zirconia Thermal Barrier Coatings Deposited by the Solution Precursor Plasma Spray Process,” J. Therm. Spray Technol., vol. 24, no. 6, pp. 895–906, Jul. 2015.

[13] R. G. Wellman and J. R. Nicholls, “A review of the erosion of thermal barrier coatings,” J. Phys. Appl. Phys., vol. 40, no. 16, p. R293, 2007.

[14] J. R. Nicholls, M. J. Deakin, and D. S. Rickerby, “A comparison between the erosion behaviour of thermal spray and electron beam physical vapour deposition thermal barrier coatings,” Wear, vol. 233–235, pp. 352–361, Dec. 1999.

[15] N. Curry, K. VanEvery, T. Snyder, J. Susnjar, and S. Bjorklund, “Performance Testing of Suspension Plasma Sprayed Thermal Barrier Coatings Produced with Varied Suspension Parameters,” Coatings, vol. 5, no. 3, pp. 338–356, Jul. 2015.

[16] C. S. Ramachandran, V. Balasubramanian, and P. V. Ananthapadmanabhan, “Erosion of atmospheric plasma sprayed rare earth oxide coatings under air suspended corundum particles,” Ceram. Int., vol. 39, no. 1, pp. 649–672, Jan. 2013.

[17] N. A. Fleck and T. Zisis, “The erosion of EB-PVD thermal barrier coatings: The competition between mechanisms,” Wear, vol. 268, no. 11–12, pp. 1214–1224, May 2010.

Microstructural control on yttria stabilized zirconia coatings by suspension plasma spraying

J. Kitamura, K. Fujimori, T. Wada, Tokyo / JPN, J. Liu, NY / USA, B. Callen, Alberta / CAN

Yttria-stabilized zirconia (YSZ) coatings are widely applied to aerospace and industrial gas turbine industries as thermal barrier to protect base alloys against high temperature. Several kinds of microstructures, such as porous, dense vertically cracked (DVC) and columnar, are intentionally designed and used properly depending on requirement and environment. Suspension plasma spraying (SPS) has been actively studied recently because it has possibility to realize much better properties than atmospheric plasma spraying (APS) as well as reduction of cost and process time than physical vapor deposition.

In this study, influence of particle size and solid concentration in ethanol base suspension as well as spray parameters, such as plasma gases, plasma power, injector size, spray distance (SD) and suspension feeding rate, have been investigated to find out how to control microstructure of YSZ coatings using SPS with cascaded arc torch.

Suitable spray distance is necessary to have dense structure when using relatively coarser particle of 0.6 μm. Columnar like structure tends to be produced when using fine particles of 0.2 μm. Using Ar/He plasma is effective to make DVC. Ar/H_2 tends to have columnar, however, increasing total gas flow seems to be effective to make DVC.

1 Introduction

Atmospheric plasma spraying (APS) is one of the well-established technologies for surface modifications. It has been improved continuously in particular about plasma gun technology from the invention about a half century ago and has applied to various industries nowadays [1]. Although improvement of the gun technology has driven the expanding market and applications, microstructure in the coatings is apparently governed by the feedstock, in particular by the particle size, when feeding powder. In conventional APS, the size is typically ranging from 10 to 100 μm. Minimum size is limited due to its flowability because it must be fed at dry conditions. Oxide ceramic is much disadvantageous to feed fine particles stably due to its lower theoretical and apparent density than metals and carbides. In order to overcome the under limit of the particle size, the use of liquid feedstock as suspension is actively studied recently to feed from sub-micron to nanometer materials very stably [2-4].

Yttria stabilized zirconia (YSZ) coatings are widely applied to aerospace and industrial gas turbine industries as thermal barrier to protect base alloys against high temperature [5]. A wide variety of coating microstructures, such as porous, dense vertically cracked (DVC) and columnar, has been applied to each part with specific requirement in the jet engines and gas turbines. It is well known that microstructures and properties of thermal sprayed coatings are significantly affected by feedstock characteristics, mainly particle size distribution and morphology of the particles, as well as spray processes. Suspension plasma spraying (SPS) with YSZ is actively studied recently because it has possibility to realize much better properties than atmospheric plasma spraying as well as reduction of cost and process time than physical vapor deposition [6-7].

Considering that productivity of the SPS, it is clear that high concentration suspension is quite valuable to increase the powder feeding rate. Also, it can be easily considered particle size in the suspension will affect coating microstructure drastically. In this paper, influence of particle size and solid concentration in ethanol base suspension as well as spray parameters, such as plasma gases, plasma power, nozzle size, spray distance (SD) and suspension feeding rate have been investigated to find out how to control microstructure of YSZ coatings using SPS with cascaded arc torch of TriplexPro™.

2 Experimental

Ethanol based suspensions of yttria stabilized zirconia (8YSZ, ZrO_2-8wt%Y_2O_3) used in this study are summarized in Table 1. Fine of 1.1 μm in D90 were dispersed into ethanol with dispersant. Suspension concentration of F25 and F35 is 25 and 35wt%, respectively. F25 (Metco 6608) has been studied in ref. 8 previously. Same preparation was done for coarser particles of 2.1μm in D90, labelled as C25 and C35.

Table 1. Ethanol based 8YSZ suspensions

Name	Particle size			Ratio [wt%]
	D10 [μm]	D50 [μm]	D90 [μm]	
F25 [8]	0.3	0.6	1.1	25
F35	↑	↑	↑	35
C25	0.4	0.7	2.1	25
C35	↑	↑	↑	35

TriplexPro-210™ from Oerlikon Metco, Switzerland was applied because of its advantageous features of stable plasma both of partially and lower pulsation [9, 10]. They are considered to be beneficial for SPS in order to vaporize the solvent and melt the particles homogeneously [11]. As for suspension feeder, test

Table 2. Spray conditions in this study

No	Mater.	Plasma gas [NLPM]	Plasma power [kW]	Injector diameter [mm]	Injector pressure [psi]	Suspension feeding rate [g/min]
1	F25 F35	Ar 40/ H_2; 5	52	0.20	50	30
2	**C25** **C35**	Ar 40/ H_2; 5	52	**0.305**	**20**	**35**
3	F25 F35	**Ar 15/ He; 50**	**47**	0.20	50	30
4	**C25** **C35**	**Ar 15/ He; 50**	**47**	0.20	50	30
5	**C25** **C35**	**Ar 15/ He; 50**	**47**	**0.305**	**20**	**35**
6	F25	Ar 40/ H_2; 5	52	**0.305**	**20**	**35**
7	**C25**	Ar **65**/ H_2; **10**	**62**	**0.305**	**50**	**58**

type product of 5MPE-SF was used also from Oerlikon Metco [12]. All suspensions were easily redispersed by pot mill prior to putting into the container of the feeder. Suspension was injected into plasma jet vertically from the injector placed at downstream from the exit of the gun nozzle.
Spray conditions are summarized in Table 2. At No. 1, F25 and F35 were fed at same spray parameters with Ar/H_2 gas. Suspension was changed to the coarse ones in No. 2 without changing plasma gas and power. Injector was changed to larger and adjusted injector pressure to become comparable suspension feed rate. Plasma gas was changed to Ar/He at No.3 from No. 1. Materials were changed to coarse ones at No. 4 from No. 3. At No. 5, injector diameter was large and can be compared with No. 4 to see the influence of injector. No. 6 is also to study the influence of injector size in fine F25. No. 7 is high gas flow rate of Ar/H_2, which can be compared with Nos. 2 and 5 directly.

Low carbon steel (JIS SS400) plates (25 × 100 × 3^t mm) were used as substrates, which were undercoated with CoNiCrAlY by APS. Spray distances for SPS were 50, 70 and 90 mm. Anode nozzle was standard with 9 mm in inner diameter in all conditions. Application rate (AR) defined as μm/pass, coating microstructure of cross section and micro Vickers hardness (HV0.3) were investigated.

3 Results and discussion

3.1 Influence of suspension concentration

Coating cross sections from No. 1 (F25 and 35 with Ar/H_2 plasma at SD of 70 mm) are shown in Fig. 1, where hardness with sigma (s) and application rate (AR) are described. High AR was attained from high concentration of F35 but porous microstructure than F25. No significant difference was seen in the hardness. Both structures look like columnar.

(a) F25, 355 (s: 36), 4.6 μ m/pass (b) F35, 317 (s: 44), 7.5 μ m/pass

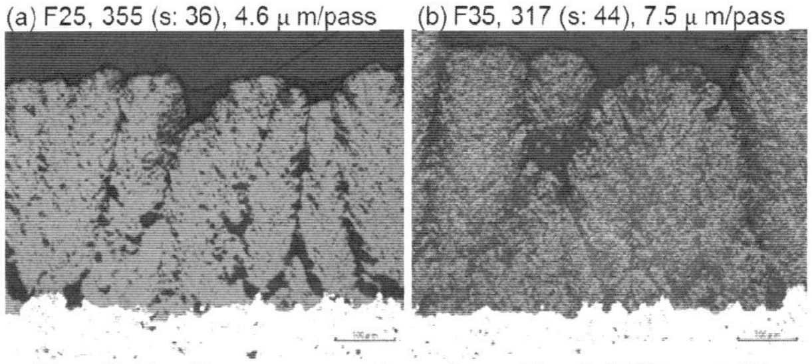

Fig. 1. Coating cross sections from No. 1 (F25 and 35 with Ar/H_2 plasma at SD of 70 mm)

Coating cross sections from No. 2 (C25 and 35 with Ar/H_2 plasma at SD of 70 mm) are shown in Fig. 2. Columnar like structures are also observed and look denser than ones in Fig. 1. High AR at 35wt% is same as fine suspension. As for hardness, in contrast, C35 coating is harder than C25 and fine ones. Comparing among 4 specimens, F25 and C35 tend to be denser than C25 and F35.

(a) C25, 392 (s: 56), 4.7 μ m/pass (b) C35, 505 (s: 38), 7.0 μ m/pass

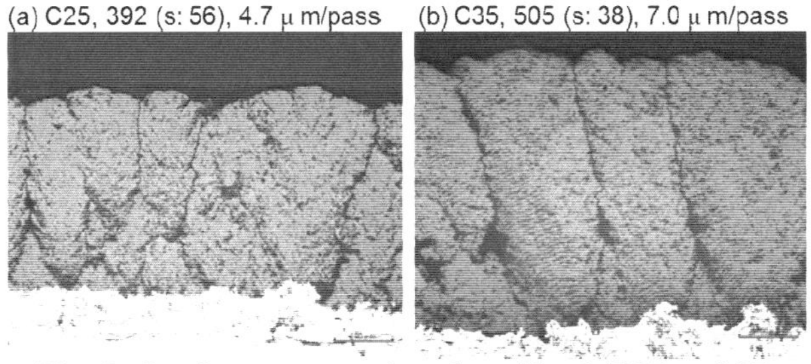

Fig. 2. Coating cross sections from No. 2 (C25 and 35 with Ar/H_2 plasma at SD of 70 mm)

Coating cross sections from No. 3 (F25 and 35 with Ar/He plasma at SD of 70 mm) are shown in Fig. 3, where porous DVC is observed. AR tended to be lower compared with Ar/H_2 plasma. In spite of much porous structure in Fig. 3(b), hardness shows relatively higher than others in Fig. 3(a) and Fig. 1.

In this SPS study, injected liquid suspension vertically into the plasma flame will be atomized by high velocity flame at first and the direction will be changed to horizontal to the direction of the flame. Liquid in the atomized droplet will be vaporized after that. Therefore, droplet size will be dominated mostly by the first atomization. High concentration suspension having relatively higher viscosity might be possible to generate large droplet that can keep at high temperature with molten state for a long time. Therefore, coating tends to be harder by depositing such a molten droplet, which would be effective to make hard coating. Relatively higher velocity with Ar/He than Ar/H_2 would contribute to keep particle molten at longer distance.

Coating cross sections from No. 4 (C25 and 35 with Ar/He plasma at SD of 70 mm) are shown in Fig. 4, observing much denser DVC. AR was almost comparable to fine ones in No. 3. Coating hardness is, in contrast, apparently higher than ones in Figs 1~3. From them, coarser particle will be effective to prepare dense and hard coatings within the experimental conditions. In particular, high velocity Ar/He plasma seems to be better for coating densification.

(a) F25, 342 (s: 74), 2.9 μ m/pass (b) F35, 417 (s: 31), 6.1 μ m/pass

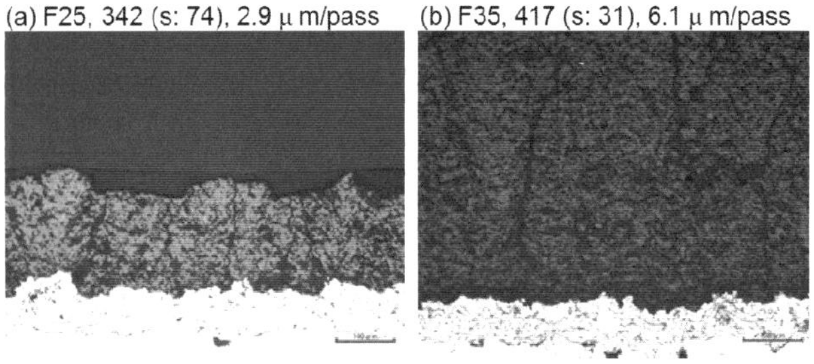

Fig. 3. Coating cross sections from No. 3 (F25 and 35 with Ar/He plasma at SD of 70 mm)

(a) C25, 593 (s: 22), 2.9 μ m/pass (b) C35, 539 (s: 34), 4.8 μ m/pass

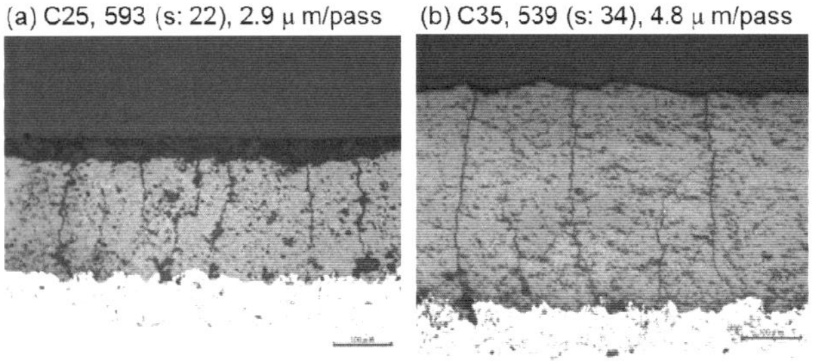

Fig. 4. Coating cross sections from No. 4 (C25 and 35 with Ar/He plasma at SD of 70 mm)

3.2 Influence of diameter of the injector nozzle

Suspension feeder of 5MPE-SF carries suspension by pressurized Ar gas simply from the container to the injector through very narrow tube. In this case, inner diameter of the injector nozzle will play important role to balance between feeding rate and intensity of injection into the plasma flame. When lowering injector pressure, both lowering feed rate and weakening injection are occurred at the same time and they cannot be controlled individually. Large diameter nozzle can be realized to lowering intensity of injection keeping same suspension feeding rate, for example.

Coating cross sections from No. 5 (C25 and 35 with Ar/He plasma at SD of 70 mm and wide injector nozzle of 0.305 mm) are shown in Fig. 5, whose microstructures in (a) C25 and (b) C35 are porous and dense DVCs, respectively. Higher AR for C35 than C25 is same tendency as narrow injector conditions shown in Fig. 4. In both cases (Nos. 4 and 5), high concentration suspension of 35wt% produced denser microstructure. When comparing with Fig. 4(a) and 5(a), as well as Fig. 4(b) and 5(b), influence of injector size is directly studied and No. 5 always shows higher AR than No. 4. The condition of suspension injection using narrow nozzle with the feeding rate in No. 4 seems to be too strong and many ratio are passed through the plasma flame reducing AR. In wide nozzle, on the hand, mild and appropriate injection can put a high portion of suspension into center of the plasma flame to increase AR. This indicates injection conditions should be optimized in each condition because plasma density and velocity are also changed by gas selection, mixing ratio, total flow and electric current.

Coating cross sections from F25 of Nos. 1 and 6 at SD of 70 mm are shown in Fig.6. Injector diameter is (a) 0.20 and (b) 0.305 mm, respectively. Fig. 6(a) is same as Fig. 1(a). Both structures are close to columnar. Coating at No. 1 in Fig. 6(a) is denser but quite lower hardness compared with Fig. 6(b). Both AR are almost same that is different from C25 and 35 as already shown in Figs 4 and 5. In spite of unclear mechanism, suitable injection pressure and feeding rate will be different when the size of particles in suspension is changed. From all results, it has a tendency that coarser particle and/or Ar/He plasma gives DVC structure easily. Columnar tends to be generated easier from finer particles and/or Ar/H_2. In some case as shown in Fig. 2(b), it will be difficult to define whether DVC or columnar. No porous structure with less vertical cracks generally made by conventional APS is prepared within this study.

(a) C25, 360 (s: 22), 4.5 μ m/pass (b) C35, 613 (s: 61), 8.4 μ m/pass

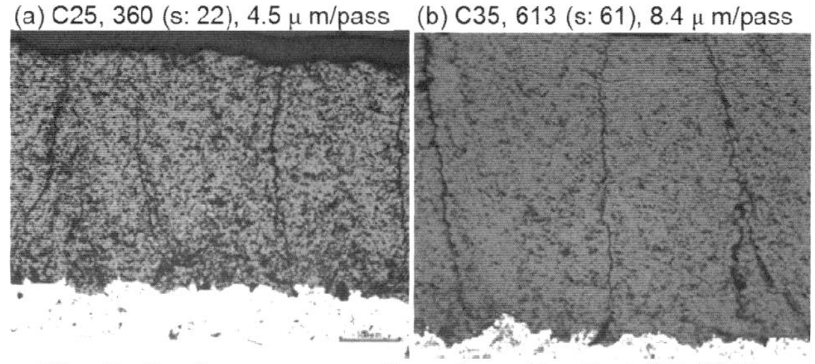

Fig. 5. Coating cross sections from No. 5 at SD (C25 and 35 with Ar/He plasma at SD of 70 mm and wide injector nozzle of 0.305 mm)

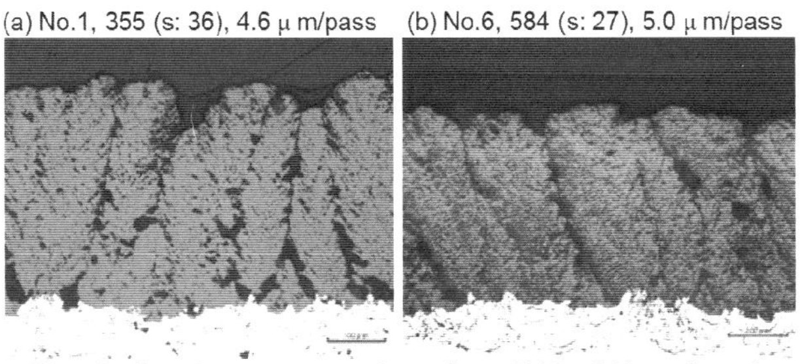

Fig. 6. Coating cross sections from F25 of Nos. 1 and 6 at SD of 70 mm. Injector diameter is (a) 0.20 and (b) 0.305 mm, respectively. Fig. 6(a) is same as Fig. 1(a).

3.3 Influence of gas flow rate of Ar/H_2

In terms of wide cross section of the nozzle with 9 mm in diameter, it may say that total gas flow rate of 45 or 60 NLPM tends to be low. In-flight particles tend to have high temperature and low velocity in general under such a low plasma gas flow rate.

Fig. 7 shows coating cross sections from C25 of Nos. 2 and 7 at SD of 70 mm, where plasma gas flow rate is low in No. 2 (Fig. 7(a)) and high in No. 7 (Fig. 7(b)), respectively. With increasing total gas flow rate, hard and dense coating has been achieved with almost same AR. As for the microstructure, the coating in Fig. 7(a) looks columnar. In contrast, the coating in Fig. 7(b) seems to be much closer to DVC but irregular surface morphology still shows the feature of columnar structure.

High velocity plasma will be attained with increasing total gas flow rate in this case. And thus, velocity of the particle will be much higher retaining molten state at longer spray distance. Therefore, dense coating can be produced in this case. This suggest denser coating could be made using at much optimized Ar/H_2 gas conditions, which must be quite economical than using Ar/He.

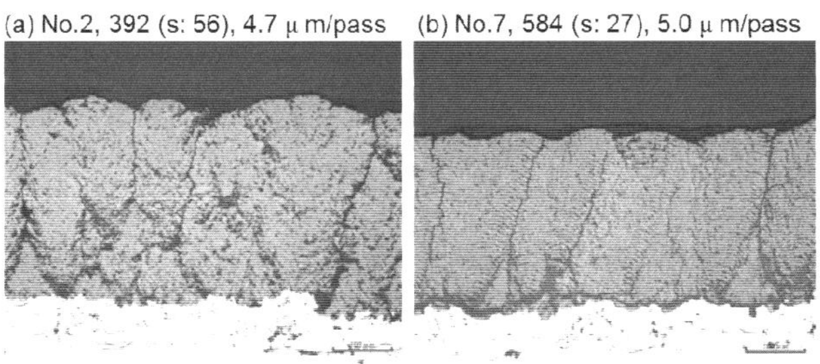

Fig. 7. Coating cross sections from C25 of Nos. 2 and 7 at SD of 70 mm. Fig. 7(a) is same as Fig. 2(a).

3.4 Influence of spray distance (SD)

Influence of spray distance (SD) is also investigated as shown in Figs. 8 and 9. In Fig. 8, cross sections are shown by F25 suspension at No. 6 using wide injector and Ar/H_2 plasma, meaning that Fig 8(b) is same as Fig. 6(b). At shorter SD of 50 mm in Fig. 8(a), coating looks like dense DVC different from columnar like in Fig. 8(b). No significant difference is seen in both hardness and AR. It is unclear now the change of the surface morphology from smooth at short SD to rough at long SD. Soft pressurize powder body, which is easily knife scratched, is obtained at longer SD of 90 mm.

Coating cross sections from No. 5 are shown in Fig. 9, with C35 using wide injector and Ar/He plasma. As for density and uniformity, coating at 70 mm in Fig. 9(b) seems to be higher than others with relatively dense DVC structure. Coating at 50 mm in Fig. 9(a) seems to be porous and looks intermediate of DVC and columnar. Since hardness is much higher, denser structure can be expected. At longer distance of 90 mm, coating has been successfully prepared. Coarse particles having higher heat capacity could contribute to slower cooling during in-flight and impact onto the substrate. Fine particles, on the other hand, rapid quenching might result in soft pressurize powder body.

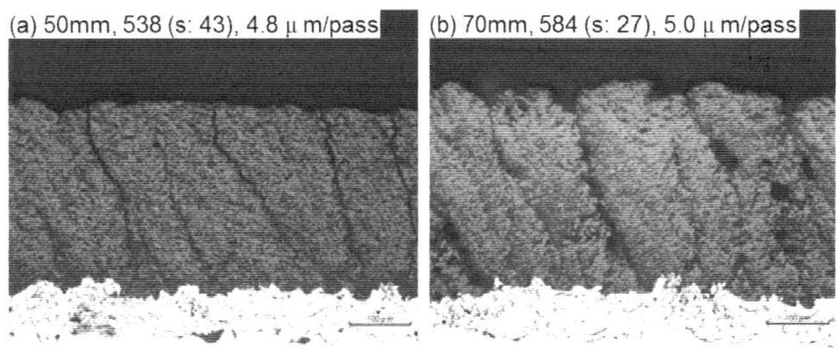

Fig. 8. Coating cross sections from F25 at No. 6 using Ar/H_2 plasma with SD of (a) 50 and (b) 70 mm. Fig. 8(b) is same as Fig. 6(b).

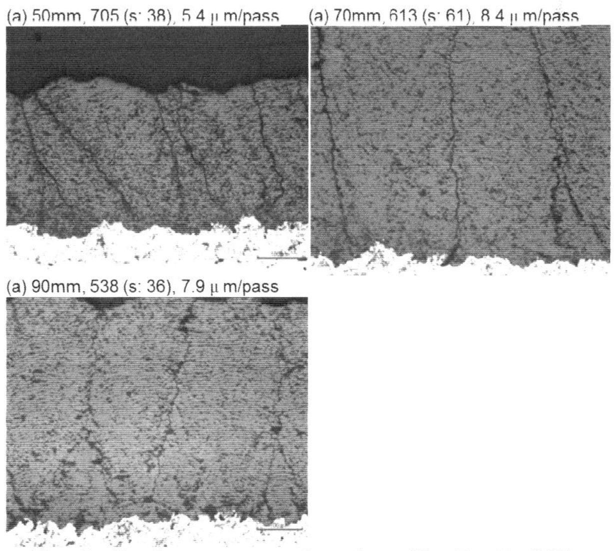

Fig. 9. Coating cross sections from No. 5 with C35, using wide injector and Ar/He plasma.

4 Conclusions

Influence of particle size and solid concentration in ethanol base suspension as well as spray parameters have been investigated to find out how to control microstructure of YSZ coatings using SPS with cascaded arc torch. Findings are as follows.

1. Coarse particles in suspension is effective to prepare hard coatings as well as make denser coatings still at longer spray distance.
2. Suspension concentration affects not only application rate but also coating microstructure.
3. Injection pressure should be optimized on each size of particle, suspension concentration and spray parameters.
4. Adjustment of injector diameter is effective method to achieve both of desired coating microstructure and desirable application rate.
5. High concentration suspension with coarser particles is suitable to prepare hard coating stably.

Future studies can be considered, for example, as follows. Trial using much coarser particles will be important to understand the influence of the size comprehensively. Developing higher concentration suspension with stable flowability will be also challenging. As for plasma parameters, higher plasma gas flow will be interesting for coating densification at longer spray distance. Higher suspension feeding rate is an interest mainly from economics. Influence of the spray angle is quite important in actual applications.

References

[1] Heimann, R.B., Plasma-Spray Coating, Principles and Applications, VCH Verlagsgesellschaft mbH, Germany (1996).

[2] Fauchais, P., Etchart-Salas, R., Rat, V., Coudert, J.F., Caron, N., Wittmann-Ténèze, K: Parameters Controlling Liquid Plasma Spraying: Solutions, Sols, or Suspensions, J. Thermal Spray Technol., Vol. 17(1), (2008), pp. 31/59.

[3] Fauchais, P., Rat, V., Delbos, C., Coudert, J.F., Chartier, T., Bianchi, L: Understanding of Suspension dc Plasma Spraying of Finely Structured Coating for SOFC, IEEE Trans. Plasma Sci., Vol. 33(2), (2005), pp. 920/930

[4] Fauchais, P., Montavon, G: Latest Developments in Suspension and Liquid Precursor Thermal Spraying, J. Thermal Spray Technol.,Vol. 19(1-2), (2009) pp. 226/239.

[5] Padture, N.P., Gell M., Jordan E. H.: Thermal Barrier Coatings for Gas-Turbine Engine applications. Science 296, (2002) pp. 280/284.

[6] Mauer, G., Guignard, A., Vaβen, R., Stöver, D.,: Process Diagnostics in Suspension Plasma Spraying, Surf. Coat. Technol. 205, (2010), pp. 961/966.

[7] Ganvir, A., Curry, N., Markocsan, N., Nylén, P., Toma, F-L.,: Comparative Study of Suspension Plasma Sprayed and Suspension High Velocity Oxy-fuel Sprayed YSZ Thermal Barrier Coatings, Surf. Coat. Technol. 268, (2015), pp. 70/76.

[8] Chen, D, Dambra, C, Sabouni, O.: Influence of Process Parameters and Feedstock Selection on Coating Microstructures with a Cascaded SinplexPro™ Plasma Torch, Proc. ITSC2014, electric file, (2014).

[9] Bernecki, T. Varley, K. Rusch, W. Wlodarczyk, J. and Klein, J., Plasma Gun with Adjustable Cathode, US Patent 4,780,591, October 25, 1988.

[10] Hawley, D., Improved Quality, Reduced Cost and Cycle Times. Sulzer Technical Review, 2007(3), pp. 14/16.

[11] Sulzer Technical Review 2/2011, Liquid suspension spray – New challenges, new opportunities, (2011), pp. 26.

[12] Sulzer Metco Flyer, “5MPE-SF Material Feed System for Suspension Plasma Spray” (2013)

Microstructure Formed by Axial Suspension Plasma Spraying: From YSZ Splat to Coating

Dachuan Li, Jingwei Feng, Huayu Zhao, Chenguang Liu, Linlin Zhang, Fang Shao, Yuexing Zhao, Shunyan Tao*, Shanghai Institute of Ceramics, Chinese Academy of Sciences / CHN
Correspondence to: sytao@mail.sic.ac.cn (S. Tao)

In the current work, yttria stabilized zirconia splats and coatings were deposited onto superalloy substrates with varied temperatures using axial suspension plasma spraying (ASPS). The splats were characterized by scanning and transmission electron microscopes (TEM). As the substrates were heated from room temperature to 300 degree C, the morphology of ASPS splats changed rapidly. However, when the substrate temperature further increased, the variation of morphology was found less conspicuous. Integrated ASPS splats were peeled off from the substrate using a "plating and etching" approach for TEM examination. The observation on morphological and crystallographic features of splats revealed that the cooling rate during impacting and spreading of ASPS splats might be lower than that of splats deposited by conventional atmospheric plasma spraying. Based on the characteristic of ASPS splats, a possible mechanism consisting shadowing effect was proposed to explain the so-called cauliflower microstructure. In addition, as-sprayed ASPS coating was also examined by X-ray diffraction and nano-indenter to determine the phase composition and microhardness, respectively.

Nomenclature	
v	velocity of droplet, m/s
d_p	diameter of droplet, m
ρ_g, ρ_d	density of gas/droplet, kg/m^3
μ	viscosity of droplet, Pa·s
σ	surface tension of droplet, J/m^2
C	drag coefficient, dimensionless

1 Introduction

Atmospheric plasma spraying (APS) is a well-established process to fabricate yttria stabilized zirconia (YSZ) thermal barrier coatings (TBCs) owing to its high depositing rate, low cost and versatility, etc. [1]. With splats piling up, the as-sprayed coating exhibits a major microstructural feature of intersplat cracks and spheroidal/ellipsoidal pores. In spite of its lower thermal conductivity, such architecture inevitably leads to unfavorable thermal cycling life and reliability [2]. Therefore, fine YSZ coatings with nano-/submicro-sized microstructure have attracted increasing attention in the community.

Currently, three different routes based on APS have been successfully carried out to prepare advanced fine YSZ coatings, namely:

i) spraying agglomerated nanometer particles with conventional APS equipment;

ii) spraying solution precursors with liquid carrier (known as solution precursor plasma spraying, SPPS);

iii) spraying suspension with liquid carrier (known as suspension plasma spraying, SPS);

Over the last decade, a number of studies and overviews [3-9] have been published, focusing on the influences of feedstock/ suspension/ solution characteristic and spraying parameters on the coating microstructure. In order to further improve the coating performance, collective efforts of the community have been continuously made to have a more comprehensive understanding of this relationship.

Earlier, another paper of our group reported the differences in thermal shock behavior and failure mechanism between thick TBCs prepared by axial SPS (ASPS) and conventional APS [10]. The aim of our current work is to compare the different manners in splat formation and detailed coating microstructure. Since the flattening and solidification behaviors of APS splats have been extensively reported [11-14], this paper is primarily devoted to splats and coatings fabricated by ASPS.

2 Experimental procedures

Nickel-based superalloy GH3128 (Ni-20Cr-8Mo-8W-0.5Al-0.5Ti, wt.%) substrates with dimensions of Φ10 mm×8 mm were used to accommodate individual splats. They were mirror-polished via a general multi-step process and ultrasonically cleaned in ethanol before spraying. An inductive heater along with a pyrometer and a programmable digital controller were used to control the substrate temperature. In order to obtain distinguishable splats, a stainless steel shield with a Φ2 mm hole was placed in front of the substrate to reduce the amount of deposited splats.

As for complete coatings, the substrates were sand-blasted GH3128 with dimensions of 25 mm×25 mm×2.7 mm. Bond coating were fabricated using commercial CoNiCrAlY powders (Co-32Ni-21Cr-8Al-0.5Y, wt.%). Detailed information on the raw material, plasma gun and spraying recipes could be found in literature [10].

Field emission scanning electron microscope (FESEM) JSM-6700F (JEOL, Japan) was employed to observe the ASPS splats. The microstructure of as-sprayed coatings were examined using a scanning electron microscope (SEM) TM3000 (HITACHI, Japan) attached with a 3D device. Transmission electron microscopy (TEM) images of splats and cross sections of coatings were acquired on JEM-2100F (JEOL, Japan). The preparation of TEM specimens are demonstrated in **Fig.1**.

The phase composition of as-sprayed coatings were identified by X-ray diffraction (XRD) employing D/Max

2550V (Rigaku, Japan) with nickel-filtered Cu Kα radiation (λ=0.154 nm) at a scan rate of 0.5 °/min^{-1}. Continuous stiffness measurements were carried out using a nano-indenter G200 (Agilent Technologies, USA) on the polished cross section of as-sprayed coatings. A strain rate of 0.05 s^{-1} and a maximum indentation depth of 5,000 nm were applied. The mean value of 10 results obtained at the depth of 500 nm was viewed as the microhardness value.

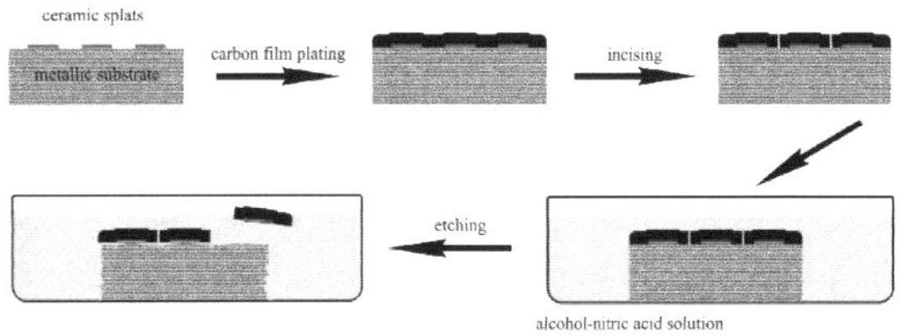

Fig. 1. Illustration of the preparation of ASPS splat specimen for TEM study. Concentration of the nitric solution was 20 vol.%.

3 Results and discussion

3.1 Splats

3.1.1 General remarks

Images of ASPS splats are shown in **Fig.2**. Morphologies of a typical APS splat are also displayed for comparison. Apart from the sizes, three major differences were identified:

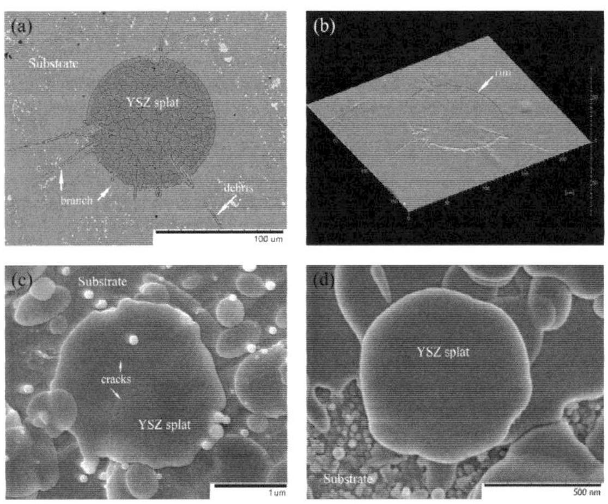

Fig. 2. (a) SEM and (b) 3D image of APS splat. SEM image of ASPS splat (c) with and (d) without intrasplat crack. The substrate temperature is 300 degree C in (a)-(c), and 900 degree C in (d).

Firstly, the ASPS splats were, by comparison, circular or elliptic with no obvious branch, while the APS splats were usually extensively fingered. Formation of the splashed branches or debris has been widely studied. The Sommerfeld parameter *K*, which is defined as

$$K = We^{0.5} \times Re^{0.25} = \frac{v^{1.25} \cdot d_p^{0.75} \cdot \rho^{0.75}}{\mu^{0.25} \cdot \sigma^{0.5}} \quad (1)$$

where *We* and *Re* are the Weber and Reynolds numbers, respectively, is of great significance to predict the splashing of splats. It is more the rule than the exception that enlarged *K* contributes to more intensive splashing [11]. The comparatively regular shape of ASPS splats might be mainly attributed to their much smaller d_p.

Secondly, the network of intrasplat cracks was frequently noticed on APS splats (**Fig.2a**), while most ASPS splats were crack-free. In the current work, the density of intra-splat crack on APS splats was measured, and the acquired result was 0.10-0.25 μm^{-1}. It was suggested that the intrasplat cracks were related to the fierce release of quenching stresses [15]. A small number of ASPS splats with cracks, as shown in **Fig.2d**, were also carefully studied. Some general features were refined as follows: i) they were with greater diameters, usually larger than 2 μm; ii) they were deposited onto non-smooth area (e.g. a former deposited splat), hence developed into a step-like or peak-like shape. The cracks were usually located near the area with small curvature radius.

Last but not least, the rim-structure was not observed on ASPS splats. As for a disk-shaped APS splat, it usually contained a rim which was much thicker than the central part [16], as highlighted in **Fig.2b**. According to the previous study [11], the curling of advancing edges caused by surface tension energy was closely related to the formation of the thick splat boundary. The substrate roughness, and more importantly, the initial solidified part at the bottom of splat periphery, could change the direction of liquid flow and therefore lead to thicker edge, or sometimes splashing. Formation of the rim-free feature of ASPS splats will be discussed in section 3.1.3.

3.1.2 Effect of substrate temperature

Fig.3 displays the splats deposited onto substrates at different temperatures. Since ASPS splats generally exhibited disk shape, the study on splat morphology was limited to the diameters. The changes of splat diameter with substrate temperature are summarized in **Fig.4**. The differences between non-heated group and heated groups were much more obvious, while under the temperature condition ranging from 300 degree C to 900 degree C, the behavior of splat formation was relatively stable.

M. Fukumoto et. al. [17] pointed out that the connecting rate (expressed as the ratio of the "truly connected" area to the total area of a splat) dominated in the splashing behavior of APS splats. Full connecting would contribute to faster solidification hence less splashing. As a result, a slower cooling rate could be inferred from the relatively irregular shape of ASPS splat on substrate at room temperature (RT). Since the initial velocity of the droplets was considered equal, the diameter of splats on substrate at RT should be the largest. Conversely, such conjecture was at variance with the appearance.

To explain the anomaly, the huge amount of tiny splats observed on non-heated substrate (**Fig.3a**) should be taken into consideration.

Literature 8 suggested that small individual particles were resulted from the disruption of poorly melted aggregates. However, considering the fact that axial injection was conducive to more efficient heating of the powder, the observed phenomenon was likely against this viewpoint. Besides, the decreased amount of tiny splats on heated substrates was also contradictious with the improved accommodating capability of pre-heated substrate [6].

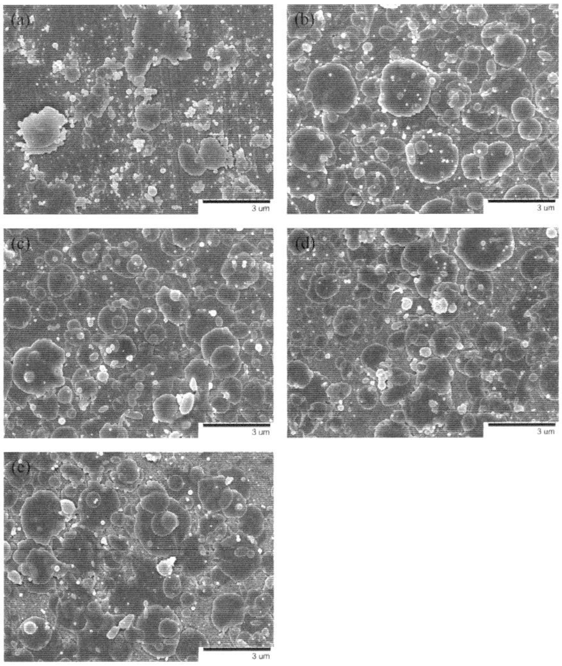

Fig. 3. Images of splats deposited onto substrates heated to different temperature. (a) RT; (b) 300 degree C; (c) 500 degree C; (d) 700 degree C; (e) 900 degree C.

Another possible reason would be that the droplet broke into smaller pieces while flying through the boundary layer of gas near the substrate shortly before impacting. Some fragments inherited the movement of "mother droplet" and hit the substrate, forming smaller splats. A rough model learned from study on the size of raindrop was proposed as follows:

$$F = \frac{1}{2} C \cdot \rho_g \cdot S \cdot v^2 \quad (2)$$

where F is the air resistance, and S is the projected area of a droplet.

To break the droplet, the F should be larger than the contractile force caused by surface tension:

$$F \geq \sigma \cdot d_p \quad (3)$$

Consequently, the critical density of gas could be presented as:

$$(\rho_g)_{critical} = \frac{8\sigma}{\pi C \cdot d_p \cdot v^2} \quad (4)$$

When the gas density is larger than the critical value, the droplet would likely to break, and vice versa. Based on equation 4, the densest gas a YSZ droplet (with a size of several microns and a velocity of hundreds of meters per second) could survive is about 10^2 kg/m^3, which is obviously much larger than the density of air (~1.3 kg/m^3 at RT). Accordingly, the above mentioned mechanism could not occur under the present situation.

Thus, it was suggested that the tiny splats were resulted from splashing during the flattening of droplets. Conversely, no evident of directional shaping was observed. This issue will be discussed in the next section.

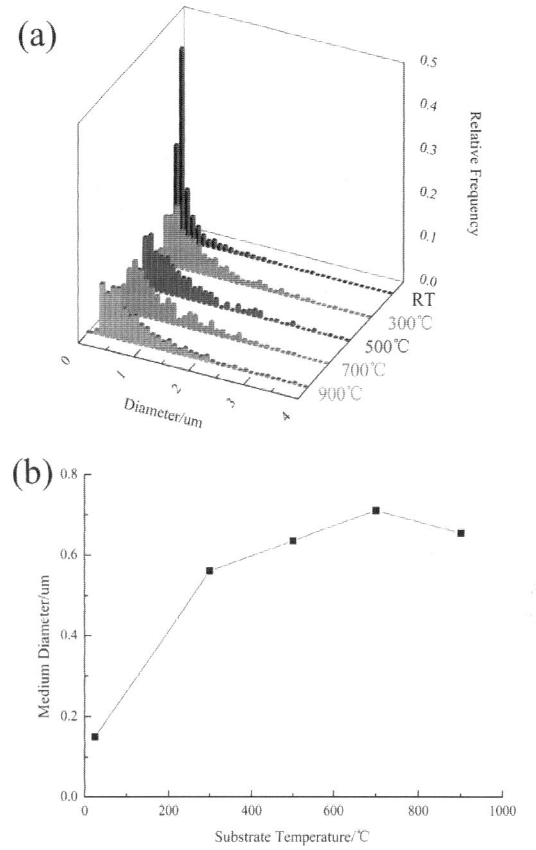

Fig. 4. (a) Function of substrate temperature on the diameter distribution of ASPS splats. (b) Median diameter of splats deposited onto substrates heated to different temperatures.

The bonding between droplets and substrate would reach a high and stable level when the substrate was heated above a certain critical temperature known as transition temperature (T_t) [17]. Fig.4(b) revealed that T_t under the current spraying condition was lower than 300 degree C. With the increase of substrate temperature, temperature gradient between the bottom of the droplet and the surface of the substrate became smaller, resulting in slower solidification and therefore enlarged splat diameter. However, when the temperature further increased, substrate roughening induced by oxidation could not be neglected. Visible change in substrate topography was observed, as shown in **Fig.2d** and **Fig.3e** (the heating period of time was about 10 min). The size of splats became smaller because of the increased connecting rate due to the oxidation of substrate surface. Besides, the obstructive effect of roughened substrate during the flattening process of a droplet could also play an important role in this phenomenon.

Additionally, the density of splats was also calculated by counting the splats within the field of view. The RT substrate had a splat density of 2.5 splats per square micrometer. However, on the other substrates which were heated to 300 degree C, 500 degree C, 700 degree C and 900 degree C, the density of splats were about 1.4 μm^{-2} in all cases. If the enlarged diameter was taken into consideration, the total area of splats in heated groups was about 2 times larger than that of the non-heated group. Enlightenment arisen from the experimental results was that slight preheating of the substrate (e.g. 200-300 degree C) would be good enough to achieve a relatively high depositing efficiency and bonding in the initial procedure, while further increase of substrate temperature would be less cost effective and was likely to cause unfavorable oxidation of the substrate or bond coating (BC) surface.

3.1.3 TEM observation

Fig.5 shows the TEM images of several ASPS splats. The splats were deposited onto mirror-polished metallic substrates then peeled off using the method illustrated in **Fig.1**. Based on the contrast, the central part appeared to be thicker than the periphery, resembling the cap of mushroom. Several factors could lead to the observed appearance of ASPS splat, and some of which even conflicted with each other. In case of rapid solidification, i.e. when a droplet was poorly heated, or the thermal resistance between droplet and substrate was insignificant, the droplet froze after the impact immediately, and the corresponding splat could assume the shape similar with a hat. Meanwhile, on condition of slow cooling, the impacted droplet could also turn a similar feature due to the surface tension force. The puzzle could be solved with the test results of selected area electron diffraction (SAED). **Fig.5** showed that the observed ASPS splats consisted of only one crystal grain with tetragonal crystal structure. Since fast cooling rate was inevitably conductive to fine-grain crystal structure, the mechanism of immediate solidification was excluded. Theoretically, there should be a critical size, above which multiple spontaneously formed grains could be observed within a splat. However, no statistics have yet been obtained in our current work.

It was worth noticing that the diameter of ASPS splat was larger than that of the columnar crystal in APS-prepared YSZ splat reported in literature 18, indicating that the cooling rate of ASPS splats was not as high as that in APS. This conclusion, however, was not in keeping with general understanding: why would the ASPS droplet whose size was much smaller possess slower cooling rate? To clarify this issue, the characteristics of ASPS must be taken into consideration. As a result of the high power output and short spraying distance in ASPS, the heat flux imposed by the gas flow was considerably large [6]. It was reasonable to assume that the heat played an overwhelming role in the formation of splats. Cooling of droplet was slowed down by the heat. As for the tiny splats mentioned in section 3.1.2, this mechanism would explain the absence of elliptic splats. Hence the rim-free and crack-free feature could be understood as well.

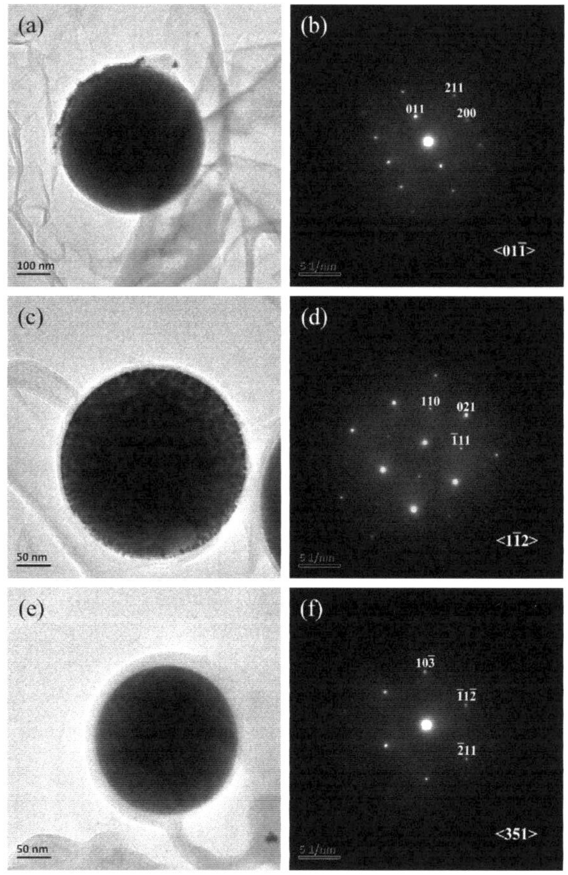

Fig. 5. TEM and selected area electron diffraction results of ASPS splats separated from substrate heated to 300 degree C. (b), (d) and (f) corresponds with the entire splat in (a), (c) and (e), respectively. Background in TEM images are carbon films.

Though of larger diameter, the YSZ splats fabricated by ASPS processed smaller grain volume owing to the height/thickness. As discussed above, it was resulted from the size of raw material instead of the cooling rate. The smaller crystal grain was assumed to be closely related to the superior phase stability of ASPS coatings reported by Y. Zhao [19]. Zirconia phase transformation from tetragonal (T) to monoclinic (M) was associated with the surface energy [20]. According to previous studies, as for 7 wt.% YSZ, the

phase transformation T→M could not occur in ceramics whose grain size was smaller than 0.9 μm [21, 22].
Additionally, the diffraction pattern in **Fig.5f** was obtained applying minimal rotation (Rotation X=-3.1 °, Rotation Y=0.6 °) of the specimen holder. Since the carbon film holding the splat was basically flat, the zone axis of <351> could be used as the orientation parameter of the very splat, or at least this direction was close to the actual normal direction of the crystal. However, since only a small number of splats were studied, the statistical preferred orientation of ASPS splats could not be firmly decided yet. Considering the coating performance was closely linked with both its microstructural and crystallographic characteristics, more attention might be paid to this topic in the future.

3.2 As-sprayed coating

Fig.6 shows the cross-sectional microstructures of as-sprayed ASPS coating. It exhibited a columnar layer which was a few microns thick at the bottom of coating (addressed as "columnar zone" in **Fig.6a**). However, above this layer, spheroidal grain took a considerably large share in the main body. Inter-columnar gaps were observed in as-sprayed ASPS coating. Unlike those in APS coatings, **Fig.6b** revealed that the segmentation cracks were resulted from regularly ranged pores (type I). Near the segmentation cracks, there was another pattern of pore arrangement (type II). The former was generally perpendicular to the substrate, while the latter was with larger angles towards the main column axis.
The transition of splat morphology was attributed to the change of cooling rate according to the coating thickness. At the beginning of deposition, since the substrate temperature was not high, the solidification of impacting droplet was accomplished within a comparatively longer period of time. Consequently, good intersplat bonding between successive splats could be achieved, resulting in the columnar feature. Later when the coating thickness reached a dozen of microns, the heat transference to BC was drastically reduced owing to the low thermal conductivity of ceramic material. The temperature of newly-formed surface increased rapidly with the growth of coating thickness. As revealed in the observation on splats, the cooling rate of YSZ droplets was then significantly improved. Additionally, considering the wettability of melted YSZ on the surface of solidified YSZ was better than on foreign material (e.g. metallic substrate), the increase was assumed to be even larger. Accordingly, splats with smaller flatness, i.e. spheroidal splat, was formed. As the spraying process continued, the increase of surface temperature became slower and slower. No great shift in splat forming took place in the following stages. It should be pointed out that a low speed or kinetic energy was suggested as a necessity in the whole process.
High resolution TEM images of the as-sprayed ASPS coating are displayed in **Fig.7**. Amorphous crystallite could be easily observed in the bonding area between splats. The hypothesis on the formation of spherical splat was proved by the continuous lattice between successive splats, as presented in **Fig.7b**. However, this mechanism was not always the case. **Fig.7c** shows two splats with different crystal orientation. The lattice in the intersplat boundary was basically consist with the splat underneath. This phenomenon suggested that the droplet corresponded with the splat above might be not well-melted as it flew in the jet. More extreme example of such case was that the primary particles were reserved in the coating, as displayed in **Fig.7d**. Nevertheless, such extreme case was not quite common within the current observation.

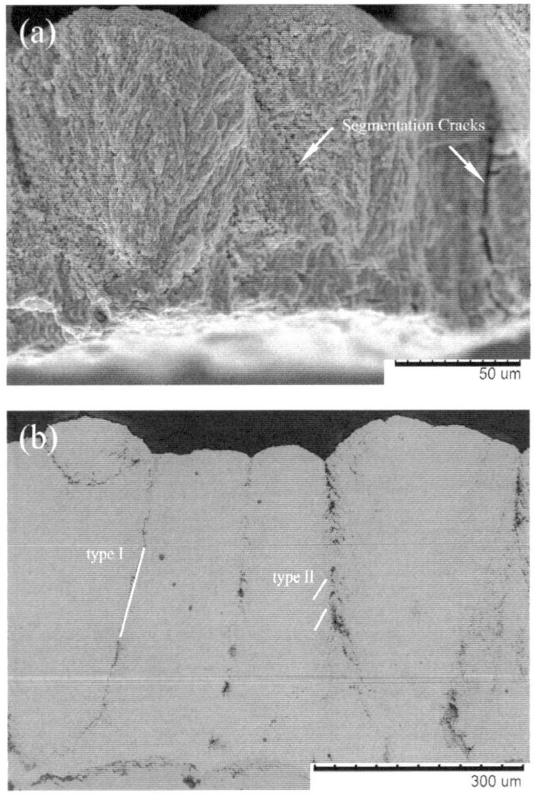

Fig. 6. (a) Fracture and (b) polished cross-section of as-sprayed ASPS coatings.

Presence of segmentation cracks were of great value for applications such as TBCs because the strain tolerance of coatings could be significantly improved. As a result, it was of great value to understand the formation of pore for the coating fabrication. Microstructures of SPS coating have been widely studied in recent years. P. Fauchais et. al. [6] have depicted a duel-featured microstructure and explained that the sticking of poorly treated particles in the beam edges was responsible for the formation of porous layer. But this explanation was obviously not the predominating law in this work, because the arrangement of pores was not in consist with the torch movement. In addition, the axial liquid injection was conductive to adequate heating to the suspension since the stream was directly injected into the hottest zone of the plasma. As a matter of fact, there were some similarities between the microstructure of coatings fabricated by ASPS and electron beam-physical vapor deposition (EB-PVD). The microstructure of EB-PVD coatings had been systematically elucidated by B. Movchan and A. Demchishin. The columnar structure of EB-PVD coatings was attributed to surface diffusion which took place at a substrate temperature of $0.3T_m$ (T_m stands for the melting point of the coating material) [23]. As a result, the forming mechanism of the patterned pores in ASPS was presumed analogous. Rationality of such analogy might lie in the presence of directional flying of impacting droplet and spheroidal feature of splat. Previous work suggested that though the temperature was lower than the conventional sintering temperature, so-called "flash sintering" could occur due to the large specific surface of ASPS splats [6]. This theory was consistent with the surface diffusion in EB-PVD.

Accordingly, the microstructure of ASPS coating could be explained using the mechanism learned from EB-PVD [24]: segmentation cracks type I originated fundamentally from the macroscopic shadowing of the columns crowns. Shadowing effect was a result of the specimen rotation. While the hefty "feather arms" (type II in **Fig.6b**) might be a consequence of shadowing of the growth steps on the center of column tips. Within this process, spheroidal splats, especially those large ones, which were randomly and occasionally deposited were considered as important origin for the initial shadowing.

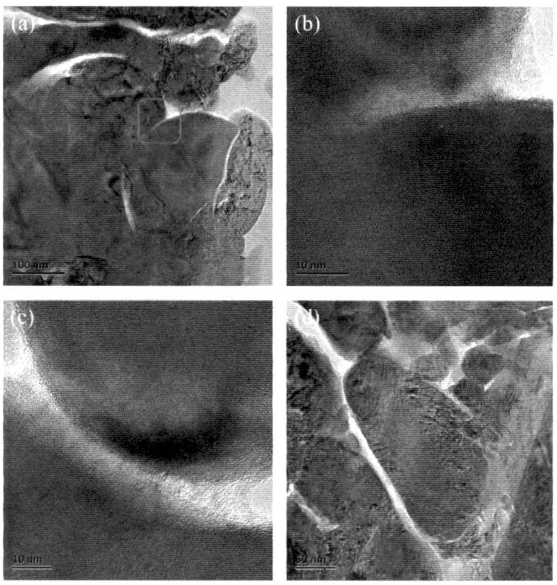

Fig. 7. High resolution TEM images of the cross-section of as-sprayed ASPS coatings. (b) is enlarged image of the highlighted area in (a).

Based on the discussion, proper roughness of the substrate was conjectured to be favorable for the formation of segmentation cracks. Work focused on this issue has been published recently [25]. Other parameters related to the temperature and velocity of droplets (e.g. torch power, size distribution and mass load of suspension) or related to the substrate temperature (e.g. feeding rate, rotation speed, air cooling) were all of pronounced influence on the detailed structure and the properties of as-sprayed ASPS coatings. This part of work was still ongoing in our group.

Under our current spraying conditions, though the porous microstructure could lead to improved strain tolerance hence notably longer cycling life [10], its unfavorable aspect should also be concerned. The large contact angle between splats, as displayed in **Fig.7a**, could be conductive to local increase of stress, resulting in formation of microcracks parallel to the substrate. In addition, because of the low kinetic energy, indeposited splats in ASPS would stick or be trapped within the coating. It was noticed that the ethanol would turn turbid during ultrasonic treatment of the as-sprayed coating samples. Premature failure of the poorly bonded branches or small columns was possible especially under the condition of thermal shock or/and exterior load. This might explain the malformed morphology of ASPS coating surface in literature 10.

Microhardness of the as-sprayed ASPS coating was 10.6±0.7 GPa (7.4±0.2 GPa for as-sprayed APS coatings). Compared with coatings fabricated by conventional APS, the size of voids in ASPS coatings were found much smaller and more uniform. Such microstructure was postulated favorable to its improved hardness. **Fig.8** shows the XRD patterns of as-deposited APS and ASPS coatings. Both coatings were primarily composed of metastable tetragonal (T') phase [26]. It was also noticed that the diffraction angle diversity between (004) at 2θ~73.2 ° and (400) at 2θ~74.2 ° in ASPS was larger than that in APS. Accordingly, the *c/a* (*c* and *a* are the unit cell parameters) ratio in ASPS was believed higher. The difference in unit cell parameters might be related to the varied phase stability and mechanical property between as-sprayed APS and ASPS coatings as well.

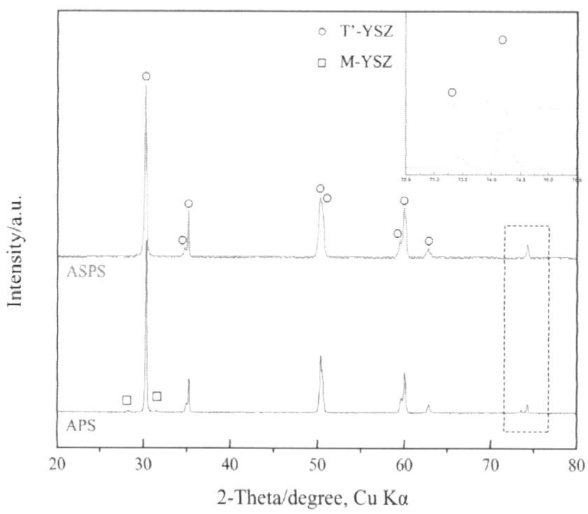

Fig. 8. X-ray diffraction patterns of as-sprayed APS and ASPS 8YSZ coatings.

4 Conclusions

In the current work, the morphology of YSZ splats and microstructure of as-sprayed YSZ coatings deposited by ASPS were experimentally investigated. The influence of substrate temperature on flattening behavior of splats were characterized by (FE)SEM. Individual ASPS splats separated from metallic substrate using a "plating and etching" approach and the as-sprayed coatings were also studied using TEM. The following conclusions could be drawn from this work:

- Compared with APS splats which usually contained long branches, ASPS splats deposited onto mirror-polished GH3128 substrate predominately exhibited near-perfect circular shape. Density of intrasplat cracks was about 0.10-0.25 μm^{-1} for APS splat, while almost no intrasplat cracks were observed on ASPS splats. In addition, the rim-structure was not detected on ASPS splats, either. Lower cooling rate during formation of ASPS splats was suggested from these characteristics.
- A drastic enlargement of deposition efficiency occurred when the substrates temperature climbed up to 300 degree C. However, further increase of substrate temperature from 300 degree C to 900 degree C exerted no significant influence on splat morphology.
- TEM study revealed that 8YSZ splats fabricated by ASPS were single crystals with tetragonal structure. A rough preferred orientation of zone axis <351> perpendicular to the substrate was also detected.
- Fundamental unit of the main as-sprayed ASPS coating was spheroidal grain, and a columnar layer at the bottom was also observed. The transition was related to the varied cooling rate of droplets. ASPS coatings exhibited a cauliflower structure resulted from two different kinds of pore arrangement. This architecture was attributed to the shadowing effect.
- Improved microhardness of 10.6±0.7 GPa was obtained in the as-sprayed ASPS coating. However, the large contact angle between spheroidal splats observed in TEM was conductive to premature failure in local areas.

5 Acknowledgements

The current work was supported by the Special Project for Army-Civilian Combination of Shanghai Municipal Commission of Economy and Informatization (Grant No. JMJH2014053).

6 Literature

[1] N.P. Padture, M. Gell, and E.H. Jordan, Materials science - Thermal barrier coatings for gas-turbine engine applications. Science 296 (2002), Issue 5566, pp. 280/4.

[2] P.L. Fauchais, J.V.R. Heberlein, and M.I. Boulos, Thermal Spray Fundamentals: From Powder to Part, Springer, (2014).

[3] C. Delbos, J. Fazilleau, V. Rat, J.F. Coudert, P. Fauchais, and B. Pateyron, Phenomena involved in suspension plasma spraying part 2: Zirconia particle treatment and coating formation. Plasma Chem Plasma P 26 (2006), Issue 4, pp. 393/414.

[4] J. Fazilleau, C. Delbos, V. Rat, J.F. Coudert, P. Fauchais, and B. Pateyron, Phenomena involved in suspension plasma spraying part 1: Suspension injection and behavior. Plasma Chem Plasma P 26 (2006), Issue 4, pp. 371/91.

[5] D. Chen, E.H. Jordan, and M. Gell, Effect of solution concentration on splat formation and coating microstructure using the solution precursor plasma spray process, Surf Coat Tech 202 (2008), Issue 10, pp. 2132/8.

[6] P. Fauchais, and G. Montavon, Latest Developments in Suspension and Liquid Precursor Thermal Spraying, J Therm Spray Techn 19 (2010), Issue 1-2, pp. 226/39.

[7] P. Carpio, Q. Blochet, B. Pateyron, L. Pawlowski, M.D. Salvador, A. Borrell, and E. Sanchez, Correlation of thermal conductivity of suspension plasma sprayed yttria stabilized zirconia coatings with some microstructural effects, Mater Lett 107 (2013), pp. 370/3.

[8] A. Joulia, G. Bolelli, E. Gualtieri, L. Lusvarghi, S. Valeri, M. Vardelle, S. Rossignol, and A. Vardelle, Comparing the deposition mechanisms in suspension plasma spray (SPS) and solution precursor plasma spray (SPPS) deposition of yttria-stabilised zirconia (YSZ), J Eur Ceram Soc 34 (2014), Issue 15, pp. 3925/40.

[9] P. Fauchais, M. Vardelle, A. Vardelle, and S. Goutier, What Do We Know, What are the Current Limitations of Suspension Plasma Spraying? , J Therm Spray Techn 24 (2015), Issue 7, pp. 1120/9.

[10] Y.X. Zhao, D.C. Li, X.H. Zhong, H.Y. Zhao, L. Wang, F. Shao, C.G. Liu, and S.Y. Tao, Thermal shock behaviors of YSZ thick thermal barrier coatings fabricated by suspension and atmospheric plasma spraying, Surf Coat Tech 249 (2014), pp. 48/55.

[11] P. Fauchais, M. Fukumoto, A. Vardelle, and M. Vardelle, Knowledge concerning splat formation: An invited review, J Therm Spray Techn 13 (2004), Issue 3, pp. 337/60.

[12] H.B. Parizi, L. Rosenzweig, J. Mostaghimi, S. Chandra, T. Coyle, H. Salimi, L. Pershin, A. McDonald, and C. Moreau, Numerical simulation of droplet impact on patterned surfaces, J Therm Spray Techn 16 (2007), Issue 5-6, pp. 713/21.

[13] A.T.T. Tran, M.M. Hyland, K. Shinoda, and S. Sampath, Influence of substrate surface conditions on the deposition and spreading of molten droplets, Thin Solid Films 519 (2011), Issue 8, pp. 2445/56.

[14] S. Goutier, M. Vardelle, and P. Fauchais, Comparison between metallic and ceramic splats: Influence of viscosity and kinetic energy on the particle flattening, Surf Coat Tech 235 (2013), pp. 657/68.

[15] H.B. Guo, H. Murakami, and S. Kuroda, Effect of hollow spherical powder size distribution on porosity and segmentation cracks in thermal barrier coatings, J Am Ceram Soc 89 (2006), Issue 12, pp. 3797/804.
[16] A.A. Syed, A. Denoirjean, B. Hannoyer, P. Fauchais, P. Denoirjean, A.A. Khan, and J.C. Labbe, Influence of substrate surface conditions on the plasma sprayed ceramic and metallic particles flattening, Surf Coat Tech 200 (2005), Issue 7, pp. 2317/31.
[17] M. Fukumoto, and Y. Huang, Flattening mechanism in thermal sprayed nickel particle impinging on flat substrate surface, J Therm Spray Techn 8 (1999), Issue 3, pp. 427/32.
[18] K. Shinoda, M. Demura, H. Murakami, S. Kuroda, and S. Sampath, Characterization of crystallographic texture in plasma-sprayed splats by electron-backscattered diffraction, Surf Coat Tech 204 (2010), Issue 21-22, pp. 3614/8.
[19] Y.X. Zhao, Plasma sprayed thick thermal barrier coatings: spraying process and failure mechanism, Master Thesis, University of Chinese Academy of Sciences, (2014), in Chinese.
[20] R.C. Garvie, Occurrence of Metastable Tetragonal Zirconia as a Crystallite Size Effect, J Phys Chem-Us 69 (1965), Issue 4, pp. 1238/43.
[21] J. Moon, H. Choi, H. Kim, and C. Lee, The effects of heat treatment on the phase transformation behavior of plasma-sprayed stabilized ZrO2 coatings, Surf Coat Tech 155 (2002), Issue 1, pp. 1/10.
[22] B. Liang, Study on Phase Stability and Thermophysical Properties of Plasma-Sprayed Nanostructured Zirconia Coatings, Ph. D. Thesis, University of Chinese Academy of Sciences, (2005), in Chinese.
[23] B.A. Movchan, and Demchish.Av, Study of Structure and Properties of Thick Vacuum Condensates of Nickel, Titanium, Tungsten, Aluminium Oxide and Zirconium Dioxide, Phys Metals Metallog 28 (1969), Issue 4, pp. 83/90.
[24] U. Schulz, B. Saruhan, K. Fritscher, and C. Leyens, Review on advanced EB-PVD ceramic topcoats for TBC applications, Int J Appl Ceram Tec 1 (2004), Issue 4, pp. 302/15.
[25] N. Curry, Z.L. Tang, N. Markocsan, and P. Nylen, Influence of bond coat surface roughness on the structure of axial suspension plasma spray thermal barrier coatings - Thermal and lifetime performance, Surf Coat Tech 268 (2015), pp. 15/23.
[26] G. Witz, V. Shklover, W. Steurer, S. Bachegowda, and H.P. Bossmann, Phase evolution in yttria-stabilized zirconia thermal barrier coatings studied by rietveld refinement of X-ray powder diffraction patterns, J Am Ceram Soc 90 (2007), Issue 9, pp. 2935/40.

Tribology and Nanoindentation Study of Suspension HVOF Thermally Sprayed Alumina Coating

T. Hussain, E.C. Shaw, Z. Pala, F. Xu Nottingham/ UK, A. S. M. Ang, Melbourne / Australia

The suspension high velocity oxy fuel (SHVOF) spraying of alumina has attracted much attention in the recent years for various tribology applications. The coatings processed via this route show unique microstructural features and the thickness of the coatings can be tailored to bridge the gap between the traditional thin film (~1-2 µm) and the conventional HVOF thermal sprayed coatings (~200 µm). In this study, a water based corundum suspension was used to deposit ~ 60 µm thick coatings onto carbon steel substrates. The aim was to develop thin coatings with superior wear properties. Hydrogen was used as a fuel gas and the effect of various processing parameters on the coating properties was examined. The coating microstructure was examined in a high resolution SEM to identify the degree of particle melting and particle morphology. XRD was used as the primary route to study the phase transformation of the feedstock suspension. At higher combustion flame energy, the coating transformed dominantly to gamma alumina and at lower flame energy the coating was composed mainly of a mixture of alpha and gamma alumina. The nanoindentation technique was applied to measure the hardness and elastic modulus of the individual phases. The nanoindentation tests, essentially, measured the sub-micrometre scale properties of the suspension HVOF sprayed alumina coatings, which are considered as "near-intrinsic" properties. The ball-on-plate wear test was performed to investigate the relationship between the wear performance and the alpha/ gamma ratios in the coatings.

1 Introduction

High Velocity Oxy-Fuel (HVOF) with liquid feedstock instead of conventional powder is a promising new technology that has the potential to produce coatings with improved coating-substrate adhesion, low in porosity and well inter-particle bonding. The motivation of suspension spraying is to process sub-micron and nanometre range particles which can provide superior properties in industrial applications. The process can deposit coating thicknesses which can fill the gap between thin-film coating and powder/ particulate based conventional thermal spraying [1, 2].

Alumina is an important engineering ceramic which is used in many key industrial applications including, insulating coatings for high temperatures, electrostatic chuck, dielectric coatings etc. [3]. Alumina is also a low cost material which makes it suitable for various wear-resistant applications. The friction and wear behaviour of alumina is of great importance for contact-mechanical applications (e.g., bearing, valves, nozzles, armour, prosthesis). In sintered form alumina contains thermodynamically stable alpha-alumina; however, when thermally sprayed the coatings consist of alpha alumina and metastable gamma and delta alumina [4,5]. The phase change from thermodynamically stable alpha to metastable alumina phases are of great interest to the suspension thermal spray community since the properties of the metastable phases differ significantly from that of the alpha alumina.

In this paper, a corundum feedstock powder in water based suspension was sprayed with hydrogen fuelled direct injection HVOF thermal spray setup. The combustion flame energy was varied to alter the ratio of alpha- gamma alumina in the coating. The coatings were examined in an SEM and XRD and the Rietveld refinement was applied to identify the fraction of phases. A comprehensive nanohardness indentation method was used to identify the mechanical properties of the individual phases. Finally, a ball on flat reciprocating wear test was performed to study the wear performance of these coatings.

2 Experimental methods

2.1 Materials and coating deposition

A commercially available corundum (CR 1, Baikowski, La Balme de Sillingly, France) was prepared in a water based suspension. The D50 particle size was 1 µm with a relative solid content of 35 wt.%. The pH value of the suspension was 8.5 and ultrapure water was used to prepare the suspension. The SHVOF coating was deposited onto the carbon steel substrate with a dimension of 60 mm by 25 mm by 2 mm. The substrates were grit blasted and cleaned with alcohol before spraying.

The suspension was sprayed with a modified UTP TopGun (Luckenbach, Germany) with a 0.3 mm diameter nozzle at the centre of the combustion chamber, i.e. axial suspension feeding was applied. A 22 mm long cylindrical combustion chamber was used in all spray conditions. Hydrogen was used as a fuel in these experiments as hydrogen results in a cleaner combustion without any traces of organic species in the coating. The samples were mounted on a rotating carousel at 73 rpm (surface speed of 1 m/s) and the gun was mounted on a traverse unit travelling at 5 mm/s which resulted in an interpass step of approximately 4 mm. Suspension was fed from a pressurised container with a mechanical stirrer (Fraunhofer IWS, Germany). The suspension was fed at a pressure of 3 bar and the suspension was stirred with a mechanical stirrer for 2 hours before spraying. In all spray runs the stand-off distance was fixed at 85 mm. Four passes of the gun was used to deposit a ~60 µm thick coating on the carbon steel substrate. Two spray parameters were chosen in this study to vary the theoretical combustion flame energy and their effect on the phase compositions. The samples were cooled using compressed air jets during the spray. Table 1 shows the detailed spray parameters used in this study.

Table. 1. SHVOF spray parameters for corundum.

Sample	O_2 Flow rate (scfh)	H_2 Flow rate (scfh)	Total gas flow (scfh)	Stoichi ometry (%)	Energy (kW)
TG 101	650	1300	1950	100	101
TG 72	466	932	1398	100	72

2.2 Sample characterisation

The samples were cross-sectioned using a precision cutting SiC saw and mounted in a conductive resin before grinding and polishing up to 1 µm diamond surface finish. The samples were further polished with a colloidal silica solution and cleaned before any microstructural investigation. A FEI XL 30 (FEI Europe, The Netherlands) SEM operated at 20 kV was employed to investigate the powder, coating and wear debris. The samples were coated with carbon before any SEM investigation.

The powder and the coatings were also investigated in a powder X-ray diffractometer (XRD) equipped with a scintillation counter detector. The instrument was operated at 40 kV and 20 mA to generate Cu *Kα* radiation with wavelength of 0.154 nm. The XRD scans were performed with a step size of 0.02◦ and 2s counting time in a 2 θ range of 20 to 90◦. The Rietveld method [6] was used on the XRD data to identify the fraction of present phases. This method is based on a least-squares fit between step-scan data of a measured diffraction pattern and a simulated diffraction pattern. The simulated XRD pattern is refined from a large number of parameters, including the background, crystal-structure parameters of each component phase, a scale factor for each constituent phase to adjust the relative intensities of the reflections. Microstructural parameters affecting the reflection profile such as crystallite size, microstrains (2nd order residual strains) and preferred orientation, are also refined.

2.3 Nanohardness

Nanoindentation on the coating was performed using an instrumented indentation platform (Hysitron TI Premier Ti, Hysitron Inc., Minneapolis, MN), which is able to record simultaneously the applied load, (P), and penetration depth, (h_i), curve for the complete loading and unloading cycle. A diamond Berkovich tip was used and the applied load of 2500 µN was found to be suitable to generate an indent impression with onset plastic deformation. Prior to indenting the samples, a scanning probe microscopy (SPM) was used to generate an image to aid the choice of indents locations to measure the individual phases of the coating. Post indent SPM imaging was also done to ensure that the accuracy of the individual indent impression was within the region of interest and that no cracking had developed. A minimum of 20 indents was carried out for each phase.

2.4 Wear testing

The friction and wear performance was carried out in linear reciprocating ball-on-flat testing using a micro tribometer (UMT-2, CETR, USA), with Si_3N_4 ball of φ 6.3 mm (Dejay Ltd., UK) for a duration of 30 mins. Samples were tested in the as sprayed condition with a fixed sliding speed of 10 mm/ s, a reciprocating stroke of 10 mm and a normal load of 10 N. The wear tracks were measured by using a focus variation profilometer (InfiniteFocus, Alicona, Austria), and the volume loss for each sample were calculated by using MountainsMap® Imaging Topography (Digital Surf, France) software. The specific wear rates (SWR) were calculated using W= V /LF, where L is sliding distance, F is applied load, and V wear volume. The wear tracks were also examined in a SEM in secondary electron mode.

3 Results and discussion

3.1 Coating microstructure

Fig.1 shows the cross-section microstructure of the alumina coatings sprayed at 101 and 72 kW combustion flame energy. The coatings appear to be well-bonded to the substrate and they do not show any sign of delamination along the coating-substrate interface. There was no significant difference on the deposition efficiency of the coatings in relation to the flame energy. The microstructure shows vertical cracks which originate mostly due to thermal and relaxation stresses generated during the spraying process of SHVOF. Clear interpass porosities were observed on both samples. This interpass porosity is a unique feature of SHVOF sprayed coating which has been reported in numerous articles. In both cases, a single pass of the gun resulted in a 15 µm coating onto carbon steel substrates.

High magnification images of the coating in **Fig. 2** show a specific microstructure. The coating sprayed at 101 kW shows mostly flattened lamellae and well bonded inter-lamellar interface. There are porosities present in the coating which are in the range from 10 nm to 2-3 µm. The flattened lamellae suggest better heat transfer from the combustion flame to the particles (longer dwell time) and increased momentum of the particles. There are bands of light grey phases and in between the light grey bands there exists a dark grey matrix. Gamma alumina has a lower density and is intrinsically vacancy rich [7], thus appears darker in the BSE image; on the other hand, alpha alumina appears light grey in the BSE image.

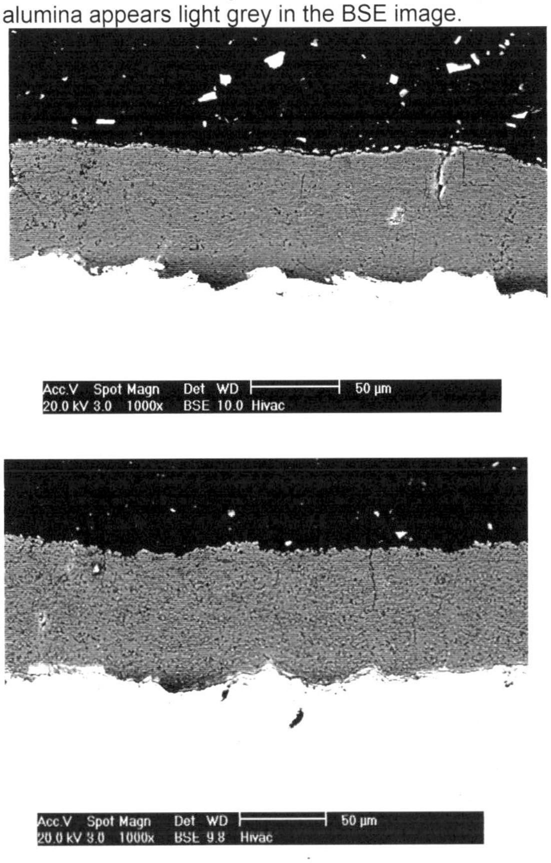

Fig. 1. BSE images of the coatings sprayed at 101 and 72 kW flame energy.

With the decrease of the total gas flow rate and the flame energy the light grey level flatted lamellae have largely decreased. Typically, in a SHVOF process the liquid carrier cools the flame as it is evaporated. By increasing the total gas flow rate or the energy of the flame the evaporation of the suspension will have a lesser cooling effect on the flame [1]. This means more heat and momentum will be transferred to the particles and the particles will arrive in molten or semi molten state. The very high impact velocity of the molten particles allows the material to flow adapting to flattened lamella. The central region of the 101 kW sample shows largely flattened lamellae of 5- 10 µm. Better particle melting also allows high inter-lamellar cohesion. The stoichiometry of the flame was kept constant in TG101 and TG72 coatings, as the excess oxygen in the flame has less influence on the oxidation of alumina. The microstructure of the TG 72 coating is different, as it contains circular unmolten particles which appear brighter in the BSE image.

Alpha phase was mostly preserved in these particles. These circular particles can be up to 2-3 µm in size. Agglomerates are probably formed in the spray as these particles are generally larger that the initial feedstock particles. There is a lot more porosity of 2-3 µm in the TG 72 coating. These porosities are circular in nature and these are closely associated with the light grey (alpha) phase.

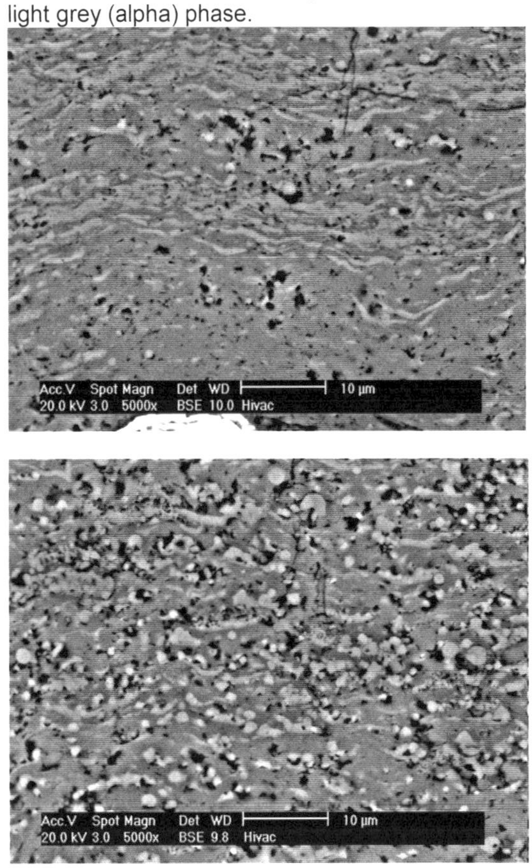

Fig. 2. High magnification images of TG 101 and TG 72 coatings.

Fig. 3 shows the top morphology of the as-sprayed alumina coating in SE imaging. The morphology of the particles show flattened lamellae and micro and nanosized spherical particles. There are less porosities on the TG 101 samples. On the other hand, many spherical particles are observed on TG 72, which could be due to the incomplete melting of the particles. It should be noted that the top surface particles did not go through subsequent impact of the incoming molten/ semi-molten particles. The particles on TG72 appear to be loosely sintered and not flattened.

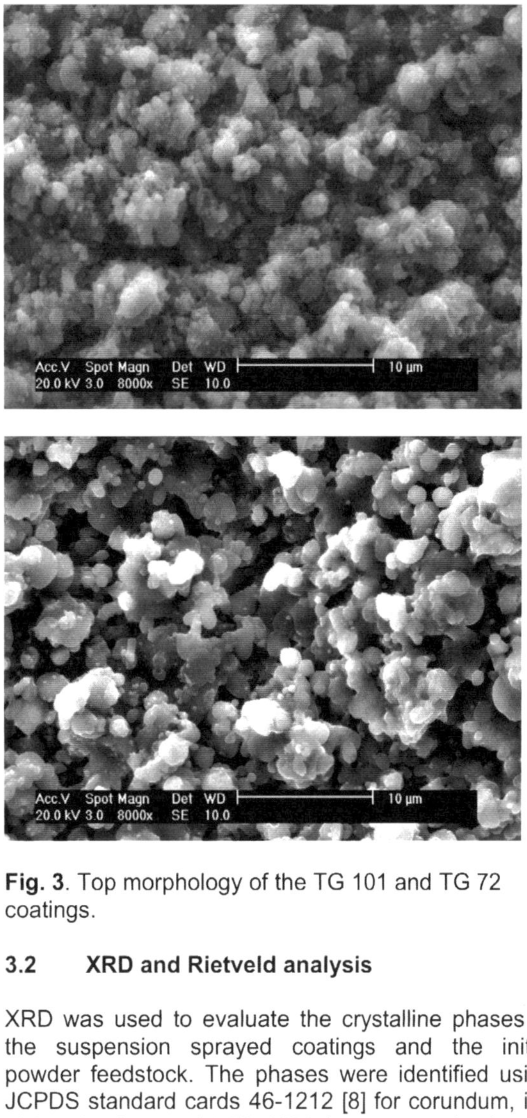

Fig. 3. Top morphology of the TG 101 and TG 72 coatings.

3.2 XRD and Rietveld analysis

XRD was used to evaluate the crystalline phases of the suspension sprayed coatings and the initial powder feedstock. The phases were identified using JCPDS standard cards 46-1212 [8] for corundum, i.e. alpha alumina, and 1-79-1558 for gamma alumina [9] which has defect spinel structure. The powder (not shown here) was composed of 100 % alpha alumina (refined lattice parameters a = 4.7582 Å, c = 12.9875 Å). **Fig. 4** shows the XRD patterns of the as-sprayed coatings. In TG 101 coating's pattern, reflections of gamma alumina are observed together with broad humps from 25 to 42° and from 57 to 65° in 2θ. However, these humps couldn't be modelled by simple pseudo-Voigt function due to finer features at (36; 39)° and (60; 62)° in 2θ. According to literature, these are manifestation of amorphous/ nanocrystalline phase(s) whose structure would be very close to mixture of tetragonal delta alumina [4] and nanocrystalline cubic gamma alumina. However, it is clear that the crystalline reflections are only from gamma alumina. On the other hand, TG72 contains a large amount of corundum (refined lattice parameters a = 4.7587 Å, c = 12.9934 Å), which was calculated to be ~70 % from the Rietveld analysis as demonstrated in **Fig. 5**. The remaining ~30 % was identified as gamma alumina. TG 72 is also the coating with the largest amount of un-melted particles and spherical splats. The percentages are relevant for the crystalline part since even for TG 72, broad amorphous humps are observed, this time without the finer features.

A wide selection of literature exists on the alpha to gamma transition of alumina during thermal spray. Sokolova et al. [11] reported two sources of alpha alumina in thermal sprayed coatings: un-melted alpha phase from the feedstock particles, which tend to increase with the decrease of input power in plasma spray. Also, secondary alpha phase can form as a result of substrate heating. The high levels of alpha alumina in C_2H_4 based SHVOF was also reported by Toma et al. [3]. It was also reported that in conventional powder based spraying alpha alumina formed from deposits only by nucleation from un-melted seeds, whereas the gamma is the result of the solidification droplet without those nuclei. In plasma spraying, the complete melting in plasma jet and subsequent rapid solidification on substrate with temperature < 300 °C results in mostly gamma phase. Coating's annealing resulted in gamma transforming firstly to tetragonal delta, then to monoclinic theta and eventually to alpha. In our study, it is believed that the decrease in total flow rate and combustion energy resulted in unmelted alpha phase from the feedstock particles. Alpha alumina is thermodynamically stable and only melts above 2015 °C. It is interesting to notice that by decreasing the total flow rate and flame energy 70% alpha phase can be retained.

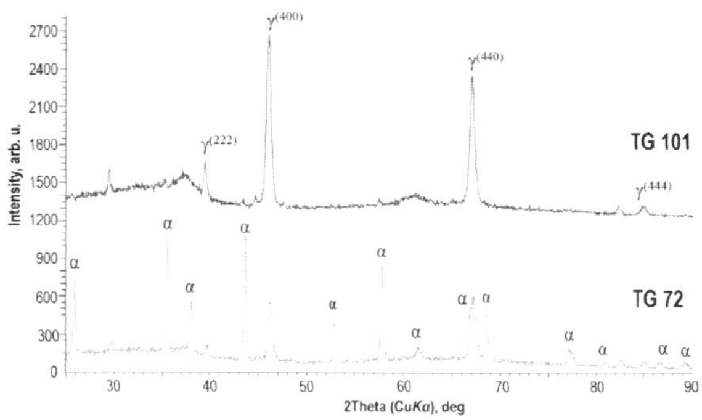

Fig. 4. XRD scans of TG 101 and TG 72 coatings.

3.3 Nano-indentation study

Oliver and Phar method [11] was used to calculate the elastic modulus of the individual phases. The apparent elastic modulus of the indenter specimen system (also known as reduced modulus, *E**) can be defined as:

$$E^* = \frac{1}{c^*\sqrt{A}}\left(\frac{dP}{dh_i}\right)$$

where A is the contact area of the indenter, dP/dh_i is the slope of the loading and displacement curve of the initial stage of unloading from applied load, P_{max}. The constant $c^*= 1.167$ for the Berkovich tip was used. Subsequently, the elastic modulus of the coating phases was determined by accounting for the elastic effects of the non-rigid indenter. Mathematically;

$$\frac{1}{E^*} = \frac{(1-\nu^2)}{E} + \frac{(1-\nu_{in}^2)}{E_{in}}$$

The characteristic material properties of the diamond indenter was E_{in}= *1141 GPa and* ν_{in}= 0.07, while the Poisson's ratio values for this current work was assumed to be ν= 0.2.

Fig. 5 shows the SPM images of TG 101 and TG 72 coating and the pattern/ grid of the nanoindentation on the polished cross-section of the samples. The SPM images are 30 by 30 µm in dimensions and two distinctive phases: white and dark are noticed in the microstructure. **Table 2** summarises the nanoindentation results of the white phase and the dark matrix on the two samples. The load was chosen such as to identify the intrinsic feature of individual splats. Typically, 2500 µN load resulted in a small material interaction volume (e.g., 177,000- 800,000 nm^2). Cross-referencing these images with the BSE images (**Fig. 2**) suggests the white phase is the light grey phase (alpha) and the dark phase is the dark grey phase (gamma). In TG72, the white phase is very hard and the nanohardness was measured to be around 20.3 GPa, whereas the matrix was only 9.1 GPa. These are the un-melted spherical particles which were found in the TG 72 coating cross-section. The higher magnification images of the nanoindentation on the matrix (10.7 GPa) and the white phase (27.6 MPa) are also shown in **Fig. 5**. There was no crack propagation along the edges of the indents. The cohesion of the SHVOF splats was much stronger and hence no cracking was observed when the lamellae were being indented. Cracking of the material can decrease the elastic modulus value of the depth sensing indentation method. The young's modulus of matrix in TG72 coating was 155 GPa and that of the white phase was 237 GPa.

The white flatted lamellae in TG 101 were slightly lower in nanohardness value (17.5 GPa). These white lamellae were flattened during impact and they were in molten state. The matrix of TG 101 is harder than that of TG72 and this phase also has a higher young's modulus. Bolelli et al. [2] reported a nanoindentation study of the SHVOF alumina coating and reported an average value of 10.5-12 GPa under various different load conditions. They reported a young's modulus of 180-190 GPa for the whole coating. The values reported in this study are higher than Bolelli's findings. It should be noted that this is the first time the intrinsic properties of the SHVOF alumina coating was reported which distinguished the different phases of alumina in the coating.

Table. 2. Nanohardness and Young's Modulus of the coatings.

	Locations	H (GPa)	E (GPa)
TG 101	White	17.5 ± 4.4	242 ± 34
	Matrix	11.2 ± 1.9	178 ± 26
TG 72	White	20.3 ± 5.9	237 ± 44
	Matrix	9.1 ± 3.3	155 ± 42

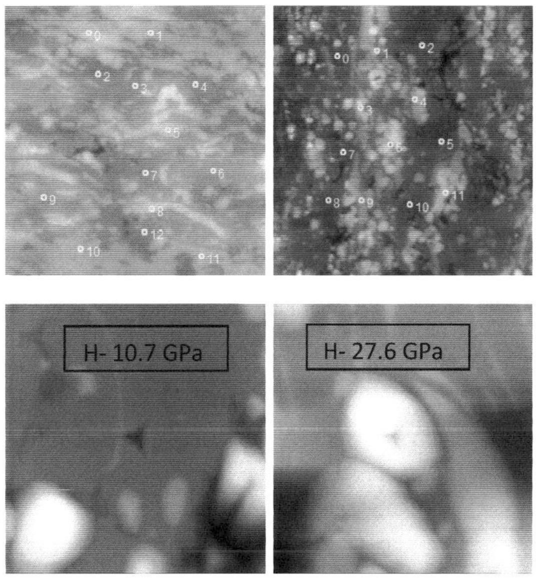

Fig. 5. Grid pattern nano-indentation of TG 101 and TG 72 coatings. Bottom images show nano-indentation of the matrix and the white phase (TG 72).

3.4 Wear performance

The specific wear rates (SWR) of both the coatings are reported in **Fig. 6**. The graph clearly shows a very small wear rate for SHOVF sprayed alumina coatings. A SWR of 3- 7 $x10^{-5}$ mm^3/ Nm value was reported by Bolelli et al [2] using sintered alumina as a counter body. The wear rates reported in this study are similar to their findings, although our study shows a remarkable difference in the wear performance of TG 101 and TG 72. TG101 coating which was composed dominantly of gamma alumina shows a lower SWR. Similarly, the wear on the counter-body was also lower in TG 101 coating (not shown here). This is rather surprising as it is generally perceived that higher fraction of alpha alumina in the coating is beneficial for wear resistance.

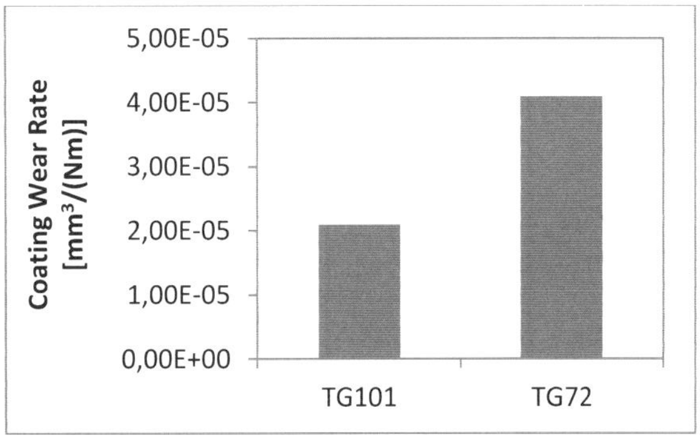

Fig. 6. Specific wear rate of the TG 101 and TG 72 coatings measured by profilometry.

The morphology of the wear track of TG 101 is shown in **Fig. 7** and the track shows flattening of asperities and loose debris on the side of the track (black arrow). A tribofilm also appears on the worn surface which is cracked at certain locations (dotted arrow). In the valleys inside the wear track, the original unaltered coating surface appears (white arrow). This tribofilm has formed possibly due to microscale plastic deformation of the coating material.

Fig. 7. SEM images of the wear track of TG 101.

Fig. 8. Wear track on TG 72 showing localise cracking

The wear track on TG 72 (shown in **Fig. 8**) appears to be wider and again a presence of a tribofilm is noticed. However, this tribofilm appears to be cracked in several places (black arrow) and there is evidence of grain pull-out in the coating (white arrow). These are probably due to fatigue cracking of the film leading to limited delamination. In conventional thermal sprayed coatings, the contact pressure is always high which can lead to brittle fracture. Also, cracks can propagate along the intra and inter-lamellar defects which leads to removal of the lamella. It was reported that SHVOF coatings possess better cohesion therefore the contact pressure reduction due to levelling of the asperities is sufficient to prevent brittle fracture [2]. In this study, both the coatings had similar surface roughness. TG101 had a surface roughness (Sa) of 3.3 µm and TG 72 had a Sa value of 3.1 µm. The higher wear rate in TG 72 is believed to be due to loosely bonded circular particles and the higher level of porosity within the coating. Further work is underway to study the inter-particle bonding and the wear mechanisms.

4 Conclusions

In this research, two different alumina coatings were deposited using SHVOF by altering the combustion flame energy. At 101 kW the coating was composed of mostly gamma alumina and at 70 kW the coating retained around 70% of alpha alumina. The alpha phase had a nanoindentation hardness between 17.5-20.3 GPa and the gamma matrix had a nanoindentation hardness of 9.1-11.2 GPa. The wear rate of the TG72 sample was higher compared to that of TG101, despite having larger amount of hard alpha alumina phase dispersed within the coating. This higher wear rate was attributed to the loosely bonded un-molten particles and the higher level of porosity within the coating.

5 References

[1] Toma, F-L. et al., Comparison of the microstructural characteristics and electrical properties of thermally sprayed Al_2O_3 coatings from aqueous suspensions and feedstock powder. J. Therm. Spray Technol. 21 (2012), Issue 3-4, pp. 480.
[2] Bolelli, G. et al., Microstructural and tribological investigation of high-velocity suspension flame sprayed (HVSFS) Al_2O_3 coatings. J. Therm. Spray Technol. 18 (2009), Issue 1, pp. 35.
[3] Toma, F-L. et al., Microstructures and functional properties of suspension-sprayed Al_2O_3 and TiO_2 coatings: an overview. J. Therm. Spray Technol. 19 (2010), Issue 1-2, pp. 262
[4] Chraska, P. et al., Alumina-base plasma-sprayed materials part I: Phase stability of alumina and alumina-chromia. J. Therm. Spray Technol. 6 (1997), Issue 3, pp.320.
[5] Ilavsky, J. et al., Alumina-base plasma-sprayed materials—Part II: Phase transformations in aluminas. J. Therm. Spray Technol. 6 (1997), Issue 4, pp. 439.
[6] Rietveld, HM., Line profiles of neutron powder-diffraction peaks for structure refinement. Acta Crystallographica. 10 (1967), Issue 1, pp.151
[7] Smrčok Ľ et al., γ-Alumina: a single-crystal X-ray diffraction study. Acta Crystallographica Section C: Crystal Structure Communications. 62 (2006), Issue 9, pp. 83.

[8] Maslen, E.N. et al., Synchrotron X-ray study of the electron density in α-Al2O3. Acta Crystallographica Section B: Structural Science. 49 (1993), Issue 6, pp. 973.
[9] Zhou, R.S. et al., Structures and transformation mechanisms of the η, γ and θ transition aluminas. Acta Crystallographica Section B: Structural Science. 47 (1991), Issue 5, pp. 617.
[10] Sokolova, T.V. et al., How certain physiochemical properties of plasma-deposited aluminium oxide depend on the deposition conditions, Inorg. Mater., 9 (1973), Issue 4, pp. 551
[11] Oliver, W.C. et al., Improved technique for determining hardness and elastic modulus using load and displacement sensing indentation experiments. J. Mat. Research, 7 (1992), Issue 6, pp.1564.

Thermal Conductivity in Suspension Sprayed Thermal Barrier Coatings: Modelling and Experiments

A. Ganvir, C. Kumara, M. Gupta, and P. Nylén, University West, Trollhättan, Sweden

Axial Suspension Plasma spraying (ASPS) can generate microstructures with higher porosity and pores in the size range from submicron to nanometre. ASPS Thermal Barrier Coatings (TBCs) have already shown a great potential to produce low thermal conductivity coatings for gas turbine applications. It is important to understand the fundamental relationship between microstructural defects in the coating such as grain boundaries, porosity etc. and thermal conductivity. Object Oriented Finite element analysis (OOF) has been shown to be an effective tool for evaluating thermal conductivity for conventional TBCs as this method is capable of incorporating the inherent microstructure as an input to the model.

The objective of this work was to analyse the thermal conductivity of ASPS TBCs using experimental techniques and also to evaluate a procedure where OOF can be used to predict and analyse the thermal conductivity for these coatings. Verification of the model was done using experimental thermal conductivity. Results showed that the varied scaled porosity has a significant influence on the thermal conductivity. Smaller grains, higher overall porosity content and lower columnar density resulted in lower thermal conductivity. It was shown that OOF could be a powerful tool to predict and rank thermal conductivity of ASPS TBCs.

1 Introduction

Columnar structured thermal barrier coatings (TBCs) are of great interest in a gas turbine industry due to their better strain tolerance ability than conventional TBCs [1]. Conventionally these columnar structured TBCs were produced for many years by electron beam vapour deposition technique (EBPVD) [2]. Suspension plasma spraying (SPS), which is relatively a cheaper technique than the EBPVD has also shown a great potential to produce a similar type of columnar structured TBCs [3], [4].

It has been a challenge for researchers to obtain lower thermal conductivity columnar structured TBCs using either of the above mentioned techniques [3], [5]. Axial Suspension Plasma Spraying (ASPS) has shown to be a promising technique to produce low thermal conductivity columnar coatings. In previous work, it was shown that ASPS technique can produce both columnar structured as well as low thermal conductivity coatings [6], [7].

SPS coatings have shown to consist of a microstructure with very fine grains, submicron or nano-sized pores, segmented cracks or inter-columnar spacing, inter-pass porosity bands etc. [4], [6]–[8]. These varied features makes the SPS coating microstructure very distinct than the conventional APS microstructure. Since thermal conductivity of a TBC is significantly dependent on the coating microstructure [6], [7], heat conduction in SPS TBCs can be significantly different. Presence of fine (submicron or nano size) microstructural defects enhances the phonon scattering in these coatings; which can lower the overall thermal conductivity. Also, since the pores present in SPS coatings are in submicron or nano size range, conductivities in the gas within the pores significantly reduces and can go below that of free gas [9], [10]. However, presence of wide enough segmented cracks or columnar spacing can increase the thermal conductivity due to the easy flow of hot gases through them [7].

Therefore, fundamental understanding of nature of heat flow through a SPS coating microstructure is necessary to achieve a low thermal conductivity coating. The objective of this work was to study the influence of various microstructural defects present in columnar SPS TBCs such as grain boundaries, pores, inter columnar spacing, etc. on thermal conductivity using experimental and modelling techniques.

2 Experimental

2.1 Sample Production

Five different types of coatings designated as S1, S2, S3, S4 and S5 were produced using the Axial III high power plasma torch (Northwest Mettech Corp., Vancouver, Canada) and Nanofeed 350 suspension feed system. Ethanol based suspension of 8 wt. % yttria stabilised zirconia (YSZ) of powder size (D50) around 500 nm and solid loading of 25 wt. % was used (Innovnano, Coimbra, Portugal). A complete TBC system had a Hastelloy®X substrate with a metallic bond coat of CoNiCrAlY material, AMDRY 9951 (Oerlikon Metco, Wholen, Switzerland) as a middle layer which was deposited using a high velocity air-fuel (HVAF) spray system (Uniquecoat, Richmond, USA); and the top layer was 8 wt. % YSZ ceramic layer. Bond coat thickness was kept constant for all samples at around 200 µm. Both, square plates (25 mm × 25 mm × 1.6 mm) and buttons (25 mm diameter and 6 mm thickness) were used as substrates.

All coatings were then prepared for further microstructural characterization. Metallographic sample preparation involved sectioning using diamond cutting blade, mounting the cross-sections in a low viscosity epoxy resin using a vacuum impregnation technique, grinding, and finally polishing semi automatically using with a Buehler PowerPro 5000 (Buehler, USA) machine.

2.2 Thermal Conductivity

Measurement for thermal conductivity was done with the help of laser flash analysis technique. The equation used [11] to calculate the thermal conductivity was:

$$\lambda = \alpha\, C_p\, \rho$$

Where $\lambda\,(W/mK)$ is thermal conductivity of the coating; $\alpha\,(m^2/s)$ is thermal diffusivity; $C_p\,(J/kg\,K)$ is specific heat capacity of the material and $\rho\,(kg/m^3)$ is the density of the coating.

A Netzsch Laser Flash Apparatus LFA 427 system (Netzsch Gerätebau GmbH, Germany) was used to measure thermal diffusivity of the coatings. Thermal diffusivity was measured both at room temperature as well as for a range of higher temperature (up to 1000°C) in air and argon atmosphere respectively.

Samples for laser flash analysis were prepared by water jet cutting 10 mm round coupons from the 25 mm square plates as a complete TBC system. The detailed description about the LFA experiment is given elsewhere [4].

2.3 Porosity

The porosity measurement was carried out by the water infiltration method. Ceramic coatings were first removed from the substrates by placing the coating in an aqua regia, a mixture of concentrated HNO_3 and HCl in the ratio (by volume) of 1:3 respectively. The samples were kept in the solution for around 6 to 12 hours and removed soon after the coatings got separated from the substrate. Over etching of the coating can increase the nano porosity in the coating, hence it was made sure that the coating was removed from the solution as soon as it was separated. The removed ceramic coating was cleaned by alcohol, dried and weighed. The ceramic coatings were then placed in a beaker containing water and kept in a vacuum for 10 min to force the water into the pores. It was then slowly infiltrated with air and vacuumed again for 10 minutes. This procedure was repeated five times to ensure the proper filling of water inside the pores. All the wet samples were then weighed one by one with the help of an electronic balance. The procedure was repeated on the same sample for three times. The detailed description about the porosity measurement is given elsewhere [4].

2.4 Column Density and Grain Size

The measurement of column density was done using scanning electron microscope (SEM) micrographs which were analysed by ImageJ, a public domain software [12]. 35 SEM micrographs were taken for each sample at 300x magnification across the 25 mm sample's cross-section. A straight line of fixed length was drawn at half of the coating thickness parallel to the interface. Each inter columnar spacing (vertical spacing between two adjacent columns) intercepting the line in the micrograph were counted as one columnar spacing. Detailed column density calculation can be found in the previous work [7].

X-ray Diffraction (XRD) was used to analyse and measure the grain size or more precisely average sizes of coherently scattering domains which were determined according to the Fundamental Parameters Approach [13]. Details about the XRD equipment and the experiments can be found in an earlier work [7].

3 Modelling

Thermal conductivity of SPS coatings was predicted by using the Object Oriented Finite element method (OOF) [14]. In this method, microscopic images taken using SEM were used to create a finite element mesh for the analysis work. OOF provides an advantage over other methods since this method can capture the detailed microscopic features like cracks, pores, different phases etc. Thus it is suitable for thermal conductivity prediction of TBCs and has been utilised extensively in previous works [15], [16]. For this study, OOF2 public domain software package [17] has been used.

3.1 Modelling Procedure

Step 1: Acquiring SEM images at two magnifications

The SPS samples were observed under the SEM in backscattered mode and images were taken at two different magnification levels, 1000x and 10000x from each sample, Fig. 1(a) and 1(b). These magnification levels were selected in order to capture the two different size scale features that can be seen in all the SPS samples; which were sub-micron/nano level pores and micron level features like large pores, vertical cracks, columnar spacing etc. At 1000x magnification level, columnar spacing can be seen easily but not the sub-micron/nano level pores (refer Fig. 1(a)). So, 10000x magnification level was selected to capture the sub-micron/nano level pores (refer Fig. 1(b)). At 10000x, SEM images were captured from an area away from the columnar spacing so that they contain only small scale pores. At each magnification level, 20 images were taken across the coating cross section (25 mm) to represent the coating microstructure. This two magnification technique was successfully used in previous work to analyse SPS TBC microstructures [7].

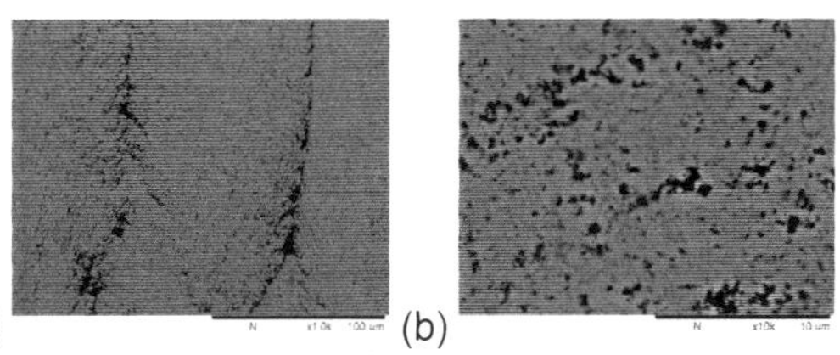

(a) (b)

Fig. 1. SEM image of a coating cross section of sample (S2) at (a) 1000x and (b) 10000x magnification level.

Step 2: Binary Image generation for 10000x images

First, 10000x images were taken into consideration. Images were converted to binary images with ImageJ using threshold value technique. Due to the binary conversion, pores in the SEM images were converted to white colour and the YSZ material areas were converted to black colour, Fig. 2(b). The threshold value was selected carefully so that the generated binary image pore distribution closely matches the actual distribution of pores. After the conversion, a finite element mesh was generated in OOF2 as described in step 3.

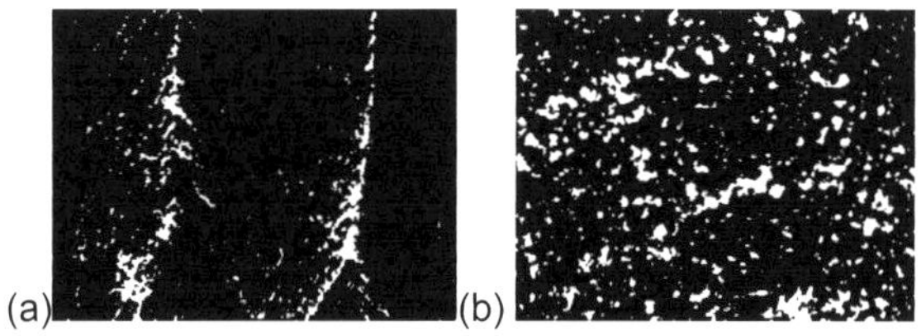

Fig. 2. Binary Image of the (a) figure 1(a), and figure 1(b). The scale bar area of the image was removed.

Step 3: Modelling of 10000x images in OOF2

The generated 10000x binary images were then imported to OOF2. Two pixel groups were produced in the OOF2, namely YSZ and PORE. Black colour pixels in the images were assigned to YSZ material group and white colour pixels were assigned to PORE material group. Thermal conductivity value of 2.5 W/Mk [10] was assigned to the YSZ material group and thermal conductivity of atmospheric air, 0.025W/mK [15], was assigned as the thermal conductivity for the PORE material group assuming that pores contain air. An adaptive meshing procedure was used to create a finite element mesh based on the image features and colours. Finer elements were generated near the interface boundaries of the PORE and YSZ pixel groups in order to account for higher thermal gradients, Fig. 3(a).

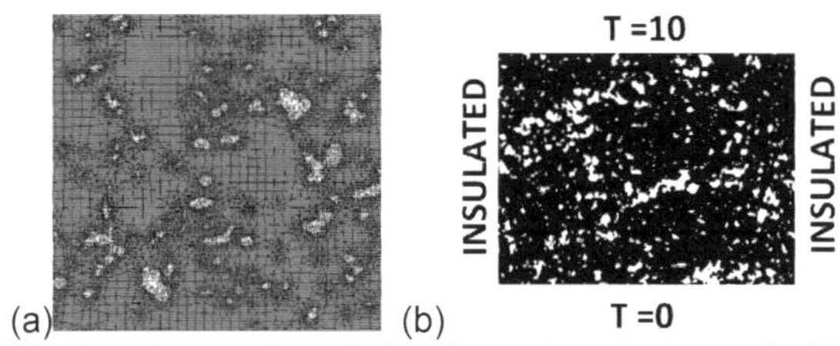

Fig. 3. (a) Part of the finite element mesh generated in OOF2 (YSZ material area is shown in Red colour for the visualization purpose) and (b) boundary conditions

Thereafter, boundary conditions were applied as shown in Fig. 3(b) for analysis. Models were then analysed and the heat flux across the cross section was calculated with OOF2. The thermal conductivity value was calculated by solving the Fourier's law of heat conduction which is:

$$\dot{Q} = -\lambda\, A\, \Delta T/\Delta X \mathrm{Q}$$

Here $\dot{Q}$ (W) is the heat flux across the cross section, λ (W/mK) is thermal conductivity of the coating area considered, A (m^2) is the cross-sectional area parallel to the top and bottom boundaries, ΔT (K) is the temperature gradient and ΔX (m) is the thickness of model perpendicular to the coating surface.

Step 4: Binary Image generation for 1000x images

The 1000x magnification images were then taken into consideration. These images were also converted into binary images with ImageJ as described in step 2 above. After converting the SEM images to binary images, it was observed that some small scale features which can be seen and already considered in 10000x images also appear in 1000x. Even though their area fraction was considerably lower than the area of large scale features, these small scale features were filtered out from the 1000x binary images so that the error due to the repetition of the features in thermal conductivity analysis can be minimized.

Image analysis porosity value was calculated by using the following equation:

$$\text{Image Analysis Porosity} = \text{Average Porosity at 1000x} + \text{Average Porosity at 10000x}$$

Step 5: Modelling of 1000x images in OOF2

1000x binary images were then imported into the OOF2. The calculated thermal conductivity at 10000x was used as the YSZ thermal conductivity instead of bulk YSZ as the fine scale porosity was not considered in the 1000x images. PORE conductivity was kept unchanged. Same boundary conditions were given as described in step 3 and the model was analysed in OOF2. Thereafter, the overall thermal conductivity was calculated using the heat conduction equation as described earlier and it was taken as the thermal conductivity of the coating.

Following major assumptions were made for the modelling calculation.

1. Heat transfer through the coating thickness happens only by means of conduction. Convection and radiation heat transfer were considered to be negligible.
2. Heat transfer only happens through the coating thickness direction. Transverse heat transfer in the coating can be neglected due to significantly larger dimensions in transverse directions.
3. Knudsen effect [18] was not taken in to account when considering the thermal conductivity of air inside the pores.

4 Results and Discussions

4.1 Microstructure and porosity

Fig. 4 shows the cross-section of all coatings and a simplified schematic of a columnar structure highlighting the important defects in the microstructure. It can be observed from Fig. 4 that all coatings have a columnar structure. An important thing to be noted here is that no fine cracks as in conventional APS coatings were observed in these columnar SPS coatings. This was confirmed by looking at higher magnifications for all coatings in SEM, Fig. 1(b).

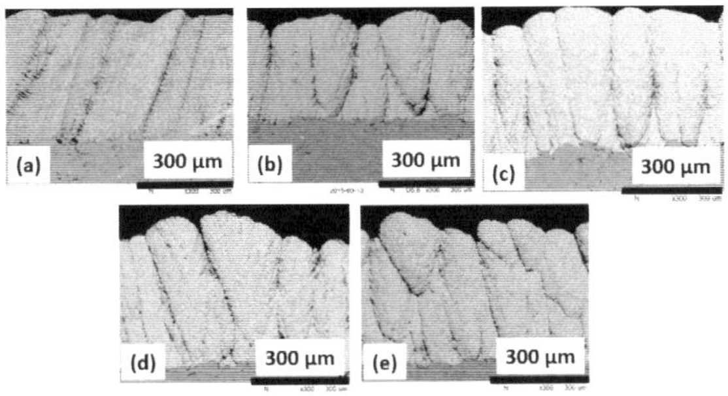

Fig. 4. Cross-sections of SPS TBCs showing columnar structure for all five types S1, S2, S3, S4 and S5 respectively shown in figure (a), (b), (c), (d) and (e)

Important defects in TBCs which can influence the thermal conductivity of the coating are porosity at a larger scale and grain boundaries at a smaller scale [6], [7]. These defects act as a barrier to the heat flow which lowers the overall thermal conductivity of the coating. In conventional APS coatings, the important microstructural defects are delamination, globular pores, fine cracks and fine scaled porosity (in the micron range) [19]. Unlike conventional APS coatings, porosity in SPS coatings includes micron, submicron or nano pores within the columns and vertical cracks or columnar spacing between the columns (refer Fig. 1 at 1000x and 10000x). Distribution of these varied scaled porosity in SPS columnar coatings hence becomes an important issue when it comes to analysing the thermal conductivity in such coatings.

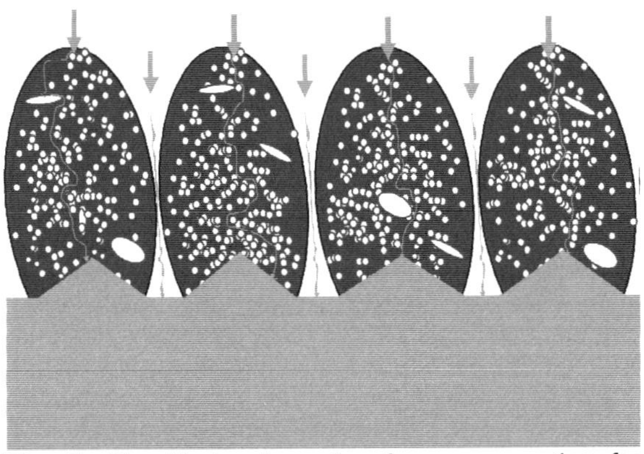

Fig. 5. Simplified schematic of a cross-section for columnar structure in SPS TBC showing columnar spacing, small and large scaled pores. Also showing the heat flow between the columns and within the columns

Simplified schematic shown in Fig. 5 shows the important microstructural features in a columnar structured coating. These features are columnar spacing (a spacing between two adjacent columns separating each other), small and large scaled pores within the column. Also, grain boundaries which are considered to be an important microstructural defect are not visible at such low magnification and hence not shown in the schematic. Fig. 5 helps in understanding the role of different scaled porosity in columnar structured coatings on thermal conductivity. It can be seen from the schematic shown, that the possible paths for heat flow through these columnar structures are firstly in between the columns (through the spacing) and secondly within the columns. In addition, it can be observed from the schematic that there are lot of smaller and larger pores within the columns. These pores can enhance the scattering of phonons and disturb the heat flow within the columns and hence help in reducing the overall coating thermal conductivity. However, inter columnar spacing can allow easier paths for the heat flow during service conditions due to pressure difference and hence increase overall coating thermal conductivity.

4.2 Measured thermal Conductivity and Grain Size

Table 1 shows the experimental data measured for all coatings. Comparing overall experimental porosity of all five coatings S1, S2, S3, S4 and S5, it can be seen that S1 showed highest (25.2%) followed by S2 (22.7%) whereas S5 showed lowest (14.2%) porosity. Coatings S3 and S4 on the other hand showed similar porosity around (15.5%) which was lower than S1 and S2 but higher than S5. Also, comparing the column density among these coatings, S5 showed the highest (13.0 cracks/mm) whereas S1 and S3 to be lowest (9.8 and 9.1 cracks/mm respectively), however coatings S2 and S4 showed 11.6 and 11.3 cracks/mm respectively.

It can be seen that coating S1 has the lowest thermal conductivity among all five coatings when comparing the experimental thermal conductivity results. The reasons for the low thermal conductivity can be highest overall porosity and smallest grain size among all coatings which means higher number of grain boundaries compare to all other coatings. Apart from this, S1 has also showed lower column density relative to other coatings. As explained earlier in section 1.2 and 4.1 that the presence of higher grain boundaries, higher small scaled pores and lower column density helps in reducing the overall thermal conductivity of the coating. These effects explain the lowest thermal conductivity of the coating S1. On the other hand, coating S5 has showed the highest thermal conductivity. Again, it can be seen that S5 has lowest porosity, large grain size and highest column density which explain the higher thermal conductivity in coating S5.

The effect of column density on thermal conductivity is however still not fully understood as concluded from a previous detailed characterisation work done by the author [6]. Moreover, the effect of enhanced heat flow in inter columnar spacing is not considered in experimental measurements due to absence of the pressure difference usually present in TBC service conditions. This would mean that thermal conductivity of these coatings would be a bit higher than measured values in actual service conditions.

Table 1. Experimental data of the of the samples

Sample Designation	S1	S2	S3	S4	S5
Thickness (μm)	388 ± 21	326 ± 22	398 ± 15	416 ± 21	423 ± 28
Column density (cracks/mm)	9.8 ± 2.3	11.6 ± 2.0	9.1 ± 2.7	11.3 ± 2.6	13.0 ± 2.8
Experimental porosity (vol. %)	25.2 ± 1.5	22.1 ± 0.5	15.7 ± 2.3	15.8 ± 1.1	14.3 ± 1.9
Grain size (nm)	100 ± 1	128 ± 3	116 ± 3	123 ± 3	122 ± 3
Experimental Thermal Conductivity (W/mK)	0.97	1.1	1.32	1.43	1.53

4.3 Modelling of Thermal Conductivity

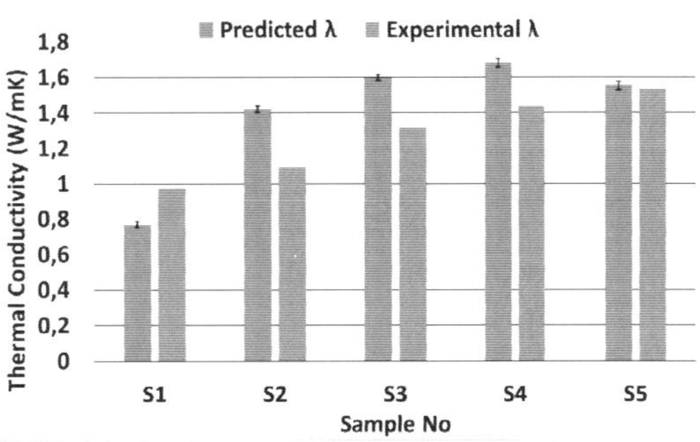

Fig. 6. Thermal conductivity results

The predicted thermal conductivity values are shown in Fig. 6 with their experimental thermal conductivity values. Out of the five samples, four samples showed same trend as the measured value for thermal conductivity of the samples. This indicates that the method that has been used for this thermal conductivity prediction is capable enough to catch the different microstructural features and rank the coatings according to their thermal conductivity value. Except the sample S1, in all other samples the predicted thermal conductivity value was above the experimental value (see Fig. 6). The reasons for over prediction of thermal conductivity by modelling could be attributed to the limitation of 2D geometry of the model and other limitations stated in section 3.1, step 5.

There are a few reasons which can explain the under prediction of the thermal conductivity for S1. Fig. 7 shows the experimental and image analysis porosity values of these coatings. The Image Analysis porosity value was 35% higher than the experimental porosity value (Refer Fig. 7). SEM images of the coating S1 showed many closed pores compared to a SEM images of a sample S2 which had the experimental porosity value close to S1 (Refer Fig. 8). Also this coating didn't show clear columnar crack like features (Refer Fig. 9). Larger closed pores may result in lower experimental porosity values since these pores may not be accessible by water in the water impregnation technique. This effect would result in higher experimental density values than reality. From equation 1, it can be concluded that experimental thermal conductivity value will then be higher than reality due to the higher experimental density. This could be a reason for the predicted thermal conductivity value to be less than the experimental value for coating S1.

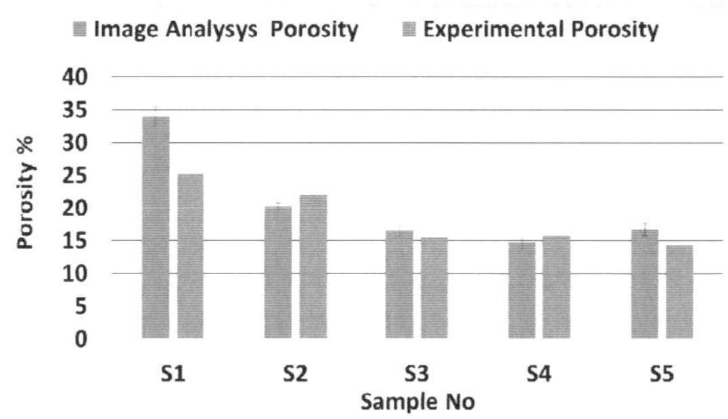

Fig. 7. Porosity results by image analysis and experimental water impregnation technique

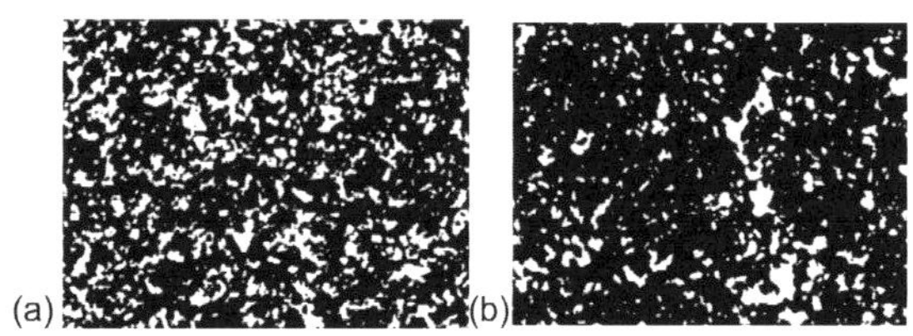

Fig. 8. Binary image of coating S1 at 10000x (a) and coating S2 at 10000x (b) having close experimental porosity levels

These results show that the modelling technique used in this work is capable of capturing the microstructural features and predict the thermal conductivity of the coatings in a reasonable way. Additionally, modelling is a more effective technique than experiments when comparing and ranking thermal conductivity of different coatings as it effectively captures the effect of closed pores on density and thermal conductivity.

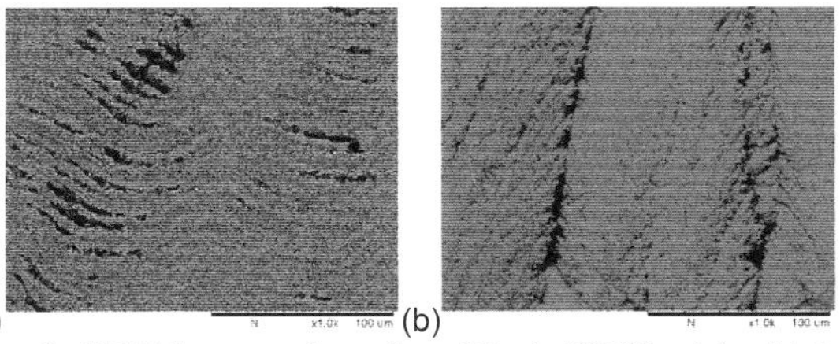

(a) (b)

Fig. 9. SEM image of coating S1 at 10000x (a) which does not have any clear columnar crack like features when compared to SEM image of a SPS coating S2 (b)

5 Conclusions

In this study, five SPS TBC samples' thermal conductivity values were analysed by using both modelling with OOF2 and standard experimental techniques. Two different magnification level SEM images were used for the modelling work. The experimental thermal conductivity values were calculated by LFA. Grain size and overall porosity were observed to be the most significant factors affecting experimentally measured thermal conductivity. The modelling method developed in this work was able to predict and rank the SPS coatings' thermal conductivity values efficiently. This method was also able to capture the effect of closed pores on thermal conductivity, which was not included in the experimental method due to the limitation in measurement of density using water impregnation technique. The results from this work show that OOF seems to be a powerful tool to predict and rank thermal conductivity of SPS TBCs.

Acknowledgements

The authors would like to acknowledge Nicholas Curry and Stefan Björklund for spraying the coatings.

References

[1] N. Curry, Z. Tang, N. Markocsan, and P. Nylén, "Influence of bond coat surface roughness on the structure of axial suspension plasma spray thermal barrier coatings — Thermal and lifetime performance," *Surf. Coat. Technol.*, vol. 268, pp. 15–23, Apr. 2015.

[2] U. Schulz, C. Leyens, K. Fritscher, M. Peters, B. Saruhan-Brings, O. Lavigne, J.-M. Dorvaux, M. Poulain, R. Mévrel, and M. Caliez, "Some recent trends in research and technology of advanced thermal barrier coatings," *Aerosp. Sci. Technol.*, vol. 7, no. 1, pp. 73–80, Jan. 2003.

[3] K. VanEvery, M. J. M. Krane, R. W. Trice, H. Wang, W. Porter, M. Besser, D. Sordelet, J. Ilavsky, and J. Almer, "Column Formation in Suspension Plasma-Sprayed Coatings and Resultant Thermal Properties," *J. Therm. Spray Technol.*, vol. 20, no. 4, pp. 817–828, Jun. 2011.

[4] A. Ganvir, N. Curry, N. Markocsan, P. Nylén, and F.-L. Toma, "Comparative study of suspension plasma sprayed and suspension high velocity oxy -fuel sprayed YSZ thermal barrier coatings," *Surf. Coat. Technol.*, vol. 268, pp. 70–76, Apr. 2015.

[5] T. R. Kakuda, A. M. Limarga, T. D. Bennett, and D. R. Clarke, "Evolution of thermal properties of EB-PVD 7YSZ thermal barrier coatings with thermal cycling," *Acta Mater.*, vol. 57, no. 8, pp. 2583–2591, May 2009.

[6] A. Ganvir, N. Curry, N. Markocsan, P. Nylén, S. Joshi, M. Vilemova, and Z. Pala, "Influence of Microstructure on Thermal Properties of Axial Suspension Plasma-Sprayed YSZ Thermal Barrier Coatings," *J. Therm. Spray Technol.*, vol. 25, no. 1–2, pp. 202–212, Nov. 2015.

[7] A. Ganvir, N. Curry, S. Björklund, N. Markocsan, and P. Nylén, "Characterization of Microstructure and Thermal Properties of YSZ Coatings Obtained by Axial Suspension Plasma Spraying (ASPS)," *J. Therm. Spray Technol.*, vol. 24, no. 7, pp. 1195–1204, Jun. 2015.

[8] H. Kassner, R. Siegert, D. Hathiramani, R. Vassen, and D. Stover, "Application of Suspension Plasma Spraying (SPS) for Manufacture of Ceramic Coatings," *J. Therm. Spray Technol.*, vol. 17, no. 1, pp. 115–123, Mar. 2008.

[9] I. O. Golosnoy, S. A. Tsipas, and T. W. Clyne, "An analytical model for simulation of heat flow in plasma-sprayed thermal barrier coatings," *J. Therm. Spray Technol.*, vol. 14, no. 2, pp. 205–214, Jun. 2005.

[10] I. O. Golosnoy, A. Cipitria, and T. W. Clyne, "Heat Transfer Through Plasma-Sprayed Thermal Barrier Coatings in Gas Turbines: A Review of Recent Work," *J. Therm. Spray Technol.*, vol. 18, no. 5–6, pp. 809–821, Dec. 2009.

[11] R. Taylor, "Thermal Conductivity Determinations of Thermal Barrier Coatings," *Mater. Sci. Eng. A*, vol. 245, no. 2, pp. 160–167, 1998.

[12] "ImageJ Software," *Image Processing and Analysis in Java*. [Online]. Available: http://imagej.nih.gov/ij/. [Accessed: 17-Nov-2014].

[13] R. W. Cheary, A. A. Coelho, and J. P. Cline, "Fundamental Parameters Line Profile Fitting in Laboratory Diffractometers," *J Res Natl Inst Stand Technol*, vol. 109, no. 1, p. 25, 2004.

[14] P. Michlik and C. Berndt, "Image-based extended finite element modeling of thermal barrier coatings," *Surf. Coat. Technol.*, vol. 201, no. 6, pp. 2369–2380, Dec. 2006.

[15] M. Gupta and P. Nylén, "Design of Low Thermal Conductivity Thermal Barrier Coatings by Finite Element Modelling," presented at the SMT24, Dresden, September 7-9, 2010, 2011, pp. 353–365.

[16] M. Gupta and P. Nylén, "Structure-property Relationships in Thermal Barrier Coatings by Finite Element Modelling," presented at the The Twenty Fifth International Conference on

Surface Modification Technologies, 2012, pp. 175–184.
[17] "OOF: Finite Element Analysis of Microstructures," *OOF: Finite Element Analysis of Microstructures*. [Online]. Available: http://www.ctcms.nist.gov/oof/oof2/. [Accessed: 17-Nov-2014].
[18] J. H. Qiao, R. Bolot, H. L. Liao, and C. Coddet, "Knudsen Effect on the Estimation of the Effective Thermal Conductivity of Thermal Barrier Coatings," *J. Therm. Spray Technol.*, vol. 22, no. 2–3, pp. 175–182, Jan. 2013.
[19] N. Curry, N. Markocsan, X.-H. Li, A. Tricoire, and M. Dorfman, "Next Generation Thermal Barrier Coatings for the Gas Turbine Industry," *J. Therm. Spray Technol.*, vol. 20, no. 1–2, pp. 108–115, Jan. 2011.

The main issues to address in modeling plasma spray torch operation

C Chazelas, A Vardelle / University of Limoges (Limoges, France), and J P Trelles / University of Massachusetts Lowell (Massachusetts, U.S.A.)

The modeling of plasma torch operation has advanced greatly in the last fifteen years due to a better understanding of the underlying physics, development of commercial, open-source computational fluid dynamics (CFD) software, and access to high performance and cloud computing. However, the operation mode of the electric arc in plasma torches is controlled by dynamic, thermal, electromagnetic, acoustic and chemical phenomena that take place at different scales and whose interactions are not completely understood yet. Even though no single model of plasma torch operation fully addresses these phenomena, most of these models are useful tools for parametric studies, if their use is underpinned by knowledge of torch operation and the model predictions are validated against experimental data. To increase the level of predictability of the current models, several further steps are needed. This study examines the issues remaining to address in the modeling of plasma spray torch operation and the current critical aspects of these.

1 Introduction

Plasma spray is one of the most versatile techniques available for the application of protective and functional coatings. The processing medium is characterized by a high energy density (10^6-10^7 Jm^{-3}) and high heat flux (10^7-10^9 Wm^{-2}) that make it well adapted to ceramics deposition. A recent thermal spray coating market analysis by Grand View Research Inc., San Francisco predicts that ceramics coatings will be the fastest growing segment in terms of revenue at a rate of 7% for the next seven years [1]. Therefore, the use of plasma spray should expand further in particular for high temperature and extreme wear and corrosion applications.

Although plasma torches are now well established tools in industry, they suffer from two inherent electrode effects, arc instability and electrode erosion. Recent torch designs involve cascaded anode and multi electrodes to limit the movement of the arc attachment at the anode and decrease the thermal load to electrodes. However, electrode effects still affect the process stability and economics [2]. A better control of these effects requires improved understanding of the arc behavior and its effects on the plasma jet. As measurements inside plasma torches are limited to the time-variation of arc voltage, inner pressure, torch thermal efficiency and acoustic emission, mathematical modeling is essential for further improvement of plasma torches.

A large body of literature deals with plasma torch modeling [3]. However, its operation is controlled by dynamic, thermal, electromagnetic, acoustic and chemical phenomena that are not fully addressed yet by the current models and several further steps are needed to achieve calculations with a high level of predictability so that everyone, and not only the authors, believe them.

This paper discusses some important issues in the modeling of arc-cathode interactions, arc column and arc-anode interactions. It assumes that the model is 3D and time dependent in order to capture the stochastic behavior of the arc.

2 Some remarks about predictive models

The ideal plasma torch model should involve the electrodes, cathode and anode regions and, arc column. The latter takes up a large part of the gas computational domain and can be considered in local thermodynamic equilibrium (LTE). Most of the current models are based on this assumption and, therefore, require some artificial treatment of the electrode-arc regions to allow the passage of the electric current. However, this assumption decreases the level of reliability and predictability of the model. A more realistic approach should take into account the departure from thermal equilibrium in the regions where the cold gas and plasma interact and, close to the electrodes. It requires at least a two-temperature model with the electron temperature (T_e) and heavy species temperature (T_h), considering electrons and heavy species with a specific Maxwell speed probability distribution. The departure from thermal equilibrium is, then, characterized by the ratio $\theta=T_e/T_h$. which can be higher than 10-15 as shown by the calculations of Trelles and al [4] (**Fig. 1**). Near electrodes, the departure from ionization equilibrium and quasi neutrality can be modeled by using sheath models that also make it possible to get rid of artificial boundary conditions at electrode walls.

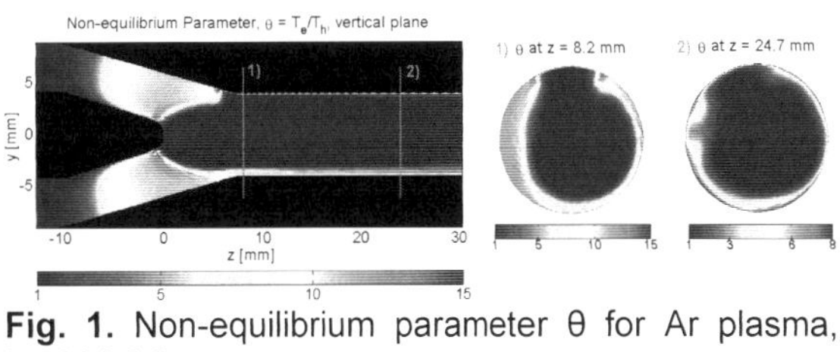

Fig. 1. Non-equilibrium parameter θ for Ar plasma, I=400A [4]

If models coupling electrodes and arc and using 2T models with electrode sheaths have been reported in transferred arc plasma torch, e.g. [5], their application to non-transferred arc plasma spray torch is much trickier because of the stochastic movement of the arc attachment at the anode wall (see section 5). In this paper, the electrode sheath models are not presented (see for example [6]), even if they are believed to be a necessary step for a fully predictive model.

3 Arc-cathode interactions

A large part of the current plasma spray torch numerical models consider only the gas domain with the electrodes included as boundary conditions. They impose a current density profile and a temperature on the cathode tip. The current profile has generally a Gaussian-like form:

$$J = J_0 \exp\left(-\left(\frac{r}{R}\right)\right)^n \qquad \textbf{Eq. 1}$$

Where J is the current density over the cathode surface, r the radial coordinate measured from the torch axis and R the radius of the spot on which the current density is imposed .

The adjustable parameters J_0, R and n of Eq.1 control the shape of the profile. However, under the operating conditions of plasma spray torch the arc radius and central temperature are largely dominated by the flow driven by the pinch pressure of the self magnetic field of the arc. The latter is determined by the current density at the cathode which in turn is controlled by the geometry of the cathode and arc current. Therefore, the arc properties are largely determined by the boundary conditions that are assumed at the cathode surface as already discussed in the early 90's for stationary 2-D models of transferred arcs [7].

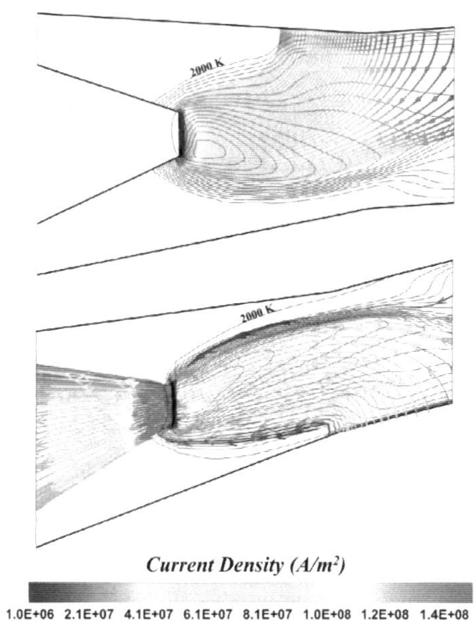

Fig. 2. Current density distribution (vector color scale) close to the cathode tip and isotherms (black lines). Top: with imposed current density profile at the cathode surface. Bottom: with arc-cathode coupling. (Conditions: gas: Ar-H_2; total current: 600 A; flow rate: 60 slpm [8).

A way to calculate the arc behavior without imposing current density and temperature at the cathode is to include the cathode in the calculation domain, impose the arc current (j=I/S where I is the arc current and S the cathode rear cross sectional section) and temperature at the rear face of the cathode, calculate the heat and current density distribution in the cathode and implement a cathode sheath model that predicts the electron and ion density and, temperature at the outer boundary of the sheath as well as the cathode voltage drop.

A first step as proposed in [8] for a LTE model is to solve the electromagnetic and energy equation in the cathode to predict the current distribution at cathode tip and calculate the heat flux between the cathode and the arc by the following equation:

$$-\lambda_c \frac{\partial T_c}{\partial z} = -\lambda_p \frac{\partial T_p}{\partial z} + Q_{rad} + Q_e \qquad \textbf{Eq. 2}$$

Where the first term of the right side represents the heat transferred by conduction by the gas, the second the net radiative flux between the plasma and the cathode and the last term corresponds to a heat flux characteristic of the thermionic cathode operation. If the cathode does not evaporate, this r term can be calculated from:

$$Q_e = -J_{em}.\left(\frac{2k_b}{e}T_c + \Phi_c\right) + J_i.\left(\frac{2k_b}{e}T_c + U_c + \Phi_i - \Phi_c\right) \qquad \textbf{Eq. 3}$$

Where the first term corresponds to the heat flux carried away by the thermionic electrons and the second the heat flux brought by the ions formed in the ionization zone that follows the cathode sheath, assuming that the heat flux brought by the back diffusion electrons and carried away by the secondary emission electrons are negligible.

In **Eq. 3**, T_c is the cathode surface temperature, Φ_c the cathode material work function, U_c the cathode voltage drop and Φ_i the first ionization energy. The electron current density J_{em} can be calculated from the Richardson-Dushman law.

Fig. 2 shows the current streamlines when a current density profile is imposed on the tip of the cathode (A) and when cathode is coupled with the arc (B). The predicted profile of current density at the cathode tip is close to that imposed at this location when the cathode is not coupled with the arc. However, this profile results from calculations, the only input data, in that case, being the current value.

The magnetic coupling of the cathode and arc allows projecting the magnetic field both in the cathode and arc with, as expected from the Biot-Savart law, a better description of the magnetic field close to the cathode tip that directly affects the Lorentz ($\vec{J} \wedge \vec{B}$) force and, thus, the arc velocity.

4 Arc column

The description of both the arc column and plasma jet are based on the coupling of fluid and electromagnetic equations and requires the thermodynamic (i.e., density, enthalpy and specific heat) and transport properties (viscosity, thermal and electrical conductivities and diffusion coefficients) of the gas mixture as well as its radiative properties.

4.1 Thermodynamic and Transport properties

The calculation methods of thermodynamic and transport properties of LTE plasmas are now well established. The former require the equilibrium composition of the gas mixture, the internal partition functions and enthalpies of formation of the different species that make up the plasma gas. Some corrections (i.e., virial, Debye–Huckel) to the ideal gas law are generally taken into account.

The transport coefficients are calculated by using the Chapman-Enskog method. It should be noted that inelastic collisions are taken into account for the gas thermal conductivity ($k_{tot} = k_h + k_{int} + k_r + k_e$) expressed as the sum of four terms corresponding to the contribution of heavy particle translation (k_h), electron translation (k_e), internal thermal conductivity k_{int}, and reaction thermal conductivity (ionization and dissociation) (k_r) [9].

Finally, the diffusion coefficients are now classically calculated with the method proposed by Murphy that uses combined diffusion coefficients [10]. It should also be noted that the metallic vapor issued from the electrodes can significantly modify the gas properties but taking them into account requires predicting the electrode phase changes (melting, vaporization) caused by the electric arc heat load. Also the vapor can recondensate and so make the gas properties calculation difficult [11].

LTE gas properties are available for practically all gases used in plasma spraying (Ar, H_2, N_2, H_2O, He, CH_4) and their binary and ternary mixtures (Ar–H_2, Ar–He, Ar-O_2, Ar-N_2, air-N_2, air-Ar, air-O_2, air-CH_4, Ar–H_2-He, Ar-H_2-Cu) [12-17]. However, the values of the interaction potentials and collision integrals that are the basis of calculations are not always well known, and may lead to uncertainties in transport coefficients. Also the data must be known up to at least 30,000K.

Contrary to LTE plasmas, the 2T plasma composition and properties are still discussed in the plasma community [11]. 2T plasma compositions are calculated by different models and use different relations to describe the ionization and dissociation reactions. These models include the Potapov's method, Van de Sanden et al.'s method, Gibbs free energy minimization method, and kinetics model. Annaloro recently showed that the Van de Sanden's method is the best adapted to 2T plasma composition calculation [18]. If there is now a generally agreed opinion on the methods of calculation of the 2T thermodynamic functions, there is still some discrepancy about the specific heat which is either calculated as a unique coefficient or divided in two coefficients for the electron and heavy particle contributions, respectively.

The 2T transport properties are calculated using essentially two approaches: the Devoto's theory that does not consider the coupling between the electrons and heavy particles in the Boltzmann equation and, the Rat's theory that considers it. The Devoto's approach cannot satisfy mass conservation and the simplified expressions predict transport properties that can be quite different (especially the reactive thermal conductivity) from those obtained with the Rat's approach. Meher [19] And Colombo [20] compared the two theories and concluded that Devoto's approach is yet adequate to determine the properties of nonequilibrium plasma. At present, the transport properties of NLTE plasmas calculated with the Devoto's model are available for some pure gases (Ar, O_2, N_2, H_2) and their mixtures (Ar–H_2, Ar–O_2, Ar–N_2, N_2–O_2) with electron temperature up to 45 000 K and θ >15 for some gas mixtures [19-24] while those calculated with the Rat's approach are available for Ar, Ar–He, Ar–Cu and Ar–H_2–He plasmas but generally for θ<3. [25-27].

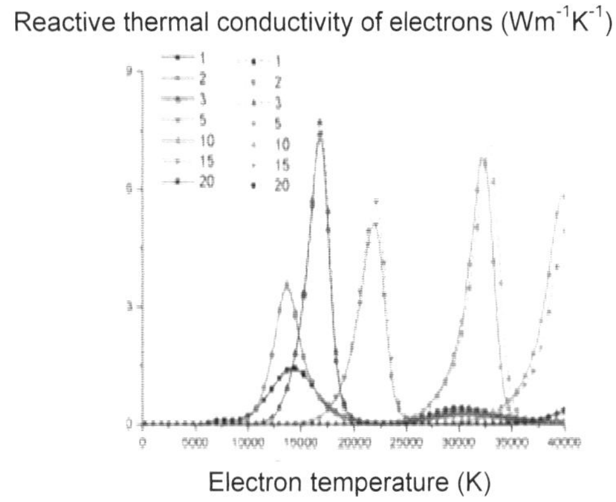

Fig. 3. Temperature dependence of the reactive thermal conductivity of nitrogen plasmas under different degrees of nonequilibrium (solid line and symbols: Debye length including only electrons; dashed line and symbols: Debye length including electrons and ions) [21].

Recently, Wang et al. [21] showed, for a nitrogen plasma, that the ionization potential reduction due to the Coulomb interaction of charged species cannot be ignored for $\theta \geq 3$ as it affects the partition functions and, thus, the plasma composition and, thermodynamic and transport properties, as illustrated in **Fig 3** for the thermal conductivity of electrons.

Most of the current works agree on the methods of calculation of viscosity, electrical conductivity and internal and translational thermal conductivities for 2T plasmas. However a still challenging question concerns the calculation of one or two reactional thermal conductivities, i.e. one for electrons $k_{re} = k_{ioni}$, and one for heavy specie $k_{rh} = k_{disso}$ and, consequently, the existence of a unique total thermal conductivity.

4.2 Modelling

The models limited to the gas domain solve the fluid conservation equations coupled with the Maxwell's equations while the models that include the electrodes in the computational domain solve also the energy and electromagnetic equations in electrodes. The equations are expressed in conservative form as a balance between accumulation, net flux, and production as follows :

$$\frac{\partial \rho\varphi}{\partial t} + \mathrm{div}(\rho\vec{u}.\varphi) = \mathrm{div}(\Gamma \mathrm{grad}\varphi) + S_\phi \qquad \textbf{Eq. 4}$$

With φ the variable to solve, ρ the mass density, υ the velocity, Γ the diffusion coefficient and S_Φ the source term.

Table 1 exemplifies the energy conservation equation. The first row shows the energy equation under LTE approximation; it corresponds to a fluid in chemical equilibrium with an internal energy characterized by a single temperature T. The second and third rows show this equation for a 2T model.

Conservation equation of Energy for a Local Thermal Equilibrium plasma (LTE)			
Conservation	Accumulation	Net flux	Net production
Int, energy	$\frac{\partial \rho h}{\partial t}$	$\nabla \cdot (\mathbf{u}\rho h + \mathbf{q}')$	$\frac{Dp}{Dt} - \boldsymbol{\tau} : \nabla \mathbf{u} + \mathbf{J}_q \cdot (\mathbf{E} + \mathbf{u} \times \mathbf{B}) - \dot{Q}_r$
Conservation equation of Energy for a Non Local Thermal Equilibrium plasma (NLTE)			
Int, energy électrons	$\frac{\partial \rho h_h}{\partial t}$	$\nabla \cdot (\mathbf{u}\rho h_h + \mathbf{q}'_h)$	$\frac{Dp_h}{Dt} - \boldsymbol{\tau} : \nabla \mathbf{u} + \dot{Q}_{eh}$
Int, energy Heavy species	$\frac{\partial \rho h_e}{\partial t}$	$\nabla \cdot (\mathbf{u}\rho h_e + \mathbf{q}'_e)$	$\frac{Dp_e}{Dt} + \mathbf{J}_q \cdot (\mathbf{E} + \mathbf{u} \times \mathbf{B}) - \dot{Q}_r - \dot{Q}_{eh}$

Table 1: Plasma energy conservation equations in LTE and non-LTE conditions. q' are the diffusive flux calculated from the Fourier's law; Q_r is the net emission coefficient and Q_{eh} the energy exchange term between electrons and heavy particles [3,4,28].

The LTE and NLTE equations have to be consistent that is the NLTE model should get reduced to the LTE model if thermal and chemical equilibrium are imposed (i.e. $T_h = T_e$.). The sommation of the two last rows of Table 1 must, therefore, leads to the LTE equation given in the first row. In the NLTE model, the electron and heavy particle energy equations involve well-known terms, but the expressions of these terms rise some questions: What is the proportion of radiation due to electrons and heavy particles for a given temperature range? What is the role of the electrons on the reactional thermal conductivity? What is the role of the heavy particles in dissociation process? These questions lead to different formulations of energy equations [29] that differ in the distribution of reactive thermal conductivity and ionization term in the two equations. The right formulation should result in departure from thermal equilibrium with $\theta \geq 1$, and T_e should be higher than T, the predicted LTE temperature while T_h should be lower than T.

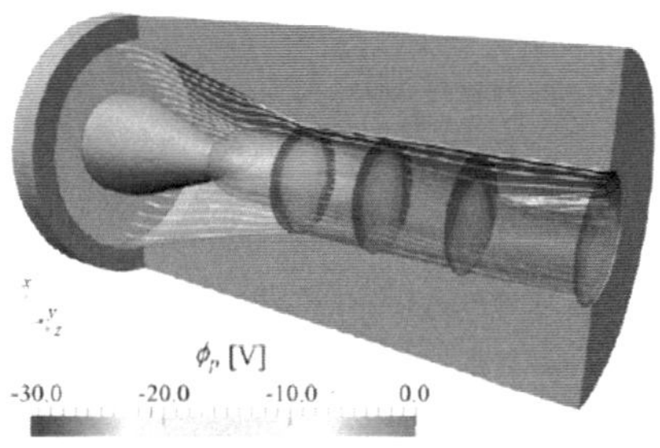

Fig. 4. Snapshot of voltage distribution over heavy species temperature slices and iso-surface 8, 12 and 16kK approximately predicted with a 2T model for argon plasma at 400 A The arc voltage is around 30V compared to 50V for LTE predictions under the same plasma conditions [30]

An example of the arc shape predicted by a 2T model together with the distribution of electric potential is given in **Fig. 4** [4]. The use of a NLTE model allows a better description of plasma flow, especially closed to the electrodes and in the arc column fringes, where steep gradients of temperature take place. It results in a better evaluation of the electrical conductivity in this zone leading to a decrease in the arc electrical resistivity and thus in the predicted arc voltage.

The solution of the set of conservation equations requires appropriate boundary conditions whose some are not straightforward to choose as the temperature of the electrons at the torch inlet in a 2T model or the value of the magnetic vector potential A when the potential vector formulation is used to compute the electromagnetic fields and electromagnetic variables [31] Also, if the computational domain is limited to the torch itself, a special care is also needed at the nozzle exit to ensure the uninterrupted progress of the flow characteristics out of the computational domain. A recommended method is the use of absorbing layers or sponge-zones to attenuate the outgoing waves and decay any reflected wave as it travels through the absorbing layer [3].

Finally an issue that has still to be addressed is the effect of acoustic waves on the arc column geometry and fluctuation. Actually some compressibility effects of the plasma-forming gas in the cathode cavity can bring about acoustic resonance phenomena (Hemholtz mode) because of a reciprocal dependence of arc voltage and pressure inside the cathode cavity that contains the cold plasma-forming gas. When the position of the anode arc attachment varies, the gas pressure and flow rate change, and thus the plasma flow [32]. To predict the effect of acoustic waves on torch operation mode, the numerical model should take into account the actual inner geometry of the torch, compressibility of the gas and couple the fluid and acoustic equations. To the best of our knowledge, this effect has not been numerically studied yet, despite the use of compressible plasma models.

5 Arc-anode interactions

The modelling of the interaction of the plasma inside the torch with the surrounding anode has to address three main aspects: (1) the displacement of the arc-anode attachment, (2) the arc reattachment process, and (3) the phenomena at the plasma-anode interface.

5.1 Displacement of the arc-anode attachment

The dynamics of the arc inside the torch can be discerned through two main features: the movement of the arc-anode attachment and the process of formation of a new arc-anode attachment, what is called the arc reattachment process [3,33,34]. The dynamic nature of the arc is evidenced by the temporal variation of the voltage drop between the electrodes, as well as by fluctuations in the temperature, pressure, and velocity at the torch exit.

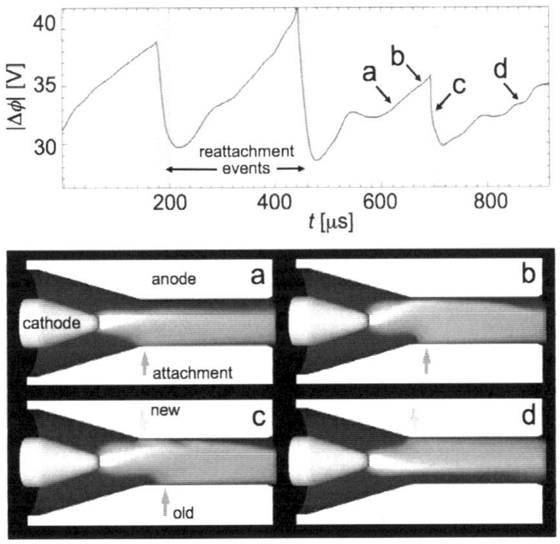

Fig. 5. Arc dynamics and arc re-attachment events. (Top) Time- evolution of the voltage difference ($\Delta\phi$) across the torch depicting the occurrence of three arc re-attachments events. (Bottom) temperature field snapshots at representative instants: **a** – single established anode attachment; **b** – dragging of the attachment along the flow direction; **c** – formation of a new attachment, which is more thermodynamically favourable due to its proximity to the cathode; and **d** – predominance of the new attachment and extinguishment of the old one. (Conditions: gas: Ar; total current: 400 A; flow rate: 60 slpm; diameter at torch exit: 8 mm. [28])

Representative simulation results of the voltage drop signal ($\Delta\phi$) together with snapshots of the temperature distribution inside a typical DC arc plasma torch are shown in **Fig. 5**. The simulations were based on a thermodynamic nonequilibrium (NLTE) plasma flow model. Details about the model, its numerical implementation, and validation, including its comparison against a thermodynamic equilibrium (LTE) model, are found in [4,28].
The results in **Fig. 5** depict the displacement of arc-anode attachment and arc reattachment process. The displacement of the anode attachment is a direct consequence of the imbalance between the drag force due to the interaction of the working gas flow on the plasma and the electromagnetic (Lorentz) force caused by the local curvature of the arc. The dominance of the drag force causes the displacement of the arc-anode attachement along the direction of the flow, linearly in the case of straight injection or following a corckscrew trajectory for swirl injection. The displacement of the arc-anode attachment is depicted in **Fig. 5 (top)** by the variation in $\Delta\phi$ between the points a and b, as well as in the temperature distribution snapshots in **Fig. 5 (bottom)** by comparing the locations of the anode attachment (indicated by the arrows) in frame **a** with respect to that in frame **b**.

The description of the displacement of the anode attachment requires the accurate and concurrent solution of the equations of the fluid dynamic model (incompressible or compressible flow, in chemical and thermodynamic equilibrium or non-equilibrium) together with those describing the evolution of the (self-induced) electromagnetic field (Maxwell's equations) in a time-dependent and three-dimensional domain, despite the constancy and axi-symmetry of operating conditions and torch geometry

5.2 Arc reattachment and electrical breakdown

A major challenge for the modelling of plasma spray torch operation is the capture of reattachment events. These events consist on the formation of a new electrical connection between the arc and the anode, and is manifested externally by a relativey sudden decrease in voltage drop, as observed in **Fig. 5**. Although the exact mechanisms driving the reattachment process are not completely understood, the relatively high electric fields, together with the abundance of excited species around the arc (by UV excitation), and the short time scale of the process, suggests that the arc reattachment is initiated by a streamer-like breakdown. This streamer connects the arc with the anode in a region somewhere upstream of the existing arc attachment, creating a conducting channel that allows the establishment of a new arc attachment. The detailed modeling of a streamer-like reattachment event is practically unfeasible with actual computational methods and computer power, especially whithin a realistic simulation of plasma spray torch operation. Therefore, approximate models have been used to describe it; they mimic the relatively spontaneous formation of the new conducting channel, e.g., [35,36].
Alternatively to breakdown-like events, a new anode attachment can form whenever a section of the arc column gets in close proximity to the anode surface. In such situation, the high temperatures, and hence high electrical conductivity, characteristic of the regions near the arc promote the establishment of new electrically conducting path. Such events have also been captured by plasma torch simulations; i.e., by the use of an artificially high electrically conductivity all along the anode surface in the case of LTE models, or by relying in the more diffusive nature of the electron temperature field in NLTE models [5].

5.3 Plasma-anode interface

An accurate description of the plasma-anode interface is very involved and encompasses complex thermal, electromagnetic, and material transport phenomena (i.e., heat transfer, current transfer, and potential material transport in the case of anode erosion [33], respectively).
For example, the heat transfer to the anode q_a can be given by:

$$q_a = -k_h \partial_n T_h - k_e \partial_n T_e + q_e + J_{qe} W_a + J_{qi}(E_i - W_a) + q_r \quad \textbf{Eq.5}$$

where ∂_n represents the derivative in the direction normal to the anode, k_h and k_e are the heavy-species (atoms, ions) and electron thermal conductivity, respectively; T_h and T_e the heavy-species and electron temperatures; J_{qe} and J_{qi} are the electron and ion current densities, q_e is the transport of energy by the electron flux often defined as $q_e = J_{qe}(h_e/e - U_a)$, where h_e is the electron enthalpy, e the elementary charge, and U_a the voltage drop across the anode sheath; W_a is the work function of the anode material, E_i is the ionization energy of the ion, and q_r is the radiative heat flux to the anode [3].
Typically, the first, third and fourth terms are the most important ones, and hence q_a depends significantly on J_{qe}. For NLTE models, the boundary conditions for T_h and T_e should be obtained by splitting **Eq. 5**, i.e.,: the terms involving electrons specify the boundary condition for T_e, and the others the condition for T_h. The expression for q_a in **Eq. 5** has to be extended to include anode material modifications, such as erosion or surface reactions, which could significantly affect the overall heat transferred.
Additionally, the accurate description of the arc-anode interface requires an adequate account of the heat transfer and current transfer through the anode domain, e.g., to describe the removal of heat from the plasma by the cooling water and the transport of current from the arc to the electric terminals connecting the anode. Such description requires incorporating, at least partially, part of the anode within the computational domain of the simulation, which can significantly increase model complexity and the computational cost of simulation

6 Conclusion

If the current LTE spray torch models are useful for parametric studies, they have to be developed further (in a sequential approach) to yield reliable and predictive calculations. 2T models proved to be necessary to better describe the arc column fringes and electrode regions and predict more realistic arc voltage. However, some questions remains about the 2T transport properties and in particular the thermal reactive conductivity and radiation of electrons and heavy particles over specific temperature ranges.
The main improvements could be the coupling of electrodes and arc with the solution of electromagnetic and energy equations in both electrodes and implementation of electrode sheaths. The current developments in close fields like plasma welding, cutting and HID lamps can help to develop such models. However, a specific issue in plasma spray torch modeling that is not addressed in transferred arc models, is the movement of the arc attachment at the anode wall that makes the implementation of the anode sheath much more complex. In a first step, the model could be developed for cascaded anode geometry. At the end, such models could predict the phase change in electrodes and time-erosion.
Finally, the coupling of the arc dynamics with the propagation of the acoustic waves generated by the compressibility effect of the plasma-forming gas in the cathode cavity should allow for better control of the arc instabilities. This coupling could be investigated first with the current compressible plasma models limited to the gas domain.

7 Literature

[1] Thermal Spray Coating Market Analysis, Grand View Research Inc., San Francisco (2015), ISBN Code: 978-1-68038-514-4
[2] A. Vardelle, C. Moreau C, N.J. Themelis, C. Chazelas A perspective on plasma spray technology. Plasma Chem Plasma Process (2015) vol 35, N°, pp 491-509
[3] J. P. Trelles, C. Chazelas, A. Vardelle and J. V. R. Heberlein: Arc Plasma Torch Modeling. Journal of Thermal Spray Technology (2009) Vol. 18, No. 5/6, pp. 728/752.
[4] P. Trelles, E. Pfender and J. V. R. Heberlein, Non-Equilibrium Modeling of Arc Plasma Torches, Journal of Physics D: Applied Physics (2007) Vol. 40, pp. 5937-5952.
[5] J. Wendelstorf, Ab initio modelling of thermal plasma gas discharges, PhD Thesis, (2000) Carolo-Wilhelmina University, Germany
[6] M. S. Benilov, “Understanding and Modelling Plasma-Electrode Interaction in High-Pressure Arc Discharges: A Review,” Journal of Physics D: Applied Physics (2008) Vol. 41, No. 14, p. 144001
[7] P. Zhu, J. J. Lowke and R.Morrow, A unified theory of free burning arcs, Cathode sheaths and cathodes, J. Phys. D: Appi. Phys. (1992) vol 25, pp 1221-1230
[8] M. Alaya, C. Chazelas, G. Mariaux, A. Vardelle, J. Thermal Spray Technol., (2015), vol 24
[9] Y. Cressault, Basic knowledge on radiative and transport properties to begin in thermal plasmas modelling, AIP Advances (2015) Vol 5 , 057112
[10] A.B. Murphy, Phys. Rev. E (1993), Vol 48, 3594
[11] A. Gleizes, Perspectives on Thermal Plasma Modelling, Plasma Chemistry and Plasma Processing (2015) Vol 35, N° 3
[12] A. B. Murphy, Transport coefficients of hydrogen and argon-hydrogen plasmas, Plasma Chem. and Plasm. Proc., (2000) vol 20, pp 279-297
[13] A. Murphy, Transport coefficients of air, argon-air, nitrogen-air, and oxygen-air plasmas, Plasma Chem. And. Plasm. Proc., (1995), vol 15, pp 279-307
[14] A. B. Murphy, Transport Coefficients of argon, nitrogen, oxygen, argon-nitrogen, and argon-oxygen plasmas, Plasma Chem. And Plasm. Proc., (1994) vol 14, pp 451-490,
[15] B. Sourd, J. Aubreton, M. F. Elchinger, M. Labrot and U. Michon: High temperature transport coefficients in e/C/H/N/O mixtures, J. Phys. D: Appl. Phys., (2006) vol 39, pp 1105-1119.
[16] A. B. Murphy, Transport coefficients of hydrogen and argon-hydrogen plasmas, Plasma Chem. and Plasm. Proc., 20, (3), 2000, 279-297
[17] Y. Cressault and A. Gleizes, Thermodynamic properties and transport coefficients in Ar-H_2-Cu

plasma, J. Phys. D:Appl. Phys., (2004),vol 37, pp 560-572.
[18] J. Annaloro, P.Teulet, A. Bultel, Y. Cressault, A. Gleizes, Calculation of composition and thermodynamic properties in thermal non-equilibrium conditions. 13th Int. Conf. High Tech Plasma (2014)
[19] K. C. Meher, N. Tiwari, S. Ghorui, Thermodynamic and Transport Properties of Nitrogen Plasma Under Thermal Equilibrium and Non-equilibrium Conditions, Plasma Chem. Plasma Process (2015) vol 35, pp 605–637
[20] V. Colombo, E. Ghedini, P. Sanibondi, Thermodynamic and transport properties in non-equilibrium argon, oxygen and nitrogen thermal plasmas., Prog. Nucl. Energy, (2008), vol 50, pp 921–933
[21] W. Z. Wang, M. Z. Rong, J. D. Yan, A. B. Murphy, and J. W. Spece: Thermophysical properties of nitrogen plasmas under thermal equilibrium and non-equilibrium conditions, Physics of Plasma (2011) vol 18, 113502
[22] V. Colombo, E. Ghedini, P. Sanibondi, Thermodynamic and transport properties in non-equilibrium argon, oxygen and nitrogen plasmas, Prog. Nucl. Energy, (2008) vol 50, pp 921–933
[23] S. Ghorui, J. Heberlein and E. Pfender, Thermodynamic and transport properties of two-temperature oxygen plasmas, Plasma Chem. Plasma Proc. (2007) vol 27, pp 267–291
[24] V Colombo, E Ghedini, P Sanibondi, Two-temperature thermodynamic and transport properties of argon–hydrogen and nitrogen–hydrogen plasmas, Journal of Physics D: Applied Physics, (2009) Vol 42, 055213
[25] J. Aubreton, et al., Thermodynamic and transport properties of a ternary Ar–H_2–He mixture out of equilibrium up to 30 000 K at atmospheric pressure. J. Phys. D: Appl. Phys. (2004.) vol 37, pp 2232–2246
[26] J. Aubreton, et al., Two-temperature transport coefficients in argon–helium thermal plasmas. J. Phys. D: Appl. Phys. (2004.) vol 37, pp 34–41
[27] J. Aubreton, M.F. Elchinger, Transport properties in non-equilibrium argon, copper and argon–copper thermal plasmas. J. Phys. D: Appl. Phys. (2003) vol 36, pp 1798–1805
[28] J. P. Trelles: Computational Study of Flow Dynamics from a DC Arc Plasma Jet. Journal of Physics D: Applied Physics (2013) Vol. 46, No. 25, 255201.
[29] P Freton, J J Gonzalez, Z Ranarijaona and J Mougenot, Energy equation formulations for two-temperature modelling of 'thermal' plasmas, Journal of Physics D: Applied Physics, (2012) Vol. 45
[30] J. P. Trelles, S. M. Modir Khazeni, Variational Multiscale Method for Nonequilibrium Plasma Flows, Computer Methods in Applied Mechanics and Engineering (2014) No. 282, pp. 87-131
[31] P. Freton, J. J. Gonzalez, M. Masquère, F.Reichert, Magnetic field approaches in dc thermal plasma modelling, Journal of Physics D: Applied Physics, (2011), vol 44,pp.345202
[32] Rat and J.F. Coudert, Acoustic stabilization of electric arc instabilities in non-transferred plasma torches, Appl. Phys. Lett. (2010) Vol 96, pp 101503,
[33] C. Baudry, A. Vardelle and G. Mariaux: Numerical Modeling of a DC Non-Transferred Plasma Torch: Movement of the Arc Anode Attachment and Resulting Anode Erosion. High Temperature Plasma Processes (2005) Vol. 9, pp. 1-15.
[34] C. Chazelas, E. Moreau, G. Mariaux and A. Vardelle: Numerical Modeling of Arc Behavior in a DC Plasma Torch. High Temperature Materials Processing (2006) Vol. 10, pp. 393-406.
[35] E. Moreau, C. Chazelas, G. Mariaux and A. Vardelle: Modeling of the Restrike Mode Operation of a DC Plasma Spray Torch. Journal of Thermal Spray Technology (2006) Vol. 15, No. 4, pp. 524-530.
[36] J. P. Trelles, E. Pfender and J. V. R. Heberlein: Modeling of the Arc Reattachment Process in Plasma Torches. Journal of Physics D: Applied Physics (2007) Vol. 40, pp. 5635/5648.

LPC - Novel Process for the Production of Photocatalytic Surfaces

B. Robinson[a], A. Tabecki[a], S. Paul[a], G. Shi[a], P. Carmichael[b], I. P. Parkin[b], J. A. Darr[b], H. L. de Villiers Lovelock[a]
[a]TWI Granta Park, Great Abington, Cambridge, CB21 6AL, UK
[b]Department of Chemistry, University College London, 20 Gordon Street, London, WC1H 0AJ

The photocatalytic applications of titanium dioxide are widely reported and include water purification, air purification and self-cleaning, however, due to the large band gap (*ca.* 3.2 eV) of titania, coatings have a limited efficiency. To therefore take full advantage of this material, low cost deposition processes are required. Titania is often incorporated into paints and concrete building materials to passively purify contacting air and remove harmful NOx and VOCs (volatile organic compounds). Incorporating titania in this manner takes advantage of the large exposed area of a building surface while minimising the cost relative to deposition processes such as PVD. Here, we report a novel deposition technique LPC (low pressure coating) for the inclusion of photoactive titania in ductile surfaces. LPC is a low cost, easily scalable technology with potential application in the production of air purifying building materials.

1 Introduction

Titania is a widely researched photocatalyst with a number of potential applications including energy production *via* photolysis and air purification.[1, 2] However, due to the large band gap of titania (ca. 3.2 eV) this photocatalyst has a limited efficiency (ca. 3%), only absorbing the UV portion of the solar spectrum.[3, 4] This limits the application of titania coatings and is further exacerbated by what can be high deposition costs depending on the coating techniques used (e.g. CVD, PVD).[5] As such, there have been a number of attempts to produce titania coatings *via* inexpensive and scalable technologies e.g. suspension spray, dip coating etc.[6, 7] Another approach to harness the photocatalytic properties of titania, is inclusion in building materials (e.g. concrete and paint) to act as an air purifier or as a self-cleaning surface.[8] Titania particles on the surface of these building materials have been shown to be capable of breaking down harmful emissions such as NO/NO_2, with a view toward improving air quality in urban areas.[9] This approach takes advantage of using low cost titania, allowing this method to be more cost effective, despite titania's limited efficiency.

Herein, a new, inexpensive, technique for the deposition of titania material, low-pressure coating (LPC), is demonstrated. In this process, angular titania feed material is accelerated into a substrate using compressed air. This results in embedment of the titania particles into the substrate, giving a photoactive surface, **Fig. 1**. This technique allows the deposition of an intermittent, adherent photocatalytic film. LPC is very inexpensive because deposition only requires the basic materials (e.g. powder and substrate) and a compressed air supply. A range of LPC titania coatings were produced varying nozzle stand-off distance, air pressure and substrate exposure time. Samples were characterised by techniques including SEM, EDX, photocurrent generation and NO_X removal testing.

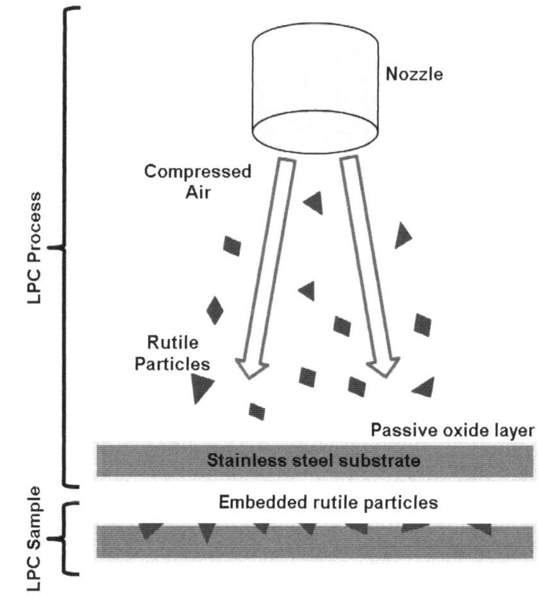

Fig. 1. A diagram showing the process of LPC and resulting sample.

2 Experimental

2.1 Materials

Fused and crushed rutile phase titania thermal spray powder (-25+5 µm, Saint Gobain, Northampton, USA) was applied to 25 x 25 x 0.5 mm AISI 304 stainless steel plates by LPC using a Guyson 200/20 blast cleaning cabinet. A small quantity of alumina impurity was present in the feedstock powder (2.35%), which was confirmed by EDX, **Fig. 2**. Samples were prepared using an 8 mm boron nitride nozzle using two pressures (30 and 60 psi), two exposure times (20 and 40 s) and a range of stand-off distances (30-60 mm).

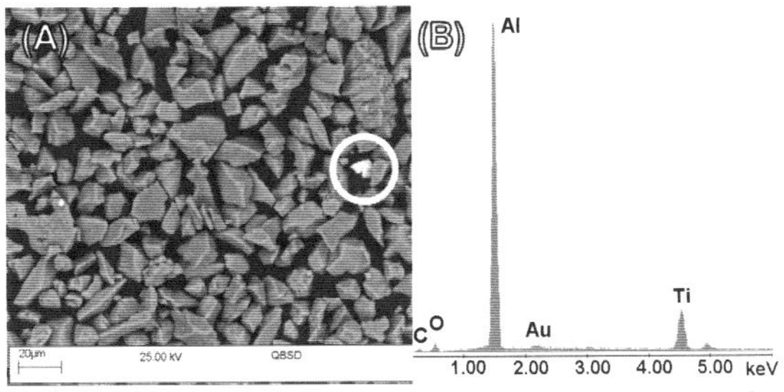

Fig. 2. SEM image of the rutile thermal spray powder used (A) with EDX spectrum highlighting an Al_2O_3 impurity.

2.2 Coating Characterisation

SEM analysis was performed using a LEO 1550 Field Emission Scanning Electron Microscope (FESEM). EDX analysis was performed using an EDAX spectrometer with Pegasus software. Area coverage was determined by volume fraction point counting (BS 7590:1992) using backscattered SEM images of the sample surface.[10] A 25 point grid (5 x 5) was used with 30 fields measured. Area coverage was calculated as the percentage of points which were located on an embedded particle with boundaries scored as 0.5. Surface profilometry was performed using an Alicona Infinite Focus SL, using a 50x objective lens. Selection of λ_c filters was carried out in agreement with standards (ISO 4288:1998).[11] Photocurrent tests were carried out using a 75 W Xenon lamp with UG320 and UG5 band pass filters to simulate the sun's UV profile. 0.1 M KNO_3 was used as the electrolyte. Measurements were performed using a three electrode glass cell using an $Ag/AgNO_3$ reference electrode and Pt wire counter electrode. The potential range was -0.4-1.2 V. Confirmation photocurrent tests were performed at Queens University Belfast using a 200 W Xenon lamp with WG320 and UG5 band pass filters to simulate the sun's UV profile. Measurements were performed using a Ag/AgCl reference electrode and Pt wire counter electrode with 0.1 M KNO_3 was used as the electrolyte. NO_x removal testing was performed in agreement with ISO 22197-1:2007.[12] A larger sample (50 x 50 x 0.5 mm) was produced for the test. The sample was irradiated with a UV lamp (1 mW cm^{-2}) with NO test gas flowed over the sample at a rate of 3 L min^{-1} for 5 hours. NO_x concentration was observed using a Enviro Technology Service plc NO_x analyser Model T200. The sample was compared with a highly active dip coated anatase phase titania coating produced from a sol-gel paste heat-treated at 450 °C.

3 Results and Discussion

The films had a dull grey colour similar to that of the powder feedstock used, indicating the inclusion of material into the surface, **Fig. 3**. The dull finish was also due to the surface being randomly roughened in all directions, scattering light. The films were robust, surviving ultrasonic cleaning with no observable change to the surface.

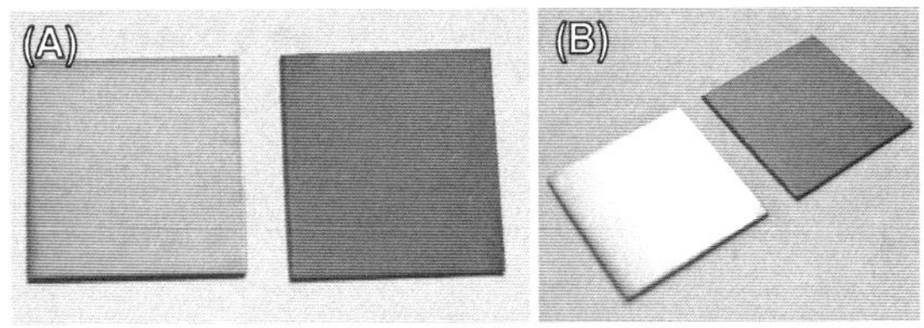

Fig. 3. Photographs of an LPC sample (40 mm 60 psi 40 s, right) relative to the pristine steel substrate (left) demonstrating the change in colour (A) and shine (B).

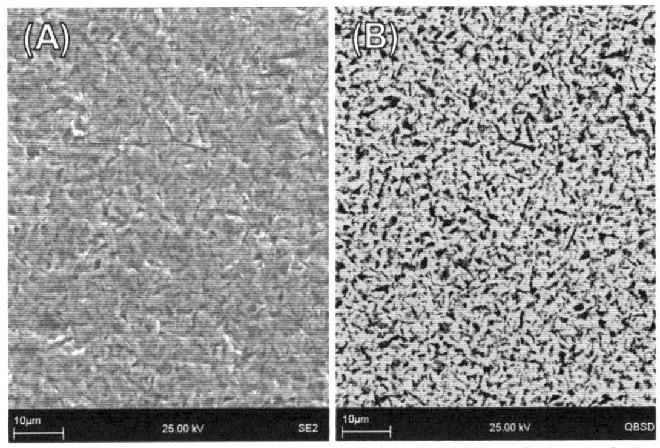

Fig. 4. SEM images showing the surface finish of the substrates after LPC (A) and a back scattered images showing the location of titania material (black) in particle impact craters (B).

SEM analysis showed that upon collision with the surface, titania powder particles shattered, depositing a segment of material lodged in the substrate, **Fig. 4**. The original particle size of the rutile powder used was -25+5 µm, significantly greater than the <1-5 µm of the particles embedded in the surface. This demonstrated the shattering of impacting particles on collision with the steel substrate. To confirm the presence of titania in these crater zones EDX and chemical mapping was performed, **Fig. 5**. Analysis showed greater concentrations of titanium and oxygen within the craters relative to the surrounding area, demonstrating the deposition of titania in these regions. Embedded particles were often observed as thin "blade-like" structures due to the breaking off of impacting particle edges, **Fig. 4**. The titania embedded in the surface did not coalesce to form a conventional coating, but rather was observed to be an intermittent film on the surface. This was due to the brittle nature of titania preventing layering of material, an effect previously observed in cold spray.[13] Area coverage of the surface by the embedded particles was determined by point counting of back scattered SEM images with the effect of running pressure, stand-off distance and exposure time noted, **Fig. 6**.

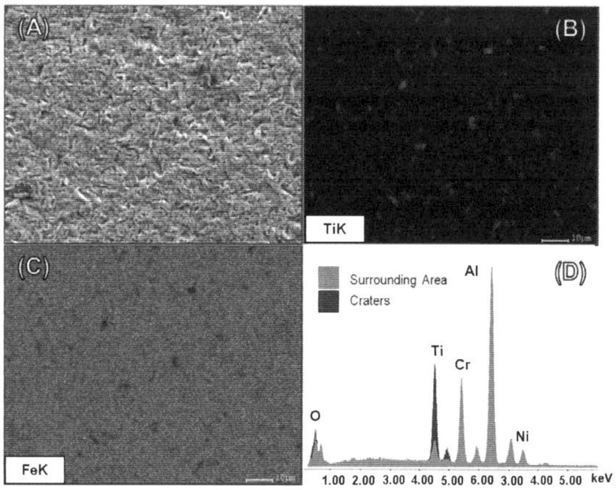

Fig. 5. Chemical mapping and localised EDX spectra of an LPC titania sample. SEM image showing the craters in an LPC sample (A). Chemical maps showing the locations of titanium (B), as titania, in the craters of the surface and Iron where the titania is blocking the iron signal (C). EDX spectra of a crater in an LPC sample and the surrounding area showing the large difference in titanium concentration (D).

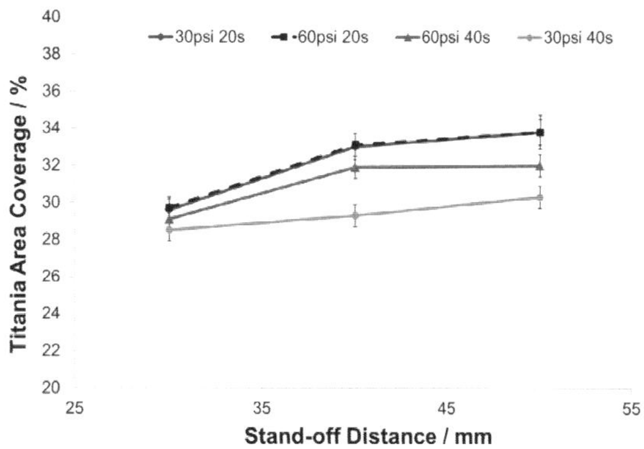

Fig. 6. The effect of stand-off distance on area coverage of embedded rutile particles

Area covered by titania particles was in the range 28.5(±0.6) to 33.8(±0.9)%, varying with process parameters, **Fig. 6**. Stand-off distance was shown to have the greatest effect on area coverage with values increasing by up to 4% (a 14% relative increase) with stand-off distance. Exposure time was shown to have an inverse effect on area coverage with values decreasing up to 3.7% (for the higher pressure setting). The running pressure used was shown to have little effect on the area covered by titania, for almost all conditions. These trends suggest that the more severe conditions typically lead to a decrease in embedded titania. This is thought to be due to greater shattering, leading to smaller particles and less material remaining post process. With greater exposure time, the arrival of particles colliding with a saturated surface leads to shattering of the previously embedded brittle material partially removing it from the surface. However, all changes in coverage observed were small with all values remaining approximately 30% of the total surface independent of parameter changes. This means that tight process control would not be necessary in a production setting which could reduce cost, particularly in on-site application.

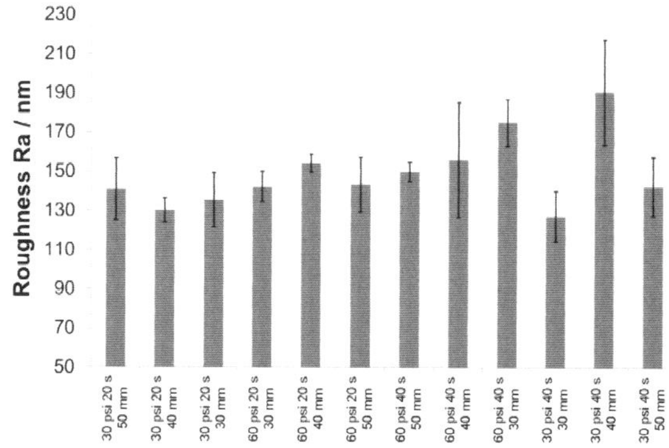

Fig. 7. Surface roughness (Ra) for all LPC coatings produced.

Surface roughness of a coating has been demonstrated to have a significant impact on coating photoactivity. To determine if coating conditions or quantity of embedded particles had any effect on surface roughness, 2D non-contact profilometry was performed. Coatings roughness Ra values varied between 127(±13) nm and 175(±11) nm, slightly below that of the initial substrate finish (202±12 nm). This fine roughness was as a result of the small titania particles used (-25+5 µm). Ra was found to be largely indistinguishable between the samples and showed no correlation with quantity of embedded titania (as did the other surface roughness parameters), **Fig. 7**. This is due to the fact that sample surface roughness is composed of two components, particles protruding from the surface and distorted/removed steel substrate, such that the quantity of embedded titania does not necessarily have a direct correlation with roughness. The lack of relationship between surface roughness and changing pressure or exposure time is in disagreement with published results for grit blasting, a very similar process to LPC, which has shown increasing surface roughness with static pressure and traverse speed.[14] However, differences in behaviour are likely as a result of the significantly different size of impacting material (alumina grit used varied in size between 125-500 µm) between these two processes. This may have led to different particle, process and surface interactions. Alumina grit is also significantly harder than titania (alumina *ca.* 9 Mohs vs rutile titania *ca.* 6.5 Mohs), which would result in different surface interactions.[15]

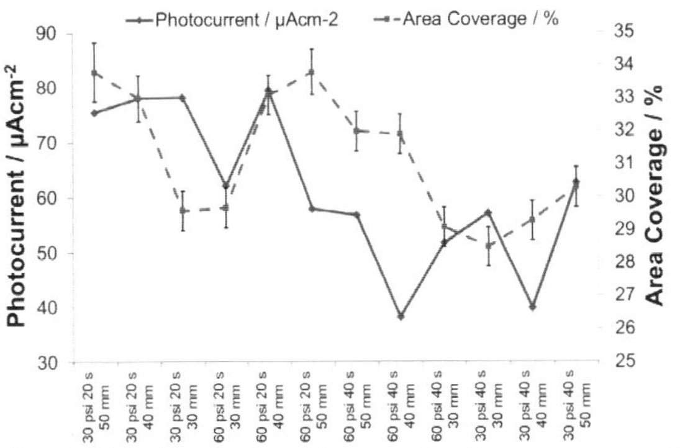

Fig. 8. The trend of photocurrent with area covered by embedded titania particles for all LPC samples produced.

To determine photocatalytic activity of the sample, photocurrent measurements were carried out in a PEC cell, with samples exposed to a Xenon lamp. Photocurrent generated was in the range 38-79 μAcm^{-2} (at 0.3 V, $Ag/AgNO_3$), with limited trend observed with titania area coverage. This lack of significant trend with area coverage may be due to alternate effects such as particle penetration into the substrate, **Fig. 8**. Photocurrent generation was also independent of coating roughness. The most active sample was sent for conformational photocurrent testing at Queens University Belfast (so that results could be compared with previous SFS coatings) as well as NO_X removal testing (ISO 22197-1: 2007) to assess the air purification properties of LPC samples.[12,16] The most photoactive sample had a photocurrent generation of 275 μA compared with 62 μA for a SFS titania coating (corrected for size differences). The relatively high photocurrent observed, was thought to be due to the intimate contact of the rutile powder and the steel substrate allowing rapid transfer of electrons into the substrate. The high velocity of the powder in this process allows the angular material to penetrate the insulating passive oxide layer that builds on the surface of stainless steel. This allows electrons generated in the titania particles to travel more readily into the substrate providing beneficial electron hole separation. The resulting film is also very thin, approaching the thickness of coatings produced by techniques such as CVD/PVD. This allows for ready separation of the photogenerated electrons and holes decreasing recombination and increasing photoactivity.

During NO_x removal testing, samples were exposed to UV light with a continuous flow of nitrogen oxide (NO) gas over the surface. Nitrogen oxide is a harmful air pollutant resulting from combustion processes such as vehicle engines.[17] The degradation of NO over TiO_2 takes place as described in the scheme below, **Eq 1**.[18-22]

$$4NO + 3O_2 + 2H_2O \overset{hv}{\rightarrow} 4HNO_3$$

Equation 1. The degradation of NO by TiO_2.

However, it has been shown that as nitric acid collects on the surface of titania a disfavourable side reaction is promoted, producing nitrogen dioxide (NO_2), **Eq 2**.

$$2HNO_3 + NO \overset{hv}{\rightarrow} 3NO_2 + H_2O$$

Equation 2. The competitive side reaction which results in production of harmful NO_2.

Nitrogen dioxide is toxic when inhaled and has been linked with increased hospitalisation and mortality rates.[23, 24] As such, to assess the air purification capabilities of samples tested, NO_x removal was measured whereby NO_x removed is defined as:

$$NO_{x\ removed} = NO_{removed} - NO_{2\ generated}$$

Equation 3. The rate of NO_x removal, taking into account NO_2 production.

The LPC sample was found to degrade 0.17 μmol of NO_x over the duration of the test, **Table 1**. Despite the existence of a standard (ISO 22197-1: 2007) for the NO_x removal testing, comparison with published values in the literature was difficult. This was due to variations in sample size, lamp power, humidity and NO flow rates, all of which have significant effects on NO_x removal rates.[17, 25] As such, a highly active anatase dip coated titania coating on a glass substrate was used for a direct comparison, dip coating being an alternative coating technology. The dip coated titania analogue was found to have degraded 0.21 μmol of NO_x under the same conditions within the same time frame. The LPC coating therefore has a comparable ability for NOx removal to that of the dip coating analogue so can be considered significantly photoactive.

4 Basic Cost Analysis

One of the significant benefits of LPC is its relatively low cost. A basic cost analysis was done. As the mass of titania could not be determined accurately due to the continuous recycling of material during the process, mass of titania used is determined from the fraction embedded in the surface of the substrate. Conservative estimates of this mass can be calculated by assuming the particles are approximately 5 micron spheres (the larger end of the particle distribution with this shape selected to artificially raise the total calculation). Average area coverage through experimentation was *ca.* 31%. This gives a mass of 4.4 g m^{-2} which at the price quoted for the titania used equates to ca. £0.15/$0.21 per m^2. In a production setting, this process would be automated using a system similar to an automated grit blaster, to improve reproducibility and production rates. As such, labour is not considered.

Table 1. NO_X removal test results for optimal LPC sample with dip coated analogue.

Sample	NO In / µmol	NO Removed / µmol	NO Unreacted / µmol	NO_2 Generated / µmol	Net NO_x Removed / µmol
LPC	39.72	1.21	38.50	1.05	0.17
Dip Coating	36.86	5.59	31.26	5.38	0.21

As previously stated, this technique is carried out as a non-loss system with recycling of powder. However, as powder size decreases, after collision with the surface, a larger and larger fraction of the material will be removed by the extraction system. This will therefore add a significant wastage to the process, which should be considered. Mass lost can be estimated to be approximately 20%, which results in an approximate price of £0.18/$0.26 per m^2 for LPC deposition.

5 Technology Limitations

LPC offers many benefits over existing technologies, namely its significantly low cost, however, this process also has disadvantages which should be considered. To achieve embedded particles, an angular feedstock material is required, in this case fused and crushed rutile powder. This limits the feedstocks available. To achieve greater photoactivities than those reported here anatase phase titania material (often reported as the more active of the two phases) would typically be used.[26] However, anatase powder cannot be produced in the required format due to its metastable nature, transforming to rutile at temperatures of *ca.* >400 °C.[27] Anatase powders are typically produced by methods such as spray drying and are therefore rounded so cannot be readily embedded in the surface by LPC. Nanopowders also cannot be used due to the low momentum of the particles, preventing penetration of the surface. Another disadvantage of this process, at its current stage of development is its inconsistent mass flow. Due to the use of fine powders (5-25 µm), flowability of the powder is poor. The blasting cabinet which was used for this process would typically use much coarser materials with greater flowability. As such, substantial mechanical force was required to facilitate the continuous feeding of the powder into the air jet. Despite this applied force, mass flow of the feedstock powder was inconsistent. To overcome this, minor modifications should be made to the cabinet to allow improved flow of powder such as altering the hopper shape and installing an agitator.

6 Potential Applications

Due to the limitations in material options, this process is unlikely to be used for the production of photolysis devices. Rutile titania cannot split water due to the location of its band gap and LPC cannot be used with nanopowders, typically used in photolysis research. However, the inexpensive production costs, scalability and high photoactivity of the surfaces produced, does lend them to application as antibacterial and air/water purifying surfaces. Films produced are robust and could be readily used in construction for external building faces for the breakdown of harmful substances such as NO_x (as demonstrated herein), particulate matter and VOCs, which reach significant levels in city areas. Films may also be used as antibacterial surfaces in hospitals preventing transmission of harmful microbes *via* contact surfaces such as counters. As no nanopowders are used in the process, this also removes possible health concerns from exposure, with the long term effects of titania nanoparticles still not fully understood.[28]

7 Conclusions

Samples were produced by LPC, resulting in rutile particles embedded in the stainless steel surface, as determined by SEM analysis. The mechanism for deposition was shown to be break-off from impacting particles, leaving smaller titania particles embedded in surface craters. Samples were produced under varying process parameters, changing running pressure, stand-off and exposure time. Stand-off distance and exposure time were observed to affect the area covered in titania particles, but had no significant effect on the photocurrent generated by the samples. This discrepancy was attributed to alternate sample parameters, such as particle penetration. The high PEC photocurrents generated are thought to be due to the thin nature of the coatings and the strong contact between the titania particles and steel substrate following their penetration through the insulating passive oxide layer. The most active sample was tested for air purification capabilities *via* a NO_x removal test. The LPC sample was found to have comparable activity to that of a highly active dip coated analogue, highlighting its significant capability for air purification. It has therefore been demonstrated that LPC is a highly inexpensive technique for the production of photoactive surfaces with potential applications as antibacterial or air purifying surfaces.

7 Acknowledgements

The authors would like to thank the Queen's University Belfast School of Chemistry and Chemical Engineering for their assistance with the PEC and NO_x removal testing. This research was partially funded by the UK's Technology Strategy Board (TSB) and the Engineering and Physical Sciences Research Council (EPSRC) under Project 100898 "Nanocrystalline Water Splitting Photodiodes II;

Device Engineering, Integration & Scale-up". The associated ESPRC EngDoc sponsorship for B Robinson is co-funded by TWI Limited under UCL's Industrial Doctorate Centre in Molecular Modelling & Materials Science.

8 References

[1]Ao, C.H. and S.C. Lee, Indoor air purification by photocatalyst TiO_2 immobilized on an activated carbon filter installed in an air cleaner. Chemical Engineering Science (2005), 60, 1, pp. 103/109.

[2] Paulose, M., *et al.*, Visible light photoelectrochemical and water-photoelectrolysis properties of titania nanotube arrays. Journal of Photochemistry and Photobiology A: Chemistry (2006), 178, 1, pp. 8/15.

[3] Wang, C., *et al.*, Preparation of cerium and nitrogen co-doped titania hollow spheres with enhanced visible light photocatalytic performance. Powder Technology (2011), 210, 3, p. 203/207.

[4] Gale, J., Astrobiology of Earth: The Emergence, Evolution and Future of Life on a Planet in Turmoil (2009) Oxford University Press.

[5] Bennett, H., *et al.* Laser Induced Damage in Optical Materials: Proceedings of a Symposium (1990), Boulder, Colorado: The International Society for Optica; Engineering.

[6] Moon, B.-H., Y.-M. Sung, and C.-H. Han, Titanium oxide Films Prepared by Sputtering, Sol Gel and Dip Coating Methods for Photovoltaic Application. Energy Procedia (2013), 34, pp. 589/596.

[7] Toma, F.-L., et al., Nanostructured photocatalytic titania coatings formed by suspension plasma spraying. Journal of Thermal Spray Technology (2006), 15, 4, pp. 587/592.

[8] Chen, J. and C.-s. Poon, Photocatalytic construction and building materials: From fundamentals to applications. Building and Environment (2009), 44, 9, pp. 1899/1906.

[9] Demeestere, K., *et al.*, Heterogeneous photocatalytic removal of toluene from air on building materials enriched with TiO_2. Building and Environment (2008), 43, 4, pp. 406/414.

[10] BS 7590:1992 Method for statistically estimating the volume fraction of phases and constituents by systematic manual point counting with a grid (1992).

[11] BS EN ISO 4288:1998 Geometric product specification (GPS). Surface texture. Profile method: Rules and procedures for the assessment of surface texture. (1998), BSI.

[12] BS ISO 22197-1:2007 Fine ceramics (advanced ceramics, advanced technical ceramics) — Test method for air-purification performance of semiconducting photocatalytic materials, in Part 1: Removal of nitric oxide (2007).

[13] Burlacov, I., *et al.*, Cold gas dynamic spraying (CGDS) of TiO_2 (anatase) powders onto poly(sulfone) substrates: Microstructural characterisation and photocatalytic efficiency. Journal of Photochemistry and Photobiology a-Chemistry (2007), 187, 2-3, pp 285-292.

[14] Begg, *et al.*, Mechanisation of the Grit Blasting Process for Thermal Spray Coating Applications, Proceeding of ITSC 2015 (2015), Long Beach, CA.

[15] Reade Advanced Materials presents: Mohs' Hardness of Abrasives. [24/08/15]; Available from:http://www.reade.com/Particle_Briefings/mohs_hardness_abrasive_grit.html.

[16] B. W. Robinson, *et al.*, 'The Effect of Fuel Ratio on Photoactivity of Suspension Flame Sprayed Coatings', Proceeding of ITSC 2015 (2015), Long Beach, CA.

[17]Boonen, E. and A. Beeldens, Recent Photocatalytic Applications for Air Purification in Belgium. Coatings (2014), 4, 3, pp. 553/573.

[18] Ibusuki, T. and K. Takeuchi, Removal of low concentration nitrogen-oxides through photoassisted heterogenous catalysis. Journal of Molecular Catalysis (1994), 88, 1, pp. 93/102.

[19] Ohko, Y., *et al.*, Photocatalytic oxidation of nitrogen monoxide using TiO_2 thin films under continuous UV light illumination. Journal of Photochemistry and Photobiology a-Chemistry (2009), 205, 1, pp. 28/33.

[20] Devahasdin, S., *et al.*, TiO_2 photocatalytic oxidation of nitric oxide: transient behavior and reaction kinetics. Journal of Photochemistry and Photobiology a-Chemistry (2003), 156, 1-3, pp. 161/170.

[21] Ballari, M.M., *et al.*, NO_x photocatalytic degradation employing concrete pavement containing titanium dioxide. Applied Catalysis B-Environmental (2010), 95, 3-4, pp. 245/254.

[22] Bowering, N., *et al.*, Photocatalytic decomposition and reduction reactions of nitric oxide over Degussa P25. Applied Catalysis B-Environmental (2006), 62, 3-4, pp. 208/216.

[23] Mills, I.C., *et al.*, Quantitative systematic review of the associations between short-term exposure to nitrogen dioxide and mortality and hospital admissions. Bmj Open (2015), 5, 5, pp. 8.

[24] Dai, J.P., *et al.*, Ambient air pollution, temperature and out-of-hospital coronary deaths in Shanghai, China. Environmental Pollution (2015), 203, pp. 116/121.

[25] Faraldos, M., et al., Photocatalytic hydrophobic concrete coatings to combat air pollution. Catalysis Today, 2015.

[26] Luttrell, T., et al., Why is anatase a better photocatalyst than rutile? - Model studies on epitaxial TiO2 films. Scientific Reports, 2014. 4: p. 8.

[27] Gouma, P.I. and M.J. Mills, Anatase-to-rutile transformation in titania powders. Journal of the American Ceramic Society, 2001. 84(3): p. 619-622.

[28] Iavicoli, I., V. Leso, and A. Bergamaschi, Toxicological Effects of Titanium Dioxide Nanoparticles: A Review of In Vivo Studies. Journal of Nanomaterials, 2012: p. 36.

Improving atmospheric plasma sprayed electrolyte layers of solid oxide fuel cells by use of powder separator

T. Liu[1] *and J. Arnold*

Institute of Engineering Thermodynamics, German Aerospace Center,

Pfaffenwaldring 38-40, 70569 Stuttgart, Germany

Abstract

In order to increase the performance of solid oxide fuel cells (SOFCs) the property of the electrolyte deposited by atmospheric plasma spraying (APS) has to be optimized concerning surface roughness and density. In this work, a dense YSZ coating with homogenous structure was realized on a NiO-YSZ anode which was supported by a porous metal substrate made of Crofer 22 H. A powder separator, which was able to diminish fluctuations of the in-flight particle stream, was used the first time to stabilize the APS process. During powder feeding a fully yttria-stabilized zirconia powder was on-line separated into two parts: a fraction of fine particles which were fed directly to the plasma, and a fraction of coarse particles which were collected and reused. The resulting coating exhibited a high homogeneity both in cross section and on surface. Validated by SEM images, the optimized coating did not possess any cracks and the porosity ratio was close to that found by vacuum plasma spraying (VPS) when two separators were employed. The gas leak rate shows an obvious increase of electrolyte performance compared to results without any powder separator.

Keywords: plasma spraying; powder separator; SOFC; coating property; electrolyte

[1] Corresponding author, Fax: +49 711 6862 747, E-Mail: taikai.liu@dlr.de.

1 Introduction

SOFCs are reliable choose for clean energy generation with high efficiency and low emission. Pure hydrogen or hydrocarbon or bio-gas can be directly fed to SOFCs to produce electricity and water with negligible emission. The size of SOFCs can be very small while keeping high efficiency which may make the fuel cells a power solution for portable electronic devices. In addition, SOFCs are very quiet during operation comparing with conventional power generator, such as turbine generator. It is possible to improve the total efficiency up to 95% by a co-generation [1]. SOFCs are based on the electrochemical oxidation and reduction reaction occurring on the anode side and cathode side respectively. A dense ceramic electrolyte is often employed to conduct oxygen ions and prevent electrons transport and to sever gas between anode and cathode. Due to the attractive conductivity of oxygen ions and robustness in high temperature environment, ceramic zirconia stabilized by yttria (YSZ) is one of the most favorite candidates for electrolyte of SOFCs. To prepare a dense YSZ electrolyte, several processes can be mentioned: physic vapor deposition (PVD), screen printing (SP) and plasma spray (PS).

Many works have been reported to successfully prepare dense YSZ electrolyte for SOFCs with PVD [2-5], and with SP process [6-8], and with PS process [9-12]. PVD is able to deposit quite dense coating with various types of material while the cost is high and the deposit efficiency is low. The SP process only can be carried on flat substrate and special post treatment as sintering is unavoidable that makes the SP process very low efficient. While PS process was proved suitable for SOFCs preparation with flexible operation conditions, such as the geometric shape of substrate, the choice of feedstock, the spraying ambiences (atmosphere, vacuum, neutral gas protect…). In addition, with PS process, it's also possible to deposit cells with structure of support-anode-electrolyte-cathode with one-step deposition. Although, necessary optimizations have been proposed to improve PS performance, the as-sprayed coatings are unavoidable porous which can help gas diffusion at two electrodes while may lead current loss at the electrolyte [9, 12-14]. Therefore, the main principle for improving SOFC performance is to enhance the density of electrolyte.

The performance of PS process is characterized by numerous factors, such as the spraying pressure, the ambient gas, the carrier medium for powder feeding, substrate surface state and etc. With different spraying pressure, the PS process can be carried under vacuum state (VPS) or atmosphere state (APS). With VPS, the plasma jet is largely extruded to promise higher impact temperature of particles and less bubbles trapped in coating. Thus, a dense coating is achieved for electrolyte of SOFCs with OCV (open circuit voltage) of 1.1 V [15]. However, both of the time cost and economic cost of VPS are unfavorable comparing with APS due to the decreasing and increasing procedure of pressure. Therefore, it's meaningful to improve the performance of APS process for achieving high dense coating. Works based both on liquid carrier (water, ethanol, ethylene glycol and etc.) [9, 12, 16] and gas carrier (mainly argon) [15, 17] have been reported to achieve OCV close to Nernst potential which is the equilibrium voltage based on balance between ion charge and ion concentration gradient. With utilization of liquid powder carrier, the APS process is denoted as suspension plasma spray (SPS) which is often used for very fine powders, e.g. Nano powders. In SPS process, feedstock is pre-dispersed in water/ethanol/... to form suspension and the settlement of particles should be prevented. That means special feeding system is necessary and the cost is higher than normal APS process.

For improving coating density, the cause of trapped pores in coating should be mentioned. Experimental and numerical works have proved that pores mainly arise from the unmolten/semi-molten state of particles before impact and the strong solidifying effect of substrate on droplets after impact [18-23]. Normally, a preheating step of the substrate is carried out to extenuate the solidifying effect to prevent droplets from splashing into long fingers/fragments where voids exist and form pores in coating [22, 23]. Due to the uneven granularity of powders, each time particles injected into plasma jet may have different size distribution: more coarse particles or more fine particles. Moreover, it's obvious that coarse particles are unable to achieve the same molten state as small particles in APS process with similar exposing time in plasma flame, especially when the feeding is accompanied by fluctuations. That will bring about pores with uneven sizes and with uncontrollable uneven distribution overall the coating.

Thus, the aim of this work is to diminish the uneven distribution of pores. In prior work, a powder separator was proposed to filter and collect coarse powder and let fine particles pass by. In this work, the powder separator was employed for depositing YSZ electrolyte on as-sprayed anode based on porous substrate of interconnects material (ITM). The microstructure of cross section and surface of electrolyte was validated by SEM (Scanning Electron Microscopy) and OM (Optical Microscopy). Gas leakage rate was measured to check the performance of half-cell.

2 Experimental conditions

Anode and electrolyte were prepared in a VPS chamber which can also serve as APS by connecting to filter pump system instead of vacuum system. A F4-VB type torch (Sulzer, Metco, Wohlen, Switzerland) with internal diameter of 6 mm was employed and held by a 2D movement motor. Powder was fed to plasma jet through an injector with internal diameter of 1.8 mm which was fixed 3 mm upstream to the nozzle exit by aligning the out port to the torch internal wall. Powder feeder stood about 3 meters away from the torch and out of the chamber. Powder separator with one inlet port and two outlet ports was employed to realize powder selection and fluctuation reduction. Two powder separators were connected inbetween the powder feeder and the torch through a carrier tube of about 5 meters. The feedstock used was Yttria-stabilized-Zirconia (YSZ, $(ZrO2)_{92}(Y2O3)_8$, Metco 6613) with size of d_{50}=14μm and d_{10}/d_{90}=7μm/23μm. All parameters mentioned in this work were provided in Tab. 1. Three square samples (S1~S3) with size of 45mm x 45mm were prepared by scanning substrate layer-by-layer with plasma flame. And two more samples (F1 and F2) were prepared by statically focusing the plasma flame on substrate (*Φ*40mm x 5mm) for different durations.

Table 1 operating parameters

parameters	samples				
	F1	F2	S1	S2	S3
I (A)	600	500	700	700	700
Ar (L/min)	40	40	50	50	50
H_2 (L/min)	5	5	5	5	5
He (L/min)	0	0	11	11	11
distance (mm)	100	300	100	100	300
feed rate (g/min)	15	15	15	15	15
Ar_{car} (L/min)	4.5	2	4.5	4.5	2
separator(s)	2	0	0	2	0
APS/VPS	APS	VPS	APS	APS	VPS
torch speed (mm/s)	0	0	400	50	400
layers/time	300 s	15 s	10	15	10

3 Results and discussion

3.1 Pores distribution

Image analysis was carried out based on optical micrographs. As shown in Fig. 1, series fractions of dark phase were obtained by controlling the gray value (threshold) which decides the type of phase: dark as void and gray as dense coating. With this threshold control, it is able to extinguish the influence of operation difference. The solid lines with hollow marks denote the change of dark phase fraction with increasing the threshold from 30 to 110. When the gray value decreased from 110 to 105, the curve 'S3' dropped from 71% to about 39% which means the region with gray value between 105 and 110 has a share of 32% over the whole cross section area, as presented by the curve 'S3_wgt' with X axis value of 110. Evident reduction is found in the fraction of dark phase by controlling threshold from 110 to 90. With threshold of 95, the fraction of dark phase in sample S2 is almost the same with the amount of dark phase in sample S3. And in the region between 30 and 95, the maximal difference between curve 'S2' and curve 'S3' is smaller than 5% with same threshold. It's obvious that S2 and S3 have comparable performance in porosity ratio. While the curve 'S1' decreases from 31% to 15% even with the same threshold as sample S2 and S3. Therefore, sample S1 is evidently poor in density than S2 and S3.

Labeled as 'S1_wgt', 'S2_wgt' and 'S3_wgt', aligning with the right Y-axis the dashed lines with filled marks depict the share of selected phase over the whole cross section. The value of weights is obtained by estimating the difference between two nearby selected gray values in the range of 30 and 110. Each data point on these lines gives a range of gray value and a value of its weight in percentage. For example, 'S3_wgt' provides a weight of 32% with gray value of 110 that means the fraction of gray phase between 105 and 110 is 32%. And 'S3_wgt' gives a weight of 14% with gray value of 105 that means the gray phase between 100 and 105 has a portion of 14% in total section area. 'S3_wgt' decreases from 32% to 2% by changing the gray value from 110 to 90. Continuously decreasing the gray value toward 35, 'S3_wgt' returns 'weight' close to 0.17% with small fluctuation. Similar tendency also presents in 'S1_wgt' and 'S2_wgt': with gray value bigger than 90, the phase weights have big variation; with gray value smaller than 90, the fluctuation of phase weights is quiet small. Therefore, it's reasonable to appraise coatings performance with a threshold of 85 for these three samples. According to the value of phase fraction in **Fig. 1** (left Y-axis), the porosity ratio of samples is: 27%, 12% and 8% for S1, S2 and S3 respectively. An obvious improvement is found from S1 to S2. It's clear that regarding the performance of porosity ratio, S2 is close to S3.

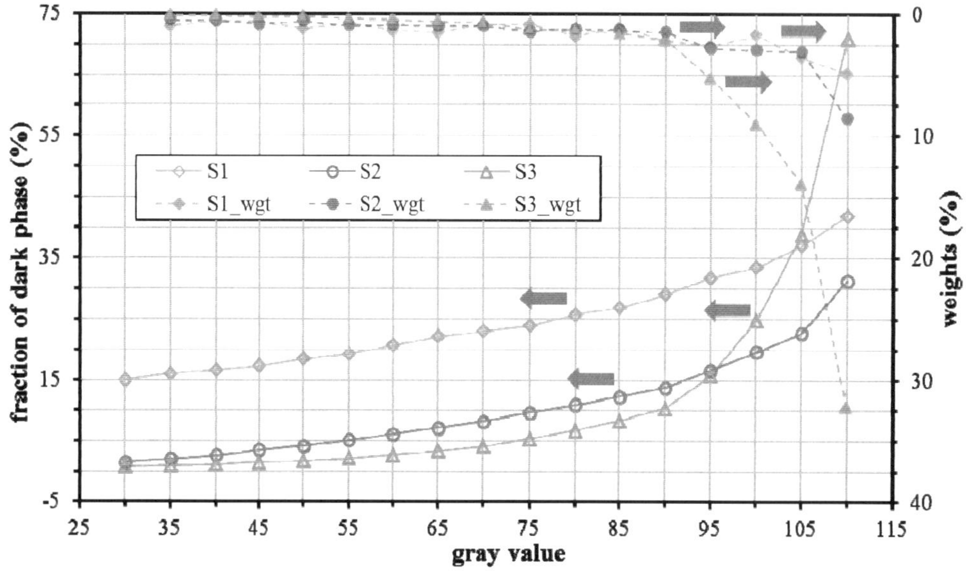

Fig. 1 fraction of dark phase and the weights

Visible improvement in pores distribution was also found comparing F1 with F2, as displayed in Fig. 2. Both F1 and F2 were prepared by stationing the torch at the center of substrate. The spraying duration is different because the different ability to melt particles between APS and VPS. Powder injector was connected to the powder feeder through a tube with length about 5 m. In order to enhance the fluctuation in powder feeding, the feeding tube was spiraled 5 times with diameter about 0.3 m both for F1 and F2. Powder separator was employed to moderate the fluctuation in powder feeding for reducing unmolten particles in plasma plume during the preparation of sample F1 [24]. F2 was prepared through normal VPS without powder separator for estimating the influence of feeding fluctuation on coating structure.

Partial cross sections of samples were presented in **Fig. 2** with gray color signifying the dense phase and black color signifying the void phase (pores). By counting the number of black pixel in each sample, the porosity ratio can be estimated as ratio between black pixel number and total pixel number. Result shows that with value of 3.7% for F1 and 5.4% for F2, the difference in porosity ratio between F1 and F2 is only 1.7%. While F1 exhibits a microstructure more homogenous than the obvious two-layer structure of F2. Calculating porosity ratio layer-by-layer, the local average porosity of the two-layer structure is 2.2% for the upper layer and 8.1% for the lower layer. This two-layer structure of F2 was induced by the serious fluctuation in powder feeding that the number of injected into plasma jet can suddenly soar up and quickly subside down [24]. Obviously, more unmolten particles will present in in-flight particle stream during such a fluctuation which will further induce pores and holes in coating. Pores with big size noticeably enrich in the lower layer because of the sharp fluctuation. According to prior works, fewer particles will be injected into the plasma jet due to the absorbing effect from the new-born slugs after the fluctuation [24]. Therefore, a dense layer will present covering the porous layer to form multilayers structure. It's evident that coatings will be inhomogeneous in microstructure with surge-like powder feeding. That can result in extreme poor reproducibility of coatings even strictly controlling the operating parameters. With powder separator extra particles are removed from the carrier gas during the fluctuation that it is always able to obtain dense coating like the upper layer of sample F2.

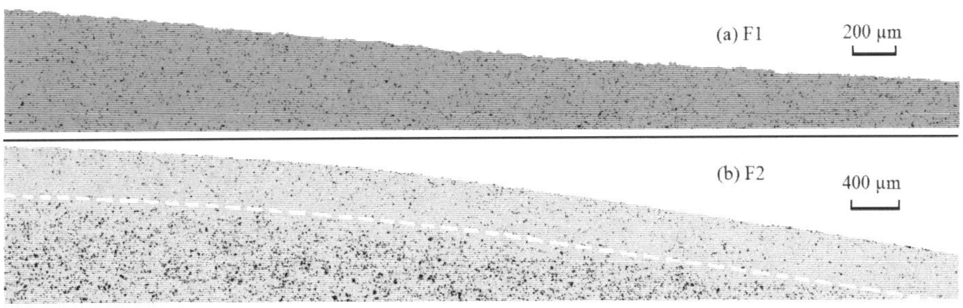

Fig. 2 partial cross section of F1 and F2

3.2 Surface morphology

The surface morphology of an electrolyte also gives contribution to the cell performance. High roughness obviously raised the contact area between electrolyte and electrodes while the possibility to have segmentation cracks and concentrated porosity in electrolytes is unavoidable increased [25, 26]. The electrolyte should be gas-tight and be thin to have low ionic resistance. A suitable surface state will be favorable for both contact area and gas-tight consideration.

Under a low magnification (x10), the surface morphology of coatings was checked with optical microscopy, as shown in Fig. 3. S1, S2 and S3 were prepared with similar anode layer which was directly sprayed on the porous substrate ('ITM'). For checking the surface morphology, samples were illuminated by a single beam with angle of 10° between the surface and the beam to project the knoll-like protrusions into shadows. In such a way, an obvious difference of surface morphology was observed from S1 to S3 in Fig. 3. Without powder separator, coating prepared under APS is comparatively rough and with evident raggedness originated from the irregular porous structure of substrate. In VPS, the coating sprayed without separator is also found uneven and with visible outcrops which are produced by over-deposition on the summits shown with arrows in subgraph 'anode'. And concavities, mainly induced from the valleys between summits in ITM, are also found to draw down the surface performance. While, fortunately, with the employment of powder separator, even in APS, the coating exhibited a high uniform topography and no distinct outcrops and summits were found, as shown in S2 of Fig. 3.

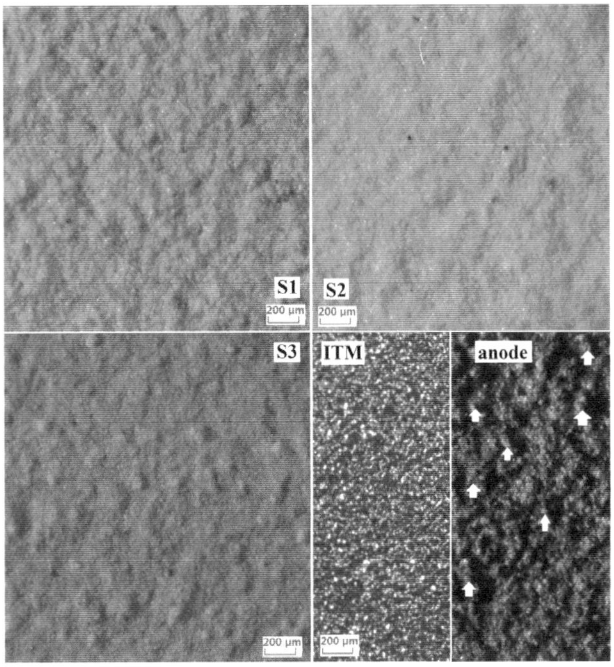

Fig. 3 surface morphology of coatings

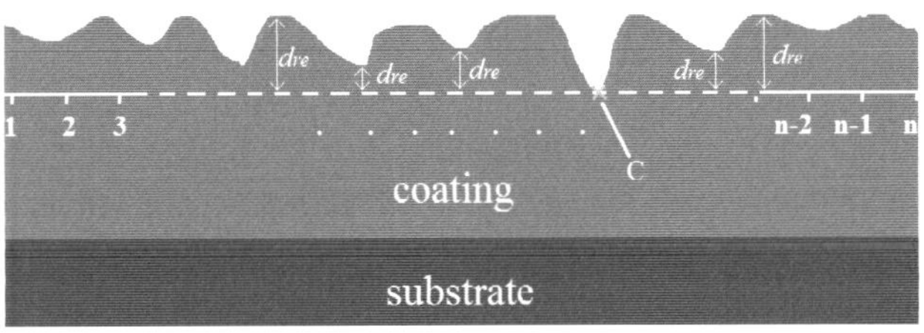

Fig. 4 profile roughness definition

For identifying the surface roughness of coatings, 40 profiles of cross section as shown in Fig. 4 were captured for each samples (S1, S2 and S3). A ground line was firstly defined to separate the rough surface from the coating by the lowest point on the surface, noted as 'C' in Fig. 4. Then relative thickness (d_{re}) of coating at each point of the ground line was defined and calculable by counting the gray pixel. The roughness of coating, R_a, was obtained by averaging the sum of d_{re} over the total number of pixels on the ground line: $R_a = 1/n\sum_1^n d_{re}$. And R_{rms}, the root mean squared roughness, was defined to describe the variance of surface roughness: $R_{rms}^2 = 1/n\sum_1^n d_{re}^2$. As displayed in Fig. 5, roughness R_a and R_{rms} of coatings are obviously decreased from S1 to S2 and to S3. Without powder separator, coating prepared under APS exhibits the roughest surface between these three samples, about 7.4 μm and 6 μm for R_a and R_{rms} respectively. While with introduction of powder separator the two roughness decrease to 6.4 μm and 4.9 μm respectively which is close to the performance of S3 (5.7 μm and 4.6 μm for R_a and R_{rms} respectively). In VPS, even without powder separator, coating established always exhibits the best performance in roughness and in porosity ratio.

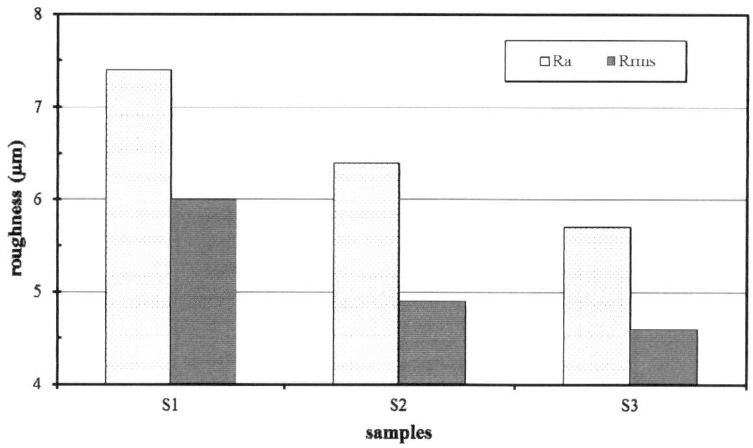

Fig. 5 roughness R_a and R_{rms}

More details about the roughness of cross section were provided in Fig. 6 with axis X signifying the relative thickness and axis Y signifying the weights of each d_{re} in the whole profile. For each sample (S1, S2 and S3), profile with width of 80 000 ~100 000 pixels was obtained from optical micrograph for weights calculating in **Fig. 6**. Bars filled with different patterns are obtained by counting the number of selected region of d_{re}. For example, d_{re} = 10 means the region between 9 μm and 10 μm, and d_{re} = 50 means the region between 30 μm and 50 μm in **Fig. 6**.

As indicated in **Fig. 6**, the selected d_{re} ranges from 1 to 50 with different interval dividing into three regions (region A, region B and region C). Noticeable difference is found when the selected d_{re} smaller than 4 μm (region A) between three samples. With summed weight of 46% in region A (sum of all data points), sample S3 obviously presents the highest weight in this region comparing with sample S1 and S2 who have weight of 35% and 38% respectively. While in region B (between 5 μm and 10 μm), there is a steady fall in the weights with increasing the relative thickness d_{re} for both of three samples. The summed weights in this region of three samples are 37%, 42% and 36% for S1, S2 and S3 respectively. It's clear that sample S2 exhibits the highest weight in region B. And in region C, the summed weights dip to 28%, 20% and 18% for S1, S2 and S3 respectively. As depicted in **Fig. 6**, sample S3 has a cross section with profile 46% in region A and 36% in region B and 18% in region C. Sample S2 has profile 38% in region A and 42% in region B and 20% in region C. Meanwhile, sample S1 has a profile that 35% in region A and 37% in region B and 28% in region C. Therefore, with employment of powder separator, the coating surface roughness is obviously improved.

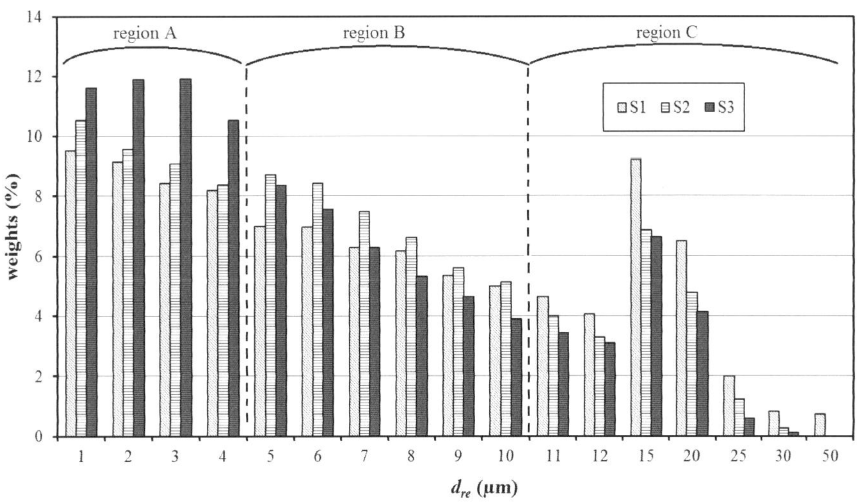

Fig. 6 distribution of relative thickness d_{re}

3.3 Electrolyte density

With high enough density, the electrolyte can completely isolate the fuel gas from oxygen. In this work electrolyte was prepared by accumulation of 8YSZ particles which have been heated up and accelerated by the plasma jet. Typical lamella structure of thermal spray was thus obtained for anode and electrolyte in this work. That means gap may exist between layers producing reduction of coating density. In addition, due to the uneven particle size and instable trajectory, particles may be partially melted or unmolten before impacting upon the substrate surface that can induce flaws cracks to decrease the coating density. Cracks also may be produced during plasma spraying due to high temperature gradient between hot layer and cold layer [27]. Therefore, even without visible holes on the surface, coating may be also permeable comparing with dense material.

Gas leak rate is widely employed to indicate the density of coating used as electrolyte for SOFC [12, 15]. Gas leak rate is dominated by the pressure difference between the two sides of electrolyte and the porosity of electrolyte. Normally, the measurement is carried out by exposing one side of electrolyte in normal atmosphere (1 atm) while pressing another side to the port of pumping system that vacuum can be formed between sample (electrolyte) and the port. If the electrolyte is dense enough, the vacuum produced can be held for a long time. Otherwise, a decay rate can be estimated according to the pressure increase of the 'vacuum'.

In this work, electrolytes of sample S1, S2 and S3 were prepared with thickness about 32 μm, 80 μm and 35 μm respectively. It's hard to control the thickness of sample S2 due to the employment of powder separator. The electrolyte of sample S2 is obviously thicker than other two electrolytes of samples S1 and S3. However, sample S3 exhibits the lowest gas leakage with rate of 19 Pa.m/s comparing with 700 Pa.m/s of sample S1 and 50 Pa.m/s of sample S2. Evidently, the electrolyte of sample S1 is completely porous for gas which is unable to generate electric power. The performance of sample S2 is much better than S1 but still not a reasonable electrolyte for SOFC. It's hard to conclude that the improvement of gas leak rate from S1 to S2 is totally dedicated by the employment of powder separator.

Regarding to the surface state of these three samples, the gas leak rate is mainly influenced by two features of coating: existence of open holes and the interval between layers. As displayed in **Fig. 7** (a) and (b), open holes are found penetrating into the porous anode layer from the electrolyte surface of sample S1 which allows the direct contact between fuel gas and air (oxygen). The pores in substrate are devoted to guide fuel gas to anode function layer (AFL) while the existence of extreme big pores is unavoidable and unforeseeable, as shown in **Fig. 7** (c) and (d), that can induce open pores in electrolyte. While, no visible open pore was found in sample S2 and sample S3. That means the open pores can be sealed either by establishing thicker coating under atmospheric condition or by spraying in vacuum.

The plasma flame of APS is obviously shorter than that in vacuum spray process. Therefore, it's difficult for particles to achieve high temperature to completely melt before impacting. While with employment of powder separator, the number of particles injected to plasma flame is reduced that will lead a higher impacting temperature for particles. Moreover, coarse particles are partially filtered out to diminish void defect in final coating. As shown in Fig. 8, sample S1 is anticipated to have the most pores and cracks/segments between these three samples (S1, S2 and S3). And the size of and pores and cracks/segments is significantly bigger than that of sample S2 and S3. These big pores and bold cracks may connect to each other to result in very porous structure of sample S1 that gives a gas leak rate of 700 Pa.m/s. However, remarkable improvement is found from sample S1 to sample S2 both in gas leak rate and coating density. With employment of powder separator, sample S2 is obviously denser than sample S1 with smaller pores and finer cracks that give a gas leak rate of 50 Pa.m/s. Therefore, it can conclude that powder separator is able to improve the density of coating. Only small pores and micro cracks are found in sample S3 that makes S3 the densest coating with gas leak rate of only 19 Pa.m/s. Regarding the performance of S3, even in VPS, the distribution of void defects is also uneven that induces poor reproducibility of coating, especially for the electrolyte of SOFC which is sensitive to the microstructure. With proper combination of powder separator and VPS process, it will be possible to establish super dense coating. Thus for robustly improving the plasma-sprayed electrolyte performance, a powder separator is helpful.

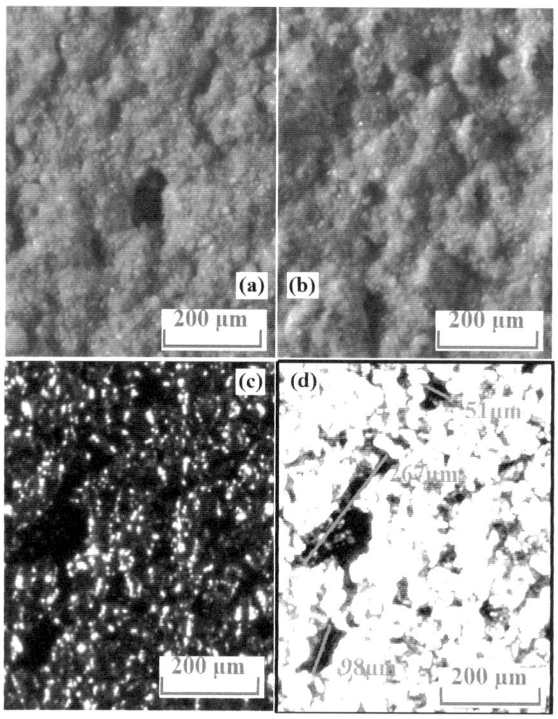

Fig. 7 open holes of sample S1 (a, b) and the substrate (c, d)

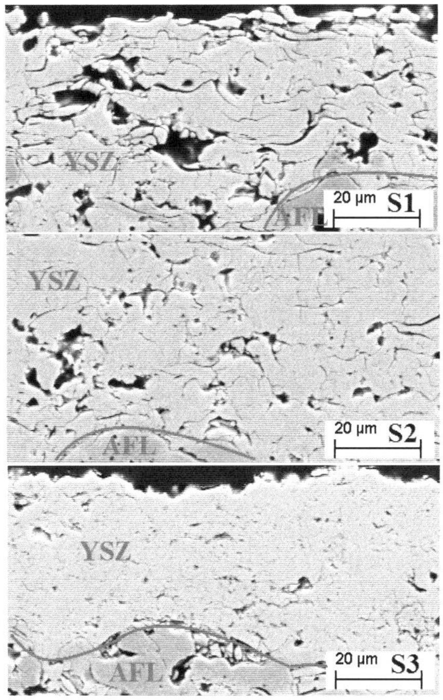

Fig. 8 electrolyte microstructure of samples (AFL: anode; YSZ: electrolyte)

4 Conclusions

Powder separator was the first time used to try to improve the density of APS-sprayed electrolyte of SOFC. Two types of samples were prepared: standard flat samples (S1, S2 and S3) by scanning substrate with plasma and center-enhanced samples (F1 and F2) by stationing torch focusing at the substrate center. Both OM and SEM were used to analyze coating microstructure. Two roughnesses (R_a and R_{rms}) were defined to estimate the surface state of coatings. Gas leak rate was measured to evaluate sample density. The results proved that the powder separator can help to homogenize coating microstructure and it's also able to optimize surface roughness. Inter-lamellar micro-cracks/segments were found obviously reduced by employing two powder separators in APS. However, with current process, VPS is still necessary for preparing high dense electrolyte coating.

Acknowledgements

Authors would like to acknowledge Deutsches Zentrum für Luft-und Raumfahrt (DLR)/Deutscher Akademischer Austauschdienst (DAAD) Research Fellowship Program for Postdocs with grant number 50019752. Authors also would like to express thanks to Grünter Roth and Ina Plock for technical support.

References

[1] A. Faes, A. Hessler, A. Zryd and J. Van Herle, A Review of RedOx Cycling of Solid Oxide Fuel Cells Anode, Membranes, 2012, **2(3)**, p 585-664

[2] A.A. Solovyev, N.S. Sochugov, S.V. Rabotkin, A.V. Shipilova, I.V. Ionov, A.N. Kovalchuk and A.O. Borduleva, Application of PVD Methods to Solid Oxide Fuel Cells, Appl. Surf. Sci., 2014, **310**, p 272-277

[3] X. He, B. Meng, Y. sun, B. Liu and M. Li, Electron Beam Physical Vapor Depositon of YSZ Electrolyte Coatings for SOFCs, Appl. Surf. Sci., 2008, **254**, p 7159-7164

[4] R. Nédélec, S. Uhlenbruck, D. Sebold, V.A.C. Haanappel, H.P. buchkremer and D. Stöver, Dense Yttria-stabilised Zirconia Electrolyte Layers for SOFC by Reactive Magnetron Sputtering, J. Power Sources, 2012, **205**, p 157-163

[5] G. Laukaitis, J. Dudonis, A.F. Orliukas and D. Milcius, Properties of YSZ Thin Films Deposited by E-beam Technique, Solid State Ionics, 2008, **179**(1-6), p 182-187

[6] R. Ried, C. Lorenz, A. Brönstrup, T. Graule, N.H. Menzler, W. Sitte and P. Holtappels, Processing of YSZ Screen Printing Pastes and the Characterization of the Electrolyte Layers for Anode Supported SOFC, J. Eur. Ceram. Soc., 2008, **28**(9), p 1801-1808

[7] D. Rotureau, J.P. Viricelle, C. Pijolat, N. Caillol and M. Pijolat, Development of a Planar SOFC Device Using Screen-printing Technology, J. Eur. Ceram. Soc., 2005, **25**(12), p 2633-2636

[8] Y. Zhang, X. Huang, Z. Lu, X. Ge, J. Xu, X. Xin, X. Sha and W. Su, Effect of Starting Powder on Screen-printed YSZ Films Used as Electrolyte in SOFCs, Solid State Ionics, 2006, **177**(3-4), 281-287

[9] D. Waldbillig and O. Kesler, Effect of Suspension Plasma Spraying Process Parameters on YSZ Coating Microstructure and Permeability, Surf. Coat. Technol., 2011, **205**(23-24), p 5483-5492

[10] C. Metcalfe, E. Lay-Grindler and O. Kesler, Characterization of Ni-YSZ Anodes for Solid Oxide Fuel Cells Fabricated by Solution Precursor Plasma spraying with Axial Feedstock Injection, J. Power Sources, 2014, **247**, p 831-839

[11] S. Wolf, R. Adam, L. Fuhrmann, S. Ansar and K.A. Friedrich, Characterization of Electrolyte Layers of Plasma-sprayed Metal Supported Solid Oxide Fuel Cells, Solid State Ionics, 2013, **243**, p 30-35

[12] M. Marr, J. Kuhn, C. Metcalfe, J. Harris and O. Kesler, Electrochemical Performance of Solid Oxide Fuel Cells having Electrolytes Made by Suspension and Solution Precursor Plasma Spraying, J. Power Sources, 2014, **245**, p 398-405

[13] R. Maric, R. Neagu, Y. Zhang-Steenwinkel, F. Van Berkel and B. Rietveld, Reactive Spray Deposition Technology-An one-step deposition technique for Solid Oxide Fuel Cell barrier layers, J. Power Sources, 2010, **195**(24), p 8198-8201
[14] G. Schiller, R. Henne, M. Lang, R. Ruckdäschel and S. Schaper, Development of Vacuum Plasma Sprayed Thin-film SOFC for Reduced Operating Temperature, Fuel Cells Bull., 2000, **21**(3), p 7-12
[15] A. Ansar, Z. Ilhan and J. Arnold, Plasma Sprayed Metal Supported SOFCs having Enhanced Performance and Durability, Thermal spray 2009: Expanding Thermal Spray Performance to New Markets and Applications, on CD-Rom, B.R. Marple, M.M. Hyland, Y.C. Lau, C.J. Li, R.S. Lima and G. Montavon, Ed., May 4-7, 2009 (Las Vegas, Nevada), DVS-German Welding Society – ASM International, 2009.
[16] D. Waldbillig and O. Kesler, Electrochemical testing of suspension plasma sprayed solid oxide fuel cell electrolytes, J. Power Sources, 2011, **196**(13), p 5423-5431
[17] J. Arnold, A. Ansar, H. Weckmann and S. Fiebig, Investigation of Time-Dependent Instabilities of Plasma Spraying Process Using Online Diagnostic Systems, Thermal Spraying 2006: Building on 100 Years of Success, B.R. Marple, M.M. Hyland, Y.C. Lau, R.S. Lima and J. Voyer, Ed., May 15-18 2006 (Seattle, USA), ASM International, 2006,396 p.
[18] S. Matthews, Shrouded plasma spray of Ni–20Cr coatings utilizing internal shroud film cooling, Surf. Coat. Technol., 2014, **249**, p 56-74
[19] R. Mcpherson and B.B. Shafer, Interlamellar Contact within Plasma-sprayed Coatings, Thin Solid Films, 1982, **97**(3), p 201-204
[20] X.C. Zhang, B.S. Xu, Y.X. Wu, F.Z. Xuan and S.T. Tu, Porosity, mechanical properties, residual stresses of supersonic plasma-sprayed Ni-based alloy coatings prepared at different powder feed rates, Appl. Surf. Sci., 2008, **254**(13), p 3879-3889
[21] P. Fauchais, M. Vardelle, A. Vardelle, L. Bianchi and A.C. Léger, Parameters Controlling the Generation and Properties of Plasma Sprayed Zirconia Coatings, Plasma Chem. Plasma Process., 1996, **16**(1), p 99S-125S
[22] A. Abedini, A. Pourmousa, S. Chandra and J. Mostaghimi, Effect of substrate temperature on the properties of coatings and splats deposited by wire arc spraying, Surf. Coat. Technol., 2006, **201**(6), p 3350-3358
[23] R. Dhiman, A. McDonald and S. Chandra, Predicting splat morphology in a thermal spray process, Surf. Coat. Technol., 2007, **201**(18), p 7789-7801
[24] T. Liu and J. Arnold, Study of In-flight Particle Stream and Particle Behaviour for Understanding the Instability Phenomenon in Plasma Spraying Process, Surf. Coat. Technol., 2016, **286**, p 80-94
[25] M. Marr, C. Metcalfe, E.S.C. Fan and O. Kesler, Improved Electrolyte Performance in Plasma Sprayed SOFCs by Electrode Modification, ECS Trans. 2013, **57**(1), p 1025-1035
[26] Y. Kawagoe, S. Namie, T. Kumakura and K. Shiozaki, New Fabricating Method of Solid Oxide Fuel Cell and Performances, Bull. M.E.S.J., 1999, **27**(1), p 14-21
[27] M. Marr and O. Kesler, Influence of plasma heat flux on segmentation cracking and permeability of thin suspension plasma sprayed coatings, Surf. Coat. Technol., 2013, **216**, p 289-296

Low temperature sintering assistant-preparation of LSGM electrolyte for solid oxide fuel cells deposited by vacuum cold spray

Lishuang Wang, Chengxin Li*, Guanjun Yang, Xiaotao Luo, Shanlin Zhang, and Changjiu Li
State Key Laboratory for Mechanical Behavior of Materials, School of Materials Science and Engineering Xi'an Jiaotong University, Xi'an Shaanxi, 710049 China.

Intermediate solid oxide fuel cell (IT-SOFC), which operates at lower temperature to reduce cost and improve efficiency, had received certain attention in recent years. Vacuum cold spray (VCS), facilitating in forming thin ceramic film based on the impact adhension of fine particles at room temperature, is therefore a new and promising method to fabricate electrolyte membrane of SOFC. However, a dense VCS-prepared coating is primarily dependent on a optimized combination of powder size range and particle velocity, which limits the application of VCS on SOFC. In this study, the VCS-prepared LSGM electrolyte was further densified assistant with low temperature sintering. LSGM electrolytes with different thickness, were fabricated by vacuum cold spray following by sintering at 1200 °C to further densify the coating. After sintering, the coating become densier in most regions due to the grain growth from one hundred nanometers to micro-scale accompanying with the appearance of some cracks in local regions of the coating surface due to thermal mismatch and constrained sintering. Therefore, the second layer was deposited on the annealed coatings to block those cracks. Then the two-layer structure was co-sintered to further densify the coating. The test results of gas permeability of the obtained coatings demonstrate that the films are dense enough to act as the electrolyte of SOFC. The output performance of cell was also tested at 600-800 °C with LSCF cathode layer deposited by air plasma spray (APS).

1 Introduction

Solid oxide fuel cell (SOFC) has attracted considerable attention in recent years due to its high-energy conversion efficiency, fuel flexibility and low pollution emisssions as well. However, the high operating temperature of SOFC brings about many serious problems, such as limited material choices of electrode and electrolyte, cell sealing and high cost. Consequently,certain efforts have been made to decrease the operating temperature of SOFC. Towards this orientation, the intermediate temperature-solid oxide fuel cell (IT-SOFC) appears to be an alternative choice for the development of SOFC in the future.. The challenges to develop IT-SOFC is twofold. First is the choices of material for both electrode and electrolyte. As a traditional electrolyte material choice, YSZ (yttria-stabilized zirconia) with ionic conductivity of 0.14 S/cm at 1000 °C, , is widely used at high temperature [1, 2]. However, , the ionic conductivity of YSZ decreases dramatically below 1073 K.As a consequence, it is of great importance to find a new electrolyte material with higher ionic conductivity at intermediate temperature. Gadolinia/samaria-doped ceria(GDC/SDC) [3, 4], stablized Bi_2O_3 and lanthanum doped with strontium and magnesium (LSGM) [5, 6] have been established as potential electrolyte candidates for SOFC applied at low temperature. As reported before, one main limitationfor CeO_2-based electrolyte wass its mixed ionic and electronic conduction (MIEC) caused by reduction of ceria from Ce^{4+} to Ce^{3+} when exposed to reduced atmosphere [3]. Although Bi_2O_3-based material presents high ionic conductivity at low temperature, phase instability in reduced atmospheres and evaporation of bismuth oxide at operating temperature are still difficult problems to overcome [7]. Recently, some attentions has been paid to LSGM with perovskite structure due to its high ionic conductivity (about 0.15 S/cm at 800 °C), negligible electron conductivity and good stability over a wide oxygen partial pressure range [5, 8]. The second challenge for IT-SOFC is the fabrication method for electrolyte. The conventional fabrication methods for LSGM electrolyte includes dip-coating [9], screen printing, sol-gel method [10] and air plasma deposition [11, 12]. Nevertheless, the aforementioned methods often are aided with high temperature sintering at 1350~1400 °C to obtain dense LSGM films. Unfortunately, undesirable chemical reaction between electrolyte with other cell components at high temperature significantly deteriorites the cell performance. Yamaji et al. [13] had reported that Ga depletion happened when LSGM exposed to reduced atmosphere. Consequently, Pt-Ga alloy formed due to the fact that Pt is used as the electrode. Additionaly, the interface reaction between Ni-based anode and LSGM at high temperature also results in unexpected degradation of cell performance [14, 15].

In our previous reports, LSGM electrolyte with a thickness of about 55 μm was successfully fabricated by air plasma spraying. The evaporation of Ga resulted in different ionic conductivity in plasma-sprayed LSGM electrolyte [12]. However, APS is relatively difficult to prepare thin electrolyte coatings applied for IT-SOFC with low ohmic resistance.Therefore, it is necessary to develop a low temperature deposition process to fabricate electrolyte for avoiding the deleption of Ga with decreased thickness.

Vacuum cold spray(VCS), also known as aerosol deposition, have been widely used for ceramic film preparation. The ceramic particles are firstly injected into the vacuum chamber by the pressure difference. Then those particles deposites on the substrate through impact and adhesion of fine particles at room temperature [16]. Although the mechanism was still unknown, numerous investigations about preparing films by VCS have been documented, such as PZT [17], Al_2O_3 [18], SiC [19], TiO_2 [20] and $Li_4Ti_5O_{12}$ [21]. Advantages of VCS include high deposition rate, controllable thickness of coatings and low cost. Meanwhile, the VCS-prepared coatings often present small grain size . Moreover, the initial powder size and morphology play critical role on the obtained film quality [22] (powder requirement). Lakshmi [23] obtained dense bulk ScSZ at 1250 °C through sintering nanocrystalline powders (the sintering temperature of ScSZ is 1400~1500 °C [24, 25]), indicating that sintering temperature of ceramic material can be lower by using nano powders.

In the present study, the VCS-prepared LSGM coating was further densified assistant with lower sintering temperature.The further sintering at 1200 °C (the sintering temperature of LSGM is above 1400 °C [8])enhanced the densification of obtained LSGM coatings. In addition, the cell performance was evaluated using GDC as buffer layer on anode-supported SOFC with LSGM as electrolyte membrance prepared by multilayer method.

2 Experimental procedure

2.1 Materials

Commercially available powder $La_{0.8}Sr_{0.2}Ga_{0.8}Mg_{0.2}O_3$ (LSGM) (Tian-yao. Inc., Qingdao, China) with a diameter of about 2 μm was used as starting powder. The starting powders were further prepared by wet milling for 2 hours by using zirconia balls as milling media and ethanol as solvent. Following on, the powders were dried and grinded to obtain fine grain size and improve deposition rate. The morphology of the powder was shown in Fig. 1. The particle size of ball-milled powder was the same as raw powder. Anode supported SOFC was used as substrate to investigate the power generation properties. The disk NiO-8YSZ (50%NiO-50%YSZ) with a diameter of 13 mm and a thickness of 0.4 mm was fabricated by tape casting processes. In order to avoid the chemical reaction between Ni and LSGM, gadolinium-doped ceria (GDC) layer (2~3 μm) was interveningly coated on NiO/YSZ substrate by tape casting.

Fig. 1. SEM micrographs of LSGM powder: (a) as-received (b) ball-milled for 2 h.

2.2 Preparation of LSGM coating

LSGM coatings were deposited using a VCS system developed in our laboratory. The schematic diagram was shown in Fig. 2. As soon as the pressure in the vacuum chamber reached 40 Pa, helium gas carried with powders was injected into the vacuum chamber. Then the nozzle scan the substrate back and forth through movement of X-Y stage. The deposition parameters were shown in table 1. The different thickness of LSGM coating were obtained through changing the scan numbers [20, 26]. Two spray pass was named as C2. In our previous reports, VCS-prepared TiO_2 films in Dye-sensitized solar cells (DSC) were porous and the porosity was 43%~49% with pore size ranging from about several nanometers to one-hundred nanometers [27-29]. Obviously, coatings were not dense enough for using as electrolyte in SOFC. Therefore, the obtained LSGM coatings were further sintered at 1200 °C for 5 hours. Microstructure of surface and fractured cross-section of films were examined by scanning electron microscope (SEM, TESCANMIRA 3 LMH, Czech). The phase structure was characterized by X-ray diffraction (XRD, Xpert PRO,PANalytical, The Netherlands).

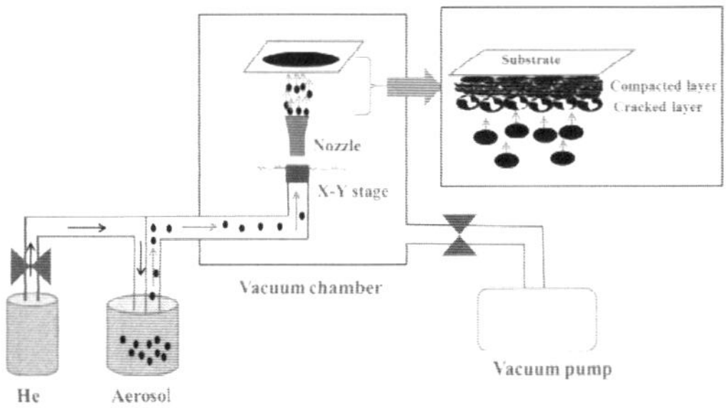

Fig. 2. Schematic of vacuum cold spray apparatus.

Table 1 Deposition parameters of vacuum cold spray

Parameter	Unit	Value
Chamber pressure	Pa	250
He gas flow rate	L/min	6.5
Distance from nozzle exit to substrate	mm	5
Nozzle traversal speed	mm/s	30
Nozzle orifice size	mm×mm	2.5×0.2
spray passes	times	2,4,6,8

2.3 Cell preparation

Commercially available agglomerated $La_{0.6}Sr_{0.4}Co_{0.2}Fe_{0.8}O_3$ (LSCF, Sulzer-Metco, USA) powder with a size range of 5~20 μm was deposited on the electrolyte using air plasma spray (APS, GP-80 KW, Jiujiang) with an area of 0.8 cm^2 and thickness of 10~20 μm. For electrochemical testing, the silver paste was employed at sides of anode and cathode as current collectors. Cell was reduced in 850℃ for 5 hours in H_2 atmosphere, then output performance of cell (I-V and I-P) was measured at 500-800 °C under humid H_2 flow of 0.05 slpm with 3% water and an air flow of 0.15 slpm.

3 Results and discussions

3.1 Microstructure of the as-deposited coatings and thermally-treated coatings

The thickness of as-deposited coatings and annealed coatings are statistically demonstrated in table 2. As a representative presentation, the cross-sectional morphology of coatings C2 is shown in Fig. 3. As can be seen that the thickness of coatings can be tripled as the spray pass increased from 2 to 8. It can also be found that the obtained LSGM coatings were not dense enough for using as the electrolyte of SOFC due to the fact that the velocity of particles was insufficient for compacting previous fragment particles [30]. In order to obtain dense coatings, coatings were annealed at 1200℃ for 5 hours. It can be found from Fig. 3(c) and (d) that grains grow from nano-size to micro size and the thickness of coatings decreased obviously. For example, the thickness of coating C2 decreased to about 1 μm after thermal exposure at 1200 °C for 5 hours, resulting in a shrinkage as much as about 50%. While for C8, the shrinkage is about 30%, as can be seen in table 3. On one hand, the VCS-prepared coating is not dense enough before heat treatment. On the other hand, the thicker film like C8 in vacuum cold spray shrinked much more as more layers result in more compressive stress due to the fact that particles bonding in VCS is derived from sufficient successive impacts of particles [30].
It is worth noting that some cracks appeared in the surface of sintered films due to the thermal mismatch between coating and the substrate, as shown in Fig. 4. According to the equation 1, strain energy release rate is proportional to the thickness of film, resulting in more cracks in thicker layers [32]. This is consistent with the results in Fig. 4 that as the thickness of coating increased to about 10 μm, large cracks appeared in the coating to release stress.

$$G_i = \frac{Eh(\Delta\alpha\Delta T)^2}{2(1-\nu)} \quad (1)$$

where h is the coating thickness, G_i is the strain energy release rate and E is the elastic modulus [32].

Table 2 Thickness of as-deposited coatings and annealed films

Spray pass	As-deposited coating thickness/μm	Annealed coating thickness /μm
2	3.03±0.38	1.52±0.32
4	4.97±0.44	3.73±0.57
6	6.21±0.56	5.03±0.21
8	8.31±0.78	5.85±0.95

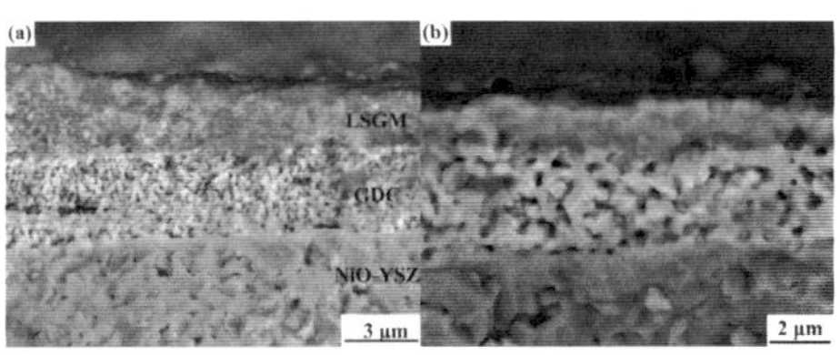

Fig. 3 Coating C2 before (a)(b) and after(c)(d) thermal exposure at 1200 °C for 5 hours: (a)(c) for surface and (b)(d) for fracture cross-section.

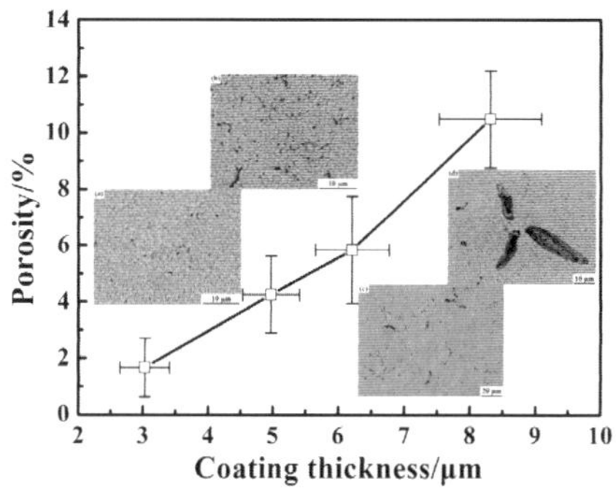

Fig. 4 Effect of VCS-prepared coating thickness on the surface porosity.

3.2 Blocking surface cracks by multilayer deposition

In order to block surface cracks, multilayers are deposited on the annealed coatings. Then the two-layer structure is co-fired to obtain dense coating. Two different combinations are used to investigate the effect of thickness of each layer on the performance of cell. Besides the C2 pattern, C4 coating is also used as the first layer. Then C2 coating was deposited on annealed C4 coating. The diagram of the method is shown in Fig. 5.

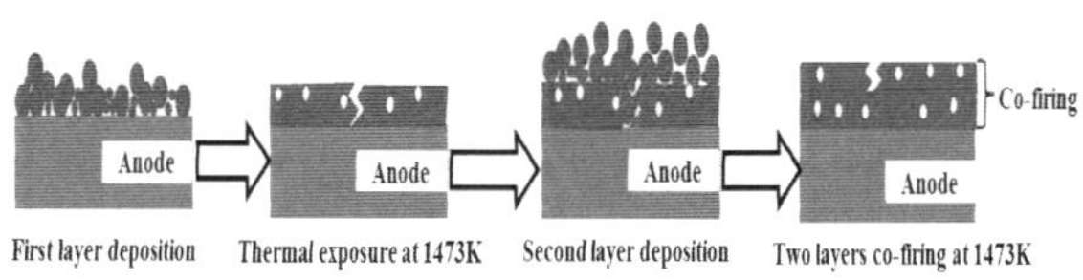

Fig.5 The diagram of co-firing multi-layer method.

The morphology of C4-2 before and after thermal exposure was shown in Fig. 6. From the coating surface shown in Fig. 6(a), there is no cracks appeared in the second layer. From the cross-sectional morphology shown in Fig. 6(b), the two layers (C4 and the following C2) before annealing are bonded well, suggesting that the surface roughness of first annealed layer is suitable for the following VCS coating [33]. This two-layer structure was co-fired at 1473K in a temperature-controlled electric furnace. Fig. 6(c) shows the cross-sectional morphologies of C4-2. It is hardly to distinguish the interface of two layers, indicating that the co-firing method is effective to merge multilayers to a single layer and thus the interfacial contact resistance between layers could be ignored.

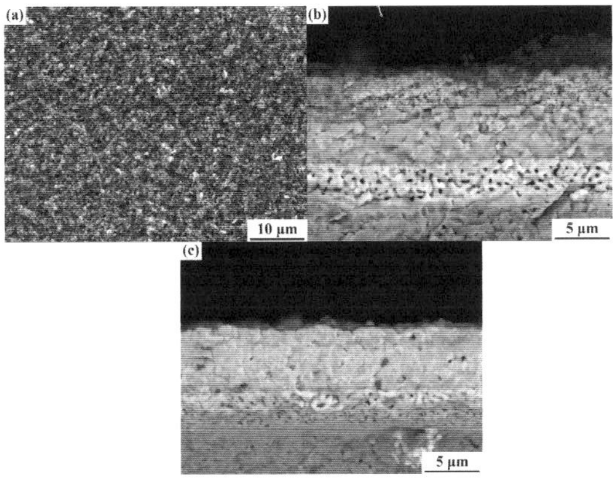

Fig.6 Morphology of C4-2 coating. (a) SEM for surface morphology before annealing; (b) back-scatter electron (BSE) for cross-sectional morphology after second layer deposition; (c) BSE for cross-sectional morphology after co-firing at 1200 °C for 5 hours.

3.3 Cell performance

The open circuit voltage (OCV) and the output performance of cell varying with test temperatures is shown in Fig. 7(a) and (b), respectively. The OCV value was about 0.8V, which was lower than the theoretical voltage [34]. The maximum power density was 250 mW·cm^{-2} at 800 °C, which was greatly lower than APS LSGM electrolyte in our previous report [12]. The gas-leakage rate of coating C4-2 was 5.08×10^{-8} cm^{4}gf^{-1}s^{-1}. Due to the fact that the coating was dense enough by the test of gas-leakage rate, it is necessary to find other reasons accounting for the relatively low OCV and maximum power density. According to the previous reports [35], the interfacial reaction between LSGM and GDC thermally exposed at temperature above 1300 °C were observed. The resistivity of the obtained reaction products $LaSrGa_3O_7$ was about 1 M ohm·cm at 800 °C. In addition, Kim [36] obtained OCV of about 0.7 V with GDC as buffer layer and dense LSGM as electrolyte. Due to the fact that the GDC layer was unable to prevent the diffusion of Ni at temperature of 1450 °C. Consequently, an electronic conduction path was generated, leading to a lower OCV. Meanwhile, they maintained that the reaction might be avoided as temperature lower than 1300 °C.

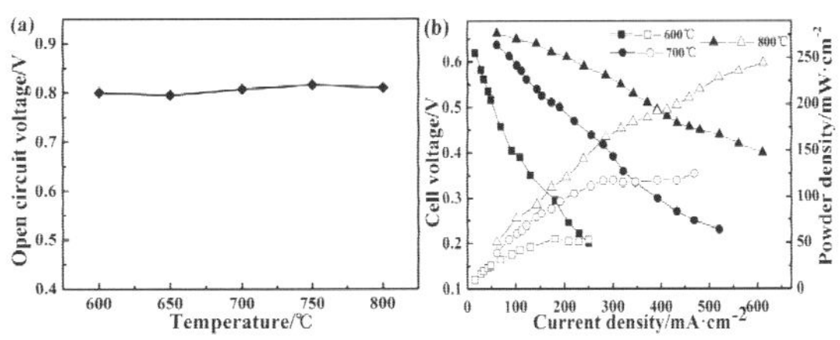

Fig.7 Cell performance: (a) open circuit voltage and (b) I-P and I-V curve.

The X-ray diffrraction analysis was used to analyze the interface reaction between LSGM and GDC after cell performance test. Fig. 8 shows the XRD pattern of LSGM after performance test. As for comparison, the phase structure of as deposited LSGM coating was also shown in Fig. 8. It can be found that the Ni and $LaSrGa_3O_7$ appeared in the interface, resulting in lower OCV value and poor output performance. Therefore, interfacical reaction between LSGM and GDC is inevitable at 1200 °C. Meanwhile, the GDC can not effectively prevent the diffusion of Ni element.

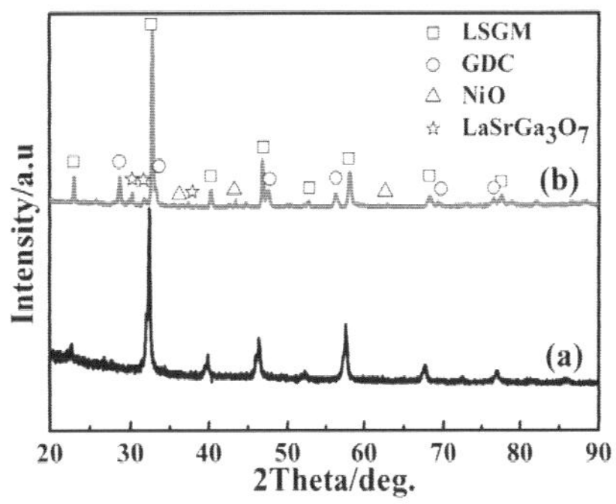

Fig.8 X-ray diffraction pattern of (a) LSGM powder (b) interface of LSGM and GDC after cell performance test.

4 Conclusions

Dense VCS-prepared LSGM electrolyte with thckness of 5 μm was successfully fabricated assistant with post-sintering at low temperature of 1200 °C using multi-layer stacking method. The optimized as sprayed thickness of first layer and second layer was about 5 μm and 3 μm, repectively, which showed a gas leakage rate of 5.08×10^{-8} cm^{4}gf^{-1}s^{-1}. The new method applied to vacuum cold spray, which overcomes the limit of particle size on the fabrication of dense coating by vacuum cold spray.

5 Acknowledgment
The present study was partially supported by the National Basic Research Program (Grant No. 2012CB625100).

6 Reference
[1] O. Yamamoto. Solid oxide fuel cells: fundamental aspects and prospects. Electrochim Acta, 45 (2000), Issue (15-16), pp. 2423-2435.
[2] M.F. Han, X.L. Tang, H.Y. Yin, and S.P. Peng: Fabrication, microstructure and properties of a YSZ electrolyte for SOFCs. J Power Sources, 165 (2007), Issue 2, pp. 757-763.
[3] S. Cho, Y. Kim, J.H. Kim, A. Manthiram, and H.Y. Wang: High power density thin film SOFCs with YSZ/GDC bilayer electrolyte. Electrochim Acta, 56 (2011), Issue 16, pp. 5472-5477.

[4] Y.J. Leng, S.H. Chan, S.P. Jiang, and K.A. Khor: Low-temperature SOFC with thin film GDC electrolyte prepared in situ by solid-state reaction. Solid State Ionics, 170 (2004), Issue (1-2), pp. 9-15.
[5] T. Ishihara, H. Matsuda, and Y. Takita: Doped $LaGaO_3$ Perovskite-Type Oxide as a New Oxide Ionic Conductor. J Am Chem Soc, 116 (1994), Issue 9, pp. 3801-3803.
[6] Raghvendra, R.K. Singh, and P. Singh, Synthesis of $La_{0.9}Sr_{0.1}Ga_{0.8}Mg_{0.2}O_{3-\delta}$ electrolyte via ethylene glycol route and its characterizations for IT-SOFC. Ceram Int, 40(2014), Issue 5, pp. 7177-7184.
[7] V.V. Kharton, F.M.B. Marques, and A. Atkinson, Transport properties of solid oxide electrolyte ceramics: a brief review. Solid State Ionics, 174(2004), Issue (1-4), pp.135-149.
[8] L.G. Cong, T.M. He, Y.A. Ji, P.F. Guan, Y.L. Huang, and W.H. Su: Synthesis and characterization of IT-electrolyte with perovskite structure $La_{0.8}Sr_{0.2}Ga_{0.85}Mg_{0.15}O_{3-\delta}$ by glycine-nitrate combustion method. J Alloy Compd, 348 (2003), Issue (1-2), pp. 325-331.
[9] N.Q. Zhang, K.N. Sun, D.R. Zhou, and D.C. Jia: Study on properties of LSGM electrolyte made by tape casting method and applications in SOFC. J Rare Earth, 24 (2006), pp. 90-92.
[10] S.G. Kim, S.P. Yoon, S.W. Nam, S.H. Hyun, and S.A. Hong: Fabrication and characterization of a YSZ/YDC composite electrolyte by a sol-gel coating method. J Power Sources, 110 (2002), Issue 1, pp. 222-228.
[11] C.X. Li, Y.X. Xie, C.J. Li, and G.J. Yang: Characterization of atmospheric plasma-sprayed $La_{0.8}Sr_{0.2}Ga_{0.8}Mg_{0.2}O_3$ electrolyte. J Power Sources, 184 (2008), Issue 2, pp. 370-374.
[12] S.L. Zhang, T. Liu, C.J. Li, S.W. Yao, C.X. Li, G.J. Yang, and M.L. Liu: Atmospheric plasma-sprayed $La_{0.8}Sr_{0.2}Ga_{0.8}Mg_{0.2}O_3$ electrolyte membranes for intermediate-temperature solid oxide fuel cells. J Mater Chem A, 3 (2015), Issue 14, pp. 7535-7553.
[13] K. Yamaji, T. Horita, M. Ishikawa, N. Sakai, and H. Yokokawa: Chemical stability of the $La_{0.9}Sr_{0.1}Ga_{0.8}Mg_{0.2}O_{2.85}$ electrolyte in a reducing atmosphere. Solid State Ionics, 121(1999), Issue (1-4), pp. 217-224.
[14] X.G. Zhang, S. Ohara, R. Maric, H. Okawa, T. Fukui, H. Yoshida, T. Inagaki, and K. Miura: Interface reactions in the NiO-SDC-LSGM system. Solid State Ionics, (133) 2000, Issue (3-4), pp. 153-160.
[15] X.G. Zhang, S. Ohara, R. Maric, K. Mukai, T. Fukui, H. Yoshida, M. Nishimura, T. Inagaki, and K. Miura: Ni-SDC cermet anode for medium-temperature solid oxide fuel cell with lanthanum gallate electrolyte. J Power Sources, 83 (1999), Issue (1-2), pp. 170-177.
[16] J. Akedo, Room temperature impact consolidation (RTIC) of fine ceramic powder by aerosol deposition method and applications to microdevices. J Therm Spray Techn, 17 (2008), Issue 2, pp. 181-198.
[17] J. Akedo, and M. Lebedev, Microstructure and electrical properties of lead zirconate titanate Pb(Zr-52/Ti-48)O-3) thick films deposited by aerosol deposition method. Jpn J Appl Phys 1, 38(1999), Issue 9B,pp. 5397-5401.
[18] J. Akedo, Aerosol deposition of ceramic thick films at room temperature: Densification mechanism of ceramic layers. J Am Ceram Soc, 89 (2006), Issue 6, pp. 1834-1839.
[19] Y. Liu, Y.Y. Wang, G.J. Yang, J.J. Feng, and K. Kusumoto: Effect of Nano-Sized TiN Additions on the Electrical Properties of Vacuum Cold Sprayed SiC Coatings. J Therm Spray Techn,19(2010), Issue 6, pp. 1238-1243.
[20] G.J. Yang, C.J. Li, S.Q. Fan, Y.Y. Wang, and C.X. Li: Influence of annealing on photocatalytic performance and adhesion of vacuum cold-sprayed nanostructured TiO_2 coating. J Therm Spray Techn, 16 (2007), Issue (5-6), pp. 873-880.
[21] R. Inada, K. Shibukawa, C. Masada, Y. Nakanishi, and Y. Sakurai: Characterization of as-deposited $Li_4Ti_5O_{12}$ thin film electrode prepared by aerosol deposition method. J Power Sources, 253 (2014), pp. 181-186.
[22] J. Akedo, and M. Lebedev, Powder preparation in aerosol deposition method for lead zirconate titanate thick films. Jpn J Appl Phys 1, 41(2002), Issue 11B, pp. 6980-6984.
[23] V.V. Lakshmi, R. Bauri, A.S. Gandhi, and S. Paul: Synthesis and characterization of nanocrystalline ScSZ electrolyte for SOFCs. Int J Hydrogen Energ, 36 (2011), Issue 22, pp. 14936-14942.
[24] C. Pascual, J.R. Jurado, and P. Duran: Electrical Behavior of Doped Yttria Stabilized Zirconia Ceramic Materials. J Mater Sci, 18(1983), Issue 5, pp. 1315-1322.
[25] Y.H. Choi, S.H. Lee, J. Wackerl, D.H. Jung, D.S. Suhr, S.Y. Choi, and D.H. Peck: Fabrication of scandia-stabilized zirconia electrolyte with a porous and dense composite layer for solid oxide fuel cells. Ceram Int, 38 (2012), pp. S485-S488.
[26] S.Q. Fan, C.J. Li, C.X. Li, G.J. Liu, G.J. Yang, and L.Z. Zhang: Preliminary study of performance of dye-sensitized solar cell of nano-TiO_2 coating deposited by vacuum cold spraying. Mater Trans, 47 (2006), Issue 7, pp. 1703-1709.
[27] S.Q. Fan, C.J. Li, G.J. Yang, L.Z. Zhang, J.C. Gao, and Y.X. Xi: Fabrication of nano-TiO_2 coating for dye-sensitized solar cell by vacuum cold spraying at room temperature. J Therm Spray Techn, 16 (2007), Issue (5-6), pp. 893-897.
[28] G.J. Yang, C.J. Li, K.X. Liao, X.L. He, S. Li, and S.Q. Fan: Influence of gas flow during vacuum cold spraying of nano-porous TiO_2 film by using strengthened nanostructured powder on performance of dye-sensitized solar cell. Thin Solid Films, (519) 2011, Issue 15, pp. 4709-4713.
[29] G.J. Yang, K.X. Liao, C.J. Li, S.Q. Fan, C.X. Li, and S. Li: Formation of Pore Structure and Its Influence on the Mass Transport Property of Vacuum Cold Sprayed TiO_2 Coatings Using Strengthened Nanostructured Powder. J Therm Spray Techn, (21) 2012, Issue (3-4), pp. 505-513.
[30] D.M. Chun, and S.H. Ahn: Deposition mechanism of dry sprayed ceramic particles at room temperature

using a nano-particle deposition system. Acta Mater, (59) 2011, Issue (7), pp. 2693-2703.

[31] Y.B. Lin, and S.A. Barnett: Co-firing of anode-supported SOFCs with thin $La_{0.9}Sr_{0.1}Ga_{0.8}Mg_{0.2}O_{3-\delta}$ electrolytes. Electrochem Solid St, 9 (2006), Issue 6, pp. A285-A288.
[32] M. Shinozaki, and T.W. Clyne, A methodology, based on sintering-induced stiffening, for prediction of the spallation lifetime of plasma-sprayed coatings, Acta Mater, 61(2013), Issue 2, pp. 579-588.
[33] C.W. Kim, J.H. Choi, H.J. Kim, D.W. Lee, C.Y. Hyun, and S.M. Nam: Effects of interlayer roughness on deposition rate and morphology of aerosol-deposited Al2O3 thick films, Ceram Int, 38 (2012), Issue 7, pp. 5621-5627.
[34] S.L. Zhang, C.X. Li, C.J. Li, G.J. Yang, and Z.H. Han: Scandia-stabilized zirconia electrolyte with improved interlamellar bonding by high-velocity plasma spraying for high performance solid oxide fuel cells, J Power Sources, 232 (2013), pp.123-131.
[35] M. Hrovat, A. Ahmad-Khanlou, Z. Samardzija, and J. Holc: Interactions between lanthanum gallate based solid electrolyte and ceria, Mater Res Bull, 34 (1999), Issue (12-13), pp. 2027-2034.
[36] K.N. Kim, B.K. Kim, J.W. Son, J. Kim, H.W. Lee, J.H. Lee, and J. Moon: Characterization of the electrode and electrolyte interfaces of LSGM-based SOFCs, Solid State Ionics, 177 (2006), Issue (19-25), pp. 2155-2158.

Thermal Plasma Synthesis of Mg-Ni Nanoparticles

Burak AKTEKİN, Tayfur ÖZTÜRK Middle East Technical University, Ankara

Nanoparticles are of great interest in energy area for a wide range of applications. A variety of techniques are available for their synthesis and among these methods, thermal plasma approach offers continuous production of high purity nanoparticles in large scale where the particle size is usually less than 100 nm. In this technique, precursors, which are normally powders, are fed to the thermal plasma where they are evaporated and nanoparticles are derived from condensation of this vapor in the quenching zone. In this study, a 30 kW RF thermal plasma system was used. The system consist of a torch and a nanopowder reactor, but when needed could be replaced by a coating chamber or a spheroidization unit. The system was configured with injection probes which were located axially in the reactor, one from the top and the other from the bottom. This allowed injection of precursor powders into different temperature zones of the plasma. In the current study, nanoparticles in Mg-Ni binary system were produced. This involved Ni, Mg and Mg2Ni. Ni nanoparticles can be produced at ease yielding particles less than 100 nm in size. Moreover, when Ni was fed together with methane, Ni particles could be encapsulated with 6-10 layers of graphite yielding a core-shell structure. Mg2Ni intermetallic nanoparticles could be synthesized from elemental powders when Ni was fed from the top injector and Mg was fed through the bottom injector which was positioned just below the quenching zone in the reactor. The weight fraction of Mg2Ni could be as high as 55 wt%. The use of thermal plasma in the synthesis of metallic nanoparticles was discussed especially with regard to intermetallic alloys.
.

1 Introduction

There is a considerable interest in sub-micron sized metallic powders as they may offer some specific size dependent properties mostly resulting from their high surface to volume ratio. One approach to producing sub-micron metallic powders is to use thermal plasma torches. Although D.C. plasma could also be used for this purpose, RF inductively coupled plasma (RF-ICP) is more common in the synthesis of nanopowders [1]. Here, the torch is electrodeless and it provides continuous production ability forming contamination free nanoparticles. Properties such as high heat capacity, high chemical reactivity and large plasma volume as well as ability of central injection and long residence time for precursors in the plasma make RF-ICP torches advantageous for the synthesis of sub-micron powders [2,3]. In this technique, plasma is basically formed by the coupling of the energy provided by an RF power to the streaming plasma gas via electromagnetic fields [4]. Initial powders are melted and evaporated in the high temperature region and condense into fine particles at the exit of the plasma. Induction thermal plasma torches have been used for the synthesis of nanoparticles of a variety of materials. These materials can be listed as mainly pure metals, carbides, nitrides, oxides, borides, nanostructured carbon-based materials and metallic alloys [5,6,7,8,9,10,11]. Metal nanoparticles include Al, Cu, Ag, Ni, Fe, Co, Mo, W, etc. [3]. Differing from pure metals, the synthesis of alloys and intermetallics which involve more than one component has its difficulties [12]. In such a system, nucleation usually starts with one of the component and other component(s) may nucleate with the existing particles, or they may form new particles [13].

The present work was undertaken to produce metallic nanoparticles in Mg-Ni binary system. Here the interest is Ni as well as Mg to produce them in sizes less than 100 nm as well as the intermetallic nanoparticles namely Mg_2Ni again in sizes smaller than 100 nm. The nanopowders reactor in use has two injectors located axially in the torch one from above and the other from the bottom. This provides flexibility in the synthesis of compounds such as Mg_2Ni as well as producing nanoparticles encapsulated with carbon.

2 Experimental

The apparatus used for the nanoparticle synthesis is a 30 kW thermal induction plasma system (Tekna Plasma Systems Inc.). The system consists of a high RF power terminal with a nominal frequency of 2-5 MHz and the inlet power of 50 kVA, Figure 1.

Induction plasma system was operated at 25 kW under 0.97 bar pressure. Central gas (plasma forming gas) used in the experiments was pure argon (Linde, 99.995% purity) and provided at a rate of 15 standard liters per minute (slpm). Sheath gas was a mixture of pure argon (60 slpm, Linde, 99.995% purity) and pure hydrogen (6 slpm, Linde, 99.995% purity). As quenching gas, nitrogen gas was used with flow rate of 150 slpm.

Materials used as precursors included methane (CH_4, 99.5% pure, Linde), nickel (99.9% pure, -325 mesh, Höganäs), magnesium (99.8% pure, -300 mesh, Alfa Aesar) and pre-alloyed Mg_2Ni (induction melted, -300 mesh). Typical duration of the production runs lasted 10-15 minutes. The torch incorporated two probes for feeding of these precursors, Figure 1(b). They were fed into the plasma using either the top or the bottom injection probe, or both of them. Carrier gas was Ar (Linde, 99.995% purity) and flow rate was kept at 5 slpm.

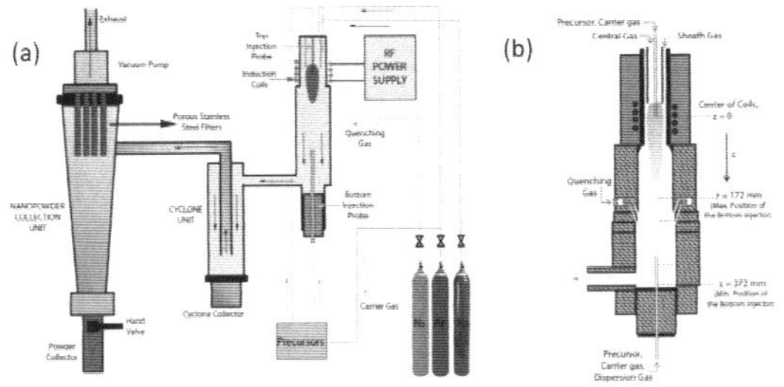

Fig. 1. A general view of the thermal plasma system (a) and the schematic view of the reactor incorporating two injection probes (b).

3. Results and Dicussion

Preliminary experiments with thermal plasma synthesis were carried out with Ni. This is one of the constituent in Mg_2Ni. Ni was preferred over Mg since it has a higher melting - boiling point and also it is less reactive. The unit was operated for 15 minutes with Ni powder at a feed rate of 3 g/min. Powders collected from bottom, cyclone and powder collector were removed and examined with SEM. Powders collected from the bottom collector are given in Figure 2(b). These powders are quite comparable to the starting powder given in Figure 2(a) except for their shape. Powders collected from the cyclone unit were also spherical in shape but much smaller (around several microns) in size and were decorated with nanoparticles, Figure 2(c). These have probably originated from incompletely vaporized liquid droplets, or highly coagulated nanoparticles. Another reason could be the relatively high exposure of some nuclei to moderate temperatures so that more time was available for the particle growth. Large agglomerates of fine particles were also present in the cyclone. SEM image of powders collected from the powder collector are shown in Figure 2(d). Here the particles are quite fine and they are again spherical in shape. Mean particle size was 80 nm with a standard deviation of 15 nm (determined over nearly 50 particles analyzed from SEM images). These were obviously derived from the evaporation-condensation process.

While metallic nanopowders may have desired properties, there are also some important drawbacks related to their size such as agglomeration and higher sensitivity to oxidation. These drawbacks usually limit their application and necessitate protective measures to be taken during their handling. One solution to this problem is seen as the stabilization of metallic nanoparticles by covering them by an inert material. For this purpose, we also fed methane (as carbon source) together with nickel to encapsulate nanopowders with graphitic layers.
Upon feeding Ni and methane to the plasma reactor, the product collected from the powder collector was mostly carbonaceous as they were black in color. In TEM images obtained from these powders, structures were mostly graphitic, but Ni particles could easily be located. Figure 3(a) and 3(b) show typical TEM

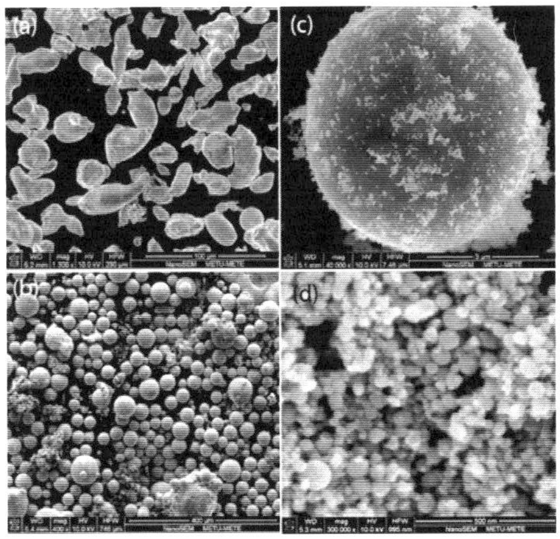

Fig. 2. SEM images of initial powders (a) and synthesized powders collected from bottom collector (b), cyclone (c), powder collector (d).

images of such particles. As seen in these images, nickel nanoparticles were encapsulated by several outer layers. Interplanar spacing measured between these layers derived from the digital diffractogram (FFT image obtained with ImageJ software) was 0.33 nm which is in well agreement with the value reported for graphite in the literature (0.3347 nm, ICSD #031829). Also, lattice fringe separation inside the core yielded a value of 0.204 nm. This value is in agreement with the reported interplanar distance for (111) plane of Ni, which is 0.2041 nm (NIST #23929). Thus carbon encapsulated Ni particles were successfully synthesized where Ni particles of 5-200 nm in size were covered by 3-9 graphitic layers. The diameter of particles was 56 nm with a standard deviation of 35 nm (measured over 35 particles). It can be concluded that carbon encapsulation also hinders the further particle growth by confining the particles within carbon layers resulting in finer particle size. These values together with TEM images (Figure 3) refer to the case where Ni was fed from the top and methane was fed from the bottom injector. Similar observations were made for the case of reverse feeding.

Following the preliminary experiments with nickel, two additional experiments were conducted with the pre-alloyed Mg_2Ni powders. They were either injected using the top injection probe or injected from the bottom probe. Samples collected were characterized with X-ray diffraction. Phases present in this sample were mainly Ni (ICDD# 4-0850) and MgO (ICDD# 45-0946). No significant Mg_2Ni was present in this sample. This implies that Mg_2Ni disintegrated within the plasma into its elements. Moreover, Mg formed in this way was oxidized to MgO. When the bottom probe was used (located 372 mm below induction coils, z=372 mm), similarly, no Mg_2Ni formation was observed. These results indicate that when Mg and Ni follow similar pathways in the reactor, they prefer to nucleate separately, which is probably due to the large differences in their nucleation temperatures.

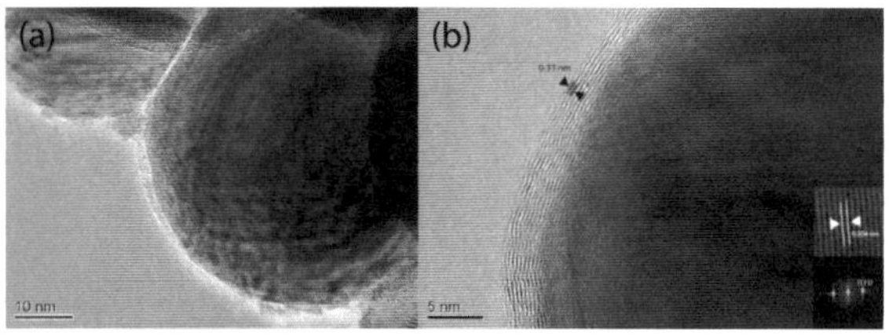

Figure 3. TEM images of carbon encapsulated nickel nanoparticles.

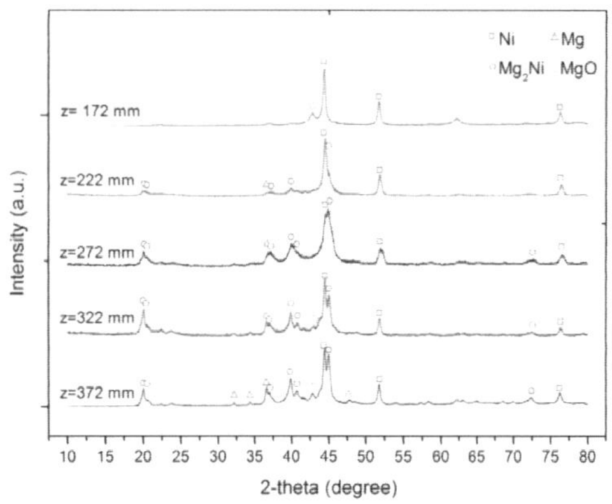

Figure 4. XRD patterns of powders when Mg was fed via bottom and Ni via top injector.

As an alternative approach, Mg and Ni were sent into the plasma separately. Since boiling and melting temperatures of Mg are lower than those of Ni, it was decided that Ni could be fed through the top probe as it provides a more effective heat transfer between the plasma and the particles, while Mg could be fed through the bottom probe (positioned at z = 372 mm). XRD pattern of nanopowders obtained in this geometry is given in Figure 4. Considerable fraction of the sample was Mg_2Ni (ICDD# 35-1225). The other phases present were Ni (ICDD# 4-0850), Mg (ICDD# 35-0821) and MgO (ICDD# 45-0946). The amount of Ni in the sample was quite comparable to that of Mg_2Ni. As done in an additional test, when excessive Mg was fed, there was only a slight increase in the Mg_2Ni content.

So as to further increase Mg_2Ni yield, an approach was adopted in which the position of Mg probe was changed in the reactor. It was anticipated that, moving the injector towards the plasma torch would affect the precursor penetration into the different regions of the plasma and could lead to alterations in the volume fractions of the phases formed. Experiments involved the use of five locations; z= 372, 322, 272, 222, 172 mm. When the bottom injector was positioned at z = 272 mm, an increase in the Mg_2Ni yield was observed; but still there was a significant amount of Ni (ICDD# 4-0850). When injector tip was moved further towards the plasma (z = 222 mm), the same trend did not continue. The fraction of Mg_2Ni was now decreased while MgO peaks became comparable to those observed in the earlier positions. It might be worth mentioning that at z = 272 mm, the tip of the injector is located just below the inlet of the quenching gas.

The resulting nanopowders gave maximum Mg_2Ni yield at z=272 mm, these powders were analyzed with SEM-EDS and also Rietveld analysis was made on XRD pattern. Mg_2Ni amount was determined as 53.9 wt% while other phases were Ni (11.6 wt%), $MgNi_2$ (11.1 wt%), MgO (17.6 wt%) and Mg (5.8 wt%). Elemental analysis result was in accordance with the Mg/Ni ratio calculated from these phase fractions.

Figure 5. SEM image of powders obtained when bottom probe was located at z=272 mm.

This sample with maximum Mg_2Ni content was further analyzed with regard to its particle size. BET analysis yielded a specific surface area of 16 m^2/g. This has yielded a value of 98 nm particle size assuming spherical shape. This value was quite close to values obtained with SEM measurements which yielded an average particle size of 89 nm (with a standard deviation of 39 nm) determined over 110 particles, Figure 5.

4 Conclusions

In this study, it was shown that the induction plasma method is an effective way of synthesizing nanoparticles in large-scale. Particles were generally spherical in shape and less than 100 nm in size. Ni nanoparticles were successfully synthesized in one-step without being oxidized. Additionally, the method of induction plasma by feeding Ni together with methane can be used successfully for the synthesis of carbon encapsulated nickel nanoparticles. Feeding methane with precursors also reduces the size of the resulting particles. Secondly, it is also possible to synthesize multicomponent materials via induction plasma method. Intermetallic Mg_2Ni compound with a significant yield was synthesized; but it was seen that the injector configuration of the precursors was quite decisive for the relative content of products. Feeding Ni from the top of the plasma and Mg from the bottom of the plasma was found to be the most effective feeding configuration.

It can be concluded that the synthesis of pure metals can be achieved easily with thermal plasma as shown for nickel in this study and as can be seen for other metals in the literature. In the case of multicomponent materials, synthesis of more stable phases such as oxides, carbides, borides with thermal plasma seems more convenient compared to metallic alloys or intermetallics. As we show in this study for Mg_2Ni, even melting and evaporation points of components are quite different from each other, significant yields of desired multicomponent materials can be achieved with a careful control of precursor feeding into the plasma.

Acknowledgement

Support for this work was provided by TUBITAK (Project number 112M193) in conjunction with COST MP1103, which we gratefully acknowledge.

5 Literature

[1] Bloggs, J. and A.N. Other: Literature example for manuscripts in DVS volumes of reports. Welding & Cutting 46 (1999), Issue 13, pp. 1156/57.

[2] Boulos, M. I. (1985). The inductively coupled RF (radio frequency) plasma. Pure&Appl. Chem., Vol. 57, No. 9,pp. 1321—1352.

[3] Cheng, Y., Shigeta, M., Choi, S., & Watanabe, T. (2012). Formation mechanism of titanium boride nanoparticles by RF induction thermal plasma. *Chemical Engineering Journal, 183*, 483-491.

[4] Jiayin, G., Xiaobao, F., Dolbec, R., Siwen, X., Jurewicz, J., & Boulos, M. (2010). Development of nanopowder synthesis using induction plasma. *Plasma Science and Technology, 12*(2), 188.

[5] Hopwood, J. (1992). Review of inductively coupled plasmas for plasma processing. *Plasma Sources Science and Technology, 1*(2), 109.

[6] Yoshida, T., & Akashi, K. (1981). Preparation of ultrafine iron particles using an rf plasma. *Trans.Jpn.Inst.Met.,22*(6), 371-378.

[7] Károly, Z., Mohai, I., Klébert, S., Keszler, A., Sajó, I., & Szépvölgyi, J. (2011). Synthesis of SiC powder by RF plasma technique. *Powder Technology, 214*(3), 300-305.

[8] Baba, K., Shohata, N., & Yonezawa, M. (1989). Synthesis and properties of ultrafine AlN powder by rf plasma. *Applied Physics Letters, 54*(23), 2309-2311.

[9] Wang, X., Li, J., Kamiyama, H., & Ishigaki, T. (2006). Fe-doped TiO2 nanopowders by oxidative pyrolysis of organometallic precursors in induction thermal plasma: Synthesis and structural characterization. *Thin Solid Films, 506*, 278-282.

[10] Szépvölgyi, J., Mohai, I., Károly, Z., & Gál, L. (2008). Synthesis of nanosized ceramic powders in a radiofrequency thermal plasma reactor. *Journal of the European Ceramic Society, 28*(5), 895-899.

[11] Cota-Sanchez, G., Soucy, G., Huczko, A., & Lange, H. (2005). Induction plasma synthesis of fullerenes and nanotubes using carbon black–nickel particles. *Carbon, 43*(15), 3153-3166.

[12] Çakmak, G., Károly, Z., Mohai, I., Öztürk, T., & Szépvölgyi, J. (2010). The processing of Mg–Ti for hydrogen storage; mechanical milling and plasma synthesis. *International Journal of Hydrogen Energy, 35*(19), 10412-10418.

[13] Aktekin, B, Çakmak G, and Öztürk, T. "Induction thermal plasma synthesis of Mg2Ni nanoparticles." International Journal of Hydrogen Energy 39.18 (2014): 9859-9864.

[14] Gurav, A., Kodas, T., Pluym, T., & Xiong, Y. (1993). Aerosol processing of materials. *Aerosol Science and Technology, 19*(4), 411-452.

Rolf Huber, Annegret Ohliger-Volmer
Gesellschaft Schweißtechnik International – GSI mbH, Branch SLV Munich

Education, training and certification for thermal spraying: experiences and future development

1. Introduction

Experience has shown that qualified supervisors and qualified workers are undisputedly one of the most important prerequisites for manufacturing premium-quality products.
This paper describes the current training situation in line with EWF guidelines and addresses prevailing curricula, changes and reforms/enhancements.
The EWF guidelines are directly linked to product-related quality management in accordance with DIN EN ISO 14922 and to the approval testing of thermal sprayers in accordance with ISO Norm 14918. Changes and reforms with respect to the standards just mentioned will likewise be presented. The final section of the paper will deal with future developments emerging in the field of thermal spray training and certification.

2. Current status

Achieving and maintaining a specific product quality depends to a great extent on the level of staff qualification.
For thermal spraying, there are two recognized training levels available for spray staff. These EWF guidelines (European Federation for Welding, Joining and Cutting) are standardized throughout Europe.
The International Institute of Welding (IIW), which currently has a membership of 40 countries worldwide, and EWF have been working together for several years now. The IIW qualification guidelines for welding inspectors have already been adopted and it is now possible to obtain IIW diplomas after successfully completing their training courses, for example, the IWE – International Welding Engineer. The proposition by EWF to adopt the EWF-ETSS and ETS guideline as an IIW/EWF guideline has already been fulfilled by IIW. In the meantime, a comprehensive set of thermal spray standards to assure the quality of spray products has been compiled along with product-related application standards (Fig. 1). It defines the requirements the spray coating needs to fulfil and the qualification which spray staff must have.

Figure 1: QM and application standards for thermal spraying

2.1. Education, training and examination of spray staff

Qualifying thermal spray staff continues to vary greatly on an international level. However, this situation has changed since EWF started to harmonize staff qualification in the field of welding and related processes including thermal spraying. These efforts began in Europe first of all and continued on a worldwide basis as well in conjunction with IIW.
The following EWF training guidelines have been successfully implemented since they were compiled and published in the period 1995/1999:

- Spray specialist
  (ETSS) European Thermal Spraying Specialist
- Thermal sprayer
  (ETS) European Thermal Sprayer

Figure 2 illustrates how the EWF training levels are linked.

The framework allows for growth and flexibility, enabling a thermal sprayer to attain the level of a thermal spraying specialist.

Figure 2: Access to EWF-ETSS education and training

If an EWF guideline is not practically implemented over many years, EWF will demand the withdrawal of this training guideline. In the last EWF General Assembly this was the case for ETSP (European Thermal Spraying Practitioner).

2.1.1. ETSS training (European Thermal Spraying Specialist)

In 2014, the education, training and examination guideline for what was the highest qualification level for spraying specialists up to then was revised based on previous experiences and proposals by the EWF working group.
With a total of 108 teaching hours, the EWF program now offers very comprehensive training for spraying specialists. By streamlining the total number of theoretical hours, double-ups can be avoided. The 3-week course primarily focuses on well-founded and practice-oriented technical knowledge.
Thermal spraying specialists trained and examined in accordance with the EWF guideline can then take on a variety of responsible positions in coating businesses, for example as a spraying supervisor/coordinator.
Pursuant to the EWF guideline, three routes are available for entry to the ETSS training program, Fig. 2:

- Specific technical qualifications as a master craftsman in a trade or in industry including 2 years job-related experience, or a trained technician or engineer.
- ETS qualification and a minimum of 2 years' thermal spray experience after successfully passing the entrance test.
- 5 years' job experience as a skilled worker in the metalworking sector and a minimum of 3 years' experience in a thermal spray-related field after successfully passing the entrance test.

Prior to the above-mentioned entrance test, candidates who feel they lack the necessary technical education may take the preparatory ETSS part 0 course, Fig. 2.
The module ETSS 0 is aimed at imparting basic technical knowledge so that candidates without a master qualification can follow and grasp the contents of the thermal spray curriculum.
In a total of 40 teaching hours, basic technical calculation, technical drawing, electrotechnics, chemistry, material properties, metal products and the machining thereof are dealt with.

2.1.2. ETS training (European Thermal Sprayer)

The professional training substantially increases the sprayer's level of knowledge and facilitates a better understanding of the relevant quality aspects.
To take part in the ETS training module, normal physical and mental capabilities are sufficient. Course participants should have a good spoken and written command of the language the course is conducted in so that they can follow the instructions in the practical section and sit the theoretical tests.
Many years of experience have shown that ETS participants definitely require basic metalworking skills. If not, a basic practical course is recommended. For candidates who have no practical experience in the spray process being tested, special prior training as prescribed by the guideline has proved to be very effective.
Module A 0 – basic theory – comprises 13 teaching hours and covers thermal spray principles and processes, health and safety, coating properties and bond mechanisms, materials, pre-treatment and preparation, spraying, maintenance and calibration, post-treatment, measuring and testing, handling

and storing coated parts, mechanization and automation as well as environmental aspects and occupational safety issues.
Module B 0 – basic practical training – is intended for participants who have no or little practical experience in thermal spraying. It comprises 14 hours of demonstrations and practical exercises in a spray job shop.
The practical training covers the following areas: surface preparation, spraying, testing the spray coating, post-treatment and occupational health and safety.
After completing this module and passing a theoretical intermediate examination, the participant can choose to attend a specific training course for one or more spray processes which, in turn, are subdivided into process-specific theoretical modules (A1-4) and practical training modules (B1-4).
The theoretical training modules A1 to A4 are aimed at imparting basic knowledge of the processes in the areas of system equipment, sprayable materials, spray process steps, coating properties, the respective areas of application as well as knowledge of occupational health and safety rules.
The duration of each specific practical training module in flame (B1), arc (B2), plasma (B3) and HVOF (B4) spraying can be determined individually as the trainer sees fit.
The duration of the practical section depends on the existing experience and the ability of the sprayer.
The practical training modules B1 to B4 are intended to take the participant through the entire thermal spray procedure, i.e. from the preparation phase through to examination and post-treatment of the coated component. For insurance reasons, the participant is, of course, supervised and supported by the trainer at all times.
The test pieces correspond to those specified in ISO 14918 "Approval testing of thermal sprayers".
After successfully completing the course and passing the examination, the candidate receives: (1) the ETS diploma pursuant to the EWF guideline, which has lifelong validity, and (2) a certificate valid for three years stating that the holder has qualified as a thermal sprayer pursuant to the prevailing standard.
The high quality awareness of staff helps to ensure that serious mistakes do not occur in the first place, thus reducing cost-intensive processing of claims and complaints (Fig. 3).

Figure 3: Flaw due to the insufficient quality awareness of staff: delivery of incorrectly coated test pieces for bond strength tests

3. Certification for coating businesses and TS staff

In the course of globalization, demand from the market for documented proof of product quality and quality management procedures in companies is increasing. Especially when high safety requirements need to be fulfilled because of the product being manufactured, it is imperative that companies comply with the standards specified. More often than not, suppliers or customers insist on such proof of quality, also with the aim of enforcing these high quality expectations in their own company.

3.1. Standards relevant to QM

In smaller enterprises in particular, quality management should not be regarded as a necessary evil. It is not only important to implement the requirements demanded by standards and directives, but also to involve and motivate staff to take personal responsibility and be accountable for product quality. (Fig. 4).

Figure 4: Quality management pillars and corresponding standards

Process-oriented quality management is particularly important for the company.

3.1.1. ISO 9001:2015

Quality management standards help to assure product quality and describe ways in which companies can establish their own QM system or how they can adapt and improve their existing QM system. This is an important basis for building trust on the part of the customer.
To date, more than 1.1 million certificates in accordance with ISO 9001 have been issued worldwide, around half a million in Europe and more than 50,000 in Germany. Approximately 178 countries are holders of one or more such certificates (ISO 9001, ISO 3834, ISO 14001, etc.).
ISO revises its standards at regular intervals, usually every 5 years, which means they are constantly adapted to the prevailing social and market requirements.
The International Standards Organization (ISO) is currently revising the standard ISO 9001 for QM systems; publication is planned for the third quarter of 2015. The current official draft contains noticeable changes in comparison to the present standard ISO 9001:2008. It is recommended that companies familiarize themselves with the new version as early as possible.
It can be assumed that after the revised ISO 9001:2015 is published, there will once again be a transition phase of 3 years in which companies can adapt their QM system to the new requirements. The new standard incorporates modern aspects of corporate management and control (Fig. 5). Users who have already been practising these measures for many years and not only require the QM certificate for marketing reasons will benefit from the additional demands.

Figure 5: Current information on ISO 9001:2015

Typical problems affecting corporate QM systems in small and medium-sized enterprises (SMEs) are:

- Customer-complaint rate and/or customer dissatisfaction is very high or often not known
- Only exceptions or one-offs are combatted – fundamental improvements are not carried out however
- Staff are not sufficiently motivated
- Decisions taken in meetings are often inadequately implemented
- Scrap and rework costs are too high and turnaround time for orders is too long
- Responsibilities and business processes are not clearly defined
- Growing cost pressure combined with increasing quality requirements and shorter lead times

Incorporating workflow processes, responsibilities and testing in a QM system with process and work instructions inevitably involves time and money. QM documents should be compiled in a team comprising the relevant staff to ensure that the contents are correctly presented and approved by all. It must be absolutely clear to everyone that the DIN EN ISO 9001 series of standards represents a universally valid model intended as a generic quality management system which is not directly aimed at assuring the quality of a particular product; i.e. it is not product-specific.
Consequently, it makes sense to incorporate the quality assurance and quality management of thermally sprayed products in the QM system. To this end, DIN EN ISO 14922 T1-4 "Quality requirements of thermally sprayed structures" can be consulted. In this way, a product-specific QM procedure can be compiled for the so-called "special process thermal spraying" compliant with ISO 9000. The term "special process" refers to processes in which quality is not exclusively tested on the end product, but in which monitoring is necessary during the manufacturing process, e.g. to ensure that the work is performed according to the prescribed parameters.

3.1.2. ISO 14922 – comprehensive quality requirements of spray businesses (T2)

In the standard EN ISO 14922 T 1-4 "Quality requirements of thermally sprayed structures", general specifications are applied to the spray field to assure product quality.
The standard defines basic principles for the manufacturer's capability and the production of spray coatings. This includes the requirements of the manufacturing process itself and the qualification of the staff employed in the manufacturing and monitoring process.

An independent authority can check compliance with these requirements and certify the suitability of the staff and the manufacturing process.
The standard comprises three levels of quality requirements pertaining to coating operations, namely comprehensive standard and elementary requirements, Fig. 6.
The requirements are listed according to their level and degree of fulfilment and described in more detail in parts 2 to 4 of the standard.
Comprehensive requirements may, for example, be imposed when spray coatings need to meet higher demands. This may be the case in the aviation, railway vehicles or chemical industry where coating failure could lead to extremely high costs or even present a risk to the public and/or the environment.
In this case, the spray company must fully fulfil the QM elements starting with the review and verification of the contract through to the quality records.

Figure 6: Interrelationship of ISO standards and EWF training for spray personnel

For manufacturing processes involving spray technology, the quality factor and its documentation is of increasing importance. The quality requirements demanded of spray businesses are specified in detail in the standards series DIN EN ISO 14922. These requirements cover the commercial offer, the conclusion of the contract through to the design, manufacture and testing of the thermally sprayed product. Maintenance and documentation pertaining to quality assurance and fulfilment of the product requirements are also included. Certification in accordance with DIN EN ISO 14922 supports compliance with the quality assurance measures specified in the standard when using thermal spray processes.

3.1.3. Standard and elementary quality requirements (DIN EN ISO 149223 T3 and 4)
On the quality level "standard quality requirements", lower quality requirements are demanded for some elements, see Fig. 4.
These elementary requirements for spray coatings are applied when no QM system is demanded, when proof of spray monitoring plays a subordinate role in the quality of the end product and when a malfunction of the coating would not cause any appreciable damage.
In this case, the QM elements need only fulfil the minimum requirements, with two exceptions. Namely, system maintenance and storage and handling of the spray materials must fully meet the requirements. If not, the quality of the spray coatings may be considerably impaired.

3.1.4. Qualification of the spray process
Assuring quality is now integral to the manufacture of technical products. Here, it is important to ensure that thermal spraying is performed in the most effective way and that suitable monitoring is provided for the entire operation. To achieve this, a written process instruction (TSPS – Thermal Spray Procedure Specification) is an important instrument, typically also a requirement of quality assurance standards and found in DIN EN 15648 "Thermal spraying – component-related procedure qualification".
Proof that the spray process and selected parameters are suitable must be established prior to the actual coating operation during the manufacturing process. It is assumed that the spray staff are trained and qualified in accordance with the requirements.

3.1.5. GTS QM – voluntary system for the certification of coating businesses and their staff
The inspection and certification of coating businesses in accordance with GTS guidelines has served as a functional and well-established QM tool for many years.
The GTS-QM guideline combines the QM elements of ISO 9001 with procedure qualifications and the theoretical and practical qualification of spray staff. In comparison to ISO 9000, certification pursuant to the GTS-QM guideline also includes testing the technical knowledge of personnel by way of

theoretical and oral examinations and observing them as they manufacture test specimens and evaluating these test specimens.
The company can also choose to undergo an optional product or coating certification for which the business being certified specifies the examination criteria. Supervised by the auditor, the parts are coated and previously specified properties are then inspected by an independent certified test laboratory - a further confidence-building QM measure which can also be used for advertising purposes.
Worldwide, GTS is the only organization of its kind in the thermal spray sector which has introduced and implemented its own transparent quality assurance network that extends far beyond DIN EN ISO standards.

3.2. Qualification requirements for spray staff

In part 2 of the standard 14922 – comprehensive requirements of thermally sprayed structures and part 3 (standard requirements), both a qualified thermal spraying specialist according to ISO 12690 "Thermal spray coordination" and a sprayer qualified in accordance with ISO 14918 "Approval testing of thermal sprayers" are required.
Part 4 (elementary requirements) simply stipulates that staff must be present who can professionally monitor the proper execution of spray operations and provide thermal spray instruction.

3.2.1. Thermal spray coordination

DIN EN ISO 12960 "Thermal spray coordination – tasks and responsibilities" specifies the quality-related responsibility and tasks including the coordination of thermal spray activities. Accordingly, the manufacturer must designate at least one thermal spray coordinator. Several spray coordinators can also be designated within the manufacturer's organization.
This standard provides the user with a guideline for specifying the quality-related tasks and the responsibility of the spray coordinator. Each activity can be linked to a number of tasks and responsibilities such as:

- stipulating specifications or their preparation
- coordinating
- checking tasks and areas connected to the manufacture of thermal spray coatings.

In accordance with the requirements of DIN EN ISO 14922 parts 2 and 3, the European Thermal Spraying Specialist – ETSS – qualified according to the corresponding EWF guidelines – or persons with comparable knowledge and experience are recognized as authorized thermal spray coordinators.

Figure 7: Preparing a ceramic spray coating in line with quality assurance: 1. incorrectly prepared, 2: correctly prepared

3.2.2. Approval testing of thermal sprayers

Thermal sprayers and also operators of thermal spray systems should either be suitably qualified and/or have industrial experience in the field. They should be able to follow oral and written instructions in order to assure the quality demands of the thermally sprayed products. This ability is verified in an examination which tests the manual dexterity of thermal sprayers and in an examination which tests operators on the thermal spray equipment.
Uniform rules for the examination are specified in the standard DIN EN ISO 14918.
In accordance with DIN EN ISO 14918, the spray qualification is valid from the day on which the practical examination is passed. The certificate is valid for three years provided that the thermal spray coordinator or the staff responsible in the employer's company confirm, every six months, that the sprayer has worked within the scope of application / with the spray process tested.

After successful approval testing of the thermal sprayer(s) in accordance with ISO 14918, the coating business receives a certificate stating that the spray operations have been carried out by trained and tested, i.e. qualified staff.
The approval test confirms that the thermal sprayer has adequate knowledge and practical experience of the spray process which he performs at work and for which he has sat the test. Testing differentiates between manual and mechanized thermal spraying and between the various spray processes.
In the first test in addition to the practical test, technical knowledge of the respective spray process is checked for which the sprayer will sit the examination.
In the technical test, knowledge of the equipment, spray process including spray parameters, base materials and spray materials, test methods for thermally sprayed coatings as well as occupational health and safety and accident prevention are examined.
The practical test assesses the blasted surface, the masking process, operating the equipment, applying the coating and occupational health and safety.
The sprayed test specimens undergo visual inspection before and after machining. Furthermore, material and process-related bond strength and metallographic and micro-hardness tests are carried out to evaluate the spray results.
The latest revision of ISO 14918 is currently awaiting approval by the ISO committee. It provides for more flexibility with respect to the spray process being tested, the spray materials used and the way the material assessments are carried out. For both coating businesses as users of the standard and certified inspection authorities, this revision is long overdue. For all stakeholders, it represents a reliable instrument in the qualification of spray staff, also from an economic perspective.
As has long been the case in the railway vehicle and power-plant sectors, experience shows that this standard is enjoying increased application for TSA – thermal sprayed aluminium – in off-shore operations and in chemical plants.

4. Future developments

For over 20 years, the GSI branch SLV Munich has been offering courses for "thermal spraying".
In addition to the exchange of experiences and metallography seminars, these focus above all on the qualifications ETSS (European Thermal Spray Specialist) and ETS (European Thermal Sprayer).
Business relations on the international stage have created increasing demand for globally recognized qualifications. Requirements are particularly high in sensitive areas such as aerospace and railway vehicles. The German Railways (DB), for example, demands from its suppliers that their thermal spray coordinators have the EWF qualification ETSS. This is now a standard part of the contract. Based on its members' decision, GTS has likewise adopted the requirement of at least one ETSS and one ETS per certified company. Certification in accordance with the GTS guideline is also in increasing demand internationally.

SLV Munich has also been offering the courses ETS and ETSS in English for the past two years. In 2014, with the approval of the Spanish ANB (Authorized National Body), three thermal sprayers gained the ETS qualification and two engineers the ETSS qualification / 3; 4/. In 2013, four thermal sprayers from Jordan took part in the ETS course in English /5/. The course feedback from the companies and participants was excellent.
The EWF training guidelines for thermal spraying specialists and thermal sprayers have facilitated the practical implementation of a qualification for thermal spraying personnel which now enjoys European-wide recognition and which will possibly enjoy worldwide recognition in the foreseeable future as well. This set of guidelines is directly linked to product-related quality management pursuant to DIN EN ISO 14922 and the approval testing of qualified thermal sprayers pursuant to ISO Norm 14918. A complete and cohesive system has thus been created which enables the manufacturer, i.e. the coating business, to increase and assure the quality of his products. He then has documented proof that he has carried out the assigned work according to the QM system described, giving the customer a

reliable means of assessing the manufacturer's ability to produce quality. This builds confidence on both sides.

5. Conclusion

In the future, customer demands vis-à-vis coating businesses will continue to increase. In addition to application standards for spray materials, spray equipment and methods for testing thermally sprayed coatings, standards have also been devised by the technical committee CEN TC 240 "Thermal spraying and thermally sprayed coatings" which provide job shops with tools to produce according to quality requirements. They also help to establish a QM system for spray technology.
Modern joining and surface technologies, especially welding and thermal spraying, are characterized by manual skills, engineering knowledge and innovative processes.
Without a doubt, consistently implementing and applying the standards and guidelines described above will visibly give the field of thermal spraying the necessary impetus for future applications.

References:

/1/ Cramer, Huber, Ohliger-Volmer: Conference Proceedings 9th Colloquium HVOF 2012, Erding
/2/ GTS-Strahl in Thermal Spray Bulletin 2/2014
/3/ GTS-Strahl in Thermal Spray Bulletin 1/2015
/4/ GTS-Strahl in Thermal Spray Bulletin 2/2013

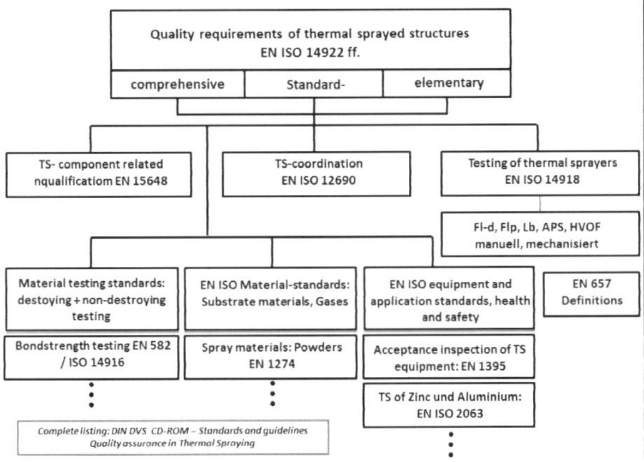

Fig. 1: QM and application standards for thermal spraying

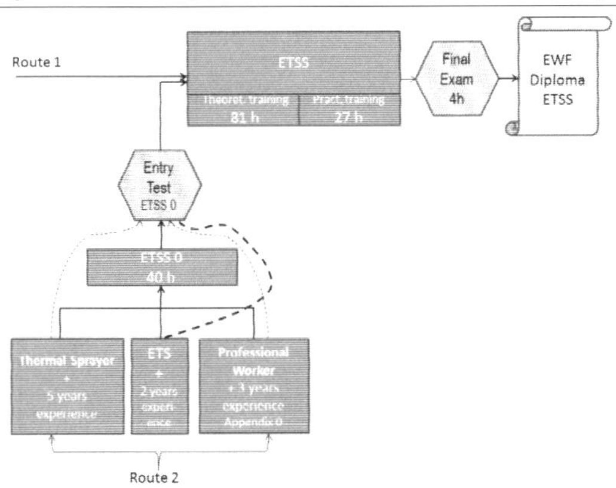

Fig. 2: Access to EWF-ETSS education and training

Fig. 3: Flaw due to the insufficient quality awareness of staff: delivery of incorrectly sprayed test pieces for bond strength testing

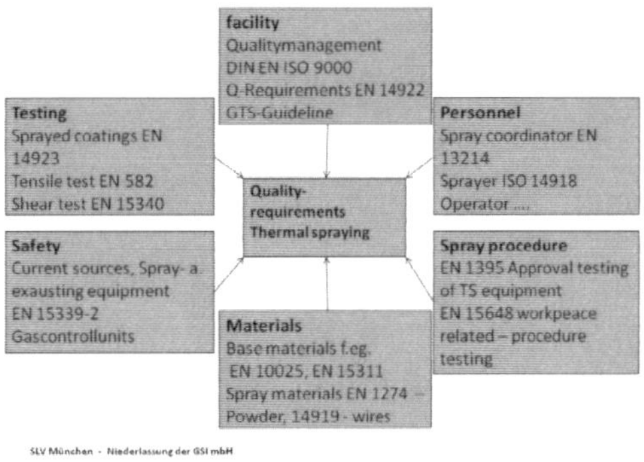

Fig. 4: Quality management pillars and corresponding standards

ISO 9001:2015

Main changes:

- greater focus on the customer
- risk-based thinking
- aligning QMS policy and objectives with the strategy of an organization
- greater flexibility with documentation
- stronger responsibility of QMS on managment level (no QM representative required)

Supporting documents:

comparison of ISO 9001:2008 and 9001:2015

Fig. 5: Current information on ISO 9001:2015

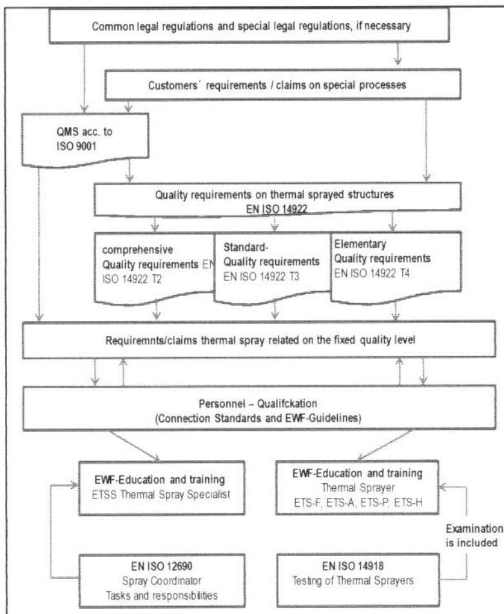

Fig. 6: Interrelationship of ISO standards and EWF training for spray personnel

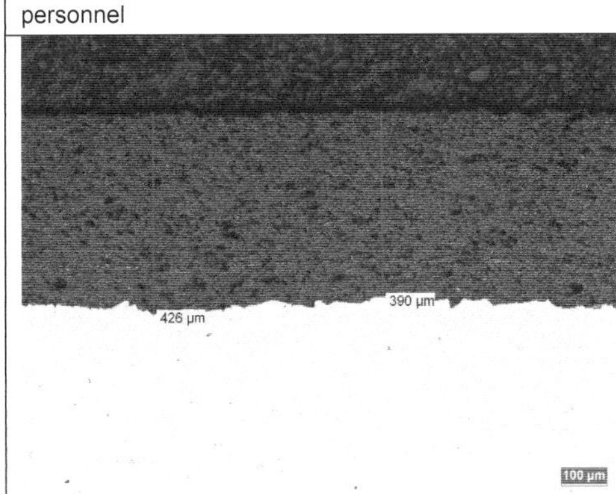

Fig. 7a: Preparation of a ceramic spray coating in line with quality assurance: incorrectly prepared

Fig. 7a: Preparation of a ceramic spray coating in line with quality assurance: correctly prepared

Investigations on the Nature of Ceramic Deposits in Plasma Spray-Physical Vapor Deposition

W. He, G. Mauer, R. Vaßen, O. Guillon, Forschungszentrum Jülich GmbH, IEK-1, Jülich / D
M. Gindrat, Oerlikon Metco AG, Wohlen / CH
R. Wäger, Helbling Technik AG, Aarau / CH

In Plasma Spray-Physical Vapor Deposition (PS-PVD) process, major fractions of the feedstock powder can be evaporated so that coatings are deposited mainly from vapor phase. In this work, Computational Fluid Dynamics (CFD) results indicate that such evaporation occurs significantly in the plasma torch nozzle and even nucleation and condensation of zirconia is highly possible there. Experimental work has been performed to investigate the nature of the deposits in the PS-PVD process, in particular coatings from condensated vapor and nano-sized clusters produced at two spraying distances of 1000 mm and 400 mm. SEM images show that, at long spraying distance, many small nano-sized particles (<20 nm) which are significantly smaller than primary particles in the feedstock powder cover the coating surface in the center of the plasma jet. However, at short spraying distance, such nano-sized particles were hardly found. Besides, in the center area of the plasma jet, XRD patterns indicate only limited preferential growth orientation at long spraying distance which might be a consequence of the low mobility of nano-sized cluster deposits. Meanwhile, at short spraying distance, XRD patterns show obvious preferred growth orientations of (110) and (002). The results give a further insight in the formation mechanisms of high quality columnar structured PS-PVD thermal barrier coatings which have already shown excellent performance in cyclic life time tests.

1 Introduction

Plasma Spray-Physical Vapor Deposition (PS-PVD) is a novel technology, named for vapor phase deposition by using plasma spraying [1]. At a very low working pressure of 50 to 200 Pa, the plasma jet is less cooled and decelerated, which leads to a plasma jet more than 2 meters in length and a diameter ranging from 200 mm to 400 mm [2]. Therefore, PS-PVD has been developed for deposition of thin and uniform coatings with large area coverage [3]. With enhanced electric input power up to 180 kW, the feedstock material even can be evaporated so that coating deposition mainly comes from vapor phase [4]. The interaction of plasma gas and vapor phase of feedstock makes non-line-of-sight deposition possible to deposit high quality columnar structured coating not only on the front side of substrate but also on the shadowed parts of substrates [5].

Currently, columnar structured thermal barrier coatings (TBCs) on aero-engines with the highest load are mainly manufactured by electron beam physical vapor deposition (EB-PVD). At PS-PVD conditions, suitable process parameters and specific feedstock powder enable manufacturing columnar structured ceramic topcoats as well. And recently, such columnar structured TBCs produced by PS-PVD have shown improved thermal cycling lifetimes more than two times higher than conventionally sprayed TBCs [6]. Moreover, the deposition rate of PS-PVD is significantly higher than that of EB-PVD, which makes PS-PVD be the most promising alternative process for depositing columnar structured TBCs. Therefore, to obtain a better understanding of deposition mechanisms of high quality columnar structured PS-PVD coatings is now a focus issue.

In the previous work, a structure zone model for magnetron sputtered coatings proposed by Thornton was transferred to PS-PVD to illustrate coating characteristics because it generally corresponds well with the characteristics of PS-PVD coatings formed by shadowing, surface diffusion, and bulk diffusion [7]. Further calculations indicated that nano-sized clusters can be formed in the boundary layer of the substrate by homogeneous nucleation and condensation from super-saturated vapors [8]. Thus, deposition in PS-PVD takes place not only from liquid splats but also from nano-sized clusters or directly from the vapor phase. In this work, Computational Fluid Dynamics (CFD) results confirm former calculations indicating that such evaporation occurs significantly in the plasma torch nozzle. Experimental work has been performed to investigate the nature of the deposits in the PS-PVD process, in particular coatings from nano-sized clusters.

2 Experimental procedures

2.1 Coating deposition

The coatings were carried out on an Oerlikon (formerly Sulzer) Metco Multicoat system, which can achieve the low working pressure of 200 Pa and a maximum input power of 150 kW. An O3CP torch was applied to generate the plasma jet with 35 slpm argon and 60 slpm helium as plasma gases plus 2x16 slpm argon carrier gas.

The feedstock powder was monoclinic zirconia agglomerated with 7 wt.% cubic yttria (YSZ) produced by Oerlikon Metco designated as M6700. The particle size distribution was about: d_{10}=6.6 µm, d_{50}=11.5 µm, d_{90}=19.0 µm.

In order to investigate the nature of vapor phase deposition in PS-PVD process, a high current of 2750 A and a low powder feed rate of 5 g/min was used in coating processes at two spraying distances of 1000 mm (test A) and 400 mm (test B). The substrate temperature (T_s) of coating at 400 mm was estimated about 2000°C, therefore graphite was utilized as substrate material in all coating processes. Spraying parameters are shown in Table 1.

Table 1. Plasma spray parameters.

Test	A	B
Spraying distance (mm)	1000	400
Plasma gas (slpm)	Ar 35 / He 60	
Current (A)	2750	
Net power (kW)	60	
Powder feed rate (g/min)	5	
Carrier gas (slpm)	32	
Spraying duration (min)	5	2

*slpm: standard liter per minute

In test A, the substrate temperature was recorded at two positions P1 and P2 by thermocouples (Type K) inserted into the substrate from backside. P1 is the center of plasma jet and P2 is 40 mm offset from center. During spraying, plasma torch and samples were not moved. The positions of substrate and thermocouples are shown in **Fig. 1**.

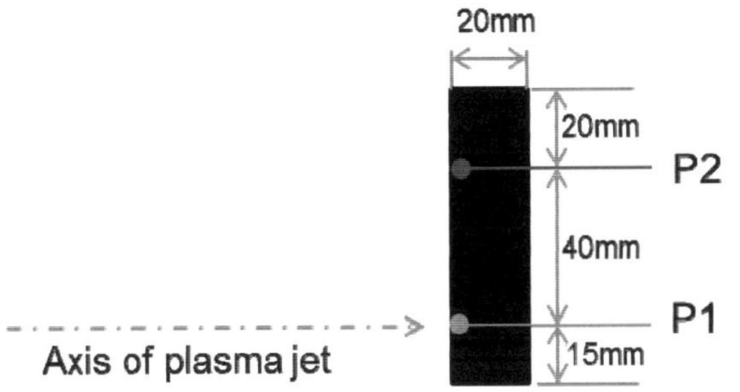

Fig. 1. Schematic drawing of positions of substrate and two thermocouples during coating in test A.

2.2 Simulation methods

CFD is the art of fluid mechanics which can solve and analyze or even provide a qualitative (and sometimes even quantitative) prediction of fluid flows by means of mathematical modeling, numerical methods, and software tools. In present work, the model and simulation were carried out with the latest version of Ansys CFX. The O3CP torch was full 360° simulated by hexahedral mesh. The outlet of the torch was set with a parabolic shaped boundary.

2.3 Characterization methods

Coating morphologies was investigated by scanning electron microscopy (SEM, Ultra 55, Zeiss - Germany) and Hitachi TM3000 tabletop microscope (Germany). X-ray diffraction (XRD, D4 Endeavor - Bruker AXS) was carried out to determine the phase composition and preferred growth orientations in the coatings.

3 Results and discussion

3.1 Computational Fluid Dynamic simulation

CFD simulation was applied to simulate the interaction of plasma and feedstock after injection. Table 2 shows some adapted material properties, boundary conditions, and parameters during spraying delivered by PS-PVD facility.

The temperature distribution in O3CP torch is shown in Fig. 2, and the highest plasma temperature is reached at the axis of torch. Particle tracks are colored by diameter. Despite that particles are not able to reach the core of plasma, the plasma temperature is much higher than vaporization temperature of zirconia. Hence, the reducing of mean particle diameter illustrates evaporation of zirconia. Table 3 gives simulation results according to the parameter setting in Table 2. In this case a significant amount of zirconia feedstock, about 56.8 wt.%, is evaporated in the nozzle.

Table 2. Constant values used for the computational simulation.

Latent heat of fusion of YSZ	H_{melt}= 700'000 J/kg
Melting temperature	T_m= 2950 K
Specific heat 1) T < 1478 K 2) 1478 K < T < T_m 3) T_m< T	1) c_p= 1000*(16.64 + 1.8$*10^{-3}$*T- 3.36$*10^5*T^2$) * 0.033977423 J/kg K 2) c_p= 0.605 J/kg K 3) c_p= 0.714 J/kg K
Cathode back wall temperature	T = 300K
Anode voltage	U = 50 V
Arc attachment on anode fixed (rel. to x-axis)	α = 135 deg
Plasma gas	Ar 35 slpm /He 60 slpm
Carrier gas	32 slpm
Carrier gas temperature	300 K
Powder (YSZ)	0.333 g/s
Current	2600 A
Voltage	48 V
Input power	124.8 kW
Net power	60 kW

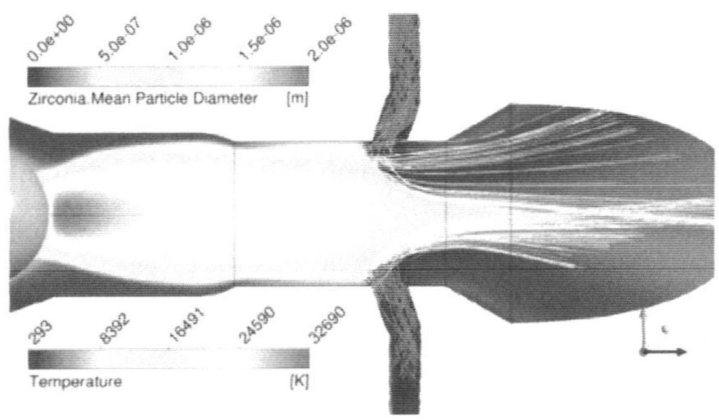

Fig. 2. Plasma temperature with particle tracks colored by diameter.

Table 3. Summary of simulation results

Power absorbed by powder	38.3 kW
Zirconia mass transfer to gas	0.189 g/s
	56.8 %

3.2 Experimental results

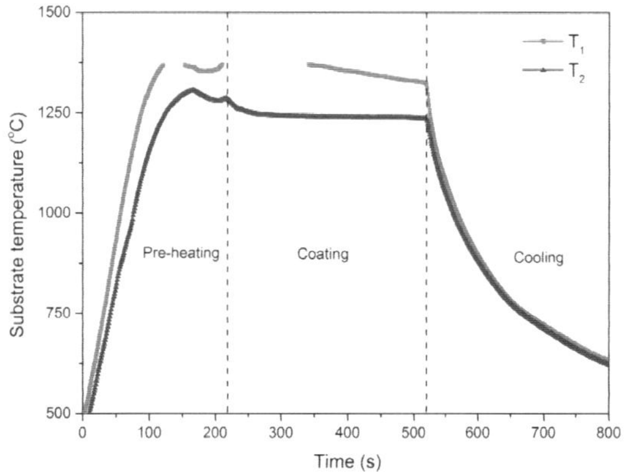

Fig. 3. Substrate temperature measurement in test A: T_1 and T_2 are temperatures at P1 and P2.

In test A, during spraying process, two thermocouples were used to monitor the substrate temperature at two positions: P1 the center of plasma jet and P2 the edge (40 mm offset from center) of plasma jet. Fig. 3 shows the temperature measurement results. The data loss of T_1 is because substrate temperature is out of measuring range of the thermocouple (1370°C). Before coating, the substrate was heated up to obtain a better adhesion between substrate and YSZ coating. During pre-heating, if only the heat transfer from plasma to substrate is considered, $T_1 > T_2$ means that the plasma temperature in the center is higher and it is assumed that the temperature difference between the coming plasma and substrate is the same for the whole substrate. The temperature drop during preheating is caused by current adjustment.

On coating onset, T_2 decreases a little due to the formation of thermal barrier coating. Then it keeps constant while T_1 keeps reducing until the end of coating. The result basically illustrates: T_1 is higher than 1324°C and T_2 is higher than 1235°C. Therefore, the ratio between T_s and melting point of zirconia (T_m = 2715°C), T_s/T_m, is between 0.45~0.55.

Columnar structured coating was deposited on the substrate. Fig. 4 shows a fracture surface of the coating in test A and test B. In test A, the coating A1 (Fig. 4A1) deposited in the center of plasma jet, a typical column starts to grow up first along the normal direction of substrate with large diameter and then branches into many finer columns. At the edge of plasma jet, coating A2 (Fig. 4A2) has limited thickness so that from the bottom to the top of columns show relatively uniform diameter. While in test B, the coating B1, also deposited in the center of plasma jet, shows a very high deposition rate. It reduces from A1 to A2 as well as from B1 to A1, indicating high concentration of deposition species either in the center of plasma jet or at short spraying distance. Besides, in the inset of Fig. 4A2, some micron-sized spherical particles are noticeable. Additionally, in Fig. 2, the particle tracks show that some particles do not reach the core of plasma and still keep their particle sizes at the exit of torch. These kinds of particles are not easy to be evaporated, and they will be directly solidify and incorporated in the coating.

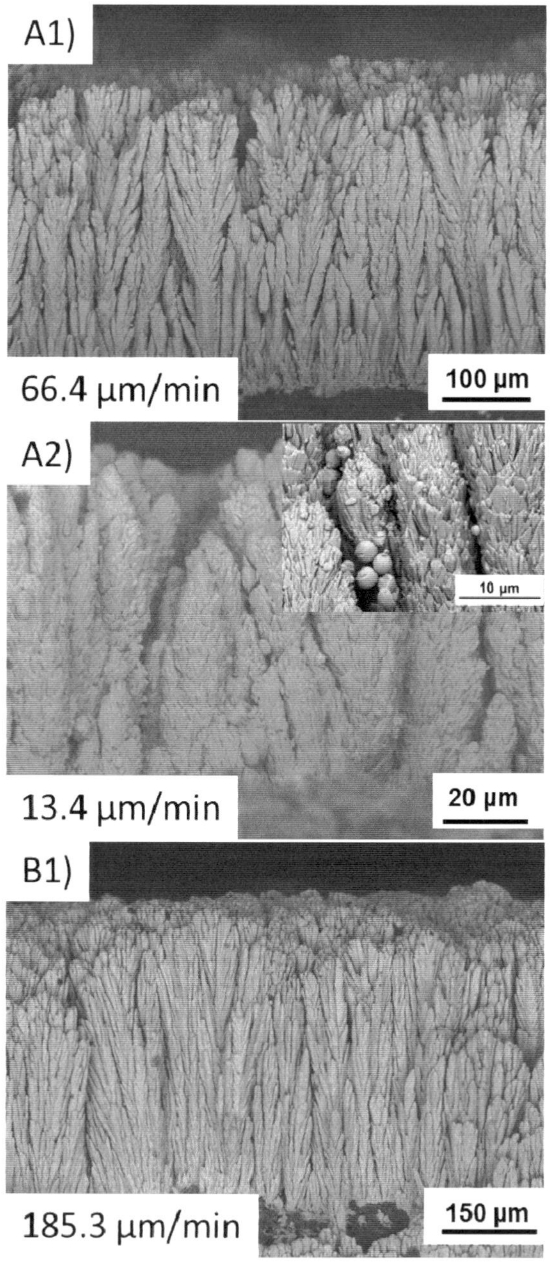

Fig. 4. SEM (back-scattered electron) images of fracture surfaces: A1), A2) and B1) are corresponding to position P1, P2 in test A (1000 mm) and P1 in test B (400 mm), respectively. An approximate growth rate in unit of μm/min is given in the images.

Fig. 6 shows the tops of columns of the coatings sprayed in test A and B. It is obvious that the coatings sprayed at 1000 mm (Fig. 6A1 and A2) have pyramid shaped top and faceted structure. But it is also visible that the faceted structure of coating A2 is not as sharp as that of coating A1. Besides, it is noteworthy that some strips and many nano-sized particles cover on every small facet of coating sprayed at 1000 mm. And the size of these particles is about or even smaller than 20 nm, which is significantly smaller than the original nano-particles in the feedstock powder (seen in Fig. 5). Hence, such kind of nano-particles must be the result of nucleation and growth of vapor phase, which could be an evidence of nano-sized cluster deposition in PS-PVD. With reducing spraying distance to 400 mm, columns in the coating B1 (Fig. 6B1) have flat tops and do not show faceted structure any more. And on the top of coating B1, nano-sized particles are not visible but some regular shaped bulges. The different microstructures of the coatings indicate different crystal growth patterns.

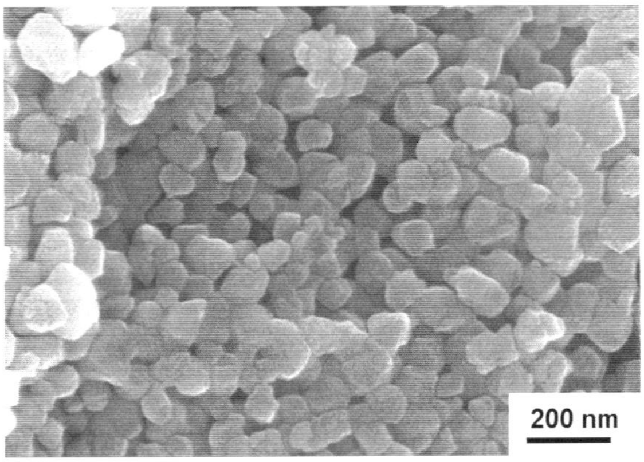

Fig. 5 SEM (secondary electron) images of original nano-sized particle in the feedstock powder.

XRD was carried out to determine the phase compositions and crystal structures of the coatings. The XRD patterns in Fig. 7 show that the main phase of coating produced by PS-PVD is tetragonal YSZ. But a small amount of monoclinic phase also exists in the coating. Because the feedstock powder is an agglomeration of monoclinic zirconia and cubic yttria, the monoclinic phase should come from unmelted or unevaporated feedstock powder particles as seen in Fig. 4A2. Meanwhile, major fraction of powder feedstock was evaporated so that the deposition of vapor mixture of zirconia and yttria leads to tetragonal phase at room temperature. Due to very high substrate temperature at 400 mm, small amount of ZrC formed in the coating. Comparing to coating A1, the narrower peaks in coating A2 mean larger crystallite size.
According to Rietveld analysis of the coatings (seen in Table 4), there is no indication of preferred orientation (PO) in coating A1. But PO of crystallographic plane (110) is found in coating A2 and coating B1 has crystallographic planes (002) and (110) as POs. Chen et al. compared that the surface free energy density of (002) is larger than that of (110) [9], which means under the same growth conditions, the growth of (110) is more thermodynamically stable than (002).

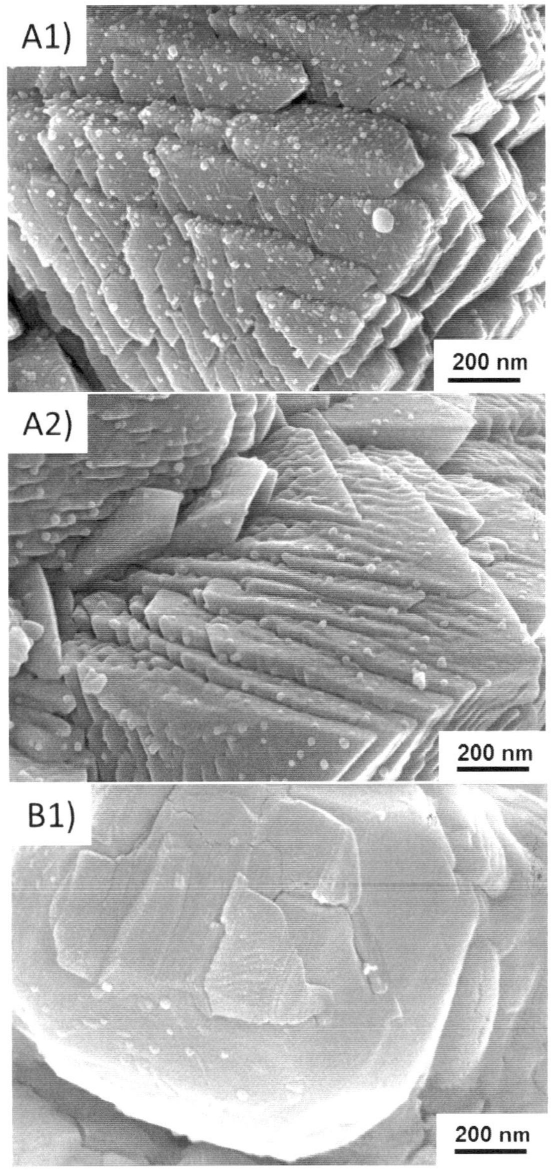

Fig. 6. SEM (secondary electron) images of column tops: A1), A2) and B1) are corresponding to position P1, P2 in test A (1000 mm) and P1 in test B (400 mm), respectively.

Table 4. Rietveld analysis of the coatings.

Coating	Amount of monoclinic phase	Preferred orientations
A1	~5%	-
A2	~3%	(110)
B1	~5%	(002) (110)

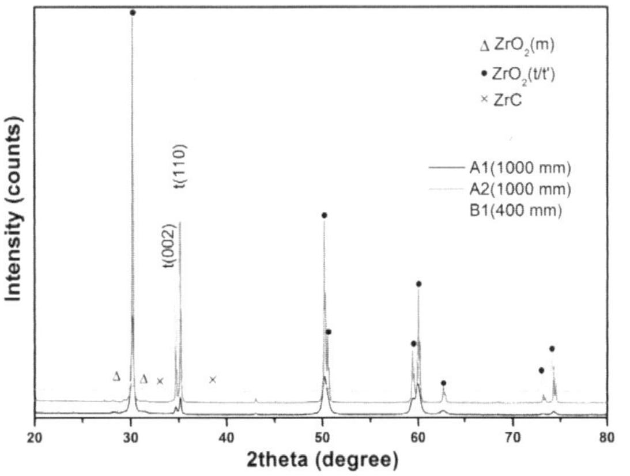

Fig. 7. XRD patterns of coatings: A1), A2) and B1) are corresponding to position P1, P2 in test A (1000 mm) and P1 in test B (400 mm), respectively.

3.3 Discussion

In a previous work, a preliminary PS-PVD structure zone model (SZM) was built up to illustrate the characteristics of PS-PVD coatings formed by shadowing, surface, and bulk diffusion [7]. It is suggested that substrate temperature (relating to the activation energy) and impingement rate are two significant parameters in relation to the characteristic features of the PS-PVD coating microstructures.

According to the CFD simulation results, evaporation of YSZ occurs significantly in the nozzle. Due to large temperature gradient between the hot plasma and the cool nozzle wall, supersaturation can be achieved in the nozzle. Likewise, the temperature gradient between substrate and the coming plasma jet is also huge, which would also lead to the formation of supersaturated vapor in the boundary layer of the substrate. In a supersaturated vapor, the critical nucleus size of homogeneous nucleation for a stable nucleus can be written as [10]

$$d^* = \frac{4\sigma v_1}{k_B T \ln S} \quad \text{(Equ. 1)}$$

wherein σ is the surface tension, v_1 is the monomer volume, k_B is the Boltzmann constant, T is the absolute temperature, S is the supersaturation ratio which is defined as n_1/n_s (n_1 is the concentration (m^{-3}) of monomer, n_s is the equilibrium monomer concentration). Once the $S > 1$, a minimum diameter for nucleation can be achieved. Therefore, the formation of nano-sized clusters in the boundary layer should also be taken into consideration.

According to S. L. Girshick's calculation [11], the nucleation burst is followed by steady growth by coagulation and the particle size distributions are broadened by non-uniform temperature, velocity profiles, and by particle deposition to substrates. The study also pointed out that the size and number of nucleated particles can be significantly affected by two factors, the cooling rate at the location of nucleation and the monomer concentration n_1 (assumed only one condensable species in vapor). Lower local cooling rates and higher concentrations both favor the nucleation of larger (and fewer) particles.

In present work, a higher concentration of deposition species in the center of plasma jet was demonstrated by the higher deposition rate. And from the temperature measurement result, at the same spraying distance an almost same cooling rate can be expected for the vapor species. Therefore, in the center of plasma jet (P1), higher concentration of deposition species will lead to formation of larger and fewer clusters.

A previous report gives experimental evidences for cluster deposition shows that the mobility of the particles strongly decreases as their size increases [12]. In other words, larger sized cluster will have higher activation energy.

At spraying distance of 1000 mm, even though the substrate temperature at P1 was higher than at P2, due to the formation of large size clusters at P1, the mobility of the clusters are inhibited and therefore they are difficult to move to each other on the substrate surface, which can be confirmed by the absence of preferred orientation. While at P2, the relatively smaller clusters are able to diffuse on the substrate surface to find a low Gibbs' energy spots to build up coating with preferred orientation of (110). And due to the diffusion of clusters, they could have chance to coalesce with each other to grow as large crystallites.

If the spraying distance is reduced to 400 mm, the substrate temperature is estimated around 2000°C. T_s/T_m is about between 0.7~0.75. The very high concentration of deposition species will lead to larger clusters, which could result in high difficulty of clusters mobility. However, due to the extremely high substrate temperature, surface diffusion or even bulk diffusion possibly occurs so that the coating B1 shows preferred orientation of (110) and (002). The growth of (002) crystal plane must be the reason for the flat column top (Fig. 6B1).

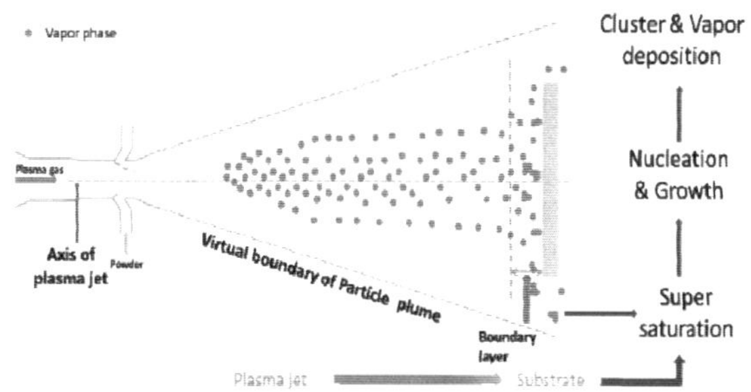

Fig. 8 Schematic drawing of the proposed cluster and vapor deposition in PS-PVD.

According to the above analysis, a proposed coating build-up of vapor and cluster deposition in PS-PVD is depicted schematically in Fig. 8. From the exit of plasma torch, the shape of particle plume is divergent, therefore high concentration of deposition species can be expected in the center of the plasma jet. In the chamber, the plasma jet is laminar and the interaction of the plasma jet with the surrounding atmosphere is weak [13]. Therefore, the plasma jet is less the plasma jet is less cooled and decelerated. As the hot plasma jet comes close to the relatively cool substrate, a boundary layer is formed due to the rapidly decreasing of temperature and velocity. In the boundary layer, nano-sized clusters can be formed as a result of supersaturation of vapor species. The size of clusters is mainly dependent on the local concentration of vapor species and also influenced by cooling rate. At the same time, vapor species nucleate and grow likely directly on the substrate. Further build-up of coating is mainly governed by the substrate temperature. In summary, the coating produced by PS-PVD is a mixture of cluster and vapor deposition.

4 Conclusions

In this work, CFD simulation for the O3CP torch gives indication of significant evaporation of YSZ already in the plasma torch. Columnar structured TBCs were successfully manufactured by PS-PVD. The microstructures and morphologies of the different areas of the coatings show formation of nano-sized particles on the top of columns. Preferred orientations of the coatings were investigated, which gives an explanation to the different microstructures of the coatings.

Based on the results, a cluster deposition in PS-PVD was proposed. The huge temperature gradient from hot plasma to cool substrate results in supersaturation of vapor phase formed in the boundary layer of substrate. Nucleation and growth of vapor species can occur in it forming clusters. The size of cluster is mainly influenced by the local concentration of vapor phase and cooling rate. After the clusters and vapor species impact on the substrate surface, the build-up of coating is significantly influenced by substrate temperature.

5 Acknowledgements

The authors would like to express their thanks to Mr. Ralf Laufs for his help to operate the PS-PVD facility, to Dr. Doris Sebold for her work on the SEM investigation and Dr. Yoo-Jung Sohn for the help of XRD analysis. The author, Wenting He, would like to acknowledge the support of China Scholarship Council.

6 Literature

[1] Von Niessen, K., M. Gindrat, and A. Refke. Vapor phase deposition using plasma spray-PVD™. Journal of thermal spray technology 19 (2010), Issue 1-2, pp. 502/09.

[2] Goral, Marek, Slawomir Kotowski, and Jan Sieniawski. The Technology of Plasma Spray Physical Vapour Deposition. High Temperature Materials and Processes 32 (2013), Issue 1, pp. 33/39.

[3] Hospach, A., et al. Columnar-structured thermal barrier coatings (TBCs) by thin film low-pressure plasma spraying (LPPS-TF). Journal of thermal spray technology 20 (2011), Issue 1-2, pp. 116/20.

[4] Mauer, G., et al. Novel opportunities for thermal spray by PS-PVD. Surface and coatings technology 268 (2015), pp. 52/57.

[5] Mauer, G., Robert V., and Detlev S. Thin and dense ceramic coatings by plasma spraying at very low pressure. Journal of thermal spray technology 19 (2010), Issue 1-2, pp. 495/501.

[6] Rezanka S., Mauer G., and Vaßen R. Improved thermal cycling durability of thermal barrier coatings manufactured by PS-PVD. Journal of Thermal Spray Technology 23 (2014), Issue 1-2, pp. 182/89.

[7] Mauer, G., et al. Process conditions and microstructures of ceramic coatings by gas phase deposition based on plasma spraying. Journal of thermal spray technology 22 (2013), Issue 2-3, pp. 83/89.

[8] Mauer, G., Rezanka, S. Hospach, A., Vaßen, R. and Guillon, O. The Role of Nucleation and Growth in Plasma Spray-Physical Vapor Deposition, to be published in Plasma Chem. Plasma Proc. 2016.

[9] Chih-Chiang, C. et al. Growth of zirconia and yttria-stabilized zirconia nanorod arrays assisted by phase transition. CrystEngComm 12 (2010), Issue 11, pp. 3664/69.

[10] Thomson, W. 4. On the Equilibrium of Vapour at a Curved Surface of Liquid. Proceedings of the Royal Society of Edinburgh 7 (1872), pp. 63/68.

[11] Girshick, S. L., and C-P. Chiu. Homogeneous nucleation of particles from the vapor phase in thermal plasma synthesis. Plasma Chemistry and Plasma Processing 9 (1989), Issue 3, pp. 355/69.

[12] Melinon, P. et al. Comparison of molecular and cluster deposition: Evidence of different percolation processes. Physical Review B 44 (1991), Issue 22, pp. 12562.

[13] Mauer, G. Plasma Characteristics and Plasma-Feedstock Interaction Under PS-PVD Process Conditions. Plasma Chemistry and Plasma Processing 34 (2014), Issue 5, pp. 1171/86.

Analysis of the High Cyclic Lifetime of Suspension Plasma Sprayed Yttria-Stabilized Zirconia Columnar Coatings

Benjamin Bernard, Aurélie Quet, Monts/ F
Luc Bianchi, Aurélien Joulia, Magny-les-Hameaux/ F
André Malié, Châtellerault/ F
Benjamin Rémy, Vandoeuvre-lès-Nancy/ F

Suspension Plasma Spraying (SPS) have been recently identified as a promising technology for enhancement of Thermal Barrier Coatings (TBCs). Particularly, Yttria-Stabilized Zirconia (YSZ) coatings presenting columnar microstructures have shown some improved capabilities as TBCs. Low thermal conductivity (< 1 $W.m^{-1}.K^{-1}$) and high thermal lifetime are demonstrated for such coatings. This study presents results of thermo-cyclic fatigue (TCF) performed at 1100 °C for SPS columnar coatings differing by the compaction of columns (from a well-separated to a compact pattern of columns). The effect of bond coat material, surface preparation and YSZ columnar microstructure are studied. Lifetime can exceed 2000 cycles to failures for YSZ coatings performed by SPS.

1 Introduction

Over many years, Thermal Barrier Coatings (TBCs) have been studied in order to improve turbine gas efficiency. Increasing thermal insulation performances of the ceramic top layer of TBCs, mainly composed of Yttria-Stabilized Zirconia, is identified as a main issue [1]. On the one hand, the common deposition process, used for YSZ ceramic layer since the 80's is the Electron Beam Physical Vapor Deposition (EB-PVD). It leads to high thermal compliant columnar structure with high thermal lifetime but also strong heat conduction paths which limit the capability to design low thermal conductivity coatings [1,2,3]. On the other hand, Atmospheric Plasma Spraying (APS), widely used before the 80's, presents a lamellar microstructure which promote a low value of thermal conductivity but which is very sensitive to thermal expansion mismatch leading to relatively low thermal lifetime also considered as a main issue [4,5].

From the 2000's, Suspension Plasma Spraying (SPS) have been proposed in order to develop nanostructured coatings. Nanosized particles are injected through the plasma jet using a liquid as carrier medium [6]. The process allows to perform coatings with a multi-scaled porosity covering 4 orders of magnitude [7]. Moreover, different microstructures are reached using SPS such as columnar structure which is uncommon for thermal sprayed coatings [8]. The low size of particles (< 5 µm) used for SPS process, allows them to be deviated by the plasma flow at the vicinity of the substrate leading to a combined normal and lateral growth of columns around substrate asperities [8,9]. YSZ columnar structures are able to significantly decrease thermal conductivity in the range 0.6 – 1 $W.m^{-1}.K^{-1}$ compared to typical values of YSZ EB-PVD coatings (~ 1.5 $W.m^{-1}.K^{-1}$) [10,11]. Some studies also report good thermal compliance of YSZ SPS columnar structure [11].

The aim of this study is to present and analyze thermal lifetime properties of YSZ columnar structures performed using SPS. Impacts of the used Bond Coat (BC), surface substrate preparation, and microstructure of YSZ columnar structure are discussed.

2 Experimental Procedure

Ethanol-based 7wt.% YSZ commercial suspensions are used for SPS coatings. Suspensions are provided by Treibacher Industry AG™ and particles are characterized by a mean particle diameter lower than 300 nm. Coatings are performed in a ventilated spray booth at atmospheric pressure using a F4-VB plasma gun with a 6 mm internal diameter nozzle provided by Oerlikon-Metco. Spraying distance is fixed at 50 mm and cooling is ensured by combination of compressed air and cryogenic liquid CO_2 jet. AM1 superalloy circular substrates of 1 inch in diameter are used. Two different BC are tested: β-NiAlPt and γ/γ'. A pre-heating step can be performed at 1 100 °C during 1 h under air to provide the growth of a thermally grown oxide (TGO) prior to YSZ deposition. Surface asperities of BC are tailored using sand blasting with two different size of corundum (size 1: 63-106 µm and size 2: 300-425 µm).

Table 1: Spraying conditions used for coating production of columnar and compact columnar structures.

	Condition	
	A	B
Gas Mixture	$Ar/He/H_2$	Ar/He
Plasma Enthalpy	2.2×10^7 J/kg	1.5×10^7 J/kg
Suspension feed rate	25 g/min	44 g/min
Deposition efficiency	58 ± 2 %	37 ± 1 %

Two different spraying conditions, differing by injection feed rate and plasma gas mixture, are investigated leading to columnar structure or compact columnar structure. Main differences between the two conditions are listed in Table 1. Deposition efficiency which corresponds to the ratio between the mass of resulting YSZ coating and the mass of YSZ injected into the plasma jet is presented for each spraying condition in Table 1.

Microstructural investigation is performed using Scanning Electron Microscopy (SEM). SEM device is a Leo 435VPi working with a secondary electron detector for SEM surface analysis and a back scattered electron detector for cross sections. Image analyses are performed using the public free domain ImageJ software.

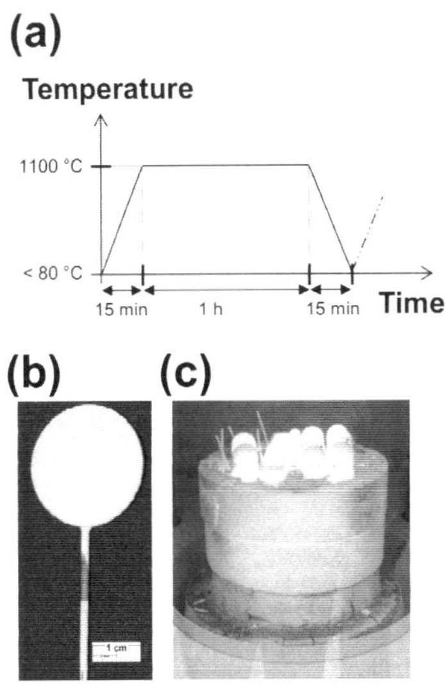

Fig. 1. (a) Description of one cycle of TCF test performed, (b) sample of YSZ coating before submitted to TCF test and (c) samples during TCF test at 1 100 °C.

Thermo-cyclic fatigue (TCF) tests are performed on SPS coatings. TCF tests consist in cycles where samples are heated in 15 min to 1 100 °C and maintained at this temperature during 1 hour before cooling in 15 min below 80 °C. The failure criterion is fixed when at least 20 % of spallation of coating area occur. Fig. 1 presents one cycle of TCF test as well as samples, before and during TCF test.

3 Results and Discussion

3.1 Microstructures

The two different microstructures investigated under TCF test are presented in Fig. 2. A typical SPS columnar structure is obtained when spraying condition A is used, similarly to those observed elsewhere [8,10,11].A compact columnar structure is obtained for spraying condition B close to recent published work [10,11].

The columnar structure obtained using spraying condition A is characterized by columnar features separated by inter-space voids as shown in Fig. 2 (a). Columnar features are characterized in top surface by cauliflowers illustrated in Fig. 2 (b). As described in previous study, image analysis is used to reach column diameter by the estimation of cauliflower size [12]. the column width is supposed to be linearly dependant to the coating thickness and initially defined by the size of asperities (therefore related to the substrate roughness). As the substrate roughness is increased the column features are wider.

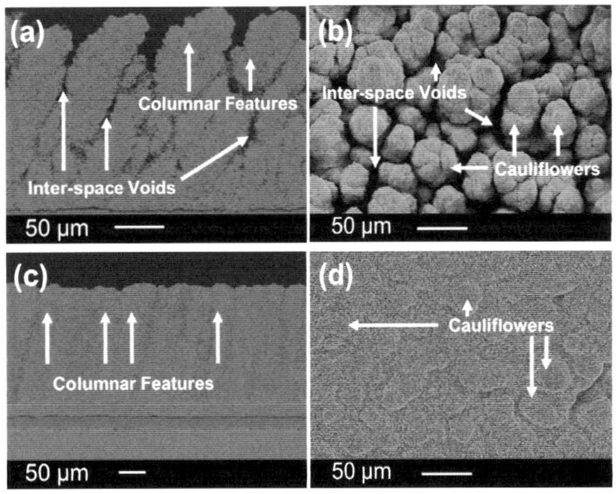

Fig. 2. SEM cross section (a) and top surface (b) of YSZ coating performed using condition A and SEM cross section (c) and top surface (d) of YSZ coating performed using condition B.

Another important aspect has to be considered to control growth of columns. Indeed, for this structure, the cone shape of column leads inevitably to perturbations during the coating build-up such as coalescence and shadowing effects resulting in the formation of big cluster and new secondary smaller columns close to them. Distribution of column diameters can be strongly modified when these effects are either delayed or accentuated. It has been shown that the distribution of column diameters can be managed using growth development speed or substrate asperity size considerations [12]. In the present study, only substrate roughness is tailored from an initial relatively smooth substrate to a rougher surface, using sandblasting. An increase of substrate roughness, favorable to the growth of biggest columns, leads to early perturbations between neighboring columns which significantly decrease the uniformity of column size as shown in Fig. 3. It leads to a broader distribution of column diameters in case of rough substrate surface when spraying condition A is employed to performed coating at 3 different thicknesses (35, 65 and 140 µm) using two levels of substrate roughness: relatively smooth for Ra_{low} and rougher for $Ra_{high} = 2.5 \times Ra_{low}$.

spraying condition A [12]. It leads to a rapid overlapping of columns from each other so called compact columnar structure.

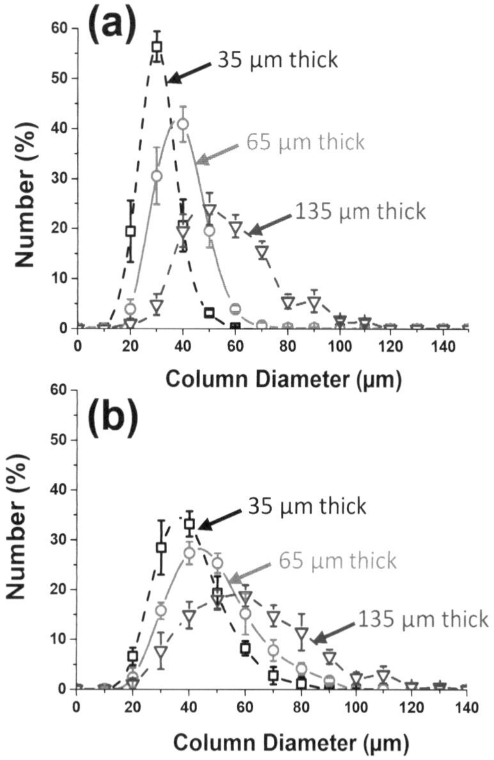

Fig. 3. Column diameter distribution for 3 coating thicknesses obtained for spraying condition A with two substrate roughness (a) Ra_{low} and (b) $Ra_{high} = 2.5 \times Ra_{low}$.

In case of spraying condition B, a compact columnar structure is observed. It can be described by columnar features in SEM cross-section and by cauliflowers in top surface, as shown in Fig. 2 (c,d). Columns are so close that inter-space voids tend to disappear. The lower plasma enthalpy associated to the higher feed rate of suspension induce a significant fall of the deposition efficiency, from 58 % for plasma A to 37 % for plasma B. In the second case, the conditions are unfavorable to the full melting of larger particles or agglomerates into the plasma core during the thermokinetic treatment leading to lower deposition efficiency. Another effect of these difficulties to melt coarse particles is the increase of the ratio of small particles which are molten and which participate to the construction of the coating. It is well known that the small particles can experience by SPS an uncommon trajectory into the plasma core compared to coarse particles. Indeed, at the vicinity of the substrate where the plasma jet can be almost orthogonal to the substrate for the short standoff distances used in SPS, the smallest particles can be deviated by the plasma flow leading to their lateral impingement onto substrate asperities [8,9]. It is the most admitted theory to explain columnar structure obtained by SPS. In our case, the increase of the ratio of well molten small particles in spraying condition B is supposed to enhance the lateral growth of columns compared to spraying condition A [12]. It leads to a rapid overlapping of columns from each other so called compact columnar structure.

Both of these microstructures, with a different level of compaction of the pattern of columns, are promising for the thermal lifetime. Indeed, inter-space voids are supposed to provide a sufficient thermal compliance for the coating in order to well resist in TCF test.

3.1 Thermo-Cyclic Fatigue Performance

Results of TCF test are summarized in Fig. 4, for all the samples prepared in this study. Columnar structures appear in blue and compact structures are plotted in red. Each number correspond to different substrate preparation.

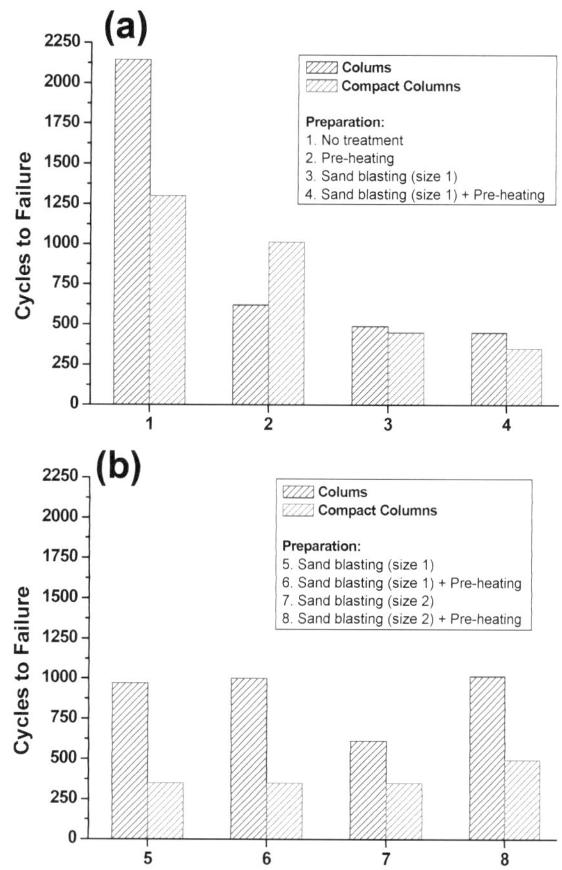

Fig. 4. Thermo-cyclic lifetimes obtained by TCF tests on YSZ SPS coatings performed on AM1 with γ/γ' (a) and β-NiAlPt (b) bond coats. Indexes 1 to 8 describe the preparation of substrate.

Samples performed using spraying condition A and leading to columnar structures with clearly visible inter-space voids seem to be more resistant to TCF tests than compact structures obtained using spraying condition B. SEM pictures of failed samples with γ/γ' bond coat, presented in Fig. 5, illustratethis assumption. Pronounced vertical cracks appear between columnar features in case of compact columnar coatings leading to the creation of sharp inter-columnar voids. The rapid development of these cracks during the stress release could weaken the interface. On the contrary, for columnar structures, intercolumnar voids appear to be more efficient for stress accommodation, leading to a soft and limited development of inter-columnar cracks. With regards to interface where debounding occurs underneath the TGO for both samples, the sharp and violent development of inter-columnar cracks for compact columns could lead to prematured debounding by weakening of the TGO as attested byinternal cracks development when its thickness increases(see Fig. 5). Preliminary observations suggest that failure remains mainly driven by rumpling of the TGO as described by Evans *et al.* [13] and already observed by Curry *et al.* for similar SPS YSZ coatings [11].

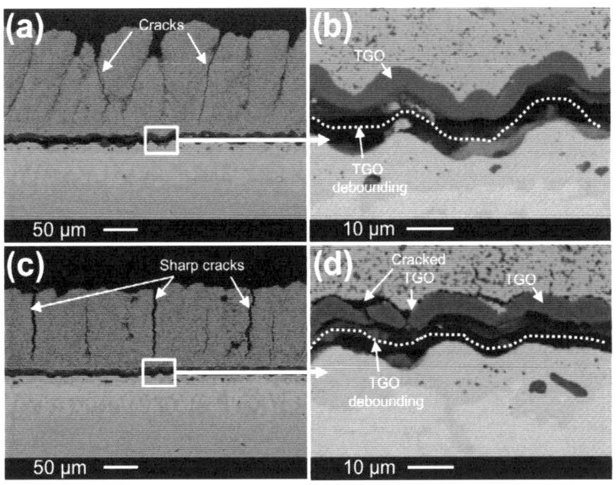

Fig. 5. Microstructures and ruptures of columns (a,b) and compact columns (c,d) after TCF test on γ/γ' samples.

The higher roughness of γ/γ' bond coat, compared to β-NiAlPt (for condition 1 and 5: $Ra_{\gamma/\gamma'} \approx 2\times Ra_{\beta\text{-NiAlPt}}$) seems to be relevant for thermal lifetime. Higher roughness is supposed to ensure more developed columnar features in case of compact columns [12]. Thus, thermal compliance can be improved by decreasing the occurrence of dramatical sharp vertical crack, development leading to higher TCF performances for samples with γ/γ' bond coat (see Fig. 4).

Considering results for samples where YSZ coatings are performed using spraying condition A without substrate heat treatment , it appears that TCF performances are higher for the smoother ones (condition 1 compared with 3 for γ/γ' and 5 compared with 7 for β-NiAlPt in Fig. 4). In fact, the higher uniformity of column sizes for the lower substrate roughness (see Fig. 3), probably enhanced the capability of the coating to release thermally induced stresses all along the layer. On the contrary, for rough surface, big columns and/or clusters can generated a disparity in the release of stress leading to a lower local thermal compliance and hences to a lower global thermal lifetime.

SEM pictures (a) and (b) of Fig. 6 illustrate the increase of substrate roughness for columnar coatings on β-NiAlPt. Columns appear to be larger and the release of stresses induces more pronounced inter-columnar cracks which probably weaken the interface and the adhesion of YSZ. In the particular case of β-NiAlPt, pre-heating seems to attenuate the drawbacks induced by a higher substrate roughness. The development of the TGO, prior to YSZ deposition, probably allows to change surface topography. It leads to a more efficient stress release by inter-columnar voids which attenuates the development of detrimental sharp inter-columnar cracks. However, for other samples, pre-heating seems not to affect drastically the TCF performances. Two assumptions are proposed to explain these observations. First, the growth of a TGO prior to YSZ deposition, does not significantly change the surface asperities excepted for a severe sand blasting using coarse corundum. Then, for non pre-heated samples, the growth of the TGO during cycling and the associated stresses, do not affect the adhesion of the YSZ coating.

The higher TCF resistance is obtained for a sample performed using condition A on a γ/γ' bond coat without any surface preparation. In that case, lifetime reach 2145 cycles to failure which is twice of typical value reached for common YSZ coatings performed using EB-PVD process. The higher TCF resistance of compact columnar structure, 1300 cycles to failure, is also reached with γ/γ' bond coat, and also exceeds the one of EB-PVD coatings.

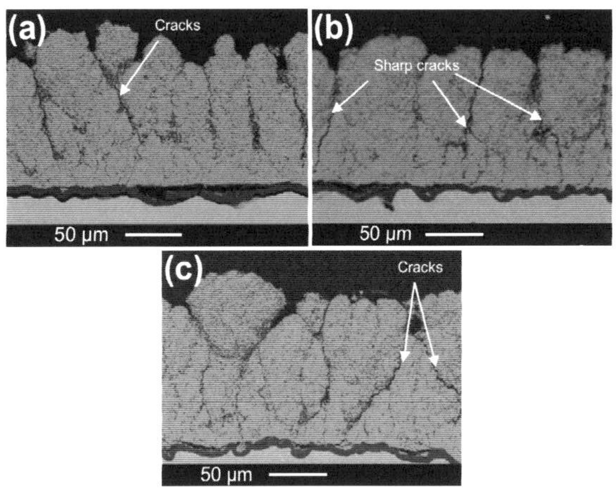

Fig. 6. Microstructures of columnar structure after TCF test on β-NiAlPt-BC for substrate preparation 5 (a), 7 (b) and 8 (c) as reported in Fig. 4. Typical rupture for these structures are illustrated by pictures (d) and (f)

4 Conclusion

Suspension plasma spraying is used to perform YSZ thermal barrier coating presenting columnar structures through the control of process parameters,. columnar structure can be varied from a well-separated to a more compact pattern of columns. A low substrate roughness is identified as a main parameter to control and ensure low-sized and well-organized columns.

Thermal lifetime, estimated by thermal cycling fatigue test, is optimized for SPS columnar structure performed on γ/γ' bond coat without any surface preparation. For columnar structure with well-separated features, 2150 cycles to failure are reached compared to 1300 cycles to failure for compact columns ones. The apparent lower TCF resistance for compact columns is attributed to a loss of thermal compliance inducing the development of sharp inter-columnar cracks. Failures seem linked to a debounding at the TGO / substrate interface.

Used bond coat , γ/γ' or β-NiAlPt,has a mojor influence on thermal lifetime. Moreover, the surface topography and preparation need to be considered carefully in order to ensure an adequate columnar structure. Particularly, roughness coupled with pre-heating considerations are helpful to improve adhesion.

Results are promising for TBC enhancement presenting higher thermal lifetime than EB-PVD structures. Even if columnar structure with well-separated columns presents higher TCF resistance compared to compact ones, the last ones are promising for thermal insulation.

Further works will be focused on the understanding of the debounding mechanism of SPS coating. Evolution of microstructure will be investigated at different cycles of TCF until failure.

5 Literature

[1] Schultz, U. and B. Saruhan, K. Fritscher, C. Leyens: Review on Advanced EB-PVD Ceramic Topcoats for TBC Applications in International Journal of Applied Ceramic Technology 1 (2004), pp. 302/15

[2] Schultz, U. and C. Leyens, K. Fritscher, M. Peters, B. Saruhan-Brings, O. Lavigne, J-M. Dorvaux, M. Poulain, R. Mévrel, M. Caliez: Some Recent Trends in research and technology of Advanced Thermal Barrier Coatings" in Aerospace Science and Technology 7 (2003) pp.73/80

[3] Schultz, U. and, M. Menzebach, C. Leyens, Y.Q. Yang: Influence of substrate material on oxidation behavior and cyclic lifetime of EB-PVD TBC systems in Surface & Coatings Technology 146 – 147 (2001) pp. 117/23

[4] Tan, Y. and V. Srinivasan, T. Nakamura, S. Sampath, P. Bertrand, G. Bertrand: Optimizing Compliance and Thermal Conductivity of Plasma Sprayed Thermal Barrier Coatings via Controlled Powders and Processing Strategies in Journal of Thermal Spray Technology 21 (2012) pp. 950/62

[5] Padture, N.P. and M. Gell, E. H. Jordan: Thermal Barrier Coating for Gas-Turbine Engine Applications in Science's Compass 296 (2002) pp. 280/84

[6] Fauchais, P. and R. Etchart-Salas, V. Rat, J.F. Coudert, N. Caron, and K. Wittmann-Ténèze: Parameters Controlling Liquid Plasma Spraying: Solutions, Sols, or Suspensions in Journal of Thermal Spray Technology 17 (2008) pp. 31/59.

[7] Bacciochini, A. and J. Ilavsky, G. Montavon, A. Denoirjean, F. Ben-ettouil, S. Valette, P. Fauchais, K. Wittmann-ténèze: Quantification of void network architectures of suspension plasma-sprayed (SPS) yttria-stabilized zirconia (YSZ) coatings using Ultrasmall-angle X-ray scattering (USAXS) in Materials Science and Engineering: A 528 (2010) pp. 91/102

[8] Van Every, K.J. and M.J.M. Krane, R.W. Trice, H. Wang, W. Porter, M. Besser, D. Sordelet, J. Ilavsky, J. Almer: Column Formation in Suspension Plasma-Sprayed Coatings and Resultant Thermal Properties in Journal of Thermal Spray Technology 20 (2011) pp. 817/28

[9] Oberste Berghaus, J. and S. Bouaricha, J.G. Lagroux, C. Moreau, Injection conditionsand in-flight particles states in suspension plasma spraying of alumina andzirconia nano-ceramics, Basel, Switzerland, ASM International, May 2–4, in: C.Berndt, E. Lugsheider (Eds.), Proceeding of the International Thermal SprayConference, 2–4, 2005.

[10] Ganvir, A. and, N. Curry, N. Markocsan, P. Nylén, F.-L. Toma: Comparative study of suspension plasma sprayed and suspension high velocity oxy-fuel sprayed YSZ thermal barrier coatings in Surface & Coatings Technology 268 (2015) pp. 70/76

[11] Curry, N. and K.J. VanEvery, T. Snyder, N. Markocsan: Thermal Conductivity Analysis and Lifetime Testing of Suspension Plasma-Sprayed Thermal Barrier Coatings in Coatings 4 (2014) pp. 630/50.

[12] Bernard, B. and L. Bianchi, A. Malié, A. Joulia, B. Rémy: Columnar suspension plasma sprayed coating microstructural controlfor thermal barrier coating application in Journal of the European Ceramic Society 36 (2016) 1081/89

[13] Evans, A.G. and D.R. Mumm, J.W Hutchinson, G.H. Meier, F.S. Pettit: Mechanisms controlling the durability of thermal barrier coatings in Progress in Materials Science 46 (2001) pp. 505/53

Cold spray of metal-polymer composite coatings onto Carbon Fiber-Reinforced Polymer (CFRP)

V. Bortolussi, F. Borit, A. Chesnaud, M. Jeandin, Evry /F, M. Faessel, B. Figliuzzi, F. Willot, Fontainebleau /F, K. Roche, G. Surdon, Argenteuil /F

The growing use of Polymer-Matrix Composite (PMC) materials within transport industry raises new security concerns, especially those due to lightning. To protect these electrically insulating materials, conductive coatings can be applied. Due to the high level of required properties, cold spray is believed to be an effective way to achieve these coatings. Recent studies showed that obstacles remained to be overcome when cold spraying metallic particles onto Carbon Fiber-Reinforced Polymer (CFRP). These are rather due to a poor adhesion of metallic particles onto carbon fibers, which prevents coating build-up. This study therefore developed the use of PEEK addition to metallic powder. The goal is to promote the coating-substrate bond strength provided that the coating could be conductive. The work focusses on composite coatings made of PEEK thermoplastic polymer powder and copper powder. The influence of spraying parameters, powder shape, size and distribution were investigated for various polymer ratios. Depending on these parameters, various microstructures could be achieved, which resulted in different electrical and bonding properties. Coating cross-sections were studied using morphological operations coupled with quantitative image analysis. Relevant parameters to feed a model for simulation of the sprayed microstructure were thus determined. This model as an effective optimization tool allowed to carefully select powders and mixture composition to produce adherent and electrically conducting coatings. Electrical properties were assessed using the Van der Pauw method with dedicated device and samples. The latter were specifically designed to lead to significant and accurate resistance measurements. These measurements showed the feasibility of PEEK-copper composite coating of CFRP. Incidentally, beyond electrical applications, this type of cold-sprayed composites can be claimed to be suitable for a large range of applications.

1 Introduction

The manufacturing of most recent commercial aircrafts involves lighter materials such as Carbon Fiber-Reinforced Polymer (CFRP). These materials often combine stacks of carbon fibers embedded in a matrix made of a thermoplastic polymer such as PEEK (Poly-Ether-Ether-Ketone). Unlike aluminum alloys, CFRP shows a high electrical resistivity due to the polymer matrix which is not suitable for protection against lightning impact. To promote this protection, aircraft manufacturers use lightning screens, which involve complex manufacturing processes. Amongst thermal spray processes, cold spray is believed to be an elegant solution to coat CFRP with electrically-conductive materials. The past 2 years showed an increasing number of articles dealing with metallic cold spray coating onto polymers and composite materials. In 2006, Sturgeon et al. [1] investigated for the first time cold spray coating of carbon fiber/PEEK with aluminum. They achieved adherent and dense coating using helium as a spraying gas and observed the behavior of the coating under thermal cycling. However, it must be mentioned that the PEEK matrix was reinforced with short carbon fibers, which means a microstructure significantly different from that with continuous fibers. Short fibers globally and locally reinforced PEEK on which the coating was sprayed. A similar study was presented by Zhou et al. [2] in 2011, still using short carbon fibers/PEEK composite on which they could spray aluminum and copper. However, to be deposited successfully, copper had to be sprayed onto an aluminum bond coat. A more recent study focused on epoxy matrix composite which exhibited a different thermomechanical behavior in the cold spray process. Affi et al. [3] stated the difficulty to grow an aluminum coating onto epoxy. They used a plasma-sprayed aluminum underlayer on which the cold spray coating could grow. They also investigated the effect of the granulometry of the aluminum powder. In 2014, Che and Yue [4] used the cold spray to coat a continuous carbon fiber/epoxy composite with aluminum, copper and tin. Only scarce particles of tin were reported to be adherent to the substrate, whereas they could not obtain a coating with copper and aluminum. These papers showed the difficulty to achieve dense and adherent coatings onto both carbon fibers and epoxy [5]. It appears that thermoset polymers are hardly suitable for cold spray while metallic coating can be obtained onto thermoplastic polymers. Hence some authors only focused on non-reinforced polymer substrates, with good results [5-7].

This paper focuses on the coating of continuous carbon fibers/PEEK composites with copper using cold spray. Previous studies showed that copper particles rebound on carbon fibers and break them at the impact. In this work to limit fiber failure and obtain an adherent coating, a method was proposed i.e. spraying a mixture of PEEK powder with copper powder. This method can be claimed to be original. One must refer only to a somewhat similar experiment focused on spraying various ratios of polymer powder with copper and aluminum in order to obtain frictionless but electrically conductive coatings [8].
In the present paper, the influence of cold spray parameters was investigated as well as powder granulometry and particle shape. The concept of critical velocity was not valid in this specific case.

Then the coating microstructure was observed and analyzed. A morphological model of the coating was then developed from quantitative image analysis of the same microstructure.

Morphological models are very useful to determine numerically physical and mechanical properties of multiphased materials exhibiting periodic or random structures. They allow to perform numerous numerical simulations on representative volumes. Varying the spatial distribution and size parameters of the phases can thus be easily done to investigate their influence on materials properties, primarily those of coatings [9-13]. In this study, the model would be used to numerically perform a morphological optimization of the microstructure, allowing to select the best powder to lead to optimized properties.

2 Materials and Processing

Two copper powders were sprayed: the first powder was a high-purity (≥99.95 wt%) spherical copper powder. Particles size ranged from 10.6 µm (d_{10}), to 33.1 µm (d_{90}) with a d_{50} of 19.6 µm as obtained by laser granulometry. The second powder was an irregular and coarser copper powder, the particle size of which ranged from 11.4 µm (d_{10}), to 114 µm (d_{90}) with a d_{50} of 57.6 µm. Copper powders were mixed with Vicote 702 PEEK powder, the particle size of which ranged from 26.5 µm (d_{10}), to 88.5 µm (d_{90}) with a d_{50} of 52.8 µm. The mixture of PEEK and copper was observed with a SEM. Spherical copper particles appeared in bright grey whereas irregular PEEK particles appeared darker, **Fig. 1.**

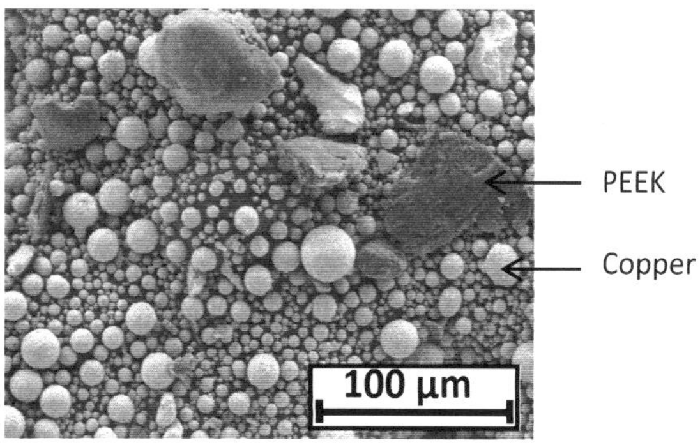

Fig. 1. SEM view of the powder mixture

This paper focuses on only one mixture of powder i.e. 80 vol.% of copper and 20 vol.% of PEEK, termed as 80/20, **Fig.1.** However, tests were performed using various ratios. The substrates were small samples of 25x25x2 mm^3 made of aeronautical grade continuous carbon fibers plies embedded in a PEEK matrix. Plain Victrex PEEK 450G samples of the same dimensions were also coated. PEEK in the composite, Victrex 450G and powdered PEEK, were all three different. However, a common glass transition of 140 °C and a melting temperature of 330 °C were considered. The CFRP surface presented an original rugosity with small bump of PEEK and valley with carbon fibers at the bottom.

Spraying experiments were carried out using a KINETICS 3000M cold spray system by CGT-Gmbh with a "MOC" nozzle perpendicular to the substrate. Spraying pressure and temperature were optimized as well as the powder flow rate, passing speed and passing step. Stand-off distance was kept at 80 mm from the substrate and all samples were sprayed in one pass only.

Coatings were cross-sectioned, polished, and metallized using a Cressington sputter coater to be observed with a Leica optical microscope. Metallization greatly enhanced the contrast between the various phases in the microstructure, which improved image analysis. Analyses were performed using ImageJ and mostly SMIL [14] in python scripts.

The electrical resistance of the coatings was checked with a simple digital ohmmeter. Coating samples for electrical conduction measurements were sprayed onto thinner (<1 mm) substrates of PEEK for their electrical behavior being investigated using the Van der Pauw method [15]. Contrary to CFRP, PEEK is insulating, which thus allowed to measure the conductivity of the coating only. These coatings were sprayed through a mask to obtain a cloverleaf shape with a diameter of 20 mm. This shape is optimal for the Van der Pauw method because it concentrates current lines in the middle of the samples, thus reducing measurement errors. The resistance of the coating were therefore measured in the various current ranges of 1 mA, 10 mA, 100 mA, 500 mA and 1 A in a HFS600E-PB4 Linkam system chamber with a Keithley 2450 Sourcemeter.

3 Results

Coatings were first optimized based on visual inspection of surface coating homogeneity and their surface mass. The first goal was to identify spraying parameters for the best coating build-up onto the substrate. Investigations were made on process gas parameters i.e. pressure and temperature. The particle surface flow rate seen by the substrate which depends on the passing speed, the passing step and the powder flow rate was also investigated.

3.1 Pressure and temperature influence

The influence of temperature and pressure of the carrier gas on coating build-up was investigated. All sprayings were performed with the same particle surface flow rate (i.e. using given passing step, speed and powder flow rate). Pressure and temperature ranged from 300 to 500 °C and 1.0 to 2.0 MPa, respectively. Coatings made of spherical copper particles tended to grow easily compared to those made of irregular particles. In the case of spherical copper particles, the optimized parameters seemed to be 1.5 MPa and 400 °C. In the case of irregular particle, thickest coatings were obtained at 1.5 MPa and 500 °C. However, these graphs do not show the coating homogeneity, **Fig. 2.** Under certain conditions, the coatings exhibited a high surface mass with many large defects at the surface such as craters or uncoated zones.

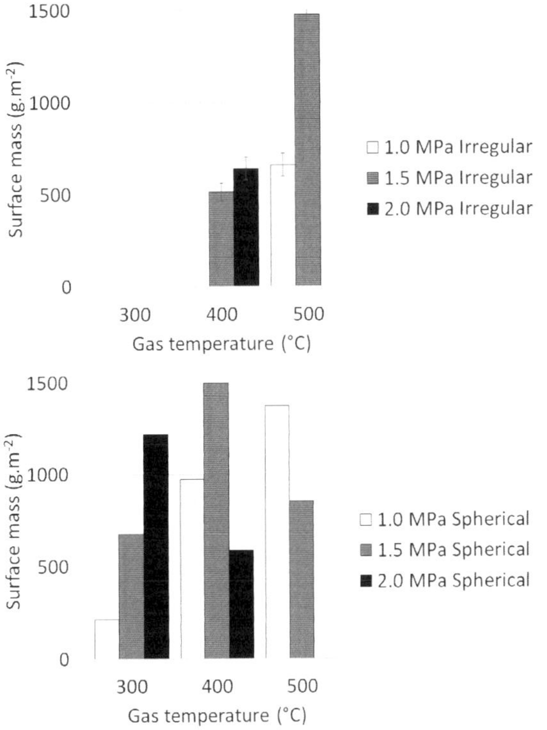

Fig. 2. Influence of the gas parameters on the surface mass of the coatings

3.2 Influence of particle surface flow rate

The influence of the particle surface flow rate was investigated with gas parameters kept at 1.5 MPa and 400 °C. The passing speed and step ranged from 100 to 300 $mm.s^{-1}$ and from 1 to 3 mm respectively. First, homogeneous coatings could be sprayed with a lower speed and a smaller step with non-optimal gas parameters, as shown on sample (a), **Fig. 3.**

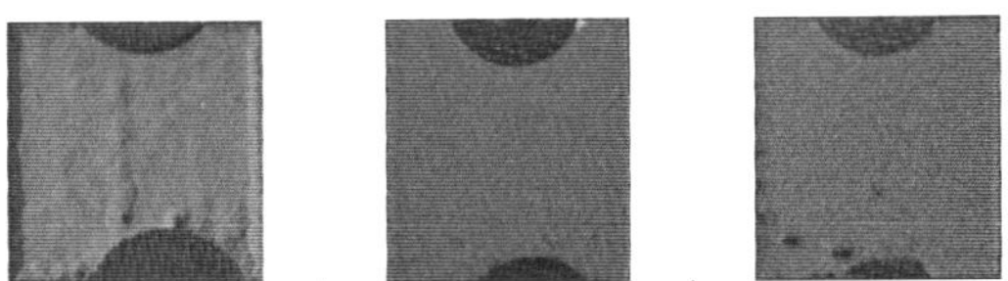

(a) Spherical copper particles	(b) Spherical copper particles	(c) Irregular copper particles
1.5 MPa, 300 °C, passing speed: 100 $mm.s^{-1}$	1.5 MPa, 400 °C, passing speed: 300 $mm.s^{-1}$	1.5 MPa, 400 °C, passing speed: 100 $mm.s^{-1}$

Fig. 3. Top views of coatings at different spraying parameters

Second, when using spherical copper particles, increasing the passing speed by 100 $mm.s^{-1}$ or the passing step by 1mm had exactly the same influence on the deposited mass. However, coating topography changed. Third, still with spherical particles, the particle surface flow rate and the gas parameters had an equal influence on coating build-up. It was possible to achieve thick and homogeneous coating by optimizing the temperature and pressure or the passing speed, step or flow rate. For irregular copper particles, the spraying window was narrower. Particle surface flow rate should be high enough to avoid surface defects but it led to thick coatings, i.e. with a high surface mass.

3.3 Coating microstructure

The microstructure of the coating showed two different phases. The PEEK powder was highly deformed, to result in a grey matrix which was surrounded by bright yellow copper particles. A third "phase" made of black holes was due to debonded copper particles. Spherical copper particles exhibited plastic deformation solely when impacting each other. They therefore mostly remained spherical, except on their impacted side.

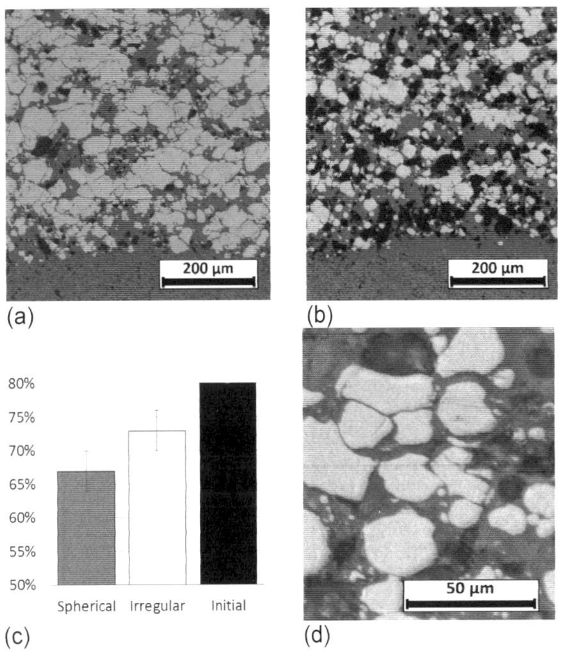

Fig. 4. Microstructure with (a) irregular particles, (b) spherical particles, (c) copper volume ratios in the coatings, (d) magnified view of (b)

Cross-section images were used for quantitative phase analysis. The surface fraction and density of copper along the vertical axis were investigated. In microstructures which involved spherical copper, the volume ratio of copper could be approximated by the surface ratio when averaged over a large number of images. The averages were computed with an automatic image segmentation method from 13 images of 2560x1920 $pixels^2$ with a resolution of 0.242 µm per pixel. The coatings made of irregular particles showed 73 vol.% of copper (a), and those made of spherical copper particles 67 vol.%, **Fig. 4** (b).

Powder shape could be responsible for the difference in copper ratio between the spherical and irregular particles. These measurements also showed that approximately 10 vol.% of copper were missing in the coating. These particles might have rebounded due to a lower velocity when impacting other copper particles. At a higher magnification, one could see that copper particles were well packed but often separated by a thin gray gap. It was believed that this gap was made of highly-deformed PEEK which surrounded the particles i.e. PEEK filled gap. This led to a microstructure which was slightly different from purely metallic coatings obtained by cold spray. PEEK prevented direct contact, diffusion and chemical bonding. Moreover, many deformed particles were still separated by PEEK even on the impacted side. The number of contacts between irregular particles was higher than that when using spherical particles.

3.4 Morphological model

The presence of PEEK filled gaps and the morphology of the microstructures could be assumed to govern the electrical conductivity of the coating. To simulate the current paths, a 3D morphological model was developed. The goal was to be able to investigate into the electrical behavior, from optimization of the metal phase morphology, and later from the selection of the best powder shape and size. The development of such model involved 3 steps:

- Segmentation of cross-sectional images to separate phases and remove visual artifacts or defects.
- Image analysis to extract image quantitative information according to 3 morphological criteria, termed correlation functions.
- Multi-scale morphological model development and optimization based on the 3 previously-mentioned morphological criteria extracted from experimental images.

The segmentation process resulted in two images: a binary image with copper particles in white and PEEK matrix in black, **Fig. 5.** (d). Debonded particles were added numerically to fill the holes. The second image showed labeled particles which allowed us to assess the efficiency of the process to segment neighboring particles, **Fig. 5.** (c). The main drawback in the segmentation process was that it could not segment every PEEK filled gap which separated close particles. The resolution and the color gradient were not good enough, so when the particles were too close, the process filled the gap with copper.

In the binary images three morphological criteria were considered to characterize the microstructure. The first criterion was the covariance, the probability that for a given point in the copper phase, if it was translated from a vector, it was still in the copper phase. This criterion corresponded to the spatial distribution of the copper numerically. The second criterion was the cumulative granulometry obtained by opening. This criterion described the size of copper phase. The third criterion related to the distribution of PEEK filled gap size. This criterion was approximate as the algorithm could not segment all the PEEK filled gap. This distribution appeared to follow a modified exponential distribution of parameters *k*.

Microstructures modeling involved two scales. The first scale was that of the copper particles. These were described by a Boolean model of spheres. The diameter of each sphere followed a gamma distribution of parameters *a* and λ. Boolean spheres were distributed spatially according to θ, the intensity of the Poisson's point process. Hence, the first scale involved three parameters which were calculated from the granulometry and the covariance through experimental images. Boolean spheres were distributed using an initial set of parameters in a 512x512xλ voxels space. The covariance and the granulometry were computed on a slice of this simulated microstructure. θ, *a* and λ were optimized using a "Nelder-Mead" algorithm to fit the experimental covariance and granulometry. The "Nelder-Mead" algorithm involved a simplex method, robust and useful in this case as there was no direct mathematical link between covariance, granulometry and these parameters. Moreover, to simplify the optimization process, a constitutive equation was derived from stereological formula linking θ, *a*, λ and the experimental surface ratio of copper.

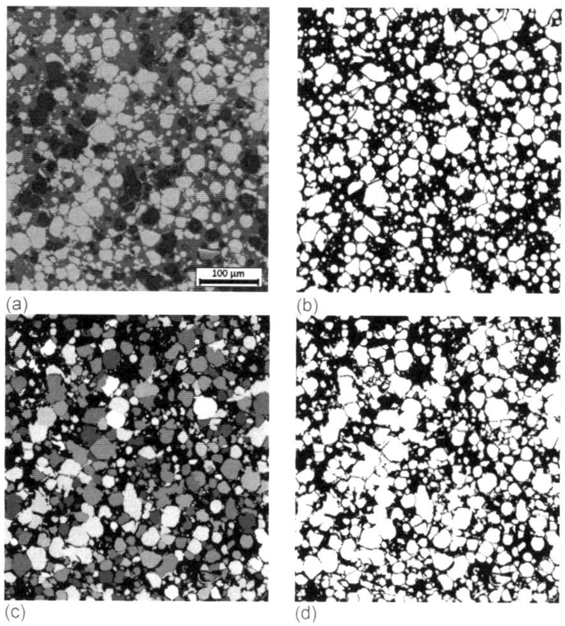

Fig. 5. (a) Original microstructure, (b) simulated microstructure, (c), (d) segmented microstructure

The second scale was that of the PEEK filled gap between neighboring particles. In the first scale, aggregates of particles were modeled by intersecting spheres, so most of them needed to be separated as in the real microstructure. In every aggregate of particles (connex spheres), germs were implanted in the center of the particles. On each germ a cell grew following a "Johnson-Mehl" tessellation. The cell nucleation and growth were time dependent. Cell boundaries were considered as PEEK filled gaps and were colored in black. They could therefore separate the particles from each other. The thickness of the boundaries was randomly computed following the exponential distribution which fitted the PEEK filled gap size distribution. Hence a simulated microstructure could be computed according to the three morphological criteria, which met the experimental microstructure, **Fig. 5** (b).

3.5 Electrical behavior of the coating

Resistance measurements showed that coatings made of spherical copper hardly conducted electricity. In contrast, when using the irregular shaped copper, one could measure a resistance which felt down to below
1 Ω, using a mere portable ohmmeter. However, the conductivity of these coatings needed to be investigated thoroughly using an insulating substrate. Coatings sprayed onto PEEK substrate were first cross-sectioned to check whether the microstructure and the copper ratio were identical to those sprayed onto CFRP. Copper ratios were measured using the previously-described segmentation method. The difference was small enough to ascertain the microstructures were similar, **Fig. 6.**

Substrate	Final average surface ratio of irregular copper	Standard deviation
CFRP	73%	3%
PEEK	68%	4%

Fig. 6. Copper ratio in coatings onto two substrates

Two thicknesses for the 80/20 coating were sprayed. Increasing the passing speed from 100 to 200 mm.s^{-1} decreased the mean thickness from 781 µm to 358 µm. Decreasing thickness also increased the number of surface defects, which consequently reduced the conductivity. The conductivity was determined from four measurements, along each side of the cloverleaf-like sample. Some slight anisotropy was showed from vertical and horizontal measurements, which might be due to defects, incorrect probe positioning or sample dissymmetry. When increasing the current intensity, a slight resistance drop was observed. Finally, regardless of thickness, under a 1 A current there was a quick damage of the coating which could burn and corrode. This left current line marks at the surface. Intensity was high enough to burn PEEK and heat up the whole sample.
Conductivity appeared to depend on coating thickness with a drop of 1500 S.m^{-1} when passing from 781 µm to 358 µm. This behavior was questionable but one can assume that the thicker the coating was, the more complex it was. This involved the paths between copper particles from one side to the other, which led to an electrical loss. At 1 A, the coating quickly burnt, which decreased the measured resistance. The presented value was not therefore obtained in an equilibrium state.
Conductivities could be compared with that obtained on similar samples of pure cold spray copper onto PEEK. The conductivity of the composite coatings was shown to be three decades below the conductivity of pure copper coatings, which, incidentally could reach the bulk copper conductivity i.e. 6.e7 S.m^{-1}, **Fig. 7.**

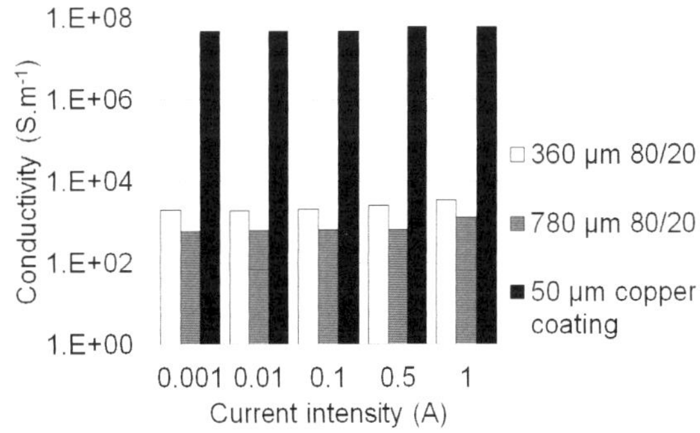

Fig. 7. Coatings conductivity

4 Discussion

4.1 Coating build-up

First, while investigating the influence of cold spray parameters, the goal was also to control the thickness and the homogeneity of the coating to obtain low-thickness, homogeneous and conductive coatings. The influence of gas pressure and temperature was studied. The optimum spraying window was shown to involve a low pressure with a rather high temperature. Moreover, the particle surface flow rate was seen to have a similar or higher influence on coating build-up when compared to that of gas parameters. A typical spraying pressure for pure copper is 2.0 MPa. One may therefore assume that the critical velocity of copper does not govern the build-up and homogeneity of the coating. These are mainly due to the presence of PEEK. This leaves bonding between deformed copper particles as a second order mechanism. This is also confirmed by the fact that the lower the pressure and the higher the temperature were, the thicker the coating was.
Second, coating build-up highly depends on the particles surface flow which is a combination of the passing speed, passing step width and powder flow rate. For example, when spraying irregular powder changing the passing step by 1mm or the passing speed by 100 mm.s^{-1} led to a similar surface mass evolution. On the other hand, this result was not valid for the spherical powder. A thorough study was carried out to compare the behavior of the two copper powders, leading to a qualitative model of the composite coating build-up. The build-up exhibits a similar non-linear behavior between the two copper powders. This behavior was particularly exhibited when reducing the particle surface flow rate with the

two powders to result in a low-thickness (as low as possible but homogeneous) coating. With the spherical powder, a 100 µm thick and homogeneous coating could be achieved, which however remained insulating. Cross-section observations showed that the coatings contained in fact only scarce copper particles. With the irregular copper, reducing the particle surface flow rate led to some large heterogeneity with many surface defects. This could be explained by the size of the particles.

Coating build-up is believed to be a two-stage process, which depends on the type of particle (fine spherical or coarse irregular), **Fig. 8.**

- Fine copper particles impact the substrate surface and rebound due to the absence of PEEK, coarse PEEK particles impact the surface at the same time as for copper particles. They adhere to the fibers, while copper particles are embedded in larger PEEK particles.
- Copper particles impact the previously-sprayed layer, deformed on copper or are embedded into PEEK.

- Coarse particles impact the substrate surface and rebound. They cannot be embedded into PEEK because of their size.
- Once a few amount of PEEK adheres to the surface with embedded smaller copper particles, larger particles finally impact, stacks or rebound. Their irregular shape can limit rebound and promotes anchoring in the already-deposited layers.

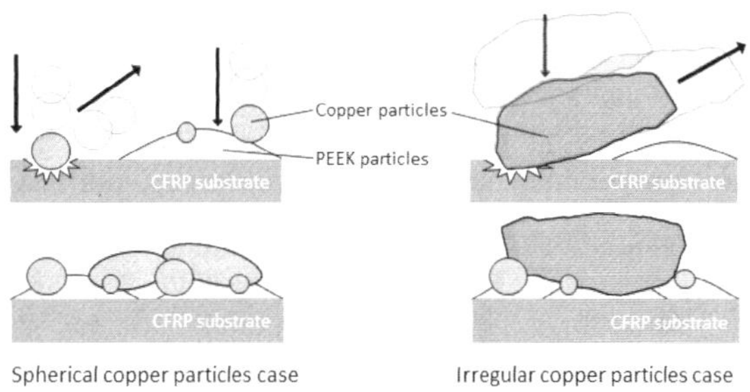

Fig. 8. Schematic illustration of the two-stage build-up process for two types of copper shape

This build-up process would explain why it was possible to obtain low-thickness coatings with spherical particles. At a lower particle surface flow rate only the first layer i.e. a sort of bond coat was sprayed. This corresponds to the first stage of the process. This layer does not contain enough copper particles to be conductive. With coarser particles sprayed at a low particle surface flow rate this type of layer cannot be obtained, since it produces a rough surface with defects. This highlights the role of the particle surface flow rate, which therefore controls PEEK deposition at the surface, allowing the coating to build-up.

The key phenomenon leading to this type of microstructure remains unknown. More precisely, the question is to know when does copper is embeded into PEEK. Is it during its flight through the nozzle or at the impact?

4.2 Electrical behavior and microstructure

The study of the microstructure led to conclude first that a significant amount of copper is lost during spraying, one may assume due to rebounds. Second, even at a high copper ratio, the metallic phase formed by the spherical particles does not percolate. The particles remain separated by thin layer of PEEK. This phenomenon seems to be unavoidable. Image analysis showed that most spherical particles were non connex, while many irregular ones were. An original morphological model was therefore developed to investigate various particle morphologies as the PEEK filled gap size distribution was approximately the same for the two types of microstructures. A higher number of contacts between irregular copper particles was assumed to govern the coating conductivity.

5 Conclusion

An original method was developed to coat Carbon Fiber-Reinforced Polymer by cold spray using polymer mixed with copper powders to result in a conductive layer. The resulting composite coatings were optimized through the study of the deposited surface mass depending on process parameters, copper particles shape and size. Interpretations of the phenomenon leading to the build-up were proposed. A morphological analysis method was developed to investigate into coating microstructures. The extraction of relevant data on the copper phase fed a morphological model of the coating microstructure. Finally, the Van der Pauw method was applied on geometrically optimized samples to assess the conductivity of the coatings. It was shown that coatings made of irregular copper particles could reach 1e3 $S.m^{-1}$ whereas those containing spherical particles remained insulating. These phenomena were mostly explained using the developed simulation tools coupled with investigation of microstructures and coating build-up.

6 Outlook

This work highlighted the interest in composite cold-sprayed coatings to achieve multi-property materials. A further goal would be to achieve gradient composite coatings in cold spray e.g. using multiple powder feeders. In this study, the polymer was shown to be an efficient coating binder. Moreover, its combination with rigid materials allowed to achieve dense coatings. Its mechanical behavior in cold spray should require

to be further studies, with mechanical simulation and/or in-situ characterization during spraying.
Last but not least, morphological models can be expected to be developed as powerful tools to optimize coatings properties which depend on particle morphology. Laborious and systematic spraying campaign could thus be avoided.

7 Acknowledgments

This work was carried out within the "C.O.MET" FUI program. The authors would like to thank all the members of the program consortium and BPI France, Pôle Astech, Pôle de la céramique, Pôle des microtechniques and DGAC for financial support.

8 References

[1] Sturgeon, A., B. Dunn, S. Celotto, and W. O'Neill: Cold Sprayed Coatings for Polymer Composite Substrate. ESA SP, 616 (2006).

[2] Zhou, X. L., A. F. Chen, J. C. Liu, X. K. Wu, and J. S. Zhang: Preparation of metallic coatings on polymer matrix composites by cold spray. Surface and Coatings Technology 206, no. 1 (2011): 132-136.

[3] Affi, J., H. Okazaki, M. Yamada, and M. Fukumoto: Fabrication of aluminum coating onto CFRP substrate by cold spray. Materials Transactions 9, no. 52 (2011): 1759-1763.

[4] Che, H. Q., and S. Yue: Cold Spray of Carbon Fiber Reinforced Polymer for Lightning Strike Protection. International Thermal Spray Conference. Barcelona: DVS, (2014).

[5] Ganesan, A., M. Yamada, and M. Fukumoto: Cold spray coating deposition mechanism on the thermoplastic and thermosetting polymer substrates. Journal of thermal spray technology 22, no. 8 (2013): 1275-1282.

[6] Lupoi, R., and W. O'Neill: Deposition of metallic coatings on polymer surfaces using cold spray. Surface and Coatings Technology 7, no. 205 (2010): 2167-2173.

[7] Giraud, D., F. Borit, V. Guipont, M. Jeandin, and J. M. Malhaire: Metallization of a Polymer using Cold Spray: Application to Aluminum Coating of Polyamide 66. International Thermal Spray Conference. DVS, (2012).

[8] Papyrin, A., Kosarev, V., Klinkov, S., Alkhimov, A. and Fomin, V. M.: Cold spray technology, Elsevier, (2006)

[9] Jeandin, M., F. Borit, V. Guipont, L. Decker, D. Jeulin, M. Suzuki, and S. Sodeoka: Lattice gas modelling in thermal spraying. Surface engineering 15, no. 3 (1999): 191-194.

[10] Amsellem, O., K. Madi, F. Borit, D. Jeulin, V. Guipont, M. Jeandin, E. Boller, and F. Pauchet: Two-dimensional (2D) and three-dimensional (3D) analyses of plasma-sprayed alumina microstructures for finite-element simulation of Young's modulus. Journal of Materials Science 43, no. 12 (2008): 4091-4098.

[11] Beauvais, S., V. Guipont, M. Jeandin, D. Jeulin, A. Robisson, and R. Saenger: Study of the porosity in plasma-sprayed alumina through an innovative three-dimensional simulation of the coating buildup. Metallurgical and Materials Transactions A 11, no. 39 (2008): 2711-2724.

[12] Bobzin, K., N. Kopp, T. Warda, and M. Öte: Determination of the effective properties of thermal spray coatings using 2d and 3d models. Journal of thermal spray technology 6, no. 21 (2012): 1269-1277.

[13] Delloro, F., M. Faessel, H. Proudhon, D. Jeulin, M. Jeandin, E. Meillot, and L. Blanchi: A morphological approach to the modelling of the cold spray process. International Thermal Spray Conference. Barcelona: DVS, (2014): 221-226.

[14] Faessel, M., and M. Bilodeau: Smil simple morphological image library. Séminaire Performance et Généricité. LRDE, (2014).

[15] Van der Pauw, L. J.: A method of measuring specific resistivity and hall effect of discs of arbitrary shape. Philips Res. Rep. 13(1958): 1-9.

Effects of in-flight change of properties on suspension plasma spraying

Amirsaman Farrokhpanah, Thomas W. Coyle, Javad Mostaghimi

A numerical study of suspension plasma spraying is presented in the current work. Atomization processes, evaporation of the fluid phase, and melting of the solid phase are included in the model. Detailed attention has also been given to the changes in physical properties of the suspension mixture as the liquid phase is evaporating. These include phenomena such as changes in heat transfer properties and changes in viscosity. Utilizing this improved model, several test cases have been studied to better evaluate effect of parameters on the final deposited surface.

1 Introduction

Using suspensions in thermal plasma coating processes has become popular nowadays. Fine ceramic powders can be used to create stable suspensions in fluids like water or ethanol. These suspensions are injected into a plasma jet. The heat from plasma will evaporate the carrier fluid and eventually melt the solid content. The suspension droplets go through several breakups before evaporation is complete. A detailed understanding of how these parameters affect the properties of the deposited coating is of great importance. Physical properties of the suspension, operating conditions of the plasma torch, along with injection parameters, i.e., position, angle, and speed, play an important role on how efficiently the fluid is evaporated and final molten drops reach the surface.

2 Effective viscosity

In suspension plasma spraying, the liquid content in the suspension is evaporating completely. This means viscosity of each drop, from injection to impacting the substrate, goes through drastic changes as concentration of solid matter varies from being dilute to highly concentrated. Here also a sample numerical example has been conducted in section 6.2 which emphasizes effects of viscosity on breakup regimes of suspension droplets.

Many studies have used analytical and empirical tools to find estimate values of suspension viscosity (μ), for different concentrations and conditions. Einstein's viscosity model is among the very first that predicts viscosity for a dilute suspension of spherical particles using [1]

$$\mu_r = 1 + 2.5\phi \quad (1)$$

With ϕ being the volume fraction for solid particles in the suspension (V_{solid}/V_{total}). μ_r, relative viscosity, is the ratio of effective viscosity of the suspension to the pure fluid viscosity, shown as

$$\mu_r = \frac{\mu_{eff}}{\mu_{liq}} \quad (2)$$

For higher concentrations, this equation was modified by Guth et al. [2] to take interaction between solid particles into account in the form of

$$\mu_r = 1 + 2.5\phi + 14.1\phi^2 \quad (3)$$

Using empirical data, Thomas enhanced this model to form [3]

$$\mu_r = 1 + 2.5\phi + 10.05\phi^2 + A\, e^{B\phi} \quad (4)$$

Where A= 0.00273 and B=16.6. For large particles at high concentrations, by taking effects of particle interactions into account, Toda el al. [4] proposed using

$$\mu_r = \frac{1 + 0.5k\phi - \phi}{(1 - k\phi)^2(1 - \phi)} \quad (5)$$

Where k is a parameter obtained using empirical results and has been reported to be in the form of $k = 1 + 0.6\phi$ for spherical yeast particles in water ($5\mu m$ radius).

Models mentioned above do not take effects of maximum packing fraction, ϕ_m, into account. ϕ_m is the largest possible volume fraction that can be achieved by adding solid particles to the suspension. Physical shape of particles along with their electric charge can affect the value of ϕ_m. Since suspension viscosity converges to infinity for solid concentrations close to ϕ_m [5], it is important to include this parameter into the model. A couple of empirical correlations have included effects of ϕ_m. Krieger et al. [6] proposed that suspension viscosity can be estimated using

$$\mu_r = \left(1 - \frac{\phi}{\phi_m}\right)^{-[\mu]\phi_m} \quad (6)$$

Here, $[\mu]$ is the intrinsic viscosity which is defined as the limiting value of $(\mu_r - 1)/\phi$ at low concentrations which dominates suspension behaviour at low concentrations [7] and is given by [8]

$$[\mu] = \lim_{\phi \to 0} \lim_{\dot{\gamma} \to 0} (\mu_r - 1/\phi) \quad (7)$$

$[\mu]$ has a value of 2.5 according to Einstein's model in equation 1.

Dabak et al. [9] also proposed a model containing another empirically adjustable variable, N, given as

$$\mu_r = \left[1 + \frac{[\mu]\phi}{N\left(1 - \frac{\phi}{\phi_m}\right)}\right]^N \quad 8$$

N here is a flow related variable and is taken to be 2 for high shear rates [10]. More recently, Senapati et al. [11] has proposed an improved model that takes effects of particles size distribution along with shear rate into account

$$\mu_r = S\left[1 + \frac{[\mu]}{\dot{\gamma}^n}\left(\frac{\phi}{\phi_m - \phi}\right)\right]^N \quad 9$$

Where $\dot{\gamma}$ is the shear rate, and n is the flow behavior index in the power law equation ($\tau = K\dot{\gamma}^n$). Parameter S is calculated from

$$S = 10\frac{C_U}{d50} \quad 10$$

Where C_U is the coefficient of uniformity calculated from

$$C_U = \frac{d60}{d10} \quad 11$$

Testing models mentioned above over a wide range of available viscosity measurements for ceramic suspensions proves lack of applicability in many cases. As a remedy, more recently, Horri et al. [12] developed an empirical model specific for ceramic suspensions for a wide range of concentrations and shear rates in the form of

$$\mu_r = 1 + 2.5\phi + K\phi\left(\frac{\phi}{\phi_m - \phi}\right)^2 \quad 12$$

Where parameter K is determined using experimental data and varies for different shear rates.

For NiO/YSZ suspension ($\rho_s = 6.1\,g/cm^3$, $\rho_l =$ *1.135gcm3, μl=5.224 mPa·s* which is chosen with regard to available experimental data of Horri et al. [12], parameter K in equation 12 can be estimated as a function of shear rate using ($R^2 = 0.9975$)

$$K = 113.1746 + \frac{4380174000 - 113.1746}{1 + (\dot{\gamma}/0.004969429)^{1.719769}} \quad 13$$

Parameter ϕ_m can also be interpolated using ($R^2 = 0.9952$)

$$\phi_m = 0.5645118 + \frac{377069.9 - 0.5645118}{1 + (\dot{\gamma}/0.001378668)^{1.495299}} \quad 14$$

Interpolations for viscosity, like the ones mentioned above, typically generate close results to the experimental data for small values of concentration. Close to ϕ_m however, as concentration should converge to infinity, results of interpolations might deviate from experimental data. These interpolations remain valid near ϕ_m as long as predicted viscosity values become very large (not necessarily infinity) and grow with a rate that is close to the rate of growth of experimental results.

For 8 mol% YSZ-water suspension using experimental results of Arevalo-Quintero et al. [13] (as plotted in **Fig. 1**, $\rho_s = 5.54\,g/cm^3$, $\rho_l = 0.997\,g/cm^3$, $\mu_l = 0.89mPa \cdot s$), under a constant shear rate with $\phi_m = 21.76$, parameter K can be estimated to be $K = 5.660117$ ($R^2 = 0.9598$).

The same procedure is applied to experimental results of Song et al. [14], which were conducted at a constant shear rate of $100\,s^{-1}$ for ceramic suspensions defined in Table 1. Fitted K values for these suspensions can be seen in Table 2.

As equation 12 produces good fits to experimental measurements of ceramic suspensions' viscosities, it has been used in this study for prediction of ceramic viscosities.

Table 1. Ceramic suspensions used in experiments of Song et al. *[14]*

Symbol	Solid Content	% wt Y_2O_3	Binder
ZSH	$Z3$	4.5-5.0	Wax-Stearic acid (13:7)
ZKH	$Z3$	4.5-5.0	Wax-KD5 (13:7)
PSH	$PSZ - Y5$	5.0	Wax-Stearic acid (13:7)
PKH	$PSZ - Y5$	5.0	Wax-KD5 (13:7)

Table 2. Fitted K values for different ceramic suspensions based on experimental measurements of viscosity

Ceramic Suspensions	K	R^2	ϕ_m	Source
YSZ	5.660117	0.9850	0.21	[13]
ZSH	651.2335	0.9972	0.59	[14]
ZKH	171.0134	0.9780	0.66	
PSH	119.3539	0.9930	0.66	
PKH	32.99419	0.8744	0.68	

3 Maximum packing fraction (ϕ_m)

Most correlations used for viscosity approximation rely heavily on a good approximation of the maximum packing fraction. Two different methods of finding ϕ_m are reviewed here. First method is applicable when experimental results for suspension viscosity at high concentrations is available. Where experimental data is missing, analytical models that follow can be used instead.

3.1 ϕ_m calculation from emperical data

For the case where experimental data for suspension viscosity is available, the method explained by

Senapati et al. [11] can be used to approximate ϕ_m. Experimental results of Arevalo-Quintero et al. [13] for viscosity of YSZ suspension in different solid contents (**Fig. 1**) is used here as an example. The suspension used in their experiment was prepared using water and 8 mol% YSZ powder with d10, d50, and d90 values close to 0.8 μm, 2.6 μm, and 6.3 μm, respectively.

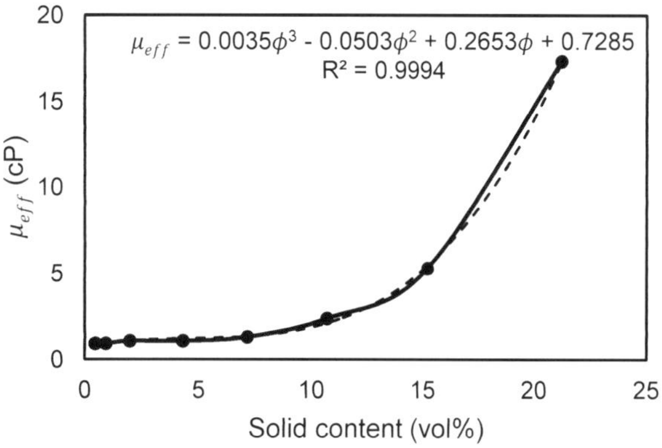

Fig. 1. Viscosity of YSZ-Water suspension under different concentrations of YSZ powder (from *[13]*). Dashed line is the trend line.

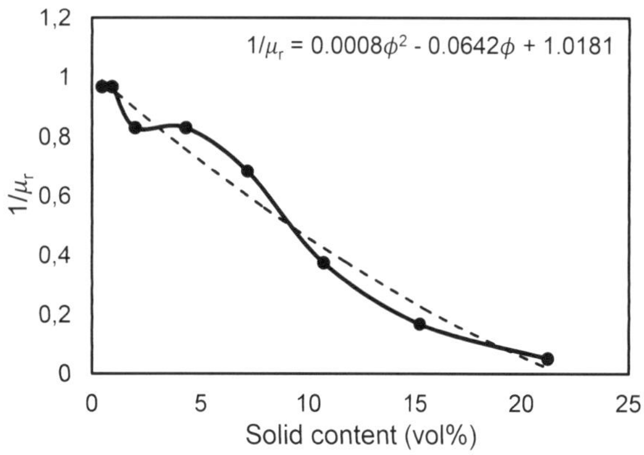

Fig. 2. $1/\mu_r$ for the YSZ-water suspension. Dashed line is the trend line.

The limiting value of $1/\mu_r$ when it goes to zero corresponds to ϕ_m. Using values presented in **Fig. *1*** for viscosity of YSZ suspensions, $1/\mu_r$ can be calculated as shown in **Fig. *2***. Using a trend line, or trial and error, the place for which $1/\mu_r$ has a very small value can be found here as $\phi_m = 21.76$.
Parameter N, appearing in equations 8 and 9, is also a suspension dependant parameter and needs to be calculated from experimental data. As indicated by Senapati et al. [11], this can be done using $(1-\mu_r)^{-1/N}$-ϕ curve. On this chart, which is shown in **Fig. *3***, the value of ϕ corresponding to upper limit of $(1-\mu_r)^{-1/N}=1$ should be identical to $\phi_m = 21.76$ which was calculated earlier. By taking N = 1, ϕ = 21.76 leads to $(1-\mu_r)^{-1/N}=1$, which is the same as ϕ_m.

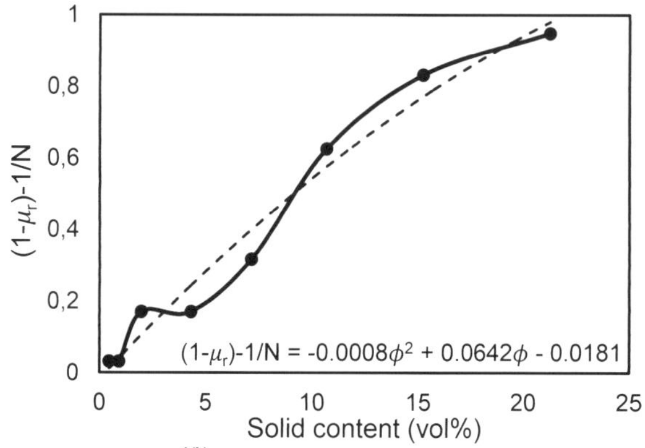

Fig. 3. $(1-\mu_r)^{-1/N}$ for the YSZ-water suspension. Dashed line is the trend line

3.2 Analytical-Empirical correlation for ϕ_m

When detailed experimental data for suspension viscosity is not available, empirically calibrated correlations can be used. With the assumption of having spherical particles with uniform diameters, Lee [15] developed an analytical model for approximation of packing of n-component mixtures. This was done by superimposing idealized packing characteristics of binary mixtures. Patton [16] later improved the model by adding effects of particle size distribution. This model assumes suspension contains three classes of particle sizes only: small (s), intermediate (i), and large (l), with volume fractions of v_s, v_i, and v_l, respectively ($v_s + v_i + v_l = 1$). In this manner, three maximum packaging concentrations for each size are calculated as

$$(\phi_m)_s = 0.639v_s + \phi_{si}v_i + \phi_{sl}(1 - v_s - v_i) \quad 15$$

$$(\phi_m)_i = \phi_{is}v_s + 0.639v_i + \phi_{il}(1 - v_s - v_i) \quad 16$$

$$(\phi_m)_l = \phi_{ls}v_s + \phi_{li}v_i + 0.639(1 - v_s - v_i) \quad 17$$

Coefficients ϕ_{si}, ϕ_{sl}, ϕ_{is}, ϕ_{il}, ϕ_{ls}, and ϕ_{li} take pairwise effects of particle size classes into account and can be obtained from

$$\phi_{ab} = 0.639 + \frac{\phi_m^* - 0.639}{1.15 - 1.017\phi_m^*} \quad ,a > b \quad 18$$

$$\phi_{ab} = 0.639 + \frac{\phi_m^* - 0.639}{1.017\phi_m^* - 0.15} \quad ,a < b \quad 19$$

Where subscripts a and b can take values of s, i, and l. The parameter ϕ_m^* is the maximum concentration for binary mixtures and can be determined using **Fig. *4***.

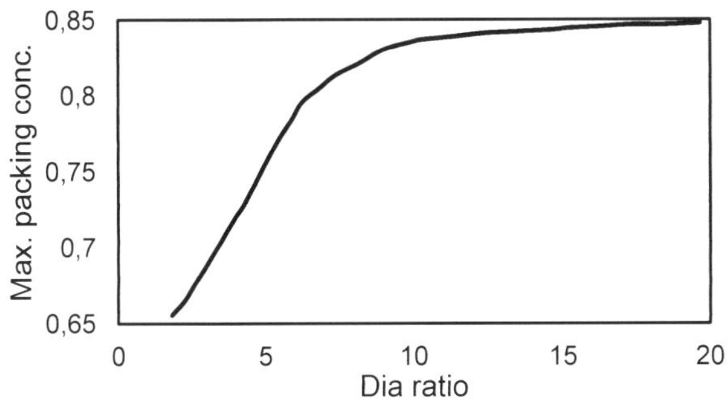

Fig. 4. Maximum concentration for binary mixtures spherical particles

The minimum of the three values obtained from equations 15-17 is the final packing factor. Dabak et al. [9] further improved this approximation by comparing results of these equations with empirical measurements. An empirically derived correction which includes effects of surface condition and effective particle shapes was introduced in the form of

$$\phi_m = \frac{1.91}{\left([\mu]\frac{N}{N_\infty}\min(\,(\phi_m)_s, (\phi_m)_i, (\phi_m)_l\,)\right)^{0.94}} \tag{20}$$

Where N and N_∞ are flow related parameters appearing in equation 8 for low and high shear rates, respectively.

4 Other physical properties

Other physical properties such as effective heat capacity and surface tension also undergo various changes during the evaporation of the liquid phase. Especially near the maximum packing factor, heat transfer mechanisms might change based on whether a shell is formed during phase change, or solid suspension particles become aggregated. The impact of these parameters is beyond the scope of current study.

5 Suspension Flight

5.1 Momentum

Plasma gas with velocity of u accelerates particles inside the domain. Particle acceleration is calculated in a Lagrangian manner using force balance between forces acting on particle and inertia in the form of [17]

$$\frac{du_p}{dt} = F_D(u - u_p) + F \tag{21}$$

With the drag force being in the form of

$$F_D = \frac{18\mu}{\rho_p d_p^2}\frac{C_D Re}{24} \tag{22}$$

Drag coefficient, is updated here based on instantaneous Re of each particle. Drag coefficient is chosen to be

$$C_{Df} = \frac{24}{Re} \tag{23}$$

for $Re < 0.01$,

$$C_{Df} = \frac{24}{Re} + 3.156\, Re^{-(0.18+0.05\log Re)} \tag{24}$$

for $0.01 < Re < 0.2$, and

$$C_{Df} = \frac{24}{Re} + \frac{6}{1+\sqrt{Re}} + 0.4 \tag{25}$$

for the range $0.2 < Re < 10^5$ based on empirical correlations for small spherical particle sizes $(5 - 100\mu m)$ [18]. Properties appearing in this equation are evaluated at particle's wall/film temperature, $(T + T_p)/2$.

For particles in moving in plasma flow, this drag coefficient needs to be modified to take effects of variable properties due to temperature gradient and Noncontinuum (Knudsen effects) into account. Influence of Basset history term is neglected for particles smaller than $100\mu m$ [19]. Effect of variable property is added using proposed method of Lee et al. [20]. Noncontinuum effects also become important for Knudsen number regime $10^{-2} < Kn < 1$ [19]. This is also considered using correction terms of Chyou et al. [21] and Chen et al. [22]. A superposition of these corrections, as proposed by Pfender [19], can be used to calculate the final drag coefficient in the form of

$$C_D = C_{Df}.f_1.f_2 \tag{26}$$

Where correction factors for strong variation of properties and noncontinuum effects are

$$f_1 = \left(\frac{\rho_\infty\mu_\infty}{\rho_w\mu_w}\right)^{-0.45}$$
$$f_2 = \left[1 + \left(\frac{2-a}{a}\right)\left(\frac{\gamma}{1+\gamma}\right)\frac{4}{Pr_w}Kn^*\right]^{-0.45} \tag{27}$$

Where the subscript w shows properties are calculated at particle's wall temperature, $(T + T_p)/2$. a is also the thermal accommodation coefficient, γ is the heat capacity ratio, and $\mathrm{Pr_w}$ is the gas Prandtl number evaluated at wall temperature. Here, Kn^* is the Knudsen number based on using effective molecular mean free path (λ_{eff}) and particle diameter (d_p). λ_{eff} is calculated from

$$\lambda_{eff} = \frac{2\bar{K}}{\rho_w\bar{V_w}\bar{C_p}}Pr_w \tag{28}$$

With $\overline{V_w}$, $\overline{C_p}$, and $\overline{K}$ being average thermal velocity, specific heat and thermal conductivity, respectively.
F in equation 21 takes thermophoretic effects into account. Thermophoretic forces are exerted on small particles travelling in gas with temperature gradient and are in the opposite direction to this gradient [17]. This force is added using proposed method of Talbot et al. [23]

$$F = -\frac{6\pi d_p \mu^2 C_s (K^* + C_t Kn)}{\rho(1 + 3C_m Kn)(1 + 2K^* + 2C_t Kn)} \frac{1}{m_p T} \frac{\partial T}{\partial x} \qquad 29$$

Where K^* is the ratio of fluid to particle thermal conductivities, m_p is the mass of particle, $C_s = 1.17$, $C_t = 2.18$, and $C_m = 1.14$. Brownian Forces are neglected since there are not many sub-micron particles involved in test cases here.

5.2 Heat transfer

Particle heating and cooling due to convection at particle surface is governed by

$$m_p C_p \frac{dT_p}{dt} = hA_p(T_\infty - T_p) \qquad 30$$

The heat transfer coefficient here is calculated with the help of Nusselt number. Based on Ranz-Marshal correlation,

$$Nu_f = 2.0 + 0.6 Re_d^{1/2} Pr^{1/3} \qquad 31$$

Similar to momentum calculations before, effects of temperature gradient and Knudsen number is included here using correction factors. Final Nusselt number is then corrected as [22, 24]

$$Nu = \frac{hd_p}{k_\infty} = Nu_f . f_{h1} . f_{h2} \qquad 32$$

Where $f_{h1} . f_{h2} = (f_1 . f_2)^{1/0.45}$. As the solvent inside particles evaporates, heat transfer mechanisms changes. If initial suspension has low ceramic concentration, a shell of molten ceramic might form imprisoning solvent inside [25]. Solvent may also evaporate completely and cloud of ceramic particles may get scattered as they melt together to form drops. Taking all these effects into account is nearly impossible due to lack of sufficient empirical correlations. Hence, it is reasonable to make simplifying assumptions.
The critical point for the suspension drop during evaporation is when concentration inside the drop reaches ϕ_m. This means that all the fluid surrounding the solid content has evaporated and solid particles are packed together at this point. The rest of solvent has to be evaporated from the liquid inside this porous and densely packed ball of solid particles. It is hence reasonable to assume from this point on that the diameter of the drop does not decrease any further as evaporation will not cause the drop to shrink any more. Surface roughness due to existence of particles is neglected as the size of each solid particle is comparably smaller than the size of the drop.

5.3 Phase change

In order to have a general inclusion of the phase change process, viscosity values are adjusted in the absence of liquid phase. This means that when the fraction of liquid content in each drop becomes negligible, the viscosity for the mixture, which now consists of only solid particles, is set to a large value. This is schematically demonstrated in **Fig. 5**. After the liquid content in drops is gone and before reaching the melting temperature of the YSZ, viscosity is chosen to have a large value (here, 1). In a 20 degree window around the melting temperature, the viscosity is chosen to be linearly dropping from this large value to its value for molten YSZ. This will also be an estimate for the mushy zone behaviour of the melting process.

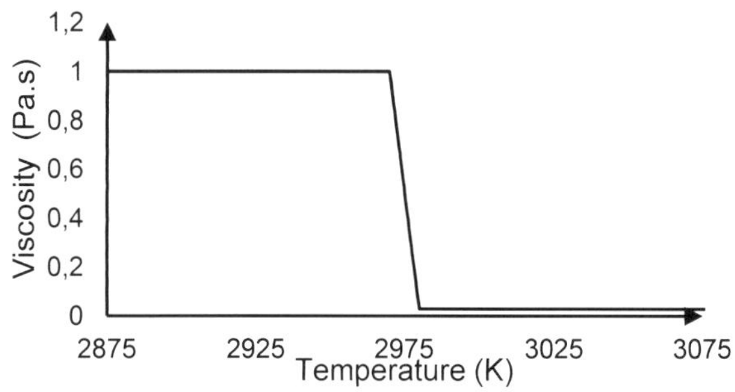

Fig. 5. Mushy zone effect on viscosity values of YSZ.

The latent heat consumed during melting process is also included by modifying C_p values of YSZ. This is done using

$$C = \begin{cases} C_s & T \leq T_1 \\ C_m + \dfrac{L}{T_2 - T_1} & T_1 < T < T_2 \\ C_l & T \geq T_2 \end{cases} \qquad 33$$

Where T_1 and T_2 mark the temperature range over which the melting occurs, and L, is the latent heat. For simplicity, C_m is chosen to be $(C_s + C_l)/2$ here. This also demonstrated in **Fig. 6**. for YSZ suspension.

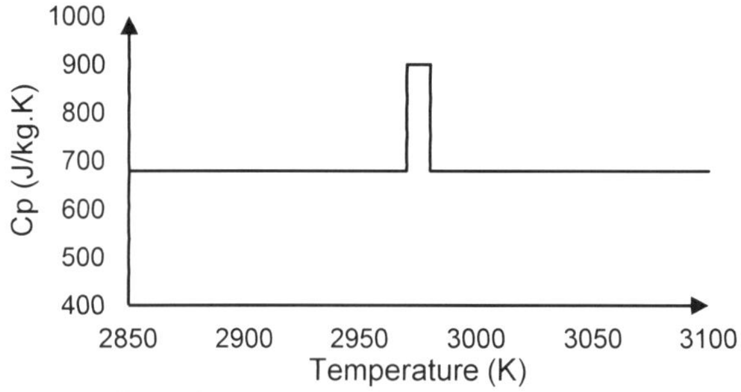

Fig. 6. Effect of latent heat of melting on C_p values.

5.4 Breakup model

Direct simulation of breakup process demands a very fine mesh resolution and also a fluid surface tracking model like VOF. Implementing these into a largescale simulation like the one studied here will make solution computationally expensive. To overcome that, suspension is initially injected into the domain in the form of drops. This means the primary breakup has already taken place. For the secondary breakup and the rest, KHRT (Kelvin-Helmholtz/Rayleigh-Taylor) model [26, 27] is used. This method, applicable to high Weber number flows [17], combines effect of aerodynamic forces which create Kelvin-Helmholtz waves along with Rayleigh-Taylor instabilities caused by acceleration of drops into the free stream. These mechanisms capture breakup using drop's surface wave growth: fastest growing instability will cause the drop to breakup.

For the suspension study here, the viscosity is updated at each iteration based on drop's concentration. Breakup model is also influenced by this change of viscosity. Breakup however is stopped when concentration reaches ϕ_m. Aggregate explosions are neglected here, hence no breakup occurs till all the solvent is evaporated and all solid content has melted. When a molten drop is formed from solid material, breakup is resumed.

6 Results and Discussion

6.1 Torch Model Validation

For the test cases here, as is demonstrated in **Fig. 7**, the plasma torch is simplified to a jet of hot fluid entering the domain. Suspension parcels are then injected into the stream in order to have evaporation, breakup, and melting. The reactions and arc influences are also simplified by adding a volumetric heat source inside the torch. Knowing torch specifications like current (I), voltage (V), and thermal efficiency (η), this volumetric heat source can be estimated as

$$\dot{Q} = \eta \frac{EI}{\Omega} \qquad 34$$

Where Ω is the volume of the anode. It is then necessary to benchmark the torch performance to make sure the simplifying assumptions generate acceptable results.

In order to avoid having a long torch entrance length, a fully developed velocity profile in the form of the power-law,

$$u(r) = u_{max} \left(\frac{r}{R}\right)^{1/n} \qquad 35$$

is taken at the Ar-H_2 inlet.

The torch modeled here has a mass flux of 35.4 slpm, using Argon with 0.1 mole H_2. Anode diameter is 6mm and gas flows in with turbulent intensity of 4.5%. Flow passing through anode is heated with a heat source to correspond to a torch with 0.47 thermal efficiency, 25.6 (V), and 600 (A).

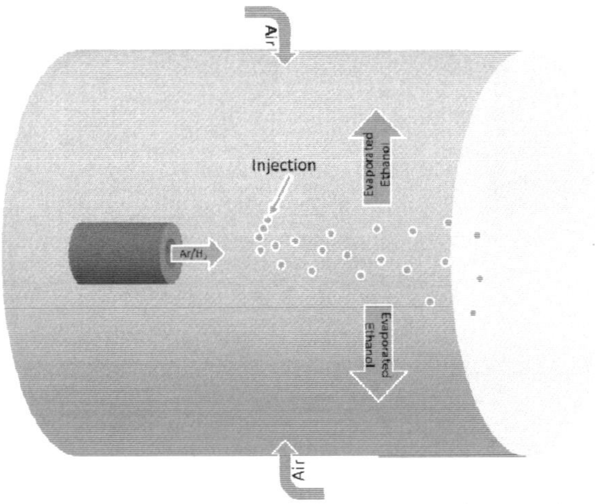

Fig. 7. Schematic drawing showing torch and the particles that are being injected downstream.

For the turbulence model, different models and resolutions have been tested here. **Fig. *8*** Shows results obtained here compared to other available test cases in literature. As is demonstrated in this figure, using k-epsilon model here with a coarse mesh resolution (384,827 nodes) provides large deviation from other results. However, by increasing resolution to a higher value of 819,952 nodes, the k-epsilon method here manages to create very good results that are a good match for experimental results of [28] and also numerical results of [29] as well as numerical RSM results of [30]. The K-epsilon method has hence been used here as the preferable turbulent model, especially due to being computationally less expensive compared to RSM [17].

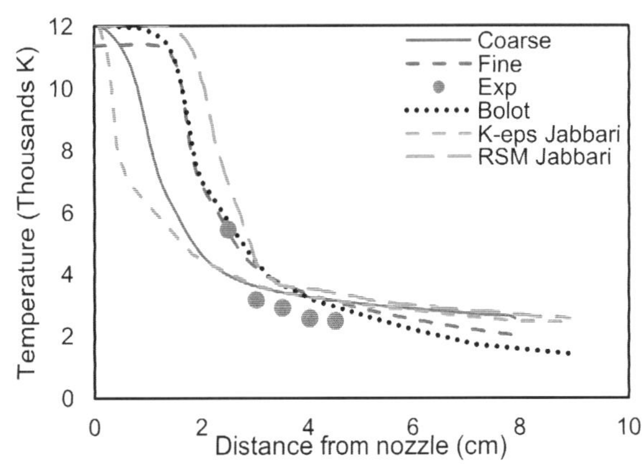

Fig. 8. Effect of turbulence model on temperature profile

The same test case above has also been performed with an extra fine mesh resolution (5,592,548 nodes)

to make sure the fine choice of resolution is acceptable. Results of these test cases are shown in **Fig. 9**. As it can be seen, the fine mesh used here produces close results to the extra fine mesh.

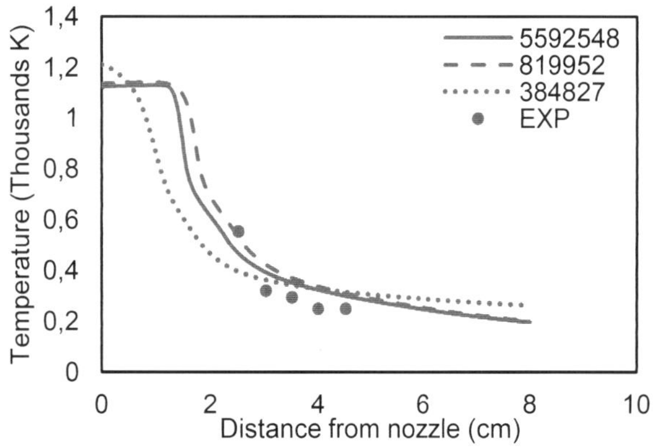

Fig. 9. Effect of mesh resolution on temperature profile

Choosing K-epsilon as the turbulence model and using 819,952 nodes, **Fig. 10** shows experimental results of Brossa [28] versus numerical results obtained here. **Table 3** shows torch conditions for each of these test cases. While the operating conditions such as flowrate and torch operating parameter are kept the same, the torch geometry used in the model here is slightly different from Brossa's experimental setup. The outer diameter of the Brossa's torch is 7.88mm while for the test here, a 6mm diameter torch has been chosen. Brossa's torch also consists of a diverging nuzzle with an angle of 95 degrees which is not modeled here.

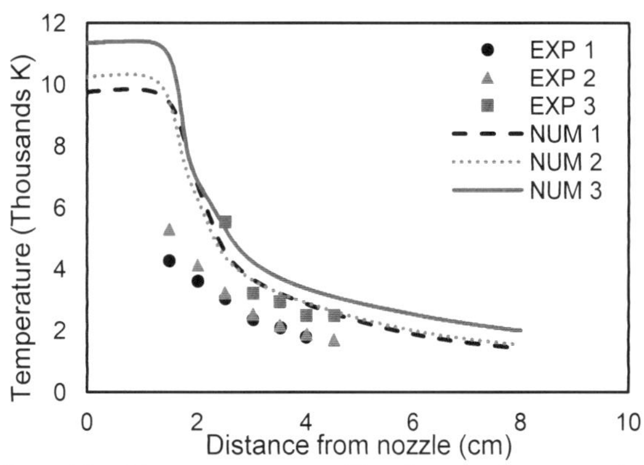

Fig. 10. Temperature profile for various torch conditions compared with experimental results.

Table 3. Torch conditions

Case #	I(A)	V(V)	$\eta(\%)$	$\dot{m}(slpm)$
1	450	27.2	51.2	47.2
2	450	24.8	47.1	35.4
3	600	25.6	47.1	35.4

6.2 Effect of viscosity in breakup

To justify the importance of using a valid viscosity model to capture suspension particle's breakup, a sample test has been conducted here. A simple torch model as discussed before has been used with mass flux of 70 slpm, flow of Argon with 0.1 mole H_2 as the gas. Anode diameter is 6mm and gas flows in with turbulent intensity of 4.5%. Particles injected are suspension of 0.1 wt% YSZ and ethanol, and have a constant viscosity. Flow passing through anode is heated with 3.2E10 W/m^3 to correspond to a torch with 0.6 thermal efficiency, 60 (V), and 500 (A).

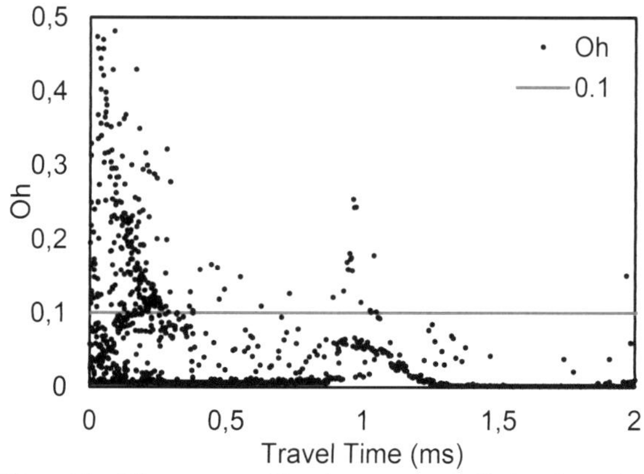

Fig. 11. Ohnesorge Number versus flight time for YSZ-Ethanol suspension particle injected into the domain

According to results of this test shown in **Fig. 11**, near 20% of particles have Oh Numbers larger than 0.1, which is the threshold from which influence of viscosity becomes significant in the break up process [31]. These particles are mostly located near nozzle exit (smaller travel times after injection) which is the place with highest shear rates and most breakups rates.

6.3 Effect of viscosity model

2As discussed in section 2, different models can be used for predicting values for effective viscosity based on instantaneous volume fraction of solids in suspension particles. Since the experimental result for viscosity of 8 mol % YSZ suspension in water at different concentrations is available (**Fig. 1**), it has been used as a benchmark for the models described in section 2. In one case, the experimental values for viscosity of this suspension have been used to update suspension viscosity at various solid concentrations. In this manner, at each time step in simulation, the volume fraction of solid content in each injected flying particle is recalculated. This value is then used to update the viscosity for the suspension mixture of that particular particle. This process is repeated for all injected particles and during all iterations.

Keeping all conditions identical, the same test case is repeated here, but instead, the viscosity of the suspension mixture is updated using equations 4, 5,

6, 8, and 12. Equation 9 is not examined as suspension powder size distribution is neglected in current framework.
Results for these test cases is summarized in **Fig. *12***. Particles traveling away from torch are here collected on a substrate located 8cm downstream of nozzle exit. This figure shows effect of viscosity model on the density distribution of particle diameters upon impact to the substrate. As evident, updating particle viscosities using equation 12 produces closer results to using experimental viscosity data of **Fig. *1***. Usage of equations 4, 5, 6, and 8 have produced larger errors.

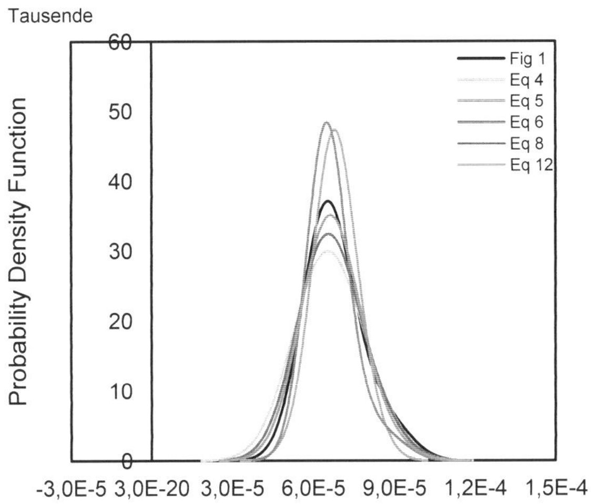

Fig. 12. Particle diameter density distribution for different viscosity models.

6.4 Test Cases

In the test cases below, the suspension of interest is YSZ in water (**Fig. *13***). Specifications of YSZ used here are given in **Table *4***.

Table 4. Pure YSZ Property

Property	Value
$\rho\ (kg/m^3)$	5560
$\mu\ (Pa \cdot s)$	0.029
$\sigma\ (N/m)$	0.43
$C_p\ (J/kg \cdot K)$	$= 1.06343\ 10^{-6}\ T^3 - 2.188953\ 10^{-3}\ T^2 + 1.709671\ T + 1.466367\ 10^2, 273 < T < 873$ $= 678.5,\ T > 873$
$\Delta H_{sf}\ (kJ/kg)$	710
$k\ (W/m \cdot K)$	2.4
$T_{mp}\ (K)$	2975

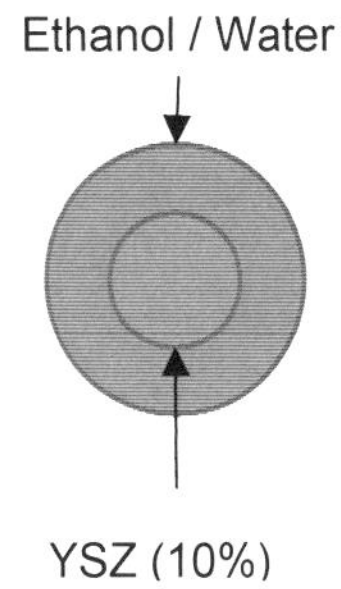

Fig. 13. Composition of a suspension drop

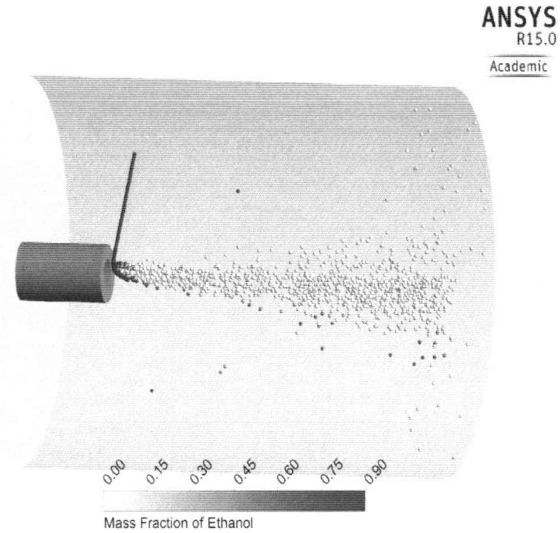

Fig. 14. Particle trajectories from torch to substrate.

Fig. *14* shows the trajectory of particles after injection. Particles here are injected with the assumption of being already atomized. For the moment, the size distribution of injection particles is neglected and is assumed all have a uniform diameter.
Injection is controlled by four parameters:

1. Velocity magnitude of particles at injection time (v),
2. Injection angle (θ),
3. Height which is injector's distance from the torch exit (H),
4. Radial distance that is injector's distance from the centre line of the torch.

After injection, particles are tracked and undergo breakups, evaporation, or might coalesce with one another. The liquid phase, ethanol or water here, evaporates during flight and then solid phase starts melting. Molten particles are carried towards the substrate. In place of the substrate, 8 cm downstream of the torch, a wall is placed to capture the particles impacting the substrate.

The torch used for the calculations here has a nozzle diameter of 6mm, thermal efficiency of 47%, and is operated under 25.6 (V) and 600 (A). Following assumptions of [22], the accommodation coefficient

(a) and γ in equation 27 are also taken 0.8 and 1.411, respectively for Ar-H_2 gas.

Table 5. Test case conditions

Case #	R (m)	H (m)	θ (deg)	v (m/s)
1				10
2			10	20
3				40
4				10
5		0.005	20	20
6				40
7				10
8			40	20
9				40
10				10
11			10	20
12				40
13				10
14	0.025	0.01	20	20
15				40
16				10
17			40	20
18				40
19				10
20			10	20
21				40
22				10
23		0.015	20	20
24				40
25				10
26			40	20
27				40
28				10
29			0	20
30				40
31				10
32		0.005	15	20
33				40
34				10
35			20	20
36				40
37	0			10
38			0	20
39				40
40				10
41		0.01	15	20
42				40
43				10
44			20	20
45				40
46		0.015	0	10
47				20
48				40
49				10
50			15	20
51				40
52				10
53			20	20
54				40

6.4.1 In-flight conditions

Study of particles during flight, including their trajectory, flight time, and rate of evaporation can give valuable information on the finished substrate coating. ***Fig. 15*** shows snapshots of particle positions during flight at 4, 5, and 6 cm from the torch exit.

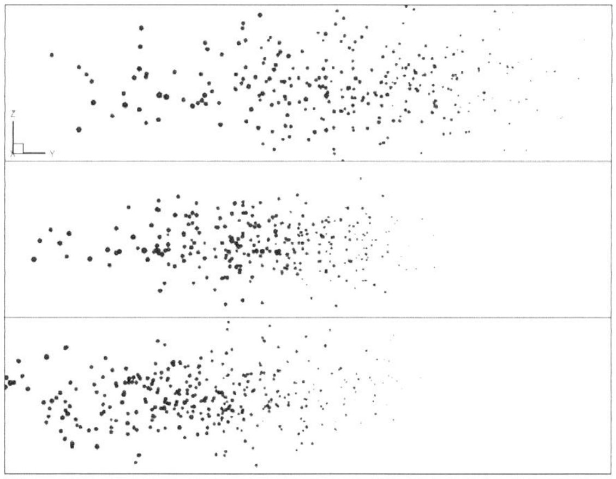

Fig. 15. Particles during flight at 0.04, 0.05, and 0.06 cm from torch exit (top to bottom).

6.4.2 Impact on substrate

Particles for each test case in **Table *5*** are captured on a substrate that is located 8cm from nozzle. The spraying process duration is approximately 25ms.

A sample particle collection is shown in **Fig. *16*** (for case #2). The size of points in this figure correspond to particle diameters upon impact. Spread of molten particles upon impact is neglected here. In this case, the torch is located at (Y=0, Z=0), and injection point is at (Y=0.025, Z=0).

The number of particles impacted the substrate during this time window of 25ms are shown in **Fig. *17***. It is clearly seen that test case #37 leaves the largest footprint on the substrate. Also it can be seen that for some test cases, based on torch flow and injection conditions, particles never reached the desired place on the substrate and were collected elsewhere. These particles have been omitted from results here as they did not meet the substrate position constraint.

Variances for coordinates Y and Z of particles hitting the substrate along with diameter and velocity magnitude for test cases shown in **Table *5*** are plotted in **Fig. *18***.

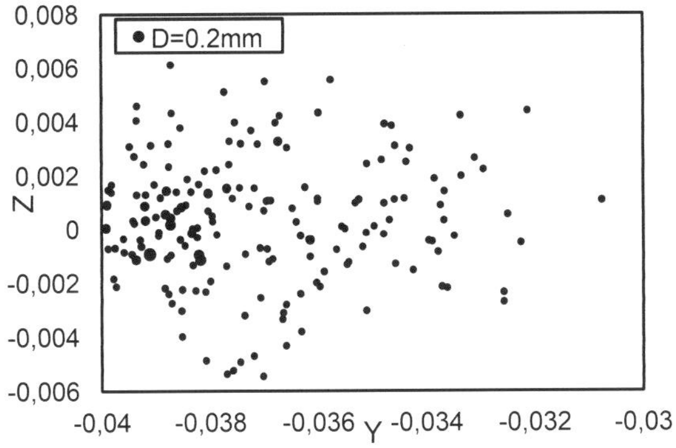

Fig. 16. Particles captured on the substrate 8cm downstream of torch for test case #52 in Table 5.

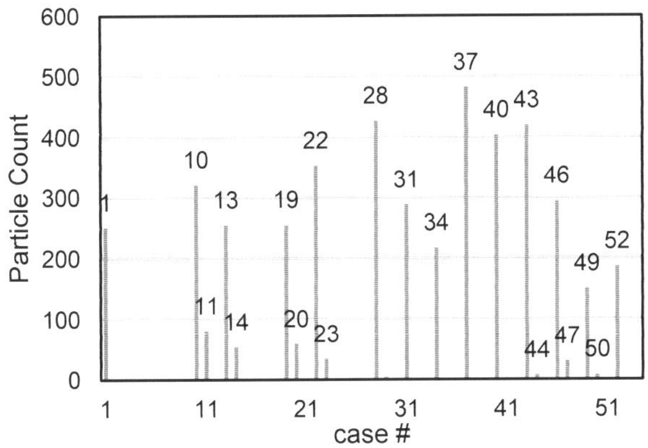

Fig. 17. Particle counts on substrate for each test case. Particles are captured during a window of 25ms.

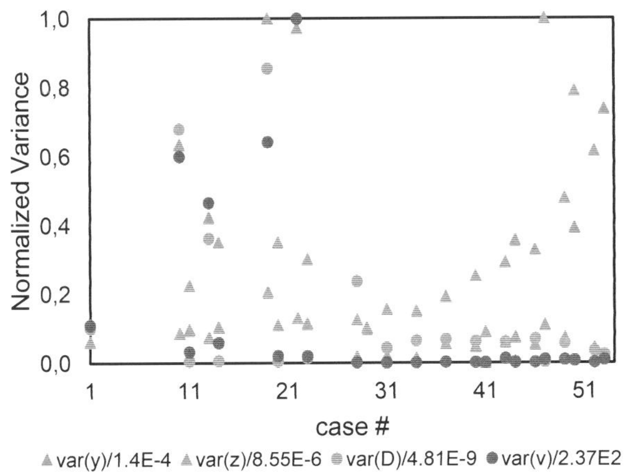

Fig. 18. Variances for coordinates Y, Z, diameter, and velocity of particles upon impact to the substrate for each test case. Values have been divided to maximum for better visualization.

For test cases with the most number of impacts, the density distribution function for particle diameters on substrate has been plotted in **Fig. *19***.

It is worth mentioning that more useful information than the ones discussed before can be obtained from the particles, which include the melting index of particles upon impact. Here, study of these parameters are left for future works.

7 Conclusion

Effects of change of properties of ceramic suspensions during flight was numerically studied here under various conditions. Results clearly indicated inclusion of a proper viscosity model for the suspension, as it goes through large concentration changes, is vital in capturing more accurate flight and breakup patterns. The choice of this viscosity model, whether from experiments or empirically derived correlations does not affect simulations enormously as long as the empirical methods follow similar growth trend to experimental viscosity-concentration data.

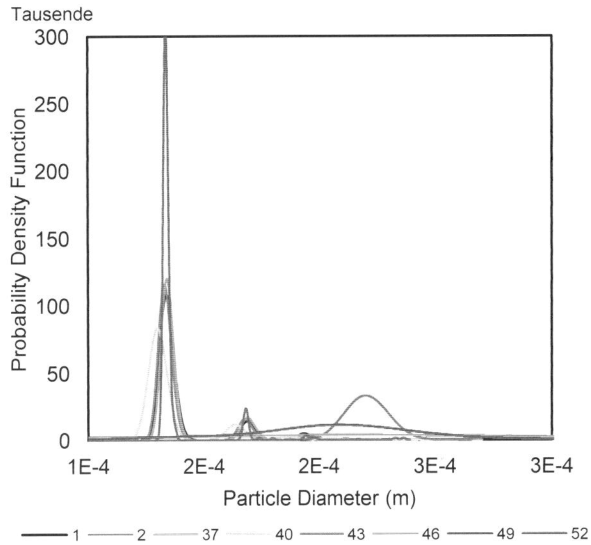

Fig. 19. Density distribution for particle diameters upon impact to the substrate

8 References

[1] A. Einstein, "Eine neue Bestimmung der Moleküldimensionen," *Annalen der Physik,* vol. 324, no. 2, pp. 289-306, 1906.

[2] E. Guth and R. Simha, "Untersuchungen über die Viskosität von Suspensionen und Lösungen. 3. Über die Viskosität von Kugelsuspensionen," *Kolloid-Zeitschrift,* vol. 74, no. 3, pp. 266-275, 1936.

[3] D. G. Thomas, "Transport characteristics of suspension: VIII. A note on the viscosity of Newtonian suspensions of uniform spherical particles," *Journal of Colloid Science,* vol. 20, no. 3, pp. 267-277, 1965.

[4] K. Toda and H. Furuse, "Extension of Einstein's viscosity equation to that for concentrated dispersions of solutes and particles," *Journal of Bioscience and Bioengineering,* vol. 102, no. 6,

pp. 524-528, 2006.

[5] J. S. Chong, E. B. Christiansen and A. D. Baer, "Rheology of concentrated suspensions," *Journal of Applied Polymer Science,* vol. 15, no. 8, pp. 2007-2021, 1971.

[6] I. M. Krieger and T. J. Dougherty, "A mechanism for non-Newtonian flow in suspensions of rigid spheres," *Transactions of The Society of Rheology,* vol. 3, no. 1, pp. 137-152, 1959.

[7] C. A. Shook and M. C. Roco, Slurry flow: principles and practice, Elsevier, 2015.

[8] L. L. Schramm, "Suspensions; Basic Principles," in *Suspensions: Fundamentals and applications in the petroleum industry*, Washington DC, American Chemical Society, 1996, pp. 3-44.

[9] T. Dabak and O. Yucel, "Shear viscosity behavior of highly concentrated suspensions at low and high shear-rates," *Rheologica Acta,* vol. 25, no. 5, pp. 527-533, 1986.

[10] H. Eilers, "Die Viskosität von Emulsionen hochviskoser Stoffe als Funktion der Konzentration," *Kolloid-Zeitschrift,* vol. 97, no. 3, pp. 313-321, 1941.

[11] P. K. Senapati, B. K. Mishra and A. Parida, "Modeling of viscosity for power plant ash slurry at higher concentrations: Effect of solids volume fraction, particle size and hydrodynamic interactions," *Powder technology,* vol. 197, no. 1, pp. 1-8, 2010.

[12] B. A. Horri, P. Ranganathan, C. Selomulya and H. Wang, "A new empirical viscosity model for ceramic suspensions," *Chemical Engineering Science,* vol. 66, no. 12, pp. 2798-2806, 2011.

[13] O. Arevalo-Quintero, D. Waldbillig and O. Kesler, "An investigation of the dispersion of YSZ, SDC, and mixtures of YSZ/SDC powders in aqueous suspensions for application in suspension plasma spraying," *Surface and Coatings Technology,* vol. 205, no. 21, pp. 5218-5227, 2011.

[14] J. H. Song and J. R. G. Evans, "Ultrafine ceramic powder injection moulding: The role of dispersants," *Journal of Rheology,* vol. 40, no. 1, pp. 131-152, 1996.

[15] D. I. Lee, "Packing of spheres and its effect on viscosity of suspensions," *Journal of Paint Technology,* vol. 42, no. 550, p. 579, 1970.

[16] T. C. Patton, Paint Flow and Pigment Dispersion, Interscience, 1964, p. 479.

[17] A. F. Ansys, "Discrete Phase," in *ANSYS Fluent Theory Guide*, 15.0 ed., ANSYS Inc., 2013, pp. 373-464.

[18] F. M. White and I. Corfield, Viscous fluid flow, vol. 3, New York: McGraw-Hill, 2006.

[19] E. Pfender, "Particle behavior in thermal plasmas," *Plasma Chemistry and Plasma Processing,* vol. 9, no. 1, pp. 167S-194S, 1989.

[20] Y. Lee, K. Hsu and E. Pfender, "Modeling of particles injected into a dc plasma jet," in *Proceedings of the 5th International Symposium on Plasma Chemistry*, 1981.

[21] Y. Chyou and E. Pfender, "Behavior of particulates in thermal plasma flows," *Plasma chemistry and plasma processing,* vol. 9, no. 1, pp. 45-71, 1989.

[22] X. Chen and E. Pfender, "Effect of the Knudsen number on heat transfer to a particle immersed into a thermal plasma," *Plasma chemistry and plasma processing,* vol. 3, no. 1, pp. 97-113, 1983.

[23] L. Talbot, R. Cheng, R. Schefer and D. Willis, "Thermophoresis of particles in a heated boundary layer," *Journal of Fluid Mechanics,* vol. 101, no. 4, pp. 737-758, 1980.

[24] X. Chen and E. Pfender, "Behavior of small particles in a thermal plasma flow," *Plasma chemistry and plasma processing,* vol. 3, no. 3, pp. 351-366, 1983.

[25] D. Chen, E. H. Jordan and M. Gell, "Effect of solution concentration on splat formation and coating microstructure using the solution precursor plasma spray process," *Surface and Coatings Technology,* vol. 202, no. 10, pp. 2132-2138, 2008.

[26] J. Beale and R. Rolf D, "Modeling spray atomization with the Kelvin-Helmholtz/Rayleigh-Taylor hybrid model," *Atomization and sprays,* vol. 9, no. 6, 1999.

[27] M. A. Patterson and R. Rolf D, "Modeling the effects of fuel spray characteristics on diesel engine combustion and emission," *SAE Technical Paper,* 1998.

[28] M. Brossa and E. Pfender., "Probe measurements in thermal plasma jets," *Plasma Chemistry and Plasma Processing,* vol. 8, no. 1, pp. 75-90, 1988.

[29] R. M. I. a. C. C. Bolot, "Mathematical modeling of a free plasma jet discharging into air and comparison with probe measurements," *Thermal Spray: A United Forum for Scientific and Technological Advances,* pp. 549-555, 1997.

[30] F. J. M. W. R. &. D. A. Jabbari, "A Numerical Study of Suspension Injection in Plasma-Spraying Process," *Journal of Thermal Spray Technology,* vol. 23, no. 1-2, pp. 3-13, 2014.

[31] N. Ashgriz, Handbook of atomization and sprays: theory and applications, Springer Science & Business Media, 2011.

Opportunities of Twin-Cathode System as new Suspension Plasma Spray System

Mohammed Shahien[1,2], Masato Suzuki[1]
[1] National Institute of Advanced Industrial Science and Technology, AIST, Tsukuba/JP
[2] Central Metallurgical Research and Development Institute, CMRDI, Cairo/EG

Using the nano-size particles in the form of suspension or precursor is the possible way to use the fine particles in thermal spray technology. It assists the formation of fine droplets, fine splats and therefore the precise control of the coating microstructure. These novel features make the suspension/solution-precursor plasma spraying (SPS/SPPS) promising solutions for several applications and research challenges.

This study investigated the feasibility and opportunities of developing a new suspension plasma spraying system with an axial feeding and low cost. The system is based on using twin-cathode plasma spray gun with three plasma torches. The main torch is in the center with reversed polarity (cathode nozzle and anode), and two sub-torches with normal polarity (anode nozzle and cathode) on the sides. The system is based on the axial feeding of the suspension, through the hollowed anode of the main torch by using a commercial dual tube nozzle. It is cost reduction system. Thus, only argon as plasma gas is required, since the enthalpy of plasma jet is high enough for melting the oxide ceramic powders. Several ceramic coatings were successfully fabricated by the new system. Moreover, the microstructure of the coatings can be controlled in wide range.

1 Introduction

Using the fine or nano-size particles is one of the promising ways to improve the functionality of the conventional thermal spray coatings. Moreover, using the nano-structured particles in the plasma spray process is considered to be a promising approach to overcome several concerns of the coating microstructure and thereby improve the coating performances and properties.

However, using the fine or nano-sized particles in the conventional thermal spray was difficult. Thus, decreasing the particle size reduce the flowability of the powders and leads to clogging of the feeding lines which makes spraying of very fine powders practically impossible. Moreover, transporting or injecting the nano-size particles in the core of plasma plume (high enthalpy flow) is difficult. It is attributed to the poor rheological properties of the fine particles, which leads to non-uniform powder transport from the feeder to the plasma plume [1].

Several efforts had been investigated to over the above concerns and assist the use of the nano-size feedstock [2-7], including: agglomeration of nano-size primary particles to form micrometer-size powders and this coating architecture is generally called as "bimodal". Using complex alloys with low critical cooling rates for metallic glass formation, and formation of amorphous plasma spray coatings. During heating such amorphous above their crystallization temperature, the nano-size structures are formed due to the low diffusion rate in the solid phase at the transformation temperature.

On the other hand, using the nano-size particles in the form of suspension (Suspension Plasma Spraying: SPS) and/or solution precursors (Solution Precursor Plasma Spraying: SPSS), had received a great attention recently.

Suspension Plasma Spraying (SPS), the process is based on suspending or dispersing the nano-sized particle (30-100 nm) into the liquid phase (water or alcohol) by means of dispersants. The process consists of mechanically injecting the suspension (feedstock particles, a liquid phase and a dispersant) in the form of a liquid stream or as drops after nebulization. Upon penetration within the plasma stream, the liquid stream rapidly fragmented due to flow shear forces and vaporized. The droplets liquid phase evaporation leads to the formation of single particle or aggregates of a few particles. After that, these particles melt and form liquid drops that impact, spread and solidify to form flattened lamellae with small or fine diameter. Finally, nano-structured coatings with controlled microstructure can be fabricated.

The first or the key issue of suspension plasma spraying, started 2o years ago (in late nineties of the last century; 1997), in Sherbrooke University in Canada, through injecting the liquid into plasma plume [7]. After that, the SPS process had received much attention in the field of thermal spray due to the promising advantages of using nano-sized powders in several applications. Several research and challenges are being investigated to develop the SPS process, including: thermal spraying system and cost, suspension Feedstock materials and particle size, suspension stability, handling and storage of the nano-suspension, injection system, stable injection, suspension flowability, plasma and liquid processing, and the coating build up mechanism.

The aim of this study is to investigate the feasibility and opportunities of developing new system with low cost, through using a twin-cathode based plasma spray system as new suspension plasma spray system. The deposition of different ceramic materials by the new SPS system was also investigated.

2 Experimental Procedures

The newly developed system is based on using a twin-cathode type plasma spray gun (Aeroplasma Corporation, Japan). The gun consists of three plasma torches as shown in schematically in Fig 1:

- P-torch (+); it is the main torch and the center of the gun. It has a reversed polarity (with a cathode nozzle and an anode).
- N-torches (-); on the right side of the main torch and it acts as sub-torch. It has normal polarity (with an anode nozzle and a cathode).
- N-torches (-); on the left side of the main torch and it acts as sub-torch. It has normal polarity (with an anode nozzle and a cathode).

Figure 2 shows the real view of the new spray system. The plasma direction is on the same gas flow of the main torch as it is clear in in Fig 2. The process starts through generating the plasma between the two cathodes of the N-torches and the anode of the P-torch and so that the plasma jet can be maintained. The plasma shape after complete generation of the plasma is shown in Fig. 3. The plasma has the hairpin-shape as significantly shown in Fig. 3 (b). Keeping the hairpin-shape of the plasma is strongly required for effective spraying.

The newly developed is based on using high voltage and low current. Therefore, the power is lower compared to other spray systems. In addition the system requires only the argon (Ar) gas as plasma gas (no need for H_2, N_2, He, ..). Thus, the voltage (enthalpy of plasma jet) is high enough (150~190 V) with Ar for melting the ceramic particles with this system, which makes it as a very cost reduction suspension plasma spray system. The operating parameters of the system vary very widely, such as plasma gas flow rate, carrier gas flow rate, and electric power inlet.

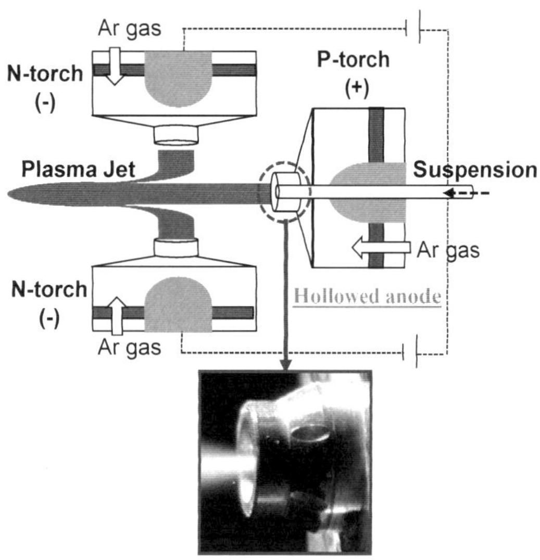

Fig. 1. Schematic diagram of new SPS system.

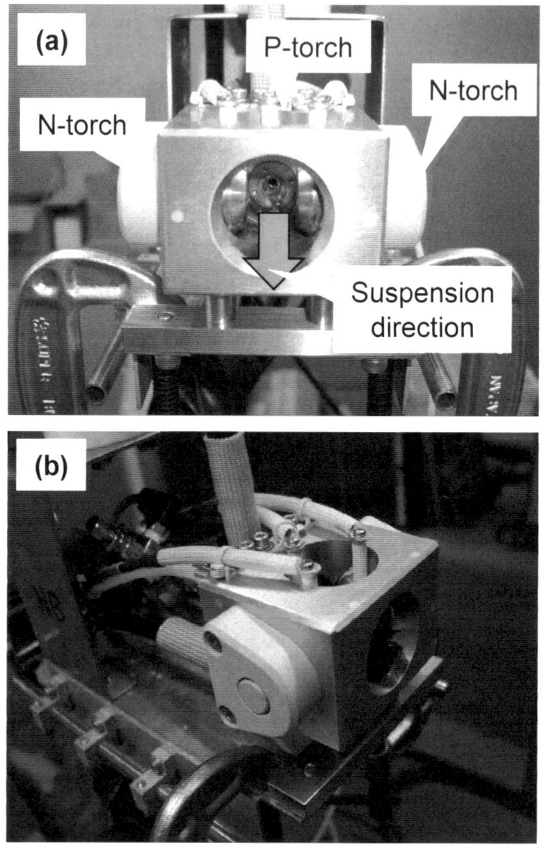

Fig. 2. The front and top views of the newly developed suspension spray system.

The newly developed system enables the axial-feeding of the suspension. Thus, the P-tourch has a hollowed electrode (anode) as shown in Fig. 1. It is well known the interesting advantages of the axial feeding in thermal spray. During the axial-feeding, the feedstock is injected into the center of the plasma jet along the axis of its flow. Therefore all of the feedstock powder will be fully entrapped inside and melted within the plasma before deposition on the substrate surface. This leads to higher Deposition efficiencies and ability to use high feed rates during spray.

In the new system a commercial dual tube nozzle, HM-6L (Fuso Seiki Co., Ltd.) was used to feed the suspension materials. It was inserted into the hollowed anode of the main torch to supply mist of suspension into the plasma jet. The real and schematic views of the dual tube nozzle are shown in Fig.4. The dual tube nozzle consists of two tubes:

- Outer tube with 0.7 mm inner and 1 mm outer diameters respectively. This tube is used for the atomizing gas (Ar gas) around the suspension.
- Inner tube with 0.2 mm inner and 0.4 mm outer diameters respectively. This inner tube is use for suspension feeding to the plasma jet.

Fig. 3. The shape of the plasma generated by the newly developed suspension spray system and the top view of its hairpin-shaped plasma.

In this study several the feasibility of deposition of several commercial feedstock materials were investigated. All the suspensions were prepared in the lab using the ethanol as solvent and using the Poly-Ethlene-Imine (PEI) as stabilizer. The mixture of the suspension was ball milled with using the ZrO_2 balls (Φ10 mm). The suspensions were prepared with various concentrations (5~15wt%).

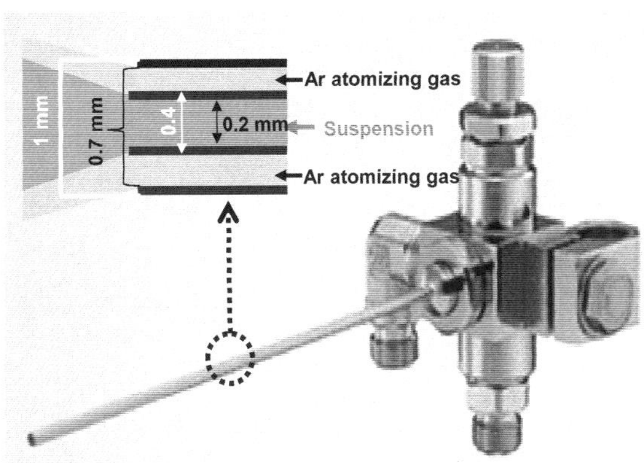

Fig. 4. The axial feeding tube (dual tube nozzle) for feeding the suspension into the plasma jet: real view and cross-section schematic diagram.

The feedstock suspensions were injected in the plasma jet using the pneumatic system (Ar gas). All the typical spray parameters are shown in table1. The grid-blasted stainless steel (SUS304) plates were used as substrates.

Table 1 The typical spray conditions

Plasma gas	Ar
The P-Torch flow rate	40 L/min
The N-Torchs flow rate	4 L/min
Power input	27 kW
Arc current	90 A
Voltage	150 V
Spray distance	15-50 mm
Atomizing gas	Ar
Atomizing gas Flow rate	5 L/min
Suspension feed rate	35 L/min
Solvent	Ethanol
Suspension concentration	5-15 wt%

3 Results and Discussion

3.1 Developing the new spray system

Figure 5 shows the microstructure of the fabricated coating using the newly developed SPS system using suspension of: 15 wt. % Y_2O_3, 5 wt. % 8YSZ and 15 wt. % TiO_2 powders. It reveals that, the temperature of the developed system with using only Ar plasma gas was enough to melt several ceramic particles in the suspensions.

It indicates that, it was possible to develop a new suspension plasma spray coating system by using a twin cathode plasma spray gun. The system requires only Ar as plasma gas which is considered as a very unique merit to decrease the cost. Beside that the system is characterized by using the axial feeding which is interesting regarding the deposition efficiency as well as controlling the microstructure of the coatings during suspension spray.
Several ceramic coatings with different melting points including: Al_2O_3, Y_2O_3, TiO_2, and YSZ were successfully fabricated by the newly developed suspension plasma spray system. This system makes it possible to vary operating parameters very widely, such as plasma gas flow rate, carrier gas flow rate, and electric power inlet. These unique novel features

of the new SPS system are showed a promising contribution in controlling the microstructure of the suspension plasma spray coatings in the future studies.

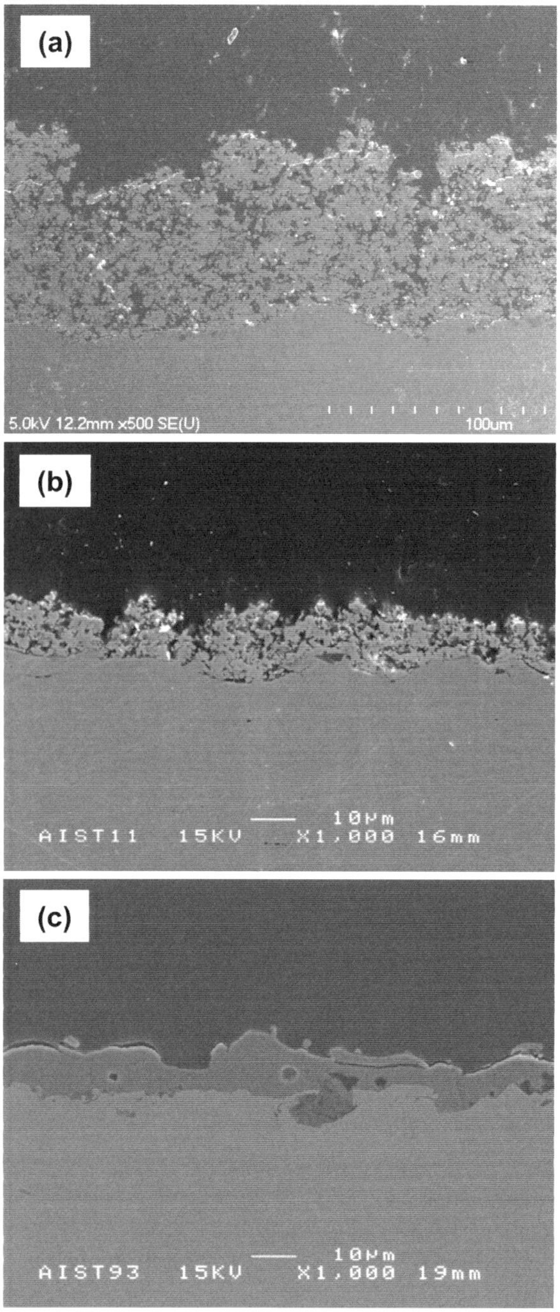

Fig. 5. The SEM cross sections of the fabricated coatings with using suspension of: (a) Y_2O_3, (b) 8YSZ and (c) TiO_2 powders.

3.2 Some novel features of the new system

Controlling the microstructure of plasma sprayed coatings is one of the big challenges for economic and industrial applications. The newly developed system showed a promising result of controlling the microstructure.

Figure 6 shows the SEM surface morphology of the fabricated coatings by using: 7.5 and 30 wt. % of Y_2O_3, 5 and 15 wt. % Al_2O_3, as well as 5 and 15 wt. % YSZ suspensions.

It reveals that, the suspension concentration strongly affected the microstructure as follow:

- With using low concentration of suspension, the coatings are consist of high amount of spherical splats and small amount of flatten splats.
- On the other hand, with using high concentration of the suspensions the coatings consist of high amount of flatten splats and small spherical splats.

Generally it is well known that, during suspension plasma spray, after injecting the suspension in the plasma plume it passes through three steps:

- Atomization of the suspension mist.
- Vaporization of the solvent and particle aggregation.
- Finally melting and then deposition on the substrate.

The influence of suspension concentration during spraying can be explained as follow:

- In case of low suspension concentration the solid content is low. During spraying after the solvent vaporization the mist tends to aggregate and forms small (light) droplet which deposit on the substrate as spherical splats and forms porous structure.
- On the other side, in case of high suspension concentration the solid content is high. During spraying after the solvent vaporization the mist tends to aggregate and forms large (heavy) droplet which flatten on the substrate and form dense coatings.

As it was very difficult to observe more details about the suspension species in the plasma during the normal plasma spray process due to the high temperature of the plasma. The influence of suspension concentration was confirmed during investigating the atomization parameters. In which the ethanol and/or suspensions were injected to the p-torch through the dual-tube nozzle without using plasma. Thus, suspension is supplied to the atomizers from pressurized liquid container.

During this experiment the suspension is supplied into the inner tube and is sprayed vertically downward into static atmospheric air. The injection pressure was also controlled. The metal halide lamp was used for observation and photograph. The laser diffraction method by using LDSA1500A (Tohnichi computer applications co. ltd.), using the histogram mode and the focus length of lens is 300 mm were applied.

Figure 7 shows the real images of spraying with using only ethanol and different concentrations of the suspension. It is clear that, increasing the suspension concentration increase the solid load gradually.

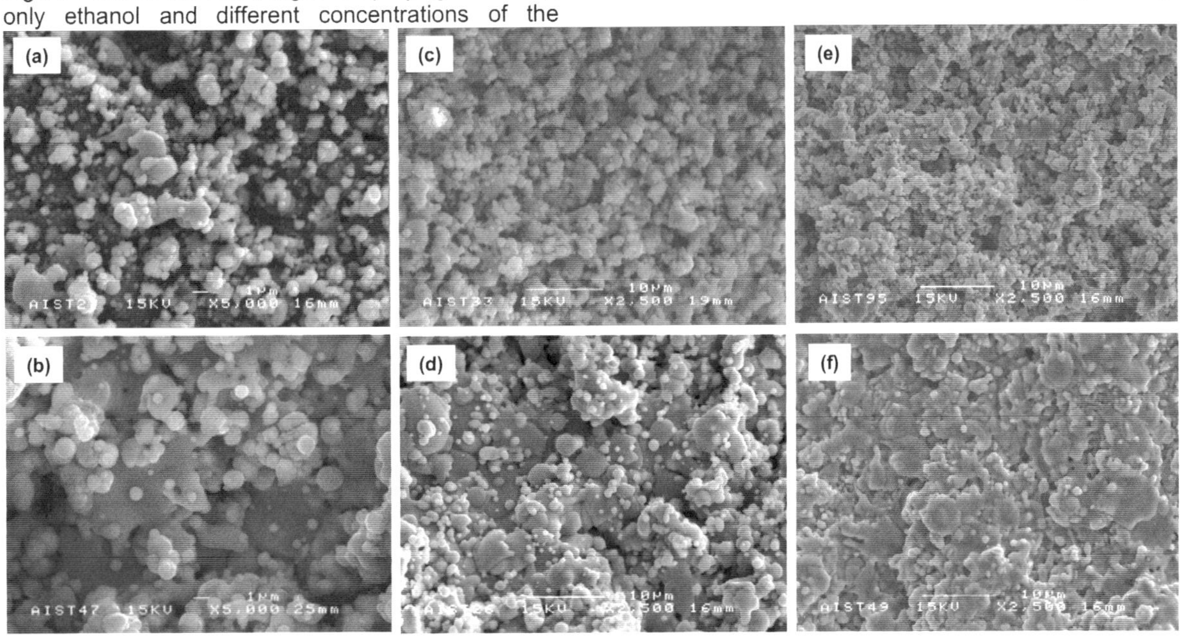

Fig. 6. The SEM surface morphology of the fabricated coatings: (a) 7.5 wt.% Y_2O_3, (b) 30 wt.% Y_2O_3, (c) 5 wt.% Al_2O_3, (d) 15 wt.% Al_2O_3, (e) 5 wt.% YSZ and (f) 15 wt.% YSZ

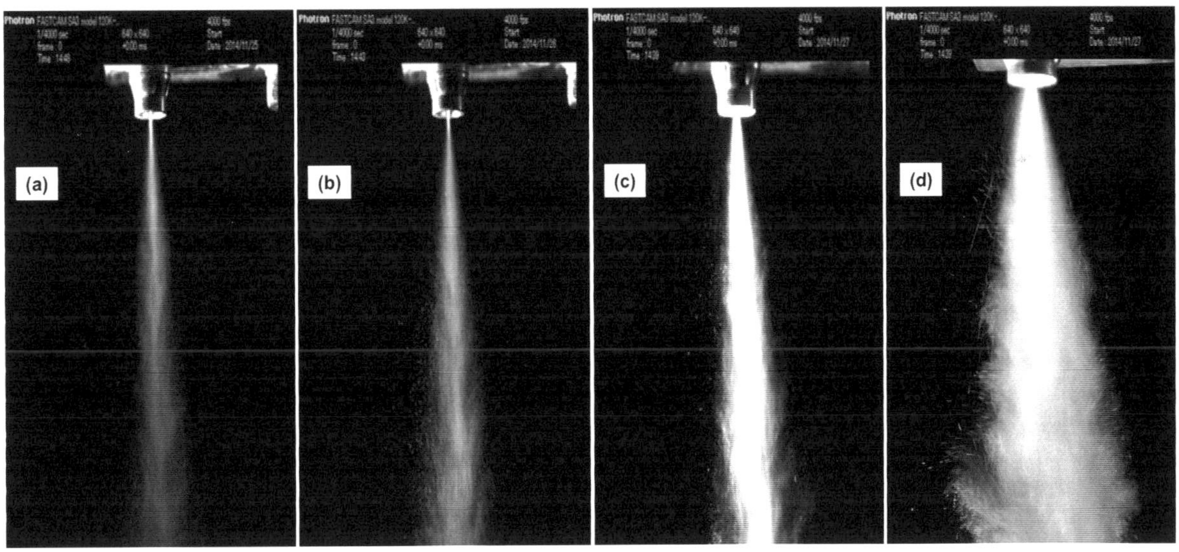

Fig. 7. The real images of spraying the only ethanol and different concentrations of the suspension vertically downward into static atmospheric air without using plasma: (a) only ethanol, (b) 5 wt. % suspension, (c) 10 wt. % suspension and (d) 15 wt. % suspension.

The reason that, the heavier droplets forms flatten splats and therefore dense coatings and the light mist forms spherical splats and form porous coatings is attributed to the movement of the droplet and its momentum near the substrate before deposition. Thus, the gas flow decrease gradually on the direction of the substrate and therefore the velocity of plasma jet decreases. This will decrease the particles velocity. The decreasing tendency of the smaller (lighter) particles is more significant than the larger ones (heavier).

It indicates that, the small (light) droplet will have very low velocity and very low momentum and therefore it will stick on the substrate as it is without flattening and forming porous coatings. On the other hand, large (heavy) droplet will have higher velocity (compared to the smaller ones) and higher momentum and therefore it will hit the substrate strongly, flatten and forming dense coatings.

Figure 8 schematically shows influence of the suspension concentration on the splat formation during spraying.

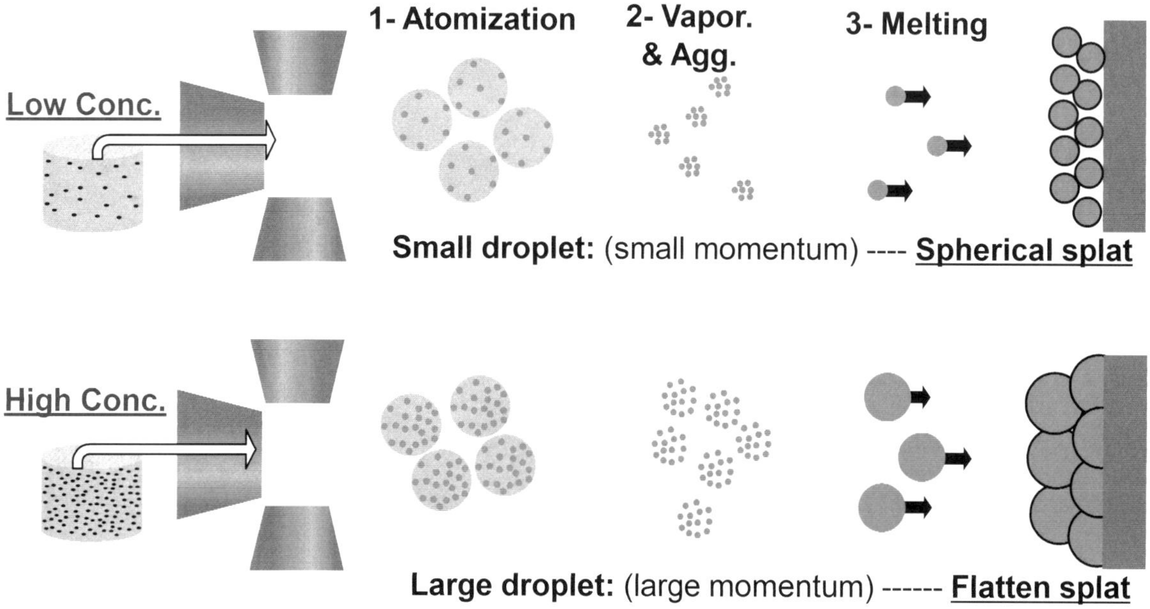

Fig. 8. The influence of the suspension concentration on the splat formation during suspension spraying.

Therefore, at low suspension concentration, the small (light) spherical droplet (with low velocity and small momentum), will stick on the substrate as it is to form porous coatings. At high suspension concentration and/or short spray distance, the large (heavy) droplet (with higher velocity and large momentum), will hit the substrate hardly and flatten to form dense coatings.

4 Conclusions

A new suspension plasma spraying system with an axial feeding and low cost was developed. It is based on using twin-cathode plasma spray gun with three plasma torches. The main torch is in the center with reversed polarity (cathode nozzle and anode), and two sub-torches with normal polarity (anode nozzle and cathode) on the sides. This system is based on the axial feeding of the suspension, through the hollowed anode of the main torch and the commercial dual tube nozzle was used for axial feeding. Furthermore, it is cost reduction system. Thus, the system requires only argon as plasma gas, since the enthalpy of plasma jet is high enough for melting the oxide ceramic powders.

Several ceramic coatings including Y_2O_3, Al_2O_3, TiO_2, and YSZ, were successfully fabricated by the new system. Furthermore, the microstructure of the fabricated coatings can be significantly controlled through controlling the suspension concentration. Using low suspension concentration assists the formation of light (small) droplets, which leads to formation of spherical splats and due to their low velocity form porous coatings. The high concentration suspension assists the formation of large (heavy) droplet which strongly hit the substrate to flatten and form dense coatings.

5 Literature

[1] Karthikeyan, J., Berndt, C.C., Tikkanen, J., Wang, J.Y., King, A.H. and Herman, H.: Preparation of nanophase materials by thermal spray processing of liquid precursors, NanoStructured Materials, (1997), 8 (1), pp. 61-74.

[2] Fauchais, P., Etchart-Salas, R., Rat, V., Coudert, J.-F., Caron, N. and Wittmann-Teneze, K.: Parameters controlling liquid plasma spraying: solutions, sols, or suspensions, J. Therm. Spray Technol., (2008), 17(1), pp. 31-59.

[3] Fauchais, P., Montavon, G., Lima, R. and Marple, B.: Engineering a new class of thermal spray nano-based microstructures from agglomerated nanostructured particles, suspensions and solutions: an invited review, J. Phys. D: Appl. Phys., (2011), 44 (no. 093001), pp. 1-53.

[4] Fauchais, P. and Vardelle, A.: Innovative and emerging processes in plasma spraying: from micro- to nano-structured coatings, J. Phys. D: Appl. Phys., (2011), 44 (no. 194011), pp. 1-4.

[5] Pawlowski, L.: Finely grained nanometric and submicrometric coatings by thermal spraying: a review, Surf. Coat. Technol., (2008), 202 (18), pp. 4318-4328.

[6] Viswanathan, V., Laha, T., Balani, K., Agarwal, A. and Seal, S.: Challenges and Advances in Nanocomposite Processing Techniques, Mat. Sc. and Eng. R, (2006) Vol. 54, pp. 121–285.
[7] Gitzhofer, F., Bouyer, E. and Boulos, M.I.: Suspension plasma spraying. US Patent 5 609 921, 3 November 1997.

Introduction of micro resin fragments with ceramic nanoparticles into plasma flame to create fine coated layers

M.Katsuki[1], H.Harada[1], and S.Kirihara[2]
(1) Hitachi Zosen Corporation, Osaka, Japan
(2) Osaka University, Osaka, Japan

Nanoparticles thermal spraying has been developed to create fine ceramic layers with higher deposition speed. Micro resin fragments including with the ceramic nanoparticles could be introduced into argon gas plasma successfully by using conventional powder feeding equipment. Zirconia nanoparticles of 200 nm in average diameter were dispersed into a thermosetting acrylic liquid resin at 57% in volume fraction. An airtight container of 150 cc in volume was revolved and rotated for degassing and mixing. Obtained pastes were solidified thorough heating at 150 °C for 30 min. Composite bulks were crushed by a vibrating milling machine. The micro fragments selected by using a sieve of 45-106 µm in mesh size. In the plasma flame, these resin fragments were burned down, and the heated nanoparticles were deposited on a metal substrate. Fine zirconia layers without micro cracks or pores were observed by scanning electron microscopy. Carbon contamination deriving from the resin material was verified by thermogravimetry for the cross section of coated layer. Vickers hardness of the ceramic phase was measured to estimate the sintered density. The fine zirconia coating will increase practical durability as heat and corrosion resistances of steam pipes in a boiler of thermal power generators inner walls of refuse incinerators.

1 Introduction

Thermal spraying is a coating process which can deposit metals or ceramics with high deposition speeds to the object. Sprayed coatings with ceramics can give the performance of heat resistance, corrosion resistance, and abrasion resistance to the object. Conventionally, as powdery spraying material have been used micro size particles about 10–100 µm and there are pores and cracks in the sprayed coatings.These pores and cracks will be disadvantage to create high dense coatings.

Recentry, thermal spraying method to use nanoparticles as spray materials has been studied for the purpose of restraing defects in the thermal spray coatings.There are some problems to use nanoparticles in plasma spraying process. Due to the nanoparticles are likely to aggregation, the transfer hose to the spray gun closes [1][2][3]. Since nanoparticles are low mass and the inertial force is weak even using the injector, it cannot be sufficiently introduced into the plasma flame [4][5][6][7]. Furthermore nano risks for human health are concerned [1].

To overcome these problems, suspension plasma spraying and pastes plasma spraying have been studied [1][3][7][8][9]. In suspension plasma spraying process, the suspension with nanoparticles which dispersed in aqueous or organic solvent are injected into the plasma or gas flame. However, the viscosity of the suspension become higher by increasing the amount of dispersion of nanoparticles and the injection will be difficult. Although there are studied about increasing the content of nanoparticles by adding dispersant, the content of nanoparticles in the solvent and deposition speeds tend to low. In addition, it is necessary to newly install a feeder and injection device of the suspension. On the other hand, pastes plasma spraying is possible to create dense sprayed coatings at higher deposition speeds by using pastes with high concentration nanoparticles. But, it is necessary to use new feeder and injection device of the pastes.

This paper shows a new nanoparticles thermal spraying process. Micro resin fragments which nanoparticles are dispersed in high concentration could be injected into the plasma flame by using conventional powder feeders and could be created fine coated layers with high deposition speeds.

Dense ceramic thermal spray coatings can be utilized in anti-corrosion coating. EfW plant (refuse incinerators) is responsible for waste disposal and energy supply. Although it is possible to improve the power generation efficiency by increasing the steam conditions of the boiler, corrosion of boiler tubes is increased [10][11][12]. Frequent replacement of the boiler tubes, increase the operating costs of the EfW plant. By the way, the thermal barrier coating (TBC) to improve the heat resistance of the structural material in a high temperature environment have been developed. TBC does not fit the purpose of blocking the structural material from the corrosive environment because the pores in the coatings enhance the thermal insulation. A dense coating of ceramics having corrosion resistance, durability of the boiler tubes can be expected.

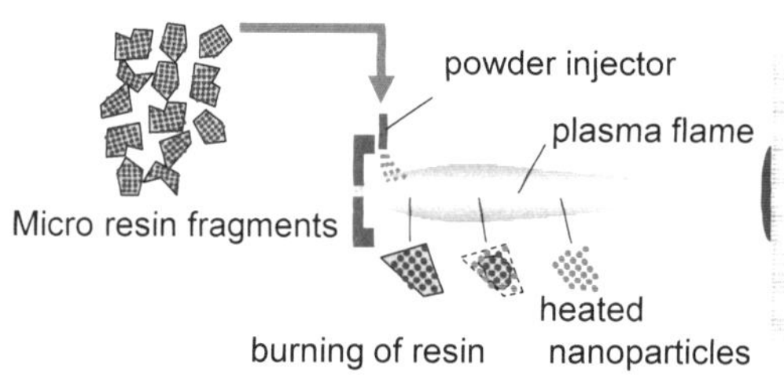

Fig. 1. The concept of nanoparticles thermal spraying process by using micro resin fragments

2 Experimental procedure

Fig.1 shows concept of nanoparticles thermal spraying process by using micro resin fragments. The micro resin fragments where nanoparticles are dispersed in high concentration are transported by using a conventional powder feeder. The resin is burned out by introducing micro resin fragments into plasma flame, and then nanoparticles are deposited on the substrate.

2.1 Preparation of micro resin fragments

The stabilized zirconia nanoparticles having the average particle size of about 200 nm (KZ-8YF, KCM Corporation, Japan) were mixed with the thermosetting acrylic liquid resin (JSR Corporation, Japan). The mixing ratio of the stabilized zirconia was 57 vol% by volume fraction (86.7 wt% by mass fraction). These were placed in a sealed container of 150 cc and mixed and degassed for 840 s by using planetary mixer SK-350T(SHASHIN KAGAKU Co.Ltd., Japan) with 1340 rpm in revolution and rotation respectively. The mixing paste was solidified by heating at 150 °C for 30 min. Composite bulks were crushed by a high speed vibrating milling machine. Micro resin fragments were screened using the sieve of 106 μm and 45 μm opening.

2.2 Plasma spray processing

The conventional spraying material (SHOCOAT K-90, SHOWA DENKO K.K., Japan), agglomerated particles (KZ-8YF, agglomerated size is 45~75μm, KCM Corporation, Japan) and micro resin fragments were sprayed using plasma spraying equipment (Sulzer Metco Inc.). Powder supply device used TWIN-120-A / H 1.0, thermal spray gun used F4 Spray Gun. The surface of SUS316 base material with dimensions 50 × 50 × 6 mm was blasting. The spray gun was traversed on this substrate and formed a coating. The spray conditions are shown in Table1. These conditions were based on a condition that can be relatively dense coating with using conventional spraying material.

Morphology of the cross section of YSZ coatings has been observed by using SEM and Vickers hardness tested by using a micro vickers hardness tester MVK-E (AKASHI, Japan). The coatings was peeled from substrate to examine the residual carbon contamination deriving from the resin and evaluated by thermogravimetry.

3 Results and discussion

3.1 About micro resin fragments

Micro resin fragments which YSZ nanoparticles are dispersed with high concentration were obtained by crashing and screening composite bulks. Fig.2 shows micro resin fragments and individual surface of higher magnification observed by SEM. It is clear that nanoparticles of 200nm in average diameter were dispersed evenly.

Table 1. Spraying experimental conditions

Parameter	Spraying material	
	Conventional, Agglomerated	Micro resin fragments
Spraying distance (mm)	100	50
Number of transitions	11	25
Power input (kW)	33.6	33.6
Plasma gases	Ar/H_2	Ar/H_2
Powder feed gas	Ar	Ar
Powder feed rate (g/min)	40	26
Transition speed (m/min)	9.9	9.9

Fig. 2. (a) YSZ nanoparticles 57 vol% micro resin fragments, and (b) Individual surface of (a) with high magunification

3.2 The introduction of micro resin fragments to the plsma frame

Micro resin fragments were able to be supplied to spray gun by using a conventional powder feeder without closure in the transfer hose. Micro resin fragments injected by the powder injector were able to be introduced to the plasma flame and formed coatings on the substrates. On visually, the length of the spray formed by micro resin fragments which being introduced to the plasma flame (distance of the particles flying at a high temperature state) was half that of conventional spraying material and the agglomerated particles.

3.3 Observation of the thermal spray coatings

Cross sectional microstructure in YSZ coated layer formed on the SUS316 substrate are shown in Fig.3. In the case of conventional spraying material, the particles were melt and deposited, but there were micro-cracks and pores. In the case of agglomerated particles, uneven morphology was formed. There were parts where unmelted particles deposited and parts where particles deposited by melting. And also there were micro-cracks and pores. The coatings of sprayed micro resin fragments were formed dense coatings which micro-cracks and pores hardly exist, although slightly unmelted particles were deposited. The deposition speeds of micro resin fragments was about 13 μm/transition.

The Vickers hardness of each coatings was compared. Test force is 100 gf and dwell time is 15s. In the case of conventional spraying material, the maximum value of the Vickers hardness was 1033 HV. The morphology was uniform and a variation of hardness was small. In the case of agglomerated particles, although vickers hardness of the part where the particles were deposited by melting is 1206 HV and exceeded the case of conventional spraying materials, vickers hardness of the part where unmelted particles were deposited is 172 HV and significantly weak. In the case of micro resin fragments, the maximum value of the Vickers hardness was 1267 HV and exceeded the case of conventional spraying materials. The vickers hardness of the other part was obtained 882 HV. It was considered that the morphology was not completely uniform because slightly unmelted particles were deposited.

The residual carbon contamination deriving from the resin was evaluated by thermogravimetry. Since the pyrolysis temperature of the thermosetting resin is about 480 °C, the sample was heated up to 1000 °C at a heating rate of 10 °C/min in Ar. Fig.4 shows the result of thermogravimetry. Weight started to decrease from 289 °C and reduced about 1wt% during heating up to 1000 °C. It was considered that the carbon deriving from the resin have been most burned out during spraying, because the micro resin fragment contain 13.3 wt% thermosetting resin.

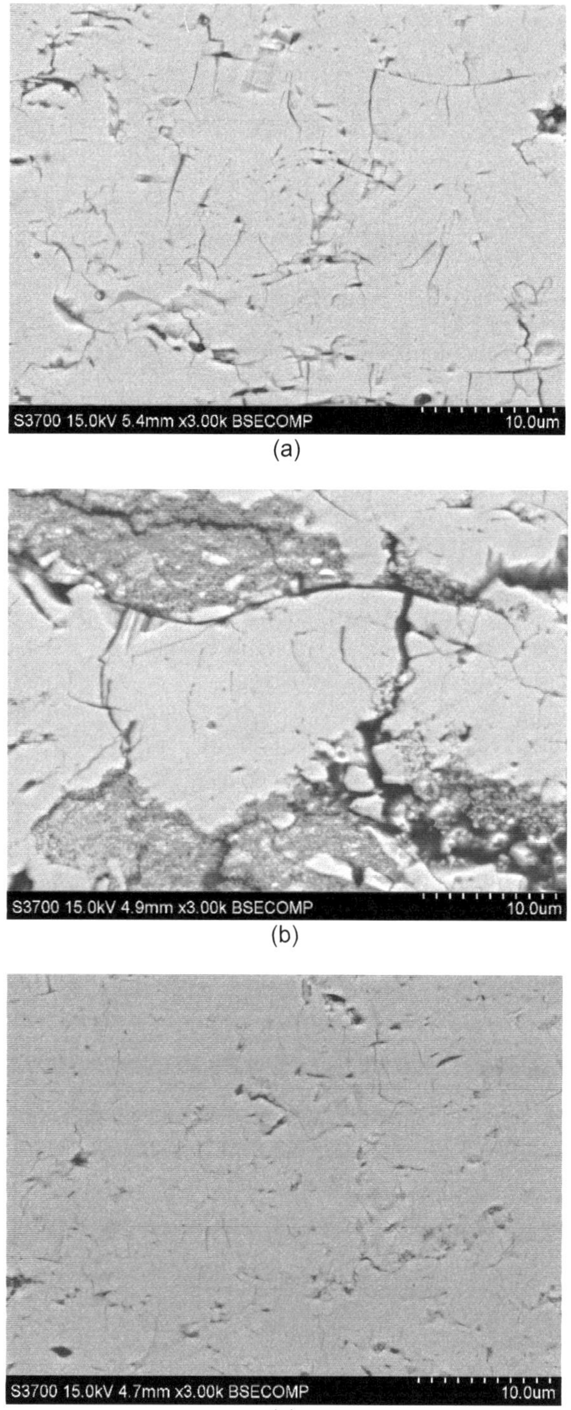

(c)

Fig. 3. Cross sectional microstructure in YSZ coated layer (a) by conventional spraying material, (b) by agglomerated particles, (c) by micro resin fragments.

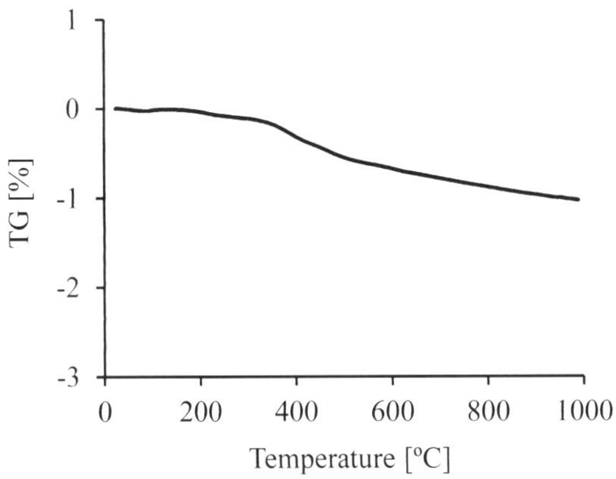

Fig. 4. TG curves of YSZ coated layer formed by micro resin fragments at a heating rate of 10 °C/min in Ar.

4 Conclusions

The thermal spraying process of nanoparticles has developed to form dense sprayed coatings with higher deposition speeds. The micro resin fragments with 57 vol% YSZ nanoparticles have been used as spraying material. The micro resin fragments could be introduced into the plasma flame with a conventional powder feeder. The resin is most burned out by introducing micro resin fragments into plasma flame, the heated YSZ nanoparticles only could be deposited on metal substrate. By using micro resin fragments, it could created dense coatings where micro-cracks and pores hardly exist, compared to the coatings using conventional spraying material. And the deposition speeds was about 13 µm/transition. The maximum value of the vickers hardness was 1267 HV and exceeded the case of conventional spraying materials. In the future, by mixing with resin and different types nanoparticles and different size nanoparticles, and by combinating them can be expected to create the composite nano morphology.

5 Literature

[1] A.Killinger, M.Kuhn, R.Gadow: High-Velocity Suspension Flame Spraying (HVSFS), a new approach for spraying nanoparticles with hypersonic speed, Surface & Coatings Technology 201, (2006), pp.1922-1929

[2] R.S.Lima, B.R.Marple: Thermal Spray Coatings Engineered from Nanostructured Ceramic Agglomerated Powders for Structual, Thermal Barrier and Biomedical Applications: A Review, Journal of Thermal Spray Technology, (2007), Volume 16(1), pp.40-63

[3] L.Pawlowski: Suspension and solution thermal spray coatings, Surface & Coatings Technology 203, (2009), pp.2807-2829

[4] R.S.Lima, A.Kucuk, C.C.Berndt: Evaluation of microhardness and elastic modulus of thermally sprayed nanostructured zirconia coatings, Surface and Coatings Technology 135, (2001), pp.166-172

[5] Y.Wang, S.Jiang, M.Wang, S.Wang, T.D.Xiao, P.R.Strutt: Abrasive wear characteristics of plasma sprayed nanostructured alumina/titania coatings, Wear 237, (2000), pp.176-185

[6] R.S.Lima, A.Kucuk, C.C.Berndt: Integrity of nanostructured partially stabilized zirconia after plasma spray processing, Materials Science and Engineering A313, (2001), pp.75-82

[7] M.Suzuki: Suspension Plasma Spraying, Jornal of the japan welding society, Vol. 83, (2014), No.2, pp.26-29, (in Japanese)

[8] D.Waldbillig, O.Kesler: The effect of solids and dispersant loadings on the suspension viscosities and deposition rates of suspension plasma sprayed YSZ coatings, Surface & Coatings Technology 203, (2009), pp.2098-2101

[9] S.Kirihara, Y.Itakura: Effective Injection of Ceramics Nanoparticle Pastes into Plasma Spray for Speedy Layer Formation, Quarterly jornal of the japan welding society, Vol. 33, (2015), No.2, pp.148-151

[10] S.Lee, N.J.Themelis, M.J.Castaldi: High-Temperature Corrosion in Waste-to-Energy Boilers, Journal of Thermal Spray Technology, (2007), Volume 16(1), pp.1-7

[11] Y.Kawahara: Application of High Temperature Corrosion-Resistant Materials and Coatings Under Severe Corrosive Environment in Waste-to-Energy Boilers, Jornal of Thermal Spray Technology, (2007), Volume 16(2), pp.202-213

[12] A. Schütz, M.Günthner, G.Motz, O. Greißl, U.Glatzel: High temperature (salt melt) corrosion tests with ceramic-coated steel, Materials Chemistry and Physics 159, (2015), pp.10-18

Fracture Characteristics of High-Velocity Suspension Flame-Sprayed Aluminum Oxide Coatings

J. Kiilakoski[1], A. Lutoschkin[2], M. Plachetta[2], M. Apostol[1], H. Koivuluoto[1], P. Müller[2], A. Killinger[2], P. Vuoristo[1]
[1]Department of Materials Science, Tampere University of Technology, Tampere/FIN
[2]Institute for Manufacturing Technologies of Ceramic Components and Composites, University of Stuttgart, Stuttgart/DE

A recent trend of development in thermally sprayed ceramic coatings is to decrease their most defining attribute, brittleness. One possible strategy to improve ductility is to alter the coating towards a finer microstructure: Due to the smaller size of the splats, the amount of crack-arresting interfaces increases. High-velocity suspension flame-spraying (HVSFS) is an emerging thermal spraying technique, which utilizes a liquid as the powder carrier instead of gas, as in the more traditional spraying techniques. This allows for nano- or submicrometer-sized granular structures to be achievable with HVSFS. The aim of this work is to evaluate the brittleness of suspension sprayed aluminum oxide (Al_2O_3) –coatings with various methods, including Vickers indentation fracture toughness, 4-point bending and High-Velocity Particle Impactor. Coatings were sprayed with HVSFS using suspensions of isopropanol and water solvents. Additionally, high-velocity oxy-fuel (HVOF) –sprayed commercially available Al_2O_3-powder feedstock was used as a reference. Differences in the melting degrees of the HVSFS-coatings were found, leading to different cracking behaviors in all tests. Both HVSFS-coatings showed improvements in strain-to-fracture,while one showed improvements in energy absorbance and fracture toughness compared to traditional HVOF-coating.

1 Introduction

Thermally sprayed ceramic coatings are commonly used in applications against corrosion, wear and for thermal insulation. Such applications are found in eg. paper- and process industries. [1–4] The wear resistance of these ceramic coatings is usually limited by their failure which happens - as in traditional bulk ceramics as well - by brittle fracture. [5–8] This is due to the nature of ceramics themselves, but also to weak links in the structure of the coating. These structural weak links include pores, inclusions and cracks already present in the as-sprayed coating, or from thermal cycling. [1,7]

However, in some studies the wear mechanism of these coatings has been altered by the change of primary particle size and spray method; Bolelli et al. have found increases in both fracture toughness [9] and wear resistance [8] when spraying Al_2O_3 and Cr_2O_3 powders with a High-Velocity Oxygen-Fuel –torch (HVOF) compared to an Atmospheric Plasma Spray torch (APS). Also Lima et al. [10] have found a consistent increase in the Weibull modulus of TiO_2 coatings when sprayed with HVOF as opposed to many other materials and methods from literature. Later the authors found that the spraying of nanostructured TiO_2 [11,12] and Al_2O_3-13TiO_2 [13] lead to superior wear resistance owing to the crack deflection effect of the semi-molten particles. Also using a HVOF-torch enhanced the improvement.

During the last decade a viable method of spraying coatings from nanosized feedstock has been studied. This method utilizes a colloidal suspension of fine powder as feedstock in HVOF-spraying and is befittingly called High-Velocity Suspension Flame-Spraying (HVSFS) [14], or suspension-HVOF (S-HVOF) [15]. Various materials, such as Al_2O_3 [16], Cr_2O_3 and YSZ [17], TiO_2 [18] as well as hydroxyapatite and bioactive glass [19] have been investigated. Dongmo et al. [20] have modelled different combustion chamber and nozzle geometries and Bolelli et al. [21] have done experimental tests on coatings prepared with different set-ups. The research has led to the technology being commercialized and some suspensions as well as equipment is already available. [15,22]

The mechanical properties of HVSFS-sprayed Al_2O_3-coatings have been studied by Bolelli et al. [21], who found that these coatings exhibited a compressive residual stress state regardless of the spray system used. Furthermore, the bond strength of some of the coatings was up to 70 MPa and sliding wear behaviour superior to that of the reference HVOF-coating. The authors also found that the optimal range of particle size to use in the suspensions is 1-2 µm, since smaller particles, <500 nm, do not attain the required momentum and have a tendency to re-solidify during flight, while larger particles remain partly unmolten in the coating. Regarding advances in the phase composition of the coatings, Toma et al. [23] have found high α-Al_2O_3 –content for coatings sprayed with HVSFS while retaining the hardness of HVOF-Al_2O_3 –coatings.

The structure and sliding wear properties of thermally sprayed ceramic coatings have been studied extensively, but so far no effort aside from indentation fracture toughness experiments has been made to evaluate an improvement in the toughness of the coatings. This study aims to bridge the gap between the toughness characteristics of the coating and the properties of the feedstock and spray system. Novel toughness measurements, such as instrumented four-point bending and high-velocity impact, are included along with traditional fracture toughness measurement.

2 Experimental

The coatings were prepared from commercially available powder materials. For the HVOF-sprayed Al_2O_3-coating, Amperit 740.008 (-25+5 µm) from H.C. Starck GmbH (Germany) was used. The Al_2O_3 in the suspensions was MR52 (d_{50}=1-1.7 µm) from Albemarle GmbH (Germany). Two suspensions were prepared, one with isopropanol (IPA) as solvent and one with water. The HVOF spraying, and the manufacturing and spraying of the water-based suspension was performed at Tampere University of Technology (TUT), and the manufacturing and spraying of the isopropanol-based suspension was performed at the Institute for Manufacturing Technologies of Ceramic Components and Composites (IFKB) at the University of Stuttgart. The spray torch in all cases was TopGun from GTV (Germany). The spray parameters are presented in **Table 1.**

Table 1. The spray parameters used in the manufacturing of the coatings.

Spray	Ethene flow [l/min]	Oxygen flow [l/min]	Oxygen-ethene ratio	Feedstock flow [ml/min]	Spray distance [mm]
HVOF	93	270	2.90	40	150
HVSFS-H_2O	75	230	3.07	70	80
HVSFS-IPA	100	300	3.00	50	120

The characterization of the coatings was performed by SEM (XL30, Philips, The Netherlands) and the surface roughnesses were measured as an average of three measurements with the SJ 301 from Mitutoyo (Japan). Coating thicknesses were measured from the SEM-cross-section as an average of ten measurements.

The fracture testing was performed with three different techniques: Vickers Indentation Fracture Toughness testing (VIFT), High-Velocity Particle Impact (HVPI) and four-point bending with acoustic emission measurement. The VIFT measurements were done in IFKB, while the two other in TUT. The VIFT tests were carried out with a Vicker hardness tester (Bareiss, Germany). The length of the resulting cracks was measured with an optical light microscope (MEF4M, Leica, Germany). A Fischerscope H100 (Helmut Fischer GmbH, Germany) was applied to measure the hardness and Young's modulus of the coatings. For the HVOF- and HVSFS-IPA-coatings 9.81 N of force was utilized, while for the HVSFS-H_2O-coating 4.80 N gave a more clear indentation mark.

The HVPI test has been developed at Tampere Wear Center (TWC) and has been utilized previously to test steels [24,25], hybrids [26,27], MMC-coatings [28] and rubbers [29]. In the test, samples of 40x50 mm were prepared by grinding with SiC-papers up to P1200 and consequently polishing up to 1 µm diamond slurries. The samples were shot with a chromium steel sphere of 5 mm diameter. Three separate repetition were done. The impact velocity was 50 m/s and the impact angle 30°. The event was recorded with a high-speed camera and the absorbed energy was calculated according to

$$E_d = E_{inci} - E_{refl} = \frac{1}{2} m_p (v_{incit} - v_{exit})^2 \qquad (1)$$

where m_p is the mass of the projectile, and v_{incit} the incident and v_{exit} the exit velocities of the projectile. When calculating the absorbed energy from the impact and relating it to the crater volume as measured by an optical profilometer (InfiniteFocus, Alicona, Austria) an estimation was made of the ability of the system (coating and substrate) to absorb impact energy. The resulting impact craters were further characterized with an SEM from the top and the cross-section.

The four-point bending setup was similar to what has successfully been used previously by other research groups [30-32]. The bending was performed with a universal testing machine (Instron, USA). Samples were grit blasted on both sides before spraying, and their dimensions were 180x25x3 mm. The samples were bent with the coating on the side of tension, with a strain gauge (KYOWA, Japan) in the middle of the sample roughly on the point of highest tension and an AE-sensor (Brüel&Kjær, Denmark) on the substrate side of the sample. The inner and outer span of the pins of the four-point bending jig was 45 mm and 110 mm, respectively. The deflection rate was 10 mm/min and the total displacement was 10 mm. The AE-signal was measured with LabView SignalExpress and analysed with DIAdem (both from National Instruments, USA).

3 Results and discussion

3.1 Coating microstructures

The microstructures of the coatings were studied with SEM (**Fig. 1**). The HVOF-sprayed Al_2O_3 coating consists of well-melted and flattened splats, and the coating is dense and has a homogeneous microstructure. Some pull-outs can be observed owing likely to domains of some unmolten spray

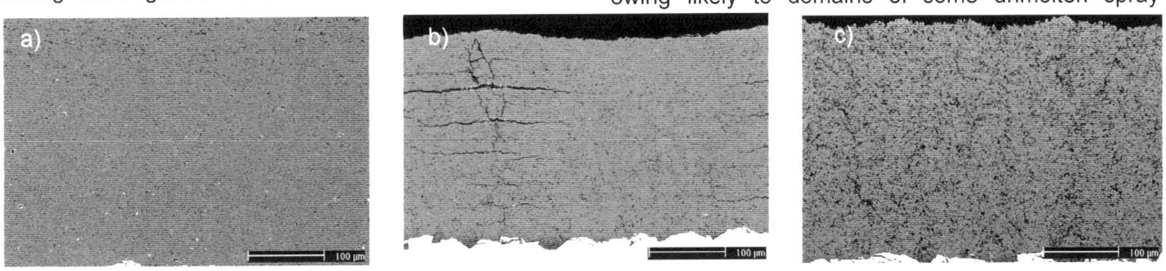

Fig. 1. SEM-images (BSE) of the cross-sections of the coatings a) HVOF-Al_2O_3, b) HVSFS-H_2O-Al_2O_3, and c) HVSFS-IPA-Al_2O_3.

powder particles. The surface of the coating is smooth with R_a=2.02 µm and R_z=13.84 µm. The coating thickness was 306 µm.

The HVSFS-H_2O-Al_2O_3–coating was generally dense and consisted of well-melted and flattened lamellae. Small, unmolten primary oxide particles were clearly visible in a matrix of molten material. However, the coating was severely cracked in many locations with tree-like cracks going vertical, and occasionally horizontally between lamellae. Similar defects in suspension-HVOF –sprayed Al_2O_3-coatings were found and characterized by Toma et al. [33]. They supposed that the occurance of cracks is due to high cooling rates between passes leading to residual stresses from thermal cycling during spraying. The respective coating thickness was 231 µm, R_a=3.48 µm and R_z=19.74 µm.

The HVSFS-IPA-Al_2O_3 -coating exhibited somewhat more apparent porosity than HVSFS-H_2O-Al_2O_3. This is likely due to pull-outs of unmolten particles which, according to fracture surface studies, seemed to be more frequent. However, the coating was more uniform without severe cracks. Likely the thermal cycling was greatly reduced by the longer spray distance, which also resulted in the greater amount of unmolten particles. However, such an effect was not observed in previous studies using the same material and parameter sets.[21] The respective coating thickness was 273 µm, R_a=3.20 µm and R_z=19.64 µm.

3.2 HVPI results and characterization

Impact craters were produced on the coatings during the test, and no coating was left intact without cracking. The crater profiles depicted material piling on the exit side of the crater in all cases. The behaviour is similar to what has been observed in steel specimen [24], but not quite as smooth. Maximum peak heights and crater depths for the coatings were HVOF: 21.53 µm and -13.59 µm, HVSFS-H_2O: 23.14 µm and -10.38 µm, and HVSFS-IPA: 62.43 µm and -44.54 µm, respectively.

SEM-images of the top-views and cross-sections of the craters are presented in **Fig. 2**. The cross-sections show that the differences in these coatings mainly arise from different coating thicknesses after sample preparation, where the HVSFS-IPA coating was abraded the most. Its profile is the smoothest and has clearly taken the form of the substrate while the coating has deformed along. Indeed, the coating showed no major cracks, but the damage was distributed quite evenly in the coating. In higher magnifications it was evident that the cracks were either arrested or deflected by unmolten primary particles. The deflection of a crack absorbs a lot of the cracks energy and therefore reduces its length. Similar behaviour was seen in the HVSFS-H_2O -coating, but to a far lesser extent. In fact, the HVSFS-H_2O –coating behaved very similarly with the HVOF-coating. The results of energy absorbance calculations are shown in **Fig. 3**. It can be seen that the greater melting degree of the water-based solution makes its behaviour similar to the HVOF-coating also in this case, while the HVSFS-IPA –coating has a lot smaller energy absorbance, likely due to a larger part of the impact being distributed into to the substrate.

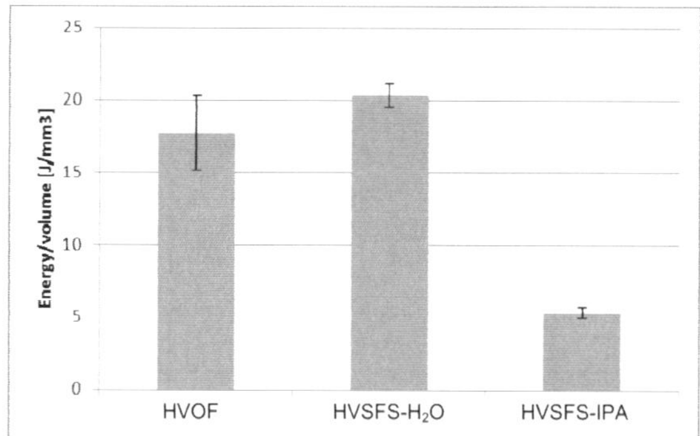

Fig. 3. The energy absorbance of the coating-substrate -system as normalized to the crater volume.

3.3 Four-point bending

The strain-to-fracture values are shown in **Fig. 4**. The first thing to mention is that generally the scattering of the values has been noted to be an indication of inhomogeneous coating quality. Therefore the HVOF-sprayed Al_2O_3 coating depicted quite a repeatable behaviour in the test, which presumably arose from a homogeneous coating structure. The average strain-to-fracture of the HVOF-coating was 0.38 %. More scattering can be seen for both HVSFS coatings, more so for the one from water-based suspension. This seems reasonable since the cross-sections of the

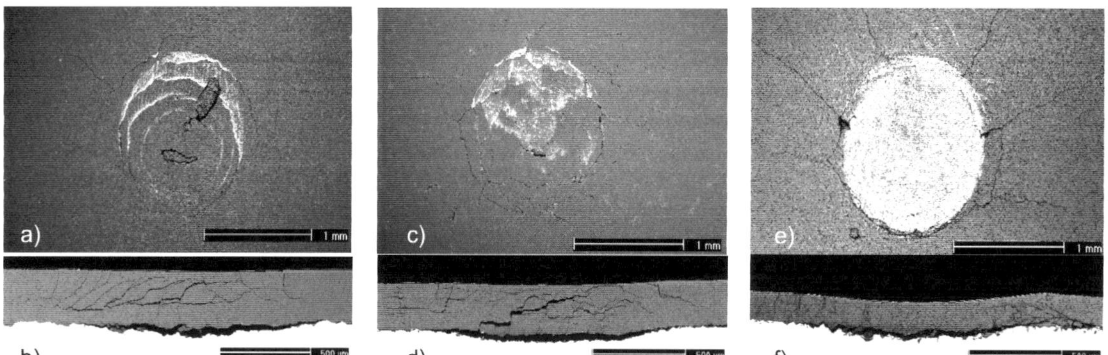

Fig. 2. SEM images of the impact craters and their cross-sections from the HVPI-test a), b): HVOF-Al_2O_3, c), d): HVSFS-H_2O-Al_2O_3, and e),f): HVSFS-IPA-Al_2O_3.

coatings quite clearly showed the inhomogeneous structure with lots of cracking in the HVSFS-H_2O – coating, which had an average strain-to-fracture of 0.45 %. The HVSFS-IPA –coating had the largest strain-to-fracture of approximately 0.7 %, which is a sign of greater formability and therefore toughness. The strain is greatly affected by the residual stress-state of the coating, with compression aiding to achieve higher values. [30,31] A noteworthy observation was that the amount of AE counts, which means peaks above the threshold value in the signal, was a lot lower for the HVSFS-IPA-coating (150) than HVSFS-H_2O (10000) and HVOF (28691). This implies small-scale cracking where probably cracks are deflected and terminated by the unmolten primary particles.

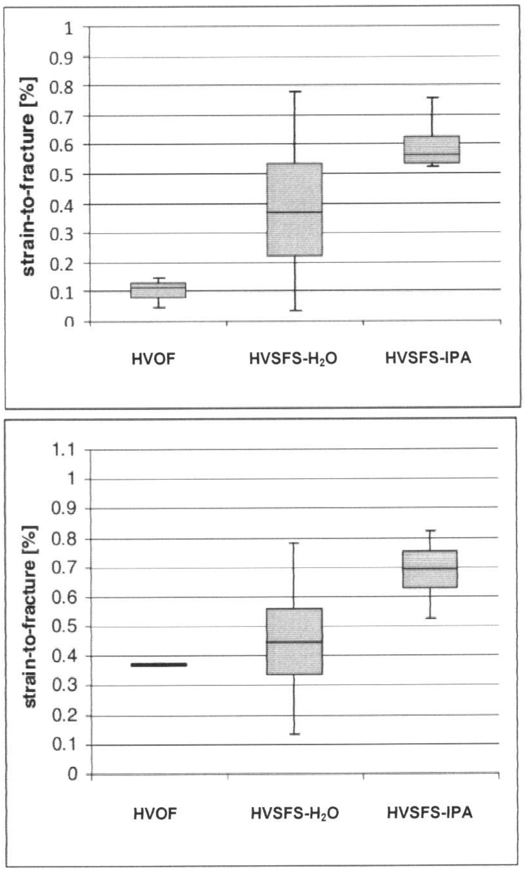

Fig. 4. Box plot of strain-to-fracture obtained for the coatings by means of a) the first occurrence of a peak in voltage reaching a defined threshold value b) the arbitrary AE energy limit.

3.4 Vickers indentation

The fracture toughness values of the coatings are presented in **Table 2.** Fracture toughnesses of the coatings calculated with various methods.While there was a significant scatter between different methods, the order of the coatings remained the same.

Table 2. Fracture toughnesses of the coatings calculated with various methods.

Coating	Fracture toughness [$MPam^{-0.5}$]			
	Niihara (Palmqvist)	Niihara (Median)	Anstis	Evans
HVOF	2.64	3.56	2.34	2.87
HVSFS-H_2O	3.05	4.2	2.81	3.22
HVSFS-IPA	1.61	2.13	1.41	1.66

The fracture toughness values were higher in the HVSFS-H_2O –coating than in the HVOF-coating, which falls in line with the results from the 4-point-bending results as well. However, unlike in the bending test, the HVSFS-IPA coating had a lower fracture toughness than the other coatings. This can be explained by their lower hardness, which is a term in all equations for calculating fracture toughness. The hardnesses were as follows: HVOF 970 $HV_{0.3}$, HVSFS-H_2O 800 $HV_{0.3}$ and HVSFS-IPA 690 $HV_{0.3}$.

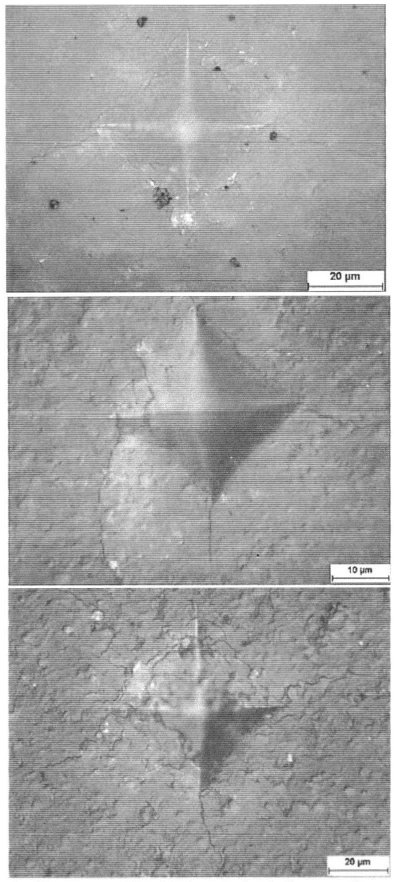

Fig. 5. Optical images of the Vickers indentations on the cross-sections of the coatings. a) HVOF-Al_2O_3 (9.81 N), b) HVSFS-H_2O-Al_2O_3 (4.90 N), c) HVSFS-IPA-Al_2O_3 (4.90 N).

Images of the indentations are presented in **Fig. 5**. The indentation of the HVOF-coating revealed an anisotropic behaviour where cracks in the horizontal direction exhibited significantly greater lengths than vertical ones. This underlines the weakness of the interfaces between lamellae of coating. Similar behaviour was not seen in the HVSFS-coatings. However, similar cracking occurred in a similar fashion to all directions. In HVSFS-H_2O –coating the cracks are not easily measurable as their path is not straight. This might be due to the phenomenon witnessed in the HVPI test, where cracks deflected from small unmolten primary particles in the coating. Similar crack behaviour occurred also in the HVSFS-IPA –coating.

4. Conclusion

In this study, Al_2O_3-coatings prepared by HVSFS from water- and isopropanol-based suspensions were compared with a HVOF-sprayed Al_2O_3-coating. Three different experimental techniques were applied to get closer to understanding the factors behind fracture behaviour of thermally sprayed ceramic coatings.

Microstructural characterization revealed large-scale cracking in the HVSFS-coating from the water-based suspension, while the isopropanol suspension produced a homogeneous coating with generally well molten particles. However, the coating also contained a high number of unmolten particles. This can be explained by thermal cycling during spraying of the water-based suspension owing to a shorter spraying distance, as well as the larger heat flux from isopropanol and longer spray distance during the spraying of the HVSFS-IPA –coating. Further investigation of this finding is needed, since previous studies to some extent reported contrary observations.

In the HVPI test, the coatings behaved quite differently. The HVSFS-IPA coating had vast small-scale cracking but no critical cohesive failure. The HVSFS-H_2O and HVOF coatings both had severe cohesive failure under impact. In all cases the substrate deformed under the impact, and caution should be exercised when comparing coatings of different thicknesses. The test does, however, give a unique visual perspective on the internal cracking behaviour of coatings under high-speed impacts.

The bending test partially confirmed the small-scale cracking seen in the HVPI test for HVSFS-IPA, as the AE counts were several magnitudes fewer than for HVSFS-H_2O and HVOF –coatings. Also, the energy of the cracks was small and highest strain-to-fracture was achieved with the isopropanol-based suspension. This might suggest a kind of "quasi-plasticity", as suggested by Lawn [34], where instead of real ductility, the ceramic could deform by stress relief via non-critical micro-scale cracking.

The indentation fracture toughness measurement of the coatings proved difficult, since the resulting cracks did not have clear straight paths. What was seen, hovewer, is that both HVSFS-coatings behaved isotropically, while the HVOF coating preferred to crack along lamella interfaces. In the case of these coatings, hardness was seen to decrease from moving from the HVOF-coating to the HVSFS-sprayed coatings.

Acknowledgements

The work has been done within FIMECC Ltd and its HYBRIDS and BSA programmes. The authors gratefully acknowledge the financial support from Tekes (Finnish Funding Agency for Technology and Innovation) and the participating companies. The authors would like to thank Mr. Mikko Kylmälahti, of Tampere University of Technology, Department of Materials Science,for spraying the HVOF and HVSFS-H_2O -coatings.

References

[1] Vuoristo, P.: Thermal spray coating processing. In: HASHMI, S. (ed.): *Comprehensive Materials Processing*. 4. First Edit. ed. : Elsevier Ltd, Oxford, UK, (2014), pp. 229/76.

[2] Nakahira, H.: Blast-furnace tuyere having excellent thermal shock resistance and high durability, United States Patent, Publication Number US3977660. Toyo Calorizing Ind. Co. Ltd.,(1976).

[3] Di Girolamo, G., E. Serra: Thermally sprayed nanostructured coatings for anti-wear and TBC applications: State-of-the-art and future perspectives. In: ALIOFKHAZRAEI, M. (ed.): *Anti-Abrasive Nanocoatings - Current and Future Applications*. Cambridge, UK : Woodhead Publishing Limited, (2015), pp. 513/41.

[4] Pawlowski, L.: *The Science and Engineering of Thermal Spray Coatings*. Second Edi. ed. West Sussex, England : John Wiley & Sons, (2008).

[5] Lawn, B.: *Fracture of Brittle Solids* : Cambridge University Press, (1993).

[6] Xie, Y., H.. Hawthorne: The damage mechanisms of several plasma-sprayed ceramic coatings in controlled scratching. *Wear* 233–235 (1999), pp. 293–305.

[7] Erickson, L.C., R. Westergård, U. Wiklund, N. Axén, H.M. Hawthorne, S. Hogmark: Cohesion in plasma-sprayed coatings — a comparison between evaluation methods. *Wear* 214 (1998), Issue 1, pp. 30/7.

[8] Bolelli, G., L. Lusvarghi, T. Manfredini, F.P. Mantini, E. Turunen, T. Varis, S.P. Hannula: Comparison between plasma- and HVOF-sprayed ceramic coatings. Part II: tribological behaviour. *International Journal of Surface Science and Engineering* 1 (2007), Issue 1, pp. 62–79.

[9] Bolelli, G., L. Lusvarghi, T. Manfredini, F. Pighetti Mantini, R. Polini, E. Turunen, T. Varis, S.-P. Hannula: Comparison between plasma- and HVOF-sprayed ceramic coatings. Part I : microstructure and mechanical properties. *International Journal of Surface Science and Engineering* 1 (2007), Issue 1, pp. 38–61.

[10] Lima, R., B. Marple: High Weibull Modulus HVOF Titania Coatings. *Journal of Thermal Spray Technology* 12 (2003), Issue 2, pp. 240/9.

[11] Lima, R.S., B.R. Marple: From APS to HVOF spraying of conventional and nanostructured titania

feedstock powders: A study on the enhancement of the mechanical properties. *Surface and Coatings Technology* 200 (2006), Issue 11, pp. 3428/37.
[12] Lima, R.S., B.R. Marple: Enhanced ductility in thermally sprayed titania coating synthesized using a nanostructured feedstock. *Materials Science and Engineering: A* 395 (2005), Issue 1-2, pp. 269/80.
[13] Lima, R.S., C. Moreau, B.R. Marple: HVOF-Sprayed Coatings Engineered from Mixtures of Nanostructured and Submicron Al2O3-TiO2 Powders: An Enhanced Wear Performance. *Journal of Thermal Spray Technology* 16 (2007), Issue 5-6, pp. 866/72.
[14] Killinger, A., M. Kuhn, R. Gadow: High-Velocity Suspension Flame Spraying (HVSFS), a new approach for spraying nanoparticles with hypersonic speed. *Surface and Coatings Technology* 201 (2006), Issue 5, pp. 1922/29.
[15] Toma, F.-L., A. Potthoff, L.-M. Berger, C. Leyens: Demands, Potentials, and Economic Aspects of Thermal Spraying with Suspensions: A Critical Review. *Journal of Thermal Spray Technology* 24 (2015), Issue 7, pp. 1143/52.
[16] Bolelli, G., J. Rauch, V. Cannillo, A. Killinger, L. Lusvarghi, R. Gadow: Microstructural and Tribological Investigation of High-Velocity Suspension Flame Sprayed (HVSFS) Al2O3 Coatings. *Journal of Thermal Spray Technology* 18 (2008), Issue 1, pp. 35–49.
[17] Rauch, J., G. Bolelli, A. Killinger, R. Gadow, V. Cannillo, L. Lusvarghi: Advances in High Velocity Suspension Flame Spraying (HVSFS). *Surface and Coatings Technology* 203 (2009), Issue 15, pp. 2131/38.
[18] Toma, F.-L., L.-M. Berger, I. Shakhverdova, B. Leupolt, A. Potthoff, K. Oelschlägel, T. Meissner, J.A.I. Gomez, Y. de Miguel: Parameters Influencing the Photocatalytic Activity of Suspension-Sprayed TiO2 Coatings. *Journal of Thermal Spray Technology* 23 (2014), Issue 7, pp. 1037/53.
[19] Bolelli, G., D. Bellucci, V. Cannillo, R. Gadow, A. Killinger, L. Lusvarghi, P. Müller, A. Sola: Comparison between Suspension Plasma Sprayed and High Velocity Suspension Flame Sprayed bioactive coatings. *Surface and Coatings Technology* 280, Elsevier B.V. (2015), pp. 232/49.
[20] Dongmo, E., R. Gadow, A. Killinger, M. Wenzelburger: Modeling of Combustion as well as Heat, Mass, and Momentum Transfer During Thermal Spraying by HVOF and HVSFS. *Journal of Thermal Spray Technology* 18 (2009), Issue 5-6, pp. 896–908.
[21] Bolelli, G., V. Cannillo, R. Gadow, A. Killinger, L. Lusvarghi, T. Manfredini, P. Müller: Properties of Al2O3 coatings by High Velocity Suspension Flame Spraying (HVSFS): Effects of injection systems and torch design. *Surface and Coatings Technology* 270 (2015), pp. 175/89.
[22] Killinger, A., P. Müller, R. Gadow: What Do We Know, What are the Current Limitations of Suspension HVOF Spraying? *Journal of Thermal Spray Technology* 24 (2015), Issue 7, pp. 1130/42.
[23] Toma, F., L. Berger, C. Stahr: Microstructures and functional properties of suspension-sprayed Al_2O_3 and TiO_2 coatings: an overview. *Journal of Thermal SprayTechnology* 19 (2010), Issue 1-2, pp. 262/74.
[24] Lindroos, M., M. Apostol, V. Kuokkala, A. Laukkanen, K. Valtonen, K. Holmberg, O. Oja: Experimental study on the behavior of wear resistant steels under high velocity single particle impacts. *International Journal of Impact Engineering* 78 (2015), pp. 114/27.
[25] Lindroos, M., M. Apostol, V. Heino, K. Valtonen, A. Laukkanen, K. Holmberg, V.-T. Kuokkala: The Deformation, Strain Hardening, and Wear Behavior of Chromium-Alloyed Hadfield Steel in Abrasive and Impact Conditions. *Tribology Letters* 57 (2015), pp. 1–11.
[26] Sarlin, E., M. Apostol, M. Lindroos, V.-T. Kuokkala, J. Vuorinen, T. Lepistö, M. Vippola: Impact properties of novel corrosion resistant hybrid structures. *Composite Structures* 108 (2014), pp. 886/93.
[27] Sarlin, E., M. Lindroos, M. Apostol, V.T. Kuokkala, J. Vuorinen, T. Lepistö, M. Vippola: The effect of test parameters on the impact resistance of a stainless steel/rubber/composite hybrid structure. *Composite Structures* 113 (2014), pp. 469/75
[28] Waudby, R., T. Varis, T. Suhonen, K. Holmberg, M. Apostol, M. Lindroos, V. Kuokkala: High velocity impact testing of thermal spray hard carbide coatings on steel substrates. In: *Proceedings of 5th World Tribology Congress WTC 2013* (2013), pp. 3–6.
[29] Molnar, W., S. Nugent, M. Lindroos, M. Apostol, M. Varga: Ballistic and numerical simulation of impacting goods on conveyor belt rubber. *Polymer Testing* 42 (2015), pp. 1–7.
[30] Cox, L.C.: The four-point bend test as a tool for coating characterization. *Surface and Coatings Technology* 36 (1988), Issue 3-4, pp. 807/15.
[31] Voyer, J., H. Kreye: Determination of cracking resistance of thermal spray coatings during four-point bend testing using an acoustic emission technique. *Journal of Thermal Spray Technology* 12 (2003), pp. 416/26.
[32] Ma, X.Q., S. Cho, M. Takemoto: Acoustic emission source analysis of plasma sprayed thermal barrier coatings during four-point bend tests. *Surface and Coatings Technology* 139 (2001), Issue 1, pp. 55–62.
[33] Toma, F., S. Langner, M.M. Barbosa, L. Berger, C. Rödel, A. Potthoff, D. Dresden: Influence of the suspension characteristics and spraying parameters on the properties of dense suspension-HVOF sprayed Al2O3 coatings. In: *International Thermal Spray Conference & Exposition (ITSC) 2011*. Hamburg: DVS Media GmbH, Düsseldorf, Germany, (2011), pp. 421/26.
[34] Lawn, B.: Indentation of ceramics with spheres: a century after Hertz. *Journal of the American Ceramic Society* 94 (1998), pp. 1977/94.

Thermal cycling behavior of suspension plasma sprayed yttria partially-stabilized zirconia thick thermal barrier coating

Xinghua ZHONG, Jinxing NI, Huayu ZHAO, Fang SHAO, Chenguang LIU, Shunyan TAO
(Key Laboratory of Inorganic Coating Materials, Shanghai Institute of Ceramics, Chinese Academy of Sciences, Shanghai 201899, China)

Wenzhong TAO, Yunsheng WANG
(Jiangsu Lida Hi-tech Special Material Co., Ltd., Changshu, Jiangsu 215506, China)

The present paper focused on evaluating the cyclic thermal shock performance of the suspension plasma sprayed (SPS) 7.5wt% Y_2O_3 partially stabilized ZrO_2 (7.5YSZ) thick thermal barrier coatings (TBCs). The SPS 7.5YSZ coating in thickness of approximately 720 microns was deposited on a 150 micron-thick CoNiCrAlY bond coat. The vertically segmented cracks were observed in the SPS 7.5YSZ coatings. The phase analysis shows that the coating crystallized in non-transformable tetragonal phase of zirconia. The thermal cycling test of the SPS 7.5YSZ coatings was carried out on a burner rig with propane-oxygen flame which heated the coating surface at temperature of 1523 K, followed by quenching to temperature of 423 K by compressed air. The results obtained shows that the SPS 7.5YSZ coatings performs longer thermal shock life relative to the atmospheric plasma sprayed (APS) counterparts under the same test conditions, which is probably attributed to the improvement in strain tolerance through the vertically segmented cracks therein. The failure behaviour of the SPS coatings is distinctly different from that of the APS coatings, which is supposed to be related to the special microstructure.

1 Introduction

Thermal barrier coatings (TBCs) have been widely used in advanced gas turbines and diesel engines in order to improve operating temperature of hot-section metal components as well as the efficiency of engines.
The state-of-the-art TBCs in thickness of 100-400 microns are double layered systems, consisting of an thermal insulated yttria partially-stabilized zirconia (YSZ) top coat and an underlying metallic bond coat. This typical TBCs can only provide a temperature reduction of 190 K during service at high temperature [1-6]. Therefore, achieving the improvements in the process efficiency and performance of engines will necessitate further increase in the thermal insulation of the hot section components.

During the past few years, an interest has been taken in increasing the thickness of the currently used YSZ top coat, and several studies upon atmospheric plasma sprayed thick TBCs (TTBCs) in thickness of more than 700 μm for lower thermal conductance application have been published [7-9]. However, some unavoidable reliability deteriorations, such as poor adhesion and low thermal shock resistance, will arise with increasing the top coat thickness. It is well known that the increased thickness of TTBCs leads to an improved temperature gradient through the coating, which result in the higher thermal mismatch stresses generated inevitably under the service conditions. In order to overcome this problem, various microstructural modification processes have been developed to improve the strain tolerance of the APS TTBCs, including manufacturing graded or multilayered structure coatings, using different types of post treatments and introducing porosity or segmentation cracks into the top coat via spray parameters controlling [10-12]. The TTBCs with segmentation cracks is found to have superior thermal shock resistance, which is attributed to that these cracks can compensate tensile stress with an opening during tensile loading. This manner is similar to the behavior of the columnar microstructure of electron beam-physical vapor deposition (EB-PVD) coatings [13-16]. The APS segmented TTBCs are fabricated through controlling spray parameters generally, such as improving the preheating temperature of the substrate, and also a higher passage thickness during the deposition [9, 16]. The density of the top coat is therefore higher than 88% that of the theoretical density, which increases the thermal conductivity and thus offsets the improved strain tolerance to some extent [17].

Recently, some experimental studies have indicated suspension plasma sprayed (SPS) and solution precursor plasma sprayed (SPPS) TBCs exhibit a unique microstructure of high segmentation cracks density with relatively high porosity levels simultaneously, which contribute to their superior thermal cyclic durability as well as a low thermal conductivity [18-20]. These results indicate a promising potential to fabricate high strain tolerated TTBCs actually. The aim of the present work is to study the possibility of the application of the suspension plasma spraying process for advanced TTBCs with high thermal cycling shock resistance. Therefore, YSZ top coat with segmentation cracks were produced via SPS, and the thermal cycling shock resistance was investigated through the thermal cycling tests. The thermal cycling behavior and the failure mechanism of the SPS YSZ TBC are studied.

2 Experimental procedure

GH3128 Ni-based superalloy plates (Beijing CISRI-GAONA Materials & Technology Co., Ltd., China) with a dimension of δ5mm×Φ30mm were used as substrates, which were grit blasted using alumina powder, followed by ultrasonically cleaning in ethanol before the spraying processes. A bond coat was deposited by the MultiCoat™ spraying system equipped with a DJ2600 gun (High Velocity Oxy-Fuel (HVOF) gun, Sulzer Metco AG, Wohlen, Switzerland) using a commercial CoNiCrAlY powders (AMDRY 9951, Sulzer Metco, USA). Segmentation-cracks structured YSZ top coatings were deposited by SPS system (Northwest Mettech Corp., North Vancouver, Canada) using the Axial III™ plasma torch with a NanoFeed™ models 350 suspension feeder. The Axial III™ plasma torch is a three cathode plasma torch, and featured by axially injecting the atomized suspension feedstock into the central core of the plasma. A commercial aqueous suspension of 7.5wt.% yttria-stabilized zirconia (7.5YSZ) submicron-particles (Jiangsu Lida Hi-tech Special Material Co., Ltd., China) was used as feedstock. The median particle diameter (D50) is 0.06 μm and the fraction of solid phase in the suspension is approximately 20 vol.%. The thickness of the as-sprayed CoNiCrAlY bond-coats were about 150 μm, whereas the thickness of the YSZ top-coat was about 720 μm. Table 1 and Table 2 present the detailed spraying parameters used for each coating sample.

Table 1. Spraying parameters for the bond coat

Item	Parameters
Hydrogen pressure (MPa)	0.8-1.0
Oxygen pressure (MPa)	1.0-1.2
Hydrogen flow rate (SLPM)	650-730
Oxygen flow rate (SLPM)	160-215
Carrier gas N_2 (SLPM)	20
Spray distance (mm)	270-290
Powder feed rate (g/min)	30-35

Table 2. Spraying parameters for the top coat

Item	Parameters
Plasma gas mixture	$Ar+N_2+H_2$
Plasma gas flow rate (SLPM)	Ar: 170-176 N_2: 48-53 H_2: 20-25
Carrier gas (SLPM)	14.5-15.2
Spray distance (mm)	70-90
Nozzle diameter (mm)	9.5
Powder inject diameter (mm)	1.5
Suspension feed rate (ml/min)	40

The phase identifications of the as-sprayed coatings were analyzed by X-ray diffraction (XRD) using Rigaku D/Max 2550 diffractometer with Cu-Kα (λ=0.15406 nm) radiation at a scan rate of 6°/min. The morphologies of coatings were investigated with scanning electron microscope (SEM, TM3000, HITACHI, Japan), Three-dimensional images of the coating surface were obtained by a 3D-View device attached to the SEM, and the elemental distribution was analyzed by means of energy dispersive spectroscopy (EDS).

A gas burner-rig test (BRT) facility with high-temperature propane/oxygen flame was constructed by ourselves in order to simulate the condition that TBCs always undergo. The cyclic thermal shock performance of the TBCs was carried out on the disk-shaped samples via the burner rig test. The flame nozzle was specially designed in order to achieve a uniform flame with low radial temperature gradient. Therefore the temperature of the whole coating surface is uniform during the BRT process, and then prevent the TBC samples from partial overheating. During the heating stage of the BRT process, the coating surface was exposed to the flame and the substrate backside was cooled by compressed air jet so that constant temperatures of surface and backside maintained. During the cooling stage, the coating surface and the substrate backside were both cooled by compressed air jet. Each cycle consists of several-minute heating stage and several-minute cooling stage, e.g., the steady-state temperature of coating surface is approximately 1523 K for 2 minutes during the heating stage, and then cooled to temperature below 423 K within 3 minutes during the cooling stage. Thermal insulation performance of the TBC samples was estimated by the difference between the temperature values of the coating surface and substrate backside. The temperatures of the coating surface and substrate backside were monitored by infrared pyrometers. At the same time, the acoustic emission technique was combined to inspect the occurrence and propagation of the cracks in the coating real-timely during the BRT. The propagation behavior of the cracks is studied via the analysis the relation between the frequency characteristic and the propagation trends of the cracks. This is beneficial to track the failure process of the TBCs real-timely. The comprehensive understanding of the degradation behavior of TBCs combined with the analysis of the failure mode can be achieved. Once the spallation area surpass the 10% of the whole coating surface, the thermal cycling test stopped. Then the number of cyclings is the thermal cycling lifetime of the coating. The photograph of the BRT for TBC sample is shown in Fig.1.

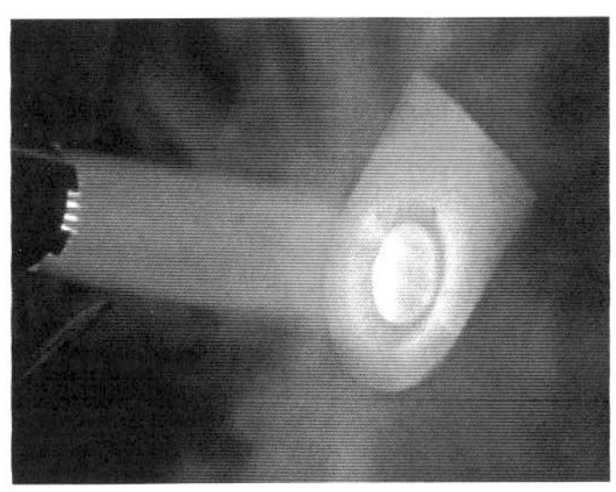

Fig.1 Photograph of the BRT for TBC sample.

3 Results and discussion

The surface morphologies of the as-sprayed SPS coating are shown in Fig. 2. It can be found that the SPS coating shows pebble-shaped protrusions in original surface with inhomogeneous size, which is typical characteristic for columnar segmented structure. The SPS coating shows an improved columnar segments structure which may result in a superior strain tolerance compared to the APS coating.

Fig. 2. Surface morphology of the as-sprayed SPS coating.

Fig. 3 shows the cross-sectional SEM images of the as-sprayed SPS coating. From the microstructure analysis, it can be seen that the SPS coating shows lack of typical splats boundaries and lamellar interfaces which can be usually seen in the conventional APS thin TBCs. Instead, the coating shows a typical structure of vertically oriented cracks running perpendicular to the coating surface and horizontally oriented branching cracks running parallel to the coating surface. One important difference from the APS coating is the lacking of splats and columnar grain structure, instead of evenly spaced horizontal cracks in the segments similar to the APS lamellar interface, which indicates a significantly different formation mechanism of segmentation cracks. As pointed out and discussed in many publications, the formation mechanism of segmentation cracks caused by tensile stress relaxation require a good intersplat bonding. On the contrary, the vertical microcracks formed in each splat could not propagate from splat to splat to form segmentations when a large amount of horizontal cracks are present. So the segmentation cracks of the SPS coating may formed in a significantly different mechanism, the columnar structures in the SPS coatings may form through the preferred deposition of feedstock, and the segmentation cracks can be the gaps of each column. And segmentation cracks are these vertical cracks penetrating almost of the coating thickness. Segmentation crack densities (Ds) were calculated by dividing the number of such cracks in a cross-section with the horizontal length of the cross-section. The estimated Ds of the SPS YSZ coating is about 4 cracks mm^{-1}. The higher Ds value of SPS coatings maybe favorable to its thermal shock resistance.

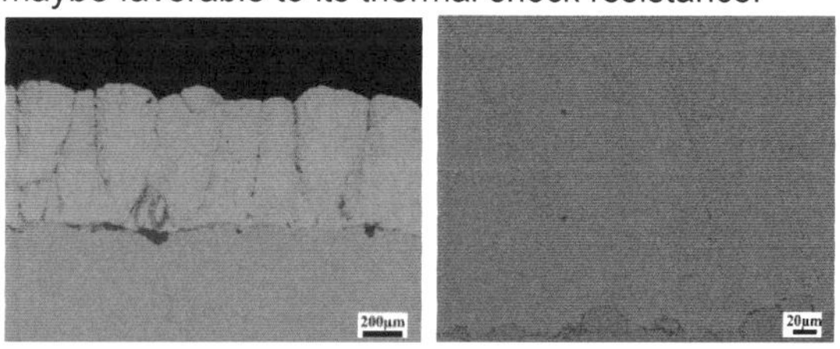

Fig. 3. Cross-sectional microstructures of the as-sprayed SPS coating.

Thermal barrier coatings (TBCs) are commonly used under a rigorous condition which is a cyclic shock by high-temperature gas current and cold gas current. The degradation behavior of TBCs is complex when they undergo such a harsh process. It is of significance for evaluating the in-service behavior and predicting the lifetime of the TBCs based on the investigation of the failure mechanism of the TBCs used under the cyclic thermal shocked condition. During the thermal cycling tests, incipient spallation was observed at the coating edge after 2508 cycles, and the spallation extended to more than 10% area of the whole coating surface after 2711 cycles. The SPS YSZ TTBC exhibits a significant improvement in the cyclic thermal shock resistance, approximately 4 times higher compared to the APS coating. Lifetime extension of the SPS coating as compared to the APS sample can be attributed to the increased Ds and superior segments microstructure. As the segmentation cracks can open during tensile loading in a manner similar to the EB-PVD coatings, the thermal mismatch stress is reduced significantly at a higher Ds condition. A reduction in thermal stress facilitates an improvement in the strain accommodation capability of the SPS TTBC in the thermal cycling test.

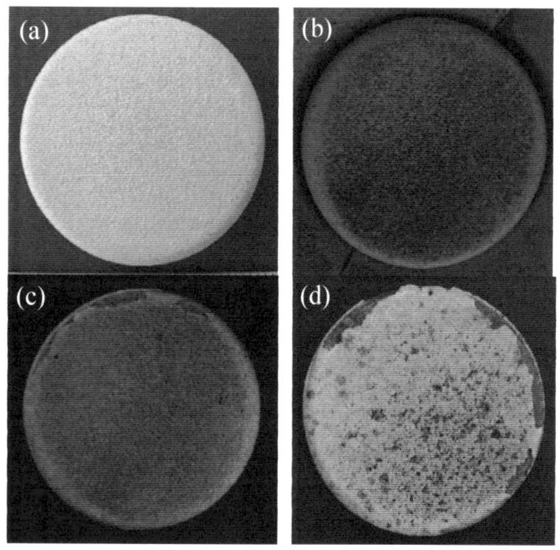

Fig. 4. The photographs of the as-spayed SPS coating (a) and after 1126 cyclings (b), 2508 cyclings (c), 2711 cyclings (d).

The surface SEM image of the failed SPS YSZ TTBC sample and the corresponding surface topography produced by 3D-View device are shown in Fig. 5. As can be seen in Fig. 5, the surface of the SPS coating appeared to have a honeycomb structure. Further inspection showed the fracture surface of the SPS coating have three different types of morphologies which can be defined as follows: (i) sockets (columnar hollows running perpendicular to the coating surface caused from segments partial spallation); (ii) pillars (segments protuberance on the failed surface); (iii) segment gaps (vertical cracks broadened occurring between individual columnar segments). All these could cause a destructive reduction to the coating temperature capability, and therefore result in local overheating of the substrate.

The distinctly different failed surface morphology of the top coat from the APS coating indicates different failure modes, which can be demonstrated in the cross-sections of the failed coatings (as shown in Fig. 6). And all the metallographical sections were selectd near the regions where failure took place. The SPS coating shows a distinctly different failure modes. In addition to fringe segments spallation (Fig. 6a) similar to that of the APS coatings, severe cracking and bridging could be seen in the center of the SPS coatings (Fig. 6b). Further microscopical observations showed that the segments spallation could be related to a crack bridging and segments pull-out phenomenon which can be seen in the previous report about the dense vertically cracked microstructure. During the heating and cooling thermal cycling conditions, some horizontal cracks in the segments will be the most prone area for cracks propagation and the prolonged cracks will join with the open gaps of the adjacent segments, which form bridges across the crack. As the crack continues to propagate, bridging will generate at the whole edge of the segment and lead to segment cracking. Eventually, the driving force for spallation in any given bridging segment exceeds the frictional resistance of the rough segments boundaries due to the residual stresses accumulated during the thermal cycling test, and therefore those bridging segments gradually pull out from their original position with left sockets, pillars and segment gaps structure to the failed surface (as can be seen in Fig. 5).

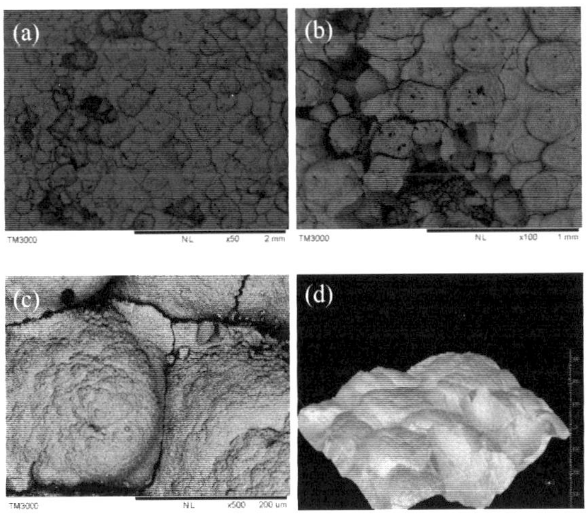

Fig. 5. Surface morphology of the failed SPS coating.

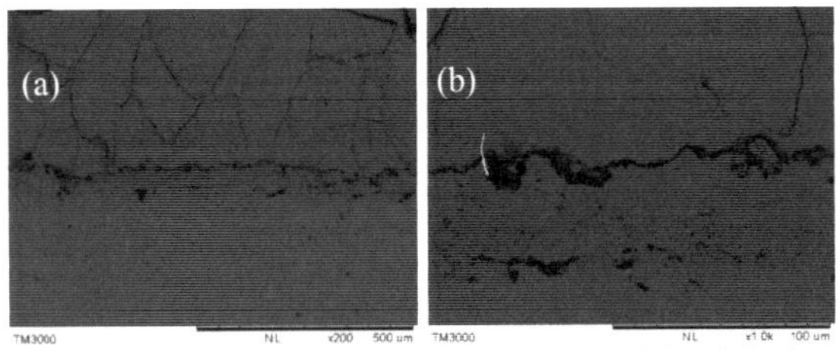

Fig. 6. Cross-sectional microstructures of the failed SPS coating.

The XRD patterns of the as-sprayed and the failed SPS YSZ TTBC are shown in Fig. 7. A small quantity of monoclinic phase of ZrO_2 (m-ZrO_2) was observed after the thermal cycling test according to the XRD patterns. The Raman spectra results also showed that the m-ZrO_2 is fouond within the failed coating (Fig. 8). It is well known that tetragonal to monoclinic phase transformation of 6-8YSZ often occurs during high-temperature service process conditions, which results in a significant volume expansion of up to 5%. The volume expansion could generate thermal stresses and eventually result in degradation and failure of the coating.

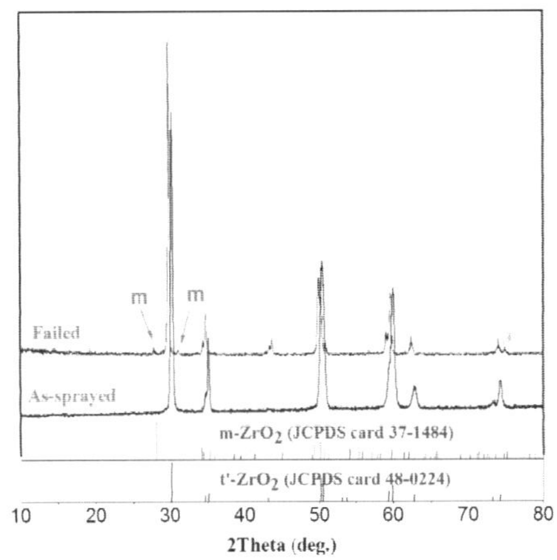

Fig. 7. XRD patterns of the as-spayed and thermal cycling failed SPS coating.

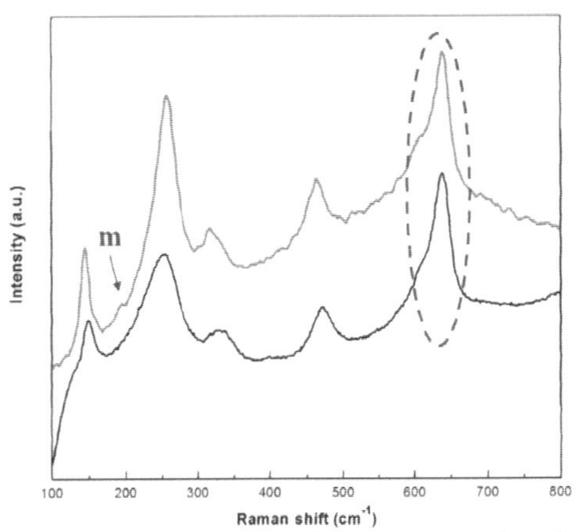

Fig. 8. Raman spctra of the as-spayed and thermal cycling failed SPS coating.

Besides the phase stability, some other factors including the substrate and bond coat factors may affect the durability of TBCs. In previous studies, major factors causing degradation of thin TBCs have been shown to be oxidation of the bond coat and thermal stress resulted from the thermal expansion mismatch between the topcoat and thermally grown oxide (TGO) /bond coat/substrate. It has also been shown that degradation of TTBCs was mainly related to the thermal expansion mismatch between the ceramic top coat and metallic substrate. Fig. 9 presents the cross-sectional SEM image of the TGO formed on the bond coat during the thermal shock test. As evident from the micrograph of the present study, there is an obviously internal oxidation in the bond coat of the coating after the thermal cycling tests. Besides, a thin, continuous TGO scale (dark black colored) formed at the bond coat/top coat interface with the thickness of 3-5 μm. However, another irregular and continuous TGO layer (light grey colored) can also be detected in the SPS TTBCs, indicating a severe oxidation of bond-coat occurred in the SPS TTBC. The severe oxidation of bond coat is due to the following two aspects: (i) SPS TTBC with higher Ds provides more interconnected cracks in YSZ top coat, which increase the channels for oxygen transportation through the top coat and (ii) SPS TTBC with higher Ds facilitates an improvement in thermal shock resistance, which increase the oxidation time. It can be concluded that the failure was most possibly dominated by thermal stresses generated due to the different thermal expansion coefficient between the TGO/bond coat/substrate and top coat during the heating and cooling thermal shock conditions. However, in addition to the thermal stress, spallation of SPS YSZ TTBC sample may be also related to the severe oxidation of the bond coat in a mechanism similar to the thin TBCs.

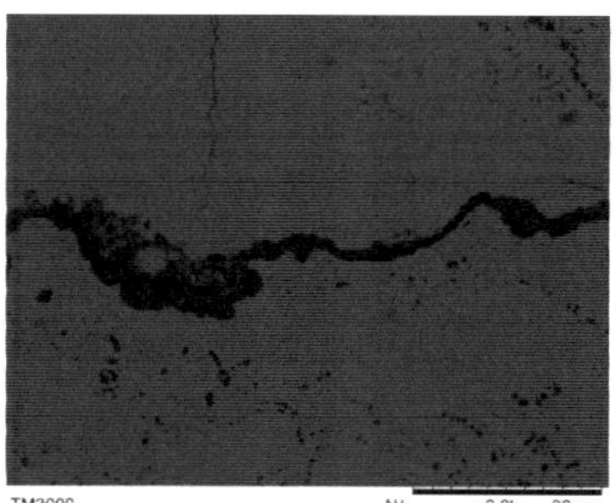

Fig. 9. Cross-section (at high magnification) of the thermally grown oxide formed on the bond coat after thermal shock test.

4 Conclusions

In conclusion, the segmented thick thermal barrier coating (TTBC) has been fabricated via suspension plasma spraying (SPS). The thermal cycling behavior was investigated through the thermal cycling test. The main conclusions can be drawn as follows:

(i) The SPS process produced a network of segmentation cracks perpendicular to the coating surface. The measured segmentation cracks density of SPS TTBC is 4 cracks mm^{-1}.

(ii) The cyclic thermal shock resistance of the SPS TTBC (2711 thermal cycles) was improved and is about five times compared to that of the APS TTBC, which may due to the increased segmentation cracks density and superior segments structure of SPS TTBC. The failure mode of the SPS coating was segments partial spallation initiated from both the edge and center of the coatings, while the APS coatings failed for the bridging of interface delamination close to the bond coat/top coat interface and the intrinsic segmentation cracks. The reason can be attributed to the different formation mechanisms of the columnar segments.

(iii) The failure mechanism of the SPS TTBC is most possibly dominated by thermal stresses generated due to the thermal expansion mismatch between the TGO/bond coat/substrate and top coat during the heating and cooling thermal shock conditions. In addition to the thermal stress, failure of the SPS TTBC is also connected to the severe oxidation of the bond coat and phase transformation of the YSZ top coat .

Acknowledgement

The authors acknowledge the financial support from the Major Program for Basic Research of Shanghai Science and Technology Committee (Grant No.12DJ1400402) and the Special Project for Army-Civilian Combination of Shanghai Municipal Commission of Economy and Informatization (Grant No. JMJH2014053).

5 Literature

[1] N.P. Padture, M. Gell, E.H. Jordan, Materials Science - Thermal barrier coatings for gas-turbine engine applications, Science, 296 (2002): 280-284.

[2] U. Schulz, C. Leyens, K. Fritscher, M. Peters, B. Saruhan-Brings, O. Lavigne, J.-M. Dorvaux, M. Poulain, R. Mévrel, M. Caliez, “Some recent trends in research and technology of advanced thermal barrier coatings”, Aerospace Sci. Technol., 7 (2003): 73–80.

[3] Y. Zhao, L. Wang, J. Yang, D. Li, X. Zhong, H. Zhao, F. Shao, S. Tao. Thermal aging behavior of axial suspension plasma sprayed yttria-stabilized zirconia thermal barrier coatings. J. Therm. Spray Technol. 24(3)(2015): 338-347.

[4] H.B. Guo, R. Vaßen, D. Stöver, Atmospheric plasma sprayed thick thermal barrier coatings with high segmentation crack density, Surf. Coat. Technol. 186 (2004): 353-363.

[5] C.G. Levi, Emerging materials and process for thermal barrier systems, Curr. Opin. Solid State Mater. Sci., 8 (2004): 77–91.

[6] Y.S. Tian, C.Z. Chen, D.Y. Wang, Q.M. Ji, Recent developments in zirconia thermal

barrier coatings, Surf. Rev. Lett., 12 (3) (2005): 369-378.
[7] H.-D. Steffens, Z. Babiak, M. Gramlich, Some aspects of thick thermal barrier coating lifetime prolongation, J. Therm. Spray Technol., 8 (1998): 517-522.
[8] P. Bengtsson, T. Ericsson, J. Wigren, Thermal shock testing of burner cans coated with a thick thermal barrier coating, J. Therm. Spray Technol., 7 (1997): 340-348.
[9] W.A. Nelson, R.M. Orenstein, TBC experience in land-based gas turbines, J. Therm. Spray Technol., 6 (1997): 176-180.
[10] D. Schwingel, R. Taylor, T. Haubold, J. Wigren, C. Gualco, Mechanical and thermophysical properties of thick PYSZ thermal barrier coatings: correlation with microstructure and spraying parameters, Surf. Coat. Technol., 108 (1998): 99-106.
[11] S. Rangaraj, K. Kokini, Estimating the fracture resistance of functionally graded thermal barrier coatings from thermal shock tests, Surf. Coat. Technol., 173 (2003): 201-212.
[12] S. Ahmaniemi, P. Vuoristo, T. Mäntylä, C. Gualco, A. Bonadei, R. Di Maggio, Thermal cycling resistance of modified thick thermal barrier coatings, Surf. Coat. Technol., 190 (2005): 378-387.
[13] U. Schulz, B. Saruhan, K. Fritscher, C. Leyens, Review on advanced EB-PVD ceramic topcoats for TBC applications, Int. J. Appl. Ceram. Technol., 1 (4) (2004): 302–315.
[14] H.B. Guo, S. Kuroda, H. Murakami, Microstructures and properties of plasma-sprayed segmented thermal barrier coatings, J. Am. Ceram. Soc., 89 (2006):1432-1439.
[15] K. Bobzin, E. Lugscheider, N. Bagcivan, Thermal cycling behavior of yttria stabilized zirconia and lanthanum zirconate as graded and bilayer EB-PVD thermal barrier coatings, High Temp. Mater. Proc., 10(2006): 103–116.
[16] A. Scrivani, G. Rizzi, U. Bardi, C. Giolli, M.M. Miranda, S. Ciattini, A. Fossati, F. Borgioli, Thermal fatigue behavior of thick and porous thermal barrier coatings systems, J. Therm. Spray Technol., 16 (2007): 816-821.
[17] T.A. Taylor, Thermal properties and microstructure of two thermal barrier coatings, Surf. Coat. Technol., 54/55 (1992): 53-57.
[18] M. Matsumoto, N. Yamaguchi, H. Matsubara, Low thermal conductivity and high temperature stability of ZrO2-Y2O3-La2O3 coatings produced by electron beam PVD, Scrip. Mater., 50 (2004): 867–871.
[19] L. Xie, D.Y. Chen, E.H. Jordan, A. Ozturk, W. Fang, X.Q. Ma, B.M. Cetegen, M. Gell, Formation of vertical cracks in solution-precursor plasma-sprayed thermal barrier coatings, Surf. Coat. Technol., 201 (2006): 1058-1064.
[20] H. Kassner, R. Siegert, D. Hathiramani, R. Vassen, D. Stoever, Application of suspension plasma spraying (SPS) for manufacture of ceramic coatings, J. Therm. Spray Technol., 17 (2008): 115-123.

SUBSTRATE PREPARATION & THERMAL SPRAY COATING PERFORMANCE-A REVIEW

S. S. Aulakh[1], G.Kaushal[2]
[1] Lecturer, Shinas College of Technology, Shinas, Oman
[2] *Assistant Professor*, Mechanical Engineering Section, Yadavindra College of Engineering, Punjabi University, Guru Kashi Campus, Talwandi Sabo, Punjab, India

Thermal spray coating is being increasingly used to enhance the performance of industrial components. New improved coatings and newer thermal spraying processes are being invented to improve the performance of the coatings. One of the biggest factor responsible for the performance of the coating is the adhesion of the coating with the substrate. Without good adhesion strength any good coating will prove ineffective. Grit blasting, laser surface treatment, dry ice blasting are some of the techniques being used or researched for good coating adhesion strength. This paper reviews the state of the art in the area of substrate preparation techniques.

Key words – substrate preparation, adhesion bond strength, grit blasting, laser surface treatment, dry ice blasting.

1 Introduction

Thermal spraying is the method of producing corrosion, wear and heat resistant coatings for various industrial applications. It is the process of applying layers of metallic and non metallic materials on a surface prepared for the process. In thermal spray processes, the coating material is melted by some sort of thermal energy like, gases, plasma, electric arc etc. The molten particles are impinged on the substrate surface, on impact with substrate surface, the particles flattens and produces adhesion to the substrate by interlocking of molten and semi molten particles with asperities of roughened surface [1]. A great number of interactive parameters have to be controlled in order to obtain good coatings and promote substrate adhesion [2]. Coating adhesion with the substrate surface is one of the fundamental factors behind the performance of the coating. To create good conditions to anchor coatings on the substrate, appropriate surface preparations are required [1]. Specific chemical, physical and mechanical characteristics are required for the substrate surface for good coating adhesion. [3]. For good coating adhesion the substrate surface should be free from any adsorbed molecules. The adsorbed molecules evaporate by the heat of molten particles and promote splashing. The substrate wettability has to be as high as possible to improve splat spreading [2]. In order to remove organic contaminants from the surface, chemical degreasing is used. Degreasing improves chemical behaviour of the surface resulting in good spreading of the particles. But to improve interlocking of the coating particles with surface further surface treatment is required. Conventionally grit blasting has been used to enhance the asperities on the surface to improve surface roughness and mechanical interlocking. The surface roughness, surface temperature, residual stress on surface, cleanliness of the surface, presence of oxide film are some of the important factors which affect the coating adhesion and hence the bond strength and subsequently the coating performance. A number of researcher have experimented with different ways of preparing the substrate surface for thermal spray coating, studying the effect of various parameters on the coating bond strength, an indicator of coating adhesion. In the following sections of the paper, these surface preparation techniques and the findings of research conducted on these techniques will be discussed.

2 Methods of Substrate Preparation

On the basis of the literature surveyed from the year 2000, different methods of substrate preparation are,

- Grit Blasting
- Laser ablation
- Laser texturing
- Dry Ice blasting

2.1 Grit Blasting

In the grit blasting process, abrasive particles are propelled onto the surface of the part to be coated. This roughens and cleans the surface for thermal spray coating process. A large number of process and other variables like the grit material and size, blasting pressure and time, stand off distance, blasting angle and hardness of the component/surface being grit blasted determine the magnitude and nature of roughening of the surface [4]. Practical grit size ranges are -10 +30 mesh (i.e., coarse), -14 40 mesh (Le., medium), and -30 + 80 mesh (Le., fine). Typically, coarse grit is used for coatings exceeding 10 mils for best adherence; medium grit is used for smoother finishes of coatings less than 10 mils with fair adherence; and fine grit grit is used for the smoothest finishes on coatings less than 10 mils to be used in the as-sprayed condition. Alumina, silica sand, crushed steel, chilled iron, copper slag, and silicon carbide are often used as abrasive grit. Grit blasting air pressure varies from 210 to 620 kPa (30 to 90 psi), with typical working distances of 50 to 150 mm (2 to 6 in.). Grit blast nozzle openings are generally 6 to 10 mm (0.25 to 0.375 in.) in diameter.
Although grit blasting is a very effective and simple method of roughening and cleaning the surface but it is an environmental hazard, there is a problem of grit inclusions, subsurface cracks specially in ductile materials like aluminium and magnesium alloys.

James day et al [5] used statistical design of expermients approach to study the effect of different grit blasting parameters like blasting pressure, no. of passes, working distance, blast angle and grit size number on substrate roughness, bond strength and grit contamination. The control variables were set at three/two levels. Multiple Linear regression and average response curve approach was used to study the correlation between independent and dependent variables. From the average response curve as shown in **Fig. 1.**, it was concluded that surface roughness is significatnly affected by grit size and blasting pressure

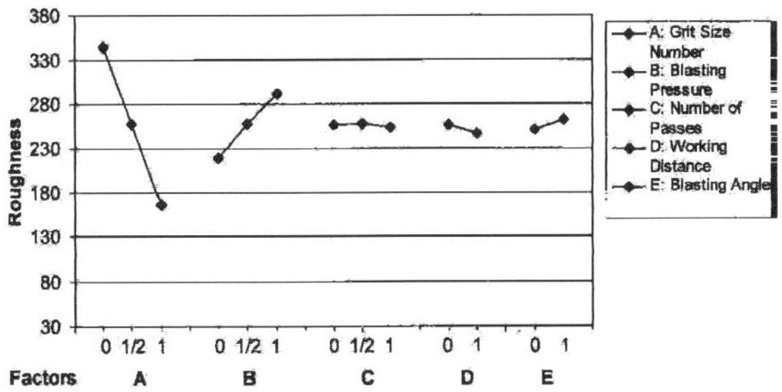

Fig. 1. Average effect of factors on surface roughness[5].

ASTM C633-79 bond strength test was conducted. The results reported were as shown in **Fig. 2.** It can be concluded that bond strength is influenced by blasting pressure, number of passes, grit size and working distance but blasting angle has little influence on bond strength.

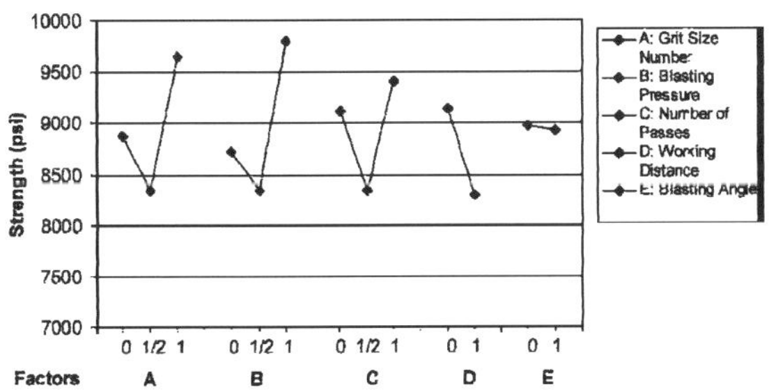

Fig. 2. Average effect of factors on bond strength [5].

It was observed (**Fig. 3.)** that grit size number, blasting pressure and number of passes could have most significant influence on grit contamination.

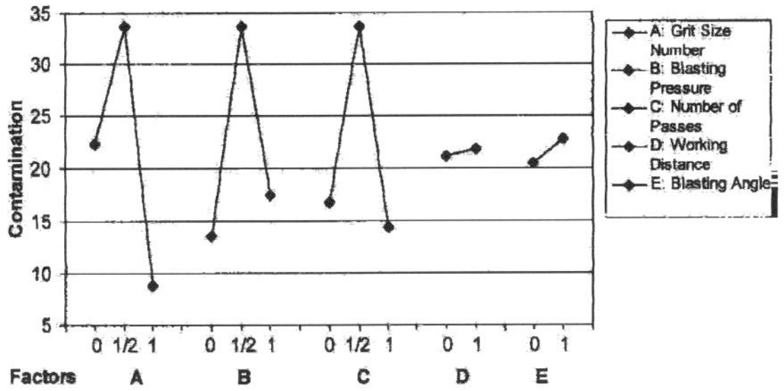

Fig. 3. Average effect of factors on surface contamination [5].

James day further concludes that bond strength can be increased by incresing blasting pressure, Increasing number of passes and increasing grit size number. In order to reduce contamination, reducing pressure, increasing grit size number and reducing number of passes is suggested. So, there seems to be a trade off redarding pressure and passes.

D. Sen et al [4] studied the influence of grit blasting pressure and alumina grit size on the roughness of mild steel substrate and bond strength of detonation sprayed coating. It was found that surface roughness doesn't vary substantially within increasing blasting time as can be seen from the **Fig. 4.**

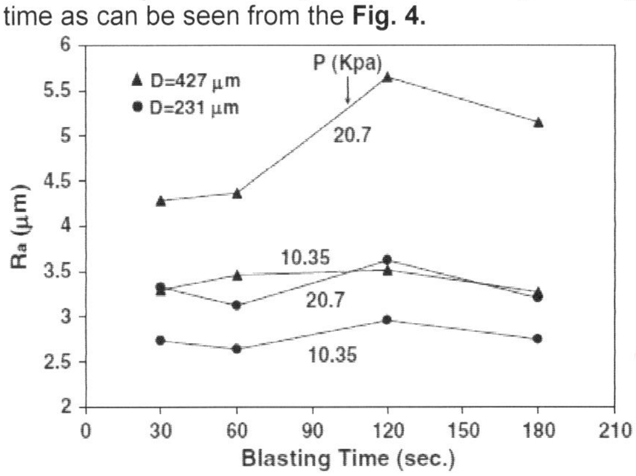

Fig. 4. Effect of blasting time on surface roughness for various combinations of blasting pressure and grit size [4].

As can be seen from **Fig. 5**. surface roughness was reported to increase continuously with increase in blasting pressure and grit size which agrees with the finding of James day too.

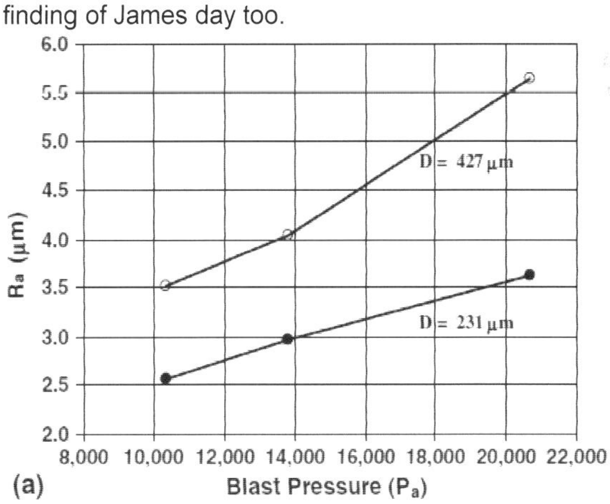

Fig. 5. Variation of surface roughness as function of blast pressure with two grit sizes [4].

There seems to be a strong correlation between bond strength and substrate roughness. From the **Fig. 6.** it can be seen that for all the coatings Cu, WC-Co and Al_2O_3, the bond strength increases from low value to high value beyond roughness value of 3.5 μm to 4 μm. So, a minimum roughness of 4 μm seems to be necessary for good bond strength.

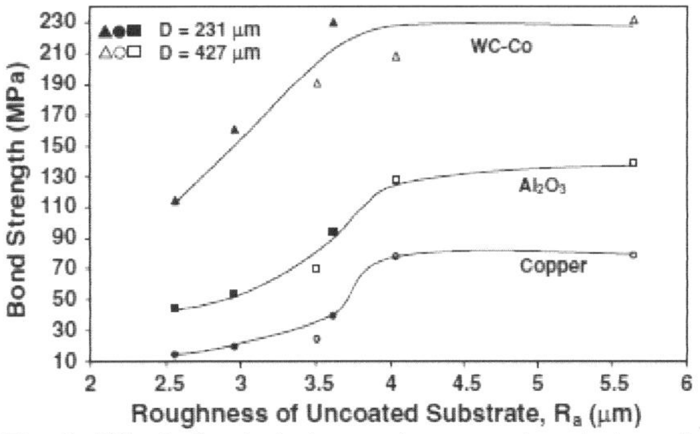

Fig. 6. Effect of substrate roughness on bond strength [4].

M H. Staia et al [6] studied the effect of surface roughness on bond strength for WC-17 Co coating HVOF sprayed on alumina #20 grit blasted AISI 1020 Steel. Three different sand blasting pressures were used to obtain three different substrate roughness values. It can be observed from **Table 1.** that roughness increases with increase in sand blasting pressure.

Table 1 Roughness values for AISI steel substrate as a function of grit blasting pressure [6].

		Sand Blasting pressure (MPa)		
		0.345	0.483	0.621
R_a	Average	9.74	10.99	12.57
R_a	S.D.	0.25	1.13	0.2

Table 2. shows the variation of bond strength with blasting pressure which in turn is correlated to surface roughness. The samples having lower surface roughness have lower value of bond strength than samples having high roughness. It indicates that a rough surface allow for a good bonding.

Table 2 Average failure loads for the tested samples as a function of grit blasting pressure [6].

Sand Blasting Pressure (MPa)	Maximum strength (MPa)
0.345	32.70
0.483	44.19
0.621	52.50

Coating failure analysis done through image analysis of frature surface also indicates the ahdesion improvement of coating with increasing surface roughness as area indicating failure thorugh bonding agent increased from 44.56% for blasting pressure of 0.345Mpa to 73.88% for 0.621 MPa blasting pressure.

2.2 Laser Surface Treatments

Conventional substrate preparation techniques consist of degreasing the surface with some solvents and subsequent Grit blasting. Although it is simple and cost effective way of preparing the surface for thermal spraying but it has some limitations [7]. Chemical solvents are not good for envvironemnt and pose health hazard to operators and grit blasting cause some damage to the surface and some grits get embedded into the surface, specially for ductile materials like aluminum and magnesium alloys.

To overcome these limitations some of the researchers tried to develop laser surface treatments as an alternative to grit blasting process. In the following section, some of the advancements in the use of laser technology for substrate preparation will be discussed.

2.2.1 Laser Ablation

S. Costil et al [8] studied the effect of flame preheating and laser preheating combined with laser ablation on splat morphology to clean the TA6V surface prior to plasma spraying of NiAl. A flame was used to preheat the surface prior to laser ablation and followed by plasma spraying. The author reported the improvement in splat shapes and reduction in splashing as the temperature of the surface is increased from 25°C to 360°C, refer to **Fig. 7.** Similar results were also achieved with preheating done thorugh laser followed by laser ablation. Combined action of heating and cleaning promotes substrates wettability and adhesion between the particles and substrate.

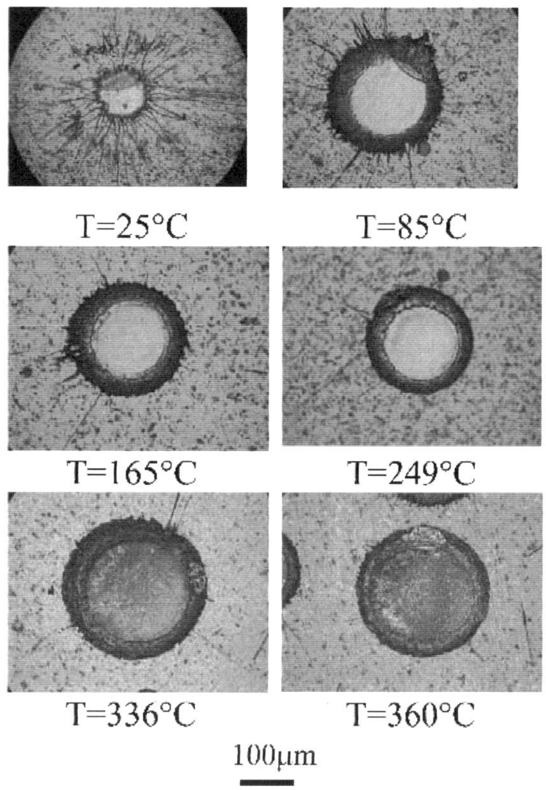

Fig. 7. Splat Morphology as a function of pre heating substrate temperature with laser ablation [8].

Y Danlos et al [2] investigated the relation between surface modifications induced by laser ablation and adhesion of the coatings. Two different materials Ti-6Al-4V and Al 2017 were plasma sprayed with NiAl coating. Two Nd:YAG laser beams in conjunction with a thermal spray gun were used in order to modify the surface prior to thermal spraying as illustrated in the **Fig. 8.**

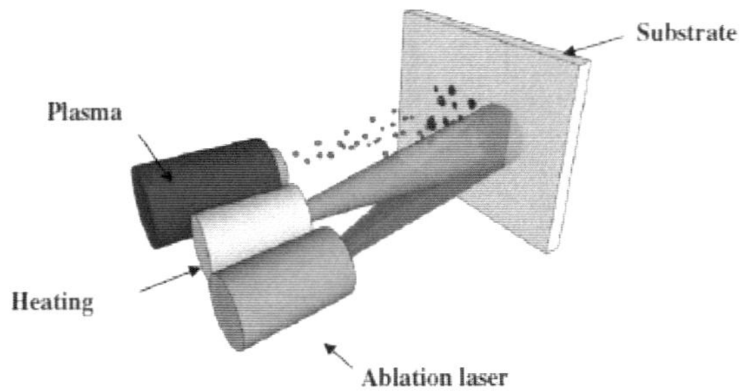

Fig. 8. Expermiental setup used by Danlos et al [2].

The authors reported that morphology of laser ablated samples is very different than that of sand blasted samples. The sandblasted samples were condsiderably rougher than the polished samples. Interestingly, the morphology of the laser ablated titanium samples is also different than that of aluminium alloy samples.

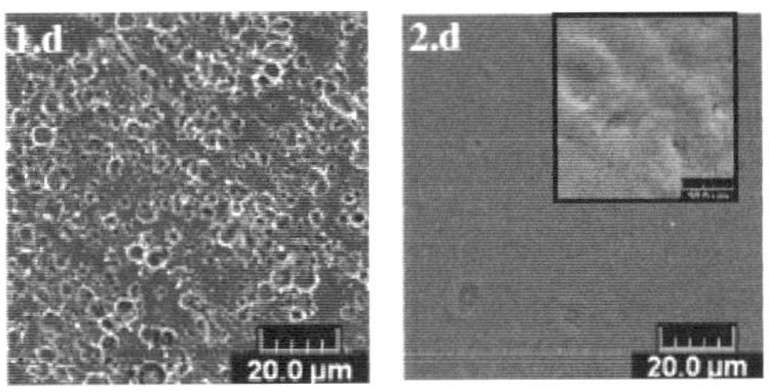

Fig. 9. SEM of AL2017(1d) and TA6V(2d) ablated with 2.3j cm^{-2} energy. [2]

As it can be observed from the **Fig. 9.** that Aluminum alloy develops some craters on the suface after laser ablations but no such craters can be seen on laser ablated titanium alloy which has a kind of melted surface. The authors attribute this phenomenon to the difference in the thermal diffusivity of both the materials. So, it can be concluded that the surface morphology obtained after laser treatment is a function of nature and composition of the substrate material. Further in this study the authors investigated the effect of laser treament on coating adhesion measured in terms of interfacial toughness. As reported by [2], the interfacial toughness is equal to 2 MPa m1/2 (σ=0.1 MPa m1/2) for the coating made on sandblasted surface whereas it is equal to 1.61 MPa m1/2 (σ=0.08 MPa m1/2) and 0.81 MPa m1/2 (σ=0.07 MPa m1/2) for those realized on surface heated by laser with an energy density of 24.8 J cm-2 and then ablated by the ablation laser with an energy density of 2.3 J cm−2 and 1.7 J cm−2 respectively. The adhesion follows the level of roughness in the samples. In general, the difference in the toughness value is much less than the difference in roughness so, the author proposes that apart from physical anchoring responsible for adhesion of coating the chemical affinity also seems to be repsonsible. The previous point was reasserted by the fact that titanium alloy has more interfacial toughness than Al 2017 for comparable roughness level, pointing towards role of chemical anchoring in coating adhesion.

M Perton et al [9] studied the effect of laser ablation treatment in case of cold spraying. They investigated the effect of pulsed laser ablation and continuous laser heating on the adhesion and cohesion of cold sprayed Ti-6Al-4V coatings. It has been reported that scratch test indicates to the reduction in cohesion strength in case of PLA samples. Although the cleaning effect of PLA was expected to produce better oxide free bonding between particles. This result is explained by generation of compressive wave inside coating during PIA process might have damaged the particle/particle bonds due to stress concentration near to pores. However when PLA was combined with laser heating the cohesion is reported to have improved compared to only PLA process. The possible explanation is given as increased plasticity due to partial heating of the coating particles.

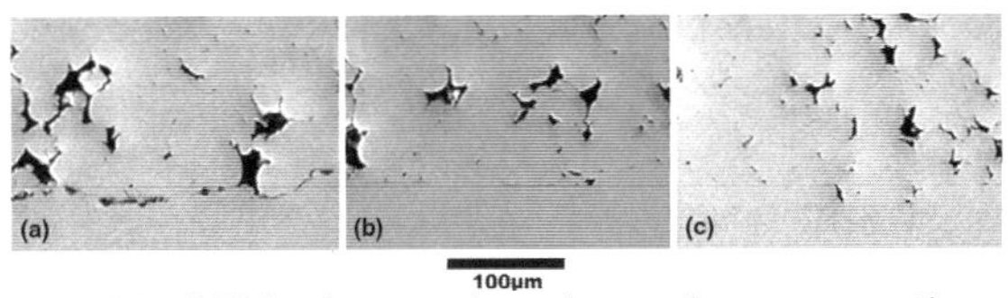

Fig. 10. SEM micrographs of sample cross-sections around the coating/substrate interface for different surface finishes: (a) grit 24 alumina blasted, (b) grit 400, ground (c) with PLA at 2.2 J/cm2 [9].

Contrary to Plasma spray or HVOF spray methods the finer surfaces appear to have better interfacial bonds in cold spray process, as can be noticed in **Fig. 10.** It is explained by the fact that in cold spray process particles are not in the molten state and need to be heavily deformed to fill the bigger asperities in case of grit blasted substrate comapred to finer ones in case of polished or PLA substrate. So, particles do not completely fill the asperities, resulting in poor bonding. This observation was further corroborated by LASAT adhesion test as illustrated in **Fig. 11.** Further, effect of laser pre heating was investigated for samples grounded with grit 400 under different laser ablation condtions. **Fig. 12.** shows that laser preheating increases the adhesion strength for PLA samples. But this is not true for non ablated samples. The authors proposes that this is because laser heating helps to maintain temperature and pressure of plasma generated during PLA process, causing it to melt the surface and produce smoother surface like polished samples, enhancing contact area and hence adhesion. Authors reccommend use of PLA process for cold spray combined with preheating.

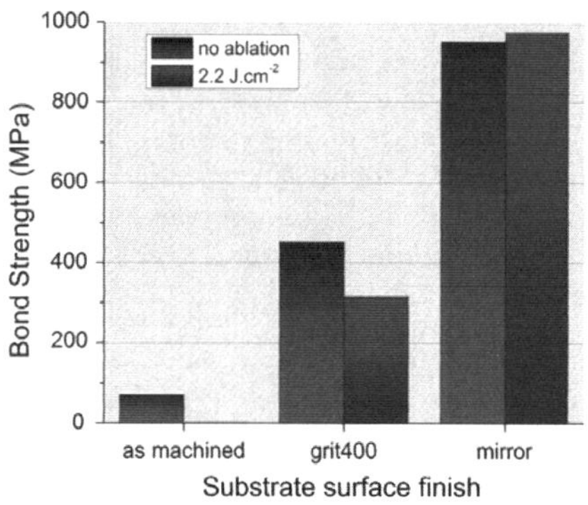

Fig. 11. Adhesion strength measured by LASAT for three roughness conditions without preheating [9].

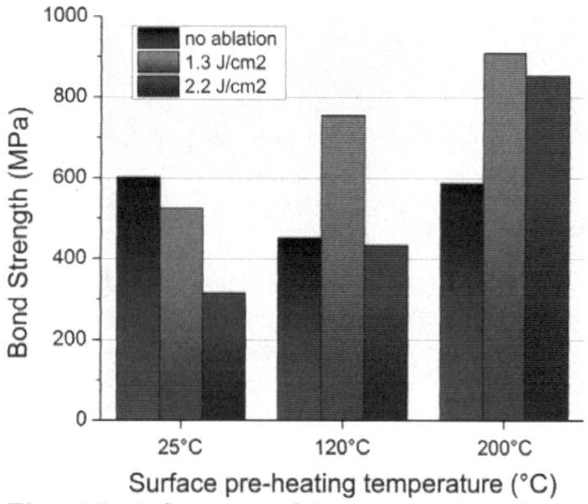

Fig. 12. Influence of Laser treatment on Adhesion strength [9].

2.2.2 Laser Surface Texturing

Laser surface texturing is a realtively new surface pretreatment technique that involves ceating precise diameter holes following preset patterns throughout surface. The accuracy of texturing is achieved by controlling laser and scanner parameter that influence the quality of surface to be coated by thermal spraying. This technique produces physico-chemical and mechanical modification on the surface.[10].
Lamraoui et al [10] introduced laser surface texturing as a substrate preparation technique. They used pulsed Nd:YAG laser(1.064µm radiation ,120ns) laser with repititive scans to texture AL 2017. The effct of several parameters like scanning rate, distance between laser impacts, frequency of pulses and incident power was studied to optimise the laser texturing process to achieve good NiAl coatings.

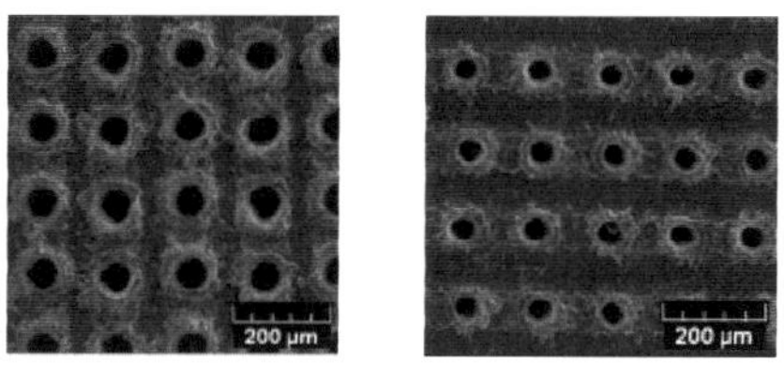

Fig. 13. SEM observations of an aluminum surface textured with (a) P=17.3 W,27 shot per hole,20 KHz, (b) P=10W, 48 shots per holes, 20 KHz [10].

Fig. 13. shows the effect of these paprameters on the texture. The relationship between morpohology of the holes and power of laser was shown as in **Fig. 14.**

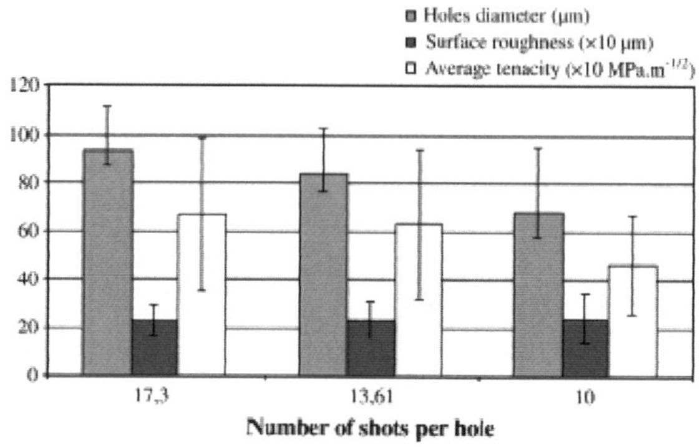

Fig. 14. Evolution of depth of the holes, the surface roughness and toughness of the textured surface as a function of the laser power.[10].

2.3 Dry Ice Blasting

Dry ice blasting is an environment friendly cleaning technique which is used for industrial cleaning [11]. This technique uses solid CO_2 pallets which are blasted on the surface to be cleaned with a high pressure. Combined thermo-mechanical effect removes any dirt, oxides etc from the surface.
Shujan dong [12] used dry ice blasting technique in combination with grit blasting to produce plasma sprayed Al_2O_3 coating on Aluminum substrate. The expermiental setup was as used in **Fig. 15.**

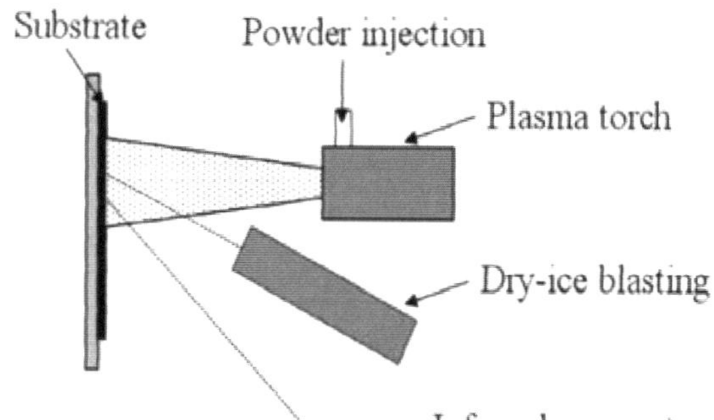

Fig. 15. Schematic diagram of experimental setup for dry ice blasting [12].

It was reported that dry ice blasting can smoothen the surface slightly. But there is not much change in topography of the surface. Roughness value S_a =6.07 µm for grit blasted aluminum sample and roughness value S_a =5.39 µm for grit blasting followed by dry ice blasted surface. Surface roughness of coated surface also follows the same trend, with coating done along dry ice balsting is slightly smoother than only grit blasted and coated surface. The authors also report that porosity of coating done with dry ice pre-treatment is 6.5% as compared to 9.1% in case of coating done with plasma spray with compressed air cooling. The decrease in porosity is attributed to cooling effect of dry ice. It is worth mentioning here that dry ice blasting insitu with plasma spray is affecting the process in two ways, one is to clean the substrate and the other efffect is to cool the substrate during plasma spray. To investigate the chemical character or cleanliness of surface wettability, contact angle of substrate samples was measured before and after dry ice blasting as shown in **Table 3.**

Table 3. Contact angle of substrate under different Conditions [12].

Specimen	Details	Contact Angle ^{0}
A-P	Initial polished Al surface.	90.5
A-PG	Al surface polluted by glycerol.	26.8
A-PGD	Al surface polluted by glycerol and then treated by CO_2.	87.3
S-P	Initial polished stainless steel surface.	72.6
S-PG	Stainless steel surface polutted by glycerol.	10.8
S-PGD	Stainless steel surface polutted by glyerol and then treated by CO_2 pallets.	69.9

Authors also report that dry ice treatment time seems to affect the splat formation. A critical processing time results in best shaped splat. Dry ice blasting also leads to increased compressive stress due to peening effect of dry ice pallets.

3 Concluding Remarks

Coating adhesion with the substrate surface is one of the fundamental factors behind the performance of the coating. A number of researcher have experimented with different ways of preparing the substrate surface for thermal spray coating, studying the effect of various parameters on the coating bond strength an indicator of coating adhesion. There seems to be a strong correlation between bond strength and substrate roughness. Grit blasting, laser surface treatment, dry ice blasting are some of the techniques being used or researched for good coating adhesion strength. Quite a few researchers have worked on the area and the study in the prospective field is still to be explored with more conclusive reports.

4. References

[1] R.S.C. Paredes, S.C. Amico, A.S.C.M. d'Oliveira: The effect of roughness and pre-heating of the substrate on the morphology of aluminium coatings deposited by thermal spraying. Surface & Coatings Technology 200 (2006), pp. 3049 – 3055.

[2] Y. Danlos, S. Costil, H. Liao, C. Coddet: Influence of Ti–6Al–4 V and Al 2017 substrate morphology on Ni–Al coating adhesion — Impacts of laser treatments. Surface & Coatings Technology 205 (2011) 2 pp 2702–2708.

[3] R.S.C. Paredes, S.C. Amico, A.S.C.M. d'Oliveira, Surf. Coat. Technol. 200 (2006), pp. 3049.

[4] D. Sen, Naveen M. Chavan, D.S. Rao, and G. Sundararajan: Influence of Grit Blasting on the Roughness and the Bond Strength of Detonation Sprayed Coating. Journal of Thermal Spray Technology Volume 19(4) June 201o, pp. 805-815.

[5] James Day, Xiao Huang, and N.L. Richard: Examination of a Grit-Blasting Process for Thermal Spraying Using Statistical Methods. Journal of Thermal Spray Technology Volume 14(4) December 2005, pp. 471-479.

[6] M.H. Staia, E. Ramos: Effect of substrate roughness induced by grit blasting upon adhesion of WC-17% Co thermal sprayed coatings. Thin Solid Films 377-378 (2000), pp. 657-664.

[7] B.J. Griffiths, D.T. Gawne, and D. Dong: The Role of Grit Blasting in the Production of High Adhesion Plasma Sprayed Alumina Coatings, Proc. Inst. Mech. Eng., Vol 211 (No. 1), part B, 1997, p 1-9.

[8] S. Costil, H. Liao et al: Influence of Surface Laser Cleaning Combined with Substrate Preheating on the Splat Morphology. Journal of Thermal Spray Technology Volume 14(1) March 2005, pp. 31-39.

[9] M. Perton, S. Costil, et al: Effect of Pulsed Laser Ablation and Continuous Laser Heating on the Adhesion and Cohesion of Cold Sprayed Ti-6Al-4V Coatings. Journal of Thermal Spray Technology, Volume 21(6) December 2012 pp. 1322-1334.

[10] A. Lamraoui, S. Costil: Laser surface texturing (LST) treatment before thermal spraying: A new process to improve the substrate-coating adherence. Surface & Coatings Technology 205 (2010), pp. S164–S167.

[11] G. Spur, E. Uhlmann, F. Elbing: Dry-ice blasting for cleaning: process, optimization and application. Wear 233–235 _1999, pp. 402–411.

[12] Shujuan Dong et al: Study on the Mechanism of Adhesion Improvement Using Dry-Ice Blasting for Plasma-Sprayed Al2O3 Coatings. Journal of Thermal Spray Technology Volume 22(2-3) March pp. 2013—213.

Wear behavior of post-treated arc sprayed cermet coatings by means of Hammer Peening

W. Tillmann, L. Hagen, Dortmund / D

Abstract

Within the field of wear protection, the utilization of cast tungsten carbide (CTC) reinforced cored wires in Twin Wire Arc Spraying (TWAS) processes provides a method for the manufacturing of cermet coatings with improved mechanical and tribological characteristics. In the field of surface engineering, machine hammer peening (MHP) is a novel surface treatment technology, which enhances the surface properties, especially for surfaces undergoing tribological contact, due to local hardness increase and insertion of compressive residual stresses below the surface.

In this study the wear behavior of WC-W_2C FeCMnSi arc sprayed coatings derived from conventional cored wires with a different chemical composition, relative to the hard phase content, are investigated. According to this, hammer peened surfaces are compared to as-sprayed and polished samples by means of metallographic investigations. With the use of ball-on-disk and dry rubber wheel tests, dry sliding and rolling wear effects on a microscopic level are scrutinized. It has been shown that under hammer peened conditions the corresponding coatings obtain lower wear resistances, but lower friction coefficients than the conventional coatings. Strain hardening effects were revealed by mechanical response using nanoindentation. However, the hammer peening process has caused a cracking of embedded carbides, which favor break-outs, leading to advanced third-body wear.

1 Introduction

Arc spraying is a long-established process to produce wear- and corrosion-resistant coatings [1]. For arc sprayed coatings, the use of cored wires with hard materials as a filling is a widely used practice to apply wear resistant coatings onto components. Following complex stress profiles, different research projects have already investigated the use of cored wires with different hard material fillings [2-4]. Employing cored wires with fine-grained hard material powder fillings can lead, for example, to an enhanced wear behavior, lower porosity, and a finely lamellar coating structure [5]. The finely lamellar coating structure thus can benefit the tribological coating properties [6].
Arc sprayed coatings exhibit properties, which are mainly dependent on the particle characteristics (particle size, -temperature, and –velocity) [7-9].
The droplet formation as well as the atomization behavior, and thus the particle characteristics are extremely dependent on the electrode phenomena as well as on the flow behavior of the atomization gas that is used [8, 10]. Additionally, the disintegration of the extruded metal sheets and the droplet formation are influenced by the selection of the feedstock material (material, wire construction) [9, 11-13]. Finally, with regard to the deposits, a heterogeneous microstructure formation, consist of intermetallic phases, eutectic mixtures and various oxides, is typical. Phase transformation processes, due to the inherent process characteristics [13, 14], also among other things, influence the formation of brittle phases, thus causing degradation of the mechanical properties [15].
Only few studies can be found in the literature, which address the use of tungsten carbide reinforced iron-based feedstock materials (FeCSiMn – WC / W_2C, Fe_3Al-WC, Fe-FeB-WC) [16-19].
The tribological behavior of tungsten carbide reinforced coating systems by means of arc spraying was examined in previous studies [6, 17, 19]. For reinforced surfaces, the wear resistance of the composites depends, inter alia, on the amount of hard materials and the embedding behavior of the hard phases [20, 21].
Although it is shown that tungsten carbide reinforced coatings represent a suitable solution improve the abrasive wear resistance of tribologically stressed surfaces, the deposit features a high roughness and inhomogeneous distribution of pores. Hence, the high roughness of the coated surfaces has to be reduced. In order to achieve the required surface quality, the protruding peaks of the topographies have to be removed by means of machining.
With regard to mechanical post-treatment processes, roller burnishing enables the compression of the rough surface topography [22]. For bulk materials, it was found that incremental cold forming processes such as roller burnishing and machine hammer peening (MHP) increase the microhardness and creating residual compressive stresses within the subsurface of the treated surface [23, 24].
This study presents the mechanical post-treatment of arc sprayed tungsten carbide coatings by MHP. In order to characterize the machined surfaces, the wear behavior was analyzed before and after the peening process. Therefore, the wear behavior of as-sprayed and polished as well as of machine hammer peened WC-W_2C FeCMnSi arc sprayed coatings derived from different cored wires is investigated. The wear behavior is determined by ball-on-disk and dry rubber

wheel tests. The stressed surfaces are analyzed by means of light and electron microscopy. Mechanical properties of the treated surfaces are measured by nanoindentation.

2 Experimentals

2.1 Substrate and feedstock material

Round C45 (diameter = 40 mm) and rectangle C45 (20 x 65 x 10 mm) steel specimens were used as substrate materials. In preparation for the thermal spray experiments, the surfaces were grit-blasted with corundum (EKF 14) and cleaned in an ultrasonic ethanol bath.

With regard to the feedstock material, a FeCMnSi cored wire, also referred to as (1.6) (Type Durmat AS-850, Fa. *Durum Verschleißschutz, Germany*), with cast tungsten carbides CTC (eutectic $WC\text{-}W_2C$ mixture) as a filling material was used. In addition, a tailor made cored wire (2.0) with a different chemical composition, relative to the hard phase content, is investigated. For all wire types, the band width and thickness of the wire sheath was 10 mm, and 220 µm respectively. **Table 1** shows the chemical composition of both wire types used.

Table 1: Chemical composition of the feedstock material

in wt.%

Sample	C	Fe	Mn	Si	W	(CTC)
1.6	2.0	Bal.	1.4	1.1	48	50
2.0	2.3	Bal.	1.4	1.0	55	56

The grain size fraction of the CTC filling for both cored wires was between -106 + 63 µm. The particle size distribution was verified by light scattering method, whereupon the 50th percentile of sampling was nearly 105 µm.

The coatings were deposited on the substrates by the Smart Arc 350 PPG spraying system (Fa. *Oerlikon Metco, Switzerland*) using compressed air as an atomization gas. The process parameter settings were kept on a constant level for all samples (atomization gas pressure = 0.6 MPa, voltage = 30 V, current = 220 A, spray distance = 110 mm, gun velocity = 200 mm / s, Y-Step = 5 mm, two overruns).

2.2 Machine hammer peening process

The subsequent mechanical post-treatment was conducted on the five-axis machining center DMU50 eVolution (Fa. *Deckel Maho, Germany*) using the pneumatic MHP tool Forge Fix Air (Fa. *3S-Engineering, Germany*). A spherically shaped carbide metal tips (tool diameters d_k = 12 mm) was used in order to smooth and compact the coatings. The MHP coating systems are referred to as 1.6 H and 2.0 H. The operating air pressure of the tool was set on 0.6 MPa, resulting in an oscillating frequency of approximately f = 220 Hz. The feed velocity v_f and the line pitch a were kept on constant levels (v_f = 1000 mm min^{-1}, a = 0.2 mm). **Fig. 1** depicts the process kinematics of MHP and shows the experimental setup on a coated and rectangle C45 (20 x 65 x 10 mm) steel specimen.

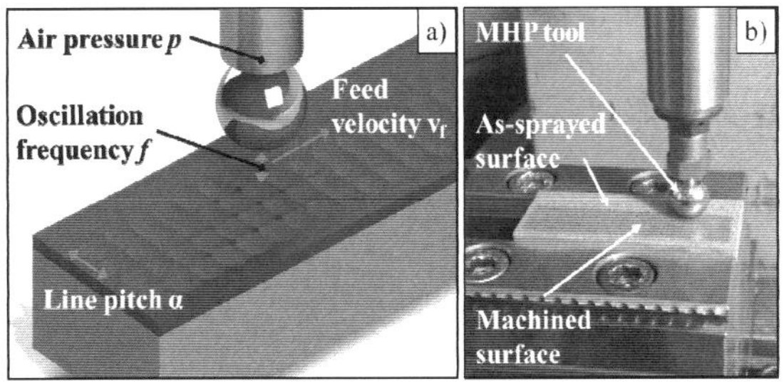

Fig. 1: a) Process kinematics of the MHP process, and b) experimental setup on a coated rectangle substrate

2.3 Analytic methods

Regarding the cross-section analyses, the coated samples were prepared by using diamond grinding discs and polishing cloths with a diamond suspension (9 µm, 6 µm, 3 µm, and 1 µm). Images of the microstructure of the coatings as well as across the wear tracks were taken, using a field emission scanning electron microscope (FE-SEM) type JSM-7001F (Fa. Jeol, Germany). EDX analyses, using the commercial software INCA (Fa. Oxford Instruments, United Kingdom), were utilized to examine the distribution of elements. Nanoindentation tests (Fa. Agilent Technologies, USA) were carried out on the surface in order to analyze the hardness and Young's modulus.

In order to examine the tribological behavior of the coatings, sliding tests, using a ball-on-disk tribometer, (BOD) (Fa. CSM, Switzerland) were carried out. A load of 10 N was applied without a lubricant supply. Towards all samples, alumina was used as a counterbody. Worn samples were analyzed within 3D-profilometry (Fa. Alicona, Austria). Accordingly, the wear coefficient was calculated, based on the average volume loss at four positions for each wear track (three out of four samples) on a constant radius.

For the abrasion tests, the dry sand rubber wheel (DSRW) technique was employed. Therefore, the relevant ASTM standard [25] was considered. A rounded silica sand with a particle size ranging from 200 to 425 µm was used. A rubberized tyre (shore hardness A58-62) with an external diameter of 227.6 mm, a breadth of 12.7 mm, and a thickness of 10 mm was utilized. The rotating speed was 195 min-1. The feedrate for the abrasive material was kept on a constant level of 340 g / min. The applied load with which the specimens are pressed to the rubberized tyre was 130 N. Five specimens of non-MHP and MHP states were employed. The mass loss of each sample was measured before and after testing using 1004 revolutions, which mean a distance of approximately 718 m.

3 Results

3.1 Machine Hammer peened surfaces

The tool impacts lead to a flattening of the surface asperities of the as-sprayed surface. The non-MHP sample provides a surface roughness of Ra = 15.1±2.6 µm for sample 1.6, and a roughness of Ra = 13.8±1.7 µm for sample 2.0. As opposed to that, the MHP samples 1.6 H and 2.0 H are planed to Ra = 6.4±0.1 µm for sample 1.6 H, and Ra = 8.1±0.2 µm for sample 2.0 H. The produced surfaces are shown in **Fig. 2**.

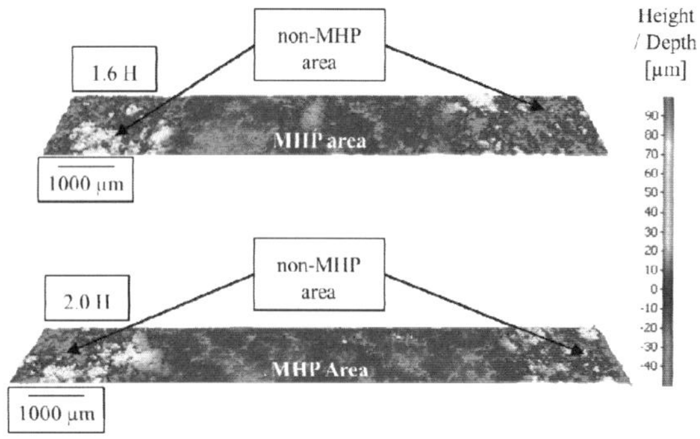

Fig. 2: 3D reconstructed images showing the topography of different samples

Moreover, the high frequently peening process leads to a compression of pores within the arc sprayed coating layer. Cross section images taken by light microscopy evince the effect of the MHP process on the coating morphology (**Fig. 3**). In view of the transition zone across the non MHP and MHP area for sample 2.0 H, the topographical effect and the change of the morphology can be clearly seen.

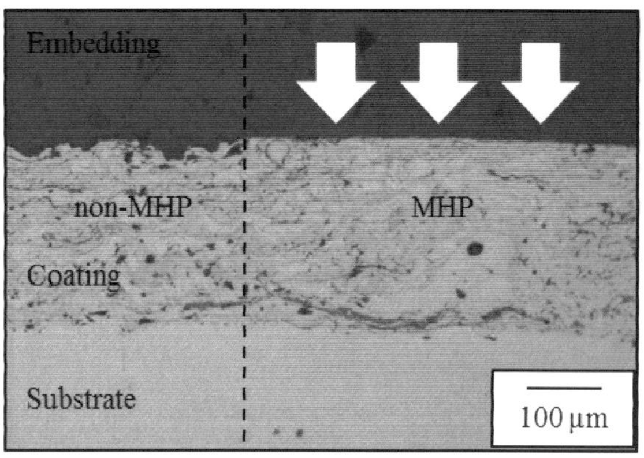

Fig. 3: Cross section images taken by light microscopy showing the coating morphology across the non-MHP and MHP area (sample 2.0)

3.2 Tribological investigations

The tribological behaviors of non-MHP, as-sprayed and polished samples (1.6 and 2.0) as well as the MHP samples (1.6 H and 2.0 H) were examined. Two different ratios of the metallic binder related to the hard phases are provided by the feedstock material (see **Table 1**). **Fig. 4** illustrates the wear tracks after BOD testing. In view of the 3D reconstructed images of the surfaces, it is obvious that both hammer peened samples (1.6 H and 2.0 H) exhibit a higher level of wear, whereas the samples 1.6 and 2.0 show a reduced wear. This is reflected by the increased width of the wear marks.

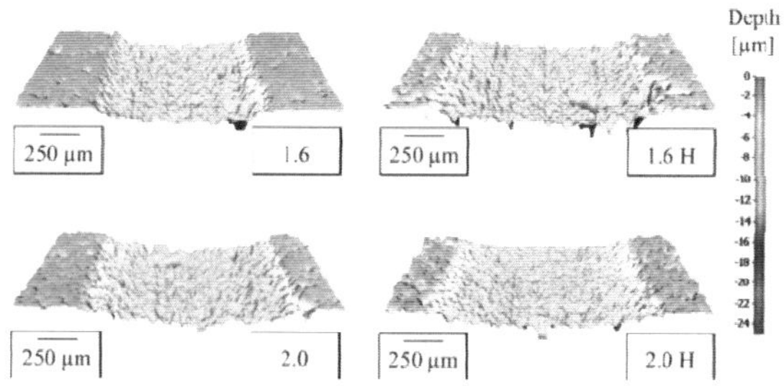

Fig. 4: 3D reconstructed images showing the worn samples after ball-on-disk testing

Regarding the wear coefficient k, which was determined by the volume loss, differences are more visible. The samples 1.6 and 2.0 possess an enhanced wear resistance compared to the MHP samples 1.6 H and 2.0 H as demonstrated by a reduced k ($k_{non\text{-}MHP}$ = 2.8±0.4 · 10^{-6} mm^3 Nm^{-1} for sample 1.6, and $k_{non\text{-}MHP}$ = 2.6±0.3 · 10^{-6} mm^3 Nm^{-1} for sample 2.0 compared to k_{MHP} = 3.9±0.7 · 10^{-6} mm^3 Nm^{-1} for sample 1.6 H, and k_{MHP} = 3.0±0.4 · 10^{-6} mm^3 Nm^{-1} for sample 2.0 H) (**Fig. 5a**).

In view of the counterbody, the results obtained from the 3D-profilometry show a similar wear behavior, determined by the volume loss of the ball material, thus resulting in almost the same tendency concerning the wear coefficient k_{Pin} (**Fig. 5b**).

When compared the non-MHP vs. MHP state, the tribological investigations reveal a higher wear resistance for the non-MHP state. For the counterbody, which is into contact towards the surface, it is assumed that its wear coefficient towards the non-MHP must be significantly increased. In contrast to that, the wear coefficient for the counterbody sliding towards sample 1.6 and 2.0 is more than half as high as for the sample 1.6 (compared to 1.6 H), and 14 % reduced for sample 2.0 (compared to 2.0 H).

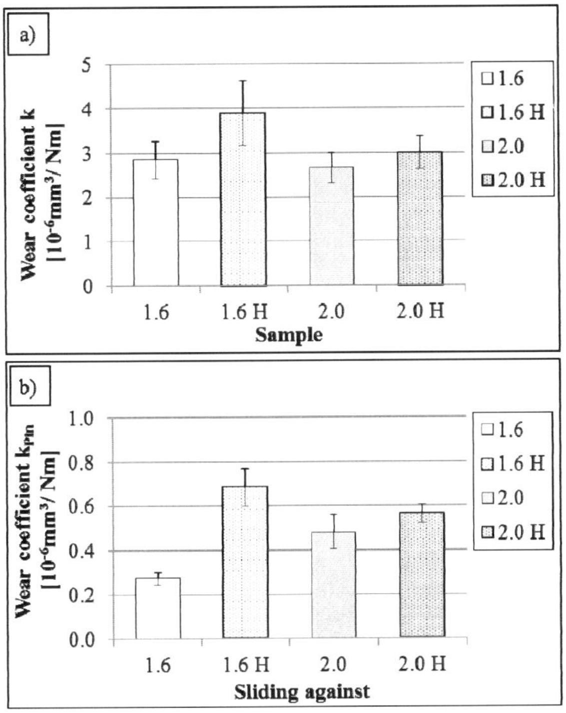

Fig. 5: (a) Wear coefficient k for the non-MHP (1.6 and 2.0) and MHP (1.6 H and 2.0 H) samples; (b) Wear coefficient k_{Pin} of the alumina counterbody towards different samples

In terms of the friction behavior, the samples 1.6 H and 2.0 H possess an enhanced friction behavior when compared to the samples 1.6 and 2.0, resulting in a reduced coefficient of friction (COF) (**Fig. 6**). For the reduced hard phase content (1.6 → 1.6 H), the COF drops significantly of about 23 % from 0.99 to 0.76, whereas for the samples, consist of a higher hard phase content (2.0 → 2.0 H), the COF only slightly decrease (about 9 % from 0.86 to 0.78). In terms of the samples 1.6 H and 2.0 H, it is conspicuous that increased deviations of the tangetial force were observed, resulting in higher fluctuations of the COF.

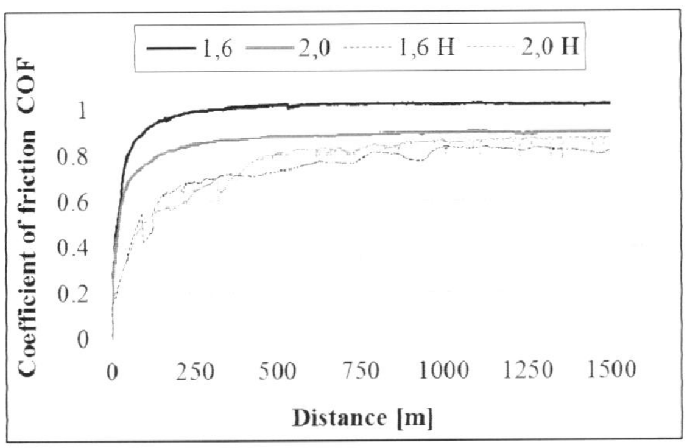

Fig. 6: Coefficient of friction COF for different samples

The results of the abrasion resistance tests are summarized in **Fig. 7**. Here, the coating's mass loss after testing due to abrasion, while corundum serves as abrasive material, is plotted. When compared to the non-MHP surfaces (1.6 and 2.0), the results for the abrasion test reveal a trend for an increase in abrasion resistance of the MHP samples (1.6 H and 2.0 H). Regarding those coating systems, which were produced with a higher amount of the metallic binder (1.6 and 1.6 H), the mass loss decreased about 23 % for the MHP sample 1.6 H. Regarding the experiments using a reduced metallic binder content (2.0 and 2.0 H), the mass loss due to abrasion for the sam-ple 2.0 H is only slightly lower (approx. 4 %), when compared to the non MHP state (2.0).
Since the ratio of the metallic binder related to the CTC hard phases varies for the different coating systems, the wear behavior of the samples 1.6 and 1.6 H is compared to the samples 2.0 and 2.0 H. It is obvious that the 1.6 series, manufactured with a cored wire providing a reduced hard phase content, obtains an advanced wear resistance when compared to the 2.0 series. This fact corresponds to a reduced mass loss for the samples 1.6 and 1.6 H as seen in **Fig. 7**.

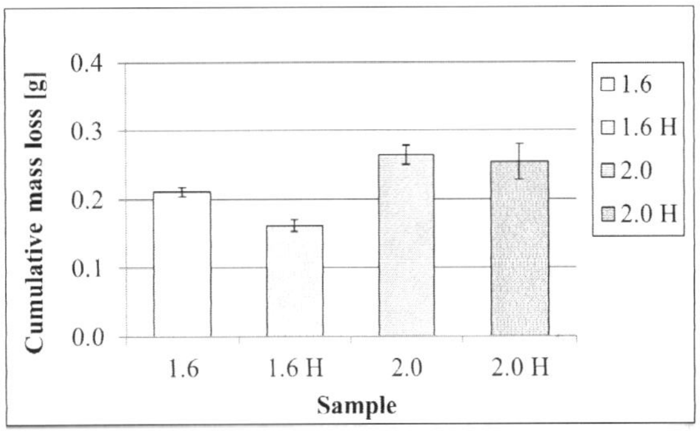

Fig. 7: Cumulative mass loss of different samples after dry sand rubber wheel testing

4 Discussion

Since an iron-based coating was applied and tested during dry sliding friction, a film caused by tribochemical reactions can be observed. The formation of oxides can be clearly seen at the contact area using EDX analyses. However, scarcely abrasive wear can be seen with most of the samples due to sliding friction. Thus, the occurrence of grooves, which had their origin in third-body wear, was detected by electron microscopy. Debris from the alumina counterbody, caused by break outs at certain inhomogenities of the coating, and break outs of the coatings' hard phases itself gets involved and affects the sliding at the contact area.
In closer view at the wear tracks, the MHP surfaces (samples 1.6 H and 2.0 H) showed a significantly high amount of break outs at the sliding contact area when compared to the non-MHP samples. Large-scaled particles were pulled out during sliding contact as observed after testing (**Fig. 8**).

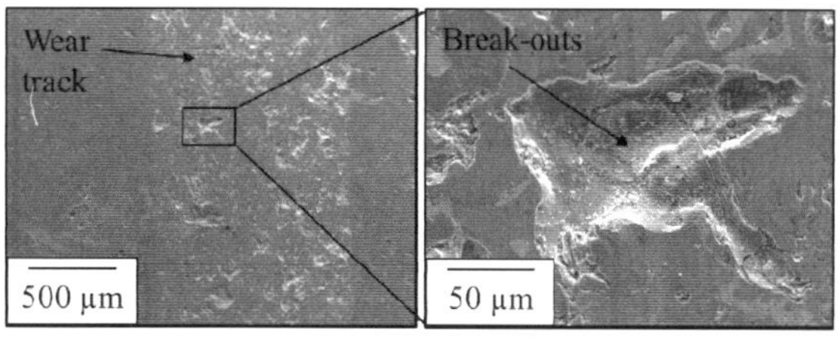

Fig. 8: SEM images showing the wear track for a machine hammer peened sample (1.6 H)

It is assumed that these particle lead to a significant third body wear as they cannot go in recess, and thus, resulting in an increased wear rate. This behavior is in good accordance with the results presented in **Fig. 5a** and **Fig. 5b** (as shown by an increased wear coefficient k for the samples 1.6 H and 2.0 H). Therefore, it was found that the MHP samples provide recognizable surface damages. With regard to cross section images of MHP samples, cracks across splat boundaries as well as transitional cracks of carbides are visible, which increase the risk of spalling (**Fig. 9**)

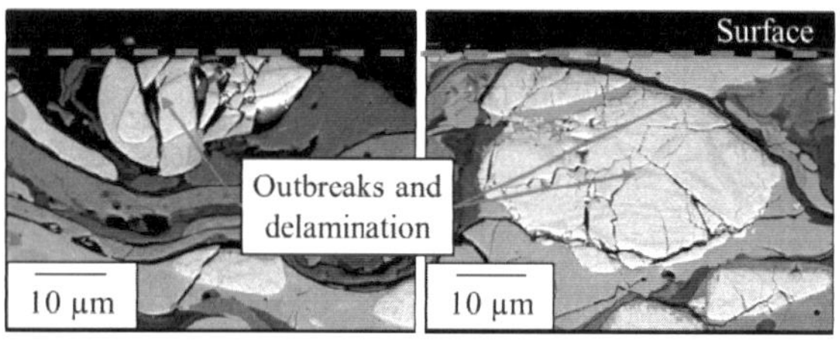

Fig. 9: Cross section images taken by electron microscopy showing surface damages due to machine hammer peening

For the dry sliding testing, all coated samples obviously exhibit the beginning of micro-pittings, in particular across the embedded carbides. Such a contact fatigue and stress cracking is a direct consequence of changing mechanical stresses. However, it is assumed that the pre-damaged state of the brittle hard phases, caused by the MHP process, is detrimental for the coating characteristics with respect to the sliding tests.

Since the amount of hard phases, related to the binder, was changed for some samples, the distribution of elements across the coating was determined by means of EDX analyses. Different spots (five out of six) across the coating have been investigated. Thus, the element concentration of tungsten was detected. Results taken from the EDX analyses were considered to calculate the mean value for the tungsten content. For the samples 1.6 and 1.6 H the mean value was 41.9±0.4 wt.%, whereas the samples 2.0 and 2.0 H consist of 44.8±0.7 wt.% of tungsten. Based on these results, it is assumed that the samples 2.0 and 2.0 H obtain a higher content of W-rich hard phases. However, the phase composition needs to be examined in more detail.

Due to the relation between the mechanical response of the coating and the ratio between hardness and Young's Modulus as demonstrated in [26, 27], nanoindentation were carried out. Nanoindents (seven times seven, depth d = 2000 nm) were performed on the different surfaces. In case of impingement onto carbides, the values were skipped for calculation. The mechanical properties of the different surface modifications are summarized in **Table 2**.

Table 2: Mechanical properties of different samples

Sample	MH [HV0.3]	H [GPa]	E [GPa]	H / E	H^3/E^2
1.6	563±82	5.5±2.2	94±35	0.0588	0.0191
1.6 H	637±171	5.9±2.0	128±25	0.0463	0.0127
2.0	456±56	4.8±2.5	87±29	0.0551	0.0146
2.0 H	560±95	4.4±1.7	106±18	0.0419	0.0078

MH = Vicker microhardness; H = Hardness; E = Young's modulus

Besides a high H / E ratio, the H^3/E^2 ratio is a significant factor governing sliding wear behavior. Thus, high H^3/E^2 ratios provide increased resistances to abrasion. Regarding the results obtained from the nanoindentation tests, the non-MHP samples (1.6 and 2.0) showed a slightly increased H^3/E^2 value when compared to the MHP samples (1.6 H and 2.0 H), which correlates to the wear coefficient k determined by the dry sliding experiments (**Fig. 5a**). Moreover, samples produced with reduced hard phase content possess higher values with respect to hardness, Young's modulus, and the ratios H / E or H^3/E^2 respectively. For thermal sprayed coatings, it is reported that the Young's modulus, inter alia, gets influenced by the porosity level and shape of pores [28, 29], thus, affecting the obtained values. However, Vickers microhardness and some of the values for the hardness taken by nanoindents reveal higher values regarding the MHP state. It is expected that strain hardening effects in the near-surface area at the metallic layer were generated due to the MHP process. It can therefore be stated that the MHP process, on the one hand, led to surface damages (e.g. by cracks of carbides), which favor break-outs, leading to advanced third-body wear. On the other hand, such post-treatment can provide enhanced mechanical properties, since the hardness of the sourrounded metallic matrix increases.

Acknowledgement

The authors gratefully acknowledge the financial support of the DFG (German Science Foundation) within the Collaborative Research Center SFB 708, subproject A1.

6 Literature

[1] Wielage, B.; Pokhmurska, H.; Student, M.; Gvozdeckii, V.; Stupnyckyj, T.; Pokhmurskii, V.: Iron-based coatings arc-sprayed with cored wires for

applications at elevated temperatures. Surf. Coat. Technol. 220 (2013), pp. 27-35.
[2] Dallaire, S.; Levert, H.: Development of Cored Wires for Improving the Abrasion Wear Resistance of Austenitic Stainless Steel. J. Therm. Spray Technol. 6 (1997), Issue 4, pp. 456-462.
[3] Fang, J.Y.; Zhuo-xin, L.; Jian-min, J.; Yao-wu, S.: Difference in particle characteristics and coating properties between spraying metallic and ceramic powder cored wires. Trans. Nonferrous Met. Soc. China 17 (2007), pp. 537-542.
[4] Liu, S.G.; Wu, J.M.; Zhang, S.C.; Rong, S.J.; Li, Z.Z.: High temperature erosion properties of arc-sprayed coatings using various cored wires containing Ti–Al intermetallics. Wear 262 (2007), pp. 555–561.
[5] Wilden, J.; Bergmann, J.P.; Jahn, S.: Einfluss der Strom-Spannungs-Charakteristik auf die Schichtbildung und -eigenschaften beim Lichtbogenspritzen. Schriftenreihe Werkstoffe und werkstofftechnische Anwendungen 18 (2004), pp. 95-100.
[6] Tillmann, W.; Luo, W.; Nebel, J.: Einfluss der Hartstoffkorngröße auf die tribologischen Eigenschaften gewalzter und geschliffener WSC-FeCSiMn-Schichten. Thermal Spray Bulletin 4(1), (2011), pp. 56-63.
[7] Newbery, A.P.; Grant, P.S.; Neiser, R.A.: The velocity and temperature of steel droplets during electric arc spraying. Surf. Coat. Technol. 195 (2005), pp. 91-101.
[8] Hussary, N.A.; Heberlein, J.V.R.: Effect of system parameters on metal break-up and particle formation in the wire arc spray process. J. Therm. Spray Technol. 16 (2007), Issue 1, pp. 1-13.
[9] Pourmousa, A.; Mostaghimi, J.; Abdini, A.; Chandra, S.: Particle size distribution in a wire-arc spraying system. J. Therm. Spray Technol. 14 (2005), Issue 4, pp. 502-510.
[10] Hussary, N.A.; Heberlein, J.V.R.: Atomization and particle-jet interactions in the wire-arc spraying process. J. Therm. Spray Technol. 10 (2001), Issue 4, pp. 604-610.
[11] Tillmann, W.; Vogli, E.; Abdulgader, M.; Gurris, M.; Kuzmin, D.; Turek, S.: Particle Behavior During the Arc Spraying Process with Cored Wires. J. Therm. Spray Technol. 17 (2008), Issue 5-6, pp. 966-973.
[12] Newbery, A.P.; Rayment, T.; Grant, P.S.: A particle image velocimetry investigation of in-flight and deposition behaviour of steel droplets during electric arc spray forming. Mat. Sci. Eng. 383 (2004), pp. 137-145.
[13] Tillmann, W.; Abdulgader, M.: Wire Composition, Its Effect on Metal Disintegration and Particle Formation in Twin-Wire Arc-Spraying Process. J. Therm. Spray Technol. 22 (2013), Issue 2-3, pp. 352-362.
[14] Newbery, A.P.; Grant, P.S.: Oxidation during electric arc spray forming of steel. J. Mater. Process. Technol. 178 (2006), pp. 259-269.
[15] Sheppard, P.; Koiprasert, H.: Effect of W dissolution in NiCrBSi–WC and NiBSi–WC arc sprayed coatings on wear behaviors. Wear 317 (2014), pp. 194–200.
[16] He, D.J.; Fu, B.J.; Jiang, J.M.; Li, X.J.: Microstructure and Wear Performance of Arc Sprayed Fe-FeB-WC Coatings. J. Therm. Spray Technol. 17 (2008), Issue 5, pp. 757-761.
[17] Tillmann, W.; Klusemann, B.; Nebel, J.; Svendsen, B.: Analysis of the Mechanical Properties of an Arc-Sprayed WC-FeCSiMn Coating: Nanoindentation and Simulation. J. Therm. Spray Technol. 20 (2011), Issue 1-2, pp. 328-335.
[18] Xu, B.; Zhu, Z.; Ma, S.; Zhang, W.; Liu, W.: Sliding wear behavior of Fe–Al and Fe–Al/WC coatings prepared by high velocity arc spraying. Wear 257 (2004), pp. 1089–1095.
[19] Tillmann, W., Luo, W.; Selvadurai. U.: Wear Analysis of Thermal Spray Coatings on 3D Surfaces. J. Therm. Spray Technol. 23 (2014), Issue 1-2, pp. 245-251.
[20] Li, Z.; Jiang, Y.; Zhou, R.; Lu, D.; Zhou, R.: Dry three-body abrasive wear behaviour of WC reinforced iron matrix surface composites produced by V-EPC
infiltration casting process. Wear 262 (2007), pp. 649-654.
[21] Zhou, R.; Jiang, Y.; Lu, D.: The effect of volume fraction of WC particles on erosion resistance of WC reinforced iron matrix surface composites, Wear 255 (2003), pp. 134-138.
[22] Hiegemann, L.; Weddeling, C.; Ben Khalifa, N.; Tekkaya, A.E.: Prediction of roughness after ball burnishing of thermally coated surfaces. J. Mater. Process. Technol 217 (2015), pp. 193-201.
[23] Scheil, J.; Müller, C.; Steitz, M.; Groche, P.: Influence of process parameters on surface hardening in hammer peening and deep rolling. Key Eng. Mater. 554-557 (2013), pp. 1819-1827.
[24] Wied J.: Oberflächenbehandlung von Umformwerkzeugen durch Festklopfen. Dissertation TU Darmstadt (2011).
[25] ASTM international: Standard test method for measuring abrasion using the dry sand/rubber wheel apparatus. ASTM stand-ards G65-04 (2004).
[26] Musil, J.; Kunc, F.; Zeman, H.; Poláková, H.: Relationships between hardness, Young's modulus and elastic recovery in hard nanocomposite coatings. Surf. Coat. Technol. 154 (2002), pp. 304-313.
[27] Cai, F.; Huang, X.; Yang, Q.: Mechanical properties, sliding wear and solid particle erosion behaviors of plasma enhanced magnetron sputtering CrSiCN coating systems. Wear 324–325 (2015), pp. 27-35.
[28] Portinha, A. et al.: Residual stresses and elastic modulus of thermal barrier coatings graded in porosity, Surf. Coat. Technol., 188–189 (2004) pp. 120-128
[29] Celik, E.; Sarikaya, O.: The effect on residual stresses of porosity in plasma sprayed $MgO–ZrO_2$ coatings for an internal combustion diesel engine, Mater. Sci. Eng., A, 379 (2004), Issue 1-2, pp. 11-16

The influence of heat treatment presence and sequence on oxidation resistance of YSZ thermal barrier coatings

Mrs. Alsu Akhmetgareeva, TSPC Ltd., Moscow, Russia

Dr. Sergey Baldaev, TSPC Ltd., Moscow, Russia

Dr. Ivan Mazilin, TSPC Ltd., Moscow, Russia

Dr. Lev Baldaev, TSPC Ltd., Moscow, Russia

Dr. Tat'yana Il'inkova, Kazan National Research Technical University named after A.N.Tupolev - KAI (KNRTU-KAI), Kazan, Russia

Abstract.

Well known that oxidation resistance of thermal barrier coatings (TBC) depends on chemical composition, thickness, the rate of formation of TGO layer.

In this paper we studied the influence of presence and sequence of air and vacuum heat treatment on properties of TBC particularly on oxidation resistance. The parameters (temperature, pressure, duration and etc.) of air and vacuum heat treatment were unchanged. TBC system were sprayed on Ni-based substrate and consists of HVOF NiCo-based bond coat (BC) and APS yttria-stabilized zirconia (YSZ) top coat. Isothermal heating test was used to determine effect of heat treatment presence and sequence on interdiffusion between bond layer and substrate and microstructural characteristics of bond coat and ceramic top coat. Before and after isothermal heating test cross-section of specimens were studied by SEM and EDX analyzes. Preliminary results demonstrated that oxidation resistance of TBC with APS-top layer and HVOF-bond layer can be improved by applying suitable sequence of heat treatment. Due to developed sequence of heat treatment TBCs service life increased up to 500 hours at 1100°C.

Effect of Pre and Post Processes on Cold Spray Coating Properties

P.K. Koh, H.W. Ng, P. Cheang, K. Loke

Cold spray process is a viable alternate repair method to address corrosion issues for gas turbine engine fan case. However, being a relatively new coating process, the effect of pre and post coating processes on the properties of the coatings needs to be better understood. The effect of pre coat grit blasting of the substrate on the mechanical properties and microstructures of the cold spray coatings was first investigated. The results showed that grit blasting may not enhance the microhardness or adhesive bond strength of the coating and thus, may be eliminated as a pre-coat process. The effect of post coating heat treatment was also evaluated. The results indicated that both stress relief and annealing processes reduced the microhardness of the coatings. On the other hand, the adhesive bond strength was unaffected by stress relief but improved significantly when subjected to annealing.

1 Introduction

Cold gas dynamic spraying (more commonly known as cold spray) is a comparatively new coating process [1]. It falls under the broad family of thermal spray processes but is unique in that it can produce a coating at relatively low temperatures. Essentially, cold spray process uses high pressure carrier gases to accelerate particles to high enough velocities such that upon impact, the particles plastically deform and bond to the surface of the substrate, thus forming a dense coating or structural deposit. On subsequent passes, the particles bond to the previously deposited layer, forming a thick coating. Preliminary study demonstrates the technical viability of cold spray process as an alternate repair method to address corrosion issues for gas turbine engine fan case. However, being a relatively new coating process, the effect of pre and post coating processes on the properties of the coatings needs to be better understood.

The positive effect of substrate surface roughness on conventional thermal spray processes has been extensively researched and validated. The effect of substrate surface roughness on the mechanical properties of cold sprayed coating is, however, not as well established. The impact of substrate surface roughness on coating bond strength was examined by Wu et al. [2]. Using cold sprayed Al-Si coating coated over a range of particle velocities on mild steel, the study evaluated the relative coating bond strength achieved. The experimental results show that the adhesive bond strength of roughened specimen is close to or slightly better than the as-polished one when the impact velocity of the coated particles is near the critical velocity of the feedstock material. However, as the impact velocity increases, the bond strength of the grit blasted specimen becomes less than that of the as-polished one. Using micrographs at high magnification, the authors identified micro-pores and defects at the interface of the coating and roughen substrate surface that they attributed to causing the lower bond strength in the grit blasted specimen. Based on these observations, the authors concluded that the effect of surface roughness on the interface bonding is minimal. In a separate study, Richer et al. [3] investigated the impact of substrate roughness on cold spray coating efficiency. Nanocrystalline Al-Mg powder was cold spray deposited on aluminium substrates that are roughened by coarse silica grit (to roughness of 316 ± 57 μm) and fine glass bead grit (to roughness of 134 ± 20 μm). Marginally better coating efficiency was observed for the rougher substrate. The authors explained that deposition efficiency is enhanced by greater substrate surface roughness which promotes more deposited mass. The authors attributed this to the greater surface area of the rougher substrate that allows larger number of particles to come into direct contact with the substrate during the spraying of the initial layer. The effect on coating adhesion strength was not investigated. In a study conducted by Kumar et al. [4] to shed more light on the effect of roughened substrates on cold spray coating properties, copper powder was deposited onto smooth and grit blasted copper and aluminium substrates and evaluated. By evaluating the flattening ratio (defined as the ratio of the diameter of the bonded particle to that of a spherical particle of the same volume), individual particle deposition characterization done in the study indicates that substrate roughness whose crest size is the same as the particle size and the trough size is the half particle size creates better bonding.Though the authors suggested that grit blasting of the substrate enhances the bond strength of the coating, the results are inconclusive. Higher adhesion strength was measured for roughened substrates at lower impact velocities (<570 m/sec) but at higher impact velocities (600 m/sec), the bond strength of the copper coating on both smooth and roughened copper and aluminium substrates are almost identical.

In conventional thermal spray processes, a post coating operation in the form of heat treatment is usually performed to enhance certain material properties such as metallurgical bonding, ductility and fracture toughness of the coating [5]. As cold spray is a solid-state coating process that takes place in a relatively lower temperature, the equivalent benefits of post coating heat treatment need to be evaluated.

In a study done by Hall et al. [6], cold sprayed aluminium specimens were annealed (300 °C, 22 hours air) and tested for its ultimate tensile strength. The experimental results show that the anneal samples demonstrated a decrease in ultimate strength and significant increase in ductility (up to

10% elongation) when compared to as-sprayed samples. Recrystallisation within the cold sprayed splats and a reduction in dislocation density in the aluminium crystals are believed to be the reasons for the alteration in the mechanical properties.

Li et al. [7] in a study on cold sprayed copper coatings, looked into the effect of annealing treatment on the microstructure, electrical resistivity and microhardness of the coatings. The study presented evidence of significant microstructure changes when the copper coatings were subjected to 350 °C and 650 °C heat treatment for various durations. The experimental results also showed that the microhardness of the cold sprayed copper coating decreased with increasing annealing temperature and at elevated annealing temperature, longer duration of annealing (12 hours) will bring the microhardness of the coating to that of annealed bulk copper. The authors also suggested that the anisotropy microstructure of the cold spray coatings can be eliminated through annealing treatment but did not provide quantitative microhardness data to support the claim. In a similar study on cold sprayed copper coatings, Cella et al. [8] obtained two sets of copper coatings sprayed at 298 and 523 K respectively using helium as the carrier gas. The experiment demonstrated a reduction of microhardness of the coating when subjected to annealing treatment at various temperatures. In addition, the specimen obtained at lower temperature exhibited recrystallisation at annealing temperature as low as 373 K with a corresponding reduction in microhardness. However, for the specimen coated at higher temperature, recrystallisation occurred at a much higher annealing temperature of 773 K. The study proposed that the specimen obtained at 523 K shows a higher dislocation density microstructurally and the low stacking fault energy of copper prevents recovery during annealing and, thus, changes in microstructure and microhardness values during annealing only start when recrystallisation begins.

Sundararajan et al. [9] found that the elastic modulus of cold sprayed coatings of a wide variety of metallic materials (Cu, Ag, Zn, Nb, Ta, Ti and 316L stainless steel) are significantly lower than the values reported for bulk cast samples. The reduction in elastic modulus of the cold sprayed coatings was related to the extent of the inter-splat boundary cracking. However, the study also found that the elastic modulus of the coatings increases significantly after undergoing heat treatment processes. The authors attributed this observation to the decrease in the number of inter-splat cracking and porosity after the heat treatment process.

Similar findings were reported by Al-Mangour et al. [10] who found that heat treatment was able to reduce porosity and improve inter-particle bonding of cold sprayed 316L stainless steel coating. The study also found increase in the ultimate tensile strength and ductility of the nitrogen-sprayed coatings after annealing. The report also showed that the microhardness of the coatings transitions four distinct stages associated with the corresponding presence (or absence) of recovery and recrystallisation of the coating microstructure.

The review found that though research work that has been reported on the pre and post coating process for cold spray, specifically surface roughening of the substrate and heat treatment after cold spray, quantitative results of their effect on coating mechanical properties such as adhesion bond strength and coating microhardness are lacking. It is the intent of this study to quantify the impact of grit blasting on the cold spray coating microhardness, tensile adhesion strength and microstructure and the effect of stress relief and full annealing on the coating mechanical properties.

2 Experimental Procedure

2.1 Feedstock powder

Commercially available gas atomized Al-6061 powder (Valimet Inc., USA) was used for the experiment. **Fig. 1** shows the scanning electron microscope (SEM) images of the Al-6061 feedstock powder. The particles are mainly spherical in shape and the diameter of the particles ranges from 5 µm to 50 µm with the average diameter measured at 28 µm.

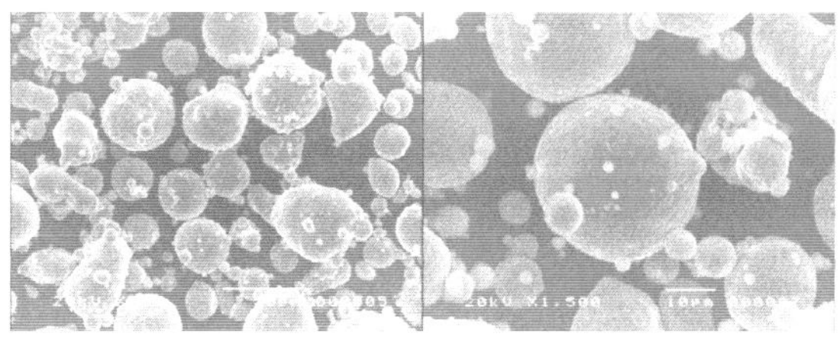

Fig. 1. SEM images of the Al-6061 feedstock powder at (a) 500X and (b) 1500X.

2.2 Cold Spray Parameters

The cold spray process was carried out using the PCS-1000 cold spray system (Plasma Giken Co. Ltd., Japan). Nitrogen was used as the carrier gas and the flow rate was set at 250 standard litre per minute (SLM). The gas pressure was set at 4 MPa and the temperature at 400 ºC. The speed of the traversing robot was 200 mms^{-1} with a 1 mm step. The standoff distance was set at 25 mm and a total of two passes were made.

2.3 Microhardness test

Two aluminium Al-6061 substrate samples 30 mm x 30 mm x 10 mm (length x width x thickness) were used. These samples were machined from a commercially available (Metal Establishment Pte Ltd, Singapore) Al-6061 square tube of same cross-sectional area. The first was grit blasted with alumina grits (#150, average size 75 μm) at 4 MPa and standoff distance of about 25 mm using a custom built grit blasting machine (Abrasive Engineering Pte Ltd, Singapore) for 2 minutes. The second was not and the substrate surface condition was as-machined. The Taylor Hobson (Leicester, UK) Surtronic Duo Surface Roughness Tester was used to measure the mean arithmetic surface roughness of the grit blasted and as-machined surfaces and the results were Ra 2.51 μm and Ra 0.10 μm respectively. Both samples were subsequently cleaned with acetone prior to the cold spray process. The coated samples were sectioned, mounted and polished. The microhardness measurements were subsequently made using the Matsuzawa (Akita, Japan) MMT-X7 tester. Five indentation measurements were taken from each sample using a 100 grams load (HV_{100g}) with a dwell time of 15 seconds along the mid-plane of the polished transverse section which is approximately 150 μm from the substrate coating interface.

2.3 Adhesive bond strength test

Two sets of experimental data each consisting of five samples was collected in accordance to the ASTM C633-01 standards (Standard Test Method for Adhesion or Cohesion Strength of Thermal Spray Coatings). The first set of substrates were grit blasted whereas the second set of substrates were left in their as-machined condition. The mean arithmetic surface roughness of the grit blasted and as-machined substrate surfaces prior to the cold spray process was measured to be Ra 2.21 μm and 0.18 μm respectively using the Taylor Hobson Surtronic Duo Surface Roughness Tester. The samples were subsequently cleaned with acetone and cold sprayed with Al-6061 powder (Valimet Inc., USA). The coated samples were glued together using epoxy glue Klebbi (Oerlikon Metco, Winterthur, Switzerland) and cured at 200 °C for 2 hours. The tensile adhesion strength of the specimens was tested using the MTS/SINTECH 65/G Universal Testing Machine with crosshead speed at 1 $mmsec^{-1}$.

2.4 Microstructural evaluation

The microstructure of the coatings was examined using a field emission scanning electron microscope (FESEM). The coating samples were sectioned, mounted, etched and prepared according to ASTM E3-11 standard. The samples were then examined and the results were analysed according to ASTM E487-85 standard.

2.5 Heat treatment of coating samples

The coatings were subjected to the following heat treatment processes:

- Stress relief: First set was subjected to 150 °C of heat treatment in an air furnace for 2 hours and air-cooled to room temperature.
- Full annealing: Second set of samples was subjected to 535 °C for 2 hours followed by 177 °C for 8 hours before air-cooled to room temperature.

3 Results and Discussion

3.1 Effect of grit blasting on coating microhardness

From **Table 1**, the average microhardness value for the coating with grit blasted substrate surface was 99.8 ± 4.1 while the measurement for the coating on the non-grit blasted substrate was 101.0 ± 8.7. The results indicate that the average microhardness values of both samples and standard deviation were relatively small, indicating the uniformity of the coating. In addition, the effect of grit blasted surfaces on the microhardness of the coating is minimal. Incidentally, the mean microhardness value for the aluminum Al-6061 substrate material was measured to be 99.4 HV_{100} and 99.2 HV_{100} for the grit blasted and non-grit blasted substrates respectively. The microhardness measurement of the substrate showed little changes in both the transverse and vertical directions. The results indicate little deviation in microhardness characteristics between both as-sprayed coating and the substrate. The constant and repeated impact of the impinging cold sprayed particle causes a gradual compaction of the coatings, thus resulting in such relatively high microhardness values.

Table 1. Microhardness measurement of cold sprayed Al-6061 coating on grit-blasted and non-grit blasted substrates.

	Specimen which was grit blasted (HV_{100g})		Specimen which was not grit blasted (HV_{100g})	
	Coating	Substrate	Coating	Substrate
1	102.8	91.4	95.1	97.1
2	104.6	100.1	98.4	96.2
3	100.2	104.1	116.0	102.1
4	94.6	97.8	95.3	108.8
5	97.0	103.8	100.2	92.0
Ave	**99.8 ± 4.1**	**99.4 ± 5.2**	**101.0 ± 8.7**	**99.2 ± 6.4**

The variation in microhardness of the coating in the vertical direction was also investigated for coatings on both grit blasted and non-grit blasted substrates. A total of five indentation measurements spread evenly were taken along the vertical direction from the top of the coating to the substrate-coating interface of each sample. The total coating thickness was approximately 250 μm for the coating on grit blasted substrate and 380 μm for the coating on the non-grit blasted substrate. The distance in between microhardness measurements was computed from the middle of the indentation to the same point on the following indentation. The results of the microhardness value of the Al-6061 coatings along the vertical direction are shown on **Fig. 2**. The measurements taken on the substrate in the vertical direction showed little variation from the mean value of 99.4 HV_{100} and 99.2 HV_{100} for the grit blasted and non-grit blasted substrates respectively. **Fig. 2** also suggests very little variation in the microhardness of the top layer of the coating between the 2 sets of samples. The maximum microhardness achieved for both samples was also almost identical at slightly more than 100 HV_{100}. This again demonstrated the minimal impact of pre-coating grit blast process on microhardness properties of the cold spray coating. In both cases, the result also indicates a progressive increase in microhardness of the coating from the top layer to the coating-substrate interface for both coatings. This is likely due to the continuing compaction of the sprayed layer as more impinging particles impact on top of the previous layer. The result also shows that the microhardness will reach a maximum value about 200 to 250 µm from the top coating layer before tapering off to the maximum value of about 100 HV_{100}. This suggests that any further compacting of the sprayed particles will have little effect on the microhardness of the underlying coating.

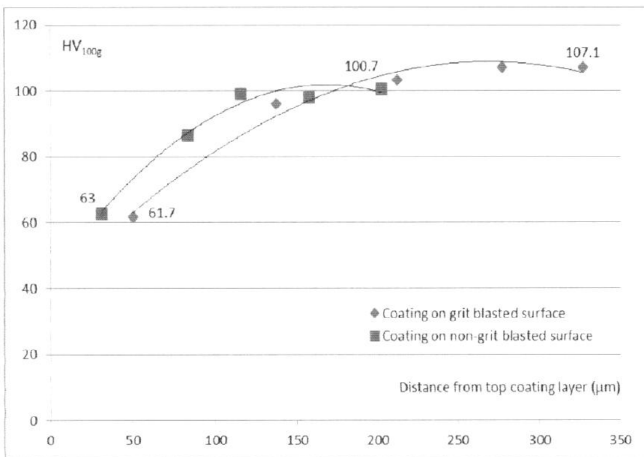

Fig. 2. Microhardness value of Al-6061 coating along the vertical direction from the top surface of the coating.

3.2 Effect of grit blasting on coating adhesive bond strength

From **Table 2**, the average tensile bond strength of the Al-6061 coating on the grit blasted and non-grit blasted substrates were 11.77 ± 4.84 MPa and 40.09 ± 3.58 MPa respectively. The failure mode for all the tests was at the coating and base material interface.

The results show that the tensile adhesion strength of the Al-6061 coating on the non-grit blasted substrate is significantly higher than the coating on the grit blasted substrates. This deviated from similar tests on conventional thermal spray coatings (plasma, HVOF, arc, etc) where the contrary is usually true. The results also seem to suggest that a pre-coating grit blast process may not be beneficial in enhancing the tensile adhesion strength of the coating. The results from this work suggest that grit blasting the substrate may actually be detrimental. In order to investigate further the cause for this anomalous result, the microstructure of the interface between the substrate and the coating for both specimens with and without grit blasting treatment were examined by SEM.

Table 2: Tensile adhesion test results for cold sprayed Al-6061 coating on grit blasted and non-grit blasted Al-6061 substrates.

	Coating with substrate grit blasted (MPa)	Failure Mode	Coating with substrate not grit blasted (MPa)	Failure Mode
1	10.36	Failure at coating/base metal interface.	36.43	Failure at coating/base metal interface.
2	13.97		39.33	
3	7.695		45.99	
4	19.10		40.14	
5	7.715		38.55	
Ave	**11.77 ± 4.84**		**40.09 ± 3.58**	

3.3 Effect of grit blasting on coating microstructure

The SEM micrographs show that the coatings prepared on different substrates have similar properties except for a few layers which are deposited initially. From **Figs. 3** and **4**, we can observe that the geometry of the particles are relatively elongated, indicating the compressing effect of the particles as they are deposited layer by layer. Plastic deformation of the overlying particles ensures that surface irregularities are completely filled, minimising inter-particle porosity. Furthermore, the high velocity impinging particles have a peening effect on the previously deposited underlying material. This action closes any gaps or small pores in the primary material. Porosity in the coating was found to be quantitatively similar for coatings prepared on grit blasted and non-grit blasted substrates. The distribution of such defects is also almost identical.

The difference between the two coatings can be seen in the interface between the coating and the substrate. **Figs. 3** and **4** suggest that the presence of a layer of impurities between the substrate and the coating on the grit blasted surface. This is the most probable explanation for the significantly reduced tensile adhesion strength for the coating on grit blasted substrate compared to the coating on non-grit blasted substrate presented on **Table 2**. **Fig. 3** shows that the grit blasting process has created a layer of contamination that prevents the larger cold sprayed particles from plastically deforming into the substrate.

Fig 3. SEM of Al-6061 coating on grit blasted substrate at 200x magnification.

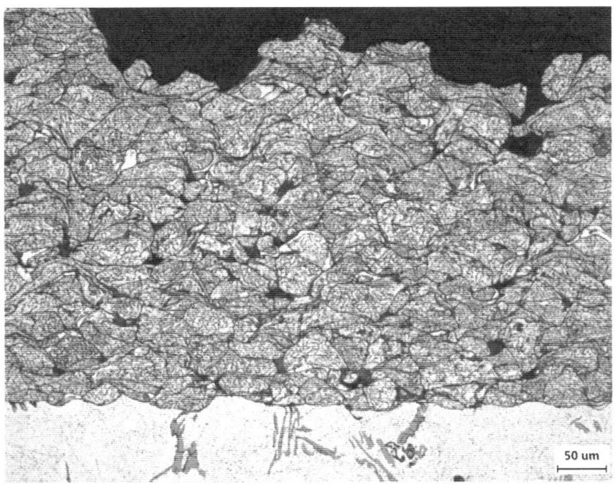

Fig 4. SEM of Al-6061 coating on non-grit blasted substrate at 200x magnification.

For effective adhesion to take place, substrate surfaces should be clean and subjected to high contact pressure. This condition is enhanced by the adiabatic shear instability and the associated localization of the plastic flow into the interfacial jets. The angular contact surfaces on the substrate may have also prevented the formation of bonds due to hydrodynamic flow instability of the adiabatic shear instability. The most likely cause of the substrate jagged surface is the alumina grit material (average size 75 µm). When blasted against the substrate, these angular particles, which are much harder than the substrate, chipped away the surfaces, leaving sharp, pointed finishes. Such surface finishes seem to prevent the cold spray particles from bonding onto the substrate and forming a well-defined interface between the coating and substrate. Furthermore, the presence of impurities, likely from the alumina grit may have contributed to the decreased in tensile adhesion strength of the coating as well.

In the case of the coating formed on the non-grit blasted substrate, **Fig. 4** shows a well-defined interface between the coating particles and the substrate. There were no impurities, porosities or voids observed between the coating and substrate. The surface of the substrate is not perfectly flat and seems to have deformed to about the same size as the impinged coating particles. The cold spray particles can be seen to plastically deform and tightly bond into the substrate, forming a good adhesion. The tensile adhesion strength test results presented in **Table 2** validates the observation.

The cold spray process can be described as a three-in-one or triplex process, in which the grit blasting, spray coating and shot peening are all carried out in a single pass. The in-process grit blasting in this case seem to be more effective than the one using alumina grit because the grit material in the form of the cold sprayed particles is essentially of identical size and shape. This ensures that the substrate surface which is inherently roughen are indented to similar shape and diameter to the particles, unlike the case where the angular alumina grit materials leave a jagged indentation instead. As such, upon impact, the impinging particles will adhere tightly to the roughened substrate as the surface area for contact between the substrate and the particle is maximized. This is consistent with the individual particle deposition characterization study done by Kumar et al. [11] which advocated that substrate surface which is roughened with crest size the same as the particle size creates better bonding.

3.4 Effect of heat treatment on coating microhardness

From **Table 3**, The average microhardness of the two samples was identical. This implies that the different heat treatment for the coatings had identical effect on the microhardness of the coatings. When compared to the microhardness values of the coating which did not go through the heat treatment process (**Table 1**), the heat treated microhardness value was lower by about 20%. This shows that the heat treatment reduced the hardness of the coating and increased its ductility. The reason for this reduction in hardness with corresponding improved ductility is the release of residual stresses in the coating and increasing inter-particle cohesion. The variation in microhardness of both coatings in the vertical direction was also investigated. Five indentation measurements spread fairly evenly were taken along the vertical direction from the top of the coating to the substrate-coating interface of each sample. The results, together with the data from the non-heat treated samples are presented in **Fig. 5**.

Table 3. Microhardness measurement of cold sprayed Al-6061 coatings which have undergone stress relief and annealing compared with untreated specimen.

	Coating stress relieved 150 deg C, 2 hours (HV_{100g})	Coating annealed 535 deg C, 2 hours & 177 deg C, 8 hours (HV_{100g})	Coating not subjected to any heat treatment. (HV_{100g})
1	84.8	83.6	102.8
2	85.8	88.5	104.6
3	88.5	89.5	100.2
4	86.2	86.4	94.6
5	90.2	89.9	97.0
Ave	**87.1 ± 2.2**	**87.6 ± 2.6**	**99.8 ± 4.1**

The results from **Fig. 5** indicate that the heat treatment process had a homogenizing effect on the microhardness of the coating. The change in the microhardness for the coating subjected to two hours of heat treatment at 150 degrees Celsius was relatively small, with the change measured at about 7.8% from the top surface of the coating to about 400 µm of coating thickness. The change was even smaller for the coating which was subjected to full annealing, with the change measured at 3.2% from the top surface to 350 µm of the coating. This is relatively small compared to the 15% change for the non-heat treated coating. Thus, uniformity of the microhardness of the coatings can be achieved with the heat treatment processes, with variation in microhardness decreasing with increasing temperature and duration which the coating is subjected to. A reason behind this phenomenon is the release of the residual stresses across the whole cross-section of the coating will the prolonged heat treatment process. This ensures uniformity of the hardness across the whole coating.

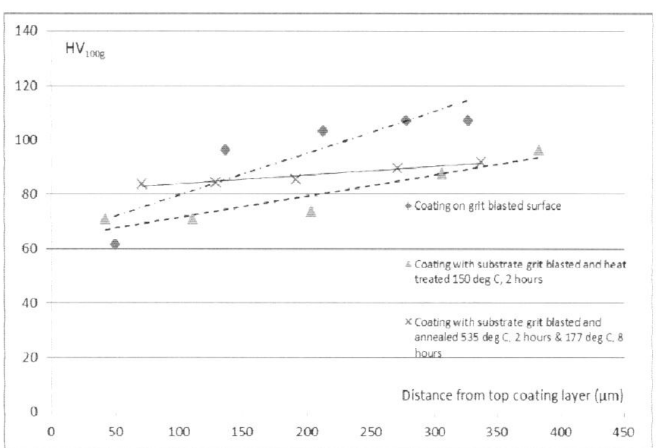

Fig. 5. Microhardness value of Al-6061 coating along the vertical direction from the top surface of the coating.

3.5 Effect of heat treatment on coating adhesive bond strength

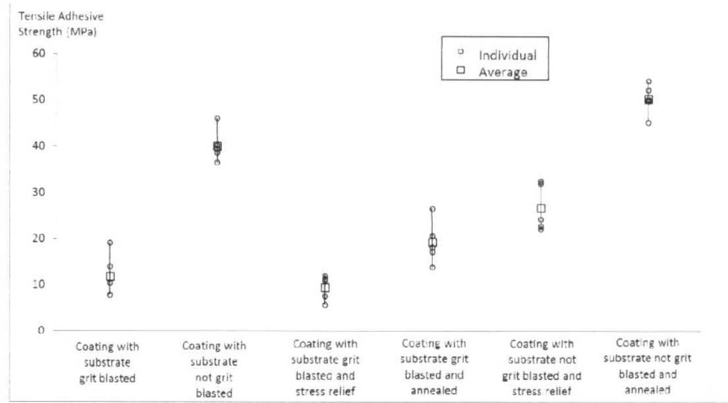

Fig 6. Tensile adhesion strength test results for cold sprayed Al-6061 coatings with and without post coating heat treatment.

From **Fig. 6**, the tensile adhesion strength for the coatings sprayed on grit blasted substrates which had undergone stress relief was not significantly better than those which did not. In fact the average tensile adhesion strength was actually lower, where the mean value of the 5 readings for non-heat treated and heat treated coatings were 11.77 MPa and 9.41 MPa respectively. This result is consistent with the data collected for the coatings which were cold sprayed on non-grit blasted substrates. The average tensile adhesion strength for the heat treated coatings was 26.72 MPa compared to 40.09 MPa for the non-heat treated samples. Both set of results indicated that unlike conventional thermal spray processes, post coating stress relief does not improve the adhesion strength of the cold spray coating. The reason could be that cold spray, being a solid state coating process, does not introduce much thermal residual stresses that may affect inter-particle cohesion. In addition, the lack of porosity in the cold spray coating may also imply minimal impact of heat treatment on enhancement of bond strength.

From **Fig. 6**, the result for the coating sprayed on grit blasted substrates which has undergone full annealing however, did suggest a significant improvement in the tensile adhesion strength. The average value of the 5 readings was 19.29 MPa, a 64% increase as compared to the coating which did not undergo any heat treatment. Similarly, for the coating sprayed on non-grit blasted substrates which has undergone full annealing, the mean tensile adhesion strength was 50.34 MPa, a 26% increase as compared to the coating which did not undergo any heat treatment. Both sets of results suggest that full annealing of the cold spray coatings contribute to the strengthening of the tensile adhesion strength of the coatings.

3.6 Effect of heat treatment on coating microstructure

Comparative examination of the microstructures of the coating which has undergone stress relief (**Fig. 7**) and that which has not (**Fig. 3**) does not reveal significant difference in grain structures or defects (voids, porosity etc) between the 2 coatings. There is also little difference observed at the substrate-coating interface for both sets of coatings. This observation is consistent with the similar tensile adhesion test results for both test coatings.

Analysis of the microstructures of the coating which has undergone full annealing (**Fig. 8**) reveals signs that recrystallisation and grain growth have taken place. In particular, the laminar particle shape of the coating in Figure 4.22c indicates that diffusion and fusion of the coating particles occurred under the elevated temperature environment. The uniformity of the change across the span of the coating also explains the fairly uniform microhardness results shown on Figure 4.17 where the change in microhardness from the top of the coating to the substrate coating interface was less than 4%.

Fig. 7. SEM of Al-6061 coating which has been subjected to stress relief at 150 °C for 2 hours (200x magnification).

Fig. 8. SEM of Al-6061 coating which has been subjected to annealing at 535 °C for 2 hours followed by 177 °C for 8 hours (200x magnification).

4 Conclusions

The effect of pre-coating substrate preparation in the form of surface roughening by grit blasting on the coating microhardness was first examined. The comparative results suggested that grit blasting of the substrate has minimal impact on the microhardness of the coating. In addition, it was also observed that the microhardness increases progressively in the vertical direction from the top of the coating to the substrate-coating interface, reaching a maximum value of about 100 HV_{100g} at between 200 µm to 250 µm from the top coating layer. The maximum microhardness value of the coating was almost identical to the microhardness of the substrate. This is attributed to the gradual compaction of the coatings as the high velocity impinging particles constantly and repeatedly impact on the previously deposited underlying coating. The cumulative effect on that process is a coating with a microhardness which is equivalent to the substrate.

The effect of grit blasting of the substrates on the coating tensile adhesion strength was also investigated. The results suggest that the tensile adhesion strength of the coating on the non-grit blasted substrate is significantly higher than that on the grit blasted substrate. Microstructure evaluation of the coating on the grit blasted substrate indicates the presence of a layer of porosity or voids between the substrate and the coating which is likely due to the grit blasting process. It is believed that the micro-voids created by the angular alumina grit could have prevented the larger cold sprayed particles from plastically deforming into the substrates as well as prevented the formation of bonds due to hydrodynamic flow instability of the adiabatic shear instability. The presence of impurities from the alumina grit may have contributed to the decrease in tensile adhesion strength as well. As for the coating formed on the non-grit blasted substrate, the microstructure evaluation suggests that the enhanced tensile adhesion strength of the coating could be the result of an inherent triplex characteristic of the cold spray process, in which the grit blasting, spray coating and shot peening are all carried out in a single pass. It can be seen that the surface of the substrate had deformed to about the same size as the impinged cold spray particles which seem to have plastically deformed and bonded tightly into the substrate.

The effect of heat treatment on the mechanical properties of the cold sprayed coatings was investigated. The average microhardness of the coatings which have undergone stress relief and full annealing was almost identical. They are however, significantly lower compared to the coating which had not undergone any heat treatment, implying that heat treatment does have an effect of increasing the ductility of the cold spray coating. The uniformity of the coating microhardness in the vertical direction from the top of the coating to the substrate-coating interface was also investigated. The results indicate that heat treatment has a homogenizing effect on the microhardness of the coating. The effect of heat treatment on the coating tensile adhesion strength was also studied. The results suggest that stress relief has little impact on the tensile adhesion strength compared to the coating which has not undergone any heat treatment. However, there was a significant improvement for coatings which has undergone full annealing. Microstructure evaluation of the latter reveals signs that recrystallization have taken place.

Acknowledgement

The authors would like to acknowledge the funding support from SIM University for this project as well as the joint research collaboration with ST Kinetics Limited (Singapore).

References

[1] Alkhimov A.P., Papyrin A.N., Dosarev V.P., Nesterovich N.J. and Shuspanov M.M., 1994, Gas-dynamic Spraying Method for Applying a Coating, US Patent 5,302,414.

[2] Wu J., Yang J., Fang H., Yoon S. and Lee C., 2006, The Bond Strength of Al-Si Coating on Mild Steel by Kinetic Spraying Deposition, Applied Surface Science, 252, p7809-7814.

[3] Richer P., Jodoin B., Ajdelsztajn L. and Lavernia E.J., 2006, Substrate Roughness and Thickness Effects on Cold Spray Nanocrystalline Al-Mg Coatings, J. Thermal Spray Technol., 15 (2), p246-254.

[4] Kumar S., Bae G. and Lee C., 2009, Deposition Characteristics of Copper Particles on Roughened Substrates through Kinetic Spraying, Applied Surface Science, 255, p3472-3479.

[5] Longo F.N., 2004, Postcoating Operations, Handbook of Thermal Spray Technology, J.R. Davis, Ed., ASM International, p128-131.

[6] Hall A.C., Cook D.J., Neiser R.A., Roemer T.J. and Hirschfeld D.A., 2006, The Effect of a Simple Annealing Heat Treatment on the Mechanical Properties of Cold Sprayed Aluminium, J. Thermal Spray Technol., 15 (2), p233-238.

[7] Li W.Y., Li C.J. and Liao H., 2006, Effect of Annealing Treatment on Microstructure and Properties of Cold-Sprayed Cu Coating, J. Thermal Spray Technol., 15 (2), p206-211.

[8] Cella E., McCartney D.G. and Shipway P.H., 2006, Effect of Deposition Conditions on the Properties and Annealing Behaviour of Cold-Sprayed Copper, J. Thermal Spray Technol., 15 (2), p255-262.

[9] Sundararajan G., Chavan N.M. and Kumar S., 2013, The Elastic Modulus of Cold Spray Coatings: Influence of Inter-splat Boundary Cracking, J. Thermal Spray Technol., 22 (8), p1348-1357.

[10] Al-Mangour B., Vo P., Mongrain R., Irissou E. and Yue S., 2014, Effect of Heat Treatment on the Microstructure and Mechanical Properties of Stainless Steel 316L Coatings Produced by Cold Spray for Biomedical Applications, J. Thermal Spray Technol., 24 (4), p188-192.

[11] Kumar S., Bae G. and Lee C., 2009, Deposition Characteristics of Copper Particles on Roughened Substrates through Kinetic Spraying, Applied Surface Science, 255, p3472-3479.

Effect of sandblasting pretreatment and heat input of APS process on property of low-temperature soldered joint of tempered glass to steel

Hong Li [1], Teng-Hai Zhu [1], Wolfgang Tillmann [2], Zhuo-Xin Li [1]
1 College of Materials Science and Engineering, Beijing University of Technology,Beijing /CN
2 Dortmund University of Technology, Dortmund/D

This research presented a new method of "glass surface pre-metallization+low temperature soldering" in order to obtain metallurgical bonding of glass to steel. Prior to the soldering process, tempered glass were deposited with Al_2O_3/Cu coatings multilayer by means of Arc Plasma Spraying (APS). Then the coated glass were soldered to steel at low temperature using Sn-based filler metal. The influence of sandblasting pretreatment of glass substrate and heat input in spraying process on microstructure and properties of ceramic coatings were investigated, as well as the mechanical properties of soldered joint. The results showed that the premetallized coating treatment effectively improved the weldability of glass with steel. The Al_2O_3 ceramic coatings obtained a maximum strength of 20.5MPa. In certain extents, with the increase of sandblasting particle size, the bonding strength of Al_2O_3 coating increased. In addition, with the increase of spraying heat input, the bonding strength of Al_2O_3 coating decreased gradually. The joints all fractured at location of interface between Al_2O_3 coatings and glass substrate. A maximum joint strength of 12.5MPa was achieved when the thickness of Cu coating is 50μm at soldering temperature 270°C and holding time 4min. However, for tempered glass substrate with high hardness and high brittleness, lots of defects, i.e. crack, chip, groove and peel-off etc. in glass surface produced by sandblasting pretreatment process were considered as the main reason for decrease of bonding strength.

1 Introduction

As typical representative of safety glass, tempered glass has high strength and good thermal stability. Its impact strength and bending strength are 3-5 times that of common glass. In addition, tempered glass can bear a temperature difference of 300°C, which is 3 times that of common glass. It is widely used in many fields, such as architectural engineering, automobile manufacturing, household utensils, instrument industry, national defense, electronics industry and so on[1]. In practical applications, the problems of joining glass to other materials are inevitably encountered.

However, glass with unique non-crystalline structure determines its significant difference between the metal in chemical bond types, physical and chemical properties, as well as mechanical properties. And the coefficient of thermal expansion of glass is so low that easily leads to residual stress when directly bonding with metal[2]. From the early 1990s, the problems of bonding glass to metal have aroused the interest of many scholars, and many new joining methods emerged. M. Rauter[3] and G. Wagner[4] performed experimental study on ultrasonic welding of glass to pure aluminum. However, ultrasonic horn has great damage on the glass surface and it is very difficult to obtain the sound joint. Hu et al.[5] have done anodic bonding test on glass/aluminum joints and glass/kovar alloy joints. They found that the ionic diffusion and anodic oxidation is the key factors influencing the bonding quality. M.Kiuchi et al.[6] obtained glass/A2011 aluminum joints with irregualr interface by using semi-solid press joining method. Sun et al.[7] brazed glass to aluminum with ultrasonic vibration assisted by using pure zinc and tin filler metal.

Anyway, considering the big difference of linear expansion coefficient between glass and metal, direct joining of glass to metal in high temperature will result in residual stress, which is easily to weaken the mechanical properties of the joints and even cause failure due to brittle fracture. As coatings produced by thermal spraying technology, such as flame, arc or cold spraying can be used as a promising solder/braze filler metal layers[8].

This paper adopted a new method of combination of surface coating technology with soldering process. At first a multilayer coating, composed of backing coating and operative coating, is deposited on the glass by thermal spraying(Arc Plasma Spraying, APS) to realize the pre-metalization of glass surface. Then low-temperature soldering is performed to realize the reliable joining of metallization glass to steel. The effect of sandblasting pretreatment and heat input of thermal spraying process on interface microstructures and properties of low-temperature soldered tempered glass/steel joints were investigated.

2 Materials and Experimental

The base materials used in the experiment are Na/Ca tempered glass and carbon steel C45, with a dimension of 70mm×50mm×10mm and Φ25mm×30mm respectively. The backing coating materials used Al_2O_3 powder and the operative coating used pure copper powder, with grain sizes -200+400 mesh and -200+325 mesh respectively. The purities of these two powders are both larger than 99.9%. Sn-3.0Ag-0.5Cu solder paste is used as filler material.

In the experiment, the brown fused alumina is used for sandblasting roughening treatment on the tempered glass surface firstly. The sandblasting process parameters are shown in **Tab. 1**. Then the roughened glass is put into the acetone solvent for ultrasonic cleaning. After drying naturally, Al_2O_3

backing coating is deposited on the glass surface using atmosphere plasma spraying, see Fig.1. Then Al_2O_3 coating deposition is succeeded by Cu operative coating deposition, finally low-temperature soldering is carried out at a temperature of 270°C and a dwell time of 4min.

Table 1. Sandblasting process parameters.

Grain size Mesh	Distance mm	Pressure MPa	Angle (°)	Time s
20, 60, 100	100	0.4	45	30

Sn-Ag-Cu solder paste is evenly painted on the laminated surface between the steel column and the metallized glass sample, and then the assembled steel-solder-glass sample is installed in the fixture and put into the T200N nitrogen lead-free reflow oven, finally complete the soldering process according to the heating curve as shown in Fig.2. After soldering, the sample is grinded and polished. The microstructure of soldered joint is observed using OLYMPUS-BX51M metallographic microscope. The coating bonding strength and the joint interface bonding strength are both measured by QBD material testing machine. Every experimental data point need testing five samples and then the average values are obtained. Finally morphology observation for fracture surface of the coating is carried out by OLYMPUS-OLS4000 Laser Scanning Confocal Microscope.

Fig. 1. Arc Plasma Spraying (APS) process.

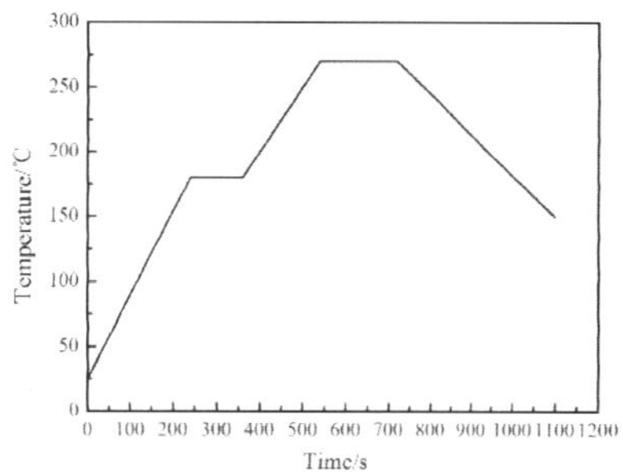

Fig. 2. Heating curve of soldering process.

3 Results and Discussion

3.1 The bonding strength of Al_2O_3 coating

According to a large number of previous experiments, it was found that the interface properties of soldered glass/steel mainly depend on the interfacial bonding strength of Al_2O_3 coating and glass substrate. While the interfacial bonding strength is mainly affected by sandblasting pretreatment process and thermal spraying process, therefore two important factors, grain size of sandblasting materials and spraying heat input are selected as the objectives in current research. **Tab. 2**. shows the relationship between spraying parameters and coating bond strength. Value range analysis was applied for the treatment of experimental data.The factor (A) grain size is the size of brown fused alumina used in sandblasting process, and the factor (B) heat input of thermal spraying process is characterized by using spraying distance and spraying power. Thus when the spray distance is relatively smaller and the spraying power is relatively larger, namely, the larger the spraying heat input, and vice versa.

It can be seen the bonding strength of coating is in the range of 5.1-20.5MPa. The primary and secondary sequence effect of the two factors on the interfacial bonding strength of Al_2O_3 coating and glass substrate is as follows: grain size>spraying heat input. Optimal levels of each factor are sandblasting grain size of 60 mesh, spraying distance of 120mm and spraying power of 33kW. The optimal combination is A2 and B1.

Table 3. The summary of value range analysis

Group	Factor	Evaluation index		
	Grain size (A) Mesh	Heat put (B)		Average Strength MPa
		Distance mm	Power kW	
1	20	120	33	11.1
2	20	100	39	8.9
3	20	80	45.5	5.1
4	60	120	33	20.5
5	60	100	39	19.2
6	60	80	45.5	16.5
7	100	120	33	18.4
8	100	100	39	16.8
9	100	80	45.5	15.6
Range	10.3	4.3		
Optimal	A2	B1		

According to the test results, the effect of grain size and spraying heat input on the bonding strength and porosity of Al_2O_3 coating is as shown in Fig.3 and Fig.4. respectively.

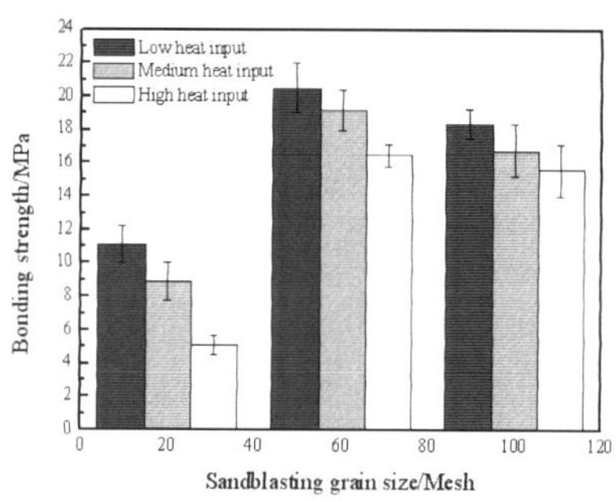

Fig. 3. Sandblasting grain size&spraying heat input v.s. bond strength of Al_2O_3 coating.

When spraying heat input is under certain condition, with the increase of grain mesh, the bond strength of Al_2O_3 coating increased first and then decreased, while porosity of Al_2O_3 coating increased. The average porosity is relatively low, between 8.79-19.6%. When sandblasting grain was 60 mesh, coating obtained a maximum bond strength of 20.5MPa. When sandblasting grain size is under certain condition, with the increase of the spraying heat input level, the bonding strength of Al_2O_3 coating showed a decreasing tendency. When the spraying heat input is low, namely when the spraying distance is 120mm and the spraying power is 33KW, the bonding strength reached the maximum value.

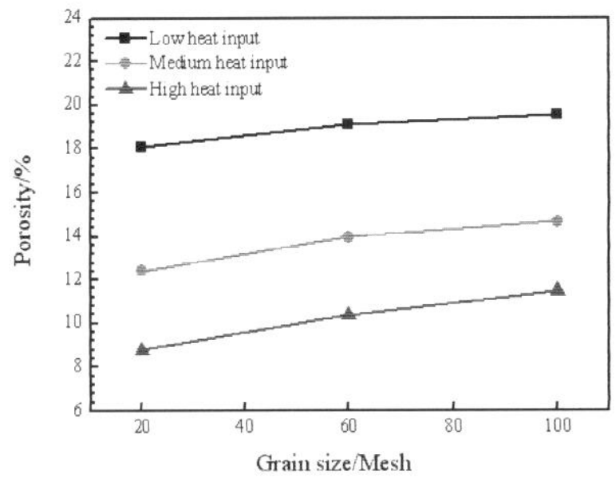

Fig. 4. Sandblasting grain size&spraying heat input v.s. porosity of Al_2O_3 coating.

Fig. 5.(a)-(c) are the bonding interface of coating and glass substrate when sandblasting grains size were 100, 60 and 20 mesh respectively. As shown in the figure, when spraying heat input is under certain condition, with the increase of sandblasting grain size, substrate surface became uneven and rough, while the bonding strength increased accordingly. But when the sandblasting grain size was furtherly increased, high-speed sand produced great impact to the surface of glass substrate, so as to induce the surface to generate more defects, such as dents, cracks, chip, groove, peeling off and etc., especially the cracks which is parallel to the surface, as shown in Fig. 5(c). These defects result in the obvious decrease of bonding strength of Al_2O_3 coating.

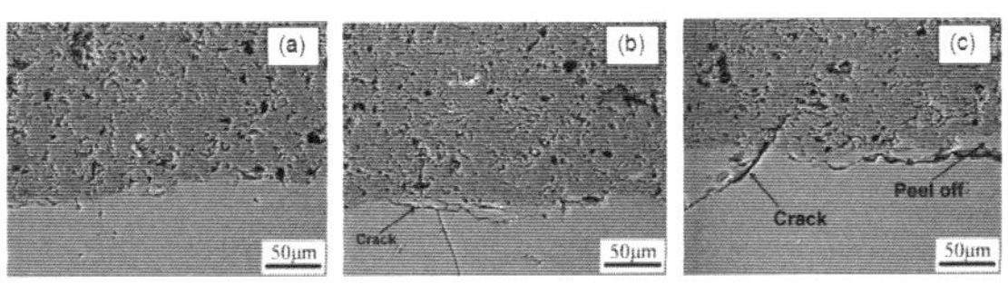

Fig. 5. SEM photograph of interface between Al_2O_3 coating and glass substrate at different sand grain size (a)100mesh, (b) 60mesh and (c) 20mesh

Fig.6(a)-(c) are the fracture surface on glass side after the bonding strength test of coating. It is analyzed that when sandblasting grain size is under certain condition, with the increase of the spraying heat input level, namely the spraying distance are reduced and the spraying power are increased, thermal shock of high temperature grain on glass substrate is increased. Due to the cracks existing on sub-surface layer of glass substrate generated by sandblasting process, the local micro zone become softened and slightly melted, or even fully peel off from the glass substrate. Furthermore, when the heat input is great, the cooling of deposited high temperature grain easily result in bigger residual stress. The damage of glass substrate surface is more serious and the bonding strength is smaller. The bright areas showed in Fig. 6(b) and (c) are caused by the failure of peeled-off local surface of glass substrate.

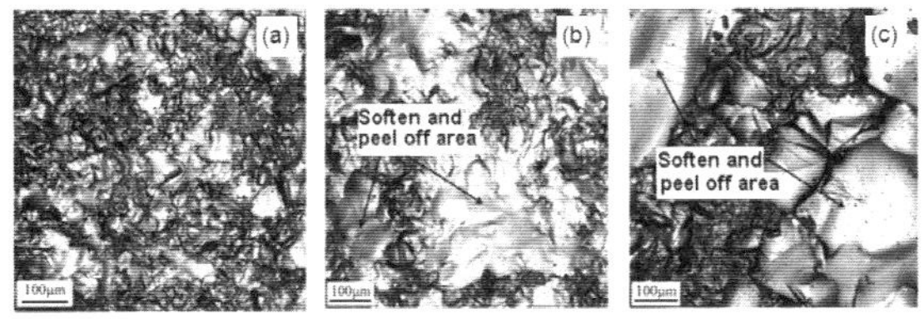

Fig. 6. SEM photograph of facture surface on glass side at different sand grain size (a)100mesh, (b) 60mesh and (c) 20mesh

When deposited Al_2O_3 coating thickness was 50μm, Cu coating thickness were 50μm, 100μm and 150μm respectively, the effect of Cu coating thickness on the tensile strength of the soldered glass/steel joints is shown in Fig.7. Along with the Cu coating thickness increases, the tensile strength of the soldered joint decreased. When the Cu coating thickness is 100μm and 150μm, the tensile strength of joint is approximate and relatively in low level. When Cu coating thickness is 50μm, the maximum joint strength is 12.5MPa.

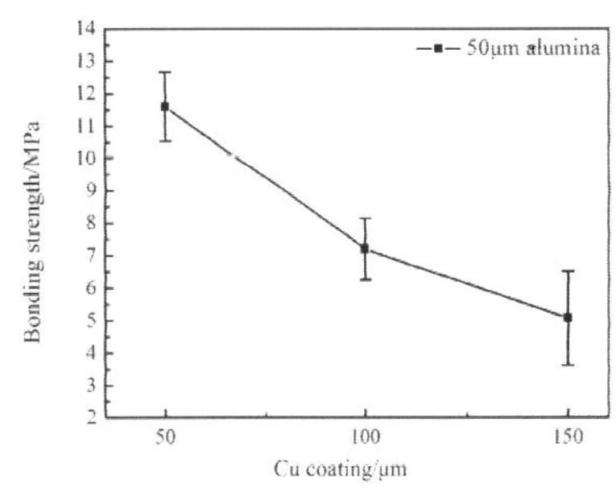

Fig. 7. Cu coating thickness v.s. strength of soldered joints

Fig.8(a)-(c) are the joint microstructure when Cu coating thickness were 50μm, 100μm and 150μm respectively.

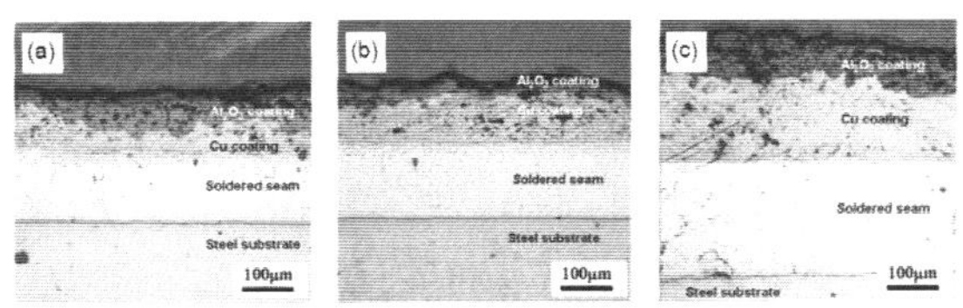

Fig. 8. Cross section microstructure of fracture joints at different Cu coating thickness (a) 50μm, (b) 100μm, and (c) 150μm

As shown, the joint fracture all occurred in the interface of Al_2O_3 coating and glass substrate. It is considered that the reason for this fracture mode is that the thermal expansion coefficient of glass, Al_2O_3 and Cu coatings are not identical, during heating and cooling stages in thermal spraying process, there is a mismatch strain between the coating and substrate, which results in residual stress in coating and internal substrate. The thicker the coating, more and more obvious the heat accumulation effect, thus the thermal stress is also larger. When Cu coating is 150µm and 100µm, the residual stress is larger and the joint strength is smaller. When Cu coating thickness is 50µm, the residual stress is smaller and the joint strength reached the maximum 12.5MPa.

4 Conclusions

In the present study, Al_2O_3 powder and pure copper powder was deposited onto tempered glass by atmospheric plasma spraying. A thick APS coating layer has been employed successfully as filler interlayer for soldering glass/steel joints. The interface bonding strength of Al_2O_3 coating is mainly affected by the sandblasting grain size and spraying heat input. Test results showed that the optimal parameters is the sandblasting grain size 60 mesh, spraying distance 120mm and spraying power 33kW. With the increase of spraying heat input, interface bonding strength decreased gradually and reached the maximum 20.5MPa. In summary, the pre-metallization coating treatment effectively improved the weldability of glass with steel, and the joints had a certain level of strength. When Al_2O_3 coating thickness is under certain condition, tensile strength test proved that the soldered joints all fractured in interface between Al_2O_3 coating and glass substrate. And the tensile strength decreases with the increase of Cu coating thickness. When Cu coating thickness is 50µm, the joint strength reached the maximum 12.5MPa.

Acknowledgements
The authors thank financial support of Beijing Municipal Natural Science Foundation (No. 3122008).

5 Literature

[1] Zhao, H.Z. and Ying, H.R. Glass technology. Chemical Industry Press (2006), Beijing
[2] Rauter, M. and Roeder, E. Ultrasonic Welding of Glass and Glass-Ceramic to Metal. Welding and Cutting(1993), Issue 4, pp26/30.
[3] Watson, J. and Overend, M: Novel connections for steel glass and glass glass structures. GLASS Performance Days(2011), pp1/4.
[4] Wagner, G. and Walther, F. Other: Glass/glass joints by ultrasonic welding. Glass Technology(2003), Issue 44(4), pp152/155.
[5] Hu, L: Investigation on the mechanism and microstructure and mechanical properties of the anodic bonding for the Pyrex glass to metals. Master thesis (2007) , pp6/57.
[6] Kiuchi, M. and Yanagimoto, J. Others. Application of mushy/semi-Solid bonding-part3. Journal of Materials Processing Technology(2003), Issue 140(1-3), pp163/166.
[7] Sun. X.L: Process and Mechanism of ultrasonic vibration assisted soldering glass. Master thesis(2009), pp9/60.
[8] Wielage, B. and Wank, A. Other: Thermally sprayed solder/braze filler alloys for the joining of light metals. Proceedings of XII. Workshop Plasmatechnik (2004), Ilmenau, D.

Closed-loop control tools for automated laser cladding processes

A. Wank, F. Beck, C. Schmengler, GTV Verschleiss-Schutz GmbH, Luckenbach / D,
J. Zierhut, K. Kloiber, Zierhut Messtechnik GmbH, München / D

As costs of laser sources have decreased strongly laser cladding and laser additive manufacturing have become fast-growing technologies due to the inherent advantages like high precision of material deposition, flexibility concerning achievable coating thickness and deposition rates, metallurgical bonding to substrate materials, low porosity and fine microstructure deposits, relatively low heat transfer to substrates and aptitude for process automation. However, there is a multitude of applications that require closed-loop control tools in order to permit complete process automation. For example change of component temperature during the cladding process, e.g. for cladding of small wall thickness components, results in need of tools to adjust laser power. And in order to fill cavities like grooves precisely despite distortion and thermal expansion of components to be clad edge recognition systems for adjustment of laser head movement are needed. GTV Verschleiss-Schutz GmbH and Zierhut Messtechnik GmbH jointly developed LPowC (Laser Power Control) and LPosC (Laser Positioning Control) systems based on digital image analyses to address these needs. Both tools are based on intelligent cameras that evaluate recordings from the cladding zone through the optical path of the laser optics. While LPowC controls local emission by the laser cladding process, LPosC detects distances between edges on component surfaces and laser spot as a base for positioning correction. Exemplary cladding of thin wall tubes with different iron, nickel and cobalt based alloys with automatic adjustment of laser power and the effect on local microstructure of the deposits are presented as well as evaluation of filling of grooves on shafts that undergo strong cooling during the cladding process.

1 Introduction

Laser additive manufacturing including laser cladding is used in more and more applications due to technological advantages on the one hand and steadily decreasing prices of laser sources [1-2] on the other hand. Laser cladding offers unsurpassed reproducibility and precision of position as well as heat transfer to component surfaces and powder feedstock. This results in exceptionally low dilution with base material without compromising metallurgical bonding, small heat affected zones and distortion as well as low porosity and homogeneous, fine grained microstructure of deposited layers.
Automation of laser cladding process is relatively easy due to typical high degree of electronic control of quality relevant process parameters like laser power, carrier and laser process gas flow rates as well as powder feed rate and handling system positions. However, full process automation sometimes does not work for control of absolute handling system positioning only. Instead precise control of actual relative positioning of powder nozzle and component surface is needed and must even cover disturbance variables like thermally induced expansion and / or component distortion that can even vary during the cladding process. Cladding of extrusion screw threads or turbine blade leading edges are examples of such jobs. Additionally typical repair jobs involve machining of damaged areas and material needs to be deposited between the edges of the machined surface area.
Also, change of component temperature during a cladding job e.g. due to small wall thickness of a component to be coated can require adaptation of laser power in order to avoid scattering of dilution with base material, coating thickness and / or microstructure.

CCD cameras are standard tools for alignment of laser beam and coaxial feed powder nozzles. But digital analyses of their images also permits determination of in-flight particle velocity [3], laser power closed-loop control [4-6] and laser cladding head positioning adaptation. The latter two features are combined in a single camera system that was jointly developed by GTV Verschleiss-Schutz GmbH and Zierhut Messtechnik GmbH. This paper describes the contained tools laser power control LPowC and laser positioning control LPosC tools both concerning their function principle and capability in use.

2 Laser Cladding Process Control Tools

2.1 LPOwC

The laser power control tool LPowC uses two analysis areas within camera images that need to be defined by the user, **Fig. 1**. The total intensity of the bigger area is used to monitor reflections from the inner surface of the powder nozzle as a means to detect impermissible pollution of the cladding head optics. If a pre-defined threshold is exceeded the respective digital signal from the interface box can be used to either shut down the laser source automatically or at least to display an error message on a control panel.
The total intensity continuously recorded within the smaller circle is used as control factor to compensate for substrate temperature change by laser power adaptation. However, LPowC cannot replace process parameter optimization. Users first need to carry out cladding tests and define optimal starting parameters including according intensity measured by LPowC at specified camera settings (exposure time and amplification).
Then users need to set minimum and maximum values around the optimal target intensity in order to

receive digital signals from the interface box to initiate laser power modification in a suitable control loop. According adjustable control loops are implemented in GTV laser cladding centre PLC programs, but can also be realized by users themselves. GTV software permits customized damping of control cycles based on number of images for averaged intensity calculation, power adaptation step and reaction time.

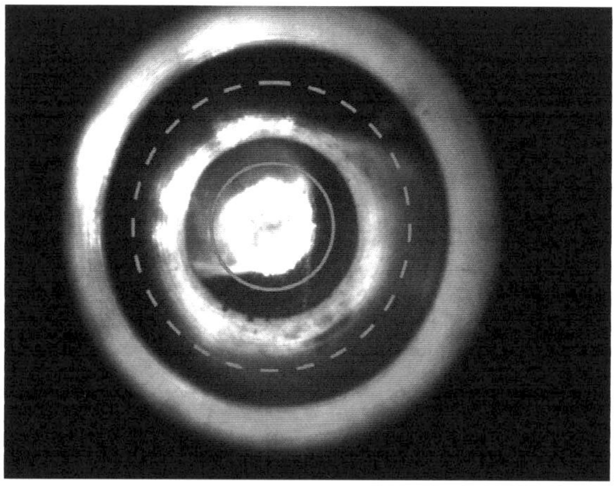

Fig. 1. Exemplary laser cladding process image with display of two observation areas; continuous line circle: image area used for process control, dashed line circle: image area for process safety check

2.2 LPosC

The laser positioning control tool LPosC detects both the actual position of laser beam centre as well as component edges and thereby permits adjustment of cladding head movement programs, **Fig. 2**. Laser beam centre point is defined by centre point of the circle formed by the powder nozzle internal bore edges on recorded images. LPosC permits considering laser beam centre position as constant or calculates it from recorded images continuously.

For actual edge detection a region of interest is specified by a rectangle with edge length ratio 2:1 (marked in yellow in **Fig. 2**). Within this area the user needs to define minimal grey scale gradients as well as their maximum length and maximum grey scale values for the system to calculate a line representing the component edge from recorded images. The software contains error correction that includes maximum permissible distance to previously detected edges as well as maximum permissible deviations from previously calculated line gradients to avoid faulty line detection and accordingly faulty cladding head movement correction. If natural emission of the cladding process is too low additional illumination e.g. by special (triggered) lasers can be applied. Such artificial illumination does not influence the general function of the edge detection software.

LPosC software by itself is not capable of defining optimal distances between component edges and laser beam centres. Optimum needs to be determined in pre-tests and depends on material combination and edge geometry as well as demands concerning permissible melting of the component edge and required metallurgical bonding degree, etc.. If the calculated distance between laser beam centre and component edge differs from a set distance more than a set value position correction in the required direction is initiated by an according digital signal to the master PLC.

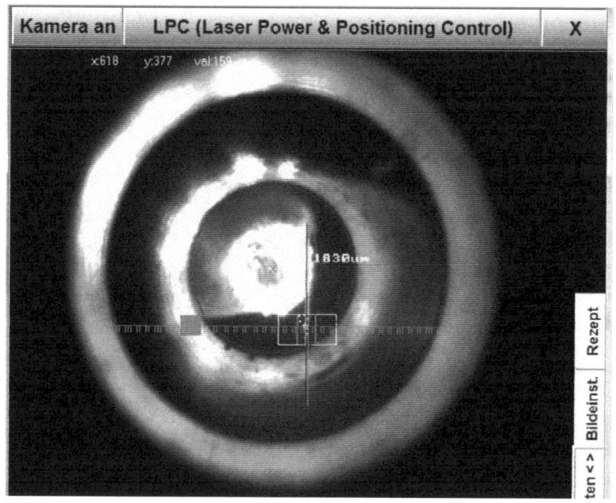

Fig. 2. Exemplary laser cladding process image with laser beam centre point too close to substrate edge as displayed by LPosC; red square indicates shift of reversal point in an oscillating head movement process to left side

3 Experimental

3.1 LPOwC tests on thin walled tubes

Three different kinds of materials are clad on 316L (1.4404) stainless steel tube with 28 mm outer diameter and 1.2 mm wall thickness on 70 mm length up to tube edges. Besides iron based hard alloy FeCrV15 (GTV 31.40.10, gas atomized, +63 -180 µm) and NiBSi 22 HRC alloy (GTV 80.10.1, water atomized, +20 -53 µm) also Stellite® 6 (GTV 31.06.10, gas atomized, +63 -150 µm) is clad. Chemical compositions of powders and cladding parameters are listed in **Table 1** and **2** respectively.

Table 1. Chemical composition of used powders

powder	31.40.10	80.10.1	31.06.10
Fe	base	0.15	1.1
Ni	-	base	1.4
Co	-	0.02	base
C	4.49	0.02	1.0
B	-	1.7	-
Si	1.18	2.76	1.2
Mn	0.97	-	-
Cr	14.5	0.1	28.4
Mo	1.13	-	-
V	15.0	-	-
W	-	-	4.4

In all three cladding tests Laserline LDF 6000-100 laser source, 80 mm collimation and 250 mm focussing length and powder nozzle GTV PN6625 with two 1-3 GTV powder splitters have been used.

Table 2. Cladding process parameters

coating	31.40.10	80.10.1	31.06.10
Fibre diameter	1,500	1,000	1,000
laser power (initial)	900	1,200	900
process gas Ar	25 l/min	25 l/min	25 l/min
carrier gas Ar	2 x 2.5 l/min	2 x 2.5 l/min	2 x 2.5 l/min
surface speed	0.5 m/min	1.2 m/min	0.8 m/min
offset / rotation	1.0 mm	1.0 mm	1.5 mm
layers	1	1	1

Laser power is recorded during the process, cross sections of claddings are prepared along the whole tube axis length and analysed metallographically by optical microscopy. Weld penetration and cladding thickness distribution are evaluated as well as hardness evolution.

3.2 LPOsC tests

Tests are carried out on tool steel X40CrMoV5-1 (1.2344) with the above mentioned powder GTV 31.40.10 as cladding material. The above mentioned laser source, optics, fibre, powder nozzle and splitters have been used. Groove depth and width are varied between 0.5 - 2.0 mm and 5 - 50 mm respectively.

4 Results

4.1 LPOwC tests on thin walled tubes

The evolution of laser power in accordance with the automatic LPosC control for the three tested cladding processes is shown in **Fig. 3**. While the combination of chosen starting laser power and working area emission ranges for FeCrV15 and NiBSi result in strong decrease of laser power after completion of the first turn of the tube, there is a slight increase of laser power in case of Stellite® 6.

Heat accumulation at the tube edge causes automatic reduction of laser power for all three tested materials. However, extent of power reduction is only small in case of FeCrV15.

Optical micrographs of claddings at the start, centre and end of the produced specimens are shown in **Figs. 4** (FeCrV15) and **5** (NiBSi). Both interface between base and clad material and cladding microstructure are homogeneous along the whole covered cladding length. While FeCrV15 claddings consist of very fine vanadium carbides dispersed in a chromium rich iron based matrix, NiBSi and Stellite®6 claddings show dendritic microstructure. All three claddings show low porosity level with small pore size typical for state-of-the-art laser cladding processes.

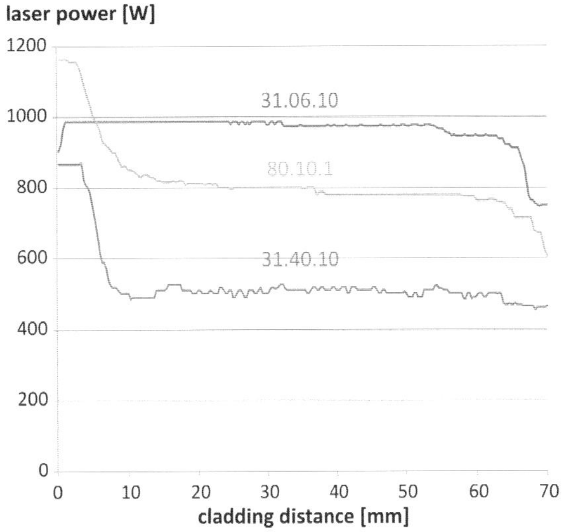

Fig. 3. Laser power evolution during cladding tests with automatic power control by LPowC

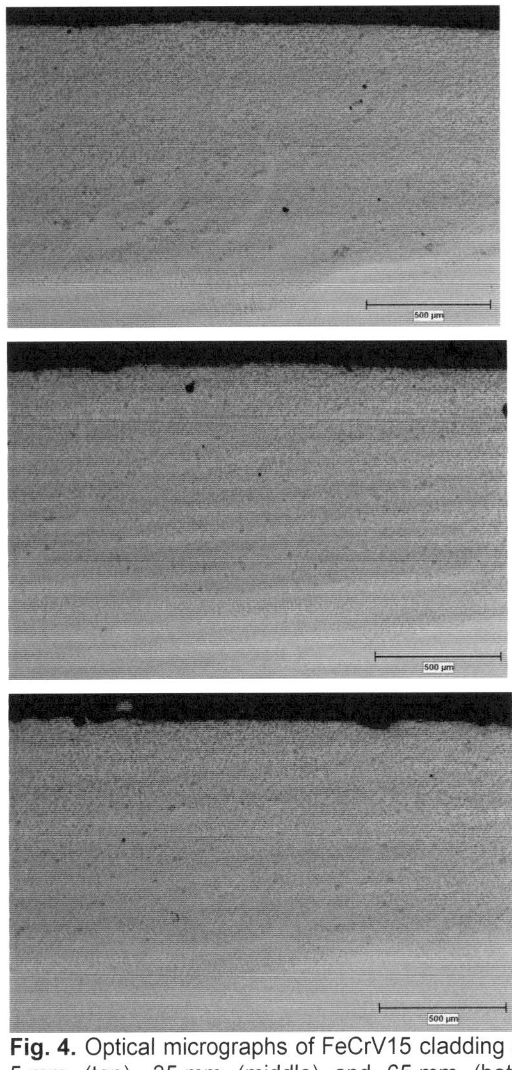

Fig. 4. Optical micrographs of FeCrV15 cladding after 5 mm (top), 35 mm (middle) and 65 mm (bottom) distance

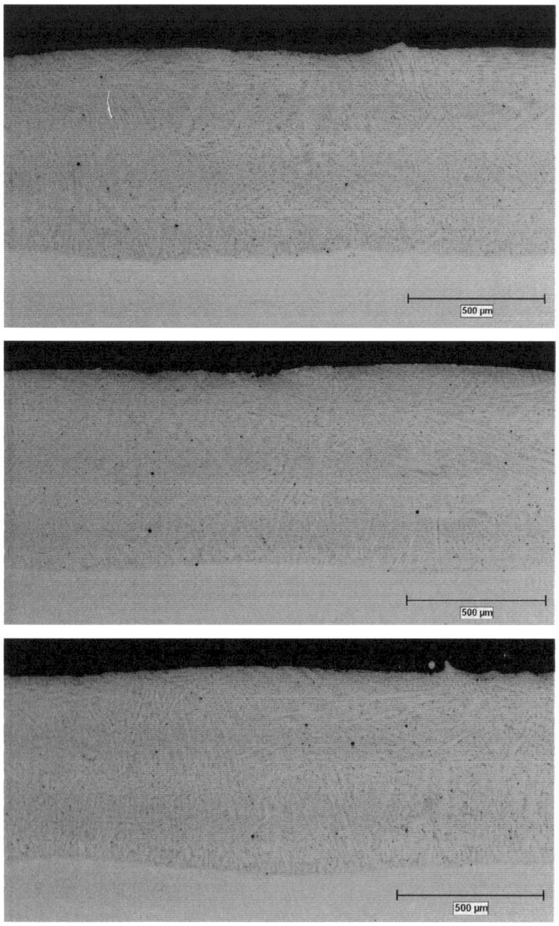

Fig. 5. Optical micrographs of NiBSi cladding after 5 mm (top), 35 mm (middle) and 65 mm (bottom) distance

The overall thickness of FeCrV15 and Stellite®6 clad 316L tubes decreases slightly along the tube axis, while the NiBSi clad tube shows perfectly constant thickness (**Fig. 6**). For FeCrV15 and Stellite®6 thicknesses decrease from 2,200 µm to 2,050 µm and from 2,050 µm to 1,950 µm respectively.

Cladding thicknesses are almost constant along the tube axis for all three kinds of cladding materials. In case of NiBSi overall and cladding thickness amount to 1,688 µm (σ: 19 µm) and 666 µm (σ: 40 µm) respectively. Accordingly weld penetration is about 180 µm, which means about 25% dilution of NiBSi coating with base material.

In the range between 5 - 65 mm cladding distance cladding thicknesses of FeCrV15 and Stellite®6 amount to about 1,050 µm and 1,500 µm respectively. So, weld penetration and base material dilution amount to 100 µm and 10% respectively for FeCrV15 and to 650 µm and 45% respectively for Stellite®6.

In contrast to FeCrV15 and NiBSi claddings that show constant cladding thickness up to the tube edge the Stellite®6 cladding shows decreasing cladding thickness as a consequence of decreased laser power towards the tube edge.

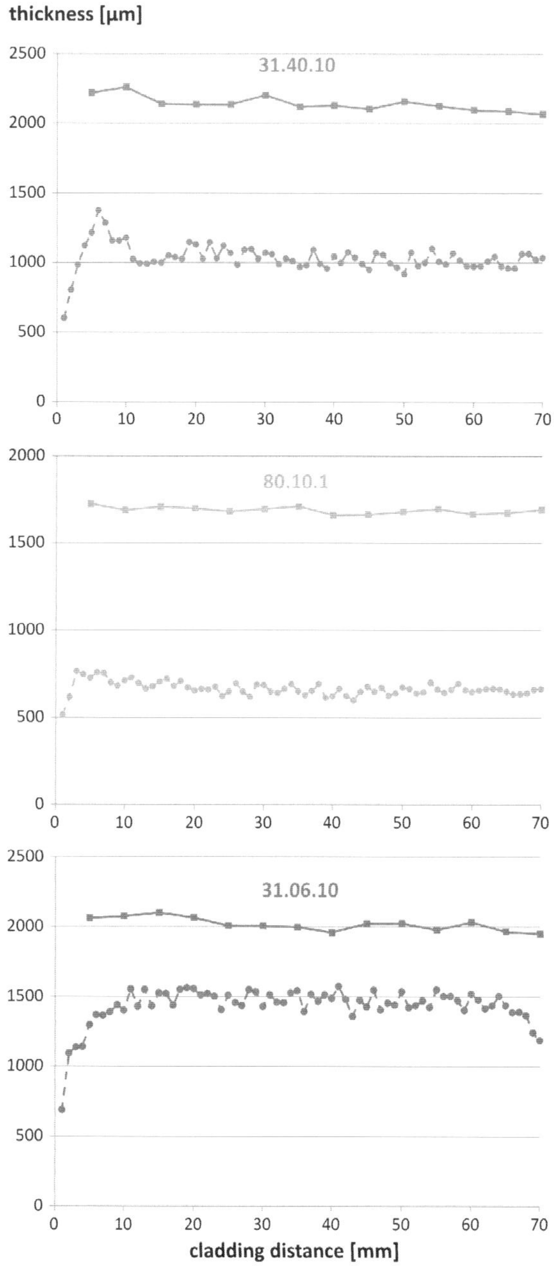

Fig. 6. Total thickness of clad tubes (continuous lines) and cladding thickness (dashed lines) along the tube axis

Despite decreased cooling rate as a consequence of increasing base material temperature during the cladding process microhardness along the tube axis remains basically constant for all three types of claddings (**Fig. 7**). Microhardness of FeCrV, NiBSi and Stellite®6 coatings amounts to 789 HV.3 (σ: 43 HV0.3), 307 HV.3 (σ: 6 HV0.3) and 308 HV.3 (σ: 13 HV0.3) respectively. Especially the NiBSi cladding shows outstanding small microhardness deviations.

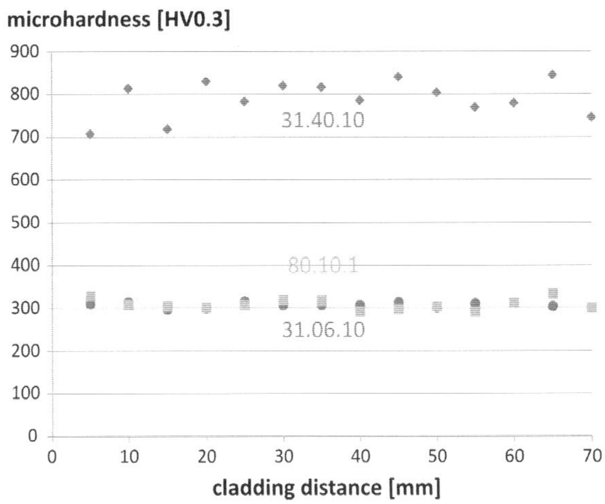

Fig. 7. Microhardness distribution of claddings along the tube axes

4.2 LPOsC tests

Besides laser power, oscillation width and speed, traverse velocity, powder feed rate and cladding head inclination to the workpiece need to be adjusted to groove geometries in order to achieve claddings that fill grooves with perfect bonding along the whole interface between base material and cladding. **Figs. 8** and **9** contain images of an exemplary 8 mm wide FeCrV15 cladding that has been deposited inside a groove with 2 mm depth and 45° inclined edges. Laser cladding head inclination to the workpiece surface was kept constant at 90°. 3,000 W laser power, 5.33 mm/s traverse speed and 10 mm oscillation width have been applied. Less than 0.2 mm had to be ground in order to achieve a cladding smooth surface at the same height like the surrounding workpiece surface.

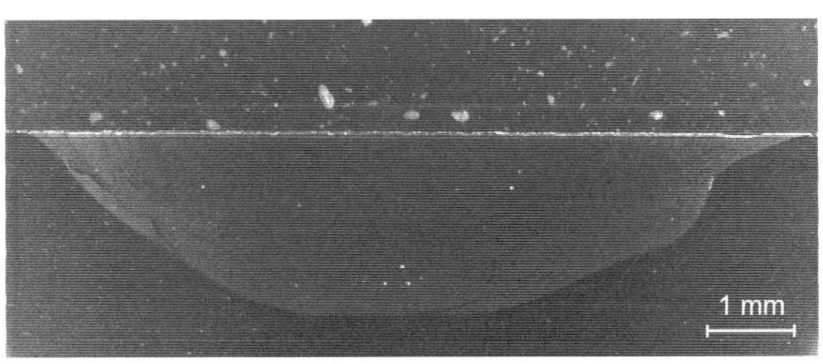

Fig. 8. Overview image of a FeCrV15 cladding that has been produced by oscillating laser cladding head movement inside a groove with 45° edges after grinding

The applied process strategy results in considerable dilution of cladding material with base material at the 45° inclined outer regions and rounded transition to the groove bottom plane. Cladding microhardness amounts to 650 HV0.3 with 20 HV0.3 standard deviation. As typical of state-of-the-art laser claddings only low porosity with small pore size (i.e. < 20 µm) is observed in cladding cross sections.

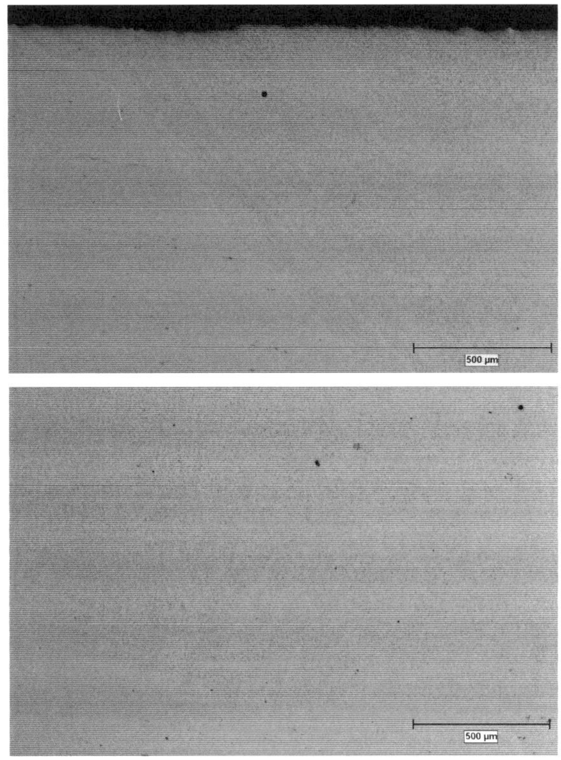

Fig. 9. Optical micrographs of a FeCrV15 cladding that has been produced by oscillating laser cladding head movement; top: groove edge, bottom: groove centre

5 Discussion

Generally, for all types of claddings produced with automatic laser power adjustment tool LPowC good homogeneity of thickness and weld penetration has been achieved even for the demanding application of relatively thick cladding of thin walled tubes. Dilution of the produced Stellite®6 specimen is certainly by far too high for typical industrial applications. Nevertheless this specimen also proves capability of LPowC to maintain cladding quality constant even in case of a process with extremely strong dilution with base material.

Despite strong increase of base material temperature in the cladding zone and accordingly decreased cooling rate all claddings show constant microstructure and even microhardness along the whole cladding distance. Low microhardness of Stellite®6 claddings must be attributed to strong dilution with the 316L base material and is not typical of Stellite®6 laser claddings.

The presented tests of the laser positioning tool LPosC are not very demanding concerning groove edge detection and positioning correction, because workpiece and groove geometry were not complex and no thermally induced workpiece expansion or shrinkage and distortion needed to be compensated.

FeCrV15 claddings inside grooves show significantly lower microhardness compared to claddings on 316L steel tubes. That can be attributed to lower content of vanadium carbides as a consequence of stronger dilution of cladding material with base material.

6 Summary and Perspectives

The high potential of closed-loop laser power control tool LPowC and laser cladding head positioning adaptation tool LPosC for complicated laser cladding tasks is confirmed. Iron, nickel and cobalt based alloys have been clad on thin walled tubes and use of LPowC permits production of claddings with homogeneous thickness, weld penetration and even microstructure as well as microhardness. LPowC still requires process parameter optimization prior to automatic process control with the aim to keep desired cladding properties constant despite workpiece temperature change. If parameters that cause excessive weld penetration and dilution with base material are chosen, the according disadvantageous cladding characteristics will also be kept constant automatically.

LPosC is capable of detecting groove edges and adjust laser cladding head positioning. Subsequent to the here presented tests the tool has successfully been applied for control of cladding processes on extrusion screws that are pre-heated prior to cladding and that show continuously changing width and inclination of screw surfaces to be clad with respect to screw axis as well as considerable distortion. Depending on the specific job demands it is necessary to combine LPosC with powerful (triggered) illumination devices or to adjust LPosC settings for different screw areas to be clad by area dedicated LPosC recipes that will be activated by an external master PLC. Thereby it is possible to use CAM data for groove machining in order to transform these automatically into CAM data for laser cladding.

LPowC and LPosC are valuable tools for automation of laser cladding processes with high complexity. Their further adaptation to specific jobs will certainly expand the field of laser cladding applications even more.

7 Literature

[1] Apter, M.M., F. Leibreich: High-power diode-laser bars come of age. Laser Focus World 42 (2006), Issue 3, pp. 91-93

[2] Bauer, A., S. Boehr: International Laser Technology Congress AKL'14: digital photonic production taking big strides forward. Fraunhofer ILT press release (2014) June 4, pp. 1-6

[3] Smurov, I., M. Doubenskaia, S. Grigoriev, A. Nazarov: Optical Monitoring in Laser Cladding of Ti6Al4V. Journal of Thermal Spray Technology 21 (2012), Issue 6, pp. 1357-1362

[4] Bia, G.J., A. Gasser, K. Wissenbach, A. Drenker, R. Poprawe: Identification and qualification of temperature signal for monitoring and control in laser cladding. Optics and Lasers in Engineering 44 (2006), Issue 12, pp. 1348-1359

[5] Hofman, J.T., B. Pathirajb, J. van Dijk, D.F. de Lange, J. Meijer: A camera based feedback control strategy for the laser cladding process. Journal of Materials Processing Technology 212 (2012), Issue 11, pp. 2455-2462

[6] Hu, D., H. Mei, G. Tao, R. Kovacevic: Closed Loop Control of 3D Laser Cladding Based on Infrared Sensing. Proceedings of the Solid freeform fabrication symposium 2001, Austin, Texas, USA, pp. 129-137

Effect of Traverse Speed on Residual Stress Distribution and Cavitation Erosion Behavior of Arc Sprayed Aluminum Bronze Coatings

M. Hauer*[1], S. Krebs*[2], W. Krömmer*[3], K.-M. Henkel*[1]; *[1]Rostock, *[2]Hamburg, *[3]Unterschleißheim / D

Within the scope of a current research project, arc spraying was used with different traverse speeds to influence the heat transfer and hence the stress state. Microstructural, chemical and mechanical analyses were realized to examine adhesive and cohesive properties. The materials used for spray experiments were the highly cavitation erosion resistant propeller alloys CuAl9Ni5Fe4Mn and CuMn13Al8Fe3Ni2. Analyses of cavitation erosion behavior were carried out to evaluate the suitability for use in marine environments. Finally the residual stress distribution was measured by modified hole-drilling method using ESPI.

1 Introduction

Severe erosion on modern containerships and ferries travelling with higher speeds endangers ship operation. One main factor is cavitation erosion, which is the formation and collapse of vapour bubbles caused by pressure oscillations. Areas exhibiting high material losses by cavitation erosion require expensive repairs by means of build-up welding and additional grinding work [1, 2]. Although thermal spray technologies like arc spraying are promising solutions for fabricating cavitation erosion resistant coatings, even in dry dock, they are not in use in the marine industry yet [3].
One reason is the lack of knowledge concerning the influence of residual stress distribution on mechanical properties and coating adhesion. Those stresses result from complex superposition of quenching, cooling and other stresses generated during the spray process, **Fig. 1** [4, 5].

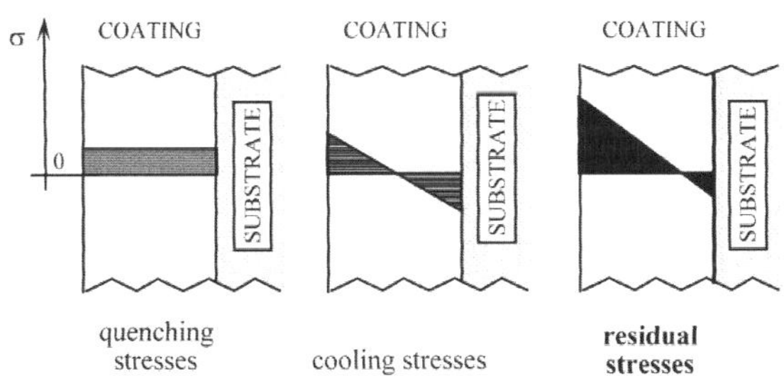

Fig. 1. Superposition of quenching (l.) and cooling (m.) stresses lead to residual stresses (r.), [4].

Cohesive and adhesive properties in combination with residual stresses of arc sprayed coatings are strongly influenced by materials, spray parameters and heat transfer in substrate and coating [6, 7]. Varying quenching stresses during spray process were considered to dominate the formation of residual stresses in arc-sprayed coatings. This leads to higher stress gradients than in plasma-sprayed coatings for equal materials [6]. Previous studies revealed that when residual stresses are accompanied by external stresses or coatings exceed a certain thickness, coating delamination occurs and thus significantly reduces service lives [8].

Therefore the focus of this work lies in potential interactions between residual stresses and microstructural, chemical and mechanical properties of the arc-sprayed coatings. The results of the study shall be evaluated in the context of thermal spray technology and its use in marine environments like repair of ship rudders.

2 Arc spray experiments and materials

The arc spray experiments were carried out using a Sulzer Metco Smart Arc (Oerlikon Metco Europe GmbH, Kelsterbach, D) in the thermal spray laboratory at Linde AG (Unterschleißheim, D). The spraying parameters were kept constant at U = 28 V, I = 180 A, p_{Air} = 4 bar and stand-off distance of 100 mm. In order to directly influence the stress state, the traverse speed was varied from 333.0 mm/s to 499.5 mm/s and 666.0 mm/s.
Materials used for spray experiments were the cavitation erosion resistant Ni-Al-Bronze CuAl9Ni5Fe4Mn (Lincoln Electric LNM CuAl8Ni6, Ø 1.6 mm, Ratingen, D) and Mn-Al-Bronze CuMn13Al8Fe3Ni2 (Bedra Bercoweld A300, Ø 1.6 mm, Heuchelheim, D). This certain materials were chosen due to their good cavitation erosion and corrosion resistance [9]. A sufficient coating thickness of at least 300 µm is needed for the following experiments; hence the number of layers needed to be increased for Ni-Al-Bronze (layers 5, 10, 10) and Mn-Al-Bronze (layers 8, 10, 15).
The substrates were grit-blasted and of type S275 mild steel (corresponding to DIN EN 10025-2) with dimensions of ca. 65 mm x 50 mm x 5 mm.

3 Results

3.1 Deposition effiency

The deposition efficiency (DE) of the coatings was determined in accordance with DIN EN ISO 17836. It rises for both materials with increasing traverse speeds, i.e. for Ni-Al-Bronze CuAl9Ni5Fe4Mn from 84.7 % to 86.2 % and 87.6 %. For the Mn-Al-Bronze CuMn13Al8Fe3Ni2 the growth from 71.8 % to 81.8 %; to finally 84.5 % was even more evident.
In general the DE of the Mn-Al-Bronze still stays below the values for Ni-Al-Bronze.

3.2 Adhesion and mechanical properties

For the purpose of quantifying the influence of traverse speed on coating adhesion, tensile adhesive tests according to DIN EN 582 were carried out using a static uniaxial testing machine (Zwick Roell Z50, Ulm, D, three specimens each), **Fig. 2**.

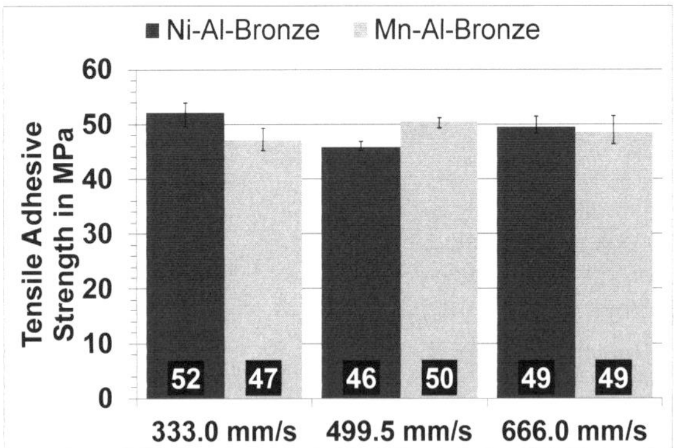

Fig. 2. Average tensile adhesive strength of the arc-sprayed bronze coatings.

It can be seen, that the average tensile adhesive strength is slightly influenced by traverse speed in the arc-spray process and that all values are within a range of ca. 10 MPa. While there is a minor tendency of rising adhesive strength values for the Mn-Al-Bronze with increasing traverse speed, the Ni-Al-Bronze coatings reveal best adhesion for a traverse speed of 333.0 mm/s. The lowest strength value for the Ni-Al-Bronze can be allocated to a traverse speed of 499.5 mm/s.

The hardness of a material is often referred to be related to cavitation erosion resistance [3]. Therefore hardness measurements on the surface of the coatings were executed by using a Wolpert 432 SVD hardness testing machine (Wilson Wolpert Instruments, Aachen, D; five measurements each), **Fig. 3**.

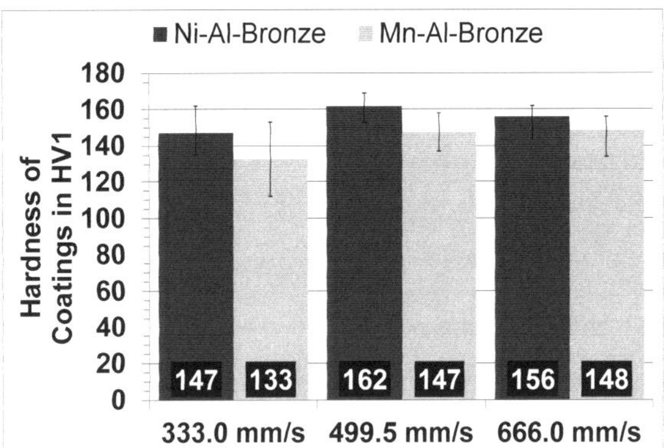

Fig. 3. Average Hardness of the arc-sprayed bronze coatings.

The results are also within a small range, but show more obvious trends. The coating hardness of the Mn-Al-Bronze coatings rise with increasing traverse speeds, analogous to the results of the tensile adhesive strengths for this material. The Ni-Al-Bronze coatings show a contrary relation, i.e. high hardness values correspond to low tensile strength values and vice versa.

3.3 Microstructure

For further information about coating quality, microstructural analyses were carried out using an optical microscope (OM) Leica DM6000M (Leica Microsystems GmbH, Wetzlar, D) and a scanning electron microscope (SEM) Zeiss Supra 25 (Carl Zeiss AG, Jena, D; acceleration voltage 20 kV). The Specimens for metallographic examinations were hot-mounted via ATM Opal 410 Hot Mounting Press (ATM GmbH, D) and gradually grinded and polished (6 µm, 3 µm suspensions, finally oxide polish). Klemm III color etchant (stock solution: sodium-thiosulphate in water; etching solution: potassium pyrosulphite) was used for enhancing contrast in the coatings and revealing different phases. The typical microstructure of the coatings is shown in **Fig. 4a**. Energy dispersive x-ray spectrometry (EDS) was performed by using a set-up of Bruker QUANTAX (Bruker Nano GmbH, Karlsruhe, D).

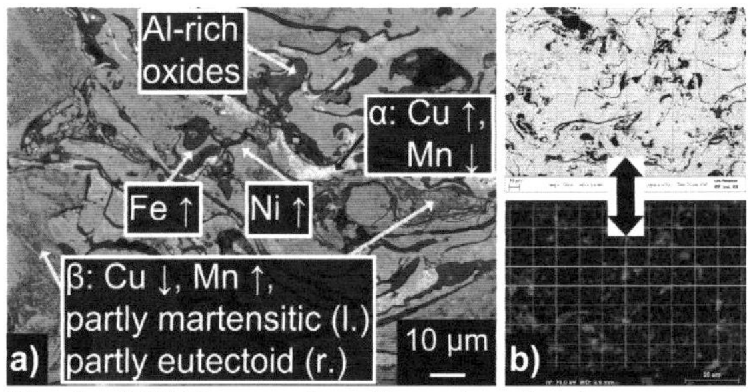

Fig. 4. a) OM image of coating microstructure **b**) SEM (top) and EDS image of O (bottom).

Comparison of OM image in **Fig. 4 a**, which displays the typical microstructure of the colour etched Mn-Al-Bronze coatings, and the same area by EDS with focus on Cu and the alloying elements Al, Ni, Fe and Mn reveals local composition variations, which are marked in **Fig. 4 a**. There are only small amounts of Al in the matrix of the coating. Instead Al forms partially large oxides, which appear dark grey. Enhanced contents of Mn and less Cu lead to a slightly brighter grey tone compared to the oxdides. These areas can be characterised as β-phase, which appear ordered and even partly martensitic and eutectoid. According to literature the ordered β-phase is stabilised by Mn, which is in good agreement with the results of EDS [9]. Areas containing more Fe or Ni are displayed in brighter grey tones, while white areas hold more Cu and less Mn. Hence these areas are identified as α-phase. It supplies ductility to the alloy and promises higher corrosion and cavitation erosion resistance than e.g. martensitic β-phases [9]. The microstructure of Ni-Al-Bronze coatings is comparable

to Mn-Al-Bronze coatings, but contains no ordered β-phase.
Further EDS images enabled a clear distinction between pores and oxides and hereby a quantitative analysis via comparing SEM and EDS images, **Fig. 4b** (99 points per image; each specimen three images).
The results of these analyses are shown in **Table 1**.

Table 1. Content of Pores / Cracks and Oxides.

Material / Traverse Speed		Content and standard deviation in %	
		Pores / Cracks	Oxides
Ni-Al-Bronze			
Coatings	333.0 mm/s	5.1	6.7
	SD	± 1.4	± 2.4
	499.5 mm/s	5.4	6.4
	SD	± 2.1	± 1.9
	666.0 mm/s	6.7	9.1
	SD	± 0.5	± 2.5
Mn-Al-Bronze			
Coatings	333.0 mm/s	5.7	4.4
	SD	± 0.5	± 1.9
	499.5 mm/s	4.7	6.4
	SD	± 1.0	± 2.1
	666.0 mm/s	4.7	4.4
	SD	± 0.5	± 1.0

The Ni-Al-Bronze coatings show rising porosity with increasing traverse speeds. During the quantitative analyses it became clear, that the Ni-Al-Bronze coatings had a more inhomogeneous pore and oxide distribution, which is evident in higher standard deviations at lower traverse speed values 333.0 mm/s and 499.5 mm/s. In contrast to this, the Mn-Al-Bronze coatings have a tendency of slightly falling porosity values with increasing traverse speed. Furthermore the coatings showed smaller standard deviations and are generally less porous compared to Ni-Al-Bronze.
Concerning the oxide formation the Ni-Al-Bronze coatings show less oxide contents for lower traverse speeds, while a higher traverse speed results in far more oxides. For Mn-Al-Bronze coatings there is no clear trend in dependence of varying traverse speeds. Specimens with both high and low traverse speed have lower oxide contents. Comparing both materials, Mn-Al-Bronze shows less oxides.

3.4 Chemical Analyses

In order to examine potential differences in chemical composition the Ni-Al- and Mn-Al-Bronze coatings were examined via optical emission spectrometry (OES) Spectromaxx (Spectro Analytical Instruments GmbH, D) using 3 measurements for each coating, **Table 2** (Cu and trace elements not displayed).
In addition the results were opposed to the chemical composition of the wire according to the manufacturer specifications. The OES results of the Ni-Al-Bronze reveal almost similar levels of Al, Ni, Fe and Mn contents for lower traverse speed values of 333.0 mm/s and 499.5 mm/s. However, for the highest traverse speed the values of Fe and Mn rise, while the contents of Al and Ni fall markedly. All values show small standard deviations; thus the measurements can be considered robust.
In comparison to the manufacturer specifications the contents of Al and Ni are higher for lower traverse speed of 333.0 mm/s and 499.5 mm/s, while for the highest traverse speed they are slightly below. All other values lie below the manufacturer specifications, which is mostly observable for Mn.

Table 2. Chemical composition of wire and coatings.

Material / Traverse Speed		Chemical composition and standard deviation in wt%			
		Al	Ni	Fe	Mn
Wire	Ni-Al-Bronze	9.0	5.0	4.0	2.5
Coatings	333.0 mm/s	9.92	5.40	3.17	0.66
	SD	± 0.24	± 0.03	± 0.03	± 0.00
	499.5 mm/s	10.00	5.37	3.15	0.66
	SD	± 0.09	± 0.02	± 0.07	± 0.01
	666.0 mm/s	8.73	4.39	3.48	1.38
	SD	± 0.04	± 0.03	± 0.01	± 0.02
Wire	Mn-Al-Bronze	7.2	2.3	2.3	11.6
Coatings	333.0 mm/s	7.87	3.12	2.47	8.74
	SD	± 0.17	± 0.00	± 0.12	± 0.21
	499.5 mm/s	8.03	3.15	2.29	8.79
	SD	± 0.08	± 0.01	± 0.02	± 0.20
	666.0 mm/s	7.60	3.12	2.46	8.56
	SD	± 0.15	± 0.04	± 0.06	± 0.06

In contrast the Mn-Al-Bronze coatings do not show any evident trend, since all measurements are within a close regime. Furthermore the contents of all alloying elements are higher than the values specified by the manufacturer except for Mn. These values are drastically lower than specified.

3.5 Cavitation erosion properties

The cavitation erosion behaviour can be considered as a measure of coating quality with respect to coating cohesion. Therefore indirect cavitation erosion tests in compliance with reference 1 were carried out for about 120 min using a Hielscher UIP1000 (frequency 19.51 kHz, peak-to-peak amplitude 50 µm, distance from tip 0.5 mm, D). The specimens were gradually grinded and polished over 6 µm to 3 µm to ensure plane surfaces, since asperities can serve as initial cracks. In further accordance to reference 1, the cumulative mass loss of the specimens (three specimens each) were converted to the cumulative thickness loss, which is designated as the mean depth of erosion (MDE). This allows comparion of materials with different densities. Another important parameter of cavitation erosion resistance is the maximum erosion rate (MER). It corresponds to the maximum slope of the MDE curve and is computed by a MATLAB routine. The MDE and MER values can be found in **Fig. 5** (no spread widths due to legibility).
All averaged cavitation erosion curves show a similar course, which is characterised on the one hand by an exponential growth to a time of approximately 60 min and on the other hand by a linear progress after 60 min. Considering the MDE curves of Ni-Al-Bronze and Mn-Al-Bronze, both diagramms show very similar trends. They emphasize that coatings sprayed at

traverse speeds below 499.5 mm/s do not differ much and result in similar MDE and MER.

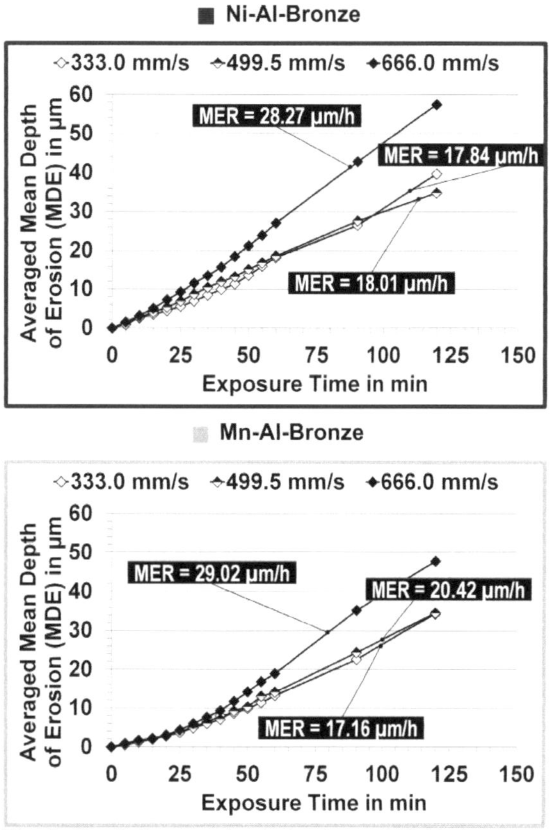

Fig. 5. Cavitation erosion behaviour of Ni-Al- (top) and Mn-Al-Bronze (bottom).

In contrast, the coatings sprayed with higher traverse speeds result in higher thickness losses which are about 1.4 - 1.7 times higher than those sprayed with slower gun velocities.

When comparing both bronze materials, the coatings sprayed with Mn-Al-Bronze obtain a lower thickness loss and erosion rate and thus show a better cavitation resistance.

3.6 Residual stress analyses

Modified hole drilling method based on "Electronic Speckle Pattern Interferometry" (ESPI) was used to identify the influence of traverse speed on residual stress state inside the Ni-Al-Bronze and Mn-Al-Bronze coatings. The ESPI technique is well documented [10, 11] and will be only shortly summarized. The contactless measurement method avoids the time consuming task of application, wiring and calibration of strain gages on a specimen. Instead, surface displacements are captured optically [10, 11]. A typical ESPI setup consists of a laser source, with its light being split by a half-mirror. One part provides a phase-stepped reference light to the CCD camera, which is used for analysis. The other part illuminates the specimen (illumination beam), which is captured on the CCD camera chip (object beam). Interference of object beam and reference light leads to speckle patterns, which vary due to surface displacements. By comparing sets of phase-stepped images before and after surface deformation, caused by the drilling process, it is possible to evaluate its sign and size at every pixel. The following analysis by an algorithm includes all pixels in a ring-shaped area around the hole and leads to the experimental displacement map. Together with elastic material parameters (Young's modulus,Poisson ratio), the stress values can be calculated by the software [10, 11].

The Stresstech PRISM system (Stresstech GmbH, Rennerod, D) was used for residual stress analyses of the coatings. The holes were machined by steps of 50 µm using high speed steel end mills (2 fluted, TiN coated, Ø 0.8 mm and 1.6 mm) up to a profile depth of 450 µm. The inner integration radius was 2.5 times the hole diameter and the outer integration radius 4.5. The Poisson ratios used for calculation were 0.3 for Ni-Al-Bronze and 0.34 for Mn-Al-Bronze (corresponding to [9]). The Young's modulus, which is needed for stress calculation, was measured using a nano indentation machine NHT (CSM Instruments, Freiburg, D) with five measurements for each coating. For both materials the results differed significantly from the values found for bulk material; i.e. for cast Ni-Al-Bronze alloys with a range of ca. 115 GPa to 121 GPa and for Mn-Al-Bronze alloys ca. 117 GPa. While the measured values decreased with rising traverse speeds for Ni-Al-Bronze (106.9 GPa, 104.0 GPa and 90.5 GPa), the modulus values varied for Mn-Al-Bronze (100.3 GPa, 108.4 GPa and 97.5 GPa).

The results of the stress measurements can be seen in **Fig. 6**. Residual Stresses σ_x correspond to stress in the dimension of specimen length; stresses σ_y to the dimension of specimen width. For each coating three measurements were carried out and evaluated. The coatings were examined as sprayed without any further pre-treatment.

In general, all Ni-Al-Bronze curves show a progress of residual tensile stresses with values up to ca. 860 MPa near the surface to a depth of at least 250 µm. Afterwards the stresses fall and later change signs, resulting in compressive stresses.

While exhibiting the lowest near surface stress value, the specimens with the highest traverse speed show an almost constant amount of residual stress of ca. 600 MPa to a depth of 200 µm (for σ_x and σ_y). Afterwards the stress first falls slightly and then in a sharp decrease between 300 µm and 350 µm markedly, resulting in a change of sign. This depth corresponds to the coating thickness, which is about ca. 360 µm.

The specimens of the different traverse speeds show higher stresses near the surface, but an instantaneously sinking progress. While the curve of the lowest traverse speed specimens has the lowest stress values of all curves, a peak with temporarily rising residual stress appears at a depth of ca. 300 µm for the curve of 499.5 mm/s in the direction of σ_x. After this peak the stress values fall again, but remain the highest stresses near the interface between coating and substrate (average coating thicknesses of ca. 340 µm for 333.0 mm/s and ca. 460 µm for 499.5 mm/s)

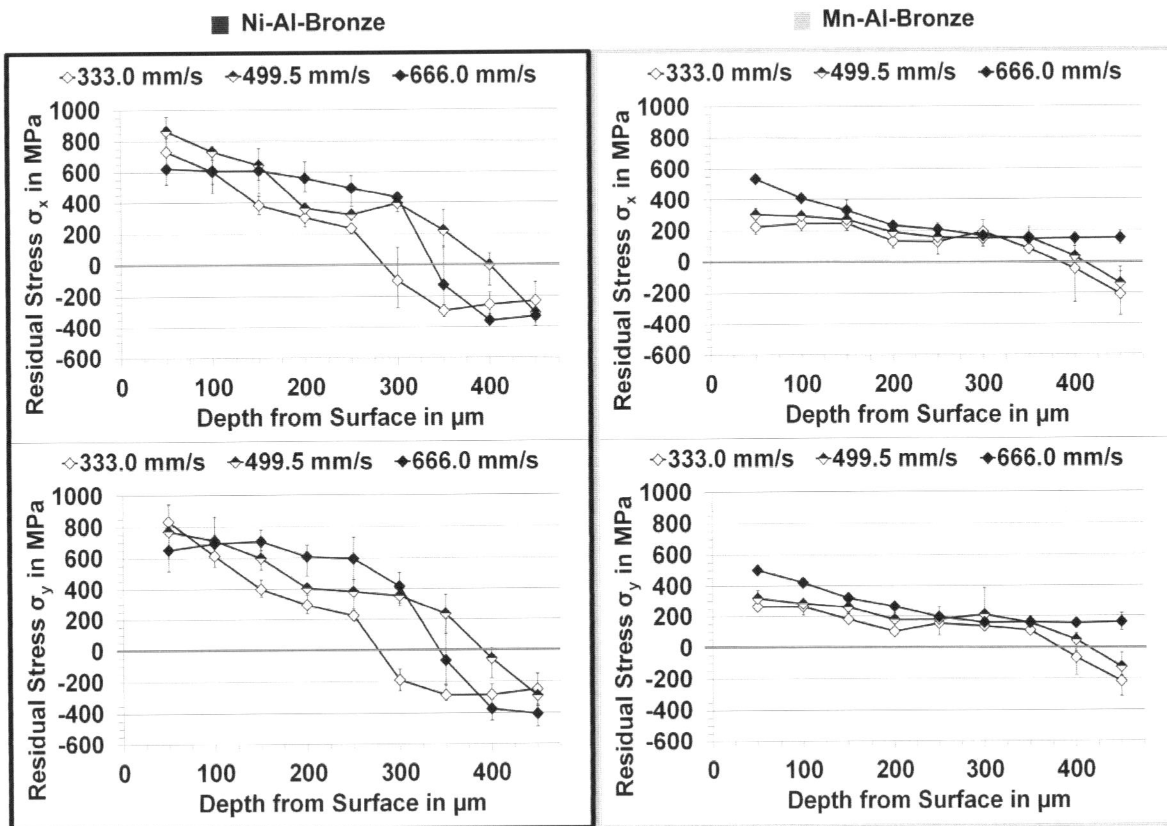

Fig. 6. Residual stresses σ_x (top) and σ_y (bottom) of Ni-Al-Bronze (left) and Mn-Al-Bronze coatings (right).

For the largest part the stress depth profile of the curve of 499.5 mm/s lies in between the curves of 333.0 mm/s and 666.0 mm/s. Additionally some variations between σ_x and σ_y are observable for all curves, especially in the tensile area, resulting in a slightly wider range of occurring stresses for σ_y.

Likewise the Mn-Al-Bronze coatings show tensile stresses to a depth of at least 350 µm and a decreasing course. However, the stresses do only vary in an lesser manner. The highest stress values occure for the highest traverse speed specimens. Those stresses reach up to 530 MPa near the surface and stay tensile throughout the further progess (average coating thickness 630 µm). The stress depth profiles of 333.0 mm/s and 499.5 mm/s are almost identical in progress. This stress values are lower than the corresponding values of 666.0 mm/s, especially near the surface. Furthermore both curves exhibit a change in sign at depth of ca. 400 µm (333.0 mm/s) and 450 µm (499.5 mm/s). The values observed do not correspond to the average coating thicknesses (ca. 590 µm for 333.0 mm/s and ca. 580 µm for 499.5 mm/s). Overall, the curves of the lowest traverse speed specimens show the lowest stresses considering Mn-Al-Bronze. Additionally there are no significant differences between σ_x and σ_y observable.

Comparison of both materials reveals far lower stresses in size for both tensile and compressive areas of corresponding curves for Mn-Al-Bronze. Besides the stresses lie in closer range and show a slower and more steadily decrease of the stress values.

4 Discussion

The results revealed two main influences,which affect coating properties. These influences were on the one hand the different traverse speeds and on the other hand the two different bronze materials investigated in the study.

Considering the different traverse speeds, the results showed higher tensile stresses for fast traverse speeds than for slow velocities with regard to both materials. Typically a trend vice versa would have been expected due to higher substrate temperatures at slowlier traverse speeds. One possible explanation might be the influence of more inhomogenities like pores and oxides for higher gun speeds. This could lead to longer heat storage within the coating and

hereby influence the cooling process and hence cooling stresses. However, porosity and oxide content is high for all experiments and also influences coating cohesion. Here the cavitation erosion tests give further information on coating properties. The results display that higher traverse speeds lead to higher erosion rates and thickness losses and therefore consequently lower cohesion. Since cavitation erosion load is comparable to fatigue, the tensile stresses in all the coatings contribute negatively to the stress regime and cause faster material loss. Thus the triangular relation between residual stress, cavitation erosion resistance and coating cohesion could verify the results. Furthermore the magnitude of the material loss seems to be dependent on the size and progress of the residual stresses near the surface, which can be seen in **Fig. 6** and **Fig. 5**. MDE curves of low and moderate speeds show comparable and more steady courses. The same fact can be seen for the stress curves. The coating adhesion does not seem to be influenced by different traverse speed for wire arc spraying, **Fig. 2**.
Comparing the different bronze materials, the Mn-Al-Bronze shows a better cavitation erosion performance and lower residual stresses, although having a slightly higher coefficient of thermal expansion. One possible approach could be the consideration of the lower melting range and consequently lower recrystallisation temperature of this bronze. The recrystallisation possibly reduces dislocations and hence coating hardness. Furthermore longer thermal influences might improve coating cohesion like typical annealing procedures. This could be proven by better coating cohesion, which is represented by cavitation erosion resistance.

5 Summary and outlook

In the present work, the influence of traverse speed on adhesive and cohesive coating properties was examined. One main objective was to determine the residual stress state for the novel material combination of Ni-Al-Bronze and Mn-Al-Bronze on mild steel substrates. It was found that higher residual stresses, especially near the coating surfaces, lead to increasing material losses in cavitation erosion test and affect coating properties and microstructure. Furthermore the stresses were far lower for Mn-Al-Bronze coatings than for Ni-Al-Bronze coatings. Besides, the varying coating formation impacted on other coating properties like hardness, rigidity and microstructure.
Further investigations will focus on reducing oxide content and more homogeneous coating microstructure. This might be achieved by the use of other spray gases than air, like nitrogen or its mixtures. Additionally, the lower temperatures of coating and substrate should result in less residual stresses.

6 Acknowledgements

The authors would like to thank all co-workers involved in the study, namely listed in alphabetical order C.-P. Homburg, D. Krüger, R. Lange, N. Néhmet, B. Ripsch, A. Schober and M. Schulze.
The research project (Nr. 18449 BG) of the research community Center of Maritime Technologies (CMT), Bramfelder Straße 164, 22305 Hamburg has been funded by the AiF within the programme for sponsorship by Industrial Joint Research (IGF) of the German Federal Ministry of Economic Affairs and Energy based on an enactment of the german parliament. This support is greatly acknowledged.

7 Literature

[1] American Society of Testing and Materials: Standard Test Method For Cavitation Erosion Using Vibratory Apparatus", G32. Annual Book of ASTM Standards (2010), pp. 98/116.

[2] Naval Surface Treatment Center: Rudder Coating Failures on Navy Ships. http://www.nstcenter.biz/wp-content/uploads/mdocs/RudderCoatingFailuresNavyShips.pdf (2003).

[3] Kim, J.-H. and Lee, M.-H.: A Study on Cavitation Erosion and Corrosion Behavior of Al-, Zn-, Cu-, and Fe-Based Coatings Prepared by Arc Spraying. Journal of Thermal Spray Technology 19 (2010), Issue 6, pp. 1224/1230.

[4] Pina, J.; Dias, A. and Lebrun, J. L.: Study by X-ray diffraction and mechanical analysis of the residual stress generation during thermal spraying. Materials Science and Engineering: A 347 (2003), Issue 1-2, pp. 21/31.

[5] Kuroda, S., T. Fukushima, and S. Kitahara. Significance of quenching stress in the cohesion and adhesion of thermally sprayed coatings. Journal of Thermal Spray Technology 1 (1992), Issue 4, pp. 325/332.

[6] Sampath, S. et al.: Role of thermal spray processing method on the microstructure, residual stress and properties of coatings: an integrated study for Ni–5 wt.%Al bond coats. Materials Science and Engineering: A 364 (2004), Issue 1-2, pp. 216/231.

[7] Chen, Y. et al.: Effect of path-plans of automatic arc spraying on coating residual stresses. Transactions of the China Welding Institution 31 (2010), Issue 8, pp. 97/100.

[8] Clyne, T. W.; Gill, S. C.: Residual stresses in thermal spray coatings and their effect on interfacial adhesion: A review of recent work. Journal of Thermal Spray Technology 5 (1996), Issue 4, pp. 401/408.

[9] H.J. Meigh: Cast and Wrought Aluminium Bronzes: Properties, Processes and Structure.
Copper Development Association / IOM Communications, London (2000)

[10] Schajer, G. S. and Steinzig, M.:
Full-Field Calculation of Hole-Drilling Residual Stresses from ESPI Data. Experimental Mechanics Vol.45 (2005),
Issue 6, pp.526/532.

[11] Steinzig, M. and Ponslet, E.:
Residual Stress Measurement using the Hole Drilling Method and Laser Speckle Interferometry: Part I.
Experimental Techniques 27 (2003),
Issue 3, pp. 43/46

Cause study to the occurrence of unmelted wire pieces during wire arc spraying

W. Mayr, E. Steiner, S. Heilmeier, L. Blau, M. Szulc, A. Atzberger, K.Hartz-Behrend, S. Zimmermann, J. Zierhut, J. Schein, Munich / D

The occurrence of process instabilities like separation of unmelted wire pieces (cold shots) or wire sticking during spraying is investigated for three different wire arc spraying equipments: An experimental spray gun, a large scale production and a small scale production equipment. Particular designed mobile diagnostics are used to record high sampling rate waveforms of quantities like arc voltage and current, voltage drop at the wire contact sleeves and finally the wire feed velocity at the feeder and just in front of the contact sleeves for push systems. A spectrum analysis based strategy is developed to predict the occurrence of process instabilities with a view to quality assurance. In addition a competitive visual inspection system of a camera and a pulsed LED illumination is designed as a tool for manual inspection of droplet detachment and spray jet generation. For each of the investigated spraying equipments, the applied diagnostics are presented and the results are interpreted with respect to the physical cause of instabilities and how to potentially avoid them.

1 Introduction

Wire arc spraying is a very cost efficient process because of its easy handling, rugged design and mobility [1]. To make it accessible to applications with higher requirements, improvements are necessary. So the emission of finest particles and the oxidation due to the use of air for atomizing has to be reduced. It is shown, that using inert gases for atomizing can lead to significant advances [2]. Inherent voltage fluctuations, caused by the high pressure atomizing air stream at the wire tips in combination with distortions of the wire feeding can lead to instabilities and even to the displacement of cold shots as shown in [3] and [4]. Kelkar et. al. showed in [5] that the physical properties of the atomizing gas influence the gas dynamics and hence the spraying process. In [6] Pourmousa et al. measure different size distributions for particles from anode and cathode.

To improve the wire arc process with view to the above mentioned limitations, a research project, funded by the BFS (Bayerische Forschungsstiftung) was established. The objective of the project is to enhance process performance by use of new atomizing gases, modified spray guns and by minimizing instabilities using process control. The use of e.g. PFI (Particle Flux Imaging) for process control is outlined in [7]. A correlation of voltage and current waveforms with high speed camera information is given in [4], where the occurrence of cold shots is interpreted and their velocities are measured.

In this paper the following quantities are accessed to better understand the occurrence of undesired instabilities and cold shots:

- arc voltage and current
- voltage drop between the wire and the contact sleeves
- wire velocity at feeder and at contact sleeves
- visual inspection of droplet formation by means of a competitive camera

Individual diagnostic setups are applied to three different wire arc spraying equipments as follows:

- experimental spraying equipment
- equipment for large scale production line
- equipment for small scale production line

The spraying equipments, diagnostic setups and results are presented and discussed in separate chapters for each of the investigated spraying equipments.

2 Investigation of the Experimental Spraying Equipment

2.1 Features of the spraying equipment

This equipment, described explicitly in [4], has been constructed with respect to easily modify parameters like wire angle and wire length. It is an open system, which means that the wire feed units are accessible during the spraying process as well as the wire contact sleeves. The wire tips at the arc are not hidden (e.g. by a shield) and can be viewed from a perpendicular position. Thus all used diagnostic units can easily be adapted. The spraying parameters for the measurements are:

Parameter	Value
Wire feed rate	4 m/min
Arc voltage	29.8 V
Arc current	190 A
Gas pressure	6 bar
Wire diameter	1.6 mm
Atomizing gas	Air

The equipment uses a push wire feed unit which feeds into a blank wire transport guide.

2.2 Diagnostic for the experimental spraying equipment

In **Fig. 1** a schematic diagram of the diagnostic system is given. It is capable of high dynamic sampling of all measured quantities. The components are subsequently described.

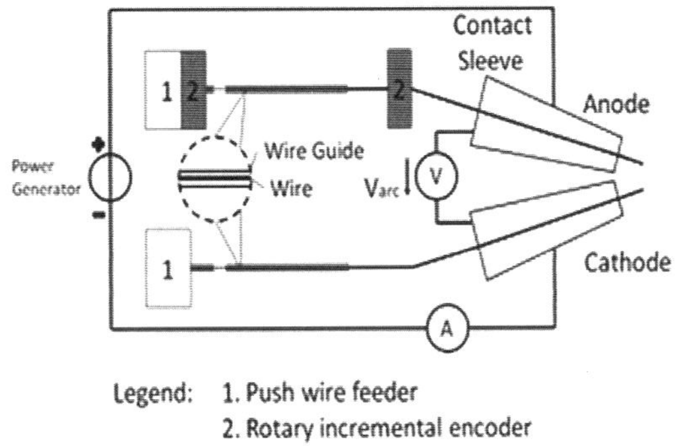

Fig. 1 Schematic diagnostics for experimental spraying equipment

- **The ammeter A** is represented by the current sensor LEM LF 1005-S and has a range of ±1.5 kA. Its signal is fed to a cRIO system by National Instruments for real time processing. The sampling rate of the used analog input is 1 MS/s.

- **The voltmeter V** corresponds to the LEM CV 3-200 sensor has a range of ±200 V. Its signal is sampled by the cRIO as well.

- **The two rotary incremental encoders** have a resolution of 1024 increments per rotation and are adapted to
  - the capstan of one push wire feeder to measure the wire velocity before the wire guide
  - just in front of the contact sleeve by a particular designed adapter.

  The signals are fed to the digital inputs of the cRIO. The sampling rate of the wire velocity for a feed rate of 4 m/min is about 620 S/s at the feed unit and about 3600 S/s at the contact sleeve.

The software of the cRIO allows triggering on the occurrence of a configurable sequence of states like open, short condition or specified slew rates. A pre trigger is adjustable. Moreover digital signals are generated to control e.g. a camera system or signalize the recognized state of the spraying process.

2.3 Results and discussion for the experimental spraying equipment

The upper graph of **Fig. 2** shows voltage and current fluctuations during normal operation of the spray gun, while the lower graph shows the wire speed, measured directly in front of the contact sleeves as well as at the wire feed unit.

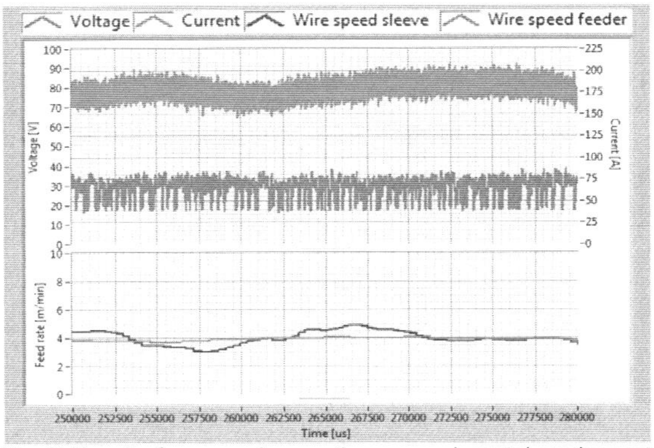

Fig. 2 Results for normal operation - experimental equipment

The duration of the waveforms is 30 ms and the x Axis is scaled in µs.

The spraying process of the experimental equipment, however, couldn´t be sustained long enough to capture an instability at normal operation. So the voltage setting is reduced to about 26 V to induce the occurrence of cold shots on which the measuring equipment triggers. In **Fig. 3** the results for the cold shot case are presented. The physical quantities are the same as in **Fig. 2**, but the duration is 60 ms instead of 30 ms.

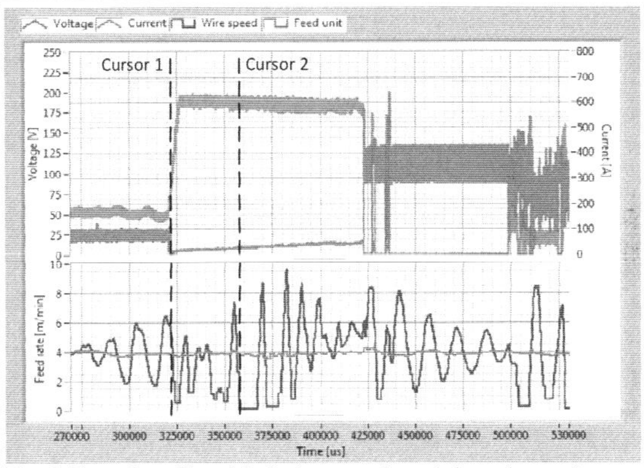

Fig. 3 Results with cold shot - experimental equipment

As can clearly be seen from **Fig. 3**, the wire speed starts a strong oscillation before the short in position of cursor 1. During this short at position of cursor 2 the wire even stops.
Since the feed unit continues feeding the wire at the beginning of the wire guide, mechanical stress builds up in guide, leading to a spurt of the wire, so that sufficient material for a cold shot accumulates. A process like this has been high speed filmed and published in [4], were the correlation between voltage and current fluctuations with high speed camera records are given.

3 Large scale Production line Equipment

3.1 Features of the Large scale Production line Equipment

The arrangement is particularly designed for spraying 3.2 mm diameter aluminum wires for high volume production in three layer operation. It is also an open system, and the used pull wire feed units are placed directly in front of the wire contact sleeves. However, due to continuous motion of the gun and restricted place in the cabin, a camera couldn´t be adapted to it.

3.2 Diagnostic for the large Scale Production line Equipment

One specific challenge for this diagnostics is to discover if process instabilities like wire sticking in the contact sleeves could be caused by the appearance of micro arcs between the contact sleeve and the wire. As a micro arc causes an enhanced voltage drop ΔV this quantity is measured for the equipment. So the sleeves and the moving wire must be contacted. For a stable contact to the moving wire, two different options are successfully tested:

- A contact brush, the wire is fed through and thus is contacted
- The housing of the pull wire feeder as its metal rolls are contacting the wire and they are connected with the housing by metal as well.

The schematic diagram of the diagnostic is given in **Fig. 4.** The waveforms of the voltage drops ΔV are measured by contacting the wire by the feeder for the anode and cathode side and the corresponding mounts of the contact sleeves. The measuring device is the handheld oscilloscope RTH1004 by Rohde & Schwarz with 4 individual isolated analog input channels [9]. It also features a wireless control which is particularly suited for application in industrial environment.

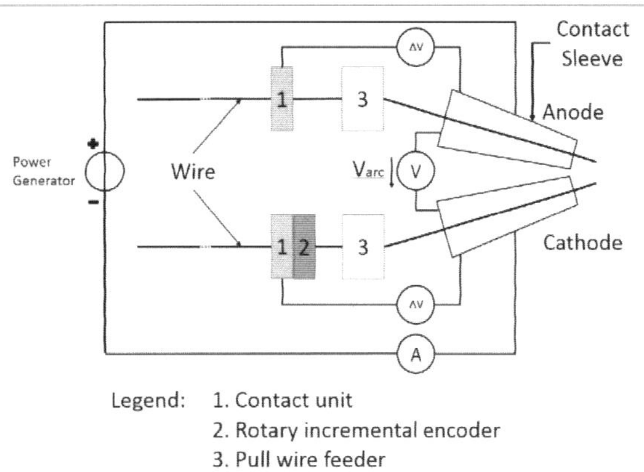

Fig. 4 Schematic diagnostics for large scale production equipment

For storage however, the scope uses a mere of 125 kSampels per channel when 4 analog channels are used. The input range is set to -0.4 V to +1.2 V. This range is, however, too small as turns out later on in chapter 3.3.

Take into account that the burner voltage is the sum of the arc voltage and the two voltage drops at the contact sleeves as can be seen form **Fig. 4**.

The rotary incremental encoder is placed directly in front of the wire feeder.

The other components (A and V) are identical to those described in chapter 2.2 with the exception of using an additional current sensor LEM.LF 2010-S (± 3kA range) for higher current values.

3.3 Results and Discussion for the large Scale Production line Equipment

Fig. 5 shows a signal section of normal operation with a duration of 70 ms.

Fig. 5

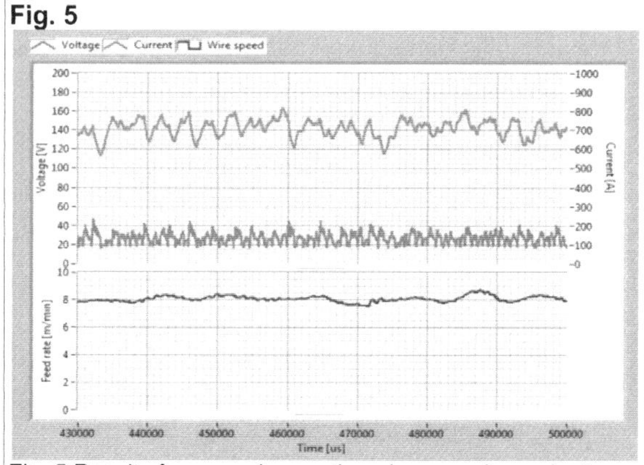

Fig. 5 Results for normal operation - large scale production

Here the voltage and current fluctuations as well as the wire speed, measured directly in front of the contact sleeves are given. The rim speed of the wire feed capstan could not be measured due to the mechanical structure of the spray gun. Notice that the relative fluctuations in wire speed, shown in **Fig. 5**, is significantly lower than for the experimental facility in **Fig. 2**. This is caused by the fact that the production facility uses a pull wire feeder, while the experimental spray gun uses a push device.

In **Fig. 6** the results are given for an instability, occurring when using worn contact sleeves.

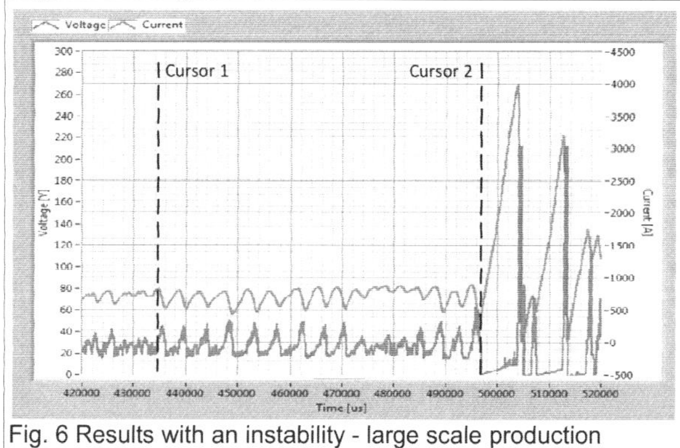

Fig. 6 Results with an instability - large scale production equipment

The diagnostic system is triggered on a short and the total measuring time is 1 s, of which 100 ms are given in the graph. The pre trigger is 500 ms, which means the first short starts at 500000 µs (time scale is in µs). At the end of the measuring time a sticking of the wire occurred. Unfortunately the wire speed could not be recorded because of a defect in the wire speed measurement unit (2 in **Fig. 4**).

Just to prevent an overlapping of the two curves, the minimum of the current axis was set to -500 amps. Notice that the current fluctuation is getting similar to periodic and coming more sinusoidal at the position of cursor 1, which is about 60 ms before the occurrence of the short of the wire tips at cursor 2 (**Fig. 6**). The short is indicated by a current rise up to 4000 A and the voltage reduction to very low values. A short may result in a cold shot [4] or sticking of wire in the contact sleeve, as was the case during this measurement. Note that recorded shorts within other measurements are also preceded by an oscillation for several 10 ms. But not all shorts lead to a sticking of the wire. This oscillation feature is also observed for another spray equipment used for the same coating application.

With view to quality assurance, a spectrum analysis based automatic detection algorithm of such oscillations is developed in [8] and applied to the recorded measurement. The principle is to continuously Fourier transform the AC component of 20 ms sub waveforms which are shifted for small time intervals. In **Fig. 7** a comparison of the results for sub waveforms between normal operation and oscillation is given. The spectrum of the sub waveform in the oscillation case significantly peaks at cursor 1 in **Fig. 7**. At normal operation no peak is observed.

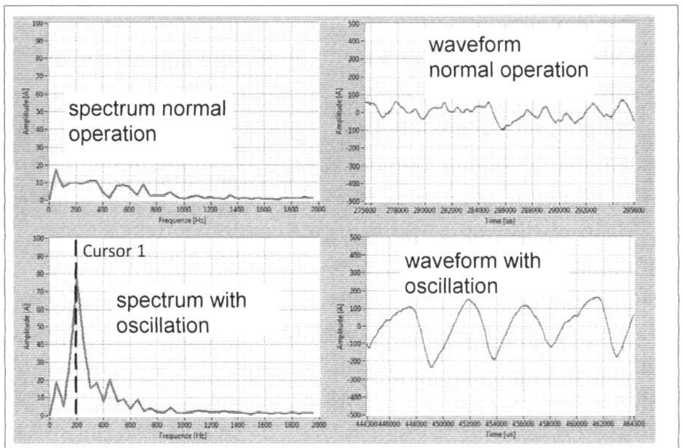

Fig. 7 Sub waveforms of 20 ms time gap and correlated spectrum (normal → upper graphs, oscillation → lower graphs)

Applying the algorithm, the point at which the oscillation is recognized the first time could be evaluated and is given in **Fig. 8** by position of cursor 1. This can be identified to be more than 40 ms before the occurrence of the short. If this procedure is deployed to a real time system, it looks promising for quality assurance, as it may enable to influence the spray process to prevent an approaching instability like wire sticking.

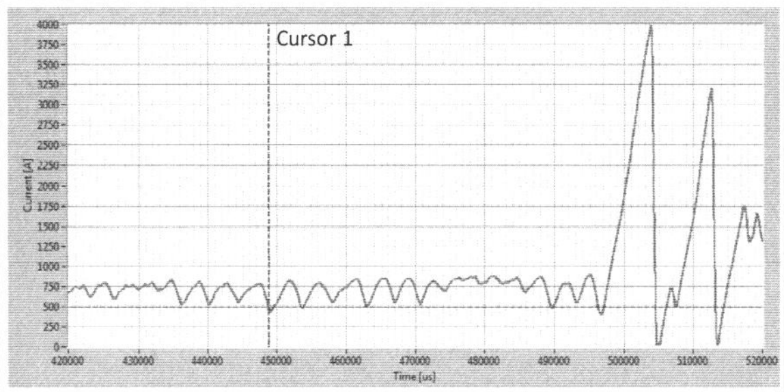

Fig. 8 Position of the first possible recognition of an oscillation

Subsequently a potential explanation of the sticking mechanism based on the ΔV measurement is given. **Fig. 9** shows the current rise during a short of the wire tips (lower graph of **Fig. 9**). In the upper graph the voltage drops ΔV for both contact sleeves are given.

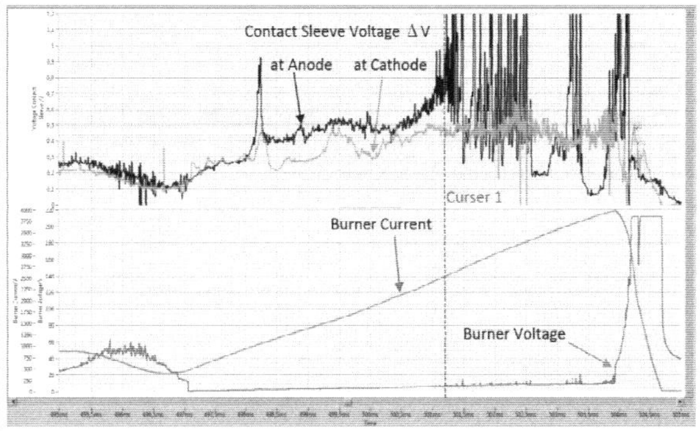

Fig. 9 Waveforms of ΔV´s, burner current and burner voltage

On the left of cursor 1, there are no over ranges in the ΔV measurements and on the burner voltage no spikes are present. To the right of cursor 1, however, the ΔV measurement at the anode sleeve runs into over range which means that ΔV is greater than 1.2 V and little spikes exists on the burner voltage graph. By zooming with the original evaluation software, it can be realized that a strict coincidence of the over ranges of ΔV and the spikes on burner voltage is present. Taking into account, that the burner voltage is the sum of the arc voltage and the two ΔV´s, it comes obvious that the spikes are the same like in the ΔV-curves but on a different scale and hence not over ranged. Most of the observed peak values of the burner voltage which coincide with anode sleeve ΔV over ranges are in between 5 V and 15 V. For the cathode sleeve ΔV reaches values up to 20 V and more.

In [10], the process of opening a switch is described and a voltage threshold V_{min} is introduced above which a light arc establishes during opening a contact. For contact metals V_{min} is in the range of 12 V to 16 V and at values below V_{min} a short instable arc without plasma can occur. Of course there are different conditions if a contact opens or if a wire is separated from a contact sleeve. Nevertheless the similar range of V_{min} and the range of the voltage drops give additional rise to the assumption that micro arcs in the sleeve are an explanation of the wire sticking.

At normal operations, no over range of ΔV could be observed. However, as there is no triggering on ΔV over ranges, only a few random selected seconds of an industrial coating process without instabilities for many hours of coating is analyzed. So one cannot excluded that micro arcs appear also during normal operation.

4 Small scale production line equipment

4.1 Features of the small scale production line equipment

The equipment is a Praxair-TAFA CoArc used by ABLER GmbH & Co. KG. It is applied to miscellaneous applications. It is a closed system, which means that the components (pull wire feed unit, contact sleeves and wire in front of sleeves) are actually not accessible during spraying process. The wire tips look out of the air cap, but a perpendicular view to it is hidden by the shield. Thus to produce a shadow image the shield was cut off.
The following spraying parameters are used:

Parameter	Value
Wire feed rate	12 ft/min (3.66 m/min)
Wire	Copper coated iron
Arc voltage	32 V
Arc current	150 A
Gas pressure	5 bar
Wire diameter	1.6 mm
Atomizing gas	Air

4.2 Diagnostic for the small scale production line equipment

For controlling standard spraying processes and for developing new applications, it is frequently desirable to manually inspect the formation of the droplets and the spray jet by vision techniques. To that behalf, a competitive visual inspection system of a camera, a pulsed LED illumination and a control unit is introduced. The optical components are arranged to yield shadow images. Special focus was put on an adjustable suppression of the luminosity of the arc, because normally it is too intense for inspecting the process.

The experimental setup was build up with:

- the LED-Spot-Illumination Lumimax Spot5WFL, that works with a flashed illumination, which is triggered by a TTL Pulse. The duration and intensity of the flash is adjusted manually.
- the camera Crevis MV-BV20G with an optical bandpass filter mounted on the front lens of the objective. The maximum frame rate is 120 fps at a resolution of 640 x 480 pixels. The minimum exposure time is 2 µs.
- item profiles for flexible fixing of the Camera and the Illumination. The optical components are adjustable by spherical joints and the item profiles can be shifted easily.
- a National Instrument MyRIO 1900, which can perform real time IO for analog and digital signals.

To produce a shadow image, the camera and the Illumination were built up as shown in **Fig. 10** and the burner is placed in between the two components. To control the recording, the MyRIO was programmed to control camera and illumination by sending TTL pulses. The camera frame rate is controlled by the frequency and the exposure time is controlled by the duration of the camera pulse.

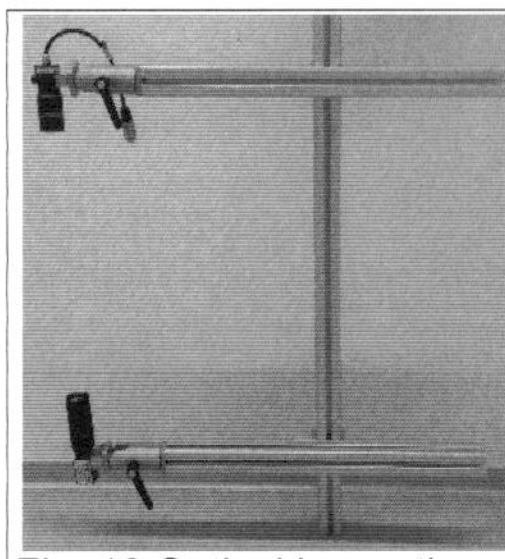

Fig. 10 Optical inspection setup

The luminosity of the arc can generally be reduced by a strong illumination and band pass filtering the exposure light of the camera. Flashed illumination allows higher light flux during exposure than a continuous one. As the luminosity of the arc is to be reduced gradually by software, the restriction that the illumination intensity must be controlled manual has to be eliminated. This is performed by controlling the effective illumination time by introducing an adjustable time shift between the camera and the illumination pulses. The overall intensity is controlled by manually setting the iris of the objective.

4.3 Results and discussion for the small scale production line equipment

In **Fig. 11** three subsequent images are given, which are taken with a frame rate of 100 fps. The parameters are set to maximum fading out of the luminosity of the arc.

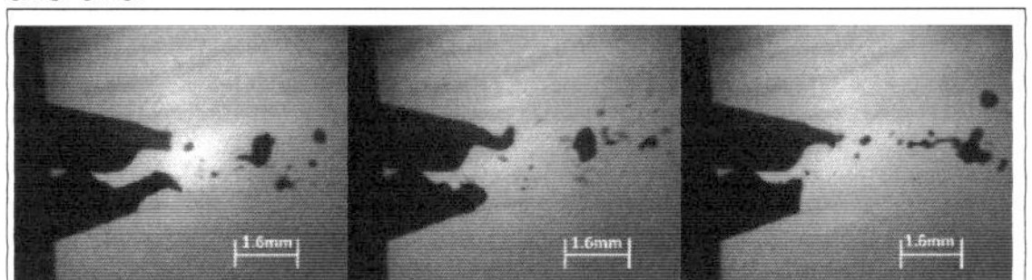

Fig. 11 Images with maximum fading out the luminosity

As the arc is actually invisible in **Fig. 11**, these settings are the best to inspect the formation of droplets and spray jet. Possible movements of the wire tips during spraying, which may precede a short of the tips, can be inspected.
In many cases, however, it is desirable to see the arc too, which is achieved easily by adjusting the appropriate parameters. In **Fig. 12** the images are given for the case of an intense visible arc.

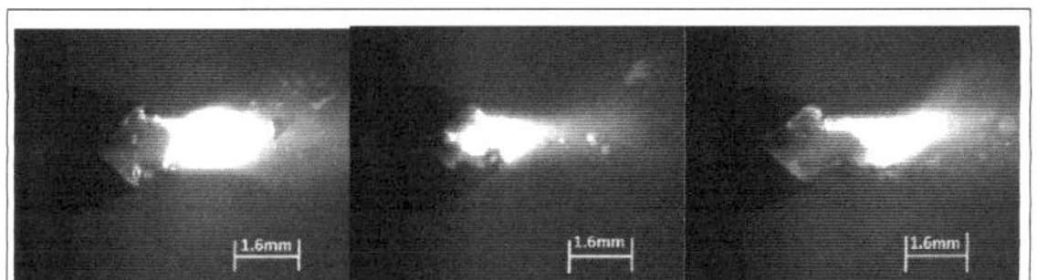

Fig. 12 Images with weak fading out the luminosity

Finally the results for an intermediate parameter setting is given in **Fig. 13**

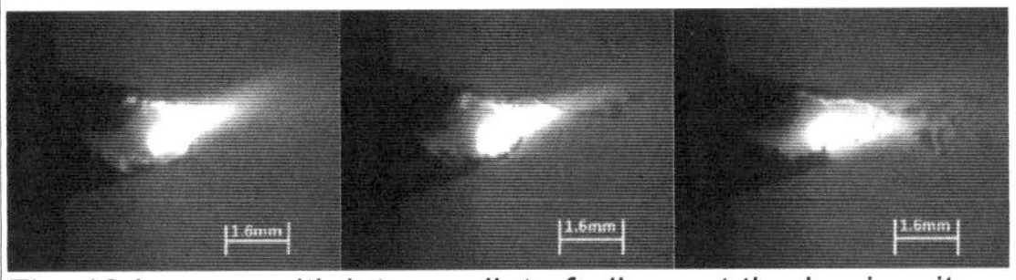
Fig. 13 Images with intermediate fading out the luminosity

It can be seen from the results, that, however, the more intense an arc is visible, the more the visibility of the wire tips is reduced.

Beside the visual inspection of the wire arc spraying process, the system can also be applied to inspect other thermal spraying processes. As with full fading out of the luminosity of the arc, a good contrast is available and the introduction of automatic image analysis looks promising.

5. Conclusions and work for the future

The experimental spraying facility has strongly fluctuating wire velocities at the contact sleeves and hence tends to instabilities. The cause of this is the use of a push wire feeder which feeds into a blank wire transport guide. No further investigations of this equipment are planned.

The large scale production equipment runs stable in general. Rarely occurring instabilities like wire sticking or cold shots are preceded by a short of the wires, which in turn is preceded by an oscillation of the arc current. With a view to quality assurance a spectrum analysis based strategy is developed to predict the occurrence of instabilities.
A physical explanation of wire sticking is probably the incidence of micro arcs in the contact sleeves, which are indicated during a short of the wire tips. This indication is based on the occurrence of peaks of the voltage drop between contact sleeve and wire.
It is intended to further investigate the physical nature of the oscillation and to deploy the strategy for prediction to a real time target in order to introduce appropriate remedies.

The presented optical inspection system is successfully applied to a small scale production line. A further improvement of the optics is intended to enable the use of image analysis software. Thus automated quality assurance can be introduced for e.g. automatic inspection of the position of the wire tips or calculating the size distribution of the spray droplets on the basis of many recorded frames.

4 Acknowledgment

The presented results were conducted in course of the project "Energy efficient production of thermally sprayed coatings through the use of a modified wire arc spraying process" funded by the Bayerische Forschungsstiftung (funding number AZ-1054-12).

3 Literature

[1] Z. Babiak and T. Wenz, „Grundlagen der thermischen Spritztechnik, Flamm- und Lichtbogenspritzen," in Moderne Beschichtungsverfahren, Weinheim, Wiley-VCH, 2005, pp. 131-149.

[2] F. T. Lang, Einflüsse des Zerstäubergases auf Umwelt und Beschichtung beim Lichtbogenspritzen, Diploma thesis, Hochschule München, 2009.

[3] N. A. Hussary and J. V. R. Heberlein, „Effect of system parameters on metal breakup and particle formation in the wire arc spray process," Journal of Thermal Spray Technology, Vol. I, No. 16, pp. 140-152, 2007.

[4] S. Eichler, S. Kirner, S. Zimmermann, J. Schein, W. Mayr and J. Zierhut, „Detection and avoiding of unmelted wire pieces during wire arc spraying," in Proceedings of ITSC 2015, Long Beach, 2015.

[5] M. Kelkar and J. Heberlein, „Wire-arc spray modelling," in Plasma Chemistry and Plasma Processing, No. 22, pp. 1-25, 2002.

[6] A. Pourmousa, J. Mostaghimi, A. Abedini and S. Chandra, „Particle size distribution in a wire-arc spraying system," Journal of Thermal Spray Technology, No. 14, pp. 502-510, 2005.

[7] J. Schaup, A. Atzberger, G. Huismann, K. Hartz-Behrend, S. Zimmermann, S. Kirner, M. Szulc, G. Forster and J. Schein, „The journey with modern power sources and torch systems to optimized arc spraying," in 10. Kolloquium HVOF Spraying, Erding, Germany, 2015

[8] E. Steiner, Further development of a real time measurement system for detecting and recording of process relevant parameters for analysis and quality assurance of the wire arc spraying process, Master thesis, Hochschule München, 2016.

[9] http://www.rohde-schwarz.de/product/rth.html

[10] E. Vinaricky "Elektrische Kontakte, Werkstoffe und Anwendungen," Springer-Verlag Berlin Heidelberg New York, 2002, pages: 44-46

Alternative industrial surface solutions

Peter Tommy Nielsen, Steen Erik Nielsen, FORCE Technology Denmark

The demand for high quality coatings, which are able to withstand the extreme circumstances, are permanent. The span goes from equipment used in food production and pharmaceutical industry to more harsh environments, such as, offshore and power plant applications. The motivation is to expand service life, reduce shut down time and minimize potential migration of component material to products. Often a need for combining of several material properties in one unit needed, which could be translated to a non-existing material.

This paper will introduce the laser cladding technique as well as laser fusing as a method of combining material properties. The focus will especially be at industrial applications, where practical methods of solutions are sought after. These solutions should replace existing techniques and should at the same time allow simple methods of repair of coated parts in place, for instance in a production line.

The paper will illuminate the manufacturing process of the surfaces on the specific applications, interpret the corrosion and wear properties of the produced coatings, exemplify real time usage and showing the benefit of repair friendly surface.

1 Introduction

Often the industry use their specific experienced based knowledge to make and manufacture components for special applications. This approach rarely fails, but sometimes it will fail, which might cause production stop, damage to machinery or large economic consequences.

To some extent, future and specific requirements to methods of applying coatings should also be considered, meaning the need to apply coatings with a somewhat more environmental friendly approach are wanted. The alternatives to the established technologies should as a minimum reach the same level of quality or have a similar or better lifetime in the service environment. This will also lead to an investigation for every intended use of components and coatings. Several aspects of coating production should be evaluated, but often the main interest lie in service value versus component lifetime and cost.

The focus of possible applications for this paper spans from offshore constructions and power plants to pharmaceutical and food production industry. All of these sectors have major issues with material losses due to corrosion during service. For offshore constructions and power plants, it is the result of extreme conditions and the damage that is sought to be minimized. The pharmaceutical and food production industry are dealing with similar challenges, all though in a smaller degree, the material losses can be quite critical because these will end up in the product. This means that the industries are seeking to design their constructions and process equipment to keep it running for a maximum amount of time with a minimum material loss, which gives longer lifecycle and shorter periods of shut down time.

The palette of coating processes suggests many solutions; however, this paper will focus on two related processes, laser cladding and laser fusing. The processes are related to welding technology and will achieve coatings with full metallurgical bonding to the substrate. The technique is applied to specific real industrial problems where, for instance, product contamination from coating or failure is not accepted.

Commonly used cobalt based coating materials represent the functional surfaces in this industrial problem solving study. The study will to some extend take advantage of previous work on the materials behavior under standard testing, specifically revealing corrosion, wear and tribocorrosion properties under simulation of real usage.

Specific industry applications such as ice producing rollers and scraped heat exchanger tubes are discussed and preliminary results are presented. Based on the mentioned applications it implicates that the coating should have no risk scalling in large flackes or contaminate the final product in an unacceptable degree. Furthermore the coating should prolong the service life and make onsite repair work possible.

2 Techniques

2.1 Introduction to laser cladding

Laser cladding is typically used for improving surface properties as wear and corrosion resistance. This is done by 'in situ' adding of a specific material which is melted and welded to the surface by the laser beam. The single layer thickness is typically in the range of 0,5-2 mm with a limited dilution of base material, carried out with a relatively low heat input. Thicker layers can be performed by multi layer techniques [10].

Surface materials are typically added either as solid wire or as fine grained powder which is then blown into the laser beam. Figures 1 and 2 illustrates the laser cladding process based on powder and wire material.

Using wire based material the loss of material as spatter is nearly zero. The availability of materials is however restricted to standard materials on the market [10].

For powder based laser cladding the loss of material varies dependent on the type of nozzle used. Co-axial nozzles may have a higher material efficiency compared to traditional methods. A material loss in the range of 20-40% may be foreseen. However, powder cladding gives the possibility of tailoring the chemical composition of the surface layer specificly by mixing different powder materials [10].

Fig. 1. Left: Laser powder cladding nozzle set up. Right: Proces in action at FORCE Technology. [1]

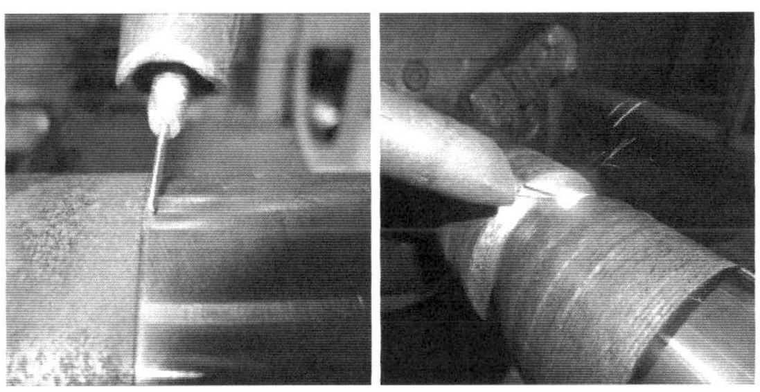

Fig. 2. Left: Laser wire cladding nozzle set up. Right: Proces in action at FORCE Technology. [1]

2.2 Introduction to laser fusing

The laser fusing technique is based on the processing of a pre-placed material on the component surface.

One method is to use the thermal spraying techniques, plasma spraying or HVOF. These thermally sprayed surface coatings is characterised by being porous and only mechnically adherent to the surface, figure 3 (left). During fusing by a laser beam the coating is melted and hereby sealed / welded to the surface, figure 3 (right).

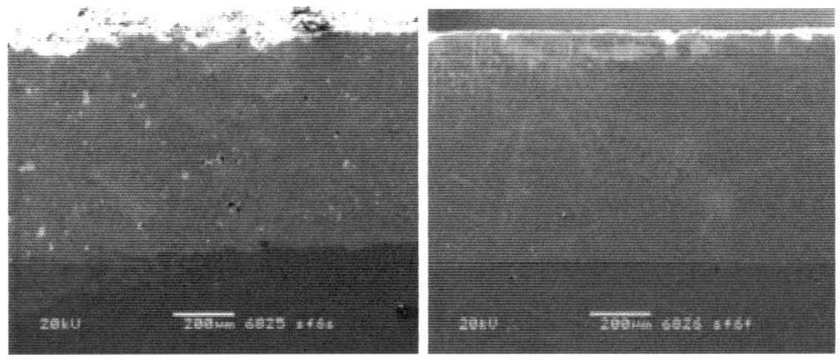

Fig. 3. Left: Plasma sprayed surface coating. Right: Laser fused surface coating.
Material; Stellite6, layer thickness; 0.5 mm. [1]

In order to laser fuse the thermally sprayed coating, the laser beam typically scans the surface in a track-overlapping pattern. Compared to the standard laser cladding process, wider tracks can be generated during laser fusing and thereby achieving higher productivity. Hereby, larger areas to be coated becomes of industrial interest. Different optical means for achieving a required uniform power distribution within the fused track can be considered [10].

Figure 4 shows a photo of an ø200 mm shaft being laser fused after sprayed with stellite6. The total surface is processed by overlapping tracks performed during a spiral movement. By optimising the degree of overlapping, 'weave heights' between tracks only in the order of 0.05 mm are achieved. This means that only a minimum of post machining might be involved. Typically, a laser fused surface just requires honing or polishing, to obtain a mirror polished surface quality.

Fig. 4: Laser fusing of a ø200 mm shaft. Cross section illustrating overlapping tracks. [1]

The laser fusing productivity is mainly dependent on the following parameters:

- coating composition,
- coating thickness,
- laser power,
- the track width involved.

As the dilution with the base material in general should be low, power level and speed are adjusted to a condition, where the wetting of the surface is adequate to perform the required surface quality for the involved layer thickness [10].

A very important aspect regarding proper wetting conditions is related to the cleaning and to the sand blasting procedure before plasma spraying. Always use clean sand, as dirt or other moisture particles blown to the surface will affect the laser fusing quality remarkably [10].

The laser fusing technique is also found suitable for multi layer techniques where several layers of the same – or different coating materials can be fused together. This gives following advantages;

- thicker layers can be produced,
- 'sandwich' layers with different metallurgy can be produced,
- a 'tailored' metallurgical layer composition can be performed.
- repairing of local areas of the surface can be carried out by plasma spraying locally followed by a local laser fusing.

2.3 Materials and expected coating properties

The well known material of Stellite 6 is sellected to form the functional coating, as previous studies has shown good in relation af laser cladding as well as laser fusing. The applications is expected to experience an enviroment similar to the state of tribocorrosion. I.e. wear, corrosion and the synergy effect of the destructive mechanisms [3] [4] [5] [6] [9].

As illustrated figure 5 the result of a combined wear and corrosion test carried out on a laser fused Stellite 6 surface compared with a cast Stellite 6 and a stainless steel material, AISI 316 L. As seen there is a reduced material loss in the range of 60-70% for the laser fused material compared with the stainless steel and about 50% less compared to the cast Stellite 6 [7].

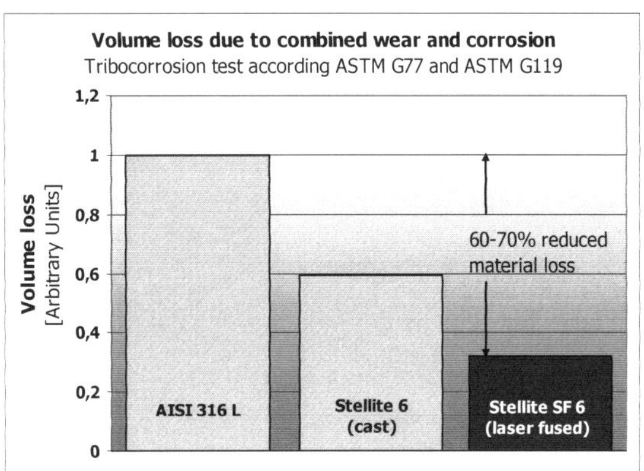

Fig. 5 Combined wear and corrosion test comparing laser fused Stellite 6 with cast Stellite 6 and stainless steel, AISI 316 L [7]

In the case of laser cladding a commersial availible Stellite 6 wire material is used. But in the case of laser fusing, the commersial availible Stellite 6 powder chemical composition is modified by adding 20% chromium powder. This modification is conduted by mixing [2].

3 Application cases and results

Two specific industrial problem statements will in this chapter be presented, discussed and solved by the means of laser fusing and laser cladding respectively.

3.1 Ice producing roller

A company making industrial machines for ice production. The ice machine, which weigh several ton and more than 3 meters high, consist of a roller that is cooled internally. On the outside, fresh water are sprayed onto the roller resulting in instantaneously freezing of the water. The ice is the scraped of and collected for intended use. See figure 6, left picture for illustration.

Sometimes the company is experiencing a problem with their product at customer production line. The problem is related to partly or full failure of protective coating, leading to severe corrosion attack and polluted ice product. It is fairly simple to exchange the roller, but it some extreme cases, for instance installation on a fishing trawler, the economic aspects can be unwanted.

The ice-producing roller needs the following:

- A wear and corrosion protective coating with full metallurgical bonding to the substrate.
- A method of applying the coating without distorting the rollers geometry – internal as well.
- A relatively cost effective coating solution.
- A repair friendly surface by the means of local weld repair.

The chosen solution consist of a traditional laser cladding approach, cladding wire Stellite 6 in 10 mm wide tracks keeping the production cost of the coating at a minimum. Figure 6, right pictures is showing the actual roller and the ongoing cladding process.

The result was a Stellite 6 coating with a full metallurgical bonding to the roller substrate in 1 mm effective coating thickness. As the laser heat input was relatively low, the roller was not distorted and the dilution rate minimized.

The coating allows local weld repair if needed, meaning uninstallation is not necessary in most cases if local damages is experienced.

Full and acute breakdown is not expected so service can be scheduled according to relevant conditions of usage.

Fig. 6 Left: example of large ice machine. [8] Right: setup of ice producing roller undergoing laser cladding. [1]

3.2 Scraped heat exchanger tube

A company making industrial machines for production of various food products. This machine, which can weigh several ton and volume wise, be very large, is basically a small chemical production unit. The heart of the machine is a tube, where typically, a crystalline process takes place – this involve heat exchange and scraping of the tube surface. As this tube treat food products, the cleaning of the machine is something that takes place regularly using more or less aggressive cleaning agents.

Usually these tubes are coated by a plating approach because the resulting surface will achieve high wear and corrosion resistance in a thin layer. A thin layer is desirable, as the thermal conductivity of the substrate should not be reduced. However due to different food products and cleaning agent – and maybe even inappropriate execution problems do occur. The protective coating has been seen to fail partly or fully, leading to severe corrosion attack and polluted food product excessing the permissible limits migration to the food.

It is not a simple maneuver to replace the heat exchanger tube, and full stop of production have unwanted economical aspects.

The heat exchanger tube needs the following:

- A wear and corrosion protective coating with full metallurgical bonding to the substrate.
- A relatively thin coating so the thermal conductivity is kept best possible
- A method of applying the coating leading to minimal post treatment.
- A long term durable coating which tolerate cleaning in place.

The chosen solution for a tube (2 m long and an internal diameter of 180 mm) consist of the two-step process, laser fusing. A self-fluxing Stellite 6 powder was modified by adding 20 % chromium and sprayed by atmospheric plasma spray process in a 0.5 mm coating thickness. Next step is to introduce the laser fusing process, by remelting the coating and forming a metallurgical bond to the substrate. The heat input is kept low and dilution rate is approximately 20%. After laser fusing a homogeneous layer thickness of 0.4 mm is achieved needing very little post treatment. Consult figure 7.

The coating allows "sandwich" structures or combination with other coating technologies, resulting in a yet unknown prolonged lifetime under production and cleaning in place.

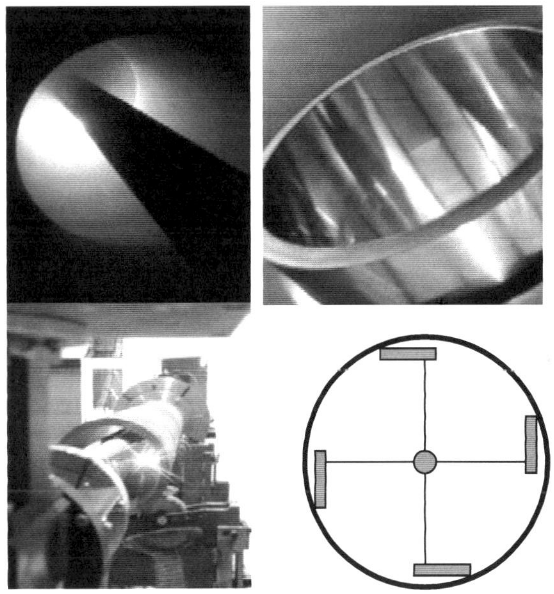

Fig. 7 Left: showing the atmospheric plasma spraying in the top (step one) and the laser fusing process in the bottom (step two). Right: upper picture is illustrating the final product with combined material properties and the lower picture is a schetch of crosssection scraping funtional tube. [1]

4 Summery and conclusions

Surface treatment of industrial components based on the laser cladding process is typically used for producing a wear- and corrosion resistant surface related to a certain application / component. The laser based surface treatment techniques are beneficial as they introduce a limited and well controlled heat input which makes it possible to produce thin coating thicknesses of high quality only introducing limited component distortion. This paper have given an example of succesfully industrial implemation of laser cladding, where a good technical solution combined with easy repair on site was higly valued for ice produsing rollers.

Also a technique where the surface material is preplaced by thermal spraying described. The thermally sprayed coating is characterised by being porous and only mechnically adherent to the surface. During fusing by a laser beam the coating is melted and hereby sealed / welded to the surface. In order to laser fuse a larger area the laser beam typically scans the surface in a track-overlapping pattern. Compared to the standard laser cladding process much wider tracks can be used for laser fusing, thereby achieving much higher productivity. Hereby, also larger coated areas becomes of industrial interest.

A widely used coating material is Stellite 6. The coating is typically plasma sprayed in a layer thickness of app. 0.5 mm. After laser fusing a homogeneous layer thickness of app. 0.4 mm is achieved. Overlapping track widths can be e.g.15-20 mm wide. The dilution with the base material should be low and typically in the range 20%. The laser fusing technique is also found suitable for multi layer techniques where several layers of the same – or different coating materials can be fused together. This gives the advantages of ; thicker layers can be produced, 'sandwich' layers with different metallurgy can be produced, a 'tailored' metallurgical layer composition can be performed, repairing of local areas of the surface can be carried out by plasma spraying locally followed by a local laser fusing procedure.

A relavent case study of scraped heat exchanger tubes has been presented, where non stop production and cleaning in place are important. The laser fusing process has shown promissing results through limited heat imput, high quality thin coatings, possible build up by sandwich laver and combination with other coating technologies.

5 Outlook

Even though laser cladding and laser fusing both are gaining ground in the industry, it could very well also result in a limiting factor.

The specific use of an application should always be considered carefully by understanding the circumstances. Then the problem should be solved by best coating material choise under thoughts of applying method.

However laser cladding and laser fusing offers exelent answers to many problems in etablished industry, where local upgrading or repair work are needed in high value components or where service life is needed prolonged. This is allready done today, but the much more optimization are waiting. This work are influenced by standards and how thing useally are done - and this need to be challenged. By doing this the argumentation, among other, lies in accelerated simulated laboratory testing for a number of relavant sectors.

6 Acknowledgements

This work is the result of an opened minded effort of numerous coworkers at FORCE Technology, division of Energy, Materials and Welding. Also results from the large project "Production in Denmark", finansed by The Danish Agency for Science, Technology and Innovation, has contributet to the work, by introdusing new technologies in the producing industry in Denmark.

7 Literature

[1] www.forcetechnology.com

[2] www.stellite.com

[3] ASTM Standard G 5

[4] ASTM Standard G 61

[5] ASTM Standard G 77

[6] ASTM Standard G 119

[7] Nielsen, P, Tribocorrosion properties of laser fused coatings, Technical University of Denmark,2008.

[8] www.buus.dk

[9] Mathiesen, T et. al, Corrosion Properties of Selected Coatings Produced by Atmospheric Plasma Spraying, HVOF and Laser Remelt Methods, Proceedings ITSC 2013.

[10] Nielsen, S, Surface Improvements Using Laser Technology, Proceedings SMT Conference 2015.

Simultaneously Spraying of Different Wires by Twin Wire Arc Spraying Using Adapted Air-Caps and Shroud Effect

W. Tillmann, M. Abdulgader and D. Bezerra, Dortmund / D

Surface engineering, particularly thermal spraying techniques, can combine the positive properties of several basis elements in a tailored coating or even give new not inherent properties like super-elasticity. Twin wire arc spraying (TWAS) stands out as a coating technique due to its low investment cost, low running cost and the simplicity as well as onsite use of the spraying apparatus. The feedstock in TWAS is introduced into the arc in form of two consumable arc electrodes "wires". The in-flight particles from the individual wires will normally not mix properly with each other in the way that a desired composition can be produced. Providing a sufficient probability for the in-flight particles from the different wires to collide, improves the mixing of compounds and thus enhances the controlling over the desired coating composition. In this work different air-cap attachments as well as shroud designs are studied to reveal their effect on the microstructure of coatings produced by spraying different materials simultaneously. Also the effect of polarity was revealed regarding the microstructure and the distribution of the splats of the individual wires. The shroud principle shows a better compound mixing and denser microstructure compared to the modified secondary atomization air-caps.

1. Introduction

Developing new coating systems is a widespread topic spectrum in thermal spraying technique with focus on wear and corrosion protection, as well as thermal insulation. In classical tribo-systems, increasing the hardness and controlling the size and arrangement of the porosity of an applied coating will minimize the adhesive and abrasive wear. This type of coating material and arrangement does not work well for all wear mechanisms. Cavitation erosion is a very complex wear mechanism that occurs mainly in many hydro dynamic systems. This kind of wear causes a severe material pitting and surface breakdown. The damage is caused by the so-called micro-jets that arise when the small cavitation bubbles collapse [1]. The generated pressure peaks damage the surface of the components and parts by removing the material adjacent to the collapsed bubbles. In the shipbuilding industry, the cavitation erosion causes damages to the propellers and ship attachments. The resulting high costs due to repairs and total failure leads to great financial losses for the shipping industry. Similar damages are also observed in closed hydro-dynamic systems like in valve seats and pipe elbows [2, 3]. The economic interest of a solution to this problem is accordingly high.

Hard coatings offer, compared to common bronzes that are used in the casting of ship propellers, hardly a benefit rather than increasing the production cost. In contrast, materials with pseudo-elastic properties like, e.g., NiTi offer a significantly higher cavitation resistance. Due to the pseudo elasticity, these materials are in the position to bounce the high pressure peaks of the micro-jets. To achieve the pseudo elasticity, a restricted alloying between the compounds is required. Therefore, producing a coating out of these compounds is a very complicated task and needs to be studied thoroughly.

In order to enable the TWAS-process to produce coatings with pseudo elastic behavior, a complete knowledge about the droplet formation and trajectory is required. Controlling and predicting the particle's in-flight behavior is very important to produce the desired coating.

In TWAS-process, one wire is connected as the anode (+) and the other as the cathode (-). When the wires are fed together an arc ignites at the shortest gap between the tips of the approaching wires. The particle size, size distribution and trajectory are directly affected by the feedstock material, adjusted voltage and current as well as the flow conditions of the atomization gas. A jet of spraying particles is formed through atomization of a molten feedstock bath, which is caused by the impingement of fast and continuous flowing air jet. **Fig. 1** shows a schematic overview of the particle formation and trajectory in TWAS-process.

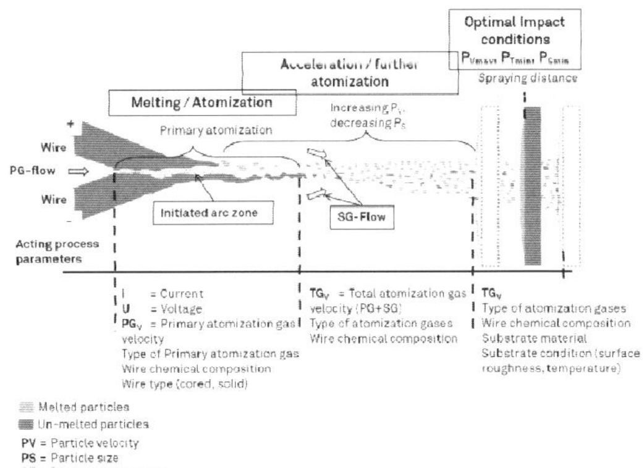

Fig. 1. Schematic view of the twin wire arc spraying (TWAS) process showing the in-flight particle initiation and trajectory.

In addition, particle size and size distributions were found to be different for the particles originating from the anode side and the cathode side as shown in **Fig. 2**. The dissimilarity was observed by simultaneously spraying similar and different feedstock materials [4-7]. Pourmousa et al. [8] found that the particles from the anode are almost twice as large in diameter as those from the cathode. The variation in particle size is due to the asymmetrical melting behavior of the anode and cathode wires [4-9]. The arc is attached at the anode over a larger area than the cathode. This heats up the anode over a bigger surface and leads to the formation of large particles from the anode side. At the cathode side, the arc is localized and more constricted, which causes the formation of small particles [4-7].

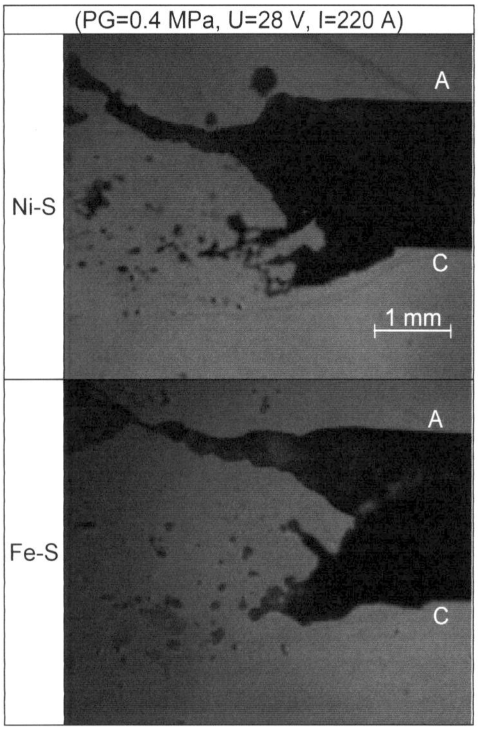

Fig. 2. High speed camera images revealing the asymmetry of the melting behavior in TWAS-process between anode (A) and cathode (C) for nickel solid wires (Ni-S) and iron solid wires (Fe-S).

Another feature in TWAS-process is the particle crossover effect, which means that the droplet originated from the anode side is found at the cathode side and vice versa as shown in **Fig. 3**. This behavior was observed in case of massive wires [5, 9-11] as well as in cored wires [10-11]. Tillmann et al. [11] reported that the adhesion layers between substrate and coating as well as between the over runs depend on the gun movement direction.
Nevertheless, TWAS-process provides an economic alternative for the production of Ti-Al intermetallic composites [12] or the creation of intermetallic alloys by the simultaneously use of different wire materials. Chen et al. [13] proclaim in their work, that almost no alloy reactions between the materials of anode and cathode are found in these processes, and thus the coatings are just mechanically mixed with the components of individual wires. Varacalle et al. [14] sprayed pseudo-alloy out of 85/15 and 70/30 Zn/Al wires for corrosion applications. Chen et al. [15] used Al und Fe to produce Iron-aluminide for the use in high temperature environments. Furthermore, other composite coatings were successfully prepared by the combination of Ti/Al wires [13, 16], carbon steel/Cu wires [5, 17], Zn/Al wires [18] stainless steel/Al wires [19]. To obtain better droplet dispersion, Watanabe et al. and Pasandideh-Fard et al. [16, 20] suggested to pole the wires regarding their melting point. The wire with higher melting temperature should be connected to the anode and the one with lower melting temperature to the cathode.

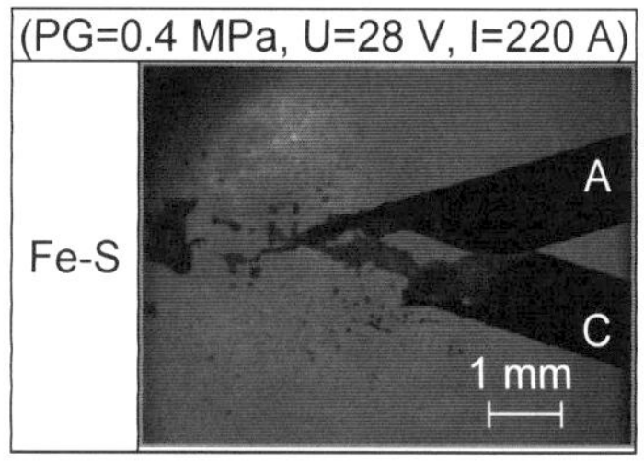

Fig. 3. High speed camera images revealing the crossover effect in TWAS-process.

Developing a cost-effective and efficient coating with pseudo-elastic properties using TWAS-process is therefore a big challenge and needs a deep understanding of the above described process phenomena. In order to understand and thus to control the coating build-up by the simultaneously spraying of different materials using TWAS-process, nickel and iron are used as feedstock materials. Different air-cap configuration were developed and tested to increase the collision possibility of the in-flight particles and thus the alloying the different materials of the sprayed wires.

1 Experimental work

2.1 Materials

Smart Arc 350 PPG (Oerlikon Metco) was used to investigate the coating microstructure by simultaneously spraying of wires with two different materials, each of which was connected onetime as anode and other time as cathode. A single overrun was deposited by each experimental setup. To study the crossover effect, a layer was prepared by a vertical movement of the gun. The wires used for this study are iron solid wire (Sprasteel11, Co. Oerlikon Metco) with the following composition in wt.% (Fe Bal., C 0.07%, Si 0.96%, Mn 1.63%) and nickel solid wire 99.0 % Ni (Metco Nickel, Co. Oerlikon Metco). Steel (1.0503) in plate form (70x50x10mm) was used as substrate. All specimens were corundum blasted to roughen the surface. The blasting of specimens was performed with alumina EKF-14 (1700-1180µm grain size). The applied gas pressure was 0.45 MPa. The blasting gun was positioned at a distance of 100 mm and an angle of 75° to the specimens. The blasted parts were cleaned with ultrasonic in an acetone bath.

2.2 Experimental Setups

The coatings were produced by moving the spray gun in the cathode direction as well as in vertical direction (in respect to the intersection between the anode and the cathode) as shown in Fig. 4. The layers were deposited by only one overrun for the different setups. The arc voltage and arc current were set constant at 28 V and 180 A respectively. The primary atomization gas was adjusted at 0.4 and 0.8 MPa and the secondary atomization gas at 0.2 and 0.4 MPa. All spraying experiments were carried out at a stand-off distance of 120 mm.

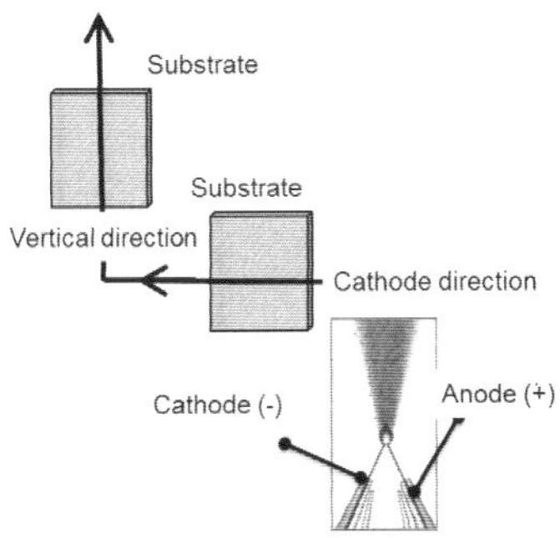

Fig. 4. Movement pattern of the spraying gun.

Our study is focused only on the charging effect of the wires, and the air-cap arrangement on the coating microstructure and the distribution of the compounds. Scanning Electron Microscopy (SEM) in form of map and line scan are used to determine the compound distribution at the different setups. The vertical moving direction was performed to investigate the crossover effect in the spraying plume by using simultaneously two different wires. **Fig. 5** shows the different air-caps and shroud attachments, which are used to perform the spraying experiments.

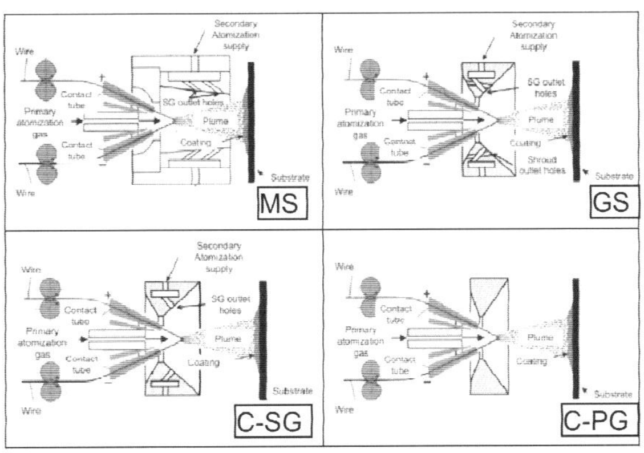

Fig. 5. Schematic overview about the air-caps and the developed shroud attachments.

To improve probability of compound mixing in TWAS-process, two types of shrouding were developed, gas shroud (GS) and massive shroud (MS). The obtained coatings were compared with the conventional primary atomization gas air-cap (C-PG) and the conventional secondary atomization gas air-cap (C-SG). The secondary gas outlet holes in GS are arranged in two arrows as can be seen in **Fig. 5** (GS) the lower one is pointing to the wires intersection point with the intention to enhance the mixing probability directly by particle initiation. The second row uses the primary atomization gas supply and is arranged to cover the spraying plume and limiting the interaction with the surrounding atmosphere. The spraying plume in MS is covered with a solid air-cap attachment, which ends directly at the prepared substrate surface "**Fig. 5** MS". The outlet holes of the secondary atomization gas are arranged to the main spraying stream with 30° and distributed along MS-attachment.

2 Results

Spraying in only one direction was performed in order to understand the coating buildup by the simultaneous spraying of wires with different materials. However, the air-cap arrangements play a decisive role in the characteristics of obtained lamellas as well as the distribution of the compounds in the produced overruns. Even though the difference in melting point between nickel and iron is not high, the lamellas structure is different for a changed poling.

2.1 Effect of the polarity reversal

At left side of the following figures appear the Fe as dark and the Ni as bright lamellas. **Fig. 6** shows the effect of polarity reversal on the distribution of the layer compounds. The influence of the secondary atomization gas is excluded in this figure by using C-PG "see **Fig. 5**". Fe-lamellas are longer and wider in case of Iron as anode (+Fe). The variation in the size of the lamellas can be explained by the dependency of the sizes of the initiated particles on the polarity. This confirms the statement of Pourmousa et al. [8] that the particles initiated from the anode side are almost twice as large in diameter as those from the cathode.

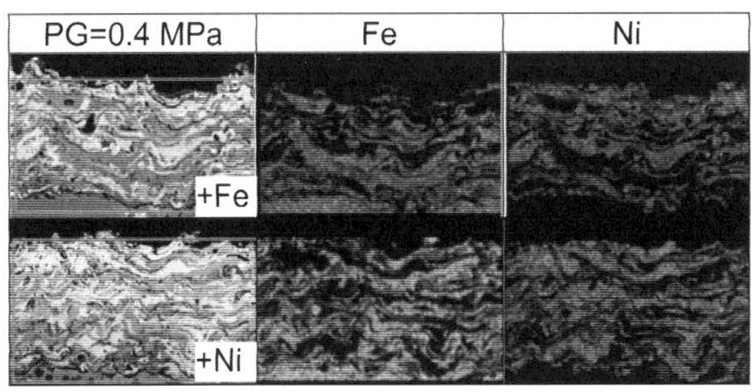

Fig. 6. Smart-maps showing the lamellas structure by changed polarity using conventional air-cap for only primary atomization gas outlet.

The crossover effect of the initiated particles was clearly observed by revealing the microstructure of the vertical sprayed layers. This phenomenon is directly related to the fact that the wires are fed with an angle to each other. The distribution of the layer compounds is shown in **Fig. 7**. Bigger portion of Iron (dark) layers can be recognized in the image in the column "left-side" for iron as anode "+Fe". The designation right-side (RS) and left-side (LS) is given with respect to the vertical direction of the gun movement see Fig. 4. Therefore, RS is the side of the substrate surface facing the anode side.

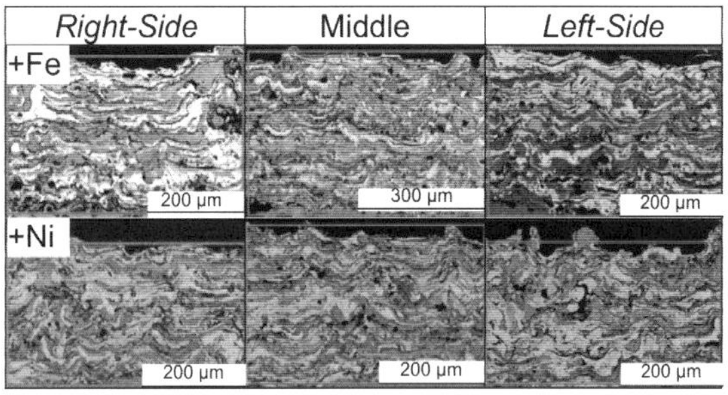

Fig. 7. Crossover effect and the distribution of the layer compounds by vertical spraying.

At the right-side, the portion of bright lamellas is larger and thus the Ni content. The size of these lamellas in general is smaller and thinner. The reason could be that the initiated Ni-particles are overall smaller than those of Fe. In addition, the viscosity and surface tension of the overheated Ni-melt is lower than the Fe-melt. That means, in turn, that the impacting of generally smaller Ni-particles will spread easier than Fe-particles. Therefore. it leads in general to a thinner and shorter Ni-Lamellas.

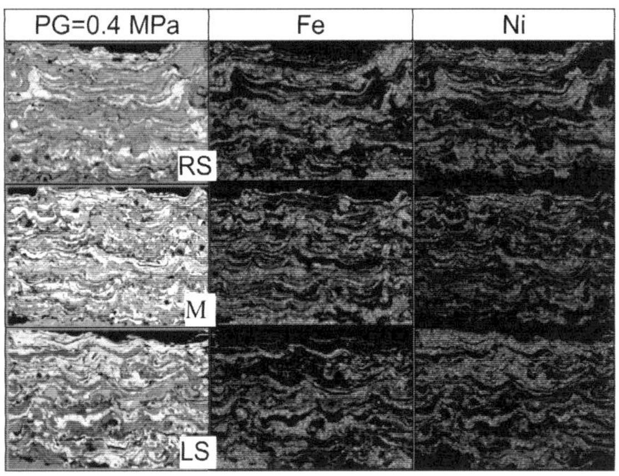

Fig. 8. Compounds distribution in vertical sprayed layers with polarity Fe-anode.

Fig. 8 shows the distribution of Ni and Fe-lamellas in vertically sprayed layer with Fe as anode. The amount and size of the Fe-lamellas are different, which reveals the crossover effect. At the middle part of the sprayed layer the distribution is more or less homogenous.

2.2 Effect of the secondary atomization gas

The flow of secondary atomization gas intersect the spraying plume in general at a certain standoff distance, which is directly related to the tilting angle of the SG-outlet holes. The intersection in most cases is directly behind the wires intersection point, where the in-flight particles are already initiated. The main purpose of the SG is to decrease the divergence of the spraying plume, enhance the secondary particle atomization and thus to increase particle velocity. The lamellas are in general finer and shorter. The compound distribution is more homogenous as can be seen in **Fig. 9**.

Increasing the SG-pressure leads to more fine lamellas. Particle temperature and the exerted aerodynamic forces control the further atomization of the initiated particles. Besides the melting temperature, surface tension and viscosity of the overheated material-melt should be also taken in the consideration by assigning the polarity of the materials to be sprayed. Materials with higher viscosity and surface tension as well as higher melting temperature should be connected as anode. If the melting temperature is about the same, then the material with the lowest viscosity and surface tension should be connected as cathode.

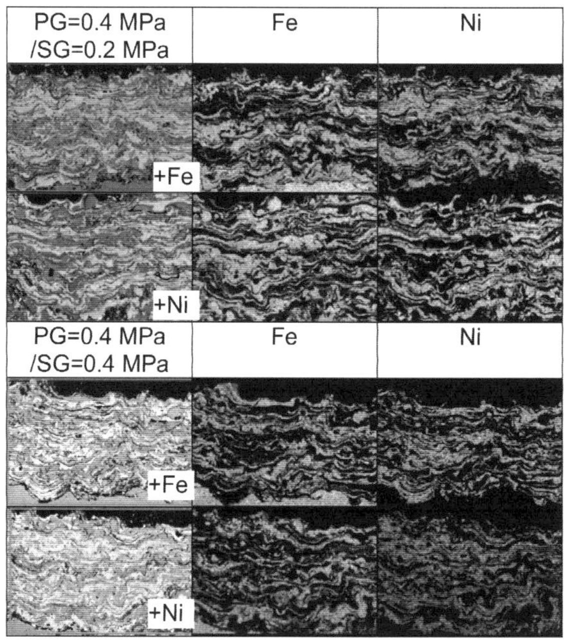

Fig. 9. Effect of the secondary atomization gas pressure (SG) on the obtained lamellas.

In case of the conventional SG air-cap the outlet holes are pointing towards the spraying plume from upper and lower side. The idea behind that is to compensate the deflection of the airflow caused by the intersecting wires. That means, the sides of spraying plume are not affected by the SG gas jet. This is obvious by revealing the layers produced by the vertical movement. The lamella distribution at the middle of the sprayed layer is almost homogenous and shows equal portion of the compounds as seen in **Fig. 10**.

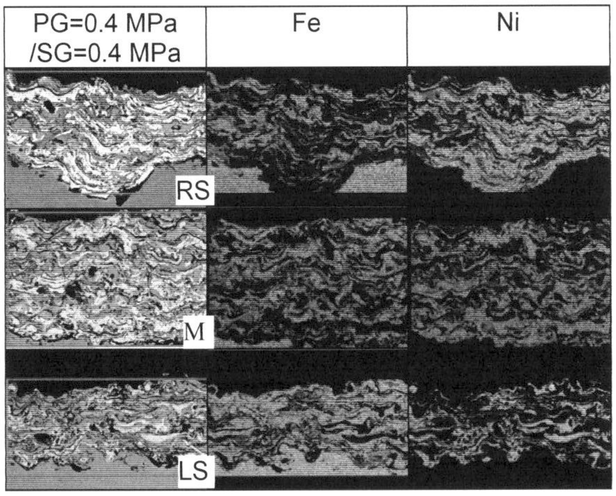

Fig. 10. Crossover effect and lamellas distribution by increased secondary atomization gas pressure using conventional air-cap

2.3 Enhancing the lamella distribution

Two different shroud principles are developed in this work to improve the lamella distribution. The design and work principle of the developed shrouds are described in chapter 2.2. The arrangement of the SG outlet holes should bring the in-flight particles to collide with each other. The gas shroud (GS) is designed in a way that should allow the collision directly at the initiation phase of the in-flight particles. However, the massive shroud (MS) is built in a way that the out streaming SG should cause a deflection in the trajectory of the in-flight particles. **Fig. 11** shows a comparison between the different SG air-cap attachments used in this study. In case of C-SG the layer arrangement is more or less a defined lamellar structure. The lamellas are bigger compared to those produced by GS and MS.

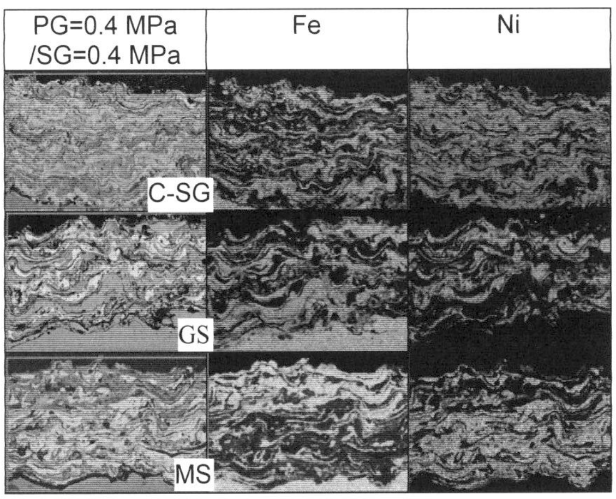

Fig. 11. SG air-cap design effect on the lamella distribution by Fe as anode

In case of GS, the Fe lamellas are thicker and shorter. The Ni particles are spread some how not in a nice lamellas form "Fig. 11". The mixing of particles is probably not good enough. This can be explained by the dominating higher aerodynamic forces cased by PG-flow.

The SG outlet holes in MS are arranged in such a way that allows the SG-flow to interact with the spraying plume at all location along the spraying distance. Therefore, the MS image reveals a better mixing, as expected, due to the enhanced collision probability. Increasing the PG to 0.8 MPa improves the mixing of the compounds as can be seen in Fig. 12

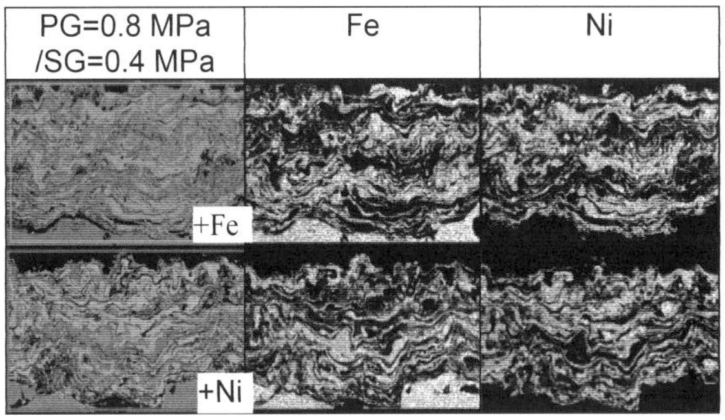

Fig. 12. Effect of the increasing the PG on the mixing of the compounds.

The enhanced mixing may lead by compounds with higher salability for each other to initiation of the desired alloy. Therefore, supplementary investigations are planed to study the effect of further increasing both PG and SG as well as to improve the design of MS principle on the mixing probability.

3 Conclusion

Different air-cap configuration were developed and tested to increase the collision probability of the in-flight particles and thus to enhance the alloying of the sprayed wires. To improve the air-cap efficiency a correlation was done between the gun movement, the polarity reversal and the adjustable spray parameter (particularly the combination of primary and secondary atomization gas). Furthermore the crossover effect and the right polarity with respect to melting point, viscosity and surface tension together should be taken into consideration by determining a suitable coating pattern.

The lamella distribution of the sprayed layers shows a relative dependency on the polarity reversal of the sprayed wires. Crossover-effect is a main feature in the TWAS-process, which effects the lamellas distribution and controls the microstructure of the obtained layers. Chen et al. stated in [13], that almost no alloy reactions between the materials of anode and cathode are found by simultaneously spraying different materials and thus the layers are built up just by the mechanically mixing. This has its validity only in case of using conventional air-caps in general, and in particular the PG air-caps. Even though the in-flight time of the spraying particles is very short, using MS increases the collision probability and allows the alloying reaction to take place. Increasing the collision probability and preserving the in-flight particles from the interaction with the surrounding atmosphere are the main goals by the development of MS used in this study.

The use and further improvement in MS principle is essential in the development of a cost-effective and efficient coating with a certain alloying limitations. The restrictions in alloying could be required to give the layers a specific property like the pseudo-elasticity, which is needed for example in case of cavitation protection. In order to control the coating build-up and material alloying, there is a substantial need for further research in the field of simultaneous spraying of different materials using TWAS-process.

4 Acknowledgment

The authors gratefully acknowledge the financial support of The German Federation of Industrial Research Associations "AIF" (IGF-Nr.: 18788 N/1)

5 Literature

[1] Kling, C. L. and F. G. Hammitt: A Photographic Study of Spark-Induced Cavitation Bubble Col-lapse, Journal of Basic Engineering 94(1972), Issue 4, pp. 825/32.

[2] Berger, J. Pohl, M. and L. Sitnik: Einfluß des Werkstoffes und seines Oberflächenzustandes auf die Kavitationsbeständigkeit, ö + p, Ölhydraulik + Pneumatik 27 (1983), pp. 714/18 in German.

[3] Pohl, M. Kast, E. Burchard, W.-G. Kleinbreuer, W. and W. Backé: Untersuchungen zur Werk-stoffzerstörung durch Kavitation in ölhydraulischen Systemen, Beitr. elektronenmikr. Direk-tabb. Oberfl. 14 (1981), S. 203/12. (German)

[4] Hussary, N.A. and J.V.R. Heberlein: Effect of system parameters on metal breakup and particle formation in the wire arc spray process, Journal of Thermal Spraying Technology 16, (2007), Issue 1, pp. 1/13.

[5] Liao, H.L. Zhu, Y.L. Bolot, R. Coddet, C. and S.N. Ma: Size distribution of particles from individual wires and the effects of nozzle geometry in twin wire arc spraying, Surface and Coating Technology 200 (2005), pp. 2123/30.

[6] Newbery, A.P. Rayment, T. and P.S. Grant: A particle image velocimetry investigation of in-flight and deposition behaviour of steel droplets during electric arc spray forming, Material Science Engineering A 383 (2004), pp. 137/45.

[7] Mohanty, P.S. Allor, R. Lechowicz, P. Parker, R. and J. Craig: Particle temperature and velocity characterization in spray tooling process by thermal imaging technique, Thermal Spray 2003, (ED) C. Moreau and B. Marple, ASM International, Ohio (2003), pp. 1183/90.

[8] Pourmousa, A. Mostaghimi, J. Abdini, A. and S. Chandra: Particle size distribution in a wire-arc spraying system, Journal of Thermal Spraying Technology 14 (2005), Issue 4, pp. 502/10

[9] Zhua, Y.L. Liaob, H.L. Coddet, C. and B.S. Xu: Characterization via image analysis of cross-over trajectories and inhomogeneity in twin wire arc spraying, Surface and Coating Technology 162 (2003), pp. 301/08.

[10] Tillmann, W. Vogli, E. and M. Abdulgader: Partikelinteraktion (Cross Over) in der Flugphase beim Lichtbogenspritzen von Fülldrähten „Particle interaction (crossover) during the in-flight phase using cored wires", SFB708, 2. Öffentliches Kolloquium Dortmund (2008), pp. 43/51 in German.

[11] Tillmann, W. Vogli, E. and M. Abdulgader: The correlation between the coating quality and the moving direction of the twin wire arc spraying gun, Journal of Thermal Spraying Technology 19 (2009), Issue 1-2, pp, 409/21.

[12] Liu, S.G. Wu, J.M. and S.C. Zhang: High temperature erosion properties of arc-sprayed coatings using various cored wires containing Ti-Al intermetallics, Wear (2007), Issue 5-6, pp. 555/61.

[13] Chen, Y. Liang, X. and S. Wei: Heat treatment induced intermetallic phase transition of arc-sprayed coating prepared by the wires combination of aluminum-cathode and steel-anode, Applied Surface Science 19 (2009), pp. 8299/304.

[14] Varacalle, D.J.JR. Zeek, D. P. and V. Zanchuck: Experimental Studies of Twin-Wire Electric Arc Sprayed Zinc/Aluminum Alloy Coatings, Journal of Thermal Spray Technology 4 (1998), pp. 513/20.

[15] Chen, Y.X. Xu, B.S. and Y. Liu: Structure and sliding wear behavior of 321 stainless steel/Al composite coating deposited by high velocity arc spraying technique, Transactions of Nonferrous Metals Society of China (English Edition) 3 (2008), pp. 603/09.

[16] Watanabe, T. Sato, T. and A. Nezu: Electrode phenomena investigation of wire arc spraying for preparation of Ti-Al intermetallic compounds, Thin Solid Films 1-2 (2002), pp. 98/103.

[17] Zhu, Y. L. Liao, H.L. and C. Coddet: Characterization via image analysis of cross-over trajectories and inhomogeneity in twin wire arc spraying, Surface and Coatings Technology 2-3 (2003), pp. 301/08.

[18] Toparli, M. Celik, E. and I. Birlik: Tribological Properties of Electric Arc–Sprayed CuSn Coating for Bearing Elements, Tribology Transactions 3 (2009), pp. 389/94.

[19] Wang, R. Song, D. and W. Liu: Effect of arc spraying power on the microstructure and mechanical properties of Zn–Al coating deposited onto carbon fiber reinforced epoxy composites, Applied Surface Science 1 (2010), pp. 203/09.

[20] Pasandideh-Fard, M. Pershin, V. and S. Chandra: Splat shapes in a thermal spray coating process: Simulations and experiments, Journal of Thermal Spray Technology 2 (2002), pp. 206/17.

Nanostructure Development of Low Pressure Plasma Sprayed Coatings with Nano-Agglomerated YSZ Feedstock

Yang Gao, Deming Yang, Dalian / CHINA

In this study a low pressure plasma spray (LPPS) was utilized to deposit nano-agglomerated feedstock of 8%Y_2O_3-stabilized ZrO_2 (YSZ). The coating having a novel fully nano-equiaxed structure was successfully fabricated in the condition of low plasma current of 300A and with plasma gas flow of 20Ar-30He-5H2 SPLM. The microstructure of the coatings ordinal change from a fully nano-equiaxed-structure to a mixing grain structure and finally to form a relative large grain structured coating with enhancing the plasma currents. Experimental result suggested that the nano-scale equiaxed structure in the coating was derived from the original agglomerated nano-particles they had undergone melting while inside the nozzle of the plasma torch and were subsequently solidified or sintered in the coating. Experiment also indicated that fully nanostructured coating had lower hardness compared with the large grain structured coating. The hardness and elastic modulus of the coating associated with the microstructure. In addition the thermal shock resistance test of the coatings on the substrate of Inconel 718 alloys was investigated.

1 Introduction

Thermal sprayed yttria-stabilized-zirconia coating has demonstrated outstanding performance and it has been applied in the varied application; particularly the coating was successful applied in the hot sections of aerospace and land-based gas turbines against the high temperature environment [1]. For the thermal barrier coating the low thermal conductivity and thermal shock resistance are most important factors and which was determined by the microstructure of the coating. It is well known that the microstructure of the thermal sprayed coating depends on the melting degrees of the feedstock powder and subsequent solidification on the substrate. The thermal spray process is intrinsically associated with the melting of particles, without some particle melting it is extremely difficult to produce coatings, particularly with ceramic materials, some degree of melting is necessary to achieve a sufficient level of particle adhesion and cohesion [2]. Plasma spraying is considered an alternative method for deposition of ceramic coating due to its high temperature of the plasma jet.

Recently, the nanostructured YSZ thermal barrier coating has shown particularly attention due to their desirable combination of properties such as high durability and lower thermal conductivity [3-6]. Need to be pointed out this nanostructured coating is not full nanostructure instead is a so-called bimodal structured coating which was currently fabricated by controlling the amount of the melting agglomerated feedstock result in semi-molten particles embedded in the coating microstructure [3-6]. The bimodal structured coating was mainly fabricated by means of the APS. Despite some initial positive research results with regard to the nanostructured YSZ coatings, there are still fundamental questions to be answered on the nanostructured YSZ coatings. These questions are related to the real role of the nanostructure in the bimodal structured coating. Lima et al. [4] have investigated the variation of the bimodal structured YSZ coating at 1400 ℃ for 1 to 20 hours, they found that the coarse porosity levels in the coating rapid increased in the heat process, for which the reasons was attributed the shrinking of the nanostructure during the heat treatment. Although thermal barrier coating with bimodal structure has showed a low thermal conductivity and high thermal shock resistance, however the reasons of this good performance are not clear yet, namely, what is the role of nanostructure in the coating? Is it caused by the nanostructure or by porosity levels in the coating?

Many literatures have been documented that formation of the bimodal structured coating is relative to the different melting degrees of the nano-agglomerated powder in the plasma jets which lead to some agglomerated powder were fully melted and others were partially melted or unmelted. The bimodal structured coating was produced in the APS process, which is related to the fluctuation of plasma jet during the ambient air entangled into the plasma plume. In contrast APS, the low pressure plasma spray is operated at low pressure of 5000-8000 Pa, which was much decreased the gas entangling and leads to small fluctuation of the plasma jet. Moreover because low pressure in the surroundings which also leads to the plasma jet to achieve a supersonic speed and it was beneficial to accelerate small sized particles to form a coating. Such reason the different microstructured coating may be expected by LPPS nano-agglomerated feedstock. With this knowledge, the goal was to attempt to create the nanostructured coating with expect be different from the bimodal structured in the microstructure of the coating. In this study the LPPS methods were adopted to spray the nanostructured yttria-stabilized-zirconia (YSZ) feedstock, and the microstructure of the coating was investigated.

2 Experiment Methods

2.1 Feedstock Materials and LPPS

The nanotructured agglomerated feedstock powder of ZrO_2-(7-8) wt. %Y_2O_3 was supplied by Beijing Sun Spraying Scientific & Technologies Development Co.,

Ltd (Beijing, China). **Fig.1a** shows the individual nanoparticles in the agglomerated feedstock that the size distribution ranging from approximately 70 to 200 nm. **Fig.1b** shows the morphology of the agglomerated feedstock powder observed by scanning electron microscope (SEM) image that the size distribution ranging from approximately 40 to 70μm and the apparent density is 1.2 g/cm^3, this sized powder can inject to the plasma plume by using a regular powder feeder.

Fig. 1 The morphology of agglomerates feedstock, (a) the original nano-particles, and (b) agglomerated feedstock.

The equipment of LPPS used in this experiment was developed in-house at the Thermal Spray Centre of the Dalian Maritime University [7] with the chamber size of Φ1.2 x 4 m. The pumping system enables the chamber pressure to vary during operation between 1000 and 10000 Pa. The ambient pressure was controlled at ranges of 5000-6000 Pa in this experiment. The goal of this study is to deposit nanostructured coating, the distinctive internal injection plasma torch was engineered and operated at plasma currents of 300-700 A were adopted. The configuration of the internal injection plasma torch was depicted in the previous report [7,16]. The compositions of plasma working gases have significant influence on the heating and the accelerating behavior of the feedstock powders [8-9]. Considering the characteristics of heating and accelerating for the feedstock powders, the plasma gases with mass flows of 20Ar-30He-5H_2 SLPM was selected for the deposition of the nano-YSZ coatings and deposition parameters are listed in **Table 1**. Low carbon steels strips, each with width of 40 mm and a thickness of 5 mm were used as substrates for the LPPS. Sand blasting was conducted before the YSZ coating. To easily remove the YSZ coatings from the substrates, a number of the YSZ coating was directly deposited onto the low carbon steels strip without any undercoat.

Table 1. Parameters for LPPS

Item	LPPS
Chamberpressure	5000-8000Pa
Primary gas Ar	20slpm
Secondary gas H_2	5slpm
Secondary gas He	30slpm
Carrier gas Ar	3 slpm
Arc current (A)	300-700 A
Standoff (mm)	350
Powder feed rate	10-12 g/min

2.2 Characterization of the Microstructures

To demonstrate microstructure of the coating, the thermal etching method was performed in a box furnace at 1200℃ for 5 min. The coating/substrate system was immersed in hydrochloric acid (HCl) to dissolve the steel substrate, which led to freestanding coatings that were subsequently used for the thermal etching and observation. The microstructure of the coating was identified using a Carl Zeiss SUPRA™ 55 field emission scanning electron microscope (FESEM). The phase structures of the nanostructured YSZ feedstock and the as-sprayed coatings were examined by X-ray diffraction (XRD) with a Cu target operating at 40 kV and 40 mA. The Knoop hardness of the coatings on cross-section was measured by a MH-6 Vickers indenter (Shanghai, China) at a load of 300 g and a dwell time of 15s. Based on the measurement of elastic recovery of the in-surface dimensions of Knoop indentations [10], the Knoop hardness (H_k) and elastic modulus (E) for brittle materials can be calculated as following equations,

$$\mathrm{Hk} = 14.22\frac{P}{a^2} \qquad (1)$$

$$E = \frac{(-\alpha H_K)}{(b'/a' - b/a)} \qquad (2)$$

Where the P is the applied load (kg), a is the long diagonals of indenter (mm), *HK* is the Knoop microhardness, b'/a' is the ratio of the short to long indention diagonals after elastic recovery, b/a is the ratio of the known Knoop indenter dimensions or geometry (1/7.11), and α is a constant having a value of 0.45.

3. Results and Discussions

3.1 Microstructure of the Coating

Fig.2 shows the cross-sectional microstructures of the coatings deposited via LPPS from arc currents of 300 to 700 A with a step of 100 A, the nanosized grain could be identified by FESEM. **Fig.2(a)** shows the cross-sectional microstructures of the coatings deposited with 300 A, a novel fully nano-equiaxed coating was fabricated, the microstructure similar to the nanostructured bulk materials made of compressed and sintered of the nano-particles [5-6].

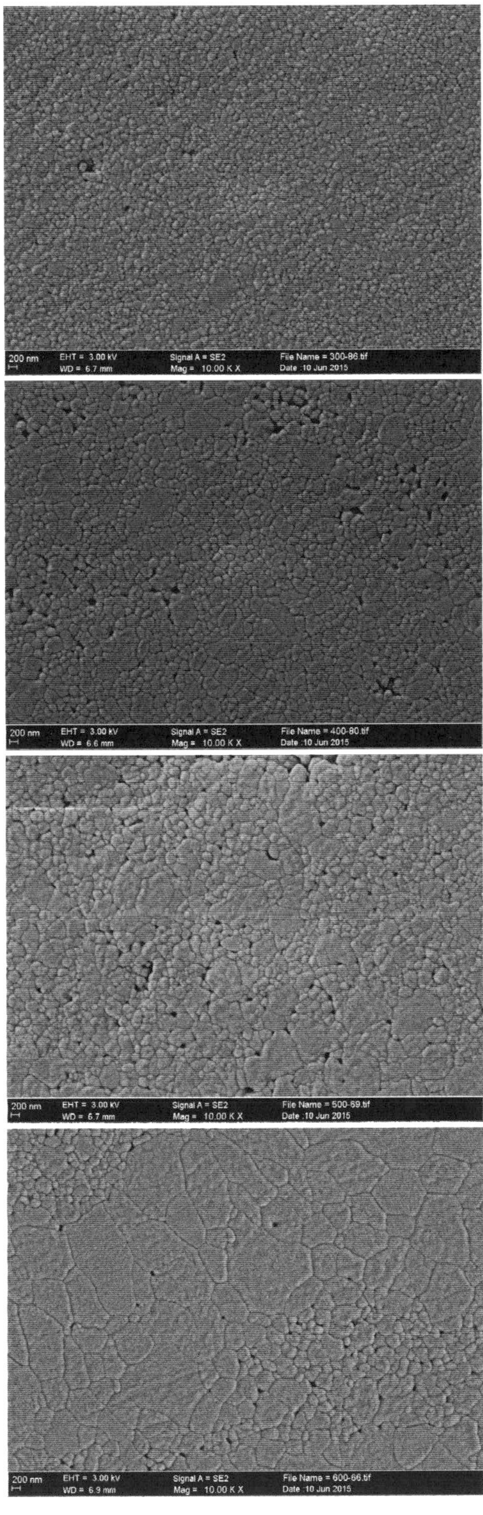

300A

400A

500A

600A

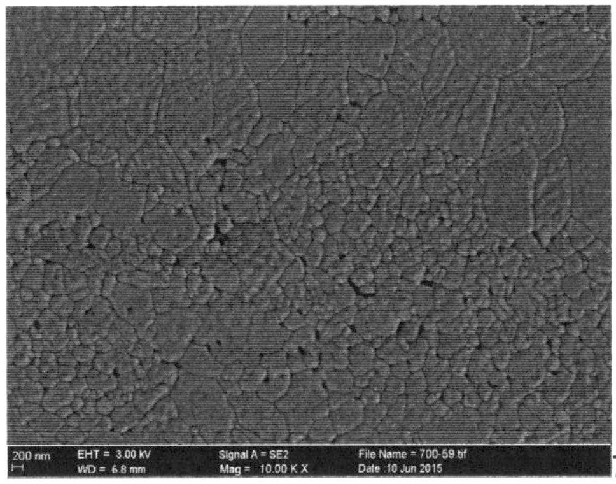

700A

Fig.2 SEM images of the microstructure of the coatings deposited by LPPS with 20Ar-30He-5H_2 working gases and at plasma currents of 300-700A.

From FESEM of the coating shows that the sizes of nano-grains in the coatings are approximately 200 nm, which is approximate to original nano-particles in the agglomerates. However, the configuration and density between the nano-feedstock and coating are difference, which suggests that the original nano-particles had undergone molten or semi-molten in the plasma jet and subsequently solidified or sintered in the coating. In this experiment the unique fully nano-equiaxed structureis only obtained by using LPPS nano-agglomerated powders and depositing at lower plasma currents of 300A. The coating with fully nano-equiaxed structure is distinct different from so-called nanostructured coating with bimodal structure that deposited by APS and the conventional APS coating with lamellar microstructure, rich of pores, splat boundaries and microcracks deposited using a fused-crushed feedstock [13-15].

The microstructure of the coatings varies with the increase of the plasma currents, i.e. a mixing grain structure that consists of a several hundred nano-scale grains and micrometers level grains appeared in the coatings with 400 A, and the amount of micrometer grains increase with increasing the plasma current. The microstructure of the coatings with 500 A, 600 A consist of larger grains with sized micrometers level. While the size of micrometers level grains are principal structures for the coatings with 700 A. In the conventional plasma spray (APS), the splat-quenched structure in the coating usually was formed by the large size particles, they partially or fully melted in the plasma plume and then quenched in the coating with typically several hundreds of microns long. In this experiment, however formation of micrometers level grains in the coating was formed when the plasma current exceed 400 A that the nano-particles were partially or fully melted and they underwent solidification or sintering in the coating. The most important feature of these coatings is the absence of horizontal, strength-degrading "splats" boundaries that always present in conventional plasma-sprayed YSZ coating.

Previous work has demonstrated that the bimodal structure is a visible feature for the nanostructured coating produced by APS that was attributed to the different malting degrees of the agglomerated feedstock during plasma spraying. The primary difference between APS and LPPS is the density of the plasma jets. APS has higher density of the plasma jets, while the LPPS has low density and temperature gradient, but high velocity of the plasma jets, which may results slow heating and high accelerating for the agglomerated feedstock. Such feature leads to the probability of APS the nano-agglomerates melting and collision being higher than that LPPS, especially when increasing the plasma current. In the plasma spraying, the properties of the plasma-forming gas largely control the characteristics of the plasma jets and momentum, heat, and mass transfer to the particles. Helium and hydrogen as second plasma gases (argon always primary gas) they increase the energy capacity of plasma jet and heat transfer to the particles. When the temperature of plasma jets below 10000 K, increases hydrogen rather than helium is more effective because of the increase of the enthalpy and thermal conductivity to the plasma gas [17]. Increasing the content of helium gas rather than hydrogen enables to achieve the higher velocity of plasma jets [18]. Thus, plasma gases of 20Ae-30He-5H_2 SLPM have high velocity and enthalpy for the plasma jets; therefore it is beneficial to create a denser coating. Our previous experiment has demonstrated that the agglomerated feedstock was primarily melted inside the nozzle for the LPPS with internal injection plasma torch [7]. When the agglomerates feedstock was injected inside the nozzle with lower plasma currents, the agglomerates would partially melt and deposited on the substrate to form a fully nano-equiaxed structured coating. The melting degree of the agglomerated feedstock depends on the plasma current, they would be fully and semi-melt melted with increase plasma current over 400A, as result to form a mixing grain structured coating. By adjusting arc current, the relative proportion of partially melted regions and the percentage of nano-sized grains in the coating can be controlled.

Fig.3 shows the XRD patterns of the powder and part of the coatings scanned at the range of 2θ=20-70°and 2θ=72-76°. The XRD patterns at the range of 2θ=20-70° is tended to detect monoclinic(m) and tetragonal(t) phase in the YSZ coating, while the XRD patterns at the range of 2θ=72-76° can detect the intensity of (004) and (220) for tetragonal phase. According to XRD patterns, tetragonal (004) peak and (220) peak should be corresponding to 73.202° and 74.230°, respectively. Careful examination of the angle (2θ) zoomed from 72° to 76° did not demonstrate the existence of cubic (400) peak in the original nanocrystalline powder, indicating that the original nanocrystalline zirconia powder consisted of monoclinic phase and tetragonal phase. As shown in Fig.3(b), the monoclinic, tetragonal and cubic peaks from the original nano-powder and the coating have changed, indicating that phase transformation occurred during the plasma spraying.

3.2 Hardness and Elastic Modulus of the Coating

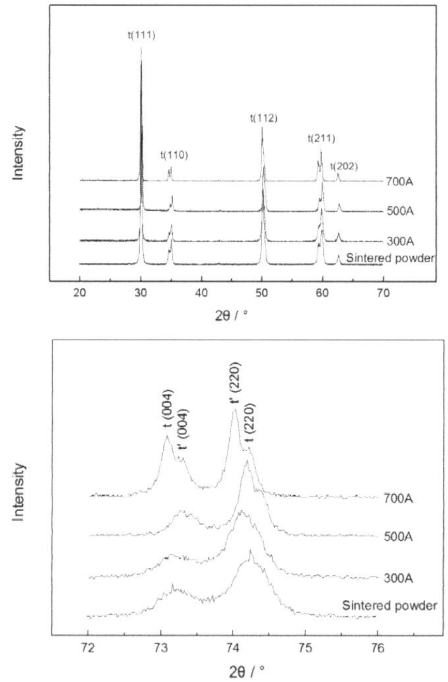

Fig.3 XRD patterns for the agglomerated feedstock and coatings.

Fig.4 exhibits the Knoop hardness of the coatings vs. the plasma currents. The hardness of the coating with 300A is 285Hv.3, that has lower hardness for thr YSZ coating. However, the hardness of the coating with 400A saltatory increased to 677Hv.3, in spite of only increased the plasma current of 100A, the hardness of the coatings increased linearly from 677 to 942Hv with increase of the plasma currents from 400 to 700 A. Lower hardness value of 285Hv for the coating with 300 A was attributed to the weak connection of the nano-grains in the coating. The increase of hardness from 677 to 942Hv for the coating from 400A to 700A was relativity to the increase content of the large grains in the coatings. The results suggest that the hardness is associated with the microstructure of the coatings. In this experiment the microstructure of the coatings can be summarized as following. The microstructure of the coating is a fully nano-equiaxed structure when the hardness is less than 300Hv. The microstructures of the coating are nano-grains mainly but contain large sized grains when the hardness is at range of 600-700Hv and the microstructure of the coatings with large sized grains are principal when the hardness exceeds 800 Hv.3.

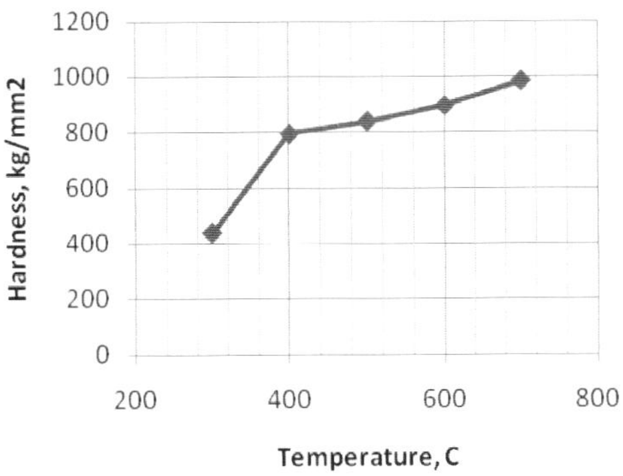

Fig.4 Knoop hardness of the coatings vs. the plasma currents.

Fig.5 shows the elastic modulus of the coating measured by Knoop indentation method. The elastic modulus of the coating increased from 35 to 198 GPa accompany with the changed microstructure from a fully nano-equiaxed structure to mixing structure and to denser large grains structure. Elastic modulus values of the sintered YSZ is 210 GPa measured by Knoop indentation method [10]. This value should be the highest value for the YSZ materials. Lima et al. have measured elastic modulus of the as-sprayed nanostructured and conventional YSZ coatings by laser-ultrasonics method [4, 15]. The elastic modulus values were 70-100 GPa for the conventional YSZ coating and 40-60 GPa for the nanostructured coating (bimodal structure). The elastic modulus value (E) of a material is directly proportional to its density (ρ) and the velocity (v) of the acoustic wave to the square travelling in the material ($E=\rho v^2$) [15]. Fully nano-grains structured coating has lower elastic modulus values (35-40 GPa), which attributes to the micropores or loosely bound of the grains in the coating. Similar elastic modulus value for the bimodal structured coating (40-60 GPa) and fully nano-equiaxed coating (35-70 GPa) may result in the similar mechanical perpetrates for both coatings.

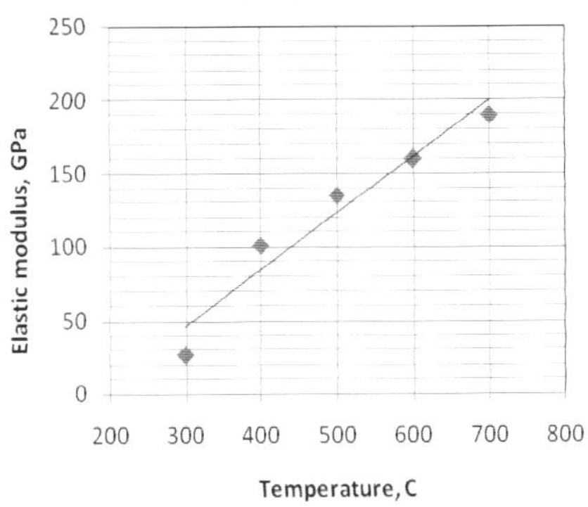

Fig.5 The elastic modulus of the coating vs. the plasma currents.

Y_2O_3-stabilized ZrO_2materials have demonstrated excellent properties and its coating have been applied in a broad area. According to the purpose of application, different microstructures of the coatings are required. For the thermal barrier coating, a low thermal conductivities and thermal shock resistance coating with certain porosity are required. For high temperature sealing systems used in jet engines, low hardness and abradable YSZ coating is expected [19]. The low hardness and fully nano-grain structured coatings may apply in this purpose. For the high temperature erosion resistance denser and high hardness coatings expected, the coating with 600-700A may suitable for this application.

4. Conclusions

Nano-agglomerated feedstock powders of Y_2O_3-stabilized ZrO_2 were plasma sprayed at an ambient pressure of 5000-10000 Pa and 100-150 Pa and at plasma current of 300-700A. Based on the results presented above, the following conclusions were drawn:

(1) A novel fully nano-equiaxed coating has been successfully deposited by LPPS or VLPPS at lower plasma current of 300-400 A and using nano-agglomerates feedstock. This fully nano-equiaxed coating is similar to the microstructure of the nano-bulk materials from the compressed and sintered of the nano-powders.

(2) The microstructures of the coatings ordinal change from a fully nano-equiaxed-structure to mixing grains and finally to a large grains structured coating along with enhancing the plasma currents.

(3) The fully nano-equiaxed coating has a lower hardnessin comparison with other structured coatings made of the same feedstock material. Low hardness attributes to loose bound among the nano-grains.

(4) The microstructures of the YSZ coatings can be predicted by the hardness measured. The microstructure of the coating is a fully nano-equiaxed structure when the hardness is lower than 500Hv. The microstructure of the coating is nano-grains but contains a few large sized grains when the hardness is at range of 600-700Hv and the microstructure of the coatings with large grained are primarily structure when the hardness exceeds 800 Hv.3.

References:

[1] Canan U. Hardwicke and Yuk-Chiu Lau, Advances in Thermal Spray Coatings for Gas Turbines and Energy Generation: A Review, *J. Thermal Spray Technol.,* 2013,22(5), p564-576.

[2] R.S. Lima, and B.R. Marple, Nanostructured YSZ thermal barrier coatings engineered to counteract sintering effects, *Materials Science and Engineering,*2008,A 485, p182–193.

[3] M. Gell, E.H. Jordan, Y.H. Sohn, D. Goberman, L. Shaw, T.D. Xiao, Development and implementation of plasma sprayed nanostructured ceramic coatings, *Surface and Coatings Technology,* 2001, 146 –147, p48–54.

[4] R.S. Lima, A. Kucuk, C.C. Berndt, Integrity of nanostructured partially stabilized zirconia after plasma spray processing, *Mat. Sci. Eng.,*2001, A313,p75-82.
[5] R.S. Lima and B.R. Marple, Thermal Spray Coatings Engineered from Nanostructured Ceramic Agglomerated Powders for Structural, Thermal Barrier and Biomedical Applications: A Review, *J. Thermal Spray Technol.,* 2007,16(1), p40-63.
[6] R.S. Lima, A. Kucuk, and C.C. Berndt, Bimodal distribution of mechanical properties on plasma sprayed nanostructured partially stabilized zirconia, *Mater. Sci. Eng.* 2002, A327, p224-232.
[7] Yang Gao, De Ming Yang, and Jianyi Gao, Characteristics of a Plasma Torch Designed for Very Low Pressure Plasma Spraying, *J. Thermal Spray Technol.,*2012, 21(3-4), p740-744.
[8] P. Eichert, M. Imber, and C.Coddet, Numerical study of an Ar-H_2 gas mixture flowing inside and outside a dc plasma torch, *J. Thermal Spray Technology,*1998, 7(4), p505-512.
[9] B. Pateyron, and P. Fauchais, Plasma spraying using Ar-He-H_2 gas mixtures, *J. Thermal Spray Technol.,* 1999, 8(4), p545-552.
Procedures, Wiley, New York, 1974, p 618–620.
[10] D.B. Marshall, Tatsuo Noma, and A.G. Evans, A Simple method for determining elastic-modulus-to-hardness ratios using Knoop indentation measurements, *Communications of the American ceramic society*, October 1982, p175-176.
[11] G.Q. Tong, K.C. Chan, L. Gao, Z.R. Lin, The grain size dependency of compressive deformation of a superplastic 3 mol% yttria stabilized tetragonal zirconia, *Materials Science and Engineering,*2002, A336, p263–269.
[12] Jie Lian, Javier E. Garay and Junlan Wang, Grain size and grain boundary effects on the mechanical behaviorof fully stabilized zirconia investigated by nanoindentation, *Scripta Materialia*, 2007, 56, p1095–1098.
[13] Guan-Jun Yang, Zheng-Long Chen, Cheng-Xin Li, and Chang-Jiu Li, Microstructural and Mechanical PropertyEvolutions of Plasma-Sprayed YSZ CoatingDuring High-Temperature Exposure:Comparison Study Between 8YSZ and 20YSZ, *Journal of Thermal Spray Technology*, 2013,Vol. 22(8), p1294-1302.
[14] W. Ma, W.X. Pan, C.K. Wu, Preliminary investigations on low-pressure laminar plasmaspray processing, *Surface & Coatings Technology,* 2005, 191, p166– 174.
[15] R.S. Lima, S.E. Kruger, G. Lamouche, B.R. Marple, Elastic modulus measurements via laser-ultrasonic and knoop indentation techniques in thermally sprayed coatings, *J. Therm. Spray Technol.,*2005,14 (1),p 52–60.
[16] Yang Gao, De Ming Yang, chengqi Sun and zhenyu Chen, Deposition of YSZ Coatings in a Chamberat Pressures below 100 Pa Using Low-PowerPlasma Spraying with an Internal Injection Powder Feeding, *J. Thermal Spray Technology,* 2013, 22(7), p1253-1258.
[17] P. Eichert, M. Imber, and C.Coddet, Numerical study of an Ar-H_2 gas mixture flowing inside and outside a dc plasma torch, *J. Thermal Spray Technology,*1998, 7(4), p505-512.
[18] A. Joulia, W. Duarte, S. Goutier, M. Vardelle, A. Vardelle, and S. Rossignol, Tailoring the Spray Conditionsfor Suspension Plasma Spraying, *Journal of Thermal Spray Technology*, 2015,Vol. 24(1-2) , p24-29.
[19] Dieter Sporer, Scott Wilson, Iacopo Giovannetti, Arno Refke and Massimo Giannozzi, On the potential of metal and ceramic based abradables in turbine seal application, *proceedings of the thirty-sixth turbo machinery symposium*, 2007, p79-86.

An Investigation on the Optimization, Reliability and Repeatability Assessment of Plasma Spray Coatings Deposited using Twin Opposing Injectors

Kwang yong Park[1,2], Ramachandran Chidambaram Seshadri[2], Sanjay Sampath[2], Byung il Yang[1], In soo Kim[1]

[1]Corporate R&D Institute, Doosan Heavy Industries & Construction, Chang won, Korea 51711
[2]Center for Thermal Spray Research, Department of Materials Science and Engineering, State University of New York at Stony Brook, Stony Brook, New York 11794-2275

In plasma spray process, the injection conditions crucially affect the particle properties and thus the characteristics of the deposited coating. The carrier gas flow has to be adjusted carefully depending on particle size and density, injector geometry and position, as well as on plasma jet characteristics to achieve optimum heat and momentum transfer. The variability of coating deposition due to parameter variations can be reduced by optimizing the injection. The situation becomes complicated when twin-opposing injectors are used to supply powder particles into the plasma jet. Further, the crossover of the two streams of particles makes it quite difficult to finalize the optimum injection set point. Hence, in this investigation a closed loop injection method was adopted to optimize the injection parameter of two opposing streams of particles. Three different opposing jet configurations namely the upstream (75°), orthogonal (90°) and downstream (105°) were employed and the methods to optimize injection were investigated systematically. The particle state - first order process map and the deposition state - second order process map were constructed using ensemble particle diagnostics method and in-situ beam curvature measurements respectively. The process property correlations were also fortified using reliability, repeatability and microstructure assessments of the coating deposits.

1 Introduction

Atmospheric Plasma Spray (APS) is a complex process with a multitude of interdependent variables that control the characteristics of the deposited material. The stochastics of the process are associated with the generation of the molten particle spray, for the formation of the deposit through successive impingement of the droplets, which can affect the deposit characteristics and properties in numerous ways. These complexities have resulted in significant challenges in our ability to manipulate and control the plasma spray process to reliably achieve desired properties and performance **[1]**. Among the different APS parameters, the injection/ carrier gas parameter is the most crucial one, which can greatly affect the reproducibility of the coating if the external injection is considered. The carrier gas flow rate (typically 2-9 lpm) is not insignificant compared with the plasma gas flow rate (typically 30-60 lpm); therefore, it may cool and perturb the plasma jet. The transverse injection velocity of particles into the plasma may considerably alter the particle behavior. Depending on the carrier gas flow rate, which determines the particle injection velocity and the momentum acquired by the particles, the particles may experience the following: penetrate into the plasma plume emerging at the opposite side, reside in the central core of the plasma plume, or rebound without entering the plume. In addition, the mixing of plasma gas with carrier gas would alter the transport properties of the plasma plume and substantial reports are available in the open literature on this topic **[2-4]**. The thermal spray industry is moving towards high throughput deposition and nowadays APS torches are fitted with two, three and six injector ports. The Injection/ carrier gas optimization becomes complicated when twin/ multiple opposing injectors are used to supply powder particles into the plasma jet. Hence, in this investigation a closed loop injection method was adopted to optimize the injection parameter of two opposing streams of particles. Three different opposing jet configurations namely the upstream (75°), orthogonal (90°) and downstream (105°) were employed and the methods to optimize injection were investigated systematically. The spray stream was characterized using, ensemble particle sensor and the real-time deposition characteristics of the coating deposits were evaluated using an In-situ coating property (ICP) sensor.

2 Experimental Method

In this, study an Yttria (8 wt. %) Zirconia powder (SG # 204, Saint-Gobain, USA) was sprayed using a 60kW DC plasma spray torch F4-MB XL (Oerlikon Metco, USA). The torch was controlled by 9MC control console (Oerlikon Metco, USA). The powder was supplied using a Twin -10-C powder feeder (Oerlikon Metco, USA). The powder feeder and the carrier gas flow rate were controlled using an independent 3D-PlumeOpt closed loop injection optimization system (Reliacoat Technologies LLC, USA). This system captures the intensity of the spray plume under different carrier gas flow rates (automatic closed-loop sweep) for a set powder feed rate and plasma spray condition using two high-speed and high-resolution cameras. The results of the carrier gas-sweep are fed into the system and it automatically sets the carrier gas corresponding to the flow rate at which maximum intensity was observed. This is how the optimum injection set point was acquired using 3D-PlumeOpt closed loop injection optimization system. The injection optimization procedure was repeated for upstream (75°), orthogonal (90°) and downstream (105°) injectors under single (top-down, bottom-up) and twin opposing jet configurations. The in-flight particle surface temperature and velocity were measured using Accuraspray-g3 (Tecnar Automation, Canada). The coatings were deposited on grit blasted on AA 6061 aluminium beam samples (230 X 25 X 2.3 mm^3) and the curvature of the beam during coating deposition was monitored using a laser based in-situ coating property (ICP sensor) developed at the Centre for Thermal Spray Research (CTSR). The integrated spray stream diagnostics adopted in the study is displayed schematically in **Figure 1**. The method for calculating the coating stress and modulus from the ICP time-curvature data can be referred elsewhere **[5, 6]**. The spray-foot print temperature during coating deposition was also monitored using a high-resolution Infrared camera FLIR-Max (FLIR systems, USA). The coating spray conditions are summarized in **Table 1**.

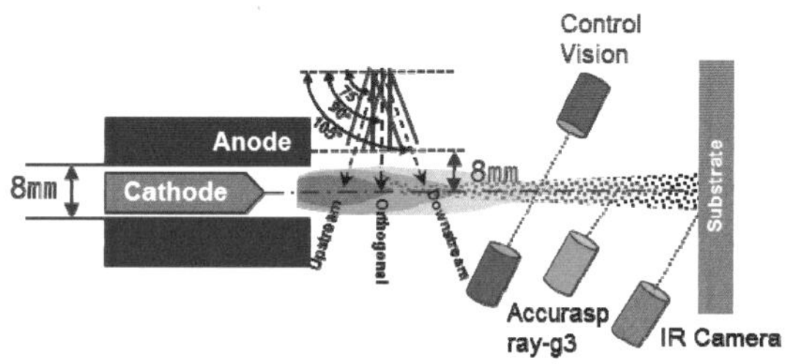

Figure 1. Schematic of the injector configurations and integrated sensors considered for spray stream diagnostics.

Table 1 Spraying conditions

Condition	Value
Powder size (μm)	10-75
Torch	F4 MB-XL
Anode internal nozzle diameter (mm)	8
Injector diameter (mm)	1.8
Current (A)	550
Primary gas flow rate(Ar) (l/min)	45
Secondary gas flow rate (H2) (l/min)	6
Injection Angles (top-down, bottom-up and twin opposing)	75°, 90°, 105°,
Optimised carrier gas flow rate (Ar in l /min) for different injector orientations	2.5 lpm (for 75°, 90°), 5 lpm (for 105°)
Powder feeder (g/min)	40
Spray distance (mm)	100
Robot speed (mm/sec)	700
Step size (mm)	3
Number of Preheating passes	4
Number of Coating passes	20

3 Results and Discussions

3.1 Injection Optimization

During radial injection since the particles do not necessarily follow an axial flow path, the interaction of the trajectory with plasma jet becomes important. Clearly the location of the particles in the plasma jet will affect the heat and momentum transfer from the plasma to the particles. The injector angle is used as a design variable to change the particle injection location in plasma sprays. It allows sampling different locations of the plasma jet for particle injection. The optimum carrier gas flow rate for 75° and 90° injection angles was found to be 2.5 lpm and for 105° injection angle, the optimum carrier gas flow rate is 5 lpm as shown in **Figure 3**. The difficulty to inject particles down stream at 105° may be related to the high drag forces acting against the particles and thus requiring higher carrier carrier gas to enter the centerline of the high velocity plasma down the stream **[7]**.

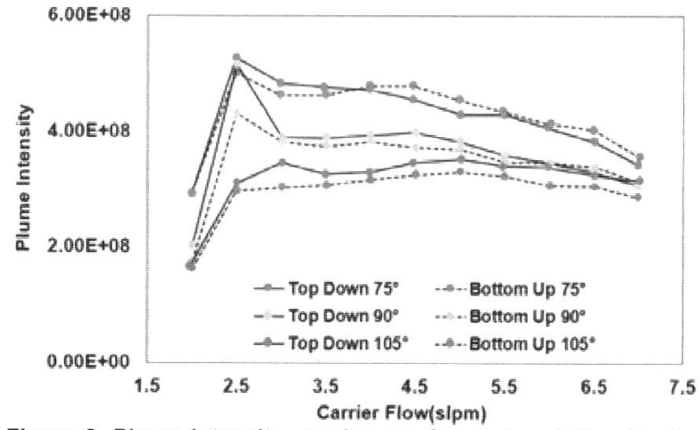

Figure 2. Plume intensity- carrier gas flow rate relationship for different injector orientations.

3.2 Inflight temperature and velocity of the particles

The surface temperature and velocities of the particles sprayed using different injector configurations are shown in **Figure 3**. The temperature and velocity of the upstream injected particles is higher compared to the orthogonal and down stream injection configurations. This is due to the higher residential time of the upstream injected particles in the plasma plume, which helps the particles to gain high heat and momentum.

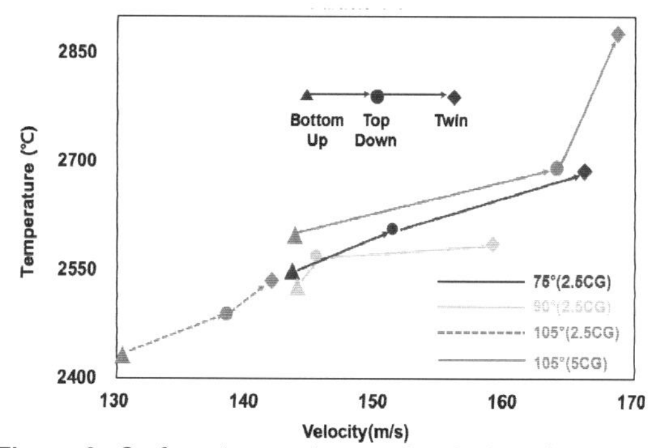

Figure 3. Surface temperature and velocity of particles sprayed under different injection configurations.

3.3 In-situ coating property analysis of the deposits

The coatings were deposited on aluminum beams mounted on an ICP sensor. The powder was spray deposited under 70°, 90° and 105° single injection (both from top-down and bottom up seperately) as well as dual opposing injections. When using single injection 40 gpm was used directly and when using twin injection the feed rate was split into half and 20 gpm of powder was injected into plasma from each one of the opposing injector. The deposition under each condition was repeated twice to check the repetability. The deposition of coatings on ICP sensor allows simultaneous measurement of substrate curvature and substrate temperature. This approach allows extraction of deposit stress evolution during deposition as well as elastic properties of the coating. The time-curvature plot derived form the ICP sensor for the twin injection configurations under consideration are shown in **Figure 4**.

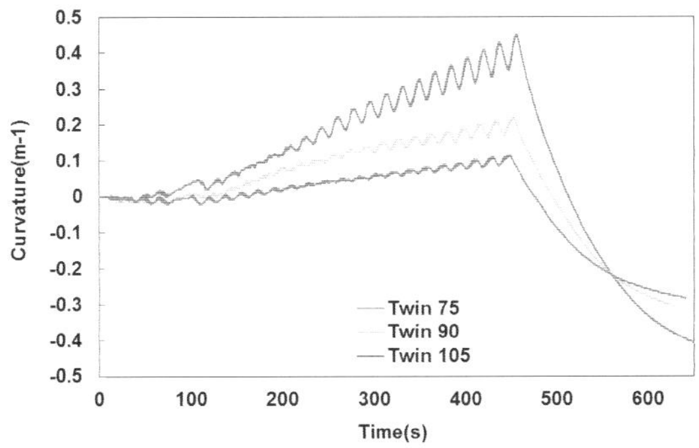

Figure 4. ICP Time-Curvature plot for coatings deposited using twin opposing jets with different orientations.

The positive values of curvatures in **Figure 4** suggest that the evolving stresses are dominanated by quenching over peening for all of the coatings which is typical for all ceramic coatings. Among the deposited coatings the evolving stress seems to be high for the upstream injection condition. This result is analogous with the temperarure velocity plot shown in **Figure 3**, which clearly indicates that the upstream injected particles are better melted. The stress-modulus relationship is pretty linear among the coatings deposited in the current study, which could be inferred from **Figure 5**. It is worth mentioning here that the top-down injection deposited coatings showed higher stress and modulus values compared to the twin opposing injection coatings, even though the loading effect and plasma quenching effect are lower in the twin configuration. May be the real benefits of the twin injection is realisable when the feed rate exceeds above the saturation feed rate levels.

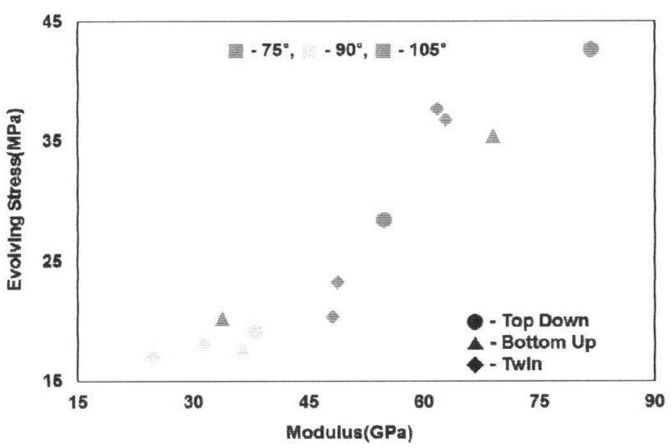

Figure 5. The Stress-Modulus relationship among the coatings

4 Summary

A closed-loop injection optimization tool 3D-Plumeopt was sucessfully utilized to optimise the injection parameter of the particles supplied though with single as well as twin oppsing injectors oriented with upstream, downstream and orhogonal configurations. The upstream injection showed improved melting which resulted in the creation of higher stress and in-plane modulus in the coatings. Though the loading effect and plasma quenching effect are lower in the twin injection configuration. The real benefits of the twin injection is realisable when the feed rate exceeds above the saturation feed rate levels. Highthroughput injection optimization with single as well as twin injectors and 3D plume profiling of the particle stream are being carriedout in our center to fortify the aforementioned phenomenon.

5 Acknowlegements

This work was supported by Korea Institute of Energy Technology Evaluation and Planning(KETEP) grant funded by Korea government Ministry of Knowledge Economy (No. 2013101010170C). The author would like to thank CTSR for collaborating. The authors would like to thank Jose Colmenares-Angulo and Jonathan Gutleber of Oerlikon metco for their help and support with the supply of torch and injector spares.

6 Literature

[1] G. Dwivedi, T. Wentz, S. Sampath, and T. Nakamura : Journal of Thermal Spray Technology(2010), 19(4), pp. 695-712

[2] T. Streibl, A. Vaidya, M. Friis, V. Srinivasan, and S. Sampath : Plasma Chemistry and Plasma Processing(2006), 26(1)

[3] A. Vaidya, V. Srinivasan, T. Streibl, M. Friis,W. Chi, S. Sampath : Materials Science and Engineering (2008), pp. 239/53

[4] V. Srinivasan M. Friis A. Vaidya T. Streibl S. Sampath : Plasma Chem Plasma Process(2007), 27, pp. 609–623.

[5] K. Shinoda, J. Colmenares-Angulo, A. Valarezo, and S. Sampath : Journal of Thermal Spray Technology 21(2012), 6, pp. 1224/33.

[6] T. Nakamura, Y. Liu, International Journal of Solids and Structures(2007), 44, pp 1990/2009

[7] W. Zhang L. L. Zheng H. Zhang S. Sampath : Plasma Chem Plasma Process(2007), 27, pp. 701/16.

Preparation and Properties of Cr_3C_2-25NiCr Thermal Spraying Powders and Coatings with Nano-sized, Micro-sized and Bimodal Structures

XIE Hui-cai, WANG Shao-yi, LIAO Xing-wen, LIU Cheng-hai, LIU Fu-ming

(Ganzhou Achteck Tool Technology Co., Ltd, Ganzhou 341300, China)

Abstract: Three types of Cr_3C_2-25NiCr thermal spraying feedstock powders with nano-sized, micro-sized and bimodal structures were prepared using a spray drying and agglomeration sintering process, and the corresponding coatings were fabricated using a high-velocity oxygen fuel spraying technique. The microstructure and phase composition of the powders and corresponding coatings were studied via metalloscopy, scanning electron microscopy and X-ray diffractometry, respectively. In addition, their microhardness, crack toughness and abrasive wear behavior were also investigated. The results indicate that all three powders are nearly spherical in shape, but the bimodal powder exhibits the best sphericity, the most uniform composition distribution, and the highest deposition efficiency, i.e., greater than 60%. All three coatings are primarily composed of Cr_3C_2 and NiCr phases, and a small amount of Cr_2O_3 was detected in the nano- and microcoatings due to oxidation. Compared with the micro-sized coating, the nano-sized coating has higher toughness but lower microhardness and lower abrasive resistance. However, the bimodal Cr_3C_2-25NiCr coating integrates the advantages of both the nano- and microcoatings, has low porosity and a dense microstructure, and is tightly combined with the substrate, thus displaying the best comprehensive performance.

Keywords: bimodal Cr_3C_2-25NiCr coating, agglomeration sintering, HVOF, deposition efficiency

---

XIE Hui-cai, Ganzhou Achteck Tool Technology Co.,Ltd, Ganzhou, China. E-mail: xiehuicaiwww@126.com

Cermet coatings fabricated using the high-velocity oxygen fuel (HVOF) spray technique dependably exhibit high density, uniform composition distribution and high bonding strength, making HVOF one of the most commonly used methods in the thermal spraying field [1]. Thermal spray Cr_3C_2-NiCr coatings have been widely used in industries such as ferrous metallurgy, petroleum chemicals, shipping, power generation, and machine manufacturing due to their excellent resistance to oxidation, abrasive wear and corrosion under high temperature conditions [2].

The properties of HVOF coatings are determined by their microstructure [3-6], which is closely associated with raw material size [7]. It was reported that coatings with significantly higher density and lower porosity could be obtained by adding a certain amount of submicron-sized or nano-sized raw material, which could also improve the microhardness and corrosion resistance of the coatings [8]. Currently, Cr_3C_2-NiCr feedstock powders are commonly prepared using micron-Cr_3C_2 ingredients with a size of approximately 2-6 μm, poor sphericity, mobility and microstructure homogeneity, and the corresponding coatings always exhibit low microhardness, crack toughness and deposition efficiency. In recent years, certain reports have addressed the nanocrystalline Cr_3C_2-25NiCr coating [9,10], and results showed that nanocrystalline coatings had significantly better properties than conventional coatings. For example, literature [9] reported that the hardness of nanocrystalline coating was 20% higher than that of the conventional coating having similar composition. However, reports on the preparation and properties of bimodal Cr_3C_2-25NiCr coatings are scarce. In the current study, three types of Cr_3C_2-25NiCr thermal spraying feedstock powders and corresponding coatings were fabricated using nano-sized, micro-sized and bimodal Cr_3C_2 as raw materials, and the effect of Cr_3C_2 size on the properties of the Cr_3C_2-25NiCr powders and coatings was investigated.

Experimental Procedure

Raw materials and powder preparation

The nano-sized Cr_3C_2 used as a raw material in this study was produced by Ganzhou Achteck Tool Technology Co., Ltd, and micro-sized Cr_3C_2 and other ingredients were sourced from commercially available raw materials. Figure 1 and figure 2 show the XRD pattern and SEM image of the nano-sized Cr_3C_2, respectively. After mixing according to table 1, a certain amount of binder and additives were added to the raw material powders to obtain a stable and homogeneous slurry. Three types of Cr_3C_2-25NiCr thermal spraying feedstock powders were prepared using a spray drying and agglomeration sintering process.

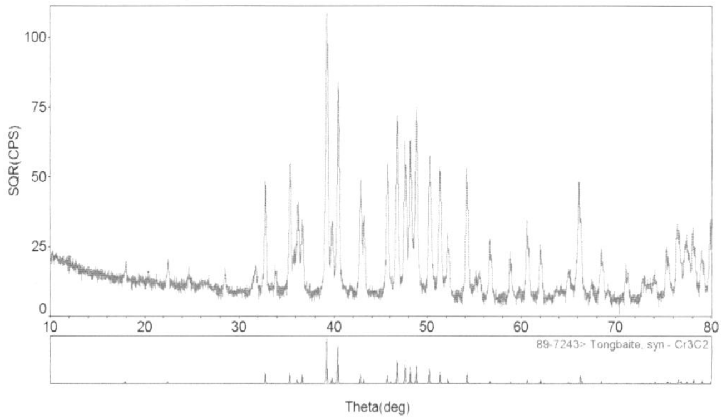

Figure 1. XRD patterns of the nano-Cr_3C_2 powder

Figure 2. Morphology of the nano-Cr_3C_2 powder

Table 1 Composition of three powder samples (mass fraction, %)

Sample	Micron-Cr_3C_2	Nano-Cr_3C_2	Ni	Cr
N	0	75	20	5
B	52.5	22.5	20	5
M	75	0	20	5

Coating fabrication

The 45 steel was selected as matrix material, and the above three types of Cr_3C_2-25NiCr powders were used as spraying feedstock powders. Initially, the substrates were de-greased, de-rusted and sandblasted prior to deposition of the corresponding coatings on the surface of the substrates using a model JP-8000 supersonic flame spraying system produced by American Praxair Corporation. The spray parameters are shown in table 2.

Table 2 Spraying parameters for HVOF

Items	Kerosene flow/GPH	O_2 gas flow /SCFH	Spray distance /mm	N_2 gas flow /SCFH
Parameters	6	1900	380	23

Characterization

The apparent densities of the Cr_3C_2-25NiCr feedstock powders were determined using a Holzer flow meter according to the ASTM B213-03 standard. The microstructures of the feedstock powders and corresponding coatings were observed by an OLYMPUS metallographic microscope and a

German Ziess scanning electron microscope SUPRA 55, and their phase compositions were analyzed using an X-ray diffractometer (XRD). The abrasive wear behavior of the coatings was tested with a MLS-225 type wet-sand rubber-wheel abrasion tester using a load of 100 N and a speed of 188 r/min. The coatings were ground for 500 rotations, and the sample weight was treated as the initial weight. The samples were abraded three times with 2000 rotations each, and the average of the three weight loss values was noted. The microhardness and crack toughness of the coatings were measured with an automatic micro-Vickers hardness tester and a HVS-10 type Vickers hardness tester, respectively. The crack toughness was calculated according to the Wilshaw formula [11]:

$$K_c = 0.079 \frac{P}{a^{3/2}} \log\left(\frac{4.5a}{c}\right)$$

where “P” represents the testing load (unit: N), “a” is half the length of the indentation diagonal (unit: μm) and “c” is the distance from indentation center to the crack tip (unit: μm).

Results and Discussion

Microstructure and properties of feedstock powders

The physical properties of the three types of Cr_3C_2-25NiCr powders are shown in table 3, and their metallographs and SEM micrographs are shown in figure 3 and figure 4, respectively. According to table 3, the powders display the same particle size distribution and nearly equal medium diameters. Both powder N and powder B have high apparent densities of approximately 2.7 g/cm^3, but powder M has an apparent density as low as 2.45 g/cm^3. The carbon loss of the feedstock powders relative to the initial carbon content can be arranged in the order N>B>M, which indicates that powders prepared using finer Cr_3C_2 undergo a significantly larger carbon loss, which means that the effect of Cr_3C_2 size on the carbon content of Cr_3C_2-25NiCr powder must be considered in the preparation process to ensure a qualified carbon content. The oxygen content of all three types of spraying

Table 3 Physical properties of the Cr_3C_2-25NiCr feedstock powders

Sample	Apparent density g/cm^3	Size distribution μm	Medium diameter μm	Loss of carbon %	Oxygen content %
N	2.67	15~53	32.1	2.56	0.08
B	2.70	15~53	31.6	1.94	0.11
M	2.45	15~53	32.5	1.62	0.09

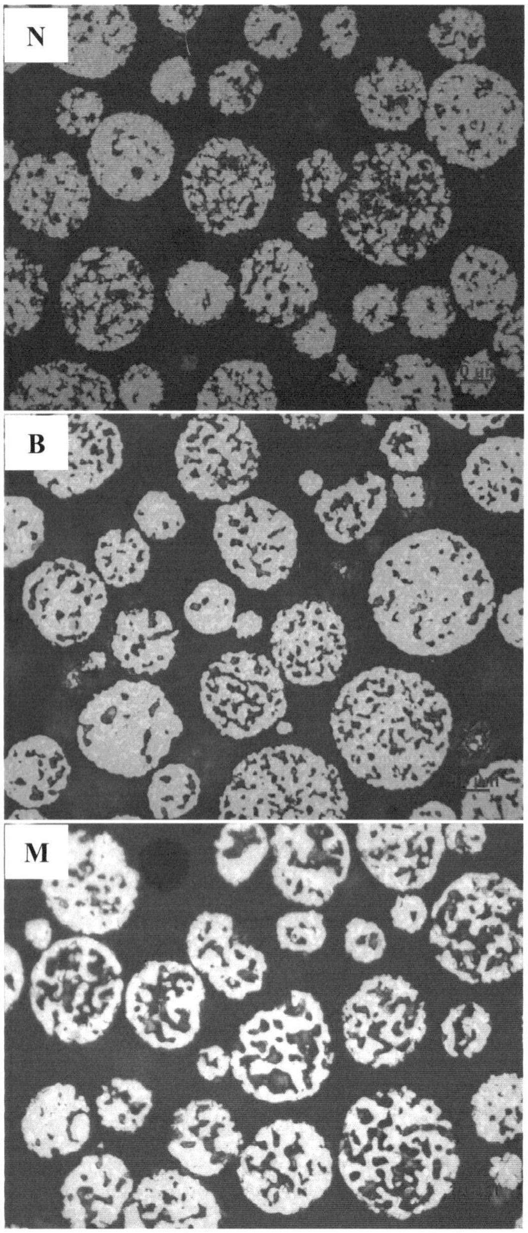

Figure 3. Metallographs of three feedstock powders

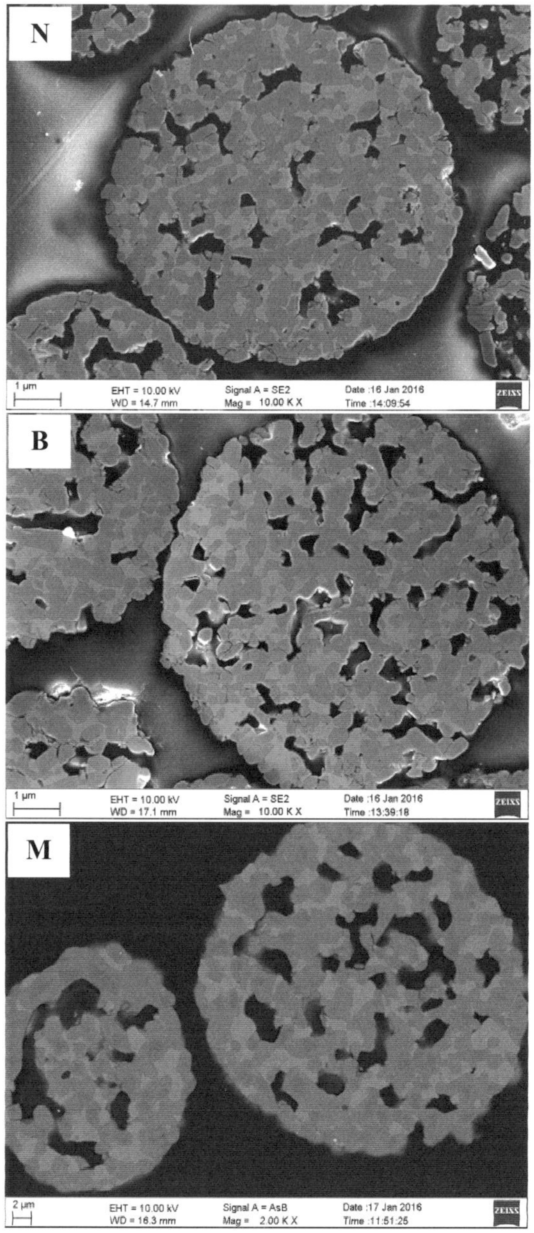

Figure 4. SEM micrographs of three feedstock powders

powders is as low as 0.1%, which is satisfactory for obtaining high quality coatings with low porosity and non-inclusion.

Figure 3 and figure 4 show that all of the Cr_3C_2-25NiCr powders are nearly spherical in shape. Non-uniform densification took place in powder N during the sintering process, which rendered certain particles loose and porous and others compact. The average grain size of Cr_3C_2 is approximately 300 nm. It is known that denser particles more easily rebound during HVOF spraying because of their low melting degree, leading to a lower deposition efficiency. Both powders B and M have a loose and porous structure, which is conducive to heat absorption and transfer of the powder in the spray flame flow and facilitates full fusing of the powder particle, resulting in higher quality coatings. The powder B particles consist of many chromium carbide grains of different sizes ranging from 100 nm to 500 nm, and the internal pores are fine and evenly distributed, which contributes to a higher apparent density. The uniformly dispersed fine grains and pores are beneficial for obtaining a uniform and dense coating. The powder M particle, however, consists of relatively thick chromium carbide grains with a size distribution of 1 μm to 5 μm, and conversely, the internal pores are large and unevenly distributed, causing a lower apparent density of only approximately 2.45 g/cm^3. Theoretically thick and unevenly distributed pores and grains are harmful to heat absorption and transfer in flame flow, and thus, the powder particles cannot be fully fused, which decreases the deposition efficiency and coating quality. During our experiment, we also found that for powder N and powder B, a high apparent density could be obtained at a relatively low temperature of 1100°C, but for powder M, an apparent density of 2.7 g/cm^3 was difficult to achieve even at a temperature as high as 1200°C. Our explanation for this observation is that the finer raw material granularity and the higher powder sintering activity reduce the sintering temperature of the powder significantly.

Figure 5 shows the XRD patterns of the three types of Cr_3C_2-25NiCr powders that display consistent diffraction curves and are all composed of Cr_3C_2 and NiCr phases. The sharp diffraction peaks indicate that each component of the powder is completely crystallized. In these Cr_3C_2-25NiCr powders, nickel and chromium exist in the form of a NiCr alloy, and the results show that the two

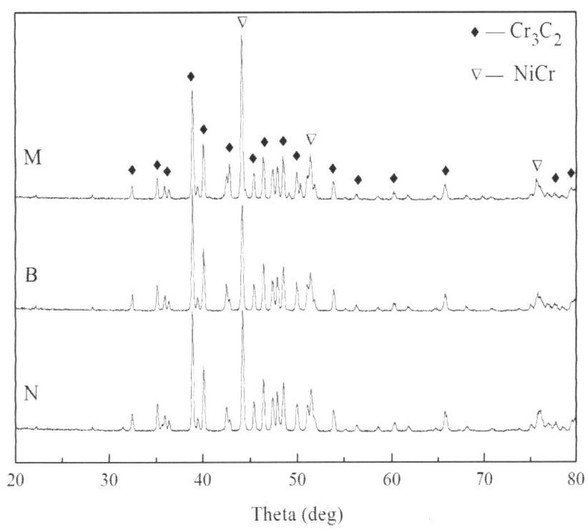

Figure 5. XRD patterns of the three feedstock powders

components can completely dissolve into each other during the sintering process, thus forming a NiCr solid solution alloy. We note that the diffraction peak heights of the NiCr phase at 44 degrees fall in the order of M>N>B; in other words, the peak of powder M is maximal and significantly higher than that of the other two and that of powder B is the lowest. We conclude that this result occurs because of the coarser grains and unevenly distributed components of powder M, which make it easier to form a NiCr alloy enrichment zone. However, the raw material granularities of powder N and B are finer, and all of their components were evenly dispersed after ball milling, which could effectively prevent the formation of a NiCr alloy enrichment zone during the sintering process, thus refining the grains and obtaining homogeneous powders. Therefore, we believe that addition of nano-sized Cr_3C_2 can inhibit formation of the NiCr alloy enrichment zone, refine the grains and aid in obtaining Cr_3C_2-25NiCr thermal spraying powders with homogeneous microstructures.

Microstructure and performances of coatings

The properties and their variation trends for the three types of Cr_3C_2-25NiCr spray coatings are shown in table 4 and figure 6, respectively. According to table 4, the deposition efficiency of the three coatings are ranked as B>N>M, i.e., the bimodal coating has the highest deposition efficiency, which is greater than 60%. The nano-sized coating follows, and these values are obviously higher than that of the micro-sized coating. As previously mentioned, the difference in deposition efficiency among the three types is mostly caused by the different microstructures of the feedstock powders.

Table 4 Properties of the Cr_3C_2-25NiCr coatings

Sample	Deposition efficiency %	Hardness $HV_{0.3kg}$ $kg \cdot mm^{-2}$	Crack toughness $Mpa \cdot m^{1/2}$	Wear weight loss g
N	57.24	849	4.5	0.0477
B	60.19	919	4.3	0.0167
M	48.94	936	3.5	0.0175

According to figure 6, the average microhardness values of the three coatings fall in the order of N<B<M, and their fracture toughness values are ordered as N>B>M, i.e., the larger the raw material Cr_3C_2 granularity, the higher the coating microhardness and the lower the crack toughness will be. The average weight losses of the three coatings due to abrasive wear after 2000 rotations are ranked as N>M>B, which means that the nano-sized coating displays the greatest weight loss, which

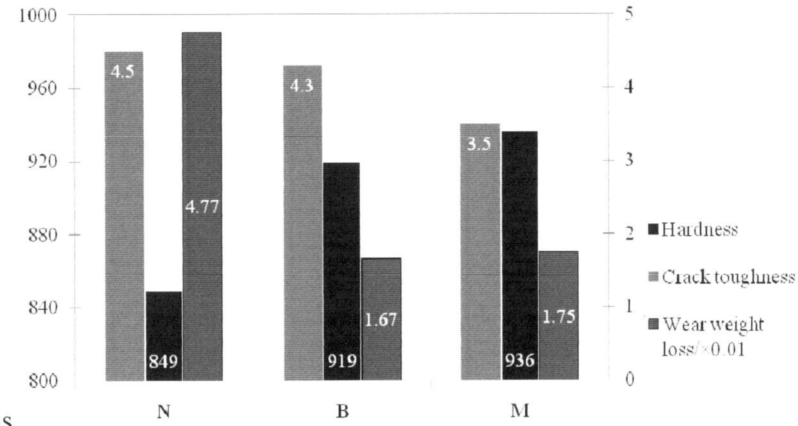

is

Figure 6. Performance comparison of the coatings

0.0477 g and is significantly greater than that of the micro-sized coating (0.0175 g), and that of the bimodal coating is the smallest. If the raw material consists of single-sized Cr_3C_2, the coarser the Cr_3C_2 grain size, the better the coating wear resistance will be, similar to the result reported by NI Ji-liang that WC-10Co-4Cr coatings made of coarser WC particles display better abrasive wear resistance [12]. In addition, coatings made of composite-sized Cr_3C_2 exhibit better abrasive wear resistance than single-sized coatings, probably because of the special bimodal structure of the coating. In summary, the micro-sized coating has a higher microhardness and better abrasive resistance but lower toughness. In contrast, the nano-sized coating has a higher toughness but lower microhardness and lower abrasive resistance. However, the bimodal coating integrates the advantages of both the nano- and micro-sized coatings and has the best comprehensive mechanical performance.

The XRD patterns of the three Cr_3C_2-25NiCr coatings are shown in figure 7, and the figure shows

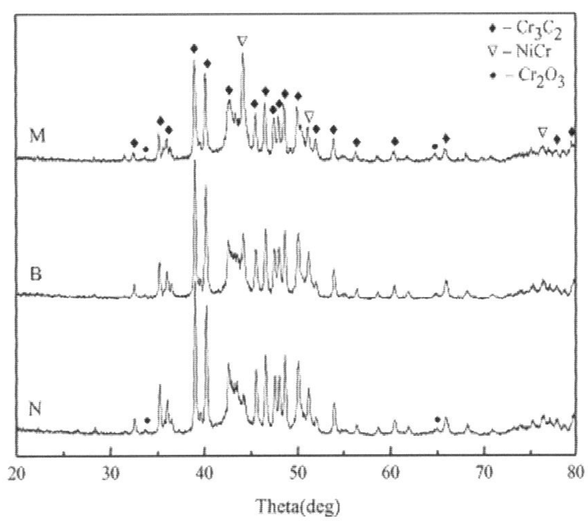

Figure.7 XRD patterns of the Cr_3C_2-25NiCr coatings

that the three coatings are primarily composed of Cr_3C_2 and NiCr phases. The Cr_7C_3 diffraction peaks were not detected in the XRD patterns, indicating no obvious decarburization phenomenon in the coatings. Nevertheless, selected weak Cr_2O_3 diffraction peaks were detected in the XRD patterns of the nano- and micro-sized coatings, indicating that the chromium in these two coatings has been significantly oxidized.

Metallographs and SEM micrographs of the coatings are shown in figure 8 and figure 9, respectively. The figures show that all three coatings present a typical layer structure and have no obvious cracks. Coatings N and B were tightly combined with the substrate, but for coating M, many defects were found on the interface of the coating and matrix because the coarse grains in powder M prevented the particles from melting during spraying and prevented them from close combination with the substrate. Many tiny pores pervade coating N, and those pores might be derived from loose and porous powder particles with light weight and poor fluidity, which caused slack powder feeding and low particle velocity and eventually left many tiny pores in the coating. Although relatively fewer pores are observed in coating M (except for a subset of dispersed large pores that might originate from original feedstock powder particles), because of the coarse grains, powder M cannot be completely fused in the spray flame flow, and the inner original large pores were not effectively removed and remained inside the coating. Compared with coatings N and M, coating B has a lower porosity and higher density, and we believe it benefited from fine and uniformly distributed grains and tiny pores inside the original feedstock powder particles, which are conducive to uniform heating and full fusion and eventually form a uniform and dense coating.

According to figure 9, coating N has fine grains on the surface but displays high porosity, and coating M consists of many coarse grains. However, coating B consists of composite Cr_3C_2 grains of different sizes that are uniformly dispersed and bonded together by the NiCr alloy to form a dense coating with low porosity. Theoretically, Cr_3C_2 grains must bear the continuous shearing force exerted by abrasive grains in the abrasive wear process because the abrasive particles (always with a size of 240-250 μm) are much larger than the Cr_3C_2 particles, and the finer Cr_3C_2 particles tend to fall off more easily due to stress concentrations interacting with the abrasive grains [13], resulting in a coating with poor abrasive wear performance. According to figure 9, the Cr_3C_2 grains are fine in coating N, and a large number of pores provide intrusion channels for abrasives such that the Cr_3C_2 particles around the pores are removed more easily due to stress concentrations during the abrasive

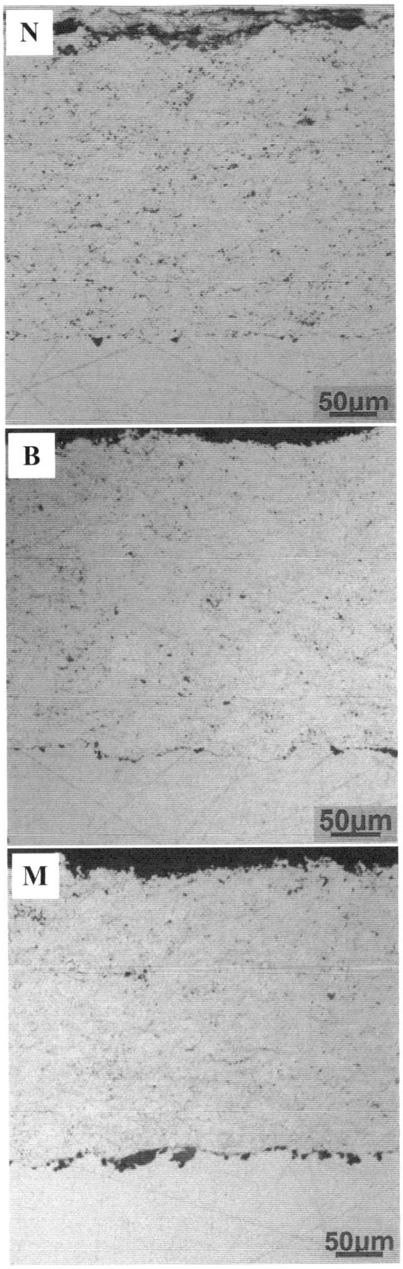

Figure.8 Metallographs of the coatings

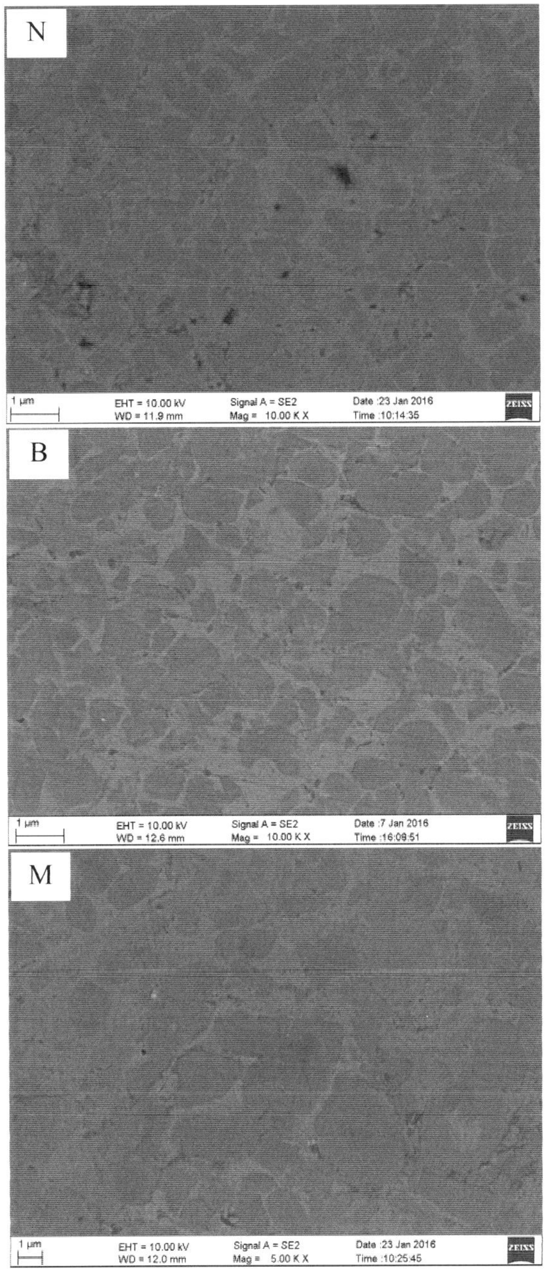

Figure.9 SEM micrographs of the coatings

wear process, and therefore, this coating has poor abrasive resistance. However, the Cr_3C_2 grains are much coarser in coating M and contain relatively fewer pores, and thus the coating has better abrasive resistance. As for coating B, the inner Cr_3C_2 particles are fairly uniformly distributed, and the coarser particles could provide protection for the finer ones. In addition to its compact microstructure, this coating is able to prevent intrusion of abrasives and thus displays the best abrasive resistance.

Conclusions

Three types of Cr_3C_2-25NiCr thermal spraying feedstock powders with nano-sized, micro-sized and bimodal structures were prepared a using spray drying and agglomeration sintering process, and the corresponding coatings were fabricated with a high-velocity oxygen fuel spraying technique. The effect of Cr_3C_2 size on the microstructure and properties of Cr_3C_2-25NiCr powders and coatings were investigated. The results showed the following:

(1) All three Cr_3C_2-25NiCr powders are nearly spherical and have the same phase composition, but the bimodal powder exhibits the best sphericity, the most uniform composition distribution, and the highest deposition efficiency, which is higher than 60%.

(2) The three Cr_3C_2-25NiCr coatings are mostly composed of Cr_3C_2 and NiCr phases, and a small amount of Cr_2O_3 was detected in the nano- and micro-sized coatings due to oxidation.

(3) The micro-sized coating has higher microhardness and better abrasive resistance but lower toughness. In contrast, the nano-sized coating has higher toughness but lower microhardness and lower abrasive resistance. However, the bimodal coating integrates the advantages of both the nano- and micro-sized coatings, has low porosity and dense microstructure, and is tightly combined with the substrate, thus displaying the best comprehensive performance.

References

[1] Tahar Sahraoui, Nour-Eddine Fenineche, Ghislain Montavon, and Christian Coddet. Structure and wear behaviour of HVOF sprayed Cr_3C_2-NiCr and WC-Co coatings [J]. Materials and Design, 2003, 24: 309-313.

[2] Mitra Akhtari Zavareh, Ahmed Aly Diaa Mohammed Sarhan, Bushroa BintiAbd Razak, and Wan Jeffrey Basirun. The tribological and electrochemical behavior of HVOF-sprayed Cr_3C_2-NiCr ceramic coating on carbon steel [J]. Ceramics International, 2015, 41: 5387-5396.

[3] S. Matthews, B. James, and M. Hyland. The role of microstructure in the mechanism of high velocity erosion of Cr_3C_2-NiCr thermal spray coatings: Part 1 — As-sprayed coatings [J]. Surface & Coatings Technology, 2009, 203: 1086-1093.

[4] S. Matthews, B. James, and M. Hyland. The role of microstructure in the mechanism of high velocity erosion of Cr_3C_2-NiCr thermal spray coatings: Part 2 — Heat treated coatings [J]. Surface & Coatings Technology, 2009, 203: 1094-1100.

[5] J.K.N. Murthy, K. Satya Prasad, K. Gopinath, and B. Venkataraman. Characterization of HVOF sprayed Cr_3C_2-50(Ni20Cr) coating and the influence of binder properties on solid particle erosion behaviour [J]. Surface & Coatings Technology, 2010, 204: 3975-3985.

[6] S. Matthews, B. James, and M. Hyland. High temperature erosion of Cr_3C_2-NiCr thermal spray coatings — The role of phase microstructure [J]. Surface & Coatings Technology, 2009, 203: 1144-1153.

[7] Gang-Chang Ji, Chang-Jiu Li, Yu-Yue Wang, and Wen-Ya Li. Microstructural characterization and abrasive wear performance of HVOF sprayed Cr_3C_2-NiCr coating [J]. Surface & Coatings Technology, 2006, 200: 6749-6757.

[8] YANG Xue, YE Fu-xing, CUI Chong, and WANG Xi-bao. Phase transformation and wear resistance of submicron structural WC-12Co coatings deposited by HVOF [J]. Thermal Spray Technology, 2009, 1(2): 53-57.

[9] M. Roy, A. Pauschitz, J. Bernardi, T. Koch, and F. Franek. Microstructure and Mechanical Properties of HVOF Sprayed Nanocrystalline Cr_3C_2-25(Ni20Cr) Coating [J]. Journal of Thermal Spray Technology, 2006, 15(3): 372-381.

[10] Manish Roy, A. Pauschitz, J. Wernisch, and F. Franek. The influence of temperature on the wear of Cr_3C_2-25(Ni20Cr) coating — comparison between nanocrystalline grains and conventional grains [J]. Wear, 2004, 257: 799-811.

[11] Evans A G, and Wilshaw T R. Quasi-static solid particle damage in brittle solids — Observations analysis and

implications [J]. Acta Metallurgica, 1976, 24:939.

[12] NI Ji-liang, CHENG Tao-tao, DING Kun-ying, and WANG Zhi-ping. Effect of WC size on abrasive wear resistance of WC-10Co-4Cr coatings [J]. Material Protection, 2013, 46(1): 19-21.

[13] Li Lin, Guo-Lu Li, Hai-Dou Wang, Jia-Jie Kang, Zhong-Lin Xu, and Hai-jun Wang. Structure and wear behavior of NiCr–Cr_3C_2 coatings sprayed by supersonic plasma spraying and high velocity oxy-fuel technologies [J]. Applied Surface Science, 2015, 356: 383-390.

Thermal cycling behavior of quasi-columnar YSZ coatings deposited by PS-PVD

Shunyan TAO, Fang SHAO, Jiasheng YANG, Huayu ZHAO, Chenguang LIU, Yin ZHUANG, Jinxing NI, Xinghua ZHONG
Laboratory of Inorganic Coating Materials, Shanghai Institute of Ceramics, Chinese Academy of Sciences, Shanghai 201899, PR China
E-mail address: sytao@mail.sic.ac.cn (S. Tao)

Abstract: Columnar structured thermal barrier coatings, owing to their high strain tolerance, are expected for their potential possibilities to substantially extend turbine lives and improve engine efficiencies. In this paper, plasma spray-physical vapor deposition (PS-PVD) process was used to deposit yttria partially stabilized zirconia (YSZ) coatings with quasi-columnar structures. Thermal cyclic tests on burner rigs and thermal shock tests by heating and water quenching method were involved to evaluate the thermal cycling and thermal shock behaviors of such kind of structured TBCs. Evolution of the microstructures, phase composition, residual stresses and failure behaviors of quasi-columnar YSZ coatings before and after the thermal tests were investigated. The quasi-columnar coating obtained before process optimization also had an average life of around 623 cycles as the spallation area reached about 10% of the total coating surface during burner rig tests with a coating surface temperature of ~ 1250℃. Failure of the coating is mainly due to the center columnar segments pull-out spallation.

Keywords: Plasma spray - physical vapor deposition (PS-PVD); Quasi-columnar structure; Thermal barrier coating; Thermal cyclic test; Thermal shock test

1 Introduction

Thermal barrier coating (TBC) systems are widely used in stationary turbines of power plants and aircraft turbines to improve the performance and efficiency of the engine [1,2]. A typical TBC system consists of an yttria partially stabilized zirconia (YSZ) topcoat, an intermediate thermally grown oxide (TGO) scale, a metallic bondcoat and a superalloy substrate [3,4]. The main properties and long-term stability of the individual TBC components are deeply depend not only on their chemical composition but also on their morphological feature. The latter can vary greatly with fabricate process. Currently, atmospheric plasma spraying (APS) and electron beam-physical vapor deposition (EB-PVD) are the two main processes used for TBC deposition, which results in very different microstructures of the coatings. An APS coatings built up of solidified droplets and produces lamellar-like splat morphology, whereas in EB-PVD, the materials condensates from the vapor phase and produces feather-like column structures [5-8]. The result is that APS TBCs have lower manufacturing costs and a lower thermal conductivity, whereas EB-PVD TBCs have a higher strain tolerance.

A new spray process is called plasma spray-physical vapor deposition (PS-PVD) [9-17], which was primarily named as low pressure plasma spraying-thin film process (LPPS-TF), enables vaporization of the injected spraying powder to a large extent and production of EB-PVD-like structures. This morphological characteristics combine the low costs of a plasma spray process with the superior strain tolerance of column structures. Improved efficiency of the vacuum pumps in PS-PVD devices leads to pressures between 50 and 200 Pa in the working chamber [10-11]. Because of the lower pressure during the process, the plasma plume expands to a length reach over 2 m, and to 200-400 mm in diameter. Therefore, the plasma jet footprint covers the sample and allows coating of regions which are not in the line of sight of the plasma torch ("non-line of sight" coating). Although the pressure in the PS-PVD process is higher than in the conventional PVD process, a high plasma jet temperatures enable fine powder evaporation in the plasma plume, and the morphology of the coatings depends on the deposition parameters [13-15,17]. Additionally, it also allows formation of the unique microstructures and enables coating of complex substrate geometries. It has been shown by Detlev Stöver et al. that different kinds of YSZ coating microstructures depending on the application, including dense coatings, PVD-like columnar coatings and nano-sized solid clusters columnar coatings (quasi-columnar), could be applied by PS-PVD process through adjusting the spraying parameters [9,10]. Thereby, this relatively new technique can fill the gap between traditional thermal spray processes and EB-PVD deposition.

In this paper, thermal cyclic tests on burner rigs and thermal shock tests by heating and water quenching method were involved to evaluate the thermal cycling and thermal shock behaviors of such kind of PS-PVD quasi-columnar structured YSZ coatings. Evaluation of the microstructures, phase composition, residual stresses and failure behaviors of the coatings before and after thermal tests were investigated.

2 Experiment procedures

2.1 Coating preparation

Ni-based superalloy (GH3128, Beijing CISRI-GAONA Materials & Technology Co., Ltd., China) was used as substrates, which were plate shape with a dimension of 25 mm × 25 mm × 2.7 mm and disk shape with a diameter of 30mm and a thickness of 5mm. Before the spraying processes, the substrate was sand blasted using alumina powder and then followed by ultrasonically cleaning in ethanol. About 100 μm thick CoNiCrAlY bond coat was deposited by VPS (VPS / LPPS-PVD / CVD hybrid system, Oerlikon Metco, Wohlen, Switzerland).

All ceramic coatings were fabricated by a PS-PVD system (Oerlikon Metco, Wohlen, Switzerland). To achieve the low working pressure of 10 - 100 Pa and a power input of 180 kW, the chamber was equipped with a vacuum pump unit consisting of a high suction capacity blower and two sets of pump stands, including screw pump and roots pump, and power sources. Since the standard F4-VB plasma torch cannot be operated at such higher power, the O3CP torch was applied which generates a supersonic DC plasma jet. The samples presented here were sprayed with a power input of 110 - 150 kW and a plasma gas flow of 95 -120 slpm (using helium and argon as well as helium argon and hydrogen) at 100 - 150 Pa chamber pressure. The feedstock material used for the coatings was an agglomerated 7 wt.% yttria partially stabilized zirconia powder (7YSZ) produced by Oerlikon Metco (Switzerland) designated as M6700 and the grain size distribution with d_{10}=2 μm, d_{50}=8 μm, and d_{90}=18 μm, respectively. This agglomerated powder consisted of submicron ZrO_2 particles and nano-sized Y_2O_3 particles. Prior to spraying, the substrate plates were preheated by the plasma jet and the temperature was controlled by sweep movements of the plasma jet over the substrate. Before the deposition of the TBC, the surface of the bondcoats were polished to smooth roughness. The improvement of lifetime by a decreased roughness has been observed to apply for PS-PVD coatings [10]. During the coating deposition process, feeding additional oxygen into the vacuum chamber was used to prevent a loss of oxygen due to the high temperatures in the plasma plume (up to 15000 K) in combination with the low oxygen pressure plasma plume, the oxide content of the spray powder can be depleted apparently by this spray process.

2.2 Thermal shock and cycling test

The thermal shock test of the PS-PVD coatings was performed by heating and the water quenching method. The plate shape samples were heated at an evaluated temperature of 1100℃ in a tube furnace (SLG 1400-60, Shanghai Shengli Test Instruments Co., Ltd., China) for 20 min and then followed by directly water quenching to room temperature. Before being placed into the furnace once again, the coating samples were dried by compressed air. Thermal cycling tests of the samples were performed in a flame shock facility (Shanghai Institute of Ceramics, Chinese Academy of Sciences, Shanghai, China). The coatings were heated for 3 min at 1250±15 ℃. During the heating stage, the backside of the samples was cooled by compressed air to maintain a temperature gradient through the coatings thickness. During the cooling stage, the samples were cooled by compressed air from both side for 3 min to below 200 ℃. When visible spallation of the coating reached about 10% of the total coating surface, the thermal cycling test was stopped and the number of the cycles was defined as the lifetime of the TBCs.

2.3 Coating characterization

The phase composition of the as-sprayed coating was characterized by X-ray diffractometer (XRD, D/max 2550 V, Rigaku Industrial Corporation, Japan) with filtered Cu Kα (40 kV, 40 mA) radiation at a scan rate of 4°/min in the 2θ range of 20°-80°. The microstructures of the samples were examined using a scanning electron microscopy (SEM, TM3000, HITACHI, Japan) operated in backscattered electron image mode.

2.4 Raman spectroscopy and Cr^{3+} photoluminescence piezo-spectroscopy (PLPS)

The micro-Raman spectroscopy technique and PLPS method was used to measure the residual stresses distribution in the YSZ ceramic topcoat and the TGO scale, respectively [18-21]. The measurements were made at room temperature in ambient air using 532 nm laser excitation in a confocal Raman microprobe (Renishaw Invia Confocal micro-Raman Microscope, Glouchestershire, UK). The diameter and spatial resolution of the laser beam at the specimen surface were about 1 and 3 μm, respectively. Resolution of a spectrometer in the spectroscopic wavenumber range used (100-800 cm^{-1}) was less than 0.2 cm^{-1}. It was well known that Raman shifts towards higher wavenumbers corresponding to an increase in the in-plane compressive residual stresses, and vice verse. The frequency shift could be converted into a biaxial stress depending on the relation for a bi-axially stressed stabilized tetragonal ZrO_2.

$$\Delta\upsilon_{TBC}=\Pi_{TBC}\,\sigma_{TBC} \quad (1)$$

where Δυ referred to the frequency shift (cm^{-1}) from the stress-free state, σ_{TBC} was the residual stress in the topcoat, Π_{TBC} was the piezo-spectroscopic coefficient, and the coefficient for plasma-sprayed TBC was reported to be 25 cm^{-1} GPa^{-1} [19].

The PLPS method was used to measure stress in the TGO through the topcoat layer non-destructively. Resolution of a spectrometer in the spectroscopic wavenumber range used (14300-14500 cm^{-1}) was less than 0.2 cm^{-1}, which gave the measured stress resolution of ~40 MPa. The frequency shift of the luminescence in the TGO layer could be converted to the residual stress by the following expression:

$$\Delta\upsilon_{TGO}=\Pi_{TGO}\sigma_{TGO} \quad (2)$$

where Π_{TGO} was the piezo-spectroscopic coefficient and for plasma-sprayed TBC was set to be 5.07 cm^{-1} GPa^{-1} [21].

3 Results and discussion

3.1 Morphology of the as-sprayed coatings

The photographs of a PS-PVD coating, **Fig. 1**. An overview of the microstructure of a specimen, the typical characteristic structure of PS-PVD coatings with its feather-like columns, **Fig. 2**. At the bottom of the YSZ topcoat, initial columns with widths of few μm were deposited which make the coating look very dense. During the further deposition, some columns grew faster than the others and the slower growing columns stopped. Thereby the feather-like columns structure which lead to the increased tolerance to strain were formed.

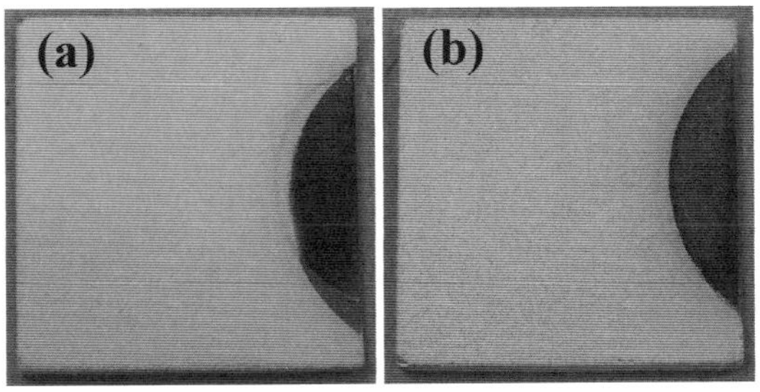

Fig.1. Front-side (a) and back-side (b) photographs of the PS-PVD coating

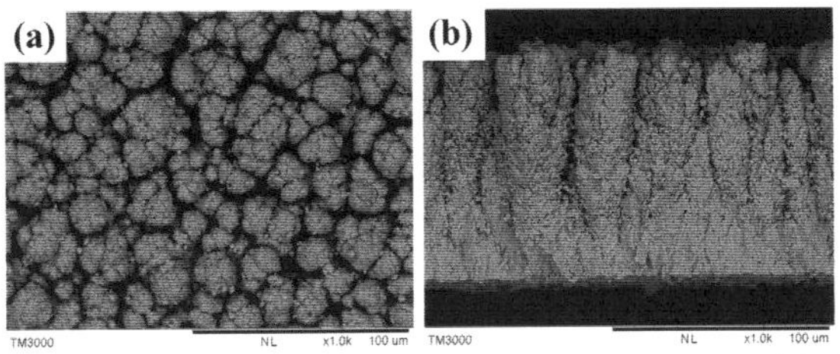

Fig. 2. Surface (a) and fracture surface (b) SEM micrographs of YSZ topcoat

The XRD patterns and Raman spectra of as-sprayed coating, **Fig. 3**. The XRD patterns show that the coatings consist of tetragonal phase zirconia. The micro-Raman spectroscopy was a powerful technique to study the structure changes in the coatings. Fig. 3(b) gives micro-Raman spectra of the topcoat. The typical Raman bands for the tetragonal phase of yttria stabilized zirconia coatings were 148, 260, 320, 470, 612 and 640 cm^{-1}, respectively. No monoclinic bands (at about 182 and 191 cm^{-1}) were observed which confirmed the results obtained by XRD.

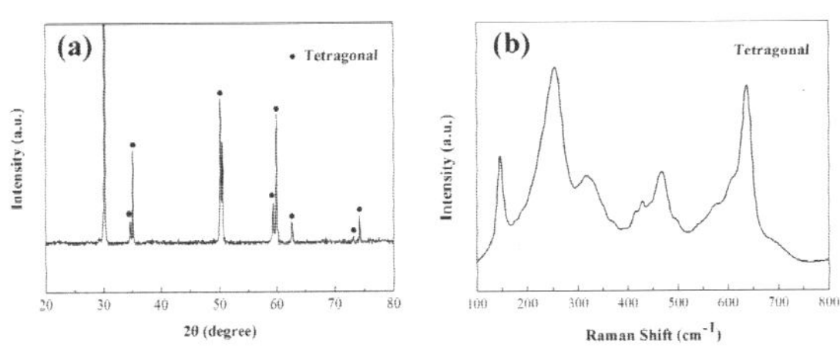

Fig. 3. XRD patterns (a) and Raman spectra (b) of as-sprayed coating

3.2 Thermal cycling behavior of PS-PVD coatings

Photographs of YSZ TBCs (here Φ30 mm × 5.8 mm substrate was used) before and after thermal cycling test, **Fig. 4**. Apparent spot spallations were observed mainly occurred at the center of the YSZ topcoat surface of the sample. With the thermal cycling going on, the spalled area enlarged. After 377 cycles, the spallation area reached to ~10% of the total surface area of the specimen and thermal cycling test was stopped. Note that after thermal cycling, the column structure retain their original appearance and inter-columnar gaps were still exist, indicating that the YSZ coating had good sintering resistance, which is to say that the sintering of YSZ ceramic layer was not the major reason for leading to the spallation of the PS-PVD coatings. What needs to be stressed was that the quasi-columnar structured TBCs failed at the latest after 623 cycles (not including here) by adjusting the process parameters. However, many more promising approaches have to be tested further in order to improve the durability of the PS-PVD coatings.

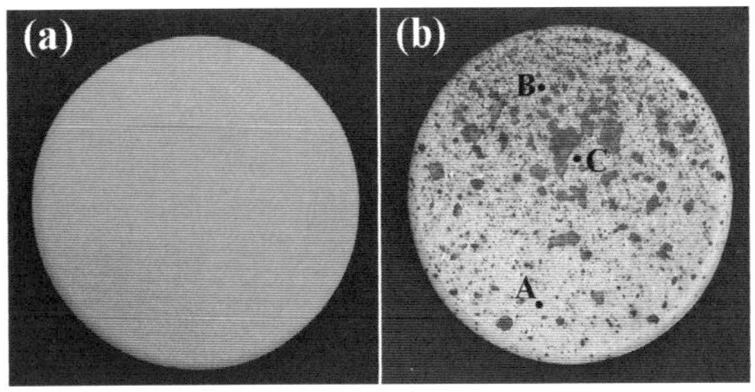

Fig. 4. Photographs of YSZ TBCs before (a) and after thermal cycling 377 cycles (b)

In order to have a better understanding of the failure mechanism. Raman spectra detected at three different locations on the specimen after thermal cycling test (shown in Fig. 4(b)), **Fig. 5**. At the edge of the coating (position A), the spectra peaks correspond to vibrational modes of tetragonal zirconia, confirming the absence of monoclinic phase. However, at the center region of the coatings (position B and C), the phase transformation from tetragonal to monoclinic was indicated by the increase in intensity of the characteristic bands of monoclinic zirconia at 182 and 191 cm^{-1}. The Raman spectra also illustrate another feature of the transformation to monoclinic, the volume fraction of the monoclinic phase could be quantified from the intensities of the tetragonal bands at 148 and 263 cm^{-1} and the monoclinic bands at 182 and 191 cm^{-1}. Therefore, it was obvious that close to the center region (Position C) of the coatings had a increasing extent of phase transformation than relatively far away from the center area (Position B) during thermal cycling. It was well known that tetragonal-monoclinic phase transformation in 6-8 wt.% YSZ was accompanied essentially by a significant volume expansion of up to 5%, indicating that the phase transformation might be the main reason for resulting in degradation and spallation of the PS-PVD coatings.

3.3 Thermal shock behavior of PS-PVD coatings

The thermal shock cycling lifetimes and corresponding macroscopic surface micrographs of the PS-PVD YSZ sample by optimizing the process parameters are graphically presented, **Fig. 6**. As can be seen, the spallation of coatings samples started at the edges of the topcoat after 20 cycles, and then more edge spallation started associated with propagation to the adjacent areas on further cycling, and less than 5% spallation occurred even after 200 cycles. Obviously, the PS-PVD TBCs exhibit a significant improvement in the thermal shock resistance compared with the APS coatings. Lifetime extension of the PS-PVD coatings, as compared with the APS samples, can be attributed to the quasi-columnar structure. As the segmentation cracks can open during tensile loading in a manner similar to the EB-PVD coatings, the thermal mismatch stress was reduced significantly and a reduction in thermal stress facilitates an improvement in the strain accommodation capability of the PS-PVD TBCs in the process of thermal shock.

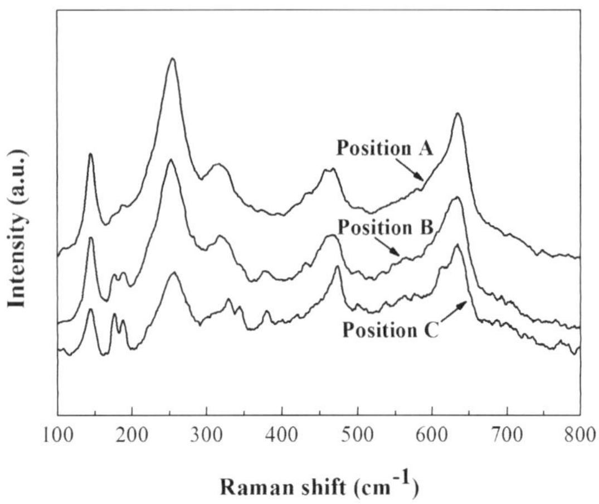

Fig. 5. Raman spectra of different positions in PS-PVD coating after thermal cycling in Fig. 4 (b)

Fig. 6. Macroscopic images of PS-PVD samples during thermal shock tests (a) as-sprayed, (b) 20, (c) 70 and (d) 200 cycles.

Micro-Raman spectra of the topcoat before and after thermal shock test, **Fig. 7**. No monoclinic bands (182 and 191 cm^{-1}) were observed which confirmed that no phase transformation occurred. Raman scattering depends on electron-phonon interactions was very susceptible to local environments, thus it was also a suitable technique to investigate the changes in the residual stresses of ceramic topcoat. The principle Raman bands of zirconia tetragonal phase were at 260 and 640 cm^{-1}, which related with stretching modes of Zr-O_{II} and Zr-O_I. In previous Raman piezo-spectroscopy measurements distribution of residual stress in tetragonal-zirconia TBCs, the shift of the Raman peak at around 640 cm^{-1} was used to measure stress distribution in the YSZ topcoat since it has a good signal to noise ratio (S/N). Considering the Raman band at about 640 cm^{-1}, some shifts were observed due to the presence of different in-plane stress states [19].

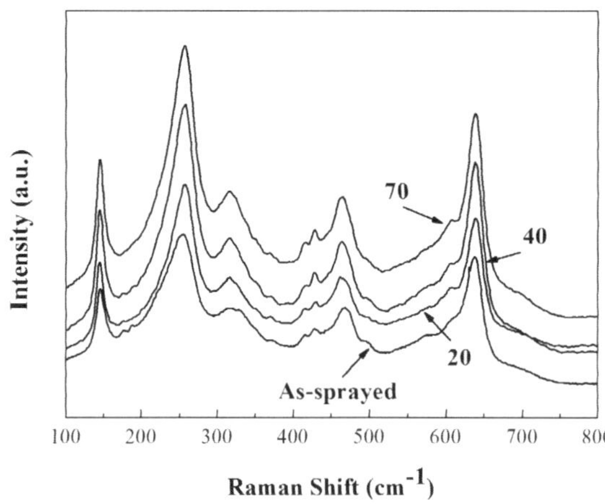

Fig. 7. Raman spectra of as-sprayed coating and the coatings after different thermal shock cycles

The variation of position of the Raman peak at around 640 cm^{-1} and the average residual stresses in the topcoat layer before and after thermal shock test, **Fig. 8**. The Raman peak of the samples after thermal shock for 70 cycles shifted to a higher wavenumber (638.86 ± 0.45 cm^{-1}) compared with the as-sprayed one (637.28 ± 0.36 cm^{-1}) and the powder (637.35 ± 0.11 cm^{-1}). According to the formula (1), the conversion of Raman peak frequency shifts measured the in-plane residual stress in the YSZ topcoat were gained. For the as-sprayed coatings, a low-level tensile stress (2.56 ± 14.66 MPa) was present in the topcoat. After thermal shock for 20 cycles, compressive stress was obtained. The average topcoat compressive stress was found to increase with thermal shock cycles and the value was σ_{TBC}= -51.04 ± 12.21, -49.28 ± 8.66, -60.09 ± 18.27 MPa for the 20, 40 and 70 cycles of thermal shock test, respectively. The tensile stress developed in the as-sprayed coatings which was due to the lower cooling rate under a pressure of 100 to 200 Pa at the end of the spraying process. After annealing, stress state was changed from tensile to compressive which was mainly due to the different coefficients of thermal expansion between the component of the TBC system ($\alpha_{Topcoat}$=11~13 × 10^{-6} /℃, $\alpha_{Bondcoat}$=13~16 × 10^{-6} /℃) [22]. Attributed to stress relaxation in the ceramic topcoat at high temperature exposure, more compressive stress would accumulated when the coatings cooling from the high temperature to ambient temperature.

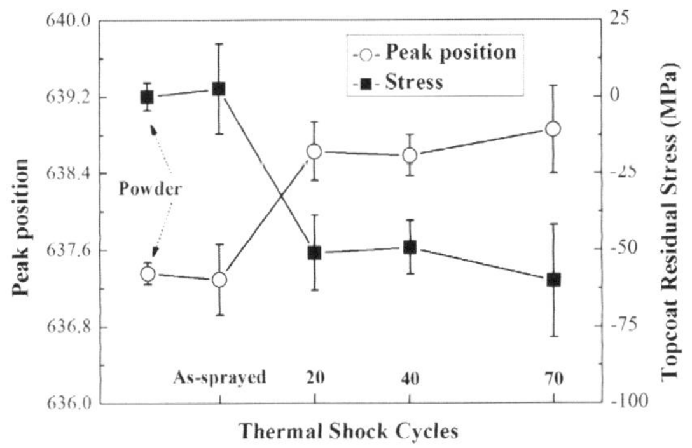

Fig. 8. Variation of the residual stresses in the YSZ topcoat as a function of thermal shock cycles

The distribution of residual stress in the TGO scale with various thermal shock cycles, **Fig. 9**. The measured in-plane stress through the topcoat surface (from -1.95 ± 0.17 to -3.93 ± 0.35 GPa) varied with the thermal shock cycles. At the beginning of thermal shock test, the measured in-plane TGO compressive stress firstly rapidly increased and then essentially maintain at a relatively stable value with further test. and the value of compressive stresses reached to σ_{TGO}= -3.94 ± 0.28, -3.99 ± 0.39, -3.93 ± 0.35 GPa for the 20, 40 and 70 cycles of thermal shock test, respectively. Nevertheless, it should be emphasized that the apparently decreasing of the TGO stress had not been observed. The decreased residual stress was mainly considered due to cracking generated in the topcoat layer or at interface between the topcoat and the TGO layer, and no obvious spallation in the PS-PVD coating was found even after thermal shock for 200 cycles, as shown in Fig. 6(d).

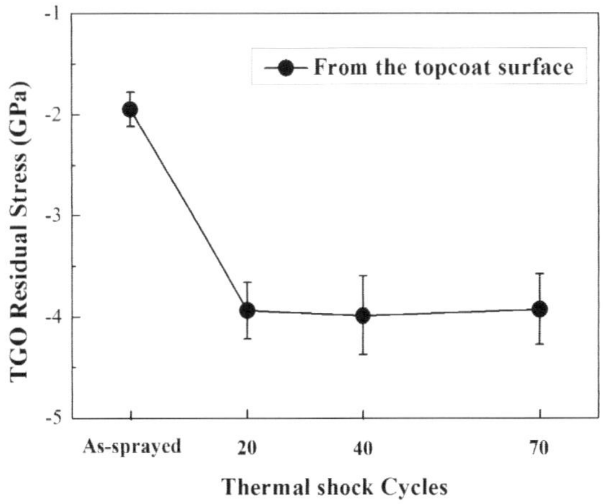

Fig. 9. Average stresses measured from the ceramic topcoat surface in the TGO as a function of the thermal shock cycles

4 Conclusions

In the present paper, quasi-columnar structured thermal barrier coatings were fabricated by plasma spray-physical vapor deposition (PS-PVD) process. Microstructure, phase composition and residual stresses of the coatings before and after thermal cycling and thermal shock test were investigated. Based on the experimental results, the following conclusions can be drawn:

(1) The columnar-structured TBCs failed at the latest after 623 cycles during flame shock testing.

(2) The as-sprayed quasi-columnar coatings consisted of tetragonal phase. Monoclinic phase was detected in the coatings after thermal cycling test especially nearby the spot spallation region.

(3) The accumulated tensile stress in the as-sprayed ceramic topcoat changed to compressive state after thermal shock for 20 cycles, and then gradually increased. In addition, the compressive stresses in the TGO measured through the ceramic topcoat surface increased rapidly at first and then essentially maintain at a relatively stable value with further test.

References

[1] Miller R.A.: Thermal barrier coatings for aircraft engines: history and directions, Journal of Thermal Spray Technology 6 (1997), Issue 1, pp. 35/42.

[2] Bose, S., J. DeMasi-Marcin: Thermal barrier coating experince in gas turbine engines at Pratt

& Whitney, Journal of Thermal Spray Technology 6 (1997), Issue 1, pp. 99/104.
[3] Vassen, R., A. Stuke, D. Stöver: Recent development in the field of thermal barrier coatings, Journal of Thermal Spray Technology 18 (2009), Issue 2, pp. 181/86.
[4] Clarke, D.R., C.G. Levi: Materials design for the next generation thermal barrier coatings, Annual Review of Materials Research 33 (2003), pp. 383/417.
[5] Kakuda, T.R, A.M. Limarga, T.D. Bennett, D.R. Clarke: Evolution of thermal properties of EB-PVD 7YSZ thermal barrier coatings with thermal cycling, Acta Materialia 57 (2009), pp. 2583/91.
[6] Cipitria, A., I.O. Golosnoy, T.W. Clyne: A sintering model for plasma-sprayed zirconia TBCs. Part I: Free-standing coatings, Acta Materialia 57 (2009), pp. 980/92.
[7] Cipitria, A., I.O. Golosnoy, T.W. Clyne: A sintering model for plasma-sprayed zirconia TBCs. Part II: Coatings bonded to a rigid substrate, Acta Materialia 57 (2009), pp. 993/1003.
[8] Lughi, V., D.R. Clarke: Low-temperature transformation kinetics of electron-beam deposited 5 wt.% yttria-stabilized zirconia, Acta Materialia 55 (2007), pp. 2049/55.
[9] Hospach, A., G. Mauer, R. Vaßen, D. Stöver: Columnar-structured thermal barrier coatings (TBCs) by thin film low-pressure plasma spraying (LPPS-TF), Journal of Thermal Spray Technology 20 (2011), Issue 1-2, pp. 116/120.
[10] Hospach, A., G. Mauer, R. Vaßen, D. Stöver: Characteristics of ceramic coatings made by thin film low pressure plasma spraying (LPPS-TF), Journal of Thermal Spray Technology 21 (2011), Issue 3-4, pp. 435/40.
[11] Rezanka, S., G. Mauer, R. Vaßen: Improved thermal cycling durability of thermal barrier coatings manufactured by PS-PVD, Journal of Thermal Spray Technology 23 (2012), Issue 1-2, pp. 182/89.
[12] Mauer, G., A. Hospach, R. Vaßen: Process development and coating characteristics of plasma spray-PVD, Surface & Coatings Technology 220 (2013), pp. 219/24.
[13] Mauer, G., M.O. Jarligo, S. Rezanka, A. Hospach, R. Vaßen: Novel opportunities for thermal spray by PS-PVD, Surface & Coatings Technology 268 (2015), pp. 52/57.
[14] Niessen, K.V., M. Gindrat, A. Refke: Vapor phase deposition using plasma spray-PVD, Journal of Thermal Spray Technology 19 (2010), Issue 1-2, pp. 502/09.
[15] Goral, M., S. Kotowski, A. Nowotnik, M. Pytel, M. Drajewicz, J. Sieniawski: PS-PVD deposition of thermal barrier coatings, Surface & Coatings Technology 237 (2013), pp. 51/55.
[16] Niessen, K.V., M. Gindrat: Plasma sprayed-PVD: a new thermal spray process to deposit out of the vapor phase, Journal of Thermal Spray Technology 20 (2011), Issue 4, pp. 736/43.
[17] Gao, L.H., H.B. Guo, L.L. Wei, C.Y. Li, H.B. Xu: Microstructure, thermal conductivity and thermal cycling behavior of thermal barrier coatings prepared by plasma spray physical vapor deposition, Surface & Coatings Technology 276 (2015), pp. 424/30.
[18] Limarga, A.M., R. Vaßen: Stress distribution in plasma-sprayed thermal barrier coatings under thermal cycling in a temperature gradient, Journal of Applied Mechanics 78 (2011), pp. 011003 1-9.
[19] Tanaka, M., M. Hasegawa, A.F. Dericioglu, Y. Kawawa: Measurement of residual stress in air plasma-sprayed Y_2O_3-ZrO_2 thermal barrier coating system using micro-Raman spectroscopy, Materials Science & Engineering A 419 (2006), pp. 262/68.
[20] Christensen, R.J., D.M. Lipkin, D.R. Clarke, K. Murphy: Nondestructive evaluation of the oxidation stresses through thermal barrier coatings using Cr^{3+} piezospectroscopy, Applied Physics Letters 69 (1996), Issue 24, pp. 3754/56.
[21] Schlichting, K.W., K. Vaidyanathan, Y.H. Sohn, E.H. Jordan, M. Gell, N.P. Padture: Application of Cr^{3+} photoluminescence piezo-spectroscopy to plasma-sprayed thermal barrier coatings for residual stress measurement, Materials Science & Engineering A 291 (2000), pp. 68/77.
[22] Rabiei, A., A.G. Evans: Failure mechanisms associated with the thermally grown oxide in plasma sprayed thermal barrier coatings, Acta Materialia 48 (2000), pp. 3963/76.

Flexure Strength of Metal Mesh Reinforced Ceramic Matrix Composite Fabricated Using Plasma Spray Method

Teymoor Mohammadi and Xiao Huang
Department of Mechanical and Aerospace Engineering
Carleton University, Ottawa, Ontario, Canada

Structural materials with higher temperature capability and mechanical integrity are sought after in gas turbine industry as the turbine inlet temperature continues to rise. While many ceramic materials are microstructurally stable at high temperatures, they exhibit poor mechanical properties under loads. In this study, Hastelloy X (HX) mesh is used as a reinforcement to strengthen 7YSZ ceramic material and improve the damage tolerance. A low cost fabrication process based on plasma spray is developed to manufacture the metal mesh reinforced 7YSZ. The mechanical test results of the HX reinforced 7YSZ CMC, based on a three-point bend test, show that HX/7YSZ CMC surpasses the monolithic 7YSZ in terms of peak load and displacement before fracture. More specifically, the incorporation of HX mesh provides over 7 times increase to the flexure strength. Heat treatment of the CMC further increases the load at the onset of crack initiation due to the enhanced bonding between 7YSZ and HX mesh.

1. Introduction

Currently the materials for hot section gas turbine components are primarily nickel- or cobalt-based superalloys. These alloys have a unique combination of high temperature strength, toughness, and resistance to degradation in corrosive and oxidizing environments. Since higher turbine inlet temperature helps improve the efficiency of the turbines and reduce the fuel cost and emissions, there is an on-going need to develop materials with greater strength and higher temperature capabilities. Although the overall performance of Ni-base superalloys have improved significantly over the past decades, the maximum operating temperatures of these superalloys are still limited. As temperature and stress distribution on a given part vary from location to location, the design and material selection for turbine components (such as blades and vanes) are primarily based on the highest localized temperature [1]. In extreme cases, air foils in modern gas turbines can reach as high as 1200°C, approaching 90% of the melting points of many superalloys. Such extreme temperature condition requires the use film cooling and thermal barrier coating on metallic components.

In addition to superalloys, ceramic materials are possible candidate for hot section components of gas turbines due to their high temperature capability. However, to withstand extreme temperature and stresses seen in gas turbines, alternative materials must also have sufficient strength, damage tolerance and environmental resistance. Monolithic ceramic materials are known to have poor tensile and impact properties and as such need to be improved before they can be considered for critical structural applications. Furthermore, materials for turbine applications must meet other requirements such as low component and life-cycle costs and less environmental impact. To meet these demands, the combination of superalloys with ceramics, in the form of metal-ceramic composites, may offer a compromise between performance and cost requirement [1].

The main objectives in incorporating metal reinforcement in ceramic-matrix composites (CMCs) are to increase the toughness through crack deflection/bridging and increase the strength by load transfer [2-4]. In traditional fiber reinforced composites, crack deflection forces the crack out of plane rather than through the fibres, i.e., the crack plane is no longer normal to the applied stress. Reinforcement of ceramic matrix with particulates and fibres can lead to an increase in toughness; but the increase is more significant when continuous fibres are employed. Both monolithic ceramic material and particulate reinforced ceramic material have been found to fail in a catastrophic manner, while the continuous fibre reinforced composite exhibited a considerable load carrying capacity even after cracks have been initiated.

This study was initiated to fabricate a low cost metal mesh reinforced ceramic matrix composite (CMC) materials and to study their mechanical properties and high temperature oxidation resistance. The manufacturing process and the test results from 3-point bend test will be reported here.

2. Materials And Experimental Procedure

2.1. Materials

Hastelloy X (HX) wire (1 mm dia.) mesh was selected in this study as the reinforcement. The composition of HX is given in **Table. 1**. This alloy was developed specifically for high temperature applications and it possesses a combination of good oxidation resistance, up to 1200°C [5], excellent formability and weldability. It is widely used in gas turbines as transition ducts, combustor cans, spray bars and flame holders.

Table. 1. Nominal chemical composition of Hastelloy X (wt.%)

Ni	Cr	Fe	Mo	Co	W	C	Mn	Si	B
Bal.	22	18	9	1.5	0.6	0.10	<1	<1	<0.008

The ceramic matrix used in this used is 7YSZ (7.5 wt.% yttria-stabilized zirconia). 7YSZ has a high melting point and can be used at temperatures up to 2,200°C in neutral or oxidizing atmospheres. In particular, yttria stabilized zirconia, within the composition in the range of 6-8 wt.% Y_2O_3, has shown to have prolonged spallation life at temperatures above 980°C [6]. As the control of interfacial strength between the reinforcement and the matrix is critical for optimizing the toughness of CMC, an interfacial layer between HX and 7YSZ was considered. Since the metal reinforced CMC will also be subjected to various oxidation/corrosion conditions, the interfacial layer must provide environmental protection to HX reinforcement. Known for its ability to prevent oxidation and diffusion of aggressive gases into metallic substrate, NiCrAlY type bond coat was chosen and applied onto HX mesh by plasma spray.

2.2. CMC Fabrication Procedure

The HX/7YSZ CMC specimens were manufactured in three steps: mesh fabrication, bond coat application, and final ceramic layer deposition. The HX mesh was manufactured by first cutting and inserting strips of wire into a jig and then spot welding the wires together. To produce a repeatable reinforcement mesh structure, a jig was created to weave the mesh and also hold the wires in place during spot welding. To maintain an even and uniform mesh aperture size during spot welding, a spacer was used. During resistance spot welding (equipment from Miyatchi Unitek, CA, USA), a flowing argon gas was used to prevent oxidation of the HX wires.

After the mesh was released from the jig, it was then mounted onto a stainless steel support fixture for plasma spray coating, **Fig. 1. (a)**. They were grit blasted with No. 46 alumina (ALODUR) for 2 minutes to create a clean, rough surface prior to plasma spraying. NiCrAlY bond coat (Ni-264-4, Praxair) and 7YSZ (ZRO-271-3, Praxair) ceramic matrix were deposited onto the HX mesh using the Mettech Axial III plasma spray system (Northwest Mettech Inc., Vancouver, Canada) with AT-1200HP powder feeder.

The plasma spray process was carried out in consecutive runs with brief cooling before each run. NiCrAlY bond coat was deposited to an average thickness of 0.6 mm on each side of the HX mesh; and about 1.5 mm of 7YSZ ceramic matrix was applied to both sides of NiCrAlY coated HX mesh after a total of 14 runs. The images after 3 and 4 runs are shown in **Fig. 1. (b) and (c)**. After eight runs (four run on each side), all openings between the mesh wires were filled, and subsequent runs continued to build up the required 7YSZ matrix thickness and lessen the ripples on the surface. The final thickness of the CMC specimens was 6.15 mm, **Fig. 2**.

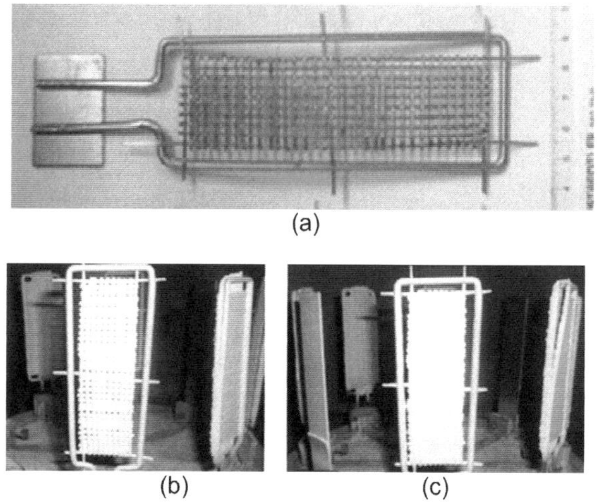

Fig. 1. (a) Hastelloy X wire mesh (10 wires per in.) mounted in a frame. (b) and (c) 7YSZ coating on HX mesh after 3 and 4 runs on each side.

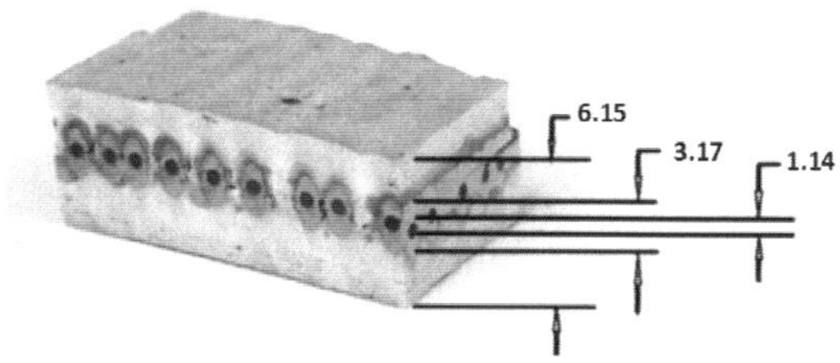

Fig. 2. Final HX/7YSZ CMC specimen dimensions (in mm).

Monolithic 7YSZ specimen (without HX mesh) was also prepared at the same time as the CMC samples and served as a baseline for this study. To produce the specimen, 7YSZ powder was plasma sprayed onto a mild steel plate. After the plasma spray process was completed, the 7YSZ coating was removed from the plate to form a free-standing specimen. The thickness of 7YSZ specimen (about 3 mm) was thinner than HX/7YSZ due to the lack of HX mesh and NiCrAlY bond coat.

When being applied as thermal barrier coating, 7YSZ often cracks and spalls off from the interface between 7YSZ and MCrAlY bond coat during isothermal and cyclic oxidation tests. It has been found that a vacuum heat treatment at 1080°C for 4 hours can delay TBC cracking and spallation [7] as it promotes the formation of a uniform Al_2O_3 (TGO) layer at the interface between 7YSZ and MCrAlY. For this reason, half of the HX/YSZ CMC specimens were heat treated in a OXY-GON vacuum furnace ($2x10^{-5}$ torr) for 4 hours at 1080°C. These specimens are referred to as heat treated specimens, whereas the other half of the CMC specimens are termed as-sprayed ones.

To prepare specimens for 3-point bent test, two rectangular test specimens, measured 50 mm x 12.5 mm x 6.15 mm, were prepared for both as-sprayed and heat treated conditions. Three-point bending test was carried out to investigate the effect of HX reinforcement on load carrying capacity and cracking resistance of 7YSZ. The bend test was carried out using a MTS 810 system with maximum load capacity of 25 kN. An extensometer (model MTS 634.12E-24) was used to record the displacements during testing. Axial force (in N) and associated axial displacement (in mm, along the thickness direction) were recorded during testing.

3. Results

3.1 Visual Inspection

The as-sprayed and heat treated HX/7YSZ CMC specimens after bend test are shown in **Fig. 3.** and **Fig. 4.** Both longitudinal and transverse cracks were observed on the as-sprayed HX-7YSZ specimens while only transverse cracks were observed in the heat treated HX/7YSZ specimens. Small fragments of 7YSZ were also found to spall off on the compression side of the as-sprayed specimens, **Fig. 3 (b)**.

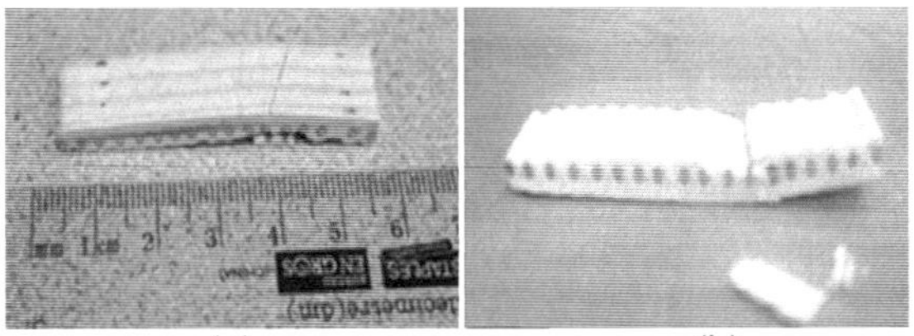

(a) (b)

Fig. 3. Images of as-sprayed HX/7YSZ CMC after bend test. (a) both longitudinal and transverse cracks observed on the tension side. (b) spallation on the compression side.

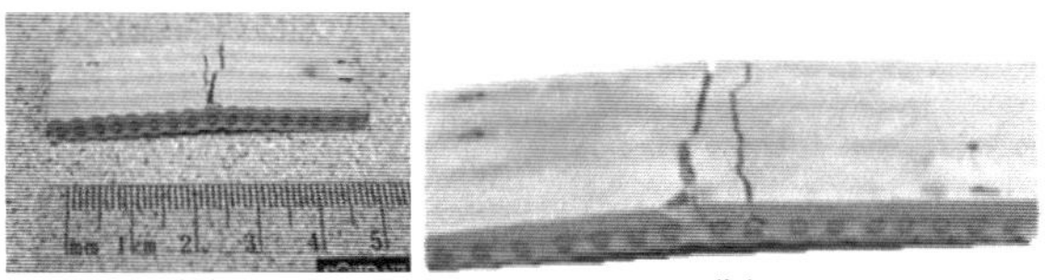

(a) (b)

Fig. 4. Transverse cracking in heat treated HX/7YSZ specimen after bend test. (a) one specimen showing three large transverse cracks and (b) another specimen with two large transverse cracks.

3.2 Bend Test Results

The load and displacement curves for the as-sprayed and heat treated HX/7YSZ CMC specimens are shown in **Fig. 5.** Overall, the CMC specimens exhibited significantly improved failure resistance (peak load and displacement prior to fracture) relative to monolithic 7YSZ. Moreover, based on the results, **Fig. 5.**, the as-sprayed HX/7YSZ CMC started to show signs of cracking at an approximate load of 516 N (540 N for the 2nd set) while monolithic 7YSZ has completely fractured under a load of 55 N. The peak load for as-sprayed HX/7YSZ CMC before fracture was 1637.4 N (and 1629 for the 2nd set), about 30 times greater than that seen in monolithic 7YSZ; its displacement at the peak load was about 6 times greater than 7YSZ.

The heat treated HX/7YSZ CMC has a larger load (40% greater) and displacement at the onset of cracking than that observed in the as-sprayed HX/7YSZ CMC, **Fig. 5.** The peak load of the heat treated HX/7YSZ CMC, however, was about 50% of the as-sprayed HX/7YSZ CMC; and the axial displacement of the heat treated CMC at this point was greater than as-sprayed HX/7YSZ (2.3 mm vs. 1.25 mm). The changes in peak load and displacement of the heat treated CMC specimens could be attributed to the modification of bond strength between NiCrAlY and 7YSZ and that of HX and NiCrAlY as a result of post spray vacuum heat treatment. A stronger bonding between 7YSZ and NiCrAlY for heat treated CMC delayed the onset of ceramic cracking, however, it also prevented effective strengthening of CMC due possibly to the hindrance of metal-wire pull out.

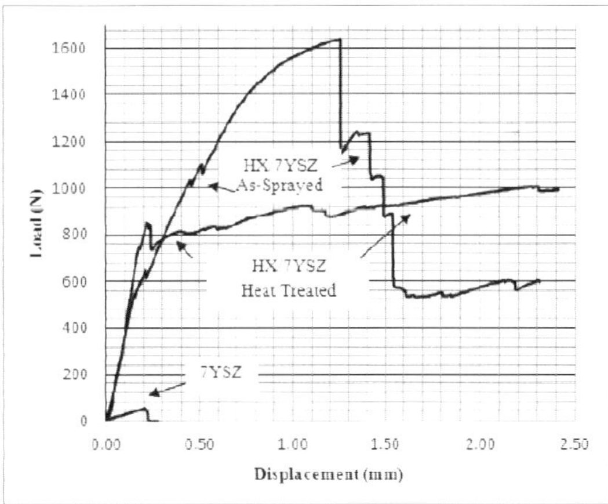

Fig. 5. Load-displacement curves for HX/7YSZ (as-sprayed and heat treated) and monolithic 7YSZ.

3.3 Microstructural Analysis

Scanning electron microscope (SEM) (Tescan Vegall XMU) was used to investigate the microstructure and fracture surface of the CMC specimens. Examination of the cross sections of the specimens before bend test showed that plasma spray process left behind internal cavities and small voids in specimens. Cavities from plasma spraying process are present between HX wires in both specimens, **Fig. 6.** (a) and (b), although the heat treated HX/7YSZ CMC contains less cavities than the as-sprayed HX/7YSZ CMC due to sintering effect. The reduced cavity in 7YSZ may also have contributed to the higher crack initiation load in the heat treated HX/7YSZ CMC specimens.

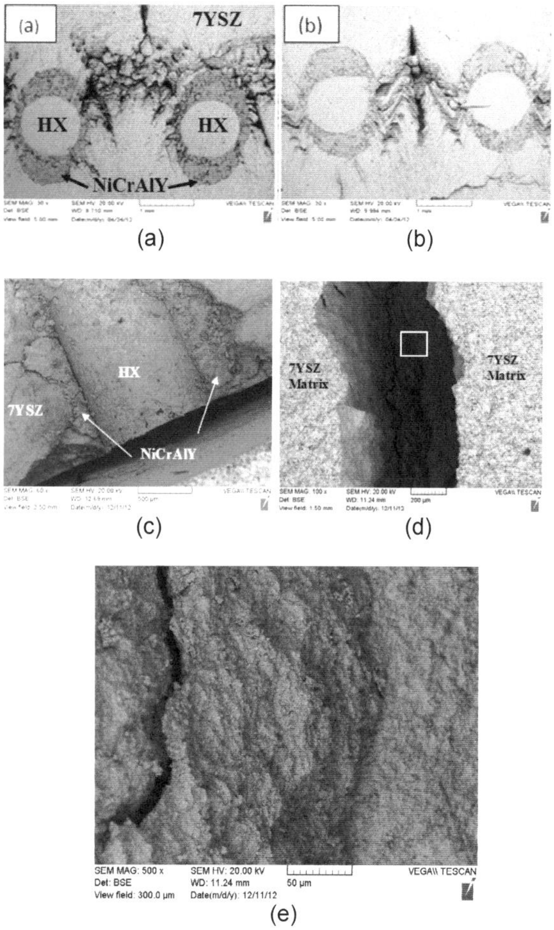

Fig. 6. Cavities between HX wires in (a) as-sprayed, (b) heat treated HX/7YSZ CMC specimens, (c) as-sprayed HX/7YSZ (compression side), (d) a transverse crack on the heat treated HX/7YSZ after bend test and (e) NiCrAlY covered HX inside of the crack (the boxed region of b).

During bend test, multiple micro-fissures along the longitudinal and transverse directions were observed in the ceramic matrix of the as-sprayed HX/7YSZ CMC specimens before reaching the peak load. As the load was increased, some of the micro-fissures developed into visible cracks, leading to large cracks and ceramic spallation, **Fig. 4.** After bend test, multiple longitudinal and transverse cracking (on tension side) and spallation (on compression side) were observed on the as-sprayed HX/7YSZ specimens; whereas the heat treated HX/7YSZ specimens only had a few transverse cracks. By examining the fracture surface of the as-sprayed specimens on the compression side, bare HX wire is clearly visible, **Fig. 6. (c)**, suggesting weak bonding between HX wire and NiCrAlY. Since no spallation was found on the heat treated HX/7YSZ CMC specimen, SEM analysis was conducted in the vicinity of a transverse crack after bend test. Inside the crack **(Fig. 6.** (d) and (e)), a transverse HX wire was found to be fully covered with NiCrAlY bond coat, an indication of stronger bonding (due to heat treatment) between NiCrAlY and HX wire. Further investigation is needed to study the effect of bond strength on the load-displacement characteristics of the fabricated CMCs using plasma spray.

4. Discussion

Although conventionally used as a coating application technique, plasma spray has the potential to provide a more economic and viable means to produce ceramic components than conventional methods such as high temperature sintering or HIPping [8]. Plasma spray process generates a typical lamellar structure in the form of layered splats. Between the splats there are numerous voids and within the splats, grains grow in columnar manner due to the large temperature gradient during solidification. Cracks are also observed along the columnar growth direction. The presence of voids between splats and cracks provides low energy path for crack propagation and delamination between splats; as such, as-sprayed materials have low strength and toughness, and are highly anisotropic [9 and 10]. Most plasma sprayed ceramic materials exhibit a porosity of 5-20%, low elastic moduli compared to conventionally processed ceramic materials (4 vs. 20 GPa) and low flexural strengths of 14-40 MPa [11 and 12]. More specifically, a sintered alumina can reach a fracture strength between 280-550 MPa, and YSZ between 800-1500 MPa [13] while free-standing plasma sprayed alumina has its flexure or tensile strength ranging from 7-22 MPa [8 and 14]. And 23 MPa tensile strength was reported for as-sprayed 8YSZ [14]. To improve the strength of as-sprayed ceramic material, post heat treatment was used [8]; the strength of alumina after heat treatment has been observed to increase by a factor of 3.

Based on the 3-point bend test results, the flexure strength of 7YSZ and HX/7YSZ CMCs was calculated using the following equation [13]:

$$\sigma_f = \frac{3PL}{2wb^2} \qquad (1)$$

where P is the fracture force, L is the outer span of the 3-point bend fixture, *w* is the specimen width and *b* is the specimen thickness. The results of the calculated flexure strength are summarized in Table 2. The as-sprayed 7YSZ, as anticipated, exhibits very low strength

DVS 324

comparing to the published values of the sintered 7YSZ. However, over 7 times of improvement was resulted by the incorporation of HX mesh into the 7YSZ matrix. This improvement is much greater than that from particulate reinforcement [15].

TABLE. 2. Summary of Flexure Strength of 7YSZ and HX/7YSZ

	P (peak load before fracture) (N)	L (m)	w (m)	b (m)	σ_f (MPa)
7YSZ	55	0.040	0.0125	0.003	29.3
HX/7YSZ	1681	0.040	0.0125	0.0062	213
HX/7YSZ heat treated	988	0.040	0.0125	0.0062	125

5. Conclusion

In this study, a new process to manufacture low cost metal mesh reinforced ceramic matrix composite (CMC) was developed. The mechanical properties of the manufactured CMC specimens were evaluated using three-point bend test. Based on the peak load before fracture, the plasma sprayed 7YSZ exhibited extremely low flexure strength. However the incorporation of HX mesh provided over 7 times increase to the strength. Heat treatment after spraying helped reduce the extend of cavity and increase the interfacial bond between HX and 7YSZ As such, the heat treated HX/7YSZ had higher load and displacement at the onset of cracking.

6. Acknowledgements

The authors thank Natural Sciences and Engineering Research Council of Canada for providing funding for this research. We also thank Dr. R. Kearsey at National Research Council of Canada for providing spot welding machine for this work.

7. Literature

[1] Matthews, F. L. and R.D. Rawlings: Composite materials - engineering and science, NW, USA, Woodhead Publishing Limited and CRC Press LLC (1999).
[2] Warren, R.: Ceramic matrix composites, New York, NY, USA, Chapman and Hall Inc. (1992).
[3] Rice, R.W.: Mechanical properties of ceramics and composites, Marcel Dekker, New York (2000).
[4] Hilling, W.B.: Strength and toughness of ceramic matrix composites. Annual Reviews of Materials Scince (1987), Issue 17, pp. 341/83.
[5] http://www.haynesintl.com/pdf/h3009.pdf (accessed 2015).
[6] Dorfman, M.R., L. Xie, A. Cipitria, S. Paul, I. Golosnoy, T.W. Clyne and W. Chi: Properties and performance of high purity thermal barrier coatings. Journal of Thermal Spray Technology (2007), Issue 16(5-6), pp. 804/8.
[7] Chen, W. R., X. Wu, B. R. Marple, R.S. Lima, R. Archer, X. Huang, and P. Patnaik: Influence of post-spraying heat treatments on the oxidation and cracking behaviour of thermal-sprayed thermal barrier coatings, 12th International Conference on Fracture ICF 2009, Ottawa, Canada.
[8] Damani, R.J., D. Rubesa and R Daner: Fracture toughness, strenghth and themal shock behaviour of bulk plasma sprayed alumina – effect of heat treatment. Journal of the European Ceramic Society (2000), Issue 20, pp. 1439/52.
[9] Lutz, E. H.: Plasma ceramic, Metallurgical International (1993), Issue 25, pp.131/7.
[10] Wanner, A., and E. H. Lutz: Elastic anisotropy of plasma sprayed, free-standing ceramics, Journal of American Ceramic Society (1998), Issue 81, pp. 2706/8.
[11] Lutz, E. H.: Microstructure and properties of plasma ceramics, Journal of American Ceramic Society (1994), Issue 77, pp. 1274/80.
[12] Pawlowski, L.: The Science and Engineering of Thermal Spray Coatings, Wiley and Sons (1995).
[13] Callister, W. D.: Materials Science and Engineering - An Introduction, Wiley and Sons (2007).
[14] Kishitake, K., H. Era and S. Baba: Tensile strength of plasma-sprayed alumina and zirconia coatings on titanium, Journal of Thermal Spray Technology (1995), Issue 4(4), pp. 353/7.
[15] Kim, S., W. Chung, K. Sohn, C. Son, S. Lee: Improvement of flexure strength and fracture toughness in alumina matrix composites reinforced with carbon nanotubes, Materials Science and Engineering A (2009), 517, pp. 293-299.

Stress analysis and failure mechanisms of plasma-sprayed thermal barrier coatings

Jiasheng YANG [a,b], Liang WANG [a], Dachuan LI [a,b], Xinghua ZHONG [a], Huayu ZHAO [a], Shunyan TAO [a*]

[a] The Key Laboratory of Inorganic Coating Materials, Shanghai Institute of Ceramics, Chinese Academy of Sciences, Shanghai 201899, China
[b] University of Chinese Academy of Sciences, Beijing 100039, China
* Corresponding author: Tel.: +862169906321; Fax: +862169906322.
E-mail address: sytao@mail.sic.ac.cn (S. Tao)

Abstract: Yttria stabilized zirconia coatings were deposited by plasma spraying and heat treated at 1100 ℃ for 50, 100, 150 and 200 h in air, respectively. The microstructure and phase composition of coatings were analyzed. Residual stresses in the ceramic topcoat and the thermally grown oxide (TGO) layer before and after thermal exposure were measured using Raman spectroscopy and photoluminescence piezo-spectroscopy techniques, respectively. It was found that the accumulated tensile stress in the as-sprayed topcoat changed to compressive after thermal exposure, and the compressive stress in the topcoat increased with an increase of thermal exposure time up to 150 h. The average compressive stresses in the TGO layer were higher than that of the cross-sectional topcoat, and the measured in-plane compressive stress increased firstly and then gradually decreased with increasing exposure time.The local interface geometry strongly affect the nature and evolution of hydrostatic stresses in the TGO. Finally, The crack initiation and propagation in the ceramic topcoat and at the topcoat / TGO / bond coat interface has been discussed with respect to the residual stresses in the plasma-sprayed TBC system.

Keywords: Thermal Barrier Coatings (TBCs); Plasma Spraying; Residual Stresses; Raman Spectroscopy; Photoluminescence Piezo-Spectroscopy (PLPS)

1 Introduction

Thermal barrier coatings (TBCs) are widely used in advanced gas-turbine engines to protect and insulate hot-section components. The use of TBCs can result in a temperature reduction of 100-200 ℃ at the superalloy surface, thereby increasing the durability of the component and improving engine performance [1,2]. Typically, a plasma-sprayed TBC system consists of nickel-based superalloy substrate, metallic bondcoat (MCrAlY, M=Ni, Co or Ni&Co), thin thermally grown oxide (TGO, primarily α-Al_2O_3) and yttria partially stabilized zirconia (YSZ) ceramic topcoat [3,4]. Research works have been done on the microstructure change and damage evolution in TBCs subjected to isothermal and cyclic oxidizing environments [5-9]. TBCs under oxidation are highly complex process which involves several phenomena such as sintering of the porous ceramic topcoat, developed of thermal expansion mismatch stress, growth of the TGO, creep and plastic deformation of the bondcoat and so on. It has been shown that the failure of plasma-sprayed TBCs occurs within the TGO or the topcoat just above the TGO due to microcracks link-up under complex stress [10-12]. Large residual stresses in the topcoat and at the topcoat/TGO/bondcoat interface develop as a result of mismatch coefficient of thermal expansion (CTE) between the components and the growth of TGO at the undulating interface. A number of models to analyze stress evolution have been developed and lifetime prediction modeling of TBCs has been conducted [13-16]. Therefore, measurement of the actual stresses distribution in the topcoat and the TGO layer and correlation of these measured stresses to those existing models are vital for providing further understanding of failure mechanisms of TBCs.

Raman spectroscopy and photoluminescence piezo-spectroscopy (PLPS) [17-20], based on a stress-induced frequency shift of spectroscopy, are powerful nondestructive techniques for measuring the residual stresses in the topcoat and the TGO, respectively. Spectroscopy methods offer a significant advantage that it can be localized by using a focused probe, thus can be used to obtain the stress distribution in the coating due to its high spatial resolution. What more important is that the PLPS technique can directly measure the residual stress in the TGO layer, which is thought to be the key critical weakness position of the TBCs.

The present work seeks to: (i) investigate the residual stresses change in the ceramic top coat and the TGO layer before and after isothermal exposure using spectroscopy techniques. (ii) obtain an understanding on how the oxidation influences the stress distribution in coatings and in turn, the causes of the observed cracking. (iii) finally, the results and the analysis performed in this study further help to understand the failure mechanisms of plasma-sprayed TBC system.

2 Experiment procedure

2.1 Coating preparation

The TBC samples consisted of a Co-32Ni-21Cr-8Al-0.5Y (wt.%) bondcoat and a ZrO_2-8wt.% Y_2O_3 topcoat, deposited by the APS method on the nickel-based superalloy plates (GH3128, Beijing CISRI-GAONA Materials & Technology Co., ltd., China) with a dimension of 30 mm × 20 mm × 1.5 mm. Detailed spray parameters were given in Refs [21]. Different layers were clearly distinguished in the section, i.e., topcoat, bondcoat and Ni-based superalloy substrate. The thickness of the topcoat and bond-coat layers was about 300 and 120 μm, respectively, **Fig. 1**.

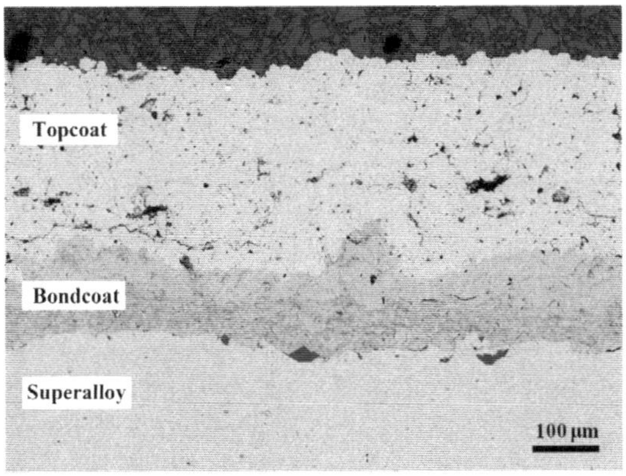

Fig. 1. Cross-sectional micrograph of the APS-TBCs in its as-sprayed condition

2.2 Heat treatment

When the TBC system was exposed to high temperature oxidation environment, a thin TGO layer was formed between the topcoat and bond coat as a result of oxidation of the bond coat. To examine the effect of thermal exposure time, the specimens were exposured in a tube furnace (SLG 1400-60, Shanghai Shengli Test Instrument Co., Ltd., China) at a temperature of 1100 ℃ with dwell time 50, 100, 150 and 200 h, respectively. Before measurement, the plan- and cross-sections of all specimens were polished with diamond paste up to 1 μm through a convectional metallographic process. The morphologies of the topcoat and the TGO were examined using a scanning electron microscopy (SEM, TM3000, HITACHI, Japan) operated in backscattered electron image mode. The accelerated voltage was set to be 15 kV.

2.3 Raman Spectroscopy for local stress measurement in topcoat

The measurements were made at room temperature in ambient air using 532 nm laser excitation in a confocal Raman microprobe (DXR, Thermo Fisher Co., USA). The diameter of the incident green laser beam at the specimen surface was ~1 μm. Because of dispersion of the laser beam within the specimen, the spatial resolution in the measurement system was ~3 μm. The measured wave number range of spectrum was from 200 to 800 cm^{-1} with the irradiation time at ~20 s. To obtain the stress distribution along the coating thickness, Raman spectra were measured at 50 μm intervals from the topcoat surface to the topcoat/bondcoat interface. The shift of the Raman peak at around 640 cm^{-1} was used to measure stress distribution since it has a good signal to noise ratio. The frequency shift could be converted into a biaxial stress depending on the relation for a bi-axially stressed stabilized tetragonal ZrO_2.

$$\sigma=\Pi I\ (639.5-\upsilon) \qquad (1)$$

where σ was the biaxial stress (MPa), and υ was the Raman shift (cm^{-1}). To calculate the stress, this technique require a previous calibration and for the YSZ topcoat each cm^{-1} corresponds to 40 MPa (ΠI=40 MPa) [18].

2.4 Cr^{3+} photoluminescence piezo-spectroscopy (PLPS) for stress measurement in TGO

The PLPS Method was used to measure stress in the TGO non-destructively. During growth of the TGO layer (typically polycrystalline α-Al_2O_3), a very small amount of Cr in the bondcoat was trapped within it and the local stress in the TGO layer was measured by the shift of the luminescence spectroscopy Cr^{3+} from a green laser (wave length of 532 nm) radiation. The measurement was done at room temperature in ambient air using a micro-focus laser luminescence spectroscopy system (Renishaw Invia Confocal micro-Raman Microscope, Glouchestershire, UK). Diameter of the laser beam at the specimen surface was ~1.5 μm. Resolution of a spectrometer in the spectroscopic wavenumber range used (14300-14500 cm^{-1}) was less than 0.2 cm^{-1}, which gave the measured stress resolution of ~40 MPa. The frequency shift of the luminescence in the TGO layer could be converted to the residual stress by the following expression:

$$\Delta\upsilon=\Pi II\ \sigma_{TGO} \qquad (2)$$

where ΠII was the piezo-spectroscopic coefficient (ΠII=5.07 cm^{-1} GPa^{-1}) [22]. The R2 peak was chosen because it has been successfully used to evaluate residual stress in this layer. A minimum of eight different points were tested for each measurement.

3 Results and discussion

3.1 Morphology of topcoat

The morphologies of the cross-section of the topcoat before and after isothermal treatment at 1100 ℃, **Fig. 2**. Two types of defect were observed: one was a globular type pore, and the other was a planar crack-like defect. The isolated pores were distributed at random, and even after exposure of 200 h, no noticeable changes in the distribution of these pores were observed. As for a comparison, the amount of horizontal and vertical cracks obviously increased with thermal aging. After exposure for 150 h, a number of larger size cracks were formed as a result of the growth as well as the linking of the horizontal and vertical cracks. With further exposure, the configuration and amount of the horizontal cracks were not significantly changed. However, larger scale vertical cracks were obviously observed. This indicated that thermal exposure resulted in a great increase in the defect area that the planar crack-like defects occupy in the direction perpendicular to the substrate than in the horizontal direction.

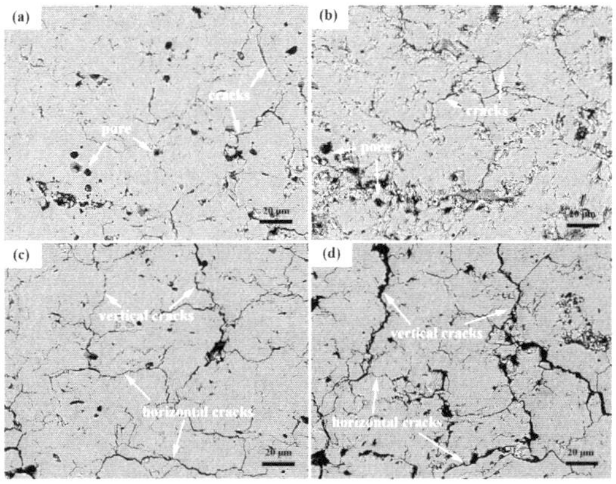

Fig. 2. Cross-section view of the topcoat after exposure at 1100 ℃ for different time: (a) 50, (b) 100, (c) 150 and (d) 200 h

3.2 TGO growth behavior and morphology

The polished transverse section of the entire TBCs after thermal exposure, **Fig. 3**. An oxide layer of predominantly Al_2O_3 formed along the interface between the topcoat and the bondcoat. Meanwhile, same mixed oxide clusters of alumina, spinel, and nickel oxide (ASN) and chromia, spinel, and nickel oxide (CSN) were also formed [23,24]. As thermal exposure continue, the TGO layer grew thicker, and cracking in the topcoat layer was limited as the exposure time increase (Fig. 3 (a) and (b)). The microcrack formation behavior was dependent on the thickness and the morphology of the TGO layer. Fig. 3 (c) and (d) shows typical microcrack formation behavior observed in the specimen after heat exposure for 150 and 200 h. It was obvious further exposure led to extensive crack coalescence and formation of macroscopic cracks located within the TGO layer. Growth and shrinkage of the formed cracks became significant after heat exposure for 200 h, some of formed cracks grown towards the topcoat and linked between the TGO layer and topcoat.

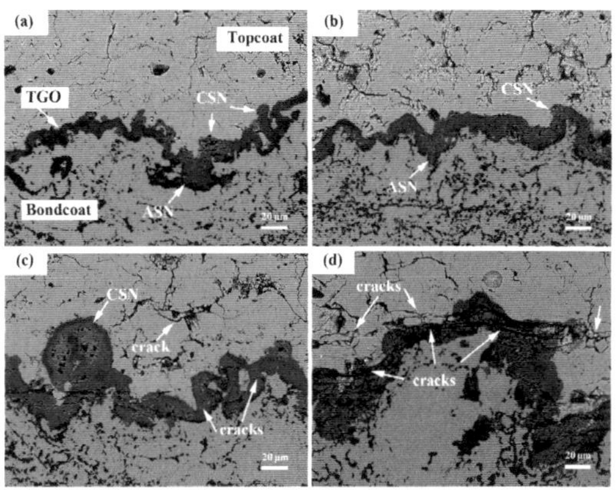

Fig. 3. Cross-section view of the TGO after exposure at 1100 □ for different time: (a) 50, (b) 100, (c) 150 and (d) 200 h

3.3 Residual stress in YSZ topcoat layer

The variation of the average residual stresses in the topcoat layer before and after thermal exposure, **Fig. 4**. For the as-sprayed coatings, a tensile stress (~70.8 ± 29.2 MPa) was present in the topcoat, after thermal exposure for 50 h, compressive stress (~ -92.4 ± 17.2 MPa) was found. The average topcoat compressive stress was found to increase with thermal exposure time from 50 to 150 h and compressive stress changed from -92.4 ± 17.2 to -126.4 ± 21.6 MPa. With further exposure, compressive stress was found slightly decreased, -123.6 ± 18.4 MPa for 200 h exposure. The distribution of residual stress in the cross-section of the topcoat layer along the through-thickness direction before and after thermal exposure, **Fig. 5**. No obvious stress changes were found on the distance from the topcoat surface to bondcoat.

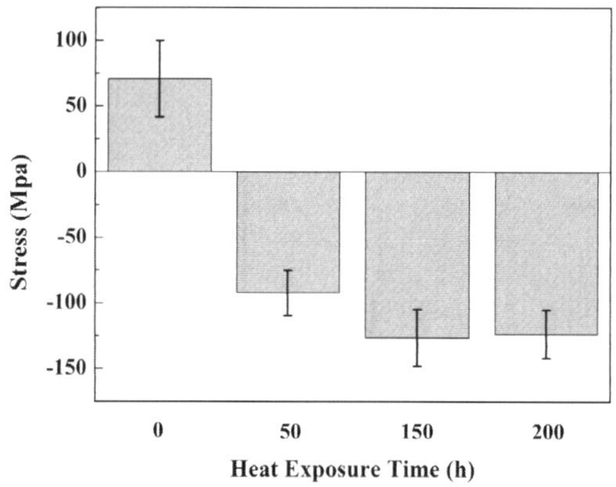

Fig. 4. Changes in the residual stresses as a function of thermal exposure time

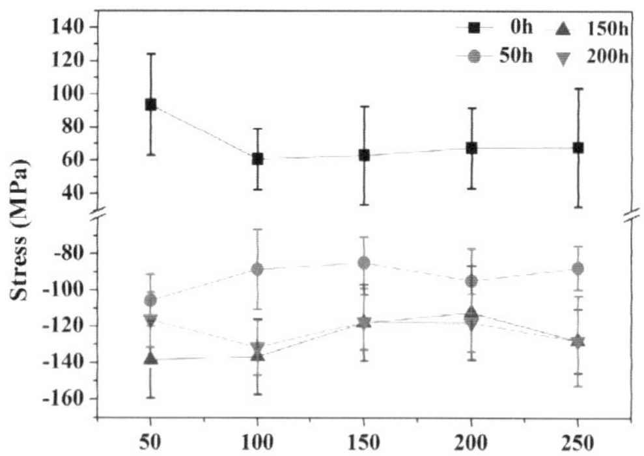

Fig. 5. Residual stress evaluation along the cross-section of the topcoat layer before and after thermal exposure

The tensile stress generated in the as-sprayed coatings was due to the rapid quenching effect of the sprayed droplets, arriving to the substrate and to the surface of the coating in the spraying. After thermal exposure, the tensile stress in the topcoat has changed from tensile state to the compressive which mainly was attributed to that at higher temperature, tensile residual stress relaxation on the ceramic topcoat occurred through the cracking of the brittle topcoat (As shown in Fig. 2). Due to this stress relaxation at high temperature, a higher compressive stress was then again developed within the coatings during the cooling procedure due to the differences between the coefficients of thermal expansion of substrate and coatings, and the growth thickness of the TGO, which had a lower thermal expansion coefficient than both of the topcoat and bondcoat.

3.4 Residual stress in TGO layer

The distribution of residual stress in the TGO with various isothermal exposure time, **Fig. 6**. The measured in-plane stress through the topcoat surface (from -2.5 ± 0.2 to -1.3 ± 0.2 GPa) was higher than those measured at the polished cross-section (form -0.7 ± 0.2 to -0.3 ± 0.1 GPa). The magnitude of the average compressive stresses measured through the topcoat surface was ~3 times of the corresponding values measured at the cross-section of the TGO layer. This mainly because of the stresses perpendicular to the polished edge were relaxed. Therefore, the measured cross-section compressive stresses were expected to be lower. At the beginning of thermal exposure, the measured in-plane TGO compressive stress firstly increased and then gradually decreased with increasing time. This compressive stress reached ~2.5 ± 0.2 GPa after 50 h heat exposure. However, after heat exposure for 150 h, the measured stress was gradually decreased. The decreased residual stresses after heat exposure for 50 h were considered due to cracking generated in the topcoat layer or at interface between the topcoat and the TGO layer. As oxidation time increases, oxides growth were reduced, and the creep in the bondcoat and topcoat layer also led to stress relaxation in the TGO layer.

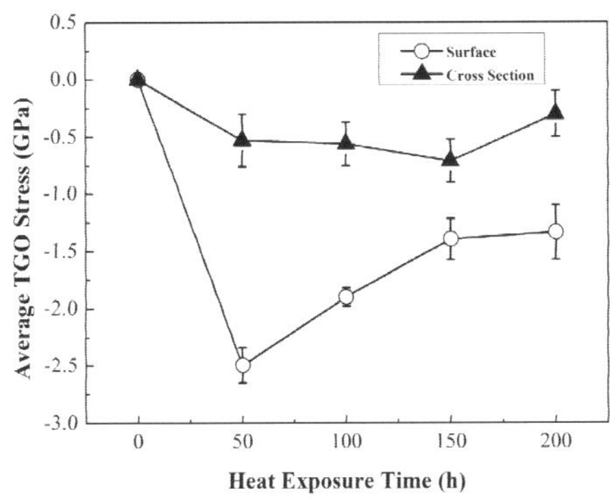

Fig. 6. Compressive stresses in the surface and cross-section of the TGO as function of thermal exposure time

Although the residual stress measured at most location was compressive state, it should be noticed that there could be components of the stress within the irregular geometry of the interface (concave and convex shape) between topcoat/TGO/bondcoat that were tensile and could initiated and promoted the microcracks in the TGO layer. Micro-fluorescence spectra was obtained from various locations. The residual stresses measured at these points, **Fig. 7**. The generated residual compressive stresses were approximately -157.8, -256.4 and -315.6 MPa at the position 1, 3 and 5, respectively, tensile stresses were approximately 631.2 and 300.1 MPa at the position 2 and 4, respectively. The compressive stresses at the position from 6 to 10 were approximately -710.0, -394.5, -1222.8, -100.3 and -433.9 MPa, respectively. These measured results clearly indicated that residual stress in the TGO layer has a strong dependent on local geometry morphology. The generated stresses at the peak region of the convex protrusion, the off-valley and valley region of the concave protrusion were compressive, with the maximum compressive stress observed at the valley of the concave protrusion (point 8). However, tensile stresses could be found in the off-peak region of the convex protrusion, and these tensile stress could initiate new microcracks generated in the TGO and promote the growth of damage within the TGO and nearby topcoat.

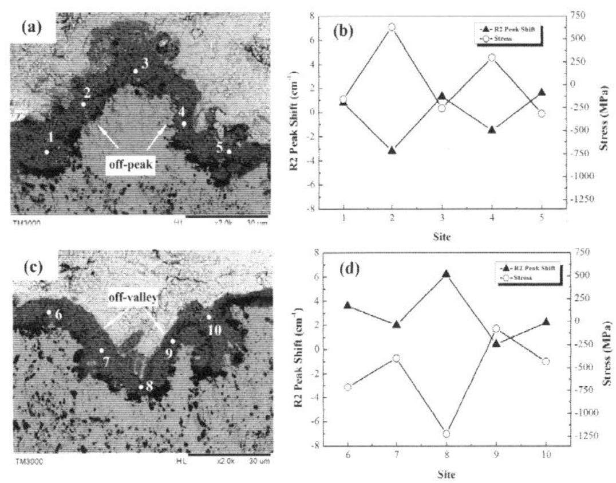

Fig. 7. Photomicrograph showing (a) bondcoat protrusions in topcoat (convex shape) and (c) topcoat protrusions in bondcoat (concave shape) in specimen isothermal exposure for 100h. (b) and (d) shows the residual stresses measured at locations indicated in (a) and (c), respectively.

3.5 Failure modes of the entire coating

From the above discussion, it was obviously that the undulation geometry of both the topcoat/TGO/bondcoat interface and the growth TGO thickness were important parameter to affect the residual stress distribution in the TBCs. The symbolic representation of the residual stresses state after the different oxidation exposure time, **Fig. 8**.

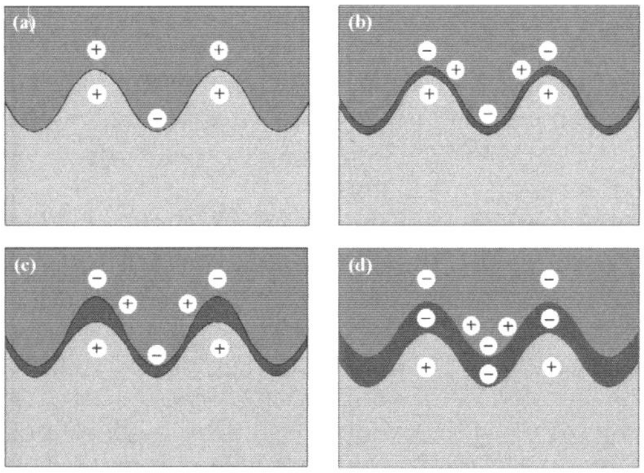

Fig. 8. Symbolic representation of the residual stresses state after the different oxidation exposure time were discussed. (a) with no or thin TGO; (b) thick TGO; (c) CSN-TGO; (d) thicker TGO. "+" and "-" signs denote tensile and compressive stresses.

At the beginning of the isothermal exposure, when the thickness of the TGO was thin, the residual stress was mainly the result of the thermal mismatch between the topcoat and the bondcoat. As the bondcoat had the larger coefficient of thermal expansion (CTE), the contract was more strongly on cooling from high temperature and this would lead to the generation of a tensile state stress within the bondcoat and topcoat "peak" region while the "valley" was under compression (shown in Fig. 8(a)). The microcracks might initiate in the peak region of the bondcoat and topcoat. However, cracking over the valley locations were still difficult due to the compressive stresses. With further oxidation and the growth of the TGO thickness, the residual stresses distribution at the interface would be changed gradually because the CTE of the TGO was smaller and the Young's modulus was higher than both the topcoat and bondcoat. Thus, TGO could partly shields the topcoat from the high shrinkage strains of the bondcoat compared to no or thin TGO thickness. Tensile stresses started to shift away from "peak" to "off-peak" location and would gradually moved to the "valley" region with further TGO growth (Fig. 8 (b) and(d)). Meanwhile, the stress in the bondcoat increased because of the high TGO stiffness and CTE mismatch to the bondcoat. As a result, cracking over the peaks of the topcoat became less favorable while cracking at the off-peak and valley locations became easier, and initiation of the cracks in the peak region of the bondcoat occured due to the generated more tensile stresses. Oxidation protection of the TBCs was obtained through the formation of a continuous α-Al_2O_3 layer on the bondcoat surface. The continuing growth of the Al_2O_3 would deplete the underlying bondcoat of Al as a result of the selective oxidation and the non-protective oxide such as CSN started to form at the apex of the protuberance of the topcoat/TGO interface after an exposure period. The α-Al_2O_3 layer ("primary" TGO) formed during thermal exposure to produce a uniform layer across the bondcoat surface. The non-α-Al_2O_3 layer (the "secondary" TGO) formed over the α-Al_2O_3 layer would obviously thicker at the apex of the protuberance (shown in Fig. 3 (c) and (d)). The out-of-plane tensile stresses also developed within the bondcoat itself at protuberances after cooling and these stresses might sufficient to induce transverse cracking of the undulation interface (shown in Fig. 8(c)).

The overview of the TBCs revealing the crack patterns after isothermal exposure for 200 h, **Fig. 9**. Some vertical branching cracks, linking splat boundaries of different sprayed layers were observed. These were caused by a primary crack initiating and propagating in the YSZ topcoat due to the generated in-plane tensile stresses as a result of different CTE of the ceramic topcoat and the metal matrix upon heating to high temperature. The widening of these vertical cracks, and their growth from the bondcoat to the surface, could be associated with cracking in the TGO. This would eventually lead to spallation of the topcoat. The propagation and linkage of the microcracks in the TGO layer, topcoat/TGO/bondcoat interface can be detected. Crack initiation and propagation mechanisms were complex as they mainly depended on the residual stress distribution in different layers and the irregular interface morphology. The growth of the TGO would promote the initiation of the microcracks in the topcoat and the TGO. The roughness at the interface would generate out-of-plane and shear stresses leading the cracks linkage and propagation.

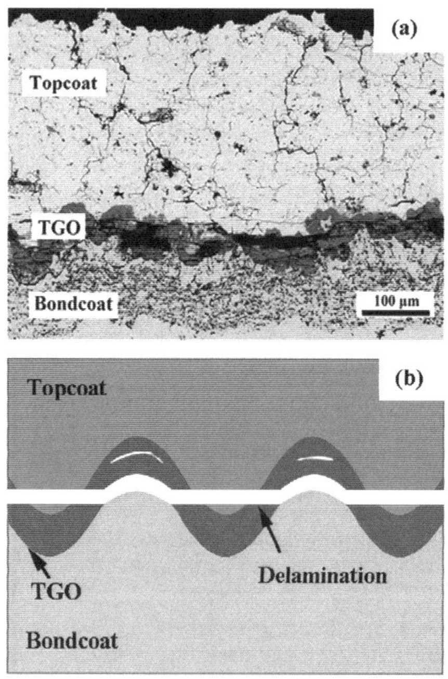

Fig. 9. (a) Overviews of the TBC system revealing the crack patterns after exposure at 1100 ℃ for 200 h,

and (b) Schematic diagram showing possible failure mode in the TBC system

From the above discussion, the figure of the failure mechanism resulting from the residual stresses distribution in the undulation interface could be concluded as follows: At the early stage of oxidation when there was no or thin TGO layer, cracking initiated at the topcoat and bondcoat peaks, driven by tensile peak stresses generated after cooling. However, the generated cracks were difficult to propagate due to the compressive stresses regions over the valleys. As the TGO thickness growth with the further oxidation, tensile stresses started to shift away from "peak" to "off-peak" location and would gradually moved to the "valley" region. Moreover, The non-α-Al_2O_3 layer formed over the α-Al_2O_3 layer would obviously thicker the total thickness of the TGO layer, and the generated out-of-plane tensile stresses within the YSZ topcoat near the flanks of the bondcoat protuberances obviously increased. All these tensile stresses could led cracks initiated in the TGO layer and along or transverse the topcoat/TGO/bondcoat interface. Finally, the delamination would occur when the linking cracks size reached to a critical level. The possible failure mode in the TBC system during the oxidation exposure is given in Fig. 9(b).

4 Conclusions

Damage evolution in the ceramic topcoat of plasma-sprayed thermal barrier coating system subjected to isothermal exposure was closely related to stresses development of the TBCs. Young's modulus and hardness were monitored by nano-indentation techniques. Residual stresses in the topcoat and TGO layer were evaluated through the spectroscopy techniques. Damage and stress evolution were investigated as a function of several parameters, such as time of oxidation, TGO thickness and local interface geometry. Finally, the figure of the failure mechanism resulting from this evolution of residual stresses was established. The following conclusions may be drawn from this work.

1) Tensile state stress developed in the as-sprayed topcoat was changed to compressive after thermal exposure. The generated compressive stress was found to increase with thermal exposure time and then slightly decreased with further exposure. No obvious changes was found that stresses distribution in the cross-section of the topcoat after thermal exposure.
2) The magnitude of the average compressive stresses measured through the topcoat surface were obvious higher than from the cross-section of the TGO layer, and the measured in-plane compressive stress increased firstly and then gradually decreased with increasing exposure time.
3) The local interface geometry strongly affect the nature and evolution of hydrostatic stresses in the TGO. After thermal exposure, the measured stresses at the peak region of the convex protrusion and the valley region of the concave protrusion were compressive, but tensile stresses could be found in the off-peak region of the convex protrusion.

References

[1] Miller R.A.: Thermal barrier coatings for aircraft engines: history and directions, Journal of Thermal Spray Technology 6 (1997), Issue 1, pp. 35/42.

[2] Bose, S., J. DeMasi-Marcin: Thermal barrier coating experince in gas turbine engines at Pratt & Whitney, Journal of Thermal Spray Technology 6 (1997), Issue 1, pp. 99/104.

[3] Vassen, R., A. Stuke, D. Stöver: Recent development in the field of thermal barrier coatings, Journal of Thermal Spray Technology 18 (2009), Issue 2, pp. 181/86.

[4] Clarke, D.R., C.G. Levi: Materials design for the next generation thermal barrier coatings, Annual Review of Materials Research 33 (2003), pp. 383/417.

[5] Rabiei, A., A.G. Evans: Failure mechanisms associated with the thermally grown oxide in plasma sprayed thermal barrier coatings, Acta Materialia 48 (2000), pp. 3963/76.

[6] Dwivedi, G., V. Viswanathan, S. Sampath, A. Shyam, E. Lara-Curzio: Fracture toughness of plasma-sprayed thermal barrier ceramics: Influence of processing, microstructure, and thermal aging, Journal of the American Ceramic Society 97 (2014), Issue 9, pp. 2736/44.

[7] Evans, A.G., D.R. Mumm, J.W. Hutchinson, G.H. Meier, F.S. Pettit: Mechanisms controlling the durability of thermal barrier coatings, Progress in Materials Science 46 (2001), pp. 505/53.

[8] Singh, J.P., B.G. Nair, D.P. Renusch, M.P. Sutaria, M.H. Grimsditch: Damage evolution and stress analysis in zirconia thermal barrier coatings during cyclic and isothermal oxidation, Journal of the American Ceramic Society 84 (2001), Issue 10, pp. 2385/93.

[9] Thompson, J.A., T.W. Clyne: The effect of heat treatment on the stiffness of zirconia top coats in plasma-sprayed TBCs, Acta Materialia 49 (2001), pp. 1565/75.

[10] Trunova, O., T. Beck, R. Herzog, R.W. Steinbrech, L. Singheiser: Damage mechanisms and lifetime behavior of plasma sprayed thermal barrier coating systems for gas turbines-Part I: Experiments, Surface & Coatings Technology 202 (2008), pp. 5027/32.

[11] Shillington, E.A.G., D.R. Clarke: Spalling failure of a thermal barrier coating associated with aluminum depletion in the bond-coat, Acta Materialia 47 (1999), pp. 1297/305.

[12] Schlichting, K.W., N.P. Padture, E.H. Jordan, M. Gell: Failure modes in plasma-sprayed thermal barrier coatings, Materials Science & Engineering A 342 (2003) , pp. 120/30.

[13] Rösler, J., M. Bäker, K. Aufzug: A parametric study of the stress state of thermal barrier coatings Part I: creep relaxation, Acta Materialia 52 (2004), pp. 4809/17.
[14] Bäker, M., J. Rösler, G. Heinze, A parametric study of the stress state of thermal barrier coatings Part II: cooling stresses, Acta Materialia 53 (2005), pp. 469/76.
[15] Rösler, J., M. Bäker, M. Volgmann: Stress state and failure mechanisms of thermal barrier coatings: role of creep in thermally grown oxide, Acta Materialia 49 (2001), pp. 3659/70.
[16] Busso, E.P., H.E. Evans, Z.Q. Qian, M.P. Taylor: Effects of breakaway oxidation on local stresses in thermal barrier coatings, Acta Materialia 58 (2010), pp. 1242/51.
[17] Limarga, A.M., R. Vaβen: Stress distribution in plasma-sprayed thermal barrier coatings under thermal cycling in a temperature gradient, Journal of Applied Mechanics 78 (2011), pp. 011003 1-9.
[18] Tanaka, M., M. Hasegawa, A.F. Dericioglu, Y. Kawawa: Measurement of residual stress in air plasma-sprayed Y_2O_3-ZrO_2 thermal barrier coating system using micro-Raman spectroscopy, Materials Science & Engineering A 419 (2006), pp. 262/68.
[19] Christensen, R.J., D.M. Lipkin, D.R. Clarke, K. Murphy: Nondestructive evaluation of the oxidation stresses through thermal barrier coatings using Cr^{3+} piezospectroscopy, Applied Physics Letters 69 (1996), Issue 24, pp. 3754/56.
[20] Schlichting, K.W., K. Vaidyanathan, Y.H. Sohn, E.H. Jordan, M. Gell, N.P. Padture: Application of Cr^{3+} photoluminescence piezo-spectroscopy to plasma-sprayed thermal barrier coatings for residual stress measurement, Materials Science & Engineering A 291 (2000), pp. 68/77.
[21] Zhao, Y.X., D.C. Li, X.H. Zhong, H.Y. Zhao. L. Wang, F. Shao, C.G. Liu, S.Y. Tao: Thermal shock behaviors of YSZ thick thermal barrier coatings fabricated by suspension and atmospheric plasma spraying, Surface & Coatings Technology 249 (2014), pp. 48/55.
[22] Liu, D., M. Seraffon, P.E.J. Flewitt, N.J. Simms, J.R. Nicholls, D.S. Rickerby: Effect of substrate curvature on residual stresses and failure modes of an air plasma sprayed thermal barrier coating system, Journal of the European Ceramic Society 33 (2013), pp. 3345/57.
[23] Chen, W.R.: Degradation of a TBC with HVOF-CoNICrAlY bond coat, Journal of Thermal Spray Technology 23 (2014), Issue 5, pp. 876/84.
[24] Chen, W.R., X. Wu, D. Dudzinski, P.C. Patnaik: Modification of oxide layer in plasma-sprayed thermal barrier coatings, Surface & Coatings Technology 200 (2006), pp. 5863/68.

The evaluation of mechanical properties of suspension plasma sprayed zirconia coatings having various microstructures

Paweł Sokołowski[1,2], Alberto Mejias[3,4], Didier Chicot[3], Radek Musalek[5], Per Nylen[6], Lech Pawłowski[2]

1 – Faculty of Mechanical Engineering, Wroclaw University of Technology, ul. Łukasiewicza 5, 50-371 Wrocław, Poland
2 – Laboratory of SPCTS, UMR CNRS 7315, University of Limoges, 12 rue Atlantis, 87068 Limoges, France
3 – Laboratoire de Mécanique de Lille, LML UMR 8107, Université Lille1, IUTA GMP, BP 90179 - 59 653 Villeneuve d'Ascq, France
4 – Universidad de Carabobo, Facultad de Ingeniería, Laboratorio de Materiales, Valencia, Venezuela.
5 – Department of Materials Engineering, Institute of Plasma Physics, Za Slovankou 3, 18200 Prague, Czech Republic
6 – Department of Engineering Science, University West, Gustava Melins Gata 2, 461 86 Trollhättan, Sweden

Abstract

The paper describes the investigations of mechanical properties of various zirconia coatings having different microstructure, phase content and chemical composition. The coatings were deposited using Suspension Plasma Spraying technology. The suspensions were formulated using three different powders: (i) ZrO_2 + 8 wt.% Y_2O_3, (ii) ZrO_2 + 14 wt.% of Y_2O_3 and (iii) ZrO_2 + 24 wt.% of CeO_2 + 2.5 wt.% of Y_2O_3. Moreover, three different plasma torches were to spray: (i) SG-100 of Praxair; (ii) Axial III of Mettech and (iii) hybrid Water-Stabilized Plasma torch of IPP Prague. The coatings had different microstructures: a *columnar-like* and a *two-zone- microstructure.*
The analysis of mechanical properties was performed using instrumented indentation technique (IIT) using a multi-cyclic loading to obtain the in-depth variation of the mechanical properties. The hardness (H) and Young's modulus (E) of coatings were found out. The measurements were made on the surface of coatings. The influence of the coating morphology on their mechanical properties was researched and discussed.

1 Introduction

Thermal Barrier Coatings are still one the key and strategic subject in the area of aviation and aerospace industry. The TBC systems have been continuously studied and developed over several last decades [1]. But there is still a desire to increase the service temperature and the lifetime of coatings what allows to decrease the fuel consumption in jet engines but also to reduce the emission of harmful substances in exhaust gases [2]. However the main difficulty is related to the fact that TBC systems operate in a very complex and demanding conditions. Thermal cyclic exposure, oxidation, hot corrosion, erosion, CMAS attack - these are some of many various loads that affect Thermal Barrier Coatings [3-5]. Following in order to ensure the proper quality, the TBC coatings should meet high requirements on thermophysical, thermomechanical and mechanical properties.
In this paper selected mechanical properties of zirconia coatings having various microstructures were studied by means of Instrumental Indentation Test. IIT allowed to determine Young modulus and Martens hardness of coatings. The indentation method was selected as a very versatile one. The analysis of load-displacement curve contains extensive information of coating behavior under the load. The main attention was paid on the influence of coating morphology on their mechanical properties, which was found to be crucial in terms of elastic modulus and hardness.

2 Experimental methods

2.1 Materials

In order to deposit the coatings three different, commercially available, zirconia powders were used:
1 – YCSZ, both yttria and ceria stabilized zirconia (ZrO_2 + 24 wt.% CeO_2 + 2.5 wt.% Y_2O_3, Metco 205NS, Oerlikon Metco),
2 – YSZ, yttria partially-stabilized zirconia (ZrO_2 + 8 wt.% Y_2O_3, Metco 204NS, Oerlikon Metco),
3 – Tosoh-YSZ, yttria fully-stabilized zirconia (ZrO_2 + 14 wt.% Y_2O_3, Tosoh TZ-8YS, Tosoh).
The YSZ and YCSZ powders were produced by HOSP method and both of them had a micrometer size in the range of -125 + 11 µm (according to powder manufacturer). The powders were initially coarse and an additional treatment was performed to crush powders particles and decrease their size. Mechanical ball milling was applied, which allowed achieving fine grained powders having unimodal distribution of particle sizes (see more details in [6]). The main particle size of milled YCSZ powder was equal to d_{n50}=0.86µm whereas for YSZ powder the mean number diameter was decreased up to d_{n50}=0.99µm. Tosoh-YSZ powder was manufactured by spray drying method. This powder had submicrometer size as-produced (d_{n50}=487nm) so there was no need to perform any additional powder preparation prior to spraying. The morphology of powders is presented on Fig. 1.

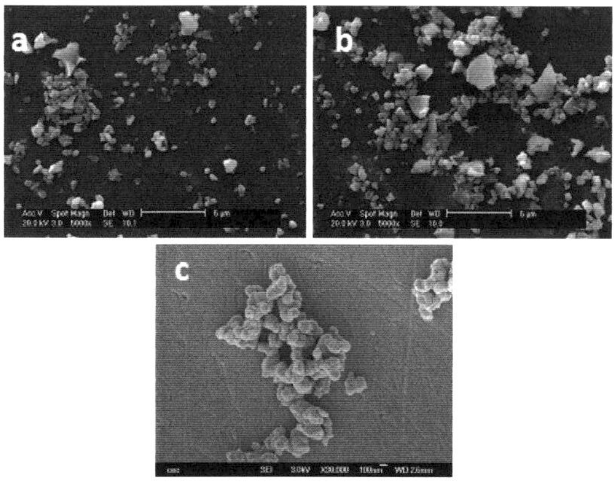

Fig. 1. The morphology of powders observed by SEM (SE detector) microscopy: YCSZ (a), YSZ (b) and Tosoh-YSZ (c).

Based on described zirconia powders the suspensions were formulated. As a solvent the mixture of distilled water and pure ethanol with a ratio of 1:1 was used. The various suspensions were prepared with three different contents of solid phase for each single powder. The amount of powder depended each time on the plasma torch. The suspension concentrations were as follows:
- 2.5 wt. %, 5 wt. %, 10 wt. % of powder for SG-100,
- 10 wt. %, 20 wt. %, 30 wt. % of powder for Axial III and h-WSP.

The stability of each suspension was assured by the addition of a dispersant agent (Beycostat C213, CECA, France). Further details of suspension preparation and stabilization were described elsewhere [6, 7].

2.2 Coating deposition process

Suspension Plasma Spraying method was applied to deposit zirconia coatings. The coatings were produced using three spray set-ups equipped with various plasma torches: (i) SG-100 of Praxair (Indianapolis, USA), (ii) Axial III of Mettech (Vancouver, Canada) and (iii) h-WSP of IPP in Prague. The set-ups were different also in the term of the suspension injection mode (injector inside/outside of the torch and axial/radial injection of liquid feedstock), electric power as well as arc stabilization principle. The main characteristics of mentioned plasma torches are summarized in the Table 1.

Table 1 The comparison of plasma torches used for spraying

Torch	Injector	Injection angle	Electric power (kW)	Arc stabilization mode
SG-100	Internal	Radial	Low (40kW)	Gas
Axial III		Axial	High (120kW)	Gas
h-WSP	External	Radial	High (150kW)	Water

Consequently the spray process parameters were optimized separately for each kind of plasma torch and the optimization is described elsewhere [6]. Prior to spraying all substrates were sand-blasted and cleaned in the ultrasonic bath in ethanol. Finally, all spray runs are synthesized in the Table 2 (samples highlighted in red color correspond to the coatings that were selected for IIT).

Table 2 The spray process design and the nomenclature of coatings

Suspension concentration	Powder	SG-100	Axial III	h-WSP
Small	YSZ-Tosoh	ST21	AT101	WT101
Medium	YSZ-Tosoh	ST51	AT201	WT201
High	YSZ-Tosoh	ST101	AT301	WT301
Small	YCSZ	SC21	AC101	WC101
Medium	YCSZ	SC51	AC201	WC201
High	YCSZ	SC101	AC301	WC301
Small	YSZ	SY21	AY101	WY101
Medium	YSZ	SY51	AY201	WY201
High	YSZ	SY101	AY301	WY301

2.3. Microstructure analysis

First, the thickness of coatings was estimated on metallographically prepared cross-section of samples under the light microscope (Nikon Eclipse LV100). Then the morphology of coatings was characterized using SEM microscopy (Philips XL30, secondary electron mode). The cross-section as well as coating top surface were investigated. Furthermore, based on SEM micrographs the calculation of a porosity was performed. For this purpose the images were edited using *ImageJ* free software. The total area of pores was calculated in different regions of samples and then the average value and standard deviation of porosity were defined.

2.4. Instrumented indentation test and theory of indentation

To investigate the mechanical properties of suspension plasma sprayed zirconia coatings the Instrumented Indentation Test (IIT) was carried out. The studies were performed using a micro-hardness tester (CSM 2-107, Peseux, Switzerland) equipped with a diamond pyramid Berkovich indenter. The measurements were realized at the surface of the zirconia coatings. The samples were characterized in as-sprayed condition. The following parameters for micro-indentation studies were applied:

- tests performed using a multicycle protocol,
- 100 cycles per one test, loading and unloading rates (mN/min) as twice the value of the maximum applied load to guarantee that all cycles last same time [8],
- increasing load value, starting from 200mN (first cycle) up to 5000mN (last cycle),
- increasing unloading values set as 20% of load value,
- dwell time equal to 15s for each maximum and minimum load.

The parameters were set according to the standard ASTM E92-82 [9]. The multicycle loading-unloading protocol is presented on Fig. 2.

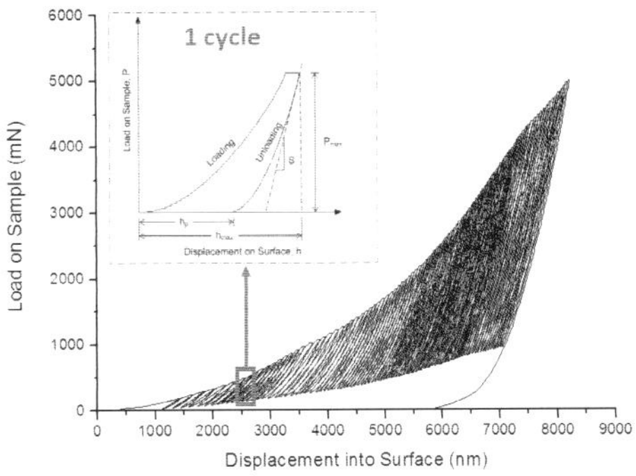

Fig. 2. The multicycle protocol realized during IIT

Elastic Modulus by IIT

In origin, the methodology of determination of reduced elastic modulus using IIT was proposed by Olivier and Pharr [10].It was assumed that elastic modulus can be calculated based on the analysis of unloading part of the "load-indentation depth" curve by using of the following formula:

$$E_R = \frac{1}{\beta}\frac{\sqrt{\pi}}{2}\frac{1}{\sqrt{A}}\frac{1}{C} \quad (1)$$

where:
E_r – reduced elastic modulus,
β – geometrical correction factor,
A – contact area,
C – total compliance.

However, many studies were performed to specify the calculation methodology as different effects should be considered when analyzing different kinds of materials, using various indenters etc. The correction factor (β) values were specified for different indenters geometries and independent on the material [11-13]. The contact area, *A*, was proposed to be calculated based on the experimental indentation depth. But the value was found to be strongly influenced by the indentation depth itself, tip defect and the type of the indenter [14-15]. Furthermore the deformation mode around the indentation was also considered as the key factor in order to define appropriate value of the contact area [10, 16]. Two different mechanism were mentioned, called *sinking-in* and *piling-up*. Yetna et al. [17] proposed a way to identify the mechanism of deformation around the indenter as a ratio between residual indentation depth and maximum indentation depth at maximum load.
Afterwards the total compliance (C) was assumed as the reciprocal of the contact stiffness that can be defined from the slope of the unloading curve at the maximum load/depth (see Fig.3). However the displacement of the indenter given by the machine was found to be overestimated as the deflection of the load frame may be registered also. Consequently the depth measurements should be corrected by the instrument compliance (C_f). Furthermore Hay et al. [18] proved that Oliver and Pharr's method may overestimate the elastic modulus when massive material are analyzed. The authors suggested to take into account a geometrical correction factor (γ) which is a function depending of the Poisson's coefficient. Finally, the following form of Olivier and Pharr expression was applied to calculate elastic modulus:

$$E_R = \frac{1}{\beta\gamma}\frac{\sqrt{\pi}}{2}\frac{1}{\sqrt{A}}\frac{1}{C-C_f} \quad (2)$$

where:
C_f – instrument compliance,
γ – geometrical correction factor depends on the Poisson's coefficient

Hardness by IIT

In order to characterize hardness of zirconia coatings, the Martens hardness, HM, was defined according to the following mathematical equation:

$$\mathrm{HM} = \frac{\mathrm{P}}{26.43\mathrm{h}^2} \quad (3)$$

where:
HM – Martens hardness,
P – applied load,
h – displacement of the indenter.

Actually, the Martens hardness was assumed to be a ratio between applied load and the actual contact area between the indenter and the material (herein defined using Berkovich indenter geometry).

3. Results and discussion

3.1 Microstructure

On the SEM microscopy image it is possible to see that the obtained coatings are characterized by various microstructures. The coatings are very different in terms of their morphology, porosity and surface topography.
The use of the uniform and submicrometer sized YSZ-Tosoh powder together with a low concentration of the suspension (ST21, WT101) allows achieving columnar-like microstructures (see Fig. 3a, b). This type of coatings is characterized by very porous microstructure where long vertical pores between each column and fine spherical pores inside each column occur. Furthermore the typical cauliflower-like surface can be observed.
On the other hand the use of coarser and more irregular powders of YSZ / YCSZ results in the deposition of much more dense zirconia coatings. Another important factor is the concentration of suspension. When the amount of solid in the suspension is greater than the coatings are more uniform and the coating' surface smoother. Due to this

change of the spray conditions the structure become to be typical *two-zones* microstructure (SC51, SY51, Fig. 3c, d) containing well melted lamellas and also fine sintered particles. The distribution of porosity is not very homogeneous and the size of pores varies between hundreds of nanometers up to few microns. Finally the coatings deposited by Axial III presents very regular lamellar microstructure having very uniform distribution of fine pores (see Fig. 3e, f). The surface of coatings AY301, AC201 and AT101 is much more smooth comparing to the other coatings. The coatings build-up mechanisms and detailed analysis of coatings' microstructures is described elsewhere [19].

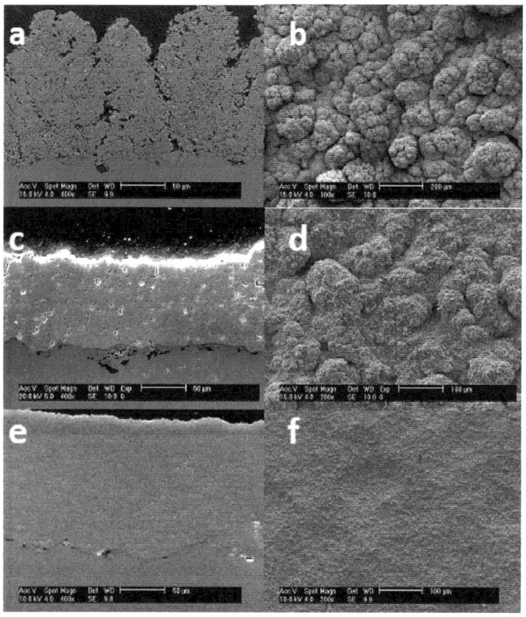

Fig. 3. The coatings' microstructure: WT101 (a, b), SC51 (c, d) and AY301 (e, f)

Table 3 Thickness and porosity of coatings selected for the instrumental indentation test

Material	Sample	Thickness [μm]	Standard deviation σ	Porosity [%]	Standard deviation σ
Tosoh-YSZ	ST21	50.4	10.8	21.4	3.7
	AT101	139.5	2.5	10.3	1.7
	WT101	176.0	30.8	21.0	3.2
YCSZ	SC51	105.2	3.8	15.5	1.5
	AC201	168.6	4.5	5.2	0.8
	WC101	145.4	11.3	14.0	1.3
YSZ	SY51	93.9	7.2	12.1	2.3
	AY301	150.7	3.0	6.4	0.6
	WY101	97.7	7.8	12.4	0.9

The morphology and porosity of coatings are the key factors that were considered in order to select the samples for mechanical properties investigations by means of IIT. The thickness of coatings chosen for indentation test is in the range of 50-176 μm whereas the porosity was calculated to be between 5 and 21%. The values of the thickness and porosity determined based on the light and SEM microscopy observations are collected in the Table 3.

3.2 Mechanical properties

Elastic modulus

To determine the value of elastic modulus in accordance to the mentioned methodology the inverse of the contact stiffness versus the inverse of the square root of the contact area is plotted as shown on Fig. 4. Then the values of Young modulus can be calculated using the method suggested by Olivier and Pharr based on the unloading part of the indentation curve.

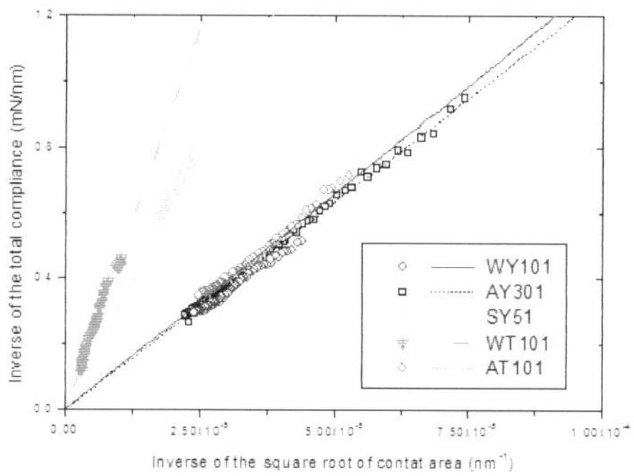

Fig. 4. The inverse of the contact stiffness vs reciprocal of contact area for zirconia coatings

The values of elastic modulus are presented in the Table 3. It can be clearly seen that elastic modulus is strongly dependent on the coating microstructure. The coatings characterized by the highest porosity and most irregular morphology (ST21, WT101) presents the lowest values of Young modulus. On the other hand samples described as dense, uniform and having very homogeneous distribution of fine pores shows the greatest values (AY301, AC201) of elastic modulus. The values of Young modulus are few times bigger than that for irregular shaped coatings (porous, columnar-like).

As presented on Fig. 4 the slope related to the elastic modulus remains constant in each measuring point when the indentation depth increases. Furthermore, each measuring point consists of 100 cycles. Therefore, the standard deviation of elastic modulus being less or around 10% in each single measuring point can be considered that the indentation test did not cause local fracture of the coating material. But the careful analysis of the load-displacement curve combined with the microscopy observations of the indentations should be performed to confirm this findings.

Table 3 Elastic modulus values of zirconia coatings investigated by IIT

Point	Sample								
	Tosoh-YSZ			YCSZ			YSZ		
	ST21	AT101	WT101	SC51	AC201	WC101	SY51	AY301	WY101
1	9.3 ± 0.2	26.2 ± 0.8	8.3 ± 0.1	7.1 ± 0.2	119.2 ± 2.9	--	129.0 ± 2.2	210.2 ± 8.1	43.3 ± 1.4
2	33.5 ± 1.1	52.3 ± 1.3	34.0 ± 0.7	40.7 ± 0.8	147.3 ± 4.9	24.5 ± 0.5	154.3 ± 6.3	61.4 ± 1.2	194.3 ± 10.7
3	13.0 ± 0.2	59.4 ± 3.8	2.8 ± 0.0	9.2 ± 0.1	167.8 ± 4.8	27.2 ± 0.6	51.7 ± 1.6	190.8 ± 5.8	66.2 ± 1.2
4	6.4 ± 0.2	90.1 ± 2.5	4.6 ± 0.0	27.2 ± 0.5	225.9 ± 18.4	--	109.0 ± 4.1	117.3 ± 2.2	89.9 ± 2.7

It should be highlighted also that the values of elastic modulus strongly varies between different measuring points. This effect can be explained by the general feature of all plasma sprayed coatings that are highly disordered in the term of their microstructures. The bimodal distribution of mechanical properties determined by IIT method was observed also by Carpio et al. [20]. However, in order to deeply examine the mechanical properties of coatings more measuring points will be analyzed as well as the indentation test will be performed at the cross-section of coatings.

Hardness

The calculation of HM hardness of coatings leads to the values presented in the Table 4, which are between 0.04 and 6.98 GPa. The same trend as in the case of elastic modulus can be observed – the more uniform coating is then the higher values of HM hardness represents. Following the most irregular coatings in the term of their morphology and porosity shows the lowest hardness values. The standard deviation is around 15% when analyzing single measuring points. But the results varies also when different measuring points are taken into account. The values of microhardness for homogeneous and relatively dense coatings corresponds to the values presented by Łatka et al [21]. On the other hand the lowest values of hardness that were achieved for irregular coatings can be a result of weak splat interfaces and high internal porosity as suggested by Lima et al. [22].

4 Conclusions

The investigations of elastic modulus and Martens hardness of suspension plasma sprayed zirconia coatings having various microstructures and chemical composition were presented.
The coatings were deposited by means of Suspension Plasma Spraying technology and using different plasma torches. Furthermore various suspension were prepared in the term of the solid phase content and zirconia powder material. This allowed achieving *columnar-like* coatings as well as coatings characterized by the well-known *two-zone-microstructure*. The coating porosity, homogeneity and surface topography varied also.
The coatings were then analyzed by Instrumental Indentation Test (IIT) using Berkovich indenter. Afterwards the elastic modulus and HM hardness of 9 different coatings were evaluated. Based on IIT test many conclusions were drawn:

- the morphology highly influence elastic modulus and hardness of thermally sprayed coatings

Table 4 Martens hardness values of zirconia coatings investigated by IIT

Point	Sample								
	Tosoh-YSZ			YCSZ			YSZ		
	ST21	AT101	WT101	SC51	AC201	WC101	SY51	AY301	WY101
1	0.14 ± 0.01	0.89 ± 0.06	0.24 ± 0.01	0.11 ± 0.01	3.97 ± 0.08	--	4.2 ± 0.2	6.31 ± 0.05	0.87 ± 0.03
2	0.14 ± 0.02	1.48 ± 0.13	2.03 ± 0.09	0.77 ± 0.12	4.26 ± 0.07	0.69 ± 0.05	4.0 ± 0.37	2.07 ± 0.10	3.34 ± 0.17
3	0.12 ± 0.01	1.99 ± 0.13	0.04 ± 0.00	0.65 ± 0.02	5.82 ± 0.35	0.35 ± 0.04	1.42 ± 0.7	5.69 ± 0.37	1.85 ± 0.04
4	0.47 ± 0.01	2.95 ± 0.13	0.15 ± 0.001	1.86 ± 0.02	6.98 ± 0.91	--	1.95 ± 0.27	4.26 ± 0.11	1.92 ± 0.08

- when the homogeneity of coating increase (morphology, porosity, size and pore distribution etc.) the mechanical properties also increase
- columnar-like coatings are characterized by much lower values of elastic modulus and HM hardness comparing to uniform and dense coatings.

5 Acknowledgements

The part of work was financed by the research project Sonata, UMO-2013/11/D/ST8/03400, awarded to the Faculty of Mechanical Engineering at the Wroclaw University of Technology by the National Science Centre (Poland). French Embassy in Poland enabled the stay of Paweł Sokołowski at University of Limoges in frame of joint PhD "BGF Co-tutelle" program. The authors would like to acknowledge the help of Dr Stefan Björklund, Dr Stefan Kozerski and Dr Leszek Łatka with the coating deposition process.

6 Literature

[1] J.T. Demasiand, K.D. Sheffler, Thermal barrier coating life prediction model development, NASA Contract Rep. 175087, 1986 (National Aeronautics and Space Administration)

[2] A. Saini, High-temperature materials increase efficiency of gas power plants, MRS Bulletin, Vol. 37, 2012

[3] N. P. Padture, M. Gell, E. H. Jordan, Thermal barrier coatings for gas-turbine engine applications, Science, vol. 296, 2002

[4] W. J. Brindley and R. A. Miller, Thermal Barrier Coating Evaluation Needs, NASA Technical Memorandum 103708, 1990

[5] L. Li, N. Hitchman, J. Knapp, Failure of Thermal Barrier Coatings Subjected to CMAS Attack, Journal of Thermal Spray Technology, 2010, Vol. 19, pp. 148-155

[6] P. Sokołowski, S. Kozerski, L. Pawłowski, A. Ambroziak, The Key Process Parameters Influencing Formation of Columnar Microstructure in Suspension Plasma Sprayed Zirconia Coatings, Surface and Coatings Technology, 2014, Vol. 260(15), pp. 97-106

[7] P. Sokołowski, L. Łatka, L. Pawłowski, A. Ambroziak, S. Kozerski, B. Nait-Ali, Characterization of Microstructure and Thermal Properties of YCSZ Coatings Obtained by Suspension Plasma Spraying, Surface and Coatings Technology, 2015, Vol. 268(25), pp. 147-152

[8] G. Quinn, P. Patel, I. Lloyd, Effect of loading rate upon conventional ceramic microindentation hardness, Journal of Research of the National Institute of Standards and Technology, 2002, vol. 107, pp. 299-306

[9] ASTM E92-82, Annual Book of ASTM Standards, 1.03 ASTM, Easton, MD, ASTM, 1996.

[10] W. Oliver, G. Pharr, An improved technique for determining hardness and elastic modulus using load and displacement sensing indentation experiments, Journal of Materials Research, 1992, vol. 6, pp. 1564–1583

[11] R. King, Elastic analysis of some punch problems for layered medium, International Journal of Solids and Structures, 1987, vol. 23, pp. 1657–1664

[12] M. Dao, N. Chollacoop, K. Van Vliet, T. Venkatesh, S. Suresh, Computational modeling of the forward reverse problems in instrumented sharp indentation, Acta Materialia, 2001, vol. 49, p. 3899–3918

[13] J. Antunes, L. Menezes et J. Fernandes, Three-dimensional numerical simulation of Vickers indentation tests, International Journal of Solids and Structures, 2006, vol. 43, 784–806

[14] M. Troyon, H. Liye, Correction factor for contact area in nanoindentation measurements, Journal of Materials Research, 2005, vol. 20, pp. 610–7

[15] D. Chicot, P. d. Baets, M. Staia, E. Puchi-Cabrera, G. Louis, Y. P. Delgado, J. Vleugels, Influence of tip defect and indenter shape on the mechanical properties determination by indentation of a TiB2–60%B4C ceramic composite, International Journal of Refractory Metals and Hard Materials, 2013, vol. 38, pp. 102-110

[16] S. Bec, A. Tonck, J.-M. Georges, E. Georges et J.-L. Loubet, Improvements in the indentation method with a surface force apparatus, Philosophical Magazine A, 1996, vol. 74, pp. 1061-1072

[17] M. Y. N'jock, D. Chicot, J. Ndjaka, J. Lesage, X. Decoopman, F. Roudet, A. Mejias, A criterion to identify sinking-in and piling-up in indentation of materials, International Journal of Mechanical Sciences, 2015, vol. 90, pp. 145-150

[18] J. Hay, A. Bolshakov, G. Pharr, Critical examination of the fundamental relations used in the analysis of nano-indentation data, Journal of Materials Research, 1999, vol. 14, pp. 2296–2305

[19] Paweł Sokołowski, Lech Pawłowski, Dagmar Dietrich, Thomas Lampke, David Jech, Advanced Microscopic Study of Suspension Plasma-Sprayed Zirconia Coatings with Different Microstructures, Journal of Thermal Spray Technology, 2016, Vol. 25, pp 94-104

[20] P. Carpio, E. Rayón, L. Pawłowski, A. Cattini, R. Benavente, E. Bannier, M.D. Salvador, E. Sánchez, Microstructure and indentation mechanical properties of YSZ nanostructured coatings obtained by suspension plasma spraying, Surface and Coatings Technology, 2013, vol. 220, pp. 237-243

[21] L. Łatka, D. Chicot, A. Cattini, L. Pawłowski, A. Ambroziak, Modeling of elastic modulus and hardness determination by indentation of porous yttria stabilized zirconia coatings, Surface & Coatings Technology, 2013, Vol. 220, pp. 131–139

[21] R.S. Lima, A. Kucuk, C.C. Berndt, Bimodal distribution of mechanical properties on plasma sprayed nanostructured partially stabilized zirconia, Materials Science and Engineering A, 2002, Vol. 327, pp. 224-232

Influence of powder characteristics on the microstructure and bond strength of cold-sprayed aluminum coating

P.E. Léger, F. Borit, N. Fabrègue, M. Jeandin, Evry / Fr
M. Ducos, Mornas / Fr

Cold spray aluminum coatings are today developed for a wide range of applications such as the repair of aircraft components, or additive manufacturing. However, powders specifically tailored for cold spray, are yet not available due to a rather low amount of results on the effect of powder characteristics on cold spray coatings properties. The influence of powder morphology on the microstructure and bond strength of cold-sprayed aluminum coating is shown in this study. Aluminum powders with spherical and irregular particle shapes were selected and cold-sprayed onto shot-peened steel. These powders were mixed with two alumina powders with either a spherical or an angular particle shape to improve coating properties. The effect of powder characteristics on coating microstructure was studied using X-Ray diffraction, optical and scanning electron microscopy. In addition, coating adhesion properties were determined using conventional pull-off testing. Aluminum coatings with extremely high adhesion were successfully cold-sprayed on steel. Alumina addition reduced the porosity and increased the hardness of the aluminum coating. Aluminum-alumina mixtures with the same particle shape seem to be more suitable to obtain dense coatings on steel with a high bond strength.

1 Introduction

Cold spray is a very attractive technology to produce coatings with a very high deposition efficiency in a wide range of applications [1]. In this process, high pressure gas (nitrogen or helium) at a temperature of less than 800°C, accelerates powder particles (typically between 5 and 50 µm in diameter) to supersonic velocities. These particles impact the substrate and deform with a high plastic strain to produce a dense coating. On the contrary of other thermal spray techniques, the powder in this process is not melted. Consequently, cold spray coatings are free of oxides and grain growth. The powder microstructure is retained by this process.
Aluminum and its alloys can achieve good bonding with various types of substrates [1]. Because of their specific properties, they are commonly used for numerous automotive and aircraft applications. In particular, they are expected to be beneficial for the coating of steel using cold spray. However, to the knowledge of the authors, there is a low amount of results from studies of the aluminum coating-steel substrate bond strength because conventional pull-off adhesion testing [2] is really difficult to use and to interpret. Aluminum coatings properties can be improved by mixing ceramic particles such as TiN [3], SiC [4] or Al_2O_3 [5], with aluminum powder. Alumina is known to reduce coating porosity and increase deposition efficiency and hardness. It also enhances mechanical properties of the coating such as wear resistance and bond strength [6, 7].
Powders specifically tailored for cold spray, are not yet available due to a lack of results on the effect of powder characteristics on cold spray coatings properties [8]. Powder parameters are numerous, e.g. grain size, particle shape, particle microstructure or oxide content. A few studies only focused on powder morphology and its effect on cold spray properties. The well-known influence of the powder particle shape a higher in-flight velocity of irregular particles compared to that of spherical particles [9]. Wong et al. [10] cold sprayed five morphologies of titanium powders (three spherical, one sponge and one irregular morphologies). They studied the role of particle morphology on coating microstructure and hardness. A medium-sized spherical powder was shown to produce the best coatings. Shockley et al. [11] investigated the influence of alumina morphology on the cold spray coating microstructure and dry-sliding wear behavior. They concluded that coating needs a spherical alumina content lower than irregular alumina content to reach a similar tribological behavior.
The present study is part of an ongoing investigation to specify some criteria to help cold spray users in selecting powders for their applications. In this article, the influence of powder morphology on microstructure and bond strength of cold-sprayed aluminum and aluminum-alumina coatings is studied. The same cold spray parameters are used for all the powders and mixtures of powders to focus on the effect of the powder particle shape. These parameters were not optimized deliberately to be sure that the influence of powder particle shape could be observed. Two aluminum and two alumina powders with either spherical or irregular shape were selected and characterized (particle shape, powder size distribution and range). Then, they were successfully cold sprayed onto a shot-peened steel substrate. Deposition efficiency, hardness and adhesion strength were then quantified to compare coating microstructures and properties. Porosity and alumina weight content were calculated using image analysis.
This study succeeded in optimizing aluminum and alumina morphologies to improve cold spray coatings properties onto steel. Some explanations about the powder morphology influence on coating properties are proposed. Numerical simulation of particle impact was undertaken to ascertain these.

2 Experimental

2.1 Materials

Two commercial aluminum powders, the one with a spherical particle shape (Poudres Hermillon, France), the other with an irregular particle shape (ECKA Granules, Germany), were cold-sprayed onto a shot-peened steel. They were mixed with two alumina powders (Saint-Gobain, France) using a Turbula® T2A blender for five hours. The Al_2O_3 content in powder mixtures was fixed at 15wt.%. The first type of alumina powder showed angular particles. The second type showed a spherical shape.

2.2. Cold spray process

Al powders and Al-15wt.%Al_2O_3 mixtures were cold-sprayed using CGT Kinetics 3000M facilities. Nitrogen was used as the spraying gas. Pressure was 2.5 MPa for a gas temperature of 400°C. The powder feed rate was fixed at 7.1 $g.min^{-1}$. The stand-off distance between the nozzle exit and the substrate, was kept at 30 mm. All the operating conditions were exactly the same for aluminum powders and aluminum-alumina mixtures. Consequently, the effect of powder morphology on cold spray coating microstructure could be studied. Deposition efficiency was determined using four samples (25x25x6 mm^3) which were weighed before and after deposition.

2.3. Microstructural characterization

Particle size distributions of the aluminum and alumina powders were obtained with a laser particle size analyzer Mastersizer 3000® (MALVERN). Powder particle shapes and coating microstructures were studied using a scanning electron microscope (SEM) Leo© 1450VP. Ten images per coating were processed using the ImageJ® software. Average porosity and thickness were calculated from ten SEM pictures for every coating. XRD measurements (copper cathode, λ=1,540 Å) were used to determine the alumina weight content in the starting powder mixtures and in the coatings.

2.4. Mechanical characterization

A Vickers indenter applied during a hold time of 10 s, was used for microhardness measurements. A 10 g and a 100 g loads were applied to measure powder and coating microhardnesses respectively. The hardness value was an average of at least ten indents taken at a random location. The bond strength of each coating was determined by pull-off testing which was conducted according to ASTM C633. Five test specimens for each aluminum powder or aluminum-alumina powder mixtures were assembled by glue DELOMONOPOX© AD286. They were pulled on a pull machine Zwick 1474© (Zwick/Roell).

3 Results

3.1 Feedstock powder characterization

Fig. 1. showed the particle shape of the four powders used in this work. Irregular Aluminum powder (**Fig. 1.a**) was atomized with nitrogen which explains its spheroidal stretched shape. Spherical Al powder (**Fig. 1.b**) was atomized with an inert gas. For the two aluminum powders, some small particles were agglomerated on large ones. These small satellites were smaller in the spherical Al powder than in the irregular Al powder. Microhardness of the irregular and spherical powders was respectively 28 ± 1.8 HV_{10} and 28 ± 2.3 HV_{10}.

Angular Al_2O_3 (**Fig. 1.c**) is the classical alumina used in cold spray [7], [11], [12]. Spherical Al_2O_3 powder was obtained by the plasma spheroidization of angular Al_2O_3 powder. Some angular shape alumina particles can therefore be found in **Fig. 1.d**. Some of the spherical alumina particles were hollow. Due to powder production, transport and handling, some of these hollow particles were broken and fragments could be observed in **Fig. 1.d**. The two alumina powders had the same hardness (more than 2000 HV10).

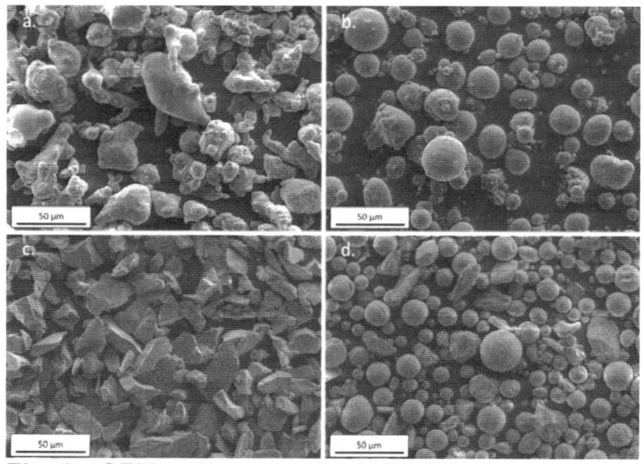

Fig. 1. SEM micrographs of feedstock aluminum powders with irregular (a.) and spherical (b.) particle shapes and alumina powders with angular (c.) and spherical (d.) particle shapes.

Fig. 2. shows the powder size distribution and the powder size range of the four powders used in this work. All the powders had almost the same powder size range (<45µm), which is rather conventional in cold spray [1]. The two Al powders size ranges were very close ([-45 + 11 µm] for irregular aluminum powder and [-32 + 11 µm] for spherical aluminum powder). The powder size distribution of the irregular Al powder was large and asymmetrical with a high fraction of coarse particles. The powder size distribution of the spherical Al powder was rather narrow and symmetrical with a low fraction of very fine particles (<5µm) which were fine and detached satellites (**Fig. 1.b**). The two Al_2O_3 powders had also very close particle size ranges, i.e. [-34 + 9 µm] for angular Al_2O_3 and [-26 + 10 µm] for spherical Al_2O_3. The powder size distribution of spherical Al_2O_3 was narrow and symmetrical to the mean particle size value (**Fig. 2.**). The powder size distribution of angular alumina was less symmetrical with a higher fraction of coarse particles. The two Al_2O_3 powders exhibited very small particles (<5µm). For spherical alumina, these small particles were assumed to be probably fragments of hollow particles which broke off.

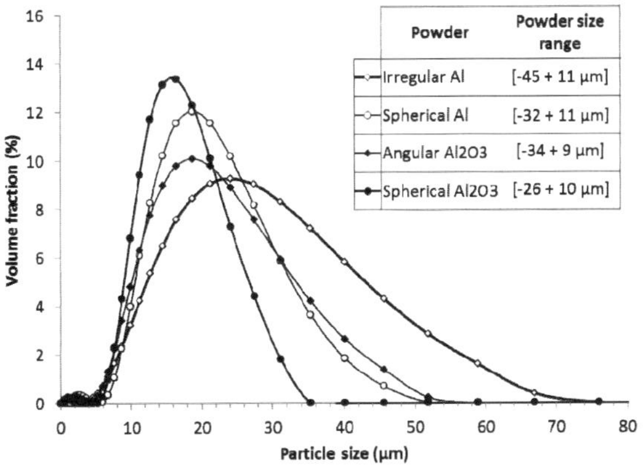

Fig. 2. Powder size distributions and powder size ranges of the four tested powders.

3.2 Process

All the powders were successfully cold-sprayed onto the steel substrate.

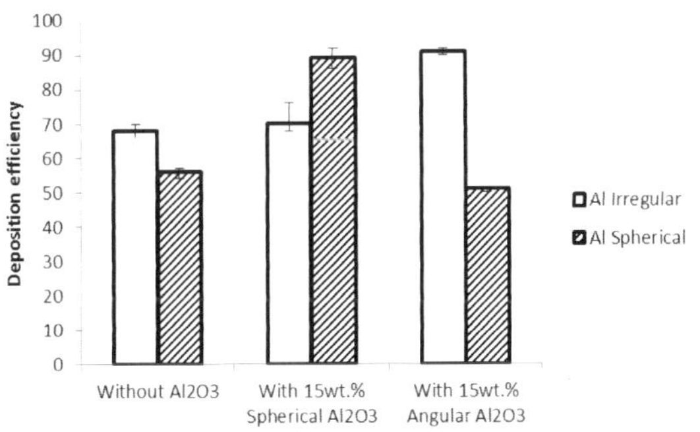

Fig. 3. Deposition efficiency for Al and Al-Al_2O_3 cold-sprayed coatings.

Despite the use of similar spraying conditions, the deposition efficiency varied depending on the type of aluminum and alumina powder (**Fig. 3.**). Irregular aluminum powder showed a higher deposition efficiency than spherical aluminum. Spherical Al_2O_3 addition increased the deposition efficiency of aluminum powders. Angular Al_2O_3 addition increased the deposition efficiency of irregular aluminum powder but decreased that of spherical aluminum powder. The higher deposition efficiency was achieved for two powder mixtures, that made of irregular aluminum powder with angular Al_2O_3 and that of spherical aluminum powder with spherical Al_2O_3. Similar particle shapes for aluminum and alumina powders led to cold spray coatings with a very high deposition efficiency.

3.3 Microstructure

Coatings made of irregular aluminum powder and those made of spherical aluminum powder will be termed respectively as "irregular aluminum coating" and "spherical aluminum coatings" in the rest of the article.

Irregular aluminum coatings (**Fig. 4.a**) were slightly thicker in average than those made of spherical aluminum powder (**Fig. 4.b**). No actual difference in porosity can be brought into light from micrographs of the two types of pure aluminum coatings.
Spherical alumina addition in the starting powder increased the thickness of the coating, regardless the aluminum powder particle shape (**Fig. 4.c** and **Fig. 4.d**). Angular alumina addition increased the coating thickness of irregular aluminum but not that of spherical aluminum as shown in **Fig. 4.e** and **Fig. 4.f**. Spherical aluminum mixed with spherical alumina resulted in the thickest and the more dense coating.

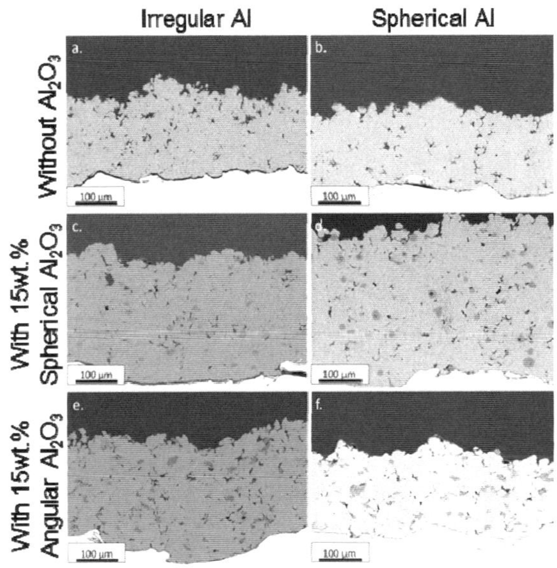

Fig. 4. Cross-sectional SEM backscattered electron micrographs of aluminum and alumina-aluminum coatings : (a.) Irregular Al, (b.) Spherical Al, (c.) Irregular Al +15wt.% Spherical Al_2O_3, (d.) Spherical Al +15wt.% Spherical Al_2O_3, (e.) Irregular Al +15wt.% Angular Al_2O_3, (d.) Spherical Al +15wt.% Angular Al_2O_3.

Spherical alumina and angular alumina were homogenously dispersed in the coatings. Some hollow particles and fragments could be observed in coatings made of spherical alumina (**Fig. 4.c** and **Fig. 4.d**). Angular alumina particles were fully dense (**Fig. 4.e** and **Fig. 4.f**). No fragments of broken particles could be observed in coatings with angular alumina addition. The content of spherical Al_2O_3 in

aluminum coatings seemed to be smaller than that of angular Al_2O_3.
Coating porosity results are summarized in **Fig. 5.** After optimization of spraying conditions, cold spray coating porosity was classically less than 1%. The high porosity values obtained in this work were due to non-optimized spraying parameters.
Aluminum coatings without alumina addition showed almost the same porosity (i.e. 7% for irregular Al coating and 6% for spherical Al coating). However, irregular aluminum coating was made of low porosity regions (4-6%) and high porosity regions (10-13%). Dense regions were more numerous in the coating. Spherical aluminum coating porosity was less discrepant from the average value (from 5% to 9%). Alumina addition reduced the porosity of aluminum coatings. Irregular aluminum coating porosity decreased with alumina addition (down to around 5% for the two aluminas) whatever the alumina particle shape. Porous regions of irregular aluminum coatings vanished with alumina addition. Spherical alumina addition decreased more the porosity of spherical aluminum coating (from 6% to 2%) than angular alumina addition (from 6% to 4%).

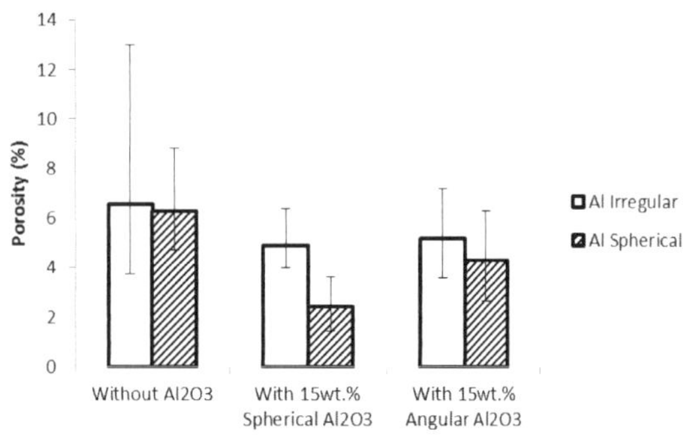

Fig. 5. Porosity of Al and Al-Al_2O_3 cold-sprayed coatings.

Alumina weight contents were determined using XRD analysis (**Table 1**). No alumina was found in pure aluminum coatings. There was neither alumina pollution nor oxidation of the aluminum powders during cold spraying.
Despite an initial alumina content of 15wt.% in all initial powder mixtures, the alumina content in the coatings was less than 15wt.%. This result was already reported [5], [7], [12]. Spherical alumina content in the coating was smaller than angular alumina content, regardless of the aluminum powder used in the mixture. Spherical alumina content was more difficult to assess because of small alumina fragments in the coating which were difficult to detect with image analysis. Alumina contents in coating were also smaller when using spherical aluminum rather than irregular aluminum as starting powder. The highest alumina content in coatings was achieved with a mixture of irregular aluminum with angular alumina.

Table 1 Alumina weight content in pure aluminum and aluminum-alumina coatings.

	Without Al_2O_3	With 15wt.% Spherical Al_2O_3	With 15wt.% Angular Al_2O_3
Irregular Al	0	3.5 ± 1	8.1 ± 1
Spherica l Al	0	2.7 ± 1	5.3 ± 1

3.4 Mechanical properties

Hardness differences (**Table 2**) were not significant (only 10 HV_{10} between the softest and the hardest coating).
Despite the same initial powder hardness (28 HV_{10}), irregular aluminum coating was harder (60 HV_{100}) than spherical aluminum coating hardness (55 HV_{100}). Alumina addition hardened aluminum coatings, whatever the aluminum powder used. Microhardness of coatings with angular alumina addition was higher than that of coatings made of spherical alumina.

Table 2 Hardness (HV100) of Al and Al-Al_2O_3 coatings.

	Without Al_2O_3	With 15wt.% Spherical Al_2O_3	With 15wt.% Angular Al_2O_3
Irregular Al	60 ± 5	62 ± 5	65 ± 3
Spherica l Al	55 ± 2	60 ± 3	62 ± 4

Bond strength tests according to ASTM C633 [2] were performed on five specimens for each coating. Specimens exhibited three types of fracture surfaces (**Fig. 6.**).
Most of the specimens broke at the coating-substrate interface (**Fig. 6.a**). The coating remained mostly bounded to the pull-off bar on which the glue was applied. This failure type is characteristic of an adhesive rupture between the coating and the substrate. These specimens could be considered to study the bond strength of the coating.
Some of the specimens broke in the glue (**Fig. 6.b**). Glue failure was the sign that the bond strength of these specimens was higher but could not be determined exactly due to the glue. These specimens therefore led to a bond strength threshold only.
A few specimens exhibited failures that occurred partly in the glue and partly at the coating-substrate interface. After SEM examination, the coating which remained on the substrate showed some glue penetration (arrowed in **Fig. 6.c**). These glue penetration islands seemed to increase the coating bond strength. However, coating thickness in this work was lower than that recommended by ASTM C633. This could explain the glue penetration in some

specimens. These samples were not kept for assessing bond strength.

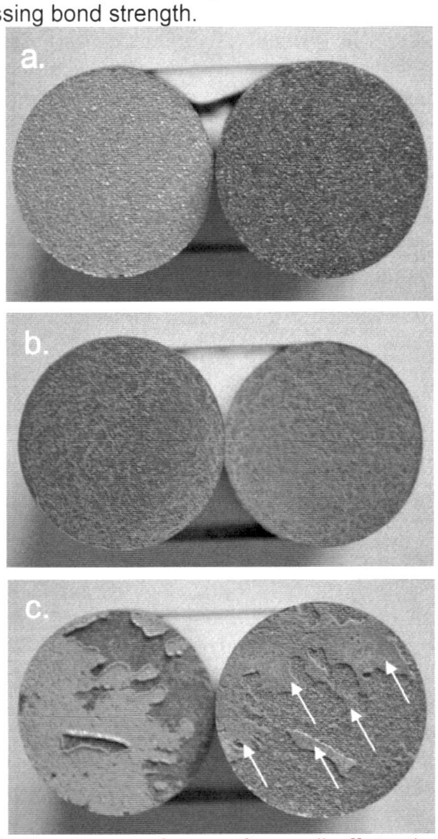

Fig. 6. Fracture surfaces after pull-off testing (the diameter of each specimen is 25mm). Glue penetration islands are arrowed.

Fig. 7. gives the bond strength of every coating. The bond strength of the glue was also tested separately. It was measured at 81±13MPa, which is really high for a glue [2].

Bond strengths of the two aluminum coatings were very high (more than 60MPa). The bond strength of the irregular aluminum coating was higher (77MPa) than that of the spherical aluminum coating (65MPa). Most of the specimens broke in the coating-substrate interface as shown in **Fig. 6.a**. Ruptures were adhesive, which was in keeping with the bond strength measurements.

Specimens with a coating made of mixtures of irregular aluminum and alumina powders, broke partly at the coating-substrate interface and partly in the glue. Bond strengths (**Fig. 7.**) have therefore to be considered as lower limits only of the real bond strength of these two powders. Consequently, these two powders could not be compared with the others.

Alumina addition increased the bond strength of spherical aluminum coating (**Fig. 7.**). Spherical alumina addition increased more the bond strength (from 65MPa to 78MPa) than angular alumina addition (from 65MPa to 70 MPa). Most of the specimens broke at the coating-substrate interface (**Fig. 6.a**). However, some of them broke partly at the coating-substrate interface and partly in the glue, due to glue penetration into the substrate (**Fig. 6.c**).

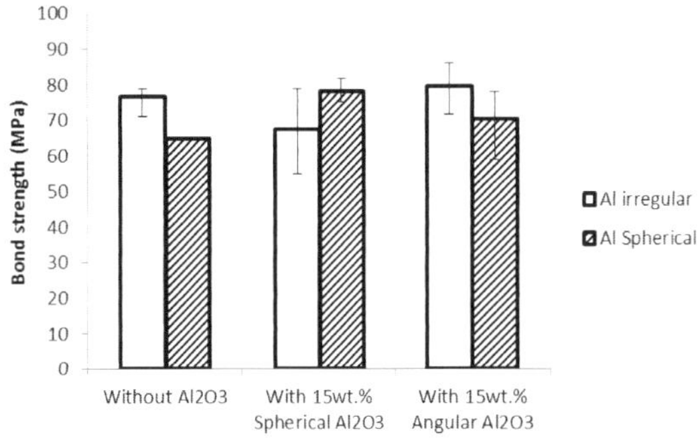

Fig. 7. Bond strength of the pure aluminum and aluminum-alumina coatings onto steel.

4 Discussion

Two aluminum powders were cold-sprayed successfully onto a shot-peened steel substrate. The coating-substrate bond strength was high compared to conventional values [1]. This is attributed to the shot-peening of the substrate. This process resulted in specimens with a high roughness and removed all the oxides from the steel. Substrate roughness and cleanliness, in that case, are supposed to improve the bond strength for this coating-substrate system.

4.1 Effect of aluminum powder morphology

In cold spray [1], [9], [10], irregular or angular particles showed higher velocities than spherical particles due to a higher drag coefficient. These higher velocities led to a higher deformation at impact therefore to higher deposition efficiency and hardness.

Irregular aluminum coating exhibited dense and highly-porous regions. These two types of regions could exist due to differences between the two types of particles (i.e. irregular and spherical) in impingement on the already-deposited particles. For suitable impinging, sprayed particles highly deformed at impact and adhered to the coating. Successive impacts compacted the previously-deposited layer, which further densified the coating. This resulted in so-called "dense regions". For unsatisfactory impinging conditions, sprayed particles do not deform and can rebound. This left "porous regions" in the coating. For all spherical aluminum particles, impinging conditions were nearly the same. They depended on particle diameter range, which was narrow in this study (**Fig. 2.**). Consequently, "dense regions" and "porous regions" could not co-exist when spraying spherical aluminum powders. However, these two types of regions existed in irregular Al coating and more dense regions were observed than porous regions. These dense regions can be assumed to be more adhesive to the substrate due to anchoring effect, which improved the coating-substrate bond strength. Consequently, for a given steel substrate, the bond strength of pure irregular Al coating was higher than that of pure spherical Al coating.

4.2 Effect of alumina powder morphology

The particle shape of additionned alumina had a strong effect on deposition efficiency for aluminum powder. When aluminum and alumina particles did not exhibit the same particle shape, the deposition efficiency of the mixture was close to that of pure aluminum powders. However, when powder mixtures involved aluminum and alumina particles of the same shape (spherical-spherical or irregular-angular), coating showed a very high deposition efficiency.
When alumina particles impact the coating, they compact the previously-deposited layer, which decreases the overall porosity of the coating. Consequently, aluminum-alumina coatings had a lower porosity than pure aluminum coatings. When angular alumina particles impact the coating, they can either rebound on already-deposited particles or be entrapped. Spherical alumina particles could be dense, hollow or made of fragments from broken hollow particles (**Fig. 1.d**). At impact, dense spherical alumina particles have the same behavior as angular alumina particles. Small alumina fragments are also entrapped in the coating during spraying but compact less the previously-deposited layer due to their small size. A few hollow spherical particles had been found in the coatings (**Fig. 4.c** and **Fig. 4.d**). During the impact, these spherical hollow particles can explode. Their fragments move in every direction and are not entrapped in the coating. Consequently, spherical alumina content was smaller than that of angular alumina entrapped in Al- Al_2O_3 coatings (**Table 1**).
Hardness of the aluminum and alumina coatings (**Table 1** and **Table 2**) was independent of the alumina morphology and only depended on its weight content. The more numerous entrapped alumina particles are, the harder the coating is. The content of spherical alumina in the coating was smaller than that of angular alumina. Consequently, coating with spherical alumina was softer than coating with angular alumina, whatever the aluminum powder used in the mixture.

5 Conclusion

High-bond strength aluminum coatings could be successfully cold-sprayed onto steel. Deposition efficiency was higher when using irregular aluminum compared to spherical aluminum. Alumina addition decreased porosity in the aluminum coating, which consequently increased hardness. Hardness of aluminum-alumina coatings depended on the alumina content only. Alumina morphology had no effect. An aluminum-alumina mixture with the same particle shape seems to be the most suitable to obtain dense coatings on steel with very high bond strength.

6 Acknowledgements

The authors would like to acknowledge D. Billières (Saint-Gobain Coating Solutions/ Cavaillon, France) for help in providing alumina powders.

7 Bibliography

[1] J. Villafuerte, Ed., *Modern Cold Spray*. New York City: Springer International Publishing, 2015.

[2] ASTM C 633-1, 'Standard Test Method for Adhesion or Cohesion Strength of Thermal Spray Coatings'. ASTM International.

[3] W.-Y. Li, G. Zhang, H. L. Liao, and C. Coddet, 'Characterizations of cold sprayed TiN particle reinforced Al2319 composite coating', *J. Mater. Process. Technol.*, vol. 202, no. 1–3, pp. 508–513, Jun. 2008.

[4] E. Sansoucy, P. Marcoux, L. Ajdelsztajn, and B. Jodoin, 'Properties of SiC-reinforced aluminum alloy coatings produced by the cold gas dynamic spraying process', *Surf. Coat. Technol.*, vol. 202, no. 16, pp. 3988–3996, May 2008.

[5] K. J. Hodder, J. A. Nychka, and A. G. McDonald, 'Comparison of 10 µm and 20 nm Al-Al2O3 Metal Matrix Composite Coatings Fabricated by Low-Pressure Cold Gas Dynamic Spraying', *J. Therm. Spray Technol.*, vol. 23, no. 5, pp. 839–848, Jun. 2014.

[6] K. Spencer, D. M. Fabijanic, and M.-X. Zhang, 'The use of Al–Al2O3 cold spray coatings to improve the surface properties of magnesium alloys', *Surf. Coat. Technol.*, vol. 204, no. 3, pp. 336–344, Oct. 2009.

[7] E. Irissou, J.-G. Legoux, B. Arsenault, and C. Moreau, 'Investigation of Al-Al2O3 Cold Spray Coating Formation and Properties', *J. Therm. Spray Technol.*, vol. 16, no. 5–6, pp. 661–668, Dec. 2007.

[8] M. Jeandin, G. Rolland, L. L. Descurninges, and M. H. Berger, 'Which powders for cold spray?', *Surf. Eng.*, vol. 30, no. 5, pp. 291–298, Mar. 2014.

[9] H. Fukanuma, N. Ohno, B. Sun, and R. Huang, 'The Influence of Particle Morphology on In-flight Particle Velocity in Cold Spray', presented at the International Thermal Spray Conference, Seattle, 2006, vol. 5, pp. 97–101.

[10] W. Wong, P. Vo, E. Irissou, A. N. Ryabinin, J.-G. Legoux, and S. Yue, 'Effect of Particle Morphology and Size Distribution on Cold-Sprayed Pure Titanium Coatings', *J. Therm. Spray Technol.*, vol. 22, no. 7, pp. 1140–1153, Oct. 2013.

[11] J. M. Shockley, S. Descartes, P. Vo, E. Irissou, and R. R. Chromik, 'The influence of Al2O3 particle morphology on the coating formation and dry sliding wear behavior of cold sprayed Al–Al2O3 composites', *Surf. Coat. Technol.*, vol. 270, pp. 324–333, May 2015.

[12] Q. Wang, N. Birbilis, H. Huang, and M.-X. Zhang, 'Microstructure characterization and nanomechanics of cold-sprayed pure Al and Al-Al2O3 composite coatings', *Surf. Coat. Technol.*, vol. 232, pp. 216–223, Oct. 2013.

AUTHOR INDEX

AUTHOR INDEX

AUTHOR INDEX

AUTHOR INDEX

AUTHOR INDEX

AUTHOR INDEX

International Thermal Spray Conference & Exposition (ITSC 2016)

Shanghai, China
10 - 12 May 2016

Volume 2 of 2

ISBN: 978-1-5108-4400-1

Printed from e-media with permission by:

Curran Associates, Inc.
57 Morehouse Lane
Red Hook, NY 12571

CURRAN ASSOCIATES INC. proceedings.com

Some format issues inherent in the e-media version may also appear in this print version.

Printed by Curran Associates, Inc. (2017)

For permission requests, please contact DVS Media GmbH
at the address below.

DVS Media GmbH
P. O. Box 10 19 65
40010 Dusseldorf
Germany

Phone: 49 (0)211 / 1591-0
Fax: 49 (0)211 / 1591-150

media@dvs-hg.de

Additional copies of this publication are available from:

Curran Associates, Inc.
57 Morehouse Lane
Red Hook, NY 12571 USA
Phone: 845-758-0400
Fax: 845-758-2633
Email: curran@proceedings.com
Web: www.proceedings.com

TABLE OF CONTENTS

VOLUME 1

AVIATION INDUSTRY III

SUSPENSION SPRAYING III

PRE- & POST-TREATMENT

ARC SPRAYING & LASER SPRAYING

POWER GENERATION II

CHARACTERIZATION & TESTING METHODS

VOLUME 2

HVOF-/HVAF-SPRAYING

NEW PROCESSES II

POWER GENERATION III

CORROSION PROTECTION

EQUIPMENT / CONSUMABLES & POWDERS, WIRES, SUSPENSIONS

IRON-BASED COATINGS

POWER GENERATION IV

MEDICAL INDUSTRY

PLASMA SPRAYING

AMORPHOUS COATINGS

POSTERS

APPLICATIONS – AUTOMOTIVE INDUSTRY

APPLICATIONS – AVIATION INDUSTRY

APPLICATIONS – GENERAL PLANT ENGINEERING

APPLICATIONS – MEDICAL INDUSTRY

PROPERTIES – CORROSION PROTECTION

PROPERTIES – METAL COATINGS

PROPERTIES – POLYMER COATINGS

PROPERTIES – PRE- & POST-TREATMENT

PROPERTIES – PROPERTIES CHARACTERIZATION & TESTING METHODS

PROPERTIES – TRIBOLOGICAL COATINGS

PROPERTIES – WEAR PROTECTION

THERMAL SPRAY PROCESSES – PLASMA SPRAYING

YOUNG PROFESSIONALS

Residual stresses measurement in plasma sprayed TBCs with cold sprayed bond coats by photoluminescence piezospectroscopy

C. R. C. Lima, S. Dosta , J. M. Guilemany, D. R. Clarke, São Paulo/ BR

A non-destructive inspection technique known as photoluminescence piezospectroscopy (PLPS) has been used to measure the residual stresses in the thermally grown oxide (TGO) from thermal barrier coatings (TBC) submitted to furnace thermal cycling at 1121 °C. The ceramic layer (YSZ) was applied by atmospheric plasma spraying onto a cold sprayed bond coated Ni superalloy. The residual stresses are analyzed as a function of the thermal exposure. The changes in stress were used to indicate the spallation and damage degree of the coating system.

1. Introduction

Thermal barrier coatings (TBC) are the state of the art for the protection of gas turbines from increasingly higher working temperatures [1, 2, 3]. A typical TBC system consists of a superalloy substrate, a metallic bond coat and an insulating ceramic top coat. In working conditions, there is also a presence of a thermally grown oxide layer, called TGO, formed as an oxidation reaction product from the bond coat oxidation in cyclic service heating [4, 5]. The ceramic layer is normally 7-8 % yttria partially stabilized zirconia (YSZ) [5-8] applied by Atmospheric Plasma Spray (APS) or Electron Bean assisted Physical Vapor Deposition (EB-PVD) [3, 8].
The bond coat usually consists of a MCrAlY alloy, where M stands for Ni, Fe, Co or some combination of them. Some alloys also include other elements like Y, Hf, Ta or Re [9, 10]. In EB-PVD TBCs, electroplated and diffusion (Ni, Pt)-Al has been also a common bond coat choice. The bond coat has the function of protecting the substrate from oxidation and hot corrosion by forming the TGO at the interface, preferentially α-alumina, which protects and helps in the anchorage of the ceramic layer. The coefficient of thermal expansion of the bond coat is between that of the metallic substrate and that of the ceramic, reducing further thermal stresses.
The traditional techniques for metallic bond coat deposition are air plasma spraying (APS), vacuum plasma spraying (VPS) or low pressure plasma spraying (LPPS) [3, 11]. HVOF spraying has been included as an alternative deposition method based mainly in its high quality deposits, low high temperature oxidation rates at the same level than those of similar VPS or LPPS coatings [10, 11, 12, 13]. More recently, cold gas spraying (CGS) has been proposed to apply metallic bond coats based on its very low oxidation perspective and high performance [14, 15, 16]. Cold Gas Spray (CGS) utilizes the kinetic energy of the accelerated particles in a supersonic flow leading to a severe plastic deformation at the impact.
When comparing CGS with conventional thermal spray techniques, the main benefits are related to the absence of thermal energy (no grain growth and no particle surface oxidation), resulting in a coating free of oxides, allowing for more free aluminum for diffusion to the surface to form the protective α-Al_2O_3 layer, one of the main functions of a bond coat in the TBC system [16, 17].
The TGO is subjected to large compressive stresses on cooling due to thermal expansion mismatch with the superalloy substrate. These stresses are accepted as the main source of stored energy that would lead to spallation of the coating and consequent function loss [1–4]. Then, a non-destructive way to access the extent of damage accumulation mainly caused by the bond coat oxidation is a very useful tool for life prediction and performance control of TBCs. This is exactly the purpose for the use of photoluminescence piezospectroscopy (PLPS).
A thermal barrier coating that is illuminated with a green or blue laser (colors corresponding to the absorption bands of Cr^{3+} in aluminum oxide) beam produces a photo-stimulated luminescence. This luminescence originates from the Cr^{3+} dopants in the aluminum oxide (TGO) formed by high-temperature oxidation of the bond coat and is characterized by an intense spectral doublet at 14400 cm^{-1} (694 nm). The frequency of the individual lines in the doublet and the shape of the doublet are sensitive to the stress, the stress distribution, and damage within the TGO [18, 19]. If the oxide is stressed, the frequency of the Cr^{3+} luminescence shifts from its stress-free value, the piezospectroscopic effect.
The frequency shift is related to the stress by a general equation and measuring this shift the biaxial stresses can be obtained after some calculation [20, 21]. PLPS is well accepted for residual stress measurements in EB-PVD TBCs. But some restrictions have been discussed when applying the technique for APS TBCs since the resulting lamellar structure is stated to excessively spread the laser beam then resulting in feeble photoluminescence signals [22].
The objective of this work was to assess residual stresses in TGOs formed on APS TBCs with cold sprayed bond coats by PLPS experiments. The results are discussed as a function of the PLPS process parameters, coating thickness and number of thermal cycles.

2 Experimental Procedure

The substrate used was an Inconel 625 Ni-based superalloy (25 x 20 x 5 mm) partially coated with just a 150 µm thick bond coat or a full TBC. The feedstock bond coat powder was a new gas atomized CoNiCrAlY powder, designation AE10086, with spherical morphology, and particle size range from 35+11 µm. The ceramic powder feedstock was an YSZ (203 NS - Oerlikon Metco) with spherical morphology, and particle size range from -125+11 µm.
The substrates were degreased whit acetone and grit blasted with white corundum immediately before spraying. The grit blasted surface roughness was 4.5 µm (Ra).
The samples were placed in a circular sample holder, moving at 500 mm s^{-1} in the substrate surface, sprayed simultaneously for every material (bond and top coat). Ceramic deposition was carried out assisted by cooling with air jet directed to back side of the sample. The chemical compositions of the metallic and ceramic powders are shown in Table 1.

Table 1. Chemical composition of the materials used in the experiments

Material	Co	Ni	Cr	Al	Y	Mo	Fe (max)	Nb+ Ta	ZrO_2	Y_2O_3	Other (max)
Inconel 625	-	bal.	20-23	-	-	8-10	5-11	3.1–4.1	-	-	-
CoNiCrAlY	bal.	29-35	18-24	5-11	0.1-0.8	-	-	-	-	-	1.0
YSZ 203 NS	-	-	-	-	-	-	-	-	92	8	-

Kinetiks® 4000 cold gas spray system (CGT GmbH) was used to apply the bond coats, using a maximum operating pressure of 40 bar and nitrogen as the propellant gas. A F4-MB plasma spray gun (Plasma Technik, Sulzer Metco, Westbury, USA) was used to apply the ceramic top coats. Table 2 shows the thermal spraying parameters used for metallic and ceramic coatings deposition. Ceramic top coatings were prepared with a thickness of 100 µm.

Table 2. Equipment parameters adjusted for thermal spraying of the coatings.

CGS	p (bar)	T (ºC)	Feed rate (Kg/h)	v (mm/s)	Passes	SOD (mm)
CoNiCrAlY	50	1000	2	550	1	20
APS	Air Flow Rate (l/min)	H_2 Flow Rate (l/min)	Amperage (A)	Voltage (V)		
ZrO_2-8%Y_2O_3	40	12	630	68	20	120

SOD= Standoff Distance; v= Particle velocity

As sprayed coatings characterization and thermal cycled coating microstructure analysis included cross-sectional Optical Microscopy (OM-Leica DMI5000 M) and Scanning Electron Microscopy (ZEISS - ULTRA 55), as well as phase analysis by Energy Dispersive Spectroscopy- EDS- (Quantum, Kevex). Coating surface roughness was measured in an Olympus LEXT Confocal Microscope - OLS 4000. Values quoted for roughness are an average of 20 measurements for each coating.
Thermal cycling was performed in an automatic system using a Thermo-Scientific Lindberg Blue M tube furnace. The thermal cycling set up can be seen in Figure 1. The cycle consisted of a 15 min period at 1121 ºC in air environment, followed by cooling to 200 ºC in ambient air. Sets of samples were systematically extracted from the holder for PLPS experiments after 50, 200 and 500 cycles. Selected samples were prepared to be observed by SEM after residual stress measurements, and the thermal grown oxide (TGO) layer was measured and analyzed.

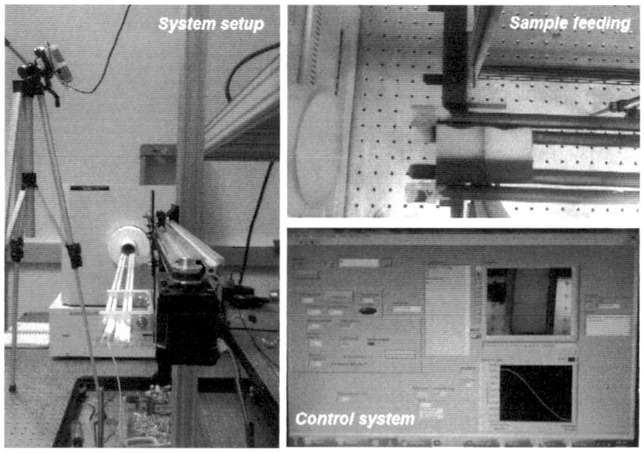

Figure 1. Thermal cycling equipment and setup

Cr^{3+} photoluminescence piezospectroscopy (PLPS) was used for residual stress measurements by using a Horiba LabRAM ARAMIS system. The laser employed was a solid state laser at 514 nm and 10% power (Filter D-1), 10X objective, grating 1800, accumulation number 10, exposition time of 5 s. A NGS LabSpec 5 software carried out the peak fitting and deconvolution. An average of 40 individual measurements was assumed for residual stress computation.

3 Results and Discussion

Representative images of the TBC coating systems after thermal cycling are shown in Fig. 2 for the different number of thermal cycles.

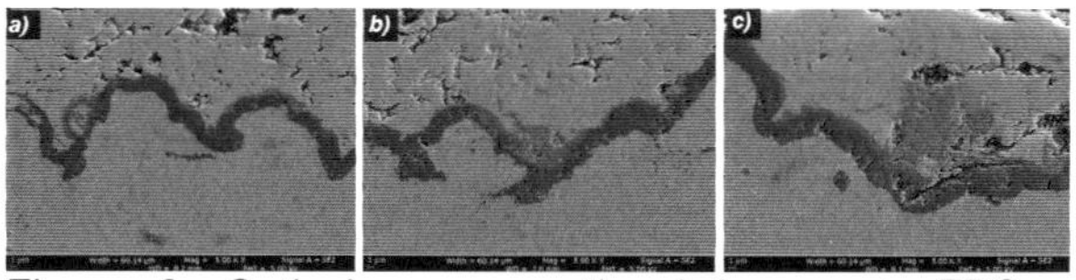

Figure 2. Optical cross sectional images a TBC coating system with a 100 µm thick ceramic over a 150 µm bond coat after thermal cycling at 1121 ºC by a) 50; b) 200; and c) 500 cycles.

The roughness of the as sprayed bond coat was measured as 3.8 µm (Ra) thought the grit blasted substrate had a roughness of 4.5 µm which shows a surface roughness reduction after bond coated by cold spray. Rough surfaces result in lower R2 frequency shifts due to tensile contributions of the stress over convex prominences of the surface.
There is no significant increase of TGO thickness after thermal cycling from 50 to 500 cycles that was measured around 4 µm. Several distinct values of

TGO thicknesses from 2 to 10 µm have been reported from other authors mainly due to differences in heating or thermal cycling procedures, feedstock material, technique for measurement, among others [23]. The maximum measured thickness was 5 µm after 500 thermal cycles, just 1 µm more than the thickness after 200 thermal cycles.

The formation of some different oxides, probably Cr and Ni oxides or spinels, can be observed in the upper part of the TGO (light phase) presenting higher incidence with further thermal cycling. The lighter phase present in the TGO was tested by PLPS experiments along to each phase, α-alumina (dark grey) and mixed oxides (mainly from Cr and Ni) of a region similar to that presented in Figure 3. The results of 40 distinct measurements are averaged in Table 3.

The residual stress in the second phase region is almost half of that of the α-Al_2O_3 region (0,77 GPa) what could justify the reduction in residual stress when increasing number of cycles as that second phase is increasingly observed for higher number of cycles. However, it should be considered that the PLPS measurement in the cross section plane of the sample is not a correct way for such practice since just a incorporated alumina oxide beneath the light phase is proved which could not reflect the correct signal emission and consequent residual stress assessment.

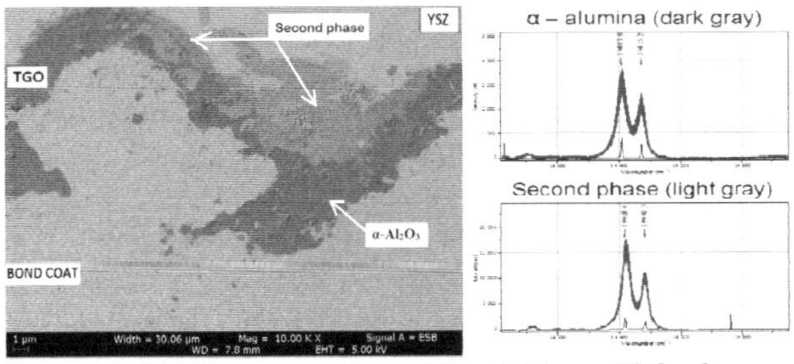

Figure 3. SEM cross section of the 100 µm TBC after 200 thermal cycles at 1121°C for 15 min showing the second phase formed over the α-Al_2O_3 TGO scale. The general spectrum from the photoluminescence piezospectroscopy (PLPS) for the two phases is added at the right side of the coating image.

Table 3. Cross sectional PLPS measurements of residual stress on different TGO phases

Thermally Grown Oxide	Average compressive residual stress (GPa)	Standard Deviation
α -alumina	1.24	0.17
Second phase	0.77	0.11

As sprayed TBCs have shown no luminescence signals in PLPS measurements thus meaning that there was no alumina traces in the coatings before thermal cycling. The results of residual stresses for the different number of thermal cycles are presented in Figure 4.

A trend of compressive stress reduction while increasing the number of thermal cycles is observed in Fig.5. The stresses are in the range of 2.0 - 2.5 after 50 cycles, then go to a range of 1.8 – 2.3 GPa after 200 cycles and finally to 1.5 to 2.4 GPa after 500 cycles. Some sporadic results are beyond such limits reaching over 3.5 GPa after 500 cycles. The dispersion of results is mainly attributed to the influence of TGO roughness and additionally to the microstructural heterogeneity of plasma sprayed coatings. The coating failure was defined as the spalling or detachment of more than 40 % of the coated area. No such failure has occurred until 500 thermal cycles.

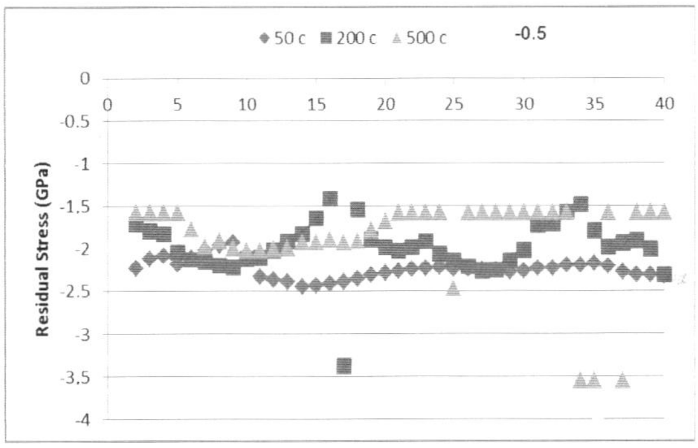

Figure 4. Compressive residual stresses measured after 50, 200 and 500 thermal cycles.

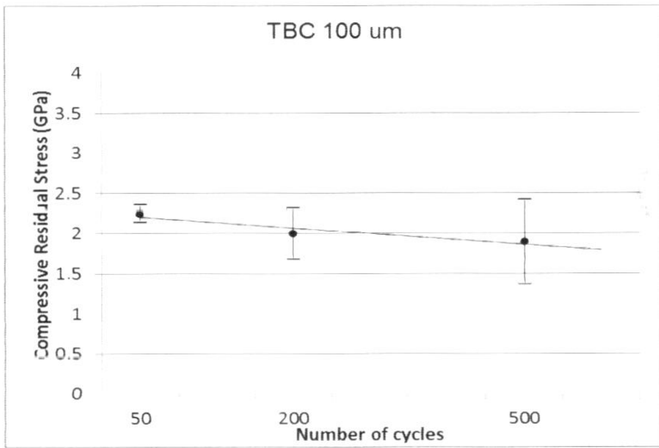

Figure 5. Residual stresses for the different thermal cycling exposition of coatings at 1121 °C. Average values and standard deviation for 40 individual measurements.

4 Conclusion

- Cr^{3+} photoluminescence piezospectroscopy was efficient to evaluate the residual stresses at the thermally grown oxide formed in air plasma sprayed TBCs with cold gas sprayed bond coats after thermal cycling at 1121 °C in static air.
- Inconel 615 superalloy substrate was bond coated by cold gas spray with a new CoNiCrAlY powder with composition Co32Ni21Cr8Al0.5Y. The coatings showed a dense microstructure with surface roughness lower than the grit blasted substrate.

- The TGO formed on the bond coat presented a thickness around 4 μm with very little variation after additional thermal cycling.
- The residual compressive stresses measured in the TGO were found to be around 2.3 GPa after 50 cycles of 15 minutes at 1121 °C, decreasing for further thermal cycles.
- A trend of decreasing compressive stress with higher number thermal cycles wa observed for the tested coatings and it is believed to be related to the formation of a second phase in the TGO changing the local thermal expansion mismatch status.

Acknowledgements

The authors would like to thank the technical help from Mary Gurak, Samuel Shian, and Peter Kjeer at SEAS, Harvard University, and Victor Crespo at CPT, University of Barcelona.
This study was conducted with support from CNPq Brazil.

3 Literature

[1] R.A. Miller, Thermal barrier coatings for aircraft engines: History and directions. Journal of Thermal Spray Technology 6 (1997) 35–42.

[2] D.R. Clarke, C. G. Levi. Materials design for the next generation thermal barrier coatings. Annual Review of Materials Research, 33 (1) (2003) 383-417.

[3] N.P. Padture, M. Gell, and E.H. Jordan, Thermal barrier coatings for gas-turbine engine applications. Science 296 (2002) 280-284.

[4] R. Vaßen, M.O. Jarligo, T. Steinke, D. E. Mack, D. Stöver. Overview on advanced thermal barrier coatings. Surface and Coatings Technology 205(4) (2010) 938-942.

[5] T.S. Hille, S. Turteltaub, A.S.J. Suiker, Oxide growth and damage evolution in thermal barrier coatings. Engineering Fracture Mechanics 78 (2011) 2139–2152.

[6] D. Stöver, C. Funke. "Directions of the development of thermal barrier coatings in energy applications." Journal of Materials Processing Technology 92 (1999): 195-202

[7] Y. Itoh, M. Saitoh, and M. Tamura, Characteristics of MCrAlY coatings sprayed by high velocity oxygen-fuel spraying system, J. Eng. Gas Turbines Power, 122 (2000), 43-49

[8] M. Peters, C. Leyens, U. Schulz, W.A. Kaysser. EB-PVD Thermal Barrier Coatings for Aeroengines and Gas Turbines. Advanced Engineering Materials 3(4) (2001) 193-204.

[9] T.A. Dobbins, R. Knight and M.J. Mayo, HVOF thermal spray deposited Y_2O_3- Stabilized ZrO_2 coatings for thermal barrier applications, J. Therm. Spray Technol., 12 (2) (2003) 214-225.

[10] P. Vuoristo, S. Ahmaniemi, S. Tuurna, T. Mantyla, E. Cordano, F. Fignino, and G.C. Gualco, Development of HVOF sprayed NiCoCrAlYRe coatings for use as bond coats of plasma sprayed thermal barrier coatings. Proceedings of ITSC 2002 – International Thermal Spray Conference and Exposition, Essen, Germany, ASM-International/DVS, v.1 (2002) 470-475

[11] W.O. Soboyejo, P. Mensah, R. Diwan, J. Crowe and S. Akwaboa, High temperature oxidation interfacial growth kinetics in YSZ thermal barrier coatings with bond coatings of NiCoCrAlY with 0,25% Hf. Mater. Sci. Eng. A 528 (2011) 2223-2230.

[12] L. Ajdelsztajn, J.A. Picas, G.E. Kim, F. L. Bastian, J. Schoenung, V. Provenzano. Oxidation behavior of HVOF sprayed nanocrystalline NiCrAlY powder. Materials Science and Engineering A, 338(1) (2002) 33-43.

[13] B.A. Pint, M.A. Bestor, J.A. Haynes, Cyclic oxidation behavior of HVOF bond coatings deposited on La- and Y-doped superalloys. Surf. Coat. Technol. 206 (2011) 1600–1604.

[14] W. R. Chen, E. Irissou, X. Wu, J.G. Legoux, B.R. Marple. The oxidation behavior of TBC with cold spray CoNiCrAlY bond coat. Journal of thermal spray technology, 20(1-2) (2011) 132-138.

[15] P. Richer, M. Yandouzi, L. Beauvais, B. Jodoin, Oxidation behaviour of CoNiCrAlY bond coats produced by plasma, HVOF and cold gas dynamic spraying. Surf. Coat. Technol., 204 (2010) 3962–3974.

[16] Q. Zhang,C.J. Li, C.X. Li, G.J. Yang, S.C. Lui. Study of oxidation behavior of nanostructured NiCrAlY bond coatings deposited by cold spraying. Surface and Coatings Technology, 202(14) (2008) 3378-3384.

[17] H. Assadi, F. Gärtner, T. Stoltenhoff, H. Kreye. Bonding mechanism in cold gas spraying. Acta Materialia, 51(15) (2003) 4379-4394.

[18] F. H. Wang, Y. Zhang, H. Wang, Residual Stress and Damage Evolution in TBCs by Optical Method, Key Engineering Materials 324-325 (2006) 1047-1050.

[19] A. Bennett, F.C. Roriz, A.B. Thakker. A philosophy for thermal barrier coating design and its corroboration by 10 000 h service experience on RB211 nozzle guide vanes. Surf. Coat. Technol. 32 (1-4) (1987) 359-375.

[20] D.R. Clarke, R.J. Christensen, V. Tolpygo. The evolution of oxidation stresses in zirconia thermal barrier coated superalloy leading to spalling failure. Surface and Coatings Technology 94-95 (1997) 89-93.

[21] Q. Ma, D.R. Clarke. Piezospectroscopic Determination of Residual Stresses in Polycrystahe Alumina. Journal of the American Ceramic Society 77 (2) (1994) 298-302.

[22] K.W. Schlichting, K. Vaidyanathan, Y.H. Sohn, E.H. Jordan, M. Gell, N.P. Padture. Application of Cr3_ photoluminescence piezo-spectroscopy to

plasma-sprayed thermal barrier coatings for residual stress measurement. Materials Science and Engineering A 291 (2000) 68–77.

[23] A. Rabiei, A. G. Evans. Failure mechanisms associated with the thermally grown oxide in plasma-sprayed thermal barrier coatings. Acta mater. 48 (2000) 3963–3976

Microstructure and performance of WC-10Co-4Cr coating with ultrafine / nanocrystalline structures

Xuezheng Wang, Haibin Wang, Xiaoyan Song *, Xuemei Liu, Deqiang You, Tao Yang
College of Materials Science and Engineering, Key Laboratory of Advanced Functional Materials, Education Ministry of China, Beijing University of Technology, Beijing 100124, China
*Corresponding author, Email: xysong@bjut.edu.cn

Abstract: The WC-10Co-4Cr composite powder was synthesized firstly. Then the composite powder was agglomerated to prepare the thermal spraying feedstock. The ultrafine/nanostructured WC-10Co-4Cr coating was prepared by high velocity oxygen fuel (HVOF) thermal spraying. The phase constitution, elemental distribution and microstructure of the coating were characterized by the X-ray diffraction (XRD) and transmission electron microscope (TEM), respectively. The wear resistance and corrosion resistance of the prepared composite coating were tested. The results show that the main phases of the coating include WC, binding phase with partial amorphous structure, with a little W2C and Co(Cr) coexisting. The distributions of Co and Cr elements from the phase boundary to the eutectic area then to Co zone were analyzed quantitatively. The mechanisms for the formation of the microstructure and effects of Cr on the performance of the composite coating are proposed.
Key words: in situ synthesis; high velocity oxygen fuel; ultrafine/nanostructure; composite coating; phase constitution

1 Introduction

The thermal sprayed WC-based cermet coatings are widely used in aerospace, automotive, petroleum, machinery manufacturing, electricity, printing, and etc.[1-3]. However, because of the poor corrosion resistance, and its applications are restricting in many fields. The studies [4-6] found that WC-Co coatings with a certain amount of Cr not only had high hardness, high bonding strength and high wear resistance, but also had excellent corrosion resistance.
In the studies of wear and corrosion mechanisms of WC-CoCr coatings, Han et al [7] suggested that a Cr-rich phase was formed during HVOF spraying process, and the WC powder was completely coated with the Cr-rich phase, so that the coating has higher corrosion resistance. Soua et al [8] found that the coating with Cr in the surface could preferentially form dense Cr oxide, which had a spinel structure and hindered the etching process, so the WC-CoCr coating had good corrosion resistance. However, the studies are lack of strong experimental evidence. In this paper, the nanostructured WC-CoCr thermal spray feedstock powder was firstly prepared using the in situ synthesized WC-Co nanocomposite powder [9-12]. Then, a series of sliding wear and corrosion tests were performed on the prepared WC-CoCr coating so as to investigate the mechanisms of wear and corrosion for the coating.

2 Experimental

2.1 Materials preparation

Commercially available tungsten oxide, cobalt oxide and carbon black powders were used as the raw materials. The raw powders were mixed by ball milling and then were subjected to in situ reactions in a vacuum furnace. In this work, the WC-10.4 wt.% Co composite powder was firstly synthesized using the above-mentioned method [9]. Then a commercially available Cr powder was added into the WC-10.4 wt.% Co composite powder by milling for 20 h with a ball-to-powder ratio of 3:1. Then the WC-10 wt.%Co-4 wt.% Cr powder was obtained . The spray drying was carried out after the WC-CoCr powder was mixed with 2.0 wt.% polyvinyl alcohol (PVA),1.0 wt.% polyethylene glycol (PEG) and 30.0 wt.% distilled water to form stable slurry. Then the spray-dried particles were heat-treated at 1215 ℃ for 5h in a vacuum tubular furnace with the pure argon as the protective gas. After physical grinding and air-classification, the WC–Co–Cr feedstock powder with particle sizes of 10-45 μm was obtained. The HVOF-K2 spraying system (GTV in Germany) was used to deposit the WC-Co coatings onto the carbon steel (0.45 wt.% C) substrates. The fuel flow rate, oxygen flow rate, spraying distance, powder feeding rate and carrier gas (Ar) flow rate were 24 L/h, 900 L/min, 340 mm, 98 g/min, and 7.5 L/min, respectively.

2.2 Characterizations

The phases in the WC–Co powders and HVOF sprayed coatings were analyzed by X-ray diffraction (XRD) using Cu Kα radiation in a Rigaku D/max-3c diffractometer. The morphologies and microstructures of the powders and coatings were examined by scanning electron microscopy (SEM) and transmission electron microscopy (TEM) with an energy dispersive spectroscopy. Microhardness tests were performed with a 300g load and a dwell time of 15 s using a Vickers indenter on the polished cross sections of the coatings. Fifteen indentations were made for each sample. The sliding wear test was carried out using a reciprocating sliding tribometer. The silicon nitride ceramic balls with a radius of 5 mm were used as the counter material. Before tests, all the coating samples were polished. The sliding wear tests were conducted at the load of 80 N, the sliding speed of 5 m/min and the total time of 15 min. The wear volume was estimated by the section profiles at three different positions of the wear scar. Three tests were performed on the coating for each load. The friction coefficient and sliding time were recorded automatically during the tests. The particle size of the powder and the grain size of the coating were measured by the linear intercept method on the SEM and TEM micrographs, respectively.

The corrosion behavior of the coatings was evaluated by electrochemical measurements. Potentiodynamic Polarization test and electrochemical impedance spectroscopy (EIS) test were carried out to compare the corrosion resistance of HVOF sprayed nanostructured WC-Co-Cr coatings. Electrochemical tests were conducted via a IVIUM advanced electrochemical system on a three-electrode test cell, with the specimen as working electrode, a platinum wire as counter electrode, and saturated calomel electrode as reference electrode. Specimens for electrochemical tests were closely encapsulated in a non-conducting epoxy resin with the rear side of the specimen soldered to a wire, leaving only an end-surface (with a surface area of about 1 cm^2) exposed to the naturally aerated 3.5 wt.% NaCl aqueous solution.

Specimens were cold mounted using epoxy resin, polished on the diamond resin polishing discs with the diamond particle sizes of 20 μm and 9 μm and then were subjected to a cloth polishing using the diamond polishing paste of 1.5 μm, then degreased in acetone in an ultrasonic bath and dried in warm air. The tests commenced after 100 min of free corrosion to allow full wetting of the surface of the coating and open circuit potential (OCP) stabilization. As the OCP became steady, potentiodynamic polarization curves were measured with a fixed sweep rate of 3 $mV \cdot s^{-1}$. The corrosion current density (i_{corr}) and corrosion potential (E_{corr}) were obtained as the intersection point of linear fits to the anodic and cathodic polarization curves, according to the Tafel extrapolation technique. EIS test was conducted at the OCP by applying a sinusoidal potential excitation of 10 mV amplitude over frequency range from 100 kHz to 10 mHz. Each test was repeated at least thrice to make sure a good repeatability of the experiment result.

3 Results and discussion

3.1 Powder characterization

Fig. 1 shows the morphology of the in situ synthesized WC-Co composite powder, which has a particle size of about 186 nm. Fig. 2 shows the cross-sectional microstructure of a single feedstock particle (Fig. 2(a)) and morphology of the spray powder (Fig. 2(b)), respectively. It is clearly shown that the powder particles exhibited nearly spherical morphology without adherent satellite particles in the given range of 15-45 μm, indicating good fluidity and stability during spraying.

The bulk density and liquidity of spray powder is 3.92 g/cm^3 and 17.47 s / 50g, respectively. The total carbon content of the feed powder was 5.23% by LECO CS844 carbon and sulfur analyzer, and the determination of its elements in chemical method, in which the content of Co is 9.47wt.%, Cr is 3.83wt.%, the balance being W, meet WC-10Co-4Cr spray powder component design requirements.

Fig.1 Morphology of the in situ synthesized WC-Co composite powder (a) and its particle size distribution (b)

Fig.2 SEM images of the WC-10Co-4Cr feedstock powder,(a)Cross-sectional microstructure of a single feedstock particle(b)morphology of the spray powder

3.2 Coating characterization

Fig. 3 shows the microstructure of the as-sprayed coating on a polished cross-section. It is evident from Fig. 3(a) that the coating is dense and well-bonded to the substrate with an average thickness of 350 μm. The average porosity of the coating is less than 1%. From Fig. 3(b) it can be noted that a high volume fraction of tungsten carbide are uniformly incorporated into the coating. The binder matrix with different grayscale is observed and regions with higher brightness are associated with significant dissolution of WC into the binder phase. Besides, some pores are observed from the microstructure. The coatings have an average hardness more than 1400 $HV_{0.3}$.

Fig. 4 shows the XRD pattern for the as-sprayed coating. It can be identified from Fig. 4 that the coating is composed of a main WC and minor W2C phases. This is because of the decarburization of phase.The Co phase in the coating is an amorphous state of Co-W-C and for this reason is not detected by XRD.

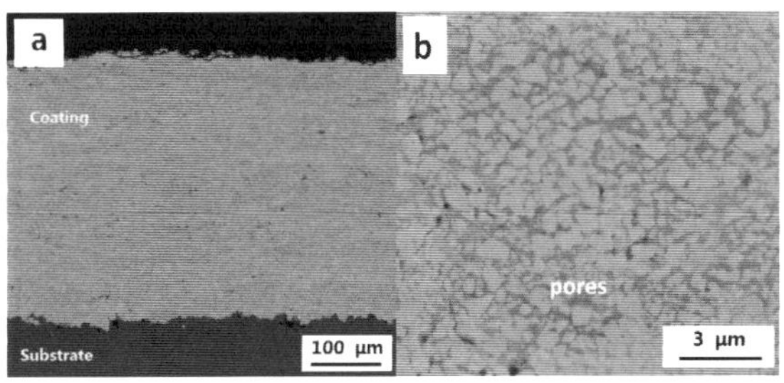

Fig.3 SEM images of a transverse section of the as-sprayed coating (a) SEM microstructures of WC-10Co-4Cr coating (b)

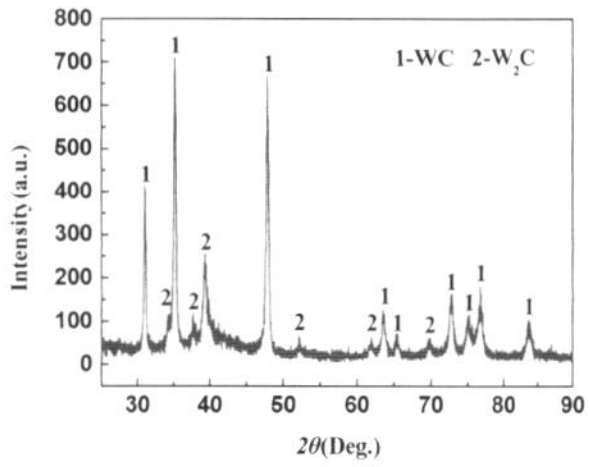

Fig.4 XRD pattern of WC-10Co-4Cr coating

3.3 Friction coefficient and wear rate

A commercially available WC-10Co-4Cr powder spray coating (hereinafter referred to as Starck) and this group of composite powders prepared homemade WC-10Co-4Cr coating (hereinafter referred to as BJUT) were reciprocated friction test.

Fig. 5 shows the WC-10Co-4Cr (Starck) coating is in the stable friction phase after 6 minutes wear, and the friction coefficient in the range of 0.40 to 0.45. The WC-10Co-4Cr (BJUT) coating is in the stable friction phase after 4 minutes wear, and the friction coefficient in the range of 0.33 and 0.35. Table 1 is a comparison of the wear rate of the two coatings, the wear rate of WC-10Co-4Cr (BJUT) coating than WC-10Co-4Cr (Starck) decreased 48.6% coating.

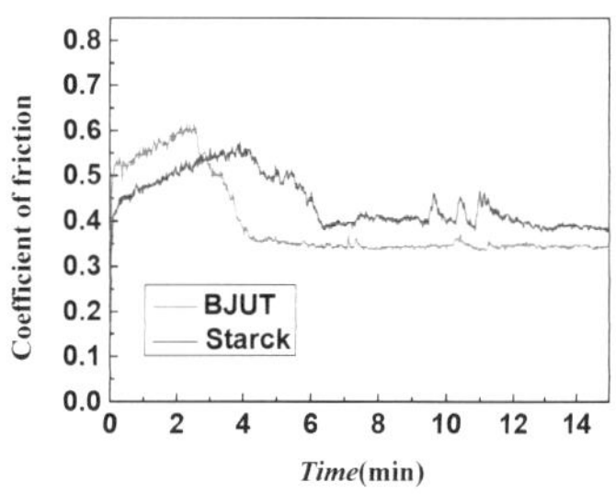

Fig.5 Sliding wear friction coefficients of coatings as a function of test time

Table 1 Wear rate of WC-10Co-4Cr coating

Sample	WC-10Co-4Cr(Starck)	WC-10Co-4Cr(BJUT)
Wear rate mm^3/mm	2.80×10^{-6}	1.44×10^{-6}

3.4 Electrochemical corrosion behavior

The potentiodynamic polarization curves of the as-sprayed WC-10Co-4Cr(BJUT) coating and WC-10Co-4Cr(Starck) coating in 3.5 wt.% NaCl solution are shown in Fig. 6(a). The corrosion potentials (E_{corr}) of the nanostructured WC-10Co-4Cr(BJUT) coating and the WC-10Co-4Cr(Starck) coating are about -0.30 V and -0.52 V, respectively. The corrosion current density (i_{corr}) of the nanostructured WC-10Co-4Cr(BJUT) coating is 3.162×10^{-6} A·cm^{-2}, which is obviously bigger than that of the WC-10Co-4Cr(Starck) coating (2.512×10^{-6} A·cm^{-2}).

The superior corrosion resistance of the nanostructured WC-10Co-4Cr(BJUT) coating is further verified by the EIS results. Fig. 6(b) shows the EIS curves of the as-sprayed WC-10Co-4Cr(BJUT) coating and the WC-10Co-4Cr(Starck) coating. It is seen from the EIS curves (Fig. 6(b)) that both impedance spectra revealed as capacitive semicircles at whole frequency, although the diameter of the capacitive semicircle for the as-sprayed WC-10Co-4Cr(BJUT) coating is larger than that for the WC-10Co-4Cr(Starck) coating. This indicates that the electrochemical mechanism is same but the corrosion rate is different. The diameter of the capacitive semicircle is related to the charge transfer resistance ,i.e., the corrosion resistance in the corrosion process; a larger diameter of the capacitive semicircle means a slower corrosion rate. Thus, it can be seen that the as-sprayed WC-10Co-4Cr(BJUT) coating has a slower corrosionrate ,i.e., a better corrosion resistance than the WC-10Co-4Cr(Starck) coating.

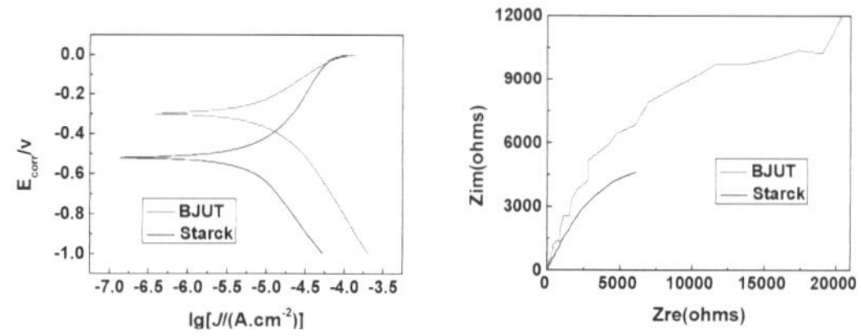

(a) (b)

Fig.6 Tafel curves of the coatings(a), EIS curves of the coatings(b)

3.5 Microstructure of the Coating

Fig. 7(a) shows a typical field TEM image of the WC-10Co-4Cr coating, which contains WC phase and amorphous phase. Through the PDF card of WC and electron diffraction spectra of the sample, Fig. 7(b) shows that the diffraction spectra of WC phase is hexagonal crystal structure (Fig. 7(b) at ①). Co area is amorphous phase (Fig. 7(c) at ②). Fig. 7(d) shows the complex phase is an amorphous phase of WC+Co(Cr). By the parameters of crystallography, the spacing and lattice parameters are that, d=2.5169(Actual measurement , d=2.60061) , and the corresponding crystal planes are $(10\bar{1}0)$ and $(1\bar{1}00)$; d=1.8829(Actual measurement , d=1.96136/1.95427), and the corresponding crystal planes are $(10\bar{1}1)$ and $(1\bar{1}01)$.

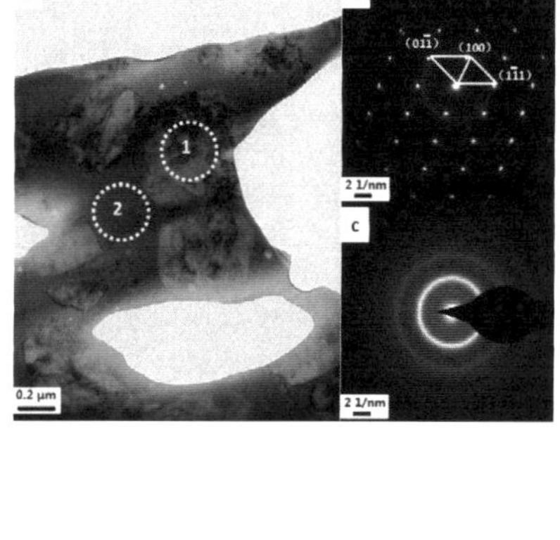

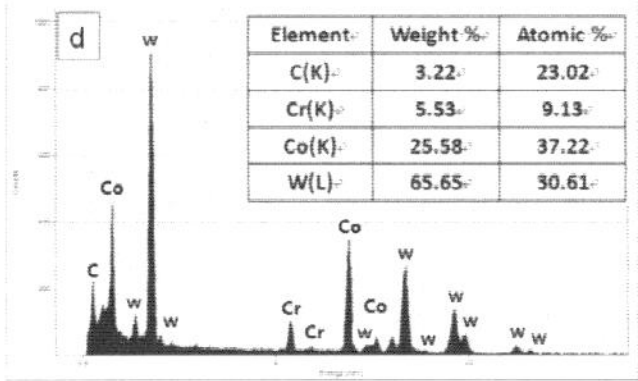

Fig.7 TEM micrograph of the coating(a) and diffraction pattern(b) at ①, and diffraction pattern(c) and EDS analysis (d) at ②

Fig. 8(a) shows a typical field TEM image of the WC-10Co-4Cr coating, too. The borders of grain are arcuate . Because in the process of spraying and under the action of high temperature and oxygen flow, the temperature of particle is very high, the tiny powder of WC-10Co-4Cr lead to the WC grain oxidation decarburization reaction and generate carbon phase deficiency. With the PDF card of W2C and electron diffraction spectra of the sample, Fig. 8(b) shows that the diffraction spectra of W2C phase is hexagonal crystal structure (Fig. 8(b) at ③). From the parameters of crystallography, with the spacing d=1.7477, the corresponding crystal planes are $(10\bar{1}2)$ and $(1\bar{1}02)$, with d=2.2752, the corresponding crystal planes are $(10\bar{1}1)$ and $(1\bar{1}01)$. Fig. 8(c) shows it is rich in Cr at ④ with the EDS analysis .

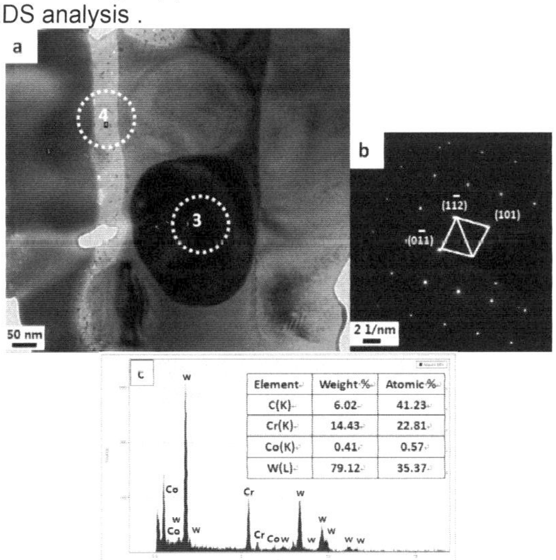

Fig.8 TEM micrograph of the coating (a) and diffraction pattern(b) at ③, and EDS analysis (c) at ④

Fig. 9(a) and (c) shows the distribution of Co and Cr by EDS analysis. From the phase boundary of WC grain to the eutectic Co area, the content of Co is gradually increased; and the content of Cr is gradually increased between the WC grains and white banded region , reached the highest value in the white banded regions and decreased between the white ribbon and Co district. Because the flame temperature of HVOF process exceeds 2700℃. During the spraying, lower melting Co (melting point 1495℃) and Cr (melting point 1857℃) [7] first begins to melt. While the powders are flying , the higher melting Cr first precipitate from the liquid phase to form a solid phase which exists in the liquid phase of Co. The solid-liquid interface lies between the liquid phase of Co and the solid phase of WC grain, and the liquid phase of Co and the solid phase of Cr. During the diffusion process, the surface tension of Cr (σ = 1628mN / m) is less than that of Co (σ = 1870mN / m) [7], resulting Cr easy to enrich the formation of Cr-rich zone at the solid phase of WC grain and Cr particle surface, and the content of Cr which is in the Co region near the Cr-rich area is less than the average content in the spraying powder (Fig.9 (b)).

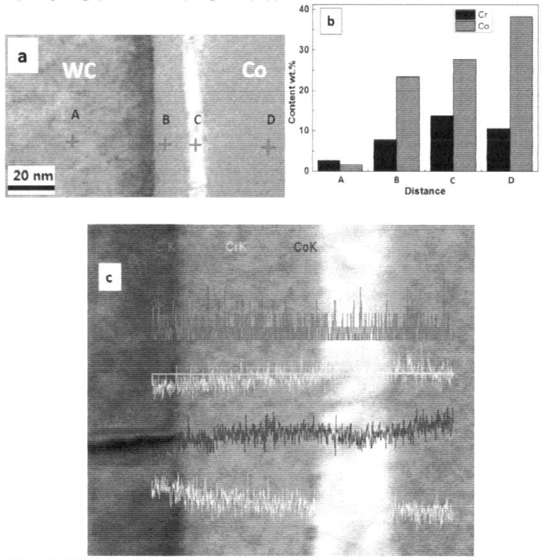

Fig.9 TEM microstructures of the coating (a) , distribution of Co and Cr (b) and EDS images of the coating (c)

4 Conclusion

The main conclusions are as follows:

(1) The main phases of the WC-10Co-4Cr coating is WC phase of hexagonal crystal structure, binding phase of amorphous structure is Co(Cr), and there is a little W_2C phase of hexagonal crystal structure and W2C+Co(Cr) multiphase of amorphous coexisting.

(2) The change rule was obtained from the phase boundary of WC grain to the eutectic Co area by the quantitative analysis of distribution of Co and Cr element in coating. The content of Co is gradually increased. The content of Cr is gradually increased between the WC grains and white banded region , reached the highest value in the white banded regions and decreased between the white ribbon and Co district.

(3) Combined WC-10Co-4Cr composite powder and process characteristics of HVOF, explain the causes of Cr in the WC-10Co-4Cr coating distribution.

Acknowledgments

This work was supported by the National High-tech R&D Program of China(2013AA032001), the Beijing Key Program of Natural Science Foundation (2131001) and the Beijing Natural Science Foundation (2154045).

5 References

[1] Bloggs, J. and A.N. Other: Literature example for manuscripts in DVS volumes of reports. Welding & Cutting 46 (1999), Issue 13, pp. 1156/57.

[1] Ding Zhangxiong, Wan Wenchen, Zhao Hui, Liu Ji Research progress and prospect of WC-Co composite coatings prepared by thermal spraying. J Therm Spray Techn. 4 (2012), Issue 2, pp.1-5+15.

[2] Song Xiaoyan. Low-cost & short-term preparation and applications of ultrafine and nano-scaled WC-Co composite powders. China Tungsten Industry 25 (2010), Issue 3, pp.23-29.

[3] Zhou Kesong, Deng Chunming, Liu Min, Song Jinbing, Deng Changguang. Characterizations of fatigue and salt spray corrosion resistance of HVAF sprayed WC-17Co and WC-10Co4Cr coatings on the substrate of 300M steel. Rare Met. Mater. Eng. 38 (2009), Issue 4, pp.671-676.

[4] Lutz-Michael Berger. Application of hardmetals as thermal spray coatings. Int. J. Refract. Met. Hard Mater. 49 (2015) , pp.350-364.

[5] Li Xuewei, Meng Yin, Wang Peng, Cheng Haoran. Research status of WC-10Co4Cr coatings prepared by thermal spraying. China Tungsten Industry 29 (2014), Issue 2, pp.28-31.

[6] Yang Weihua, Cui Xiangzhong. Anti-fretting performance of WC10Co4Cr coatings on 30CrMnSiNi2A.Rare Met. Mater. Eng. 42 (2013), Issue S2, pp.197-200.

[7] Han Tao, Deng Chunming, Liu Min, et al. Fracture toughness of normal high velocity oxygen-fuel flame and low temperature high velocity oxygen-fuel sprayed WC-10Co-4Cr coating. J. Chin. Ceram. Soc. 42 (2014), Issue 3, pp.409-415.

[8] Souza V A D,Neville A. Mechanisms and kinetics of WC-Co-Cr high velocity oxy-fuel thermal spray coating degradation in corrosive environments .J. Therm. Spray Technol. 15 (2006), Issue 1, pp.106-116.

[9] Zhang Jiuxing, Song Xiaoyan, Liu Wenbin. A simple and rapid method for the preparation of ultrafine WC-Co composite powder: China [P],200610165554.2.

[10] Liu Wenbin, Song Xiaoyan, Zhang Jiuxing, Zhou Meiling. Study on preparation of ultrafine WC-Co composite powder by mixed oxide powder situ reaction. China Powder Industry (2008), Issue 1, pp.19-23.

[11] W. B. Liu, X. Y. Song, J. X. Zhang, et al. Thermodynamic analysis for in situ synthesis of WC–Co Composite Powder from Metal Oxides . Mater. Chem. Phys. 109 (2008), Issue 2-3, pp.235-240.

[12] W. B. Liu, X. Y. Song, J. X. Zhang, et al. Preparation of ultrafine WC-Co composite powder by in situ reduction and carbonization reactions. Int. J. Refract. Met. Hard Mater. 27 (2009), Issue 1 , pp.115-120.

Performance of HVOF sprayed Ni-20 Cr+1%Zr Coating in Na_2SO_4-60%V_2O_5 Environment at 900°C

G. Kaushal
Yadavindra College of Engineering, Punjabi University, Guru Kashi Campus, Talwandi Sabo, Bathinda, Punjab 151302, India
Email: gagankaushal@pbi.ac.in Tel.91-98727-05486

H. Singh
Indian Institute of Technology Ropar, Rupnagar, Punjab, India

S. Prakash
Indian Institute of Technology Roorkee, Roorkee, Uttarankhand, India

Abstract

It has been learnt that there is a need to develop high performance coating compositions to be deposited by thermal spray processes for various industrial applications especially for the power plant boilers. Here-in, HVOF sprayed Ni-20Cr+1%Zr on A213 TP347H boiler steel has been assessed by severe corrosion tests at high temperature in molten salt environment. Corrosion kinetics followed by FE-SEM/XRD analysis are recorded to evaluate the performance and the outcome is discussed in the present study.

Keywords: High temperature Corrosion, HVOF, Ni-20Cr, Zirconium, Boiler Environment

Introduction

Thermal spray processes are highly favorable in surface engineering industry due to their, relatively low cost and their capacity to deposit almost any material followed by high production rate. These are considered technically-effective coating technology that is applied to many industrial applications [1,2]. Among the various techniques of thermal spray processes, high velocity oxy-fuel (HVOF) thermal spray process process has gained popularity for benefits in wear and corrosion management applications [3-5]. Extensive studies have been reported on the said process and are considered highly successful because of mainly due to relatively lower particle temperature which reduces the formation of undesirable phases, and high particle velocity, which improves the intersplat bond strength and produces high density coatings with favorable hardness and toughness.

Recently, there has been significant interest to produce and investigate the performance of various coating compositions developed by HVOF spraying for applications where resistance to wear and oxidation/corrosion is required. Moreover role of rare earth elements as beneficial additions in the coating compositions is also significantly studied. Zirconium (Zr) provides additional improvement in adherence through the formation of an extensive oxide peg network beneath the scale, which mechanically locks the scale onto the substrate [6]. The aim of the current study is to assess the performance of commercially available coating composition namely Ni-20Cr blended with 1% Zr by weight and deposited by HVOF thermal spray process for high temperature applications. It is anticipated that the blended composition will provide a better Corrosion resistant coating.
The investigations are performed on a boiler steel ASTM A213 TP347H (347H) steel under cyclic thermal loading conditions at an elevated temperature in molten salt environment at 900°C.

Experimental Procedure

Development of Coating
Substrate Material:347H boiler steel tube material with chemical composition C- 0.04-0.1, Mn -2.0, P - 0.04 max, S - 0.03 max, Si - 0.75, Cr -17-19, Ni -9.0-13 and Fe- 65.08 (mass%) has been used as a substrate material in the current study. Specimens, each measuring 20mm x 15mm x 5mm approximately, were cut from the fresh boiler tubes. The specimens were polished down to 180 grit SiC paper finish and were grit blasted with Al_2O_3(grit 60) before the deposition of the coating.
Coating Formulation: The specimens each measuring 20mm x 15mm x 5mm approximately were cut from the fresh boiler tube. The specimens were polished down to 180 grit SiC paper finish, and before the deposition of the coating. The Ni-20Cr+1%Zr coating was coated on the steel substrates using commercially available HVOF-spray (HIPOJET-2100) thermal spray process. The coating was deposited at Metallizing Equipment Company Private Limited, Jodhpur, India, operating with oxygen and liquid petroleum gas (LPG) as input gases. The coating parameters adopted are reported in **Table 1**[7].
2.2 Characterization of as-sprayed coating

The thickness of as-sprayed coating was examined along cross-section. Apparent surface porosity of the as-sprayed Ni-20Cr+1%Zr coating was measured by Image analyzer system from Chennai Metco Pvt. Ltd (India) using Envision 3.0 software based on ASTM 276. X-ray diffraction (XRD) analysis of the coated samples was carried out using Bruker AXS D-8 Advance Diffractometer (Germany) with CuKα radiation and nickel filter at 20 mA under a voltage of 35 kV. FE-SEM/EDS analysis (FE-SEM, FEI, Quanta 200F) was done to characterize the surface morphology of the coating. After the surface characterization, the samples were sectioned, mounted in epoxy powder by using hot mounting method and mirror-polished. The mounted samples were characterized to obtain their cross-sectional morphology and compositions by using the FE-SEM/EDS analysis. X-ray mappings of the various elements were also taken by the same FE-SEM/EDS machine.

Table.1 *Spray parameters employed for HVOF coating [7]*

Oxygen flow rate	*200 SLPM*
Fuel (LPG) flow rate	*50 SLPM*
Air-flow rate	*900 SLPM*
Spray distance	*20 cm*
Powder feed rate	*25-30 gm/min*
Fuel pressure	*6.00 kg/cm^2*
Oxygen pressure	*8.00 kg/cm^2*
Air pressure	*6.00 kg/cm^2*

Corrosion Studies in Molten Salt Environment

The studies were performed in a molten salt environment consisting of a Na_2SO_4-60%V_2O_5 mixture for 50 cycles. Each cycle consisted of 1 hour of heating at 900°C in a silicon carbide tube furnace followed by 20 minutes of cooling at room temperature. The corrosion studies were performed on the uncoated as well as coated specimens for the purpose of comparison. Prior to the corrosion testing, the specimens were polished with SiC papers down to 1200 grit, which was followed by cloth wheel polishing with 1 μm alumina powder. This was done to obtain a consistent surface preparation for all the specimens. Salt mixture of Na2SO4-60%V2O5 was applied on the preheated specimens (250°C) with a camel hairbrush. Attempts were made to achieve a uniform thickness of salt weighing 3 to 5 mg/cm2. Themass-change measurements were taken at the end of each cycle with the help of an electronic balance machine CB-120 (Contech, Mumbai, India) having a sensitivity of 10-3 g. The spalled scale was also included at the time of measuring mass-change to determine the total rate of corrosion. After the exposure, surface, analysis of the corroded samples was done by X-ray diffraction (XRD) and fieldemission scanning electron microscopy/energy dispersive spectroscopy (FE-SEM/EDS)..

The aim of cyclic loading by laboratory tests was to create conditions for the accelerated testing. After exposure the corroded samples were tested by the XRD and FE-SEM/EDS techniques.

Results

Characterization of As-Sprayed Coatings

Thickness and Porosity: Average cross-sectional thickness of Ni-20Cr+1%Zr coating was 225 μm and apparent surface porosity of coating was 1.3-1.6 % . Micro-hardness of the coating on the selected boiler steel has been measured along the cross-section.

XRD Analysis of As-Sprayed Coating: The XRD diffractogram of the HVOF-spray of Ni-20Cr+1%Zr coating on 347H boiler steel in as-sprayed condition is shown in Figure 1. The analysis indicates the formation of Ni as prominent phase as per the powder composition of deposited coating.

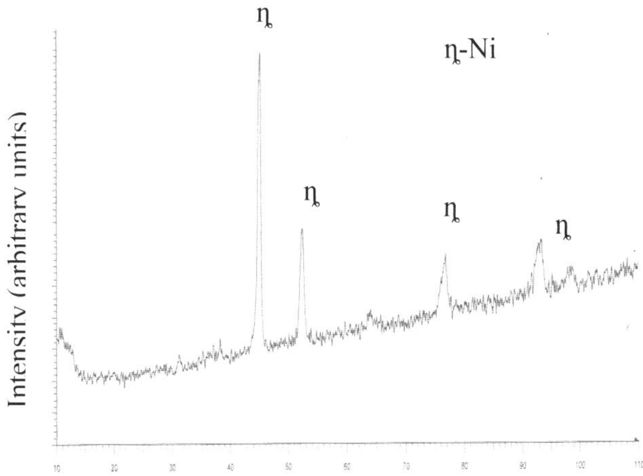

Figure 1: X-ray diffraction profiles of the HVOF-spray Ni-20Cr +1%Zr coating on A213 TP347H boiler steel

FE-SEM/EDS Analysis of As-Sprayed Coating: The microstructure of coating [Fig. 2a] is observed as fine-grained, with minimal porous appearance. The EDS analysis of the coating again shows the dominance of Ni in the coating composition alongwith the significant presence of Cr and Zr. Zr concentration in found to be maximum 24% at point 1 whereas it is absent at points 2. Small amount of O is also observed at points 1 and 3. Cross-sectional morphology alongwith variations of elemental composition of the HVOF-sprayed steels have been shown through Fig. 2b . The coating, in general, has splat-like laminar morphology, the splats are clearly distinguishable with inter-connecting splat boundaries. There is a consistent presence of Zr along the thickness of this coating for the substrate steels. Fe is mainly restricted to the coating area in the HC2 coated steel .

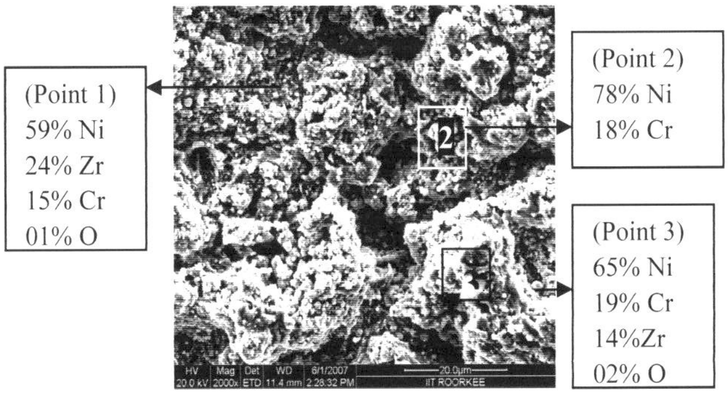

(a)

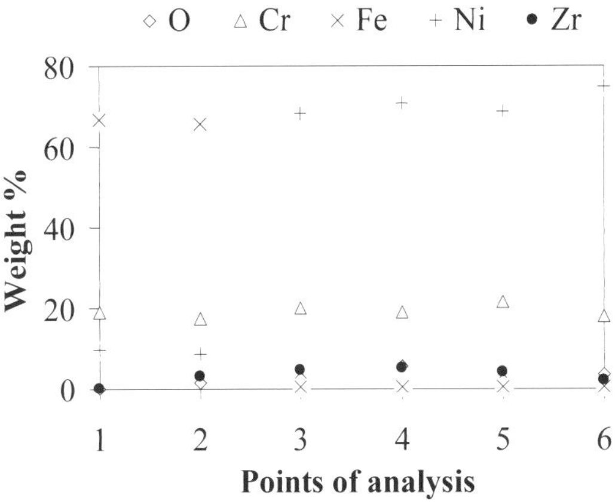

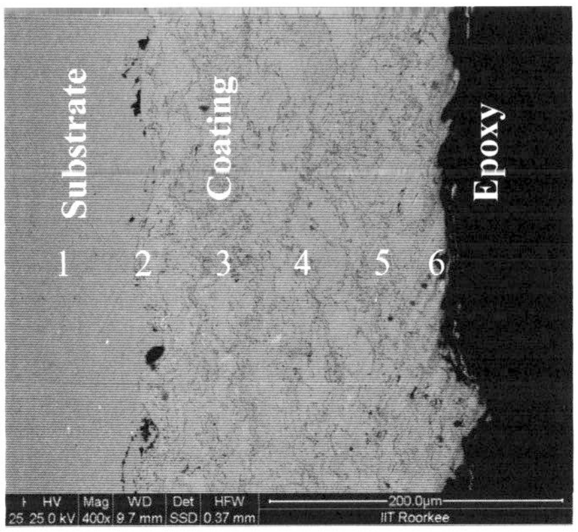

(b)

Figure 2: FE-SEM /EDS analysis of the HVOF-spray Ni-20Cr +1%Zr coating on A213 TP347H boiler steel (a) surface(b) cross-section

Corrosion tests in Molten Salt Environment

Mass Change Data: The performance of the HVOF spray blended coating has been recorded by mass change value. Mass change (mg/cm^2) versus number of cycles plots for 347H corroded steel at 900°C in molten salt environment upto 50 cycles have been shown in Fig. 3. It has been observed that only a slight mass loss has been observed for the first 14 cycles, then mass gain was observed for the next 16 cycles and subsequently mass loss was recorded. The overall mass change shown by the coated steel was -7.209 mg/cm^2. It is clear from the plot that coating showed minimum mass change values after exposure to severe aggressive environment. Some scale was seen in the boat by the completion of cyclic studies and apart from that no visible cracks were observed.

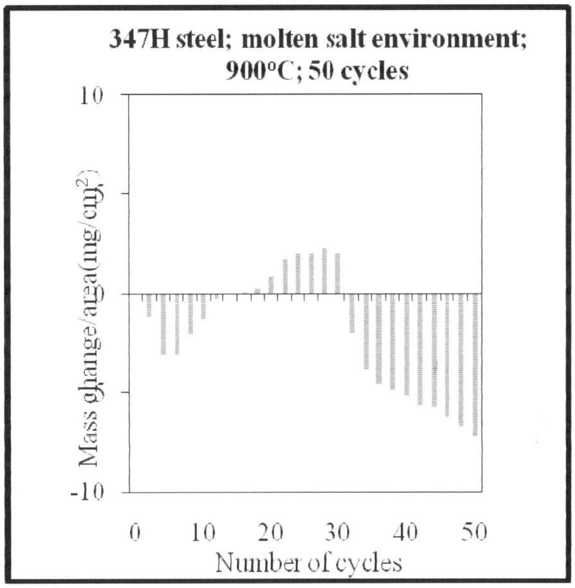

Figure 3: Mass change/area vs. number of cycles plot for the HVOF-spray Ni-20Cr +1%Zr coating on A213 TP347H boiler steel subjected to after corrosion in molten salt (Na_2SO_4-60%V_2O_5) environment at 900°C for 50 cycles

XRD Analysis of Exposed Specimen: The XRD diffractogram of the HVOF-spray of Ni-20Cr+1%Zr coating on 347H boiler 347H corroded steel at 900°C in molten salt environment upto 50 cycles have been shown in Fig. 4. The XRD analysis of HC2 coated steel indicated NiO as main phase alongwith Cr_2O_3 and $NiCr_2O_4$ as minor phases. Some relatively weak phases of $CrVO_4$ were also observed. Furthermore minor peaks of ZrO has also been revealed in the XRD analysis.

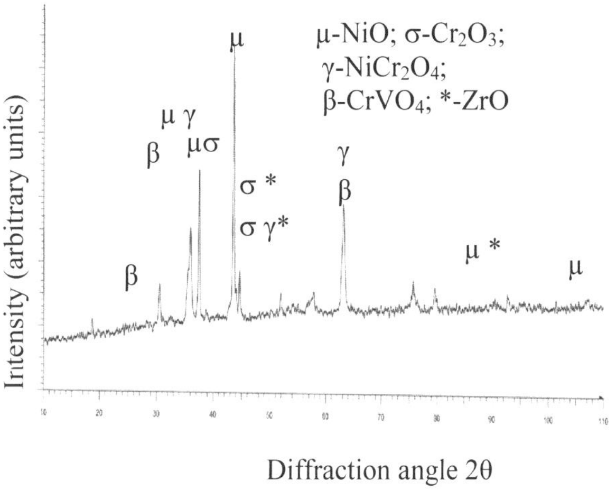

Figure 4: X-ray diffraction profiles of the HVOF-spray Ni-20Cr +1%Zr coating on A213 TP347H boiler steel subjected to after corrosion in molten salt (Na_2SO_4-60%V_2O_5) environment at 900°C for 50 cycles

FE-SEM/EDS Analysis of Exposed Specimen: The oxide scale of coating [Fig. 5a] exhibited very fine oxidized nodules dispersed in the matrix as evidenced by EDS analysis at point 1, the scale has Ni as major constituent and O in appreciable quantity probably indicating the formation of NiO layer. The EDS analysis at point 2 revealed Ni (57%), Cr (23%), O (12%), Zr (5%) and Fe (2%) in the scale region. A similar analysis for the corroded 347H steel along cross-section after the exposure is shown in Fig. 5 (b). The scale appears to have a splat-like laminar morphology and seems well intact even after the exposure. The EDS analysis of the cross-section of the coated steel at point 1 reveals that the substrate consists mainly of iron and Cr The composition of Fe is drastically reduced from point 3 onwards. Ni concentration is increased after point 3 and is maximum at point 6 and Cr has maxima at point 5. Zr has shown its presence from point 4 outwards in outer scale.

Discussion

The HVOF -spray process provides the possibility of developing blended composition of Ni-20Cr –Zr coatings on the 347H boiler steel with the given process parameters. Average cross-sectional thickness of 225 μm was successfully attained of the blended weight percent composition of Ni-20Cr and 1%Zr. Porosity of the coating has a significant role to play as far as the hot corrosion resistance of thermal sprayed spray coatings is concerned. Dense coatings usually provide

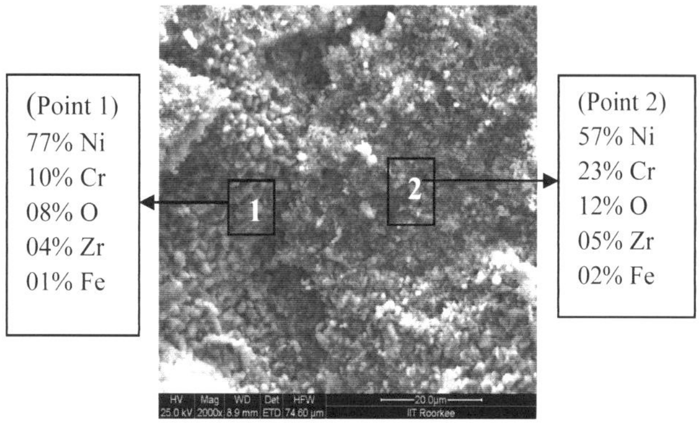

(a)

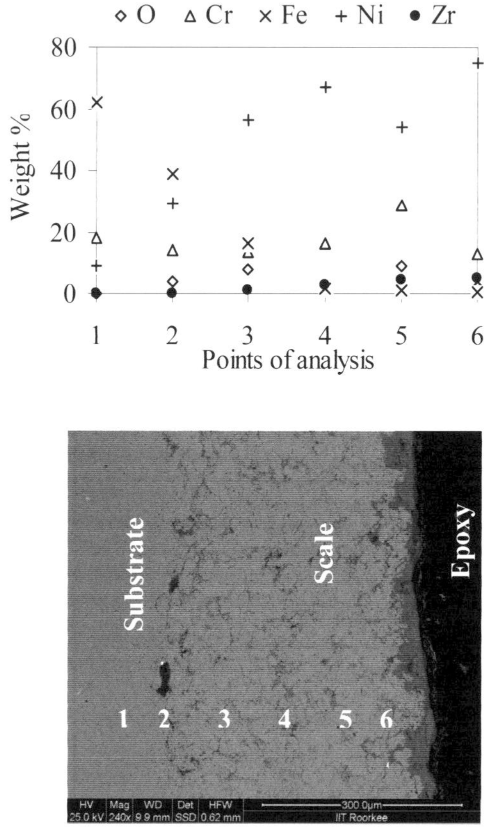

(b)

Figure 5: Figure 2: FE-SEM /EDS analysis of the HVOF-spray Ni-20Cr +1%Zr coating on A213 TP347H boiler steel (a) surface(b) cross-section subjected to after corrosion in molten salt (Na_2SO_4-60%V_2O_5) environment at 900°C for 50 cycles better corrosion resistance than the porous coatings. The measured apparent surface porosity value was found to be less than 2.0 percent and is in good agreement with earlier studies on HVOF spray coating [7-9]. The XRD and FE-SEM/EDS analysis of the as-sprayed coating is as per the feed stock powder composition of respective blended coating and there is a consistent presence of Zr along the thickness of this coating for the substrate steel (Fig 2b). The overall mass change shown by the coated steel subjected to after corrosion in molten salt (Na_2SO_4-60%V_2O_5) environment at 900°C for 50 cycles was quite negligible indicating that the blended coating has been successful in imparting corrosion resistance even in the severe accelerated environment. The XRD analysis of coated steel after corrosion testing indicated NiO as main phase alongwith Cr_2O_3 and $NiCr_2O_4$ as minor phases. These phases are reported to be protective oxides [10, 11]. These mixed oxides are known for providing superior corrosion resistance, as has been reported by several investigators [12,13]. Some relatively weak phases of $CrVO_4$ were also observed. Presence of $CrVO_4$ phase depicts the penetration of salt elements (Vanadium) through the scale. Furthermore minor peaks of ZrO has also been revealed in the XRD analysis, which could well predict the formation of oxide pegs that are beneficial to provide adherence to the coating and slower the oxide scale growth. It has been clearly observed in the FE-SEM/EDS analysis along cross-section that even after the severe corrosion testing a adherent and intact scale is observed after the completion of studies., moreover percentage of Ni and Cr has increased towards the top layer of scale along with uniform presence of Zr and O in outward scale clearly endorsing the XRD analysis. Hong et al [6] reported that Zr-rich pegs are generated by Zr addition during cyclic oxidation at high temperature and provide better resistance.

Based upon the performance of coating on 347H steel, it may be concluded that blending of minor addition of Zr in the commercially available the Ni-20Cr coating could be useful in corrosion environment.

Conclusions

- The Ni-20Cr coating blended with 1%Zr composition could be successfully deposited on on A213 TP347H boiler steel by HVOF spraying.

- The corrosion resistance of Ni-20Cr coating could be enhanced by minor addition of Zr due to additional improvement in adherence, which mechanically locks the scale formed of protective phases like NiO, Cr_2O_3 and $NiCr_2O_4$.

References

1. J. R. Davis, Handbook of Thermal Spray Technology, ASM International, Materials Park, OH, USA., 2004.
2. Bach, F. W., Laarmann, A., Wenz, T. (Eds.), ModernSurface Technology, Wiley-VCH Verlag GmbH & Co. (Germany, 2006).
3. T.S. Sidhu, S. Prakash, R.D. Agrawal, Surf. Coat. Technol. 201 (2006) 792.
4. Kaushal, G. Singh, H. and Prakash, S., (2010) "High Temperature Erosion-Corrosion Performance of HVOF Sprayed Ni-20 Cr Coating in Actual Boiler Environment," Metallurgical and Materials Transactions. A, Vol. 42, No. 7, pp 1836-1846.
5. Abu-Aesh, M., (2008) "Dependence of Sliding Wear Resistance and Microhardness of Al-Spray Coating Layers on Substrate Conditions Using High-Velocity Oxygen Fuel (HVOF)," Materials and Manufacturing Processes, Vol. 23, pp 723-728.
6. S.J. Hong, G.H. Hwang, W.K. Han, S.G. Kang, Intermetallics, 2010, 18, p 864–870.
7. Kaushal, G. Singh, H. and Prakash, S., Proc. of International Thermal Spray Conference and Exposition. (2008), The Netherlands. 1351.
8. Sahraoui, T. Fenineche, N.E. Montavon, G. and Coddet, C., Mater. Des. 24 (2003) 309.
9. Gil, L. and Staia, M.H., Thin Solid Films., Vol. 420-421, (2002), 446.
10. Miguel, J.M. Guilemany, J.M. and Vizcaino, S., Tribol. Int., Vol. 36, (2003), 181.
11. Sundararajan, T., Kuroda, S., Itagaki, T. and Abe, F., (2003), "Steam Oxidation Resistance of Ni-Cr Thermal Spray Coatings on 9Cr-1Mo Steel. Part 1: 80Ni-20Cr," ISIJ Int., Vol. 43, No. 1, pp. 95-103.
12. Ul-Hamid, A., (2003), "Diverse Scaling Behavior of the Ni-20Cr Alloy," Maters. Chem. Phys., Vol. 80, pp. 135-142.
13. Sidhu,T.S. Agrawal, R.D. and Prakash, S., J. Mater. Engg. Perform., Vol.15, No.1,(2006), 130.
14. S.S. Chatha, H.S. Sidhu, B.S. Sidhu., Surf. Coat. Technol., 2012, 206, 19–20, 3839-3850

Effects of powder structures and HVOF processes on erosion wear performance of WC-10Co4Cr coatings

Xiang Ding[1], Xu-Dong Cheng[1], Jin Shi[2]*, Chao Li[2], Zhang-Xiong Ding[2]
1. State Key Laboratory of Advanced Technology for Materials and Processing, Wuhan University of Technology, Wuhan 430070,China;
2. School of Energy and Power Engineering, Wuhan University of Technology, Wuhan 430063, China
*Corresponding author. E-mail shijinwhut@163.com

Abstract: Conventional and multimodal WC-10Co4Cr cermet coatings were sprayed respectively by High Velocity Oxygen Gas Fuel spraying (HVOGF) and High Velocity Oxygen Liquid Fuel spraying (HVOLF), and the coating structures were analyzed by OM, SEM, XRD. Porosity, microhardness and fracture toughness of the coatings were measured. Erosion resistance of the coatings to solid sand was tested and the failure mechanisms were explored. Results showed that the mechanical properties of WC-10Co4Cr coatings deposited by HVOLF were much more superior to those deposited by HVOGF. Multimodal WC-10Co4Cr coating deposited by HVOLF had the highest microhardness and fracture toughness, the lowest porosity, the most excellent resistance of sand solid erosion wear. Its solid sand erosion resistance was enhanced above 15% and 40% respectively than conventional coatings deposited by HVOLF at 30°and 90°impact angles, and was approximately 3 times than that of multimodal coatings deposited by HVOGF at same testing conditions.

1 Introduction

Erosion wear exists widely in industries including ship, machinery, metallurgy, energy, construction materials, aviation and aerospace, and has become one of the most serious issues for the material damage or equipment failure. Since erosion behavior occurs only at the surface of the equipment, surface engineering techniques, especially the nano–structured coating, have become the most effective solutions for the erosion problems [1-3]. WC based cermet coatings are well-known for their outstanding wear resistance, which has drawn much attention in the research of the erosion resistant coating materials in recent years [4]. Adding Cr into WC-Co coatings can further enhance the strength and corrosion resistance, so excellent erosion resistance is expected to be achieved for WC-CoCr coatings in corrosion medium [5,6].

The WC particle size in the WC-CoCr coating significantly affects the microstructure and properties of the coating. Generally, mechanical properties and wear resistance of the coating enhance with the decreasing WC particle size [7]. Recently, several research have showed that the hardness and toughness of nano-structured materials can be improved simultaneously, which granted nano WC–CoCr coatings unique wear resistance [8,9]. Multimodal WC based coatings, in which WC is composed of both micro and nano sized particles, are proposed in order to restrain the decarburization of nano WC and reduce the high cost of nano coatings [10,11]. The multimodal coatings are expected to obtain dense structure and excellent erosion wear resistance. Among all spraying approaches, high velocity oxy-fuel spraying (HVOF) possesses high flame velocity and moderate flame temperature, which hinders the decarburization of WC during the heating by flame. Therefore, HVOF is considered as an ideal method to prepare various structured WC-CoCr cermet coatings [12,13].

In this work, multimodal and conventional WC-10Co4Cr coatings were deposited by High Velocity Oxygen Liquid Fuel spraying (HVOLF) and High Velocity Oxygen Gas Fuel spraying (HVOGF). The microstructure, mechanical properties and the erosion behavior in dry sand of various coatings were studied. The research results will provide important reference for WC-CoCr anti-erosion coating design and optimization of HVOF process.

2 Experimental procedure

2.1 Materials

Multimodal and conventional WC-10Co4Cr composite powders were produced by Ganzhou Achteck Tool Technology Co., Ltd. The characteristics of the two WC-10Co4Cr composite powders are listed in **Table 1**. 304 stainless steel is used as substrate for spraying.

Table1. Characteristics of two WC-10Co4Cr composite powders

Powder No.	Powder type	Manufacturing method	Nano WC size (nm)	Micron WC size (μm)	Nano WC rate (%)	Powder size (μm)
P1	Multimodal	Agglomeration	100-200	2.5	30	20-53
P2	Conventional	Agglomeration	-	1.5	-	10-45

2.2 Methods

JP8000 HVOLF (using kerosene as fuel) and ZB2000 HVOGF (using propane as fuel) systems were used to prepare WC-10Co4Cr coatings. Prior to spraying, the substrates' surface was cleaned with acetone and blasted with 60 mesh corundum. The thickness of the coatings was controlled about 0.5mm. All the samples were ground and polished to an average surface roughness Ra≤0.2μm before any characterizations.

The morphology and microstructures of the powders and the coatings were observed with VHX-2000 digital optical microscope, FEI Quanta 250 scanning election microscope (SEM). Phase identification for the powders and the coatings was carried out by a D/max-2550 diffraction meter (XRD) using Cu-Ka radiation with λ = 0.154nm.

Microhardness measurements of the coatings were performed via HVS-1000 microhardness tester at a load of 200g for 15s. The porosity was measured with metallographic photos by Axiovet 40 MAT metallographic microscope, followed by porosity calculation using IQ materials software. The fracture toughness was calculated according to the Wilshaw equation [14] and the result was the average value of ten measurements.

The erosion test was operated in GP-1 dry sand blasting machine with nozzle diameter of 8 mm. At 30° and 90° impact angle, 6 min and 4 min erosion time, 120 mm and 150 mm erosion distance were used respectively. Abrasives were 60 mesh brown fused alumina and air pressure is 0.4 MPa. TG328 electronic balance with a sensitivity of 0.1mg was used to weigh mass losses.

3 Results and discussion

3.1 Microstructure of WC-10Co4Cr powders and coatings

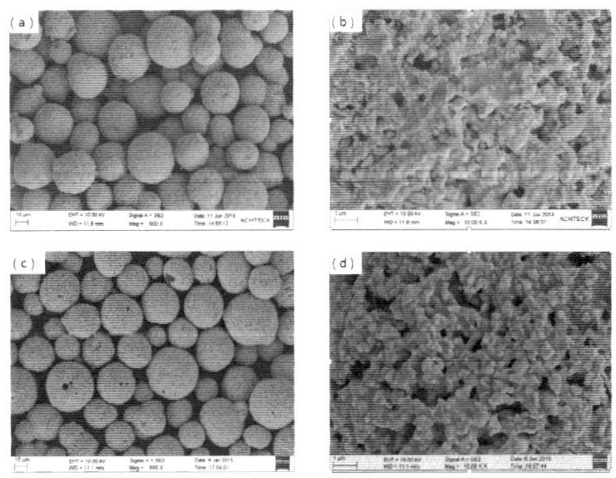

Fig. 1. Micrographs of WC-10Co4Cr powders at different magnification. a), b) P1 powder and c), d) P2 powder.

The SEM micrographs of multimodal WC-10Co4Cr powder (P1) and conventional powder (P2) are shown in **Fig. 1**. It can be seen that both powders are highly spherical. A large number of pores exist in the surface of P2 powder, demonstrating that P2 powder is packing much looser than P1 powder. In the micrograph of P1 powder with higher magnification, micro WC particles wrapped by the finer nano-particles and the CoCr binding phase can be clearly observed.

The cross-sectional microstructures of the multimodal and conventional WC-10Co4Cr coatings deposited by HVOF are shown in **Fig. 2**. All coatings possess dense structure due to the high speed which WC-10Co4Cr particles obtain during HVOF process, and the high kinetic energy causes particle deformation when particles collide into the substrate and provides strong binding strength between adjacent particles. It can be further observed that coatings sprayed by HVOLF are denser than HVOGF sprayed coatings, as a result of distinct configuration of two spraying systems. JP8000 has a separated combustion chamber and longer injector compared to ZB2000, which can produce higher pressure due to more completely combustion. After contacting with the flame, WC-10Co4Cr powders can be melt more homogeneously and possess higher velocity in JP8000, thus leading to the formation of denser coatings.

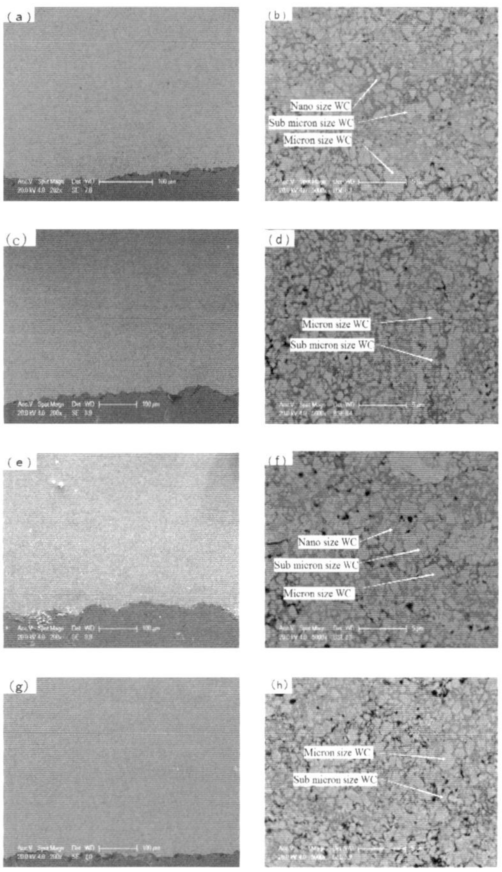

Fig. 2. Cross-sectional micrographs of WC-10Co4Cr coatings of a), b) M1; c), d) C1; e), f) M2; g), h) C2.

In case of the multimodal WC-10Co4Cr coatings, nano WC particles can be easily melt due to their high specific surface area as well as CoCr binding phase because of its low melting temperature. This melted mixture will fill in the space between molten or semi-molten micro WC particles, forming the multimodal structure with very low porosity, as shown in **Fig. 2** (a).

It can be observed in **Fig. 2** (b) and (f) that the nano, submicron and micron sized WC particles exist in multimodal WC-10Co4Cr coating, while in conventional WC-10Co4Cr coating only the submicron and micron sized WC particles can be observed. The submicron sized WC particles in the multimodal and conventional WC-10Co4Cr coatings are produced by micron WC particle breakage during the power ball mill process.

The coating porosity is shown in **Table 2**. Porosity values of two HVOLF sprayed coatings (<0.5%) are apparently much less than those of HVOGF sprayed coatings (> 1.0%), demonstrating the influence of HVOF spraying processes. Moreover, porosity values of multimodal coatings are less than conventional ones using the same HOVF, since the gap between micro WC particles is filled with melted nano WC and CoCr binding phase.

3.2 Phase composition of WC-10Co4Cr powders and coatings

Fig. 3 represents the XRD analysis of WC-10Co4Cr powders and coatings. The XRD patterns of multimodal and conventional WC-10Co4Cr powders show no obvious difference, which are mainly composed WC, Co, and Co_4W_2C crystalline phases. The combination of a small amount of Cr and C may occur during the sintering due to the high affinity, leading to decarburization in partial WC-10Co4Cr powders, thereby forming a small amount of Co_4W_2C. Co_4W_2C phase disappears in the XRD patterns of the coatings due to the decomposition in spraying process, meanwhile a small amount of decarburization phase W_2C is observed. Co transforms to amorphous phase during spraying, so it can neither be observed in the XRD spectra of the coatings. Conventional WC-10Co4Cr coating sprayed by HVOLF has the maximum proportion of W_2C, while its counterpart gets the minimum proportion. Therefore, the selection of spray powder's structure and HVOF process has a significant influence on the phase composition of the coatings.

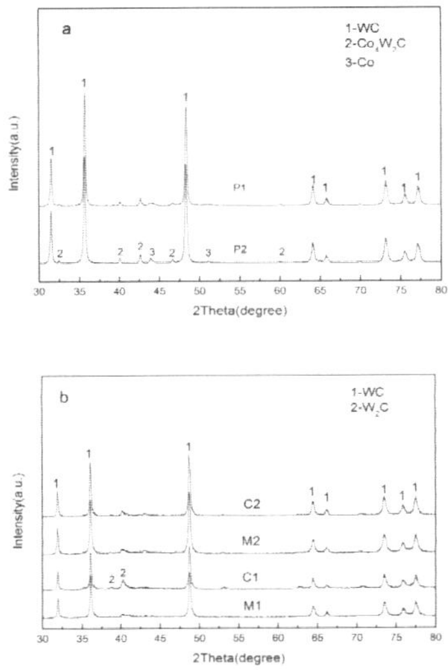

Fig.3. XRD patterns of a) WC-10Co4Cr powders and b) coatings

3.3 Mechanical properties of WC-10Co4Cr coatings

Table 2. Porosity, microhardness and fracture toughness of WC-10Co4Cr coatings

Specimen	Porosity (%)	Microhardness ($HV_{0.2}$)	Fracture toughness ($MPa \cdot m^{1/2}$)
M1	0.26±0.07	1345±136	5.56±0.50
C1	0.43±0.12	1322±181	3.85±0.52
M2	1.04±0.25	1141±200	3.57±0.87
C2	1.83±0.31	1069±217	3.38±0.48

Mechanical properties of multimodal and conventional WC-10Co4Cr coatings including microhardness and fracture toughness are listed in **Table 2**. HVOLF sprayed coatings possess higher microhardness and fracture toughness due to the greater inter-particle bonding force generated by higher velocity of the flame. HVOLF sprayed WC-10Co4Cr coating has the highest microhardness of ~1345 $HV_{0.2}$ and also obtains an extraordinary fracture toughness of ~5.56 $MPa \cdot m^{1/2}$, which is 45% higher than that of other three coatings. This outstanding property can provide obstacles to the formation and propagation of cracks during erosion.

3.4 Erosion wear performance of WC-10Co4Cr coatings

The erosion rates of multimodal and conventional WC-10Co4Cr coatings at 30° and 90° impact angle are shown in **Fig. 4**. The erosion rates of WC-10Co4Cr coatings sprayed by HVOGF are much higher than those sprayed by HVOLF. The HVOGF sprayed conventional and HVOLF sprayed multimodal coatings have the highest and lowest erosion rate respectively. All coatings suffer more erosion at impact angle of 90° than 30°, even though at 90° the erosion distance increases by 25% and the erosion time shortens by about 30%. The erosion rate distinctions of four coatings at two erosion angle have the same tendency with the erosion rate, with minimum value obtained by the HVOLF sprayed multimodal coating and maximum value achieved by the HVOGF sprayed conventional coating. Therefore the erosion wear performance of WC-10Co4Cr coatings depends not only on the structure of feedstock powder, but also on the HVOF process and the impact angle. The multimodal WC-10Co4Cr coating sprayed by HVOLF process has the most excellent erosion wear resistance.

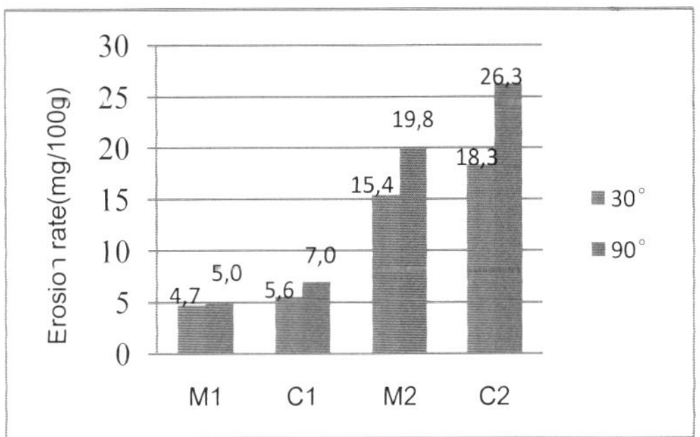

Fig.4. Erosion rate of WC-10Co4Cr coatings at a) 30° and b) 90°

3.5 Erosion wear mechanisms of WC-10Co4Cr coatings

The erosion surface morphologies of multimodal and conventional WC-10Co4Cr coatings at 30° and 90° impact angle are illustrated in **Fig. 5** and **Fig. 6**. In **Fig. 5**, obvious long furrow shaped grooves can be observed on the surface of the coatings at 30°. The morphology characteristics of the erosion surface of multimodal and conventional WC-10Co4Cr coatings sprayed by HVOLF are similar, only the amount of scratches of the conventional WC-10Co4Cr coating is relatively higher. On the other hand, micro cracks can be observed around the scratches on the erosion surface of multimodal and conventional WC-10Co4Cr coatings sprayed by HVOGF. The micro damage zone is formed with the micro cracks growth, and the partial material is stripped from the surface, leaving a clear pit, which causes the damage of the coating. At 90° impact angle, the coating surface displays wedged erosion pits with honeycomb shape, and material is extruded around the pits. HVOLF sprayed multimodal WC-10Co4Cr coating has the most shallow erosion pits, while conventional WC-10Co4Cr coating sprayed by HVOGF suffers most serious particle loss.

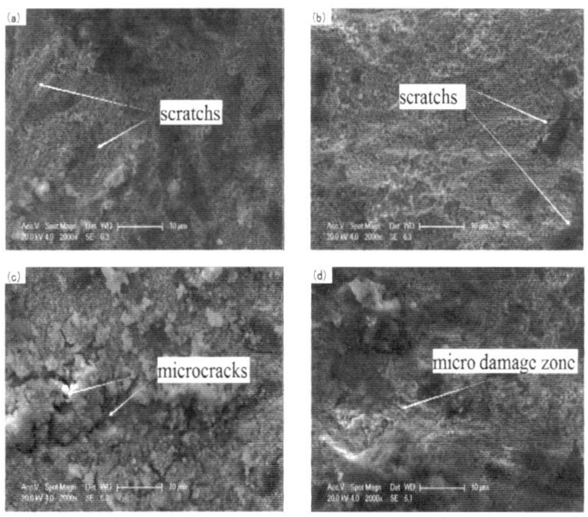

Fig.5. Surface micrographs of WC-10Co4Cr coatings of a) M1, b) C1, c) M2 and d) C2 after erosion at 30°

Fig.6. Surface micrographs of WC-10Co4Cr coatings of a) M1, b) C1, c) M2 and d) C2 after erosion at 90°

The erosion wear performance and worn surface morphology of WC-10Co4Cr coatings sprayed by HVOF are mainly determined by the structures, properties and wear mechanism of the coatings. When the impact angle is 30°, the erosion rate of the HVOLF sprayed multimodal WC-10Co4Cr coatings is 15% lower than that of the HVOLF sprayed conventional coating and 70% lower than that of the multimodal coating sprayed by HVOGF. When the impact angle is 90°, the erosion rate of the HVOLF sprayed multimodal WC-10Co4Cr coating is reduced by 30% and 75% compared to the HVOLF sprayed conventional coating and the HVOGF sprayed multimodal coating respectively. In the erosion wear, the main wear mechanism is micromachining at 30°, and the most critical factor related to the wear resistance is the coating microhardness. At 90° the damage of the coating is mainly affected by fatigue, and the main factor that affects the coating erosion performance is the fracture toughness. It shows that the optimization of spraying process is particularly important to improve the erosion resistance of the coating.

In WC-10Co4Cr coatings, fine WC particles and CoCr binding phase commonly exist in the grain boundary, which greatly improve the coatings' microhardness, fracture toughness and erosion resistance due to the high binding strength between the particles. The binding strength enhances in the multimodal because of the evenly distributed nano sized WC particles, further improving the coating erosion resistance. At the same time, HVOLF technique can achieve WC-10Co4Cr coatings with denser microstructure and lower porosity than HVOGF. The microhardness and fracture toughness of HVOLF sprayed multimodal coating are considerably enhanced, and these increased the coating erosion resistance.

4 Conclusions

The main results can be summarized as followed in this study.:

- The WC-10Co4Cr coatings deposited by HVOLF are mainly composed of WC, amorphous CoCr, and a small amount of W2C, no obvious WC decarburization can be detected.
- The WC-10Co4Cr coatings sprayed by HVOLF

exhibit better mechanical properties than those sprayed by HVOGF, in which the HVOLF sprayed multimodal coating exhibits the lowest porosity, highest hardness and fracture toughness (40% higher than others).

- The HVOLF sprayed WC-10Co4Cr coatings, especially the multimodal WC-10Co4Cr coating, possess better erosion resistance than those sprayed by HVOGF.
- The erosion resistance of WC-10Co4Cr coatings depends not only on the structure of the feedstock powders but also on HVOF spraying process. The superior erosion resistance of the HVOLF sprayed multimodal WC-10Co4Cr coating is attributed to its low porosity, high hardness and toughness.

5 Acknowledgments

This research was supported by the National Natural Science Foundation of China under the project (51422507).

6 Literature

[1] Santa J F, Baena J C, Toro A: Slurry erosion of thermal spray coatings and stainless steels for hydraulic machinery. Wear, 263 (2007), Issue 11, pp. 258/264.

[2] Padhy M K, Saini R P: A review on silt erosion in hydro turbines. Renewable & Sustainable Energy Reviews, 12 (2008), Issue 7, pp. 1974/87.

[3] Kim J H, Joo H G, Kang Y L: Simulation of solid particle erosion in WC-Ni coated wall using CFD. Journal of Materials Processing Technology, 224 (2015), pp. 240/245.

[4] Kim HJ, Yang HS, Baik KH: Development and properties of nanostructured thermal spray coatings. Current applied physics, 6 (2006), Issue 6, pp. 1002/1006.

[5] Zhou KS, Deng CM, Liu M: Characterizations of Fatigue and Salt Spray Corrosion Resistance of HVAF Sprayed WC-17Co and WC-10Co4Cr Coatings on the Substrate of 300M Steel. Rare metal materials and engineering, 38 (2009), Issue 4, pp. 671/676.

[6] Hong S, Wu YP, Zhang JF: Cavitation Erosion Behavior and Mechanism of HVOF Sprayed WC-10Co-4Cr Coating in 3.5 wt% NaCl Solution. Transactions of the Indian Institute of Metals, 68 (2015), Issue 1, pp. 151/159.

[7] Armstrong RW: The hardness and strength properties of WC-Co composites. Materials, 4 (2011), Issue 7, pp. 1287/1308.

[8] Zhao XQ, Zhou HD, Chen JM: Comparative Study of the Friction and Wear Behavior of Plasma Sprayed Conventional and Nanostructrured WC-12%Co Coatings on Stainless Steel. Materials Science and Engineering, 431 (2006), Issue 1-2, pp. 290/297.

[9] Chen H, Gou GQ, Tu MJ: Characteristics of Nano Particles and Their Effect on the Formation of Nanostructures in Air Plasma Spraying WC-17Co Coating. Surface & Coatings Technology, 203 (2009), Issue 13, pp.1785/89.

[10] Skandan G, Yao R, Sadangi R: Multimodal coatings: A new concept in thermal spraying. Journal of thermal spray technology, 9 (2000), Issue 3, pp. 329/331.

[11] Guilemany JM, Dosta S, Miguel JR: The enhancement of the properties of WC-Co HVOF coatings through the use of nanostructured and microstructured feedstock powders. Surface & Coatings Technology, 201 (2006), Issue 3-4, pp. 1180/90.

[12] Wang Q, Chen ZH, Li LX, Yang GB: The parameters optimization and abrasion wear mechanism of liquid fuel HVOF sprayed bimodal WC-12Co coating. Surface & Coatings Technology, 206 (2012), Issue 8-9, pp. 2233/41.

[13] Wang Q, Tang ZX, Cha LM: Cavitation and Sand Slurry Erosion Resistances of WC-10Co-4Cr Coatings. Journal of materials engineering and performance, 24 (2015), Issue 6, pp. 2435/43.

[14] Tahar Sahraoui, Sofiane Guessasma, M: Ali Jeridane et al: HVOF Sprayed WC-Co Coatings: Microstructure, Mechanical Properties and Friction Moment Prediction. Materials and Design, 31 (2010), Issue 3, pp. 1431/37.

Process Simulation & Modeling in HVOF Qualification and Optimization

Liang Zeng & Brentt Ramsey;
RBC-Sonic Industries, 20030 Normandie Ave., Torrance, CA 90502

Recently, High Velocity Oxy-fuel (HVOF) technique was widely used in aerospace industry due to its excellent combination of wear resistance, fatigue, and environmental performances. However, due to the intrinsic nature of aerospace industry, process qualification could take a few years to accomplish. This is primarily due to the qualification and optimization of coatings properties for a given specification needs a careful control of the plenty of operational spray parameters and high requirements.

In this work, process models with combination of statistics, materials science and mechanics were developed to qualify and optimize HVOF process. Two different models were built. The first processing model was trying to correlate the processing variables with typical specimens' metallurgical properties, such as tensile strength, Oxide and porosity, residual stress, micro-hardness. The second model was focused on fatigue properties vs. HVOF processing variables. The processing model adopted response surface methodology (RSM) with five parameters over three levels for each factor. A total of 34 runs were used. By combining both models together, statistical models were developed to predict parts from microstructural to mechanical properties, including specimen spalling cycles. In addition, the most optimized process-operating parameters were developed.

1 Introduction

Thermal spray coatings are widely applied for better corrosion and wear resistance. High velocity processes — especially HVOF (High velocity oxy-fuel) spraying—are the preferred methods for producing coating with low porosity and high adhesion. Most aerospace customers require detailed qualification, ranging from basic microstructure and residual stress, to more complicated damage tolerant system such as fatigue/spalling, wear or corrosion, etc. Furthermore, HVOF spraying is a very complex process, which has a large variety of variables affecting the deposit formation and hence coating properties. These variables include hardware characteristics, such as nozzle geometry and spraying distance, and process parameters, including fuel, gas flow density, and powder feedstock. In the spray process, the powder particles experience very high speed combined with fast heating up to its melting or semi-melting point or even above. This high temperature may cause evaporation of the powder or some components of it, dissolution, and phase transformations. Due to this complex nature of HVOF technique, the control and optimization of the process in order to achieve coating with desired properties is a highly challenging task. There are different ways of optimizing and analyzing the HVOF processes and deposit formation. These include statistical methods such as design of experiments (DOE), numerical modeling and simulation, and Finite Element Methodology (FEM) [1-2]. Among all the optimization tools, DOE was widely used since the optimization of coating properties for a given specification needs a careful control of the operational spray parameters and statistical methods can be useful in preparation of the experimentation and productions. However, DOE alone may not be sufficient. In most cases, the resuts in DOE, especially Taguchi approach, is dependent on the design and selection of varibles and the result may be misleading. It would be good to combine DOE with other approaches.

Thus, the objective of this study was to find the new way of HVOF process qualification and optimization using process models with combination of statistics, materials science and mechanics. This approach can also be used to identify most optimized process-operating parameters or process input variables to meets all customer requirements. In this evaluation, we choose most common aerospace customer qualification requirements including typical microstructural and mechanical requirements per AMS2448, and additional fatigue requirement. In addition, the process-structure-property relation was also developed by process maps, which was used as design tool for coating processing.

2 Microstructure DOE Experimental setups

2.1 Equipment descriptions

The HVOF Equipment used is a TAFA JP8000 with a TAFA 1264 Powder Feeder Closed Loop System. The set points were preset based on trials that demonstrate all values within will achieve acceptable coating application. The coating system provides fixtures for mounting the HVOF torch and holding the work piece, so that torch-to-work distance and transverse rate variations can be minimized. In this process qualification development, a well-defined 1350VF/WC-731 powder per specification AMS 7882.

Micro-hardness testing was conducted using Mitutoyo Hm-112 at 300 gram load. All specimens were cold mounted, followed by fully polished and analysized using Zeiss Microscopy.

2.2 DOE methodology

As described above, HVOF is a complicated process containing many process input factors which could impact coating properties. The actual choice of parameters needs some understanding of the process as there are as many as 50 process variables [3]. To limit the amount of experimental runs, the chosen inputs for experimentation of this DOE was be Oxygen Flow, Fuel flow, Powder Feed Rate, Stand-off Distance, Surface Speed, and Surface preparation roughness, for a total of 6 factors (Table 1). Since the relationship between the experimental response and factors are not always linear and the factors at multiple levels have to be used to determine quadratic or even cubic relationship using Response Surface

Factors	Low (-1)	High (+1)	Center Point
Oxygen Flow, SCFH	1800	2000	1900
Fuel flow, GPH	5.6	6.4	6.0
Powder Feed Rate, GFM/RPM	64	90	77
Stand Off Distance, inch	12	16	14
Surface speed, SF/min	200	300	250
Surface roughness, µin	79	157	118

Table 1: Setting of Factors and Two Levels for DOE of HVOF Coating

Methodology (RSM). The D-Optimal design is made usually constructed and evaluated to calculate the matrix for a given mathematical model, number of experimental runs, etc [3]. Statistical analysis software JMP was used for DOE and analysis. Desirability profiling and optimization within JMP were used to identify optimal first and second order processing map development. In desirability profiling, all response have a specified desirability function. Then, the overall desirability was defined as the geometric mean of the desirability for each response.

Based on our extensive scientific literature study and our initial experience [5-6], the low, high and center points for each factor settings were chosen as listed in Table 1. All other process input variables were remained constant in all runs. A total of 34 experimental runs were made.

2.2 DOE responces

For DOE responces, it is important to understand the difference between the general output of the process and the characteristics/properties of the final coating deposit. As stated earlier, the final goal of the coating was maximized microstructure and fatigue performance. However, when the coating is initially sprayed, only a set of simple measurements can be used for quality control of the process, as follows:

- Microstructure (primarily measurement of porosity, un-melted particles, and oxides)
- Micro-hardness
- Almen Strip (residual stress)
- Substrate Temperature (during coating)

Table 2 indicates information of test coupons used in DOE for each run. The coupon size of each type meets the requirement in accordance with AMS 2448. Prior to abrasive blasting, the parts were solvent cleaned by wiping the part with isopropyl alcohol and a lint free cloth. The parts were then visually inspected to ensure complete removal of all surface contaminants. Abrasive blasting was performed in a manner so as not to remove and/or distort the base material ensuring that there is no abrasive blast pulsation affecting the grit blast process.

Testing	Requirements	Material	Coating Thickness, inch
Bending	N crack	4130 steel	0.001-0.003"
Micro-hardness test	950 min @ HV_{300}	4130 steel	0.009-0.012"
Microstructure	Porosity, Oxide <1%	4130 steel	0.009-0.012"
Tensile bond strength	10ksi min,	4130 steel	0.009-0.012"
Residual stress	0.003-0.012"	Almen N strip	0.0045-0.0053"

Table 2: Test coupons used in DOE for each run

3 Microstructure model & results

DOE order runs 1 through 34 were tested. Table 3 represents all the results that were analyzed in the DOE study.

All of the 34 DOE orders exhibited no failure in the bending test. There were no evidences of peeling, cracking or separation at neither the bent coating surface nor the edges of the alloy strip. It is safe to say that adhesion quality of the coating is not a concern for each of the 34 DOE orders. For microstructure, there were multiple orders which showed oxide and/or porosity over 1%. The variation of the porosity and oxide in the microstructure of the HVOF coating is controlled by numerous variables in the parameters. The main controlling factor in microstructure results is the oxygen/fuel ratio and stand-off distance. This ratio controls the flame temperature which effects how the powder dissolute and melts onto the substrate. As expected, lower stand-off distance tends to have lower porosity.

Of the 34 DOE orders, 3 of them did not meet the required tensile bonding strength of 69 MPa. The other 31 orders only failed due to adhesive adhesion failure. In order to fully understand and analyze the results from this test, it is important to take note of how much of the coating separation exists after the two test specimens separate. The orders that have failed or had a lower tensile bond breaking point are shown to have parameters that led to a higher substrate temperature and high powder feed rate. This is mostly attributed to closer standoff distances,

powder grams per minute as well as the fuel/oxygen ratio.
The minimum Vickers hardness of HV300g - 950 was surpassed in each of the 34 DOE orders. Statistical model has 93% R^2, indicating statistically significant. Factors that produced a hard coating include a closer stand-off distance and higher fuel. Although a harder coating seems optimal for the aerospace field, a softer coating seems logical for the ideal HVOF parameters. Albeit better corrosion and wear resistance, a harder coating generally tends to be more brittle. Consequently, a more brittle coating has more of a chance to exhibit spallation. The parameters that produced a softer coating are chosen in order to eliminate the chance of spallation and to increase fatigue life.

Figure 1 shows the developed microstructural model. Based on the model, first order process map represents relationship between the spray stream measured responses with coating properties, such as hardness, microstructural properties, residual stress, etc was identified. The overall process map was precisely determined. . This process map can be used to establish manufacturing robustness and the process window for a reliable process.

Runs	Porosity	Oxide	Almen Strip Arc Height	Hardness	Tensile-Bond
	(%)	(%)	(um)	HV300	(MPa)
1	0.39	1.32	313	1374	76
2	0.61	0.86	178	1365	82
3	0.71	1.11	130	1338	76
4	1.32	1.60	110	1247	71
5	0.45	0.70	343	1343	73
6	0.34	0.93	257	1314	70
7	0.41	1.30	227	1361	81
8	0.25	0.55	401	1347	81
9	0.18	0.64	371	1427	38
10	0.26	0.57	217	1304	82
11	0.36	0.67	219	1320	82
12	0.26	0.56	270	1318	78
13	0.23	0.73	246	1309	85
14	0.45	1.38	175	1353	81
15	0.41	1.19	217	1319	84
16	0.21	0.73	308	1342	86
17	0.53	0.94	207	1333	80
18	0.52	0.62	223	1342	77
19	0.89	1.25	68	1260	77
20	0.77	1.45	204	1272	82
21	0.52	0.57	172	1291	81
22	0.35	1.23	241	1360	79
23	0.31	0.60	277	1417	69
24	0.37	0.86	235	1402	88
25	0.32	0.64	178	1394	38
26	1.24	1.31	64	1272	77
27	0.89	0.82	130	1288	83
28	0.22	0.66	262	1465	85
29	0.89	0.69	180	1369	83
30	0.65	1.10	266	1388	81
31	0.60	0.56	139	1370	79
32	0.57	0.86	209	1321	77
33	0.25	0.36	347	1458	40
34	0.28	0.54	272	1359	83

Table 3. Mcrostructural DOE results.

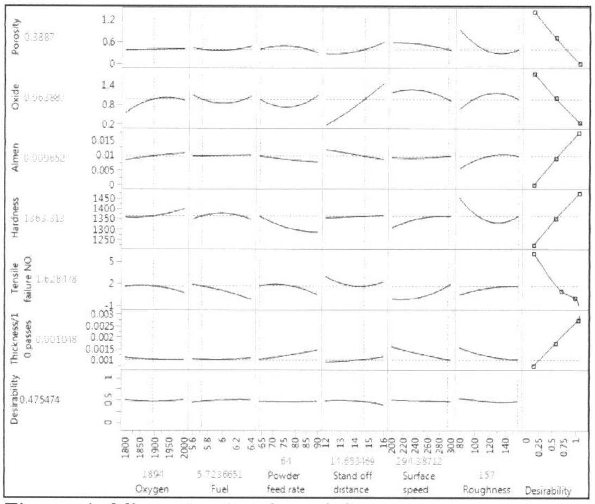

Figure 1. Microstructural model

4 Fatigue DOE setups

Fatigue is a very critical property in the aerospace industry, because of the repeated cyclic loading for landing gear, actuators, airframe parts, and gas turbine engine components. The purpose of fatigue testing is to evaluate the effect of the coating on the fatigue of the underlying material [4]. In addition, coatings must maintain their integrity under expected service conditions (i.e. not delaminate during testing at stresses seen in service, although delamination may occur on final failure failure once specimens break).

It is almost impossible to conduct another 34 runs of fatigue testing. Thus, fatigue testing DOE variable selection was based on microstructural model analysis, plus fundamental understanding of materials science including fatigue science and thermal spray process. From microstructural model, low powder feed rate provides decent bonding strength that expected to provide high fatigue/spalling cycles, too. Thus, the powder feed rate was fixed at 64 gpm. The specimen surface speed was chose at 200 SF/min due to relative low sensitivity. Parts surface roughness was chosen to be 79 Ra. At both surface speed 200 SF/min and roughness of 79Ra, specimens demonstrated very low porosity and substrate temperature that supposed to be good for fatigue and will be discussed later. Therefore, fatigue testing DOE focused on oxygen, kerosene and stand-off distance factors (Table 4). The corresponding response was fatigue spalling cycles. Under ideal situation, the spalling occurred during final fatigue failure. Plus, the spalling location would be exactly the same location of final fatigue failure location. In non-ideal case that spalling started prior to fatigue failure or spalling location was away from fatigue failure, it is generally considered not acceptabel.

Factors	Low (-1)	High (+1)	Center Point
Oxygen Flow, SCFH	1800	2000	1900
Fuel flow, GPH	5.6	6.4	6.0
Stand Off Distance, inch	12	16	14

Table 4. Fatigue DOE design table.

4.1 Fatigue test specimens & setup

Fatigue testing conditions was based on load-controlled constant amplitude axial fatigue testing in accordance with ASTM E466-96:

- Fatigue testing Load: 14.2 Klbf
- Specimen gage diameter (prior to coating): 0.315 inch
- Max Fatigue stress: 182 ksi
- Testing frequency: 4 Hz
- Fatigue testing ratio: R= -1
- Fatigue specimen surface finish: As HVOF coated
- Fatigue specimen geometry: Figure 2
- Specimen material type: 300M with tensile strength 280-300 ksi (1930-2068 Mpa)
- During fatigue testing, video cameras were used to precesely record the start of coating spalling. If spalling occurred during final fatigue failure, then spalling cycles is the fatigue cycle.

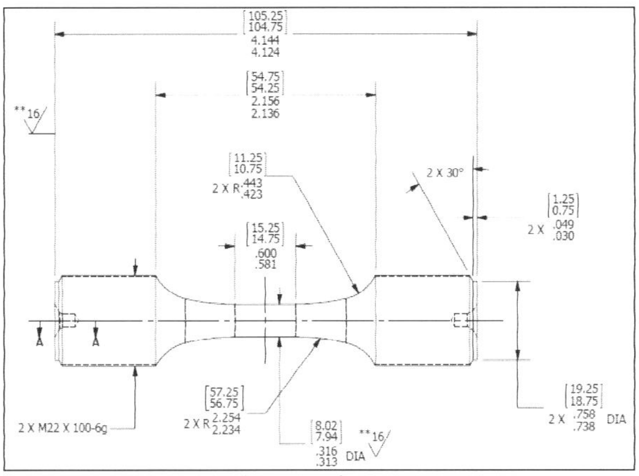

Figure 2. Fatigue testing specimens geometry

All fatigue specimen preparation involved four major steps: rough machining, vacuum heat treatment, finish machining, and shot peening. The heat treat for 300M was in accordance with MTL-1201. All specimens to be tested were prepared in the same fashion to ensure a common starting condition prior to any surface preparation techniques used for the coatings.

5 Fatigue model & results

Figure 3 shows the fatigue model based on testing results. With R^2 of 72% and F ratio of 2.6, this fatigue model is statistically significant. The most significant factors are the oxygen and kerosene contents.

Apparently, for our application, low oxygen and kerosene and high distance is preferred to have maximum fatigue/spalling cycles. Table 5 showed one of the best fatigue testing results. For both specimens, all fatigue specimens show no sign of spalling until final fatigue failure (figure 4). After failure, fatigue tested specimens were put back together to measure maximum spalling distance. The measured distance is between 2-4 mm.

Specimen No.	Sapllling cycles	Failure cycles
NT-1	N/A	10305
NT-2	N/A	11136

Table 5. Fatigue test results.

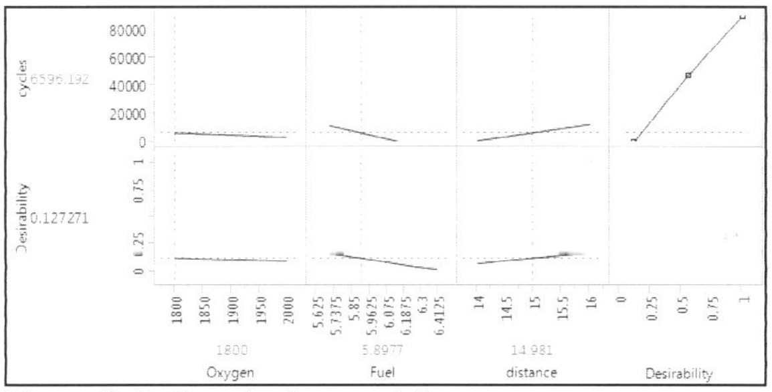

Figure 3. Fatigue testing model

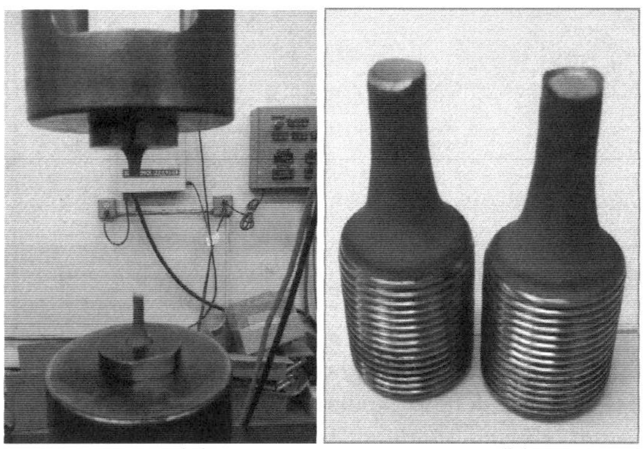
(a) (b)

Figure 4. Fatigue testing and tested specimens

6 Combinted models

After generating both mistructural and fatigue models, it is time to combine both models together to identify appropriate conditions to meet all customer requirements. Interestingly enough, the best conditions to meet each model did not coinside with each other. In other words, the best spray processing conditions to generate lowest porosity did not have the highest fatigue/spalling cycles. Correspondingly, the best spray processing conditions to generate highest spalling cyles did not have the lowest porosity, eithor. However, it is not difficult to identify spray processing conditions that could meet both microstructural and fatigue/spalling conditions.

Table 6 showed final fatigue results with decent microstructure in figure 5.

Specimen No.	Sapling cycles	Failure cycles
M6-1	N/A	8967
M6-2	N/A	10736
M6-3	N/A	10712
M6-4	N/A	11324
M6-5	N/A	11356

Table 6. Fatigue test results.

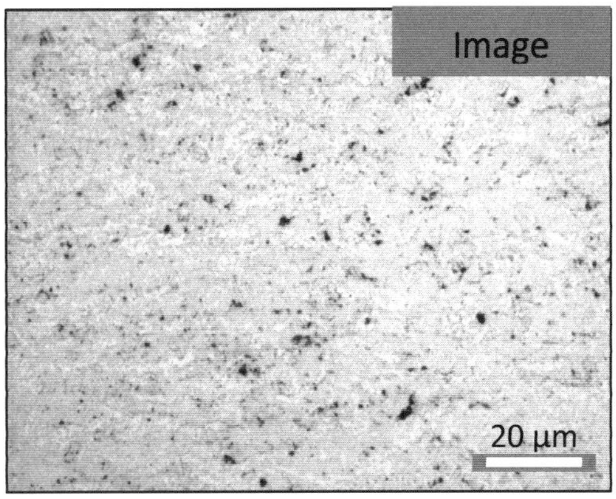

Figure 5. Microstructure of fatigue tested specimen

7 Discussion

Current example demonstrated that HVOF process qualification using combination of statistics, materials science and mechanics models is an useful and effective approach to successfully meet customer requirements within short time frame. It is interesting to discuss some important aspect of HVOF models.

7.1 Pre-liminary model

One of the most fascinating feature of HVOF models are that there is more than one "good" processing condition for the coating process. Due to its' complexity, the desired coating properties cannot be achieved by just changing a single variable since there are complicated interactions. There are so many factors in the process that even in altering one parameter might have a substantial effect in the coating deposition. That is also the reason statistical could be very helpful. In addition, the varible selection became highly critical to preveent any potential misleading information. This has to come from fundamental understanding of material science and thermal spray process. In aerospace thermal industry, each equipment is alway slightly different. Thus, finding the right nitch for individual equipment is critical. The authors would suggest to conduct pre-liminary DOE to identify the most critical varibles based on final applications. In this example, the first model, microstructure is the pre-liminary DOE.

7.2 Final model

For second model, fatigue model, is the most important application model. It has to based on final customer applications. The most critical fatigue model varible selection were determined based off of key trends seen from the pre-liminary DOE data. For a dynamic environment, such as the aerospace industry, fatigue life and performance should be considered as the most critical factor. So in order to select the most ideal HVOF spraying condition, fatigue was used as the primary guideline for the model varible selections. Therefore, the primary determinant in the selection is controlling the temperature of the substrate. Thus, these parameters prioritized in keeping the temperature as low as possible. The combustion gas ratio of oxygen/fuel is another determinant in the selection process. The low points for the oxygen and fuel flow settings kept the porosity and oxide low, while keeping fatigue life high. Therefore, only oxygen, kerosene and stand-off distance were chosen to be the fatigue model varibles.

7.3 Statistics

Due to the non-linear relationship, it is highly receommend that Response Surface Methodology or other optimal design be used. There are many commercially available software on the market.

8 Conclusions

This evaluation outlines new way of High Velocity Oxy-Fuel (HVOF) spraying qualification and optimization using combined statistical, materials science and mechanics. It has been proved to be an very effective way to be customer qualified successfully wiin extremely short time frame.

The approach can also be used to identify and demonstrate the ideal setting of HVOF process-operating parameters, with which the deposited coating meets customer requirements, including fatigue. All these models adopted response surface methodology.

This approach includes developing two DOEs. The first processing model, pre-liminary DOE, was trying to identify following goals based on final applications:
1. Equipment nitches;
2. Processing map: correlate the processing variables with typical specimens' metallurgical properties, such as tensile strength, oxide and porosity, residual stress, micro-hardness, etc.

The second model was focused on final applications, fatigue properties vs. HVOF processing variables. The varible selections were based on pre-liminary model and fundamental of material science. In this example, it would be the substrate temperature temperature.

9 Literature

[1] J. Mostaghimi, S. Chandra, R. Ghafouri-Azar, A. Dolatabadi, *Surf. Coat. Technol. 163-164 (*2003) 1.
[2] E. Dongmo, M. Wenzelburger, M. Gadow, *Surf. Coat. Technol. 202 (*2008) 4470.
[3] C. Pierlot, L. Pawlowski, M. Bigan, P. Chagnon, *Surf. Coat. Technol.* 202 (2008) 4483
[4] M. Oksa, E. Turunen ,T. Suhonen , T. Varis, S. Hannula, *Coatings* 1(2011) 17-52
[5] R.S. Lima , J. Karthikeyan , C.M. Kay , J. Lindemann , C.C. Berndt, *Thin Solid Films* 416 (2002) 129
[6] W. Fang, T.Y. Cho, J.H. Yoon, K.O. Song, S.K. Hur, S.J. Youn, H.G. Chun, *journal of materials*

Characterization of YSZ coatings deposited by conventional Dc and CO_2/CH_4 Torches

H.R. Salimijazi[1], R. Ghasemi[2], J. Mostaghimi[3], L. Pershin[3], [1]Isfahan University of Technology, Isfahan, Iran, [2] Developing Advanced Testing Equipment for Materials and Coating (DATMC), Materials and Energy Research Center, Karaj, Iran, [3]University of Toronto, Toronto, Canada

Usually, argon is used for plasma generation in conventional DC plasma torches. Its low thermal conductivity and enthalpy, and erosion of tungsten cathode limit the application and durability performance of the spray process. In the current research, hollow sphere yttria stabilized zirconia powders were deposited using a new patented DC plasma gun which uses molecular gases; carbon dioxide-hydrocarbons mixture (CO_2+CH_4). Compared to Ar and Ar+H_2 mixture, the enthalpy and thermal conductivity of the CO_2+CH_4 mixture is much higher, resulting in higher torch power and heat transfer to the sprayed particles. The plasma plume was scanned by DPV 2000 and in-flight particles temperature and velocity distributions upon the impact were collected. Moreover, foot prints of single splats were collected on mirror polished stainless steel substrates. Morphology of single splats and microstructure of coatings were characterized and compared to single splats and coatings deposited by F4 plasma torch under the same torch power. The majority of the in-flight particles temperature was at least 100 ºC higher than the YSZ melting point. SEM results showed that almost all single splats in the foot print were completely melted. The coating surface is smoother than the one deposited by the conventional torch indicating less unmelted and partially melted splats in the coating. Coating cross section characterization showed less porosity and denser coating compared to the coatings deposited by F4.

1 Introduction

Thermal barrier coatings (TBCs) are widely used for high temperature corrosion protection of high temperature components which need long time performance. These coatings were designed to be used on combustion chambers. At begging of 1980 TBCs were applied on aero-engine blades. In recent years TBCs are applied on combustion chambers, vanes and blades of gas turbines and piston crown of diesel engines [1].

Thermal barrier coatings typically consist of two layers: a metallic bond coat which is usually MCrAlY and a ceramic top coat that is based on zirconia [1, 2]. The bond coat increases the adhesion between ceramic top coat and the substrate and it has high oxidation resistance [3]. The top coat is expected to have high melting point, low density, low thermal conductivity, chemically inert, conformity of thermal expansion with substrate, good adhesion to the substrate and high erosion resistance. Many studies have been done to apply a coating that has the best properties of mentioned parameters. Between these parameters, the most essential one for TBCs is low thermal conductivity. Among ceramics, zirconia has a low thermal conductivity (2 W/mk). Furthermore thermal conductivity of zirconia is not temperature dependent. So, it is one of the best candidates for using as a top coat in thermal barrier coatings [4].

There are some methods that are used for applying TBC. Thermal spraying, Electron beam-Physical vapor deposition (EB-PVD) and Chemical Vapor Deposition (CVD) are very practical [5]. Nowadays, TBCs are deposited by using atmospheric plasma spraying and PVD [6, 7]. Due to higher cost and lower deposition rate of PVD, plasma spraying are used widely [8].

During plasma spraying, a stream of molten or solid particles strikes to substrate with a high velocity and then it deforms and rapidly solidifies like disk shaped splats. Splats are joined together by chemical or mechanical bonding. Depending on process parameters, some splat splashing and pores are formed. Therefore, physical and mechanical properties of deposits are strongly deped on feature of splats, cohesion strength between splats, adhesion strength between deposit and substrate, porosity shape, size and distribution, cracks, and defects [9, 10].

The majority of plasma spray torches use argon (Ar) for plasma generation. However the low thermal conductivity and enthalpy of argon limit the heat conduction to the in-flight particles. To overcome this limitation, a secondary gas such as hydrogen and helium are most commonly used is mixed with argon. Introduction of these gases increases the thermal conductivity of the mixture. Additionally, molecular gases must dissociate before ionization which requires larger energy input, resulting in higher enthalpy of the plasma and subsequently higher temparature of the in-flight particles (Ref. 5, 6). Use of carbon dioxide-hydrocarbons mixture has a number of advantages. Compared with Ar and Ar+H_2, the enthalpy and thermal conductivity of the CO_2+CH_4 mixture is much higher, leading to a higher torch power and improved heat transfer to the sprayed particles (Ref. 1, 5). In this paper, we compared the results of YSZ coatings deposited by Conventional DC plasma torch (F4) and a new patented CO_2+CH_4 torches.

2 Experimental Procedure

A conventional DC plasma torch, F4-MB, Sulzer Metco AG, Switzerland, and a new patented CO_2+CH_4 torch developed by the Centre for Advanced Coating Technology (CACT), University of Toronto, were used to deposit hollow sphere yttria stabilized zirconia (YSZ) on stainless steel substrates. The feedstock powders, HOSP ZrO_2 $8Y_2O_3$, (204NS) were supplied by Oerlikon metco, Switzerland, with particle size distribution of +11-125 µm. SEM micrographs of the powders are shown in Fig. 1.

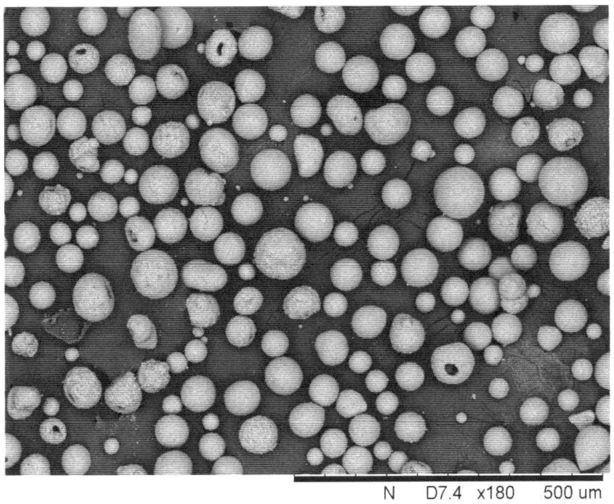

Fig. 1. SEM image of HOSP YZS powder particles.

It was tried to kept the power of plasma torches almost the same during the YSZ spraying by controlling the process parameters. The plasma spray process parameters are tabulated in Table 1.

Table 1. Plasma spray process parameters.

Gun	Gas flow, (lpm)	Current (A)	Voltage (V)	Spray Distance, mm	Thermal efficiency
F4-MB	40 (Ar) 20(H2)	600	70	75	35%
CACT	35 (CO2) 15 (CH4)	300	150	100	56%

The DPV-2000 scan of the YSZ in-flight particles in the CO_2+CH_4 plasma plum cross section upon the impact are shown in Fig. 2. The majority of in-flight particles has a velocity of above 90 m/s and temparature of above 3100 °C.

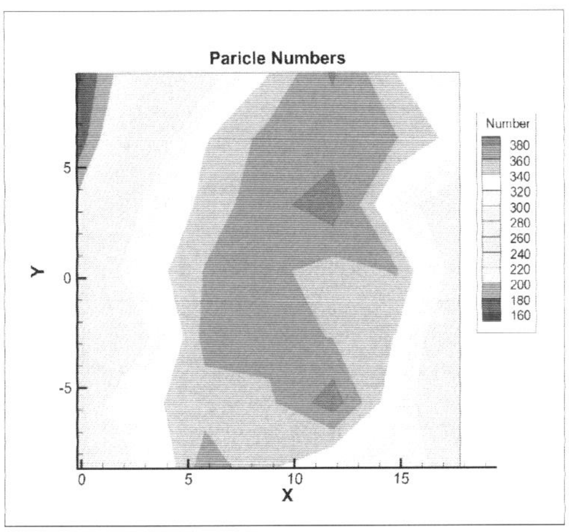

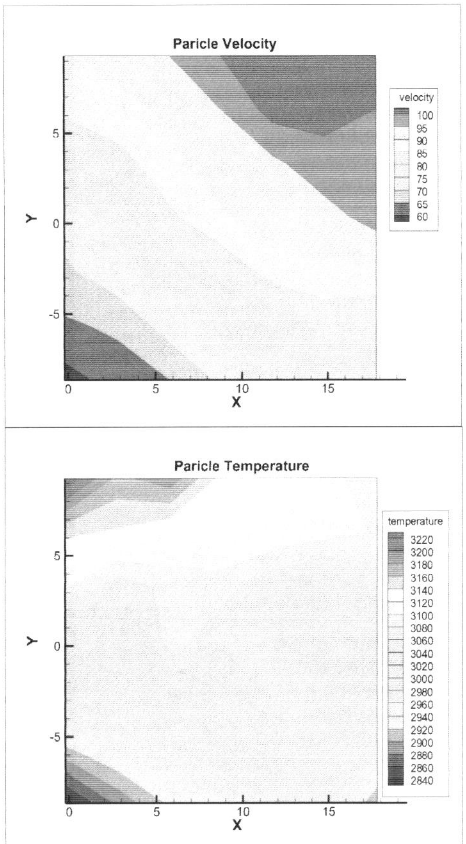

Fig. 2. DPV scan of in-flight particles characteristics at stand off distance of 100 mm.

3 Results and Discussion

Fig. 3 shows the morphologies of single splats deposited by CACT and F4-MB plasma gun. The splats morphologies in footprints indicate the degree of particle melting in the plasma jet. Increasing the enthalpy and efficiency of the plasma jet by using CO_2+CH_4 gas mixture in the CACT gun compared to use Ar+H2 mixture in F4-MB gun results in transfering more heat into the in-flight particles. Therefore, a high degree of flattening would be expected. The diameter of the splat in the coating deposited by CACT gun (Fig. 3a) is range from 70-100 µm; whereas, the diameter of the splat in the coating deposited by F4-MB (Fig. 3b) is in the range of 30-70 µm.

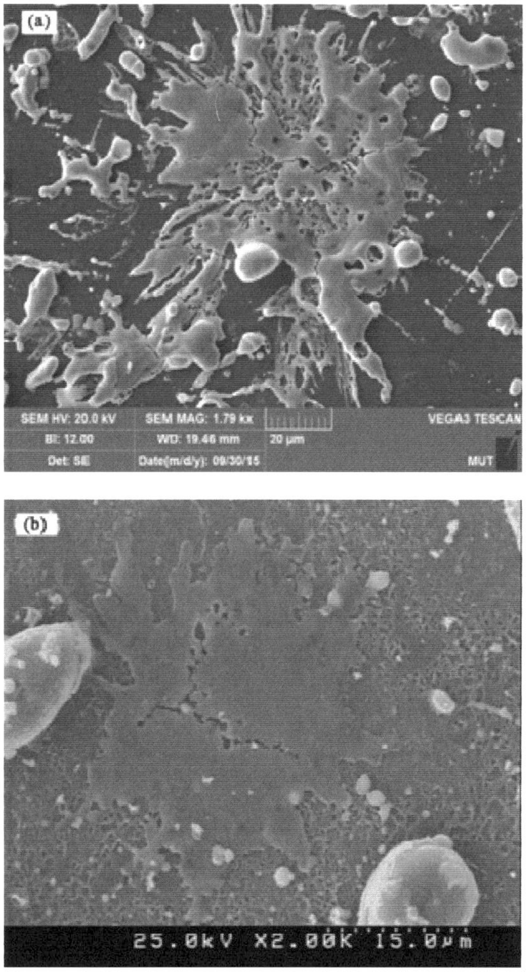

Fig. 3. SEM migrograph of YSZ single splat deposited on stailess steel by (a) CACT gun and (b) F4-MB gun.

Fig. 4 shows the fracture surface morphologies of the as-sprayed coating deposited by F4-MB and CACT plasma gun. The microstructure features of both coatings consist of typical lamellar structure with columnar grains which enclose pores, voids, unmelted particles, inter splat-cracks, and intra-splat cracks. Vertical intra-splat crack are formed since high residual stresses due to shrinkage during rapid cooling and thermal expansion mismatch of zirconia and metallic superalloy. The cohesion strength between splats is considerably lower than the adhesion strength betwen coating and metallic substrate. Therefore, horizontal inter-splat cracks are formed at splat boundaries due to the low cohesion strength. The formation of pores and voids is due to accidental deposition, since splats are not fully covered by adjacent splat as well as entrapment of the air during the plasma spraying.

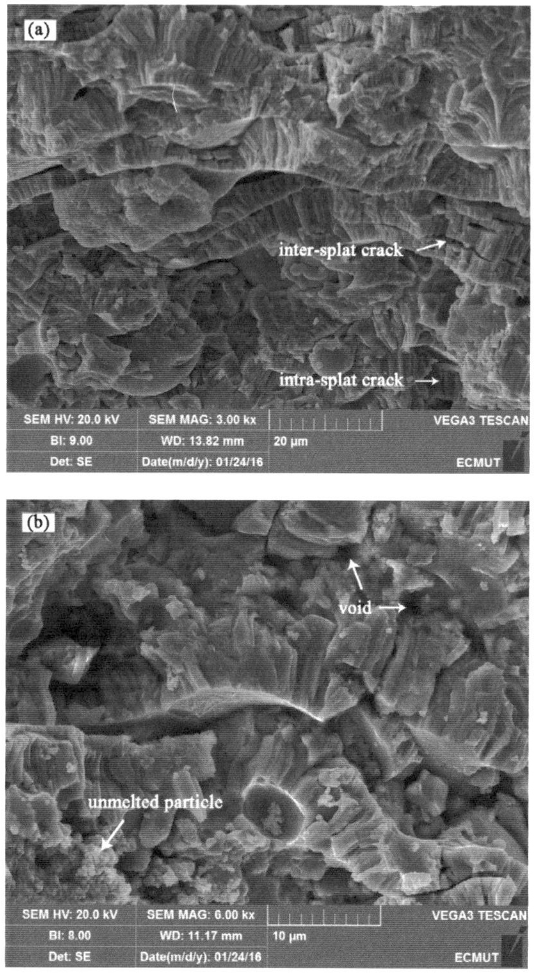

Fig 4. SEM microstructure of the fractured surface of YSZ coating deposited by (a) CACT gun (b) F4-MB gun.

The unmelted particles volume, size of lamellae, voids between lamellae and splats degree of flattening in the coating deposited by CACT gun (Fig. 4a) is different than the coating deposited by F4-MB gun (Fig. 4b). Because of the higher particles temperature in the CACT gun, the molten droplets flatten and spread to a much higher degree than in the case of F4-MB gun. This high degree flattening has led to reduction of lamellae thickness. In coating deposited by CACT gun, due to most of the sprayed particles reach above their melting point, the impact energy spread the liquid interface into surface void and ultimately reduces the voids. The high temperature also breaks up unmelted particles, leading to improve inter-splat cohesion and lower porosity.

4 References

[1] N. P. Padture, M. Gell, L. Eric and H. Jordan, "Thermal Barrier coatings for gas turbine engine applications", Material science 296(2002)280-285.

[2] U. Schulz, C. Leyens, K. Fritscher, M. Peters, B. Saruhan-Brings, O. Lavigne, J.M. Dorvaux, M. Poulain, R. Mévrel and M. Caliez, "Some recent trends in research and technology of advanced thermal barrier coatings", Aerospace Science & Technology 7(2003)73-80.

[3] C.H. Lee, H. K. Kim, H. S. Choi and H. S. Ahn, "Phase transformation and bond coat oxidation behavior of plasma-sprayed zirconia thermal barrier coating", Surface & Coatings Technology 124(2000)1-12.

[4] X.Q. Cao, R. Vassenb and D. Stoeverb, "Ceramic materials for thermal barrier coatings", The European Ceramic Society 24(2004)1-10.

[5] R. Taylor and J.R. Brandon, "Microstructure, composition and property relationships of plasma sprayed thermal barrier coatings", Surface and Coatings Technology 50(1992)141-149.

[6] O. Altun and Y.E. Boke, "Effect of the microstructure of EB-PVD thermal barrier coatings on the thermal conductivity and the methods to reduce the thermal conductivity", Archives of Materials Science 29(2008)7-16.

[7] G. Moskal, L. Swadzba and T. Rzychot, "Measurement of residual stress in plasma-sprayed TBC with a gradient of porosity and chemical composition", Journal of Achievements in Materials and Manufacturing Engineering 23(2007)3l-34.

[8] Y. Wang and G. Sayre, "Commercial thermal barrier coatings with a double-layer bond coat on turbine vanes and the process repeatability", Surface & Coatings Technology 203(2009)2186-2192.

[9] J. Mostaghimi and S. Chandro, "Splat formation in plasma-spray coating process", Pure Applied Chemistry 74(2002)441-445.

[10] H.R. Salimijazi, L. Pershin, T.W. coyle, J. Mostaghimi, S. chandra, Y.C. Lau, L. Rosenzweig and E. Moran, "Effect of droplet characteristics and substrate surface topography on the final morphology of plasma-sprayed zirconia single splats", Journal of Thermal Spray Technology 16(2007)291 -299.

High velocity spray deposition of WC cermets by an air-oxygen controlled combustion process

M. Parco*, I. Fagoaga, G. Barykin, C. Vaquero, Fundación Tecnalia Research & Innovation, IRUN, Spain

HVOF processes represent the state of art for the spray deposition of wear and corrosion resistant coatings since their supersonic gas velocities in combination with moderate flame temperatures allow the deposition of optimal coatings with very high bond strengths, fine surface finishes and low oxide levels. However, new generation coating materials (fine powders), stringent quality requirements and the high productivity demanded by the industry, push the HVOF technologies to their limits. Recently, a novel air-oxygen controlled high velocity combustion process has been development by Tecnalia. The system operates within the supersonic regime using a broad range of fuel/oxidant ratios thanks to the use of air-oxygen mixtures and a carefully designed gun. Extremely low flame temperatures can be achieved while keeping a supersonic flow of combustion products, thus allowing the solid state deposition of almost all industrially relevant metal alloys with superior deposit qualities. This work deals with the development of superhard cermet coatings using different types of conventional and fine WC-Co(Cr) powders. The process parameters and the system configuration have been optimised looking for the best coating properties and microstructure. Results are compared in terms of ascertained coating properties (microstructure, microhardness and toughness) and the functional performance under sliding wear.

1 Introduction

Because of their high hardness, high wear-resistance and good fracture resistance, hardmetal coatings are attractive and successful composite materials that have been widely used in many industrial applications. In particular, High Velocity Oxy-Fuel (HVOF) spray coatings based on tungsten carbide (WC) have been introduced some years ago as an alternative to electrolytic hard chrome (EHC). Such HVOF sprayed coatings have supposed a clear improvement from a technical and environmental perspective. In applications where the coatings will be used in an erosive-corrosive environment, HVOF WC-CoCr coatings are frequently recommended and are, in fact, the current industrial standard as an alternative to EHC [1].

HVOF sprayed WC-Co coatings are formed by the successive deposition of a stream of solid–liquid two-phase droplets. However, the thermal cycling of WC-Co spray particles up to a temperature over the melting point of the binder phase during thermal spraying leads to the decarburization of carbide, which significantly influences wear performance of the resultant coating. The decarburization of carbide occurs through the dissolution of carbide into molten binder cobalt alloy, thermal decomposition of WC, oxidation of dicarbide and rebounding of large carbide particles within droplet. As a result, the phases present in the thermally sprayed WC-Co may include main phase WC carbide, certain content of W_2C (or WC_{1-x}), minor phases of M_6C (Co_3W_3C, Co_2W_4C) and W, trace of Co, and even $M_{12}C$ (Co_6W_6C). The experimental results have shown that the thermal decarburization of WC is limited to the thermal decomposition of WC to W_2C and carbon. The formation of metallic tungsten in the coating results from the oxidation of W_2C. Therefore, the effective suppression of thermal decomposition of WC becomes effective to suppress the formation of tungsten. However, in order to reach a sufficient deformability of the cermet particles upon high velocity impact, it is necessary to heat the feedstock just to melting of the binder phase through spray condition control [2,**[3**].

On the other side, the use of fine feedstock powders (< 10 μm) can extend the feasibility and scope of HVOF coatings to new fields of applications. Especially for the purpose of near-net-shape, these powders facilitate homogeneous layer morphologies, very low porosities (<1%) and smooth coating surfaces (Rz < 10 μm) [4]. However, fine powder feedstocks have showed to feature a very high sensitivity to the gas flow, velocity, and temperature of the spray jet. In principle, there is a small range between sufficient particle heating and overheating [5].

Within the industry, all this has motivated a growing interest in finding thermal spray processes able to operate under "colder" conditions and/or with higher gas velocities. Recently, High-Velocity Air-Fuel (HVAF) processes and even cold spraying have challenged HVOF deposited WC-CoCr coatings. HVAF processes, for instance, have proved to be a viable technology with many practical advantages over conventional HVOF systems [6]. Activated Combustion HVAF (AC-HVAF) sprayed WCCoCr coatings were found to outperform the wear resistance of their HVOF counterparts sprayed with conventional HVOF systems like JP5000 and DJ2600, performing even 20-30 times better than electroplated hard chrome and spray-fused coatings [7]. Moreover, liquid fuelled HVOF systems can be modified by the addition of an inert gas (N_2) for cooling purpose, together with the reduction of the effective flame power by changing the design of the combustion chamber, like the so-called nanoHVOF® technology [8].

* *Corresponding author: M. Parco, Industry and Transport Division, TECNALIA, Mikeletegi Pasealekua 2, 20009 San Sebastián-Donostia, Spain.*
E-mail address:maria.parco@tecnalia.com

Comparing with kerosene fuelled HVOF systems (~200kw), gas fuelled HVOF systems work at lower power levels (~100kW), which in combination with the implementation of low temperature fuel gases, as natural gas, and the substitution of O_2 by air, can further reduce the flame temperature while keeping a stable performance in the supersonic regimen. In the practice, the oxidant mixture can be adjusted from pure O_2 to full compressed air, leading to a wide range of energetic conditions in the flame.

In line with this research trend, the thermal spray group from TECNALIA has recently developed a High-Velocity Oxy/Air-Fuel (HVO/AF) system able to operate within a wide range of process conditions. Low gas temperatures are essentially reached by using natural gas as fuel and by injecting compressed air and oxygen to the combustion mixture. This report aims to present the results collected in relation to the development of wear resistant WC-Co(Cr) coatings, using both conventional and novel fine feedstock powders.

2 Process description

In this work, a novel High-Velocity Oxy/Air-Fuel (HVO/AF) system has been implemented. This system has been under development during the last 10 years at TECNALIA and has proved to be a cost effective thermal spray technology for the deposition of high quality metal [9] and cermet base coatings, **Fig. 1**.

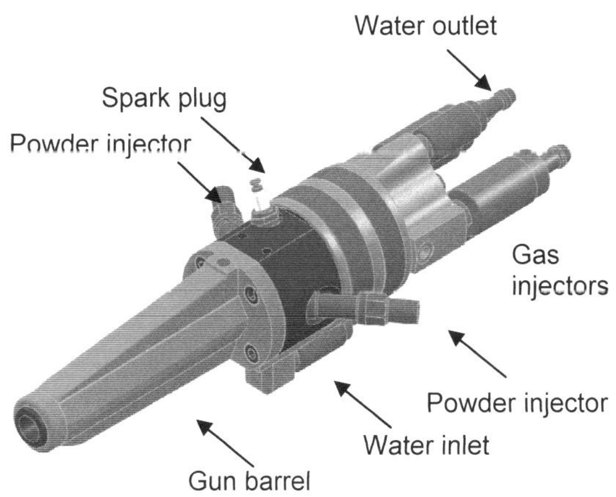

Fig. 1. Overview of the HVO/AF gun (proprietary technology of TECNALIA).

Basically, thanks to the use of air-oxygen mixtures and a carefully designed combustion chamber and gas injection module, the system is able to operate within the supersonic regime using a broad range of fuel/oxidant ratios. This fact allows the operation of the system with extremely low flame temperatures while keeping a supersonic flow of combustion products. Moreover, following a modular design, further critical gun components as the gun barrel and the powder injection module, can be easily exchanged to tailor the gun configuration for the processing of materials with very different physical natures, thus allowing the deposition of almost all industrially relevant metal base and cermet materials with superior deposit qualities. The system is operated using natural gas as fuel. Among the investigated coating materials, the temperature of the flame and thus the process conditions are mainly adjusted by varying the natural gas flow rate between 150 and 250 NLPM. The feedstock powders are injected radially using two feeding lines, allowing the processing of up to 6 kg/h of powder (for tungsten carbide). The whole gun is water cooled.

3 Experimental procedure

3.1 Feedstock powders

As feedstock materials, two conventional WCCo(Cr) powders were employed: Amperit® 526.074 (WC-17Co, agglomerated and sintered with coarse WC) and Amperit® 554.074 (WC10Co4Cr, sintered and crushed with medium WC). These feedstocks are commercially available from H.C. Starck with a size distribution of 15−45 µm. On the other side, a fine WC-10Co-4Cr powder from Thermico GmbH & Co. KG was investigated (article Nr. SJA 179/080 B, spheroidized powder with a size distribution between, 1-12 µm, with sub-micron WC ~400-800 nm).

3.2 Coating deposition

Coatings were sprayed on flat mild steel substrates. Prior to the coating deposition, substrates were face grinded and degreased. No grit blasting pre-treatment was employed.

Three different conical expansion nozzles with 70, 100 and 150 mm in length were investigated to assess the influence of the resulting coating properties..

As a starting point, a broad parameter window was screened to define the most suitable parameters for each coating material. Subsequently, the process parameters were further optimized to produce a dense and homogeneous coating microstructure with different melting degrees. It can be noted that several nozzles of varying geometry and length were investigated; however, this paper narrows the discussion to three nozzles. **Table 1** summarizes the optimized process parameters. The feedstock powders are injected radially using two feeding lines. A relative gun velocity of 1.4 m/s was used in all cases.

The deposition efficiency (DE) has been calculated as a ratio of the mass of the coating deposited on a flat mild steel substrate over the amount of feedstock material fed into the system during time spend to deposit the coating. The mass of the coating has been calculated by weighting the sample before and after the coating deposition. Taking sample A3 as a reference for WC-10Co-4Cr coating qualities (i.e. microstructure and hardness as reported in the following section) typically achieved with conventional HVOF systems [1], it can be stated that a decent DE is achieved with the investigated system (~55%).

Indentation fracture toughness was also assessed by performing indentations into the polished cross-sections of the coatings using a Vickers-shaped indenter. A peak load of 5 kgf (≈49 N, load period 10 s) was employed in order to induce crack propagation from the indent corners. Fracture toughness was determined by measuring crack lengths on optical micrographs (200× magnification). Only cracks parallel to the substrate starting at the left and right tip of indents, respectively, were considered according to the method described in [12] , **Fig. 2**. The critical stress intensity (K_{IC}) was calculated according to Shetty et al. based on the Palmqvist crack model:

Table 1. Optimized process parameters

Ref.	Feedstock	Natural gas [NLPM]	O_2 [NLPM]	Air [NLPM]	Carrier gas [NLPM]	Feeding rate [g/min]	Standoff distance [mm]	Nozzle length [mm]	DE [%]
A1	Amperit 554.074	250	500	1000	70	~64	200	70	~22
A2	Amperit 554.074	250	500	1000	70	~64	200	100	~36
A3	Amperit 554.074	250	500	1000	70	~64	200	150	~55
A8	Amperit 526.074	250	500	1000	70	~60	200	70	~44
A4	Amperit 526.074	250	500	1000	70	~60	200	100	~50
A5	Amperit 526.074	250	500	1000	70	~60	200	150	~50
T1	Thermico WCCoCr	250	500	1000	70	70-75	150	70	~68
T2	Thermico WCCoCr	250	500	1000	70	70-75	150	100	~70

In general, a sharp decrease in DE down to 22% is noticed when using shorter gun nozzles. This is explained by the reduced dwell time within the system and thus by the reduced thermal transfer to the spray particles, which results in lower melting degree of the metal binder. However, under similar process conditions, the deposition efficiency of the fine spray powder is quite higher (~68-70%), without noticeable variations as a function of the gun nozzle length under the investigated process parameters. For the WC-17Co (coarse) powder, also a moderate decrease in DE is observed when moving from the 100 mm to 70 mm long nozzle.

3.3 Coating characterization

X-ray diffraction (XRD) measurements were carried out on a Bruker D8 Advance diffractometer operating at 40 kV and 40 mA using a Cu radiation (Kα= 1,54184 Å). The integral intensities of the (100) peak of WC (I_{WC}) and of the (101) peak of W_2C (I_{W2C}) were assessed by pattern fitting using pseudo-Voigt functions and the index of carbide retention (I), quantifying the retention of the WC phase against the formation of W_2C, was computed as $I = I_{WC} / (I_{WC} + I_{W2C})$, according to the definition in [10].

Coatings were cut, mounted, ground, and polished and cross sections by were examined by optical microscopy (OM) and scanning electron microscopy (SEM: JSM-5010LV, JEOL) equipped with energy-dispersive X-ray (EDX) microanalysis (D8 Advance, Bruker). Coating microhardness (Struers Duramin A-300 micrometer) was evaluated by Vickers indentation using a test load of 300g and a dwell time of 15 s, following an indentation pattern that has been optimized in previous works [11]. The reported hardness values are the average of 10 measurements performed on the cross section of the coating.

$$K_{IC} = 0.0889(HW)^{\frac{1}{2}} \quad (1)$$

Where H is the hardness and W is the crack resistance parameter (also quoted as Palmqvist toughness). W is calculated based on the following equation:

$$W = \frac{(P - P_0)}{4\bar{a}} \quad (2)$$

Where $\bar{a}$ is the mean radial crack length, P is the indentation load and P_0 is the threshold indentation load for inducing a crack.

Fig. 2. Microhardness indentation pattern, where a: 100 µm and b: 50 µm.

3.4 Wear testing

The sliding wear and friction performance of the coatings was evaluated under dry and lubricated (with P3 oil) conditions in a pin-on-disk test (MicroTest MT 60/NI/HT) using WC balls (diameter 6 mm) as counterpart. Prior to the test, the surface coating was grinded (Ra~ 0,2 – 0,3μm). In order to study the wear differences on ceramic coatings, the test was performed for dry conditions using a load of 60 N with linear speed of 50 mm/s for 3 h. Likewise, to study the variations in the friction coefficient, the test was performed for dry and lubricated conditions using a load of 10 N with a linear speed of 20 mm/s for 8 hours. The friction coefficient was recorded on-line by the instrument. The coating wear rate was measured by contact profilometry on the wear track (Dektak 150).

4 Results and discussion

4.1 Microstructural characterization

Representative cross sectional micrographs of the coatings produced under conditions summarized in Table 1 are showed in **Fig.3** - **Fig. 5**. Coating produced with the conventional WC10Co4Cr powder (Amperit 554) at conditions A1, A2 and A3 are shown in **Fig. 3.** They exhibit macroscopically dense microstructures with very low porosity detectable at these magnifications. At greater magnifications, some pores (producing black contrast in backscattered electron imaging mode) become visible (**Fig. 6**,a,b). Even in the case of coating A3, where the feedstock material is melted to a greater extent (**Fig. 6**b), still some microporosity mainly related to the original internal porosity of the spray particles is evident.

Coatings produced with the WC-17Co composition at conditions A4, A5 and A6 (**Fig. 4**) feature also macroscopically dense microstructures with low porosity detectable at these magnifications. In the coating produced with the longest gun nozzle (A5), the spray particles are melted to a greater extend, as evidenced by the metal binder differenced from the hardmetal phase on some isolated locations. Detailed inspection of this sample (**Fig. 6c**), revealed the existence of a bright phase in the matrix which may correspond to η-phasesM_6C (e.g. Co_3W_3C) according to the EDX analysis.

Representative cross sectional micrographs of the coatings produced with the fine WC10Co4Cr powder at conditions T1 and T2 are shown in **Fig. 5.** Some residual porosity can be recognized at this magnification for both coatings produced under these conditions. The observed micropores are, however, more uniformly distributed in comparison to the coatings produced with the coarse WC10Co4Cr powder (**Fig. 3**). Nevertheless, at higher magnifications, when comparing the WC10Co4Cr coatings produced with gun nozzles having either 70 or 100 mm in length, coatings produced with fine powder (T1 / T2, **Fig. 6**d) exhibit a similar structure of those produced with the coarse powder (A1 / A2, **Fig. 6**a), but with finer carbides and interlamellar volume defects (pores) as it could be expected. It is well known that during flight spray particles melt to different extent depending on their size and dwell times.

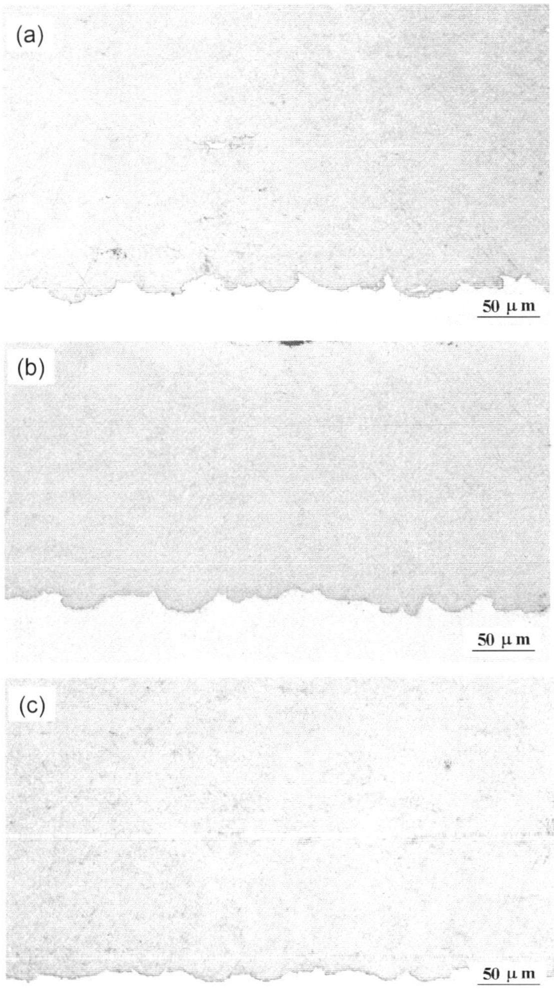

Fig. 3. Cross sectional micrograph of WC10Co4Cr (Amperit 554) coatings produced at different process conditions: a) sample ref. A1, b) A2 and c) A3.

The carbon loss and typical metallurgical reactions observed in HVOF sprayed WC-Co(Cr) coatings are perceivable through XRD patterns (**Fig. 7**). For the WCCoCr composition, WC, the η-phase $(Co,Cr)_3W_3C$, and a (Co,Cr,W) alloy binder are expected to be in thermodynamic equilibrium. Thus, chromium is contained both in the metallic binder and the hard phase. Basically the chromium addition intensifies the effect of *η*-phase formation compared to plain WC–Co [1]. For the coating produced with the coarse WC10Co4Cr powder, the XRD pattern evidences the formation of W_2C when increasing the nozzle length (comparison of A1, A2 and A3). Assuming the XRD technique has a detection threshold of about 2%, the index of WC retention is ≥98% for those coatings produced with gun nozzle having 70 or 100 mm in length (**Table 2**). The coating produced with 150 mm long nozzle undergoes a higher decarburization as result of the higher particle temperature reached inside the nozzle. Nevertheless, the resulting Index of WC retention is moderate in comparison to values reported for WC10Co4Cr coatings produced with conventional gas-/liquid-fuelled HVOF systems [6].

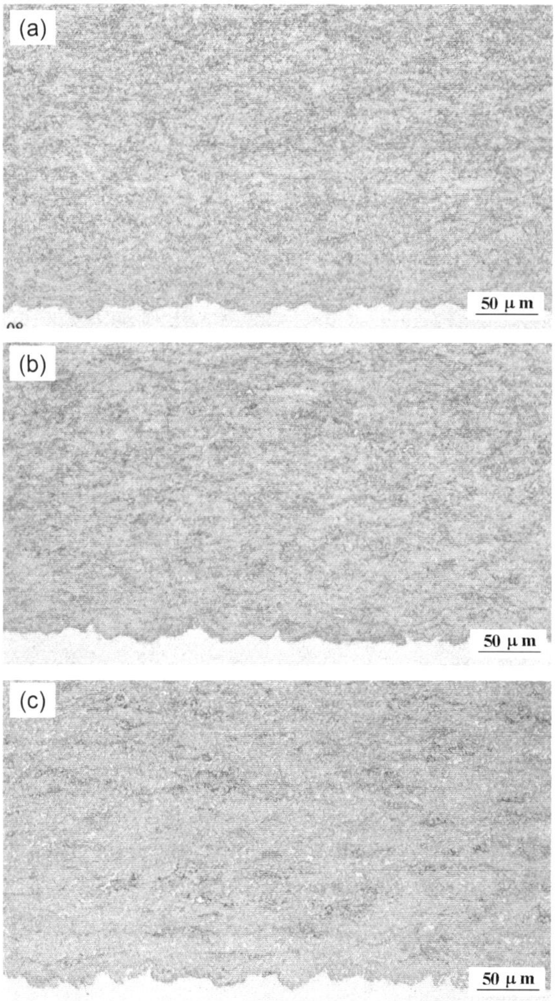

Fig. 4. Cross sectional micrograph of WC17Co (Amperit 526) coatings produced at different process conditions: a) Sample reference A8, b) A4 and c) A5.

Those coatings produced with the conventional WC-17Co powder (bigger carbide sizes), exhibit a high retention of the WC phase. The low value calculated for the coating produced with the 100 mm long nozzle (A4), can be explained by the difficulty in identifying the main W_2C reflection peak (101), which is quite close to the second peak from the M_6C phase. The XRD pattern evidences the formation of amorphous and/or nanostructured phases around the main peak of the M_6C phase in all these coatings, which can be attributed to the higher binder content of this material. It is well known that carbides that have been dissolved in the binder during spraying often remain in significant amounts in the binder phase of the coatings. Due to the rapid solidification rate upon impact, the binder phase in the coatings is typically in an amorphous or nanocrystalline state giving rise to broad, diffuse peaks in XRD patterns [1].

In the case of the WC10Co4Cr coatings produced with the fine spheroidized powder (T1 and T2), the material undergoes a remarkably higher decarburization in comparison to the other coatings, exhibiting an index of WC retention of around 90 and 78% for samples T1 and T2, respectively. Dissolution of WC into the matrix of powders with such sub-micron sized carbides has been proved to be more sensitive to spraying parameters than conventional powders (WC size 1-3 ~μm) [13]. Additionally, on small size particles, the whole particle can be affected from melting and decarburization, making the control of the particle temperatures an issue.

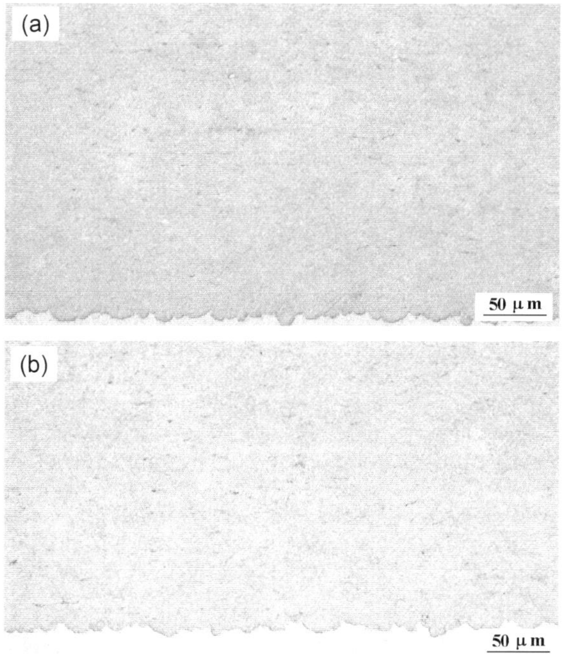

Fig. 5. Cross sectional micrograph of WC10Co4Cr (Thermico) coatings produced at different process conditions: a) Sample reference T1 and b) T2.

Table 2. Measured properties for WC-Co(Cr) coatings

Ref.	Coating material	Index of WC retention, %	Hardness HV 0,3
A1	WC10Co4Cr	>98	1790 ± 93
A2	WC10Co4Cr	>98	1728 ± 148
A3	WC10Co4Cr	>95	1309 ± 181
A8	WC-17Co	>98	1185 ± 82
A4	WC-17Co	>96	1250 ± 119
A5	WC-17Co	>98	1431 ± 119
T1	WC10Co4Cr	>90	1674 ± 99
T2	WC10Co4Cr	>78	1653 ± 93

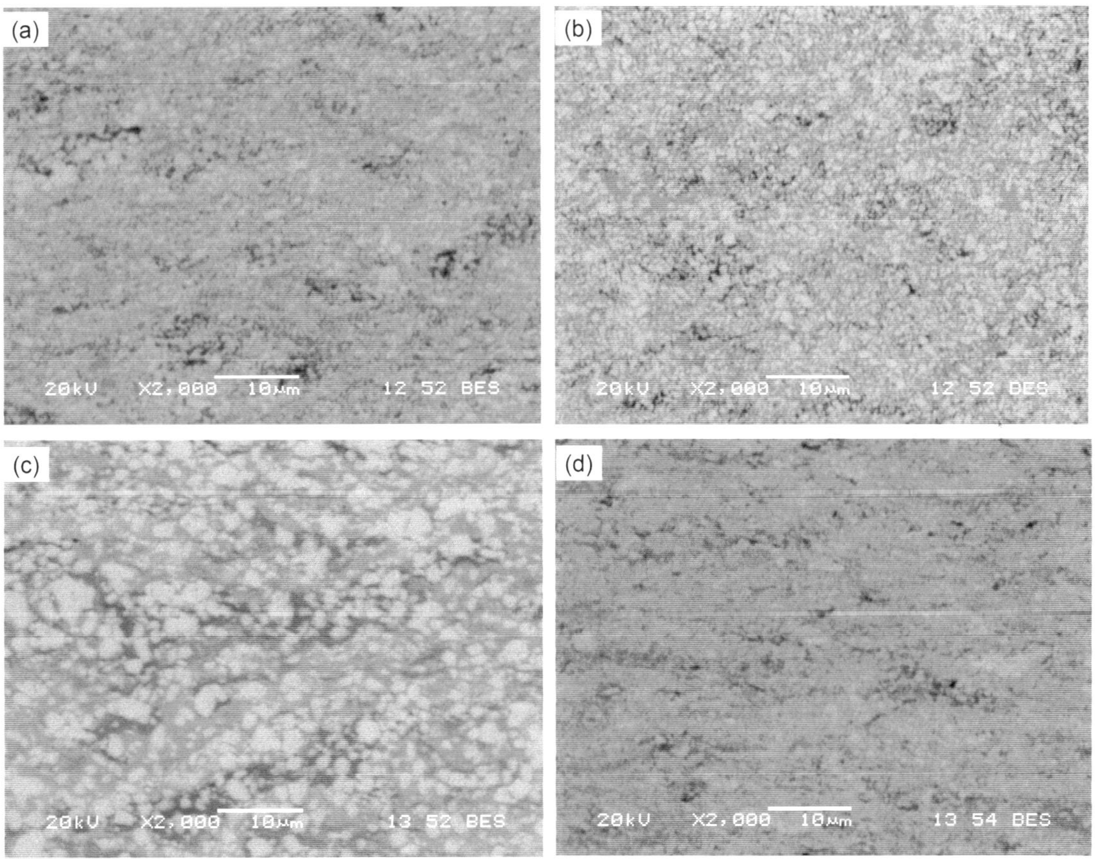

Fig. 6. Representative SEM micrograph of WC-Co(Cr) coatings produced with different process conditions: a) Sample reference A1, b) A3, A5 and b) T2.

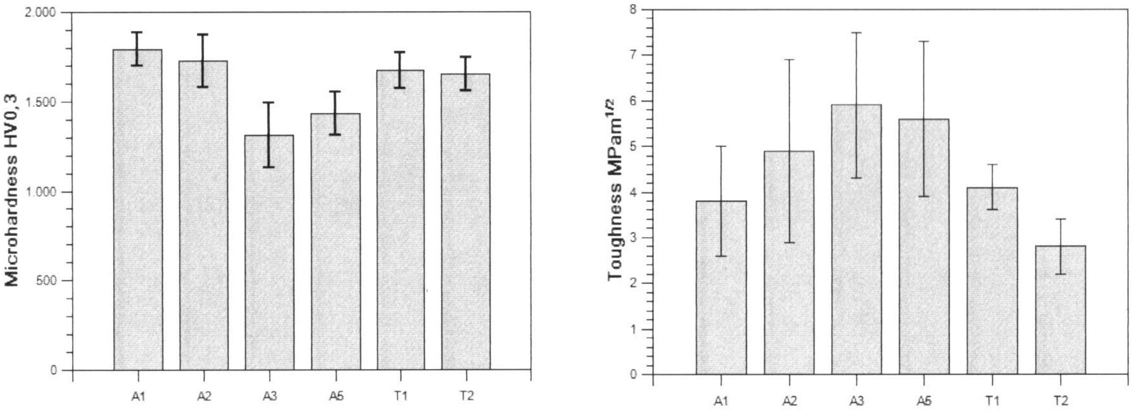

(a) (b)

Fig. 8. (a) Vickers microhardness and (b) Indentation fracture toughness (K_{IC}) values of selected coatings.

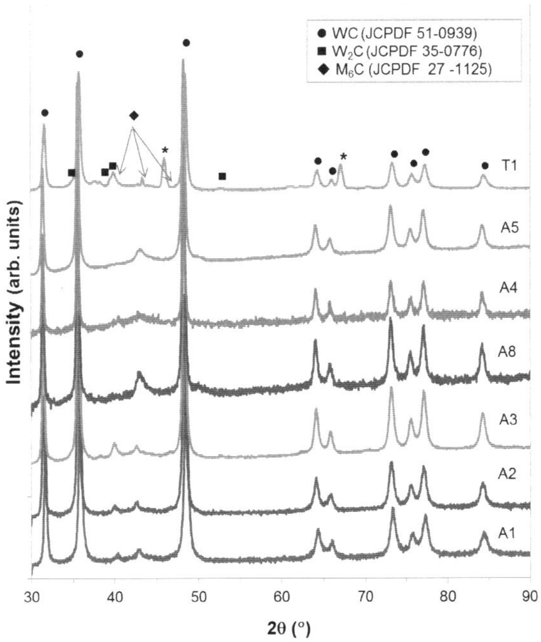

Fig. 7. XRD patterns of all WC-Co(Cr) coatings (* indicates some contamination coming from a protective Al_2O_3 layer and an organic sealer applied to protect sample for metallographic preparation).

4.2 Mechanical properties

For comparison purposes, the microhardness values of selected coatings are plotted in parallel to the calculated indentation fracture toughness (K_{IC}) values in **Fig. 8**. Even if having comparable mean microhardness values, the WC10Co4Cr coatings produced with the conventionally sized powder using the 70 and 100 long nozzles shows a clear influence of the gun nozzle length on the resulting coating toughness. In general, it can be noticed that the indentation facture toughness of these coatings increases with the increase of the gun nozzle length, what can be attributed to the higher melting degree of the binder matrix and thus, to an apparently higher interlamellar cohesion (deformed and re-solidified spray particles upon impact on the substrate).

In previous studied it was found that the cracks initiating from the edge of the Vickers indenter tip follows the path of lamella boundaries. According to Varis et al. [3], the increased particle temperature improves the lamella bonding and makes more difficult for crack to grow until the toughness of the lamella boundary region has decreased excessively due to carbide dissolution effects. Fracture toughness, which is measured from the cross section of the coating, was found to be therefore connected to the properties of the lamella interfaces.

It explains the comparably lower fracture toughness observed by the coatings produced with the colder parameters (short barrel lengths 70 and 100 mm) in comparison to the more melted coatings, deposited with the longer barrel of 150 mm.

4.3 Wear testing

Collected results (wear rates and Coefficient of Friction, CoF) are plotter in **Fig. 9**. At room temperature, the dry sliding wear behaviour of WC10Co4Cr coatings has been reported to be dominated by two concurrent mechanisms: ductile flow, which causes the formation of "wavy" lips with protruding WC grains, and brittle fracture, which causes direct detachment of small portions of material from the coating surface. In principle, the relative incidence of the two mechanisms is mainly controlled by intra-lamellar features, such as decarburisation. This is probably related to the small size of the contacting asperities, which matches with the intra-lamellar length scale [6]. This can explain the higher wear rates measured on the coating produced with the fine WCCoCr powder and the WC17Co powder at a test load of 60N.

A similar explanation could be employed to explain the tendency observed among the WCCoCr coatings produced with the conventional powder (A1, A2, A3). By the coating produced with the shortest gun nozzles (A1), a lower intra-lamellar strength could be attributed to the lower melting degree of the binder matrix. On the other side, by the coating produced with the longest gun nozzle (A3), the decarburization could be negatively affecting the coating performance. Conditions reached with the intermediate gun nozzle length (A2) might be a good compromise to obtain a good intra-lamellar cohesion strength while keeping a low decarburization degree. With mean values roughly around 0,5, the friction coefficients are mild for all coatings (**Fig. 9**b). For comparison, **Fig. 10** shows the plots of the recorded friction coefficients under dry (at 60 N) and lubricated (at 10 N) conditions.

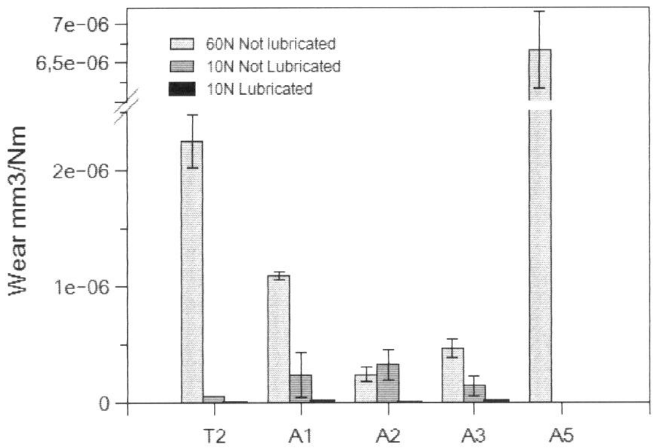

(a)

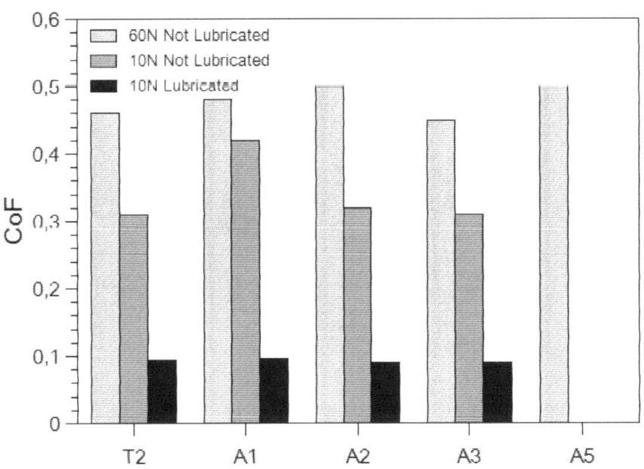

(b)

Fig. 9. (a) Wear rates and (b) friction coefficients obtained by ball-on-disk under dry and lubricated sliding wear tests at room temperature (RT).

At a test load of 10 N, the sliding wear rates of all WC–10Co4Cr coatings (**Fig. 9**a) are below 3E−07 mm^3/(Nm), thus corresponding to a mild wear regime [6]. The selected WC-17Co coating (A5) was not tested under this condition. Due to the large scattering of the results, no clear tendencies could be found among the tested specimens. It is however noticeable, that the wear rate of the coating produced with the fine powder is almost negligible under selected test conditions.

Under lubricated conditions at a test load of 10 N, the profile of the wear track could not be detected on any of the samples. The mean values of the friction coefficients are closely around 0,91 and 0,96 for all tested WCCoCr coatings (**Fig. 9**b). Nevertheless, some minor deviations in the plots of the recorded values over the test durations are observed, **Fig.10**. Further characterization of the wear tracks and wear debris is required to better understand the differences imposed by the inherent coating properties.

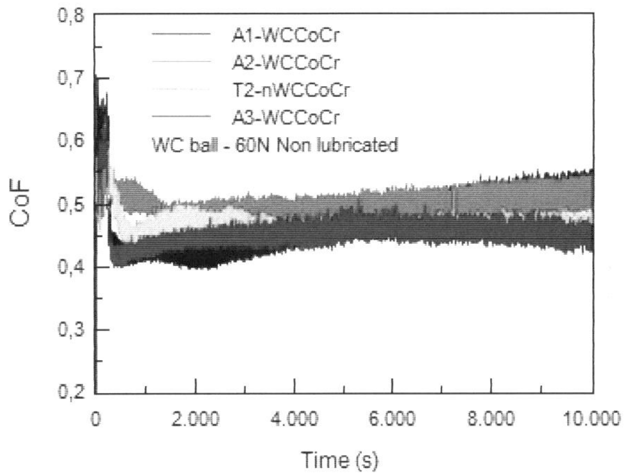

(a)

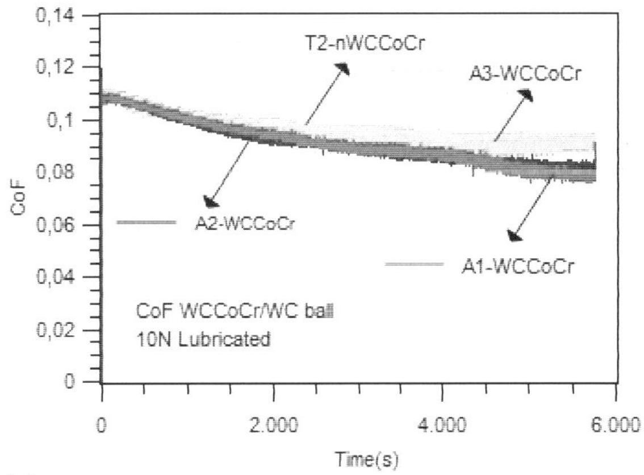

(b)

Fig. 10. Friction coefficients obtained by ball-on-disk under dry (a) and lubricated (b) sliding wear tests.

5 Conclusions

Superhard cermet WC10Co4Cr and WC17Co coatings have been deposited with the new High Velocity Oxy/Air-Fuel based system developed by Tecnalia. For comparison, a conventional (15-45 μm) as well as a new fine WC10Co4Cr powder (particle sizes between 1-12 μm with sub-micron WC ~400-800 nm) were investigated. The influence of the gun nozzle length on ascertained coating properties like the microstructure, phase composition, microhardness and toughness has been explored, allowing the selection of the most suitable gun configuration for each feedstock material. Three different conical expansion nozzles with 70, 100 and 150 mm in length were investigated. Collected results led to the conclusion that for the processing of conventional (coarse) WC-17Co feedstock powders, the longest gun nozzle (150 mm) is the most appropriate to achieve a hard (~1431 HV0,3) and dense coating with negligible decomposition of the tungsten carbides into undesirable phases like W_2C. Relatively tough (~5,6 MPa*m$^{1/2}$) coatings are achieved with this material.

For conventionally sized WC-CoCr powders, the best approach to achieve a sufficient melting degree of the binder matrix, while limiting the decarburization of the coating, is provided by the 100 mm long nozzle, thus leading to the best combination of microhardness (~1728 HV0,3), indentation fracture toughness (~ 5 MPa*m$^{1/2}$) and sliding wear resistance (under dry conditions at 10N load ~3E-07 mm^3/Nm).

When using the fine powder of the same composition (WC10Co4Cr), the spray particles are evidently more sensitive to decarburization. Coatings with a relatively high retention index of WC (>90%), even high microhardness (~1674 HV0,3) and moderate toughness (~ 4 MPa*m$^{1/2}$) are achieved with the 70 mm long nozzle. With a longer gun nozzle (100 mm), ascertained properties like the microhardness are not significantly affected, while decarburization mechanisms seem to be accelerated as result of the prolonged dwell time within the gun. The sliding wear performance of resulting sub-micron structured WC10Co4Cr coatings overpass the performance of coatings produced with conventional coarse powders at low test loads (10 N). Nevertheless, the wear performance under high loads (60 N) seems to be compromised by insufficient intralamellar particle cohesion and/or the formation of brittle phases.

It should be noticed that further research works are required to fully explore the potential of this technology for the thermal spray processing of fine/sub-micron structured powders. On the other side, further investigation of the functional performance of the developed coatings under sliding and abrasive wear conditions will be required to better understand the link to the mechanical properties and key microstructural features of the coatings. These works are in progress and will be reported in the future.

6 Acknowledgements

The authors would like to thanks Dr. José Angel Sánchez García from the *Materials for Energy and Environment Department* (Energy Division, TECNALIA) for his contribution to the wear characterization of the coatings and the analysis of results.

7 Literature

[1] L.-M. Berger. Application of hardmetals as thermal spray coatings. Int. Journal of Refractory Metals and Hard Materials 49 (2015), pp. 350–364.

[2] Chang-Jiu, L. and G.-J. Yang: Relationships between feedstock structure, particle parameter, coating deposition, microstructure and properties for thermally sprayed conventional and nanostructured WC–Co. Int. Journal of Refractory Metals and Hard Materials, 39 (2013), pp. 2-17.

[3] T. Varis et al. Formation Mechanisms, Structure, and Properties of HVOF-Sprayed WC-CoCr Coatings: An Approach toward Process Maps. J. Therm. Spray Technol. Volume 23(6) August 2014, pp.1009 - 1018.

[4] G. Matthaeus, W. Brandl, I. F. Secosan, G. Marginean, V. Chirila, and D. Ackermann, Standard HVOF Process Compared to the HVOF Process for Internal Coating with Fine Powders, Thermal Spray 2008: Crossing Borders, on CD-ROM, E. Lugscheider, Ed., May 2-4, 2008 (Maastricht), DVS Media, 2008, pp. 485-488.

[5] W. Tillmann et al. Influence of Fine Powder Feedstock (210 + 2 μm) on the HVOF Spraying Characteristics, Coating Morphology, and Properties of WC-CoCr 86-10-4. Journal of Thermal Spray Technology, Volume 22 (2-3) March 2013, pp. 242–249.

[6] G. Bolelli et al. Tribology of HVOF- and HVAF-sprayed WC–10Co4Cr hardmetal coatings: A comparative assessment. Surface & Coatings Technology 265 (2015) 125–144.

[7] A. Verstak and V. Baranovski: AC-HVAF Sprayed Tungsten Carbide: Properties and Applications. Building on 100 Years of Success: Proceedings of the 2006 International Thermal Spray Conference, B.R. Marple, M.M. Hyland, Y.C. Lau, R.S. Lima and J. Voyer, Eds., May 15-18, 2006 (Seattle, WA, USA), ASM International, Materials Park, OH, 2006.[

[8] V. Wesling, R. Reiter, L. Lau and G. Matthäus. nanoHVOF – Process technology for and properties of a new high density protective coating. Mat.-wiss. u. Werkstofftech. 2014, 45, No. 6. (In german)

[9] M. Parco, G. Barykin and I. Fagoaga. HVOF spraying of metal alloys with Natural Gas and Oxygen-Air mixtures. Thermal Spay Bulletin 7 (2014) No.1. Specialist article.

[10] C. Bartuli, T. Valente, F. Cipri, E. Bemporad and M. Tului. Parametric Study of an HVOF Process for the Deposition of Nanostructured WC-Co Coatings. J. Therm. Spray Technol. 14 (2) (2005) 187–195.

[11] G. Barykin and M. Parco. The Oxy-Fuel Ionisation (OFI) Spray Process. Thermal Spray 2007: Global Coating Solutions. B. R. Marple, M. M. Hyland, Y. -C. Lau, C. -J. Li, R. S. Lima and G. Montavon, Eds., May 14-16, 2007 (Beijing,

China), ASM International, Materials Park, OH, 2007.
[12] M.M. Lima et al. Toughness evaluation of HVOF WC /Co coatings using non-linear regression analysis. Materials Science and Engineering A357 (2003), pp. 337-345.
[13] A. Ghabchi et al. Behavior of HVOF WC-10Co4Cr Coatings with Different Carbide Size in Fine and coarse Particle Abrasion. J. Therm. Spray Technol. Volume 19 (1-2) January 2010, pp. 368 – 377.

An in-situ cooling solution for thermal spray coating process

Feng Qin*, Tim Bao, and Liang He
Air Products and Chemicals, Inc., Shanghai 201203, China
*Email: qinf1@airproducts.com

Thermal spray is widely used to deposit protective coatings on a metal surface, by directly spraying melted or heated metal precursors. The coating materials are often heated by plasma or oxy-fuel combustion means. During spraying, the substrate temperature must be cooled down and well controlled to avoid coating quality issues related to overheating. A cryogenic nitrogen gas-based cooling solution has been developed to cool the part simultaneously while it is being spray coated, and therefore improves productivity by eliminating the inter-pass cooling breaks (wait for the parts to cool down) associated with traditional air cooling methods.

The system does not only cool but also controls the work piece temperature within a specified range, with infrared sensors and a close-loop control scheme. The real-case experiments demonstrated thermal spray cooling resulted in 50% shorter process time, improved deposition efficiency, much tighter temperature control (±20F), smaller micro-porosity density (0.05%), higher substrate hardness, less coating oxidation or carbon loss, and minimized residual stresses in coatings.

1 Background

The metal industry constantly demands for high precision, high hardness, high wear-resistance, and high corrosion-resistance surface coatings for materials and parts used in automotive, aerospace, military defence, construction and mechanical applications. While the conventional surface treatment processes have their own limitations, thermal spraying coating process has become a more and more popular alternative. High performance materials like WC-CO, WC-CoCr, Stellite, even ceramics can be coated onto conventional materials by thermal spraying process, which can be significantly improved surface hardness, wear resistance, corrosion resistance, insulation property, or combination of these properties. Thermal spraying coating is especially useful for repair and restoration of wearable parts.

However, several limitations are found with the thermal spraying process: temperature control and thermal management, low productivity, and cost competitiveness.

Improper temperature control during thermal deposition frequently leads to coating and workpiece overheating, thermal deformation and degradation of substrate material, and damaging stresses due to a mismatch of thermal expansion coefficients between the coating and substrate. When damage occurs through overheating and thermal stress, delamination of the coatings can occur in service.

An effective cooling method must be applied to avoid overheating. Among cooling techniques, compressed-air and liquid CO_2 cooling methods are not fast enough to remove the generated heat from thermal spray. This causes the inter-pass "cooling breaks", which result in productivity losses and waste in feed powder and fuel or energy. Consequently, it is often difficult for a thermal spraying production line to achieve full capacity.

In addition, CO_2 and air are prone to oxidation metallic coating surfaces, which would decrease deposition efficiency of the sprayed powders on to the substrate.
Low coating rate with low productivity compromises the cost competitive advantage of thermal spray compared to traditional protection methods such as hard Chromium plating. Therefore, a cryogenic nitrogen gas-based cooling solution has been developed to mitigate the limitations of current thermal spraying processing.

2 Introduction to TSC system

2.1 Description of the system

The thermal spray cooling (TSC) system is an in-situ cooling solution to maintain part temperature during thermal spray coating operations using liquid nitrogen (LIN) and cold nitrogen gas (GAN) as the refrigerants. This makes high-quality thermal spray coatings possible by maintaining part temperature within a predefined range, even for heat-intensive spraying processes. It can enable users to apply coatings faster and at a lower cost than traditional cooling methods. A complete TSC system consist of the following components shown in Fig. 1,

- Liquid nitrogen storage tanks are typically stored outside the facility and delivered to the TSC control cabinet in the form of liquid nitrogen (LIN) and nitrogen gas (GAN).
- Control cabinet contains a regulating valve that controls the liquid and gas mix, based on the temperature feedback from a IR sensor pointing at the work piece.
- Cooling nozzle to direct the nitrogen coolant onto the work-piece.
- Ventilation and oxygen monitoring devices.

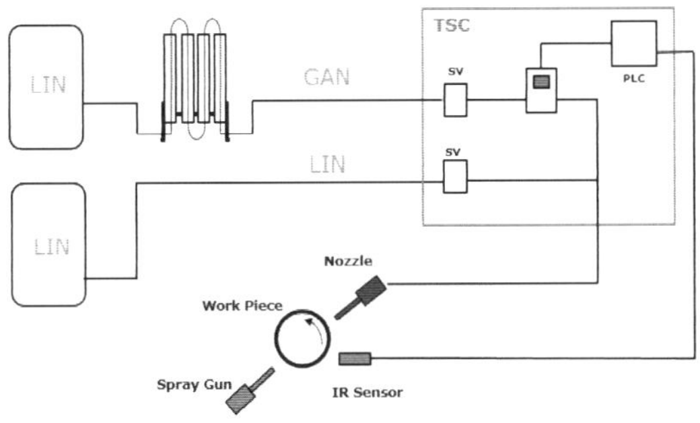

Fig. 1. Schematics of TSC system

2.2 Advantages of the system

The TSC system is develop to work together with any thermal spray equipment of any thermal source, regardless of high velocity oxy-fuel (HVOF) or plasma. It is easy and flexible to mount onto a robot arm, and it can also be combined with the existing cooling system to further improve the cooling efficiency.

A recent study [1] showed that the TSC system not only enhances the cooling capability, but also improves the deposition efficiency (DE). As shown in Fig. 2, the LIN-GAN-Air combination coolant can offer higher cooling efficiency and lower surface temperature, compared to the traditional air cooling or LCO2-Air cooling. The LIN-GAN temperature was pre-set at 310C in Fig. 1 and actually one can set it even lower. However, the LCO2 and compressed air temperature was the result of the maximum cooling capacity. For LIN-GAN cooling, the faster cooling speed not only maintains a lower surface temperature, but also generates a more uniform and stable temperature profile with a minimal temperature fluctuation.

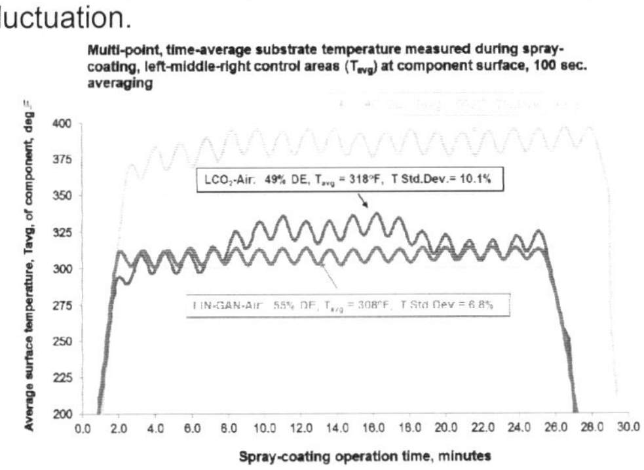

Fig. 2 Substrate surface temperature trend during a spray coating test [1]

In the meantime, LIN-GAN cooled coatings are less oxidized and less decarburized than coatings cooled by LCO2 or compressed air. With nitrogen as inert atmosphere protection, carbon loss on the coating surface is decreased. Fig. 3 shows that the TSC process with LIN-GAN helps maximize the content of the desired, tough, and hard WC phase. Another benefit of TSC is to minimize metallic oxidation, which increases the powder deposition efficiency and also reduces the coating porosity levels, by comparing the three cooling media.

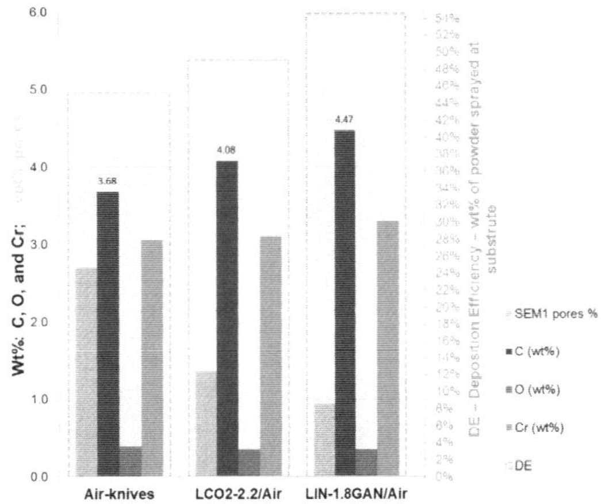

Fig. 3. Effect of cooling method on powder DE (deposition efficiency), coating composition (C-carbon, O-oxygen, and Cr-chromium), and cross-section porosity.

Air Products [2] reported that LIN-GAN cooling solution had been successfully applied in combination with high spray rate guns to deposit metallic and ceramic coatings (tungsten carbide chrome cobalt, WC-CoCr) powder to repair surfaces of aircraft landing gears in the aerospace industry.

The TSC cooling solution is also proven beneficial for many different coating materials. See Table 1 for examples on hard metal coatings. Table 2 lists major benefits for ceramic coatings with TSC cooling.

The TSC technique helps increase thermal spray productivity and brings savings in consumables from improved cooling efficiency, the inert nitrogen atmosphere can also minimize the oxidation and improve the coating quality. Table 3 shows the coating quality improvement from TSC applications.

Table 1 Benefits of TSC on hard metal coatings

Major Benefit	Process saving	Applicability
Elimination of intercoolin g Breaks	- 30 – 60% savings in coating deposition time - 15 – 60% powder savings - 30 – 60% process gas savings (H , O , N)	HVOF and high-spray rate plasma systems
Increased deposition efficiency	10 – 30% savings in coating deposition time, powder and process gases	HVOF and high-spray rate plasma systems
Rapid, Flexible, Masking	Savings in setup time and masking material –use of silicon-rubber masking materials possible with LIN cooling	All parts that require masking regardless of the spraying process used

Table 2 Benefits of TSC on ceramic coatings

Majo Benefit	Process saving	Applicability
Local cryo-jet cooling	- Direct intensive cooling behind the coating flame - close mount to coating flame - regulation of cooling intensity	HVOF and high-spray rate plasma systems
Elimination of intercooling breaks	- deposition time saving - powder saving - process gas savings (H2, O2, N2)	HVOF and high-spray rate plasma systems
Increased coating thickness	- Improved coated product quality - Increased product performance - Increased deposition efficiency	HVOF and high-spray rate plasma systems
System is quick to install and masking is flexible	Savings in setup time and masking material –use of siliconrubber masking materials possible with LIN cooling	All parts that require masking regardless of the spraying process used

Table 3 Benefits of TSC to improve product quality

LIN Cooling Benefit	Quality Enhancement	Applicability
Preservation of Substrate Properties	Heat-treated and/or shot peened parts do not soften during spray coating due to elimination of overheating	All aerospace parts and the parts with alternate/variable loading
Minimized Coating Oxidation	Nitrogen inerting reduces coating oxidation, resulting in less brittle coatings	All metallic, carbide, and composite coatings except for oxide ceramic (e.g. TBC and HPA) coatings
Minimized Residual Stresses in Coatings	Reduction in stress gradient across the coating/substrate interface, resulting in coatings that are more resistant to flexing, without delamination or flaking	All coatings characterized by lower coefficient of thermal expansion than substrate, e.g.: WC-Co on steel or Al, TBC on any substrate, etc.

3 Conclusion

The LIN/GAN-based cooling solution has been demonstrated to be an effective technique to control the substrate temperature for thermal spraying coating processing. As a result, productivity, coating quality, material cost are dramatically improved, which alleviates many limitations associated with traditional cooling methods. Therefore, without the limitation of cooling issues, thermal spray could be more widely used for all kinds of industry applications, including aerospace, aircraft, defense, automotive, and paper making. The LIN/GAN-based cooling solution can be also used in welding industry and the emerging industry 3D-printing. Table 4 lists the industry segments which TSC can be applied.

Table 4 Areas of applications for TSC cooling

Industry	Coating targeted properties	Examples of parts coated
Aerospace/ Aircraft	Thermal Barrier, wear, oxidation & corrosion resistant, clearance control	Combustion Chamber High Pressure Nozzles, Leading gears, compressor Stators, seals & blades, bearing parts
Heavy machinery & Defence	Wear & corrosion resistant, corrosion protection	Bolts, rods, polis, Propeller Shaft Bearing, Rolls and other mills parts
Power Generation	Wear, oxidation & corrosion resistant, oxygen diffusion (fuel cell)	Gas turbine parts, Hydrofoils, injection nozzles, Sputting targets, SOFC
Oil and Gas	Wear & corrosion resistant, corrosion protection, oxidation	Valves and its components: e.g.gate valves, balls, etc
Automotive	Wear & corrosion resistant Oxygen sensing, electrically insulating, electrically conductive, EMI/RFI shielding	Brake discs, fuel injection nozzles, turbo charger parts, cylinders,Crankshafts, Exhaust,Mufflers,reinforced polymer
Paper Making	Wear & corrosion resistant release(non-stick), traction, coating metering	Rolls, bearing parts

4 Literature

[1] Z. Zurecki, Air Products and Chemicals, Inc,. Influence of Substrate Cooling Method on Compositional Shift and Deposition Efficiency of HVOF-sprayed WC-Co Type Coatings. ITSC 2012.

[2] Thomas Mebrahtu, Air Products and Chemicals, Inc,. Cooling Hot Coatings with Nitrogen. Finishing Today magazine, March 2008, 330-08-007-US.

Performance and economic characteristics of multi-chamber detonation sprayer used in thermal spray technology

Yu. Tyurin, O. Kolisnichenko, Kyiv/UA, Jeff Jia, Beijing/CN, N. Vasilik, Moscow/RU, M. Kovaleva, M. Prozorova, M. Arseenko, V. Sirota, Belgorod/RU

Distinctive feature of the Multi-chamber detonation sprayer (MCDS) is an availability of several chambers where there is combustion of gas mixtures with frequency up to 30 Hz. The pulsed flow of the combustion products acquires a high velocity (up to 2100 m/s) and pressure (up to 3.2 MPa).Powder is accelerated by using the combustion products, which are formed in the MCDS chambers and are converged before entering the nozzle, where they interact with the two-phase gas-powder cloud. These allowed to accelerate the powder sprayed to the velocity up to 1000 m/s. MCDS has replaceable barrels with the diameter from 16 to 20 mm and length from 300 to 500 mm. MCDS completed with a feeder The exact dosing and feed of powder into the barrel provide high deposition efficiency before the initiation of detonation. The deposition efficiency is 80-90% and 50-70%, respectively for cermet and ceramic powders. Flow rate of fuel mixture components are 0.6-1.6 m^3/h of propane and 3.0-8.0 m^3/h of oxygen with productivity up to 3 kg/h (metal powders), up to 2 kg/h (cermet powders) and up to 1 kg/h (ceramic powders).

1 Introduction

With the availability of a wide variety of methods for thermal spraying of coatings, equipment to implement them, workpieces for spraying and a wide range of spraying materials the choice of an economically sound coating method involves difficulties.

In practice, the wide use is made of such mass produced HVOF units as JP-5000/8000 (USA), DJ 2600/2700 (USA), Intelli-Jet (USA), HV 50 HVOF (The Netherlands), etc. Of interest among the reported new developments is a high-frequency detonation device intended for thermal spraying of coatings [1]. Torch TOPGUN® AIRJET [2], which is capable of spraying both powder and wire, is an example of successful upgrading of the HVOF unit. The most interesting device is Torch HVAF (Russia) [3], where the air flow rate is more than 300 m3/h and propylene flow rate is up to 16 m3/h. The torch is cooled by fuel mixture components, this increasing its efficiency.

Company "Sulzer Metco (US) Inc." conducted comprehensive analysis of the costs of coatings produced by using vacuum-plasma LVPS unit and HVOF units Diamond Jet-DJ9W and JP5000 [4]. The analysis showed that spraying one kilo of a coating costs $ 80-110.

The "cold" methods for spraying of coatings are being developed now. The principle of operation here is as follows: the flow of a heated gas via a very narrow nozzle is provided at the required rate. The equipment has a high power intensity of 40-80 kW and provides deposition of coatings of ductile metals.

The Pulsed-Cold Gas Dynamic Spraying Process is being developed to increase the efficiency of formation of the high-velocity gas jet [5].

Drawbacks of the known thermal spraying devices and technologies include the high power intensity and the use of gases under a pressure of above 4 MPa. This makes control of gases more difficult, increases requirements to safety and reduces efficiency.

2 Technology and equipment

In this connection, it is necessary to develop technologies and equipment characterized by simplicity in operation, low cost and high efficiency of utilization of materials and power. These devices should be versatile and provide high-quality coatings. The authors developed a new design of a multi-chamber detonation sprayer (MCDS) unit for deposition of coatings. The essential difference of the MCDS from the detonation one is that it sums up the energies of products of detonation combustion of fuel gas mixtures from several specially profiled detonation chambers. The rate and temperature of the combustion products depend only on the parameters of filling of each chamber with the fuel gas mixture.

The MCDS equipment (**Fig. 1**) consists: 1 - MCDS device; 2 – powder feeder; 3 – gas panel; 4 – technology control system; 5 – manipulators for displacement of the device; 6 – manipulators for workpiece.

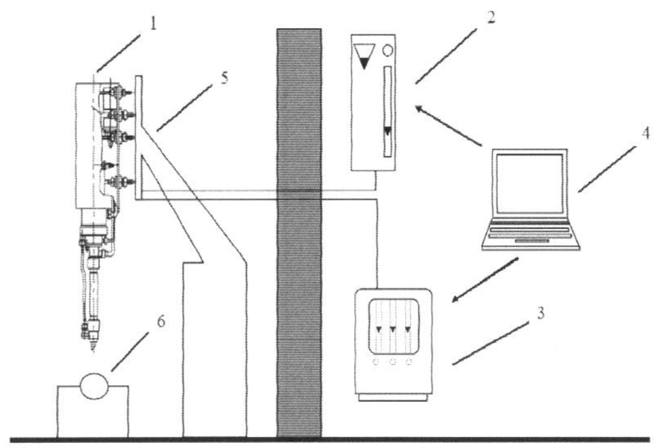

Fig. 1. Equipment for MCDS deposition of coatings.

The velocity parameters of the powder jet were estimated by using a two-channel optical measurement system.

The high combustion initiation frequency (20-30 HZ) and MCDS provide the possibility of implementing the quasi-continuous coating technology. This allows employing standard devices for feeding gases and powder under a low pressure.
The device consists (**Fig. 2**) of detonation chamber - 1, where the detonation combustion mode is implemented. Furthermore, the device comprises a ring detonation chamber - 2, which operates using combustible mixtures at any concentration. A cylindrical nozzle (diameter is 16 to 20 mm) - 3 is used for heating and accelerating powder materials. Also, MCDS has powder delivery unit - 4, automobile spark plugs - 5 for initiating the detonation regime of combustion, and the piping system - 6 for input component combustible gas mixture.
Low thermal power of the combustion products permits formation of coatings from a small distance (20-80 mm), this providing a substantial increase in spraying efficiency and decrease in oxidation and losses of a spraying material. Coatings can be deposited on small-size pieces with thin walls (0.3-1.0 mm), this widening the application field for the technology. The technology is of the pulsed type. The combustion products contact the walls of the combustion chambers for $1\text{-}2\cdot10^{-3}$ s, this decreasing the MCDS thermal intensity and cooling costs. In addition, the use of initial low-pressure gases (0.1-0.3 MPa) allows the control panels to be fitted with simple standard devices, thus reducing the price of the equipment and, what is most important, improving the operational safety. The use of the low-pressure process gases considerably increases the efficiency of utilization of gas bottle systems.
The Investigations of the velocity and amplitude values of shock waves in MCDS confirmed the possibility of achieving the high velocities of the gas flow. The investigations showed that in the stoichiometric propane + butane -oxygen-air mixture at an initial pressure of 0.1 MPa the average measured values of the velocity of a detonation wave in the nozzle were 2300 ± 100 m/s at a pressure of 3.0 ± 0.15 MPa.
Radiation from the powder jet was fed to fast-response germanium photodiodes FD287 via fiber-optic cables, which were placed at a preset distance from each other. Spectral characteristics of the photodiodes allowed minimizing the effect of the detonation products on integrated radiation of the gas-powder jet. Therefore, the velocity parameters of a heated powder were determined from the time shift of the radiation spikes.
As shown experimentally, the velocity of the powder jet depends on the characteristics of the powder, filling of the cumulative-detonation chamber with the fuel gas mixture and its composition. For example, maximal value of the velocity of the powder Al_2O_3 AMPERIT® 740.0 near nozzle exit section may range to 1200 ± 200 m/s.
MCDS device can be used horizontally, vertically as well as at an angle to the sprayed surface for deposition of the coating on the inner surface of the product (**Fig. 3**).

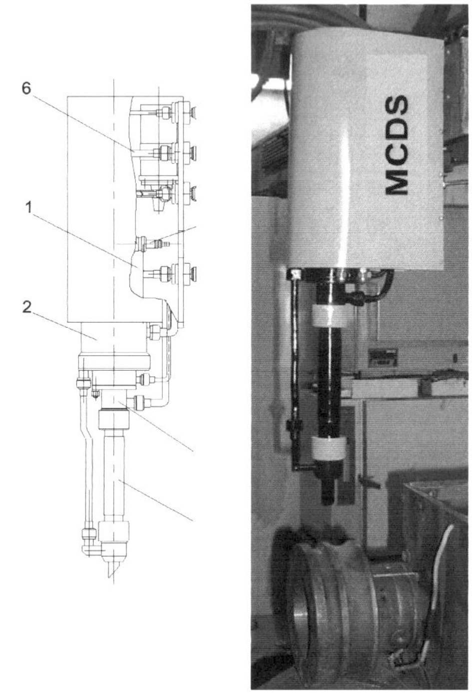

(a)

(b)

Fig. 2. Multi-chamber detonation sprayer (MCDS) (a), Control panel of the MCDS(b).

3 Comparative analysis

The efficiency of the MCDS was compared to that of the best samples of the HVOF units, which provide a high velocity of the spraying powder (up to 1000 m/s) and minimal porosity of the resulting coating (1-4 %). Analysis of the consumption of power and gases in thermal spraying units shows **(Table 1)** that over 10 m^3 of the fuel gas mixture components is spent per kilo of a coating when using the HVOF technology.

Fig. 3. Coating the inner surface with MCDS.

Almost all coating deposition devices consume 30-50 kW·h of the electric power for operation of the cooling system or heating of gases and powder.
The costs of materials and electric power are compensated for by the high efficiency of the coatings. But the problem of emissions of fumes and hothouse gases into the atmosphere remains.
It is a known fact that to provide quality coatings it is necessary to avoid overheating of the material deposited and provide it with the kinetic energy required for plastic deformation, as well as high adhesion and cohesion strength.
With the MCDS technology the energy of the fuel gas mixture is transformed into a high-velocity jet with a low thermal power, this providing formation of a high-quality layer with no overheating of a workpiece. The possibility of formation of the high-velocity and low-thermal power jet leads to a substantial decrease in consumption of the fuel gas, and allows formation of the high-quality coatings on thin-walled parts made from low-melting point materials. Depending on the parameters of filling the detonation combustion chambers, it is possible to deposit bronze, ceramic or hard-alloy coatings.
For example, **Fig. 4** shows the transverse section of a coating from different powders. As proved by the analysis results, the deposited coatings are hard, dense, has thickness up to 400 µm. Some modes of formation of coatings from different powders are represented in **Table 2.**

Table 1. Basic parameters of devices for spraying cermet coatings

Parameters	Devices				
	Jet-DJ9W [4]	JP5000 [4]	HVAF [3]	DZ9000E [6]	MCDS
Consumption, m^3/h:					
Nitrogen, m^3/h	0,8	0,8	-	-	0,8
Oxygen, m^3/h	18	60	-	20	6,4
Air, m^3/h	-	-	300	2	3,0
Propane, m^3/h	-	-	-	5,4	1,4
Methane, m^3/h	12	-	-	-	-
Kerosene, l/h	-	23	-	-	-
Propylene, kg/h	-	-	30	-	-
Electric power, kW*	47	40	12	28	3
Consumption of nozzles, pieces/h	0,12	0,12	0,1	0,1	-
Material utilization, %	55-70	40-60	70	60-70	70-90
Productivity, kg/h	3,2	3,2	10	3	1-3
Spraying distance, mm	250-300	300-400	100-200	150-200	30-80

*Common electric power of compressor, air drier, cooling system, control panel.

To form the high-velocity gas jet in the HVOF unit nozzle, it is necessary to create a high pressure of the fuel mixture in the combustion chambers. But this gas has an excessive thermal power, which exerts the negative effect on the coating formation process, decreases the spraying efficiency and increases emissions of harmful materials into the atmosphere. Moreover, the high energy density in the nozzle substantially limits its service life (to 5-10 h). Attempts to extend service life of the nozzles and the unit as a whole make the cooling system more complicated and increase its power intensity. Analysis of the cost of a coating deposited by using the HVOF units shows that current material and power expenditures for spraying are commensurable with the cost of a powder and make up $ 80-110 per kilo of a coating [4].

Table 2. Modes of formation of coatings from powders of ceramics, metal-ceramic and metal

Mode No.	Flow rate of fuel mixture components, m^3/h			Nozzle length, mm	Powder feeding, g/h	Frequency of detonations (Hz)
	Oxygen	Propane - butane	Air			
Powder AMPERIT® 740.0 - Al_2O_3 (H.C. Starck) Powder ZrB_2+12%$MoSi_2$						
1	*4.4 / **4.0	*0.8 / **0.82	*0.14 / **0.25	500	720-1000	20
Powder AMERIT 554.074 WC-Co-Cr (H.C. Starck)						
2	*3,13 / **3,25	*0,71/ **0,75	*1,56/ **1,4	300	1200-2000	20
Powder Ti, (Raymor Industries Inc.)						
3	*2.22/ **1.82	*0.57/ **0,4	*1.18/ **1.08	300	1100	18
*Cylindrical combustion chamber. **Ring combustion chamber						

Properties of Coatings. Examinations of the coating layer were conducted by using scanning electron-ion microscopes Quanta 200 3D and Quanta 600 (SEM) equipped with an X-ray detector of the PEGASUS 2000 system.

Porosity was determined by the metallographic method with elements of the qualitative and quantitative analyses of the geometry of the pores by using optical inverted microscope Olympus GX51. The hardness tests were carried out with the automatic micro-hardness analysis system DM-8 by the Vickers method with the indenter loads of 300 g.

These coatings contain a large volume fraction of finely, uniformly dispersed WC grains in the Co-based metallic binder phase. The higher hardness 967±230 $HV_{0.3}$ (powder feed rate 2000 g/h) is most likely associated with a greater degree of decarburization of WC particles and the presence of a higher level of the harder W_2C phase in the coating.

The specific wear rate decreases with a decrease in the powder feed rate from 2000 to 1200 g/h and the densest coatings with a specific wear rate of $1.171 \cdot 10^{-5}$ mm^3 $(m \cdot N)^{-1}$ are produced at a powder feed rate of 1200 g/h.

Coating from powder Al_2O_3 AMPERIT® 740.0 (**Fig. 4 a**) was implemented at a maximal velocity of the powder cloud. Increasing the barrel length to 500 mm provided the formation of a dense ceramic layer with hardness of 1320 ±25 $HV_{0.3}$ and porosity below 1 % In a multilayer coating deposited in 8-12 passes the value of micro-hardness monotonically decreased from the maximal value on the surface to lower values

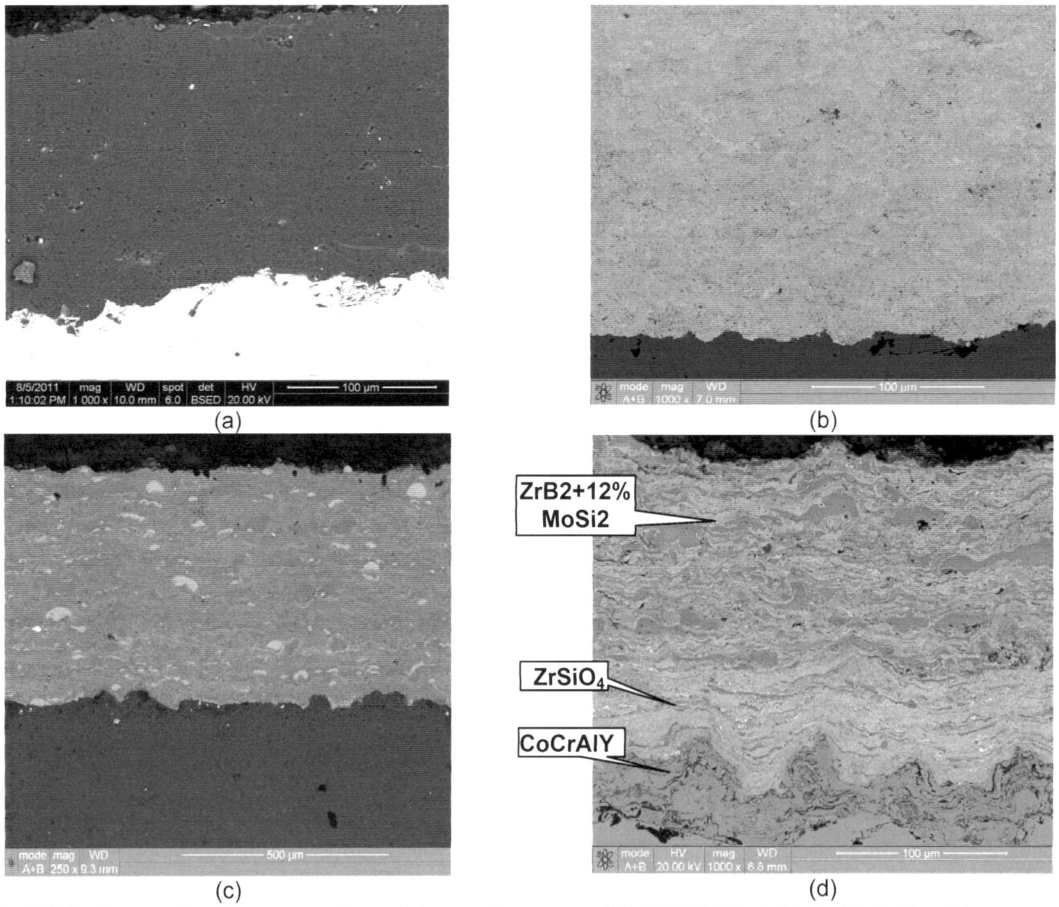

Fig. 4. SEM micrographs of cross-sections of the coatings from: AMPERIT 740.0 Al_2O_3 (H.C. Starck), powder size 5.6-22.5 μm (a), AMERIT 554.074 WC-Co-Cr 86%-10%-4% (H.C. Starck), powder size 15-45 μm (b), titanium (Raymor Industries Inc.), powder size 30-60 μm (c), Tree-layer coating (from the bottom to the surface) - CoCrAlY, $ZrSiO_4$, ZrB_2+12%$MoSi_2$ (d).

deeper in the coating [7].
The corundum crystalline form (α-Al_2O_3) is the one having the highest values of hardness and density, and content of crystalline form α-Al_2O_3 in the coating reaches up to 47%.
As a result, it was established that the visible interface between the Al_2O_3 ceramic layer and the steel substrate was free from defects, i.e. from regions with weakened bonds.
A high velocity of the dispersed materials leads to their deformation, chemical reactions and formation of a sufficiently thick (5-15 μm) transition layer, which provides adhesion of the ceramic layer to the steel substrate.
Coatings from AMPERIT 554.074 WC-10Co-4Cr powders (**Fig. 4 b**) on corrosion resistance steel have been obtained by multi-chamber detonation sprayer with powder feed rates 1200-2000 g/h and using barrel length of 300 mm. Microstructure and properties of coatings were studied. MCDS application allows the formation of a dense hard layer with a porosity of less than 0.5%. A slight decarburization of the WC particles to hemicarbide (W_2C) was observed in all coatings [8].
The microstructure of the coatings deposited using barrel length of 500 mm is dense and homogeneous. In this case the coatings had porosity of 0.24%.
The titanium-based coatings (**Fig. 4 c**) were deposited with different powder feed rate 800–1100 g/h and barrel length (300-500 mm) [9]. The analysis performed allows us to make a conclusion that the fine powder particles were heated and deformed to a state of fine lamellae and that they filled the spacing between the coarse particles to form a dense coating. Coating material deposited using barrel length of 300 mm is not deformed; it consists of the on deformed discrete particles and has the porosity of 13.49%. The coating layer deposited using barrel length of 400 mm was denser and had a porosity of 0.92%. However, a lot of coarse inclusions in the form of no deformed discrete particles were detected. The phase analysis shows that the coating fabricated with a barrel length of 300 mm is mainly composed of unreacted Ti, TiO, and Ti_2O_3. With the increase of the barrel length up to 500 mm, the coating is composed as a new component, TiO_2 (anatase, rutile) was found in the coating. Consequently, the microhardness increases from 685 to 1300 ± 250 $HV_{0.05}$ with the increase of barrel length in the range of 300–500 mm.
Experiments have been conducted as well for deposition of multi-layer coatings of different types of powders. For example tree-layer coating (from the bottom to the surface) - CoCrAlY, $ZrSiO_4$, ZrB_2+12%$MoSi_2$ (**Fig. 4 d**) was obtained with using MCDS.

4 Conclusion

MCDS is a versatile unit. It allows depositing coatings of low-melting point and refractory materials. Depending on filling the chamber and design features of MCDS it is possible to widely vary the velocity (up to 1200 m/s) and temperature of the powder material. For example, one and the same device allows depositing metal powder and ceramics. The MCDS is advantageous over HVOF in that it requires a much lower consumption of fuel gases and electric power at same quality of the coatings. Pressure in the gas lines is not in excess of 0.4 MPa. Design peculiarities of the unit allow the spraying process to be performed without overheating of devices parts and without damaging them. The results of this work open up new prospects for the further elaboration of new technologies for making protective coatings that can enhance the resistant properties of machinery parts.

Literature

[1]. Parco M., Barykin G., Fagoaga I. and Vaquero C. Development of wear resistant ceramic coatings by HFPD. ITSC 2008, pp.140/4.

[2]. Wielage B., Rupprecht C., Paczkowski G., Menzen R., Weissenfels G., Bernhardt H.-U. and Runkel M. A new way in HVOF technology – CFD optimized TOPGUN® AIRJET for powder and wire. ITSC 2008, pp. 145-150.

[3]. Korobov Y. and Filippov M. HVAF coating application in steel-making process. ITSC 2008, pp. 1362/65.

[4]. Meyer P., Rusch W. Production Coating Cost Comparison. ITSC 2003, pp. 123/8.

[5]. Jodoin B., Richer P., Bérubé G., Ajdelsztajn L., Yandouzi M., Erdi A. Pulsed-Cold Gas Dynamic Spraying Process: Development and Capabilities. ITSC 2007, pp.19-24.

[6]. Wang R.J., Xu L., Zhang T.J., Huang X.O., Luo H.J. Development and Application of DZ9000E Portable HVOF System. ITSC 2007, pp. 577/81.

[7]. Kovaleva M., Tyurin Yu., Vasilik N., Kolisnichenko O., Prozorova M., Arseenko M. Deposition and characterization of Al_2O_3 coatings by multi-chamber gas-dynamic accelerator. Surface and Coatings Technology (2013), Volume 232, pp. 719/25.

[8]. Kovaleva M.G., Tyurin Y.N., Vasilik N.J., Kolisnichenko O.V., Prozorova M.S., Arseenko M.Y., Sirota V.V., Pavlenko I.A. Properties and peculiarities of WCCoCr coatings formed by multi-chamber detonation sprayer. Applied Mechanics and Materials (2015), Volume 752-753, pp. 752/3.

[9]. Kovaleva M., Tyurin Yu., Kolisnichenko O., Prozorova M., Arseenko M. Properties of detonation nanostructured titanium-based coatings. Journal of Thermal Spray Technology (2014), Volume 22, No. 4, pp. 518/24.

Influence of Ce and Ru on the Behaviour of HVOF NiCoCrAlX Coatings in High Temperature Oxidation

Pimin Zhang [1], Ru Lin Peng[1], Xin-Hai Li [2], Sten Johansson[1]

[1] Department of Management and Engineering, Linköping University, 581 83, Linköping, Sweden
[2]Siemens Industrial Turbomachinery AB, SE-61283 Finspång, Sweden

Abstract: MCrAlX (M: Ni or Co or both, X: minor elements) have been widely used to protect hot components in gas turbines against oxidation and hot corrosion at high temperatures. Understanding the influence of the X-elements on oxidation behaviour is important in the design of durable MCrAlX coatings. In this study, NiCoCrAlX coatings doped with Y+Ru and Ce, respectively were deposited on Inconel 792 substrate by HVOF. The samples were subjected to oxidation test in laboratory air at 900, 1000, 1100 ºC and cyclic oxidation test between 100 ºC and 1100 ºC with 1 hour dwell time at 1100 ºC. It was observed that the coating with Ce shows a much higher oxidation rate than the coating with Y+Ru under both isothermal and cyclic oxidation tests. Besides, β-depletion due to interdiffusion between coating and substrate was significantly lower when Ru was present. This paper presents a detailed study of the influence of Ce and Ru on oxidation behaviour including investigation of the microstructure evolution influenced by oxidation time at coating surface and coating-substrate interface. Also simulation using an established oxidation-diffusion model was undertaken to give a further understanding of the oxidation behaviour.

1 Introduction

MCrAlX coatings (M for Ni and/or Co, X for minor elements) are widely used to protect the base materials (superalloys) against high-temperature oxidation and hot corrosion in gas turbine engines. The protection offered by MCrAlX coatings against high-temperature oxidation relies on the performance of a thermally grown oxide (TGO) scale formed at the coating-gas interface [1,2]. For a MCrAlX-coated superalloys at high temperature, the growth and adhesion of TGO at the coating surface, and the interdiffusion-induced Al depletion at the coating-substrate interface are important for the durability of the coating. The overall oxidation resistance is closely related to the coating's composition and a large number of publications have been dedicated to studying the effect of minor addition of alloying elements [3–7].

Cerium (Ce) is a typical reactive element, it's effect in improving oxidation performance has been reported in recent years. In [8], Saito et al. reported that the adherence of the oxide film formed on Ni-20Cr alloys was improved by small additions of Ce and Si. The combination of Ce and Si doping aids the formation of the root shape protrusions which inhibits the crack propagation at TGO and increases the spallation resistance [9]. However, no efforts are made to compare the capabilities of improving oxidation resistance between Ce and other reactive elements (REs) like Y.

The interdiffusion between coating and substrate is dependent on coating composition, substrate composition, and temperature [10,11]. The inward diffusion of Al from coating into substrate also causes depletion of Al, then an inner β-depletion zone forms on the coating side [10]. The reduction of inwards diffusion of Al from coating to substrate may significantly extend the life of MCrAlX coatings. According to [12], the formation of a (Ru,Ni)Al layer effectively slowed down the inward diffusion of Al. The addition of Ru in Ni-base superalloys is also beneficial since it decreases topologically close-packed (TCP) phase precipitation rate and growth rate, which prevent the degradation of mechanical properties [13]. But, the effect of Ru on interdiffusion between MCrAlX coating and superalloys substrate has rarely been discussed.

The current work investigates the oxidation behaviour of two modified MCrAlX coatings, one doped with Ce and the other with Ru+Y. While Ce and Y are both reactive elements for enhancing the adhesion of the protective oxide scale, the addition of Ru is for studying its effect on interdiffusion in the coatings. A comparative research is carried out on coatings sprayed by high velocity oxygen fuel (HVOF) technique at isothermal oxidation tests and cyclic oxidation tests. This paper aims at contributing to increased understanding the effects of Ru and Ce on the behaviour of MCrAlX coatings at high temperatures oxidation tests.

2 Experiment and simulation

The material system studied in this work was NiCoCrAlX coatings deposited on Inconel 792 substrate (composition by wt.%: 12.5 Cr, 9 Co, 4.2 W, 4.2 Ta, 4 Ti, 3.4 Al, 1.9 Mo 0.08 C, 0.03 Zr+B, balanced by Ni), using the HVOF spraying technique. The chemical composition of the as-heat treated MCrAlX coatings is shown in **Table 1**. It was measured by combination of energy dispersive spectroscope (EDS) and wavelength dispersive spectroscope (WDS). Prior to coating deposition, the substrate surfaces were grit blasted. After the deposition, the coated specimens were subjected to a heat treatment, including solution annealing at 1120 °C for 2 hours in vacuum, followed by aging treatment at 845 °C for 24 hours in air and air cooling to room temperature. Such heat treatment, which is intended for hardening the substrate, also increases the bond strength between the coating and substrate.

Isothermal oxidation tests at different temperatures and cyclic oxidation test on the coated specimens were conducted in laboratory air, details see **Table 2**. One cyclic oxidation consists of heating at 1100 °C for 1 hour and 10 min forced-air cooling to 100 °C. The surface and the cross-sections of the specimens were examined using SEM for phase and chemical composition determination. The thickness of the TGO and the layer containing β-phase (β-left zone) within coatings were quantified by image analysis software.

Table 1 Chemical composition of as-heat treated NiCo15Cr12.5AlX coatings [in wt.%].

MCrAlX with X=	Ru	Ce	Y	Others
C1	3.3	-	0.3	0.3Si, Mo,Ta<0.6
C2	-	0.2	-	

Table 2 Conditions of oxidation test.

Condition	Time/Cycles
ISO-900 °C	500, 1500, 3000, 5000 h
ISO-1000 °C	500, 1500, 5000 h
ISO-1100 °C	50, 300, 500, 800 h
Cyclic oxidation	50, 300, 500, 800 cycles

To have a better understanding of the effect of elements on the coating behaviour in the oxidation test, the experiment work is supplemented with computational simulation, using Thermocalc software package. Phase of corresponding composition was confirmed by Thermocalc, interdiffusion between coating and substrate during oxidation was performed in homogenization model by DICTRA [11,14–16].

3 Results and Discussion

3.1 Isothermal oxidation rate

Since the formation and growth of TGO play an important role in its subsequent thickening [17], the alumina scale growth rate were evaluated on the specimens subjected to the isothermal tests. **Fig. 1** compares the evolution of surface oxide morphology of the Y+Ru doped C1 and Ce doped C2 coatings after the initial heat treatment and after oxidation at 1000 °C for 500 h and 5000 h, respectively. From **Fig. 1a** and **d**, it can be observed that no visible transient aluminium-oxide developed on the surface of both coating during the heat treatment. On the specimens subjected to 500 h isothermal exposure, aluminium oxides mainly in needle shape were observed. They could be transient aluminium oxide [12] or stable alumina with the inherent morphology of transient aluminium oxide. As the oxidation test continued to 5000 h, the needle-like oxides of C2 coating grew larger, but the oxide morphology remained the same (**Fig. 1e** and **f**). On the other hand, the oxide scale on C1 coating exhibited an obvious morphological evolution (**Fig. 1c**) from 500 h to 5000 h: an evenly distributed nodal-like oxides without the presence of needle-like oxides on the coating surface, indicating that most of the needle-like oxides had transformed into nodal-like morphology.

Cooling from the oxidation temperature caused local TGO spallation in the coatings exposed to 1000 °C for 5000 h. The fractured surface of the TGO in such a spalled region is shown in **Fig. 2** for both coatings. The examination revealed that the oxide scale of C2 contains two layers (**Fig. 2b**): equaxial grains (inner layer) and needle-like Al_2O_3 (outer layer). However, on the C1 coating surface it mainly consisted of

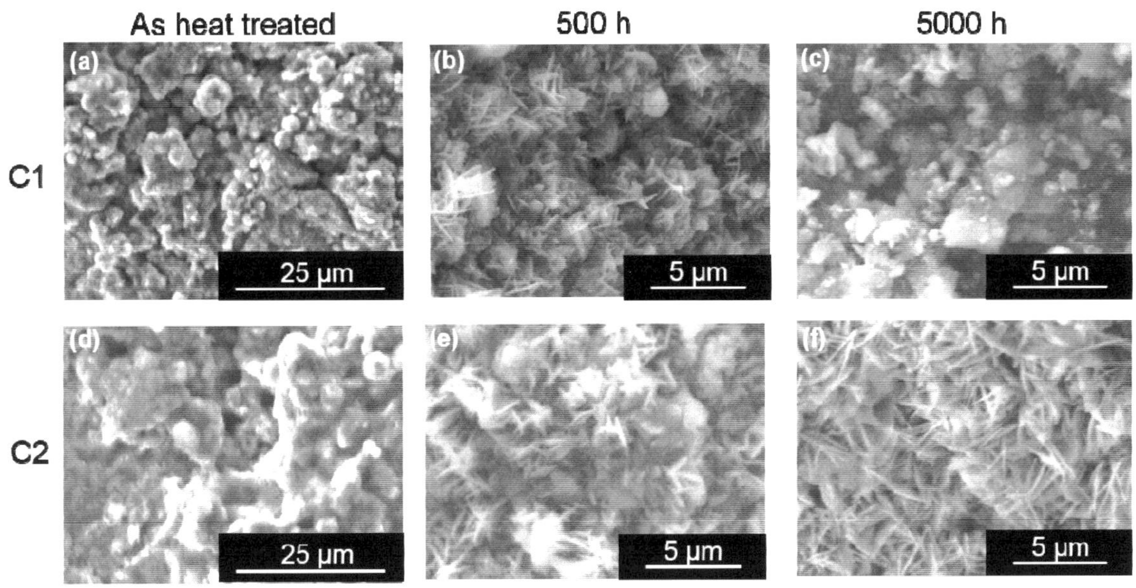

Fig. 1. SEM images of the Y+Ru doped C1 coating (a-c) and Ce-doped C2 coating (d-f) surface after heat treatment, 500 h and 5000 h oxidation tests at 1000 °C.

equaxied α-Al_2O_3 grains in the outer layer (**Fig. 2a**) and columnar grain in the inner layer. As reported by Brady et al. [18], doping with REs such as Y, Hf , La and Zr have some significant effects on the thickening of oxide scale: a tendency for thick oxide scales to exhibit a columnar morphology. However, such effects are not observed for Ce in C2 coating.

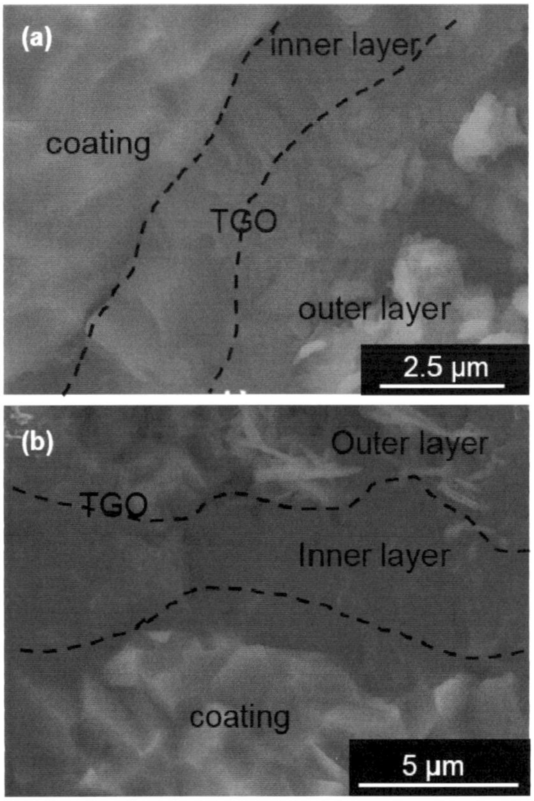

Fig. 2. SEM images for the oxide scales fractured sections of Y+Ru doped C1 coating (a) and Ce-doped C2 coating (b) after 5000 h isothermal oxidation tests at 1000 °C. (Noted that the images are in different magnifications.)

As is recognized, the growth of TGO is assumed to be diffusion-controlled and the TGO thickening follows a parabolic oxidation law:

$$h = (kt)^{1/2}, \quad (1)$$

with k following the Arrhenius relationship:

$$k = k_0 e^{-\frac{Q}{RT}}, \quad (2)$$

where h represents TGO thickness, *k* is the oxidation rate constant, *t* denotes oxidation time, *Q* is the activation energy, *T* is temperature and *R* is gas constant 8.314 $Jmol^{-1}K^{-1}$.

Fig. 3a compares the TGO thickness growth of C1 and C2 coatings in isothermal oxidation tests at 1100 °C. As can be seen, the oxidation rate at the initial stage (below 50 h) is similar for both coatings and is higher than the stage followed, which is attributed to the formation of a fast growing, transient Al_2O_3 oxide scale. After 50 h oxidation, the C1 coating shows a lower TGO growth rate than the C2 coating, indicating a slower Al depletion and therefore a better oxidation resistance of C1 coating.

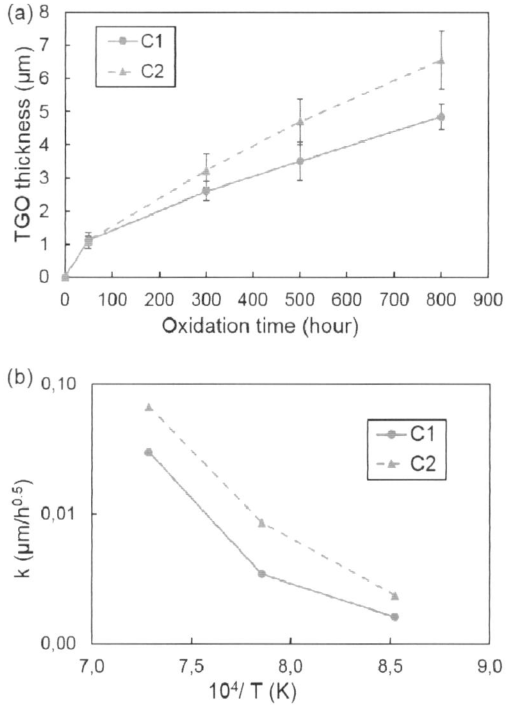

Fig. 3. (a) TGO thickness growth of C1 and C2 coatings in isothermal oxidation tests at 1100 °C, (b) Arrhenius relationship: *ln(k)* of C1 and C2 coatings for the isothermal oxidation at 900, 1000 and 1100 °C vs. $10^4/T$.

By fitting Eq. (1) to experimentally determined TGO thickness evolution with oxidation time (e.g. **Fig. 3a**), the oxidation rate constant *k* was calculated for the isothermal tests at 900, 1000 and 1100 °C. Then, the Arrhenius relationship (see Eq. (2)) between *k* and *T*, *Q* was derived. In **Fig. 3b** the logarithm of the oxidation rate constants *k* of C1 and C2 coatings were plotted against the inverse of temperature $10^4/T$. The TGO growth rate indicated by oxidation rate constant *k* increases with the increasing oxidation temperature and is lower for C1 coating at all test temperatures. One possible explanation for the lower TGO growth rate of C1 coating can be the effect of Y which retards the aluminium oxide transition from transient θ to stable α [19]. When the θ-Al_2O_3 is allowed to well develop in the initial stage, the finally obtained α-Al_2O_3 grains would be larger leading to a reduction of the TGO growth rate, since the scale grows predominantly by the diffusion of oxygen along the α-Al_2O_3 grain boundaries. Further work is required to explore the microstructure evolution of Al_2O_3 in the initial stage.

Fig. 3b reveals a near linear *ln(k)* versus $10^4/T$ relationship for the C2 coating. This means a similar diffusion mechanism controlling the TGO growth for all the three testing temperatures. In the case of C1 coating, however, the slope decreased largely at low temperature (900-1000 °C). Such a change could be related to enhanced diffusion of oxygen. The existence of columnar grain boundaries promotes fast diffusion of Y into the scale, $YAlO_x$ oxide forms and provides extra high-diffusivity paths for oxygen [20]. At higher temperatures, the tendency for thick oxide scales to exhibit a columnar morphology was weakened due to a less profound doping effect of C1 coating [18].

3.2 Comparison of oxide scale adhesion

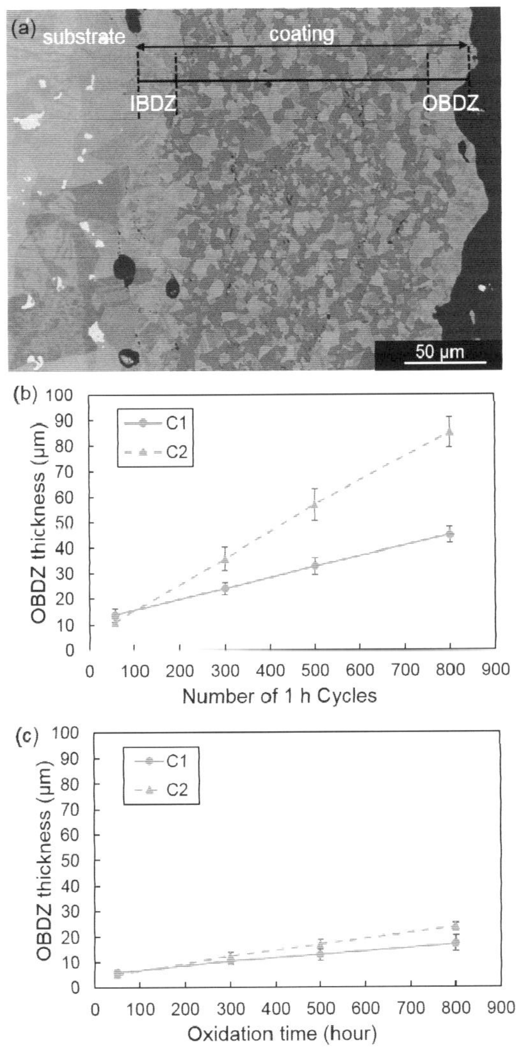

Fig. 4. Microstructure of the C1 coating (BSE image) of iso-1000 °C specimen after 5000h (a), The OBDZ thickness of C1 and C2 coating of cyclic oxidation testing at 1100 °C (b), and isothermal oxidation testing at 1100 °C (c).

Due to the oxidation of the MCrAlX coating at the coating-gas interface, an outer beta phase depletion zone (OBDZ) appeared in the coating below the coating-oxide interface (**Fig. 4a**). **Fig. 4b** and **4c** show the OBDZ thickness changes of Y+Ru doped C1 coating and Ce-doped C2 coating during cyclic oxidation testing at 1100 °C (b), and isothermal oxidation testing at 1100 °C (c), respectively. The results were obtained by measuring the thickness evolution of OBDZ with oxidation time. The spallation of oxide scale will accelerate the consumption of Al, which leads to the fast depletion of β-phase in the coating, and thus the OBDZ growth rate can be used to evaluate the spallation resistance of the oxide scale.

From **Fig. 4**, it can be seen that for the C2 coating the OBDZ thickness after 800 cycles in cyclic oxidation test (85.1±5.9 μm) is 3.6 times that after 800 h in isothermal oxidation test (23.6±1.5 μm). For the C1 coating, the ratio is much smaller, 2.6 times (44.7±3.3 μm against 17.2±3.0 μm). As no obvious TGO spallation was observed under isothermal oxidation test, the enhanced OBDZ growth rate observed in cyclic oxidation test for both coatings is mainly related to TGO spallation due to thermal stresses induced by the tempearture changes. The much larger OBDZ thickness of C2 coating, twice of that of C1, after 800 cycles indicates a poorer spallation resistance of the Ce-doped C2 coating. There is no doubt that a significant difference existing between Y+Ru and Ce dopants in improving the scale adhesion.

It was noted from [20] that the slow aluminium oxide θ-α transition could lead to an extended life of the θ phase layer, which was proposed to be beneficial to the relaxation of the stresses induced by the growth of α-Al_2O_3. The Y addition in C1 coating retards the θ-α transition, which could partly explain the coating's better spallation resistance.

According to Ogawa et al. [21], root-like TGO has been found in coating containing Ce, which can improve TGO adhesion due to anchor effect, and the influence became more pronounced when the amount of Ce increased. However, no such root-like TGO were found in Ce-doped C2 coating, which can be resulted from too small amount of Ce (0.2 wt.%) addition in C2 coating compared with Ogawa's work (0.5, 1, 1.5 wt.%).

Therefore, considering the poor isothermal oxidation and cyclic oxidation resistance of Ce-doped coating, it seems that an equal amount of Ce (0.2 wt.% in C2 coating) as REs addition in MCrAlX coating is not sufficient to provide good oxidation resistance as Y (0.3 wt.% in C1 coating).

3.3 Comparison of interdiffusion

Interdiffusion between the coating and substrate can result in β-depletion near the coating-substrate interface in the coating side (**Fig. 5**) as the aluminium move from the coating to the substrate. The β-phase depletion rate, controlled by diffusion, is expected to follow Eq. (1).

Fig. 5a shows the simulation results of β-phase depletion of Y+Ru doped C1 coating at 1100 °C, based on the oxidation model described in [6,9,10]. Simulation was also performed for the C2 coating but the results are not presented here. **Fig. 5b** shows the isothermal 1100 °C oxidation test data of IBDZ (inner β-depletion zone, **Fig. 4a**) thickness against square root of time. The IBDZ growth rate of C1 and C2 coating are 1.44 and 2.17 µm/h$^{0.5}$, respectively, which are consistent with the simulation results (1.52 µm/h$^{0.5}$ for C1 coating and 2.05 µm/h$^{0.5}$ for C2 coating), and the Y+Ru doped C1 coating has a lower interdiffusion rate of Al than Ce-doped C2 coating.

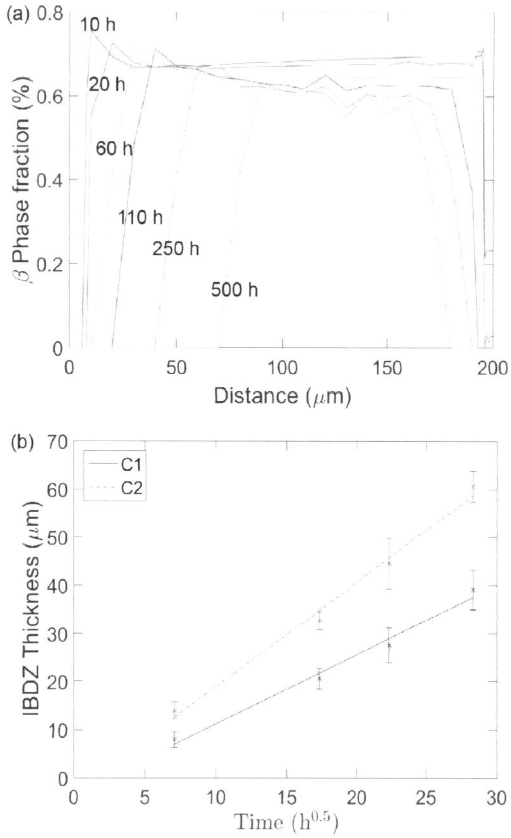

Fig. 5. Simulation results of β-phase depletion of Y+Ru doped C1 coating at 1100 °C (a), IBDZ thickness data of C1 and C2 coating from isothermal oxidation test at 1100 °C vs. square root of time (b).

The effect of Ru on the interdiffusion has been further investigated by equilibrium calculation using Thermocalc, see **Fig. 6**. Several different alloy systems with similar compositions as C1 coating were chosen including: NiCo15Cr12.5Al+1Ru, 1.5Ru, 2Ru, 2.5Ru, 3Ru, 3.5Ru (in weight percent). It's clear that the β-phase fraction decreased with the decreasing Ru content in the alloy, indicating that Ru stabilizes β-phase in the alloy system. This result is consistent with [7], which claimed that Ru greatly stabilizes β-phase by largely extending β-phase field in the phase diagram.

In summary, both simulation and experimental analyses showed that the addition of Ru in C1 coating decreased interdiffusion induced β-phase depletion. And the fact that Ru are β-phase stabilizer contributes to the decreasing of inward diffusion of Al.

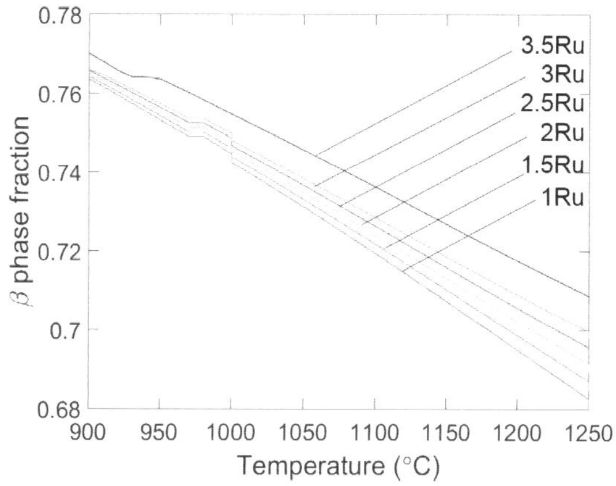

Fig. 6. β-phase fraction for different alloy system from 900 to 1250 °C, derived from equilibrium calculation by Thermocalc. The composition of alloy systems are NiCo15Cr12.5Al +1Ru,1.5Ru,2Ru,2.5Ru,3Ru,3.5Ru.

4 Conclusion

Oxidation behaviour of NiCoCrAl coatings with Ce and Y+Ru additions was compared at several test conditions. Conclusions can be drawn as follows:

1. In comparison with Ce-doped coating, the Y+Ru-doped coating showed a lower growth rate of the outer β-depletion zone (below the protective scale) in both isothermal and cyclic oxidation tests, indicating a better resistance to oxide growth and a stronger adhesion of TGO. The difference can be partly attributed to the positive effect of Y. Its addition may relax the growth stress of α-Al_2O_3 by retarding the θ-α transformation, and may reduce the diffusion of oxygen through the oxide scale by promoting a coarse grain structure of α-Al_2O_3.
2. The Y+Ru-doped coating exhibited a lower growth rate of inner β-depletion zone (near the coating-substrate interface) induced by interdiffusion. This is attributed to that the addition of Ru stabilize the β-phase and reduce the lose of Al by interdiffusion.
3. While the addition of about 3.3 wt% Ru seems to significantly increase resistance against interdiffusion, the substitution of Y (0.3 wt %) by a similar amount of Ce (0.2 wt%) decreases greatly the oxidation resistance at the coating surface.

5 Acknowledgment

The auther would like to thank the Siemens Industrial Turbomachinery AB (Finspång, Sweden) and Swedish Energy Agency through KME consortium - ELFORSK are greatly acknowledged for their financial support. The research school of Agora Materials and Strategic Faculty Grant SFO-MAT-LiU#2009-00971 and project teams at Linköping University are also acknowledged for their support on this research.

6 Literature

[1] Brandl W, Toma D, Krüger J, Grabke HJ, Matthäus G. The oxidation behaviour of HVOF thermal-sprayed MCrAlY coatings. Surf Coatings Technol 1997;94-95:21–6.

[2] Chen LC, Zhang C, Yang ZG. Effect of pre-oxidation on the hot corrosion of CoNiCrAlYRe alloy. Corros Sci 2011;53:374–80.

[3] Taylor TA, Bettridge DF. Development of alloyed and dispersion-strengthened MCrAlY coatings. Surf Coatings Technol 1996;86-87:9–14.

[4] Nijdam TJ, Sloof WG. Effect of reactive element oxide inclusions on the growth kinetics of protective oxide scales. Acta Mater 2007;55:5980–7.

[5] Czech N, Schmitz F, Stamm W. Improvement of MCrAlY coatings by addition of rhenium. Surf Coatings Technol 1994;68-69:17–21.

[6] Luthra KL, Briant CL. Mechanism of adhesion of alumina on MCrAlY alloys. Oxid Met 1986;26:397–416.

[7] Yuan K, Lin Peng R, Li XH, Johansson S, Wang YD. Some aspects of elemental behaviour in HVOF MCrAlY coatings in high-temperature oxidation. Surf Coatings Technol 2015;261:86–101.

[8] Y. Saito, K. Kiriu, T. Kimura TA and SY. Spalling of the Surface Oxide Formed on Ni-20Cr Alloys with Small Additions of Cerium and Silicon. Japan Inst Met 1975;39:1110–7.

[9] Tanno M, Ogawa K, Shoji T. Effect of Cerium and Silicon Additions to MCrAlY on the High-Temperature Oxidation Behavior and Bond Strength of Thermal Barrier Coatings. Key Eng Mater 2004;261-263:1061–6.

[10] Itoh Y, Tamura M. Reaction Diffusion Behaviors for Interface Between Ni-Based Super Alloys and Vacuum Plasma Sprayed MCrAlY Coatings. Trans ASME 1999;121:476–83.

[11] Yuan K, Eriksson R, Lin Peng R, Li XH, Johansson S, Wang YD. MCrAlY coating design based on oxidation–diffusion modelling. Part II Lifing aspects. Surf Coatings Technol 2014;254:79–96.

[12] Wang Y, Guo HB, Peng H, Peng LQ, Gong SK. Diffusion barrier behaviors of (Ru,Ni)Al/NiAl coatings on Ni-based superalloy substrate. Intermetallics 2011;19:191–5.

[13] Sato A, Harada H, Yokokawa T, Murakumo T, Koizumi Y, Kobayashi T, et al. The effects of ruthenium on the phase stability of fourth generation Ni-base single crystal superalloys. Scr Mater 2006;54:1679–84.

[14] Yuan K, Eriksson R, Lin Peng R, Li XH, Johansson S, Wang YD. MCrAlY coating design based on oxidation-diffusion modelling. Part I: Microstructural evolution. Surf Coatings Technol 2014;254:79–96.

[15] Yuan K, Eriksson R, Lin Peng R, Li XH, Johansson S, Wang YD. Modeling of microstructural evolution and lifetime prediction of MCrAlY coatings on nickel based superalloys during high temperature oxidation. Surf Coatings Technol 2013;232:204–15.

[16] Sundman B, Jansson B, Andersson JO. The Thermo-Calc databank system. Calphad 1985;9:153–90.

[17] Toscano J, Vaβen R, Gil A, Subanovic M, Naumenko D, Singheiser L, et al. Parameters affecting TGO growth and adherence on MCrAlY-bond coats for TBC's. Surf Coatings Technol 2006;201:3906–10.

[18] Brady, M. P. and Pint, B. A. and Tortorelli, P. F. and Wright, I. G. and Hanrahan RJ. High-Temperature Oxidation and Corrosion of Intermetallics. Mater. Sci. Technol., Wiley-VCH Verlag GmbH & Co. KGaA; 2006, p. 229–325.

[19] Peng X, Clarke DR, Wang F. Transient-alumina transformations during the oxidation of magnetron-sputtered CoCrAl nanocrystalline coatings. Oxid Met 2003;60:225–40.

[20] Peng H, Guo H, He J, Gong S. Microscale lamellar NiCoCrAlY coating with improved oxidation resistance. Surf Coatings Technol 2012;207:110–6.

[21] Ogawa K, Ito K, Shoji T, Seo DW, Tezuka H, Kato H. Effects of Ce and Si Additions to CoNiCrAlY Bond Coat Materials on Oxidation Behavior and Crack Propagation of Thermal Barrier Coatings. J Therm Spray Technol 2006;15:640–51.

The correlation of the TBC lifetimes in burner cycling test with thermal gradient and furnace isothermal cycling test by TGO effects

Chang-Jiu Li, Hui Dong, Hang Ding, Guan-Jun Yang
State Key Laboratory for Mechanical Behavior of Materials, School of Materials Science and Engineering, Xi'an Jiaotong University, China / E-mail: licj@mail.xjtu.edu.cn

Two types of typical thermal cycling tests are used for the evaluation of thermal cycling lifetime of thermal barrier coatings. Those are burner cycling tests with a thermal gradient and isothermal furnace cycling test. There are diverse explanations to test results up to now. Although certain correlations should exist between the results obtained by two types of the tests, no evident parameters in two tests were directly related possibly due to large range of difference test conditions. In this investigation, a series of TBC samples with carefully prepared Al_2O_3-based TGO of different thicknesses were used for both burner cycling and furnace cycling tests. The relationships between thermal cycling lifetime and TGO thickness were obtained for two types of the tests. It was found that TGO thickness presents the same influence tendency despite different thermal cycling tests. The results reveal the existence of the critical TGO thickness by which the transition of failure mode takes place. Moreover, the values of the critical TGO thickness for two tests are comparable. The results evidently suggest that the lifetimes during different thermal cycling tests can be correlated by TGO effects on failure behavior. However, it is clear that the apparent dominant driving factor to TBC failure is different in two type tests. Thus, the burner cycling test could be used for optimizing the durability of ceramic top coat, while the furnace cycling test results represent the integrated TBC durable performance of the bond coat and top ceramic coating.

1. Introduction

Thermal barrier coatings (TBCs) are widely used to protect parts in the hot-sections of mechanical equipment, especially parts in the gas turbines and diesel engines. The applications of TBCs can either significantly enhance the reliability of engines by lowering working temperature of superalloy parts or significantly increase the engine efficiency by allowing use of higher temperature combustion gas [1-2]. Therefore, TBCs are characterized by two important performance parameters, i.e., thermal barrier effect and durability [2]. With a plasma-sprayed TBC, a desirable thermal barrier effect could be realized by selecting ceramic coating materials, designing coating structure including coating thicknesses and tailoring coating microstructures [3]. On the other hand, the durability is crucial to TBC applications since it is directly related to its failure. The durability is usually limited by spalling of the top ceramic coat. For plasma sprayed TBCs, the lamellar structure of ceramic coating benefits thermal barrier effect but lowers cracking resistance due to low toughness [4, 5].

The durability of TBCs is directly related to the evolution of microcracks network within ceramic coating [6]. It has been pointed out that the durability is a complex function of materials properties including superalloy substrate, bond coat and top ceramic coat, coating design and service conditions [2]. For example, although the coating adhesion is supposed to be associated to the durability of TBCs, the investigation revealed no direct relation between coating adhesion and its durability [7]. Up to now, it is still a difficult challenge to predict reasonably lifetime of a TBC [8]. Moreover, for land-based gas turbines, the expected long operation time up to several years makes full scale part field test impossible due to unreasonably high cost. Therefore, it is necessary to make reliable durability tests for ranking the materials selected and optimizing coating processing parameters and microstructures [9].

Usually, two types of durability tests are employed for TBCs: furnace thermal cycling and rig-burner cycling [2, 9]. Each of the tests possesses certain advantages [2]. The typical features of the furnace tests are generally low cost, accurately controlled test temperature, and easy monitoring of specimen weight gains with less contaminants. While the burner rig tests (Fig. 7) can be used to simulate the conditions under which the TBCs serve. For example, by allowing the cooling at the substrate surface a thermal gradient over ceramic top coat with a high heating or cooling rate can be applied, so that short-duration cycle durability representative of subsonic engine takeoff and climb can be simulated. The greatest problem with rig-burner tests involves the difficulty in measuring temperature although it can force hot spots thereby avoiding edge effect failures [2]. Moreover, the specimens can be exposed to salt contaminants conditions. The variety of the rig-burner tests, being referred to as burner cycling test in the present study, is to use a coin-type sample with evenly heating flame to control the surface temperature and thermal gradient within TBCs. Although two types of tests are usually adopted for evaluating the performance of TBCs samples, there are no data to clearly illustrate the correlation between the results obtained by two types of the tests possibly due to lack of standardized conditions which make the tests in the literature be performed in larger variety of conditions.

One typical feature inherent to YSZ TBCs is that YSZ coatings used for thermal barrier are usually permeable to oxygen at high temperature in operation environment [10], which causes the gradual oxidation of superalloy bond coat, leading to thermally growing of oxides (TGO) at the interface between the bond coat and ceramic top coat [11]. Since TGO growth is inevitable and its growth along with other factors leads usually to the spalling of ceramic top coat and thus failure of TBCs, the TGO growth is supposed to be one of the important factors dominating TBCs durability [7, 12]. The previous investigations have revealed that using either burner thermal cycling test or furnace test the thickness of TGO prior tests significantly influences the lifetime of TBCs [12]. Therefore, in the present study, based on the results of thermal cyclic lifetime dependency on TGO thickness obtained by two types of thermal cycling tests which were performed using a series of the same TBC samples with different TGO thickness [12, 13], the correlation between the representative results from two types of the tests was examined to provide fundamentals to guide the introduction of thermal cycling tests.

2. Experimental procedure

2.1 Materials and Coating Deposition Process

The previous investigation showed that the thermal cycling lifetime of TBCs consisting of cold-sprayed MCrAlY bond coat is higher than that with LPPSed bond coat due to easy formation of a conformal Al_2O_3-based TGO [14]. Thus, in the present study, the cold spray system (CS-2000, Xi'an Jiaotong University, China) was used to deposit MCrAlY of ~150 μm on a disk-shaped nickel-based superalloy Inconel 738 substrate (Φ25.4×3 mm). The nominal compositions of MCrAlY powder used, being in a spherical shape, are Ni23Co20Cr8.5Al4.0Ta0.6Y (Sulzer-Metco Amdry 997, -37 μm, +9 μm). Table 1 shows the cold spray conditions. The well established $8Y_2O_3$ stabilized ZrO_2 (8YSZ) was used at the top coat material. A commercial atmospheric plasma spray (APS) system was used to deposit 8YSZ coating on the bond coat. To ensure sufficient melting of YSZ particles during plasma spraying, a hollow spherical 8YSZ (Metco 204B-NS, -75 μm, +45 μm) was employed. The thickness of topcoat was controlled to be approximately 250 μm. The APS spray parameters are shown in Table 2.

Table 1: Cold spray parameters

Processin g gas	Gas pressur e (MPa)	Gas Temp. (°C)	Gun travers e speed (mm/s)	Spray distanc e (mm)
He	2.0	730±2 0	150	20

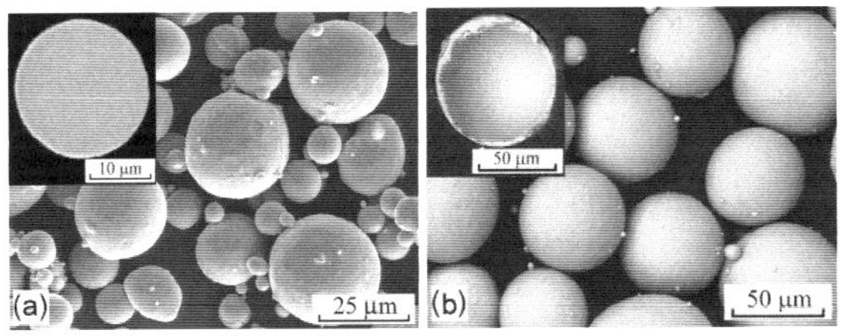

(a) (b)

Fig. 1 SEM images of powders: (a) NiCoCrAlTaY powder for bond coat in a spherical shape and (b) hollow and spherical 8YSZ powder for topcoat. The inset in (b) exhibits the hollow feature of YSZ powder by cross section.[12]

Table. 2: Plasma spray conditions

Parameters	Unit	Valu e
Arc power	kW	39
Arc current	A	650
Arc voltage	V	60
Primary plasma gas (Ar)	slpm	60
Secondary plasma gas (H_2)	slpm	4
Spray distance	mm	80

2.2 Pretreatment of TBC samples and thermal Cycling Test for TBCs

To produce a uniform α-Al_2O_3-based TGO, all TBC samples were pre-treated in an argon atmosphere (O_2 content< 10ppm) at a temperature of 1000 °C for 4h and then 1080 °C for 4h. Through such treatment, the TGO mainly contains α-Al_2O_3, and its thickness is 1.3±0.3 μm. Subsequently, in order to fabricate TBCs with TGOs of different thicknesses, the samples were further isothermal-oxidized at 1080°C for 20h, 50h, 105h, 225h, 300h and 350h, respectively, before thermal cycling test. A scanning electron microscope (SEM, VEGA II-XMU, TESCAN, Czech) image in the backscattered electron image (BEI) mode was employed to confirm TGO thickness and examine microstructure of TBCs. 100 measurements per sample were made for the estimation of mean TGO thicknesses after isothermal oxidation. After isothermal oxidization, the TGO still contains α-Al_2O_3, while its thicknesses became 2.7±0.2 μm, 3.9±0.3 μm, 5.0±0.4 μm, 6.5±0.4 μm, 7.0±0.6 μm and 7.7±0.6 μm, respectively, which have been reported in the previous report [12].

Isothermal furnace cycling test was performed by using atmospheric furnace. One thermal cycling includes 26 min holding in the furnace and 4 min cooling out the furnace. In this study TBC samples were rapidly put into the furnace being kept at a temperature of 1120°C, hold for 26 minutes and then taken out furnace for rapidly cooling down to a temperature about 80°C in 4 minutes by blowing air [13]. Thermal cycling test was terminated when the area of YSZ spalling visually reaches 50% of YSZ topcoat surface. The thermal cyclic lifetime of TBCs is an average of three TBC specimens.

The burner thermal cycling test was performed by using a gas burner test facility operating with propane and oxygen by 70s heating, 50s holding and 120 s cooling. The samples were heated by columned flame with a diameter of about 20 mm, which can give a homogeneous temperature distribution on the surface of sample. The change of the surface temperatures of both the substrate and YSZ coating during thermal cycling test is shown in Fig.2. YSZ surface was heated to a maximum temperature of 1150°C by a flame and the sample was cooled simultaneously by compressed air jet from the surface of substrate. The surface temperature of the substrate was maintained at approximately 935°C, when the YSZ surface reached to the peak temperature of 1150°C. Then, the samples were rapidly cooled down in 120s from both the surfaces of the substrate and YSZ coating by compressed air jets. The peak temperature of the bond coat and YSZ interface was estimated to be ~1000°C [12].

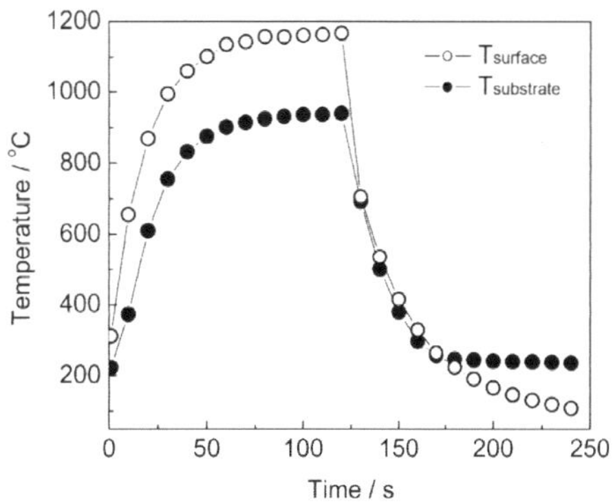

Fig.2 Temperature change of the substrate and YSZ surfaces during each thermal cycling test [12]

During thermal cycling test, the surface temperature of YSZ coating was monitored by a non-contact infrared thermometer with a wavelength of 8-14 μm. Simultaneously, the substrate temperature was also monitored by a non-contact infrared thermometer with the wavelength of 1.6 μm. Thermal cycling test was terminated when a visible spallation of the YSZ was observed. The lifetimes of TBCs were an average of the three specimens at each TGO thickness.

Metallographic cross section was prepared from the failure samples to examine the microstructure features and failure modes. After failure, the samples were mounted with epoxy adhesive by vacuum infiltration before cutting to avoid any artifact cracks. A scanning electron microscope (SEM, VEGA II-XMU, TESCAN, Czech) in the backscattered electron image (BEI) mode was employed to examine microstructure of TBCs.

3. Result and discussion

3.1 Microstructural features of the TBCs for thermal cycling tests

Figure 3 shows a typical microstructure of TBC sample at the as-sprayed conditions consisting of the cold-sprayed MCrAlY bond coat and APS YSZ top coat. The cold-sprayed bond coat presents a dense microstructure without apparent oxides, which is contributed to low temperature features of cold spraying. The top YSZ coating presents a classical lamellar structure with a porosity of 10% which was estimated by image processing of cross-sectional images. The different types of pores including global pores, micro-cracks in individual splats, and non-bonded interface voids are present in the coating. Fig. 4 shows the cross-sectional SEM micrographs of the TBCs with the TGOs of different thicknesses after isothermal oxidation. The α-Al_2O_3-based TGO with a black contrast in Fig. 4 was uniformly distributed at the interface between YSZ and bond coat for all samples. Including the TBC samples with only pre-treatment, TBCs with seven TGO thicknesses were used for thermal cycling tests accordingly. A close examination revealed that the cracks appeared at the TGO/bond coat interface or TGO /YSZ interface when TGO grew to a thickness of 6.5 μm (as marked by arrow in Fig. 4 d).

Fig. 3 The typical microstructure of TBCs [12]

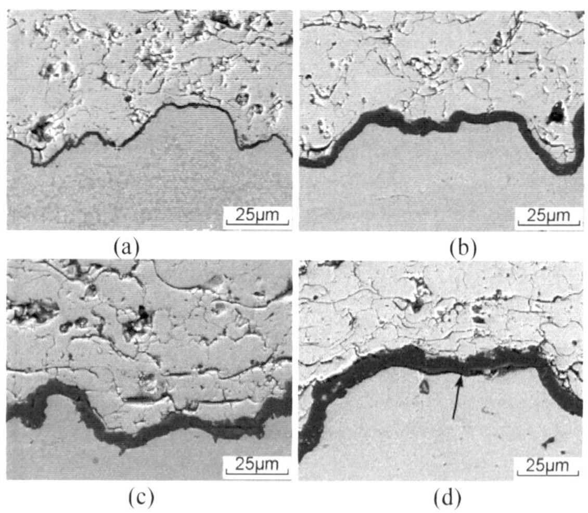

Fig. 4 Cross sectional microstructure of TBCs with the TGOs of different thicknesses: (a) 1.3 μm, (b) 3.9μm, (c) 5.0 μm , (d) 6.5 μm

3.2 Dependency of thermal cycling lifetime of TBCs on TGO thickness

Figure 5 shows the effect of initial TGO thickness on burner thermal cyclic lifetime of the TBC. It is clear that the lifetime of the TBC is significantly influenced by the initial TGO thickness. When the initial TGO thickness is 1.3μm, the mean thermal cyclic lifetime of the TBC reached higher than 9000 cycles. With the increase of TGO thickness, the cyclic lifetime decreases significantly. At a TGO thickness of 5.0 μm, the mean cyclic lifetime decreases to less than 2000 cycles. In particular, with the further increase of TGO over 5μm, the thermal cyclic lifetime presents a more rapid decrease. When the TGO thickness grew to 7.7 μm, the mean lifetime decreased to less than 400 cycles. Thus, when the initial TGO thickness increases from 1.3μm to 7.7 μm, the mean lifetime of the TBC reduces by a factor of more than 20. At small TGO thickness region, the lifetime of TBC, N_{bc}, can be expressed through the regression of experimental data as:

$$N_{bc} = \frac{1.2 \times 10^4}{\delta} \quad (1)$$

Where N_{bc} is burner thermal cycling lifetime of the TBC, and δ is TGO thickness.

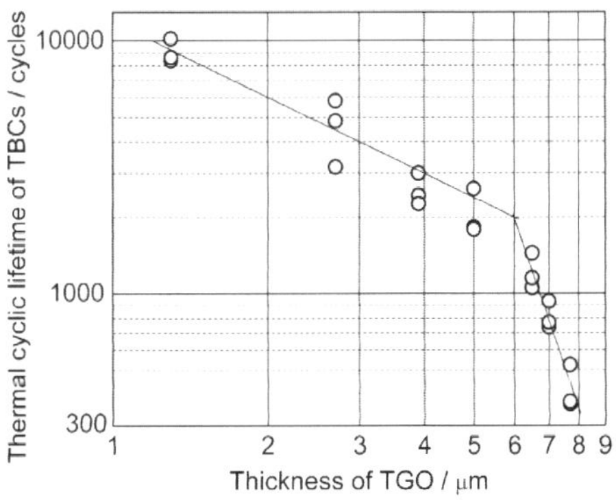

Fig.5 Effect of TGO thickness on the thermal cyclic lifetime under burner cycling test with a gradient thermal [12]

Fig.6 shows the effect of TGO thickness on the lifetime of TBCs under isothermal furnace thermal cycling. With the increase of TGO thickness the lifetime of TBCs presents a decreasing tendency, which is clearly similar to that observed for burner thermal cycling test as shown in Fig.5. In the same way, at the small TGO thickness region, the thermal cycling lifetime can be obtained according to the regression result of the experimental data shown in Fig.6 as:

$$N_{fc} = \frac{144}{\sqrt{\delta}} \quad (2)$$

Compared with the results shown in Fig.5 and Fig.6 it can be clearly recognized that either under the burner thermal cycling or furnace thermal cycling TGO in TBC presents the effect with the comparable tendency on the lifetime. The relationship between thermal cyclic lifetime can be defined by two similar power functions with different negative powers, which can be divided by the same characteristic TGO thickness despite burner thermal cycling or furnace thermal cycling. As discussed later such characteristic TGO thickness corresponds to a critical TGO thickness over which the failure mode of TBCs changes.

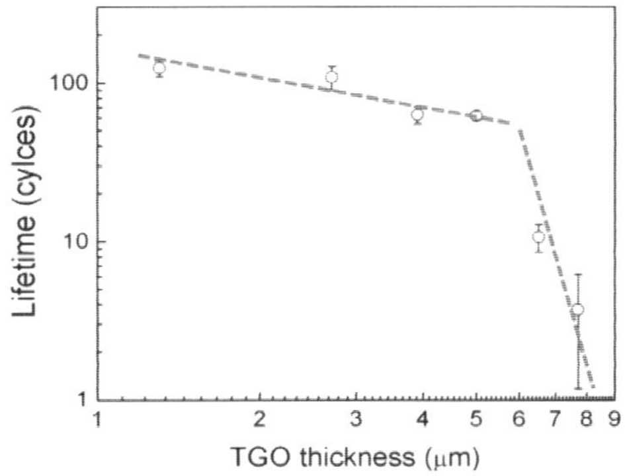

Fig.6 Effect of TGO thickness on the thermal cyclic lifetime under furnace isothermal cycling test

3.3 Consistence of TBC failure modes

The previous examination showed that with TGO thickness increasing TBCs present two typical failure modes [12]. At the TGO thickness less than 5 μm the spalling of top YSZ coat occurs through propagation of cracks along lamellar interfaces within YSZ above the TGO. Fig.7a shows the typical crack observed in the TBC with 1.3 μm thick TGO where YSZ still survived after TBC sample failure was observed at furnace cycling test. It illustrated that the cracks propagated in YSZ several tens of micrometers away from the interface between the bond coat and YSZ. On the other hand, Fig.7c shows the cross sectional microstructure observed in the region where YSZ spalling occurred after burner thermal cycling. It was found that YSZ of several tens of micrometers covered over substantial bond coat surface. This fact means that under both furnace thermal cycling and burner cycling tests the TBC failure occurs at the same mode. Moreover, when the initial TGO thickness becomes larger than 5 μm, it was recognized that TBC spalling occurs mainly from the interface between TGO and YSZ and/or interface between TGO and bond coat as shown in Fig.7b and d for failure TBC samples which experienced burner and furnace cycling tests, respectively. Thus, after YSZ spalling little YSZ adhered to the bond coat as shown in Fig.7d. Apparently, two types of failure modes correspond to two different TGO thickness regions in the same fashion for two types of thermal cycling tests. This may the essential cause that under two different types of thermal cycling tests the lifetime presents the comparable dependency against TGO thickness.

3.4 The critical TGO thickness for failure mode transition

Although for the TBC with an initial TGO of 1.3μm thick the TBC withstood up to 10,000 cycles under the burner cycling before failure the increment of TGO thickness after the test was observed to be limited. This is because the bond coat surface was exposed to a low temperature of 1000°C for much limited accumulative duration. On the other hand, since the furnace thermal cycling test was carried out at a temperature of 1120°C, TGO grows remarkably compared with the bond coat surface temperature of 1000°C under burner cycling test. Thus the thickness of TGO in the TBC samples was measured after TBC failure was observed for the furnace cycling test. Fig.8 shows the comparison of TGO thickness after TBC failure with those in the the initial TBC samples. It was found that with the TBC samples of TGO less than 5 μm thick after failure the TGO grew to a comparable thickness of ~ 5μm. Moreover, from Fig.6 it was clearly shown that when TGO thickness increases to over 5 μm the lifetime decreases rapidly with the increase of TGO thickness. Therefore, the TGO thickness corresponding to the point where two straight lines intercept in Fig.6 can be defined as the critical TGO thickness.

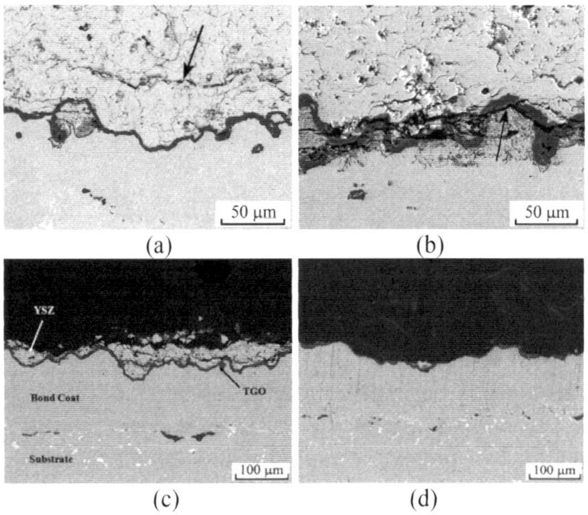

Fig.7 Cross sectional microstructures of TBCs with different initial TGO thicknesses after failure: (a) 1.3 μm TGO under burner cycling,(b) 6.5 μm TGO under burner cycling,(c) 1.3 μm TGO under furnace cycling,(d) 6.5 μm TGO under furnace cycling

As discussed above over the critical TGO thickness the failure mode transforms from cracking within YSZ to TGO interfaces. With the present TBC system the critical TGO thickness for failure mode transition becomes about 5 μm. From the results shown in Figs.5 and 6, when the TGO thickness reaches to over 8 μm, the lifetime of the present TBC decreases rapidly towards zero. Those factors reveal the critical TGO thickness at spalling of APS YSZ. This thickness of about 8 to 9 μm is consistent with the data reported in the literature [7]. Therefore, two conceptual critical TGO thicknesses can be introduced to explain the failure behavior. The one small critical TGO thickness of about 5μm to 6 μm corresponds to the transition of cracking path from inside of YSZ to two interfaces across TGO. Another larger critical TGO thickness corresponds to final critical one for complete spalling of top ceramic coating. All those factors indicate that for long durable TBCs the TGO growth plays a dominant role on TBC failure.

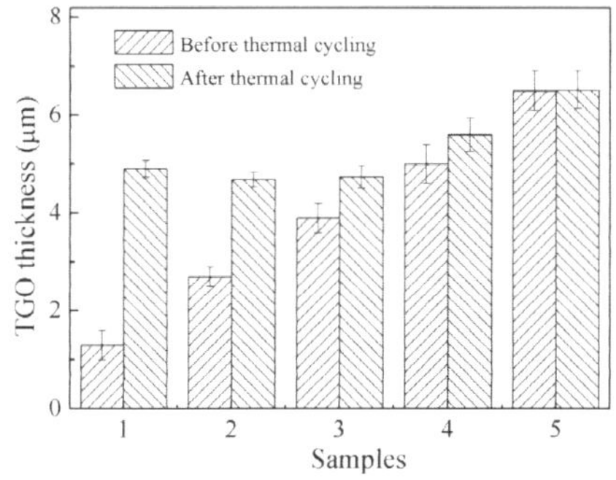

Fig.8 Comparison of TGO thickness at TBC failure after the furnace cycling test with initial TGO thickness for TBC samples with different initial TGO thicknesses (No.1: 1.3 μ m, No.2: 2.7 μm, No.3: 3.9 μm m, No.4: 5.0 μm, No.5: 6.5 μm) [13]

3.5 Comparison of apparent driving factors for TBCs failure

Many factors influence the durability and thus lifetime of TBCs during thermal cycling tests, including substrate properties, coating materials and microstructures, and test conditions. In the present study, the same TBC system was tested under different TGO thicknesses by both the burner and furnace cycling tests. It is clear that the test condition and TGO thickness influence the lifetime significantly. However, as discussed above the failure modes and its dependency on TGO thickness are the same. Those facts suggest that using two different tests mainly alters crack evolving rates with cracks propagating at different rates. On the other hand, it is clear that the thermal cycling lifetime presents a remarkable difference by two orders of magnitude for two type tests. Comparing such large difference in the lifetime, it is evident that the dominant driving factors for cracks evolution leading to TBC failure are different in two tests.

Since lower temperature of the bond coat surface and limited accumulative exposure time to high temperature under burner cycling test, TGO growth over whole test duration is limited under burner cycling test. Thus, frequent cycling is the driving factor rather than TGO growth. The lifetime will depends on cooling rate during thermal cycling and thermal gradient within TBC during tests besides TGO thickness [15,16]. According to the dependency of the lifetime on TGO thickness, the effect of TGO on cracking at a given TGO thickness can be underlined. On the other hand, during the furnace cycling test since the lifetimes are much limited the TGO growth during the test becomes driving factor. Of course, the quenching stress associating with rapid heating and/or cooling during thermal cycling affects the absolute lifetime under certain test conditions. Therefore, the results of the burner cycling test, especially under such test conditions that during whole test duration TGO growth is limited, represent the resistance of top ceramic coating to crack evolution and propagation. On the other hand, the furnace cycling test results represent the integrated effects of the bond coat oxidation resistance and the top ceramic coat cracking resistance. Therefore, although the results of the lifetime against TGO thickness for two type tests are presented with similar dependency on TGO in the present study, using the burner cycling test under the present conditions the performance of top ceramic coatings can be characterized and ranked. On the other hand, using the furnace cycling tests the integrated performance of a top coat and a bond coat is evaluated. Thus, for the optimization of top ceramic coatings the burner cycling test would be more suitable.

4. Conclusions

Using APS 8YSZ TBC with well controlled TGOs of different thicknesses the failure behavior of TBCs under both the burner and furnace cycling tests was examined. Results revealed that the thermal cyclic lifetime of the TBC under two type tests decreased following the power functions with the change of the TGO thickness. The same failure modes were observed for TBCs under two types of thermal cycling tests. Both the tests showed that there exists a critical TGO thickness from 5 to 6.0 μm for the transition of the TBC failure mode from cracking within YSZ, i.e. cohesive failure mode to cracking from the interfaces around TGO. Although based on the same failure modes and the same dependency of the lifetime on TGO thickness either of the tests can be used to evaluate the integrated performance of TBCs, it is clear that by adjusting test condition thc burncr cycling test results can be more suitable for evaluating the performance of top ceramic coating in TBCs. While the furnace cycling test results integrate the effects of both the bond coat properties, especially oxidation resistance, and ceramic top coating cracking resistance. Thus, the burner cycling and furnace cycling tests could be employed for examining dominant effect of typical factors controlling TBC failure.

Acknowledgements

The present project is supported by National Basic Research Program (Grant No. 2012CB625100).

Literature

[1] C. U. Hardwickre and Y. C. Lau, Advances in Thermal spray coatings for gas turbines and energy generation: A Review, J. Therm. Spray. Technol., 22[5] (2013) 564-576
[2] R.A. Miller, Thermal barrier coatings for aircraft engines: history and directions, J. Thermal Spray Technol., 6[1] (1997) 35-42
[3] N. P. Padture, M. Gell, and E.H. Jordan, Thermal barrier coatings for gas-turbine engine applications, Science, 296 (2002) 280-284
[4] C. J. Li and A. Ohmori, Relationships between the microstructure and properties of thermally sprayed deposits, J. Thermal Spray Technol., 11[3] (2002) 365-374
[5] C.-J. Li, G.-J. Yang, C.-X. Li, Development of the particle interface bonding in thermal spray coatings: A Review, J. Thermal Spray Technol., 22[2] (2013) 192
[6] C-J Li, Y. Li, G. J. Yang, and C. X. Li, A Novel plasma-sprayed durable thermal barrier coating with a well-bonded YSZ interlayer between porous YSZ and bond coat, J. Therm Spray Technol, 21 (2012) 383–390
[7] R.S. Lima, D. Nagy, and B.R. Marple, Bond coat engineering influenceon the evolution of the microstructure, bond strength, and failure of TBCs subjected to thermal cycling, J. Therm Spray Technol, 24 (2015)152-159
[8] R. Vassen, S. Giesen, D. Stover, Lifetime of plasma-sprayed thermal barrier coatings: Comparison of numericaland experimental results, J. Therm Spray Technol., 18 (2009) 835-845
[9] R. Vassen , Y. Kagawa , R. Subramanian, P. Zombo, and D. Zhu, Testing and evaluation of thermal-barrier coatings, MRS Bulletin, 37 (2012) 911-916
[10] Y.Z. Xing, C.-J. Li,Q. Zhang, C.X. Li, G.J. Yang, Influence of microstructure on the ionic conductivity of plasma-sprayed yttria-stabilized zirconia deposits, J. Am. Ceram. Soc., 91 [12] (2008), 3931-3936
[11] Y. Li, C.-J. Li, Q. Zhang, G.-J. Yang, C.-X. Li, Influence of TGO composition on the thermal shock lifetime of thermal barrier coatings with cold-sprayed MCrAlY bond coat, J. Therm Spray Technol., 19[1-2] (2010) 168-177
[12] H. Dong, G.-J. Yang, C.-X. Li, X.-T. Luo, C.-J. Li, Effect of TGO thickness on thermal cyclic lifetime and failure mode of plasma-sprayed TBCs, J. Am. Ceram. Soc., 97 [4] (2014) 1226-1232
[13] H. Ding, H. Dong, C.-J. Li, G.-J. Yang, Effect of TGO thickness on isothermal cyclic lifetime of plasma-sprayed YSZ thermal barrier coatings, Proceedings of ITSC'2015, Long Beach, USA, May 11-14, 2015, ASM International, 801-805
[14] Y. Li, C.-J. Li, G.-J. Yang, L.-K. Xing, Thermal fatigue behavior of thermal barrier coatings with the MCrAlY bond coats by cold spraying and low-pressure plasma spraying, Surf. Coat. Technol., 205 (2010) 2225–2233
[15] H. Dong, G.-J. Yang, H.-N. Cai, H. Ding, C.-X. Li, C.-J. Li, The influence of temperature gradient across YSZ thickness on thermal cyclic lifetime of plasma-sprayed thermal barrier coatings, Ceram. Int., 41[9] (2015)11046-11056
[16] H. Dong, C.-J. Li, H. Ding, G.-J. Yang, C.-X. Li,

Effect of MCrAlY bond coat temperature on thermal cyclic lifetime of plasma-sprayed thermal barrier coatings, Proceedings of ITSC'2015, Eds., A. McDonald, A. Agarwal, G. Bolelli, A. Concustell, Y.-C. Lau, F.-L. Toma, E. Turunen, C. Widener, Long Beach, USA, May 11-14, 2015, ASM International, 795-800

Performance of HVOF Sprayed Ni based Coating on T22 Steel in Actual Boiler Environment

Niraj Bala, BBSBEC, Fathegarh Sahib,Punjab, India
Harpreet Singh, IITR, Ropar, Punjab, India
Satya Prakash, IITR, Roorkee, Uttrakhand, India

High Velocity Oxy Fuel (HVOF) spray process represents a significant part of overall thermal spray business. Ni based alloys have been used for thermal spraying for various applications. In the present work commercially available Ni-20Cr blended with TiC powder was successfully deposited on a boiler steel (SAE 213-T22). To evaluate the performance of the coating in actual conditions the bare as well as the coated steels were subjected to cyclic exposures, in the superheater zone of a coal fired boiler for 15 cycles. The weight change and thickness loss data were used to establish kinetics of the erosion-corrosion. X-ray diffraction and FE-SEM/EDS techniques were used to analyse the as-sprayed and eroded-corroded specimens. The better erosion-corrosion resistance of the coating in comparison to the uncoated steel may be attributed to the formation of oxides and spinels of nickel and chromium.
Keywords: T22 boiler steel; coating; HVOF spray

1 Introduction

High temperature corrosion accompanied by erosion is a severe problem, which may result in premature failure of the boiler tubes. One countermeasure to overcome this problem is the use of thermal spray protective coatings. In recent years, high velocity oxygen fuel (HVOF) spraying, a novel variation of combustion spraying has had a dramatic influence on the field of thermal spray. Due to the high velocity and high impact of the sprayed powder, the coating produced is less porous and has higher bond strength than that produced by other methods [1]. The HVOF spray process represents the state-of-the-art for thermal spray metallic coatings and results in very dense, tightly adherent coatings with little or no oxidation during the application and low residual stresses [2]. Ni based alloys represent a significant part of overall thermal spray business. These materials are widely used as bond coats and as top coats in number of applications requiring combination of properties, such as good wear resistance and corrosion resistance at the same time [3-6].

Recently, there has been significant interest in the production of TiC based cermet coatings by HVOF spraying for applications where resistance to wear and oxidation or corrosion is required [7-9]. In the present work, peformance of the erosion-corrosion behavior of two coatings Ni-20Cr and Ni-20Cr blended with 24wt% TiC (Ni-20Cr-TiC) deposited by HVOF thermal spray technique on T22 boiler steel has been studied in the actual industrial environment of a coal fired boiler plant.

2. Experimental Procedure

2.1 Substrate and Feedstock Powder

The substrate material selected for the study was 2.25Cr-1Mo steel, designated ASTM-SA213-T-22 (T22), with chemical composition C 0.15, Mn 0.3-0.6, P 0.03 max, S 0.03 max, Si 0.5, Cr 1.9-2.6, Mo 0.87-1.13 and Fe 94.66 (weight %), procured from Guru Gobind Singh Super Thermal Power Plant, Ropar (India). The specimens measuring approximately 20mm x 15mm x 5mm, were cut from the fresh boiler tubes and were polished down to 180 grit SiC paper finish and grit blasted with Al_2O_3 (grit 60) before the deposition of the coating. Two types of coating powder compositions were deposited on the steel, which included commercially available Ni-20Cr and Ni-20Cr blended with 24wt% TiC (Ni-20Cr-TiC) powder.

2.2 Deposition Technique and Equipment

HVOF spray coatings were performed using commercially available HVOF-spray (HIPOJET- 2100) thermal spray process at Metallizing Equipment Company Private Limited, Jodhpur, India, operating with oxygen and liquid petroleum gas (LPG) as input gases. The average coating thickness obtained was 200µm. The substrate steels were cooled by compressed air jets during and after spraying. The coating parameters adopted are reported in Table Table 1 [10].

Table 1 Spray parameters employed for HVOF coating

Oxygen flow rate	200 SLPM
Fuel (LPG) flow rate	50 SLPM
Air-flow rate	900 SLPM
Spray distance	20 cm
Powder feed rate	25-30 gm/min
Fuel pressure	6.00 kg/cm^2
Oxygen pressure	8.00 kg/cm^2
Air pressure	6.00 kg/cm^2

2.3 As-Sprayed Coating Characterization

X-ray diffraction (XRD) analysis and Field Emission Scanning Electron Microscope/Energy Dispersive Spectroscopy (FE-SEM/EDS) of the coated samples was carried out, the procedure of the same has been elaborated elsewhere [10, 11]. After the surface characterization, the samples were sectioned, mounted in transoptic powder and subjected to polishing, using emery papers of 220, 400, 600 grit and subsequently 1/0, 2/0, 3/0 and 4/0 grades. Fine polishing was carried out to obtain a mirror finish using a 0.3 μm diamond paste. The cross-sectional microhardness of the coating was measured with a load of 2.94 N using the Digital Micro Vickers Hardness tester (SHV-1000, Chennai Metco Pvt. Ltd, India). The polished samples were characterized to obtain their cross-sectional morphology and compositions by using the FE-SEM/EDS analysis.

2.4 Erosion-corrosion (E-C) studies in coal-fired boiler

The coated, as well as, uncoated samples were hung through soot blower dummy points in the middle zone of the low temperature superheater of the Stage-II Boiler of Guru Gobind Singh Super Thermal Power Plant (GGSSTPP), Punjab (India) for erosion-corrosion studies under cyclic conditions. In this zone the temperature was about 700±10°C and volumetric flow of flue gases was around 700 tonnes/hr. Flue gases contain 16% CO_2 and 3% O_2 by volume. The chemical analysis of the coal and that of ash and flue gas inside the said boiler is reported elsewhere [10]. The studies were conducted for 15 cycles; each cycle consisted of 100 hours exposure to the boiler environment followed by 1 hour cooling at ambient conditions. The weight change data were taken after each cycle along with physical observations to approximate the kinetics of E-C. The thickness was measured using a Sylvac micrometer screw gauge (Swiss make, resolution 0.001) and represented value was taken as the average of 3 measured values [10].

2.5 Analysis of the exposed samples

The samples, after exposure to erosion-corrosion were analysed for the characterization of corrosion products. The analysis was performed for surface and cross-section of the exposed samples. Corroded samples were subjected to XRD, surface and cross-sectional FE-SEM/EDS analysis. However, in this case, the weight change data could not be of much use for predicting corrosion behaviour due to suspected spalling and ash deposition on the samples. Furthermore, the weight change data might also have been affected by the erosion due to flyash. In view of above facts, the extent of E-C has also been evaluated by measuring the thickness of the scale lost due to erosion, spalling or evaporation by finding the difference in thickness of the specimens before and after 1500 hours exposure to the actual environment. From the thickness loss data corrosion rates in mils/year (mpy) were evaluated as per the following example:

If total thickness loss is y mm for 1500 hours. In a year there are 8760 hours. 1 mm is equal to 39.37 mils. So mpy is y x 39.37 x 8760/1500 [10].

3. Experimental Results

3.1. Microhardness

The microhardness of HVOF sprayed Ni-20Cr coating has been found to be in the range of 175 to 351 Hv, with an average value of 267 Hv for the coating, while the substrate steel has an average microhardness of 166 Hv. Evidently, the microhardness of the coating is higher than that of the substrate steel. Similarly, in case of Ni-20Cr-TiC coating the microhardness lies in the range of 258-394 Hv with an average value of 326 Hv. Obviously, the microhardness of the Ni-20Cr-TiC sprayed coating is found to be higher.

3.2 XRD analysis of as-sprayed coatings

The XRD diffractogram of the HVOF spray Ni-20Cr coating on T22 boiler steel in as-sprayed condition indicated the formation of Ni and NiO phases. Whereas, for Ni-20Cr-TiC coating Ni and TiO_2 phases were observed

3.3 FE-SEM/EDS analysis of the as-sprayed coatings

3.3.1. Surface Morphology

The FE-SEM for Ni-20Cr-TiC coating has been depicted in **Fig. 1**, which represents a micro-structure consisting of well-formed splats at most of the regions. At some other locations, fine grained structure has been revealed. The coating at point 1 contains 42% C, 27% O, alongwith 15% Ni, 11% Ti, and 5% Cr, which indicates the probable formation of oxides and carbides of the said elements at certain locations. At point 2, significant presence of Ni along with Cr can be seen. The FE-SEM micrograph of HVOF sprayed Ni-20Cr coating in as-sprayed condition has been discussed elsewhere [10].

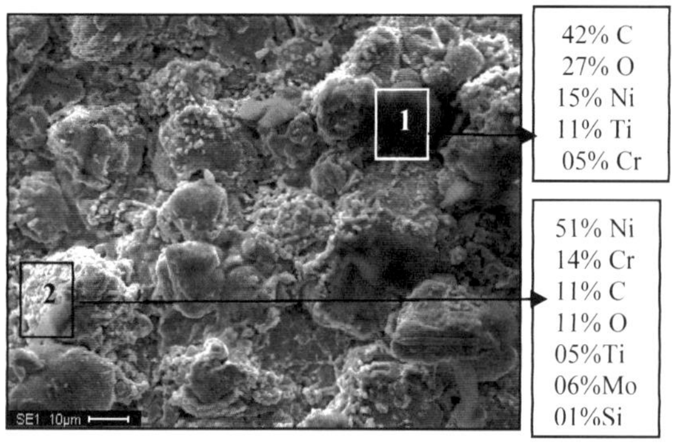

Fig.1 Surface morphology and EDS analysis of HVOF sprayed Ni-20Cr-TiC coating on T22 boiler steel showing elemental composition (%) at selected points.

3.3.2. Cross-sectional Analysis

The FE-SEM of the cross-section of Ni-20Cr-TiC coating on T22 steel represents a dense structure with layered morphology with only a marginal presence of micropores **(Fig. 2).** The coating/substrate interface is faultless. The concentrations of Ni and Cr are found to be constant from points 3 to 6. Similarly, carbon is present in significant amounts along the cross-section. 4% of Ti is seen at point 3.Though Fe shows its presence upto the coating/substrate interface (point 2), whereas minor amounts of Mo have been indicated at points 1, 2 and 5. The FE-SEM cross-sectional analysis of HVOF sprayed Ni-20Cr coating in as-sprayed condition has been discussed elsewhere [10].

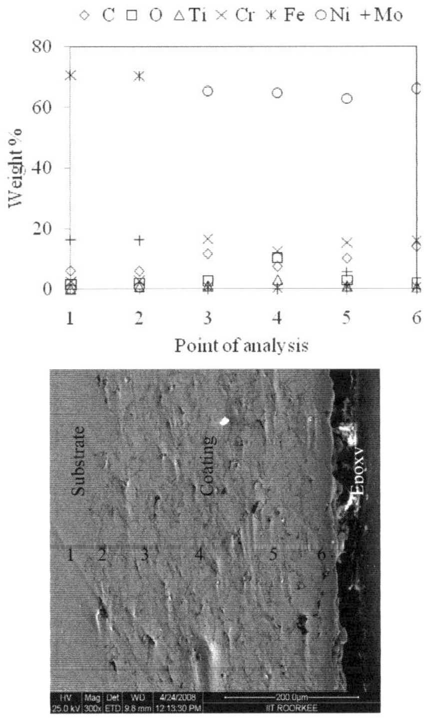

Fig. 2 Surface scale morphology and variation of elemental composition across the cross-section of HVOF sprayed Ni-20Cr-TiC coating on T22 steel.

3.4 Weight change and thickness lost data for erosion-corrosion studies

Weight change per unit area expressed in mg/cm^2 as a function of time expressed in hours for bare and coated T22 steel in actual boiler environment has been depicted in **Fig. 3**. The uncoated T22 steel showed a rapid weight gain at the end of 1st cycle. The plot represents a decrease in weight gain from the 2nd to 4th cycle. Since no spallation was observed till the end of the 4th cycle the weight loss may be attributed to the erosion of the sample by ash particles. Else, there is a formation of some volatile species, which escaped from the steel surfaces. However, the probability of the latter is marginal at the working temperature of 700°C. Followed by this an increase in weight was observed from 4th to 8th cycle. A sharp decline in weight was observed at the end of 8th cycle which might have been due to the spallation of the scale. After the 9th cycle not much weight changes have been observed. The steel showed an overall weight gain of 1.596 mg/cm^2 during erosion-corrosion in the actual boiler environment, which indicates that the process might have taken place by the growth of oxide scale on the surface. Else, there is the possibility of deposition of ash particles on the sample. The thickness of metal lost due to the exposure is 0.54 mm. The corrosion rates have been calculated based upon this thickness data in term of mils per year (mpy). Corrosion rate indicated by T22 steel is 124.16 mpy. After the deposition of the HVOF sprayed coatings on T22 steel, overall weight losses were observed during the early cycles of exposure. However, after an initial weight loss, the weight change became insignificant with the progress of experiment for both the coated steels. The overall weight loss has been found to be 4.472 and 4.893 mg/cm^2 for Ni-20Cr and Ni-20Cr-TiC coating. The thickness of metal lost after 1500 hours of exposure to boiler environment is measured as 0.1 mm and 0.13 mm for the Ni-20Cr and Ni-20Cr-TiC coating respectively. Corrosion rate has been calculated as 22.992 and 29.89 mpy for the the Ni-20Cr and Ni-20Cr-TiC coating respectively

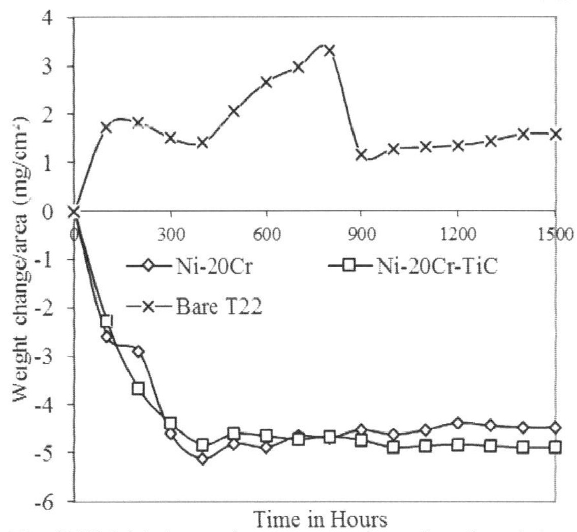

Fig. 3 Weight change/area vs. number of cycles plot for the bare and coated T22 steel subjected to actual boiler environment at 700°C for 1500 hours.

3.7. XRD analysis of exposed samples

In the given environment the bare steel indicated the formation of Fe_2O_3 as the main constituent of scale along with peaks of Al_2O_3 and Cr_2O_3 phases. The prominent phases revealed by the analysis for the Ni-20Cr coating were Cr_2O_3, Al_2O_3 and Fe_2O_3. Similar phases have been indicated for the scale of the Ni-20Cr-TiC coated steel; in addition NiO, $NiCr_2O_4$, TiO_2 and TiO phases have also been indicated.

3.8. FE-SEM/EDS Analysis

3.8.1. Surface analysis

The FE-SEM micrograph for the eroded-corroded surface of Ni-20Cr-TiC coated T22 base steel subjected to actual boiler environment at 700°C for 1500 hours is represented in **Fig. 4**. The oxide scale for the eroded-corroded Ni-20Cr-TiC coating represents low lying area embedded with ash particles. The composition of scale at point 1 is showing the presence of flyash particles, whereas the top scale is mainly having Cr, O and Ni. The surface FE-SEM analysis of exposed Ni-20Cr coating has been discussed elsewhere [10].

3.8.2 Cross-sectional Analysis

The morphology along the cross-section of Ni-20Cr-TiC coating (**Fig. 5**) a laminar layered adherent scale, having the presence of some oxides and or carbides. Small amount of Al (3%) is seen at point 6. Concentration of Cr is almost constant from point 2 to 5. Whereas, Ni has an enhanced presence at point 3, 5 and 6. Ti and C represent a maxima at point 4. High amount of O has been observed at the scale-substrate interface. Fe shows its presence upto point 2. The FE-SEM cross-sectional analysis of exposed Ni-20Cr coating has been discussed elsewhere[10].

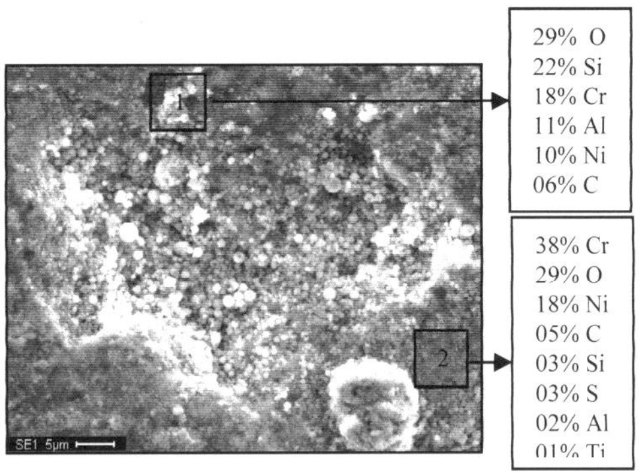

Fig. 4 Surface morphology with EDS analysis for the bare and Ni-20Cr coated T22 boiler steel subjected to actual boiler environment at 700°C for 1500 hours.

4. Discussion

4.1. As- sprayed coatings

Ni-20Cr and Ni-20Cr-TiC powder was successfully deposited on T22 boiler steel by the HVOF spray process. The hardness of Ni-20Cr-TiC coating was found to be more in comparison to the conventional Ni-20Cr counterpart in the present investigation.

Ni was observed as the main phase in both the coatings. Ni as the main phase for as-sprayed Ni-20Cr has also been observed by Singh [12], and Singh [13] for plasma sprayed Ni-20Cr coatings. The phases identified by XRD analysis for as-sprayed coatings are further confirmed by the surface EDS analysis. The presence of O as indicated by the EDS analysis in the case of HVOF sprayed coatings indicated the formation of oxide inclusions. Moreover, an additional NiO phase has been identified for the HVOF sprayed Ni-20Cr coating and further TiO_2 phase was also identified for the Ni-20Cr-TiC coating. The FE-SEM analysis indicated a typical splat-like morphology for HVOF sprayed coatings. The EDS observations represented the presence of oxygen in substantial amounts which confirmed the formation of oxide phase in the microstructure of the coatings. Similar findings of the presence of oxide phases in the microstructure of HVOF coatings have been reported by Dent et al [14], Kong et al [15], and Zhang et al [16].

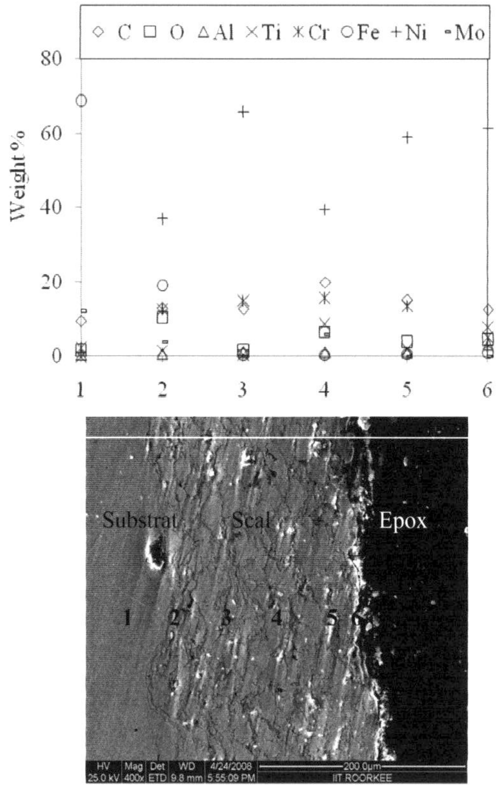

Fig. 5 Oxide scale morphology and variation of elemental composition across the cross-section of HVOF sprayed Ni-20Cr-TiC coated T22 boiler steel subjected to actual boiler environments at 700ºC for 1500 hours.

The cross-sectional FE-SEM images of HVOF sprayed coating represented a dense appearance with layered morphology. The coating/substrate interfaces were found to be intact and continuous in general. The coating elements were restricted to the coating region.

4.2. Erosion-corrosion studies

The HVOF sprayed Ni-20Cr, Ni-20Cr-TiC coating on T22 steel showed good resistance to erosion-corrosion (E-C) in the actual boiler environment. No visual crack was seen and the scales in general were intact in all the cases. This trend of weight change may be due to the erosion of micro- hills present on the surfaces of the specimen as the coal used in Indian power stations has large amounts of ash (about 50%), which contain abrasive mineral species such as hard quartz (up to 15%) which increase the erosion propensity of coal [17]. However, once the surfaces became smooth, no further erosion-corrosion was noticed. This indicates towards the useful nature of the coatings to control erosion-corrosion during longer durations of exposure.

The corrosion rate of Ni-20Cr coating was found to be lower in comparison to Ni-20Cr-TiC. The XRD revealed the formation of Cr_2O_3, Al_2O_3 and Fe_2O_3 phases for the eroded-corroded scale of Ni-20Cr coating. Nicholl and Wahl [18], Stroosnijder et al [19], and Toma et al [20] suggested that Cr_2O_3 inhibits oxidation of the substrate alloys by blocking the diffusion of reacting species. Moreover, Cr_2O_3 being significantly harder phase can also provide erosion resistance alongwith the corrosion resistance. The formation of Fe_2O_3 phase might have been due to diffusion of Fe from the substrate as has been suggested by Singh [12], and Sundararajan et al [21]. Else it might have come from the fly ash present in the environment of the boiler. The Al_2O_3 phase might be due to the ash depositions. The EDS analysis further confirmed the formation of Cr_2O_3 at the outer region,.

Although the weight gain data as well as corrosion rate data show that Ni-20Cr-TiC coating provides relatively less protection to the base steel in the given boiler environment as compared to Ni-20Cr coating, but the coating was successful to reduce the corrosion rate of steel by 76%. The cross-sectional FE-SEM analysis showed that the oxidized coating was in good contact and adherent to the substrate and free from cracks/defects **(Fig.5).** NiO, Cr_2O_3, $NiCr_2O_4$, TiO_2, TiO and Al_2O_3 and Fe_2O_3 phases were identified for the oxide scale of coating. NiO and Cr_2O_3 phases are protective in nature against E-C, especially against corrosion. Presence of spinel $NiCr_2O_4$ in the surface oxides also helps in developing corrosion resistance as spinel phases usually have much smaller diffusion coefficients for the cations and anions than those in their parent oxides [22]. Formation of TiO_2 phase was also identified by Zeng et al [23], in their air oxidation studies on Ni-Ti alloys at 650°C. According to them the TiO_2 phase acts as a diffusion barrier to the outward diffusion of Ni.

5. Conclusions

1. Ni-20Cr and Ni-20Cr-TiC powder was successfully deposited on T22 boiler steel by the HVOF spray technique at the given process parameters.
2. In actual boiler environment the uncoated steel showed an overall weight gain, while both HVOF sprayed coatings represented an initial weight loss followed by only marginal weight change.
3. Both HVOF sprayed Ni-20Cr and Ni-20Cr-TiC coatings were successful to reduce the erosion-corrosion rate of T22 steel by 82% and 76 % respectively in actual boiler environment. The coatings were found to retain their surface contact with the steels without any signs of scale spallation.

Ni-20Cr > Ni-20Cr-TiC >Bare T22 ste

6. References

[1] Tan, J.C., Looney, L. and Hashmi M.S.J.: Component Repair Using HVOF Thermal Spraying. J. Mater. Processing Technol. 92-93 (1999), pp. 203/208.

[2] Fedrizzi, L., Rossi, S., Cristel, R. and Bonora, P.L.: Corrosion and Wear Behaviour of HVOF Cermet Coatings Used to Replaced Hard Chromium. Electrochimica Acta 49 (2004), pp. 2803/2814.

[3] Hidalgo, V.H., Varela, F.J.B. and Rico, E.F.. Erosion Wear and Mechanical Properties of Plasma-sprayed Nickel and Iron-based Coatings Subjected to Service Conditions in Boilers. Trib. Int. 30 (1997), Issue 9, pp. 641/649.

[4] Wang, B.Q.: Erosion-corrosion of Thermal Sprayed Coatings in FBC Boilers. Wear, 199 (1996), pp. 4/32.

[5] Koivuluoto, H. and Vuoristo, P.: Effect of Ceramic Particles on Properties of Cold-sprayed Ni-20Cr + Al_2O_3 Coatings. J. Therm. Spray Technol. 18 (2009), Issue 4, pp. 555/562.

[6] Koivuluoto H. and Vuoristo, P. Structural Analysis of Cold-Sprayed Nickel-Based Metallic and Metallic-Ceramic Coatings. J. Therm. Spray Technol., 19 (2010), Issue 5, pp. 975/989.

[7] Paco, J.M., De Nutting, J., Guilemany, J.M., Sanchez, F.J., Miguel, J.R. Smith, P.: Structure-property relationships of Ti-C-Ni+Ti and (Ti,W)C-Ni powders manufactured by SHS process, and the result- and HVOF-sprayed coatings, Thermal Spray: A United Forum for Scientific and Technological Advances, C.C. Berndt, Ed. Sept 15-18, 1997 (Indianapolis, IN), ASM International, (1997) pp. 935/942.

[8] Blatchford, M.T. Horlock, A.J. Shipway, P.H. McCartney, D.G. Wood, J.V.: Production and characterization of HVOF sprayed Ni(Cr)-TiC coatings using SHS powder feedstock, Thermal Spray: Surface Engineering via Applied Research, C.C. Berndt, Ed. May 8-11, 2000 (Montreal, Quebec, Canada), ASM International, 2000, pp. 515/521.
[9] Guilemany, J.M.. De Paco, J.M Miguel, J.R Sanchez, J. Smith, P. Corrosion resistance HVOF coatings based upon TiC+ NiTiC and (Ti, W)C + Ni, Thermal Spray: Meeting the Challenges of the 21st, Century, C. Coddet, Ed., May 25-29, 1998 (Nice, France), ASM International (1998) pp. 57/61.
[10] Bala, N., Singh, H., and Prakash, S.: High Temperature Corrosion Behavior of Cold Spray Ni-20Cr Coating on Boiler Steel in Molten Salt Environment at 900°C. Journal of Thermal Spray Technology, 19 (2010), pp. 110/118.
[11] Bala, N., Singh, H., and Prakash, S.: High Temperature Oxidation Studies of Cold Sprayed Ni-20Cr and Ni-50Cr Coatings on SAE 213-T22 Boiler Steel. Appl. Surf. Sci. 255 (2009), Issue15, pp. 6862/6869.
[12] Singh, Buta, (2003), Studies on the Role of Coatings in Improving Resistance to Hot Corrosion and Degradation, Ph.D. Thesis, Met. & Mat. Eng. Dept., Indian Institute of Technology Roorkee, Roorkee, India.
[13] Singh, Harpreet, (2005), Hot Corrosion Studies of Plasma Sprayed Coatings Over Some Ni- and Fe-Based Superalloys, Ph.D. Thesis, Met. & Mat. Eng. Dept., Indian Institute of Technology Roorkee, Roorkee, India.
[14] Dent, A.H., Horlock, A.J., McCartney, D.G. and Harris, S.J.: Microstructural Characterisation of a Ni-Cr-B-C Based Alloy Coating Produced by High Velocity Oxy-Fuel Thermal Spraying. Surf. Coat. Technol. 139 (2001), pp. 244/250.
[15] Kong, G., Zhang, D., Brown, P.D., McCartney, D.G. and Harris, S.J.: Microstructural Characterisation of High Velocity Oxy-Fuel Thermally Sprayed Stellite-6. Mater. Sci. Technol. 19 (2003), pp. 1003/1011.
[16] Zhang, D., Harris, S.J. and McCartney, D.G.: Microstructure Formation and Corrosion Behaviour in HVOF-Sprayed Inconel 625 Coatings. Mater. Sci. Eng. A 344 (2003), pp. 45/56.
[17] Krishnamoorthy, P.R. and Seetharamu, S.: The Wear Life of Pipe Bends and other Components under Pulverized Coal Erosion, Proc. of LIPREX-89, Hyderabad, India, January 1989.
[18] Nicoll, A.R. and Wahl, G.: The Effect of Alloying Additions on M-Cr-Al-Y Systems-An Experimental Study. Thin Solid Films 95, (1983) pp. 21/34.
[19] Stroosnijder, M.F., Mevrel, R. and Bennet, M.J.: The Interaction of Surface Engineering and High Temperature Corrosion Protection. Mater. High Temp. 12, (1994), Issue 1, pp. 53/66.
[20] Toma, D., Brandl, W. and Koester, U.: Studies on the Transient Stage of Oxidation of VPS and HVOF Sprayed MCrAlY Coatings. Surf. Coat. Technol. 120-121 (1999), pp. 8/15
[21] Sundararajan, T., Kuroda, S., Itagaki, T. and Abe, F.: Behaviour of Mn and Si in the Spray Powders during Steam Oxidation of Ni-Cr Thermal Spray Coatings. ISIJ Int. 44 (2004), pp. 139/144.
[22] Chatterjee, U.K., Bose, S.K. and Roy, S.K., (2001), 'Environmental Degradation of Metals,' Pub. Marcel Dekker, 270 Madison Avenue, New York
[23] Zeng, C.L., Li, M.C., Liu, G.Q. and Wu, W.T.: Air Oxidation of Ni-Ti Alloys at 650- 850°C. Oxid. Metals 58 (2002), Issue 1-2, pp. 171/184

Performance of Thermally Sprayed Corrosion Resistant Alloy (CRA) Coatings on Carbon Steel in Supercritical CO_2 Environments at Different Temperatures

S Paul, Cambridge / UK

This paper reports the performance of HVOF-sprayed corrosion resistant alloy (CRA) coatings in an aqueous solution containing supercritical CO_2. 316L stainless steel, Ti, alloy C-276 and alloy 625 were procured in power form and sprayed onto carbon steel with a JP500 HVOF system and 8 mm holidays were drilled in the coating to expose the steel substrate. These were then exposed to 3.5 wt.% NaCl solution for 30 days, bubbled with 10MPa CO_2. Tests were conducted at two different temperatures, 40°C and 80°C. Post-test microstructural characterization revealed that the coatings protected the steel substrate from CO_2 corrosion when intact. The bare steel in the exposed region formed a siderite scale, while no such scale was seen in the case of undamaged CRA-coated steel. The substrate close to the coating showed accelerated corrosion due to galvanic effects. It can be concluded that thermally sprayed CRA coatings are a candidate to provide cost-effective corrosion mitigation in wet supercritical CO_2. However, care must be taken to ensure that the thermally sprayed CRA layer does not have any through porosity or damage; else, such coatings may accelerate corrosion of the underlying steel due to galvanic interactions.

1 Introduction

High pressure CO_2 is often encountered in several industries such as oil and gas exploration, chemical and petrochemical plants, power generation etc. Often the temperatures and pressures are such that the CO_2 is in supercritical[1] condition.

It has become common industrial practice to use carbon steels for several industrial applications due to its low cost, weldability and availability. Inspite of its obvious advantages, the application of carbon steels is often limited to areas to low corrorosivity. It is well documented that carbon steels corrode in the presence of low pressure wet CO_2 [1-5], and high pressure and supercritical CO_2 [6-10]. The corrosion rate of carbon steels in some environments have been reported to be more than a few millimeters per year [4,9,10]. Such high corrosion rates could render the component unfit for service. In such situations, costly corrosion resistant alloys (CRAs) is often recommended. However, the cost of CRAs can be prohibitive for emerging technologies, such as CCS which needs to prove economic viability before its wider adoption [11]. If carbon steel is coated with a CRA by a cost-effective industrial method where the coating would provide the corrosion resistance, and the carbon steel, the strength then such systems can be used economically as an alternative to CRAs as structural components in CCS and other applications handling wet supercritical or 'dense phase' CO_2. If the coating is damaged then it can be repaired and the component can be reused without costly replacement. This concept has been researched recently; however, the likelihood of accelerated corrosion of the underlying carbon steel when the CRA coating is damaged has raised concerns [12-14]. Further, the application of such coatings at different temperatures has not been explored.

To address these concerns and technology gaps, several CRAs such as 316L stainless steel, Ti, alloy C-276 and alloy 625 were sprayed onto carbon steel using a JP5000 HVOF system, holidays were drilled to expose the steel substrate, and tested in deaerated 3.5 wt.% NaCl solution bubbled with 10MPa CO_2 at two different temperatures, 40°C and 80°C in separate autoclaves for 30 days. After testing for 30 days, the specimens were taken out and microstructural evaluation of cross-sections was carried out to assess the performance of CRA coatings.

2 Experimental

2.1 Materials

Carbon-Manganese steel plates of dimensions 40×40×5mm were selected as the substrate material conforming to specification S355J2+N(2004). Titanium (Crucible Research, USA; now ATI Metals), alloy C-276 (Sulzer Metco, USA; now Oerlikon), alloy 625 (Anval, Sweden, now Carpenter Powder Products) and 316L (Sandvik Osprey, Sweden) were sourced in powder form for spraying. The composition of the substrate and powders is given in **Table 1**.

Table 1. Chemical composition of substrate and coating material in wt. %.

Material	C	Mn	Cr	W	Ti	Fe	Nb	Ni	Mo
Carbon steel	0.12	1.39	-	-	-	Bal.	-	-	-
Ti	0.08	-	-	-	Bal	0.2	-	-	-
Alloy C-276	0.01	-	16.0	3.8	-	5.5	-	Bal.	15.5
Alloy 625	0.05	-	21.5	-	-	1.5	3.6	Bal	9.0
316L	0.02	2.0	17.0	-	-	Bal.	-	12	2.5

[1] Phase at a temperature and pressure above the critical temperature and pressure. The critical point beyond which CO2 exists in the supercritical phase is 31.1°C and 7.39 MPa. At high pressure CO2 may exist as a liquid if the temperature is below the critical temperature. When CO2 exists above the critical pressure it is often termed 'dense phase' fluid.

2.2 Coating production

The steel plates were degreased, grit blasted with F60 alumina grit and given a final air blast to produce a clean, rough surface. Spraying was carried out using a JP5000 high velocity oxy-fuel (HVOF) system (Praxair, Indianapolis, USA). Only one surface was coated with CRAs to a nominal thickness of 300 μm (except 316L which was ~220 μm) by using the spray parameters given in **Table 2**.

Table 2. Spray parameters used for preparing CRA coatings.

Coating material	Power size (μm)	Powder feed rate (g/min)	Spray distance (mm)	Gas flow rate (scfh)		Kerosene flow rate (US gal/h)
				O_2	Ar	
Ti	44-74	200	380	1850	25	6
Alloy C-276	33-45	410	380	1825	21	6
Alloy 625	16-44	330	380	1850	25	6
316L	15-45	270	355	1800	20	6

2.3 Corrosion testing in supercritical CO_2

As only one surface was sprayed, the other surfaces were coated with a curable polymer (henceforth referred to as the sealant), Belzona 1591 (Belzona Polymerics Ltd., Harrogate, UK) to protect the edges and back surface and expose only the surface coated with a CRA to the test environment, **Fig. 1**.

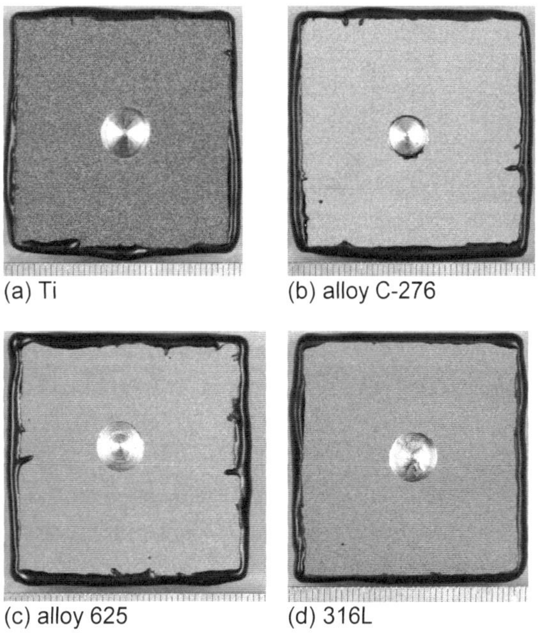

Fig. 1. Photographs of the specimens before testing showing (a) Ti, (b) alloy C-276, (c) alloy 625, and (d) 316L coatings. The specimens were 40×40×5 mm. This set was tested at 40°C. The other set tested at 80°C is not presented to save space. The 8 mm diameter holiday in the centre of the specimen exposed the underlying carbon steel.

The test specimens were devided into two sets- one tested at 40°C and the other at 80°C. Specimens were placed in an autoclave and purged with N_2. 3.5wt% NaCl solution, previously purged with N_2, was introduced in the autoclave and heated to the test temperature. The test pressure was raised to 10 MPa by pumping CO_2. The test duration was 30 days and once the test was completed the specimens were taken out, washed with de-ionised water and dried in air for further analysis.

2.5 Microstructural characterisation

After the autoclave exposure, changes to the coating appearance were recorded (i.e. the samples were photographed). The surfaces of the specimens were examined by light optical microscopy and SEM (samples were gold coated prior to this), including element mapping obtained using energy dispersive X-ray spectroscopy (EDX; Carl Zeiss, Oberkochen, Germany).

3 Results

3.1 Visual examination

The photographs of the specimens tested at 40°C and 80°C clearly show evidence of environmental damage due to supercritical CO_2 exposure, **Fig. 2** and **Fig. 3**.

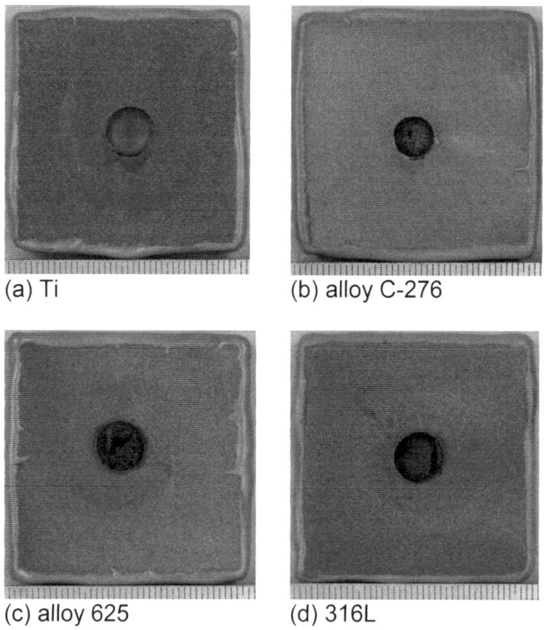

Fig. 2. Photographs of the CRA-coated specimens after testing in a 3.5wt% NaCl solution bubbled with 10MPa CO_2 at 40°C.

It is evident that the sealant used to protect the side and the back surfaces has undergone some changes in supercritical CO_2. The discoloration is evident, however, the edge retention is acceptable in all cases. Some rust spots were seen on the surface of the Ti coating specimen. These were formed during drying process and not in the de-oxygenated test environment. The holiday region (exposed steel substrate) was covered with dark grey/black deposits in most cases, presumably $FeCO_3$ (product of carbon steel corrosion in CO_2-water system).

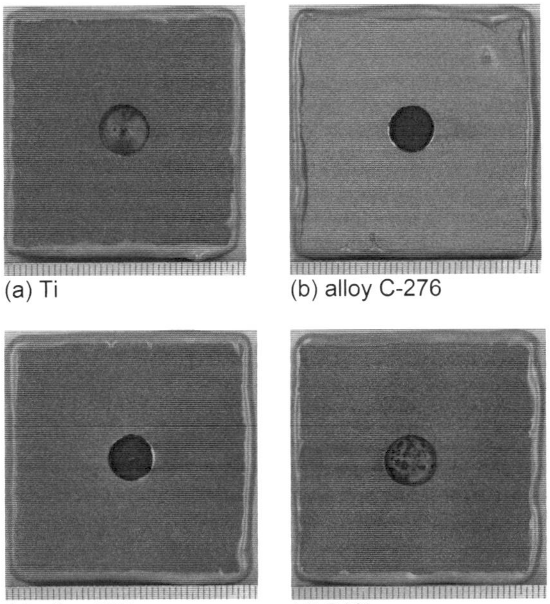

(a) Ti (b) alloy C-276

(c) alloy 625 (d) 316L

Fig. 3. Photographs of the CRA-coated specimens after testing in a 3.5wt% NaCl solution bubbled with 10MPa CO_2 at 80°C.

3.2 Microstructure of coating cross-section

When sectioned along the holiday region, the specimens showed corrosion products on the surface of the steel substrate, **Fig. 4** and **Fig. 5**. However, in the region close to the CRA coating undercutting was seen.

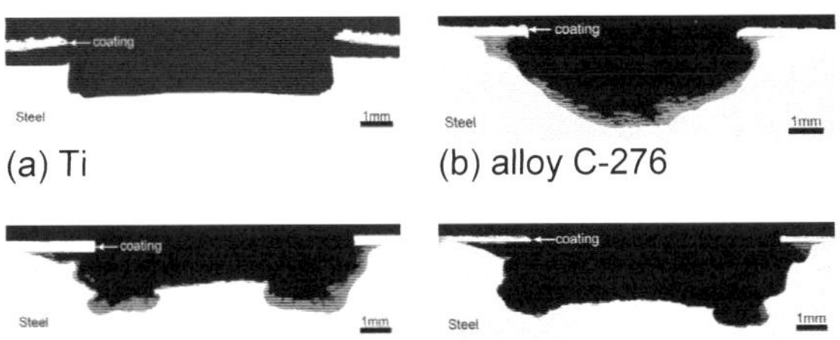

(a) Ti (b) alloy C-276

(c) alloy 625 (d) 316L

Fig. 4. SEM micrographs of coated specimen cross-sections after testing in 3.5wt% NaCl bubbled with 10MPa CO_2 for 30days at 40°C.

The presence of a CRA coating seems to have accelerated the corrosion of the carbon steel substrate close to the holiday. Ti coating has disbonded from the substrate when testes at 40°C. The reason for this is not entirely clear. However, residual stresses during thermal spraying is expected to have some influence. The holiday region in this case is intact probably due to the disbondment of the coating. Some corrosion product was seen in the region close to the interface between the coating and substrate, **Fig. 4a**.

The shape of the holiday region post exposure depends on the type of CRA and test temperature, **Fig. 4**. Alloy 625 and 316L coating specimens showed similar behavior in terms of the nature of corrosion pits formed in the holiday region. There was severe undercutting adjacent to the holiday region of these coatings at both temperatures. Some scale was also seen in the holiday region. The scale was more pronounced for coatings tested at 80°C, possibly due to conditions more conducive to scale formation. Scales are more uniform in specimens tested at 80°C.

Alloy C-276 coating also showed some undercutting with the formation of a scale on the steel surface in the holiday region. However, the nature of the undercutting is different for alloy C-276 coating at different temperatures. In most cases, the CRA coatings result in the formation of undercutting or corrosion close to the coating on the carbon steel side. However, for the alloy C-276 coating the undercutting is different and more in the central region away from the coating. This is unexpected and possibly an artefact of the drilling process to prepare holidays in specimens.

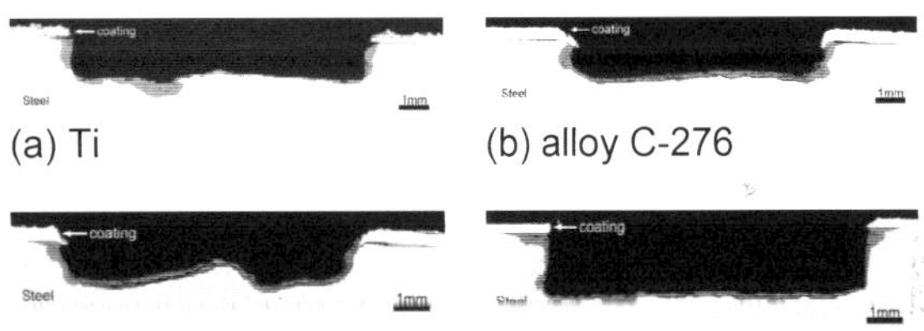

(a) Ti (b) alloy C-276

(c) alloy 625 (d) 316L

Fig. 5. SEM micrographs of coated specimen cross-sections after testing in 3.5wt% NaCl bubbled with 10MPa CO_2 for 30days at 80°C.

Fig. 5 shows the cross-section of CRA coatings tested at 80°C with evidence of corrosion at the steel-coating interface. The region adjacent to the defect shows corrosion products on the carbon steel side. It would appear that the carbon steel has been preferentially corroded and the coating has been left intact. The EDX analysis of the corrosion product (not shown) confirms the presence of Fe, C, O, Au in the scale which indicate $FeCO_3$ scale. The Au seen in the EDX spectra is from the sputtered coat of Au deposited to make the scale conductive and avoid any charging during SEM studies.

The coatings tested at 80°C showed some porosity which is expected in thermally sprayed coatings, **Fig. 6.** The region away from the defect shows detailed microstructural features such as porosity. Thermally sprayed coatings are built-up by impingement of molten, and sometimes, semi-molten particle, initially on the substrate and later on the previous deposited splats. As the particles impinge and cool, they form pores of various morphologies. These could be sharp crack-like to more regular, globular pores. In the specimens examined, most of this porosity was of inter-splat type. Some of these pores could also be surface-connected. However, when seen under a microscope very little evidence of corrosion was seen at the coating-substrate interface in most cases. Alloy C-276 (not shown) and alloy 625 coatings were very dense and this seems to have hindered the ingress of corrosive species through the coating, **Fig. 6b**. Only Ti coating showed some evidence of corrosion at the interface, even in the regions away from the holiday, **Fig. 6a**. The coating microstructure shows more inter-splat porosity with clear splats boundaries and in some cases gaps between the splats. These gaps can be a few microns in places. This could be the reason for the formation of some corrosion products at the coating-substrate interface. Hence, the coating was examined in detail using EDX, **Fig. 6c**. The detailed EDX of the corrosion product at the interface gave peaks corresponding to Fe, O, Ti, Au and C indicating the possible presence of $FeCO_3$. The Ti peak is was given off by the coating and the Au peak was the residual Au from the sputtering process. These corrosion products were only present at certain locations where the surface-connected porosity was present. None of the CRA coatings showed corrosion of the CRA material itself, as expected.

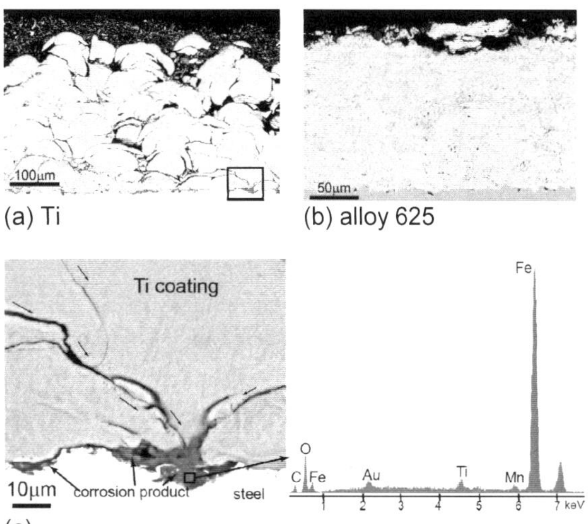

Fig. 6 SEM micrographs of coated specimen cross-sections after testing in 3.5wt% NaCl bubbled with 10MPa CO2 for 30days at 80°C: (a) Ti, (b) alloy 625, and (c) EDX of the regin highlighted in (a).

The uncoated carbon steel specimen showed a dark scale with a crystalline appearance, containing Fe, C and O (not shown). The crystallites comprising the scale had a faceted structure. This information when combined with the composition suggests the presence of siderite. The presence of siderite ($FeCO_3$) was confirmed by x-ray diffraction (XRD, not shown). The XRD pattern revealed that the scale was entirely comprised of siderite, a crystalline form of iron carbonate.

A corrosion rate could not be determined as it varied and depended on the region examined. The region in the carbon steel substrate close to the CRA coating showed more corrosion than the region away from it. In case of 316L coating a 1.5mm loss of substrate (conversion to a carbonate scale) close to the interface was recorded in 30days.

4 Discussion

4.1 Mechanism of corrosion of steel in CO_2

Corrosion in CO_2-containing environmentshas been intensively studied by the oil and gas industry for a few decades While the mechanism of corrosion of steel has been studied for over 60 years, the most significant advances were made only after the work of de Waard and Milliams [2]. Further advances were made in the last few decades by various researchers [1,3,15]. They argue that CO_2 corrosion of steel is an electrochemical process involving anodic dissolution of iron and cathodic evolution of hydrogen.

When CO_2 is dissolved in water it is hydrated to form carbonic acid (H_2CO_3).

$$CO_2(aq) + H_2O(\ell) = H_2CO_3(aq) \qquad (1)$$

This weak acid can dissociate to form a bicarbonate which can further form a carbonate depending on the pH .

$$H_2CO_3(aq) = H^+ + HCO_3^-(aq) \qquad (2)$$

$$HCO_3^-(aq) = H^+ + CO_3^{2-}(aq) \qquad (3)$$

The anodic dissolution of iron is affected by the presence of CO_2. The formation of a Fe-CO_2 complex is argued to be responsible for catalysing the dissolution of iron [15]. Several multi-step models were proposed, but the overall reaction is:

$$Fe(s) = Fe^{2+}(aq) + 2e^- \qquad (4)$$

Fe^{2+} (aq) can further oxidise to Fe^{3+}(aq) if dissolved oxygen or oxidants are present. From the above equation it would appear that the anodic dissolution of iron in CO_2-containing solutions is the same as in other acids, i.e. pH dependent. This, however, is refuted recently and it is assumed that the anodic dissolution of iron in CO_2-containing solution proceeds with little influence of pH (at $pH > 4-5$).

The cathodic reduction, on the other hand, usually involves two main pH-dependent reactions namely proton reduction and carbonic acid reduction. Many have assumed that these two reactions are independent and the net cathodic current is the sum of the currents of the two reactions. At $pH < 4$, the proton reduction to form hydrogen is the dominant cathodic reaction:

$$2H^+(aq) + 2e^- = H_2(g) \tag{5}$$

At a $pH > 4$, the dominant reaction is the reduction of carbonic acid.

$$2H_2CO_3(aq) + 2e^- = H_2(g) + 2HCO_3^-(aq) \tag{6}$$

Bicarbonate ion can be further reduced to give hydrogen

$$2HCO_3^-(aq) + 2e^- = H_2(g) + 2CO_3^{2-}(aq) \tag{7}$$

There is some debate in literature regarding the direct reduction of bicarbonate. At currents above the limiting current density of H^+ or H_2CO_3 reduction (and $pH > 5$, and low p_{CO_2}), hydrogen evolution from the direct reduction of water becomes the dominant cathodic reaction:

$$2H_2O(l) + 2e^- = H_2(g) + 2OH^-(aq) \tag{8}$$

The overall reaction for aqueous CO_2 corrosion of iron or carbon steel is:

$$Fe(s) + CO_2(g) + H_2O(\ell) = FeCO_3(s) + H_2(g) \tag{9}$$

The electrochemical reactions are often accompanied by the formation of iron carbonate and its deposition in the form of scales particularly at elevated temperatures, which can be protective or non-protective depending on the conditions under which they are formed. Some other oxides of iron may also form depending on the availability of dissolved oxygen in the system. From the present work it is evident that a scale is formed at a temperature of 40°C, which is low when compared to some of the other reports on siderite formation [15]. Formation of a tenacious scale at such low temperatures is likely to be due to the quiescent experimental conditions and the high CO_2 partial pressure. At high CO_2 partial pressure, the aqueous phase gets saturated with CO_2 very quickly. This may results in accelerated kinetics of formation of carbonic acid and subsequent dissolution of Fe which is likely to form corrosion products when its solubility limit is exceeded. Thus, under these conditions the formation of siderite scale may be facilitated.

4.2 Corrosion performance of damaged CRA-coated steel in supercritical CO_2 environments

CRA such as Ti, alloy C-276, 316L and alloy 625 are all cathodic to carbon steel in most industrial environments. In the presence of holidays the CRA coating is expected to accelerate the corrosion of the underlying carbon steel substrate due to galvanic interactions. The conductive solution (3.5 wt% NaCl) acts as the electrolyte which completes the galvanic cell comprising the CRA coating (acting as the cathode) in electrical contact with the carbon steel substrate (acting as the anode). The difference in the corrosion potential of the anode (steel) and cathode (CRA coating) acts as the driving force for galvanic corrosion.

The relative area of the cathode and anode exposed to the corrosive environment has a pronounced effect on the rate of corrosion of the anode. A small anode (exposed steel in the holiday region) joined to a larger cathode (CRA coating) results in a high current density on the steel, and hence accelerated dissolution of steel giving Fe^{2+} in solution. The dissolved CO_2 forms carbonic acid which gives carbonate and bicarbonate ions by reactions given in equations 6 and 7. This also helps with the cathodic reduction reactions (equations 5, 6 and 7) which take place at the CRA coating surface in contact with the electrolyte, ultimately generating hydrogen. The anodic reaction is the dissolution of iron (equation 4). Thus, the steel at the coating-substrate interface gets attacked first resulting in preferential dissolution of Fe causing visible undercutting. As the dissolution of iron continues along the interface the concentration of dissolved Fe^{2+} increases which leads to its saturation (and possibly supersaturation) in the vicinity of the holiday close to the coating-steel interface due to high surface of steel to volume of conductive CO_2-containing solution ratio. Owing to the saturation of Fe^{2+}, precipitation of $FeCO_3$ takes place on the steel surface close to the interface first. With time the precipitation of $FeCO_3$ continues along the steel surface ultimately covering the entire holiday in most cases.

To avoid such circumstances a dense, pore-free coating is essential. If such coatings cannot be practically accomplished, one can at least endeavour to produce CRA coatings free from surface-connected porosity or use a sealant to plug the pores. The sealant should be inactive in supercritical CO_2.

The mechanism of corrosion described earlier was based mainly on gaseous species. However, in the presence of supercritical CO_2 the actual concentration of the reacting species i.e. CO_2 can be significantly higher [16]. If scaling conditions are present then supersaturation may be reached quickly and the scale development may be accelerated. If scaling conditions are not reached and the concentration of CO_2 is high then this may lead to accelerated dissolution of the carbon steel substrate. As the probability of corrosion of carbon steel is high one may mitigate the risk by using monitoring systems. Non-destructive evaluation and condition monitoring of steel components coated with a CRA can help reduce the risk of corrosion initiated failure. If any defect in the coating is detected they can be repaired very easily if the damaged area can be accessed.

5 Conclusions

Coatings of Ti, Alloy C-276, Alloy 625 and 316L were thermally sprayed onto carbon steel using a JP5000 system. Holidays were drilled to expose underlying substrate and the specimens were exposed to 3.5 wt.% NaCl solution saturated with 10 MPa CO_2 in two separate autoclaves; at 40°C and 80°C. After testing for 30 days, the specimens were taken out and microstructural evaluation of the cross-sections was carried out.

Microstructural characterization revealed that the thermally sprayed layer protected the steel substrate from CO_2 corrosion when intact. The bare steel formed a siderite scale. However, in a non-scaling condition this would not be protective and under loading conditions cracking of the scale might be of concern. No scale was seen in the case of CRA-coated steel in the regions away from the holiday. It can be concluded that dence coatings of Ti, Alloy C-276, Alloy 625 and 316L can provide a cost-effective corrosion mitigation method for infrastructure likely to be in contact with wet supercritical CO_2. The same coatings can possibly be used as inner lining of pipes for transport of impure CO_2. However, care must be taken to ensure that the thermally sprayed layer does not have any through porosity; else, such coatings may accelerate corrosion of the underlying steel due to galvanic interactions, as evidenced from regions close to the holiday. In the presence of through-thickness porosity or holidays (defects), sealants not affected by wet supercritical CO_2 should be used to close the surface-connected pores and hence create a coating system capable of providing a cost-effective corrosion mitigation solution.

6 Acknowledgements

The work was funded by TWI. The author would like to thank TWI staff, especially Ian Wallis, Mike Bennett, Barnaby King, Andrew Tabecki and Sheila Stevens for their technical support.

7 References

[1] Kermani, M.B. and A. Morshed: Carbon dioxide corrosion in oil and gas production- A Compendium. Corrosion 59, (2003), Issue 8, pp. 659/83.

[2] De Waard C. and D.E. Milliams: Carbonic acid corrosion of steel. Corrosion 31 (1975), pp. 177/81.

[3] A. Dugstad, “Mechanism of protective layer formation during CO_2 corrosion of mild steel. CORROSION/98, Paper No. 31, NACE, Houston, TX, 1998.

[4] Dugstad, A. et al.: Corrosion of transport pipelines for CO_2–effect of water ingress. Energy Procedia 4 (2011), pp. 3063/70.

[5] Kermani, B and T. Chevrot: Recommended practice for corrosion management of pipelines in oil and gas production and transportation” (Leeds, UK: EFC 64, 2012), pp. 109.

[6] Paul, S. et al.: Material selection for supercritical co_2 transport. The First International Forum on the Transportation of CO_2 by Pipeline, paper no. 16 (Newcastle, UK: Clarion, 2010), pp. 1/21.

[7] Paul, S. et al.: Material assessment for high pressure CO_2 transport. The Second International Forum on the Transportation of CO_2 by Pipeline, paper no. 8 (Newcastle, UK: Clarion, 2011), pp. 1/29.

[8] Paul, S. et al. Selection of materials for high pressure CO_2 transport. The Third International Forum on the Transportation of CO_2 by Pipeline, paper no. 19 (Newcastle, UK: Clarion, 2012), pp. 1/16.

[9] Seiersten, M.: Material selection for separation, transportation and disposal of CO_2.” CORROSION/2001, Paper no. 01042, (Houston, TX: NACE, 2001).

[10] Ayello, F. et al.: Effect of impurities on corrosion of steel in supercritical CO_2.” CORROSION/2010, Paper no. 10193, (Houston, TX: NACE, 2010).

[11] Metz, B. et al.: Carbon dioxide capture and storage: IPCC special report prepared by working group III of the intergovernmental panel on climate change. (Cambridge, UK: Cambridge University Press, 2005), pp. 431.

[12] S. Paul, "Thermally sprayed corrosion resistant alloy coatings on carbon steel for use in CCS," EUROCORR/14, paper no. 7456 (Pisa, Italy: EFC, 2014), pp. 1/6.

[13] Paul, S.: Thermally sprayed corrosion resistant alloy coatings on carbon steel for use in supercritical CO_2 environments. CORROSION/2015, paper no 5939 (Houston, TX: NACE, 2015), pp.1/9.

[14] Paul, S.: Mitigating corrosion of carbon steel in supercritical CO_2 environments using HVOF coatings. ITSC/2015, paper no. 38553 (Long Beach, CA: ASM, 2015), pp. 1072/77.

[15] Nesic, S.: Key issues related to modelling of internal corrosion of oil and gas pipelines – a review. Corrosion Science 49 (2007), Issue 12, pp. 4308/38.

[16] Weiss, R.F.: Carbon dioxide in water and seawater: the solubility of a non-ideal gas. Marine Chemistry 2 (1974), pp. 203/15.

High temperature protective performance of LT-HVOF sprayed TiAl3-Al coating for Ti-6Al-4V substrate

Chunming Deng*, M. Liu, Wenpu Yang, Jifu Zhang, Jie Mao, Changguang Deng, Kesong Zhou

Guangdong Institute of New Materials, National Engineering Lab for Modern Surface Engineering Technology, The Key Lab of Guangdong for Modern Surface Engineering Technolog, Guangzhou 510650, P.R.China

**E-mail: denghans@126.com*

Abstract

Ti alloys found their wide applications as important structural materials in aviation, aerospace and chemical industry for the high specific strength and excellent corrosion resistance performance. However these applications are only restricted at low temperature because of the terrible oxidation resistance performance at over 350 °C and the great degradation of mechanical strength at high temperature. Protective coatings are needed in order to improve the service temperature of Ti alloys.

Conventional protective MCrAlY coating leads to great degradation of mechanical strength due to the co-diffusion of metallic element. TiAl3 intermetallic is considered as ideal oxidation resistant coating for Ti-6Al-4V because of the similar chemical composition. Unfortunately as an intermetallic material, TiAl3 is a very brittle material, and metallic material is added to increase the toughness of the coating. In this paper, TiAl3-Al composite coating is prepared on Ti-6Al-4V substrate via Low temperature HVOF process, the microstructure for as sprayed coating is analyzed by means of SEM. Finally the oxidation resistance of LT-HVOF sprayed TiAl3 Al composite coating on Ti 6Al 4V substrate is characterized holding different time at the temperature of 700 °C. The protective mechanism of the coating is briefly discussed in this paper.

Introduction

Titanium alloys are widely considered as ideal structural materials in wide range of aerospace, chemical, petroleum industries and etc. for the high specific strength and excellent corrosion resistance in most aggressive media. However the maximum engineering application temperature for Titanium alloy is within the range of 550-600 °C[1]. Of the great research focuses in the recent years have been TiAl based alloys. These alloys own excellent high temperature strength up to 750-800 °C, but are far from engineering applications as a result of the rather poor ductility at room temperature[2,3]. In general titanium alloys perform awful oxidation resistance due to the non-protective oxide layer formed on the alloys. What's more, the oxidation of the alloys will also lead to the high chemical impurities (O, N) into titanium alloys, which greatly degrade the mechanical performance. In general the high temperature strength was only 1-10% strength of titanium alloy at room temperature[4] because titanium alloys (especially ultimate strength) are all highly sensitive to chemical impurities.

Most of titanium alloys are applied in relatively low temperature environment, which greatly restrict their wide applications. For example, most of titanium alloys serve as low pressure compressor in gas turbine which operates at the maximum service temperature of about 350 °C[1]. If titanium alloys can withstand higher service temperature, lightweight of facilities is probably further improved.

One of the effective approaches to improve service temperature is to apply surface engineering except the high temperature TiAl based alloys mentioned above. However the chemical impurity is the key issue for Titanium alloys. The conventional oxidation resistance coatings, like NiCrAlY, CoCrAl, and etc. used for Ni- and Co-based high temperature alloys cannot achieve satisfactory results. One of the reasons lies in the diffusion of coating with titanium alloys as chemical impurities during high temperature service, which leads to the great reduction in mechanical strength of Titanium alloys[5].

Due to the stringent impurity requirement of titanium alloys, Ti-Al based coatings were generally recommended. These coatings own not only satisfactory oxidation resistance performance but also the high adhesive strength due to the similar CTE to titanium alloys substrate[6]. A dense protective layer was formed on TiAlCr coating surface during service only when Cr content was over 10at.%. Unfortunately the protective layer seemed to be fragile and easily detached from the titanium alloys substrate due to the relatively higher Cr content[7]. Addition of Ag into the TiAlCr coating could result in the improved oxidation resistance even when the Cr content was lower than 10at.%[8]. Besides most of these coatings were prepared via PVD technology, the low deposition efficiency and component size requirement restricted the wide applications of these coatings.

Thermal sprayed MCrAlY coatings were attempted as the oxidation resistance coatings for titanium alloys, and the results were not so good. The Ti-Al system is still the base material selection as oxidation resistant coating for the impurity requirement. Of the system, TiAl3 intermetallic material owns the best oxidation resistance material in the Ti-Al intermetallic system due to the formation of α-Al2O3 layer on the coating surface, but the material also has terrible ductility and is far from application[9].

Xiong tianying et al. [10] prepared a dense Al-Ti composite layer by means of cold spraying with a mixture of Al and Ti powder, afterwards heat treatment was applied in vacuum or inert gas shrouding environment to achieve TiAl3 layer. The results revealed that a layer of TiAl3-Al composite material was achieved and exhibited excellent protection performance from the high temperature oxidation for Titanium alloys. However the process is complicated due to the high purity shrouding gas system applied. Basing on the excellent oxidation resistance performance of TiAl3, Al material is added to improve the ductility of TiAl3 material so as to enhance the oxidation resistance performance of Titanium alloys.

In this paper TiAl3-Al composite coating were prepared on Ti-6Al-4V by means of LT-HVOF (Low Temperature HVOF) with mechanically mixed powder as feedstock, the microstructure, fundamental properties and the oxidation resistance performance for the composite coating on Titanium alloys were characterized.

Experiments and characterizations

Ti-6Al-4V with a dimension of Φ6mm * 40mm is selected as substrate. Prior to deposition, the substrate was ultrasonically cleaned in ethanol, and then grit blasted to achieve a roughened surface. TiAl3 and Al powder with a size of about D50=20μm and D50=50μm were chosen as thermal spray powders. Both powders had a globular microstructure. The morphologies and particle size distribution for both powders are shown in Fig. 1. Mechanically mixed powder with a ratio of 92wt.% TiAl3 and 8 wt.% Al powder were selected as feedstock.

Table.1 Deposition parameters of LT-HVOF TiAl3-Al coating

Process	Oxygen (L/min)	Fuel (L/h)	Chamber pressure (bar)	Spraying distance (mm)
LT-HVOF	780	13	15.4	100-310

Modified HVOF torch (K2, GTV, Germany), called Low Temperature HVOF (LT-HVOF) process, was used to deposit the composite coating, and the deposition parameters are presented in table 1. LT-HVOF process is characterized by lower plume temperature and higher plume velocity than conventional HVOF process through increasing combustion chamber pressure (15.4bar, almost twice of convention HVOF)) reducing fuel consumption, therefore dense metallic coatings with low oxygen content could be manufactured. This process excels on the preparations of low melting point metallic coatings, like Al, Cu and their alloys[11]. The coating thickness was about 30μm.

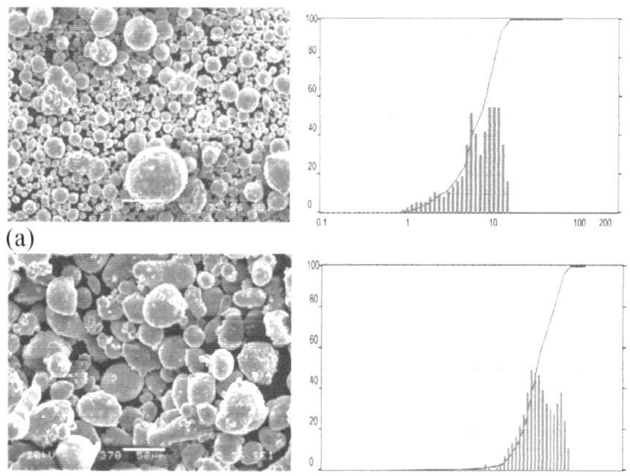

(a) (b)

Figure 1: SEM images and particle size distributions for TiAl3 (a) and Al (b) powder

The microstructure for as sprayed TiAl3-Al composite coating was characterized by Field Emission Gun SEM (Sirion 200, FEI, USA). The coated sample was put into a muffle furnace (MFL—2301K type, Yixing Electrical Furnace factory, China) with different holding time at the temperature of 700°C. The mass variations of sample were obtained after various oxidation duration using BS224S electronic balance with an accuracy of 0.1mg.

Results and Discussion

Coating microstructure

Cross-sectional image for LT-HVOF sprayed TiAl3-Al composite coating on the substrate of Ti-6Al-4V alloy is presented in Fig. 2. The layer with a thickness of about 30μm is well bonded on the titanium alloy substrate, meaning the good adhesion with substrate. The as sprayed coating has highly dense microstructure and no big pore was found in the coating. In addition, the coating exhibits clear laminar microstructure, indicating the full melting and flattened state of the particles. Al particles are randomly deposited in the coating. It should be noted that the addition of Al into TiAl3 material just was used to increase the ductility, but not only acted as adhesive phase of TiAl3. In fact, the authors have successfully deposit both very dense Al coating with oxygen content (0.5wt.%) and dense TiAl3 coating. So in the coating, oxidation layer on the flattened particles is not observed due to the applied LT-HVOF process, while this is often detected in the HVOF coating. The reason is also for the lower plume temperature and higher plume velocity of LT-HVOF process.

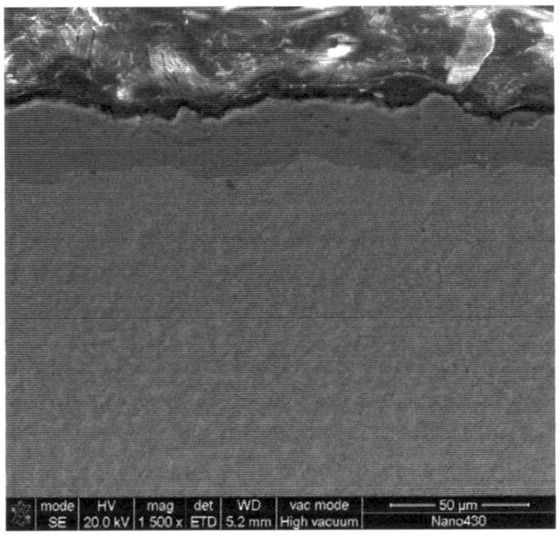

Figure 2: SEM images for cross section of TiAl3-Al composite coating

Oxidation test

The images for the coated Ti-6Al-4V sample experiencing various oxidation times at 700 °C are presented in Figure 3. The sample displayed different color on the surface. In the center it is brown, while it appeared to be dark color at the both ends. This was probably attributed to the different microstructure resulted from the twice deposition. But the sample kept the good integrity and no any spallation was observed on the coating surface even oxidation test for 1000h. The results showed the excellent corrosion resistance performance of the coating. While bare Ti-6Al-4V alloy had terrible oxidation resistance performance, clear spallation was occurred after oxidation test only for 96h.

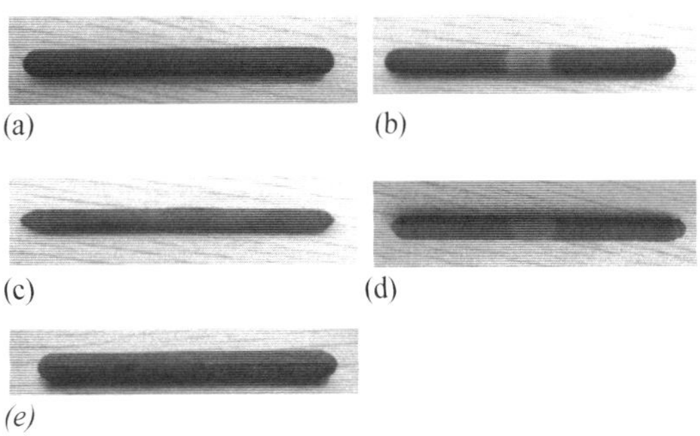

Figure 3: photos of TiAl3-Al coated Ti-6Al-4V with various oxidation times at 700 °C, (a) 8h; (b) 100h; (c) 800h; (d) 1000h and Ti-6Al-4V oxidation for 96h at 700 °C (e)

The oxidation dynamic curve for coated Ti-6Al-4V is shown in Figure 4 after oxidation test for 1000h at 700 °C. At the beginning (about 200h) obvious mass gain was detected for the coated sample, afterwards the mass gain almost remained to be constant till the oxidation test for 1000h. This indicates the excellent oxidation resistance of TiAl3-Al coating for Ti-6Al-4V.

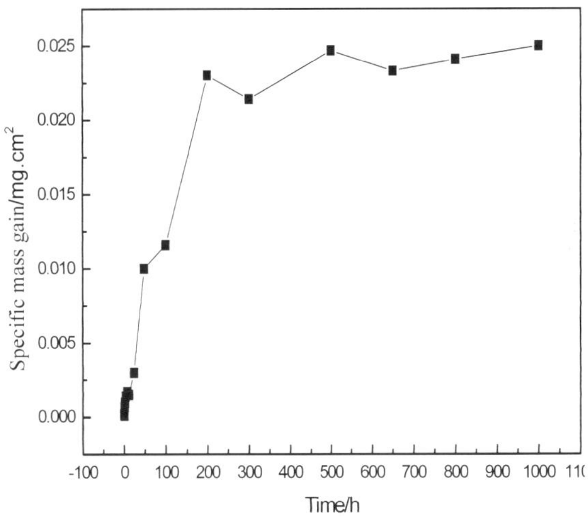

Figure 4: Oxidation dynamic curve of TiAl3-Al coated Ti-6Al-4V at 700 °C

Discussion

The cross-sectional image for TiAl3-Al coated Ti-6Al-4V is presented in Figure 5 after oxidation test for 1000h, the elemental distribution is also attached in the image. There was no variation in coating thickness after oxidation, and the diffusion of coating and substrate was occurred during the oxidation. From the EDS analysis, high oxygen content was detected at the subsurface (about 10µm), while oxygen content dropped quickly. This indicated that a protective oxide layer was probably formed on the coating surface. Unfortunately the stable and dense oxide layer was not clearly observed and the phase composition was still unknown. Further work regarding the oxidation resistant mechanism for the TiAl3-Al composite coating will be performed, such as phase composition of the protective oxide layer, high magnification of coating. Besides the effects of applied coating on the mechanical performance of titanium alloy will be done in the later work.

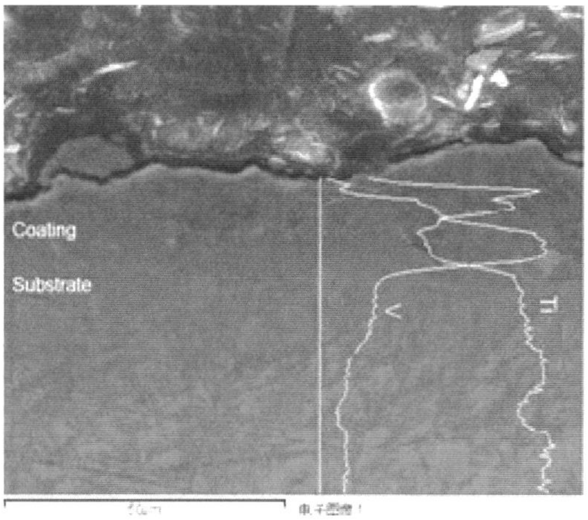

Figure 5: cross sectional image and the elemental EDS results of TiAl3-Al coated Ti-6Al-4V at 700 °C after oxidation for 1000h

Conclusions

In order to improve the service temperature of Ti alloys, protective coating should be applied. In this paper, a TiAl3-Al composite coating is manufactured on Ti-6Al-4V substrate via Low temperature HVOF process. The results showed that the as-sprayed coating had highly dense microstructure and was well bonded with substrate. Obvious mass gain was observed for at the starting 200h, afterwards the mass gain was almost constant till the oxidation test for 1000h for the coated sample. This indicates the excellent oxidation resistance of TiAl3-Al coating for Ti-6Al-4V. The protective oxide layer is the reason for the excellent oxidation resistance performance of the coating. Further work regarding the oxidation resistance mechanism should be performed later.

Reference

1. Valentin N. Moiseyev, Titanium alloys: Russian aircraft and aerospace applications. *Taylor & Francis*, 2006
2. JP Lin, XJ Xu, YL Wang, et al. High temperature deformation behaviors of a high Nb containing TiAl alloy. *Intermetallics*, 2007, 15(5):668–674
3. T Kruml, K Obrtlik. Microstructure degradation in high temperature fatigue of TiAl alloy. *International Journal of Fatigue*, 2014, 65(65): 28-32
4. VM Fedirko, OH Luk, Yanenko, VS Trush, Influence of the Diffusion Saturation with Oxygen on the Durability and Long-Term Static Strength of Titanium Alloys. *Materials Science*, 2014, 50(3): 415-420
5. D.W. McKee , K.L. Luthra, Plasma-sprayed coatings for titanium alloy oxidation protection. *Surface and Coatings Technology*, 1993, 56(2): 109–117
6. Z Tang, L Niewolak, V Shemet, et al. Development of oxidation resistant coatings for γ-TiAl based alloys. *Materials Science and Engineering: A*, 2002, Volume 328(1–2): 297-301
7. Zhaolin Tang, Fuhui Wang, Weitao Wu, Effect of a sputtered TiAlCr coating on the oxidation resistance of TiAl intermetallic compound. *Oxidation of Metals*, 1997, 48(5): 511-525
8. Y. J. Xi, F. H. Wang, L. L. He, Oxidation and Hot-Corrosion Behavior of Sputtered Ti-48Al-8Cr-2Ag Nanocrsytalline Coating, *Materials Science Forum*, 2004, 461-464: 1181
9. J.L. Smialek. Oxidation behaviour of TiAl3 coatings and alloys. *Corrosion Science*, 1993, 35(5–8): 1199–1201, 1203–1208
10. Li SHEN, Ling-yan KONG, Tian-ying XIONG, et al. Preparation of TiAl3-Al composite coating by cold spraying, *Transactions of Nonferrous Metals Society of China*, 2009, 19(4): 879–882
11. C. M. Deng, C. G. Deng, M. Liu, et al., Corrosion of Ti coating prepared by modified HVOF process, *Proceedings of the International Thermal Spray Conference of 2010*, Singapore: Thermal spray: Global solutions for future application.

Acknowledgement

We would like to acknowledge the financial support from National Basic Research Program of China ("973" program) (Grant No. 2012CB625100) and Guangdong Science and Technology programming (No. 2013B050800031, No. 2014B050502008) and Guangzhou Science and Technology programming (No. 201508030001).

From nano-particles to nano-structured thermal spray coatings

Y. Wang*, Y. Liu, S.Y. Liu, L. Wang, W. Tian and Z.Y. Pan
Laboratory of Nano Surface Engineering, School of Materials Science and Engineering, Harbin Institute of Technology, Harbin 150001, China

Abstract: Thermal Spray is the most effective method for surface modification in industry. When the nanomaterials and nanotechnology is applied in thermal spray, a novel technology of thermal spray is developed. Because the common nanoparticles are small in size, light in weight, they are easily to blow away by gas flow and ablate off by high temperature flame, thus they are not suitable for thermal spray. Therefore, to fabricate nanostructured feedstocks suitable for thermal spray through reconstitution process of common nanoparticles becomes the key point. The paper summarized the research results of our group on reconstitution process of nanoparticles and nanostructured coating technology during the last decade, from which the necessity and importance of reconstitution process is explained.
Keywords: reconstitution, thermal spray, nanostructured coatings
Abstract, abridged version

1. Introduction

Thermal spray is the most effective method for surface modification in industry, which can manufacture all kinds of coatings including metal, alloy, ceramic and polymer. According to the market research report on new technology of BBC company in 2012, the sale of ceramic coating reach up to 1.9 billion dollars in the North American market, in which the market shares is above 65% for thermal spray coatings [1].

Nanomaterials are experiencing a rapid development in recent years due to their existing and/or potential applications in a wide variety of technological areas such as electronics, catalysis, ceramics, magnetic data storage and structural components etc. The study of nanomaterials also has been extended to coatings process using the thermal spray technique, which marks the coming of a new era to design the surfaces of real parts using nano-surface engineering. The common nanoparticles are small in size, light in weight, they are easily to blow away by gas flow and ablate off by high temperature flame, therefore they are not suitable for thermal spraying [2-3]. The reconstitution process becomes the key point for spraying nanostructured coatings by thermal spray method.

In the beginning of 21st century, US navy declared that a revolutionary novel nanostructured coating fabricated by thermal spray method had passed the all kinds of tests. And these coatings are widely applied to around hundreds of components in US navy warship, submarine, minesweeper and aircraft carrier (including inlet and exhaust valves of submarine, cabin door strut of submarine, motor and oil pump shaft of aircraft carrier, main propulsion rod of minesweeper, etc.) [3-9]. This first application of nanostructured coatings creates a new era of nano-surface engineering [8,9].

In the technology of thermal spray nanostructured coatings, the reconstitution process is to fabricate the sprayable feedstock from the individual nanoparticles by ball-milling, spray drying aggregation, high temperature sintering, etc. The comprehensive properties of coatings depend on the composition, microstructure, density and flowability of feedstock. Before the emergence of liquid feedstock, the reconstitution process of nanostructured particles is the only way to thermal spraying nanostructured coatings. Moreover, it can control the composition and microstructure of sprayable feedstock through the reconstitution process, subsequently to fabricate different coatings by thermal spray method.

The paper introduces the reconstitution process in thermal spraying technology by three examples, which is as follows.

2. Effect of reconstitution process on nanostructured Al2O3-TiO2 coating [10-14]

Because of the better strength and chemical stability, ceramic materials are widely used based on the wear and corrosion resistance. They are also used for coating materials to improve the surface properties. However, the brittleness and poor thermal shock resistance limit the application of ceramic materials for coatings. Besides, the adhesive strength of coating/matrix interface and coating densities should be considered if the ceramic materials are used for coatings.

In order to improve the adhesive strength, wear resistance, corrosion resistance and thermal shock resistance of Al_2O_3-TiO_2 coatings, the original Al_2O_3-TiO_2 nanoparticles were reconstituted. The process is as follows: (a) nanostructured materials are uniformly dispersed in liquid medium by ultrasonic or ball-milling treatment; (b) adding organic adhesive to the liquid medium in (a) to obtain solution; (c) spray drying the solution to obtain the nanostructured aggregate; (d) high temperature sintering to wipe off residual water, oxygen in aggregate, also increase the sinter process to obtain the nanostructured feedstock suitable for thermal spray. The reconstitution process is shown in Figure 1.

Table 1 and Figure 2 show the comparative nanostructured Al_2O_3-TiO_2 feedstocks fabricated by three different reconstitution processes. After spray drying and loose sintering, plasma treatment is employed to reconstitution process. It can be seen that the density and flowability of reconstitution particles after plasma treatment are obviously increased.

Because spray drying particles show the worst density and strength, the comparative aspect of loose sintering (Undensified nanostructured coating) and plasma sintering (Densified nanostructured coating) coatings are shown here. The microstructured Metco 130 coating is also introduced. And the mechanical properties of the three coatings are displayed in Table 2. It can be seen that the mechanical properties of densified nanostructured coating is obviously increased.

Table 3 shows the wear rate of three coatings tested in a three-body testing. It can be seen that the wear resistance of both nanostructured coatings is better than that of the Metco 130 coating. And the densified nanostructured coating display the best wear resistance, which is about 3 times than Metco 130 coating.

Table 4 shows the results of electrochemistry parameters of different coatings corroded in 5%HCl and 3.5% NaCl solutions. The corrosion potential of densified nanostructured coating is obviously higher than that of the other two coatings, which implies the corrosion tendency of densified nanostructured coating is the lowest one.

Table 5 showed the thermal cycling life of three coatings at 650℃ and 850℃. It can be seen that the thermal shock resistance of densified nanostructured coating is the best one among those three coatings. The thermal shock resistance of Metco 130 coating is worst.

In all, the densified nanostructured coating shows the best properties including adhesive strength, hardness, spalltion resistance in bend, wear resistance, corrosion resistance, and so on. Those improved properties are attributed to the network structure in densified nanostructured coating [10-14].

3. Effect of reconstitution on SiCp/Al2O3-20wt.%ZrO2 coating [15-18]

Adding micro-sized and nano-sized SiC particles into the Al_2O_3-20wt.%ZrO_2 particles to obtain mixed particles. After aforementioned reconstitution process, the mixed particles can be fabricated as feedstock which suitable for thermal spraying. Six kinds of coatings are fabricated by plasma spray method. AZ is the coating of Al_2O_3-20wt.%ZrO_2 fabricated by loose sintered particles. AZP is the coating of Al_2O_3-20wt.%ZrO_2 fabricated by plasma sintered particles. AZSm is the coating of micro-sized SiC particles enhanced Al_2O_3-20wt.%ZrO_2 fabricated by loose sintered particles. AZSmP is the coating of micro-sized SiC particles enhanced Al_2O_3-20wt.%ZrO_2 fabricated by plasma sintered particles. AZSn is the coating of nano-sized SiC particles enhanced Al_2O_3-20wt.%ZrO_2 fabricated by loose sintered particles. AZSnP is the coating of nano-sized SiC particles enhanced Al_2O_3-20wt.%ZrO_2 fabricated by plasma sintered particles.

Table 6 shows the comparative properties of the six coatings. It can be seen that the properties of coatings fabricated by nano-sized SiC particles enhanced Al_2O_3-20wt.%ZrO_2 are better than that fabricated by micro-sized SiC particles enhanced Al_2O_3-20wt.%ZrO_2. And the plasma sintering of feedstock obviously improved properties of coatings.

Figure 3 shows the thermal shock resistance of the six coatings at 1000 ℃. It can be seen that AZSnP coating shows the best thermal shock resistance.

Figure 4 showed the erosion wear rate of the six coatings. The AZSnP coating displays the excellent erosion wear properties.

4. Effect of reconstitution on thermal barrier coating [19-24]

Nanostructured 8YSZ and zirconate (Aluminate)-based are expected to be the new type of candidate thermal barrier coating materials (TBCs) in the future. It has a great significance to prepare the TBCs with excellent performance in order to improve the durability and reliability of high temperature components, such as aero engine, turbine blade and so on. From the viewpoint of materials design, the double ceramic layer $La_2Zr_2O_7$ (LZ)/8YSZ coating, nanostructured 8YSZ coating and conventional 8YSZ coating have been prepared, and their mechanical properties, thermal insulation ability, thermal shock resistance and high temperature oxidation resistance have been compared and investigated systematically.

Nanostructured 8YSZ and LZ feedstock has been prepared by Nano-particle reconstitution process. The conventional 8YSZ coating, nanostructured 8YSZ coating, double ceramic layer LZ/8YSZ TBC has been prepared onto the GH4169 super-alloy substrate with the NiCoCrAlY as the bond-coating. The bonding strength testing result indicate that the nanostructured 8YSZ and DCL LZ/8YSZ has higher bonding strength compared with the conventional 8YSZ TBC. The difference of the bonding strength between the nanostructured 8YSZ and DCL LZ/8YSZ is not evident. The thermal insulation effects of the three as-sprayed coatings have been measured by the oxygen flame assisted with the device designed by us. The investigation results indicate that the DCL LZ/8YSZ TBCS has the best thermal insulation effect and the conventional 8YSZ has the worst thermal insulation effect, as shown in Figures 5 and 6[19-20].

5. Conclusion

Nano-thermal spray technology creates a new era of nano-surface engineering. While the reconstitution process becomes the necessary and important key technology during thermal spray. How to effective control the composition and nanostructure of feedstock is depended on the reconstitution process, by which different coatings can be fabricated.

References

[1] Market Research Report of Business Communications Co. at December 2004.

[2] P.R. Strutt, R.F. Boland and B.H. Kear: US Patent 6,025,034, (2000).

[3] L.T. Kabacoff, Nanoceramic Coatings Exhibit Much Higher Toughness and Wear Resistance than Conventional Coatings. AMPTIAC, 2002, 6(1): 37- 42.

[4] Oral Remarks of Dr. James Murday for the Department of Defense before the Senate Committee on Science. Commerce and Transportation, http://www.nano.gov/html/res/MurdaySenTest03.htm.

[5] More Products, http://www.nano.gov/html/facts/MoreProds.htm.

[6] U. S. Navy Finding New Applications For Advances In Nanotechnology, http://www.Nationaldefensemagazine.org/issues/2004/oct/US_Navy_Finding.htm.

[7] J.S. Murday, The Coming Revolution: Science and

Technology of Nanoscale Structures, AMPTIAC, 2002, 6(1): 5-10.
[8] Nanostructured coating approved for use on Navy ships, http://www.onr.navy.mil/onr/newsreel/nr000405.htm.
[9] All Benefit From DoD-Industrial Dual-Use Partnerships, http://www.vnis.com/vetnews/usdefense/usdefense2000/usdefense2000-045.htm.
[10] Y. Wang, W. Tian, T. Zhang, Y. Yang, Microstructure, Spallation and Corrosion of Plasma Sprayed Al_2O_3-13%TiO_2 Coatings, Corrosion Science, 2009, 51(12): 2924~2931.
[11] Y. Wang, W. Tian, T. Zhang, Y. Yang, Electrochemical Corrosion Behavior of Plasma Sprayed Al_2O_3-13%TiO_2 Coatings in Aqueous Hydrochloric Acid Solution, Materials and Corrosion, 2010, 61(7): 611-617.
[12] W. Tian, Y. Wang, T. Zhang, Y. Yang, Sliding Wear and Electrochemical Corrosion Behavior of Plasma Sprayed Nanocomposite Al_2O_3-13%TiO_2 Coatings, Materials Chemistry and Physics, 2009, 118(1): 37~45.
[13] Y. Wang, S. Jiang, M.D. Wang, et al., Abrasive wear characteristics of plasma sprayed nanostructured alumina/titania coatings, Wear, 2000, 237, 176-185.
[14] Y. Wang, W. Tian, G. Liu, Thermal spray nanostructured Al_2O_3/TiO_2 coatings and their applications, Materials Science and Technology, 2006, 14(3): 254- 257.
[15] Z.Y. Pan, Y. Wang, X.W. Li, C.H. Wang, Fabrication and characterization of heat and plasma treated SiC/Al_2O_3-YSZ feedstocks used for plasma spraying, Vacuum, 2012, 86(10):1558-1567.
[16] Z.Y. Pan, Y. Wang, C.H. Wang, X.G. Sun, L. Wang, The effect of SiC particles on thermal shock behavior of Al_2O_3/8YSZ coatings fabricated by atmospheric plasma spraying, Surface and Coatings Technology, 2012, 206(8-9):2484-2498.
[17] Z.Y. Pan, Y. Wang, X.W. Li, C.H. Wang, Effect of submicron and nano SiC particles on erosion wear and scratch behavior of plasma sprayed Al_2O_3/8YSZ coatings, Journal of Thermal Spray Technology, 2012, 21(5).995-1010.
[18] Z. Y. Pan, Y. Wang, X. G. Sun, Fabrication and characterization of spray dried Al_2O_3-ZrO_2-Y_2O_3 powders treated by calcining and plasma, Powder Technology, 2011, 212(2):316-326.
[19] L.Wang, Y. Wang, X.G. Sun, J.Q. He, Z.Y. Pan, C.H. Wang, A novel structure design towards extremely low thermal conductivity for thermal barrier coatings: Experimental and mathematical study, Materials and Design, 2012, 35(3):505-517.
[20] L. Wang, Y. Wang, X.G. Sun, J.Q. He, Z.Y. Pan, L.L. Yu, Preparation and characterization of nanostructured $La_2Zr_2O_7$ feedstock used for plasma spraying, Powder Technology, 2011, 212(1):267-277.
[21] L. Wang, Y. Wang, X.G. Sun, J.Q. He, Z.Y. Pan, C.H. Wang, Finite element simulation of stress distribution and development in 8YSZ and double-ceramic-layer $La_2Zr_2O_7$/8YSZ thermal barrier coatings during thermal shock, Applied Surface Science, 2012, 258(8):3540-3551.
[22] L. Wang, Y.Wang, X.G. Sun, J.Q. He, Z.Y. Pan, C.H. Wang, Thermal shock behavior of nanostructured 8YSZ and $La_2Zr_2O_7$7/8YSZ thermal barrier coatings fabricated by atmospheric plasma spray, Ceramics International. 2012, 38(5): 3595-3606.
[23] L.Wang, Y. Wang, X.G. Sun, Z.Y. Pan, J.Q. He, Y. Zhou, P.L.Wu, Microstructure and surface residual stress of plasma sprayed nanostructured and conventional ZrO_2-8wt%Y_2O_3 thermal barrier coatings, Surface and Interface Analysis, 2011, 43(5):869-880.
[24] L.Wang, Y. Wang, X.G. Sun, Preparation of 8YSZ and Double-ceramic-layer $La_2Zr_2O_7$/8YSZ Thermal Barrier Coatings on GH4169 Superalloy Substrates and Their Static High Temperature Oxidation Behavior, Materials Science Forum, 2013, 749: 617-632.

*Corresponding author: Y. Wang,
Email address of the corresponding author:
wangyou@hit.edu.cn

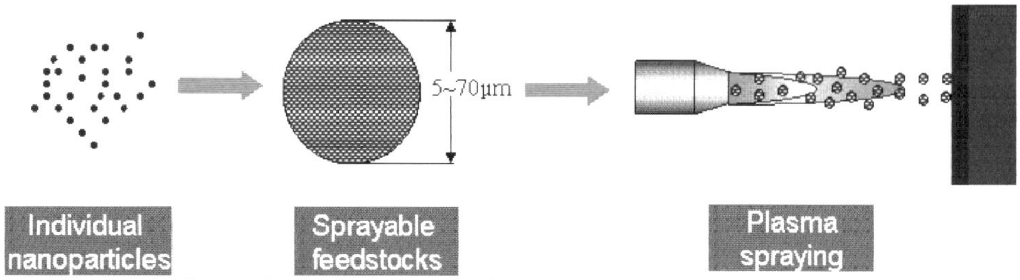

Figure 1 sketch map of reconstitution process of nanostructured feedstock suitable for thermal spray

Table 1 The density and flowability of particles after different treatment

Properties	Spray drying	Loose sintering	Plasma treatment
flowability（g/min）	13.61	9.9	57.32
density（g/cm^3）	1.15	1.2	2.24

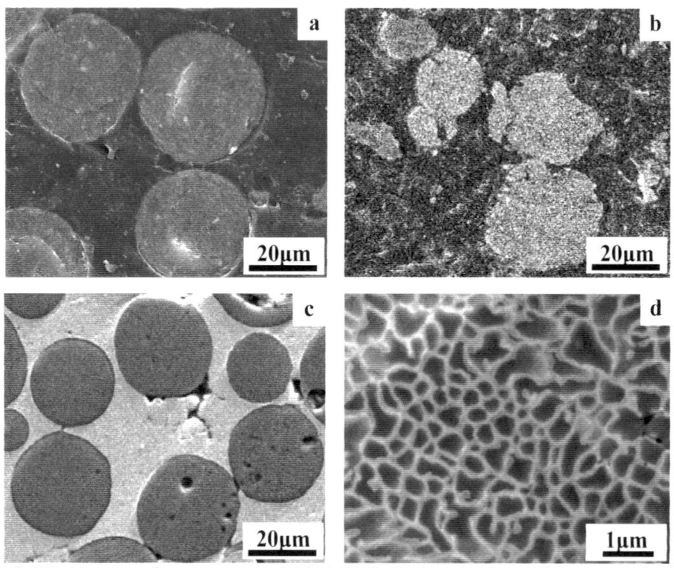

Figure 2 SEM images of nanostructured Al_2O_3-TiO_2 feedstock after different constitution process
(a) spray drying; (b) loose sintering; (c) plasma sintering; (d) high magnification image of (c)

Table 2 Mechanical properties of Al_2O_3-TiO_2 coatings

Coatings	Adhesive strength（MPa）	Hardnes($HV_{0.3}$) (GPa)	Resistance to crack extension $(1/\mu m)\times10^{-3}$	Spalltion resistance in bend (degree)
Metco 130 coating	25.0±5.	8.32±0.85	3.31±0.64	35.7±4.9
Undensified nanostructured coating	20.6±4.6	7.94±1.03	3.58±0.49	54.0±8.7
Densified nanostructured coating	31.0±2.1	9.20±0.56	4.44±0.50	63.3±4.5

Table 3 Wear rate of three coatings tested in a three-body testing

Coatings	Wear rate（$mm^3/N \cdot m$）$\times 10^{-4}$
Metco 130 coating	17.9
Undensified nanostructured coating	11 4
Densified nanostructured coating	5.9

Table 4 Results of potentiodynamic corrosion test of different coatings in 5%HCl and 3.5% NaCl solution.

Coatings	5% HCl		3.5% NaCl	
	E_{corr}（mV）	i_{corr} ($\mu A/cm^2$)	E_{corr}（mV）	i_{corr} ($\mu A/cm^2$)
Metco 130 coating	-281.7	489.0	-263.7	522.4
Undensified nanostructured coating	-280.5	217.0	-184.3	2647.9
Densified nanostructured coating	-219.0	81.5	-209.2	126.8

Table 5 Thermal cycling life of three coatings at 650 and 850

Coatings	Thermal cycling life	
	650	850
Metco 130 coating	41	3
Undensified nanostructured coating	56	5
Densified nanostructured coating	69	6

Table 6 Properties of Al_2O_3-20wt.%ZrO_2-based coating

Coatings	Hardness ($HV_{0.3}$)	Density (g/cm^3)	Porosity (%)
AZ	911.673±100	3.26415	7.65
AZP	1054.25±160	3.83784	6.39
AZSm	867.803±110	3.57895	5.12
AZSmP	1006.71±1500	3.91304	6.14
AZSn	918.253±210	3.66667	5.37
AZSnP	1200.41±80	4.04762	3.79

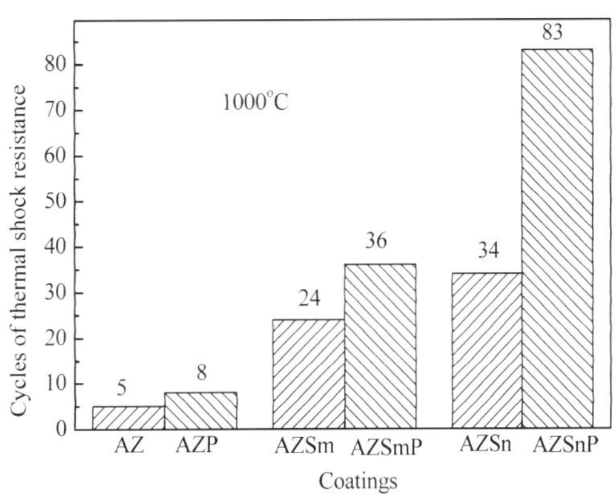

Figure 3 Thermal shock behaviors of Al_2O_3-20wt.%ZrO_2-based coating

DVS 324

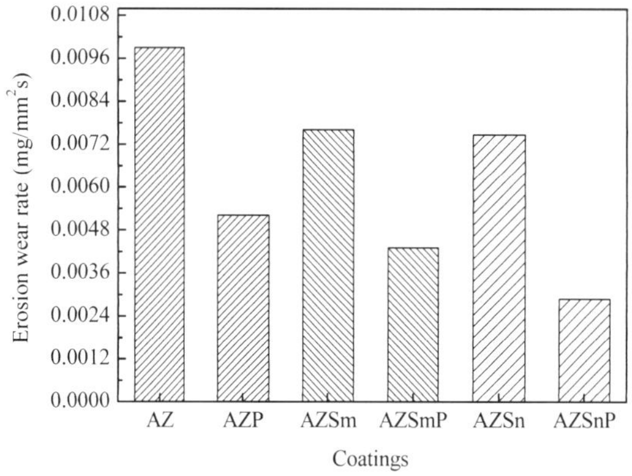

Figure 4 Erosion wear properties of Al_2O_3-20wt.%ZrO_2-based coating

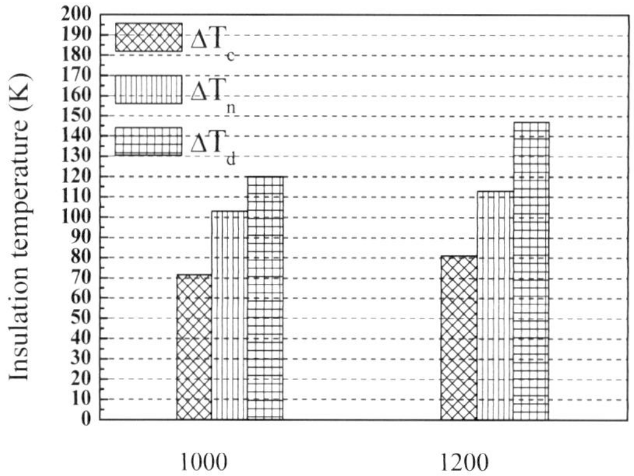

Figure 5 The insulation temperatures of the three different coatings at different coating surface temperatures

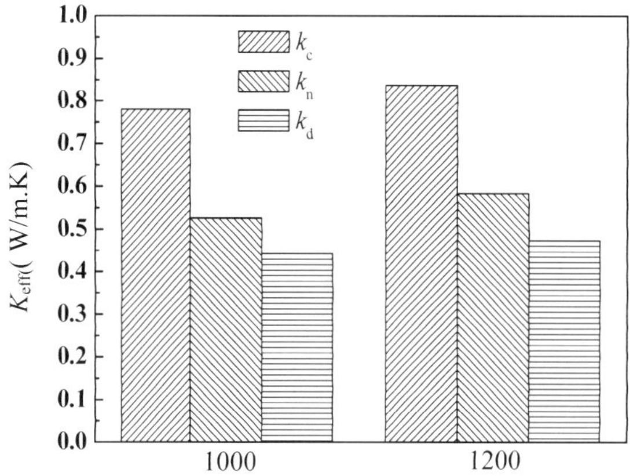

Figure 6 Effective thermal conductivity of three different as-sprayed coatings at different coating surface temperature

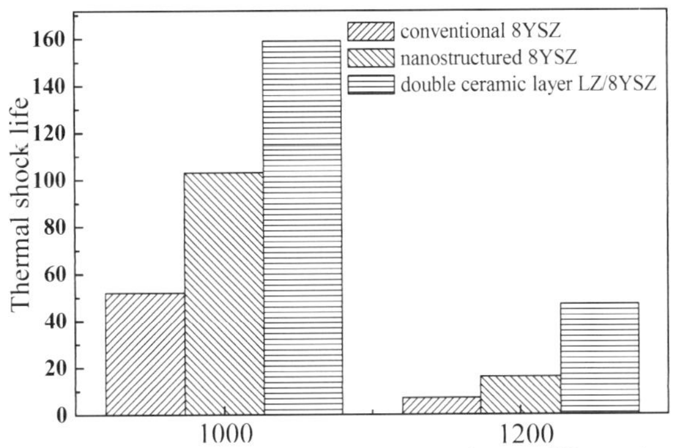

Figure 7 Thermal cycle life of different coatings at different thermal shock temperature

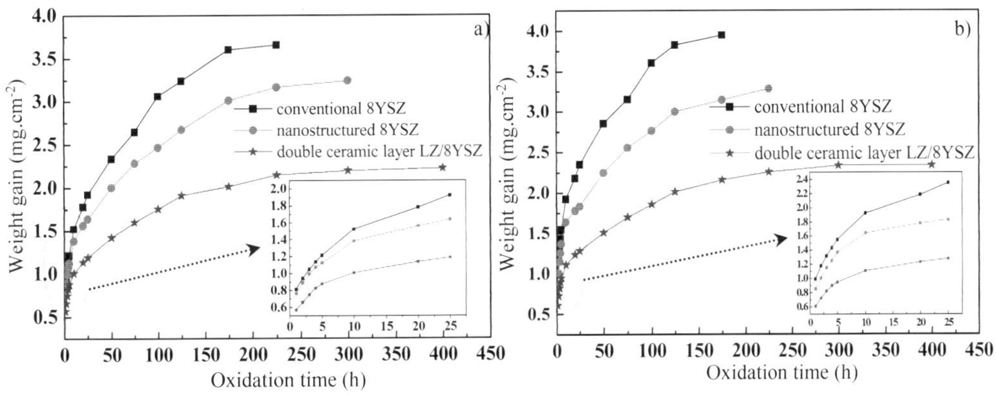

Figure 8 The weight gain as the function of time at 1000 a) and at 1200 b)

Structural and phase state of the material between the ceramic layer and the metallic substrate formed by high-speed jet of powder

M. Prozorova, M. Kovaleva, M. Arseenko, V. Sirota, Belgorod/RU, Yu. Tyurin, O. Kolisnichenko, Kyiv/UA, Jeff Jia, Beijing/CN, N. Vasilik, Moscow/RU.

In the present study a new multi-chamber detonation sprayer (MCDS) was employed to deposit Al_2O_3 ceramic coatings on substrate of alloys based on iron and titanium. MCDS provides a working gas rate up to 3000 m/s with low thermal power excepting overheating of the powder material and substrate. The ceramic powders were fed into the cylindrical chamber barrel. Each of discrete particles had the same conditions for the acceleration and heating in the high-speed jet of product of detonation combustion. The studies have shown that ceramic micropowders accelerated up to 2000 m/s. Generally, only the surface layer of ceramic particles has time to warm up at this speed. When a particle strikes a solid material of substrate a surface layer receives additional energy and cooperates with the metal elements forming the intermetallic compounds. Electron microscopic studies of the layer between the coating and substrate have shown the presence of the intermetallic compound type of FeAl, TiAl formed during plastic deformation and thermal diffusion of the coating and the substrate. The intermetallic materials of this layer were formed in a nanocrystalline state. This increases their plasticity and conducive the relaxation stresses generated in the formation of the ceramic layer.

1 Introduction

The ceramic coatings from the powder of Al_2O_3 are usually used to protect the details of which work in the hard conditions. They show a high resistance to corrosive and oxidizing environments and provide protection for details of which work at high temperatures. In addition, powders based on aluminum oxide have a low value, which determines the cost-effectiveness of their application. A new direction in this area is modification of surface by plasma-detonation technology that allows obtaining the ceramic coatings with a definite composition and thickness of reduction of the cost, which determines the cost-effectiveness of their application. Plasma-detonation technology is based on the use of pulsed plasma jets, accelerated by the electromagnetic field up to speed 5000...8000 m/s [1]. All this makes it possible to obtain high quality coatings of ceramic and metal on metal substrates [2,3]. The aim of this work was to study the structure and the physical and mechanical properties of high-quality coatings of Al_2O_3 obtained by a new multi-chamber detonation sprayer.

2 Methods of sample manufacturing and analysis

2.1 Materials

Powder AMPERIT® 740.0 Al2O3 was used to form a dense ceramic layer on the steel STE255 and titanium alloy OT 4 sv (Fe: ~ 0.3%; C:~0.1%; Si: ~0.12%; Zr: ~0.9%; O: ~ 0.15; Mn: 0.8-2.1%; N: ~0.05%; Al: 3.5-5%). The powder consisted of crushed particles with a maximal size of up to 50 μm (5-10%), the main fraction being 5.6-22.5 μm, the phase composition is mainly represented by γ-Al_2O_3 (92.34%), α-Al_2O_3 (3.83%), SiO_2 (3.83%) (Fig. 1). Flat specimens of hot-rolled carbon steel (Fe-0.25;C-0.90;Mn-0.04;P-0.05;S-0.20;Cu, all in wt pct) were used as substrates and they were sandblasted using alumina grits 25A F360 prior to spraying. The dimensions of the samples were of 30 x 30 x 5 mm. The sample was prepared by grinding with abrasive SiC paper (gradation 200, 500, 800, 1000), followed by polishing with 1 μm diamond slurry according to the procedure recommended by Struers company for ceramic coatings.

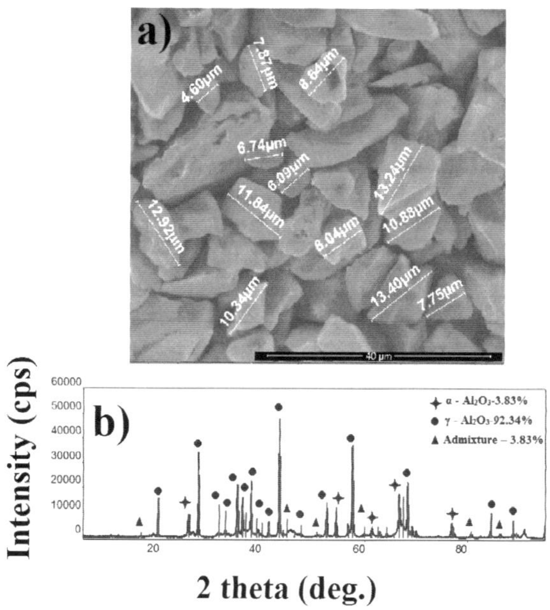

Fig. 1. The morphology (a, SEM) and the phase composition (b, X-ray) of the Al_2O_3 powder

2.2 Apparatus and Procedures

The cumulative-detonation technology and equipment is created for deposition of nanocomposite powder coatings on samples. The device provides a high velocity of the powder materials (>2000 mm/min) without its overheating. Productivity was 0.72÷1 kg/hour, gas mixture consumption (C_3H_8+O_2) –

5.08 m^3/hour, frequency – 20 Hz, the nozzle section to a specimen - 60 μm.

Morphological and compositional characterization of powder of Al_2O_3 and surface of ceramic coatings were performed by electron ion microscope Quanta 600 FEG and Quanta 200 3D equipped with integrated microanalysis system Pegasus 2000 (SEM-EDS). The microhardness of coatings was determined on transverse sections of system «coating/substrate» by an automatic analysis system microhardness DM-8 according to the method of micro-Vickers. X-ray diffraction studies of coatings were performed on a diffractometer Ultima IV Rigaku.

The evaluation of wear resistance of coatings was carried out by the methods of tribometer using the an automated machine friction (Tribometer, CSM Instruments, Switzerland) under the standard scheme of test «disc/ball» on the air under a load of 6 N. The porosity of the coatings from the powder of Al_2O_3 was determined by metallographic method using an optical inverted microscope Olympus GX51.

3 Results and Discussion

A series of Al_2O_3 coatings (200 μm) have been prepared on steel and titanium substrate by a new multi-chamber detonation sprayer with a nozzle length of 500 mm. Velocity parameters of the gas-powder flow were estimated over a length of 20-80 mm from the exit section of the nozzle by using a two-channel optical measurement system, which comprised two photodiodes FD287 (spectral sensitivity range: 0,5 - 1,7 μm), plastic optical fibres (POF), amplifiers and a device for real-time recording of signal waveforms. The photodiode spectral sensitivity range and POF spectral window provided filtration of the gas-powder stream radiation by cutting off radiation of combustion products. The boundaries of the heated powder cloud were registered by two measurement systems installed at a distance of 20 mm from each other. The precise time of visualisation in two planes makes it possible to calculate the velocity of the frontal area of the heated powder cloud and its length [4]. 20 measurements of exact time visualization of the front explosion simulation at each point were averaged, and characteristics of the velocity of the frontal area of the clouds were obtained from the experimental results.

Electron microscopic investigation of transverse sections of the «coating/substrate» has shown that the coating of powder of Al_2O_3 characterized by the alternation of compressed flakes and deformed discrete particles of the oxide. Coating formed by stacking of a plurality of deformed particles with different temperature , velocity , mass and state of aggregation.

Thick coatings with good adhesion to the substrate and the low porosity less than 1% obtained by plasma-detonation device from the alumina powder. The quality of coatings was achieved by the presence of α and θ - Al_2O_3, their share to 87% (Fig. 2). Analysis of the structural features of the interface «coating/substrate» has shown that the detonation powder coatings have three visually distinguishable zones (substrate, transition layer and main coating) (Fig. 3). The main components of the matrix are the Al and O, and their concentration at different points changes insignificantly. Impurity elements in the transition region of the substrate are the elements of the substrate. The presence of impurity elements in the coatings in the transition region has been caused by mixing of detonation gases, air and sputtered particles of the substrate.

Analysis of the structure of the interface «coating/substrate» after etching has shown that the coating has three visually distinct layers (Fig. 4). Therefore, the material of boundary was a mix of elements of the substrate and coating.

At the boundary between the transition layer and the main coating (Fig.5a) the concentration of atoms of aluminum was about 50 wt.%, and oxygen ~24 wt.%. It was established that the content of iron in the transient layer is 20 wt.% what can be attributed to its mass transfer from the substrate. It was found the presence of iron, metallic aluminum and intermetallic AlFe on the border between the coating and substrate. It was established that the content of titan in the transient layer (titanium substrate) is 19 wt.% that also explains the phenomenon of mass transfer from the substrate. It was found the presence of titan, metallic aluminum and intermetallic AlTi on the border between the coating and substrate (Fig.5b). This favors the relaxation of stresses arise during the deposition of coatings [5]. Therefore, the iron and titan should be considered as an impurity, which is a favorable effect on the mechanical properties of coatings.

Thus, to improve the performance properties was selected regime of the deposition of the coating in an oxidizing atmosphere (with excess oxygen). Oxidizing atmosphere also prevented the hydrogenation coatings, as it is known that alumina adsorbs hydrogen well in a heated state. Thus, the results of microanalysis of the transition layer and separate sections covering confirm the mixing of the coating material - Al_2O_3 with the substrate material, indicating that a strong connection of the resulting system «coating/substrate» interface [6].

The microhardness of the substrate of steel STE255 under the coating was changes to a depth of 200 μm from 550 $HV_{0.3}$ to medium hardness of the substrate material - 246 $HV_{0.3}$.

The microhardness of the substrate of titanium alloy OT 4 sv under the coating was changes to a depth of 200 μm from 630 $HV_{0.3}$ to medium hardness of the substrate material - 400 $HV_{0.3}$.

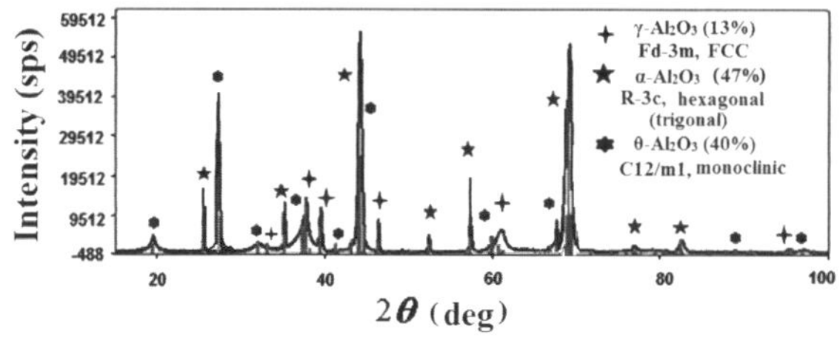

Fig. 2. Results of X-ray analysis of Al_2O_3 coating.

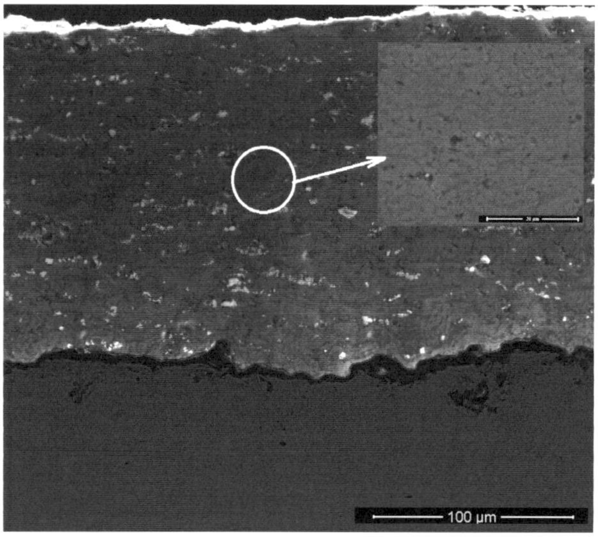

Fig. 4. SEM images of cross section of the Al_2O_3 powder coating

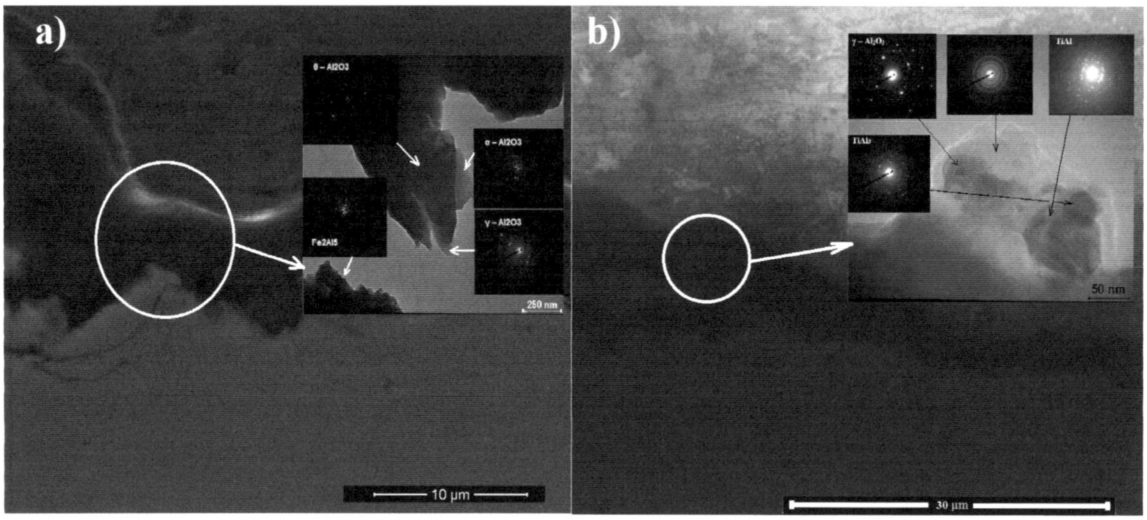

Fig. 5. Microstructure and elemental composition of transverse sections of the sample (SEM) and image and diffraction (TEM) photographs of a transition layer «coating/substrate»: a – steel substrate; b – titanium substrate

The figure 6 demonstrates that the microhardness of the substrate decreases by 1.5 times at a depth of 15 microns from the boundary «coating/substrate» and reached the minimal value at a depth of 180 microns. It means explains the presence of work hardening near the boundary with the surface, was formed as a result of impact in the process of abrasive blasting and coating. The microhardness of coating was measured from the surface to the substrate gradually decreases from the surface of the substrate from

1380 to 1000 $HV_{0.3}$. This decrease is due to the mechanism of formation of layered coating and heating below the underlying layers. These conditions are the result of the formation of a soft plastic and aluminum oxide phases (Fig. 6).

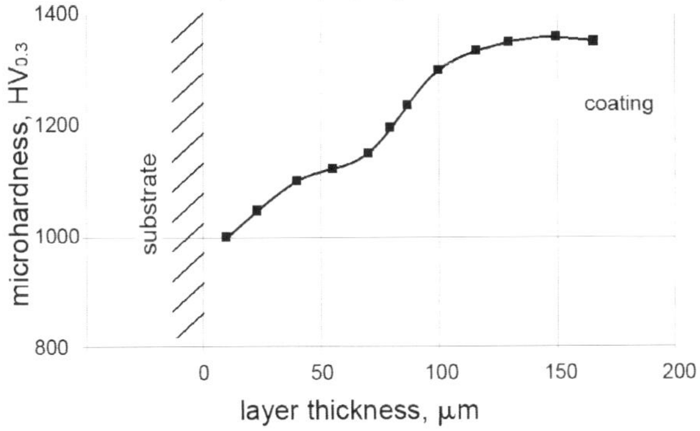

Fig. 6 The microhardness of the coating

Wear resistance tests performed on air at load of 6 N, linear speed is 15 cm / sec, the radius of curvature of 5 mm of wear, sliding distance of 1200 meters. From the data of table 1 shows that the investigated coating significantly increases the wear resistance of the sample STE255 and titanium alloy OT 4 sv, also provides a low ability to wear out the coating.

Investigation of the topology of friction surface coating shows that the initial friction coefficient of coated steel was 0.038, which may be due to the high roughness of the sample. Found that despite the increase in counterface of wear, track wear and tear, however, without destroying the coating, which is associated with good wear resistance of the coating [7].

Table 1. – Tribological characteristics

Sample	The coefficient of friction (μ)		The wear factor, $\mu m^3 \cdot N^{-1} \cdot m^{-1}$	
	initial	test	counter body (x10^{-5})	sample (x10^{-5})
Steel (STE255)	0.204	0.674	0.269	35.36
Titanium alloy OT 4 sv	0.154	0.643	0.82	69.7
Steel/Al_2O_3	0.038	0.959	1.61	19.39
Titanium alloy OT 4 sv /Al_2O_3	0.162	0.825	1.07	2.15

4 Conclusion

Ceramic multifunction dense coatings from the powder of Al_2O_3 were obtained by a new multi-chamber detonation sprayer. The microhardness of coatings de-creases smoothly from the surface of the substrate from 1380 to 1000 $HV_{0.3}$.

Analysis of the transition layer «coating/substrate» interface was showed that the layer is well adjacent to the substrate. Is formed transition layer, consisting of the dissociation products of elements of the substrate and the oxide. The hardness of the substrate layer is adjacent to the coating, increasing to 550 $HV_{0.3}$ at a hardness of the basis 246 $HV_{0.3}$. (steel substrate) and increasing to 630 $HV_{0.3}$ at a hardness of the basis 400 $HV_{0.3}$ (titanium alloy). The results of this work open up new prospects for the further elaboration of new technologies for making protective coatings that can enhance the microhardness and the wear resistant properties of industrial iron and titanium alloy machinery parts.

Acknowledgements

All of studies were carried out on the equipment of the Joint Research Center of Belgorod National Research University «Diagnostics of structure and properties of nanomaterials».

5 Literature

[1] E. Turunen, T. Varis, S.-P. Hannula, J. Keskinen, P. Lintunen, Thermal Spray 2007: Global Coating Solutions, Materials Park (USA, Ohio).

[2] M.A. Riley, A.J. Sturgeon, Thermal Spray 2007: Global Coating Solutions, Materials Park (USA, Ohio).

[3] Toma, D., Brandl, W., Marginean, G. Wear and corrosion behaviour of thermally sprayed cermet coatings. Surface and Coatings Technology 138 (2001), pp. 149/58.

[4] Vasilik, N., Tyurin, N., Kolisnichenko, O. Method for gas-dynamic detonating speedup of powders and device for its implementation, RU Patent 2506341, 2012.

[5] Yu.N. Tyurin, M.A. Zhadkevich., A.D. Pogrebnyak, O.V. Kolisnichenko, 6th International Conference "Films and coatings" (2001, Jule) (Russia, St. Petersburg).

[6] Kovaleva, M., Tyurin, Yu., Vasilik, N., Kolisnichenko, O., Prozorova, M., Arseenko, M. Deposition and characterization of Al_2O_3 coatings by multi-chamber gas-dynamic accelerator. Surface and Coatings Technology 232 (2013), pp. 719/25.

[7] Kovaleva, M., Tyurin, Yu., Vasilik, N., Kolisnichenko, O., Prozorova, M., Arseenko, M., Sirota, V., Pavlenko, I. Effect of heat treatment on the microstructure and microhardness of detonation nanostructured Al_2O_3 coatings. Journal of Thermal Spray Technology 23 (2014), Issue 7, pp. 1199/209.

Characterization of WC-Co coating modified by nano-sized TiC addition using electron microscopy methods

H. Myalska, R. Swadźba, R. Rozmus, G. Moskal, J. Wiedermann, Katowice, Gliwice, PL

Abstract

Scanning electron microscopy and scanning- transmission electron microscopy were employed to microstructure characterization of WC-Co coating modified by 5 % wt. nano-sized TiC and WC-Cocoating as a referential obtained by high velocity air fuel spraying (HVAF) on a steel substrate. Formation of a new phase containing tungsten in cobalt matrix that was identified as $Co_{0.9}W_{0.1}$by selected area diffraction was revealed. Observed diffusion effect was less intensive in case of TiC application. Any crystalline phases with titanium except of TiC were not found. Obtained results suggests that nano-sized TiC may increase the stability of metallic matrix microstructure in a WC base coating.

1 Introduction

Thermally sprayed hardmetal coatings are widely applied in many industrial applications in order to resist the elements working under very different service conditions such as aggressive media and high temperatures [1-4].

Tungsten carbide (WC) is a well-known material used for that conditions due to its high hardnessand corrosion resistance. The overall performance of WC is improved by placing in ductile metals, such as cobalt or nickel (with the range from 3 to 30 wt.%). The tungsten carbide cermet powder can be deposited using few different spraying technologies such as plasma spraying, high velocity oxygen fuel (HVOF) and high velocity air fuel (HVAF) spraying processes [5-8].

Cermet coatings obtained by HVOF and HVAF spray processes usually are characterized with low porosity, microstructure homogeneity, high hardness, high adhesion and compression stress [9-11]. The main difference between those systems is that in HVAF compressed air and fuel gas are mixed in the combustion chamber instead of pure oxygen. This provides lower temperature and reduces oxidationduring spraying process. Moreover, the air application instead of oxygen for combustion and a high spray rate are expected to lower the cost of the coating and promote its competitiveness over other processes if the coating quality is comparable or better [12-14].

It is well known that during thermal spraying of WC-Co coatings two negative phenomena such as decomposition and decarburization of carbides appear. Following phases such as W_2C, W, and amorphous or nano-crystalline Co-W-C are formed in the process of decarburization which is based on the reduction of WC content in coating. The overall performance of thermal sprayed WC-Co coatings is significantly affected by the decarburization of the deposited powder. Controlling the decarburization of the carbide is therefore crucial towards achieving the cermet coatings with super mechanical properties [15-20].

Different concepts of WC-Co coating properties improvement may be consider and an application of nanocomponents being the mechanical strengthening additions is one of them [21-22]. In case of that solution both nano-sized WC was as well as another nano-sized carbides were examined [23]. Own research works on WC-Co coatings with nano-sized TiC particles confirmed an increase of mechanical properties comparing with conventional WC-Cocoating that was deposited at the same conditions of HVAF process [24]. Because the effect of nano-sized TiC applied in the initial powder mixture on a final coating microstructure was not known, the detailed examinations with high resolution were necessary.

The aim of the work is a characterization of a thermally sprayed WC-Co coatings modified by 5wt.% of nano-sized TiC on a steel substrate, by different electron microscopy methods including phase identification. This allows to answer whether the used compounds are stable in the process of coating deposition and to characterize the function of nano-sized TiC in WC-Co coating microstructure formation.

2 Experiment

The WC-Co coating with nano-sized TiC and without that addition for comparison were obtained by HVAF technique and the processing parameters are shown in Table 1.

Tab.1. Parameters of coatings deposition.

Air	Pressure [bar]	8.5
	Flow [l/min]	7.5
Propane	Pressure [bar]	9.2
	Flow [l/min]	85
Nitrogen	Pressure [bar]	8.4
	Flow [l/min]	30
Powder feed rate [rpm]		3-7
Surface speed [mm/s]		500

The coatings were formed from pure WC-Co standard powder (Amperit 526.074, FST) and mechanically blended mixture of WC-Co powder with 5wt.% of nano-sized TiC (Hefei EvNano Technology). The particle size of applied powders was respectively 45/15 µm for WC-Co and 40-100 nm for TiC. The initial morphology of powders used in experiment and their configuration in obtained mixture observed with different magnification, using field emission scanning electron microscope (HITACHI S-4200), was shown in Fig. 1.

The SEM observations revealed that the agglomerates of WC-Co standard powder were built of irregular WC grains surrounded by the cobalt matrix(Fig. 1a and b). A slight tendency for the agglomeration of TiC nano-sized particles as manufactured was observed, however the agglomerates were sized of approximately of 300 nm (Fig. 1c and d). After applied mixing procedure of WC-Co powder with nano-sized TiC powder a characteristic configuration of components was formed (Fig.1d and e). The globular particles of WC-Co powder were regularly covered with the disintegrated grains of TiC powder and the isolated agglomerates of TiC typical for initial TiC were not detected in obtained powder mixture. That configuration of TiC on WC-Co powder should ensure greater spraying efficiency as well as more homogenous distribution of TiC in a metal matrix of the deposited WC-Co-nano-sized TiC coating.

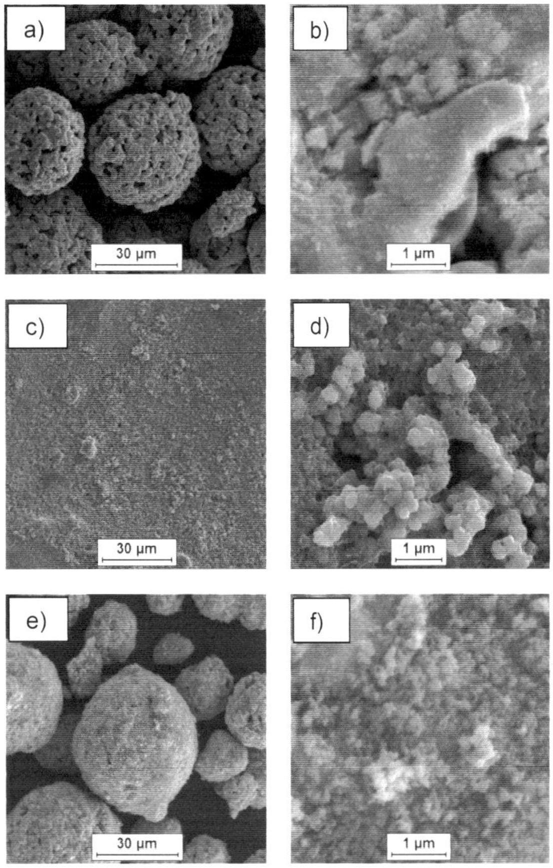

Fig.1. SEM micrographs of a) and b) WC-Co powder, c) and d) nano-sized TiC powder, e) and f) WC-Co + 5wt.% nano-sized TiC mixture.

The examined WC-Co-nano-sized Ti coating and a referential WC Co coatingdeposited on steel surface were cut across, grinded andpolished.Then the scanning electron microscope (HITACHI S-4200) with X-ray energy dispersive spectrometer (EDS, Noran System NSS 2.2) was employed for characterizationof microstructure andthe chemical composition in microareas.

For coatings characterization with a higher magnification and their phase composition identification the thin foils were prepared by a focused ion beam (FIB) method with Quanta 200i DualBeam by FEI. The samples were examined with theUltrahigh Resolution Scanning-Transmission Electron Microscope (S/TEM) TITAN 80-300 by FEI Company, and images in bright field (BF), dark field (DF)techniques as well as selected area diffraction patterns (SADP) were obtained.

3 Results and discussion

The SEM micrographs registered in secondary electrons mode and the elements mapping of cross-sectioned coatings are shown in Figs 2-3.

The analysis of elements distribution in the coating formed from WC-Co powder confirmed a presence of phases containing tungsten with carbon and pure cobalt, and titanium with carbon in case of WC-Co-nano-sized Ti mixture application. In addition the microareas, where a presence of cobalt and tungsten were found, independently on nano-sized TiC content. However, the amount of the regions containing cobalt and tungsten was lower in the case of the coatings modified with nano-sized TiC. In order to ensure a correct characterization of that microareas the use of transmission electron microscopy methods with selected area electron diffraction was required.

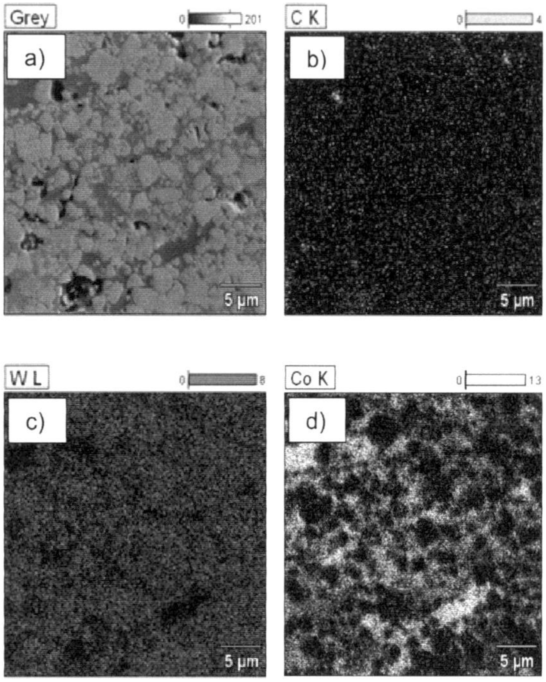

Fig.2. Microstructure (a) and element maps of carbon (b), tungsten (c) and cobalt(d) of coating obtained from WC-Co powder, SEM, EDS.

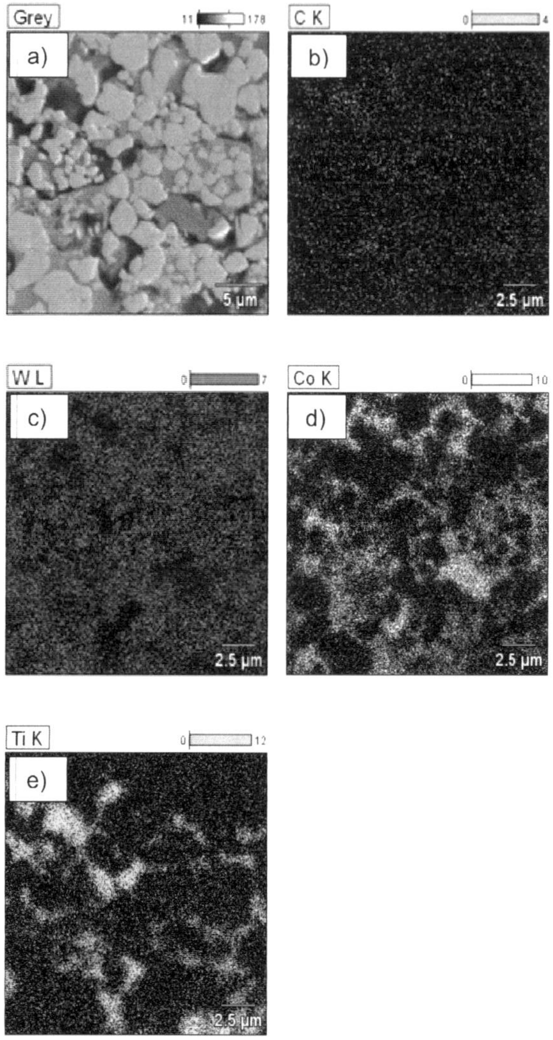

Fig.3. Microstructure (a) and element maps of carbon (b), tungsten (c), cobalt(d) and titanium (e) of coating obtained from a powder mixture of WC-Co + 5wt.% nano-sized TiC, SEM, EDS.

The S/TEM images of WC-Co coating carried out in bright field and dark field and EDS analysis confirmed a presence of the grains with sharp edges, and tungsten with carbon content (WC). They were surrounded with the cobalt matrix where the microareas enriched in tungsten were detected (Fig. 4). That microareas were located in vicinity of a WC grains and their edges at the border with pure cobalt were rounded. Observed the cobalt matrix microstructure change isa result of tungsten diffusion and indicates a new phase formation because of evident sharp change of W and C concentration at it borders both with WC and pure Co microareas.

Results of the same STEM examination procedure applied for WC-Co-nano-sized TiC coating are presented in Fig. 5 and they exhibited similar effect of cobalt matrix enrichment in the tungsten (Fig. 5c).

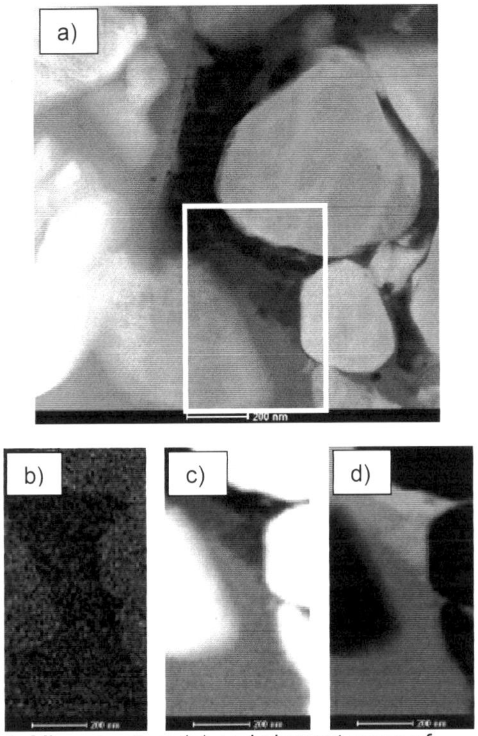

Fig.4. Microstructure (a) and element maps of carbon (b), tungsten (c) and cobalt (d) of coating obtained from WC-Co powder, STEM, HAADF, EDS.

Selected area diffraction patterns (SADP) obtained for both investigated coatingsand presented in Fig. 6 for WC-Co and in Fig. 7 for WC-Co-nano-sized TiC system. Theyconfirmed in obtained materials the presence of phases used in initial powders as WC and Co, and TiC respectively. The microareas of Co matrix enriched with W were identified as $Co_{0.9}W_{0.1}$ phase independently on nano-sized TiC application (Fig.6 e and 7 e).

Neither phases containing titanium, except titanium carbide applied as the mechanical properties improving component, nor evident changes in the nano-sized TiC morphology were found. This fact indicates the stability of TiC structure during the coating microstructure formation. However, it should be mentioned that the amount of a new $Co_{0.9}W_{0.1}$ intermetallic phase formed in vicinity of the WC grains was noticeably lower in the coating with nano-sized TiC. It allows to suppose that in examined material the nano-sized TiC increases the cobalt matrix structure stability. That effect can also appear in the coating service conditions and be more effective because of lower temperature and less intensive diffusion processes, but it has to beverified experimentally. The influence of the identified $Co_{0.9}W_{0.1}$ phase with irregular shape and size on coating properties must be examined, but one may expect that it presence can be disadvantageous because the intermetallic phases are brittle usually.

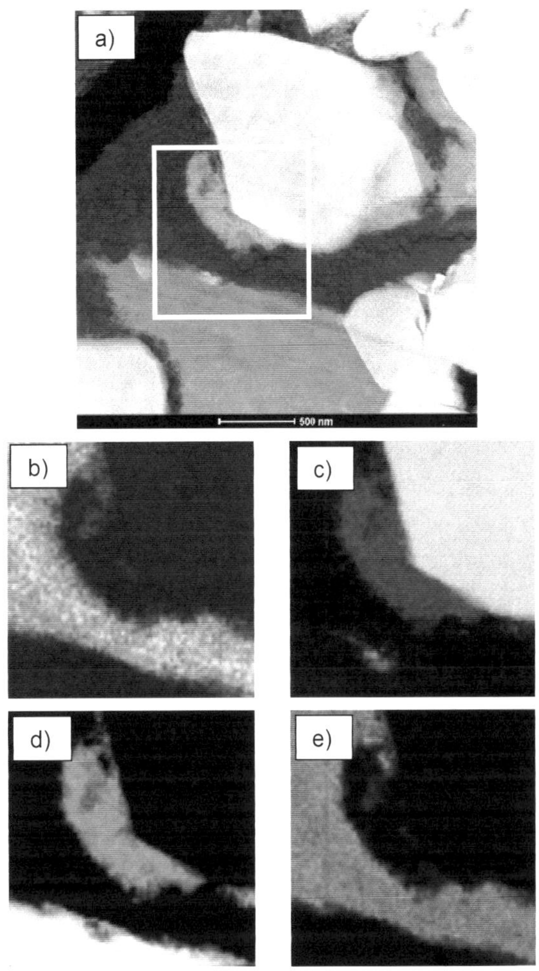

Fig.5. Microstructure (a) and element maps of carbon (b), tungsten (c), cobalt (d) and titanium (e) of coating obtained from WC-Co + 5 wt.% nano-sized TiC powder mixture, STEM, HAADF, EDS.

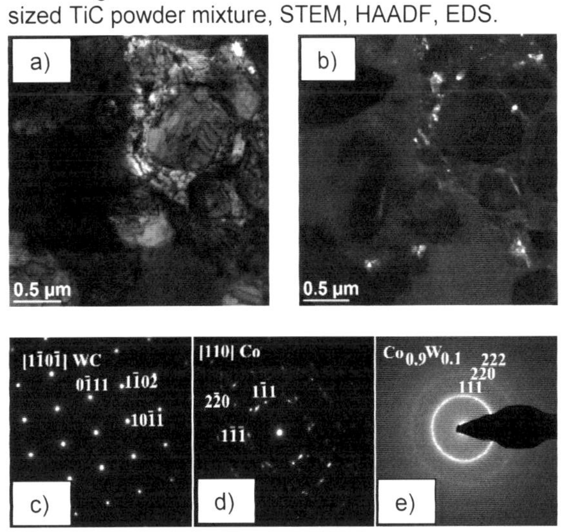

Fig.6.TEM micrographs in BF (a) and DF (b) of WC-Co coating and SADP of identified phases WC (c), Co (d) and $Co_{0.9}W_{0.1}$ (e).

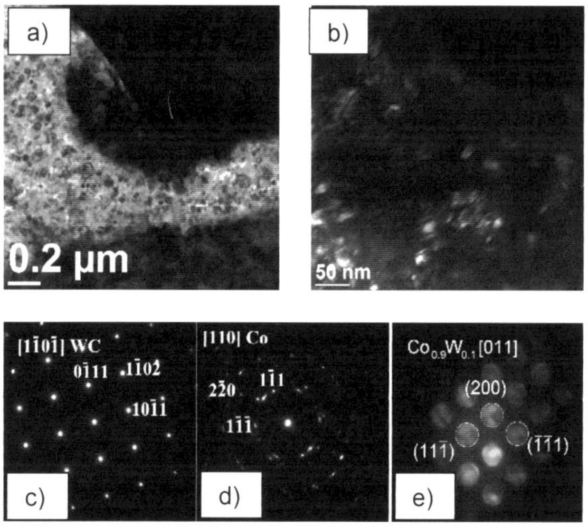

Fig.7. TEM micrographs in BF (a) and DF (b) of WC-Co-nano sized TiC coating and SADP of identified phases WC (c), Co (d) and $Co_{0.9}W_{0.1}$ (e).

Generally, the literature data suggest [20] that a formation of new phases induce a decrease a tungsten content in tungsten carbide is an unfavorable effect during WC-Co coating deposition. That problem is regardingalso acoating stability in service conditions where the intensive diffusion may appear. Conducted experiment indicates that the nano-sized TiC except a strengthening function in the WC-Co coating may also decrease the intensity of structure transformation.

4 Conclusions

The WC-Co coating with the nano-sized TiC addition and WC-Co coating as a referential material obtained by HVAF method were characterized by different electron microscopy methods and as follows conclusions were formed

- In cobalt matrix the microareas enriched in tungsten were detected at the vinicity of the WC grains by EDS and they were identified as $Co_{0.9}W_{0.1}$ phase by SADP method.
- A new phases containing Ti were not found in the examined coating, which indicates the stability of nano-sized TiC during the deposition process.
- An effect of the new $Co_{0.9}W_{0.1}$ phase formation was detected in both types of examined WC-Co coatings but it was weaker when a nano-sized TiC addition was applied.
- The purposefulness of nano-sized TiC adding to WC-Co coating, which was previously revealed by the mechanical properties increase, was additionally confirmed by it influence on microstructure stability observed in presented experiment.

Acknowledgement

We gratefully acknowledge the financial supportof Polish National Science Centre (decision No. 2014/15/N/ST8/03588).

5 Literature

[1] Davis J.R. (Ed.), Handbook of Thermal Spray Technology, ASM International, Materials Park, OH, USA, 2004.

[2] Pawlowski L., The Science and Engineering of Thermal Spray Coatings, second ed., J. Wiley and Sons Ltd., Chichester, UK (2008), pp. 85/89.

[3] Bolleli G., Berger L.-M., Bonetti M., Lusvarghi L., Comparative study of the dry sliding wear behaviour of HVOF-sprayed WC–(W,Cr)2C–Ni and WC–CoCr hardmetal coatings, Wear 309 (2014), pp. 96/111.

[4] Szymański K., Hernas A., Moskal G., Myalska H., Thermally sprayed coatings resistant to erosion and corrosion for power plant boilers - A review, Surface & Coatings Technology 268 (2015) pp. 153/164.

[5] Lidong Z, Matthias M, Falko F, Robert D, Erich L. Influence of spray parameters on the particle in-flight properties and the properties of HVOF coating of WC– CoCr, Wear 257 (2004), pp.41/46.

[5] Murthy JKN, Venkataraman B. Abrasive wear behaviour of WC–CoCr and Cr3C2-20(NiCr) deposited by HVOF and detonation spray processes. Surface & Coatings Technology 200 (2006), pp. 2642/2652.

[6] Picas J.A., Xiong Y., Punset M., Ajdelsztajn L., Forn A., Schoenung J.M., Microstructure and wear resistance of WC–Co by three consolidation processing techniques, Int. Journal of Refractory Metals & Hard Materials 27 (2009), pp. 344/349

[7] Hulka I, Şerban V.A., Secoşan I., Vuoristo P., Niemi K., Wear properties of CrC–37WC–18M coatings deposited by HVOF and HVAF spraying processes, Surface & Coatings Technology 210 (2012),pp.15/20.

[8] Koutsky J., High velocity oxy-fuel spraying, Journal of Materials Processes Technology 157–158 (2004), pp. 557/560.

[9] Wielage B., Wank A., Pokmurska H., Grund T., Rupprecht C., Reisel G., Friesen E., Development and trends in HVOF spraying technology Surface & Coatings Technology 201 (2006), pp. 2032/2037.

[10] Guo R.Q., Zhang C., Chen Q., Yang Y., Li N., Liu L., Study of structure and corrosion resistance of Fe-based amorphous coatings prepared by HVAF and HVOF, Corrosion Science 53 (2011), pp. 2351/2356.

[11] Wang A.P., WangZ.M., ZhangJ., WangJ.Q., Deposition of HVAF-sprayed Ni-based amorphous metallic coatings, Journals of Alloys and Compounds 440 (2007), pp. 225/228.

[12] Gorlach I.A., The application of high velocity processes for the deposition coatings, R&D Journalof the South African Institution of Mechanical Engineering 24 (2008), pp. 9/16.

[13] Guo R.Q., Zhang C., Chen Q., Yang Y., Li N., Liu L., Study of structure and corrosion resistance of Fe-based amorphous coatings prepared by HVAF and HVOF Corrosion Science 53 (2011), pp. 2351/2356.

[14] Zeng Z., Sakoda N., Tajiri T, Kuroda S., Structure and corrosion behavior of 316L stainless steel coatings formed by HVAF spraying with and without sealing, Surface & Coatings Technology 203 (2008), pp. 284/290.

[15] Guilemany J.M., Depaco J.M., Nutting J., Miguel J.R., Characterization of the W2C phase formed during the high velocity oxygen fuel spraying of a WC + 12 pct Co powder, Metallurgical and Materials Transactions A 30 (1999), Issue 8, pp 1913/1921.

[16] Verdon C., Karimi A., Martin J.L., A study of high velocity oxy-fuel thermally sprayed tungsten carbide based coatings. Part 1: Microstructures, Materials Science and Engineering: A 246 (1998), Issue 1-2, pp. 11/24.

[17] Stewart D.A., Shipway P.H., McCartney D.G., Abrasive wear behaviour of conventional and nanocomposite HVOF-sprayed WC–Co coatings, Wear 225-229 (1999), pp. 789/798.

[18] Schwetzke R., Kreye H., Therm J., Microstructure and properties of tungsten carbide coatings sprayed with various high-velocity oxygen fuel spray systems, Spray Technology 8 (1999), pp. 433/439.

[19] Chen H., Hutchings I.M., Abrasive wear resistance of plasma-sprayed tungsten carbide–cobalt coatings, Surface & Coatings Technology 107 (1998), pp. 106/114.

[20] Yuan J., Zhan Q., Huang J., Ding S., Li H., Decarburization mechanisms of WCeCo during thermal spraying: Insights from controlled carbon loss and microstructure characterization, Materials Chemistry and Physics 142 (2013), pp. 165/171.

[21] Poblano-Salas C.A., Cabral-Miramontes J.A., Gallegos-Melgar A., Ruiz-Luna H., Aguilar-Escobar J.D., Espinosa-Arbelaez D.G., Espinoza-Beltrán F., Trapaga-Martínez G., Muñoz-Saldaña J., Effects of VC additions on themechanical properties of bimodalWC–Co HVOF thermal sprayed coatings measured by nanoindentation, Int. Journal of Refractory Metals and Hard Materials 48 (2015), pp. 167/178.

[22] Myalska H., Szymański K., Moskal G., Microstructure and properties of WC-Co HVAF coatings obtained from standard, superfine and modified by sub-micrometric carbide powders, Archives of Metallurgy and Materials 60 (2015), Issue 2, pp. 759/766.

[23] Yuan J., Ma Ch., Yu Z., Li H, Improving the wear resistance of HVOF sprayed WC-Co coatings by adding submicronsized WC particles at the splats interphases, Surface & Coatings Technology 285 (2016), pp. 17/23.

[24] Myalska H., Moskal G., Szymański K., Microstructure and properties ofWC–Co coatings,modified by sub-microcrystalline carbides, obtained by differentmethods of high velocity spray processes, Surface & Coatings Technology 260 (2014), pp. 303/309.

An Investigation of the Corrosion and Fouling Resistance of Zn-Ni/Cr_2O_3-TiO_2-CuO-SiO_2/PTFE Composite Ceramic Coatings

LI Qing-yu[1], CHENG Xu-dong[1], YE Wei-ping[2]

(1.State Key Laboratory of Advanced Technology for Materials Synthesis and Processing, Wuhan University of Technology, Wuhan 430070, China

2. School of Materials science and Engineering, Wuhan University of Technology, Wuhan 430070, China)

Abstract: In this study, Zn-Ni/Cr_2O_3-TiO_2-CuO-SiO_2/PTFE composite coatings were deposited on stainless steel (SS) substrates by Atmospheric Plasma Spraying (APS), aiming at possessing the capability of wear, corrosion and stain resistance. The microstructures and phases of the composite coatings were revealed by X-Ray Diffraction (XRD) and Scanning Electron Microscopy (SEM). The corrosion resistance of the coatings was characterized by Neutral Salt Spray (NSS) test, polarization curve, weightlessness experiment and immersion corrosion test. The fouling resistance of the coatings was investigated by wear test under water-lubrication condition. The reciprocating parts with the coatings were relatively intact after 5000h immersion test in realistic ocean environment. The research revealed that the cyclical changes of the coatings weight were influenced by the weight fluctuate of dissolution of oxide film and the accumulation of secondary products. Furthermore, the PTFE sealed the porosity of the coatings surface which applied the surface of coatings less micro-porosity. With sealing reagent and TiO_2 particle penetrating into the pores, leading to the decline of porosity in the coatings, the low-porosity made the composite ceramic coatings effective in reducing and retarding the corrosion and wear of reciprocating parts.

Key words: APS; Anti-corrosion; Composite ceramic coatings; Corrosion mechanism; Fouling-resistance

1 Introduction

Several reciprocating parts of cylinder in ocean environment, such as piston rods, are notoriously subjected to fast degradation at typical service conditions, because of the working environment as well as of the severe attack promoted by oxygen and molten salts [1]. The application of overlay composite ceramic coatings allows improving their lifetime, by defending their surface against marine corrosion, wear and friction.

The tradition protecting method by plating hard chromium on surface is effective in resisting marine corrosion and mechanical wear. However, after long-term application of its corrosion and wear mechanism, it revealed the fact that the trivalent chromium produced by electroplating has a significant impact on human health and other living organisms in t he environment [2]. Otherwise, without consideration for its environmental problems, the surface of steel structure with chromium coating begin to rust and occur pitting corrosion after immersion in the marine for 5-7 years. For the higher potential of chromium than steel, the expansion speed of corrosion part is relatively high. And as friction pairs of hydraulic moving parts, three-body abrasion is easily caused by gravel, iron chips and corrosion products embedded on the surface of chromium coating by its limited hardness. In addition, the microbe contacted with the surface of friction piece and transmission device deteriorate the working condition. It is well recognised that ceramic coatings with high hardness, low thermal conductivity, and low friction coefficient are able to reduce friction of the surface as well as the damaging promoted by corrosion, wear and friction [3].

Atmospheric Plasma Spraying (APS) is a usual method to apply ceramic coatings. The friction performance is an important index to evaluate the ceramic coating; many scholars have conducted massive research to the friction and wear performance of APS ceramic coatings, obtained rich empirical study material and formed several very influential theories. In 1981, professor Brookes [4-6] predicted that ceramic materials have a wild field of application marine machinery, and successfully developed an axial plunger pump in which plunger piston and valve plate was made by ceramic materials. In wear and friction performance, research made by S. Matthews and B. James [7] has shown that WC-Co, Ni-Cr-Mo and Cr_2O_3 coatings produced by APS possess excellent friction and wear resistant. Similar study has also been reported by Wassemrna. Ch [8], this research demonstrates that WC-Co, Ni-Cr-Mo and Cr_2O_3 coatings keep stable in salt-spray test, which is of better ability in corrosion resistant than the chromium plating. The Japanese DSRV water pump uses the Al_2O_3 and Al_2O_3-TiO_2 coatings by APS, exhibiting a effective erosion-corrosion property [9]. In 1996, Pushpvanam.M [10] has obtained Zn-Ni alloys by electro-deposition in citrate bath solution; this discovery promotes the replacement of chromium plating in anti-corrosion. In our laboratory, Qi.Zh has got the Zn-Ni powder by ball grinding mill.

However, some technical restrictions have been remarked for single layer coatings-their relatively simple function. Therefore, further investigations are needed in order to obtain composite ceramic coatings to resist corrosion, wear and friction. In this paper, ZnNi/Cr_2O_3-TiO_2-CuO-SiO_2/PTFE composite ceramic coatings were deposited on stainless steel (SS) substrates by APS. The phase composition, microstructure, wear, friction and corrosion ability were investigated, aiming to the development of reciprocating parts in seawater.

2 Material and methods

2.1 Coating preparation

The Zn-Ni/Cr_2O_3-TiO_2-CuO-SiO_2/PTFE composite ceramic coatings were deposited by APS system equipped with GP-80 plasma torch (Jiujiang Spray Equipment Corporation, China) on the 304 stainless steel (25mm×25mm). The powders were prepared by spray drying in this research. Several raw materials, including metal oxides, organic binder, and demonized water, were mixed in proportion to be well-ground slurry; the slurry was sprayed to be powders in spray dry tower. The component (wt%) of the binding layer is made up by 70% Zn and 30% Ni, the ceramic coating powder is composed by 75% Cr_2O_3, 17% TiO_2, 2.5%CuO and 5.5% SiO_2. **Table. 1** shows the operating parameters for preparing spray-dried spinel powders. The sprayed powders have been chosen with a granularity of 40-90μm and the powder of Cr_2O_3-TiO_2-CuO-SiO_2 has been baked from 1100 to 1200°C. The deposition efficiency was calculated from the ratio between the coating mass and the total feedstock mass delivered to the plasma torch, which was determined from the powder feed rate and the spraying time on the substrate [1].The average thickness of the composite ceramic coatings was about from 300 to 500μm. Finally, the composite ceramic coatings Zn-Ni/Cr_2O_3-TiO_2-CuO-SiO_2 were sealed by Polytetrafluoroethylene (PTFE).

2.2 Characterization

The phase compositions of the coatings were identified by XRD (D/MAXIII) with a scanning increment of 0.02°/step and a scanning speed of 5°/min over the 2θ range of 20-90°, X-ray diffraction was carried out using a Cu Kα radiation generated at 40kV and 40mA. Micro structural observation of the coatings was performed by SEM (JSM-5610LV). The porosity was measured by the method of thermogravimetric volatilization of liquids [11] which was calculated from weight measurements according to

$$P = \frac{W_1 / d_1}{W_s / d_s + W_1 / d_1} \quad (1)$$

Where W_1 and W_s are the weight of the liquid and solid, d_1 and d_s are the densities of them, respectively.

2.3 Immersion corrosion test

The immersion corrosion test of the composite ceramic coatings were evaluated immersing the coatings into 3.5 wt% NaCl solution for Neutral Salt Spray (NSS) test, taking out the sample every 24 hours and cleaning it in deionized water by ultrasonic method. Then, the sample was weighted on electronic scales after constant temperature of 60°C for an hour. The mass loss was calculated by the following equation, presenting the corrosion resistance of the coating. The mass loss (M) was defined as:

$$M = M_1 - M_2 \quad (2)$$

Where M_1 (g) and M_2 (g) were the mass of the sample before and after immersion test.

2.4 Electrochemical measurements

The corrosion behavior of the composite coatings was investigated by potentiodynamic polarization and electrochemical impedance spectroscopy (EIS) in 3.5 wt% NaCl solution by Potentiostat (PARSTAT-2773). C-view software was used to measure the potentiodymic polarization curves of the sample. The potentiodyic polarization behavior was recorded after immersion every day. In this test, a three-electrode system was used, the sample was applied as the working electrode, a saturated calomel electrode (SCE) and a platinum plate was the reference electrode and counter electrode, respectively. The scan range of the polarization curve was from -200 to +800mV with a scan rate of 0.333mv/s. The frequency range was from 10kHz to 0.01Hz and 10mV peak-to-peak voltage amplitude versus open circuit potential (OCP) was used to measure EIS, and the data of it was analyzed by Zsimp Win software. It should be noted that the test were performed for two replicate tests at least.

3 result and discussion

3.1 Structure characterization

The XRD patterns of as-sprayed Zn-Ni and Cr_2O_3-TiO_2-CuO-SiO_2 powders and coatings are shown in **Fig. 1a and b**, respectively. As shown in **Fig. 1a,** the diffraction peak of Ni is non-existance after deposition as coating, for the atom radius of Ni is larger than that of Zn and the Ni dissolved in Zn, the phase pattern of this coating is composed by Zn-Ni supersaturated solid solution. As similar, the content of CuO and SiO_2 is relatively low which result to the disappearance of the corresponding peaks **(Fig. 1b).** The powder is mainly composed of escolaite Cr_2O_3 with a small amount of CuO and SiO_2. The original particle size of TiO_2 is small, resulted the possession of size effect and large specific surface area. This phase is not present in the XRD patterns of the composite coatings. Indeed, during the plasma spraying unbalance condition, the TiO_2 powder dissolved in Cr_2O_3, producing the supersaturated solid solution-$(Cr_{0.88}Ti_{0.12})_2O_3$, while the diffraction peak of it not finding because of its tiny content. Furthermore, the plasma flame flow leads the oxygen loss of TiO_2, the structure diffraction peaks vanished for the formation of TiO_{2-x} [12]. While, no obvious CuO and SiO_2 XRD patterns are detect in the composite coatings for its minute amounts. As a result, only the phase of Cr_2O_3 is observed in the coating.

The microstructure of the particle of two powders is demonstrated in **Fig.1c-f** The Zn-Ni powder particle is composed of spherical hollow particles with a average diameter of 50μm as shown in **Fig. 1c**. **Fig. 1d** shows the detailed morphology of the Zn-Ni grains with well detecting of the metal particle size between 300 to 500nm. As shown in **Fig. 1e**, there are some tiny holes in the surface of as-sprayed powders. Their porous microstructure is characterized by the powders feedstock of micro-size Cr_2O_3 and nanoscale TiO_2, CuO and SiO_2, with the disparity of size, the surface of hollow sphere agglomerated powder is unable to be completely densified. **Fig.1f** presents the detailed morphology of the surface of Cr_2O_3-TiO_2-CuO-SiO_2 hollow grains, the nanoscale additions TiO_2, CuO and SiO_2 fill among the rubbly Cr_2O_3 which is typical blocky or angular morphology. Although during the period of baking, the surface of particle has formed holes by the volatile of binder, the micro pore hardly affect the mobility and stability of the powder when spraying coatings.

Fig. 1g-i reveals the surface microstructure of the as-sprayed coatings. The molten state of the coating is well without cracks and holes, which is associated to the sphere nanoscale powder and melting evenly in the plasma flame, getting an adequate spread on the substrate (Fig. 1g). Whereas, Fig. 1h presents few discharge channels and open pores on the surface, which might be attributed to the internal residual stresses resulting from the rapid cooling of the coatings. Certain particles failed to melt completely, which also resulted in the particle stack and voids [13]. However, this defect of voids and cracks might be offset by the pore sealing agent which is mentioned and described below. The Fig. 1i reveals the cross sectional image of the coatings. Voids were found in two coatings, while no obvious cracks were viewed between the Zn-Ni coating and the substrate, Zn-Ni coating and Cr2O3-TiO2-CuO-SiO2 coating as the same, which indicated firm combination between them. The typical lamellar microstructure of cross section is observable in two coatings, those voids are produced by the partially melted particles, and the incomplete fill between the particles and rough surface as well as the inadequate combination between particles.

3.2 Cavitation erosion behavior

Fig. 2a displays the variation of mass loss of ZnNi/Cr_2O_3-TiO_2-CuO-SiO_2 coatings and ZnNi/Cr_2O_3-TiO_2-CuO-SiO_2/PTFE coatings after immersion in 3.5 wt% NaCl solution for 18 days, under a test temperature of 20°C. The variability of the two coatings is similar in 3.5wt% NaCl solution, the mass loss occur in initial stage, while the weight of two coatings increases with the proceeding of corrosion, the total mass of the coatings circulate the change. However, the mass variation of day decreases and the rangebility of mass change have a diminution over corrosion time. For oxide film, ion Cl^- is strong destructive and penetrates into oxide film, resulted in the disintegration of oxide film [14-16]. In addition, the total loss (g) and mass change rate (g/d) of the coatings without PTFE are 1.49 and 0.75, while the coatings with PTFE; those have decreased to 1.04 and 0.63, respectively. According to the disaffinity, it is obviously observed that the total mass loss and mass change rate of ZnNi/Cr_2O_3-TiO_2-CuO-SiO_2/PTFE coatings is less than that without pore seal agent- PTFE, indicating that PTFE has a passive influence on the erosion resistance of this kind of coating. The porosity of the two coatings has been calculated according to equation (2) and is to be 0.47 and 0.65 without PTFE coatings, respectively. Which predict that the PTFE is passive in reducing the porosity of ZnNi/Cr_2O_3-TiO_2-CuO-SiO_2 coatings and effective influence on erosion resistantance.

As well observable with the cross-sectional morphologies of two coatings in **Fig. 2b, c,** the coatings protect the substrate from cavitation erosion and no deep pits which penetrate into the substrate are found. Although, compared to the cross section of coatings before erosion, the quantity of pores increases in Cr_2O_3-TiO_2-CuO-SiO_2 coating and it have an impact on bonding laye, for the coatings with PTFE, the pores merely added in Cr_2O_3-TiO_2-CuO-SiO_2 coating. The comparison implies again that PTFE is resultful in cavitation erosion resistance in 3.5 wt% NaCl solution. The holes and pores still exist although with pore sealing for the reason of PTFE might not spread well on the cross section of the coatings.

The surface morphologies of two coatings after cavitation erosion for 48h in 3.5 wt% NaCl solution are shown in **Fig. 2d and e** to investigate the cavitation damage. From **Fig. 2d**, although been immersed in 3.5 wt% NaCl for 48h, the most pores and holes have been sealed by PTFE, and the lamellar structure has no obvious change. As can be seen in **Fig. 2e**, unlike the composite coatings with PTFE, it possesses a considerable number of discharge channels and open pores on its surface. More pores producing might result in more discharge channels on its surface, furthermore, the corrosion rate speeds up and corrosion depth increases. Meanwhile, for the mass-transfer characteristics of corrosion fluid and the gradient of the concentration in coatings, the corrosion product diffuses from the discharge channels, led to the failure of ceramic coatings. There are two ways of ceramic coatings failure, as corrosion product contact with the substrate through the discharge channels leading to the shed of ceramics coatings; the existence of unbalance phases, porosity and residual stress leading to chemical corrosion in coatings. However, the corrosion rate slow down gradually, might be for the reason of the corrosion product sealing the holes or channels, exerting a brake on erosion.

3.3 Corrosion behaviour

Potentiodynamic polarization and EIS measurement were used to investigate the corrosion behaviour of the ZnNi/Cr_2O_3-TiO_2-CuO-SiO_2/PTFE coatings in 3.5 wt% NaCl solution at 20°C. The result is demonstrated in **Figs. 3a**, depicting the potentiadynamic polarization of the coatings after 1h and 120h, and the corrosion potential of immersion after 120h has a trend to upper shift compared with that of 1h. After immersion for 1h, the corrosion current increases sharply with the increasement of voltage anodic polarization voltage in 0.307 V - 0.183 V. While when the scanning voltage continues to increase between - 0.103 V to 0.373 V, the changes of current tend to be stable and the coatings are being in a passivation stage, after coming into a short transpassive region the coatings enter a passivation zone again. Compared to it, for the curve of 120h, the coatings are in a polarized stage during the scanning voltage of 0.263 ~ 0.162 V, however, when voltage adds to 0.123 V and the current surge again, the coaings entered a passivation region and transpassive zone, respectively. The polarization resistance valves (R_p) of the coaings were calculated by Sten-Geary equation as follows [17]:

$$R_p = \frac{b_a b_c}{2.303 i_{corr}(b_a + b_c)} \quad (3)$$

Where i_{coor} is corrosion current density and b_a and b_c are anodic and cathodic Tafel constants, respectively, which can be extracted from polarization curves using Tafel extrapolation method [18]. The fitting result of R_p is presented in **Table. 2**. As can be seen, the self-corrosion voltage of the 120h after soaking having a tiny rise in corrosion potential, while the self-corrosion current increased more than 20 times and the polarization resistance sharply reduced more than 50 times, which means the oxide film has formed and the increase of i_{coor} means more pores and holes has appeared. **Figure 3b, c** is the surface morphology of 1 h and 120 h after immersion test. Visually, the ceramic coating surface morphology of sample after 120h immersion has changed and the number of pores increases significantly as corrosion time pasted. For parts of the corrosion products transfer through the coatings from inside to outside delivered by corrosion liquid; when the liquid contact with the surface of the coatings, the surface appear many irregular arrangement holes of about 1μm in diameter. This shows that with the extension of corrosion time the number of holes and pores increase, so does the corrosion discharge channel. The oxide film produced by the binding sites between metal substrate with ceramic coating, resulted in the rising of the corrosion potential. At the same time, the micro fracture, blind hole and inclusions form the through-hole by the erosion of fluid inclusions. With the increase of corrosion channel, the self-corrosion current rises and the polarization resistance reduces. Due to the blocking effect of the ceramic coating lamellar structure, the corrosion fluid permeates along the disordered cracks and reactive particles transfer unparallel along the direction of the fluid concentration, which causes the nolinear variational change of the corrosion current and the polarization resistance.

The measured EIS for different corrosion time coatings is shown as Nyquist curves and Bode plots in **Fig.5a** and **b**, respectively. It is clearly seen that the impedance curves are overlap at higher frequency, for a saturated electrolyte solutions of permeating; the capacitance is in a stable range which would not increase with the permeating of the saturated electrolyte solutions. As shown in **Fig. 5b**, the Nyquist cures of the 1h are similar with the 2h and 4h are similar with 72h, which illustrates that the corrosion mechanism is variational in different immersion time. Therefore, the equivalent circuit is given in early immersion stage for 1h, 2h and the result is fitted (**Table. 3**) according to **Fig. 4a**. Where R_1 is the capacitance of 3.5 wt% NaCl solution, R_2 and R_3 are the capacitances which are related to the ceramic coatings and the binding layer, respectively. According to it, in the initial corrosion stage, the capacitance of the coatings obviously decreases attributed to the appearance of corrosion discharge channels by the penetration of the micro fracture, blind hole and inclusions in the corrosion fluids.

In the later immersion stage when after 4h, the fitting results of the comparison between Nyquist and Bode curves are listed in **Fig.5e** and **f** according to the equivalent circuit (**Fig. 4b**). The fitting result is also shown in **Table.3** which displays the electrochemical data in the later immerse stage, where C and R_4 are the electrochemical transfer resistance of double electrode layer and the double electrode layer capacitance between binding layer and substrate. After immersed in the 3.5 wt% NaCl solutions for 4h, the solutions has touched the substrate through the porosity and formed oxide films on the substrate. However, after 24h corrosion, by the self-closing behaviour of the ceramic coatings, the corrosion products penetrate into the micro-holes and block the holes, the corrosion solution is hard to touch the substrate as the corrosion channel

has changed, which can be concluded by the increase of R_3. The increase of C value at 24h exhibits the dense change and thickening of the passivation film, while the decrease of it in 72h means the passivation membrane has dissolved by the corrosion of ion Cl^-, and the R_4 adds to 165.4 in 120h which demonstrates that the oxide film has formed again. The conclusion can give a specific explanation to the previous mass loss test.
It is assumed that the corrosion mechanism can be implied according to the change of electrochemical data in **Table.3**, this process can be classified by whether oxide layer has formed or not. In the early corrosion stage (1, 2h), the capacitances related to ceramic coatings and binding layer are in a decline trend, causing by the increment of the number of pores and holes, which means the corrosion fluid infiltrate the coatings but not touch the substrate, the R_2 value decreases all along which reveals the holes and pores are increasing in the whole corrosion process. In the later corrosion stage (4, 24, 72h), the corrosion fluid begin to contact with the substrate, compared the data between those after 4h and 24h, the thickening of passivation membrane adds by the increase of C value; the corrosion product has sealed the holes by the increase of R_3. Then, that film and corrosion product dissolves in corrosion fluid gradually by the decrease of C value and R_3 in 72h. At last, the oxide film has formed again according to the augment of R_4 in 120h.Therefore, the mechanism can be simplified as that the corrosion process occurs repeatedly according to the achieve rules above.

4 Conclusion

The formation of ZnNi/Cr_2O_3-TiO_2-CuO-SiO_2/PTFE composite coatings by Atmospheric Plasma Spraying (APS) method was investigated in this research. The result revealed that the addition of pore seal agent, PTFE has decreased the porosity with a value of 0.47 from 0.65, and the coatings with PTFE lose less weight has decreased to 1.04g from 1.49g without PTFE coating after 18days in the mass loss test and been similar to the weight change rate come from 0.75g/d to 0.63g/d which also means lower porosity is essential for corrosion resistance.
The mass change of the coatings can be explained by the formation and dissolution of oxide film in corrosion fluid. According to the electrochemical measurements, the corrosion stage can be concluded as 1. The corrosion fluid diffuses go through from the pores and reach the boundary interface between binding layer and substrate. 2. The corrosion products appear on the interface and the oxide passivation layer has formed. 3. The Cl- ion contacted with the oxide layer, leading to the disintergration of it and the fresh interface of the surface is presented in corrosion fluid. Furthermore, for the accumulation of corrosion products, it clogs the channel and holes as well as slows down the rate of corrosion; therefore, it improves the resistance of the coating to some extent.
The reciprocating parts with the coatings were relatively intact after 5000h immersion test in realistic ocean environment, the pictures are not shown in this paper for the workpieces protected from the public viewing.

References

[1] G. Di Girolamo et al. Microstructural, mechanical and thermal characteristics of zirconia-based thermal barrier coatings deposited by plasma spraying. Ceramics International 41 (2015), pp. 11776/11785.
[2] Shubo Deng. Removal of trivalent and hexavalent chromium with animated polyacrylonitrile fibers: performance and mechanism. Water Research 38 (2004), Issue 9, pp. 2424/2432.
[3] P. Mohan, T. Patterson, V.H. Desai, Y.H. Sohn. Degradation of free-standing air plasma sprayed CoNiCrAlY coatings by vanadium and phosphorus pentoxides. Surf Coat Technol 203 (2008), pp. 427/431

[4] Brookes. C. A. The tribological application of ceramic materials in water hydraulics potential and problems. 1st Bath Int on Fluid Power (1988), University of Bath, UK.
[5] Brookes. C. A. The selection and performance of ceramic components in seawater pump. The 3th JHPS Int Symp on Fluid Power (1996), Yokohama, Japan.
[6] Brookes. C. A. The development of water hydraulic pumps using advanced engineering ceramics. Proc of 4th Scandinavian International Conference of Fluid Power (1995), Tampere, Finland.
[7] S. Matthews, B. James. Review of thermal spray coating applications in the steel industry: part 1-hardware in steel making to the continuous annealing process. Thermal Spray Technology 19 (2016), Issue 6, pp. 1267/1276.
[8] Wassemrna. Ch, Roland. B, Gustasfson. S. Planting in printing maehineyr, Thermal Spray 2001: New surface on new millennium. Proceeding the International Thermal Spray Conference (2001), Ohio, USA.
[9] Ter. V.J, Kuikko. T. V, Development of seawater hydraulic power pack. Proceedings of 4th Scandinavian International Conference on Fluid Power (1995), Tampere, Finland.
[10] Pushpvanam. M. Electro-deposit high corrosion resistant of Zn-Ni alloy. Trans SAEST 31 (1996), Issue 2, pp. 35/40.
[11] Bradley. D. Measurement of porosity in ceramic coatings by thermogravimetric volatization of liquids. Am Ceram soc 79 (1996), Issue 12, pp. 3317/20.
[12] R.M. Li. Radiation properties modelling for plasma-sprayed-alumina-coated rough surfaces for spacecrafts. Materials Science and Engineering (2006), Issue 132, pp. 209/214.
[13] Hui pu. Ablation of vacuum plasma sprayed TaC-based composite coatings. Ceramics International (2015), Issue 41, pp. 11387/11395.
[14] B. P. Lochel and H. H. Strehblow, On The Mechanism Of Breakdown Of Passivity Of Iron For Instationary Conditions. Werks . Korros 31 (1980), Issue 5, pp. 353/358.
[15] C. Y. Chao, L. F. Lin, and D. D. Macdonald, A point-defect model for anodic passive films. I. Film growth-kinetics, J. Electrochem. Soc 128 (1981)，Issue 6, pp. 1187/1194.
[16] D. D. Macdonald, The point-defect model for the passive state. J. Electrochem. Soc 139 (1992)，Issue 12, pp. 3434/3449.
[17] D. Sreekanth. Effect of K_2TiF_6 and $Na_2B_4O_7$ as electrolyte additives on pore morphology and corrosion properties of plasma electrolytic oxidation coatings on ZM21 magnesium alloy. Surf. Coat. Technol 222 （2013）. pp.31/37.
[18] A. Bahramian. An investigation of the characteristics of Al_2O_3/TiO_2 PEO nanocomposite coating. Applied Surface Science 351 (2015), pp. 12/26.

Microstructure and properties of FeCrNbBSiC coating prepared by wire-arc spraying process

Z.H. Qian, Z. Zhou, H.H. Yao, X. Wu, Y .M. Wang, D.Y. He, Beijing/CN

A new FeCrNbBSiC-cored wire was designed to produce protective coatings on ASTM 4032 aluminum alloy substrate by wire-arc spraying process. The phase composition, microstructure, microhardness, bonding strength and thermal conductivity of the coating were investigated in comparison with a commercial 316L stain steel coating under the same condition. The results show that the FeCrNbBSiC coating exhibited a dense and homogeneous lamellar with much less oxide inclusions than 316L coating. The phase component of Fe-based coating was consisted of amorphous phase and α-Fe nanocrystalline phase with a crystallization temperature 625 °C. The microhardness of FeCrNbBSiC coating was higher than that of 316L coating, reaching to 1020.9$HV_{0.1}$, which mainly attributes to the composite structure with amorphous and crystalline. Meanwhile the bonding strength of the designed coating is 31.8MPa slightly higher than that of 316L coating. The thermal conductivities of both coatings were measured at room temperature to evaluate the ability of heat transfer. It was observed that the thermal conductivity of designed coating approached 2.16W/mK less than the commercial 316L coating with 3.95 W/mK, which is much low in the typical metal-based coatings by arc spraying process. Therefore the wire-arc sprayed FeCrNbBSiC coating could be an effective and economic candidate for thermal insulation protective coating in engineering applications.

1 Introduction

The automobile industry is facing a serious challenge to improve vehicle fuel efficiency and reduce exhaust emission [1]. Therein, reduction vehicle weight is considered to be one of the most effective ways by using light weight alloy instead of traditional steel parts. However, light alloy, like Aluminium alloy, has lower hardness and melt point compared to steel, which cannot satisfy the requirements of application. One of the effective approaches to protect the light alloy components is depositing a functional coating on the surfaces of components [2-3].

During the last decades, thermal barrier coatings (TBCs) technologies have been enormously improved and enhanced by many researchers. For instance Yttria-Stabilized Zirconia (YSZ) is widely used as a thermal insulator at elevated environment. Some researchers have brought YSZ coatings into the protection for hot section of metal components in advanced gas-turbines or internal combustion engines [4-5]. Results indicate that the use of TBCs can contribute to a temperature reduction of as much as 200°C at the metal surface, thereby improving the durability of the metal component as well as enhancing the fuel efficiency [5]. However, TBCs using the ceramic materials have a poor toughness and unpredictable life-time or failure upon mechanical loading or thermal shock, due to their inherent brittle behavior. As regarding the nanostructured or amorphous metals are considered to exhibit low thermal diffusivity and conductivity [1, 3]. Therefore, using metal-based TBCs (MBTBCs) with nanostructured or amorphous structure have been proposed. The thermal conductivity of an Fe-based amorphous coating prepared by induction plasma-sprayed process is as low as 2 W/mK [3], which is close to traditional YSZ coatings.

Recently, significant efforts have been made to prepare amorphous and nanoscale coatings, because of their superior mechanical, physical and chemical properties [6]. By far, there are a lot of methods to prepare amorphous metallic coatings, such as High velocity oxygen fuel spraying, Plasma spraying, and Arc spraying. Among of these Thermal spraying technologies, wire-arc spraying has attracted increasing interest because of following advantages compared to other deposition methods: low cost, high spray rate, high deposition, simplicity and flexibility [7-9].

Many Fe-based amorphous alloys coatings prepared by wire-arc spraying were studied the wear and erosion performance in high-temperature components. The NanoSteel Company reports two typical iron-based nanocomposite coatings with conventional wire-arc spraying technology, SHS 7170 [10]. The as-deposited microstructure is characterized by a glass matrix containing crystallites in sizes from 60 to140 nm, and the coatings demonstrate high performance during elevated-temperature erosion test for boiler application. J.B. Cheng et al. developed FeCrBSiMnNbY and FeCrBSiNbW metallic glass coatings [11-12], which show high hardness, low wear rate and improved corrosion resistance. However, regarding heat protection in wire-arc spraying Fe-based amorphous coatings is almost a blank.

In this study, an FeCrNbBSiC cored wires were designed to deposit coating by wire-arc spraying, and explore the potential as heat protective coating in transportation field. The microstructure was characterized and the microhardness, bonding strength and thermal conductivity were also examined. A commercial 316L coating was studied simultaneously under the same experimental conditions.

2 Experimental Procedure

2.1 Materials selection and preparation

The ASTM 4032 aluminum alloy (with dimensions of 150×100×10 mm^3) was employed as substrates in this study. FeCrNbBSiC -cored wires with diameter 2.0 mm (the nominal composition of the cored wire is shown in **Table1**) was designed as the feedstock material, and deposited on the substrate by wire-arc spraying. The 0.3 mm thick wire cover was made from stainless steel tape (Cr: 17wt. % and Fe: balance). The filling coefficient of the cored wire was 30~32%. The filling powders were mixture of industrial ferro-boron (B: 18 wt.% and Fe: balance), ferro-silicon (Si: 75 wt.% and Fe: balance), ferro-niobium (Nb: 65 wt.% and Fe: balance), boron carbide(B:78 wt.% and C balance), pure chromium(≥99 wt.%) and iron(≥99 wt.%). The wire-arc spraying was performed using the 8835 arc spray gun (Praxair) mounted on a robot, with an arc voltage of 30-32 V, a current of 180-200 A and a stand-off distance of 200 mm. For comparison, a commercially available solid wire 316L with a diameter 1.6 mm (the nominal compositions are also shown in **Table 1**) was applied to deposit coating under the same conditions.
Two kinds of specimens were deposited for different experiments: (1) the coatings were deposited on the top of surface of the ASTM 4032 aluminum for 400-500μm. These specimens were used to investigate the microstructure of the coatings (cutting with 15×10×10mm^3); (2) the coatings were deposited on the substrates for nearly 1mm and then separated from the substrates. Circular discs with a dimension of Φ11mm×1mm^3 were for density and thermal diffusivity, respectively. Before spraying the substrates were degreased by acetone, dried, and then grit-blasted.

2.2 Microstructure Characterization

The phase identification of the coatings was used to perform by X-ray diffraction (XRD, D8 ADVANCE, BRUKER/AKS) with 2θ diffraction angle ranging from 20°to 85°. The microstructure of as-deposited coatings was characterized by scanning electron microscopy (SEM, Quanta-200, FEI) and transmission electron microscopy (TEM, JEM-2100, JEOL). Thin coating slices for TEM were first peeled from substrates by mechanical method and ground to about 50 μm, and then thinned by ion-beam milling technique until perforation. Percentage of the porosity in the coatings was evaluated using image analysis on optical microscopy (OM, BX51M, OLYMPUS). Oxygen content of the powders and detached coatings were analysed by an Oxygen/Nitrogen Determinator (G8 GALILEO, BRUKER). The thermal stability and amorphous fraction of the samples was examined by differential scanning calorimeter (DSC, STA 449F3, NETZSCH) in a continuous heating mode at a rate of 0.33 K/s from room temperature to 1000°C under argon atmosphere.

2.3 Coating properties

Vickers microhardness measurements were conducted on the cross section of the wire-arc spraying coatings with an applied load of 100 gf and indentation time of 10s, and calculated as an average of 15 measurements. The bonding strength was measured according to ASTM C633-01 standard using a universal mechanical test machine (QBD-100, Chembound, China). At least three tests were done for each sample and the averaged value was used as the indicator of the bonding strength.
The apparent density (ρ) of the sprayed coatings was measured by the Archimedes principle. Thermal diffusivity (α) measurement of the coatings was carried out at room temperature by using the laser flash thermal constant analyzer (LFA427, NETZSCH). Specific heat (Cp) of the coatings was also measured using the Differential scanning calorimetry (STA 449F3, NETZSCH) at room temperature. Then, the Thermal conductivity (k) was calculated from the relation $k = a\rho Cp$.

3. Result and Discussion

3.1 Microstructure Characterization

Fig. 1 presents the X-ray diffraction patterns of wire-arc spraying FeCrNbBSiC coating and 316L coating. A broad diffraction halo appeared at 45° with a small amount of crystalline diffraction peaks, indicating that the coating is primarily amorphous matrix with some nanocrystals, as shown in **Fig. 1(a)**. It manifests the alloy chosen here, satisfies the three empirical rules for the stabilization of the super cooled liquid, leading to highly dense random packed atomic configurations, higher viscosity, and lower atomic diffusivity [13]. The major crystalline phase in the FeCrNbBSiC coating was α-Fe. Meanwhile, it was identified from **Fig. 1(b)** that 316 L coating consisted of γ-(Fe, Ni), α-(Fe, Cr) and $FeCr_2O_4$ phases.

Fig. 2 shows typical SEM images of wire-arc spraying FeCrNbBSiC coating and 316L coating. It is recognizable that both coatings present relatively dense and homogeneous lamellar microstructures. No evident cracks or spallation are observed, and the coatings are bonded the substrate well with a relative compact interface. The average porosity of the FeCrNbBSiC coating is about 6.3% by image analysis. While the porosity of 316L coating is lower than designed Fe-based coating, approximate 4.4%. A large amount of dark-contrast oxide inclusions are noticeable between the flattened particles in as-deposited 316L coating (**Fig. 2 b**). The inclusions were identified Fe, Cr and O elements by EDS. According to the XRD analysis result, the inclusions are iron-chromium oxide ($FeCr_2O_4$). In contrast, the amount of oxide inclusions in FeCrNbBSiC coating, as shown in **Fig. 2 a**, is significantly fewer than that in the 316L coating. To confirm this observation, the oxygen content of FeCrNbBSiC coating and 316L coating were measured using O/N determinater, ascertained 1.74% and 10.23% respectively, thus it could be concluded that severe oxidation occurred during the spraying 316L solid wire. It further proved that because of the addition of boron and silicon, the oxidation of molten particles decreased enormously during spraying.

In order to obtain more detailed information on microstructure of FeCrNbBSiC coating, the TEM was undertaken as shown in **Fig. 3**. Fig. 3 (a) is a bright field TEM image of as-deposited coating. The selected area electron diffraction (SAED) is also shown in the inset of **Fig. 3 (a)**. It presents diffused halo rings, which means the coating is basically composed of amorphous phase, which is according with the results of XRD examination well. Nevertheless, some nanocrystalline grains can also be seen clearly in the coating in **Fig. 3 (b).** The coexistence of amorphous phase and crystalline grains were verified by the diffused halo rings and diffraction spots in the SAED pattern, as shown in the inset of **Fig. 3 (b)**. It is also notable that the nanocrystalline grains are embedded in the amorphous matrix with the maximum dimension 100nm. On the other hand, it is inferred that the generation of nanocrystalline caused by heat accumulated inside the coating.

Table 1 nominal chemical composition of the wires

Materials	Cr	Nb	B	Si	C	Mn	Ni	Mo	S,P	Fe
Fe-based cored wire	≤16	≤7	≤4	≤2.5	≤1	≤0.5	...	...	...	Balance
Commercial 316L solid wire	≤18	...	...	≤1	≤0.03	≤2	≤15	≤3	≤0.07	Balance

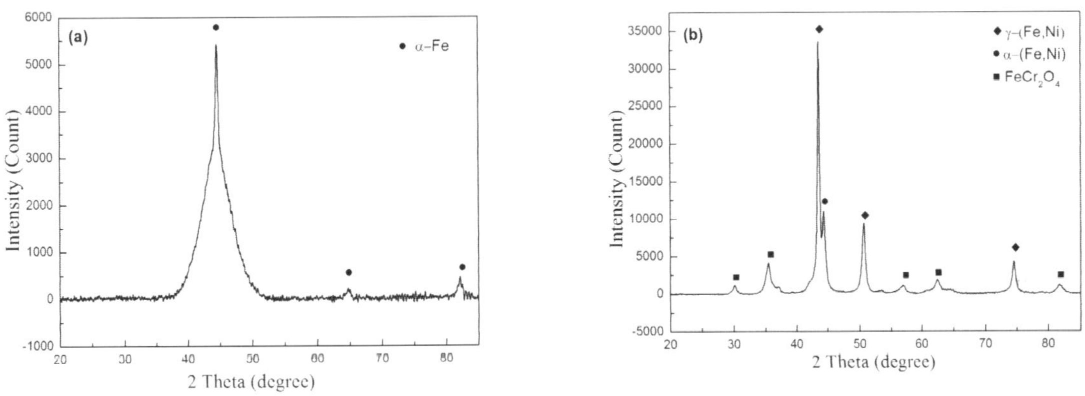

Fig. 1 XRD patterns of as-sprayed coatings: (a) FeCrNbBSiC coating, (b) 316 L coating

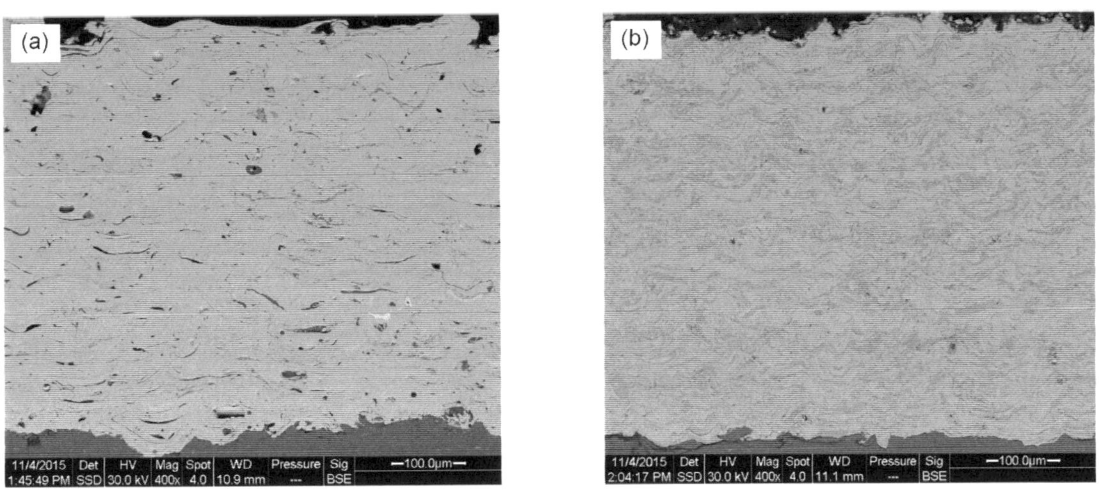

Fig. 2 SEM images of as-sprayed coatings: (a) FeCrNbBSiC coating, (b) 316 L coating

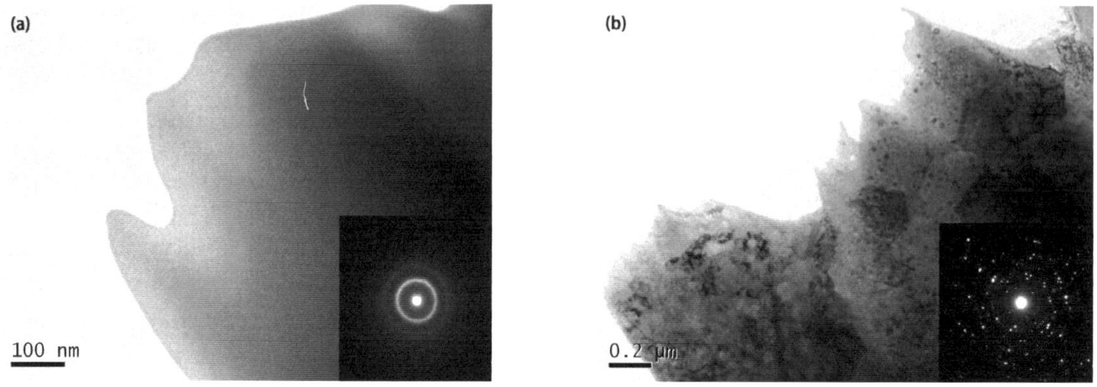

Fig. 3 Bright-field TEM images and selected area electron diffraction (SAED) patterns of FeCrNbBSiC coating: (a) an image of amorphous phase; (b) coexistence of amorphous phase and nanocrystalline grains.

Fig. 4 illustrates the DSC traces of the FeCrNbBSiC coating. An exothermic solid-state transformation is observed from the curve. The onset temperature of the crystallization Tx is 625°C, indicating that the coating has good thermal stability maintained less than this temperature. XRD and TEM patterns manifest that the FeCrNbBSiC coating is in the form of amorphous phase. It is attributed to the large glass formation ability (GFA) of this alloy system firstly, which meets the three empirical rules proposed by Inoue et al [13]. In this work, the addition of Fe, Cr, Nb, B, Si and C causes the sequential change in the atomic size, leading to the formation of random and packed structure in the super cooled liquid. The high thermal stability of super cooled liquid against crystallization, an inherent characteristic of the alloy as illustrated in Fig. 4 is also beneficial to the large GFA [14]. Secondly, the rapid cooling rate of the individual splat during arc spraying reaches 10^{-5}K/s [15], which can satisfy to form an amorphous phase. However, the temperature in arc can reach 5000°C during the spraying process [15-16], which is much higher than the crystallization temperature. Therefore, the amorphous and nanocrystalline structure are emerged in FeCrNbBSiC coating, as shown in **Fig. 3**.

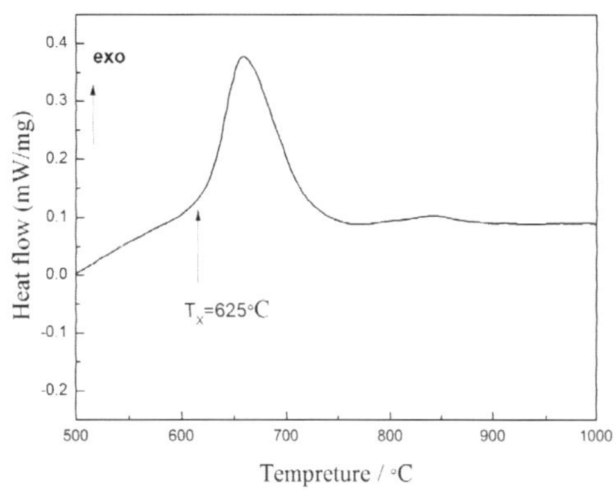

Fig. 4 Typical DSC trace of FeCrNbBSiC coating

3.2 Microhardness and bonding Strength of the Coatings

Fig. 5 shows the distribution of microhardness along the cross-sections of coatings via distance from the coating and substrate interface. It can be seen that the microhardness of the aluminum alloy substrates is less than $100HV_{0.1}$. The increases in microhardness of both substrates are observed at the coating/substrate interface. It is because of the work hardening effect of sand blasting before the processing, and the impact of the molten particles during the wire-arc spraying.

The microhardness of the designed Fe-based coating is in the range of 940-1100 $HV_{0.1}$, which is two times higher than 316L coating with an average microhardness value of $458.2HV_{0.1}$. This suggests that the composite microstructure with amorphous and nanocrystalline in FeCrNbBSiC coating is contributed to the high hardness of coating.

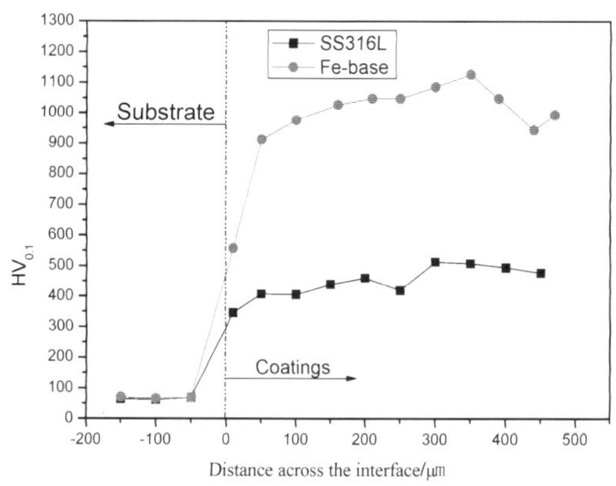

Fig. 5 Microhardness profiles of the as-deposited coatings

Meanwhile, the bonding strength of as-deposited coatings was also examined. As a result, the bonding strength of the designed Fe-based coating is 31.8 MPa, higher the average value of 316L coating, 29.9 MPa. The bonding strength is most depending on the spraying method, for instance the bonding strength of arc spraying coating is lower than HVOF spraying coating. That is because, the flying particles with high speed can produce compressive stress on the surface of substrate, which is beneficial to coatings more compact mechanical bite with the substrate as well as a lower porosity. However, the bonding strength of FeCrNbBSiC may be able to meet the requirements of vehicles engine.

3.3 Thermal Conductivity

The thermal conductivity directly indicates the ability of material to conduct heat [1]. The room temperature thermal conductivity (*k*) of the FeCrNbBSiC and 316L coatings were calculated based on the measurement of densities (ρ), thermal capacities (*Cp*) and thermal diffusion coefficients (α) of coatings, presented in **Table 2**. The result shows that, both arc spraying coatings are kept in typical low level in metal-based coatings, less than 4 W/mK. Therein, the thermal conductivity of the designed Fe-based coating is lower than commercial 316L coating, reached 2.16 W/mK, which is close to conventional YSZ TBCs. The reason could be that the amorphous structure and grain refinement are disadvantaged to heat transport in coatings [17-19]. Furthermore, the pores and oxide inclusion can also hinder the radial heat transport in coating via scattering the heat-transfer electron or phonon [20-21]. Therefore, the designed Fe-based coating exhibits better ability of insulation heat owing to the synergistic effects from microstructural character.

Table 2 The densities, thermal capacity and thermal diffusion coefficient and thermal conductivities of coatings

Coatings	ρ (g/cm^3)	*Cp* (J/g·K) $\times 10^{-1}$	α (cm^2/s) $\times 10^{-2}$	*k* (W/mK)
Fe-based	6.667	5.212	0.624	2.168
316L	6.471	5.441	1.142	3.948

4 Conclusions

(1) The designed FeCrNbBSiC-cored wire and commercial solid wire stain steel 316L were produced and deposited on the ASTM 4032 aluminum substrate by wire-arc spraying process. The phase component of Fe-based coating was consisted of amorphous and α-Fe nanocrystalline phase with a crystallization temperature 625 °C. Both as-deposited coatings are composed of dense and homogeneous lamellar. The oxide inclusions in FeCrNbBSiC coating are significantly fewer than 316L coating.
(2) The microhardness of FeCrNbBSiC coating is higher than that of 316L coating, reaching to 1020.9HV0.1, which mainly attributes to the composite structure with amorphous and crystalline. Meanwhile the bonding strength of the designed coating is 31.8MP slightly higher than that of 316L coating.
(3) The thermal conductivities of coatings were measured at room temperature to evaluate the ability of heat transfer. The thermal conductivity of designed Fe-based coating approaches 2.16W/mK less than the commercial 316L coating with 3.95 W/mK, which is much low in the typical metal-based coatings by arc spraying process. Therefore the wire-arc sprayed FeCrNbBSiC coating could be an effective and economic candidate for thermal insulation protective coating in engineering applications.

5 References

[1] Poirier, D., Lamarre, J.M. and Legoux, J.G.: Thermal cycling assessment of steel-based thermal barrier coatings for Al protection. Journal of Thermal Spray Technology 24(2014), Issue 1-2, pp. 175/84.

[2] Lugscheider, E., Dicks, R., Kowalsky, K., Cook, D., Nassenstein, K. andVerpoort, C.: A Materials System and Method of Its Application for the Wear Protection of Aluminium Engine Cylinder Bore Surfaces, Thermal Spray 2004: Advances in Technology and Applications, May 10-12, 2004, Osaka, Japan.

[3] Shin, D., Gitzhofer, F., Moreau, C.: Properties and induction plasma sprayed iron based nanostructured alloy coatings for metal based thermal barrier coatings. Journal of Thermal Spray Technology 16 (2007), Issue 1, pp. 118/27.

[4] Slifka, A.J., Filla, B.J., Phelps, J.M., Bancke, G. and Berndt, C.C.: Thermal Conductivity of a Zirconia Thermal Barrier Coating. Journal of Thermal Spray Technology 7 (1998), Issue 1, pp. 43/46.

[5] Schlichting, K.W., Padture, N.P., and Klemens, P.G.: Thermal Conductivity of Dense and Porous Yttria Stabilized Zirconia. Journal of Materials Science 36 (2001), Issue 12, pp. 3003/10.

[6] Zhou, Z., Wang, L., He, D.Y. and etc.: Microstructure and wear resistance of Fe-based amorphous metallic coatings prepared by HVOF thermal spraying. Journal of Thermal Spray Technology 19 (2010), Issue 6, pp. 1287/93.

[7] Venugopal, K., Agrawal, M.: Evaluation of arc sprayed coatings for erosion protection of tubes in atmospheric fluidised bed combustion (AFBC) boilers. Wear 264 (2008), Issue 1-2, pp. 139/45.

[8] He, D.Y., Fu, B.Y., Jiang, J.M., Li, X.Y.: Microstructure and Wear Performance of Arc Sprayed Fe-FeB-WC Coatings. Journal of Thermal Spray Technology 17 (2008), Issue 5-6, pp. 757/61.

[9] Li, R., Zhou, Z., He, D.Y.: Microstructure and high-temperature oxidation behavior of wire-arc sprayed Fe-based coatings. Surface & Coatings Technology 251 (2014), pp. 186/90.

[10] Branagan, D.J., Breitsameter, M., Meacham, B.E. and Belashchenko, V.: High-performance nanoscale composite coatings for boiler applications. Journal of Thermal Spray Technology 14 (2005), Issue 2, pp. 196/204.

[11] Cheng, J.B., Wang, Z.H., Xu, B. S.: Wear and corrosion behaviors of FeCrBSiNbW amorphous/nanocrystalline coating prepared by arc spraying process. Journal of Thermal Spray Technology 21 (2012), Issue 5, pp. 1025/31.
[12] Cheng, J.B., Liang, X.B., Chen, Y.X.: High-temperature erosion resistance of FeBSiNb amorphous coating deposited by arc spraying for boiler applications. Journal of Thermal Spray Technology 22 (2013), Issue 5, S. 820/27.
[13] Inoue, A. High strength bulk amorphous alloys with low critical cooling rates. Materials Transactions JIM 36 (1995), Issue 7, pp. 866/75.
[14] Inoue, A., Takeuchi, A. and Zhang, T.: Ferromagnetic bulk amorphous alloys, Metallurgical and Materials Transactions A 29 (1998), Issue 7, pp. 1779/93.
[15] Newbery, A.P., Grant, P.S., and Neiser R.A.: The velocity and temperature of steel droplets during electric arc spraying, Surface and Coatings Technology 195 (2005), Issue 1, pp. 91/101.
[16] Lin, J.R., Wang, Z.H., Lin, P.H. and et al: Microstructure and cavitation erosion behavior of FeNiCrBSiNbW coating prepared by twin wires arc spraying process, Surface and Coatings Technology 240 (2014), pp. 432/36.
[17] Wu, J., Guo, H.B., Zhou L.: Microstructure and Thermal Properties of Plasma Sprayed Thermal Barrier Coatings from Nanostructured YSZ. Journal of Thermal Spray Technology 19 (2010), Issue 6, pp. 1186/94.
[18] Shin, D., Gitzhofer, F., Moreau, C.: Thermal property evolution of metal based thermal barrier coatings with heat treatments. Journal of Material Science 42 (2007), Issue 15, pp. 5915/23.
[19] Yang, H.-S, Eastman, J.A., Thompson, L.J. and Bai, G.R.: Grain-Size-Dependent Thermal Transport Properties in Nanocrystalline Yttria-Stabilized Zirconia. Materials Research Society Symposium Proceedings 703 (2002), pp. 179/84.
[20] Shin, D., Gitzhofer, F., Moreau, C.: Properties of Plasma Sprayed Iron Based Nanostructured Alloy Coating for Metal Based Thermal Barrier Coatings, Journal of Thermal Spray Technology 16 (2005), Issue 1, pp. 118/27.
[21] W.G. Chi. Thermal Transport Properties of Thermally Sprayed Coatings: An Integrated Study of Materials, Processing and Microstructural Effects [D]. Stony Brook University, 2007.

An Influence of Strain-Induced Nucleation of Martensitic Transformations on Tribological Properties of Sprayed and Surfaced Depositions

Yu. Korobov, V. Verkhorubov, S. Nevezhin, M. Filippov, G. Tkachuk, A. Makarov, I. Zabolotskikh,
Ekaterinburg, Russia

As shown by previous studies there are differences in structural features of metastable austenite type coatings and overlays due to Fe-based alloys. In the present work tribological properties of the depositions produced from Fe-C-Cr-Ti-Al feedstock were analyzed. Coatings were deposited by arc spraying and overlays were deposited by arc surfacing. Abrasion and adhesive wear tests, structural and phase analyses were performed. A strain-induced nucleation of martensite under external load was confirmed by structure changes. The difference in the conditions of formation of the depositions was resulted in varying of tribological properties at abrasive wear, dry and lubricated friction.

1 Introduction

Alloy of metastable austenite structure (MAS) is a proper material to withstand various mechanical loads. When the external load is applied to such material, synergetic effects of transformation from austenite to martensite take place. First, rise of martensitic phase share leads to hardness increase. Second, energy of external load, which is applied to the working surface, is dissipated due to strain-induced nucleation of martensitic transformation. As a result, wear resistance under different conditions (e.g. abrasive, hydro and gas-abrasive, erosion, cavitation, adhesion, and fatigue) is improving [1-4]. Steel 1.5Cr8Ti2Al, which is studied here, belongs to metastable austenite type alloys. It is used for increasing wear resistance as by arc surfacing [5] and by arc spraying [6].
Comparison of overlays' and coatings' properties is of interest for researchers. Lima C.R.C. et al. showed differences in abrasive wear resistance between surfaced layers, arc sprayed and HVOF coatings. The wear resistance of arc sprayed coatings is 40 % less comparing to surfaced layers. HVOF coatings' wear resistance is 1.5 - 2 times as large comparing to the surfaced ones. However, these results are of limited use because different types of materials (cored wire, solid wire and powder) and different alloying systems (140MCX, stainless steel, WC-Co and Cr_3C_2-NiCr) were applied [7]. A. Pelz studied the wear resistance of coatings from powders and cored wires of various Fe-based composition (FeCrC, FeCrNbC, FeCrVC). The feedstocks were deposited by HVOF, arc spraying and plasma-transferred-arc welding (PTA). Weight loss of arc sprayed coating from FeCrC wire increases by 2.2 and 5.5 times comparing to HVOF and PTA coatings respectively. PTA coating from FeCrC and FeCrNbC showed the lowest wear in all test conditions, even lower than the WC/Co and NiCr-CrC HVOF coatings [8]. A. G. M. Pukasiewicz et al studied arc sprayed coatings from Fe-Mn-Cr-Si-Ni cored wires. They marked that after load by cavitation martensite strain induced phase transformation occurs [9].
Lack of direct compare of wear resistance of the same metastable austenite structure due to coatings and surfaced layers takes place. It obstructs to ground rational application areas of such depositions. The aim of this study was to fill that space.

2 Experimental Details

Sprayed coating (hereinafter "coating") and a surfaced layer (hereinafter "layer") were performed on the working surface of samples. Coating of 1.0 - 1.3 mm thickness was done by activated arc spraying gun ADM-10 (UWI, Ekaterinburg, Russia) [10] from cored wires. Its distinctive feature is using of propane-air combustion products as a carrier gas. Spraying parameters were the following: air pressure 3.4 MPa, propane pressure 3.2 MPa, spraying distance 100 mm, traverse velocity 200 mm/s, current 180 A, voltage 28 V. Core wire of ∅1.6 mm was used as a feedstock. Layer was done by manual metal arc welding using electrode of ∅3 mm, welding current 100 A, voltage 28 V. Weld metal from 4th layer was investigated. A low-carbon wire was used as electrode rod. High-carbon ferrochrome and ferrotitanium were used as electrode coating components. Rutile was a slag-forming base of electrode coating. Steel alloy 1.5Cr8Ti2Al was taken as a feedstock in both cases.
Two-body wear tests were carried out by pin-on-plate reciprocating technique, **Table 1**.

Table 1: Operating modes of tribological tests.

Mode	1	2	3
Friction condition	dry	dry	lubricant[1]
Counter sample material	Abrasive[2]	D2 ASTM, HRC 63	D2 ASTM, HRC 63
Full friction path, m	60	160	320
Velocity, m/s	0.16	0.07	0.07
Specific load, MPa	1	6	16 and 30
Work surface area, mm	10×10	7×7	7×7

[1] - Industrial oil I-30A GOST 20799-88; [2] - Corundum abrasive paper with grain size up to 150 μm was fixed on the plate

In modes 2, 3 of the tests a friction force was measured by tensometric gauges installed on the elastic spring. The coatings were examined in as-sprayed state and after additional surface rolling down, which was carried out by "ball-on-plate" reciprocating technique. Test parameters were the following: ball diameter 10 mm, ball steel E52100 ASTM (HRC 65), trav-

erse speed 0.16 m/s, stroke 125 mm, a friction path 5 m, axial load 100 N which corresponds to a pressure of 2000 MPa [11].
Micro hardness was measured by PMT-3 device (LOMO, St. Petersburg, Russia) under 100 g load; coating structure was studied by microscope MET 2 (Altami, St. Petersburg, Russia); x-ray diffractometer DRON-3.0 (Bourevestnik Inc., St. Petersburg, Russia) in the Fe- and Co- K_α radiation; scanning electron microscope Philips SEM 525 (FEI, Hillsboro, USA)".

3 Results and Discussion

Structure study of the layer
Metallographic analysis of the layer showed that the microstructure is of dendritic - cellular morphology, **Fig. 1**.

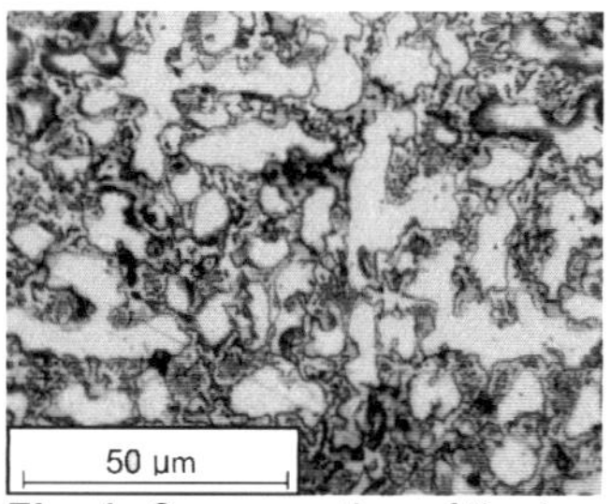

Fig. 1: Cross-section of the layer

Light areas represent the matrix metal in the form of slightly elongated ellipses, average size of 15×10 μm. For the qualitative evaluation of phase changes in the layer we applied equilibrium diagram of Fe-Cr-C (Cr wt. 8%) [12] at C wt. 1.5 %. An approximate estimate of phase changes is possible due to relatively low cooling rate 10-50 deg./sec. [13, 14].
Neglecting the titanium influence, formation of the structure can be represented as follows. At first formation of primary composite carbides like (Cr, $Fe)_3C$ and (Cr, $Fe)_7C_3$ takes place, due to the highest melting temperature, then crystals of austenite dendrite structure are formed. Between axes of the dendrites a eutectic "austenite + carbide (Cr, $Fe)_7C_3$ + (Cr, $Fe)_3C$" is crystallized. During the subsequent cooling process of excessive carbides precipitation is developing. Its degree of completion depends upon cooling velocity in the solid state. In addition, a disintegration of overcooled austenite takes place due to its depletion by carbon and chromium. If a cooling rate of overcooled austenite exceeds critical value, its decomposition is finished by the martensitic transformation at lower temperatures. Thus, by analyzing the Fe-Cr-C diagram we can conclude that the metal matrix is of martensite - austenite mixture with globular shape inclusions of dispersed carbide particles like (Cr, $Fe)_7C_3$. A metal matrix in a form of ellipses is bordered by mixed colored grid of austenite - carbide eutectic of ~3 μm thickness. Weld metal hardness is 630-640 $HV_{0.1}$.
The X-ray spectral analysis of layer allows us to refine the results of an analysis of the Fe-Cr-C diagram. It indicates the presence of primary titanium carbides like TiC. Their presence will change an order of structure formation. Refractory carbides TiC allocate at first during cooling due to highest melting point. More fusible carbides of cementite type like (Cr, $Fe)_7C_3$ are present in the eutectic. Quantitative metallography calculation showed that amount of metal base and allocated carbides within the weld is about 75 vol. %. The ratio of residual austenite and martensite in the matrix was 70/30 vol. %. The presence of chromium in solid solution increases the amount of retained austenite by lowering the martensite start temperature M_s. Titanium alloying indirectly increases the rate of formation of strain-induced martensite. Causal relationship is the following. A part of a carbon is taken by titanium from a solid solution to form titanium carbide TiC. As the carbon is the main stabilizer of austenite, its content lowering leads to acceleration of the strain-induced martensite nucleation.
Structure study of the coating
The coating has a typical lamellar porous character, which is seen in a cross-section, **Fig. 2**. The lamellas thickness is basically 5-10 μm. Etching reveals several distinctive colors and shades of structural components. The dominant light color fragments alternate with dark gray and black inclusions of various sizes and configurations.

a)
b)

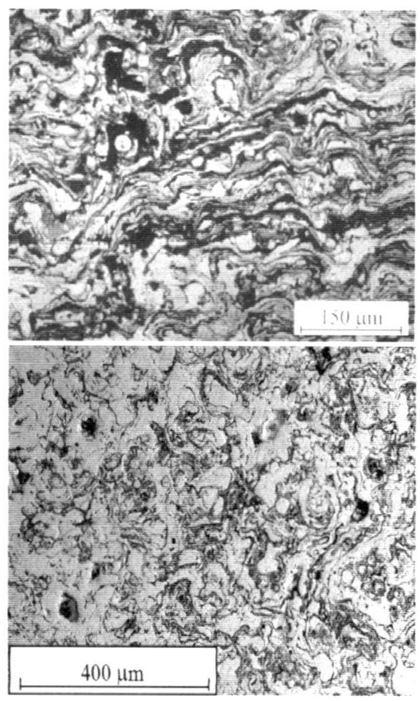

Fig. 2: Coating microstructure: a) cross-section; b) surface view

The microstructure of the coating surface is of the same coloring map as described in the cross section, **Fig. 2b**. It represents an irregular mosaic set of disperse structural components of 10-50 μm in size. The microstructure reflects the typical flattening behavior of droplets [15]. The average microhardness of the coating is 580–600 $HV_{0.1}$.
SEM analysis of the coating surface showed that light fields are the metal which contains Fe (bal.), Cr (6-10 wt.%) and negligible quantity of titanium. That metal fragments represent a solid solution of chromium in austenite and in martensite. Gray color fragments are of maximum concentration of alloying elements, wt. %: Cr (10-12); Ti (2-3), Fe (bal.). These gray areas are

likely complexes of carbides and oxides, due to spraying process.
X-ray diffraction analysis of the coating surface showed that the phase structure is based on metal (70 vol. %). It consists of martensite and retained austenite in a 50/50 ratio, vol. %. The share of martensite in the coating is increased comparing to the layer, because of specific features of the coating formation. As known, droplets at spraying face with already solidified splats [16, 17]. So strain-induced martensite nucleation is facilitated by impacts of droplets during coating formation.
According to X-ray diffraction analysis, other share of the structure includes titanium carbide, chromium carbide and oxides. So light fragments comprise metal base; gray ones comprise carbides, i.e. Ti, Cr; dark ones comprise Fe oxides.

Comparing of the structures

The coating structure is formed at rapid cooling of the sprayed particles on the substrate. The cooling rate of the coating it is 10^4-10^5deg./sec. [16, 18]. It is some orders greater than in case of the layer. It changes the conditions of phase formation. In the coating crystallization corresponds to high rate type which occurs by diffusionless way. That type of crystallization is characterized by delayed precipitation of carbides from the solution [19]. So, primary carbides and eutectic are absent in the structure of as-sprayed coating. Only secondary carbides, allocated from the austenite after crystallization, are represented in the structure.
Thus, the layer structure is more homogeneous and dense than the coating one. However, the structure of coating is more dispersed and transformation of austenite to martensite has passed in a greater degree comparing to the layer. There are primary carbides and eutectic carbide in the layer, which are not observed in the microstructure of the coating.
Such differences in the structure of the depositions are originates from features of their formation at surfacing and spraying:

- at spraying only sprayed material is in molten state while substrate remains solid. At surfacing as the substrate and the layer are melted. This causes mixing of both materials and changing of their properties;
- the cooling rate of the layer is higher by several orders comparing to the coating. It changes the conditions of phase formation;
- at spraying specific square of the droplets (square/volume ratio, 1/m) is higher on some orders of magnitude than at surfacing. In addition, lifetime of the sprayed particles in liquid state drops per order of magnitude comparing to surfaced ones. It significantly effects on their oxidation in the liquid state [20].
- Content of austenite in the as-sprayed coating it lower comparing to the initial layer due to difference in conditions of formation.

Coatings and layers after loading

Measurements of the coating surface showed that microhardness rises from 600 $HV_{0,1}$ before loading to 880 $HV_{0,1}$ after ball-on-plate test, 2^d rolling. Microhardness after ball-on-plate test, 1st rolling (770 $HV_{0,1}$) is the same as after pin-on-plate test, **Table 1, mode 1**. It indicates the changes in the structure of the coating and of the layer. Reducing the amount of austenite in the metal matrix after multiple loading by ball rolling was confirmed by X-ray analysis, **Table 2**.

Table 2: Phase change in the deposition after surface loading

Type of test		Austenite/martensite ratio, vol. %	
		Layer	Coating
Before loading		70/30	50/50
Ball-on-plate	1 rolling	-	40/60
	2 rolling	-	30/70
Pin-on-plate, mode 1		20/80	20/80

During shifting of coating over the abrasive paper the abrasive particles are intruded into coating and microcutting takes place. Border areas and pores in the coating are damaged primarily. Thereby wear of the coating is accelerated in comparison with the weld layer.
Wear tests at different conditions showed the following results. At "pin-on-plate" technique using fixed abrasive, Table 1, mode 1, weight loss (Δm) in case of layer is 7 times lower than that of low carbon steel like C1020 ASTM, and wear resistance of the coating is half as much again than layer, **Fig. 3**.

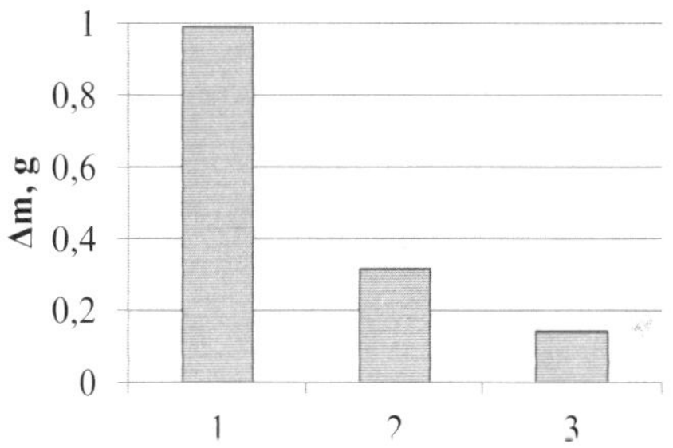

Fig. 3. Weight loss at pin-on-plate test using fixed abrasive. 1 - steel like C1020 ASTM; 2 – coating; 3 - layer.

The worn surface for layer and coating according to SEM results show different width of scars and destruction type. The width of scars of coatings is an order of magnitude larger than in case of layer. In addition, the separation of fragments from the friction surface is observed, **Fig. 4**.

a) 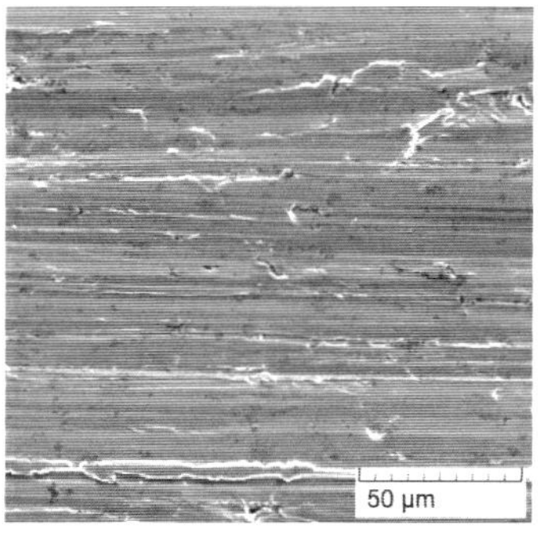

b) 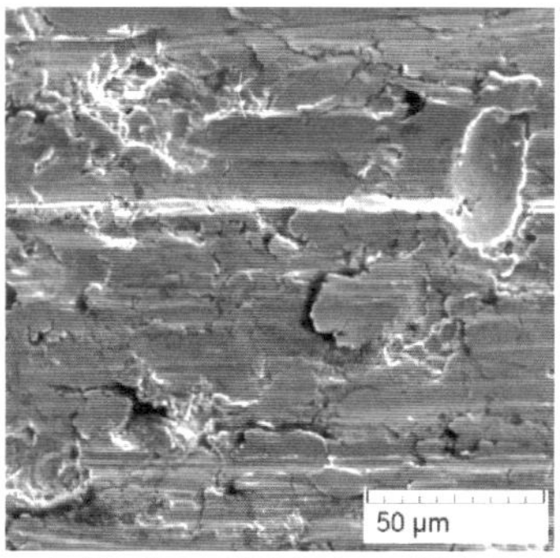

Fig. 4. Surface SEM view after wear, mode 1 of table 1, view field 166 μm: a) layer, b) coating

After abrasive wear thin foils were made from a working surface of the layer by electro-polishing of the opposite side of the working surface. They were studied by transmission electron microscopy (TEM), **Fig. 5**. According to decryption of pointed electronograms the structure of the metal base represents crystals of deformation martensite with high dislocation density (**Fig. 5a**) and larger twin crystals of cooling martensite (**Fig. 5b**).

a) 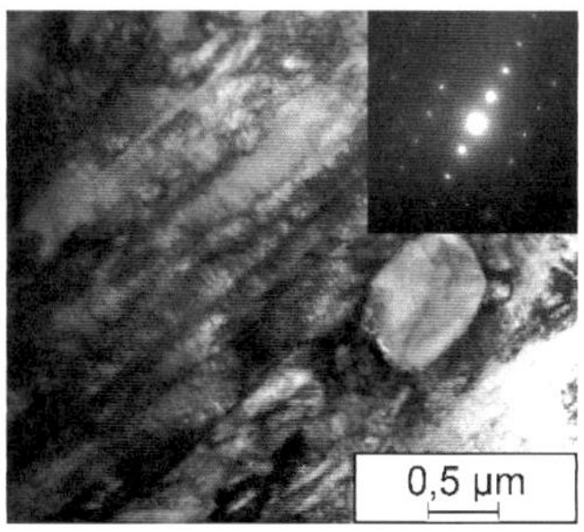

b) 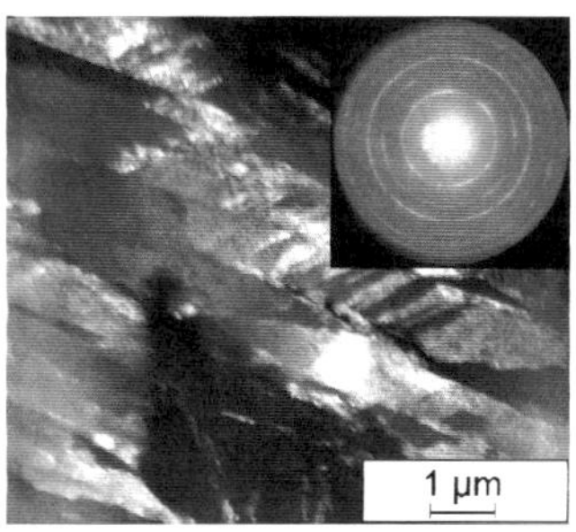

Fig. 5. Electronograms of the opposite side of the layer working surface. a) Crystals of deformation martensite b) Large twin crystals of martensite

TEM and X-ray diffraction analysis showed that the structure of the layer includes about 35% of carbides. They in turn consist of large primary carbides of titanium, carbide eutectic and dispersed secondary carbide of (M, $Fe)_7C_3$ type.
The dark field image showed that the areas of residual austenite are arranged in streaks between the crystals of martensite. Such morphology of martensite crystals indicates that the carbon content in them is not less than 0.7% [21].
At dry pin-on-plate test according to Table 2, mode 2, the coating wear rate is significantly higher compared to the layer, **Fig. 6**. In case of layer wear rate is approximately constant on the entire experimental friction path from 8 to 160 m, **Fig. 6**, curve 2. Oppositely, the coating wear rate differs along the test duration, **Fig. 6**, curve 1. Until mark of 40 m a stage of break-in period is clearly denoted. At sliding distance from 40 to160 m a steady wear of less intensity takes place. So, at dry wear the advantage of layer before coating decreases with increasing friction path: at the end of break-in period, mark 40 m, the difference in the values of the wear rate of layer and coating is 13 times, at the stage of steady wear the difference drops to 7 times. During the test, the friction coefficients of coatings ranged within 0.45-0.61, while in case of layer it ranged within 0.53-0.61, which is typical in both cases for adhesive wear of steel surfaces [20].

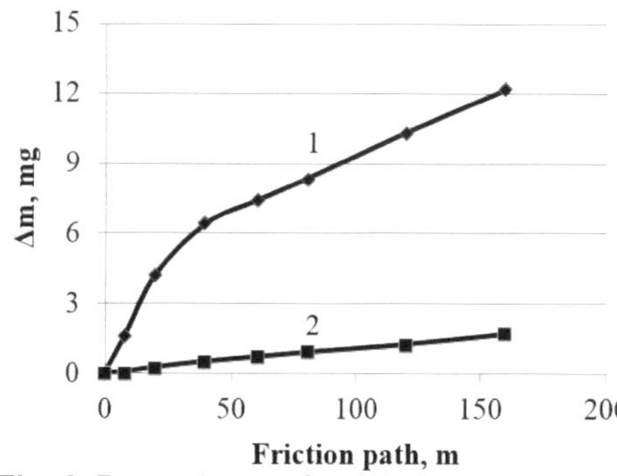

Fig. 6. Dependence of weight loss at dry pin-on-plate test. 1-coating; 2-layer.

Reduced coating wear resistance under adhesive wear is obviously connected with the presence of pores and layer boundaries between lamellas in the form of oxides, which promote accelerated separation of fragments from the friction surface of the coating. The same conclusion was done applying to cavitation wear of arc sprayed metastable austenite type coating from Fe-Cr-Mn-Si-Ni alloy [9].
Lubricated sliding friction tests, Table 1, mode 3, showed an inverse picture. At specific load 16 MPa a zero wear is fixed in case of layer, while the coating weight increase by 1.2 mg takes place due to void's filling by lubricant at absence of wear. Then specific pressure was increased up to 30 MPa during the subsequent testing. The weight loss of layer was 0.4 mg, while the coating remained at the "negative" wear. Friction coefficients (f) in the coating and layer ranged within 0.11-0.13, which corresponds to the regime of boundary friction [22]. Thus, contrary to dry friction, in case of boundary friction coating is of higher wear resistance due to its good oil-retaining ability by pores. Differences in the wear resistance of sprayed coatings and surfaced layers, primarily is caused by features of their formation. Contrary to the welded layer, sprayed coating is formed from a plurality of metal particles, which contact as with the substrate, and with mated parts. On the spraying distance the oxidation of particles occurs and the resulting oxide film is located in the coating. It leads to the layering structure of the coating. Subsequently, this layered structure and the presence of pores influence on resistance of the coating against abrasive and adhesive wear.

Thus, basing on this analysis, rational method of strengthening or restoration of surfaces may be selected depending on the operating conditions of the details.

4 Applications

Basing on investigation results developed metastable austenite structure feedstocks were successfully applied for Arc Spraying and Surfacing during last decade. Using developed Active Arc Spraying guns restoration of heavy-duty parts and producing new ones was done in various industrial sectors:

- Vehicle repair enterprise – engine parts, including crankshafts;
- Metallurgy – processing rolls hydro press plungers, piston rods and shaft journals;
- Gas-oil – bearing caps and rotor's necks of electric motors

Arc surfacing is also used for typical applications were resistance against shock-sliding, cavitation, abrasion is necessary.

6 Conclusions

1. The layer represents a heterogeneous structure. Its main component is an austenite-martensite-carbide matrix in forms of slightly elongated ellipses of an average size of 15 × 10 mm. It is surrounded by carbide eutectic with inclusions of martensite and residual austenite. In case of the coating, heterogeneous structure consists of martensite-austenite metal base, dispersed carbides, oxide layers and pores.
2. The tribological properties of arc sprayed coatings and surfaced layers of metastable austenite type structure differ due to conditions of their formation.
3. At abrasive wear, the surfaced layer resistance is 2 times as large comparing to the sprayed coating. At the stage of steady wear in conditions of dry friction, the resistance of the layer is 7 times larger. Under conditions of lubricated friction, the inverse dependence is observed. Namely, wear resistance of the sprayed coating in 2 times larger than the surfaced layer one.
4. Surfacing alloys with metastable austenite structure is preferable for parts operating under conditions of abrasive wear and dry friction and arc spraying for units operating in friction with lubrication.
5. Metastable austenite type materials are were successfully applied for Arc Spraying and Surfacing in base industrial sectors.

References

[1] Bogachev, I.N., and R.I. Mintz. *Cavitation Destruction of Iron-Carbon Alloys*, Mashgiz: Moscow-Sverdlovsk, 1959, p. 111 (in Russian).
[2] Bogers, A. J., and W. G. Burgers. *Acta Met.,* 12 (1964) 255 p.
[3] Olson, G. B., and M. A. Cohen, Mechanism for the Strain-Induced Nucleation of Martensitic Transformation, *J. of the Less-Common Metals*, 28, (1972), pp. 107-118.
[4] Schastlivtsev, V. M., and M. A. Filippov. Role of the Bogachev - Mints Concept of Metastability of Austenite in Choosing Wear-Resistant Materials, *Metal Science & Heat Treatment.* 2005, Vol. 47 Issue 1/2, pp. 3-5 (in Russian).
[5] Kulishenko, B. A., A.N. Balin, and M.A. Filippov. Electrodes for Hardfacing Components Subjected to Abrasive and Impact-Abrasive Effects, *Welding International*, 19 (4), 2005, pp. 326-329.
[6] Korobov, Yu., M. Filippov, A. Belozertsev, S. Neveszin., and V. Shymiakov. Metastable Austenite Type Core Wire for Arc Spraying - Modeling of Heat Transfer in End Face and Coating Structure Analysis, *Thermal Spray 2010: Global Solutions for Future Applications,* B.R. Marple, A. Agarwal, M.M. Hyland, Y.-C. Lau, C.-J. Li, R.S. Lima, G. Montavon, Ed., May 3-5, 2010 (Singapore), ASM International, pp. 631 – 635.
[7] Lima, C.R.C., B.V. Bette, F. Camargo, H.C. Fals, and V.A. Ferraresi. Wear Performance of Thermally Sprayed and Welded Coatings Using Conventional and Nanostructured Materials. *ITSC 2013: Innovative coating solutions for the Global Economy*, ASM International, 2013, pp. 424 – 428.
[8] Pelz, A., Analysis of Fe-Base Materials and Evaluation of their Suitability for Wear Protection Coatings, *Thermal Spray 2010: Global Solutions for Future Applications,* B.R. Marple, A. Agarwal, M.M. Hyland, Y.-C. Lau, C.-J. Li, R.S. Lima, G. Montavon, Ed., May 3-5, 2010 (Singapore), ASM International, pp. 751-757.
[9] Pukasiewicz, A. G. M., A. R. Capra, and R. S. C. Paredes, Cavitation Erosion Mechanism in Fe-Mn-Cr-Si-Ni Arc Thermally Sprayed Coatings, *Thermal Spray 2011: Proceedings of the International Thermal Spray Conference,* B.R. Marple, A. Agarwal, M.M. Hyland, Y.-C. Lau, C.-J. Li, R.S. Lima, and A. McDonald, Ed., September 27-29, 2011 (Hamburg, Germany), ASM International, pp. 902-907.
[10] Korobov, Yu. S., Efficiency of Using Activated Arc Metallization for the Deposition of Protective Coatings. *Welding International*, 19(7), 2005, pp. 580 - 582.
[11] Elagina, O.Yu., *Technological Methods to Improve the Wear Resistance of Machine Parts*, University book, Logos: Moscow, Russia, 2009, p. 488 (in Russian).
[12] Houdremont, E. *Handbuch der Sonderstahlkunde*, Springer-Verlag: Berlin, 1966, p. 734, (In German).
[13] Bagryansky, K.V., Z.A. Dobrotina, and K.K. Khrenov. *Theory of Welding Processes*, Vischa Shkola: Kiev, 1976, p. 424, (in Russian).
[14] Petrov, G.L., and A.S. Tumarev. *Theory of Welding Processes*, Vischaya Shkola: Moscow, 1977, p. 392, (in Russian).
[15] Solonenko, O.P., M. Ushio, and A. Ohmori, Comprehensive Investigation of Metal Drop-Substrate Interaction. *Thermal Spray Coatings: Research, Design and Applications*, C.C. Berndt and T.F. Bernecki, Ed., June 7-11, 1993 (Anaheim, CA), ASM International, pp. 55-60.
[16] Kudinov, V., *Plasma Sprayed Coatings*, Nauka: Moscow, 1977, p. 184, (in Russian).
[17] Dykhuizen, R.C. Review of Impact and Solidification of Molten Thermal Spray Droplets, *J. of Therm. Spray Tech.*, 3 (4) 1994, pp 351-361.

[18] Boronenkov, V., and Yu. Korobov, *Fundamentals of Arc Spraying. Physic-chemical Regularities*. (Switzerland), Springer International Publishing, 2015, 256 p.
[19] Popov, A.A. *Phase transformations in metallic melts*, Metallurgizdat: Moscow, 1963, p. 311, (in Russian).
[20] Korobov, Yu., and V. Boronenkov. Modeling of Oxygen Intake by Particles at Arc Spraying, *Thermal Spray 2008: Crossing Borders,* on CD-ROM, E Lugscheider, Ed., June 2-4, 2008 (Maastricht, The Netherlands), DVS-German Welding Society, 2008, pp. 317-322.
[21] Schastlivtsev, V.M., and V.I. Zel'dovich. *Physical fundamentals of metallurgy*. Ekaterinburg, Ural Branch of Russian Academy of Sciences, 2015. 224 p (in Russian).
[22] Bowden, F.P., and D. Tabor. *The Friction and Lubrication of Solids*, Oxford University Press: New York, 1950, p. 428.

Highly Economical Fe-based Protective Coating System by Novel Material Design and Full Utilization of the Atomized Powder

Kirsten Bobzin, Lidong Zhao, Mehmet Öte, Tim Königstein
Surface Engineering Institute, RWTH Aachen University, Aachen, Germany

In this study, a novel economical Ni- and Mo-free wear and corrosion resistant FeCrMn alloy was designed. Spray powders of this alloy were produced through gas atomization. The total atomized powder was fractioned to -90 +45 µm, -45 +25 µm, -25 +10 µm and -10 µm. These powders were characterized in terms of their size distributions and morphologies. Subsequently, they were sprayed by a three-cathode plasma generator TriplexPro^{TM}-210. It was found that fine powders with fractions of -25 +10 µm and -10 µm showed a significantly higher affinity to oxidation than coarser ones. Nevertheless, dense coatings with low oxide contents could be realized using suitable parameters even with the finest powder. The results show that a full utilization of the atomized powder can be realized by three-cathode plasma spraying due to its wide parameter window. The average deposition efficiency is more than 70 %. In this way, the atomized powder can be used very efficiently. This contributes additionally to the high economical potential of this coating system which exhibits higher wear resistance than conventional stainless steel coatings and significantly higher corrosion resistance compared to carbon steels as well as cast irons.

1 Introduction

Main advantages of thermally sprayed Fe-based coatings compared to conventional Ni- and Co-based coatings, are their favourable lower costs as well as their environmental behavior. Besides classic coating systems based on 400 series and 300 series stainless steels [1], high performance Fe-based wear and corrosion resistant coatings have been developed by HVOF-spraying in recent years [1-11]. Their wear resistance is mainly attributed to precipitated fine borides as well as carbides in the coatings or to their amorphous/nanocrystalline composite contents. High Cr-contents in a range of 17-33 wt. % are applied to provide the desirable corrosion resistance of the coatings. Furthermore, alloying elements such as Ni and Mo are added to increase the corrosion resistance and W and Nb are applied to improve the glass-forming ability in amorphous/nanocrystalline composite coatings. Since these coatings were developed using HVOF-spraying, their applicability in plasma spraying has not been investigated invensively yet. An increased sensitivity to cracking was reported when being sprayed using APS process [11].

Three-cathode plasma generators such as TriplexPro™-210 with three cathodes and one anode (Oerlikon Metco Europe GmbH, Kelsterbach, Germany) improve the plasma stability compared to conventional one-cathode plasma generators significantly, leading to more stable and uniform treatment of particles [12-14]. Metallic fine powders can be processed using three-cathode spraying system [15, 16].

This study aims at developing a highly economical Fe-based wear and corrosion resistant coating system by APS spraying. The high cost-effectiveness can be attributed to low material costs as well as to the high application efficiencies of the total atomized powder. Developed coating systems should exhibit a higher wear resistance and comparable corrosion resistance in comparison to conventional stainless steel coatings.

2 Experimental procedure

The chemical composition of the designed alloy is FeCr25Mn10BC. As this composition does not contain cost-intensive alloying elements, significantly lower material costs can be realized compared to the above mentioned Fe-based coating systems which are developed for HVOF spraying. A high chromium content of 25 wt. % similar to those in duplex stainless steels was used to provide high corrosion resistance. A relatively high manganese content of 10 wt. % was applied to increase the ductility of this alloy and to reduce the losses of free chromium caused by oxidation during spraying. Boron and carbon (altogether less than 2. wt. %) were added to enable formation of borides and carbides for increasing wear resistance. This feedstock material was gas atomized. The total atomized powder was fractioned to -90 +45 µm, -45 +25 µm, -25 +10 µm and -10 µm by air separation (Nanoval GmbH & Co. KG, Berlin, Germany). The different fractions should favour the development of spray parameters to produce dense coatings with low oxide contents. The phase composition of the atomized powder was investigated by X-ray diffraction analysis using XRD 3000 (GE Inspection Technologies GmbH, Hürth, Germany) with a Cu target and the morphology of the powders was observed using scanning electron microscopy (SEM). Their size distributions were ascertained by a particle analysis system Morphologi G2 (Malvern Instruments GmbH, Herrenberg, Germany). Plates made of grey cast iron EN-GJL-250 were used as substrates. Due to the above mentioned advantages of three-cathode plasma spraying, the plasma generator TriplexPro^{TM}-210 was utilized to produce the coatings. The spray parameters are listed in Table 1. A conventional 316L-coating produced by a plasma generator F4MB-XL (Oerlikon Metco Europe GmbH, Kelsterbach,

Germany) with a powder fraction of -106 +45 µm was used as reference coating.

Table 1: Spray parameters for TriplexPro™-210.

Fraction	-90 +45 µm	-45 +20 µm	-25 +10 µm	-10 µm
Nozzle (mm)	9	9	9, 6.5	6.5
Current (A)	360-400	300-360	250-360	300-360
Ar (SLPM)	50-70	70-100	70-140	70-140
N_2 (SLPM)	0-4	0-4	-	-
Distance (mm)	100	100	100	100

The microstructures of the coatings were investigated by means of optical light microscopy and scanning electron microscopy (SEM). Their phase compositions were investigated using XRD 3000. Microhardnesses were measured on the cross-sections of the coatings using a Vickers hardness tester. A microhardness value was given as an average of ten measurements. Pin-on-Disc tests were conducted to investigate the wear behavior of selected coatings using a tribometer (CSM Instruments, Freiburg, Germany). The tests were carried out at room temperature without additional lubricants. Al_2O_3 balls with a diameter of Ø=6 mm were used as counterpart. The normal load was set to F=2 N. The sliding velocity was v=100 mm/s. The sliding distance was d=500 m. The volume losses were determined using a confocal laser-scanning microscope (CLSM) VK-X200 (Keyence Deutschland GmbH, Neu-Isenburg, Germany). Salt spray test according to ISO9227 was used to evaluate the corrosion resistance of the coatings comparatively.

3 Results and discussion

The particle analysis proved that the size distributions correspond to the given fractions above. The diameter corresponding to the 10th percentile was D=38.60 µm, D=19.12 µm,D=11.11 µm and D=4.43 µm for the four powders, respectively. The diameter corresponding to the 90th percentile was D=75.43 µm, D=39.28 µm, D=25.60 µm and D=16.41 µm, respectively. The shape factor "aspect ratio" amounted to 0.82, 0.85, 0.89 and 0.87 for the powders, indicating the finer particles exhibited a pronounced sphericity. **Fig.1** shows SEM-micrographs of the powders. A high number of spheric particles can be recognized. Consistent with the results of the particle analysis, coarser powders exhibit less sphericity. More satellites are recognizable on particles of coaser powders, indicationg that the atomization process was not set up properly. The X-ray diffraction analysis of the powders identified both ferrite and boride in the powders. Austenite was not detected despite the high Mn-content in this alloy.

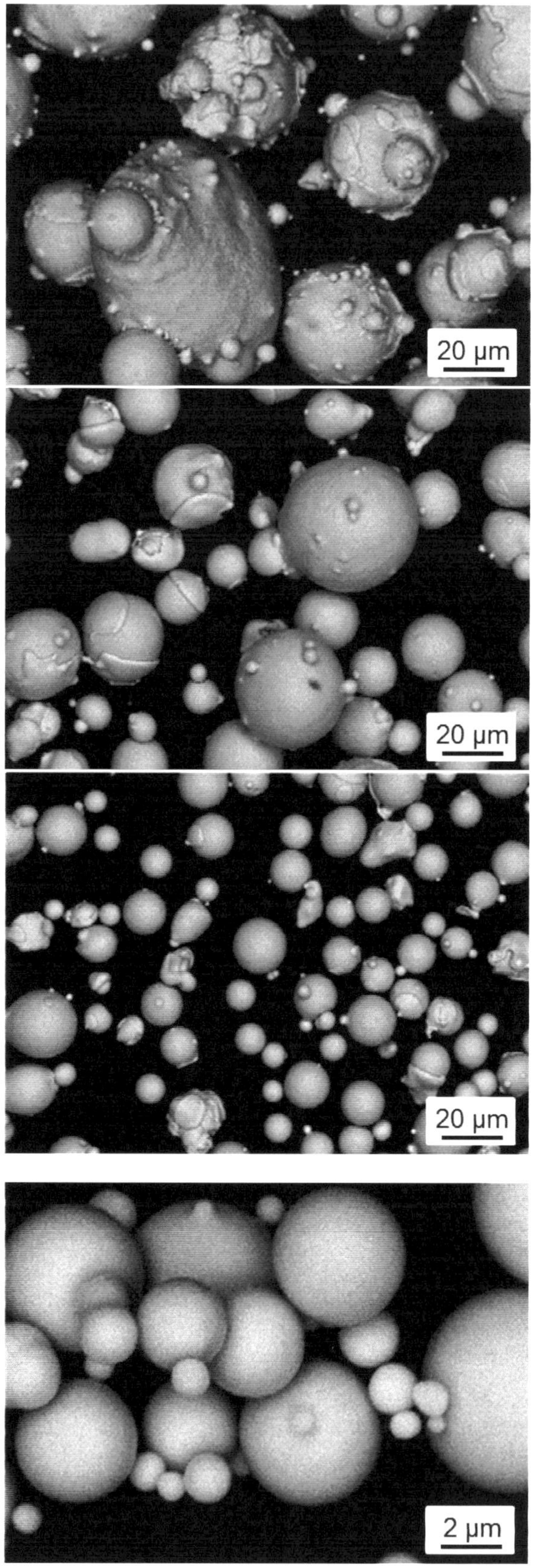

Fig. 1. SEM-micrographs of the spray powders.

The spray experiments revealed that coarser powders exhibit a high suitability for APS spraying. Dense coatings with low oxide contents were attained in a relatively wide parameter range. The chemical analysis of selected coatings revealed that coatings exhibit a low oxygen content of less than 1.2 wt. %. The powder with the fraction of -90 +45 µm was less sensitive to oxidation compared to the powder with the fraction of -45 +20 µm. This can be attributed to lower specific surface areas. This effect is consistent with works dealing with the application of conventional Fe-based coatings by APS spraying. In order to minimize the oxidation of feedstock material during spraying, usually coarse powders, for example, with typical fractions of -106 +45 µm or -90 +45 µm are employed [1].

Figs. 2 and 3 show cross-sections of a coating deposited with the fraction of -90 +45 µm and a coating deposited with the fraction of -45 +20 µm. It is recognizable that both coatings exhibit typical lamellar microstructures, indicating that the particles were mostly molten during impacting the substrate. Due to its higher dimensions of the particles, the lamellae in the coating deposited with the fraction of -90 +45 µm were generally thicker than those in the coating deposited with the fraction of -45 +20 µm. Both coatings were quite dense. Their porosities were about 2.5 %. The oxide contents which can be recognized as dark particles and thin lamellae in the pictures were low. The lowest hardness measured was more than 630 HV0.1. This value is significantly higher than the hardness value of conventional stainless steel coatings. High hardnesses of deposited coatings can be mainly attributed to borides and carbides in the coatings. The deposition efficiencies for the coatings shown in **Figs. 2 and 3** were 75 % and 77 % for the fraction of -90 +45 µm and -45 +20 µm, respectively. The high deposition efficiencies show that the powders are very suitable for APS spraying, contributing to the desirable high cost-effectiveness of the coating system.

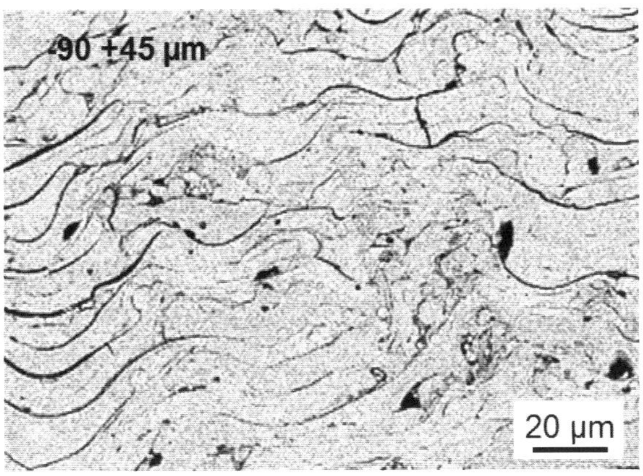

Fig. 2. SEM-micrograph of a cross-section of a coating deposited with the powder -90 +45 µm.

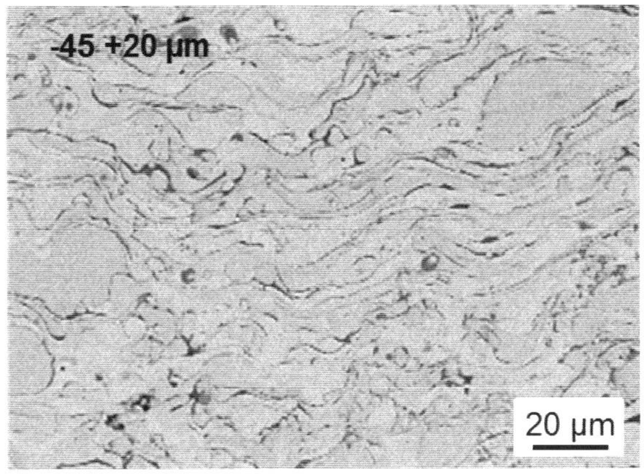

Fig. 3. SEM-micrograph of a cross-section of a coating deposited with the powder -45 +20 µm.

The spray experiments showed that finer powders with the fraction of -25 +10 µm and that of -10 µm were less suitable for APS spraying compared to coarser powders, because they were significantly more sensitive to the spray parameters regarding their oxidation behavior. **Fig. 4** shows for example a micrograph of a cross-section of a coating deposited with the fraction of -25 +10 µm. Numerous oxides are recognizable. The oxides are identified as $(Fe,Cr,Mn)_3O_4$ by means of X-ray diffraction analysis.The lamellar microstructure indicates that fine particles were well molten during impacting the substrate. Due to their higher specific surface areas, fine particles are significantly oxidized.

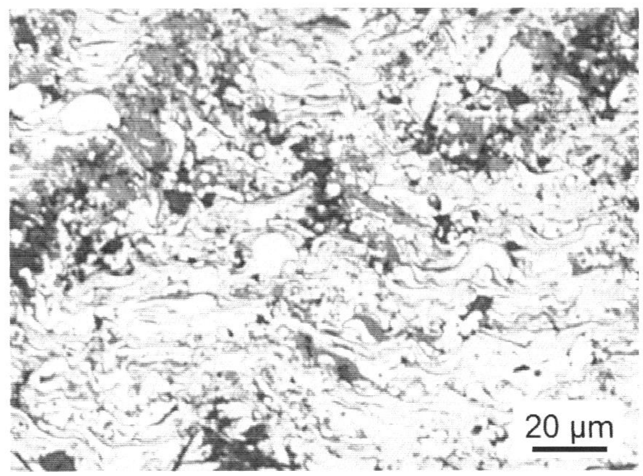

Fig. 4. Micrograph of a coating containing numerous oxides deposited with the powder -25 +10 µm.

The oxidation behavior of particles can be controlled by increasing particle velocities and decreasing the melting degree of particles. This was realized by using the smaller nozzle with a diameter of Ø=6.5 mm, increasing argon flow rate and reducing the current. **Figs. 5 and 6** show SEM-micrographs of two coatings with low oxide contents. Both coatings do not show typical lamellar microstrucutres. Instead, they exhibit a microstructure which is typical for HVOF-sprayed metallic coatings. Numerous semi-molten particles were included in the coatings. This can be attributed to high particle velocities and relatively low melting degree of them. The oxides as shown in **Fig. 4** are hardly recognizable in case of modified process parameters (see **Fig. 5 and Fig. 6**). The particle diagnostic measurements by SprayWatch 4s (Oseir Ltd., Tampere, Finland) showed that particle velocities higher than 650 m/s can be realized using modified parameters which are employed to deposit the coating shown in **Fig. 5**. The X-ray analysis of the coatings identified ferrite and boride. No oxides were detected. The deposition efficiency of the powder with the fraction of -25 +10 µm was 75 %. In contrast, the deposition efficiency of the fraction of -10 µm was about 50 %. This can be attributed to lacking penetration of small particles into the high temperature regions of the plasma jet. Because the powder with the fraction of -10 µm was only about 20 % of the total atomized powder, the average deposition efficiency of the total amount of the atomized powder was still more than 70 %. The high average deposition efficiency means a high application efficiency for the total atomized powder. The lowest hardness of the coatings deposited with the fraction of -25 +10 µm amounted to 541±37 HV0.1. It was 561±53 HV0.1 for the coatings deposited with the fraction of -10 µm.

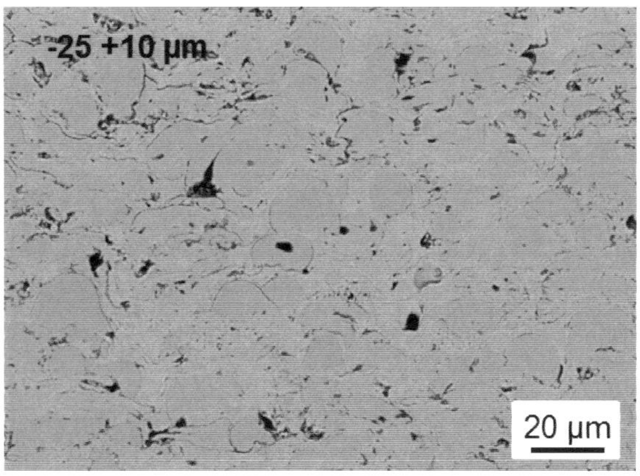

Fig. 5. SEM-micrograph of a cross-section of a coating with a low oxide content deposited with the powder -25 +10 µm.

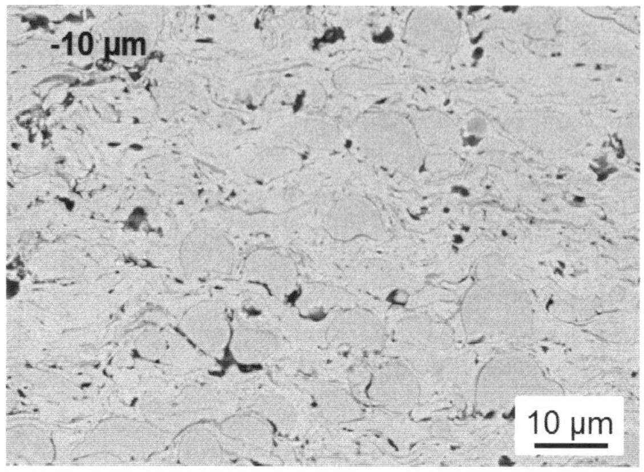

Fig. 6. SEM-micrograph of a cross-section of a coating with a low oxide content deposited with the powder -10 µm.

The pin-on-disc test showed that the wear rate of the most wear resistant FeCrMnBC-coating amounts to 4351 $\mu m^3/N.m$. This is only a quarter of that of the reference 316L-coating, verifiying that the FeCrMnBC-coatings exhibit singnificantly higher wear resistance in comparison to conventional stainless steel coatings. The higher hardness, borides and carbides in the coatings contribute to measured high wear resistance.

Fig. 7. Graphs of a sample of uncoated grey cast iron EN-GJL-250, a FeCrMnBC-coating and the reference 316L-coating after salt spray tests.

Fig. 7 shows pictures of uncoated EN-GJL-250, the FeCrMnBC-coating in **Fig. 3** and the reference 316L-coating prior to being tested and after different test durations. After 6 h, EN-GJL-250 was completely corroded. Therefore, it was taken out of the test. Both the FeCrMnBC-coating and the reference 316L-coating showed significantly better behavior compared to EN-GJL-250. After 30 h, small rust traces were visible on their surfaces, indicating both coatings were attacked selectively. The FeCrMnBC-coating exhibited slightly less corrosion products compared to the reference 316L-coating. No significant differences between both coatings were recognizable after 120 h. The salt spray test revealed that the FeCrMnBC-coating exhibits a comparable corrosion resistance to the reference 316L-coating under the same test condition. **Fig. 8** shows micrographs of cross-sections of both coatings after salt spray test. It is evident that selective corrosion occurred in the surface areas of both coatings. In addition, undercorrosion at the interface to the substrate are recognizable for both coatings.

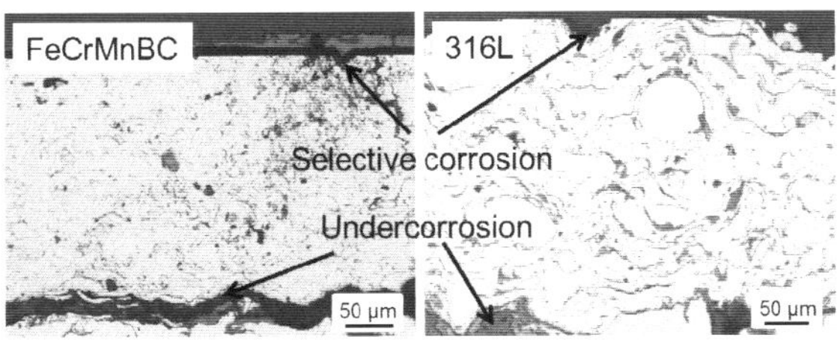

Fig. 8. Micrographs of cross-sections of the FeCrMnBC-coating and the reference 316L-coating after salt spray test.

4 Conclusions

In this study, a novel Fe-based coating system was developed. It is highly economical, because it does not contain cost-intensive alloying elements. The total amount of the atomized powder can be utilized for APS spraying with a very high average deposition efficiency. The coatings exhibit higher wear resistance compared to conventional stainless steel coatings with comparable corrosion resistance. The developed coating system and process promise a highly economic alternative to conventional systems. These can be employed in various applications to protect carbon steel and cast iron substrates.

5 Acknowlegements

This work is based on IGF-project 17727 N. The IGF-project of VDMA has been, on the basis of a resolution of the Lower House of the German Parliament, promoted by the German Ministry of Economic Affairs and Technology via AiF within the frame work of the program for the promotion of joint industrial research and development (IGF). The authors would like to thank all parties involved for funding and support.

3 Literature

[1] Oerlikon Metco AG: Thermal Spray Materials Guide V2015.04 (EN)

[2] A. Milanti, H. Koivuluoto, P. Vuoristo, G. Bolelli, F. Bozza and L. Lusvarghi, Microstrucutral Characteristics and Tribological Behavior of HVOF-Sprayed Novel Fe-Based Alloy Coatings, Coating, 4 (2014), pp. 98/120.

[3] G. Bolelli, B. Bonferroni, J. Laurila, L. Lusvargi, A. Milanti, K. Niemi and P. Vuoristo, Micromechanical Properties and Sliding Wear Behaviour of HVOF-Sprayed Fe-Based Alloy Coatings, Wear, 276-277 (2012), pp. 29/47.

[4] G. Bolelli, T. Börner, A. Milanti, L. Lusvargi, J. Laurila, H. Koivuluoto, K. Niemi and P. Vuoristo, Tribological Behaviour of HVOF- and HVAF-Sprayed Composite Coatings Based on Fe-Alloy + WC-12% Co, Surface and Coatings Technology, 248 (2014), pp. 104/112.

[5] A. Schwenk, HVOF-K2-Spraying of Iron-Based Materials for Applications in Printing Industry. Proc. of 8.Colloquium High Velocity Oxy-Fuel Flame Spraying, Erding, 2009, pp. 191/198

[6] A.M. Osmanda, A.J. Battenbough and A. M. Staines, Fe-Cr and Fe-Cr-Ni Based HVOF Coating Systems Customised for Wear and Corrosion Protection, Thermal Spray Bulletin 5 (2012), pp. 117/122.

[7] R.Q. Guo, C. Zhang, Q. Chen, Y. Yang, N. Li and L. Liu, Study of Structure and Corrosion Resistance of Fe-Based Amorphous Coatings by HVAF and HVOF, Corrosion Science, 53 (2011), pp. 2351/2356.

[8] H.S. Ni, X.H. Liu, X.C. Chang, W.L. Hou, W. Liu and J.Q. Wang, High Performance Amorphous Steel Coating Prepared by HVOF Thermal Spraying, Jornal of Alloys and Compounds, 467 (2009), pp. 163/167.

[9] D. Zois, A. Lekatou, M. Vardaoulias, Preparation and Characterization of Highly Amorphous HVOF Stainless Steel Coatings, Journal of Alloys and Compounds, 504 (2010), pp. 283/287.

[10] W.H. Liu, F.S. Shieu and W.T. Hsiao, Enhancement of Wear and Corrosion Resistance of Iron-Based Hard Coatings Deposited by High-Velocity Oxygen Fuel (HVOF) Thermal Spraying, Surface and Coatings Technology, 249 (2014), pp. 24/41.

[11] V. Varadaraajan, R. K. Guduru and P.S. Mohanty, Synthesis and Microstructural Evolution of Amorphous/Nanocrystalline Steel Coatings by Different Thermal-Spray Processes, Journal of Thermal Spray Technology, 22 (2013), pp. 452/462.

[12] K. Landes, Plasma Generators for Thermal Plasma Processes, International Journal of Materials Research,102 (2011), pp. 959/963.

[13] K. Bobzin, N. Kopp, T. Warda, M. Schäfer, M. Öte, A Numerical Investigation: Influence oft he Operating Gas on the Flow Characteristics of a Three-cahtode Air Plasma Spraying System, Proc. of ITSC 2013, Busan, Korea, 2013, pp. 400-405

[14] K. Bobzin, M. Öte, Eine numerische Untersuchung: Atmosphärisches Plasmaspritzen mittels eines Dreikathoden-Plasmabrenners, Thermal Spray Bulletin, 8(2015), pp. 118-123.
[15] K. Bobzin, F. Ernst, J. Zwick, M. Bruehl and G. Matthaeus, Analysis of In-flight Particle Properties of Ultrafine Powders, Thermal Spray 2007: Global Coating Solutions, B.R. Marple, M.M. Hyland, Y.-C. Lau, C.-J. Li, R.S. Lima and G. Montavon, Ed., May 14-16, 2007 (Beijing, China), pp. 582/587.
[16] K. Bobzin, T. Schläfer, T. Warda, I. Petkovic, M. Schäfer, K. Landes, G. Forster, S. Zimmermann, J. Marques-Lopez, S. Kirner, M. Kauffeld and J. Schein, APS TriplexPro-200/-210 and HVOF K2: Comparison of Two Modern Thermal Spray Systems with Regard to Particle in-Flight and Coating Properties Using Iron Based Feedstock Materials, Thermal Spray 2008: S.R. Lima, A. Argawal, M.M. Hyland, Y:C- Lau, C.J. Li, A. McDonald and F. L. Thoa, Ed., May 21-24, 2012, (Houston, Texas, USA), pp. 956/963.

Fe-based Powder Alloys Deposited by HVOF and HVAF for Sliding Wear Applications

Senad Dizdar, Höganäs AB, Sweden

Three Fe-based powder alloys have been deposited by HVOF- and HVAF-spraying onto EN 355 MC steel plates. Test blocks machined from these plates are tested for dry sliding adhesive wear. The wear testing has a base approach from ASTM G65 norm but wear-mechanism mapping method is applied in order to avoid risks with localized presentation of wear data. Wear maps for all three Fe-based powder alloys are similar, show regions of plasticity dominated wear, wear transition and oxidational wear. Achieved wear rate and friction coefficient are lower than for EHC-plated reference blocks.

1. Introduction

This paper is a third one in a series [1,2] discussing feasibility of replacement of electrolytic hard chrome (EHC) plating by HVOF/HVAF spraying for a number of structural components in aerospace, automotive, printing, process and oil & gas applications. A widely used wear/corrosion remedy for these components is electrolytic hard chrome (EHC) plating with a layer thickness of about 0.25 mm with hardness 900 to 1100 HV. However, EHC plating baths include hexavalent chromium, both health and environmental hazard. Accordingly, health/environmental government agencies demand substitutes for EHC coatings [3].

Fe-based powder alloys for HVOF/HVAF spray coating applications open for cost effective consumables that retain high performance due to a range of alloys developed for corrosion and wear resistance. However, depending on the specific alloy system, powder may oxidize/corrode during high temperature spray deposition, 3000°C and 1900°C flame temperature, for HVOF- respective HVAF spraing. Choice of a proper alloying system prevents powder from oxidation. In addition, Fe-based powder alloys are less harmful in comparison to WC-Co-Cr, Co- and Ni-based alloys [1].

The components are often simultaneously exposed to a number of wear and corrosion damaging mechanisms. Often one of the damaging mechanisms is critical and taken as the main design criterion. However, a sudden change in component loading conditions and working environment can occur and alter the critical damaging mechanism. Therefore, it is important to understand a components performance for basic wear and corrosion mechanisms. Previous papers in this series showed potential of three Fe-based powder alloys to meet EHC-plating performance in abrasion [1] and erosion [2] exposed applications. However, these alloys are so far not classified according to EN 14700 norm [11], which widely summarizes use of powders and other consumables for hard facing against wear and corrosion and even review the critical damaging mechanisms for some typical hard facing applications.

This investigation aims to generate wear maps for dry sliding adhesive wear of test samples HVOF/HVAF sprayed by using three Fe-based powder alloys, compare with wear maps for EHC-plated samples and correlate wear with the microstructure. The applications of interest are industrial valves and other dry sliding wear applications.

2. Method

Structural components, which in their operation come into metal-to-metal sliding contact with each other or with metallic wear particles, can suffer from adhesive wear damages and premature failures. Fully functioning lubrication by oil, grease or solid lubricants prolongs normally the component service life. However, fully functioning lubrication may fail or cannot be established under increased temperature, dirty or aggressive environments. These situations are in common called as unlubricated – dry friction and wear. Here, friction and wear control is based on selection of metals with limited/low solid solubility to each other or prevention of direct metal-to-metal contact by self-lubricated phases or selection of one contact member from non-metallic material.

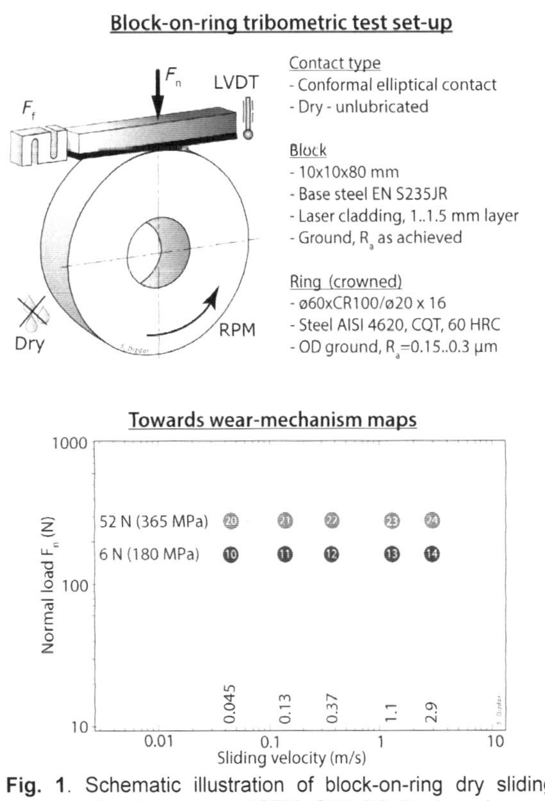

Fig. 1. Schematic illustration of block-on-ring dry sliding wear testing based on ASTM G77 [5] (top) and wear-mechanism mapping method (bottom).

Fig. 1 (top) illustrates schematically block-on-ring dry sliding wear testing based on ASTM G77 [5] standard. A commercial tribometer - a multipurpose friction and wear tester is used. An AC-geared motor drive rotates the crowned test ring. A test block is pressed onto the ring by using a load arm and dead weights. The test block is by friction force pushed against a strain gauge load transducer, while an LVDT (linear variable displacement transducer) registers total linear wear in the contact. Fig. 1 (bottom) illustrates wear-mapping method used in this investigation. The wear-mapping method eliminates risks with localized presentation of wear data (Bowden [6], Lim and Ashby [7]). Instead of a normal load/velocity point, wear testing is performed at preselected and stepwise increasing normal loads and sliding velocities. A general picture over friction and wear mechanisms is by this generated for the preselected range. Sliding velocities in this investigation are 0.045, 0.13, 0.37, 1.1 and 2.9 m/s while normal loads are 6 and 52 N (corresponding 180 respective 365 MPa for a block and a ring made of steel). This preselection is a good loading description for applications such as industrial valves etc.

Edrisy et al. [8] constructed wear mechanism maps for dry sliding of sprayed disk samples against AISI M2 cylindrical pins. The discs were contact members being discontinuously in the contact, what makes the results comparison more complex. In addition, the relative humidity was low, only 10%. The test discs were produced by PTWA (plasma-transferred wire-arc) and HVOF-spraying by using AISI 1020 mild steel wire as pure and as alloyed with 2.5% Al. They did not report friction coefficient from the wear mapping. Wear rates and SEM analysis of surface oxides on the wear track were base for depicting the wear mechanism regions.

3. Experiment

Test blocks in this investigation were manufactured at the same time as blocks from [1,2]. Table 1 lists Fe-based powder alloys used in this investigation, test blocks and their manufacturing routes. Fig. 2 shows a bar chart with two-dimensional surface roughness parameters. The top coating layer sprayed with HVOF/HVAF technology is known to be less homogeneous in comparison to the layers below due to the lack of so called "peening effect". The top layer was therefore removed by plane grinding. This grinding removed about 15 µm from the top of the coating but resulted in no additional densification or homogenization of the coated sample surface. The top sprayed layer of sharp components is normally super-finished in some way for the same reason.

Roughness in Fig. 2 shows that surface of all test blocks has deep surface pores. The valley depth component, R_v, is substantialy larger than the peak height component R_p. and substantially exceed 50% of the ten-point-height R_z. FeSP586, both HVOF and HVAF, and 6AB HVAF blocks have fine surface roughness with average roughness R_a<1 µm. Reference EHC-plated blocks do not diverge from the sprayed blocks.

Table 1: Powder grade composition, test blocks and their manufacturing routes.

	1	2	3	4
Grade	FeSP529	FeSP586	6AB	EHC
C	1.9	0.7	2.1	
Mo	5.8	-	-	
Cu	1.5	-	-	
Ni	4.2	12.3	5.9	
V	0.2	-	-	
Fe (bal.)	(70.7)	(52.3)	(70.9)	
Mn	0.2	-	0.3	
Cr	9.7	30.6	14.4	
Si	2.1	0.6	3	
B	3.7	3.7	3.4	
Coating spraying	HVOF-DJH, sieve cut -45+20µm Tampere Univ. of Techn. Finland; HVAF-M3, sieve cut -36+20µm University West, Sweden			EHC Bath Plating
Steel substrate	EN S355 MC 8 mm thick plates			EN C45
Coating thickness	About 350 µm (by multiplying 15 µm thick single spray layers			250 µm
Test surface finish	Coating's top 15 µm ground away.			As delivered
Hardness	$HV_{0.3}$	$HV_{0.3}$	$HV_{0.3}$	$HV_{0.3}$
HVOF	560	750	800	950
HVAF	850	810	920	

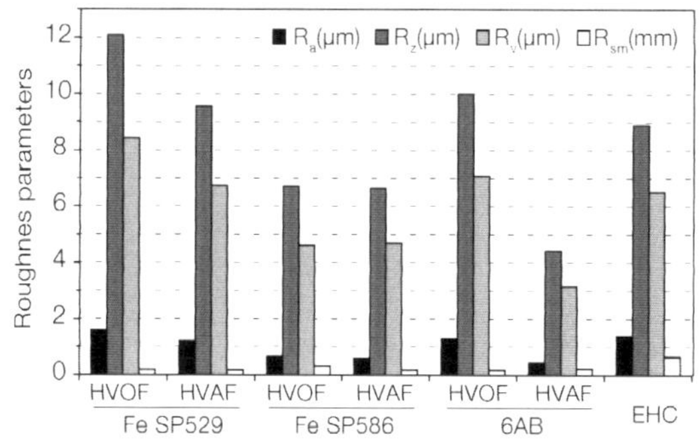

Fig. 2. Surface roughness parameters of test blocks.

Microstructure of test blocks is described by microstructure cross cut of the blocks and XRD analysis of the blocks (Fig. 3). The microstructure shows typical voids and oxide stringers for HVOF FeSP529, FeSP586 and 6AB samples. HVAF samples of the same powder show the same features but to a lower extent. FeSP586 blocks show finest overall microstructure. XRD chart shows matrix and hard phase constituents of the powder and sprayed coatings. Outcome of a Thermocalc simulation gives an indication of hard phase percentage in the matrix. [1].

Table 2 Outcome of Thermocalc simulation, at.%

	FeSP529	FeSP586	6AB
Austenite	26.1	60.7	6.8
Ferrite	6.2		28.5
Carbides	21.9	7.4	24.8
Borides	44.3	30.6	38.8
Graphite			0.5
Total	98.5	98.7	99.4

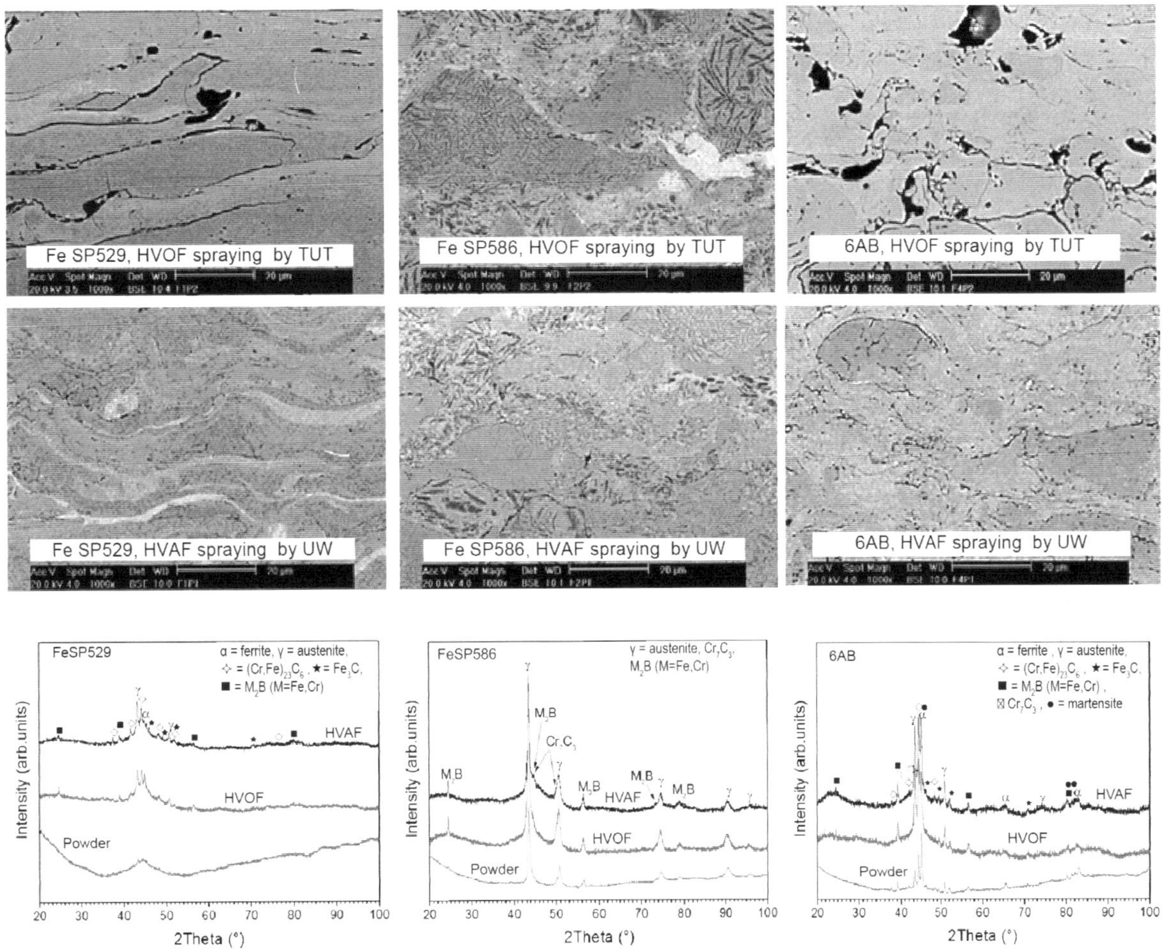

Fig. 3. Typical SEM BSE microstructure cross cut photographs of test samples (top) and XRD analysis of the powder and sprayed blocks [1].

4. Results and discussion

Wear-mechanism maps generated for three sprayed blocks and reference EHC-plated blocks are shown in Fig. 4. In addition, Fig. 5 illustrates some friction and wear logs for FeSP529 blocks at 6N in normal load and 0.13, 1.1 and 2.9 m/s in sliding velocity. Finally, Fig. 6 shows details of wear track appearance for test blocks corresponding test logs in Fig. 5.

A general observation valid for all test runs was that wear particles generated the test dry sliding contact were fine reddish powder, likely a mixture of iron oxides. Lab room temperature was 20°C and relative humidity was 30 to 35%. A partition of wear particles generated in the contact adhered to the counter rings and biased wear logging up to some extent.

Wear rate shown in Fig. 4 is block wear rate in $mm^3/(Nm)$ calculated based on size of the wear track ellipse on the test blocks.

For all sprayed blocks at both 6 and 52 N in normal load and 0.045, 0.13 and 0.37 m/s in sliding velocity, friction coefficient is 0.7 to 0.9. Wear rate of the sprayed blocks was at a level of 10^{-6} $mm^3/(Nm)$. Fig. 6 top shows appearance of the wear track for FeSP529 samples at 0.13 m/s in sliding velocity. Longish vertical wear marks show evidence of plastic deformation of contacting asperities and adhered (previously detached and transferred) wear particles. Based on presented evidence, wear mechanism is plasticity dominated wear.

A further increase in sliding velocity causes changes in friction and wear behavior. At 1.1 m/s in sliding velocity, all sprayed test blocks achieved a highest wear rate while fluctuation of friction coefficient around the mean level decreased somewhat (Fig. 5 middle). Increased sliding velocity to 1.1 m/s smooths out surface of vertical wear marks of adhered material, likely due to increase in contact flash temperature with sliding velocity (Fig. 6 middle). However, the wear marks still show evidence of plastic deformation up to some limited extent. Simultaneous presence of both wear track smoothening and plastic deformation indicate that the wear is in a transition between plasticity dominated and oxidational wear.

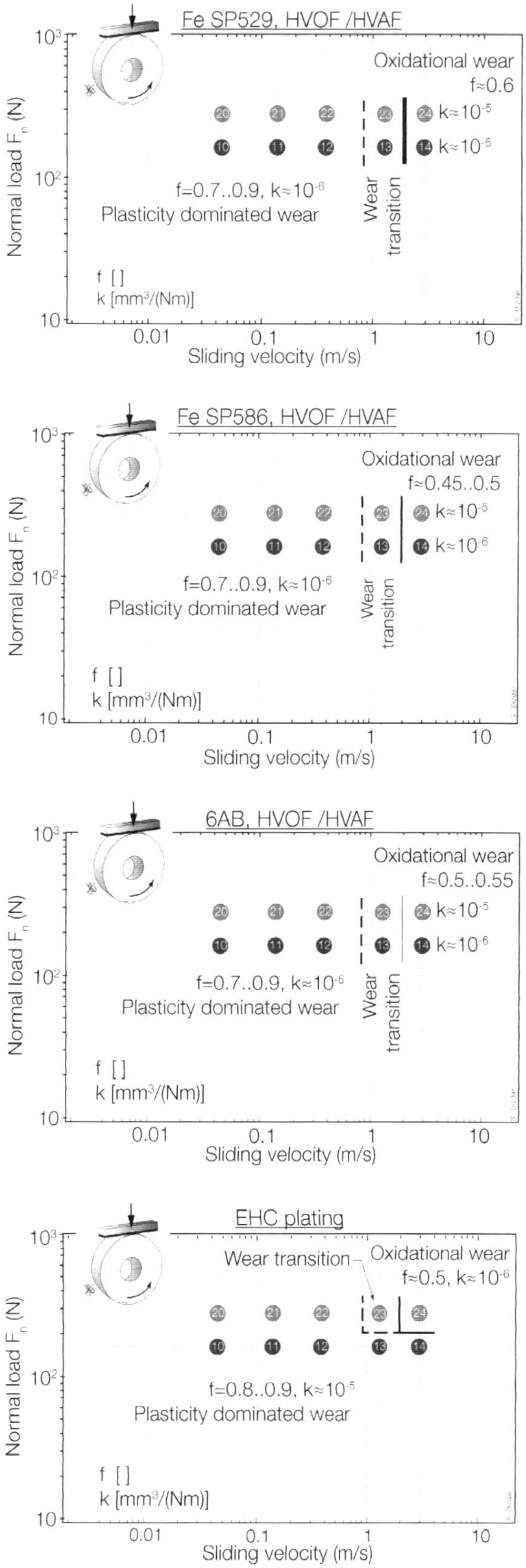

Fig. 4. Wear-mechanism maps with wear rate of ground sprayed test blocks.

Finally, at the highest test sliding velocity of 2.9 m/s, friction coefficient drops for all test materials except for EHC-plating. The drop is highest for FeSP586, down to 0.45 to 0.5, then for 6AB, 0.5 to 0.55 and FeSP529, 0.6. The wear rate, however, remains unaltered at 6 N but increase, for one power of ten, to 10^{-6} $mm^3/(Nm)$ at 52 N in normal load. Fig. 6 middle shows thinnest and smoothest vertical wear marks of adhered material in comparison to the marks for previous sliding velocities. The contact surface is in overall flat and smooth, with exception for transferred wear particles. This evidence suggests dominance of oxidational wear mechanism. The contact flash temperature likely exceeds 700°C when layers of surface oxides, such as magnetite Fe_3O_4 forms.

Friction coefficient drops from 0.7 to 0.9 for plasticity dominated wear to e.g. 0.45 to 0.5 for oxidational wear of FeSP586 alloy is a large one. One would normally expect drop in wear rate too when approaching mild oxidational wear. However, further work needs to depict mild/severe sub regions of oxidational wear region.

For EHC-plated blocks, the friction and wear behaviors do not change for both test normal loads and for sliding velocities of 0.045, 0.13 and 0.37 m/s. The friction coefficient is at 0.8 to 0.9 level while the wear rate is one power of ten higher than for all sprayed blocks, i.e. 10^{-5} $mm^3/(Nm)$. Axis length of the wear track ellipse is consecutively larger for EHC-plated blocks than for equivalent data points for other test materials. First at 1.1 m/s in sliding velocity, friction coefficient curve fluctuation around the mid value decreases. This is a sign of plasticity dominated wear transition to oxidational wear. When the sliding velocity reaches 2.9 m/s, the friction coefficient curve achieves 0.5-level, becomes smooth and the wear rate drops to 10^{-6} $mm^3/(Nm)$. This proves occurrence of oxidational wear.

Described friction and wear behaviors follow main wear-mechanism mapping depiction for dry steel-to-steel contacts presented by Lim and Ashby [7]. Of course, materials tested here are thermally sprayed coatings containing metal particle splats, oxide stringers, voids and other inclusions. Another point is that the sprayed materials are in fact steels highly alloyed with C, Cr and Ni, B and Si (two last are not that common in common steels).

In published literature, there are not many papers on dry sliding wear and Fe-based HVOF/HVAF-sprayed coatings. Bolelli et al. [10] investigated dry sliding wear of HVOF-sprayed Fe-based coatings. However, they used ball-on-disc, a different type of tribological set-up and set the sprayed test plate as the contact member being discontinuously in the contact. This makes the results comparison more complex. Anyway, friction coefficient recorded was higher for EHC-plating in comparison to Fe-based HVOF sprayed discs when dry slid against EN 100Cr6 steel balls.

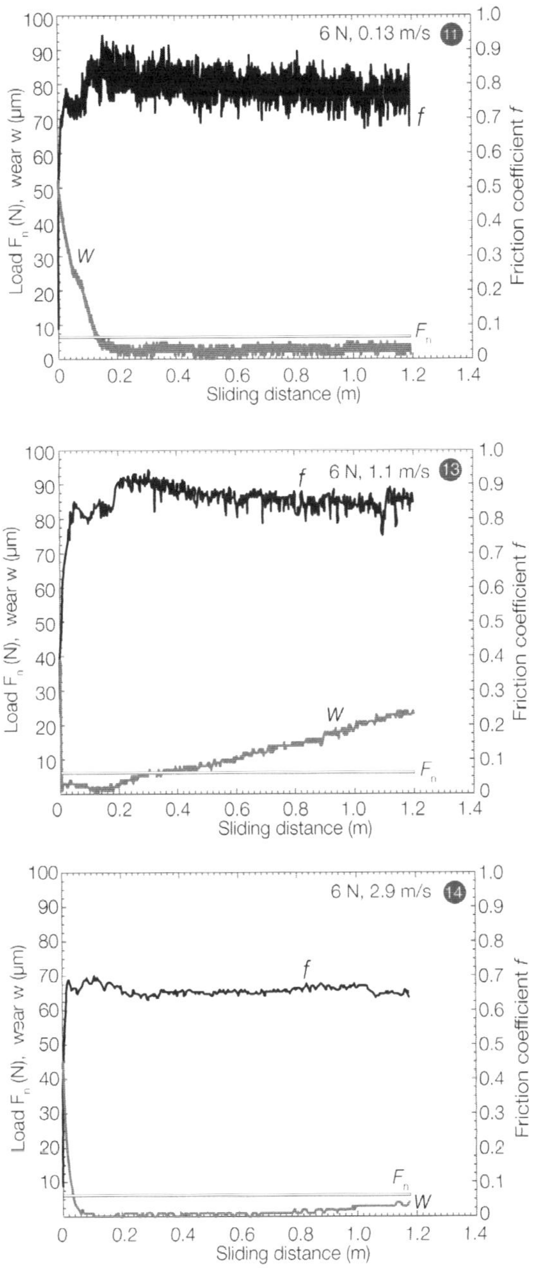

Fig. 5. Friction and wear logs for FeSP529 samples

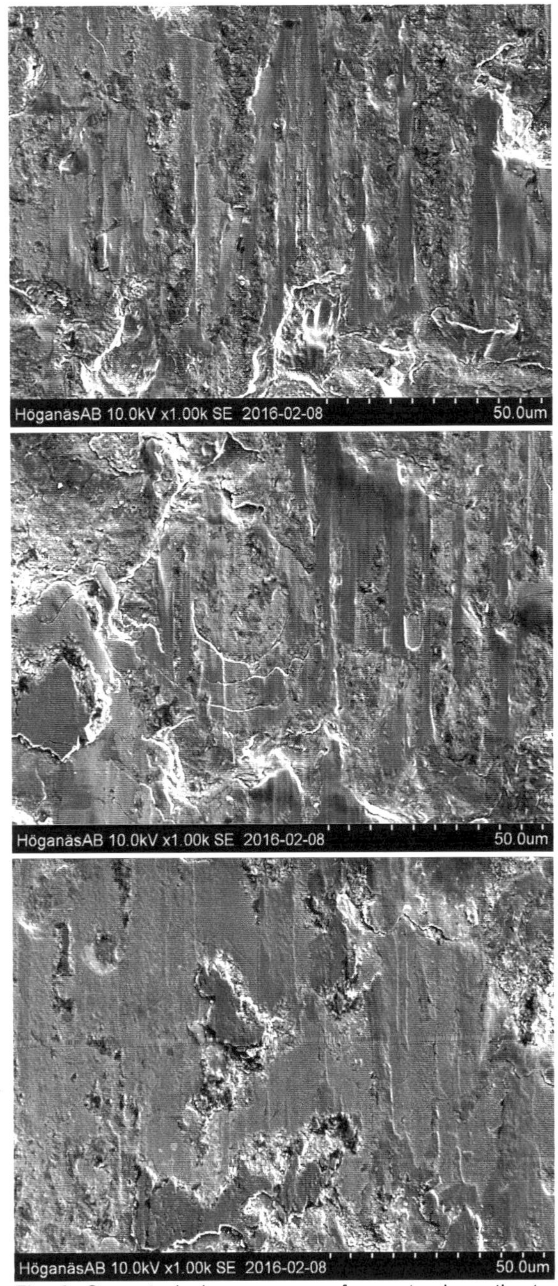

Fig. 6. Some typical appearance of wear track on the test blocks.

Publications on dry sliding wear of EHC-plating are not many too. Budinsky [9] reported ASTM G77 dry sliding wear testing of EHP-plated blocks against AISI 4620 steel rings as CQT/ground at 0.6 m/s in sliding velocity and 45 N in normal load. The friction coefficient was 0.7 while the wear rate was around $4 \cdot 10^{-7}$ $mm^3/(Nm)$ (calculated from data in [9]). These values are lower than those present in this investigation, in particular the wear rate is about two powers of ten lower. The friction and wear data for EHC-plated blocks was already reported to vary substantially [1] depending on variation of the EHC-plating technology and processing at particular plating workshops.

As already mentioned, friction and wear in dry sliding contacts is based on selection of metals with limited/low solid solubility to each other or prevention of direct metal-to-metal contact by self-lubricated phases or selection of one contact member from non-metallic material. Alloys used in this investigation includes substantial amounts of carbides and borides, Following indication from the Thermocalc simulation, total carbides and borides content is approximately at 60% level for FeSP529 and 6AB, and 40% level for FeSP586. FeSP529 has an austenitic/ferritic matrix, 6AB ferritic/austenitic and FeSp586 austenitic matrix. Friction and wear behavior for plasticity dominated wear region was similar for blocks from all three alloys. Oxidational wear region shows differences, in particular in friction coefficient drop due to oxide layer formation. However, lowest friction of 0.45 to 0.5 shows FeSP586 blocks with austenitic matrix and least hard phase content. A completion work with deeper analysis of the wear track and the oxides formed could provide more understanding here.

The mapping presented here was in strictly dry conditions, i.e. the worst case. However, surface roughness of the test blocks includes deep surface pores i.e. their lubricant or fluid retention ability is high. Any presence of water or other operating fluid will for sure substantially change the friction- and the wear behaviors. Accordingly, repetition of wear mapping as performed here, but with test blocks impregnated with lubricant or lubricated with water or other powering fluid, can be of interest for industrial valves and other typical dry wear applications.

5. Conclusions

Three Fe-based powder grades have been HVOF- and/HVAF-sprayed onto EN S335MC structural steel plates. Sprayed and ground test block made of FeSP529, FeSP586 and 6AB powder grades were tested for dry sliding wear against AISI 4620 steel CQT ground counter rings by using wear-mapping method. Following conclusions were drawn:

- Plasticity-dominated adhesive wear occurs for all sprayed test blocks at sliding velocities lower than 0.37 m/s disregarding normal load up to 52 N (about 365 MPa in Hertzian max. contact pressure) followed with wear rate of about 10^{-6} $mm^3/(Nm)$ and friction coefficient 0.7 to 0.9.
- Wear transition occurs when sliding velocity increases to 1.1 m/s.
- Oxidational wear occurs when sliding velocity increases to 2.9 m/ s, friction coefficient drops below 0.6 but wear rate stays unchanged, about 10^{-6} $mm^3/(Nm)$.
- Lowest friction coefficient in oxidational wear showed FeSP586-, 0.45 to 0.5, then 6AB-, 0.5 to 0.55, and finally FeSP529-sprayed blocks, 0.6.
- EHC-plated reference blocks achieved similar friction coefficient but one power of ten higher wear rate compared to sprayed blocks, 10^{-5} $mm^3/(Nm)$. Wear transition and oxidational wear occur also at 1.1m/s respective 2.9 m/s in sliding velocity but only at 52 N normal load. Then the friction coefficient drops to 0.5 for oxidational wear.
- In overall, HVAF/HVOF sprayed blocks performed better than EHC-plated reference blocks.

6. Acknowledgements

The author appreciates efforts of Christophe Lyphout, University West, Trollhättan, Sweden, and colleagues from Tampere University of Technology, Tampere, Finland Andrea Milanti, Heli Koivuluoto and Petri Vuoristo, for cooperation in manufacture of test and valuable discussions.

The author appreciates work on Thermocalc simulations performed by Sven Bengtsson and experimental work performed by Henrik Holmqvist.

7. Further work

Further work suggested:

- A completion of this paper by deeper analysis of the wear track and the oxides formed.
- A repetition of the wear mapping performed here but in lubricated conditions

8. Literature

1. Dizdar, S, Lyphout, C., Milanti, A., Koivuluoto H., Vuoristo P., Fe-based Powder Alloys Deposited by HVOF and HVAF for Abrasive Wear Applications, ITSC 2014 Int. Therm. Spray Conf., 21-23 May 2016, Barcelona, Spain
2. Dizdar, S., Kumar, M., Fe-based Powder Alloys Deposited by HVOF and HVAF for Applications Exposed to Solid Particle Erosion, ITSC 2015 Int. Therm. Spray Conf., 11-14 May 2016, Long Beach CA, USA
3. European chemical agency (ECHA), http://echa.europa.eu.
4. EN 14700-2014: Welding consumables - Welding consumables for hard-facing.
5. ASTM G77 - 2005: Standard Test Method for Ranking Resistance of Materials to Sliding Wear Using Block-on-Ring Wear Test.
6. Tabor, D., Status and direction of tribology as a science in the 80's. Understanding and prediction, Proc. Int. Conf. “Tribology in the 80's”, **1** (1983) 1-17, NASA Lewis Research Center.
7. Lim S.C., Ashby M.F., Wear-mechanism maps, Acta Metall. 35 (1987)1–24.
8. Edrisy A., Perry T., Alpas A.T., Wear mechanism maps for thermal-spray steel coatings, *Metall. Mater. Trans. A*, 36A (2005) 2737-2749.
9. Budinski K.G., The wear resistance of diffusion treated surfaces, *Wear* **162-164** (1993) 757-762.
10. Bolelli G., Bonferroni B., Laurila J., Lusvarghi L., Milanti A., Niemi K., Vuoristo P., Micromechanical properties and sliding wear beahviour of HVOF-sprayed Fe-based alloy coatings, *Wear* **276-277** (2012) 29-47.

A Comparison between Nanostructured YSZ Coatings Fabricated by LPPS and APS Method Using Unsintered Porous Agglomerates

***Yan Zhao*[1] *Yang Gao*[1]**

1 Thermal Spray Center of the Dalian maritime University, Dalian, Liaoning, China

Email: gaoyang@dlmu.edu.cn, Tel and Fax: +8641187426895

Abstract

The nanostructured zirconia-7-8 wt.% yttria agglomerates were fabricated by low pressure plasma spraying (LPPS) method at 5000 Pa with the current of 350 A, 500 A and 650 A. The coatings were composed of tetragonal, cubic phase zirconia and small amount of monoclinic phase zirconia. The Weibull plots of Knoop microhardness under a 10gf load indicated that the coatings possessed bimodal characters when the current below 650 A. In contrast with the air plasma sprayed nanostructured coatings, the LPPS coatings exhibited fine equiaxed grain structure, including consolidated coarser grains and loose packed nano granules. Especially with the current of 350 A, larger amount of crystalline grain in the full molten part fell into the nano scale, and significantly reduced the average grain size of the coating. Further increased the arc current to 650 A, the coating exhibited a uniform fine equiaxed grain structure of ~200 nm. The mechanical properties (including average microhardness and elastic modulus) were investigated via indentation method. The LPPS coatings showed slightly higher microhardness, but the elastic modulus merely halved.

Introduction

The nanostructured ceramic coatings are generally achieved using conventional plasma spray method with nanostructured agglomerated powder [1]. The as-sprayed coatings are formed by a mixture of agglomerates that were fully molten and semi-molten, exhibiting a "bimodal microstructure". Through carefully controlling the particle surface temperature, as many as 24 % ~ 35 % [2, 3] nano zones were reported to be remained in a bimodal structured coating, which led to ~ 41 % decrease in thermal diffusivity in contrast with that of the conventional coating (at 1400 °C, 20 h) [2]. Significant improvements on coating properties make the nanostructured coatings more attractive and, further investigations on the structure-property relationship of nanostructured coatings elucidated a necessity in preserving the fine nano structure of less than 100 nm as much as possible [4-6]. However, the nature of thermal spray technology requires enough melting degree of the feedstock powder to supply sufficient adhesion/cohesion strength, which constraints the amount of nano zone in a bimodal structured coating.

Racek et al. [7] achieved coatings with mere full nanostructure of granule morphology (less than 1 μm) by spraying nanostructured YSZ agglomerates (~ 25 μm) via TTPR (triple torch plasma reactor) method with input power of 33 kW at 100 torr (~ 13 kPa). The as-sprayed coatings presented a multi-modal pattern evolving from porous nano granules to highly consolidated particles, elucidating the inhomogeneous heating status before deposition. Correspondingly, the mechanical properties was also expressed the same distribution modal [8]. Though the particles behavior within the plasma jet was not mentioned, through comparing the fine granules in the as-sprayed coatings with the original ones, it can be acknowledged that a sever disintegration occurred in the plasma jet; besides, no sign of flattening was identified among the consolidated granules, which is to say that the droplets were of a reduced size of less than 1 μm. It is worth noting that most of the results reported a denser lamellae structure formed by molten splats [9, 10] during depositing sintered nanostructured agglomerates under low pressure conditions, therefore, it can be concluded that the realization of finer droplets is a premise to develop the uniform equiaxed granule microstructure. It is known that the Knudsen effect is significant at low pressure, and the transport of momentum, heat, and mass in the vicinity of finer particles is governed no longer by a continuum transport mechanisms, but a gas-particle collision, as suggested by Joshi et. al [11]. In consequence, the behavior of disintegrated fragments might be quite different from those at ambient pressure, which provides the possibilities of a novel coating microstructure by controlling the processing conditions. In this study, the experiments were fulfilled with the conventional low pressure plasma spray (LPPS) method at 5000 Pa with a gas mixture of Ar-He-H_2. And Lower arc currents of 350 A, 500 A and 650 A were selected to observe the coating microstructure. The microstructures, especially the nano zone characters were compared with those obtained with air plasma spray method. At last the mechanical behaviors (microhardness, elastic modulus, and fracture toughness) corresponding to different coating microstructures were evaluated and compared.

Experimental procedure

Feedstock powder preparation

The 7-8 wt% yttria-stabilized zirconia (YSZ) nano powder (~ 60 nm) was obtained through a commercial source (Guangzhou Hongwu Nano-Science and Technology Co., Ltd., China). The morphology of the original nano powder is shown in Fig. 1a. The starting nano powder was dispersed into polyvinyl alcohol (PVA, Aladdin®) aqueous solution, following the weight ratio (powder to PVA) of 11:1, and then fed into a 6000 Y experimental spray dryer (Shanghai, China) to produce hollow spherical agglomerates (10~100 μm) with apparent density of 0.95 g/cm^3, shown in Fig.1 b (calculated by measuring the weight of the free falling agglomerates within a fix volume container).

Fig.1 The morphology of (a) the original nano powder and (b) spray dried agglomerates.

Coating deposition

The stainless steel plates (8 mm × 50 mm × 160 mm) were used as substrates and the NiCrAlY bond coat was pre-deposited using APS method. The YSZ top coats were fabricated using air plasma spraying and low pressure plasma spraying respectively. The spray gun was made by Dalian Maritime University with internal powder feeding along radical direction. And the low pressure plasma system was also homemade. The cylindrical working chamber is 1.2 m in diameter and 4 m in height. The plasma gun (maximal current, 800 A) is mounted on a fixed support vertically, the motion of which is controlled by a 3-dimensional axis of transformation unit. The chamber pressure could be reduced to as low as 50 Pa by using a pumping system with a high power mechanical pump and a rotary piston pump. The coating thickness was controlled by the number of spraying passes. The specific spray parameters are listed in Table 1.

Table 1 Spray parameters

Parameters	APS	LPPS
Primary gas Ar (slpm)	40	20
Secondary gas H_2 (slpm)	5	5
Secondary gas He (slpm)		30
Carrier gas Ar(slpm)	5	3
Arc current (A)	350	350/500/650
Arc voltage (V)	60	50
Stand-off (mm)	100	400
Powder feed rate (g/min)	10	10

Microstructure characterization

The coating microstructure was analyzed using field emission scanning electron microscopy (FESEM, Carl Zeiss SUPRA™ 55). The cross-section samples were sectioned and mounted with epoxy resin, wet-grounded with silicon carbide paper, then polished. Prior to observation, the samples were thermally etched at 1200 °C for 5 min in a box furnace. The phase structure of the original powder and the as-sprayed coatings were examined using X-ray diffraction (XRD, D/max-IIIA, Rigaku, Japan) with Cu K_α (λ=1.54056 Å). Slow scan of 71 - 76 ° was carried out by a step width of 0.01 and a time per step equal to 3 s.

Mechanical properties

Vickers` microhardness measurements were performed under a load of 500 gf for 15 s on the polished cross-sections of as-sprayed coatings. The brittle ceramic materials exhibit a decrease in microhardness with increasing indentation load, as it is called the indentation size effect (ISE), and a constant hardness is typically reached at loads between 5 and 100 N depending on materials [12-14]. In current experiment, a load of 500 gf was chosen. Besides, a total of 15 measurements were carried out to evaluate the average microhardness of coating samples. Another set of 25 points was taken under a load of 200gf with a Knoop indenter and the results were further statistically analyzed with Weibull plots. The variations in materials properties could be told through statistically analysis of microhardness obtained with a lower load, as mentioned by Lin et al. [15]. Lima [16] and Zhou [17] also used this method to analyze the bimodal character of nanostructured YSZ coatings fabricated with APS.

The Knoop indentation method was used to determine the hardness-to-modulus ratio of YSZ coatings by measuring the elastic recovery of the indentation diagonals as [18]:

$$\frac{b}{a}-\frac{b`}{a`}=\alpha\frac{H}{E}$$

Where H is the microhardness (Pa), $b`/a`$ is the ratio of the short to long indention diagonals after elastic recovery, b/a is 1/7.11 for Knoop indenter geometry, and α is a constant having a value of 0.45. In this work, the Knoop indentation tests were performed under 300 gf load for 15s.

Results

The reconstituted agglomerates were successfully deposited at 5000 Pa and the cross-sections of the coating samples after thermal etching are displayed in Fig. 2a-c. it could be seen that when applying a lower current of 350 A, the coating presented a typical bimodal structure, however, it was quite different from that engineered at atmospheric pressure as displayed in Fig. 2d. The coating was composed of alternating layers of loose nano granules to compacted coarser equiaxed particles which is quite different from the well-known bimodal structured nano coating, composed by unmelted nano granules and columnar grains. Increasing arc current caused a structure transformation at the layer interfaces. The size differences between the loose nano granules and the grains at the splat interface was reduced the current to 500 A, and uniform fine grain structure was found as the current increased to 650 A. The coating was composed of fine equiaxed granules who have similar size along three directions. However, a larger portion of the granules were beyond 100 nm and the largest amount of nano granules resembling to the original powder was found in the coatings fabricated with 350 A (Fig. 2 a).

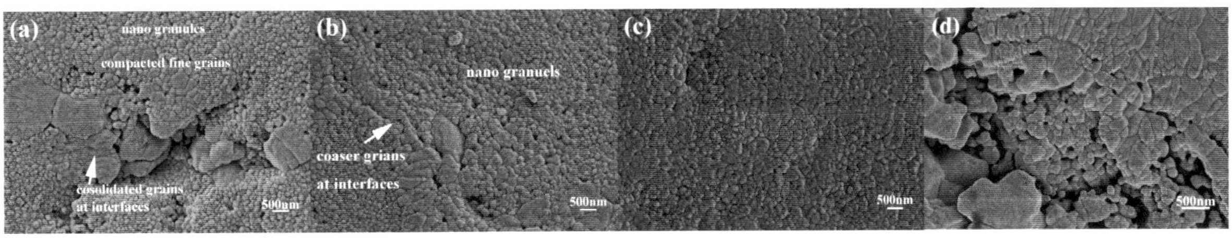

Fig. 2 The cross-sections of coating engineered by LPPS method with the current of (a) 350A, (b) 500A and (c) 650A as well as the cross-section of the coating fabricated with APS method with the current of 450 A.

X-ray diffraction patterns of YSZ as-sprayed coatings fabricated by APS (Fig. 3a) and LPPS (Fig. 3b-d) are shown in Fig. 3. No monoclinic phase zirconia is noticed in the APS coating, while all the coatings engineered by LPPS contain monoclinic phase zirconia, even with the current of 650 A. And a slow scan at high angle region (71 - 76°) is displayed in Fig. 3e. With increasing the arc current, the patterns show slight fluctuations of peak intensity or peak shift. It has been acknowledged that the nano granules experienced a rapid grain growth during deposition [19]. In present study, the heating might be impeded due to the low apparent density of 0.95 g/cm^3 which promoted the retention of the original nano granules in larger amount, even at higher arc current.

The structure feature was further investigated through statistical analysis of Knoop microhardness with 200 gf load, and the Weibull plots are displayed in Fig. 4. It is known that in the nanostructured coatings deposited by air plasma sprayed method contain regions of molten and non-molten particles which also may be considered as two phases, correspondingly, the Weibull modulus present bimodal distributions. The LPPS coatings engineered with the current of 350 A and 500 A also showed bimodal distributions of Weibull modulus, as shown in Fig. 4a, the same distribution as that of the APS coating (Fig. 4b). While further increased the current to 650 A, the Weibull modulus regressed to single line distribution. Though the coatings are composed by fine equiaxed particles, most of them were larger than 100 nm. However, it needs to point out that despite a single line distribution of Weibull modulus, the coating microstructure is quite different from the conventional lamellae coating.

The Vickers` indentations were applied under a load of 500 gf, and the results are shown in Fig. 5. The microhardness of LPPS coatings were 305 ± 41 HV, 513 ± 107 and 649 ± 130 HV respectively, arranged in ascending order of arc currents.

Micro-hardness is usually considered to be an index of penetration, deformation, and a reflection of the splat-to-splat cohesion of a thermal sprayed coating. When the input power were close, the LPPS coating (500 A) showed lower microhardness than the APS coating, and it might be attributed to the novel coating microstructure, especially the full molten area. The values of elastic modulus obtained through indentation tests are shown in Fig. 5b. With increasing the arc current, the elastic modulus of the LPPS coatings are 23±15 GPa, 31±19 GPa, 34±18 GPa respectively, remarkably lower than the APS coating(125 ± 55 GPa). The overall value of the elastic modulus of the LPPS coatings were also lower than the related reports on nanostructured coatings, and even that of the coatings with similar structure [20]. It is well known that relatively high thermo-mechanical stability of the plasma-sprayed zirconia top coats is to a larger extent attributed to the low macro-scopic stiffness, providing high strain tolerance [21].

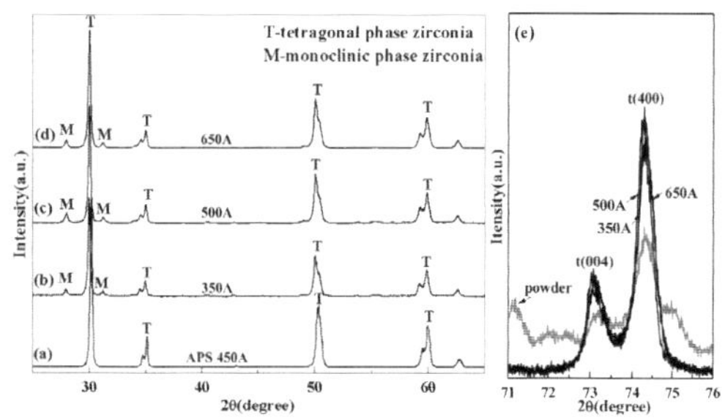

Fig.3 (a) Shows the XRD patterns of APS nanostructured YSZ coating, (b-d) shows the patterns of LPPS coatings and (e) the patterns at selected angle range of 71- 76 °.

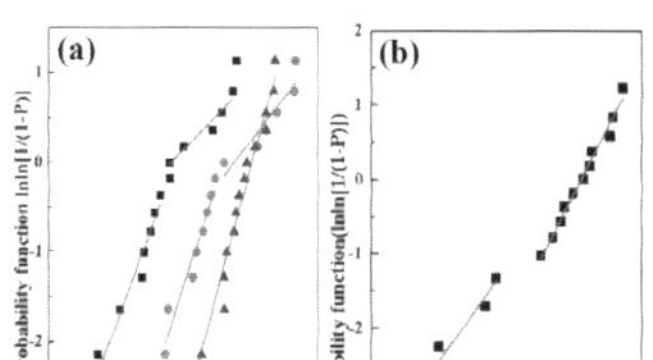

Fig. 4 Weibull plots of Koop microhardness under a lower load of 200 gf for (a) LPPS coatings and (b) APS coating engineered with 450 A.

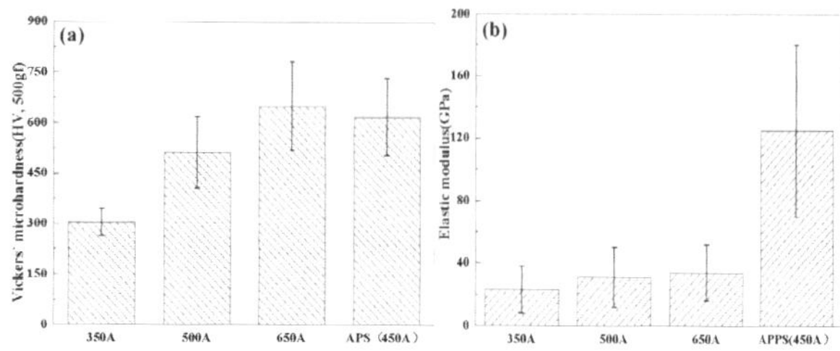

Fig. 5 (a) Vickers` microhardness and (b) elastic modulus of both LPPS and APS coatings

Discussion

The coating microstructure, to a large extent is dependent on the heat trajectories and velocity of individual droplets as well as the heat conduction between the adjacent layers when the substrate conditions were kept constant. At low pressure, the plasma jet exhibit unconventional features related to weak collisionality and propagated geometry which would significantly affect the particles heating and accelerating. The jet temperature mainly depends on gas composition, input power, chamber pressure and gun performance, etc. An appropriate combination of chamber pressure and input power level during LPPS was reported to be capable of evaporating the feedstock, forming a composite structure of solid splats and condensates [22]. Previous OES (optical emission spectroscopy) measurements on a loaded flow of 20Ar-20He-10H_2 at 100 Pa revealed that pretty low evaporation of YSZ occurred when the arc current lower than 700 A [23]. And according to the enthalpy probe measurements with varied chamber pressures below 1000 Pa by Dorier et al. [24], the jet length and the temperature along axis decreased as the chamber pressure increased to 1000 Pa. Consequently, it is reasonable to consider that the vaporization of YSZ can be neglected for the current gas composition, chamber pressure and input power level.

The specific constitution of gas mixture determine the plasma jet properties and affect the momentum, heat, and mass transfers with the particles. The addition of hydrogen or helium could enhance particle melting, and the former is prominent due to high thermal conductivity [25]. Previous results [26] involving a gas mixture of Ar-H_2 showed that with a lower input power, a typical bimodal microstructure was developed at low pressure of 1000-5000 Pa, and further raising input power or reducing chamber pressure would lead to full melting or even evaporation of particles, forming the conventional lamella structure with micronsized grains or a combination with fine nano condensates. Helium possess higher viscosity due to a higher ionization potential and Janisson et al. [27] pointed out that at a constant total gas flow rate, a higher volume percentage of helium resulted in lower jet temperature and slow decrease in viscosity as the temperature exceeding 11000 K, in consequence enhanced the plasma jet stability. Simultaneously adding them as second gas can take advantages of them both. What is more, a gas mixture with helium can yield a dense nano-granule structure solidified from fine liquid droplets at very low pressure (less than 100 Pa) [28] which implied that a ternary gas mixture with higher portion of helium is beneficial for accelerating the finer body.

The agglomerates used in the experiments are actually hollow spherical which made them behaving differently when being subjected into the plasma jet in contrast with the sintered sprayed dried particles and a modeling work of Gulyaev et al. [29] is helpful in understanding the specific process. When the porous agglomerates entered the plasma jet, the solid and the entrapped gas would be heated. The gas expanded to the surroundings at low pressure, and then recaptured by the molten shell to form a single cavity, as displayed in Fig. 6. As soon as the completed melting achieved, a pressure balance was build up between the inner gas and the additional surface pressure, leading to a notable reduction in droplets size. The dynamic process was also revealed in the work of Fogarassy [30] et al. It needs to draw attention that the model is limited by a porosity level of less than 60 % which might fail in describing the behavior of highly porous agglomerates. It is also known that the pressure within the gun channel is higher than the chamber pressure and, the agglomerates are mainly heated and accelerated within the nozzle. The high porosity of the feedstock powder would cause a larger populations of agglomerates to crowd into the gun channel in a unit time which could promote the collisions against the channel wall and between particles, and accelerate the integration process, as drafted in Fig. 6 b. However, it is hard to know the specific motion and heating conditions of particles within the gun channel. According to the pioneer works, the porous agglomerates (not hollow shell) possess higher heat resistance, therefore it can be perceived that the unsintered agglomerates with low apparent density might result in its sluggish grain growth during deposition with higher arc current. A low chamber pressure would cause a rarified gas environment, the relationship between the mean free path of molecules (λ) and the chamber pressure (P) could be expressed as:

$$\lambda = \frac{RT}{\sqrt{2}\pi d^2 N_A P}$$

where R is a constant of 8.3145 J/mol K; N_A is the Avogadro`s number; T is temperature, K; d is molecular diameter, Å. Reduction in chamber pressure and the size of agglomerates would cause higher Knudsen number (defined as λ/L, L is the characteristic length of the body in question), meaning the heat and momentum transportation to small particles would be hindered due to the strong Knudsen effect. The lower content of full molten part in the as-sprayed coating even at higher current also evident a vast disintegration process in turn.

When the droplets impinged on the substrate, the substrate temperature played an important role in developing the coating microstructure and it promoted the formation of fine equiaxed grain structure as the current went to 650 A. Although a higher substrate temperature was reported to affect the grain growth behavior during solidification [31], it was also found to be a necessity in depositing finer granules and developing equiaxed grains [20, 32]. The relationship between coating microstructure and substrate temperature can be understood through an interlayer mixing model suitable for low pressure plasma spray [33]. In brief, if each depositing layer had sufficient enthalpy to reheat the underlying layer to the solidus temperature, even allowing the presence of some liquid between the layers, in consequence, an equiaxed crystalline microstructure can be obtained. The transitory presence of some liquid phase played a role in mixing the interlayer boundaries and were proposed to depend on the interlayer time, the average mass deposited per unit area and the substrate temperature. Low temperatures promoted an inhomogeneous banded microstructure with partially solidified and unmelted particles. But the molten parts in the coatings engineered with a lower current of 350 A showed dense and fine equiaxed grains, as displayed in Fig. 2a. But their interfaces with the neighboring unmelted nano granules were clearly identified, therefore, the crystallization of fine equiaxed grains was more likely attributed to large undercooling [34] and gas contamination [35]. The retarded grain growth during deposition might also be a result of the Knudsen effect which lengthened the time interval between the subsequent layers, and the droplets might endure a large undercooling in a longer traveling distance of 350 mm.

Conclusions

In summary, a novel coating microstructure can be obtained using easy disintegrated agglomerates with low pressure plasma spray method. However, the random breakup of agglomerates and significant Knudsen effect distinguished the as-sprayed nanostructured coatings from the nanostructured coating engineered at atmospheric pressure. Increasing the operating current could improve the melting degree of agglomerates and their velocity, leading to a compacted coating microstructure. The Weibull modulus of microhardness also displayed bimodal distribution when the current was lower than 650 A. In contrast with the air plasma sprayed nano coating, the as-sprayed LPPS nanostructured coating contained monoclinic phase zirconia, even for the coating with higher melting degree. And when the input power was closer, the microhardness of LPPS coating was slight lower than that of the APS coating, but the elasticmoduls was remarkable lower. Though the coating deposited with 650 A had single Weibull modulus, the coating microstructure was quite different from the conventional microstructure with full molten splats. The coating microstructure was controlled by the LPPS processing technology and the feedstock properties.

Reference

[1] R. S. Lima and B. R. Marple, Thermal spray coatings engineered from nanostructured ceramic agglomerated powders for structural, thermal barrier and biomedical applications: A review, *J. Therm. Spray Technol.*, 2007, **16**(1), p40-63.

[2] H. Jamalin, R. Mozafarinia, R. S. Razavi, R. A. Pidani, Comparison of thermal shock resistances of plasma-sprayed nanostructured and conventional yttria stabilized zirconia thermal barrier coatings, *Ceram. Int.*, 2012, **38**(8), p6705-6712.

[3] R. S. Lima, B. R. Marple, Nanostructured YSZ thermal barrier coatings engineered to counteract sintering effects, *Mater. Sci. Eng. A.*, 2008, **485**(1), p182-193.

[4] P. J. He, H. Sun, Y. F. Gui, F. Lapostolle, H. Liao, C. Coddet, Microstructure and properties of nanostructured YSZ coating prepared by suspension plasma spraying at low pressure, *Surf. Coat. Technol.*, 2015, **261**, p318-326.

[5] M. Aliofkhazraei, Nanocoatings-size effect in nanostructured films, Springer, New York, 2011, p132-133.

[6] A. Keyvani, M. Saremi, M. H. Sohi, Z. Valefi, A comparison on thermomechanical properties of plasma-sprayed conventional and nanostructured YSZ TBC coatings in thermal cycling, *J. Alloys Compd.*, 2012, **541**(22), p488-494.

[7] O. Racek, C. C. Berndt, D. N. Guru, J. Heberlein, Nanostructured and conventional YSZ coatings deposited using APS and TTPR techniques, *Surf. Coat. Technol.*, 2006, **201**(1), p338-346.

[8] O. Racek, C. C. Berndt, Mechanical property variations within thermal barrier coatings, *Surf. Coat. Technol.*, 2007, **202**(2), p362-369.

[9] M. F. Smith, A. C. Hall, J. D. Fleetwood, P. Meyer, Very low pressure plasma spray-A review of an emerging technology in the thermal spray community, *Coat.*, 2011, **1**(2), p117-132.

[10] M. Žadvydas, S. Tamulevicus, T. Grinys, Relation between the YSZ powder properties and vacuum plasma spray deposited layers, *Mater. Sci.*, 2004, **10**(4), p349-352.

[11] S. V. Joshi, J. Y. Park, P. R. Taylor, L. S. Richardson, Knudsen effect on plasma- particle mass transfer. I. Formulation and application to self-diffusion, *Plasma Chem. Plasma Proc.*, 1986, **6**(3), p281-298.

[12] P. J. Patel and J. J. Swab. Indentation Size Effect (ISE) of Transparent AlON and $MgAl_2O_4$, *Army Res. Lab.* (2006) ARL-TR-3852.

[13] H. Li, R. C. Bradt, The Microhardness Indentation size-load effect (ISE) in hard ceramic materials, *J. Hard Mat.*, 1992, **3**(3-4), p403-419.

[14] H. Li, R. C. Bradt, The effect of indentation-induced cracking on the apparent microhardness. *J. Mat. Sci.*, 1996, **31**(4), p1065-1070.

[15] C. K. Lin, C. C. Berndt, Statistical analysis of microhardness variations in thermal spray coatings, *J. Mater. Sci.*, 1995, **30**(1), p111-117.

[16] R. S. Lima, A. Kucuk, C. C. Berndt, Bimodal distribution of mechanical properties on plasma sprayed nanostructured partially stabilized zirconia, *Mater. Sci. Eng. A*, 2002, **327** (2), p224-232.

[17] H. Zhou, J. Wang, B. He, F. Li, J. Ma, B. Sun, Mechanical properties of nanostructured zirconia coatings fabricated by air plasma spraying, *Key Eng. Mater.*, 2008, **368-372**, p1280-1283.

[18] D. B. Marshall, T. Noma, A. G. Evans, A simple method for determining elastic-modulus-to-hardness ratios using Knoop indentation measurements, *J. Am. Ceram. Soc.*, 1982, **65**(10), C175-176.

[19] T. Mostafa, S. Morteza, S. Mehdi, The effect of heat treatment and thermal spray processes on the grain growth of nanostructured composite CoNiCrAlY/YSZ powders, *J. Alloys Compd.*, 2015, **646**, p372-379.

[20] L. D. Xie, X. Q. Ma, E. H. Jordan, N. P. Padture, D. T. Xiao, M. Gella, Identification of coating deposition mechanisms in the solution-precursor plasma-spray process using model spray experiments, *Mater. Sci. Eng. A*, 2003, **362**(1), p204-212.

[21] J. A. Thompson and T. W. Clyne. The effect of heat treatment of the stiffness of zirconia top coatings in plasma-prayed TBCs, *Acta Mater.*, 2001, **49**(9), p1565-1575.

[22] A. Hospach, G. Mauer, R. Vaßen, D. Stover, Columnar-structured thermal barrier coatings (TBCs) by thin film low pressure plasma spraying (LPPS-TF), *J. Therm. Spray. Technol.*, 2011, **20**(1-2), p116-120.

[23] Y. Gao, D. Yang, C Sun, Z. Chen, Deposition of YSZ coatings in a chamber at pressures below 100 Pa using low-power plasma spraying with an internal injection powder feeding, *J. Therm. Spray. Technol.*, 2013, **22**(7), p1253-1258.

[24] J. L. Dorier, M. Gindrat, C. Hollenstein, Plasma jet properties in a new spraying process at low pressure for large area thin film deposition, Proceedings of international thermal spray conference, Singapore, 2001, p28-30.

[25] B. Pateyron, M. F. Elchinger, G. Delluc, P. Fauchais, Thermodynamic and transport properties of Ar-H and Ar-He plasma gases used for spraying at atmospheric pressure. I: properties of the mixtures, *Plasma Chem. Plasma Proc.*, 1992, **12**, p421-448.

[26] C. Q. Sun, Y. Gao, D. M. Yang, Plasma torch designed for low pressure plasma spraying and formation of yttria stabilized zirconia coatings, ITSC 2014, Barcelona, DVS302, p979-984.

[27] S. Janisson, A. Vardelle, J. F. Coudert, E. Meillot, B. Pateyron, P. Fauchais, Plasma spraying using Ar-He-H_2 gas mixtures, *J. Therm. Spray Technol.*, 1999, **8**(4), p545-552.

[28] Y. Gao, Y. Zhao, C. Sun, J. Gao, D. M. Yang, Fully nano-equiaxed-structured thermal barrier coating deposited using very-low-pressure plasma spray, ITSC 2015, California, p427-33.

[29] I. P. Gulyaev, O. P. Solonenko. Modelling of the behavior of hollow ZrO_2 particles in plasma jet with regard to their thermal expansion, *Thermophys. Aeromech.*, 2013, **20**(6), p769-782.

[30] P. Fogarassya, D. Gerdayb, A. Lodini, Agglomerated nanostructured particles disintegration during the plasma thermal spraying process, *Mech. Res. Commun.*, 2005, **32**(2), p221-239.

[31] E. J. Yang, X. T. Luo, G. J. Yang, C. X. Li, C. J. Li. M.

Takahashi, Epitaxial grain growth during 8YSZ splat formation on polycrystalline YSZ substrates by plasma spraying, *Surf. Coat. Technol.*, 2015, **274**, p37-43.
[32] K. H. Baika, P. S. Grantb, B. Cantor, The equiaxed-banded microstructural transition during low pressure plasma spraying, *Acta Mater.*, 2004, **52**(1), p199-208.
[33] B. Cantor, K. H. Baik, P. S. Grant, Development of microstructure in spray formed alloys, *Prog. Mater. Sci.*, 1997, **42**(1-4), p373-392.
[34] J. H. Perepezko. Nucleation controlled phase selection during solidification, *Mater. Sci. Eng. A*, 2005, **413-4**14, p389-397.
[35] K. Lu. Nanocrystalline metals crystallized from amorphous solids: nanocrystallization, structure, and properties, *Mater. Sci. Eng.*, 1996, **16**(4), p161-221.

Comparative Investigation on the Mechanical Properties of Thermal Barrier Coatings Made by Atmospheric Plasma Spaying and Suspension Plasma Spraying

M.Gizynski[1, 2], X.Chen[1], M.Watanabe[1], H.Araki[1], S. Kuroda[1], Z.Pakiela[2]

1) National Institute for Materials Science, 1-2-1 Sengen, Tsukuba, 305-0047 Ibaraki, Japan

2) Faculty of Materials Science and Engineering, Warsaw University of Technology, Wołoska 141, 02-507 Warsaw, Poland

Abstract

In the present work fundamental mechanical properties of both SPS and APS coatings were investigated. Four point bending test of coated samples revealed lower values of Young's modulus for as-sprayed SPS than APS coatings. Measurements of interfacial toughness between the YSZ top coating and the CoNiCrAlY-based bond coating showed higher values for SPS coatings. Low stiffness of SPS coatings limits elastic energy stored in the ceramic coating. This coupled with good interfacial toughness might make SPS coatings less prone to delamination under various thermomechanical loading conditions such as these encountered during thermal cycling.

1. Introduction

Introduction of Thermal Barrier Coatings (TBC) in the 70s of the 20th century allowed increasing the operating temperature of gas turbines and jet engines by 100-150°C [1]. Strong increasing trend in turbine efficiency and inlet gas temperature requires new solutions in the TBC systems. One of such solutions can be application of nanosized yttria-stabilized zirconia (YSZ) powders. Application of fine powders (in a range of single µm) has been reported to improve certain properties of TBCs [2]. Nevertheless, problems with flowability and injection of such powder in the plasma jet require use of suspension as the carrying medium. Suspension Plasma Spray (SPS) is a spraying technique in which a suspension containing nanosized powder is injected into a plasma stream, where it atomizes to fine droplets, from which the liquid phase evaporates. Heated by plasma the solid particles melt and impact the substrate to build up a coating. TBCs made by SPS are expected to have several merits as compared with conventional coatings, e.g., lower thermal conductivity, high strain tolerance and longer lifetime in thermal shock tests [3,4]. Thus SPS coatings seem to merge the benefits of both coatings deposited by Electron Beam Physical Vapor Deposition (EB-PVD) - low thermal conductivity and compliance, and that of conventional Atmospheric Plasma Spray (APS) i.e., lower thermal conductivity with undemanding fabrication process and relatively low cost.

Although numerous studies were made on SPS process and coating formation [5,6], very few works have reported mechanical properties of SPS TBCs. In this manuscript results of investigation on basic mechanical properties such as Young's modulus and interfacial toughness of SPS coatings and conventional coatings made by APS are reported.

2. Materials and Experiments

The substrate material used for coating deposition was Ni-based superalloy a Inconel-718. Prior to spraying the substrate had been cut into beams with dimensions of 100 mmx10 mmx2.65 mm, roughened by grit blasting and cleaned in acetone. Interfacial toughness was measured on specimens with additional ca. 250 µm thick CoNiCrAlY bond coating. The bond coating was sprayed by modified 2 stage High Velocity Oxygen Fuel (HVOF) technique also called as Warm Spray [7,8]. Details of which are presented in Table 1.

6 various kinds of ceramic coatings were sprayed - 3 by APS and 3 by SPS, the parameters of the spraying processes are listed in **Table 2**. Two YSZ powders were used as the feedstock material in APS process: 204NS (Metco, USA) and K90 (Showa Denko, Japan). 204NS is hollow-oxide sphere powder with the particle size distribution of -125 µm +11 µm, while K90 is dense fussed and crushed powder with the particle size distribution of -45 µm+10 µm. In the SPS processing ethanol based suspension of submicron

(D_{50}≈400 nm) YSZ powder was employed. The solid phase concentration in the suspension was 25 wt. %. Suspension was delivered to a plasma jet through a Oerlikon Metco 5MPRE-SF feeding system.

Table 1. List of spraying parameters for bond coating

Method	Warm Spray
Powder	AMDRY 9954
Feed rate *g/min*	70
Spraying distance *mm*	200
Gun velocity *mm/s*	200
Fuel/O_2 *slm*	0.391/805
Nitrogen flow *slm*	500

Table 2. List of spraying parameters for YSZ coating

Coating	A1	A2	A3	S1	S2	S3
Method	APS	APS	APS	SPS	SPS	SPS
Feedstock	204NS	K90	K90	Susp.	Susp.	Susp.
Spraying distance *mm*	100	100	80	50	70	70
Plasma forming gases flow *slm*	Ar/He 80/20	Ar/He 80/20	Ar/He 50/20	Ar/He 44/6	Ar/He 44/6	Ar/H_2 40/10
Current *A*	800	800	1000	480	480	480
Gun Power *kW*	24	24	34	45.2	45.2	58.3

In order to reveal the microstructure of the sprayed coatings some specimens were cut using a slow rotating saw, molded in epoxy resin and subjected to a standard polishing procedure, where the final stage was polishing with 0.2 µm alumina suspension. Observations was done by a scanning electron microscope (SEM) JEOL JSM-6010LA in the secondary electron mode.

Mechanical tests carried out on the coatings include Young's modulus measurements, hardness and interfacial toughness evaluation. Stiffness was investigated by two methods- indentation-based test and 4-point bending test. Indentation method uses the Oliver and Pharr model [9] where Young's modulus is proportional to the slope of the initial part of an unloading curve. Instrumented ultra-microharndess tester Shimadzu W201S was used in this study. The tester is equipped with a 115° triangular pyramidal indenter. APS coatings were tested at a load of 1,961 mN, while coatings fabricated by SPS were subjected to a load of 100 and 1,961 mN. At least 16 indents for 100 mN and 10 for 1,961 mN were done in order to obtain representative information about the coatings properties. Keeping in mind that the indentation method gives information about a limited volume of the tested material, 4 point bending test was carried out. The presented data were obtained in the arrangement where the ceramic coating was in compression. The tests were conducted on a Shimadzu EZ Graph universal testing machine equipped with four point bending arrangement with supporting and loading span of 40 and 20 mm, respectively. Displacement in the midspan was measured with a displacement sensor. Known dimensions of the substrate and coating and the elastic properties of the substrate allow using equations of the elastic beams theory to recalculate forces acting on a specimen into stresses. The Young's modulus can be calculated using two equations expressing "bending modulus" (EI) of a composite beam [10],

$$(EI)^* = \frac{2D^3}{3}\frac{\Delta P}{\Delta \delta} \qquad (1)$$

$$(EI)^* = \frac{wE_{sub}}{3}[(t_{sub} - t_{NA})^3 - t_{NA}^3] + \\ + \frac{wE_C}{3}[(t_{sub} + t_C - t_{NA})^3 - (t_{sub} - t_{NA})^3] \qquad (2)$$

where: w is the width of a specimen, E- Young's modulus, I- area moment of inertia, D-distance between supporting and loading spans, $\Delta P/\Delta\delta$- the slope of the elastic range of load-displacement curve, t- thickness; subscripts sub, C and NA refer to the substrate, coating and the neutral axis, respectively. The position of the neutral axis in the composite beam is given by:

$$t_{NA} = \frac{E_{sub}t_{sub}^2 + E_C t_C^2 + 2E_C t_{sub} t_C}{2(E_{sub}t_{sub} + E_C t_C)} \qquad (3)$$

A hardness tester Akashi MVK-E with a Vickers indenter was employed for the hardness measurements with a load of 3 N. In order to provide statistically relevant data, at least 10 indentations were performed.

Interfacial toughness was determined using an indentation method- several indents with Vickers indenter under various load (P) ranging from 1 N to 300 N. One of the imprint's diagonal had been aligned

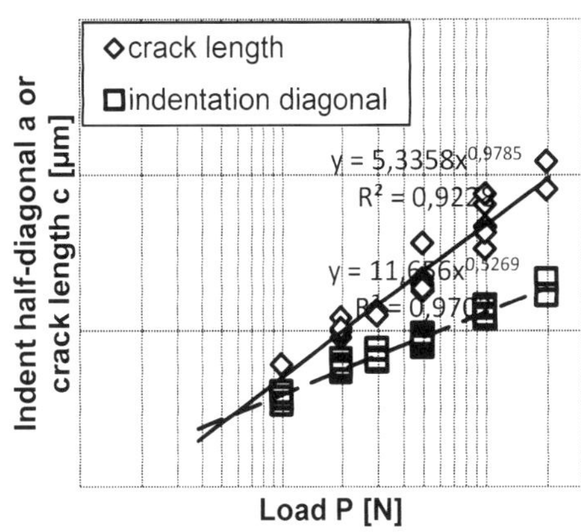

Figure 1. Method of evaluation of critical load and crack length in indentation-based apparent toughness

DVS 324

technique. Presented data were obtained for coating S2.

with the bond coating/ceramic coating interface, then the diagonal of indent (*2a*) and crack length (*2c*) were measured using SEM. Results of the assessment were plotted on a bilogarithmical graph log P vs. log a or log c, then the critical load and crack length were estimated as the intersection of "hardness line" and "crack line",**Fig. 1**. The critical values of the crack length (c_c) and load (P_c) permit to calculate the interfacial toughness [11]:

$$K_{af} = \alpha \frac{E^{\frac{1}{2}} P_c}{H^{\frac{1}{2}} c_c^{\frac{3}{2}}} \qquad (4)$$

where α is a constant equals 0.016, H stands for the apparent hardness of the bond coat and the ceramic coating. More detailed description of the method was reported by Sniezewski et al. [12]. All the specimens subjected to the indentation based tests were prepared as specimens for microstructure evaluation and indents were located on the cross section.

3. Results and Discussion

The microstructure of fabricated coatings is showed in **Fig. 2**. Coating A1 inherited porous microstructure from the HOSP- type powder. **Fig. 2a** reveals a few micrometer sized porosity, myriads of intersplat cracks and voids. Utilization of fused powder resulted in the denser microstructure of coating A2 with very limited number of pores. Shorter spraying distance and higher deposition temperature led to formation of vertical cracks in coating A3,**Fig. 2c**, which have been often reported to have beneficial effect on strain accommodation and lifetime under thermal loads [13].

Figure 2d shows transitional microstructure of S1 coating with features of both dense vertically cracked (DVC) and columnar microstructures. Numerous fine pores are located in coating's segments, that is known to reduce thermal conductivity of the coating. In mid-thickness of S1 coating long horizontal discontinuity is visible; this defect was formed when spraying process had to be paused due to overheating problems. Longer spraying distance resulted in columnar microstructure of coatings S2 and S3. Large voids and plenty of unmelted/resolified particles were found inside columns.

Spraying conditions and microstructure features corresponded with hardness measurements, **Fig. 3**. Coatings A1 and A2 despite different levels of porosity manifest similar hardness around 6.5 GPa. Smaller spraying distance and higher processing temperature of coating A3 resulted in vertical cracks and higher hardness. Short spraying distance and high power applied during deposition of coating S1 resulted in dense fine microstructure with submicrometric features (such as splat thickness, crystal size), good intersplat bonding and low porosity. To these can be attributed in the highest measured harness- 8.7 GPa. Strong intersplat bonding and high hardness restrain permanent deformation during indentation and the net work of indentation which is displayed by the smaller area covered by indentation curve as compared with curve for coating A2, **Fig. 4a**. Longer stand-off distance resulted in much more porous microstructure containing poor bonded resolidified particles, which, in turns, manifested as the lowered hardness of coating S2 and S3. The wide indentation curve of S2 corresponds to the large work of plastic deformation by indentation, which indicates that the coating has high ability to accommodate strain by permanent deformation.

Similar tendency was observed for the indentation-based Young's modulus values, especially for the indentation load of 1,961 mN, **Fig. 4b**. The reason of higher values obtained for the lower load of 100 mN may be the indentation size effect [14] or smaller

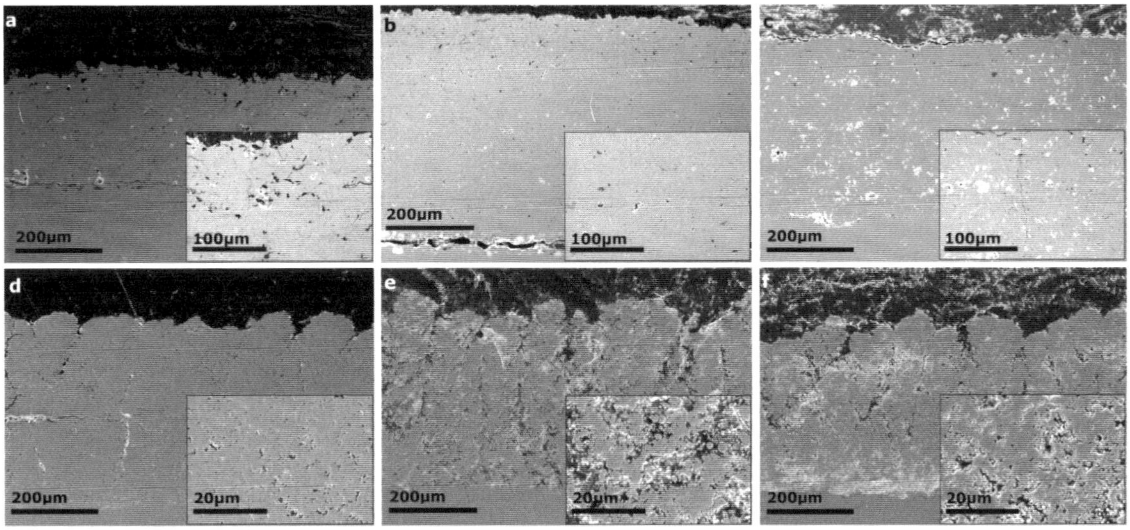

Figure 2. Cross sections of YSZ coatings fabricated by APS and SPS methods. (a) Porous microstructure of A1 coating formed from HOSP powder, (b) dense A2 coating made from fused and crushed K90 powder, (c) dense microstructure of A3 coating with a few vertical cracks. (d) Dense structure of S1 coating with vertical cracked/columnar transitional microstructure, higher magnification picture reveals porosity with size of around 1 µm and smaller; porous columnar microstructure of coatings (e) S2 and (f) S3.

volume of material under the indenter. In the latter case, the contribution of porosity and defects to the final value might be smaller than for the higher loads, so values of modulus would be higher. Since the indentation-based segmented parts rose. More pronounced drop of S1 coating's modulus is a consequence of a greater number of vertical cracks as compared with coating A3, **Fig. 2c** and **Fig. 2d**. Microstructure containing porosity, compliant columns and wide inter-columnar spaces lead to even less stiff coatings S2 and S3, which Young's moduli are about 65 and 35 GPa, respectively.

Interfacial toughness of TBCs is frequently reported as one of the parameters determining their reliability and lifetime under thermal loads [15]. All the coatings investigated in this work coatings manufactured by APS have similar interfacial toughness around 1.6 MPa m$^{1/2}$, **Fig. 5**, which is a typical values for as-sprayed APS coatings [16]. SPS coatings show significantly higher values 3.8 MPa m$^{1/2}$ for S1 and about 2.1 MPa m$^{1/2}$ for S2 and S3. The superior interfacial toughness of S1 coating could be an outcome of high processing temperature and high temperature of impacting ceramic droplets. While impacting onto a substrate, hot droplets are immediately cooled down, solidify and anchor to the bond coating's asperities. Longer spraying distance used for S2 and S3 coatings processing resulted in

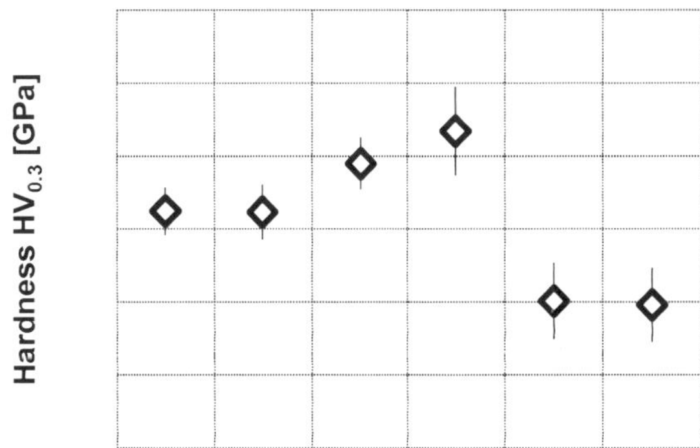

Figure 3. Hardness of studied coatings measured on cross sections by Vickers indenter.

lower temperature of the substrate and ceramic droplets, slower impacting velocity and, in turn, weaker bonding between the substrate and ceramic layer.

One of the most important modes of TBC failure is delamination of the ceramic coating. The strain energy (*U*) stored in coating is found as driving force of the delamination. The *U* can be given by:

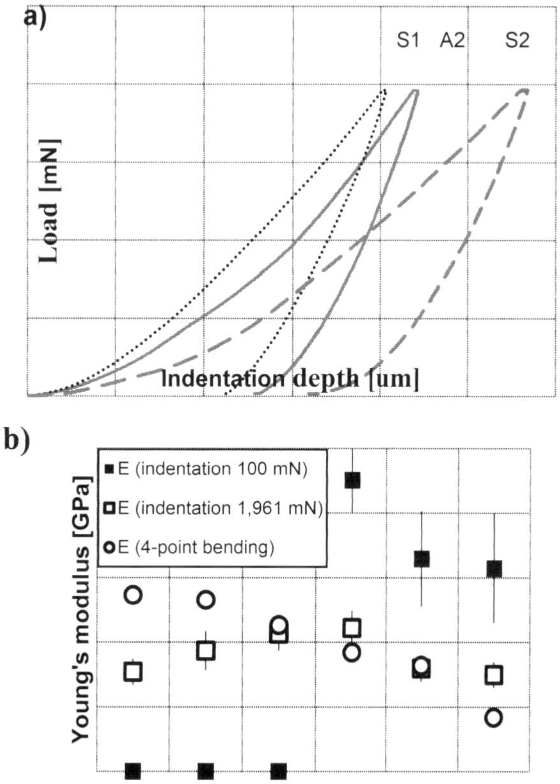

Figure 4. (a) Indentation curves of A2, S1 and S3 coatings recorded at load of 1,961 mN. (b)Young's modulus values measured by ultra-micro indentation and 4-point bending test

$$U = t_C \int_0^{\varepsilon} \sigma(\varepsilon) d\varepsilon \qquad (5)$$

Where: ε stands for the coating's strain and σ is stress acting on the coating. Assuming for simplicity that coating is a linear-elastic material and strain is the uniform through coating's thickness ($t_{sub} >> t_C$), one can notice that strain energy is proportional to:

$$U \approx \frac{1}{2} t_C E \varepsilon^2 \qquad (6)$$

On the other hand, energy required to generate unit area of crack (Γ) is given by:

$$\Gamma = \frac{K_{af}^2}{E} \qquad (7)$$

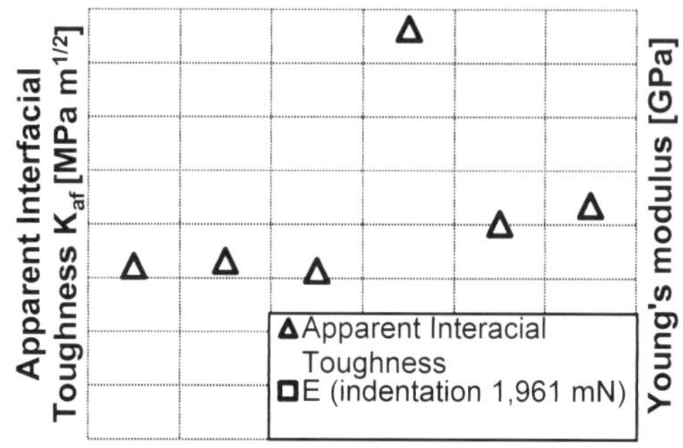

Figure 5. Apparent Interfacial Toughness of YSZ coatings with CoNiCrAlY bond coatings. Young's modulus values of respective coatings are also plotted

If delamination failure would be considered as competition between two mentioned factors (strain energy and "bonding" energy- Γ), it is clear that coatings showing low Young's modulus and high interfacial toughness can withstand larger strains without failure. Among the investigated coatings, ones manufactured by SPS have these properties and may be less prone to failure by delamination and make promise of improved performance in such applications as TBC for gas turbines and jet engines.

4. Conclusions

Coatings manufactured in SPS process have distinctive microstructural features and mechanical properties comparing with conventional APS coatings.

Investigated SPS coatings showed:

- Vertically cracked/columnar mixed microstructure (S1) or columnar microstructure (S2 and S3),
- Microstructure features and defects in size of single micrometers and smaller,
- Dense SPS coating (S1) showed large Young's modulus of slabs, though global stiffness of the coating was lowered by numerous vertical cracks and intercolumnar gaps,
- S1 coating due to high temperature deposition had superior interfacial toughness,
- Coatings S2 and S3 had porous columnar microstructure, Young's modulus and hardness measured in columns were low, global stiffness of the coatings was even lower because wide intercolumnar spaces,

- Despite interfacial toughness of S2 and S3 coatings is lower than for S1 coating, its value is significantly higher than values typical for APS coatings.

It appears that the mechanical properties of SPS-fabricated coatings could be designed in a wider range, which provides new opportunity to tailor the coating behavior to specific application. Particularly, enhanced interfacial toughness coupled with high compliance (**Figure 5**) shows a potential of improved SPS coatings' performance in such applications as TBC for gas turbines.

Literature

[1] T.E. Strangman, Thermal barrier coatings for turbine airfoils, Thin Solid Films 127 (1985) 93-105

[2] O.Tingaud, A.Bacciochini, G.Montavon, A.Denoirjean, O.Fauchais, Suspension DC plasma spraying if thick finely-structured ceramic coatings: Process manufacturing mechanisms, Surface & Coating Technology 203 (15) (2009) 2157-2161

[3] H.Kassner, R.Siegert, D.Hathiramani, R.Vassen, D.Stoever, Application of Suspension Plasma Spraying (SPS) for Manufacure of Ceramic Coatings, Journal of Thermal Spray Technology 17 (2008) 115-123

[4] E.Meillot, R.Vert, C.Caruyer, D.Damiani, M.Vardelle, Manufacturing nanostructured YSZ coatings by suspension plasma spraying (SPS): Effect of injection parameters, Journal of Applied Physics 44 (19) (2011) 194-208

[5] L.Pawlowski, Suspension and solution thermal spray coatings, Surface & Coating Technology 203 (2009) 2807-2829

[6] K.VanEvery, M.J.M.Krane, R.W.Trice, H.Wang, W.Porter, M.Besser, D.Sordelet, J.Ilavsky, J.Almer, Column Formation in Suspension Plasma-Sprayed Coating and Resultant Thermal Properties, Journal of Thermal Spray Technology 20 (2011) 817-828

[7] S.Kuroda, M.Watanabe, K.H.Kim, H.Katanoda, Current status and future prospects of warm spray technology, Journal of Thermal Spray Technology 20 (2011) 653-76

[8] S.Kuroda, J.Kawakita, M.Watanabe, H.Katanoda, Warm spraying- a novel coating process based on high-velocity impact of solid particles, Science and Technology of Advanced Materials 9 (2008) 1-17

[9] W.C.Oliver, G.M.Pharr, An improved technique for determining hardness and elastic modulus using load and displacement sensing indentation experiments, Journal of Materials Research 7 (1992) 1564-1583

[10] A.Kucuk, C.C.Berndt, U.Senturk, R.S.Lima, C.R.C.Lima, Influence of plasma spray parameters on mechanical properties of yttria stabilized zirconia coating. I: Four point bending test, Materials Science and Engineering A 284 (2000) 29-40

[11] G.R.Anstins, P.Chantikul, B.R.Lawn, D.B.Marshal, A Critical Evaluation of Indentation Techniques for Measuring Fracture Toughness:I, Direct Crack Measurements, Journal of American Ceramic Society 64 (1981) 533-538

[12] J.Sniezewski, V.Vidal, P.Lours, Y. Le Maoult, Thermal barrier coatings adherence and spallation: Interfacial indentation resistance and cyclic oxidation behaviour under thermal gradient, Surface & Coating Technology 204 (2009) 807-811

[13] H.Guo, H.Murakami, S.Kuroda, Thermal Cyclic Behavior of Plasma Sprayed Segmented Thermal Barrier Coaitngs, Materials Transactions 47 (2006), 306-309

[14] S.Verpek, S.Mukherjee, P.Karvankova, H.D.Mannling, J.L.He, K.Moto, J.Prochazka, A.S.Argon, Limits to the strength of super- and ultrahard nanocomposite coatings, Journal of Vacuum Science and Technology A 21 (2003) 532-544

[15] Y.Wang, H.Guo, S.Gong, Thermal shock resistance and mechanical properties of $La_2Ce_2O_7$ thermal barrier coatings with segmented structure, Ceramics International 35 (2009) 2639–2644

[16] Y.Zhao, A.Shinmi, X.Zhao, P.J.Withers, S. Van Boxel, N.Markocsan, P. Nylen, P.Xiao, Investigation of interfacial properties of atmospheric plasma sprayed thermal barrier coating with four-point bending and computed tomography technique, Surface & Coating 206 (2012) 4922-4929

Nanotopographical features of plasma sprayed coatings on bone implant surfaces

Guocheng Wang[1]*, Xiaoxia Song[1], Xiaobing Zhao[2]
[1]Research Center for Human Tissues & Organs Degeneration, Shenzhen Institute of Advanced Technology, Chinese Academy of Sciences, Shenzhen, Guangdong 518055, China
[2]School of Materials Science and Engineering, Changzhou University, 213164 Changzhou, China

Plasma spraying, a well-known surface modification technique has been widely used in many industry fields. In biomedical field, it has been commercialized to produce hydroxyapatite coated orthopaedic implants. In the last decades, a large amount of studies have been carried out to develop new coating products with plasma spraying technique, aiming to overcome the drawbacks of the currently used hydroxyapatite coated implants. Most of the studies have been conducted using coating materials with different chemical compositions, however, seldom of them focus on the exploration of how to make innovative application of this traditional surface technique for better serving the development of new biomedical coatings. The effects of nanotopography on responses of bone-related cells have been well-documented, making "how to generate and refine nanotopography of plasma sprayed biomedical coatings as well as how they influence their biological performance" especially important in the modern biomedical coating design. In our previous work, attempts were made to address these questions by several independent studies. In this proceeding, we are going to review these studies with a purpose to address the following three questions: 1. the effects of nanotopography on bone cell adhesion, 2. the effects of nanotopography on bone-like apatite formation in simulated body fluid. 3. How to refine the nanotopography of plasma sprayed coatings.

1 Introduction

Titanium and its alloy (Ti-6Al-4V) are promising metal materials and have been widely used in orthopedic applications, such as artificial hip joint, bone plates and dental implants. Although they have good mechanical properties, their suboptimal osseointegration resulting from lack of surface bioactivity of their surfaces are causing a high rate of revision surgery, limiting their further wide application. Various surface modification techniques have been applied on the Ti and its alloy implants to enhance their bioactivity through alternation of their chemistry and physical properties.
Surface chemistry and topography are two of the most important properties that determine the biological performance of the implants.[1,2] Chemically, the bioactivity of the implant surface can be enhanced by incorporation of "bone-seeker" trace element ions, such as Sr, Si, and Mg etc[3,4]. or embedding functional groups (i.e. hydroxyl -OH) on the surfaces[5]; physically, it can be enhanced by endowing the implant surface with topographic features at both micro and nanoscale[2,6]. The beneficial effects of those ions and topographic features on the bone formation have been widely demonstrated and reviewed in a variety of papers[3,6,7].
Plasma spraying technique is possibly the most widely used surface technique nowadays utilized in commercial bone implants. It has the potential to achieve both chemical and physical modification in one single process due to its high temperature characteristic and superhigh cooling rate. Plasma spraying yields temperatures up to 12 000 K in the core region of a plasma jet, with which various ceramic powders can be melted down and deposited onto the substrates. Therefore, it provides a wide selection range of chemical composition for the implant surface. A cooling rate up to 1×10^6 to 1×10^7 $K \cdot s^{-1}$ during plasma spraying process is able to cause the occurrence of ultrarapid quenching of the melts during solidification[8]. Under such thermal conditions, the growth of crystallites is strongly suppressed, leading to the formation of some nonequilibrium phases and unique fine structures. In our previous study, we succeeded in the production of nanostructured coatings using conventional powders based on this specific thermal history of plasma spraying[4,9–11]. We also demonstrated that the obtained hybrid nano/microstructures enhanced the cellular activity and promoted bone healing.
Given the importance of the surface nanotopographic features, we in our recent studies made an attempt to refine the nanotopography of plasma sprayed coatings[8]. In this paper, we are going to review our recent progress in refining the surface nanotopographic features of metallic implants by plasma spraying.

2 Effects of the nanostructure of plasma sprayed coatings on bone cell adhesion

Bioactive trace elements, such as strontium (Sr), silicon (Si) and zinc (Zn), have been proved to be essential for normal bone metabolism. Their beneficial effects on bone formation at fracture sites have been well-documented. Sr is one of the important essential trace elements in human body. Its deficiency can lead to poor bone mineralization. It has been shown to be a promising therapeutic agent to heal bone disease [12,13]. For example, strontium renalate has been used in clinics for long time to treat osteoporosis. Sr has been incorporated in engineered biomaterials to make a "bioactive" materials for bone tissue engineering applications, with encouraging results [14,15]. Si ions have been proved to be able to promote both osteogenesis and angiogenesis, making silicate-based bioglass and/or bioceramic very popular for bone regeneration and replacement applications[16,17]. Zn is another element that have been widely studied for bone frature healing. It can enhance bone cell proliferation and differentiation[18,19].

Therefore, it seems of great interest to combine all these beneficial elements in one surface and meanwhile produce nanotopographic features. In our previous work, we used plasma spraying technique to deposit Sr-$Ca_2ZnSi_2O_7$ (Sr-HT) coatings on Ti alloy implant surface and evaluated their bone formation ability by both *in vitro* cell culture experiments and *in vivo* animal studies[4].

The $Ca_2ZnSi_2O_7$ (HT) and Sr-HT coating have a hybrid micro/nanostructure with a surface roughness of about 8μm. Compared to the hydroxyapatite coating that used as a control group showing a smooth surface at nanoscale (**Fig. 1A**), nanosized grains with around 50nm in size were observed to embed in the microstructures, as shown in **Fig. 1B** and **C**. Since the feeding stock powders are more than 100μm in size, the nanograins are supposed to be formed due to the superfast cooling during the solidification of the coating. This nanostructure has been proved to have great influence on the cellular adhesion.

We investigated the adhesion of primary human osteoblasts (HOBs) on the coatings, compared to those cultured on the cell culture plates using the extracts of the coatings as cell culture medium. No significant difference in expression of cell-adhesion related integrin β1 gene in the cells cultured in the extracts of the different coatings (hydroxyapatite (HAp), HT and Sr-HT) (**Fig. 2, right**), however, significant upregulation of the integrin β1 expression was found for the cells cultured on the surface of the Sr-HT and HT coatings, compared to those cultured on the HAp coating (**Fig. 2, left**). Since all these three coatings share a similar surface micro-roughness and differ in their nanotopography: HT and Sr-HT coatings have a similar nanotopography while no obvious nanotopography can be found on the HAp coating. Therefore, it is plausible to conclude that the nanotopography played critical influence on the initial adhesion of the HOBs on the coating surface.

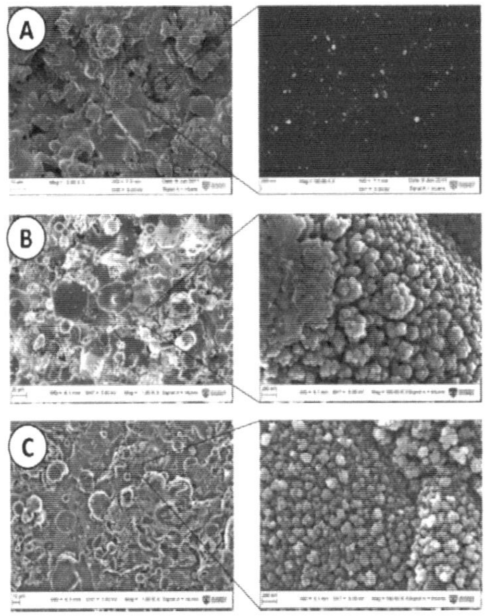

Fig. 1. Surface morphology of plasma sprayed hydroxyapatite (A), hardystonite (B) and Sr-hardystonite (C). The right image is a magnified view of the labelled areas in the left images. All the coatings were deposited on the biomedical grade Ti alloy substrates with sand-blast pre-treatment. The plasma sprayed hydroxyapatite was produced using vacuum plasma spraying technique with parameters used for clinical application[4].

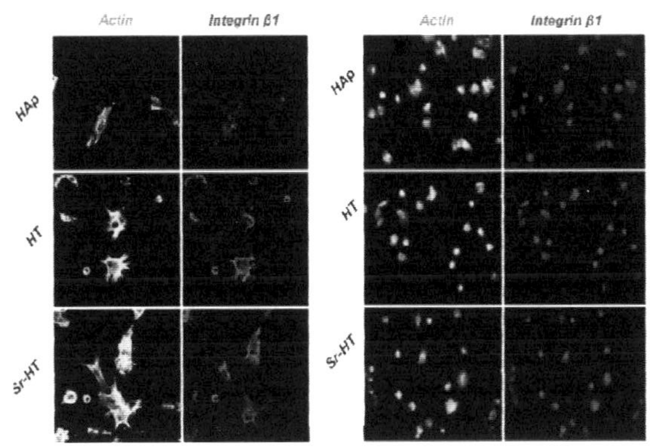

Fig. 2. The expression of integrin β1 (red) and F-actin (green) was detected by immunofluorescence assay. Left: cells were directly cultured on coating surfaces; Right: cells were cultured in the extract solution of the different coatings.

3. Effects of the micro/nanotopography of plasma sprayed coatings on bone-like apatite deposition

Besides the ability to enhance cell adhesion, the nanotopography can also promote the bone-like apatite formation in simulated body fluid (SBF). In previous work, we produced as-sprayed zirconia coatings, polished coating and heat-treated coating[20]. The as-sprayed coating has hirechical micro/nanostructure (**Fig. 3A** and **B**), the polished coating has a smoother surface with neither microstructure nor nanostructure (**Fig. 3C** and **D**), while the heat-treated coating has only microstructure, as shown in **Fig. 3E** and **F**.

SBF immersion tests showed that, the surface of the as-sprayed coating was almost fully covered by bone-like apatite (**Fig. 4A**) after immersion in SBF for 14 days. And the apatite layer became denser and thicker after immersion in SBF for 28 days. However, no precipitates are observed on the polished and the heat-treated coating surface. This results indicate the significant effects of the nanotopography of plasma sprayed coatings on *in vitro* mineralization in SBF.

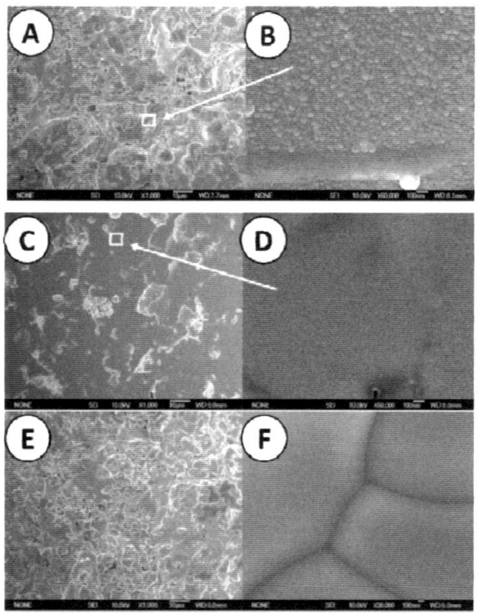

Fig. 3. Surface morphology of plasma sprayed calcium stabilized zirconia coatings with different post-treatments: (A and B), no treatment; (C and D) polishing and (E and F) heat-treatment. The as-sprayed coating has hybrid nano/microstructures, the polished coating has neither nano nor microstructure, whereas the heat-treated coating has the similar microstructure to that on the as-sprayed coating, but doesn't have obvious nanostructure as seen on the as-sprayed coating[20].

Cell proliferation on the as-sprayed, the polished and the heat-treated coatings was evaluated by alamarBlue^{TM}. The cells used are MG63 osteoblast-like cells. From the results shown in **Fig. 5**, Cells cultured on the as-sprayed coating surface proliferated faster, compared to those on polished and the heat-treated coatings. Considering the topographic differences among the three groups of coatings, the higher proliferation rate for the cells cultured on the as-sprayed coatings should be ascribed to the nanostructures on their surfaces.

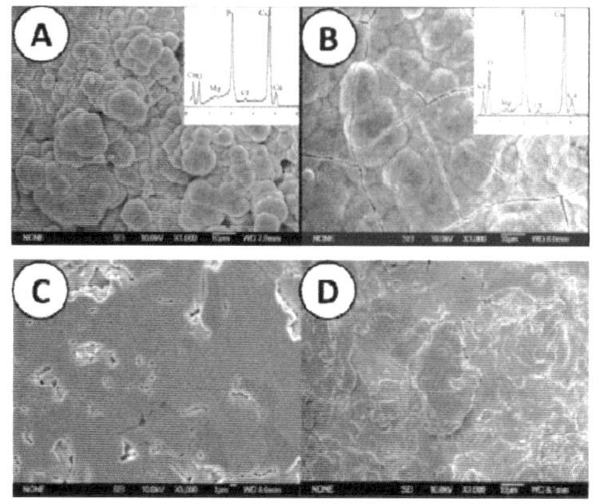

Fig. 4. Surface morphology of plasma sprayed calcium stabilized zirconia coatings with different post-treatments after immersion in simulated body fluid: (A) the as-sprayed coatings without any treatment after immersion in simulated body fluid for 14 days (B) the as-sprayed coatings without any treatment after immersion in simulated body fluid for 28 days; (C) the polished coating after immersion in simulated body fluid for 28 days and (D) the heat-treated coating after immersion in simulated body fluid for 28 days.

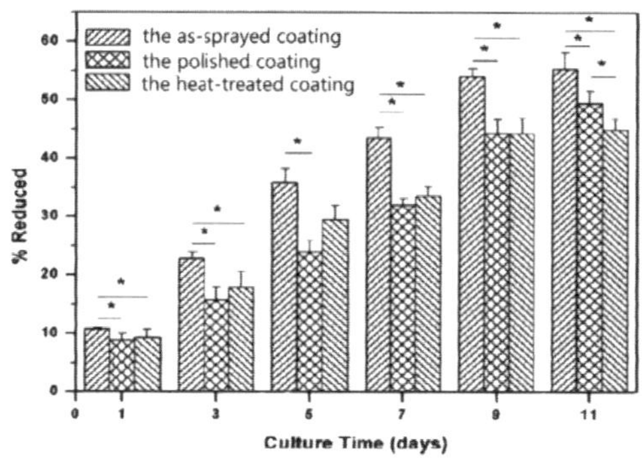

Fig. 5. Graph of percentage alamarBlue^{TM} reduction for MG63 cells cultured on the surface of the as-sprayed, the polished and the heat-treated coatings for various periods. (Cell density of the cell suspension was 5×10^5 cell/ml). *p value < 0.05.

4. Refining nanotopography of plasma sprayed coatings

Despite that various nanotopographic features have been proved to enhance bone cell activity and promote new bone formation, their effects are not equal. Therefore, it is of great importance of being able to alter the nanotopographic features. Plasma spraying, as mentioned before, is characterized by its high temperature and superfast cooling rate, which leads to a rapid solidification of the molten feeding stock. During the solidification process of the plasma sprayed coatings, the melt recrystallizes under non-equilibrium condition. This recrystallization process is influenced by the thermal physical properties of the coating materials and any physical and chemical reactions during the solidification being able to cause changes in the heat. Therefore, it is possible to refine the nanotopographic features of plasma sprayed coating surface through adjusting the thermal chemical reactions of the feeding stocks. In our previous work, we produced Nb_2O_5/TiO_2 composite coatings, and adjust the content of Nb_2O_5 in the composite coating to tailor the thermal chemical reactions[21]. We found that the Nb_2O_5 and TiO_2 coatings have similar nanostructures composed of nanosized grains with a similar size. As the content of Nb_2O_5 in the coating increases, the nanostructure slightly changed until it reached up to 50%. Differently from the nanostructure of the 10%Nb_2O_5 and 30%Nb_2O_5-TIO_2 coatings, 50%Nb_2O_5-TiO_2 coatings show a nanosheet structure, as shown in **Fig. 6**. XRD data demonstrated it was mainly composed of a solid solution $Ti_{0.95}Nb_{0.95}O_4$. The formation of solid solution is the result of the atom migration, which requires energy adsorption and in turn leads to the change in heat, which influences the recrystallization process during the solidification. This possibly accounts for the specific fine structure of the Nb_2O_5 doped TiO_2 oatings.

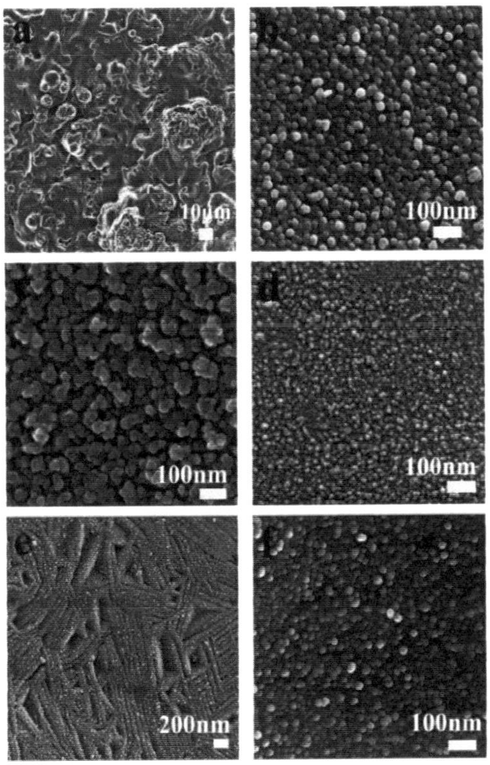

Fig. 6. SEM images of the surface of the TiO_2 coatings doped with different percentage of Nb_2O_5: (a) Low magnification image of the pure TiO_2 coating showing the typical rough surface of plasma sprayed coatings; High magnification views of: (b) the pure TiO_2 coating, (c) 10%Nb_2O_5/TiO_2; (d) 30%Nb_2O_5/TiO_2; (e) 50%Nb_2O_5/TiO_2 and (f) Nb_2O_5 coatings.

5. Literature

[1] Dohan Ehrenfest, D.M., Coelho, P.G., Kang, B.S., Sul, Y.T., Albrektsson, T. Classification of osseointegrated implant surfaces: materials, chemistry and topography. Trends in Biotechnology 28 (2010), Issue 4, pp. 198/206.

[2] Bettinger, C.J., Langer, R. J., Borenstein, T. Engineering substrate topography at the micro- and nanoscale to control cell function. Angewandte Chemie International Edition 48 (2009),Issue 30, pp. 5406/15.

[3] Hoppe, A., Güldal, N.S., Boccaccini, A.R. A review of the biological response to ionic dissolution products from bioactive glasses and glass-ceramics. Biomaterials 32 (2011), Issue 11, pp. 2757/74.

[4] Zhang, W., Wang, G., Liu, Y., Zhao, X., Zou, D., Zhu, C. et al. The synergistic effect of hierarchical micro/nano-topography and bioactive ions for enhanced osseointegration. Biomaterials 34 (2013), Issue 13, pp. 3184/95.

[5] Hong, Y., Yu, M., Lin, J., Cheng, K., Weng, W., Wang, H. Surface hydroxyl groups direct cellular response on amorphous and anatase TiO_2 nanodots. Colloids and Surfaces B: Biointerfaces 123 (2014), pp. 68/74.

[6] Wennerberg, A. And Albrektsson, T.Effects of titanium surface topography on bone integration: a systematic review. Clinical Oral Implants Research 20 (2009), Issue s4, pp. 172/84.

[7] Lavenus, S., Louarn, G., Layrolle, P. Nanotechnology and dental implants. International Journal of Biomaterials 2010 (2010), pp. 915327.

[8] Zhao, X., Wang, G., Zheng, H., Lu, Z., Zhong, X., Cheng, X. et al. Delicate refinement of surface nanotopography by adjusting TiO_2 coating chemical composition for enhanced interfacial biocompatibility. ACS Applied Materials & Interfaces 5 (2013), Issue 16, pp. 8203/09.

[9] Wang, G., Lu, Z., Liu, X., Zhou, X., Ding, C., Zreiqat, H. Nanostructured glass-ceramic coatings for orthopaedic applications. Journal of the Royal Society Interface 8 (2011), Issue 61, pp. 1192/203.

[10] Wang, G., Liu, X., Gao, J., Ding, C. In vitro bioactivity and phase stability of plasma-sprayed nanostructured 3Y-TZP coatings. Acta Biomaterialia 5 (2009), Issue 6, pp. 2270/8.

[11] Wang, G., Meng, F., Ding, C., Chu, P.K., Liu, X. Microstructure, bioactivity and osteoblast behavior of monoclinic zirconia coating with nanostructured surface. Acta Biomaterialia 6 (2010), Issue 3, pp. 990/1000.

[12] Marie, P.J., Ammann, P., Boivin, G., Rey, C. Mechanisms of Action and Therapeutic Potential of Strontium in Bone. Calcified Tissue International 69 (2001), Issue 3, pp. 121/9.

[13] Saidak, Z. and Marie, P.J. Strontium signaling: molecular mechanisms and therapeutic implications in osteoporosis. Pharmacology & Therapeutics 136 (2012), Issue 2, pp. 216/26.
[14] Wu, C., Ramaswamy, Y., Kwik, D., Zreiqat, H. The effect of strontium incorporation into CaSiO3 ceramics on their physical and biological properties. Biomaterials 28 (2007) Issue 21, pp. 3171/81.
[15] Zhao, L., Wang, H., Huo, K., Zhang, X., Wang, W., Zhang, Y. et al. The osteogenic activity of strontium loaded titania nanotube arrays on titanium substrates. Biomaterials 34 (2013), Issue 1, pp. 19/29.
[16] Li, H., Xue, K., Kong, N., Liu, K., Chang, J. Silicate bioceramics enhanced vascularization and osteogenesis through stimulating interactions between endothelia cells and bone marrow stromal cells. Biomaterials 35 (2014), Issue 12, pp. 3803/18.
[17] Sun, J., Wei, L., Liu, X., Li, J., Li, B., Wang, G. et al. Influences of ionic dissolution products of dicalcium silicate coating on osteoblastic proliferation, differentiation and gene expression. Acta Biomaterialia 5 (2009), Issue 4, pp. 1284/93.
[18] Seo, H.J., Cho, Y.E., Kim, T., Shin, H.I., Kwun, I.S. Zinc may increase bone formation through stimulating cell proliferation, alkaline phosphatase activity and collagen synthesis in osteoblastic MC3T3-E1 cells. Nutrition research and practice 4 (2010), Issue 5, pp. 356/61.
[19] Aina, V., Malavasi, G., Pla, A.F., Munaron, L., Morterra, C. Zinc-containing bioactive glasses: Surface reactivity and behaviour towards endothelial cells. Acta Biomaterialia 5 (2009), Issue 5, pp. 121/22.
[20] Wang, G., Liu, X., Zreiqat, H., Ding, C. Enhanced effects of nano-scale topography on the bioactivity and osteoblast behaviors of micron rough ZrO2 coatings. Colloids Surfaces B Biointerfaces 86 (2011), Issue 2, pp. 267/74.
[21] Zhao, X., Wang, G., Zheng, H., Lu, Z., Zhong, X., Cheng, X. et al. Delicate refinement of surface nanotopography by adjusting TiO_2 coating chemical composition for enhanced interfacial biocompatibility. ACS Applied Materials & Interfaces 5 (2013), Issue 16, pp. 8203/9.

Preparation and *in vitro* evaluation of plasma sprayed cerium oxide coatings

Kai Li, Youtao Xie, Mingyu You, Liping Huang, Xuebin Zheng*
Key Laboratory of Inorganic Coating Materials, Shanghai Institute of Ceramics, Chinese Academy of Sciences, 1295 Dingxi Road, Shanghai 200050, PR China
*Email address: xbzheng@mail.sic.ac.cn

Abstract: In our work, protective effects of cerium oxide (CeO_2) coatings on the response of osteoblast to H_2O_2-induced oxidative stress were investigated. CeO_2 coating significantly increased osteoblast growth when compared to control. H_2O_2-induced reduction of differentiation marker ALP was recovered in the presence of CeO_2. The above effects were attributed to the fact that CeO_2 coating could decrease reactive oxygen species production in H_2O_2-treated osteoblasts. Taken together, CeO_2 coatings with antioxidant property would be promising for bone regeneration.

1 Introduction

Titanium (Ti)-based metallic biomaterials have gradually become the most widely used metal implants in clinics [1]. However, their osseointegration capacities may dramatically decrease in some disease conditions. For example, osteoporosis has been proved to adversely affect the biological performance of Ti implants, associated with the much slower bone regeneration [2]. Recent studies have shown that osteoporosis could dramatically increase the level of reactive oxygen species (ROS) in both mitochondrial and cytoplasma [3]. Elevated production of ROS cause oxidative stress, which is known to be able to disrupt the cellular oxidant/antioxidant balance leading to the slow tissue regeneration and wound healing. A growing body of evidence has shown that oxidative stress induced by exogenous stimuli such as hydrogen peroxide (H_2O_2) is able to inhibit osteoblast differentiation and induce apoptosis of osteoblastic cell line [4]. Hence, we hypothesize that the ROS may play a critical role in the adverse effect of osteoporosis on the biological performance of Ti implant, and imparting the Ti implant with antioxidant capability may attenuate the effects of oxidative stress and improve post-operative outcomes.

Cerium oxide (ceria) ceramics have been gaining considerable attention due to its free radical scavenging properties [5]. Work has demonstrated that the addition of CeO_2 nanoparticles in bioactive glass scaffolds increased the proliferation and downstream collagen production of human mesenchymal stem cells, by acting as free radical scavenger [6]. Therefore, nanoceria has been demonstrated as a potent material for treatment of disease conditions caused by ROS. However, its use as orthopedic coating material has yet to be investigated.

In this work, CeO_2 powders were coated onto Ti-6Al-4V substrate using plasma spraying techniques. We aimed to investigate CeO_2 coating on the changes of osteoblasts function and oxidative injury in osteoblastic MC3T3-E1 cells after exposure to H_2O_2.

2 Materials and methods

2.1 Preparation and characterization of CeO_2 coatings

CeO_2 powders were sprayed onto Ti-6Al-4V substrates applying an atmospheric plasma-spraying system (F4-MB, Sulzer Metco, Switzerland). The detailed parameters are described in our previous study [7]. The phase composition of the as-sprayed coating was measured using an X-ray diffractometer.

2.2 Cell culture and analysis of cell viability

MC3T3-E1 cells were cultured in DMEM supplemented with 10 % FBS and 1 % antibiotics in a humidified atmosphere of 5% CO_2 at 37 °C. 1 ml of cell suspension supplemented with a 10% 1mM H_2O_2 were seeded on each sample. For cell viability test, the cells were cultured for 1, 4, and 7 days. Cells without H_2O_2 treatment served as control. At the predetermined time point, the medium was removed and replaced with 0.9 ml culture medium and 0.1 ml Cell Counting Kit-8. After additional culture for 3 h, 100 μl of the reacted reagent was transferred and the absorbance was read at 450 nm using a microplate reader.

2.3 ALP activity

1 mL of cell suspension supplemented with 10% 1mM H_2O_2 were seeded on each sample. The cells were cultured for 4, 7 and 14 days. Cells without H_2O_2 treatment served as control. At the predetermined time point, culture medium was removed and coating samples were washed gently twice times with PBS, and then cells were lysed in 150 μl of 0.1% Triton X-100. Lysates were sonicated after being centrifuged at 12,000 rpm for 15 min at 4 °C. 50 μl of supernatant was mixed with 150 μl working solution according to the manufacturer's protocol (Beyotime, China). The relative ALP activity was expressed as nmol/min/mg protein.

2.4 Determination of oxidative damage in MC3T3-E1 cells

Cellular oxidative damage was evaluated by the levels of ROS. Relative changes of intracellular ROS within H_2O_2-treated cells were monitored using a fluorescence probe DCFH-DA as previously described. Briefly, MC3T3-E1 cells on each coating sample were exposed to H_2O_2 for the indicated time (1, 4 and 7 days) and then stained with 5µg/ml DCFH-DA for 20 min at room temperature. The ROS production was measured with a fluorescence reader.

2.5 Statistical Analysis

Results were expressed as means ± standard deviation (SD) for all experiments and were analyzed using one-way ANOVA with a post-hoc test. A *p*-value < 0.01 was considered statistically significant.

3 Results

3.1 Characterization of CeO_2 coating

Figure 1 shows the XRD patterns of the CeO_2 powder and the as-sprayed coating. It can be seen that the powder was mainly composed of CeO_2 phase (JCPDS No. 34-0394). There was no obvious change in phase composition between the powder and the coating.

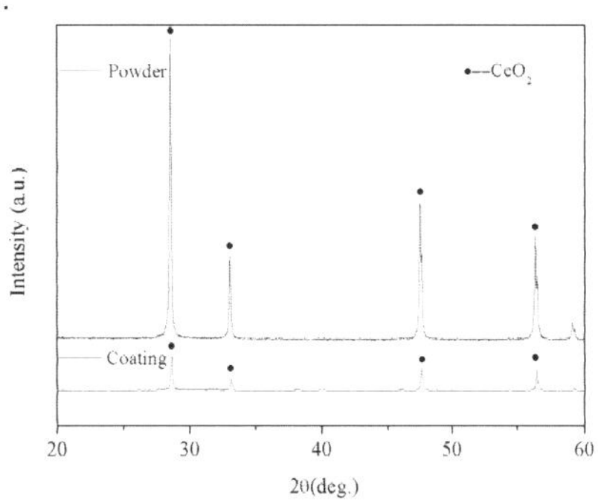

Fig. 1. XRD patterns of the CeO_2 powder and the as-sprayed coating.

3.2 CeO_2 coating alleviates H_2O_2-induced cytotoxicity in MC3T3-E1 osteoblastic cells

In our study, treatment of cells with H_2O_2 for 1-7 days significantly decreased cell viability in time-dependent manner, as shown in Fig. 2. The cell viability of the H_2O_2 group was reduced to 30.9 ± 1.1% (day 1), 25.4 ± 1.1% (day 4) and 23.8 ± 0.56% (day 7) respectively compared with the control group. However, considerable improvement in the cell proliferative capability was observed for the CeO_2+H_2O_2 group with respect to the H_2O_2 group. The cell viability of the CeO_2 +H_2O_2 group approached 53.2 ± 1.2% (day 1), 43.3 ± 1.7% (day 4) and 43.2 ± 1.5% (day 7) under the same treatment.

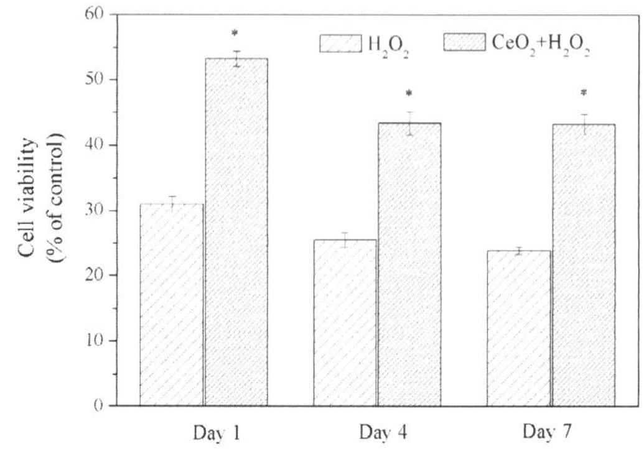

Fig. 2. Cell viability of MC3T3-E1 cells in response to H_2O_2. The viability of control cells without H_2O_2 challenge was defined as 100%. *$p < 0.01$ vs the H_2O_2 group.

3.3 CeO_2 coating attenuates H_2O_2-suppressed ALP activity

We examined the activity of ALP to evaluate the effect of CeO_2 coating on osteoblast differentiation. As shown in Fig. 3, incubation of cells with H_2O_2 in the absence of CeO_2 coating markedly reduced the level of ALP activity in MC3T3-E1 cells. The ALP activity of MC3T3-E1 cells for the CeO_2+H_2O_2 group even showed higher levels of expression than the control group without H_2O_2 treatment, indicating that CeO_2 coating significantly attenuated H_2O_2-downregulated level of ALP activity.

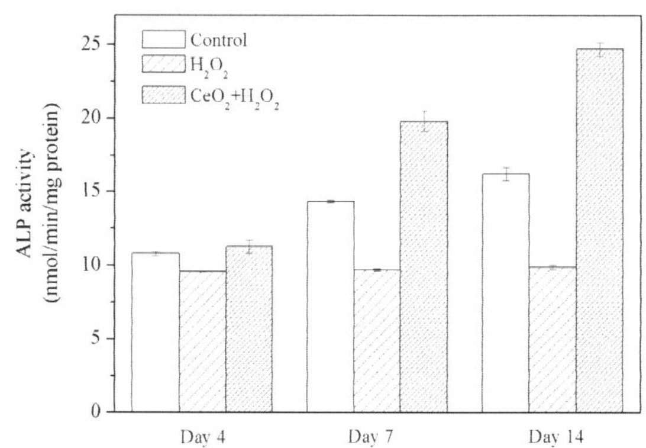

Fig. 3. Effects of the CeO_2 coating on activity of ALP in MC3T3-E1 cells exposed to H_2O_2.

3.4 Antioxidant effects of CeO_2 coating in MC3T3-E1 cells

To investigate the antioxidant effects of CeO_2 coating, we examined the effect of the CeO_2 coating on H_2O_2 mediated oxidative damage in MC3T3-E1 cells by monitoring intracellular ROS generation. As shown in Fig. 4, the levels of ROS in MC3T3-E1 cells for H_2O_2 group increased 4.7- (day 1), 4.9- (day 4), 5.9-fold (day 7) respectively after treatment with 0.1 mM H_2O_2 for 1, 4, and 7 days. Such H_2O_2-evoked ROS production was eliminated by CeO_2 coating for 31.1% at day 7.

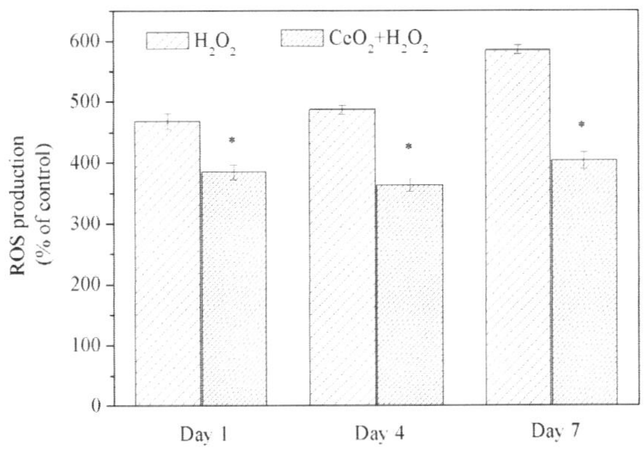

Fig. 4. Effects of the CeO_2 coating on H_2O_2–induced intracellular ROS production in MC3T3-E1 cells. *$p < 0.01$ vs the H_2O_2 group.

4 Discussion

Increases in oxidative stress occur acutely in major surgical procedures including orthopedic surgery, especially when ischaemia is followed by reperfusion [8]. ROS are known to cause damaging oxidative stress if produced in excessive quantities [9]. H_2O_2, one of the main ROS, may diffuse across biological membranes and produce a wide range of injury. It has been reported that H_2O_2 induces apoptosis or necrosis of various types of cells [10].

In the present study, we demonstrated that CeO_2 coating alleviated these deleterious effects of H_2O_2 on osteoblastic proliferation and differentiation. At least 70% increase in cell viability was observed for the CeO_2+H_2O_2 group when compared to the H_2O_2 group under the same treatment. Moreover, the cells on the CeO_2 coating exhibited the higher survival rate than those on the control. We continued to investigate the effect of CeO_2 coating on osteoblastic differentiation under the condition of oxidative stress. We found that in this study H_2O_2 considerably suppressed the ALP activity MC3T3-E1 cells for H_2O_2 group. However, CeO_2 coating significantly attenuated H_2O_2-downregulated level of ALP activity. Among the cells treated with H_2O_2, the ones on the CeO_2 coating exhibited the highest levels of ALP activity. These data suggested that the CeO_2 coating could protect osteoblastic cells against oxidative stress and promote their proliferation and differentiation.

We further examined the underlying mechanism of the protective effects of CeO_2 coating. At physiological low levels, ROS function as redox messengers in intracellular signaling and regulation, whereas excess ROS induce cell damage and apoptosis. Previous studies reported that the deleterious effects of H_2O_2 on bone marrow stem cells were partially mediated through ROS production from H_2O_2. In our study, after treatment of MC3T3-E1 cells with H_2O_2, much higher levels of intracellular ROS were observed for H_2O_2 group coating than CeO_2+ H_2O_2 group. It suggested that CeO_2 coating reduced H_2O_2-stimulated intracellular ROS production. Taken together, CeO_2 coating may scavenge excessive ROS derived from oxidative damage, ameliorating H_2O_2-impaired cell survival and proliferation.

5 Conclusions

In this study, we prepared CeO_2 coating via plasma spraying technology. CeO_2 coating suppressed intracellular ROS generation in osteoblastic MC3T3-E1 cells, protecting the proliferation and differentiation cells from oxidative stress induced by H_2O_2. The use of CeO2 incorporated calcium silicate biomedical coatings can promote bone recovery associated with increased oxidative stress.

Acknowledgments

This work was supported by the Natural Science Foundation of China (Grant No. 51172264, No. 51502328).

References

1. M. Browne and P.J. Gregson, Effect of Mechanical Surface Pretreatment on Metal Ion Release, Biomaterials, 2000, 21(4), p 385-392
2. Kinov, P., A. Leithner, R. Radl, K. Bodo, G.A. Khoschsorur, K. Schauenstein, and R. Windhager, Role of free radicals in aseptic loosening of hip arthroplasty. Journal of Orthopaedic Research, 2006. 24(1): p. 55-62.
3. Rosenfeldt, F., M. Wilson, G. Lee, C. Kure, R.C. Ou, L. Braun, and J. de Haan, Oxidative stress in surgery in an ageing population: Pathophysiology and therapy. Experimental Gerontology, 2013. 48(1): p. 45-54
4. Lee, D.H., B.S. Lim, Y.K. Lee, and H.C. Yang, Effects of hydrogen peroxide (H2O2) on alkaline phosphatase activity and matrix mineralization of odontoblast and osteoblast cell lines. Cell Biology and Toxicology, 2006. 22(1): p. 39-46
5. Inaba, H. and H. Tagawa, Ceria-based solid electrolytes - Review. Solid State Ionics, 1996. 83(1-2): p. 1-16
6. Nicolini, V., E. Gambuzzi, G. Malavasi, L. Menabue, M.C. Menziani, G. Lusvardi, A. Pedone, F. Benedetti, P. Luches, S. D'Addato, and S. Valeri, Evidence of Catalase Mimetic Activity in Ce3+/Ce4+ Doped Bioactive Glasses. Journal of Physical Chemistry B, 2015. 119(10): p. 4009-4019.
7. K. Li, J. Yu, Y. Xie, L. Huang, X. Ye, and X. Zheng, Chemical Stability and Antimicrobial Activity of Plasma Sprayed Bioactive Ca2ZnSi2O7 Coating, J. Mater. Sci. Mater. Med., 2011, 22(12), p 2781-2789.
8. Cheng, Y.J., C.T. Chien, and C.F. Chen, Oxidative stress in bilateral total knee replacement, under ischaemic tourniquet. Journal of Bone and Joint Surgery-British Volume, 2003. 85B(5): p. 679-682.
9. Hosoya, S., H. Suzuki, M. Yamamoto, K. Kobayashi, and Y. Abiko, Alkaline phosphatase and type I collagen gene expressions were reduced by hydroxyl radical-treated fibronectin substratum. Molecular Genetics and Metabolism, 1998. 65(1): p. 31-34.

10. Gardner, A.M., F.H. Xu, C. Fady, F.J. Jacoby, D.C. Duffey, Y.P. Tu, and A. Lichtenstein, Apoptotic vs nonapoptotic cytotoxicity induced by hydrogen peroxide. Free Radical Biology and Medicine, 1997. 22(1-2): p. 73-83.

Osteogenic behavior of bone marrow mesenchymal stem cells on calcium silicate based coatings by hydrothermal treatment

Dandan Hu, Kai Li, Youtao Xie, Liping Huang, Xuebin Zheng*

Key Laboratory of Inorganic Coating Materials, Shanghai Institute of Ceramics, Chinese Academy of Sciences, 1295 Dingxi Road, Shanghai 200050, PR China

*Email address: xbzheng@mail.sic.ac.cn

Abstract: In our study, a surface modification strategy encompassing the use of bioactive trace elements together with nanotopographical modifications was employed to enhance the osteogenic abilities of calcium silicate (Ca-Si) coatings. We developed strontium-loaded nanolayer on plasma sprayed Ca-Si coating via hydrothermal treatment. Original Ca-Si (CS) coating and Ca-Si coating modified with similar nanolayer (NT-CS) were parallel investigated for comparison. Bone marrow mesenchymal stem cells (BMSCs) cultured on NT-CS and Sr modified NT-CS coating surfaces spread more and exhibited polygonal morphology with a relatively larger size compared to those on CS coating. BMSCs differentiation ability of Sr-NT-CS was superior to that of NT-CS and CS coatings. These results suggested that Sr-NT-CS combined the advantages of nanotopography and strontium to enhance osteogenic ability of BMSCs.

1 Introduction

Calcium silicate (Ca-Si) ceramics have attracted great attention due to their bioactivity and osseointegration property [1-2]. Plasma sprayed Ca-Si coatings on Ti substrate, which combine the bioactivity of bio-ceramics and the mechanical properties of titanium, are potential bio-materials for orthopedic implant application. However, the high degradation rate of Ca-Si coatings causes an increase in pH, which can be detrimental for cell growth in this region and thus for the osseointegration.

It has been reported that the incorporation of Mg, Zn or Sr into Ca-Si coatings resulted in enhanced stability and better biological properties such as higher cell proliferation and differentiation, gene expression, and bone regeneration compared to pure Ca-Si coating [3-5]. As a member of these modified Ca-Si based coating, Sr-incorporated coating has been widely considered and comprehensive studies about its in vitro and in vivo osteogenic properties have been performed by numerous research groups [6-7].

Although the biological properties of Sr-$CaSiO_3$ ceramics and coatings are promising, osteogenic abilities of these materials could be improved by architecture of surface topography. Surface nano-scale topographies are supposed to directly influence the cell morphology and cytoskeleton distribution, which have been considered to be closely related with the cell fate orientation and osteogenic tendency [8-9].

In this study, nano-structured topography and Sr-loaded surface were prepared. The impact of the nano-structured topography and Sr ion release on cell morphology and cytoskeleton distribution as well as the following cell differentiation and mineralization was evaluated.

2 Materials and methods

2.1 Coating fabrication and surface modification

The $CaSiO_3$ powders were synthesized by the chemical precipitation method using $Ca(NO_3)_2$ and Na_2SiO_3 solutions as raw materials. The obtained $CaSiO_3$ powders were sieved through 60 meshes and sprayed onto the treated Ti substrates (ø 10 mm × 1 mm) by an atmosphere plasma spray (APS) system (F4-MB, Sulzer Metco, Switzerland).

The nano-structured topography on $CaSiO_3$ coating were fabricated via hydrothermal treatment with deionized water. Briefly, the sample was immersed in 50 mL deionized water in a Teflonlined reaction vessel at 180°C for 24 h. Then, the modified samples were rinsed with deionized water and dried.

The Sr-loaded surface on $CaSiO_3$ coating were modified via hydrothermal treatment with 1 mol/L $Sr(NO_3)_2$ aqueous solutions.

2.2 Surface characterization

The surface topography of the coatings were characterized by field-emission scanning electron microscopy (FESEM, Magellan 400, FEI, USA). Surface roughness (Ra) measurement was completed using a stylus profiler (T8000, HOMMEL, Germany). The contact angle of the fabricated samples was evaluated with Automatic Contact Angle Meter Model (SL200B, Solon information technology Co., Ltd., China)

2.3 Cell adhesion observations

The BMSCs were seeded on each sample at a density of 3×10^4 cells per well in 24-well culture plate. Cells were fixed with 4 % paraformaldehyde (PA) solution for 15 min at room temperature, washing twice, and premeabilized with 0.1 % Triton X-100 for 15 min. After washed twice, cells were treated with 4, 6-diamino-2-phenylindole (DAPI, Sigma) at a concentration of 4 μg / mL for 5 min in dark. After rinsing for three times, the cell cytoskeleton were stained with 5 μL / mL rhodamine phalloidin (Molecular Probes) for 40 min without light. After through washing with PBS, cells on samples were visualized with confocal laser scanning microscope (CLSM, Leica).

2.4 Extracellular matrix collagen (COL)

To determine the cell differentiation and mineralization, the BMSCs were seeded on each sample in 24-well plate at a density of 5×10^4 cells per well. Extracellular matrix collagen (COL) was observed with sirius-red-staining at 10 days culture in osteogenic induction.

2.5 ALP activity

Alkaline phosphatase (ALP) activity, cells were lysed in 500 μL of 0.2 % Triton X-100 and then centrifuged at 10^4 rpm for 5 min at 4℃. 30 μL of the supernatants was mixed with 150 μL of the working solution, and enzyme activity was determined using the ALP assay kit (BioAssay Systems, Hayward, CA). Total protein values were measured using the Micro-BCA kit (Pierce, Thermol Scientific, USA) at 630 nm. The results were normalized and presented as OD values per mg of total protein

3 Results

3.1 Surface characterization

Surface morphology of the Ca-Si based coatings was presented in Figure 1. The original coating has a rough surface and a few of micro-cracks built by randomly stacking of fully melted and semi-melted particles. After hydrothermal treatment of the coatings with deionized water or $Sr(NO_3)_2$ solution, nanoplate-like structures with thickness of 20-50 nm were obtained on the coating surface. The surface average roughness (Ra) of the original coating, NT-CS coating and Sr-NT-CS coating was 16.93 ± 1.28, 14.72 ± 1.07 and 15.60 ± 0.84 μm, respectively.

Fig. 1. SEM images of the surface topography at low and high magnification. CS, the original $CaSiO_3$ coating; NT-CS, nanolayer structure on the $CaSiO_3$ coating surface; Sr-NT-CS, nanolayer structure and Sr-loaded on the $CaSiO_3$ coating surface

3.2 Cell morphology

As shown in Figure 2, BMSCs stained with DAPI-phalloidin to label the nuclei-actin cytoskeleton were visible by confocal microscopy after 24 h incubation. The morphology of the cells on CS coating surface presented round morphology with less-spread body and few filopodia, while they were more spread and exhibited a polygonal morphology with a relatively larger size by the introduction of nanolayer structure (NT-CS and Sr-NT-CS).

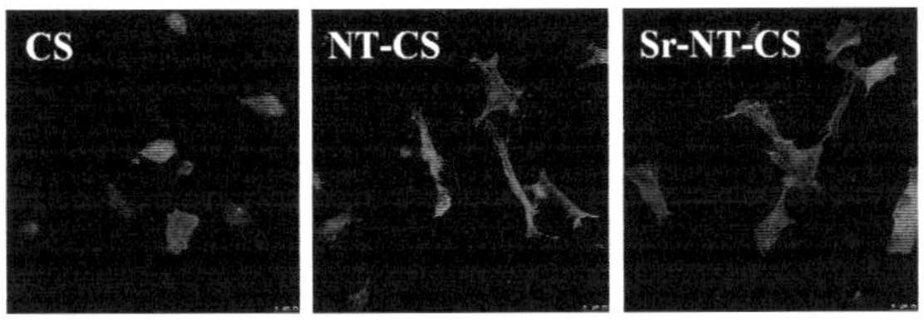

Fig. 2. Confocal micrographs of BMSCs adhering to various surfaces. Surfaces are as indicated in the legend for Fig. 1.

3.3 Extracellular matrix collagen (COL)

Figure 3 showed the extracellular matrix collagen (COL) staining. Compared to CS coating surface, higher extracellular matrix collagen was observed in cells grown on the NT-CS surface, and no significant improvement was observed on the Sr-NT-CS surface

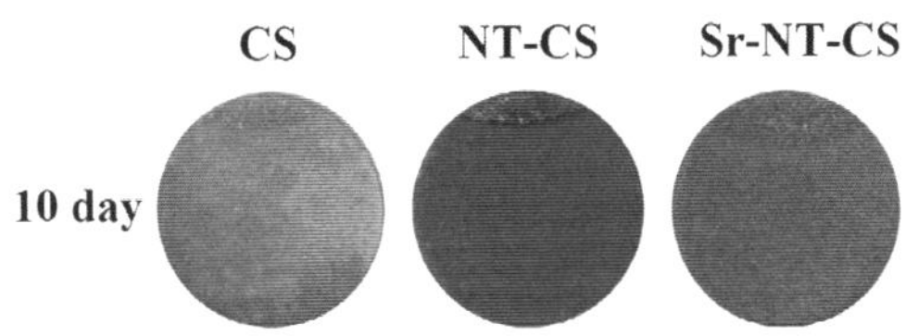

Fig. 3. Extracellular matrix collagen (COL) staining of BMSCs cultured on the surfaces at day 10. Surfaces are as indicated in the legend for Fig. 1

3.4 ALP activity

ALP activity was measured at day 10 to assess the osteogenic differentiation potential of BMSCs on the orginal and modified coatings (Figure 4). Sr-NT-CS had the highest ALP activity, followed by NT-CS and CS coating surfaces.

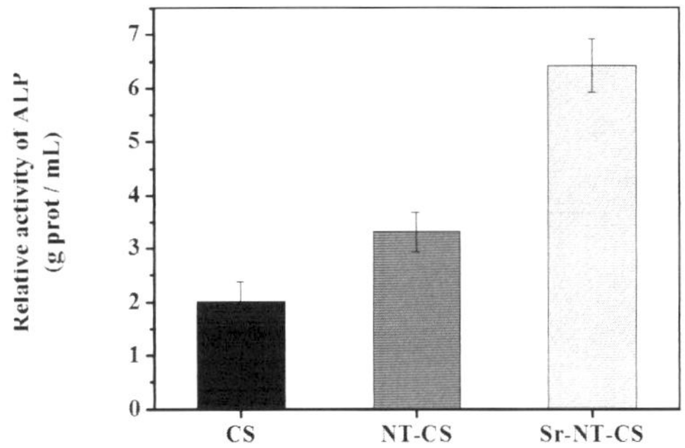

Fig. 4. Alkaline phosphatase (ALP) activity of BMSCs cultured for 10 days on various surfaces. Surfaces are as indicated in the legend for Fig. 1.

4 Discussion

The success of bone substitution materials largely depends upon the formation of bio-active interface between materials and bone tissue. A series of processes occur through the initial adhesion of cells with bone substitution materials. Cellular adhesion and subsequent cellular responses on the materials, therefore, are critical and prerequisite parameters for osteointegration and osteoconduction.

Nanolayer structured surface (NT) showed two obvious characteristics as compared to surface without nano-structured topography. Firstly, it brought an increase in crystallinity, which might contribute to the improvement of stability. Secondly, BMSCs on the nano-structured coating surfaces (NT-CS and Sr-NT-CS) were seen adhered and spread well.Sr-loaded surface (Sr-NT-CS) induced greater ALP activity compared to that observed with the NT-CS surface, consistent with studies demonstrating that Sr induced osteoblastic differentiation in cultured BMSCs. These results suggested that Sr-NT-CS combined the advantages of hierarchical nanotopography and Sr to enhance osteogenic ability

5 Conclusion

In this study, Sr-loaded calcium silicate coatings with nano-topography surface (Sr-NT-CS) were fabricated by plasma spraying and hydrothermal method. Nanolayer structured surface led to better cell spreading. When Sr was introduced, greater ALP activity was oberserved with the Sr-NT-CS surface. These results suggested that Sr-NT-CS combined the advantages of hierarchical nanotopography and Sr to enhance osteogenic ability.

References

[1] Ni S, Chang J, Chou L, Zhai W. Comparison of osteoblast-like cell responses to calcium silicate and tricalcium phosphate ceramics in vitro. Journal of biomedical materials research Part B, Applied biomaterials. 2007;80(1):174-83.

[2] Xu S, Lin K, Wang Z, Chang J, Wang L, Lu J et al. Reconstruction of calvarial defect of rabbits using porous calcium silicate bioactive ceramics. Biomaterials. 2008;29(17):2588-96.

[3] Wu CT, Chang J. A novel akermanite bioceramic: Preparation and characteristics. J Biomater Appl. 2006;21(2):119-29.

[4] Li K, Yu J, Xie Y, Huang L, Ye X, Zheng X. Effects of Zn Content on Crystal Structure, Cytocompatibility, Antibacterial Activity, and Chemical Stability in Zn-Modified Calcium Silicate Coatings. Journal of Thermal Spray Technology. 2013;22(6):965-73.

[5] Wu CT, Ramaswamy Y, Kwik D, Zreiqat H. The effect of strontium incorporation into $CaSiO_3$ ceramics on their physical and biological properties. Biomaterials. 2007;28(21):3171-81.

[6] Zhu YF, Zhu M, He X, Zhang JH, Tao CL. Substitutions of strontium in mesoporous calcium silicate and their physicochemical and biological properties. Acta Biomaterialia. 2013;9:6723–6731.

[7] Lin KL, Xia LG, Li HY, Jiang XQ, Pan HB , Xu YJ et al. Enhanced osteoporotic bone regeneration by strontium-substituted calcium silicate bioactive ceramics. Biomaterials. 2013;34:10028-10042

[8] Rolando A. Gittens, Taylor McLachlan, Rene Olivares-Navarrete, Ye Cai, Simon Berner et al. The effects of combined micron-/submicron-scale surface roughness and nanoscale features on cell proliferation and differentiation. Biomaterials. 2011;32:3395-3403

[9] Pan HH, Xie YT, Li K, Hu DD, Zhao J, Zheng XB and Tang TT. ROCK-regulated synergistic effect of macropore/nanowire topography on cytoskeletal distribution and cell differentiation. RSC Adv. 2015;5:101834–101842

Hierarchical macro/nano titanium coating promotes cell differentiation via ROCK-regulated synergistic cytoskeletal distribution

Houhua Pan[1], Youtao Xie[1], Xuebin Zheng[1*], Tingting Tang[2**]

[1]Key Laboratory of Inorganic Coating Materials, Shanghai Institute of Ceramics, Chinese Academy of Sciences, 1295 Dingxi Road, Shanghai 200050, PR China
*Email address: xbzheng@mail.sic.ac.cn

[2]Shanghai Key Laboratory of Orthopaedic Implants, Department of Orthopaedics, Shanghai Ninth People's Hospital, Shanghai JiaoTong University School of Medicine, Shanghai 200011, PR China
**Email address: tingtingtang@hotmail.com

Abstract: Plasma sprayed titanium coating (TPS) with rough macroporous structure is one of the modification technology commonly used on orthopedic implants. Considering their poor early cell responses, nano-modified plasma sprayed titanium coating (NTPS) has been considered as an effective approach to induce more rapid osteogenesis. However, less attention has been paid to the detailed process of the cell-surface interaction of osteoblasts on the NTPS surface. In this study, focus has been put on the detailed mechanism between surface topography and cell differentiation performance in terms of both MC3T3 and BMSC cells. Results showed that synergistic effect of cytoskeleton distribution was obtained on hierarchical NTPS surface, with better cell spreading induced by nanotexture, and meanwhile multi-dimentional cytoskeleton distribution brought by rough macroporous structure. Western blotting results showed higher ROCK activity on NTPS, which meant greater cytoskeleton reorganization and higher intracellular tension. Consequently, better MC3T3 and BMSC differentiation performance were both achieved on NTPS surface, which were further down-regulated by the inhibition of ROCK. It can be concluded that the better performance of NTPS in promoting cell differentiation attributes to the synergistic effect of macro/nano-topography on cytoskeletal distribution and the concomitant higher intracellular tension.

1 Introduction

Plasma sprayed titanium coating (TPS) is one of the modification technology commonly used and its unique macroporous structures effectively encourages the bone tissue in-growth [1]. Considering the poor early cell responses [2], nano-modified plasma sprayed titanium coating (NTPS) has been developed and it exhibits more pronounced effects to improve the early osseointegration [3-4]. Nonetheless, the mechanism mediating the response of cells to these NTPS surfaces are largely unknown, which thereby precludes precise understanding of its biological effect. Thus, it is necessary to carry out the detailed process and mechanism of cell-surface interaction on NTPS structures.

Surface topographies are supposed to directly influence the cell morphology and cytoskeleton distribution, which have been considered to be closely related with the cell fate orientation and osteogenic tendency. Mcbeath et al. found that well spread stem cell undergoes osteogenesis, while the poorly spread one becomes adipocytes [5]. This attributed to the cytoskeleton tension caused by morphology change and the following intracellular mechanotransduction, which have been considered to play a critical role in osteogenesis inducement. Meanwhile, Rho-associated kinase (ROCK) is known to play a pivotal role in cytoskeletal reorganization and is essential to cytoskeletal tension generation [6]. Considering this, it provides an insight to the problem mentioned above by studying the status of cell morphology and cytoskeleton distribution as well as ROCK activity on different designed surfaces and find the bridge between surface topography and cell differentiation.

In this study, both MC3T3 and BMSC cells were used for better evaluation. The purpose of this study was to focus on how the hierarchical structures influence cytoskeleton distribution as well as cell differentiation, and to investigate the distinct role of nano and macrotopography in regulating the osteogenic fate.

2 Materials and methods

2.1 Sample preparation and surface characterization

Pure Ti substrates were treated by plasma spraying method and/or hydrothermal treatment. In brief, samples were firstly polished and then treated by 2.5M NaOH in a Teflon-lined vessel at 150°C for 15h and finally annealed at 450°C for 2h to obtain the anntase nanowire surface (NT). Meanwhile, vacuum plasma spraying was utilized to fabricate the plasma sprayed titanium coating (TPS). The TPS samples were further treated by the above-mentioned method to form the NTPS surfaces. The topographies of the samples were examined by field-emission scanning electron microscopy (Hitachi S-4800, Japan).

2.2 Cell culture

Mouse osteoblastic cell line MC3T3-E1 was obtained from the Cell Bank of the Chinese Academy of Sciences (Shanghai, China). Rat bone mesenchymal stem cells (rBMSCs) were also used for evaluation. In brief, four-week-old male Sprague Dawley rats were obtained from the Ninth People's Hospital Animal Center (Shanghai, China). Their BMSCs were obtained by flushing the bone marrow from their tibias and femur bones using high-glucose Dulbecco's modified Eagle's medium (DMEM, Gibco) supplemented with 10% fetal bovine serum (FBS, Sigma) and 1% penicillin/streptomycin (Invitrogen). MC3T3 cells and BMSCs were seeded in culture dishes and placed in an incubator containing 5% CO_2 at 37°C. The culture medium was replaced every three days and cells were passaged when they reached 80% - 90% confluence. BMSCs at passage 3 - 5 were used. All the experiments in this study were performed using MC3T3 cells and BMSCs respectively.

2.3 Cell morphology

Cells were seeded onto each specimen at a density of 2×10^4 cells/well in a 24-well. After 24h of incubation, the samples were washed twice with PBS, fixed in 2% glutaraldehyde overnight, and washed twice again. Then cells were dehydrated in a graded series of ethanol (30, 50, 70, 90, 100%) to prevent deformation of cytoskeleton, and dried in 37 °C overnight. Finally the morphology of cells were observed by the SEM.

2.4 Western blotting

3×10^5 cells/well was seeded in a 6-well plate. Cells on the surface were washed by PBS and carefully collected by Tyrisin after 3 days. After centrifugation and thorough cleaning three times with PBS, cells were treated by Ripa lysis buffer (Beyotime) with 1% protease inhibitor. After protein concentration of all samples were quantified, necessary volumes were calculated and equal protein of lysates were loaded into acrylamide gels for electrophoresis, and transferred onto a PVDF membrane (Millipore). Membranes were blocked with blotto (5% nonfat milk powder in PBS plus 0.1% Tween 20) for 1 hour at room temperature and incubated with specific pMYPT1 and GAPDH primary antibodies (CST) at 4 °C overnight. Membranes were washed three times with PBS plus 0.1% Tween 20 and incubated with secondary antibody conjugated with HRP (CST) and subsequently incubated in chemiluminescent reagents and detected (Thermo Scientific, Waltham, MA).

2.5 RT-qPCR assay and kinase inhibition

The osteogenesis-related gene expression was assayed by real-time quantitative polymerase chain reaction (RT-qPCR). Cells were seeded on the samples in a 6-well plate at an initial density of 3×10^5 cells/well. After incubation for 14 days, total RNA was isolated using trizol (Life Technology), and quantitative real-time PCR was performed using an ABI 7500 Real-Time PCR System (Applied Biosystems, USA) with a PCR kit (SYBR Premix EX Taq, TaKaRa). The relative expression of type I collagen (COL I), osteopontin (OPN) and osteocalcin (OC) were evaluated and β-actin was used as the internal control for normalization. Futhermore, ROCK was inhibited using ROCK-specific inhibitor Y-27632 (Sigma, 10μM). After 14 days, osteogenic gene expression was also detected by real-time PCR as previously described. All of the Gene expression of other three groups were presented relative to the corresponding results of polished titanium.

2.6 Statistical analysis

All experiments were conducted in triplicate and repeated three times, and all the data were expressed as the means ± standard deviation (SD). Statistical differences were determined by methods of *T-test*. Values of $p< 0.05$ were considered to be statistically significant.

3 Results

3.1 Sample preparation and surface characterization

The surface morphologies of the fabricated samples were characterized by SEM, as shown in **Fig. 1**. The polished titanium (PT) was used as control. After hydrothermal treatment, an anisotropic nanowire-like texture (NT) appeared with the wire diameters between 20-50 nm. The TPS surface was composed by molten and flattened particles, which on the whole presented an irregular rough macroporous morphology. After hydrothermal treatment, hierarchical structures with similar nanowires were found for NTPS surface while no obvious changes could be seen in the original micro-scale texture.

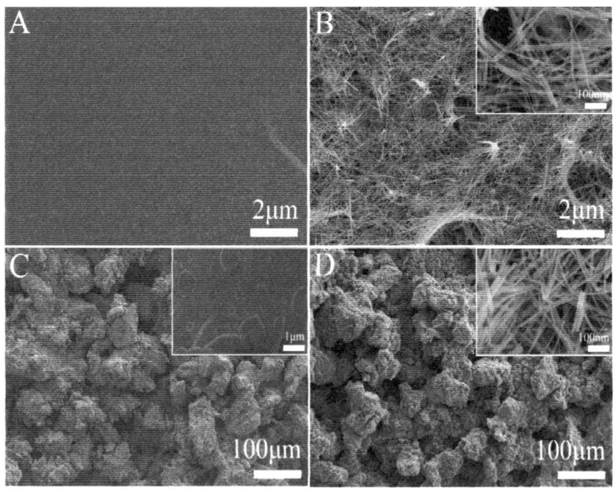

Fig. 1. SEM pictures of surface topography (A) PT; (B) NT; (C) TPS; (D) NTPS

3.2 Cell morphology

The morphology of MC3T3 and BMSC on different substrates were displayed, **Fig. 2-3**. Similar cytoskeletal status could be observed for these two cell species. On PT surfaces, most of the cells presented spindle or round morphology with less-spread body and few filopodia. With the introduction of nanowire, cells were more spread and exhibited a polygonal morphology with a relatively larger size. For the TPS surface, cells showed an elongated body which adapted themselves to the complicated rough topography of coating. However, poor cell spreading all around with few filopodia could be seen. On the hierarchical NTPS surface , synergistic cytoskeletal distribution could be observed. Cells went along with the coating's 3-D topography and spread well all around simultaneously. Large amounts of filopodia could also be seen on the edge of skeleton, presenting a good interaction with the nanowire texture.

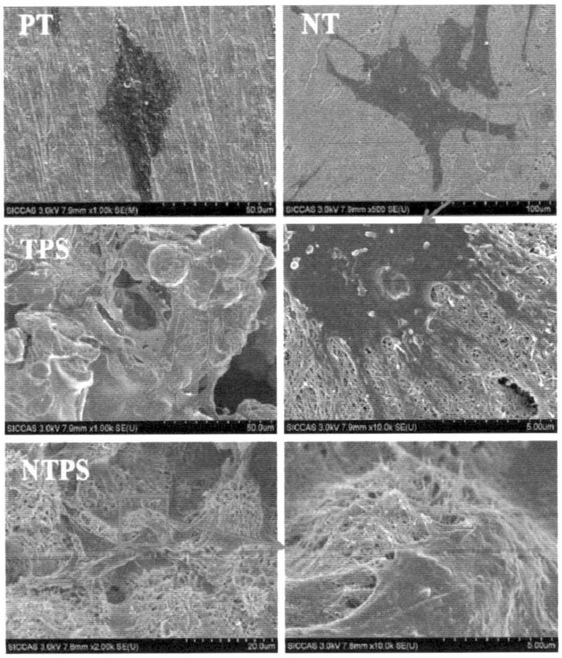

Fig. 2. SEM pictures of MC3T3 morphology

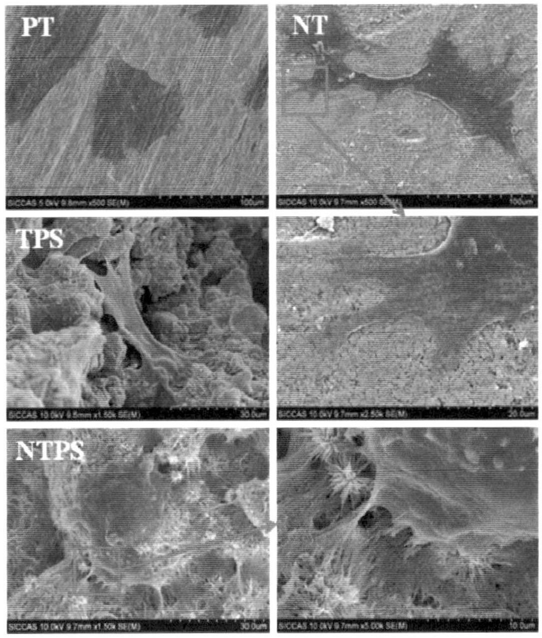

Fig. 3. SEM pictures of BMSC morphology

3.3 Western blotting

To examine whether the specific cell morphology and cytoskeleton distribution aforementioned were related with the cytoskeleton tension, the ROCK activity was detected by phosphorylation of MYPT1, one of ROCK's substrate. Results showed that the ROCK activity of MC3T3 and BMSC cells on NT and TPS was greater than that on PT, while cells on NTPS presented higher ROCK activity than NT and TPS, **Fig. 4-5.**

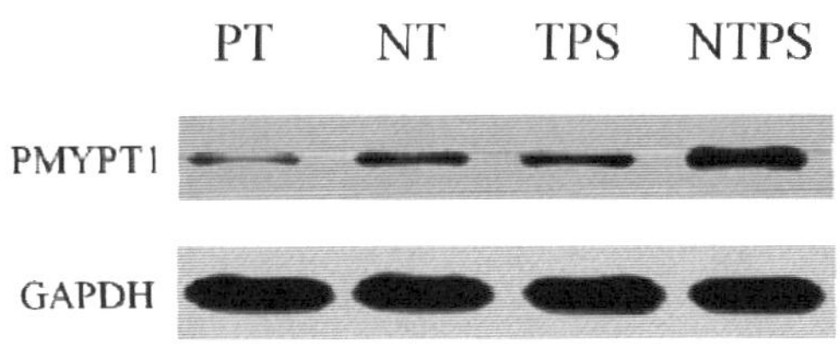

Fig. 4. Western blotting of pMYPT1 of MC3T3 cells

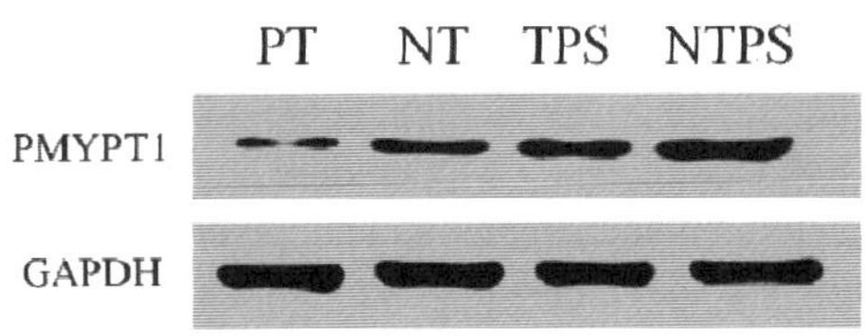

Fig. 5. Western blotting of pMYPT1 of BMSCs

3.4 Real-time PCR of osteogenic genes and ROCK inhibition

The osteogenic gene expressions, including COL1, OPN and OC after 14 days, were quantified by real-time PCR. NT and TPS facilitated all of them in different degrees, while comprehensively higher expressions on the NTPS surface could be observed. In order to further confirm the relationship among cytoskeleton tension and cell differentiation, ROCK was further inhibited with Y27632. Results showed that after ROCK was inhibited, all the three osteogenic gene expression underwent a slide for all four groups. The gaps between different samples diminished significantly, **Fig. 6-7.**

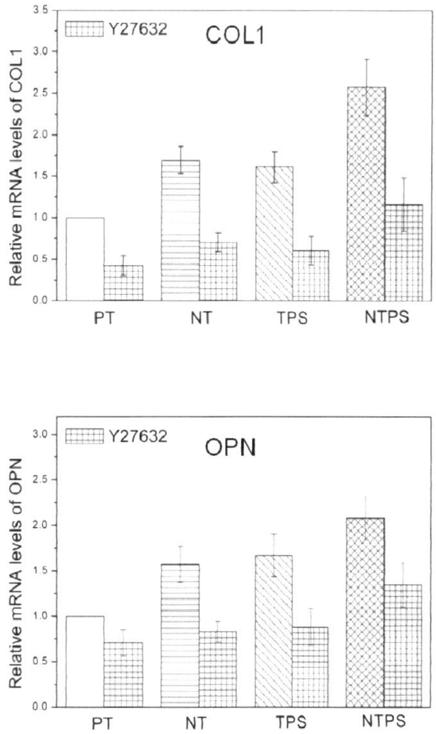

Fig. 6. Osteogenic genes of MC3T3 cells after 14 days

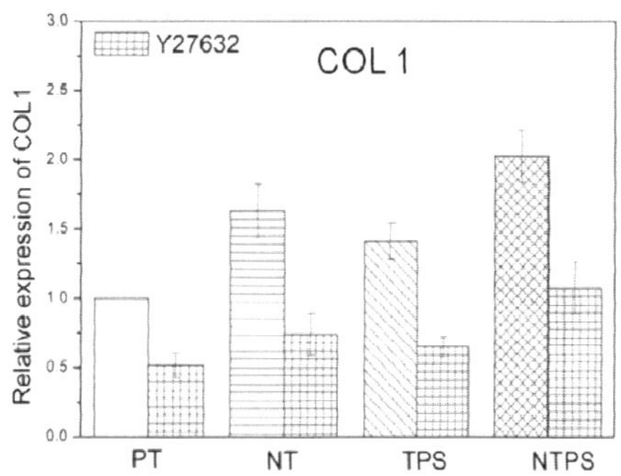

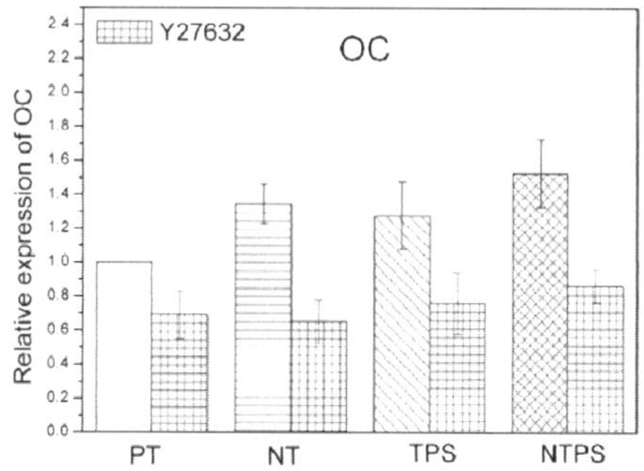

Fig. 7. Osteogenic genes of BMSCs after 14 days

4 Discussion

Nano-modified plasma sprayed titanium coating (NTPS) has been considered as a promising approach for enhancing ealrly osseointegration in recent years. However, the detailed mechanism mediating the response of cells to these NTPS surfaces was poorly investigated. Thus, in this study, further exploration about detailed mechanism has been processed. MC3T3 cells and BMSCs were simultaneously researched for better illustration. Our results suggested no obvious difference between MC3T3 and BMSC behaviors as regards the topography-differentiation mechanism. The NT and TPS structure did play different roles in cytoskeleton distribution respectively, and the combination of the two could lead to a synergistic effect, which resulted in higher ROCK activity and enhanced cytoskeleton tension and finally made a positive effect on cell differentiation.

Nanostructures have been believed to have significant effect on early cell response. In this study, when seeded on nanowire surface, cells were outstretched and exhibited a polygonal morphology with extended filopodia. It is widely accepted that nano-scale structures can contribute to cell spreading [7]. This probably attributes to its high surface energy and high hydrophilia, which will benefit the specific protein adsorption and integrin expression and leads to better cell spreading [8]. cells on the TPS surface showed an elongated body and extended along the rough up-and-down topography formed by melted particles, but poor cell spreading accompanying with little filopodia could also be observed, which means poor adaptation to the native rough surface topography and inferior cytoskeleton development. These phenomenons were common on TPS surface as reported [3-4]. Cells on the hierarchical NTPS surface spread well all around while went along with the coating's rough topography. It seemed that the macro/nano topography could provide multi-dimensional cytoskeleton deformation alongwith well-spread cytoskeleton in the same time, which meant synergistic effect was obtained.

Cytoskeleton tension has been considered to accompany with cell spreading and cytoskeleton reorganization. ROCK is known to play a pivotal role in cytoskeleton reorganization. In our study, Moderate ROCK activity were found on NT and TPS and higher ROCK activity could be observed on NTPS. These were in line with the cell morphology aforementioned. Despite the excellent cell spreading on NT surface, small roughness meant cytoskeleton deformation in a single dimension , which meant relatively small cytoskeleton tension. Similarly, TPS provided rough macroporous surface but cells just took on elongated bodies rather than spread all around on the rough surface, accompanied by less intracellular tension. As a matter of course, when good cell spreading met rough surface or marcopores, three-dimensional well-developed cytoskeleton distribution was formed and additional tension was produced.

Cytoskeletal tension are believed to influence cell differentiation through intracellular mechanotransduction, namely signal pathways [9]. Focal adhesions (FAs) play a important role in the transduction of information of extracellular environment and amounts of kinases locate in these areas [10]. Besides, Focal adhesions are linked to actin through specific protein [6,11]. Hence, the elevated cytoskeleton tension by NTPS might made a effect to focal adhesion conformation and the phosphorylation of specific kinases could also be regulated. Thus signaling pathways concerning the osteogenic lineage could be activated. The synergistic effect of NTPS on cytoskeletal distribution finally turned into better differentiation performance. Further inhibition of ROCK impeded expression of osteogenic genes on different surfaces, which verified the pivotal role of ROCK and intracellular tension in cell differentiation.

5 Conclusion

In this paper, synergistic effect of NTPS on cytoskeleton distribution was observed for both MC3T3 and BMSC. Better cell spreading and three-dimensional cytoskeleton deformation were obtained on NTPS simultaneously, which led to higher ROCK activity and finally enhanced differentiation. It points out that the cell morphology and cytoskeleton and the concomitant intracellular mechanotransduction might be decisive for the cell osteogenic fate of MSCs on NTPS.

Acknowledgments
This work is financially supported by the Opening Project of Shanghai Key Laboratory of Orthopaedic Implant (KFKT2014002).

References

[1] A. Wisbey, P. J. Gregson, L. M. Peter and M. Tuke. Effect of surface treatment on the dissolution of titanium-based implant materials. Biomaterials 12 (1991), Issue 5, pp. 470-473.

[2] D. A. Puleo and A. Nanci. Understanding and controlling the bone–implant interface. Biomaterials 23-24 (1990), Issue 20, pp. 2311-2321.

[3] Y. T. Xie, X. B. Zheng, L. P. Huang and C. X. Ding. Influence of hierarchical hybrid micro/nano-structured surface on biological performance of titanium coating. Journal of Materials Science 47 (2012), pp. 1411-1417.

[4] Y. T. Xie, H. Y. Ao, S. G. Xin, X. B. Zheng and C. X. Ding. Enhanced cellular responses to titanium coating with hierarchical hybrid structure. Materials Science & Engineering C 38 (2014), pp. 272-277.

[5] R. McBeath, D. M. Pirone, C. M. Nelson, K. Bhadriraju and C. S. Chen. Cell shape, cytoskeletal tension, and RhoA regulate stem cell lineage commitment. Dev Cell 6 (2004), Issue 4, pp. 483-495.

[6] C. H. Seo, K. Furukawa, K. Montagne, H. Jeong and T. Ushida. The effect of substrate microtopography on focal adhesion maturation and actin organization via the RhoA/ROCK pathway. Biomaterials 32 (2011), Issue 36, pp. 9568-9575.

[7] K. von der Mark, J. Park, S. Bauer and P. Schmuki. Nanoscale engineering of biomimetic surfaces: cues from the extracellular matrix. Cell and tissue research 339 (2010), Issue 1, pp. 131-153.

[8] A. Klymov, L. Prodanov, E. Lamers, J. A. Jansen and X. F. Walboomers. Understanding the role of nano-topography on the surface of a bone-implant. Biomaterials Science 1 (2013), pp. 135-151.

[9] L.A. Turner and M. J. Dalby. Nanotopography – potential relevance in the stem cell niche. Biomaterials Science 2 (2014), Issue 11, pp. 1574-1594.

[10] B. K. K. Teo, S. T. Wong, C. K. Lim, T. Y. S. Kung, C. H. Yap, Y. Ramagopal, L. H. Romer and E. K. F. Yim. Nanotopography modulates mechanotransduction of stem cells and induces differentiation through focal adhesion kinase. Acs Nano 7 (2013), Issue 6, pp. 4785-4798.

[11] Y.R.V. Shih, K. F. Tseng, H. Y. Lai, C. H. Lin and O. K. Lee. Matrix stiffness regulation of integrin-mediated mechanotransduction during osteogenic differentiation of human mesenchymal stem cells. Journal of Bone and Mineral Research 26 (2011), Issue 4, pp. 730-738.

Oxidation Behavior of Titanium Carbonitride Coating Deposited by Atmospheric Plasma Spray Synthesis

L. Zhu, J. He, D. Yan, Tianjin/CN, N. Zhang, Shenyang/CN

As a high-hard and anti-frictional material, titanium carbonitride (TiCN) thick-coatings or thin-films are increasingly used in many industrial fields. In the present study, TiCN coatings were obtained by atmospheric plasma spray synthesis or reactive plasma spray. In order to promote the reaction between Ti particles and reactive gases, a home-made gas tunnel was mounted on a conventional plasma gun to perform spray process. The oxidation behavior of the TiCN coatings under different temperatures at static air was carefully investigated. As a result, when the temperature was over 700 °C, the coatings suffered a serious oxidation, and finally they were entirely oxidized to be TiO_2 phase at 1100 °C. The principal oxidation mechanism was clarified, indicating that the oxygen can permeate into the defects and react with TiCN at high temperature.In addition, concerning the use of TiCN coating at high temperature conditions, the microhardness of the oxidized coatings at different treatment temperatures were also evaluated.

1 Introduction

Plasma spray synthesis technique, or reactive plasma spray (RPS), has been well established and developed in recent decades. Until now, a large number of RPS coating materials has been successfully synthesized contributing to different application needs [1-5]. Generally, RPS can be considered a combination of conventional plasma spray and self-propagating high-temperature synthesis (SHS) process, which makes coating deposit and chemical reaction into a single step with low cost and high deposition rate [6]. In contrast to conventional plasma spray process, RPS coatings possess many excellent characteristics, including high bond strength, further refined coating microstructure and good flattening [7, 8]. Thus, as an advanced coating method, RPS provides a potential opportunity to elaborate varied compounds or metastable phases by an exothermal reaction between selective powders and surrounding reactive gaseous matters in the plasma jet [9], such as nitrides, carbides, boride, carbonitrides, etc.

It is well-known that TiCN cermet is an attracted coating material applicable for various structural and functional component surfaces due to its excellent wear, corrosion and oxidation resistance, high hardness, high thermal conductivity, and relatively good toughness [10-13]. In the previous studies, TiCN coatings were frequently obtained either by physical vapour deposition (PVD) or by chemical vapour deposition (CVD) methods [14] and usually used as tool materials in cutting and forming applications [15]. Although these methods can fabricate high-quality and defect-free coating products, some inherent features of film processes certainly lead to ultra-thin coating thickness, low deposition rate, as well as costly investment [16], which cannot meet some special applications. For example, such thin coatings difficultly serve under extremely heavy load conditions. Therefore, in order to seek an effective and economical solution, RPS is involved to elaborate TiCN thick coatings with high deposition rate and economical cost.

In the present study, a gas-solid chemical reaction between Ti powder and reactive gases of nitrogen and acetylene (N_2 and C_2H_2) was used to synthesize TiCN coating. However, it is known that high-temperature in-flight melted particles are vulnerable to be oxidized under atmospheric environment, and the ultra-fast SHS reaction rate probably results in an incompletely reactive product. Based on these two points, a home-made gas tunnel filled with N_2 and C_2H_2 gases was thereby proposed and mounted in the front of conventional plasma gun promoting SHS reaction and decreasing the oxidation of in-flight melted particles.

In most cases, TiCN coatings remain exposed to elevated temperatures in oxidizing environments [17] in use and they can also suffer a serious oxidization under wear conditions due to high friction heat. Thus, the oxidation resistance or thermal stability of the coatings under such conditions plays an important role in influencing coating lifetime. Accordingly, the objective of this work is to investigate the oxidation behavior of RPS TiCN coatings under different temperature conditions in static air. Phase compositions and microstructure characteristics of the oxidized TiCN coatings were carefully examined. The oxidation mechanism and the hardness of the oxidized coatings were also studied.

2 Experimental methods

The starting feedstock was a commercially available titanium (Ti) powder (GRINM, Beijing, China) with an irregular shape and a particle size distribution of 15 to 45 µm as shown in Fig. 1, and its element ingredients are listed in Table 1. The ASTM 1015 mild steel substrate samples were firstly degreased and grit-blasted prior to RPS process. The GP-80B atmospheric plasma spray equipment (Jiujiang, China) was employed to deposit TiCN coating. A home-made reactive gas tunnel were assembled with a normal DC (direct current) BT-G3 plasma torch as a reactive spray system [5, 18], as shown in Fig. 2. The Ni/10 wt%Al alloy powder was firstly pre-sprayed onto the substrate surface as a bond coat to improve the adherence between coating and substrate. Afterward, the pure Ti powder was introduced into the plasma jet reacting with N_2 and C_2H_2 gases inside the gas tunnel, and finally TiCN coatings with a thickness of more than 200 µm were obtained. The used RPS parameters are displayed in Table 2.

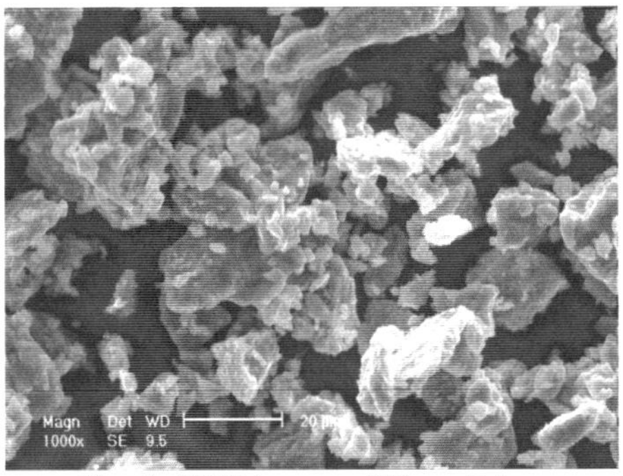

Fig. 1. SEM image of Ti particles.

Table 1. Ingredients of raw Ti powder.

Element	Ti	Fe	Si	Cl
wt%	>99.4	<0.06	<0.02	<0.01
Element	C	N	O	H
wt%	<0.02	<0.04	<0.4	<0.02

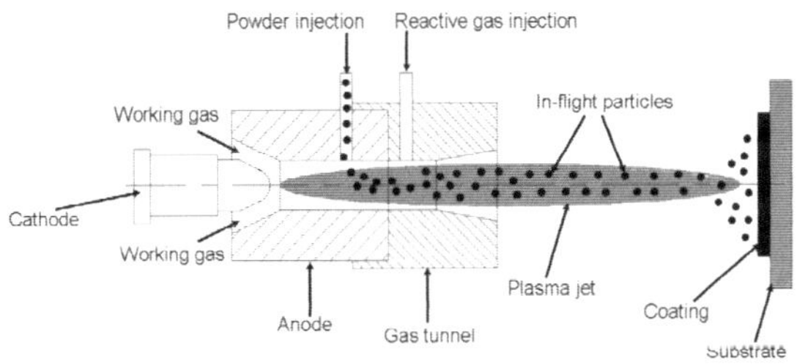

Fig. 2. Experimental configuration of the gas tunnel mounted on a conventional plasma gun.

Table 2. The used RPS parameters.

Parameters	Value
Standoff distance (mm)	110
Current (A)	500
Voltage (V)	70
Power (kW)	35
Ar (slpm)	40
N_2 (slpm)	60
C_2H_2 (slpm)	4

The obtained TiCN coatings were peeled off from the substrates (without bond coat). The oxidation experiments of the coatings were performed in a high-temperature furnace under atmospheric ambience for ten cycles, and five oxidation temperatures of 300 °C, 500 °C, 700 °C, 900 °C, and 1100 °C were selected in this experiment. In each cycle, the coating was held for one hour followed by 15 min cooling at room temperature. The increase of oxidation weight were measured at the end of each cycle with help of electronic balance with a sensitivity of 0.1 mg.

The phase composition of the oxidized coatings was determined by mean of an X-ray diffractometer (XRD, JEOL 2500 PC) with a CuKα radiation. A scanning electron microscopy (SEM, PHILIPS XL30-TMP) was used to determine coating microstructure (with substrate and bond coat).

In order to evaluate the properties of the TiCN coatings after oxidation, the microhardness was measured at the polished cross-section of the coating by means of a digital Vickers microhardness tester HXD-100 (Shanghai, China). Load of 100 g and duration time of 15 s were selected. 10 indentations were randomly impressed for each load and the average microhardness values were used to estimate the coating hardness.

3 Results and discussion

Fig. 3 shows the correlation of the oxidation weight increase versus thermal cycle number of the TiCN coatings oxidized under different temperatures (300 to 1100 °C). It is obvious that the coatings suffered an oxidation under all temperature conditions. Especialy, the TiCN coatings oxidized under relatively low temperatures (300 and 500 °C) display less weight increase in comparison with those oxidized under higher temperature conditions (more than 700 °C). The total oxidation weight increases after 10 cycles are measured to be 2.9, 6.7, 67.6, 112.5, 161.4 mg/cm^2 corresponding to 300, 500, 700, 900, and 1100 °C, respectively. At the oxidation early stage, the oxidatation weight rapidly increases as the temperature increased more than 700 °C, and then this increase is slowed down, which suggests that the coatings undergo more serious oxidation at high temperature.

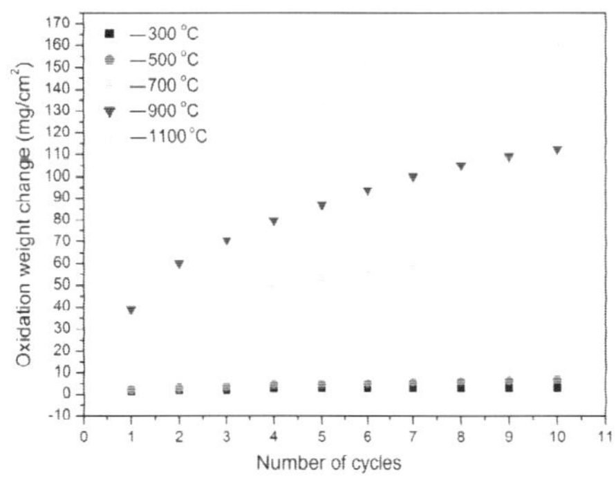

Fig. 3. Oxidation weight change of the TiCN coatings oxidized under different temperatures.

The XRD patterns of the TiCN coatings oxidized under different temperatures (20 to 1100 °C) are displayed in Fig. 4. It is found that the synthesized coating is mainly composed of $TiC_{0.2}N_{0.8}$ phase (see JCPDS: # 76-2484) with a small amount of Ti_3O at 20 °C, and displays a face centered cubic (FCC) structure. Also no elementary titanium is detected in the coating. The similar result has been obtained in our previous research [18], and it is also in good agreement with other researchers' studies [20, 21]. Similarly, when low oxidation temperatures (300 and 500 °C) are used, the coating takes place a slightly oxidation. Note that the content of Ti_3O phase increases and the $TiC_{0.2}N_{0.8}$ phase keeps comparatively high stability during the oxidation. As the increase of oxidation temperature (700 and 900 °C), TiO_2 phase appears in the coating, which means that $TiC_{0.2}N_{0.8}$ phase undergoes a serious oxidation. When the temperature continuously elevates at 1100 °C, the coating is fully corroded to be TiO_2 phase.

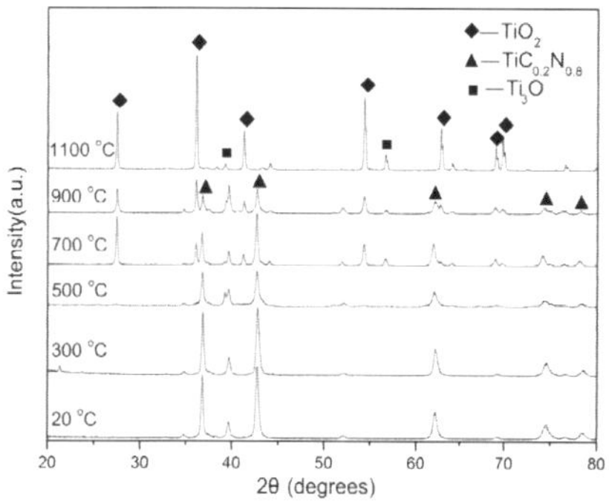

Fig. 4. XRD patterns of the TiCN coatings oxidized under different temperatures.

In order to further understand the oxidation process of TiCN coatings, the cross-section microstructures of the coatings oxidized under different temperatures after 10 cycles are presented in Fig. 5. According to the observation of the coating cross-section morphologies, the coating shows a lamallar microstructure with some pores and microcracks before the inoxidation (Fig. 5 (a)). It is also found that the coating has been oxidized slightly at 300 °C from the back scattering electron (BSE) image (Fig.5 (b)), and the cracks and pores increase and a little of oxidized phases (gray phase) can also observed in the coating. Further, when the temperature increases to 700 °C (Fig.5 (d)), more oxidized phases can be observed in the coating. Based on the XRD analysis, these oxidized phases are TiO_2. Moreover, the pores and cracks are filled by these oxidized phases making the coating more dense, which implys that the oxidation initially occurs at these defect position. As the temperature increases, the oxidation becomes more serious. The new pores appear again and their size gradually enlarges. Finally, the coating is fully oxidized at 1100 °C, a large amount of cracks and pores are presented in the coating, which causes that the coating peels off from the substrate.

The degradation of the TiCN coatings is thought to be due to the formation of oxidized phases (main TiO_2) at their surfaces and defects causing delamination. It can be seen that the oxygen can gradually penetrate into the TiCN coating to form some defects, lowering the mechanical properties. During the oxidation process, the nitrogen and carbon atoms of the TiCN coating can be substitute by the oxygen atoms, and the gas phases forms by reducing the carbon or nitrogen atoms. Furthermore, these atoms released in the oxidation process tend to be trapped within the oxide layer or at the interface, and therefore cause pores or delamination [22].

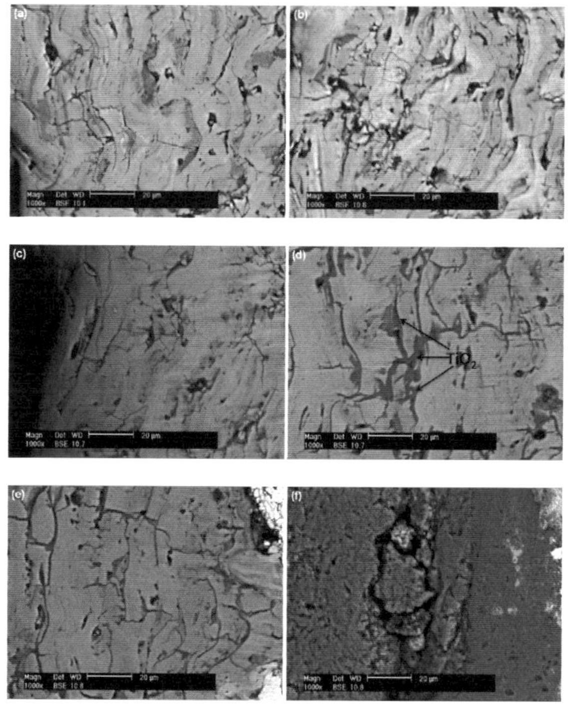

Fig. 5. BSE images of cross-sectional microstructures of the oxidized TiCN coatings after 10 cycly: (a) 20 °C; (b) 300 °C; (c) 500 °C; (d) 700 °C; (e) 900 °C and (f) 1100 °C.

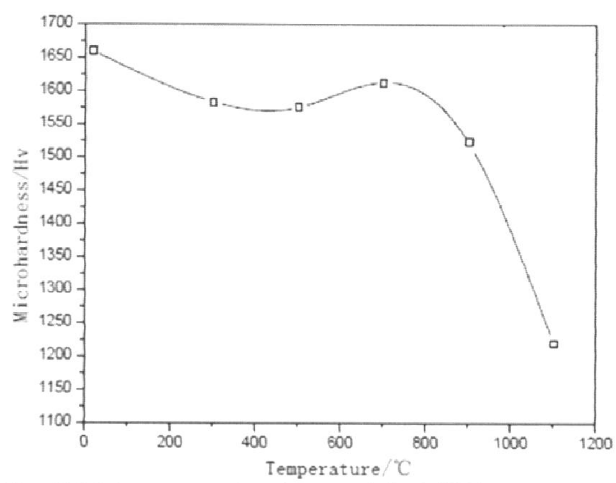

Fig. 6. Vickers microhardness of the TiCN coatings as a function of oxidation temperatures.

Based on the application consideration of TiCN coating, the properties of the oxidized coatings are evaluated by microhardness tests. A comparison study of microhardness measurements on the oxidized and inoxidized coatings is shown in Fig. 6.

The micrhardness of the inoxidized coating present the highest value compared to those of the oxidized catings. As the oxidation temperature increases, the microhardness progressively slows down. However, it is remarkable that the microhardess value of the oxidized coating show slightly increases at 700 °C, because the TiO_2 phase formed at the defect position improves the density of the coating. Afterward, the microhardness value is severely decreased because of the change of microstructure and phase and more pores and cracks formed, thus the coating shows the extremely low microhardness after the oxidation at 1100 °C. It is indicated that the coating can be applied under 700 °C

4 Concluding remarks

In summary, TiCN cermet coatings were successfully synthesized through RPS process using a home-made gas tunnel plasma spray torch in this work. According to the experimental results of oxidation and hardness measurements, the following important viewpoints have to be pointed out:

(i) The oxidation weight change and XRD analysis at the different temperature presented that the TiCN can be slowly oxidized under 700 °C. As temperature exceeded 700 °C, the oxidation weight was strongly increased, and a large amount of oxidized phase formed. The coating was fully oxidized to be TiO_2 at 1100 °C in final.

(ii) The cross-sectional morphologies of the oxidized coatings at different temperature displayed that the TiO_2 was formed at the defect position. As temperature increased, the fraction of oxidized phase in the coating increased. The oxidation caused the appearance of more defects, which resulted the degradation of the coatings.

(iii) The microhardness value of the coating was progressively decreased as the temperature increased due to the oxidized phase formed.

Acknowledgment

This research is financially supported by Natural Science Foundation of Hebei province, China (Grant Nos. E2009000052 and E2005000056). The authors would like to thank Mr. SUN Weichang (Tianjin University of Technology, Tianjin, China) for help in XRD analysis.

References

[1] P. Fauchais, A. Vardelle, A. Denoirjean, Reactive Thermal Plasmas: Ulltrafine Particle Synthesis and Coating Deposition, Surf. Coat. Technol., 1997, 97: pp 66/78.

[2] C. Tekmen, Y. Tsunekawa, M. Okumiya, In-Situ TiB2 and Al2O3 Formation by DC Plasma Spraying, Surf. Coat. Technol., 2008, 202(17): pp 4170/4175.

[3] C. Tekmen, Y. Tsunekawa, M. Okumiya, Effect of plasma spray parameters on in-flight particle characteristics and in-situ alumina formation, Surf. Coat. Technol., 2008, 203: pp 223/228.

[4] G.J. Yang, C.J. Li, Y.Y. Wang, Phase formation of nano-TiO2 particles during flame spraying with liguid feedstock, J. Therm. Spray Technol., 2005, 14(4): pp 480/486.

[5] L. Zhu, J.N. He, D.R. Yan, L.S. Xiao, Y.C. Dong, D.C. Xue, D.L. Meng, Titanium Carbonitride Thick Coating Prepared by Plasma Spray Synthesis and its Tribological Properties, Chin. Sci. Bull., 2007, 57: pp 1849/1855.

[6] M. Yamada, Y. Kouzaki, T. Yasui, and M. Fukumoto, Fabrication of Iron Nitride Coatings by Reactive RF Plasma Spraying, Surf. Coat. Technol., 2006, 201: pp 1745/1751.

[7] F. Casadei, R. Pileggi, R. Valle, A. Matthews, Studies on a Combined Reactive Plasma Sprayed/Arc Deposited Duplex Coating for Titanium Alloys, Surf. Coat. Technol., 2006, 201: pp 1200/1206.

[8] D. Zhao, F. Luo, W. Zhou, D. Zhu, Microwave absorption properties and complex permittivity of Fe/FeAl2O4 coatings deposited by reactive plasma spraying Al/Fe2O3 powders, Surf. Coat. Technol., 2011, 205: pp 4254/4259.

[9] C. Tekmen, M. Yamazaki, Y. Tsunekawa, and M. Okumiya, In Situ Plasma Spraying: Alumina Formation and In-Flight Particle Diagnostic, Surf. Coat. Technol., 2008, 202(17): pp 4163/4169.

[10] L. Zhu, J. He, D. Yan, Y. Dong, L. Xiao, Titanium Carbonitride Coatings Prepared by Reactive Plasma Spraying Ti powders, Surf. Rev. Lett., 2007, 14: pp 171/177.

[11] O. Zgalat-Lozynskyy, M. Herrmann, A. Ragulya, Spark plasma sintering of TiCN nanopowders in non-linear heating and loading regimes, J. Eur. Ceram. Soc., 2011, 31: pp 809/813.

[12] Y. Mu, M. Wang, D. Yu, Synthesis of Ti(CN) powders by combustion reaction from Ti powder and a novel carbon–nitrogen precursor, Int. J. Refract. Met. Hard Mat., 2011, 29: pp 326/328.

[13] M. Stueber, H. Holleck, H. Leiste, K. Seemann, S. Ulrich, C. Ziebert, Concepts for the design of advanced nanoscale PVD multilayer protective thin films, J. Alloys Compd., 2009, 483: pp 321/333.

[14] E. Gustenau-Michalek, P. Herzig, A. Neckel, Titanium carbonitrides, Ti(C,N): electronic structure and chemical bonding, J. Alloys Compd., 1995, 219: pp 303/306.

[15] B. Gómez, A. Jiménez-Suarez, E. Gordo, Oxidation and tribological behaviour of an Fe-based MMC reinforced with TiCN particles, Int. J. Refrac. Metals Hard Mater., 2009, 27: pp 360/366.

[16] Y. Yang, D.R. Yan, Y.C. Dong, L. Wang, X.G. Chen, J.X. Zhang, J.N. He, X.Z. Li, J. In Situ Nanostructured Ceramic Matrix Composite Coating Prepared by Plasma Spraying Composite Powders, Alloys Compd., 2011, 509: pp L90/L94.

[17] F. Monteverde, A. Bellosi, Oxidation behavior of titanium carbonitride based materials, Corros. Sci., 2002, 44: pp 1967/1982.

[18] L. Zhu, J. He, D. Yan, L. Xiao, Y. Dong, J. Zhang, H. Liao, Synthesis and Microstructure Observation of

Titanium Carbonitride Nanostructured Coatings Using Reactive Plasma Spraying in Atmosphere, Appl. Surf. Sci., 2011, 257: pp 8722/8727.
[19] W.R. Feng, D.R. Yan, J.N. He, X.Z. Li, Y.C. Dong, Reactive plasma sprayed TiN coating and its tribological properties, Wear, 2005, 258: pp 806/811.
[20] X. Feng, L.Y. Shi, Novel chemical metathesis route to prepare TiCN nanocrystallites at low temperature, Mater. Chem. and Phys., 2005, 94: pp 58-61.
[21] J.C. Caicedo, C. Amaya, L. Yate, W. Aperador, G. Zambrano, M.E. Gómez, J. Alvarado-Rivera, J. Muñoz-Saldaña, P. Prieto, Effect of Applied Bias Voltage on Corrosion-Resistance for TiC1-xNx and Ti1-xNbxC1-yNy Coatings, Appl. Surf. Sci., 2010, 256: pp 2876/2883.

[22] I. Milošev, H.-H. Strehblow, B. Navinšek, XPS in the study of high-temperature oxidation of CrN and TiN hard coatings, Surf. Coat. Technol., 1995, 74-75, pp 897/902.

Comparison of ZrB_2-$MoSi_2$ composite coatings fabricated by atmospheric and vacuum plasma spray processes

Yaran Niu, Zhong Wang, Xuebin Zheng, Chuanxian Ding
Key Laboratory of Inorganic Coating Materials, Shanghai Institute of Ceramics, Chinese Academy of Sciences, Shanghai 200050, China

In this work, ZrB_2 - 20vol.% $MoSi_2$ (named as ZM) composite coatings were fabricated by atmospheric plasma spray (APS) and vacuum plasma spray (VPS) techniques, respectively. Phase composition and microstructure of the composite coatings were characterized. Their oxidation behaviours and microstructure changes at 1500 °C were comparatively investigated. The results showed that VPS-ZM coating was composed of hexagonal ZrB_2, tetragonal and hexagonal $MoSi_2$. While certain amount of ZrO_2 existed in APS-ZM coating. The oxide content, surface roughness and porosity of VPS-ZM coating were apparently lower than those of APS-ZM coating. The mass gain of APS-ZM coating was maximum at the beginning (1500 °C, 0 h) and then decreased with extending the oxidation time. While the mass of VPS-ZM coating gradually increased with increasing the oxidation time from 0 h to 10 h. VPS-ZM coating had much better oxidation resistance than APS-ZM coating, presenting potential application as high-temperature protecting materials. The possible reasons for the difference of the two kinds of coatings were analysed.

1 Introduction

Ultra-high temperature ceramics (UHTCs) are potential thermal protective materials due to their extremely high melting points [1-2]. Zirconium diboride (ZrB_2) is typical among UHTCs owing to its unique combination of properties, including low density, high thermal conductivity, good thermal shock resistance and relatively low cost [3]. However, the poor oxidation resistance of ZrB_2 above 1100 °C limits it for extreme environment applications. It is well accepted that addition of silicon-containing materials is vital for the improvement of the oxidation resistance of ZrB_2. Therefore, several silicon-containing ceramics, such as SiC, $MoSi_2$, and $TaSi_2$, have been incorporated into ZB_2 [4-5]. Among them, $MoSi_2$ has excellent oxidation-resistant property at high temperature and presents quite high thermal stability during fabrication processes [6-7]. The addition of $MoSi_2$ was considered to be an effective way to promote the high temperature oxidation resistance of ZrB_2.

Plasma spray technique is characterized by ultrahigh temperature of flame and high velocity of flying particles, which has been widely applied in diverse industries. In previous work in our laboratory, ZrB_2-SiC, ZrB_2-$MoSi_2$ and ZrB_2-$TaSi_2$ composite coatings have been fabricated by vacuum plasma spray (VPS) [8-9]. It is found that ZrB_2-based coatings with relatively dense microstructure could be obtained, which displayed excellent high temperature oxidation resistance. Commonly, the coating fabrication can be processed in air and inert gas environments by air plasma spray (APS) and vacuum plasma spray (VPS), respectively. The fabrication processes would have great influence on the microstructure of the resultant coatings and then their properties. However, to the best of our knowledge, no studies have been done on the comparison between the ZrB_2-based coatings prepared by APS and VPS technologies. And few researches have been carried out on the influence of oxide on the microstructure of the composite coatings and their oxidation behaviors.

In this work, ZrB_2-$MoSi_2$ composite coatings were fabricated by APS and VPS techniques, respectively. The phase compositions and microstructures of the composite coatings were characterized. The high temperature oxidation resistant property was evaluated at 1500 °C. The microstructure evolutions of the two kinds of composite coatings were comparatively studied. The possible explanations on the differences of oxidation behaviors between APS-ZM and VPS-ZM coatings were discussed.

2 Experimental procedure

2.1 Coating preparation

Commercially available ZrB_2 powder (Shanghai St-nano Science And Technology Co., Ltd., China) and $MoSi_2$ powder (Qinghuangdao Eco Material Co. Ltd., China) were used as feedstocks. ZrB_2 with 20 vol.% $MoSi_2$ (denoted as ZM) powders were mechanically mixed for 6 h to ensure the uniform distribution of the particles before spraying. The coatings deposition was carried out by a plasma spray system (A-2000, Sulzer Metco, Switzerland) equipped with a F4-MB torch for APS and a F4-VB torch for VPS, respectively. Argon and hydrogen were used as plasma forming gases. Free-standing coatings with thicknesses of 600-800 μm were deposited on graphite substrates and then mechanically peeled off for characterization.

2.2 Oxidation tests

Experiments of the oxidation resistance were carried out using a high temperature box furnace in air. The samples were placed in alumina crucible, heated at a rate of 5 °C /min from room temperature to 1500 °C and held for different time of 0 h - 10 h in stagnant air, and then naturally cooled down. The masses of the samples before and after the oxidation tests were measured by an electronic balance with accuracy of 0.1 mg. The mass gain (W_m) was defined as,

$$W_m = (m_2 - m_1) / m_1 \times 100\% \quad (1)$$

Where m_1 (g) and m_2 (g) were the masses of the samples before and after oxidation tests, respectively.

2.3 Composition and microstructure characterization

The compositions of the as-sprayed and oxidized coatings were analyzed by X-ray diffraction (XRD, RAX-10, Rigaku, Japan) operating with Cu Kα (λ = 1.5406 Å) radiation. The surface and polished cross-sectional morphologies of the as-sprayed and oxidized coatings were investigated by field emission scanning electron microscope (FESEM, Magellan, FEI, USA) with X-ray energy dispersive spectrometry (EDS, INCA ENERGY, Oxford, UK) which was appended to FESEM for chemical analysis. The surface roughness of the coatings (Ra) were measured by a profilometer (Hommelwerke T8000-C, Germany). The porosity of the composite coatings was measured by an image analysis system on the polished cross sections using image analysis (IMAGE, National Institute of Health, Bethesda, USA). The oxygen content in the coatings was characterized by Nitrogen/Oxygen Determinator (TC-600C, Leco, USA). All the values were averaged from 3-5 measurements.

3 Results and discussion

3.1 Compositions and microstructures

The XRD pattern of the as-received ZM powder is given in Fig. 1. It can be seen that the powder was composed of hexagonal ZrB_2 and tetragonal $MoSi_2$ phases. The APS-ZM and VPS-AM composite coatings are shown in Fig. 2. After the VPS deposition process, the intensity of tetragonal $MoSi_2$ in the composite coating decreased and some peaks corresponding to hexagonal $MoSi_2$ appeared. It is thought that the high cooling rate of the coating contributed to the phase transformation of $MoSi_2$ [10]. After the APS deposition process, certain amount of monoclinic ZrO_2 formed in APS-ZM coating. The oxygen content in APS-ZM coating was about 3.98 wt.%, which was six times higher than that of VPS-ZM coating (0.59 wt.%). The oxygen content in the as-received powders was also measured to be about 0.58 wt.%. It can be concluded that VPS process greatly reduced the introduction of oxide.

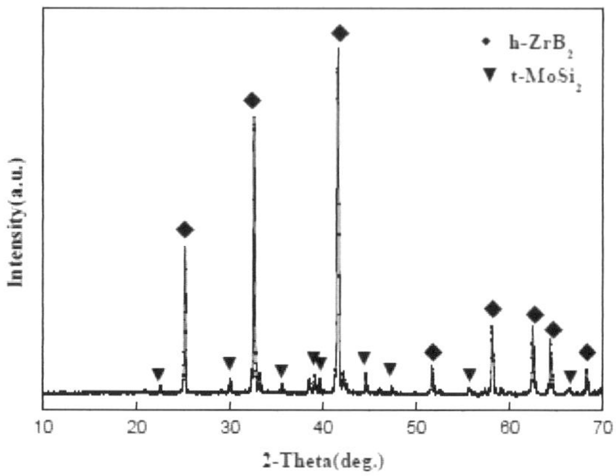

Fig. 1. XRD pattern of ZM powder.

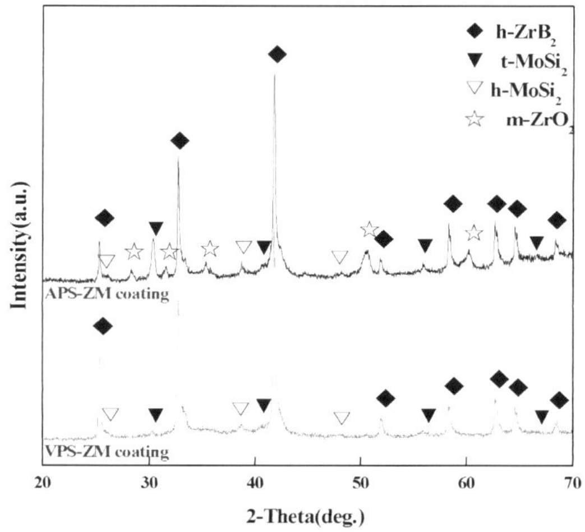

Fig.2. XRD patterns of APS-ZM and VPS-ZM coatings.

Fig. 3. Surface morphologies of APS-ZM coating (a) and VPS-ZM coating (b); inserts corresponding to cross-sectional morphologies.

The surface and cross-sectional morphologies of the composite coatings are illustrated in Fig. 3. It can be seen that both the coatings were composed of melted or semi-melted particles. The surface roughnesses (Ra) of APS-ZM and VPS-ZM coatings were about 9.36 μm and 7.42 μm, respectively. The roughness of APS-ZM coating was obvious higher than that of VPS-ZM coating. Moreover, there were more pores formed on the surface of APS-ZM coating (Fig. 2a). The cross-sectional images show that there were many pores with large size in APS-ZM coating. Moreover, it is found that some of the pores were connective, which would supply channels for oxygen and worsen the oxidation resistance. While the pores in VPS-ZM coating were of relatively small size and contactless. The measured porosities of APS-ZM and VPS-ZM coatings were about 9.3 % and 6.8 %, respectively.

3.2 Property characterization

The coatings were tested in a box furnace at 1500 °C for 0 h – 10 h to evaluate the oxidation resistance. The macro-photographs after oxidation for different hours are presented in Fig. 4. It can be seen that the color of APS-ZM coating have changed after heating to 1500 °C and then cooling down to room temperature (1500 °C, 0h). And it totally turned to white after 3 h oxidation. However, there was no obvious change for VPS-ZM coating even after 10 h oxidation.

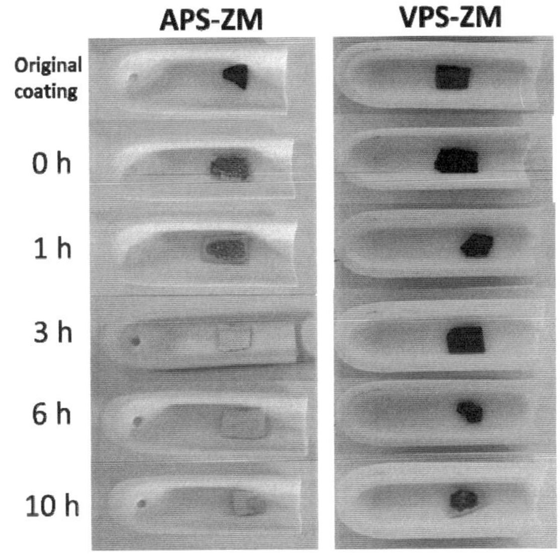

Fig. 4. Macro-photographies of APS-ZM and VPS-ZM coatings after oxidation at 1500 °C for different hours.

Fig. 5 illustrates the mass changes of the two kinds of coatings, from which, it can be seen that the oxidation behaviors were quite different for APS-ZM and VPS-ZM coatings. The mass of VPS-ZM coating kept gradually increasing during the tests. While the mass gain of APS-ZM coating was maximum at the beginning (1500 °C, 0h) and then decreased with prolonging the oxidation time. After 6 h treatment, the mass of APS-ZM did not change any more, indicating that it had been totally oxidized.

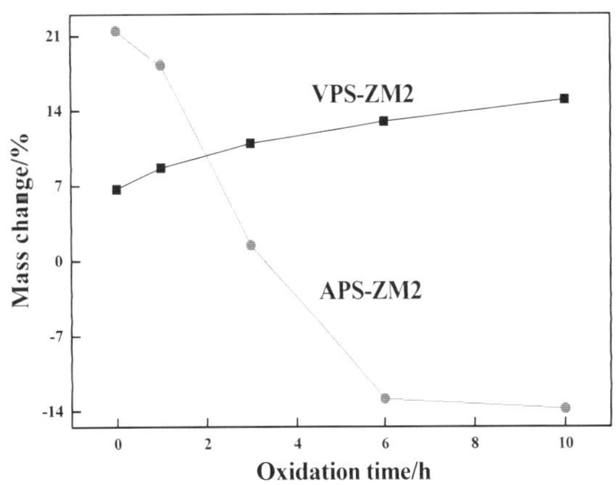

Fig. 5. Mass-changes of APS-ZM and VPS-ZM coatings after oxidation at 1500 °C for different hours.

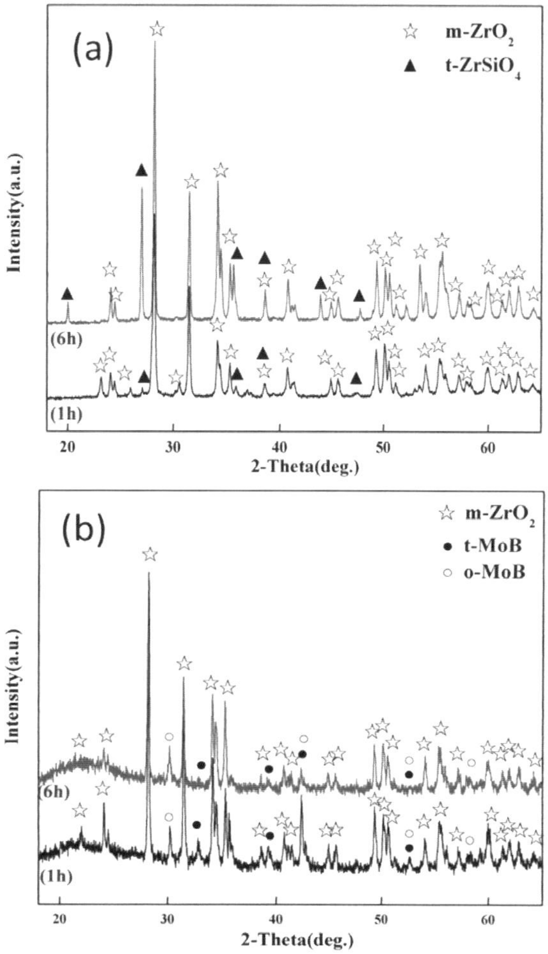

Fig. 6. XRD patterns of APS-ZM coating (a) and VPS-ZM coating (b) after oxidation at 1500 °C for 1 h and 6 h.

The phase compositions of APS-ZM and VPS-ZM coatings after oxidation for 1 h and 6 h are presented in Fig. 6. It is found that monoclinic ZrO_2 and tetragonal $ZrSiO_4$ were formed for APS-ZM coating after 1 h oxidation, which increased after 6 h oxidation. While for VPS-ZM coating, tetragonal and orthorhomb MoB was detected besides ZrO_2, however, no peaks corresponding to $ZrSiO_4$ appeared.

The surface morphologies of APS-ZM coatings after oxidation for 1 h and 6 h are illustrated in Fig. 7. A thin film covered the APS-ZM coating surface after 1 h oxidation, which was proved to be silica by EDS which was not shown here (Fig. 7a and 7b). However, the silica film disappeared after 6 h oxidation and polycrystalline grains with large size (Fig. 7b) covered the coating surface. The element maps prove that the grains were mainly composed of Si, O and Zr (Fig. 8). Combined with the XRD result, it can be concluded that the grains were ZrO_2 and $ZrSiO_4$.

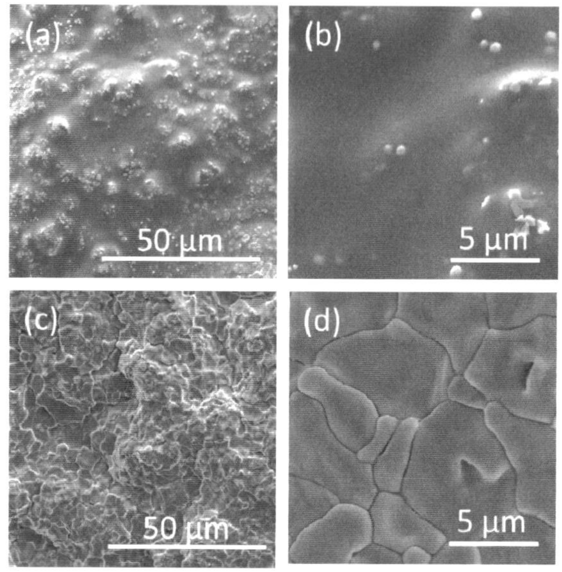

Fig. 7. Surface morphologies of the APS-ZM coatings after oxidation at 1500 °C for (a, b) 1 h and (c, d) 6 h.

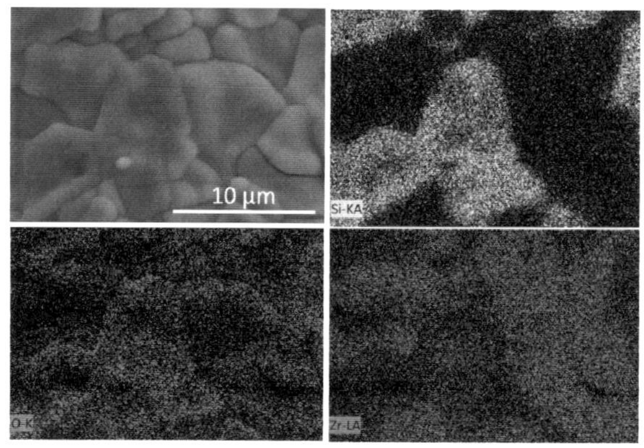

Fig. 8. Element maps of APS-ZM coating after oxidation at 1500 °C for 6 h.

The surface morphologies of VPS-ZM coatings after oxidation for 1 h and 6 h were illustrated in Fig. 9. It can be seen that a relatively dense silica film covered the surface of VPS-ZM coating after 1 h treatment. After 6 h oxidation, the surface become smoother and some new substance in white color appeared. The EDS results (Fig. 10) further confirmed that the white substance was ZrO_2.

It is well known that the formation of dense silica rich glassy layer on ZrB_2 surface is vital to its high temperature oxidation resistance. The oxidation process of ZrB_2-$MoSi_2$ composite which could take place in ambient air under high temperature (like 1500 °C) following the reactions:

$$5MoSi_2(s)+7O_2(g) \rightarrow Mo_5Si_3(s)+7SiO_2(s) \quad (2)$$

$$ZrB_2(s)+5/2O_2(g) \rightarrow ZrO_2(s)+B_2O_3(g) \quad (3)$$

$$2Mo_5Si_3(s) + 21O_2(g) \rightarrow 10MoO_3(g) + 6SiO_2\,(l) \quad (4)$$

$$2MoSi_2(s) + B_2O_3(g) + 5/2O_2(g) \rightarrow 4SiO_2(s) + 2MoB\,(s) \quad (5)$$

$$2MoB(s) + 9/2O_2(g) \rightarrow 2MoO_3(g) + B_2O_3\,(g) \quad (6)$$

$$ZrO_2(s) + SiO_2(l) \rightarrow ZrSiO_4(s) \quad (7)$$

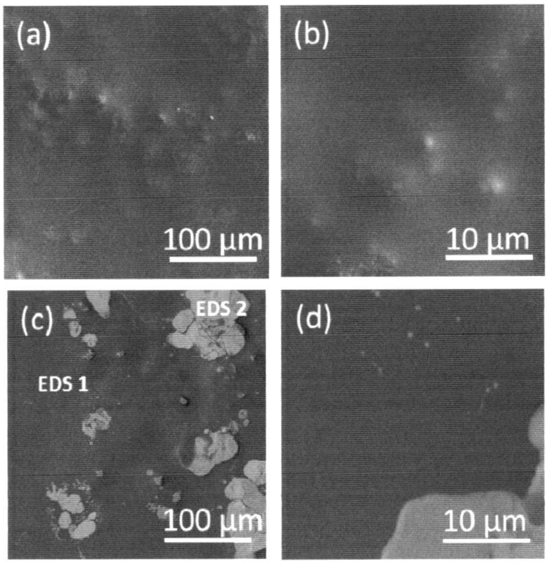

Fig. 9. Surface morphologies of the VPS-ZM coatings after oxidation at 1500 °C for (a, b) 1 h and (c, d) 6 h.

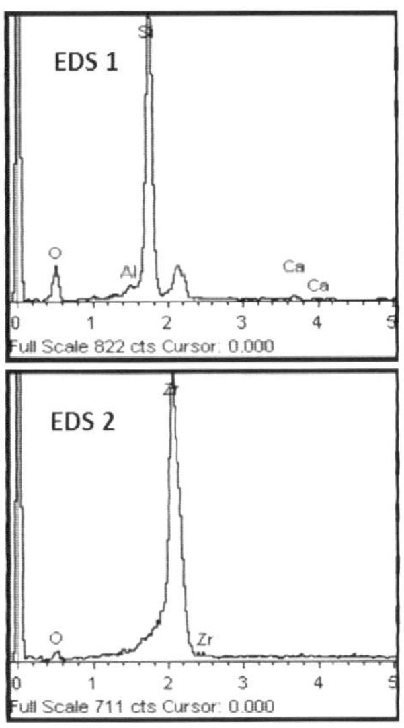

Element	Weight%
O K	34.48
Al K	1.28
Si K	62.70
Ca K	1.53
Tatals	100.00

Element	Weight%
O K	10.42
Zr K	89.58
Tatals	100.00

Fig. 10. EDS results of VPS-ZM coating after oxidation at 1500 °C for 6 h.

The XRD results confirmed that certain amount of $ZrSiO_4$ was formed for APS-ZM coating during the oxidation process even after 1 h oxidation. For APS-ZM coating, the resource of ZrO_2 can be divided into two parts, one came from the already existed ZrO_2 which was formed during the fabrication process, the other one came from the oxidation of ZrB_2 during the oxidation test. Previous work from our labratory have confirmed that $ZrSiO_4$ was formed in ZrB_2-$MoSi_2$ composite coating after oxidation at 1200 °C for 1 h. Increasing the content of $MoSi_2$ in the composite coating would result in producing more $ZrSiO_4$ [10]. Therefore, it is supposed that the formed silica would be consumed to synthesize $ZrSiO_4$ with the pre-existed ZrO_2. Besides that, the formation of $ZrSiO_4$ (ZrO_2) would result in thermal stresses or microstructure defects, due to the mismatch of thermal expansion coefficients among ZrB_2, ZrO_2 and $ZrSiO_4$. This would provide more channels for oxygen diffusion and then accelerate the oxidation process. At the same time, the large pores and high surface roughness of APS-ZM coating also inhibited the formation of continuous silica film.

For VPS-ZM coating, it had a relatively dense microstructure. A dense continuous silica film was formed on its surface and the thickness of which increased with extending the oxidation time. The silica film greatly reduced the diffusion of oxygen and prevented the further oxidation of underlying coating. The formation of MoB was also good for reducing the evaporation of B_2O_3 and MoO_3, defensing the integrity of silica film. Therefore, VPS-ZM coating presented excellent oxidation resistance in high temperature enviorment.

4. Conclusions

ZrB_2-$MoSi_2$ composite coatings were fabricated by air and vaccum plasma spray techniques, respectively. Certain content of ZrO_2 was formed during the APS fabrication process. APS-ZM coating had higher surface roughness, porosity and oxygen content campared with those of VPS-ZM coating. The oxiation tests at 1500 °C showed that the oxidation behaviors of APS-ZM and VPS-ZM coatings were quite different. A thin silica film was formed on the surface APS-ZM coating after 1 h treatment, while it disappeared after 6 h treatment. A thick dense silica film kept maintained on the surface of VPS-ZM coating after 6 h treatment. A lot of $ZrSiO_4$ was formed for APS-ZM coating after 1h oxidation. No $ZrSiO_4$ was detected in VPS-ZM coating after 6 h oxidation. It was supposed that the ZrO_2 phase and loose microstructure led to the poor oxidation resistance of APS-ZM coating.

Acknowledgements

This work was supported by the National Natural Science Foundation (for Young Scholar) of China under grant 51102267.

References

[1] K. Upadhya, J.M.Yang and W.Hoffman: Advanced materials for ultra-high-temperature structural

applications, American Ceramic Society Bulletin. 76 (1997) pp. 51/56.
[2] D.D. Jayaseelan, Y. Xin, L. Vandeperre, P. Brown, and W.E. Lee: Development of multi-layered thermal protection system (TPS) for aerospace applications. Composites Part B, 79 (2015) pp. 392/405.
[3] S.Q. Guo: Densification of ZrB_2-based composites and their mechanical and physical properties: a review. Journal of the European Ceramic Society, 29 (2009) pp. 995/1011.
[4] M.M. Opeka, I.G.Talmy, E.J.Wuchina, J.A.Zaykoski and S.J.Causey: Mechanical, thermal and oxidation properties of refractory Hafnium and Zirconium compounds. Journal of the European Ceramic Society 19 (1999) 2405/14.
[5] L. Silvestroni, G. Meriggi and D. Sciti: Oxidation behavior of ZrB_2 composites doped with various transition metal silicides. Corrosion Science, 83 (2014) 281/91.
[6] K. Bundschuh and M. Schütze: Oxidizing atmospheres Part I: basic considerations on materials selection. Materials and Corrosion, 52 (2001) 204/12.
[7] Y. Niu, X. Fei, H. Wang, X. Zheng and C. Ding: Microstructure characteristics and oxidation behavior of molybdenum disilicide coatings prepared by low pressure plasma spraying. Journal of Thermal Spray Technology, 22 (2013) 96/103.
[8] Y. Niu, H. Wang, H. Li, X. Zheng and C. Ding: Dense ZrB_2-$MoSi_2$ composite coating fabricated by low pressure plasma spray (LPPS). Ceramics International, 39 (2013) 9773/77.
[9] Z. Wang, Y. Niu, C. Hu, H. Li, Y. Zeng, X. Zheng, M. Ren and J. Sun: High temperature oxidation resistance of metal silicide incorporated ZrB_2 composite coatings prepared by vacuum plasma spray. Ceramics International, 41 (2015) 14868/75.
[10] Y. Niu, H. Wang, Z. Liu, C. Hu, X. Wang, X. Zheng and C. Ding: Microstructure evolution of ZrB_2-$MoSi_2$ composite coatings at middle and high temperatures. Surface & Coatings Technology, 273 (2015) 30/38.

Plasma-sprayed composite coatings for severe conditions in nuclear power plants

A. Werry, C. Chazelas, A. Denoirjean, S. Valette, A. Vardelle ; Univ. Limoges, CNRS, ENSCI, SPCTS, UMR 7315, F-87000 Limoges, France.

E. Meillot, CEA, DAM, Le Ripault, F-37260 Monts, France

The Sodium-cooled Fast Reactor (SFR) operates in fast neutron spectrum with a closed fuel cycle; it makes an efficient use of uranium resources and has the ability to burn actinides and so achieve sustainable development of nuclear energy. The liquid sodium coolant exhibits superior thermal properties, good neutronic characteristics, low activation and low density allowing a low pumping power. In addition the stainless steel corrosion rates in pure liquid sodium are limited. However, liquid sodium removes the protective oxide layer that forms on the surface of stainless steel components in the primary coolant circuit and, thus, facilitates the self-welding of metal components subjected to high-pressure contact.
The idea developed in this work was to protect the stainless steel components from self-welding by using composite coatings made from cobalt-free nickel-base alloy and submicronic alumina particles. The coatings were prepared by plasma spraying combining powder of metallic matrix and suspension of ceramic particles. The effect of the addition of ceramic particles in the matrix was investigated by observing the microstructure of coatings, and comparing their mechanical behavior (micro hardness and tribological tests representative of the application) for different ceramic particle contents (from 0 to 13 wt%).

1 Introduction

The 4th-generation of nuclear reactors is nearly entering the demonstration phase. The claimed benefits are sustainability, safety and reliability, economic competitiveness and proliferation resistance [1]. This reactor generation also aims to minimize the formation of long-lived radioactive elements that contaminate the cooling circuits.
The Sodium cooled Fast Reactor (SFR) is one of the six designs considered by The Generation IV International Forum. It utilises depleted uranium as fuel matrix and has a coolant temperature of 500-550°C ; its expected operation lifetime is 60 years.
It uses liquid sodium as reactor coolant, allowing high power density with low coolant volume, at low pressure because of high boiling point (881.4 °C), low heat capacity (1279 J/kg.K), high thermal conductivity (71.6 W/m.K) and low density (828 kg/m^3 at 500°C).
For liquid sodium, the oxygen concentration has to be controlled as low as possible (< 5 ppm) because the corrosion rate by liquid sodium increases significantly with oxygen concentration [2-3-4-5-6]. However, high-purity sodium cleans metal surface by dissolving the oxides and surface inclusions. It, thus, favors the self-welding tendency of metal components in static contact, under load, for extended periods while immersed in high-temperature sodium [7].
316LN stainless steel in some components, such as pumps, valves, and core support structures are sensitive to this phenomenon [8]. Thick coatings of stellite could be used to protect them as they have good resistance to sodium corrosion up to temperatures of 600°C and high anti-wear properties thanks to the cobalt matrix [9-10].
Nevertheless this element [11] is activated under neutron fluence to ^{60}Co, a radioactive element with a half-life of 5.27 years that contaminates the primary circuit.
Iron-based and nickel-based hardfacing alloys have been considered as substitutes to stellites but without success [12-19]: they have a higher hardness than stellites up to 400°C but also high friction coefficient and they cannot be used under contact pressure higher than 40 MPa as they lose their integrity under such conditions.
An option to reduce the self-welding tendency of metal coatings in contact and thus reduce friction between metals could be the addition of ceramic particles to the metal matrix. This could be achieved by a mechanical alloying of the metal alloy and ceramic powders [20-21] or a dry mixing of both powders before coating deposition [22]. Another solution could be to integrate the ceramic particles in the metal matrix during the coating deposition phase [23]. This is the approach developed in this study that deals with the combination of plasma spraying of metal alloy powder and suspension plasma spraying of submicrometric ceramic particles. It should be noted that the plasma spray technology is well adapted for hardsurfacing of metal components as it does not generate shape deformations if an adequate cooling of substrate is used. However, this approach requires to use conditions of powder and suspension injection so that the spray jets of the large metal particles and submicrometric ceramic particles coincide on the same spot on the target surface.
The metal powder is a cobalt-free nickel-base alloy (Colmonoy) and the submicronic particles are alumina; they are deposited on 316L stainless steel substrates. The coatings are intended to be used in the Advanced Sodium Technological Reactor for Industrial Demonstration (ASTRID) [24].
The study investigates the effect of the addition of the submicronic alumina particles to the Colmonoy matrix

on the microstructure, micro-hardness and wear resistance of the composite coatings.

2 Experimental Procedure

2.1 Materials and Method

The metal alloy powder was a nickel-based alloy that contains chromium borides and carbides (Colmonoy 52H). The particle size distribution ranged between 22 µm and 45 µm. The chemical composition of this powder is presented in **Table 1**.

The alumina powder (AKP30, Sumitomo Chemical Corp.) used to prepare the suspensions had an average particle size of 300 nm. The metal and ceramic powders were both oven dried at 80°C for 5 hours before plasma spraying.

The suspensions consisted of alumina powder dispersed in pure ethanol with Polyethyleneimine (PEI) as dispersant. PEI also helped to decrease the suspension viscosity. The compositions of the various suspensions are shown in **Table 2**.

Table 1: Chemical composition (in wt%) of 316L sStainless steel and Colmonoy 52H.

Elements (wt %)	316L	Colmonoy 52H
Fe	Balance	3.8
Cr	17.4	12.2
Ni	13.6	Balance
Co	0	-
Mo	2.1	-
Si	0.5	3.7
B	0.0025	2.2
C	0.024	0.5
Others	N-S-P-Al	-

Table 2: Composition of alumina suspensions.

Suspension	Solvent	Solid content (wt%) of Al_2O_3	Weight% of PEI per dry powder weight
S1	Ethanol	10	0.5
S2	Ethanol	20	0.5

2.2 Sample Preparation

The coating deposition was performed using an Oerlikon-Metco Triplex Pro plasma torch. The plasma operation parameters are listed in **Table 3**. The dried metal alloy powder was injected through a 1.8-mm diameter injector with 7 SLM Argon gas flow. The alumina suspension was injected through a 150-µm diameter injector under a pressure of 5 bars. A schematic drawing of the plasma process is given in **Fig. 1**.

Coatings were deposited on sand-blasted 316L stainless steel cylinders of 25 mm diameter x 15 mm thickness, mounted onto a turntable rotating on a horizontal axis, in front of which the torch moved horizontally. The substrate mean roughness ranged

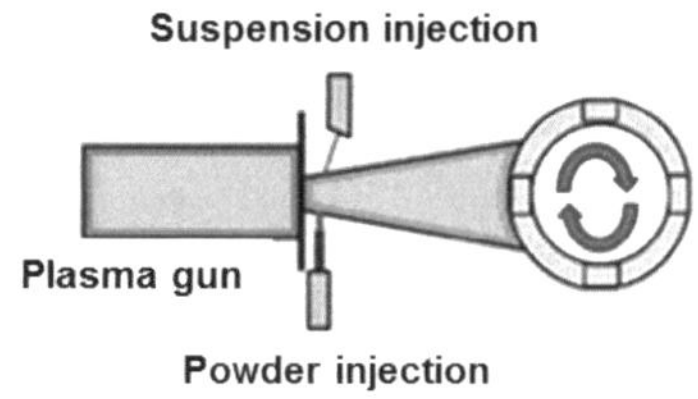

Fig. 1: Hybrid Plasma spray process combining powder and suspension, adapted from [23].

Table 3: Process parameters.

Plasma parameters	Value
Arc current (A)	450
Plasma gas flow rate, Ar (SLM)	50
Arc voltage (V)	93.4
Plasma specific enthalpy (MJ/kg)	16
Substrate stand off distance (mm)	70

between 6 and 8 µm. The substrates were heated to 250°C prior to deposition with the plasma jet; they were maintained at this temperature during the deposition process thanks to air cooling. The mass ratio between the injected metal powder and suspension was varied according to the parameters listed in **Table 4**. This made it possible to achieve different concentrations of submicronic alumina particles in the metallic matrix.

Table 4: Powder and suspension injection parameters

Sample	Metallic powder feed rate (g/min)	Suspension	Suspension feed rate (g/min)	Alumina/Colmonoy weight ratio
PI_0	40.3	none	0	0
PI_2	40.3	S1	24	0.06
PI_10	40.3	S2	29.5	0.15
PI_13	24.7	S2	29.5	0.24

All coatings were deposited with 50 spray cycles and after deposition cooled down in ambient air.

2.3 Sample Characterization Procedure

The polished cross sections of coatings were examined using optical microscopy and scanning electron microscopy (Philips XL 30 SEM) in conjunction with energy dispersive X-ray spectrometry (EDS). BackScatter Electron (BSE) imaging and energy dispersive X-ray analysis provided the distribution of alumina particles in the samples while the coating alumina content was determined using X-ray fluorescence (ARL SMS-Omega from Thermo Scientific).

Coatings were attacked in an electrolytic mix of hydrochloric and acetic acid for 10 seconds with a voltage of 2.4V to reveal the microstructure.

Micro hardness tests were performed using a Buehler Micromet 6040 micro-indenter with a load of 300g and a dwell time of 15 seconds. Measurements were conducted on the polished cross-sections of coatings. The open porosity was measured by the Archimedes method. The adhesion strength of coatings was tested following the ASTM C633-01 Standard test method [25].
The tribological behavior of coatings was evaluated by using a tribometer with a flat-on-flat configuration under inert atmosphere. The tests used a 6-mm diameter 316L cylinder in contact with the coating polished surface. They are representative of the contacts of 316L components in service life operation. The parameters of the tests are shown in **Table 5**.

Table 5: Flat-on-flat test parameters.

Parameter	Value
Temperature (°C)	5
Linear speed (mm/s)	2.7
Wear distance (m)	1.6
Test duration (min)	10
Atmosphere	Argon
O_2 content (ppm)	< 2
Load (N)	1414
Starting surface	Mirror polished

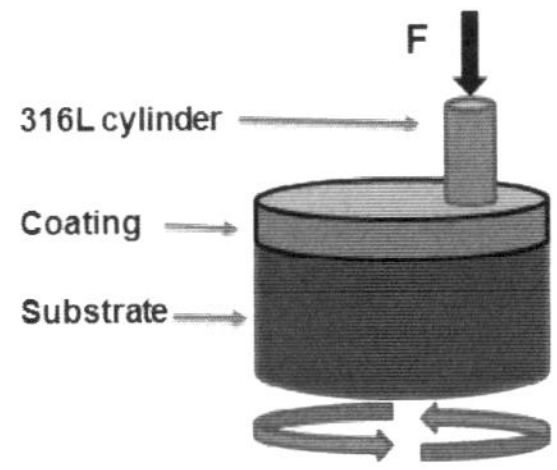

Fig. 2: Tribological flat-on-flat test.

A drawing of the tribological set-up is presented in **Fig. 2**. Samples were weighted before and after the wear test to quantify the material transfer between the two materials from the weight gain and loss of both sample and cylinder. The worn surfaces were also observed with optical microscope.

3 Results and discussions

3.1 Coating Microstructure

The composite coatings exhibited the typical plasma-spray microstructure composed of lamellae (fully melted and flattened particles), unmelted large particles and pores. The substrate-coating interface did not show cracks or delamination as observed in **Fig. 3**.
At low magnification, agglomerates of alumina fine particles were also observed (**Fig. 4)** while at higher magnification, the alumina particles appeared to be homogeneously distributed in the metal matrix between metallic lamellae (**Fig. 5**) for the coatings prepared with an alumina/Colmonoy weight ratio of 0.15 (Pl_10) and 0.24 (PL_13), respectively. The alumina was easily detected from the X-ray map of coatings as Colmonoy matrix is aluminum free.
The mass percentage of alumina in each coating quantified by X-ray fluorescence is presented in**Table 6**.
The integration of alumina particles in the metallic matrix is expected to affect coating properties such as microhardness, adhesion and tribological behaviour as it will be discussed in the next section.

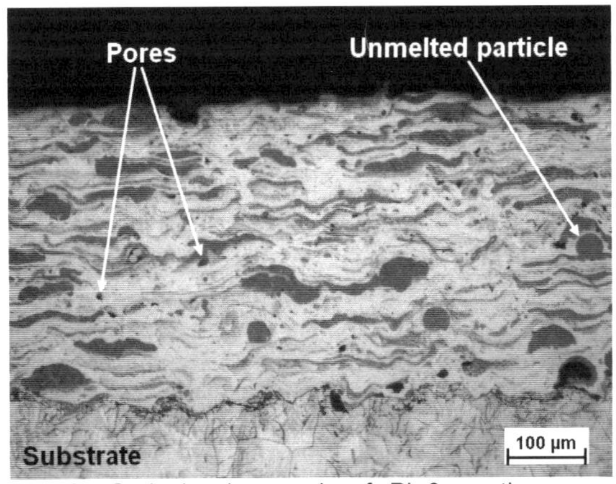

Fig. 3: Optical micrograph of Pl_0 coating cross section.

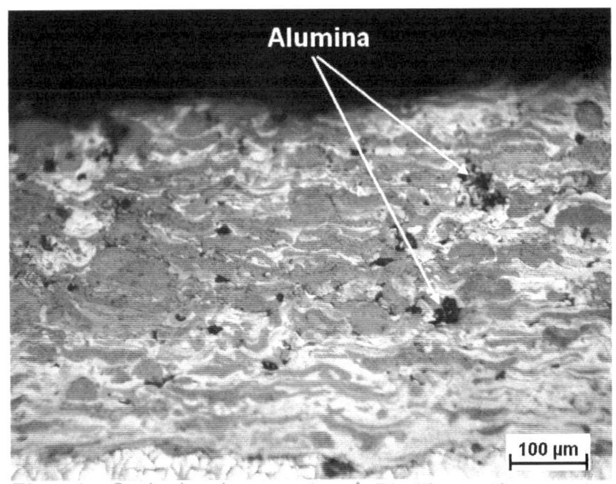

Fig. 4: Optical micrograph of Pl_10 coating cross section.

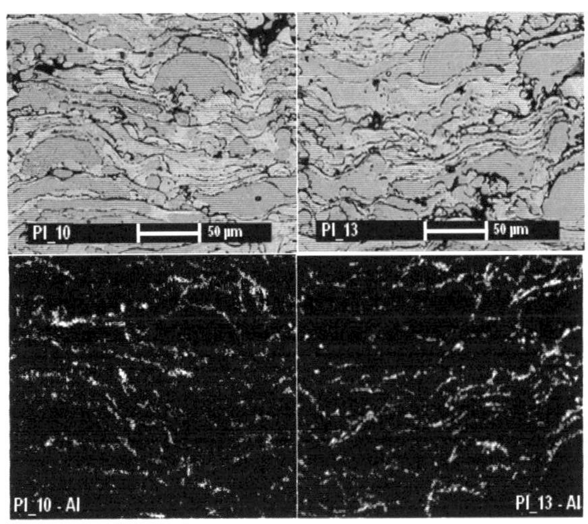

Fig. 5: SEM images of Pl_10 and Pl_13 coatings (top pictures) and corresponding Aluminum X-ray maps (bottom pictures) .

Table 6: Alumina weight content in coatings measured by X-ray fluorescence.

Sample	Powder feed rate (g/min)	Alumina load-Suspension feed rate (g/min)	Alumina wt% in final coating
PI_0	40.3	0	0
PI_2	40.3	10% - 24	2.4
PI_10	40.3	20% - 29.5	10.2
PI_13	24.7	20% - 29.5	13

3.2 Microhardness, porosity and adhesion

Microhardness measurements performed on the polished coating cross sections are presented in **Fig. 6**. Surprisingly, the microhadness decreased as the alumina content increased.

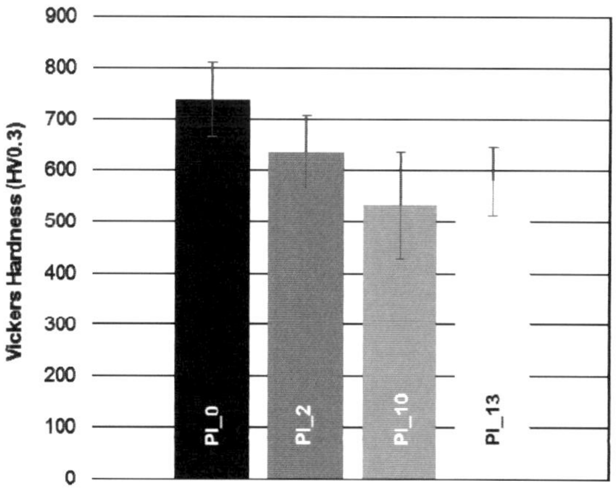

Fig. 6: Microhardness of plasma coatings as a function of alumina content.

The alumina particles were actually integrated at the lamellae interfaces and so decreased the coating cohesion and, cracks tended to appear at the indentation edges during microhardness testing. This decrease in microhardness could also be linked to an increase in the coating open porosity as shown in **Fig. 7**.

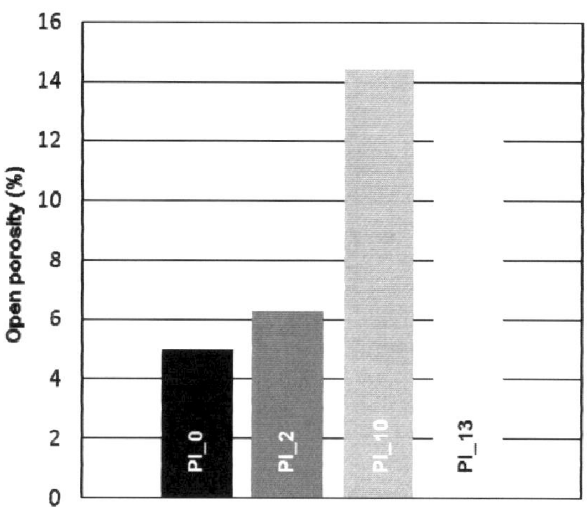

Fig. 7: Coating open porosity as a function of alumina content

As the alumina content was increased, the metal lamella surface was covered by alumina particles, creating voids more or less filled by alumina particles and cracks could, then, be easily initiated in these zones.
The coating adhesion strength showed the same tendency with the alumina content increase but to a lesser extent as presented in **Table 7.**

Table 7: Breaking loads of plasma coatings.

Sample	Breaking load (MPa)	Standard deviation (MPa)
PI_0	37.8	2.0
PI_2	40.4	1.0
PI_10	35.4	2.3
PI_13	33.7	0.9

These results show that increasing the alumina content in coatings tend to reduce the breaking load which is explained by a loss of coating integrity. All coatings exhibited a cohesive breaking and no delamination at coating-substrate interface; the coating adhesion strength seems thus to be higher than coating cohesion. This is a positive point for the use of the flat-on-flat tribological test because coatings showed that they have a low tendency to eventually delaminate during wear test.

3.3 Coatings sliding wear test

Although the microhardness and cohesion of coatings were negatively affected by the addition of alumina particles in the metal matrix, their tribological behavior exhibits an opposite tendency. The results on material transfer between the 316L cylinder surface and coated flat surface are shown in **Fig. 8** and **Fig. 9**.

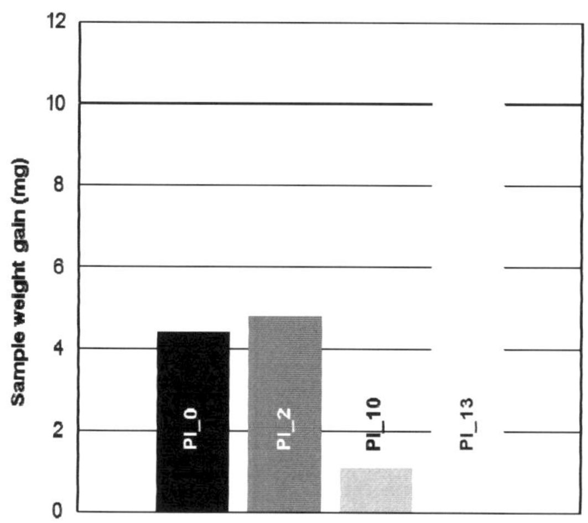

Fig. 8: Weight gain of coated samples after wear test as a function of coating alumina particle content.

The 316L cylinder surface is much softer than coating materials (about 160-170 HV0.3 against 500-700 HV0.3) and so, during the wear tests it should lose

weight while the sample coating should gain weight. At first sight, the integration of alumina particles reduced this phenomenon driven by adhesion wear (weight loss of Pl_0 > Pl_2 > Pl_10). The wear of the 316L
cylinder

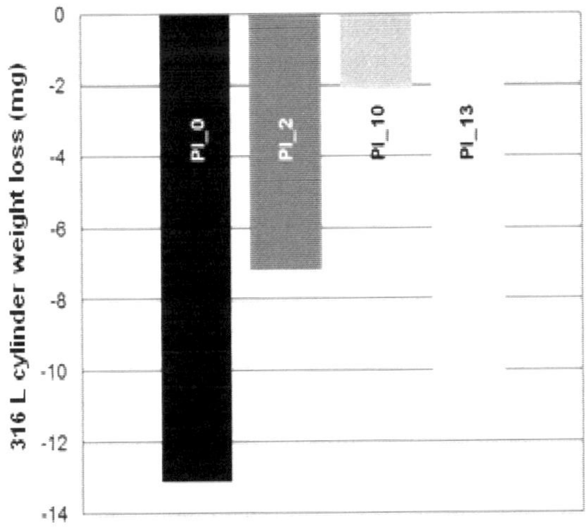

Fig. 9: Weight loss of 316L (SS) cylinder after wear test as a function of coating alumina particle content.

surface appeared to be drastically reduced for a coating alumina content of about 10% . The presence of alumina particles limited the zones where the transfer between the two rubbing surfaces occured; it also disminished the adhesion of these surfaces and self-welding tendency of the metallic parts in contact as shown in **Fig. 10.** However, when the alumina content was increased above 10% (PL_13 coating), the coating integrity was seriously affected and an increase in the wear of the two opposing surfaces was observed.

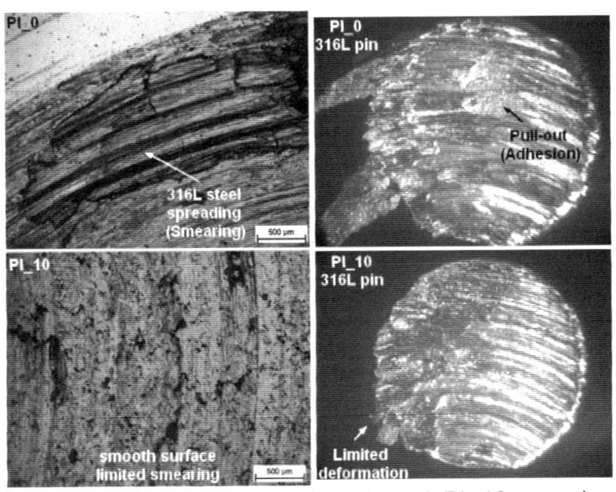

Fig. 10: Wear surfaces of Pl_0 and Pl_10 samples (left pictures) and corresponding 316L cylinder surfaces (right pictures).

The worn surface of the Pl_10 coating exhibited a smoother surface than that of the Pl_0 coating which was covered by large 316L spread bands. Some parts of the Pl_0 coating were pulled out from the surface and stuck onto the 316L cylinder surface. Also the 316L cylinder showed less deformation with the addition of alumina fine particles.

4 Conclusions

This study dealt with plasma-sprayed coatings of nickel-based alloys reinforced with alumina submicrometric particles deposited on 316L stainless steel substrates. These coatings are intended to be used as hardfacing coatings in Sodium-cooled Fast Reactors as substitute to stellite coatings. The main results are the following:

- An homogeneous dispersion of alumina particles in the coating has been achieved.
- Alumina addition decreases the coating micro-hardness and increases the open porosity level.
- Adhesion strength is higher than coating cohesion and is reduced by alumina addition.
- Wear is reduced by alumina addition because the adhesion and self-welding tendency are limited.

Nevertheless, the coating cohesion has to be improved and the optimal alumina content has to be found to obtain better coating performances.

5 References

[1] Technology Roadmap Update for Generation IV Nuclear Energy Systems, January 2014, Issued by the OECD Nuclear Energy Agency for the Generation IV International Forum

[2] M.V. Polley, G. Skyrme, , An analysis of the corrosion of pure iron in sodium loop systems, Journal of Nuclear Materials 66 (1977) 221–235 .

[3] Roy, M., Mécanisme de dissolution des aciers austénitiques dans l'eutectique plomb-bismuth à 500°C, PhD thesis, in French, University of tcchnology of Compiègne, France, 2012.

[4] Weeks, J. R., et al, "Corrosion and deposition of steels and nickel-base alloys in liquid sodium", Metallurgy and Materials Science Division, New York, United States, 1971.

[5] Fukushima, K., et al, "Electronic structure calculation studies on corrosion of stainless steel in liquid metal", Japanese Journal of Applied Physics, Vol.33 (1994), pp. 2613-2618.

[6] Zhang, J., et al, "Models of Liquid Metal Corrosion", Journal of Nuclear Materials, Vol. 404 (2010), pp. 82–96.

[7] Chander, S., et al., "Experimental Study of Self-Welding of Materials in High Temperature Liquid Sodium", Wear, Vol. 162-164 (1993), pp. 458-456.

[8] Meikandamurthy, C., et al, "Effect of cold-work on sefl-welding susceptibility of austenitic stainless steel (alloy D9) in high temperature flowing sodium", Journal of Nuclear Materials, Vol. 407 (2010), pp. 165-170.

[9] Iordache, L., "Analyse tribo-métallurgique des rechargements base cobalt par fusion soudage d'outillages à chaud", PhD thesis, Ecole nationnale supérieure d'arts et métiers, 2006.

[10] Revel, P., "Etude du comportement et de l'endommagement d'un superalliage base cobalt soumis à la fatigue thermique", PhD thesis, University technologic of Compiègne, France, 1991.

[11] Müller, R.A., et al., "Liquid Metal Cooled Fast Breeder Reactor Development and Its Mechanical-Structural Requirements and Problems", Nuclear Engineering and Design, Vol. 19 (1972), pp. 85-98.

[12] Cachon, L., "Tribological Qualification of Cobalt-Free Coatings for Pressurized Water Reactor Primary-Circuit Gate Valve Applications", Surface and Coatings Technology, Vol. 85 (1996), pp. 163-169.

[13] Airey, G.P., "Qualification of Cobait Free Hardfacing Alloys for PWR Valve Applications", Environmental Degradation of Materials in Nuclear Power Systems -- Water Reactors, G.J. Theus and J.R. Weeks (eds.), (The Metallurgical Society, 1988), pp. 755-760.

[14] Benhamou, C., "Testing of Cobalt-Free Alloys for Valve Applications Using a Special Test Loop", Nuclear Science and Technology (1992), pp. 1-19.

[15] Burdett, W.B., "Development of Cobalt Free Wear Resistant Alloys for Nuclear Applications", Swf Eng., Vol.8 (2) (1992), pp. 131-135.

[16] Ocken, H., "Wear Tests of Low-Cobalt Alloys for Hardfacing Nuclear Components", The American Society of Mechanical Engineers, (Palo Alto, CA), (1985), pp. 529-538.

[17] Shiels, S.A., et al., "Laboratory Evaluation of Low Cobalt Wear Materials for Nuclear Applications", Proc. Int. Syrup. Fontevraud III, Vol. 1, (Société Française d'Energie Nucléaire), (1994), pp. 99-106.

[18] Vikström, J., "Galling Resistance of Hardfacing Alloys Replacing Stellite", Wear, Vol. 179 (l994), pp. 143-146.

[19] Yonezawa, T., et al., "Applicability of Cobalt Free Resistant Materials to Valves", Proc. Int. Syrup. Fontevraud III, Vol. 1, (Société Française d'Energie Nucléaire), (1994), pp. 116-123.

[20] Zhao, L., "Wear Behavior of Al2O3 Dispersion Strengthened MCrAlY Coating", Surface and Coating Technology, Vol.184 (2004), pp. 298-306.

[21] Ageorges, H., et al., "Plasma Spraying of Stainless-Steel Particles Coated with an Alumina Shell", Thin Solid Films, Vol. 370 (2000), pp. 213-222.

[22] Werry, A., et al, "Multi-Scale-Structured Composite Coatings by Plasma-Transferred Arc for Nuclear Applications", *Journal of Thermal Spray Technology,* Vol. 25 (2016),_pp 375-383 .

[23] Ravaux, A., Réalisation et étude de dépôts composites multi-échelle élaborés par projection plasma pour des applications tribologiques à hautes températures, Ph D.thesis University of Limoges, France, December, 4th 2014.

[24] 4th Generation Sodium–Cooled Fast Reactor: The ASTRID Technological Demonstrator, CEA, December 2012, Direction de l'énergie nucléaire, Centre de Saclay, 91191 Gif-sur-Yvette cedex, www.cea.fr.

[25] ASTM C633-01 – Standard Test Method for Adhesion or Cohesion Strength of Flame-Sprayed Coatings; http://www.astm.org/DATABASE.CART/HISTORICAL/C633-01.htm ; [Feb 2016].

Microstructure and mechanical property of Al_2O_3/YAG wear-resistant composite coating deposited by atmospheric plasma spraying

Kai Yang [a,*], Jian Rong [a,b], Chenguang Liu [a], Shunyan Tao [a], Chuanxian Ding [a]
[a] The Key Laboratory of Inorganic Coating Materials, Shanghai Institute of Ceramics, Chinese Academy of Sciences, Shanghai 201899, PR China
[b] University of Chinese Academy of Sciences, Beijing 100039 , PR China

Plasma sprayed Al_2O_3 ceramic coatings show great engineering potential in the combined working conditions , such as high specific pressure, high temperature, serious oxidation and thermal shock. Strengthening and toughening from grain refinement or solid solution may be beneficial to the improvement of the strength and toughness of the Al_2O_3 coating. However, poor high-temperature microstructure stability restricts some practical applications. Therefore, an interesting work has been taken in this article. The spray-dried Al_2O_3/Y_2O_3 composite powder was prepared using commercially available nano-sized Al_2O_3 and Y_2O_3 feedstock. The $Al_2O_3/Y_3Al_5O_{12}$(YAG) eutectic structure in the coating is expected to be obtained due to high enthalpy, large temperature gradient and rapid solidification in plasma spraying process. Interestingly, the as-sprayed coating showed obvious amorphous feature, which may be conducive to the anti-corrosion performance. After heat treating, the coating is composed of the crystalline phases of α-Al_2O_3 and YAG. The phase composition, microstructure and mechanical property of Al_2O_3/YAG composite coating were investigated.

1 Introduction

The Al_2O_3 coatings are considered to be the typical representative of oxide ceramic coatings [1-4]. The previous studies were focused on the strength-toughness enhancement of the Al_2O_3 coatings, including porosity control [5], metal phase addition [6], other oxide solution [7-9], structure nano-crystallization [10], component or structure gradient [11] and post processing [12]. Directionally solidified Al_2O_3-based eutectic ceramic in situ bulk composites with inherently high melting point, excellent microstructure stability, low defect, outstanding resistance to creep, corrosion and oxidization at elevated temperature, have attracted great interest as promising candidate for high-temperature applications [13]. Binary and pseudo-binary eutectics, ternary eutectics and even some off-eutectic compositions of the ternary system Al_2O_3–ZrO_2–Y_2O_3 were explored in detail [14], including Al_2O_3–ZrO_2, Al_2O_3–YSZ (yttria-stabilized zirconia), Al_2O_3–$Y_3Al_5O_{12}$ (YAG), Al_2O_3–YAG–ZrO_2 and Al_2O_3–YAG–YSZ eutectic oxide ceramics. In particular, binary Al_2O_3–YAG eutectics have been reported to exhibit excellent mechanical performance, thermal conduction and creep resistance at high temperatures [15], which could be attributed to large area fraction of clean and strong interfaces without glassy phases and continuous interpenetrating networks formed by nano-scale or sub-micron eutectic phases [16]. From the point of view of processing, the key to obtaining the homogeneous microstructure of Al_2O_3-based eutectic ceramics is to keep flat solid-liquid interfaces during growth at microscopic and macroscopic levels, and this requires large thermal gradient in the solidification direction. High-performance directionally solidified Al_2O_3-based bulk eutectics have been successfully produced in small cylindrical geometry by means of various techniques [17], such as the Bridgman, laser floating zone (LFZ), edge-defined film-fed growth (EFG), micro-pulling-down (μ-PD), laser engineered net shaping (LENS), and selective laser melting (SLM) methods. Based on the analysis of process conditions, there are three similarities existing in these eutectic ceramic preparation procedures: ①high enthalpy; ②steep temperature gradient; ③rapid solidification. The three features may be critical to fabricate Al_2O_3-based eutectic ceramics. Furthermore, whether the distinct performance superiority of Al_2O_3-based eutectic bulk ceramics could be realized in the form of corresponding eutectic ceramic coatings?

Plasma spraying is a thermal spraying process used to produce high quality coatings by a combination of high temperature, high energy heat source, a relatively inert spraying medium, and high particle velocities. In plasma spraying devices, the arc is formed between two electrodes in plasma forming gas, which usually consists of either argon/hydrogen or argon/helium. As the plasma gas is heated by the arc, it expands and is accelerated through a shaped nozzle, creating velocities up to MACH 2. Temperatures in the arc zone approach 20000K. Plasma sprayed coatings are made up of flattened particles, known as splats. The splats constitute layers in the coating and these layers in turn create the lamellar structure of the deposit. The high quench rate of the molten particles during plasma spraying is about $10^6 K \cdot s^{-1}$ [18]. The estimated temperature gradients in splats approach $10^4 K \cdot cm^{-1}$. According to the above analysis, plasma spraying possesses three basic features, which may be considered to be three critical factors of Al_2O_3-based eutectic bulk ceramic preparation. This indicates the corresponding eutectic

* Corresponding author. Tel: +86 021 69906320; Fax: +86 021 69906322; E-mail address: kaiyang@mail.sic.ac.cn (K. Yang).

ceramic coatings might be fabricated. In this paper, we proposed a bold hypothesis that Al_2O_3–YAG eutectic ceramic coating may be obtained via atmospheric plasma spraying. The objective of this study is to investigate the microstructure characterization of plasma-sprayed Al_2O_3–YAG composite coating and to verify the feasibility of expected eutectic ceramic coating preparation.

2 Experimental procedure

2.1 Sprayable feedstock preparation

Commercially available nano-sized Al_2O_3 and Y_2O_3 powders were used as feedstocks. The average sizes (D50) of Al_2O_3 and Y_2O_3 nanopowders are 76.4nm and 74.2nm, respectively. The nano-sized particles had to be granulated to micron-sized granules since they could not be used as sprayable feedstocks directly. The Al_2O_3 and Y_2O_3 nanopowders with the eutectic composition (82mol% Al_2O_3, 18mol% Y_2O_3) were put into the ball milling tank, which contained alumina based balls (the weight ratio of balls to powers was 4:1). After 10h ball milling, as s typical mixture of the deionized water, binder, dispersant, and nanopowders were stirred homogenously for 4h. When the corresponding slurry was prepared, the spray drying process was continued directly in order to avoid the precipitation of slurry. The main controlled operating parameters were the air temperature at the entry (220 ~ 240°C), at the exit (120 ~ 130°C), inside the chamber (170°C), the rotation speed of atomizer (12000rpm), and slurry flow rates (1 ~ 1.2kg/h). In order to increase the density of granules and eliminate all the organic additives, the granulated particles obtained from the spray drying process, were calcined. The sintering temperature (1000 ~ 1100°C) was determined by the thermo gravimetric analysis (TGA) method. After that, small (<15μm) and large particles (>65μm) were removed by passing them through sieves to meet the demands of sprayable feedstocks.

2.2 Coating deposition

The Multicoat atmospheric plasma spraying system equipped with a F4-MB plasma gun (Sulzer Metco AG, Switzerland) was applied to produce the coating. Prior to spraying, the graphite substrates were degreased ultrasonically in acetone and then grit blasted with corundum. A mixture of argon and hydrogen was used as plasma gas. The important plasma spraying parameters were arc current (645 ~ 650A), Ar flow rate (45slpm), H_2 flow rate (7slpm), carrier gas flow rate (3slpm), powder feed rate ($35g{\cdot}min^{-1}$) and spray distance (120mm). After plasma spraying, heat treatments were carried out at 1200°C for 24h and 96h.

2.3 Coating characterization

The phase compositions of the feedstock and coating were identified by X-ray diffraction (XRD) using a Rigaku D/Max2550 Diffractometer with nickel-filtered Cu Kα radiation (λ=0.15406nm). The cross-sectional morphologies of the coating were characterized by a Nova NanoSEM 230 scanning electron microscope (FEI, Hillsboro, USA). Transmission electron microscopy (TEM) images were taken on JEM 2100F to characterize the microstructure of the coating. Vickers microhardness measurements were carried out on the polished cross-section of the coating using an Instron Wilson-Wolpert Tukon 2100B Hardness Tester under the load of 200g with a dwell time of 15s.

3 Results and discussion

Fig. 1 exhibits the XRD patterns of sprayable feedstock, as-sprayed coating and heat-treated coating. The XRD measurements were performed in the 2θ range from 10° to 90° at a scanning speed of $2°{\cdot}min^{-1}$. The sprayable feedstock consists of Al_2O_3/Y_2O_3 granules obtained from the spray drying and the corresponding XRD pattern presents two phases of α-Al_2O_3 and c-Y_2O_3. Interestingly, the obvious amorphous feature appears in the XRD pattern of the as-sprayed coating (shown in Fig. 1b). Simultaneously, α-Al_2O_3 and YAG phases were also detected. This may indicate that the amorphous phase is dominant in the as-sprayed coating. Furthermore, a few α-Al_2O_3 and YAG crystalline phases exist in the amorphous coating matrix. The as-sprayed coating was heated at 1200°C for 24h. The heat treatment facilitates complete crystallization of the coating (shown in Fig. 1c). The heat-treated coating is composed of α-Al_2O_3 and YAG. Consequently, the Al_2O_3–YAG composite coating was obtained via atmospheric plasma spraying.

Fig. 2a shows the TEM image of the as-sprayed coating. The coating matrix is mainly comprised of these zones with no boundary feature (marked by the yellow circle). The corresponding selected area electron diffraction (SAED) pattern displays diffuse halo, which confirms the formation of amorphous phase in plasma spraying process. Moreover, a small amount of crystalline grains were dispersively distributed in the amorphous coating matrix. The conspicuous changes in the coating microstructure due to heat treatment were presented in Fig. 2b. As is clearly seen, the heat-treated coating almost completely consists of closely arranged crystalline grains. The corresponding coating possesses few porosity, compact structure and strong grain boundary bonding.

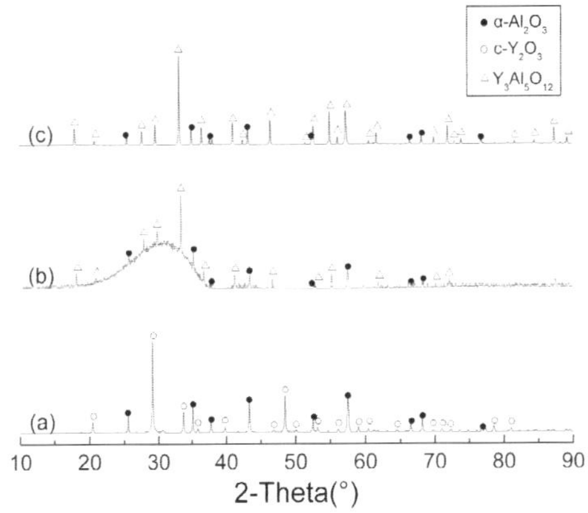

Fig. 1. XRD patterns of powder and coating: (a) sprayable feedstock; (b) as-sprayed coating; (c) heat-treated coating at 1200℃ for 24h.

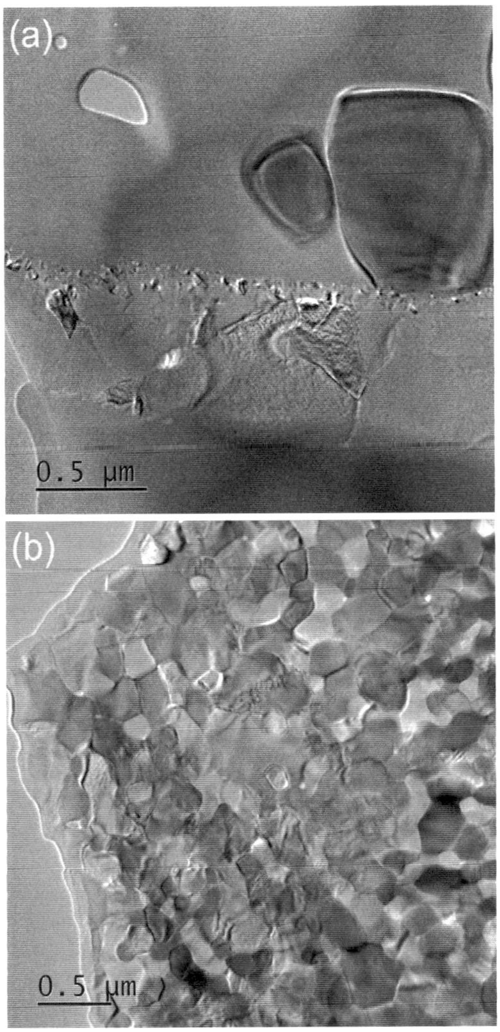

Fig. 2. TEM images of (a) as-sprayed coating and (b) heat-treated coating at 1200℃ for 24h.

The cross-sectional morphologies of as-sprayed and heat-treated Al_2O_3–YAG composite coatings are revealed in Fig. 3. There are no observable two phase characteristics in the SEM morphological image of the as-sprayed coating (shown in Fig. 3a). It can be obviously seen from Fig. 3b that nano-scale amouphous clusters constitute the as-sprayed Al_2O_3–YAG composite coating. Compared with the as-sprayed coating, the heat-treated coating possesses significant structure changes. Fig. 3c displays typical interpenetrating network microstructure of the coating, which consists of α-Al_2O_3 phase (the black) and YAG phase (the gray) without any other phases. The phase volume ratio is Al_2O_3 : YAG = 46 : 54, which coincides well with that expected for eutectic composition. The heat-treated Al_2O_3–YAG composite coating (1200℃ for 24h) presents an interpenetrating network as "Chinese Script" type structure, in which the α-Al_2O_3 and YAG phases are homogenously distributed and interconnected with each other. Accordingly, the microstructure of Al_2O_3–YAG composite coating with heat treatment is very similar to that of Al_2O_3–YAG eutectic bulk ceramic. As the holding time was extended to 96h, the Al_2O_3–YAG composite coating exhibited this pattern with great similarity consisting of entangled network of α-Al_2O_3/YAG eutectic phases (shown in Fig. 3d). Furthermore, the interphase spacing slightly increases due to the prolonged heat treatment time.

Fig. 4 reveals vickers microhardness of the coatings. The coating microhardness represented the average of 15 indentations. Vickers hardnesses of Al_2O_3 coating, as-sprayed and heat-treated Al_2O_3–YAG composite coatings are 9.26GPa, 7.23GPa, 15.98GPa and 15.48GPa, respectively. Compared with pure Al_2O_3 coating, the hardness of the as-sprayed Al_2O_3–YAG composite coating decreases, which may be attributed to the major amorphous phase. Vickers hardness of heat-treated Al_2O_3–YAG composite coating (1200℃ for 24h) increases by 72% compared with that of pure Al_2O_3 coating. As the heat treatment time was prolonged to 96h, the coating hardness slightly lowered (about 3%). The hardnesses of heat-treated Al_2O_3–YAG composite coatings are consistent with those of reported Al_2O_3–YAG eutectic oxide bulk ceramics [14]. Accordingly, plasma-sprayed Al_2O_3–YAG composite coatings with heat treatment possess representative three-dimensional interpenetrating (TDI) structure of eutectic ceramics. With increasing the heat preservation time, the microhardness of Al_2O_3–YAG composite coating basically maintained stable, which may be due to high temperature microstructure stability of the coating with typical TDI structure of eutectic ceramics.

4 Conclusions

In summary, this paper presents the results of an experimental study of the microstructure and mechanical property of plasma-sprayed Al_2O_3–YAG composite coating. The amorphous phase is dominant

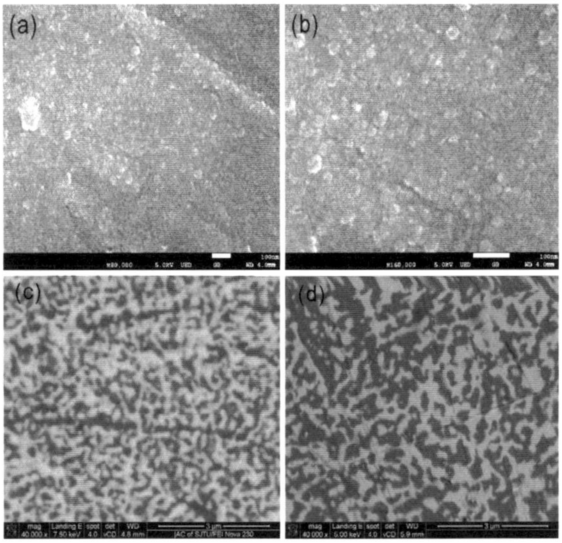

Fig. 3. Cross-sectional morphologies of the coatings: (a) as-sprayed coating; (b) Magnified image of (a); (c) heat-treated coating at 1200°C for 24h; (d) heat-treated coating at 1200°C for 96h.

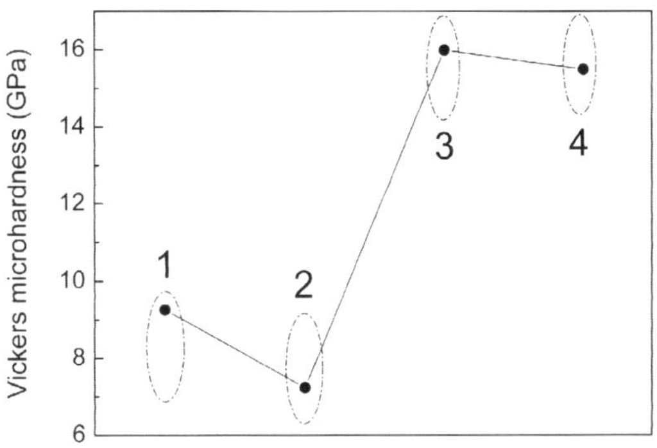

Fig. 4. Vickers microhardness and thermal conductivity of the coatings. 1: Al_2O_3 coating; 2: as-sprayed Al_2O_3–YAG composite coating; 3: heat-treated Al_2O_3–YAG composite coating (1200°C for 24h); 4: heat-treated Al_2O_3–YAG composite coating (1200°C for 96h).

in as-sprayed Al_2O_3–YAG composite coating. After heat treatment, the eutectic Al_2O_3–YAG ceramic coating was successfully obtained, which exhibited the deposition microstructure as "Chinese Script" pattern consisting of three-dimensional interpenetrating network of α-Al_2O_3/YAG eutectic phases. The plasma-sprayed Al_2O_3–YAG composite coating with the heat treatment shows excellent microstructure and microhardness stabilities under high temperature.

Acknowledgments

The study is jointly supported by National Nature Science Foundation of China (51302299), Shanghai Nature Science Fund Project (13ZR1446000) and Youth Innovation Promotion Association, Chinese Academy of Sciences.

References

[1] D. W. Lee, H. J. Kim, Y. H. Kim, Y. H. Yun, and S. M. Nam, "Growth Process of alpha-Al_2O_3 Ceramic Films on Metal Substrates Fabricated at Room Temperature by Aerosol Deposition," J. Am. Ceram. Soc., **94** [9] 3131-3138 (2011).

[2] S. R. Bakshi, K. Balani, and A. Agarwal, "Thermal conductivity of plasma-sprayed aluminum oxide-multiwalled carbon nanotube composites," J. Am. Ceram. Soc., **91** [3] 942-947 (2008).

[3] N. P. Padture, M. Gell, and E. H. Jordan, "Thermal Barrier Coatings for Gas-Turbine Engine Applications," Science, **296**, 280-284 (April 2002).

[4] D. Chen, E. H. Jordan, M. Gell, and X. Ma, "Dense alumina-zirconia coatings using the solution precursor plasma spray process," J. Am. Ceram. Soc., **91** [2] 359-365 (2008).

[5] Z. Yin, S. Tao, X. Zhou, and C. Ding, "Particle in-flight behavior and its influence on the microstructure and mechanical properties of plasma-sprayed Al_2O_3 coatings," J. Eur. Ceram. Soc., **28** [6] 1143-1148 (2008).

[6] Z. Yin, S. Tao, X. Zhou, and C. Ding, "Tribological properties of plasma sprayed Al/Al_2O_3 composite coatings," Wear, **263** [7-12] 1430-1437 (2007).

[7] Z. Y. Pan, Y. Wang, C. H. Wang, X. G. Sun, and L. Wang, " The effect of SiC particles on thermal shock behavior of Al_2O_3/8YSZ coatings fabricated by atmospheric plasma spraying," Surf. Coat. Technol., 206 **[8-9] 2484-2498 (2012).**

[8] K. Yang, J. Feng, X. Zhou, and S. Tao, "Microstructural characterization and strengthening-toughening mechanism of plasma-sprayed Al_2O_3–Cr_2O_3 composite coatings," J. Therm. Spray Technol., **21** [5] 1011-1024 (2012).

[9] A. Rico, C. J. Múnez, and J. Rodríguez, "On the relevance of microstructure in the fracture toughness of nanostructured alumina–13% titania coatings," Surf. Coat. Technol., **243**, 46-50 (March 2014).

[10] D. Zois, A. Lekatou, M. Vardavoulias, and A. Vazdirvanidis, "Nanostructured alumina coatings manufactured by air plasma spraying: Correlation of properties with the raw powder microstructure," J. Alloy. Compd., **495** [2] 611-616 (2010).

[11] L. R. Krishna, P. S. V. N. B. Gupta, and G. Sundararajan, "The influence of phase gradient within the micro arc oxidation (MAO) coatings on mechanical and tribological behaviors," Surf. Coat. Technol., **269**, 54-63 (May 2015).

[12] Y. Wang, C. Li, L. Guo, and W. Tian, "Laser remelting of plasma sprayed nanostructured Al_2O_3–TiO_2 coatings at different laser power," Surf. Coat. Technol., **204** [21-22], 3559-3566 (2010).
[13] J. Zhang, H. Su, K. Song, L. Liu, and H. Fu, "Microstructure, growth mechanism and mechanical property of Al_2O_3-based eutectic ceramic *in situ* composites," J. Eur. Ceram. Soc., **31** [7] 1191-1198 (2011).
[14] J. LLorca and V. M. Orera, "Directionally solidified eutectic ceramic oxides," Prog. Mater. Sci., **51** [6] 711-809 (2006).
[15] A. Larrea, V. M. Orera, R. I. Merino, and J. I. Pena, " Microstructure and mechanical properties of Al_2O_3–YSZ and Al_2O_3–YAG directionally solidified eutectic plates," J. Eur. Ceram. Soc., **25** [8] 1419-1429 (2005).
[16] E. C. Dickey, C. S. Frazer, R. R. Watkins, and C. R. Hubbard, " Residual stresses in high-temperature ceramic eutectics," J. Eur. Ceram. Soc., **19** [13-14] 2503-2509 (1999).
[17] F. Niu, D. Wu, G. Ma, J. Wang, M. Guo, and B. Zhang, " Nanosized microstructure of Al_2O_3–ZrO_2 (Y_2O_3) eutectics fabricated by laser engineered net shaping," Scripta Mater., **95**, 39-41 (January 2015).
[18] K. Yang, X. Zhou, H. Zhao, and S. Tao, "Microstructure and mechanical properties of Al_2O_3–Cr_2O_3 composite coatings produced by atmospheric plasma spraying," Surf. Coat. Technol., **206** [6] 1362-1371 (2011).

Evaporation enhancement of YSZ spray particles for PS-PVD by modifying the nozzle in conventional 80 kW plasma spray system

Qing-Yu Chen, Chang-Jiu Li*, Cheng-Xin Li, Guan-Jun Yang, State Key Laboratory for Mechanical Behavior of Materials, School of Materials Science and Engineering, Xi'an Jiaotong University, Xi'an, Shaanxi, China/ E-mail: licj@mail.xjtu.edu.cn

In this paper, a modified nozzle with a large orifice was used to reduce the velocity of plasma jet in order to prolong the dwelling time of spray particle in plasma jet. Optical emission spectrometer (OES) was employed to characterize the temperature, electron number density, also the quantity of Zr atoms and Zr^{1+} ions in plasma jet. The deposition characteristics of YSZ coatings by PS-PVD using 80 kW plasma spray system in vacuum at the condition of droplet filtering were examined. Field emission-scanning electron microscope (FE-SEM) was used to characterize the microstructure of the YSZ coatings. The results show that the temperature and electron number density of plasma jet with the modified nozzle of large orifice are larger than that with the normal nozzle. The cross sectional morphology of the fracture coating deposited through droplet filtering reveals a columnar-structure. The surface morphology of coating exhibits a typical island-structure which is composed of nano-particles. Through droplet filtering, the thickness of columnar-structured coating deposited with modified nozzle is much larger than that with the normal nozzle. The higher temperature and electron number density enhance the intensity of YSZ particle evaporation, resulting in a higher deposition rate with modified nozzle than that with normal nozzle.

1 Introduction

Atmospheric plasma spray (APS) is a conventional thermal spray technology with characterization of high temperature and high velocity [1-3]. The plasma jet has enough energy to melt and even evaporate any powder particles since the center temperature of plasma jet can reach above 15000 K [4]. However, the complete evaporation of ceramic powder with high melting point such as YSZ is almost impossible through APS because the high velocity of plasma jet results in insufficient heat transfer to YSZ ceramic particles [5]. Thus, in atmospheric plasma spray process, YSZ particles is usually heated to a just fully molten or even semi-molten state by plasma jet generated by conventional 80 kW class plasma system [6].

Plasma spray-physical vapor deposition (PS-PVD) is an emerging technology which is operated in low pressure below 200 Pa [7-9]. In PS-PVD process, plasma jet is significantly expanded both in radial and axial directions due to very low pressure atmosphere which results in plasma jet with a velocity higher than sound and low temperature [10-12]. In expanded plasma jet, powder particles have a longer dwelling time for heating, which would make it possible to evaporate most kinds of the powder particles before deposition. The evaporation of ceramic powder with high melting point such as YSZ is a challenging work since the high velocity of plasma jet makes evaporation of powder particles insufficient. To enhance the evaporation of YSZ particles, one approach is to use high power plasma system (up to 180 kW) with highly intensive heating ability since velocity of plasma jet is much higher [13]. With conventional 80 kW low pressure plasma spray (LPPS) system, due to lower plasma heating ability, plasma temperature control is essentially important which directly influences evaporation of spray materials. In order to control plasma temperature for enhancing evaporation of YSZ powder particles, one effective approach is to use a shrouded plasma torch to restrict the expansion of the plasma jet. The previous study showed that the plasma temperature can be effectively increased in a restricted zone which results in an effective enhancement of the evaporation of YSZ powder particles [14]. However, there still exist substantial droplets in the plasma jet due to lower plasma heating ability of low power plasma [15]. Despite the existence of droplets in plasma jet, by introducing droplet filtering, it has been demostrated in the previous study a columnar structured YSZ coating can be deposited by full vapor phase through PS-PVD using conventional 80 kW plasma system at a power level from 45 kW to 78 kW [16].

Another approach for enhancing the evaporation of YSZ ceramic powder is to increase the dwelling time of particles in the plasma jet which can be realized through decreasing the velocity of plasma jet. Reducing the flow rate of primary gas and secondary gas is an impractical method since it can reduce the plasma power simultaneously. Since anode nozzle geometry has great influence on the characteristic of plasma jet such as enthalpy, temperature and velocity etc. [17-19], in this paper, a modified nozzle with a larger oriffice was used to modify plasma jet characteristics in order to improve the particle heating in plasma jet and thus the deposition rate of YSZ coating via vapor phase.

2 Materials and Experiments

A commercially available agglomerated 7-8 wt. % YSZ powders (Metco 6700, Sulzer-Metco, Switzerland, 5-25 μm) was used as the feedstock. The sintered YSZ with dimensions of Φ18× 2 mm (diameter × thickness) was employed as a substrate for coating deposition. Prior to spraying, the substrate was ground by diamond disk and degreased in an ultrasonic acetone bath. The substrate positioning with respect to plasma torch and torch geometry could be referred in the literature [16]. The plasma torch was fixed during the spraying. Three samples were placed along plasma jet axis at an interval of 80 mm in order to constitute three-stage shielding. The first sample was placed at a spray distance of 260 mm with a shielding pole of 25 mm in diameter 80 mm ahead of the first sample. Thus the second and third samples were deposited at spray distances of 340 mm and 420 mm. The detailed setting of the samples could be referred in the previous publication [16]. In order to investigate the comformity of deposited coating, superalloy 738 substrate deposited with cold-sprayed MCrAlYTa bond coat of a rough surface was used.

Plasma system (MF-P 1000 APS/VPS, GTV, Germany) equipped with an <80 kW direct-current (dc) plasma system was used in this experiment to realize PS-PVD process. The plasma torch was fixed on a six-axis robot which was place in a 10 m^3 chamber with a controlled Ar atmosphere. The main deposition conditions are summarized in Table 1. The power of plasma torch was set at 60 kW. The chamber pressure was maintained at 100 Pa. For comparison, the coating was also deposited at a power of 78 kW.

Table 1. The main deposition conditions of PS-PVD

Parameters	Unit	Value
Chamber pressure	Pa	100
Plasma arc power	kW	60
Arc current	A	750
Ar gas flow	slpm	60
H_2 gas flow	slpm	12
Spray distance	mm	260/340/420
Powder feed rate	g/min	3

Optical emission spectroscopy was used to characterize the properties of plasma jet. For data acquisition, a grating of 300 grooves/mm was selected to diagnose the light emission from plasma, and the resolution of the spectrometer was 0.058 nm. The wavelength band selected was from 300 nm to 1000 nm.

The electron number density was estimated by using Stark broadening method which is suitable for the plasma at either equilibrium or nonequilibrium [20, 21]. In the present paper, the spectrum line (at a wavelength of 486.1 nm) of hydrogen balmer series spectra was used to calculate electron number density [22] by following equation.

$$\log n_e = 1.452 \log \Delta\lambda + 14.017 \quad (1)$$

where n_e (cm^{-3}) is the electron number density, $\triangle\lambda$ (Å) is the full width at the half maximum (FWHM). In the calculation, the spectral line was fitted by Voigt function which is a convolution of a Lorentz and a Gauss function. Since only Stark broadening causes a significant Lorentzian contribution [23]. Lorentzian half widths were used to calculate the electron number density.

The double line method (also known as intensity comparison method) can be used to measure the plasma temperature by following equation [24].

$$\frac{I_{ji}}{I_{kl}} = \frac{\lambda_{kl} A_{ji} g_j}{\lambda_{ji} A_{kl} g_k} \exp\left(\frac{E_k - E_j}{kT}\right) \quad (2)$$

where I_{ji} (a.u.) is the intensity of spectra generated via energy level transition from j to l, I_{kl} (a.u.) is the intensity of spectrum generated via energy level transition from k to l, λ_{ji} (nm) is the wavelength of spectra generated via energy level transition from j to l, λ_{kl} (nm) is the wavelength of spectra generated via energy level transition from k to l, A_{ji} (s^{-1}) is the Einstein spontaneous transition probability of energy level from j to l, A_{kl} (s^{-1}) is the Einstein spontaneous transition probability of energy level from k to l, g_j is the statistical weight of energy level j, g_k is the statistical weight of energy level k, E_j (eV) is the excitation energy of energy level j, E_k (eV) is the excitation energy of energy level k. k is the Boltzmann constant. In these data, the light intensity, which was selected H_β (486.1 nm) and H_α (656.3 nm) [25] corresponding to I_{ji} and I_{kl}, can be measured in the experiment. The other parameters can be obtained from the NIST database [26].

The quantity of Zr atoms and Zr^{1+} ions in the plasma jet can be calculated by the following formula [20]:

$$I_{ji} = A_{ji} h\nu_{ji} \frac{g_j}{Z} N \exp(-E_j / k_B T) \quad (3)$$

$$I^+_{ji} = A^+_{ji} h\nu^+_{ji} \frac{g^+_j}{Z^+} N^+ \exp(-E^+_j / k_B T) \quad (4)$$

where I_{ji} and I_{ji}^+ is the intensity of Zr atom and ion spectra generated via energy level transition from j to l, A_{ji} and A_{ji}^+ (s^{-1}) is the Zr atom and ion Einstein spontaneous transition probability of energy level from j to l, v_{ji} andv_{ji}^+ is the frequency of the Zr atom and ion, E_j and E_j^+ (eV) is the excitation energy of energy level j, N and N^+ is the quantity of Zr atom and ion, k_B is the Boltzmann constant, h is the Planck constant.

The more detailed test parameters of OES could be referred in the literature [14]. The surface morphology and cross sectional morphology of fractured coating were examined by field emission-scanning electron microscopy (FE-SEM).

3 Results and discussion

3.1 Characteristics of plasma jet

The electron number density of plasma jet generated by a modified nozzle in comparison with that by normal nozzle at a spray distance of 260 mm is shown in Fig.1. The electron number density of plasma jet from the modified nozzle is higher than that with normal nozzle. Generally, the electron number density may be decreased with the increased diameter of nozzle for the dilution of the electron number density in an expanded area. With the modified nozzle, although the orifice of nozzle is larger, the electron number density is higher at the same spatial possible due to reduced plasma jet velocity. As the velocity of plasma jet is reduced, the amount of electrons remaining at the fixed area in a

unit time may be increased which results in the increase of the electron number density. It can be also seen that the change of electron number density for two nozzles with or without powder feeding is the same. Thus, the powder feeding has little influence on the change of the electron number density [14].
The measurement of the temperature of plasma jet showed that the temperature ranged from 6000K to 7000K with limited change by nozzle modifying and powder feeding at a spray distance of 260 mm. Fig.2 shows the quantity of Zr atoms and Zr^{1+} ions in the plasma jet generated by two types of the nozzles. It is clear that the quantity of either Zr atoms or Zr^{1+} ions in the plasma jet by using modified nozzle was significantly increased. These results are attributed to the enhanced evaporation of YSZ particles in plasma jet. Since plasma temperature was not increased significantly by increasing oriffice of nozzle, such enhanced evaporation is evidently contributed to lower velocity of plasma jet.

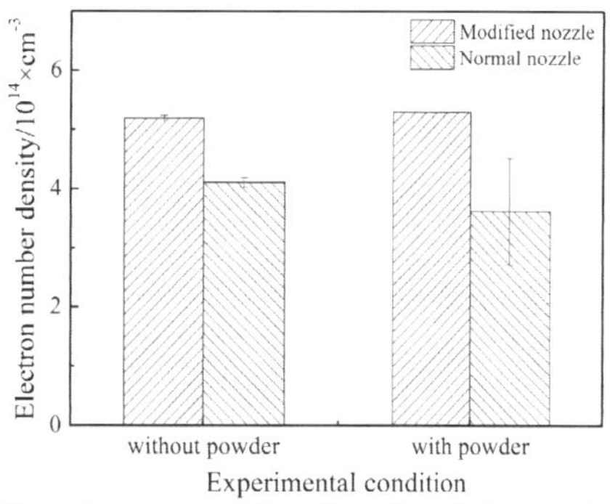

Fig.1. The electron number density of plasma jet by the modified nozzle compared with normal nozzle at a spray distance of 260 mm

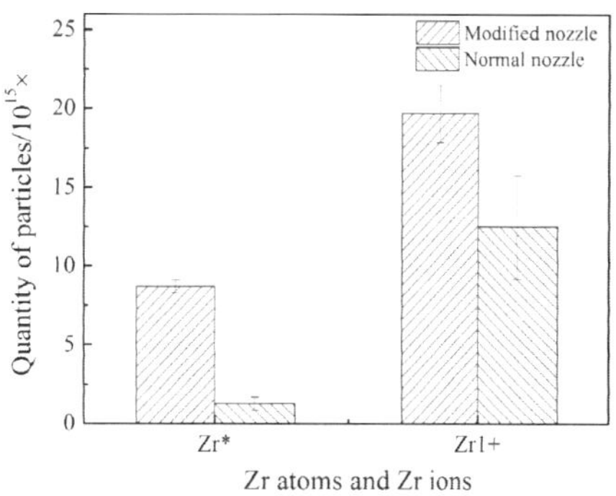

Fig.2. Comparison of the numbers of Zr atom and Zr^{1+} ion in the plasma jet generated by the modified nozzle in comparison with the normal nozzle at a spray distance of 260 mm

3.2 Microstructure of deposited coatings

The surface morphology of deposited coatings is shown in Fig.3. The surface of all coatings presents an island-like structure. The size of island-like structure decreased with the increase of the spray distance. From high magnification images, it can be seen that the unit of island-like structure is composed of nano particles with angular shape. The size of nano particles is also decreased with increasing the distance.

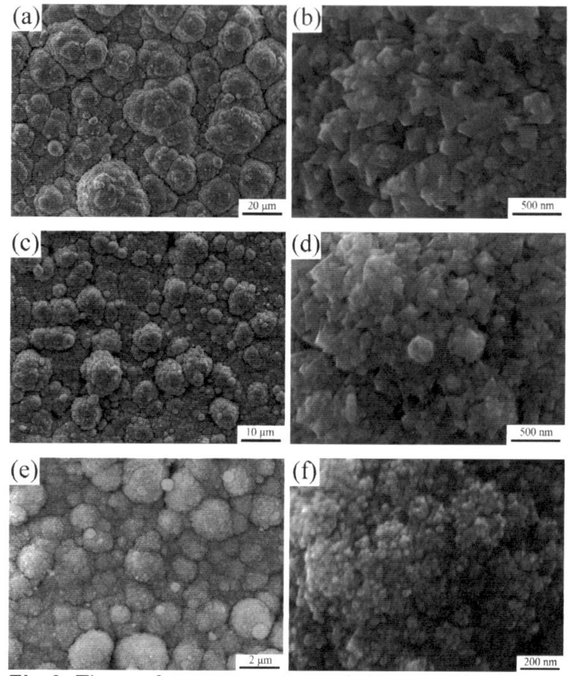

Fig.3. The surface morphology of YSZ coatings deposited at different spray distances: (a), (b) 260 mm; (c), (d) 340 mm; (e), (f) 420mm

From the cross sections of fracture coatings shown in Fig.4, it is evident that the coatings present a columnar structure. Moreover, it was observed that the size of of the unit cell in the columnar structure decreased with increasing the distance.

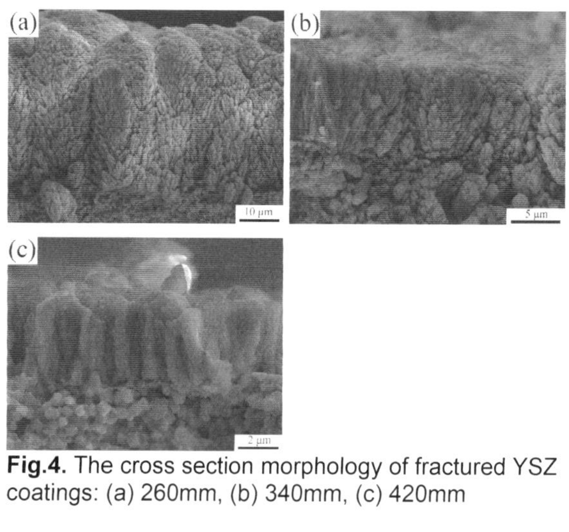

Fig.4. The cross section morphology of fractured YSZ coatings: (a) 260mm, (b) 340mm, (c) 420mm

It was observed that the growth of the columnar cell grows into a "V" shape possibly due to the preferred growth orientations. Some columnar cells grow fast and limit the growth of other columnar cells arround them. Then, the surface becomes island structure.

3.3 The effect of modified nozzle on the deposition rate of YSZ coatings

The measurement showed that the thickness of deposited coatings at the same plasma conditions with the modified nozzle is larger than that with the normal nozzle (Fig.5). The coating thickness with the modified nozzle at 260 mm and 340 mm under 60 kW is even larger than those deposited with the normal nozzle under 78 kW. This can be attributed to enhanced evaporation of YSZ particles in plasma jet. The thickness of coatings deposited with modified nozzle at 420 mm distance became comparable to that with normal nozzle by the same plasma power. This is probably due to the fast attenuation of the plasma temperature at long plasma jet axis due to the low velocity.

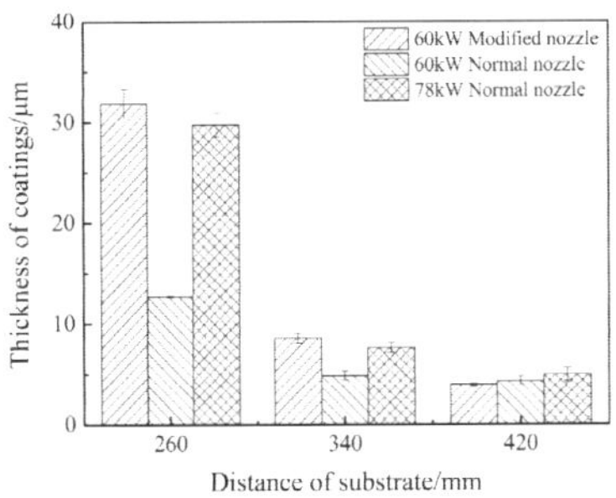

Fig.5. The thickness of deposited coatings with modified in comparison to that of the normal nozzle at two different power levels

The size of the nano-particles observed from the surface of island-structured coatings was estimated based on the surface images. Despite a larger size of islands as shown in Fig.6a, the size of the nano-particles in the coating deposited with the modified nozzle is a little smaller than that with the normal nozzle as shown in Fig. 6b at the same plasma conditions. On the other hand, it can be found that the size of the nanoparticles in columnar cells depends more on spray distance. This is because the size of islands correspond to the size of local column cell size which is closely related to the length of column cell due to its V-shape like morphology.

3.3 Comformity of Deposited Coating

The microstructure of YSZ coating deposited on a rough cold-sprayed MCrAlY surface was shown in Fig.7 at a spray distance of 260 mm. It can be seen from the image at a low magnification that the coating presents a columnar structure on the rough surface and columnar cells grow perpendicular to the substrate surface over either vally, hill or top of the curved bond coat surface. The thickness of the columnar structured coating is uniform. The columnar structured coating is bonded well to the substrate as shown in Fig.7b. Therefore, the columnar structured coating with good comformity over the substrate surface can be deposited under enhanced evaporation of YSZ condition by PS-PVD through droplet filtering.

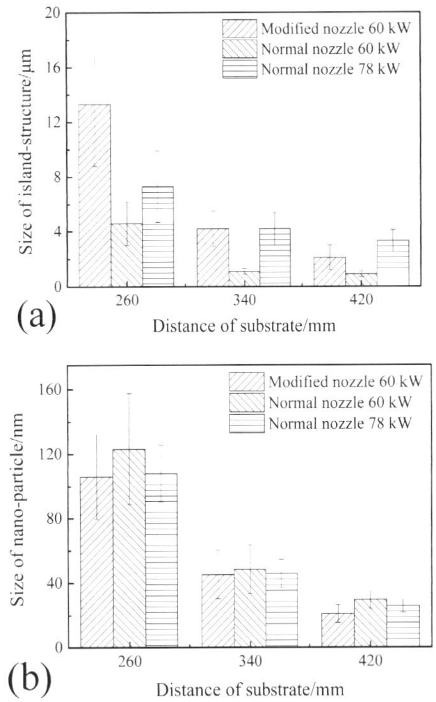

Fig.6. the sizes of unit structure in the deposited coatings with the modified nozzle compared with the normal nozzle: (a) size of Island structure, (b) size of nano-particle

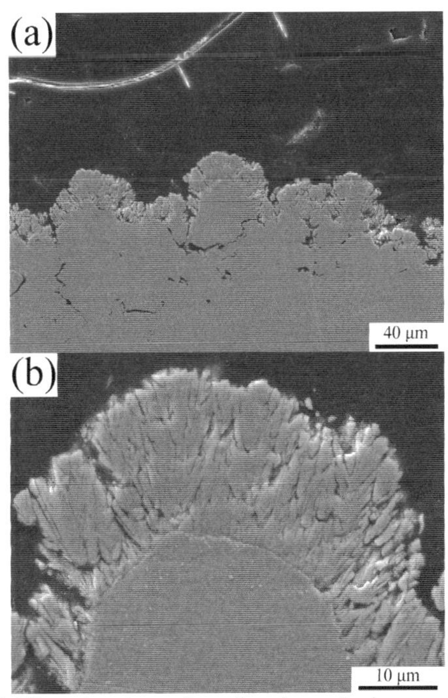

Fig.7. Microstructure of coating deposited on rough surface: (a) low magnification, (b) high magnification

4 Conclusions

By modifying the plasma torch nozzle through increasing oriffice diamerter, it was found that the electron density numbers and consequently the quantity of Zr gas phase including Zr atoms and Zr^{1+} ions in the plasma jet can be significantly increased. Accordingly, the evaporation of YSZ particles was evidently enhanced by modifying plasma nozzle. As a result, using conventional 80 kW class low pressure plasma spray systems, the columnar structured YSZ coating was realized at a higher deposition rate through PS-PVD process by using the modified nozzle. Moreover, it was revealed that the columnar structured YSZ coatings present good comformity on the rough surface of cold-sprayed McrAlY substrate.

Acknowledgements

This research work was sponsored by National Basic Research Program of China (2012CB625100) and National Natural Science Foundation of China (51342003).

Literature

[1] W. Xu, easurement of Plasma ion temperature and flow velocity from chord-averaged emission line profile, J. Thermal Spray Technol., 32 (2011), Issue 1, pp. 303/05.

[2] S. C. Snyder, G. D. Lassahn and J. D. Grandy, Direct determination of gas velocity and gas temperature in an atmospheric-pressure argon–hydrogen plasma jet. J. Quant. Spectrosc. RA, 107 (2007), Issue 2, pp. 217/25.

[3] Q. B. Fan, L. Wang and F. C. Wang, Numerical simulation of temperature and velocity fields in plasma spray. J. Cent. South Univ. T., 14 (2007), Issue 4, pp. 496/99.

[4] J. Cizek, I. Dlouhy and F. Siska et al., Modification of plasma-sprayed TiO_2 coatings characteristics via controlling the in-flight temperature and velocity of the powder particles. J. Therm. Spray Technol., 23 (2014), Issue 8, pp. 1339/49.

[5] T. Streibl, A. Vaidya and M. Friis et al., A critical assessment of particle temperature distributions during plasma spraying: experimental results for YSZ. Plasma Chem. Plasma P., 26 (2006), Issue 1, pp. 73/102.

[6] C. Zhang, A. F. Kanta and C. X. Li et al., Effect of in-flight particle characteristics on the coating properties of atmospheric plasma-sprayed 8 mol% Y_2O_3–ZrO_2 electrolyte coating studying by artificial neural networks. Surf. Coat. Technol., 204 (2009), pp. 463/69.

[7] A. Hospach, G. Mauer and R. Varen et al., Columnar-structured thermal barrier coatings (TBCs) by thin film low-pressure plasma spraying (LPPS-TF). J. Therm. Spray Technol., 116 (2011), Issue 1-2, pp. 116/20.

[8] M. Goral and J. Sieniawski, The technology of plasma spray physical vapour deposition. J. Achieve. Mater. Manufact. Eng., 55 (2012), Issue 2, pp. 689/93.

[9] M. Goral, S. Kotowski and A. Nowotni, PS-PVD deposition of thermal barrier coatings. Surf. Coat. Technol., 237 (2013), pp. 51/55.

[10] K. von Niessen and M. Gindrat, Plasma spray-PVD: A new thermal spray process to deposit out of the vapor phase. J. Therm. Spray Technol., 20 (2011), Issue 4, pp. 736/43.

[11] K. von Niessen, M. Gindrat and A. Refke, Vapor phase deposition using plasma spray-PVDTM. J. Therm. Spray Technol., 19 (2010), Issue 1-2, pp. 502/09.

[12] K. von Niessen and M. Gindrat, Vapor phase deposition using a plasma spray process. J. Eng. Gas Turb. Power, 133 (2011), pp. 061301-1/7.

[13] A. Refke, M. Gindrat and K. V. Niessen et al., LPPS thin film: A hybrid coating technology between thermal spray and PVD. Thermal Spray 2007: Global Coating Solutions, OH, USA, May 2007, pp. 705/10.

[14] Q. Y. Chen, X. Z. Peng and G. J. Yang et al., Characterization of plasma jet in plasma spray-physical vapor deposition of YSZ using a <80 kW shrouded torch based on optical emission spectroscopy. J. Therm. Spray Technol., 24 (2015), Issue 6, pp. 1038/45.

[15] Q. Y. Chen, C. J. Li and C. X. Li et al., A preliminary microstructure investigation of YSZ coatings deposited by plasma spray–physical vapor deposition using a shrouded plasma torch. International Thermal Spray Conference & Exposition 2014, Barcelona, Spain, May 2014, pp. 268/72.

[16] Q. Y. Chen, C. X. Li and T. Wei et al., Controlling grain size in columnar YSZ coating formation by droplet filtering assisted PS-PVD processing. RSC Adv., 5 (2015), Issue 124, pp. 102126/33.

[17] J.A. Browning, Plasma spray torch with high velocity hot gas particle supply|extends circumferential anode life. U.S. Patent 4,916,273, 1990-4-10.

[18] T. H. Hwang, D. U. Kim and J. M. Park et al., Effects of anode nozzle geometry on thermal plasma characteristics generated by non-transferred torches for material processing. Proceedings of IEEE 29th International Conference on Plasma Sciences, Banff, Alta, Canada, May 2002, pp. 117/17.

[19] S. Choi, T. H. Hwang and J. H. Seo, Effects of anode nozzle geometry on ambient air entrainment into thermal plasma jets generated by nontransferred plasma torch. IEEE Transaction on Plasma Science 32 (2004), Issue 2, pp. 473/78.

[20] Y.K. Ke and H.R. Dong, Handbook of analytical chemistry third fascicle spectrum analysis, Chemical Industry Press, Beijing, China, 1998.

[21] S. Pellerin, K. Musiol, B. Pokrzywka, and J. Chapelle, Stark width of 4s-4p ArI transition

(696.543 nm), J. Phys. B-At. Mol. Opt. 29 (1996),Issue 17, pp. 3911/24.
[22] A.A. Ovsyannikov and M.F. Zhukov, Plasma Diagnostics, Cambridge International Science Publishers, Cambridge, British, 1995.
[23] G. Mauer and R. Vaßen, Plasma spray-PVD: plasma characteristicsand impact on coating properties, J. Phys. Conference Ser. 406 (2012), Issue 1, pp. 1/12.
[24] D.R. Qiu, The atomic spectrum analysis, Fudan University Press, Shanghai, China, 2002.
[25] G.A. Odintsova and E.P. Vaulin, Excitation temperature and gas temperature in supersonic low-density flows, J. Appl. Spectrosc. 7 (1967), Issue 5, pp. 507/08.
[26] http://physics.nist.gov/PhysRefData/ASD/lines_form.html

Corrosion behaviors of amorphous coating under simulated deep sea environment

Cheng Zhang, Zhi Wei Zhang, Lin Liu
State Key Lab for Materials Processing and Die & Mould Technology, School of Materials Science and Engineering,Huazhong University of Science and Technology, Wuhan 430074, People's Republic of China

In-situ corrosion behaviors of a Fe-based amorphous coatings under simulated deep sea environment (80 atm) are first investigated using a serial of electrochemistry analysis including potentiodynamic polarization, pitting incubation tests and corrosive morphologies observations, as compared to 316L stainless steel alloys. Structure and mechanical properties of passive films formed under different pressures were directly analyzed by transmission electron microscopy (TEM) and nanoindentation measurements. The results demonstrate that amorphous coating exhibited high pitting resistance than 316L SS alloys under high hydrostatic pressures, evidenced by higher pitting potential, longer pitting incubation time and smaller pitting growth ability. In addition, passive films formed on amorphous coating were found to be thicker, more uniform and harder than those on the SS alloys, indicating that the former is more difficult to be mechanically break-down and penetrated by Cl- ions. The findings may pave the way for exploring the amorphous coatings for deep-sea applications.

1 Introduction

The demand for exploring vast amounts of oil and gas resources in deep sea has increased. The deep sea presents a harsh corrosive environment for metallic materials because of the combined effects of low temperature, variable dissolved oxygen content and high hydrostatic pressure. Among these effects, hydrostatic pressure has emerged to play an increasingly important role in pressure-accelerated corrosion and failure for most industrially relevant metals, by strongly degrading their performances and design lives.

Due to the high susceptible to pitting corrosion of the stainless steel alloys, they are generally required surface coatings in the practical deep-sea applications. Although polymer coatings offers appreciable corrosion protection for steels under the hydrostatic pressure conditions, they always suffered from poor durability because of accelerated water absorption and penetration [1]. Recently, HVOF-sprayed Fe-based amorphous coatings have attracted tremendous attentions owing to their exceptionally high pitting resistance in various corrosive media [2]. However, corrosion behavior of amorphous coatings at high hydrostatic pressures have not been reported so far.

2 Experimental proceedures

$Fe_{48}Mo_{14}Cr_{15}Y_2C_{15}B_6$ (at. %) amorphous powders produced by gas atomization was used as the feedstock powder. Amorphous coating was fabricated by means of HVOF on a 316L stainless steel substrates. In-situ electrochemical experiments were performed in a deep sea simulation system. Potentiodynamic polarization tests were performed in a standard three-electrode system in a 3.5 wt% NaCl. For pitting incubation measurement, the potentiostatic technique was applied to measure the anodic current trace as a function of immersion time at various hydrostatic pressure. In order to directly observe the structure and thickness of passive films, an aberration corrected TEM (G2 60-200)was carried out. The TEM samples were cut out by *in-situ* lift-out focus ion beam (Quanta 3D FEG) system. The mechanical properties of the passive films were measured on a nano-indentation system with 50 nm Berkovich tip radius. The loading rate and the peak load depth is 50 μN s^{-1} and 500 μN.

3 Results and discusstions

3.1 Potentialdynamic polarizations

Fig. 1 shows potentiodynamic polarization curves of Fe-based amorphous coating and SS alloy in a 3.5 wt. % NaCl solution at atmospheric pressure (1 atm) and at high hydrostatic pressure (80 atm). It discloses that the pitting poentail (E_{pit}) values of the SS alloy at all pressures are significantly lower than that of the amorphous coating, indicative of an inferior stability and protectiveness of the passive film on the SS alloys. More interestingly, the E_{pit} is increased for the coating (ΔE_{pit}=4.2%) while E_{pit} dramatically decreases by approximately 11.9% for the SS alloy at 80 atm. Accordingly, the hydrostatic pressure might have different mechanisms in affecting the pitting resistance on the two alloys, i.e., improvement for the amorphous coating while deterioration for the 316L SS alloy.

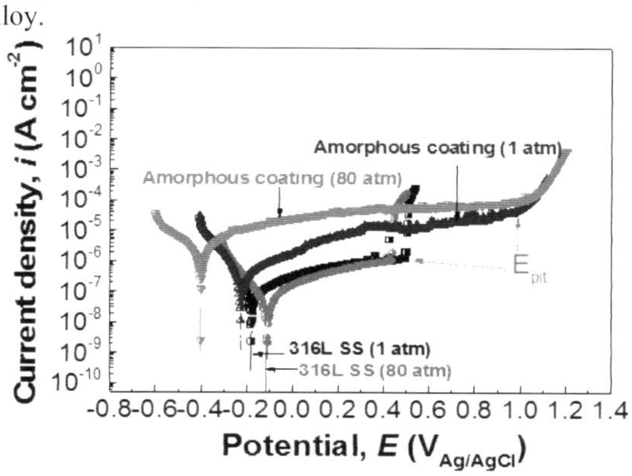

Fig.1. Potentiodynamic polarization curves of amorphous coatings and 316L stainless steel (SS) alloys at the pressures of 1 atm and 80 atm

3.2 Pitting incubation measurements

Fig. 2(a) displays the log *i*-log *t* curves of amorphous coating and 316L SS alloy tested at 0.45 $V_{Ag/AgCl}$ at 1 atm and 80 atm, respectively. It highlights that the pitting incubation time for the coating is noticeably longer than that of the 316L SS alloy at both pressure conditions, indicating that the passive film formed on the amorphous coating has much lower rupture probability at the deep sea environment. Furthermore, both the pitting initiation and stable growth stages for the 316L SS alloy are obviously accelerated by the high pressure, evidenced by short pitting incubation time and higher current at the second stage. However, at this potential, the stable growth for pitting on the amorphous coating is impossible to be observed. The pitting incubation tests were additionally carried out at a higher constant potential of 0.95 $V_{Ag/AgCl}$, which is 0.05 $V_{Ag/AgCl}$ below the E_{pit} of the coating. As expected, there is a substantial increase of τ at the pressure of 80 atm, compared to that of 1 atm, as shown in Fig. 2(b).

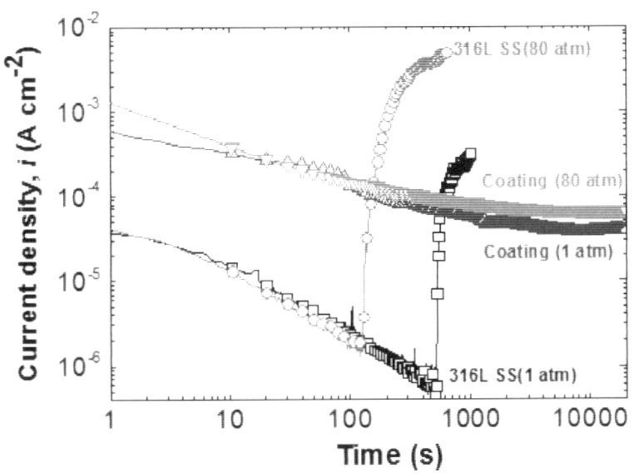

(a)

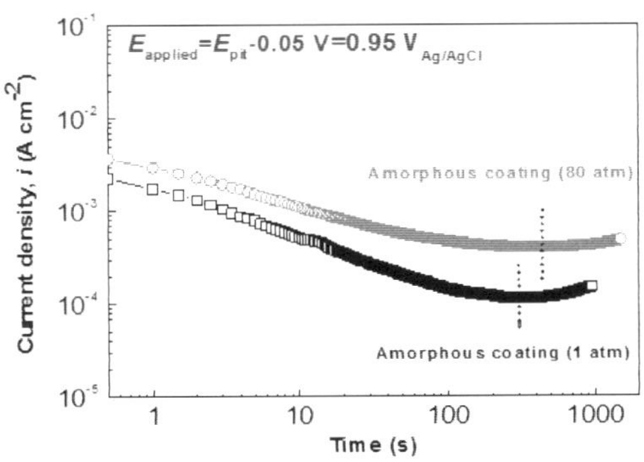

(b)

Fig.2. log (*i*) *vs* log(*t*) plots for the amorphous coating and the 316L SS alloy under the applied potential of (a) 0.45 $V_{Ag/AgCl}$ and (b) 0.95 $V_{Ag/AgCl}$ at 1 atm and 80 atm.

3.3 Observatin of nanoscale passive films

Using FIB and HRTEM techniques, we directly examined the atomic structure and film thickness of the passive films. Figs. 3 shows the HRTEM images of passive films. The HRTEM image and its diffraction patterns clearly reveal an entirely amorphous structure of the passive films formed under both pressures. The passive films on large scale are found to be very uniform, continuous and have a thicknesses of 2-7 nm depending on the pressures. Surprisingly, the passive film formed increases from ~2 nm at 1 atm to 6~7 nm at 80 atm, supporting the fact that pressure promoted the dynamic processes of oxide film growth-up. The EELS technique which has an ultrahigh spatial resolution for light elements such as O was further carried out to identify the cross-sectional composition of passive films. Fig. 3(e) is the corresponding background-subtracted EELS spectrum that was collected from the passive films, Cr, Fe peaks and clear O peak are highlighted. As compared to 1 atm, the passive film formed 80 atm exhibits a significantly higher O, Cr and Fe intensity, implying the higher Cr-oxides and Fe-oxides content of the oxide film indeed formed with the aid of pressure.

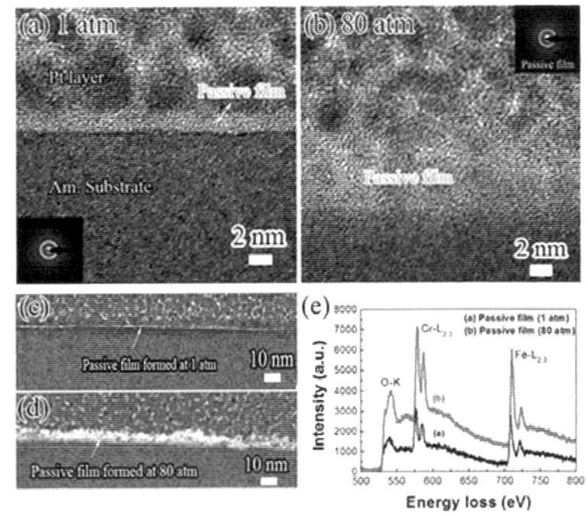

Fig.3 (a, b) High-resolution bright-field TEM of passive films and (c, d) Colored passive films formed on the amorphous coating at different pressures, showing the thickness and uniformity. (e) electron energy loss spectroscopy (EELS) from the passive films.

3.4 Nanoidentation measurement

In order to further verify the fact that the passive film on the amorphous coating formed at high hydrostatic pressure is superior than the one formed at atmospheric pressure, nano-mechanical properties such as critical breakdown load (P_{crit}) and nano-hardness of the passivated surface are evaluated by using the nano indentation tests. The nanohardness presented in Fig.4 indicates that the passivated amorphous sample at 80 atm is relatively harder than that at 1 atm, i.e., H_{80atm}=16.3±0.40 GPa and H_{1atm}=15.0±0.30 GPa; however, the passivated 316L SS alloy at 80 atm is softer (as H_{80atm}=3.5±0.12 GPa and H_{1atm}=3.8±0.07 GPa). The tendency of P_{crti} and nanohardness with the pressure for the two materials is coincident with the electrochemical results and TEM observations.

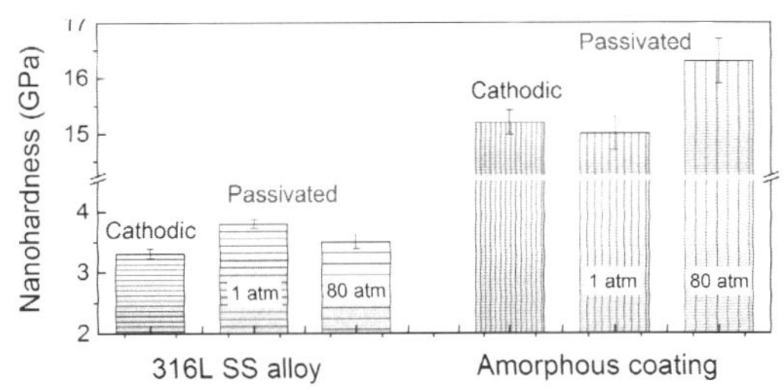

Fig.4. Summary of the surface hardness for cathodic and passivated samples at various pressures.

4 Literatures

[1] Y. Liu, J. Wang, L. Liu, Y. Li, F. Wang, Corros. Sci., 74 (2013) 59-70.

[2] J. Farmer, J.S. Choi, S. Cheng, J. Haslam, D. Dan, P. Hailey, T. Lian, R. Rebak, J. Perepezko, J. Payer, D. Branagan, B. Beardsley, A. D'Amato, L. Aprigliano. Mater. Trans. A, 40 (2008) 1289-1305.

Improvement of bonding strength and impact resistance of Fe-based amorphous coatings by structure modifications

Lin Liu*, Cheng Zhang, Wei Wang, M. Yasir
State Key Lab for Materials Processing and Die & Mould Technology, School of Materials Science and Engineering,Huazhong University of Science and Technology, Wuhan 430074, People's Republic of China

Amorphous coatings are well-known for their high strength, high hardness and outstanding corrosion and wear resistance. During the past decades, despite the fabrication, structure, corrosion and wear properties of amorphous coatings prepred by thermal spraying have achieved grear advances, the relatively poor bonding strength and impact resistance have became new problems that restricing the wide applications of such coatings. In this paper, we reviewed the recent progress about amorphous coatings made in our group via a serial of structure modifications. In the first route, ductile metals, ceramic particles and polymer phases were introduced into amoprhous coatings; while in the second route, laminar structures were designed to form homogenous amorphous composite coatings. It will be demonstrated that those amorphous composite coating exhibited significantly improved bonding strength, impact resistance, as well as wear/friction resistance over pure amorphous coatings.

1 Introduction

Amorphous coatings, an alternative form of BMGs, have recently attracted increasing interest in many industrial applications, because of the possibilities of synthesizing a large-scale surface protection layer with an excellent combination of properties, such as high hardness and wear resistance, superior corrosion resistance and neutron-absorption ability. While a great deal of research on amorphous coatings has been focused on corrosion and tribological properties, recent studies reveals very limited bonding strength (~30 MPa) and impact resistance of the amorphous coatings, which has largely restricting their industrial applications.

In the presnet paper, the recent progresses about amorphous coatings made in our group via a serial of structure modifications are reviewed. In the first route, ductile metals, ceramic particles and polymer phases were introduced into amoprhous coatings; while in the second route, laminar structures are designed to form homogenous amorphous composite coatings. It will be demonstrated that those amorphous composite coating exhibited significantly improved bonding strength, impact resistance, as well as wear/friction resistance over pure amorphous coatings.

2 Results and discusstions

2.1 By adding stainless steel particles

From the viewpoint of bonding mechanism in thermally sprayed coatings, the soft metallic phases might be effective to absorb impact energy and impede the crack propagation. A few Fe-based amorphous matrix composite coatings reinforced by various volume fractions of ductile 316L stainless steel (SS) powders have been directly synthesized by HVOF. Fig. 1a illustrats clearly a mixed structure containing the 316L SS phase (the dark-contrast region) and the amorphous matrix. The interfaces between the two phases are continuous and free of porosity. Additionally, it were composed of both amorphous phases and nano-sized crystals, demonstrating the occurrence of atomic diffusion and chemical reactions at the interfaces between amorphous powder and stainless steel powders.
The bonding strength of the coatings has been assessed by "pull-off" tensile test according to ASTM C633 standard. Fig. 1(c) shows that the addition of 316L SS powders is very effective in improving the bonding strength of the coatings. For example, the composite coatings with 8% SS had a significantly higher bonding strength of about 45 MPa, which is approximately 170% higher than that of the monolithic coating. The stress-strain curves for various free standing coatings under bending are shown in Fig.1(d), which further reveals that the fracture strain of the composite coatings greatly increases with the increase of the volume fraction of stainless steel phase. This result implies that the optimum addition of 316L SS phases is around 8 vol. %, in achieving substantial improvement in toughness of the coatings.

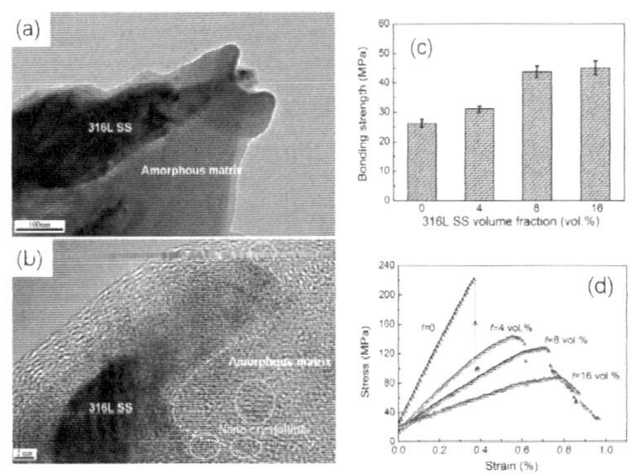

Fig.1. (a) Bright-field TEM image of the interfacial regions between the 316L stainless steel particle and the amorphous matrix for sample with v=16 vol. %; (b) the enlarged HRTEM image on the SS/amorphous interface. (c) Bonding strengths for the coatings with different volume fractions of 316L SS and (d) Stress-strain curves of free-standing coatings with different volume fractions of 316L SS under 3-point bending.[1]

2.2 By adding ceramic particles

The impact resistance of the coatings was evaluated by drop-weight impact test with different impact energies. Figs. 2(a, b) illustrate the cross-sectional morphologies of the amorphous coating and the composite coating after being impacted with 14.2 J energy (corresponding to 80 cm height), respectively. Clear delamination at the interface between NiCrAl layer and the substrate, and penetrating cracks were observed in the amorphous coating. In contrast, both horizontal and vertical cracks were substantially suppressed in the composite coating due to the presence of Al_2O_3 particles. In order to study the toughening mechanism, the composite coating was further subjected to a higher energy impact. Fig. 2(c) displays the cross-section image of the composite coating after being impacted with 21.2 J, no interface delamination happened although large plastic deformation of the coating/substrate occurred at such high impact energy. Figs. 2(d, e) show the enlarged fracture morphologies of the composite coating after impact test, which reveal that the Al_2O_3 particles have two roles during impacting: (1) Al_2O_3 particles block the crack propagation in the coating; (2) Al_2O_3 absorb the impact energy by the formation of micro cracks inside the particles. Based on the cracking behavior and fracture morphologies, two possible toughening mechanisms are proposed here, as illustrated schematically in Figs. 2 (f, g). First, the hard Al_2O_3 particles can hinder the crack propagation from the amorphous matrix and additionally deflect cracks from the main crack path. Second, the brittle nature of Alumina particles allow micro cracking to occur inside them during higher stress concentration at the crack tips.As a result, the impact energy could be highly consumed, which, in turn, enhance the impact resistance of the composite coating.

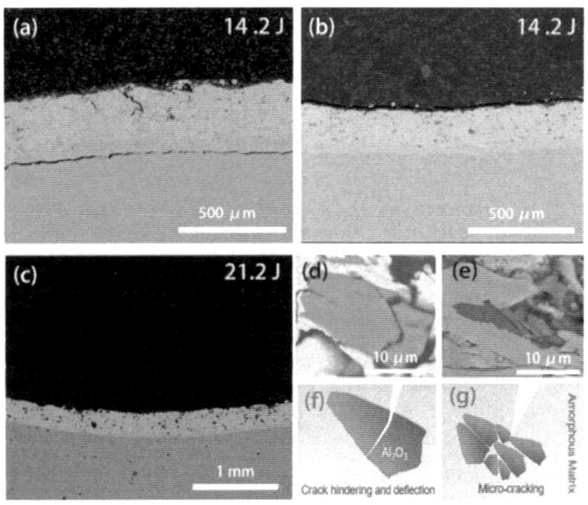

Fig.2. The cross-sectional fracturing structure of the amorphous coating and the composite coating reinforce with 20% Al_2O_3 particles under different impact energies. (a) The amorphous coating after impact with the energy of 14.2 J; (b, c) the composite coating after impacts with the energy of 14.2 J and 21.2 J, respectively (d, e) the enlarged images from (c) showing the cracking behaviors of Al_2O_3 particles and (f, g) show two kinds of toughening mechanism, i.e., crack deflection and micro-cracking.

2.3 By laminar structure designing

In this section, using the similar structural design concept, we have developed a few Fe-based amorphous composite coatings with laminar structure *via* high velocity oxygen fuel (HVOF) thermal spraying technique. The resulting coatings comprise of thick amorphous layers by alternate thin, adhesive and oxidation resistant NiCrAl crystalline layers. Figs. 3 (a) shows the cross-sectional SEM images of as-sprayed laminar composite coatings consisting of alternate amorphous and NiCrAl layers. The composite coatings display periodically laminated architecture with the amorphous layer being the outmost surface. The interfacial structure between the NiCrAl and the amorphous layer was further characterized by cross sectional TEM observations. As shown in Fig. 3a, the uniform image constrast of the amorphous layer and the corroponding SAED pattern indicate a fully amorphous structure in this region. The SAED pattern taken from the NiCrAl region shows a polycrystalline structure, from which the crystalline phases were indexed as fcc-Ni and Ni_3Al. These results verified the formation of metallurgical bonding between the NiCrAl and amorphous phases.

Fig. 3(c) shows the bonding strength of the coatings measured via tensile test. It was found that the monolithic amorphous and NiCrAl coatings have bonding strengths of 26±3 and 44±6 MPa, respectively. The bonding strength of the 4-layered and 6-layered composite coatings is 36±4 MPa and 40±3 MPa, respectively, which are 38% and 54% increase over the monolithic amorphous coating at similar thickness (~640 μm). The impact resistance of the coatings was evaluated by drop-weight impact test. Fig. 3(d) illustrates the comparison of the impact resistance for various coatings. The monolithic amorphous coating is very sensitive to impacting with very low impact energy (< 1.8 J). However, the laminated composite coatings possess higher impact resistance with the energy of 1.8-3.5 J for the 4-layered coating and 17.6-21.2 J for the 6-layered coating, respectively.

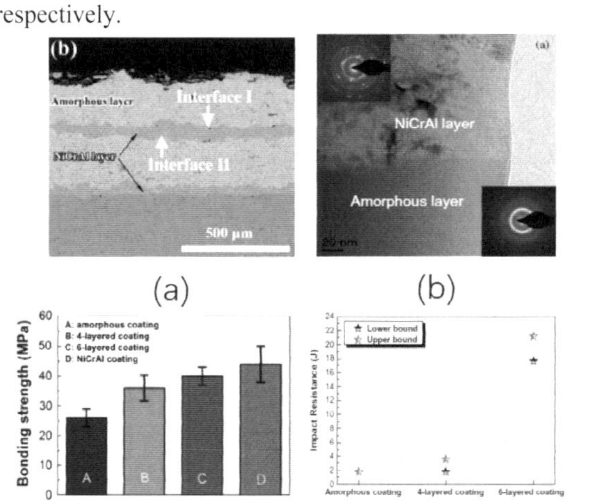

Fig.3. (a) The cross sectional structure of a 4-layered coating containing two layers of amorphous coating and two layers of NiCrAl coatings; (b) Cross-sectional bright-field TEM image of an interfacial region between the NiCrAl and the amorphous layer. Insets are the SAED patterns taken from the NiCrAl and amorphous regions, respectively; (c) Bonding strength values for various coatings; (d) Comparison of the impact resistance of various coatings. [2]

3 Literatures

[1] H. Zhou, C. Zhang, W. Wang, M. Yasir, L. Liu. J Mater Sci Technol. 31 (2015) 43.

[2] C.Zhang, H.Zhou, L. Liu, Acta Mater. 72 (2014) 239.

Enhancement of wear and impact resistance of Fe based amorphous coatings by the addition of Alumina particles

Muhammad Yasir, Cheng Zhang, Wei Wang, Peng Xu, Lin Liu*

State Key Lab for Materials Processing and Die & Mould Technology, School of Materials Science and Engineering, Huazhong University of Science and Technology, Wuhan 430074, People's Republic of China

Abstract

The influence of the addition of Al_2O_3 particles on the wear and impact properties of Fe-based amorphous coatings has been studied. The wear behaviors were evaluated in air and in 3.5 wt. % NaCl solution. The homogeneously distributed hard Al_2O_3 particles resulted in improved wear resistance and reduced coefficient of friction (COF) in both conditions. The composite coating reinforced with 20 wt.% Al_2O_3 particles exhibit the best wear performance, having extremely low COF (<0.2) and high wear resistance (2-3 times higher than monolithic amorphous coating). Detailed SEM analysis indicated that the wear is dominated by oxidative delamination in air and by corrosion wear in 3.5% NaCl solution. Furthermore, the impact resistance/damage tolerance of amorphous coatings was also improved three times by the addition of alumina particles. The deformation and fracture behavior under impact loading were investigated via scanning electron microscopy observations as well as finite element simulation. The hard second phase particles present in the amorphous coating matrix disperses the residual stresses generated during the impact loading. The brittle alumina particles absorb the impact energy by breaking itself which stop the initiation of cracks and also play a vital role in the crack arresting and blocking of the crack propagation.

Keywords: Fe-based amorphous coating; Al_2O_3 particle addition; Wear, corrosion wear, Impact resistance, crack initiation.

*Corresponding author, Tel.: +86-27-87556894; Fax: +86-27-87554405.

E-mail address: lliu2000@mail.hust.edu.cn (L. Liu).

High temperature corrosion of selected thermally sprayed coatings deposited by HVOF technology

Z. Česánek, J. Schubert, Š. Houdková, Pilsen/CZ, V. Matikainen, H. Koivuluoto, P. Vuoristo, Tampere/FI

Corrosion resistance of coatings deposited by High Velocity Oxygen Fuel (HVOF) thermal spraying technology requires high density in coating and good adhesion to the substrate. The majority of thermally sprayed materials meet the requirements for high corrosion resistance in terms of their composition. However, the porous structure and also the inhogeneous microstructure raise doubts about the performance of thermally sprayed coatings in providing of sufficient protection to the base material. In fact, corrosion protection is one of the basic coating functions. Anyway, in comparison with wear resistance research, no sufficient attention is often dedicated to the issue of component protection against corrosion attack using HVOF sprayed coatings.

This paper summarizes the results of high temperature corrosion and erosion tests for Cr_3C_2-25%NiCr, Cr_3C_2-25%CoNiCrAlY, TiMoCN-29%Ni (experimental powder), Stellite 6, NiCrBSi and Hastelloy C-276 coatings. The coatings were applied on W.Nr. 1.4923 steel substrate material, which is also used in nitrided state as an effective constructional material for components in power equipment. All evaluated coatings were deposited by HVOF hermal spray technology. In hot corrosion testing the salt mixture with composition of 59% Na_2SO_4, 34.5% KCl and 6.5% NaCl in an environment with 30% humidity was chosen as an aggressive corrosive environment. Furthermore, two temperatures of 525 °C and 575 °C were studied. The test duration was in both cases 168 hours. After the testing, all evaluated coatings were analyzed using scanning electron microscope (SEM) and afterwards elemental composition (EDAX) was conducted. Alloy based coatings showed very similar corrosion mechanism in the selected aggressive environment.

1 Introduction

High temperature corrosion is corrosion in the presence of molten salts. The corrosion process involves deposition of salts on the material surface. Certain salts are at operating temperatures in liquid state or form complex salt mixtures in the presence of sulfur containing gases. Salt mixtures melt at much lower temperatures than are the melting temperatures of their individual constituents. The material attack accompanied by the deposition of salt mixtures on the surface is caused not only by the impurities deposited in coal or oil, but also by the penetration of salts to the material surface, e.g. from sea water [1].

High temperature corrosion is nowadays a very serious problem in the construction of power generation plants. Various solutions are used to face high temperature corrosion, but the majority prevents or limits the high temperature corrosion only partially and temporarily. The application of protective coatings deposited using HVOF technology seems to be one of the most effective options to protect components against corrosion attacks. However, no coating is universally suitable for every type of environment, and therefore it is very important to investigate which coatings are appropriate for particular environment [2], [3].

Moreover, even some kind of erosive stress on materials can occur in the environment with high temperature corrosion. It is clear that synergic effect of high temperature corrosion with erosion differs from separate evaluation of above mentioned phenomena. On the other hand, the results also have a specific added value and are for this reason also included in the paper.

Erosion is characterized by particle removal and damage of functional surface. Erosive wear occurs e.g. in machine parts that transmit suspension (components of pumps, water turbines, pipes, fittings and nozzles), is caused by particle dispersion in air or gas (boilers, components of fans, cyclones, air-blasting equipment and pipes) or is generated by air, steam, droplet or gas flow (steam fittings, components of gas and steam turbines). Dealing with erosive wear caused by particles, several ways of particle impact angle on material surface need to be differentiated. If a particle with inclined direction to the surface has sufficient energy, it penetrates into the exposed surface and may dislodge or divide material. If a particle with almost parallel direction to the surface has small impact energy, it loads the surface only within the limits of elastic deformation, and therefore causes no damage. Another case is perpendicular impact of particles on the surface resulting in elastic or plastic deformation. After the perpendicular impact on surface, the particles are usually rebounded back [4].

The main methods of erosive wear assessment are based on particle flow. The particles are transmitted in the air, gas or on the basis of so-called centrifuge. While assessing erosion using air transmission, the particles are accelerated by the air flow under the certain pressure towards the specimen and impact the surface perpendicularly, or under specific angle obtained by specimen positioning. In the second case, the erosion testing is conducted using centrifugal force produced by rotation and acceleration of abrasive particles which are distributed from the channels in rotating disk. Deng et al. [5] evaluated differences between the above mentioned methods, particularly in relation to size, shape, concentration and velocity of abrasive particles. Regarding speed, the relationship between air pressure and particle velocity in a gas device is almost linear for all shapes, sizes and concentrations of abrasive particles. In centrifugal device, linear dependency was also proved between angular speed of rotating disk and velocity of particles distributed from channels. In case of gas device, it was found that small and sharp-edged particles are accelerated to higher velocities. Regarding concentration and velocity of particles, it was found that higher concentrations lead to reduction of particle velocity as a result of collisions which decelerate particles and reduce their kinetic energy. On the contrary, the relationship between particle properties and their velocity is minimal in centrifugal device. The expansion of particle stream after leaving the accelerating tube is another important effect occurring only in gas equipments. The expansion increases with increasing velocity and concentration of particles. This phenomenon was attributed to the aerodynamic association of particles and air flow, to the mutual interaction of particles and to the interaction of particles with acceleration tube wall. So-called "dead zone" in the middle of particle flow is formed as a result of gas flow expansion. Finer grains have higher tendency to copy the gas flow in comparison with rougher abrasive grains. Furthermore, the effect of abrasive particle properties on erosive wear of materials was mentioned. The shape of abrasive particles is another important factor. Sharp-edged particles cause higher wear resulting in surface cutting, especially at smaller impact angles. On the contrary, oval particles result in material removal caused by means of plastic deformation and grooving.

2 Experiment

Steel W.Nr. 1.4923 was used as a base material and coatings were deposited by HVOF thermal spraying technology using TAFA JP 5000 system. **Fig. 1a** shows specimens after application of salt mixture.
Nowadays, several methods of high temperature corrosion testing are used, but their principles differ in many respects. The basic evaluation method for high temperature corrosion is application of specific salt amount from saturated salt solution on specimen surface with subsequent measurements of weight loss or gain. Conditions of the tests performed in this paper were following. The test was based on application of certain amount of salt mixture with specific composition (59% Na_2SO_4 34.5% KCl and 6.5% NaCl) on the surface of evaluated coatings. The specimens were placed on ceramic plate in an autoclave, see **Fig. 1b**. The environment with 30% humidity was maintained throughout the test using a device for humidity generation. Two measurements in the time interval of 168 hours were performed. Besides different temperatures, other parameters were identical for all tests. The temperature of the first measurement was 525 °C and of the second was 575 °C.
After the exposure time, the specimens were then removed from the autoclave and cold moulded using a resin. This procedure prevents from losing of corrosion products that were formed on specimen surfaces during the corrosion test. Specimens were afterwards cut and moulded again in order to conduct cross-sectional metallographic observation for all evaluated coatings. Metallographic evaluation was conducted using scanning electron microscopy (SEM). The exposed specimens were further evaluated using the change of elemental composition. This evaluation was performed by elemental analysis EDAX. In conclusion, all the results were analyzed in terms of coatings suitability in chosen environment.

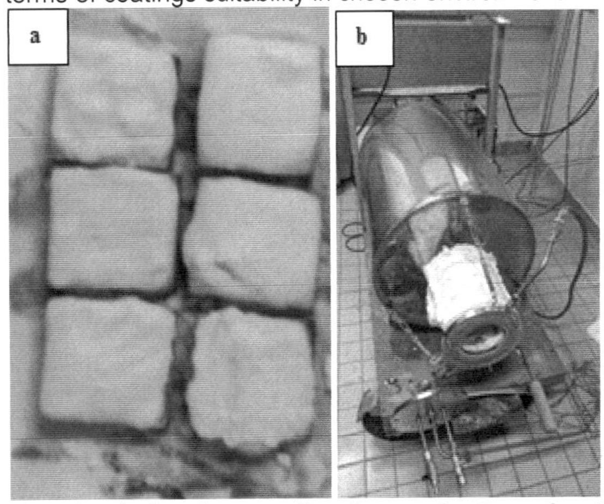

Fig. 1. a) Specimens after the application of salt mixture. b) Equipment for evaluation of high temperature corrosion resistance at Tampere University of Technology.

In this study, the attention was paid to the evaluation of solid particle erosion of selected thermally sprayed coatings. The influence of solid particle impact angle on the wear mass loss and wear mechanism was also studied and discussed. For solid particle erosion wear testing, the equipment was designed and manufactured. It uses the centrifugal construction for generation of kinetic energy of solid particles impinging the surface of coated samples (**Fig. 2**). The sample holders enable to adjustment of the particles impact angle [6].

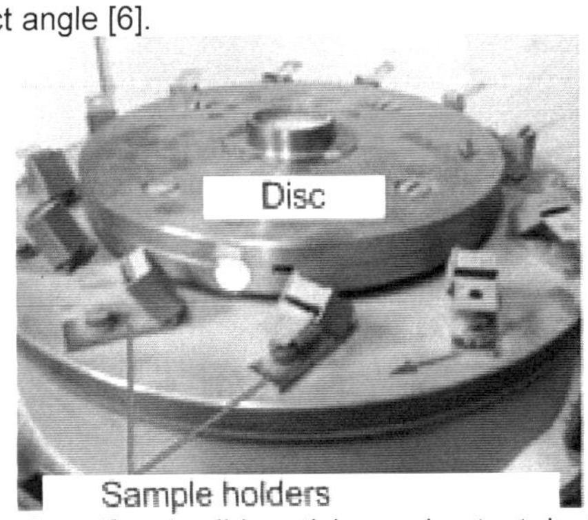

Fig. 2. Centrifugal solid particle erosion test device.

Based on a previous study [7], the test parameters suitable for HVOF coatings were chosen. Al_2O_3 (212-250 μm size range) abrasive particles were used, 47 m/s circumferential speed of rotating disc was set, the exposition time was 2 min. Four different angles of abrasive particles impact were set: 90°, 60°, 30°, 15°. The erosion wear mass loss was determined by samples weighing before and after the test, and was converted into the volume loss using specific density of coating materials [6].

3 Results

3.1 High temperature corrosion

Pictures made using scanning electron microscopy in **Fig. 3** show the structure of HVOF **Hastelloy C-276** coating in cross-section before and after the exposure at 525 °C and 575 °C. The pictures show evident occurrence of thin most probably oxide layer formed on the coating surface during the corrosion tests. It was found that increasing temperature leads also to increase in thickness of oxide layer, which is evident from **Fig. 4**. **Fig. 4** moreover proves that the increase in temperature led to the formation of compact oxide layer covering the entire specimen surface. Thanks to the creation of oxide layer, the penetration of aggressive corrosive environments was prevented and the coating was protected with this layer throughout the whole testing period.

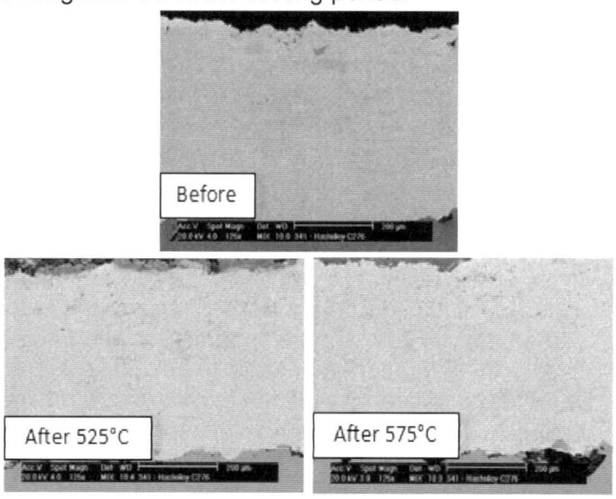

Fig. 3. HVOF Hastelloy C-276 coating above before exposure to corrosion testing, left after the exposure at 525 °C and right after the exposure at 575 °C, magnification 125x, SEM images.

Fig. 4. HVOF Hastelloy C-276 coating after the exposure at 575 ° C, right 500x magnification and left 1000x magnification, SEM images.

Pictures in **Fig. 5** show HVOF **Stellite 6** coating before and after the exposure to the corrosive environment at temperatures of 525 °C and 575 °C. The pictures after the corrosion test also show a thin oxide layer which was not influenced by the temperature increase compared with Hastelloy C-276 coating. According to the results of EDAX analysis, it can be argued that the protective oxide layer on the coating surface was composed mainly of Cr oxides, see **Fig. 6**. The results of EDAX analysis revealed the penetration of oxides and salt remains up to 150 µm below the surface at the exposure temperature of 575 °C. These salt remains and oxides penetrated into the coating along splat borders, see **Fig. 7**.

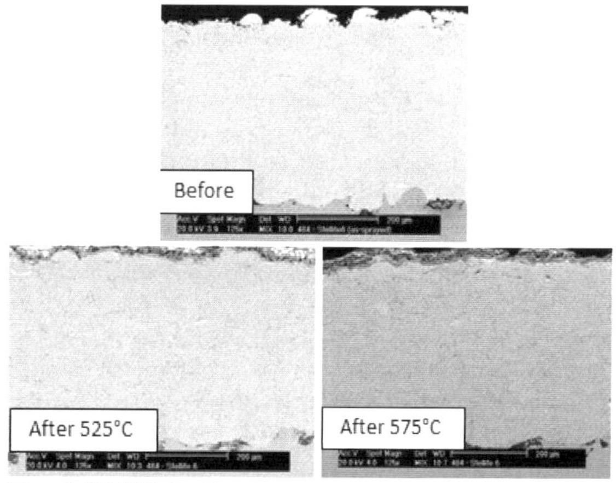

Fig. 5. HVOF Stellite 6 coating above before exposure to corrosion testing, left after the exposure at 525 °C and right after the exposure at 575 °C, magnification 125x, SEM images.

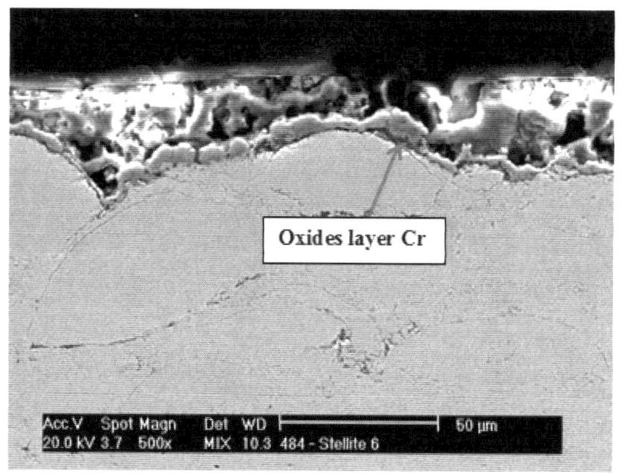

Fig. 6. HVOF Stellite 6 coating, after the exposure at 525 °C, magnification 500x, SEM images.

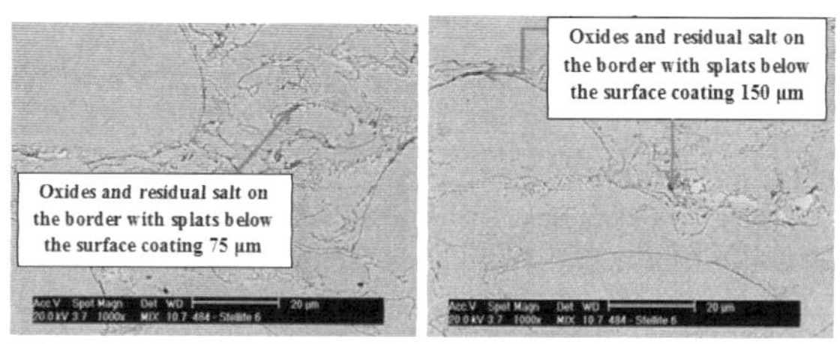

Fig. 7. HVOF Stellite 6 coating after the exposure at 575 °C, magnification 1000x, SEM images.

HVOF **NiCrBSi** coating behaved similarly in the corrosive environment of salt mixtures like above described alloy based Hastelloy C-276 and Stellite 6 coatings. However, the picture in **Fig. 8** shows that at the oxide layer became significantly thicker at higher exposure temperature. According to pictures in **Fig. 9**, the corrosion mechanism includes the formation of thin, mainly Ni oxide layer on the coating surface. The corrosion further continued with gradual separation of coating matrix from the surface. EDAX analysis proved thin layer of Ni and Si oxides after the exposure at the temperature of 575 °C. Moreover, there was measured growth and coarsening of Cr-Ni phase formed as consequence of coating exposure to elevated temperature. This measurement is consistent with the results of previous study by (Schubert et al.) [8], which states that elevated temperature leads to progressive increase of Cr-Ni phase and its formation into needle-shaped structures.

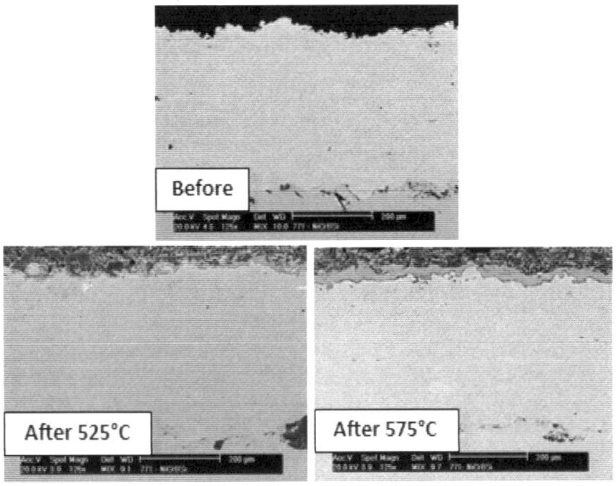

Fig. 8. HVOF NiCrBSi coating above before the exposure, left after the exposure at 525 °C and right after the exposure 575 °C, magnification 125x, SEM images.

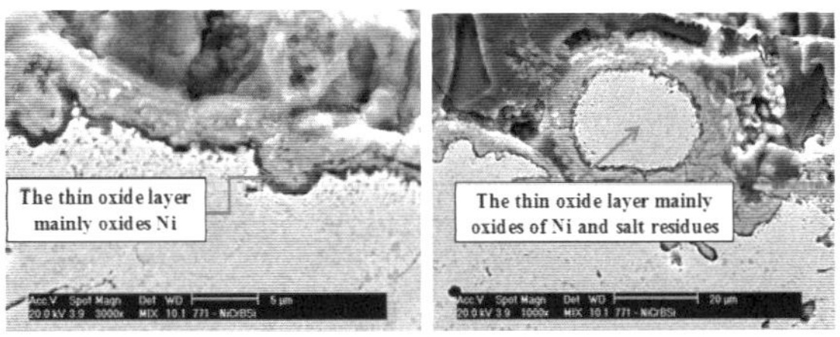

Fig. 9. HVOF NiCrBSi coating after the exposure at 525 °C, right magnification 3000x and left magnification 1000x, SEM images.

In the case of carbide coatings, HVOF **Cr_3C_2-25%NiCr** coating exhibited at the exposure temperature of 525 °C relatively good corrosion resistance properties, which is also evident from **Fig. 10**. However, at the temperature of 575 °C, the rapid reduction in coating thicknesses indicated a significant degradation. **Fig. 11** shows a layer composed of Cr oxides and salt remains on the coating surface. The corrosion mechanism of exposed coating is clearly visible in **Fig. 11**. Corrosion attacked matrix and lead to delaminating of individual carbides from the coating surface. It is very interesting that the corrosion attack is not so significant at the exposure temperature of 525 ° C in comparison with the exposure temperature of 575 °C, which is "only" about 50 °C higher, see **Fig. 10**. No tests have been conducted to explain this coating behavior. For this reason, it would be good to continue in this research.

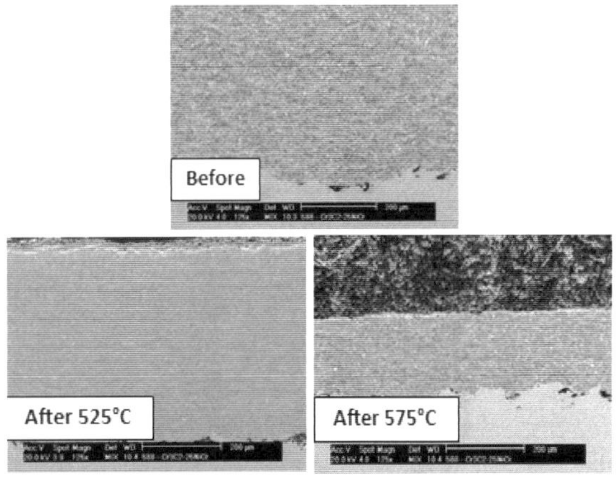

Fig. 10. HVOF Cr_3C_2-25% NiCr coating above before the exposure, left after the exposure at 525 °C and right after the exposure at 575 °C, magnification 125x, SEM images.

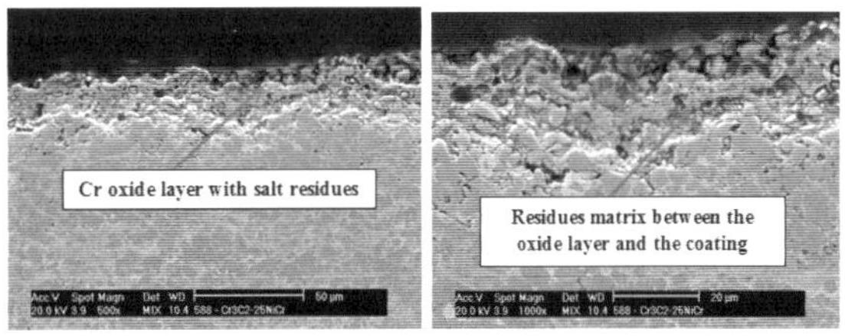

Fig. 11. HVOF Cr_3C_2-25% NiCr coating after the exposure at the temperature of 525 °C, right 500x magnification and left 1000x magnification, SEM images.

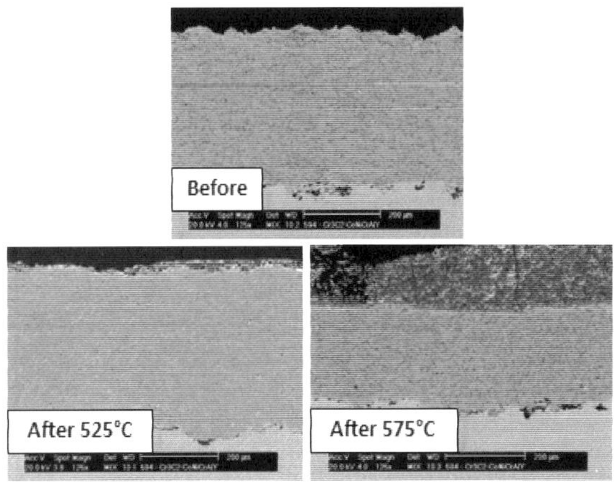

Fig. 12. HVOF Cr_3C_2-25% CoNiCrAlY coating above before the exposure, left after the exposure at 525 °C and right after exposure at 575 °C, magnification 125x, SEM images.

HVOF **Cr_3C_2-25%CoNiCrAlY** carbide coating behaved similarly like the previously described Cr_3C_2-

25% NiCr carbide coating. Almost no corrosion attack occurred at the exposure temperature of 525 °C. At the temperature of 575 °C, the speed of corrosion attack rapidly increased and resulted in the destruction of almost half of the coating thickness, see **Fig. 10**. **Fig. 13** clearly depicts the mechanism of coating corrosion damage leading to gradual corrosion of matrix and delaminating of individual carbides. The corrosion proceeded in the direction from the surface. As already mentioned several times above, this process increases in speed at the exposure temperature of 575 °C. **Fig. 14** shows the structure of Cr_3C_2-25% CoNiCrAlY coating in the cross-section. It clearly presents the composition of oxide layer covering the entire surface of the evaluated coating.

Fig. 13. HVOF Cr_3C_2-25%CoNiCrAlY coating after the exposure at the temperature of 525 °C, right magnification 1000x and left magnification 3000x, SEM images.

Fig. 14. HVOF Cr_3C_2-25%CoNiCrAlY coating after the exposure at the temperature of 525 °C, right magnification 3000x and left magnification 1000x, SEM images.

The final evaluated coating was an experimental HVOF **TiMoCN-29%Ni** (T10) coating. The results of high temperature corrosion testing proved that this coating is absolutely unsuitable for use at high temperatures, see also **Fig. 15**. The results apparently show that higher temperatures lead to total coating destruction, which is most likely caused by the formation of new phases between splats. Ni phases with larger volume were formed in this coating, and their formation resulted in the increase of internal stress and consequent relaxation, which means coating delaminating. This phenomenon was observed after the tests at both selected exposure temperatures. The exact reasons of this coating behavior are still under the investigation.

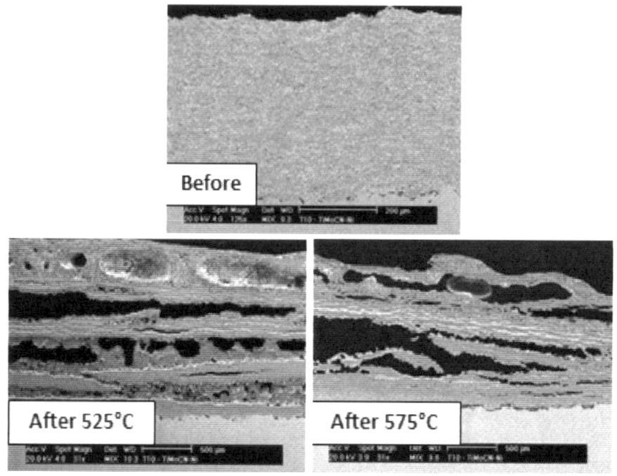

Fig. 15. HVOF TiMoCN-29%Ni coating above before the exposure, left after the exposure at 525 °C and right after the exposure at 575 °C, magnification 125x, SEM images.

Fig. 16 shows probably Ni phase formed as a result of coating exposure to high temperature.

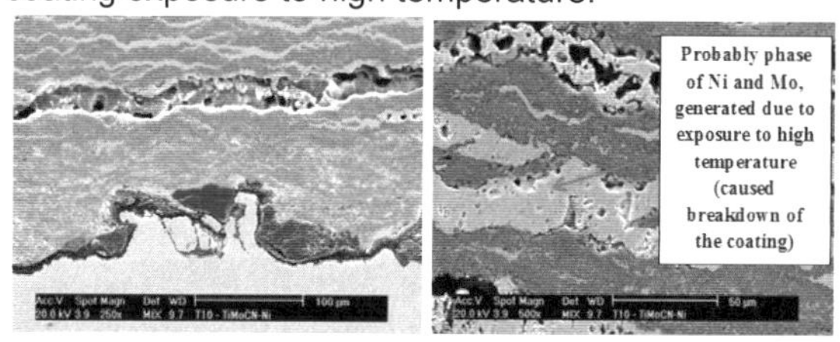

Fig. 16. HVOF TiMoCN-29%Ni after the exposure at the temperature of 575 °C, right magnification 250x and left magnification 500x, SEM images.

3.2 Erosion

The weight loss measurement results for evaluated coatings under four impact angles of erosive particles are summarized on the graph in **Fig. 17**.

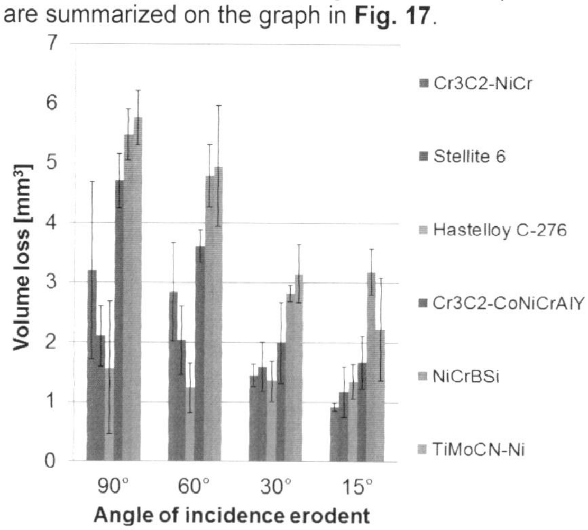

Fig. 17. The results of coatings erosion evaluation.

Alloy based Hastelloy C-276 coating showed the highest erosion resistance under the particle impact angle of 90°. Carbide coatings and alloy based NiCrBSi coating exhibited the highest values of volume loss. These results are not surprising due to the crucial role of erosive wear mechanism, which depends not only on hardness, but also on microstructure [9, 10]. The obtained results enable the prediction that the greater is the impact angle, the higher is the effect of microstructure in carbide coatings. Most likely, hard particles (carbides) beak out of the coating, which leads to significant weight loss. A similar trend can be observed also by other coatings. The greater is the angle value, the higher is the weight loss, but the weight loss of alloy based coatings is less significant. The lower weight loss is caused by lower material removal in comparison with carbide coatings due to different microstructure. The erosive media impact angle of 60° shows identical trend like the impact angle of 90°. The differences between most evaluated coatings balanced at the impact angle of 30°. However, the situation suddenly slightly changed for the sharpest impact angle of 15°. Cr_3C_2-25% NiCr coating exhibited the highest erosion resistance due to strong bonding of carbide particles in matrix. The obtained results proved that other evaluated coatings do not have particles so well bonded in matrix like Cr_3C_2-25% NiCr coating. Low resistance to erosive stress of NiCrBSi coating is most likely affected by its lower hardness compared with the other evaluated coatings.

4 Conclusion

Based on the conducted experiment, it can be argued that HVOF thermally sprayed coatings, if they are appropriately chosen, protect functional surfaces operating at high temperatures in corrosive-aggressive environment.
It was proved that the ability of a coating to protect the substrate material decreases with increasing temperature of the environment.The obtained results show that carbide coatings exhibit inferior protective properties compared with alloy based coatings in selected corrosive environment. The corrosion mechanism of carbide-based coatings involves the degradation of matrix and subsequent release of individual carbides from the coating surface. The degradation of carbide-based coatings is strongly dependent on the exposure temperature. The obtained results evidently show that the small increase in temperature by 50 °C lead to dramatically different degradation rate. The question remaining to answer is, if such degradation is caused by changes of coating material or by changes of salt mixture phase composition, e.g. from solid to liquid state at higher temperature. Out of all evaluated coatings, the HVOF Hastelloy C-276 coating showed the best protection against high temperature corrosion.
The alloy based Hastelloy C-276 coating exhibited also the highest erosion resistance under the erosive particle impact angle of 90°. This result is most likely dependent on the mechanism of erosive wear. Alloy based NiCrBSi coating showed the highest values of volume loss.It was also found that the greater is the erosive particle impact angle, the higher is the effect of microstructure by carbide coatings. The trend is identical for the impact angles of 60° and 90°. The differences between evaluated coatings balanced at the impact angle of 30°. However, the situation suddenly slightly changed for the sharpest impact angle of 15°. HVOF Cr_3C_2-25% NiCr coating exhibited the highest erosion resistance due to strong bonding of carbide particles in matrix.

Acknowledgement
The paper was prepared thanks to the project of Technology Agency of the Czech Republic no. TE01020068.

3 Literature

[1] Khanna, A. S.: High Temperature oxidation and Corrosion. Delhi: ASM International, 2002, ISBN 0-87170-762-4.
[2] Česánek, Z., Schubert, J., Houdková, Š., Prantnerová, M.: Vysokoteplotní koroze vybraných žárově stříkaných povlaků vytvořených technologií HVOF – High Velocity Oxygen Fuel, In: All for Power, ročník 9, 2/2015, pp. 30-36, ISSN 1802-8535.
[3] Česánek, Z., Smazalová, E., Schubert, J., Houdková, Š., Prantnerová, M.: High temperature hot corrosion behavior of selected thermally sprayed coatings on X22 stainless steel in an aggressive environment at 750 °C. In: Metal 2015 Conference proceedings, pp. 131, ISBN 978-80-87294-58-1.
[4] Bushan, B.: Introduction to Tribology, John Wiley & Sons, Inc., New York, 2002.
[5] Deng, T., Bingley, M.S., Bradley, M.S.A., Silva, S.R.: A comparison of the gas-blast and centrifugal-accelerator erosion testers: The influence of particle dynamics, Wear, 265 (2008) p. 945-955.
[6] Houdková, Š., Česánek, Z., Polach, P.: Solid Particle Erosion of Selected HVOF Sprayed Coatings. Applied Mechanics and Materials, Applied Methods of the Analysis of Static and Dynamic Loads of Structures and Machines II (Padevět, P. ed.), Proceeding of Trans Tech Publications Ltd, Vol. 827, February 2016, pp. 39-46.
[7] Česánek, Z., Houdková, Š.: Hodnocení erozní odolnosti vybraných žárově stříkaných povlaků, VYZ-1519/2013, VZÚ Plzeň Ltd., Pilsen, 2013.
[8] Schubert, J., Smazalová, E., Houdková, Š., Česánek, Z., Prantnerová, M.: Standard Remelting Methods of Thermally Sprayed NiCrBSi based Coatings, In: Metal 2014 Conference proceedings, pp. 137, ISBN 978-80-87294-52-9.
[9] Sidhu, H.S., Buta, B.S., Prakash, S.: Solid particle erosion of HVOF Sprayed NiCr and Stellite-6 coatings, In: Surface & Coatings Technology 202, 2007, pp. 232-238.
[10] Kumar, A., Boy, J., Zatorski, R., Stephenson L.D.: Thermal Spray and Weld Repair Alloys for the Repair of Cavitation Damage in Turbines and Pumps: A Technical Note, J. Therm. Spray Technol., 14(2), (2005) pp.177-182.

Influence of heat-treatment methods on mechanical and erosion properties of thermally sprayed NiCrBSi based coating

J. Schubert, Z. Česánek, Š. Houdková, Pilsen / CZ

Heat treatment of thermally sprayed coatings is a post-treatment, which is capable of sealing surface porosity without using any other necessary additives. This advantage is mainly beneficial in applications in corrosive environments. The benefits are derived from the positive influence on coating microstructure and removing inhomogeneities introduced into a coating during thermal spraying process. Heat treatment allows to reduce inhomogeneities and the amount of porosity in the resulting coating or to eliminate the porosity entirely. The paper presents the results of mechanical and tribological properties evaluation made on specimen of NiCrBSi alloy based coatings and describes differences between heat treatment methods. The specimens were deposited using Flame spray (FS) and High-velocity-oxygen-fuel (HVOF) thermal spraying technologies and heat treated by acetylene flame. The HVOF samples were prepared as the reference ones for properties comparison. The resulting coatings were tested using the erosion resistance test along with measurements of basic mechanical properties (HR15N, HV0.3, roughness). The microstructure of the specimens was evaluated using optical microscopy.

1 Introduction

Now, thermal spraying of coatings is widely used and available technology that provides the components with superior mechanical and anti-corrosion properties. These technologies produce functional coatings on the components surfaces providing them with the desired enhanced properties. This research is focused on the way how to enhance the properties further more. The improvement is achieved by different methods of coatings heat-treatment. The quality of the final coating deposited using thermal spraying methods is determined by many parameters such as particle size, deposition temperature, fuel type, feed rate, spray angle, spray distance, substrate temperature, chamber pressure, spraying technique used and many others. Precise adjusting of these parameters is necessary in order to achieve the best results. This research consists in comparing coatings prepared by HVOF (High Velocity Oxy Fuel) and FS (Flame spray), which are widespread and commonly used thermal spraying technologies.

Both technologies produce coatings with a certain degree of porosity and both use similar powder materials. The coating prepared using the FS technology is characterized by higher porosity and weaker adhesion bond to the substrate. This paper presents results showing that the properties can be substantially enhanced by heat-treatment methods. The NiCrBSi coating was chosen for this study because it shows good results while subjected to heat treatment. This coating also exhibits excellent corrosion resistance due to the selective oxidation mechanism. The resulting oxides in pores block the pores and prevent the further penetration of corrosive environment and further diffusion of corrosive media to the substrate. Very low porosity of this coating and microstructure composed of flat splats also contribute to the increased resistance to high temperature corrosion at elevated temperatures (guaranteed especially by high content of Cr). These advantageous properties can be further improved by subsequent heat-treatment of the produced coating, which is ensured by high content of B and Si (both elements lower the melting temperature) [1, 2]. For this reason, the NiCrBSi coating is widely used in applications requiring corrosion and oxidation resistance and mechanical durability. The resulting coating structure before the heat-treatment is determined especially by the technology of the coating preparation. A basic structure of commonly prepared NiCrBSi coatings contains pores, oxides and unmelted particles. Thanks to the unique composition of this coating it is possible to eliminate these structural deficiencies by heat-treatment techniques. Actual heat-treatment of the NiCrBSi coating further improves particular anti-corrosion and anti-oxidation properties (due to the structure homogenization and the sealing of pores). Carbon contained in the coating allows the formation of carbides and improves the wear resistance [2]. Heat-treatment by flame [3, 4] and recently incoming heat-treatment by laser [5] are classic and most commonly used methods for heat-treatment of coatings.

The aim of this research is to evaluate the effect of different heat-treatment methods on microstructure and mechanical properties of NiCrBSi coatings applied on steel substrate. Furthermore this study deals with comparison between the depositions methods (HVOF and FS) and examines if there are any significant differences and if the process is economical. This paper is based on previous results and is continuing the research of heat-treatment of coating [6, 12]. The main reason for continuing this research was to increase deposition efficiency using different powder and to observe the mechanical and other properties of this newly prepared coating with properties of previously prepare coatings [12].

The NiCrBSi coating was applied on the substrate using two different methods (HVOF and FS). The coatings were then heat-treated using acetylene flame technology. The impact of this technology on the final properties of the coating was further evaluated in order to find the relationship between the heat-treatment technology and the microstructure or mechanical properties.

2 Experimental procedure

Experimental samples were prepared using two technologies (TAFA HP/HVOF and Praxair 6PII). The prepared samples were then heat-treated using the acetylene flame method. This method was selected due to the previous results [12]. Samples were subsequently tested and evaluated in terms of mechanical properties and their microstructure was evaluated together with the erosion resistance. The surface roughness was measured using the Mitutoyo SJ-201P surface tester according to DIN EN ISO 4287. Hardness (HR15N), micro hardness (HV0.3) and roughness were included in the mechanical properties testing. Preparation of samples, heat-treatment and subsequent testing was performed in the laboratories of the Research and Testing Institute Plzen (VZU Pilsen) applying the same procedure as in the previous research [6, 12].

2.1 Deposition parameters

Standard optimized parameters were used for both HVOF and FS spraying technologies optimized in previous research [6, 12]. The deposition parameters of newly sprayed FS are shown in **Tab. 1**. The NiCrBSi powder with a particle size distribution +20 – 53 mm was selected as a powder material for HVOF. Powder with a particle size distribution +45 – 106 mm designated 771.91 (its chemical composition: 15%Cr, 4%Fe, 4.25%Si, 3%B, and 0.7% C, Ni rest) was selected for the flame spray technology. Comparison was made with the FS sprayed NiCrBSi coating (designated 771.33) with a particle size distribution +20 - 53mm [12]. Coatings were sprayed onto carbon steel, with dimensions designed according to the requirements of relevant testing technologies. The substrate surface was degreased and grit blasted before spraying (brown corundum (Al_2O_3) F22 grit 0.8 to 1.0mm was used as an abrasive medium).

FLAME SPRAY DEVICE: GTV 6P-II	
Oxygen	39.1 l/min
C_2H_2	19.6 l/min
Spray distance	200 mm
Air cap	6P4
Barrel	P76-M
Siphon plug	6P-605
Traverse speed	300 mm/s
Feed rate	45 g/min
Offset	3 mm
Number of passes	4

Tab. 1. Deposition parameters of the FS NiCrBSi coating 771.91.

2.2 Measurement conditions

Microstructure evaluation was conducted using the FEI Quanta 200 scanning electron microscope in NTC of the University of West Bohemia and using the Nikon Epiphot 300 optical microscope in VZU Plzen. Surface hardness HR15N was evaluated applying the Rockwell HT 8003 hardness test on unpolished coating surface. Results were calculated as an average of five measurements conducted in the Laboratories of VZU Plzen. Micro hardness HV0.3 was evaluated in the coating cross section. Results were calculated as an average of seven measurements conducted similarly in VZU Plzen on the LECO DM–400A device. The influence of solid particle erosion was tested by a device designed in our laboratories. It is constructed as a centrifugal device allowing testing the erosion resistance for different impact angles [9]. The device design is shown in **Fig. 1** and **Fig. 2**. The principle of this test is thoroughly described in the paper written by T. Deng [13]. Parameters of erosion testing are presented in the following table (**Tab. 2**). The test was conducted at 22 ± 2 °C. Artificial white corundum, grain size of 212-250 microns, was used as an abrasive medium.

Angle of particle impact	**90°, 60°, 30°, 15°**
Exposure time	2 min
Speed of rotational disc	4500 rev/min
Diameter of diaphragm	20 mm

Tab. 2. Test parameters of erosion testing

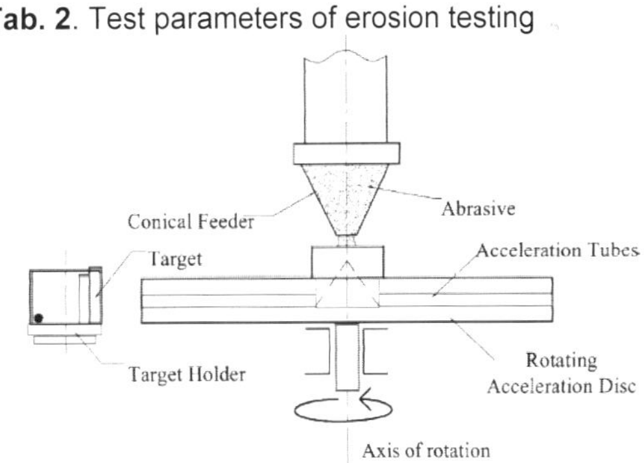

Fig. 1. Erosion test device schematic design [13].

Fig. 2. Detail of sample holders of erosion tester.

2.3 Heat-treatment method

Based on previous results [6, 12] the heat-treatment by flame was selected as the best candidate for comparing the influence of heat-treatment on the resulting mechanical properties. The heat-treatment by flame was carried out using an acetylene torch with a constant feed rate and using a high temperature output torch (**Fig. 3**). This torch was selected considering the high temperature output and a wide diameter of flame. The reason for this choice was to quickly deliver necessary heat into the whole sample/part and simultaneously reduce the heat transferred into the substrate material.
The heat-treated coating was heated up to optically evident melted state (glass as a surface effect). Heat-treated coating was then cooled in air without additional cooling. This procedure was chosen due to the impossibility of precise coating temperature control during this method.

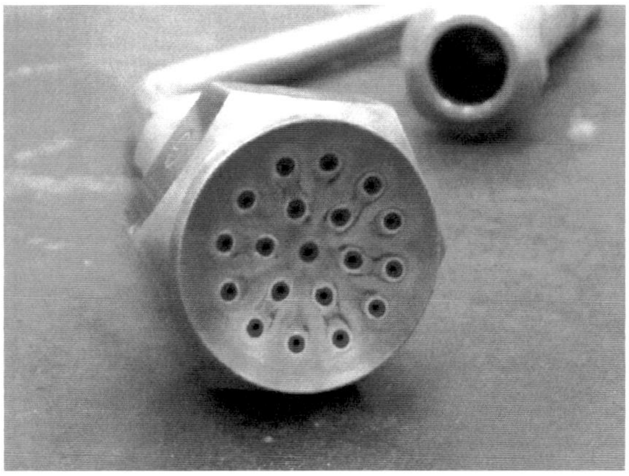

Fig. 3. Acetylene flame torch used for heat-treatment of thermally sprayed coatings.

3 Results

3.1 Microstructure of tested coating

The microstructure of tested samples is shown in the following figures (**Fig. 4** and **Fig. 5**). The first figure focuses on characterizing the change between the as-sprayed coatings prepared using different powders. **Fig. 4 a)** represents the microstructure of a newly used coating 771.91 of much coarser grain size. The microstructure of this coating is typical for the FS sprayed coatings. **Fig. 4 b)** [12] on the other hand captures microstructure of previously prepared coating 771.33 with finer grain size. Both a) and b) represent the as-sprayed state of coating and it is possible to see that the new powder is able to produce much finer final microstructure than the second one, despite the coarser grain size of the powder. **Fig. 5** represents the microstructure of both coating after heat-treatment. There is possible to see the complete change in coating microstructure. **Fig. 5 a)** shows the microstructure of heat-treated coating 771.91. **Fig. 5 b)** shows the microstructure of previously sprayed FS coating heat-treated by flame (771.33) [12]. Both coatings exhibit a clear reduction of porosity (remaining pores are closed). The microstructure of coatings appears to be considerably finer and with obvious homogenization of structure, which is the crucial factor for final coating properties [10]. Recrystallization, which occurs during this process, is typical for heat-treatment. Magnitude of recrystallization is determined by initial microstructure of the heat-treated coating. The finer is the initial microstructure after deposition the coarser will be the final microstructure after the heat-treatment process. This assumption was proved in previous research [6]. It is assumed that the recrystallization occurs particularly in the nickel stage [11]. The heat-treated coating also forms a transition layer composed of Ni phases with the base material lying in the area between the coating and the base material [11].

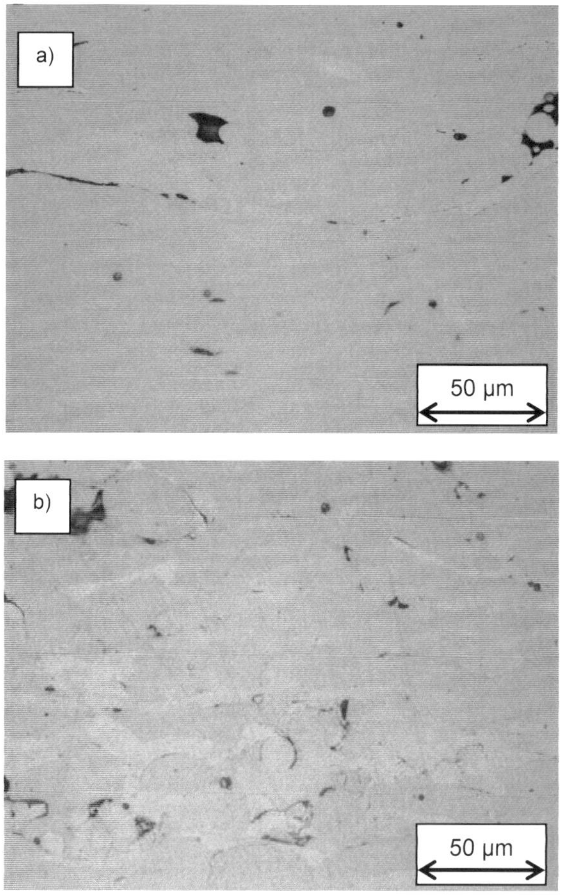

Fig. 4. Microstructure of coatings in as-sprayed state (magnification 500); a) a new coating 771.91, b) previously used coating 771.33.

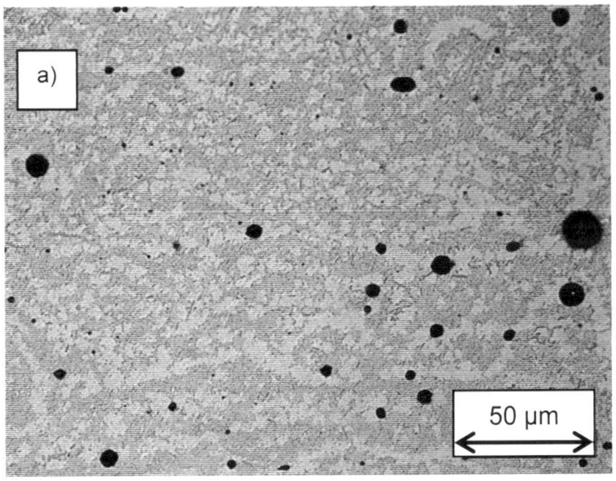

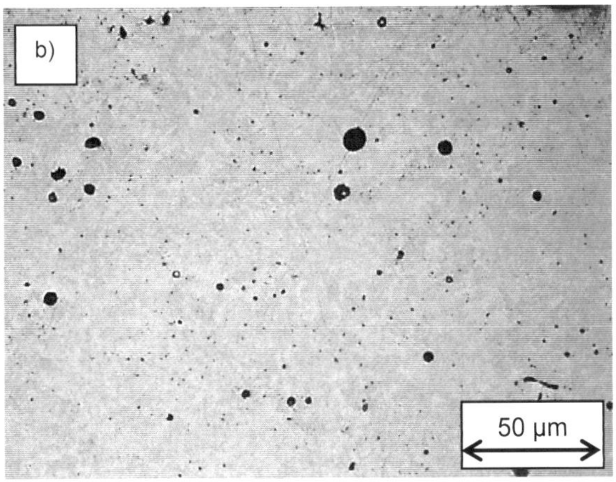

Fig. 5. Microstructure of heat-treated coatings (magnification 500); a) a new coating 771.91, b) previously used coating 771.33.

3.2 Mechanical properties

For the accuracy of comparison the measured values of HVOF and FS 771.33 coatings were used from previous research [12]. The measured values (see Tab. 1) indicate that the reference HVOF sprayed coatings have higher micro and macro hardness compared with the FS sprayed coatings. Previous measurements confirmed that after heat-treatment the mechanical properties such as macro hardness and micro hardness decrease at coatings deposited using the HVOF technology [6, 7, 8, 12]. This phenomenon is attributed to the influence of relaxation of internal stresses, which are present in the coating prior to the heat-treatment procedure. The increase in hardness of coating prepared by the FS technology is probably caused by the homogenization of the coating itself, because prior to the heat-treatment the coating contains a plenty of pores, which lower the overall hardness.
Furthermore, high deviations are probably caused by the measurement inaccuracy, because hardness is very difficult to measure at the coating with large pores content. In accordance with assumption, the roughness of the coating was decreased after the application of the heat-treatment method. This phenomenon is caused by the homogenization and the increased coating compactness. The difference between the as-sprayed coating and the heat-treated one can be seen in **Fig. 6**. This figure provides comparison between surface quality and roughness before and after the heat-treatment process. This figure also shows partial heat-treatment of a real product (compressor plunger). Application of the heat-treated NiCrBSi coating on this part is very beneficial because of the enhancement of mechanical properties and other characteristics in connection with a low cost of the method. In some cases there is no need for subsequent gridding due to such a low roughness. It is very important to note that the combination of much cheaper technologies is able to produce the coating with almost similar mechanical properties to the HVOF sprayed coating and much finer roughness.
When comparing the powders prepared using the FS technology (powders 771.33 and 771.91) it is obvious that the results are not significantly different. Slight deviation between the values can be caused by the measurement error, which is evident from the standard deviations of measurement mentioned in **Tab. 3**.

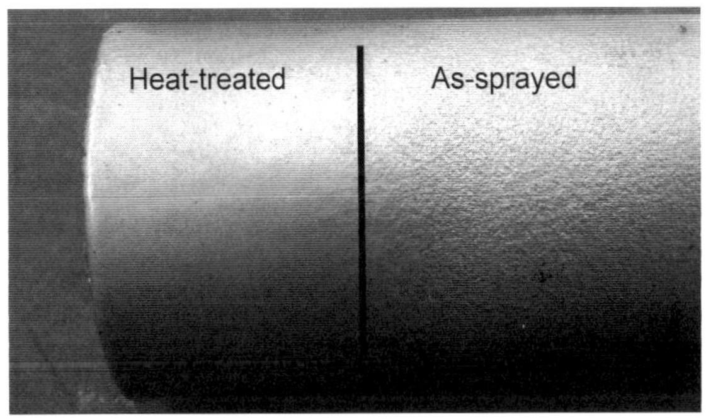

Fig. 6. Heat-treatment of the NiCrBSi 771.91 coating applied on a real part (compressor plunger).

Powder	HR15N	HV0,3	Roughness
A-S 771.33	73.75 ± 4.91	474.7 ± 47.08	8.03 ± 0.78
A-S 771.91	72.25 ± 3.41	438.28 ± 34.82	14.6 ± 0.75
H-T 771.33	83.88 ± 1.65	748.43 ± 34.13	2.38 ± 0.34
H-T 771.91	83.1 ± 3.03	734.25 ± 35	2.31 ± 0.11
A-S (HVOF)	87.78 ± 1.83	815 ± 58.76	6.72 ± 0.95

Tab. 3. Mechanical properties (heat-treated, as-sprayed and HVOF coatings). HVOF and as-sprayed 771.33 values are used from previous research for the accuracy reasons [12]; A-S (As-Sprayed), H-T (Heat-Treated).

3.3 Erosion properties

Erosion test was conducted on all as-sprayed and heat-treated coatings. The measurements were performed on the HVOF coating and the FS sprayed coating 771.91 in as-sprayed and heat-treated state. Results of this test are shown in **Fig. 7,** which presents the overall results of the specific coating. Dependency of erosion particle impact angle on the performance of coating is displayed in **Fig. 8**. The FS coating shows a great reduction of erosion volume loss after the heat-treatment. The heat-treated FS coating also exhibit significantly lower erosion volume loss comparing to the HVOF sprayed coating.

Especially the dependency on the particle impact angle is almost negligible. This is a very important fact, because the sensitivity to the impact angle is a main problem when it comes to the selection of the right coating for erosion protection.

In this way the heat-treated NiCrBSi coating has outperformed the expectations and can be considered suitable for the erosion resistant coating for many applications.

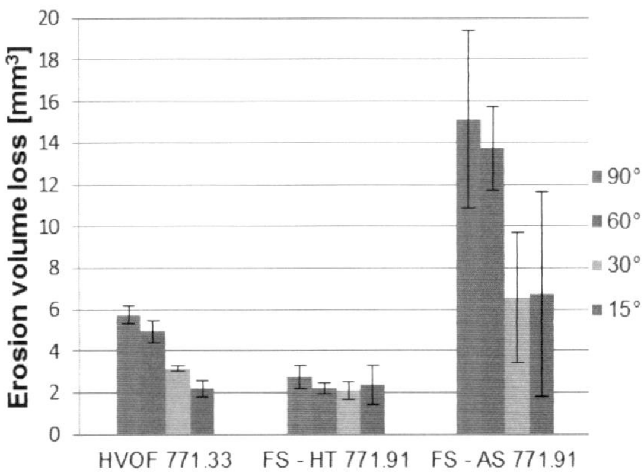

Fig. 7. Erosion test results of the overall coating erosion resistance (FS – Flame Sprayed; HT – Heat-Treated).

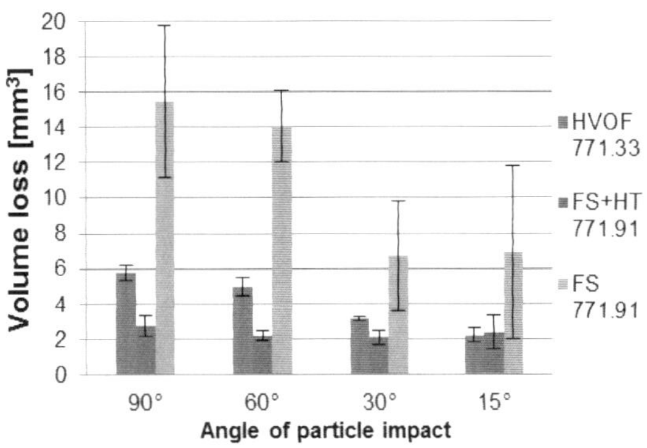

Fig. 7. Erosion test results of as-sprayed and heat-treated coating with dependency on different particle impact angles (FS – Flame Sprayed; HT – Heat-Treated).

4 Discussion

Heat treatment was proved to be a suitable method for enhancing mechanical properties of coatings produced by flame spray. Using flame spray and subsequent heat-treatment by the acetylene flame is a commonly available technology compared to the HVOF systems. Furthermore the combination of flame spraying and heat-treatment is more cost effective than the HVOF technology in terms of preparing the NiCrBSi functional coatings. As it was stated above this method has also many advantages in enhancing final mechanical and tribological properties further than those achievable using the HVOF systems. Using heat-treatment a complete change of microstructure was achieved and confirmed. It was also previously confirmed that the microstructure of coatings is influenced by the type of the heat-treatment method [12].

Using the flame heat-treatment method provides a significant increase in mechanical properties, erosion and abrasion resistance and also ensures homogenization of the coating microstructure [12]. This assumption was also confirmed by herby conducted research. The resulting mechanical properties of coatings prepared by HVOF (as-sprayed) and FS with heat-treatment are quite identical and in some ways the FS heat-treated coatings exceed the performance of the HVOF coatings. Enhancement of erosion resistance of the heat-treated NiCrBSi coating, which is significantly higher than at HVOF sprayed coating, is most surprising. Sensitivity to the particle impact angle, which is unique among carbide or alloy based coatings, is also interesting. It is assumed that this phenomenon is caused by the coating homogenization and by the elimination of residual pores together with recrystallization and formation of new Ni phases in the vicinity of the coating surface.

The process of recrystallization taking place in the NiCrBSi coating at elevated temperatures [11] is also interesting. This process was observed during preliminary testing [12] and confirmed during this study. The process of recrystallization is probably significantly responsible for the increase in mechanical properties and for increase of erosion and abrasion resistance together with homogenization of microstructure. During the heat-treatment the recrystallization process takes place together with the growth of Ni phases. If the microstructures of the flame as-sprayed coating and of the heat-treated one (**Fig. 4** and **Fig. 5**) are compared significant reduction of the size of particles and much finer microstructure of the heat-treated coating can be seen.

Previous studies confirmed that the recrystallization and homogenization of coating microstructure change the mechanism of abrasion damaging [12] and it is assumed that this is the same reason for the increased erosion resistance. Further study will be conducted to uncover the cause of this enhancement and to precisely document the erosion mechanism of the heat-treated coating itself.

As mentioned above coating 771.91 deposited using FS and heat-treatment showed increase in its hardness and this increase is equivalent to coating 771.33. Differences between heat-treated coatings 771.33 and 771,91 are especially evident in the resulting coating microstructure varying in the position and the size of particles and phases.

The process of heat-treatment was fully mastered as proved on real parts of the compressor plungers. Therefore, this heat-treatment technique for the NiCrBSi coating is suitable for such applications either from the perspective of mechanical and tribological properties but also from economic perspective.

This study also proved that the NiCrBSi powder (771.91) is suitable for commercial use on real parts, is substitutable with powder 771.33 and meets the requirements on mechanical and other parameters together with lowering the cost of the process by increasing the deposition efficiency.

For future research it is important to recommend focusing on areas of residual stresses and the influence of heat-treatment on the stress decomposition. It would also be advisable to focus on corrosion properties of the heat-treated NiCrBSi coating, which would allow considering this post-processed coating to be a complex protection against mechanical, tribological, erosion and corrosion damage.

Acknowledgement
The paper has originated in the framework of institutional support for the long-time conception development of the research institution provided by Ministry of Industry and Trade of the Czech Republic.

3 Literature

[1] RODRIGUEZ J., MARTIN A., FERNÁNDEZ R., FERNÁNDEZ J.E., An experimental study of the wear performance of NiCrBSi thermal spray coatings, Wear 255 (2003), 950–955

[2] SERRES, N., HLAWKA, F., COSTIL, S., LANGLADE, C., MACHI, F., An investigation of the mechanical properties and wear resistance of NiCrBSi coatings carried out by in situ laser remelting, *Wear 2011*, Vol. 270, p. 640-649

[3] NAVAS, C.,COLACO, R., de DAMBORENEA, J., VILAR, R. Abrasive wear behavior of laser clad and flame sprayed-melted NiCrBSi coatings , *Coat. Surf. Technol.*,2006, Vol. 200, p. 6854-6862

[4] GONZÁLES, R., CADENAS, M.,FERNÁNDEZ, R., CORTIZO, J.L., RODRÍGUEZ, E. Wear behavior of flame sprayed NiCrBSi coating remelted by flame or by laser, *Wear 2007*, Vol. 262, p. 301-307

[5] MIGUEL,J. M.,GUILEMANY, J.M., VIZCAINO, S. Tribological study of NiCrBSi coating obtained by different processes, *Tribology International*, 2003, Vol. 36, p. 181-187

[6] SCHUBERT, J., SMAZALOVÁ, E., HOUDKOVÁ, Š., ČESÁNEK, Z., PRANTNEROVÁ, M.; Standard Remelting Methods of Thermally Sprayed NiCrBSi based Coatings, In: Metal 2014 Conference proceedings, 21.-23.5.2014, Brno, ČR, pp. 137, ISBN 978-80-87294-52-9.

[7] ASTM Standard G65 - 04, 2010, "Standard Test Method for Measuring Abrasion Using the Dry Sand/Rubber Wheel Apparatus," ASTM International, West Conshohocken, PA, 2010, DOI: 10.1520/G0065-04R10

[8] HOUDKOVÁ, Š., VOSTŘÁK M., HRUŠKA M., ŘÍHA J., SMAZALOVÁ E., ČESÁNEK Z., SCHUBERT J., COMPARISON OF NiCrBSi COATINGS, HVOF SPRAYED, RE-MELTED BY FLAME AND BY HIGH-POWER LASER, 15. - 17. 5. 2013, Brno, Czech Republic, EU, *conference Metal 2013*

[9] HOUDKOVÁ, Š., ČESÁNEK, Z., POLACH, P.: Solid Particle Erosion of Selected HVOF Sprayed Coatings. Applied Mechanics and Materials, Applied Methods of the Analysis of Static and Dynamic Loads of Structures and Machines II (Padevět, P. ed.), Proceeding of Trans Tech Publications Ltd, Vol. 827, February 2016, pp. 39-46.

[10] Z. BERGANT, J. GRUM, Quality Improvement of Flame sprayed, heat treated, and remelted NiCrBSi coatings, *Journal of Thermal Spray Technology*, 18 (3) (2009) 380-390

[11] GRUZDYS E., MEŠKINIS Š., TAMULEVIČIUS S., GRINYS T., Effects of Heating by Arc Discharge on Structure and Chemical Composition of Coatings Formed Using High Velocity Oxygen Fuel and Flame Spraying Processes, *Materials Science*, ISSN 1392–1320, Vol. 14, No. 3. 2008

[12] HOUDKOVA, Š, SMAZALOVA, E., VOSTRAK, M., SCHUBERT, J.; Properties of NiCrBSi coating, as sprayed and remelted by different technologies, Surface and Coating Technology 253, 14-26, Elsevier, 2014

[13] DENG, T., BINGLEY, M.S., BRADLEY, M.S.A., Influence of particle dynamics on erosion test conditions within the centrifugal accelerator type erosion tester, Wear 249, 1059-1069, Elsevier, 2001

Influence of multiple guide vane geometry on characteristics of TBCs deposited by PS-PVD method

J. Mao*, M. Liu, C. G. DENG, C. M. DENG and K. S. ZHOU
Guangdong Institute of New Materials, Guangzhou; National Engineering Laboratiry for Modern Materials Surface Engineering Technology; The Key Lab of Guangdong for Modern Surface Engineering Technology, Guangzhou, 510651, P.R.China.
* Corresponding e-mail address: jmao0901@163.com.cn

Abstract

The influence of the multiple guide vane geometry on the thickness and microstructure of thermal barrier coatings was presented in this article. The assessment of model vane made of DZ40M alloy used in the first stage of turbine engines was performed. The multicomponent NiCoCrAlTaY coating as bond coat as well as the zirconia stabilized by yttrium oxide coating as top ceramic coat was deposited on the vane surface by PS-PVD process. The thickness measurement and microstructure observation was performed with a use of optical and scanning microscopy. It has been proven that the entire surface of double vane was covered by NiCoCrAlTaY and YSZ coatings. The thickest coating was found close to the leading edge and trailing edge of the vane. In those places the coating was approximate 80-100% thicker than in the other areas on double vane. The obtained results indicate that it is possible to manufacture the thermal barrier coatings including metallic bond coat and top ceramic coat by single PS-PVD process on the first stage turbine blades of the aircraft engine. It indicates the possibility of application of this new method for TBCs production in the industrial practice.

Keywords

Multiple guide vane; Thermal barrier coatings; PS-PVD; Thickness distribution; Microstructure

1 Introduction

Plasma spray-physical vapor deposition (PS-PVD) is a very promising technique to manufacture coatings of different characteristic features and unique properties, combining the high deposition efficiency of thermal spray process and characteristic advantages of thin PVD coatings [1-3]. Developed on the basis of low pressure plasma spray (LPPS), PS-PVD operates in considerable lower pressure, ranging from 50 to 200 Pa, resulting in an extended plasma jet of more than 2 m in length and 200-400 mm in diameter. Under 200Pa condition, the supersonic plasma jet possesses high velocity about 3000 m/s and high temperature more than 12000°C [1]. Although the PS-PVD plasma jet is significantly larger compared to the standard vacuum plasma spray (VPS), the distribution of temperature and particle velocity is more homogenous because of the high power capability of plasma torch (> 120 kW) and efficient feedstock method[4]. Particularly at pressure below 2 kPa, PS-PVD can result in a vaporization of injected ceramic powder and a uniform deposition onto target substrate out of the vapor phase. Consequently, PS-PVD allows for deposition of splat-like coatings significantly thinner than APS and condensation of vaporized material similar to PVD method by changing the processing parameters [2, 4, 5]. This technology has been reported to coat large and thin areas as well as one square meter with a 10μm layer of Al_2O_3 in less than one minute [6]. The microstructure variety, feedstock flexibility, and high deposition rate make PS-PVD an attractive processing technique for a variety of industries, including gas-sensing membranes, fuel cell components, wear or electrically resistive coatings, diffusion barrier layers and so on.

The initial application aim and the most important research efforts of PS-PVD process have focused on gas turbine engine materials. The thermal barrier coatings (TBCs) used on hot section components of the aircraft engine are an ideal solution to resist dangerous oxidation and serious hot corrosion and to improve efficiency and life of aircraft engine. Typical TBCs generally comprises two layers, an external ceramic layer made of zirconium oxide stabilized by 7-8% yttrium oxide and an internal MCrAlY layer which provides anti-corrosion protection and ensures appropriate adhesion of the ceramic layer to the substrate. The TBCs deposited with the use of the physical vaporation method (EB-PVD) has a column structure, which results in less efficiency of the coating, a higher thermal conductivity coefficient, and improved thermo-mechanical properties [7, 8]. The coating microstructure formed by APS is noticeably porous and parallel splat to the base, what results in a low thermal conductivity coefficient [9]. Nevertheless, neither EB-PVD nor APS can not fully guarantee the uniformity of the deposited coating on the entire surface of the parts having complex geometry and shadowed areas. What's more, partial shadowed location can not even cover any coatings. The new process of PS-PVD enables vaporization of the injected powder materials to a great extent, leading to another outstanding advantage, non-line-of-sight deposition, which is not achievable using PVD or other conventional plasma spray techniques [2, 10, 11]. The vaporized material incorporated in the supersonic gas stream can completely cover the parts having complex geometries and shadowed areas such as multiple nozzle guide vanes (NGV), and acquire good uniformity in thickness and structure.

In this paper, the influence of the double guide vane geometry on the thickness and microstructure of the multicomponent NiCoCrAlTaY coating as bond layer as well as the zirconia stabilized by yttrium oxide coating as top layer was presented. The non-line-of-sight capability of PS-PVD process was demonstrated and indicated the possibility to manufacture the metallic bond layer and top ceramic layer on complex components in a single production step.

2 Experimental

All experiments were carried out on a MultiCoat™ hybrid plasma-coating system, which was built by Guangzhou Research Institute of Non-ferrous Metals (GRINM) in 2014 as shown in Fig. 1. The very low working pressure of 50 Pa and high power input of 150 kW are achieved with the strengthened vacuum pump systems, increased power supplies and the use of O3CP plasma gun. But the system can also be operated with a F4-VB torch.

Fig. 1 MultiCoat™ hybrid plasma-coating system at GRINM.

The spraying powder for bond coating is spherical NiCoCrAlTaY powder (-38+5μm, Amdry 997, Oerlikon Metco) and the powder for the top ceramic coating is agglomerated 7.5 wt.% yttria partially stabilized zirconia powder (-30+1μm, Metco 6700, Oerlikon Metco), respectively. The substrates and bond-coated substrates were preheated immediately before coating by the plasma jet up to temperatures above 650°C and 950°C, respectively.

The simplified double vane model with parallel platforms and solid airfoils made of DZ40M alloy is used as substrate, as shown in Fig. 2. It has a cubic dimension of 170×75×55 mm. The thickness distribution measurement was performed at three profiles with different height from the bottom platform, marked as 20% airfoil, 50% airfoil and 80% airfoil. In Fig. 3, it's shown that 10 measuring points were performed on eath profile of both trailing vane (vane A) and leading vane (vane B).

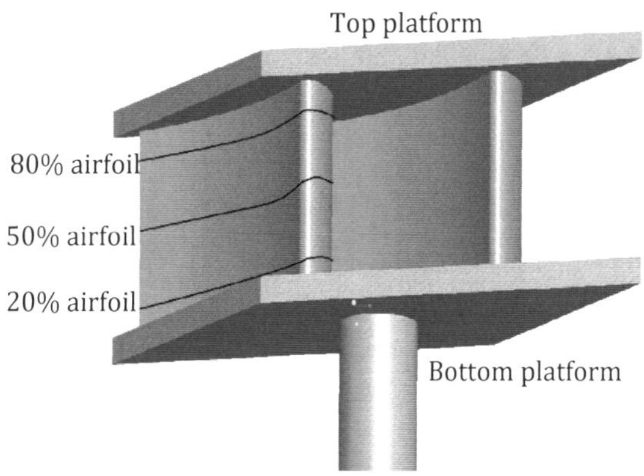

Fig. 2 3D sketch of simplified double vane model with parallel platforms and solid airfoils made of DZ40M alloy.

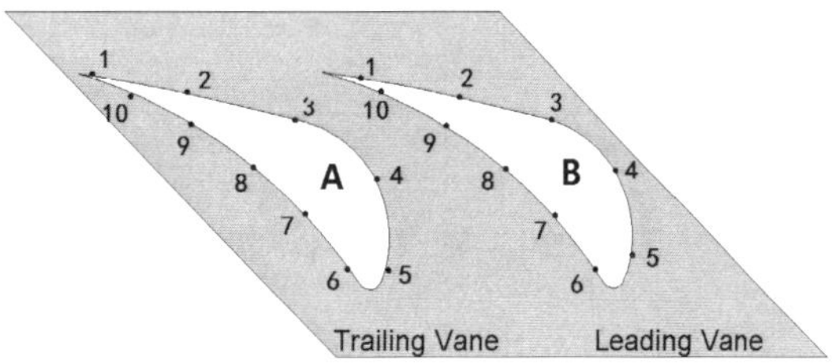

Fig. 3 Measuring point distribution on the double vane model.

Additional oxygen gas was fed to compensate for the oxygen deficiency of YSZ during coating. The spray process parameters used to deposit the NiCoCrAlTaY and YSZ coatings are listed in Table 1. Particularly, it's important to point out that NiCoCrAlTaY coatings were prepared under the vacuum chamber pressure in the range of LPPS condition, but used the O3CP plasma gun instead of F4-VB gun and operated with much higher current and power, much higher plasma gas flow rate. The spray distance increased correspondingly, adapting to extended plasma jet. Of course, the degree of plasma expansion is much lighter than that under typical PS-PVD process condition.

Table 1 Spraying parameters used for the NiCoCrAlTaY and YSZ coating deposition.

Spray Parameter	Value	
	NiCoCrAlTaY	YSZ
Chamber pressure, Pa	4000	150
Gun current, A	1650	2600
Torch power, kW	85	129
Ar gas flow rate, slpm*	110	35
He gas flow rate, slpm*	-	60
H_2 gas flow rate, slpm*	6	-
Spray distance, mm	350	950
Feed rate, g/min	40	20
Oxygen flow, slpm*	0	2

*slpm, standard liters per minute

During the spray process, the part rotates on its axis with the speed of 20 rpm, and, at the same time, the gun performs a rapid reciprocating motion of the plasma jet over the part between the two platforms. The velocity of the gun movement is 1000 mm/s. The reversal spot is outside of the platforms to prevent overheating. The part is coated at a temperature of approx. 700 and 1000°C for the NiCoCrAlTaY and YSZ coatings, respectively. The coated part is shown in Fig. 4.

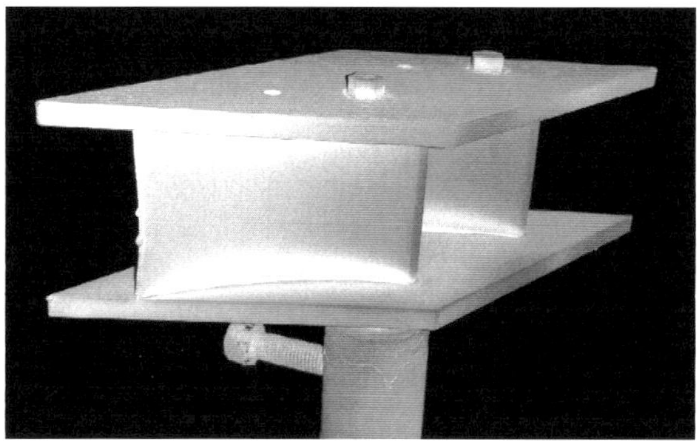

Fig. 4 Double vane model with YSZ coating deposited by PS-PVD method

3 Results and Discussion

3.1 Thickness and microstructure results of NiCoCrAlTaY coating

The research has proven that the entire surface of double vane was covered by NiCoCrAlTaY and YSZ coatings. As shown in Fig. 5, typical lamellar NiCoCrAlTaY coatings with different thickness deposited onto DZ40M metallic substrate. Fig. 5a-5d respectively shows the cross-section microstructure of NiCoCrAlTaY coatings in selected spots 1, 3, 6 and 8 on vane A marked in Fig. 3. The coatings have a high as-sprayed density with only very fine and homogenously distributed interlaminar porosity. Between the coating and substrate surface, there is no obvious porosity, indicating a good adaptation to the substrate surface. But in Fig. 5b a few porosity is visible in bonding position because of the dramatic decrease of coating thickness.

The Fig. 5e is a high magnification of the rectangle in Fig. 5a. It's clearly shows that a dramatic deformation, flattening and cumulation behavior of thermal sprayed particles occurred in deposition period, indicating that powder particles underwent a sufficient heating and acceleration because of the high plasma velocity and temperature. The particles melt sufficiently and achieved very high velocity before impacting onto the substrate surface. As a result, few porosities remained and a high density presented in coating. Fig. 5b corresponds to the spot 3 on vane A as marked in Fig. 3, which is a shadowed area of double guide vane. Due to the mutual obstruct effect of airfoils, the deposition rate decrease evidently. A few porosities formed and couldn't be filled up because of poor spray angle and relative low impact velocity of particles.

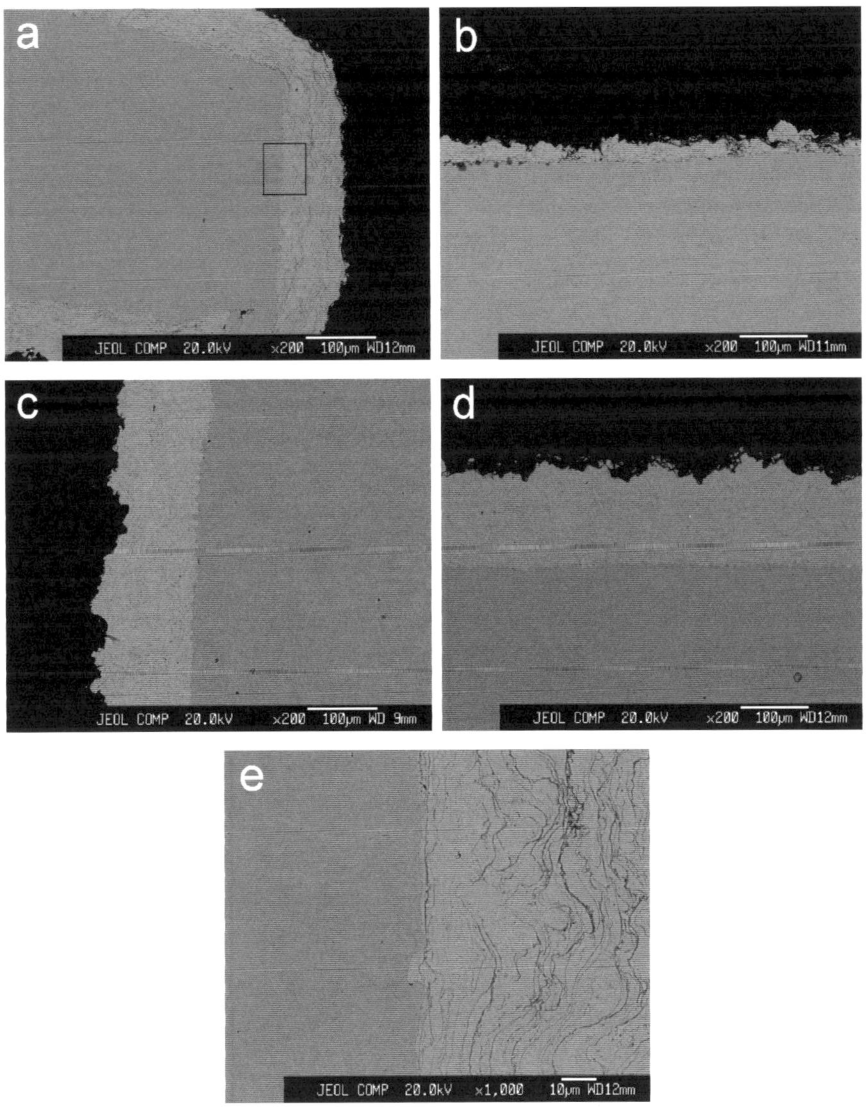

Fig. 5 Typical lamellar NiCoCrAlTaY coatings deposited at different positions.

Thickness measurements have shown significant differences in the NiCoCrAlTaY coating's thickness of the blade profile, as shown in Fig. 6. It shows that the coating thickness varies in range of 20-150 μm on trailing vane and 50-140 μm on leading vane. For leading vane (vane B) as shown in Fig. 6b, the coating thickness has a more uniform distribution. The thickest coating up to 140 μm has been formed near spot 1 and the thinnest coating about 50 μm has been formed near shadowed spot 10. More noteworthy is that the coating thickness distribution on the trailing vane (vane A) has a wider range as shown in Fig. 6a. The thickest coating up to 150 μm has been formed near spot 9 and 10 and the thinnest coating about 20 μm has been formed near shadowed spot 3.

Although the coatings are thin near shadowed positions, for example spot 10 on leading vane or spot 3 on trailing vane, but using conventional low pressure plasma spraying (LPPS), these shadowed positions are hardly deposited any coatings because the LPPS plasma jet is too short and small in diameter. But in this experiment, the O3CP plasma gun was used instead of F4-VB gun and operated at a much higher current and power level. Compared with LPPS, the plasma jet was also extended to some extent. Altrough, the degree of plasma expansion is much lighter than that under typical PS-PVD process condition. In this sense, the use of PS-PVD system can make metallic coating entirely cover the blade surface and has a function-required thickness in the shadow positions, demonstrating clearly an important technical characteristic of non-line-of-sight deposition. In addition, the bond coats obtained by this process is very dense, that is beneficial for TBC coatings system to improve property of high temperature oxidation resistance. Of course, by adjusting the parameters of spraying process, the thickness uniformity of NiCoCrAlTaY coating can be further improved, which is helpful and promoted for real practicality and feasibility of PS-PVD system to produce metallic coatings.

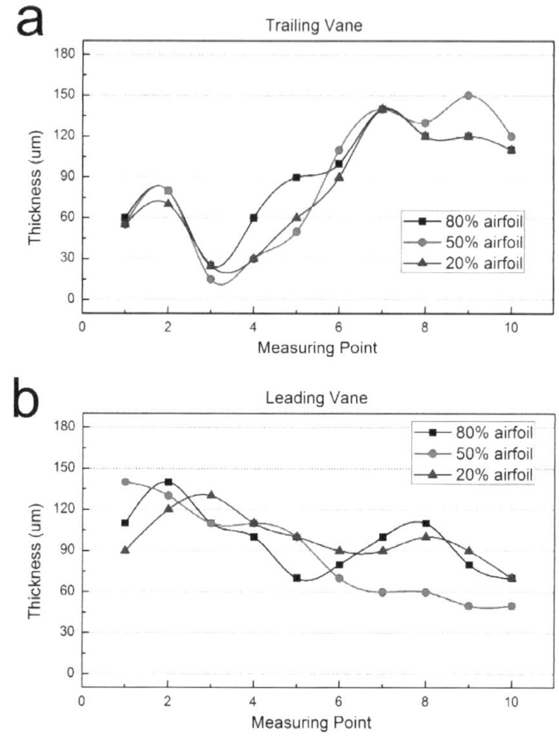

Fig. 6 NiCoCrAlTaY coating thickness distribution of a PS-PVD coated double vane model

3.2 Thickness and microstructure results of YSZ coating

As shown in Fig. 7, feather-like YSZ coatings were deposited onto NiCoCrAlTaY bond coating at different positions. Fig. 7a-7d respectively shows the cross-section microstructure of YSZ coatings in selected spots 1, 4, 6 and 9 on vane A marked in Fig. 3. In different positions, the YSZ coatings are all feather-like morphology, possessing different thickness and microstructure. The coating in spot 4 shows a higher density. In contrast, the coatings in spots 1, 6 and 9 have loose feather structure and many gaps between feather-like crystals. It's clearly shown by the Fig. 5e and Fig. 5f, which are high magnification of the rectangle in Fig. 5a and Fig. 5b respectively.

It's noticed that there are a large number of fine spherical particles in the gaps in the coating at spots 1, 6 and 9. Most of these spherical particles are in size of 3 μm or less and EDS show its chemical composition has no difference from the composition of feather-like grains. So it's reasonable to inferred that the YSZ particles (d50 = 10μm) fed into the expanded plasma jet were evaporated from the particle surface, but larger particles in heating process is not effectively changed into vapours completely, leading to an condensation and bonding each other in the feather grain gaps in the form of small liquid droplets.

The position in spot 4 is a shadowed position of double vane. The larger particles even if not fully evaporated, also due to the shielding effect were blocked and can not be solidified directly on the substrate. Only fully vaporized material can bypass the mutual occluded airfoils and deposit on the shadowed positions of double vane. Therefore, the coating at spot 4 has a more obvious feather-type structure, and the structure is denser and no fine spherical particles distributed in gaps.

A looser feather-type structure and a larger porosity help to reduce the thermal conductivity of thermal barrier coating and improve thermal cycle life. Of course, a denser feather structure or columnar structure can be obtained by adjusting the process parameters. There are three aspects to consider. The first is to increase the evaporation ratio of the feedstock material, including increase input power and/or the proportion of high-enthalpy gas. The second is to improve the substrate temperature. In general, a higher substrate temperature is advantageous to the formation of a well-defined columnar microstructure. The third is to reduce the substrate surface roughness. The YSZ columns grow perpendicular to the substrate surface if the substrate was polished to very low surface roughness such as EB-PVD process. In this experiment, the YSZ coatings were deposited directly onto the as-sprayed NiCoCrAlTaY bond coating with an average surface roughness of about 10 μm. As a result, it is visible from Fig. 7 that the feathers do not really grow perpendicular to the substrate surface. The reason is when the coating deposits on rough surface, the growing features interfere with each other and do not form a homogeneous structure. Reducing the surface roughness is favorable for structural uniformity. The surface roughness of substrate to obtain a well-defined feather or columnar microstructure by PS-PVD should not exceed a Ra value of 2 μm [4].

The fast reciprocating motion ensures a homogeneous temperature distributed on the substrate surface. Deposition efficiency of about 0.15 μm/pass calculated according to the average coating thickness could be achieved. This corresponds to a deposition rate on real turbine blades in the range of 5-10 μm/min.

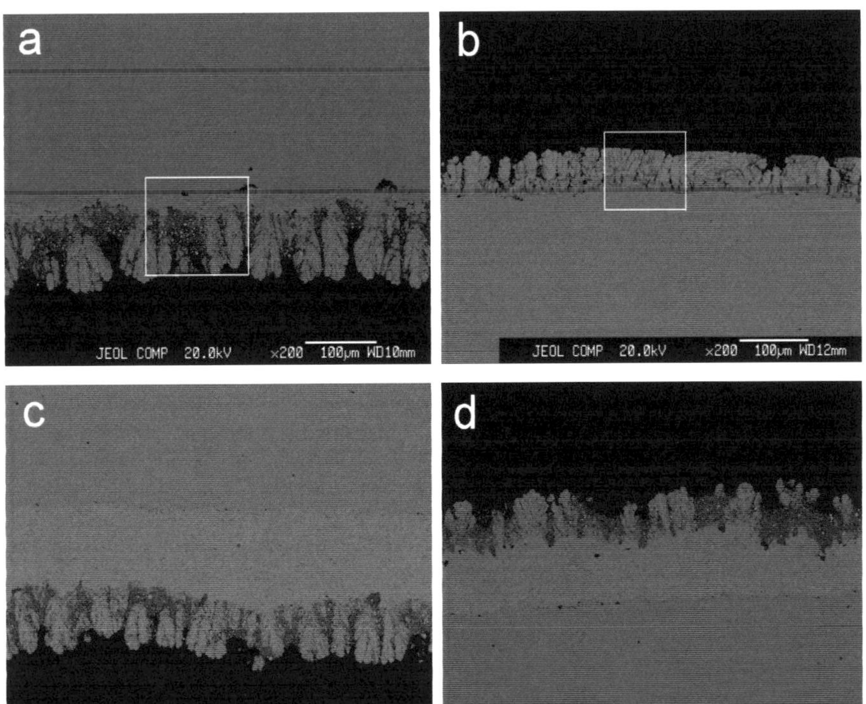

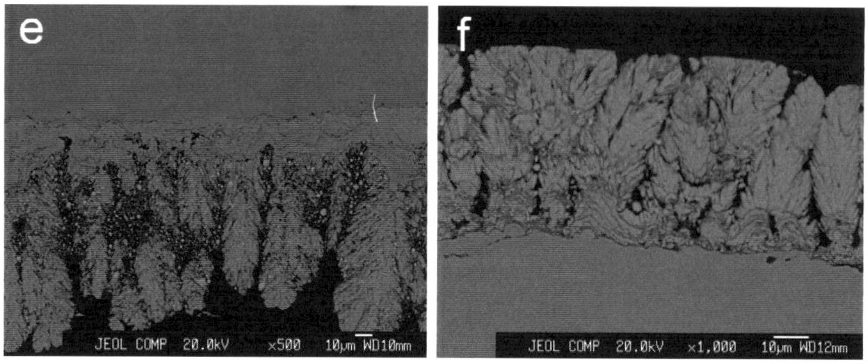

Fig. 7 Feather-like YSZ coatings deposited at different positions by PS-PVD.

Obvious differences in the YSZ coating's thickness of the blade profiles were shown in Fig. 8. As shown in the profile of 50% airfoil, the coating thickness varies in range of 100-190 μm on trailing vane and 90-180 μm on leading vane. On trailing vane, the thickest YSZ coating up to 190 μm has been formed near spot 10 and the thinnest coating about 100 μm has been formed near shadowed spot 4 and fringe spot 9. On leading vane, the thickest coating up to 180 μm has been formed near spot 4 and the thinnest coating about 90 μm has been formed near shadowed spot 9. It has been proven that the thickest coating was close to the leading edge and trailing edge of the vane and the thinest coating was located in shadowed core place. In those edge places the coating thickness is approximate 80-100% larger than in the other areas on double vane. In ref. 4, the thickest coating is 2~3 times as the thinest coating, though the overall coating thickness is larger than the experiment in this article.

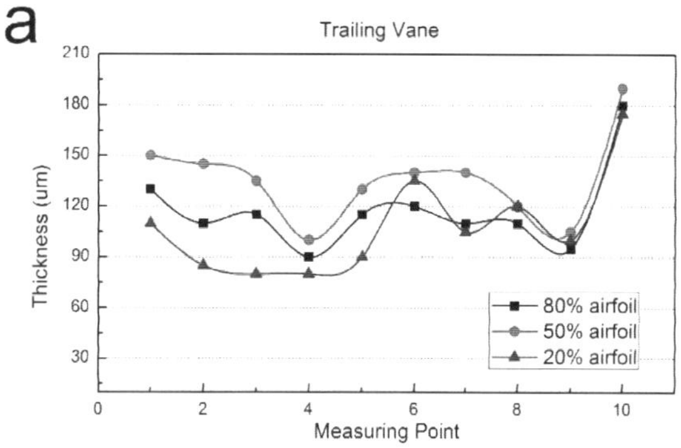

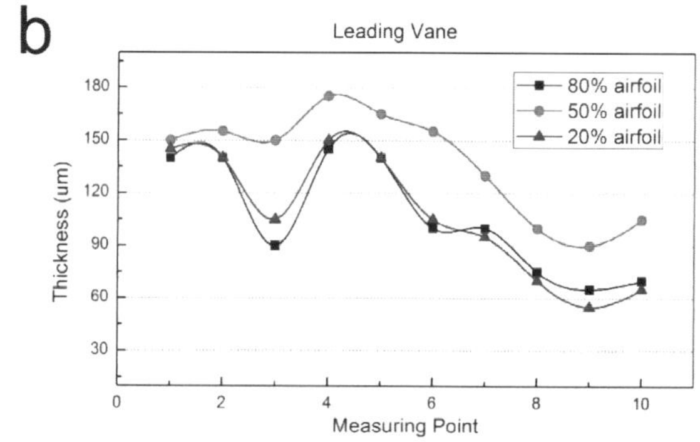

Fig. 8 YSZ Coating thickness distribution of a PS-PVD coated double vane model

Application of thermal barrier coating can decrease the base material temperature and increase the engine operating temperature, improving the turbine efficiency and reducing fuel consumption and harmful gas emission. Application of thermal barrier coating in connection with internal and external air cooling system can lower the blade base temperature about 165℃[7]. Generally, the EB-PVD (Electron beam physical vapour deposition) and APS (Air plasma spraying) are classical alternative technologies are used for deposition of YSZ coatings, depending on property requirements. As same as microstructure and property, the thickness of TBCs has a strong dependence on coating deposition method, especially on multiple airfoil vane with complex geometry. Although the coating is thin near shadowed position like spot 3 on trailing vane and spot 9, 10 on leading vane, but using conventional EB-PVD and APS method, it's hard to deposit any YSZ coatings in these shadowed positions because of line-of-sight deposition restriction. Application of PS-PVD technology can make YSZ coating completely to cover the entire blade including the shadow positions. In contrast to EB-PVD and APS, PS-PVD incorporates the vaporized coating material into a supersonic plasma stream. The high velocity gas stream forced the incorporated vaporized material into complex shaped parts. Even shadowed areas which are not in line of sight can be coated homogeneously. According to the result mentioned above, it's easy to deduce that when the thickest YSZ coating reached up to 300 μm, the thinnest coating on double vane can at least deposit over 150 μm. Such a thickness of YSZ coating is sufficient to meet the thickness requirement and ensure the thermal barrier effect.

It's worthy to note that the YSZ coating's thickness distribution is more uniform than the NiCoCrAlTaY coating's. The reason is the NiCoCrAlTaY coating builds up the layer by liquid splat because of the lower power of PS-PVD process and the superior thermal conductivity of metal. The NiCoCrAlTaY particles hardly vaporized and deposited around the shadowed position. In contrast, as a result of the high power input up to 130 kW, the poor thermal conductivity of ceramics and the addition of high enthalpy gas He. PS-PVD can vaporize feedstock material and produce coatings when the vapor phases go around the barrier and reach shadowed position. So the thickness distribution of YSZ coatings is superior to this of NiCoCrAlTaY coatings.

Unfortunately, although the thickness on airfoil has the same distribution tendency at three profiles with different height, marked as 20% airfoil, 50% airfoil and 80% airfoil, the coating thickness along each profile has an obvious difference, that is, the coating along 50% airfoil profile is thicker than the others. The most important reason that affecting coating thickness along each profile is spraying angle. During the spray process, the part rotates on its axis, and, at the same time, the gun performs a rapid reciprocating motion with a particular spraying angle defined the angle between the plasma jet axis and the horizontal line. The angle 25° is chosen in this experiment. The coating thickness difference between 50% airfoil profile and the other two profiles enlarged when the spraying angle decreased to 15°, which provided an instructive metheod to decrease thickness difference on double vane with different height.

4 Conclusions

Dense multicomponent NiCoCrAlTaY coatings and feather-structured YSZ coatings were deposited by PS-PVD system on double nozzle guide vane. The coating thickness varies in a wide range although the YSZ coating's thickness distribution is more uniform than the NiCoCrAlTaY coating's. The geometry and arrangement of parts showed a big influence on the coating thickness and microstructure. The thickest coating was found close to the leading edge and trailing edge of the vane. In those places the coating thickness was approximate 80-100% larger than in the other areas on turbine blade. Inspiringly, PS-PVD technology can make NiCoCrAlTaY and YSZ coating completely cover the entire double vane. In contrast to EB-PVD and APS, PS-PVD can obtain a relatively homogeneous coating, even on shadowed areas which are not in the line of sight.

On multiple nozzle guide vane, manufacturing the thermal barrier coatings including metallic bond coating and top ceramic coating by single PS-PVD process seems to be possible. It demonstrates the possibility of application of this innovative PS-PVD method for TBCs production in the industrial field.

Acknowledgments

The authors gratefully acknowledge the financial support from the National Program on Key Basic Research Project (Grants No. 2012CB625100), the Science and Technology cooperation project of Guangdong province, and of Guangzhou city (Grant Nos. 2013B050800027, 201508030001).

Reference

1. J.L. Dorier, M. Gindrat, C. Hollenstein, M. Loch, A. Refke, A. Salito and G. Barbezat: Plasma Jet Properties in a New Spraying Process at Low Pressure for Large Area Thin Film Deposition. *Proceedings of the ITSC*, Singapore, May 28-30, 2001, p 759-764.
2. K. von Niessen, M. Gindrat and A. Refke: Vapor Phase Deposition Using Plasma Spray-PVD. *J. Therm. Spray Technol.*, 2010, **19**(1-2), p 502-509.
3. A. Hospach, G. Mauer, R. Vaßen and D. Stöver: Characteristics of Ceramic Coatings Made by Thin

Film Low Pressure Plasma Spraying (LPPS-TF). *J. Therm. Spray Technol.*, 2012, **21**(3-4), p 435-440.
4. K. von Niessen and M. Gindrat: Plasma Spray-PVD: A New Thermal Spray Process to Deposit Out of the Vapor Phase. *J. Therm. Spray Technol.*, 2011, **20**(4), p 736-743
5. S. Rezanka, G. Mauer, and R. Vaßen: Improved Thermal Cycling Durability of Thermal Barrier Coatings Manufactured by PS-PVD. *J. Therm. Spray Technol.*, 2014, 23(1-2), p 182-189
6. A. Refke, M. Gindrat and K. von Niessen: LPPS Thin Film: A Hybrid Coating Technology between Thermal Spray and PVD for Functional Thin Coatings and Large Area Applications. *Proceedings of the ITSC,* Beijing, China, May 2007, p 705-710
7. M. Hetmanczyk, L. Swadzba and B. Mendala: Advanced materials and protective coatings in aero-engines application. *J. Achi. Mater. Manuf. Eng.,* 2007, **24**(1), p 372-381.
8. J. Singh, D.E. Wolfe and J. Singh: Architecture of thermal barrier coatings produced by electron beam-physical vapor deposition (EB-PVD). *J. Mater. Sci.,* 2002, **37** (15), p 3261-3267.
9. M.O. Jarligo, D.E. Mack, R. Vassen and D. Stöver: Application of plasma-sprayed complex Perovskites as thermal barrier coatings. *J. Therm. Spray Technol.*, 2009, **18**(2), p 187-193.
10. M. Góral, J. Sieniawski, S. Kotowski, M. Pytel and M. Masłyk: Influence of turbine blade geometry on thickness of TBCs deposited by VPA and PS-PVD methods. *Archi. Mater. Sci. Eng.*, 2012, **54**(1), p 22-28
11. Q. Y. Chen, C. J. Li, C. X. Li, G. J. Yang and J. Z. Zhao: A preliminary microstructure investigation of YSZ coatings deposited by plasma spray - physical vapor deposition using a shrouded plasma torch, *Proceedings of the ITSC*, Barcelona, Spain, May 2014, p 268-272

Effect of nitric acid corrosion on the luminescent intensity of plasma sprayed YAG:Ce Coatings

Peng Zeng, Hehui Wang, Weize Wang, Jingye Yu, Liangmin Wu
Key Lab of Pressure System and safety, Ministry of Education, East China University of Science and Technology, Shanghai 200237, PR China
Corresponding author: wangwz@ecust.edu.cn

Abstract: In order to monitor the corrosion condition of components in chemical containers, YAG: Ce^{3+} Coating is prepared, and the effect of acid corrosion on the luminescence intensity of coating is studied. The powder is synthesized by the high-temperature solid phase synthesis technique. The coatings are prepared by air plasma spraying. Nitric acid solution is used as the typical acid. Microstructure and phases are characterized through using SEM and XRD, respectively. Effect of immersion duration in the acid is studied on the luminescence intensity. It is found that the luminescence intensity shows the fluctuation tendency with the immersion duration, which relates with phase composition.

1 Introduction

Corrosion exists in many parts of the industry, resulting in the reduction and degradation performance of materials. Corrosion causes a lot of economic losses annually. The corrosion of oil and petrochemical equipment is more prominent in many industries. Among them, the condensate system of the refinery tower is one of the most serious corrosion equipment in the refinery due to the direct effect of various corrosive media. With the sulfur content in crude oil increasing, the corrosion of equipment is becoming more and more serious [1]. The safety problems caused by corrosion are gradually increasing. It is very important to monitor the corrosion condition for the equipment, so as to grasp the changing information in time.

The purpose of corrosion detection is to obtain the change of the structure and properties of the material by measuring the corrosion rate or corrosion condition etc. Detection methods include non-destructive testing [2-4], physical and mechanical means [5], chemical and electrochemical analysis [6-7] etc. Fluorescence method for characterizing the material loss is intuitive, which is currently used for measuring the wear [8], peeling-off [9], and stress state [10] of coatings. It is also reported that fluorescence method has been used to monitor the corrosion of aluminum alloy matrix [11] with high sensitivity.

Ce doped YAG can effectively absorb blue light, emit yellow light and thus form white light. The preparation and properties of YAG:Ce powder and the related materials have been studied [12]. In this study, YAG:Ce coating was prepared by plasma spraying, and the influence of corrosion on the fluorescence property of coating was studied. It is expected to provide a new way for the monitoring of the corrosion condition.

2 Experimental

YAG:Ce powder is synthesized through using high temperature solid-state reaction method. Cerium ion concentration is 0.06. The chemical reaction equation for the synthesis at high temperatures is shown as follows:

$$5Al_2O_3 + 3Y_2O_3 + 0.06CeO_2 \xrightarrow{1200\sim1600^{\circ}C} 2Y_{2.94}Ce_{0.06}Al_5O_{12}$$

After accurately weighting the Al_2O_3, Y_2O_3 and the activator CeO_2, they are put into the zirconia ceramic ball mill. 1.15% of the total mass of the reactants, BaF_2 and H_3BO_3 (the mass ratio of BaF_2 and H_3BO_3 is 1: 1), is added to jar and milled for one hour to make the powders fully mixed. After that, the powders are transferred to the corundum crucible, heated to 1300 °C in the reducing atmosphere for 3 hours. The obtained YAG:Ce powder is ground and mixed with PVA (polyvinyl alcohol) and the deionized water. Spray granulation is carried out by a high-speed centrifugal spray dryer. Eventually the YAG:Ce powders with good liquidity are obtained.

YAG:Ce coating are deposited on the preheated substrate by plasma spraying. Argon is used as the primary plasma operating gas and hydrogen is selected as an auxiliary gas. The pressure of argon and hydrogen is fixed at 0.4 MPa and 0.25 MPa, respectively. Spraying power is 36kW. The traverse speed of spray gun is kept as 150 mm/s with a spray distance of 85 mm.

The coating samples are immersed in 1.5 mol/L nitric acid solution at 25 °C for 5, 10, 15, 20 and 25 hrs, respectively. After corrosion, the samples are cleaned and dried for the further examination. In addition, another one sample without corrosion is used as the blank data.

The phase compositions of powders and coatings is examined by X ray diffraction (XRD, D/MAX RIGAKU 2550 VB/PC). The cross section morphology of the coating was observed by scanning electron microscopy (EVO MA15 ZEISS). The luminescence properties of the powders and coatings were analyzed by fluorescence spectrometer (Jobin Yvon Fluorolog-3-P).

3 Results and discussion

3.1 Microstructure and luminescent properties of YAG:Ce powders

The microstructure of YAG:Ce powder prepared by high

temperature solid-state reaction is shown in Fig. 1. After the granulation, the particle size of the powder is mainly between 40-60 microns. The phase composition of the powder (Fig. 2) is dominated by YAG ($Al_5Y_3O_{12}$). Small amount of $YAlO_3$ (YAP) is also detected.

Fig. 3 is the excitation and emission spectra of YAG:Ce powder. It can be found the excitation spectra of YAG powder is of the nominal two-peak structure. An excitation peak at 340 nm is observed in the near ultraviolet region, and a broadband excitation peak in the visible region. In the YAG crystals, the ground state of Ce^{3+} split to a double state of $^2F_{7/2}$ and $^2F_{5/2}$ by the effect of spin coupling. At room temperature, the former is occupied by electrons, and the *4f* energy level is shielded in the inner layer, which can be affected a little by the crystal field. While the radial wave function of *5d* is located outside the $5s^25p^6$ shell, the *5d* state is strongly influenced by the crystal field. When Y^{3+} (r_{Y3+}=10.19 nm) is replaced by Ce^{3+} (r_{Ce3+}=11.43 nm) in YAG crystal, the crystal field splitting is intensified, due to the distance between the central ion and the ion pair becomes shorter. Therefore, Ce^{3+} can obtain a larger *5d* energy gap in YAG crystal field [13]. In the excitation spectra, the lowest *5d* energy band can be brought to 491 nm and 342 nm, the former corresponding to the energy level transition from $4f^{12}F_{5/2}\Gamma_{8u}$ to $5d^{12}E_g\Gamma_{8g}$, and the latter to the energy level transition from $2F_{5/2}\Gamma_{7u}$ to $5d^{12}E_g\Gamma_{7g}$. The broadband excitation peaks at 466 nm corresponds to the energy level transition from $4f_{12}F_{5/2}\Gamma_{7u}$ to $5d^{12}E_g\Gamma_{8g}$. The excitation peaks between 440-480 nm are of the multiple structure, besides the peak at 466 nm, which are not due to the level transition, but the test instrument, and have no other effects on the analysis [13]. Fig. 3 (b) is the emission spectrum of YAG:Ce powder at the excitation wavelength of 466 nm. The excited electrons transit from the *4f* ground state to the *5d* excited state, and then most of electrons from the *5d* excited state transit further to the $2F_{5/2}$ and $2F_{7/2}$ levels. The two transitions combine to form the wide emission bands.

3.2 Microstructure of coatings

After the rare earth ions replace yttrium ion locating in the center of the YAG dodecahedral crystal, the difference between the radius of yttrium and rare earth ion causes the formation of the defect and also the variation in the corresponding emission intensity. At the same time, due to the difference between the radius of the cerium ion (0.1283nm) and yttrium ion (0.1159nm) is relatively large [14], the defects in the YAG materials doped with cerium ions are much more than with other ions. Therefore, this kind of material is easy to react in the corrosive environment.

Fig. 1 Morphology of YAG: Ce^{3+} powders

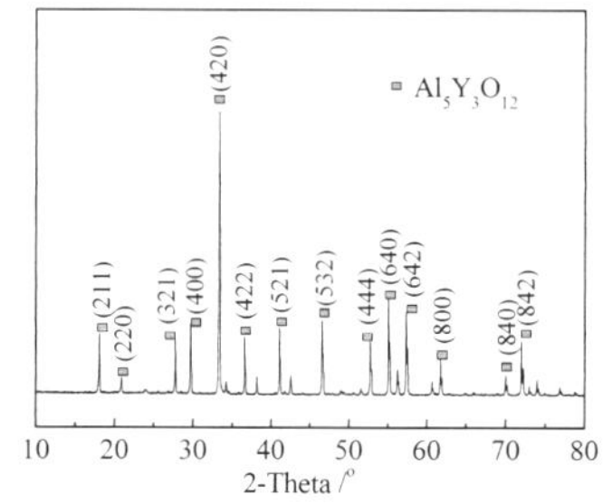

Fig. 2 XRD patterns of YAG: Ce^{3+} powder

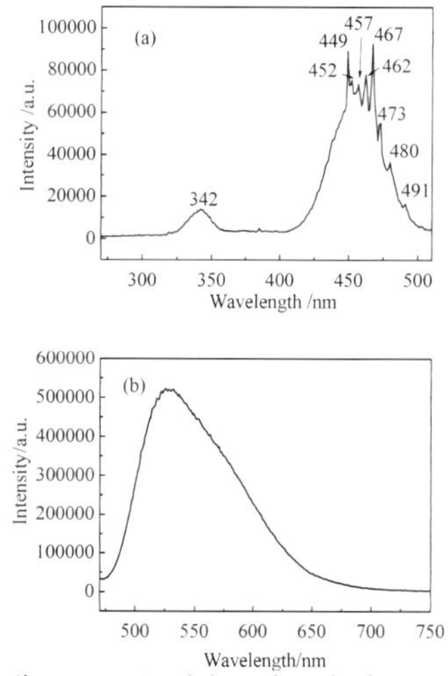

Fig. 3 Excitation spectra (a) and emission spectra (b) of YAG: Ce^{3+} powders

The coatings are corroded in the 1.5 mol/L nitric acid solution for different time. With the increment of corrosion time in acid solution, the size of pores in the coatings gradually increases (Fig.4). From the microscopic structure, the corrosion occurs mainly between the lamella of coatings. When the molten particles are impacting on the substrate or the formed layer during spraying, the depositing surface temperature is lower than that of the droplets, which is the easy condition for forming the amorphous phase in coatings [15]. In the process of corrosion, the amorphous phase reacts with the corrosion liquid [16]. In addition, there are non-bonded interfaces between the lamella in coatings. Combining with the vertical cracks in coatings, they compose the immersing access for the corrosion solution. The effect of armorphous phase and pore results in the continued corrosion.

The sprayed coating mainly includes the phases of YAG and YAP, and also a small amount of Y_2O_3. After the coating is corroded by nitric acid, both the intensity and the position of the (420) peak changes. When the corrosion time is 0, 5, 10, 15, 20 and 25 hrs, the peak intensities of the coatings are 6120, 6110, 6600, 7200,

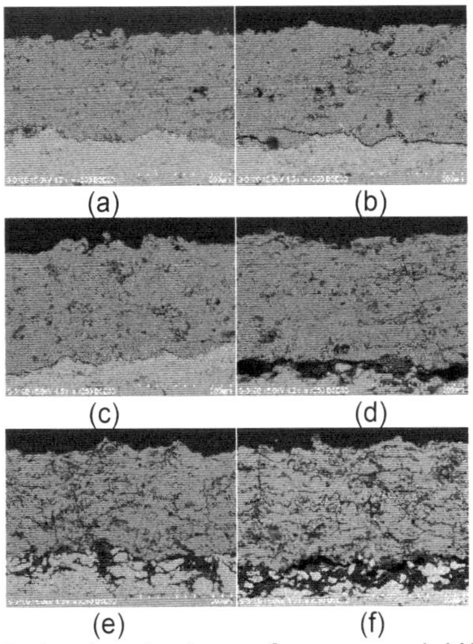

Fig. 4 Typical microstructure of as-sprayed YAG:Ce^{3+} coatings (a) and after immersion in nitric acid solution for 5 (b)、10 (c)、15 (d)、20 (e) and 25 (f) hours

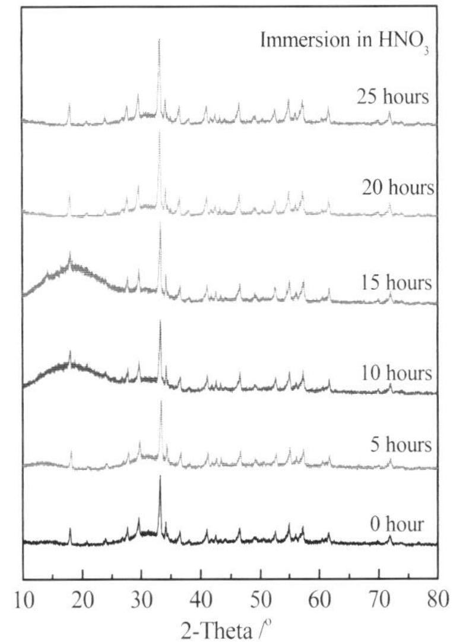

Fig. 5 XRD patterns of YAG: Ce coatings after immersion in nitric acid solution for different durations

7500 and 7500 respectively. With the increment of corrosion time, the peak intensity of the coating gradually increases, and finally maintains unchangeable, which is related to the continuous dissolution of amorphous phase in the coating. And with the corrosion time increasing from 0 to 25 hours, the peak positions are respectively 33.20°, 33.41°, 33.29°, 33.29°, 33.28° and 33.23°. The main peak position shifts firstly to the large angle, and then remains unchangable, and finally to the small angle. The changes of peak location relate to the phase composition and the doped elements. Though the main composition of coatings is the phases of YAG and YAP, the proportion of phases changes, which may cause variations in the main peak position of XRD patterns.

3.3 Luminescent properties of coatings

The luminous intensity of the coating is shown in Fig. 6 after the plasma sprayed YAG:Ce^{3+} coating has been corroded for different durations in the nitric acid solution. With the increase in corrosion time, the luminiscent intensity increases first and then decreases, and finally increases again. The luminous intensity of the coating shows the fluctuating trend.

YAG:Ce^{3+} belongs to the luminescent material having the discrete luminescence center. Luminescence is due to the Ce^{3+} ion transition between the energy levels of 4*f* level and 5*d* level. The energy level of the discrete Ce^{3+} ion is affected by the surrounding crystal field. From the microstructure of the coating after the corrosion, it can be seen that the porosity of the coating increases with the corrosion time. Because of the appearance of pores, the surface dangling bonds of Ce^{3+} ion increase correspondingly, which may lead to the phenomenon of fluorescence quenching. Therefore, with the increase of the coating porosity, the luminescent property of the coating decreases.

Moreover, with the increase of corrosion time, the peak position of XRD patterns shifts to the large angle, and then remains unchanged, and finally increases. In the process of corrosion, the pores are generated in the coating, the tensile stress is released, which causes the decrease of crystal plane spacing. At the same time, due to the larger radius of the doped Ce^{3+} ion than that of the Y^{3+} ion, it will have an effect on the crystal field, resulting in the expansion of the crystal field. Therefore, the diffraction peak can be expected to move to the small angle. In addition, the coating phase composition is mainly of YAG and YAP, but the proportion is different. The main peak position of YAG locates at 33.317°, while that YAP at 33.064°. If the content of YAG increases, the main peak would shift to large angle. Therefore, with the increase of the corrosion time, the diffraction peaks of the coating show the change tendency as above. And the appearance of YAP phase can cause the decrease of the emission intensity of the coating.

The shift of the diffraction peak has a significant effect on the emission intensity of coatings between 0 and 15 hrs of corrosion time, comparing the influence of the porosity and crystallinity. With corrosion time increasing to higher than 20 hrs, the influence of crystallinity of the coating is dominant. The luminous intensity of coatings is affected by the porosity, the intensity of the diffraction peak and the shifting of the diffraction peak etc, which has a fluctuated

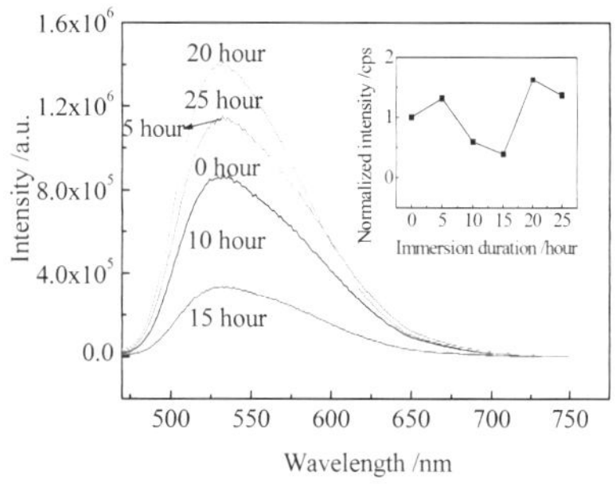

Fig. 6 Effect of immersion duration in nitric acid solution on the luminescent intensity of the YAG: Ce^{3+} coatings

tendency. When the YAG:Ce^{3+} coating is used to monitor the corrosion state, the coating structure and other factors should also be considered.

4 Conclusions

In order to expand the application of the fluorescent element in the field of corrosion monitoring, the influence of nitric acid corrosion on the microstructure and luminescence properties of plasma sprayed YAG:Ce^{3+} coating is studied. The results show that the YAG:Ce^{3+} powder formed by high temperature solid-state reaction is mainly of YAG phase, and the coating is YAG and YAP. After the coating is corroded in nitric acid solution for different durations, the porosity of the coating increases and the intensity of diffraction peaks shows the similar trend with the porosity. And the peak position of XRD patterns of coatings shifts to large angle firstly, then remains unchanged, and finaly move to small angle direction. When the corrosion time increase sfrom 0 to 25 hours, the fluorescence intensity of the coating increases first, then decreases, and finaly increases. The fluorescence intensity of the coating is influenced by the porosity, the intensity of the diffraction peak and the offset of the diffraction peak etc.

Acknowledgements

The authors gratefully appreciate the support of National Natural Science Foundation (No. 51275172), Science and Technology Commission of Shanghai Municipality Project (No. 14DZ2261205), Aviation Fund (2015ZES7001, 2013ZFS7001) and Shanghai Pujiang Program (15PJD009).

References

[1] Alvisi. et al.: An overview of naphthenic acid corrosion in a vacuum distillation plant. Engineering Failure Analysis, 18 (2011), pp.1403/6.

[2] Foudazi Ali. et al.: Characterization of Corroded Reinforced Steel Bars by Active Microwave Thermography. IEEE Transactions on Instrumentation and Measurement, 64 (2015), pp.2583/5.

[3] Buck Jeremy A. et al.: Pulsed Eddy Current Inspection of Support Structures in Steam Generators, IEEE Sensors Journal, 15 (2015), pp.4305/12.

[4] Farhidzadeh Alireza. et al.: Reference-free corrosion damage diagnosis in steel strands using guided ultrasonic waves, Ultrasonics, 57(2015), pp.198/208.

[5] Puls RW. et al.: Long-term performance monitoring for a permeable reactive barrier at the US Coast Guard Support Center, Elizabeth City, North Carolina, Journal of Hazardous Materials, 68(1999), pp.109/124.

[6] Lenartova I. et al.: Hydrogen embrittlement for austenitic alloys: behaviour of microstructure and segregation of sulphur and phosphorus impurities, Revue De Metallurgie-Cahiers D Informations Techniques, 96 (1999), pp.243/251.

[7] Cao Xianlong. et al.: Electrochemical investigation on atmospheric corrosion of carbon steel under different environmental parameters, Anti-Corrosion Methods and Materials, 60 (2013), pp.199/205.

[8] Muratorea C. et al.: Smart tribological coatings with wear sensing capability, Wear, 256(2008), pp.913/920.

[9] Jeffrey I Eldridge. et al.: Monitoring delamination of plasma-sprayed thermal barrier coatings, Surface and Coatings Technology, 201 (2006), pp.3926/3930.

[10] Wang, W, Z. et al.: Effect of Plasma Spraying Parameters on the Luminescent Intensity of YSZ: Tm Coatings, Rare Metal Materials and Engineering, 43(2014), pp.481/4.

[11] Li, S, M. et al.: Preparation and performance of fluorescent sensing coating for monitoring corrosion of Al alloy 2024, Trans. Nonferrous Met. Soc. 16(2006), pp.159/164.

[12] Zhang, K. et al.: Co-precipitation synthesis and luminescence behavior of Ce-doped yttrium aluminum garnet (YAG:Ce) phosphor: The effect of precipitant, Journal of Alloys and Compounds, 453(2008), pp.265/270.

[13] Zhao, B, Y, et a: Studies on Optical Properties and Ce Concentration of Ce:YAG Single Crystal for WLEDs, Chemical Journal of Chinses Universities, 25(2014), pp.230/6.

[14] Zeng, G, P. et al.: Defeats in Doped YAG Crystal, Journal of Synthetic Crystals. 4 (1999), pp.354/8.

[15] Gross K A, Saber Samandari S S, Heemann K S: Journal of Biomedical Materials Research B, 93(2012), pp. 1-8.

[16] Wang W Z, Liang J C, Guo X P et al. Mechanical properties and dissolution behavior of plasma sprayed wollastonite coatings deposited at different substrate temperatures, Journal of Thermal Spray Technology, 21(2012), pp. 496/504.

Sulfidation behavior of plasma-sprayed Al-Mo coatings at high temperature

Liangmin Wu, Weize Wang, Jingye Yu
Key Lab of Safety Science of Pressurized System, Ministry of Education, East China University of Science and Technology, Shanghai 200237, Peoples Republic of China.
Corresponding author: wangwz@ecust.edu.cn

Abstract: Gas burner is the key component of coal-water slurry gasifier system. However, due to the sulfidation at high temperature, gas burner often fails in advance, which affects the stable operation of the system. In this work, sulfidation corrosion resistance of plasma-sprayed Al-Mo coating with NiCrAlY and Mo bonding layer is investigated. The sulfidation behavior of Al-Mo coatings is compared with Mo coatings and Inconel 600 alloy at temperatures of 973K, 1073K and 1173K, respectively. the The sulfidation tests show that Al-Mo coatings protect the substrate better than Mo coatings when surfering high-temperature corrosion in sulfur vapor. In addition, the oxidation resistance of Al-Mo coatings is also high than that of Mo coatings. Plasma sprayed Al-Mo coatings are promising candidates for the corrosion protection for steels under the coal gasification condition.

1 Introduction

As a core component of the coal-water slurry gasifier, burner can atomize the coal-water slurry with the use of high-velocity oxygen. But during the actual operation, some radial radioactive cracks often appear at the end of burner, which has a significant influence on the production continuity and results in a huge economic loss. It is generally considered that the radial radioactive cracks are mainly caused by high temperature sulfidation. In the meantime, the high temperature oxidation, erosive wear, thermal shock and stress concentration also have influences on the burner failure[1].
The corrosion of metals and alloys in sulfur-contain atmospheres is usually much more severe than in purely oxidizing environment owing to the fact that common metal sulfides have larger deviation from stoichiometry and higher density of lattice defects, in comparison with the corresponding oxides [2]. It has been reported that Mo coating deposited by ion plasma deposition could protect the substrate well from high temperature sulfidation owing to the compact molybdenum sulfide layer formed on the coating surface in sulfidation atmosphere [2]. It is also reported that sputter-deposited amorphous Al-Mo alloys have excellent resistance to highly sulfidizing environments up to about 1273K. Due to the formation of Al_2O_3 on Al-Mo coatings in oxidizing atmosphere which could diminish the evaporation of volatile molybdenum, the addition of aluminum could improve the oxidation resistance of Mo coatings [3]. However, sputter deposition is not very economical and convenient. It is valuable to develop a practical way to deposite Al-Mo coatings on the substrate for the industrial application.The present study aims to improve the sulfidation resistance of Inconel 600 by plasma-sprayed Al-Mo coatings.

2 Experimental

2.1 Power and coating preparation

Pure elemental aluminum and molybdenum were mixed to yield an average starting composition of Al-27at.%Mo. Mechanical alloying was carried out in a ND8-L mixer/mill with zirconia ceramic balls of 6mm in diameter in Ar atomosphere. The ball-to-power weight ratio was 10:1. About 1wt% stearic acid was added as a process control agent in order to prevent excessive sticking of the power samples to the balls and to the vial walls. After milled up to 31h in the stainless-steel vial, the mixed powders were annealed for 2h at 1273K in the tube vacuum furnace. To obtain powders in the range of 200 to 500 mesh, sintered powders were milled in the vial for the appropriate period. Inconel 600 alloy was used as substrate. Cylindrical specimens φ16×3mm were cut from the steel, then the specimens were ultrasonically cleaned in acetone and grit blasted using 60-80 mesh Al_2O_3. Mo and Al-Mo coatings were plasma-sprayed on the substrates with a thickness about 300 μm. Mo and NiCrAlY were used as the bonding layer with a thickness about 130 μm.

Table 1. Spraying parameters for the coatings

	Mo	NiCrAlY	Al-Mo
Spraying current (A)	600	500	500
Spraying voltage (V)	60	60	60
Spraying distance (mm)	120	100	100

2.2 High temperature oxidation and sulfidation test

The sulfidation tests were carried out in a Ar-S_2 atmosphere with the sulfur pressure of 1 kPa in a tube vacuum furnace at 973 K, 1073 K and 1173 K, respectively. The oxidation tests were carried out in the air atmosphere in a muffle furnace at temperatures of 973 K, 1073 K and 1173 K, respectively.
The exposure time was 12h for 6 cycles. After each cycle, the weight gains of the specimens after sulfidation were measured using a balance with an accuracy of 10^{-5}g. The microstructure of cross-section and macromorphology of coatings after sulfidation for 12h were observed by Scanning Electron Microscope (SEM).

3 Results and discussion

3.1 Powders and coatings

Fig.1 shows that after milled for 31h in Ar atmosphere, the phases composition Al-27at.%Mo powders shows no obvious change. Fig.2 shows that after milled for 31h and heat treated for 2h at 1273K in Ar atmosphere, Al-27at.%Mo powders completely turn into intermetallic compounds of Al_8Mo_3 and $Al_{3+X}Mo_{1-X}$. After milled in the vial after the appropriate period, the typical microstructure of sintered powders is shown in Fig.3. Lots of tiny debris bonds together to compose the powder particle. Though the particles are not granular, the feed rate during spraying can be accepeted. The typical microstructure of Al-Mo coatings after plasma spraying is shown in Fig. 4, which is comparetively dense.

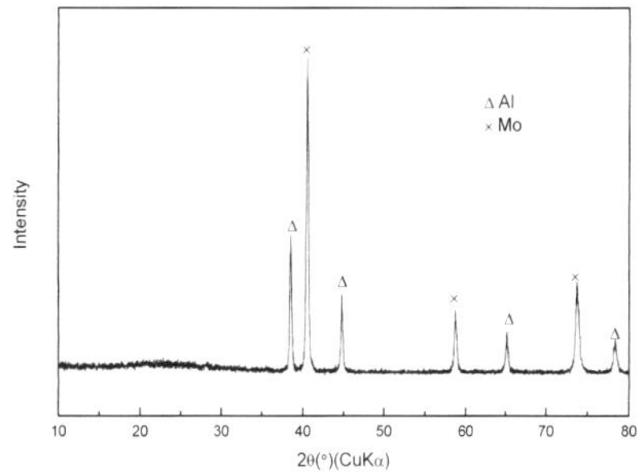

Fig. 1. X-ray diffraction pattern of the Al-27at.%Mo powders milled for 31h

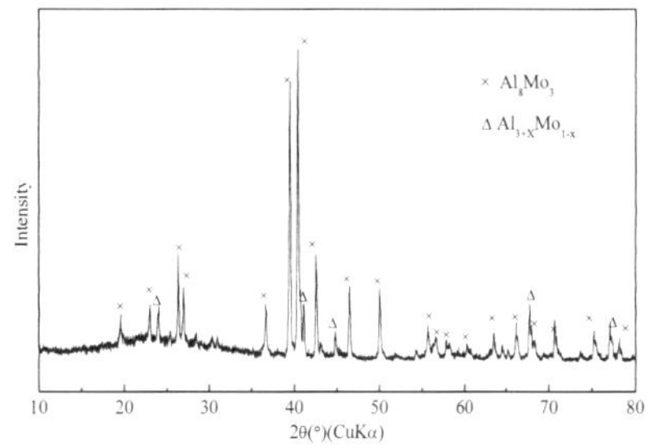

Fig. 2. X-ray diffraction pattern of the Al-27at.%Mo powders milled for 31h and heat treated for 2h at 1273K

Fig. 3. Microstructure of sintered Al-Mo powders

3.2 Sulfidation behavior of Mo and Al-Mo coatings with NiCrAlY bonding layer

The macro morphology of Inconel 600, Mo coating and Al-Mo coating with NiCrAlY bonding layer is shown in Fig. 5 after sulfidation at 973K, 1073K and 1173 K, respectively. As shown in Fig. 5, there is no visible cracks and peeling-off on the suface of Mo coating and Al-Mo coating after sulfidation at 973K for 12h. However, both Mo coating and Al-Mo coating fail after sulfidation at 1073K for 10h. Al-Mo coating separates from the substrate at the edge of the coating. Mo coating detaches from the substrate overall after sulfidation at 1073K for 8h.

Sulfidation kinetics at 973K and 1073K are shown in Fig.6. The weight gains of Inconel 600 alloy are much higher than those of coated specimens. For the coated specimens, the weight gains of Al-Mo coatings are

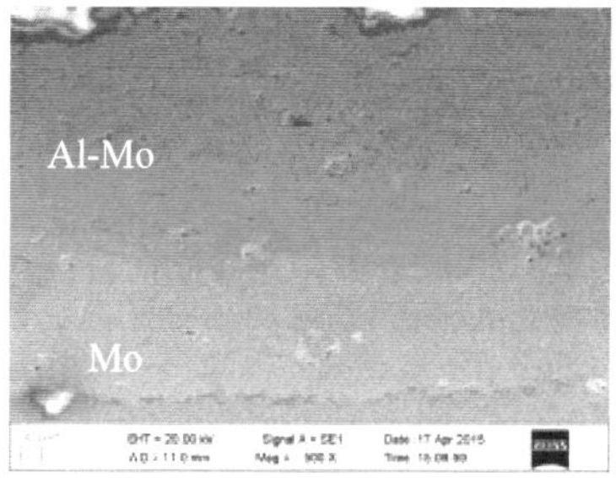

Fig. 4. Microstructure of plasma-sprayed Al-Mo coatings

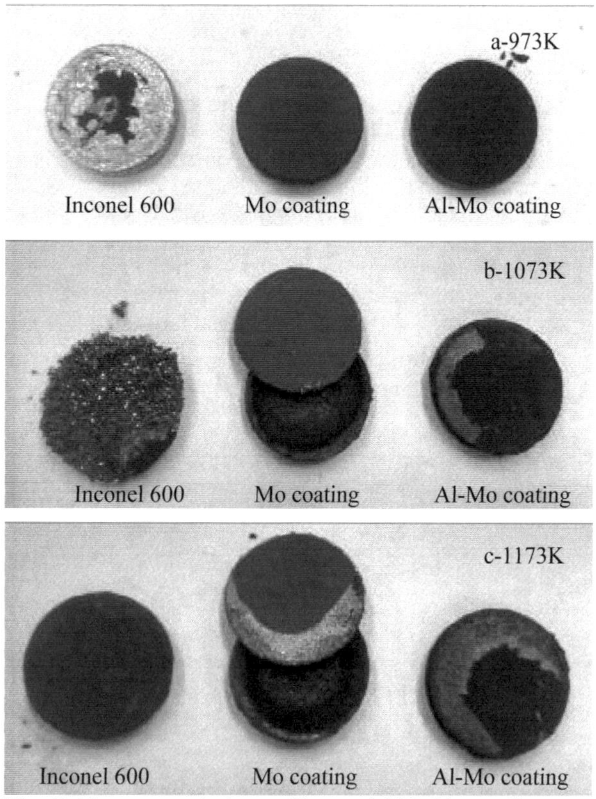

Fig. 5. Macro morphology of Inconel 600, Mo coating and Al-Mo coating with NiCrAlY bonding layer after sulfidation at 973K (a), 1073K (b) and 1173 K (c)

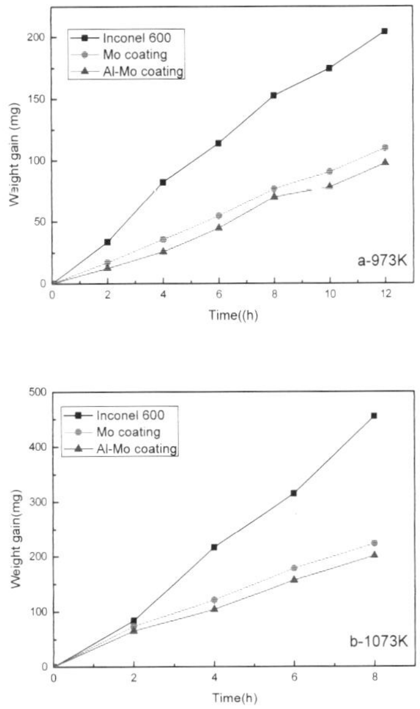

Fig. 6. Sulfidation kinetics of Inconel 600 alloy, Mo coating and Al-Mo coating with NiCrAlY bonding layer at 973K (a) and 1073K (b)

lower than that of Mo coatings. The failure modes of Mo coating and Al-Mo coating at 1073K and 1173K are very similar. Large area of Al-Mo coating and overall Mo coating faile after sulfidation for 2h and 1h, respectively.

It can been seen that Mo and Al-Mo coatings with NiCrAlY bonding layer could protect the substrate from sulfidation at 973K. However, the sulfidation rates increase significantly at higher temperature especially at 1173K. Sulfur could pass through Mo and Al-Mo coating, reacts with NiCrAlY layer and generates low melting point eutectic Ni_3S_2, resulting in the rapid failure of coatings.

3.3 Sulfidation behavior of Mo and Al-Mo coatings with Mo bonding layer

As shown in Fig. 7, at 973K and 1073K, the surface appearances of all the coated specimens are uniform and compact after sulfidation for 12h. However, a large part of Mo coating detaches from the substrate after sulfidation at 1173K for 8h. A yellow corrosion products layer is formed on the surface and thus the coating loses its integrity after sulfidation. Although Al-Mo coating also separates from the substrate at the edge of the coupon, the most area of the coating surface connects with the substrate after sulfidation at 1173K for 12h. Further test shows that most area of Al-Mo coating connects the substrate well after sulfidation at 1173K for 109h.

Sulfidation kinetics of Al-Mo coatings, Mo coatings and Inconel 600 alloy at temperatures of 973K, 1073K and 1173K are shown in Fig.8 with Mo bonding layer. The weight gains of Inconel 600 alloy are much higher than those of coated specimens because corrosion products formed on metal could not protect specimens from continuing corroding. The weight gains of coated specimens are about half of that of bare specimens at 973K and 1073K. For the coated specimens, the weight gains of Al-Mo coatings are lower than those of Mo coatings.

The microstructure of Al-Mo coating with Mo bonding layer is shown in Fig. 9 after sulfurdation at 1173K for 12h. It can be seen that no visible large cracks generate in the cross section of Al-Mo coating after

Fig. 7. Macro morphology of Inconel 600, Mo coating and Al-Mo coating after sulfidation at 973K (a), 1073K (b) and 1173 K (c) for 12h

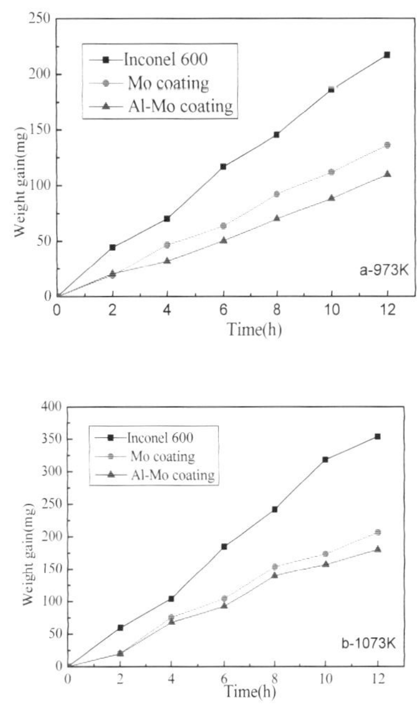

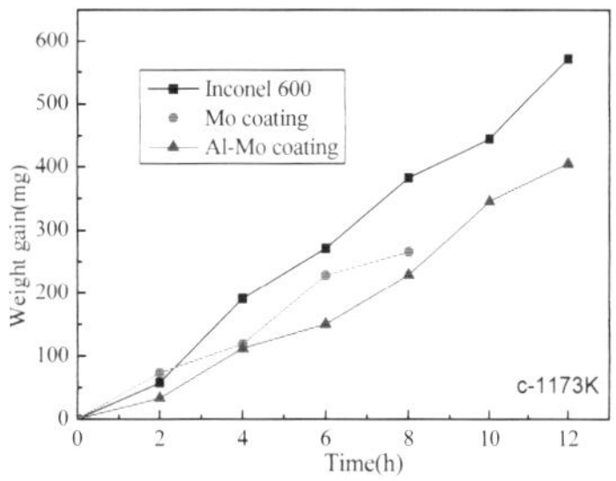

Fig. 8. Sulfidation kinetics of Al-Mo coatings, Mo coatings and Inconel 600 alloy with Mo bonding layer at 973K (a), 1073K (b)and 1173K (c)

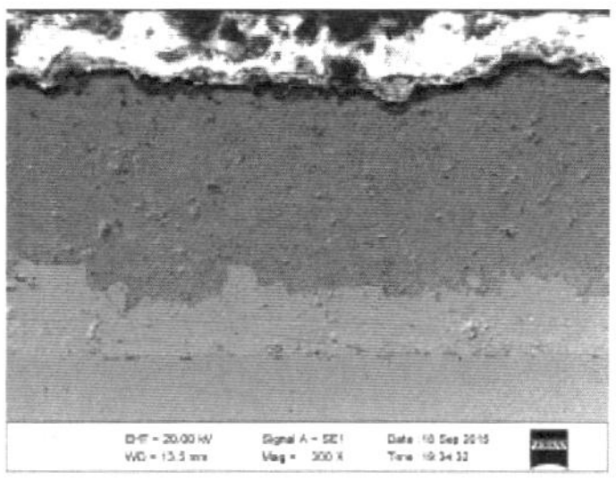

Fig. 9. Microstructure of Al-Mo coating with Mo bonding layer after sulfidation at 1173 K for 12h

the sulfidation. From the detailed examination for the coating microstructure, a compact corrosion products layer forms on Al-Mo coating surface, which could protect the specimen from continuing corroding.

3.4 Oxidation behavior of Mo and Al-Mo coatings with Mo bonding layer

After oxidation at 973K for 12h in the muffle furnace, loose faint yellow corrosion products generate on the surface of Mo coating. However, the surface of Al-Mo coating after sulfidation is compact and integrated. At the temperatures of 1073K and 1173K, after putting Mo coating in the muffle, due to the evaporation of volatile molybdenum oxide, massive smoke raises from the surface. Mo coating transforms into powders when it is taken out. As for Al-Mo coating, it separates from the substrate at the edge of the coating at 1073K, with a white corrosion products layer consisting of Al_2O_3 which could slow down the oxidation rate generating on the surface. The failure rate at the edge of Al-Mo coating also accelerates at 1173K.

4 Conclusions

Al-Mo and Mo coatings are deposited by plasma spraying with NiCrAlY and Mo coatings as the bonding coat. The sulfurdation resistance of the coatings are compared. The oxidation behavior is also investigated considering the gas atmosphere of gas burner. It is found that NiCrAlY is not compatibly used as bonding layer because of the formation of low melting point eutectic Ni_3S_2. The weight gains of plasma-sprayed Al-Mo coatings are lower than those of Mo coatings and Inconel 600 alloy in sulfur vapor especially at 973K and 1073K, which could be caused by the compact corrosion products layer. The oxidation resistance of the Al-Mo coatings with Mo bonding is higher than that of Mo coatings because of the fomation of protective Al_2O_3. Correspondingly, Al-Mo coatings with the Mo bonding layer are promising candidates for the corrosion protection for steels under typical coal gasification condition.

Fig. 10. Macro morphology of Inconel 600, Mo coating and Al-Mo coating after oxidation at 973K (a), 1073K (b) and 1173 K (c)

Acknowledgements

The authors gratefully appreciate the support of National Natural Science Foundation (No. 51275172), Science and Technology Commission of Shanghai Municipality Project (No. 14DZ2261205), Aviation Fund (2015ZES7001, 2013ZFS7001) and Shanghai Pujiang Program (15PJD009).

References

[1] Tian X, Qin C, Xu H, Cao J: Cracking causes of the burner made of Inconel 600 in a gasifier. Oxidation and protection, 35(2014), pp. 956/8.

[2] Mrowec S: The problem of sulphur in high-temperature corrosion. Oxidation of Metals, 44 (1995), pp. 177/209.

[3] Wendler B, Danielewski M, Dabek J: Modern refractory AlMo and AlMoSi coatings on steels with diffusion barrier.Thin Solid Films, 459 (2004), pp. 178/82.

Effect of Substrate Preheating on the Bonding Strength of Detonation Sprayed WC10Co4Cr Coatings

Yang Gao, Yunlei Hou and Jianyi Gao, Dalian / CHINA

Although detonation thermal spraying is considered to be a good method for the preparation of WC-10Co4Cr coating, the depositing parameter especial the oxygen-fuel ratio observably influenced the properties of the coating. In this study a detonation thermal sprayed WC-10Co4Cr coatings were deposited on the low carbon steels substrates with no preheated and preheated 300°C, respectively. The microstructure, hardness, bonding strength, and fracture behavior of the coating were investigated by means of Field Emission Scanning Electron Microscope (FESEM), X-ray diffraction (XRD) and the pull-off pin test. The results demonstrated that with increases the temperature of substrate result in the hardness of the coating increased. The low strength of binder CoCr phase in the coating plays a decisive role in the fracture toughness of the coating. Preheating decreased the hardness of the substrate, which makes the WC particles more easily embedded in the substrate, as well as reduce the stress between the interface of the substrate and the coating which is beneficial to enhance bonding strength. Furthermore the fracture surface and particle impacting and flattening on the substrate were analyzed.

1 Introduction

Tungsten-carbide based cermets coating WC-10Co4Cr with metallic cobalt-chromium binders are frequently used in the surface modification for the metallic mechanical parts to enhance wears and appropriate corrosion resistance [1-3]. The hard WC particles in the coatings play the role of high hardness and wear resistance, while the metallic binder CoCr supplies the necessary coating toughness. However, the WC phase in the powder easy undergoes decarburization during the thermal spraying process [4, 5], which may lead to degrade properties of the coating. Therefore the deposition method was limited to the flame having high jet speed and suitable temperature. Among thermal spraying methods the high-velocity oxygen-fuel (HVOF) and/or detonation spray with a high deposition efficiency and has demonstrated to be most efficient techniques to deposit high performance coatings due to its supersonic jets and suitable flame temperature of 2800-3200°C [4-7].

Detonation thermal spray is an intermittent spraying, a particles impact onto the substrate with a certain frequency pulse. The feedstock powder was heated and accelerated in the tubular barrel caused by explosion energy. Detonation spraying can achieve supersonic jet by flammable gas mixture of fuel and oxygen ignites in the tubular barrel combustion chamber (usually with tube length of 1300 mm and diameter of 25-30mm) and near the exit of the barrel. The characteristic of detonation thermal spray is that its supersonic jet is not produced by high pressure and high flow fuel and oxygen combustion, instead detonation thermal spray used a low pressure and low flow gas fuel and oxygen, for example, the oxygen and acetylene flows were 30-50 SLPM enough for detonation spraying. Because of such characteristics, the total enthalpy of the detonation spraying was much low than that HVOF and additional cooling purge gas make the enthalpy of the detonation wave are little influence on the substrate, usually the temperature of substrate would below 150°C during the process, such it was beneficial for spraying small or thin wall parts. But on the other hand, since the detonations thermal spray usually adopts acetylene as a fuel that having high combustion temperature than HVOF using propane or propylene as a fuel. Moreover, detonation thermal spraying can be deposited coating in the varied broader range of oxygen-fuel (OF) ratio, therefore the decarburization of WC particle were consistently observed in the detonation thermal spray, especially at high OF ratio [5] or using a small sized WC particles of feedstock [4]. The WC-Co or WC-10Co4Co coating deposited by the detonation spray may consists of multiphase such as hard phase has WC, W_2C and the metallic binder matrix contain CoCr or solid solution of W and C in Co that was determined by the extent of the decarburization of WC particle in the powder during the coating process. Diversiform microstructure of the coating would influences properties, such as hardness, wear resistance, fracture toughness and adhesion strength. In the various properties of the coating, the adhesion strength of coating is particularly important to evaluate the performance and reliability of coatings in engineering applications. Many studies has demonstrated that adhesion is not a constant in practical applications, but rather a complicated property depend on loading conditions, coating thickness [7-8] and different parameters such as grit blasting to roughen the substrate surface [8]. Furthermore, the residual stresses due to the mismatch in thermal and mechanical properties between coatings and substrate also is an important factor [9]. In the process of detonation spraying, the molten or semi-molten particles collide on the surface of substrate and its flattening behavior could affect the performance of the coating particular adhesion strength between the coating and substrate, while the preheating substrate would change the flattening and solidify behavior of the molten particles [10-11]. However, there are only few literatures about the influence of the preheating substrate on the bonding strength of tungsten carbide coating, in particular, the fracture interface and morphology of coatings. In this article the OF ratio was controlled in a certain range to fabricate the coating by change the substrate preheating temperature, and the microstructure, hardness and adhesion of the coating were investigated. Furthermore the fracture surface and particle impacting and flattening on the substrate were analyzed.

2. Experimental Materials and Methods

2.1 Feedstock Powder and Coatings

It is well known that the tungsten carbide bases feedstock powder such as WC-10Co4Cr for the thermal spray used has many fabricating methods, for example the fused-crashed, agglomerated-sintering and sintered-crashed etc. Taking into account the decarburization and deposition efficiency of the powder used for detonation spray, commercially available sintered-crashed WC-10Co4Cr powder was utilized in this investigation. The decarburization of the WC particles in the powder is related to the size of WC particle, small size WC particle easily lead to decarburization to form a phase of W_2C, or even to form W phase during thermal spraying process [12-14]. In this experiment, commonly the WC grains size of 0.5-2μm was selected. **Figure 1** shows the morphologies of the feedstock powder and its cross sectional structure in interior of the particle, the cuboids WC grains can be observed.

Fig. 1. the morphologies of the feedstock powder fabricated by fused- crashed. (b) High magnification of the cross-sectional of the powder.

Ukraine DNB-3 detonation equipment was adopted. Acetylene and oxygen was employed as flammable gas, while nitrogen as carrier gas to introduce the sprayed powders into the gun, as well as purge gas. Samples of low carbon steels with size of 45mm×45mm×5mm were prepared as a substrate. Prior to the detonation spray, the substrate was grit-blasted with brown alumina grit of #36 mesh size to increase the substrate surface roughness to approximately 5.5μm Ra. The WC-10Co4Cr coatings with approximate thickness of 0.2-0.3mm were directly deposited on grit-blasted substrate using the detonation spraying process and spraying parameter is shown in Table 1. In consideration of the decarburizing of WC grains in the detonation thermal spraying, especially in the case of a high OF ratio, a lower ratio of OF=1.1 was mainly selected in this experiment, in addition to compare the microstructure of the coating also a high ratio of OF=2.0 spraying parameters was carried out. To evaluate the effect of preheating substrate on the bonding strength, the coatings were deposited on the substrate at room temperature and 300℃ preheated, respectively.

Table 1 Parameters for detonation spray WC10Co4Cr

Items	Setting values
Explosive frequency	6 times/second
Fuel flow (acetylene)	30 SLPM
Oxygen flow	32 SLPM
Stand off	180 mm
oxygen/fuel ratio	1.1
Feed rate of powder	100 g/min
Moving speed of gun	1000 mm/min

2.1 Bonding Strength Tensile Test

In general, the detonation sprayed WC-Co coatings has very high bonding strength which probably exceed 50MPa [15-16], thus high bonding strength value would lead to measure it become difficult if by using the tensile test with standardized sample were joined with the cylindrical counter parts using an adhesive of epoxy resin [17]. In this study the bonding strength of the WC-10Co4Cr coatings was measured by the pull off conical cylinder pin method with the application of a force directly on the coating–substrate interface without using epoxy adhesive. The schematic diagram of this tensile test is shown in **Figure 2**. The coating was deposited on the surface of a conical pin (d=4mm) and fixed inserted a substrate made by low carbon steels. To avoid the coating was torn, a piece of steel as a reinforcement bolster was adhered on the top of the coating by the epoxy resin. Three measurements were carried out for each spraying condition. After tensile test the fracture surface was investigated by using SEM image to estimate fractured location.

2.3 Characteristics and Hardness of the Coating

Olympus GX51 metallographic microscope and Carl Zeiss SUPRA™ 55 field emission scanning electron microscope (FESEM) were used to observe the microstructure of the coatings. The crystalline structures of WC-10Co4Cr powder and resulting coatings were investigated by using a Rigaku D/MAX - 3A (Japan) X-ray diffractometer operating with Cu Kα (λ = 0.154056 nm) radiation. The analyzed range of the diffraction angle 2θ was between 20° and 85° by a step width of 0.02°. microhardness of the WC-10Co4Cr coatings was obtained by using a Wolpert Wilson Instruments 401 WVA hardness tester (China) with Vickers indenter; microhardness was measured on the cross-section of the coatings with varied load from 10 to 1000 gf and hold time for 5s.

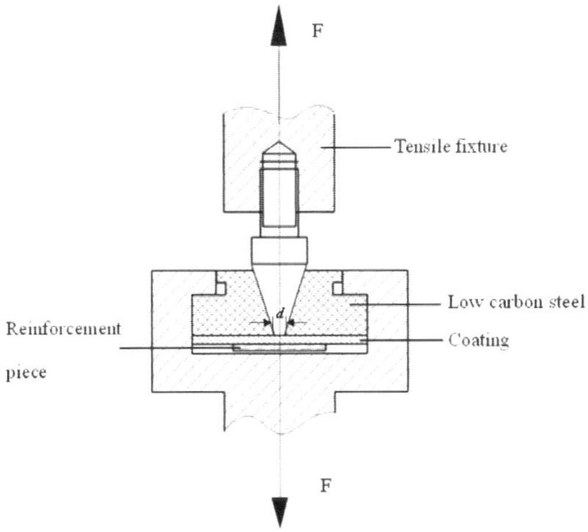

Fig. 2. Schematic diagram of the direct pull- off pin test.

3. Results and Discussions

3.1 Phases of the Powder and Coatings

Figure 3 shows the XRD patterns of the WC-10Co4Cr powder and its coatings deposited on the low carbon steel substrate with initial non-preheated and 300°C preheated, respectively. It can be found that the feedstock powder is composed of WC phase while no W2C phase. At the same time, the cobalt peak could be found but very weak. From XRD patterns of the coatings also can be found that WC phase is major phase, while a little W2C phase was also observed in the coatings, that indicating decarburization of WC were little in this detonation spray condition and preheated substrate was not influence the phase structure of the coatings. WC has high melting point (2867℃) and easy decarburizing in the high temperature flame when the particles excess temperature of 1250 °C [16]. Although the temperature of detonation flame is excess the 1250 °C, the lower OF parameter result in decarburization of WC was restrained. The metallic tungsten is not detectable in both coatings also relative to the lower OF. Notable quantity of binder CoCr alloy was not present in the coating, possibly because the binder of CoCr is easy to form nanocrystalline and/or amorphous phase in the coatings, which was been reported in the literature [14]. Formation of the coating also depended on the molten degrees of CoCr alloy and dissolves a fraction of the tungsten carbide. With increasing temperature of detonation flame by change the ratio of fuel/oxygen would form intermetallic Co_3W_3C or Co_6W_6C and precipitates W2C around the original WC grains [4, 5]. It should be noted that the degree of decarburization of D-gun sprayed WC-10Co4Cr coating in this parameter is much lower than that other thermal spraying technique [1, 4] and even same detonation sprayed [5], it was attributed to the low OF ratio and the sintered-crashed powders adopted.

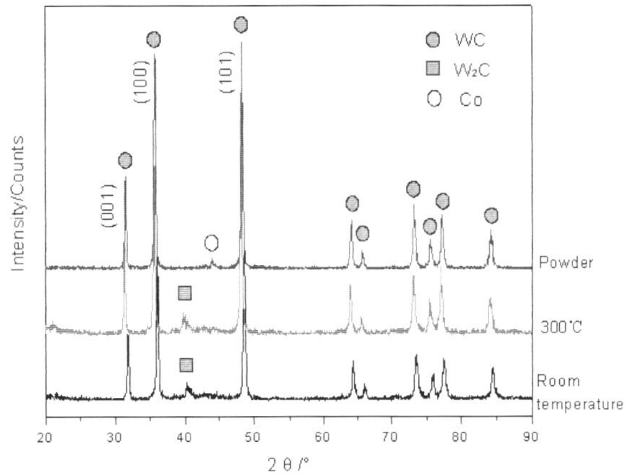

Fig. 3. XRD patterns of powder and coatings.

3.2 Microstructure and Hardness of the Coatings

The microstructure of WC-10Co4Cr coatings deposited on the low carbon steel substrate with non-preheated and 300℃ preheated were sectioned views obtained using SEM and results are provided in **Figure 4 and Figure 5.** It can be observed that the detonation sprayed WC-10Co4Cr coatings are dense, and the both microstructure of the coating is almost the same. However, detailed observe the microstructure of the coating from **Figure 5**, in the vicinity of the bottom of the substrate, the binder phase in the coating shows a relatively lighter color compared to the binder phase at the top of the coating with deposited on 300℃ preheated substrate. Corresponding to this porosity of the coating deposited on substrate with 300 ℃ preheated is slightly low. Others the WC grains in the coating retain their cuboids edges which similar to the powder feedstock (**Figure 1**). Some literature [5,12] have shown that decarburized W_2C particles in the coating showed a smooth edges and around it is bright background represents the decarburized region with different from dark background cobalt rich region that can easy identified by Back scattered SEM images. Combined with the results of XRD (**Figure 3**) it was showed that the WC particles in the powder were seldom decarburized in this process.

Figure 6 shows the Vickers hardness of the coating measured along the vertical direction from near the substrate to the surface of the coating with a measure step of 50 μm and at load of 300 gf. The hardness increased from 800 kg/mm^2 near the substrate-coating interface to 1050 kg/mm^2 near the surface of the coating, and then a

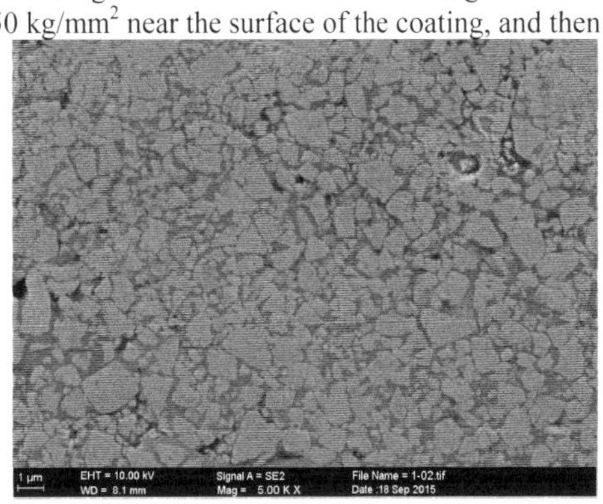

Fig. 4. Morphologies of the coating deposited on the no heating substrate.

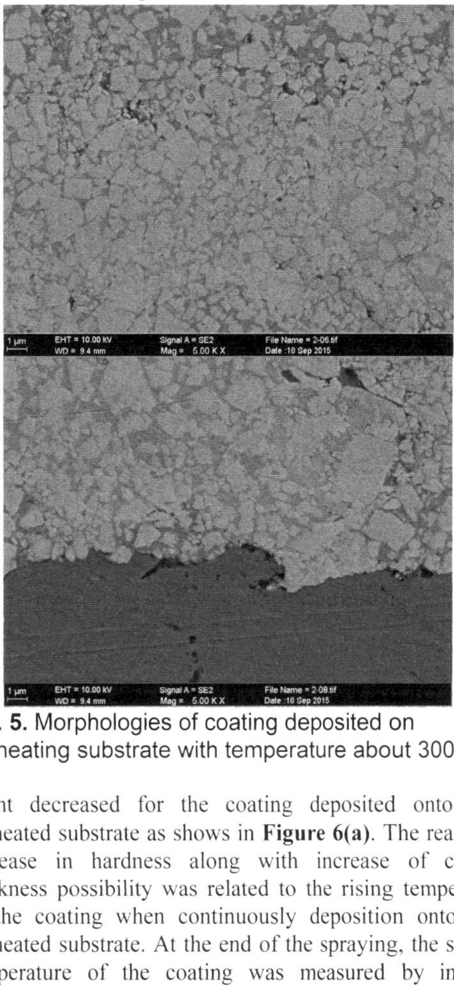

Fig. 5. Morphologies of coating deposited on preheating substrate with temperature about 300°C

slight decreased for the coating deposited onto non-preheated substrate as shows in **Figure 6(a)**. The reason of increase in hardness along with increase of coating thickness possibility was related to the rising temperature of the coating when continuously deposition onto non-preheated substrate. At the end of the spraying, the surface temperature of the coating was measured by infrared method, it was about 110 ℃. **Figure 6(b)** shows the distribution of hardness in the coating deposited onto the substrate with 300℃ preheated, compared with the **Figure 6(a),** the hardness continuously decreased from 1130 kg/mm^2 near substrate to 820 kg/mm^2 near the surface of the coating, the reason also can consider related to the decreasing temperature of coating from 300℃ preheated to final about 110 ℃, due to the detonation spraying is accompany with the cooling purge gas and a low enthalpy of the flame.

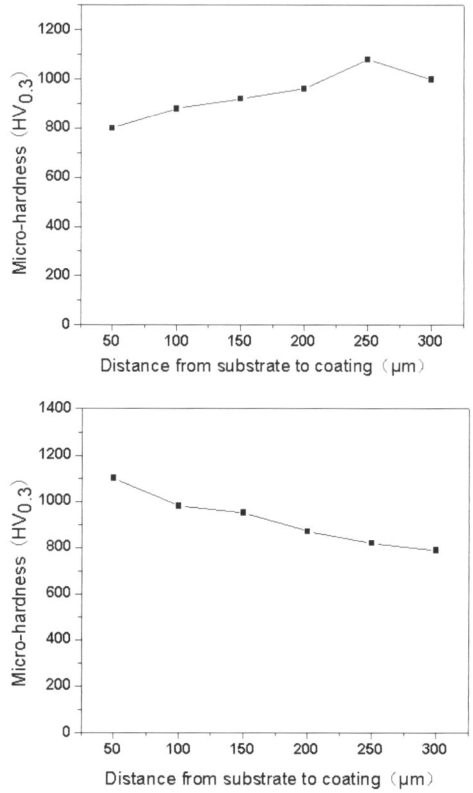

Fig. 6. Hardness along the thickness direction of coating. (a) Substrate temperature about 25°C (b) Substrate temperature about 300°C

The Vickers hardness number is defined as the load divided by the surface area of the indentation which is determined from the following equation.

$$DPH = 1.854p / L^2 \qquad (1)$$

Where p is applied load, kg, L is average length of diagonals, mm. The indentation on the coatings was estimated by taking indents on the polished cross-section of the coatings. The hardness of the coating was calculated from equation of (1) by measuring indentation length of diagonals. **Figure 7** displayed the hardness calculated at load of 10, 25, 50, 100, 200, 300, 500, and 1000 gf, respectively. The hardness number rapidly decreased with increase applied load at range of 10, 25, 50 gf indicated that small loads result in the deformed indentation included the partial elastic deformation. While the hardness number slow decreased from 1000 kg/mm^2 to 900 kg/mm^2 with

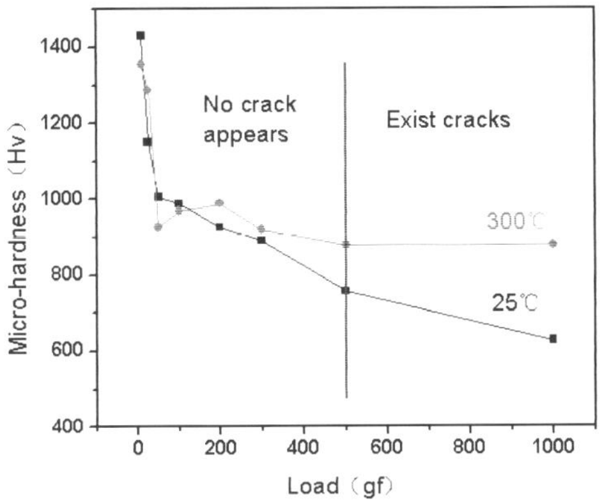

Fig.7. Hardness number is related to the measured load.

increasing the load at range of 100, 200 and 300 gf and no corner cracks were discernable in this range. And the corner cracks presented at a load higher than 500 gf. SEM observations of crack paths generated from indentation test shown in **Figure 8** indicate that cracks preferentially propagated through the CoCr binder regions indicating lower fracture toughness of these phases. The fracture toughness of the coating was related to the crack propagated length, the microstructure of the coating, Co content, WC phase size and extent of decarburization may influence the crack propagation [18]. It should be noted that since thermally spray coatings generally have an anisotropic microstructure, typically characterized by the so called lamellar structure cracks produced under indentation test mainly develop in the direction parallel to lamellae. The WC particles has high hardness and modulus of elasticity (E= 680 GPa) compare with the metal binder CoCr (E=211 GPa) [19], consequently the fracture almost present in the binder CoCr phase. Another reason of fracture almost present in the binder CoCr phase generally that the stress is compressive in the WC phase and tensile in the CoCr phase. This originates from the thermal stresses when the material is sintered between 1650 and 1720 K and cooled to room temperature due to the differences in coefficients of thermal expansion of WC and CoCr. At room temperature, WC is in compression with a stress of about 500 MPa and Co is under tensile stress of about 2500 MPa [20, 21].

3.3 Bonding Strength of the Coatings

The bonding strength of the detonation sprayed WC-10Co4Cr coatings on the low carbon steel substrate with no-preheated and preheated 300℃ were measured by the pull off method and results are shown in **Figure 9**. The values of bonding strengths are 58-74 MPa for the coatings deposited on the non-preheated substrate, while the bonding strength on the substrate with 300℃ preheated are 75-80 MPa that high than deposited on the non-preheated substrate.

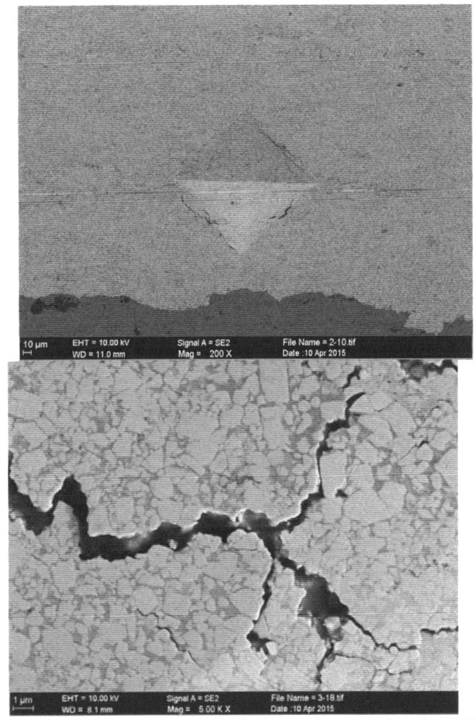

Fig.8. The corner cracks indentation at a load of 1000 gf on the cross-sectional coating (OF=1.1).

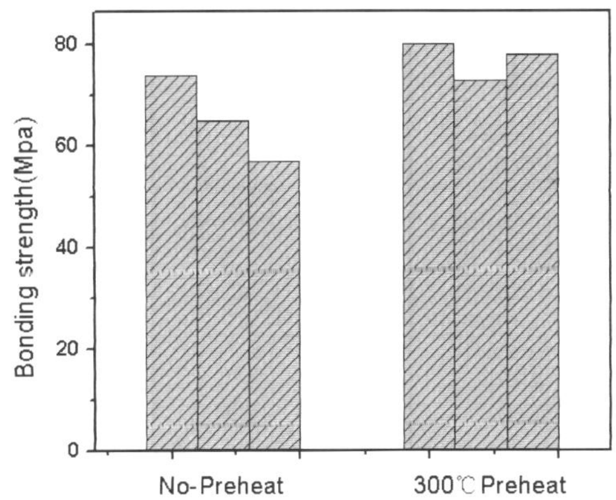

Fig.9. Bonding strengths of the coatings on the low carbon steel substrate with non-preheated and 300℃ preheated.

4. Conclusions

In this study WC-10Co4Cr coatings were deposited onto non-preheated and 300℃ preheated low carbon steel substrate by the detonation thermal spray with low OF ratio. The effect of substrate temperature on the microstructures, hardness, fracture morphology of the coating, bonding strengths and particle impacted on the low carbon steel substrate were investigated. The conclusion is as follows:

(1) The microstructure of the coating especially the WC grains were not change whether substrate is preheated to 300℃ or not. However, the on the near substrate the

porosity of the coating deposited on substrate with non-preheated is slightly high.

(2) The hardness number of the coating is related to the temperature of substrate or previous coating, with increases the temperature the hardness of the coating increased.

(3) When the indentation test was conducted and applied load exceed a criticality value, the crack propagation in the coating could be observed along the CoCr binder matrix, and the strength of the coating was determined mainly by the metallic binder matrix.

(4) Preheating substrate to 300℃ increases the bonding strength between the coating and the substrate. In this case, the fracture occurs mainly in the interior of the coating.

(5) Preheating reduced the hardness of the low carbon steel substrate, which makes the WC particles more easily embedded in the substrate, as well as reduce the stress between the interface of the substrate and coating which is beneficial to enhance bonding strength.

References

[1] Andrew Siao Ming Ang, Hugo Howse, Scott A. Wade, and Christopher C. Berndt, Development of Processing Windows for HVOF Carbide-Based Coatings, *Journal of Thermal Spray Technology, 2016,25 (1-2), 28-35.*

[2] L.M. Berger, Application of Hard metals as Thermal Spray Coatings, *Int. J. Refract Metal Hard Mater., 2015, 49(1), p 350-364.*

[3] J.A. Picas, M. Punset, M. Teresa Baile, E. Marti´n, and A. Forn, Tribological Evaluation of HVOF Thermal-Spray Coatings as a Hard Chrome Replacement, *Surf. Interface Anal., 2011, 43(10), p 1346-1353.*

[4] S.Y. Park, M.C. Kim, C.G. Park, Mechanical properties and microstructure evolution of the nano WC–Co coatings fabricated by detonation gun spraying with post heat treatment, *Materials Science and Engineering A 449–451 (2007) 894–897.*

[5] Suresh Babu, Bikramjit Basu, G. Sundararajan, Processing–structure–property correlation and decarburization phenomenon in detonation sprayed WC–12Co coatings, *Acta Materialia 56 (2008) 5012–5026.*

[6] Jerzy Robert Sobiecki, Janusz Ewertowshi, Tomasz Babul, Tadeusz Wierzchon, Properties of alumina coatings produced by gas-detonation method, *Surface and Coatings Technology 180 -181 (2004) 556–560.*

[7] Tong Yul Cho, Jae Hong Yoon, Jae Young Cho, Yun Kon Joo, Jin Ho Kang, Shihong Zhang, Hui Gon Chun, Soon Young Hwang, Sik Chol Kwon, Surface properties and tensile bond strength of HVOF thermal spray coatings of WC-Co powder onto the surface of 420J2 steel and the bond coats of Ni, NiCr, and Ni/NiCr, *Surface & Coatings Technology 203 (2009) 3250–3253.*

[8] M.H. Staia, E. Ramos, A. Carrasquero, A. Roman, J. Lesage, D. Chicot, G. Mesmacque, Effect of substrate roughness induced by grit blasting upon adhesion of WC-17% Co thermal sprayed coatings, *Thin Solid Films, Vol. 377-378, (2000), 657-664.*

[9] D. Mari, A.D. Krawitz, J.W. Richardson, W. Benoit, Residual stress in WC-Co measured by neutron diffraction, *Mater. Sci. Eng. A209, (1996), 197-205.*

[10] H.R. Salimijazi, L. Pershin, T.W. Coyle, J. Mostaghimi, S. Chandra, Y.C. Lau, L. Rosenzweig, and E. Moran, Effect of Droplet Characteristics and Substrate Surface Topography on the Final Morphology of Plasma-Sprayed Zirconia Single Splats, *Journal of Thermal Spray Technology, 2007,16 (2),291-299.*

[11] Yongang Zhang, Margaret Hyland, Anh Tuyet Tran, and Steven Matthews, Effect of Substrates Temperatures on the Spreading Behavior of Plasma-Sprayed Ni and Ni-20 wt.% Cr Splats, *Journal of Thermal Spray Technology, 2016,25 (1-2), 71-81.*

[12] Suresh Babu Pitchuka, Bikramjit Basu, and G. Sundararajan, A Comparison of Mechanical and Tribological Behavior of Nanostructured and Conventional WC-12Co Detonation-Sprayed Coatings, *Journal of Thermal Spray Technology, 2013, 22(4), 478-490.*

[13] J.K.N. Murthy, D.S. Rao, B. Venkataraman, Effect of grinding on the erosion behaviour of a WC–Co–Cr coating deposited by HVOF and detonation gun spray processes, *Wear, (2001), 249, 592–600.*

[14] M.S. Mahdipoor, F.Tarasi, C.Moreau, A.Dolatabadi, M.Medraj, HVOF sprayed coatings of nano-agglomerated tungsten-carbide/cobalt powders for water droplet erosion application, *Wear, 330-331(2015)338–347.*

[15] Hao Du, Weigang Hua, Jiangang Liu, Jun Gong, Chao Sun, Lishi Wen, Influence of process variables on the qualities of detonation gun sprayed WC-Co coatings, *Materials Science and Engineering A 408 (2005) 202–210.*

[16] M. Hadad, M. Hockauf, L.W. Meyer, G. Marot, J. Lesage, R. Hitzek, S. Siegmann, Adhesion evaluation of multilayered based WC–Co–Cr thermally sprayed coatings, *Surface & Coatings Technology 202 (2008) 4399–4405.*

[17] D.J. Greving, J.R. Shadley, and E.F. Rybicki, Effect of coating thickness and residual stresses on the bond strength of ASTM C633-79 Thermal spray coating test specimens, *J. Therm. Spray Technol., 1994, Volume 3(4), p371-378.*

[18] Pornthep Chivavibul, Makoto Watanabe, Seiji Kuroda, Kentaro Shinoda, Effects of carbide size and Co content on the microstructure and mechanical properties of HVOF-sprayed WC-Co coatings, *Surface & Coatings Technology 202 (2007) 509–521.*

[19] Onur Sayman, Faruk Sen, Erdal Celik, Yusuf Arman, Thermal stress analysis of WC–Co/Cr–Ni multilayer coatings on 316L steel substrate during cooling process, *Materials and Design 30 (2009) 770–774.*

[20] Andrew M. Venter, Thilo Pirling, Thomas Buslaps, O. Philip Oladijo, Axel Steuwer, Tshepo P. Ntsoane, Lesley A. Cornish, Natasha Sacks, Systematic investigation of residual strains associated with WC-Co coatings thermal sprayed onto metal substrates. *Surface & Coatings Technology 206 (2012) 4011–4020.*

[21] D. Mari, A.D. Krawitz, J.W. Richardson, W. Benoit, Residual stress in WC-Co measured by neutron diffraction, *Mater. Sci. Eng. A209, (1996), 197-205.*

Corrosion testing of hydroxyapatite and hydroxyapatite – silicon oxide coated titanium

Gurpreet Singh, Hazoor Singh, Buta Singh Sidhu

In the current investigation, plasma spray technique was used for depositing hydroxyapatite (HA) and hydroxyapatite – silicon oxide (SiO_2) coatings on titanium (Ti) substrate. In HA-SiO_2 coating, 10 wt% SiO_2 was mixed with HA. The corrosion resistance of the uncoated, HA coated and HA + 10 wt% SiO_2 coated Ti was investigated by potentiodynamic polarization test after immersion in simulated human body fluid (Ringer's solution) for 24 hours. The as-sprayed and dipped samples were analyzed by X-ray diffraction (XRD), scanning electron microscopy (SEM) and energy-dispersive X-ray spectroscopy (EDX) analyses. The results of electrochemical study reveals that the HA coating improves the corrosion resistance of Ti. The addition of SiO_2 in HA further enhances the corrosion resistance HA coating on Ti. The results of the Scherrer's equation confirm the rise in crystallite size (τ) of the coated samples after corrosion testing in Ringer's solution.
Keywords: Hydroxyapatite; Silicon oxide; Corrosion; Plasma Sprayed

1 Introduction

Metallic bio-implants are usually fabricated from light metals or alloys such as titanium (Ti) and Ti alloys, because of their suitable mechanical properties and excellent biocompatibility [1, 2]. A bioimplant should have very high corrosion and wear resistance in highly corrosive body environment which includes blood and other constituents of the body fluid which encompass several constituents like water, sodium, chlorine, proteins, plasma and amino acids [3, 4]. Bio-implant should possess sufficient fatigue strength and fracture toughness to sustain the varying loading conditions to which they are subjected. A minimum service period of 15 to 20 years in older patients and more than 20 years for younger patients is expected from a bio-implant. The dissolution of surface oxide film and corrosion are the two mechanisms which introduce additional ions into the body. The excess release of ions from prosthesis can generate undesirable biological reactions leading to mechanical failure of implant [5]. A bio-implant should have excellent biocompatibility with human body so that it should not cause any adverse effects like allergy, maturation, inflammation and toxicity. The clinical success of bio-implant is not only achieved with excellent implant material, the other factors such as implant design, surgery techniques, host bone quality, load bearing and surface quality plays important role [6].

The surfaces of metallic implants are usually coated with bioactive material to improve the mechanical, chemical, and biological properties of implant material for biomedical applications [7]. Hydroxyapatite HA:$Ca_{10}(PO_4)_6(OH)_2$) is the most established bioactive material which has been used in the orthopedic and dental fields in clinical applications due to its excellent osteoconduction and chemical composition similar to that of the natural bone in comparison with other bioactive materials [8-11]. Osteoconductive HA coatings promote bone healing leading to the rapid biological fixation of implants [12]. However, HA have poor mechanical properties such as fretting fatigue, toughness, abrasive wear and adhesive strength. But in some high load bearing bones such as femoral and tibia bones adhesive strength between HA and metallic implants is important to ensure good bone bonding ability and high fracture toughness [13, 14]. The bonding ability of bone and muscle as well as crosslinking agent in connective tissues in bioactive materials has been improved by addition of silicon oxide [15]. One of the main functions of silicon is to partake in cellular development and gene expression. The silica content has a chemical function as the nature of bonding to bone of bioglasses relates to the in vivo solubility of these glasses [16]. The formation of apatite (bone like tissues) was found to be enhanced when plasma sprayed silicon substituted hydroxyapatite coated bio-implants were immersed in simulated body fluid [17]. The mechanical function of silica particle is to improve the strength of a hydroxyapatite coating by enhancing the crack arrest or crack deflection [16].

Among the numerous experimental deposition processes plasma spray process is the most commercially, well established preferred technique to deposit HA on metallic implants [18, 19]. The clinical success of biomaterials for their use as surgical implants has been investigated by performing various in vitro and in vivo studies. In vitro studies are performed to give an overview of the behavior of the material in simulated body fluid such as Hank's solution or Ringer' s solution. The in vivo tests are performed on animal models to evaluate the actual performance of the materials as per the standards and guidelines approved by FDA (Food and Drug Administration, USA). Furlong and Osborn who were the first to perform the clinical trials of HA-coated implants has been reported that HA coatings can successfully enhance clinical success [20].

In this work, Atmospheric Plasma Spray technique was used to deposit hydroxyapatite and hydroxyapatite – silicon oxide coatings on titanium substrate. The as-sprayed coatings were characterized by x-ray diffraction (XRD), scanning electron microscopy (SEM) and energy-dispersive X-ray spectroscopy (EDX) techniques. In vitro study has been performed to investigate the corrosion resistance of uncoated and coated titanium specimens by the Tafel extrapolation method in simulated body fluid (Ringer's solution). After the electrochemical corrosion testing the crystallinity and morphology of the exposed specimens was investigated by XRD, SEM and EDX. The outcomes of this work attempts to mark new dimensions for the reinforcement of HA with SiO_2 to control the corrosion of biomaterials in human body.

2 Experimental procedure

2.1 Materials

Medical grade HA powder (IFGL Bio Ceramics Limited, Kolkata, India) and silicon oxide (SiO_2) powder (Metroblue Industries, Madurai, India) with particle distribution of 57 – 200 µm and 15 – 150 µm respectively were used in this study. HA and SiO_2 powders were mechanically stirred and mixtures of HA + 10 wt% SiO_2 were prepared by stirring the powders in a ceramic pot for 20 min. Commercial available biomedical material unalloyed titanium (Ti-ASTM-F-67-00) was used as substrate. Specimens of Ti of size 15 x 15 x 3 mm were coated at 500 A arc current and 50 V arc voltage with HA and HA + 10 wt% SiO2 powders using plasma spray (Miller Spray System) at Anod Plasma, Kanpur, India. Before spraying the substrate surface was grit blasted with alumina of particle size 50 – 60 µm at a pressure of 5 bars for 2 minutes to roughen the surface and subsequently air blasted to remove any residual grit. Because a highly roughened substrate surface exhibited higher bond strength as compared to a smooth substrate surface [21]. The argon which was the primary gas was supplied at 57 slpm (standard liter per minute) and the flow rate of secondary gas hydrogen was 8 slpm. The powder was flown at 11 slpm and the spraying distance between the plasma torch and substrate was 75 mm.

2.2 Characterization of coatings

The phase structure of the both feedstock powders and coatings were analyzed by XPERT-PRO Xray diffractometer system. In the phase analysis, the radiation source was Cu Kα; the operating generator setting was 45 kV / 40 mA. The feedstock powder and coated samples were scanned over the 2θ range of 10° to 60° and 20° to 60° respectively. The crystallite size (τ) of the coated samples before and after corrosion testing in Ringer's solution has been calculated using Scherrer's equation [22]:

$$\tau = \frac{K.\lambda}{W.\cos(\theta)} \qquad (1)$$

where: K is dimensionless shape factor dependent on crystallite shape, $0.8 < K < 1.1$; $K = 0.94$ for FWHM (full width at half max) of spherical crystals with cubic symmetry; λ is the x-ray wavelength of monochromatic radiation ($\lambda_{\alpha 1} = 1.5405$ Å). W is the defined as the full width of peak from the intensity distribution pattern measured at half of the maximum intensity value. This value is the difference between two 2 θ values, in radians; and θ is the Bragg angle, in degree.

Microstructural investigation was carried out on the surfaces and polished cross-sections of the coatings by SEM (EVO MA 15 ZEISS) coupled with EDX. As-sprayed coatings were cut with a low speed precision saw at 75 rpm speed and mounted in hot resin using a hot mounting press, followed by polishing with emery papers of 220, 320, 400, 600, 800, 1000, and 2000 grades, and finally mirror finished by buffing using an alumina slurry solution on napped cloth. To achieve the desired conductivity for observation in SEM the gold sputter coating were applied to samples. Elemental analysis of the coatings was carried out using an EDX to display the distribution of elements in the coatings and to evaluate Ca/P ratios of both coatings as well as to provide evidence of the presence of silica particles in HA + 10 wt% SiO_2 coating.

2.3 Surface Roughness

The surface roughness of uncoated, HA coated and HA-SiO_2 coated Ti samples was measured by a roughness tester (SJ-201; MITUTOYO), using a filter of Gaussian type for a cut-off wavelength of 0.8 mm. Each reported value of roughness is the average of four measurements at various positions on the surface of the coated samples. The measured roughness parameters are Ra (the arithmetic mean of the departures of the roughness profile from the mean line) is the most commonly used roughness parameter.

2.4 Electrochemical Corrosion Studies

Electrochemical techniques are preferably used to measure the corrosion resistance of the implant material because of their quick response and ability to measure very slow corrosion rate. In in-vitro metallic corrosion testing, under wet conditions corrosion generally occurs in cell at the metal surface. A potentiodynamic polarization test was conducted to investigate the electrochemical corrosion resistance of the uncoated, HA-coated and HA-SiO_2 coated Ti specimens. A Potentiostat / Galvanostat (Series G-750; Gamry Instruments, Warminster, PA), interfaced to a computer and supported with Gamry electrochemical software DC105 was used to conduct the potentiodynamic polarization test. Ringer's solution with chemical composition (in g/L) as 9-NaCl, 0.24-$CaCl_2$, 0.43-KCl, and 0.2-$NaHCO_3$ at pH 7.2 was used as the electrolyte for simulating human body fluid conditions. Each specimen was stabilized by dipping just below the surface of Ringer's solution in the electrochemical cell for 24 hours. The Ti specimen and a saturated calomel electrode (SCE) were used as the working and reference electrode respectively whereas a graphite rod served as the counter electrode. The instrument measures and controls the potential difference between a non-current carrying reference electrode and one of the two current carrying electrodes (the working electrode). All tests were performed at a scan rate of 1 mV/sec and fresh solution was used for each experiment. Polarization curves were initiated at −250 mV to +250 mV relative to open circuit potential. The parameters for conducting the potentiodynamic scan to calculate the corrosion resistance of uncoated, HA coated and HA + 10 wt% SiO_2 coated Ti are listed in Table 1.

3 Results and discussion

3.1 XRD Analysis

The XRD pattern of HA powder, SiO_2 powder and HA + 10 wt% SiO2 powder published elsewhere confirms that all the powders are composed of crystalline phases and the major peaks of HA powder and SiO2 powder match the JCPDS cards 74-0566 and 86-1629 respectively [23]. In XRD pattern of HA + 10 wt% SiO2 powder H and S denotes the HA and silicon oxide peaks respectively. XRD patterns of plasma sprayed pure HA and HA + 10 wt% SiO2 coating on Ti are shown in Figure 1. The structure of pure HA and HA + 10 wt% SiO2 coatings is composed of HA phase. The degree of crystallinity of HA + 10 wt% SiO2 coating on Ti is less in comparison to HA coating on Ti as the peaks became less sharp and broadened. Because of the presence of high melting point silica particles near or in contact with HA particles facilitate the melting degree of HA particles. According to the Scherrer's equation the crystallite size of HA and HA + 10 wt% SiO2 coatings were 19.268 nm and 18.268 nm respectively. These results are in agreement with previous XRD patterns of HA and HA-SiO2 coatings [24].

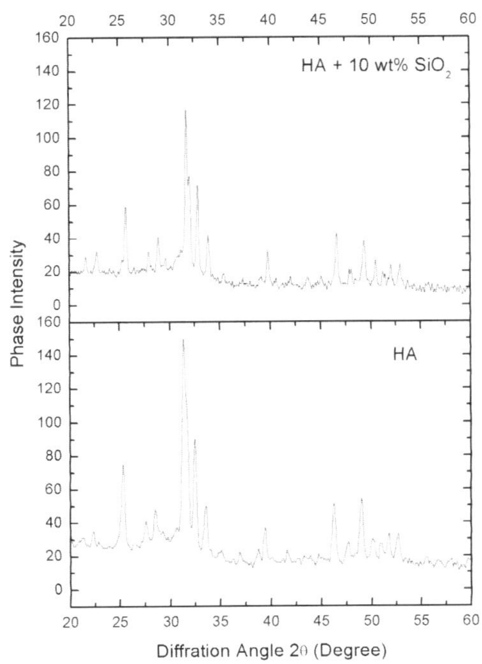

Fig. 1. XRD pattern of plasma sprayed HA coating and HA + 10 wt% SiO_2 coating on Ti

3.2 SEM / EDX analysis

3.2.1 Surface Analysis

The micrograph of HA and silicon oxide powders published elsewhere confirm that the HA powder has the spherical shape and silica powder has the angular and irregular shape [23]. The size distribution range in both the powders seems to be wider. It is worthwhile to mention that the particles size and shape influence the final microstructure of the plasma spray coating. The SEM micrograph along with EDX point analysis of an HA-coated specimen is shown in Figure 2 (a). The microstructure consists of well formed interconnected splats with uneven surface. Many pores were observed on the coating surface but no micro crack was observed on surface of as-sprayed sample. The EDX analysis HA coated specimen at three different positions confirmed the presence of significant amounts of calcium (Ca), phosphorous (P) and oxygen (O) elements which are the main elements of HA powder.

SEM micrograph of HA + 10 wt% SiO_2 coating surfaces in Figure 2 (b) shows that both powders melted enough to form a dense surface. The numbers of observed pores / voids were less in comparison to HA coated surface. The presence of silica particles decreases the porosity and makes the structure denser as clear from micrograph because the silica particles raise the temperature of nearby HA particles which decrease the number of un-melted particles of HA [15, 24, 27]. It is well known that porosity is less if the un-melted particles are less. No micro crack was observed on the surface of as-sprayed coating. The EDX analysis proves the presence of silica particles. The average Ca/P ratio was 1.68 and will remain in agreement with the guidelines described by the Food and Drug Administration and in the ISO standards [25, 26].

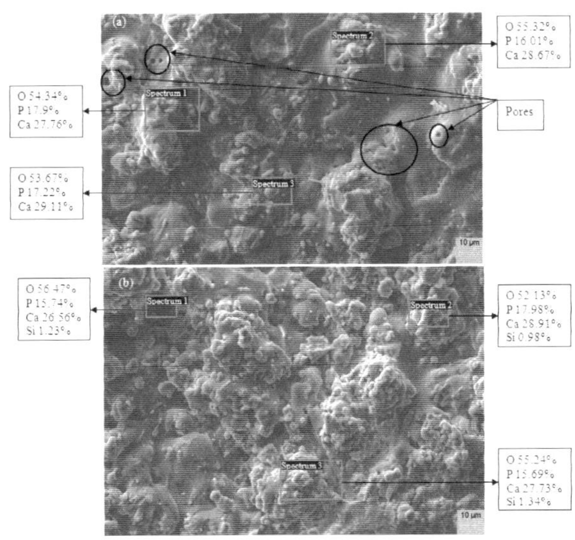

Fig. 2. SEM micrograph and EDX Analysis of Plasma Sprayed (a) HA coating (b) HA + 10 wt% SiO_2 coating on Ti

3.2.2 Cross-sectional analysis

SEM and EDX analysis at the cross-section of HA coated and HA + 10 wt% SiO_2 coated Ti is shown in Figure 3. The HA coating (Figure 3 (a)) has a typical splat-like laminar morphology where HA + 10 wt% SiO_2 coating (Figure 3 (b)) revealed a laminar morphology and appears smoother then HA coating. Average thickness of the coating is measured at four positions from the cross-sectional micrograph and is 100 ± 15 µm. The coating shows the good adhesive strength with substrate and no crack appears at the interface or within the coating. The EDX analysis show the presence of Ca, P and O elements in both coatings and presence of silicon in HA + 10 wt% SiO_2 coating.

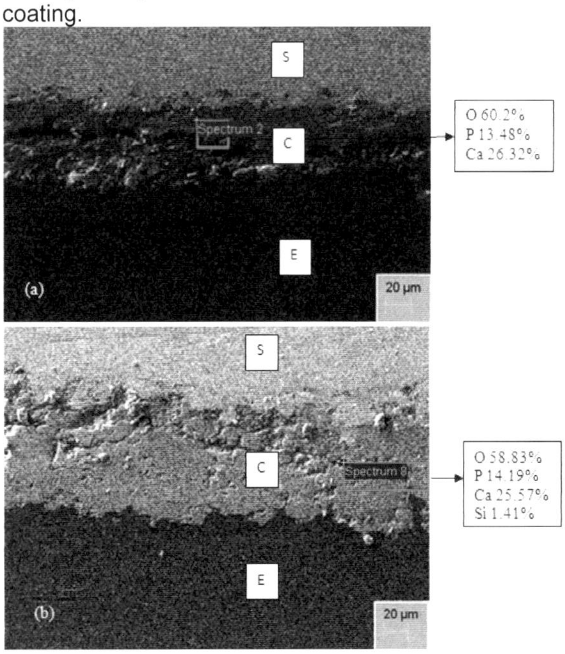

Fig. 3. SEM and EDX along the cross-section of Plasma Sprayed (a) HA coating (b) HA + 10 wt% SiO_2 coating on Ti (S, C, E, represents the Substrate, Coating and Epoxy respectively)

3.3 Surface Roughness

The average surface roughness (Ra) value for the uncoated, HA coated and HA + 10 wt% SiO_2 coated Ti substrates were 0.934 ± 0.2 µm, 5.87 ± 0.3 µm and 5.61 ± 0.3 µm respectively. Gross and Babovic [28] reported that plasma sprayed HA coating give a surface roughness of 4-6 µm while incorporating the partially melted particles. The surface roughness of HA coating was in agreement as reported by Gross and Babovic. High surface roughness of plasma sprayed HA coating leads to the increase in dis-solution rate and apatite precipitation [29]. Surface roughness, surface topography, surface energy and chemical composition have been reported as very important factors for implant tissue interaction and to affect the biocompatibility in clinical use [29, 30]. The bone anchoring and bio mechanical stability is found to be more in roughened surface implants [12]. High surface roughness will increase the coating adhesion and body-fluid interface, and thus increase the dissolution rate and apatite precipitation [30, 31].

3.4 Corrosion Behavior

The potentiodynamic scan of the uncoated, HA coated and HA + 10 wt% SiO_2 coated Ti samples in Ringer's solution was conducted. The Tafel extrapolation method was used to determine the corrosion parameters such as anodic Tafel slope (βa), cathodic Tafel slope (βc), corrosion potential (E_{Corr}), and corrosion current density (I_{Corr}) from the potentiodynamic curves. The results of these corrosion parameters are shown in Table 1. The chance of corrosion in a material depends upon the corrosion current density (I_{Corr}) at a given potential, materials with higher value of I_{Corr} will be more prone to corrode. The result of Tafel slope values shows that corrosion current density of uncoated Ti specimen in Ringer's solution is (I_{Corr} = 2.12 µA, E_{Corr} = -73.10 mV) is higher than HA coated (I_{Corr} = 1.6 µA, E_{Corr} = -113.0 mV) and HA + 10 wt% SiO_2 coated (I_{Corr} = 1.43 µA, E_{Corr} = -246.0 mV) Ti sample in Ringer's solution. This dissolution is related to the chemical composition and crystallographic characteristics of the coatings. Corrosion of HA coated substrate is concerned with incursion of chloride ion and water into the coating and then subsequent electrochemical reactions at the interface of HA and metallic substrate [32]. Zhang et al reported the corrosion mechanism of HA coated Ti substrate having pores and cracks in two steps[61]: Hydrogen ions (H+) are produced at the interface area where corrosion occurs as shown in equation 2. It is then followed by the dissolution of HA in the high concentration area as shown in equation 3.

$$Ti + 2H_2O \rightarrow TiO_2 + 4H + + 4H^{e-} \quad (2)$$

$$Ca_{10}(PO_4)_6(OH)_2 + 2H^+ \rightarrow 10Ca^{2-} + 6PO_4^{3-} + H_2O \quad (3)$$

So the analysis of Tafel slope values indicates that the HA + 10 wt% SiO_2 coated specimens with lower I_{Corr} values are more corrosion resistant than uncoated and HA coated Ti specimens in Ringer's solution. The earlier corrosion studies on HA coatings shows the same kind of results [33-36]. The results are in agreement with previous studies [37-38]. Gallardo et al reported that the SiO_2 coating have shown very good electrochemical behavior to prevent corrosion [39]. These results demonstrates that the presence of SiO_2 in HA act as a protective agent to act against corrosion and diffusion of ions. Gibson et al. demonstrated that the mixing of silicon in HA improves the osteoblast cell activity and apatite formation in SBF in comparison to the pure HA [40]. Patel et al. conducted a in vivo study and showed a significant rise in the bone apposition on the surface of HA-Si implant in comparison to pure HA when implanted in rabbit [41].

After electrochemical corrosion testing in Ringer's solution, the XRD scans of HA coated and HA + 10 wt% SiO_2 coated Ti (Figure 4) showed an interesting behavior. Both the coatings appeared more crystalline and the intensity of XRD peaks found to be increased. The crystallite size of HA coating and HA + 10 wt% SiO_2 coatings on Ti were calculated to using Scherrer's equation and were increased to 21.23 nm and 19.13 nm respectively. The sharp peaks after immersion indicates the dissolution of the amorphous phases. It is reported that amorphous phases are more soluble then crystalline HA as they encourage the early bone growth [42]. The dissolution of HA was favorable for the early stages of transformation of biological equivalents that act as mediator between osteoclast and osteoblast differentiation [43]. The higher crystalline coating leads to longer implant life, while some implant manufacture prefer a faster dissolving coating to enhance the bone growth [44]. It has been reported in in-vivo studies that the phase purity, crystallinity and microstructure of HA coatings affects the biological response of HA coating [45].

The compositional changes, if any, on the surfaces of the exposed specimens were further examined by SEM and EDX. The surface morphology of the HA coated samples (Figure 5 (a)) appears to be changed after 24 hour immersion in Ringer's solution. The size of pores / voids appears to be increase but no crack was observed. Each pore acts as active site for entrapping surrounding corrosive impurities [46]. However the porous coatings appear to be prospective implant materials for tissue engineering application because as the cell in growth propagates in to the pores it will assist the tissue regeneration [47]. The morphology of HA + 10 wt% SiO_2 coating (Figure 5 (b)) changed to flattened particles of irregular shape after exposure to the corrosion testing in Ringer's solution. No cracks were found HA + 10 wt% SiO_2 coated specimens after 24 hour immersion in Ringer's solution. The preferential site of corrosion on the substrate surface depends on the site of pores because the pore provides the conducting path between the corrosive medium and the metallic substrate. The formation of denser and less porous coatings shifts the corrosion current density to lower values [48].

EDX analysis confirms the presence of Ca, P, and O elements in both coatings and the presences of silicon in HA + 10 wt% SiO_2 coated Ti specimens. EDX analyses of HA and HA + 10 wt% SiO_2 coated specimen shows that the percentage of oxygen increases and the Ca and P (at %) decreases by taking the average of elemental composition at three spectrums after 24 hour immersion in Ringer's solution. The earlier literature reported that the decrease in the values indicates that phosphate accumulate on the surface which suggests that incongruent dissolution of the HA has taken place [34]. No constituent of substrate was found on the surface of any coatings after the immersion for 24 hour in ringer's solution.

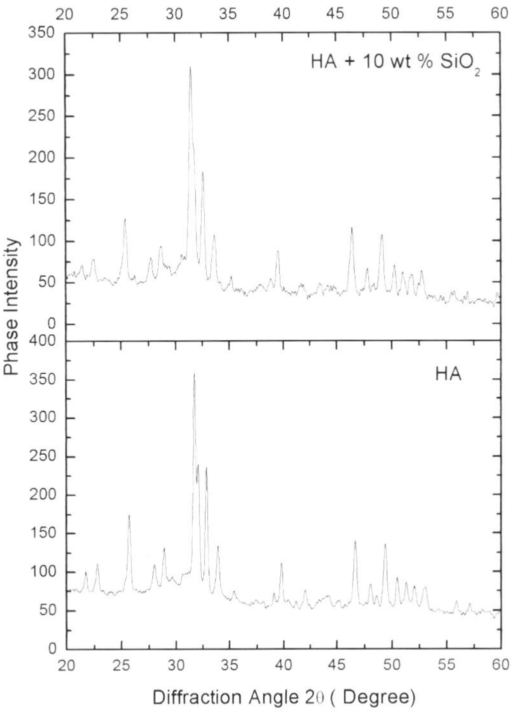

Fig. 4. XRD pattern of plasma sprayed Pure HA coating and HA + 10 wt% SiO_2 coating on Ti after corrosion testing in Ringer's solution

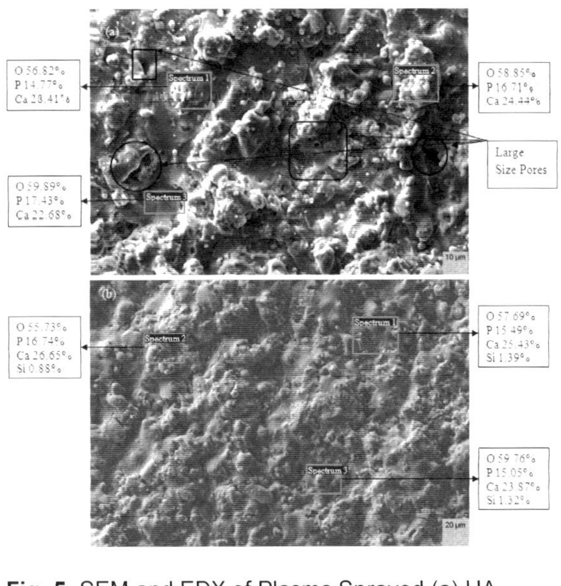

Fig. 5. SEM and EDX of Plasma Sprayed (a) HA coating (b) HA + 10 wt% SiO_2 coating after immersion in Ringer's solution

4 Conclusions

In the present study, plasma spray technique was used to deposit HA and HA + 10 wt% SiO_2 coatings on Ti implant material. The plasma sprayed HA coating exhibited higher surface roughness (R_a = 5.87 ± 0.3 μm) than HA + 10 wt% SiO_2 Coating (R_a = 5.61 ± 0.3 μm). The electrochemical study showed an improvement in corrosion resistance of Ti after the deposition of plasma sprayed HA + 10 wt% SiO_2 and HA coating in Ringer's solution. After electrochemical corrosion testing the intensity of XRD peaks of both coatings increased and owing to the dissolution of amorphous phase during immersion in Ringer's solution. The crystallite size of HA coating and HA + 10 wt% SiO_2 increased from 19.26 nm to 21.23 nm and 18.26 nm to 19.13 nm respectively. A comparison of the SEM micrographs of both plasma sprayed coatings before and after corrosion testing in Ringer's solution showed that the sizes of pores / void on the surface of HA coating appears to be increased and making it more porous where as the morphology of HA + 10 wt% SiO_2 coating changes to flattened particles of irregular shape results in the small grain boundaries, hence making the coating less porous and more corrosion resistant. The decrease in the value of Ca and P confirms the accumulation of phosphate on the coating surface that incongruent the dissolution of HA. No crack was observed on the surface of HA and HA +10 wt% SiO_2 coated Ti even after immersion in Ringer's solution.

Future in-vivo studies of plasma sprayed HA and HA + 10 wt% SiO_2 coated Ti and a complete interpretation of these results can help in assessing their use in clinical applications.

References

[1] Bardos DI (1990) Titanium and titanium alloys. In: William D, editor. Encyclopaedia of medical and dental materials. Oxford: Pergamon Press, pp.360–365.

[2] Boyer R, Welsh G, Colling EW. (Eds.) (1994) Materials properties handbook: titanium alloys. Ohio: ASM International pp. 181–219.

[3] Pourbaix M (1984) 'Electrochemical corrosion of metallic biomaterials', *Biomaterials*, Vol. 5, pp. 122–134.

[4] Lawrence SK and Gertrude Shults M (1925) 'Studies on the relationship of the chemical constituents of blood and cerebrospinal fluid', *Journal of Experimental Medicine,* Vol. 42 No. 4, pp. 565-591

[5] Davis JR (2003) Handbook of materials for medical devices, ASM International.

[6] Albrektsson T, Branemark PI and Hansson BO (1981) 'Lindstrom Journal of Osseointegrated titanium implants', *Acta Orthopaedica Scand,* Vol. 52, pp.155–170.

[7] Liu X, Chu PK and Ding C (2004) 'Surface modifications of titanium, titanium alloys and related materials for biomedical applications', *Mater Sci Eng,* Vol. 47, pp. 49–121.

[8] Edwards JT, Brunski JB and Higuchi HW (1997) 'Mechanical and morphologic investigation of the tensile strength of a bonehydroxyapatite interface', *Journal of Biomedical Material Research,* Vol. 36, pp. 454–468.

[9] Hench LL and Polak JM (2002) 'Third generation biomedical materials', *Science,* Vol. 295, pp. 1014–1017.

[10] Paul W and Sharma CP (1999) 'Development of porous spherical hydroxyapatite granules: Application towards protein delivery', *Journal of Material Science: Materials in Medicine,* Vol. 10, pp. 383–388.

[11] Vijayalakshmi U, Balamurugan A and Rajeshwari S (2005) 'Synthesis and Characterization of Porous Silica Gels for Biomedical Applications', *Trends Biomaterial: Artificial Organs*, Vol. 18, No. 2, pp.101–105.

[12] Guehennec LL, Soueidan A, Layrolle P and Amouriq Y (2007) 'Surface treatments of titanium dental implants for rapid osseointegration', *Dental Materials*, Vol. 23, No. 7, pp. 844–854.

[13] Kato H, Nakamura T, Nishiguchi S, Matsusue Y, Kobayashi M, Miyazaki T, Kim HM and Kokubo T (2000) 'Bonding of alkali- and heat treated tantalum implants to bone', *Journal of Biomedical Materials Research,* Vol. 53, pp. 28–35.

[14] Kokubo T, Miyaji F and Kim HM (1999) 'Spontaneous formation of bonelike apatite layer on chemically treated titanium metals', *Journal of American Ceramic Society,* Vol. 79, pp. 1127–1136.

[15] Morks MF (2008) 'Fabrication and Characterization of Plasma-Sprayed HA/SiO_2 Coatings for Biomedical Application', *Journal of the Mechanical Behavior of Biomedical Materials,* Vol. 1, pp. 105 – 111.

[16] Hijon N, Victoria Cabanas M, Juan, Pena, ValletRegi and Maria (2006) 'Dip Coated Silicon Substituted Hydroxyapatite Films', *Acta Biomaterials* Vol. 2, No. 5, pp. 567–574.

[17] Gross KA and Berndt CC (1998) 'Thermal Processing of Hydroxyapatite for Coating Production', *Journal of Biomedical Materials Research*, Vol. 39, pp. 580–587.

[18] Yang CY, Lin RM, Wang BC, Lee TM, Chang E, Hang YS and Chen PQ (1997) 'In Vitro and In Vivo Mechanical Evaluations of Plasma-Sprayed Hydroxyapatite Coatings on Titanium Implants: The Effect of Coating Characteristics', *Journal of Biomedical Material Research* , Vol. 37, pp. 335-345.

[19] Gu YW, Loh NH, Khor KA, Tor SB and Cheang P (2002) 'Spark Plasma Sintering of Hydroxyapatite Powders', *Biomaterials*, Vol. 23, pp. 37–43.

[20] Furlong RJ and Osborn JF (1991) 'Fixation of Hip Prostheses by Hydroxyapatite Ceramic Coatings', *Journal of Bone Joint Surgery*, Vol. 73 (B), pp. 741–746.

[21] Nimb L, Gotfredsen K and Steen JJ (1993) 'Mechanical Failure of Hydroxyapatite-Coated Titanium and Cobalt-Chromium-Molybdenum Alloy Implants', *An Animal Study. Acta Orthopaedica Belgica*, Vol. 59, pp. 333–338.
[22] Bouyer E, Gitzhofer F and Boulos MI (2000) 'Morphological Study of Hydroxyapatite Nanocrystal Suspension', *Journal of Materials Science: Materials in Medicine*, Vol. 11 No. 8, pp. 523–531.
[23] Singh G, Singh H and Sidhu BS (2013) 'Characterization and Corrosion Resistance of Plasma Sprayed HA and HA–SiO_2 Coatings on Ti–6Al–4V', *Surface and Coatings Technology*, Vol. 228, pp. 242-247.
[24] Fernandez J, Gaona M and Guilemany JM (2004) 'Tribological Study of Plasma Hydroxyapatite Coatings', *Bioceramics 16, Key Eng Mat*, Vol. 254-256, pp. 383-386.
[25] FDA (1992) Calcium Phosphate (Ca-P) Coating Draft Guidance for Preparation of FDA Submissions for Orthopedic and Dental Endosseous Implants, Washington, DC: Food and Drug Administration, pp. 1–14.
[26] International Organization of Standards (1996) Implants for Surgery: Coating for Hydroxyapatite Ceramics, ISO, pp.1–8.
[27] Singh G, Singh H and Sidhu BS (2013) 'Corrosion behavior of plasma sprayed hydroxyapatite and hydroxyapatite-silicon oxide coatings on AISI 304 for biomedical application', *Applied Surface Science*, Vol. 284, pp. 811-818.
[28] Schwartz Z and Boyan BD (1994) 'Underlying Mechanisms at the Bone-Biomaterial Interface', *Journal of Cell Biochemistry*, Vol. 56, No. 3, pp. 340-347.
[29] Lee TM, Tsai RS, Chang E, Yang CY and Yang MR (2002) 'The Cell Attachment and Morphology of Neonatal Rat Calvarial Osteoblasts on the Surface of Ti6Al4V and Plasma Sprayed HA Coating: Effect of Surface Roughness and Serum Contents', *Journal of Material Science: Materials in Medicine*, Vol. 13, pp.341-350.
[30] Mohammadi Z, Ziaei-Moayyed AA and Sheikh-Mehdi Mesgar A (2007) 'Grit blasting of Ti-6Al-4V alloy: Optimization and its effect on adhesion strength of plasma-sprayed hydroxyapatite coatings', *Journal of Materials Processing Technology*, Vol. 194, pp. 15–23.
[31] Gross KA and Babovic M (2002) 'Influence of Abrasion on the Surface Characteristics of Thermally Sprayed Hydroxyapatite Coatings', *Biomaterials*, Vol. 23 No. 24, pp. 4731-4737.
[32] Zhang Z, Dunn MF, Xiao TD, Tomsia AP and Saiz E (2002) 'Nanostructured hydroxyapatite coatings for improved adhesion and corrosion resistance for medical implants Nanotech', *Biotechnology Convergence, Stamford, CT* , pp. 291–296.
[33] Sousa SR and Barbosa MA (1996) 'Effect of Hydroxyapatite Thickness on Metal Ion Release from Ti6Al4V Substrates', *Biomaterials,* Vol. 17, No. 4, pp. 397-404.
[34] Fathi MH, Salehi M, Saatchi A, Mortazavi V and Moosavi SB (2003) 'In Vitro Corrosion Behavior of Bioceramic, Metallic and Bioceramic-Metallic, Coated Stainless Steel Dental Implants', *Dental Materials*, Vol. 19, No. 3, pp.188-198.
[35] Yen SK, Chiou SH, Wu SJ, Chang CC, Lin SP and Lin CM (2006) 'Characterization of Electrolytic HA/ZrO2 Double Layers Coatings on Ti-6Al-4V Implant Alloy', *Material Science Engineering*, Vol. 26, No. 1, pp. 65-77.
[36] Cachinho SCP and Correia RN (2008) 'Titanium Scaffolds for Osteointegration: Mechanical, In Vitro and Corrosion Behavior', *Journal of Material Science: Materials in Medicine*, Vol. 19, No. 1, pp. 451-457.
[37] Jeong YH, Brantley WA and Choe HC (2012) 'AC Impedance Behavior of Silicon-hydroxyapatite Doped Film on the Ti–35Nb–xZr Alloy by EB-PVD Method', *Surface & Coating Technology* [doi:10.1016/j.surfcoat.2012.04.075.].
[38] Vijayalakshmi U and Rajeswari S 'Development of Silica Glass Coatings on 316L SS and Evaluation of its Corrosion Resistance Behavior in Ringer's Solution', *Metallurgical and Materials Transactions A*, [DOI: 10.1007/s11661-012-1283-5].
[39] Gallardo J, Galliano P and Duran A (2001) 'Bioactive and Protective Sol-Gel Coatings on Metals for Orthopaedic Prostheses', *Journal of Sol-Gel Science and Technology*, Vol. 21, pp. 65–74.
[40] Gibson IR, Huang J, Best SM and Bonfield W (1999) 'Enhanced in Vitro Cell Activity and Surface Apatite Layer formation on Novel Silicon Substituted Hydroxyapatites. In: Ohgushi H, Hasting GW, Yoshikawa T, editors. 12th International Symposium on Ceramics in Medicine, Japan. Nara, pp. 191-194.
[41] Patel N, Best SM, Bonfield W, Gibson IR, Hing KA, Damien E and Revell PA (2002) 'A Comparative Study on the in Vivo Behavior of Hydroxyapatite and Silicon Substituted Hydroxyapatite Granules', *Journal of Material Science: Materials in Medicine*, Vol. 13, pp. 1199–1206.
[42] Maxian SH, Zawadsky JP and Dunn MG (1993) 'Mechanical and Histological Evaluation of Amorphous Calcium Phosphate and Poorly Crystallized Hydroxyapatite Coatings on Titanium Coatings', *Journal of Biomedical Materials Research*, Vol. 27 No. 6, pp. 717-728.
[43] Xue W, Tao S, Liu X, Zheng XB and Ding C (2004) 'In Vivo Evaluation of Plasma Sprayed Hydroxyapatite Coatings Having Different

Crystallinity', *Biomaterials*, Vol. 25, pp. 415-442.
[44] Gross KA and Saber-Samandari S (2007) 'Nano-Mechanical Properties of Hydroxyapatite Coatings with a Focus on the Single Solidified Droplet', *Journal of Australian Ceramic Society*, Vol. 43, No. 2, pp. 98-101.
[45] Yang YC and Chang E (2001) 'Infuence of Residual Stress on Bonding Strength and Fracture of Plasma-Sprayed Hydroxyapatite Coatings on Ti-6Al-4V substrate', *Biomaterials*, Vol. 22, pp. 1827-1836.
[46] Sarao TPS, Sidhu HS and Singh H (2012) 'Characterization and In Vitro Corrosion Investigations of Thermal Sprayed Hydroxyapatite and Hydroxyapatite-Titania Coatings on Ti Alloy', *Metallurgical and Materials Transactions A*, Vol. 43 A, pp. 4365-4376.
[47] Gopi D, Indira J, Kavitha L and Ferreira JMF (2013) 'Hydroxyapatite Coating on Selectively Passivated and Sensitively Polymer-protected Surgical Grade Stainless Steel', *Journal of Applied Electrochemistry*, Vol. 43, pp. 331-345.
[48] Roland T, Pelletier H and Krier J (2013) 'Scratch Resistance and Electrochemical Corrosion Behavior of Hydroxyapatite Coatings on Ti6Al4V in Simulated Physiological Media', *Journal of Applied Electrochemistry*, Vol. 43, pp. 53-63.

Fabrication and characterization of graphene oxide/fluorhydroxyapatite composite materials for biomedical applications

Yulong Bai, Jingjun Gao, Wen Ma, Yu Bai*

School of Materials Science and Engineering, Inner Mongolia University of Technology, Hohhot, Inner Mongolia, P. R. China

Abstract

Hydroxyapatite (HA) has been currently used for bone repair or regeneration due to its similar chemical composition, crystallinity and morphology to bone minerals. However, its mechanical strength is insufficient for major load bearing implants. Additionally, the higher dissolution of HA in physiological environment also limits its application. Thus in order to overcome these disadvantages, the fabrication of HA composite materials should be one of possible solutions. In this work, the graphene oxide (GO) as a nanofiller is introduced to fabricate the HA composite material. Meanwhile, OH^- groups of HA are partially substituted by F^- ions using an *in situ* synthesis method. The graphene oxide/fluorhydroxyapatite (GO/FHA) powder is consolidated through spark plasma sintering technology. FTIR, XPS, XRD, SEM and TEM are used to characterize the obtained GO/FHA composite material. Compared with HA, the GO/FHA displays an increased mechanical strength and decreased solubility. Moreover, the addition of GO and the partial substitution of F^- also increase the bioactivity of HA based on cell viability tests. Therefore, the GO/FHA composite is expected to be a promising material for orthopedic implants.

Keywords: Biomedical implants, Graphene oxide, Fluorhydroxyapatite, Mechanical property, Bioactivity, Stability

1. Introduction

Biomaterials based on calcium phosphate, especially hydroxyapatite (HA, $Ca_{10}(PO_4)_6(OH)_2$), have attracted a significant amount of attention for use in bone and teeth implants due to their unique properties, such as excellent biocompatibility, bioactivity and osteoconductivity. However, the poor mechanical properties of HA compared with bone make it unsuitable for use in major load-bearing positions [1, 2]. Low physic-chemical stability of HA during long-term contact with biological fluids/tissues is another drawback for its successful application, especially in oral and dental fields [3, 4]. Therefore, in order to overcome these disadvantages, the fabrication of HA composite materials should be one of possible solutions. Over past several decades, some conventional reinforcing phases, such as TiO_2 [5], ZrO_2 [6], bioglass [7] and polyethylene [8], have been used to combine with HA for achieving a desired property. Recently, graphene and its derivatives have found wide acceptance as reinforcement or filling materials in ceramic matrices due to their extraordinary high mechanical properties such as Young's modulus (up to 1TPa) and intrinsic strength (approximately 130 GPa)[9, 10]. Among graphene derivatives, graphene oxide (GO) shows desirable dispersion behavior due to the reactive oxygenic groups present on the edges and planes. These groups can not only stabilize the dispersion of GO in aqueous phase, but also increase the interfacial bonding within the components and facilitate the stress transfer in the composites.

Although GO, as a reinforcing agent, can improve the mechanical properties of HA, the chemical stability of HA after implantation is not still satisfying. Compared to pure HA, however, fluoroapatite (FA, $Ca_{10}(PO_4)_6F_2$) formed by fluoride ion substituted for hydroxide ion in the OH^- lattice position, is found to possess an apparent stability, such as an increased acid resistance and antimicrobial activity to caries bacteria. Indeed, fluorine itself is also an essential trace element in bone tissues and teeth, which can promote the mineralization and crystallization of calcium phosphate in the process of bone formation.

In this work, we fabricated GO/FHA nanocomposite materials with GO component and fluorine element incorporated simultaneously into the nanocomposite via *in-situ* synthesis method to achieve the long-term biostability and the improvement of mechanical properties.

2. Experimental procedures

2.1. Materials

The chemicals of $NaNO_3$, $KMnO_4$, H_2O_2, $Ca(NO_3)_2 \cdot 4H_2O$, $(NH_4)_2HPO_4$ and NH_4F were purchased from Sinopharm Chemical Regent Co., Ltd.. The graphite powder was obtained from Aladdin Industrial Co., Ltd.. Osteoblast cells (MC3T3-E1), α-Minimum Essential Medium (α-MEM), dimethyl sulfoxide (DMSO) and phosphate buffered saline (PBS) solution were purchased from Procell Biotech. Co., Ltd.. 3-(4, 5-dimethylthiazol-2-yl)-2, 5-diphenyl tetrazolium bromide (MTT) were provided by Sigma-Aldrich Co..

2.2 Preparation of GO

GO was prepared from graphite powder according to the Hummers and Offeman method [11]. Initially, 23 mL of H_2SO_4 was added to a beaker containing 0.5 g graphite powder and 0.5 g $NaNO_3$ at room temperature. Then, 3.0 g $KMnO_4$ was gradually added to the mixture. The mixture was stirred at 35 ℃ for 1 h. After adding 40 mL distilled water (DW), the temperature of mixture was increased to 90 ℃ and then kept the temperature for 35 min. Finally, the suspension was diluted with 100 mL DW. In order to ensure the reduction of the residual permanganate into soluble manganese ions, H_2O_2 (30 %) solution was added. After the synthesis, the GO suspension was filtrated and washed repeatedly with DW. The resulting product was separated from the suspension by centrifugation operation at 10000 rpm.

2.3 Synthesis of GO/FHA composites

The GO/FHA composite was prepared using *in-situ* synthesis method. 10 mg of GO was dispersed in 100 mL of 0.01 mol/L aqueous $Ca(NO_3)_2 \cdot 4H_2O$, and sonicated for 3 h to form solution A. 0.93 g $(NH_4)_2HPO_4$ and 0.26 g NH_4F were dissolved in 70 mL DW to form solution B. Prior to reaction, the pH of solution A and solution B was adjusted to 9 by $NH_3 \cdot H_2O$, respectively. Subsequently, Solution B was added into solution A dropwise with continuous stirring. After dropping, the stirring was maintained for 3 hours. The mixture was then filtrated and washed with DW. Finally, the resulting product [GO/$Ca_{10}(PO_4)_6(OH)F$ composite] was vacuum-dried at 60 ℃ for 24 h.

The consolidation process was performed by SPS technology. The resulting powders were loaded into a graphite die (15 mm inner diameter), in which graphite papers (3-4 layers) were placed between the powders and die for easy specimen removal. The SPS (SPS-20T-10, Shanghai Chenhua electric Co., China) sintering process was conducted under a vacuum of 5 Pa and applied a uniaxial pressure of 50 MPa throughout the sintering cycle. The sintering temperature was set at 1100 ℃ at a rate of 150 ℃/min and a holding time was 5 min during the sintering.

2.4 Characterization

The morphology of the fabricated sample was examined by a field emission scanning electron microscopy (FE-SEM, FEI Quanta FEG 650, USA) and transmission electron microscopy (TEM, JEM 1200EX, Japan). Atomic force microscopy (AFM, MutiMode 8, Bruker, USA) was employed in tapping mode to investigate the GO sheets. The GO sheets were deposited on a freshly cleaved mica substrate using a spin-coating method. The structure and phase composition of the samples were characterized by X-ray diffraction (XRD, Rigaku D/Max-2500/PC, Japan) using Cu Kα radiation with a scanning rate of 3°/min at a range from 20° to 50°. Fourier transform infrared spectroscopy (FT-IR) was performed using a Spectrum GX spectrometer (Nicolet 870, USA) in KBr pellets. The spectra were recorded from 4000 cm^{-1} to 450 cm^{-1}. Raman spectra analyses were carried out with a Renishaw Invia Raman Microscope using a laser with a wavelength of 514 nm. X-ray photoelectron spectroscopy (XPS, Thermo escalab 250Xi, USA) was employed to indentify the composition and elemental state of the samples.

2.5 Mechanical property measurements

The analysis of the mechanical properties was carried out using Micro Vickers Hardness Tester (HXD-1000TM/LCD, Shanghai Taiming Optical Instrument Co., China) with the load of 100 mN and 300 mN (a dwell time of 10 s) for hardness and fracture toughness, respectively. The indentation and crack patterns were used to calculate the fracture toughness (K_{IC}) using following equation [12]:

$$K_{IC} = 0.16 H_V a^2 C^{-\frac{3}{2}}$$

where H is the hardness, a is the half length of the indent diagonal, C is the crack length.

2.6 Solubility measurements

The dissolution behavior of the samples was investigated by immersing them into individual centrifuge tubes with an aqueous solution (45 mL, pH = 4.0) at 37℃. The solution was prepared by mixing PBS solution and 1 mol/L citric acid solution (used to adjust the pH value). The concentration of the released Ca^{2+} in the immersion solution was measured at different time intervals (1, 2, 3, 4, 5, 7, 9, 11, 13 and 15 days) using an inductively coupled plasma atomic emission spectrometer (ICP-OES) (Perkin Elmer Optima 7300V, USA).

2.7 *In vitro* biocompatibility assessment

MC3T3-E1 cells were cultured in α-MEM supplemented with 10 % fetal bovine serum (FBS) and 500 unit/mL penicillin-streptomycin in a humidified atmosphere with 5 % CO_2 at 37 ℃. When the cultured cells reached confluence, they were detached to provide adequate numbers of cells for the following studies.

Cell proliferation: Proliferation of the cultured cells was evaluated using a MTT assay. 18 mL of α-MEM without FBS was mixed with 2 mL of a MTT solution (5 mg/mL). 1 mL of the MTT mixture solution was then added to the aspirated wells and incubated for 4 h. The medium was removed, and the dark blue crystals remaining in the wells were then dissolved in 1 mL of DMSO. The solution was then transferred to a new 96-well microplate, and the optical density (OD) of the solution in each well was measured at a wavelength of 570 nm using a microplate reader (Epoch 2, BioTek, USA).

2.8 Data analysis

All experiments were carried out in triplicate and were expressed as the mean ± standard deviation (SD). Single-factor analysis of variance was used to assess the statistical significance of the results. A P value < 0.05 was considered significant.

3. Results and discussion

3.1 Preparation and characterization of GO and GO/FHA nanocomposites

GO nanosheet was prepared from pristine graphite according to the modified

Hummers and Offeman method. The as-synthesized GO shows wrinkled-paper-like morphology [Fig. 1(a)]. Most of the GO nanosheet exists as single or few layers and the thickness of the GO nanosheet is around 1.4 nm, as shown in Fig. 1(b). According to the XRD pattern with a characteristic peak centered at 11.2° [Fig. 1(c)], layer-to-layer distance of the GO nanosheet is calculated. The detailed value is 7.8 Å for the d_{001} plane. The d-spacing of GO nanosheet is larger than that of graphite (3.4 Å) due to the presence of oxygen-containing functional groups such as carboxyl (-COOH), hydroxyl (-OH), and epoxy groups as well as other structural defects. The presence of these groups not only facilitate a homogeneous dispersion of GO in aqueous phase, but also can provide numerous reactive sites for the subsequent formation of FHA nanoparticles. At the initial stage of reaction, Ca^{2+} is absorbed on the GO through the coordination interaction of the C-O-C groups or through ions exchange with H^{+} from COOH and OH groups. Then, the absorbed Ca^{2+} becomes the nucleation site for the following absorption of PO_4^{3-} and OH^- groups. Because the diameter of F^- ions (1.32 Å) is smaller than that of OH^- ions (1.68 Å), it is possible that OH^- is substituted by F^- ions during the formation of HA crystal growth. By controlling the pH value and adjusting the adding amount of F^- ions, partially substitution of OH^- by F^- ions can be achieved on the basis of the stoichiometric ratio.

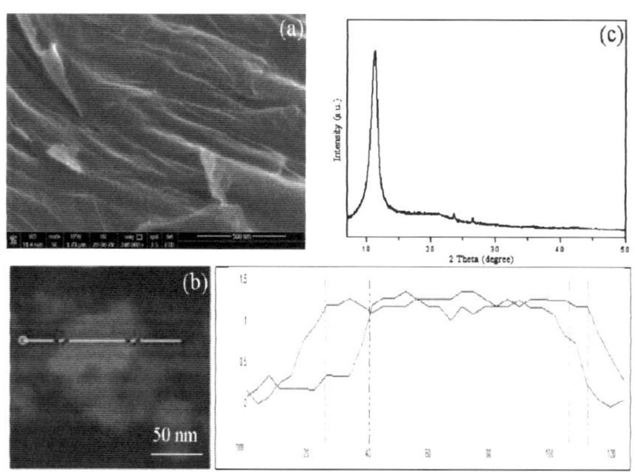

Fig. 1 (a) SEM image of the GO surface morphology, (b) AFM analysis of GO showing the height of around 1.4 nm, (c) XRD pattern of GO.

Fig. 2(a) and (b) shows the TEM and SEM images of GO/FHA composites. It can be observed that FHA nanoparticles are grafted on the GO sheet through a strong chemical bonding. The wrinkled surface of GO can be clearly discernible in the image as the shown from the arrow in Fig. 2(b). Fig. 2(c) shows the result from EDS that is performed on the composite, where the atomic Ca to P ratio is around 1.56, which is similar with the stoichiometric ratio for the natural HA.

Fig. 2 TEM (a) and SEM (b) images of the GO/FHA composites. (c) shows the EDS spectrum of the GO/FHA composites.

The FTIR spectra of GO, FHA and GO/FHA composites are shown in Fig. 3. The broad bands ranging from 3400-3500 cm^{-1} correspond to the O-H stretching vibration of adsorbed water molecules. The spectrum of FHA shows characteristic bands of phosphate species. The band that appear at 1031 cm^{-1} belong to the asymmetric stretch vibration of PO_4^{3-} groups, while the bands at 565 and 600 cm^{-1} originate from *v*2 stretching symmetric vibration. The oxygen-containing functional groups of GO are revealed by the bands at 1730, 1621, 1400, 1116, and 1049 cm^{-1}. These bands can be assigned to the C=O stretching vibration of COOH group, C=C stretching vibration, C-O-H deformation vibration, C-OH stretching vibration, and C-O stretching vibration, respectively. After *in-situ* synthesis of FHA on the surface of GO, the bands associated with oxygen-containing functional groups decrease significantly, and some of the bands disappear.

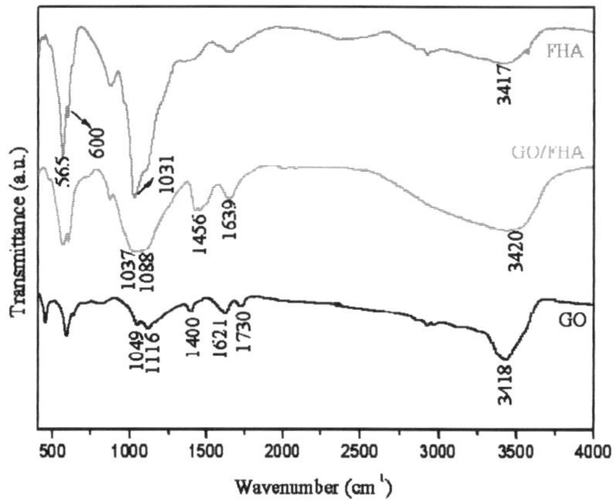

Fig. 3 FTIR spectra of GO, FHA and GO/FHA composites.

Although FTIR spectra can clearly indicate some functional groups existed in GO/FHA composites, it is not enough to demonstrate the presence of GO in the composites. Hence, Raman analysis is performed. Fig. 4 shows the Raman spectra of GO and GO/FHA composite with and without SPS processing. Both the typical D band (1350 cm^{-1}) and G band (1593 cm^{-1}) of GO can be observed. It is well known that the G band is a characteristic feature of graphic carbon layers corresponding to tangential vibration of carbon atoms and the D band is a typical sign of presence of detective graphitic carbon [13, 14]. D band and G band can also be clearly seen in the as-synthesized and sintered GO/FHA composite, suggesting the existence of GO in composite. However, the D band and G band for as-synthesized GO/FHA powder displays a small red-shift to 1360 and 1596 cm^{-1}, respectively, compared with that of GO. The spectral shift could be ascribed to the disturbing of the GO structure caused by the chemical interactions between the carbon/oxide atoms of GO and the reactive sites of FHA. With the sintered GO/FHA composite, shifting of D and G band to higher frequency can be found as compared to the GO, especially D band shows a ~ 20 cm^{-1} blue-shift. The compressive stress applied on GO during SPS might be the underlying reason for the blue-shift.

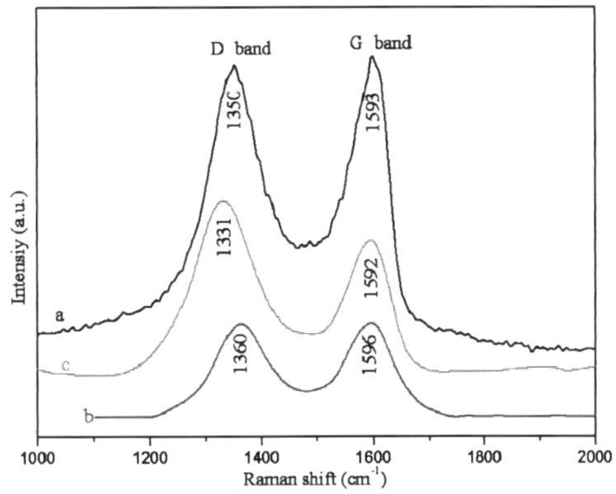

Fig. 4 Raman spectra of GO (a), as-synthesized (b) and sintered GO/FHA composites (c).

Theoretically, F^- ions easily replace OH^- groups in HA on thermodynamic ground (ΔE_x = -0.4-0.6kJ/mol). The lowest energy of the crystal structure is reached at x =1.0 ($Ca_{10}(PO_4)_6(OH)_{2-x}F_x$) with a configuration that HA crystal consists of alternating planes of F^- ions and OH^- groups, where they are alternating in the *c*-axis but collocated within the *a/b*-plane. Therefore, it is necessary to actually confirm the substitution of F^- ions into HA structure, which can be studied from XRD. The XRD patterns of as-synthesized GO/FHA and HA powders are shown in Fig. 5. The diffraction peaks of all the samples are assigned to apatite phase. With reference to the pure HA (JPCDS 09-432), there is no other apatite phase found in GO/FHA. However, all the main diffraction peaks of GO/FHA have a slight shift toward higher degree, suggesting that some of the OH^- groups are substituted by the F^- ions to form FHA, which is in accordance with the results obtained in other studies [15-17]. Unfortunately, it is difficult to found the diffraction peak of GO in GO/FHA, likely due to the small content of GO (1 wt. %) or the presence of strong FHA peaks. Nevertheless, the existence of GO has been confirmed by Raman spectrum characterization.

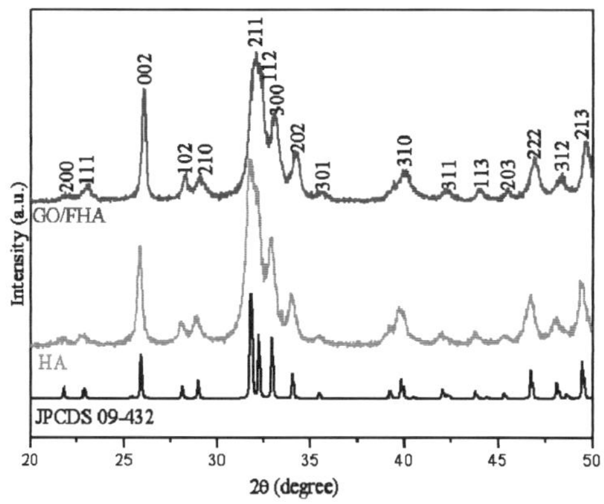

Fig. 5 XRD patterns of HA and as-synthesized GO/FHA composites.

XPS is another effective way to identify the F^- incorporation into the HA structure as well as the presence of GO in GO/FHA composite. Fig. 6 represents the XPS spectra of GO and GO/FHA composites. Fig. 6(a) shows that the spectrum of GO contains only two elements of C and O. By contrast, the signal peaks of Ca, P and F emerge in the XPS survey spectrum of GO/FHA, indicative of the formation of FHA. The narrow scans of F1s are shown in Fig. 6(b), and the binding energy of F1s centered at 684.3 eV reveals that F ions have been incorporated into the HA structure. This is because F 1s at 684.3 eV is the fingerprint of F in FHA, which has been reported by other researchers [18-20]. Chemical state of the O atoms within the GO and GO/FHA composite are identified in Fig. 6(c) and (d), respectively. In GO spectra, the deconvolution of the O 1s band is composed of two peaks. The peak with smaller binding energy is observed at 531.8 eV, which is usually connected to the presence of double bond between C and O atoms. And another peak has a binding energy of 532.9 eV. According to the literature [21-23], the peak is most often ascribed to the O in a single C-O bond. In GO/FHA spectra, the O 1s band is resolved into four peaks. The peaks centered at 531.3 and 532.3 eV are attributed to the O in PO_4^{3-} and OH^- groups, whereas another two peaks at 531.8 and 532.9 eV correspond to the O in C=O and C-O groups which is also consistent with the deconvolution of O 1s band in GO. The XPS results further demonstrate the presence of GO and FHA, which agrees with the FTIR and XRD results.

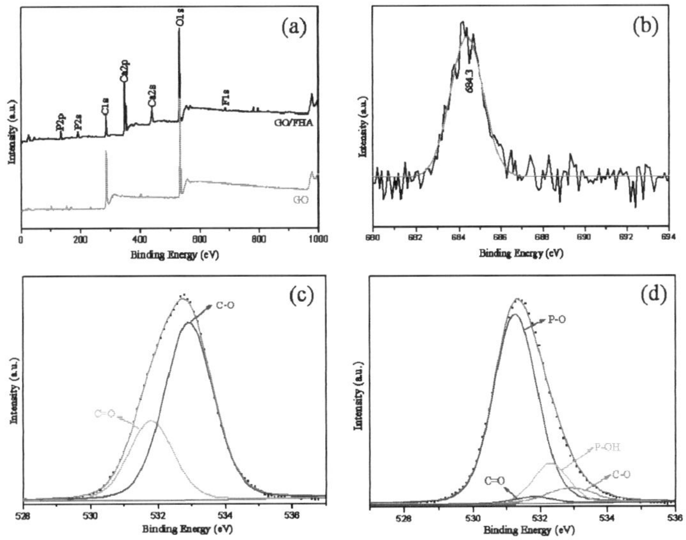

Fig. 6 XPS survey spectra of GO and GO/FHA composites (a), F1s XPS spectrum of GO/FHA (b), O1s spectra of GO (c), O1s spectra of GO/FHA composites (d).

3.2 Mechanical property

The hardness and fracture toughness of sintered samples are denoted in Table 1. It can be seen that the hardness and fracture toughness of GO/FHA composite increase ~ 86 % and ~137 %, respectively, as compared to pure HA. And improvements of ~ 41 % and ~20 % are observed compared with FHA corresponding to hardness and fracture toughness. The result indicates that the introduction of GO and F^- ions significantly enhance the mechanical property of pure HA.

Table 1 Mechanical properties of different samples

Samples	HA	FHA	GO/FHA
H_V (GPa)	4.25 ± 0.44	5.60 ± 0.49	7.92 ± 0.40
K_{IC} (MPa·m$^{0.5}$)	0.43 ± 0.15	0.85 ± 0.10	1.02 ± 0.10

It is well known that graphene, as a toughened phase, has a large surface area, which provides a more contact between graphene and FHA particles. And oxygen-containing functional groups existed on graphene surface provide a reactive site for the formation of FHA crystal, causing a close bonding and mechanical interlocking between graphene and FHA matrix. The close interaction enhances the load transfer efficiency of graphene toward the FHA matrix and thus increases the mechanical property of FHA.

On the other hand, from the results in table 1, it is also seen that the hardness and fracture toughness of FHA are higher than that of pure HA. The positive effect of F^- ions on the mechanical property of HA can be explained by considering the crystal structure. In HA, H^+ in an OH^- can bond with O^{2-} from different directions along the c-axis. When an OH^- is replaced by a F^-, a stronger hydrogen bond forms between the F^- and the nearby OH^- because of the higher affinity of the fluoride in respect to the oxygen, producing a more ordered structure, which cause the increase of HA hardness. Besides, the substitution of F^- ions into HA crystal structure also causes the changes of lattice parameters, namely there is a significant decrease in the a-axis of HA. The change of lattice parameters is bound to cause the lattice distortion. The production of lattice distortion is thought to enhance the fracture toughness of HA. In all, the incorporation of both F^- ions and GO into HA crystal structure produces a combination of their individual functions to obtain a high mechanical property.

3.3 Dissolution behavior

The solubility of the samples is usually measured by immersing them in covered containers with buffered solutions. The appearance of metal ions and anions after immersion and the variation of their concentrations can be considered as the result of the dissolution of the samples and thus can be utilized to evaluate the relative stability. It is known that the collapse of the apatite structure is mainly due to the release of Ca^{2+} and PO_4^{3-} ions caused by their low stability in physiological environment [24, 25]. Therefore, in this work, we choose the Ca^{2+} ion release as an index to evaluate the dissolution behavior of samples. Fig. 7 compares the Ca^{2+} ion release of pure HA and GO/FHA composite after immersing them in a citrate modified PBS solution with different times. It is seen that the two dissolution curves exhibit a similar varying trend and can be divided into two distinct stages: the initial sharp increase in Ca^{2+} concentration and the sharp decrease in the dissolution rate during the first 7 days, followed by a slow increase in Ca^{2+} concentration and finally a nearly constant dissolution rate. Importantly, it is also seen from the Fig. 7 that the substitution of F^- ions into HA crystal structure has a positive effect on the dissolution resistance of HA. At the same immersion time, the Ca^{2+} ion release of pure HA is higher than that of GO/FHA composites. The main reason is that the fluoride substitution stabilizes the crystal structure of HA and improves crystallization.

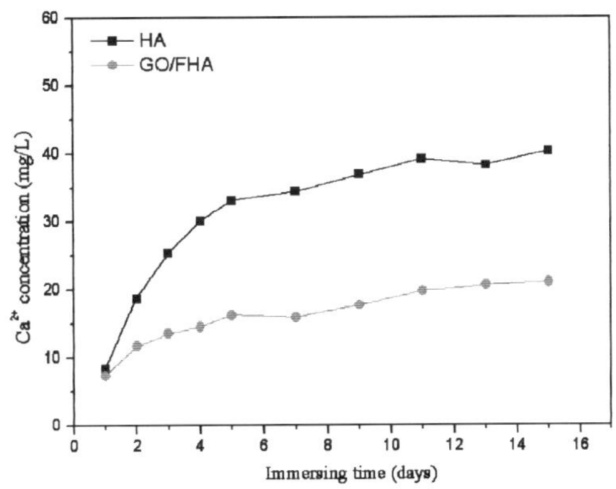

Fig. 7 Dissolution behavior of HA and GO/FHA composites in a citrate modified PBS solution: Ca^{2+} concentration as a function of immersing time.

3.4 *In vitro* cytotoxicity evaluation

The living cell numbers are evaluated by an MTT assay. Fig. 8 shows the cell density as a function of culture time for the HA and GO/FHA composites with the same initial cell implanting density of 5×10^4 cell/cm^2. It is apparent that the absorbance values increase on both samples with the extension of culture time, meaning the number of living cells increased, i.e., the cells proliferated. It is also seen that the absorbance values of GO/FHA composites is higher than that of HA, especially after 3 days of incubation, suggesting the GO/FHA composites have a positive effect on osteoblastic proliferation.

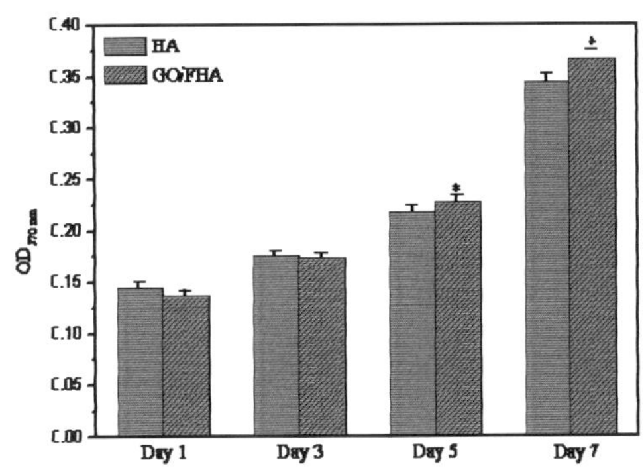

Fig. 8 MTT assay of osteoblast cells cultured in HA and GO/FHA composites for 1, 3, 5 and 7 days. Significant difference $*p < 0.05$ compared with HA.

DVS 324

Conclusions

GO and F^- ions are simultaneously incorporated into HA structure via *in-situ* chemical synthesis method. A consolidation procedure is performed by SPS technology at 1100   and 50 MPa. XRD patterns, FTIR and XPS spectra suggest that the OH^- groups of HA are partially substituted by F^- ions. Raman spectra evidence that GO retains its original structure after experiencing harsh processing conditions. Compared with HA, the fabricated GO/FHA composites display 86% and 137% improvement in hardness and fracture toughness, respectively. Solubility measurement result shows that the incorporation of F^- ions increases the dissolution resistance of HA. *In-vitro* bioactivity studies reveal that the co-effects of GO and F^- ions promote osteoblast proliferation. These results will pave the way for GO-based FHA biocomposite to be used as dental implant material in practical clinical applications with combined improvement in comprehensive properties.

Acknowledgments

This work was financially supported by "The Key Laboratory of Inorganic Coating Materials, Chinese Academy of Sciences (KLICM-2014-05)" and The Natural Science Foundation of Inner Mongolia Autonomous Region (2014BS0504)".

References

[1]Q. Tan, K. Zhang, S. Gu, J. Ren, Mineralization of surfactant functionalized multi-walled carbon nanotubes (MWNTs) to prepare hydroxyapatite/MWNTs nanohybrid, Appl. Surf. Sci. 255 (2009) 7036–7039.

[2]C. Kealley, M. Elcombe, A.V. Riessen, B. Ben-Nissan, Development of carbon nanotube-reinforced hydroxyapatite bioceramics, Physica B 385-386 (2006) 496–498.

[3]Y. Chen, X. Miao, Effect of fluorine addition on the corrosion resistance of hydroxyapatite ceramics, Ceram. Int. 30 (2004) 1961–1965.

[4]A.S. Furqan, Fluoride-containing bioactive glasses: Glass design, structure, bioactivity, cellular interactions, and recent developments, Mater. Sci. Eng. C 58 (2016) 1279–1289.

[5]A. Pittrof, S. Bauer, P. Schmuki, Micropatterned TiO_2 nanotube surfaces for

site-selective nucleation of hydroxyapatite from simulated body fluid, Acta Biomater. 7 (2011) 424–431.

[6]S. Peyman, A. Abdollah, Effects of zirconia content on characteristics and corrosion behavior of hydroxyapatite/ZrO_2 biocomposite coatings codeposited by electrodeposition, Surf. Coatings Tech. 262 (2015) 166–172.

[7]Z. Yazdanpanahn, M.E. Bahrololoom, B. Hashemi, Evaluating morphology and mechanical properties of glass-reinforced natural hydroxyapatite composites, J. Mech. Behave. Biomed. Mater. 41 (2015) 36–42.

[8]H. Fouad, R. Elleithy, Y. Othman, O.Y. Alothman, Thermo-mechanical, Wear and Fracture Behavior of High-density Polyethylene/Hydroxyapatite Nano Composite for Biomedical Applications: Effect of Accelerated Ageing, J. Mater. Sci. Technol. 29(6) (2013) 573-581.

[9]I.W. Frank, D.M. Tanenbaum, A.M. Van der Zande, P.L. McEuen, Mechanical properties of suspended graphene sheets, J. Vac. Sci. Technol. B 25(6) (2007) 2558–2561.

[10]L. Zhang, W. Liu, C. Yue, T. Zhang, P. Li, Z. Xing, Y. Chen, A tough graphene nanosheet/hydroxyapatite composite with improved in vitro biocompatibility, Carbon 6 (2013) 105-115.

[11]H.N. Lim, N.M. Huang, S.S. Lim, I. Harrison, C.H. Chia, Fabrication and characterization of graphene hydrogel via hydrothermal approach as a scaffold for preliminary study of cell growth, Int. J. Nanomed. 6 (2011) 1817–1823.

[12]A.G. Evans, E.A. Charles, Fracture Toughness Determinations by Indentation, J. Am. Ceram. Soc. 59(7-8) (1976) 371-372.

[13]S. Kim, S.H. Ku, S.Y. Lim, J.H. Kim, C.B. Park, Graphene–Biomineral Hybrid Materials, Adv. Mater. 23 (2011) 2009–2014.

[14]M.N. Gururaj, O. Aderemi, L. Zhiping, In situ deposition of hydroxyapatite on graphene nanosheets, Mater. Res. Bull. 48 (2013) 175–179.

[15]M. Esnaashary, M. Fathi, M. Ahmadian, The Effect of the Two-Step Sintering Process on Consolidation of Fluoridated Hydroxyapatite and its Mechanical Properties and Bioactivity, Int. J. Appl. Ceram. Technol. 11(1) (2014) 47–56.

[16]R. Rojaee, M. Fathi, K. Raeissi, A. Sharifnabi, Biodegradation assessment of nanostructured fluoridated hydroxyapatite coatings on biomedical grade magnesium alloy, Ceram. Int. 40 (2014) 15149–15158.

[17]Q. Rao, X. Gu, X. Fan, B. Dai, J. Wu, M. Gong, C. Wu, Fluoridation of synthetic apatite: Effect on the formation of calcium-deficient hydroxyapatite and the properties of porous scaffold, Ceram. Int. 42(2) (2016) 3442-3451.

[18]C.C. Wu, S.T. Huang, T.W. Tseng, Q.L. Rao, H.C. Lin, FT-IR and XRD investigations on sintered fluoridated hydroxyapatite composites, J. Mol. Struct. 979 (2010) 72–76.

[19]K. Cheng, S. Zhang, W. Weng, The F content in sol–gel derived FHA coatings: an XPS study, Surf. Coat. Technol. 198 (2005) 237– 241.

[20]T. Hidekazu, Y. Akemi, K. Kazuhiko, I. Tatsuo, Surface structure and properties of fluoridated calcium Hydroxyapatite, Colloids Surf. A 204 (2002) 251–259.

[21]Z.B. Zhang, J.J. Wu, Y. Su, J. Zhou, Y. Gao, H.Y. Yu, J.S. Gu, Layer-by-layer assembly of graphene oxide on polypropylenemacroporous membranes via click chemistry to improveantibacterial and antifouling performance, App. Surf. Sci. 332 (2015) 300–307.

[22]C.G.K. Yue, M.N. Ging, H.A.H. Cheng, Identification of functional groups and determination of carboxyl formation temperature in graphene oxide using the XPS O 1s spectrum, Thin Solid Films 590 (2015) 40–48.

[23]Y.Q. Wang, H. Viswanathan, A.A. Audi, P.M.A. Sherwood, X-ray Photoelectron Spectroscopic Studies of Carbon Fiber Surfaces. 22. Comparison between Surface Treatment of Untreated and Previously Surface-Treated Fibers, Chem. Mater. 12 (2000) 1100-1107.

[24]J.T. Christopher, M.Y. Anne, A.A.N. Ensanya, G. George, C.K. Jonathan, Hydroxyapatite, fluor-hydroxyapatite and fluorapatite produced via the sol–gel method: dissolution behaviour and biological properties after crystallization, J. Mater. Sci.: Mater. Med. 25 (2014) 47–53.

[25]Y. Cai, S. Zhang, X. Zeng, D. Sun, Effect of fluorine incorporation on long-term stability of magnesium-containing hydroxyapatite coatings, J. Mater. Sci.: Mater. Med. 22 (2011) 1633–1638.

Corrosion resistance and apatite forming ability of plasma sprayed CP-Ti/Bioglass composite coating

S. Bsat, B. De Snoo and X. Huang
Department of Mechanical and Aerospace Engineering
Carleton University

The bio-functionality and corrosion resistance of plasma sprayed CP-Ti and CP-Ti/bioglass coatings were studied. The bio-functionality of the coatings were evaluated using simulated body fluid (SBF) immersion testing. After four weeks of immersion the surface chemistry of the coatings were examined using energy dispersive spectrometry (EDS) elemental analysis and X-ray diffraction (XRD). The corrosion resistance of the coatings were evaluated by measuring the mass change following SBF immersion. After four weeks of immersion in SBF the Ca content on the CP-Ti/bioglass coatings increased from 4 to 16 wt. % while no noticeable increase in Ca was observed on CP-Ti coatings. HA formation was observed on both coatings, although the relative intensity of HA on the XRD spectrum was higher for CP-Ti/bioglass coatings. The results suggest that the addition of bioglass enhances bio-activity and promotes HA formation. It was found that CP-Ti experienced a mass gain while CP-Ti/bioglass underwent a mass loss following SBF immersion. The mass loss of CP-Ti/bioglass is likely due to the dissolution of calcium and phosphate.

1.0 Introduction

Traditionally, implants were fixated using polymethylmethacrylate (PMMA) bone cement to provide a rigid fixation between the implant and the bone. While PMMA bone cement remains in use as a method for implant fixation, osseointegration as a form of implant fixation is widely used as well. Osseointegration requires the growth of the adjacent bone into the porous structure of the implant thus providing a rigid fixation. For osseointegration to take place, an appropriate implant surface chemistry and surface topography are required. The surface chemistry of the implant in particular must be bio-functionalized to help promote the initial formation of hydroxyapatite (HA), later leading to bone growth and integration [1-5].

In addition to bio-functionality, the corrosion resistance, is another important factor to consider for implant performance. The three main metallic biomaterials currently used are stainless steel, titanium, and cobalt-chrome, all of which exhibit excellent corrosion resistance in the environment of the human body [6] and adequate mechanical properties. Each of these metals is able to form a nano-scale oxide film that acts as a corrosion barrier preventing direct contact between body fluid and the metal implant surface, making them ideal biocompatible materials with limited corrosion [7].

To bio-functionalize the surface of implants, a wide range of coatings with varying composition and manufacturing methods have been researched in order to maximize the formation of HA and bone. The current work evaluates the bio-functionality of plasma sprayed commercially pure titanium-bioglass (CP-Ti/bioglass) coatings in contrast to CP-Ti coatings by examining their HA forming ability using simulated body fluid (SBF) immersion testing. The surface chemistry of CP-Ti and CP-Ti/bioglass coatings were examined before and after immersion in SBF using energy dispersive spectroscopy (EDS) and X-ray diffraction (XRD). Furthermore the corrosion rate of CP-Ti and CP-Ti/bioglass coatings were evaluated using the weight change method based on ASTM G 31.

2.0 Materials and Experimental Procedure

2.1 Commercially Pure Titanium and Bioglass

Commercially pure titanium powder (Raymor, Quebec, Canada) and 45S5 bioglass powder (MO-SCI Co., Missouri, USA), with compositions shown in **Table 1**, were used in this study. The particle size of the CP-Ti powder ranged between 15-45 microns and the average particle size of the bioglass was -53 microns. Commercially pure titanium plates were used as substrate material to apply the coatings.

Table 1. Composition of 45S5 bioglass

	SiO_2	CaO	Na_2O	P_2O_5
Mol %	46.1	26.9%	24.4	2.5

2.2 Composite Powder Processing

CP-Ti/bioglass composite powder was prepared by mixing 15wt. % of 45S5 bioglass powder with 85wt. % CP-Ti powder. The powders were mixed using a ball mill (Fritsch GmbH Planetary Mono Mill Pulverisette 6). The powders were mixed for 20 minutes clockwise and 20 minutes counter clockwise to avoid grinding.

2.3 Plasma Spraying Process

CP-Ti plates of size 5 x 2 x 0.1 cm^3 were used for the deposition of the coatings. Prior to plasma spraying, the substrate surface was ultrasonically cleaned in acetone for 15 minutes and grit blasted with alumina oxide particles. The plasma spraying process was conducted in air using the Axial III plasma spray system (Mettech Corp, Vancouver, Canada). The substrates were pre-heated to avoid thermal stresses in the coating. The parameters used during plasma spraying include a total gas flow of 225 l/min (35 % hydrogen and 10 % nitrogen in argon plasma gas), a spray distance of 150 mm and a nozzle size of 1.4 cm. The CP-Ti powder and CP-Ti/bioglass composite powder were fed at a rate of 40 g/min and 42 g/min, respectively, using argon as the carrier gas.

Elemental analysis of the coated samples was conducted using EDS (Tescan Vega-II XMU VPSEM and Oxford Inca Energy 250X EDS).

2.4 SBF Immersion

Hank's balanced salt solution (Bishop, Canada) was use for SBF immersion testing [2,4,5]. The composition of the solution is provided in **Table 2**. The ion concentration in the SBF closely resembles the ion concentration in human blood plasma, **Table 3**. The solution was buffered to a pH of approximately 7.4, matching that of human blood plasma. Before being immersed in SBF, the samples were rinsed in acetone and ultrasonically cleaned for 15 minutes. The samples were then dried in a furnace for 6 hrs at 60°C to ensure there was no moisture in the samples. The sample's weights were measured using a scale accurate to ±0.0001 g. After weighing, each sample was then immersed in 50 mL SBF separately. The beakers were sealed to remain sterile and placed in a water bath kept at 37°C, the physiological temperature of the human body.

Table 2. Composition of Hank's balanced salt solution.

Chemicals	Amount (g/L)
NaCl	8
$CaCl_2$	0.14
KCl	0.4
$MgCl_2 \cdot 6H_2O$	0.10
$MgSO_4 \cdot 7H_2O$	0.06
$NaHCO_3$	0.35
$Na_2H_2PO_4$	0.10
$Na_2HPO_4 \cdot 2H_2O$	0.06
Glucose	1

After immersion for four weeks, the samples were removed from SBF solution, gently rinsed with distilled water and dried for 6 hrs at 60°C. The solutions changed from red (pH=7.4) to yellow (pH=6.5-6.8) with slightly increased acidity. Elemental analysis of the coated samples was conducted using EDS (Tescan Vega-II XMU VPSEM and Oxford Inca Energy 250X EDS) and the formation of a HA phase was determined using XRD (Phillips 1710 X-ray Diffractometer).

Table 3. Ion concentration of blood plasma and Hank's balanced salt solution.

Species	Blood Plasma (mmol/L)	Hanks Balanced Salt Solution (mmol/L)
Ca^{2+}	2.5	2.2
HPO_4^{2-}	1.0	0.8
Na^+	142.0	140.3
Cl^-	148.8	148.0
Mg^{2+}	1.5	1.3
K^+	5.0	5.3
SO_4^{2-}	0.5	0.5
HCO_3^-	4.2	4.0

After four weeks of immersion in the SBF each sample was weighed in order to calculate the mass change. To account for the uncoated sides of the samples which were also in contact with the SBF, two samples of commercially pure titanium substrates were also immersed SBF and weighed to calculate the average weight change. This weight change was then subtracted from the weight changes measured for the coated samples. All mass changes given in the results section were normalized to the surface areas. The specimen preparation, apparatus set up, test conditions, methods of cleaning specimens, and evaluation of results were based on the methods and procedures described by the ASTM G 31 - Standard Practice for Laboratory Immersion Corrosion Testing of Metals.

3.0 Results and Discussion

3.1 Bio-functionality of plasma sprayed CP-Ti/bioglass and CP-Ti coatings

After four weeks of immersion in SBF, EDS elemental analysis was used as a preliminary method for evaluating the bio-functionality of CP-Ti/bioglass coatings. HA, the main mineral in bone contains both Ca and P, hence the change in weight percent of Ca on each coating surface before and after SBF immersion was measured. The average weight percent of Ca before and after immersion in SBF for both plasma sprayed CP-Ti (**Fig. 1.**) and CP-Ti/bioglass (**Fig. 2.**) coatings are presented, **Table 4**. After immersion in SBF the CP-Ti/bioglass coatings showed a 12 wt. % increase in Ca whereas negligible Ca was detected on CP-Ti coatings, suggesting that the addition of bioglass to plasma sprayed CP-Ti enhances bio-functionality.

Table 4. wt. % of calcium before and after four weeks of immersion in SBF.

Coating	Approximate wt. % of calcium before immersion	Approximate wt. % of calcium after immersion
CP-Ti	0	0
CP-Ti/bioglass	4	16

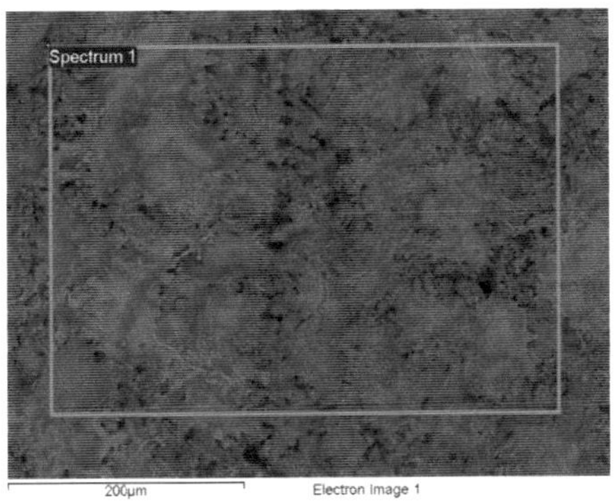

Fig.1. CP-Ti plasma sprayed coating

Fig. 2. CP-Ti/bioglass plasma sprayed coating

To furthur study the bio-functionality of CP-Ti/bioglass coating, XRD was used to detect HA formation following immersion in SBF. From XRD patterns shown in **Fig. 3.** it can be concluded that HA has formed on both the CP-Ti and CP-Ti/bioglass coatings. The diffraction peaks labeled in the figures are similar to those found on plasma sprayed titanium coatings after in vitro testing in other studies [2,4,5]. Several identical peaks were observed on XRD spectra for both the CP-Ti and CP-Ti/bioglass coatings. They were identified as titanium and titanium oxide (rutile). The rutile was formed during plasma spraying when titanium powder was exposed to oxygen and high temperatures [5]. Other phases in bioglass were not detected on CP-Ti/bioglass sample during XRD analysis possibly due to their amorphous nature [5].

The relative intensities of HA peaks (to Ti peaks) formed on the two coatings were different. The relative intensity of HA formed on the CP-Ti/bioglass coating was greater than that of HA on the CP-Ti coating. This is clearly due to the addition of bioglass in the composite coating assisting the HA formation. It has been shown that when bioglass is subjected to SBF, dissolution of calcium and phosphorous ions occurs helping to supersaturate the solution and increase HA precipitation on the coating surface [4]. The HA formation on CP-Ti is believed to be very superficial as EDS could not detect the presence in calcium. The XRD results coincide with EDS, suggesting that the addition of bioglass to plasma sprayed CP-Ti coatings enhances bio-functionality and promotes HA formation on the coaitng surface.

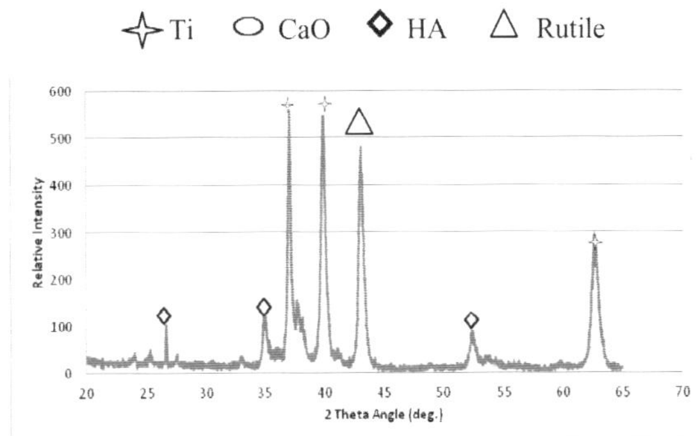

(a) CP-Ti

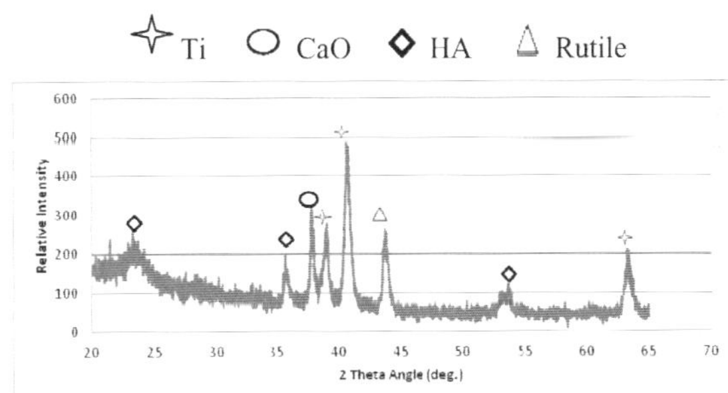

(b) CP-Ti/Bioglass

Fig. 3. (a) XRD pattern of CP-Ti coating after four weeks in SBF and (b)XRD pattern of CP-Ti/bioglass coating after four weeks in SBF

In addition to the difference in HA intensities of the two coatings, CP-Ti/bioglass coating also contained a prominent crystalline calcium oxide peak (Figure 4). This was attributed to the high temperatures reached during plasma spraying of bioglass, converting some amourphous bioglass to crystalline calcium oxide [4]. This peak will likely decrease in intensity with prolonged exposure to SBF as calcium oxide continues to dissolve in SBF, adding more calcium ions to the SBF solution.

3.2 Corrosion resistance of CP-Ti and CP-Ti/bioglass coatings

The mass of each coating sample was measured before and after immersion in SBF to detect any mass gain or loss. The averaged mass change per unit area, after subtracting the weight change on the backside of the coating sample, is given in **Table 5**. The results showed that CP-Ti coating experienced a mass gain while CP-Ti/bioglass underwent a mass loss. The mass gain for CP-Ti can be attributed to the precipitation of HA onto the coated surface (as was detected by XRD). The mass loss for CP-Ti/bioglass was a result of bioglass dissolution in SBF. This dissolution of bioglass helps supersaturate the SBF with sufficient calcium and phosphorous ions for HA formation. However, while this mass loss may promote increased HA formation it can be detrimental as excessive calcium and phosphorous ions may be deposited into the surrounding tissues [9].

Table 5. Average change in mass after immersion in SBF for CP-Ti and CP-Ti/bioglass

Coating	Average Mass Change (mg/cm^2)
CP-Ti Substrate	0.307
CP-Ti Coating	0.753
CP-Ti/bioglass Coating	-1.703

4.0 Conclusion

Based on the observation of increased Ca and HA following SBF immersion this study suggests that the CP-Ti/bioglass composite coating is superior to the CP-Ti coating in terms of bio-functionality. However, in terms of corrosion resistance, the CP-Ti/bioglass coating had substantial weight loss when compared to CP-Ti coating. Further studies are needed to address the weight loss issue associated with plasma sprayed CP-Ti/bioglass coatings.

5.0 Acknowledgements

The following research could not have been completed without the help of several individuals who were kind enough to contribute to the progressions of our study. We would like to acknowledge Colin Miyata for helping with the preparation of the samples, Fred Barrett for his patience in conducting several plasma spraying trials, Jianqun Wang for being flexible and helpful in booking time for using the SEM/EDS and George Dicks for expanding our knowledge on XRD.

6.0 Literature

[1] Paital S.R. and N. B. Dahotre: Calcium phosphate coatings for bio-implant applications: materials, performance factors, and methodologies. Materials Science and Engineering 66 (2009), Issue 1-3, pp.1/70.
[2] Lee Y. C. Wang and T. Huang: In vitro characterization of postheat-treated plasma-sprayed hydroxyapatite coatings. Surface and Coatings Technology 197 (2005), pp.367/74.
[3] Morks M.F.: Plasma spraying of zirconia-titania-silica bio-ceramic composite oating for implant application. Materials Letters 64 (2010), pp. 1968/71.
[4] Gu Y.W and K. A. Khor: Activity of plasma sprayed yttria stabilized reinforced hydroxyapatite/Ti-6Al-4V composite coatings in simulated body fluid. Biomaterials 25 (2004), pp. 3177/85.
[5] Gu Y.W and K.A. Khor: In vitro studies of plasma-sprayed hydroxyapatite/Ti-6Al-4V composite coatings in simulated body fluid SBF. Biomaterials 24 (2003), pp. 1603/11.
[6] Gilbert J.L.: Mechanically assisted corrosion of metallic biomaterials. ASM Handbook- Corrosion: Environments and Industries, ASM International 13C (2006), pp. 827/35.
[7] St. John K.R.: Corrosion effects on the biocompatibility of metallic materials and implants. ASM Handbook-Corrosion - Environments and Industries, ASM International 13 C (2006), pp. 820/5.
[8] Callister W.D and D. G. Rethwisch: Materials Science and Engineering: An Introduction (8th ed.). John Wiley and Sons Inc (2010).
[9] Hoffman A.S., B. D. Ratner and F. J Schoen: Biomaterials Science: An Introduction to Materials in Medicine (2nd ed.). Elsevier Academic Press (2004).

Plasma Sprayed Hydroxyapatite Coatings on 3D Printed Polymer Substrates for Orthopedic Applications: A Preliminary Study.

L. Barillas[1,2], J. M. Cubero-Sesin[3], T. Guillén-Girón[3]

[1]Plasma Laboratory for Fusion Energy and Applications, Instituto Tecnológico de Costa Rica, Cartago, Costa Rica
[2]School of Electromechanical Engineering, Instituto Tecnológico de Costa Rica, Cartago, Costa Rica
[3]School of Materials Science and Engineering, Instituto Tecnológico de Costa Rica, Cartago, Costa Rica.

New manufacturing technologies such as 3D printing, have allowed wide range of medical devices from generic "one size fits all", expensive devices to the customized, more affordable options. Polymers can play an important role in customized medical devices such as orthopedic implants, due to their lower weight, cost and performance when compared with other materials. Additionally, polymers are widely sought for this kind of devices to support bone in-growth and to enhance the osseointegration. In order to further enhance their biocompatibility, these medical devices can be coated with bioactive materials, such as hydroxyapatite (HA). Among different thermal spray techniques, atmospheric plasma spraying (APS) can be used to obtain HA coatings on implants. Nevertheless, plasma sprayed HA coatings have been primarily applied to metallic substrates, and unfortunately, only a few studies have been performed on polymers and composite materials. This work aims to present a preliminary study of plasma sprayed HA coatings on biopolymer and metallic substrates by using a modified plasma cutting equipment. Furthermore, the HA coatings were characterized by thermal imaging, visual inspection and scanning electron microscopy (SEM).

1 Introduction

Biomaterials are substances from synthetic or natural origin that interact temporarily or permanently in the body, which augment or replace partially or totally any affected tissue, organ or bodily function, in order to maintain or improve the quality of life of the individual [1]. Within these biomaterials, polymers (bio-inert or bio-absorbable) are often used for medical devices due to their lower weight, cost and improved performance in comparison with other materials [2]. Along with these advantages, new and improved additive manufacturing technologies –like 3D printing– have allowed biopolymers to present new applications with great potential in the customized orthopedics industry and tissue engineering. Implants that were impossible to manufacture with traditional machining methods due to their complex geometries, are now being 3D printed to deliver fairly fast and affordable well-fitted solutions.

Therefore, there are good reasons to develop polymer composite implants, especially when compared to metallic counterparts. First, because of their nonmetallic nature, the generation of particles and ions formed at metallic implant surfaces is avoided. Secondly, because of their anisotropic properties and versatility in design, they can exhibit iso-elastic properties, and similar stiffness to their surrounding host tissues, which has been shown to promote bone ingrowth and reduce stress shielding [3]. Nonetheless, these biopolymers often lack adequate bioactivity to achieve proper osseointegration between the tissue and the implant surface [4,5]. Previous works [3–8] have suggested that one way to promote the desired osseoconductivity is with the functionalization of the implant's surface by coating it with bioactive materials, such as hydroxyapatite.

Hydroxyapatite is a ceramic biomaterial derived from calcium phosphate which is chemically similar to the primary mineral content of bones and teeth [6], and has proven excellent biocompatibility, bioactivity, and osseoconductivity, supporting bone ingrowth with the remaining bone or tissue structure when used in implants [3–8].

Atmospheric plasma spraying (APS) is a well-known deposition technique for bioactive hydroxyapatite coatings [5, 7, 10]. Nevertheless, plasma sprayed HA coatings have been primarily applied to metallic substrates, and only a few studies have been performed on polymers [3, 7].

This work shows a preliminary study of plasma sprayed HA coatings on three biopolymer substrates: Polylactic acid (PLA), Poly(oxy-1,4-phenyleneoxy-1,4-phenylenecarbonyl-1,4-phenylene) (PEEK) and Poly(vinyl alcohol) (PVA). As a proof-of-concept, the coatings were performed using a modified plasma cutter equipment instead of a conventional plasma spray system. Deposition over metallic substrates of Ti-6Al-7Nb alloy and commercial purity Al, was carried out for comparison purposes.

2 Materials and Methods

2.1 Test Materials

Hydroxyapatite ($HCa_5O_{13}P_3$) used in this work had a purity of ≥90% and an average grain size of 5 µm (distributed by Sigma-Aldrich).

The polymers used in this work were Poly(vinyl alcohol) (PVA), Mw: 146 000-186 000, 99+%hydrolyzed and Poly(oxy-1,4-phenyleneoxy-1,4-phenylenecarbonyl-1,4-phenylene) (PEEK), Mw: 20

800 (distributed by Sigma-Aldrich). Additionally, a 1.75 mm diameter Polylactic acid (PLA) filament (distributed by Hacthbox) was used for 3D printing of specimens.

PLA discs of 30 mm diameter and 2 mm thickness, with equally distributed through holes of 2 mm and 4 mm diameter (as shown in **Fig. 1**), were 3D printed using fused deposition modeling (FDM) technology (printing parameters: 0.2 mm layer height, 0.8 shell thickness, 0.4 mm bottom/top thickness, 70% fill density, 45 mm/s print speed, 210 ºC printing speed). This geometry was selected to emulate the design of a fixation plate used in orthopedic applications.

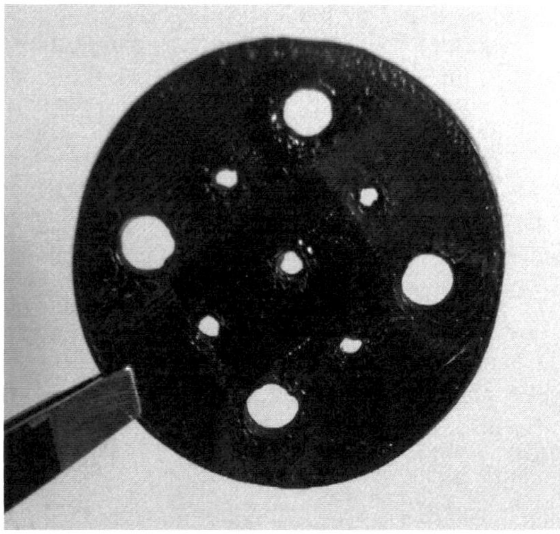

Fig. 1. PLA specimen to be coated with HA fabricated by 3D printing.

PEEK specimens of approximately 30x20x3 mm^3 were obtained by a hot-pressing method to melt pellets and form a compacted substrate with the aforementioned dimensions. Similar PVA specimens were obtained using the technique described for PEEK specimens. Previous to the HA deposition, the surfaces of the polymer samples were ground with 240-grit SiC polishing paper. One PLA sample was left to be coated in an as-printed condition.

Aluminum sheet specimens of 40x40x1.5 mm^3 were also ground with 240 grit grade SiC sandpaper. A 25 mm diameter and 1 mm thickness disc of Ti-6Al-7Nb extracted from a hot-extruded bar by electrical discharge machining (EDM), was also used. The surface of the disc was ground to eliminate the oxidized layer from the EDM process. All metallic specimens were pre-heated with the plasma gun for 5 seconds prior to the HA deposition.

All specimens were cleaned with ethylic alcohol 80% v/v prior to the coating process. After coating, the samples were air blown at a pressure of 8 bar and swept with a soft brush, in order to discard large amounts of unmolten particles on the surface of the samples.

2.2 Equipment and Experiment Setup

Since no commercial plasma spray equipment was available, a Thermal Dynamics model Pak Master 50XL plasma cutting equipment was used. A custom made bronze nozzle was fabricated to serve as an attachment to the plasma torch (model PCH/M-28) in order to inject the hydroxyapatite powder, as shown in **Fig. 2**. The original parts of the torch head were kept, including the crown shield cup. The nozzle had two injection ports perpendicular to the plasma jet at different distances from the torch outlet, but only the closest injection port to the torch outlet (at 7.3 mm) was used. **Fig 3.** shows the CAD schematic of the nozzle.

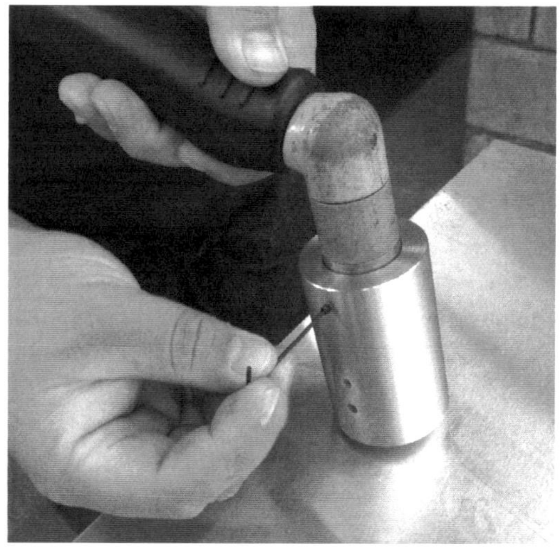

Fig. 2. Attachment of nozzle to plasma cutter torch.

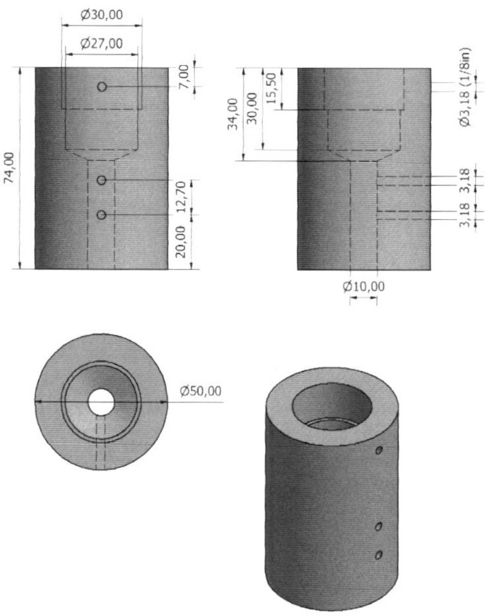

Fig. 3. Schematic of nozzle designed as attachment to plasma cutter torch.

All samples were coated using a power set of 208 V_{ac} single phase input voltage and a current output of 28±2.5 A, using compressed air at 8 bar as the plasma gas. The plasma gun was placed vertically, as well as the specimens, at a stand-off distance of 70 mm from the torch outlet. **Fig. 4** shows the experiment set up used in this study.

Since an automatic or autonomous powder deposition chamber was not available at the time of the experiment to inject the HA, a bulb syringe was used, placing the powder inside the bulb and manually squeezing it. For all polymer and aluminum specimens, the bulb was squeezed once, whereas for the Ti-6Al-7Nb substrate it was squeezed 3 times (in approximately less than 1.5 second intervals). Immediately after injection all specimens were removed from the heat source and left to cool in air.

2.3 Characterization

Scanning electron microscopy was performed using a Hitachi TM-3000 under an acceleration voltage of 15 kV to evaluate splat configuration, coating uniformity and thickness. The polymeric samples were coated with a thin layer of Au-Pd by physical vapor deposition. A cross-sectional specimen from the Ti-6Al-7Nb sample was prepared by embedding the sample in epoxy resin before cutting it with a high-speed diamond disk.

A thermal imaging camera (FLIR i50) was employed in order to estimate the temperature of the polymeric substrates immediately after coating.

Fig. 4. Experimental setup used to apply HA coating to polymeric and metallic specimens.

3 Results

The results were sought in terms of surface macro and microstructure. Initially, the coated specimens were visually inspected for thermal damage or signs of burning, particularly in the polymers. The microstructure of the coating, regarding morphology, splat configuration and uniformity were analyzed by SEM.

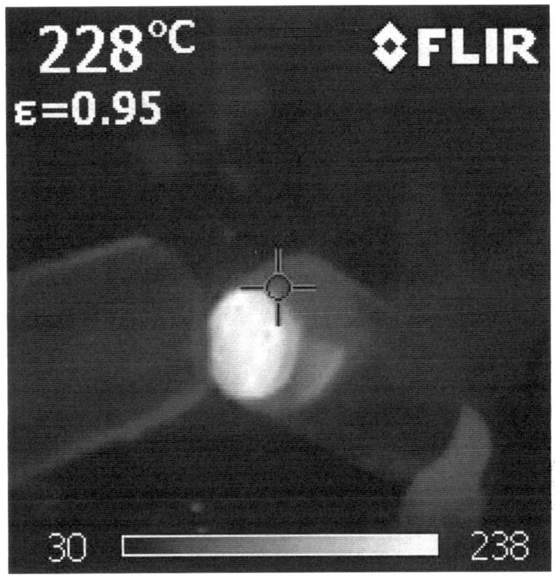

Fig. 5. Thermal image of PLA substrate, represented by brightest circular object.

3.1 Visual Inspection

Some degree of thermal damage to the polymer surfaces was visible, represented mainly by fine blisters and slightly deformed edges (mainly in the holes of the PLA samples). However, they still maintained their original shape. Although some regions of the polymer samples reached approx. 230 - 290 °C according to the thermal image shown in **Fig. 5**, no signs of severe burning or discoloration were observed. Metallic substrates were not affected by the plasma.

3.2 SEM Analysis

On the polymeric substrates, widespread regions of deposited HA were not found by SEM observations. However, splats were detected randomly across the surface of PLA samples, as shown in **Fig. 6,** which corresponds to the surface-ground specimen in **Fig. 6(a)** and as-printed specimen in **Fig. 6(b)**. **Fig. 7** shows similar images for (a) the PEEK and (b)the PVA samples.

Figure 8 shows SEM images of Ti-6Al-7Nb samples, which display a more uniform coated surface, when compared to the polymeric samples. Visible pores, cracks and splat stacking can be observed by close inspection of the surfaces of PEEK in **Fig. 8(a)** and PVA in **Fig 8(b)**. **Figure 8 (c)** shows the cross section of the sample where a layer of HA of nearly 50 µm can be identified in bright contrast. **Figure 9** shows SEM images of the surface of the aluminum sample. **Figure 9(a)** shows a region with several splats of different sizes can be seen. **Figure 9(b)** shows a magnified view of one of the splats with some porosity visible in it.

4 Discussion and Conclusions

In the plasma spraying technique several conditions must be met to achieve uniform coatings, according to Fauchais [9]. Especially, all particles that are injected should be uniformly heated and reach the substrate with a temperature over their melting point (but not overheated to reach vaporization, for example), with velocities as high as possible but compatible with a fully molten state, since high velocities decrease particle residence time and thus affecting their heating.

As shown in the results of SEM observations for PLA, PEEK and PVA specimens, it was not easy to find widespread zones where it was clear that the HA reached a semi-molten state to properly cover a large area of the surface. This could be due to several reasons. First, the way the specimens were exposed to the plasmatron was such that it would not physically affect the polymers by the high temperatures (PLA glass transition (Gt)=60-63 °C, melting point (Mp)=170-180 °C; PEEK Gt=~143 °C, Mp=~322 °C; and PVA Gt=75-85 °C, Mp= 180-228 °C). In this sense, the spraying was performed intentionally as a 'wipe' test, where the substrate is moved quickly across the loaded plasma jet to collect individual particle splats. This method serves as a useful technique to estimate the approximate spray parameter settings, in terms of plasma energy and spray distance [8].

It is known that substrate roughness controls mechanical properties and residual stresses of coatings [8], which could partially explain the differences between splat morphologies seen in **Fig. 6 (a)** and **(b)** for PLA specimens with different surface preparations. It can be seen in **Fig. 6(b)** that the energy supplied to the HA particles was not sufficient to achieve melting in all of them. Also, if **Fig. 6(a)** is compared to **Fig. 6(b)**, one could state that in the first one more particles achieved a semi-molten state, than in the second one, even causing some rifts on the surface, as pointed by arrows on **Fig 6 (a).** These rifts were observed in the PEEK sample on **Fig. 7(a)** as well. In the case of the PVA specimen (**Fig. 7 (b)**), a tiny splat with almost no pores and a few cracks represents a state close to ideal melting [8]. Although some degree of porosity is highly desired to favor implant bioresorption and osseoconductivity [5, 9], the case of the PVA represents perhaps the best deposition on polymers in this study. More tests are required to further improve the coatings substrates, but according to the SEM observations in **Fig. 6** and **Fig. 7**, there is potential for application of widespread coated areas if instrumental variables can be further controlled.

Deposition on metallic specimens was carried out to compare the capabilities of the equipment setup used in this study, with a well-known technology.

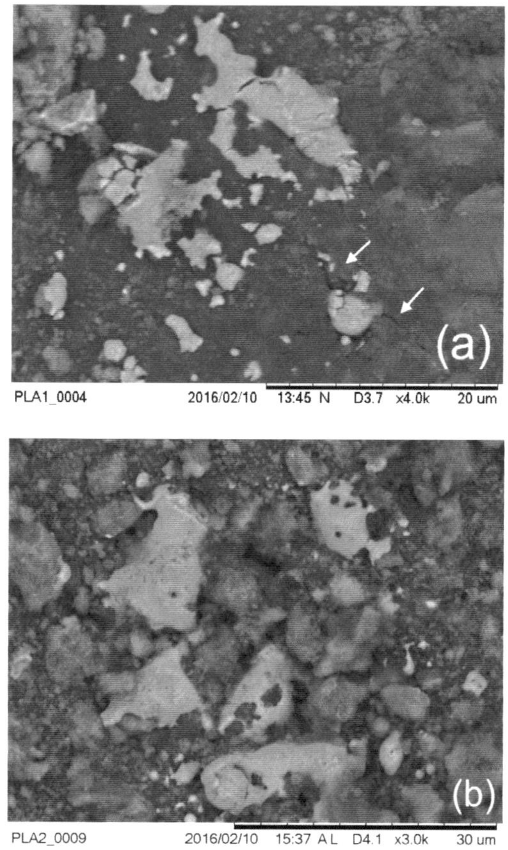

Fig. 6. PLA specimens with HA splats over (a) grinded surface and (b) as-3D printed surface.

The results are satisfactory in the metallic samples, concerning number of splats, morphology and coating uniformity. A widespread region with HA coating was obtained in the Ti-6Al-7Nb specimen, as shown in **Fig. 8**, which shows characteristics comparable with successful depositions in different studies regarding porosity, cracks and splat stacking [5, 6, 8, 10]. One factor that could play an important role is the pre-heating of the surface, since surface heating or cooling influences the splat configuration, residual stresses of coatings and morphology of splats. It is also possible to control the splat shape by enhancing surface diffusion; therefore preheating leads to improved coating integrity, microstructure and other mechanical and thermal properties [8]. This was not attempted for the polymeric substrates and it could be considered as a contributing factor to the difference in the characteristics of the splats. The cross-section SEM image of **Fig. 8(c)** resulted in a layer thickness ranging from 20 - 50 μm, approximately. In most cases, a thin HA coating (<50 μm) yields better adhesion compared with thicker coatings, owing to reduced residual coating stresses. However, it will quickly reabsorb in the course of bone integration, on the one hand. On the other hand, thicker coatings are needed to stimulate bone reconstruction over longer periods to ensure a more permanent bond [8].

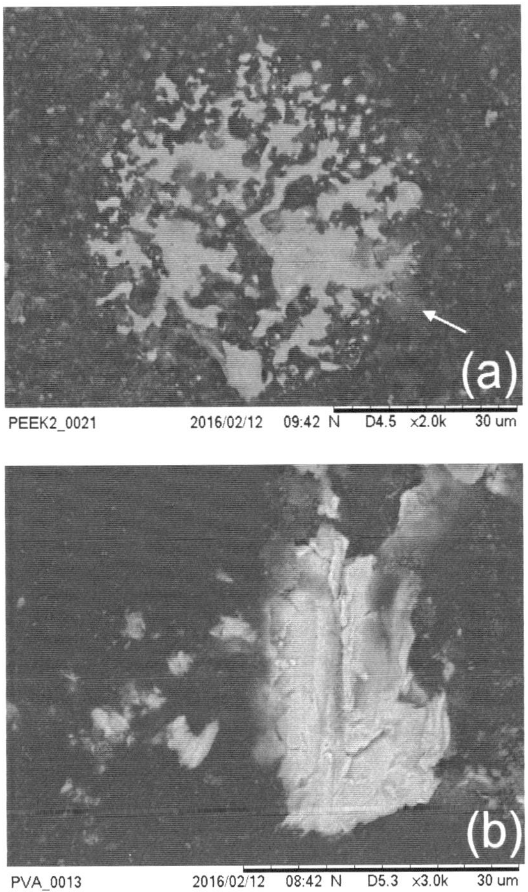

Fig. 7. Polymer specimens with HA splats, where (a) corresponds to PEEK, and (b) corresponds to PVA.

As for the aluminum specimen shown in **Fig. 9**, splat configuration displays what is known as 'frozen-in-time' traces of severely overheated HA particles, resulting in 'exploded' splat geometries [8]. This usually happens when plasma power is increased and/or stand-off distance is decreased. The latter could be the most probable cause since the power was not changed for any specimen, however stand-off distance was measured but not meticulously controlled (human dependent).

The results of the SEM observations carried out in this work demonstrate the successful deposition of HA to a certain extent on polymer and metallic substrates, with a non-conventional plasma spray device. Dey and Mukhopadhyay [6], performed successful HA coatings on metals using a non-industrial plasma spray setup, consisting of a micro-plasma spraying equipment from a commercial welder, using a low plasmatron power below 1.5 kW and a powder feeder chamber.

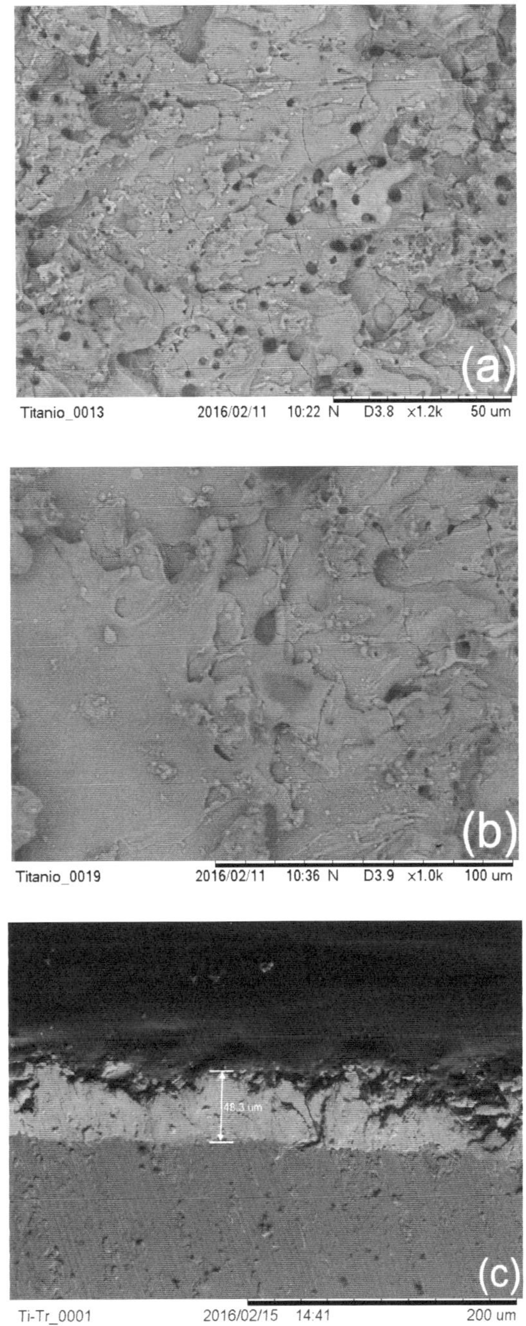

Fig. 8. Ti-6Al-7Nb coated with HA. (a) and (b) are surface images, while (c) is a cross-section image showing approximate HA layer thickness (brightest).

Based on a similar strategy, there were good reasons for using a modified plasma gun to attempt deposition of HA coatings on different substrates. This alternative represents a more economical equipment to apply coatings, especially on polymeric substrates due to their potential application for orthopedic implants. Even though more tests must be performed on the proposed technique, the results obtained show a promise that uniform coatings can be achieved if critical parameters, such as power input, arc gas type and pressure are controlled. Additional variables such as power feed rate and gas, grain size and shape, injection angle, surface roughness and substrate pre-heating, as well as spray distance and divergence, are considered in a future study, with a proper DOE to collect significant data on the conditions for plasma-sprayed HA on polymeric substrates.

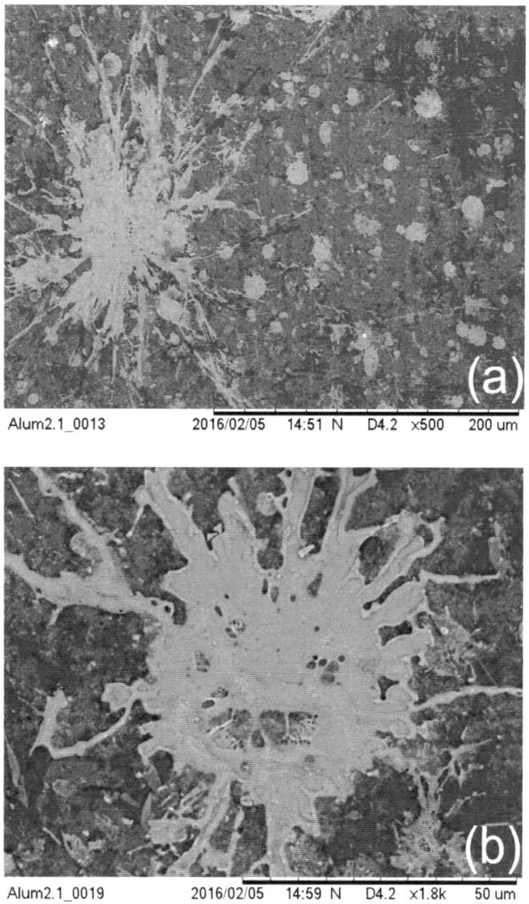

Fig. 9. Aluminum specimens coated with HA. Image (a) shows several splats of different sizes, and (b) a close up to one of the splats.

Acknowledgements

The authors would like to thank Mr. Alejandro Medaglia, Mr. Fernando Alvarado and Ms. Daniela Murillo from the Microscopy Laboratory for their support with the SEM observations. Also, Mr. Pablo Campos from the Electromechanical Workshop, Ms. Diana Morales, Ms. Dayana Méndez, Ms. Verónica Valverde, Mr. Allan Carmona, Mr. Johel Fernández, Mr. Alexander Gatica and Mr. Augusto Arce for their aid with the setup of the plasma equipment, all from the *Instituto Tecnológico de Costa Rica*.

References

[1] C. P. Bergmann and A. Stumpf, "Biomaterials," in Dental Ceramics, 1st ed., Springer-Verlag Berlin Heidelberg, 2013, pp. 9–14.
[2] L. W. McKeen, "Plastics Used in Medical Devices," in Handbook of Polymer Applications in Medicine and Medical Devices, Sina Ebnes., Elsevier Inc., 2014, pp. 21–53.
[3] C. Auclair-Daigle, M. N. Bureau, J. G. Legoux, and L. Yahia, "Bioactive hydroxyapatite coatings on polymer composites for orthopedic implants," J. Biomed. Mater. Res. - Part A, vol. 73, no. 4, pp. 398–408, 2005.
[4] R. Ma and T. Tang, "Current strategies to improve the bioactivity of PEEK," Int. J. Mol. Sci., vol. 15, no. 4, pp. 5426–5445, 2014.
[5] P. Robotti, G. Zappini, and S. M. Kurtz, "Thermal Plasma Spray Deposition of Titanium and Hydroxyapatite on Polyaryletheretherketone Implants," in PEEK Biomaterials Handbook, 1st ed., Massachussets: Elsevier Inc., 2012, pp. 119–143.
[6] A. Dey and A. K. Mukhopadhyay, Microplasma Sprayed Hydroxyapatite Coatings. Kolkata, India: CRC Press Taylor & Francis Group, 2015.
[7] S. Beauvais and O. Decaux, "Plasma Sprayed Biocompatible Coatings on PEEK Implants," Therm. Spray 2007 Glob. Coat. Solut., pp. 371–377, 2007.
[8] R. B. Heimann, Plasma Spray Coating: Principles and Applications, 2nd ed. Weinheim: Wiley-VCH Verlag GmbH & Co. KGaA, 2008.
[9] P. Fauchais, "Understanding plasma spraying," J. Phys. D. Appl. Phys., vol. 37, no. 9, pp. R86–R108, 2004.
[10] R. Gadow, a. Killinger, and N. Stiegler, "Hydroxyapatite coatings for biomedical applications deposited by different thermal spray techniques," Surf. Coatings Technol., vol. 205, no. 4, pp. 1157–1164, 2010.

Liquid plasma-sprayed nanonetwork $La_{0.4}Sr_{0.6}Co_{0.2}Fe_{0.8}O_3/Ce_{0.8}Gd_{0.2}O_2$ composite cathodes for intermediate-temperature solid oxide fuel cells

Shan-Lin Zhang, Chang-Jiu Li, Cheng-Xin Li, Guan-Jun Yang, State Key Laboratory for Mechanical Behavior of Materials, School of Materials Science and Engineering, Xi'an Jiaotong University, Xi'an, Shaanxi, China, E-mail: licj@mail.xjtu.edu.cn

Here, we investigated the feasibility of using a liquid plasma spray process as a novel method for the cost-effective fabrication of a nanonetwork of $La_{0.4}Sr_{0.6}Co_{0.2}Fe_{0.8}O_3$ (LSCF) and $Ce_{0.8}Gd_{0.2}O_2$ (GDC) composite as a high-performance cathode for intermediate-temperature solid oxide fuel cells (IT-SOFCs). A suspension containing well-dispersed nanosized GDC particles in an LSCF precursor solution was designed as the feedstock. Particularly, the effects of GDC concentration in the suspension on the phase composition, microstructure, and electrochemical performance of the resulting composite cathode were studied. The GDC content in the final cathode increased with GDC concentration. When the GDC concentration increased to 15 g L^{-1}, the nanosized GDC particles distributed uniformly and continuously on the LSCF backbone to form a porous and network structure. The electrochemical studies further indicate that the cathode polarization decreased with the increase in GDC concentration from 0 g L^{-1} to 15 g L^{-1}, whereas a further increase in the GDC concentration increased the cathode polarization instead because of the increased size of GDC clusters and decreased number of active sites. At 750 °C, the cathode prepared using the suspension with a GDC concentration of 15 g L^{-1} exhibited an impressive area-specific polarization resistance of 0.009 Ω cm^2.

1 Introduction

Solid oxide fuel cells (SOFCs) convert chemical energy stored in a fuel to electricity with many advantages such as fuel flexibility, high efficiency, low emission, and environmental friendliness [1-4]. To commercialize the SOFC technology, a major research effort in recent years has been to reduce the operating temperature to below 700 °C [5], where cost and reliability, the two main challenges of SOFC technology, can be best met [4, 6-10]. However, lowering the operating temperature would not only increase the electrolyte resistance, but also decrease the kinetics of electrode reactions, resulting in an appreciable increase in the cell resistance [11, 12]. With the use of thin-film electrolytes, the performance of intermediate temperature SOFCs (IT-SOFCs) has become more limited by cathode polarization. Therefore, a critical step towards IT-SOFCs is the development of new materials [13-15] or novel microstructures with a low polarization loss [16, 17].

$La_{0.6}Sr_{0.4}Co_{0.2}Fe_{0.8}O_3$ (LSCF)-based cathodes have been extensively investigated for IT-SOFCs because of significantly higher mixed ionic and electronic conductivity in the temperature range 600–800 °C [18, 19]. $Ce_{0.8}Sm_{0.2}O_2$ (SDC) [20], $Ce_{0.8}Gd_{0.2}O_2$ (GDC) [21], and a precious metal such as Pt, Pd, and Ag [22-24] have been commonly blended with LSCF to further increase the performance by increasing oxygen reduction reaction (ORR)-active triple-phase boundary (TPB). Many studies have clearly indicated that the polarization resistance of LSCF/GDC composite cathode is several times lower than that of the pure LSCF cathode [25, 26]. Moreover, the cathode performance is not only affected by the material itself, but also by the microstructure, mainly controlled by the fabrication process. The most common fabrication methods for LSCF-based composite cathodes follow either a screen printing or an infiltration approach [27-29]. In particular, the nanostructures obtained by the infiltration method yield a low polarization resistance in the IT-range. In this method, a porous LSCF backbone is first prepared by the screen printing or tape casting method. Then, the backbone is infiltrated with the GDC or SDC precursor solution. After the firing, the precursor decomposes into GDC or SDC nanoparticles on the surface of the LSCF backbone. One disadvantage of this method is the high temperature (>900 °C) required for sintering the porous LSCF backbone in an oxidation atmosphere, which is not suitable for metal-supported SOFCs. Moreover, the infiltration process is tedious and production unfriendly, requiring many steps to achieve the desired mass loading. Therefore, alternative fabrication processes should be developed to prepare high-performance LSCF-based composite cathodes at low temperatures and with a low cost.

Suspension plasma spraying (SPS) and solution precursor plasma spraying (SPPS) techniques have been recently developed for the cost-effective fabrication of SOFCs. Plasma spraying (PS) is a widely proven low-cost, large-scale industrial coating process. One of the advantages of PS for SOFCs is the fact that the sprayed cells are ready for assembly, and further processing of the coatings at higher temperatures is not required after the deposition [30, 31]. SPS and SPPS are two derivatives of conventional PS where a liquid solution is used instead of solid particles as the feedstock. The advantage of SPS and SPPS over the conventional PS is the ability to fabricate nanostructures; this is critical for increasing the ORR activity of IT-SOFCs. We have previously demonstrated that SPS can be used to deposit a high-performance composite anode [28]. However, to the best of our knowledge, only a few studies have been reported on the fabrication of LSCF-based composite cathode by SPS. Particularly for SPPS, impure phases are frequently formed during the spraying process when two types of liquid precursors are mixed, thus making it difficult for SPPS deposition for composite cathode application.

We recently observed that high-performance LSCF/GDC composite cathodes can be deposited by a combination of SPS and SPPS. In this study, nanosized GDC particles were added to the LSCF precursor solution, affording a suspension as the feedstock. The precursor solution may react with LSCF under heating a plasma gas, whereas GDC particles adhere to LSCF, forming composite agglomerates in the final deposits. Compared to the high melting point of GDC (~2500 °C), LSCF with a much lower melting point (~1890 °C) is easier to sinter and melt during the spraying. Therefore, the LSCF may form backbones with larger connected grains; nanosized GDC particles adhere to them, forming a network structure, and provide more TPBs.

In this study, the phase composition, microstructure, and electrochemical performance of SPS–SPPS-derived LSCF/GDC composition cathodes for IT-SOFCs were systematically characterized.

2 Experimental

Stoichiometric amounts of $La(NO_3)_3 \cdot 6H_2O$ (99.99%, SCRC), $Sr(NO_3)_2 \cdot 6H_2O$ (99.99%, SCRC), $Co(NO_3)_2 \cdot 6H_2O$ (99.99%, SCRC), and $Fe(NO_3)_3 \cdot 9H_2O$ (99.99%, SCRC) were dissolved in a water/alcohol (99.7%, Ante) solvent (1:1 vol.%), affording a nitrate solution of 0.05 M concentration. Nanosized GDC particles (~50 nm, Fuel cell materials, USA) were then added to the precursor solution, affording a suspension. The GDC concentration of the suspension was controlled at 0, 5, 10, 15, and 20 g L^{-1}. The prepared suspension was sprayed using a conventional air plasma spraying system (GDP-80, Jiujiang, China). Fig. 1 shows the schematic diagram of the spraying process. The suspension was injected into the plasma jet using a peristaltic pump (Qdos 30, England). The parameters for spraying LSCF/GDC cathodes are shown in Table 1.

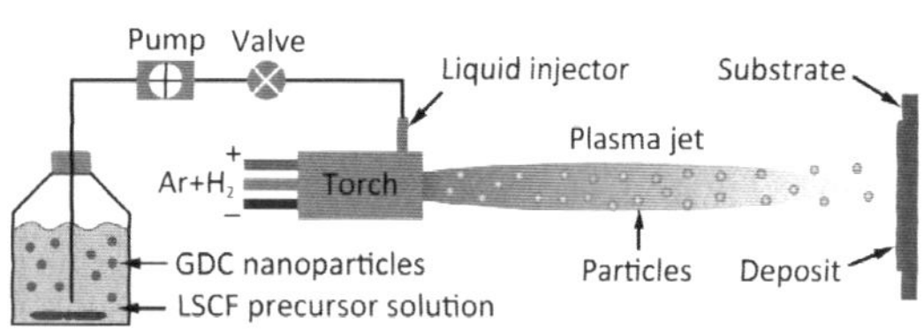

Fig. 1. Schematic diagram of the spray process.

Sc_2O_3-stabilized ZrO_2 (ScSZ) pellet was used as the substrate for deposition; ScSZ served as the electrolyte because of its high ionic conductivity at intermediate temperatures (600–800 °C). To prepare ScSZ substrates, ScSZ powders (~300 nm, TERIO, Qingdao, China) were pressed into pellets with a diameter of 15 mm under a pressure of 100 MPa. The pellets were then densified at 1400 °C for 5 h. The final products had a diameter of 13 mm and a thickness of ~1 mm.

To characterize the cathode performance, symmetric cells of the LSCF/GDC|ScSZ|LSCF/GDC were prepared by depositing approximately 20–30 μm thick LSCF/GDC composite on both sides of the ScSZ substrate with a diameter of 8 mm. Prior to the electrochemical characterization, silver paste/meshes were attached as the current collectors for the symmetric cells and annealed at 700 °C in air for 2 h. The electrochemical impedance spectroscopy (EIS) of the symmetrical cells was then carried out using a Solartron SI 1260/1287 impedance analyzer in ambient air under open-circuit voltage (OCV) conditions. The sweeping frequency range was 0.1–10^5 Hz at an amplitude of alternating circuit (AC) voltage of 20 mV.

The phase compositions of the prepared cathodes were investigated by X-ray diffraction (XRD) analysis (X'Pert PRO, PANalytical, Netherlands). The microstructures of the cathodes were analyzed using a scanning electron microscope (SEM, TESCAN MIRA 3 LMH, Czech). The distribution and content of GDC in the final cathodes were quantified by energy-dispersive X-ray (EDX) analysis.

Table 1 Spray parameters for making LSCF/GDC composite cathode

Parameter	Unit	Value
Plasma arc power	kW	36
Arc voltage	V	60
Arc current	A	600
Flow rate of Ar	slpm	50
Flow rate of H_2	slpm	4.5
Solution feed rate	ml min^{-1}	50
Deposition temperature	°C	200–400
Spray distance	mm	80
Thickness	μm	20–30

Sc_2O_3-stabilized ZrO_2 (ScSZ) pellet was used as the substrate for deposition; ScSZ served as the electrolyte because of its high ionic conductivity at intermediate temperatures (600–800 °C). To prepare ScSZ substrates, ScSZ powders (~300 nm, TERIO, Qingdao, China) were pressed into pellets with a diameter of 15 mm under a pressure of 100 MPa. The pellets were then densified at 1400 °C for 5 h. The final products had a diameter of 13 mm and a thickness of ~1 mm.

To characterize the cathode performance, symmetric cells of the LSCF/GDC|ScSZ|LSCF/GDC were prepared by depositing approximately 20–30 μm thick LSCF/GDC composite on both sides of the ScSZ substrate with a diameter of 8 mm. Prior to the electrochemical characterization, silver paste/meshes were attached as the current collectors for the symmetric cells and annealed at 700 °C in air for 2 h. The electrochemical impedance spectroscopy (EIS) of the symmetrical cells was then carried out using a Solartron SI 1260/1287 impedance analyzer in ambient air under open-circuit voltage (OCV) conditions. The sweeping frequency range was 0.1–10^5 Hz at an amplitude of alternating circuit (AC) voltage of 20 mV.

The phase compositions of the prepared cathodes were investigated by X-ray diffraction (XRD) analysis (X'Pert PRO, PANalytical, Netherlands). The microstructures of the cathodes were analyzed using a scanning electron microscope (SEM, TESCAN MIRA 3 LMH, Czech). The distribution and content of GDC in the final cathodes were quantified by energy-dispersive X-ray (EDX) analysis.

3 Results and discussion

3.1 Phase composition and microstructure

Fig. 2 shows the XRD patterns of the cathodes deposited by different suspensions. When GDC particles were not added to the LSCF precursor solution, the pure LSCF cathode exhibited a single phase of perovskite structure; the XRD pattern was similar to the standard XRD pattern of an $La_{0.6}Sr_{0.4}Co_{0.8}Fe_{0.2}O_3$ (PDF 48–0124), indicating that the cathode had a pure phase of LSCF. When GDC particles were added to the LSCF precursor solution, the cathode showed a mixture of LSCF and GDC. The XRD patterns show that the intensity of the diffraction peak for GDC increased with GDC concentration, indicating an increase in the GDC content in the cathode. Moreover, impurity phases were not observed, indicating that the GDC did not react with the precursor or LSCF during the spraying.

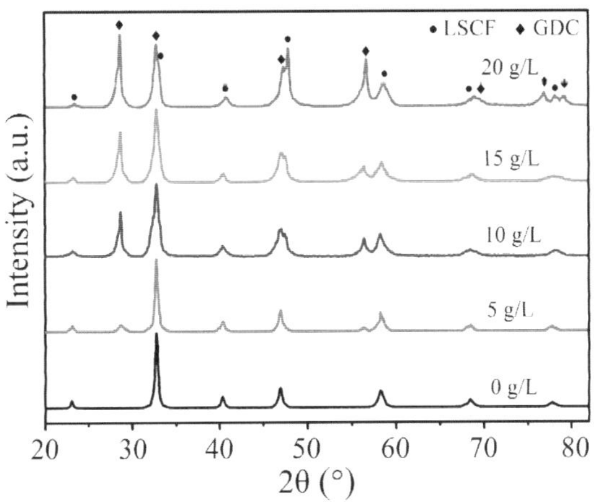

Fig. 2. XRD patterns of the cathodes deposited by the different suspensions containing different contents of GDC particles.

Fig. 3 shows the morphologies of the cross-section of the cathodes. The lower magnification images clearly show that the all the cathodes have a thickness of 20–30 μm and are well bonded with the ScSZ electrolyte. Fig. 3 (b) shows that the pure LSCF

Fig. 3. Fractured surface morphologies of LSCF/GDC cathodes deposited by different concentration suspension; (a) and (b) 0 g L^{-1}, (c) and (d) 5 g L^{-1}, (e) and (f) 10 g L^{-1}, (g) and (h) 15 g L^{-1}, (i) and (j) 20 g L^{-1}.

cathode has a high porosity and is composed of nano and submicron particles. The LSCF particles are well connected by the sintering-like necks. For the LSCF/GDC composite cathodes, LSCF formed the backbone for adhering GDC particles. This also indicates that GDC particles are distributed along the LSCF surfaces as well as interfaces between the LSCF particles. When the cathode was deposited by the 15 g L^{-1} suspension, the GDC particles on the surface of LSCF backbone connected to form a porous network. Besides, the results indicate that the network structure of GDC bonded well with the LSCF backbone. Furthermore, more GDC particles were observed at a higher suspension concentration. When a 20 g L^{-1} suspension was used, almost all the LSCF surface was covered with the GDC particles.

3.2 Eeffect of suspension concentration on GDC content

To determine the GDC content and its distribution in the final cathodes, the cross-sections of the polished cathodes were analyzed. To prevent the removal of particles in the porous cathodes during the sample preparation, the samples were sealed with epoxy adhesives by vacuum infiltration before the sectioning. Figs. 4 (a) and (b) show the microstructure of the cathode deposited by the 15 g L^{-1} suspension. The lower magnification, Fig. 4 (a), shows that the cathode has a porous and island-like structure. Microsized channels are present between the "microislands", facilitating oxygen transport in the cathode. Furthermore, as indicated by the black arrows, the attachment of cathode to the electrolyte is excellent. Except for the large channels, based on the distribution of the infiltrated resin, it can be inferred that most of the pores are present in the stacking gaps between the small particles. Fig. 4 (b) shows that the cathode has a porosity of >40%. Fig. 4 (c) shows the corresponding EDX element mapping for Fig. 4 (b), indicating that the LSCF and GDC are uniformly distributed across the cathode. Besides, according to the EDS analysis, the GDC content in different cathodes can be obtained.

Fig. 5 shows the effect of GDC concentration in the suspension on the resulting GDC content in the composite cathodes. This indicates that with the increase in the GDC concentration, the GDC concentration in the final cathodes also increased as expected. The GDC content (wt.%) in the final cathodes deposited by the 5, 10, 15, and 20 g L^{-1} suspensions were 8.6%, 20%, 32%, and 46%, respectively. Notably, the final GDC content is much lower than that in the feedstock suspension (red line), indicating a loss of GDC to the surrounding during the spraying. This can be attributed to two factors. First, the GDC particles have a very small diameter and thus showed a small inertial. When they are not attached to the LSCF particles, they easily flow with the plasma gas and do not deposit on the substrate. Second, GDC has a much higher melting point than LSCF, with a difference of ~800–900 °C. During the spray process, it is more difficult to melt the GDC particles. Moreover, even when the GDC particles are melted, they easily cool down before reaching the substrate due to their small size. Therefore, most of the GDC particles deposited onto the final cathodes adhered to the LSCF particles. Therefore, the GDC content in the final cathodes was lower than that in the starting solution. Therefore, a loss of GDC during the spraying was expected.

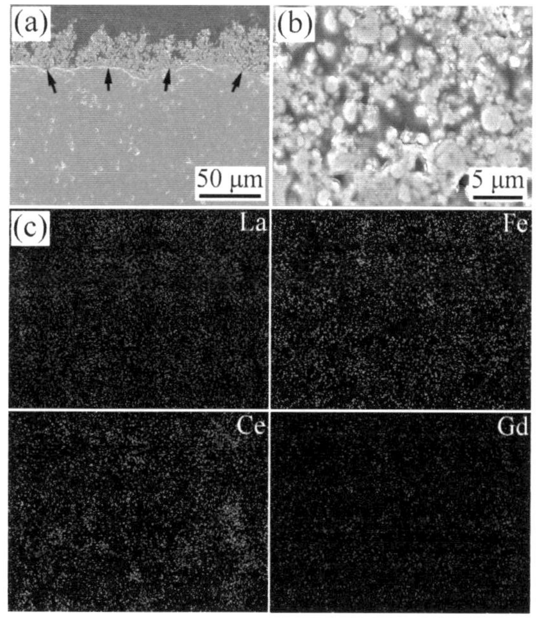

Fig. 4. Microstructure of polished cross-section for LSCF/GDC cathodes deposited by the 15 g L^{-1} suspension, (a) lower magnification, (b) higher magnification, (c) the corresponding EDX elemental mapping for (b).

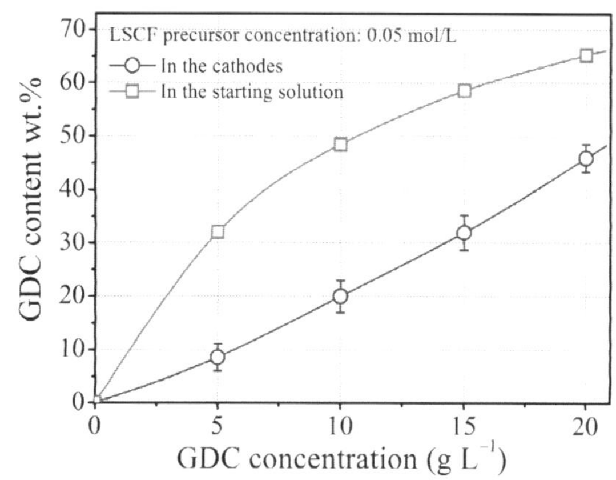

Fig. 5. Effect of suspension concentration on the GDC content in the final cathodes.

3.3 Electrochemical performance

Fig. 6 (a) shows some of the typical impedance data obtained from symmetric cells comprising LSCF/GDC cathodes deposited with different concentrations of suspensions. To easily compare the cathode polarization resistance (R_p), the grain bulk resistances form the electrolytes were removed from the impedance data. At 450 °C, as shown in the lower section of Fig. 6 (a), the left semicircle at the high frequency region in the spectrum shows the grain-boundary resistance (R_{gb}) from the electrolyte. The overlap of the left circles indicates that all the samples have a similar R_{gb} because the electrolyte substrates were prepared under the same conditions. Similarly, the right semicircles at the low frequency region in the spectrum can be attributed to R_p. However, at 700 °C, the left semicircle is too small and cannot be distinguished from the spectrum because of the high grain-boundary conductivity at high temperatures. Therefore, the R_p at high temperatures can be obtained from the intercept of the semicircle. Fig. 6 (b) shows the R_p for different types of cathodes at different temperatures. The R_p decreased rapidly with the increase in the concentration of GDC suspension. Interestingly, a minimum polarization resistance was observed for all the temperatures at a GDC concentration of 15 g L^{-1}. However, with the further increase in suspension concentration, the cathode polarization increased. At 600 °C, the cathodes deposited by 0, 5, 10, and 15 g L^{-1} suspension had a polarization resistance of 0.76, 0.52, 0.38, and 0.11 Ω cm^2, respectively. However, when the suspension concentration increased from 15 g L^{-1} to 20 g L^{-1}, the cathode polarization increased from 0.1 Ω cm^2 to 0.28 Ω cm^2. When the measurement temperature was increased to 700 °C, the polarization resistances were 0.19, 0.12, 0.095, 0.023, and 0.06 Ω cm^2 for the cathodes deposited by 0, 5, 10, 15, and 20 g L^{-1} suspensions, respectively. Particularly, the R_p was <0.01 Ω cm^2 at 750 °C when the cathode was deposited by the 15 g L^{-1} suspension.

Moreover, the activation energy (E_a) for the electrode polarization can be calculated from the linear relationship $\ln(R_p)$–1/T shown in Fig. 6 (b) using Arrhenius law. The E_a for the pure LSCF cathode was calculated to be 1.02 eV; the E_a values for 5 g L^{-1}, 10 g L^{-1}, 15 g L^{-1}, and 20 g L^{-1} suspensions were 1.05 eV, 1.05 eV, 1.12 eV, and 1.09 eV, respectively. In this study, the E_a for the pure LSCF cathode was significantly lower than that reported in the literature [25, 32]. This can be attributed to the smaller grain size of a thermal-sprayed deposit because of the rapid cooling process. Therefore, the ionic conductivity of grain boundary increased at low temperatures, causing a higher oxygen diffusion rate. Therefore, the cathode showed a lower polarization resistance at low temperatures and lower E_a. Besides, the E_a almost did not vary when GDC was added, indicating that the ORR reaction kinetics and electrocatalytic activity of LSCF cathode did not change in this study. However, extensive studies are needed to obtain an accurate analysis of the dominant mechanism for cathode reaction.

Fig. 6 (c) shows the polarization resistance of the cathode prepared by the 15 g L^{-1} suspension during the isothermal dwelling at 650 °C for 350 h. The R_p remained practically constant (0.05 Ω cm^2), indicating that the liquid plasma-sprayed LSCF/GDC composite cathode demonstrates an impressive stability at 650 °C.

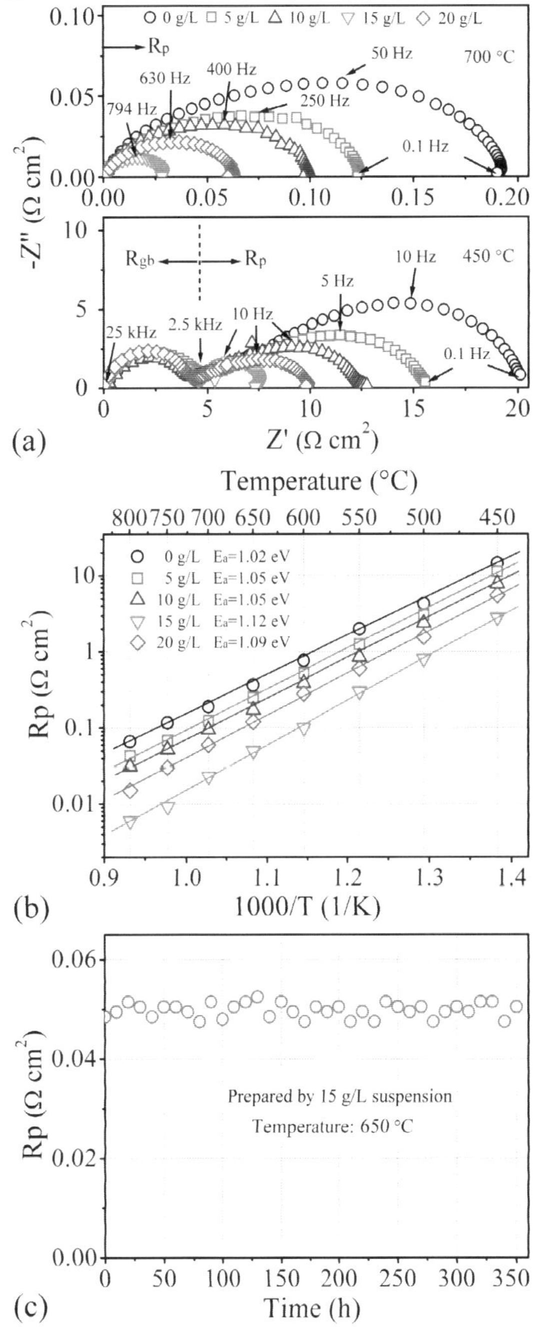

Fig. 6. (a) Nyquist plots for LSCF/GDC cathodes at different temperatures, (b) Arrhenius plots of polarization resistances for the LSCF/GDC cathodes deposited by different suspensions, (c) Polarization

resistance for the cathode prepared by 15 g L^{-1} suspension at 650 °C for 350 h.

4 Conclusions

Liquid plasma spraying was used as a novel method for the preparation of high-performance nanonetwork SOFC cathodes. The effect of the GDC in LSCF on the polarization resistance of LSCF/GDC composite cathodes was studied. The deposited composite cathodes showed a homogeneous island-like microstructure with a porosity of >40%. The GDC content in the final cathodes increased with the GDC concentration in the feedstock suspension. Moreover, the resulting LSCF/GDC composite cathodes outperformed the pure LSCF counterpart. The nanosized GDC on the LSCF backbone showed a network structure, endowing the cathode with a high ionic conductivity and increasing the number of active sites. Among all the composite cathodes fabricated, 15 g L^{-1} GDC was the optimal concentration of the suspension; at this concentration, the polarization resistances of the cathode were 0.1 Ω cm^2 and 0.009 at 600 °C and 750 °C, respectively. The stability measurements showed a good long-term stability of the optimal cathode with a polarization resistance of 0.05 Ω cm^2 at 650 °C for more than ~350 h. Overall, it was demonstrated that liquid plasma spraying is a viable technique to deposit high-performance composite cathodes for IT-SOFCs.

5 Literature

[1] N.Q. Minh, J. Am. Ceram. Soc. 76(1993), pp. 563/68.

[2] S.C. Singhal, Solid State Ionics 152(2002), pp. 405/10.

[3] M. Liu, M. E. Lynch, K. Blinn, F. M. Alamgir and Y. Choi, Materials Today 14(2011), pp. 534/46.

[4] L. Yang, S.Z. Wang, K. Blinn, M.F. Liu, Z. Liu, Z. Cheng and M.L. Liu, Science 326(2009), pp. 126/29.

[5] S. Ha, P.C. Su and S.W. Cha, J. Mater. Chem. A 1(2013), pp. 9645/49.

[6] C.W. Kwon, J.W. Son, J.H. Lee, H.M. Kim, H.W. Lee and K.B. Kim, Adv. Funct. Mater. 21(2011), pp. 1154/59.

[7] Y.B. Kim, T.P. Holme, T.M. Gür and F.B. Prinz, Adv. Funct. Mater. 21(2011), pp. 4684/90.

[8] Z. Gao, E. C. Miller and S.A. Barnett, Adv Funct Mater 24 (2014), pp. 5703/09.

[9] L. Yang, Y. Choi, W. Qin, H. Chen, K. Blinn, M. Liu, P. Liu, J. Bai, T.A, Tyson and M. Liu, Nat. Commun. 2(2011), pp. 357.

[10] D. Ding, X. Li, S. Lai, K. Gerdes and M. Liu, Energy and Environ. Sci. 7(2014), pp. 552/75.

[11] J. M. Ralph, A.C. Schoeler and M. Krumpelt, J. Mater. Sci. 36(2001), pp. 1161/72.

[12] S. Jiang, J. Solid State Electrochem. 11(2007), pp. 93/102.

[13] X. Chen and S.P. Jiang, J. Mater. Chem. A 1(2013), pp. 4871/78.

[14] Z. Du, H. Zhao, Y. Shen, L. Wang, M. Fang, K. ´Swierczek and K. Zheng, J. Mater. Chem. A 2(2014), pp. 10290/99.

[15] B. Qian, Y. Chen, M. O. Tade and Z. Shao, J. Mater. Chem. A 2(2014), pp. 15078/86.

[16] N. Ortiz-Vitoriano, I.R. de Larramendi, S.N. Cook, M. Burriel, A. Aguadero, J.A. Kilner and T. Rojo, Adv. Funct. Mater. 23(2013), pp. 5131/39.

[17] M. Mamak, N. Coombs and G. A. Ozin, Adv. Funct. Mater. 11(2001), pp. 59/63.

[18] L. Blum, L.G. J. de Haart, J. Malzbender, N.H. Menzler, J. Remmel and R. Steinberger-Wilckens, J. Power Sources 241(2013), pp. 477/85.

[19] Z. Wang, R. Peng, W. Zhang, X. Wu, C. Xia and Y. Lu, J. Mater. Chem. A 1(2013), pp. 12932/40.

[20] H.J. Hwang, J.-W. Moon, S. Lee and E.A. Lee, J. Power Sources 145(2005), pp. 243/48.

[21] E. Perry Murray, M.J. Sever and S.A. Barnett, Solid State Ionics 148(2002), pp. 27/34.

[22] M. Sahibzada, S.J. Benson, R.A. Rudkin and J.A. Kilner, Solid State Ionics 13-115(1998), pp. 285/90.

[23] J. Chen, F. Liang, B. Chi, J. Pu, S.P. Jiang and L. Jian, J. Power Sources 194(2009), pp. 275/80.

[24] Y. Sakito, A. Hirano, N. Imanishi, Y. Takeda, O. Yamamoto and Y. Liu, J. Power Sources 182(2008), pp. 476/81.

[25] Y. Leng, S.H. Chan and Q. Liu, Int. J. Hydrogen Energy 33(2008), pp. 3808/17.

[26] V. Dusastre and J.A. Kilner, Solid State Ionics 126(1999), pp. 163/74.

[27] L. Nie, M. Liu, Y. Zhang and M. Liu, J. Power Sources 195(2010), 4704/08.

[28] X. Zhang, H. Zhang and X. Liu, J. Power Sources 269(2014), pp. 412/17.

[29] M. Liu, D. Ding, K. Blinn, X. Li, L. Nie and M. Liu, Int. J. Hydrogen Energy 37(2012), pp. 8613/20.

[30] R. Hui, Z. Wang, O. Kesler, L. Rose, J. Jankovic, S. Yick, R. Maric and D. Ghosh, J. Power Sources 170(2007), pp. 308/23.

[31] R. Henne, J. thermal spray tech. 16(2007), pp. 381/403.

[32] L. Zhao, S. Amarasinghe and S.P. Jiang, Electrochem. Commu. 37(2013), pp. 84/7.

Study of High Velocity Arc Sprayed heat resistant coatings from FeCrAlBY cored wire

Yu. Korobov, S. Nevezhin, M. Filippov, A. Makarov, I. Malygina, Ekaterinburg, Russia
D. Fantozzi, A. Milanti, H. Koivuluoto, P. Vuoristo, Tampere, Finland

Arc sprayed coatings are widely used in industry applications due to their high output and good quality. Fe-based wire is an attractive feedstock as it is much cheaper than Co- and Ni-based spraying materials. In this study, the influence of yttrium content in a FeCrBAl core wire on the coating heat- and wear resistance was examined. The coatings were produced by Active Arc Spraying. The process allows improving the coating quality by using propane-air combustion products as a carrier gas. The coating was characterized in terms of its surface morphology, microstructure, and phase composition by SEM and XRD analysis, and microhardness. The coating was evaluated by two-body abrasive wear test, and tests of heat resistance at 700°C in air, and under salt deposit at 550°C.The tests showed high temperature oxidation resistance when compared to typical wrought steels for boilers and high abrasive wear resistance compared to alternatives. The corrosion products thickness on the surface of the coating is quite low and of the same magnitude of the other metal coatings deposited with thermal spray technologies and studied in the same conditions.

1 Introduction

Arc sprayed (AS) coatings from Fe-based cored wires (CW) are one of the most cost effective solutions for protection against wear and hot corrosion [1]. Fe-Cr-Al alloy is a traditionally used material for protection against gas corrosion in form as of bulk material and of coatings [2, 3]. It is an attractive substitute of significantly more expensive Co- and Ni-base materials. Alloying the Fe-Cr-Al by boron increases the wear resistance of the alloy by forming reinforcing carboboride phases [4, 5].
Heat resistance of the Fe-Cr-Al alloy is provided by a surface alumina layer, which forms at high operating temperatures. This oxide layer is characterized by high melting temperature, chemical and thermal stability and low rate of growth.
Addition of yttrium is favorable for protection against gas corrosion and wear by the following reasons. First, Y-containing Fe-Cr-Al alloys showed good oxide adherences on the arc sprayed coatings at high temperature operation under thermal cycling [6, 7]. Second, during the AS process an alloying elements burnout can take place. It leads to a decrease of alloying elements concentration reducing the coating heat-resistance. Yttrium affinity to oxygen is higher than that of other alloying elements. Therefore, yttrium protects them from oxidation at transition from the feedstock to the coating. And the elements save their ability to resist against gas corrosion and wear.
Active Arc Spraying (AAS) was applied for coating deposition in the present study. The process differs from typical AS process by the use of propane-air combustion products as a carrier gas instead of compressed air. This increases the particle velocity, temperature, and forms a reducing atmosphere. These features can decrease the burnout of spraying metal in flight, increase adhesion strength of the coating and lower porosity of the coating [8, 9].
The aim of this work is to evaluate the influence of Y addition into FeCrBAl core wire on heat-, wear- and corrosion resistance of active arc sprayed coatings.

2 Experimental Methods

The following parameters of AAS Gun ADM-10 (UWI, Ekaterinburg, Russia) were employed: current 250 A, voltage 30V, spraying distance 100 mm, input pressure of gases: air - 0.42 MPa, propane - 0.41 MPa.
Two Fe-based cored wires Cr10.8Al5.6B4.8 of 2.0 mm in diameter were used. The composition of the second wire differs by alloying of yttrium (0.7 wt. %).
The Y content in the wire was designated according to neural network model, which has been developed with respect to arc spraying [10].
The samples of ASME 1020 steel, with the dimensions of 30x20x3 mm and 10x10x50 mm were used. The coating thickness was 0.4 mm at the 30x20 mm surface and at 1.5 mm at 10x10 mm surface correspondingly. The first type of coatings was used for evaluation of the heat resistance and the second type was used for wear test. The amount of oxygen in the coating was measured by melting the coating in inert gas flow by analyzer ON-900 (Eltra, Haan, Germany).
The porosity of the coatings was determined by metallographic method on the transverse sections using commercial hardware and software SIAMS 700 (SIAMS, Ekaterinburg, Russia) to estimate the distribution of pores by size, area, and volume.
The adhesion strength of the coatings was determined by the method of separation of a conical pin [11]. A schematic view of the test is presented in **Fig.1**.

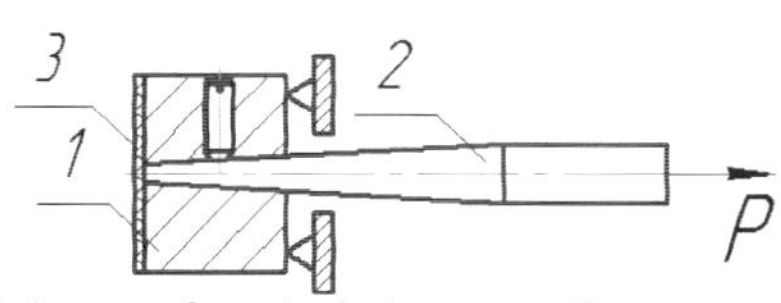

Fig. 1 Scheme of conical pin separation.

Here the washer 1 serves as the basis; pin 2 is inserted into its hole so that its end surface is flush with the external plane of the washer. The total surface of the pin and the washer after preparation is coated with the coating 3.The test consists of pulling the pin by applying a force (*P*). The results were averaged for three samples.

Heat resistance of the coatings from these wires was evaluated in accordance with GOST 9.312 [12]. The gravimetric method was used according to weight-loss measurements of the samples exposed for 100 h in the air at 700°C. To decrease measuring error the 30x20x3 mm samples was deposited by 10 μm galvanic nickel. Before spraying the layer from 30x20 mm surface was removed.

The high temperature corrosion test was performed in an alumina tube furnace at 550°C for duration of 168 h. The samples of ASME 1020 steel was cut of the size of 20x20 mm and an area of about 255 mm^2 (a 18mm-diameter circle) was covered with approximately 1 g of KCl salt. The salt was finely grinded in a ceramic mortar and mixed with ethanol to create a paste and facilitate the deposition. The furnace environment was purged with 1.5 l/min of air with the addition of water vapor to reach 12% of specific humidity.

Two-body abrasive wear studies were carried out by pin-on-plate reciprocating technique. Corundum abrasive paper with grain size up to 150 μm was fixed on the plate and the sprayed sample was fixed in a holder. Test conditions were the following: full friction path 60 m, pin velocity 0.16 m/s, specific load 1 MPa, work surface area 10×10 mm.

Microhardness measuring was performed by Leica VMHT device (Walter Uhl, Aßlar, Germany) under 50 and 300 g loads on the polished surface of the sample.

A microstructural study of the coating from Y-containing wire was performed. Phase composition on the coating surface was determined by X-ray diffractometer XRD-7000 (SHIMADZU, Kyoto, Japan, Cr-$K_α$ radiation). Microchemical structure and composition of the coatings was studied by scanning electron microscope VEGA II XMU (Tescan, Brno, Czech Republic). It was equipped with a wavelength dispersive X-ray spectrometer INCA Wave 700 and energy-INCA Energy 450XT microanalyzers (Centre "Plastomertia", the Institute of Engineering Science, Ural Branch of RAS, Russia).

The microstructure was investigated by Scanning Electron Microscope (SEM), Philips XL30 equipped with Energy Dispersive X-ray (EDX) microanalysis on the cross section, whereas phase composition was assessed by X-Ray Diffractometry (XRD, Empyrean, PANAnalytical, Co-$K_α$ radiation) of the surface of the coating. Experimental conditions include 2θ range 20°-120°, step size 0.02°, beam mask 20 mm, programmable divergent slit fixed at ½ degree, Fe-filter and PANAnalytical PIXcel 3D detector. Phase identification was performed using the PANAlytical X'Pert High Score Plus software using the ICDD JCPDF-2 database (International Centre for Diffraction Data, Newtown Square, PA, USA).

3 Results and discussion

3.1 Oxygen inflow during the spray process

The studies showed that the oxygen content in AAS-Y coating (1.3 wt. %) is 1.8 times less compared to the AAS coating. This result coincides with the side data [6], according to which the oxygen content in the AS coatings is cut in half by introduction of rare earth metals into CW. Obviously, it is influenced by high affinity of yttrium to oxygen [13]. It prevents burnout of alloying elements during spraying process preserving their proper content for further operation.

3.2 Porosity and adhesion strength of the coatings

The porosity of the coating from the FeCrBAlY wire (AAS-Y) is 2.8%, the maximum pore size is 24.2 μm, the minimum - 1.2 μm, a number of pores less than 10 microns is 97.4%, and their integral square and volume are 73.1% and 50.4% correspondingly. In the coating from Y-free FeCrBAl wire, porosity is 3.5 % and with similar pore distribution.

In AAS-Y coating, adhesion strength is 43 MPa, which is 1.2 times higher than that of the coating prepared from the FeCrBAl Y-free wire (AAS). Reduced porosity and increased adhesive strength of AAS-Y coating is reasoned by lower degree of oxidation of the sprayed particles and improved interaction in contact "particle-substrate".

3.3 Structure of the as-sprayed coatings

According to surface analysis, the coating with Y alloying consists of iron *α*-phase (solid solution of Fe-Cr-B, enriched with carbon) with reinforcing phases: borides Fe_2B, $(Fe,Cr)_2B$ and carboborides $(Fe,Cr)_2(B,C)$, **Fig. 2**. Yttrium-aluminum oxides were found in the coating from Y-containing wire at the borders and within the structural components, see point 1 at **Fig. 3**.

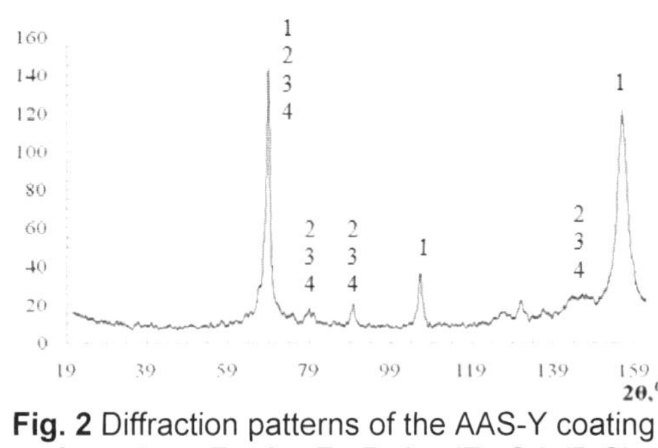

Fig. 2 Diffraction patterns of the AAS-Y coating surface. 1 – α-Fe, 2 – Fe_2B, 3 – $(Fe,Cr)_2(B,C)$, 4 – $(Fe,Cr)_2B$.

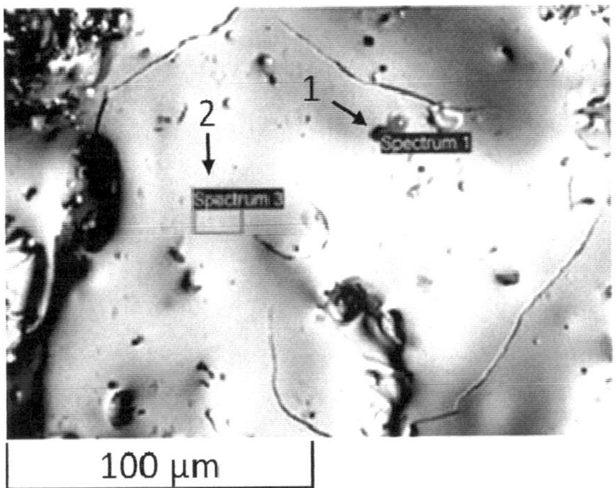

Element, content		O	Al	Si	Cr	Y	Fe
Spectrum 1	wt.%	38.2	37.4	0.3	4.6	11.1	8.3
Spectrum 2	wt.%	-	2.4	0.7	9.9	11.1	87.0

Fig. 3 SEM image of the coating surface and quantitative EDS elemental analyses of details 1 and 2. (Si was not specially added, it exists only as a charge component).

Cross section of the AAS-Y coating is shown in **Fig. 4**.

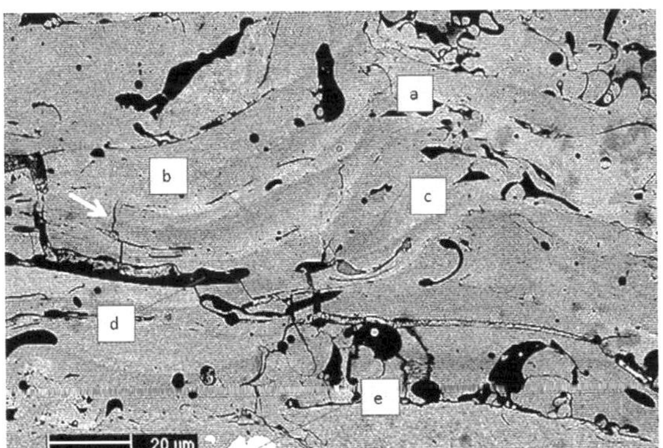

Element, content		*B*	*O*	*Al*	*Y*	*Cr*	*Ni*	*Fe*
Spectrum a	*wt.%*	*4.68*	*1.37*	*3.79*	*0.39*	*8.57*	*--*	*81.19*
Spectrum b	*wt.%*	*--*	*20.8*	*37.96*	*25.48*	*2.75*	*--*	*13.01*
Spectrum c	*wt.%*	*--*	*13.62*	*13.03*	*67.38*	*1.10*	*0.76*	*4.12*
Spectrum d	*wt.%*	*--*	*23.31*	*38.52*	*30.93*	*1.44*	*1.27*	*4.53*
Spectrum e	*wt.%*	*--*	*--*	*--*	*3.79*	*--*	*92.94*	*3.27*

Fig. 4 Microstructure of the AAS-Y coating and EDS quantitative elemental analyses of the indicated microstructural details.

Zones pointed as (a-e) were detected by EDX microanalysis. Different grey scale colored striations (a, b) differ by the content of Al and Y. In particular darker colored striations have major amount of Al and Y. Small cracks were detected along the boundary of such zones (white arrow). Some small dark grey drop-like inclusions (c) are rich in Y and Al. Pores (d) are present in the coating. At the coating-substrate interface, several white, long inclusions of the size of about 100 µm were observed (e). These big inclusions are Ni rich that may come from the surface preparation method.

3.4 Heat resistance of the coating

The measured specific weight-loss for the Y-added coating was 0.54 g/(m^2·h). It is an order of magnitude smaller than those of T11 and T122 steels and similar to the heat resistance of austenite steel, **Table 1.**

Table 1 Heat resistance of commercial boiler materials [14, 15]

Grade of steel	Corresponding ASM grade	Class of steel	Specific mass loss, g/(m^2·h)
12Cr1MoV GOST20072	T11 ASME A213	pearlite	80
1Cr12W2MoV GOST5632	T122 ASME A213	martensite-ferrite	10
1Cr18Ni12T GOST5632	TempaloyA-1 ASME SA312	austenite	0.4
Cr23Ni18 GOST5632	TP310 ASME SA312		0.1

3.5 Corrosion resistance of the Y-containing coating

An overall image of the coating after test and corrosion products is presented in **Fig. 5**.

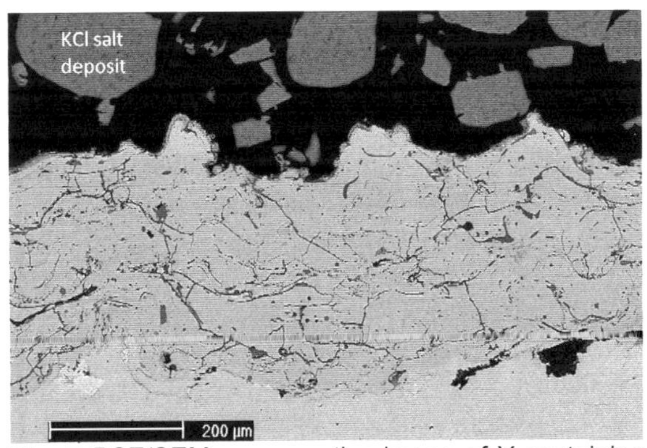

Fig. 5 BSE/SEM cross section image of Y-containing coating after the test.

A magnification of the corrosion products is given in **Fig. 6.** EDX microanalysis showed the following. A homogeneous thin corrosion products layer was detected on top of the coating. The topmost corrosion products layer (**a**) is rich in metal oxides probably in the form of chromates, ferrates and aluminates of potassium. A thin porous layer lies between the corrosion products and the coating (**b**). The EDX spot analysis in the pores (**c**) indicates the presence of Cr, Y and Al. The high amount of Fe probably belongs to the coating. This porous morphology might be a precipitation zone of borides, which was subjected to selective corrosion of Cr, as the XRD analysis suggests.

This type of morphology may be caused by chlorine induced corrosion. Chlorine can selectively react with alloying elements and precipitates like borides and carbides, forming volatile chlorides that leave the surface in form of vapor, resulting in a porous area. The formed chlorides then migrate toward the surface where the higher partial pressure of oxygen makes them deposit in the form of oxides [16]. The chromium oxide in the layer lying on the top of the porous area (**c**) was probably originated in such way. Several inclusions of different shapes and sizes were detected (**d**). The presence of Al and Y oxides is revealed in these inclusions.

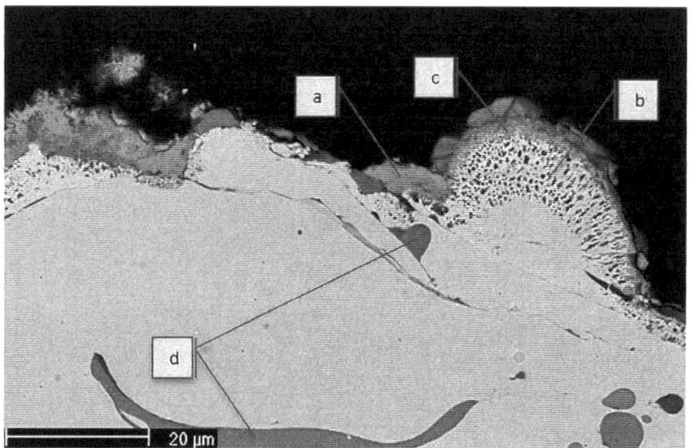

Element, content		O	Al	Cl	K	Cr	Y	Fe
Spectrum a	wt.%	22.6	2.00	0.33	3.41	8.69	--	62.97
Spectrum b	wt.%	2.92	1.94	0.23	0.33	1.54	0.55	92.49
Spectrum c	wt.%	22.67	0.51	0.41	43.46	28.86	--	4.08
Spectrum d	wt.%	20.15	35.3	--	--	0	0.97	5.53

Fig. 6 Cross section of coating AAS-Y after high temperature corrosion test. Corrosion product areas and microstructural details (a-d) are described in the text.

Along the coating cross section the multiple cracks are filled with Fe, Cr and Al oxides (**Fig. 7).**

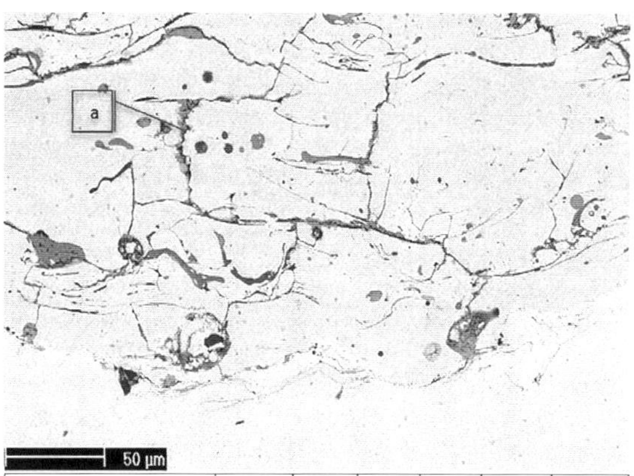

Element, content		*O*	*Al*	*Cl*	*K*	*Cr*	*Fe*
Spectrum a	*wt.%*	*20.47*	*7.00*	*0.21*	*6.21*	*13.42*	*52.69*

Fig. 7 Cracks and oxides trough the cross section of the coating and at the coating/substrate interface of sample AAS-Y

The coating is well bonded to the substrate. However, some cracks are visible and metal oxides including Fe, Cr, Al, and very little amount of Cl were detected (**Fig. 7, area a**). The presence of chlorine along the coating cross section indicates that chlorine penetrated the coating, probably through open porosity and cracks in the coating microstructure. Even though only a little amount of chlorine was found along the coating, the presence of corrosion products at the interface coating-substrate is a sign that corrosive agents (such as chlorine) could reach the substrate ad thus, it can be stated that the coating is not fully protective against gases such as chlorine. However, it performed similarly to other compared Ni-based thermal spray coatings showing very thin corrosion products layer.

3.6 Wear resistance of the Y-containing coating.

Microhardness of the AAS-Y coating is $HV_{0.05}$ 1222±52 and $HV_{0.3}$ 1111±64. The weight loss of the AAS-Y samples after tribological tests was in a range of the (0.15-0.19) g. AAS-Y wear was compared with AAS, AAS coating from 150Cr8Ti2Al cored wire, and ASME 1020 solid steel. Coatings of Fe-based metastable austenite type structure from 150Cr8Ti2Al cored wire are applied for wear resistant applications [17].
Ratio of weight losses of AAS-Y coating and alternatives is the following: (a) it is lower by 20 % compared to the AAS coatings from the yttrium-free wire. That result coincides with the side data concerning effect of REM additives on wear resistance of Fe-based coating at two-body wear [6]; (b) it is more than two times less than that of 150Cr8Ti2Al coating due to formation of complex carbides like $(Fe,Cr)_2(B,C)$ in the coating; (c) it is an order of magnitude smaller than in case of ASME 1020 steel.

4 Conclusions

1) Oxygen content in the Active Arc Sprayed coating from FeCrBAl wire is 1.8 times as much comparing to coating from the same wire with added 0.7 wt. % yttrium. That result is caused by high affinity of yttrium to oxygen. It prevents burnout of other alloying elements during spraying, which allows keeping their proper content for further operation. In addition, adhesion strength rise and porosity drops drastically.
2) Heat resistance of the coatings from FeCrBAl type core wires was evaluated in air at 700°C, 100 h. It is shown that the addition of yttrium into the wire leads to 15 % increase of the coating's oxidation resistance. Coatings from the wires are of the same heat resistance level as the austenite steels and 1-2 orders of magnitude higher as compared to the pearlite and ferrite-martensite steels, which are widely used as boilers materials. It is associated with the formation of complex oxides like $(Fe, Al, Y)_2O_3$.
3) Two-body abrasive wear of Active Arc Sprayed Coatings from FeCrBAlY core wires was evaluated. Wear resistance of Arc Sprayed coatings from the wires is an order of magnitude higher compared to ASME 1020 steel and more than twice as high compared to Arc sprayed wear-resistant metastable austenite Fe-based coating. It is caused by formation of complex carbides like $(Fe,Cr)_2(B,C)$.The introduction of yttrium in the core wire increases the wear resistance by 20 % compared to the coatings from the yttrium-free CW.
4) Chlorine induced corrosion took place mainly on the surface of the coating from the yttrium-added wire.

However, it was detected with minor intensity also along the cross section of the coating. The corrosion products thickness on the surface of the coating is quite low and of same magnitude of other Ni-based metal coatings deposited with thermal spray technologies and tested.

Acknowledgements

The present study was partially supported by program of OEMMPU RAS (project No.12-T-1-1010) and program of FASIE (projects No.11906r/21605).

Additionally, the authors would like to thank V. Verkhorubov (UrFU) for his support at spraying samples.

The work has been done partly within FIMECC Ltd and its HYBRIDS and BSA programs. The authors gratefully acknowledge the financial support from Tekes (Finnish Funding Agency for Technology and Innovation) and the participating companies.

References

1. Wilden, J., V. Drescher, and M. Djahanbakhsh. Wire Arc Sprayed Iron-based Anti-Corrosion Coatings for Waste Incineration Plants. *Thermal Spray 2010: Global Solutions for Future Applications,* B.R. Marple, A. Agarwal, M.M. Hyland, Y.-C. Lau, C.-J. Li, R.S. Lima, and G. Montavon, Ed., May 3-5, 2010 (Singapore), Springer, 2011, pp. 309-312.
2. David, J.Y. *High Temperature Oxidation and Corrosion of Metals.* (Netherlands), Elsevier, 2008, 574 p.
3. Korobov, Yu., M. Filippov, S. Nevezhin, and M. Karabanalov. High-Temperature Oxidation Behavior of Fe-25Cr-5Al Arc Spraying Coatings. *Thermal Spray 2012: Proceedings of the International Thermal Spray Conference, Air, Land, Water, and the Human Body: Thermal Spray Science and Applications,* Basil R. Marple, Arvind Agarwal, Laura Filofteia-Toma, Margaret M. Hyland, Yuk-Chiu Lau, Chang-Jiu Li, Rogerio S. Lima, and André McDonald, Ed., May 21-24, 2012 (Houston, Texas), pp. 729-733.
4. Borisova, A.L., I.V. Mits, T.V. Kaida, and A.Sh. Kleiman, Structure and properties of coating based on ferrochrome and ferrochrome aluminum alloys obtained by arc spraying flux-cored wires. *Avtom svarka*, [in Russian], 1995, 6, pp. 3-6.
5. Wielage, B., H., Pokhmurska, M. Student, V. Gvozdeckii, T. Stupnyckyj, and V. Pokhmurskii. Iron-based coatings arc-sprayed with cored wires for applications at elevated temperatures. *Surface & Coatings Technology*, 2013, 220, pp. 27-35.
6. He D., and J. Jiang. Effect of Rare Earth Elements on the Wear Resistance of Iron-Based Thermal Sprayed Coatings. *Thermal Spray 2004: Advances in Technology and Application,* ASM International, May 10-12, 2004 (Osaka, Japan), ASM International, 2004, pp. 61-64.
7. Amano, T. Rare earth application for heat-resisting alloys. *J. of Rare Earth*, 2010, 28 (Spec. Issue), pp. 12-21.
8. Boronenkov, V., and Yu. Korobov. *Fundamentals of Arc Spraying. Physic-chemical Regularities.* (Switzerland), Springer International Publishing, 2015, 256 p.
9. Korobov, Yu. Efficiency of using activated arc metallization for the deposition of protective coatings. *Welding international*, 2005, 19 (7), pp. 580-582.
10. Korobov, Yu., S. Nevezhin, V. Verkhorubov, G. Rimer, and A. Zhilin. Optimization of alloying of heat resistant core wires for arc spraying by neural network modeling. *Thermal Process Modeling and Computer Simulation,* June 16-19, Orlando, USA, 2014, pp. 322-327.
11. Tyshinski, L.I. A.V. Plokhov, O.A. Tokarev, and V.I. Sindeev. *Methods of study the materials.* Mir: Moscow [in Russian], 2004, 384 p.
12. "Unified system of corrosion and ageing protection. Protection coatings. Methods of heat resistance determination" GOST 9.312, USSR State Committee on Standards, 1989, 11 p.
13. Kulikov I.S. *Metal deoxidation.* [in Russian], Metallurgy: Moscow, 1975, 504 p.
14. Nikitin V.I. *Calculation of heat resistance of metals.* [in Russian], Metallurgy: Moscow, 1976, 207 p.
15. Davis J.R. *ASM Specialty Handbook: Heat-Resistant Materials.* ASM, 1997, 591 p.
16. Grabke, H. J., E. Reese and M. Spiegel. The effects of chlorides, hydrogen chloride, and sulfur dioxide in the oxidation of steels below deposits. *Corrosion Science* 1995, 37 (7), pp. 1023-1043.
17. Korobov, Yu., M. Filippov, A. Makarov, I. Zabolotski, Yu. Khudorozhkova, and G. Rimer. Comparison of metastable austenite type structure of Fe-based coatings produced by Arc Spraying and Surfacing. *Thermal Spray 2014: Not Fiction: Thermal Spray the Key Technology of Modern Life!*, May 21-23, 2014 (Barcelona, Spain), ASM International, 2014, pp. 874-878.

Porous nanostructured hydroxyapatite coatings deposited by suspension plasma spray

H. Xu, X. Geng, G. Liu, P. Zhu, C. Zhang, Yangzhou / CN

Abstract
Hydroxyapatite (HA) coatings were successfully deposited by suspension plasma spray (SPS) method on stainless steel substrates. The suspension was prepared by a wet chemical method using $Ca(NO_3)_2$ and $(NH_4)_2HPO_4$ as source materials. The TEM micrographs show that the HA particles are needle-like nanowires. FE-SEM images confirm that the coatings are nanostructured. X-ray Diffraction analysis shows that the as-sprayed coatings are HA and the crystal size is estimated to be less than 30 nm. Fourier transform infrared spectroscopy and MicroRaman spectroscopy indicate the significant presence of OH^- and CO_3^{2-} in coatings. In addition, the effect of hydrogen flow rate in plasma forming gas on the coating properties was also studied. The SPS process proved its simplicity and reliability in fabricating porous and nanostructured HA coatings.

1. Introduction

Atmospheric plasma spray (APS) is the most widely used coating deposition process for HA coatings in commercial application due to its high deposition efficiency and low cost [1-3]. In APS process, individual nanoparticles are not able to be distributed into plasma flame by carrier gas due to the nanoparticles easily block the powder feeding tube. In order to solve this problem, the nanoparticles can be agglomerated into micrometer-sized powder (>5μm) and then be sprayed [4]. However, this is not a straightforward process since the preparation of agglomerated feedstock powders is needed and the nanoparticles cannot be totally retained after a stay in an ultra-high temperature plasma flame.
Moreover, owing to the high temperature of plasma plume and the rapid cooling rate of molten particles, the phase and microstructure of the obtained coatings are changed compared with those of the feedstock powder [1-3]. The decomposition phases in APS coatings deteriorate the dissolution behavior of the HA coatings. Besides, high temperature atmosphere often results in the loss of hydroxyl ions. The concentration of OH^- ions is believed to play an important role in biological properties. The reduction of OH^- will lead to the formation of amorphous phase and aggravate the decomposition of HA [5]. In order to fabricate nanostructured HA coatings with less impurity phases, it is essential to develop a novel process. Suspension plasma spray (SPS) may be an alternative process. SPS process is a novel method for depositing nanostructured ceramics coatings in which a liquid feedstock is injected into a plasma flame [6].
In this study, the suspension prepared by a wet chemical synthesis method was directly used as the liquid feedstock and SPS process was used to prepare nanostructured HA coatings. Then, the morphology and phase constitutions of the synthesized powders and coatings were characterized by the transmission electron microscope (TEM), field-emission scanning electron microscopy (FE-SEM), X-ray diffraction (XRD) analysis, Fourier transform infrared spectroscopy (FTIR) and MicroRaman spectroscopy (MRS). The effect of hydrogen flow rate in plasma forming gas on coating properties was investigated.

2. Experimental method

The HA suspension was prepared by a wet chemical method on the basis of the following reaction:

$$10\ Ca(NO_3)_2 + 6\ (NH_4)_2HPO_4 + 8\ NH_4OH \rightarrow Ca_{10}(PO_4)_6(OH)_2 + 20\ NH_4NO_3 + 6\ H_2O$$

$(NH_4)_2HPO_4$ aqueous solution (0.24 mol/L) was added into stirred $Ca(NO_3)_2$ aqueous solution (0.4 mol/L) according to a Ca/P atomic ratio of 1.67. The pH value of the solution was adjusted to 10 with ammonia (NH_4OH) solution. The reaction was kept at 95℃ for four days. Finally, the solution was aging for two days and then dialyzed with the deionized water to remove ammonia.
The particle size and morphology of the HA precursor were observed by TEM (Tecnai 12, Philips, Netherlands). Then, the freeze-drying method was used to obtain as-synthesized HA powder. The phase structure of the powder was determined by XRD (D8 Advance, Bruker AXS, Germany) operated at 40 KV and 40 mA using a CuKα radiation. The 2θ range was 20-60° with a scanning rate of $1^\circ\ min^{-1}$. Meanwhile, the powder structure was identified by FTIR (Cary610/670, Varian, USA) at 400-4000 cm^{-1}. The as-synthesized powder was compressed into a potassium bromide (KBr) plate for FTIR analysis. Finally, the powder was characterized by MRS (In

via, Renishaw, England) with a 50× objective in the optical microscope. An excitation wavelength of 532 nm and a spectral resolution of 1 cm^{-1} were used in this work. The spectra were recorded in the range of 200-3800 cm^{-1}.

HA coatings were deposited on stainless steel substrates by SPS with a F4MB plasma spray torch (Oerlikon Metco, Switzerland) attached to a six-axis ABB robot arm (ABB, Sweden). The suspension was directly injected into the plasma flame by a nozzle with an inner diameter of 0.26 mm driven through a peristaltic pump. The flow rate of the liquid was controlled at 40 mL/min. The primary plasma gas was argon with a flow rate of 40 L/min, and the secondary plasma gas was hydrogen. In order to study the effect of hydrogen flow rate on coating properties, the flow rate was chosen as 2, 4, 6 and 8 L/min, respectively. Compressed air at a pressure of 0.3 MPa was used as cooling gas for the substrates during spray. The surface morphology were observed by field-emission scanning electron microscope (FE-SEM, S4800Ⅱ, Japan). The XRD, FTIR and MRS characterizations of the obtained HA coatings were performed using the same parameters for the powder. For FTIR, the coatings were firstly removed from the substrate and then grinded into powder.

3. Results and discussion

3.1 Characteristics of HA suspension

Fig. 1 presents the particle size and morphology of the as-synthesized HA suspension. The HA is composed of needle-like nanowires with a diameter of c.a 5nm and a length of c.a. 100 nm.

Fig. 2(a) shows the XRD pattern of the as-synthesized powder obtained by freeze-drying method. It can be found that the result is in accordance with hexagonal HA corresponding to PDF 09-0432. Moreover, it should be noticed that the peaks locating at 31.77, 32.20, 32.90 and 34.05° are seriously broadening. Similar results were also found in [7-9], which are assigned to small crystal size. In order to further determine the phase structure, the four broadening peaks are deconvoluted with Lorentz fitting functions by software Origin (OriginLab Corp.), and the coefficient of determination R^2 is 0.99, as shown in Fig. 2(b). It can be believed that the synthesized powder was a HA.

Fig. 3 shows the FTIR spectra of the HA powder. The absorption bands located at 630 and 3570 cm^{-1} are associated with the stretching and vibration of the OH^- groups. The absorption band observed at 3300-3500 cm^{-1} is assigned to the presence of H_2O in the specimens, and the absorption band at 3137 cm^{-1} is linked with the NH_4^+ groups remaining in the samples from the HA suspension. The bands at 875 and 1400-1500 cm^{-1} were seen in the spectra, which are attributed to CO_3^{2-}. All the band position was coincide with the data in [10].

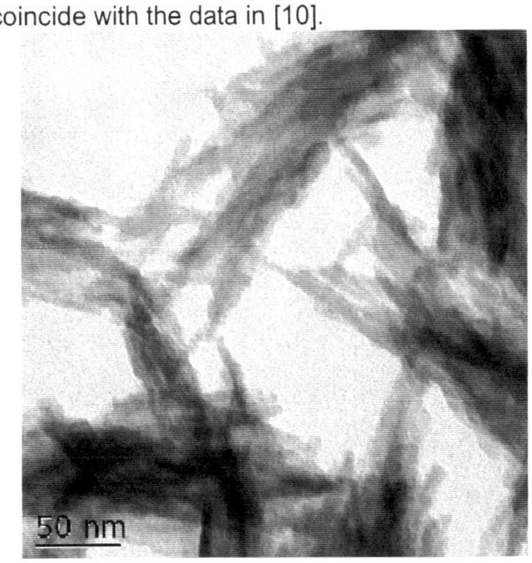

Fig.1. TEM micrographs of the HA liquid precursor

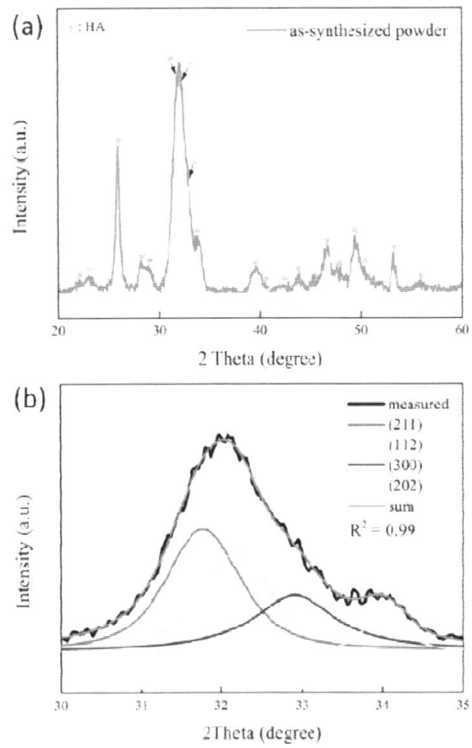

Fig.2. X-ray diffraction patterns of powders obtained by freeze-drying method, (a) as measured, (b) fitting patterns.

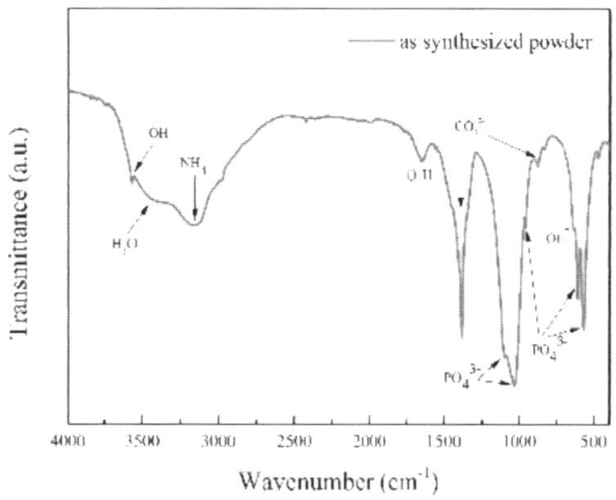

Fig.3. FTIR spectra of powders in the range of 400-4000 cm^{-1}.

As displayed in Fig. 4. The v1 frequency, located at 962 cm^{-1}, is the most intense peak corresponding to symmetric stretching of the P-O bond [11]. The v2 frequency represents the doubly degenerate O-P-O bending mode, which are situated at 430 cm^{-1} [12]. The v3 frequency at 1046 cm^{-1} arises from the triply degenerate T_2 mode involving a symmetric P-O stretching and also P motion [13]. Finally, the v4 frequency attributed to the triply degenerate T_2 modes of mainly O-P-O bending character is located at 590 cm^{-1}. In addition, the hydroxyl stretch at 3572 cm^{-1} for the powder and coatings are displayed in the spectra, which confirms the strong presence of OH^- [12].

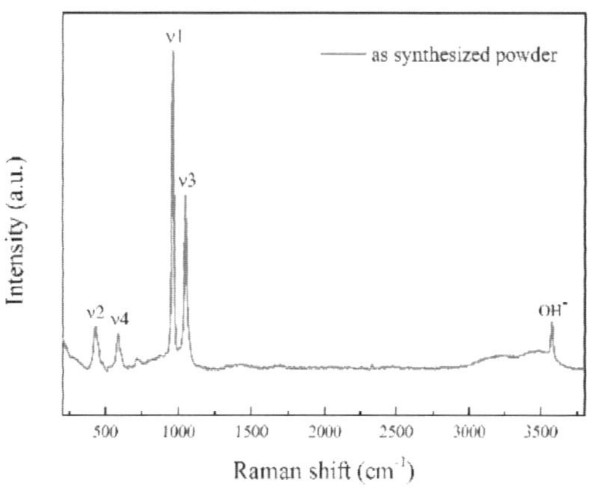

Fig.4. Raman spectra of powders in the range of 200-3800 cm^{-1}.

3.2 Characteristics of HA coatings

Fig. 5 presents FE-SEM images of the four as-sprayed HA coatings deposited with 2, 4, 6 and 8 L/min H_2 (specimens are abbreviated as 2, 4, 6 and 8 L coating in this paper). It can be found that the thermal conductivity of plasma flame (H_2 flow rate) plays an important role in coating microstructure. The coatings always consist of needle-like HA nanowires. As H_2 flow rate increases, the coatings (6 and 8 L coating) present a rough and porous surface. Moreover, the nanowires gradually disappear and agglomerates form.

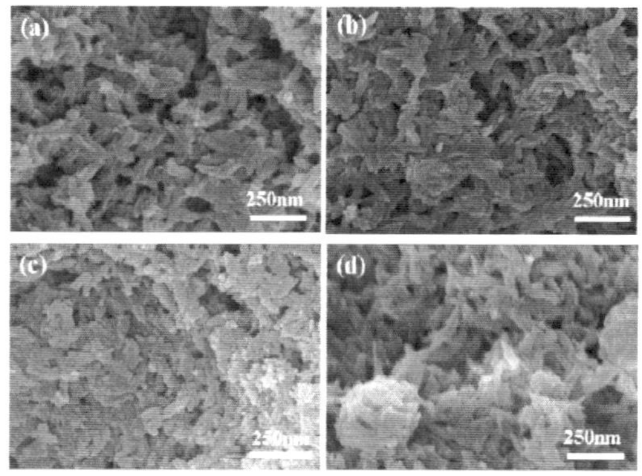

Fig.5. FE-SEM images of 2, 4, 6 and 8 L coatings.

Fig. 6(a) shows XRD patterns of the four as-sprayed HA coatings. Then, the patterns are fitted with the four peaks as for the powder, with R^2 equal to 0.99. The results demonstrate that the as-sprayed HA coatings are also well in line with hexagonal HA (PDF 09-0432), as displayed in Fig. 6(b). Similar to the pattern of powder, the four peaks for coatings are also broadening, which indicates that the crystal size in the as-sprayed coatings was also very small.

By means of the identification with the Jade 6.0 software, it can be found that the 2 L coating is mainly composed of pure HA. With increase in hydrogen flow rate, some decomposition phases such as α-TCP, β-TCP, TTCP and CaO appear, and the amount of decomposition phases increases with hydrogen flow rate. The diffraction peak of α-TCP, β-TCP, TTCP and CaO are in agreement with PDF 09-0348, 09-0169, 25-1137 and 82-1690, respectively. The phenomenon is different from that in APS. According to literature [3,14,15], even though using low plasma power, the HA coating deposited by APS consists of all the decomposition phases, e.g., CaO, α-TCP, β-TCP and TTCP . Compared with that in APS, the decomposition phases and their amount in SPS are less.

Fig. 7 presents the FTIR spectra of the four HA coatings. The band position is the same as that of the powder. The as-synthesized HA powder is mainly composed of the absorption bands related to the OH^-,O-H, H_2O, NH_4^+ and PO_4^{3-} groups. The intensity of the NH_4^+ group adsorption band in the as-sprayed coatings remarkably decreases when compared to HA powder. As H_2 flow rate increases, the NH_4^+ group disappears. In addition, the adsorption band of the H_2O and O-H band gradually reduces.

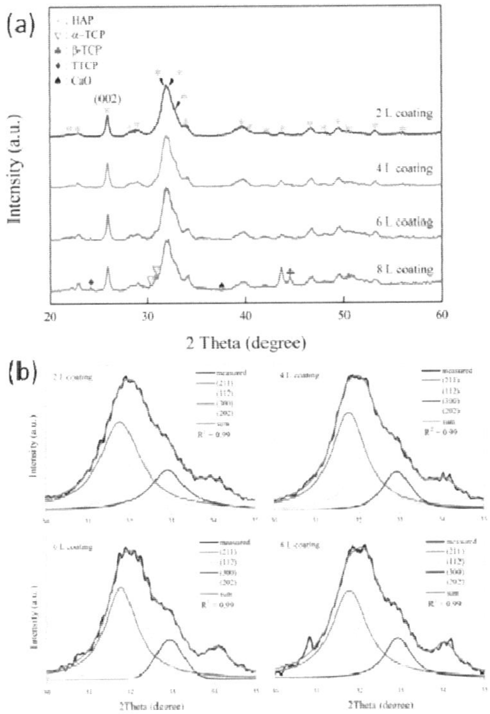

Fig.6. X-ray diffraction patterns of 2, 4, 6 and 8 L coating, (a) as measured, (b) fitting patterns..

From the FTIR spectra, it can be seen that the strong presence of the main adsorption band corresponding to CO_3^{2-}. The carbonation during the SPS process is desirable for HA coatings in orthopedic and dental applications.

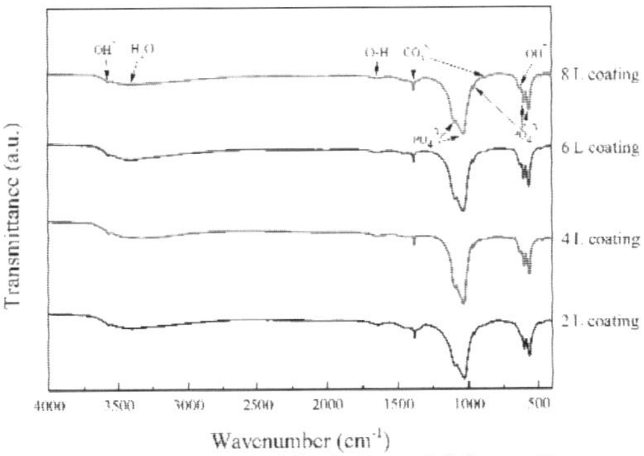

Fig.7. FTIR spectra of 2, 4, 6 and 8 L coatings.

Fig. 8 shows the MRS spectra of the as-sprayed HA coatings. The spectra bands referring to OH^- and PO_4^{3-} groups are also observed and the band position is in line with the powder. It is pointed out that the peak band at 962 cm^{-1} is assigned to crystalline HA and the amorphous phase is located at 952 cm^{-1} [11,12,16,17]. In this study, the peak located at 962 cm^{-1} without any shoulder around 952 cm^{-1} reveals that the powder and the coating are well-crystallized HA.

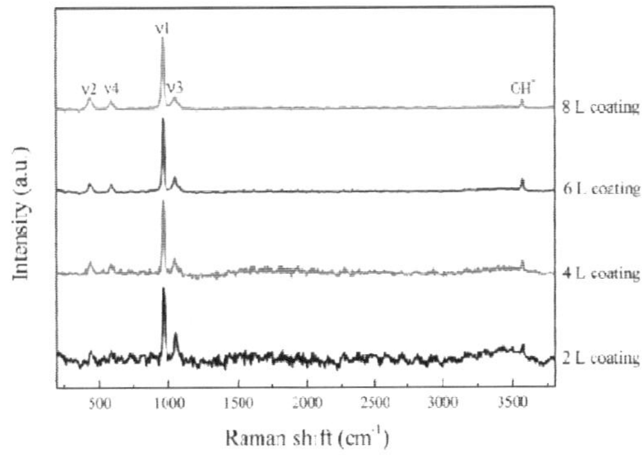

Fig.8. Raman spectra of 2, 4, 6 and 8 L coatings ranging from (a) 200 to 3800 cm^{-1} and (b) 920 to 1000 cm^{-1}.

4. Conclusions

In this work, nanostructured HA coatings have been successfully prepared by suspension plasma spray (SPS). The formation of nanostructured coatings ascribes to the suppressed grain growth by using suspension as feedstock. The SPS coatings with low H_2 flow rate are HA without any impurity phases. When high H_2 flow rate is employed, the coatings contain decomposition phases. The significant presence of OH^- leads to the formation of well crystallized HA which will enhance the biocompatibility and bioactivity. The SPS process paves a straightforward and effective routine for fabrication of HA nanostructured coatings.

5. Literature

[1] L. Sun, C.C. Berndt, C.P. Grey. Phase, structural and microstructural investigations of plasma sprayed hydroxyapatite coatings. Materials Science and Engineering: A, 360 (2003) 70-84.

[2] R.A. Surmenev. A review of plasma-assisted methods for calcium phosphate-based coatings fabrication. Surface and Coatings Technology, 206 (2012) 2035-2056.

[3] S. Dyshlovenko, L. Pawlowski, B. Pateyron, et al. Modelling of plasma particle interactions and

coating growth for plasma spraying of hydroxyapatite. Surface and Coatings Technology, 200 (2006) 3757-3769.
[4] C. Zhang, M. Debliquy, A. Boudiba, et al. Sensing properties of atmospheric plasma-sprayed WO3 coating for sub-ppm NO2 detection. Sensors and Actuators B: Chemical, 144 (2010) 280-288.
[5] J. Weng, X. Liu, X. Li, etc. Intrinsic factors apatite influencing its amorphization during plasma spray coating. Biomaterials, 16 (1995) 35-44.
[6] L. Pawlowski. Suspension and solution thermal spray coatings. Surface and Coatings Technology, 203 (2009) 2807-2829.
[7] S.H. Teng, M.H. Liang, P. Wang, et al. Biomimetic composite microspheres of collagen/chitosan/nano-hydroxyapatite: In-situ synthesis and characterization. Materials science & engineering C, Materials for biological applications, 58 (2016) 610-613.
[8] C. Zhang, X. Zhang, C. Liu, et al. Nano-alumina/hydroxyapatite composite powders prepared by in-situ chemical precipitation. Ceramics International, 42 (2016) 279-285.
[9] E. Park, A. Condrate, D. Lee, etc. Characterization of hydroxyapatite: before and after plasma spraying. Journal of materials science: materials in medicine, 13 (2002) 213-218.
[10] R.T. Candidato, P. Sokołowski, L. Pawłowski, et al. Preliminary study of hydroxyapatite coatings synthesis using solution precursor plasma spraying. Surface and Coatings Technology, 277 (2015) 242-250.
[11] R. D'haese, L. Pawlowski, M. Bigan, et al. Phase evolution of hydroxapatite coatings suspension plasma sprayed using variable parameters in simulated body fluid. Surface and Coatings Technology, 204 (2010) 1236-1246.
[12] S. Saber-Samandari, K. Alamara, S. Saber-Samandari, et al. Micro-Raman spectroscopy shows how the coating process affects the characteristics of hydroxylapatite. Acta biomaterialia, 9 (2013) 9538-9546.
[13] Y. Yang, K.H. Kim, C.M. Agrawal, et al. Interaction of hydroxyapatite-titanium at elevated temperature in vacuum environment. Biomaterials, 25 (2004) 2927-2932.
[14] Y. Huang, L. Song, X. Liu, et al. Hydroxyapatite coatings deposited by liquid precursor plasma spraying: controlled dense and porous microstructures and osteoblastic cell responses. Biofabrication, 2 (2010) 045003.
[15] R.B. Heimann. Thermal spraying of biomaterials. Surface and Coatings Technology, 201 (2006) 2012-2019.
[16] B.R. Gligorijević, M. Vilotijević, M. Šćepanović, et al. Substrate preheating and structural properties of power plasma sprayed hydroxyapatite coatings. Ceramics International, 42 (2016) 411-420.
[17] S. Saber-Samandari, K. Alamara, S. Saber-Samandari. Calcium phosphate coatings: Morphology, micro-structure and mechanical properties. Ceramics International, 40 (2014) 563-572.

Characterization and properties of cermet coatings formed by a new multi-chamber detonation sprayer

Yu. Tyurin, O. Kolisnichenko, Kyiv/UA, Jeff Jia, Beijing/CN, N. Vasilik, Moscow/RU, M. Kovaleva, M. Prozorova, M. Arseenko, V. Sirota, Belgorod/RU

The work reported presents the results of a design of experiment study carried out on a new multi-chamber detonation sprayer (MCDS) used to prepare cermet coatings. A series of cermet coatings WC-10Co-4Cr, Cr_3C_2-NiCr and Cr_3C_2-TaC-NiCr have been prepared on a steel substrate. The microstructures and properties of cermet coatings were examined with the use of optical microscopy, scanning electron microscopy, X-ray phase analysis, and a Vickers hardness tester at a test load of 300 g. The wear behavior was carried out by the ball-on-disk tests. The thickness of cermet coatings was varied in the range of 80 to 375 microns. MCDS application allows the formation of a dense layer with porosity of less than 1.0 % and hardness of 974, 980, 780 (±200 $HV_{0.3}$), respectively WC-10Co-4Cr, Cr_3C_2-NiCr and Cr_3C_2-TaC-NiCr. The cermet coatings show the low specific wear rate 1.2, 1.1, $2.34 \cdot 10^{-5}$ mm^3 $(m*N)^{-1}$. The wear rate of cermet coating was found to be lower than that of steel. The results of this work open up new prospects for the further elaboration of new technologies for making protective cermet coatings that can enhance the protective and wear-resistant properties of industrial iron and iron alloy machinery parts.

1 Introduction

Thermally sprayed cermet coatings have excellent wear-resistant properties and have been widely used in aerospace, chemical and machinery industries [1]. Thermally sprayed cermet coatings consist of ceramic particles mainly in form of carbides reinforced by metallic binder exhibit unique microstructural and mechanical characteristics. Such structure brings in a novel combination of hardness and toughness enabling application of this class of material in wear resistant surfaces. The microstructure of a thermally sprayed coating is usually inhomogeneous. Discontinuities, such as pores, oxide lamellas or incompletely molten spray particles are typically present in the sprayed materials [2]. The deposition methods for the cermet coatings are atmospheric plasma spraying; high velocity processes, such as high velocity oxy-fuel (HVOF) processes; and detonation gun spray (DS) processes [2], [3], [4]. During the spraying processes and the cooling of deposits complex chemical transformations of the materials occur. Structure and properties of thermal spray cermet coatings are very sensitive to spray method, input parameters, powder morphology, and environment. Higher particle velocities in high velocity processes lead to better coating properties, such as a higher bond strength, density and lower oxide content [5]. A critical need exists for transitioning from manufacturing technologies involved pollution disposal to technologies that allow for pollution avoidance and energy savings while producing materials with superior combinations of microstructure and properties [6]. In this study to some cermet coatings spraying, it has been proposed to use a new multi-chamber detonation sprayer (MCDS) [7], [8].

2 Experimental procedure

2.1 Materials

The WC–10Co–4Cr powder (H.C. Starck, AMPERIT®554.074) with grain size in the range of -45+15 µm; Cr_3C_2–25(NiCr) powder (H.C. Starck, AMPERIT®584.054) with grain size in the range of -45+10 µm and 50Cr_3C_2-20TaC-30NiCr powder (Toshiba Materials Co.) with grain size in the range of -53+15 µm (**Fig. 1**) were coated on a corrosion-resistant steel substrates (Fe-0.12C-0.90Mn-0.025P-0.01S-16Cr-0.20Cu, all in wt pct).

Fig. 1. SEM-micrographs of powders: (a) WC–10Co–4Cr; (b) Cr_3C_2–25(NiCr); (c) 50Cr_3C_2-20TaC-30NiCr.

SEM micrographs of the feedstock powders are shown in **Fig. 1**. The WC-10Co-4Cr powder consists of particles have a irregular morphology with sharp facets (**Fig. 1a**). SEM micrographs of Cr_3C_2–25(NiCr) and 50Cr_3C_2-20TaC-30NiCr powders show that the powder particles have spherical, porous, constituents homogenously distributed (**Fig. 1b**) and angular shape with angular carbide particles aggregated densely in NiCr alloy matrix (**Fig. 1c**). Phase composition of the feedstock powders is shown in **Table 1**.

Table 1. Phase composition of the feedstock powders and sprayed coatings

	Powder	Coating
WC–10Co–4Cr	WC, Co_6W_6C, Cr_7C_3	WC, W_2C, W, Co_6W_6C
Cr_3C_2–25(NiCr)	Cr_3C_2, Cr_7C_3 and NiCr matrix	Cr_3C_2, Ni, Cr_2O_3
50Cr_3C_2-20TaC-30NiCr	Cr_3C_2, TaC, $Al_{0.14}Ni_{0.86}$, Fe_3C	OTa, Cr_7C_3, NiO, $Cr_{23}C_6$

2.2 Thermal spray tests

In the present study, a multi-chamber, vertically mounted, detonation sprayer (MCDS) [7], [8] was employed to deposit the cermet coatings on a corrosion-resistant steel substrate. The automated equipment consists of device for spraying; standard powder feeder with a feed rate of up to 3 kg/h; a standard low-pressure (max. 0.3 MPa) gas panel for feeding oxygen, propane-butane and air; an automated control system for the technological process; an automated manipulators for moving the MCDS; and a specimen holder. The pulsed (frequency 20 Hz) flow of the combustion products acquires a high velocity (up to 2100 m/s) and pressure (up to 3 MPa), which shortens the time of the effect on the nozzle walls and surface that are treated ($1x10^{-3}$ s). MCDS was an equipped with several combustion chambers serially connected by channels. These channels were intended for the expiration of the combustion products. They were converged in the central chamber - nozzle. Filling of all chambers by the combustible mixture was carried out simultaneously. MCDS was equipped with a standard powder feeder (by Metco). Continuous gas-powder jet was supplied along the conduit (4-8 m), divided into portions and inserted into the nozzle.

Flat specimens of corrosion-resistant steel were sandblasted by 25A F360 alumina grits prior to spraying. The various sets of parameters used in these experiments are shown in **Table 2**.

Table 2. Spraying parameters and properties of coatings

Parameters		**Coatings**		
		WC-10Co-4Cr	**Cr_3C_2–25(NiCr)**	**50Cr_3C_2-20TaC-30NiCr**
Flow rate of fuel mixture components (m^3/h)	Oxygen	*2.9 / **2.64	*3.43 / **3.04	
	Propane (30%) + butane (70%)	*0.66 / **0.60	*0.83 / **0.75	
	Air	*1.56 / **1.20	*1.16 / **0.99	
Barrel length (mm)		300		
Barrel diameter (mm)		18		
Spray distance (mm)		55		
Frequency (Hz)		20		
Powder feed rate (g/h)		1200	840	800
Thickness (μm)		355 ± 20	100 ± 20	210 ± 20
Porosity (%)		0.15	0.98	0.15
Microhardness, ±200 $HV_{0.3}$		974	980	780
Specific wear rate·($x10^{-5}$)* ($mm^3(m·N)^{-1}$)**				
	Counterbody	0.12	0.15	0.29
	Specimen	1.17	1.10	2.34

*Cylindrical combustion chamber **Disk combustion chamber.
***Surface roughness (not polished surface) R_a = 7.5 ± 0.01 μm.

2.3 Powder and coating characterization

To determine the microstructure and elemental composition of the powders and coatings scanning electronic microscopes (SEM, Quanta 200 3D) were performed. To identify the cross-sectional details, the samples were cut across the cross section, mounted in transoptic powder, and subjected to mirror polishing. The coating thickness was measured with a scanning electron microscope. Porosity was determined by the metallographic method with elements of the qualitative and quantitative analysis of the geometry of the pores using an optical inverted Olympus GX51 microscope [9]. Phase composition of the powders and coatings was determined by the X-ray phase analysis method (diffractometer Rigaku Ultima IV). XRD spectrum for phase analysis was determined by shooting the scheme θ-2Θ scan focusing by Brega-Brentano in the angular range of 5 ... 95 deg. 2Θ. Investigations were carried out in the θ/2θ step scan mode at a step of 0.02 degrees in the 2θ range at a rate of 0.6 deg./min. Crystalline phases were identified by the ICDD PDF-2 (2008) powder diffraction database.

Micro-hardness of the coatings was determined by an automatic micro-hardness tester DM-8B (Affri) by Vickers's test at a test load of 300 g. Ten samples were obtained by different operation cycles under the same conditions.

The tribological evaluation of the coated substrates under dry conditions was performed using a ball on disc tribometer that was manufactured by CSM Instruments (Switzerland) according to ASTM wear testing standard G-99. All tests were performed at 25ºC. A 6 mm in diameter aluminum oxide (Al2O3) ball was used as a counterbody. Specimens from all series were tested under a 10 N normal load, 0.10 $m \cdot s^{-1}$ sliding speed and a total sliding distance of 1000 m. The total wear volume was calculated by measuring the track cross-sectional area with a stylus profilometer (Taylor-Hobson) at ten different locations along the wear track. The ASTM G-99 standard determines the amount of wear by measuring the appropriate linear dimensions of both specimens (ball and disk) before and after the test [10]. All of the obtained samples are characterized by almost the same microstructures, microhardness and wear resistance. Arbitrary selected data are presented in the paper.

3 Results and discussions

Fig. 2 presents high-magnification SEM micrographs (back-scattered electron mode) of the polished cross-sections of the sprayed coatings. The thickness of observed sprayed coatings was varied in the range of 80 to 375 microns.

The coating is formed through the successive layering of a set of deformed particles with different temperatures, speeds and masses. The bulk of the sprayed coating material is deformed and closely packed. It was clearly recognized that the coatings consisted of lamellae elongated in the direction parallel to coating surface. A close examination of the microstructure reveals that carbide particles were distributed in the coating evenly The dense coating was formed through the intensive plastic deformation, as a result from the impact of a particle moving with high velocity and the tamping effect on the source side of following particles [11]. The interface between the sprayed coating and the substrate had no visible macro defects. The coatings presented very good bonding to the substrate. The particles are distributed uniformly in the coating. The sprayed coatings consist of remelted splats of "flattened powder particles", small amounts of porosity and unmelted particles (**Fig. 2**). Porosity is one of the most important characteristics evaluating the coating quality. It degrades coating cohesion and reduces coating hardness, which is undesirable for wear resistance. The porosity measurements that were calculated from the calculation of the quantitative imaging analysis for the obtained coatings are given in **Table 2**. The metallographic porosity levels in the coatings were very low.

Phase composition of the sprayed coatings is shown in **Table 1**. All of the coatings underwent a phase transformation during spraying.

Comparing the phase composition of the original powder and sprayed coatings WC-10Co-4Cr (**Table 1**), it appears that decomposition of WC, to W_2C and W, has occurred. The XRD trace showed a broad maximum in the 38 to 46 two-theta range in the XRD, for all of the sprayed coatings, which is indicative that the cobalt–chromium binder material may be in an amorphous/nanocrystalline form, most likely as a result of the high cooling rate that is commonly associated with the spray process [12], [13].

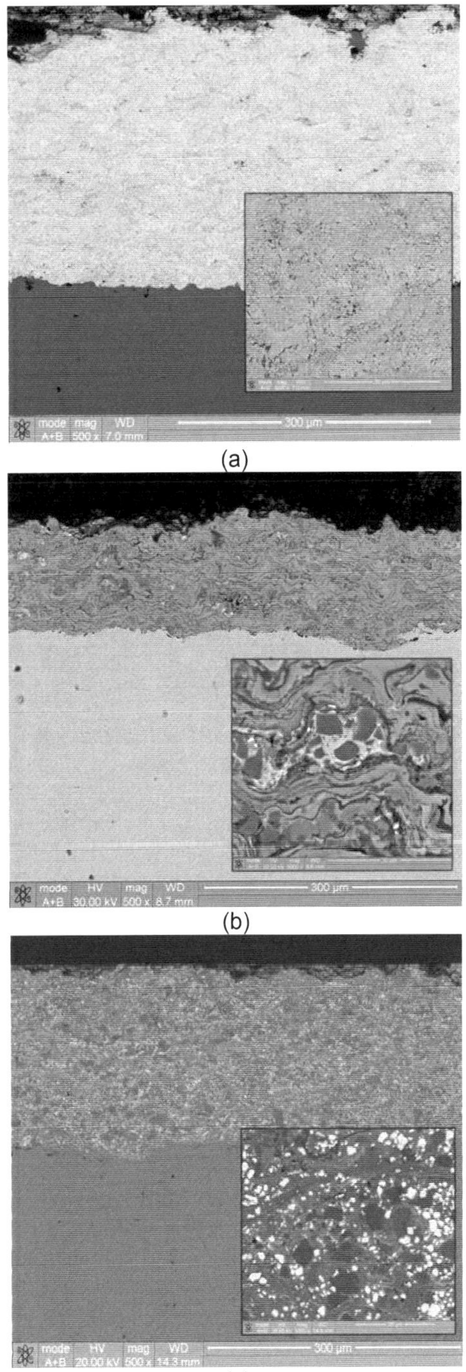

Fig. 2. Coatings cross-section SEM micrographs: (a) WC-10Co-4Cr; (b) Cr_3C_2–25(NiCr); (c) 50Cr_3C_2-20TaC-30NiCr (back-scattered electron mode, Quanta 200 3D).

In the $50Cr_3C_2$-20TaC-30NiCr sprayed coating, the carbide Cr_7C_3 was present. It is generally considered that this carbide is formed through decarburization of Cr_3C_2 carbide.
Additionally, the Cr_3C_2–25(NiCr) and $50Cr_3C_2$-20TaC-30NiCr sprayed coatings contain Cr_2O_3 and NiO, respectively, which is formed by the oxidation of the powder during the spraying process. This indicates that although the spraying conditions are not very oxidizing, the powder was oxidized during processing.The measured hardness values and specific wear rate of the sprayed coatings are reported in Table 2. A wide range of hardness values in sprayed coatings is caused by their different phase composition.
A specific wear rate of sprayed coatings was much less than a specific wear rate of steel ($35.99 \cdot 10^{-5}$ $mm^3(m \cdot N)^{-1}$). The main reason that the wear resistance of the coated samples was much better than that of the uncoated is that there exist many types of carbide and few oxides in the dense coating, with good bonding to the substrate.

4 Conclusion

The cermet coatings on the flat specimens of the corrosion-resistant steel were produced by a multi-chamber detonation sprayer. It was established that MCDS has provided the conditions for formation of a dense layer with low porosity of less than 1.0%, high microhardness and low wear rates. The results of this work open up new prospects for the further elaboration of new technologies for making protective cermet coatings that can enhance the protective and wear-resistant properties of industrial iron and iron alloy machinery parts.

Acknowledgements

All of studies were carried out on the equipment of the Joint Research Center of Belgorod National Research University «Diagnostics of structure and properties of nanomaterials».

5 Literature

[1] Chuanxian, D., Bingtang, H., and Huiling, L. Preparation and characterization plasma-sprayed wear-resistant ceramic and cermet coating materials. Thin Solid Films 118 (1984), pp. 485/93.
[2] Toma, D., Brandl, W., Marginean, G. Wear and corrosion behaviour of thermally sprayed cermet coatings. Surface and Coatings Technology 138 (2001), pp. 149/58.
[3] Li, S.Q., Gong, S.L., Duan, Y.P., Liu, S.H. Studies of laser hybrid plasma spraying WC–10Co4Cr coatings. Surface Engineering 30 (2014), Issue 1, pp. 1/5.
[4] Nikolaev, Yu.A., Vasil'ev, A.A., Ulianitsky V.Yu. Gas detonation and its application for technique and technologies. Combustion, Explosion, and Shock Waves 39 (2003), Issue 4, pp. 22/54.
[5] Murthy, J.K.N., Venkataraman, B. Abrasive wear behaviour of WC-CoCr and Cr_3C_2-20(NiCr) deposited by HVOF and detonation spray processes. Surface and Coatings Technology 200 (2006), pp. 2642/52.
[6] Picas, J.A., Xiong, Y., Punset, M., Ajdelsztajn, L., Forn, A., Schoenung, J.M., Microstructure and wear resistance of WC–Co by three consolidation processing techniques. International Journal of Refractory Metals and Hard Materials 27 (2009), pp. 344/9.
[7] Vasilik, N., Tyurin, N., Kolisnichenko, O. Method for gas-dynamic detonating speedup of powders and device for its implementation, RU Patent 2506341, 2012.
[8] Kovaleva, M., Tyurin, Yu., Vasilik, N., Kolisnichenko, O., Prozorova, M., Arseenko, M. Deposition and characterization of Al_2O_3 coatings by multi-chamber gas-dynamic accelerator. Surface and Coatings Technology 232 (2013), pp. 719/25.
[9] Moskal, G.J. The porosity assessment of thermal barrier coatings obtained by APS method. Journal of Achievements in Materials and Manufacturing 20 (2007), Issue 1-2, pp. 483/6.
[10] Standard test method for wear testing with a Pin-on-Disk apparatus, ASTM G99-05(2010), Book of ASTM Standards, American Soc. for Metals 03.02 (2010), 6 p.
[11] Kovaleva, M., Tyurin, Yu., Vasilik, N., Kolisnichenko, O., Prozorova, M., Arseenko, M., Sirota, V., Pavlenko, I. Effect of heat treatment on the microstructure and microhardness of detonation nanostructured Al2O3 coatings. Journal of Thermal Spray Technology 23 (2014), Issue 7, pp. 1199/209.
[12] Picas, J.A., Xiong, Y., Punset, M., Ajdelsztajn, L., Forn, A., Schoenung, J.M. Microstructure and wear resistance of WC–Co by three consolidation processing techniques. International Journal of Refractory Metals and Hard Materials 27 (2009), pp. 344/49.
[13] Kovaleva, M., Tyurin, Yu., Vasilik, N., Kolisnichenko, O., Prozorova, M., Arseenko, M., Yapryntsev, M., Sirota, V., Pavlenko, I. Effect of processing parameters on the microstructure and properties of WC-10Co-4Cr coatings formed by a new multi-chamber gas-dynamic accelerator. Ceramics International 41 (2015), Issue 10, pp. 15067/74.

Failure behavior of the plasma-sprayed yttria stabilized zirconia thermal barrier coatings under the three point bending test via acoustic emission technique

Liang WANG[1,*], Jiasheng YANG[1,2], Jinxin NI[1], Xinghua ZHONG[1], Huayu ZHAO[1], Fang Shao[1], Chengguang LIU[1], Shunyan TAO[1], Chuanxian DING[1]
[1]Shanghai Institute of Ceramics, Chinese Academy of Sciences, Shanghai 200050, PR China;
[2]University of the Chinese Academy of Sciences, Beijing 100039, PR China
* Corresponding author. Tel.:+86-021-69906320 Fax: +86-021-69906322
E-mail address: L.Wang@mail.sic.ac.cn

Abstract: In this paper, the failure behavior of plasma-sprayed yttria stabilized zirconia thermal barrier coatings fabricated by atmospheric plasma spraying(APS-TBCs) under three-point bending (3PB) has been characterized via acoustic emission (AE) technique. Linear Position Sensing (LPS) method has been adopted to monitor dynamic failure process of the APS-TBCs under 3PB. The investigation results indicate that the variation of AE parameters (ring down counts, amplitudes and cumulative energy) correspond well with the change of stress-strain curve of the loading processes. The failure mechanism can be analyzed systematically based on the characteristics of AE parameters. The distribution of frequency of crack propagation has been revealed. The AE signals came from plastic deformation of substrates and cracking of coatings. The different types of cracks and modes of cracks propagation have been distinguished. LPS method has indicated that many segmentation cracks initiate at the inner of the top-coat. And some main cracks tend to propagate toward the interface. The actual failure mechanism of the APS-TBCs under 3PB results from debonding of ceramic coating from the substrates and then horizontal crack propagate along the bond-coat/top-coat interface under action of flexural moment. The additional digital image technique (DIC) has been used and further testified variation of deformation process of the APS-TBCs.
Keywords: Thermal barrier coatings; Plasma spraying; Failure; Three-point bending; Acoustic emission

1 Introduction

In the past decades, the thermal barrier coatings (TBCs) have been widely used in the aero-engine and land-base turbine. It is one type of the most important coating materials in the industry. The TBCs play an important role in protecting the supealloy substrates from being oxidized, corrosion and worn[1-4]. The TBCs are usually composed of double or multilayers. The common TBCs are composed of single ceramic layer and the metallic layer which is often called the bond-coat. The role of the bond-coat has two aspects. Firstly, it will relieve the misfit of the coefficient of the thermal expansion between the top-coat and the substrate. Secondly, it will improve the anti-oxidation ability of the substrate. And the yttria stabilized zirconia is usually used as the top-coat due to its relatively lower thermal conductivity and higher thermal expansion coefficient, especially high fracture toughness[5-8]. Till now, there are many techniques to fabricate the TBCs, as for the bond-coat of the TBCs, it can be prepared via high velocity oxygen flame(HVOF) spray[9-11], vacuum plasma spray (VPS)[12], low pressure plasma spray (LPPS)[13] and atmospheric plasma spraying(APS), the top-coat of the TBCs can be fabricated by vacuum plasma spraying, LPPS,VPS, plasma spraying-physical vapor deposition (PS-PVD)[14-16], electron beam-physical vapor deposition (EB-PVD)[17], and solution (precursor) plasma spraying (SPS,SPPS) etc.[18-20].

Failure is inevitable when the TBCs was endured with the actual service condition. Generally, the failure modes are very complicated due to the irregular microstructure of the TBCs and the varied external conditions[21-24]. It can be expressed as the following function:

$$\text{Failure modes} = f\,(\text{microstructure, external service condition}) \qquad (1)$$

The microstructure of the TBCs can be viewed as the function of thickness of each layer, the micro-pores, micro-cracks, interfaces etc. While the service condition include the external temperature condition, heat radiation, heat convection, oxygen concentration, erosion media, such as CAO-MgO-Al_2O_3-SiO_2 (CMAS) and so on[25,26].

In fact, the microstructure is also changed with the actual service condition changing. So the failure of the TBCs is a highly non-linear and dynamic process. Many previous work focused on the investigation of the failure of the TBCs via the experimental and simulation methods. As for the simulation aspect, most work focused on the finite element simulation of the residual stress of the as-sprayed TBCs and the development and distribution of the residual stress in the TBCs under the actual service condition. It belongs to static method for characterizing the failure behavior of the TBCs, and these work tend to judge the failure modes based on the calculation of the maximum residual stress[27, 28].

As for the experimental methods, the main characterization method is using the SEM observation. The morphology of the cross-section of the TBCs after failure can be obtained, but the preparation of the coating sample has often experienced cut, grind and polish process. These mechanical process may introduce new defects in the TBCs, and further cover up the failure information of the TBCs after failure, such as the distribution and geometrical characteristic of the cracks. It is a destructive method. In addition, there are many other methods to characterize the failure of the TBCs from different aspects. As for monitoring the grow of the

TGO and measuring the residual stress, Photo-stimulated Luminescence piezo spectroscopy (PLPS) and AC Impedance Spectroscopy(IS)are two common methods. The PLPS has often been used to measure the residual stress in the TGO layer, and the ACIS is often used to characterize the growth of the TGO layer especially the variation of the composition and thickness of the TGO layer [29-34].

As for to monitor the crack existence and crack propagation behavior, the Infrared Thermograph and acoustic emission (AE) are very effective methods. The Infrared Thermograph can only detect the static defects, i.e. the existence of the cracks. The AE which was regarded as a non-destructive method can monitor the crack initiation, grow and propagation timely and dynamically. The failure information can be obtained via the analysis of the AE parameters[35-41].

In the current work, the three-point bending (3PB) test has been performed on the as-sprayed yttria stabilized zirconia TBCs. The failure behavior of the TBCs has been monitored by the AE technique dynamically. The failure information, especially the distribution of the amplitude and the frequency of the crack propagation has been obtained via the analysis of AE signals. The failure mechanism of the TBCs under the 3PB test has also been proposed.

2. Experimental procedures

2.1 Coating fabrication

The coating samples which were used for 3PB test were fabricated by atmospheric plasma spraying (APS). The 304 stainless steel substrates with the dimension 80×6×4mm were immersed in the ethanol for ultrasonic cleaning with the dwell time 30min in order to remove the oil stain before spraying. Then the substrates were grit-blasted with 24 mesh white corundum sand at a pressure of 6 bar in order to increase the bonding strength. The sprayed-powder(feedstock) was dried in an air oven at 120°C for 2h. Then CoNiCrAlY, YSZ coating were deposited onto the substrates in turn to fabricate the as-sprayed TBCs. All the feedstocks were thermally sprayed using an Ar/H_2 APS torch (F4-MB plasma gun, Sulzer Metco, Switzerland). The feedstocks which were used to deposit the coatings were given externally. The nominal thicknesses of the bond-coat and top-coat for the as-sprayed TBCs are 100μm and 300μm, respectively. The obtained roughness of the grit blasted substrate prior to bonding is about 6~10μm. The detailed APS preparation parameters for the TBCs can be seen in the published papers of our group[42,43].

2.2 Three-point bending (3PB) test

The three-point bending (3PB) test for the coating samples has been performed on the compression test machine(Instron-5566) as shown in Fig.1. The linear displacement loading has been imposed onto the coating samples, the maximum load can reach to 10kN, and the loading rate is 0.005mm/min. The plot of the force as the function of the loading displacement is displayed timely in the process of the 3PB test. When the coating separate from the substrate, and the bending moment reach to a certain value, the 3PB test stopped, but the substrate has not failure, if the failure happened, larger force and displacement should be imposed onto the coating samples. As the purpose of our current work is to find some relationship between the failure modes and the AE parameters, we only performs a certain force and load displacement. The schematic illustration of three point bending test for the thermal barrier coatings afflicted with AE system is shown in Fig.2.

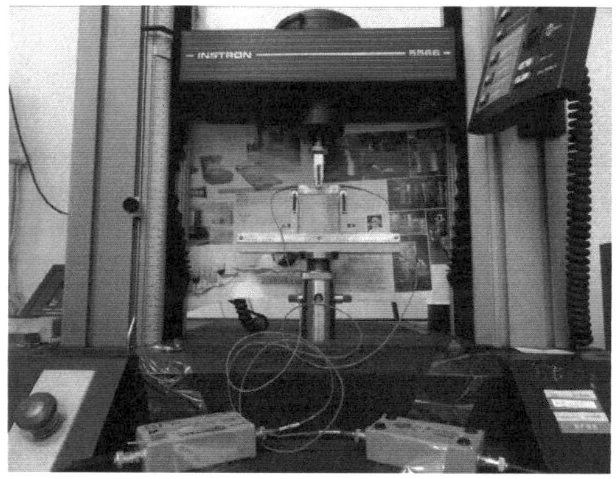

Fig.1 The macro image for the 3PB test afflicted with the acoustic emission system

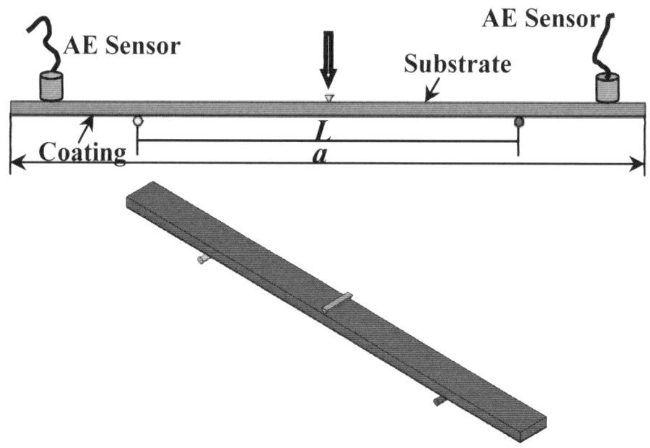

Fig.2 Schematic illustration of three point bending test for the thermal barrier coatings afflicted with AE system

2.3 AE monitor during the 3PB test

In the current work, the 3PB test was performed and accompanied by the monitor of the AE parameters, the nano30 sensor was attached to the coating samples, positing test has been implemented. Before the test, the hardware parameters has been set properly, the threshold value was set to be 40db in order to exclude the disturb of the environmental noise. In order to analyze the waveform, the waveform stream was acquired during the AE test, the relationship plots, correlation plots and 3D plots have been displayed timely. In order to reflect the deformation characteristics as the function of the time, the digital image correlation (DIC) has been used which was used as the attachment of testing system. The variation of the stain of the coatings system can be obtained via the DIC.

3. Results and discussion

3.1 Dynamic behavior of the TBCs under 3PB test

In this section, the failure mechanism of the coating samples under 3PB test has been analyzed systematically, the AE signal has been also analyzed in detail in order to find the failure modes and failure characteristic of the TBCs under 3PB test. Previous work have also done some analysis of AE signals about the TBCs under the bending[44,45]. In this paper, four different coating system have been further compared and discussed. Fig.3 shows the dynamic deformation characteristics of the coating specimen with the loading displacement increasing via the digital image technique (DIC). It can seen that the metallic substrate has experienced the elastic deformation and plastic deformation, and the bending moment increase continuously with the increasing of the loading displacement. But the ceramic top-coat of the TBCs was usually regarded as the linear-elastic. So only the elastic deformation occurred, when the bending stress of the ceramic layer exceed its fracture strength, the coating will peel from the supported substrate directly. The detailed failure mechanism of the TBCs under 3PB test will be further discussed in the following section.

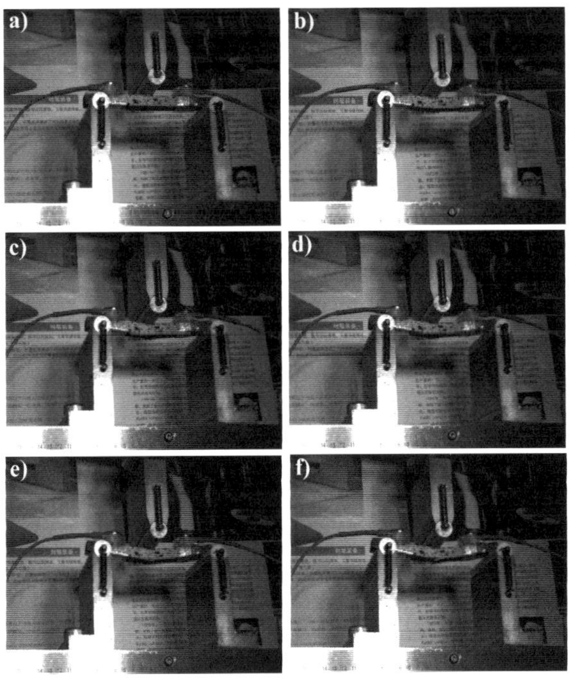

Fig.3 The dynamic failure process of the TBCs under 3PB test using DIC technique

Fig.4 shows the macro-image of profile side and front side of the coating specimen after 3PB test. As for the coating specimen with only the bond-coating, it can be seen that the coatings has detached from the substrate, there are several branched cracks which can be clearly seen on the surface, this indicates that the metallic bond-coat has a certain plasticity and toughness. While for the coating specimen with the bond-coat/top-coat, only a large and deep crack can be visible on the surface of the coatings which indicated the brittle intrinsic characteristic of the ceramic coating.

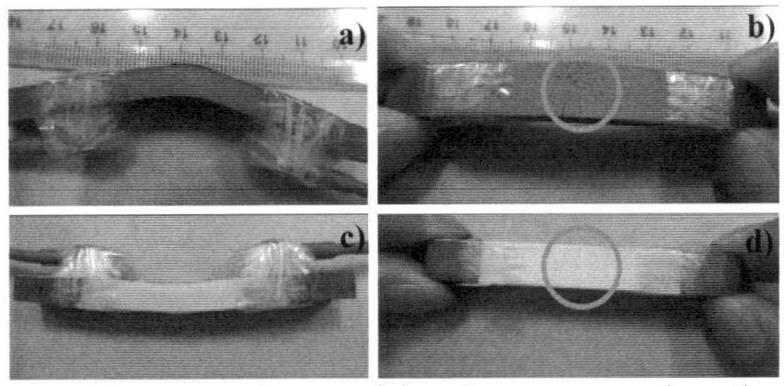

Fig.4 The macro-image of the plasma sprayed coating after 3PB test
a) profile side and b) surface of the bond-coating
c) profile side and d) surface of the bond-coat/top-coat

Fig.5 shows the plot of the flexure stress as the function of the flexure strain for the different types of coating systems. It can be clearly seen that the plot of the flexure stress as the function of the flexure strain for the superalloy and the superalloy/bond-coat is nearly the same, this is because that the bond-coat is metallic layer, the thermal expansion coefficient of the bond-coat is near to that of the metallic substrate, and the bond-coat is relatively thinner compared with the ceramic top-coat and the substrate, and the bonding strength between the substrate and the bond-coat is usually very high which even reach to 50MPa. When the applied load is not large enough, the bond-coat and the substrate will deform together. But once the ceramic top-coat was deposited onto the surface of the bond-coat, the curve will deviate at the initial stage of the plastic deformation. The corresponding flexure stress will increase when the flexure strain is the same compared with the superalloy/BC coating system, this is because that the ceramic top-coat has more higher elastic modulus than that of the metallic substrate, the plastic deformation will need higher stress and energy. While for the stainless steel substrate, it is usually more softer compared with the superalloy substrate, the elastic modulus and hardness of the 304 stainless steel is usually lower than that of the superalloy substrate, so it will need lower stress level when produce the same strain for the stainless steel/BC/YSZ coating system.

3.2 Analysis of AE signals

Fig.6 shows the typical AE signals which correspond to the plot of flexure strain vs. flexure stress for the four coating systems. It can be seen that all the stress-strain curves for the four coating systems can be divided into two stages, i.e. elastic deformation and plastic deformation stage. As for the standing superalloy substrates, it usually has high elastic/elastic deformation ability, only the bending phenomenon has occurred in our current experiments, because the compress strain has not reach to its fracture stain, the AE signals usually came from the slip of the dislocation and the grain boundaries, so the AE signals are very weak cor

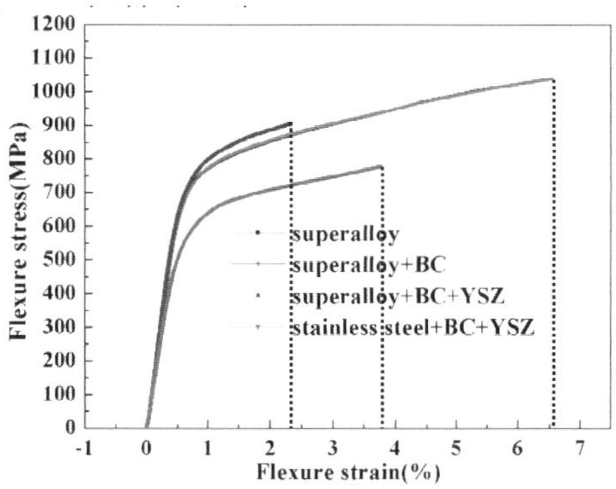

Fig.5 The plot of the flexure stress as the function of the flexure strain for the different types of coating systems

coating systems(Fig.6a)). As for the superalloy/bond-coat system, it can be easily seen that when the plastic deformation of this system occurred, large amounts of AE signals have appeared(Fig.6b)). As for the superalloy/BC/YSZ system, it can be seen from the value of the vertical coordinates, there are large amounts of AE signals, the ceramic YSZ top-coat is brittle compared with the metallic substrate and the bond-coat. Once the ceramic YSZ top-coat began to initiate and grow, the large amounts of AE signals will appear(Fig.6c)). As for the stainless steel/BC/YSZ system, the large amounts of AE signals appear at the plastic deformation stage. As the deformation ability of the stainless steel will release some stress concentration at the inner of the ceramic top-coat, so the event counts will be lower compared with the superalloy/BC/YSZ substrate.

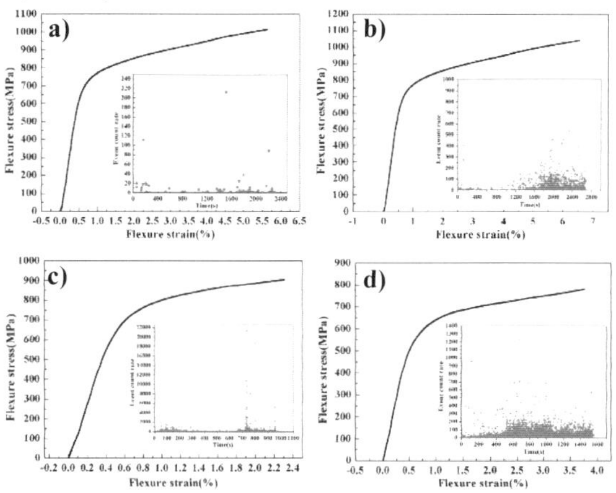

Fig.6 The typical AE signal which correspond to the plot of flexure strain vs. flexure stress for the four coating systems
a) standing superalloy substrate b) superalloy/BC
c)superlaloy/BC/YSZ c)stainless steel/BC/YSZ

Fig.7 shows the plot of amplitude as the function of the time for the four coating systems. It can be further seen that the AE signals are very weak for the standing superalloy substrates. The maximum value of the amplitude only reach to 64db. As for the superalloy/BC system, the relation plot of amplitude vs. time can be divided into two distinct stages, which corresponds to the two stages of the stress-strain curves. During the stage of the plastic deformation, large amounts of AE signals will appear. The maximum value of the amplitude can reach to 85db. As for the superalloy/BC/YSZ system, the relation plot of amplitude vs. time can be also divided into two distinct stages, which corresponds to the two stages of the stress-strain curves, during the stage of the plastic deformation stage, large amounts of AE signal will also appear, the maximum value of the amplitude can reach to 95db. The stainless steel/BC/YSZ has the same trend with that of the superalloy/BC/YSZ system.

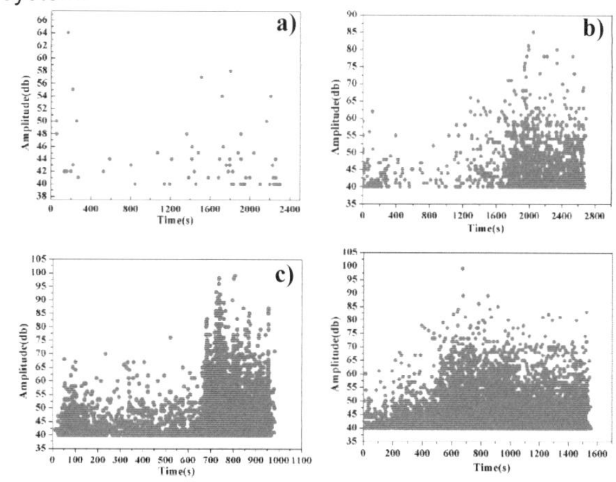

Fig.7 The plot of amplitude as the function of the time
a) standing superalloy substrate b) superalloy/BC
c) superlaloy/BC/YSZ c)stainless steel/BC/YSZ

Fig.8 shows the plot of energy as the function of the time for the four coating systems, it can be seen that the energy of the AE signals is very low for the standing superalloy substrate (Fig.8a)). As for the superalloy/BC system, it can be seen that the energy of AE signals at this plastic deformation stage is much higher than that of the elastic deformation stage(Fig.8b)). As for the superalloy/BC/YSZ system, it can be seen from the value of the vertical coordinates that the value of the energy of the AE signals is very high, this is because the brittle YSZ ceramic top-coat will launch large amounts of cracks during the 3PB test and release large energy(Fig.8c)). As for the stainless steel/BC/YSZ system, when the strain exceed a certain value, a lot of cracks will initiate at the inner of the ceramic YSZ layer, and large amounts of AE signals will appear, but the low elastic modulus of the stainless steel compared with the superalloy substrates, the deformation will be easier compared with the superalloy substrates. It will release some stress concentration in the coating, so the energy of the AE signals will be lower than that of the superalloy/BC/YSZ system(Fig.8d)).

Fig.9 shows the plot of peak frequency as the function of the amplitude for the four coating systems.

As for the standing superalloy substrates, the peak frequency is in the range of 140-170KHz which corresponds to the frequency of the slip of the dislocation and the grain boundaries or plastic deformation of the

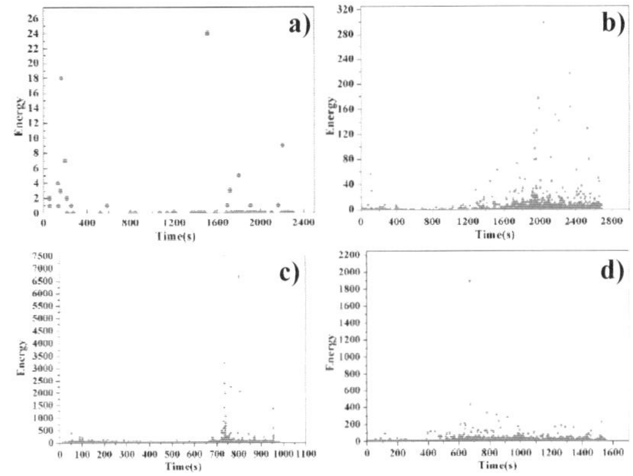

Fig.8 The plot of energy as the function of the time
a) standing superalloy substrate b) superalloy/BC c)superlaloy/BC/YSZ c)stainless steel/BC/YSZ

alloy(Fig.9a)). As for the superalloy/BC system, the plot of peak frequency as the function of the amplitude can be divided into three typical zones, the peak frequencies are in the ranges of 150-200KHz, 200-300KHz, 300-400KHz(Fig.9b)). As for the superalloy/BC/YSZ system, the plot of peak frequency as the function of the amplitude can be divided into four typical zones, the peak frequencies are in the ranges of 125-175KHz, 175-225KHz, 225-300KHz and 300-400KHz(Fig.9c)). As for the stainless steel/BC/YSZ system, the plot of peak frequency as the function of the amplitude can be also divided into four typical zones, the peak frequencies are in the ranges of 130-170KHz, 180-200KHz, 210-270KHz and 280-400KHz(Fig.9d)).

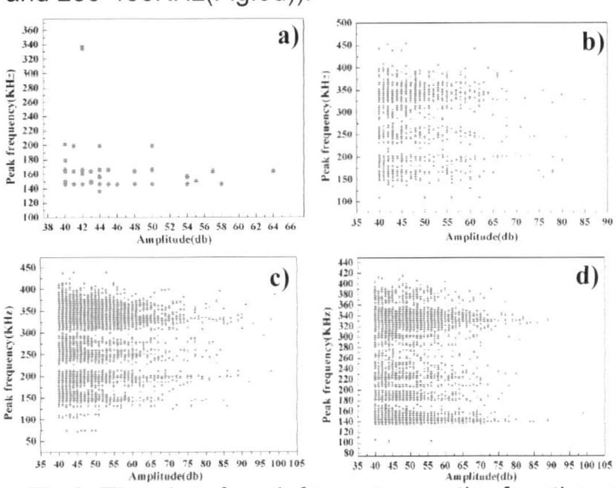

Fig.9 The plot of peak frequency as the function of the amplitude
a) standing superalloy substrate b) superalloy/BC c)superlaloy/BC/YSZ c)stainless steel/BC/YSZ

Fig.10 shows the plot of center frequency as the function of the amplitude. It can be seen that the center frequency of the AE signals for the standing superalloy is in the range of 235KHZ-260KHz(Fig.10a)). As for the superalloy/BC system, the center frequency of the AE signals is in the range of 240KHZ-360KHz(Fig.10b)). As for the superalloy/BC/YSZ system, the center frequency of the AE signals is in the range of 220KHZ-360KHz(Fig.10c)). As for the stainless steel/BC/YSZ system, the center frequency of the AE signals is in the range of 220KHZ-350KHz(Fig.10d)).

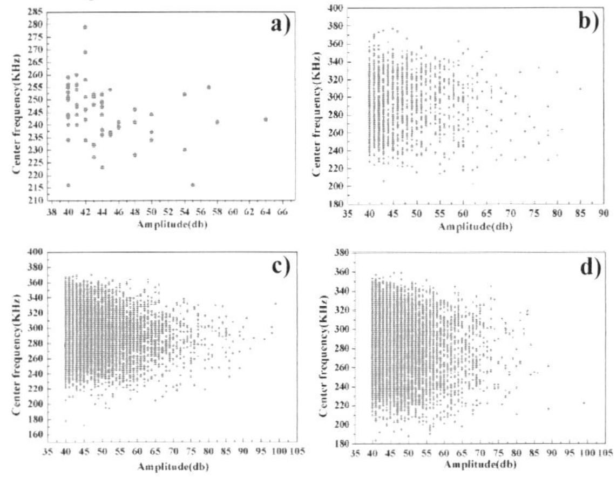

Fig.10 The plot of center frequency as the function of the amplitude
a) standing superalloy substrate b) superalloy/BC
c) superlaloy/BC/YSZ d) stainless steel/BC/YSZ

3.3 Failure mechanism

Fig.11 shows the schematic illustration of the failure process of the substrate/BC/YSZ system under the three point bending test.

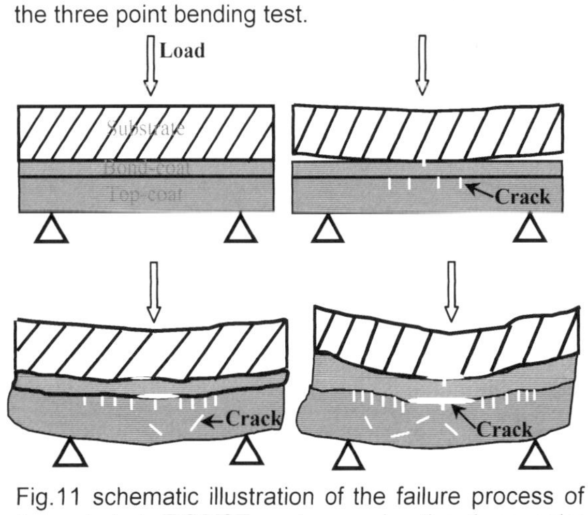

Fig.11 schematic illustration of the failure process of the substrate/BC/YSZ system under the three point bending test

Fig.11a) shows the initial state. With the applied load increasing, the metallic substrate firstly began to deform, and some micro-cracks will initiate at the substrate/bond-coat interface, and some micro-cracks will also initiate at the inner of the ceramic layer (Fig.11b)). With the applied load further increasing, the horizontal crack will initiate at the bond-coat/topcoat interface, and a certain number cracks will also initiate at the inner of the ceramic top-coat due to the deformation of the brittle ceramic top-coat(Fig.11c)). With the applied load further increasing, the horizontal crack will propagate, linked along the bond-coat/top-coat interface and the propagating crack will detached from the substrate, the coating totally failure eventually (Fig.11d)).

Conclusions

In this paper, the failure behavior of the plasma sprayed yttria stabilized zirconia thermal barrier coatings under 3PB test has been characterized via acoustic emission techniques. The following occlusions can be drawned:

(1) The in-situ AE technique can monitor the deformation and failure behavior of the thermal barrier coatings dynamically, and the digital image correlation technique can examine the realistic displacement and strain for the bending coating specimen.

(2) The flexure stress as the function of flexure strain can be divided into two stages, i.e. elastic deformation and plastic deformation, which corresponds well to the plot of AE signals of event count rate vs. time or amplitude vs. time.

(3) The brittle fracture of the ceramic YSZ top-coat can induce large amounts of AE signals. The AE signals of the standing superalloy is very limited, which mainly came from the slip of the dislocation and the grain boundaries during the plastic deformation. The failure of the TBCs under three point bending is attributed to the formation and propagation of the cracks at the top-coat/bond-coat interface.

Acknowledgment

We express our gratitude to Senior engineer Li lei and Gao Jianhua for their experimental assistance. This work was jointly supported by the National Natural Science Foundation of China (NSFC) under the Grant No.51202277, Young Scholar Project (No.12ZR1452000) supported by the Shanghai Science and Technology Committee, and we also give our thanks to the special Project for Army-Civilian Combination of Shanghai Municipal Commission of Economy and Informatization (Grant No. JMJH2014053).The current work was also supported by the Major Program for Basic Research of Shanghai Science and Technology Committee (No.12DJ1400402), the Key Research Program of Chinese Academy of Sciences (No. KGZD-EW-T06) and the research grant (No.14DZ2261203) from Shanghai government.

References

[1] Padture NP , Gell M, Jordan EH. Materials science- Thermal barrier coatings for gas-turbine engine applications.Science.2002, 296(5566): 280-284

[2]John H. Perepezko. The Hotter the Engine, the Better.Science.2009, 326(5956):1068-1069

[3] Kristen A. Marino, Berit Hinnemann, and Emily A. Carter. Atomic-scale insight and design principles for turbine engine thermal barrier coatings from theory. PNAS, 2011,108(14):5480–5487

[4] L. Yang, Y. C. Zhou, C. Lu. Damage evolution and rupture time prediction in thermal barrier coatings subjected to cyclic heating and cooling: An acoustic emission method. Acta Mater, 2011, 59: 6519-6529.

[5] S. Uwe, L. Christoph, F. Klaus, P. Manfred, S.B. Bilge, L. Odile, Some recent trends in research and technology of advanced thermal barrier coatings, Aerosp. Sci. Technol.7 (2003) 73-80.

[6] D.R. Clarke, M. Oechsner, N.P. Padture, Thermal-Barrier Coatings For More Efficient Gas-Turbine Engines. MRS Bulletin. 2012,37 (10) :891-902.

[7] C.G. Levi, J.W. Hutchinson, M.-H. Vidal-Setif, C.A. Johnson, Environmental degradation of thermal-barrier coatings by molten deposits. MRS Bulletin. 2012, 37(10): 932-941

[8] A.G. Evans, D.R. Clarke, C.G. Levi, The influence of oxides on the performance of advanced gas turbines. J. Eur. Ceram. Soc. 2008, 28 (7) 1405-1419.

[9] Myoung Sang-Won, Lu Zhe, Jung Yeon-Gil, Jang Byung-Koog, Paik Ungyu. Control of bond coat microstructure in HVOF process for thermal barrier coatings. Surface & Coatings Technology. 2014, 260: 63-67

[10] Ni L Y, Liu C, Huang H, Zhou C. G. Thermal Cycling Behavior of Thermal Barrier Coatings with HVOF NiCrAlY Bond Coat. Journal of Thermal Spray Technology.2011, 20(5): 1133-1138

[11] Ma Kaka, Schoenung Julie M. Influence of cryomilling on the microstructural features in HVOF-sprayed NiCrAlY bond coats for thermal barrier coatings: Creation of a homogeneous distribution of nanoscale dispersoids. Philosophical Magazine Letters. 2010, 90.(10)739-751.

[12] Rudolphi M, Renusch D, Zschau H.–E, Schuetze M. The effect of moisture on the delayed spallation of thermal barrier coatings: VPS NiCoCrAlY bond coat plus APS YSZ top coat. Materials at high Temperatures. 2009,26(3): 325-329.

[13] Hospach Andreas, Mauer Georg, Vassen Robert, Stoever Detlev. Columnar-Structured Thermal Barrier Coatings (TBCs) by Thin Film Low-Pressure Plasma Spraying (LPPS-TF).Journal of Thermal Spray Technology. 2011, 20(1-2): 116-120.

[14] Song Jinbing, Zhang Xiaofeng, Deng Chumming, Liu Min, Zhou Kesong, Tong Xin. Research of in situ modified PS-PVD thermal barrier coating against CMAS (CaO-MgO-Al_2O_3-SiO_2) corrosion. Ceramics International. 2016, 42,(2): 3163-3169

[15] Rezanka S, Mauer G, Vassen R, Improved Thermal Cycling Durability of Thermal Barrier Coatings Manufactured by PS-PVD. Journal of Thermal Spray Technology.2014, 23(1-2):182-189.

[16] Goral Marek, Kotowski Slawomir, Nowotnik Andrzej, Pytel Macie, Drajewicz Marcin, Sieniawski Jan. PS-PVD deposition of thermal barrier coatings. Surface & Coatings Technology. 2013, 237: 51-55.

[17] Xu Zhenhua, Wang Zhankao, Huang Guanghong, Mu Rende, He Limin. Morphology, bond strength and thermal cycling behavior of (Ni, Pt)Al/YSZ EB-PVD thermal barrier coatings. Journal of Alloys and Compounds. 2015, 651: 445-453.

[18] Ajay A, Raja V S , Sivakumar G , Joshi S. V. Hot corrosion behavior of solution precursor and atmospheric plasma sprayed thermal barrier coatings. Corrosion science. 2015,98:271-279

[19] Jiang Chen, Jordan Eric H, Harris Alan B, Gell Maurice, Roth Jeffrey. Double-Layer Gadolinium Zirconate/Yttria-Stabilized Zirconia Thermal Barrier Coatings Deposited by the Solution Precursor Plasma Spray Process. Journal of Thermal Spray Technology. 24(6):895-906

[20] Yuexing Zhao, Liang Wang, Jiasheng Yang, Dachuan Li, Xinghua Zhong, Huayu Zhao, Fang Shao, Shunyan Tao. Thermal aging behavior of axial suspension plasma sprayed yttria-stabilized zirconia (YSZ) thermal barrier coatings. Journal of Thermal Spray Technology. 2015,24(3):338-347

[21] Evans AG, Mumm DR , Hutchinson JW, Meier GH , Pettit FS. Mechanisms controlling the durability of thermal barrier coatings. Progress In Materials Science. 2001, 46 (5):505-553

[22] Cao XQ , Vassen R , Stoever D . Ceramic materials for thermal barrier coatings. Journal of The European Ceramic Society. 2004,24(1): 1-10

[23] Rabiei A, Evans AG. Failure mechanisms associated with the thermally grown oxide in plasma-sprayed thermal barrier coatings. Acta Materialia. 2000,48(15): 3963-3976

[24] L. Wang, D.C. Li, J.S. Yang, F. Shao, X.H. Zhong, H.Y. Zhao, K. Yang, S.Y. Tao, Y. Wang. Modeling of thermal properties and failure of thermal barrier coatings with the use of finite element methods: A review. Journal of the European Ceramic Society. DOI:10.1016/j.jeurceramsoc.2015.12.038

[25] Krause Amanda R, Li Xing, Padture Nitin P. Interaction between ceramic powder and molten calcia-magnesia-alumino-silicate (CMAS) glass, and its implication on CMAS-resistant thermal barrier coatings. Scripta Materialia. 2016, 112: 118-122

[26] Su Luochuan, Chen Xi ,Wang T J. Numerical analysis of CMAS penetration induced interfacial delamination of transversely isotropic ceramic coat in thermal barrier coating system. Surface & Coatings Technology. 2015, 280: 100-109.

[27] Liu D, Kyaw S T, Flewitt P E J, Seraffon M, Simms N J, Pavier M, Jones I A. Residual stresses in environmental and thermal barrier coatings on curved superalloy substrates: Experimental measurements and modeling. Materials Science and Engineering A. 2014, 606: 117-126

[28] Ranjbar-Far M, Absi J, Mariaux G, Shahidi S. effect of Residual Stresses and Prediction of Possible Failure Mechanisms on Thermal Barrier Coating System by Finite Element Method. Journal of Thermal Spray technology.2010,19(5): 1054-1061

[29] Vakhtin Andrei B, Peterson Kristen A, Kane Daniel J, Jordan Eric H, Hansen Geoffrey, Teicholz Matthew. Combination of Fourier-domain optical coherence tomography and photo-stimulated luminescence piezo-spectroscopy as an NDE tool for thermal barrier coatings. Proceedings of the ASME Turbo Expo. 2007, Vol 1,pp. 275-279

[30] Liu Dong, Lord Oliver, Stevens Oliver, Flewitt Peter E. J. The role of beam dispersion in Raman and photo-stimulated luminescence piezo-spectroscopyof yttria-stabilized zirconia in multi-layered coatings. Acta Materialia. 2013, 61(1): 12-21

[31]Anderson PS, Wang X, Xiao P. Impedance spectroscopy study of plasma sprayed and EB-PVD thermal barrier coatings. Surface & Coatings Technology. 2004,185(1):106-119

[32] Huang Hui, Liu Chao, Ni Liyon, Zhou Chungen. Evaluation of TGO growth in thermal barrier coatings using impedance spec troscopy. Rare Metals. 2011,30(1): 643-646

[33] Gomez-Garcia J, Rico A, Garrido-Maneiro M A, Munez C J, Poza P, Utrilla V. Correlation of mechanical properties and electrochemical impedance spectroscopy analysi s of thermal barrier coatings. Surface & Coatings Technology.2009,204(6-7):812-815

[34] Yang Fan, Xiao Ping. Nondestructive Evaluation of Thermal Barrier Coatings Using Impedance Spectroscopy. International Journal of Applied Ceramic Technology.2009,6(3):381-399

[35] W. Zhu, L. Yang, J. W. Cuo, Y. C. Zhou, C. Lu. Determination of interfacial adhesion energies of thermal barrier coatings by compression test combined with a cohesive zone finite element model. Int. J. Plasticity, 2015, 64: 76-87.

[36] X.Q. Ma, S. Cho, M. Takemoto. Acoustic emission source analysis of plasma sprayed thermal barrier coatings during four-point bend tests. Surface and Coatings Technology 139Ž2001.55_62

[37]Daniel Nies, Birgit Rehmer, Birgit Skrotzki and Robert Vaβen. Damage Characterization of Thermal Barrier Coatings by Acoustic Emission and Thermography. Advanced Engineering Materials 2012, 14, No. 9,790-794

[38] Kaita Ito, Hitoshi Kuriki, Hiroshi Araki, Seiji Kuroda and Manabu Enoki.Detection of segmentation cracks in top coat of thermal barrier coatings during plasma spraying by non-contact acoustic emission method.. Sci. Technol. Adv. Mater. 15 (2014) 035007 (11pp)

[39] L. Yang*, H. S. Kang, Y. C. Zhou, C. Lu. Frequency as a key parameter in discriminating the crack modes of thermal barrier coatings: cluster analysis of acoustic emission signals. Surf. Coat. Technol., 2015, 264: 97-104.

[40] K. Ito, H. Kuriki, H. Araki, S. Kuroda, and M. Enoki. Evaluation of Generation Mechanism of Vertical Cracks in Top Coat of TBCs During APS Deposition by Laser AE Method. Volume 24(5) June 2015 Journal of Thermal Spray Technology.848-856.

[41] L. Yang, Y. C. Zhou, W. G. Mao, C. Lu. Real-time acoustic emission testing based on wavelet transform for the failure process of thermal barrier coatings. Appl. Phys. Lett., 2008, 93: 231906.

[42]L. Wang, C.G. Liu, X.H. Zhong, Y.X. Zhao, H.Y. Zhao, J.S. Yang, S.Y. Tao, Y. Wang. Investigation of crack propagation behavior of thermal barrier coatings fabricated by atmospheric plasma spraying in the process of uniaxial tension using acoustic emission technique. Journal of Thermal Spray Technology.2015, 24(3):296-308

[43]Yuexing Zhao, Dachuan Li, Xinghua Zhong, Huayu Zhao, Liang Wang, Fang Shao, Chenguang Liu, Shunyan Tao*. Thermal shock behaviors of YSZ thick thermal barrier coatings fabricated by suspension and atmospheric plasma spraying. Surface and Coatings Technology. 2014, 249, 48-55

[44] L. Yang, H. S. Kang, Y. C. Zhou, L. M. He, C. Lu. Intelligent discrimination of failure modes in thermal barrier coatings: wavelet transform and neural network analysis of acoustic emission signals. Exp. Mech., 2015, 55: 321-330.

[45] L. Yang, Z. C. Zhong, J. You, Q. M. Zhang, Y. C. Zhou, W. Z. Tang. Acoustic emission evaluation of fracture characteristics in thermal barrier coatings under bending. Surf. Coat. Technol., 2013, 232: 710-718.

Influence of laser glazing on the Characterization of plasma-sprayed YSZ coatings

Yan WANG[a*], Jiang-wei LIU[a], Han-lin LIAO[a], Geoffrey DARUT[a], Jorge STELLA[c], Thierry POIRIER[b], Marie-Pierre PLANCHE[a]
[a]Univ. Bourgogne Franche-Comté, UTBM, LERMPS, BELFORT France
b Institut de Céramique Française, LIMOGES France
c Universidad Simón Bolívar, Departamento de Ciencia de los Materiales, CARACAS, Venezuela
Tel: +33 7834360 Mail:yan.wang@utbm.fr

In this study, 8wt. % yttria-stabilized zirconia (8YSZ) powder was deposited on the substrates by atmospheric plasma spray (APS). The coatings were post treated by laser glazing under different parameters in order to densify them. The characterization of the laser molten pools under different laser treatment conditions were studied. Preheating processes were also employed. Scanning electron microscopy (SEM) observations of the surface and cross-section of as-sprayed and laser-glazed coatings were carried out to investigate the influence of laser glazing on the microstructure on laser glazed coatings. The results show that preheating processes improve the coating in terms of deepening the laser-glazed layer, reducing the number of vertical cracks and surface density of cracks and widening the molten pool. Finally, the linear energy density influences the characterization of the glazed layer.
Keywords: YSZ coating, APS, laser-glazing, SEM observation, laser molten pool

1 Introduction

8wt. % yttria-stabilized zirconia (8YSZ) is mostly used to manufacture thermal barrier coating (TBC) as a ceramic layer [1,2]. TBC is widely applied to the combustion chamber of gas turbine engines. Because the coating can provide high thermal gradient with cooling backside and hot outside. So it enables the gas turbine to work in relative high temperature environment. Thus it increases the thermal efficiency significantly [3]. In addition, the stress induced phase transformation from tetragonal phase to monoclinic phase in YSZ increases its fracture toughness [4]. Moreover, YSZ material is also used in applications like solid oxide fuel cells (SOFC) and oxygen sensors for its advantages like adequate ionic conductivity , stability in dual environment especially [5], but the required microstructure is much denser than in TBC. However, after plasma spraying, volume shrinkage and tensile stress on the top of the fabricated coating lead to crack networks on the surface of YSZ coatings. These cracks and pores are detrimental for mechanical properties. In addition, the oxidation of the bonding layer and the corrosion of the components lead together to the failure of TBCs[6] . In the case of the SOFC application, plasma spraying induces pores which cause diffusion of fuel or air through the electrolyte [7]. For all these reasons, it is necessary to heal and densify the coatings by post treatment. And a laser glazing post treatment is a useful way to heal and densify coatings[8–10].
The aim of present study is to densify the as-sprayed coating by laser remelting. Different parameters of laser treatment were adjusted to observe the influence of laser glazing on the microstructure of plasma sprayed YSZ coating, particularly the effect of the pre-heating process.

2 Experimental procedures

2.1 Preparation of the coating deposition

Commercial ZrO_2-8 wt. % Y_2O_3 powder (Plasmatex-powder) was employed as feedstock. Carbon steel disks were used as substrates with the specific dimensions of Ø25×9mm.The substrates were first cleaned in an ultrasonic acetone bath during 10 min and second grit-blasted with alumina. A Sulzer-Metco F4 plasma gun was used to deposit 8YSZ coating. A plasma gas mixture composed of Argon and Hydrogen was employed. Argon was also employed as carrier gas. **Table. 1** illustrates the detail information of spraying parameters.

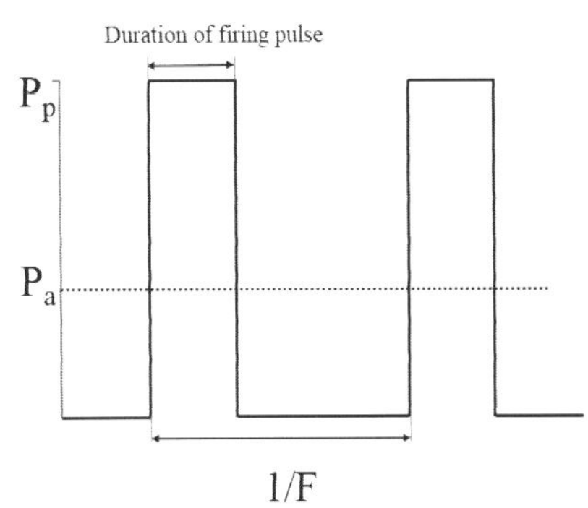

Fig. 1. Schematic signal of a pulsed laser

2.2 Laser gazing conditions

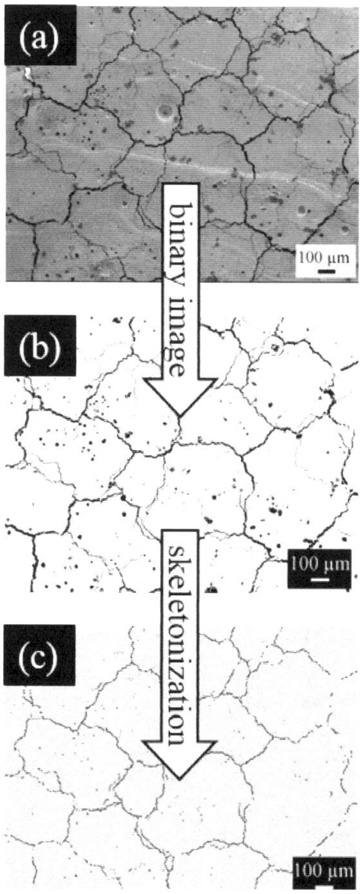

Fig. 2. Conversion of binary image and skeletonization during crack measurement

The post laser treatment process was achieved by a Nd-YAG laser source designed by the Cheval company (Cheval, Pirey, France). This laser can supply a strong power with a maximum average power of 1100 W. The pulse duration can be adjusted from 0.5 to 10 ms. Wavelength and maximum frequency are 1064 nm and 200 Hz respectively. Off-set distance is fixed to 340mm and the diameter of the spot is 3mm at this focalization point.

Table. 1. Spraying parameters

Parameters	value
Electric arc current (A)	620
Electric arc voltage (V)	60
Plasma gas (NL/min) Ar	32
H2	10
Carrier gas (NL/min)	3.5
Liner speed (m/s)	1.2
Stand-off distance (mm)	115
substrate	Ø25mm*9mm
powder	ZrO2-8%Y2O3

During laser-glazing process, three factors were mainly used to control the linear density of laser energy, hence to get different microstructures and morphologies of the glazed coatings. They concern duration of laser firing pulse, the scanning speed and the energy density. The duration of each laser cycle is calculated from the ratio 1 / F, F corresponding to a given frequency. The laser firing period is the heating time required by the laser during each cycle. In this model, it was assumed that the laser power is constant during each pulse as showed in **Fig. 1**. **Table. 2** depicted the parameters of the laser treatment.

To evaluate the effect of laser parameters, ten unique tracks remelting were carried out with different scanning speeds, different durations for each fire pulse, and different intensities of the laser (the laser intensity of the imposed for varying the peak power laser). A calorimeter (Labmaster 4DOPTICAL USA) was used to measure the average power of the laser as defined by the following relationship:

$P_a = E \cdot F$ (1)

Where P_a is its average power (W), E its energy (J) and F its frequency (Hz)

The energy E can be obtained by the following relationship:

$E = P_p \tau$ (2)

Where P_p is the peak power (W) and τ the pulse duration of firing a pulse (ms)

The laser power density can be calculated by the following relationship:

$\varphi(r) = \frac{P}{2\pi R^2} \exp(-\frac{r^2}{2R^2})$ (3)

where $P/2\pi R^2$ is the power density of laser at a pulse ($W.mm^{-2}$) and R represents the radius of dispersion of the pulsed laser(1.5mm)

In addition, the as-sprayed coatings were pre-heated by 4, 8 and 12 laser passes with a spot of 10mm and a scanning speed of 50mm/s before laser glazing. Then, coatings were laser glazed following the parameters matrix described **in tables. 2** and **3** varying the firing duration, the scanning speed and the energy density.

Table. 2. Parameters of laser treatment

Energy density=480 W/mm2, v=5mm/s, f=10Hz	Energy density=480 W/mm2 Duration=5ms, f=10Hz	v=5mm/s, f=10Hz, Duration=5ms,
Duration of laser firing	Velocity	Energy density
3ms	5mm/s	348W/mm2
4ms	10	480
5ms	20	570
6ms	30	688

2.3 Characterizations of the samples

Laser diffraction method (Maserter2000, Malvern, England) was used to measure the size distribution of the 8YSZ powder. Feedstock, coating cross and its surface morphology were characterized by scanning electron microscopy (SEM) and optical microscopy (OM).). During the laser glazing process, the temperature of the coatings were measured by infrared thermal imaging camera (SC 5210) developed by FLIR company.

Table. 3. Energy densities for different parameters

Case	Duration (ms)	Velocity (mm/s)	Average power (W)	Power pulse (W)	Energy density (W/mm2)	Liner energy density (J/mm)
1	3	5	170	4700	332	28.2
2	4	5	261	6525	462	52.2
3	5	5	339	6780	480	67.8
4	6	5	424	7067	500	84.8
3	5	5	339	6780	480	67.8
5	5	10	339	6780	480	33.9
6	5	20	339	6780	480	16.95
7	5	30	339	6780	480	11.3
8	5	5	246	4920	348	49.2
3	5	5	339	6780	480	67.8
9	5	5	402	8060	570	80.60
10	5	5	486	9727	688	97.27

Image J software was employed to estimate the porosity and the surface density of cracks. For the case of porosity tests, 10 SEM pictures were used obtained at random places of polished cross-sections at a magnification of 300. While with respect to surface density of cracks, first a threshold value was set to distinguish cracks from pores and dense areas in the coatings. Second, the pictures were converted to binary mode as shown in **Fig. 2 (b)**. Black pixels represent both cracks and pores while white pixels corresponded to dense areas. Finally, the binary image was skeletonized (**Fig. 2 (C)**) in such a way that the width of cracks and pores were ignored. The black pixels lines were skeletonized as cracks. This way, the length of cracks could be estimated as the numbers of black pixels. The value of surface density of cracks is finally deduced from the ratio between the numbers of black pixels and the total number of pixels [11].

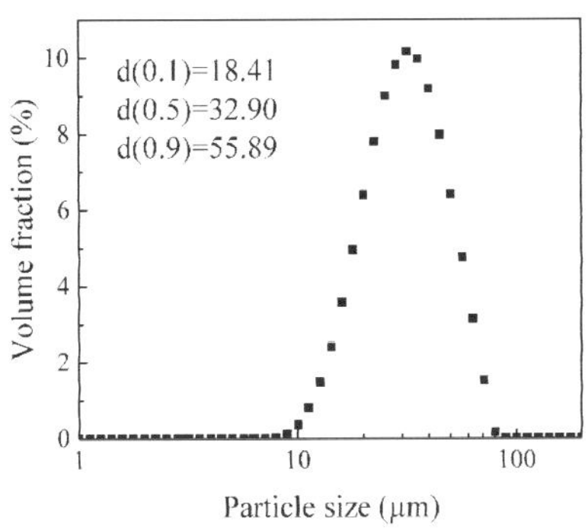

Fig. 3. Powder morphology and size distribution

Fig. 4. Cross-section morphology of as-sprayed YSZ coating

3 Results and discussions

3.1 Characterization of the feedstock powder and the as-sprayed coatings

From the results shown in **Fig. 3**, average size of YSZ powder is 35.267μm with d(0.1)=18.41, d(0.5)=32.90, d(0.9)=55.89. **Fig. 4** displays the cross-section morphology of as-sprayed YSZ coating. It corresponds to a typical thermal spray structure with a porosity of 8.5±1.2%.

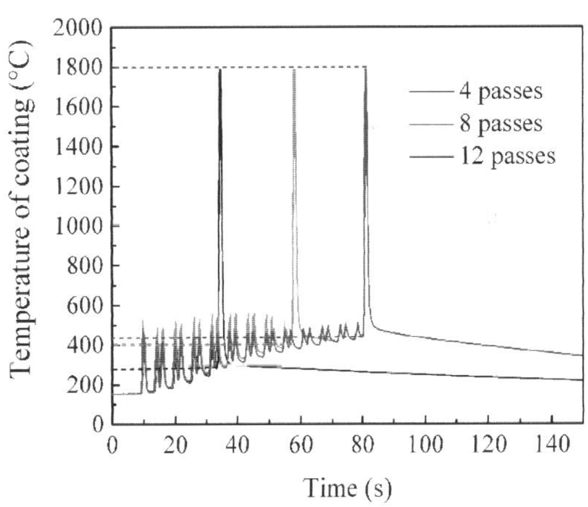

Fig. 5. Evolution of the coating temperature during the preheating process

3.2 Characterization of laser-glazed coatings

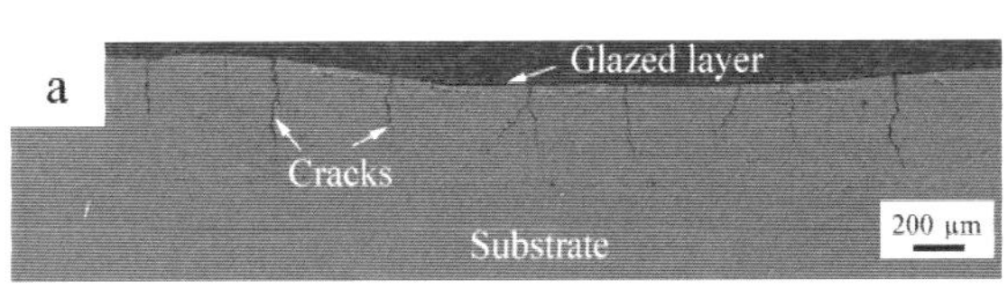

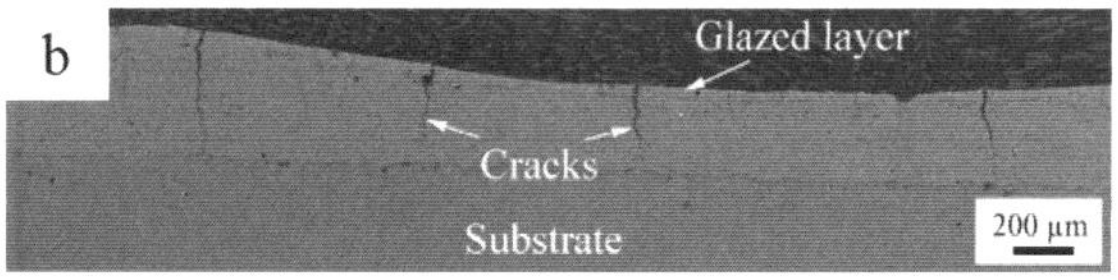

Fig. 5 Cross-section microstructure of laser glazed coatings under parameters of case 3 (pulse laser duration: 5 ms, scanning speed: 5mm/s, energy density: 480 W/mm2) without preheating (a), preheating to 420°C

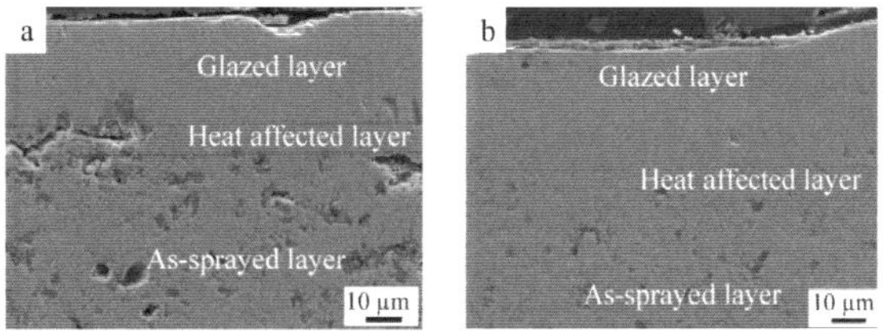

Fig. 7. Close view of cross-section microstructure of laser glazed caoting under parameters case 3 (a) without preheating, (b) preheating to 420°C

Fig.6 illustrates the surface temperature of coating during preheating process. Coatings were preheated to 285 °C, 400°C and 420°C respectively before laser glazing. And temperature of coating increases to 1800°C during laser process.

Cross-section microstructure of laser glazed coatings are showed in **Fig. 6**. The coatings exhibit a glazed layer with numerous vertical cracks. The quantity of the vertical cracks is much less for the preheated coating than for the one without preheating. Because the energy density of the laser spot is a Gaussian distribution (see equation 3), the energy provides to the central area is higher than that to the border, on the contrary, the cooling rate is much higher for border areas compared to the central areas. Therefore, the solidification arises from the border to the central zone and induces tension stresses to the central areas due to the volume contraction. As a consequence, vertical cracks could be generated if the stress can't be released properly. As a result, **Fig. 6(a)** exhibits that a number of vertical cracks formats to release stresses. Preheating process makes the temperature of the coating increasing to 420°C, Inconsistent cooling rate between different areas is reduced. Hence, numbers of vertical cracks decreases significantly as can be observed in **Fig. 6(b)**.

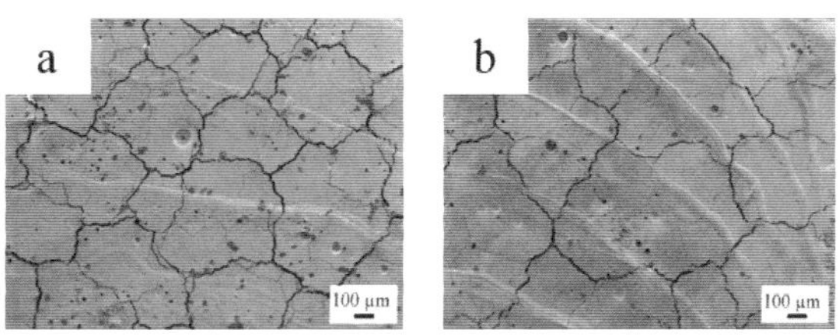

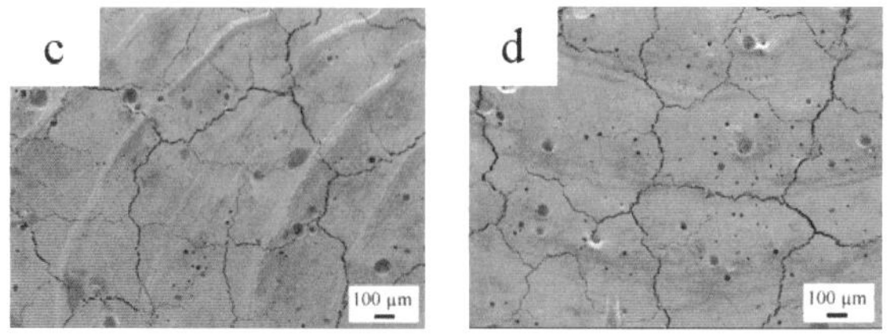

Fig. 8. Surface morphology of laser glazed coatings without preheating (a), preheated to 285 °C(b), 400°C (c), and 420°C (d)

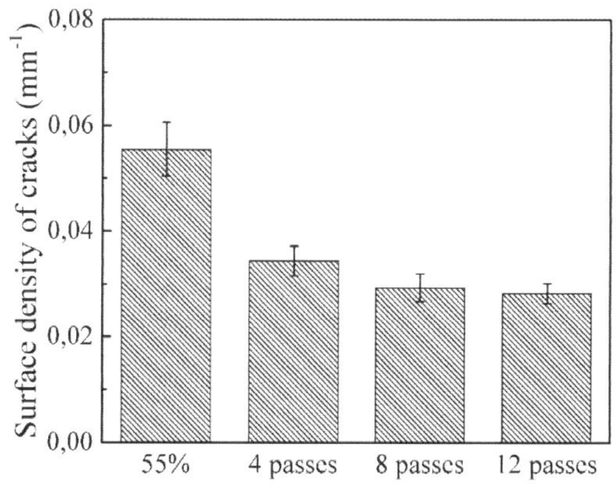

Fig. 9. Surface density cracks in function of the parameters of laser glazing process

Looking at the cross-section microstructure of laser glazed coating (**Fig. 7**), three parts can be distinguished in the laser glazed coating:

- → The first one corresponds to a glazed layer. This part looks like to bulk material with dense microstructure and only a few defects.
- → The second one concerns the heat affected layer with the presence of a transformation area including small pores and defects.
- → The third and last one is as-sprayed layer. The microstructure of this layer is identical to the as-sprayed coating.

Comparing **Fig. 7(a)** with **Fig. 7(b)**, it can be noted that the depth of glazed layer preheated to 420°C is larger than that without preheating and the defects in the second zone are less numerous. After the preheating of coating, it could be supposed that the thermal gradient between molten layers and unmolten ones is small leading to reduce stresses caused by cooling rate. As a consequence, cracks in this specific zone are not need to release stress **Fig. 7(a)**.

Fig. 8 illustrates surface crack nets in function of different process parameters. Cracks combine one with another in the whole surface of laser-glazed coating. While preheating is applied, the number of cracks decreases sharply as well as the width of cracks. The surface density of such cracks changes from 0.055mm^{-1} to 0.028 mm^{-1} (**Fig.9**).

3.3 Linear energy density and microstructure

During the laser glazing process, different parameters (**Table.2**) are adjusted to get a proper microstructure. The influence of the linear energy density's effect on the microstructure characteristics of laser-glazed coatings is visible.

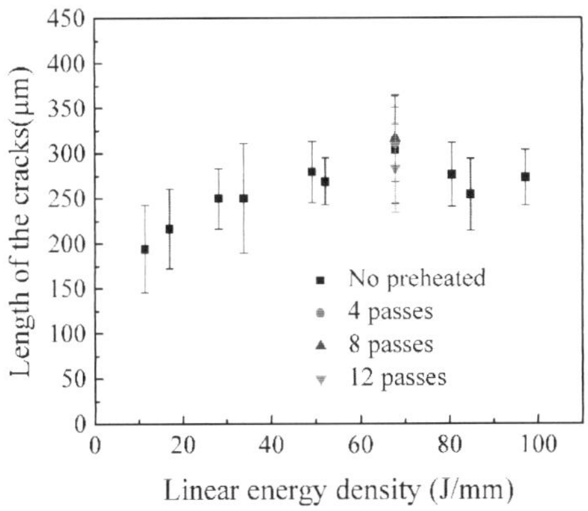

Fig. 10. Length of cracks for different linear energy densities and different parameters of laser glazing process

Fig. 10 displays the evolution of the length of the vertical cracks as a function of linear energy density. It can be observed, that linear energy density doesn't influence the length of cracks very much. In fact, when the linear energy density increases from 11.3 J/mm to 97.3 J/mm, the length of vertical cracks varies from 194.5μm to 316.7 μm. The corresponding results include two distinct stages:

- At first, the length of cracks increases linearly with the increase in linear energy density.
- At second, the length of cracks still constant around 280μm regardless of change of linear density. Also, the length of cracks of coating preheated don't differ from that without preheating.

The width of molten pool in function of different linear energy densities is also investigated. Increasing the linear energy density results in the linearly increase of width of molten pool as it is proved by the fitting function shown in **Fig.11**, excepted for case 1. In this specific case, the duration of laser pulse is 3ms and the linear density is 28.2 J/mm. Preheating processes also widen the molten pool. Indeed, width of molten pool increase from 4.42mm to 5.1mm as preheating temperature increases from 0°C to 420°C.

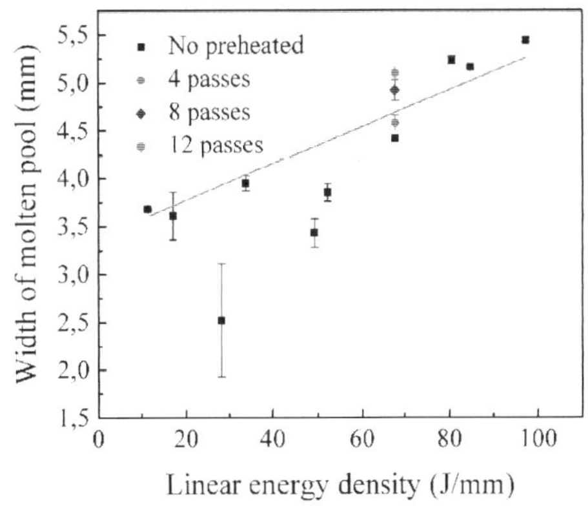

Fig. 11. Width of molten pool for different linear energy densities and different parameters of laser glazing process

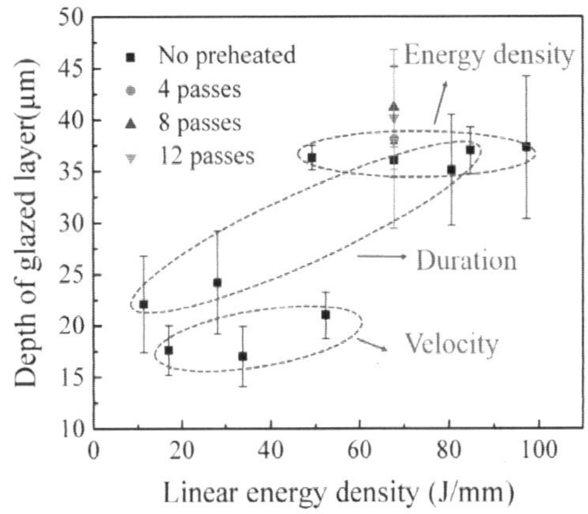

Fig. 12. Depth of glazed layer for different linear energy densities and different parameters of laser glazing process

Fig12 shows the relationship between depth of glazed layer and linear energy density. It seems that no evident correlation could be found between the two. Further investigation reveals that velocity, pulse duration and energy density should be also taken into account to estimate the depth of glazed layers marked with different cycles in **Fig. 12**. It is noticed that the change in scanning speed and energy density, do not influence so much the depth of glazed layer. However, depth of glazed layer is very sensitive to the laser pulse duration. As the duration increases from 3ms to 6ms, the depth of glazed layer increases consequently from 21mm to 36mm. Therefore, the preheating process increases the depth of glazed layer (**Fig.7**) as well as it decreases the thermal gradient and cracking in the YSZ coatings.

4 Conclusion.

In the study, different post laser parameters were applied to plasma sprayed coating to obtain a dense glazed layer. The influence of laser glazing parameters on the microstructure of coating were investigated. It can be concluded that the preheating process makes a significant difference on the laser glazing coating quality, since it deepens the glazed layer, widens the molten pool and reduces the cracks.

5 Acknowledgement

The author WANG Yan shows great thanks to China Scholarship Council for the financial support.

6 Literature

[1] Antou, G., Montavon, G., Hlawka, F., Cornet, A., Coddet, C. and Machi, F. Surf. Coat. Technol. **172**(2003), Issue2-3, pp. 279/90.
[2] Wolfe, D. E., Singh, J., Miller, R. A., Eldridge, J. I. and Zhu, D. M. Surf. Coat. Technol. **190**(2005). Issue1, pp. 132/49.
[3] Clarke, D. R Oechsner, .M. and Padture, N. P. MRS Bull. **37**(2012). Issue10, pp. 891/98.
[4] Stella, J., Poirier, T. and Pohl, M. Wear **300**(2013). Issue1-2, pp. 163/68.
[5] Minh, N. Q. and Takahashi, T. Science and Technology of Ceramic Fuel Cells (Elsevier, 1995).
[6] Evans, A. G., Mumm, D. R., Hutchinson, J. W., Meier, G. H.and Pettit, F. S. Prog. Mater. Sci. 46, (2001). Issue5, pp. 505.
[7] Khor, K. A., Chen, X. J., Chan, S. H. and. Yu, L. G. Mater. Sci. Eng. A **366** (2004). Issue1, pp. 120/26.
[8] Ghasemi, R., Shoja-Razavi, R., Mozafarinia, R.and Jamali, H.Ceram. Int. **39**2013). Issue8, pp. 9483/90.
[9] Tsai H. L.and P. C. Tsai, P. C.J. Mater. Eng. Perform. **7**(1998). Issue2, pp. 258/64.
[10] Tsai, H. L. Tsai, P. C and Tu, D. C.Mater. Sci. Eng. A **161**, 145 (1993). Issue1, pp. 145-155.
[11] Liu, Q., Danlos, Y., Song, B., Zhang, B., Yin, S. and. Liao, H.L., J. Mater. Process. Technol. **222** (2015). pp. 61/74.

Impedance Analysis of 7YSZ Thermal Barrier Coatings during High-Temperature Oxidation

CHEN Wen-Long[a,b], LIU Min[b], ZHANG Ji-Fu[b]
[a].School of Materials and Energy, Guang dong University of Technology, Guangzhou 510006, China;
[b].Guangdong Institute of New Materials, National Engineering Lab for Modern Materials Surface Engineering Technology,The Key Lab for Guangdong for Modern Surface Engineering Technology, Guangzhou, 510650, China

Abstract:The ZrO2-7wt%Y2O3 (7YSZ) thermal barrier coatings (TBCs) were prepared by Atmospheric-Plasma-Spray (APS). High-temperature oxidation of 7YSZ-TBCs was accomplished at 1000 ℃characterized by impedance spectroscopy and scanning electron microscopy (SEM) equipped with Energy Dispersive Spectrometer (EDS) .The results indicate that the thermally grown oxide (TGO) mainly contain alumina. The increase of TGO layer thickness appears to follow the parabolic law. The impedance analysis demonstrates that the resistance of TGO increases with oxidation time increasing and also follows the parabolic law, where the TGO thickness characterized by the resistance of TGO in fitting results according to the equivalent circuit is feasible. The resistance of the YSZ grain boundary increased due to the cracks increase within YSZ coating when the oxidation time is less than 50 h. While more than 150 h, the resistance of YSZ grain boundary slightly decrease mainly due to the YSZ coat sintering during oxidation process.

Key words: thermal barrier coatings; Impedance spectroscopy; High-temperature oxidation; TGO

1 Introduction

Thermal barrier coatings (TBCs) as a surface protection technology are widely used in the hot-component of gas turbines like blades and vanes [1-3]. A typical TBC system consist of a heat-insulating ceramic (usually YSZ) as top coating and an oxidation-resistance alloy (usually MCrAlY) as bond coating on the Ni-based superalloy substrate. During high temperature environment, the top coating and bond coating are to be sandwiched by a thermally grown oxide（TGO）layer. It has been establish that the growth of the TGO layer is an essential factor in influencing the degradation of TBCs [4-5].Therefore, it is very important to characterize the growth and composition change of the TGO in the high temperature environment.
Non-destructive evaluation (NDE) techniques have been used to examine the degradation of TBCs, such as infrared thermography [6-7] which was used to evaluate the delamination of TBCs, acoustic emissions [8-10] has been used for the examination of cracking in TBCs, photoluminescence [11-12] which can check TGO residual stress. Despite these non-destructive techniques can be used to examine the formation of deficiency in TBCs as cracking occurs, hardly did these techniques evaluate the TGO thickness and composition change. Impedance spectroscopy is a kind of non-destructive technique, which has been used widely to examine the electrical properties of materials [13-15]. In the past decade, the impedance spectroscopy has been used to reflect the growth and composition changes of TGO during oxidation and effect of cracks propagation of YSZ layer on YSZ electrical properties. For example, Zhang et al. [16] have also used impedance spectroscopy measurements the relationship between microstructure of top coat and its electrical properties; thickness and composition change of the TGO in different oxygen partial pressure environment in EB-PVD (electron beam physical vapor deposition) TBCs at 1050 ℃. Wang et al. [17] have studied the TGO growth in APS (atmospheric-plasma-spray) TBCs after oxidation in air at 1100 ℃. Ali et al. [18] also adopted impedance spectroscopy investigated the relevance between microstructure of the TGO and electrical properties of TGO in APS-TBCs oxidation in air at 1150 ℃.
The great mass of former studies were concentrated on the oxidation temperatures of TBCs above 1000 ℃， which is excessed the operation temperature for superalloy assembly in engines. The aim of the present work is to evaluate the oxidation of the TBCs at a lower temperature of 1000 ℃ using impedance spectroscopy. Impedance analysis of TBCs in this work showed that three relaxation processes, where two of them correspond to YSZ grain boundaries, the TGO, and the metal electrode effect. As is well know, the impedance of polycrystalline zirconia is determined by the donation of both the grain boundary and the grain, and should gain two semicircles [17]. However, only one semicircle corresponding to the YSZ is examined in this work. The main reason is that the conductivity of YSZ at low temperature is determined by the grain boundary, but the conductivity at high temperature is fully dominated by the grain. The conversion process happens at the temperature ranging from 323 ℃ to 1223 ℃. Thus the impedance measurement of YSZ coating at 400 ℃ in this work gains impedance spectra with one semicircle.

2 Experimental

2.1 Specimen preparation

Nickel based super-alloy K417 with composition (wt%) of 14-16Co, 8.5-9.5Cr, 4.8-5.7Al, 4.5-5.0Ti, 2.5-3.5Mo, 0.13-0.22C, rest Ni was cut into columnar specimen with a dimension of Φ25.3×6 mm as substrates, which were grit-blasted before air-plasma spray. The NiCoCrAlYTa bond coating (~100um thick) with composition (wt%) of 22-24Co, 19-21Cr, 7.5-8.5Al, 0.4-0.8 Y, 3.5-5.5 Ta, rest Ni was deposited by low pressure plasma spray (LPPS) on the surface of superalloy. The 7YSZ top coating (~150 um thickness) was prepared APS on the surface of bond coating. The spraying parameters of 7YSZ are given in Table 1.

Table 1.Spray parameters of 7YSZ coating prepared by APS

Material	Ar	H2	Feed rate	Current	Distance
7YSZ	40 slpm	9 slpm	2×1.5 g.min-1	630 A	110 mm

2.2. High-temperature oxidation testing

The specimens were heated in air at 1000 ℃ with a 10 ℃/min and isothermal oxidized for 5 h, 10 h, 50 h, 100 h, 150 h, and 250 h, respectively. Five specimens were used for impedance measurements for different oxidation condition. Meanwhile, the microstructure and TGO thickness of the cross-section of the coating were observed by SEM equipped with Energy Dispersive Spectrometer (EDS), and the phase transition were examined by X-ray diffraction (XRD).

2.3. Impedance spectroscopy measurement

For impedance measurements, the K417 alloy substrate without TBCs was welded on stainless steel wires acting as an electrode. The TBC surface was coated with a platinum paint with an area of π×52 mm2, which served as the other electrode (as shown in Fig.1). In order to consolidate the platinum paint and enhance its adhesion to the specimen surface, the paint was sintered at 800 ℃ for 10min. Impedance measurements were implemented at 400 ℃ using an Ametek Parstat 4000 electrochemical impedance spectroscopy (EIS) analyzer, which is computer controlled. Spectra analysis was performed using Zview impedance analysis software to obtain the electrical properties of TBCs. In this measurement, alternating current (AC) amplitude of 100 mv was employed and the AC frequency was in the range 0.1 to 1×10^6 Hz.

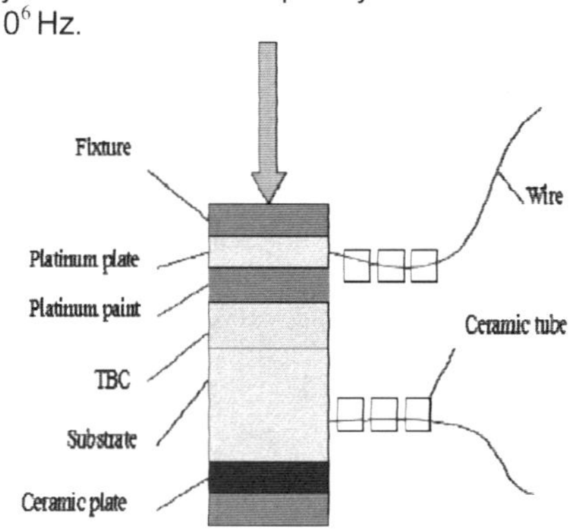

Fig. 1 Impedance spectroscopy samples and fixtures

3. Results and discussion

3.1. Microstructure and oxidation kinetic

Fig. 2 shows the cross-sections of the specimens after oxidation. In this specimen, after oxidation of 5 h, a continuous TGO layer has formed at the YSZ/bond coat interface. When the time of oxidation is less than 100 h, the SEM images of the TGO layer showed that the thickness of the TGO occurs apparent thicken with increasing oxidation time, demonstrating the bond coats had a rapid oxidation. After oxidation of 100 h, the thickness had no apparent change but color deepened of TGO layer, indicating the TGO growth slowed down. The TGO layer of specimen after oxidation 250 h has been analyzed by the XRD, revealing that the TGO layer mainly consists of alumina (as shown in Fig. 3). Due to the aluminum content of bond coating is relatively high in oxidation of initial stage, the increased rapidly of TGO thickness when the time of oxidation is less than 100 h. The thickness of TGO increased with increasing oxidation time, which result in the greater diffusion resistance of oxygen into the bond coating, thereby the growth rate of TGO layer decreased.

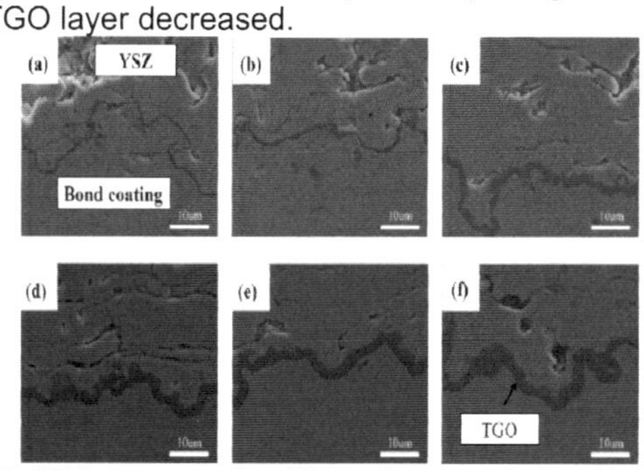

Fig. 2 SEM of cross-sectional TBCs after atmospheric oxidation at 1000 ℃ for (a) 5 h, (b)10 h, (c)50 h, (d)100 h, (e)150 h, (f) 250 h.

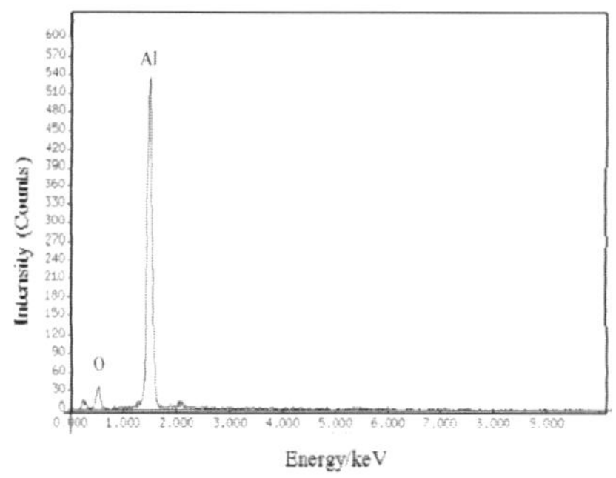

Fig. 3 EDS analysis of TGO layer after oxidation of 250 h

The Fig.4 shows oxidation kinetics of TGO layer showing linear relationship between TGO thickness and square root of time, indicating that the oxidation of bond coating follows a parabolic law. The oxidation kinetics may be characterized by

$$\delta^2 = k_c t \quad (1)$$

Square root on both sides of the equation:

$$\delta = \sqrt{k_c t} \quad (2)$$

Where δ is the thickness of the TGO layer, t is the oxidation time, and kc is the reaction rate parameter. The figure shows that the line slope for oxidation times less than 100 h is greater than that for oxidation more than 100 h, this like that above microstructure analysis. This indicates that the reaction rate parameter is larger during the initial oxidation stage than the oxidation time longer than 100 h because the square root of the kc equals to the slope of the line.

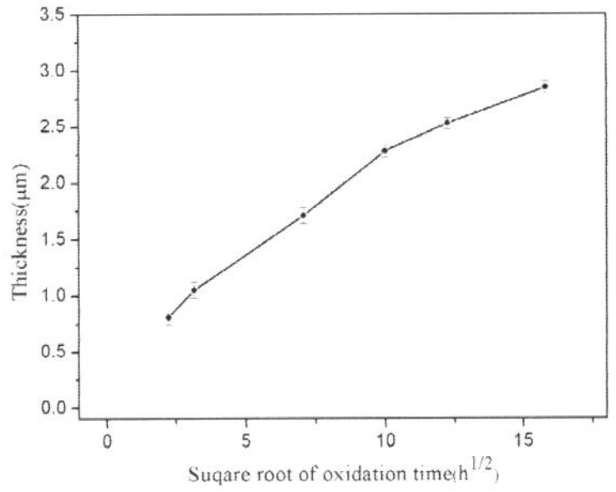

Fig. 4. Oxidation kinetics of the TBCs at 1000 ℃

3.2. Impedance spectra of TBCs after oxidation and equivalent circuit

The impedance spectra of specimens with different oxidation time have been achieved using impedance measurement. There are two types of impedance diagrams, namely Nyquist plots and Bode plots which are most frequently used. In a Nyquist plot, the impedance is represented by a real part Z' and an imaginary part Z" with the formula Z (ω)= Z' +j Z" ,where ω is the angular frequency and j= $\sqrt{1}$. In the Bode plot, the modulus of the impedance and the phase angle are both plotted as a function of frequency. For a simple resistor-capacitor (R-C) circuit, the Nyquist plot is characterized by a semicircle. The Nyquist plot is normally used to determine the major parameters, such as resistance and capacitance corresponding to an electrochemical system. If the system is complicated, the spectra have to be fitted with an equivalent circuit model, which is representative of the microstructural features of the materials [19].

Typical Nyquist plots and Bode plots for TBCs high-temperature oxidation in air at 1000 ℃ for different oxidation time are represented in Figs. 5(a-b), respectively. Two semicircles are apparently presented in each Nyquist plot in Fig. 5(a), the larger and small semicircle correspond to TGO layer and YSZ coat, respectively. It has been verified that small semicircle at high frequency corresponds to the ceramic coating and larger semicircle at low frequency corresponds to the TGO layer [20]. In the Bode plots, there are four relaxation processes which were attributed to a metal electrode response (e), a TGO layer, YSZ grain boundary (GB), and YSZ grain (G).

An equivalent circuit model is established for fitting the measured impedance spectra, so the electrical properties of the YSZ coating and TGO layer can be obtained. The AC equivalent circuit of APS-TBCs system is presented in Fig. 5(c). Due to the constant phase element (CPE) is more suitable to describe the behavior of a non-ideal, i.e. not a pure capacitor, caused by the dispersion of electrode and chemical inhomogeneity, using of CPE instead of capacitance in the equivalent circuit. The impedance of CPE is given by:

$$Z_{CPE} = A^{-1}(j\omega)^{-n} \qquad (3)$$

Where A is a parameter independent of frequency, ω is angular frequency and j= $\sqrt{1}$, and n is an exponential index which represents a dispersion of relaxation frequency. When the n=1, the CPE functions as an ideal capacitor and A is therefore equal to the capacitance C. However, in most cases, n is less than 1. In the case of no ideal capacitive response, the value of A cannot be used to represent the capacitance of the TBC system. This work adopt an equivalent capacitance C, which may be acquired by [21]

$$C = R^{(1-n)/n} A^{1/n} \qquad (4)$$

where R is the resistance.

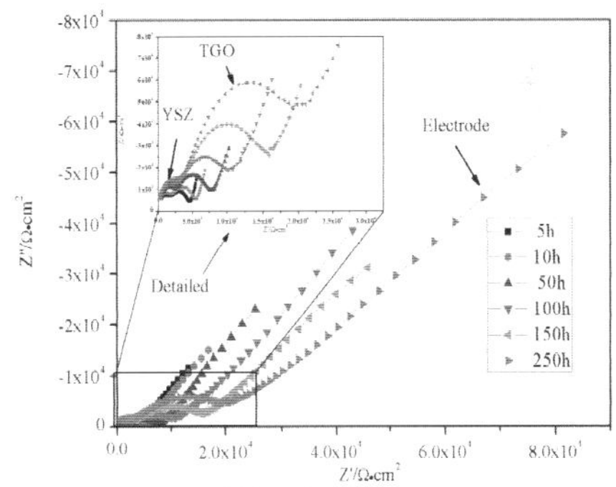

(a) Nyquist plot

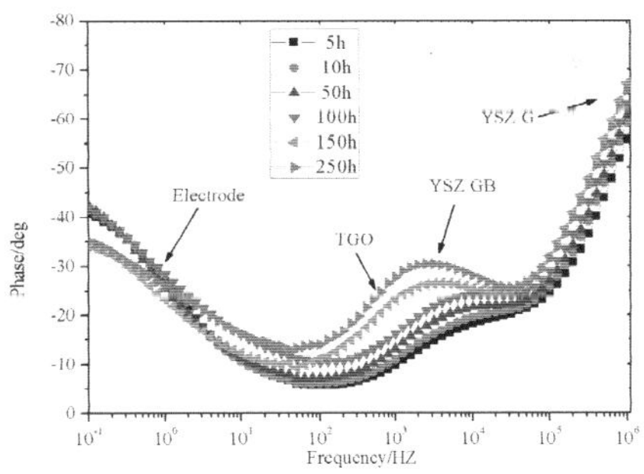

(b) Bode plot

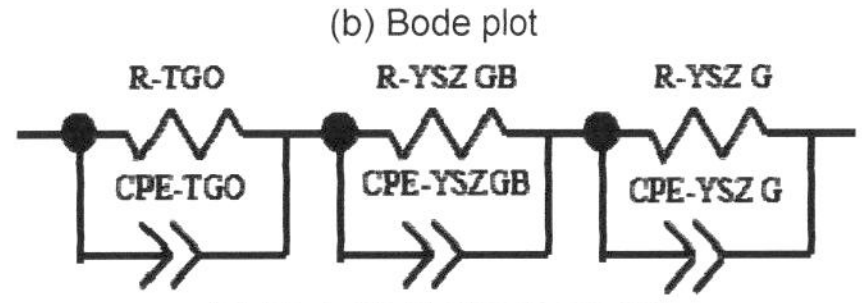

(c) Equivalent circuits model

Fig. 5 Impedance spectroscopy of TBCs oxidation for different time, (a) Nyquist plot, (b) Bode plot, (c) Equivalent circuits model

3.3. Relation between electrical properties and TGO growth

According to the equivalent circuits, Ogawa et al. [22] obtained the relation between the TGO thickness and the resistance of TGO and Wang et al. [18] calculate the thickness of TGO from the diameter of modulus spectra. In this work, due to the TGOs layer fully consist of alumina, it is possible to calculate the TGO thickness according to the relationship between resistance and thickness as shown in the following equation:

$$R = \frac{\rho d}{A} \qquad (5)$$

where R is the resistance, d is the thickness of electrode, namely the thickness of TGO in this study. A is the area of electrode and ρ is the resistivity (specific resistance) of electrode material, for alumina, its value of resistivity is constant at the preset temperature. As is shown in Fig. 6, the resistance of TGO is fitted according to the equivalent circuit plotted as a function of the square root of oxidation time. Clearly, there is a linear relation between the TGO resistance and the square root of oxidation time, indicating that the oxidation of 250 h in air at 1000 °C follows a parabolic law, at the same time, characterizing the TGO thickness by the resistance of TGO in fitting results according to the equivalent circuit is feasible.

Moreover, Fig. 7 shows the resistance and capacitance of TGO as a function of TGO thickness. Obviously, the resistance of the TGO increase and the capacitance decreases with increasing TGO thickness. However, the above current are not absolutely linear and the slope increase slightly after oxidation of 100 h. According to the equation (5) and following equation (6) analysis indicating that the resistivity (ρ) increasing and relative dielectric constant (ε) decreasing during oxidation.

$$C = \frac{\varepsilon \varepsilon_0 A}{\delta} \qquad (6)$$

Where C is capacitance, ε is the relative dielectric constant, ε_0 is the space dielectric constant, and A is the area of electrode. Increase of resistivity may be caused by TGO densification after oxidation. As is well known, Al2O3 is a p-type semiconductor. The resistivity is determined by electron hole concentration and inverse proportion relationship. The electron hole concentration decreases with decreasing oxygen partial pressure. The denser alumina will lead to a decrease in oxygen partial pressure within the TGO layer resulting in an increase of the resistivity. The reason for relative dielectric constant decreasing is ill-defined and needs to be researched continually.

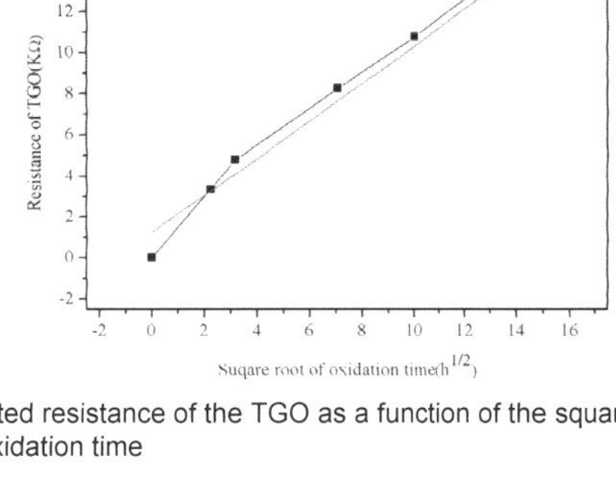

Fig. 6 Fitted resistance of the TGO as a function of the square root of oxidation time

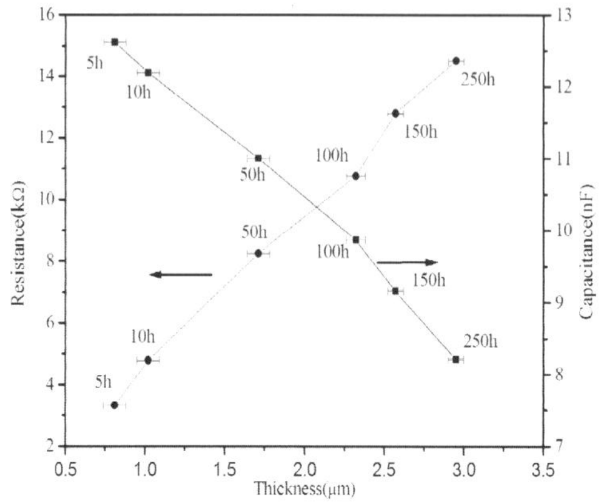

Fig. 7 Fitted resistance and capacitance of the TGO as a function of TGO thickness

3.4. Relation between electrical properties and microstructure of the YSZ

The microstructure of the YSZ coating after oxidation for 5 h, 50 h, 150 h, and 250 h is showed in Fig. 8(a-d), respectively. There is an apparent increase of micro-cracks within the YSZ coating after oxidation time from 5 h to 50 h (seen in Fig. 8(a-b)). The Fig. 8(b-c) show that the micro-cracks are no obvious propagation in the internal YSZ coating along with oxidation time increase from 50 h to 150 h. After oxidation 250 h, there is no apparent micro-cracks in the YSZ coating mainly due to the sintering which cause the micro-cracks healing together and form a more dense YSZ coating.

Zhang et al. [16] had studied the relation between micro-cracks of YSZ and the resistance of the YSZ variation, indicating the resistance of the YSZ increases with micro-cracks increasing. At the same time, the resistance decreases of the YSZ mainly result from increase of the YSZ grain boundary conductivity due to the sintering of YSZ coating during oxidation process. In this work, the resistance and capacitance of the YSZ coating are plotted in Fig. 9 as a function of oxidation time. As depicted above, the resistance and capacitance of YSZ coating mainly contributed to the YSZ grain boundary because the impedance test temperature is 400 °C. Clearly, there is no apparent change in the capacitance of YSZ grain boundary for different oxidation time. However, the grain boundary resistance increases quickly to 2663Ω after 50 h oxidation because some small micro-cracks in the YSZ coating increased during the initial oxidation that caused the decrease of YSZ conductivity. Then the resistance of the YSZ grain boundary has no apparent change because no apparent propagation of cracks for oxidation times shorter than 150 h. Finally the resistance of the YSZ grain boundary decreases slightly after 250 h oxidation due to sintering of the YSZ coating during oxidation.

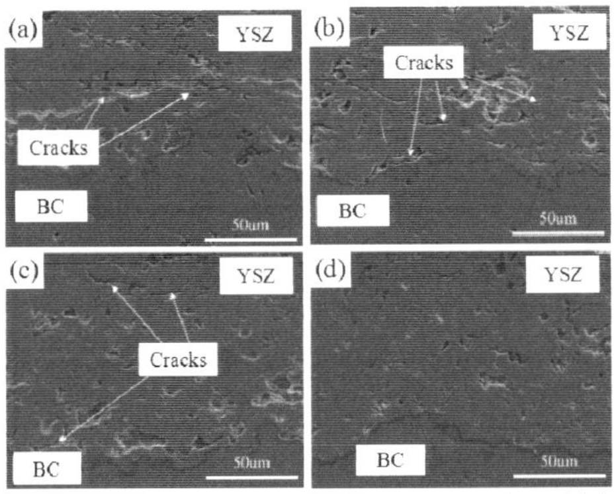

Fig. 8 Microstructure of the YSZ coating after oxidation of (a) 5 h, (b) 50 h, (c) 150 h, and (d) 250 h.

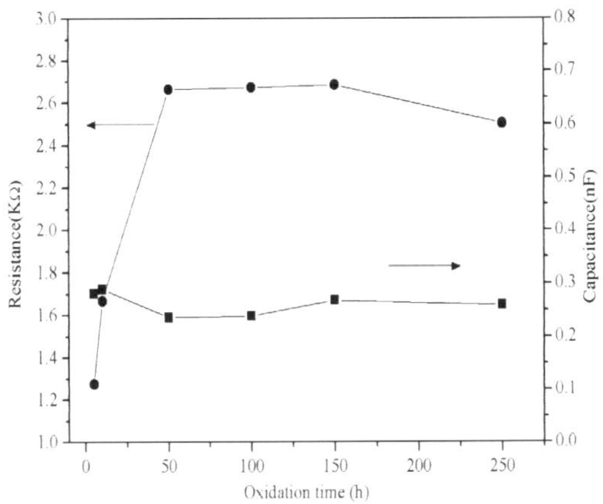

Fig. 9 Fitted resistance and capacitance of the YSZ as a function of oxidation time

4. Conclusions

High-temperature oxidation of TBCs prepared by APS was accomplished at 1000 °C and was investigated using impedance spectroscopy assisted with SEM. The TGO mainly consists of alumina with the increase of TGO layer thickness following the parabolic law. The relation between TGO growth, microstructure of YSZ coating and their corresponding electrical properties are summarized respectively as follows :

(1) The resistance and resistivity of TGO layer is related to the thickness and porosity of TGO layer, respectively. The resistance of TGO layer increases with increasing oxidation time due to the increase of TGO thickness. The resistivity increased of TGO layer after oxidation of 100 h due to the porosity decreased of TGO layer. This implies that the protective capability of the TGO is strengthened with increasing oxidation time.

(2) The resistance of the YSZ grain boundary increased due to the cracks increase within YSZ coating when the oxidation time is less than 50 h. While more than 150 h, the resistance of YSZ grain boundary slightly decrease mainly due to the YSZ coat sintering during oxidation process.

References

[1] Gurrappa I, Rao A S. Thermal barrier coatings for enhanced efficiency of gas turbine engines. Surface and Coatings Technology 2006; 201(6):3016-3029.

[2] Zhang X F, Zhou K S, Xu W, et al. Reaction Mechanism and Thermal Insulation Property of Al-deposited 7YSZ Thermal Barrier Coating. Journal of Materials Science & Technology 2015; 31(10):1006-1010.

[3] Vaßen R, Jarligo, M O, Steinke T, et al. Overview on advanced thermal barrier coatings. Surface and Coatings Technology 2010; 205:938-942.

[4] Huang W, Zhu C, Liu X Y, et al. A comparative study of the microstructure and oxidation behavior in supersonic and conventional air plasma sprayed thermal barrier coatings. Surface and Coatings Technology 2013, 235:853-859.

[5] He M Y, Evans A G, Hutchinson J W, Effects of morphology on the decohesion of compressed thin films. Materials Science and Engineering A 1998, 245(2):168-181.

[6] Chaki S, Marical P, Panier S, et al. Interfacial defects detection in plasma-sprayed ceramic coating components using two stimulated infrared thermography techniques. NDT & E International 2011, 44(6):519-522.

[7] Schweda M, Beck T, Offermann M, et al. Thermographic analysis and modelling of the delamination crack growth in a thermal barrier coating on Fecralloy substrate. Surface and Coatings Technology 2013, 217:124-128.

[8] Yang L, Zhou Y C, Lu C. Damage evolution and rupture time prediction in thermal barrier coatings subjected to cyclic heating and cooling: An acoustic emission method. Acta Materialia 2011, 59(17):6519-6529.

[9] Yang L, Zhong Z C, You J. Acoustic emission evaluation of fracture characteristics in thermal barrier coatings under bending. Surface and Coatings Technology 2013, 232:710-718.

[10] Yao W B, Dai C Y, Mao W G. Acoustic emission analysis on tensile failure of air plasma-sprayed thermal barrier coatings. Surface and Coatings Technology 2012, 206(18):3803-3807.

[11] Wang X, Wu R T, Atkinson A. Characterisation of residual stress and interface degradation in TBCs by photo-luminescence piezo-spectroscopy. Surface and Coatings Technology 2010, 204(15):2472-2482.

[12] Wang X, Lee G, Atkinson A. Investigation of TBCs on turbine blades by photoluminescence piezospectroscopy. Acta Materialia 2009, 57(1):182-195.

[13] Ai J H, Chen Y Z, Urquidi-Macdonald, M. Electrochemical impedance spectroscopic study of passive zirconium. Journal of Nuclear Materials 2008, 379(1-3):162-168

[14] McIntyre J M, Pham H Q,

Electrochemical impedance spectroscopy: a tool for organic coatings optimizations. Progress in Organic Coatings 1996, 27(1-4):201-207.
[15] Amirudin A, Thieny D, Application of electrochemical impedance spectroscopy to study the degradation of polymer-coated metals. Progress in Organic Coatings 1995, 26(1):1-28.
[16] Zhang C X, Zhou C G, Gong S K, et al. Evaluation of thermal barrier coating exposed to different oxygen partial pressure environments by impedance spectroscopy. Surface and Coatings Technology 2006, 201(1-2):446-451.
[17] Wang X, Mei J F, Xiao P. Non-destructive evaluation of thermal barrier coatings using impedance spectroscopy. Journal of the European Ceramic Society 2001,21(7):855-859.
[18] Ali M S, Song S H, Xiao P. Evaluation of degradation of thermal barrier coatings using impedance spectroscopy. Journal of the European Ceramic Society 2002, 22(1):101-107.
[19] Fletcher J G, West A R, Irvine J T S. The AC impedance spectroscopy response of the physical interface between yttria-stabilized zirconia and YBa2Cu3O7-x. Journal of The Electrochemical Society 1995, 142(8):2650-2654.
[20] Anderson P S, Wang X, Xiao P. Impedance spectroscopy study of plasma sprayed and EB-PVD thermal barrier coatings. Surface and Coatings Technology 2004, 185(1):106-119.
[21] Song S H, Xiao P, Weng L Q, Evaluation of microstructural evolution in thermal barrier coatings during thermal cycling using impedance spectroscopy. Journal of the European Ceramic Society 2005, 25(7):1167-1173.
[22] Ogawa K, Minkov D, Shoji T, et al. NDE of degradation of thermal barrier coating by means of impedance spectroscopy. NDT & E International 1999, 32(3):177-185.

Preparation and cytocompatibility of nanoporous SiO_2/TiO_2 coatings by plasma spraying and HF etching

Xiaobing Zhao
School of Materials Science and Engineering, Changzhou University, Changzhou, Jiangsu, 213164

Abstract:
SiO_2/TiO_2 coating with nanoporous structures were prepared by atmospheric plasma spraying technology and subsequently hydrothermal etching method in hydrogen fluoride solution. The effects of hydrothermal time and temperature on the microstructures and osteoblast behavior of the SiO_2/TiO_2 coatings were investigated. Results demonstrated that the as-sprayed SiO_2/TiO_2 coating is mainly composed of rutile and quartz phases. After etching, the phase structures almost do not change. Nanoporous topographics are formed on the surface of the coatings and the hydrothermal parameters have important influences on the size and shape of the pores. The interconnected network pores on the coating surface are observed at the appropriate hydrothermal conditions (the hydrothermal time and temperature are 60 min and 100 ℃, respectively). Compared to TiO_2 and SiO_2/TiO_2 coatings, nanoporous SiO_2/TiO_2 coatings exhibit enhanced osteoblast adhesion and promoted cell proliferation. Our results suggest the potential application of the porous coatings for enhancing the biological performance of the currently used dental and orthopaedic implant materials.
Keywords: Surface topography, porous materials, plasma spray, osteoblast

1 Introduction

Titanium (Ti) and its alloy have been successfully used as biomedical implants due to their excellent mechanical properties such as low specific weight, high corrosion resistance and good biocompatibility [1, 2]. However, the main drawback of Ti implants is their insufficient bioactivity which results in poor interfacial bonding between the implant and bone [3]. Nevertheless, the existence of harmful ions (such as V) in Ti alloys may enhance the expression of proliflammatory factors and osteolytic mediators, resulting in failure of the implants [4]. To overcome these drawbacks, coating technology on the surface of Ti and Ti alloy is an efficient method and has been successfully applied in orthopedic field. However, there are still some problems of the coated implants in the long term stability and safety, such as the dissolution and the release of metallic ions in biological environment [5, 6]. TiO_2 coating has been proposed to be a substitute for the currently used biomedical implant coatings due to its chemical stability, nontoxicity and biocompatibility [7, 8]. However, TiO_2 coating is bioinert and cannot induce new bone formation at the implant surface, rendering its potential application in orthopedic implants. In order to enhance the bioactivity of the implant surfaces, surface modification approaches including surface chemistry and/or topography adjustments have been utilized. Silicon (Si) as a biocompatible trace element has been proved its beneficial effect on the improvement of biological performance of biomaterials [9, 10]. More importantly, Surfaces with micro- and nano-topography have been proved to enhance the cell behaviors, such as adhesion, activation, orientation, morphology and movement [11]. The effects of nanotopographies on the cell behavior have been investigated by using nanoscale patterns including nanograins [12], nanotubes [13], nanopits [14], nanopost and nanograte [15], nanopores [16]. Recently, porous topography is particularly attractive for enhancing cell activation, adhesion, orientation and movement of implant materials. Karlsson et al. [17] reported that nanoporous alumina could provide a good surface for osteoblast rapidly spreading, flattening and adhesion. Liu et al. [18] prepared Mn-MAO coatings on cp-Ti and the results indicated that the formation of a porous 3D topography was capable to improving osteoblastic cell differentiation. An investigation has been shown that mesoporous TiO_2 films could be applied in implant coatings without decreasing the biomechanical stability of the implant [19]. Methods to produce nanoporous surfaces can be classified as subtractive (i.e. etching) and additive (i.e. surface deposition). In this work, we firstly deposit SiO_2/TiO_2 coatings on Ti substrates by plasma spraying and then nanoporous topographies are formed on the SiO_2/TiO_2 coating by etching with HF-hydrothermal treatment through adjusting hydrothermal condition. The phase composition, topography and cytocompatibility of the coatings were investigated.

2. Experimental

2.1. *Coating fabrication*

Commercial TiO_2 (P25, Deggusa; 30 nm, ≥99.6 %) and SiO_2 (Shanghai Ruiyu optical and electrical materials Co., Ltd, China; 20 μm, ≥99.2 %) powders were used as feedstocks. To improve the flowability of the composite powders, TiO_2 and SiO_2 powders with weight ratios of 1:1 were thoroughly mixed by ball-milling and reconstituted using the following way: 100g composite powders, 50 mL deionized water and 10 mL poly(vinylalcohol) (PVA) solution (5wt%) were mixed in the ball milling tank for 2 hours, dried at 80 °C for 24 hours and then sieved by 80 mesh sieves. Only powders below 80 meshes were used to deposit the coatings.

An atmospheric plasma spraying (APS) system (9M, Sulzer Metco, USA) was used to produce the coatings on commercially pure titanium (cp-Ti) discs with a dimension of 10 mm×10 mm×1 mm. Prior to plasma spraying, the cp-Ti discs were ultrasonically cleaned in absolute ethanol and grit-blasted with brown corundum. The spraying parameters are as follows: the power was 42 kW, primary gas (Ar) flow rate was 40 slpm, auxiliary gas (H2) flow rate was 12 slpm, spraying distance was 100 mm and the feeding rate of powders was 20 g/min. After plasma spraying, the coatings were ultrasonically cleaned in acetone and deionized water.

2.2. *Coating etching*

The as-sprayed coatings were immersed in hydrogen fluoride (HF) solution (50 mL, 0.4%). The hydrothermal carving experiments were conducted in a Teflon-lined stainless steel autoclave at different treated time (30, 60 and 90 min) and temperature (80, 100 and 120 ℃). After cooling to room temperature, the coatings were ultrasonically cleaned in deionized water for three times.

2.3. *Coating characterization*

The phase structures of the coatings before and after hydrothermal treatment were examined by X-ray diffraction (XRD) employing a diffractometer of Shimadzu 6000 with Cu Kα radiation (λ=1.5418 Å). The surface morphologies were observed by field emission scanning electron microscopy (FESEM, Zeiss SUPRA-55). The surface roughness (Ra) of the coatings was determined by measuring four randomly chosen tracks on each coating using a surface profilometer (Hommelwerke T8000-C, Germany). Data were presented as mean ± S.D. For ion release experiment, the coatings were incubated for 1, 4, 7, 14 and 21 days in 10 ml saline water at 37 ℃ without stirring. The amounts of released ion were determined by analyzing the resulting solutions by inductively-coupled plasma optical emission spectrometry (ICP-OES).

2.4. *Cell culture*

The cytocompatibility of the coatings were evaluated by an *in vitro* assay. The procedures of cell culture in this work are same as our previous work [20]. Briefly, osteoblast precursor cells (MC3T3-E1, Cells Resource Center, Shanghai Institutes for

Biological Sciences) were seeded on the coatings at a density of 5.0×10^4 cells per well using 24-well tissue culture plates and holders (Costar, USA). After cultured for 1, 4 and 7 days in a 5%CO_2 incubator at 37 ℃, the cell proliferation was determined using the alamarBlue™ (AbD Serotec Ltd., UK). For fluorescence microscopy observation, 1 mL cell suspension was seeded onto the coatings. After incubating for 1, 4 and 24 h, the cells were rinsed with phosphate buffered saline (PBS) three times and fixed in a 3% glutaric dialdehyde diluent for 20 min at room temperature, and then rinsed three times with PBS. The cells were permeabilized with 0.2% (v/v) Triton X-100 (Amresco, USA) for 4 min at room temperature followed by three rinses with PBS. After stained with 4, 6-diamidino-2-phenylindole (DAPI, Invitrogen), the morphologies of the cells cultured on the coating surfaces were observed by a fluorescence microscope (Olympus GX71, Japan).

3. Results and Discussion

3.1. *Phase composition*

Figure 1 and 2 show the XRD patterns of the SiO_2/TiO_2 coatings before and after HF-hydrothermal treatment with different time and temperature. The as sprayed SiO_2/TiO_2 coating is mainly composed of rutile and quartz phases. After hydrothermal treatment, there are no obvious changes in the phase composition.

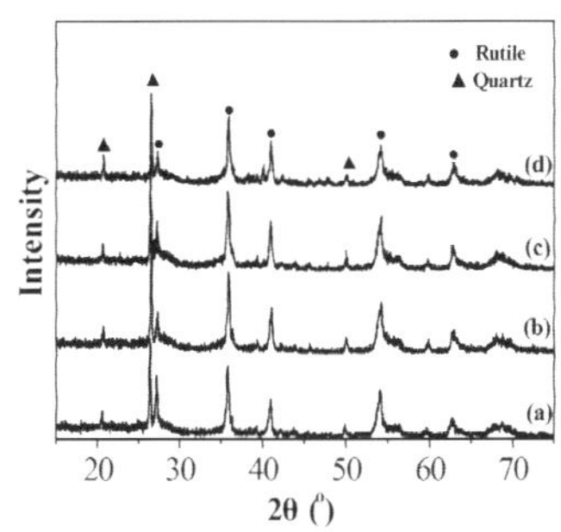

Fig. 1 XRD patterns of the (a) SiO_2/TiO_2 and etched SiO_2/TiO_2 coatings for different hydrothermal time: (b) 30 min, (c) 60 min, (d) 90 min

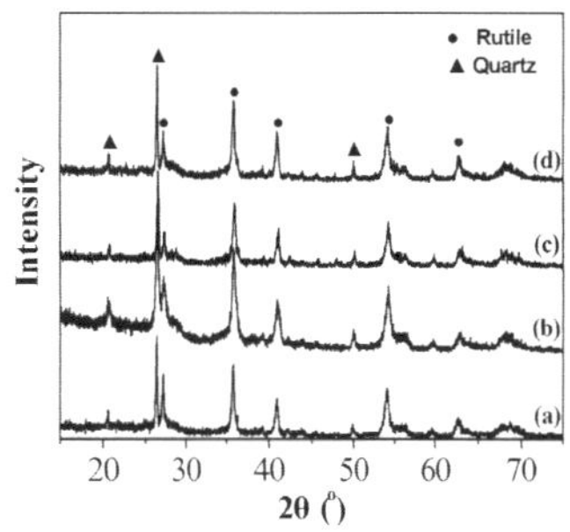

Fig. 2 XRD patterns of the (a) SiO_2/TiO_2 and etched SiO_2/TiO_2 coatings at different hydrothermal temperature: (b) 80 ℃, (c) 100 ℃, (d) 120 ℃

3.2. *Surface morphologies*

Figure 3 and 4 display the surface morphologies of the SiO_2/TiO_2 coatings carved by HF-hydrothermal treatment at various time and temperature. Compared to the as-sprayed SiO_2/TiO_2 coating, as shown in our previous work [21], HF-hydrothermal treated SiO_2/TiO_2 coatings have the typical characteristics of plasma sprayed coatings, exhibiting rough surface with many splats (Fig. 3 a, b, c and Fig. 4 a, b). However, prominent differences in the structures at nanoscale can be observed among the HF-hydrothermal treated SiO_2/TiO_2 coatings (Fig. 3 a1, b1, c1 and Fig. 4 a1, b1).

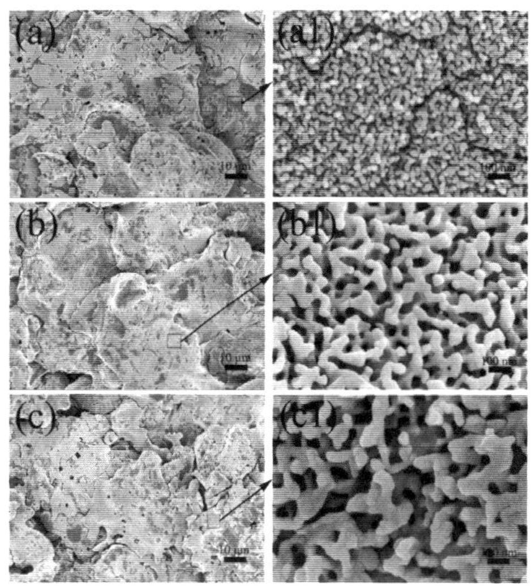

Fig. 3 Surface morphologies of the SiO_2/TiO_2 coatings after etching for (a) 30 min, (b) 60 min and (c) 90 min

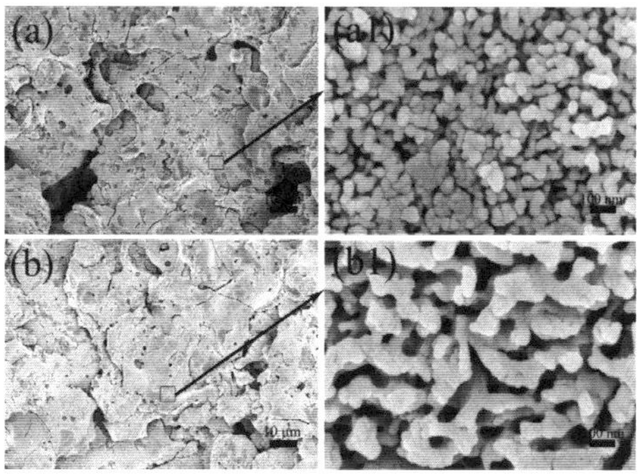

Fig. 4 Surface morphologies of the SiO_2/TiO_2 coatings after etching at (a) 80℃ and (b) 120 ℃

Our previous work showed that the as-sprayed SiO_2/TiO_2 coating displayed nanotopography with irregular hairy protrusions [21]. For the SiO_2/TiO_2 coating after HF-hydrothermal treatment for 30 min, the nanograins on the surface are stacked forming small aggregates due to the dissolve a small amount of SiO_2 particles in HF solution (Fig. 3 a1). With the elongation of hydrothermal time to 60 and 90 min, the nanoporous netlike structures are formed on the surfaces of the coatings, and the nanoporous structure of the coatings treated for 60 min seems more regular and well-defined than that for 90 min (Fig. 3 b1 and c1). In addition, the nanotopographies of the SiO_2/TiO_2 coatings treated at low (80 ℃) and high (120 ℃) hydrothermal temperature (Fig. 4 a1 and b1, respectively) present irregular and less-defined structures compared to the coating treated at 100 ℃. The results of SEM observation indicate that HF-hydrothermal treatment does change the nanotopographies of the SiO_2/TiO_2 coatings, but the influences are based on hydrothermal conditions. At appropriate hydrothermal time and temperature, nanoporous netlike structures are formed on the surface of the SiO_2/TiO_2 coatings. It is noted that the SiO_2/TiO_2 coating are serious breakage when HF-hydrothermal time is 90 min (Fig. 3c), thereby the subsequent experiments ignore this kind of coating.

3.3. *Surface roughness*

Surfaces with micrometer-sized toughness not only have positive effects on the cell attachment, proliferation and differentiation [22], but also provide a large surface area for the anchorage of new bone, which can enhance the osseointegration of the implant [23, 24]. Fig. 5 and Fig. 6 present the surface roughness of the as-sprayed TiO_2 and SiO_2/TiO_2 coatings before and after HF-hydrothermal treatment. After treatment, the surface roughness (Ra) values of the SiO_2/TiO_2 coatings slightly change compared to the as-sprayed TiO_2 and SiO_2/TiO_2 coatings, whereas all coatings exhibit micrometer-sized roughness, suggesting a potential application in implant materials.

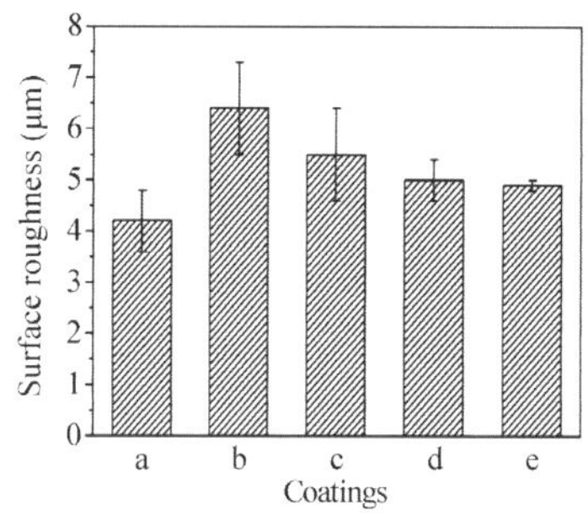

Fig. 5 Surface roughness of the (a) TiO_2, (b) SiO_2/TiO_2 and etched SiO_2/TiO_2 coatings for different hydrothermal time: (c) 30 min, (d) 60 min, (e) 90 min

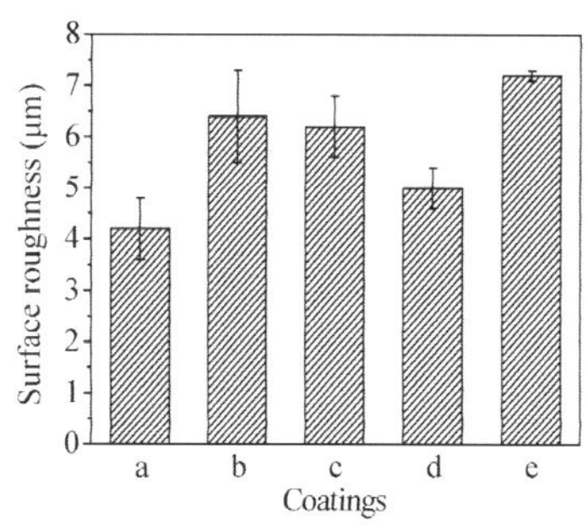

Fig. 6 Surface roughness of the (a) TiO_2, (b) SiO_2/TiO_2 and etched SiO_2/TiO_2 coatings at different hydrothermal temperature: (b) 80 ℃, (c) 100 ℃, (d) 120 ℃

3.4. *ion release*

The beneficial effects of Si ion releases on the biological properties of implants have been reported by many researches [25, 26]. In the present study, the concentrations of Si ion of the as-sprayed TiO_2 and SiO_2/TiO_2 coatings before and after HF-hydrothermal treatment were evaluated by ICP-OES, shown in Fig. 7 and Fig. 8. The concentration of Si ions released form the as-sprayed SiO_2/TiO_2 coating is only 3.86μg/mL (ppm) after soaking in saline water for 21 days. The concentrations of Si ions released form the HF-hydrothermal treated SiO_2/TiO_2 coatings are increased, but the Si ion concentrations from all coatings are all lower than 11 ppm.

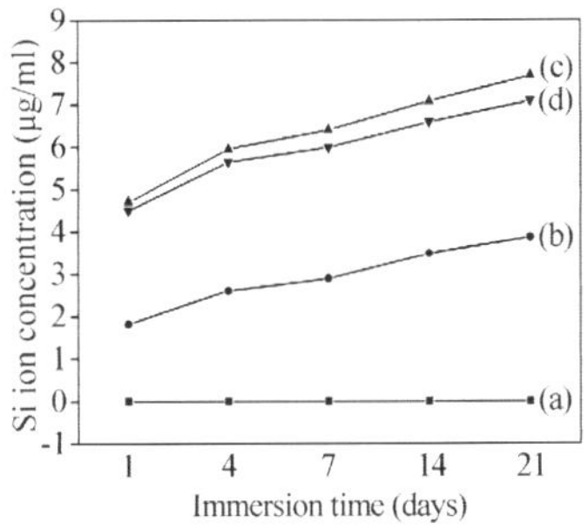

Fig. 7 Si ion concentrations after immersion in saline water of the (a) TiO_2, (b) SiO_2/TiO_2 and etched SiO_2/TiO_2 coatings for different hydrothermal time: (c) 30 min, (d) 60 min

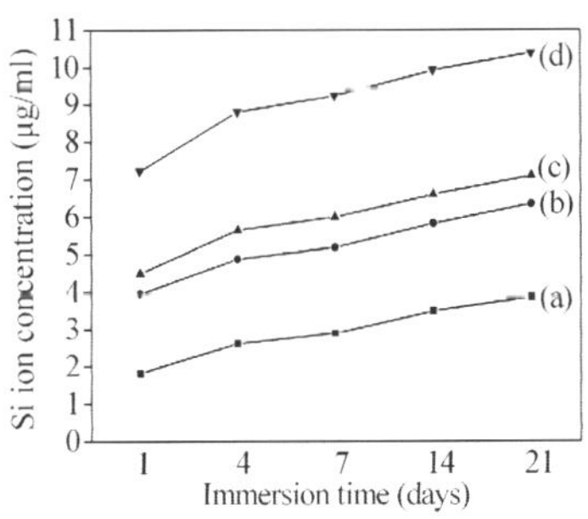

Fig. 8 Si ion concentrations after immersion in saline water of the (a) SiO_2/TiO_2 and etched SiO_2/TiO_2 coatings at different hydrothermal temperature: (b) 80 ℃, (c) 100 ℃, (d) 120 ℃

3.5. *cell culture*

Fig. 9 and 10 depict the fluorescence microscope images of the MC3T3-E1 cells seeded on the surfaces of the TiO_2, SiO_2/TiO_2 before and after hydrothermal treatment. After 1 and 4 h of culture, compared to those on the as-sprayed TiO_2 and SiO_2/TiO_2 coatings, MC3T3-E1 cells seeded on the surfaces of hydrothermal treated SiO_2/TiO_2 coatings are larger and more spread out with abundant lamellae and filopodia extending form the cell bodies. After culture for 24 h, the cells seeded on the surfaces of HF-hydrothermal treated SiO_2/TiO_2 coatings become more voluminous and stretched than those on the as-sprayed TiO_2 and SiO_2/TiO_2 coatings.

Fig. 11 and 12 present the proliferation of MC3T3-E1 cells cultured on the surfaces of the as-sprayed TiO_2, SiO_2/TiO_2 and HF-hydrothermal treated SiO_2/TiO_2 coatings for 1, 4 and 7 days. The proliferation rates of MC3T3-E1 cells on the surfaces of all coatings increase with the increase in the culture time. In the initial 1 day of culture, the MC3T3-E1 cells proliferation rates are similar. After culturing for 4 days, the cells proliferation rates on the treated SiO_2/TiO_2 coatings are higher than those on the as-sprayed coatings. At day 7, significantly higher proliferation rates are observed for the cells cultured on the treated SiO_2/TiO_2 coatings. And the proliferation rate on the SiO_2/TiO_2 coating carved by 60 min-100 ℃ hydrothermal treatment is highest. These results suggest that the enhanced bioactivities of SiO_2/TiO_2 coatings by HF-hydrothermal treatment, especially regular nanoporous structures are formed on the surface of the SiO_2/TiO_2 coating.

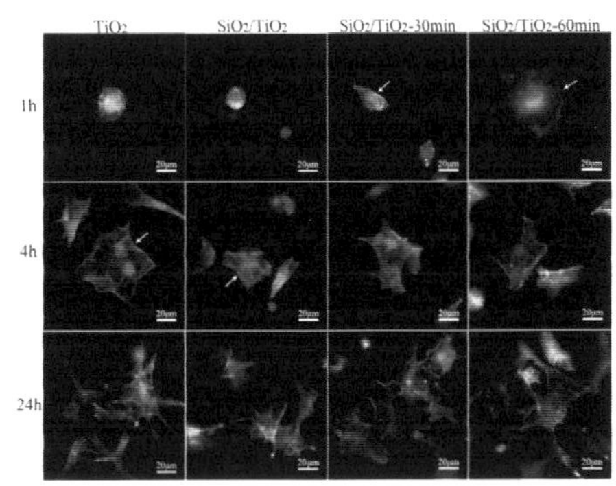

Fig. 9 Fluorescence images of MC3T3-E1 cells seeded on the surface of the TiO_2 and SiO_2/TiO_2 coatings before and after etching for different hydrothermal time

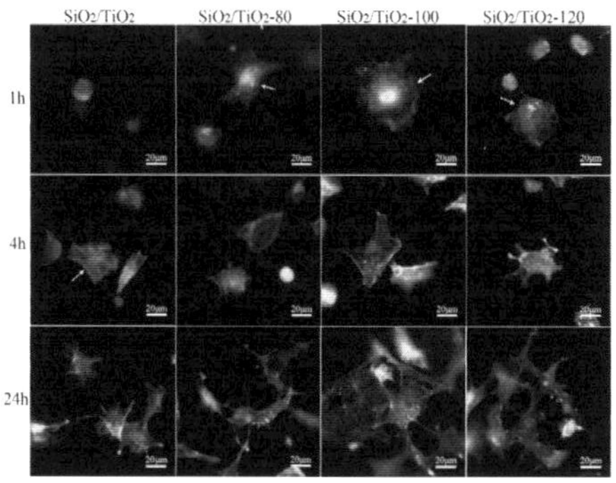

Fig. 10 Fluorescence images of MC3T3-E1 cells seeded on the surface of the SiO_2/TiO_2 coatings before and after etching at different hydrothermal temperature

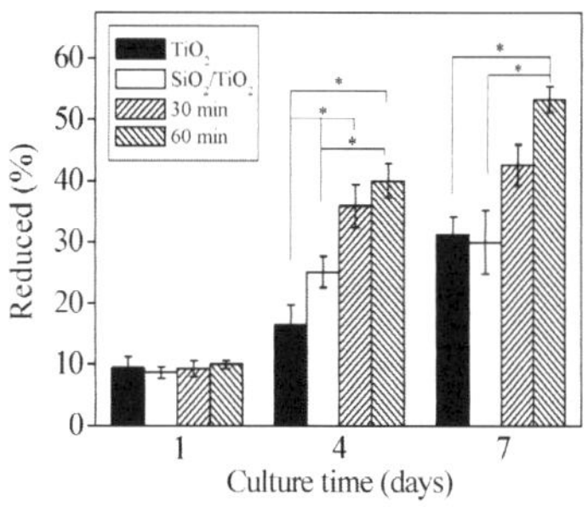

Fig. 11 Proliferation of MC3T3-E1 cells seeded on the surface of the TiO_2 and SiO_2/TiO_2 coatings before and after etching for different hydrothermal time, *$p<0.05$

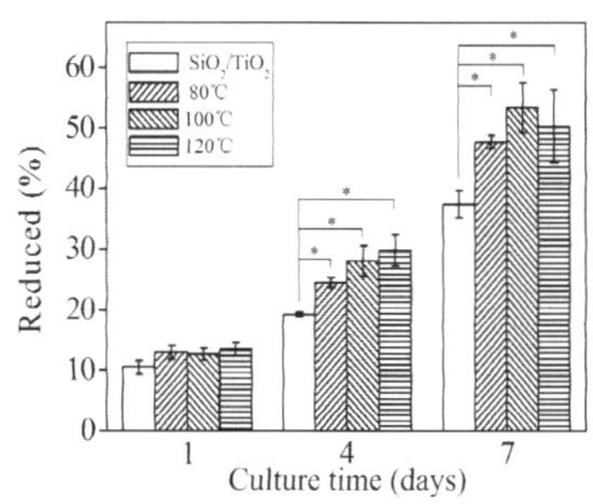

Fig. 12 Proliferation of MC3T3-E1 cells seeded on the surface of the SiO_2/TiO_2 coatings before and after etching at different hydrothermal temperature, *$p<0.05$

4. Conclusion

In summary, the SiO_2/TiO_2 composite coatings were deposited onto cp-Ti substrates by atmospheric plasma spraying technique. To enhance the cytocompatibility of the SiO_2/TiO_2 coatings, nanoporous structures were formed on the surfaces of the SiO_2/TiO_2 coatings with hydrothermal method in HF solution. The SiO_2/TiO_2 coatings were composed of rutile and quartz phases. All coatings presented micro-sized rough surfaces, but differed in their nanoporous structures at various hydrothermal conditions. A appropriate hydrothermal condition (60 min, 100℃) could induce interconnected and well-defined nanopores to form on the coating surface. The topographies with nanoporous structures of SiO_2/TiO_2 coatings exhibited enhanced cell adhesion and proliferation, suggesting a crucial dependent on the topography. These results indicate that nanoporous SiO_2/TiO_2 coatings have potential for applications as biomedical coatings in orthopedic and dental implants and can expand the application scope of plasma spray in the biomedical field.

References

[1] X. Zhu, J. Chen, L. Scheideler, R. Reichl, J. Geis-Gerstorfer, Effects of topography and composition of titanium surface oxides on osteoblast responses, Biomaterials 25 (2004) 4087-4103.

[2] M. Long, H.J. Rock, Titanium alloys in total joint replacement-a materials science perspective, Biomaterials 19 (1998) 1621-1639.

[3] J.M. Gomez-Vega, E. Saiz, A.P. Tomsia, Glass-baded coatings for titanium implant alloys. J. Biomed. Mater. Res. 46 (1999) 549-559.

[4] S.D. Rogers, D.W. Howie, S.E. Graves, M.J. Pearcy, D.R. Haynes, In vitro human monocyte response to wear particles of titanium alloy containing vanadium or niobium, J. Bone Joint Surg. Br. 79-B (1997) 311-315.

[5] W. Xue, X. Liu, X. Zheng, C. Ding, Dissolution and mineralization of plasma-sprayed wollastonite

coatings with different crystallinity, Surf. Coat. Technol. 200 (2005) 2420-2427.
[6] B.S. Ng, I. Annergren, A.M. Soutar, K.A. Khor, A.E.W. Jarfors, Characterisation of a duplex TiO_2/CaP coating on Ti6Al4V for hard tissue replacement, Biomaterials 26 (2005) 1087-1095.
[7] M. Geetha, A.K. Singh, R. Asokamani, A.K. Gogia, Ti based biomaterials, the ultimate choice for orthopaedic implants-A review, Prog. Mater. Sci. 54 (2009) 397-425.
[8] H. Shao, C. Yu, X. Xu, J. Wang, R. Zhai, X. Wang, Influence of Ti nanocrystallization on microstructure, interface bonding, surface energy and blood compatibility of surface TiO_2 films, Appl. Surf. Sci. 257 (2010) 1649-1654.
[9] S.M.A. Shibli, S.J. Mathai, Development and bio-electrochemical characterization of a novel TiO_2-SiO_2 mixed oxide coating for titanium implants, J. Mater. Sci.: Mater. Med. 19 (2008) 2971-2981.
[10] L. Wang, L. Shi, J. Cheng, Z. Shi, L. Ren, Y. Wang, Biocompatibility of Si-incorporated TiO_2 film prepared by micro-arc oxidation, Mater. Lett. 116 (2014) 35-38.
[11] R.C. Flemming, C.J. Murphy, G.A. Abrams, S.L. Goodman, P.F. Nealey, Effects of synthetic micro- and nano-structured surfaces on cell behavior, Biomaterials 20 (1999) 573-588.
[12] T.J. Webster, R.W. Siegel, R. Bizios, Osteoblast adhesion on nanophase ceramics, Biomaterials 20 (1999) 1221-1227.
[13] L. Zhao, S. Mei, W. Wang, P.K. Chu, Z. Wu, Y. Zhang, The role of sterilization in the cytocompatibility of titania nanotubes, Biomaterials 31 (2010) 2055-2063.
[14] M.J. Dalby, N. Gadegaard, M.O. Riehle, C.S.W. Wilkinson, A.S.G Curtis, Investigating filopodia sensing using arrays of defined nano-pits down to 35 nm diameter in size, Int. J. Biochem. Cell Biol. 36 (2004) 2005-2015.
[15] C.H. Choi, S.H. Hagvall, B.M. Wu, J.C.Y. Dunn, R.E. Beygui, C.J. Kim, Cell interaction with three-dimensional sharp-tip nanotopography, Biomaterials 28 (2007) 1672-1679.
[16] A.V. Sapelkin, S.C. Bayliss, B. Unal, A Charalambou, Interaction of B50 rat hippocampal cells with stain-etched porous silicon, Biomaterials 27 (2006) 842-846.
[17] M. Karlsson, E. Pålsgård, P.R. Wilshaw, L.D. Silvio, Initial in vitro interaction of osteoblasts with nano-porous alumina, Biomaterials 24 (2003) 3039-3046.
[18] Y.T. Liu, K.C. Kung, T.M. Lee, T.S. Lui, Enhancing biological properties of porous coatings through the incorporation of manganese, J. Alloy. Compd. 581 (2013) 459-467.
[19] J. Karlsson, R. Jimbo, H.M. Fathali, H.O. Schwartz-Filho, M. Hayashi, M. Halvarsson, A. Wennerberg, M. Andersson, In vivo biomechanical stability of osseointegrating mesoporous TiO_2 implants, Acta Biomater. 8 (2012) 4438-4446.
[20] H. Cao, X. Liu, F. Meng, P.K. Chu, Biological actions of silver nanoparticles embedded in titanium controlled by micro-galvanic effects, Biomaterials 32 (2011) 693-705.
[21] X. Zhao, G. Wang, H. Zheng, Z. Lu, X. Zhong, X. Cheng, H. Zreiqat, Delicate refinement of surface nanotopography by adjusting TiO_2 coating chemical composition for enhanced interfacial biocompatibility, ACS Appl. Mater. Interfaces 5 (2013) 8203-8209.
[22] G. Zhao, A.L. Raines, M. Wieland, Z. Schwartz, B.D. Boyan, Requirement for both micron- and submicron scale structure for synergistic responses of osteoblasts to substrate surface energy and topography, Biomaterials 28 (2007) 2821-2829.
[23] H.J. Rønold, J. E. Ellingsen, Effect of micro-roughness produced by TiO_2 blasting-tensile testing of bone attachment by using coin-shaped implants, Biomaterials 23 (2002) 4211-4219.

[24] K. Suzuki, K. Aoki, K. Ohya, Effects of surface toughness of titanium implants on bone remodelling activity of femur in rabbits, Bone 21 (1997) 507-514.
[25] S. Ni, J. Chang, L. Chou, W. Zhai, Comparion of osteoblast-like cell responses to calcium silicate and tricalcium phosphate ceramics in vitro, J. Biomed. Mate.r Res. B 80 (2007) 174-183.
[26] D. Lozano, M.J. Feito, S. Portal-Núñez, R.M. Lozano, M.C. Matesanz, M.C. Serrano, M. Vallet-Regí, M.T. Portolés, P. Esbrit, Osteostatin improves the osteogenic activity of fibroblast growth factor-2 immobilized in Si-doped hydroxyapatite in osteoblastic cells, Acta Biomater. 8 (2012) 2770-2777.

DVS 324

Dense YSZ Coatings Fabricated by Plasma Spray-Physical Vapor Deposition

Fang SHAO*, Huayu ZHAO, Chenguang LIU, Xinghua ZHONG, Yin ZHUANG, Jinxing NI, Shunyan TAO

The Key Laboratory of Inorganic Coating Materials, Shanghai Institute of Ceramics, Chinese Academy of Sciences, Shanghai 201899, PR China

Plasma spray-physical vapor deposition (PS-PVD) is a novel technique which can improve coating density and surface characteristics. In the present study, dense YSZ coatings were deposited using PS-PVD with equitable process parameters. The spray distances were changed from 500 mm to 800 mm at intervals of 100 mm and the effects of spray distances and off-sets (along axial and radial directions of the torch) on the microstructure of the as-sprayed coatings were discussed. Four typical coatings were studied in detail, their microstructure development, phase composition, mechanical properties were explored. The as-sprayed coatings have dense lamellar/columnar hybrid microstructures, the porosity can be as low as 0.44%. Both monoclinic and tetragonal zirconia were detected in the coatings. Further, a part of oxygen lose during the spray process with nearer spray distance. The mechanical properties, such as hardness (*H*) and Young's modulus (*E*) of the coatings were both changed as a function of spray distance. The maximum values of *H* and *E* were 16.6 and 234.3 GPa, which were obtained in the densest coatings. This basic research might provide some useful insight for the possible application of dense PS-PVD YSZ coatings.

1 Introduction

Yttria-stabilized zirconia (YSZ) coatings have gained wide applications due to their excellent mechanical and thermal properties. Both air plasma spraying (APS) and electron-beam physical vapor deposition (EB-PVD) technologies have been applied to prepared YSZ coatings [1]. Conventionally plasma-sprayed YSZ coatings are normally formed by the stacked splats originating from high velocity melted or semi-melted particles, whereas there is a big difficulty in the deposition of uniform, dense ceramic coatings [2]. High porosity in YSZ coatings is beneficial to thermal insulation applications because it reduces the thermal conductivity of coating, however, it is deleterious for anti-corrosion properties and SOFC applications [3]. In contrast, EB-PVD processes can produce high-quality coatings, but the low deposition rates and high investment costs seriously limit their application ranges.

In recent years, plasma spray physical vapor deposition (PS-PVD) have been developed with the possibility to offer new opportunities to obtain advanced microstructures and thus to comply with the growing demands on modern functional coatings [4]. As a new promising technology, PS-PVD combines the advantages of thermal spraying (high deposition rates and cost-efficiency compared to EB-PVD). Due to the lower pressure during the process, the plasma jet length can reach over 2 m, and its diameter is 200-400 mm [5]. Further, high enthalpy plasma stream with high plasma torch output up to 180 kW can be obtained. Thus PS-PVD can apply coatings with high density and large areas at unprecedented speeds. There have been some reports on the coatings deposited by PS-PVD [6, 7]. Despite the importance of dense YSZ coatings, the reports mainly focus on columnar structures, few information is available on the detailed microstructures and property of PS-PVD dense coatings.

In this paper, dense YSZ ceramic coatings were successfully deposited by PS-PVD technology, which is expected to contribute to the applications of electrical insulation, SOFC, anti-corrosion TBCs etc. To exploit the potential of PS-PVD technology, the dependency on process conditions should be well understood. Besides spray distance, the radial location of the samples can affect the morphologies to a large extent. Thus, coatings deposited with different spray distance and radial distance were investigated. Dense coatings can be obtained at nearer spray distance and edger regions. Since porosity is one of the main microstructural features affecting the property of the coatings, microstructure and porosity details of PS-PVD dense YSZ coatings have been studied extensively. Based on the results, attempts were made to discuss the formation of dense coatings. Mechanical properties, e.g., hardness (*H*) and Young's modulus (*E*), of the coatings were investigated to evaluate the durability of coatings and get a better understanding of the relationship between coating microstructure and mechanical properties.

2 Experimental procedure

2.1 Plasma Spraying

All coating procedures were carried out on an Oerlikon Metco (Switzerland) LPPS Hybrid system. The polished graphite was used as the coating substrate. The spray procedure was illustrated in **Fig. 1**. The spray distance is the distance from the nozzle exit of spray gun to the substrate surface. Different spray distances from 500 mm to 800 mm at intervals of 100 mm were applied during the experimental process. To determine the radial location precisely, the graphite was in groups of 19 pieces during each deposition process and a piece of graphite was 50×10×9 mm^3. Metco 6700 Yittria stabilized Zirconia (YSZ) ceramic powders were used to deposit topcoats. The YSZ powders had spherical morphology with particle size ranging 1～30 μm. The deposition parameters are given in Table 1. Before coating deposition, the chamber pressure was lowered down to the working pressure (～1.5 mbar). The positions of spray gun and samples were kept constant in all cases, the obtained samples were labeled as (1～19)-(500～800).

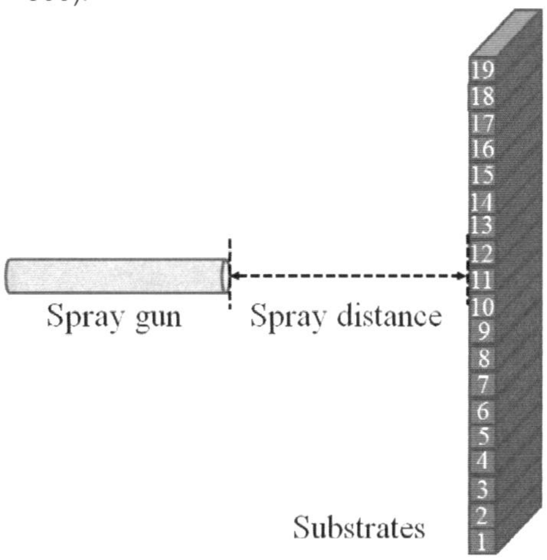

Fig. 1. Schematic diagram of the spray procedure

Table 1 PS-PVD operating parameters.

Parameters	value	Unit
Current	2600	A
Net power	65	kW
Primary plasma gas (Ar)	35	slpm
Secondary plasma gas (He)	60	slpm
Carrier gas (Ar)	8	slpm
Powder feed rate	5	g/min
Spray distance	500-800	mm

2.2 Characterization

The phases of the powders and coatings were identified by x-ray diffraction (XRD, D/max 2550 V, Rigaku Tokyo, Japan). The morphologies of the as-deposited YSZ coatings were characterized by a scanning electron microscope (SEM, TM3000, HITACHI, Japan) equipped with an energy-dispersive spectrometer (EDS). Porosity of the coatings was estimated by image analysis (IA) method. For each coating sample, 5 micrographs (magnification of 5000) at random locations on the cross section were analyzed. The roughness of the as-sprayed coatings was measured by a roughometer T8000 R120-400 (JENOPTIK Industry Metrology GmbH, Germany). Hardness (*H*) and Young's modulus (*E*) of the coatings were measured on the polished cross section by Nanoindentation technique, which were performed as our previous work [8]. A G-200 nanoindenter (Agilent Technologies, Oak, Ridge, USA) equipped with a Berkovich-type indenter was applied and the instrument was operated in a continuous stiffness mode (CSM). The maximum indentation depth was 5000 nm, the values at a penetration depth of about 500 nm were applied to represent the results. 10 indents randomly chosen on the cross sections were analyzed. A Poisson's ratio of 0.25 was used to calculate the elastic modulus.

3 Results and discussion

3.1 Microstructure

The microstructure of the coatings changed with the spray distance and radial location. Along the radial direction, the thickness of the coatings was gradually decreased from the center regions to the edge regions. The thickest coating was obtained at the jet center position for each spray distance, which is the NO.11 sample. On the same position, the deposition rate decreased with the increase of spray distance. The deposition rate is more sensitive to radial location than spray distance. For example, compared with 11-500, the deposition rate of 12-500 and 11-600 decreased 19.9% and 12.4%, respectively. 12-500 had a 52.9% lower deposition rate while 11-600 has a 34.8% lower deposition rate than 11-500. However, the morphology is more sensitive to spray distance than the radial location. The spraying distance controls the amount of molten particles in the coating layer and the heat load of the substrate.
For coatings deposited at 700 mm and 800 mm, the coatings on the center regions ((9～12)-700 and (9～13)-800) represent typical columnar structure, then changed to columnar/lamellar hybrid structures on transition and edge regions. The results are similar with other published reports [9, 10]. Since this work here mainly focused on the dense coatings, 11-500, 11-600, 13-700 and 14-800 coatings were discussed here. **Fig. 2** presents the surface morphologies of 11-500, 11-600, 13-700 and 14-800 coatings. Apart from the splats, there were protrusions on the surface of the samples. At the same position, the protrusions increased in numbers as the spray distance increased from 500 mm to 600 mm. 13-700 showed similar surface morphology while on the surface of 14-800, the protrusions were more and bigger. The mean surface roughness (*Ra*) of 11-500, 11-600, 13-700 and 14-800 coatings are 1.43, 1.29, 1.56 and 1.75 μm, respectively.

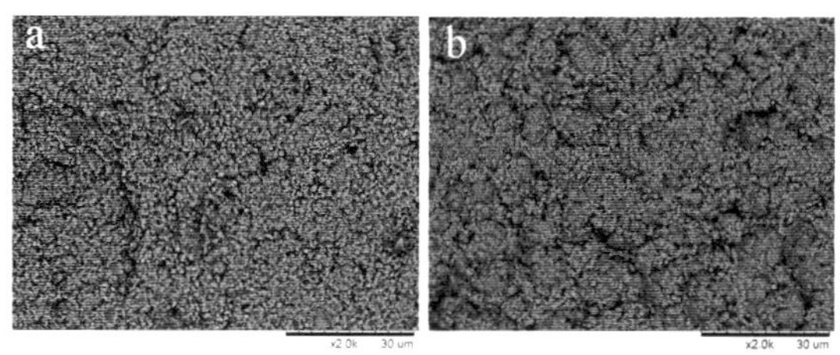

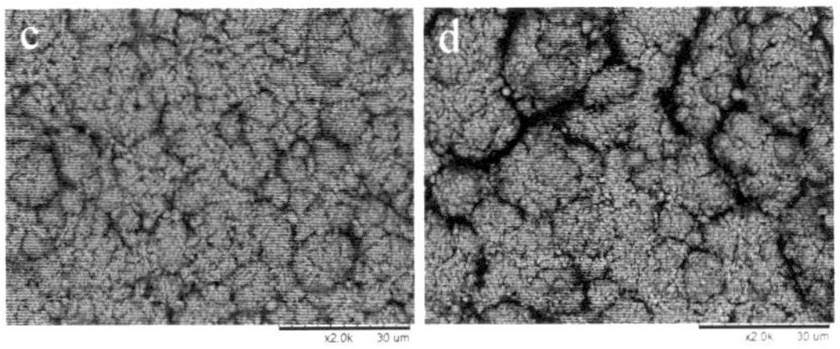

Fig. 2. Surface morphologies of the (a) 11-500, (b) 11-600, (c) 13-700 and (d) 14-800 coatings.

As shown in **Fig. 3**, the fracture images of 11-500, 11-600, 13-700 and 14-800 coatings all showed dense lamellar/columnar hybrid microstructures. The lamellar structure prevailed in 11-500, 11-600 while quasi-columnar structures can be observed in 13-700 and 14-800 coatings.

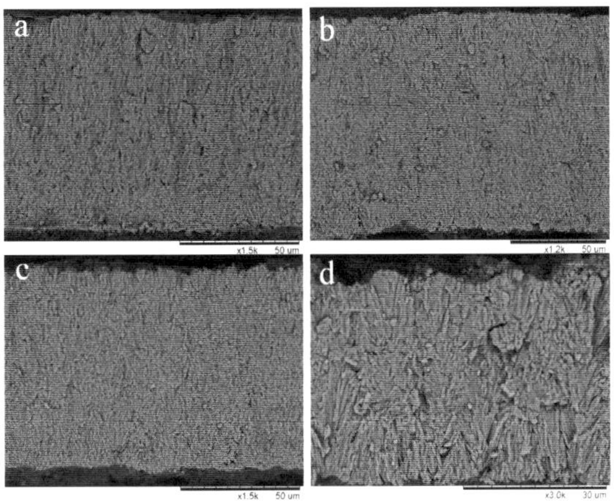

Fig. 3. Fracture cross-section SEM images of the (a) 11-500, (b) 11-600, (c) 13-700 and (d) 14-800 coatings.

High magnification cross-sectional (fractured) SEM images of the deposited coating are given in **Fig. 4** to highlight microstructural details. It indicates that the 11-500 coating is dense and consist mostly of short columnar grains. The grains are coarse and well adhered, which have a width of ~1.2 μm and a length of ~2.3 μm. There are some narrow pores between the columns and lamellar structures. These pores are formed probably as the result of incomplete inter-column and inter-splat connections. By increasing the spray distance to 600 mm, the grain adhered more closely. Thus the interface between the columns are more obscure and the voids are decreased. The closely packed columnar grains and dense splats collectively make the whole coating to obtain a compact structure with few pores. For 13-700 and 14-800 coatings, the columnar grains are smoother, which are about 0.6 μm in width. It have been demonstrated that the diameter of columns depend on the solidification time of the YSZ [11]. For a given substrate, splat thickness and speed of particle impact, the column diameter increases with increase of the solidification time [12]. At shorter spray distances, the substrates remains at higher temperatures due to the incoming heat flux from plasma flame. Therefore the columnar solidification occurs at slower growth rates. This may explain the width decrease of columns in 13-700 and 14-800 coatings compared with those obtained at shorter spray distances. As shown in **Fig. 4** c and d, the tendency of columns in 13-700 and 14-800 coatings to grow often at an angle rather than perpendicular to the substrate. The length of columns can be long, especially in 14-800 coatings, and the growth of columns can be followed in the next splat. Some columns grow faster than the others and the slower growing crystals stop when they come close to their neighbor. Thereby the gaps between the columns are formed [7].

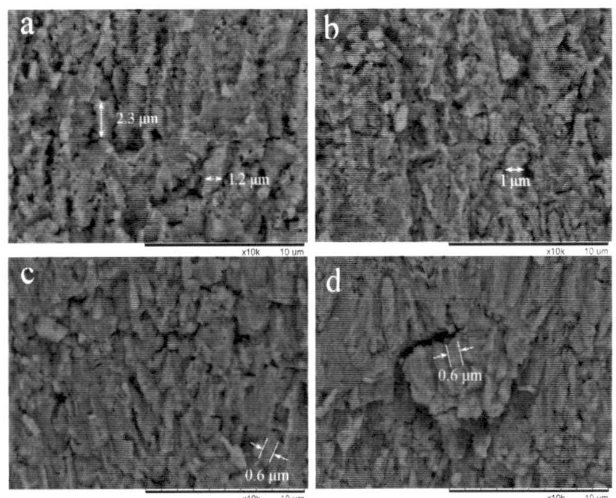

Fig. 4. High magnification SEM micrograph of the fracture surface of the (a) 11-500, (b) 11-600, (c) 13-700 and (d) 14-800 coatings

To further investigate the microstructure, the cross-sectional samples were polished with abrasive papers followed by mirror polishing. As shown in **Fig.5** and **Fig. 6**, both of 11-500 and 11-600 showed denser coatings than other coatings. The porosity can be evaluated by IA method and the results are presented in **Fig. 7**. Porosities of 11-500, 11-600, 13-700 and 14-800 are 0.89%, 0.44%, 4.99% and 5.15%, respectively. Therefore, 11-500, 11-600 coatings possess denser structures, which is consistent with the morphology analysis results. The chemical compositions of point A-E marked with white crosses in **Fig. 6** b and d were listed in **Table 2**. The contents of Y element varied a lot in all of the four coatings and there were regions with a little or even no Y element. During the spray process, ZrO_2 and Y_2O_3 may not mix sufficiently, thus the dispersion of Y was not uniformly. The usual deposition mode for conventional plasma spraying is that liquid droplets form splats piling up to build the coating [13]. In LPPS-PVD conditions, the deposits can be formed also from clusters and/or vaporized atomic species besides liquid splat depostion. For the investigated 11-500 and

11-600 coatings, there may be only a small amount of the feedstock particles that can be evaporated. The preheated substrate can increase the contact temperature, and reduce the viscosity of the impinging melted droplets and the thermal stress. The coatings are obtained mainly from the highly molten, well flowable splats, developing large flattening ratios [14]. Due to the good flowability, the pores and microcracks in the deposits can be filled. Additionally, the low pressure ambience can considerably improve particle velocities and kinetic impact energies. As a result, very dense microstructures are formed. When the contents of condensates from the vapor phase further increased in the coatings, the gaps between the columns are formed. Especially, the contact of columns with different angles may lead to the large pores. It's worth noting that many pores detected in 13-700 and 14-800 coatings have a shallow pit morphology, structures can be seen in the interiors of the pores, as shown in **Fig. 6**. The gaps between the columns may be narrow and small, thus the corresponding pores are shallow and slender.

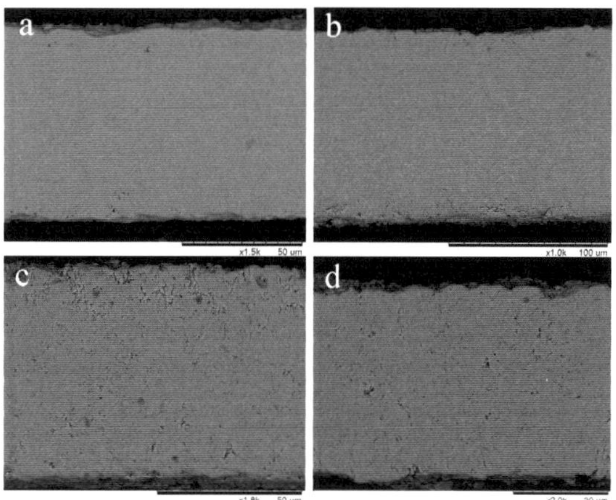

Fig. 5. The polished cross-section morphologies of (a) 11-500, (b) 11-600, (c) 13-700 and (d) 14-800 coatings.

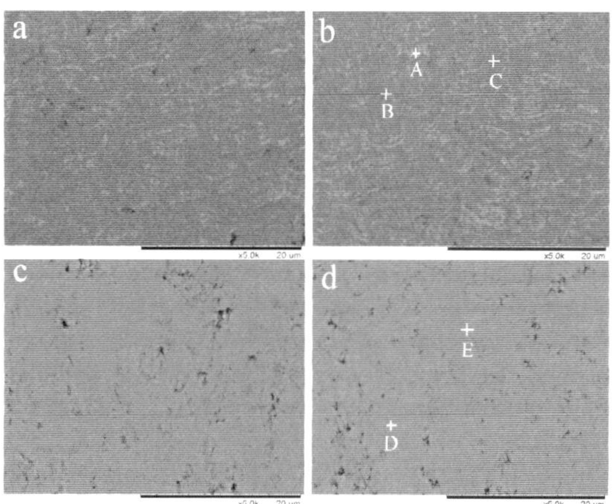

Fig. 6. High magnification SEM micrograph of the polished cross-section morphologies of the (a) 11-500, (b) 11-600, (c) 13-700 and (d) 14-800 coatings.

Table 2 Chemical composition (wt%) of point A-E mark by with white crosses in **Fig. 6** b and d.

Elements	A	B	C	D	E
O	20.89	27.58	27.92	25.63	24.38
Y	/	12.84	1.93	2.71	0.95
Zr	79.11	59.58	70.15	71.66	74.67

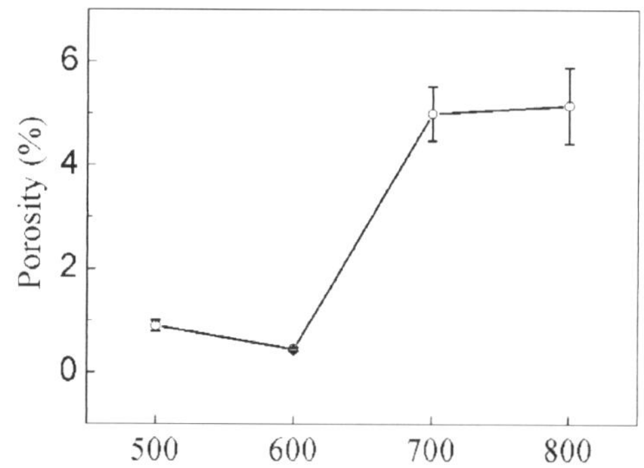

Fig. 7. Porosities of the 11-500, 11-600, 13-700 and 14-800 coatings.

3.2 Phase structure

Fig. 8 shows the X-ray diffraction patterns of the powders and as-sprayed coatings. The feedstock consists of monoclinic zirconia (M-ZrO_2) and yttria (Y_2O_3). Both M-ZrO_2 and tetragonal zirconia (T-ZrO_2) were detected in 11-600 and 14-800. Based on the analysis of EDS results, it can be inferred that in some regions of the as-sprayed coatings, the content of Y may be insufficient to stabilize the tetragonal zirconia, therefore monoclinic zirconia is detected in 11-600 and 14-800. Further, suboxides of zirconia (ZrO_{2-x}) was also present in 11-600 while 14-800 included only M- ZrO_2 and T-ZrO_2. The presence of ZrO_{2-x} indicates the oxygen loss during spraying [15].

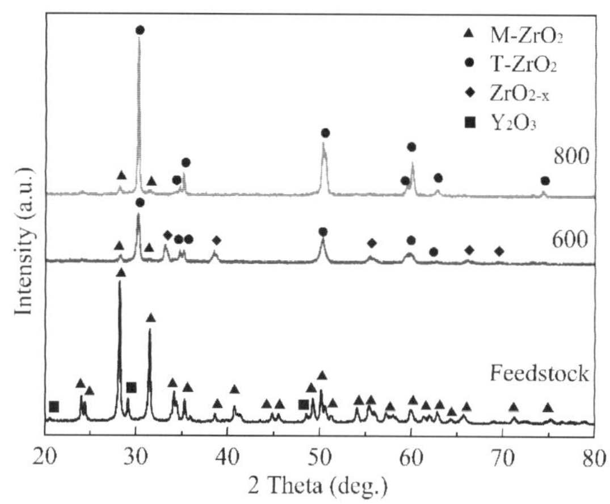

Fig. 8. XRD patterns of the feedstock and as-sprayed 11-600, 14-800 coatings.

3.3 Mechanical Properties

Fig. 9 shows the hardness (*H*) and Young's modulus (*E*) of 11-500, 11-600, 13-700, 14-800 coatings. As can be seen, the mean *H* and *E* values of 11-600 are 16.6 and 234.3 GPa, respectively, higher than those of other coatings. The *H* value was slightly larger than that of 13-700 and 14-800 while the *E* value of 11-500 was roughly equal or even lower, which was not completely consistent with the expectations. This may be due to the shallow pores in 13-700 and 14-800 coatings. The shallow pores contributes a lot to the porosity, however, they have limited effects on the hardness and Young's modulus.

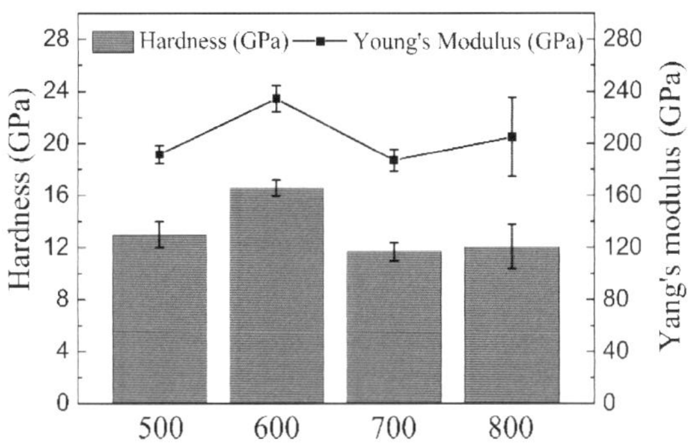

Fig. 9. Hardness (*H*) and Young's modulus (*E*) of 11-500, 11-600, 13-700, 14-800 coatings.

4 Conclusions

In this paper, dense YSZ coatings were produced by PS-PVD under varying spray distances and radial distances. 11-500, 11-600, 13-700 and 14-800 coatings were studied in detail. The microstructure, phase and mechanical properties of the coatings were investigated. Some conclusions can be drawn as follows:

(1) Apart from the splats, there were protrusions on the surface of the four coatings. The protrusions increased in numbers as the spray distance increased from 500 mm to 600 mm. 13-700 showed similar surface morphology while on the surface of 14-800, the protrusions were more and bigger. The mean surface roughness (Ra) of 11-500, 11-600, 13-700 and 14-800 coatings are 1.43, 1.29, 1.56 and 1.75 μm, respectively. The fracture images of the four coatings all showed dense lamellar/columnar hybrid microstructures. The lamellar structure prevailed in 11-500, 11-600 while quasi-columnar structures can be observed in 13-700 and 14-800 coatings.

(2) Both monoclinic and tetragonal zirconia were detected in 11-600 and 14-800 coatings. In some regions of the as-sprayed coatings, the content of Y may be insufficient to stabilize the tetragonal zirconia. Further, a part of oxygen lose during the spray process with nearer spray distance.

(3) Among the coatings investigated, 11-600 coating exhibited the lowest porosity (0.44%), the highest hardness (16.6 GPa) and Young's modulus (234.3 GPa) due to the dense microstructure. Though the porosities of 13-700 and 14-800 (4.99% and 5.15%) are much higher than 11-500 (0.89%), there is no significant decrease in the mechanical properties (*H* and *E*) of 13-700 and 14-800. The shallow pores in 13-700 and 14-800 coatings contributes a lot to the porosity, however, they have limited effects on the hardness and Young's modulus.

5 Literature

[1] M.H. Vidal-Setif, C. Rio, D. Boivin, O. Lavigne, Microstructural characterization of the interaction between 8YPSZ (EB-PVD) thermal barrier coatings and a synthetic CAS, Surf Coat Tech, 239 (2014) 41-48.
[2] L. Zhu, N.N. Zhang, B.C. Zhang, F. Sun, R. Bolot, M.P. Planche, H.L. Liao, C. Coddet, Very low pressure plasma sprayed alumina and yttria-stabilized zirconia thin dense coatings using a modified transferred arc plasma torch, Appl Surf Sci, 258 (2011) 1422-1428.
[3] D. Stender, N. Schauble, A. Weidenkaff, A. Montagne, R. Ghisleni, J. Michler, C.W. Schneider, A. Wokaun, T. Lippert, Dense zig-zag microstructures in YSZ thin films by pulsed laser deposition, Apl Mater, 3 (2015).
[4] K. Iizuka, M. Kambara, T. Yoshida, Growth of tin oxide thick films by plasma spray physical vapor deposition, Sensor Actuat B-Chem, 155 (2011) 551-556.
[5] M. Goral, S. Kotowski, A. Nowotnik, M. Pytel, M. Drajewicz, J. Sieniawski, PS-PVD deposition of thermal barrier coatings, Surf Coat Tech, 237 (2013) 51-55.
[6] L.H. Gao, H.B. Guo, L.L. Wei, C.Y. Li, S.K. Gong, H.B. Xu, Microstructure and mechanical properties of yttria stabilized zirconia coatings prepared by plasma spray physical vapor deposition, Ceram Int, 41 (2015) 8305-8311.
[7] S. Rezanka, G. Mauer, R. Vassen, Improved Thermal Cycling Durability of Thermal Barrier Coatings Manufactured by PS-PVD, J Therm Spray Techn, 23 (2014) 182-189.
[8] Y.X. Zhao, L. Wang, J.S. Yang, D.C. Li, X.H. Zhong, H.Y. Zhao, F. Shao, S.Y. Tao, Thermal Aging Behavior of Axial Suspension Plasma-Sprayed Yttria-Stabilized Zirconia (YSZ) Thermal Barrier Coatings, J Therm Spray Techn, 24 (2015) 338-347.
[9] C.Y. Li, H.B. Guo, L.H. Gao, L.L. Wei, S.K. Gong, H.B. Xu, Microstructures of Yttria-Stabilized Zirconia Coatings by Plasma Spray-Physical Vapor Deposition, J Therm Spray Techn, 24 (2015) 534-541.
[10] L.H. Gao, H.B. Guo, L.L. Wei, C.Y. Li, H.B. Xu, Microstructure, thermal conductivity and thermal cycling behavior of thermal barrier coatings prepared by plasma spray physical vapor deposition, Surf Coat Tech, 276 (2015) 424-430.
[11] L. Bianchi, A. Denoirjean, F. Blein, P. Fauchais, Microstructural investigation of plasma-sprayed

ceramic splats, Thin Solid Films, 299 (1997) 125-135.
[12] R. Sobhanverdi, A. Akbari, Porosity and microstructural features of plasma sprayed Yttria stabilized Zirconia thermal barrier coatings, Ceram Int, 41 (2015) 14517-14528.
[13] G. Mauer, A. Hospach, N. Zotov, R. Vassen, Process Conditions and Microstructures of Ceramic Coatings by Gas Phase Deposition Based on Plasma Spraying, J Therm Spray Techn, 22 (2013) 83-89.
[14] G. Mauer, M.O. Jarligo, S. Rezanka, A. Hospach, R. Vassen, Novel opportunities for thermal spray by PS-PVD, Surf Coat Tech, 268 (2015) 52-57.
[15] G. Mauer, A. Hospach, R. Vassen, Process development and coating characteristics of plasma spray-PVD, Surf Coat Tech, 220 (2013) 219-224.

Tribological performance of plasma sprayed Al_2O_3-Y_2O_3 composite coatings

Jian Rong[a,b,c], Kai Yang[a,b,*], Huayu Zhao[a,b], Chenguang Liu[a,b], Yin Zhuang[a,b], Shunyan Tao[a,b,*]

[a] Shanghai Institute of Ceramics, Chinese Academy of Sciences, Shanghai 201899, China

[b] The Key Laboratory of Inorganic Coating Materials, Chinese Academy of Sciences, Shanghai 201899, China

[c] University of Chinese Academy of Sciences, Beijing 100039, China

*Corresponding author: Tel.: +86 21 69906320; fax: +86 21 69906322.
E-mail address: kaiyang@mail.sic.ac.cn (K. Yang); sytao@mail.sic.ac.cn (S. Tao);

In the present study, Al_2O_3 and Al_2O_3-Y_2O_3 coatings were deposited by atmospheric plasma spraying. The microstructure, some mechanical properties and dry sliding wear behavior of the coatings were examined. The results showed that the mass ratio of α-Al_2O_3/γ-Al_2O_3 in as-sprayed Al_2O_3 coating was 0.08, however, the corresponding value in Al_2O_3-Y_2O_3 composite coating was 0.53, which may be beneficial to improve the mechanical properties and thermal conductivity of the composite coating. In addition, the tribological examination shows that the friction coefficients of composite coating/graphite pairs are more stable, and exhibiting better wear resistance, which may be related to the formation of more effective and stable graphite transferred film on the surface of the coating subjected to abrasion. Especially, the Al_2O_3-Y_2O_3 composite coating was proved to be a worthwhile wear resistant coating.

1. Introduction

Thermal sprayed ceramic coatings, as a good candidate, have been widely used in industries, to offer wear and corrosion protection, erosion resistance and thermal insulation [1-4]. Al_2O_3 coatings, by virtue of their inherent attrition and corrosive resistance, excellent dielectric and thermal shock resistance have attracted attention for a variety of applications [5, 6]. Whereas, alumina displays polymorphism, except for the stable rhombohedral α-Al_2O_3, other known metastable Al_2O_3 polymorphs are γ, δ, θ, k, η, χ, β and θ' [7]. However, the alumina phases in the plasma sprayed alumina coatings were turned out to be predominately γ-Al_2O_3 rather than α-Al_2O_3, sometimes δ-Al_2O_3 also existed [8], even though the feedstock powder was composed of α-Al_2O_3 alone. The hardness and thermal conductivity of α-Al_2O_3 phase were higher than γ-Al_2O_3. As to severe wear conditions, the invalidation of the coatings were often attributed to the thermal stress generated by massive friction heat. Therefore, attempts to improve the thermal conductivity of the plasma sprayed alumina coatings were considered as an effective method to improve the wear resistance. The enhanced wear resistance was proved by the improved thermal conductivity of alumina coatings through proper control of porosity or moderate addition of metal Al under the effort of Zhijian Yin [9, 10]. Therefore, the reservation of more α-Al_2O_3 was also an effective path to enhance the thermal conductivity of coatings. Heat treatment can realize the transition of metastable alumina into α-Al_2O_3 at a temperature over 1100℃ [8]. Unfortunately, this is often unattainable on account of the possible substrate failure. Laser surface modification of plasma sprayed alumina coatings is also an effective routine to contribute the transition of metastable alumina to α-Al_2O_3 [11]. Whereas, the process is complex and not cost-effective. In addition, the laser surface modification may also produce large residual stress in the coating. Other attempts have been conducted to stabilize the α-Al_2O_3 of plasma sprayed alumina coatings via the addition of oxide [12, 13] or non-oxide constituent, such as SiC [14]. Researches performed by Kozlova [12] have uncovered that the addition of ZrO_2 and MgO could increase the α-Al_2O_3 content in the as-sprayed coatings, and other oxides, such as Fe_2O_3, Co_2O_3, NiO, CaO, SiO_2, displays the opposite way. These results were attributed to the effects of these oxides on melting temperature and corresponding change in the quantity of unmelted α-Al_2O_3 seeds. R. Venkataraman, et al [15] have studied the plasma sprayed alumina-13 wt.% titania coatings and found an increased α-Al_2O_3 content in contrast with pure alumina coatings, which may be due to the interplay of the powder in the plasma flame resulting in phase separation of titanium cations from the feedstock. Kai Yang, et al [16] focused on the addition of Cr_2O_3 into Al_2O_3 coatings and realized the stabilization of α-Al_2O_3 through the effects of heterogeneous nucleation and partial solid solution.

Rare earth elements are extensively used as additives to refinement of microstructure owing to its high chemical activity. However, very few researchers have focused on the rare earth oxides on the effect of stabilization of α-Al_2O_3 and corresponding tribological performance. In the present research, we studied the addition of Y_2O_3 into Al_2O_3 coatings and found an enhanced α-Al_2O_3 phase content, the corresponding tribological behaviors compared with pure alumina coatings were further studied.

2. Experimental procedures

2.1 Materials and preparation

Stainless steel (1Cr18Ni9Ti) plates with a dimension of 16mm×27mm×1.3mm were used as substrates, which were sand blasted using corundum, followed by ultrasonically cleaning in ethanol before the spraying process. Plasma spraying was performed through a MultiCoat^{TM} spraying system equipped with a F4MB-XL plasma gun (Sulzer Metco AG, Switzerland). NiCr powder was chosen to deposit the bond coat. Commercially available Al_2O_3 and Y_2O_3 powders were used as feedstock. Al_2O_3 were fused and crushed powders, showing angular and irregular morphology. While, Y_2O_3 were agglomerated and sintered powders, showing spherical shape, **Fig. 1.** The average sizes (D_{50}) of Al_2O_3 and Y_2O_3 feedstock powders are 20.2 μm and 39.0 μm, respectively. The mole ratio of Al_2O_3 and Y_2O_3 was taken as 82:18. The Al_2O_3 and Y_2O_3 powders were fed into a jar proportionally and mechanical mixed for 120h and then sprayed for the preparation of composite coatings. Pure Al_2O_3 coatings were also deposited for comparison. The plasma spraying parameters are shown in **Table. 1.**

2.2 Coating characterization

The phase compositions were determined by X-ray diffraction using a Rigaku D/Max2550 Diffractometer with nickel-filtered Cu Kα radiation (λ=0.15406 nm). The XRD data were collected in the 2θ range from 10° to 90° at a scanning speed of 1° min^{-1}. The micro-structure of the coatings was analyzed by a Hitachi TM3000 scanning electron microscope. Image analysis (IA) method was employed to estimate the porosity of the coatings. Ten random selected SEM micrographs (magnification of 1000×) were quantified for each coating sample.

Information of the thermal diffusivity (α), together with the specific heat (c_p) and bulk density (ρ), allows the determination of thermal conductivity (λ) from reference [17]. Pure ceramic coatings without bond coating were deposited on graphite substrate (dimension Φ30mm ×3.5mm) for thermal diffusivity measurement. Coating specimens were peeled off from the graphite substrate and further machined to measure the thermal diffusivity using the laser-flash technique [17]. Specific heat capacity measurements were carried out using a modulated differential scanning calorimeter (Diamond DSC, PE, USA). Vickers microhardness was measured with an Instron Wilson Wolpert Tukon 2100B Hardness Tester (200 g load for 10 s) on the polished cross- section of as-sprayed coatings. Ten indentations were tested and averaged for the hardness.

2.3 Tribological Testing

Friction and wear tests were carried out on a MM-200 (Shanghai Research Institute of Materials, China) tribological tester using a block-on-ring layout at room temperature. The dimensions of stainless steel ring and graphite block were Φ_{outer} 40 mmΦ $\times_{inner}$ 16 mm× 10 mm and 30 mm× 7 mm× 6 mm, respectively. The thickness of top ceramic coatings on the stainless steel ring surfaces after grinding were about 450 μm. Before the wear testing, the surface roughness (Ra), determined by a HOMMEL-ETAMIC T8000 roughness tester of coated ring treated by grinding was 0.8 μm. The wear tests were taken at following conditions: relatively high loads of 1000 N; a rotational speed of 400 rpm (equivalent to a sliding velocity of 0.84 m/s). Friction coefficient was acquired from friction torque, which was directly manifested on the tester. Wear rate was calculated by wear mass loss, which was performed by weighing the samples before and after each of the wear tests, being divided by wear time (h). The specimens were cleaned in an ultrasonic bath with acetone for 15 min and then dried in an oven at 120℃ for 60 min before weighing. Three repeated tests were conducted in order to ensure the reliability of the measurements.

3. Results and discussion

3.1 Phase composition analysis

The Al_2O_3 feedstock is composed of α-Al_2O_3, whereas, the resulting coating consists of γ-Al_2O_3 (major) and α-Al_2O_3 (minor), **Fig. 2(a)**. The residual α-Al_2O_3 content was estimated from X-ray analysis by using the direct intensity comparison method [15]. It is generally accepted that the preferential formation of γ-Al_2O_3 is attributed to the easier nucleation of γ-Al_2O_3 than α-Al_2O_3 owing to its lower interfacial energy [18-20]. It can be noted that the composite powders is composed of two phases, namely, α-Al_2O_3 and c-Y_2O_3 (cubic yttria), whereas, the composite coating consists of four phases, γ-Al_2O_3 and m-Y_2O_3 (monoclinic yttria) appeared, **Fig. 2(b)**. As to the pure Al_2O_3 coating, the maximum diffraction peak intensity ratio of α-Al_2O_3 to γ-Al_2O_3 ($I_{(113)}$(α-Al_2O_3)/$I_{(440)}$ (γ-Al_2O_3)) is equal to 0.08. However, the corresponding value of Al_2O_3-Y_2O_3 composite coating was 0.53. The dramatic variation may be due to the existence of Y_2O_3 in the corresponding composite coating. Thus, we can deduce that the addition of Y_2O_3 is conducive to the stabilization of α-Al_2O_3. Since the melting point of Y_2O_3 (2410℃) is over 360℃ higher than alumina (2050℃), it is likely that it raises the amount of unmelted seeds resulting in the formation of greater amount α-Al_2O_3 phase in the composite coatings. The enhanced content of α-Al_2O_3 in composite coatings may be beneficial to improve the mechanical properties and thermal conductivity. In addition, the existence of metastable m-Y_2O_3 phase was attributed to the high pressure and ultrafast quenching process involved in the plasma spray opera-

Table. 1.
Plasma spraying operating parameters

Parameters	Value	Unit
Arc current	640-650	A
Primary plasma gas (Ar)	45-50	splm
Secondary plasma gas (H_2)	7-9	splm
Carrier gas (Ar)	3.0-4.0	splm
Powder feed rate	40	g min^{-1}
Spray distance	100-120	mm

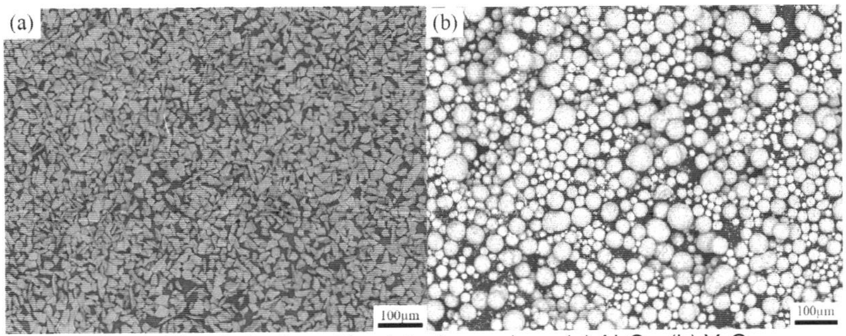

Fig. 1. Morphologies of feedstock powders: (a) Al_2O_3, (b) Y_2O_3.

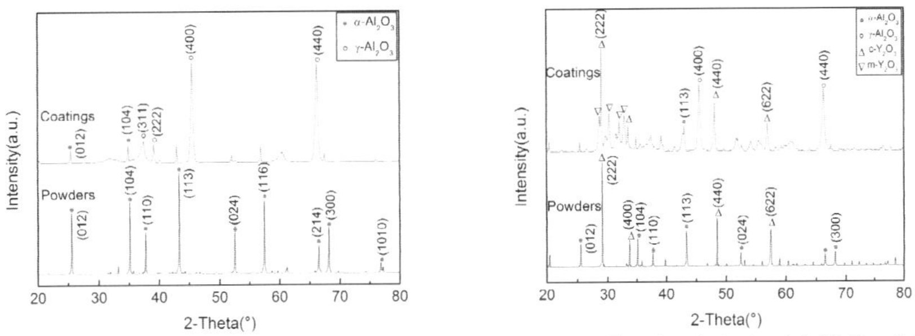

Fig. 2. XRD patterns of the as-sprayed coatings and corresponding feedstock: (a) Al_2O_3, (b) Al_2O_3-Y_2O_3.

tions. Similar results were also reported by other researchers [21].

3.2 Microstructure of as-sprayed coatings

As to the surface of Al_2O_3 coatings, plenty of smaller splashed droplets (elucidated by rectangle) can be clearly seen, **Fig. 3a**. Thus, a great deal of asperities were formed, which easily contributed to the formation of defects, such as pores. With regard to the composite coatings, some smaller splashed droplets (elucidated by arrows) of Al_2O_3 were perfectly covered by the Y_2O_3 droplets and formed a relatively flat surface with less pores and scarcely any new smaller splashed Y_2O_3 droplets were formed, **Fig. 3b**. Which effectively facilitated to densify the microstructure and alleviate the generation of defects by shadow effects. The borders (shows by rectangular) between Al_2O_3 and Y_2O_3 splats were much closer than Al_2O_3 and Al_2O_3 splats, which may be beneficial for the composite coatings to possess favorable interlamellar cohesion and thermal conductivity, **Fig. 3b**. From the analysis of the cross-sectional morphologies, it may be noted that there were plenty of micro-pores as well as some interlamellar cracks inside the alumina coatings, **Fig. 4a**. However, less micro-pores with smaller volume were manifested in Al_2O_3-Y_2O_3 composite coatings. In addition, the interface between Al_2O_3 and Y_2O_3 phases were closely integration without pores and cracks, which shows better interlamellar cohesion, **Fig. 4b**. It could be observed that the bonding at NiCr bond coating/steel substrate and top ceramic coating/NiCr bond coating interfaces is favorable. No obvious coating defects, including crack, pore and inclusion are found in the above-mentioned binary interfaces. The porosity of the composite coating was estimated to be 0.84±0.32 %, whereas, the corresponding value of alumina coating was 2.72±1.11 %, much larger than the composite coating.

3.3 Hardness and thermal conductivity

Mechanical properties and thermal conduction property have salient effects on the coating anti-wear performance under severe conditions. The hardness of Al_2O_3 and Al_2O_3-Y_2O_3 and Y_2O_3 composite coatings were 9.26±0.57 GPa、8.16±0.091 GPa and 4.12±0.21 GPa, respectively. Owing to the existence of the low hardness Y_2O_3 phase, the hardness of the composite coating was slightly lower than alumina coating but comparable. Meanwhile, the standard deviation of the composite coating was smaller than the alumina coating, indicating a uniform structure. The severe wear process easily generates huge friction heat and external stress, which may lead to the coating failure owing to the thermal expansion coefficient mismatch between the coating and the metal substrate. The coatings showing higher thermal conductivity usually possess greater wear resistance [22]. It can be found that the thermal diffusivity and thermal conductivity of composite coatings were higher than Al_2O_3 coatings, **Fig. 5**. Higher content of α-Al_2O_3、lower porosity、larger intersplat adhesion and better inter-lamellar cohesion along with high thermal conductivity of Y_2O_3 may be responsible for the enhanced thermal conductivity of the Al_2O_3-Y_2O_3 composite coatings.

3.4 Tribological behaviors

As to tribological behaviors, the friction coefficients of both coating/graphite pairs increase first, following with decrease and then gradually stabilize, **Fig. 6a.** The increase of the friction coefficients were attributed to the

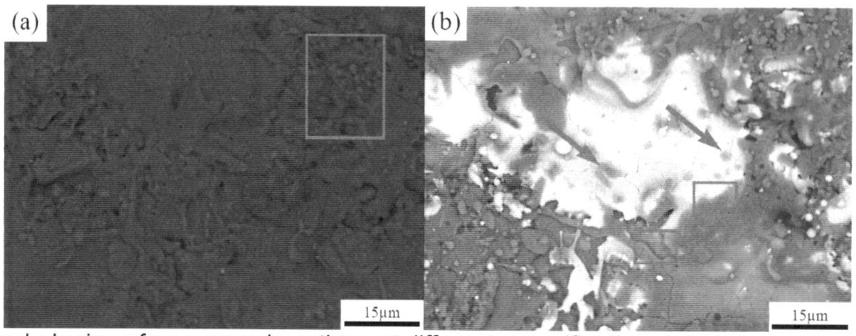

Fig. 3. Surface morphologies of as-sprayed coatings at different magnification: (a) Al_2O_3 coating and (b) Al_2O_3-Y_2O_3 composite coating.

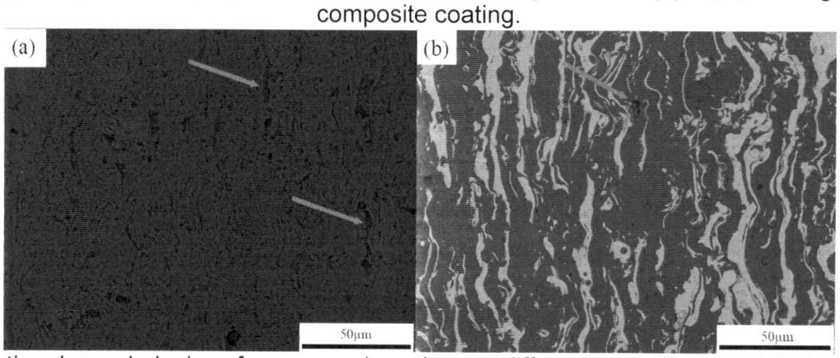

Fig. 4. Cross-sectional morphologies of as-sprayed coatings at different magnification: (a) Al_2O_3 coating and (b) Al_2O_3-Y_2O_3 composite coating.

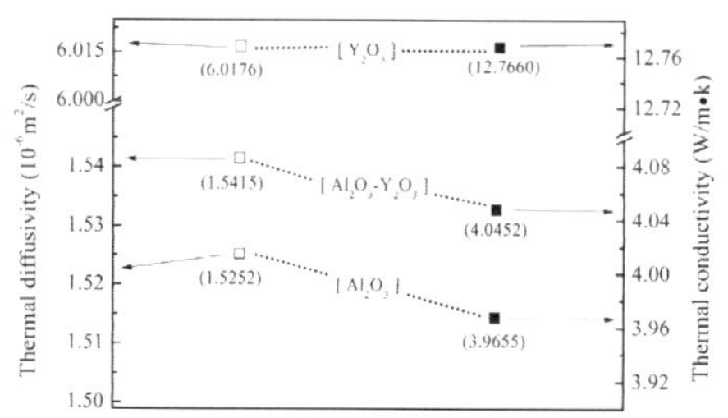

Fig. 5. The thermal diffusivity and corresponding thermal conductivity of Al_2O_3, Al_2O_3-Y_2O_3, Y_2O_3 coatings at room temperature.

unstable of worn surface topography in the initial wear stage. Among the wear process, multifold reaction may occur on the real contact surfaces of the friction pairs under the combined effects of the contact stress and frictional heat [23, 24]. The fragmentation and detachment of the worn surface in the initial stage may result in the unstability of the tribological behaviors. With sliding time going on, the generation of graphite transferred films on the coating wear ring surfaces facilitates to the lubrication effect, which coincide with the descending trend of the corresponding friction coefficients. The phenomenon of transferred film are ubiquitous in the process of wearing [10, 25]. The average friction coefficients of Al_2O_3 and Al_2O_3-Y_2O_3 composite coatings were 0.078±0.012, 0.044±0.006, respectively. Compared to the Al_2O_3 coating/graphite pair, the composite coating/graphite pair owns greater stability in the friction coefficient. Which may be related to the formation of more effective and stable graphite transfer film on the surface of the coating subjected to abrasion. Meanwhile, the composite coating/graphite pair demonstrates lower wear rates of ring and block, **Fig 6b**. The corresponding values are -0.0016 (negative value means the increment of weight by the adhered graphite) and 0.0018 g/h, respectively. Therefore, the composite coating exhibits better wear resistance compared with alumina coating.

In order to uncover the failure process and mechanism of the coatings, the worn surfaces of Al_2O_3 and Al_2O_3-Y_2O_3 composite coatings were analyzed. As is often the case, both of the coatings have formed graphite transferred film (the black phases were proved to be transferred graphites by EDS (not shown here)), **Fig. 7**. As to the former, some stripe-type plowing traces appeared on the Al_2O_3 surface subjected to abrasion. Moreover, some smaller wear debris can be found on the worn surface, which severs as abrasive and facilitates to the abrasive wear. Meanwhile, several cracks were clearly seen on the worn surface, which may be owing to the fatigue-induced brittle fracture. During the wear test, by the effects of wear debris along with the elevated cyclic and thermal stress, the pre-existed cracks may easily propagated along the splat boundaries and led to extensive cracking of the surface

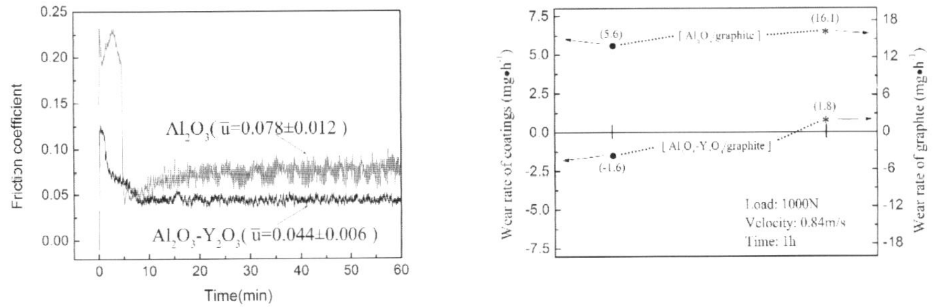

Fig. 6 Friction coefficients (a) and wear rate (b) of Al_2O_3/Graphite and Al_2O_3-Y_2O_3/Graphite friction pairs.

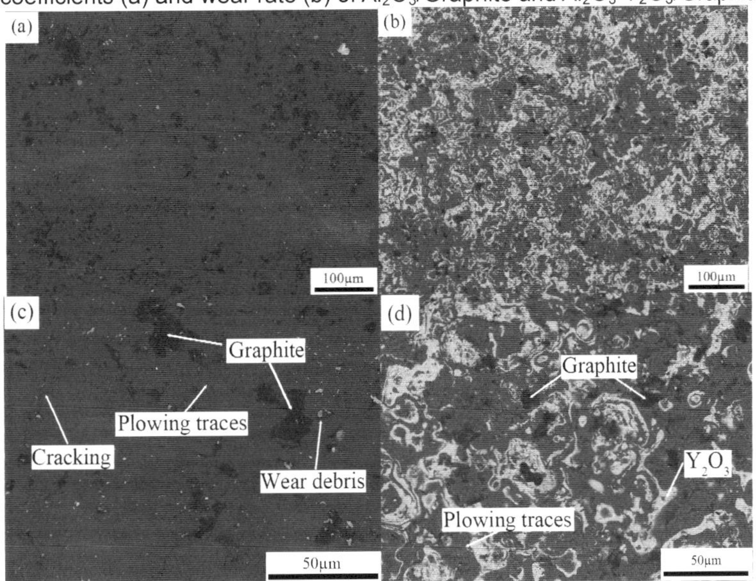

Fig. 7. The worn surface morphologies of coatings at different magnification: (a) (c) Al_2O_3 coating and (b) (d) Al_2O_3-Y_2O_3 composite coating.

splats and resulted in the splats spallation or detachment of the transferred film. According to the above analysis, the main failure mechanism of alumina coating was plastic deformation along with some abrasive wear and fatigue-induced brittle fracture. As to the latter, very few stripe-type wear topography is found on the worn surfaces of Al_2O_3-Y_2O_3 composite coatings. Meanwhile, scarcely any wear debris and cracks were displayed on the worn surface. These phenomenon may be attributed to the addition of the relatively soft yttrium oxide phase, which improved the α-Al_2O_3 content along with increased toughness, densified the microstructure and enhanced interlamellar cohesion of the composite coating, thus, protected the splats from spallation. The content of the graphite on the surface of composite coating was slightly lower than alumina coating, which may be correlated to the lower adhesive efficiency of the low hardness Y_2O_3 phase (graphite transferred film apt to adhere on the harder alumina phase). Even though the graphite transfer film has the ability of lubrication and alleviate the abrasion, the large amount of wear debris generated on the alumina coating worn surface which acted as abrasive and seems to have overwhelming effect than graphite. In addition, the high thermal conductivity of the composite coating, which effectively alleviated the friction heat generated at the contact surface and conducive to the formation of more effective and stable transferred film subjected to abrasion.

Conclusions

In the present work, Al_2O_3 and Al_2O_3-Y_2O_3 composite coatings were fabricated by plasma spraying. The microstructure and tribological properties of the coatings were investigated. The following conclusions arising from this work can be drawn:

1. The α-Al_2O_3 phase content of the composite coatings were enhanced compared with pure alumina coatings, which indicated that the addition of Y_2O_3 can effectively contribute to the stabilization of α-Al_2O_3.
2. Compared with Al_2O_3 coating/graphite pair, the friction coefficients of Al_2O_3-Y_2O_3 composite coating/graphite pair are more stable. In addition, the corresponding wear rates of the coating and graphite are lower. The composite coating possess pleasurable wear-resistance performance.
3. The composite coating/graphite pair exhibits better tribological performance, which may be owing to its greater thermal conductivity of the coating. High thermal conductivity contributes to lower worn sur-

face temperature, which may be beneficial to the formation of more effective and stable graphite transfer film on the surface of the coating subjected to abrasion.

References

1. Y. Jin, and Y. Yang. Tribological behavior of various plasma-sprayed ceramic coatings. Surf. Coat. Technol. 88 (1996), pp. 248/54.
2. Y. Dianran, H. Jining, and W. Jianjun. The corrosion behaviour of a plasma spraying Al_2O_3 ceramic coating in dilute HCl solution. Surf. Coat. Technol. 89 (1997), pp. 191/95.
3. T. Branco, R. Gansert, and S. Sampath. Solid Particle Erosion of Plasma Sprayed Ceramic Coatings, Materials Research 7 (2004), Issue 1, pp. 147/53.
4. T.M. Yonushonis. Overview of thermal barrier coatings in diesel engines. J. Therm. Spray Technol. 6 (1997), Issue 1, pp. 50/56.
5. L. Pawlowski. The relationship between structure and dielectric properties in plasma-sprayed alumina coatings. Surf. Coat. Technol. 35 (1988), pp. 285/98.
6. C.J. Li, G.J. Yang, and A. Ohmori. Relationship between particle erosion and lamellar microstructure for plasma-sprayed alumina coatings. Wear 260 (2006), Issue 11-12, pp. 1166/72.
7. M.G. Gee. The formation of aluminium hydroxide in the sliding wear of alumina. Wear 153 (1992), pp. 201/27.
8. R. Mcpherson. On the formation of thermally sprayed alumina coatings. Journal of Materials Science 15 (1980), pp. 3141/49.
9. Z.J. Yin, S.Y. Tao, X.M. Zhou, and C.X. Ding. Particle in-flight behavior and its influence on the microstructure and mechanical properties of plasma-sprayed Al_2O_3 coatings. J. Eur. Ceram. Soc. 28 (2008), Issue 6, pp. 1143/48.
10. Z. Yin, S. Tao, X. Zhou, and C. Ding. Tribological properties of plasma sprayed Al/Al_2O_3 composite coatings. Wear 263 (2007), Issue 7-12, pp. 1430/37.
11. R. Krishnan, S. Dash, R. Kesavamoorthy, C. Babu Rao, A.K. Tyagi, and B. Raj. Laser surface modification and characterization of air plasma sprayed alumina coatings. Surf. Coat. Technol. 200 (2006), Issue 8, pp. 2791/99.
12. K. I.R. Structure transformations in deposited aluminium oxide. Neorg. Mater. 7 (1971), Issue 8, pp. 1372/76.
13. L.L. Shaw, D. Goberman, and R. Ren. The dependency of microstructure and properties of nanostructured coatings on plasma spray conditions. Surf. Coat. Technol. 130 (2000), pp. 1/8.
14. M. Uma Devi. On the nature of phases in Al_2O_3 and Al_2O_3–SiC thermal spray coatings. Ceram. Int.30 (2004), Issue 4, pp. 545/53.
15. R. Venkataraman, and R. Krishnamurthy. Evaluation of fracture toughness of as plasma sprayed alumina–13wt.% titania coatings by micro-indentation techniques. J. Eur. Ceram. Soc.26 (2006), Issue 15, pp. 3075/81.
16. K. Yang, X. Zhou, H. Zhao, and S. Tao. Microstructure and mechanical properties of Al_2O_3–Cr_2O_3 composite coatings produced by atmospheric plasma spraying. Surf. Coat. Technol. 206 (2011), Issue 6, pp. 1362/71.
17. H.J. Rätzer-Scheibe, U. Schulz, and T. Krell. The effect of coating thickness on the thermal conductivity of EB-PVD PYSZ thermal barrier coatings. Surf. Coat. Technol. 200 (2006), Issue 18-19, pp. 5636/44.
18. R. Mcpherson. Formation of metastable phases in flame- and plasma-prepared alumina. *Journal of* Materials Science 8 (1973), pp. 851/58.
19. N.N. Ault. Characteristics of Refractory Oxide Coatings Produced by Flame-Sparying. J. Am. Ceram. Soc. 40 (1957), Issue 3, pp. 69/74.
20. P. Fauchais, M. Fukumoto, A. Vardelle, and M. Vardelle. Knowledge Concerning Splat Formation: An Invited Review. J. Therm. Spray Technol. 13 (2004), Issue 3, pp. 337/60.
21. V. Gourlaouen, G. Schnedecker, and A.M. Lejus. Metastable phases in yttrium oxide plasma spray deposits and their effect on coating properties. Mater. Res. Bull.28 (1993), pp. 415/25.
22. D. Chuanxian, H. Bingtang, and L. Huiling. Plasma-sprayed wear-resistance ceramic and cermet coating materials. Thin Solid Films 118 (1984), pp. 485/93.
23. X. Lin, Y. Zeng, and C. Ding. Tribological behaviour of nanostructured Al_2O_3-3 wt% TiO_2 coating against steel in dry sliding. Tribology Letters 17 (2004), Issue 1, pp. 19/26.
24. V. Fervel, B. Normand, and C. Coddet. Tribological behavior of plasma sprayed Al_2O_3-based cermet coatings. Wear 230 (1999), pp. 70/77.
25. B. Normand, V. Fervel, and C. Coddet. Tribological properties of plasma sprayed alumina–titania coatings role and control of the microstructure. Surf. Coat. Technol. 123 (2000), pp. 278/87.

Tubular asymmetric $La_{0.6}Sr_{0.4}Co_{0.2}Fe_{0.8}O_{3-\delta}$ oxygen transport membranes prepared by plasma spraying-physical vapor deposition (PS-PVD) and supersonic air-gas plasma spraying (SAPS)

Shaopeng Niu[a,b], Kesong Zhou[a,b], Liping Xu[b], Changguang Deng[b], Min Liu[c,b], Jie Mao[b]
[a] School of Materials Science and Engineering, South China University of Technology, Guangzhou 510 640, China
[b] Guangdong Institute of New Materials, National Engineering Laboratory for Modern Materials Surface Engineering Technology, The Key Lab of Guangdong for Modern Surface Engineering Technology, Guangzhou 510 651, China.
[c] Guangdong Academy of Sciences, Guangzhou 510075, China.

Abstract
Tubular asymmetric $La_{0.6}Sr_{0.4}Co_{0.2}Fe_{0.8}O_{3-\delta}$ (LSCF) oxygen transport membranes were successfully prepared by SAPS and PS-PVD on porous austenitic stainless steel for the first time. Microstructure and oxygen permeablity of the two kinds of membranes were mainly studied. The oxygen permeation flux measurements were carried out on a self-designed test device at atmosphere pressure in an air-helium gradient. Oxygen permeation fluxes of the two tubular membranes (6.1×10^{-4}ml/cm^2·min for PS-PVD membrane, 0.34×10^{-4}ml/cm^2·min for SAPS membrane at 900 ℃) are reported, the extremely low permeation is believed to be caused by the oxidation of metallic porous supports.

1 Introduction

Dense oxygen transport membranes (OTM) made from mixed oxygen ionic and electronic conductors have attracted intense research interest owing to their potential applications in oxygen separation, SOFC and oxygen-related catalytic processes, such as partial oxidation of hydrocarbons to value-added products. Since the membrane transfers oxygen in a way of ionic diffusion, its oxygen permeability can be effectively improved by reducing thickness, but doing so also reduces the mechanical strength of membranes. However, asymmetric membranes comprising a thin dense film over a relatively thick porous support have been proposed to balance the contradiction [1-3]. Asymmetric membranes generally include two main configurations, planar and tubular. Compared with planar membranes, tubular membranes possess more oxygen permeable area and have more advantages in high temperature sealing and connection. Easier componentized tubular asymmetric membranes will accelerate the sizable application of OTM. However, because of the complicated preparation process, the commonly used wet chemical methods still have difficulty in manufacturing asymmetric membranes, especially the tubular membranes with large scale [4-5].
Newly developed plasma spraying-physical vapor deposition technique (PS-PVD) provides an effective and more efficient avenue for asymmetric membranes preparation without requiring subsequent sintering. By means of high-powered plasma gun and operating pressures down to 200Pa and below, the spray system produces a hot supersonic gas stream about 2000m/s with temperatures higher than 6000℃. A dense and well-adherent ceramic coating can be deposited rapidly [6-9]. And successful planar membrane preparation has already been reported by researchers [10-12]. High energy supersonic air-gas plasma spraying technique is also a good option for preparing OTM. Due to the special three-electrode spray gun with enhanced capacity of arc compression and fire gas explosion effect, an ultrahigh temperature and supersonic plasma jet can also be achieved. The process does not require a vacuum or expensive plasma gas (only use compressed air plus minor fire gas additions), so if practical, SAPS might be more promising for actual application because of its equivalent ability to produce rapidly dense ceramic coatings but cost less [13].

2 Experimental

2.1 Powder and porous supports

Metco 6830A powders (Oerlikon Metco, Switzerland) were used as the LSCF feedstock material. The mean powder particle size d50 is 21±4μm detected by laser particle sizer (BT-9300st, Bettersize, China). Commercial Commercial porous AISI 316L disks and tubes were used as the membranes' supports.porous tubes which actually applied to the filter elements have 5-10μm pore size and 35-50% porosity (GaoQ Functional Materials, China). The size of all the tubes is 250mm length, 20mm diameter and 2mm wall thick.

2.2 Membrane deposition

PS-PVD preparation work was carried out in an Oerlikon Metco PS-PVD Multicoat facility using a current-controlled O3CP gun with power capability of up to 150kW and nozzle diameter of 12mm. Argon and Helium were used as plasma gases, and additional oxygen was pumped into the vacuum chamber to avoid the reduction of LSCF membrane during spraying. The spray parameters used are shown in Table 1.

Table 1. Deposition parameters of PS-PVD

Parameter	Value
Chamber pressure, Pa	150
Current, A	2400
Torch power, kW	56-57
Ar, NLPM	100
He，NLPM	15-20
Spray distance, mm	1000
Carrier gas Ar, NLPM	9.0
Powder feed rate, g/min	20
Deposition efficience	80μm/100sec

SAPS preparation work was proceeded in PLAZER 180 PL-S system (PLAZER LTD, Ukraine) using a special three-electrode spraying gun with power up to 180kW. The system operated in air, compressed air and propane were used as the working gases. The spray parameters are listed in Table 2.

Table 2. Deposition parameters of SAPS

Parameter	Value
Current, A	250-260
Voltage, V	480-500
Power, kW	120-125
Compressed air, MPa	0.3-0.4
Propane, L/min	2.0-2.5
Spray distance, mm	230
Carrier gas N_2, L/min	12.0
Powder feed rate, g/min	8
Deposition efficience	130μm/53sec

2.3 Microstructures and oxygen permeation measurement

The membranes in cross-sections were analyzed using scanning electron microscopes (JEOL 8100, Japan). A specially designed test device was made to measure the oxygen permeation fluxes of the tubular membranes. A tubular furnace with 150mm length flat-temperature zone was used for heating in tests. Schematic diagram of the device is shown in Fig.1a, Fig. 1b is the physical picture. There are two gas paths as can be seen in the Fig.1a. Gas path A consists of the quartze tube, the sealing flange components at the tube's two ends and the gas pipes connect to the flanges. In the quarze tube, the tubular membrane with two sealing bite type fittings and the connecting gas pipes constitutes the gas path B.

It is based on the membranes' high mechanical strength that the bite type fittings can be used for membrane sealing. And because of the tubular geometry, the fittings can be placed out of the heating zone and therefore low temperature seal can be accomplished.

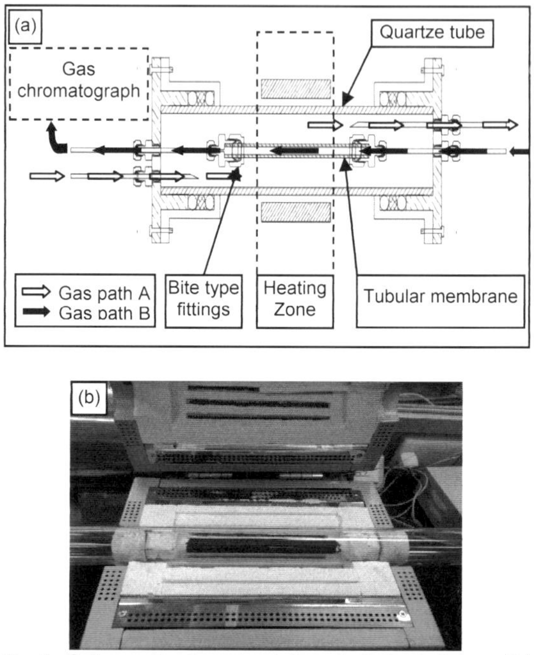

Fig.1. Schematic diagram (a) and phyical picture (b) of the self-designed test device

In the test, oxygen permeation was measured at two temperatures 700℃ and 900℃, each temperature was held 1h from the moment the oxygen permeation became stable. Dry air was used as feed gas (300ml/min) which was led into gas path A. Helium selected as sweep gas (50ml/min) was led into gas path B. Both gases were fed at atmosphere pressure. Because of oxygen concertration difference formed by the two gas paths, oxygen can pass through the tubular membrane from its surface to the porous support side at high temperatures. The permeated oxygen was carried by sweep helium to the gas chromatograph (Agilent 7610, US) where the oxygen flow rate can be analyzed every three minutes. Considering extra oxygen will be brought in due to the potential air leakage, oxygen permeation flux must be refined by removing the leak oxygen, which can be calculated by measuring the leak nitrogen and using the knowing O_2/N_2 ratio of the feed air.

3 Experimental

3.1 Microstructure

Fig.2 shows the cross-section backward-scattering images of tubular membranes deposited by PS-PVD and SAPS. Both membranes show mainly laminar microstructures. A few unmelted powder particles can be observed in PS-PVD membrane (Fig.2c), which led to the formation of pores and cavities (Fig.2b), and more distinct interlaminer cracks also can be found in the membrane. It's thought that these unmelted particles and interlaminer cracks were caused by the use of oversized powder feedstock, because a big particle is not easy to be accelerated and well molten in the expanded plasma jet with lower viscosity and wider temperature distribution. By comparison, SAPS membrane shows a much denser microstructure due to the well spreading of molten droplets with very high impact energy, only very few closed micropores can be observed at low magnification (Fig.2e). Different with PS-PVD membrane, there are few interlaminer cracks but much more dispersed vertical microcracks existing in the SAPS membrane (Fig.2f). It is thought that the vertical microcracks were attribute to the very fast cooling of individual droplet in deposition, the spreading droplet experienced solidification and shrinkage in a very short time and would suffer a high tensile stress which caused the splats break up.

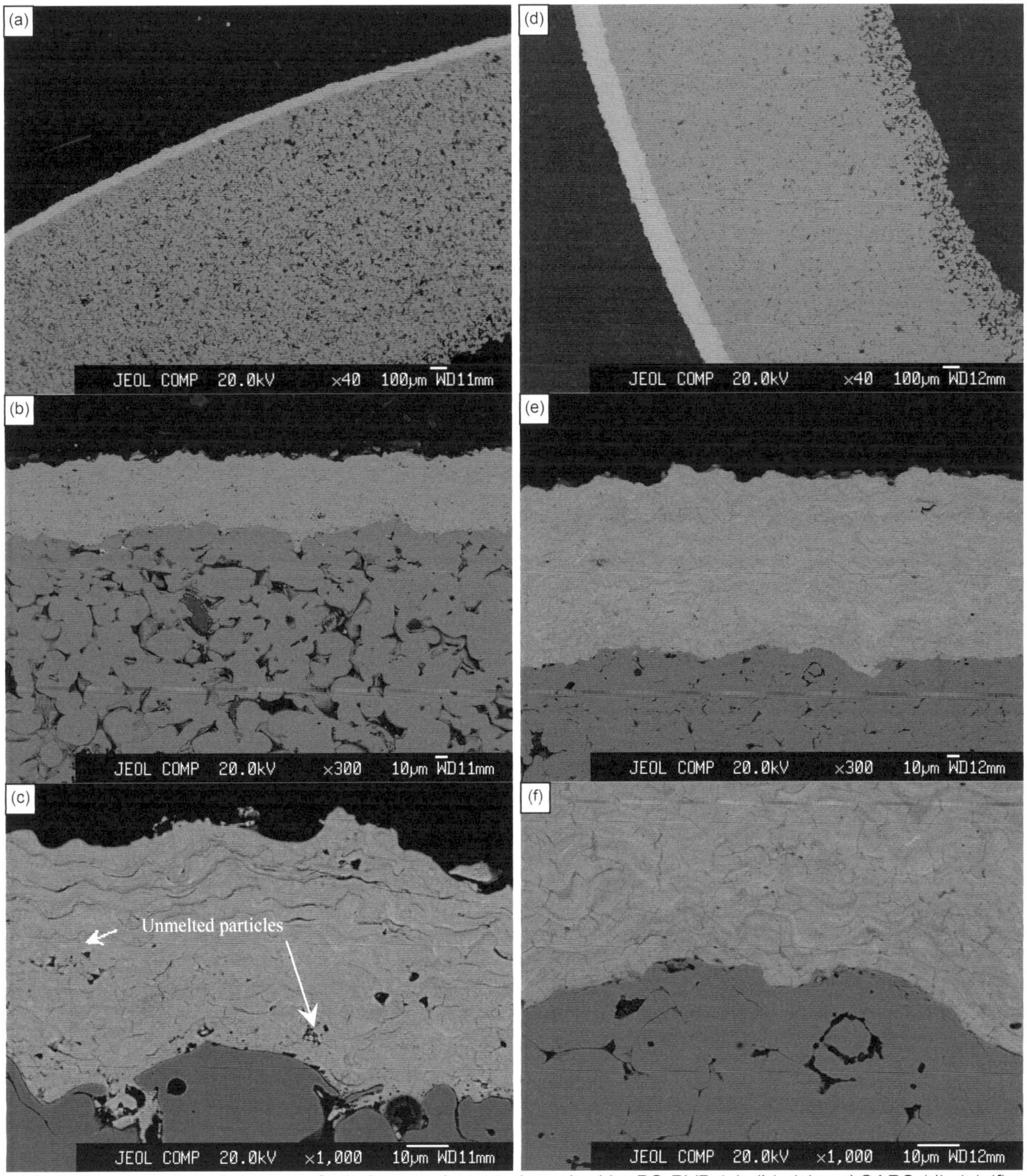

Fig.2. Cross-sectional images of tubular membranes deposited by PS-PVD (a), (b), (c) and SAPS (d), (e),(f)

3.2 Oxygen permeation

Fig.3 shows the oxygen permeation flux and leak gas flow rates of the two membranes at 700℃ and 900℃. The oxygen permeations of both membranes increase as temperature rises. From 700℃ to 900℃, oxygen permeation flux of PS-PVD membrane increases from 2.3×10^{-4}ml/cm^2·min to 6.1×10^{-4}ml/cm^2·min, and that of SAPS membrane also shows a minor increase from 0.21×10^{-4}ml/cm^2·min to 0.34×10^{-4}ml/cm^2·min. Actually the measured oxygen permeation datas are far lower than the LSCF membranes as-sintered in the literature (about 0.2-0.5×10^{-4}ml/cm^2·min at 900℃ [2, 14]). The extremely low oxygen permeation is thought to be caused by the oxidation of metallic supports. As shown in Fig.4, despite swept by helium, the substrates of the two membranes still got oxidized after oxygen permeation test., and the more serious substrate's oxidation of SAPS membrane (Fig.4c, 4d) seems to explain its lower oxygen permeation than PS-PVD membrane. Studies is going on to figure out the much easier oxidation of substrate for SAPS membrane, a lower oxygen permeation temperature for SAPS membrane is given as the initial guess, which means perhaps serious oxidation has occurred before reaching the test temperature.

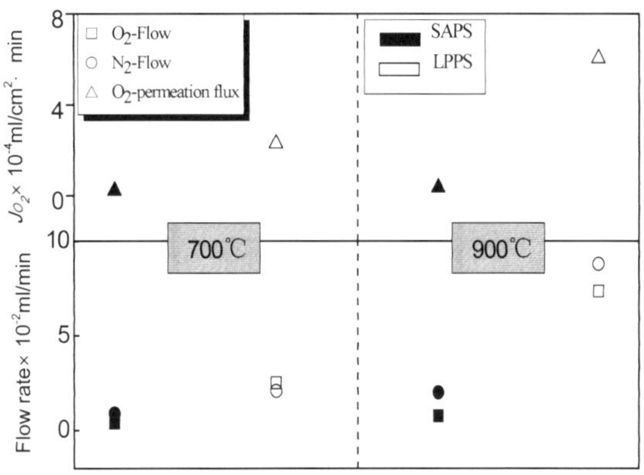

Fig.3. The oxygen permeation flux and leak gas flow rates of tubular membranes at 700℃ and 900℃

The gas leakage of the two membranes also increase with a rise of temperature as seen in Fig.3. From 700 ℃ to 900 ℃, N_2 flow rate of SAPS membrane increases from 0.91×10^{-2}ml/min to 2.0×10^{-2}ml/min, and that of PS-PVD membrane increases from 2.1×10^{-2}ml/min to 8.8×10^{-2}ml/min. The great leakage growth of PS-PVD membrane may be relate to the vertical cracks which is believed to come into existence in heating period because of the tensile thermal stress applied by the metal substrate. For SAPS membrane, few vertical cracks can be observed as the thermal stress can releases by the original formed vertical microcracks in splats, and owing to this, the membrane can keep a lower and relatively constant gas leakge. Truthfully, that's pretty low gas leakages for the two kinds of tubular membranes with such a large size, and this also shows the promising feasibility of the fitting-type sealing.

4 Conclusion

successful preparation of large-size tubular porous metallic supported OTM along with the oxygen permeation tests using a self-designed device were firstly carried out in the work. Although the test results were not high in reliability because of the oxidation of porous substrates, the easier sealing and modular tubular membranes with high strength can still contribute to improving the current bottlenecks of OTMs in application if suitable substrate can be used. In fact, some porous super alloy and ceramic supports have been trying to be used to deposit OTM, and respective investigation is underway.

5 Acknowledgements

The authers acknowledge the financial support from the Natural Science Foundation of China (51174251).

6 Literature

[1] Y Teraoka, T Fukuda, N Miura, et al. Development of oxygen semipermeable membrane using mixed conductive perovskite-type oxides (part 2). J. Ceram. Soc. Jpn, 97(1989), pp. 533/38.

[2] W.Q. Jin, S.G. Li, N.P. Xu, et al. Preparation of an asymmetric perovskite-type membrane and its oxygen permeability. J. membr. Sci., 185(2001), pp. 237/43.

[3] S Baumann, J.M Serra, M.P Lobera, et al. Ultrahigh oxygen permeation flux through supported $Ba_{0.5}Sr_{0.5}Co_{0.8}Fe_{0.2}O_{3-\delta}$ membranes. J. Membr. Sci., 377(2011), pp. 198/205.

[4] M Salehi, F Clemens, E.M Pfaff, et al. A case study of the effect of grain size on the oxygen permeation flux of BSCF disk-shaped membrane fabricated by thermoplastic processing. J. Membr. Sci., 382(2011), pp. 186/93.

[5] S Baumann, W.A Meulenberg, H.P Buchkremer. Manufacturing strategies for asymmetric ceramic membranes for efficient separation of oxygen from air. J. Euro. Ceram. Soc., 33(2013), pp. 1251/61.

[6] K. von Niessen, M. Gindrat, and A. Refke, Vapor phase deposition using plasma spray—PVD, J. Therm. Spray Technol., 19(2010), pp. 502/09.

[7] A. Hospach, G. Mauer, R. Vaßen, et al, Characteristics of ceramic coatings made by thin film low pressure plasma spraying (LPPS-TF), J. Therm. Spray Technol., 21(2012), pp. 435/40.

[8] K. von Niessen and M. Gindrat, Plasma spray-PVD: A new thermal spray process to deposit out of the vapor phase, J. Therm. Spray Technol., 20(2011), pp. 736/43.

[9] M. Gindrat and R. Damani, LPPS hybrid technologies for emerging energy applications—recent developments, International thermal spray conference, 27-30 September 2011, Hamburg.

[10] Zotov N, Hospach A, Mauer. G, et al. Deposition of $La_{1-x}Sr_xFe_{1-y}Co_yO_{3-\delta}$ coatings with different phase compositions and microstructures by low-pressure plasma spraying-thin film (LPPS-TF) processes. J. Therm. Spray Tech., 21(2012), pp. 441/47.
[11] Zotov N, Baumann S, Meulenberg W A, et al. La–Sr–Fe–Co oxygen transport membranes on metal supports deposited by low pressure plasma spraying-physical vapour deposition. J. Membr. Sci., 442(2013), pp. 119/23.
[12] Jarligo M.O, Mauer G, Bram M, et al. Plasma spray physical vapor deposition of $La_{1-x}Sr_xFe_{1-y}Co_yO_{3-\delta}$ thin-film oxygen transport membrane on porous metallic supports. J. Therm. Spary Tech., 23(2014), pp. 213/19.
[13] J. Mao, C.M. Deng, Z.Q. Kuang and et al: Transactions of Materials and Heat Treatment. 36(2015), pp.143/46.
[14] M.P. Lobera, J. M. Serra, S.P. Foghmoes et al. On the use of supported ceria membranes for oxyfuel process/syngas production, J. Membr. Sci., 385-386(2011), pp.154/61.

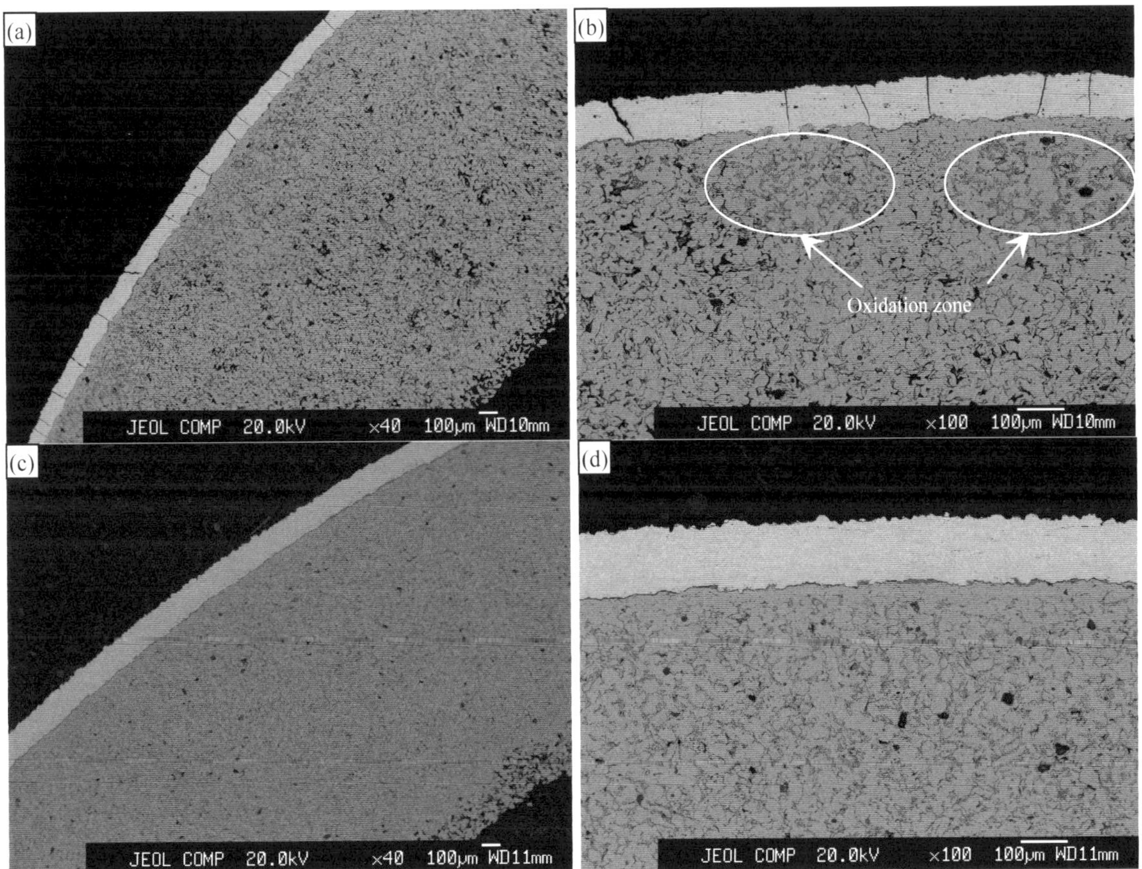

Fig.4. Cross-sectional images of LSCF membranes deposited by PS-PVD (a), (b) and SAPS (c), (d) after oxygen permeation test

Sintering and phase transformation behavior of high-purity nanocrystallized thermal barrier coatings at very high temperatures

Kang Yuan, Yueguang Yu, Xiaojuan Ji, Jie Shen
Beijing General Research Institute of Mining and Metallurgy, Beijing 100160, China

Abstract. High-purity nanocrystallized yttria-stabilized zirconia (YSZ) powders were used to manufacture thermal barrier coatings (TBCs) by air-plasma spray (APS) technique. Comparing to a commercial YSZ (Metco 234A), the high-purity YSZ had a lower content of oxide-impuries (SiO_2, TiO_2, etc.). The nanocrystallized microstructure in the high-purity YSZ poweders was found to be largely degredated after the APS process. After the spray, the coating samples were aged at very high tempertures (1200, 1300 and 1400 °C). The XRD results indicated a tremendous tetragonal-to-monoclic phase transformation in the commercial YSZ coatings aged at 1400 °C for 100 hours, while the high-purity YSZ coating showed much higher phase stability. The sintering behavior of the YSZ coatings and the influence of the MCrAlY oxidation were also discussed.

1 Introduction

The development of moden aero and industrial gas turbines urges high-performance thermal barrier coatings (TBCs) in high-temperature services. As the most commonly used TBC matetial, yttria stabilized zirconia (YSZ) has been applied and developed for several decades. YSZ TBCs may fail during the services because of ceramic sintering or unwised phase-transformation. It is now well established that the clearage of impurities such as SiO_2, TiO_2, and Al_2O_3 from the YSZ materials can significantly decrease the sintering rate therefore to improve the TBCs' toughness [1-3]. The tetragonal-to-cubic/monolic phase transformation is easier to occur with the addition of the oxide-impurities [1-3]. If a YSZ coating with higher purity is used, the coating properties can be highly improved [4].

In this paper, the ageing behaviour of TBCs made by high-purity nanocrystallized YSZ powders was studied at temperatures between 1200 °C and 1400 °C. The samples contained both YSZ layers and MCrAlY layers, therefore the effect of the MCrAlY oxidation on the sintering and phase transformation behavior of the ceramic material can be also investigated, which was rarely seen in previous literatures.

2 Experimentals

The high purity nanocrystallized YSZ powder, namely *YSZ-a*, was produced from BGRIMM (-62+25 μm); the commercial YSZ powder, namely *YSZ-b*, was from Oerlikon Metco company (-90+15 μm). The powders' composition is listed in **Table 1**; *YSZ-a* had a higher purity than *YSZ-b*. The morphology of the YSZ powders is shown in **Fig. 1**. MCrAlY powders had a composition of Ni-23Co-17Cr-12Al-0.5Y (-74+38 μm). The MCrAlY and YSZ powders were alternatively air-plasma sprayed (APS) on a flat stainless steel to form MCrAlY-YSZ layered samples; the typical morphology of the coating samples is presented in **Fig. 2**. The MCrAlY layer was about 70~100 μm thick and the YSZ layer was 300~400 μm thick. The APS spraying parameters used was: Ar flow rate as 38 $L \cdot min^{-1}$, H_2 flow rate as 13 $L \cdot min^{-1}$, plasma power as 45 kW, spraying distance as 100 mm. Some thick MCrAlY samples (~ 1 mm) were produced to get thermal weight gain data of the MCrAlY.

Table 1. Chemical composition (wt.%) of the YSZ powders.

	ZrO_2	Y_2O_3	HfO_2	SiO_2	TiO_2	Al_2O_3	Fe_2O_3	Others
YSZ-a[1]	Bal.	7-9	2.5	0.5	0.2	0.2	0.2	0.8
YSZ-b[2]	Bal.	6-9	~2	0.01	0.01	0.01	0.01	<0.05

[1]according to Metco production manual; [2]measured via X-ray fluorescence method (Fe_2O_3), gravimetric method (ZrO_2), and coupled plasma atomic emission spectroscopy analysis (Y_2O_3, HfO_2, TiO_2, Al_2O_3, and others).

Fig. 1. The cross-sectional morphology of the powders: (a, b) for *YSZ-a* and (c, d) for *YSZ-b*. Figure (b, d) show the inner structure of the powders.

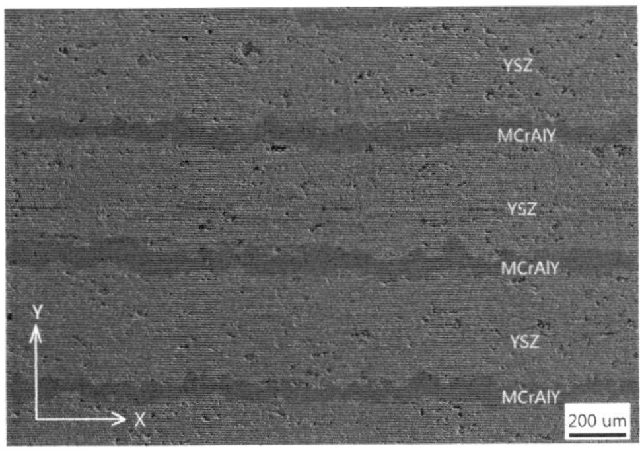

Fig. 2. The cross section of the samples as received showing MCrAlY-YSZ layered structure.

Ageing tests were carried out at 1200, 1300 and 1400 °C for 50 and 100 hours. The thick MCrAlY samples were weighed at 50 and 100 hours using an electronic balance with resolution of 0.0001 g. X-ray diffraction (XRD) measurement was performed on the YSZ sides of coating-layered samples for different ageing temperatures and times. In the XRD measurement, Cu-Ka radiation (λ=1.5406 Å) was applied with theta-2theta scanning between 10-100°. Scanning electron microscope (SEM) and energy dispersive spectroscope (EDS) were used to analyze the microstructure and chemical composition in the materials.

3 Results

3.1 Oxidation of the MCrAlY

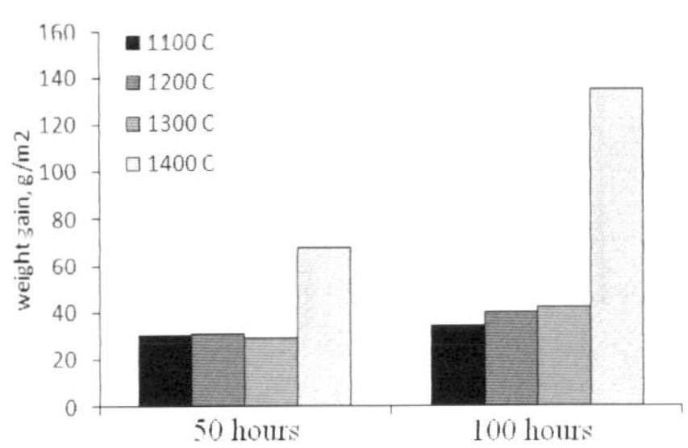

Fig. 3. The weight gain results of the MCrAlY for 50 and 100 ageing hours. The result for 1100 °C was from a measurement on a MCrAlY-coated superalloy disk, which was taken as a reference.

The weight gain results of the MCrAlY at different temperatures are shown in **Fig. 3**. At temperature lower than 1300 °C alumina was the main oxide formed on the thick MCrAlY samples upto 100 hours, while at 1400 °C spinel oxides were formed largely giving the significant higher oxidation rate at 1400 °C. The oxidation behaviour of the MCrAlY layers (~70 µm) in the MCrAlY-YSZ samples, however, was not the same as the thick MCrAlY sample. **Fig. 4** shows the typical morphology of the MCrAlY layers. At 1200 °C Al oxidation was dominant for all the MCrAlY layers (**Fig. 4b**); at 1300 °C mixture oxidation occurred in some MCrAlY layers; at 1400 °C all the MCrAlY layers got mixture-type oxidation (**Fig. 4c**).

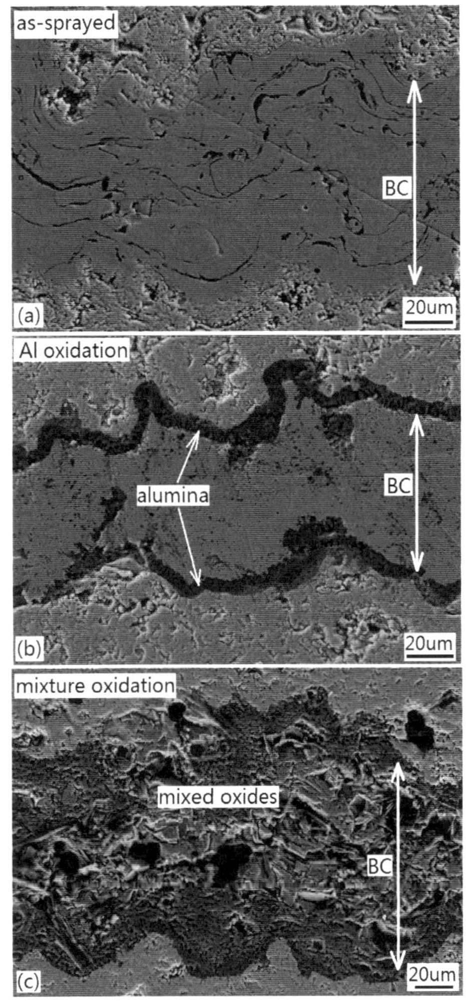

Fig. 4. The cross sections of the MCrAlY coatings: (a) before oxidation (i.e. as-sprayed), (b) at the stage when Al oxidation dominant, and (c) at the mixture oxidation stage.

3.2 Phase transformation in the YSZ

The XRD results in YSZ powders, as-sprayed YSZ coatings and aged YSZ coatings are presented in **Fig. 5-6**. Two phases could be identified in the YSZs, i.e. tetragonal and monoclic phases (t-phase, m-phase). The highest peak of m-phase is particularly denoted in **Fig. 5-6**. An example of the identification of the two phases is shown in **Fig. 7**. Accordingly, some m-phase existed in the powders of both *YSZ-a* and *YSZ-b*. After the APS spray, m-phase seemed to disappear from the coatings. The t-phase was the only detectable phase up to 1300 °C.The t-to-m phase transformation took place at 1400 °C for 100 hours especially in *YSZ-b* coating. **Table 2** lists the phase amount calculated according to the height of the XRD peaks of the two phases. The *YSZ-b* powder had a somewhat higher m-phase fraction than the *YSZ-a*. T YSZ coatings stimulated more t-to-m phase transformation in *YSZ-b* coating at 1400 °C for 100 hours. Such results coincided with the foundings in literatures [1-3], i.e. that a YSZ coating with lower impurity would have higher phase stability.

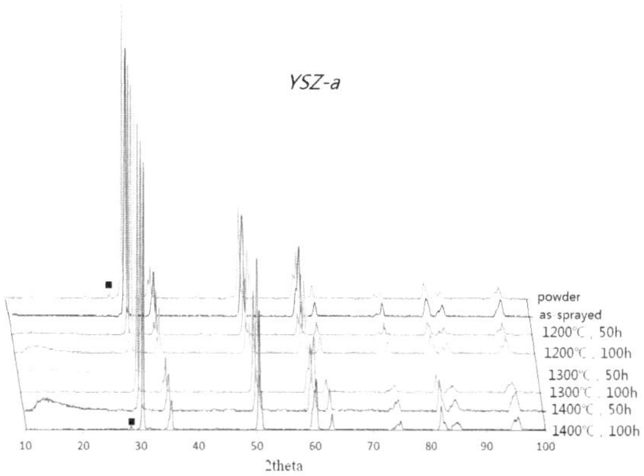

Fig. 5. XRD results of *YSZ-a* powders and coatings. Solid squares denote the highest peak of monoclinic phase.

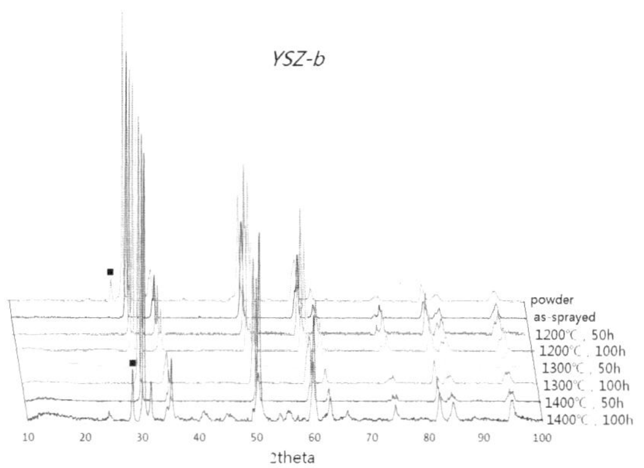

Fig. 6. XRD results of *YSZ-b* powders and coatings. Solid squares denote the highest peak of monoclinic phase.

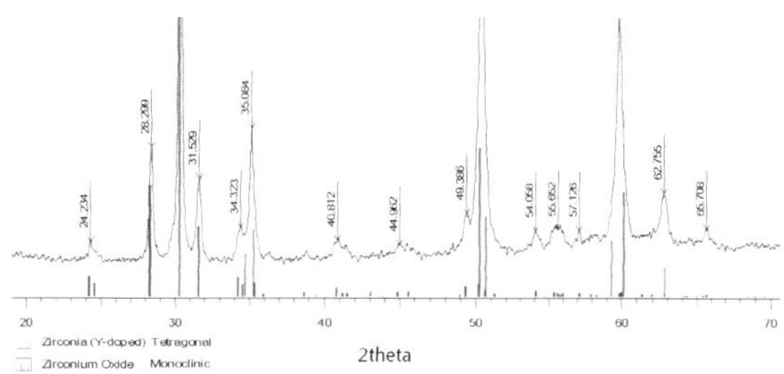

Fig. 7. The identification of tetragonal (red) and monoclinic (blue) phases in the XRD spectrum (*YSZ-b* coating aged at 1400 °C for 100 hours).

Table 2. Phase amount in the YSZ powders and coatings according to XRD results (Monoclic – m; tetragonal - t).

	YSZ-a	*YSZ-b*
Powders	~95% t, 5% m	~84% t, 16% m
As-sprayed	>98% t	>98% t
1200°C,100 h	>98% t	>98% t
1300°C,100 h	>98% t	>98% t
1400°C,50 h	>98% t	>98% t
1400°C,100 h	~96% t, 4% m	~65% t, 35% m

3.2 Sintering behavior of the YSZ coatings

Fig. 8 shows the development of the porosity (including pores and visible cracks) in the YSZ coatings. A decrease of the porosity from the as-sprayed status to the aged status is quite obvious, indicating that sintering occurred in the YSZs. Higher temperature gave heavier sintering effect; the data on 1300C-50h for *YSZ-b* is anomaly high probably due to a problem on sample preparation. The porosity in *YSZ-b* for 1400C-100h condition was high because of the occurrence of occurance of the highly t-to-m transformation and the influence by MCrAlY layers due to the mixture oxidation.

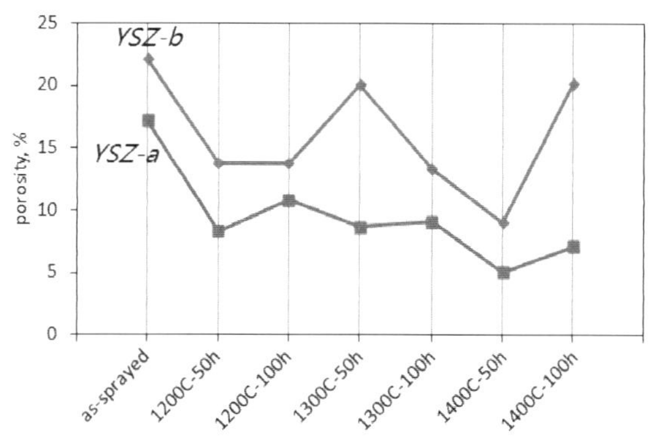

Fig. 8. The development of the porosity of the YSZ coatings. The porosity was measured by image analysing.

The morphology of pores and cracks in *YSZ-a* and *YSZ-b* coatings is presented in **Fig. 9**. *YSZ-b* showed typical splat-on-splat structures and the splats' boundaries (the horizontal cracks) can be clearly seen. In *YSZ-a*, however, horizontal splat-boundary-cracks can not be observed; instead, some network-like cracks were found. The difference between the two coatings should be due to the microstructures owned by the YSZ powders. *YSZ-a* powder had nanocrystalline structures with quite high amount of surface/interfaces. During the APS spray, such nanocrystalline structure would be degraded with extra energy released which could promote the melting of the powders. In this case, *YSZ-a* powders could be fully molten and therefore the splats' boundaries can not be easily found. The network-like cracks were formed due to a shrinkage of the molten powders.

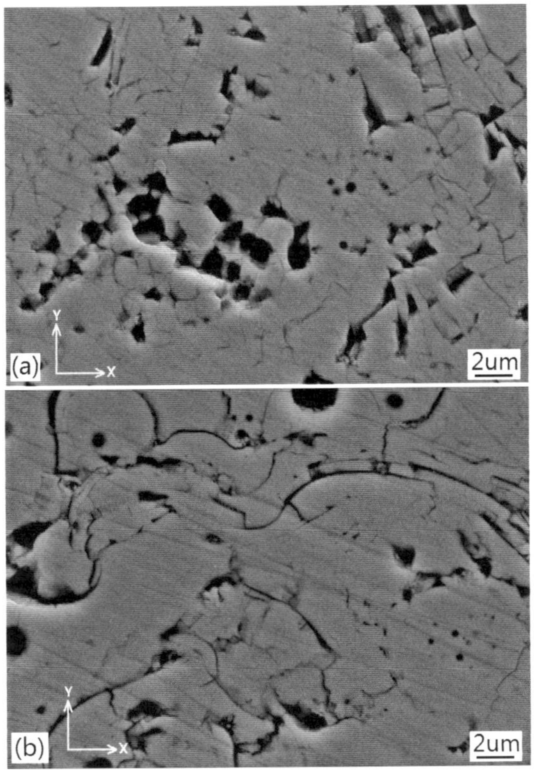

Fig. 9. The cross section of (a) *YSZ-a* and (b) *YSZ-b* coatings as-sprayed.

Fig. 10 and **Fig. 11** compare microstructures in the YSZ coatings before and after the ageing processes. It can be clearly seen that cracks and small pores disappeared due to sintering mechanism. The size of the large pores was also decreased during the ageing process for both *YSZ-a* and *YSZ-b* coatings. Based on the results in **Fig. 8**, it is hard to say the high-purity *YSZ-a* coating performed higher sintering resistance than *YSZ-b*. However, small narrow cracks indeed were found in *YSZ-a* coating while the small cracks in *YSZ-b* coating usually appeared round or pore-like (**Fig. 11**). At 1400 °C, mixture oxidation occurred (both for 50 and 100 hours); the formation of spinels and Ni/Co-oxides caused a large volumetric increase of MCrAlY layer which caused tensile stresses in the neighbouring YSZ layers (both for *YSZ-a* and *YSZ-b*). The vertical cracks or crack-like pores (along the Y direction in **Fig. 12**) should be a result due to the tensile stresses. The *YSZ-b* coating at 1400 °C for 100 hours was an exception since the t-to-m phase-transformation would cause material shrinkage which diluted the influence of the MCrAlY oxidation.

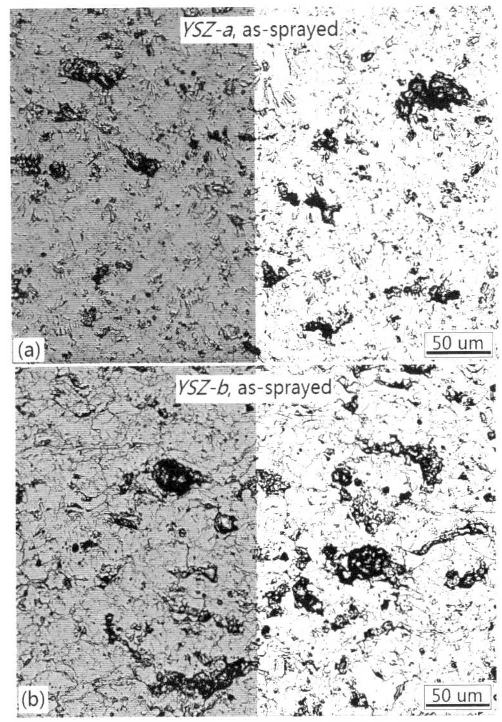

Fig. 10. The cross section of (a) YSZ-a and (b) YSZ-b coatings as-sprayed. Half of the images is transformed into black and white showing pores and cracks in black.

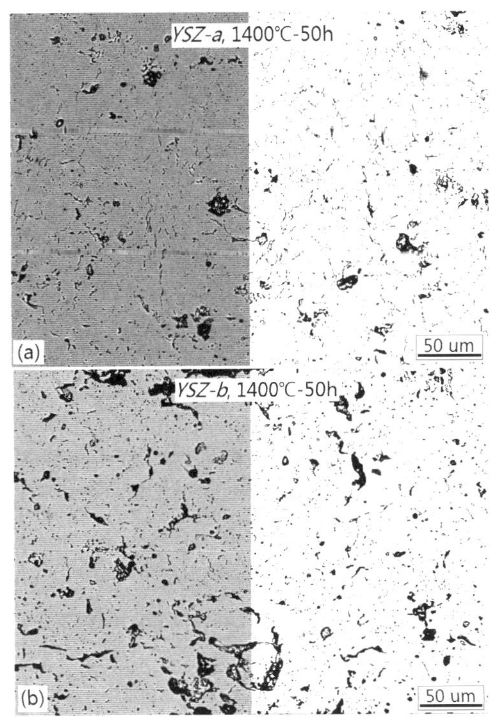

Fig. 11. The cross section of (a) *YSZ-a* and (b) *YSZ-b* coatings aged at 1400 °C for 50 hours. Half of the

images is transformed into black and white showing pores and cracks in black.

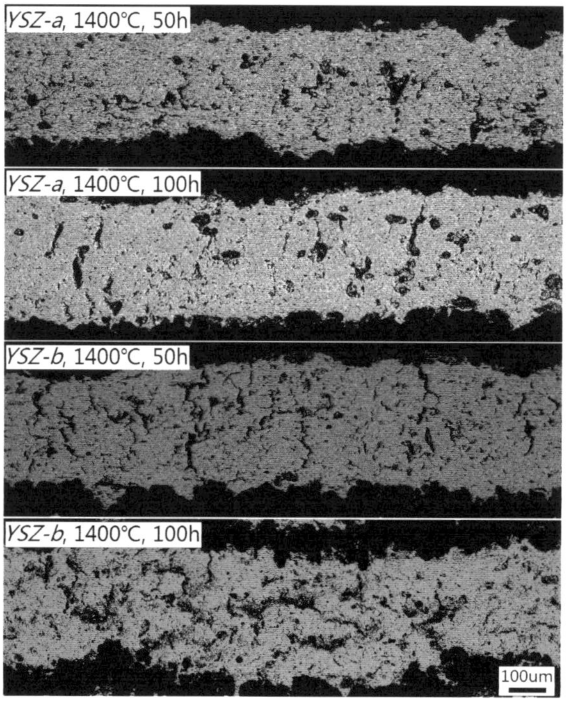

Fig. 12. Overview of the coatings' cross sections aged at 1400 °C. The large cracks or the crack-like pores were mostly along Y direction, indicating a tensile stress along X direction in the YSZ layers.

4 Discussion

The coatings made by high-purity nanocrystallized YSZ (i.e *YSZ-a*) powders showed higher resistance against tetragonal-to-monoclic phase transformation at high temperatures. That mainly attributed to the lower impurity of *YSZ-a* [1-3]. The phase transformation (> 30%) in the low-impurity *YSZ-b* coatings seemed to induce a significant shrinkage of the coating (**Fig. 8**). It had been reported that around 3-5% shrinkage would occur due to such phase transformation [5].
The oxidation of the MCrAlY coatings had a important effect on the sintering behavior of the YSZ coatings. The formation of alumina layer on MCrAlY gave no large effect. However, when the MCrAlY coating lost its Al-oxidation resistance, spinel and Ni/Co-oxides exploresively formed which brought a great volumetric increase in the MCrAlY layers. That caused tensile stresses in the YSZ layers, which promoted the formation of enlongated pores/cracks normal to the MCrAlY-YSZ interfaces.
The high-purity *YSZ-a* coating should have performed a higher sintering resistance than *YSZ-b* [1-3], which however was not clearly observed according to the porosity analysis. That because the sintering behavior of the YSZs was also highly influenced by phase transformation in the YSZs and the oxidation of MCrAlY layers in this study. Eventhough, some results indeed indicated a higher sintering resistance of the high-purity *YSZ-a*, e.g. that more narrow cracks were observed in *YSZ-a* coating (**Fig. 11**).
An interesting phenomenon observed in this study was the disappeance of the nanocrystalline structures of the powders after the manufacturing of the *YSZ-a* coatings. That indicated a potencial of the nanocrystallized YSZ powders in an application of low-energy spray for producing thermal barrier coatings. Some work will be done in future.

5 Conclusions

In this stury, high-purity nanocrystallized YSZ powders were used to manufacture MCrAlY-YSZ coatings. It was found that the high-purity YSZ coating had higher sintering resistance and phase-transformation resistance than a commercial YSZ coating. The oxidation of MCrAlY coatings can strongly affect the sintering behavior of the YSZ coatings when the MCrAlY lost its Al-oxidation resistance.

6 References

[1] S. Paul, A. Cipitria, I. Golosnoy, L. Xie, M. Dorfman, and T. Clyne, "Effects of impurity content on the sintering characteristics of plasma-sprayed zirconia," *Journal of Thermal Spray Technology,* vol. 16, pp. 798-803, 2007.

[2] S. Hodgson, J. Cawley, and M. Clubley, "The role of SiO 2 impurities in the microstructure and properties of Y–TZP," *Journal of materials processing technology,* vol. 86, pp. 139-145, 1999.

[3] S. Hodgson, J. Cawley, and M. Clubley, "The role of Al 2 O 3 impurities on the microstructure and properties of Y-TZP," *Journal of Materials Processing Technology,* vol. 92, pp. 85-90, 1999.

[4] M. Helminiak, N. Yanar, F. Pettit, T. Taylor, and G. Meier, "The behavior of high-purity, low-density air plasma sprayed thermal barrier coatings," *Surface and Coatings Technology,* vol. 204, pp. 793-796, 2009.

[5] S. Bose, *High Temperature Coatings.* Oxford: Butterworth-Heinemann, 2007.

Synthesis of Y_2O_3 films with high thermal-shock resistance from a metal–EDTA complex using flame-spraying apparatus

Keiji Komatsu*, Ayumu Toyama, Tetsuo Sekiya, Tomoyuki Shirai, Atsushi Nakamura, Ikumi Toda , Shigeo Ohshio, Hiroyuki Muramatsu, Hidetoshi Saitoh

In this study, yttrium oxide (Y_2O_3) films with high thermal-shock resistance were synthesized from a metal–ethylenediaminetetraacetic (metal–EDTA) complex by employing a H_2–O_2 combustion flame. A rotation apparatus and various cooling agents (compressed air, liquid nitrogen, and atomized purified-water) were used during the synthesis to control the thermal history during film deposition. An EDTA•Y•H complex was prepared and used as the staring material for the synthesis of Y_2O_3 films with a flame-spraying apparatus. Although thermally extreme environments were employed during the synthesis, all of the obtained Y_2O_3 films showed only a few cracks and minor peeling in their microstructures. For instance, the Y_2O_3 film synthesized using the rotation apparatus with water atomization units exhibited a porosity of 22.8%. The maximum film temperature immediately after deposition was 453°C owing to the high heat of evaporation of water.

1 Introduction

We propose here a method using a flame-spraying apparatus to synthesize metal oxides from a metal–ethylenediaminetetraacetic (EDTA) complex as a starting material [1–4]. The oxides synthesized using the metal–EDTA complex have uniform stoichiometry [5–8]. Compared to methods involving chemical vapor deposition (CVD) (e.g., atmospheric CVD [9]), physical vapor deposition (PVD) (e.g., RF magnetron sputtering [10]), and thermal spraying (e.g., plasma spraying [11,12]), the proposed synthesis is based on a unique deposition process that includes both CVD and PVD, as described below.

The deposition of metal oxide in the proposed synthesis involves two mechanisms, namely, chemical reaction and physical collision. It begins with a chemical reaction in which the starting material, metal–EDTA complex, is decomposed and oxidized to form metal-oxide particles. Therefore, the properties of the deposited film depend on the temperature and velocity of the moving particles. It is the physical collision of the metal-oxide particles with the substrate, however, that forms a film as the accumulated layer of incident particles solidify. Although a higher substrate temperature would improve the properties of the deposited film, it tends to damage the substrate. For example, when combustion gases with higher calorific values are used to synthesize dense films with excellent properties, the combustion flame could melt the substrate. In one such synthesis, we used hydrogen gas with a flow rate of 0.048 g/s, and heat was generated at an estimated rate of 0.68 kJ/s. Because the time of flight for the moving particles was 1.3×10^{-3} s, it was difficult to melt a metal oxide. In contrast, the energy required to decompose a metal–EDTA powder is 5.30×10^{-17} kJ. Therefore, the complex could be easily decomposed by the combustion gases. In the case of a H_2–O_2 flame, this required dissipating approximately 1000 J of heat into the surroundings. Hence, the substrate temperature would be a fundamental hindrance to the use of this method in commercial applications such as thermal-barrier coatings (TBCs). In TBCs, alloy layers are set between ceramics shielding the layer and substrate. The alloy layers aid heat-stress formation between materials because of differences in their thermal-expansion coefficients [13,14]. As the changes in temperature of a material result in cracking, destruction, or peeling, the effects of thermal history on a metal-oxide film synthesized from a metal–EDTA complex using a flame-spraying apparatus are determined by fundamental parameters, as in the case of TBC application.

In this study, we designed experimental conditions for controlling the thermal history during synthesis. Yttrium (Y^{3+}) ions were selected to form the metal–EDTA complexes, and metal-oxide films were deposited on a stainless steel (SUS) substrate with a H_2–O_2 combustion flame. First, we developed a rotation apparatus to avoid overheating the substrates during deposition. Next, various cooling units—compressed air, liquid nitrogen, and atomized water—were used during the synthesis to reduce the substrate temperature. To control the microstructure of the deposited films, we attempted to manipulate the films' thermal history by changing the cooling units. This is because the deposition processes (cooling and solidification) during the synthesis could be varied according to experimental conditions. First, we measured the in situ temperature on a substrate during deposition for digitized records of the thermal history. The microstructures, elemental distributions, crystal structure, and organic chemical composition of the deposited films were then investigated.

2 Experimental

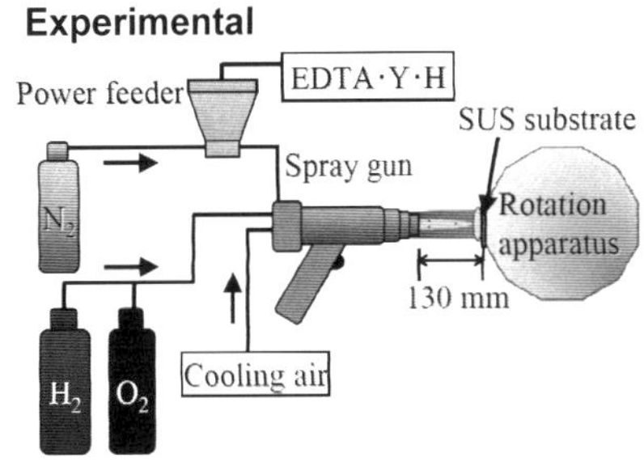

Fig. 1. Schematic illustration of steps involved in the deposition of oxide films onto substrates placed in a rotation apparatus. A metal–EDTA complex is used as the source material.

We first prepared EDTA•Y•H (Chubu Chelest Co., Ltd.) for the deposition of Y_2O_3 films. A conventional flame-spraying system consisting of a feeding unit (5MPE, Sulzer Metco) and a spray gun (6P-II, Sulzer Metco) was used to perform reactive spraying. Figure 1 shows a schematic illustration of the steps involved in the deposition of oxide films by using a rotation apparatus and the metal–EDTA complex. The rotation apparatus with the capability of depositing twelve samples at a time was fabricated in-house. Mortar was used in the apparatus for rolling. The rotation velocity of the apparatus was kept within the range of 45–90 rpm by controlling the applied voltage of a variable autotransformer. EDTA•Y•H was placed in the feeding unit and transported by N_2 gas to the spray gun; at this point, the vibrator in the feeding unit was turned on. A mixture of H_2 and O_2, with the gases flowing at rates of 32.6 and 43.0 L/min, respectively, was used as the flame gas. EDTA•Y•H powder was introduced into the flame, resulting in the thermal decomposition of EDTA and oxidation of the metal yttrium, thereby forming its oxide.

Next, we introduced four different cooling units, which all cooled the rotation apparatus from above. Table 1 lists the experimental conditions for Samples 1–4. To investigate the effects of the cooling units, we used only the rotation apparatus at 45 rpm for the control, Sample 1. For Sample 2, the rotation apparatus at 45 rpm and a compressed air unit were used. Compressed air was driven over the sample by an air nozzle. The distance between the air nozzle and substrate was 50 mm, and the airflow rate was 100 L/min. For Sample 3, a hand-made SUS container (350 × 240 × 240 mm3) filled with liquid nitrogen was used. The rotation apparatus at 45 rpm was dipped into the liquid nitrogen container. For Sample 4, purified-water atomization units were used with the rotation apparatus at 45 rpm as cooling units. The deposition time for each sample was 330 s.

We measured the substrate temperature under various experimental conditions, as described below. A K-type thermocouple was used to measure the in situ temperature of the substrates. First, the thermocouple was combined with a data logger (EL-USB-TC-LCD, MK Scientific). Next, the thermocouple was inserted in a hole bored into the substrate. The thermocouple was fixed to the substrate by a bolt to measure the in situ substrate temperature. Temperature profiles were obtained in each experiment using analysis software.

X-ray diffractometry (XRD; M03XHF22, MAC Science) was conducted using Cu Kα radiation to observe the crystal structure of the samples. The particle-size distribution analysis for raw metal–EDTA complexes was conducted using a particle-size analyzer (MT-3300, Nikkiso). The surface and cross-sectional morphologies of the films were observed by field-emission scanning electron microscopy (FE-SEM; JSM-6700F, JEOL). The elements contained in the samples and their distributions were estimated by energy-dispersive X-ray spectroscopy (EDX) combined with FE-SEM.

Table 1. Experimental conditions employed during the synthesis of Samples 1–4.

Sample ID	Experimental condition
Sample 1	45 rpm rotation
Sample 2	45 rpm rotation +compressed air
Sample 3	45 rpm rotation +liquid nitrogen
Sample 4	45 rpm rotation + purified water atomization

3 Results and discussion

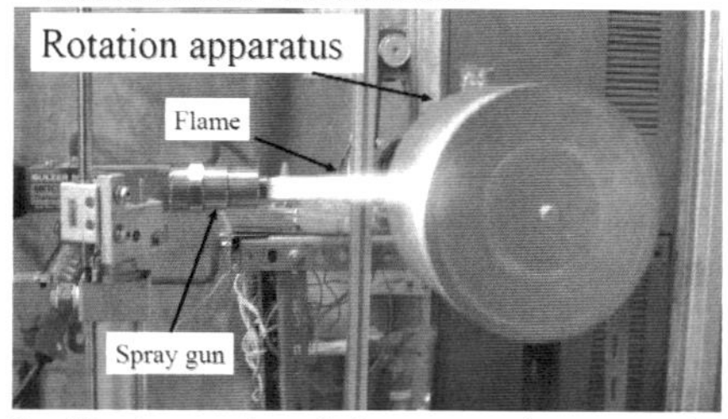

Fig. 2. Photograph of the experiment using the rotation apparatus with a commercial flame-spraying apparatus and a H_2–O_2 flame.

Figure 2 shows a photograph of the experiment performed using the rotation apparatus with a commercial flame-spraying apparatus and a H_2–O_2 flame. The color of the H_2–O_2 flame was orange. The torch length was approximately 30 cm. After the introduction of EDTA•Y•H in the H_2–O_2 flame, the flame became clear and white. The measured feeding rate for EDTA•Y•H was approximately 5.0 g/min. During deposition, the combustion flame came into contact with the rotation apparatus. Thereafter, the red heat phenomenon did not occur on the substrate.

We measured the temperature profiles on the substrate during deposition using the thermocouple system. The measured temperature profiles were equivalent to the thermal history of the samples, and we attempted to estimate the thermal history for each sample by a quantitative approach. For the temperature measurement of objects, the thermo-couple system has a higher spatial resolution than that of a radiation thermometer [15]. However, the radiation thermometer system has an advantage in terms of the surface area of objects in that the reading can be taken with the naked eye [16]. Figure 3 shows the temperature profiles that displayed no dependence on the rotational speeds of 45, 60, and 90 rpm. The substrate temperature (Ts) was increased at a rate of 10°C/s until 40 s after deposition, and thereafter increased at a rate of 0.5°C/s. Subsequently, the substrate reached the thermal equilibrium state, as shown by the flattening of the temperature profile. At 330 s after deposition, the substrate temperature was approximately 620°C. The air-cooling effect resulting from the increased number of rotations during deposition was small compared to the heating effect of the H_2–O_2 flame. As a result, Ts was constant at rotational speeds of 40–90 rpm.

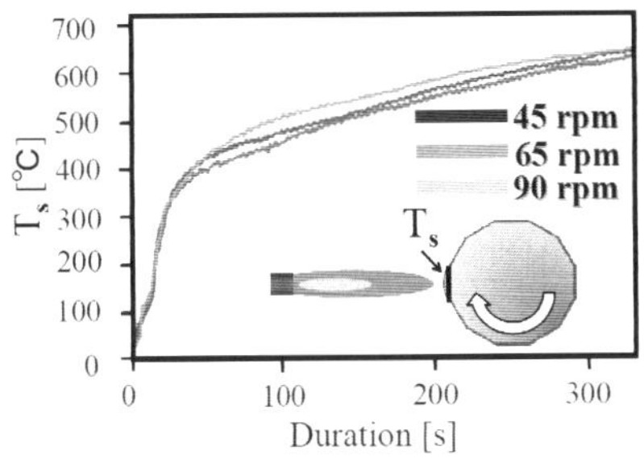

Fig. 3. Temperature profiles of samples prepared using various rotational speeds: 45, 60, and 90 rpm..

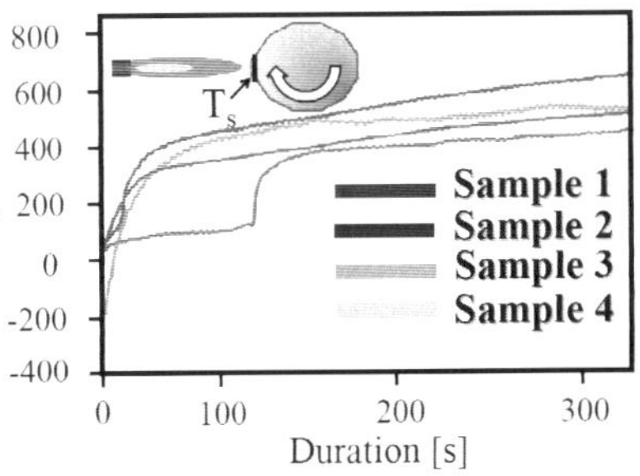

Fig. 4. Temperature profiles of Samples 1–4 prepared under various experimental conditions.

Figure 4 shows the temperature profiles under various experimental conditions for Samples 1-4. In the high-velocity oxy-fuel (HVOF) deposition process, liquid-nitrogen fluxes (refrigeration) are used to cool the deposited films on samples [17]. We introduced four different cooling units in this study, and the obtained temperature profiles of the substrates were different for each cooling unit. Furthermore, the temperature gradients and substrate temperature at 330 s (T_{330s}; just after deposition) were different. First, Ts of Sample 1 increased at a rate of 10°C/s until 30 s after deposition and at 1°C/s thereafter; T_{330s} of Sample 1 was 646°C. Next, the rotation apparatus at 45 rpm and compressed air were introduced for Sample 2. The Ts value of Sample 2 also increased at a rate of 10°C/s until 30 s after deposition and at 1°C/s thereafter; T_{330s} of Sample 2 was 523°C. The heating rate for Sample 3 was 10°C/s until 30 s after deposition, and it was 1°C/s when the rotation apparatus at 45 rpm and liquid nitrogen were used. The substrate temperature was -189°C at the start of deposition; T_{330s} of Sample 3 was 519°C. The temperature of Sample 4 gradually increased at a rate of 1°C/s until 100 s after deposition. In addition, Ts rapidly increased to approximately 300°C in 100 s after deposition and increased at 1°C/s thereafter; T_{330s} of Sample 4 was 453°C. Figure 5 shows the substrate temperatures of Samples 1–4 at 330 s (just after film deposition). The highest T_{330s} value was obtained during the synthesis using the rotation apparatus. In contrast, the lowest T_{330s} value was obtained during the synthesis using the rotation apparatus and atomization of purified water. These results could be attributed to the Leidenfrost effect on substrates. A significant amount of water in contact with a substrate can produce an insulating vapor layer that would prevent the liquid from boiling rapidly during deposition. In other words, water with a high heat of evaporation (2250 kJ/kg) can absorb heat on a substrate quickly [18]. Consequently, the thermal histories for Samples 1–4 were different under different experimental conditions.

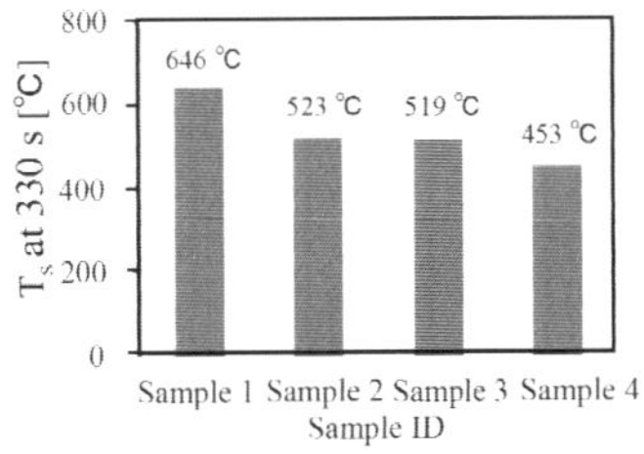

Fig. 5. Substrate temperature of Samples 1–4 at 330 s (just after film deposition).

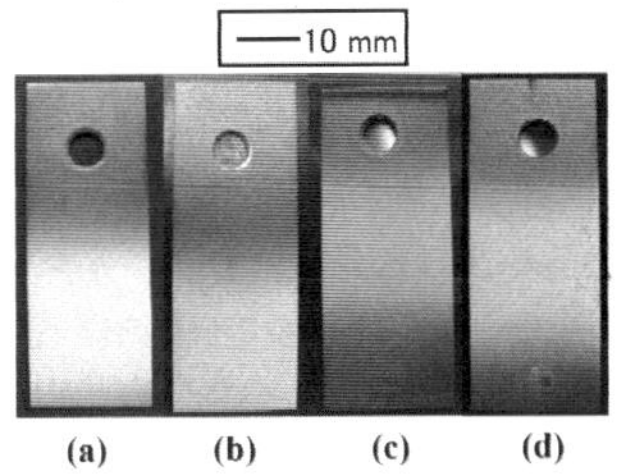

Fig. 6. Photographs of films synthesized on SUS304 substrates using the flame-spraying apparatus: (a) Sample 1; (b) Sample 2; (c) Sample 3; (d) Sample 4.

Figure 6 shows the photographs of films synthesized on SUS304 substrates, Samples 1–¬4. The color of the blasted SUS substrate was light gray. In this study, we employed thermally extreme environments during deposition. After deposition, the substrate color changed to white without any cracking or peeling, thus confirming the existence of a film on the SUS substrate. From another perspective, for example, Sample 3 was exposed to a H_2–O_2 flame (~2700°C [19]) and liquid nitrogen (-189°C) within a cycle. These different temperatures are equivalent to those obtained in cyclic thermal-exposure testing with hard temperature swings. The temperature range in this test was approximately 20–1100°C (where 20°C ≈ room temperature). After thermal cycling, materials such as those for TBCs show many areas with cracking and peeling in their microstructures [20–22]. The deposited films in this study did not show cracking or peeling, however, indicating that the deposited films had high thermal-shock resistance.

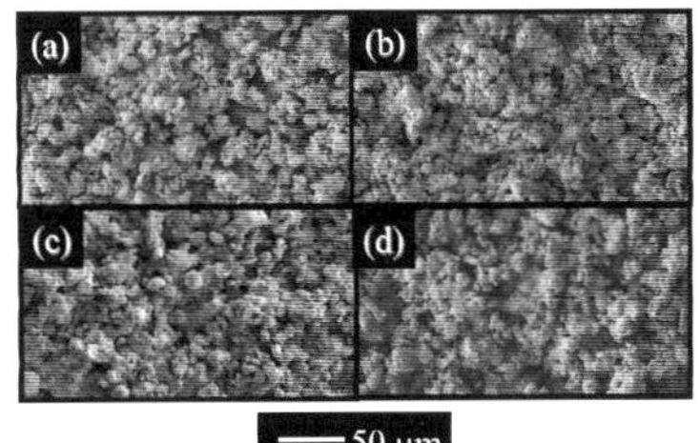

Fig. 7. Surface SEM images of films synthesized on SUS304 substrates using the flame-spraying apparatus: (a) Sample 1; (b) Sample 2; (c) Sample 3; (d) Sample 4.

Figure 7 shows the surface SEM images of films synthesized on SUS304 substrates (Samples 1–4) using the flame-spraying apparatus. Structures consisting of air bubbles amidst spherical powder particles, with feature sizes on the scale of several tens of micrometers, and mostly flat (splatting) morphologies were observed in Samples 1–4. In particular, when compared to Samples 1–3, there were fewer spherical powder particles and bubbles in Sample 4.

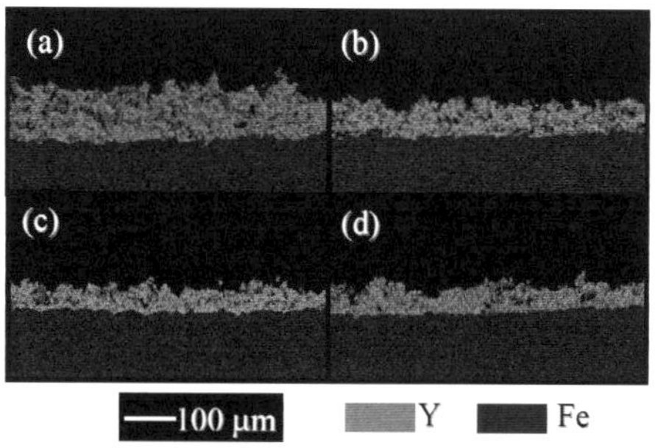

Fig. 8. Cross-sectional EDX images of films synthesized on SUS304 substrates using the flame-spraying apparatus: (a) Sample 1; (b) Sample 2; (c) Sample 3; (d) Sample 4. The white points indicate erbium, and the gray points indicate iron.; (d) Sample 4.

Figure 8 shows the cross-sectional elemental-mapping images of films synthesized on SUS304 substrates (Samples 1–4) using the flame-spraying apparatus. None of the samples showed cracking or peeling in its microstructure; however, the existence of a few pores was confirmed. We estimated the film thickness and porosity of Samples 1–4 from their EDX images. Table 2 shows the estimated values for film thickness, cross-sectional porosity, and degree of flatness for films synthesized on SUS304 substrates. Sample 3 exhibited the thinnest film (36.9 μm). In contrast, Sample 4 exhibited the densest film with a cross-sectional porosity of 22.8%. As the obtained films showed no cracking or peeling, the estimated porosities correspond to open pores existing in the deposited films. A comparison of the films revealed that Sample 3 had the highest degree of flatness, obtained when liquid nitrogen, the lowest-temperature cooling agent in this study, was used. In addition, homogeneous elemental distributions of Y atoms in the films were confirmed. These results indicate that the proposed synthesis route resulted in the formation of a dense microstructure with high thermal-shock resistance.

Table 2. Estimated values of film thickness, cross-sectional porosity, and degree of flatness for films synthesized on SUS304 substrates.

Sample ID	Film thickness [μm]	Cross-sectional Porosity [%]	Degree of flatness[%]
Sample 1	89.1	33.1	58.7
Sample 2	53.4	30.2	64.9
Sample 3	36.9	36.4	70.9
Sample 4	42.7	22.8	65.8

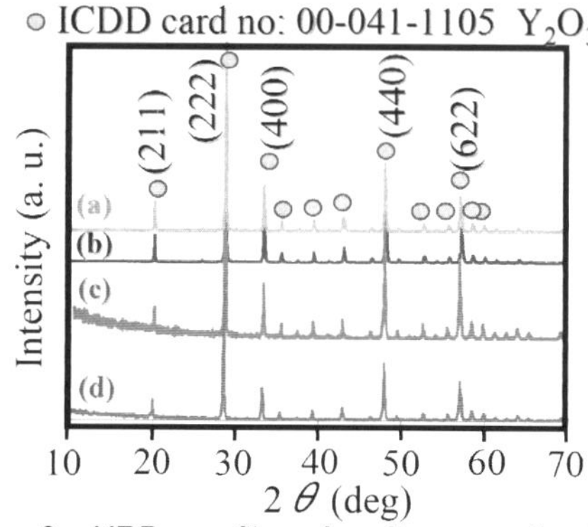

Fig. 9. XRD profiles for films synthesized on SUS304 substrates using the flame-spraying apparatus: (a) Sample 1; (b) Sample 2; (c) Sample 3; (d) Sample 4.

Figure 9 shows the XRD profiles for films synthesized on SUS304 substrates (Samples 1–4) using the flame-spraying apparatus. The diffraction peak positions and intensities were almost identical for the samples, and none of the obtained diffraction peaks showed any shift. The existence of Y_2O_3 crystals with a cubic system was confirmed by identification of the peaks against files from the International Centre of Diffraction Data (ICDD card No. 00-041-1105). The crystallite diameters estimated from the (222) diffraction peaks for Samples 1–4 are summarized in Table 3, with Sample 4 exhibiting the smallest crystallite size. These results suggest that the thermal history during deposition had a small effect on film crystallinity. In the Y_2O_3–Er_2O_3 system [23], the melting point for Y_2O_3 is 2425°C; moreover, cubic Y_2O_3 forms below 2380°C. The theoretical combustion flame temperature for the H_2–O_2 flame (~2700°C) suggests that cubic Y_2O_3 crystals were formed as a result of the decomposition and oxidation of the EDTA•Y•H complex. The enthalpy values reported in the literature for Y_2O_3 provide a proof for the formation of cubic Y_2O_3 crystals and support this theory [24]. Y_2O_3 has a polymorphic character. Both cubic and monoclinic Y_2O_3 crystals exist, with surface enthalpies of 1.66 ± 0.14 and 2.78 ± 0.49 J/m2, respectively. Thus, the formation rate of cubic Y_2O_3 was increased from a thermodynamic perspective.

Table 3. Estimated crystallite diameter for Samples 1–4.

Sample ID	Crystallite diameter [nm]
Sample 1	635.8
Sample 2	587.3
Sample 3	605.3
Sample 4	579.9

Figure 10 shows the FT-IR spectra for Samples 1–4; the spectrum for the EDTA•Y•H complex is shown for comparison. The EDTA•Y•H complex exhibited peaks at 1440 and 1500–1700 cm^{-1}, which are assigned to carbonyl functional groups. Moreover, the broad peak at 2900 cm-1 in the IR spectrum indicates the existence of carbon–hydrogen bonds and hydroxyl groups. In addition, an absorption peak was observed at approximately 3500 cm^{-1}, which is assigned to the hydroxyl (–OH) group stretching mode [1]. All of the obtained Y_2O_3 films exhibited no IR peaks, confirming that there were only a few organic constituents. These results indicate that the EDTA•Y•H complex was successfully decomposed and oxidized for the synthesis of each sample.

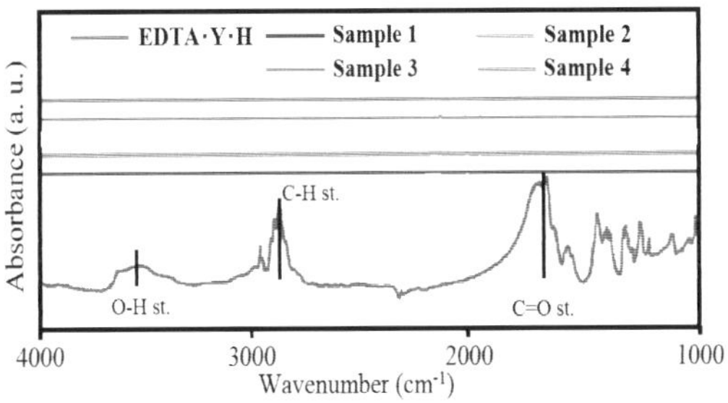

Fig. 10. FT-IR spectra for Samples 1–4, shown in comparison with that of the EDTA•Y•H complex.

The deposited Y_2O_3 films had high thermal-shock resistance, and despite their thermal histories with extreme temperature swings, none of the obtained Y_2O_3 films showed any cracking or peeling in their microstructures. We performed a quantitative analysis of the obtained results, and our observations are discussed below. First, the differences in the thermal histories of Samples 1–4 (Fig. 4) could have originated from the specific heat capacities of the samples. Using the respective specific heat capacities of 1.005, 1.034, and 4.178 kJ/(kg•K) for air, nitrogen, and water, we estimated the heat absorbed by each sample from the substrate. Then, we assumed the weight of the SUS substrate was 1 g and the temperature difference during deposition was 500°C. The estimated values of heat absorbed by Samples 2, 3, and 4 were 502, 517, and 2089 J, respectively. Indeed, T_{330s} of Sample 4 was 453°C when atomized water was used for cooling. The film properties of Samples 1–4 can be evaluated from the aspect of heat-transfer velocity. We estimated the heat-transfer velocity (q) by using the following formula [25]:

(1)

where λ is the specific thermal conductivity, A is the heat transfer area, ΔT is the temperature difference during deposition, and t is the substrate thickness. We assumed λ = 15.9 W/(m•K). Before deposition, the H_2–O_2 flame temperature was 2700°C, while the substrate temperature immediately after film deposition was 646, 523, 519, and 453°C for Samples 1–4, respectively, as shown in Fig. 5. Therefore, the estimated values of q were 7.76, 8.22, 8.24, and 8.49 kcal/s for Samples 1–4, respectively, with Sample 4 showing the highest q value. These results clearly indicate that the deposited Y_2O_3 films had high thermal-shock resistance.

Using the above estimated values of heat absorbed by each sample, we can explain the changes in the Y_2O_3 film microstructure as follows. In this study, the densest Y_2O_3 film with 22.8% porosity was synthesized on the SUS substrate that was cooled with atomized water (Sample 4). During the synthesis of this film, the metal–EDTA complex was decomposed to form metal-oxide particles. Y_2O_3 particles were then introduced in a high-temperature (>2000°C) region where they collided with the substrate surface and formed the film on the substrate upon rapid cooling (solidification). Various morphologies (e.g., splatted, spherical) were observed among the deposited metal-oxide particles, as shown in Fig. 7, suggesting that the film morphologies were related to the solidification rate during synthesis. Thus, the solidification rate is expected to be a function of the substrate temperature: a lower the substrate temperature results in a faster solidification rate; in contrast, a higher substrate temperature results in a slower rate. For instance, by increasing the solidification rate, a Y_2O_3 film with more splatted morphology was synthesized without the particles dispersing into the surroundings during a short deposition period. We can deduce from this result that Sample 4 had a dense microstructure with 22.8% porosity as a result of the fast solidification rate when atomized water was used. In other words, significant heat absorption from the substrate occurred owing to a high heat of evaporation in the cooling water. In contrast, Samples 2 and 3 showed relatively more porous microstructures with 30.2–36.4% porosity because of a lower solidification rate when other cooling units (air compression and liquid nitrogen) were used. These results indicate that the film microstructure could be varied by controlling the thermal history of the substrate during synthesis. In addition, the obtained film microstructure also affected the physical properties and behavior of the film. For instance, the surface-sweeping effect resulting from compressed air was a possible phenomenon. The air-compression pressure would have removed particles with low adherability from the deposited Y_2O_3 films and only the spherical powder particles with air bubble morphologies remained attached. Therefore, this suggests that the resulting Y_2O_3 films consisted of particles with high adherability.

4 Conclusion

Y_2O_3 films with high thermal shock resistance were synthesized from a metal–EDTA complex using a flame-spraying apparatus. The thermal histories of the deposited Y_2O_3 films were controlled by using a rotation apparatus and cooling units of compressed air, liquid nitrogen, or atomized water. The existence of Y_2O_3 crystals in the films was confirmed, and none of the obtained Y_2O_3 films showed cracking and peeling in their microstructures. It was determined that differences in the thermal history of each sample had an influence on its porosity. These results indicate that Y_2O_3 films with high thermal-shock resistance could be synthesized by the proposed method.

5 References

[1] H. Akasaka, M. Ohto, Y. Hasebe, A. Nakamura, S. Ohshio, H. Saitoh, Yttria coating synthesized by reactive flame spray process using Y-EDTA complex, Surf. Coat. Tech. 205 (2012) 3877–3880.

[2] K. Komatsu, Y. Hasebe, T. Sekiya, A. Nakamura, S. Ohshio, H. Akasaka, I. Toda, H. Saitoh, Structural analysis of TiO_2 particles synthesized from Ti-EDTA•NH_4 complex, in press.

[3] K. Komatsu, T. Tsuchiya, Y. Hasebe, T. Sekiya, Y. Hasebe, A. Toyama, A. Nakamura, H. Akasaka, H. Saitoh, Synthesis of metal oxide particles using reaction route from rare-earth metal-EDTA complexes, J. Therm. Spray Tech, (in press).

[4] K. Komatsu, T. Sekiya, A. Toyama, Y. Hasebe, A. Nakamura, M. Noguchi, Y. Li, S. Ohshio, H. Akasaka, H. Muramatsu, H. Saitoh, Deposition of metal oxide films from metal-EDTA complexes by flame spray technique, . Therm. Spray Tech, (in press).

[5] H. Saitoh, K. Kawahara, S. Ohshio, A. Nakamura, N. Nambu, Metal composition of Y_2O_3: Eu powder evaluated using particle analyzer, Sci. Tech. Adv. Mater. 6 (2005) 205–209.

[6] K. Komatsu, S. Tsuchida, H. Maruyama, A. Nakamura, S. Ohshio, H. Akasaka, H. Saitoh, Synthesis of a violet Sr-Al-O:Eu^{2+} phosphor particle using elemental Al diffusion, Inter. J. Appl. Ceram. Tech. (2012) 1–8.

[7] K. Komatsu, T. Shirai, A. Nakamura, A. Kato, S. Ohshio, N. Nambu, I. Toda, H. Muramatsu, H. Saitoh, Synthesis and luminescence properties of Eu^{2+}-doped 8-coordinated SrO phosphors, Ceram. Inter. 39 (2013) 7115–7118.

[8] K. Komatsu, A. Nakamura, S. Ohshio, I. Toda, H. Muramatsu, H. Saitoh, Optical and structural stability of blue SrO:Eu^{2+} phosphor, J. Solid State Chem. 204 (2013) 186–189.

[9] N. Tanaka, S. Ohshio, H. Saitoh, K. Uematsu, Crystal growth of highly oriented zinc oxide by laser deposition technique with metal-EDTA (ethylene diamine tetra-acetic acid) complexes, Jpn. J. Appl. Phys. 37 (1998) L1165–L1168.

[10] H.M. Ansari, M.D. Rauscher, S.A. Dregia, S.A. Akbarn, Epitaxial pore-free gadolinia-doped ceria thin films on yttria-stabilized zirconia by RF magnetron sputtering, Ceram. Int. 39 (2013) 9749–9752.

[11] P. Mechnich, W. Braue, Air plasma-sprayed Y_2O_3 coatings for Al_2O_3/ Al_2O_3 ceramic matrix composites, J. Euro. Ceram. Soc. 33 (2013) 2645–2653.

[12] H. Jamalin, R. Mozafarinia, R.S. Razavi, R.A. Pidani, Comparison of thermal shock resistances of plasma-sprayed nanostructured and conventional yttria stabilized zirconia thermal barrier coatings, Ceram. Inter. 38 (2012) 6705–6712.

[13] Z. Lu, S.-W. Myoung, H.-S. Kim, M.-S. Kim, J.-H. Lee, Y.-G. Jung, J.-C. Jang, U. Paik, Microstructure evolution and interface stability of thermal barrier coatings with vertical type cracks in cyclic thermal exposure, J. Therm. Spray Tech. 22 (2013) 671–679.

[14] R. Eriksson, H. Brodin, S. Johansson, L. Ostergren, X.-H. Li, Influence of isothermal and cyclic heat treatments on the adhesion of plasma sprayed thermal barrier coatings, Surf. Coat. Tech. 205 (2011) 5422–5429.

[15] T. Tong, J. Li, Q. Chen, J.P. Longtin, S. Tankiewicz, S. Sampath, Ultrafast laser micromachining of thermal sprayed coatings for microheaters: design, fabrication and characterization, Sensors and Actuators A 114 (2004) 102–111.

[16] C.E. Leshock, J.-N. Kim, Y.C. Shin, Plasma enhanced machining of Inconel 718: modeling of workpiece temperature with plasma heating and experimental results, Inter. J. Mach. Tools. Manufact. 41 (2001) 877–897.

[17] M. Magnani, P.H. Suegama, N. Espallargas, S. Dosta, C.S. Fugivara, J.M. Guilemany, A.V. Benedetti, Influence of HVOF parameters on the corrosion and wear resistance of WC-Co coatings sprayed on AA7050 T7, Surf. Coat. Tech. 202 (2008) 4746–4757.

[18] W.M. Grissom, F.A. Wierum, Liquid spray cooling of a heated surface, Inter. J. Heat Mass Trans. 24 (1981) 261–271.

[19] Thermal spray technology an introduction (In Japanese) p. 15.

[20] P.L. Ke, Q.M. Wang, J. Gong, C. Sun, Y.C. Zhou, Progressive damage during thermal shock cycling of D-gun sprayed thermal barrier coatings with hollow spherical ZrO_2–$8Y_2O_3$, Mater. Sci. Eng. A 435–436 (2006) 228–236.

[21] A.N. Khan, J. Lu, Thermal cyclic behavior of air plasma sprayed thermal barrier coatings sprayed on stainless steel substrates, Surf. Coat. Tech. 201 (2007) 4653–4658.

[22] E. Altuncu, E.I. Karaall, G. Erdogan, F. Ustel, A. Turk, The effect of samples geometry and thermal cycling test type on the thermal shock behaviour of plasma sprayed TBCs, Plasma Process. Polym. 6 (2009) S711–S715.

[23] E.R. Andrievskaya, Phase equilibria in the refractory oxide systems of zirconia, hafnia and yttria with rare-earth oxides, J. Euro. Ceram. Soc. 28 (2008) 2363–2388.

[24] P. Zhang, A. Navrotsky, B. Guo, I. Kennedy, A.N. Clark, C. Lesher, Q. Liu, Energetics of cubic and monoclinic yttrium oxide polymorphs: phase transitions, surface enthalpies, and stability at the nanoscale, J. Phys. Chem. C 112 (2008) 932–938.

[25] M.P. Fard, V. Pershin, S. Chandra, J. Mostaghimi, Splat shapes in a thermal spray coating process: simulations and experiments, J. Therm. Spray Tech. 11 (2002) 206–217.

Corrosion resistance of nanostructured WC-CoCr coating with Co-Cr alloy binder

Haibin Wang, Xuezheng Wang, Xiaoyan Song*, Xuemei Liu, Tao Yang, Beijing / China

College of Materials Science and Engineering, Key Laboratory of Advanced Functional Materials, Education Ministry of China, Beijing University of Technology, Beijing 100124, China

The nanostructured WC-CoCr coating was fabricated by the high velocity oxy-fuel (HVOF) thermal spraying using a new type of WC-CoCr powder, in which the Co-Cr exists in a form of metallic compound. The Co-Cr alloy powder was prepared by the induction melting and subsequent mechanical milling. Detailed characterizations were carried out on the phase constitution and microstructure of the Co-Cr powder and the cermet coating, as well as the mechanical properties and corrosion resistance of the coating. The results show that the nanostructured WC-CoCr coating has greatly improved corrosion resistance as compared with the conventional WC-Co-Cr coating where Cr exists as single metal.

1 Introduction

The thermal sprayed WC-CoCr coatings have been widely used in various industries for the excellent combination of wear and erosion-corrosion resistant properties [1-6]. The corrosion resistance mechanisms for the WC-CoCr coating are mainly attributed to the formation of passive film containing Cr which inhibits the dissolution of binder materials in the corrosion environment [3]. Therefore, whether the Cr distributes uniformly in the cermet coating has a great effect on its corrosion resistance.However, it is difficlut to obtain an homogeneous distribution of metallic Co and Cr by mechanical mixing, e.g. ball milling, due to the good ductility of pure metals. In conventional WC-CoCr coatings, the Cr exists mainly in the form of single metal and/or Co-Cr solid solution [7]. In fact, the latter is rarely demonstrated by previous experimental results. In addition, the added Cr easily reacted with carbon dissolved in the Co matrix, forming the stable Cr_3C_2 phase which not only resulted in the formation of Co_3W_3C phase due to the loss of carbon but also weakened the effect of Cr as a corrosion-resistant element [8-10].

According to the Co-Cr phase diagram [11] as shown by Fig. 1, the Co-Cr system can also form the intermetallic phase as the weight percent of Cr is between 50.5 and 64%. Generally, the intermetallic compounds have much better corrosion resistance than pure metal in spite of the increased brittleness [12, 13]. Moreover, the homogeneous mixing of Co and Cr on the atomic scale can also be obtained. Therefore, a superior corrosion-resistant property for the WC-CoCr coating can be expected by partly replacing the pure metallic Co/Cr matrix with the Co-Cr intermetallic compound.

With regards to the advantages of nanostructured WC-based cermets, the aim of the present study is to explore the effect of Co-Cr addition in the form of intermetallic compound on the mechanical properties and corrosion resistance of the nanostructured WC-CoCr coating.

2 Experimental

2.1 Materials preparation

The commercially available Co and Cr bulks (Purity>99.9%, Alpha Inc.) were used as the starting materials.The Co-Cr ingot with the Co/Cr weight ratio of 43:57 was prepared by the levitation induction melting. The ingot was subjected to a mechanical ball milling procedure and the Co-Cr alloy powder was then obtained. A nanoscale WC-8wt.%Co composite powder was synthesized by in situ reactions of tungsten oxide, cobalt oxide and carbon black powders after milling [14-16]. The WC-CoCr powder with an approximate nominal composition of 86wt.% WC, 10wt.%Co and 4wt.%Cr was subsequently obtained by mixing the in situ synthesized WC-8Co composite powder and Co-Cr alloy powder. Prior to thermal spraying, the WC-CoCr powder was agglomerated by spray drying and a heat-treatment at 1230°C for 5h to prepare the feedstock. The high velocity oxy-fuel (HVOF) spraying system was used to deposite the feedstock powder on a stainless steel. For comparsion, another nanostructured WC-10Co4Cr feedstock powder was also prepared by adding the metallic Cr in the in situ synthesized WC-Co composite powder. The procesing details can be found in our previous work [16].

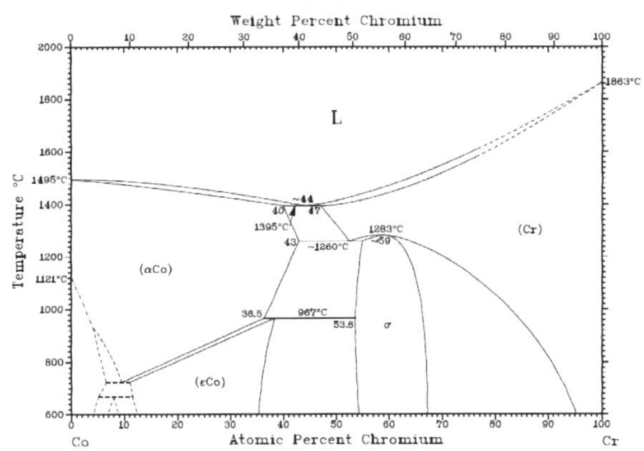

Fig. 1. Phase diagram of the Co-Cr system

2.2 Testing methods

The phases of the powders were investigated by X-ray diffraction (XRD, XRD-7000, SHIMADZU) using Cu-Kα radiation source. The morphologies of the prepared feedstock powders and coatings were analyzed by scanning electron microscopy (SEM, Nova NanoSEM 200, FEI). Nanoindentation hardness and modulus measurements were conducted on the polished top surface of the coating by Agilent Nano Indenter G200. The corrosion behaviour of the materials was evaluated by three-electrode electrochemical measurements using the Ivium CompactStat workstation (Ivium Technologies, Netherlands). In the electrochemical tests, a platinum foil, a saturated calomel electrode and the specimen were used as reference, counter and working electrodes, respectively. Potentiodynamic polarization test and electrochemical impendence spectroscopy (EIS) test were carried out to estimate the corrosion resistance of the Co/Co-Cr bulks and nanostructured WC–CoCr coatings. The Co-Cr alloy bulk with the size of 10×10×5 mm was obtained by wire electrical discharge machining on the ingot. All the specimens for electrochemical tests were mounted in a non-conducting epoxy resin with the rear side of the specimen soldered to a wire, leaving only an end surface (with a surface area of about 1 cm^2) exposed to the aerated 3.5 wt.% NaCl aqueous solution. The specimens were immersed in the saline solution for 1 day to stabilize the open circuit potential (OCP). After that, potentiodynamic polarization curves were measured with a fixed sweep rate of 1 mV/s. The corrosion current density (i_{corr}) and corrosion potential (E_{corr}) were determined as the the coordinates of the intersection point of linear fits to the anodic and cathodic polarization curves, according to Tafel's law [41]. EIS test was conducted at the OCP by applying a sinusoidal potential excitation of 10 mV amplitude over frequency range from 100 kHz to 10 mHz. Each test was repeated for three times to ensure the reproducibility of the results.

3 Results and discussion

3.1 Preparation of WC-CoCr feedstock

Fig. 2 shows the phase constitutions of the in situ synthesized WC-Co composite powder and the prepared Co-Cr alloy powder. The composite powder has only WC and Co phases. The Co-Cr powder is mainly composed of the CoCr-σ intermetallic phase which has a tetragonal structure corresponding to PDF card # 09-0052. According to the Co-Cr phase diagram, both Co and Cr have a certain solubility in the CoCr intermetallic compound, which facilitates the preparation of the powder with single phase. Fig. 3(a) shows the morphologies of the CoCr alloy powder. After milling, the alloy powder has irregular and sharp-edged structures with the particle size of several microns. The high-magnification microstructure of a single powder particle confirms the typical characteristic of brittle fracture. The EDS analysis result shows that the Co/Cr weight ratio of the alloy powder is about 44:56 which is well consistent with the initial composition design.

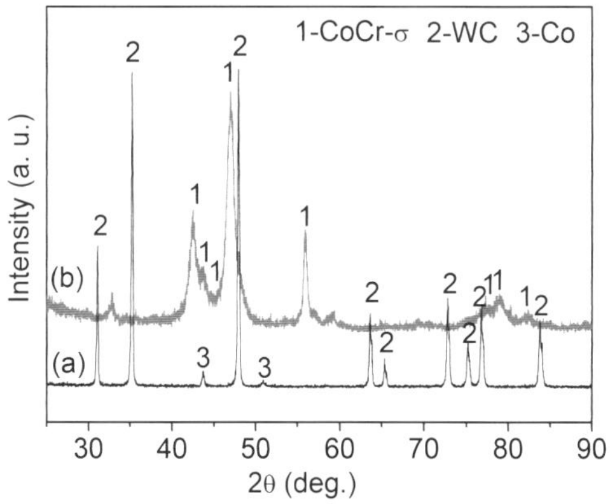

Fig. 2. XRD patterns of the prepared WC-8Co composite powder (a) and Co-Cr alloy powder (b)

Fig. 3(b) and (c) shows the surface morphologies of the WC-CoCr feedstock powder containing the intermetallic CoCr compound. The feedstock particle has nearly spherical shape and a high density which endows the powder with good flowability.

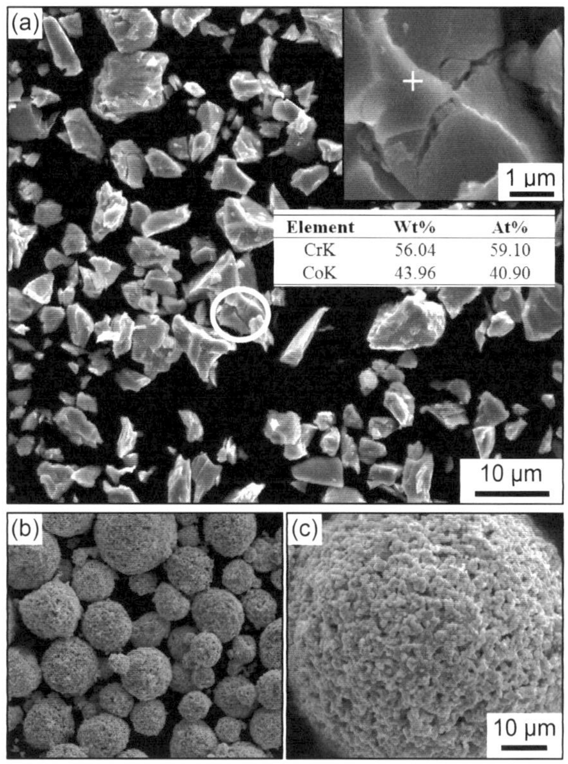

Fig. 3. (a) Morphologies of the CoCr intermetallic powder and the enlarged view of the circular region and EDS analysis result of the region marked with "+", (b) Morphologies of the WC-CoCr spraying feedstock

powder containing the intermetallic CoCr compound and (c) the enlarged view of a single particle.

3.2 Microstructure and mechanical properties

Fig. 4 shows the cross-sectional microstructures of both WC-CoCr coatings. Due to the addition of brittle CoCr intermetallic compound in the feedstock powder, the fabricated WC-CoCr coating has a higher porosity than that prepared by individual addition of Cr. However, the coating containing CoCr compound has a ratively finer WC particle size. This is mainly owing to the longer period of ball milling for the initial powder considering the homogeneity of CoCr compound and WC-Co composite powder.

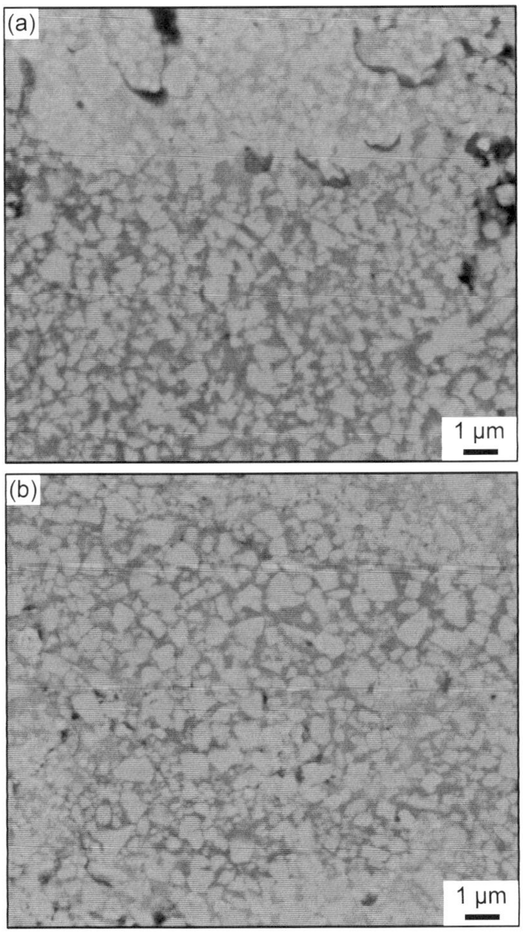

Fig. 4. Cross-sectional BSE images of both WC-CoCr coatings prepared by (a) addition of CoCr compound and (b) individual addition of Cr.

The nanoindentation hardness and modulus of both WC-CoCr coatings are listed in Table 1. For comparison, the mechanical properties of pure Co and CoCr compound are also summarized in Table 1. The CoCr compound ingot has a hardness about 14.5 GPa, which is double that of pure Co bulk. Correspondingly, the WC-CoCr coating containing CoCr compound has a relatively higher hardness than that prepared by individual addition of single Cr whereas an opposite trend for the modulus of the coatings is obtained. The reason may be attributed to the poor wettability of CoCr compound and WC phase.

Table 1 Hardness and modulus of Co, CoCr and WC-CoCr coatings

Specimen	Hardness (GPa)	Modulus (GPa)
Co	6.4±0.4	239.1±15.3
CoCr	14.5±0.6	311.8±11.1
Coating with CoCr	22.8±1.5	303.2±32.6
Coating with single Cr	19.4±1.8	394.5±33.8

3.3 Electrochemical behaviour of coatings

Fig. 5(a) and (b) shows the potentiodynamic polarization curves of the Co and CoCr ingots and as-sprayed WC-CoCr coatings in 3.5 wt.% NaCl solution, respectively. The corresponding electrochemical values are listed in Table 2. The corrosion potentials of Co and CoCr ingots are obviously higher than that of the thermal sprayed cermet coatings due to that the bulks have metallurgical bonding interfaces rather than weak mechanical bonding between splats for the coatings. Correspondingly, the ingots have much lower current densities. As compared to pure Co, the CoCr compound exhibit superior corrosion resistance, i.e. with higher corrosion potential and lower current density. As a reason, the corrosion current density of the WC-CoCr coating containing CoCr is no more than 1/3 of the cermet coating prepared with addition of single Cr in WC-Co. For the as-sprayed coating, the micro-galvanic corrosion resulting from the micro-galvanic cells between the WC phase and the Co-Cr binder is the dominant corrosion mechanism [17, 18]. Owing to the increase of corrosion potential of CoCr compound as compared to single Co, the micro-galvanic corrosion in the cermet coating with CoCr alloy binder was thus significantly inhibited.

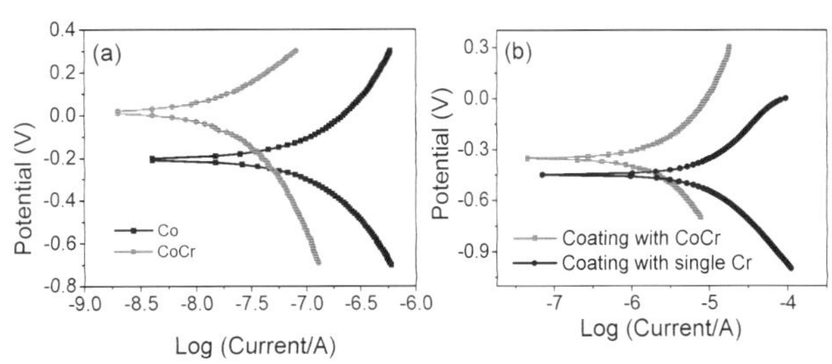

Fig. 5. Potentiodynamic polarization curves of Co and CoCr ingots (a) and two types of HVOF sprayed nanostructured WC–CoCr coatings **(b)** in 3.5 wt.% NaCl aqueous solution.

The superior corrosion resistance of the CoCr bulk and correspondingly fabricated WC–CoCr coating is further verified by the EIS results as shown by Fig. 6(a) and (b). For the Nyquist diagrams, a larger diameter of the capacitive semicircle means a higher charge transfer resistance in the corrosion process and thus a slower corrosion rate of the materials. As seen from Fig. 6(a) and (b), the CoCr ingot and the cermet coating containing CoCr binder have a larger diameter than that of the pure Co bulk and the coating prepared with addition of single Cr, respectively. This indicates the CoCr compound and correspondingly fabricated cermet coating have better corrosion resistance.

Table 2 Corrosion current densities (i_{corr}) and potentials (E_{corr}) of the coatings

Specimen	Corrosion potential (mV)	Current density (A/cm^2)
Co	-0.2	9.3×10^{-8}
CoCr	0	1.9×10^{-8}
Coating with CoCr	-0.35	2.2×10^{-6}
Coating with single Cr	-0.45	6.9×10^{-6}

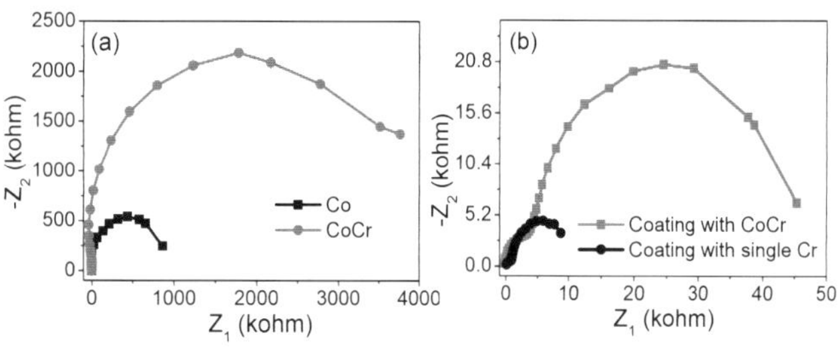

Fig. 6. Nyquist diagrams of EIS tests for Co and CoCr ingots (a) and two types of HVOF sprayed nanostructured WC–CoCr coatings **(b)** in 3.5 wt.% NaCl aqueous solution.

4 Conclusions

Based on the idea that partly replacing the single metallic Co-Cr binder in conventional WC-CoCr coating with corrosion-resistant CoCr compound , the preparation and properties of a new type of cermet coating were explored. The following conclusions can be drawn.

(1) The CoCr compound powder with several microns can be obtained by combination of induction melting and mechanical milling.
(2) The fabricated CoCr ingot has a nanoindentation hardness of 14.5 GPa and modulus of 311.8GPa, both of which are significantly higher than that of the Co bulk.
(3) As compared with pure Co, the CoCr compound has higher corrosion potential and lower corrosion current density. As a reason, the HVOF sprayed WC-CoCr coating containing CoCr alloy binder has much higher corrosion resistance than that of the cermet coating prepared by addition of single Cr.

Acknowledgments
This work was supported by the National High-tech R&D Program of China (2013AA032001), the National Science Fund for Distinguished Young Scholars (51425101) and the Beijing Natural Science Foundation (2154045).

5 Literature

[1] Thakur L. and N. Arora: A study on erosive wear behavior of HVOF sprayed nanostructured WC–CoCr coatings. J Mech Sci Technol 27 (2013), Issue 5, pp. 1461/7.
[2] Thakur L. and N. Arora: Sliding and abrasive wear behavior of WC–CoCr coatings
a) with different carbide sizes. J Mater Eng Perform 22 (2013), Issue 2, pp. 574/83.
[3] Hong S., Wu Y.P., Wang B., Zheng Y.G., Gao W.W. and G.Y. Li: High-velocity oxygen-fuel spray parameter optimization of nanostructured WC–10Co–4Cr coatings and sliding wear behavior of the optimized coating. Mater Des 55 (2014), pp. 286/91.
[4] Picas J.A., Rupérez E., Punset M. and A. Forn: Influence of HVOF spraying parameters on the corrosion resistance of WC–CoCr coatings in strong acidic environment. Surf Coat Technol 225 (2013), pp. 47/57.
[5] Bolelli G., Giovanardi R., Lusvarghi L. and T. Manfredini: Corrosion resistance of HVOF-sprayed coatings for hard chrome replacement. Corros Sci 48 (2006), Issue 11, pp. 3375/97.
[6] Balamurugan G.M., Duraiselvam M. and V. Anandakrishnan: Comparison of high temperature wear behaviour of plasma sprayed WC–Co coated and hard chromium plated AISI 304 austenitic stainless steel. Mater Des 35 (2012), pp. 640/6.
[7] Perry J.M., Neville A. and T. Hodgkiess: A comparison of the corrosion behavior of WC-Co-Cr and WC-Co HVOF thermally sprayed coatings by in situ atomic force microscopy (AFM). J Therm Spray Technol 11 (2002), Issue 4, pp. 536/41.
[8] Wang Q., Zhang S.Y., Cheng Y.L., Xiang J., Zhao X.Q. and G.B. Yang: Wear and corrosion performance of WC-10Co4Cr coatings deposited by different HVOF and HVAF spraying processes. Surf Coat Technol 218 (2013) pp. 127/36.
[9] Wang L.J., Qiu P.X., Liu Y., Zhou W.X., Gou G.Q. and H. Chen: Corrosion behavior of thermal sprayed WC cermet coatings containing metallic binders in saline environment. Trans Nonferrous Met Soc China 23 (2013), pp. 2611/17.
[10] Liu S.L., Zheng X.P. and G.Q. Geng: Influence of nano-WC-12Co powder addition in WC-10Co-4Cr AC-HVAF sprayed coatings on wear and erosion behaviour. Wear 269 (2010), Issue 5-6, pp. 362/7.
[11] ASM Handbook, Alloys Phase Diagrams, 3 (1998) pp. 9/26, 635.

[12] Verdian M.M., Raeissi K. and M. Salehi: Fabrication and corrosion resistance of HVOF-sprayed Ni_2Si intermetallic compound. Applied Surf Sci 273 (2013), pp. 426/31.
[13] Mokgalaka M.N., Popoola A.P.I. and S.L. Pityana: In situ laser deposition of NiTi intermetallics for corrosion improvement of Ti–6Al–4V alloy. Trans Nonferrous Met Soc China 25 (2015), Issue 10, pp. 3315/22.
[14] Liu W.B., Song X.Y., Zhang J.X., Zhang G.Z. and X.M. Liu: Preparation of ultrafine WC–Co composite powder by in situ reduction and carbonization reactions. Int J Refract Met Hard Mater 27 (2009), pp. 115/20.
[15] Wang H.B., Song X.Y., Liu X.M., Wei C.B., Gao Y. and J. Fu: Effect of heat-treatment of spray-dried powder on properties of ultrafine-structured WC-Co coating. Surf Coat Technol 207 (2012), pp. 117/22.
[16] Wang H.B., Wang X.Z., Song X.Y., Liu X.M. and X.W. Liu: Sliding wear behavior of nanostructured WC–Co–Cr coatings. Applied Surf Sci 355 (2015), pp. 453/60.
[17] Saha G.C. and T.I. Khan: The corrosion and wear performance of microcrystalline WC-10Co-4Cr and near-nanocrystalline WC-17Co high velocity oxy-fuel sprayed coatings on steel substrate. Metall Mater Trans A 41A (2010), Issue 11, pp. 3000/9.
[18] Cho J.E., Hwang S.Y. and K.Y. Kim: Corrosion behavior of thermal sprayed WC cermet coatings having various metallic binders in strong acidic environment. Surf Coat Technol 200 (2006), Issue 8, pp. 2653/62.

Influence of different content of free silicon on the oxidation resistance of SiC coatings on the C/SiC composites

Fang Jia[1,2], Haoran Peng[1,2], Xiaojuan Ji[1,2], Xin Zhang[1,2], Jie Shen[1,2]
(1. Beijing General Research Institute of Mining & Metallurgy, Beijing 100160, China;
2. Beijing Engineering Technology Research Center of Surface Strengthening and Repairing of Industry parts, Beijing 102206, China;)

Abstract: SiC coatings with different content of free silicon were prepared on C/SiC composites in a vacuum atmosphere by pack cementation technique to protect the substrate from oxidation. The phase and microstructure of the coatings were investigated by XRD, SEM and EDS analyses. Cyclic oxidation test at 1300 °C in air atmosphere was performed. Cracks and voids of the coatings were reduced with the increasing content of free silicon. The SiC coating with appropriate content of silicon exhibited better oxidation resistance compared to that without silicon. However, higher content of free silicon was not beneficial for SiC coating to improve anti-oxidation performance. Possible mechanisms related to oxidation were preliminarily discussed.

1 Introduction

Carbon fiber reinforced silicon carbide composites (C/SiC) are one of the extremely attractive high temperature structural materials for applications in aeronautics and astronautics fields [1]. These applications require the C/SiC composites to operate in an oxidizing atmosphere. However, the oxidation of C/SiC composites makes the performance of these materials decreased rapidly at temperatures above 723 K and the oxidation rate increases quickly with the temperature increase [2, 3]. To rectify this problem, anti-oxidation coating is a logical method to protect these composites against oxidation for their successful utilization at high temperatures.

Ceramic coatings are commonly employed to protect carbon materials from oxidation [4]. Many ceramics are promising candidate materials as oxidation protective coatings such as TiC, TaC, B_4C, NbC, SiC and ZrC, etc. Among the possible materials, SiC ceramic material exhibites attractive properties, which has been applied widely as oxidation protective coating at high temperature due to its good mechanical properties, excellent oxidation resistance and good physical–chemical compatibility with C/SiC composites [5].

Several methods such as pack cementation [6], chemical vapor decomposition (CVD) [7] and laser induced chemical decomposition (LICD) [8] have been developed to produce SiC coating. Among these methods, the pack cementation is widely used in producing oxidation protective coatings and it is a good technique to deposit SiC coatings on the surface of a carbon material [9]. However, the SiC coating obtained by pack cementation usually possesses many structure defects such as pores, cracks or pinholes, which is disadvantageous to its oxidation resistance at high temperature because oxygen can attack the substrate through the defects [10]. Therefore, it is crucial to produce a kind of dense coating to achieve an excellent oxidation resistance of the coated C/SiC composites. In pack cementation technique, it is required not only to design the chemical composition of powder but also the preparation temperature for producing the dense coating on the surface of C/SiC composites.

In the present work, dense SiC coatings with different content of free silicon were produced on the surface of C/SiC composites substrates by pack cementation method. The microstructures of the coatings were investigated. Moreover, to better understand the oxidation resistance properties of the coatings, the cyclic oxidation behavior at 1300 °C was also studied.

2 Experimental procedures

Carbon fiber reinforced silicon carbide composites (C/SiC) with the dimension of 10 mm × 10 mm × 2 mm were used as substrate materials. Before pack cementation procedure, the specimens were cleaned with ethanol and dried at 273 K for 90 min. The SiC coatings were prepared by a pack cementation process with SiC, Si, C and Al_2O_3 powders.

Pack compositions for the pack cementation process were as follow: 25–50 wt.% SiC, 20–60 wt.% Si, 7–12 wt.% graphite and 6–10 wt.% Al_2O_3. All of these powders have the granularity of below 325 mesh and the purity of analytical grade. The C/SiC composites were immersed in the powder mixture in a BN crucible with a BN lid. Then the crucibles were placed into an electric furnace that was heated in vacuum atmosphere to 1700 °C and held at the temperature for 2 h to produce SiC coatings with different content of free silicon. After deposition of the coatings, the pack was allowed to furnace cool, and the samples were obtained from the pack then ultrasonically cleaned in the acetone in order to remove the loosely embedded pack substance.

Thermal cyclic oxidation experiments of the as-deposited coatings were conducted in air furnace (KSL 1700X) in dry and static air at 1300 °C. These specimens were put inside or taken out of electrical furnace at certain period. The weight change of coated samples was recorded using an electronic balance (FA1104N) with a precision of 0.1mg.

The surface and cross sectional morphologies of the specimens were characterized by a field emission-scanning electron microscope (FE-SEM Supra 55, Germany) equipped with energy dispersive X-ray spectrometry (EDS) and back scattering electron (BSE) detector. The phases of the as-deposited coatings and oxides were identified by X-ray diffraction (XRD, Rigaku D/max 2200PC, Japan) with Cu Ka radiation.

3 Results and discussion

3.1 Microstructure of as-deposited coatings

The surface morphologies of the as-deposited coatings obtained from different pack cementation process are shown in Fig. 1a–c. Coatings produced with 25 wt. % Si in pack materials exhibit relatively loose surface and it can be seen that some pores exist in the coatings (Fig. 1a). These pores could be self-seal partially at high temperature, but C/SiC substrate would still be oxidized due to oxygen diffusing through the pores at the temperature between the pore sealing temperature and the original oxidizing temperature of C/SiC composites. When the mass fraction of Si in pack materials is 40 wt. % and 50 wt. %, it reveals the formation of dense and unsmooth coating structures without pores (Fig. 1b-c). This can be attributed to enough free silicon melting at high temperature which sealed the pores. It has been reported that the dense structure is beneficial to improving the oxidation resistance of the coating.

Fig. 1. Surface morphologies of the as-deposited coatings: (a) 25 wt. % Si, (b) 40 wt. % Si, and (c) 50 wt. % Si in pack materials.

The corresponding cross-section morphologies of three

kinds of coatings are shown in Fig. 2a–c. The coating obtained with 25 wt. % Si in pack materials is ~60μm in thickness with penetrable cracks or small holes. By EDS analysis, the coating is mainly composed of SiC Phase with no free Si. When the mass fraction of Si in pack materials is 40 wt. %, the thickness of coating is around 90μm without penetrable cracks or large holes in the coating (Fig. 2b). It can be seen that the thickness of coating, which is ~200μm with 50 wt. % Si in pack materials (Fig. 2c), increased distinctly with increasing content of Si in original pack powders. It displays a dense structure without any cracks that directly penetrated into the C/SiC substrate. This demonstrates that the density of the coating is improved with the increase of Si in pack materials which is mainly due to penetration of enough molten Si into the holes in the porous SiC coating during the heat treatment of pack cementation. On the basis of the above analyses, it is concluded that the special structures and the thickness of the SiC coating could be obtained by controlling the mass fraction of Si in pack materials. Moreover, Fig. 2b-c shows obviously that there are two kinds of crystalline particles characterized as grey SiC and white Si which are existent not only on the coating surface but also inside the coating and the content of each phase is quantitatively analyzed by XRD in following description.

3.2. Phase analysis of as-deposited coatings

Fig. 3 shows the surface XRD patterns of the as-deposited coatings with different mass fraction of Si in pack materials. It could be seen that the coating (25% Si) is basically comprised of α-SiC and β-SiC phases with no other phases detected. It is consistent with the cross-section morphology of the coating (Fig. 2a). Besidesα-SiC and β-SiC phases, peaks of silicon appeared in the coatings (40% Si and 50% Si in pack materials), which indicates that some silicon diffusing into the coating did not react with the C powder and C/SiC composites. In addition, the intensity of the Si peaks was enhanced with the increase of Si in pack materials. By Semi-quantitative XRD analysis, the content of free silicon in the three kinds of coatings is 0%, 6% and 19%, respectively.

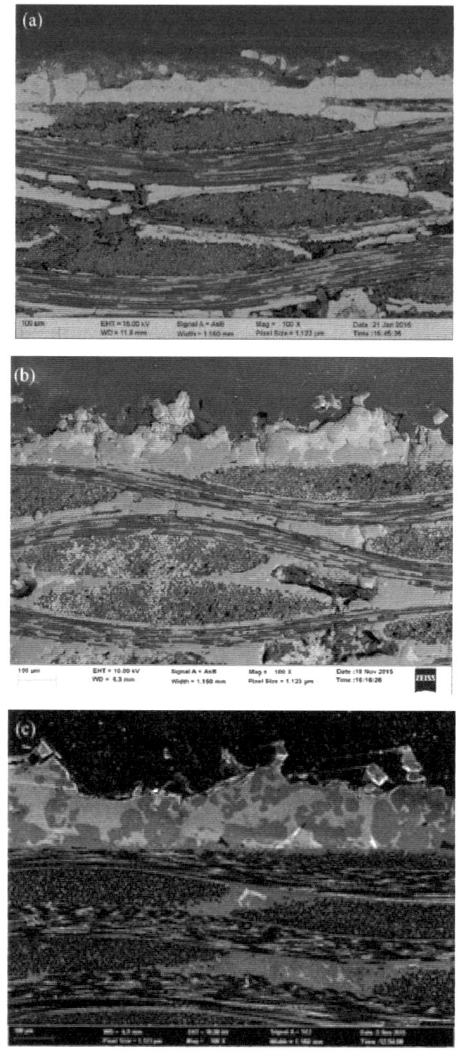

Fig. 2. Cross-section morphologies of the as-deposited coatings: (a) 25 wt. % Si, (b) 40 wt. % Si, and (c) 50 wt. % Si in pack materials.

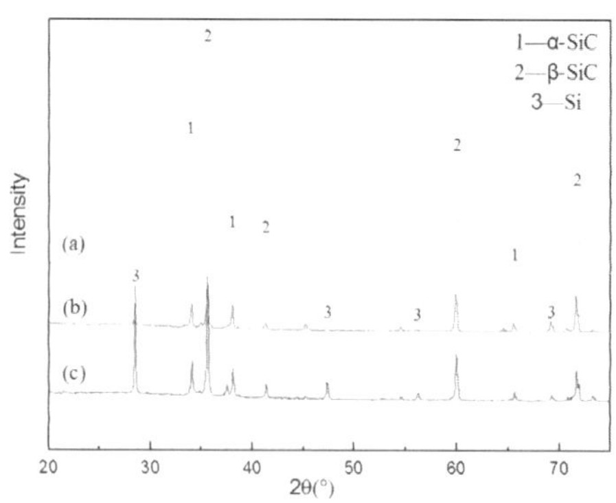

Fig. 3. XRD patterns of the as-deposited coatings: (a) 25 wt. % Si, (b) 40 wt. % Si, and (c) 50 wt. % Si in pack materials.

3.3. Oxidation resistance of the coatings

Fig. 4 shows the of the weight change of the SiC coatings with different content of free silicon in air at 1300 °C. After the oxidation in air for 10 h, the weight loss of coated sample with 25% Si in pack materials is up to ~15.1%. The coated sample with 50% Si in pack materials reveals a lower mass loss of ~9.13% after 10h. However, the coated sample with 40% Si in pack materials loses only 5.52% mass after exposure to air at 1300 °C for 10 h. This indicates that the oxidation resistance of SiC coating with appropriate free Si is better than that of coating with excessive free Si and no free Si.

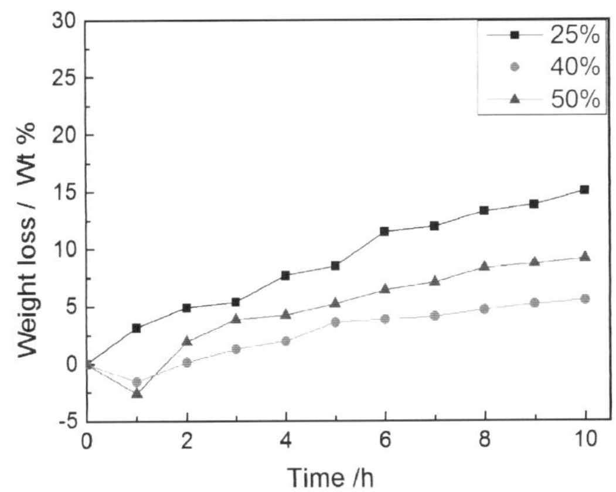

Fig. 4. The 1300 °C oxidation curves of the coated samples prepared with different Si contents in pack materials.

The oxidation behavior and the oxidation mechanism of the coated C/SiC samples are varied at different times. According to the oxidation curve (Fig. 4), it can be seen that in the initial oxidation stage, the weight of the coated samples (40% Si and 50% Si in pack materials) increased with time and this weight gain of coated samples may be caused by the formation of the vitreous SiO_2 glass layer which is due to the reactions of oxygen with Si and SiC exposed to air at high temperature. With the oxidation time extending, the surface of the coating changes entirely into dense SiO_2 glass oxide film which could reduce oxygen permeability and efficiently prevent the oxygen from permeating into the C/SiC substrate. In addition, the excessive gas and the quick cooling from 1300 °C to room temperature during the oxidation process results in the formation of pores and cracks on the SiO_2 glass oxide film and the oxygen passes through these defects diffusing to the underlying substrate which results in the weight loss of the coated samples.

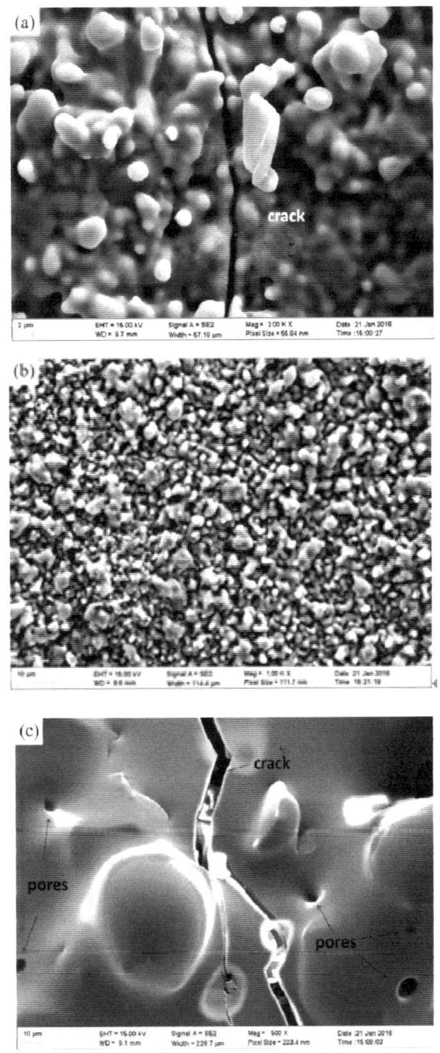

Fig. 5. Surface morphologies of coatings after 10 h oxidation at 1300 °C: (a) 25 wt. % Si, (b) 40 wt. % Si, and (c) 50 wt. % Si in pack materials.

The surface morphologies of the coatings after 10 h oxidation are shown in Fig. 5a–c. It could be observed that the surface of coatings is all covered with SiO_2 glass oxide film. The glass oxide film is not continuous with cracks existing in the coatings which were produced with 25% Si and 50% Si in pack materials, as shown in Fig. 5a, c. These cracks might form during the cooling process from 1300 °C to room temperature for weighing. In addition, there are some apparent pores in the oxide film for the coating which was produced with 50% Si in pack materials as denoted by the arrows in Fig. 5c. The formation of these pores may be attributed to the escape of gases. At 1300 °C, CO and CO_2 generated from the oxidation of SiC and the C/SiC substrate would get out through the SiO_2 film. In the Fig. 5b, relatively continuous SiO_2 glass oxide film which is advantageous to the oxidation resistance can be observed on the surface of coating which was produced with 40% Si in pack materials.

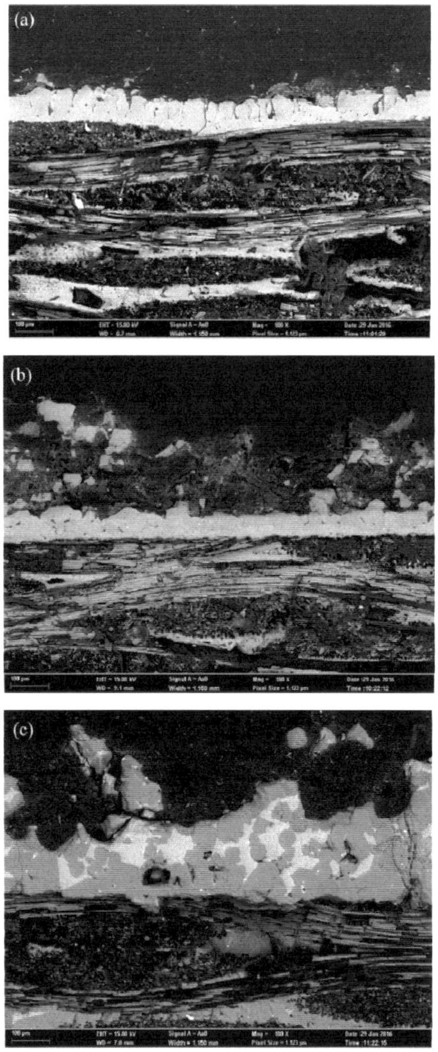

Fig. 6. Cross-section morphologies of coatings after 10 h oxidation at 1300 °C: (a) 25 wt. % Si, (b) 40 wt. % Si, and (c) 50 wt. % Si in pack materials.

Figure 6 shows the cross-sectional micrograph of the coatings after 10 h oxidation. Some pores or small cracks are found in the coatings. The volatilization of glass coatings at 1300 °C and the formation of cracks and pores are the main factors for the mass loss of the coated specimens. The thickness of the coating decreases respectively after oxidation. This also proves the volatilization of the glass coating at 1300 °C. XRD pattern of the coatings after oxidation for 10 h (Fig. 7) shows that Si phase disappears in the coating which was produced with 40% Si in pack materials and decreases in the coating which was produced with 60% Si in pack materials, while a new SiO_2 phase forms compared with XRD patterns of the coatings before oxidation. It also indicates that the above-mentioned dense glass film is SiO_2 film.

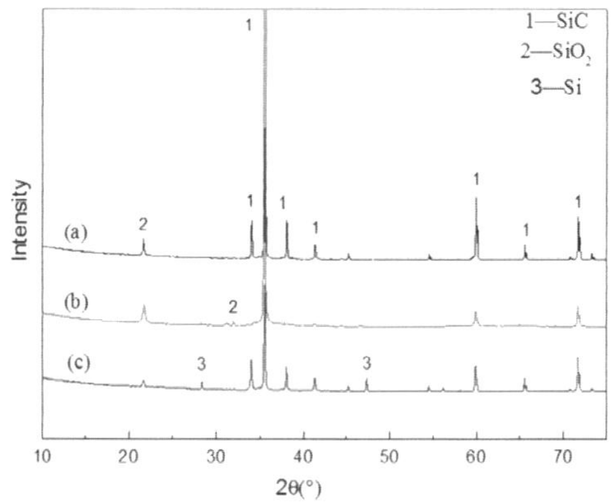

Fig. 7. XRD patterns of coatings after 10 h oxidation at 1300 °C: (a) 25 wt. % Si, (b) 40 wt. % Si, and (c) 50 wt. % Si in pack materials.

4 Conclusions

SiC coatings with different content of free silicon were prepared on C/SiC composites in a vacuum atmosphere at 1700 °C by pack cementation technique and oxidation behavior at 1300 °C was investigated. The main conclusions can be drawn as follows:

1) The chemical compositions of pack powder in a pack cementation technique have a great influence on the phase, microstructure of the coatings. Cracks and voids of the coatings are reduced and the thickness of the coatings is increased with the increasing content of free silicon.

2) The coated sample with 6% free Si in coating loses only 5.52% mass after exposure to air at 1300 °C for 10 h which demonstrates that the SiC coating with

appropriate content of free silicon exhibits better oxidation resistance compared to that without silicon. However, higher content of free silicon is not beneficial for SiC coating to improve anti-oxidation performance.
3) The weight loss of the SiC coated samples can be due to the oxidation of C/SiC substrate by the oxygen diffusing through the penetrable cracks in the coatings.

Acknowledgements

This research is sponsored by National International Science and Technology Cooperation Special Project under Grant No. 2013DFB50390.

References

[1] T.M. Besmann, B.W. Sheldon, R.A. Lowden. Vapor phase fabrication and properties of continuous filament ceramic composites. Science 253 (1991), Issue (6), pp. 1104/09.

[2] M. Imuta, J. Gotoh. Development of high temperature materials including CMCs for space application. Key Eng Mater 164–165 (1999), pp. 439/44.

[3] M.E. Westwood, J.D. Webster, R.J. Day, F.H. Hayes, R. Taylor. Oxidation protection for carbon fiber composites. J Mater Sci 31 (1996), pp. 1389/97.

[4] Q.S. Zhu, X.L. Qiu, C.W. Ma. Oxidation resistant SiC coating for graphite materials. Carbon 37 (1999), pp. 1475/84.

[5] L.F. Cheng, Y.D. Xu, L.T. Zhang, X.G. Luan. Oxidation and defect control of CVD SiC coating on three dimensional C/SiC composites. Carbon 40 (2002), pp. 2229/34.

[6] C.A.A. Cairo, M.L.A. Graca, C.R.M. Silva, J.C. Bressiani. Functionally gradient ceramic coating for carbon/carbon antioxidation protection. J Euro Ceram Soc 21 (2001), pp. 325/29.

[7] L.F. Cheng, Y.D. Xu, L.T. Zhang. Preparation of an oxidation protection coating for C/C composites by low pressure chemical vapor deposition. Carbon 38 (2000), pp. 1493/98.

[8] S. Lloyd, N. Avery, M. Pal. A novel laser technique for oxidation resistant coating of carbon–carbon composite. Carbon 39 (2001), pp. 991/99.

[9] Q.S. Zhu, X.L. Qiu, C.W. Ma. Oxidation resistant SiC coating for graphite materials. Carbon 37(1999), pp. 1475/84.

[10] H.J. Li , H. Xue, Y.J. Wang, Q.G. Fu, D.J. Yao. A MoSi2–SiC–Si oxidation protective coating for carbon/carbon composites. Surface & Coatings Technology 201 (2007), pp. 944/47.

Can thermally sprayed aluminium (TSA) mitigate corrosion of carbon steel in carbon capture and storage (CCS) environments?

B Syrek-Gerstenkorn, S Paul, Cambridge / UK

Combustion of fossil fuels generates carbon dioxide (CO_2), a greenhouse gas (GHG) whose atmospheric release is widely acknowledged as the main cause of global warming. To continue using fossil fuels whilst minimizing CO_2 release, carbon capture and storage (CCS) may be used. CCS requires capture, transport and storage of CO_2. For transport of CO_2, low cost carbon steel pipelines are selected owing to their negligible corrosion rates in dry CO_2. However, in the presence of liquid water carbonic acid forms, which is corrosive to carbon steel. In order to mitigate wet CO_2 corrosion, use of expensive corrosion resistant alloys (CRAs) such as stainless steels are recommended, but the increased cost makes such selection economically unfeasible; hence, new methods of corrosion mitigation are sought. One such method is the use of thermally sprayed aluminium (TSA) coatings, which mitigates corrosion of carbon steel in offshore service, but there are concerns regarding its suitability in CO_2-containing solutions. A 30-day test was carried out during which carbon steel specimens sprayed with aluminium were immersed in deionized water at ambient temperature in 0.1 MPa CO_2 and the acidity (pH), potential and amount of dissolved Al^{3+} ions were monitored.

1 Introduction

One of the major issues being face globally is climate change. Anthropogenic activities, such as burning of fossil fuels, which releases CO_2, a greenhouse gas, in the atmosphere is recognised as one of the causes [1-2]. The increasing amount of CO_2 in the atmosphere is linked to the rise in average temperature of the Earth's surface (a phenomenon often referred to as 'global warming') which results in a change in the statistical distribution of global weather patterns, and, consequently, exacerbates both the - frequency and intensity - of extreme weather events worldwide. At present, fossil fuels are the primary source of energy (~85% worldwide) and are likely to remain that way in the near future [1]. One way of utilising fossil fuels whilst minimising the atmospheric release of CO_2 is carbon capture and storage (CCS). CCS technology consists of three main elements: (i) collecting and concentrating CO_2, (ii) transporting it to a suitable storage location, and (iii) storing it underground in depleted oil and gas fields or deep saline aquifers [2-5]. Usually, captured CO_2 has to be transported from the point of capture to the place of storage. There are two means of transport of large quantities of CO_2: by pipelines and by ships [6]. Pipelines are more commonly used means to transport the compressed carbon dioxide due to its cost-effectiveness and safety [3,7,8]. Carbon steel pipelines are used because of their low cost and low corrosion rate in dry CO_2 environment. However, in presence of free water, carbon dioxide forms highly corrosive carbonic acid (H_2CO_3) [9]. Free water in CCS systems may come from the combustion process (especially in case of post-combustion capture systems where the possibility of concentration of water in the inlet stream is higher in comparison to the oxyfuel or pre-combustion systems) [10]. In order to mitigate corrosion, when CO_2 cannot be dried, corrosion resistant alloys (e.g. stainless steels) are used [11,12]. However, as the cost of pipelines made from stainless steels is relatively high, new methods of corrosion mitigation are being researched. One of the potential methods, that can be used, is thermally sprayed aluminium (TSA) coating. TSA has been used to mitigate corrosion of offshore steel structures for decades [13]. It is known to mitigate corrosion of carbon steel by-(i) forming a protective barrier when intact, and (ii) acting as a sacrificial layer when damaged. TSA acts sacrificially by its position in the galvanic series- being more anodic to carbon steel in seawater. Although, TSA's ability to protect steels in offshore service is proven, there is no evidence of its suitability in wet CO_2 environments. In wet CO_2 environments, corrosion of Al may ensue and TSA may not be suitable for such applications. Corrosion of Al in aqueous solution involves its anodic dissolution following (1):

$$Al \rightarrow Al^{3+} + 3e^- \qquad (1)$$

supported by the corresponding cathodic reactions (2) in aerated, (3) in anoxic/neutral, and (4) in anoxic/acidic environments:

$$O_2 + 4H^+ + 4e^- \rightarrow 2H_2O \qquad (2)$$
$$2H_2O + 2e^- \rightarrow H_2 + 2OH^- \qquad (3)$$
$$2H^+ + 2e^- \rightarrow H_2 \qquad (4)$$

Both anodic and cathodic reactions occur simultaneously at the same rate on a metal's surface, hence one cannot exist without the other.
Corrosion resistance of aluminium is associated with natural oxide layer, which forms on the metal's surface on contact with oxidising environment. A simple version of the chemical reaction involved in given below (5):

$$Al^{3+} + 3H_2O \rightarrow Al_2O_3 + 6H^+ \qquad (5)$$

The stability of this thin oxide film is strongly dependent on pH of the solution [12]. In the range of pH 4-9 the oxide layer is stable, but in more acidic and alkaline solutions it undergoes dissolution and Al^{3+}

ions and hydroxide ions are formed, respectively. Knowledge about the rate of dissolution of Al and stability of alumina in a given media is crucial to ascertain suitability of using TSA as a corrosion mitigation method. Another factor is the galvanic potential of Al and steel in the environment of interest as this defines the thermodynamic drive for one to act as an anode and protect the other, which acts as the cathode.

This paper attempts to address the above-mentioned knowledge gaps. A 30-day test was carried out during which carbon steel specimens thermally sprayed with aluminium were immersed in deionized water saturated with 0.1 MPa CO_2 at ambient temperature. The acidity (pH) and potential were monitored, and the amount of dissolved Al^{3+} ions were analysed after completion of the test. Corrosion potential (Ecorr) of carbon steel, stand-alone TSA and TSA-coated steel was also measured in a separate reactor under similar conditions to access the suitability of TSA to mitigate corrosion of steel in CO_2-containing aqueous media.

2 Results and discussion

2.1 Monitoring the acidity/alkalinity (pH) of the solution

For a 30-day test, 29 TSA-coated specimens were divided into two sets (each assigned to a separate glass reactor) in a manner that the combined surface area of aluminium coating in one reactor was twice of that in the other. pH was monitored in both reactors for the entire duration of the tests. The results are shown in the graph below, **Fig.1.**:

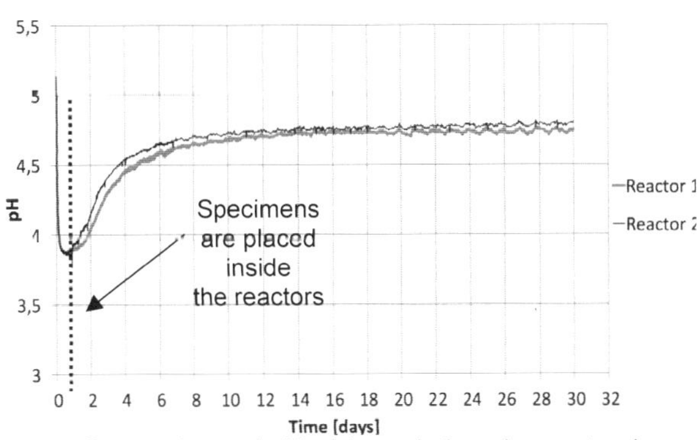

Fig. 1. Comparison of pH of the solutions in reactor 1 and 2.

One can notice that trends of both plots are very similar, with the variation being <6%. At the beginning of the test, pH dropped from 5.1 to 3.8 within the first 2 days. This change was due to the introduction of CO_2 into deionised water. The saturation level was identified when the pH reached 3.8. Once CO_2-saturation was reached, the specimens were put in the solution. This resulted in an increase in the pH. It can be noticed in **Fig. 1.** that pH in both the reactors was increasing until day 12 and after that the pH stabilised at around 4.7.

The decrease of pH from the value of around 5 to around 3.8 is associated with the dissolution of CO_2, and its subsequent dissociation in solution. When the dissolved CO_2 forms carbonic acid in water, it can dissociate to release H^+ ions and as a result, pH of the solution decreases. The sequence of reactions leading to the reduction in pH is given under. Reaction between carbon dioxide and water is expressed by the following equation:

$$CO_{2(g)} + H_2O \rightleftharpoons CO_{2(aq)} + H_2O \quad (5)$$

Some of carbon dioxide reacts with water and carbonic acid is formed following the reaction below:

$$CO_{2(aq)} + H_2O \rightleftharpoons H_2CO_3 \quad (6)$$

Next, some part of carbonic acid undergoes two-stage dissociation in accordance with the following equations (7 and 8):

$$H_2CO_3 \rightleftharpoons H^+ + HCO_3^- \quad (7)$$

$$HCO_3^- \rightleftharpoons H^+ + CO_3^{2-} \quad (8)$$

This results in two dissociation constants whose values at room temperature are given below:

$$K_{a1} = \frac{[H^+][HCO_3^-]}{[H_2CO_3]} = 4.45 \text{x} 10^{-7} \quad (9)$$

$$K_{a2} = \frac{[H^+][CO_3^{2-}]}{[HCO_3^-]} = 4.8 \text{x} 10^{-11} \quad (10)$$

Analysing the graph in **Fig. 1.**, it can be seen that the pH of CO_2-saturated deionised water started to increase after putting the specimens inside the reactors. This increase of pH was caused by consumptions of H^+ ions in the solution due to cathodic reactions. Despite different amount of aluminium in both reactors, the rate of change of pH with time was the same in both reactors due to continuous CO_2 supply to the solution. Therefore, the graph does not give any information about the kinetics of the reactions which took place inside reactors.

Analysing the Pourbaix diagram of aluminium it can be noticed that this metal shows corrosion resistivity in the pH range between ~ 4 and 9 due to formation of a passive film. pH-monitoring showed that water saturated with CO_2 reaches the pH of 3.9 which is outside the "safe zone".

Nevertheless, it should be borne in mind that the Pourbaix diagram is based merely on theoretical thermodynamic considerations. This diagram shows the potential of aluminium (immersed in distilled water at 25 °C) measured with respect to standard hydrogen electrode.

2.2 Potential measurements

In order to determine the potential of TSA in CO_2-containing solution tests on stand-alone TSA, TSA on steel, and steel were conducted. Information about the potential of TSA and steel are of great significance where those metals are in contact in the presence of an electrolyte. To protect the steel by using TSA, aluminium coating must have more electronegative potential or more active than steel in a given environment. **Fig.2.** below represents the results of potential measurements of steel, TSA on steel, and stand-alone TSA in CO_2 environment during 9 days.

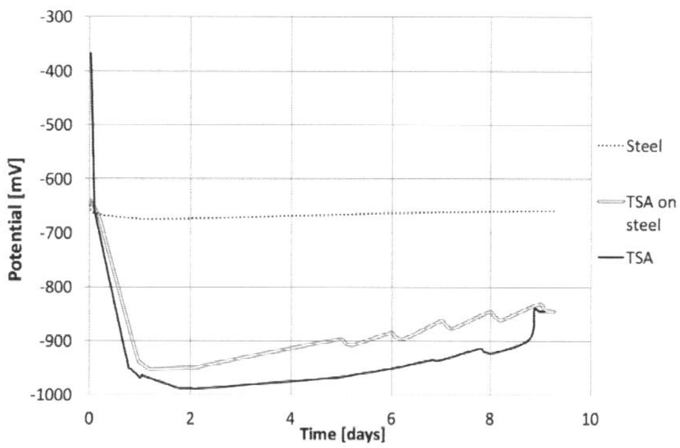

Fig. 2. Variation of potential of steel, TSA on steel and TSA over time during 9 days in CO_2-saturated water.

It can be seen that potential of steel did not change significantly during the test. At the beginning, there was a gentle decline from the value of -650 to -675 mV during the first day but then the potential of the steel flattened out at -660 mV. In contrast, potential of stand- alone TSA decreased dramatically in first day from -350 to -980 mV and it continued to fall to nearly -1000mV until day 2. Between the second and ninth day there was a gradual rise of the potential of stand-alone TSA from the value of around -1000 mV to about -850 mV with two slight drops in day 8 and 9. One can observe that potential of TSA on steel also decreased substantially at the beginning of the test from around -650 to -950mV but then it rose steadily to -900 mV in day 5. Despite an overall increase of potential of TSA on steel until the end of the test, one can observe the number of peaks and troughs over the last 4 days of the test. As can be seen from the graph, potentials of stand-alone TSA and TSA on steel showed similar trends in CO_2 environment.
However, TSA on steel reached higher values than stand-alone TSA for the entire duration of the test except for the first two hours when the potential of TSA was significantly higher. It can well be seen from the graph in **Fig. 3.**, which depicts the change of the potential during the first couple of hours of the test. Analysing the graph one can also see that during first 4 hours potential of steel was lower than potential of TSA on steel, but after this period potential of TSA on steel went down significantly, which resulted in higher values of potential of steel in comparison to TSA. Less negative value of potential of TSA during first 4 hours is critical in view of lack of the protection provided by TSA in case of damaged aluminium layer. More noble potential of TSA leads to protection of aluminium by preferential dissolution of steel, therefore the first few hours, when the potential of TSA is higher than potential of steel, are crucial.

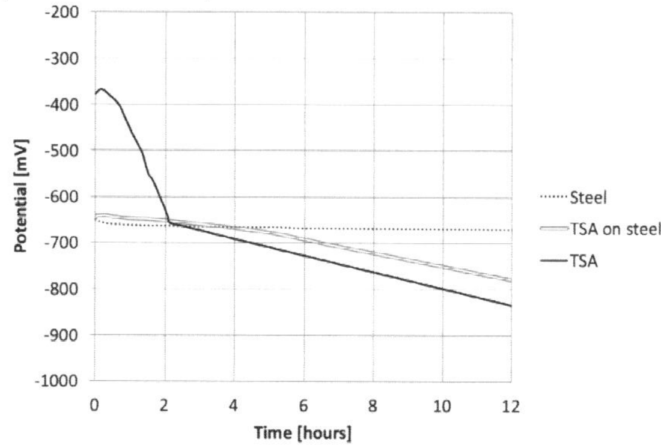

Fig. 3. Variation of potential of steel, TSA on steel and TSA over time during 12 hours.

Knowledge about how fast the potential of aluminium drops in CO_2 environment could be very useful for developing a fully effective TSA-coating. This gives an indication of how long TSA-coating should be supported by other protective layer in order to exclude the possibility of excessive corrosion of steel due to the polarity reversal phenomenon.

2.3 Analysis of amount of dissolved aluminium ions

The results of chemical analysis of aliquots of known volume from the test solution extracted from both glass reactors are shown in **Tab. 1.** below:

Tab. 1. Results of analysis of amount of dissolved aluminium ions.

Reactor 1		Reactor 2	
Surface to volume ratio [cm^{-1}]	Amount of Al^{3+} [mg/L]	Surface to volume ratio [cm^{-1}]	Amount of Al^{3+} [mg/L]
0.038	13.5	0.079	12.6

It can be observed that although the surface area of TSA present in second reactor was twice as big as surface area of TSA in first one, the amount of dissolved Al ions was almost the same in both reactors (the difference is less than 1 mg/L). This may indicate that the saturation of the test solution with aluminium was achieved in both reactors. Information about achieving the saturation can be very useful in the light of corrosion mitigation. Knowing the maximum amount of aluminium, which can dissolve per litre of the solution, can give an indication about the minimum thickness of the coating that can protect the steel in the confined volume of the solution.

3 Experimental Procedure

3.1 Specimen preparation

Prior to spraying, four carbon steel plates with dimensions of 10 by 15 cm were de-burred using a file, degreased with alcohol and then prepared by blasting (at 80psi) with F60 and NK12 alumina abrasive blasting grit to give a rough surface (Ra~10µm, Rz~53µm and Rt~94 µm).

Tab. 2. Composition of substrate and coating consumable used for specimens preparation

Substrate: S355 carbon steel								
Grade	Element wt%							
	C	Si	Mn	P	S	V	N	Fe
EN100 25 S355 J2G3	0.1 2	0.39 0	1.3 9	0.14	0.01 9	0.06 5	0.00 3	balan ce
Coating consumable: aluminium wire								
	Fe	N	Si	Mn	Cu	Al		
	0.2 1	0.01	0.0 7	<0.0 1	<0.0 1	bal.		

The substrate surface was prepared to Sa3 and sprayed with commercially pure aluminium (99.5wt%) using a standard commercial Metallisation 528 twin wire arc spray (TWAS) gun. Spray parameters recommended by the equipment manufacturer (i.e. current, arc voltage, standoff distance) were used to attain a nominal coating thickness of ~300µm. Composition of the wire used is given in **Tab. 2.** . A summary of the arc spray parameters is provided in **Tab. 3.** .

Tab. 3. Spray parameters used for preparing TSA coatings

Equipment used	Equipment Make	Wire size [mm]	Spray Distance [mm]
528	Metallisation Ltd.	2.3	98.7
Spray Distance [mm]	Increment Step [mm]	Horizontal traverse [m/s]	thickness range [µm]
95	15	0.5	250-300

Subsequently, the steel plates were sectioned into 8 parts using vertical band saw PROCUT F-280 and abrasive cutter BUEHLER ABRASIMET 2 and then sharp edges were grinded using BUEHLER PHOENIX 3000. Once completed, the back and sides of 29 specimens were protected with two layers of the lacquer - LACOMIT VARNISH – AGAR G371, to avoid exposure of bare steel to a corrosive environment. After lacquering all the specimens were left to dry overnight at 20°C and then marked (1-29) by using a black waterproof marker.

3.2 Assessment of suitable sealant system

In order to check the suitability of lacquer the following test was performed. LACOMIT VARNISH – AGAR G371 was applied on four glass slides, which were then immersed in a beaker containing deionised water at ambient temperature (19±2°C). Simultaneously, second beaker was filled with deionised water with 1g NaCl (to increase conductivity) at ambient temperature. The pH of both solutions was monitored during a 40-day exposure using the pH meter (370 pH meter JenWAY A921) with the pH probe (HANNA instruments HI1230). Every day, prior to measurements, the pH meter was calibrated with the use of buffer solutions of pH=7 and pH=4. The change of pH (of both solutions) over time was presented in form of the graph. It was observed that LACOMIT VARNISH – AGAR G371 doesn't influence the pH of the solution, therefore the conclusion was made that this lacquer is suitable for the test.

3.3 Surface measurements and coating characterisation

Since thermally sprayed aluminium is a porous material, its real surface area is greater than the apparent geometric area. Surface area was determined using an Alicona InfiniteFocus SL 3D surface profilometer and associated IF Measure Suite software. The area of 4.4506x4.4506 cm was scanned and the surface area of this region was 25.0396 cm^2. Based on the results obtained from 3D profilometry it was possible to establish the overall surface area of all specimens. The geometric surface area of lacquered specimens was determined with the use of the ruler in the following way:

1) Dimension measurements of the inner rectangle
2) Dimension measurements of the outer rectangle
3) Determination of the geometric area of the inner and outer rectangle using the following formula:

$$GA = a * b \tag{7}$$

Where "a" is the length of the shorter side of the rectangle and "b" is the length of the longer side of the rectangle.

4) Determination of the actual geometric surface area by taking the average of inner and outer surface area

3.4 A 30-day exposure test

Two 5-litre glass reactors were filled with deionised water, then closed and sealed. Next, CO_2 was bubbled into the solution with the constant flow rate of 0.03 litres per minute in reactor 1 and 0.06 in reactor 2 through a gas distributer. Next day, when the saturation was achieved (pH=0.93), both glass reactors were opened (without discontinuation of CO_2 supply) and the first set of specimens was put into the reactor 1, while the second set was placed into the reactor 2 with the pair of tongs. Afterwards, both glass reactors were closed and sealed once again. pH of both solutions was continuously monitored during the whole 30-day test using PHH- SD1 pH meters with real-time data loggers and pH probes (HANNA instruments HI1230). Before the start of the experiments the pH meter was calibrated at the test temperature (20°C) in the buffer solutions of pH=7, pH=4 and pH=10. The experimental setup is shown in **Fig.4.** below.

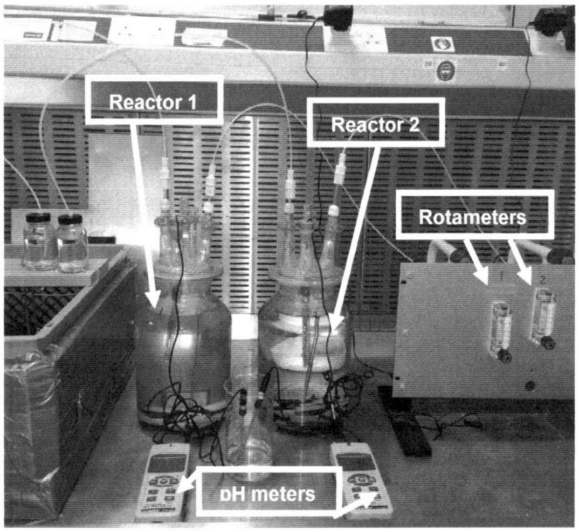

Fig. 4. Experimental setup showing the arrangement.

After 30 days, an aliquot of known volume (25 ml) was taken out from each test solution by a pipette and put in separate labelled plastic containers containing 2 ml hydrochloric acid (HCl). The acid was used to acidify the solution in order to avoid hydrolysis and hence precipitation of metal ions. Next, all specimens were retrieved from the glass reactors for analysis, dried up with the paper towel and placed in an oven at (90°C) for 24 hours to dry. After drying, samples were placed on the paper towel to cool down.
The volume of the solution from both reactors was determined from its mass obtained by weighing the reactors containing solutions, then weighing empty reactors and taking the difference. Measurements were conducted on Precisa IBK 12000D scale.

3.5 Analysis of amount of dissolved aluminium

The concentrations of dissolved Al^{3+} ions in the solution was determined by analysing an aliquot of known volume (25 ml) from the test solution using an inductively coupled plasma optical emission spectroscopy (ICP-OES). The analysis was carried out by a subcontracted laboratory. The results reported here were adjusted for a dilution factor of 1.08 as 2mls of dilute acid was added to each 25mls of sample solution before analysis.

3.6 Potential measurements

When aluminium is coupled to steel, the polarity reversal phenomenon may occur under certain conditions. Potential measurements were performed in order to check if aluminium becomes cathodic to steel in a CO_2 – containing aqueous environment, because it would lead to accelerated corrosion of steel.

For this test, two S355 carbon steel specimens with dimensions of 30 mm by 50 mm were used. One of them was thermally sprayed with aluminium as described in section 3.1. Both specimens were attached to a stainless steel rod cover with insulation as shown. Uncoated specimen was abraded with emery paper, washed with distilled water and acetone in order to guarantee a reproducible starting surface state. Then, the connection between the rod and the specimen was protected with the same lacquer, which was used in previously described test to prevent galvanic interaction between the stainless steel stud and the sample. All uncoated faces of TSA-coated specimen were also protected with 3 layers of lacquer in order to enable determination of a potential of aluminium.

Fig. 5. Electrochemical cell showing the specimens.

The potential of steel and TSA on steel was measured against a reference electrode (pH probe containing silver chloride – HANNA instruments HI1230), which was calibrated before the start of a test against a saturated calomel electrode (SCE) of a known potential. Potential measurements were performed in the same solution as used in the 30-day test (deionised, de- aerated water in 1 barg CO_2 at ambient temperature). An electrochemical cell is shown in **Fig. 5.** .The data, which was collected manually with the use of a potentiostat, was used to plot the change of potential over time.

Additionally, potential of TSA sprayed on glass was measured in a separate reactor containing the same solution under the same conditions with the use of a potentiostat. An electrochemical cell containing a pH probe as a reference electrode and TSA-specimen as a working electrode was used for this test. Data was collected manually using a potentiostat. This test was conducted in order to check the potential of TSA without the influence of the substrate (steel). Due to porosity of TSA, when potential of TSA on steel was measured it was in fact mixed potential of TSA together with steel.

4 Conclusions

Experimental studies showed that thermally sprayed aluminium is a possible candidate for corrosion mitigation in CO_2-containing water. In this work, it was shown that aluminium dissolves in deionised water saturated with CO_2 until the saturation of the solution with Al^{3+} ions is achieved. This dissolution of aluminium is accompanied with an increase of pH of the solution. Studies showed that concentration of aluminium ions in deionised water in equilibrium with 0.1 MPa CO_2 is equal to around 13 mg per litre. This information can give an indication that thermally sprayed aluminium can be possibly used to mitigate corrosion in confined volume of the solution since no more than 13 mg of aluminium will dissolve per one litre of the solution. Moreover, one of the advantages of using TSA as a corrosion mitigation method is the easy way to repair damaged or deteriorated aluminium coating on-site. However, potential measurements of TSA and steel in CO_2 environments revealed that, during the first 4 hours after immersion, undesirable polarity reversal phenomenon takes place. In the event of the TSA coating being damaged, this period is crucial, as steel would be preferentially dissolved, giving protection to the aluminium, as opposite to what is expected from TSA coating. In order to get a better understanding of this phenomenon longer-term test should be conducted.

5 Acknowledgments

The authors would like to thank TWI for funding the research, and Mike Bennett and Ryan Bellward for their technical assistance.

4 Literature

[1] IEA: CO_2 emission from fuel combustion - highlights. Paris: OECD/IEA, (2014).

[2] IPCC: IPCC Special Report on Carbon Dioxide Capture and Storage. Prepared by Working Group III of the Intergovernmental Panel on Climate Change . Metz, B., O. Davidson, H. C. de Coninck, M. Loos, and L. A. Meyer (eds.); Cambridge, United Kingdom and New York, NY, USA: Cambridge University Press: (2005).

[3] Fu, C., and T. Gundersen. Carbon Capture and Storage in the Power Industry: Challenges and Opportunities. Energy Procedia, 2012 International Conference on Future Energy, Environment, and Materials (2012), 16, Part C, pp1806/12.

[4] Bobicki, E. R., Q. Liu, Z. Xu, and H. Zeng: Carbon capture and storage using alkaline industrial wastes. Progress in Energy and Combustion Science (2012), Issue 38, pp. 302/20.

[5] Figueroa, J. D., T. Fout, S. Plasynski , H. McIlvried, and R. D. Srivastava: Advances in CO_2 capture technology—The U.S. Department of Energy's Carbon Sequestration Program. International journal of greenhouse gas control (2008), Vol. 2.

[6] Svensson, R., M. Odenberger, F. Johnsson, and L. Stromberg. Transportation systems for CO_2—application to carbon capture and storage. Energy Conversion and Management (2004), 45, pp. 2343/53.

[7] Pires, J. C. M., F. G. Martins, M. C. M. Alvim-Ferraz, and M. Simões: Recent developments on carbon capture and storage: An overview. Chemical Engineering Research and Design 89 (2011), Issue 9, pp. 1446/60.

[8] Herzog, H. J.: Scaling up carbon dioxide capture and storage: From megatons to gigatons. Energy Economics 33 (2011), Issue 4, pp. 597/604.

[9] Paul, S., R. Shepherd, A Bahrami, and P. Woollin: Material selection for supercritical CO_2 transport. Newcastle, Proceedings of the first International Forum on the Transportation of CO_2 by pipeline (2010), pp. 1/16.

[10] Bilio, M., S. Brown, M. Fairweather, and H. Mahgerefteh: CO_2 pipelines material and safety considerations. IChemE 155 (2009), pp. 423/29.

[11] Heggum, G., T. Weydahl, R. Mo, M. Mølnvik, and A. Austegaard: Chapter 15 - CO_2 conditioning and transportation. In Carbon Dioxide Capture for Storage in Deep Geologic Formations - Results from the CO2 Capture Project: Vol 2 - Geologic Storage of Carbon Dioxide with Monitoring and Verification, by D. C. Thomas, & S. M. Benson, edited by D. C. Thomas, & S. M. Benson; Elsevier: Amsterdam, (2005), pp. 925/36.

[12] Vargel, C.: Chapter B.1 - Introducton to the corrosion of aluminium. In corrosion of aluminium, by C. Vargel; Elsevier: Amsterdam, (2004); pp. 81/109. ISBN: 978-0-08-044495-6.

[13] Paul, S: Corrosion Control for Marine- and Land-Based Infrastructure Applications. In ASM Handbook, vol. 5A; Robert C. Tucker, Jr. (eds.); ASM International: (2013); pp. 248/52. ISBN: 978-1-61503-996-8.

Electrochemical Impedance Spectroscopy of Thermally Sprayed Aluminium in Synthetic Seawater

S Paul, Cambridge / UK

Commercially pure aluminium (>99.5wt% Al) was arc-sprayed on carbon steel. Electrochemical impedance spectroscopy (EIS) was carried out on TSA-coated steel samples when exposed to synthetic seawater. A simple pore network circuit model was used to analyse the data. Corrosion rate was calculated using this pore network model. The corrosion rate was calculated to be ~5-15μm/y from the charge transfer resistance value. The mechanism of protection offered by TSA is also discussed.

1 Introduction

The use of structural materials such as carbon steels for offshore applications depends on the ability to mitigate corrosion of such materials economically for long term. Coating systems are sought to mitigate corrosion in most cases where the use of corrosion resistant alloys would be economically unviable. Finding a coating system that can withstand the corrosive action of seawater, sun's UV radiation and floating debris is a challenge. Polymeric coating systems which protect the underlying structure by forming a barrier are risky to use in regions such as splash and tidal zones where damage by floating debris is likely. When the barrier layer is damaged no protection is offered by organic paint systems. However, this is not the case when a sacrificial coating such as aluminium is used.

Arc-sprayed aluminum has been used in offshore applications for several decades. Aluminium, being anodic to steel, offers cathodic protection to steel by being preferentially consumed Protection offered in such a manner is named sacrificial protection. One of the earliest recorded commercial applications of thermally sprayed aluminium (TSA) in the UK includes the Hutton tension leg platform in the North Sea in the early 1980s. Since then, several researchers have studied the behavior of these coating in saline solutions- some using seawater [1-4], natural or synthetic, and others NaCl solution [5]. The use of NaCl solution to simulate offshore immersion conditions is not appropriate as this is not representative of seawater. Seawater contains cations such as Ca^{2+} and Mg^{2+}, essential for the deposition of calcareous matter on cathodically protected steels. Without the presence of these cations the protective mechanism is altered and one is likely to get misleading data [6]. Understanding the mechanism of protection offered by TSA is important as such information can be used to improve upon existing coatings or develop novel coating systems.

The study of corrosion performance of TSA is not new. Polarization tests [7] and salt spray tests [8] to show the protective behaviour of TSA from corrosion in chloride-containing environments are often carried out in the laboratory. However, long-term field tests in the marine atmosphere to investigate the corrosion are rare [9].

The corrosion rates are often calculated using linear polarisation technique. However, this technique does not give information regarding the mechanism of corrosion. Electrochemical Impedance Spectroscopy (EIS), on the other hand, is a technique that can give insight into the corrosion mechanism and help us understand the protective function of TSA when applied on carbon steel.

In this paper, the mechanism of corrosion protection and its changes during synthetic seawater immersion investigated using electrochemical impedance spectroscopy (EIS) is reported. TSA-coated steel samples were exposed to synthetic seawater for different times to study the long-term protective performance of arc-sprayed aluminum coatings.

2 Experimental

2.1 Materials

Carbon-Manganese steel was selected as the substrate material which conforms to specification S355J2+N(2004). The coating material selected was 2.3mm diameter commercially 'pure' aluminium (99.5% Al) supplied by Metallisation Ltd. (UK). The nominal composition of the substrate and coating material is given in **Table 1**.

Table 1. Composition of steel substrate and Al wire (coating material).

Materials	Element (wt%)							
	C	Mn	Si	S	P	Fe	Cu	Al
Steel substrate	0.12	1.39	0.39	0.019	0.014	Bal.	-	-
Al wire	-	<0.01	0.07	-	-	0.21	<0.01	Bal

2.2 Coating production

The substrate was cut into 50×50×6mm flat panels and the edges were rounded off using a file prior to further preparation. Prior to coating, the substrates were degreased and then prepared by blasting with F60 and NK12 alumina to give a rough surface to which the molten and semi-molten spray particles can key onto.

Thermal spray coatings were prepared using a Metallisation 528 twin wire arc spray gun mounted on a 6-axis robot. Dry compressed air, supplied at 5.5bar, was used as the particle atomising and propellant gas. Coatings were produced using the spray parameters given in **Table 2** to attain a nominal coating thickness of 300µm.

Table 2. Spray parameters used for preparaing TSA coatings

Wire feed rate (g/min)	Spray distance (mm)	Increment step (mm)	Traverse speed (m/s)
98.7	95	15	0.5

2.3 Corrosion testing

Immersion test in ASTM D1141 synthetic seawater was carried out at 35°C [10]. The conventional 3-electrode arrangement, with the TSA-coated steel as the working electrode, standard calomel electrode (SCE) as the reference and Pt/Ti electrode as the counter, was used to gather electrochemical data for 9 months. Of the electrochemical techniques available, EIS seemed to give vital information on the corrosion mechanism and hence it was used. An ACM Gill 12 electrochemical workstation was used for potential and EIS studies. A 30mV RMS AC signal amplitude was used with frequency scans between 10kHz and 1Hz.

2.4 EIS data analysis

A suitable model was selected to fit the EIS data, **Fig. 1**. Data analysis was carried out using ZSimWin (EChem Software)

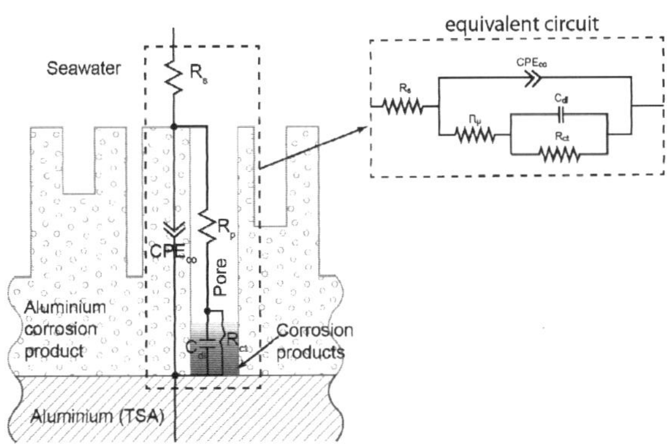

Fig. 1. Schematic of the coating showing equivalent circuit elements such as solution resistance (R_s); constant phase element (CPE_{co}), pore resistance (R_p), interface reaction (C_{dl}) and charge transfer resistance (R_{ct}).

The various circuit elements are shown in **Fig. 1**. The model was selected to represent the various processes that are likely to occur in the coating system during exposure to saline solutions [4,5]. Based on previous work, it was assumed that the coating would have an oxide layer on TSA which when exposed to seawater may be transformed into a hydrated oxide layer [6]. Thus, the aluminium corrosion product is assumed to act as a constant phase element (CPE_{co}). In addition to this, various elements have been added to represent the seawater solution resistance (R_s), resistance offered by the pore network (R_p), and R_{ct} and C_{dl} to simulate the charge transfer and double-layer capacitance of the interface between Al and seawater in the pore.

2.4 Determination of corrosion rate

The corrosion rate (CR) was calculated by taking the R_{ct} value into account and using the Tafel equation. The equation used for TSA is:

$$CR = (i_{corr} \times 3237 \times EW)/(d \times A) \quad (1)$$

Where $i_{corr} = 18/R_{ct}$ (for Al in seawater), EW is the equivalent weight of Al(9), d and A are density and area in gm/cm^3 and cm^2, respectively. The density of Al was taken to be $2.7 g/cm^3$.

2.5 Microstructural characterisation

After the corrosion experiments, changes to the coating appearance were recorded (i.e. the samples were photographed). The surfaces of the corroded specimens were examined by light optical microscopy and SEM (samples were gold coated prior to this), including element mapping obtained using energy dispersive X-ray spectroscopy (EDX).

3 Results

3.1 Visual examination

The samples were photographed before and after exposure to synthetic seawater, **Fig. 2**.

(a) (b)

Fig. 2. Photographs of samples (a) before and (b) after testing in synthetic seawater. The field of view is 25×25mm.

The surface of the sample was also analysed under SEM using a EDX analyser. The EDX spectra of the while spots showed the presence of Al and O (not presented in this paper), implying the presence of oxide and possibly hydrated oxides of alumina. The reflectivity of the coating also changed due to the corrosion product layer on the surface, as expected.

3.2 EIS model validation

Several values of the circuit elements selected in the model shown in **Fig. 1** were tried using ZSimWin to find the best fit to the experimental data. In all cases, the seawater resistance (R_s) was assumed to be between 15-45Ωcm. The data obtained after 134days of immersion are shown in **Fig. 3**.

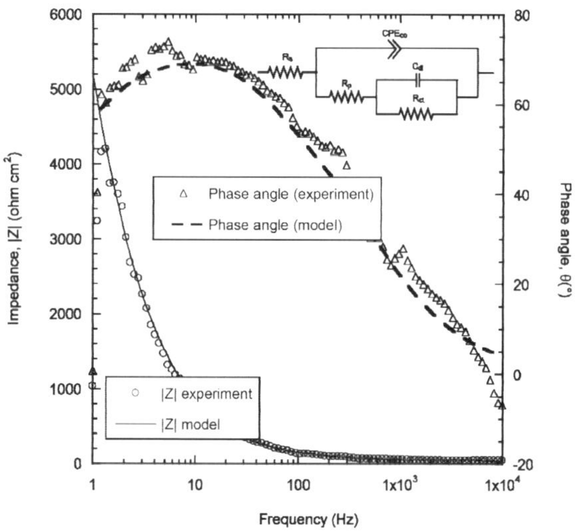

Fig. 3. Impedance and phase data after 134days of exposure showing the correlation between experimental and model data. Model parameters used are CPE_{co} = $3.5\times10^{-5}Ss^n$, n = 0.81, R_p = 64Ω, C_{dl} = 2.5×10^{-6}F and R_{ct} = 1.6×10^4Ω.

A reasonable fit was obtained by using the values shown in the figure caption.

3.3 EIS of TSA in synthetic seawater

The impedance data was used to understand the kinetic parameters for electron transfer reactions at the TSA/electrolyte interface. There are several ways of representing the EIS data. The most commonly used representations are Nyquist, Bode and modulus of Impedance plots. The representation of real vs imaginary impedance data collected as a function of frequency gives Nyquist plot. The EIS Nyquist plots obtained during the exposure of TSA in synthetic seawater shows different behaviour in the early stages of exposure when compared to long-term data, **Fig. 4**. The data in the first few days of exposure shows tendency to form a semi-circle, however, this is not the case after a month of testing, **Fig. 4** (inset). It would appear that the mechanism of protection of a TSA-coated steel is changing with the time of exposure resulting in the formation of semicircles at the start, but not so, as the corrosion reaction progresses. The data was collected from 10kHz and 1Hz so the response at very low frequencies is not available.

The intercept on the real impedance axis (Z') corresponde to the resistance (R_s) provided by the synthetic seawater. The change in the behaviour of the Z'-Z" plot with time gives us useful information regarding the change in the nature of the coating with time. The nature of the Z'-Z" plot changing with time from that of a semi-circle to that of a straight line. The real impedance values increased by several orders of magnitude with the time of exposure in synthetic seawater. The overall impedance is a function of resistive and capacitive elements as the model selected has these elements. The way these combine to give the overall impedance depends on the way these are arranged, **Fig.1, 3**. Nonetheless, useful information can be extracted from the Z'-Z" plot.

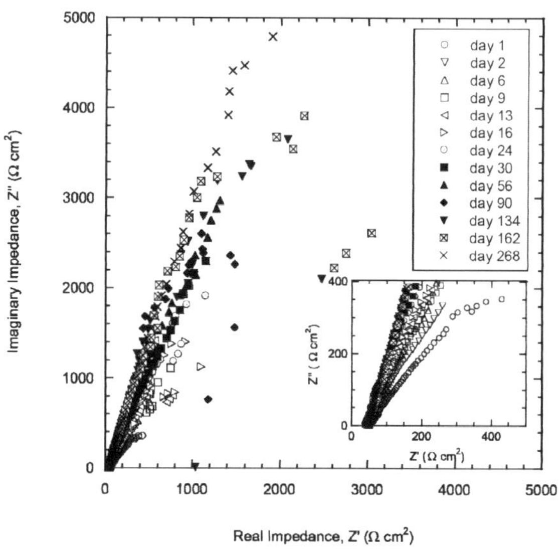

Fig. 4. Experimental Nyquist plot obtained during the exposure of TSA in synthetic seawater at 35°C. The inset shows the higher frequency Nyquist plot.

Another commonly used representation of EIS data is the phase angle plot. This is a plot of phase angle as a function of frequency. The impedance phase angle plot, also called the bode plot showed an increase in the value as the frequency decreased from 10kHz to 10Hz, **Fig. 3, 5**. However, the nature changed below 10Hz and the phase angle decreased as the frequency was reduced.

The model selected for data analysis gave a reasonable fit to the experimental data, **Fig. 3**. The constant phase element which incorporated the non-ideal capacitive part of the coating system required input exponent (n) values between 0.6-0.8 to get a good fit. The long-term phase angle plot showed higher experimental values of phase angle with exposure time, **Fig. 5**. In all cases, the angle at 1Hz was found to be between 40-70°. This implies that the circuit elements involving capacitors have a significant impact on the behaviour of the coating system when exposed to synthetic seawater.

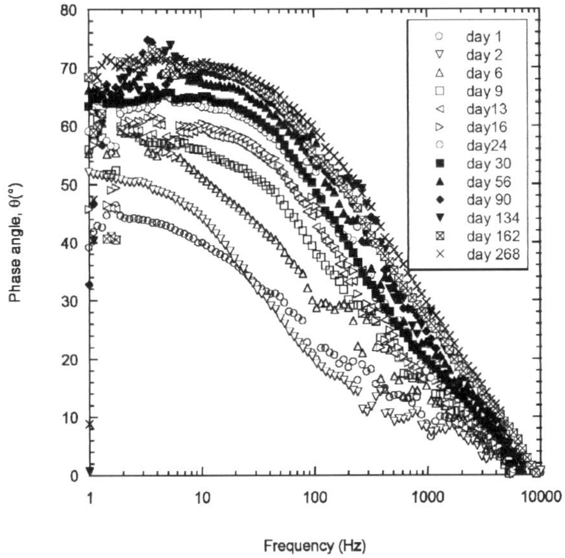

Fig. 5. Experimental long-term Bode phase plot of the TSA-coated steel immersed in synthetic seawater at 35°C.

In addition to the Nyquist and Bode phase plot, modulus of impedance is also used to represent EIS data. The modulus of impedance (|Z|) is often calculated as: $|Z| = (Z'^2 + Z''^2)^{1/2}$. Thus, the modulus is dependent on the real (Z') and imaginary (Z") impedance.

The modulus of impedance data plotted as a function of frequency, also called the Bode impedance plot is showen in **Fig. 6**.

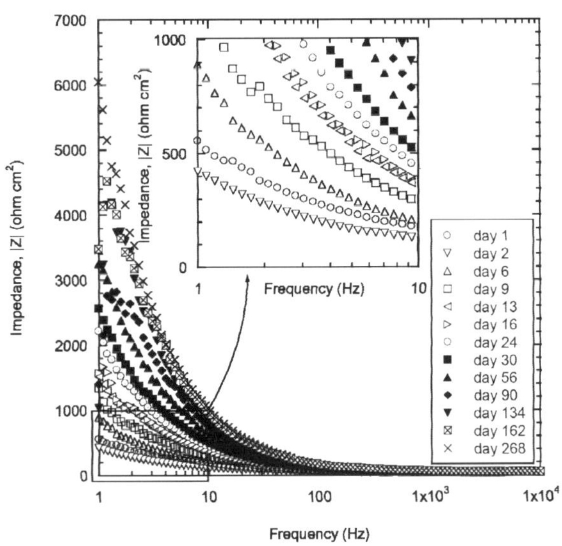

Fig. 6. Bode impedance plot showing the behaviour of TSA-coated steel during long-term exposure in synthetic seawater. Inset shows the low frequency (1-10Hz) impedance response.

In general, the impedance increases with the decrease in the applied frequency, **Fig. 6**. At 10kHz the modulus of impedance was found to be ~50Ωcm² irrespective of the exposure period, while it varied by an order of magnitude at 1Hz depending on the exposure time- increasing with the time of exposure. The modulus of impedance at 1Hz increased from $\sim 5\times10^2 \Omega cm^2$ in the first few days of exposure to $\sim 6\times10^3 \Omega cm^2$ after ~9 months in synthetic seawater. However, it would appear that the modulus of impedance after 2 days of exposure was lower than the day 1 value. Beyond this point the impedance seems to be increasing steadily with the exposure period.

3.4 Corrosion rate of TSA-coated steel

The corrosion rate was obtained by incorporating the R_{ct} values into equation 1. The R_{ct} values were obtained by finding the values of the circuit parametrs that gave the best fit to the experimental data, **Fig. 3**. The R_{ct} values were low at the start of exposure, $10^4\Omega$, and then it had values exceeding $10^6\Omega$ after a few months. These values when incorporated in equation 1 gave corrosion rate values between 1mm/y to 5µm/y. The corrosion rate of TSA decreased with time and reached values of ~5-15µm/y in the later stages of exposure. The high initial corrosion rate was only seen in the first few scans.

4 Discussion

4.1 Electrochemical Impedance Spectroscopy

Althought, EIS is a powerful technique for studying corrosion mechanism, the choice of circuit elements and models determine how effectively the data is used to understand the charge transfer mechanisms occurring at the electrode-electrolyte interface. There are several methods to determine if the model selected for data analysis is a reasonable one. In some cases these require basic understanding of the circuit behaviour, and for others, complex mathematical analysis such as the use of Kramers-Kronig relations is required [11]. For example, a device with a purely resistive element exhibits no phase shift between the voltage and current, hence, would only show a frequency-independent straight line in the Nyquist plot. Thus, the model needs to have only a resistive circuit element. This, however, is a very simple example and no real system involving coating is likely to be that simple.

In the system studied in this work, several circuit elements were required to explain the behaviour of the coatings. Each of the elements serve a function and represent a physical phenomena. For example, the high frequency loop (**Fig. 4**, inset) at the start of the exposure corresponds to the coating, while the low frequency response can be attributed to the corrosion process [12]. The seawater which penetrated into the coating through the pore network and splat boundaries caused the dissolution of aluminium inside the coating and generated capacitive loop at lower frequencies. This, however, was not obvious as the data was not collected below 1Hz. Data going well below 0.1Hz is required to see more prominent capacitive loop [5].

Capacitor, another circuit element often used to represent physical phenome involving coatings, are formed when two conducting plates are separated by a non-conducting media, called the dielectric. The non-conducting media could be air, vacuum, coating or the corrosion product. The value of the capacitance (C) depends on the area of the plates (A), the distance between the plates (d) and the dielectric constant of the material ($\mathcal{E}$) and is given by $C = \mathcal{E}A/d$. In addition to the coating or the corrosion product being a capacitor, the properties of a pore is also represented by capacitors. The environment an electrolyte sees inside a pore is often different from that of the bulk. On one side of the pore the electrolyte is experiencing the metal where corrosion reaction is taking place, thereby changing the local composition of the electrolyte and on the other side the electrolyte is extending into the bulk solution. The interface between the electrolyte in the pore and the bare metal is often modeled as a double-layer capacitor (C_{dl}). The use of pure capacitors to represent real coatings is frought with difficulty as real coating hardly have a pure capacitance in the electrochemical processes. To overcome this challenge, constant phase element (CPE) are often used. They are now commonly used to replace capacitance in coated systems to represent the inhomogeneities of the real electrode surface.

Although, the model selected in this work gave a reasonable fit to the experimental data several other circuit elements could have been used to better fit the data. From a mechanistic point of view, the use of certain elements to represent diffusion might seem appropriate. Diffusion of species can create an impedance which is often represented by a Warburg element (also called Warburg impedance). This impedance depends on the frequency of the potential perturbation. Unlike a pure resistive element, this impedance is frequency dependent. At high frequencies, the Warburg impedance is small, but at low frequencies the Warburg impedance is large. On a Nyquist plot the Warburg impedance appears as line with 45° slope. On a Bode Plot, the Warburg impedance exhibits a phase shift of 45°. This is inline with what was observed in this work implying that certain reactions occurring at the coating-seawater interface involve diffusion. In general, the accumulation of the corrosion products in the pores of coatings leads to the appearance of Warburg behaviour which causes the mass-transfer reaction [13].

The selection of a model is crucial and incorrect model may give values of circuit elements not representative of corrosion processes. It may well be argued that such kinetics-based models are too sensitive to the selected models and are prone to errors. Nonetheless, the interpretation of experimental data is not entirely dependent on the model selected.

One point mentioned earlier is the data presented in the Bode impedance plot, **Fig. 6**. Although, one may be tempted to dismiss the apparent discrepancy in the day 1 and day 2 data as experimental error, but on closer inspection it appears to be a real effect. Part of the reason for the initial reduction in impedance is the dissolution of air-formed oxide film on TSA in synthetic seawater [14]. The oxide film formed on aluminium in water at ambient temperatures is thin and amorphous and may consist of some bayerite [15]. Once the air-formed film is replaced by the water-formed corrosion product the impedance of the TSA-coated steel rises to a slightly higher value depending on the protective nature of the oxide and the morphology of the deposit formed on the exposed steel surface.

4.2 Corrosion perfromance of TSA-coated steel

Aluminium being anodic to carbon steel in seawater protects it by being preferentially consumed when electrically connected, thus providing cathodic protection. Owing to this nature, aluminium is thermally sprayed onto carbon steel to protect offshore steel structures from the corrosive marine environment. Although, aluminium is anodic to steel, the primary function of the coating is to act as a barrier. If the barrier is breached, the anodic nature of the coating becomes crucial. Even in the absence of an external damage thermally sprayed coatings contain porosity which are often surface-connected and at times reach the substrate. The presence of surface-connected porosity could result in the electrolyte (seawater) reaching the substrate, but since the substrate is cathodic to the coating (anode), corrosion of steel is not expected.

The anode (aluminum in the case of TSA-coated steel) polarises the exposed steel and results in the production of hydroxyl ions (OH^-), which increases the interfacial pH. The generation of OH^- occurs either due to the reduction of dissolved oxygen or by the reduction of water (which occurs at slightly greater negative potential) [17]. The generation of OH^- ions facilitates the precipitation of magnesium hydroxide from seawater. Moreover, the increased production of OH^- leads to changes in the stability of dissolved Ca^{2+} ions at the steel surface. The Ca^{2+} ions form the basis of calco-carbonic equilibrium which is governed by the dissolution and subsequent dissociation of CO_2 in seawater. It is generally recognised that during subsea cathodic protection, the formation of calcareous deposit on exposed steel surface is favourable since it reduces the cathode area. These deposits on steel under cathodic protection comprises $CaCO_3$ and $Mg(OH)_2$ at ambient seawater temperatures [6]. This is in line with the observation of the experimental work where an increase in impedance was observed with exposure time and a reduction in corrosion rate with time.

5 Conclusions

The following conclusions can be drawn from this work where EIS was used to understand the mechanism of long-term corrosion protection offered by TSA to carbon steel in a simulated marine immersion environment at 35°C:

- A simple model involving capacitors, resistors and constant phase elements can represent the coating behaviour. However, the use of a diffusion element such as 'Warburg' impedance may be used to further modify the model and provide better represent the coating behaviour.
- The representation of EIS data using Nyquist and Bode representations was found suitable for understanding the behaviour of TSA-coated steel in seawater. However, it is evident that data collected over a wider range of frequencies, particularly the ones below 1Hz are useful.
- The coating impedance increased with time which was inline with the microstructural changes in the coating during exposure. The sealing of surface-connected pores by corrosion products resulted in the increase of impedance.
- The charge transfer resistance obtained by fitting the selected model to the experimental EIS data. gave the corrosion rate value of TSA to be ~5-15μm/y in the later stages of exposure. After 9 months of exposure the corrosion rate was ~5μm/y.
- The low long-term corrosion rate and high integrity of TSA implies that it can provide long-term corrosion protection to carbon steel offshore structures. The anodic nature of TSA also ensure the protection of steel even when TSA is damaged in certain regions.

6 Acknowledgements

The work was funded by TWI. The author would like to thank TWI staff, especially Andrew Tabecki, Mike Bennett and Sheila Stevens for their technical support.

7 References

[1] Paul, S. et al.:Thermally Sprayed Aluminum (TSA) Coatings for Extended Design Life of 22%Cr Duplex Stainless Steel in Marine Environments. Journal of Thermal Spray Technology 22 (2013), Issue 2-3, pp328/336.

[2] Paul, S. et al.: Corrosion Testing of Thermally Sprayed Aluminium. ITSC/2015, paper no. 38552 (Long Beach, CA: ASM International, 2015), pp 964/970.

[3] Gartland, P.O. and T.G. Eggen: Cathodic and anodic properties of thermally sprayed Al- and Zn-based coatings in seawater. Corrosion/90, paper no. 367 (Houston, TX: NACE, 1990), pp. 1/16.

[4] Zhu, Q.J. et al.:Electrochemical impedance spectroscopy analysis of cold sprayed and arc sprayed aluminium coatings serviced in marine environment. Surface Engineering 28 (2012), Issue 4, pp. 300/305.

[5] Esfahani, E.A. et al.: Study of corrosion behavior of arc sprayed aluminum coating on mild steel. Journal of Thermal Spray Technology 21 (2012), Issue 6, pp. 1195/1202.

[6] Paul, S: Corrosion performance of damaged thermally sprayed aluminium in synthetic seawater at different temperatures. Thermal Spray Bulletin 8 (2015), Issue 2, pp. 139/146

[7] Han, M.S. et al.: Effects of thickness of Al thermal spray coating for STS 304. Transactions of Nonferrous Metals Society of China 19 (2009), pp. 925/929.

[8] Rodriguez, R.M: Comparison of aluminum coatings deposited by flame spray and by electric arc spray. Surface and Coatings Technology 202 (2007), pp. 172/179.

[9] Kuroda, S. et al.: An 18-year exposure test of thermal-sprayed Zn, Al, and Zn-Al coatings in marine environment. Corrosion 62 (2006), Issue 7, pp. 635/647.

[10] ASTM D1141-98(2013), Standard Practice for the Preparation of Substitute Ocean Water, ASTM International, West Conshohocken, PA, 2013.

[11] Orazem, M.E.: A tutorial on impedance spectroscopy. Corrosion/98, paper no. 302 (Houston, TX: NACE, 1998), pp. 1/11.

[12] Liu, C. et al.: EIS comparison on corrosion performance of PVD TiN and CrN coated Mild Steel in 0.5 N NaCl aqueous solution, Corrosion Science 43 (2001), Issue 10, pp 1953/1961.

[13] Y. Wang, Y. et al.:Microstructure, spallation and corrosion of plasma sprayed Al2O3-13%TiO2 coatings, Corrosion Science 51 (2009), Issue 12, pp. 2924/2931.

[14] Lorking, K.F., and J.E.O. Mayne: The corrosion of aluminium. Journal of Applied Chemistry 11 (1961), No. 5, pp. 170/80.

[15] Ghali, E.: Aluminum and Aluminum Alloys, in Uhlig's Corrosion Handbook, 3rd Edition (Ed. R. W. Revie), pp. 715/45, John Wiley & Sons, Hoboken, NJ, USA, 2011.

[16] Paul, S.: Corrosion control for marine- and land-based infrastructure applications, ASM Handbook, Vol. 5A, pp. 248/52, ASM International, Ohio, 2013.

[17] Barchiche, C. et al.: Characterization of calcareous deposits by eletrochemical method role of sulphates, calcium concentration and temperature. Eletrochimica Acta 49 (2004), Issue 17-18, pp. 2833/2839.

Effect of multi-pass coatings on surface structure of cold sprayed Si coatings

J. Nakamura, H. Saito, K, Fujita, K, Sakaki, Nagano/J

Pristine Si powders were cold sprayed on Cu foil. To thicken the coated layer, in addition to 1 pass coating, 2 pass coatings and 3 pass coatings were carried out. In regard to Ni-P coated Si powders, 2 pass coated anode shows excellent charge discharge characteristics. That means 2 pass coated layer seems to be thicker than 1 pass coated layer. Unfortunately for the pristine Si powders, even if 2 pass operation is performed, additionally attached Si mass is only 2～3 %.

1 Introduction

Li-ion secondary batteries are widely utilized for mobile electronic equipment as personal computers or smart phones. Recent years large capacity Li-ion secondary batteries are heavily used for automotive vehicles.

At present for Li-ion secondary batteries. carbon anode materials are mainly utilized. Theoretical charge discharge capacity of carbon anode is 372 mAh/g. Whereas Si anode, its capacity is 3572 mAh/g[1], that is 10 times larger than that of carbon anode. In our laboratory cold sprayed Si anodes have been studied[2] over 10 years. To obtain excellent charge discharge characteristics, about 5 μm thickness of Si is preferable[3]. Notwithstanding obtained thickness of cold sprayed Si coatings were few microns. To increase coating thickness, many trials have been studied, such as setting of Cu foil, control of particle velocity, spray gas temperature or multi-pass coatings. H. Usui et al[4] reported Ni-P coated Si particles. Its Ni-P coated Si anode has exhibited excellent performance in charge-discharge cycling number. In this report we deal with survey of Ni-P coated Si anode, using cold spray technique, and at the same time examined effectiveness of multi pass coating on charge discharge characteristics.

2 Procedure

Pristine Si powders of average 5 μm diameter and Ni-P coated Si powders were cold sprayed on Cu foils.

Starting materials

Pristine Si powders

Si powders were provided by Kishimoto Sangyo Co. jet mill grinded and sieved. The particle size distribution is shown in Fig.1.

Ni-P coated Si powders:

At first pristine Si powders were immersed in alkaline solution to degrease. After immersed in Sn sensitizing solution and Pd activating solution, then chemically Ni-P is deposited under 90℃ 2.5～8 min.

Cu foils:

For collector substrate, electrolyticaly deposited Cu foils of 35 μm thickness ,provided by Furukawa Electric are used.

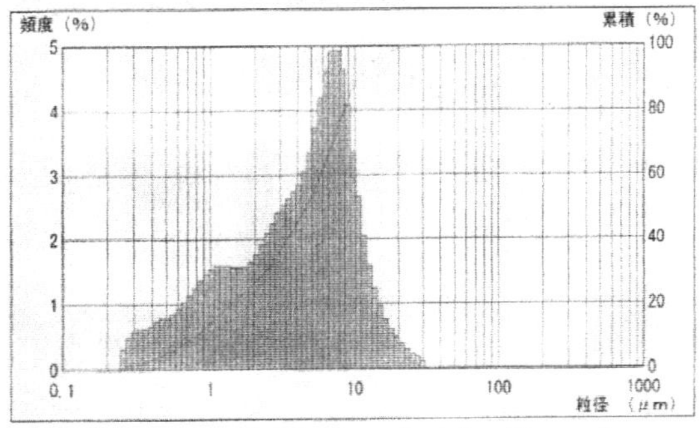

Fig.1. Particle size distribution of pristine Si powder

Cold spray equipment:

Cold spray equipment is shown in Fig.2.

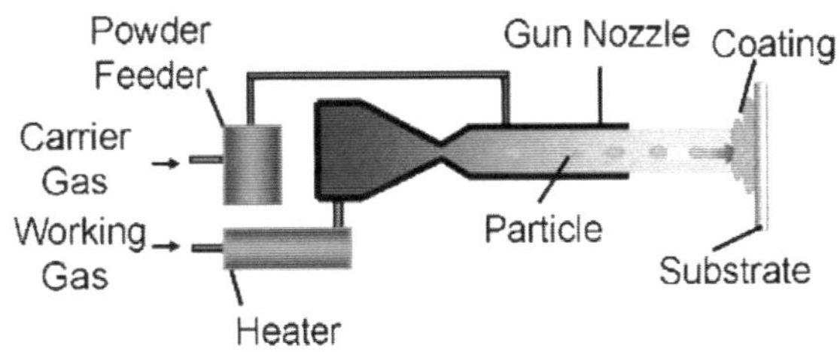

Fig.2. Schematic diagram of cold spray equipment

3 Experiment

Pristine Si powders and Ni-P coated Si powders are cold sprayed at

Inlet N_2 gas pressure: 1.5 MPa

Inlet N_2 gas temperature: 573K

Gun nozzle: 140mm x 7 φ

Spray distance : 15mm

Gun traverse speed: 20 mm/s

1 pass spray, 2 pass spray and 3 pass spray were carried out. At the moment of 2 pass spray, temperature of the collector substrate is too high so about 10 min standby was inserted to next spray operation..

Si and Si composite cold sprayed Cu substrate are then punched to φ 16 to put together CR2032 coin cell. The CR2032 coin cell is illustrated in Fig.3.

Cold sprayed Si and Si-Ni coted Cu foils were examined by FE-SEM JSM-7000F(JEOL Ltd.) and TM 3030 PLUS Miniscope(HITACHI).

Charge-discharge characteristics were measured by "RTS2004W(NAGANO Co.)"

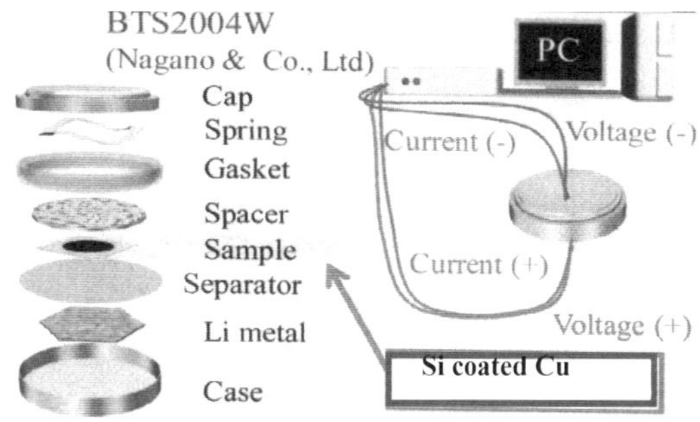

Fig.3 Schematic diagram of CR2032 coin cell

4 Results and discussions

Charge discharge characteristics of Ni-P coated Si anodes

Charge discharge characteristics of Ni-P 1pass coated anode is shown Fig.4.

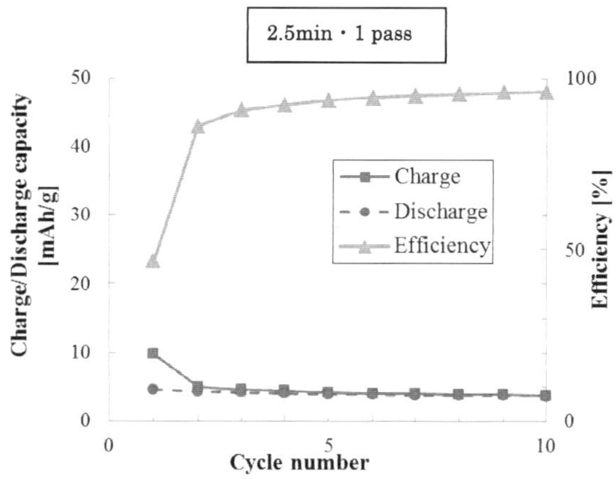

Fig. 4. Charge discharge curve of Ni-P 1 pass coated anode

Charge discharge characteristics of Ni-P 2pass coated anode is shown Fig.5.

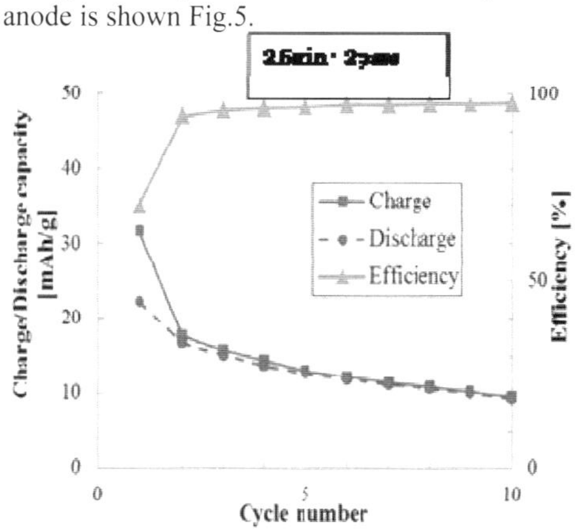

Fig.5. Charge dischsrge curve of Ni-P 2pass coated anode

Charge discharge capacity of Ni-P 2 pass coated anode is nearly twice as high as that of 1 pass coated anode. Unfortunately its value of 22 to 32 mAh/g is as low as compared to theoretical value of 4000 mAh/g.

At cold sprayed Si anodes, even though 2 pass film is coated, there appear no tendency to increase coating thickness for long years of research in our laboratory.

The 2 pass coating of Ni-P coated Si is considered to be effective to value up of charge discharge capacity. So the authors try again to multi-pass coating of pristine Si powder.

4.2 SEM observation and EDX analysis

The surface morphologies, characteristic X-ray intensity mapping of Si were examined. SEM image of 1 pass coated Si 1200x and 3000x are shown in Fig.6 and 7.

Fig.6. 1 pass coated Si 1200x

Fig.7. 1 pass coated Si 3000x

Pristine Si powders consist of 1 to 15 μ m diameter particles, and average particle size is 5 μ m. On the coated surface, only small particles of 2 to 3 μ m is observed. Electrolyticaly deposited Cu surface is not smooth and many projections around 10 μ m height are existed. In the observation of enlarged image(3000x), any large Si particles of 5 to 10 μ m is found.. For 1 pass coated substrate,characteristic X ray mappings of Si are shown in Fig.8. In the same way, for 2 pass coated substrate SEM image and X ray mapping of Si are shown in Fig.9. and Fig.10.

X ray intensity ratio of Si and Cu at 1 pass to 3 pass coated substrates is shown in Fig.11.

Considering details in EDX of Si and Cu , Si intensity ratio of 1 pass coated substrate is 46.5%,.

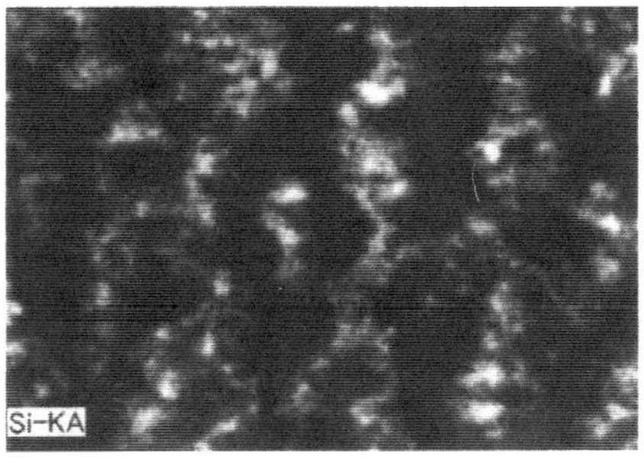

Fig.8. Si 1 pass coated 1200x
EDX characteristic X rav of Si

Fig.9. Si 2 pass coated 1200x

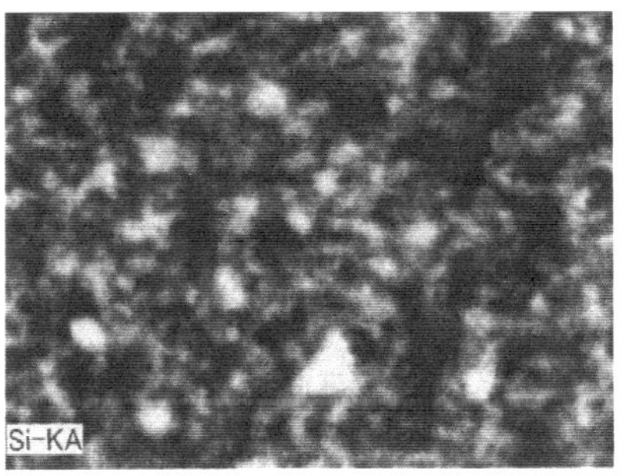

Fig.10. Si 2pass coated 1200x
EDX characteristic X ray of Si

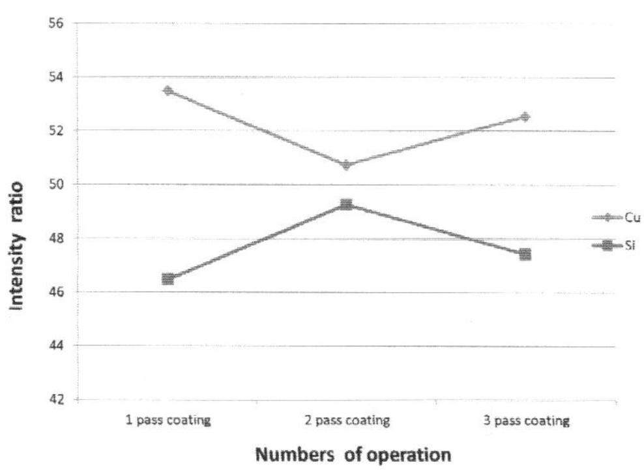

Fig.11. X ray intensity ratio of Si versus Cu

in contrast that of 2 pass coated substrate is 49.3%

At 1 pass coating, fine Si particles cover Cu foil surface and succeeding 2 pass operation, some of the fine Si particles cling to pre attached Si surface. In 3 pass coating operation, additional attach of Si particles is expected, but unfortunately any additional clings is observed .In 3pass operation, Si X ray intensity ratio falls to 47.4%,thogh it is higher than that of 1 pass operation.

5 Simulation of gas velocity and particles velocity

Gas velocity:U_g and particles velocity:U_p are simulated using ANSYS FLUENT R15.0

The shape of spray nozzle is shown in Fig.12.

Gas Velocity and particles velocities during aviation are shown in Fig.13. Particles velocities at the Cu foil are shown in Fig.14.

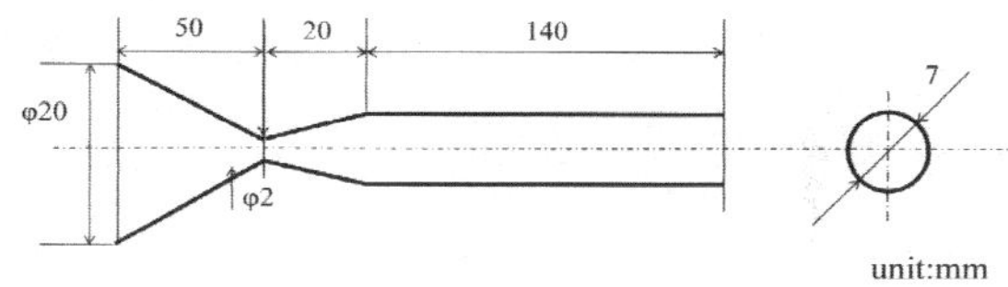

Fig.12.Dimensions of spray nozzle

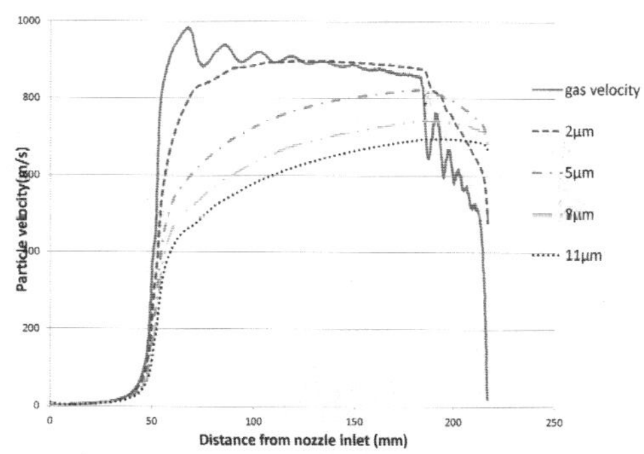

Fig.13 Particles and gas velocity during aviation

At Cu foil position, particles over 4μm size have enough speed over 650m/s. As far as considering collision speed, coarse particles have ability to deposit on the board. Actually only fine particles can deposit. When reached at the board surface, we consider that coarse Si particles which is hard, do not adhere to pre attached hard Si surface.

6 Summry

At the 2 pass coated anode with Ni-P coated Si powders, charge discharge characteristics are improved. Then the authors expected even in pristine Si powders,multi pass coatings will be effective to thicken the coating thickness. At the 2pass coated substrate, Si X ray intensity is obviously strong

than that of 1 pass coated substrate.However the quantity of additionally attached Si is estimated only 2 to 3 %,far bellow expected 50 to -100%..
In spite of using average 5 μ m diameter Si powders,the size of attached Si particles is 1to 2 μ m.Coarse Si particles do not adhere to Cu foil.

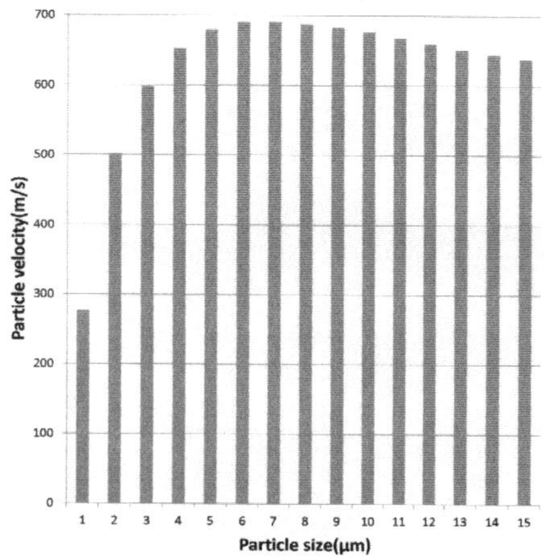

Fig.14. Particles velocity at Cu foil

Acknowledgement

The authors thank to Dr. Y. Shimizu of Shinshu University for his profitable advise, professor T. Hayashi of Shinshu University for favorably lending us charge-discharge measuring equipment, and K. Arai of Sinko Kogyo for the preparation of Ni-P coated Si powders.

literature

[1] Kasavajjula,U,; Wang,C.S.;Appleby,A.J. Nano- and Bulk-Silicon-Based Insertion Anodes for Lithium-Ion Secondery Cells. J. Power Sources.2007,163,1003-1039
[2] K.Sakaki et al: CD-ROM of Proc.ITSC(2007), p.13
[3] R.Nakanishi, K.Sakaki, J.Nakamura, A.Yumoto,: Preparations and Evaluation of Si anode for Next-Generation Li-ion Battery by Supersonic Free-jet PVD Process. Japan Thermal Spray Society, No.100 (2014 Autumn conference),101,p.1(2014)
[4] H.Usui, M.Shibata, K.Nakai, H.Sakaguti.: Anode properties of thick-film electrodes by gas deposition of Ni-coated Si particles. J. Power Sources.196(2011)2143-2148

Study on the thermal spray process and the properties of WC-12Co and WC-10Co-4Cr coatings

Wenjie ZHU, Jianming XU, Qian WANG and Xinghai TAN
Shanghai Baosteel Industry Technological Service Co., Ltd, Shanghai, 20190,China
Contact e-mail:174369@baosteel.com

Abstract: WC-12Co and WC-10Co-4Cr coating were prepared by high velocity oxygen fuel (HVOF) spraying and plasma spraying technique on Q345 steel substrate. The coating properties were analyzed by microscope, micro-hardness tester and electrochemical test. The results show that the WC-10Co-4Cr coating prepared by HVOF spraying exhibited better wear resistance especially in corrosion resistance compared with WC-12Co. The practical application of the coating effect is good.

Key words: HVOF spraying, plasma spraying, WC-12Co,WC-10Co-4Cr, Orthogonal analysis, Electrochemical test, Coating properties

1. Introduction

Cold-rolled production line in the steel industry (including hot-dip galvanizing, continuous annealing and other units) is mainly used for high quality steel strip products. These rollers exposed in the air used for conveying steel are called process rollers. With the continuous improvement of product quality requirements and production technology the rollers' surface must be able to maintain the performance of wear resistance, corrosion resistance and roughness to prevent strip steel skid. Therefore, the use of high hardness surface spraying process roller, has became the development trend of advanced production line of hot-dip galvanizing.

Thermal spraying is a kind of surface hardening technology, which is to use heat source made spraying material to melt or semi-molten state ,then high-speed jet and deposit material on a pretreated substrate surface, to form coating [1]. High-velocity oxygen fuel (HVOF) thermal spraying is one of flame spraying, its flame has high jet speed (up to 500 m/s ~ 500 m/s) and lower temperature. The higher particle velocity is helpful to obtain coating with high density and good bonding strength. The lower flame temperature can effectively inhibit or reduce the decomposition of the carbide hard phase during the spraying process, so the coating of spraying parts can get good organizational structure and high hardness and anti-wear ability [2-3]. Plasma spraying with high temperature and speed flow flame, has the advantages of well atmosphere controllability, flexible craft and the universality of coating materials, generally used for spraying ceramic materials to produce coating with high hardness, well wear resistance and corrosion resistant[4] .

WC-Co coating is one of the most applied valuable of carbide metal ceramic coating, the coating which has high density, high bonding strength with substrate [5-6]

has been widely used in aerospace, metallurgy, electric power and other fields. However, its application in corrosion environment is relatively insufficient. The addition of a certain amount of Cr in the WC-Co coating, not only can improve the corrosion resistance, but aslo can make better erosion resistance of the coating [7 -8].

This research chooses the coating porosity and the hardness as the measure to optimize the technological parameter of preparing WC-12Co and WC-10Co-4Cr coating by HVOF spraying and plasma spraying with the help of quadrature analysis. In order to obtain the coating with good wear resistance and impact resistance.

2. Experimental

2.1 The process parameters design

In order to determine the optimum parameters of the product process, often need to go through repeated experiments. Orthogonal test is an effective way of design the experiment with multifactor. Spraying process parameters are the main factors influencing the coating structure and performance. HVOF spraying (JP 8000 Type) experiment choose three orthogonal parameters: oxygen flow, powder feeding rate and the spraying distance . Plasma spraying experiment choose three orthogonal parameters: Current, powder, feeding rate and spraying distance. Considering the performance and equipment characteristics of WC-12Co and WC-10Co-4Cr powders, the parameters value of the orthogonal experiment are shown in table 1.

Table1.spraying parameter value

Powder	HVOF spraying			Plasma spraying		
	oxygen flow(scfh)	powder feeding rate(g/min)	spraying distance(mm)	Current(A)	powder feeding rate(g/min)	spraying distance(mm)
WC-10Co-4Cr	1900	80	320	650	45	90
	1950	100	360	700	50	110
	2000	120	400	750	55	130
WC-12Co	1900	80	360	650	30	100
	1950	95	380	700	35	120
	2000	110	400	750	40	140

1.2 Coating preparation

WC-12Co and WC-10Co-4Cr coating were prepared by high velocity oxygen fuel spraying (JP 8000 Type) and plasma spraying (Type F4) equipment on Q345 steel substrate. Experimental parameters set up according to the orthogonal experiment design。

1.3 The performance test

Considering the experimental conditions, the density and micro hardness of the coating are chosen as the factors to optimize the coating process parameters. The density can be measured by coating porosity, and the hardness is the ability of the material to resist the plastic deformation under the action of external force, scratch, wear or cutting. The hardness is one of the important mechanical properties of the coating , which relate to the performance of wear resistance, strength, service life and so on. After this ,testing the microstructure, microhardness, wear resistance and corrosion resistance of the coating made by the optimum parameters of the product process.

3. The results and discussion

3.1 The orthogonal experiment

According to the above mentioned parameters, the hardness and porosity of the coating were tested.

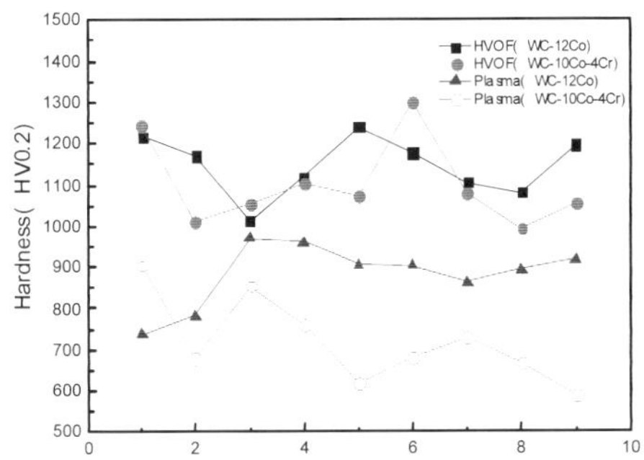

Figure 1.The coating hardness contrast figure of the orthogonal experiment

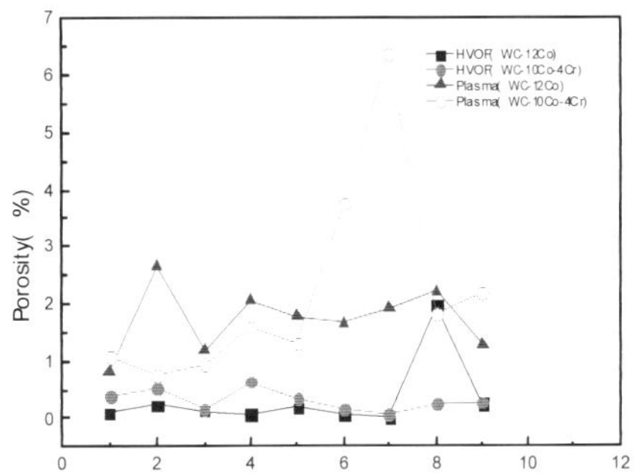

Figure 2.The coating porosity contrast figure of orthogonal experiment

As shown in Figure 1, we can see that the hardness of the HVOF coating is higher than that of the plasma sprayed coating, and the hardness of the two HVOF group is greater than that of the 1000HV. Microhardness directly affects the wear resistance of the coating, and the coating hardness on the process roll need to be above 1000.

As shown in figure 2, it can be seen that the plasma spraying coating porosity is higher than the HVOF spraying coating. Due to the layered structure, the thermal spraying coating is inevitable in the presence of porosity, and the density of the coating can be expressed by the porosity of the coating. Considering the impact resistance of the coating, the pore is likely to affect the boundary of the crack propagation, so the lower the porosity, the higher the density, the better the performance of the coating.

This analysis of the orthogonal experiment is choose the coating microhardness and porosity as an index, range R analysis. The range analysis can determine the degree of influence of each experimental parameter. Considering the importance of the two indicators in the coating application, weighting the two indicators score Y. Range R analysis was carried out on the weighted score Y result to get the optimum parameters of each process [8]. The index orthogonal analysis of the two coating by HVOF spraying are shown as Table 2 and Table 3.

Table 2 The orthogonal analysis of WC-12Co coating by HVOF spraying

The R range analysis of coating hardness				The R range analysis of coating porosity				The R range analysis of Y_i			
I	3401	3443	3476	I	0.45	0.32	0.18	I	19.04	65.3	42.05
II	3381	3385	3485	II	2.23	0.53	2.28	II	64.73	49.18	85.61
III	3537	3491	3358	III	0.31	1.14	0.43	III	68.1	84.27	61.97
R	156	106	127	R	1.92	0.82	2.10	R	82.57	45.09	42.68

The R value shows the influence of coating parameters of the two index ,and the optimization process set can get by the R value of Y_i .

As shown in table 2 ,the main factors influencing the microhardness and porosity of WC-12Co HVOF coating is spraying distance, the second is the oxygen flow rate, feeding rate is the minimal impact. Resulting from the range analysis of Y_i, the optimization process to spray WC-12Co coating by HVOF spraying is oxygen flow 2000/scfh, feeding rate 110g/min, spraying distance 380mm.

Table 3 The orthogonal analysis of WC-10Co-4Cr coating by HVOF spraying

The R range analysis of coating hardness				The R range analysis of coating porosity				The R range analysis of Y_i			
I	3306	3421	3202	I	1.07	1.08	0.75	I	-40.11	-22.58	25.08
II	3475	3408	3167	II	1.11	0.58	0.55	II	13.01	-48.76	-96.6
III	3126	3078	3538	III	0.57	1.09	1.45	III	-25.04	19.2	18.45
R	349	343	371	R	0.54	0.51	0.9	R	53.12	67.96	121.7

As shown in table 3,the main factors influencing the microhardness and porosity of WC-10Co-4Cr HVOF coating is spraying distance, the second is the oxygen flow rate, feeding rate is the minimal impact. Resulting from the range analysis of Y_i, the optimization process to spray WC-10Co-4Cr coating by HVOF spraying is oxygen flow 1950/scfh, feeding rate100g/min, spraying distance 360mm.

The optimization spray parameters to spray WC-12Co and WC-10Co-4Cr coating by HVOF spraying are shown in Table.4.

Table.4. HVOF spray parameters of WC-Co coating

Coating	Oxygen flow (scfh)	Powder feed rate (g/min)	Spraying distance (mm)	Other parameters
WC-12Co	2000	110	380	*
WC-10Co-4Cr	1950	100	360	**

3.2 Microscopic characterization of the coating

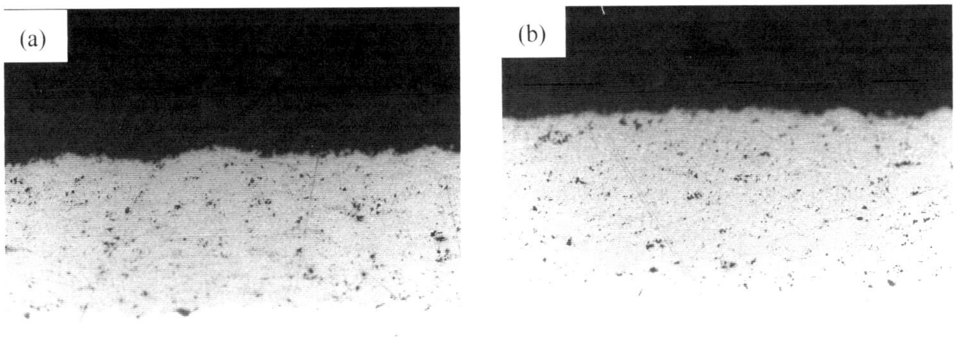

Figure.3.Microstructure of WC-12Co coating(a) and WC-10Co-4Cr coating(b)

As shown in Figure 3.The micristructure of the WC-12Co and WC-10Co-4Cr coating are dense and the porosity the Porosity is both lower than 1%. The microhardness of the WC-10Co-4Cr coating is $HV_{0.2}1385$, higher than the WC-12Co coating $HV_{0.2}1213$.

3.3 Wear resisitance test

During the work of process roller ,the roller surface continuely bear the effects of friction and wear. So the coating is required the excellent wear resistance. The wear resistance of WC-12Co coating, WC-10Co-4Cr coating and process roll substrate was tested by the pin-on-disk tribometer.

Fig. 4 The wear rate of the base material, the WC-12Co coating , the WC-10Co-4Cr coating

As shown in Fig. 4, The wear rate of the matrix material is the highest, and the wear resistance is the worst. Due to the hardness of HVOF coating is much higher than that of the base material, the wear resistance of the process roll is significantly improved

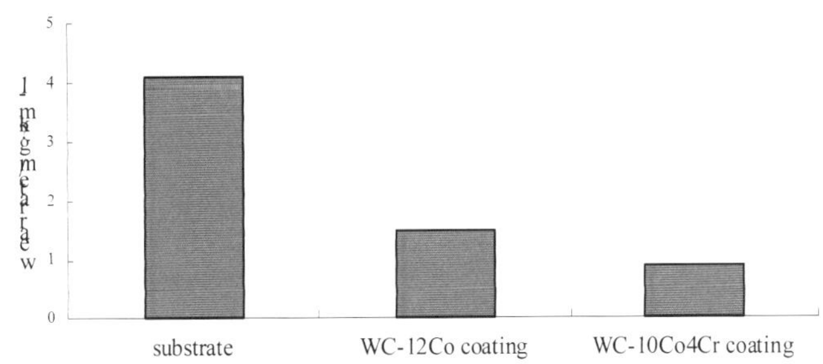

Fig. 4 The wear rate of the base material, the WC-12Co coating , the WC-10Co-4Cr coating

The wear rate of WC-10Co-4Cr coating is lower than that of WC-12Co coating, and its wear resistance is better. Because a small amount of W2C is formed during the spraying process, the hardness is higher than that of tungsten carbide, the strengthening phase is not easy to fall off, the formation of abrasive grains, reducing the wear of abrasive grain.[10,11].

3.4 Electrochemical characteristics test

The electrochemical properties of WC-10Co-4Cr and WC-12Co metal ceramic coating were tested by RST5200 electrochemical workstation.

The test choose standard three-electrode system, the sample epoxy encapsulated as a working electrode, a platinum plate as an auxiliary electrode, a saturated calomel electrode (SCE) as a reference electrode. Test the coating potentiodynamic polarization curves in 3.5% NaCl solution at 25℃. The working electrode have immersed in 3.5% NaCl solution for 0.5h before electrochemical test. Potential sweep range -100~100mV (relative to open-circuit potential), scan rate 0.5mV/s.

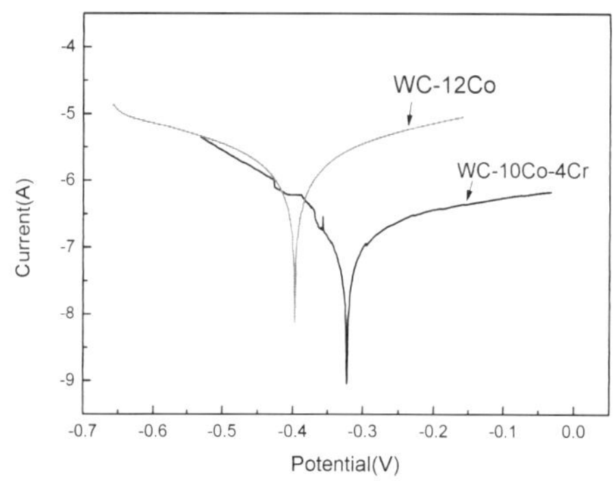

Figure 5. The polarization curves of WC-12Co and WC-10Co-4Cr coating

As shown in figure 5,the corrosion potential of WC-10Co-4Cr coating (-0.32309V) is higher than WC-12Co coating(-0.39774),this is due to the presence of Cr form a stable phase, improved resistance to electrochemical corrosion coating. The WC-10Co-4Cr coating has better electrochemical properties.

3.5 Practical application

Figure 6. The process roller with WC-10Co-4Cr coating

As shown in figure 6, the WC-10Co-4Cr coating sprayed by HVOF spray can be used on the surface of process roller. This application can improve the corrosion resistance of the roller surface, avoid the spalling of the coating caused by corrosion .

4. Conclusion

(1) The hardness and density of HVOF sprayed WC-Co coating are higher than that of plasma spraying coating, and the surface of process roll should be sprayed with

HVOF.

（2）The best spraying technology of the two kind of WC - Co coating and the factors influence of spraying process can be shown through the orthogonal analysis. The optimization process to spray WC-12Co coating by HVOF spraying is oxygen flow 2000/scfh, feeding rate 110g/min, spraying distance 380mm.The optimization process to spray WC-10Co-4Cr coating by HVOF spraying is oxygen flow 1950/scfh, feeding rate100g/min, spraying distance 360mm.

（3）The micristructure of the WC-12Co and WC-10Co-4Cr coating are dense and the porosity is both low(<1%).The WC-10Cr-4Co coating has better hardness ,wear resisitance and electrochemical properties.

（4）The WC-10Cr-4Co coating sprayed by HVOF, which has good hardness, wear resistance and electrochemical performance, can be used to improve the surface properties and extend the life of the process roll.

5. Literature

[1]Zijian Wu， Chaojun Wu， Keli Zeng， et al． Thermal Spray Technology and Application[M]．Beijing：Machinery Industry Press，2005．

[2]Steffens H，Brandl W.Moderne beschichtungsverfahren [M] .Weinheim: Deutsche Gesellschaft fur Materialkunde，2005.

[3]Niemine R，Vuoristo P，Niemi K，et al. Rolling constact fatigue failure mechanisms in plasma and HVOF sprayed WC-Co coating[J]. Wear，1997，212(1): 66-77.

[4] Jimei Xia．Research on preparation of plasma spraying wear-resistant coating and progress [J]．Modern Manufacturing Technology and Equipment，2009，(4)：23-26．

[5] Jinguo Zhang. Research on thermal spray powders and wear-resistant coating preparation of WC-17%Co[D].Changsha：Materials Processing Engineering, Central South University，2005.

[6] Xiaojing Yuan，Hangong Wang，Bolin Zha，et al. Multifunctional research on the WC10Co4Cr HVOF coating wear performance[J]. Materials Science and Engineering，2004，22（2）：204-208.

[7] Guanghua Zhang，Shu Li，Yang Liu，et al.Slurry Erosion Behavior of HVOF WC-10Co-4Cr coating[J]. China Surface Engineering，2007，20（7）：16-28.

[8] Kui Chen. Experimental Design and Analysis，Tsinghua University Press，1996

[9] Piao Zhongyu, Xu Bingshi, Wang Haidou, Cheng li. Analysis of contact fatigue mechanisms of thermally sprayed cermet coatings. Materials Protection, 2008, 41(10): 1~6.

[10] R.Ahmed, Contact fatigue failure modes of HVOF coatings. Wear, 2002, 253: 473~487.

Microstructure and wear resistance of atmospheric-plasma-sprayed AlCoCrFeNiTi high-entropy alloy coating

Wei Xiong, Li-Hui Tian*, Chuan Liu, Sheng Lu, Zhenjiang/China

An equiatomic AlCoCrFeNiTi high entropy alloy coating with a thickness of about 236 μm, a porosity of about 1.6%, and an adhesive strength of higher than 50.3 MPa was deposited by atmospheric plasma spraying. Besides BCC1 and BCC2 phases, minor FCC was detected by X-ray diffraction in the as-sprayed coating. The microhardness of the as-sprayed AlCoCrFeNiTi high entropy alloy coating was 664 Hv, which was 4 times of that of the 316 stainless steel. The volume wear rate of the as-sprayed coating at room temperature was 0.77×10^{-4} $mm^3/(N\cdot m)$, which was about a quarter of that of the substrate material (2.55×10^{-4} $mm^3/(N\cdot m)$). The volume wear rate decreased as the wear temperature increased from 25°C to 700°C and increased when the temperature increased from 700°C to 900°C. At the temperature of 700°C, the volume wear rate was 0.23×10^{-4} $mm^3/(N\cdot m)$, which was about one-ninth of that of the substrate material. The wear mechanism changed from delamination wear to oxidation wear and adhesion wear.

1 Introduction

High-entropy alloys (HEAs) are novel materials which broke the conventional conception of alloy design that based on one principal element or compound in order to gain the primary property. Compared to the conventional alloy, HEAs contain 5~13 main elements with each elemental equimolar or near-equimolar ratios. These alloys are unique since they have high configurationally entropy which is restricted to the formation of many intermetallic compounds or other complex structures, contributing to formation of simple body centered cubic (BCC) and/or face centered cubic (FCC) solid solutions, even hexagonal close packed (HCP) solid solutions and amorphous phases [1]. The HEAs possess promising properties such as good thermal stability [2, 3], excellent corrosion resistance and oxidation resistance, high hardness and strength, and wear resistance [4], which can be used as moulds, dies, tools, mechanical parts and so on [5, 6].

There are many advanced methods reported to prepare HEAs, such as arc melt casting, laser cladding, magnetron sputtering and so on. With the rapid development of modern industry, the use of harsh environment leads to higher requirements on the wear resistance coatings, especially for the coating materials with high hardness, wear resistance and high temperature oxidation resistance. The traditional wear-resisting coating materials represented by TiN are difficult to meet the demand of modern industry. Therefore, the novel HEAs coating attracted more and more attentions.

Atmospheric plasma spraying (APS), based on the advantages of the lower dilution of the substrate material, which is distinguished from the laser cladding, high temperature flame, concentration of the energy and high bonding strength, is widely used in aerospace and petrochemical industry. So far, the preparation of HEAs coating by APS is still in infancy. A.S.M. Ang [7] prepared nanostructured AlCoCrFeNi and MnCoCrFeNi HEAs coatings using mechanically alloyed powders. Results show that the Vickers hardness was 4.13±0.43 Hv and 4.42±0.6 GPa, respectively. It was also reported that there were minor oxide peaks detected. L.M. Wang [8] deposited $Ni_xCo_{0.6}Fe_{0.2}Cr_ySiAlTi_{0.2}$ coating by plasma spraying. After the annealing at 1100°C for 10 h, the Vickers hardness was parallel to that of the as-casted state (1045 Hv). T.M. Yue [9] investigated the AlCoCrCuFeNi HEA coating prepared by plasma spray processing and laser remelting. It was found that the microstructure of the coating after laser remelting exhibited an epitaxial growth of columnar dendrites.

By now, CoCrCuFeNi HEA as representative of FCC solid solution and AlCoCrFeNi HEA as representative of BCC solid solution have been widely reported [10]. In this study, CoCrFeNi HEA matrix was selected and Al and Ti elements were added to enhance the hardness and wear resistance of the coating. AlCoCrFeNiTi feedstock powder was prepared by ball milling, and AlCoCrFeNiTi HEA coating was deposited by APS. The microstructure, microhardness and wear resistance of the as-sprayed coating were investigated.

2 Experimental

2.1 Preparation of feedstock powder

Element powders of Al, Co, Cr, Fe, Ni and Ti with purities of higher than 99.0 wt.% were used to synthesize the feedstock powder via mechanical alloying (MA) in equiatomic ratio. Mechanical alloying was carried out in a high energy planetary ball mill (XM-4, Xiangtan) with 304 stainless steel balls and vials as the milling medium. The spindle speed of the mill was 800 r/min and the ball to powder weight ratio was 10:1. The powder milled for 30 h was sieved into a size range from 35 to 65 μm, which was suitable for APS due to its good flowability.

2.2 Preparation of HEA coating

Praxair 3710 high energy plasma spray system was used to deposit AlCoCrFeNiTi coating onto 316 stainless steel substrates with a dimension of Φ 25×8 mm. Prior to spraying, the surface of the substrate was sand-blasted to a roughness of R_z=82.32, using corundum (16 mesh). The detailed spraying parameters are listed in **Table 1**.

Table 1. Parameters of atmospheric plasma spraying

Parameter	Value
Plasma arc power (kW)	45
Primary gas (Ar) flow (psi)	60
Secondary gas (N_2) flow (psi)	50
Powder feed rate (r/min)	0.8
Spray distance (mm)	100
Gun traverse speed (mm/s)	200

2.3 Characterization

The phase structure of the feedstock powder and the as-sprayed coating was examined using X-ray diffractometer (XRD, XRD-6000, Shimadzu Corporation) with Cu-Kα radiation. The scanning speed was 6 °/min and the range of 2θ was from 20° to 90°. The microstructure of the feedstock powder, the as-sprayed coating and the worn surfaces was analyzed by scanning electron microscopy (SEM) equipped with energy dispersive spectroscopy (EDS). The porosity of the HEA coating was evaluated by the image analysis software Photoshop.

The adhesive strength of the coating was obtained through tensile test with a tensile speed of 1 mm/s. E-7 glue was used as the adhesive. Three samples were tested and the average tensile strength was obtained as the adhesive strength of the coating.

MH-5D microhardness tester was used to measure the microhardness of the cross-section of the HEA coating from the substrate to the surface of the coating. For each position, 10-15 points were measured. The load was 200 g and the dwelling time was 10 s.

Tribological experiments were conducted using the HT-1000 high temperature wear tester under the test temperature of 25°C, 500°C, 700°C and 900°C. The test was carried out with a spindle speed of 573 r/min, a load of 5 N, and a duration time of 20 min. The Si_3N_4 balls with a diameter of 5 mm were used as the wear pair. Prior to test, the coating and the substrate were polished to R_a=0.2. According to the wear equation, the wear rate W can be calculated as $W=V/(S\times L)$ [11], where V is the volume loss (mm^3), S is the sliding distance (m) and L is the normal load applied (N).

3 Results and discussion

3.1 Microstructure of feedstock powder

The XRD pattern of the feedstock powder is shown in **Fig. 1**. It is obvious that the ball milled powder was composed of two kinds of body centered cubic phases (BCC1 and BCC2). It was reported that the AlCoCrFeNiTi high-entropy alloy was consisted of Fe-Cr rich A2 structure and Al-Ni-Ti rich B2 structure [12]. It can also be seen that the major peak was broaden, which may due to the high lattice strain, the refined crystal size and the adverse crystallinity [13].

The surface morphology of the AlCoCrFeNiTi feedstock powder is shown in **Fig. 2**. The powder exhibited an irregular and equiaxial shape, which can result in a good flowability.

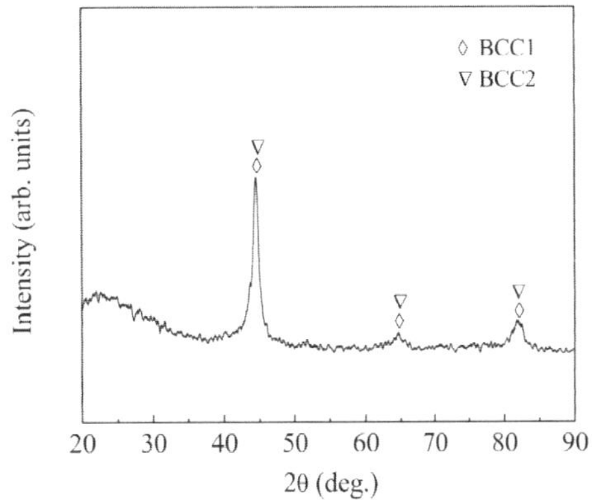

Fig. 1. XRD pattern of the feedstock powder.

Fig. 2. SEM image of the feedstock powder.

3.2 Microstructure of HEA coating

The XRD pattern of the as-sprayed AlCoCrFeNiTi coating is shown in **Fig. 3**. Comparing the XRD pattern of the coating with that of the feedstock powder, it can be found that they had similar diffraction peaks. However, after spraying, the peaks shift to left side slightly and the intensity of the peaks increased. The increment of the crystal constant and the internal energy release may be the explanation of this phenomenon at high temperature plasma sprayed states. Besides BCC1 and BCC2 phases, minor FCC was detected in the as-sprayed coating. In addition, NiCr2O4 oxide phase was detected, which indicates that high temperature oxidation of the feedstock powder occurred.

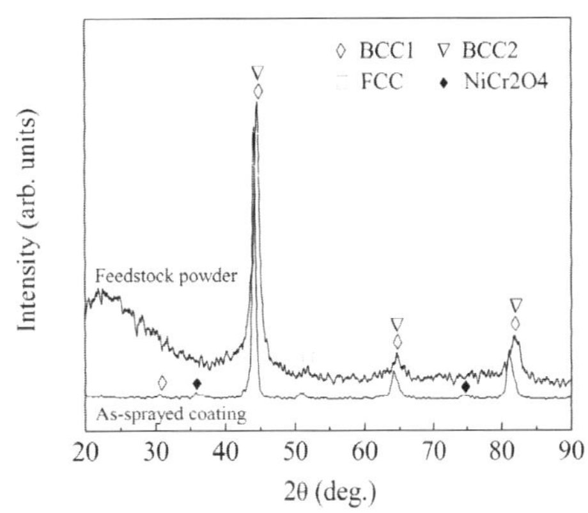

Fig. 3. XRD patterns of HEA coating and the feedstock.

Cross-sectional microstructure of the as-sprayed AlCoCrFeNiTi coating is shown in **Fig. 4**. It can be seen that a dense coating with a thickness of about 236 μm was obtained, **Fig. 4(a)**. Some voids with an irregular shape and a size less than 20 μm were presented in the coating. The porosity of the coating was about 1.6%. Besides, a good bonding was obtained between the coating and the substrate. The result of the tensile test shows that the fracture occurred in the coating with an average strength of 50.3 MPa, which indicates that the adhesive strength of the coating was higher than 50.3 MPa.

A typical lamellar structure can be seen in **Fig. 4(b)**. Most of the splats were fully melted and well flattened. Besides, some unmelted particles can also be found. Light gray regions (a and b) with some Al element maybe the BCC phase, which was caused by the large atomic radius of Al. Region c, which consisted of higher proportion Co and Ni elements but without Al element, may be the FCC phase [14]. In addition, some dark gray area (d) with oxygen content of 30.6 wt.% was due to the oxidation of the metal elements during plasma spraying. It was reported that Cr_2O_3, iron oxide and aluminum oxides was formed when CoCrFeNiTiAl high-entropy alloy was annealed at a high temperature of 900°C for 200 h [15].

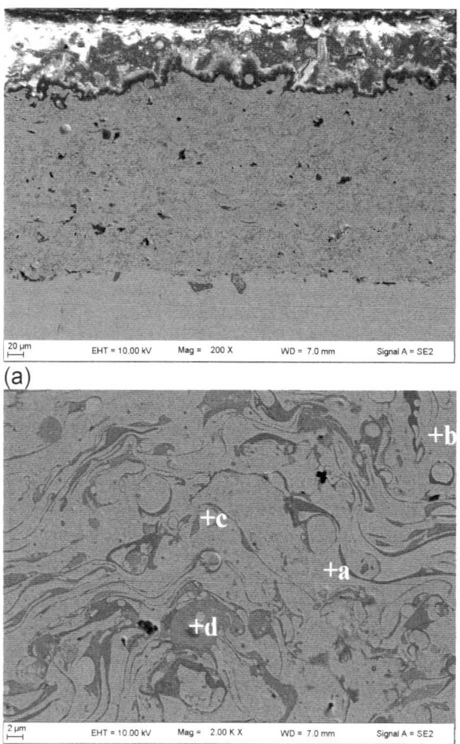

Fig. 4. SEM images of the HEA coating at lower (a) and higher (b) magnifications.

Table 2. EDS results of the HEA coating

Point	Al	Co	Cr	Fe	Ni	Ti	O
a	5.7	20.9	14.7	28.3	19.9	10.4	0.0
b	7.8	20.5	18.2	19.3	20.5	13.7	0.0
c	0.0	34.2	4.9	22.5	36.2	1.8	0.0
d	6.2	16.6	8.1	15.9	10.8	11.7	30.6

3.3 Microhardness of HEA coating

The microhardness of the as-sprayed AlCoCrFeNiTi high-entropy alloy coating as a function of the distance from the interface between the substrate and the coating is shown in **Fig. 5**. Due to the formation of the hard BCC phases, the average hardness of the coating was about 664 Hv, which was 4 times of that of the substrate (167 Hv). In addition, the hardness of the substrate near the substrate/coating interface increased, which was due to the heat treatment of the plasma arc during the high temperature spraying process.

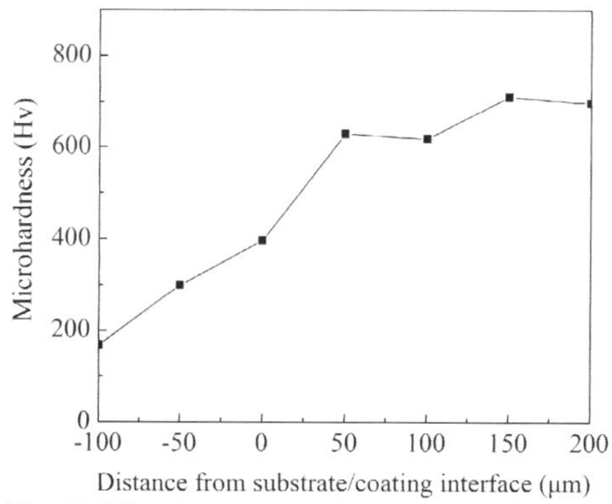

Fig. 5. Microhardness of the HEA coating.

3.4 Wear resistance of HEA coating

The worn surface morphology of the as-prayed high-entropy alloy coating at 25°C, 500°C, 700°C and 900°C is shown in **Fig. 6**. With the increase of the wear temperature, obvious differences of the worn surfaces can be seen.

At the wear temperature of room temperature, pits (region c) can be observed in **Fig. 6(a).** The worn surface undergoes a periodic delamination fracture [16]. It is obvious that soft and ductile FCC phase existed on the worn surface. In addition, shallower grooves and oxidized regions (regions a and b) can be found on the worn surface, which was due to the resist to plastic deformation of the hard BCC phase and oxides in the coating. Therefore, the wear mechanisms of the coating at room temperature were predominantly delamination wear along with adhesion wear.

For the coating at the wear temperature of 500°C, delamination of the splats was also found, **Fig. 6(b)**. Minor regions such as d and f had a higher oxygen content and most of the regions like e had a lower oxygen content. The above results indicate that the wear mechanisms of the AlCoCrFeNiTi high-entropy alloy coating were predominantly delamination wear along with oxidation wear.

Shallower and narrower grooves and rougher surface of local regions can be seen in **Fig. 6(c)**. EDS analysis shows that the average oxygen content in regions g, h and i was 30.8 wt.% and the Si content of region g was up to 3.7 wt.%. It indicates that oxidation wear and adhesion wear were the main wear mechanisms, since

AlCoCrFeNiTi high-entropy alloy can keep a high hardness at 700°C [15].

At the temperature of 900°C, smooth grooves and ductile deformation along the grooves can be seen in **Fig. 6(d)**. Besides, the whole worn surface was oxidized and the average oxygen content of regions j, k and m was 35.5 wt.%, **Table 3**. In addition, minor silicon element in region k and m was detected. During test, oxides and Si_3N_4 were detached from the surfaces of the coating and the Si_3N_4 ball. Then the detached particles were retained between the surfaces of the friction pairs and crushed into the worn surface, which was contributed to the level of smoothening effect [17]. Therefore, typical adhesion wear was the main wear mechanism at 900°C.

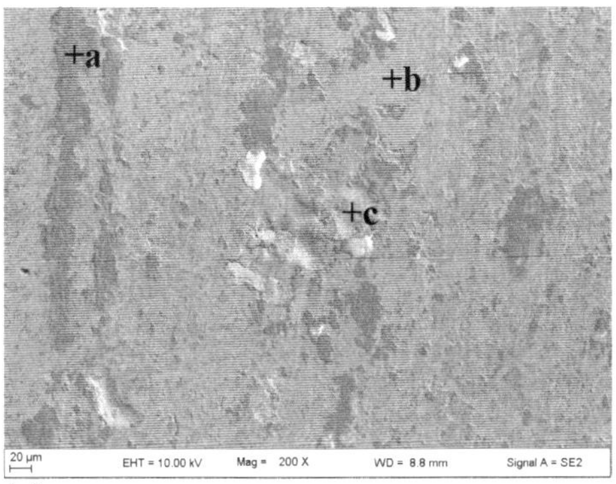

(a)

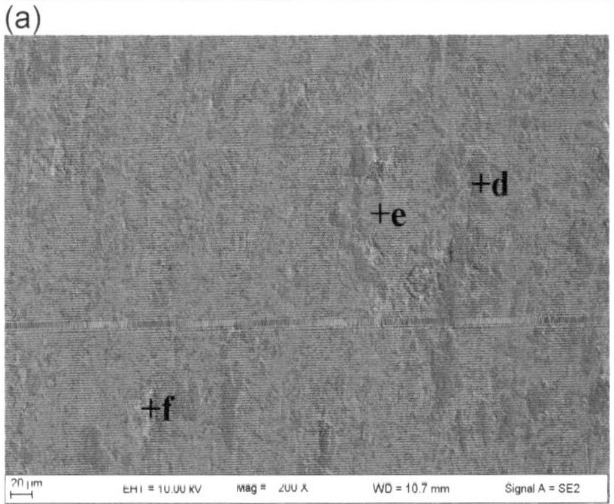

(b)

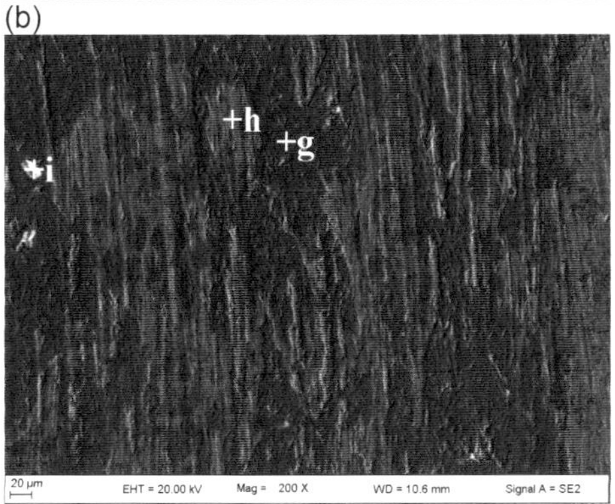

(c)

(d)

Fig. 6. SEM images of worn surfaces of the HEA coating at 25°C (a), 500°C (b), 700°C (c) and 900°C (d).

Table 3. EDS results of worn surfaces of the HEA coatings

Point	Al	Co	Cr	Fe	Ni	Ti	Si	O
a	5.4	8.7	6.9	8.1	8.9	11.3	0.0	46.8
b	8.8	17.8	15.8	17.7	20.7	14.2	0.0	5.0
c	9.0	18.1	26.7	16.1	18	12.1	0.0	0.0
d	5.0	14.3	10.9	13.4	14.1	11.4	0.0	31.0
e	9.0	20.8	13.3	16.2	21.1	15.8	0.0	3.9
f	4.6	12.9	7.9	21.5	16.6	10.7	0.0	25.9
g	4.9	12.9	8.7	12.6	12.1	10.4	3.7	34.6
h	6.2	14.8	12.3	14.4	15.0	12.1	0.0	25.2
i	3.8	11.8	8.9	11.0	11.9	14.5	1.3	32.7
j	7.4	10.5	12.9	10.7	9.6	16.7	0.0	32.1
k	3.3	16.2	7.6	16.5	11.5	8.5	0.8	35.6
m	4.9	12.1	9.6	12.5	11.1	10.6	1.3	37.7

The volume wear rate of the as-sprayed AlCoCrFeNiTi high-entropy alloy coating and the 316 stainless steel substrate is shown in **Fig. 7**. It can be found that the volume wear rate of the as-sprayed coating at room temperature was 0.77×10^{-4} $mm^3/(N\cdot m)$, which was about a quarter of that of the substrate material (2.55×10^{-4} $mm^3/(N\cdot m)$). With the increase of the wear temperature from 25°C to 700°C, the volume wear rate decreased. At the temperature of 700°C, the volume wear rate was 0.23×10^{-4} $mm^3/(N\cdot m)$, which was about one-ninth of that of the substrate material. At room temperature, a portion of ductile FCC phase existed in the coating and hard oxidation phases were formed at high temperature. And the wear mechanism changed from delamination wear to oxidation wear and adhesion wear. However, the volume wear rate increased when the wear temperature increased from 700°C to 900°C.

4 Conclusions

(1) Equiatomic AlCoCrFeNiTi high entropy alloy coating with a thickness of about 236 µm, a porosity of about 1.6%, and an adhesive strength of higher than 50.3 MPa was deposited by atmospheric plasma spraying. Besides BCC1 and BCC2 phases, minor FCC was detected in the as-sprayed coating.

(2) The microhardness of the as-sprayed AlCoCrFeNiTi high entropy alloy coating was 664 Hv, which was 4 times of that of the 316 stainless steel.

(3) The volume wear rate of the as-sprayed coating at room temperature was 0.77×10^{-4} $mm^3/(N\cdot m)$, which was about a quarter of that of the substrate material

(2.55×10^{-4} $mm^3/(N\cdot m)$). The volume wear rate decreased as the wear temperature increased from 25°C to 700°C and increased when the temperature increased from 700°C to 900°C. At the temperature of 700°C, the volume wear rate was 0.23×10^{-4} $mm^3/(N\cdot m)$, which was about one-ninth of that of the substrate material.

(4) The wear mechanism changed from delamination wear to oxidation wear and adhesion wear.

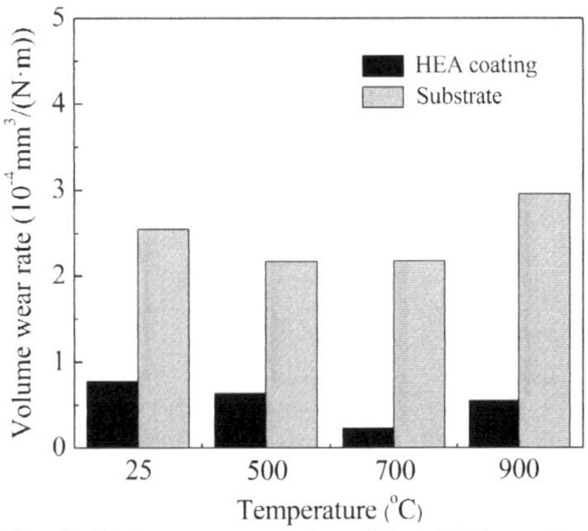

Fig. 7. Volume wear rate of the HEA coating and 316 stainless steel at 25°C, 500°C, 700°C and 900°C.

Acknowledgement

The present work is supported by the National Natural Science Foundation of China (No. 51401091).

References

[1] Y.P. Wang, B.S. Li, M.X. Ren, C. Yang and H.Z. Fu: Microstructure and compressive properties of AlCrFeCoNi high entropy alloy. Mater. Sci. Eng. A 491 (2008), Issues 1-2, pp. 154/58.

[2] J.W. Yeh, S.K. Chen, S.J. Lin, J.Y. Gan, T.S. Chin, T.T. Shun, C.H. Tsau and S.Y. Chang: Nanostructured high-entropy alloys with multiple principal elements: novel alloy design concepts and outcomes. Adv. Eng. Mater. 6 (2004), Issue 4, pp. 299/303.

[3] C.J. Tong, Y.L. Chen, J.W. Yeh, T.T. Tsau and S.J. Lin: Microstructure characterization of $Al_xCoCrCuFeNi$ high-entropy. Metall. Mater. Trans. A 36 (2005), Issue 4, pp. 881/93.

[4] M.H. Chuang, M.H. Tsai, W.R. Wang, S.J. Lin and J.W. Yeh: Microstructure and wear behavior of $Al_xCo_{1.5}CrFeNi_{1.5}Ti_y$ high-entropy alloys. Acta Mater. 59 (2011), Issue 16, pp. 6308/17.

[5] C.P. Lee, C.C. Chang, Y.Y. Chen, J.W. Yeh and H.C. Shih: Effect of the aluminum content of $Al_xCrFe_{1.5}MnNi_{0.5}$ high-entropy alloys on the corrosion behaviour in aqueous environments. Corros. Sci. 50 (2008), Issue 7, pp. 2053/60.

[6] Y. Zhang, Y.J. Zhou, P.L. Jin, G.L. Chen and K.L. Peter: Solid solution and phase formation rules for multi-component alloys. Adv. Eng. Mater. 10 (2008), Issue 6, pp. 534/38.

[7] A.S.M. Ang, C.C. Berndt, M.L. Sesso, A. Anupam, S. Praveen, R.S. Kottada, B.S. Murty: Plasma-sprayed high entropy alloys: microstructure and properties of AlCoCrFeNi and MnCoCrFeNi. Metall. Mater. Trans. A 46 (2015), Issue 2, pp. 791/800.

[8] L.M. Wang, C.C. Chen, J.W. Yeh and S.T. Ke: The microstructure and strengthening mechanism of thermal spray coating $Ni_xCo_{0.6}Fe_{0.2}Cr_ySi_zAlTi_{0.2}$ high-entropy alloys. Mater. Chem. Phys. 126 (2011), Issue 3, pp. 880/85.

[9] T.M. Yue, H. Xie, X. Lin, H. Yang and G. Meng: Microstructure of laser re-melted AlCoCrCuFeNi high entropy alloy coatings produced by plasma spraying. Entropy 15 (2013), Issue 7, pp. 2833/45.

[10] F.Y. Wang: Study of microstructure and mechanical properties of $AlCoCrFeNiTi_X$ high entropy alloy. Taiyuan University of Technology, Taiyuan, 2013.

[11] H. Liu: Friction and wear behaviors of PVD nitride coatings at elevated temperatures. Shandong University, Ji'nan, 2012.

[12] Y. Yuan, J. Wang, J.S. Li, H.C. Kou and W.M. Liu: Characterization of BCC phases in $AlCoCrFeNiTi_x$ high entropy alloys. Mater. Lett. 138 (2015), pp. 78/80.

[13] K.B. Zhang, Z.Y. Fu, J.Y. Zhang, W.M. Wang, S.W. Lee and K. Niihara: Characterization of nanocrystalline CoCrFeNiTiAl high-entropy solid solution processed by mechanical alloying. J. Alloys Compd. 495 (2010), Issue 1, pp. 33/38.

[14] K.B. Zhang, Y. Fu, J.Y. Zhang, W.M. Wang, Y.C. Wang, Q.J. Zhang and J. Shi: Microstructure and mechanical properties of $CoCrFeNiTiAl_x$ high-entropy alloys. Mater. Sci. Eng. A 508 (2009), Issues 1-2, pp. 213/19.

[15] Y.J. Chang and A. Yeh: The evolution of microstructures and high temperature properties $Al_xCo_{1.5}CrFeNi_{1.5}Ti_y$ high entropy alloys. J. Alloys Compd. 653 (2015), pp. 379/85.

[16] Y. Wang, T.Q. Lei and J.J. Liu: Tribo-metallographic behavior of high carbon steels in dry sliding: wear mechanisms and their transition. Wear 231 (1999), pp. 12/19.

[17] H.T. Duan, Y. Wu, M. Hua, C.Q. Yuan, D. Wang, J.S. Tu, H.C. Kou and J. Li: Tribological properties of AlCoCrFeNiCu high-entropy alloy in hydrogen between the friction pairs. Wear 297 (2013), Issues 1-2, pp. 1045/51.

The deposition of pure aluminium via cold spray for the corrosion protection of steel

Ildiko Peter[1a], Barry Aldwell[2b], Rocco Lupoi[2c], Mario Rosso[1d]
[1] Department of Applied Science and Technology, Politecnico di Torino, Italy
[2] Department of Mechanical and Manufacturing Engineering, Trinity College Dublin, Ireland
[a]ildiko.peter@polito.it, [b]aldwellb@tcd.ie, [c]lupoir@tcd.ie, [d]mario.rosso@polito.it

In this paper, development and characterization of cold sprayed protective pure Al coatings on a non-regular steel surface will be proposed. The research is oriented to the optimization of the deposition route selecting the most appropriate process parameters and the really applicable layer thickness, and on the study of how the defects and the imperfections of the substrate influence the coating performance and the corrosion resistance of the coated material.

1 Introduction

Due to their outstanding mechanical strength and relatively low cost, steel is widely used in our modern society. It is considered a strategic material for many industrial applications. However, the tendency to oxidize in environment containing halide ions or aggressive species can limit its applications [1-3], that can be reduced by ani corrosion coating. Cr (VI) compounds are the most commonly used anticorrosion layers, but due to their high toxicity makes this solution less attractive [4-6] and a strong effort is taking on to substitute them with other non-toxic elements [7-9]. Actually, various technological solutions are available to deposit protective layers onto steel substrate. Some of them are related to nitriding processes, physical vapour deposition, chemical vapour deposition or atmosphere plasma jet metallization [10, 11]. High-velocity oxy fuel process has attracted much attention for coating realization, due to its ability of creating coatings with lower porosity, higher hardness, superior bond strength and less decarburization compared to many of the other thermal spraying methods [12, 13].
Aluminium and its alloys are remarkable materials for different industrial applications and the driving force for their use, apart from their performance, is principally coming from the need to moderate the CO_2 gas emission, which is considered one of the most important product responsible for the generation of the greenhouse effect. Full or partial replacement of a heavier steel counterparts with lighter elements (Al, Mg, Ti based alloys) or the combined use of steel and light alloys is a growing topic [14]. The corrosion resistance of aluminium comes from its facility to develop a natural oxide film on the surface reducing the corrosion rate. Due to the hydrogen discharge, the electrodeposition of Al from aqueous solution is difficult, while high vacuum techniques, i.e. chemical vapour deposition and physical vapour deposition are time-consuming and very expensive too. Electrodeposition from molten salts constitutes a reliable solution, but the high temperature involved makes this route inadequate for some applications. Additionally, it reveals high energy consumption, pollutant emission and difficult realization of the procedures. The development of a molten salts called "room temperature molten salts", gives different occasion to the development of this process. The main difficulty using this kind of materials is related to their water and air stability [15] and the development of ionic liquids in order to reach higher stability is still now limited. The room temperature ionic liquid, namely $AlCl_3$-[EMIm]Cl has a number of attractive features such as very low vapour pressure, relatively high electrical conductivity and a wide electrochemical potential window and they can be considered as potential electrolytes for the electrolytic extraction, recycling, refining and deposition of aluminium. Anyway, carefully managing the ionic liquids, the deposition can be realized [16, 17], but it appears difficult to do it in a totally safely manner.
On the other hand, cold spray (CS) as low-temperature coating process is developed and is still to be more and more employed for the quick realization of a compact coating layer onto different substrate materials. Several material systems have already been coated using CS and generally speaking the suitability of the material used in this process is mostly governed by their deformation properties. Materials with relatively low melting point and lower mechanical strength satisfy this condition and Zn, Al, Cu and their alloys are ideal materials for this purpose [18- 20]; however the deposition of strong metals such as Ti is also possible [21]. Recently, the deposition of diamond was also demonstrated possible using powder that are cladded with a ductile phase [23].
In this context, the aim of the present paper concerns to the development and characterization of cold sprayed protective Al coatings on a non-regular steel surface. The research is carryed out with a twofold purpose: one of them is oriented on the optimization of the deposition route selecting the most appropriate process parameters (i.e. velocity orprotective environment.) and the really applicable layer thickness, while the second one is targeted to investigate how the defects and the imperfections of the substrate influences the coating performance and the corrosion resistance of the material.

2 Materials and methods

2.1 Coated samples: substrate, coating material and deposition technique

The material used as substrate in the present investigation was an AISI 1045 steel with 0.43 C, 0.80 Mn, 0.20 Si, 0.02 S, and <0.09 P, supplied in bars and

subsequently sectioned and prepared according to the specific application. Commercially pure Al (99.5%) was used for the deposition.
The nozzle used was a de-laval deisgn, with 6mm exit diameter. Samples were generated using nitrogen and helium as processing gas, with a level of temperature pre-heating.

2.2 Morphological analysis and corrosion resistance evaluation

Coated samples have been carefully extracted by a cutting procedure from the macro samples and then have been prepared by a typical metallographic technique involving mounting and polishing procedures. The microstructure of the samples has been investigated using an optical microscope, (OM, MeF4 Reichart-Jung) and Scanning Electron Microscopy (SEM, Leo 1450VP) equipped with Energy X-rays Dispersive Spectroscopy unit (EDS, Oxford microprobe) used for compositional analysis. X-ray technique (X-ray, PANanalytical tool with Cu Kα wavelength of 1.5418 Å) has been employed to verify the deposition, to identify the different phases present and to give a qualitative and comparative information about the presence of the residual stress in the interested regions.
The corrosion resistance of the samples has been investigated by the accelerated salt spray chamber test. The corrosion resistance has been monitored on the whole sample surface inserted in the test chamber containing the corrosive media.The solution used was NaCl 5wt% and the test has been carried out for 168 hours. The chamber was maintained at a temperature of 35°C for the entire test period of the test. At pre-set intervals the samples have been extracted from the chamber, they have been inspected and introduced again in the chamber till the end of the test. Following the complete immersion, the samples have been cleaned by water and ethanol washing and then have been dried. Following the visual inspection of the whole configuration, some significant areas have been observed by SEM and EDS analysis and comparison of different zones has been carried out.

3 Results and discussion

The Al powders used for the research has been analysed by SEM, EDS and X-Ray. The morphology of the powders is reported in Fig.1. Globular grain shape and a relatively regular variation of the grains size (20÷40 μm) can be evidenced. The Xray pattern will be illustrated later on together with the diffraction patterns of the substrate and the coating layer.
Some preliminary studied have been carried out using an AISI 1045 steel sheets and performing some thin layer deposition in order to identify the starting parameters for the deposition (Fig.1 right side). . In this case, N_2 was used at 30bar inlet pressure, using 78, 155, 224 and 268 C (top to bottom) as nozzle inlet temperature.

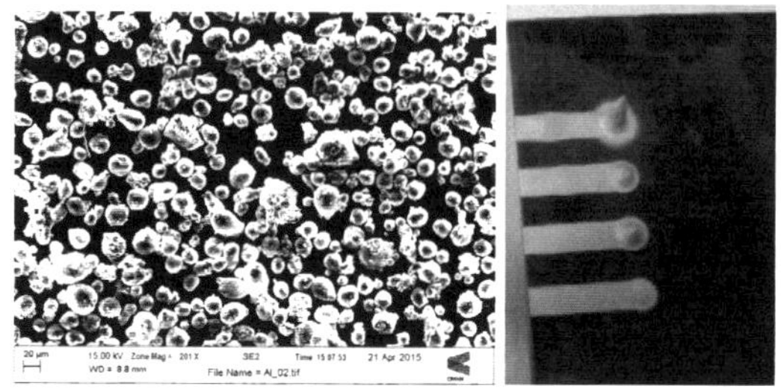

Fig. 1. SEM micrograph for the powders used for coating (left) and photos oft he preliminary coatings (right).

Using N_2 as carrier gas a porous and irregular coating layer was realized (Fig.2a) with N_2 at 30bar, 250C, and with a nozzle Transverse Speed (TS) of 15mm/s.The coating geometry is in the form of "tracks"; thus, to cover the full surface tracks were overlapped (4mm) Generally, porosity can act as crack initiation site during service compromising the real lifetime of the coated surface. To overcome this step the further coating have been realized on some steel disks using helium as carrier gas, with 10bar inlet pressure, room temperature, and 60mm/s for TS. The same tracks overlap as for the nitrogen case was used.
These conditions generate a more homogeneous coating layer with no presence of macro-porosity (Fig.2b). The samples coated with N_2 have been abandoned, while those obtained using He have been submitted to further investigations in order to assess the coating performance, the adhesion of the coating to the steel substrate and its corrosion resistance.

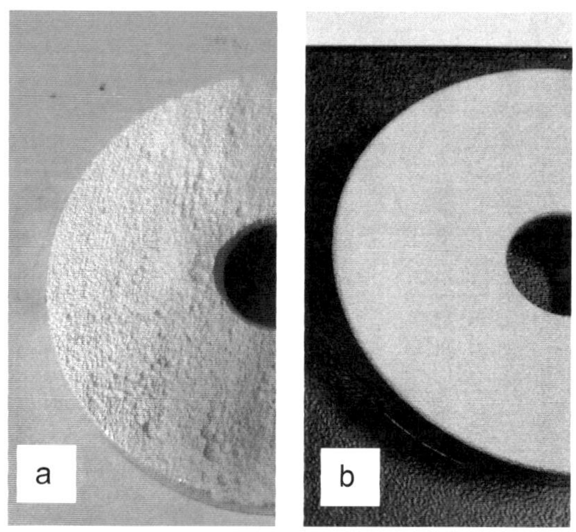

Fig. 2. Photographs of the coating realized with N_2 (a) and with He (b) as particle carrier gas. .

After cutting and standard metallographic preparation, X-Ray diffraction measurement (Fig. 3) reveals that the coating perfectly covers the steel substrate. It can also be observed that during the deposition process no phase change arise.

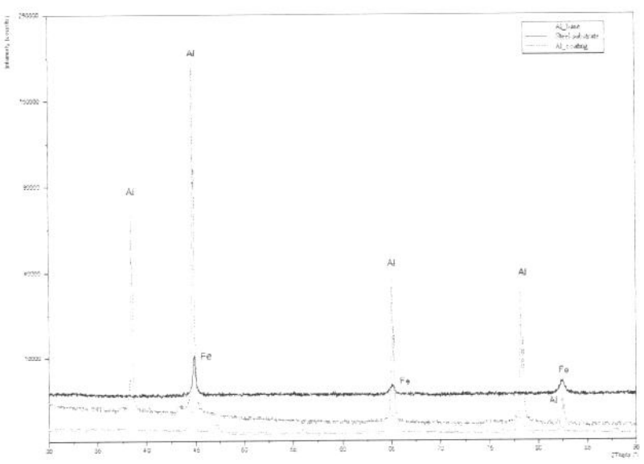

Fig. 3. X-ray pattern showing showing the diffraction signs fort he substrate, starting coating material and the coated steel.

Similar to the initial powder the coating diffraction pattern shown the presence of Al peaks, matching the JCPDS 01-089-2837 data card. There are no important oxidation and the absence of any chemical reactions during the deposition process has been detected.
SEM analysis has been carried out to evaluate the coating adhesion to the substrate, the thickness of the coating produced, the presence of internal porosity and to identify the possible defects. Fig. 4 reports the SEM microstructures for two realized thicknesses: Fig.4a, b are related to a coating thickness of about 480 μm and Fig.4 c, d with a coating thickness of about 1.2 mm, respectively. At higher magnification, it is evidenced that a continuous coating layer with good adhesion has been obtained spraying a thinner layer (Fig.4c) compared to the thicker one (Fig.4d). The interface between the substrate and the coating layer in this case is more or less defect free (Fig.4b). Contrarily, in the case of denser layer the development of some cracks and the presence of porosities have been revealed, highlighted with yellow arrows in Fig.4d.

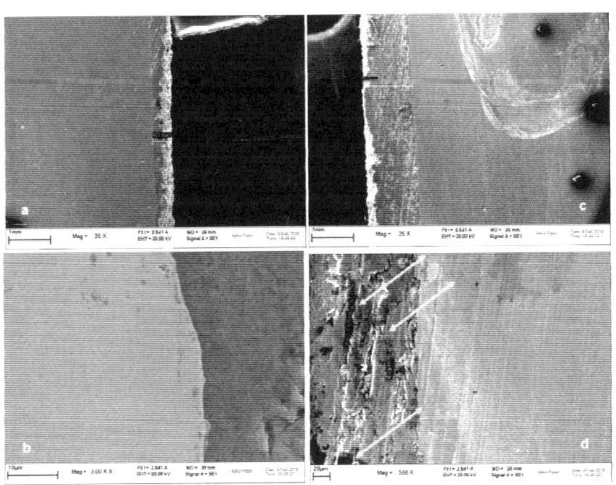

Fig. 4. SEM micrographs of the coated samples, showing: a general overview (a, c), detail on the cross sections (b, d)

Preliminary results concerning the residual stress present in the coating and on the subsequent layer are in good agreement with the data reported in [19], increasing the coating layer thickness a decrease of the shear adhesive bond strength occurs due to the accumulated residual stress in the coating layer.

The coating layer with thickness of about 1.2 mm did not provide an adequate protection to the substrate. Within a few hours of its exposure to the corrosive media the steel substrate has been highly corroded and the surface is totally covered by corrosion products principaly made of iron oxide (rust). In such situations no significative differences between the un-coated and coated substrate appear. The porosity of the coating and the large cracks growth illustrated in Fig.4d determine the corrosion performance and in this case it cause the rapid failure of the substrate with very high corrosion rate.

On the other hand, the samples with thinner thickness, as expected, has very promising corrosion resistance up to 168 h. Generally, the disposition of the samples in the salt chamber is like that reported in Fig.5, while the photos of the samples extracted and examined periodically have been illustrated in Fig.6, indicating also their performance in the corrosion media in hour.
The un-coated sample shows after a short exposure, the first sign of the corrosion as in the early experiment and it proceed very quickly. After 168 h the surface is completely covered by the corrosion product. On the contrary, even if the coated steel show some differences compared to the initial state after its contact with the salt solution, the coating survives the entire period of the test. Formation of some sediment coming principally from the salt solution and partially from the oxidation of the Al has been detected. The coating layer loss its original shininess as reported in the photos on the right side in Fig.6.
The most relevant origin of these weaknesses is related to the fact that only one side of the sample has been coated and the other sides are exposed to the corrosion in a same way as the un-coated substrate.

Fig. 5 Disposition of the samples in during accelerated salt chamber test

The development of the iron oxide on these side is easily transported by the solution on the coated part. After carefully cleaning with acetone, the EDS analysis reveal only the presence of Al_2O_3 and NaCl as major compunds as reported in Fig.7.

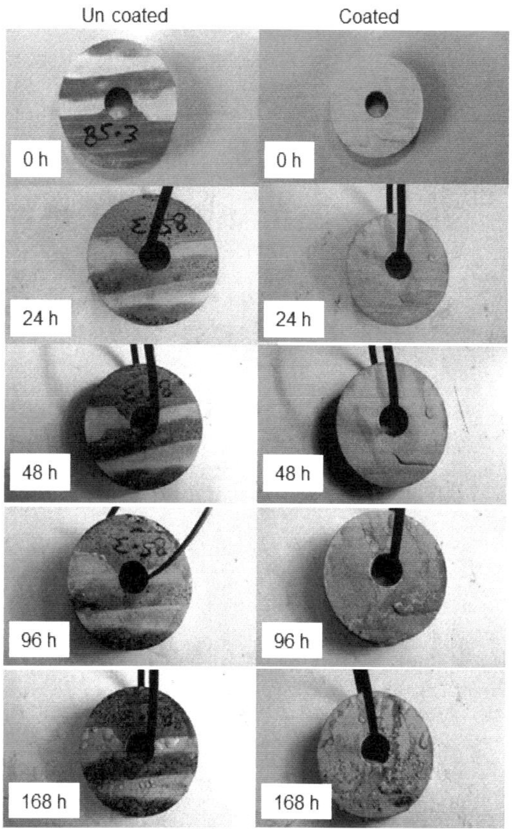

Fig. 6. Photographs of the un-coated and coated samples, taken at prefixed time, after their permanence in a salt spray chamber.

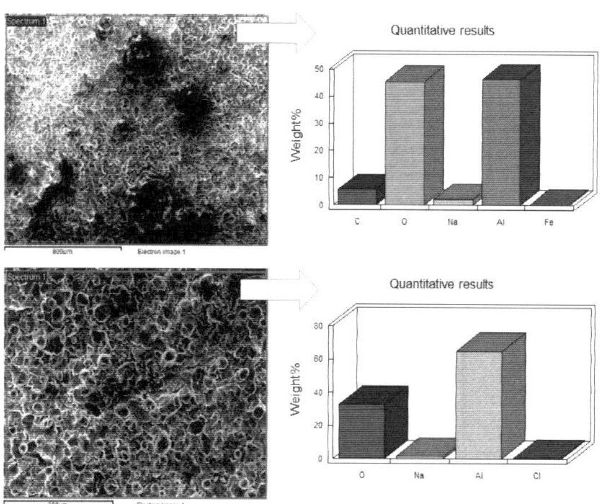

Fig. 7. Results of EDS analysis showing the area where the analysis was performed and the compositional analysis results.

As introduced, the sample shows some imperfections as indicated by the yellow arrows in Fig.8b which are more accentuated after corrosion test as reported in Fig.8c.

.

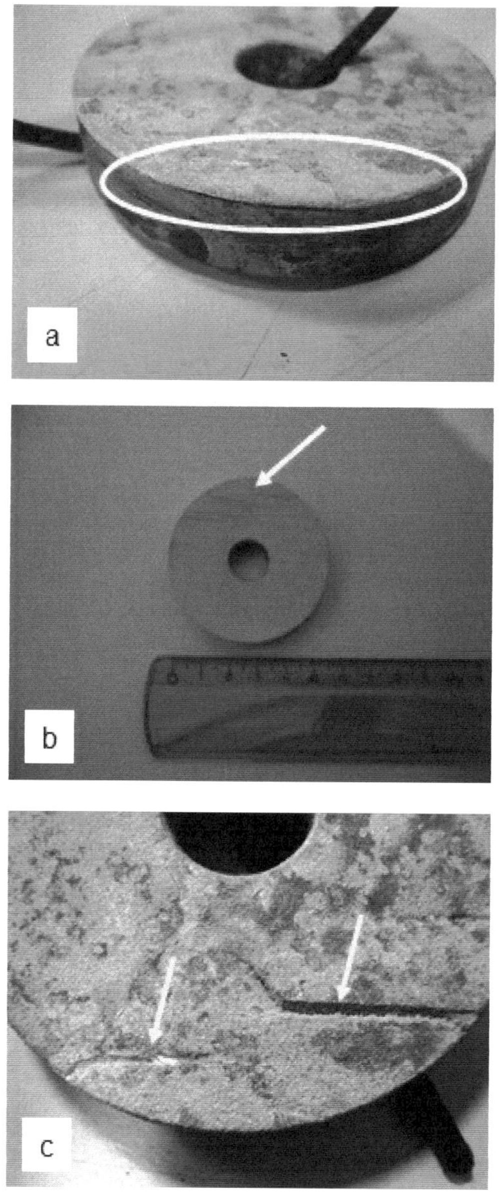

Fig. 8. SEM micrographs of the coated samples showing the zone with the presence of defect (b) and some details of the coating in the defect free area (c) and in the area with no macro defect (a).

It appears that defects are critical from corrosion resistance point of view and have a negative outcome on the final coated surface lifetime. Defects play an important role because they stop the passivation of the active pits on the surface, therefore in such cases pitting corrosion propagation rapidly occurs. Close to the defect zone, a higher amount of rust has been deposited on the top of the Al coating layer, which during the time cause a rapid corrosion of the whole surfaces.

The adhesion of the coating layer is compromised in the region close to the defect and the coating layer is detached from the substrate as reported in the highlighted area of Fig.8 a. In other regions, with no significant imperfections the coating layer adher properly to the substrate (Fig.4b) and the interface is continuous. In such areas on the top of the coating only the sign of the compounds described in the previous section has been detected. The research pointed out that the covering all the sides of the sample, if achieved with no significative defects, can lead to a good protection against corrosion

4 Conclusions

In the present paper development and characterization of cold sprayed protective Al coatings on a non-regular steel surface were performed. The research was focalized, firstly, one the optimization of the deposition route choosing the most suitable process parameters and the most appropriate layer thickness and furthermore to investigate how the defects of the substrate influence the coating performances and the corrosion resistance of the coated material.

Cold spray method is appropriate technique for Al deposition because thank to the low process temperature do not involve any phase transformation during deposition.

A coating of about 480 µm thickness was deposited successfully onto AISI 1045 steel substrate, using helium as carrier gas with 10 bar nozzle inlet pressure. The research demonstrates that defects are dangerous as corrosion resistance concerns and compromise the adhesion of the coating layer to the substrate, negatively influencing the device lifetime.

The feasibility of the cold spray deposition in case of non-conformal substrates was demonstrated. Some on-going research is related to the realization of coating with no significant defects and including all the walls of the sample by creating a barrier against corrosion.

4 Acknowledgements

The authors wish to thank ing. E. Pallavicini and D. Pezzini for their contribution to some measurements.

5 Literature

[1] S. Fajardo, D.M. Bastidas, M. Criado, J.M. Bastidas: Electrochemical study on the corrosion behaviour of a new low-nickel stainless steel in carbonated alkaline solution in the presence of chlorides, Electrochimica Acta 129 (2014) pp.160–170.

[2] S.M. Abd El Haleem, E.E. Abd El Aal, S. Abd El Wanees, A. Diab: Environmental factors affecting the corrosion behaviour of reinforcing steel: I. The early stage of passive film formation in Ca(OH)2 solutions, Corrosion Science 52 (2010) pp. 3875–3882.

[3] I.Peter, M. Rosso, F.S. Gobber: Study of protective coatings for aluminum die casting molds, Applied Surface Science 358 (2015) pp. 563–571.

[4] Pacyna JM.: Monitoring and assessment of metal contaminants in the air, Toxicology of Metals. Boca Raton, FL, CRC Press (1996) pp. 9–28.

[5] Gregory John Gibbson, Robert George Hansell: Journal of materials processing technology 2 0 4 (2008) pp. 184-191

[6] G. Bolelli, V. Cannillo, L. Lusvarghi, S. Ricco: Surface & Coatings Technology 200 (2006) pp.2995 – 3009

[7] Gulizia, S., et al.: Performance evaluation of PVD coatings for high pressure die casting, Surface and Coatings Technology, Vol. 140, No. 3 (2001), pp. 200–205.

[8] Dobrazanski, L.A., et al.: Structure and properties of wear resistance PVD coatings deposited onto X37CrMoV5-1 type hot work steel, Journal of Materials Processing Technology, Vol. 164 –165, (2005), pp. 843 – 849.

[9] I. Peter, et al.: Investigations on coating of dies for advanced squeeze casting process, Acta Metallurgica Slovaca, Vol. 20, 2014, No. 1, pp. 18-27.

[10] H.S. Sidhu, B.S. Sidhu, S . Prakash: The role of HVOF coatings in improving hot corrosion resistance of ASTM-SA210 GrA1 steel in the presence of Na_2SO_4–V_2O_5 salt deposits, Surf. Coat. Technol. 200 (2006) pp. 5386–5394.

[11] N . Bala, H. Singh, S. Prakash: An overview of characterizations and high temperature behaviour of thermal spray NiCr coatings, Int. J. Mater. Sci. 2 (3) (2007) pp. 201-218.

[12] J.A. Picas, Y. Xiong, M. Punset , L. Ajdelsztajn, A. Forn, J.M. Schoenung: Int. Journal of Refractory Metals & Hard Materials 27 (2009) pp. 344–349

[13] S. Thiele, K. Sempf, KJ. Roessler, LM Berger, J. Spatzier: J.Thermal Spray Technol 20 (1-2) 2011 pp. 358-365.

[14] I. Lichioiu, I.Peter, B.Varga, M.o Rosso, “Preparation and Structural Characterization of Rapidly Solidified Al-Cu Alloys” J. Mater. Sci. Technol., 2014, 30(4), pp. 394-400.

[15] F. Endres, et al: Electrodeposition from Ionic Liquids, WILEY-VCH Verlag GmbH & Co. KGaA, (2008), ISBN 978-3-527-31565-9.

[16] M. Rosso, F. Calosso, I. Peter: “Grain growth on galvanic deposition of aluminium”, Metalurgia International, pp. 15-18, (2011), Vol. 5.

[17] I. Peter, M. Rosso: “Simulation of electrodeposition of Al from ionic liquid”, Materials Science Forum Volume 794-796, 2014, pp 229-234.

[18] R. Ghelichi, D. MacDonald, S. Bagherifard, H. Jahed, M. Guagliano and B. Jodoin: Microstructure and fatigue behaviour of cold spray coated Al5052, Acta Materialia 60 (2012) pp. 6555–6561.

[19] M. S. Lee, H. J. Choi, J. W. Choi and H. J. Kim: Application of cold spray coating technique to an underground disposal copper canister and its

corrosion properties, Nucl. Eng. Technol. 43 (2011) pp. 557–566.

[20] P. C. King, G. Bae, S. H. Zahiri, M. Jahedi and C. Lee: An experimental and finite element study of cold spray copper impact onto two aluminum substrates , J. Thermal Spray Technol. 19 (2010) pp. 620–634.

[21] R. Lupoi: Current Design and Performance of Cold spray Nozzles: Experimental and numerical observations on deposition efficiency and particle velocity, Syrface Engineering, 30 (5) (2014) pp. 316-322.

[22] Barry Aldwell, Shuo Yin, Kevin A. McDonnell, Daniel Trimble, Tanvir Hussain, Rocco Lupoi, A novel method for metal-diamond composite coating deposition with cold spray and formation mechanism, *Scripta Materialia*, 115, 2016, pp.10 – 13.

Research of influence of technological parameters of process of spraying on properties of polymeric coatings

S. Mankovsky, L. Baldaev, S. Baldaev, S. Sheremetev, M. Fedorova, A. Rigin,
Technological Systems for Protective Coatings (TSPC), Moscow, Russia
adm@tspc.ru, +7(495)7838220

Abstract

The paper discusses the results of research on the influence of technological parameters of spraying processes on the properties and characteristics of polymer coatings.

Introduction

The application of polymer coatings on the surface of equipment and facilities is perspective direction. Using the polymer materials contributes to the formation of protective coatings with a wide spectrum of physico-chemical and mechanical properties, such as:
- low specific weight,
- wide range of environmental effects resistance,
- good electrical insulation properties,
- corrosion resistance in various aggressive environments,
- resistance to abrasion wear.

These properties allows to increase the lifetime of the protected components significantly.
Coatings of this type are used for protection of parts operating in urban, rural, marine, chemical corrosion. Also for protection of parts in contact with food.
Common methods for coating the surface by the polymer powders are flame spraying and electrostatic deposition in the electric field.
Flame spraying process is simple and versatile. As working gas is often used acetylene or propane, rarely - hydrogen. The spray material is melted by the flame of a combustible gas mixture with oxygen and sprayed with compressed air. Flame spraying process is the most economical and easily implemented methods of applying polymer coatings. Spraying and melting layer coating is in a single process operation [1].
The advantages of flame spraying of coatings are:
- the high productivity of the process with a high deposition effeciency (about 85 %);
- relatively low noise and light radiation, allowing to work without additional protection;
- the ease and simplicity of maintenance, low cost and mobility of equipment, allowing you to spray on the spot, without dismantling products [1].

This process, if necessary, can be automated to provide a constant spraying distance from the nozzle exit of the gun to the surface, application of polymer coatings on the external or internal surface of metal structures can be conducted both in the field and in the factory.
Electrostatic deposition finds wide application in almost all areas of industrial production, where protective, decorative, insulating, chemically resistant and other coatings needed. Process for producing coatings from polymer powders is the transfer of the electric charge to the powder particles of the dielectric material, depositing them under the influence of electrical forces from an aerosol cloud that surrounds the earthed surface and fixing the applied layer. Powder coating formed is melted, sintered or polymerized and is transformed into a monolithic film during heat treatment. They significantly improve the performance characteristics of the coated products. Benefits are provided by:
- exceptions to the paint technology of expensive and polluting solvents;
- the possibility of obtaining coating thickness of 40 microns and above for one cycle of spraying;
- high productivity, improve working conditions, etc.

The final stage of formation of the coating is a thermal processing required for the polymerization of the polymeric material. Common methods for thermal processing required for the polymerization of the polymeric materials are heating in the oven and heating with induction heater.
Oven heating offers uniform heating, but having the following disadvantages: limitation in size, low heating rate, large amounts of energy and the inability of local heating.
The peculiarity of method induction heating is that the induction heaters do not produce heating of the polymer. Induction heaters make the heating of metals by electric currents which are induced by the alternating magnetic field. At high frequency the eddy currents are displaced educated by them magnetic field in thin surface layers of metal products Δ (skin effect), resulting in their density increases sharply, and the product is heated. Coatings of metal powder and a polymer on its surface, heated by conduction.
Advantages: rapidity process; energy saving; the possibility of local and sequential processing sections.
Disadvantages: risk of overheating of individual sections because of the uneven distribution of fields. Overheat at junctions of a product consisting of several individual parts that are bonded together mechanically. The complexity of processing of some metals with low resistance.
This paper presents the results of research, testing and comparison of polymer coatings PA11 made by 3 different ways:
- Flame spraying;
- Electrostatic spraying followed by the heating in the oven;

- Electrostatic spraying followed by the heating by induction heater.

Experimental Methods

For application of coatings were used two types of steel samples: without coating and with a bondcoat of corrosion-resistant metal alloy. The base material – steel A107. Sample dimension is 85×50×2 mm. For surface preparation of the samples abrasive blasting was used.

The designation of samples is achieved by two digits. The first digit represents the presence of a bondcoat : 1 - without bondcoat, 2 – with bondcoat. The second digit indicates the method of heating of the polymer, as described previously: 1 – heating by induction heater, 2 – heating in the oven, 3 – flame spray heating.

For gas-flame spraying polymer powders was used self-made portable burner. It has been designed and built to carry out flame spraying of thermoplastic powders applied by melting. The equipment must be connected to a source of compressed air and LPG (liquid propane gas). The working process can be outlined as follows: compressed air coming from the air compressor and LPG fed from the special bottle are supplied to the torch (their mixing is adjusted by a special remote control that is also used to adjust the mixture outlet pressure). Thermoplastic powder is then added to the air-LPG mix inside the gun. Inside the gun, the powder, gas and air are mixed and the powder is then flame-sprayed to coat.

For electrostatic spraying polymer powders was used gun "Leader" — designed for even application of powder on the surface. Powder invades from the tank into the tube of the gun, passes near the corona electrode and receives a charge necessary for the adhesion to the surface. The conical nozzle sets a spray radius. Voltage regulation from 0 to 60 V allows to change the strength of the reported charge that affects the maximum possible thickness of the layer. Air is used as a lift gas.

For heating of the polymer powders deposited by electrostatic method was used induction heater "ELSIT".

The equipment consist of a high-frequency generator with microprocessor control system and a display displaying information, and block matching transformer with the resonant capacitors. On the front panel of the induction heater placed terminals for connection of the inductor and controls the heating process with a LCD-display to control the induction heating process. Installation and the inductor are cooled flow-through technical water. The control unit can be adjusted for the capacity in percent and set the timer.

Research and testing of the polymer coatings was carried out according to the following control methods:

- visual method control - assessment of the quality of the appearance of the coatings was carried out according to [2] (color uniformity, presence of unmelted particles, cracks, delaminations).
- optical microscopy of coatings - the study of boundary "the basis-coating" and the coating structure was carried out on the microscope-hardness tester Zwick/Roell ZHμ.
- hardness measurement of coatings - were used the hardness tester TH-320, allowing measurements by the method superficial Rockwell scale HR15Y. The tests were carried out according to the operating instructions of the equipment.
- test was conducted on a special stand with the abrasive particles (quartz sand grains 100-400 μm). The sample is fixed in an equipment and pressed to a rotating rubber circle with a force of 135 N. Sand is supplied on the sample from hopper. Rubber circle captures particles and carries them by the sample surface providing wear. Conducted qualitative and quantitative evaluation results: visual inspection, mass loss. Both samples were tested for 20 sec [3].

The scheme of test stand is shown in- method of determination of adhesion. Cross-cut test method (GOST 15140-78).

The spray materials were:

- for applying the bondcoat was used wire of Monel with diameter of 3 mm. The alloy composition is: Ni+Co (base), Cu (27-29%), Fe (2-3%), Mn (1.2-1.8%);
- for applying the polymer coatings was used polymer powder PA11 with non-spherical particles, comminuted form. Photos of particles of polymer powder are presented in Fig 1. Characteristics of polymer material PA11 are presented in Table 1.

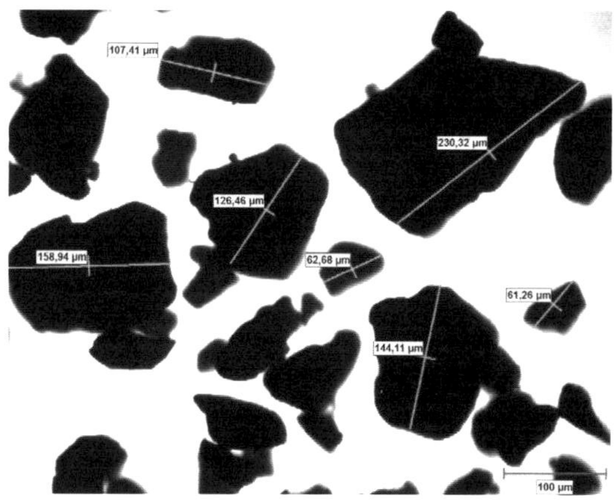

Fig. 1. Photo of particles of polymer powder

Table 1. Characteristics of polymer material PA11

Parameter	Value
Characteristic of powder	
Melting temperature,°C	190
Grain size 90%, μm	80-200
Bulk density, g/cm^3	0.49
Color	grey

The surface of the base material was cleaned from impurities and contaminants. To do this, abrasive blasting was used. This provided the activation and the surface roughness for an effective adhesion of the coating. The coating was applied within 2 hours after the surface activation.

For the spraying of metal bondcoat was used gas-flame burner with the robotic manipulator "KUKA". Mode the flame was neutral. The bondcoat thickness is 150-250 microns.

Parameters of electrostatic spraying (ES) presented in the Table 2.

Table 2. Parameters of electrostatic spraying

Parameter	Value	
	Sample 1.1, Sample 1.2	Sample 2.1, Sample 2.2
Air pressure, bar	2	
Voltage, kV	30	
Coating thickness per pass, μm	14-18	
The number of passes to obtain the desired coating thickness	10	

Parameters of induction heating presented in the Table 3.

Table 3. Parameters of induction heating

Parameter	Value	
	Sample 1.1	Sample 2.1
Frequency, Hz	55475	
Power, %	45	
Average heating temperature of the samples, °C	230	230

Parameters of the oven heating presents in the Table 4.

Table 4. Parameters of the oven heating

Parameter	Value	
	Sample 1.2	Sample 2.2
Preheating temperature of the oven, °C	220	
The time of exposure of the sample to complete melting powder, min	15	40

Parameteres of flame spraying (FS) presents in the Table 5.

Table 5. Parameteres of flame spraying of polymer

Parameter	Value	
	Sample 1.3	Sample 2.3
Air pressure, bar	4	
Powder flowrate, standard unit	2	
Distance preheating, mm	250	
Distance of spraying, mm	350-400	
Distance of fusing, mm	200-300	

Discussion

The results of the research and testing of polymeric coatings are set forth below.

Visual control showed color uniformity, the pavement surface is smooth, relatively uniform. Comparison roughness polymer coatings on the samples with metallic bondcoat and samples without metallic bondcoat revealed no significant differences. The roughness of the polymer coatings with the metallic bondcoat samples is 0,25 Ra. This is higher than the roughness of the polymer coatings on the samples without the metallic bondcoat (1.6 times, 0,15 Ra).

Optical microscopy of coatings: boundary "the basis-coating" and photos of the microstructure of polymer coatings is shown in Fig.2, Fig.3 and Fig.4.

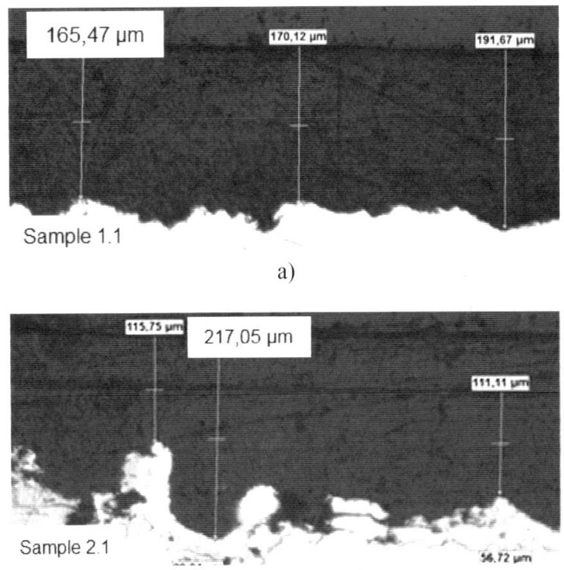

b)

Fig.2. Microstructure electrostatic spraying + induction heating coatings: a) Sample 1.1 (without metal bondcoat), b) Sample 2.1 (with metal bondcoat)

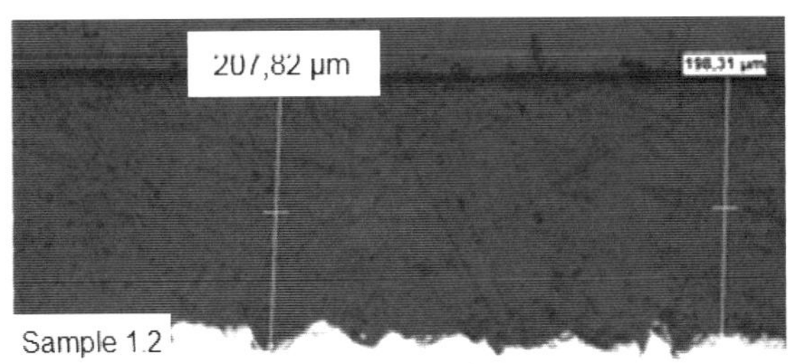

a)

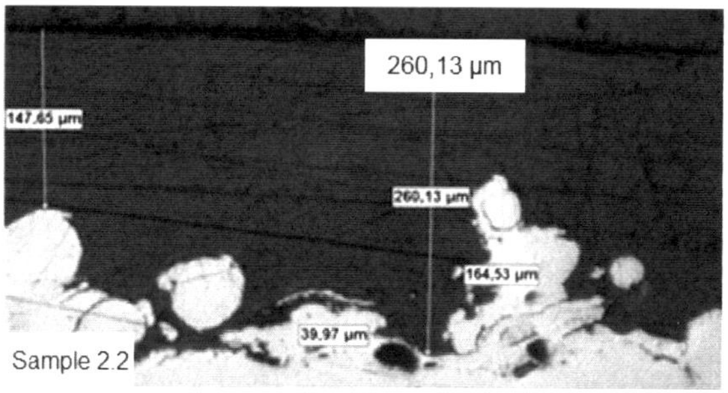

b)

Fig.3. Microstructure electrostatic spraying+oven heating coatings a) Sample 2.1 (without metal bondcoat), b) Sample 2.2 (with metal bondcoat)

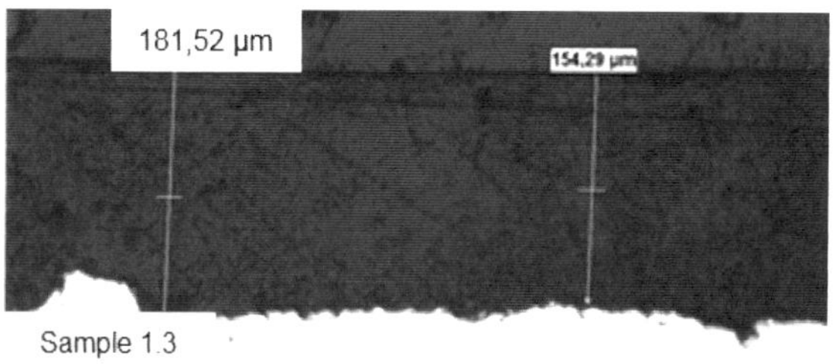

a)

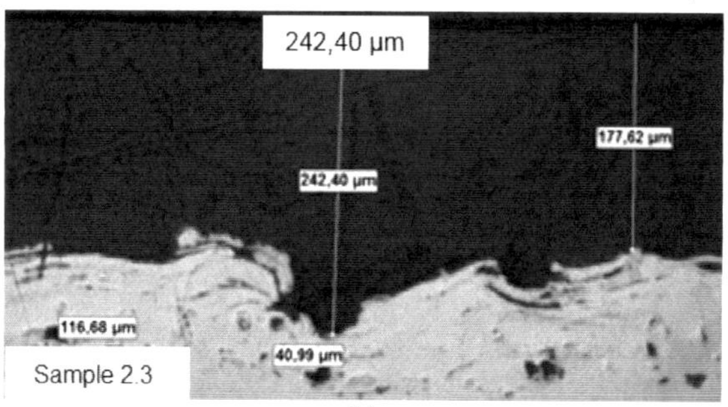

b)

Fig.4. Microstructure Flame spraying coatings: a) Sample 1.3 (without metal bondcoat), b) Sample 2.3 (with metal bondcoat)

Microstructural analysis shows: significant differences in the structure of the coating are not detected. The method of application and the heating does not have a significant impact. The coating thickness for methods ES+induction heating and FS is uniform in the range of 200-250 µm, but for method ES+oven heating the coating thickness is smaller (150-165 µm) possibly due to effects of volume of the heat source (Table 6). Border "the basis-coating": for samples without bondcoat is a clear, smooth; for samples with bondcoat is rough, with protrusions which are filled with polymer.

Table 6. Average thickness of polymer coatings

Parameter	Value, µm	
	Without bondcoat (1)	With bondcoat (2)
Sample 1(ES+induction heating)	200	224
Sample 2 (ES+oven heating)	165	151
Sample 3 (FS)	246	251

Hardness testing of coatings: the method of production of coatings and the presence of the bondcoat does not influence the hardness coatings. Hardness polymer coatings is uniform in the range 85-89 units (Table 7).

Table 7. Average hardness of polymer coatings (HR15Y scale)

Parameter	Value	
	Without bondcoat (1)	With bondcoat (2)
Sample 1(ES+ induction heating)	87	87
Sample 2 (ES+oven heating)	85,8	80,9
Sample 3 (FS)	87,9	86

The results testing of coatings for wear by a directional stream of abrasive particles showed a high protective properties, namely increased abrasion resistance on average 5 times compared with the sample without coating (Table 8).

Table 8. Abrasive resistance of polymer coatings

Parameter	Value, Δm	
	Without bondcoat (1)	With bondcoat (2)
Control sample (without coating)	0,12	
Sample 1(ES+induction heating)	0.033	0.035
Sample 2 (ES+oven heating)	0.033	0.035
Sample 3 (FS)	0.033	0.035

The results of testing shown in Fig.5

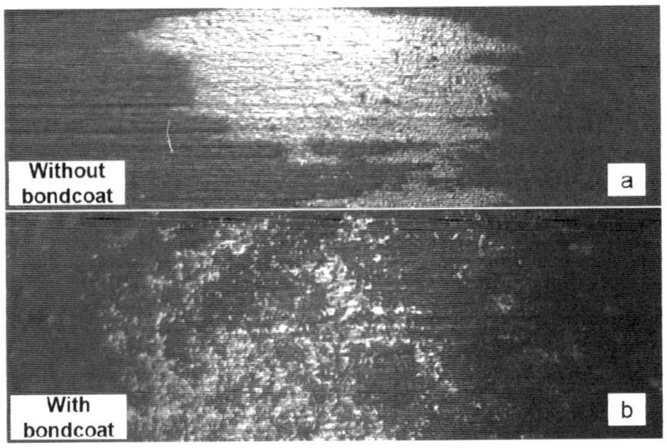

Fig.5. Surface of sample after abrasive resistance test: a) polymer is completely worn away, the steel bases are not protected from corrosion; b) polymer is partially worn, sample still protected with anti-corrosion bondcoat and the polymer remaining in the pores of the bondcoat.

That confirmed in the test of abrasion resistance. The wear of polymer layer on the sample without bondcoat leads to the interaction of the base metal with the environment. The presence of the bondcoat prevents the formation of corrosion at the same degree of wear, because on the base metal remains a layer of corrosion resistant metal with inclusions of polymer that closes the pores.
In determining the adhesion method of lattice cuts, coatings showed high adhesion – score A1 (scale 1-4), regardless of the presence of the bondcoat and a method of applying.

Table 9. Results of the adhesion testing of polymer coatings

Parameter	Value	
	Without bondcoat (1)	With bondcoat (2)
Sample 1(ES+induction heating)	1	1
Sample 2 (ES+oven heating)	1	1
Sample (FS)	1	1

Conclusions

The paper shows that the described methodes can be used for the application of polymer coatings. These methods have their advantages and disadvantages and can be used, depending on their technological characteristics.
Experiments showed that results of research of coatings are similar but have minor differences.
The flame-spray method versatile and efficient method of applying polymer coatings. This method has one-step operation of applying and heating of polymer powder, and the possibility of installing the gun on the robotic manipulator allows to achieve a high accuracy in thickness control and uniformity of the coating.
Process electrostatic deposition of polymer coatings with subsequent induction heating can have high productivity, but disadvandage is two-step operation and limited product configuration.
The process of electrostatic deposition of polymer coatings with subsequent heating in a oven required much more time for forming the coating, in comparison with the previous methods, with similar results. On this basis, we can say that the use of the oven for heating, it is advisable only for small parts. The process of electrostatic deposition of polymer coatings suitable for flat surfaces of large area, where the thickness control is not of paramount importance. Thus, for the application of polymer coatings can be used various ways, but most acceptable - flame-spray method.
When using metal coatings for corrosion protection uses thickness coatings not less than 300-400 microns.
When using a polymer coating deposited on the metal bondcoat can lower the thickness of the metal bondcoat 2 times (instead of 300-400 microns, 150-200 microns) without decrease the quality and get economic effect.
The polymer coatings has good adhesion to the metal bondcoat with the developed roughness, and when applied to the metal bondcoat penetrates the irregularities and pores and secured in them and can provide a dense coatings without open pores to protect the base metal, and prevents the penetration of external influences (chips, scratches) a base metal.
Integrated coatings based on metals and polymers of various types, have great potential use in petrochemical, chemical, food, agricultural, aerospace, nuclear and other industries.

References

[1] Baldaev L. Other: Student training manual. Thermal spraying (2007), pp. 344/121.
[2] GOST 9.304-87 Gas-thermal coatings. General requirements and methods of control (1987).
[3] Vinogradov V., M. Sorokin. Other: Training manual. Abrasive wear (1990), pp. 224/89.

Laser re-melting assisted with induction heating of yttria partially stabilized zirconia thermal barrier coatings

Rujia Wang, Xia Dong*, Zhengjie Fan, Kedian Wang, Wenqiang Duan, Xuesong Mei, Wenjun Wang, Jianlei Cui
(State Key Laboratory for Manufacturing Systems Engineering, Xi'an Jiaotong University,710049 Xi'an)

Abstract
Laser re-melting of thermal barrier coatings (TBCs) results in certain issues, such as surface segmented network cracks, surface crater and low processing efficiency. In this study, laser re-melting of ZrO2-7wt% Y2O3 using a pulsed laser with and without induction preheating was investigated. A wide variety of TBCs surface temperatures were attained by the induction heater before laser re-melting. The laser re-melted samples were investigated by scanning electron microscope (SEM) and X-ray diffraction (XRD). It was observed that the utilization of induction preheating can effectively decrease the threshold laser energy (30% of peak power, 67.9J/cm^2) of fully re-melting TBCs with relatively less crack density compared to without the application of it (35% of peak power, 79.2 J/cm^2). It was found that some longitudinal cracks distributed along the re-melted/un-melted interfaces without the application of preheating or at low surface temperatures (below 900℃) while disappeared when heated to higher temperatures (above 900℃) due to it helps developing relatively steadier melt temperatures as well as decreasing the cooling rates between molten pool and un-melted coating and consequently the thermal gradient during the subsequent re-melting process. XRD results indicated that the enhanced surface temperature can decrease the content of monoclinic phase.
Keywords: laser re-melting; thermal barrier coating; induction preheating; network crack.

1 Introduction

Plasma sprayed thermal barrier coatings (TBCs) have been widely applied to gas turbine and diesel engines to enhance thermal insulation and increase component durability, thus extending their service life and improving their efficiency [1]. Commercial TBCs system consists of yttria-stabilized zirconia (YSZ) ceramic top layer, an oxidation resistant bond layer (typically MCrAlY) and a superalloy substrate [2, 3]. However, the ceramic top layer has significant amount of microstructural defects such as voids, porosities, unmelted particles and microcracks due to the plasma spraying process[3]. The porosities can provide a low thermal conductivity[4], whereas these defects contribute to premature spallation or fracture of coatings.

Along with the increasing demands for more efficient gas turbines and diesel engines, a harsh and high-temperature environment is inevitable. Thermal interaction of laser beam with materials has gradually been put on the agenda. Laser remelting can be regarded as a strategy in this field to improve the performance of TBCs [5, 6]. Large quantities of studies have indicated that laser remelting is conducive to reduction in the surface roughness, formation of segmented network cracks and columnar grains[3, 7] perpendicular to the coating surface as well as sealing porosities beneath the surface [2, 8-11]. A smoother surface is of great benefit to enhance the hot corrosion [1, 6, 10]and erosion resistance [6]of coatings and segmented cracks are of great help for improving the thermal shock resistance [2, 6, 10, 12, 13] of coatings. However, the thermal damage and crack propagation can be caused in the laser remelting process because of its high cooling rate and elevated temperature gradient. Therefore, an increasing interest in preheating the thermal barrier coatings for surface modification has led to the development of new technologies for laser remelting procedures.

In the present study a combination of laser remelting with induction preheating is utilized. Induction heating has been widely used in laser cladding or welding [14-16]. Studies have indicated that evolution of the molten pool in laser cladding assisted with induction heating was uniform and smooth [14]. Few attempts have been conducted on laser remelting assisted with induction preheating. The various preheating temperatures are of special interest in this study due to their ability to produce local improvements in the cooling rate and consequently thermal gradient in the remelted zone. And the main intention of this paper is to understand of the effect of laser remelting assisted with induction preheating on TBCs.

2 Experimental

2.1. Material preparation

Type 718 nickle-based superalloy (Inconel 718) was acted as substrate material and prepared into samples with a dimension of 10mm×5mm×5mm. NiCoCrAlY and nanostructured yttria-stablized zirconia power (ZrO2-7wt.% Y2O3) were prepared for the bond coating and top ceramic coating, respectively.

2.2. Air plasma spraying

Plasma spraying was carried out with a GTV delta three anodes atmospheric plasma spraying system (GTV, German). Nitrogen gas was the primary plasma gas whilst hydrogen and argon gas were added as the secondary plasma gases. The substrates were grit blasted, degreased and cleaned with acetone using ultrasonic cleaning and preheated with plasma gun prior to spraying. After that, the well-prepared samples were subjected to plasma spraying process.

2.3. Induction preheating and laser re-melting

In this study, a JK 300D Nd:YAG pulsed laser (GSI, England) was employed to complete the re-melting task. It features a wavelength of 1064 nm, maximum average laser power 300 W, repetition frequency range of 1-1000 Hz, pulse duration range of 0.2-5 ms and a Gaussian distribution. Compared with directly laser re-melting, a novel method was adopted in this work. Prior to each re-melting process, samples were preheated by an induction heater and their surface temperatures were monitored by an infrared thermometer. The operating mode of the induction heater was not switched from heating to heat-insulation until the surface temperatures of coatings arrived at the scheduled values. The scheduled preheating temperatures ranged from 400 ℃ to 1000 ℃ with an interval of 100 ℃. Moreover, a heat-conducting intermediate was employed as the directly heated object rather than the samples consequently rendering a smooth and quiet heating process. And meanwhile comparative trails were conducted, wherein laser energy density is used as the independent variable of a series experiments with non-preheating. The schematic diagram of laser re-melting and induction preheating system used in this investigation is shown in **Fig. 1**. Optimum laser parameters obtained from preliminary experimental results were used to prepare high quality remelted coatings. The corresponding laser parameters mentioned above are summarized in **Table 1**.

Table 1
Parameters of laser remelting process

Parameters	
peak power	30%,32%,33%,34%,35%
scanning speed	5mm/s
pulse during	1ms
repetition frequency	50Hz
beam spot	3 mm
preheating temperature	400,500,600,700,800,900,1000 ℃

2.4. Coating characterization

The microstructures and surface topographies of laser remelted samples were examined by scanning electron microscopy (JSM-6390A, JEOL, SEM, Japan). Then the samples mounting and polishing employed following procedures consist of using a cold curing resin material such as araldite with hardener at room temperature and an automatic polisher respectively. After polishing, the cross-section surfaces were cleaned with acetone and then also determined by SEM. X-ray diffraction (XRD) studies were carried out using a X-ray diffractometer (D/MAX-2400, Rigaku, Japan) to examine the phase compositions present after laser re-melting. Cu target Kα produced at 40KV and 100 mA with scanning speed of one step/s. and the step length of 0.02° was used to determine the peak positions of the phases in the 2θ range of 20° to 90°.

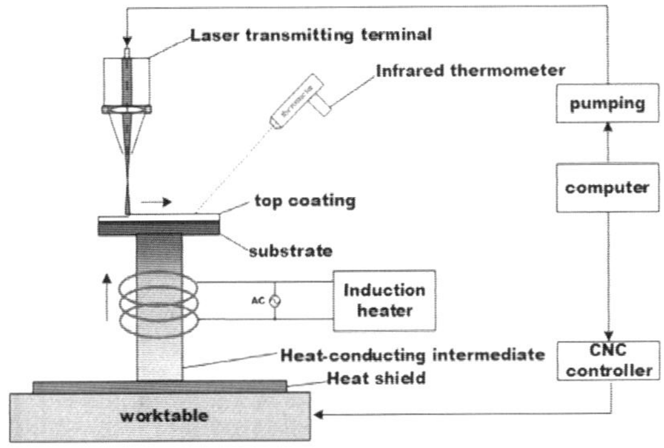

Fig. 1. Schematic diagram of laser re-melting and induction preheating system

3 Results and discussion

3.1. Microstructure

Fig. 2 shows the surface SEM topographies of the coatings subjected to laser remelting combined with previous induction preheating. Laser pulse energy is 4.8J (30% of peak power, energy density is 67.9 J/cm2), and the induction preheating temperature varies from 400 ℃ to 1000 ℃. The influence of laser remelting is considered to help produce a smoother surface in the remelted region contrary to the unmelted one. **Fig. 2a** ~ **Fig. 2f** reveal that the segmented network cracks and craters as well as pores were present on the surfaces of the remelted coatings. However, there exists a big difference among them when they are closely examined. It was observed that the crack density showed a downward trend along with the decrease of the preheating temperatures. Nevertheless, the longitudinal cracks gradually appeared at the remelted/unmelted interfaces along the laser scanning direction with the decrease of the preheating temperatures. **Fig. 2a** and **Fig. 2b** represent the remelted surface with preheating temperature at 1000℃ and 900℃, respectively. As can be seen that, the boundaries between remelted and unmelted regions were free of cracks. However, there exhibited a severer crack phenomenon in the internal of the remelted region than other samples with relatively lower preheating temperatures. Exactly the opposite, as we can see in **Fig. 2e** and **Fig. 2f**, it obtained a less crack density in the internal of the remelted region of APS coatings. Furthermore, obvious longitudinal cracks appeared at the interfaces between the remelted and unmelted region with preheating temperature at 600℃ and 400℃. While under the condition of preheating

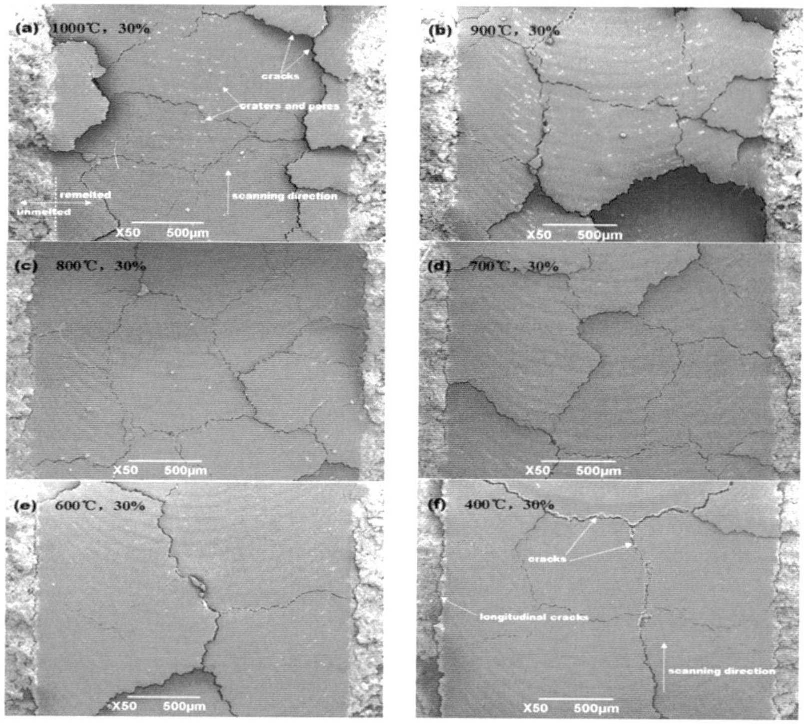

Fig. 2. Surface SEM topographies of the coatings subjected to laser remelting combined with previous induction preheating:(a)1000°C,(b)900°C,(c)800°C,(d)700°C,(e)600°C,(f)400°C

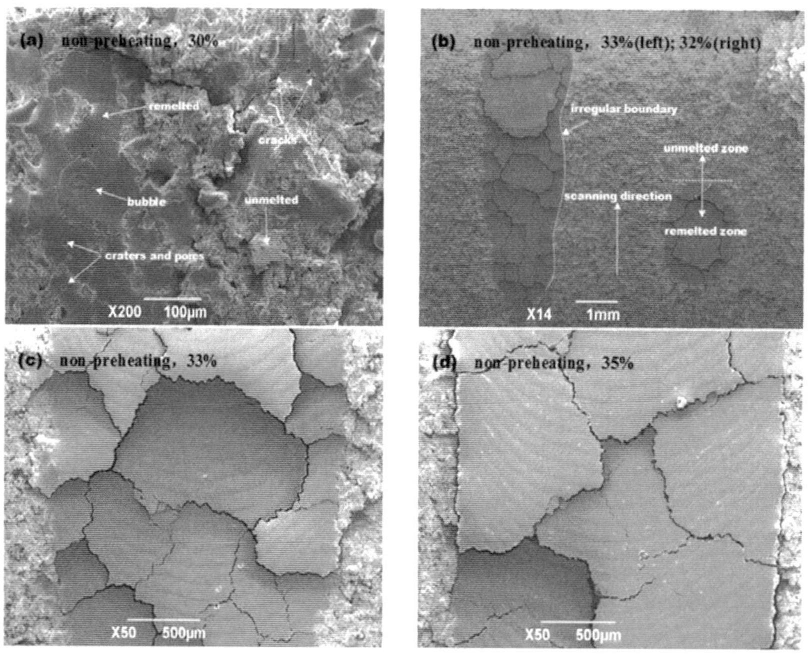

Fig. 3. Surface SEM morphologies of the remelted APS coatings:(a)30%,(b)33%&32%,(c)33%,(d)35%

temperature at 800 °C and 700 °C, there existed several tiny longitudinal cracks at the remelted/unmelted interface but a dense crack distribution in the internal of the remelted region of coatings, as shown in **Fig. 2c** and **Fig. 2d**. Therefore, a conception of 'transfer of crack' is raised to explain what it happened in the laser remelting process assisted with different preheating temperatures. In the laser remelting process, there existed a temperature gradient between the laser beam and the remelting zone because of the laser source has a Gauss distribution, contributing to a distinct difference between the center and margins of the remelting zone. When the surface of the coatings were preheated at high temperatures (900℃ and 1000℃), the center of the remelting zone exhibited a severer temperature gradient and on the contrary the margins of the remelting zone a moderate one. Thus, the cracks were initialed and consequently propagated in the internal of the remelting zone. And a small expansion and shrinkage occurred at remelted/unmelted interfaces during remelting and solidification as a result of high preheating temperatures. The residual stress [1, 3] had to be released through the propagation of internal cracks. Therefore, the remelted zone featured free of cracks at the remelted/unmelted interfaces and a dense crack distribution in the internal. While the surface of the coatings were preheated at relatively lower temperatures(400℃ and 600℃), the decrease of the internal crack density in the remelted zone may be due to the formation and propagation of remelted/unmelted cracks and a totally mild thermal gradient was acquired. On the one hand, a great expansion and shrinkage of between the remelted zone and unmelted zone caused the formation of longitudinal cracks [3, 6, 10, 11]. The other is the thermal stress caused by laser moving could be released through longitudinal cracks, thus realizing the transfer of cracks and reducing the internal cracks.

Fig. 3 presents the surface SEM morphologies of the remelted APS coatings obtained under the condition of various laser energy densities with non-preheating. It should be noted that laser energy density acted as the independent variable of such comparative experiments. The laser energy density ranged from 67.9 J/cm2 to 79.2 J/cm2, corresponding to 30% and 35% of peak power respectively. As can be seen in **Fig. 3a**, surface characteristics of coatings did not change significantly at 30% of the peak power, whose laser energy density is 67.9 J/cm2. It did not form a complete remelted microstructure. This can be considered that the heat energy input applied to the surface of the ceramic top coating was insufficient to accomplish such changes including reheating, melting and solidification. However, it provided several cues to investigate the causes of the formation of craters and pores and also threw a light upon the existence of cracks after laser remelting. A bubble-like microstructure and craters and pores as well as cracks were observed in the remelted region in **Fig. 3a**. The bubble-like microstructure was just like the left traces of suddenly burst out of bubbles. These bubbles may stem from the coalescing and raising of pores [11]located in the ceramic top coating. The craters and pores may be the consequent of failing to completely escape out from surface before re-solidifying of the molten zone [3]. **Fig. 3b** reveals that the heat input was also not enough to make a difference at 32% of the peak power and an irregular remelted shape was attained at 33% of the peak power compared to that obtained at 30% of the peak power with induction preheating. **Fig. 3c** and **Fig. 3d** give the morphologies of the remelted zone acquired at 33% and 35% of the peak power, respectively. It should be noted that only when the energy density rose to 79.2 J/cm2 (35% of the peak power) can make the surface of coatings a normal and regular remelted zone with non-preheating. However, the remelted zone shown in **Fig. 3d** was characterized by a dense crack distribution and a significant longitudinal crack on account of the aggressive energy input (79.2 J/cm2). In addition, under the condition of previous induction preheating, the crack density of the remelted coatings was less than that of the remeled coatings under the condition of non-preheating process. This is mainly due to imposing induction preheating on the coatings, which helps in developing a relatively steadier melt temperature as well as decreasing the cooling rates [14] between the molten pool and unmelted coating and consequently the thermal gradient [14] during the subsequent remelting process. Thus, providing induction preheating prior to laser remelting can make the molten pool stay in a relatively stable state and a small localized temperature gradient. The benefits embodied that it made the development of cracks in a relatively gentle process other than become so vigorous. As a result, under the condition of induction preheating it can effectively reduce the laser energy threshold (79.2 J/cm2 down to 67.9 J/cm2) and crack density of the remelted zone, thereby reducing the thermal damage of high laser energy to the as-sprayed coatings.

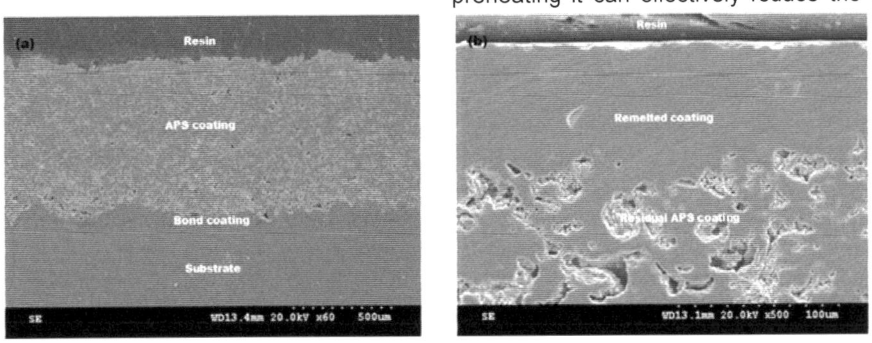

Fig. 4. Cross-sectional micrographics of as-sprayed (a) and laser remelted TBCs (b)

Fig. 4 shows the cross-sectional micrographics of as-sprayed and laser remelted TBCs. It is demonstrably obvious that there exists a big difference between **Fig. 4a** and **Fig. 4b**. APS coating, bond coating and substrate are indicated visually in

Fig. 4a. The as-sprayed coating contained mounts of pores, microcracks and unmelted powders and owned a rough
morphology. The high porosity of the as-sprayed TBCs helps in providing low thermal conductivity [7]. As can be seen in **Fig. 4b**, the laser remelting smoothened the rough surface of as-sprayed coating and rendered the top coating a dense layer. Besides, this may help in enhancing the hot corrosion and erosion resistance of coatings with a smoother and denser microstructure [10].

Fig. 5 exhibits the micrographics of crack distributions on cross-section of laser remelted TBCs. In all situations, the cracks originated from the segmented network cracks presented in **Fig. 2** and **Fig. 3** and gradually went deep into the coatings along the direction perpendicular to the coating surface. When the cracks came across the defects such as pores, microcracks as well as voids during the development of cracking, their traces would deviate from the primary direction and were affected by the defects exist in coatings [9]. As shown in **Fig. 5a** ~ **Fig. 5d**, the characteristics of cracks across the cross-section were quite different. With the decrease of preheating temperatures, the vertical cracks became wider and deeper in transverse and longitudinal direction, respectively. The crack length in **Fig. 5d** has around fourfold increase in longitudinal compared with that in **Fig. 5a**. Compared with the cracks depicted in **Fig. 5a** ~ **Fig. 5d**, the vertical cracks became worse and severer with an increased width, as shown in **Fig. 5e** and **Fig. 5f**. This was mainly due to a higher thermal gradient was obtained and consequently the solidification was faster, leading to higher thermal stresses and thus higher crack initiation and propagation in the as-sprayed coatings with non-preheating of as-sprayed TBCs. The concentration of defects may also be considered as a factor of serious cracks in cross-section of remelted coatings. What is more, the cross-sectional morphology of columnar crystals grown perpendicular to the coating surface in the form of amplification can also be seen in **Fig. 5f**. The dense columnar crystals were rooted in the bottom of the remelted layer and the formation of this microstructure was owed to the direction of heat flow[3]. The remelted depth was also

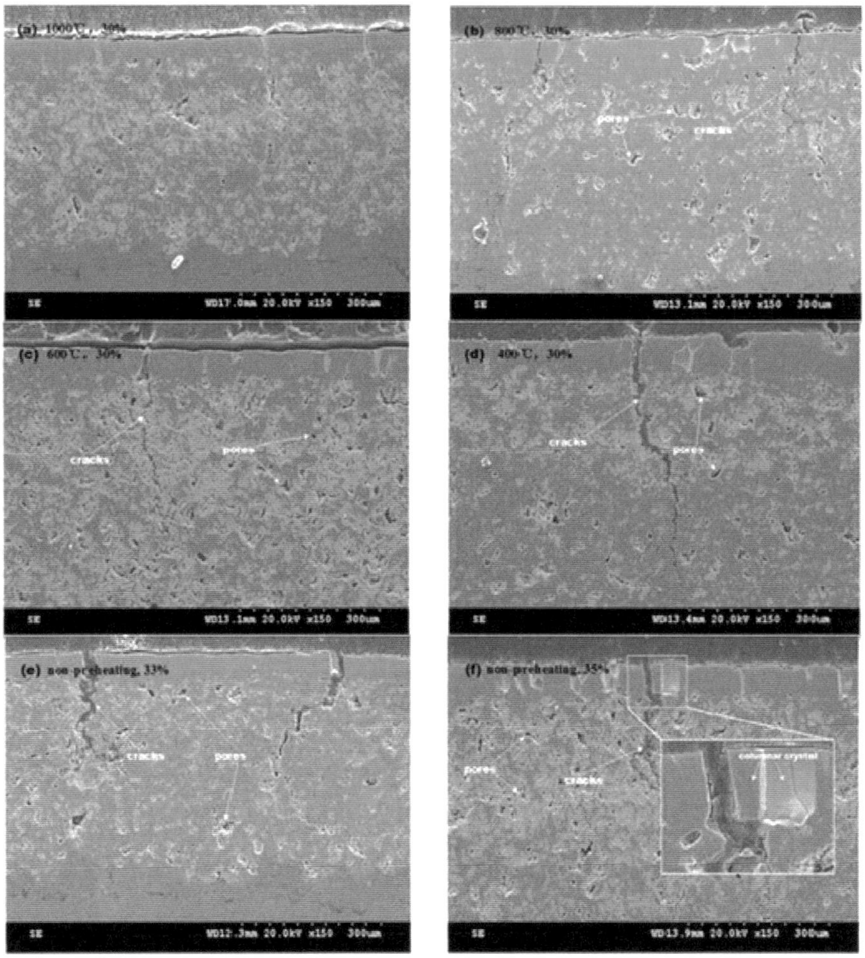

Fig. 5. Micrographics of crack distribution on cross-section of laser remelted TBCs: (a) (b) (c)(d), preheating; (e) (f), non-preheating.

measured in this study. The results were presented in **Fig. 6a** and **Fig. 6b**. The values were calculated through the method of acquiring the average of ten different positions located at the cross-section of the remelted coatings. **Fig. 6a** depicted the curve of mean remelted depth versus the preheating temperature. **Fig. 6b** presented the curve of mean remelted depth of 79.2 J/cm2. Furthermore, there is also a phenomenon that we cannot turn a blind eye. Let us back to the SEM morphologies of laser modified TBCs. There are mounts of craters and pores in the remelted region presented in **Fig. 2a**, **Fig. 2b** and **Fig. 3d**. This is mainly due to the abundant pores coalesced and rose to surface but failed to completely escape out as

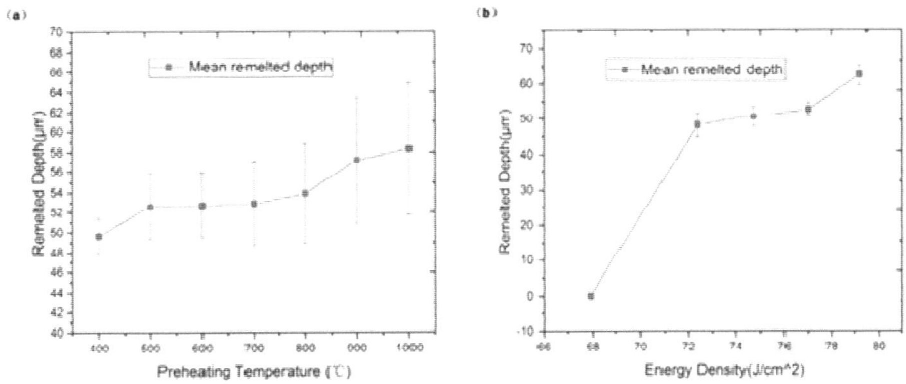

Fig. 6. The remelted depth at (a) different preheating temperatures and (b) different energy densities

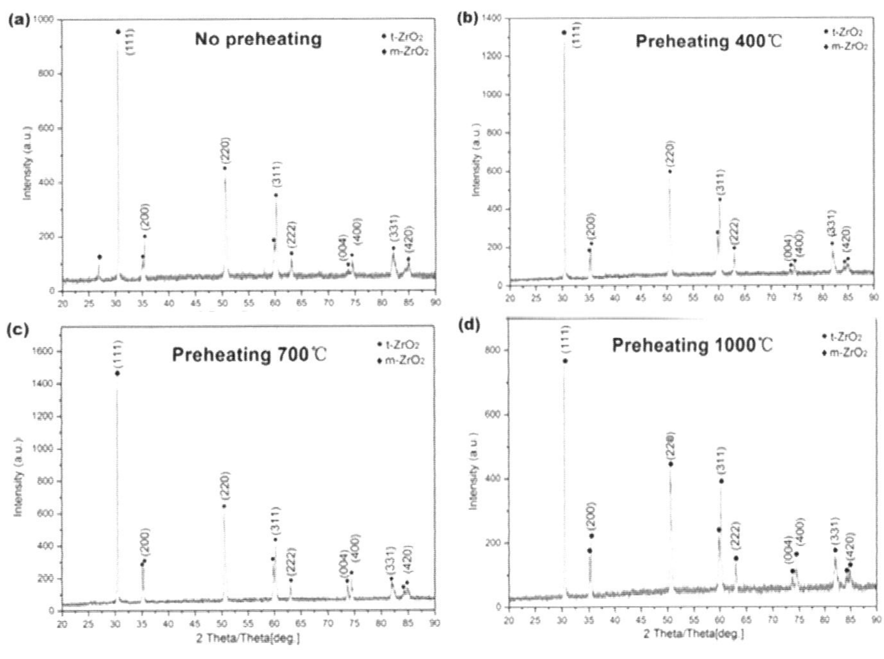

Fig. 7. XRD patterns for laser remelted coatings with and without preheating

versus the energy density. It is predicted that the mean remelted depth manifested an increasing trend along with the rise of preheating temperatures. However, there is an extremely small growth when the temperature ranged from 500℃ to 800℃. The mean remelted depth could arrived at 58.35 μm when the preheating temperature was 1000℃. Moreover, the mean remelted depth also went up with the gradually enhanced energy density. The mean remelted depth could reach to 62.49 μm under the condition of energy a result of the corresponding larger remelted depth (58.35 μm and 62.49 μm).

3.2. Phase identification

XRD patterns for laser remelted coatings with and without preheating were given in **Fig. 7**. As can be seen in **Fig. 7a**, laser remelted thermal barrier coatings without preheating consisted of tetragonal phase (T), cubic phase (c) and a small amount of monoclinic phase (m). However, there was no formation of monoclinic phase in the remelted coatings with preheating presented in **Fig. 7b**, **Fig. 7c** and **Fig. 7d**. The situations mentioned above have a relationship with the cooling rate and distribution of the Y2O3 stabilizer. The cooling rate in plasma spraying and laser remelting process was estimated as high as 10^7 °C/s and $10^3 \sim 10^4$ °C/s [1, 3, 11], respectively. This can lead to the inhomogenization of composition such as the Y2O3 stabilizer. In this case, it provided chances for transformation from tetragonal phase to monoclinic phase. This can explain the existence of monoclinic phase in **Fig. 7a**. Nevertheless, induction preheating can lengthen the retention period of molten pool by decreasing the cooling rate during laser remelting process. This was conducive to the diffusion of the stabilizer, which enhanced the inhibitory effect on the monoclinic transformation and consequently obtained the elimination of the monoclinic phase.

4 Conclusion

In the present study, a new method of induction preheating was introduced to the surface modification of plasma sprayed YSZ thermal barrier coatings. Prior to laser remelting process, the samples were preheated by an inductor at various temperatures. From the results presented in this study, it is clearly indicated that laser remelting assisted with induction preheating exhibiting a promising approach for surface modification. Imposing induction preheating on the coatings helps in developing a relatively steadier melt temperature as well as decreasing the cooling rates between the molten pool and unmelted coating and consequently the thermal gradient during the subsequent remelting process. Besides, it can effectively decrease the threshold laser energy (30% of peak power, 67.9J/cm2) of fully re-melting TBCs with relatively less crack density compared to without the application of it (35% of peak power, 79.2 J/cm2). And the crack density of the remelted coatings with preheating was less than that with non-preheating. In addition, a high preheating temperature can make the interfaces at the remelted/unmelted zone free of cracks. What is more, induction preheating can enhance the inhibitory effect on the monoclinic transformation and consequently obtain the elimination of the monoclinic phase.

Acknowledgments

The authors acknowledge the financial support provided by the National Natural Science Foundation of China (Grant no. 51375374), National High Technology Research and Development Program of China (863 Program) (Grant no. 2013AA040101), Program for ChangJiang Scholars and Innovative Research Team in University (Grant no. IRT1172).

References

[1] R. Ahmadi-Pidani, R. Shoja-Razavi, R. Mozafarinia, H. Jamali, Improving the hot corrosion resistance of plasma sprayed ceria yttria stabilized zirconia thermal barrier coatings by laser surface treatment, Materials & Design, 57 (2014) 336-341.

[2] Z. Fan, K. Wang, X. Dong, W. Duan, X. Mei, W. Wang, J. Cui, J. Lv, Influence of columnar grain microstructure on thermal shock resistance of laser re-melted ZrO2-7wt.% Y2O3 coatings and their failure mechanism, Surface and Coatings Technology, 277 (2015) 188-196.

[3] M.G. Gok, G. Goller, Microstructural evaluation of laser remelted gadolinium zirconate thermal barrier coatings, Surface and Coatings Technology, 276 (2015) 202-209.

[4] D. Wang, Z. Tian, L. Shen, Z. Liu, Y. Huang, Influences of laser remelting on microstructure of nanostructured Al2O3 13wt.% TiO2 coatings fabricated by plasma spraying, Applied Surface Science, 255 (2009) 4606-4610.

[5] Y. Ge, W. Wang, X. Wang, Z. Cui, B. Xu, Study on laser surface remelting of plasma-sprayed Al Si/1wt% nano-Si3N4 coating on AZ31B magnesium alloy, Applied Surface Science, 273 (2013) 122-127.

[6] D. Wang, Z. Tian, L. Shen, Z. Liu, Y. Huang, Effects of laser remelting on microstructure and solid particle erosion characteristics of ZrO2 7wt%Y2O3 thermal barrier coating prepared by plasma spraying, Ceramics International, 40 (2014) 8791-8799.

[7] C. Zhu, P. Li, A. Javed, G.Y. Liang, P. Xiao, An investigation on the microstructure and oxidation behavior of laser remelted air plasma sprayed thermal barrier coatings, Surface and Coatings Technology, 206 (2012) 3739-3746.

[8] C. Li, Y. Wang, L. Guo, J. He, Z. Pan, L. Wang, Laser remelting of plasma-sprayed conventional and nanostructured Al2O3 13wt.%TiO2 coatings on titanium alloy, Journal of Alloys and Compounds, 506 (2010) 356-363.

[9] C. Batista, A. Portinha, R.M. Ribeiro, V. Teixeira, M.F. Costa, C.R. Oliveira, Surface laser-glazing of plasma-sprayed thermal barrier coatings, Applied Surface Science, 247 (2005) 313-319.

[10] R. Ghasemi, R. Shoja-Razavi, R. Mozafarinia, H. Jamali, Laser glazing of plasma-sprayed nanostructured yttria stabilized zirconia thermal barrier coatings, Ceramics International, 39 (2013) 9483-9490.

[11] R. Ahmadi-Pidani, R. Shoja-Razavi, R. Mozafarinia, H. Jamali, Laser surface modification of plasma sprayed CYSZ thermal barrier coatings, Ceramics International, 39 (2013) 2473-2480.

[12] F. Chang, K. Zhou, X. Tong, L. Xu, X. Zhang, M. Liu, Microstructure and thermal shock resistance of the peg-nail structured TBCs treated by selective laser modification, Applied Surface Science, 317 (2014) 598-606.

[13] R. Ahmadi-Pidani, R. Shoja-Razavi, R. Mozafarinia, H. Jamali, Improving the thermal shock resistance of plasma sprayed CYSZ thermal barrier coatings by laser surface modification, Optics and Lasers in Engineering, 50 (2012) 780-786.

[14] P. Farahmand, S. Liu, Z. Zhang, R. Kovacevic, Laser cladding assisted by induction heating of Ni–WC composite enhanced by nano-WC and La2O3, Ceramics International, 40 (2014) 15421-15438.

[15] H. Kugler, F. Vollertsen, Inductive preheating in laser beam welding of multimaterial joints of 22MnB5 and AA6016, in: C. Emmelmann, M.F. Zaeh, T. Graf, M. Schmidt (Eds.) Lasers in Manufacturing, 2013, pp. 41-48.

[16] J. Cui, L. Yang, Y. Wang, X. Mei, W. Wang, C. Hou, Nanospot soldering polystyrene nanoparticles with an optical fiber probe laser irradiating a metallic AFM probe based on the near-field enhancement effect, ACS Applied Materials and Interfaces, 7 (2015) 2294-2300.

Characterization of CrMoBW- Fe base in-flight particles and splats fabricated by nano cored wire arc spraying

Man Tuiprae[1,a*], Kiattisak Meekhanthong[2,b], Komsanti Chokethawai[3,c], Sittichai Wirojanupatump[3,d]
[1] Faculty Engineering, Rajamangala University of Technology lanna, Chiang Mai 50200, Thailand.
[2] Advanced Surface Technology Co., Ltd. Thailand.
[3] Faculty of Science, Chiang Mai University, Chiang Mai 50200, Thailand.
[a]Tman@rmutl.ac.th, [b]kiattisak.ast@gmail.com, [c]komsanti@gmail.com, [d]wish- 2000@hotmail.com

Key words: cored wire arc spray, in-flight particle, splat

Abstract

Properties and performance of the coating fabricated by wire arc spray process strongly depend on various factors including spray parameters and characteristics of both in-flight particle and splat. In this study, relationship between in-flight particle, splat formation and microstructure of the CrMoBW-Fe based nanocomposite coating fabricated by wire arc spray process were investigated. The Focus ion beam (FIB) milling was used to prepare a transversal cross-section of the in-flight particle and splat to be characterized by scanning electron microscopy (SEM) along with an energy dispersive x-ray spectroscopy (SEM-EDS). In addition, phase compositions were also analyzed by using x-ray diffractometer (XRD). The results revealed that size of the in-flight particles and splat morphology greatly influenced microstructure and properties of the coatings. In-flight particles were revealed a spherical-shape with a broad size distribution ranging from 0.8 to 115 μm at the average of 26 μm particle diameter. In-flight particles were well flattened on impact giving flower-shape splat with the average diameter of 100 μm. Flattening degree was found to be 3.8. Arc sprayed CrMoBW-Fe base nanocomposite coating showed very thin splat with fine lamella structure which implied that particles were fully molten upon impact onto substrate.

1 Introduction

Serviceable engineering components depend not only on their bulk properties but also on surface characteristics. Twin wire arc spraying has been widely used as a quick and convenient process for producing thick coatings. It has been widely used in thermal spray industry as being an economical spraying process which combines low equipment cost and operating cost along with very high deposition efficiencies [1-4].

Arc spray process can be applied with a wide range of electrical conductive materials including metal, alloy and metal matrix composites or cermet known as cored wire. Properties of the arc spray coatings strongly affected by spray parameters and relate with droplet formation, size of in-flight particle and particle velocity.[5-9]. Many researchers reported that substrate parameters such as temperature, surface profiles and their interactions on impact influence the splat formation and coating properties [10-11]. However, understanding of the mechanisms of the splat formation is still limited. In this work, the splat formation of the arc sprayed CrMoBW-Fe base cored wire onto a polished stainless steel substrate was investigated.

2 Experimental procedures

2.1 Characterization of cored wire

CrMoBW-Fe base cored wire was studied. Characterization of the as recieved cored wire including morphology, particle size, microstructure, chemical compositions and phase compositions. Nominal compositions of the cored wire materials provided by supplier was shown in Table1. Analysis of the filler particles taken from the cored wire was prepared by putting particles on a carbon tape followed by gold sputtering just prior to SE-SEM, BSE-SEM investigation along with EDS-SEM for chemical analysis (JEOL JSM-5910, Japan). Size distribution of filler particles was analyzed by laser particle size analyzer. X-ray diffractometer (Rigaku, TTRAX III) was employed for phase analysis.

2.2 Collection of in-flight particle and Splat

CrMoBW-Fe base cored wire was sprayed by arc spray system (TAFA Arcjet 9000, USA). Arc spray parameters were shown in Table 2. In-flight particles were collected by spraying through Ø 3 mm holes of stainless steel plate into distilled water (Fig.1a). The in-flight particles were collected and cleaned by alcohol follow by air drying immediately. Splat collection was done by spraying onto polished surface of stainless steel plate (Fig. 1b).

2.3 Characterization of in-flight particle

Morphology and microstructure of the in-flight particls were revealed by SEM along with EDS-SEM for chemical analysis. Cross-sectioning of in-flight particles was prepared by FIB technique. Size distribution was also analyzed by laser particle size analyzer. Phase analysis of the in-flight particls was analyzed by X-Ray diffraction technique. The XRD was operated at 50 kV and 30 mA using Cu K_α radiation (1.54 Å). The goniometer was set with a step-width of 0.02° and a scan rate of 2.4 degree per second over a 2θ ranging from 20° to 40°. Phase compositions were identified by using JCPDS.

2.4 Characterization of a single splat

Characteristics of splat produced by CrMoBW-Fe base cored wire were investigated by OM and SEM. FIB technique was used for microstructure characterization of a single splat cross-section. Measurement of perimeter and area of individual splat were performed by using image analysis with SCENTIS software. Measurement was randomly performed on 100 splats in order to obtain the average value. Splat diameter, d_s was calculated by equation (1) using a measured splat area. The degree of flattening is the ratio of the splat diameter (d_s) to the in-flight particle diameter (d_i) as shown in equation (2), while the degree of splashing is the ratio of calculated splat area ($P^2/4\pi$) to actual splat area (A) measured by image analysis as shown in equation (3) [5].

$$d_s = (4A/\pi)^{1/2} \quad (1)$$

$$DF = d_s/d_i \quad (2)$$

$$DS = (P^2/4\pi) \times (1/A) \quad (3)$$

When P is the perimeter of splat, and A is the actual (measured) area of splat.

Table 1 Chemical compositions of studied cored wires as nominated by supplier.

Designated code	Chemical Compositions (wt%)							
	Si	B	Mn	Cr	Mo	WC	Nb	Fe
CrMoBW - Fe base	2	5	3	25	6	16	12	Bal.

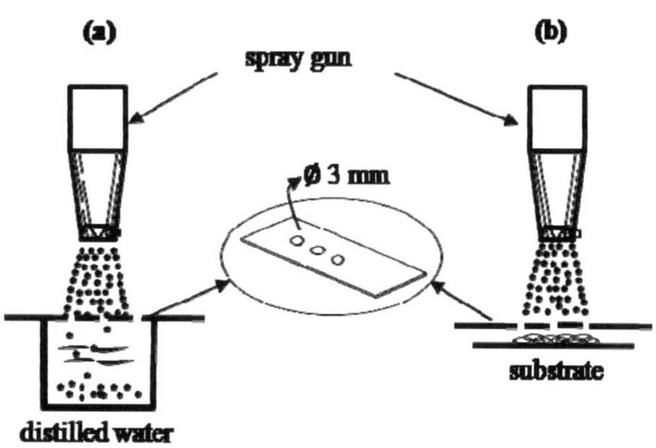

Figure 1 Schematic diagram shows collection methods (a) in-flight particle collection and (b) splat collection.

Table 2 Arc spraying parameters.

Arc Voltage (V)	35
Arc current (A)	200
Air pressure (bar)	4.2
Wire feed rate (g/min)	95
Spray distance (mm)	150

3 Results and discussion

3.1 Cored wire characterization

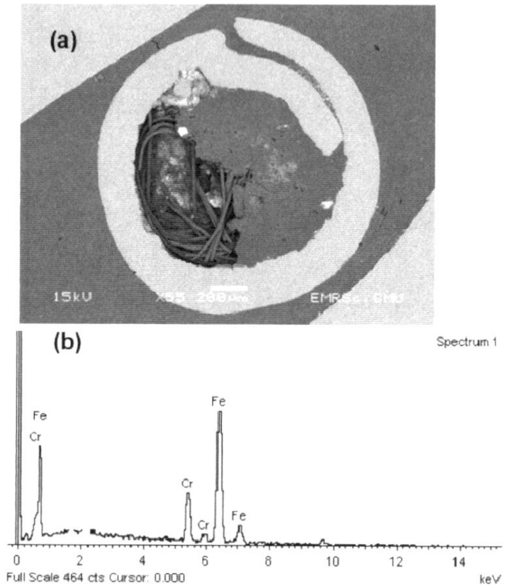

Figure 2 (a) BSE-SEM micrograph and (b) EDS spectrum of CrMoBW- Fe base shell.

Fig. 2(a) shows a cross-section of CrMoBW-Fe base cored wire illustrating shell and filler component of cored wire. The shell part was about 200 μm thick with a diameter of 1.6 mm. Chemical analysis of CrMoBW-Fe base cored wire showed Fe and Cr as can be seen in Fig. 2(b). Morphology of the filler particles showed as irregular shape as shown in Fig. 3. Filler particles had broad size distribution ranging from 0.2 to 125 μm with a mean diameter of 52 μm (Fig. 4). Phases found in CrMoBW-Fe base cored wires filler including W_2C, WC, B_4C, FeNb, Fe_2O_4, and FeCr as shown in Fig. 5.

3.2 In-flight characterization

Analysis of in-flight particles from CrMoBW- Fe base cored wire by SEM indicated that in-flight particles are spherical shape with rather smooth particle surface (Fig. 6 (a)). The average diameter of CrMoBW- Fe base in-flight particles was 26.7 μm. with particle size distribution ranging from 0.8 to 115 μm as shown in Fig. 6(b). SEM cross-section image of CrMoBW- Fe base in-flight particle (Fig. 7(a)) shows a good homogeneity of the eutectic and dendrite structure. EDS-SEM analysis of chemical compositions taken from mapping mode was shown in Fig. 7 (b). It can be observed from the EDS spectrum that all elements including Fe, Cr, Nb, Mo, W, C, Si and Mn were fully melted during arc spraying process.

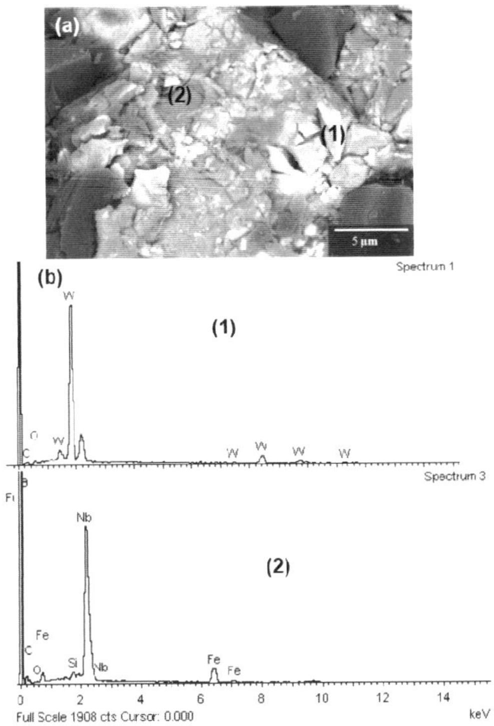

Figure 3 (a) SE-SEM micrographs show morphology of CrMoBW- Fe base filler (b) EDS spectrums of area 1 and 2 of CrMoBW- Fe base filler

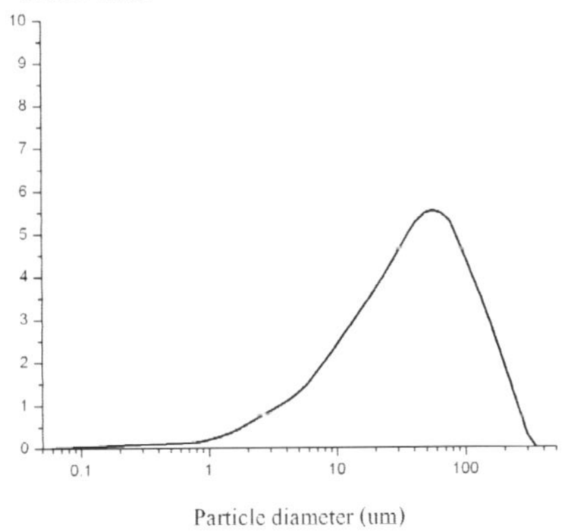

Figure 4 Size distribution of filler particle of CrMoBW- Fe base cored wire.

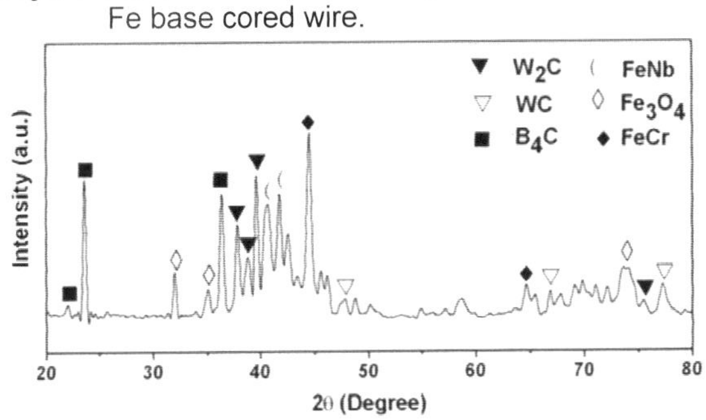

Figure 5 X-ray diffraction patterns of the filler of CrMoBW- Fe base.

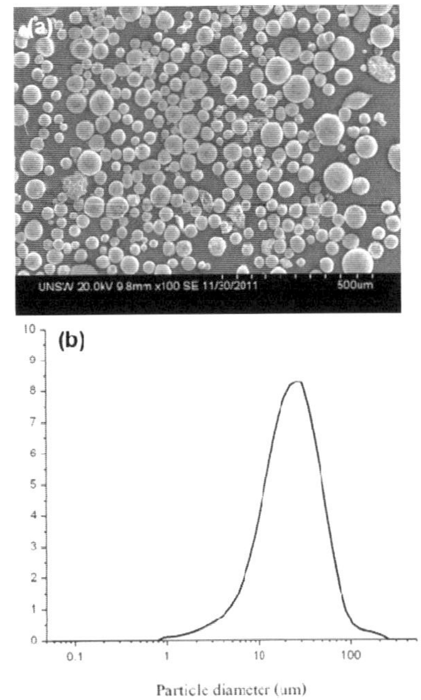

Figure 6 (a) SE-SEM micrograph of in-flight particles (b) Size distribution of in-flight particles from CrMoBW- Fe base core wire.

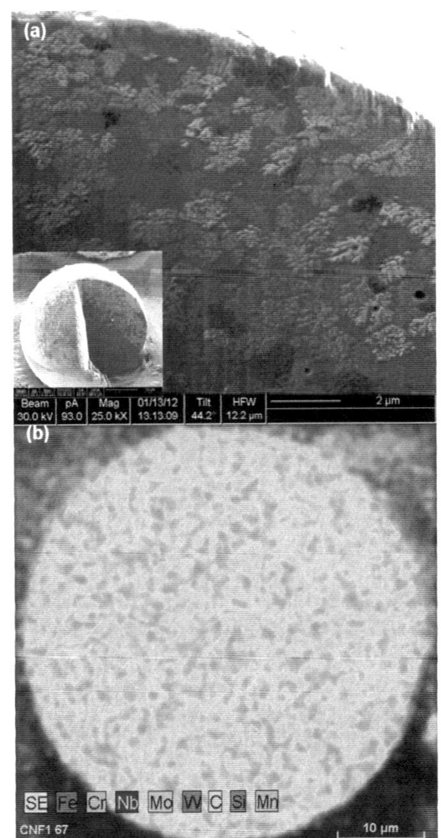

Figure 7 (a) SEM-FIB image of in-flight particles of CrMoBW- Fe base (b) EDS spectrum mapping analysis of the in-flight particle

In Fig. 8 also shows peaks of FeCr as a predominant phase formed along with carbide and oxide including W_2C, Fe_3O_4 and Fe_3C.

3.3 Splat characterization

Fig. 9 shows SEM images of the CrMoBW- Fe base splats with a disk shape like (Fig.9a) and a flower shape like (Fig.9b). The average diameter of the splat was calculated to be 100 μm. The degree of flattening and degree of splashing were 3.8 and 3.3, respectively (Table 3).

Fig.9(c) shows well bonded interface between splat and substrate with a thin splat layer (2 μm). It also revealed a typical void inside splat structure as shown in Fig. 9(d). SEM images of the coating cross-section revealed a thin splat and fine lamella structure (Fig. 10). This implied that particles were fully molten upon impact. The micro pores can also be observed.

Table 3 Characteristics of in-flight particle and splat of CrMoBW- Fe base.

Mean size (μm)			Degree of flattening	Degree of splashing
Filler particle	In-flight particle	Splat		
52	26	100	3.8	3.3

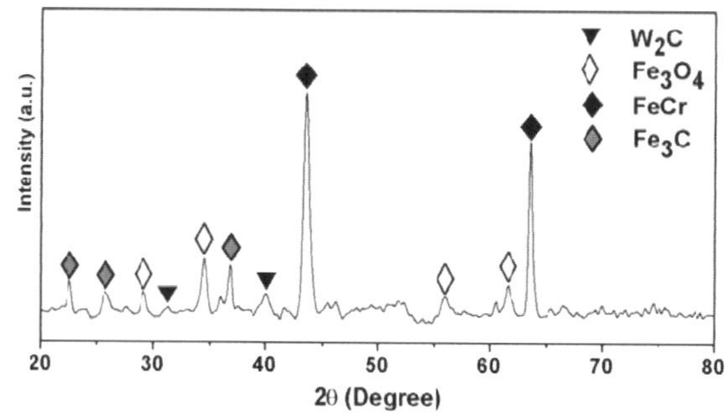

Figure 8 X-ray diffraction patterns of the in-flight particles of CrMoBW- Fe base cored wire

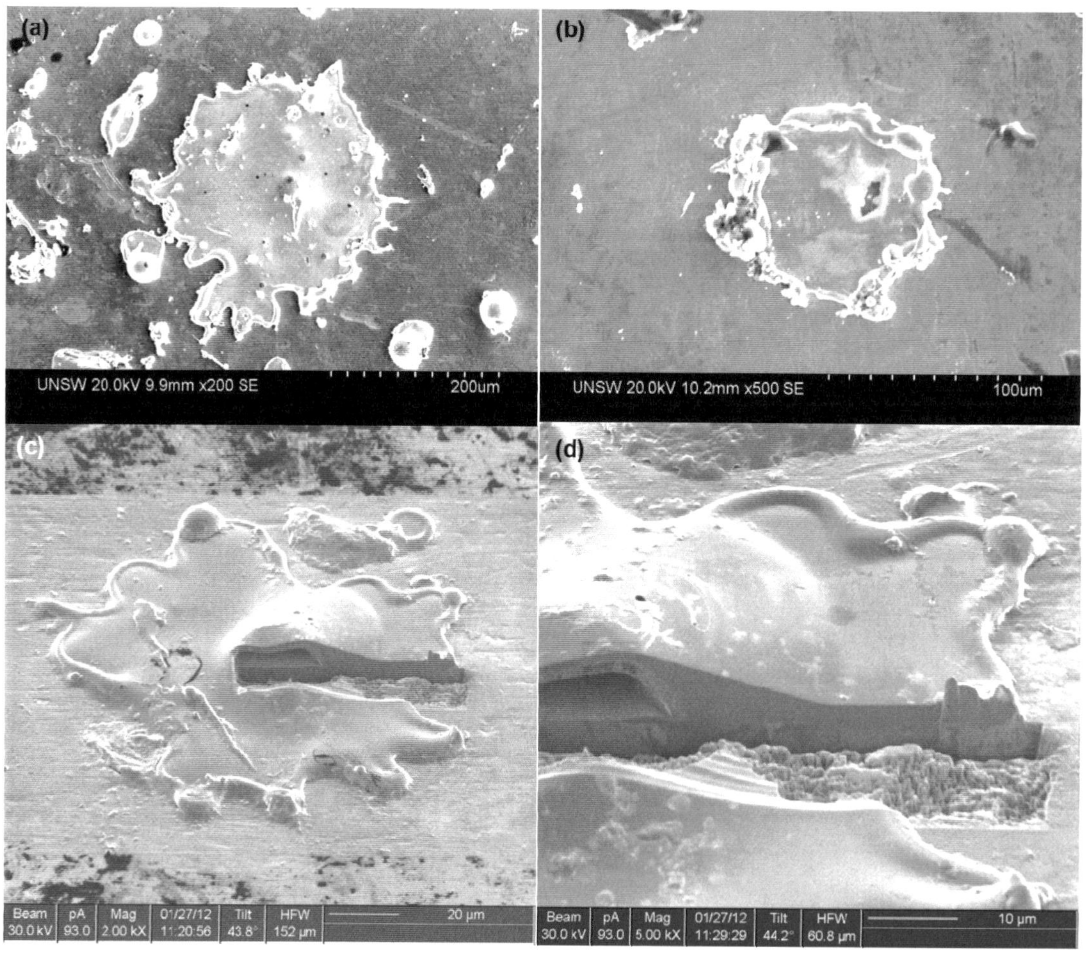

Figure 9 SE-SEM micrographs and SEM-FIB micrographs of CrMoBW- Fe base splats particles

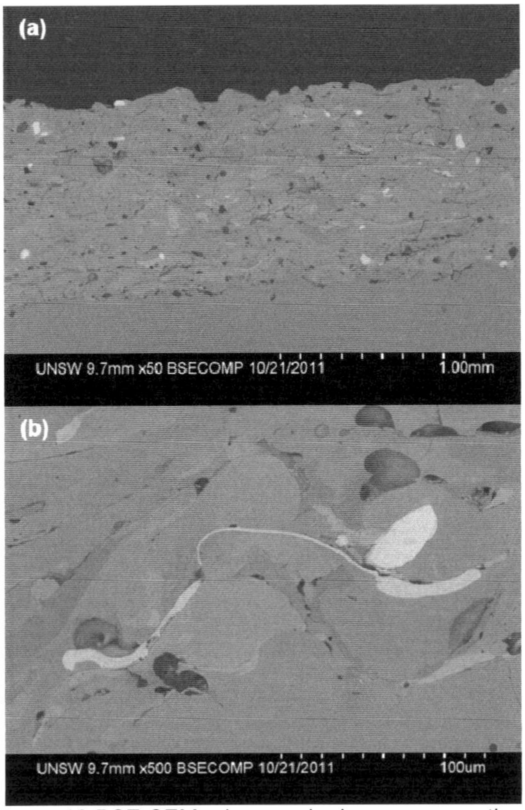

Figure 10 BSE-SEM micrograph show cross-sections of CrMoBW- Fe base coating at (a) 50X and (b) 500X magnification.

4 Conclusions

A spherical shape like of the in-flight particles of CrMoBW-Fe base nanocomposite with a narrow size distribution were successfully obtained by cored wire arc spray process. Upon impact onto the stainless steel substrate, particles were well flattened showing flower-shape splat with the average diameter of 100 μm. Microstructure of the arc sprayed CrMoBW-Fe base nanocomposite coating exhibited a thin splat and fine lamella structure. This implied that particles were fully molten.

5 Acknowledgements

The authors would like to acknowledge the financial support from Rajamangala University of technology Lanna and Chiang Mai University. The authors are also gratefully acknowledge Advance Surface Technology Co., Ltd., for facilitating spraying process.

6 Reference

[1] I. Gedzevicius, A.V. Valiulis, Analysis of wire arc spraying process variables on coatings properties, J. Materials Processing Technology 175 (2006)206-211.

[2] M.P. Planche, H. Liao, C. Oddet: Relationships between in-flight particle characteristics and coating microstructure with a twin wire arc spray process and different working conditions, Surface & Coatings Technology 182(2004)215-226.

[3] C.M. Fernandes, V. Popovich, M. Matos, A.M.R. Senos, M.T. Vieira, Carbide phases formed in WC-M (M=Fe/Ni/Cr) systems, Ceramics International 35(2009)369-372.

[4] V. Stoica, R. Ahmed, T. Itsukaichi, Influence of heat-treatment on the sliding wear of thermal spray cermet coating, Surface & Coating Technology 199 (2005)7-21.

[5] A. Pourmousa, J. Mostaghimi, A. Aabedini, S. Chandra, Particle size distribution in a wire-arc spraying system, J. Thermal Spray Technology. 14(2005)502-510.

[6] K. Cooke, G. Oliver, V. Buchanan, N. Palmer, Optimisation of the electric wire arc-spraying process for improved wear resistance of sugar mill roller shells, Surface & Coating Technology 202(2007)185-188.

[7] I. Gedzevicius, A.V. Valiulis, Analysis of wire arc spraying process variables on coating properties, Journal of Materials Processing Technology 175 (2006)206-211.

[8] M.P. Planche, H. Liao, C. Coddet, Relationships betaween in-flight particle characteristics and coating microsturcture with a twin wire arc spray process and different working conditions, Surface & Coating Technology 182(2004)215-226.

[9] K. Venugopal, M. Agrawal, Evalution of arc sprayed coating for erosion protection of tubes in atmosheric fluidsed bes combustion (AFBC) boilers, Wear 264(2008)139-145.

[10] S. Brossard, P.R. Munroe, A.T.T. Tran, and M.M. Hyland, Effects of Substrate Roughness on Splat Formation for Ni-Cr Particles Plasma Sprayed onto Aluminum Substrates, J. Thermal Spray Technology 19(5)(2010)1131-1141.

[11] S. Brossard, P.R. Munroe, A.T. Tran, and M.M. Hyland, Study of the Splat Microstructure and the Effects of Substrate Heating on the Splat Formation for Ni-Cr Particles Plasma Sprayed onto Stainless Steel Substrates, J. Thermal Spray Technology 19(5)(2010)1100-1114.

Analysis of ultrasonic-assisted laser re-melting of thermal barrier coatings

Zhengjie Fan, Xia Dong*, Rujia Wang, Kedian Wang, Wenqiang Duan, Xuesong Mei, Wenjun Wang, Jianlei Cui, State Key Laboratory for Manufacturing Systems Engineering,Xi'an Jiaotong University, Xi'an, P.R. China
*E-mail: xdongxjtu@163.com

Abstract: The effect of ultrasonic vibration, which was injected into the melt during the entire stages of re-melting, on the crack distribution, surface morphology, and grain refinement in a laser re-melted ZrO_2-7wt% Y_2O_3 thermal barrier coatings (TBCs) was researched. Prefect uniform distribution of segmented network cracks and refined microstructure were obtained with the application of ultrasonic vibration. The increased fracture toughness, resulting from greater strain tolerance of the uniformly distributed cracks in ultrasonic-assisted laser re-melted TBCs, was confirmed by the thermal shock experimental verification and theoretical analysis. Visible ribbon-like loops, which appeared at surface ripples and were caused by surface tension, were eliminated during ultrasonic-assisted laser re-melting due to enhancement in surface convection. The surface morphology results indicated that the outline of re-melted coating surface obtained by utilizing the ultrasonic vibration with low power output (~20%) was much flatter compared to that obtained without laser re-melting. Moreover, evident undulated crests and troughs geometries appeared at the re-melted surface when ultrasonic vibration with high power output (above 50%) was introduced. The results of X-ray diffraction exhibited that the transformation from tetragonal to monoclinic was prevented when the power output of ultrasonic vibration was higher than 30%.
Keywords: laser re-melting; thermal barrier coating; ultrasonic vibration; surface morphology; network crack distribution; grain refinement.

1 Introduction

Laser processing offers opportunities to modify the microstructure, phases, element distribution, residual stress and material property through laser-material interactions such as shock peening, welding, re-melting[1-3]. Laser re-melting technology is a currently recognized as promising method for improving the performance of plasma sprayed TBCs. Several studies done in the past show that laser re-melting can improve the compactness, surface roughness, micro-hardness, wear resistance, thermal shock resistance and high temperature corrosion resistance performance of plasma-sprayed TBCs [3-8]. Laser re-melting eliminates the lamellar structure of plasma sprayed coatings [9], generates compact columnar grains in the direction of heat flow and equiaxed grains at the surface [10,11]. Columnar grains in the laser-glazed coating can improve the thermal shock resistance of coatings by increasing stress tolerance of the laser re-melted coatings [11]. However, the presence of abundant segmented cracks can cause the decrease of high temperature oxidation resistance and thermal insulation capability of thelaser re-melted coatings in comparison to that of the plasma sprayed coatings[5,12]. Moreover, the inhomogeneity of the structure and voids occurred in the re-melted coatings, caused by the nonuniform heating and the rapid cooling of the laser characteristic, has negative effect on the thermal cycle life[13].
The ultrasonic vibration-assisted machining is widely investigated for manufacturing of brittle ceramic, the casting and welding fields. Most of the processes that utilize the desired acoustic streaming and mechanical stirring effects of ultrasonic vibration on slowing the cooling and solidification rates during solidifying[14,15], resulting in the pronounced effects on the microstructure refinement and improving the uniformity of the element distribution[16,17]. In this paper, a new ultrasonic vibration assisted laser re-melted approach is presented for surface modification of TBCs. Uniform distribution of segmented network cracks and uniformity microstructure were obtained using the ultrasonic vibration laser re-melting process.

2 Experimental procedure

A nickel-based superalloy (Inconel 718) was chosen as the substrate material and was cut into 10mm×5 mm×2 mm specimens. ZrO_2-7wt% Y_2O_3 powder was used to prepare the top ceramic coating in order to effectively improve the performance of the TBCs. The diameter range of the powders used was 20～40μm, **Fig. 1a**. NiCoCrAlY was chosen as the bond coating material with a particle diameter range of 50～70μm, **Fig. 1b**. In this study, the plasma-sprayed coatings were prepared by a GTV delta three anodes atmospheric plasma spraying system (GTV, Germany).

Fig.1. (a) SEM image of YSZ powder and (b) of NiCoCrAlY powder.

After preparation of TBCs sample, the ultrasonic-assisted laser re-melting experimental was carried on as a schematic of experimental system, **Fig. 2**. A JK300D Nd:YAG pulsed laser model (GSI, England) with maximum average power of 300W and with a wavelength of 1064nm was irradiated on the vibrating specimens. The laser beam with 1 mm pluse duration and 40 HZ repetition rates was irradiated perpendicular to the TBC surface with defocused diameter of 3mm and with the scanning rate of 5 mm/s. The coatings were re-melted with the laser pulse energy of 5.2J and 6J respectively. The ultrasonic power supply delivered the power output of 2600w at a fixed frequency of 20 kHz. For LRU-C, LRU-E, LRU-F samples, the processing was conducted with three ultrasonic vibration power outputs: 40%, 50%, 60%. These parameters are summarized in **Table 1**. The direction of ultrasonic vibration was perpendicular to the surface of the substrate.

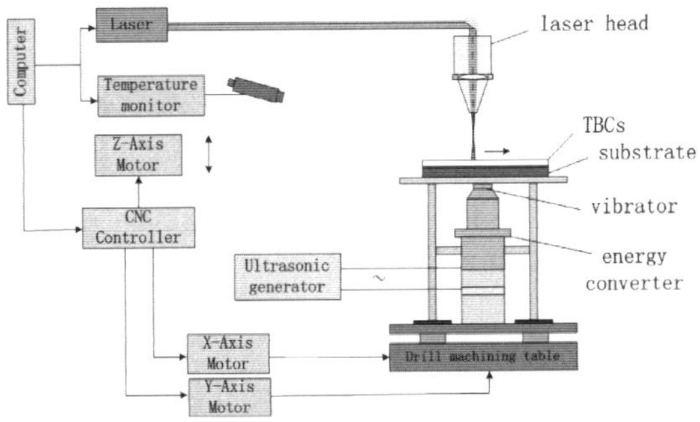

Fig.2. Schematic of the ultrasonic-assisted laser re-melting experimental system

3 Results and discussion

3.1 Microstructure

A work-piece surface melted by pulsed laser irradiation was shown in this micrograph, **Fig. 3**. The images reveal that all glazed coating surfaces have some segmented network cracks that are perpendicular to the surface. The microstructures obtained without the application of ultrasonic vibration during laser remelting were shown, **Fig. 3(a)** and **Fig. 3(b)**. It can be observed that the cracking network of the remelted layer was irregular and the width of the cracks almost reach up to 20μm. Obviously, surface cracks are relatively serious for the high laser energy (compare LR-A with LR-B).This is because the molten pool has higher temperature as well as internal thermal stress under the larger pulse energy irradiation.

Table 1 Parameters of laser re-melting

Label	Single pulse energy (J)	Ultrasonic power(w)
LR-A	6	0
LR-B	5.2	0
LRU-C	6	520(20%)
LRU-D	5.2	520(20%)
LRU-E	6	780(30%)
LRU-F	5.2	780(30%)
LRU-F	6	1300(50%)
LRU-F	5.2	1300(50%)

These figures show the surface morphology of LRU-C, LRU-E and LRU-G respectively when the pulse energy is 6J, **Fig. 3(c)**, **Fig. 3(e)** and **Fig. 3(g)**. It appears that the distribution of segmented network cracks is more regular and uniform with the application of assist ultrasonic vibration. In comparison with the coatings remelted without assist ultrasonic vibration, the cracks formed in the surface of the coatings subjected to the assist ultrasonic vibration has a smaller crack width while they have higher network crack density. This is mainly because the application of low-energy ultrasonic vibration (20%) can increase the heat transfer effect and reduce heat reunion effect of the molten pool. Further, it can eliminate the coarse cracks due to local high thermal stress, and the enhanced convection can make the pool to be heated uniformly, thus form dense and minor cracks. When the power output is more than 50%, it can enhance the mass transfer effect and increase the blocking effect of the crack. The voids and gaps due to the shrinkage of the liquid ceramic in the solidification of the molten pool can be replenished in time by the transferred ceramic. Therefore, surface crack density decreased with the ultrasonic vibration power increase. However, it can introduces molten pool too much energy and significantly increase the driving force of crack propagation with the accumulation of thermal stress when the ultrasonic vibration power output is more than 50% of the range, and the crack has been extended to the un-melting area **Fig. 4 (g)**.

These figures show the surface morphology of LRU-D, LRU-F and LRU-H when the pulse energy is 5.2J, **Fig. 3(d)**, **Fig. 3(f)** and **Fig. 3(h)**. The change trend of the crack distribution of the coating remelted by laser with pulse energy of 5.2 J is the same as that of 6J when using various ultrasonic vibrations. The difference is that the surface remelted by the laser pulse energy of 5.2J has smaller crack density than that of 6J.

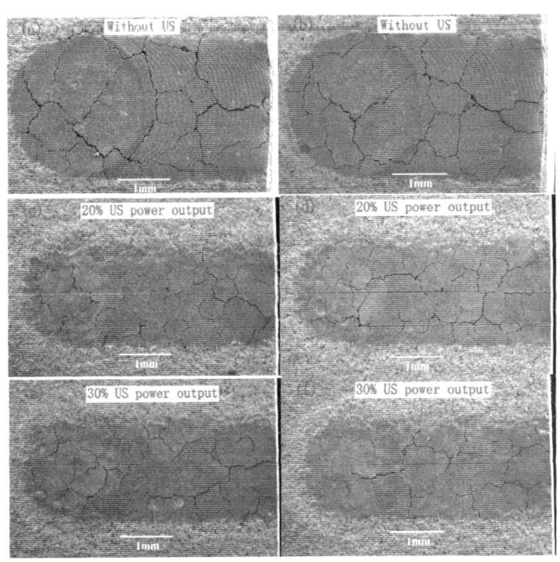

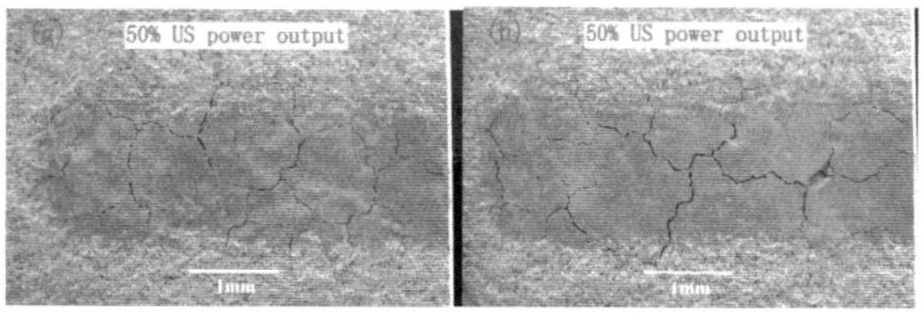

Fig.3. Surface micrographs of coatings remelted by various pulse energies with ultrasonic vibration or without ultrasonic vibration
(a),(c),(e),(g)pulse energy 6J; (b),(d),(f),(h)pulse energy 5.2J

Visible ribbon-like loops appear as surface ripples caused by the laser pulses without ultrasonic-assisted vibration, **Fig. 3(a)** and **Fig. 3(b)**. The surface ripples were supposed to caused by surface tension. These loops formed during laser pulse duration (pulse on) with ripple and wrinkle structure are brighter and more uneven in microstructure than these formed during pulse internal time (pulse off). It can be seen in Fig.4, the loops structure is similar to amorphous organization. Which is mainly due to laser irradiation continuously heating the molten pool and the solidification area during laser pulse-on period and so the cooling rate is too low to reach cooling rate of forming equiaxed grains. The formation of the ripple and wrinkle structure is mainly because of surface tension difference caused by the movement of laser spot. The surface tension difference prompt the fluid flow reverse to the laser scanning direction and flow state of the fluid is preserved in the rapid solidification of molten pool.

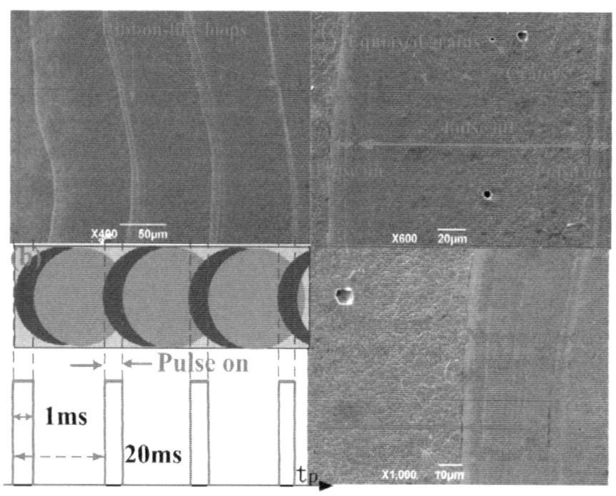

Fig.4. (a) Ribbon-like loops structure in pulse laser remelted coatings surface; (b) schematic diagram of formation in ribbon loops; (c)Microstructure obtained within a period of one laser pulse; (d)high magnification microstructure around loops.

It can be seen from that visible ribbon-like loops were eliminated during ultrasonic-assisted laser re-melting, **Fig.5(a)**. On the one hand, it enhanced the convection effect of the molten pool and reduced the surface tension difference. On the other hand, the increased convection can block the fluid flow moving towards the rear of molten pool. Thus, the ultrasonic vibration can eliminate the loops structure and contribute to develop more uniform and finer equiaxed grains. In addition, the molten pools are heated evenly and thus the thermal stress distribute more uniform with the ultrasonic vibration effect. Therefore, the network cracks formed with ultrasonic assisted vibration are finer than without of that while the cracks are more densely for the application of ultrasonic vibration, which may be due to the total energy as well as summation of the thermal stress caused by pulse laser is certain, **Fig.5**.

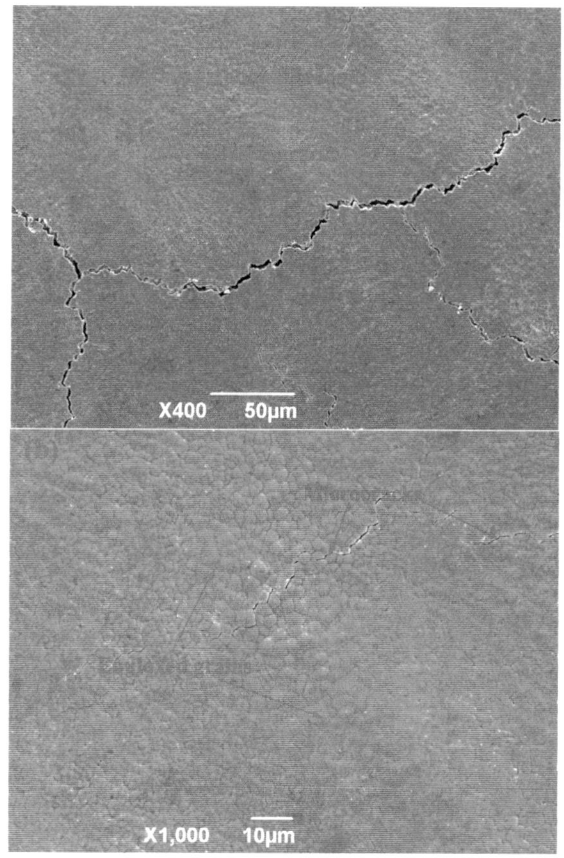

Fig.5. (a)Surface micrographs of TBCs obtained with ultrasonic assisted vibration laser remelting (b)high magnification surface micrographs

3.2 Phase composition

The changes in the phase constitution of the coatings were analyzed using XRD analysis. It presents the XRD patterns of the LRCs and the LRUCs in the 2 theta range of 20°-90°, **Fig. 6**. It demonstrates that the phase composition of coatings remelted without the ultrasonic vibration are a mixture of tetragonal phase (t), cubic phase (c) and a small amount of monoclinic phase (m). The monoclinic phase was not eliminated completely due to the cooling rates ($10^{3}\,^{\circ}CS^{-1}$~$10^{4}\,^{\circ}CS^{-1}$) in laser remelting process was estimated relatively high, combined with the heating non-uniformity characteristic of the Gaussian pulse laser. Under such conditions, local inhomogeneity in the Y_2O_3 stabilizer distribution in the remelted coatings results in tetragonal to monoclinic phase transformation was not inhibited during cooling. When the laser pulse energy is 6J, there is no monoclinic phase in the remelted coatings when ultrasonic vibration is at any power output (20%, 30%, 50%). With the ultrasonic vibration effect, the enhancement of mass and heat transfer in molten pool can increases the element transfer rate and thus leads Y_2O_3 stabilizer to distribute uniformly. When the laser pulse energy is 5.2J, the monoclinic phase was absent only when the ultrasonic vibration was at low output power (20%) and the XRD spectra appeared diffraction peak of monoclinic phase at high output power (30%, 50%). This is caused by the instable strongly convection effect produced by the high power ultrasonic vibration significantly increases the cooling rate of the molten pool. Moreover, the noteworthy cavitation phenomenon obtained at high power ultrasonic vibration causes the aggregation effect of stabilizer in molten pool. Therefore, monoclinic phase transformation cannot be suppressed in the area without stabilizers Y_2O_3. When the pulse energy is 6J, however, the effect of convection enhancement induced by ultrasonic vibration is not sufficient to release the excessive thermal energy on the molten pool irradiated by high laser intensity in a very short time, which limits the increase if cooling rate of molten pool and improves the diffusion of the stabilizer due to the molten pool staying for longer time.

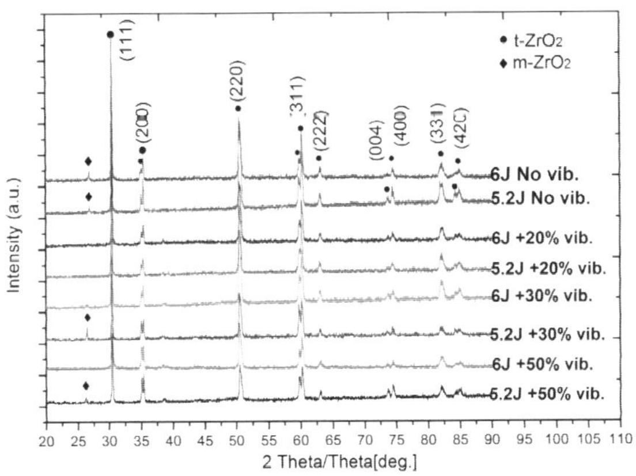

Fig. 6. XRD patterns of remelted coatings with and without ultrasonic vibration

3.3 Thermal Shock Behavior

It shows the cycle number when the spallation was initiated for the coatings re-melted in different pulse energy (5.2J and 6J) with and without ultrasonic vibration during thermal shock testing, **Fig. 7(a)**. It is obvious that the number of thermal cycle required for failure of coatings re-melted with 20% and 30% power output ultrasonic vibration is much more than that of without ultrasonic vibration respectively. However, the re-melted coatings with the application of 50% power output ultrasonic vibration endures the almost equivalent number of thermal cycle than the coatings re-melted without the ultrasonic vibration when the pulse energy is 5.2J, even less than that of the coatings re-melted with the pulse energy 6J. It proved that the segmented cracks that are more helpful in relieving stress and increasing the strain tolerance of top coatings. As previously demonstrated, the distribution and density of segmented cracks were significantly changed by the ultrasonic vibration during re-melting process. Further, it has a important factor affecting the durability of re-melted coatings.
Based on Shi et al. measurements, the typical stress-strain curves of ceramic coatings exhibit nonlinear relations and could be expressed by:

$\varepsilon=\varepsilon^{e}+\varepsilon^{ne}$

where ε^{ne} represents

$\varepsilon^{e}=\sigma/E$

Which is the nominal strain and is estimated based on linear elastic behavior, and where ε^{ne} represents the inelastic strain caused by microcracks and laminar structures. This study also showed that the fracture strain of the ceramic coatings increased with a decrease in coating thickness. In general, there is no evident difference in nominal strain among the same material, and the increase in fracture strain could be expected from increases in inelastic strain. The level of inelastic strain depends on the strain needed for microcrack extension and crack deflection and branching before catastrophic failure. Based on this, the topcoat nearly have the same transverse fracture strain (parallel to TBC surface direction) since thickness and laminar structure of melted layer re-melted with and without ultrasonic assisted vibration are almost equal to each other. However, the longitudinal fracture strain (perpendicular to TBC surface direction) are various among the coatings, which have segment cracks with different crack density. The longitudinal fracture strain of independent segment topcoat should increased with a decrease in crack spacing (treated as thickness in case of longitudinal fracture strain). Therefore the LRU-C has greatest total longitudinal fracture strain due to the topcoat have most uniform and greatest crack density than that of other coatings. The increased longitudinal fracture strain can effectively prevent longitudinal crack deflection and propagation , **Fig. 7(c)**. On the contrary, the LR-A and LRU-F have smaller total longitudinal fracture strain as the topcoat have greater crack spacing. There are observable longitudinal cracks deflection and propagation occurred at topcoat where have large crack spacing after a certain thermal cycle, **Fig. 7(b)**.

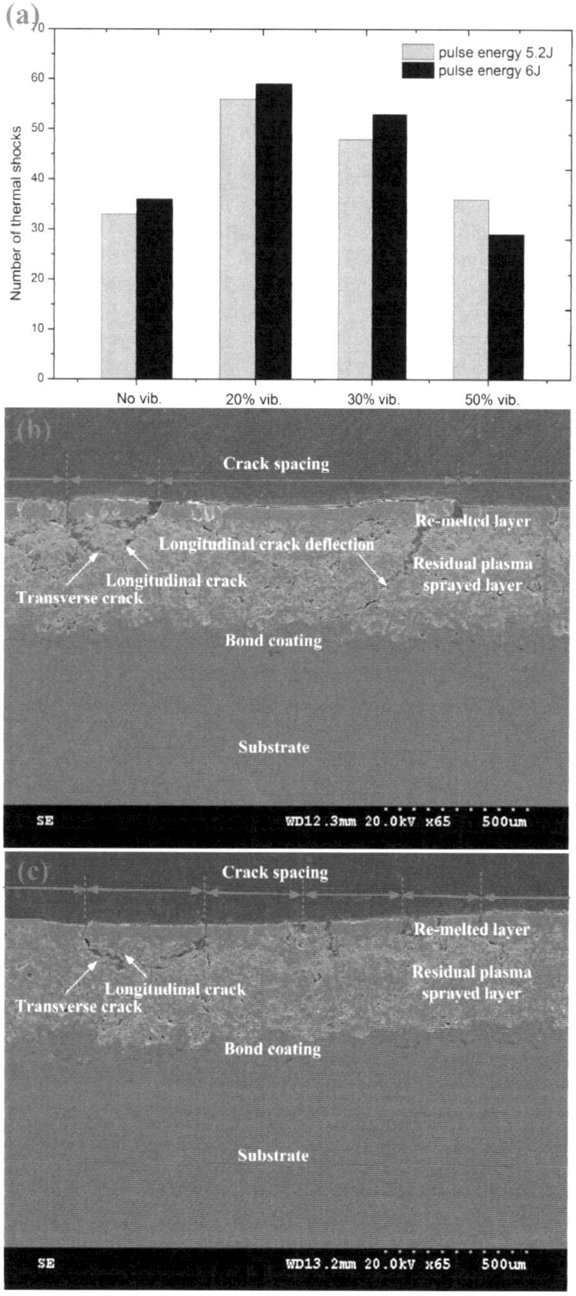

Fig.7. (a)Cyclic life time of with and without ultrasonic vibration re-melted coatings; Crack deflection and propagation after 30 thermal cycles for LR-A sample (b) and LRU-C sample (c)

4 Conclusion

In this paper ultrasonic assisted vibration laser re-melting of thermal barrier coatings was researched. Visible ribbon-like loops were eliminated during ultrasonic-assisted laser re-melting due to enhancement in surface convection. Moreover, the transformation from tetragonal to monoclinic was prevented when the power output of ultrasonic vibration was higher than 30%.

The thermal shock results demonstrated that ultrasonic assisted vibration with 20% or 30% power output can significantly improved the spallation initiation of laser re-melted coatings. Improving the longitudinal fracture strain of the topcoat through formation of uniformly distribute segmental cracks with small crack spacing is recognized as the major enhancement mechanism for extension of the ultrasonic vibration laser re-melted coatings life time.

Acknowledgments

The authors would like to thank the National Natural Science Foundation of China (grant no. 51575432), National Natural Science Foundation of China (grant no. 51375374), Program for ChangJiang Scholars and Innovative Research Team in University (grant no. IRT1172) for supporting this work.

References

[1] R.K. Nalla, I. Altenberger, U. Noster, G.Y. Liu, B. Scholtes, R.o. Ritchie, On the influence of mechanical surface treatments-deep rolling and laser shock peening-on the fatigue behavior of Ti-6Al-4V at ambient and elevated temperatures. *Mater. Sci. Eng. A,* 355.1 (2003), pp. 216/230.

[2] J.L. Cui, L.J.Yang, Y. Wang, X.S. Mei, W.J. Wang, C.J. Hou, Nanospot Soldering Polystyrene Nanoparticles with an Optical Fiber Probe Laser Irradiating a Metallic AFM Probe Based on the Near-Field Enhancement Effect, ACS Appl. Mater. Interfaces, 7(4) (2015), pp. 2294/2300.

[3] M.F. Morks, C.C. Berndt, Y.Durandet, M.Brand, J.Wang, Microscopic observation of laser glazed yttria-stabilized zirconia coatings, App. Surf. Sci., 256 (2010), pp. 6213/6218.

[4] R.Ghasemi, R.Shoja-Razavi, R.Mozafarinia, H.Jamali, The influence of laser treatment on thermal shock resistance of plasma-sprayed nanostructured yttria stabilized zirconia thermal barrier coatings, Ceram. Int., 40 (2014), pp.347/355.

[5] R. Ahmadi-Pidani, R. Shoja-Razavi, R. Mozafarinia, H. Jamali, Laser surface modification of plasma sprayed CYSZ thermal barrier coatings, Ceram. Int., 39 (2013), pp. 2473/2480.

[6] C. Batista, A. Portinha, R.M. Ribeiro, V. Teixeira, C.R. Oliveira, Evaluation of laser-glazed plasma-sprayed thermal barrier coatings under high temperature exposure to molten salts, Surf. Coat. Technol., 200 (2006), 6783/6791.

[7] P.C. Tsai, J.H. Lee, C.S. Hsu, Hot corrosion behavior of laser-glazed plasma-sprayed yttria-stabilized zirconia thermal barrier coatings in the presence of V_2O_5, Surf. Coat. Technol., 201 (2007), 5143/5147.

[8] Y. Fu, A.W. Batchelor, H. Xing, Y. Gu, Wear behaviour of laser-treated plasma-sprayed ZrO_2 coatings, Wear, 210 (1997), pp. 157/164.

[9] J.H. Lee, P.C. Tsai, C.L. Chang, Microstructure and thermal cyclic performance of laser-glazed plasma-sprayed ceria-yttria-stabilized zirconia thermal barrier coatings, Surf. Coat. Technol., 202 (2008), pp. 5607/5612.
[10] H. Chen, Y. Hao, H. Wang, W. Tang, Analysis of the microstructure and thermal shock resistance of laser glazed nanostructured zirconia TBCs, Journal of Therm. Spray Technol., 19 (2010), pp, 558/565.
[11] K. Kokini, B.D. Choules, Y.R. Takeuchi, Thermal Fracture Mechanisms in Ceramic Thermal Barrier Coatings, Journal of Therm. Spray Technol., 6(1) (1997), pp, 43/49.
[12] R. Ghasemin, R. Shoja-Razavi, Laser glazing of plasma-sprayed nanostructured yttria stabilized zirconia thermal barrier coatings, Ceram. Int., 39 (2013), pp, 9483/9490.
[13] Z.J. Fan, K.D. Wang, X. Dong, W.Q. Duan, X.S. Mei, W.J. Wang, J.L. Cui, J. Lv, Influence of columnar grain microstructure on thermal shock resistance of laser re-melted ZrO2-7 wt% Y2O3 coatings and their failure mechanism, Surf. Coat.Tech., 277 (2015), pp, 188/196.
[14] A. Ramireza, Ma Qiana, B. Davisb, T. Wilksc, D.H. StJohnd, Potency of high-intensity ultrasonic treatment for grain refinement of magnesium alloys, Scr. Mater., 59 (2008), pp, 19/22.
[15] Y. Cuia, C.L. Xua, Q. Hanb, Effect of ultrasonic vibration on unmixed zone formation, Scr. Mater., 55 (2006), pp, 975/978.
[16] D.J. Wu, M.H. Guo, G.Y. Ma, F.Y. Niu, Dilution characteristics of ultrasonic assisted laser clad yttria-stabilized zirconia coating, Mater. Lett., 141(2015), pp, 207/209.
[17] S. Habib Alavi, Sandip P. Harimkar, Melt expulsion during ultrasonic vibration-assisted laser surface processing of austenitic stainless steel, Ultrasonics, 59 (2015), pp, 21/30.

Sliding Wear Investigation of Plasma and High Velocity Oxy-Fuel Sprayed Alumina Coatings

O. Ali, R. Ahmed, School of Engineering and Physical Sciences, Heriot-Watt University, Edinburgh, EH14 4AS, UK
H. Alawadhi, M. Shameer, College of Sciences, University of Sharjah, P.O. Box: 27272, Sharjah, UAE

Alumina coatings display a multitude of properties such as resistance to chemical reactions, sliding wear and thermal barrier properties. This paper investigates the sliding wear resistance of thermally sprayed alumina coatings produced using two different techniques of High Velocity Oxy-Fuel (HVOF) and Air Plasma Spray (APS). Coatings were deposited on AISI 440C stainless steel substrates using two different types of feedstock powder, with fine powder (1-5μm) used for HVOF spraying and conventional sized powder (10-45μm) for APS spraying. Sliding wear tests were carried out according to the ASTM G133-02 standard using sintered WC-6Co ball as the counter-body. Coatings were characterised using Scanning Electron Microscope (SEM), X-Ray Diffraction (XRD) and nano-indentation technique. Both coatings showed the presence of α- and γ-phases, where APS coating had mostly γ-phase and HVOF coating had mostly α-phase. Regions of wear track were analysed using SEM to observe the wear mechanisms. HVOF coating showed an order of magnitude improvement in the wear resistance when compared to the APS coating. The relatively lower performance of the APS coating is attributed to the presence of γ-phase, coating fracture and higher porosity when compared to the HVOF coating. Results are discussed in terms of the wear mechanism and structure-property relationships.

1. Introduction

Wear and friction is a major contributor to energy loss in industrial applications. Ceramic materials display high hardness and chemical inertness, and hence are a good candidate for the wear and corrosion resistance applications. Ceramics also display thermal and electrical insulation properties. Surface coatings made of ceramics allow parts (susceptible to corrosion and high wear) to last longer and reduce costs [1, 2]. Wear resistant coatings are mostly based on transition metal carbides (WC, TiC, Cr_3C_2, etc.) and hard oxides (Al_2O_3, TiO_2, Cr_2O_3), which can be deposited using thermal spray techniques such Air Plasma Spray (APS), High Velocity Oxy-Fuel (HVOF) and flame spraying [2]. Alumina (Al_2O_3) is a high hardness and thermally stable ceramic, although during specific thermal spray techniques (APS) the stable α-phase changes to softer and lower density θ-, γ- or δ-phases, due to the lower nucleation energy of these compounds. The addition of Zirconia (ZrO_2) or Titania (TiO_2) allows for better wear resistance and higher fracture toughness [3].
Investigation of wear resistance of plasma sprayed alumina coatings started off with conventional sized feedstock and recently heading towards nanostructured coatings using suspension techniques [4]. Wear resistance of these coatings is dependent on i) their microstructure which is influenced by coating process parameters (i.e. particle size, spray distance, temperature, etc.), ii) bulk properties of counter-body materials and iii) tribological testing parameters.
This paper compares the wear resistance of alumina coatings deposited using APS and HVOF thermal spray techniques with a view to understand the wear mechanisms and structure property relationship.

2. Experimental

2.1 Test materials

Details of feedstock powder, thermal spray guns and spraying parameters are listed in **Table. 1.** Two feedstock powders were used for coating deposition, a conventional powder was used for the APS alumina coating with particle sizes ranging from 10 to 45 μm, and a fine powder for HVOF alumina coating with particle sizes from 1-5 μm. The coatings were deposited using a 9 MB Metco spray gun for APS spraying and θ-gun for HVOF spraying using optimised parameters. Coatings were 250 μm thick and deposited on AISI 440C martensitic stainless steel discs of diameter 31 mm and 8 mm thickness [5].

2.2 Coating characterisation

Coatings were characterised using Scanning Electron Microscope (SEM), Electron Dispersive X-Ray Spectrometry (EDX), X-Ray Diffraction (XRD), Vickers microhardness and nano-indentation techniques. The SEM was used to observe the microstructure of coatings which were cryogenically fractured to preserve microstructural features. XRD analysis was performed using a Bruker AXS, model D8 ADVANCE X-Ray Diffractometer operating at 40kV and 40mA, using Cu Kα radiation (wavelength λ=0.1542 nm). Vickers microhardness was measured using a calibrated Mitutoyo, MVK-H1 machine to make five indentations on the surface of each sample using a load of 1.96 N (or HV0.2) load. Nano-indentation analysis was performed using a calibrated TriboIndenter® (Hysitron Inc., MN) equipped with Berkovich indenter at the load of 100mN. These measurements were performed on coating cross-sections to measure the hardness and elastic modulus based on the Oliver and Pharr method [6].

Table. 1. Thermal spray process parameters [5, 7].

Spray Gun	Spray conditions and powders	
9-MB Metco (APS)	Spray material	Angular/crushed alumina (Al_2O_3 > 98.0%)
	Powder size	10-45 μm
	Arc current	500 A
	Arc voltage	70 V
	Primary gas	37.6 L/min (Ar)
	Secondary gas	7.1 L/min (H_2)
	Spray distance	80 mm
	Microhardness (HV0.2)	683±38
θ-Gun (HVOF)	Spray material	Angular/crushed powder alumina (Al_2O_3 > 98.0%)
	Powder size	1-5 μm
	Oxygen flow rate	893 L/min
	Kerosene flow rate	0.3 L/min
	Acetylene flow rate	43 L/min
	Spray distance	150 mm
	Microhardness (HV0.2)	632±29

2.3 Wear testing

Dry sliding wear resistance of coatings was investigated using a Ball-on-Plate reciprocating sliding wear tribometer (BLR2000 M, Bud Labs, USA) at room temperature in accordance with the ASTM G133-02 standard. The counter-body material was a sintered WC-6%Co ball of diameter 9.53 mm (3/8 inches). The total sliding distance for the wear test was 500m with a normal load of 25N and stroke length of 10mm, at a reciprocating frequency of 2Hz. Tests were repeated five times for each pair of coating and counter-body material. Coating volume loss was calculated using the average cross-sectional area of the wear track and stroke length. The cross-sectional area of the coating wear track was measured using an interferometer (Zygo New View). The volume loss of the counter-body (WC-6Co ball) was measured using a weighing scale accurate to 10^{-5}g and the average density of the ball (14.95 g/cm^3 - supplier information). Friction force was measured using a tension-compression load cell mounted on the tribometer. Wear tracks were analysed using a Scanning Electron Microscope (SEM), in Secondary electron (SE) and Back Scattered Electron (BSE) mode to study the wear mechanisms.

3. Results

3.1 Coating microstructure and XRD analysis

SEM images of cryogenically fractured coating cross-sections are shown in **Fig. 1.** These fractured cross-sections highlight subtle differences between the two coatings, the HVOF coating shows some unmolten particles and relatively lower porosity. Coating porosity was measured as 8.8% ± 2.0 and 2.8% ± 0.64 for the APS and HVOF coatings, respectively [7].

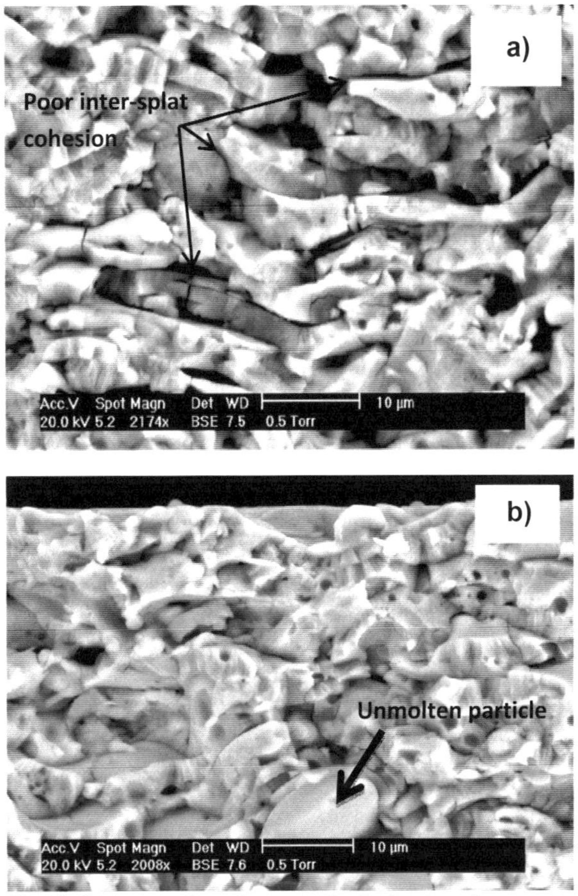

Fig. 1. Cross-section of cryogenically fractured coating of a) APS b) HVOF [8].

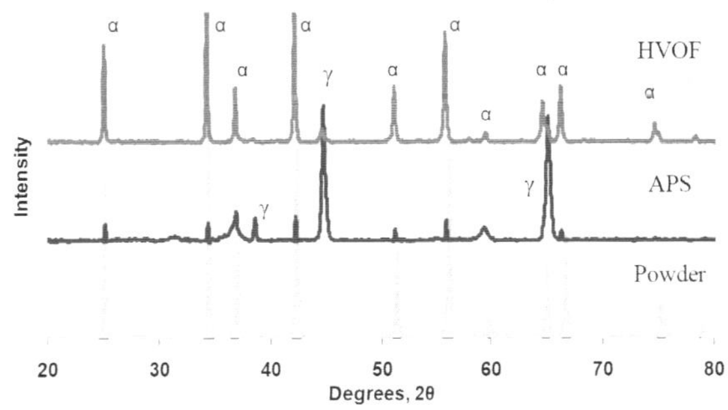

Fig. 2. XRD analysis of coatings and feedstock [7].

XRD analysis of the feedstock powders showed the presence of a predominantly *α*-phase compared to *γ*-phase. **Fig. 2** shows the XRD patterns for the two coatings and the feedstock powders. Both powders showed very similar analysis results and therefore have been summarised by a single pattern. The APS coating showed relatively higher *γ*-phase transformation when compared to predominantly *α*-phase in the HVOF coating.

3.2 Nano-indentation tests

The nano-indentation investigation of two coatings were performed in a previous study [7] and included here as **Fig. 3** to aid the discussion. In this figure it can be noted that the hardness and elastic modulus of the HVOF coating is higher than that of the APS coating. The results are represented for both the coating and substrate materials.

3.3 Sliding wear tests

The averaged coefficient of friction (CoF) values along with their standard deviation is illustrated in **Fig. 4a.** The average friction for the last 300m sliding distance for the HVOF and the APS coatings was measured as 0.47 ± 0.03 and 0.53 ± 0.03, respectively. Sliding wear performance of the HVOF coating was better than the APS coating, where the total wear rate of the HVOF and the APS coatings was 1.23×10^{-6} mm^3/Nm and 11.52×10^{-6} mm^3/Nm, respectively, as shown in **Fig. 4b**. This figure also shows the components of total wear rate (ball and coating) along with the standard deviation of the averaged wear rate results. SEM observations of the wear tracks for the APS and HVOF coatings along with the EDX map inlays of W concentration are shown in **Fig. 5**. The presence of W highlights the presence of tribofilm.

4. Discussion

4.1 Coating microstructure and composition

The microstructure of fractured coatings (**Fig. 1**) shows porosity, unmolten particles, poor inter-splat cohesion and vertical cracks in splats. The individual splats also display columnar grain feature, which is consistent with previous investigations of plasma sprayed alumina coatings [9, 10]. This feature is also present in the HVOF coating but at a relatively lower level. The columnar structure is caused by the crystallisation of alumina in highly non-equilibrium condition, as the molten content loses heat due to conduction to the substrate/ underlying splats in a direction orthogonal to the plane of the substrate/ underlying material [9]. Also the cross-section of APS coating indicated weaker inter-splat bonding in comparison to the HVOF coating.
The splat thickness of both coatings was similar despite the large differences in powder size (Table 1). This is due to the malleability of the partially molten particles and high particle velocities in the HVOF coating when compared to the APS coating. The thermal loading of powder particles influences the malleability and therefore the splat dimensions. The APS coating has a more uniform lamellar structure but splat cohesion is low as can be seen by the gaps between the splats which are attributed to the low impact velocity. The HVOF coating has a combination of lamellar structure with unmolten particles, but splat cohesion is better which is attributed to the inherent high impact velocities.
The quantity of γ-phase alumina in thermally sprayed coatings is used as an indicator to verify the extent of melting of the feedstock powder (feedstock powders are most often made of 99% α-phase alumina) [11]. **Fig. 2** indicates that there was relatively higher α-phase $\rightarrow\gamma$-phase transformation during the APS coating deposition.

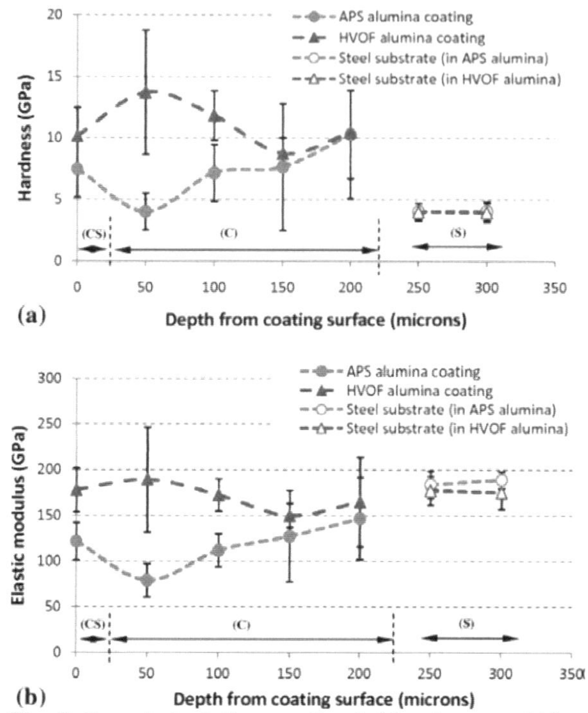

Fig. 3. Nano-indentation results of a) hardness and b) modulus of APS and HVOF coatings (CS-coating surface, C-coating, S-substrate) [7].

4.2 Sliding wear resistance

The sliding wear test results (**Fig. 4**) show that the HVOF coating wear rate was roughly ten times lower than the APS coating, whereas the ball wear rates for both coatings were similar. After the running-in stage, the CoF of HVOF coating was also relatively lower than the APS coating.

4.2.1 Coefficient of friction

In dry sliding wear, CoF is generally linked to the amount of energy dissipated during asperity interactions, both in 2-body and 3-body interactions in the contact region and the tribo-mechanical properties of the asperities. As the sliding distance increases, the averaged CoF values for the APS and HVOF coatings indicate different behaviours. The CoF for APS coating gradually decreases as the experiment continues after the peak value following the running-in stage. The near linear nature of CoF for the APS coating after the running-in stage suggests that the wear mechanism remained the same throughout the tests. The CoF of HVOF coating peaks higher than the APS coating, however after the running-in stage it drops down to a stable but relatively lower value than the APS coating; a trend which continues for the remainder of the experiment. These changes were linked to the differences in the sliding wear mechanism, as discussed later.

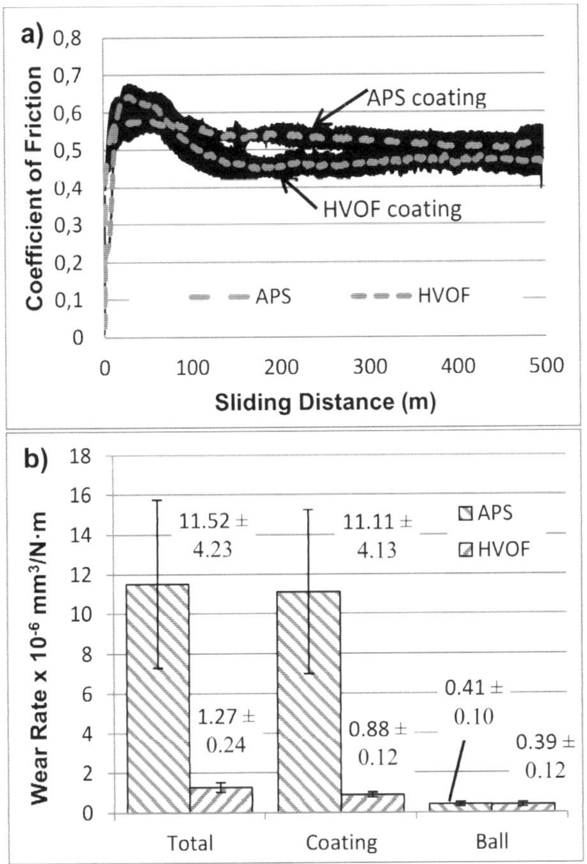

Fig. 4. a) CoF against sliding distance and b) total wear rate (ball and coating) and wear rate of the coating and ball materials.

4.2.2 Sliding wear observations

Wear track observations (**Fig. 5**) show the formation of W-rich tribofilm in coatings. The proportion of the tribofilm present in the wear tracks is significantly different in both coatings. The APS coating has areas of exposed coating, which can be seen between the flat features indicating the presence of W-rich tribofilm shown in the BSE image (**Fig. 5a**). The white patches have relatively higher concentration of W (based on EDX analysis). The APS coating also showed vertical fractures (**Fig. 1**), these fractures can propagate as the APS coating experiences cyclic stresses under sliding wear. The exposed coating shown in **Fig. 5a** is caused by the failure of relatively poor inter-splat bond leading to subsequent delamination in the APS coating. This is consistent with the investigation reported by Psyllaki et al. [12]. The debri collected after the APS coating wear tests showed agglomeration of fine particles and also larger delaminated sections of alumina coating which was identified by the EDX analysis.
Observations of the HVOF wear track showed that regions of exposed coating were much smaller when compared to the APS coating. The wear track surface is covered with large quantities of W-rich tribofilm as confirmed by the by EDX analysis shown in **Fig. 5b.** Besides fracture, plastic deformation of coating can also occur during the sliding wear process. Although this behaviour is generally considered to be a property of metals, but ceramic coatings have been observed to plastically deform and result in the formation of smooth flaky debris, as observed by Bolelli et al. [10]. The plastic deformation can occur due to sliding along "shear fault" like lamellae boundaries in thermal spray coatings. The contributing factors for this plastic deformation are the size of the lamellae and the contact pressure, i.e. the contact pressure can cause plastic deformation in ceramic coatings once it reaches a critical point. Increase in contact stress can also result in the deformation of substrate material [13].

5. Sliding Wear Mechanism

Wear resistance of alumina coatings is generally linked to their hardness and toughness, both of which are dependent on coating porosity, inter-splat bond strength, cracks and crystalline phases. These factors can also influence coating density e.g. metastable phases are less dense [14]. The relative influence of hardness and toughness is also dependent upon the tribological conditions [15]. In the current investigation, HVOF coating performed significantly better than the APS coating, both in terms of the wear rate and CoF, which is linked to the changes in the wear mechanism.
The wear mechanism for both APS and HVOF coatings was dominated by i) coating properties (elastic modulus, hardness, fracture toughness and dominant crystalline phase (α-phase and/or γ-phase)), ii) formation of W-rich tribofilm on the coating and O-,C-rich tribofilm on the WC-Co ball, and iii) the extent of wear debri entrapment in the contact region.

5.1 Role of coating properties

There were subtle differences in the elastic modulus and hardness of the APS and HVOF coatings (**Fig. 3**). Higher nano-hardness of the HVOF coating can be attributed to the dominant α-phase of the coating microstructure and the relatively lower porosity. The indentation size effects indicated a relatively higher macrohardness value for the APS coating [16]. The elastic modulus of the HVOF coating was also relatively higher, which indicates better inter-splat bonding and coating cohesion. A parametric study of the HVOF alumina coatings performed by Turunen et al. showed that coatings that had the highest densities showed the highest wear resistance [17]. Previous studies by the authors on the fracture toughness of these coatings have indicated that the fracture toughness of the APS coating was too low to be measured by the macro-indentation method, whereas the indentation fracture toughness of the HVOF coating was measured as 5.5 MPa·m$^{1/2}$ (Palmqvist model) [18]. Based on the combination of relatively higher hardness, elastic modulus, fracture toughness and lower porosity, the delamination and wear resistance of the HVOF coating significantly improved, when compared to the APS coating.

5.2 Influence of tribofilm

Tribofilms are caused by the transfer of elemental species from the interacting surfaces in sliding wear. These films serve the purpose of reducing wear and CoF by providing a relatively low shear strength interface between interacting asperities. SEM and EDX analysis of the wear tracks and the wear debris indicated that a W-rich tribofilm formed on the surface of both coatings due to asperity interactions and flash temperatures, during the sliding wear process. The presence of this tribofilm reduces CoF and coating wear rate. SEM observations shown in **Fig.5** indicate that the formation and stability of this tribofilm was relatively higher for the HVOF coating. Hence the tribofilm provided lower protection during asperity interactions for the APS coating, which also resulted in its relatively higher CoF and wear rate.

Two factors were critical in the formation and stability of this W-rich tribofilm. First, being the difference in the dominant crystalline HCP α-phase for the HVOF coating as opposed to Cubic γ-phase for the APS coating. It is postulated that the α-phase of the HVOF coating not only prompted the formation of relatively higher amount of W-rich tribofilm, but also the stability of this tribofilm was also higher, as observed in the EDX and SEM analysis. The stable tribofilm covering larger areas of the wear track therefore resulted in relatively lower CoF and wear rate for the HVOF coating.

The second factor relates to the relatively higher amount of wear debri in the wear track of the APS coating, as discussed in section 5.3. The tribofilm is formed by the asperity interaction of WC-Co ball with the Al_2O_3 coating material. As this tribofilm grows in thickness it is removed during the wear process and the tribofilm reforms due to the asperity interactions between the ball and coating materials. This is therefore a continuous process of tribofilm formation and removal. The presence of entrapped debri in the contact region will reduce this 2-body WC-Co ball/Al_2O_3 coating interaction for the APS coating, and relatively increase 3-body (ball/debri and debri/coating) interaction. This 3-body interaction is not favourable for the formation and stability of the tribofilm in the APS coating.

Similarly, C-rich and O-rich tribofilm was observed on the surface of the worn WC-Co balls. However, there were no significant differences observed between the two coatings.

5.3 Wear groove and debri entrapment

Wear track analysis of both coatings indicated that the depth of wear groove for the APS and HVOF coatings was on average 20 µm and 2.5 µm, respectively. Similarly, the width of the wear groove was on average 1200 µm and 600 µm for the APS and HVOF coatings, respectively. The relatively higher depth and width of the wear groove for the APS coating would result in a higher tendency for the coating and ball wear debri to be entrapped in the contact region. This debri entrapment modifies the wear mechanism from 2-body to 3-body abrasion. Hence the APS coating wears more, as it is abraded by the relatively higher proportion of Al_2O_3 debris entrapped in the contact region instead of the WC-Co ball. Although the wear debri can roll and slide to reduce CoF, however, they can also consume more energy as they deform and fracture within the contact region. Hence the presence of entrapped debri can lead to higher CoF, wear rate and relatively lower stability of W-rich tribofilm for the APS coating.

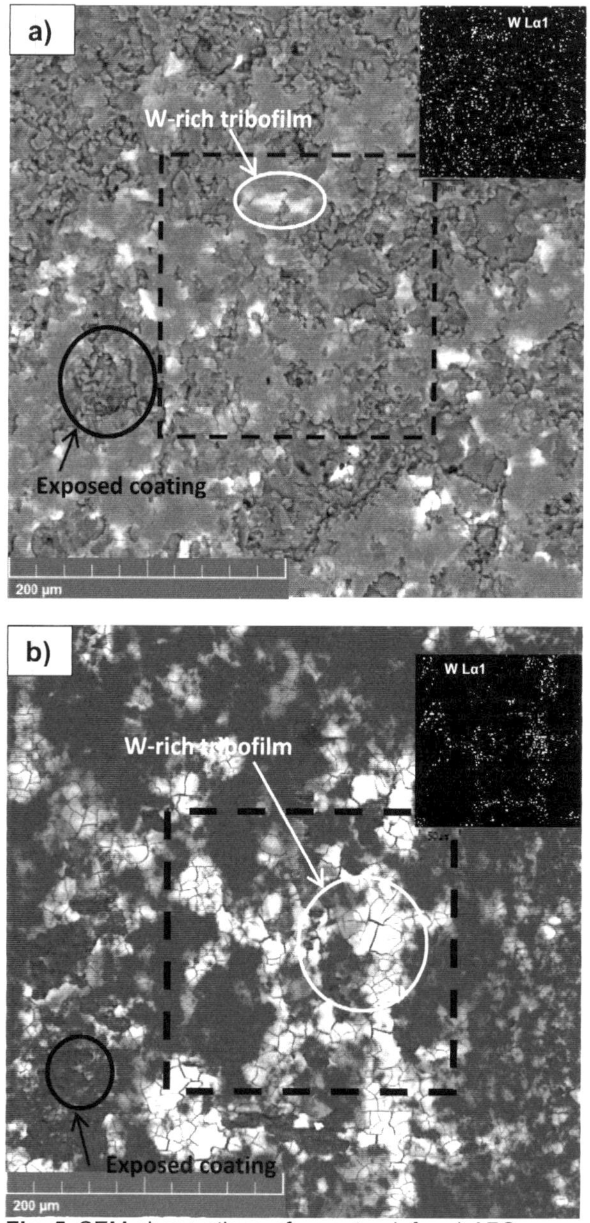

Fig. 5. SEM observations of wear track for a) APS coating b) HVOF coating, with EDX map inlays of W in the boxed section.

6 Conclusion

1) The wear resistance of the HVOF coating was an order of magnitude higher than the APS coating. Similarly, the average CoF of HVOF coating was lower than that of the APS coating.
2) A wear mechanism based on the influence of coating properties, formation of W-rich tribofilm and the extent of wear debri entrapment in the contact region is presented to account for the differences in the wear and friction behaviour.
3) The combination of relatively higher α-phase content, nano-hardness, modulus, fracture toughness and lower porosity improved the delamination and wear resistance of the HVOF coating.
4) The formation and stability of W-rich tribofilm was higher in the HVOF coating.
5) Wear debri entrapment in the contact region increased the wear rate and CoF of the APS coating.

6 References

[1] Fernandez, J.E., Rodriguez, R., Wang Y. and Rincón A: Sliding wear of a plasma-sprayed Al2O3 coating. Wear 181-183 (1995), part 1, pp. 417/25.

[2] Singh, V.P., Sil, A. and Jayaganthan, R: A study on sliding and erosive wear behaviour of atmospheric plasma sprayed conventional and nanostructured alumina coatings. Materials and Design 32 (2011), issue 2, pp. 584/91.

[3] Fervel, V., Normand, B., Liao, H., Coddet, C., Beche, E. and Berjoan, R.: Friction and wear mechanisms of thermally sprayed ceramic and cermet coatings. Surface and Coatings Technology 111 (1999), issue 2-3, pp. 255/62.

[4] Toma, F-.L., Potthoff, A., Berger, L-.M. and Leyens, C.: Demands, potentials, and economic aspects of thermal spraying with suspensions: a critical review. Journal of Thermal Spray Technology 24 (2015), issue 7, pp. 1143/52.

[5] Faisal, N.H. and Ahmed, R.: Acoustic emission analysis of Vickers indentation fracture of cermet and ceramic coatings. Measurement Science and Technology 22 (2011), issue 12, pp. 125704.

[6] Oliver, W.C. and Pharr, G.M.: An improved technique for determining hardness and elastic modulus using load and displacement sensing indentation experiments. Journal of Materials Research 7 (1992), issue 6, pp. 1564/83.

[7] Ahmed, R., Faisal, N.H., Paradowska, A.M. and Fitzpatrick, M.E.: Residual strain and fracture response of Al2O3 coatings deposited via APS and HVOF techniques. Journal of Thermal Spray Technology 21 (2012), issue 1, pp. 23/40.

[8] Faisal, N.H.: Acoustic emission analysis for quality assessment of thermally sprayed coatings. Heriot-Watt University (2009), Ediburgh, UK

[9] Bianchi, L., Denoirjean, A., Blein, F. and Fauchais P.: Microstructural investigation of plasma-sprayed ceramic splat. Thin Solid Films 299 (1997), issue 1-2, pp. 125/35.

[10] Hao, S., Li, C-.J. and Yang, G-.J.: Influence of deposition temperature on the microstructures and properties of plasma-sprayed Al_2O_3 coatings. Journal of Thermal Spray Technology 20 (2011), issue 1, pp. 160/69.

[11] Bolelli, G., Rauch, J., Valeria, C., Killinger, A., Lusvarghi, L. and Gadow, R.: Microstructural and tribological investigation of high-velocity suspension flame sprayed (HVSFS) Al2O3 coatings. Journal of Thermal Spray Technology 18 (2009), issue 1, pp. 35/49.

[12] Psyllaki, P.P., Jeandin, M. and Pantelis, D.I.: Microstructure and wear mechanisms of thermal-sprayed alumina coatings. Materials Letter 47 (2001), pp. 77/82.

[13] Osawa, S., Itsukaichi, T. and Ahmed, A.: Influence of substrate properties on the impact resistance of WC cermet coatings. Journal of Thermal Spray Technology 14 (2005), issue 4, pp. 495/501.

[14] Andersson, J.M.: Controlling formation and stability of alumina phases. Linköpings University (2005), Dissertation no. 987.

[15] Faisal, N.H., Ahmed, R. and Reuben, R.L.: Indentation testing and its acoustic emission response: applications and emerging trends. International Materials Review 56 (2011), issue 2, pp. 98/142.

[16] Ahmed, R., Ashraf, A., Elameen, M., Faisal, N.H., El-Sherik, A.M., Elakwah, Y.O. and Goosen, M.F.A.: Single asperity nanoscratch behaviour of HIPed and cast stellite 6. Wear 312 (2014), issue 1-2, pp. 70/82.

[17] Turunen, E., Varis, T., Hannula, S.-P., Vaidya, A., Kulkarni, A., Gutleber, J., Sampath, S. and Herman, H.: On the role of particle state and deposition procedure on mechanical, tribological and dielectric response of high velocity oxy-fuel sprayed alumina coatings. Materials Science and Engineering A 415 (2006), pp. 1-11.

[18] Faisal, N.H., Ahmed, R., Prathuru, A.K., Spence, S., Hossain, M. and Steel, J.A.: An improved Vickers indentation fracture toughness model to assess the quality of thermally sprayed coatings. Engineering Fracture Mechanics 128 (2014), pp. 189/204.

Influence of Ni content on the wear resistance of B_4C-Ni coatings sprayed by APS

Youcha. Zhang*, Chengchang. Jia
University of Science and Technology Beijing, Beijing, China
Peng. Jia, Jinlei. Zhang
Beijing United Coatings Technologies Co., Ltd. Beijing, China
Yongchang. Jia
Beijing Langqiao Material Technology Co., Ltd. Beijing, China
*E-mail: zhang.youcha@united-coatings.com

Abstract

Wear protection is one of the major applications of B_4C-Ni coatings. B_4C-Ni powders with Ni theoretical content of 5, 40 and 60wt% fabricated by pressurized hydrogen reduction process were characterized and deposited by air plasma spray process. The microstructure and phase structure of the powders and the as-sprayed coatings were analyzed by scanning electron microscope (SEM) and X-ray diffraction (XRD). The microhardness Hv0.3 and the adhesive strength were measured. The influence of Ni content on the coating microstructure, performance and fretting wear resistance for the as-sprayed coatings were investigated. Results indicate that the nickel content affect the fretting wear resistance of coatings, coating with nickel of 40wt% shows the best physical performance and wear resistance.

Introduction

Boron carbide (B_4C) represents one of the hardest and densest materials on earth, which is a composite of B_4C and carbon in graphitic form. It is a black crystalline powder of high melting point, low density, high hardness, excellent thermoelectric performance and excellent chemical resistance etc.[1,2].

As a result, B_4C coating as an economical and flexible existence is widely used in nuclear power industry to safely contain nuclear control rods and prevent radiation leaks, which also is used in the manufacturing industry and aerospace industry to make materials that require the highest level of durability for its high coating hardness and fine chemical stability, especially requirement of perfectly chemical resistance against potassium chlorate and nitric acid as well as the slow reaction with chlorine and oxygen below 1000 °C[3,4,5]. It is reported that B_4C coated chamber wall is used to replace the previously anodized aluminum surface in semiconductor industry [6].

The B_4C coatings could be formed by air plasma spray process, vacuum plasma spray process, low pressure plasma spray process(LPPS), a supersonic plasma jet chemical vapor deposition process [7] and electromagnetically accelerated plasma spraying (EMAPS) method[8] etc. Among which, the air plasma spray methods is the most common and economical fabricating method, while B_4C coatings face the problem of the high melting point of the boron carbide of 2400 °C, which induces a low(<20%) deposition efficiency and a high void content of the coating layer[9].

Nickel is a corrosion resistance element, which can be used as binder phase of B_4C coatings to increase the deposition efficiency. Plasma sprayed Nickel coated B_4C coatings with the feedstocks (remarked as B_4C /x wt.% Ni)fabricated by Pressurized hydrogen reduction process hasn't been reported in the past years, as well as the corresponding microstructure, performance of coatings. Therefore this study will focus on the performance of Nickel coated B_4C powders with three nickel contents (5, 40, 60 by wt.%), which are remarked as B_4C/5Ni, B_4C/40Ni and B_4C/60Ni. Furthermore, on the change rule of microstructure, physical performance and the friction rate with the nickel content of the coatings sprayed by plasma spray process.

Materials and Experimental Procedures

Materials Preparation

B_4C powder (purity>95%, Mudanjiang Hongda Boron Carbide Factory) and nickel coated B_4C powders with nickel content of 5, 40, 60 wt.% fabricated by pressurized hydrogen reduction process were prepared for air plasma spraying process, with a particle sizes in the range of 45-105μm. The morphology and phase composition of the powders were measured on SEM (type: FEI Quanta 200) and XRD (X-ray diffraction, type: DMAX-RB, Rigaku Corporation).

The Specimens

For the experiments, the substrate specimens are made of mild carbon steel (Q235 steel) were utilized. First, the pieces with the size of 20 mm×50 mm×5 mm were used as metallographic specimens. Second, bar-shaped specimens of Φ25.4×45mm were used to determine the adhesion strength. Third, the plate-taped substrate specimens (Φ10×20mm buttons) were used on the conventional ball-on-plate test.

All the substrates surfaces were grit blasted to clean and roughen before spraying by alumina grit (36 mesh) and blasting angle of about 75°. Multicoat Air plasma spraying system (Sulzer Metco) was used for depositing B_4C/5Ni, B_4C/40Ni and B_4C/60Ni powders on mild carbon steel substrate. The plasma spray parameters are summarized in Table 1.

Table 1 Process parameters of plasma spray

Voltage	Current	Gas flow rate(Ar)	Gas flow rate(H_2)	Feed rate	Spraying distance
70V	650A	55 NLPM	6.5NLPM	6g/min	140mm

The cross section metallograph and porosity were evaluated on type of OLYMPUS BX 51M optical microscope. The microhardness Hv0.3 was measured on the as-sprayed coatings cross section using hardness tester Micromet 1503(Buehler, USA). The reported values are average from at least 6 measurements. The adhesive strength was measured on Instron 3369 type material testing machine. The reported values are average from 3 measurements for each coating.

Wear test

To analyze the effect of the nickel content on the wear resistance of the as-sprayed coatings, a ball-on-plate fretting wear resistance test was carried out on a tribometer (type: SRV-4, Optimol Instruments, Germany) using the disk-shaped specimens, and the ZrO_2 ball (Φ6mm) as the upper sample. The fretting parameters used were as following: a normal force of 50N, frequency 5 Hz, duration time of 600s. Fretting tests were performed in ambient atmosphere at room temperature.

Results and Discussion

Particles Characterization

B_4C powder and nickel coated B_4C feedstock particles produced by pressurized hydrogen reduction process are observed in SEM images (Fig. 1) and EDS(Table 2). All the feedstock particles indicated 100% of diameter between 45-105μm, with irregular shape. The nickel coating integrity and the nickel layer thickness increase with the nickel content increasing from 5 to 60wt.%. B_4C/60Ni powder shows an integral nickel coating outside the B_4C core, while powders with nickel content of 5 & 40wt. % are partially coated. The spray powders still differ with each other on the powder morphology, for example, the B_4C/60Ni powder indicates a near-spherical shape, with no sharp corners, and the other two powders both have sharp corners.

The chemical composition analysis indicates the original B_4C powder has a purity of 96.63%, which also shows the actual nickel content of three coated powders are 5.03, 40.05, 60.04% by weight. This implies low nickel content of 5 and 40wt.% isn't enough to completely coat the B_4C particle, unless the nickel content increases as high as 60wt.%.

The phase composition could also be analyzed by means of an X-ray diffraction (XRD) of the B_4C and nickel coated B_4C spray powders. A typical XRD pattern is shown in Fig. 2. It shows B_4C powder had a single B_4C phase. It also indicates nickel is the main phase of the B_4C/60Ni powder, and nearly no B_4C phase, this result is conformity with the B_4C powders are coated with Ni entirely, which also is conformity with the SEM results (Fig.1 (d)). The B_4C/5Ni and B_4C/40Ni powders both show B_4C and nickel phase, which means these two powders weren't coated finely.

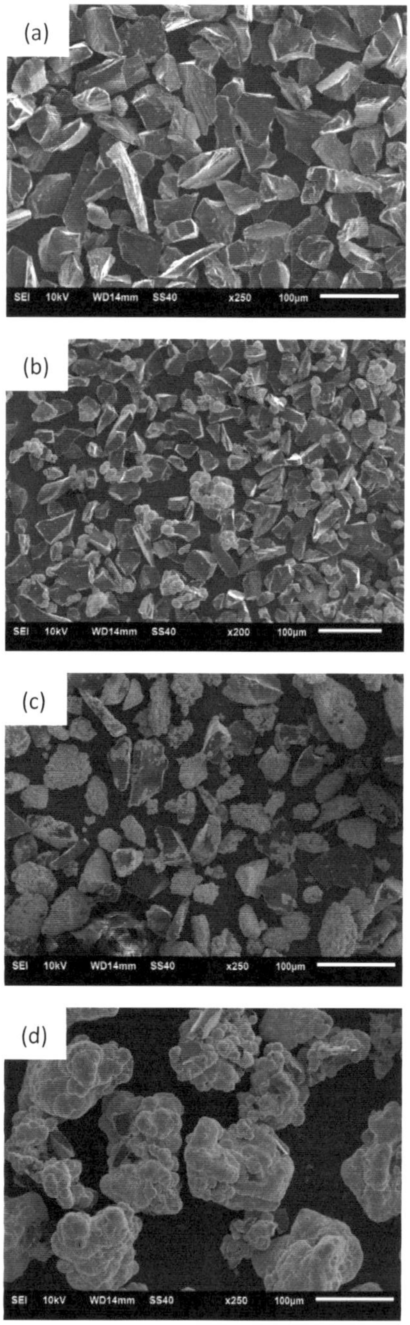

Fig.1: SEM micrograph of B_4C powder a) and Nickel coated B_4C powders: b) B_4C/5Ni, c) B_4C/40Ni, d) B_4C/60Ni

Table 2: EDS of the thermal spray powders

Powders	Elements content /wt.%					
	B	C	O	Si	Ni	Fe
B_4C	38.65	57.98	2.39	0.27	--	0.71
B_4C/5Ni	45.70	47.63	1.49	0.15	5.03	--
B_4C /40Ni	18.60	38.59	2.50	0.04	40.05	0.22
B_4C/60Ni	--	37.11	2.85	--	60.04	--

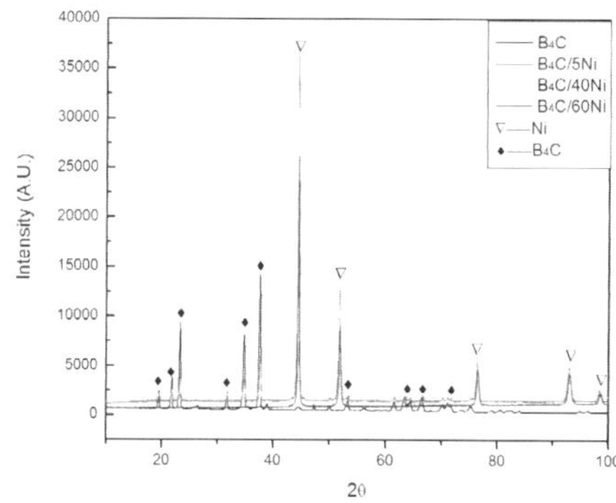

Fig.2 X-ray patterns of the thermal spray powders

Coating Characterization

A cross section metallograph of the as-sprayed B_4C/Ni coatings shows grey-dark contrast around the splats, which was identified as B_4C particles(Fig. 3). Among three coatings, B_4C/40Ni coating has the largest number of B_4C particles, the most uniform and dense microstructure in the cross section, which is consistent with the lowest porosity of 0.62% in Table 3. The cross section of the as-sprayed B_4C/5Ni and B_4C/60Ni coating show a little amount of B_4C particle.

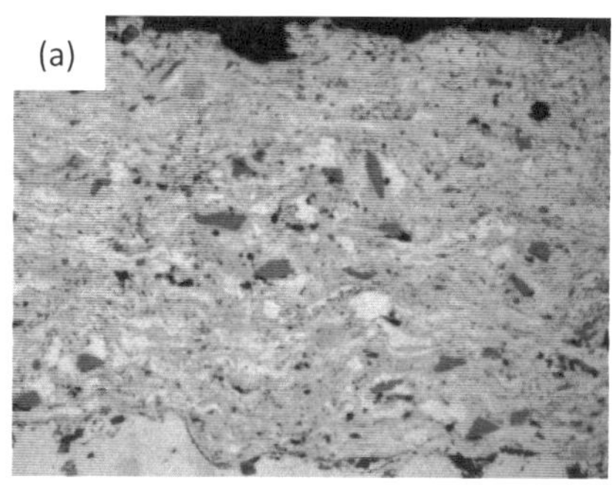

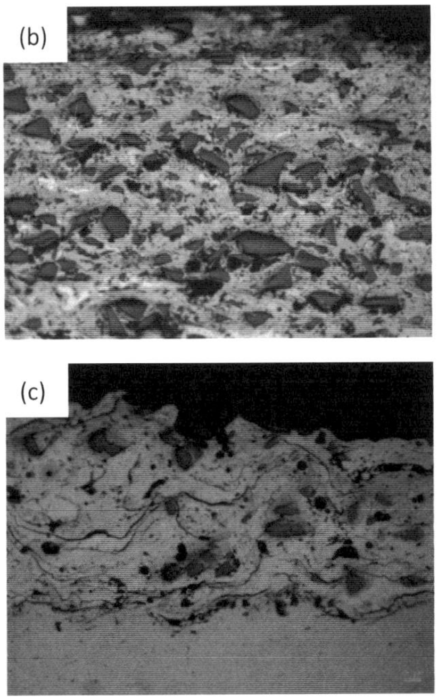

Fig.3 Optical microscopy of the fabricated coatings: a) B_4C/5Ni, b) B_4C/40Ni, c) B_4C /60Ni

As revealed by microhardness tests and adhesive strength (Table 3) on cross sections of coatings with varying nickel content, the B_4C/40Ni coating shows the highest microhardness and a much lower adhesive strength of 43.20Mpa, at the same time, the coatings with low and high nickel content(B_4C/5Ni and the B_4C/60Ni) show almost equal microhardness and adhesive strength value. This demonstrates that the hardness of composite coating was mainly dependent on B_4C particle number, this implies higher B_4C particle number led to much higher hardness because of the high hardness of B_4C hard phase particle. All the measured tested samples were broken between the coating and the substrate, with the value of tensile stress exceeding 40 MPa.

Table3 Porosity, microhardness and adhesion strength of the coatings

coating	porosity	Hv0.3	Adhesive strength(Mpa)
B_4C/5Ni	1.07%	504.3	49.08
B_4C/40Ni	0.62%	569.6	43.20
B_4C/60Ni	1.25%.	503	49.17

Wear performance

Plots of the coefficient of friction(COF) of the fabricated coatings (Fig.4)showed that the B_4C/40Ni coating exhibit a transient stage (~100s) followed by steady state values around 0.553. As to B_4C/5Ni and B_4C/60Ni coating, the COF increase with time and finally stabilize at a value of 0.747 and 0.826. Among the three coatings, B_4C/40Ni coating exhibits the lowest measured coefficient of friction.

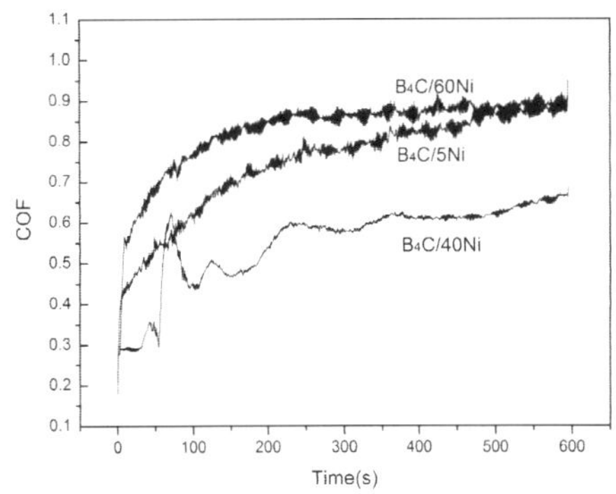

Fig.4 The COF evolution measured during ball-on-plate test of the as-sprayed B_4C/Ni coatings

The wear resistance of thermal spray coatings mainly depends on coating properties such as the microstructure, porosity and hardness, etc. of the coatings [10]. Low magnification images of the contacting surfaces of the B_4C/Ni coatings after sliding 600s cycles (Fig. 5) revealed, the fatigue loading is taking place, leading to a pronounced wear. The least worn was the B_4C/40Ni coating, reinforced by B_4C hard particles with good binding to the matrix. The wear is associated namely with the braking of hard particles, followed by their pulling out (Fig. 5 a&b). Both of B_4C/5Ni and B_4C/60Ni coating revealed a much higher wear width than coating with nickel content of 40wt.%, this implies these two coatings underwent a much higher wear, which wasthe same results with the COF rate(Fig.4). The wear of the B_4C/5Ni and B_4C/60Ni coating was realized by splat delamination in consequence of cyclic loading, but much less pronounced wear mechanism was observed.

The corresponding EDS evolutions of the worn coatings show that, as to the oxide content of three coatings(Fig.5 A), both the smooth points(Fig.5 B) and wear debris(Fig.5 C) of the wear scars were partially oxidized.

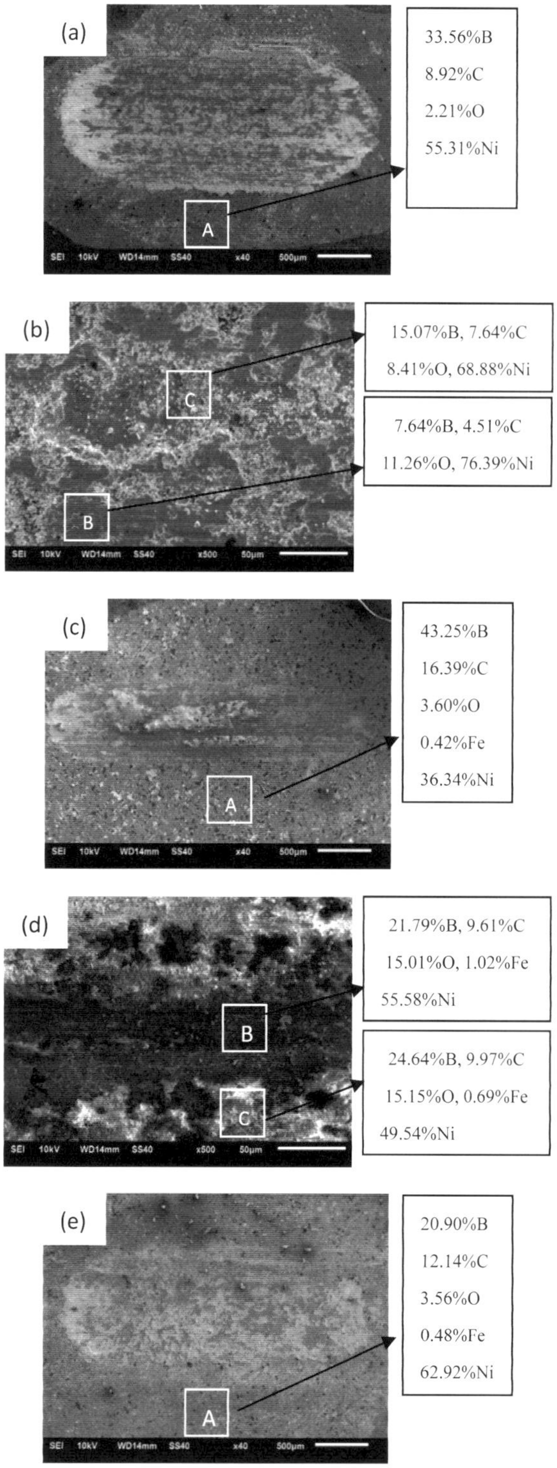

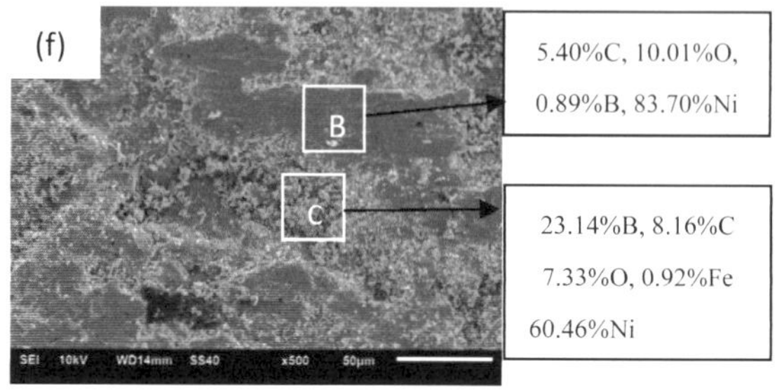

Fig. 5 SEM & EDS analysis of the worn coatings: a)&b) B_4C/5Ni. c)&d) B_4C/40Ni ,e)&f) B_4C/60Ni

Conclusions

Nickel coated B_4C powders with different nickel content(5, 40, 60wt.%)were used to gain coatings by air plasma spray process and their fretting behavior were compared. Based on the results presented above the following conclusions have been drown:

1 B_4C particles could be coated completely with nickel content of 60wt.%.

2 The B_4C/40Ni coating showed the highest microhardness, the lowest porosity, and the lowest COF of 0.553, than coating with nickel content of 5, 60wt%.

3 Among the plasma sprayed coatings with different nickel content, the B_4C/40Ni coating can gain the best fretting wear resistance than low (5wt.%) and high(60wt.%) nickel content.

4 B_4C/Ni coatings can improve the wear resistance compared with the conventional air plasma sprayed B_4C coatings.

Acknowledgements

The authors gratefully acknowledge the financial support from united coatings technology(UCT) of china. They also gratefully acknowledge help from Mrs. Bian and Mr. Chao Zhang for data analysis, Mr. Jinlei Zhang for sample preparation, and Mr. Pengfei He for fretting wear test.

References

[1] J.R. Davis, "Handbook of Thermal Spray Technology", www.asminternational.org. pp, 286.

[2] YU Xiao-He, LU Tie-Cheng, LIN Tao, et al., "Preparation of Boron Carbide Spherical Films and Hollow icrosphere", Journal of Inorganic Materials, Vol. 27(12)(2012), PP., 1325-1331.
[3] DU Xian-Wu, ZHANG Zhi-Xiao, WANG Wei-Min, et al., "Effect of Particle Size on Densification and Mechanical Properties of Hot-pressed Boron Carbide", 28(10)(2013), PP.,1062-1066.
[4] Chun-Ming Lin, "Parameter Optimisation of a Vacuum Plasma Spraying Process Using Boron Carbide", Journal of Thermal Spray Technology, volume 21(5)(2012), PP., 873-881.
[5] Fr.-W. Bach, Hannover/D, K. Möhwald et al., "Qualification of a modified triplex II plasma gun for processing of liquid precursors and wire- or powder shaped spray materials under controlled atmosphere", ITSC 2005,pp., 1177-1184.
[6] Hong Shih, "A Systematic Study and Characterization of Advanced Corrosion Resistance Materials and Their Applications for Plasma Etching Process in Semiconductor Silicon Wafer Fabrication", Corrosion Resistance, www.intechopen.com, 2012.
[7] V. Rat, E. Bouyer, R.H. Henne et al., "DC Plasma Diagnostics for Improvement of Plasma Spraying Process under Soft Vacuum Conditions", Thermal Spray,(2003),PP.1269,1278.
[8] J. Kitamura, S. Usuba, Y. Kakudate, et al., "", J. Thermal Spray Technol. 12 (2003),PP., 70-76.
[9] LIN Chu-Cheng, ZHU Hui-Ying, MASAO Matsuyama, et al., "Characterization of Tritium Retention in Plasma Sprayed B_4C/Mo Coatings", Journal of Inorganic Materials, 28(9)(2013), PP., 1040-1044.
[10] W. Tillmann, W. Luo, U. Selvadurai, "The Influence of Residual Stress on Wear Resistance of Thermal Spray Coatings", ITSC 2015, PP., 718-724.

Microstructure and tribological performance of Cr_7C_3-$(Ni,Cr)_3(Al,Cr)$ coating deposited from Cr_7C_3 in-situ synthesized alloy powder

Zhu Hongbin, Shen Jie, Gao Feng, Yu Yueguang, Beijing General Research Institute of Mining and Metallurgy, Beijing Engineering Technology Research Center of Surface Strengthening and Repairing of Industry parts, Beijing 102216, China, zhb851020@163.com
Li Changhai, Chalmers University of Technology, Gothenburg 41296, Sweden

The work is aimed at developing a new type of Cr_7C_3-$(Ni,Cr)_3(Al,Cr)$ coatings for parts used in heavy-duty diesel engine. The feedstock, in which the stripe shaped Cr_7C_3 was in-situ synthesized, was firstly prepared by vacuum melting and gas atomization, then it was subjected to high velocity oxy-fuel(HVOF) spraying to deposited coating.The element composition, microstructures and the phase composition of the powder, as well as the sprayed coating were analyzed by chemical analysis, SEM and XRD. The hardness and sliding wear performance of the sprayed coating was also tested and compared to a commercial Cr_3C_2-NiCr coating used on piston ring.The results showed that the content of carbon in feedstock was almost the same as it was designed, and the volume content of Cr_7C_3 was increased with the more carbon and chromium added. The phases of the sprayed coating was also mainly the Cr_7C_3 and the Cr doped Ni_3Al, and the ratio of them was almost similar to it in the powder. With the increased content of Cr_7C_3 in coating, the microhardness at room temprature was firstly increased to about $1000Hv_{0.3}$, but then minor decreased. When the temprature rised to 700℃, the microhardness of the coating was almost no changed, which showed its potential application at the high temprature.The coating involved 70vol.% and 77 vol.%Cr_7C_3 was showed the improved wear resisrance, lower friction coefficient and better compatibility to contourpart than other Cr_7C_3-$(Ni,Cr)_3(Al,Cr)$ coatings and Cr_3C_2-NiCr coating.

1 Introduction

Modern heavy-duty diesel engine working under a high pressure(180-195bars) and temperature(around 600℃ in exhaust gases). In this condition, the oil lubrication is failure, and the parts in it would be suffered severe wear[1].
Thermal spraying is an effective surface protection technology to deposit wear resistance coatings on the industrial parts. In the past twenty years, the conventional ceramic-metallic composited coatings, such as WC-Co and Cr_xC_y-NiCr had been applied on the wear parts in the diesel engine[2-5]. However, the WC-Co coating is proved unsuitable for wear protection above the temperature raised to 450℃[6]. Although the wear resistance of Cr_xC_y-NiCr material is relatively worse than WC-Co, it is considered to use under the temperature up to 850℃, so some research work was focused on it and its relevant update versions[4,5].
In the Cr-C carbide system, Cr_3C_2 is the most commonly used, but it is relatively more active and tends to decarburize during long time service at high temperature. That would deteriorated the interface between it and binder phase, and finally decrease the wear resistance[7]. The Cr_7C_3 has more thermal dynamic stability than Cr_3C_2 through the gibbs free energy calculation, and it has similar mechanical property as Cr_3C_2, so it has been believed as a potential chromium carbide used in high temperature wear condition[8, 9].
Ni_3Al-based intermetallics have been used as a high-temperature structural material widely for their good mechanical property and oxidation resistance[10]. However, some laboratory studies have indicated that the Ni_3Al-based intermetallics had significant potential in wear-critical application, especially in cavitation erosion and in sliding wear at temperature range of 400℃ to 650℃[11,12]. The reasons may be related to the work-harden ability and increased yield strength with temperature raised to 750℃, which is a different characteristic with NiCr alloy[10, 11].
Moreover, the Ni_3Al has a good wettability and close thermal expansion matching with Cr_7C_3[13].Therefore, the combination of the Cr_7C_3 with Ni_3Al-based intermetallic would be developed as a novel wear-resistance materials at the elevated temperature.
Prior works by had successfully prepared the 15~30vol.% Cr_7C_3 reinforced Ni_3Al composites and cladding, and they indicated they had a great abrasive and sliding wear-resistance[14,15].
However, there were few reports on the thermal sprayed Cr_7C_3-Ni_3Al coating so far as the feedstock are hard to obtain by conventional composite powder preparation method.
As a preliminary study, this work aims at developing a new type of Cr_7C_3-$(Ni,Cr)_3(Al,Cr)$ coating used in modern diesel engine. The Cr_7C_3-$(Ni,Cr)_3(Al,Cr)$ feedstock with different composition were firstly prepared by vacuum melting and gas atomization process, during which the Cr_7C_3 could be in-situ synthesized. Then the powder was deposited by HVOF spraying. Though the microstrcuture, hardness as well as tribological analysis of the coating, the optimized Cr_7C_3-$(Ni,Cr)_3(Al,Cr)$ coating with better friction and wear resistance was selected and compared with commercially Cr_3C_2-NiCr coating.

2 Experimental

2.1 The composition design for novel Cr_7C_3-$(Ni,Cr)_3(Al,Cr)$ material

In order to develop the novel Cr_7C_3-$(Ni,Cr)_3(Al,Cr)$ material with excellent performance, the composition design is very important. There are two factors needed to be considered: one is the characteristics of the material, and another is the technic realizability of the feedstock preparation.

Firstly, as is known, the single phase Ni_3Al alloy is usually brittle with inter-granular fracture, so 0.1~0.3wt.% boron is designed to add in all the materials to aviod this intrinsic brittleness[16].

Then, to obtain the in-situ synthesized Cr_7C_3 phase, the designed ratio of Cr to C should be 7:3. However, as a portion of Cr would be solid-soluted in Ni_3Al to form the $(Ni,Cr)_3(Al,Cr)$, so it designed to add another part of Cr in the material system. As the element analysis in the prior experiment, this part of Cr is about 4~7wt.%.

Moreover, as the Al-Ni phase diagram shows in **Fig.1**, during the cooling process, if slight segregation occurred in the Ni_3Al melt with 25at.% stoichiometric Al, the NiAl phase with high solid point(1638℃) is easy to be pre-precipitated and the atomized nozzle would be blocked. For this reason, the (Al,Cr) content in the $(Ni,Cr)_3(Al,Cr)$ should be designed less than 25at.%, for restraining the pre-precipitation of NiAl.

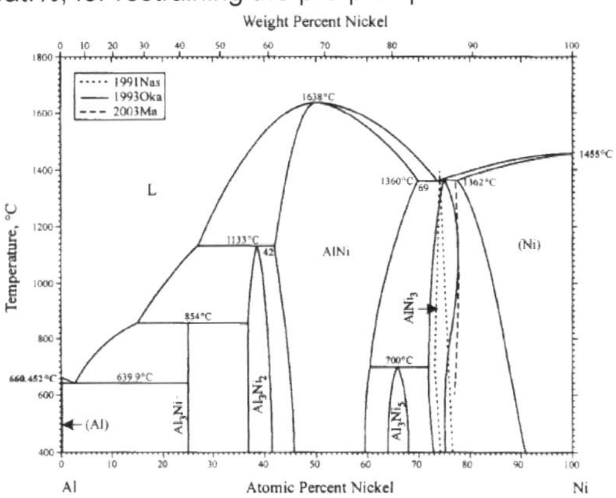

Fig. 1. Al-Ni phase diagram[17]

To optimize the microstructure and relevant property of Cr_7C_3-$(Ni,Cr)_3(Al,Cr)$ coating, five compositions with 3~7wt.% carbon were designed for this work according to the composition design principles mentioned above, which named CCNA3, CCNA4, CCNA5, CCNA6 and CCNA7.

2.1 Preparation of Cr_7C_3-$(Ni,Cr)_3(Al,Cr)$ feedstock and coating

The Cr_7C_3-$(Ni,Cr)_3(Al,Cr)$ feedstock was prepared by the vacuum melting and gas atomization. On the basis of the composition design, the raw Ni, Cr, C, Al, B material was put into the vacuum melting furnace(<30Pa) and heated to 1600-1800℃ for 30-60 min. Then the formed molten liquid was poured to a tundish, and flowed uniformly through the atomization nozzle, whose diameter was 5-7mm. At the same time, the Ar inert gas with 3-4MPa pressure was used to break up the liquid stream into droplet, and cool it to form the atomized Cr_7C_3-$(Ni,Cr)_3(Al,Cr)$ powder. Finally, all the powder was sieved to the range of 15-45μm for HVOF spraying.

The as-prepared feedstock was sprayed by GTV-K2 gun to form the Cr_7C_3-$(Ni,Cr)_3(Al,Cr)$ coating. A commercial Cr_3C_2-NiCr feedstock, prepared by BGRIMM Co. Ltd. with spray dried and sintered method was selected as a comparison. The carbon steel(0.45C%) and GH4169 were employed as the substrate respectively for room temperature and high temperature test. Prior to spraying, the substrate was grit blasted with 500-700 μm SiC particles at 4.5 bar pressure and then ultrasonic cleaned with acetone. The HVOF parameters are listed in **Table 1**.

Table 1 HVOF parameters used for the preparation of Cr_7C_3-$(Ni,Cr)_3(Al,Cr)$ and Cr_3C_2-NiCr coating

Parameters	Value
Oxygen, l/h	900
Kerosene, l/h	26
Powder feed rate, g/min	50
Spray distance, mm	380
Pre-heating,℃	300

2.2 Test and Characterization

The morphology and the cross-section of the powder as well as the coating were observed using a scanning electron microscope (SEM, ZEISS™ Gemini55, Germany), equipped with energy dispersive X-ray analysis (EDS, Oxford instruments, UK). The carbon content in powder and coating was tested by chemical analysis. The phase constituent was characterized by X-ray diffraction (XRD) employing a monochromatic Cu-Kα radiation (λ = 0.1541 nm) at 40 kV, 30 mA. The porosity and the volume fraction of the in-situ synthesized Cr_7C_3 phase in the powder and coating were estimated by imaging analysis using a commercial software package OLYCIA m3 developed by OLYMPUS. At least 10 images from different zones were taken for the measurement.

The room temperature microhardness of coating was tested using a FM-700 Vickers hardness tester with 0.3 kgf load, and the high temperature microhardness was tested on the AVK-A hardness tester (AKASHI, Japan) with 1 kgf load. Prior to testing, the specimen room was pumped to vacuum and filled Ar to positive pressure. Each coating received at least 10 indents.

The tribological performance test was performed on a SRV® IV tribo-tester(OPTIMOL, Germany). In order to approach the real wear status of the parts in the heavy-duty diesel engine, the linear reciprocation wear mode was applied, and the grey cast iron was employed as an immobile counterpart. **Fig.2** shows the schema of the testing. Before the test, the coating sample was polished to a surface finish R_a of 0.5 μm. The test load, frequency and stroke were 100 N, 50 Hz and 2 mm respectively, each coating was tested for 0.5 h. The coefficient of friction was recorded automatically. After testing, the wear mass of the coating was measured and converted to wear volume through true density testing by Archimedes' method. The morphology of the wear scar was also observed by SEM.

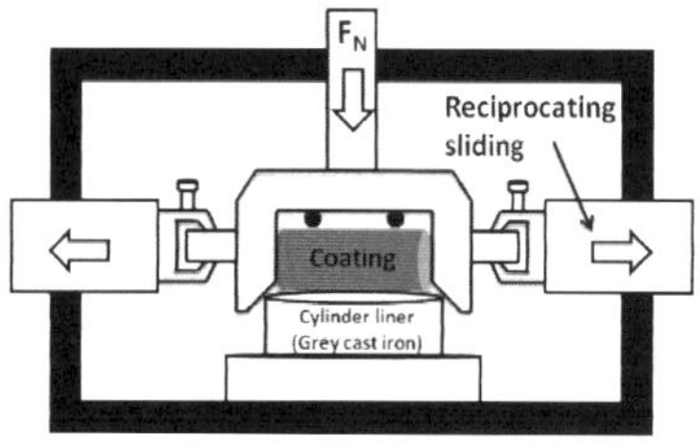

Fig. 2. Samples for linear sliding wear testing

3 Results and discussion

3.1 Characterization of the as-prepared feedstock

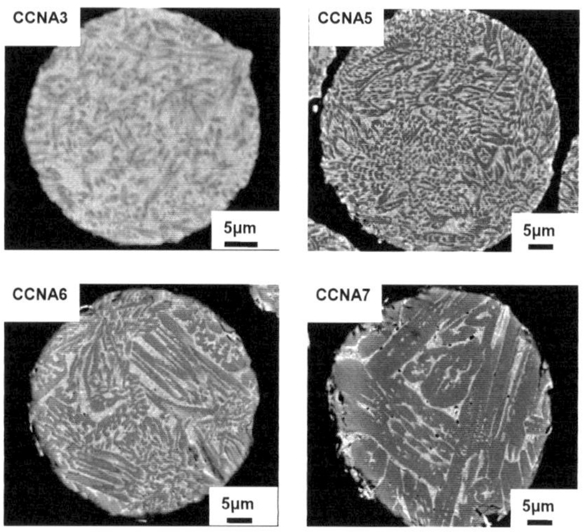

Fig. 3. Cross sectional observation of as-prepared Cr_7C_3 in-situ synthesized Cr_7C_3-(Ni,Cr)$_3$(Al,Cr) powder

Fig.3 shows the cross section morphology of representative Cr_7C_3-(Ni,Cr)$_3$(Al,Cr) powder with different carbon addition in BSE mode. All the powder has a spherical shape with smooth surface and dense structure, which is a unique characteristic of this atomized powder. Another feature could be seen was the dark and stripe shaped phase distributed uniformly in the matrix phase. Though the EDS analysis, there were mainly Cr and C detected at the dark phase, and their proportion was about 65.3at.% and 34.7at.% respectively, indicating the phase was about Cr_7C_3. Whereas at the matrix phase, the detected elements were mainly Ni, Cr and Al but few C, and the proportion of them was 72.1at.%, 11.0at.%and 16.9at.%. The Cr in matrix could be considered solid soluted in Ni and Al half each, so the matrix could be deemed as (Ni,Cr)$_3$(Al,Cr). The phase constituent of the powder could be further verified by the XRD analysis in **Fig.4**, which indicated the main phases were Cr_7C_3 and Cr-doped Ni_3Al.

In fact, other two chromium carbides, such as Cr_3C_2, $Cr_{23}C_6$ was also possible formed in the powder as their negative Gibbs free energy. However, the Cr_3C_2 (-268. 4 kJ /mol) has a much higher Gibbs free energy than Cr_7C_3 (-605. 3 kJ /mol), and the most stable $Cr_{23}C_6$ should be formed by aging treatment at lower temprature[7,9]. Thus, during the rapid solidification of the droplet, the Cr_7C_3 phase was firstly precipitated by the in-situ reaction of Cr and C and few of Cr_3C_2 and $Cr_{23}C_6$ was detected in the powder.

The in-situ synthesized Cr_7C_3 is wrapped in the matrix and has a strongly bonding interface with it, such characteristics would improve the deposition of carbide and decrease the oxidation potential of it, when comparing the conventional agglomerated Cr_3C_2-NiCr or WC-Co powder with porous structure.

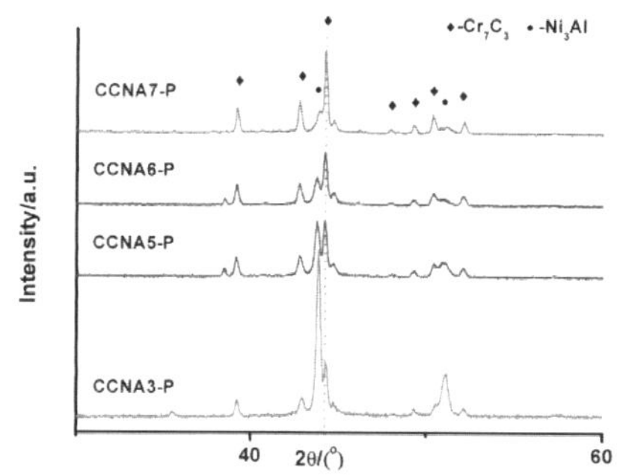

Fig.4. XRD pattern of the Cr_7C_3-(Ni,Cr)$_3$(Al,Cr) powders with varied carbon content

From the **Fig.3**, it can be seen the high carbon content and the presence of sufficient chromium resulted in a high volume percentage of in-situ synthesized Cr_7C_3 phase. The XRD pattern also shows the Cr_7C_3 peak was evidently higher in the powder prepared with high carbon content. The statistical value of Cr_7C_3 in different powder is listed in the **Table 3**, which is closed to the theoretical value. That indicated almost all the carbon was reacted with chromium, and few of it was dissolve into the matrix as its poor solubility with Ni and Al[18].

Table 3 Microstructure characteristics of the prepared Cr_7C_3-(Ni,Cr)$_3$(Al,Cr) feedstock and sprayed coating

	Carbon content (wt.%)		Cr_7C_3 content (vol.%)		Porosity (vol.%)
	powder	coating	powder	coating	coating
CCNA3	3.04	2.96	33.31	40.37	1.31
CCNA4	3.92	3.82	48.48	52.71	1.77
CCNA5	4.95	4.79	63.11	70.31	1.75
CCNA6	5.93	5.74	76.41	77.08	1.86
CCNA7	6.89	6.73	84.36	83.67	4.23

Moreover, with the increased content of Cr_7C_3, its structure is also graudually coarser at the same time. The reason may be related to the precipitated primary

Cr_7C_3 and its growth from the melt with more content of Cr and C[19].
As the addition of Cr and C in the Ni-Al-Cr-C system was increasing, the hypereutectic composition of the melt would be formed, when the temprature decreased, the Cr_7C_3 would be precipitated from the hypereutectic melt as the form of primary Cr_7C_3. The primary Cr_7C_3 would be further grown with consuming the Cr and C from the melt until the composition of the melt reach to eutectic area. Then the $(Ni,Cr)_3(Al,Cr)+Cr_7C_3$ eutectic would be formed.
The primary Cr_7C_3 is much coarser than the Cr_7C_3 in eutectic, and it grows along the [1000] direction as its anisotropic structure. Its size was dependent on the residual Cr and C could be consumed in the melt when the cooling rate was about consistently.
Though the analysis of the C-Cr-Ni phase diagram, the primary Cr_7C_3 would be precipitated when the composition of Ni lower than 40at.%[20]. The total content of Ni and Al in CCNA4 and CCNA5 powder was designed as 45.6at.% and 35.1at.%, so it could be considered that the primary Cr_7C_3 was formed in the CCNA5,6,7 powder.
However, as **Fig.3** showed, the growth of primary Cr_7C_3 was not evidently in the CCNA5 powder, this may be due to the high cooling rate of atomization processing and less Cr and C in CCNA5 powder than them in CCNA6 or CCNA7.

3.2 Microstructure and hardness of the sprayed coating

The HVOF sprayed Cr_7C_3-$(Ni,Cr)_3(Al,Cr)$ coating from different powder is displayed in **Fig.5**. As the low temperature and high velocity of the HVOF spraying process, most powder was heated to semi-melton and impacted onto the substrate with plastic deformation, so almost all the coatings presented typical lamellar and dense structure. The estimated average porosity of the sprayed coating is listed in **Table 3**. However, the porosity of CCNA7 coating is much higher than other coating, reached to 4.23%. The reason could be found in the magnified microstructure of it, in which the binder phase is too scarce to fill the pores and hold ceramic phase during deposition.
Beside the CCNA7, the statistical porosity in other coating is mainly from the interlaminar cracks and pores, which may be formed by insufficient deformation of some deposited particles. Comparing to the Co(Cr), Ni(Cr), *etc.* metallic or alloy materials, the Ni_3Al intermerallic had been proved exhibiting the relatively higher yield stress, and which was anomalously increasing with the temprature up to about 650℃~750℃[10, 21]. Therefore, for the Cr_7C_3-$(Ni,Cr)_3(Al,Cr)$ powder, a higher temprature is needed to soften the $(Ni,Cr)_3(Al,Cr)$ binder phase. For this, the optimization of HVOF spraying will be conducted in future.

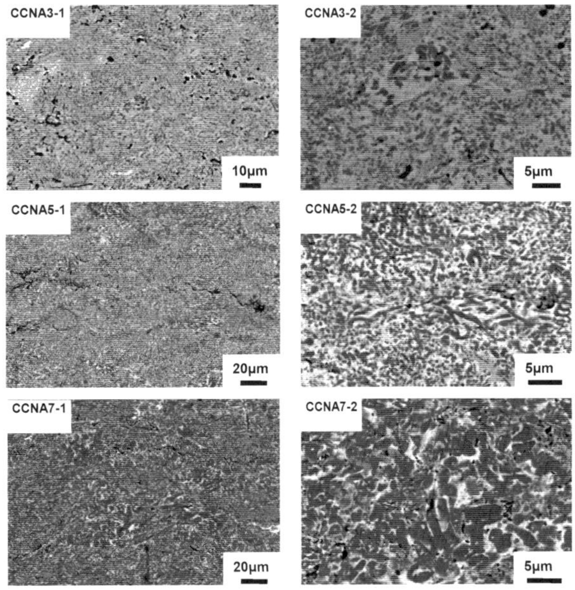

Fig.5. Cross sectional observation of the HVOF sprayed Cr_7C_3-$(Ni,Cr)_3(Al,Cr)$ coatings

Besides some pores, the microstructure of powder was well inherited to the coatings. Based on the microstructure and XRD analysis in **Fig.6**, the Cr_7C_3 and $(Ni,Cr)_3(Al,Cr)$ phase was find well retained in the coatings and few other phases was detected. That could be ascribed to the high stability of the Cr_7C_3 and its enwrapped structure in the dense powder.

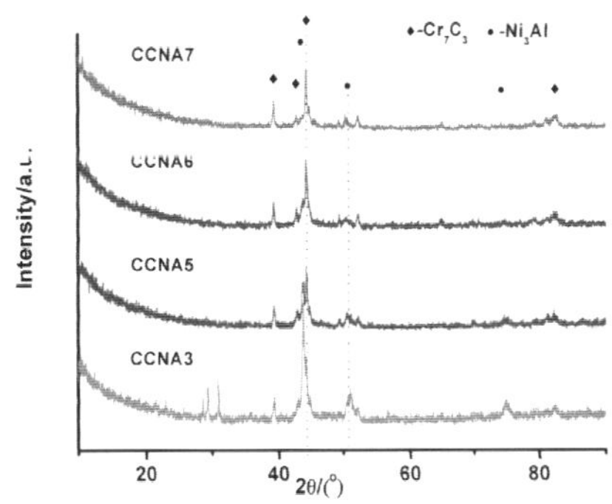

Fig.6. XRD pattern of the sprayed Cr_7C_3-$(Ni,Cr)_3(Al,Cr)$ coating deposited from powders with varied carbon content

However, the carbon content in the coating is reduced when comparing with it in powder (Table 3), and its decrement tends to increase in the coating with higher carbon content. That indicated the carbon was still lost during the HVOF spraying process, but just very slightly. As mentioned before, owing to the dense structure of the powder, the loss of carbon by the direct oxidation of Cr_7C_3 should be very few, so the main mechanism for it may be considered to the indirect oxidation of carbon in Cr_7C_3. When the powder was heating in the high temprature HVOF

plume, the C, Cr atom from Cr_7C_3 was easy to diffuse to the $(Ni,Cr)_3(Al,Cr)$ matrix at the interface. Then as the poor solubility of carbon in Ni_3Al, the C atom would diffuse to the surface of powder and reacted with oxygen to form the CO or CO_2. This indirect oxidation mechanism of carbide was also found in the thermal spraying Cr_3C_2-NiCr and WC-Co. Through the EDS analysis, the detected proportion of Cr in the matrix of the coating was about 18.3at.%, evidently higher than it in the powder, which could further verify the diffusion of Cr from the Cr_7C_3 to the matrix.
It also must be pointed out the statistical Cr_7C_3 volume fraction in the coating might be deceptive. The error may be from the different cross-section morphology between raw spherical powders and corresponding deformed lamellar. As the ceramic phase could not be deformed, this error is much smaller between the powder and coating with more Cr_7C_3.
In general, altough the highest loss of carbon in the CCNA7 coating was up to 0.16wt.%, it was still very slight when comparing with the convention Cr_3C_2-NiCr coating(0.91wt.%), because the diffusion of C andCr was reduced as the higher thermal stability of Cr_7C_3 and the direct oxidation was restrained in the dense powder.

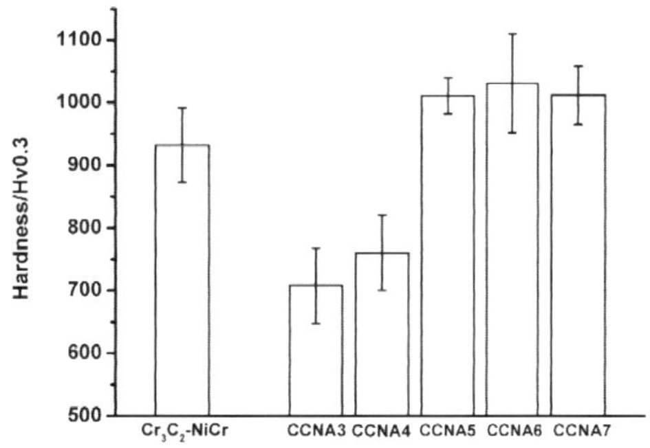

Fig.7. Micro-hardness of the sprayed coatings

Fig.7 illustrated the micro-hardness of the Cr_7C_3-$(Ni,Cr)_3(Al,Cr)$ coatings and commercial Cr_3C_2-NiCr coating at the room temprature. The microhardness was obviously improved with the increased content of C or Cr_7C_3 in the coating, but a minor decreased in the CCNA7 coating. The reason is the much higher hardness of Cr_7C_3 than the $(Ni,Cr)_3(Al,Cr)$ matrix. However, when the content of Cr_7C_3 was too much to be wholly held by the scarce matrix phase, the bonding between them would be weak and the hardness was reduced, like the hardness of CCNA7 coating showed.
With the suitable proportion of Cr_7C_3 and $(Ni,Cr)_3(Al,Cr)$, the highest hardness is obtained from the CCNA6 coating, which is about $1031Hv_{0.3}$. Comparing with the Cr_3C_2-NiCr coating, the CCNA5, CCNA6 and CCNA7 coating all presents higher hardness. Considering the similar hardness of Cr_3C_2 and Cr_7C_3, the improved hardness may be from the $(Ni,Cr)_3(Al,Cr)$ matrix and its strong bonding with the situ synthesized Cr_7C_3 phase.

The Ni_3Al intermetallic often exhibited higher yield strength than the NiCr alloy as its long-range order crystalline structure with strong oriented atomic bonding[10]. The addition of B and Cr in it could further improve its ductility. Thus the $(Ni,Cr)_3(Al,Cr)$ matrix may have a better ability to help the coating resist the indentation.
Moreover, the yield strength of the $(Ni,Cr)_3(Al,Cr)$ could be further increased with the elevated temperature below 750℃. This anomalous yield behaviour also affected the high temperature property of the Cr_7C_3-$(Ni,Cr)_3(Al,Cr)$ coatings. As the **Fig.8** shows, in the argon atmosphere, the hardness of the Cr_7C_3-$(Ni,Cr)_3(Al,Cr)$ coatings was little changed as the temperature raised to 700℃. However, the hardness of Cr_3C_2-NiCr coating was decreasd clearly. This anomalous yield behaviour is owing to the cross-slipping pinning at higher temperature. If the temperature further increased to more than 750℃, other slip system would be start-up and the strength of the $(Ni,Cr)_3(Al,Cr)$ would be decreased quickly. Therefore, comparing with the conventional WC-Co and Cr_3C_2-NiCr coating, the Cr_7C_3-$(Ni,Cr)_3(Al,Cr)$ coating may exhibit better wear or erosion property at the temperature of 450℃ to 750℃.

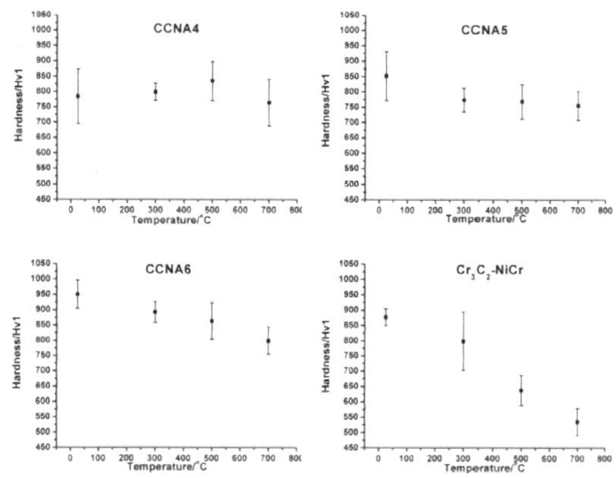

Fig.8. Comparison of high temperature hardness of Cr_7C_3-$(Ni,Cr)_3(Al,Cr)$ and Cr_3C_2-NiCr coating

3.3 Tribological performance of the prepared Cr_7C_3-$(Ni,Cr)_3(Al,Cr)$coating

Typical dynamic COF curve and corrsponding wear volumn of the Cr_7C_3-$(Ni,Cr)_3(Al,Cr)$ coatings and referred Cr_3C_2-NiCr coating at room temprature is displayed in **Fig.9**. and **Fig.10** respectively. From the carefully observation, it cound be found the steady state COF and wear volume were all initially decreasing but then gradually raising with the increased Cr_7C_3 in the coating.The CCNA5 and CCNA6 coating with moderate composition present relatively lower mean COF and wear volume, as well as smoother COF curve than the other Cr_7C_3-$(Ni,Cr)_3(Al,Cr)$ coatings and Cr_3C_2-NiCr coating.

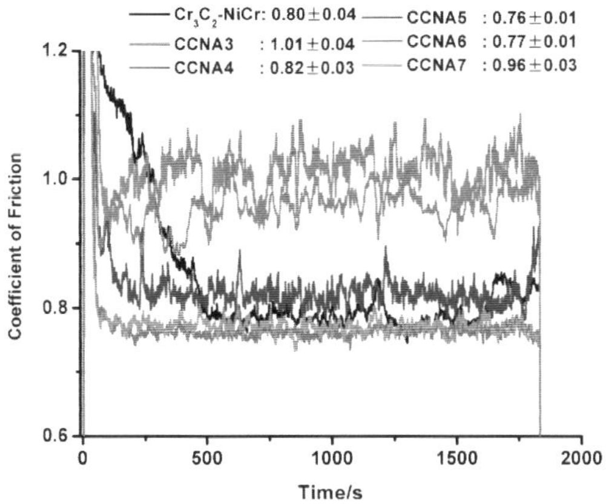

Fig.9. Friction coefficient of the prepared Cr_7C_3-$(Ni,Cr)_3(Al,Cr)$ coatings and the commercial Cr_3C_2-NiCr coating

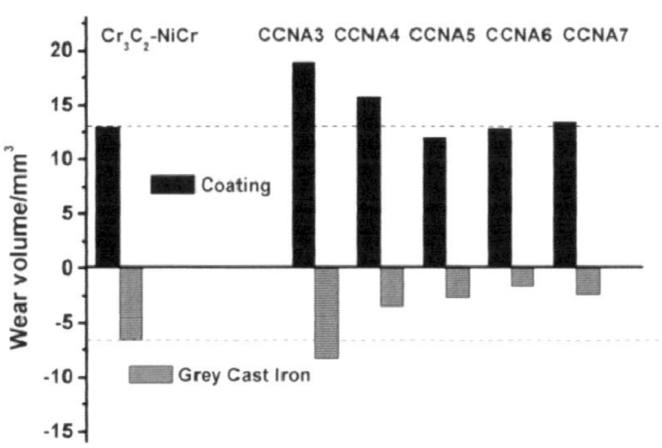

Fig.10. Wear volume of the prepared Cr_7C_3-$(Ni,Cr)_3(Al,Cr)$ coatings and the commercial Cr_3C_2-NiCr coating

This result is also well matched with the variation of hardness in the **Fig.7**, indicating the tribological performance of the coating was largely dependent on its microstructure and mechanical property.
This relationship can be related to distinct wear behaviour for different coatings. The representative worn surfaces of the coatings are showed in **Fig.11**. It could be found all the coatings presented many grooves except of CCNA7. However, the grooves on the CCNA5 and CCNA6 coatings are much milder than it on the Cr_3C_2-NiCr coating and CCNA3 coating. Moreover, on the CCNA3, the adhesion layer was also seen besides the grooves. Although the grooves were inconspicuous on the CCNA7, there were micro-cracks and large pits. This different morphology of the worn surface indicates the distinct wear behaviour for coatings.
According most research works, the main wear behavior of the ceramic-metallic material or coating during the sliding wear could be briefly summarized as adhesive wear, ploughing and abrasive wear[22-24].
The adhesive wear usually occures between two surfaces with higher and similar chemical reactivity, such as surface of metal or alloy[25]. Hence, the increasing of metallic phase in the coating would improve the adhesive wear, such as the CCNA3. The specific morphology of the adhesive is the adhesion layers and spalling left by the separation of the adhered region, like the **Fig.11(b1)** shows, which will lead to a severe wear of the counterparts and suddenly increased COF.
Besides the adhesion, the softer metallic phase is also prone to plastic deformation and extrudation from the surface under the Hertzian loading, which could be called ploughing. When the metallic phase is removal, the exposed ceramic phase would be fractured and pulled out with accumulated strain, acting as abrasive particles, thus the abrasive wear will start and the ploughing will be more serious. These ploughing-abrasive looped wear behaviour is mostly influenced by the phase compositons, hardness, strenghth of the materials and even the bonding between ceramic and binder phase. It will evidently increase the COF and wear rate of the coatings.
The $(Ni,Cr)_3(Al,Cr)$ intermetallic phase is believed had a better resistance to plastic deformation and ploughing than the NiCr alloy as its high yield strength and strong strain-harden ability. At the same time, it also had a strong bonding with the in-situ synthesized Cr_7C_3, so the Cr_7C_3-$(Ni,Cr)_3(Al,Cr)$ with proper composition, such as CCNA5 and CCNA6, would present more milder ploughing and abrasive wear, which eventually caused lower, smoother COF and lower wear volume of the counterparts.
Although the CCNA7 have a best resistance to the adhesive wear as its highest content of ceramic phase, the limited bonding with the scarce metallic matrix cause it prone to pull out after brittle facture under the cyclic stress. Hence, more large pits and cracks are seen on its worn surface(**Fig.11(c2)**) and the wear volume of the counterparts, as well as the COF is much higher than the CCNA5 and CCNA6 coating.
In summary, the main wear behaviour of the CCNA5, CCNA6 coating and the Cr_3C_2-NiCr coating is ploughing and abrasive wear. As the better resistance to plastic deformation and ploughing of the $(Ni,Cr)_3(Al,Cr)$ phase and its stronger bonding with Cr_7C_3, the CCNA5 and CCNA6 coating present better tribological performance than the Cr_3C_2-NiCr coating. The tribological performance of the CCNA3 coating is the worst as it was undergoing not only adhesive wear but also ploughing and abrasive wear.
With respected to the CCNA7 coating, the wear behaviour is specific, which is more related to the brittle facture of the ceramic phase as it had scarce metallic phase. Thus its wear volume is at relatively low level, but the COF is higher.

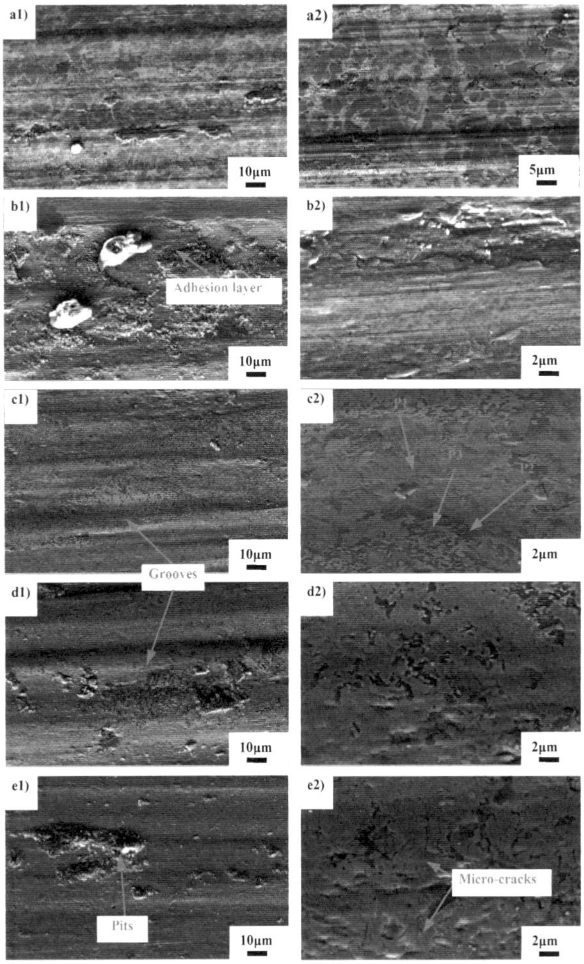

Fig.11 Morphology of the worn surface of Cr_3C_2-NiCr coating (a1,a2) and varies Cr_7C_3-$(Ni,Cr)_3(Al,Cr)$ coatings (b1,b2-CCNA3; c1,c2-CCNA5; d1,d2-CCNA6; e1,e2-CCNA7)

4 Conclusions

From the research work, several points can be concluded:

1. Spherical Cr_7C_3-$(Ni,Cr)_3(Al,Cr)$ powder with different composition was designed and successfully prepared by melting and atomization method. The Cr_7C_3, presenting stripe shape as its anisotropic crystal structure, is in-situ synthesized during the solidification process of the powder, and uniformly distributed in the Cr-doped Ni_3Al phase. With more carbon added in the material system, the content of synthesized Cr_7C_3 is correspondingly increased and its structure is coarser.
2. As the dense structure of powder and high thermal dynamic stability of Cr_7C_3, the microstructure of the feedstock was almost inherited to the HVOF sprayed Cr_7C_3-$(Ni,Cr)_3(Al,Cr)$ coating with few carbon loss. the main phases in the coating are also the Cr_7C_3 and Cr-doped Ni_3Al. With more content of Cr_7C_3 in coating, its microhardness at room temperature was also increased. However, when the Cr_7C_3 content was more than 80vol.%, the supporting effect of binder phase would be weakened, thus the microhardness was decreased. The highest hardness of Cr_7C_3-$(Ni,Cr)_3(Al,Cr)$ coating was about $1000Hv_{0.3}$, and from the room temperature to 700℃, it was almost unchanged, that could be ascribed to good mechanical properties of Cr-doped Ni_3Al and anomalous dependence of strength on temperature.
3. The friction coefficient, the wear volume of the Cr_7C_3-$(Ni,Cr)_3(Al,Cr)$ coating and counterparts were also firstly decreased but then increased with the content of Cr_7C_3. As the finer structure and appropriate content of Cr_7C_3 phase in the coating deposited from CCNA5 and NACC6 powder, they showed better Comprehensive tribological performance than other Cr_7C_3-$(Ni,Cr)_3(Al,Cr)$ coatings and Cr_3C_2-NiCr coating.
4. The main wear mechanism between the coating and counterpart are abrasive wear by extruded hard phase and the adhesive wear. The good high temperature property and work-hardening effect of $(Ni,Cr)_3(Al,Cr)$, the stripe shaped morphology of Cr_7C_3, and the strong bonding between them, were the main reasons for better wear resistance than Cr_3C_2-NiCr coating. The lower chemical affinity of $(Ni,Cr)_3(Al,Cr)$ than NiCr was also beneficial for the mild adhesive wear between the Cr_7C_3-$(Ni,Cr)_3(Al,Cr)$ coating and its counterpart.

5 Acknowledgements

The authors are grateful for the financial supported by National Natural Science Foundation of China (51541107) and International Science &Technology Cooperation Program of China(2015DFA50970).

6 Literature

[1] Herbst. D. C: Thermal Spray Solutions For Diesel Engine Piston Rings. International Thermal Spray Conference(2003) , pp.129/38.

[2] Fukutome. H, Shimizu.H, Yamashita. N et al.: The application of cermet coating on piston ring by HVOF. International Thermal Spray Conference [1995]. pp.21/26.

[3] Herbst. D. C: Thermally applied coating of mechanically alloyed powders for piston rings, Patent US6,887,585, May 3, 2005

[4] Picas. J. A, A. Forna, G.Matthaus: HVOF coatings as an alternative to hard chrome for pistons and valves. Surface& Coatings Technology(2006), Issue 261(5-6), pp. 477/484.

[5] Rastegar. F, Richardson. D.E: Alternative to chrome: HVOF cermet coatings for high horse power

diesel engines. Surface& Coating Technology(1997), Issue 90(1), pp. 156/63.
[6] Berger. L. M: Hardmetals as thermal spray coatings. Powder Metallurgy(2007), Issue 50(3): pp. 205/14.
[7] Matthews.S: Development of high carbide dissolution/low carbon loss Cr_3C_2-NiCr coatings by shrouded plasma spraying. Surface & Coatings Technology(2014), Issue 258: pp. 886/900
[8] Mayr, W., et al.: Phase Equilibria and Mulitiphase Reaction Diffusion in the Cr-C and Cr-N Systems. Journal of Phase Equilibria(1999), Issue 20(1): pp. 35/44.
[9] Small M, Ryba E: Calculation and evaluation of the gibbs energies of formation of Cr3C2 , Cr7C3 , and Cr23C6. Metallurgical Transactions A(1981), Issue 12(8): pp.1389/96.
[10] Jozwik. P, Polkowski. W, Bojar. Z: Applications of Ni3Al based Intermetallic alloys-current stage and potential perceptivities, Materials(2015), Issue 8(5): pp. 2537/68.
[11] Blau. P. J, Devore. C. E:Sliding behavior of alumina/nickel and alumina/nickel aluminide couples at room and elevated temperature. Journal of Tribology(1988), Issue 110(4): pp. 646/52.
[12] Johnson. M, Mikkola. D. E, March P. A., et al.: The resistance of nickel and iron aluminides to cavitation erosion and abrasive wear. Wear(1990), Issue 140(2): pp. 279/89.
[13] Shang. P. L, Feng. D, Luo. H. L, et al.: Development of new wear-resistant surface coating at elevated temperature. Journal of Iron & Steel Research International(2006), Issue 13(5): pp. 37/40.
[14] Gong. K, Luo. H, Feng. D, et al.: Wear of Ni3Al-based materials and its chromium-carbide reinforced composites. Wear(2008), Issue 265(11): pp. 1751/55.
[15] La. P, Xue. Q, Liu. W: Effects of boron doping on tribological properties of Ni3Al-Cr7C3 coatings under dry sliding. Wear(2001), Issue 249(1-2): pp. 93/99.

Fracture toughness of hard coatings applied by Flame Spray process using flexicord feedstock

C. R. C. Lima, M. S. Orozco; H. C. Fals; M. A. R. Mojena, V. A. Ferraresi, São Paulo/ BR

The hardness of a coating determines the depth of penetration of hard particles that are free to roll and slide against its surface. Therefore, an increase in hardness reduces abrasive wear. Additionally, the fracture toughness impacts on critical load needed to achieve the crack propagation along the coating. Then, it is crucial to present high hardness and high fracture toughness for the coating to increase wear resistance. Vickers indentation technique was employed to assess fracture toughness in hard coatings applied by flame spray using cored-wire feedstock. A comparative analysis of the values of fracture toughness obtained by experimental and theoretical methods was performed allowing to validate the experimental results and to understand the relationship of this property to the hardness of the coatings for each of the tested materials.

1. Introduction

Thermal spray processes are increasingly used in several industries, such as the automotive, aerospace, and petrochemical sectors, among others [1-3]. Wear and corrosion resistance are some of the main applications for thermal spray [4]. Thermal spray technology has been basically divided into three groups that correspond to the energy source used: those using expanding compressed gas commonly known as cold gas spray (CGS), as well as electric and combustion processes. Powder or wire flame spray (FS) is one of the most traditional and widely used combustion thermal spray processes. The recent flame spray alternative technique employing cords (Flexicord or FS/FC) allows obtaining metallic, ceramic and carbide high quality coatings at a relatively low expense [5]. Very few studies can be found related to characterization or performance of coatings obtained by FS process using cords as feedstock.

Referring to wear application, it is generally accepted that the hardness and fracture toughness of the coatings are the most important mechanical properties determining the wear resistance [2, 6, 7]. The hardness of the composites determines the depth of penetration of the hard particles free to roll and slide between two surfaces. Then, an increase in hardness reduces abrasive wear. On the other side, fracture toughness impact on critical load necessary for crack propagation along the coating. Wear resistance increases with higher hardness and fracture toughness [8].

Vickers indentation can be used for determining fracture toughness (K_c) with advantages such as simple sample preparation, the standardization of Vickers indenters and the small number of samples required. Some methods commonly applied for determining the fracture toughness by indentation are considered not truly reliable since these methods employ two common theoretical crack models, Palmvisqt and Half Penny, thought the nature of the cracks is really more complex [9]. One model that is available uses the total length of surface cracks [9], rather than the radial crack length measured in the cross section, commonly used by the earlier theoretical methods [10].

This work performs Vickers indentation to determine fracture toughness of coatings obtained by flame spray of flexicord feedstock (FS/FC). The results are compared with the results obtained using theoretical models the literature. The relationship of K_c with the hardness of the coatings is discussed.

2 Experimental Procedure

Three different materials were applied by the FS/FC technique on a low carbon steel substrate (ASTM 1020) with dimensions of 180 mm x 70 mm x 3 mm. Just before spraying, the substrate surface was degreased and grit blasted with aluminum oxide. The equipment used was a Master Jet Gun 2 (Saint Gobain, FR). The main parameters used in both processes are shown in Table 1.

Table 1 Main thermal spray parameters used for the application of the powder and cord materials

Material	Air Pressure (Bar)	Standoff Distance (mm)	Oxygen	Acetylene	Feed rate (cm/min)
HardKarb 1350	4,5	100-150	50	60	45
KhromKarb Fine 75-25	4,5	100-150	60	70	90
HardKarb 12CoF	4,5	100-150	50	60	46

The chemical compositions of the feedstock materials are shown in Table 2. For a better understanding of this work, feedstock materials will be further referenced in the text, as designated in Table 2.

Table 2 Chemical composition of feedstock materials

Material	Designation	Chemical Composition (%Wt)							
		C_{tot}	W	Co	Cr	Ni	Si	Al	Other
HardKarb 1350 (10Co 4Cr)	1350 FSFC	5,5	Bal.	10,0	4,10				
KhromKarb Fine (75CrCo/25NiCr)	7525 FSFC	12,9			Bal.	3,1	1,13		0,98
HardKarb 12CoF	8812 FSFC	5,5	80,0	12,0					2,50

Figure 1 illustrates the scanning electron microscopy images (SEM) of flexicords for FS. In general, the materials show a spherical morphology. The powder particles are bonded by a polymer, as can be observed in the cross section of the cord.

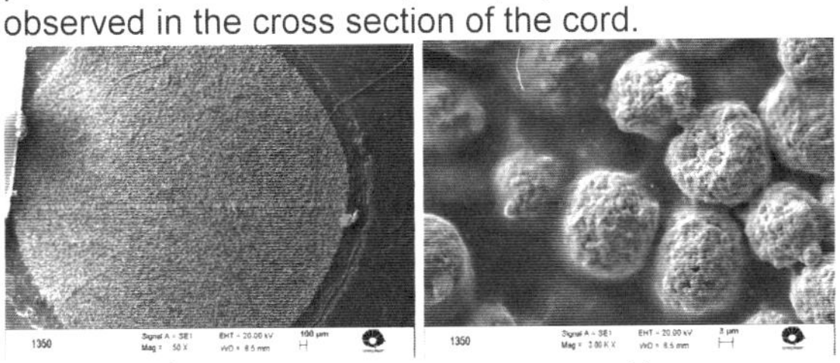

a) b)

Fig. 1 Representative SEM images of the Flexicord feedstock HardKarb 1350 for FS: a) cross section of the cord; b) Detail of the powder morphology and agglomeration

The coated samples were cut, embedded into Bakelite, polished and cleaned by ultrasonic immersion in an acetone bath. Scanning electron microscopy (SEM) was used for the analysis of the surface and cross section of the coatings. Microhardness measurements were performed with a Shimadzu HMV-G microdurometer, having a Vickers indenter, load of 300 g and a residence time of 15 s. Indentation was carried out in the middle plane of the cross section of the coating in order to minimize the edge and substrate interface effects. Ten microhardness measurements were averaged discarding the lowest and highest values.

For fracture toughness determination, experimental measurement and theoretical calculation were applied and the respective results were compared. Vickers indentation method was used for experimentally determining the fracture toughness (K_{Ic}) applying a procedure similar to measure microhardness. The load applied was 1kg, residence time was 20s. The average value of ten measured K_{Ic} values is reported, discarding the lowest and the highest values.

Some selected equations found in the literature [10, 11, 12, 13, 14] were used for coatings' fracture toughness calculation. These equations correspond to models that assume the half penny crack type since the equations assuming Palmvisqt cracks have presented no consistent results. Kc is the fracture toughness by indentation expressed in $MPa.m^{1/2}$; P is the load in N; a is the length of the mean diagonal of the Vickers indentation, in meters; c is the length of the radial crack measured from the center of the indentation, in m; E is the Young's modulus; and H_v is the Vickers hardness, both expressed in GPa.

3 Results and Discussion

In Figure 3, representative micrographs of the cross sections and surfaces of the 7525 coatings deposited by HVOF and 8812 deposited by FS flexicords are shown. It is evident that the coatings obtained have a laminar structure resulting from the accumulation of molten and semi-molten particles on the substrate (Fig. 3.a, b). It is further observed that the coatings have a dense structure, especially those obtained with HVOF, and good adhesion to the substrate. The average thickness of the coatings was 440 µm.

Figure 3 shows SEM images of the surface and cross section of the as sprayed coatings. It can be seen the irregular nature of the surface which justify the necessity of performing several readings to reduce measurement error introduced by these irregularities in the evaluation of hardness and fracture toughness of the coatings.

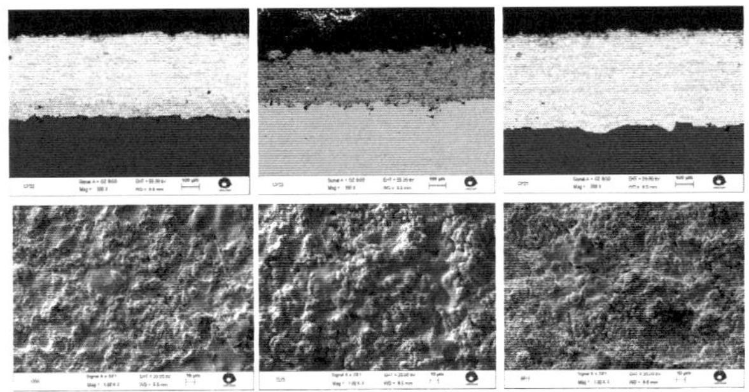

a) b) c)

Fig. 3 SEM images of the coatings in cross section and surface: a) 1350; b) 7525; and c) 8812 FSFC.

Microhardness (H_v300) and fracture toughness results are presented in Table 4. It should be noted that the H_V values obtained for FS/FC coatings averaged 626.44 ± 66.61, with a confidence level of 90%.

Table 4. Mean values of microhardness (H_v) and fracture toughness (K_c).

Materials	H_v300	K_c Experiment	K_c Eq. 1	K_c Eq. 2	K_c Eq. 3	K_c Eq. 4	K_c Eq. 5	K_c Eq. 6	K_c Eq. 7
1350 FSFC	553,02	2,388	1,227	1,160	1,789	1,306	1,501	1,472	1,333
7525 FSFC	683,01	3,321	2,763	2,221	2,393	2,480	2,850	2,892	2,756
8812 FSFC	643,27	3,594	1,872	1,681	2,318	2,121	2,438	2,361	2,098

The largest value of H_V was obtained for 7525 (base NiCr), yielding the lowest value for 1350 (base WC). This is based on the formation of undesirable phases in the WC-based coatings due to the decarburization phenomenon during FS/FC process, which leads to a deterioration of the mechanical properties.

The fracture toughness (Kc) was determined experimentally with the use of microhardness tester. First, it was necessary to estimate the Young's modulus (E) for the materials. Thus, for determining the Kc of 1350 coatings, a value of E = 282 GPa [15] was used; for 7525 coatings, the value of E was 214 GPa [16]; and for 8812 coatings, E = 300 GPa [17, 18]. Also in Table 4, it is shown the Kc values determined both experimentally and theoretically by using seven distinct equations from the literature. The comparative analysis first allowed the confirmation that the results obtained experimentally from the assumed E values, with respect to the values calculated with the equations, are within the confidence interval of 35% and considered acceptable and accurate.

Values of fracture toughness, both measured and calculated, vary between Kc = 1.16 ÷ 3.39 MPa.m$^{1/2}$ for 1350, 2.22 ÷ 3.32 MPa.m$^{1/2}$ for 7525, and 1.68 ÷ 3.59 MPa.m$^{1/2}$ for the 8812. For all cases, the values were within the confidence interval of 35% that is considered acceptable.

Figure 4 shows a trend graph of the Kc values that were experimentally obtained and calculated by using the equations of the literature. This graph shows that similar values of K_c were obtained for each of the different coating materials.

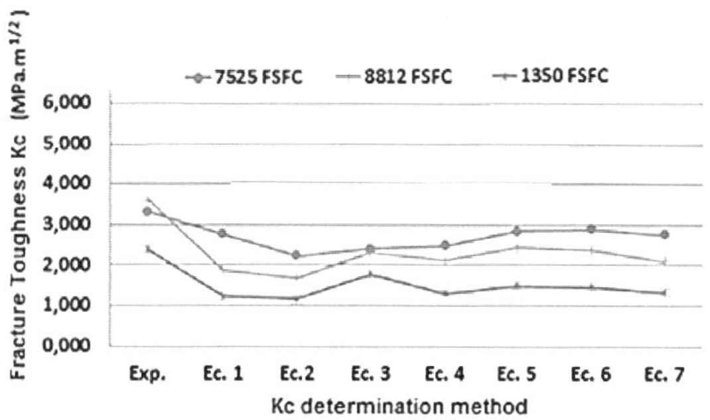

Fig. 6 Comparison of the K_c values measured and calculated.

The highest values of fracture toughness were obtained for 7525; that is, for NiCr-based material, in the same manner that was applied with respect to the microhardness values. Then, it can be stated that the materials used in the coatings under investigation that have indicated higher hardness values have exhibited superior toughness fracture as well. The lowest values of microhardness and fracture toughness were for the 1350 coating with showed the worst mechanical properties. Considering the high temperatures reached during FS/FC process and the long time that the materials were exposed to such temperatures, it is believed that this WC-based material has undergone a decarburization or dissolution of carbides, as has been previously reported by other authors [2, 19], leading to the formation of new phases and deteriorating the mechanical properties. The flame spray technique using flexicords (FS/FC) can be used satisfactorily as an alternative method to obtain coatings with good mechanical properties, especially in situations where the use of other thermal spray methods are restrict due to inaccessibility or high cost.

4 Conclusion

Fracture toughness was evaluated for coatings obtained with the use of flame spray (FS) applying flexicord feedstock. It can be concluded that:

- The coatings presented hardness values in the range of 626.44 ± 66.61 H_V and values of fracture toughness between 1.16 ÷ 3.59 MPa.m$^{1/2}$;
- In the coating obtained with the HardKarb 1350 material (WC-10Co 4Cr) using the FS/FC method, the occurrence of the phenomenon of decarburization of tungsten carbide particles was observed, then leading to the lower mechanical properties presented by this coating.
- Fracture toughness values reported in this work were determined both experimentally and theoretically with less than 35% of error, what is considered satisfactory.

Acknowledgements

This work has been supported by CAPES, Brazil, Project MES-CUBA no. 131/11.

5 Literature

[1] Longo FN (1992) Industrial Guide-Markets, Materials, and Applications for Thermal-Sprayed Coatings, J Therm Spray Techn 1(2):143-145.

[2] Xie, M, Zhang, S, Li, M (2013) Comparative investigation on HVOF sprayed carbide-based coatings, Appl Surf Sci 273:799– 805.

[3] Murugan, k, Ragupathy, A, Balasubramanian, V, Sridhar, K (2014) Optimizing HVOF spray process parameters to attain minimum porosity and maximum hardness in WC–10Co–4Cr coatings, Surf Coat Tech 247:90–102.

[4] Hermanek, FJ Thermal Spray Terminology and Company Origins (pub.) ASM International, Materials Park, Ohio, USA, 2001.

[5] Fauchais, P, Heberlein, JVR, Boulos, M (204) Thermal Spray Fundamentals. From powder to part. ISBN: 978-0-387-28319-7. 1566 p.

[6] Murthy, JKN and Venkataraman, B (2006) Abrasive wear behaviour of WC–CoCr and Cr3C2–20(NiCr) deposited by HVOF and detonation spray processes, Surf Coat Tech 200(8):2642– 2652.

[7] Ma, N, Guo, L, Cheng, Z, Wu, H, Ye, F, Zhang, K (2014) Improvement on mechanical properties and wear resistance of HVOF sprayed WC-12Co coatings by optimizing feedstock structure, Appl Surf Sci 320:364–371.

[8] Mateen, A., Saha, G.C., Khan, T.I., Khalid, F.A. (2011) Tribological behaviour of HVOF sprayed near-nanostructured and microstructured WC-17wt.%Co coatings, Surface & Coatings Technology, v. 206(6), 1077–1084.

[9] Faisal, NH, Ahmed, R, Prathuru, AK, Spence, S, Hossain, M, Steel, JA (2014) An improved Vickers indentation fracture toughness model to assess the quality of thermally sprayed coatings, Eng Fract Mech 128(C):189–204

[10] Ponton, C.B., Rawlings, R.D. (1989) Vickers indentation fracture toughness test Part 1 Review of literature and formulation of standardized

indentation toughness equations, Materials Science and Technology, v. 5(9), 865-872.

[11] Houdková, T and Kašparová, MŠ (2013) Experimental study of indentation fracture toughness in HVOF sprayed hardmetal coatings, Eng Fract Mech 110:468–476.

[12] Evans, AG and Wilshaw, TR (1976) Quasi-static solid particle damage in brittle solids—I. Observations analysis and implications, Acta Metall 24(10):939-956.

[13] Movahedi, B (2013) Fracture toughness and wear behavior of NiAl-based nanocomposite HVOF coatings, Surf Coat Tech 235:212–219.

[14] Li, H, Khor, KA, Cheang, P (2002) Young's modulus and fracture toughness determination of high velocity oxy-fuel-sprayed bioceramic coatings, Surf Coat Tech 155(1):21–32

[15] Ghabchi, A, Varis, T, Turunen, E, Suhonen, T, Liu, X, Hannula, S-P (2010) Behavior of HVOF WC-10Co4Cr Coatings with Different Carbide Size in Fine and Coarse Particle Abrasion, J Therm Spray Techn 19(1-2):368-377.

[16] Houdková, S, Bláhová, O, Zahálka, F, Kašparová, M. (2012) The Instrumented Indentation Study of HVOF-Sprayed Hardmetal Coatings, J Therm Spray Techn 21(1):77-85 [19] Jacobs, L, Hyland, MM, De Bonte, M (1998) Comparative Study of WC-Cermet Coatings Sprayed via the HVOF and the HVAF Process, J Therm Spray Techn 7(2):213-218.

[17] Jafari, M, Enayati, MH, Salehi, M, Nahvi, SM, Park, CG(2013) Improvement in tribological properties of HVOF sprayed WC–Co coatings using electroless Ni–P coated feedstock powders, Surf Coat Tech 235:310–317.

[18] Jafari, M, Enayati, MH, Salehi, M, Nahvi, SM, Park, CG (2013) Microstructural and mechanical characterizations of a novel HVOF-sprayed WC-Co coating deposited from electroless Ni–P coated WC-12Co powders, Mat Sci Eng A Struct 578:46–53.

[19] Lee, CW, Han, JH, Yoon, J, Shin, MC, Kwun, SI (2010) A study on powder mixing for high fracture toughness and wear resistance of WC–Co–Cr coatings sprayed by HVOF, Surf Coat Tech 204(14):2223–2229.

Research to the coating properties of different nickel-chromium alloy for Cr_3C_2-25%NiCr powders by high velocity oxygen fuel spray

Zhenduo Li; Yueguang Yu;
Beijing genernal research institute of mining & metallurgy, China

Cr_3C_2-25%NiCr coating by high velocity oxygen fuel spray (HVOF) have been extensively applied to resist abrase and erosion in high temperature applications such as power boiler and turbine blade.A type of ultrafine nickel-chrome alloy were used to made Cr_3C_2-25%NiCr powders with spray drying that possess excellent flow ability and appearance with agglomerated and sintered mothed. Microstructure of the ultrafine Cr_3C_2-25%NiCr powders and coatings were observed through scanning electronic microscope (SEM), and bond strength and microhardness of coatings were measured by tensile shearing test and Vickers hardness test. It is indicated that ultrafine Cr3C2-25%NiCr coatings have some outstanding properties to traditional Cr_3C_2-25%NiCr coatings by HVOF.

1 Introduction

Cr_3C_2-25%NiCr powders possess excellent flow ability and abrasion resistance that have been extensively applied to spray the coating of resist abrase and erosion in high temperature applications. Such as power boiler, roller and turbine blade[1-3]. This powders are prepared through agglomerated and sintered process as a popular method,the chrome carbide particle size are used the between of 1~4μm , but as the metal bland powder the nickel chromium alloy are withou clear size even to 1~40μm. In fact the nickel chromium-chrome carbide composite powder are research as Co-WC series powder focus on the microstructure of NiCr wetting Cr_3C_2 phase and degree of sphericity. He jianhong, Enriaue J. through agitator ball milling forming to prepare the nanometer 25%NiCr-Cr_3C_2 composite powder that research result indicated nanometer coating have more highly wear-resisting abradability[4-5]. But the process is time-consuming to the industry perpare for Cr_3C_2-25%NiCr powder. In this paper, the size of 1-1.5μm chrome carbide particles are used to prepared Cr3C2-25%NiCr powders and the different type of NiCr alloys through spray drying and vacuum sintering process, that the aim to obtain high properties and low cost the 25%NiCr-Cr_3C_2 coating.

2 experiment procedure

2.1 powders

In the paper,two type of Cr_3C_2-25%NiCr powders are prepared with the different NiCr alloy and single Ni &single Cr powder as raw material.Nickel chromium alloy (Ni80Cr20) or ultrafine single powder and chrome carbide particle were milled about 16h in the water,then the slurry was made to form the spheric powder by the sprayed drying.the two type of powder were sintered in the vacuum sintering furnace with 1250℃for 2h sinter time. The powder are sized to 45~15μm.The morphology of Cr_3C_2-NiCr and Cr_3C_2-Ni-Cr surface and cross powders were sprayed drying and sintered that showed in Fig.1, and the XRDdiffraction patterns of the powder of Cr_3C_2-NiCr and Cr_3C_2-Ni-Cr in Fig.2.

Fig.1 (a,b) Details of the Cr_3C_2-25%NiCr powders by contant from NiCr alloy (X600, X1000).

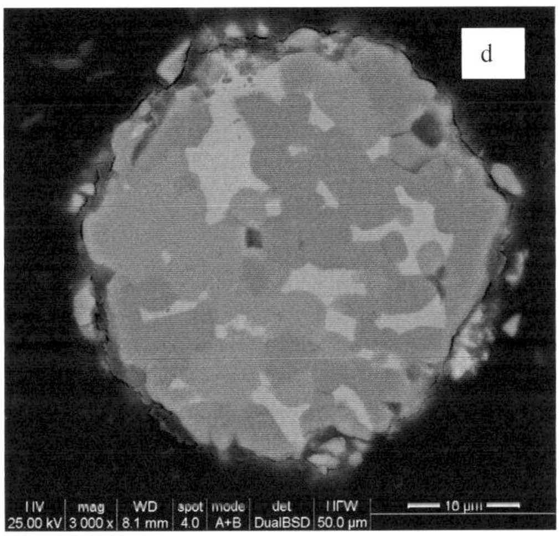

Fig.1 (c,d) Details of the Cr_3C_2-25%NiCr powders by contant from single Ni and Cr powder (X600, X1000).

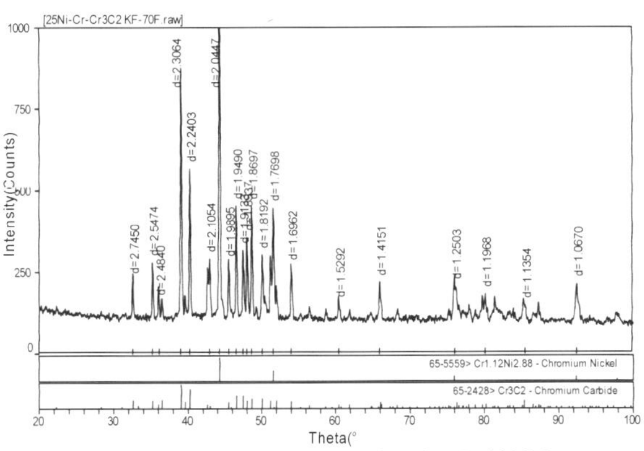

Fig.2 a,The XRD patten of the Cr3C2-25%NiCr powders by contant from NiCr alloy

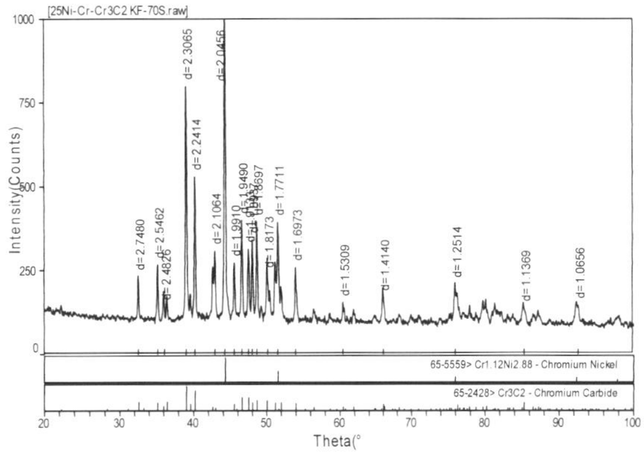

Fig.2 b,The XRD patten of the Cr_3C_2-25%NiCr powders by contant from single Ni and Cr powder

2.2 HVOF processing

Thermal spraying of the two type of powders were carried our with GTV K2 gun. The powders were sprayed on the steel substrates samples which was roughed by shut blasting with sand, than it cutted to 20×50×4mm by linear cutting machine. The coated samples were cuted by linear cutting machine, and cross section were grinded and polished to satisfy scanning electron microscope (HITACHI S-3500N) and microhardness analyse. The coatings were deposite with GTV K2 system of the spray parameters of coatings showed in **tables 1**.

Tab.1 HVOF parameters of Cr_3C_2-25%NiCr powder

parameter	value
Aviation kersoene	30L/h
oxygen	60m^3/h
Spraying distance	380mm
Carrier powder rate	80~120g/min

The coating bond strength is determined by universal test machine, tensile sample was sprayed about 0.25mm thickness, using FM1000 adhesive film as bonding materal at 190℃ for 2h solidification process. The coating mircohardness is detected with 402MVA Vickers hardness testing machine, and using 300N load to the diamond penetrator on the cross section Cr_3C_2-25%NiCr coatings.

The model for erosion was gritted blasting with sand in the coating surface, that the sand blowing angle was 30° for about 1 min, and the samples was weighed at the front and back.

3 result and discussion

Fig.1(a,b) It is the scanning electron microscope phase of Cr_3C_2-25%NiCr powders by the nickel chrome alloy, surface is more smooth and spherical , duo to fine NiCr alloy cann't completely wetting the chromium carbide particles that the cracks and proes exist in the cross of powders. **Fig.1(c,d)** show the single Ni and Cr raw of Cr_3C_2-25%NiCr powders that were same surface and degree of sphericility, and the cross section is more uniform and denser than alloy type, because single raw is more fines and easy mixing with Cr_3C_2 particles.

The XRD diffraction patterns of the powder of Cr_3C_2-25%NiCr by NiCr alloy or single rew are showed by fig.2(a,b),however it is not obvious difference between to NiCr alloy or single rew. This is only have a diffracted intensity different, the powder of Cr_3C_2-25%NiCr by NiCr alloy has a more intensity diffraction peak, that showed the powder is kept alloying in the process of sintered.

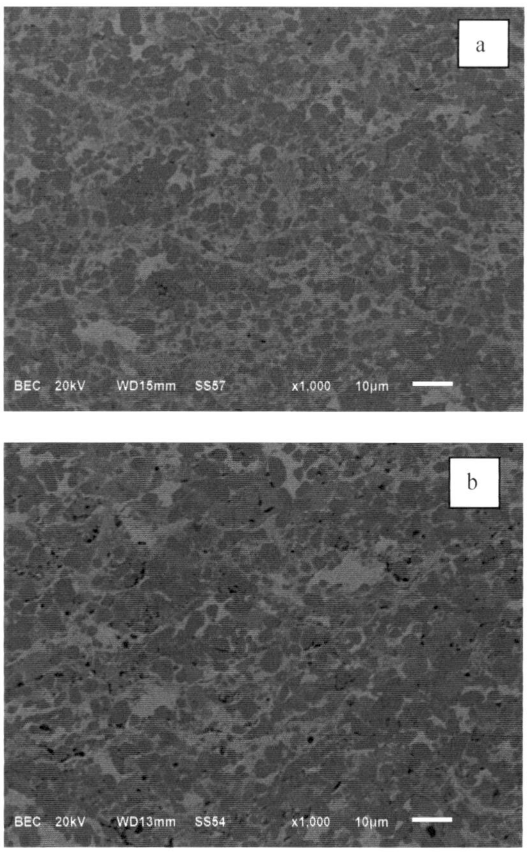

Fig.3 a)microstructure of NiCr alloy of Cr_3C_2-25% NiCr coatings by HVOF; Fig.2 b) microstructure of single raw of Ni and Cr of Cr_3C_2-25%NiCr coatings by HVOF.

Fig.3 a) The microstructure of Cr_3C_2-25%NiCr coatings by NiCr alloy indicate the NiCr alloy more easily moisten that can wetting Cr_3C_2 particles compactly. **Fig.3** b) as ultrafine single raw Ni and Cr of Cr_3C_2-25%NiCr coatings has obvious porosity compared with coatings by NiCr alloy in the HVOF process, because the melted state NiCr alloy can enough spread and moisten in the space of Cr_3C_2 particle, however single raw of Ni and Cr cannot completely deformate or melt to wetting Cr_3C_2 particle.

The average value of bond strength of Cr_3C_2-25%NiCr coating by NiCr alloy is 55Mpa, and coating by single raw of Ni and Cr average value is 52Mpa.

Tab.2 coatings mircohardness of Cr3C2-25%NiCr

	Cr_3C_2-25%NiCr coating by NiCr alloy	Cr_3C_2-25%NiCr coating by Ni and Cr
1	1216	893
2	1138	917
3	1141	725
4	986	1098
5	958	974
average value	1087	921

Tab.2 the mircohardness value of coating by NiCr alloy result isHv0.3:1087 and coating by Ni and Cr is 921. The result indicates the coating Cr_3C_2-25%NiCr by NiCr alloy has more better compactness, moreover the mircohardness of NiCr alloy is more higher hardness than single Ni and Cr exit in the coating of Cr_3C_2-25%NiCr.

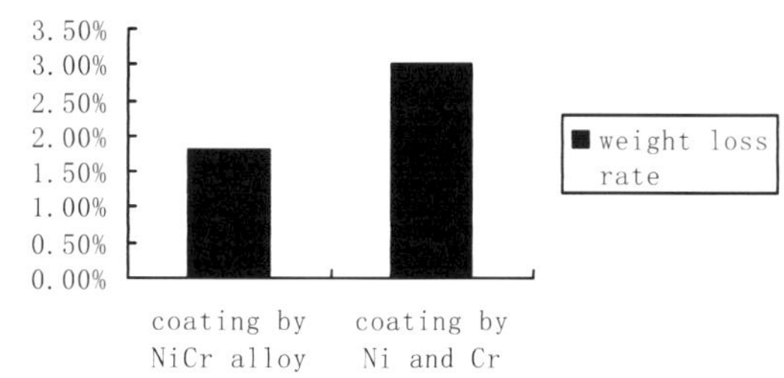

Fig.4 the weight loss rate of the coating Cr3C2-25%NiCr by NiCr alloy or single Ni and Cr.

The erosion experiment result is showed in Fig.4. It is obviously showed that the coating of Cr_3C_2-25%NiCr contanted by NiCr alloy has more better erosion resistance than the coating by single Ni and Cr as binding phase. The main reason is the NiCr alloy have higher the hardness of and melt ability in the process HVOF than the coating constituted by single raw.

4 conclusions

In this paper the Cr_3C_2-25%NiCr powder with different raw materials that include NiCr alloy or single Ni and Cr powder are prepared and sprayed, the result show the two type powder can double used to application for business. The results are showed the mircohardness and erosion resistance better of Cr_3C_2-25%NiCr by NiCr alloy in the HVOF process.

Literature

[1] He Jianhong, Enrique J.L, Precipitation phenomenon in nanostructured Cr_3C_2-25 % NiCr coating[J].Materials Science and Engineering.A301 (2001) : 69-79.

[2] G.matthäus, J.A.Picas,A.Forn. Effect of feedstock powder size on the sliding wear behaviour of thermal sprayed HVOF Cr_3C_2-25%NiCr coatings. Proceedings of 2004 international thermal spray conference, Osaka, Japan, May 2004.

[3] Li Chang Jiu,SONOYA K.,Ji Gang chang.Effect of powder type on the relationship between spray parameters and properties of HVOF spray Cr_3C_2-NiCr coatings. proceedings of the 15th international spray conference, 1998 May, Nice France, 25-29.

[4] Chang jiu Li, Gang chang Ji, Yu yue Wang, Keiji Sonoya. Dominant effect of carbide rebounding on the carbon loss during high velocity oxy-fuel spraying of Cr_3C_2-NiCr. Thin solid films 419(2002) 137-143.

[5] Stein K.J., Schorr B.S and Marder A.; Erosion of thermal spray MCr-Cr_3C_2 cermet coatings. Wear 224 (1999), pp.153-159.

Tribological behavior of arc sprayed and compressed coatings with different carbide grain size fraction

W. Tillmann, L. Hagen, P. Schröder, Dortmund / D

Due to their outstanding properties, $WC-W_2C$ iron-based cermet coatings are widely used in the field of wear protection. Regarding commonly used $WC-W_2C$ reinforced coating systems, it has been reported that their tribological behavior is mainly determined by the carbide grain size fraction. Although the manufacturing route for arc sprayed $WC-W_2C$ cermet coatings is in an advanced state, there is still a lack of knowledge concerning the performance of cored wires with fine grain-sized carbides as filling material and their related coating properties when post-treatment processes are used such as machine hammer peening (MHP).

A major objective was to characterize $WC-W_2C$ FeCMnSi coatings, deposited with different carbide grain size fillings, with respect to their tribological behavior. Thus, investigations using dry rubber wheel and ball-on-disk tests were performed at room temperature in order to analyze the wear and friction behavior. In addition, metallographic investigations were carried out to analyze their lamellar microstructure. As-sprayed and polished samples were compared to compressed samples since MHP processes lead to a densification of the microstructure formation. The effect of MHP on the tribological behavior was investigated as the utilization of conventional and fine grain-sized carbides lead to a different coating microstructure.

1 Introduction

Within the field of wear protection, arc sprayed coatings, produced by using cored wires with hard materials as a filling, are widely used to enhance the tribological and mechanical characteristics of surfaces. In terms of arc sprayed coatings, only few studies have discussed the use of tungsten carbide reinforced iron-based feedstock materials [1-4]. The tribological behavior of WC/W_2C reinforced deposits by means of twin wire arc spraying (TWAS) was investigated in previous studies [5-7]. It was found that a fine lamellar coating structure, provided by utilizing fine-grain hard material fillings, could lead to enhanced tribological coating characteristics [5]. For tribologically stressed surfaces, the wear resistance of WC-composites depends, among other things, on the hard material component and the embedding behavior of the hard phases [8, 9].

Due to overlaps of many individual spray particles, arc sprayed coatings are produced, which exhibit properties that mainly depend on the particle characteristics (particle size, -temperature, and -velocity) of the individual impinging particles [10-12]. Thus, in-flight particle characteristics are the controlling factors in the splat formation and in the determination of the morphology of the coating. The atomization behavior of the molten electrode tips depends on the electrode phenomena [11, 13]. The disintegration of the molten electrode tips, and thus, the droplet formation is significantly affected by the wire configuration [12, 14-16]. It was found that the droplet characteristics, among other things, are influenced by the grain size distribution of the powder filling [14, 16].

However, the deposits produced by arc spraying processes feature a heterogeneous microstructure formation. Moreover, such deposits have an inhomogeneous distribution of pores and a high roughness. A high surface roughness, a high level of porosity, and weak intersplat boundaries can be detrimental to the coating properties as these characteristics affect the wear resistance of tribologically stressed surfaces. In order to modify the surface properties, especially for surfaces that are exposed to tribological contact, protruding peaks across the surface of the coating have to be removed. Machining is a commonly used method in order to achieve the required surface quality.

With regard to fine surface finishing processes in the field of tool- or mould-making technologies, different surface treatments, such as shot peening, compacting, ball burnishing or MHP were already used [17-21]. For bulk materials, it was found that these methods increase the microhardness and induced residual compressive stresses within the subsurface of the treated area, which lead to an improved service life. In the field of surface engineering, MHP is a novel surface treatment technology, which enhances the surface properties. For the MHP process, a hard metal punch performs an oscillating movement via an electromagnetic operating principle [22, 23]. The surface is compacted by the pulse upon impact of the punch on the surface. A study by Tai et al. [24] showed the evidence of residual compressive stresses, induced by MHP, into the subsurface area. Moreover, Bleicher et al. [25] demonstrated in a recent study that the MHP process can be used to implemented WC hard phases, which are positively locked on a hot work steel surface.

The aim of this study is to analyze the applicability of the MHP process to arc sprayed coatings systems. A major objective was to characterize FeCMnSi arc sprayed coatings, deposited with different $WC-W_2C$ grain-sized carbides as filling material, with respect to their mechanical and tribological behavior. Thus, dry sliding experiments using ball-on-disk tests as well as abrasion resistance experiments, utilizing dry rubber wheel tests, were performed for MHP and non-MHP coating surface conditions.

2 Experimental setup

2.1 Feedstock materials

Iron-based cored wires, with a diameter of 1.6 mm (Fa. Durmat, type AS-850, Germany), were used as feedstock material. A cast tungsten carbide (CTC), which serve as a hard material, was used as a filling. The feedstock material varied in terms of its grain size fraction (**Table 1**). The chemical composition of the wire was 2% C, 1.4% Mn, 1.1% Si, 48% W and Fe bal.

Table 1. Grain size fraction of the CTC filling material ("UF" = Ultrafine)

Sample	**Carbide grain size fraction**
1,6	-106+63µm
1,6 UF	-45+15µm
1,6 H	-106+63µm
1,6 H UF	-45+15µm

The proportion of CTC (eutectic mixture of $WC\text{-}W_2C$) was kept constant for all samples (50%). Samples marked with "H" were post-treated with MHP process.

2.2 Machine hammer peening (MHP)

In order to smooth and compact the arc sprayed coatings, MHP (Fa. DMG, type DMU 50 eVolution, Germany) was used as a post-treatment with an air pressure of 6 bar, a feed velocity of 1000 mm/min, a line pitch of 0.2 mm, and a ball diameter of 12 mm. The oscillating frequency of the punch was approximately 220 Hz

Fig. 1 shows the setup of the MHP process.

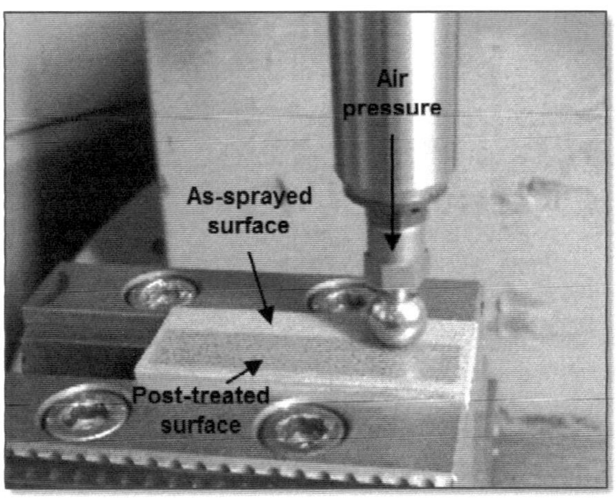

Fig. 1. Setup of MHP process

2.3 Coating characterization

The coating morphology was analyzed by means of a field emission scanning electron microscope (FE-SEM, type JSM-7001F, Fa. Jeol, Germany). Before the metallographic investigation, the coated samples were prepared by using diamond grinding discs and polishing cloths with a diamond suspension (9 µm, 6 µm, 3 µm, and 1 µm).

A ball on disk test was employed in order to determine the sliding wear and friction behavior of the different coatings. An alumina ball with a diameter of 6 mm was used during the tests with a high temperature tribometer (Fa. CSM Instruments, Switzerland), and the coefficient of friction (COF) was determined. In addition, in order to obtain the wear coefficient, a 3D optical surface profilometer (type Infinite Focus, Fa. Alicona, Austria) was employed to measure the volume loss of the wear track after 30,000 revolutions.

For the abrasion tests, the dry sand rubber wheel (DSRW) technique was employed. Therefore, the relevant ASTM standard [26] was considered. The samples were pressed against a rotating rubber wheel with a force of 130 N and a wear distance of 718 m. An abrasive material in powder form (rounded quartz sand) was added to the contact area. The particle size range was between 200 and 425 µm. For the rubber wheel, a tyre (shore hardness A58-62) with an external diameter of 227.6 mm was utilized

3 Results

3.1 Dry sliding experiment

When performing the dry sliding experiment, it is clearly visible that a reduction of the carbide grain size causes a decrease in the sliding wear coefficient k, as shown in

Fig. 2. Values of $3.88\cdot10^{-6}$ mm^3/Nm and $3.32\cdot10^{-6}$ mm^3/Nm for the samples 1,6 H and 1,6 H UF, respectively, were obtained.

Moreover, the MHP process has also an influence on the sliding wear performance since the wear coefficient k is higher for post-treated surfaces.

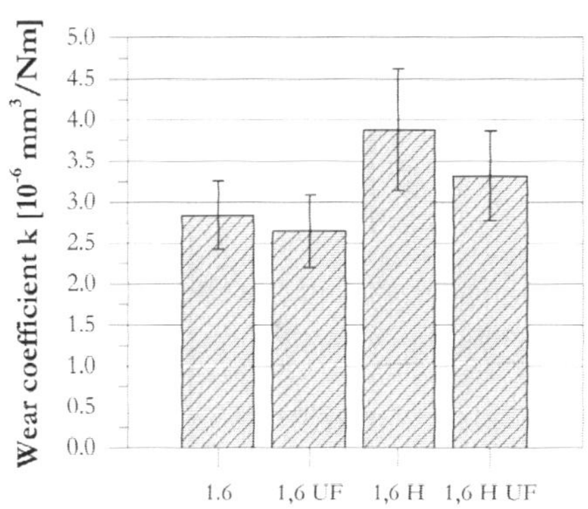

Fig. 2. Wear coefficient of the surfaces

When using the MHP process, an increase of the wear coefficient k_{Pin} was also identified for the counterpart used in the ball-on-disk-tests (
Fig. 3).

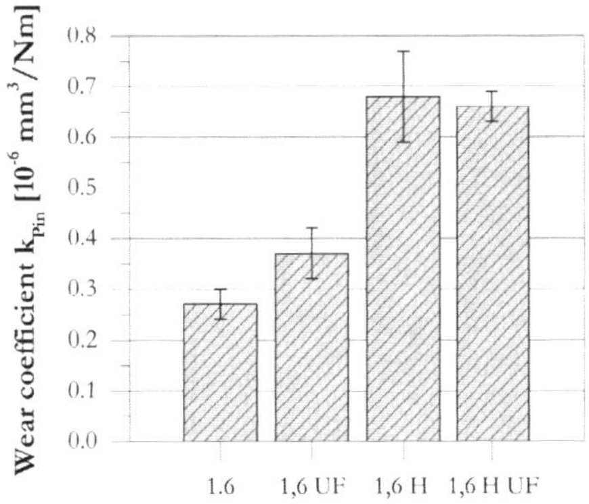

Fig. 3. Wear coefficient k_{Pin} of the counterpart towards different samples

Regarding the friction behavior, the use of MHP as well as the use of fine grain-sized coatings lead to a better friction performance induced by a decreased COF, as shown in

Fig 4. The absolute values of the average friction coefficient vary between 0.79 for sample 1,6 H and 1.0 for sample 1,6. Furthermore, small irregularities and fluctuations in the friction behavior were determined for post-treated coatings when comparing them to polished samples.

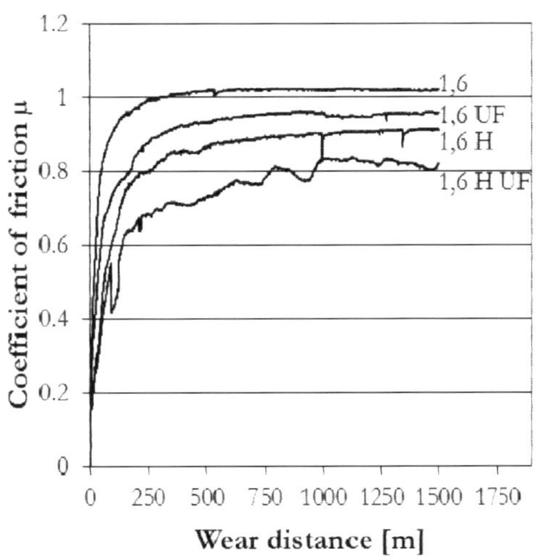

Fig 4. COF of the surfaces

By analyzing the abrasive wear caused by the dry rubber wheel test, a reduction of the cumulative mass loss when using MHP process can be observed (
Fig. 5).

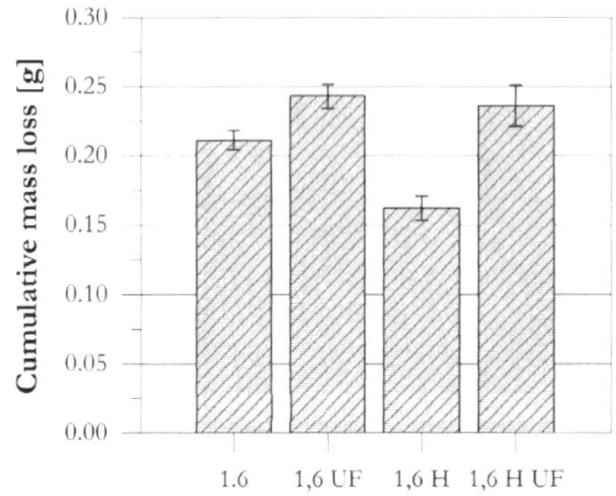

Fig. 5. Cumulative mass loss of different samples after dry rubber wheel test

MHP surfaces have a cumulative mass loss of 0.162 g (1,6 H) and 0.236 g (1,6 H UF), whereas polished coatings exhibit values of 0.211 g (1,6) and 0.243 g (1,6 UF). Furthermore, by comparing these results, a reduction of the cumulative mass loss due to an increased carbide grain size fraction can be recognized.

4 Discussion

4.1 Morphology and microstructure

By comparing the morphology of the samples, it is clearly visible that the post-treatment by MHP causes a compression of the coating and consequently a decrease of the thickness. It can be assumed that the protruding peaks and splats at the subsurface area are plastically deformed due to this compression [27]. Thus, the size of the pores is reduced. The reduction of the porosity is shown in

Fig. 6. Sample 1,6 UF shows the highest porosity (5.83%), whereas the post-treated and ultrafine coating (1,6 H UF) has the lowest porosity of 3.25%. The samples 1,6 and 1,6 H have a porosity of 5.75 % and 3,96%, respectively.

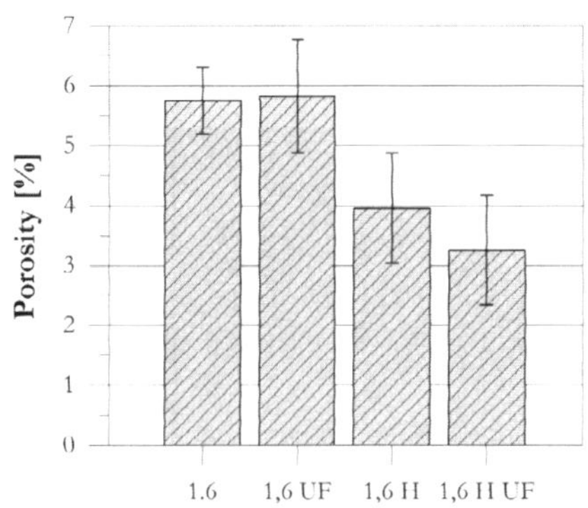

Fig. 6. Porosity of the coatings

However, in contrast to the positive effects, a formation of cracks across splat boundaries could also be observed when applying the MHP process due to high loads (
Fig. 7). Thereby, these cracks occur mainly in hard phases as the compression induces brittle fractures instead of plastic deformations.

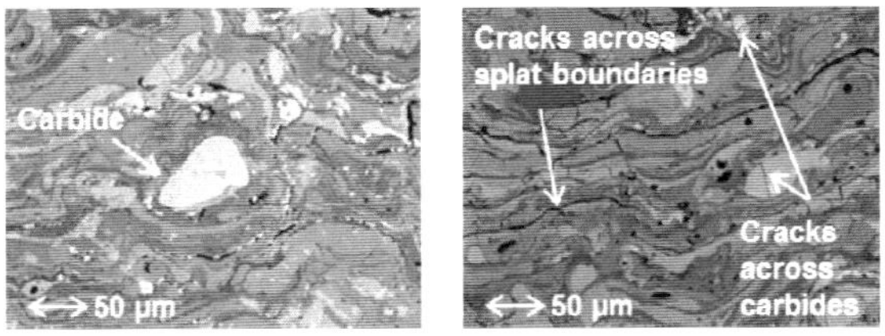

Fig. 7. Microstructure of the sample 1,6 (left) and the sample 1,6 H (right)

Furthermore, EDX-analyses showed that the tungsten content was for all samples similar. It varied between 38.4 % (1.6 UF) and 41.9 %(1.6).

4.2 Mechanical properties

Regarding the mechanical properties, the hardness and Young's modulus were analyzed (
Fig. 8). The hardness of the surface increases as a consequence of MHP process. Thus, the hardness value increases from 4.4 GPa to 5.9 GPa for the ultrafine samples and from 5.5 GPa to 5.9 GPa for the conventional samples.

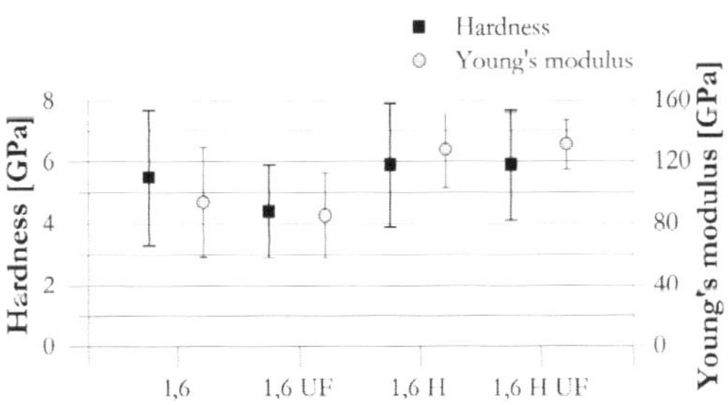

Fig. 8. Hardness and Young's modulus of different samples

The increase of hardness is caused by strain hardening effects due to plastic deformation in the material [31].

4.3 Sliding wear behavior

An enhanced wear behavior induced by a decrease of the sliding wear coefficient for coatings with fine grain-sized carbides was observed within these investigations, as shown in
Fig. 2.

Fig. 9 compares the wear tracks of non-MHP samples. It is clearly visible that fine grain-sized carbide coatings and conventional surfaces differ in terms of the intensity of the damage.

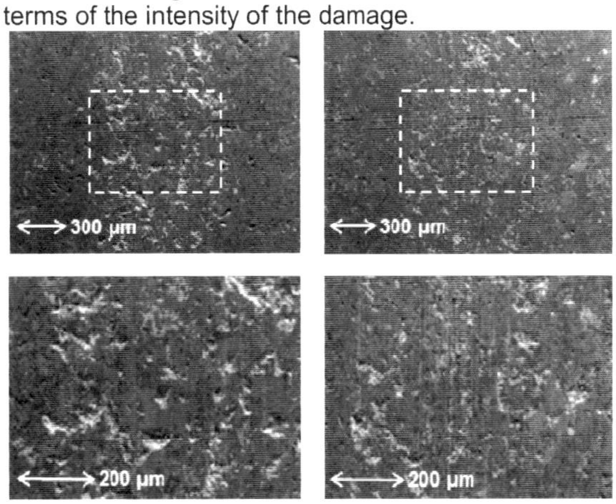

Fig. 9. Electron microscope images showing the wear track of different samples (left: 1,6; right: 1,6 UF)

The fact that a smaller carbide size is beneficial for the wear performance was already discovered by several other authors [29]. One explanation for this is that the probability of local overloads of individual carbide particles is reduced due to a distribution of the loads in the contact area to a larger amount of carbides. Therefore, the risk of cracking is smaller [30, 31]. Yang et al. [32] describe another effect. During the ball-on-disk test, the soft and ductile matrix is first deformed, whereby the carbides are exposed. Thus, a lower damage of the surface occurs during the contact between the sample and the ball when using fine grain-sized carbides.
Furthermore, the MHP has a negative effect on the wear behavior. It can be assumed that the cracks, which result from great cyclic loads of this process, are the reasons for this. Similar results were achieved in a previous study [33] when analyzing rolled surfaces. An increase of microcracks could be noticed due to the mechanical post-treatment, which leads to a poorer wear performance.

4.4 Friction behavior

The use of fine grain-sized coatings improves the friction performance, as shown in
Fig 4. This effect can be explained by the fine and homogenous microstructure with the same proportion of WC-W_2C. A finer coating structure with short and narrow flakes was also recognized in previous investigations [5] in arc sprayed coatings when using a fine grain-sized carbide fraction. This effect leads to a bigger amount of WC particles which provide a better support for the ball, so that the load is dispensed on more carbides. Thus, the friction resistance against the ball is reduced [30].
Moreover, MHP coatings show a better friction behavior. The roughness in the contact area of the sliding process increases during the ball-on-disk test

because the pores of the coatings are exposed. Due to the compression induced by MHP process, post-treated coatings have a lower porosity than conventional coatings. Therefore, a reduction of the COF is achieved for the MHP coatings.
The fluctuations determined for post-treated coatings could also be observed at the end of the ball-on-disk test in previous investigations [30] with HVOF-sprayed coatings. These fluctuations are promoted by an increasing wear [30]. Since the irregularities in the friction behavior occur mainly in post-treated coatings, it can be assumed that the cracks, which were induced previously by the MHP process, lead to a higher amount of break outs of broken carbides, and thus affecting the tangential force during the tests.

4.5 Abrasive wear behavior

The MHP process as well as the use of fine grain-sized carbides have an impact on the abrasive wear performance, as presented in
Fig. 5. The higher wear resistance of MHP coatings in dry rubber wheel tests can be explained by the increase of hardness (
Fig. 8). In addition, the abrasive wear appears first on the protruding peaks of the surface. Since the amount of protruding peaks is reduced due to a mechanical compression and a levelling of the surface caused by MHP, the abrasive wear decreases.
This leads to the assumption that the influence of the induced cracks is not as high as in the ball-on-disk test due to lower fatigue loads within this test. Therefore, MHP has a positive effect on the abrasive wear performance for dry rubber wheel test.
For different feedstock materials, the use of fine grain-sized carbides causes a poorer abrasive wear resistance. This can be explained by a higher mean free path of the metallic binder for the stressed surface, when compared to samples produced with a larger carbide size as a feedstock material. This was observed in previous investigations for HVOF-sprayed coatings [30]. Nevertheless, for the use of the fine grain-sized carbide feedstock material EDX analyses revealed a slightly lower content of tungsten at the contact area, which can be correlated to a reduced amount of tungsten carbides (WC, W_2C, M_6C, $M_{12}C$). However, those phases need to be verified by x-ray diffraction analyses.

5 Conclusion

The main focus of these investigations was to analyze the impact of the post-treatment using MHP as well as the influence of the carbide grain size fraction on the tribological behavior of arc sprayed coatings. In summary, the following results were obtained:

A finer carbide grain size fraction causes:

- A decrease in sliding wear coefficient due to a reduction of the risk of cracking.
- An increase of the abrasive wear, proven by dry rubber wheel tests.

The post-treatment of surfaces by applying the MHP process causes:

- A poorer wear behavior in the case of dry sliding experiments due to the occurrence of cracks, which are previously induced by high loads of the MHP process.
- An improved coefficient of friction
- An enhanced wear behavior towards abrasion, which was proven by dry rubber wheel tests.

Acknowledgement

The authors gratefully acknowledge the financial support of the DFG (German Science Foundation) within the Collaborative Research Center SFB 708, subproject A1.

6 Literature

[1] He, D.J.; Fu, B.J.; Jiang, J.M.; Li, X.J.: Microstructure and Wear Performance of Arc Sprayed Fe-FeB-WC Coatings. J. Therm. Spray Technol. 17 (2008), Issue 5, pp. 757-761.
[2] Tillmann, W.; Klusemann, B.; Nebel, J.; Svendsen, B.: Analysis of the Mechanical Properties of an Arc-Sprayed WC-FeCSiMn Coating: Nanoindentation and Simulation. J. Therm. Spray Technol. 20 (2011), Issue 1-2, pp. 328-335.
[3] Xu, B.; Zhu, Z.; Ma, S.; Zhang, W.; Liu, W.: Sliding wear behavior of Fe–Al and Fe–Al/WC coatings prepared by high velocity arc spraying. Wear 257 (2004), pp. 1089–1095.
[4] Tillmann, W., Luo, W.; Selvadurai. U.: Wear Analysis of Thermal Spray Coatings on 3D Surfaces. J. Therm. Spray Technol. 23 (2014), Issue 1-2, pp. 245-251.
[5] Tillmann, W.; Luo, W.; Nebel, J.: Einfluss der Hartstoffkorngröße auf die tribologischen Eigenschaften gewalzter und geschliffener WSC-FeCSiMn-Schichten. Thermal Spray Bulletin 4(1), (2011), pp. 56-63.
[6] Tillmann, W.; Klusemann, B.; Nebel, J.; Svendsen, B.: Analysis of the Mechanical Properties of an Arc-Sprayed WC-FeCSiMn Coating: Nanoindentation and Simulation. J. Therm. Spray Technol. 20 (2011), Issue 1-2, pp. 328-335.
[7] Tillmann, W., Luo, W.; Selvadurai. U.: Wear Analysis of Thermal Spray Coatings on 3D Surfaces. J. Therm. Spray Technol. 23 (2014), Issue 1-2, pp. 245-251.
[8] Li, Z.; Jiang, Y.; Zhou, R.; Lu, D.; Zhou, R.: Dry three-body abrasive wear behavior of WC reinforced iron matrix surface composites produced by V-EPC infiltration casting process. Wear 262 (2007), pp. 649-654.
[9] Zhou, R.; Jiang, Y.; Lu, D.: The effect of volume fraction of WC particles on erosion resistance

of WC reinforced iron matrix surface composites, Wear 255 (2003), pp. 134-138.
[10] Newbery, A.P.; Grant, P.S.; Neiser, R.A.: The velocity and temperature of steel droplets during electric arc spraying. Surf. Coat. Technol. 195 (2005), pp. 91-101.
[11] Hussary, N.A.; Heberlein, J.V.R.: Effect of system parameters on metal break-up and particle formation in the wire arc spray process. J. Therm. Spray Technol. 16 (2007), Issue 1, pp. 1-13.
[12] Pourmousa, A.; Mostaghimi, J.; Abdini, A.; Chandra, S.: Particle size distribution in a wire-arc spraying system. J. Therm. Spray Technol. 14 (2005), Issue 4, pp. 502-510.
[13] Hussary, N.A.; Heberlein, J.V.R.: Atomization and particle-jet interactions in the wire-arc spraying process. J. Therm. Spray Technol. 10 (2001), Issue 4, pp. 604-610.
[14] Tillmann, W.; Vogli, E.; Abdulgader, M.; Gurris, M.; Kuzmin, D.; Turek, S.: Particle Behavior During the Arc Spraying Process with Cored Wires. J. Therm. Spray Technol. 17 (2008), Issue 5-6, pp. 966-973.
[15] Newbery, A.P.; Rayment, T.; Grant, P.S.: A particle image velocimetry investigation of in-flight and deposition behavior of steel droplets during electric arc spray forming. Mat. Sci. Eng. 383 (2004), pp. 137-145.
[16] Tillmann, W.; Abdulgader, M.: Wire Composition, Its Effect on Metal Disintegration and Particle Formation in Twin-Wire Arc-Spraying Process. J. Therm. Spray Technol. 22 (2013), Issue 2-3, pp. 352-362.
[17] Kim, J.C. et al.: Residual stress relaxation and low- and high-cycle fatigue behavior of shot-peened medium-carbon steel. International Journal of Fatigue 56 (2013), pp. 114-122.
[18] Byrne, G.D., O'Neill, L., Twomey, B., Dowling, D.P.: Comparison between shot peening and abrasive blasting processes as deposition methods for hydroxyapatite coatings onto a titanium alloy. Surf. Coat. Technol. 216 (2013), pp. 224-231.
[19] Juijerm, P.; Altenberger, I.; Scholtes, B.: Influence of ageing on cyclic deformation behavior and residual stress relaxation of deep rolled as-quenched aluminium alloy AA6110. International Journal of Fatigue 29 (2007), Issue 7, pp. 1374-1382.
[20] Bleicher, F. et al.: Mechanism of surface modification using MHP technology. CIRP Annals - Manufacturing Technology 61 (2012), Issue 1, pp. 375-378.
[21] Rodríguez, A. et al.: Surface improvement of shafts by the deep ball-burnishing technique. Surf. Coat. Technol. 206 (2012), Issue 11-12, pp. 2817-2824.
[22] Scheil, J. et al. Influence of Process Parameters on Surface Hardening in Hammer Peening and Deep Rolling. Key Engineering Materials 554-557 (2013), pp. 1819-1827.
[23] Wied, J.: Oberflächenbehandlung von Umformwerkzeugen durch Festklopfen. Dissertation, TU Darmstadt, 2010.
[24] Tai, M.; Miki, C.; Suzuki, K.: Mechanism of Fatigue Strength Improvement by Applying Peening Treatment. Procedia Engineering 14 (2011), pp. 117-125.
[25] Bleicher, F. et al.: Improving the tribological characteristics of tool and mould surfaces by machine hammer peening. CIRP Annals - Manufacturing Technology 62 (2013), Issue 1, pp. 239-242.
[26] ASTM international: Standard test method for measuring abrasion using the dry sand/rubber wheel apparatus. ASTM stand-ards G65-04 (2004).
[27] Adjassoho, B.; Kozeschnik, E.; Lechner, C. et al.: Controlled surface treatment with machine Hammer Peening. Proceedings of the 22nd international conference on metallurgy and materials (2013), Czech Republic.
[28] Trauth, D.; Klocke, F.; Shirobokov, A.; Schongen, F.: Analysis of friction between stainless steel sheets and machine hammer peened structured tool surfaces: experimental and numerical investigation of the lubricated interaction gap. Production Engineering 8 (2014), Issue 3, pp. 263-272.
[29] van Acker, K.; Vanhoyweghen, D.; Persoons, R. et al.: Influence of tungsten carbide particle size and distribution on the wear resistance of laser clad WC/Ni coatings. Wear 1-4 (2005), pp. 194-202.
[30] Tillmann, W.; Baumann, I.-T.; Hollingsworth, P.S.; Hagen, L.: Sliding and Rolling Wear Behavior of HVOF-Sprayed Coatings Derived from Conventional, Fine and Nanostructured WC-12Co Powders. J. Therm. Spray Technol. 23 (2014), Issue 1-2, pp. 262-280.
[31] Baumann, I.-T.: Hochverschleißfeste und konturnahe Werkzeugoberflächen durch Hochgeschwindigkeitsflammspritzverfahren, Vulkan-Verl. (2012), Essen.
[32] Yang, Q.; Senda, T.; Ohmori, A.: Effect of carbide grain size on microstructure and sliding wear behavior of HVOF-sprayed WC-12% Co coatings. Wear 1-2 (2003), pp. 23-34.
[33] Tillmann, W.; Luo, W.; Tekkaya, A. et al.: Verschleißanalyse gewalzter und geschliffener Eisenbasislegierungsschichten mit und ohne Hartstoffverstärkung. Materialwiss. Werkstofftech. 6 (2012), pp. 485-494.

Cold spray coatings for spacecrafts low-temperature soldering application

Cui qingxin, zhang jiaqiang*, bai jingying, zhangligong, wang bingcun, ma gang , wen chen, Li li, wen qiangmiao
Beijing Spacecrafts, China Academy of Space Technology, Beijing 100190, PR China

The demands for increased onboard power consumption have resulted in an ever increasing heat dissipation capacity of radiator in spacecrafts. Low temperature soldering is a key technology to improve the thermal conductivity between heat pipe and radiator when the pipe is encapsulated with volatile working fluid. While the low temperature solder of Ga-Cu eutectic alloy has a vital drawback that may lead to corrosion of aluminum alloy. In this work, we investigated cold sprayed Cu solderable coatings on 6063 and LD10 aluminium plate to improve the wettability of aluminium substrates during the low temperature soldering process and provide a diffusion barrier to protect aluminium. The effect of the coating on the joint's interface microstructure, wetting behaviour, shear strength and soldering properties was investigated in detail.

1 Introduction

The demands for increased onboard power consumption have resulted in an ever increasing heat dissipation capacity of radiator in spacecrafts, which require improving heat transfer between heat pipe and radiator of spacecrafts. Low temperature soldering is a key technology to improve the thermal conductivity between heat pipe and radiator when the pipe is encapsulated with volatile working fluid. There are two problems need to be solved in soldering process of aluminum alloy. First, aluminum alloy is difficult to be wetted during the soldering process due to its surface oxides. Besides, the soldering process should be carried out at a low temperature no more than 80℃, as heat pipe is filled with volatile working fluid. While the low temperature solder of Ga-Cu eutectic alloy has a vital drawback that may lead to corrosion of aluminum alloy. Thus, the joint surface of aluminum substrate must be deposited a coating that can be solder and has protective ability.[1-3]

Cold gas dynamic spraying technology, or cold spraying for short, is developed in the mid 1980s at the Institute of Theoretical and Applied Mechanics of the Siberian Division of the Russian Academy of Science in Novosibirsk, which is a rapidly developing technology for the preparation of coatings or bulk materials in the solid state.[4-5] During cold spray process, powder particles, usually in the size range of 1 to 50 μm, are accelerated through a De Laval type of converging-diverging nozzle through a supersonic carrier gas flow. Typically particle speeds of 300 to 1500 m/s are attained and the powder particles are directed to impact on a substrate or previously deposited powder to form a coherent coating. Heated air or nitrogen may be used as the accelerating gas but the best quality of coatings is generally achieved by using helium at room, or slightly elevated temperature.[6-8]

Compared to traditional thermal spray process, the most distinct characteristic of cold spray is its lower processing temperature, consequently lower thermal effects to the deposited powder and substrate.[9] Therefore, cold spray is particularly suitable to prepare coatings that are sensitive to oxidation or thermal influences. Cold sprayed coatings are reported to have relatively a low oxygen or oxide content. Moreover, cold spray coatings could have extremely dense under suitable spray conditions. Dense coatings with low oxidation will also mean that they have excellent mechanical, thermal and electrical properties.[10-11] Therefore the coating deposited by cold spray is a proper candidate to apply as a solderable coating on aluminium surface to joint aluminium heat pipe and radiator for enhancing the thermal conductivity and improving the heat dissipation capacity of radiator.

The purpose of the present work was to assess the potential for employing cold sprayed Cu coatings as solder layers. The main objective of this paper is to systematically characterize the cold sprayed Cu coatings and the results of the low temperature soldering. Therefore, cold sprayed Cu coatings approximately of 40 to 60μm in thickness were deposited on 6063 and LD10 aluminium plate, respectively. The process of the cold spray, properties of the coating and soldering results were investigated in detail.

2 Experimental procedure

2.1 Materials

Pure Cu powder of nominal size range 5 to 45 μm purchased from commercial sources was used as feedstock for cold spray with a nitrogen carrier, and its morphology is shown in Fig. 1. The pure cu powder shows a near spherical morphology. Before cold spraying, 6063 and LD 10 aluminium substrate had been cut into 200mm×40mm ×2mm in size, and deceased with hot alkaline solution, ultrasonically cleaned in distilled water, and dried in air condition at room temperature. The low temperature Ga-Cu solder was prepared by Beijing Spacecrafts.

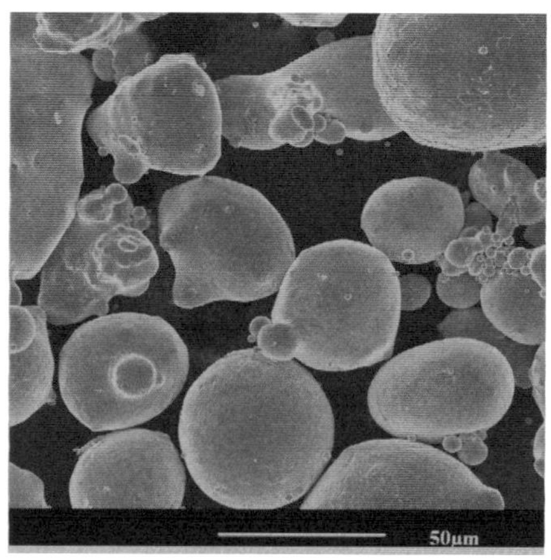

Fig. 1. SEM image of the Cu powder.

2.2 Cold spray deposition

Cold spray was carried out at a high pressure commercial cold spray system. The carrier gas used in this study was nitrogen at 2Mpa pressure and heated to 500□ temperature. The distance between the nozzle exit and the aluminium substrate was set to 40mm, the spraying process was carried out by using a speed of 100mm/s of the spray gun relative to the substrate to prepare the copper solderable coating. The coating thickness is about 40μm to 60μm.

2.3 Soldering

The material used in this work was Ga-Cu(GC)active solder paste containing amounts of noble element. Soldering experiments were carried out using the sample geometry is shown in Fig.2. Before soldering, all samples were ultrasonically cleaned using ethanol, rinsed using acetone and distilled water. The surface of the sprayed coating deposited on 6063 Al and LD10 Al plate was ground with silicon carbide grinding paper to 1200 grade. The direct active soldering process was performed on an electric hot plate with thermostatic control. Prior to joining, the specimens was preheated using the hot plate system at 36-42℃. The hot plate system was calibrated until the temperature of the specimens was steady. The GC active solder was then placed on the surface of sprayed cu coating of the specimens. The 6063 substrate was set on the LD 10 plate surface with a pressure by clamp, and the sandwich was heated to 75℃ in an oven for 72h and cooled in air.

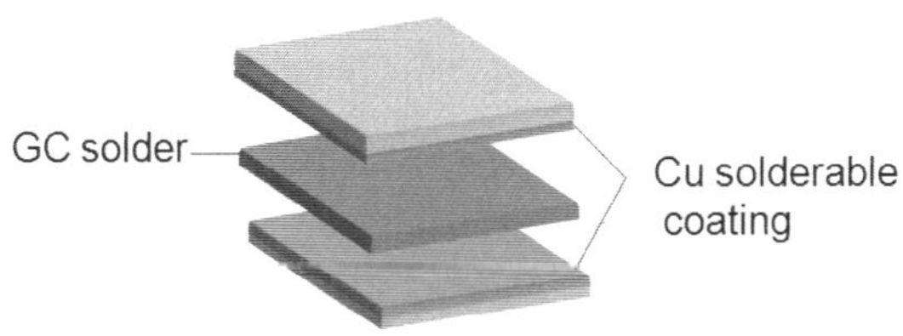

Fig.2. Schematic sample geometry for soldering experiments.

2.4 Coating and solder joint characterization

The mirror-polished cross-section and top surface of the cold sprayed Cu coating were examined with a metallographic microscope (Olympus GX51) and SEM. Cross-sections of the coatings were prepared for surface and interface measurements using the following procedure. The coating samples were first cut into pieces approximately 10×10 mm and then mounted in epoxy resin that was cured at room temperature for 24h. The mounted samples were then successively ground with 400, 800, and 1000 grit SiC papers and finally polished using diamond slurries of 3.5 μm and 1.5 μm for 5 min and 5 min, respectively.

3 Results and discussion

3.1 Microstructure

The microstructure of the cold spraying solderable coating is characterized by using metallographic microscope. Fig.3 presents the images showing the microstructures of the sprayed copper coatings. It can be seen that extremely dense cu coating with no evidence of significant inherent material defects has obtained by the cold spray process. The sprayed Cu coating material is in intimate contact with aluminum substrate forming a metallurgical bond during the high velocity impact and plastic deformation process. The surface morphology of the coating shows that Cu particle deformed intensively during the deposition process owing to the high particle velocity and its excellent ductility. The sprayed surfaces of the coating also exhibit a cellular morphology of particles that were impacted and bonded on the previously deposited particles (Fig. 3c).

Fig.3. Metallographic images of cold spray Cu coating: (a) Al substrate with copper coating, (b) Metallographic of polished cross section(×200), (c) Metallographic of top surface(×1000).

3.2 Oxide analysis

The EDS was used to analyze the oxygen content of the sprayed copper coating. The results revealed that almost no oxidation occurs in the coating, in which oxygen content is less than 1%. In thermal spraying, the conductivity of a coating is mainly reduced by non-metallic bonds at internal interfaces and trapped oxides. However, the cold spray process does not melt the sprayed particles compared to a conventional thermal spray and thus provides an excellent opportunity to produce high pure copper coatings with low porosity and low oxide content, which is beneficial to improve the thermal conductivity of the solder parts.

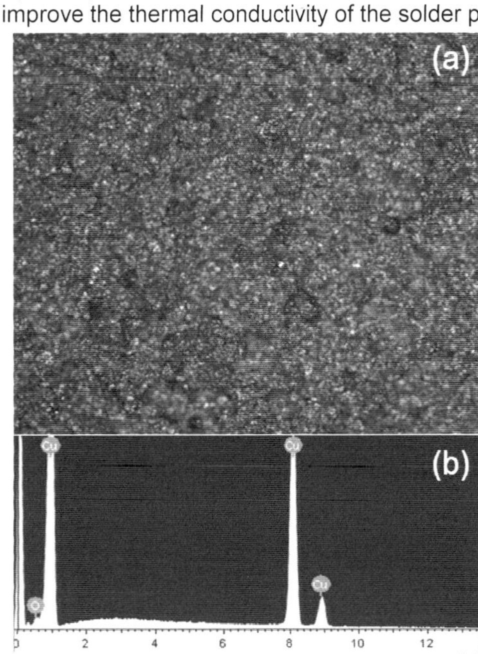

Fig.4. (a) Typical SEM image of the copper coating as sprayed, (b) corresponding EDS spectra of the coating surface.

3.3 Wetting behaviour

The wettability is a critical property that used to evaluate the bonding quality between solder and substrate. As shown in Fig.5. the measured contact angle between solder and aluminium substrate is 20° with a cold spray solderable copper coating, indicating that introduction of copper coating improved the wettability of aluminium alloy.

Fig.5. The wetting angle of the low temperature solder on the sprayed copper coating.

3.4 shear strength

Fig.6. shows the average shear strengths of the soldering samples with cold sprayed copper coating and the naked aluminium joints. The joints strength of specimens with sprayed solderable copper coating is 28.9MPa, higher than those of naked joints(12.3Mpa). These results indicate that the cold sprayed copper coating have excellent adhesive with aluminum. The copper coating can effectively improve the solder reliability of aluminum alloy.

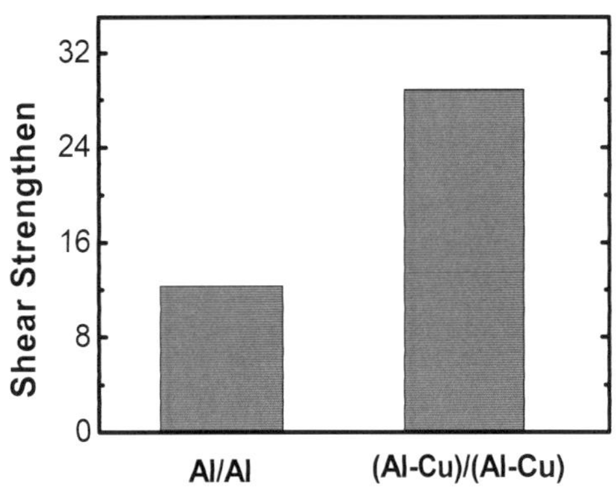

Fig.6. Shear strengths of the soldering samples.

3.5 Interfacial characterization

To get further information into the solderability of the cold sprayed copper coating, the cross-section of the soldering sample has been studied. As shown in Fig.7, the solder layer is sandwiched in two layers of the cold spray copper coating. The Cu and Ga element is diffusion each other to form a Cu and $CuGa_2$ mixed phase in the interface of copper and solder layers. But there is no Ga element could be detected in the deeper layer of the copper coating which is close to the aluminum substrate, indicating that the cold sprayed copper coating could effectively prevent the diffusion of gallium to provide a barrier layer for the protection of aluminum substrate.

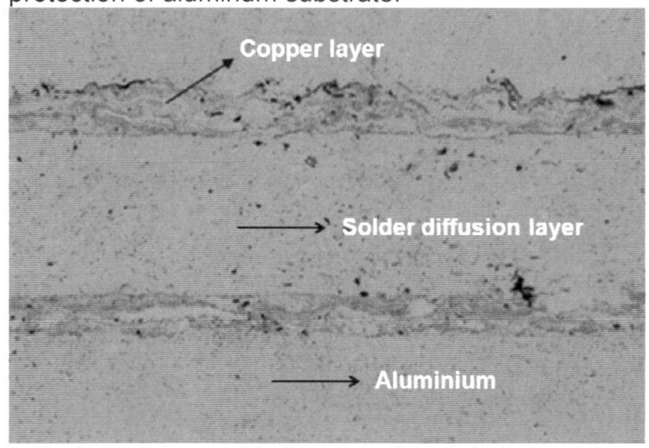

Fig.6. Metallographic images of the interface of the soldering samples with solderable coating.

4 conclusion

Solderable copper coatings have been successfully deposited onto 6063 and LD 10 aluminium substrates by a cold spray process using pure nitrogen as carrier gas in 500℃. Metallographic observation and EDS results show that the copper particles formed a very dense coating with low porosity and oxide in the deposit process. The sprayed copper coating could improve the wettability of aluminium alloy in the solder process, which has potential application in spacecrafts to joint heat pipe and radiator to improve the soldering interconnect reliability and thermal conductivity of the radiator.

5 References

[1] Lovro Gorjan, Gurdial Blugan, Manfred Boretius, Stefano De La Pierre, Monica Ferraris, Valentina Casalegno, Stefano Rizzo, Thomas Graule, Jakob Kuebler. Fracture behavior of soldered Al_2O_3 ceramic to A356 aluminumalloy and resistance of the joint to low temperature exposure. Materials & Design, 2015, 88(25), 889-896.

[2] Z. Mirski, T. Wojdat, M. Stachowicz. Soldering of aluminium with copper and steel using intermediate layer Zn-Ni. Archives of Civil and Mechanical Engineering, 2015, 15(4), 903-910.

[3] J.F. Li, P.A. Agyakwa, C.M. Johnson, D. Zhang b, T. Hussain b, D.G. McCartney. Characterization and solderability of cold sprayed Sn–Cu coatings on Al and Cu substrates. Surface & Coatings Technology, 2010, 204, 1395-1404.

[4] A.P. Alkhimov, V.F. Kosarev, and A.N. Papyrin, A Method of Cold Gas Dynamic Deposition, Sov. Phys. Dokl., 1990, 35(12),1047-1049.

[5] R.C. Dykhuizen and M.F. Smith, Gas Dynamic Principles of Spray, J. Therm. Spray Technol., 1998, 7 (2), 205-212.

[6] Xinkun Suoa, Shuo Yina, Marie-Pierre Planchea, Taikai Liu, Hanlin Liao. Strong effect of carrier gas species on particle velocity during cold spray processes. Surface & Coatings Technology, 2015, 268, 90-93.

[7] T. Schmidt, F. Gaertner, H. Kreye. New Developments in Cold Spray Based on Higher Gas and Particle Temperatures. Journal of Thermal Spray Technology, 2006, 15(4), 488-494.

[8] T. Stoltenhoff, H. Kreye, H. J. Richter, Horst.J.Richter. An Analysis of the Cold Spray Process and Its Coatings., Journal of Thermal Spray Technology, 2002, 11(4), 542-550.
[9] A.S. Khanna, W.S. Rathod. Development of CoNiCrAlY oxidation resistant hard coatings using high velocity oxy fuel and cold spray techniques. International Journal of Refractory Metals and Hard Materials, 2015, 49, 374-382.
[10]D. Seo , K. Ogawa, K. Sakaguchi, N. Miyamoto, Y. Tsuzuki. Parameter study influencing thermal conductivity of annealed pure copper coatings deposited by selective cold spray processes., Surface & Coatings Technology., 2012, 206, 2316-2324.
[11]Amirthan Ganesan, Jon Affi, Motohiro Yamada, Masahiro Fukumoto. Bonding behavior studies of cold sprayed copper coating on the PVC polymer substrate. Surface & Coatings Technology , 2012, 207, 262-269.

Effect of Nozzle Material on Downstream Lateral Injection Cold Spray Performance

D. MacDonald, S. Leblanc-Robert, R. Fernández, A. Farjam, B. Jodoin/ Ottawa, CA

In cold gas dynamic spraying, the gas nature, process stagnation pressure and temperature, and the standoff distance are known to be important parameters that affect the deposition efficiency and coating quality. This investigation attempts to elucidate the effect of nozzle material on coatings produced using a downstream lateral injection cold spray system. Through experimentation, it is shown that the nozzle material has a substantial effect on deposition efficiency and particle velocity. It is proposed that the effects are related to complex interaction between the particles and the internal nozzle walls. The results obtained lead to the conclusion that during the particle/nozzle wall contact, a nozzle with higher thermal diffusivity transfers more heat to the particles. This heat transfer results in lower critical velocities and therefore higher deposition efficiencies, despite a noticeable reduction of particle velocities which is also attributed to particle-nozzle interactions.

1 INTRODUCTION

Cold Gas Dynamic Spraying (CGDS) has been attracting interest from the thermal spray community since the original discovery of the process [1]. It has been established that the critical velocity is the fundamental physical parameter governing the CGDS process [2]–[4], with each feedstock powder material having its own critical velocity, above which the particles start to deform and adhere to the substrate to form a coating. Studies have allowed establishing the critical velocity for a wide variety of materials that can be sprayed by CGDS. It has also been established that critical velocity is a function of the substrate material as well as feedstock powder material, oxide content, size, shape, and impact temperature [5]–[8].

Since particle impact velocity is at the heart of the CGDS process, many studies have also explored the important area of nozzle design. It has been shown that the gas flow inside CGDS nozzles is largely dependent on the nozzle internal geometry. As such, studies have used classic gas dynamics principles to better understand the parameters governing proper nozzle design and also extended the design effort of such nozzles with the use of more complete three-dimensional Computational Fluid Dynamics (CFD) models, providing notable insight on nozzle design [9]–[11].

However, no study has yet looked at the effect of nozzle materials on the particle dynamics and resulting coatings, although several materials have been used, typically for nozzle fouling prevention [12], [13]. The current paper aims at reporting the effect of the nozzle material on the process performance through the limited study of the sprayability of Al 7075 feedstock powder on Al 7075 substrates using downstream lateral injection typically used in low-pressure cold spray. In particular, depositions are made using four nozzles with identical geometries but made of plastic, ceramic, steel and copper while using identical spray parameters. Relative Deposition Efficiencies (DE) are measured and compared between each nozzle, demonstrating a large variance. Exit particle velocity is measured for each nozzle in an effort to understand and explain the physical mechanisms governing the coating formation for the various nozzle material used.

2 EXPERIMENTAL PROCEDURES

2.1 CGDS System and Parameters

The CGDS system used to produce the coatings for this work was the commercially available SST Series EP Cold Spray System (Centerline Ltd, Windsor, Canada). The system has a maximum operating gas temperature of 500°C and a maximum gas pressure of 3.8 MPa. A quick connect nozzle holder with a 2 mm diameter throat was used with 120 mm long diverging nozzles. Powder was fed using a commercially available rotary powder feeder from Praxair Surface Technologies (Model 1264, Praxair Surface Technologies, Concord, USA).

In order to test the effect of different nozzle materials on deposition efficiency and quality, coatings were deposited using four different nozzle materials using the deposition parameters given in Table 1. All parameters were chosen as they are typical of low pressure cold spray.

Table 1. CGDS Parameters

Parameter Selection	Value
Gas Temperature	500°C
Gas Pressure	3.45 MPa
Gas Nature	Nitrogen
Traverse Speed	10 mm/s
Step Size	1 mm
Passes	4
Powder Feed rate	13.75g/min
Feeder Gas Flow Rate	1.7 m^3/h
Powder Feeder Gas Nature	Nitrogen
Standoff Distance	15 mm

D V3

2.2 Nozzle Materials and Geometry

The nozzles in this study are based off of the commercially available Ulti-flow nozzle (Centerline Ltd, Windsor, Canada) in which powder if fed radially after the nozzle throat, to prevent nozzle throat clogging. All nozzles, other than the polymer nozzle, were machined in house to match the internal geometry of the commercially available polymer nozzle. The nozzle geometry can be seen in Fig. 1.

In order to test the impact of the materials used for the nozzles, materials were selected for a range of thermal properties and mechanical properties. The materials must also be easily machinable, which limits the use of most ceramics. Table 2 presents the nozzle material properties (values taken from [14]–[17]).

The nozzle internal surface roughness levels were similar for all nozzles and are reported in Table 3. These values were measured using a Phase II SRG-4000 (Phase II, Upper Saddle River, NJ, USA) stylus roughness meter.

2.3 Powder and Substrate Materials

The powder used for this study is a gas atomized, aluminum 7075 powder (AA7075 - 200 mesh) (Valimet, Stockton, CA, USA). Laser diffraction analysis (Microtrac S3500, Nikkiso, Japan) was used to determine the mean particle sizes based on volume (MV), number of particles (MN), and area (MA) which are 42.10 µm, 13.46 µm, and 29.62 µm respectively. Typical relevant properties of this alloy can be found in Table 2.

The substrates used in this study were aluminum 7075-T6 flat plates. Immediately prior to CGDS deposition, the substrates were prepared through grit blasting and acetone cleaning.

2.4 Coating Analysis

Coatings and powders were investigated using a scanning electron microscope (EVO-MA10, Zeiss, UK) and an optical microscope (VHX-2000E, Keyence, Canada). Porosity measurements were quantified through analysis of the optical images over the entire coating cross-section.

2.5 Particle Velocity Measurement

Particle velocity measurements were taken using a ColdSprayMeter (CSM) eVOLUTION (Tecnar Automation Ltd, St-Bruno, Canada). This system uses a continuous 2.4 W (λ=810 nm) laser with a dual spit photomask in order to perform in-flight diagnostic on individual particles. This system provides precise velocity and diameter measurements of individual particles during flight.

2.6 Modelling of Gas Flow in Nozzle

A computational fluid dynamics (CFD) model was developed using the commercially available solver Fluent. The mass, momentum, and energy equations were solved using a coupled implicit scheme. The gas was also assumed to obey the perfect gas law and the turbulence was modeled using a k-epsilon model. The flow was assumed to be in a steady-state regime and 2D-axisymetric. All of the surrounding surfaces of the domain are taken by default as adiabatic, except for the inlet and outlet. At the domain inlet, the flow was assumed to be uniform and normal to the inlet cross section and the gas stagnation pressure and temperature are prescribed following the spray parameters. The pressure at the outlet was considered as atmospheric pressure. The effect of gravity on the flow was neglected.

For the k-epsilon model, the closure coefficients used are respectively Ce1=1.44, Ce2=1.92, Cl=0.09, rk=1.0, and re=1.3. The convergence criteria for absolute residuals were set to a maximum of 10-6. For the special cases where heat transfer to the surroundings was considered, it was modelled using a convection coefficient of 60 W/m^2K and surrounding temperature of 23°C. Grid independency was checked and it was found that a 100 by 30 mesh was sufficient to model the 2D-axiymetric nozzle flow adequately.

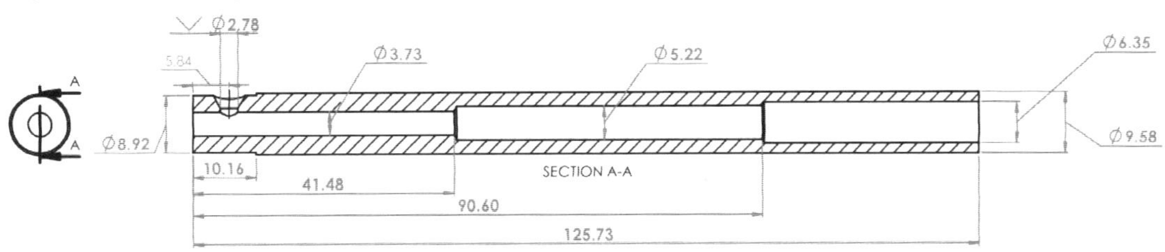

Fig. 1. Nozzle cross-section dimensions (dimensions given in mm).

Table 2. Relevant Nozzle and Powder Properties

Name	Specific Material	k	ρ	C	α	R_a
		W/m K	kg/m^3	J/kg K	m^2/s	μm
Polymer	Polybenzimidazole (PBI)	0.038	1300	1256	2.33E-08	1.8±0.6
Ceramic	Glass-Mica Ceramic	1.15	2700	1200	3.55E-07	5.5±0.9
Steel	304 Stainless Steel	20.2	8000	500	5.05E-06	5.8±0.4
Copper	C110 Copper	388	8890	385	1.13E-04	4.3±0.4
N/A	Aluminum 7075	130	2800	960	4.84E-05	N/A

2.7 Modeling of Heat Transfer between Particle and Nozzle Walls

The heat transfer module of ABAQUS/Standard was used for the heat transfer simulations involving the particle at the nozzle walls. Each simulation considered an aluminum 7075 particle in contact with the four different nozzle materials. A transient, 2D-axisymmetric model was used for a total duration of 100 microseconds where the heat transfer equation was solved. An 8 μm radial contact was set between the particle and the nozzle wall with no contact resistance, this contact zone was computed as the contact area of a 42 μm particle indenting into the wall a depth equal to the lowest surface roughness found in the nozzles. A constant temperature was set at the internal borders of the nozzle as a boundary condition to simulate the heat given by the nozzle. This temperature will be based on the temperature estimated by the CFD analysis. Finally, a quadratic-dominated mesh (DC2D4) was used with a maximum element size of 1/50 particle diameter.

3 Results and Discussion

A coating of significant thickness could not be deposited using the polymer nozzle. All other nozzles resulted in coatings with porosity levels less than 1% (See Table 3 for measured values). The results of the coating thickness can also be found in Table 3. The relative DE is a normalized value for coating thickness, taking the thickest coating as 100%. Fig. 2 shows the comparison of the coating thicknesses obtained and presents cross section images of these coatings.

Table 3. Results for Different Nozzle Materials

Nozzle Material	Porosity	Thickness	Relative DE
	%	μm	%
Polymer	N/A	0±0	0.0±0
Ceramic	0.0±0.0	290±6	25.8±2.1
Steel	0.0±0.0	725±21	64.4±2.9
Copper	0.1±0.1	1125±50	100.0±4.4

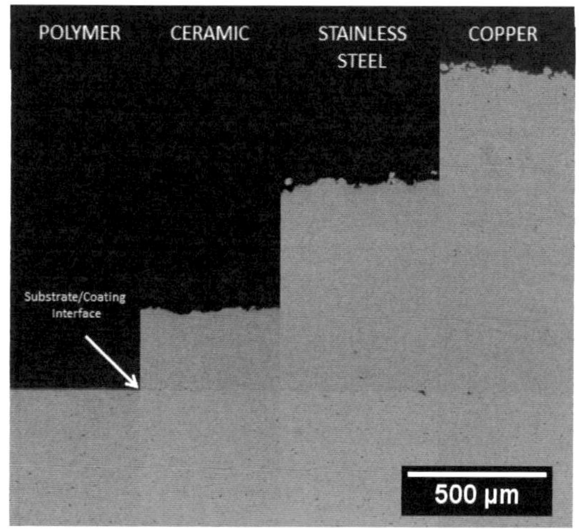

Fig. 2. Optical image of coating cross-sections demonstrating difference in coating thickness for the various nozzle materials.

A drastic change in relative DE is seen in the samples when switching between nozzle materials. Since deposition efficiency is typically affected by the particle impact velocity, particle velocity measurements were taken as a way to understand the phenomenon and are represented using a box and whisker plot in Fig. 3 (left). A range of velocities is seen for all nozzle materials, which is expected as particles of different size will have different velocities and follow different paths. From the results presented in Fig. 3 (left), it can be concluded that the nozzle material also had a significant influence on the particle velocity. However, these results reveal that deposition efficiency did not increase with measured particle velocity. In fact they appear to be inversely correlated with measured particle velocity. Furthermore a large change of relative DE is observed for similar particle velocities with the ceramic and stainless steel nozzle, again indicating that particle velocity is not the only parameter dictating the impact/adhesion behaviour of the particles. These results seem to be in contradiction with the results generally reported for the CGDS process [4], where deposition efficiency increases with particle impact velocity.

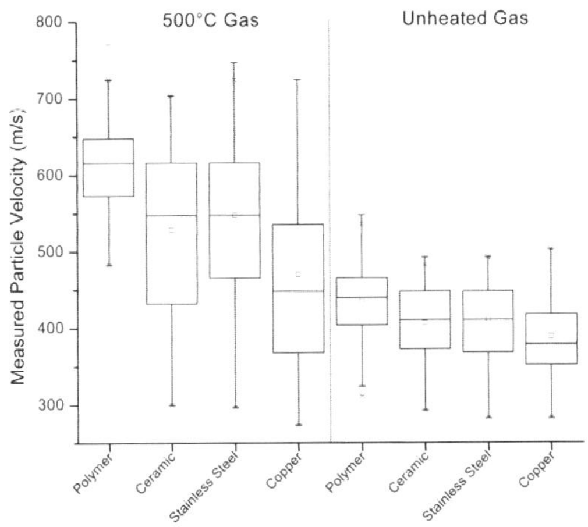

Fig.3. Particle velocity vs. nozzle material for heated gas (left) and for unheated gas (right).

In an effort to try elucidating these results, an investigation was completed in order to determine the cause of the particle velocity variation amongst the various nozzles, keeping in mind the fact that the nozzle geometries are all identical and that the variance in nozzle internal surface roughness is very small (Table 2) and does not correlate with the measured particle velocity.

The nozzle flow-particle momentum transfer can be represented by Eq 1 [18],

$$mV_p \frac{dV_p}{dx} = \frac{C_D A_D \rho (V - V_p)^2}{2} \qquad \text{(Eq 1)}$$

where m is the mass of the particle, V_P is the particle velocity, x is the position along the nozzle length, C_D is the drag coefficient, A_D is the cross sectional area of the particle, ρ is the density of the gas, and V is the gas velocity.

From Eq 1, one can observe that the particle velocity can be affected by a change in gas density or velocity in the nozzle. In this case, since the nozzle geometries are all identical gas dynamics analysis would generate the same gas density and velocity along the nozzle length for each nozzle materials. However, the gas dynamic equations would not account for heat being transferred from the nozzle gas flow to the surroundings through the nozzle wall, thus affecting the gas-particle momentum transfer. This could result in different particle velocities as the different nozzle materials would have different thermal resistance to the surroundings. A CFD model of this was completed incorporating both the nozzle flow gas dynamics and the heat transfer from the nozzle to the surroundings. The graphical results of this model are omitted from this proceeding for brevity, but the gas velocities and densities were found to nearly remain identical. Therefore, the heat transfer from the nozzle to the ambient air is minute. As such, this should not result in the large changes of particle velocity measured and reported.

To further support the modeling results obtained and ensure that the lower particle velocities are not the result of heat leaving through the nozzle walls, an experiment was done with the outside of the copper nozzle insulated with high temperature flexible ceramic insulation (aluminum silica). The median particle velocity for the copper nozzle went from 448 m/s to 483 m/s with the addition of the insulation. Although an increase in particle velocity was observed, the particle velocities were well below those of polymer nozzle (with a median velocity of 616 m/s). From this result and the results from the modelling, it is clear that the lowered particle velocities are not primarily a result of heat transfer between the nozzle and surroundings.

To investigate the role that gas stagnation temperature plays in particle velocity, an experiment was done using nitrogen gas at room temperature (as opposed to 500°C used to produce the reported coatings). These tests did not result in any coating (DE of 0%) and the relative difference in particle velocities for different nozzle materials was greatly reduced (see Fig. 3 (right)). Although the difference in particle velocities between each nozzle was reduced, a similar trend in velocities is seen when the stagnation temperature was lowered. From this it can be hypothesized that the phenomenon slowing the particles is related to and enhanced by the gas stagnation temperature.

Close observation of the flow modeling results reveals that the nozzle wall temperature is expected to be substantially higher than the gas temperature on the nozzle centerline due to flow stagnation at the walls (graphical results omitted from this proceeding for brevity). These wall/gas temperatures differ only slightly between nozzle materials, and these differences should not be able to account for such a drastic change of velocity and DE between the various nozzles.

Adhesion of particles along the nozzle walls and eventual clogging after extended use was observed in all nozzles other than the polymer nozzle. This observation has revealed that there must be interactions between the nozzle walls and the particles. It is hypothesized that the particle-wall interactions could cause particles to smear and slow down, with a different behaviour for each material and a stronger effect as the nozzle wall/gas stagnation temperature increases. During the particle/wall interactions particle velocity is expected to change as a result of friction, thermal diffusivity, material resilience, and particle temperature. All of these factors may be at play during the deposition process, resulting in the range of particle velocities seen in Fig. 3.

It is known that particle impact temperature has an effect on the CGDS deposition efficiency by reducing the critical velocity required for deposition. Assadi et al developed a relationship for critical velocity as expressed in Eq 2 [19],

$$V_{cr} = 667 - 14\rho_P + 0.08T_m + 0.1\sigma_u - 0.4T_i \quad \text{(Eq 2)}$$

where V_{cr} is the critical velocity required for deposition, ρ_P is the particle density, T_m is the melting temperature, σ_u is the ultimate strength of the particle, and T_i is the initial particle temperature.

Eq 2 suggests that an increase in particle impact temperature could lower the particle critical velocity enough to account for the increase in deposition efficiency. Since it was shown that the gas temperature and density inside the nozzle do not vary significantly between the different nozzles, the change in particle temperature that could explain the reduced critical velocity leading to increased DE despite the lower measure particle velocity can potentially be related to the interactions between the particles and the nozzle walls. As with the particle velocity, the particle temperature differences for different nozzles would be a result of the properties of the nozzle materials itself. Fig. 4 shows the relative DE plotted against the thermal diffusivity of the nozzle materials.

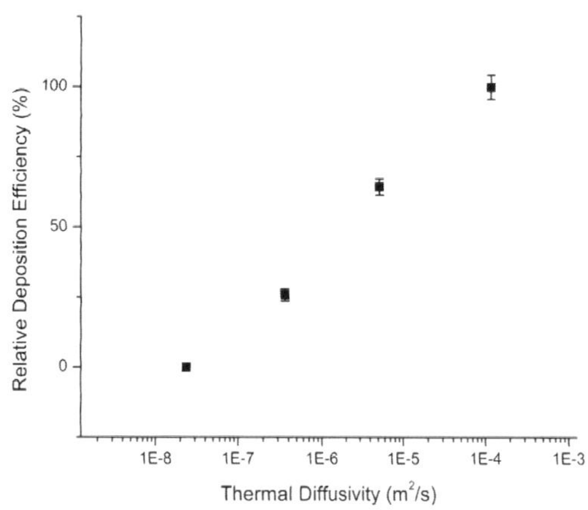

Fig. 4. Relative DE vs. thermal diffusivity.

A clear trend is observed; materials with higher thermal diffusivity resulted in a higher deposition efficiency for identical nozzle geometries and spray parameters. Following these results, it is hypothesized that nozzle material with higher thermal diffusivity would transfer heat more efficiently to the particles during the particle-wall interactions involved in cold spray using lateral powder injection. These particles being at a higher temperature after their interaction with the nozzle walls would have a lower critical velocity and produce a coating despite the lower measured velocities reported. In all cases where deposition occurred on the substrate, the nozzles had evidence of deposition/erosion on the internal walls.

To further support the hypothesis that heat is being transferred from the nozzle walls to the particles, a simple modeling of a particle in contact with the nozzle wall was done. The particle size was chosen based on the volumetric mean particle size (42 μm), the initial particle temperature was chosen as approximately the centerline gas temperature according to the model (100°C), and the wall temperature was chosen to approximately match the model (415°C). A plot of average particle temperature vs. time is shown in Fig. 5.

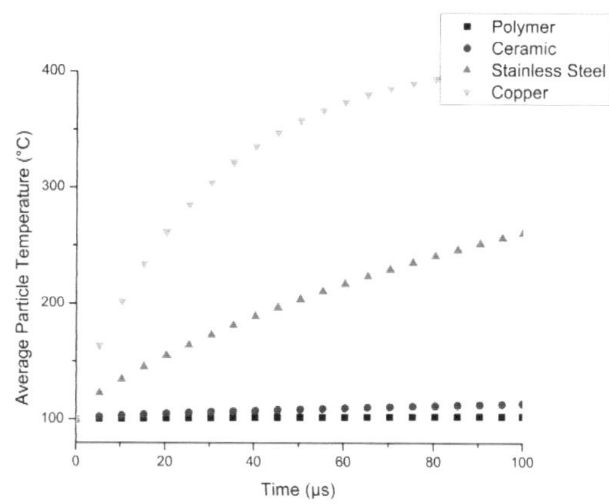

Fig. 5. Average particle temperature as function of time in contact with nozzle wall.

The minimal time a particle could be inside the nozzle can be estimated by assuming the particles are travelling the length of the nozzle at the recorded velocities. For example, the fastest particles came from the polymer nozzle and had a median velocity of 616 m/s. From this, the minimal time a particle could be inside the nozzle can be estimated to be approximately 200 microseconds.

While there are many unknowns in this problem, this model demonstrates that if a particle touches the nozzle wall for even 10% of the minimum time (20 μs) it is inside the nozzle, substantial differences in particle temperatures may occur. These differences would result in lower required critical velocities, which could be the reason for the increase in deposition efficiency seen in this study. In order to get a better understanding of the particle/wall interaction, a full CFD model of the nozzle including particles should be done, but this is beyond the scope of this research and will be investigated in a future paper.

4 CONCLUSIONS

This investigation focused on the impact of nozzle material on the coatings produced using a downstream lateral injection cold spray system. Although this paper only covers several nozzle materials, a single set of spray parameters, and one powder/substrate combination, the results demonstrate that nozzle material is a parameter that should be taken into consideration in cold spray. Substantial changes in DE can occur by changing the nozzle material. It is concluded that the change in deposition efficiency observed is related to the thermal diffusivity of the nozzle material, as a higher thermal diffusivity will transfer more heat to the particle during particle/wall interaction. These higher particle temperatures would therefore lower the critical velocity required for the particles to deposit on the substrate, resulting in an increased the deposition efficiency.

5 LITERATURE

[1] Alkhimov, "Gas-Dynamic Spraying Method for Applying a Coating," 53024141994.

[2] E. Calla, D. G. McCartney, and P. H. Shipway, "Effect of Deposition Conditions on the Properties and Annealing Behavior of Cold-Sprayed Copper," *J. Therm. Spray Technol.*, vol. 15, no. 2, pp. 255–262, Jun. 2006.

[3] F. Raletz, M. Vardelle, and G. Ezo'o, "Critical particle velocity under cold spray conditions," *Surf. Coatings Technol.*, vol. 201, no. 5, pp. 1942–1947, Oct. 2006.

[4] D. L. Gilmore, R. C. Dykhuizen, R. A. Neiser, T. J. Roemer, and M. F. Smith, "Particle Velocity and Deposition Efficiency in the Cold Spray Process," *J. Therm. Spray Technol.*, vol. 8, no. December, pp. 576–582, 1999.

[5] T. Schmidt, F. Gärtner, H. Assadi, and H. Kreye, "Development of a generalized parameter window for cold spray deposition," *Acta Mater.*, vol. 54, no. 3, pp. 729–742, Feb. 2006.

[6] C.-J. Li, W.-Y. Li, and H. Liao, "Examination of the Critical Velocity for Deposition of Particles in Cold Spraying," *J. Therm. Spray Technol.*, vol. 15, no. 2, pp. 212–222, Jun. 2006.

[7] J. Vlcek, L. Gimeno, H. Huber, and E. Lugscheider, "A Systematic Approach to Material Eligibility for the Cold-Spray Process," *J. Therm. Spray Technol.*, vol. 14, no. 1, pp. 125–133, Mar. 2005.

[8] W.-Y. Li, C. Zhang, H.-T. Wang, X. P. Guo, H. L. Liao, C.-J. Li, and C. Coddet, "Significant influences of metal reactivity and oxide films at particle surfaces on coating microstructure in cold spraying," *Appl. Surf. Sci.*, vol. 253, no. 7, pp. 3557–3562, Jan. 2007.

[9] W.-Y. Li, H. Liao, G. Douchy, and C. Coddet, "Optimal design of a cold spray nozzle by numerical analysis of particle velocity and experimental validation with 316L stainless steel powder," *Mater. Des.*, vol. 28, no. 7, pp. 2129–2137, Jan. 2007.

[10] A. P. Alkhimov, V. F. Kosarev, and S. V Klinkov, "The Features of Cold Spray Nozzle Design," *J. Therm. Spray Technol.*, vol. 10, no. February 2000, pp. 375–381, 2001.

[11] B. Jodoin, "Cold Spray Nozzle Mach Number Limitation," *J. Therm. Spray Technol.*, vol. 11, no. December, pp. 496–507, 2002.

[12] F. Gärtner, T. Stoltenhoff, T. Schmidt, and H. Kreye, "The Cold Spray Process and Its Potential for Industrial Applications," *J. Therm. Spray Technol.*, vol. 15, no. 2, pp. 223–232, Jun. 2006.

[13] E. Irissou, J.-G. Legoux, A. N. Ryabinin, B. Jodoin, and C. Moreau, "Review on Cold Spray Process and Technology: Part I—Intellectual Property," *J. Therm. Spray Technol.*, vol. 17, no. 4, pp. 495–516, Oct. 2008.

[14] R. B. Sandor, "PBI (Polybenzimidazole): Synthesis, Properties and Applications," *High Perform. Polym.*, vol. 2, no. 1, pp. 25–37, 1990.

[15] R. Steiner, "ASM Handbook Volume 1, Properties and Selection: Irons, Steels and High Performance Alloys," p. 2521, 1993.

[16] "MACHINABLE GRADES TECHNICAL DATA SHEET," Clifton NJ, 2014.

[17] R. Nunes, *ASM Handbook Volume 2, Properties and Selection: Nonferrous Alloys and Special Purpose Materials.* .

[18] V. F. Kosarev, S. V Klinkov, A. P. Alkhimov, and A. N. Papyrin, "On Some Aspects of Gas Dynamics of the Cold Spray Process," *J. Therm. Spray Technol.*, vol. 12, no. June, pp. 265–281, 2003.

[19] H. Assadi, F. Gärtner, T. Stoltenhoff, and H. Kreye, "Bonding mechanism in cold gas spraying," *Acta Mater.*, vol. 51, no. 15, pp. 4379–4394, Sep. 2003.

Influence of Length of Divergent and Parallel Section of the Rectangular Cross-Section Nozzle on Behavior of Particles of Copper in High-Pressure Cold Spray

Kazuhiko SAKAKI, Shinji ARAI and Tatuyuki TAGAMI , Nagano / J

In the cold spray process, cross-sectional shape of the nozzle has a significant effect on spray pattern of coatings. Spray pattern of the rectangular spray nozzle is wider than that of the circular spray nozzle. Accordingly, rectangular nozzle is better efficiency than circular nozzle at spray a constant width. The goal of this investigation is to establish a design for the cold spray gun nozzle in order to gain a flatter the spray pattern of coatings. We have investigated the effect of expansion ratio of rectangular nozzle on spray pattern of copper coatings in cold spraying by the computational fluid dynamics (CFD) and experiments. And the particle velocity and deposition efficiency of the copper coatings were dependent on the expansion ratio and geometry of the rectangular nozzle.
In this study, CFD and experiments investigated the influence of length of divergent and parallel section of the converging-diverging parallel-shaped rectangular cross-section (C-D-Para-Re) nozzle on behavior of the gas and copper particles in high-pressure cold spray, and C-D-Re nozzle, too. As the results, it found that there is an optimal ratio of the length of the diverging section and the parallel section for C-D-Para-Re nozzle.

1 Introduction

One of the major elements of the cold spray (CS) process [1] is the high-speed gas jet, which is governed by gas dynamics. In gas dynamics, supersonic flows are obtained with convergent-divergent (or de Laval) nozzles, which are used for rocket motors. Rocket motors, including their nozzle design, have been studied and analysed in detail. While the principal purpose of the design of a rocket motor nozzle is to maximize the thrust, in thermal spraying including cold spraying, the main purpose is to obtain better coating quality [1-16]. Previous studies on thermal spraying show the coating properties are principally determined by the thermal and kinetic energy states of particles upon impact with the substrate. To have a balance between these two states, various changes to the design of the high velocity oxygen fuel (HVOF) gun nozzle have been attempted. However, works concerning the influence of nozzle geometry on the thermal spray process are sparse [2-6]. Previously, the effects of the throat diameter and exit divergence of the gun nozzle on the HVOF process have been considered [4]. The combustion gas flow (such as pressure, velocity, temperature and expansion state of gas jet from the nozzle exit), the particle behavior and, therefore, the nature of coatings were found to be significantly influenced by these nozzle parameters. In addition, the effect of the expansion state of the combustion gas jet on the HVOF process was investigated using a diverging nozzle exit. The particle velocity reached a maximum with the correct expansion state of the gas jet due to an increased gas jet velocity. This resulted in an increase in the bonding strength of the NiCrAlY coating [5].
The nozzle geometry is also important with regard to the cold spray method. In the cold spray method, a coating is formed by exposing a substrate to high-velocity solid-phase particles, which have been accelerated by supersonic gas flow at a temperature much lower than the melting or softening temperature of the feedstock. Cold spray systems employ various kinds of gun nozzle contours, such as convergent-divergent (or de Laval) [1, 4-12, 14-15, 18, 19], convergent-divergent-barrel [8, 15, 18, 19] and convergent-barrel [9, 10] nozzle contours.
And the cross-section of divergent section of the cold spray nozzle has either circle or rectangle. The circular nozzles are one-piece construction, are machined using standard fabrication practices. Rectangle cross-section nozzle is fabricated in two pieces to active the rectangular geometry and weld and machined to final configure ration. Thus, in the commercial CS device, circular section nozzles are often found.
The particle velocity profiles are quite different comparing the circular and rectangular exit geometries in cold spray. The rectangular nozzles provide a more uniform particle velocity across the exit of nozzle. The circular exit is parabolic in shape. Both profiles are characteristic of their respective exit geometries. A particular desired spray profiles may be obtain by changing the nozzle exit geometry [12]. And there is a report [15] that deposition efficacy of copper coatings became maximum with 12 from 11 in of the expansion ratio of circular nozzle for stagnation pressure 3MPa and stagnation temperature 350℃ of nitrogen gas. On the other hand, the expansion ratio of rectangular nozzle is from 1.8 to 4.8 [11], or from 3.8 to 4.0[16]. This expansion ratio is considerably smaller than 11which is the optimum expansion ratio of the circular section nozzle.
The goal of this investigation is to establish a design for the cold spray gun nozzle in order to gain a flatter the spray pattern of coatings. We have investigated the effect of expansion ratio of rectangular nozzle on spray pattern of copper coatings in cold spraying by the computational fluid dynamics (CFD) and experiments [18, 19]. And the particle velocity and deposition efficiency of the copper coatings were dependent on the expansion ratio and geometry of the rectangular nozzle.
In this study, CFD and experiments investigated the influence of length of divergent and parallel section of the converging-diverging parallel-shaped rectangular cross-section (C-D-Para-Re) nozzle on behavior of

the gas and copper particles in high-pressure cold spray. And we study about the influence of length of divergent section of the converging-diverging shaped rectangular cross-section (C-D-Re) nozzle, to compare with the circular nozzle. As the results, it found that there is an optimal ratio of the length of the diverging section and the parallel section for C-D-Para-Re nozzle.

2 Numerical simulation of cold spray

There are several techniques that can be used to calculate the gas flow of cold spray systems. Recent analyses have used computational fluid dynamics (CFD) methods to simulate complex phenomenon in two dimensions. In the present paper, Ansys Fluent (version 15.0) commercial software [20] was used to predict the gas and particle flow. A schematic drawing of the cold spray equipment for the present study is shown in **Fig. 1**. **Fig.2** is shown a modelling of an axial-symmetric supersonic nozzle with rectangular cross-section. The nozzle throat diameter d_t is 2 mm (constant), the length of the converging section $L_{con.}$ is 50 mm (constant), the total length of divergent section $L_{div.}$ and the parallel length $L_{para.}$ of the nozzle is 200 mm(constant), and ten different ratio of the length of the diverging section and the parallel section for C-D-Para-Re nozzle are used, as shown in **Table 1**. It has been optimized in the previous report [9, 19] that the total length of divergent section $L_{div.}$ and the parallel length $L_{para.}$ of the nozzle is 200 mm.

The quarter three-dimensional models were used to reduce the computation cost. The computational domain was partitioned into a number of hexahedral cells. In order to achieve a grid independent solution, a variable number of elements about 150,000 and number nodes of about 60,000 were required, depending on the cross-sectional shape of the nozzles and nozzle outlet peripheral region.

The gas inlet was chosen as pressure inlet with a pressure value of 3.0 MPa[gage] and a temperature value of 623 K with process (nitrogen) gas. The particle velocity of 10 m/s was used in many models, thus in our model all particles are injected randomly at the surface of powder inlet with a constant velocity of 10 m/s at room temperature (300K). The outlet of the computational domain was treated as pressure outlet with an atmosphere pressure (0.0MPa [gage]). A standard no-slip condition was used for the nozzle wall and substrate surface. The heat transfer process between the gas and the wall was not considered. The fluid was computed through the ideal gas law, therefore the density was assumed constant. In order to capture the turbulent flow features accurately, the standard K–ε turbulence model available in FLUENT was used to model the turbulent flow in the simulation. The standard wall function was chosen for the near-wall flow treatment.

The trajectory of magnesium particles was simulated. The physical properties of copper particles are shown in **Table 2**. The acceleration of particles was computed using discrete phase modelling (DPM). The heat transfer between the gas and particles had been considered. Particle–particle interactions and the effect of particles on the gas phase were neglected. More than 100 particles were used to obtain reliable statistical data in each case. In order to take turbulence effects into account, the Stochastic-Tracking type model implemented in Fluent was used.

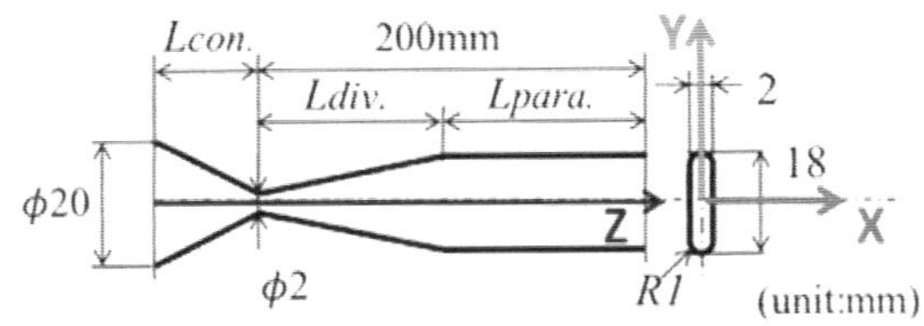

Fig. 1. Schematic cross-section diagram of converging-diverging parallel-shaped rectangular cross-section (C-D-Para-Re) cold spray nozzle used.

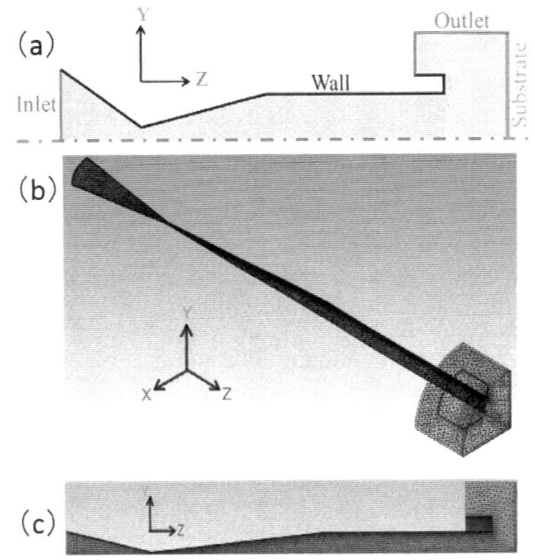

Fig. 2. Modeling of an axial-symmetric supersonic nozzle with rectangular cross-section; (a) schematic of simulation modelm, an example of a computational mesh (b) overall view(outside), (c)overall view(plane of symmetry).

Table 1 Cold splay nozzle dimensions unit: mm

No.	Nozzle Name C-D-Para(Re)-	$L_{con.}$	$L_{div.}$	$L_{para.}$	$L_{div.} : L_{para.}$	$\frac{L_{para}}{L_{div}+L_{para}}$
1	20-180		20	180	1 : 9	0.9
2	40-160		40	160	1 : 4	0.8
3	60-140		60	140	3 : 7	0.7
4	80-120		80	120	2 : 3	0.6
5	100-100	50	100	100	1 : 1	0.5
6	120-80		120	80	3 : 2	0.4
7	140-60		140	60	7 : 3	0.3
8	160-40		160	40	4 : 1	0.2
9	180-20		180	20	9 : 1	0.1
10	200		200	0	10 : 0	0

Table 2 Physical properties of Copper particles.

Diameter (μm)	10
Density (kg/m^3)	8933
Specific heat, C_p (J/(kg· K))	384.5
Thermal conductivity (W/(m· K))	399.7

In this approach, the Discrete Random Walk (DRW) model was used to predict the particle velocity and distribution. The fundamental fluid-dynamic and particle dragging equations employed in the model can be found in the reference [20].

3 Experimental equipment and methods

A schematic drawing of the cold spray equipment for the present study is shown in **Fig. 1**. Compressed nitrogen gas is introduced to a heater and a powder feeder. The pressurized gas is heated in an electric furnace. Usually, a convergent-divergent (C-D) nozzle is used in a typical cold spray device, and feedstock is injected axially into the main process gas flow at the gun nozzle intake. The cold spray equipment for this study was developed at Shinshu University [3, 4] and has a convergent-divergent-barrel (C-D-B) nozzle [18]. As described below, the results of CFD, were used C-D-Re-100-100 nozzle, of which the nozzle throat diameter d_t is 2 mm, the length of the converging section $L_{con.}$ is 50 mm, the length of divergent section L_{div} is 100 mm and the parallel length of the nozzle is 100 mm, shown in Table 1 No.5, in the experiment.

The gas pressure value in nozzle chamber is 3.0 MPa [gage] and the gas temperature is 623 K. The pressure of the carrier gas is 3.1 MPa in order to insure the powder feeding.

The Feedstock powder is delivered by a precision metering device called "powder feeder" and typically is injected axially into the center of the high-pressure side of the nozzle intake (axial injection of powder: A/I) . Copper is cold sprayed onto mild steel (JIS SS400) substrates using pure copper (Cu-HWQ 10, Fukuda Metal Foil & Power Co., Ltd., Kyoto) with averaged particle diameter of 10 µm as the starting material. Spray pattern of copper coating was measured every 2mm with a dial gauge.

The deposition efficiency was ascertained by measuring the weight gain on substrates for a spray time of 60 s taking into account the known feed-stock powder flow rate.

4 Results and discussion

4.1 Influence of the ration of length between divergent section and parallel section on the gas and Cu particles velocity by CFD

Fig. 3 shows the Influence of the ration of length between divergent section and parallel section on the process (nitrogen) gas velocity on the nozzle Z-axis in nozzle the influence of the ratio of the divergent section of the nozzle and the parallel section length on the distribution of the gas velocity on the nozzle central axis.

Gas reaches sonic velocity at the throat section becomes a supersonic accelerated by expansion in a nozzle divergent section, reaches the nozzle end slowly decreases due to pipe friction in the nozzle the nozzle cross-sectional area does not change the

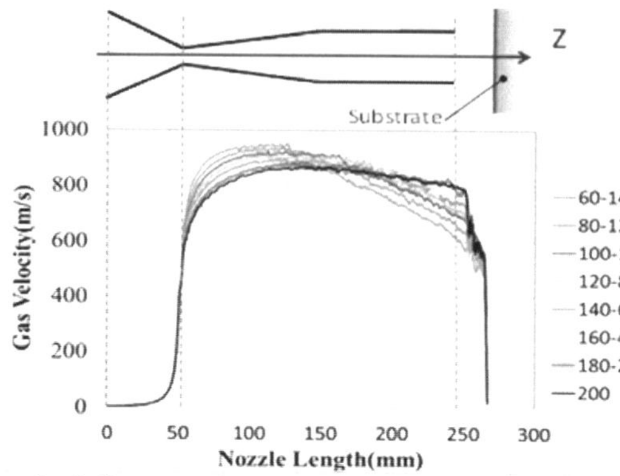

Fig. 3 Influence of the ration of length between divergent section and parallel section on the process (nitrogen) gas velocity on the nozzle Z-axis in nozzle. (P_i=3MPa, T_{gi}=350°C).

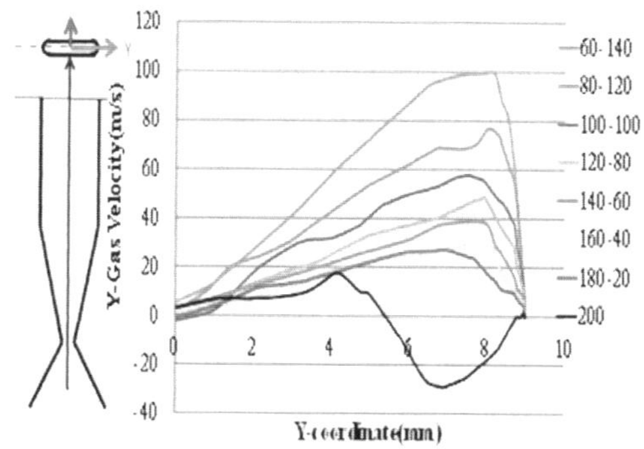

Fig. 4 Influence of the ration of length between divergent section and parallel section on the process (nitrogen) gas velocity on the nozzle Y-axis at end of divergent part in nozzle. (P_i=3MPa, T_{gi}=350°C).

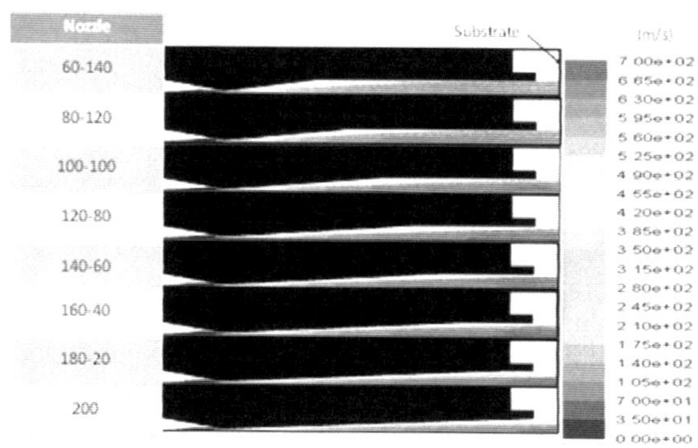

Fig. 5 Influence of the ration of length between divergent section and parallel section on Cu particles velocity. (N_2, P_i=3MPa, T_{gi}=350°C).

parallel section. Nozzle expansion ratio used in the present paper (optimum area ratio of the throat and the exit in nozzle), with respect to the nozzle inlet pressure 3.0MPa [gage], since the gas jet is the proper expansion, of the gas jet at the nozzle outlet and later rapid expansion (under expansion jet) and contraction (hyperinflation jet) are not. However, since the substrate is in a 15mm away, abruptly decelerated

from around 10 mm of the substrate immediately before. Although not shown the pressure is suddenly occurs up plate shock wave. Incidentally, C-D-Para. (Re) -20-180, -40-160 since the abruptly expanding short divergent section, since the flow occurs is peeled from the nozzle inner surface, not shown.
In addition, those nozzle divergent section is short, C-D-Para. (Re) -60-140 is, at 948m / s and the current nozzle at the nozzle divergent section end has been reached in the fastest. Further, in **Fig. 4**, it shows the distribution in the nozzle width direction of the *Y*-direction velocity of the gas flow in the nozzle divergent section end. In CD-Para. (Re) -60-140, has a significant peak compared to other nozzles, *Y* = fairly early gas of about 1/10 of 100m / s and the axial gas velocity is in the width direction in 8mm It flows, the driving force that attempts to move from the center of the particles in the width direction. However, the speed of the Y-direction becomes the spray pattern obtained by biasing the too large particles to the ends [18]. Therefore, this time, C-D-Para. (Re) is per -100-100, it considered a nozzle to appropriately distribute the particles in the nozzle width direction.
In **Fig. 5**, in the flow of each nozzle, 100 copper particles, discrete is considered random [20]. As a result of the effect of turbulence by using a random walk. As the ratio of the nozzle parallel section length is greater, the particles are widely distributed in the nozzle width (Y) direction. On the other hand, as shown in **Fig. 6**, the collision rate of the copper particles to the substrate on the nozzle central axis, parallel section length is not more than 100 mm, is greater than 660mm / s.
From the above results, the distribution and the velocity distribution of the nozzle the width direction of the particles, In this CFD results, although not sufficiently examined is obtained, the experimental result [18] be based, rectangular cross section nozzle it was decided to trial the CD-Para.- (Re) -100-100 nozzle.

4.2 Experimental results (Effect of nozzle shape on the coating spray pattern)

In Fig. 7 and 8, a rectangular cross-section converging-diverging nozzle parallel CD-Para. (Re) -100-100 the end section and the thickness of the rectangular cross-section converging-diverging nozzle CD (Re) -120 of the coating surface was used for comparison indicate distribution (thermal spray pattern), respectively.
Both nozzle outlet section is a rectangle of 2 × 18 ㎜, it is firmly deposited as shown in **Fig. 7 (a)**, the boundary of the coating is clear. This is also reflected in the coating thickness distribution of the spray pattern in **Fig. 8**, it has spread as the further 19 ㎜ and 0.6 ㎜ the width. Further, the coating cross-sectional area than the results in **Fig. 8, (a)** 1.66 ㎜ ^{2} and **(b)** a 1.26 ㎜ ^{2}, it is considered that the density of the copper coating is no great difference, deposition rate is 1.3 times as high as that become.
Therefore, to put the parallel section of the

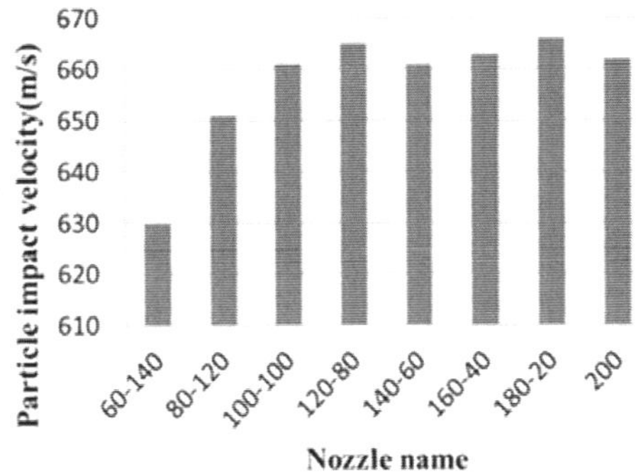

Fig.6 Influence of the ration of length between divergent section and parallel section on the impact velocity on the subetrate (N_2, P_i=3MPa, T_{gi}=350°C).

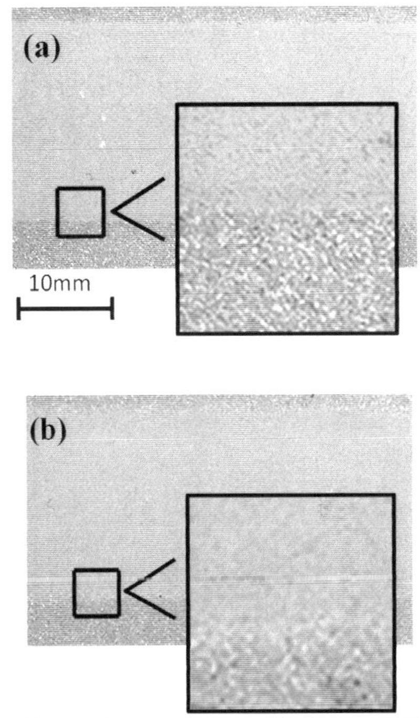

Fig. 7 Photo of spray pattern of copper coatings with (a)C-D-Para(Re)-100-100 nozzle, (b)C-D(Re)-120 nozzle [18] (N_2, P_i=3MPa, T_{gi}=350°C).

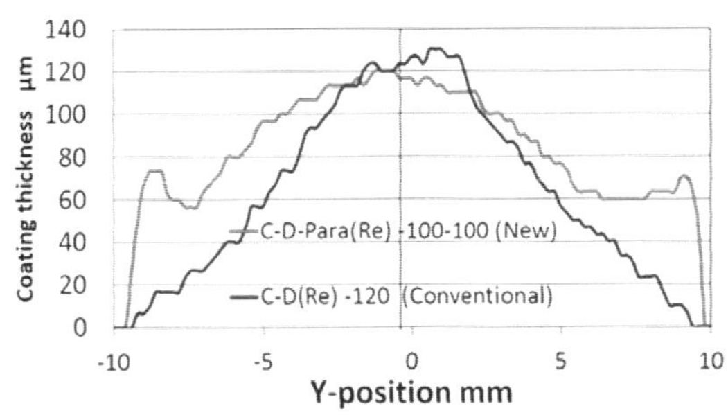

Fig. 8 Influence of width of rectangular cross-section nozzle on copper coating spray pattern (N_2, P_i=3MPa, T_{gi}=350°C).

appropriate length after the nozzle divergent section, it was found to be effective in smoothing the spray pattern. This gas conditions (N_2, P_i = 3MPa [gage], T_{gi} = 350 ℃) to the length ratio of the divergent section of the subsequent throat and the parallel section was relatively appropriate at around 0.5.

5 Summary

The divergent section and a length ratio of the parallel section of the converging-diverging parallel shaped rectangular cross-section nozzles was varied, with the following results.

a) From the results of a) CFD, parallel section is shorter it is, the maximum speed of the gas is as high as 948m / s, it occurs near the end of the divergent section. In addition, in the divergent section end, gas velocity of the rectangular width direction, still high shorter nozzle of the parallel part, fairly high, about 1/10 of 100m / s and the axial velocity.
b) At a result converging-diverging parallel-shaped nozzle was fabricated on the basis of the b) CFD, it was able to produce a film of more flat thermal spray pattern than the converging-diverging nozzle shape.

ACKNOWLEDGMENTS

Part of this study, Scientific Research Grant Program (Academic Research Grants Fund grant) (Foundation (C) (general): 2015 - 2016 fiscal year) was carried out by the issue number 15K06460.

The authors thank Fukuda Metal Foil & Power Co., Ltd. for supplying the cupper feedstock powder. The authors thank students Masatoshi Itoh, who works now in Seiko Epson Corp., of graduate school of Shinshu University for their assistance.

6 Literature

[1] A. P. Alkimov, A. N. Papyrin, V. F. Kosarev and M. M. Shushpanav, US Pat. 5302414(1994)

[2] C. M. Hackette and G. S. Settles: The Influence of Nozzle Design on HVOF Spray Particle Velocity and Temperature", Thermal Spray Science & Technology, C. C. Berndt and S. Sampath, Ed, ASM International, Material Park, OH, (1995), p.p.135/40.

[3] K.Kopiola, J. P .Hirvonen, L. Laas and F.Rossi: The Influence of Nozzle design on HVOF Exit Gas Velocity and Coating Microstructure", J. Thermal Spray Technology, l6 (4), (1997), pp.469/74.

[4] K. Sakaki, Y. Shimizu, Y. Gouda and A. Devasenapath: Effect of nozzle geometry on High Velocity Oxygen-Fuel (HVOF) Thermal spraying process," proceedings of the 15th International Thermal Spray Conference, C .Coddet, ed., ASM International, (1998), pp. 445-450.

[5] K. Sakaki, Y. Shimizu and Y. Gouda, 'Effect of Expansion State of Combustion Gas Jet on High Velocity Oxygen-Fuel Thermal Spraying Process', J. Japan Institute of Metals, 63(2),(1999), 269/76 (in Japanese).

[6] D.L. Gilmore, R.C. Dykhuizen, R.A. Neiser, T.J.Roemer, and M.F.Smith: Particle Velocity and deposition Efficiency in Cold Spray Process," J. Thermal Spray Technology, 8 (4), (1999), pp. 576/82.

[7] K. Sakaki and Y. Shimizu: Effect of Increase in Entrance Length of Gun Nozzle on HVOF and Cold Spray Process," J. of Thermal Spray Technology, 10 (3), (2001), pp.487/96.

[8] T. Stoltenhoff, H. Kreye, and H.J.Richter "An analysis of the cold spray process and its coatings," J. Thermal Spray Technology, 11 (4), (2002), pp. 542/50.

[9] K. Sakaki, N. Huruhashi, K. Tamaki and Y. Shimizu: Effect of nozzle geometry on cold spray process, International Thermal Spray Conference, Lugscheider, E., and Berndt, C.C., ed.,DVS Deutscher Verband für Schweißen, (2002),pp. 385/89.

[10] W-Y. Li, et al.:Optimal design of a convergent-barrel cold spray nozzle by numerical method," Applied Surface Science, 253(2) , (2006), pp.708/13.

[11] K.Sakaki, S.Shinkai, N.Ebara and Y.Shimizu: Effect of geometry of the gun nozzle, the increase in the entrance convergent section length and powder injection point on cold sprayed titanium coatings' Materials Science Forum Vols. 534-536, (2006), pp.413/16

[12] A. N. Papyrin ed. :Cold Spray Technology, Elsevier, Cambridge, U.K., (2007), p.119/78.

[13] A. N. Papyrin ed., Cold Spray Technology, Elsevier, Cambridge, U.K., (2007), p.216/20.

[14] V. K. Champagne, ed.: The Cold Spray Materials Deposition Process," Woodhead Publishing Materials, Cambridge, U.K., (2007), pp. 117/26.

[15] W.-Y. Li, H.L. Liao, G. Douchy, C. Coddet: Optimal design of a cold spray nozzle by numerical analysis of particle velocity and experimental validation with 316L stainless steel powder", Materials and Design, 28 (7), (2007), pp.2129/37.

[16] K. Sakaki, K. Takeda, K. Takada, T. Hosono, and Y. Shimizu: Influence of the expansion ratio of the gun nozzle and gas pressure on properties of cold sprayed copper coatings, Proceedings of the 3rd JSME/ASME International Conference on Material and Processing, (2008), MSEC_ICM&P2008-72015.

[17] X.K. Suo, T.K. Liu, W.Y. Li, Q.L. Suo, M.P. Planche, H.L. Liao: Numerical study on the effect of nozzle dimension on particle distribution in cold spraying", Surface & Coatings Technology, Vol.220, (2013), pp. 107/11.

[18] K.Sakaki, T. Akashi, and T.Hosono, "Influence of Cross sectional Shape of Cold Spray Nozzle on Spray pattern of copper coatings", DVD of Proc. of the JSME/ASME 2014 International Conference on Materials and Processing, Detroit (2014), ICMP2014-4961.

[19] S. Arai, K. Sakaki, T. Tagami and T. Tuchida: Optimization of the Shape of Rectangular Cross-Section Nozzle in High-Pressure Cold Spray (Influence of Length of Barrel and Divergent Section of Nozzle), Proc. of 2015 Japan Thermal spray society fall meeting, (2015), pp.5/6(in Japanese).

[20] ANSYS, Inc. : ANSYS FLUENT Theory guide for Release 15.0, (2015).

Effect of gas pressure on interfacial bonding features of the cold-sprayed Ni coatings on Al substrate

Yingchun XIE[1], Marie-Pierre PLANCHE[1], Rija-nirina RAOELISON[1], Christophe VERDY[1], Philippe HERVE[2], Hanlin LIAO[1]
1 IRTES-LERMPS, Université de Technologie de Belfort-Montbéliard, Belfort, France
2 LEME. Université Paris Ouest Nanterre la Défense, Ville d'Avray, France

In this study, the interface after heat treatment between Ni coating and Al substrate was analyzed depending on the gas pressures. Relationships between the interfacial bonding features and gas pressures were discussed. It was found that metallurgical bonding area is a relevant factor that induces interfacial atomic diffusion. It was also demonstrated that higher gas pressure increased intermetallic phase area leading to enhance adhesion of coating. As the gas pressure increased, the porosity decreased, thus enabling the formation of denser coating layer.
Key words: Cold spraying, gas pressure, interfacial bonding features, intermetallic phase, coatings properties

1 Introduction

Cold spray is a relatively new coating technology with a fast growing coating applications during the last decades. In this process, the powder feedstock is accelerated to high speed by a high velocity gas flow in the Laval nozzle[1]. The particles impact onto the substrate, and a coating is formed through the intensive plastic deformation at a temperature below the melting point of the considered feedstock material. So this technique can minimize the negative effect caused by a molten or semi-molten state. As a result, spray materials experience little microstructure change, little oxidation or decomposition as compared to the conventional thermal spray processes and provides the possibility of coating oxygen-sensitive materials. At the same time, the cold spray process can provides a possibility to design the coating microstructure through feedstock powders. Many metals and their alloys and composites have been successfully coated by cold spraying[2-5].
Currently, the most accepted bonding theory for cold spray are mechanical and metallurgical bonding. In this high speed impacting process , the strong plastic deformation causes disruption of the thin oxide surface and provides discontinue intimate contact between the particles and the substrate, which is believed to be a necessary condition for metallurgical bonding[2, 6]. At the same time, the thermal softening induced by adiabatic shear instabilities occurs at contact zone between particles and substrate, which leads the material mixing and extruded from the interface, to form an outward metal jet and contributes also to form metallurgical bonding at fresh surface between particles and substrate.[7, 8] Simultaneously, mechanical interlocking are generated induced by materials rollups and vortices at the interface region. The major variables of the cold spray process include many aspects as the propelling gas pressure for example. It is one of the most important factors and for now, much effort has been devoted to research how the propelling gas pressure affect the properties of cold sprayed coating[9-12]. Numerous studies, either experimental or modeling have been conducted to explore the main influence of the propelling gas pressure. In some works, it is reported that increasing propelling gas pressure can improve the deposition efficiency and the adhesion strength. Despite of many meaningful optimizations[9, 12-16], the coating properties have been obtained by increasing propelling gas pressure. For example, when the propelling gas pressures increased to 5MPa to achieve higher particle velocities. High particle velocity can generate an ultra-strong adhesive strength of Cu coatings (more than 200 MPa) even on stainless steel substrates[13]. However, the effect of propelling gas pressure on interfacial bonding features of cold sprayed coating is still not very clear. In this study, the main work pays special attention to the effect of the propelling gas pressure on the metallurgical bonding features at the interface between the coating and the substrate, trying to explain reason of increased adhesion of coating more directly. Furthermore, the effects of propelling gas pressure on the properties of cold sprayed Nickel coating were also investigated.

2 Materials and Experiments

2.1 Materials

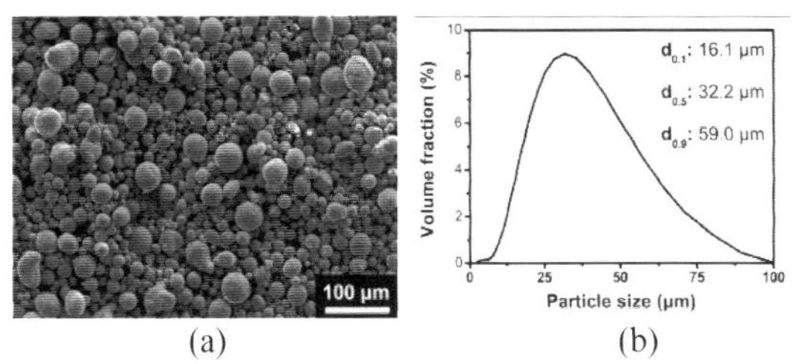

Fig. 1. Morphology (a) and diameter distributions (b) of Nickel particles.

The powder used in this study was Nickel powder. Nickel particles show a spherical morphology with a mean particle diameter of 32µm (Fig. 1 (a)). Particle size distribution was measured by the laser diffraction particle size analyser (Mastersizer 2000, MALVERN Instruments, UK) in Fig. 1(b). Aluminium cylindrical with 25 mm diameter were used as substrate. Polished and grit blasting surfaces of substrates were prepared for individual splat deposition and full coating deposition.

2.2 Coating deposition and heat treatment

Cold spray coatings and splats were produced by CS system (LERMPS, UTBM, France),which consists of computer control, powder feeder, pressure gas system, gas heater, and spray gun with a MOC nozzle. Fig. 2 illustrated the nozzle dimension, a 172.4 mm length nozzle with a 2.7 mm diameter throat, and a 7.8 mm diameter exit. The nozzle has a round cross-section with an expansion ratio of about 8.3. High pressure argon and compressed air were used as powder carrier gas and propelling gas, respectively. The temperature of the compressed air was about 600 °C, and the spray distance was 30 mm. The pressures of propelling gas were set to 2.0 MPa, 2.4 MPa, and 2.8 MPa respectively. The gun traverse speed to manufacture the individual cold sprayed spat and full coating were either 500 mm s^{-1} or 100 mm s^{-1}.

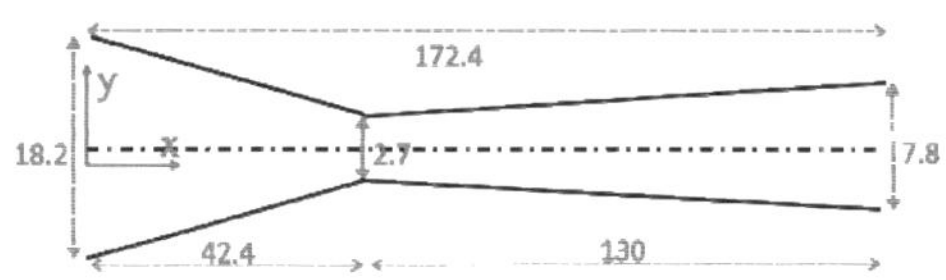

Fig. 2. Schematic diagram of the MOC nozzle geometry.

In order to directly analyze the adhesion degree of coating, the intermetallic phase must be generated at interface between Al and Ni. So parts of samples were heat treated in a furnace. The samples were heated to about 400 °C, and held at this temperature about for 15 min and then furnace cooled down to room temperature.

2.3 Characterization

Samples manufactured under different parameters were polished to observe their microstructure. Microstructures of the powder and coatings were performed using scanning electron microscope (SEM) (JEOL, JSM-5800LV, Japan). The adhesion strength between the coating and substrate were carried out using a pull-off test (ASTM C-633-01).

2.4 Numerical methodology

In this study, a computational fluid dynamics model which matches the real nozzle experimental condition was used to predict particle velocity and temperature prior to the impact. The two-dimensional axisymmetric model was used which is conductive for saving the computational time. The computational domain is meshed using a regular and structured grid of quadrilateral elements. In the simulation, three propelling gas pressures (2.0, 2.4 and 2.8 MPa) were tested, which matched also the experimental parameters. The spherical Nickel particles present at thenozzle inlet an initial velocity of 25 m/s and a temperature of 25 °C. The substrate was located 30 mm away from the nozzle exit.

3 Results and Discussions

3.1 Simulation results

Based on the simulation results, it is found that propelling gas pressure significantly affects the velocity and temperature prior to the impact. Fig. 3 shows the evolution of particle impact temperature and velocity versus the propelling gas pressures.

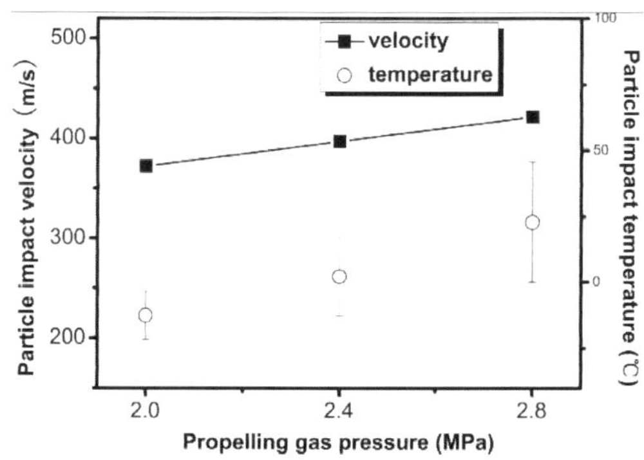

Fig. 3. Particle impact velocity and temperature in function of propelling gas pressures

As propelling gas pressure gradually increases, both particle impact temperature and velocity increase which could facilitate the deposition of cold sprayed. Higher propelling gas pressure not only improves particle impact velocity, and leads to higher impact temperature. Especially, the particle impact temperature is below 0°C under 2.0 MPa conditions. When the propelling gas pressure increases to 2.8 MPa, the corresponding particle impact temperature is about 23 °C.

3.2 Coating properties

3.2.1 Coating microstructures

Fig. 4 shows the cross section micrographs of the Nickel coating on Aluminum substrate corresponding to different spraying conditions. As can be seen from Fig. 4, different coating thicknesses were generated for the three different pressures of the propelling gas. The coating thickness exhibits an upward trend with increasing the propelling gas pressure.
These micrographs clearly indicate that denser Nickel coatings could be produced under higher propelling gas pressures. The black zones in the coating correspond to pores and microcracks. As Fig. 4 shows, many large pores and microcracks are generated in the case of 2.0 MPa pressure. As propelling gas pressure increases, it is possible to observe that the size of the pores decreases. This way, higher gas pressure leads to close the small

pores and microcracks in the coating, microcracks could be easily seen in the coating.

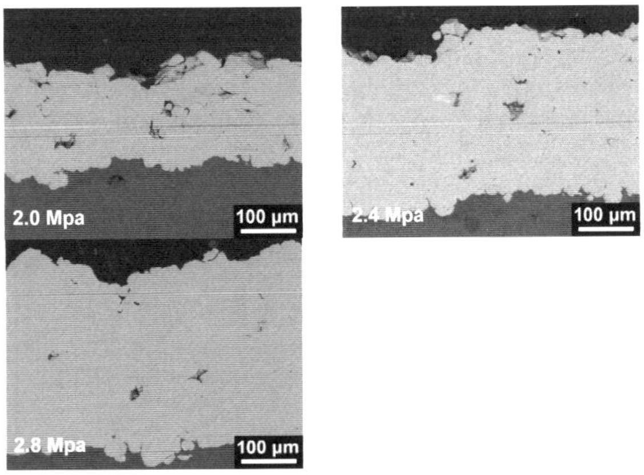

Fig. 4. Cross-sectional microstructures of coatings in function of propelling gas pressures

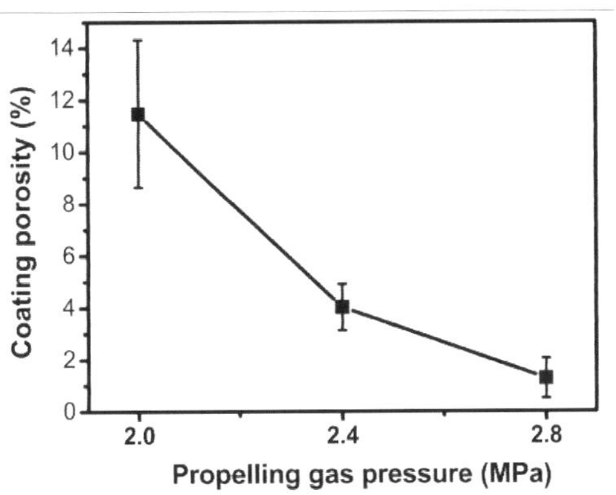

Fig. 5. Coating porosity in function of propelling gas pressures

Fig. 5 displays the evolution of coating porosities in function of propelling gas pressures. The porosity level porosity decreases from 11.4% to 1.3 with the propelling gas pressure increase.

3.2.2 Adhesion Strength of Coatings and Fracture Analyses

Generally, when the pores and microcracks of coating decrease, the adhesion strength of coating will rise. The result of adhesion strength in function of different propelling gas pressures is shown in Fig. 6. The higher the propelling gas pressure, the higher the adhesion strength is. The adhesion strength is 36.7 ± 6.6 MPa for the 2.0 MPa pressure. When the propelling gas pressure increases to 2.4 MPa, the adhesion strength increases to 41.9 ± 7.1 MPa, respectively. Finally, with the highest considered propelling gas pressure reached, 2.8 MPa, the adhesion strength continues to rise to 53.9 ± 7.5 MPa. Therefore, propelling gas pressure increase contributes to form a better bonding thus can optimize the adhesion strength of nickel coating.

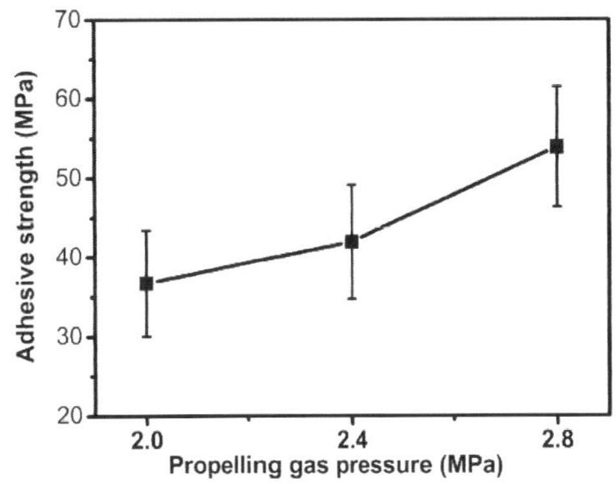

Fig. 6. Adhesive strength in function of propelling gas pressures

3.3 Discussion

From the results mentioned above, the propelling gas pressure significantly influences the adhesion strength between coating and substrate. In order to investigate a little more the influence of the propelling gas pressure on the properties of cold sprayed Ni coating, single Ni particle sprayed under different propelling gas pressures was deposited on Al substrate.

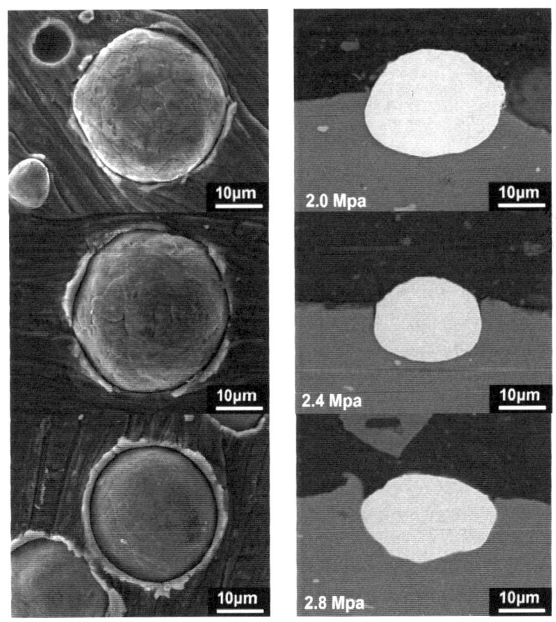

Fig. 7. Surface morphologies and Cross-sectional of a single splat in function of propelling gas pressures

Fig. 7 displays surface and cross-sectional micrographs of a single Ni particle deposited on Al substrate at different pressures. For all cases, the Al substrate and Nickel particle experience intense deformation after the high speed collision, the nickel particle embedded in the Al substrate and tightly together. It is seen that with increasing pressure from 2.0 to 2.8 MPa, the Al substrate has deformed more extensively, and a metal jet of substrate around the particle becomes more clear. When the propelling gas pressure raises, it could be found, the particle also experiences increasingly intensive deformation, becoming more flat. And the particles have obviously penetrate into the substrate when the propelling gas pressure is 2.8 MPa. These results are consistent with the reported results in literature. Because particle impact velocity increases as propelling gas pressure increases, particles have more kinetic energy resulting in more plastic deformation. At the same time, particle impact temperature is also improved as propelling gas pressure increases, higher particle impact temperature leads to softening effect that can increase the ductility of such particles, which is beneficial to cold spray deposition. The pores and micro cracks become smaller, even disappear when the degree of deformation of particles and substrate is more intense.

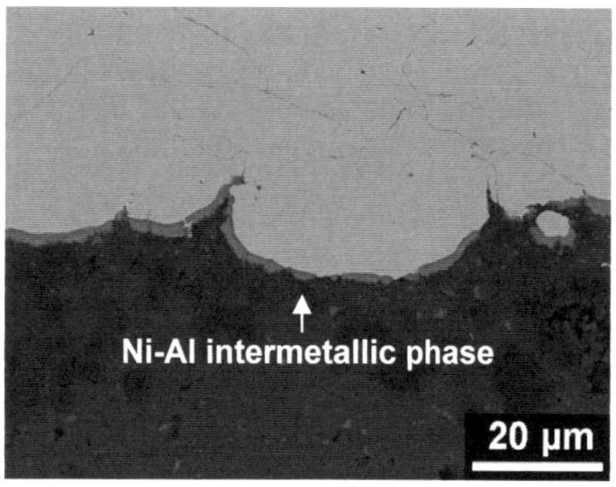

Fig. 8. Ni–Al intermetallic phase at the interfacial region

Higher propelling gas pressure leading to an increase in particle velocities. Higher particle velocity can generate stronger adhesive strength that is a consensus view[9, 12, 17, 18]. Particle-substrate and particle-coating bonding is a combination of different mechanisms. Mechanical interlocking and metallurgical bonding are believed to be the main mechanisms of bonding in cold spraying. When the propelling gas pressure gradually increases, as shown in Fig. 7, it is generally assumed that the larger deformation of particle and substrate increases the degree of mechanical interlock, simultaneously, more deformation is beneficial to improve the effective contact area between particles and substrate. In this case, more opportunities for forming metallurgical bonding are present therefore leading to higher adhesion strength. It is also admitted, fresh metal-to-metal contact between particles and substrate is very important because it influences the formation of bonding strength for cold spray coating quality. However, it is very difficult to observe by SEM directly the bonding situation at interface between coating and substrate. Although all sample interfaces look very good combination before heat treatment, in fact, the real situation is different with a lot of unbounding regions. For example, the original existing oxide film on the metal surface can be disrupted and pushed out from the interface by the outward metal jet. But there is also a large number of oxide film debris at this interface. There are some works reported on this phenomenon[19-23], the oxide film debris between individual particles insidealuminium, copper, and nickel deposits has also been analyzed using transmission electron microscopy (TEM) and simulation, part of them will be left at the interface after particle deposition. Oxide film debris remaining at the interface prevent fresh metal contact between particles and substrate, which significantly weakens the bonding strength and brings some defects inside the coating. So how much area is not binding which has been a problem for the cold spray study .
Ni-Al intermetallic phase formed by atomic diffusion during the annealing process is therefore effective to solve this problem. Fig. 8 presents high magnification SEM micrographs of interfacial bonding region between coatings and substrate, more details of Ni-Al intermetallic phase could be found. The observation suggestes that the original Ni–Al interface before annealing must be considered as interfaces oxide free to allow the direct metal-to-metal contact for atomic diffusion.

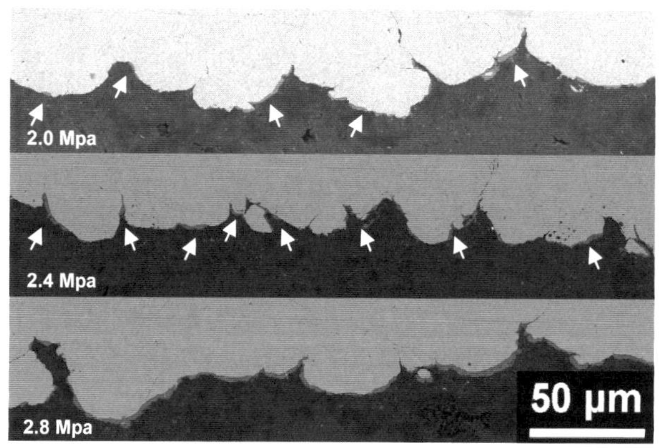

Fig. 9. interface bonding feathers with different propelling gas pressure

The evidence for metallurgical bonding is provided in Fig. 9 which presents SEM micrographs of interfacial bonding region between coatings and substrate obtained under different pressure conditions. It could be found a significant difference between interfaces at the four different pressures after the heat treatment. As propelling gas pressure gradually increases, the thickness of Ni–Al intermetallic phase becomes thicker, and the area of intermetallic phase also increased. The phase of intermetallic phase became abivous. For the specific case of 2.8 MPa, the Ni–Al intermetallic phases fille in most interface regions. That means a much area at the interface to allow the direct metal-to-metal contact exists, and more metallurgical bonding area could formed. So the size of metallurgical bonding area is more reasonable explaination why the adhesion strength of cold sprayed Ni coating increases when propelling gas pressure is higher.

4 Conclusions

In the research, cold sprayed Ni coating and individual particle splats were deposited on Al substrate with different propelling gas pressures.
Denser coating and stronger adhesion strength could be gotten by improving propelling gas pressure. Material deformed intensively can disrupt the oxide film, therefore increases the metal area allowing to achieve metallurgical bonding. Material experienced more intense plastic deformation can disrupt the surface oxide film of material under higher propelling gas pressure, which benefits the mechanical interlock and produces more oxide-free surface leading to metallurgical bonding. Combined with a post thermal method, a direct observation of bonding situation at interfacial between coating and substrate by SEM can be done, which permits to assume that the main cause of elevated adhesion strength is related to propelling gas increase.

Acknowledgments The authors would like to acknowledge the support by China Scholarship Council (CSC) and Marie Curie FP7-IPACT-268696(EU).

Literature

[1]. Alkhimov, A.P. and Kosarev, V.F. Other: A Method Of Cold Gas-Dynamic Spraying. Dokl Akad Nauk Sssr+ 315 (1990), Issue 5, pp. 1062-65.

[2]. Xie, Y. and Planche, M.-P. Other: Effect of Substrate Preheating on Adhesive Strength of SS 316L Cold Spray Coatings. Journal of Thermal Spray Technology 25 (2015), Issue 1-2, pp. 123-30.

[3]. Yang, G.-J. and Li, C.-J. Other: Low temperature deposition and characterization of TiO2 photocatalytic film through cold spray. Applied Surface Science 254 (2008), Issue 13, pp. 3979-82.

[4]. Kim, K. and Watanabe, M. Other: Bonding mechanisms of thermally softened metallic powder particles and substrates impacted at high velocity. Surface and Coatings Technology 204 (2010), Issue 14, pp. 2175-80.

[5]. Zhao, Z.B. and Gillispie, B.A. Other: Coating deposition by the kinetic spray process. Surface and Coatings Technology 200 (2006), Issue 16-17, pp. 4746-54.

[6]. Luo, X.-T. and Li, C.-X. Other: High velocity impact induced microstructure evolution during deposition of cold spray coatings: A review. Surface and Coatings Technology 254 (2014), Issue, pp. 11-20.

[7]. Assadi, H. and Gärtner, F. Other: Bonding mechanism in cold gas spraying. Acta Materialia 51 (2003), Issue 15, pp. 4379-94.

[8]. Grujicic, M. and Zhao, C.L. Other: Adiabatic shear instability based mechanism for particles/substrate bonding in the cold-gas dynamic-spray process. Materials & Design 25 (2004), Issue 8, pp. 681-88.

[9]. Lee, H. and Shin, H. Other: Effect of gas pressure on Al coatings by cold gas dynamic spray. Materials Letters 62 (2008), Issue 10-11, pp. 1579-81.

[10]. Tang, W. and Liu, J. Other: The effects of two gas flow streams with initial temperature and pressure differences in cold spraying nozzle. Surface and Coatings Technology 240 (2014), Issue, pp. 86-95.

[11]. Yin, S. and Zhang, M. Other: Numerical investigations on the effect of total pressure and nozzle divergent length on the flow character and particle impact velocity in cold spraying. Surface and Coatings Technology 232 (2013), Issue, pp. 290-97.

[12]. Yin, S. and Liu, Q. Other: Effect of injection pressure on particle acceleration, dispersion and deposition in cold spray. Computational Materials Science 90 (2014), Issue, pp. 7-15.

[13]. Huang, R. and Ma, W. Other: Development of ultra-strong adhesive strength coatings using cold spray. Surface and Coatings Technology (2014), Issue

[14]. Numerical Studies on the Effects of Stagnation Pressure and Temperature on Supersonic Flow Characteristics in Cold Spray Applications. (2011), Issue

[15]. Bae, G. and Kumar, S. Other: Bonding features and associated mechanisms in kinetic sprayed titanium coatings. Acta Materialia 57 (2009), Issue 19, pp. 5654-66.

[16]. Lee, J. and Shin, S. Other: Effect of gas temperature on critical velocity and deposition characteristics in kinetic spraying. Applied Surface Science 253 (2007), Issue 7, pp. 3512-20.

[17]. Schmidt, T. and Assadi, H. Other: From Particle Acceleration to Impact and Bonding in Cold Spraying. Journal of Thermal Spray Technology 18 (2009), Issue 5-6, pp. 794-808.

[18]. Yu, M. and Li, W.Y. Other: Finite Element Simulation of Impacting Behavior of Particles in Cold Spraying by Eulerian Approach. Journal of Thermal Spray Technology 21 (2011), Issue 3-4, pp. 745-52.

[19]. Yin, S. and Wang, X. Other: Deformation behavior of the oxide film on the surface of cold sprayed powder particle. Applied Surface Science 259 (2012), Issue, pp. 294-300.

[20]. Li, W.-Y. and Li, C.-J. Other: Significant influence of particle surface oxidation on deposition efficiency, interface microstructure and adhesive strength of cold-sprayed copper

coatings. Applied Surface Science 256 (2010), Issue 16, pp. 4953-58.

[21]. Li, C.J. and Wang, H.T. Other: Influence of Spray Materials and Their Surface Oxidation on the Critical Velocity in Cold Spraying. Journal of Thermal Spray Technology 19 (2009), Issue 1-2, pp. 95-101.

[22]. Li, W.-Y. and Zhang, C. Other: Study on impact fusion at particle interfaces and its effect on coating microstructure in cold spraying. Applied Surface Science 254 (2007), Issue 2, pp. 517-26.

[23]. <Significant influences of metal reactivity and oxide films at particle.pdf>. (2007), Issue

Improving corrosion resistance of Al6061 coating deposited by an in-situ shot-peening assisted cold spray

Ying-Kang Wei, Xiao-Tao Luo, Chang-Jiu Li
State Key Laboratory for Mechanical Behavior of Materials, School of Materials Science and Engineering, Xi'an Jiaotong University, Xi'an, China / E-mail: licj@mail.xjtu.edu.cn

Abstract
In this study, Al6061 coating was deposited on AZ31B magnesium by cold spray with a commercially Al6061 powder blended with large sized stainless steel particles (as shot-peehing particles for *in-situ* shot-peening assited densifying of the coating) using nitrogen gases at low working gas pressure and temperature. Powders and coatings were characterized by scanning electron microscopy and microhardness measurements. It was found that through introducing the in-situ tamping effect by using shot-peening (SP) particles, the plastic deformation of deposited particles is significantly enhanced, thereby resulting in a dense Al6061 coating. SEM observations revealed that no SP particle was incorporated into Al6061 coating. The effect of SP particles on the deposition efficiency of Al6061 particles was also examined. The electrochemical corrosion experiments revealed that using the novel in-situ shot-peening assisted cold spraying it is possible to deposit fully dense Al6061 coating which can provide complete protection for the magnesium materials from corrosion.

1 Introduction

Magnesium alloys have led to the increasing concern of its use as a structural component in the aerospace and automotive industry especially for die-cast AZ and AM series, since Mg exhibits low density, high specific strength, good buffer and good performance of thermal conduction [1]. However, the poor corrosion resistance of Mg alloys hampers their wider application [1, 2]. AZ31 Mg alloy, generally comprising 3 wt.% Al, 1 wt.% Zn and 0.2–0.5 wt.% Mn, has been widely exploited for commercial applications. Current research shows that the most effective way to prevent corrosion for Mg alloys is to coat a barrier layer in the alloy surface [3], such as chemical conversion coatings, electrochemical plating, anodizing, gas-phase deposition processes and thermal spraying [3-7]. However, these techniques are usually difficult, expensive and not environmentally-friendly to use in Mg alloys surface.
Cold spray coating is one of the protective coating methods that can be employed for magnesium corrosion-protective. The low processing temperature of cold spray process makes it possible to coat magnesium alloy with barrier coating without evident thermal effect on the Mg-based substrate. Furthermore, the cold spray process does not require heavy metal or toxic ions such as Cr^{6+} and CN^{-} in the electrochemical deposition process and without complex pretreatment for magnesium alloy substrate. Therefore, corrosion protection by cold spray is promising to address many of the shortcomings associated with conventinal methods for corrosion protection of magnesium alloys [8].
Aluminium alloys are regularly chosen as a matrix because of their low density, good isotropic mechanical properties, excellent corrosion resistance and reasonable cost. Amongst aluminium alloys, 6061 is an Al–Mg–Si alloy widely used for structural applications due to its good strength, corrosion resistance, immunity to stress corrosion cracking as well as heat treatability by forming precipitates that increase the strength at the cost of somewhat reduced ductility [9, 10]. Aluminum is a very active metal with a low standard potential (-1.66 V). When aluminum is exposed to air, a thin aluminum oxide layer formed in the surface that prevents further oxidation. Therefore, some researchers focus on the field of cold spraying to coat magnesium alloys with Al coating or aluminium alloy coating [11-13]. However, the previous study revealed that it is difficult to prepare completely dense aluminum or aluminium alloy coating through cold spray with nitrogen as the accelerating gas and large pores opened to coating surface are formed which affect significantly the corrosion resistance [14].
A so called tamping effect, in which impact of subsequent particles on the deposited particles is useful to densify the inter-particle pores in the previously deposited coating, was reported [15, 16]. Meanwhile, researchers in the field of composite materials coating manufactured dense aluminum composite coating by using pure aluminum powders and hard ceramic or metal particles mixtures [17-19]. Furthermore, Luo et al. [20] decreased the porosity of Ti and Ti6Al4V coatings by mixing large sized shot-peeing (SP) particles into the spraying powders through deliberate introduction of tamping effect. Such process was referred to as the in-situ shot-peening assisted cold spraying. It was found that it is possible to achieve a full dense coating by deliberately introduced SP densifying effect. Thus, in the present investigation, stainless steel SP particles were blended into the Al6061 powders to introduce an enhanced in-situ tamping effect and therefore prepare dense Al6061 coating on the AZ31B magnesium alloy. Two principle objectives were: (i) to investigate the effect of SP particles on the Al6061 coating microstructure characteristics; (ii) to investigate the corrosion resistance behavior of the prepared Al6061 coatings.

2 Materials and Methods

2.1 Materials

The commercial Al6061 powders were purchased from Beijing Xing Rong Yuan Technology Co., Ltd., with a particle size of -300 mesh. The chemical compositions of the Al6061 powders are summarized in **Table 1.** The available stainless steel SP particles (1Cr18) have a diameter range of 200-300 μm. **Fig. 1.** represents the typical morphology of the Al6061 spray powders and SP particles used in this study.
Substrate for cold spraying were AZ31B magnesium alloy plates with dimensions of 100×20×4 mm. In order to improve the adhesion of coating/substrate, the substrate surface was sandblasted using 24mesh alumina grits at compressed air pressure of 0.6 MPa before the deposition. After sandblasting, the substrate was cleaned in an ultrasonic acetone bath for 3 minutes.

Table 1. Chemical compositions of Al6061 powders used for depositing cold spray coatings (wt.%)

Al	Si	Fe	Cu	Mn	Mg	Zn	Ti
97	0.3	0.7	0.3	0.15	1.1	0.27	0.18

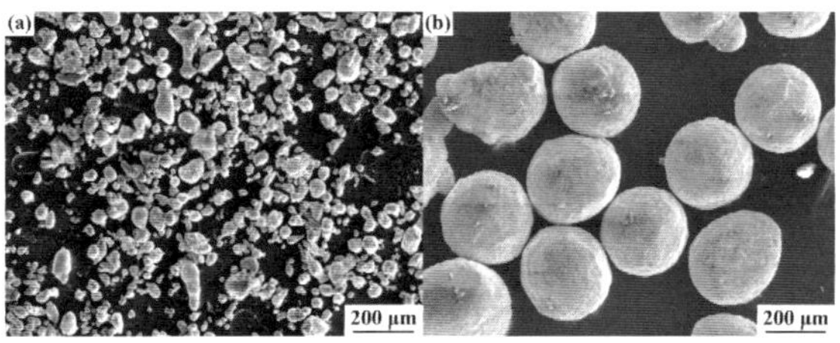

Fig. 1. SEM images of Al6061 spray powders (a) and stainless steel SP particles (b)

2.2 Cold Spraying of AL6061 coatings

In this study, all the Al6061 coatings were performed by using DWCS-2000 cold spraying system which produced by Dewei Automation Company (Xi'an, China). The mixed powders of 60vol.% SP particles with Al6061 powders were used as feedstock in *in-situ* SP assisted cold spraying. Nitrogen was used as accelerating gas with stagnation pressure and temperature of 2.5 MPa and 300 °C respectively. The nitrogen was also as the powder carrier gas at a pressure of 2.7 MPa. The coating was also deposited via routine cold spraying only using Al6061 powders at a gas pressure of 3 MPa and a temperature of 350 °C for comparison.
For all coatings, the stand-off distance and traverse speed of the spray torch were set to be 20 mm and 40 mm/s, respectively. The feeding rate of feedstock powders was about 20 g/min.

2.3 Corrosion behavior Testing

The corrosion resistance of the coatings was tested using CS310 electrochemical workstation which was manufactured by Corrtest Instruments CO., Ltd (Wuhan, China). Three-electrode electrochemical cell system was employed with a saturated calomel electrode (SCE) as the reference electrode and platinum electrode as the counter electrode. The coating sample with surface polished before the testacted as a working electrode (1 cm^2 exposed surface). Coating thickness ranged from 200 to 300μm. The electrolyte was 3.5wt.% NaCl aqueous solution and the scanning rate of potential was 0.5 mV/s from -0.6 V to +0.7 V with respect to the corrosion potential at room temperature (15 °C). Corrosion experiments were also performed on Al6061 bulk and un-coated AZ31B magnesium alloy substrate for comparison.

3 Experimental Results and Discussion

3.1 Characterization of coating microstructure

Fig. 2. shows the cross-sectional microstructures of two types of Al6061 coatings deposited by the *in-situ* SP assisted cold spraying and conventional cold spraying. The Al6061 coating prepared the in-situ SP assisted process with the mixed powders reveals a completely dense microstructure while numerous of pores can be observed from the coating prepared routinely directly with Al6061 powders. The average porosity of Al6061 coatings was estimated to be 14.4±0.6% and 0.4±0.1% for routine process and the novel *in-situ* SP assisted process, respectively.
Fig. 3. compares the surface morphology of the Al6061 coatings deposited by *in-situ* SP assisted cold spraying and routine cold spraying. Many pits caused by the impact of SP particles can be clearly observed on the surface of th Al6061 coating deposited by the novel process. The regular pits on Al6061 coating surface reveal intense forced plastic deformation by SP particles. On the other hand, from the surfaces of porous Al6061 coating many nearly spherical Al6061 particles were still clearly observed, revealing limited plastic deformation and poor contact among neighboring deposited particles.
It can be found from **Fig. 2.** and **Fig. 3.** that only relying on the impact of Al6061 particles, plastic deformation of the particles is not intensive enough to yield a dense microstructure. However, by introducing in-situ tamping by the heavy impact of SP particles the further intenisve plastic deformation of deposited Al6061 particles is obviously developed.
It is undesirable that the stainless steel SP particles are included in the Al6061 coating. The average velocity of SP particles is estimated to be ~288 m/s, lower than the critical velocity calculated from Schmidt et al. [21]. Therefore, the SP particles are not deposited in Al6061 coating. Our preliminary experiments reveal that the SP particles may be included in the coating under some high parameters

and by controlling processing gas conditions the inclusion of SP particles can be effectively prevented.

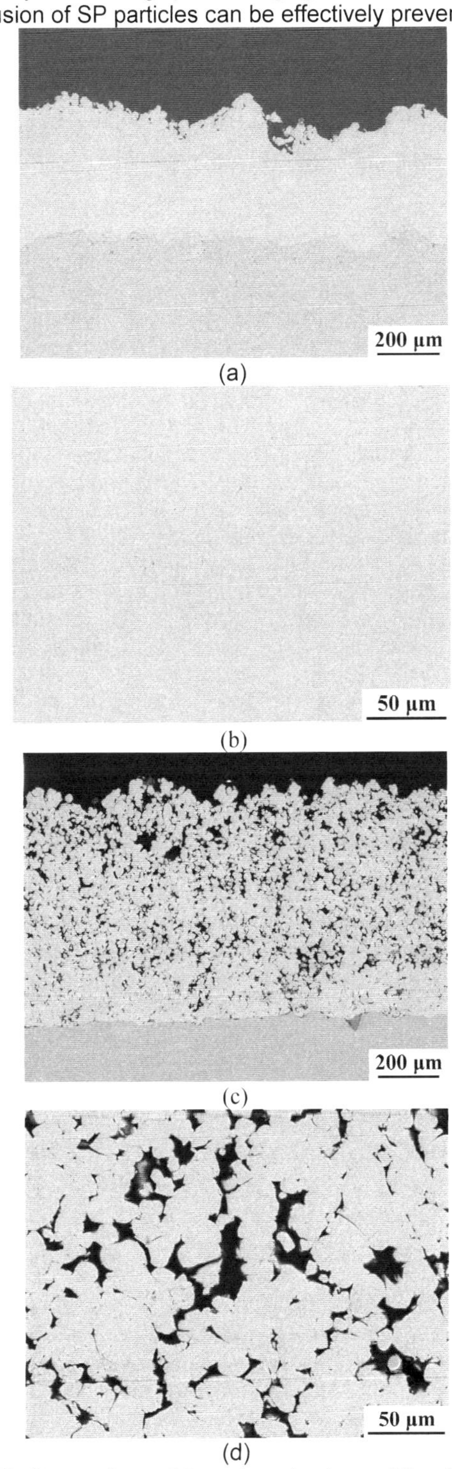

Fig. 2. Comparison of the microstructure of the dense Al6061 coating deposited by the novel in-situ SP assisted process (a) and (b) with porous Al6061 coating deposited by the routine cold spraying (c) and (d)

In this study, compared with the routine cold spraying, accelerating gas pressure and temperature of in-situ SP assisted cold spraying were decreased by 0.5 MPa and 50 °C, respectively. As a result, no stainless steel SP particles were observed in the Al6061 coating using mixed powders. However, the decrease of gas pressure and temperature and the enhanced plastic deformation by impact of SP particles on previously deposited coating are known to affect the deposition efficiency. In this study, the deposition efficiency of Al6061 powders indeed decreased from 35.7% for routine cold spraying to 15.1% for novel process. An further study is required to optimize the spray conditions.

3.2 Microhardness of Al6061 coatings

The microhardness profiles across the coating/substrate interface are plotted in **Fig. 4.** The successive impacts of the incident Al6061 and/or SP particles on the previously deposited coating causes the tamping effect. Such effect forces the impacted coating surface region to deform besides self-impact induced deformation. Since with the increase of coating thickness, the accumulative tamping effect is gradually increased. Hence, the hardness of coatings is inversely proportional to the distance to the substrate/coating interface. Moreover, the microhardness of Al6061 coating deposited by in-situ SP assisted cold spray is higher than those deposited by routine cold spraying in the same distance to interface. Such increase of the microhardness of coating is attributed to additional peening effect introduced by the SP particles to the deposited Al6061 particles. Generally, it is considered that the increased hardness at the substrate side is attributable to the work hardening effect resulting from Al6061 and SP particles impact on the surface of substrate, which is similar to the shot peening process [22]. Consequently, it can be rationally concluded that blending SP particles can substantially increase coatings microhardness.

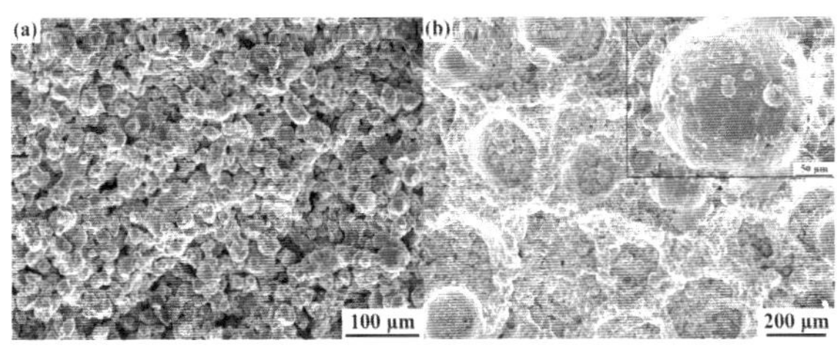

Fig. 3. Comparison of the surface morphology of the dense Al6061 coating deposited by the novel in-situ SP assisted process (a) and (b) with porous Al6061 coating deposited by the routine cold spraying (c) and (d)

3.3 Corrosion resistance

3.3.1 Open-Circuit Potential Measurements.

Open-circuit potential measurement is one of the electrochemical methods used to evaluate the corrosion performance and indicate the thermodynamically tendency to oxidation or passivation of coating in a corrosive medium. **Fig. 5** shows the open-circuit potential curves of the un-coated AZ31B, Al606l bulk and AZ31B specimens coated with Al606l coatings. With the open-circuit potential curve of the bulk Mg alloy almost no obvious variation was observed, and the potential was stabilized at -1.47 V in sodium chloride media. In the first 300 seconds immersion, a continuous increase in open-circuit potential was observed for Al606l bulk sample toward positive potential, indicating that the thermodynamic tendency of corrosion is decreased. This initial rise seems to be related to the generation of hydrous aluminum oxides in the coating surface. The potential of the Al606l bulk is stabilized at -0.59 V after 300 seconds of immersion. Therefore, it is evident that the thermodynamic tendency of corrosion of Al6061 bulk surface aluminum oxide layer is very stable in chloride solution. During the whole immersion period, the potential of AZ31B specimens coated with dense Al606l coatings remained at -0.6 V. The result shows that the Al606l particles have good cohesion after experienced heavily impacting by *in-situ* SP particles and there are no connected pores in the coatings to permit the infiltration of corrosive medium. Furthermore, this result also reveals that on the dense coating surface a relatively stable oxide film formed during immersion in NaCl aqueos solution. However, a small scale of oscillating variation of the open-circuit potential oscillation was observed in the AZ31B specimens coated with dense Al606l coatings. This potential variation may be attributed to a reaction at the interparticle interface and to the formation of new interfaces such as pitting [23].

The open-circuit potential of AZ31B specimens coated with porous Al606l coatings exhibited a gradual drop slightly at first 1000s and tended to slightly increase trend and reached to a potential of about -1.24 V after one hour. The result shows that many connected pores exist in the porous coatings and the corrosive medium has permeated to the Mg alloy substrate surface through the connected pores.

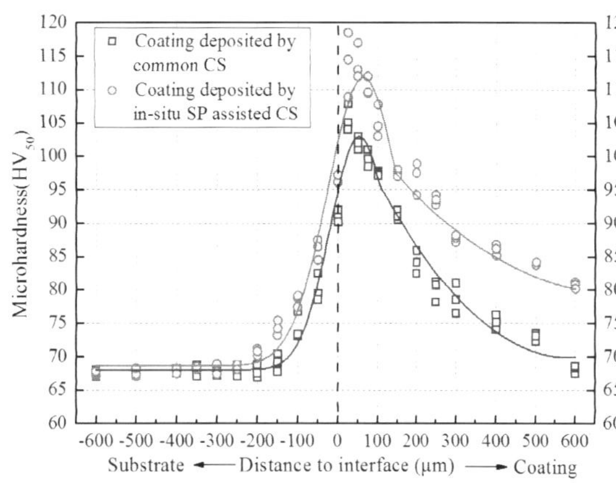

Fig. 4. Distrubution of microhardness within the substrate and coating across the substrate/coating interface of AZ31B alloys with Al6061 coatings

deposited by the routine cold spraying and novel in-situ SP assisted cold spraying

3.3.1 Evaluation by Dynamic Polarization The potentiodynamic polarization testing is used to evaluate the coating corrosion properties. **Fig. 6.** shows the potentiodynamic polarization curves of of the un-coated AZ31B, Al606l bulk and AZ31B specimens coated with Al606l coatings. It can be recognized that the polarization curve of AZ31B specimens coated with dense Al606l coatings by the in-situ SP process is comparable to the that of the Al6061 bulk. Corrosion potential (Ecorr) and corrosion current density (Icorr) which are derived from the potentiodynamic curve are shown in **Table 2.** Owing to a dense protective oxide on the surface, Al6061 bulk revealed a low corrosion current density of 0.604 μA/cm^2. On the contrary, AZ31B substrate revealed a very high corrosion current density of 96.3 μA/cm^2. The AZ31B specimens coated with porous Al606l coatings revealed a current density of 5.86 μA/cm^2. This value is lower than the result of AZ31B substrate but still much higher than the result of Al606l bulk.The corrosion current density of dense Al6061 coating is 0.753 μA/cm^2, being comparable to that of Al6061 bulk. With the heavy impact of SP particles, the Al6061 particles in the dense coating by the novel process have obviously experienced much intensive plastic deformation. Such effect induces higher strain rate at the interfaces among the particles on the coating surface and could act as a preferential reaction zone for pitting corrosion [13]. Therefore, compared with the Al6061 bulk, the Icorr value of the dense Al6061 coatings increased a little compared with Al6061 bulk and however much lower than the values of AZ31B substrate and AZ31B specimen coated with porous Al606l coatings.

Therfore, it is clear that the dense Al6061 coating coated AZ31B exhibits comaprable electrochemical behavior to Al6061 bulk. This fact reveals that the corrosion of the AZ31B can be fully prevented by applying the dense Al6061 coating through using the present novel in-situ SP assisted cold spraying.

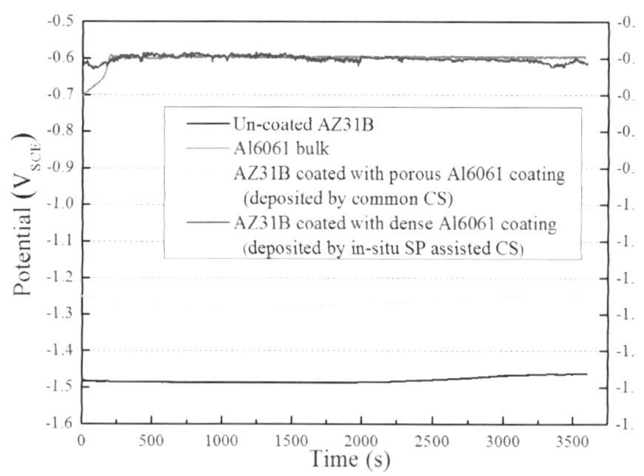

Fig. 5. Open-circuit potential curves of the un-coated AZ31B, Al606l bulk and AZ31B specimens coated with two types of Al606l coatings by the routine and novel processes

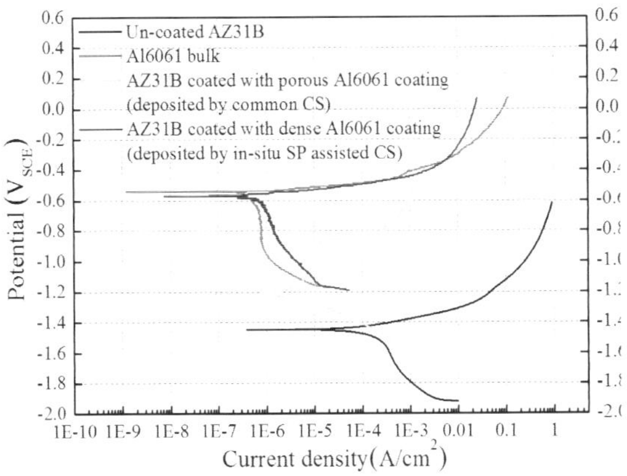

Fig. 6. Potentiodynamic polarization curves of the un-coated AZ31B, Al606l bulk and AZ31B specimens coated with two types of Al606l coatings by the routine and novel processes

Table 2 Potentiodynamic corrosion results showing the measured corrosion potential and current density of the un-coated AZ31B, Al606l bulk and AZ31B specimens coated with Al606l coatings

Samples	Ecorr (V)	Icorr (mA/cm^2)
un-coated AZ31B	-1.45	9.63×10^{-2}
Al6061 bulk	-0.54	6.04×10^{-4}
AZ31B specimens coated with porous Al6061 coatings	-1.21	5.86×10^{-3}
AZ31B specimens coated with dense Al6061 coatings	-0.57	7.53×10^{-4}

4 Conclusions

Using the novel in-situ SP assisted cold spraying by deliberately introducing SP particles into the spray powders can significantly enhance the plastic deformation and consequently cohesion of deposited Al606l particles by enhanced accumulative tamping effect. The Al606l coating deposited by the in-situ SP assisted cold spraying reveals a completely dense microstructure with a porosity of 0.4±0.1%. The electrochemical evaluation of the dense Al6061 coating coated AZ31B alloy reveals a comparable open circuit potential and a dynamic polarization behavior to that of Al6061 bulk. As a result, the dense Al6061 coatings deposited by in-situ SP assisted cold spraying can provide completely effective protection for the AZ31B magnesium material from corrosion.

Acknowledgements

This work is supported by National Nature Science Foundation (No. 50115814).

References

[1] R.L. Edgar, In: K.U. Kainer (Ed.): Magnesium Alloys and their Application, Wiley-VCH, Germany, 2003, pp. 3.

[2] I.J. Polmear, Introduction to Magnesium in ASM Specialty Handbook: Magnesium and Magnesium Alloys, ASM International, Materials Park, OH, 1999, pp. 3/15.

[3] J.E. Gray, B. Luan, Protective coatings on magnesium and its alloys — a critical review, J. Alloy & Compd., 336 (2002), pp. 88/113.

[4] G.-L. Song, A. Atrens, Recent insights into the mechanism of magnesium corrosionand research suggestions, Adv. Eng. Mater., 9 (2007), pp. 177.

[5] A. Pardo, P. Casajus, M. Mohedano, A.E. Coy, F. Viejo, B. Torres, E. Matykina, Corrosion protection of Mg/Al alloys by thermal sprayed aluminium coatings, Appl. Surf. Sci., 255 (2009), pp. 6968/77.

[6] L. Anicai, R. Masi, M. Santamaria, F. Di Quarto, A photoelectrochemical investigation of conversion coatings on Mg substrates, Corrosion Sci., 47 (2005), Issue 12, pp. 2883/2900.

[7] C. Wang, F.H. Wang, S. Zhu, F. Jiang, Cerium conversion coatings for AZ91D magnesium alloy in ethanol solution and its corrosion resistance, Corrosion Sci., 51 (2009), pp. 2916/23.

[8] G. Shayegana, H. Mahmoudia, R. Ghelichi, J. Villafuerte, J. Wang, M. Guagliano, H. Jahed, Residual stress induced by cold spray coating of magnesium AZ31B extrusion, Mater. Design, 60 (2014), pp. 72–84.

[9] Y.-C. Kang, S. Chan, Tensile properties of nanometric Al_2O_3 particulate-reinforced aluminum matrix composites, Mate. Chem. Phys., 85 (2004), Issues 2/3, pp. 438/43.

[10] C. Suryanarayana, Synthesis of nanocomposites by mechanical alloying, J. Alloys Comp., 509 (2011), pp. 229/34.

[11] Y. S. Tao, T. Y. Xiong, C. Sun, L.Y. Kong, X.Y. Cui, T.F. Li, G.L. Song, Microstructure and corrosion performance of a cold sprayed aluminium coating on AZ91D magnesium alloy, Corrosion Sci., 52 (2010), pp.3191/97.

[12] B. S. DeForce, T. J. Eden, Other: Cold Spray Al-5% Mg Coatings for the Corrosion Protection of Magnesium Alloys. Journal of Thermal Spray Technology 20 (2011), pp. 1352/58.

[13] H.Y. Bu, M. Yandouzi, C. Lu, D. MacDonald, B. Jodoin, Cold spray blended Al+Mg17Al12 coating for corrosion protection of AZ91D magnesium alloy, Surf. Coat. Technol., 207 (2012), pp 155/62.

[14] W.-Y. Li, C.-J. Li, in: X. Huang (Ed.), Proceedings of the Fifth International Seminar and the Sixth China National Thermal Spraying Conference, 2002, Anshan, China, pp. 50/55.

[15] W.-Y. Li, C.-J. Li, Deposition characteristics of titanium coating in cold spraying, Surf. Coat. Technol., 167 (2003), pp. 278/83.

[16] M. Jahedi, S. Zahiri, S. Gulizia, B. Tiganis, C. Tang, D. Fraser, Direct Manufacturing of Titanium

Parts by Cold Spray, Mater Sci Forum, 618-619 (2009), pp. 505/08.
[17] H.-T. Wang, C.-J. Li, G.J. Yang, C.X. Li, Cold spraying of Fe/Al powder mixture: Coating characteristics and influence of heat treatment on the phase structure, Appl. Surf. Sci., 255 (2008), pp. 2538/44.
[18] E. Irissou, J.G. Legoux, B. Arsenault, C. Moreau, Investigation of Al-Al_2O_3 Cold Spray Coating Formation and Properties, J. Thermal Spray Technol., 16 (2007).
[19] E. Sansoucy, P. Marcoux, L. Ajdelsztajn, B. Jodoin, Properties of SiC-reinforced aluminum alloy coatings produced by the cold gas dynamic spraying process, Surf. Coat. Technol., 202 (2008), pp. 3988/96.
[20] X.-T. Luo, Y.-K. Wei, Y. Wang, C.-J. Li Microstructure and mechanical property of Ti and Ti6Al4V prepared by an in-situ shot peening assisted cold spraying, Mater. Des., 85 (2015), pp. 527/33.
[21] T. Schmidt, F. Gartner, H. Assadi, H. Kreye, Development of a generalized parameter window for cold spray deposition, ACTA. Mater., 54 (2006), pp. 729/42.
[22] P. Zhang, J. Lindemann, Effect of roller burnishing on the high cycle fatigue performance of the high-strength wrought magnesium alloy AZ80, Scrip. Mater., 52 (2005), pp. 485/90.
[23] D. Dzhurinskiy, E. Maeva, E. Leshchinsky, R.G. Maev, Corrosion protection of light alloys using low pressure cold spray, J. Thermal Spray Technol., 21 (2012), pp. 304/13.

Feature of interfacial bonding achieved by medium-velocity particle impacts in cold spray

Yu-Juan Li, Ying-Kang Wei, Xiao-Tao Luo, Chang-Jiu Li
State Key Laboratory for Mechanical Behavior of Materials, School of Materials Science and Engineering, Xi'an Jiao tong University, Xi'an, Shaanxi, 710049, China/ E-mail: licj@mail.xjtu.edu.cn

Due to the low deposition temperature and high materials building rate, cold spraying can be used as additive manufacture technology. The inter-particle bonding in cold-sprayed coating is considered to be the most crucial factor determining its mechanical properties, physical and chemical behavior such as tensile strength, hardness, electrical and thermal conductivities, as well as corrosion resistance. Therefore, in this study Cu coatings are prepared by cold spraying and the feature of interfacial-particle bonding is investigated. Results show that an extremely dense Cu coating with very little pores is prepared. Quantitative analysis with the image analysis indicates that the porosity of coating is only 0.12%. The coating micro-hardness is about 163 $HV_{0.1}$ which is much higher than that of pure Cu bulk. The electrical conductivity of as-sprayed reaches 76 %IACS which is close to that of cold rolled pure Cu bulk. Tensile tests show that the mean ultimate tensile strength of cold sprayed Cu sample is about 65.5 ± 5.5 MPa. The elongations to failure of all the Cu samples do not exceed 0.2%. Moreover, the mean Young modulus obtained from tensile curves is merely 0.65 ± 0.5 GPa.

1 Introduction

In cold spraying, micrometer scale particles (typically 10-50 μm) are fed into high-speed gas flow and are accelerated to velocities of the order of 300-1200 m/s, and then impact on the substrate at the fully solid state [1,2]. The coatings are formed layer by layer with a high degree of plastic deformation between particle/substrate and particle/particle. Due to the relatively low deposition temperature (< 800 °C), cold spraying is particularly suitable for coating material which are sensitive to heating or oxidation and would transform or react in conventional spraying processes, such as Cu, Ti and so on [3-5]. The other advantages of cold spraying is its higher material building rate in contrast with the conventional additive manufacture technology, which makes it great potential for 3D print. Generally, particle velocity prior to the impact is an important factor that determines whether particles can deposit on the substrate or on the deposited coating. It has been widely accepted that there exists a material-dependent critical velocity for a given condition (e.g. specific particle size, temperature and material properties), only above which the bonding at the particle/substrate or particle/particle interface can be successfully formed [3,6]. Therefore, a large number of works were devoted to investigate the critical velocity of cold sprayed materials. It was reported that the critical velocity of Cu particle deposited on Cu substrate is in the range of 300-640 $m \cdot s^{-1}$ [6-8].

As for the bonding mechanism, many experimental efforts have been carried out during the past years to better understand this phenomenon. Currently, the most acceptable view of the bonding mechanism can be regarded as the occurrence of adiabatic shear instability (ASI) at the interface which results from the high strain rate and the intensive localized deformation during the particle deposition process [3,4,6,9]. At the region where ASI occurs, adiabatic heating-induced thermal softening is dominant over work hardening. Consequently, metals behave like a viscous material and are extruded from the interface, forming an outward metal jet at the rim. Such viscous-like metal jet helps to clean up the cracked native oxide film which originally exists on particle surfaces, allowing the metal-to-metal contact and thus the metallic bonding to occur [6,10-12]. Metallurgical bonding and mechanical interlocking are commonly perceived to be two mechanisms of the metallic bonding in cold spray. Metallurgical bonding is known as a result of the atomic diffusion at the oxide-free interface [10,13]. Basically, the formation of intermetallic phase at the interfacial region [14,16], dimple-like and groove-like features [10,17,18] appearing at the fracture surface or very thin amorphous layer can be recognized as a marker for oxide-free interface and thus true the metallurgical bonding. As for the mechanical interlock, it results from the extrusion and physical mixture of the soft metal materials and always happens at the coating-substrate interface [19]. The strength of mechanical-based bonding is normally much smaller compared with that of metallurgical-based bonding [20,21].

The bonding of coatings prepared by cold spraying determines its applications in the industrial field. Therefore, much effort has been devoted to improve the bonding strength and mechanical performance of cold-sprayed coatings during the past decades [22.23], while mainly focused on the adhesive strength, due to that coating-substrate bonding is mainly determined by the substrate surface condition, and hence there are various methods to improve the adhesive strength of cold sprayed coating. However, for the particle-particle bonding only determined by the original properties of the feedstock and the working parameters, most of the applied methods of improving the coating cohesive strength can merely be based on the improvement of particle conditions prior to impact, increasing particle velocity or temperature. Therefore, in the present study, a kind of Cu feedstock with low oxygen content was used to prepare coating by cold spraying with relatively high spray pressure and temperature in order to improve the inter-particle bonding and gain a coating with high cohesive strength. The feature of interfacial bonding achieved by high-velocity particle impact in cold spraying was researched.

2 Experimental procedure

2.1 Feedstock powder

Commercially available gas-atomized Cu powder was used as the raw materials. The morphology of the powder is presented in Fig. 1(a). The powder particles reveal a near spherical morphology. The particle size distribution was measured by laser diffractometry (Laser Mastersizer, Malvern Instruments, Malvern, UK) as shown in Fig.1b, indicating that the diameter of powder particles ranges from 10 to 56 μm with a volumetric average diameter of 29 μm. The oxygen content measurement of feedstock manifests that the Cu powders have a oxygen content of 0.07 wt.%.

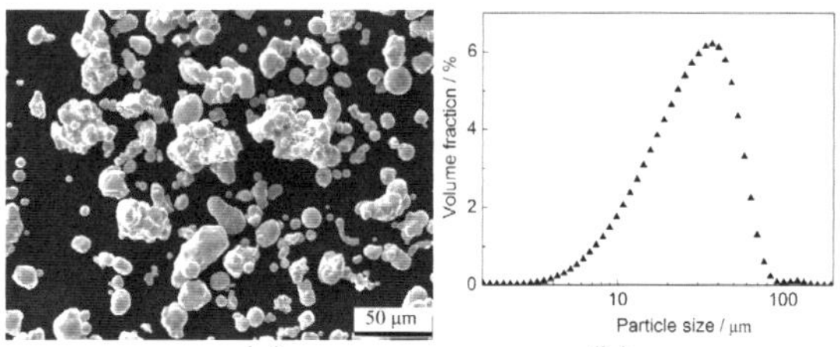

(a) (b)

Fig. 1. Morphologies (a) and particle size distributions (b) of Cu powder particles.

2.2 Cold spraying process

Commercial pure Al plates with dimensions of 120 mm × 80 mm × 6 mm were employed as the substrate. Prior to the coating deposition, the substrates were grit-blasted by 40 mesh alumina grits. An in-house developed cold spray system (CS-2000, Xi'an Jiao tong University) was used for coating deposition. In this system, a Laval nozzle was used with throat diameter of 2 mm, outlet diameter of 4.4 mm, and downstream distance of 170 mm. N_2 gas was used both as the accelerating gas and the powder carrier gas. The spray parameters are listed in Table 1.

Table 1 Cold spray parameters for coating deposition

Spray parameters	Values
Main gas pressure/MPa	3.0
Carrier gas pressure /MPa	3.2
Gas temperature/°C	400±20
Powder feeding rate/g·s^{-1}	18
Gun traverse speed /mm·s^{-1}	80
Stand-off distance /mm	20

2.3 Characterization of the coating

Cross sections of cold sprayed coatings were prepared vertically to their surfaces by a conventional mechanical polishing method with SiC paper and diamond suspensions. The mirror-polished cross-sections were observed by a field emission scanning electron microscope. The porosity was estimated based on the image analyzing with the cross section photos.

2.4 Tensile test, micro-hardness and Electrical conductivity

Cohesive strength of coating was measured by tensile test. Dimensions of the tensile specimens are shown in Fig. 2 following the standards of ASTM E8/E8M-13a. In order to understand the failure mechanism of the cold-sprayed samples, the fracture surface of the tensile specimens were also characterized by the field emission scanning electron microscope. The electrical conductivity of the coatings was measured by double bridge route. The micro-hardness of the coating was tested under a load of 100 g with a dwell time of 30 s. The hardness was evaluated by the average value of 10 indentations.

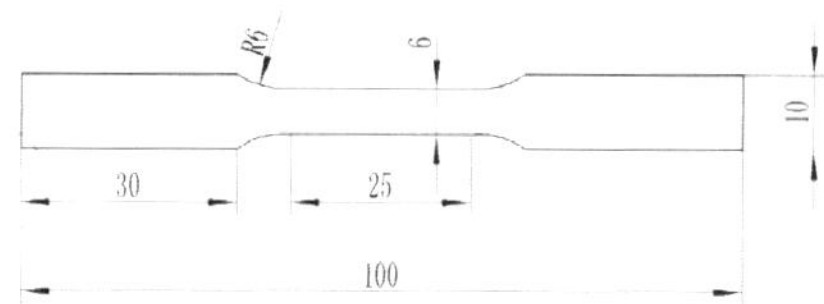

Fig. 2. Dimension of tensile specimen.

2.5 Numerical simulations

FEA was performed as an assistant method. Simulation was carried out using ABAQUS with the Lagrangian algorithm and Dynamic-explicit solver. As for saving the computational time and reducing the computational cost, a two-dimensional impacting model was employed with an adiabatic process and further simplified as a 1/2 symmetrical model. A Cu particle with the diameter of 29 μm was used to represent the average level of the feedstock diameters. It should be noted that the particle size indeed has some effects on the particle deposition because the particle impact velocity and critical velocity are both size-dependent parameters [24]. Fixed boundary condition was imposed to the substrate bottom and side walls. Based on the real working conditions and nozzle geometry, the particle impact velocity and temperature were calculated as 524 m/s and 190°C by the computational fluid dynamic (CFD) technique. The substrate temperature was given as 25°C. The material properties were described by the Johnson–Cook plasticity model which accounts for strain, strain rate hardening, as well as thermal softening. The property parameters for Cu used in the model can be found elsewhere [25].

3 Results and discussion

3.1 Microstructure and porosity

Fig.3 shows the cross section microstructures of cold sprayed coating in both directions which are parallel and perpendicular to the moving direction of the gun. As shown in Fig. 3a and Fig.3b, it can be seen that dense Cu coating was obtained by cold spray at the parallel direction. Very little pores can be observed from cross sectional microstructure. In order to quantitatively evaluate the density of the cold sprayed coating, porosity of coating is tested by means of image analysis. The measured result shows that coating porosity is extremely low, only 0.12% ± 0.09 at the parallel direction, indicating that the present spray condition is suitable to prepare dense cold sprayed Cu coating.

It has been commonly accepted that spray angle plays an adverse role on the particle bonding at deposition process [26,27], which may results in the formation of inter-particle pores, leading to a high porosity of cold sprayed coating. As it is known, particles experience normal impact during deposition process in a single track. However, the impact of particles between single tracks is oblique. In order to comprehensively evaluate the densification of as-sprayed coating, the cross section microstructures at the perpendicular direction is also observed as shown in Fig. 3c and 3d. It should be noted that the as-sprayed coating is dense at perpendicular direction as well as that at parallel direction. The porosity measured according to image analysis is 0.12% ± 0.01, further demonstrating that the coating is very dense at perpendicular direction. The reason may be that the ductility of Cu particles is so excellent at this spray condition that a certain oblique impact can be negligible.

According to above results, it can be concluded that a suitable spray condition can prepare dense Cu coating with cold spray.

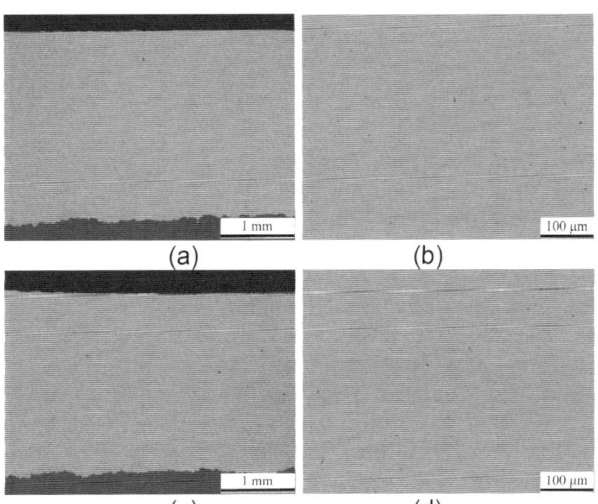

Fig. 3. Cross sections of cold sprayed Cu coating in both directions parallel (a, b) and perpendicular (c, d) to the moving direction of gun.

3.2 Micro-hardness and electrical conductivity

Micro-hardness and electrical conductivity of cold sprayed Cu coating are presented in Fig.4. As shown in Fig.4, it should be noted that due to the strain-hardening effect during particle deposition process, the micro-hardness of the as-sprayed Cu coating is about 163 $HV_{0.1}$ which is much higher than that of 100 $HV_{0.1}$ reported for pure Cu bulk. Therefore, it can be concluded that the cold sprayed coating has a higher micro-hardness than that corresponding to conventional bulk counterpart. Similar result has been reported by Li et al. [28] who cold sprayed Ti onto mild steel substrate and found that the micro-hardness of the as-sprayed Ti coating is slightly higher than that of conventional bulk Ti.

In Fig.4, it can be seen that the electrical conductivity of cold sprayed sample reaches 76 %IACS which is close to that of cold rolled pure Cu bulk. The reasons are as follows. On the one hand, the porosity in coating is the barrier for the movement of electron, and thus a higher porosity of the coating leads to an inferior electrical conductivity while a superior electrical conductivity is attributed to a lower porosity of the coating. As presented in Fig.3, the porosity of cold sprayed coating is as low as 0.12% giving rise to a relatively high electrical conductivity. On the other hand, the oxide at the interface in coating resulted from the feedstock powder or the oxidation in deposition process also blocks the movement of electron resulting in an inferior electrical conductivity. In this experiment, the oxide at the interface is reduced to the minimum due to the following two reasons. First, the oxygen content of feedstock power is low leading to very few residues in coating. Second, because of the low deposition temperature, the oxidation of particle can be effectively avoided in deposition process. Segall and Stoltenhoff [29] studied cold sprayed coatings of oxidation sensitive materials such as Ti or MCrAlY and reported that the coatings show similar, or only slightly higher oxygen contents than their respective powder feedstock. As a consequence, the cold sprayed coating exhibits a superior electrical conductivity due to above causes.

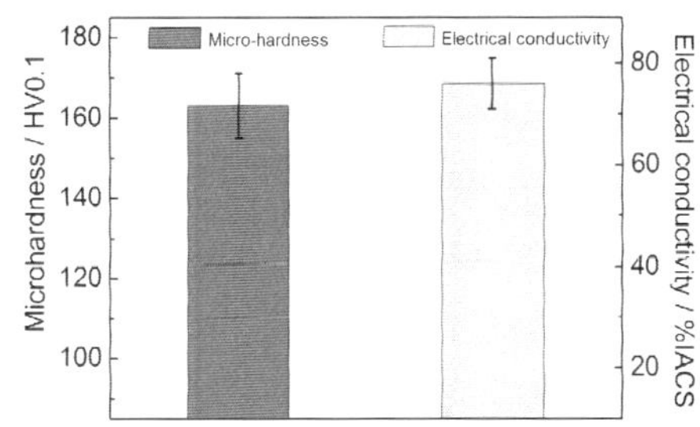

Fig. 4. Micro-hardness and electrical conductivity of cold sprayed coating.

3.3 Cohesive strength

Fig.5 presents a representative stress-strain curve of cold sprayed Cu sample as obtained from tensile test. Tensile tests show that the mean ultimate tensile strength value of cold sprayed Cu sample is about 65.5 ± 5.5 MPa. The elongations to failure of all the Cu samples do not exceed 0.2%, which reflects the brittleness of cold sprayed coatings. However, it has been reported that the ultimate tensile strength of conventional corresponding Cu bulk is at least above 200 MPa according to literatures, which is much higher than that of cold sprayed Cu coating. Moreover, the mean Young modulus of Cu samples obtained from tensile curves is merely 0.65 ± 0.5 GPa, far below the corresponding value of conventional Cu bulk (above 100 GPa). These phenomena reveal that mechanical interlocking between the deformed particles is the main bonding mechanism. Besides, there may be very little bonding while a great deal of un-combined interface exists between deformed particles in cold sprayed coating although with an extremely low coating porosity.

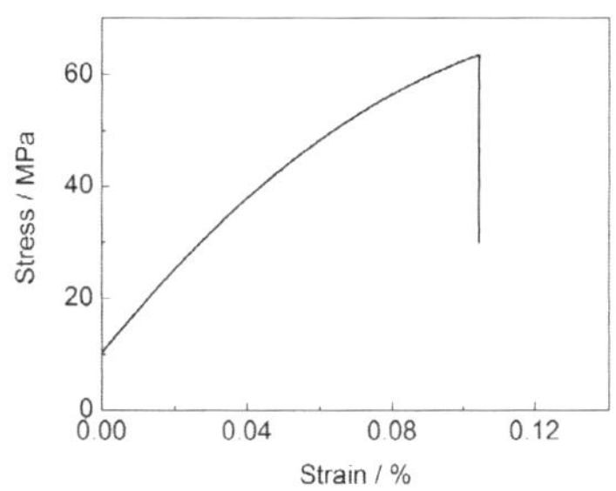

Fig. 5. Stress-strain curve of the cold sprayed Cu sample.

In order to obtain further evidences of fracture mechanism of cold sprayed coating, the fracture surfaces of the typical tensile samples were observed and are presented in Fig.6. A locally smooth fracture surface for the cold sprayed tensile sample can be seen from Fig.6. Tracking the crack propagation, it is found that it occurs along the interfaces of the deposited particles rather than through the deformed particles as shown in Fig.6. It can be rationalized that the main bonding mechanism of cold sprayed Cu coating in this experiment is mechanical bonding between deposited particles determining the cohesive strength of coating. Therefore, the tensile strength of cold sprayed sample is relatively low and the fracture morphology exhibits the brittle characteristics as obtained in tensile tests shown in Fig.5.

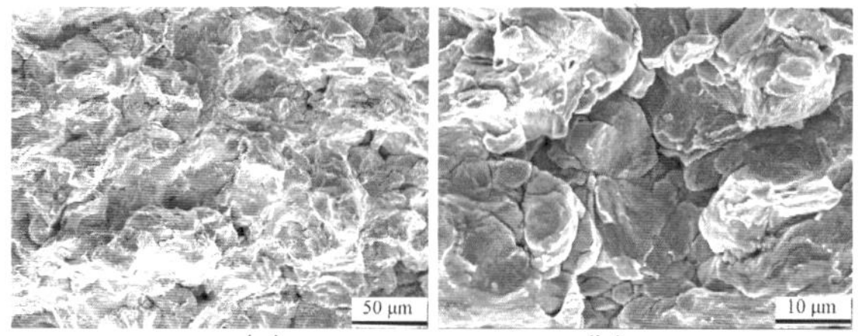

(a) (b)

Fig. 6. SEM fracture morphology of cold sprayed Cu sample.

A simplified particle impact model that a single Cu particle deposited on a Cu substrate was carried out to clarify the interfacial bonding features between particles in cold sprayed Cu coating. The simulated equivalent plastic strain (PEEQ) and temperature (TEMP) contours at the end of impact are shown in Fig.7. Obviously, because of a relatively high velocity before impact, the Cu particle and substrate experience intensive deformation at the same time during the deposition process and the metal jet formed at the rim of contact interface between the flattened particle and deformed substrate. It should also be noted that the localized intensive plastic deformation in the particle and substrate occurs at the periphery of the contact interface, where the temperature is much higher than the inner part of the particle due to plastic dissipation induced adiabatic temperature rise. These phenomena are identical with the work reported by Grujicic et al. [6] and Assadi et al. [3].

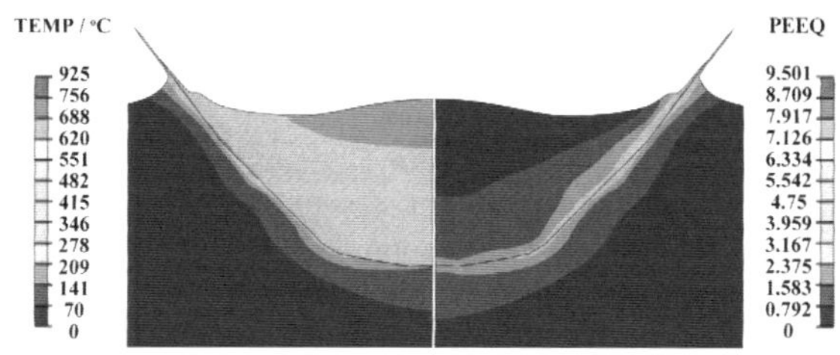

Fig. 7. Contours of the temperature and equivalent plastic strain of a single Cu particle deposited on a Cu substrate.

It has been generally recognized that bonding at the interface between deformed particle and substrate in cold spraying occurs only when the particle velocity is above the critical velocity. When a particle is deposited just at the critical velocity, there exists a so-called critical element at the particle interface which undergoes the highest amount of deformation within the particle. During the impact, the equivalent plastic strain and the temperature of the critical element show a monotonic increase to a certain particle/substrate contact time and once past this contact time, they undergo an abrupt increase, which is the significant symbol of the occurrence of adiabatic shear instability aroused by localized the plastic strain and thermal-softening effects. The temporal

development of the equivalent plastic strain and the temperature of the element on the deformed particle in this work are presented in Fig.8. It is interesting to find that the equivalent plastic strain and the temperature of the element 1 rise abruptly during the initial stage of the impact and then exhibit a much more gradual increment. This gradual increment continues until another abrupt climb in the temperature, following which the temperature eventually plateaus. While for element 2, the first element below element 1 along the interface between deposited particles, this phenomenon does not happen. This means that the adiabatic shear instability is onset at element 1. Therefore, the bonding occurs in the regions above the element 1 along interfaces. However, for regions below the element 1, bonding can not occur due to the absence of the adiabatic shear instability and a large number of un-combined interfaces are remaining. These un-combined interfaces in cold sprayed coating give rise to a relatively low cohesive strength tested in tensile test as shown in Fig.5 compared with that corresponding to conventional bulk counterpart.
Therefore, in order to gain a cold sprayed coating with a comparable or even higher cohesive strength than their bulk counterparts, it is necessary to further improve the particle velocity to reduce the un-combined regions and make metallurgical bonding occur at interfaces between deposited particles. The feature of interfacial bonding achieved by high-velocity particle impacts in cold spray will be investigated in the future.

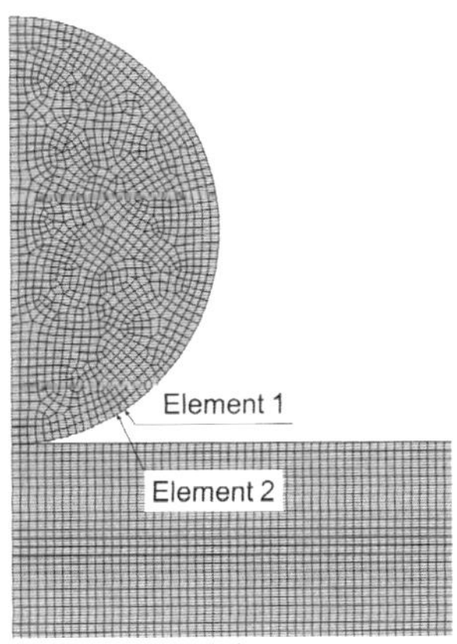

(a)

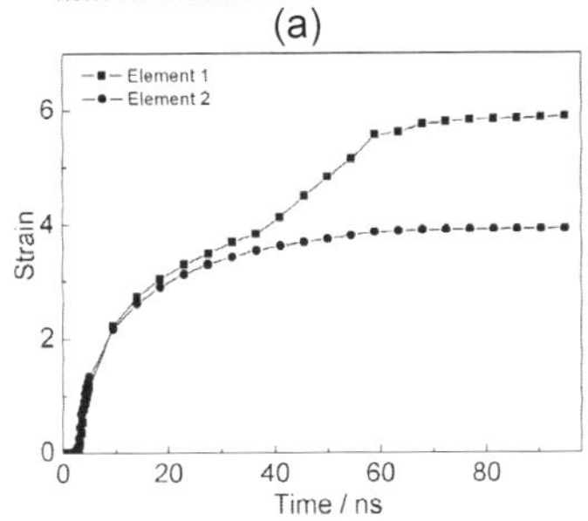

(b)

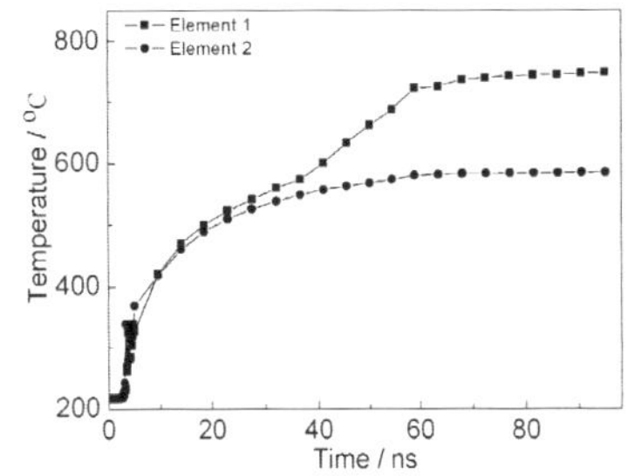

(c)

Fig. 8. Positions of element 1 and 2 (a), calculated temporal development of the equivalent plastic strain (b) and the temperature (c) at two elements marked in (a).

4 Conclusions

In this study, a Cu feedstock with low oxygen content was used to prepare coating with a medium velocity particle impact in order to deposit a coating with relatively high cohesive strength. The feature of interfacial bonding in the cold sprayed Cu coating was investigated and the conclusions are as followings. Because of a medium particle velocity before impact calculated by the computational fluid dynamic technique, the Cu coating is extremely dense with very little pores. Quantitative analysis with the image method yielded the porosity of coating of only 0.12% from the two cross sections which are parallel and perpendicular to the moving direction of gun. Due to the strain-hardening effect during particle deposition process, the micro-hardness of the as-sprayed Cu coating is about 163 $HV_{0.1}$ which is much higher than that of 100 $HV_{0.1}$ reported for pure Cu bulk. The electrical conductivity of the as-sprayed coating reaches 76 %IACS which is close to that of cold rolled pure Cu bulk on account of a low oxygen content in feedstock, a relatively low deposition temperature and a low porosity in coating. Tensile tests show that the mean ultimate tensile strength of cold sprayed Cu sample is about 65.5 ± 5.5 MPa. The elongations to failure of all the Cu samples do not exceed 0.2%, which reflects the brittleness of cold sprayed coatings. Moreover, the mean Young modulus of Cu samples obtained from tensile curves is merely 0.65 ± 0.5 GPa. All these tensile results is attributed to the main bonding mechanism of cold sprayed Cu coating in this experiment is mechanical bonding between deposited particles deduced from the fracture morphology. Besides, the deformation behavior of the Cu particle simulated by a finite element analysis employing the explicit code of ABAQUS reveals that there exists a large number of un-combined interfaces between deformed particles in coating, which is the another explanation of the lower cohesive strength in contrast with that corresponding to conventional bulk counterpart.

Acknowledgements

This work is supported by National Natural Science Foundation of China (No. 50115814).

Literature

[1] C. Borchers, F. Gartner, T. Stoltenhoff, H. Assadi, H. Kreye, Microstructural and macroscopic properties of cold sprayed copper coatings, J. Appl. Phys., 93 (2003), Issue 12, pp. 10064-10070.

[2] W.-Y. Li, C.-J. Li, H Liao, Effect of Annealing Treatment on the Microstructure and Properties of Cold-Sprayed Cu Coating, J. Therm. Spray Technol., 15 (2006), Issue 2, pp. 206-211.

[3] H. Assadi, F. Gartner, T. Stoltenhoff, H. Kreye, Bonding mechanism in cold gas spraying, ACTA Mater., 51 (2003), Issue 15, pp. 4379-4394.

[4] T. Schmidt, H. Assadi, F. Gartner, H. Richter, T. Stoltenhoff, H. Kreye, T. Klassen, From Particle Acceleration to Impact and Bonding in Cold Spraying, J. Therm. Spray Technol., 18 (2009), Issue 5-6, pp. 794-808.

[5] J. Pattison, S. Celotto, R. Morgan, M. Bray, W. O'Neill, Cold gas dynamic manufacturing: A non-thermal approach to freeform fabrication, J. Mach. Tool Manuf., 47 (2007), Issue 3-4, pp. 627-634.

[6] M. Grujicic, C.-L. Zhao, W. S. DeRosset, D. Helfritch, Adiabatic shear instability based mechanism for particles/substrate bonding in the cold-gas dynamic-spray process, Mater. Design 25 (2004), Issue 8, pp. 681-688.

[7] T. Schmidt, F. Gartner, H. Assadi, H. Kreye, Development of a generalized parameter window for cold spray deposition, ACTA Mater., 54 (2006), Issue 3, pp. 729-742.

[8] C.-J. Li, W.-Y. Li, H. Liao, Examination of the Critical Velocity for Deposition of Particles in Cold Spraying, J. Therm. Spray Technol., 15 (2006), Issue 2, pp. 212-222.

[9] K, Kim, M, Watanabe, S. Kuroda, Thermal softening effect on the deposition efficiency and microstructure of warm sprayed metallic powder, Script. Mater., 60 (2009), Issue 8, pp. 710-713.

[10] G. Bae, S. Kumar, S. Yoon, K. Kang, H. Na, H.-J. Kim, C. Lee, Bonding features and associated mechanisms in kinetic sprayed titanium coatings, ACTA Mater., 57 (2009), Issue 19, pp. 5654-66.

[11] S. Barradas, V. Guipont, R. Molins, M. Jeandin, M. Arrigoni, M. Boustie, C. Bolis, L. Berthe, M. Ducos, Laser Shock Flier Impact Simulation of Particle-Substrate Interactions in Cold Spray, J. Therm. Spray Technol., 16 (2007), Issue 4, pp. 548-556.

[12] P.C. King, M. Jahedi, Relationship between particle size and deformation in the cold spray process, Appl. Surf. Sci., 256 (2010), Issue 6, pp. 1735-1738.

[13] H.-T. Wang, C.-J. Li, G.-J. Yang, C.-X. Li, Cold spraying of Fe/Al powder mixture: Coating characteristics and influence of heat treatment on the phase structure, Appl. Surf. Sci., 255 (2008), Issue 5, pp. 2538-2544.

[14] H. Lee, H. Shin, K. Ko, Effects of Gas Pressure of Cold Spray on the Formation of Al-Based Intermetallic Compound, J. Therm. Spray Technol., 19 (2009), Issue 1-2, pp. 102-109.

[15] H.-T. Wang, C.-J. Li, G.-C. Ji, G.-J. Yang, Annealing Effect on the Intermetallic Compound Formation of Cold Sprayed Fe/Al Composite Coating, J. Therm. Spray Technol., 21 (2012), Issue 3-4, pp. 571-577.

[16] V.K. Champagne, D. Helfritch, P. Leyman, S. Grendahl, B. Klotz, Interface Material Mixing Formed by the Deposition of Copper on Aluminum by Means of the Cold Spray Process, J. Therm. Spray Technol., 14 (2005), Issue 3, pp. 330-334.

[17] P.C. King, G. Bae, S.H. Zahiri, M. Jahedi, C. Lee, An Experimental and Finite Element Study of Cold Spray Copper Impact onto Two Aluminum Substrates, J. Therm. Spray Technol., 19 (2009), Issue 3, pp. 620-634.

[18] W.-Y. Li, C. Zhang, X. Guo, C.-J. Li, H. Liao, C. Coddet, Study on impact fusion at particle interfaces and its effect on coating microstructure in cold spraying, Appl. Surf. Sci., 254 (2007), Issue 2, pp. 517-26.

[19] Y. Xiong, X. Xiong, S. Yoon, G. Bae, C. Lee, Dependence of Bonding Mechanisms of Cold Sprayed Coatings on Strain-Rate-Induced Non-Equilibrium Phase Transformation, J. Therm. Spray Technol., 20 (2011), Issue 4, pp. 860/65.

[20] I. Manna, J. Dutta Majumdar, Enhanced kinetics of diffusion coating of aluminium on copper by boundary diffusion, J. Mater. Sci. Lett., 12 (1993), pp. 920-22.

[21] F.A. Calvo, J.M. Gomez De Salazar, F. Molleda, Special features of the formation of the diffusion bonded joints between copper and aluminium, J. Mater. Sci., (1988), Issue 23, pp. 2273-2280.

[22] R. Huang, H. Fukanuma, Study of the Influence of Particle Velocity on Adhesive Strength of Cold Spray Deposits. J. Therm. Spray Technol., 21 (2011), Issue 3-4, pp. 541-549.

[23] D.J. Varacalle, D.P. Guillen, D.M. Deason, W. Rhodaberger, E. Sampson, Effect of Grit-Blasting on Substrate Roughness and Coating Adhesion. J. Therm. Spray Technol., 15 (2006), Issue 3, pp. 348-355.

[24] G. Bae, K. Kang, H. Na, J.-J. Kim, C. Lee, Effect of particle size on the microstructure and properties of kinetic sprayed nickel coatings, Surf. Coat. Technol., 204 (2010), Issue 20, pp. 3326-3335.

[25] S. Yin, X. Wang, X. Suo, H. Liao, Z. Guo, W. Li, C. Coddet, Deposition behavior of thermally softened copper particles in cold spraying, Acta Mate., 61 (2013), Issue 14, pp. 5105-5118.

[26] S. Yin, Y. Sun, X. Wang, Z. Guo, H. Liao, Effect of Spray Angle on Temperature Distribution within the Metallic Substrate in Cold Spraying, J. Therm. Spray Technol., 22 (2013), Issue 6, pp. 983-991.

[27] W.-Y. Li, S. Yin, X.-F. Wang, Numerical investigations of the effect of oblique impact on particle deformation in cold spraying by the SPH

method, J. Appl. Phys., 256 (2010), Issue 12, pp. 3725-3734.
[28] B. Al-Mangour, P. Vo, R. Mongrain, E. Irissou, S. Yue, Effect of Heat Treatment on the Microstructure and Mechanical Properties of Stainless Steel 316L Coatings Produced by Cold Spray for Biomedical Applications, J. Therm. Spray Technol., 23 (2014), Issue 4, pp. 641-6452.
[29] T. Stoltenhoff, C. Borchers, F. Gärtner, H. Kreye, Microstructures and key properties of cold-sprayed and thermally sprayed copper coatings, Surf. Coat. Technol., 200 (2006), Issue 16-17, pp. 4947-4960.

The Effect of Annealing on the Mechanical Properties of Iron - Stainless Steel composites

Rajib Barua[1], Huseyin Aydin[2], David Frost[1], Rosaire Mongrain[1], Phuong Vo[3] and Stephen Yue[2]
[1]Department of Mechanical Engineering, McGill University, 817 Sherbrooke Street West, Montreal, Quebec, H3A 0C3, Canada; [2]Department of Mining and Materials Engineering, McGill University, 3610 University Street, Montreal, Quebec, H3A 2B2; Canada; [3]NRC Boucherville, Quebec.

Stainless steel (316L) powder is mixed with commercially pure iron (CP Fe) (by weight of CP Fe in the range between 20% and 80%) and cold sprayed onto a steel substrate to produce a metal composite designed primarily to exhibit controlled properties. For these composites, porosity is very low, but annealing between 600°C to 1100°C for an hour reduces porosity; the lowest porosity was exhibited by the 50-50 composite. Annealing also 'sinters' the interparticle interfaces, leading to vastly improved fracture properties. The fully annealed single component 316L material exhibits a much higher strength compared to the fully annealed CP Fe specimen, but the addition of only 20% 316L to CP Fe leads to a composite with the same fully annealed strength level as that of the 316L.

1 Introduction

Composite metal-metal materials produced by spraying powders of two or more different metal powders can provide desirable multifunctional properties [1]. Some metals and alloys been targeted for mixed powder applications by using cold spray [1, 2-4] for biomedical applications. Hydroxyapatite/ Ti mixed powders have been sprayed to fabricate load bearing surgical implants with less cytotoxicity and better mechanical properties [5]. Mixed coatings for metallic biodegradable coating were based on cobalt-chromium L605 alloy mixed with stainless steel 316L in order to explore the possibility of micro-galvanic corrosion [6].

In this work, 'commercial purity' (CP) Fe is mixed with 316L stainless steel to nominally produce composites with CP Fe compositions of 20%, 50% and 80wt% by cold spraying onto steel substrates. The primary concern was to control the corrosion characteristics of the composite, but also of great importance are the mechanical properties of the coatings. This paper describes the mechanical properties of these composites in the as sprayed and annealed conditions.

2 Materials and Methods

Feedstock powders were purchased from Sandvik Osprey Limited, UK. The 316L stainless steel contained about 16 wt% Cr, 11 wt % Ni and 0.014 wt % C. The 'commercial purity' (CP) Fe had about 0.4 wt % Mn and 0.03 wt% C. The average powder diameters were about 44 μm and 23 μm for the 316L and CP Fe, respectively. Both powders were basically spherical (Fig. 1) but the 316L had very fine particles coating the 44 μm particles. The microhardness values of CP Fe and 316L powders are about 290 and 230 $Hv_{0.01}$, respectively, which are much higher than the bulk values of 150 and 155 Hv for Fe and 316L, respectively [7,8], and may be due to the presence of martensite.

The composites were fabricated using a KINETICS® 4000 cold spray system (Sulzer Metco, Westbury, NY) with nitrogen as the propellant gas and an MOC24 nozzle at the McGill – NRC Cold Spray facility housed in the National Research Center (NRC), Boucherville, QC. Mixed powders were prepared by using either a rolling mixer without balls or a rotating tumbler with balls for an hour. Mild steel 1020CR substrates were blasted with 24 grit alumina before spraying to increase adhesion of the particles to the substrate. The gas temperature was 700 °C, the gas pressure was 4 MPa, powder feed rate was around 20 g/min, gun traverse speed was 300 mm/s and the standoff distance was 80 mm.

316L

CP Fe

Fig. 1. As received powders

The 'coatings' were removed from the substrates by electric machine discharge prior to mechanical testing and annealing. Annealing was performed to increase the ductility of the metals. To minimize oxidation, the samples were wrapped in a steel sheet and annealed at 1100°C for 1 h in an argon gas atmosphere. After one hour had elapsed, the sample remained in the oven for a further 30 min while the oven cooled and the sample was removed and allowed to cool to room temperature.

Cross sections were ground to a typical metallographic finish and the porosity was determined using a Nikon Epiphot 200 microscopy equipped with Clemex Vision software.

Microhardness was measured with a Vickers Microhardness tester (Clark, Clemex™ CMT) at a load of 50 gram

Prior to mechanical testing, the samples were polished in order to eliminate any micro-cracks on the surface. Mechanical properties were determined by micro-shear punch testing, which is a quick and easy method to characterize limited volumes of material. The test is based on a blanking operation using a flat, cylindrical punch of 1.55 mm diameter, which is moved at a constant speed to punch a hole in a flat, thin sample,10 mm by 4 mm by 0.60-0.85 mm thick. Load-displacement curves obtained from micro-shear punch testing can be correlated with conventional tensile tests to obtain yield and ultimate tensile strength of the coatings [9].

3 Results

3.1. Compositions of the composites as-cold sprayed

The chemical compositions of the composites were measured and then used to determine the amount of CP Fe and 316L in the coatings. The results reveal that the CP Fe levels are approximately 5 wt% lower in the as sprayed composite compared to the as-mixed powders, for all mixed powders. This may be due to the lower deposition efficiency of the CP Fe compared to the 316L (approximately 30% vs 70%), but it is interesting that the absolute decrease in the CP Fe level did not change with composition of the composite.

3.2. Effect of annealing on microhardness

Figure 2 shows the change in microhardness of the 316L and CP Fe in all the composites. In the as-sprayed state, the hardnesses of both powders are quite similar and remain largely unaffected by tempering at 400 °C. At 800 °C there is a significant drop in hardness of both powders, indicating recrystallization. As well, after annealing at 800 °C and above, there is a clear separation in the hardness values of the component metals, with CP Fe now clearly softer than the 316L.

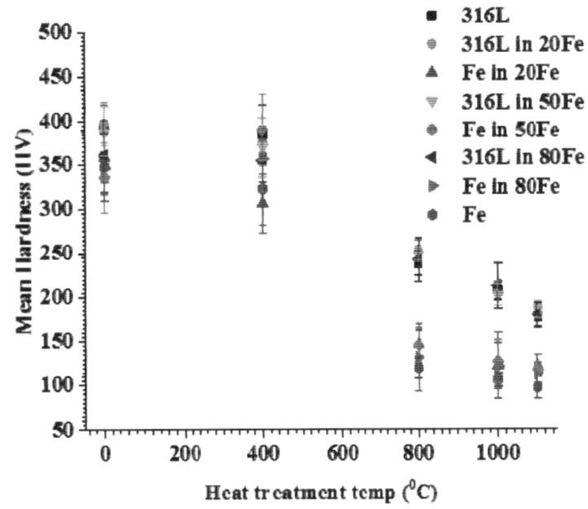

Fig. 2. Effect of annealing temperature on microhardness of 316L and CP Fe in all composites.

3.2. Effect of annealing on porosity

The change in porosity of samples exposed for 1 hour over temperatures of 400, 800, 1000 and 1100°C, was monitored. Significant reductions in porosity start to occur at 800 °C coinciding with the noticeable hardness reduction and recrystallization. Recrystallization is an indication that diffusion rates are significant and it is this transfer of material that is leading to a reduction in porosity. All composites exhibited porosities less than 0.6% after annealing at 1100°C, but the 50% Fe composite exhibited the lowest porosity, which is possibly related to having the highest CP Fe / 316L interfaces.

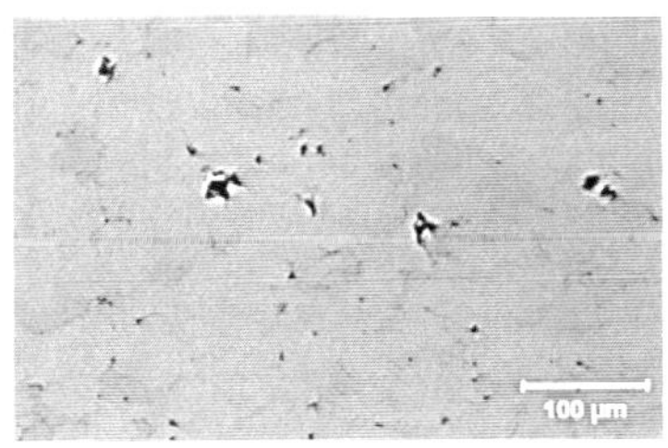

400 °C

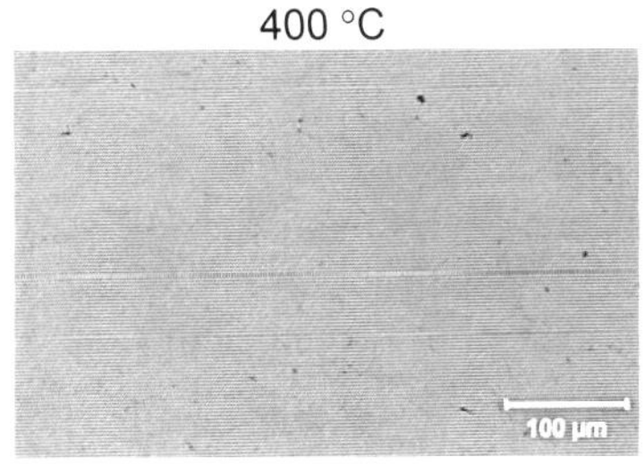

1100°C

Fig. 3 Effect of annealing temperature for one hour on porosity of 20% CP Fe composite.

3.3. Micro-Shear Punch Test of the Coatings

Figure 4 shows the effect annealing temperature on the flow behaviour of the single component cold sprayed 316L. In the as sprayed condition, the ductility is negligible, and this leads to a very low fracture strength. At 400 °C there is little change but at 800 °C there is a dramatic increase in ductility and a corresponding increase in strength. There are further increases in ductility and strength with increasing annealing temperature.

This increase in properties appears to coincide with the recrystallization observed in the microhardness results. However, recrystallization is a *softening* mechanism. The increase in strength is more likely directly related to the increase in ductility, which, in turn, is likely to be related to the reduction in porosity. However, the level of porosity even in the as-sprayed condition is very low and the observed level of porosity decrese is unlikely to lead to such a dramatic improvement. It is more probable that interparticle 'sintering' has occurred, increasing the ductility in this way. Recrystallization may lead to sintering because considerable movement of atoms takes place, mainly through grain boundary motion. This may help metallurgical bonding to take place at particle to particle interfaces.

Figure 5 shows the corresponding behaviour for single component cold sprayed CP Fe. Once again the as- sprayed properties are very low; 400 °C brings little improvement and 800 °C dramatically improves ductility and fracture strength. In contrast to 316L, however, increase the annealing temperature above 800 °C leads to virtually no further improvement. This suggests that the kinetics of 'sintering' (and/or recrystallization) is faster in CP Fe compared to 316L. Note also that the 316L reaches much higher strength levels than the CP Fe, although the ductilities are similar.

Figure 6 shows the effect of annealing temperature on the shear punch curves of the 80% CP Fe composite. Its behaviour is more similar to the 316L than the CP Fe in that the fracture strength continues to increase with annealing temperature above 400 °C in contrast to the cold sprayed CP Fe. However, there is a fundamental difference in flow behaviour between the composite and either of the single component materials, and this is discussed below.

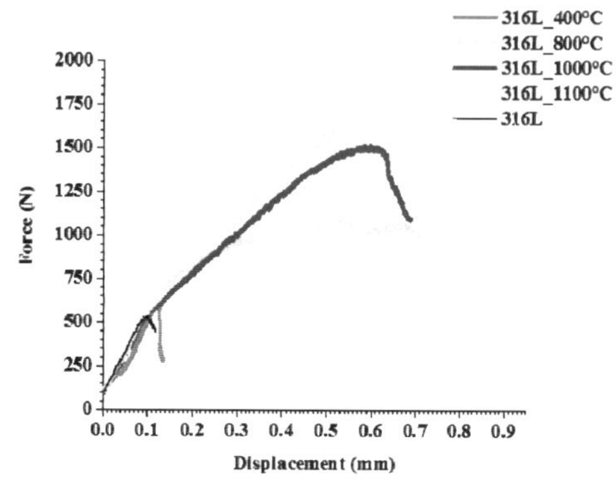

Fig. 4 Effect of annealing on cold sprayed 316L

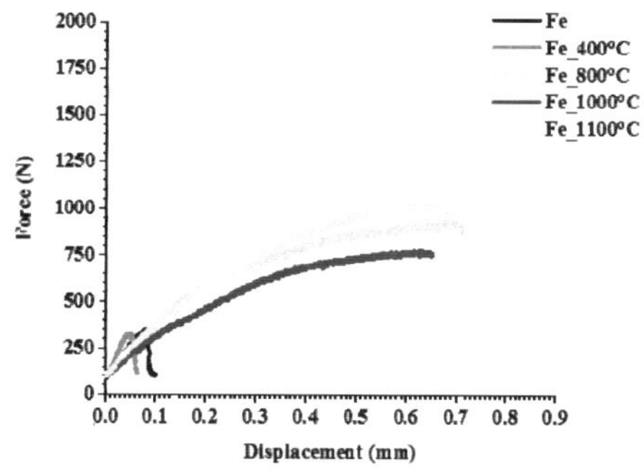

Fig. 5 Effect of annealing on cold sprayed CP Fe.

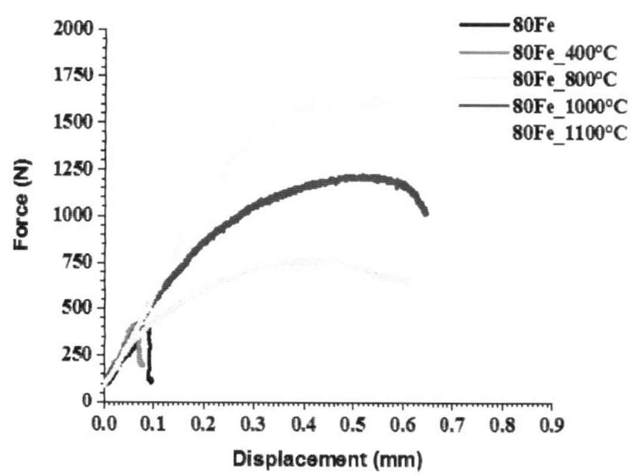

Fig. 6 Effect of annealing on 80% CP Fe composite.

4. Discussion

As noted earlier, the effect of annealing is to increase the strain to fracture, which, in turn increases the fracture strength, primarily through work hardening. This effect is well established at 800 °C for all the cold sprayed materials shown in this paper. For the 316L, increasing the temperature above 800 °C leads to further increases in ductility and strength, but note that all the curves fall on top of each other initially and deviate from each other only when fracture takes place. In other words, the *plastic flow* behaviour is identical for each annealing temperature but the *fracture* characteristics differ. Plastic flow is essentially the breaking and reforming of atomic bonds, I.e. dislocation motion, whereas fracture is breaking of bonds. The fact that the flow curves fall on top of each other at low strains suggest that the plastic flow behaviour is the same for the specimens annealed at 800 °C and above, regardless of the annealing temperature, whereas the fracture behaviour is different.

For the same microstructure, any differences in flow behaviour between the annealing conditions would be due to different amounts of work hardening. Work hardening is removed during annealing by recrystallization. Because the plastic flow behaviour is the same, it suggests that recrystallization is more or less complete at 800 °C. If it is assumed that fracture preferentially occurs at particle/particle interfaces, then the increasing fracture strain with increasing annealing temperature above 800 °C is probably due to increasing levels of 'sintering'. Normally, sintering, in classical powder metal processing, is primarily a means to reduce porosity, but at the same time metallurgical bonds are formed across the particle/particle interfaces. For these as-cold sprayed powders, porosity is low and not a first order issue, but metallurgical bonding between particles is limited. Annealing at high enough temperatures will rapidly sinter the material by increasing the metallurgical bonds.

In the case of CP Fe, annealing at above 800 °C does not significantly change the plastic flow or fracture behaviour, indicating that 'sintering' is easier in CP Fe. This may be due to the absence of the tenacious Cr oxide film on the powder particle surfaces, which is part of the reason why 316L is 'stainless'.

In the case of the composite, like the 316L, the fracture strength increases with annealing temperatures above 800 °C, but there are two differences: (i) at low strains, the curves do not fall on top of each other and (ii) the fracture strain does not increase with annealing temperature. Since it appears that 800 °C is sufficient for full recrystallization of both 316L and CP Fe, the change in fracture strength with annealing temperature is probably mainly due to sintering of the particle-particle interfaces.

There are three types of such interfaces in the composite: CP Fe - CP Fe, 316L - 316L and CP Fe - 316L. Therefore, at 800 °C, it can be assumed that the fracture strain is mainly due to the bonding of CP Fe particles to each other and the flow behaviour is due mainly to CP Fe. At 1000 °C, the mixed metal interfaces bonding is 'complete' and flow behaviour includes some 316L deformation. At 1100 °C, 316L - 316L bonding is 'complete' and all the 316L in the composite contributes to the flow behaviour. Thus the increase in flow strength of the composite is directly due to an increasing amount of 316L contributing to the load bearing of the composite, as opposed to increasing the metallurgical bonding between particles per se. Clearly, the UTS in a metal composite is not directly proportional to the volume fraction of the component metals since the UTS of the 80 % CP Fe composite is almost the same as the 100% 316L specimen after annealing at 1100 °C. However, it is difficult to explain why the fracture strain seems to be independent of the increase in particle-particle sintering. Perhaps the strain to fracture is predominantly controlled by the CP Fe – CP Fe bonding, since CP Fe constitutes 80% of the composite and the CP Fe particles are all sintered at 800 °C.

5. Conclusions

Composites of 316L and CP Fe were successfully made by cold spraying, although all composites exhibited a CP Fe content approximately 5% lower than the composite in the as-mixed powder condition.

The porosity of the as-cold sprayed coatings was low and fell to less than 0.6% for all materials after annealing at 1100 °C. The 50-50 composite exhibited the lowest porosity in the annealed condition.

Annealing at 800 °C significantly lowered porosity, drastically reducing the hardness of both the 316L and CP Fe (whether in the single component material or in a composite) and radically increased fracture strength and strain to fracture.

Much of these improvements are associated with recrystallization, which in turn is due to large scale movement of atoms. But recrystallization per se is not thought to lead to increases in fracture strain and stress. It is suggested that sintering, which is also due to large scale atom movement, leads directly to the increased fracture strain and stress.

There is considerable potential in terms of increasing mechanical properties by mixed metal cold spray strategies since a composite containing 20% 316L almost attained the fracture strength of 100% 316L, which represents a factor of 3 increase compared to the fracture strength of 100% CP Fe.

6 Literature

[1] Sova, A., Pervushin, D., and Smurov, I.: 'Development of multimaterial coatings by cold spray and gas detonation spraying', Surface & Coatings Technology, 2010, 205, (4), pp. 1108-1114.

[2] Kikuchi, S., Yoshino, S., Yamada, M., Fukumoto, M., and Okamoto, K.: 'Microstructures and Thermal Properties of Cold-Sprayed Cu-Cr Composite Coatings', Journal of Thermal Spray Technology, 2013, 22, (6), pp. 926-931

[3] Wang, H.-T., Li, C.-J., Yang, G.-J., and Li, C.-X.: 'Cold spraying of Fe/Al powder mixture: Coating characteristics and influence of heat treatment on the phase structure', Applied Surface Science, 2008, 255, (5), pp. 2538-2544

[4] Ko, K.H., Lee, H., and Choi, J.O.: 'Effects of Cu Feedstock on Intermetallic Compound Formation in Cold Sprayed Cu-Sn Coatings', International Journal of Material Science, 2013, 3, (2), pp. 45-53

[5] Choudhuri, A., Mohanty, P.S., and Karthikeyan, J.: 'Bio-ceramic Composite Coatings by Cold Spray Technology', in Marple, B.R., Hyland, M.M., Lau, Y.-C., Li, C.-J., Lima, R.S., and Montavon, G. (Eds.) (Thermal Spray 2009: Expanding Thermal Spray Performance to New Markets and Applications), pp. 391-396

[6] Al-Mangour, B., Mongrain, R., Irissou, E., and Yue, S.: 'Improving the Strength and Corrosion Resistance of 316L Stainless Steel for Biomedical Applications using Cold Spray', Surface & Coatings Technology, 2013, 216, pp. 297-307

[7] http://www.matweb.com/search/DataSheet.aspx?MatGUID=654ca9c358264b5392d43315d8535b7d, accessed Octoer 24, 2014

[8] http://www.matweb.com/search/datasheet.aspx?MatGUID=a2d0107bf958442e9f8db6dc9933fe31, accessed October 23, 2014

[9] Aydin, H.: 'Effect of Microstructure on Static and Dynamic Mechanical Properties of Third Generation Advanced High Strength Steels', Ph.D. thesis, McGill University, 2013

Acknowledgements

The authors would like to thank Frederic Belval and Jean-François Alarie from the National Research Council, Canada (NRC, Boucherville, Canada) who performed the cold spray. The authors acknowledge Dr. Olivier Bertrand from Laval University and Claude Belleville from Opsens Inc. for their financial support on this research. Last but not least, the financial contributions of NSERC (Natural Sciences and Engineering Research Council) and the CFI (Canadian Foundation for Innovation) are gratefully acknowledged.

Flame sprayed HDPE-Cu composite coatings: an innovative technique for constructing antifouling layers

Z.M. Jia, X.K. Suo, Y.F. Gong, Y. Liu, Ningbo / CN, P.P. Jin, Xining / CN, H. Li, Ningbo / CN

Biofouling and corrosion are two predominate worldwide concerns in modern marine industries. Large-scale fabrication of protective coatings with both sufficient corrosion resistance and favorable antifouling property yet remains elusive. Here we propose a new approach for fabricating corrosion-resistant high density polyethylene (HDPE) coatings with unique dispersion of copper particles for excellent long-term antifouling performances. Electroless plating was employed for designing the HDPE-Cu particles for coating deposition by flame spraying. Copper shell with a thickness of ~1 µm surrounding individual HDPE particle has been achieved, which has been retained in the coatings. Neutral salt spray testing for 14 days reveals almost intact state of the HDPE-Cu coatings. Significantly constrained attachment of Bacillus sp. bacteria on the surfaces of the Cu-containing coatings is observed, suggesting exciting antifouling performances of the coatings. The strategy of fabricating the novel HDPE-Cu coatings gives bright insight into potential antifouling applications in the marine environment.

1 Introduction

Marine biofouling is an undesirable attachment behavior of marine organisms on a surface immersed in seawater [1]. Antifouling based on biocides is a well-established approach for effectively preventing biofouling in marine industries. Copper has become the most important alternative biocide since organotin compounds were banned. As it is challenging to apply biocides on the substrate surface directly for long-term services, developing appropriate techniques for incorporating biocides into coatings has been one of the recent research efforts.

Exciting findings of thermal spray technology have shed light on its potential applications in marine industries. Thermal sprayed aluminum, zinc and aluminum-zinc alloy coatings have shown long-term corrosion protection for steel structures in the marine environment for more than 20 years [2-4]. Thermal sprayed polymer coatings like HDPE, low-density polyethylene (LDPE) [5] have proved to perform well in corrosive environment. For marine applications, however, the polymeric coatings lack antifouling properties. In fact, there have been attempts in recent years of using copper to construct antifouling coatings by thermal spraying. Antifouling performances of cold sprayed Cu/Cu_2O inorganic coatings have been reported by Ding R. et al. and increased Cu_2O content brought about enhanced antifouling performances [6]. Yet, the coatings do not offer the capability to resist corrosion. Embedding of copper particles by cold spray into polymer [7] might avoid the deficiency in corrosion resistance, while facilitated the antifouling performances by impeding attachment of marine organisms through release of copper ions from the composite structure. Further investigation on bronze and zinc embedded structures revealed similar antifouling performances [8-10]. However the thin layer of the composite structure achieved by embedding metal particles into polymer matrix might influence release efficiency and potential application of the structure against biofouling. Cost-efficient fabrication of large-scale anti-corrosion/fouling polymer-based coatings is to be further explored. Taking into account the fact that use of copper is still permitted for antifouling purpose, appropriate preparation of the starting polymer-Cu composites is essential for developing consequent antifouling coatings.

Thermal sprayed coatings for marine applications have been extensively investigated, however, few of single-layer coatings have been attempted to satisfy both sufficient anti-corrosion and antifouling properties. In this study, electroless plating was developed for making HDPE-Cu composite powder for subsequent coating fabrication by flame spraying. Microstructure, corrosion resistance and antifouling properties of the composite coatings were characterized and elucidated. Even distribution of Copper in the coatings has been achieved. Excellent corrosion resistance and promising antifouling property were revealed.

2 Experimental setup

2.1 Materials

Commercially available HDPE powder was used as the starting feedstock. The particles showed ellipsoidal shape (Fig. 1) with a particle size range of +54-140 µm. To achieve a desirable coating structure, prior to the spraying, a thin Cu wrapping layer was made covering HDPE particle by electroless plating. Pretreatment of the HDPE powder followed the procedures reported by others [11]. The HDPE powder was added into electroless plating bath at 50°C and the synthesis was maintained for 40 minites.

2.2 Deposition of the coatings

Copper, HDPE, Cu-coated HDPE particles were deposited respectively on mild steel plates using flame spraying system. Oxygen and acetylene were used as the combustion-supporting gas and fuel gas, respectively. Feed rate of the powder was set as 60 g/min. The standoff distance from nozzle exit to sample surface was 250 mm. Traverse speed of the gun was 200 mm/s. The fabrication route for making the HDPE-Cu coatings is schematically depicted in Fig. 2.

Fig. 1. Surface morphology of original HDPE powder.

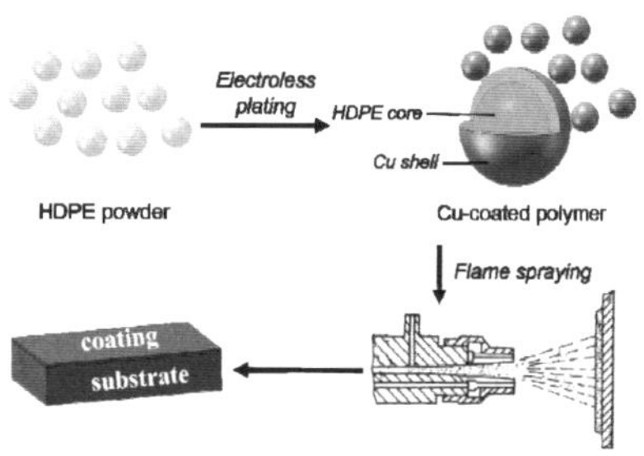

Fig. 2. Schematic illustration of the fabrication process of HDPE-Cu composite coatings.

2.3 Microstructure characterization powder and the coatings were characterized

Microstructures of the Cu-coated HDPE by scanning electron microscope. Distribution of copper component in the coatings were analyzed using energy dispersive spectrometer (EDS) equipped with FESEM. Phase composition of the powder and the coatings was examined by X-ray diffraction (XRD, D8 Advance, Bruker AXS, Germany). Structural evolution of the powder was evaluated using differential scanning calorimetry and thermogravimetric analysis (DSC/TGA, Pyris Diamond DSC/TG/DTA, Perkin-Elmer, USA) in nitrogen atmosphere with a heating rate of 10°C/min. Two heating-up processes were employed to eliminate heat history of the feedstock powder.

2.4 Assessment of anti-corrosion and antifouling properties

Corrosion resistance of the HDPE coatings, the copper coatings and the HDPE-Cu composite coatings was evaluated using neutral salt spray testing. Surface morphology, cross-sectional microstructure and corrosion products of exposed samples in salt spraying were examined periodically using FESEM. Formation of bacterial biofilm on and further antibacterial efficiency against gram-positive Bacillus sp. of the coatings in ASW was examined. The coating samples were immersed in the bacteria-containing ASW for 3 days. For FESEM observation of the bacteria colonized on the surfaces of the samples, the bacteria after incubation were fixed in 2.5% glutaraldehyde for 24 hours, followed by gradual dehydration and coating with Au.

3 Results and Discussions

It is clear that the HDPE particles are entirely wrapped by thin copper layer after electroless plating (Fig. 3). This is not surprising since copper tends to nucleate and grows on Ag catalytic sites that have already affiliated on the surfaces of HDPE particles. Cross-sectional view of individual particle and corresponding EDS analysis further evidences the composite structure (Fig. 3c,d).

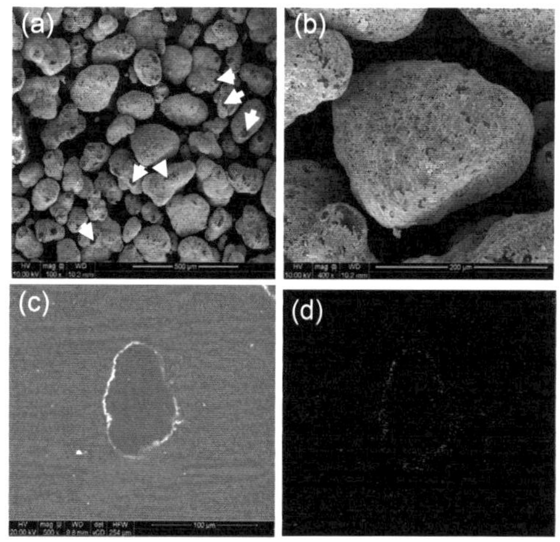

Fig. 3. Surface morphology (a,b), cross-section microstructure (c) and EDS mapping (d) of Cu-coated HDPE powder.

The coatings were successfully fabricated by flame spraying. Dense structure has been achieved (Fig. 4). No marked flaws for instance microcracks or pores are seen from either surfaces or cross-sections of the coatings. It is exciting to note even distribution of copper in composite coatings, and copper exists at the interfaces between individual HDPE splats (Fig. 4b). This on the other hand suggests that the copper layer sticks to HDPE particles regardless of fully or partially melted state of the HDPE particles during the spraying.
XRD detection suggests complete reaction of Ag catalytic sites with copper ion in the solution of electroless plating (Fig. 5a). In addition, no extra peaks are seen except those for copper and HDPE, which deduces that negligible oxidation or other undesirable chemical changes during spraying. Thermal analyses using TG and DSC (Fig. 5b, c, d, and Table 1) clearly indicate that the melting temperature of HDPE in the coatings decreases slightly as compared to that in the starting powder, which is probably due to thermal decomposition during spraying [12]. Higher $T_{5\%}$ and $T_{d\,(max)}$ of HDPE in the Cu-coated HDPE powder compared to the pure

HDPE powder is likely due to that copper shell absorbs heat during the heating-up process. Interestingly, it is worth noting that for the HDPE-Cu composites, the coatings show remarkable decrease in both $T_{5\%}$ and $T_{d\,(max)}$, suggesting that the copper shell probably promotes thermal decomposition of HDPE during the coating fabrication. This could be explained by the prolonged heating effect of metallic layer to HDPE. Pretreatment of the starting HDPE for preparing the HDPE-Cu powder already resulted in a decrease in crystallinity. However, it is surprising to note that the crystallinity of HDPE in the coatings is significantly augmented compared to the starting HDPE powder. Presence of copper obviously facilitates the increased crystallinity. This could also be attributed to the annealing effect offered by copper layer. Enhanced strength could be expected for the coatings since the increase in crystallinity of HDPE gave rise to enhanced properties[13].

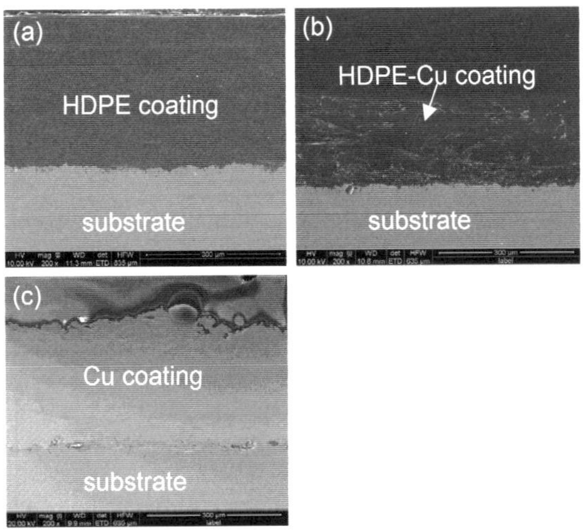

Fig. 4. Cross-section microstructure (a) and surface morphology (b) of HDPE coatings, HDPE-Cu composite coatings and Cu coatings.

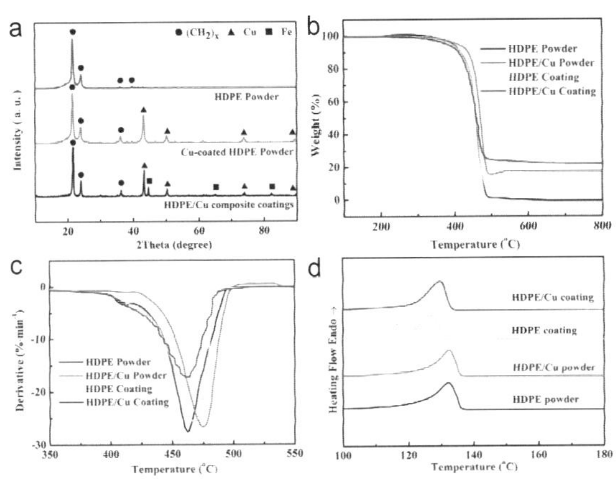

Fig. 5. XRD curves (a), TG curves (b), DTG curves (c) and DSC curves (d) of HDPE-Cu composite powders and coatings.

Table 1. Thermal properties of powders and composite coatings [a].

Samples	T_m (°C)	$T_{5\%}$ (°C)	$T_{d\,(max)}$ (°C)	ΔH_m ($J{\cdot}g^{-1}$)	χ_c (%)
HDPE powder	132.2	395.4	462.7	142.1	52.6
Cu-coated HDPE powder	132.4	408.7	474.8	103.3	38.2
HDPE coatings	126.5	390.9	467.4	168.3	62.3
Cu-coated HDPE coatings	129.6	376.3	461.5	168.8	62.5

[a] Melting temperature T_m, Temperature at 5% weight loss is determined by TGA $T_{5\%}$, Maximum decomposition temperature $T_{d\,(max)}$, Melting enthalpy ΔH_m, Degree of crystallinity of HDPE χ_c.

Anti-corrosion performance persists as one of the major challenges for nowadays available antifouling thermal sprayed coatings for marine applications. Thus the neutral salt spray test was conducted, and the coating samples were exposed in salt spraying fog for up to 14 days. It is realized that after 6 hrs exposure, corrosion product was firstly seen for the pure copper coatings. 7 days exposure of the copper coating samples already gave rise to full coverage on the surfaces of the coatings by corrosion products namely Fe_3O_4 and α-FeOOH (Fig. 6b), whereas other samples were not corroded yet. After 14 days exposure, the surfaces of HDPE coatings still exhibited intact state, suggesting excellent anti-corrosion performances.

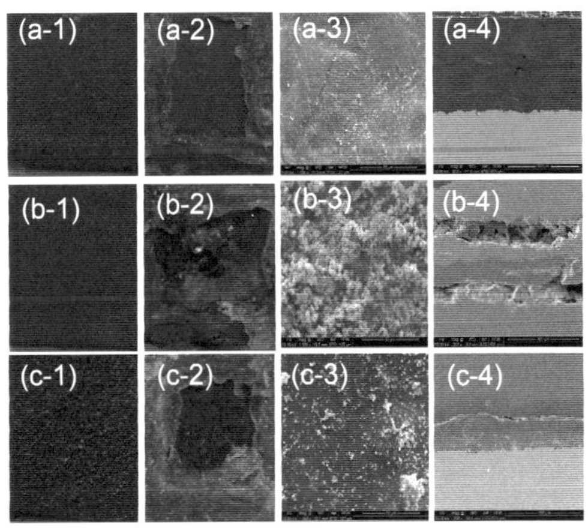

Fig. 6. Surface morphologies and cross-sectional microstructures of HDPE coatings (a), copper coatings (b) and HDPE-Cu composite coatings (c) after salt spray test for 14 days. (a-1), (b-1) and (c-1) are the macro surface morphologies of coatings before salt spray test; (a-2), (b-2) and (c-2) are the macro surface morphologies of coatings after salt spray test; (a-3), (b-3) and (c-3) are micro surface morphologies of coatings after salt spray test; (a-4), (b-4) and (c-4) are cross-sectional morphologies of coatings after salt spray test.

Assessment of the antifouling properties of the HDPE-Cu coatings reveals prohibited attachment of Bacillus sp. bacteria. It is noted that after 3 days exposure in the bacteria-containing ASW, colonization of the bacteria on the HDPE coatings are clearly seen (Fig. 7a). In contrast, the Cu-containing coatings and the pure copper coatings show non attachment of the bacteria on their surfaces, which is predominately due to release of copper ions into the solution [14-16]. Extensive investigations have proposed many possible antifouling mechanisms of copper ions [17]. Copper ions released from the Cu-containing coatings could engage with bacteria's DNA and subsequently inhibit bacteria multiplication [16]. Copper ions adsorbed on cell surface could interact with caryon and cytoplasm of bacteria cell, in turn damage cell membrane or destroy the function of bacterial enzymes [18, 19].

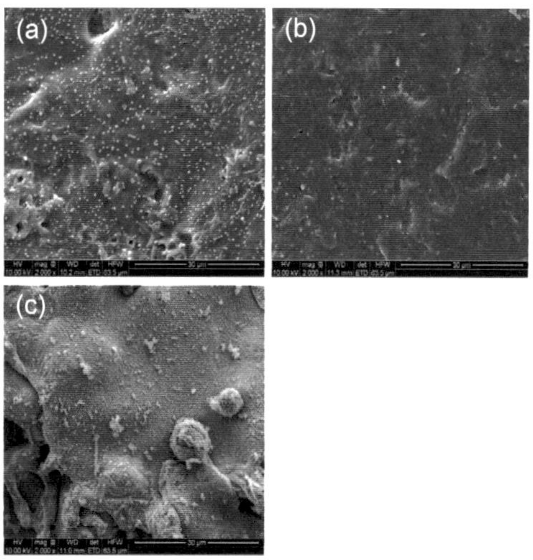

Fig. 7. Surface morphologies of HDPE coatings (a), HDPE-Cu composite coatings (b) and copper coatings (c) after immersion in bacteria-containing artificial seawater for 72 hours.

4 Conclusions

A new processing route for making HDPE-Cu powder and coatings was proposed. The flame sprayed HDPE-Cu coatings show homogenous dense microstructure with unique distribution of copper at interfaces between individual HDPE splats. Both excellent antifouling and anti-corrosion performances were revealed for the Cu-containing coatings. The special coating structure offers controllable release of copper for inhibiting attachment of bacteria. The results shed light on potential applications of the HDPE-Cu coatings for marine applications and the approach is effective for cost-efficiently making large-scale polymer-metal composite coatings for a variety of applications.

5 References

[1] Legg M. and Yücel M. K., Garcia de Carellan I., Kappatos V., Selcuk C., Gan T. H.: Acoustic methods for biofouling control: A review. Ocean Engineering 103 (2015), pp. 237/47.

[2] Kuroda S. and Kawakita J., Takemoto M.: An 18-year exposure test of thermal-sprayed Zn, Al, and Zn-Al coatings in marine environment. Corrosion 62 (2006), Issue 7, pp. 635/47.

[3] Wood R. J. K. and Speyer A. J.: Erosion-corrosion of candidate HVOF aluminium-based marine coatings. Wear 256 (2004), Issue 5, pp. 545/56.

[4] Rogers F. S.: Thermal spray for commercial shipbuilding. Journal of Thermal Spray Technology 6 (1997), Issue 3, pp. 291/93.

[5] Jothi K. J. and Santhoskumar A. U., Amanulla S., Palanivelu K.: Thermally Sprayable Anti-corrosion Marine Coatings Based on MAH-g-LDPE/UHMWPE Nanocomposites. Journal of Thermal Spray Technology 23 (2014), Issue 8, pp. 1413/24.

[6] Ding R. and Li X., Wang J., Xu L.: Study on Antifouling Effect of Cold Spray Cu-Cu2O Coating. Paint & Coatings Industry 43 (2013), Issue 9, pp. 1/6.

[7] King P. C. and Poole A. J., Horne S., de Nys R., Gulizia S., Jahedi M. Z.: Embedment of copper particles into polymers by cold spray. Surface and Coatings Technology 216 (2013), pp. 60/67.

[8] Vucko M. J. and King P. C., Poole A. J., Carl C., Jahedi M. Z., de Nys R.: Cold spray metal embedment: an innovative antifouling technology. Biofouling 28 (2012), Issue 3, pp. 239/48.

[9] Vucko M. J. and King P. C., Poole A. J., Hu Y., Jahedi M. Z., de Nys R.: Assessing the antifouling properties of cold-spray metal embedment using loading density gradients of metal particles. Biofouling 30 (2014), Issue 6, pp. 651/66.

[10] Vucko M. J. and King P. C., Poole A. J., Jahedi M. Z., de Nys R.: Polyurethane seismic streamer skins: an application of cold spray metal embedment. Biofouling 29 (2013), Issue 1, pp. 1/9.

[11] Narkis M. and Yacubowicz J., Vaxman A., Marmur A.: Eleectrically conductive composites prepared from polymer particles coated with metals by electroless deposition. Polymer Engineering and Science 26 (1986), Issue 2, pp. 139/43.

[12] Li C. and Hou T., She X., Wei X., She F., Gao W., Kong L.: Decomposition properties of PVA/graphene composites during melting-crystallization. Polymer Degradation and Stability 119 (2015), pp. 178/89.

[13] Dou G. and Dou Q.: Crystallization kinetics, spherulitic morphology, mechanical properties and heat resistance of β-nucleated impact-resistant propylene-ethylene copolymer. Thermochimica Acta 614 (2015), pp. 21/32.

[14] Brooks S. and Waldock M.: The use of copper as a biocide in marine antifouling paints. Advances in Marine Antifouling Coatings and Technologies. Woodhead Publishing Limited, Cambridge (2009), pp. 492/521.

[15] Finnie A. A. and Williams D. N.: Paint and coatings technology for the control of marine fouling. Biofouling 10 (2010), pp. 185/206.
[16] Yuranova T. and Rincon A. G., Bozzi A., Parra S., Pulgarin C., Albers P., Kiwi J.: Antibacterial textiles prepared by RF-plasma and vacuum-UV mediated deposition of silver. Journal of Photochemistry and Photobiology A: Chemistry 161 (2003), Issue 1, pp. 27/34.
[17] Hobman J. L. and Crossman L. C.: Bacterial antimicrobial metal ion resistance. J Med Microbiol 64 (2015), Issue Pt 5, pp. 471/97.
[18] Zhang X. and Aifantis K. E.: Interpreting the softening of nanomaterials through gradient plasticity. Journal of Materials Research 26 (2011), Issue 11, pp. 1399/1405.
[19] Liu P. C. and Hsieh J. H., Li C., Chang Y. K., Yang C. C.: Dissolution of Cu nanoparticles and antibacterial behaviors of TaN-Cu nanocomposite thin films. Thin Solid Films 517 (2009), Issue 17, pp. 4956/60.

Influence of HVOF spray process on performances of NiCrAl coatings

Lu Xiaoliang[1, 2], Ji Xiaojuan[1, 2], Hou Weiao[1, 2] , Shen Jie[1, 2]
1. Beijing General Research Institute of Mining & Metallurgy, Beijing 100160;
2.Beijing Engineering Technology Research Center of Surface Strengthening and Repairing of Industry Parts, Beijing 100026

Abstract: NiCrAl coatings were prepared on quenched 1045 steel substrate by HVOF technology. The effect of the spraying distance on the microstructure, porosity, bonding strength, micro hardness of the coating, and the efficiency of powder deposition was investigated. The results showed that: NiCrAl coating porosity, bond strength, micro hardness, deposition efficiency significantly affected by spraying distance, the trend of the spraying distance impact on coating porosity, bond strength, micro hardness and deposition efficiency is increases first and then decreasing, the bonding mode between the coating and the substrate was mechanical bonding. The optimal spray parameters was obtained: spray distance 360 mm, kerosene flow rate 22 L/h, oxygen flow rate 850 L/min. Using the optimal spraying parameters, the coating bonding strength is 66 MPa, microhardness is 418.56 HV0.3, porosity is 1.1%, powder deposition efficiency up to 65.3%. The coatings shows good mechanical properties, and lay a good foundation for the preparation of high temperature work coating.

1 Introduction

HVOF is a kind of thermal spray technology developed on the basis of explosive spraying. with a low flame flow temperature (2600-3000 ℃and high flame velocity (approximately 1500m / s). The metal oxidation and decomposition can be effectively inhibited in the spraying process, so the coatings show high bonding strength, low porosity, and homogeneous microstructure[1]. HVOF is widely used in preparing metal ceramic coatings, but only a few research of the application of preparing NiCrAl coatings were reported by Mahesh, who studied the HOVF sprayed NiCrAl coating on super alloy matrix about the structure and mechanical properties[2], and the NiCrAl coating corrosion behavior at 900 ℃ [3]. When the coating material is selected, the type of gun, gas flow, spray distance influence the microstructure, bonding strength, and porosity of the coating. Bonding strength, porosity and spray deposition efficiency are the main indicators for evaluating the quality of the coatings, as well as the most used is the bonding strength. Because the failure of the coating often occurs at the interface between the substrate and coating . Under certain conditions of fuel and oxygen flow, the temperature and the velocity of the powders in the high velocity flame flow will vary with the change of spraying distance, so the performances of coatings are significantly affected by the spraying distance.In this study, GTV-HVOF systems and K2 gun were used to prepare the NiCrAl coatings with different spray distances, the effects of spraying distance on bonding strength, porosity, microhardness, powder deposition efficiency were investigated .

Thermal spray coatings often used in the high-temperature part as a surface strengthen coating because of their unique chemical composition and microstructure, which, but the relative low bonding strength between the coating and substrate limit the application of surface strengthen coatings, in order to improve the bonding strength between the surface strengthen coatings and the substrate, adhesive layer often sprayed between the substrate and the surface strengthen coating. NiCrAl coatings prepared by plasma spraying technology are generally used as an adhesive layer to enhance the bonding strength of the working layer and the substrate[4-5], but the plasma sprayed coatings have relative high porosity and low bonding strength. Improving the bonding strength and reducing the porosity are directly related to the service life of the coating system, but the optimization of the HVOF spraying process for preparing NiCrAl coatings were rarely reported. Therefore, studying the influence of HVOF spraying process on NiCrAl coating properties has an important scientific significance.

2 Experiment

2.1 Preparation of Coating

The self-adhesive nickel chromium aluminum powder with the particle size of 15-45 μm (KF110, Beijing General Research Institute of Mining and Metallurgy) is used as spray powder. The substrate material is quenched 1045 steel, the size of the sample is 60 mm × 25 mm. Before spraying, the surface was washed by acetone, and sandblasted using 50-70 mesh corundum process, in order to enhance the bonding strength between the coating and the substrate. The coatings were prepared by using ASM6600 robots and K2 GTV spray systems, with the spray gun barrel 150/11. The samples were fixed on a special stainless steel fixture with the line speed of 0.15 m/s and the diameter of 200 mm. During the spraying process, the spray gun perpendicular to the axis of the fixture, the moving speed is 6 mm/s, and the temperature of samples before spraying was maintained at 150-200 ℃. Spray parameters are listed in Table 1.

Table 1 Spraying parameters of HVOF under different spray distances

Parameters	Samples 1	2	3
Material	NiCrAl	NiCrAl	NiCrAl
Spray gun	K2	K2	K2
Nozzle	150K	150K	150K
Carrier gas (Ar)	7L/min	7L/min	7L/min
Oxygene	850[L/min]	850[L/min]	850[L/min]
Kerosene	22[L/h]	22[L/h]	22[L/h]
Powder feed rate	100g/min	100g/min	100g/min
Spraying distence	340mm	360mm	380mm
Spraying velocity	0.15m/s	0.15m/s	0.15m/s
Number of passes	12	12	12
Coating Thickness (μm)	120	190	140

2.2 Coating characterization

The sample has been grinded, mechanical polished and ultrasonic cleaned before testing Quanta 200 scanning electron microscope was used to analysis the microstructure of the coatings; The microhardness was measured by using HXD-1000TM testerwith the load of 0.3kg (2.94 N)and the load time of 15s of coating the bonding strength was measured by using WE-100 electro-hydraulic servo universal testing machine; LeiCa DMIMR image analysis system sectional sample coating porosity determined randomly selected five field Determination of coating porosity and calculate the average; the roughness were tested by using TR100 surface roughness tester.

3 Results and analysis

3.1 Coating microstructure analysis

As shown in Fig. 1, the NiAlCr coating prepared by HVOF technology shows a dark grey appearance. Fig 1 (a), (b) and (c) show the surface morphologies of the NiCrAl coatings prepared using the spraying distance of 340 mm, 360 mm and 380mm, respectively. It can be seen from the figures that the coating prepared with the spray distance of 360 mm are well melted. This is because of the flattening degree of the particles decreases with the increasing spray distance, and some particles are even not melted. These unmelted particles in bonding coating decreases the bonding strength of the surface layer. The roughness values of these three coatings were measured. The roughness values of the coating using 340 mm, 360 mm and 380 mm are Ra:6.24 Rz:34.33, Ra:5.63 Rz:33.33, Ra:8.11 Rz:45.63, respectively. It can be seen that the NiCrAl coating using 360 mm has the lowest roughness and highest particle flattening degree.

Fig.2(a),(b),(c) are the cross-sectional morphologies of NiCrAl coatings perpared with different spray distances of 340mm, 360mm, 380mm, respectively. As seen in Fig.2(a), with the spray distance of 340mm, there are large amount of micro-holes, and the porosity is 2.1%. The main reason is the powders haven't been fully melted, and the flight speed of the particles are not high enough. Therefore, the combination of the matrix and coatings is weakness, and so is the adhesive strength. What has mentioned above is t correspond to adhesive strength of the spray parameter. Fig.2(b) is the cross-sectional morphology of the coating prepared with the spraying distance of 360mm, the Fig. shows that the powder spread well and the porosity is low(1.1%), as well as the bonding strength is the highest of three parameters. When the spray distance is 380mm,the porosity of the coating is 1.95%. Temperature and velocity of the powders decreased, which led to a molten state the powder when contact the substrate and the coating due to the temperature is not high enough to produce mixed with the raw material layer, then combined with the strength of the coating.

(a) (b) (c)

Fig. 1. Surface morphologies of NiCrAl coatings under different spray distances

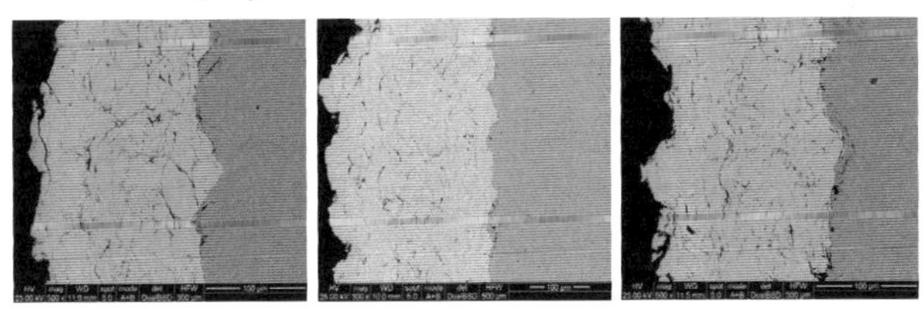

(a) (b) (c)

Fig. 2. Cross-sectional morphologies of NiCrAl coatings under different spray distances

3.2 Bonding strength of the coatings

The morphologies of the fractured coatings with three different spraying distances are shown in Fig. 3, 4, and 5. All the fracture occurred within the coatings. The tensile strength values of the coatings were about 45 MPa for spraying distance of 340 mm, 65 MPa for 360 mm, and 54MPa for 380 mm. Such firstly increase then decrease of the coating strength with spraying distance increasing was due to the melting level of the powders. At a short distance, the powders were not fully molten, while at a too long distance, molten powders were cooled, both of which resulted in a decrease of the bonding strength.

Fig. 3. Distance of 340mm cross-sectional view

Fig. 4. Distance of 360mm cross-sectional view

Fig. 5. Distance of 380mm cross-sectional view

3.3 Deposition efficiency analysis

By fixing the moving times of the spraying gun, the thickness of the coating sprayed on the substrate as well as the weight gain of the substrate by 100 gram powders were measured to characterize deposition efficiency. As shown in table 2, with the spraying distance increasing, the deposition efficiency firstly increased but then decreased. At the spraying distance of 340 mm, some large powders could not be fully molten, rebounding occurred with a phenomenon of sparking during the spraying; at 380 mm, some small powders may be solidified resulting in raw-material segments in the coating; at 360 mm, the melting of the powders got maximum and the deposition efficiency became highest. Extra test by spraying 100 gram powders onto substrate was done to measure the value of the deposition efficiency. As shown in table 2, the efficiency reached 65.3% at the spraying distance of 360 mm.

Table2 deposition efficiency test of NiCrAl coating

Parameters	Samples		
Spray distance (mm)	340	360	380
Number of passes	12	12	12
Coating Thickness (μm)	120	190	140
Powder weight before coating(g)	100	100	100
Weight of coating on the substrate(g)	41.2	65.3	49.6

2.4 Coating Mechanical Properties Analysis

Fig. 6 show the microhardness distribution curves of NiCrAl coatings at different spraying distances. It's easy to know that the hardness varies a lot with spraying distance. When the spray distance is 340mm, the average microhardness of the coating is HV0.3 392.13. When spraying distance is 360mm, the average microhardness of coating for HV0.3 418.56. When the spraying distance is 380mm, the average microhardness of the coating is HV0.3 350.30. The microhardness of the coating with the spray distance of 360mm is higher than that of spraying distance of 340mm and 380mm. It shows that the coatings have uniform structure, low porosity, without large pore. The coating can satisfy the requirements of the actual working conditions.

3 conclusion

1) The influence of spraying distance on the performances of the NiCrAl coatings prepared by

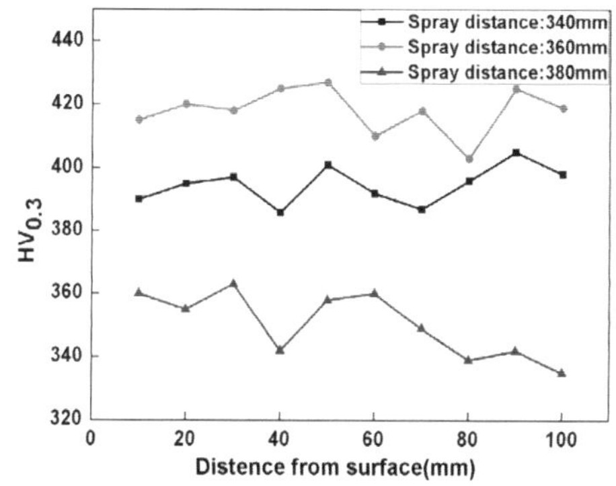

Fig. 6. Microhardness distribution curves of NiCrAl coating under different spraying distances

using HVOF technology was studied. The optimized spraying distance is 360mm.
2) When the spraying distance is 360mm, the powder sufficiently deformed spread, the coating was dense and uniform with a high bonding strength (66 MPa), high microhardness HV0.3 418.56, low porosity (1.1%),and high deposition efficiency (65.3%) .
3) Using the optimization process parameters, the coatings show excellent mechanical properties, can meet the needs of working conditions

References

[1] Wang haijun. Thermal Spray Engineer's Guide [M]. Beijing: National Defense Industry Press,2010.
[2] R.A. Mahesh, R. Jayaganthan, S. Prakash. Microstructural characteristics and mechanical properties of HVOF sprayed NiCrAl coating on superalloys[J]. Journal of Alloys and Compounds, 2009, 468:392-405.
[3] R.A. Mahesh, R. Jayaganthan, S. Prakash. Evaluation of hot corrosion behaviour of HVOF sprayed NiCrAl coating on superalloys at 900℃[J]. Materials Chemistry and Physics,2005,111:524-533.
[4] Duan zongqin, Wang zhehua. Plasma Spray Optimization of NiCrAl Coating Process [J]. Thermal processing technology and material research, 2008(5):80-82.
[5] Duan zongqin, Zhang baoxia, Wang zhehua, Lin pinghua. Plasma Spraying NiCrAl coating performance with the thickness [J]. Heat Treatment Technology, 2009, 38(8):104-109.

Influence of the segmentation crack on the stress distribution and failure behavior of the thick thermal barrier coatings fabricated by atmospheric plasma spraying: A numerical simulation study

Liang WANG[1,*], Jiasheng YANG[1,2], Dachuan LI[1,2], Huayu ZHAO[1], Xinghua ZHONG[1], Fang SHAO[1], Chengguang LIU[1], Shunyan TAO[1], Chuanxian DING[1]

[1]Key Laboratory of Inorganic Coating Materials, Shanghai Institute of Ceramics, Chinese Academy of Sciences, Shanghai 201899, PR China;

[2]University of the Chinese Academy of Sciences, Beijing 100039, PR China

* Corresponding author. Tel.:+86-021-69906320 Fax: +86-021-69906322

E-mail address: L.Wang@mail.sic.ac.cn

Abstract: In this paper, the influence of segmentation crack on the stress distribution and failure behavior of thick thermal barrier coatings fabricated by atmospheric plasma spraying (APS-TTBCs) has been investigated via finite element modeling. The density and length of segmentation cracks have been specially focused on. The investigation results indicate that the existence of the segmentation crack is not beneficial to improve thermal insulation effect of the TTBCs. But to control the appropriate segmentation crack density will be beneficial to improve thermal shock resistance of TTBCs, and the segmentation crack length also affect residual stress of the crack tip. The stress intensity factor and energy release rate (*J* integration) will increase with the increasing of segmentation crack length. The existence of segmentation crack will improve strain tolerance of TTBCs. The strain tolerance has been characterized by segmentation crack density and segmentation crack length quantitatively. The simulation results have been further verified via experimental methods. It has been found that the as-sprayed TTBCs exhibited superior thermal shock resistance when the segmentation crack density is about 4mm^{-1}. The failure mechanism of APS-TTBCs can be attributed to the propagation of segmentation crack and formation of the horizontal crack at the bond-coat/top-coat interface.

Keywords: Thick thermal barrier coatings; Segmentation crack; Residual stress; Failure; Finite element modeling

1 Introduction

Advanced thermal protection is often used in the engine, turbine blade and aircraft and so on. As a kind of very important ceramic coating, thermal barrier coatings are widely used in these components. The existence of the TBCs enables the hot-section components withstand more higher temperature and protect them from being against worn, oxidized and erosion. With the development of the superalloy materials, the single-crystal alloy can withstand the temperature 1150℃ which is the highest till now. But when the advanced TBCs were coated onto the superalloy substrates, it can make the underling substrate can operate at 1350℃-1450℃, this will improve the heat efficiency of the air-craft significantly[1-4]. But the failure problem is inevitable when the TBCs were endured long time high temperature service. Especially, the reliability and durability are the two key aspects for the TBCs, The peeling off from the substrates is a common phenomena which restrict the wide application of the TBCs. At some cases, the failure of the TBCs may be disastrous[5-7]. The TBCs often operate under the alternate cooling and heating conditions. So to improve the thermal shock resistance ability is a direct and efficient routine to prolong the lifespan of the TBCs.

As we know, the properties or performance of the materials are controlled by its microstructure. As for the TBCs, the thermal shock resistance ability of the TBCs are dependent on the microstructure and the exterior service conditions. When the exterior service conditions are constant, the microstructure will become the first factor for the improvement of the thermal shock resistance ability of the TBCs. The various microstructure can be observed for the TBCs due to the different fabrication techniques, such as atmospheric plasma spraying[8-10], electron beam-physical vapor deposition[11], solution (precursor) plasma spraying[12], vacuum plasma spraying, low pressure plasma spraying, LPPS-PVD, high velocity oxygen spaying (HVOF) and so on. The TBCs will exhibit different microstructual characteristic when they were fabricated using different techniques, even the microstructures are also very different using the different parameters but with different fabrication methods [13,14]. Laminar microstructure and columnar or feather-like microstructure are common structural characteristics. Whatever structure, pores and cracks are the basic characteristic for the TBCs. Previous investigation results indicate that pores and cracks will play an important role in affecting the thermal insulation effect and the thermal shock resistance of the TBCs fabricated by atmospheric plasma spraying. The formation of the pore and the cracks are attributed to the characteristics of thermal spraying, during thermal spraying. The sprayable powder (feedstock) was melted in the plasma flame, then it will be atomized and fly to the substrate,subsequently it impinged onto the substrate, the later splats will deposited on to the former splat, when adjacent pores were not arranged fully or the splat was nor fully melted, pores will induce among the splats. Or the inner of the splats was not sprayed fully, and air bubble will produce in the splat, when it was cooled, the bubble was excluded, and the pores will retained in the splats. The formation of the cracks are usually attributed to the quenching stress and the thermal mismatch stress[15,16]. Once the continuous splats have been arranged, the coating will be formed. Especially, except the intrinsic chemical compositions, the pores and cracks will play an important role in controlling the performance of the TBCs when they were used under the actual service conditions. As for the pores , the orientation, shape, size and porosity will affect the thermal insulation effect or the effective thermal conductivity of the TBCs[17,18].

As for the cracks, the horizontal crack, vertical crack, cracks at the inner of the top-coat or cracks at the interface, crack length will also affect the failure behavior of the TBCs[19-24], especially, as for the thick thermal barrier coatings (TTBCs) fabricated by atmospheric plasma spraying(APS), the vertical crack which is also called segmentation crack was distributed at the inner of the top-coat(ceramic layer), usually the length of the segmentation crack is beyond the half of the thickness of the top-coat, the existence of the segmentation cracks will reduce the stress concentration of the TTBCs, and thus improve the strain tolerance of the coatings[25-27], so it will ensure the thermal shock resistance ability has a significant enhancement. Although the segmentation crack is not beneficial to the decrease of the effective thermal conductivity of the TBCS, but the thickness of the TTBCs is large enough, and this will also ensure that the TTBCs still have the high thermal insulation effect[28-30].

Previous literatures have reported the fabrication, microstructure characterization and the associated thermal and mechanical properties, but most work is based on the experimental results. The description of the failure mechanisms is often based on the observation of the cross-section, it belongs to apparent explanation. In fact, the stress level and distribution play an very important role in controlling the failure modes of the TBCs during thermal shock, the special microstructure of the TTBCs will affect the stress distribution and the maximum stress level[31]. Especially, the existence of the segmentation crack at the inner of the ceramic layer of the TTBCs will release some stress concentration more or less.

The objective of the current work is to investigate the stress distribution and the failure modes of the TTBCs via finite element modeling in order to further understand the failure mechanisms deeply. Some experimental results haven been presented to testify our modeling results. In addition, the coating design method has been proposed based on the simulation results.

2. Simulation procedures

2.1 Model and mesh

The model of the TTBCs was established using commercial available software ANSYS 15.0 affilicated with APDL code[32]. The model with different density of segmentation crack (*D*s) has been established and compared. In the current work, the cylinder coating model has been used. In order to save the computational time and improve the solution efficiency, the 2D Axisymmetric plane stain model has been established, the thicknesses of the substrate, bond-coat layer and top-coat layer were 6mm, 100μm and 800μm, respectively. In fact, as for the TTBCs, the crack length should exceed the half of the thickness of the ceramic layer, the detailed model is illustrated in Fig.1. The segmentation crack density can be defined as follows:

$$Ds = N/L \quad (1)$$

Where *N* and *L* are the crack number and length along the horizontal direction of the cross-section, respectively. In order to reflect the stress concentration at the crack tip, the mesh around the crack tip has been refined. In addition, finer mesh was used at interface layers where excessive gradient of temperature and stress may occur. The partial mesh can be seen in Fig.1b).

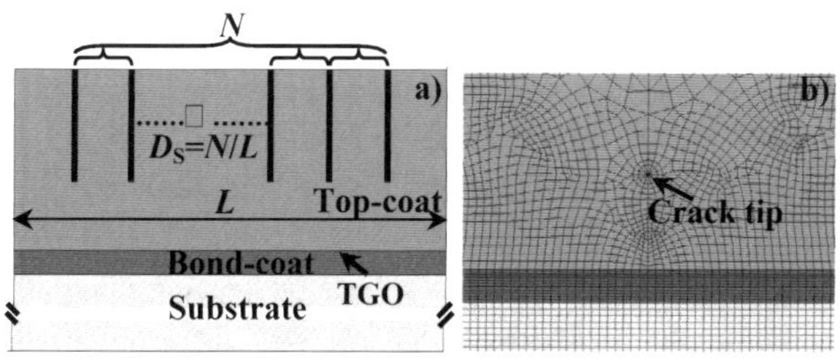

Fig. 1 Finite element model of thick thermal barrier coatings a) and mesh used in the finite element simulation b)

2. 2 Basic assumptions, initial and boundary conditions

The current finite element modeling is based on the following hypothesis:(1) the super-alloy substrate, metallic bond-coat obey the elastic-plastic constitutive behavior, the ceramic YSZ layer and the thermally grown oxide (TGO) were regarded as liner-elastic. The detailed materials properties can be seen in our published paper.(2) The geometric morphology at the TGO/TC and TGO/BC interface are flat without considering the fluctuate characteristics of the TGO layer. (3) The calculation of the creep effect was started only when the temperature exceed 600℃;(4) Only the segmentation cracks were considered, the micro-pores was not considered for all layers. The detailed properties parameters for each layer of the TTBCs can be seen in our published paper[33].

The calculation of the residual stress induced in the TBCs after plasma spraying has been performed via birth and death element technique before performing the calculation of the thermal cycling stress. The detailed simulation procedures about how to calculate the residual stress using birth and death element technique ca be seen in our published paper[34]. The calculated residual stress as the initial condition was added to the heat stage of the first thermal cycling. When the as-sprayed coating was placed into the muffle at the heating stage of a thermal cycle, the temperatures of the whole coating system including the substrate are set to be 25℃.

The boundary conditions can be described as follows: during the heating stage of a thermal cycle, the whole coating system is placed into the muffle, the bottom of the model is set to be the temperature of the muffle, the left boundary was set to be axisymmetric and UX is equal to zero, the displacement in the *y* direction of the bottom was set to be UY=0. The multi-points coupled (MPC) strain was imposed onto the right side of the model in order to make the nodes of the right side has the same displacement in the *x* direction. During the cooling stage of a thermal cycle, the geometric conditions are the same with that of the heating stage, but the boundary conditions of the temperature are absolutely different, the whole coating system cooled from the muffle temperature to the ambient temperature[35].

2.3 Experimental verification

In order to further verify our simulation results, the experimental test has been performed. Firstly, the yttria stabilized zirconia TBCs have been fabricated, the detailed plasma spraying parameters can be also seen in our published paper. The as-sprayed coatings were endured with burner rig test (BRT) in order to examine the thermal shock resistance ability. The oxygen-dimethylmethane flame has been used to heat the coating samples, the flame was vertically imposed onto the surface of the coating, the heating time last for 3min, then the coating sample will rotate and was cooled with compressed air from the surface and the backside for 3min, this is just a thermal cycle, then repeat the same process and proceeded the next thermal cycle. When 5% area of the coating surface has spalled, then the coating can be regarded as failure, and the thermal cycling numbers can be counted with the lifetime of the coating for the thermal shock test.

3. Results and discussion

3.1 calculation of thermal stress and fracture mechanic parameters of TTBCs

Fig.2 shows the distribution of the radial stress of the thick thermal barrier coatings with different segmentation crack density, it can be seen that the maximum tensile stress tend to appear at the crack tip, while the maximum compressive stress existed at the TGO layer. In fact, with the number of the segmentation crack increasing or the segmentation crack density increasing, the maximum tensile stress at the crack tip will decrease. It must be specially mentioned that when one segmentation crack existed at the inner of the ceramic top-coat, the maximum tensile stress at the crack tip is 546MPa, while there are two segmentation cracks at the inner of the top-coat, the maximum tensile stress at the crack tip decrease to 373MPa. With the number of the segmentation cracks further increasing, the maximum tensile stress at the crack tip will decrease gradually. This further indicated that when the segmentation crack density was controlled at a appropriate value, the segmentation cracks will release the stress concentration of the thick thermal barrier coating evidently, and resulting the improvement of the strain tolerance which will be beneficial to the improvement of the thermal shock resistance of the as-sprayed thick thermal airier coatings.

Fig.3 shows the distribution of the radial stress along the radial direction for the thick thermal barrier coatings with different segmentation crack density. It can be seen that when there are no segmentation cracks, the radial stress along the radial direction

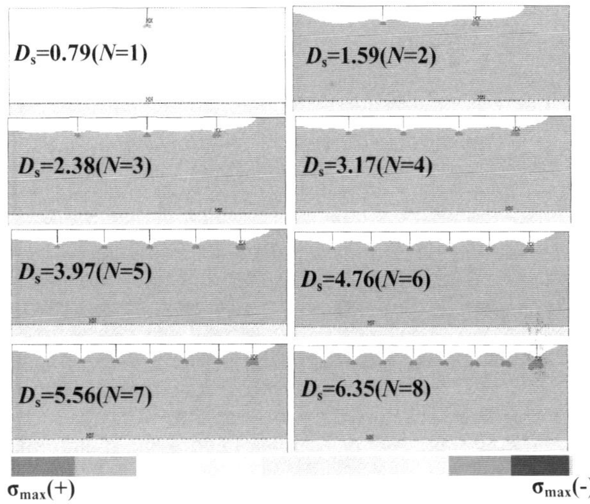

Fig.2 Distribution of the radial stress of the thick thermal barrier coatings with different segmentation crack density

which indicates in Fig.1 nearly keep the same (about 130MPa). When there existed a segmentation crack, it can be easily seen that the jump phenomena of the stress has appeared at the location around the crack tip. While at the positions far away from the crack tip, the stress is even lower than that of the TTBCs without segmentation cracks, when the number of the segmentation crack further increase to 3, i.e. the crack density is equal to 2.38, the jump phenomenon of the stress also appeared around the segmentation crack tip, while the stress at the positions where are far away from the crack tip exhibited decline tendency with the segmentation crack density decreasing. But when the number of the segmentation cracks further increase, i.e. The segmentation crack density increase to a certain value, the mutual supplementation effect will be imposed with each other for the multiple cracks, large area will have stress concentration because many cracks are near with each other, based on the simulation results, the appropriate segmentation crack density is about 3.97cracks/mm.

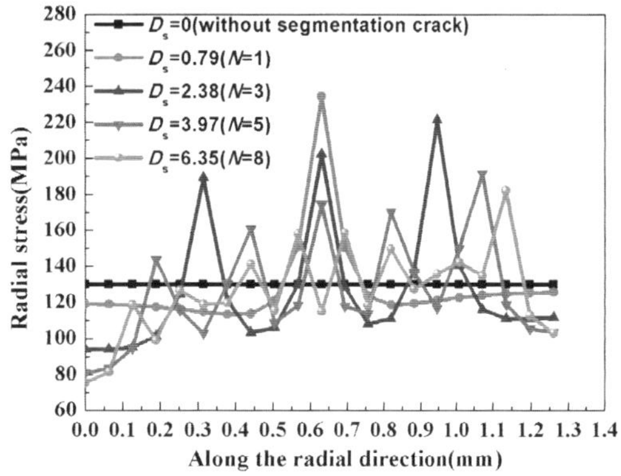

Fig.3 The distribution of the radial stress along the radial direction for the thick thermal barrier coatings with different segmentation crack density

Fig.4 shows the plot of maximum radial stress and the maximum compressive stress of the TGO layer as the function of the segmentation crack density. It can be easily seen that when the segmentation crack length keep a constant, the radial stress will increase with the segmentation crack density increasing from 0 to 1, then the radial stress at the segmentation crack tip will decease. Especially , when the number of the segmentation crack increase from 1 to 2, i.e. the segmentation crack density increase from 0.79 to 1.59, the radial stress at the segmentation rack tip decrease abruptly, then the radial stress decrease gradually with increasing the segmentation crack density from 2.38 to 6.35. Especially, when the segmentation crack density exceed 0.79, the maximum tensile radial stress at the segmentation crack tip will increase with the segmentation crack length increasing(Fig.4a)). The maximum compressive radial stress at the TGO layer also increase when the segmentation crack density increase from 1 to 0.79, then the maximum compressive radial stress at the TGO layer will decrease with the segmentation crack density further increase from 0.79 to 3.17, when the segmentation crack density increase from 3.17, the maximum compressive stress at the TGO layer nearly keep the same. In fact, the maximum compressive radial stress have not changed too much with the variation of the segmentation crack density. The segmentation crack length will not affect the maximum compressive radial stress at the crack tip too much.

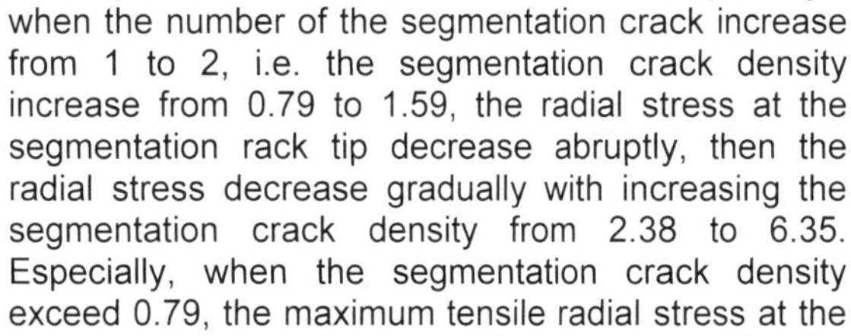

Fig.4 The plot of maximum radial stress and the maximum compressive stress of the TGO layer as the function of the segmentation crack density

Fig.5 shows the plot of stress intensity factor and the energy release rate of the segmentation crack as the function of the segmentation crack density. It can be seen that the stress intensity factor decease abruptly when the segmentation crack density increase from 0.79 to 1.59, then the stress intensity factor will increase with the segmentation crack density increasing, this is because that the multi-cracks appeared at the body of the ceramic top-coat, the field area of the stress concentration will also increase, the stress concentration effect at the crack tip may be gelegt. In addition, the stress intensity factor will also increase with the segmentation crack length increasing under the condition that the segmentation crack density keep a constant. This is consistent with the linear fracture mechanic(Fig.5a)). Fig.5b) shows the energy release rate as the function of the segmentation crack density, it has the same trend with the stress intensity factor, i.e. the energy release rate decease abruptly when the segmentation crack density increase from 0.79 to 1.59, then the energy release rate will increase with the segmentation crack density increasing, In addition, the energy release rate will also increase with the segmentation crack length increasing under the condition that the segmentation crack density keep a constant.

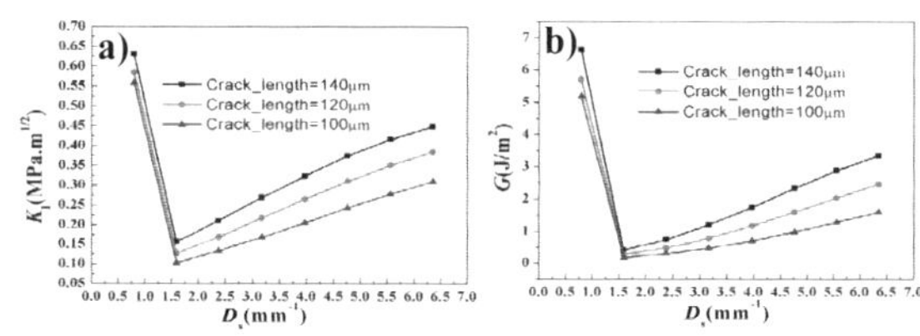

Fig.5 The plot of stress intensity factor and the energy release rate of the segmentation crack as the function of the segmentation crack density

3.2 Experimental validation

Fig.6 shows the results of burner rig test for the thick thermal barrier coatings with the different segmentation crack density. The detailed preparation parameters can be seen in the published papers of our group[36,37]. Fig.6a) shows the cross section image of the TTBCs after 172 thermal cyclings, the segmentation crack density is 4cracks/mm, the left figure shows the macro-image of the TTBC after 172 thermal cycling, it can be easily seen that the surface of the coating nearly keep completed, no peeling phenomena can be seen for this coating. Only some sintering phenomena has occurred for this coating. Fig. 6b) displays the cross section image of the TTBC after 192 thermal cycling with the segmentation crack density equal to 2.5 cracks/mm, the left figure shows the macro-image of the coating samples, it can be seen that large coating has detached form the coating, the coating has reached to its failure criteria, i.e. over 5% of the coating surface has peeled from the substrate.

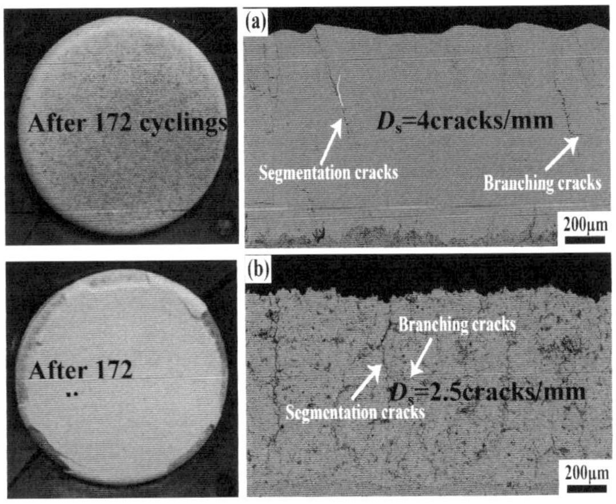

Fig.6 The results of burner rig test for the thick thermal barrier coatings with the segmentation crack density equal to 4cracks/mm a) and 2.5cracks/mm b)

3.3 Failure mechanism

In this section, a model has been proposed to explain the failure mechanism of the thick thermal barrier coatings during thermal shock. Based on the simulation results, the ceramic top-coat usually generate large tensile radial stress, as the TGO layer is usually very thin, the substrate, bond-coat and TGO layer were viewed as a whole assembly which was defined as "equivalent substrate". The thickness of the coating layer(ceramic top-coat) and equivalent substrate layer was set as h_c and h_s (Fig.7a)). The crack gap can be formed between two adjacent splats under the action of the tensile radial stress. Then the segmentation cracks can be formed at the inner of the ceramic top-coat, the stress concentration usually occurred at the segmentation crack tip, the distribution characteristic of the radial stress around the segmentation crack tip can be also seen in the figure 7b). The segmentation crack tip has the maximum tensile radial stress. In fact, when the displacement between two adjacent segmentation cracks is near enough, the stress field will be overlapped, it can be imagined that when the number of the segmentation cracks is large enough at the inner of the ceramic top-coat, there will be large tensile radial stress at the inner of the ceramic top-coat even higher than that of the TTBCs without segmentation cracks. this is not beneficial to the improvement of the thermal shock resistance ability of the TTBCs, so the segmentation crack density must be controlled at a suitable range(Fig.7b)). In addition, the branching crack can be also formed due to the shear stress between two contiguous upward and underlying splats, the generation of the breaching cracks is a result of the segmentation cracks. The branching cracks are usually the horizontal crack which can propagate along the laminar interface of the splats and lead the coating failure eventually(Fig.7c)).The critical shear stress can be expressed as follows(Fig.7d))[38,39]:

$$\tau^*=\pi h_c \sigma_c^* / \lambda^* \qquad (2)$$

here λ is the longitude length as ingle spat which can be expressed as :

$$\lambda=L/(N+1) \qquad (3)$$

The failure mechanism for a actual TTBCs can be attributed to the co-action of the multi-factors. The segmentation cracks will initiate from the surface of the as-sprayed coatings, and with the thermal cycling number increasing, the radial stress will increase, and the segmentation crack will propagate toward to the bond-coat/top-coat or top-coat/TGO interface, and maybe some branching cracks will be formed at the interface between two contiguous upward and underlying splats, the branching cracks will propagate, link at the horizontal direction and then reach to the edge of the coating eventually, then the failure of the coating will occur. In addition, the inner of the ceramic top-coat also has some kink cracks and pores, the kink cracks will be a barrier to the propagated cracks, some micro-pores will release some stress concentration of the segmentation crack and delay the propagation rate of the segmentation cracks.

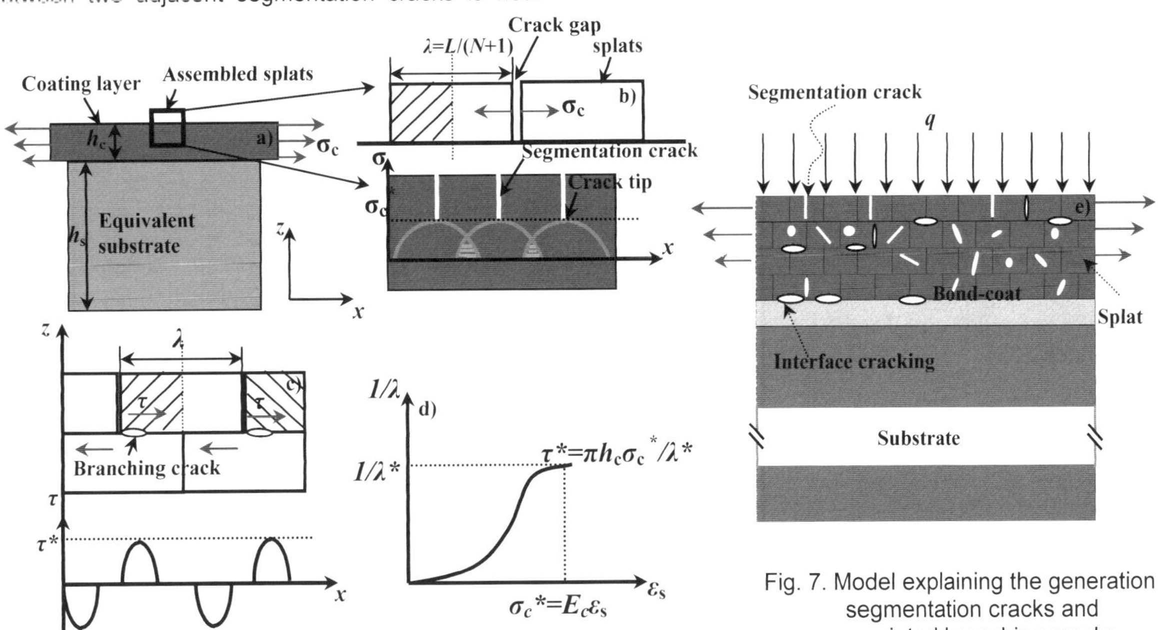

Fig. 7. Model explaining the generation of segmentation cracks and associated branching cracks

Conclusions

In the current work, the influence of the segmentation crack on the stress distribution and failure behavior of the thick thermal barrier coatings fabricated by atmospheric plasma spraying has been investigated via numerical simulation, the following conclusions can be obtained as follows:

(1) The segmentation cracks are beneficial to decrease the stress level of the ceramic top-coat and further improve the strain tolerance of the coatings and resulting in the improvement of the thermal shock resistance ability.

(2) The stress intensity factor and energy release rate (*J* integration) will increase with the increasing of segmentation crack length. The strain tolerance has been characterized by segmentation crack density and segmentation crack length quantitatively.

(3) The appropriate segmentation crack density is in the rage of 2.5-4.6mm^{-1} which will be more beneficial to improve the thermal shock resistance ability. In our work ,the appropriate segmentation crack density is about 3.79 which correspond well with our experimental results, the TTBCs with segmentation crack 4cracks/mm has superior thermal shock resitance ablity.

Acknowledgment

We express our gratitude to the Integrated Computational Materials Research Center of Shanghai Institute of ceramics, Chinese Academy of Sciences for providing software support. This work was jointly supported by the National Natural Science Foundation of China (NSFC) under the Grant No.51202277, Young Scholar Project (No.12ZR1452000) supported by the Shanghai Science and Technology Committee , and we also give our thanks to the special Project for Army-Civilian Combination of Shanghai Municipal Commission of Economy and Informatization (No. JMJH2014053).The current work was also supported by the Major Program for Basic Research of Shanghai Science and Technology Committee (No. 12DJ1400402), the Key Research Program of Chinese Academy of Sciences (No. KGZD-EW-T06) and the research grant (No.14DZ2261203) from Shanghai government.

References

[1] Padture NP , Gell M, Jordan EH. Materials science-Thermal barrier coatings for gas-turbine engine applications.Science.2002, 296(5566): 280-284

[2] John H. Perepezko. The Hotter the Engine, the Better. Science.2009, 326(5956):1068-1069

[3] Evans A G, Mumm D R , Hutchinson JW, Meier G H, Pettit F S. Mechanisms controlling the durability of thermal barrier coatings. Progress In Materials Science. 2001, 46 (5):505-553

[4] Kristen A. Marino, Berit Hinnemann, Emily A. Carter. Atomic-scale insight and design principles for turbine engine thermal barrier coatings from theory. PNAS,2011,108(14):5480–5487

[5] Cao XQ , Vassen R , Stoever D. Ceramic materials for thermal barrier coatings. Journal of The European Ceramic Society. 2004,24(1): 1-10

[6] Rabiei A, Evans A G. Failure mechanisms associated with the thermally grown oxide in plasma-sprayed thermal barrier coatings. Acta Materialia.2000,48(15): 3963-3976

[7] L. Wang, D.C. Li, J.S. Yang, F. Shao, X.H. Zhong, H.Y. Zhao, K. Yang, S.Y. Tao, Y. Wang. Modeling of thermal properties and failure of thermal barrier coatings with the use of finite element methods: A review. Journal of the European Ceramic Society. DOI:10.1016/j.jeurceramsoc.2015.12.038

[8]M. Karger, R. Vaβen, D. StÖevr. Atmospheric plasma sprayed thermal barrier coatings with high segmentation crack densities: Spraying process, microstructure and thermal cycling behavior. Surface & Coatings Technology 206 (2011) 16-23

[9]H.B. Guo, R. Vaßen, D. Sto¨ver.Atmospheric plasma sprayed thick thermal barrier coatings with high segmentation crack density. Surface & Coatings Technology 186 (2004) 353– 363

[10] Ed F. Rejda, Darrell F. Socie, Takamoto Itoh. Deformation behavior of plasma-sprayed thick thermal barrier coatings. Surface and Coatings Technology 113 (1999) 218-226

[11]F. Cernuschi, L. Lorenzoni, S. Capelli, C. Guardamagna, M. Karger, R. Vaβen, K. von Niessen , N.Markocsan, J. Menuey, Carlo Giolli,Solid particle erosion of thermal spray and physical vapour deposition thermal barrier coatings. Wear 271 (2011) 2909-2918

[12] F. Rousseau, C. Fourmond, F. Prima, M.H. Vidal Serif, O. Lavigne, D. Morvan, P. Chereau. Deposition of thick and 50% porous YpSZ layer by spraying nitrate solution in a low pressure plasma reactor. Surface & Coatings Technology 206 (2011) 1621-1627

[13]S. Ahmaniemi, J. Tuominen, M. Vippola, P. Vuoristo, T. Mäntylä, F. Cernuschi, C. Gualco, A. Bonadei, and R. Di Maggio. Characterization of Modified Thick Thermal Barrier Coatings. Journal of Thermal Spray Technology. 2004,13(3) :361-369.

[14]Hongbo Guo, Hideyuki Murakami and Seiji Kuroda. Effects of Heat Treatment on Microstructures and Physical Properties of Segmented Thermal Barrier Coatings. Materials Transactions, 2005, 46(8):1775-1778

[15] H.B. Guo, S. Kuroda, H. Murakami. Segmented thermal barrier coatings produced by atmospheric plasma spraying hollow powders. Thin Solid Films 506-507 (2006) 136-139

[16] Pyung-Ho Lee, Sang-Yup Lee, Jae-Young Kwon, Sang-Won Myoung, Je-Hyun Lee, Yeon-Gil Jung, Hyun Cho, Ungyu Paik. Thermal cycling behavior

and interfacial stability in thick thermal barrier coatings. Surface & Coatings Technology 205 (2010) 1250-1255

[17] Andrea Scrivani, Gabriele Rizzi, Christoper C. Berndt. Enhanced thick thermal barrier coatings that exhibit varying porosity. Materials Science and Engineering A 476 (2008) 1-7.

[18] Giovanni Di Girolamo, Caterina Blasi, Leonardo Pagnotta, Monica Schioppa.Phase evolution and thermophysical properties of plasma sprayed thick zirconia coatings after annealing. Ceramics International 36 (2010) 2273-2280

[19] A.K. Ray, S.C. Bose, P.K. De, D.K. Das.Lifetime evaluation of a thick thermal barrier coated superalloy used in turbine blade. Materials Science and Engineering A 527 (2010) 5474-5483

[20] H.-D. Steffens, Z. Babiak, and M. Gramlich. Some Aspects of Thick Thermal Barrier Coating Lifetime Prolongation. Journal of Thermal Spray Technology Volume 8(4) December 1999,517-522

[21] Hongbo Guo, Hideyuki Murakami and Seiji Kuroda. Thermal Cycling Behavior of Plasma Sprayed Segmented Thermal Barrier Coatings. Materials Transactions, 2006, 47 (2):306-309

[22] S. Ahmaniemi, P. Vuoristo, T. Mäntylä, C. Gualco, A. Bonadei, R. Di Maggio. Thermal cycling resistance of modified thick thermal barrier coatings. Surface & Coatings Technology 190 (2005) 378- 387

[23] A. Scrivani, G. Rizzi, U. Bardi, C. Giolli, M. Muniz Miranda, S. Ciattini, A. Fossati, and F. Borgioli. Thermal Fatigue Behavior of Thick and Porous Thermal Barrier Coatings Systems. Journal of Thermal Spray Technology. 2007,16(5-6):816-821

[24]S. Krämer, S. Faulhaber, M. Chambers, D.R. Clarke, C.G. Levi, J.W. Hutchinson, A.G. Evans. Mechanisms of cracking and delamination within thick thermal barrier systems in aero-engines subject to calcium-magnesium-alumino-silicate (CMAS) penetration. Materials Science and Engineering A 490 (2008) 26-35

[25] S. Ahmaniemi, P. Vuoristo, T. Mäntylä, F. Cernuschi, L. Lorenzoni. Modified thick thermal barrier coatings:Thermophysical characterization. Journal of the European Ceramic Society 24 (2004) 2669-2679.

[26] Manish Madhwal, Eric H. Jordan, Maurice Gell. Failure mechanisms of dense vertically-cracked thermal barrier coatings. Materials Science and Engineering A 384 (2004) 151-161

[27] H.B. Guo, R. Vaßen, D. StÖver. Thermophysical properties and thermal cycling behavior of plasma sprayed thick thermal barrier coatings. Surface & Coatings Technology 192 (2005) 48-56

[28] S. Ahmaniemi, P. Vuoristo, T. Mäntylä. Improved sealing treatments for thick thermal barrier coatings. Surface and Coatings Technology 151-152 (2002) 412-417

[29]S. Ahmaniemi, J. Tuominen, P. Vuoristo, and T. Mäntylä.Sealing Procedures for Thick Thermal Barrier Coatings. Journal of Thermal Spray Technology, 2002,11(3):320-332.

[30]F. Cernuschi, L. Lorenzoni , S. Ahmaniemi,1, P. Vuoristo, T. Mäntylä. Studies of the sintering kinetics of thick thermal barrier coatings by thermal diffusivity measurements. Journal of the European Ceramic Society 25 (2005) 393-400

[31]S.A. Tsipas, I.O. Golosnoy. Effect of substrate temperature on the microstructure and properties of thick plasma-sprayed YSZ TBCs. Journal of the European Ceramic Society 31 (2011) 2923–2929

[32]ANSYS Inc. Release15.0. Documentation for ANSYS, 2016.

[33] L.Wang, J.S.Yang, J.X. Ni, C.G. Liu, X.H. Zhong, F. Shao, H.Y. Zhao, S.Y. Tao, Y. Wang. Influence of cracks in APS-TBCs on stress around TGO during thermal cycling: A numerical simulation study. Surface and Coatings Technology, 285(2016) 98-112

[34]L. Wang, Y. Wang, X.G. Sun, J.Q. He, Z.Y. Pan, C.H. Wang. Finite element simulation of residual stress of double-ceramic-layer $La_2Zr_2O_7$/8YSZ thermal barrier coatings using birth and death element technique. Computational Materials Science, 2012, 53(1):117-127

[35] L. Wang, Y. Wang, X.G. Sun, J.Q. He, Z.Y. Pan, C.H. Wang. Finite element simulation of stress distribution and development in 8YSZ and double-ceramic-layer $La_2Zr_2O_7$/8YSZ thermal barrier coatings during thermal shock. Applied Surface Science, 2012, 258(8):3540-3551

[36] Yuexing Zhao, Dachuan Li, Xinghua Zhong, Huayu Zhao, Liang Wang, Fang Shao, Chenguang Liu, Shunyan Tao*. Thermal shock behaviors of YSZ thick thermal barrier coatings fabricated by suspension and atmospheric plasma spraying. Surface and Coatings Technology. 2014, 249, 48-55

[37] Yuexing Zhao, Liang Wang, Jiasheng Yang, Dachuan Li, Xinghua Zhong, Huayu Zhao, Fang Shao, Shunyan Tao. Thermal aging behavior of axial suspension plasma sprayed yttria-stabilized zirconia (YSZ) thermal barrier coatings. Journal of Thermal Spray Technology. 2015,24(3):338-347

[38] D. C. Agrawal and R. Raj, Measurement of the Ultimate Shear Strength of a Metal/Ceramic Interface, Acta Metall., 1989,37(4):1265-1270

[39] H. B. Guo, H. Murakami, and S. Kuroda. Effect of Hollow Spherical Powder Size Distribution on Porosity and Segmentation Cracks in Thermal Barrier Coatings. J. Am. Ceram. Soc., 2006,89 (12):3797-3804

Introduction of micro resin fragments with ceramic nanoparticles into plasma flame to create fine coated layers

M.Katsuki[1], H.Harada[1], and S.Kirihara[2]
(1) Hitachi Zosen Corporation, Osaka, Japan
(2) Osaka University, Osaka, Japan

Nanoparticles thermal spraying has been developed to create fine ceramic layers with higher deposition speed. Micro resin fragments including with the ceramic nanoparticles could be introduced into argon gas plasma successfully by using conventional powder feeding equipment. Zirconia nanoparticles of 200 nm in average diameter were dispersed into a thermosetting acrylic liquid resin at 57% in volume fraction. An airtight container of 150 cc in volume was revolved and rotated for degassing and mixing. Obtained pastes were solidified thorough heating at 150 °C for 30 min. Composite bulks were crushed by a vibrating milling machine. The micro fragments selected by using a sieve of 45-106 μm in mesh size. In the plasma flame, these resin fragments were burned down, and the heated nanoparticles were deposited on a metal substrate. Fine zirconia layers without micro cracks or pores were observed by scanning electron microscopy. Carbon contamination deriving from the resin material was verified by thermogravimetry for the cross section of coated layer. Vickers hardness of the ceramic phase was measured to estimate the sintered density. The fine zirconia coating will increase practical durability as heat and corrosion resistances of steam pipes in a boiler of thermal power generators inner walls of refuse incinerators.

1 Introduction

Thermal spraying is a coating process which can deposit metals or ceramics with high deposition speeds to the object. Sprayed coatings with ceramics can give the performance of heat resistance, corrosion resistance, and abrasion resistance to the object. Conventionally, as powdery spraying material have been used micro size particles about 10–100 μm and there are pores and cracks in the sprayed coatings.These pores and cracks will be disadvantage to create high dense coatings.

Recentry, thermal spraying method to use nanoparticles as spray materials has been studied for the purpose of restraing defects in the thermal spray coatings.There are some problems to use nanoparticles in plasma spraying process. Due to the nanoparticles are likely to aggregation, the transfer hose to the spray gun closes [1][2][3]. Since nanoparticles are low mass and the inertial force is weak even using the injector, it cannot be sufficiently introduced into the plasma flame [4][5][6][7]. Furthermore nano risks for human health are concerned [1].

To overcome these problems, suspension plasma spraying and pastes plasma spraying have been studied [1][3][7][8][9]. In suspension plasma spraying process, the suspension with nanoparticles which dispersed in aqueous or organic solvent are injected into the plasma or gas flame. However, the viscosity of the suspension become higher by increasing the amount of dispersion of nanoparticles and the injection will be difficult. Although there are studied about increasing the content of nanoparticles by adding dispersant, the content of nanoparticles in the solvent and deposition speeds tend to low. In addition, it is necessary to newly install a feeder and injection device of the suspension. On the other hand, pastes plasma spraying is possible to create dense sprayed coatings at higher deposition speeds by using pastes with high concentration nanoparticles. But, it is necessary to use new feeder and injection device of the pastes.

This paper shows a new nanoparticles thermal spraying process. Micro resin fragments which nanoparticles are dispersed in high concentration could be injected into the plasma flame by using conventional powder feeders and could be created fine coated layers with high deposition speeds.

Dense ceramic thermal spray coatings can be utilized in anti-corrosion coating. EfW plant (refuse incinerators) is responsible for waste disposal and energy supply. Although it is possible to improve the power generation efficiency by increasing the steam conditions of the boiler, corrosion of boiler tubes is increased [10][11][12]. Frequent replacement of the boiler tubes, increase the operating costs of the EfW plant. By the way, the thermal barrier coating (TBC) to improve the heat resistance of the structural material in a high temperature environment have been developed. TBC does not fit the purpose of blocking the structural material from the corrosive environment because the pores in the coatings enhance the thermal insulation. A dense coating of ceramics having corrosion resistance, durability of the boiler tubes can be expected.

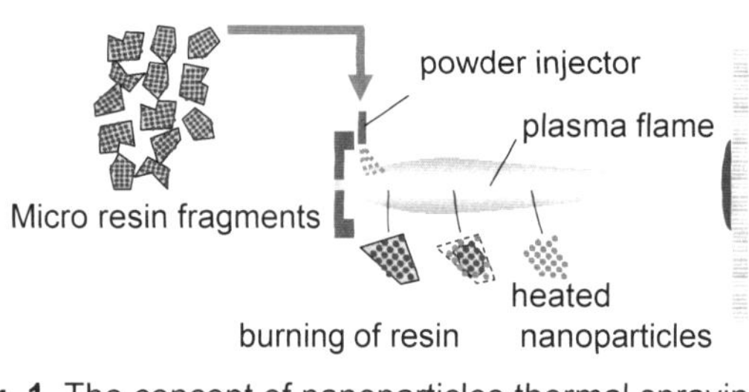

Fig. 1. The concept of nanoparticles thermal spraying process by using micro resin fragments

2 Experimental procedure

Fig.1 shows concept of nanoparticles thermal spraying process by using micro resin fragments. The micro resin fragments where nanoparticles are dispersed in high concentration are transported by using a conventional powder feeder. The resin is burned out by introducing micro resin fragments into plasma flame, and then nanoparticles are deposited on the substrate.

2.1 Preparation of micro resin fragments

The stabilized zirconia nanoparticles having the average particle size of about 200 nm (KZ-8YF, KCM Corporation, Japan) were mixed with the thermosetting acrylic liquid resin (JSR Corporation, Japan). The mixing ratio of the stabilized zirconia was 57 vol% by volume fraction (86.7 wt% by mass fraction). These were placed in a sealed container of 150 cc and mixed and degassed for 840 s by using planetary mixer SK-350T(SHASHIN KAGAKU Co.Ltd., Japan) with 1340 rpm in revolution and rotation respectively. The mixing paste was solidified by heating at 150 °C for 30 min. Composite bulks were crushed by a high speed vibrating milling machine. Micro resin fragments were screened using the sieve of 106 μm and 45 μm opening.

2.2 Plasma spray processing

The conventional spraying material (SHOCOAT K-90, SHOWA DENKO K.K., Japan), agglomerated particles (KZ-8YF, agglomerated size is 45~75μm, KCM Corporation, Japan) and micro resin fragments were sprayed using plasma spraying equipment (Sulzer Metco Inc.). Powder supply device used TWIN-120-A / H 1.0, thermal spray gun used F4 Spray Gun. The surface of SUS316 base material with dimensions 50 × 50 × 6 mm was blasting. The spray gun was traversed on this substrate and formed a coating. The spray conditions are shown in Table1. These conditions were based on a condition that can be relatively dense coating with using conventional spraying material.

Morphology of the cross section of YSZ coatings has been observed by using SEM and Vickers hardness ~~have~~ tested by using a micro vickers hardness tester MVK-E (AKASHI, Japan). The coatings was peeled from substrate to examine the residual carbon contamination deriving from the resin and evaluated by thermogravimetry.

3 Results and discussion

3.1 About micro resin fragments

Micro resin fragments which YSZ nanoparticles are dispersed with high concentration were obtained by crashing and screening composite bulks. Fig.2 shows micro resin fragments and individual surface of higher magnification observed by SEM. It is clear that nanoparticles of 200nm in average diameter were dispersed evenly.

Table 1. Spraying experimental conditions

Parameter	Spraying material	
	Conventional, Agglomerated	Micro resin fragments
Spraying distance (mm)	100	50
Number of transitions	11	25
Power input (kW)	33.6	33.6
Plasma gases	Ar/H_2	Ar/H_2
Powder feed gas	Ar	Ar
Powder feed rate (g/min)	40	26
Transition speed (m/min)	9.9	9.9

(a)

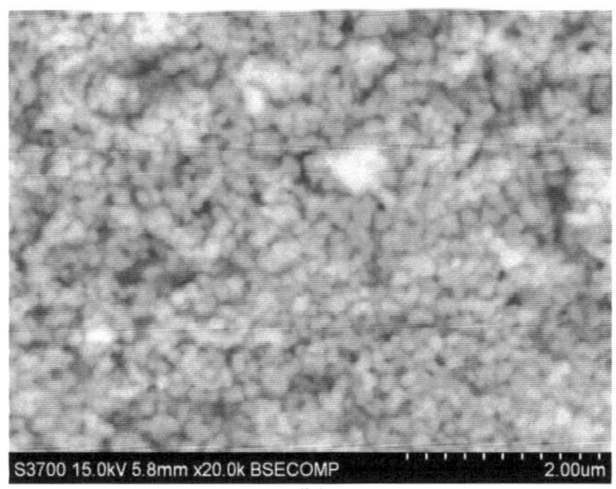

(b)

Fig. 2. (a) YSZ nanoparticles 57 vol% micro resin fragments, and (b) Individual surface of (a) with high magunification

3.2 The introduction of micro resin fragments to the plsma frame

Micro resin fragments were able to be supplied to spray gun by using a conventional powder feeder without closure in the transfer hose. Micro resin fragments injected by the powder injector were able to be introduced to the plasma flame and formed coatings on the substrates. On visually, the length of the spray formed by micro resin fragments which being introduced to the plasma flame (distance of the particles flying at a high temperature state) was half that of conventional spraying material and the agglomerated particles.

3.3 Observation of the thermal spray coatings

Cross sectional microstructure in YSZ coated layer formed on the SUS316 substrate are shown in Fig.3. In the case of conventional spraying material, the particles were melt and deposited, but there were micro-cracks and pores. In the case of agglomerated particles, uneven morphology was formed. There were parts where unmelted particles deposited and parts where particles deposited by melting. And also there were micro-cracks and pores. The coatings of sprayed micro resin fragments were formed dense coatings which micro-cracks and pores hardly exist, although slightly unmelted particles were deposited. The deposition speeds of micro resin fragments was about 13 μm/transition.

The Vickers hardness of each coatings was compared. Test force is 100 gf and dwell time is 15s. In the case of conventional spraying material, the maximum value of the Vickers hardness was 1033 HV. The morphology was uniform and a variation of hardness was small. In the case of agglomerated particles, although vickers hardness of the part where the particles were deposited by melting is 1206 HV and exceeded the case of conventional spraying materials, vickers hardness of the part where unmelted particles were deposited is 172 HV and significantly weak. In the case of micro resin fragments, the maximum value of the Vickers hardness was 1267 HV and exceeded the case of conventional spraying materials. The vickers hardness of the other part was obtained 882 HV. It was considered that the morphology was not completely uniform because slightly unmelted particles were deposited.

The residual carbon contamination deriving from the resin was evaluated by thermogravimetry. Since the pyrolysis temperature of the thermosetting resin is about 480 °C, the sample was heated up to 1000 °C at a heating rate of 10 °C/min in Ar. Fig.4 shows the result of thermogravimetry. Weight started to decrease from 289 °C and reduced about 1wt% during heating up to 1000 °C. It was considered that the carbon deriving from the resin have been most burned out during spraying, because the micro resin fragment contain 13.3 wt% thermosetting resin.

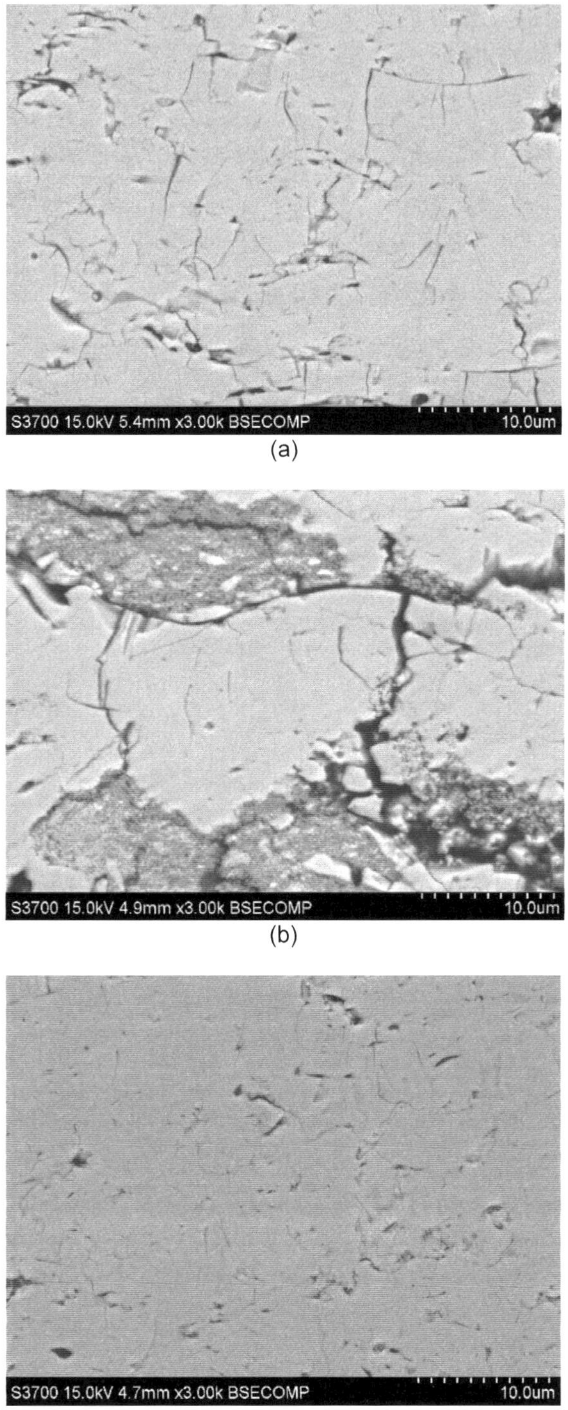

Fig. 3. Cross sectional microstructure in YSZ coated layer (a) by conventional spraying material, (b) by agglomerated particles, (c) by micro resin fragments.

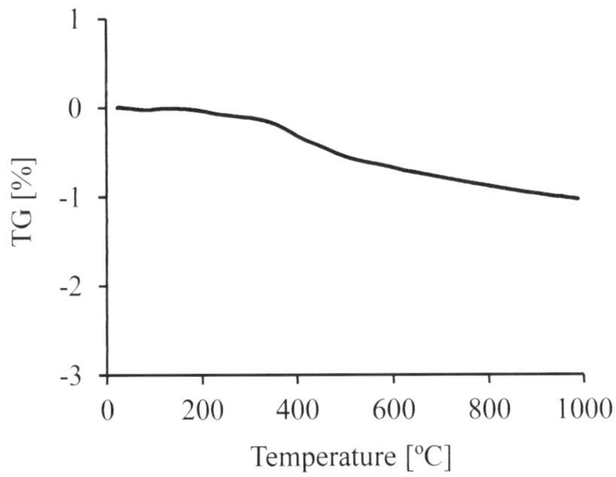

Fig. 4. TG curves of YSZ coated layer formed by micro resin fragments at a heating rate of 10 °C/min in Ar.

4 Conclusions

The thermal spraying process of nanoparticles has developed to form dense sprayed coatings with higher deposition speeds. The micro resin fragments with 57 vol% YSZ nanoparticles have been used as spraying material. The micro resin fragments could be introduced into the plasma flame with a conventional powder feeder. The resin is most burned out by introducing micro resin fragments into plasma flame, the heated YSZ nanoparticles only could be deposited on metal substrate. By using micro resin fragments, it could created dense coatings where micro-cracks and pores hardly exist, compared to the coatings using conventional spraying material. And the deposition speeds was about 13 µm/transition. The maximum value of the vickers hardness was 1267 HV and exceeded the case of conventional spraying materials. In the future, by mixing with resin and different types nanoparticles and different size nanoparticles, and by combinating them can be expected to create the composite nano morphology.

5 Literature

[1] A.Killinger, M.Kuhn, R.Gadow: High-Velocity Suspension Flame Spraying (HVSFS), a new approach for spraying nanoparticles with hypersonic speed, Surface & Coatings Technology 201, (2006), pp.1922-1929

[2] R.S.Lima, B.R.Marple: Thermal Spray Coatings Engineered from Nanostructured Ceramic Agglomerated Powders for Structual, Thermal Barrier and Biomedical Applications: A Review, Journal of Thermal Spray Technology, (2007), Volume 16(1), pp.40-63

[3] L.Pawlowski: Suspension and solution thermal spray coatings, Surface & Coatings Technology 203, (2009), pp.2807-2829

[4] R.S.Lima, A.Kucuk, C.C.Berndt: Evaluation of microhardness and elastic modulus of thermally sprayed nanostructured zirconia coatings, Surface and Coatings Technology 135, (2001), pp.166-172

[5] Y.Wang, S.Jiang, M.Wang, S.Wang, T.D.Xiao, P.R.Strutt: Abrasive wear characteristics of plasma sprayed nanostructured alumina/titania coatings, Wear 237, (2000), pp.176-185

[6] R.S.Lima, A.Kucuk, C.C.Berndt: Integrity of nanostructured partially stabilized zirconia after plasma spray processing, Materials Science and Engineering A313, (2001), pp.75-82

[7] M.Suzuki: Suspension Plasma Spraying, Jornal of the japan welding society, Vol. 83, (2014), No.2, pp.26-29, (in Japanese)

[8] D.Waldbillig, O.Kesler: The effect of solids and dispersant loadings on the suspension viscosities and deposition rates of suspension plasma sprayed YSZ coatings, Surface & Coatings Technology 203, (2009), pp.2098-2101

[9] S.Kirihara, Y.Itakura: Effective Injection of Ceramics Nanoparticle Pastes into Plasma Spray for Speedy Layer Formation, Quarterly jornal of the japan welding society, Vol. 33, (2015), No.2, pp.148-151

[10] S.Lee, N.J.Themelis, M.J.Castaldi: High-Temperature Corrosion in Waste-to-Energy Boilers, Journal of Thermal Spray Technology, (2007), Volume 16(1), pp.1-7

[11] Y.Kawahara: Application of High Temperature Corrosion-Resistant Materials and Coatings Under Severe Corrosive Environment in Waste-to-Energy Boilers, Jornal of Thermal Spray Technology, (2007), Volume 16(2), pp.202-213

[12] A. Schütz, M.Günthner, G.Motz, O. Greißl, U.Glatzel: High temperature (salt melt) corrosion tests with ceramic-coated steel, Materials Chemistry and Physics 159, (2015), pp.10-18

Nitrogen-dioxide sensing properties of solution precursor plasma sprayed zinc oxide coatings with oxygen vacancies

C. Zhang, X. Geng, J. Wang, J. You, Yangzhou / CN

Abatract

ZnO_x coatings were prepared by solution precursor plasma spray on alumina substrate equipped with interdigital gold electrodes. The field-emission scanning electron microscopy images illustrated that the as-sprayed coatings were porous and nanostructured. The X-ray diffraction pattern showed that the main peaks were hexagonal ZnO together with obvious alumina peaks emitted from the substrate. The XPS characterization gave an obvious presence of oxygen vacancies in the coatings. The UV-Vis spectra confirmed a significant absorption in visible-light range. The band-gap of the as-sprayed coating was estimated to be 2.1 eV, which was much less than that of stoichiometric ZnO. The oxygen-vacant sensors show significant responses to NO_2 even as low as 0.4 ppm, and the sensor response increases linearly with NO_2 concentration. The enhanced sensor properties were attributed to the presence of oxygen vacancies and the porous nanostructure.

1. Introduction

In the last decades, ZnO has became the most widely used sensitive material in gas sensor application [1]. Currently, typical ZnO based sensors are generally required to be operated at elevated temperatures [2], which will bring about many problems, such as excessive power consumption, growth in grain size and deterioration of sensor stability. Hence, one of the research focus in gas sensors is to develop room temperature sensors. Some researchers adopted visible light as an activation source to achieve this. In view of the insufficient photon energy that fail to activate ZnO, it is necessary to add photosensitizer [3-4] or increasing the amount of defects in sensitive layers. The native defects, such as oxygen vacancies, will form mid-gap state between valence and conduction band so that the obtained sensitive layers can be more effective [9-11]. Till now, few studies have attempted to enhance the sensing performance by preparing oxygen-deficient ZnO sensitive layer. An application of a post-treatment process with H_2O_2 and annealing is utilized to obtain oxygen-vacant ZnO for improving the sensing characteristics in [9]. However, this is not a straightforward method and too complex and tedious to be used in large-scale coating deposition process. The use of another deposition method, solution precursor plasma spray (SPPS), should be an alternative process [5,7,12]. As well known, the gas sensing response is greatly influenced by the coating microstructure [5,7,12]. The gas sensors with nanostructured and porous sensitive layers are expected to obtain a highly excellent performance [8]. SPPS is such a method to achieve this [12].
In this study, oxygen-vacant, nanostructured and highly porous ZnO sensitive layers were directly deposited by SPPS process without any post-treatment. The phase constitution and surface morphology of the obtained coatings were characterized by XRD and FE-SEM, respectively. The formation of surface oxygen vacancies were confirmed by XPS analysis. The optical properties of the obtained coatings were recorded by UV-Vis. The sensors based on the SPPS ZnO coatings have been subjected to gas sensor testing under visible light illumination at room temperature. The effect of light wavelength on the sensor characteristics were investigated.

2. Experimental method

ZnO coatings were deposited on alumina substrates equipped with a pair of interdigitated gold electrodes (C-MAC Micro Technology Company, Belgium) with a F4MB plasma spray torch (Oerlikon Metco, Switzerland) attached to a six-axis ABB robot arm (ABB, Sweden). In this work, the liquid used was aqueous solution containing 0.2 M zinc acetate, and injected into the plasma jet by a nozzle with an inner diameter of 0.26 mm through a peristaltic pump. The flow rate of the liquid was controlled at 20 mL/min. The primary plasma gas was argon with a flow rate of 40 L/min, and the secondary plasma gas was hydrogen with a flow rate of 2 L/min. The arc current was 513 A and consequently the plasma power was 25 kW. Compressed air at a pressure of 0.3 MPa was used as cooling gas for the substrates during spraying.
Surface and cross-section morphology of the as-sprayed coatings were inspected using field-emission scanning electron microscopy (FE-SEM, S4800 II, Japan). The crystal structure of the obtained coatings were determined by X-ray diffractometer (XRD, D8 Advance, Bruker,

Germany) using a CuKα radiation. The 2θ range of XRD diffraction angle was 20-70° with a scanning rate of 5°·min^{-1} and 0.1°·min^{-1} for 30-40° was used in this test. The presence of oxygen vacancies in the as-sprayed ZnO coatings were verified by several optical techniques, including X-ray photoelectron spectroscopy (XPS, ESCALAB 250Xi, Thermo Scientific, USA), Photoluminescence spectroscopy (PL, with He-Cd 325 nm laser) and electron paramagnetic resonance (EPR, A300-10/12, Bruker, Germany). Additionally, UV-Vis diffuse reflectance spectrophotometer (Cary5000, Varian, USA) using $BaSO_4$ as the reference was used to record UV-Vis absorption spectra of the coatings within light wavelength ranging from 300 to 800 nm.
Visible light illumination and electrical resistance measurement were carried out in the chamber. The ZnO sensors were connected to an Agilent data acquisition instrument to record the electrical resistances of the sensors.

3. Results and discussion

Fig. 1 shows XRD pattern of the SPPS ZnO coatings deposited by 2 L/min H_2 with a scanning rate of 5°·min^{-1}. Due to the thin thickness of the coatings, there is also Al_2O_3 peak emitted from the alumina substrate. This result demonstrates that the as-sprayed coatings are well in line with hexagonal wurtzite ZnO (PDF 36-1451).

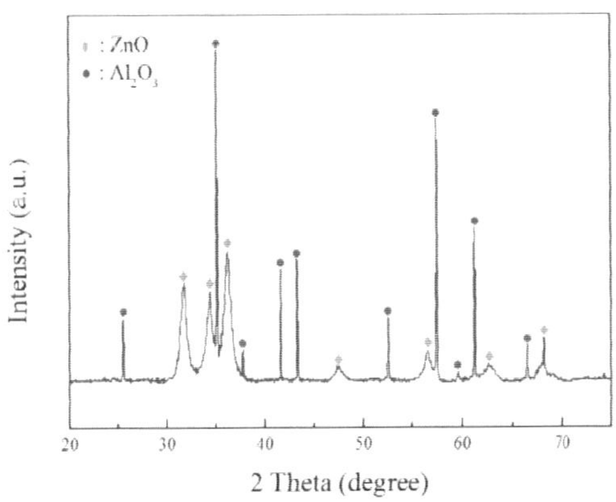

Fig. 1. X-ray diffraction patterns of the as-sprayed coatings.

Fig. 2(a) and (b) illustrate the surface morphology with low and high magnifications. It can be observed that the coatings are nanostructured with grain size of c.a. 15nm, which is consistent with the results of XRD. The plasma sprayed coatings are generally deposited by the rapid splat cooling process. A higher cooling speed over 10^5 K/s will result in the formation of smaller grain size in the as-sprayed coating. Moreover, it could also be found that the coating is highly porous microstructure. The size of pores existing in the SPPS ZnO coatings are about microns, which are attributed to the entrapped gases, cracking and premature solidification of some grains [8].

Fig. 2. Surface morphology of the as-sprayed coatings: (a) low magnification; (b) high magnification.

XPS is a useful technique to find out the surface phase composition and chemical state of the element, here it was used to provide quantitative evidence for the presence of oxygen vacancies on ZnO surface. Fig. 3 shows the surface XPS survey spectra for ZnO coatings. Through calculation, it can be found that the content of zinc atom equal to 42.56% is higher than that of oxygen atom being 41.66%, the ratio of zinc atom to oxygen atom being more than 1, which reveals that the as-sprayed coatings contain oxygen vacancies [6]. The binding energy was calibrated by taking the C1s peak at 284.8 eV.
To investigate the effect of oxygen vacancies on the coating optical properties, some of the oxygen-vacant ZnO coatings were annealed at 500℃ for 4 h to get stoichiometric ZnO. The UV-Vis absorption spectra was used to measure

the optical properties of as-sprayed and heat-treated ZnO coatings.

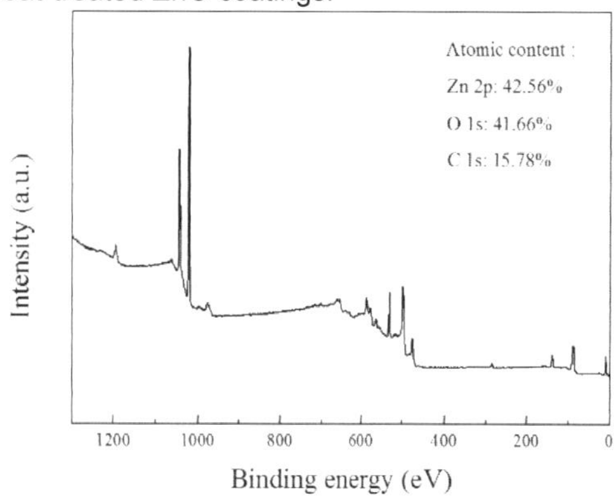

Fig. 3. XPS survey scan spectra of the as-sprayed coatings.

As displayed in Fig. 4(a), the UV-Vis absorption spectra of ZnO coating is recorded under light illumination with wavelength of 300-800 nm. It can be observed that as-sprayed coatings have absorption in the whole visible light range, while the heat-treated ZnO coatings only can absorb ultra-violet light. Optical band gap of ZnO can be calculated using the Tauc model [14]. Based on this equation, the band gap of the as-sprayed coatings is estimated to be 2.1 eV from the Tauc plot of $(\alpha h\nu)^{1/2}$ versus $h\nu$ (figure not listed). However, the band gap of heat-treated ZnO is 3.12 eV. The reduction of energy band gap of the as-sprayed coatings is mainly attributed to the presence of oxygen vacancies, which leads to a red shift of the UV-Vis absorption spectra. Similar results are found in [9-10].

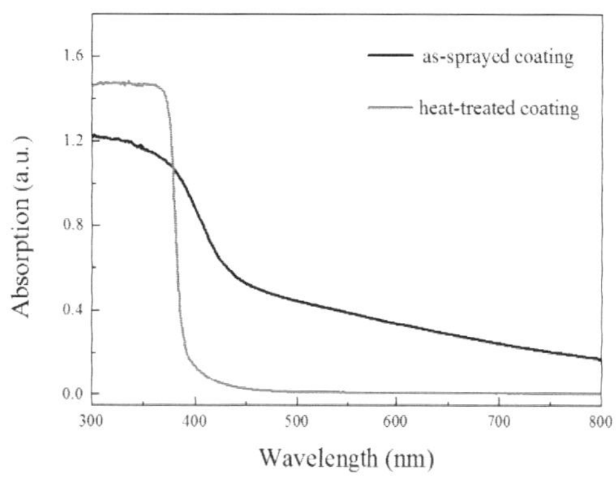

Fig. 4. UV-Vis absorption spectra of the as-sprayed and heat-treated coatings.

Fig. 5(a) illustrates the electrical resistance responses of the oxygen-vacant ZnO sensors to NO_2 gas with a concentration from 0.4 to 1 ppm under blue light illumination at room temperature. The oxygen-vacant ZnO sensors exhibit significant response to NO_2. The response is as high as 2.8 even the NO_2 concentration was as low as 0.4 ppm, indicating that the gas sensors developed in this study have a high detectability. In addition, it can also be found that the sensor response increases linearly with NO_2 concentration, as shown in Fig. 5(b).

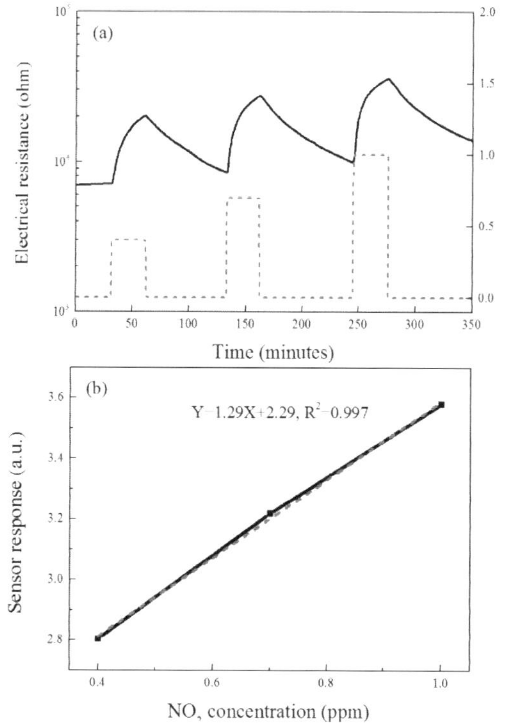

Fig. 5. The sensing response of oxygen-vacant ZnO sensors to 0.4-1 ppm NO_2 in air under blue light illumination: (a) electrical resistance response; (b) sensor response.

The light wavelength is an important parameter determining sensor properties. The electrical resistances of sensor based on SPPS ZnO coatings are plotted against time towards 1 ppm NO_2 illuminated by light with wavelength ranging from 400-640 nm. It can be noted that the bulb having different light wavelength play an important role in the sensor properties. According to Fig. 6, some results are summarized. Firstly, the base resistance R_{air} diminished when light wavelength decreased from 640 to 480 nm. The base resistance decreased from 5.63×10^4 under red light illumination to 5.48×10^3 Ω irradiated by purple light. As well known, the photon energy increase with the decrease of light wavelength on the basis of the formula: $E= hc/\lambda$, where E is the light energy, h is the Planck's constant, c is the velocity of light and λ is the light wavelength. Hence, more electron-hole pairs on ZnO surface are generated as the light wavelength decreases, which results in the reduction of electrical resistance. Secondly, it could also be observed that the sensor response increases when shorter wavelength light is used (except purple light). This is attributed to the presence of oxygen vacancies. The reason for the variation in sensor response will be discussed in the sensing mechanism section. Thirdly, the response and recovery time diminishes when using shorter wavelength light (except purple light). The role of photon energy on response and recovery rate is similar to heating [12-13]. With the light wavelength decreases, the photon energy increases, which enhances the response and recovery rate.

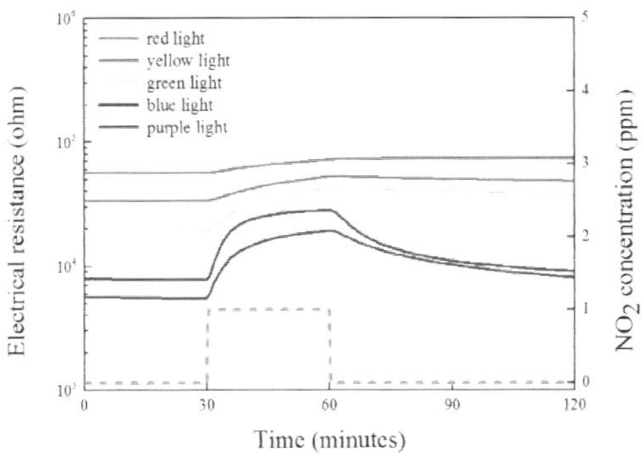

Fig. 6. Electrical resistances of sensors based on oxygen-vacant ZnO coatings with 1 ppm NO_2 under different light conditions at room temperature.

4. Conclusions

A room temperature NO_2 gas sensor has been developed based on oxygen-vacant ZnO sensing coatings. The oxygen-vacant coatings were deposited by SPPS process due to the reducing atmosphere and rapid cooling rate. SPPS has proven its simplicity and reliability in fabricating oxygen-vacant, nanostructured and highly porous ZnO sensitive layers. The presence of oxygen vacancies in the as-sprayed coatings were confirmed by the XPS characterization. The adsorption range for oxygen-vacant coatings were extended to visible light range by forming a donor level between the conduction and valence band. The oxygen-vacant sensors show significant responses to NO_2 even as low as 0.4 ppm, and the sensor response increases linearly with NO_2 concentration. The enhanced sensor properties were attributed to the presence of oxygen vacancies and the porous and nanometric microstructure. The effect of light wavelength on the sensor performance was also investigated.

5. Literature

[1] T. Chen, H. Chen, C. Hsu, et al. Characteristics of ZnO nanorods-based ammonia gas sensors with a cross-linked configuration. Sensors and Actuators B: Chemical, 221 (2015) 491-498.

[2] H. Woo, C. Kwak, J. Chung, et al. Highly selective and sensitive xylene sensors using Ni-doped branched ZnO nanowire networks. Sensors and Actuators B: Chemical, 216 (2015) 358-66.

[3] M. V. Malashchonak, E. Astreltsov, A. V. Mazanik, et al. Band-gap and sub-band-gap photoelectrochemical processes at nanocrystalline CdS grown on ZnO by successive ionic layer adsorption and reaction method. Thin Solid Films, 589 (2015) 145-152.

[4] J. Qi, W. Liu, C. Biswas, et al. Enhanced power conversion efficiency of CdS quantum dot sensitized solar cells with ZnO nanowire arrays as the photoanodes. Optics Communications, 349 (2015) 198-202.

[5] C. Zhang, X. Geng, H. Li, et al. Microstructure and gas sensing properties of solution precursor plasma-sprayed zinc oxide coatings. Materials Research Bulletin, 2015, 63:67-71.

[6] J. Zhao, C. Xie, L. Yang, et al. Enhanced gas sensing performance of Li-doped ZnO nanoparticle film by the synergistic effect of oxygen interstitials and oxygen vacancies. Applied Surface Science, 330 (2015) 126-133.

[7] C. Zhang, X. Geng, M. Olivier, et al. Solution precursor plasma-sprayed tungsten oxide coatings for nitrogen dioxide detection. Ceramics International, 40 (2014) 11427-11431.

[8] C. Zhang, M. Debliquy, A. Boudiba, et al. Sensing properties of atmospheric plasma-sprayed WO_3 coating for sub-ppm NO_2 detection. Sensors and Actuators B: Chemical, 144 (2010) 280-288.

[9] W. Kim, M. Choi, K. Yong. Generation of oxygen vacancies in ZnO nanorods/films and their effects on gas sensing properties. Sensors and Actuators B: Chemical, 209 (2015) 989-996.

[10] L.E. Halliburton, N.C. Giles, N.Y. Garces, et al. Production of native donors in ZnO by annealing at high temperature in Zn vapor. Applied Physics Letters, 87 (2005) 172108.

[11] L.S. Vlasenko, G.D. Watkins. Optical detection of electron paramagnetic resonance in room-temperature electron-irradiated ZnO. Physical Review B, 71 (2005) 125210.

[12] X. Geng, C. Zhang, M. Debliquy. Cadmium

sulfide activated zinc oxide coatings deposited by liquid plasma spray for room temperature nitrogen dioxide detection under visible light illumination. Ceramics International, 42 (2016) 4845-4852.
[13] C. Zhang, A. Boudiba, P. Marco, et al. Room temperature responses of visible-light illuminated WO_3 sensors to NO_2 in sub-ppm range. Sensors and Actuators B: Chemical, 181 (2013) 395-401.
[14] M.H. Habibi, M.H. Rahmati. Fabrication and characterization of ZnO@CdS core-shell nanostructure using acetate precursors: XRD, FESEM, DRS, FTIR studies and effects of cadmium ion concentration on band gap. Spectrochimica acta Part A, Molecular and biomolecular spectroscopy, 133 (2014) 13-18.

Densification of metal oxide films synthesized from a metal complex by flame apparatus

Keiji Komatsu*,Tomoyuki Shirai, Ayumu Toyama, Takao Iseki, D. Y. Enkin, Atsushi Nakamura, Yu Li, Shigeo Ohshio, Hiroyuki Muramatsu, Hidetoshi Saitoh

In this study, we synthesized dense metal oxide films from a metal-ethylenediaminetetraacetic acid complex using a flame spray apparatus. Erbium oxide (Er_2O_3) films were synthesized on stainless steel (SUS) substrates using N_2, N_2+O_2 (air), or O_2 as the carrier gas and a H_2-O_2 mixture as the combustion gas. When O_2 was used as the carrier gas, a 19.6-μm-thick Er_2O_3 layer was deposited on the SUS substrate. The cross-sectional porosity of this film was 3%. In contrast, a 5.2–22.0-μm -thick Er_2O_3 layer was synthesized on SUS substrates when N_2 and air were used as the carrier gases. The cross-sectional porosities of these films were 5.9–23.4%. These results indicated that the carrier gas used during the synthesis process plays an important role in determining the thickness and porosity of the resulting film.

1 Introduction

Thermal spraying is a surface-modification technique commonly used to deposit thermal barrier coatings, which are effective in improving the durability and anticorrosion (chemical resistance) and heat-shock properties of structural materials [1][2][3]. Thermal spraying techniques have numerous advantages, including a high deposition rate and the ability to deposit coatings on large areas on an industrial scale. For example, the chemical resistance of a material plays a critical role in determining the stability of the material in corrosive environments at temperatures lower than 550℃ [4]. Some of the commercial materials that exhibit high chemical resistances are zinc, aluminum, and zinc-aluminum alloys [5]. Four approaches are defined in the Japanese Industrial Standard H8250-1998 for fabricating films with high chemical resistances. These are (1) corrosion-prevention spraying to prevent the corrosion of the raw metal, (2) anti-rust spraying to prevent the formation of rust in steel, (3) corrosion-resistant spraying to protect high-resistance materials from corrosive environments, and (4) anti-high-temperature oxidation spraying to prevent the high-temperature oxidation of materials in severe environments. Ideally, the corrosion-resistant films deposited on materials to be protected should not exhibit cracks. However, a few cracks are always formed on industrially coated films. In such cases, a sealing treatment must be performed on the films to fill these cracks [6][7]. Thus, corrosion-resistant films with higher densities are desirable.

The temperature and velocity of the in-flight particles are the key factors for obtaining films with a dense morphology. For examples, plasma spraying (PS) and high-velocity oxygen fuel (HVOF) coating techniques are known to suitable for synthesizing dense films by thermal spraying [8][9][10]. In the PS technique, the in-flight particles have temperatures of 10000–15000 K, and their velocity reaches 100–200 m/s. In contrast, for the HVOF coating technique, these parameters are 2300–2900 K and 550–800 m/s. The films synthesized by the HVOF coating technique are denser than those obtained by the PS technique. Thus, it seems likely that the in-flight velocity of the coating particles determines the density of the deposited films. Recently, metal oxides were synthesized from metal–ethylenediaminetetraacetic acid (EDTA) complexes using a flame spraying technique [11][12][13]. The metal-EDTA complex was placed in a feed unit and transported by flowing N2 gas to the spray gun. As this technique allows for high deposition rates, films with thicknesses of several tens of micrometers could be obtained. The cross-sectional porosities of these films were 6–15 μm. This deposition technique involves two processes: a chemical reaction and physical collisions. During the chemical reaction, the metal-EDTA complex decomposes and oxidizes to form metal oxide particles. In the second process, these particles collide with the substrate and form a thick metal oxide layer. During the synthesis process, the supply of oxygen atoms to the metal-EDTA complex is crucial for ensuring that the chemical reaction takes place.

In this study, dense Er_2O_3 films were synthesized from an EDTA•Er•H complex through flame spraying; the process involved a chemical reaction of the complex. The Er_2O_3 films were deposited on stainless steel (SUS) substrates using various gases as the carrier gas and a mixture of H_2/O_2 as the combustion gas. The carrier gases used were nitrogen, commercial air (nitrogen and oxygen), and oxygen. First, we determined the effects of the carrier gas used on the decomposition of the EDTA•Er•H complex. Next, we examined the effects of the carrier gas on the microstructures of the deposited films. Furthermore, the effects of the carrier gas on the characteristics of the deposited films were also studied.

2 Experimental

As mentioned above, the EDTA complex EDTA•Er•H (Chubu Chelest Co., Ltd.) was used for synthesizing the Er_2O_3 films. A conventional flame-spraying apparatus consisting of a feed unit (5MPE, Sulzer Metco) and a spray gun (6P-II, Sulzer Metco) was used to perform the reactive spraying. This apparatus is used commercially for the deposition of metal or self-fluxing alloy films with an acetylene or hydrogen flame. As has been stated above, we employed different carrier gases to determine the effects on the carrier gas on the properties of the deposited films. The EDTA•Er•H complex was placed in the feed unit and transported by O_2, N_2+O_2 (air), or N_2 to the spray gun. The carrier-gas flow rate was 7.1 L/min. In this study, we selected hydrogen gas as the fuel. A mixture of H_2 and O_2, with the respective flow rates being 32.6 and 43.0 L/min, was used as the flame gas.

The EDTA · Er · H powder was introduced to the flame and made to react with oxygen, which resulted in the thermal decomposition of EDTA. The particles of the resulting product were then sprayed onto the SUS substrate (SUS304, 30 × 50 × 1 mm), which had been previously blasted with #60 alumina grit (purity of 99.7%, particle size of 212–250 μm, Fuji Manufacturing Co., Ltd.), resulting in the deposition of a metal oxide film. The distance between the spray gun and the substrate was 150 mm. The deposition duration was 4–10 s. The gun traverse rate was 50 mm/s. Next, the spray nozzle was moved in a longitudinal direction, and deposition was performed again for approximately 6 s for each area without preheating the substrate. The yield percentage (i.e., deposition efficiency) of this technique was approximately 60%, as determined by actual measurements.

The thermogravimetric (TG)/differential thermal analysis (DTA) profiles of the EDTA•Er•H complex in the N_2+O_2 and N_2 atmospheres were obtained using a thermogravimetric analyzer (ThermoPlus 2, Rigaku). X-ray diffraction (XRD) analysis (M03XHF22, MAC Science) were performed using Cu Kα radiation to observe the crystal structures of the films. The surface and cross-sectional morphologies of the films were observed using field-emission scanning electron microscopy (FESEM) (JSM-6700F, JEOL). The constituent elements in the films and their distributions were estimated through energy-dispersive X-ray spectroscopy (EDX) combined with FESEM. The porosities of the films were determined by analyzing SEM images of the films using the software Image J. The temperatures and velocities of the particles in the combustion gas were measured with an infrared camera (Accura Spray-G3: Tecnar). First, a combustion flame was formed by mixing hydrogen and oxygen. Next, the camera was set vertically at a distance of 20 cm from the center of the flame plume. The temperature and velocity of the in-flight particles were measured at nozzle-substrate distances of 110, 130, 150, and 170 mm.

3. Results and discussion

We first investigated the thermal decomposition behaviors of the EDTA•Er•H complex in the different atmospheres. The TG/DTA profile of the complex in N_2+O_2 (air) is shown in Fig. 1. An exothermic peak can be seen at approximately 365 ℃. This peak indicated that the oxidative decomposition of the EDTA•Er•H complex occurred in the N_2+O_2 atmosphere. Further, the TG/DTA profile of the complex in the N_2 is shown in Fig. 2. An endothermic peak can be seen at approximately 400℃. The peak indicated that the vaporization of EDTA•Er•H occurred in the N_2 atmosphere. Thus, the thermal pyrolysis behavior of EDTA•Er•H depended on the reaction environment.

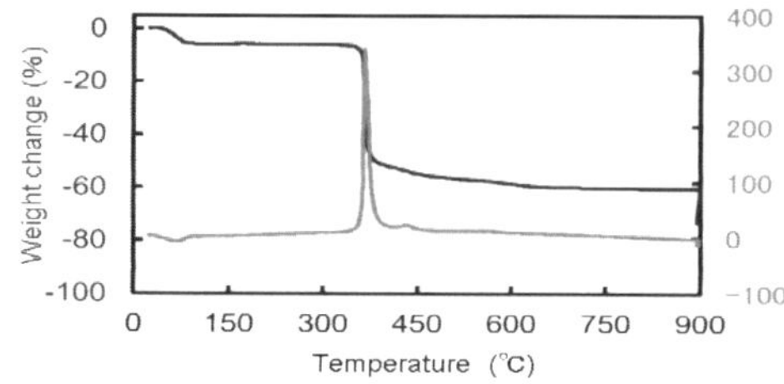

Fig. 1. TG/DTA curves for the EDTA•Er•H complex with N_2+O_2 as the carrier gas.

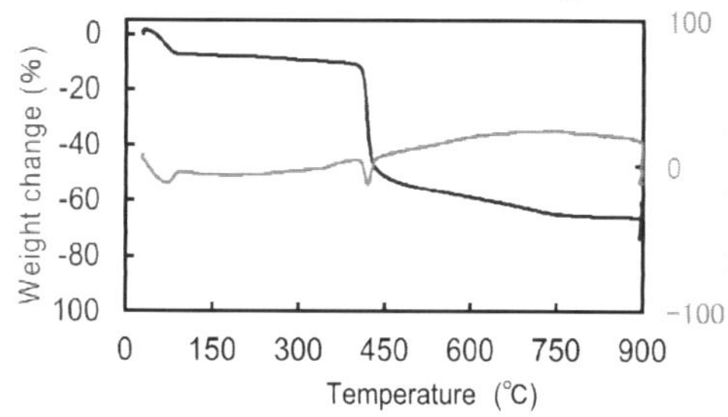

Fig. 2. TG/DTA curves for the EDTA•Er•H complex with N_2 as the carrier gas.

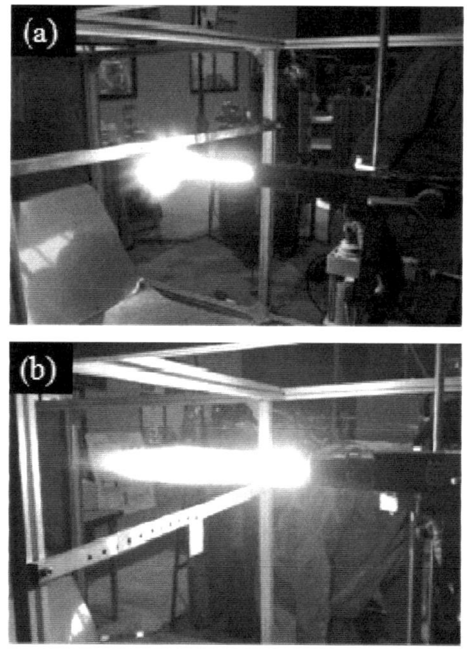

Fig.3. Photograph of the experimental setup, which included a commercial flame-spraying apparatus and a H_2-O_2 flame and used (a) O_2 as the carrier gas and (b) as N_2 the carrier gas.

Figure 3 shows photographs of the experimental setup. The commercial flame-spraying apparatus and the H_2-O_2 flame can be seen. When O_2 was used as the carrier gas, a brighter flame was obtained that that when N_2 was used as the carrier gas. To determine the crystalline structures of the deposited films, XRD analyses were performed. Figure 4 shows the XRD profiles of the three films deposited on the SUS304 substrates. The peaks in their XRD profiles could be assigned using the International Centre for Diffraction Data (ICDD) cards as references. All the films deposited using EDTA•Er•H exhibited cubic and monoclinic Er_2O_3 crystalline phases (ICDD Card Nos. 00-008-0050 and 01-077-6226, respectively). Er_2O_3 exhibits three structural polymorphisms: cubic, monoclinic, and hexagonal, which are commonly known as C-, B-, and A-type structures, respectively [14]. Thus, the choice of the carrier gas had no effect on the crystal structures of the deposited films.

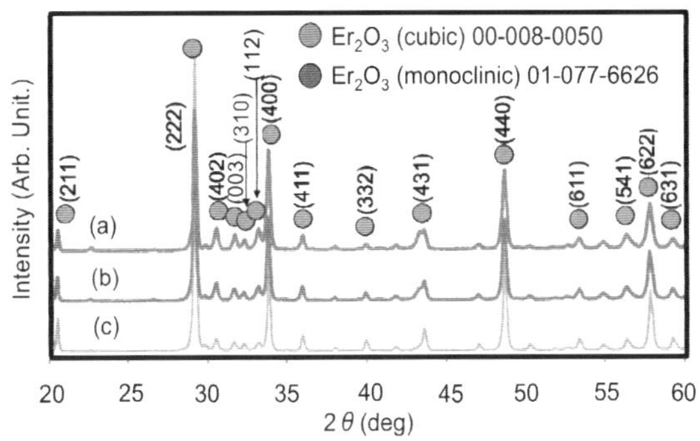

Fig.4.XRD profiles for 2θ of 10–80°for the films synthesized using (a) O_2 as the carrier gas, (b) N_2+O_2 as the carrier gas, and (c) N_2 as the carrier gas.

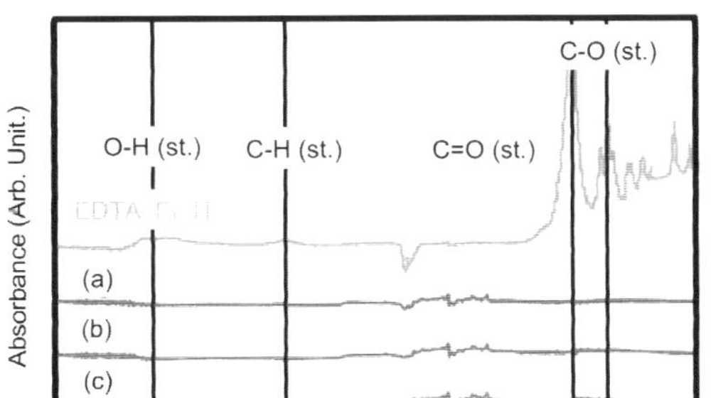

Fig.5. FTIR spectra of the metal oxide films synthesized using (a) O_2 as the carrier gas, (b) N_2+O_2 as the carrier gas, and (c) as the N_2 carrier gas.

The chemical compositions of the obtained films were assumed to be an indicator of the degree of decomposition of the EDTA•Er•H complex. The organic constituents were determined by Fourier transform infrared spectroscopy (FTIR). Figure 5 shows the FTIR spectra of the Er_2O_3 films obtained using the different carrier gases. The FTIR spectrum of the EDTA•Er•H complex is shown for comparison.

Fig.6. Surface SEM images of the films synthesized using (a) O_2 as the carrier gas, (b) N_2+O_2 as the carrier gas, and (c) N_2 as the carrier gas.

The EDTA•Er•H complex exhibited peaks at 1440 and 1500–1700 cm^{-1}, which could be assigned to the carbonyl group. Further, the broad peaks at 2900 and 3300 cm^{-1} in the FTIR spectrum indicated the presence of carbon-hydrogen bonds and the hydroxyl group [11]. In contrast, the Er_2O_3 films did not exhibit any FTIR peaks, confirming that they did not contain many organic constituents. These results indicated that the decomposition and oxidation of the EDTA•Er•H complex occurred in all the carrier gases.

Figure 6 shows surface SEM images of the three Er_2O_3 films. When N_2 was used as the carrier gas, a number of spherical (i.e., not flattened) particles with sizes of approximately 5–10 μm were also observed. Further, even though large pores were present, flattened particles were also seen in the Er_2O_3 film synthesized using N_2+O_2 as the carrier gas. Finally, a large number of flattened particles were seen in the Er_2O_3 film deposited using O_2 as the carrier gas. Particles with the highest degree of flatness (or splatting) were observed in the Er_2O_3 film obtained using O_2 as the carrier gas, for which flat Er_2O_3particles were deposited (Fig. 6c). In contrast, when N_2 was used as the carrier gas, the degree of flatness was the lowest, with spherical particles also being observed (Fig. 6a). Thus, the choice of the carrier gas caused differences in the surface morphologies of the deposited Er_2O_3 films. In general, the microstructures of the films deposited by the PS and HVOF coating techniques vary with the combustion gas used and its composition [15][16]. This is because these parameters determine the flame's temperature and the density of the plasma in the reaction system. These, in turn, determine the temperature and velocity of the in-flight particles. Thus, the chemical reaction processes (decomposition and oxidation) of the EDTA•Er•H complex can be varied by changing the carrier gas used.

Next, the cross-sectional morphologies and elemental distributions of the films were determined using SEM and EDX analyses. Figure 7 shows cross-sectional SEM images and elemental maps for the Er_2O_3 films obtained using the various carrier gases.

The films consisted of several Er_2O_3 layers with thicknesses on the micron scale. Large pores with diameters of approximately 2–5 μm were observed in the Er_2O_3 film deposited using N_2 as the carrier gas. Further, it was found that the number of pores in the Er_2O_3 film synthesized using N_2+O_2 as the carrier gas was lower. Finally, the film deposited using O_2 as the carrier gas had the fewest pore. Table 1 shows the thicknesses and cross-sectional porosities of the various Er_2O_3 films; the porosities were determined by

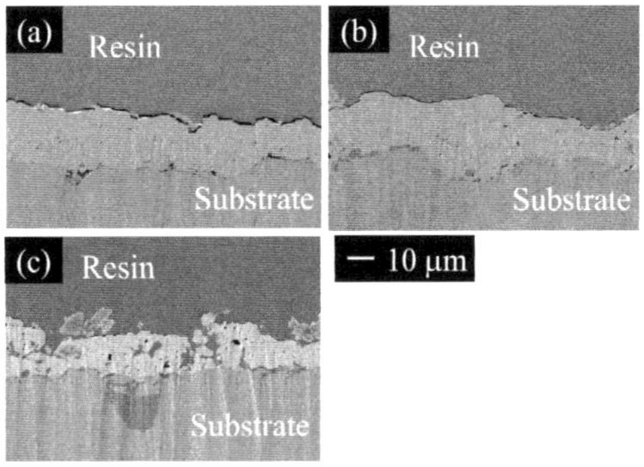

Fig.7. Cross-sectional SEM images of the films synthesized using (a) O_2 as the carrier gas, (b) N_2+O_2 as the carrier gas, and (c) N_2 as the carrier gas.

analyzing SEM images of the films. This analysis method can be used to determine the numbers of both
open and closed pores [17]. The Er_2O_3 film obtained using N_2 as the carrier gas had an average thickness of 15.2 μm and a cross-sectional porosity of 23.4%. Next, in the case of N_2+O_2 as the carrier gas, these values were 22.0 μm and 5.9%, respectively. Finally, when O_2 was the carrier gas, the resulting film was the densest, having a porosity of 3% and an average thickness of 19.6 μm. These results indicated that O_2 is the most suited carrier gas for fabricating dense metal oxide layers.

Table 1 Thicknesses and cross-sectional porosities of the films synthesized on the SUS304 substrates using the different carrier gases.

Carrier gas	Film thickness (μm)	Cross-sectional porosity(%)
N_2	15.2	23.4
N_2+O_2(Air)	22.0	5.9
O_2	19.6	3.0

With respect to thermal spraying, knowing the temperature and velocity of the in-flight particles is useful for determining their collision behavior, which affects the crystalline structure and morphology of the deposited film [18,][19]. Figures 8 and 9 show the temperature and velocity of the in-flight particles for each carrier gas as a function of the nozzle-substrate distance. At a nozzle-substrate distance of 150 mm, the temperatures were 2290℃ in N_2 gas, 2300℃ in N_2+O_2, and 2400℃ in O_2 gas. Further, the velocities of the particles were 99 m/s in N_2 gas, 96 m/s in N_2+O_2 gas, and 90 m/s in O_2 gas. The maximum temperature was in N_2 gas at a distance of 150 mm. However, when the distance was changed to 110 m, the highest temperature was obtained in N_2+O_2 and O_2 gas. The velocity in N_2 gas was the highest at a distance of 150 mm and decreased gradually as the distance was decreased. In contrast, the velocities in N_2+O_2 and O_2 gas decreased with an increase in the nozzle-substrate distance.

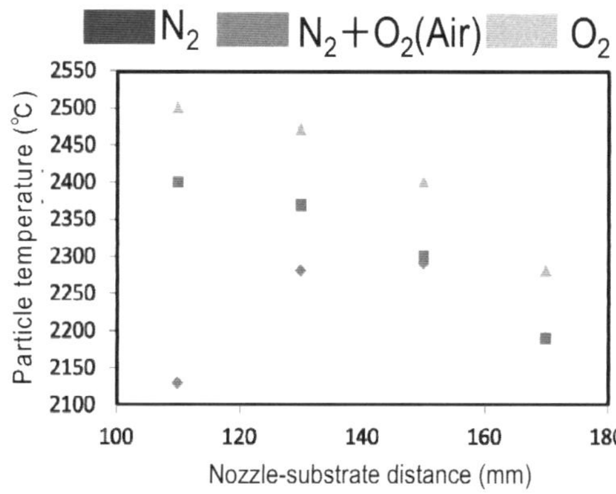

Fig.8.Temperatures of the in-flight particles propelled in the H_2-O_2 combustion gas mixture at different nozzle-substrate distances while using various gas as the carrier gas.

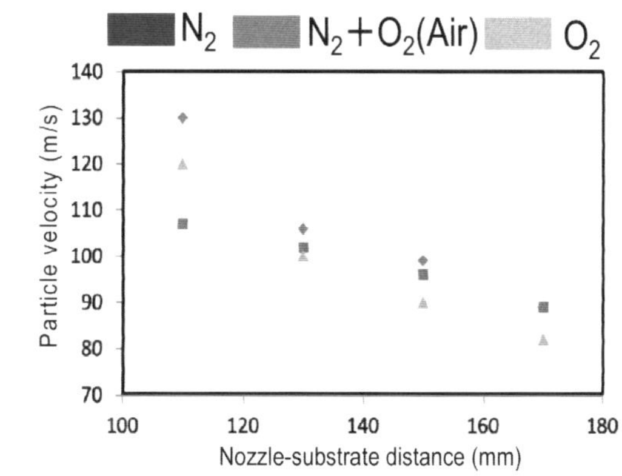

Fig.9.Velocites of the in-flight particles propelled in the H_2-O_2 combustion gas mixture at different nozzle-substrate distances while using various gas as the carrier gas.

On the basis of these results, the effects of the carrier gas on the properties of the deposited films are discussed below. The cross-sectional porosities were 23.4%, 5.9%, and 3.0% when carrier gas was N_2, N_2+O_2, and O_2, respectively. These results indicated that the carrier gas used has an effect on the decomposition and oxidation of the EDTA•Er•H complex. The chemical reaction of EDTA•Er•H (M: metal species) can be represented by Equation (1). In (1), EDTA•Er•H reacts with O_2 to form a Er_2O_3 film.

$$2EDTA \cdot Er \cdot H + 24O_2 \rightarrow Er_2O_3 + 20CO_2 + 13H_2O + 4NO_2 \quad (1)$$

Hence, Equation (1) confirms that, by varying the carrier gas used, the properties of the deposited film can be changed. Denser metal oxide films can be formed by supplying higher amounts of thermal energy to the metal oxide particles. Bae et al. reported that a higher thermal energy resulted in films with lower porosities [20]. In this study, the temperature of the in-flight Er_2O_3 particle obtained from the decomposition and oxidation of EDTA•Er•H was 2400℃ in O_2 and N_2+O_2 and 2290℃ in N_2. The thermal energy of the particles can be calculated as shown below [21]:

$$Q = mcT \quad (2)$$

where Q is the thermal energy (J/s), m is the weight of sample (kg), c is the specific heat of the material (kJ/(kg K)), and T is the temperature (K).

This equation suggests that higher the temperature, higher would be the thermal energy. Hence, the densest Er_2O_3 film was obtained using O_2 as the carrier gas because the temperature of the in-flight particles was the highest in this case. The obtained TG/DTA profiles indicated that the choice of the carrier gas determined the decomposition temperature of EDTA•Er•H, which decreased to 365℃ in N_2+O_2 from 420℃ in N_2. In addition, the TG/DTA profiles suggested that an exothermic reaction occurred in N_2+O_2 while an endothermic reaction took place in the N_2 atmosphere. In the N_2+O_2 atmosphere, the exothermic reaction involving EDTA•Er•H occurred owing to the oxygen present. In contrast, the endothermic peak indicated that a self-decomposition reaction occurred in the N_2 atmosphere. Hence, the use of O_2 as the carrier gas resulted in the synthesis of dense films with improved microstructures.

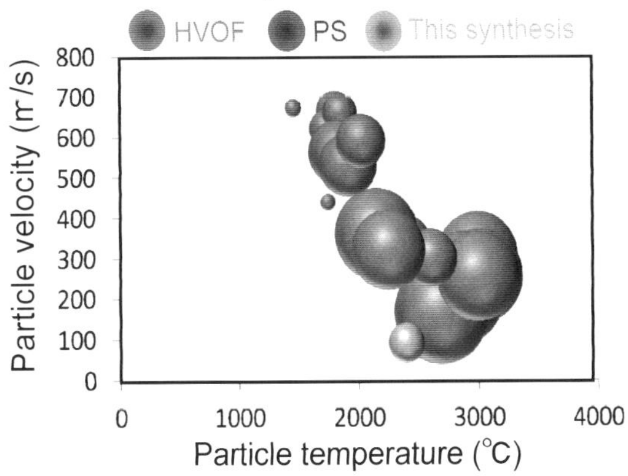

Fig.10. Relationships between the temperature and velocity of the in-flight particles and the porosities of films deposited using various deposition techniques. The sizes of the circles indicate the porosities of the deposited films.

Finally, we compared the properties of the synthesized films with those obtained by PS and HVOF coating techniques. Figure 11 shows the temperatures and velocities of the in-flight particles and the porosities of the films obtained using these three methods [8][9][10]. The temperature of the in-flight particles for the method investigated in the present study lay in the mid range, while their velocity was the lowest. Further, the porosity of the obtained films was the same as those of the films synthesized by the HVOF coating technique. This indicated that the proposed method is ideal for synthesizing dense films at low costs.

4 Conclusion

Dense Er_2O_3 films with a cross-sectional porosity of 3% were synthesized from a metal-EDTA complex using a commercial flame-spraying apparatus. O_2, air (N_2+O_2), and N_2 were used as the carrier gases, in order to investigate the effects of the carrier gas on the properties of the deposited films. By varying the carrier gas, the cross-sectional porosity of the deposited Er_2O_3 films could be varied from 3.0% to 23.4%, with the thickness being 15.2–22.0 μm. It was found that O_2 was the ideal carrier gas for fabricating dense metal oxide films.

5 References

[1] C. Zhou, Q. Zhang, Y. Li, Thermal shock behavior of nanostructured and microstructured thermal barrier coatings on a Fe-based alloy, Surf. Coat. Technol. 163 (2010) 51–59.

[2] H. Pokhmurska, B. Wielage, T. Lampke, T. Grund, M. Student, N. Chervinska, Post-treatment of thermal spray coatings on magnesium, Surf. Coat. Technol. 202 (2008) 4515–4524.

[3] R.S. Lima, B.R. Marple, Nanostructured YSZ thermal barrier coatings engineered to counteract sintering effects, Mater. Sci. Eng. A 485 (2008) 182-193.

[4] D. Toma, W. Brandl, G. Marginean, Wear and corrosion behaviour of thermally sprayed cermet coatings, Surf. Coat. Technol. 138 (2001) 149–158.

[5] A. Pardo, P. Casajus , M. Mohedano, A.E. Coy, F. Viejo, B. Torres, E. Matykina, Corrosion protection of Mg/Al alloys by thermal sprayed aluminium coatings, Appl. Surf. Sci. 255 (2009) 6968–6977.

[6] S. Liscano, L. Gil, M. H. Staia, Effect of sealing treatment on the corrosion resistance of thermal-sprayed ceramic coatings, Surf. Coat. Technol. 188-189 (2004) 135–139.

[7] S. Ahmaniemi, P. Vuoristo, T. Mantyla, Improved sealing treatments for thick thermal barrier coatings, Surf. Coat. Technol. 151-152 (2002) 412- 417.

[8] J. –G. Legoux, B. Arsenault, L. Leblanc, V. Bouyer, C. Moreau, Evaluation of Four High Velocity Thermal Spray Guns Using WC-10% Co-4% Cr Cermets, J. Therm. Spray Technol. 11 (2002) 86–94.

[9] M. Prystay, P. Gougeon, C. Moreau, Structure of plasma-sprayed zirconia coatings tailored by controlling the temperature and velocity of the

sprayed particles, J. Therm. Spray Technol. 10 (2001) 67–75.
[10] M.P. Planche, H. Liao, B. Normand, C. Codde, Relationships between NiCrBSi particle characteristics and corresponding coating properties using different thermal spraying processes, Surf. Coat Technol. 200 (2005) 2465–2473.
[11] H. Akasaka, M. Ohto, Y. Hasebe, A. Nakamura, S. Ohshio, H. Saitoh, Yttria coating synthesized by reactive flame spray process using Y-EDTA complex, Surf. Coat Technol. 205 (2011) 3877–3880.
[12] K. Komatsu, T. Sekiya, A. Toyama, Y. Hasebe, A. Nakamura, M. Noguchi, Y. Li, S. Ohshio, H. Akasaka, H. Muramatsu, H. Saitoh, Deposition of metal oxide films from metal–EDTA complexes by flame spray technique, J. Therm. Spray Technol. 23 (2014) 833–838.
[13] K. Komatsu, T. Sekiya, A. Toyama, A. Nakamura, I. Toda, S. Ohshio, H. Muramatsu, H. Saitoh, Synthesis of Y2O3 films by spray coating with milled EDTA•Y•H complexes, Int. J. Chem. Nuc. Metal. Mater. Eng. 8 (2014) 1322–1326.
[14] P. P. Bose, M. K. Gupta, R. Mittal, S. Rols, S. N. Achary, A. K. Tyagi, S. L. Chaplot, Phase transitions and thermodynamic properties of yttria, Y2O3: Inelastic neutron scattering shell model and first-principles calculations, Phys. Rev. B 84 (2011) 094301.
[15] E. Lugscheider, C. Herbst, L. Zhao, Parameter studies on high-velocity oxy-fuel spraying of MCrAlY coatings, Surf. Coat Technol. 108–109 (1998) 16–23.
[16] B. Pateyron, M.-F. Elchinger, 1 G. Delluc, P. Fauchais, Thermodynamic and transport properties of Ar-H~ and Ar-He plasma gases used for spraying at atmospheric pressure: properties of the mixtures, Plasma Chem. Plasma Process. 12 (1992) 421–448.
[17] S. Deshpande, A. Kulkarni, S. Sampath, H. Herman, Application of image analysis for characterization of porosity in thermal spray coatings and correlation with small angle neutron scattering Surf. Coat. Technol. 187 (2004) 6–16.
[18] F. Tarasi, M. Medraj, A. Dolatabadi, J. Oberste-Berghaus, C. Moreau, Effective parameters in axial injection suspension plasma spray process of alumina-zirconia ceramics, J. Therm. Spray Technol. 19 (2010) 787–795.
[19] A. Aguero, F. Camon, J. Garcıa de Blas, J.C. del Hoyo, R. Muelas, A. Santaballa, S. Ulargui, P. Valles, Phase formation and transformation in alumina/YSZ nanocomposite coating deposited by suspension plasma spray process, J. Therm. Spray Technol. 20 (2011) 1292–1309.
[20] G. Bae, S.Kumar, S. Yoon, K. Kang, H. Na, H.-J. Kim, C. Lee, General aspects of interface bonding in kinetic sprayed coatings, Acta Mater. 57 (2009) 5654–5666.
[21] K. Taheria, R. Gadow, A. Killinger, Exergy analysis as a developed concept of energy efficiency optimized processes: The case of thermal spray processes, Procedia CIRP, 00 (2014) 000–000.

7YSZ coating prepared by PS-PVD based on heterogeneous nucleation

X. F. Zhang*, K. S. Zhou*, M. Liu, C. M. Deng, J. Mao, Z. Q. Deng, C. G. Deng

(Guangdong Institute of New Materials, National Engineering Laboratory for Modern Materials Surface Engineering Technology, The Key Lab of Guangdong for Modern Surface Engineering Technology, Guangzhou 510651, China)

Abstract: Plasma spray-physical vapor deposition (PS-PVD) as a novel coating process based on low pressure plasma spray (LPPS) has been significantly used, where deposition takes place not only from liquid splats but also from nano-sized clusters and from the vapor phase in contrast to conventional technologies. As a consequence, thin, dense and porous ceramic coating for special application needs can be deposited through PS-PVD due to its high power, very low pressure, etc. This offers new opportunities to obtain different advanced microstructures and thus to meet the growing requirements in modern functional coatings. To better exploit the potential of such a gas phase deposition from PS-PVD, the deposition mechanisms and their dependency on process conditions must be understood detailedly. Therefore, coating experiments were performed at variant process conditions. Based on the observed phenomenon, a deposition mechanism of heterogeneous nucleation is presented.

1. Introduction

The plasma spray-physical vapor deposition (PS-PVD) process has been developed with the aim of depositing different structured functional coatings, such as thin, gas tight and columnar coating, etc. with large area coverage by plasma spray [1-3]. The PS-PVD is developed based on low pressure plasma spray (LPPS), where electrical current up to 3000 A and plasma gas flow up to 200 slpm, an input power level of 180 kW could be achieved. With operation pressure decreasing, the plasma plume expands to a length of more than 2.2 m and 400 mm in diameter. Using appropriate parameters, it is possible to evaporate the powder feedstock materials obtaining variant microstructures and non-line-of-sight deposition [4-5]. Li et al. [6] reported that using advanced PS-PVD process, the nanohardness and microhardness of prepared dense coatings are markedly higher than those of conventional YSZ coatings (i.e., the coating fabricated by atmospheric plasma spray or electron beam-physical vapor deposition), even comparable with the sintered YSZ polycrystal. Goral et al. [7] presented that columnar YSZ coatings were obtained by PS-PVD through evaporating powders. The microstructures of coating were affected by feed rate, chamber pressure, and sample rotation rate, plasma gas ratio (Ar, He, N_2). Hospach et al. [8] showed that columnar-structured YSZ coatings between 20 and 750 μm were deposited by PS-PVD. Thinner and thicker coatings seem to be possible. The geometry and arrangement of the sample and the sample holder indicated a big influence on the coating quality.

Despite much investigation have been done about PS-PVD in the several decades, however, the basic process technology, such as heating of powder particles to spray-deposited molten, semi-molten droplets or vapor gas phase onto substrate surface, has remained essentially the same [9-10]. And significant gaps currently exist in our understanding of the PS-PVD of processes. These areas consist of particle-plasma interactions in the rarefied plasma, particle vaporization and its affect on plasma properties, and deposition mechanisms associated with different microstructures. Now, increased understanding in each of these areas will enable the spray community to more easily apply PS-PVD technology to meet emerging coating challenges [11-12].

To exploit the potential of such as gas-phase deposition using PS-PVD technology, the deposition mechanisms and their dependency on process conditions must be better understood. The PS-PVD process can be summarized as three steps [9]: feedstock processing in plasma torch, plasma jet formation and materials transport, deposition and coating growth. During the third step, it mainly includes heterogeneous and homogeneous nucleation depending on spray distance [13]. When spray distance was set at middle area in axial direction of plasma flame, coating deposition primarily relies on heterogeneous nucleation on substrate surface. In this

*Corresponding authors: K.S. Zhou, E-mail: kszhou2004@163.com, Tel: +86-20-37238263; fax: +86-20-37238531; zxf200808@126.com (X.F. Zhang)

work, different structured coating based on heterogeneous nucleation had been obtained. And these principles were summarized in this investigation.

2. Experimental Procedure

The experimental set-up is based on a Sulzer-Metco Multicoat TM PS-PVD system together with an O3CP plasma torch mounted on a robot manipulator of ABB inside a Φ2.5×4.5 m vacuum chamber. The PS-PVD system is obtained through a comprehensive reconstruction of an existing conventional LPPS system. In particular, the system is equipped with an additional vacuum pumping unit, a large vacuum blower to provide sufficient pumping capacity at low pressures, enhanced cooling capacity, additional power sources, a new torch transfer arc system and new operational control units. In terms of the powder feeding system, two powder injectors are located in the cylindrical section of the O3CP nozzle (*D* 12.5 mm) close to divergent part.

Feedstock agglomerated 7YSZ powers (Metco 6700, Sulzer-Metco) were used and their grain size ranges from 5 to 22 μm. 7YSZ coatings were deposited on graphite, sintered zirconia, and nickel-based superalloy K417 substrate (size Φ25.4×5 mm, surface roughness < 2 μm) at spray distance 950 mm, where the Ar-He hybrid plasma was operated at 67 kW net power of O3CP under operation pressure of 1.5 mbar and the substrate temperatures were controlled at 850 ℃ and room temperature, respectively. During the pre-heating or deposition process, the substrate remained still, while the plasma gun was moved at a speed of 1000 mm/s. The microstructures of 7YSZ coatings were characterized by Field Emission-Scanning Electron Microscope (FE-SEM, Nova-Nono430, FEI) and Transmission Electron Microscopy (TEM, JEM2100F, JEOL). Additionally, before TEM characterization, the test samples were prepared by Focused Ion Beam (FIB, 450S, FEI) milling.

3. Results and discussion

3.1. Variation of coating microstructure

3.1.1. Effect of substrate material

Taking graphite as substrate without pre-heating by plasma flame, 7YSZ coatings were prepared by PS-PVD showing typical columnar microstructure, seen in Fig. 1(a). The interface of coating and substrate is good bonding without crack. Among columns, there exists different size of gap. Between columns, many small particles were formed, which results from condensation of vapor phase [14-15], as shown in the magnified Fig. 1(b). When the graphite substrate was replaced by sintered zirconia, similar columnar 7YSZ coating was obtained with the same parameters at room temperature of substrate, as presented in Fig. 2(a). Through comparison, it can be know that the horizontal width of single column is larger than the column deposited on graphite, but the coating has lower deposition rate when taking zirconia as substrate. Due to similar properties of 7YSZ material and sintered zirconia, the interface of coating and substrate has a better bonding. Between columns, there is no apparent vertical gap and also no small particle appeared in the gap. Besides, the column is made of fine grains and denser than that coating deposited on graphite.

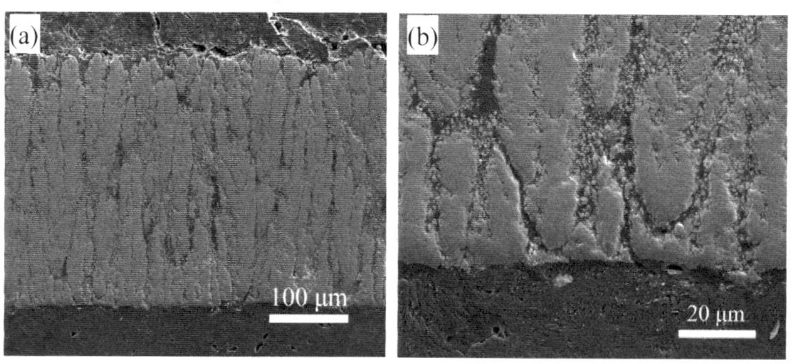

Fig. 1 7YSZ coating deposited on graphite at room temperature, (a) cross section, (b) magnified cross section.

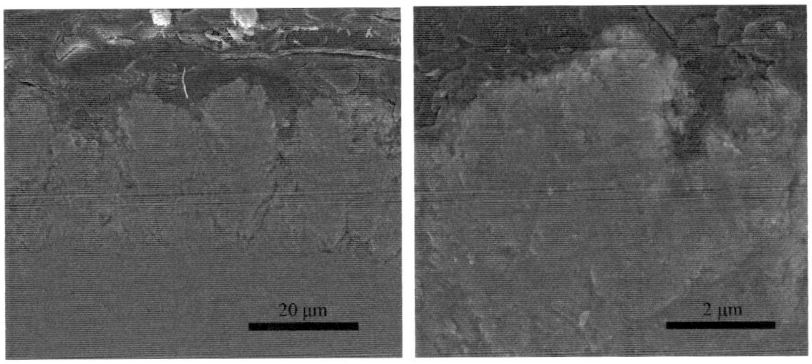

Fig. 2 7YSZ coating deposited on sintered zirconia at room temperature, (a) cross section, (b) magnified cross section.

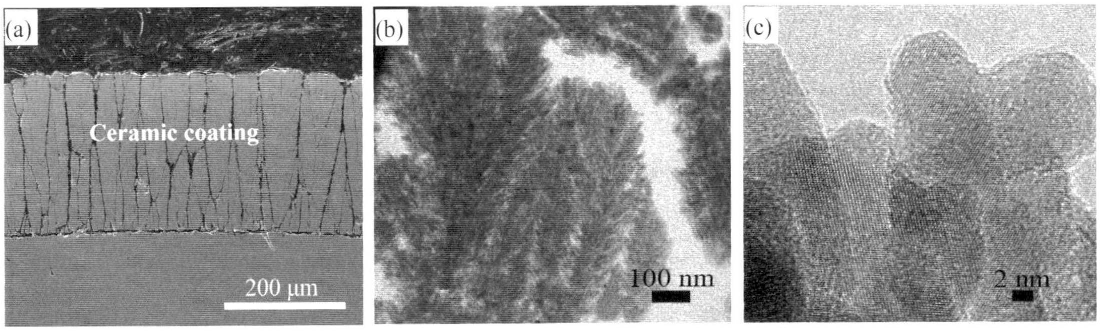

Fig. 3 7YSZ coating deposited on K417 at substrate temperature of 850℃, (a) columnar coating, (b) bright filed micrograph of TEM, (c) high resolution image of TEM

Fig. 4 7YSZ coating deposited on K417 at room substrate temperature, (a) 7YSZ coating, (b) magnified micrograph of ceramic coating, (c) bright filed image of TEM.

3.2. Heterogeneous nucleation

According to the reported literature [13], it showed that when spray distance was set at 950 mm, heterogeneous nucleation of absorbed gas particles will occur on the substrate surface. When vapor particles in plasma plume impinge on the substrate, heterogeneous nucleation occurs on the substrate surface that lower the critical free energy required to form a stable nucleus of mean dimension *r*. Fig. 5 shows a spherical cap-shaped solid nucleus on the substrate with a contact angle *θ*. The critical-sized nucleus is related to the amount of undercooling by the following relation [16-17]:

$$r^* = -\frac{2\gamma_{vn} T_m}{\Delta H_f \Delta T} \quad (1)$$

Where T_m is equilibrium freezing temperature of feedstock powder, ΔH_f is latent heat of fusion, and ΔT is amount of undercooling at which nucleus formed, γ_{vn} is specific surface free energy between vapor and nucleus. Thus, free-energy change of heterogeneous nucleation can be given by

$$\Delta G^*_{He} = \frac{16\pi n\gamma_{vn}^3}{3\Delta G_v^2}\cdot\left(\frac{2-3\cos\theta+\cos^3\theta}{4}\right) = \frac{16\pi n\gamma_{vn}^3}{3\Delta G_v^2}\cdot f(\theta) \quad (2)$$

Where ΔG_V is volume free energy of nucleus, *n* is the number of formed nucleus.

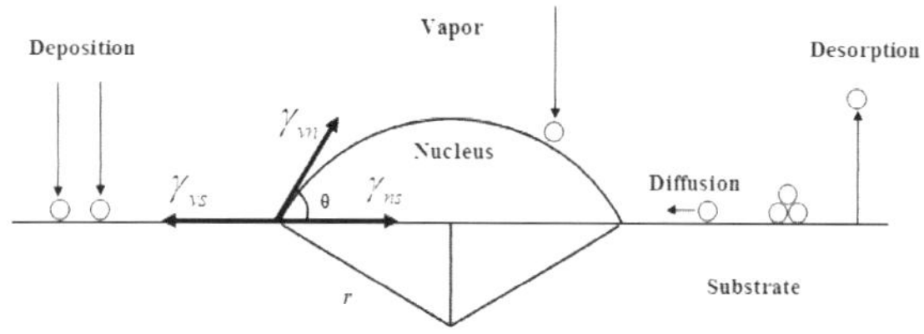

Fig. 5 Schematic of heterogeneous nucleation on substrate surface during vapor deposition [16].

In section 3.1.1, two substrates were used showing that the width of columnar grain deposited on zircona is larger than the column on graphite, as illustrated in Fig. 6. For heterogeneous nucleation to take place, the substrate must be wetted by the vapor phase. The Fig. 5 indicates that a nucleus formed on substrate surface, creating a contact angle θ between the nucleus and substrate. Importantly, the contact angle θ depends solely on the surface properties of the involved materials, and which will affect the r^* and ΔG^*_{He}. The contact angle θ between nucleus and zirconia substrate is larger than that θ between nucleus and graphite due to due to higher surface tension of zirconia at room temperature. Thus the r^* and ΔG^*_{He} will increase with a decrease of contact angle θ based on Eq. 1-2, resulting in larger width of column deposited on zirconia substrate. However, the number of column will reduce.

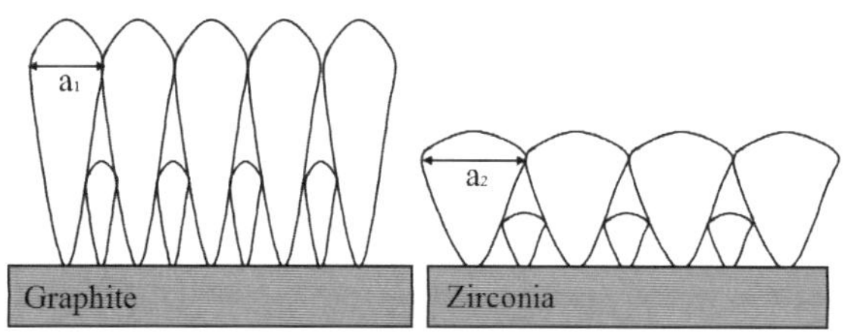

Fig. 6 Schematic of different structure of columnar coating deposited on different substrate material.

3.3. Coating growth

It can be inferred from the Eq. 1-2 reveals that with decrease of substrate temperature, the r^* and ΔG^*_{He} will decrease correspondingly. Taking superalloy K417 as substrate, the coating structure will change from oriented column to fine grain when substrate temperature vary from 850 ℃ to room temperature, as described in section 3.1.2. Due to lower surface energy than the inorganic materials such as graphite and zirconia, so the coating tends to be dense structure instead of porous columns.

Most of feedstock 7YSZ powders with rate of 18 g/min will be vaporized into vapor particles impinging on the substrate surface [13, 17-18]. Some of vapor particles were absorbed but remains were rebounded by the rigidity substrate and vortex flow. Thus, ahead of substrate surface, the number of vapor particles in unit volume C_V is variable and given by [19]

$$C_V = C_0\left(1+\frac{1-K_0}{K_0}e^{-R_x/D}\right) \quad (3)$$

Where C_0 is the number of vapor particles without rebound effect, K_0 is equilibrium partition coefficient, R_x is absorbed rate of vapor particle by substrate that is proportional to distance x ahead of substrate surface, D is a constant for certain vapor material. Thus, the critical temperature of heterogeneous nucleation can be calculated by Eq. 4, which is influenced by C_V.

$$T_V = T_A - m\cdot C_V = T_A - m\cdot C_0\left(1+\frac{1-K_0}{K_0}e^{-R_x/D}\right) \quad (4)$$

Where T_A is melting temperature of feedstock material, m is a constant. T_V is proportional to the distance x from substrate surface as plotted in Fig. 7. When x=0, the interface temperature T_i between substrate and coating given by following relation

$$T_i = T_A - m\cdot C_0 / K_0 \quad (5)$$

The practical temperature T_D of vapor phase ahead of substrate depends on gradient temperature $G(dT/dx)$ and it can be described by following equation

$$T_D = T_i + G\cdot x = T_A - mC_0/K_0 + Gx \quad (6)$$

The Eq. (6) indicates that the T_D is linear with the distance x from substrate surface. Coating structure deposited on the substrate mainly depends on the amount of undercooling when $T_D < T_V$. When substrate temperature was set at 850 ℃, the amount of undercooling $\triangle T(1)$ and practical temperature $T_D(1)$ are shown in Fig. 7(a). With substrate temperature decreasing, $\triangle T(2)$ and $T_D(2)$ also be described as Fig. 7(b). And when the substrate temperature reduces to room temperature, Fig. 7(c) indicates the $\triangle T(3)$ and $T_D(3)$. The Fig. 7 indicates that the undercooling of interface between substrate and vapor phase $\triangle T(1)<\triangle T(2)<\triangle T(3)$, and the slope of T_D is increase along with decrease of substrate temperature $T_D(1)<T_D(2)<T_D(3)$. In the Figs. 7(a) and (b), with increase of distance x, the $\triangle T$ increase first and then

decrease when $T_D<T_V$, which can be illustrated by Fig. 8(a) called negative temperature gradient distribution. As opposite to Fig. 7(c), the $\triangle T$ decreases continuously called positive temperature gradient distribution, seen in Fig. 8(b). If the $\triangle T$ is low, columnar coating can be obtained demonstrated in Fig. 3. And when the $\triangle T$ decreases, the mixed structure composed of columnar coating and fine grains can be obtained as illustrated in Fig. 4. Thus, if the $\triangle T$ reduces continuously, dense coating made of fine grains fully can be fabricated.

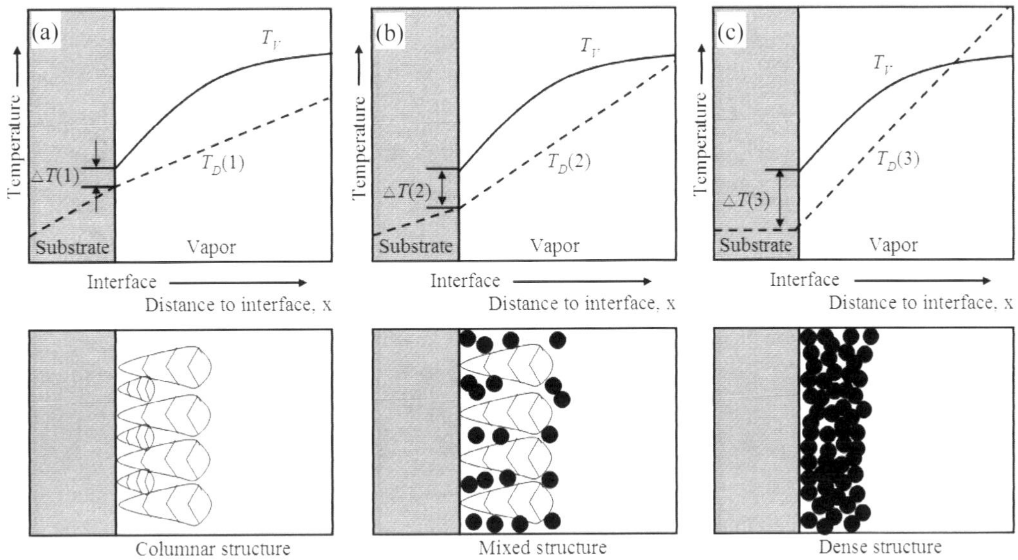

Fig. 7 Different amount of undercooling related to variant coating structure.

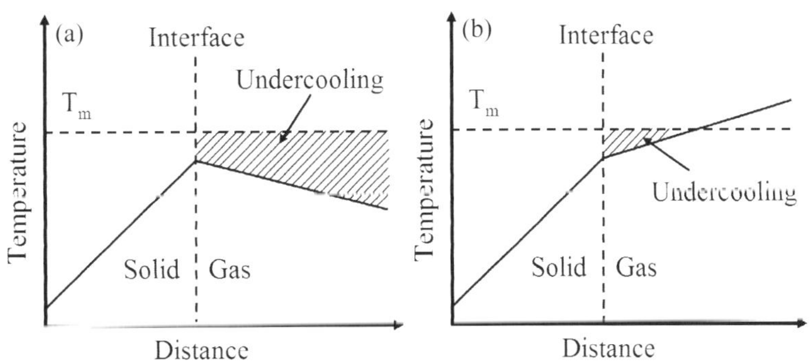

Fig. 8 (a) Negative temperature gradient distribution, (b) positive temperature gradient distribution in gas phase

4. Conclusions

Different columnar structured 7YSZ coatings were prepared by plasma spray-physical vapor deposition (PS-PVD) on graphite and zirconia substrate. Besides, the effect of substrate temperature on coating structure deposited on metallic material has been investigated. The following gas-deposition principles based on heterogeneous nucleation can be drawn as following:

(a) Without pre-heating, the coating deposited on the graphite and zirconia substrate is columnar structure. Due to higher surface energy of zirconia, the horizontal width of column is larger than that columnar grain deposited on graphite.

(b) Undercooling of interface between substrate and vapor phase plays an important role in coating structure. With substrate temperature decreasing, the temperature gradient of gas-solid interface changes from negative to positive distribution, and the typical columnar structure will transform into mixed structure made of column and fine grain.

Acknowledgements

We would like to acknowledge the financial support from National Basic Research Program of China (2012CB625100), National Natural Science Foundation of China (51174251) and Guangzhou Science and Technology Program (201510010095).

References

[1] M. Gindrat, H.M. Hohle, K.V. Niessen, P. Guittienne, other: Plasma spray-CVD: A new thermal spray process to produce thin films from liquid or gaseous precusors. J. Thermal Spray Technol. 20(2011), Issue 4,pp.882/887

[2] N. Zotov, S. Baumann, W.A. Meulenberg, other: La-Sr-Fe-Co oxygen transport membranes on metal supports deposited by low pressure plasma spraying-physical vapour deposition. J. Membrane Sci. 442 (2013), pp.119/123

[3] M.F. Smith, A.C. Hall, J.D. Fleetwood, other: Very low pressure plasma spray-A review of an emerging technology in the thermal spray community. Coatings 1(2011), pp.117/132.

[4] G. Mauer, A. Hospach, and R. Vaben, Process development and coating characteristics of plasma spray-PVD. Surf. Coat. Technol. 220 (2013), pp.219/224

[5] L. Zhu, N.N. Zhang, B.C. Zhang, other: Very low pressure plasma sprayed alumina and yttria-stabilized zirconia thin dense coatings using a modified transferred arc plasma torch. Applied Surf. Sci. 258 (2011), pp.1422/1428

[6] J.Q. Li, H.J. Huang, T. Ma, other: Twin-structured Yttria-stabilized t' zirconia coatings deposited by plasma spray physical vapor deposition: microstructure and mechanical properties. J. Am. Ceram. Soc., 90 (2007), Issue 2, pp.603/607.

[7] M. Goral, S. Kotowski, A. Nowotnik, other: PS-PVD deposition of thermal barrier coatings, Surf. Coat. Technol. 237 (2013), pp. 51/55

[8] A. Hospach, G. Mauer, R. Vaben, other: Columnar structured thermal barrier coatings (TBCs) by thin film low-pressure plasma spraying (LPPS-TF), J. Thermal Spray Technol. 20(2011), Issue1-2, pp.116-120

[9] G. Mauer, A.Hospach, N. Zotov, R. Vaben: Process conditions and microstructures of ceramic coatings by gas phase deposition based on plasma spraying, J. Thermal Spray Technol. 22(2013), Issue, 2-3, pp.83-89

[10] L.H. Gao, H.B. Guo, L.L. Wei, other: Microstructure, thermal conductivity and thermal cycling behavior of thermal barrier coatings prepared by plasma spray physical vapor deposition. Surf. Coat. Technol. 276 (2015), pp. 424/460

[11] X.F. Zhang, K.S. Zhou, M. Liu, other: Toughness and elasticity behaviors in nano-structured 7 wt.% Y2O3-stablized ZrO2 coating, Surf. Coat. Technol. 276 (2015), pp. 316/319.

[12] J.B. Song, X.F. Zhang, C.M. Deng, other: Research of in situ modified PS-PVD thermal barrier coating against CMAS corrosion, 2015, in press, DOI: http://dx.doi.org/10.1016/j.ceramint.2015.10.106

[13] X.F. Zhang, K.S. Zhou, M. Liu, other: Gas deposition mechanisms of 7YSZ coating based on plasma spray-physical vapor deposition. J. Eur. Ceram. Soc. in press, 2015, DOI:10.1016/j.jeurceramsoc.2015.10.041

[14] K. Lizuka, M. Kambara, T. Yoshida: Growth of tin oxide thick films by plasma spray physical vapor deposition, Sensor and Actuators B 155 (2011), pp. 551/556

[15] G. Mauer, R. Vaben: Plasma spray-PVD: Plasma characteristics and impact on coating properties, Journal of Physics: Conference Series 406 (2012), pp. 012005/012017

[16] M. Ohring: Materials Science of Thin Films Deposition and Structure, 2th, 2006, Elsevier (Singapore) Pte Ltd. p 357-406.

[17] W.F. Smith, J. Hashemi: Foundations of Materials Science and Engineering, 2009, 5th, Mc Graw-Hill, Inc. p136-147.

[18] L.H. Guo, H.B. Guo, L.L. Wei, other: Microstructure and mechanical properties of yttria stabilized zirconia coatings prepared by plasma spray physical vapor deposition, Ceram. Int. 41 (2015), Issue 7, pp. 8305-8311

[19] W. Kurz, D.J. Fisher, Fundamental of Solidification, 1984, Trans. Tech. Pubn., pp. 53/58

Deposition of nanostructured YSZ coating from spray-dried particles of no heat treatment

Yan ZHAO [1] *Yang GAO* [1]
1 Dalian Maritime University, Dalian, Liaoning, China
Email: gaoyang@dlmu.edu.cn; Tel and Fax: +8641187426895

Abstract

In this study, the nanostructured YSZ agglomerates with no heat treatment were internally fed into a regular water cooled anode nozzle and a lengthened gun barrel without water cooling respectively. The coatings were deposited via atmospheric plasma method and the differences in coating microstructure, especially the characteristics of the nano-zone were investigated by optical microscope (OM) and field emission scanning electron microscope (FESEM). The results show that the porous nanostructured YSZ agglomerates with no heat treatment could be effectively deposited for both conditions. The unsintered agglomerates were favorable for developing more porous nano zone with finer grains. Larger amount of nano structure that resemble to the original agglomerates were preserved when fed through the lengthened gun barrel without water cooling.

Introduction

Nanostructured coatings show more potentialities than conventional coating structure in improving thermal and mechanical properties of components for different application purpose [1-3]. The plasma spray method, including a process of partially melting the powder particles, accelerating, impingement and solidified on the substrate surface, is the most commonly used method in fabricating nanostructured ceramic coatings due to its flexibility and low cost. By controlling the surface temperature of feedstock particles, the original nano structure could be retained and embedded in the full molten area, forming a bimodal coating structure. The unmelted cold core of agglomerates correspond to the nano zone in the as-sprayed coating. There are lots of experimental [4-8] and theoretical studies [9-11] on fabricating nanostructured ceramic coatings to provide reliable technique routes to engineer nanostructured ceramic coatings from reconstructed powder. Spray parameters and feedstock properties are considered as the major impact factors of preserving the nano feature of feedstock powder as well as the characteristics of nano-zone structure which determine the excellent properties of the nanostructured coatings. For instance, a porous nano-zone would significantly reduce the thermal conductivity of nano-structured coatings [12] and supply certain strain tolerance through crack deflection [13], whereas a denser nano-structured coating might be favorable for wear-resistant coatings [14].

Generally, the reconstitution process include spray drying and sinter treatment, and the properties of the agglomerates (e.g. grain structure and porosity) would depend on the specific process conditions. The heat treatment to remove the binder and densify the particle would cause the coarser of nanocrystal which is evidenced by the work of Liu et al. [15] on the growing behavior of nanocrystalline zirconia powder at 1200~1400℃. The relationship of grain size between the starting powder and the nanostructure in the as-sprayed coating was also investigated by Wang et al. [16] and Bai et al. [17]. In order to overcome the effects of heat treatment on the nano grain, the most direct idea is to skip this procedure. However, there are many consequences, such as the loose connected powder would easily disintegrated into small pieces which could not obtain enough inertia to deposit and the porous structure might influence the flight behavior which might introduce difficulties in process control. Fogarassy et al. [18] modeled the behavior of agglomerated nanostructured particles in the plasma jet and suggested that the main cause for the gradual disintegration or explosion of agglomerated particles seems to be the high tensile stress developed inside the particles during their violent external heat. The forces exerted on particles injected into the plasma jet would be simplified as the viscous drag force and the thermophoresis. Due to the severe temperature gradient, it could be expected that a molten outer shell might developed as the disintegration occurred due to the evaporation of binder which would disturb the trajectories of agglomerates in the plasma jet. It could be speculated that if the fragments of agglomerated particles were still in range of particles that being appropriately heated and accelerated, then a coating with finer nanostructure would be obtained. Besides, it is the fluctuations of turbulent plasma jet and its dependencies on spray parameters provide the feasibility of fabricating porous structure powder which is meaningful in engineering nanostructured coating.

The object of this work is to fabricate nanostructured ceramic coatings by using spray dried agglomerated powder of no heat treatment and to characterize the differences in microstructure caused by changing of feedstock properties. The spray gun was renovate to realize the effective deposition of the unsintered agglomerates and the coatings were compared with an internal feeding spray gun with regular water cooling. The coating properties including porosity, micro-hardness and thermal conductivity of bimodal structured nano YSZ coating were examined.

Experimental Methods

The nanostructured ZrO_2-7~8 wt. % Y_2O_3 is supplied by Beijing Sun Spraying Scientific & Technologies Development Co., Ltd. The nano granules (~20nm) were reconstituted into solid sphere of 70~100 μm by spray drying method (Fig.1a). The cross-section of as-received powder are presented in Fig. 1b, showing a homogeneous structure compared to the dense shell of sintered agglomerates. The high magnification on agglomerates surface is displayed in Fig. 1c. Different from regular commercial available nanostructured powder, the spray dried agglomerates used in this study have not been calcinated at high temperature, and the apparent density is 1.2 g/mm^3. The substrates were 304 L stainless steel plates (5 mm × 20 mm × 40 mm rectangular plates). Fe-Cr-Al (45-70 μm, Beijing Yi Xin' an Technology & Development Co., Ltd,) powder was used as metallic bond coat.

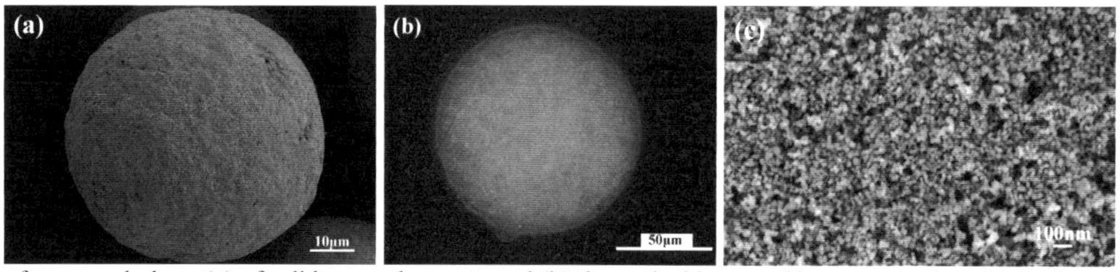

Fig. 1 The surface morphology (a) of solidary agglomerate and (b) the optical image of its cross-section. (c) Surface magnification show the fine nano components.

One of the spray gun used in this study was renovated based on the spray gun (LETS5) developed by thermal spray center of Dalian Maritime University. The anode-nozzle is 6 mm in diameter and 30 mm in length. A short copper tube of the same diameter as the nozzle was installed coaxially to the nozzle and the powder entrance is 10 mm away from nozzle exit without water cooling. The supplemented short tube is not electrically insulated from the nozzle. The sketch of the gun configuration and the loaded jet flow is displayed in Fig. 2. The other spray gun is also a homemade spray gun with internal powder feeding along radical direction. The feeding spot was on the side of the anode nozzle and cooled by circulating water. The spay guns were noticed as Gun1 and Gun2 respectively.

Fig.2 The sketch of gun configuration and the loaded jet flow.

The substrates were sand-blasted with corundum granules and then degreased with acetone. Subsequently the NiCrAlY bond coat layer of 100-150 μm was deposited by atmospheric plasma spray (APS) method. Specific spray parameters are listed in Table 1.

Table 1 Spray parameters of bond coat and YSZ

Parameter	Bond coat	YSZ	
		Gun 1	Gun 2
Power (kW)	12.5	21	30
Current (A)	250	200/250/300	250/350/450
Voltage (V)	50	70	60
Primary gas Ar (L/min)	30	28.9	28.9
Secondary gas H_2 (L/min)	5	5	5
Powder feed rate (g/min)	10	5	10
Spray distance (mm)	100	100	100

Field emission scanning electron microscopy (Carl Zeiss SUPRA™ 55) and optical microscope (Olympus-GX51) were utilized to observe the coating microstructure. The cross-section of samples were sectioned and mounted with epoxy resin, then metallurgical polished. To investigate the nano structure in the as-deposit coatings, thermal etch was carried out in a box furnace at 1200□ for 7min.The phase structure of both nanostructured YSZ feedstock and as-sprayed coatings was examined by an X-ray diffraction (XRD) with Cu target. The average size of crystallites was determined based on XRD peak broadening; details of this method could reference our former work [19]. The average grain size could be estimated by Debye Scherrer formula [20-22]. Silicon specimen was used to measure the instrumental broadening. The porosity was evaluated by image analysis using OLYCIAm^3 software with 15 pictures per-sample. Micro hardness of test samples was conducted by MH-6 Vickers indenter with a load of 200 g for a dwell time of 15 s. The Weibull statistics is utilized to analysis the hardness of nanostructure coatings.

Results

The optical images of polished cross-sections are shown in Fig. 3. The unsintered agglomerates were successfully deposited using both spray guns. By assembling a short tube, the Gun 1 provided protection for the particle flow from the ambient cold air and forced the agglomerates to immerse into the plasma jet, therefore, the deposition efficiency was measured to be about 40% with current of 200 A, and maintained this level as the current increased to 300 A. Most of the nano area were found to be band-like and parallel to the substrate surface, except for the condition of 300 A. The sever deformation of the nano zone parallel to the substrate surface implied a higher deposition speed of particles. When the arc current was lower than 300 A, the melting of the agglomerates was quite limited, as a result the original porous nanostructure was retained. By contrast, the coating prepared by Gun 2 showed higher content of full molten area, which could be attributed to the higher heating efficiency of internal power feeding method. The solid spheres of YSZ spray dried powder used in this study are bigger and more porous than those commonly used in APS, and the rate of heat conduction within individual particle would be too slow to ensure a homogeneous heating during their residence in the plasma jet [23], a condition which is favorable for preserving the original nano character. Therefore, even with the internal powder feeding method, the nano zone still can be retained until the arc current reached 450 A

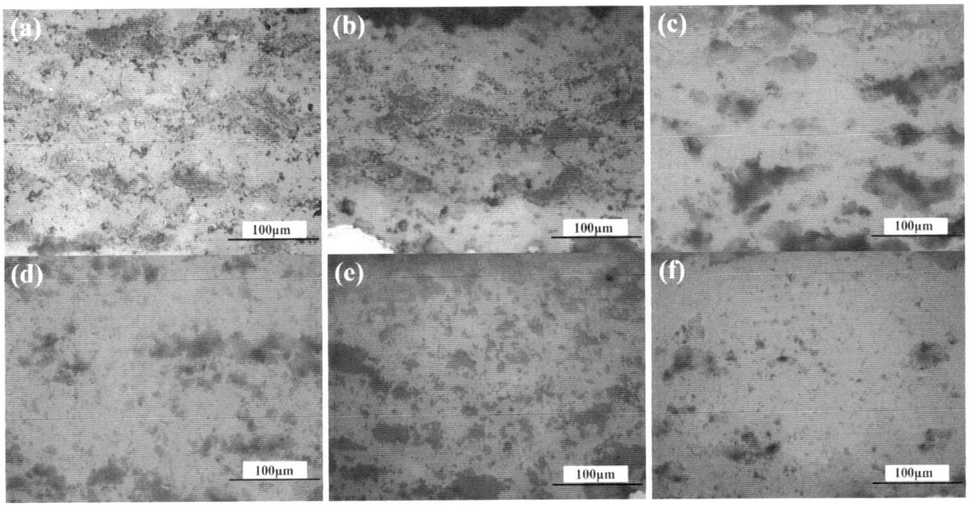

Fig.3 the cross-sections of coatings engineered with Gun 1(a-c) and Gun 2(d-f)

Higher magnifications on nano zone are displayed in Fig.4. It could been seen that highly porous and weakly bonded nano zone was developed by Gun 1 with 200 A. Fig. 4b is the higher magnification on the selected area in Fig. 4a which indicate that the original nanostructure has been well preserved. The cracks within the nano zone elucidated that severe deformation occurred during flattening. As the current increased to 250 A, the nano area reduced due to the increased melting degree. However, the input power is pretty lower than previously used (higher arc current with outward powder feeding) and the porous nature of the unsintered agglomerates might also be inherited by the as-sprayed coatings. By contrast, the nano granules retained in the coatings deposited by Gun 2 were remarkably coarser than that of the coatings sprayed by Gun 1. Due to the porous nature of the unsintered agglomerates, the coatings, especially the unmolten parts were also porous. Although the coatings presented two types of structure character, the unmelted granules grew apparently which elucidated that the agglomerates experienced remarkable grain growth during deposition by Gun 2.

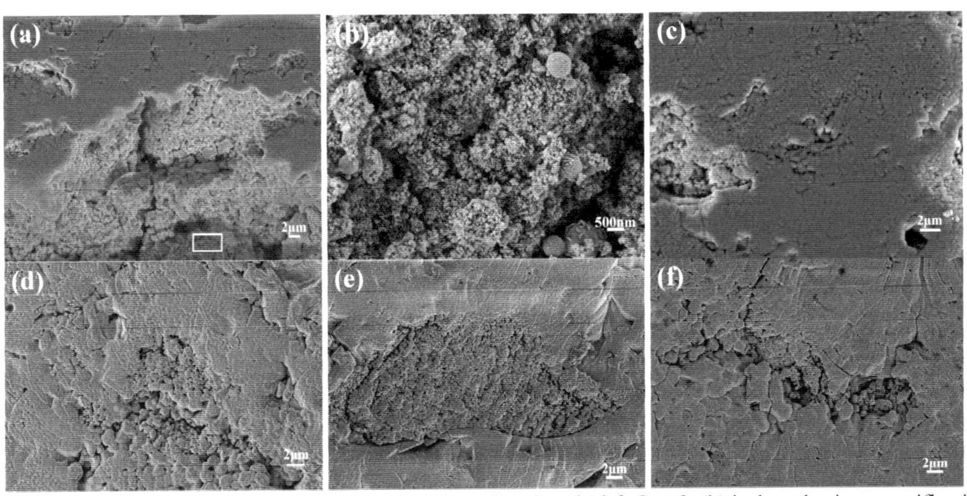

Fig.4 Higher magnifications on the nano zones produced by (a-c) Gun 1 and (d-f) Gun 2. (b) is the selective magnification marked in (a).(The figures are arranged with the increase of arc current)

Fig.5 shows the XRD patterns of the as-sprayed coatings in the selected angle range. The patterns at the low angle region of 26 - 32 ° are displayed in Fig.5a. No monoclinic phase zirconia was observed in the as-sprayed coatings. The patterns at high angle range of 71 -76 ° are displayed in Fig. 5b. It can be seen that the coating is composed of non-transformable tetragonal and tetragonal phase zirconia. In a nanostructured coating with bimodal distribution, the formation of the non-transformable tetragonal phase is promoted by the high quenching rate (10^6-10^7 K^{-1}) of the molten droplets, while the tetragonal phase is associated to the preserved unmelted particles [24]. Among the coatings fabricated with Gun1, no apparent increase of peak intensity or peak shift was observed with the increase of arc current, while among the patterns of the coatings deposited with Gun2, obvious peak shift was found. It is known that the unmelted nano zone is characterized by high porosity inherited from the original agglomerates and, the porous nano zone is favorable for the release of the thermal stress, therefore lower the residual stress. The nano zone content of the coatings engineered with Gun1 were evaluated to be about 37%, 35% and 30% respectively, which were remarkably higher than those of the coatings fabricated by Gun 2 (listed in Table 2). Average grain size of sample coatings calculated by Debye Scherrer formula is also listed in Table2. The average grain size of the coatings engineered with Gun1 were also lower than those of the coatings deposited by Gun 2. Considering the grain size of the original powder, it can be concluded that a larger amount of the nanostructure of original size was preserved in coatings from powder without heat treatment, especially when using Gun 1. And though there are many area with loose packed fine granules can be identified in the coatings engineered with Gun 2, especially with 450A, apparent grain growth occurred during deposition, most of the nano character were lost. Therefore, controlling the arc current is a necessity to obtain fine nano structure as much as possibile when using regular internal powder feeding method.

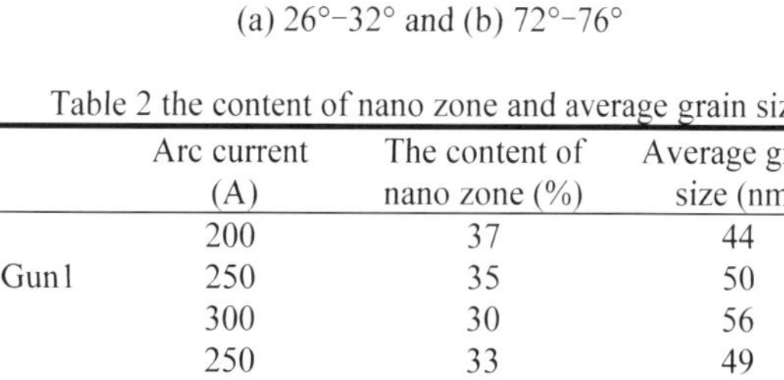

Fig.5 xrd patterns of the coatings at selective angle range of (a) 26°-32° and (b) 72°-76°

Table 2 the content of nano zone and average grain size

	Arc current (A)	The content of nano zone (%)	Average grain size (nm)
	200	37	44
Gun1	250	35	50
	300	30	56
	250	33	49
Gun2	350	25	67
	450	18	120

Fig.6 shows the Weibull plots of Koop indentations under a lower load of 10 gf. Except for the coating fabricated by Gun 1 with the current of 300 A, all of the coatings presented a bimodal distribution of Weibull modulus. It is known that the bimodal distribution of Weibull plots is attributed to the incorporation of unmelted agglomerates and a lower load can exclude the effects of architecture defects and reflect the intrinsic properties of assessed material [25]. Using lower load is intended to exclude the impacts of the structure defects (such as coarser porosity, cracks and splats boundaries) as much as possible and to assess the intrinsic properties of YSZ. The flat areas probed by the indents actually reflected the sole character of the full molten parts. Although the porous unmelted area can be identified in the coating fabricated by Gun 1 with the current of 300 A, as shown in Fig. 3c, only one Weibull modulus is presented. That is to say, the flat area had only one structure character which is quite different from that observed in the coatings fabricated by Gun 2 with similar arc current. Besides, all of the coatings fabricated by Gun 2 showed bimodal distribution of Weibull modulus, which further elucidated that nanostructured coating could be obtained with internal powder feeding method, on the premise of using unsintered agglomerates.

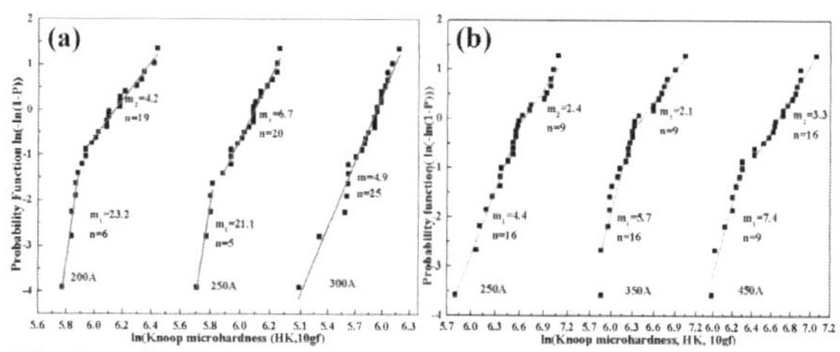

Fig.6 Weibull plots of the coatings engineered with (a) Gun 1 and (b) Gun 2.

Fig.7 shows the Vickers` indentations which performed on the cross-section of nanostructured YSZ coatings under 500gf. Micro-hardness is usually considered to be an index of penetration, deformation, scratching resistance and also a reflection of the splat-to-splat cohesion of thermal spray coating. The hardness of coatings produced by Gun 1 are 303.9 ± 84.6 HV, 371.3 ± 90.7 HV and 449.4 ± 138.4 HV respectively. A higher level of micro hardness was obtained in coatings sprayed by Gun 2, due to the higher melting degree of the agglomerates, which are 408.8 ± 85.2 HV, 548.5 ± 87 HV, 628.6 ± 48.1 HV respectively. The elastic modulus of coatings are displayed in Fig.8. It could be seen that the elastic modulus of the coatings engineered with the tow spray guns were close to each other when the arc currents were similar. It can be seen that using porous unsintered agglomerates can remarkably reduce the elastic modulus of the coatings engineered with internal powder feeding method.

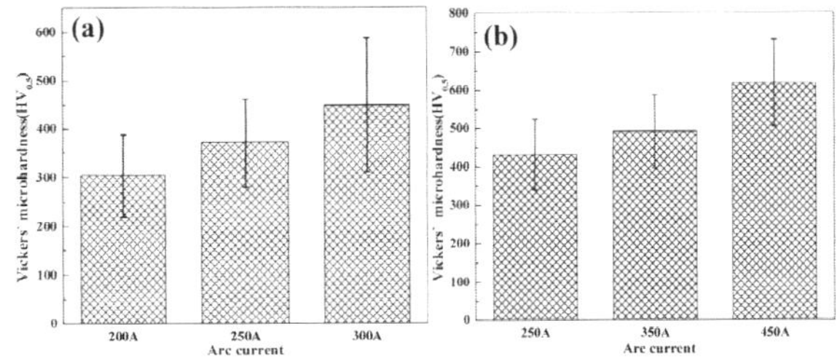

Fig. 7 The microhardness of coatings engineered with (a) Gun 1 and (b) Gun 2.

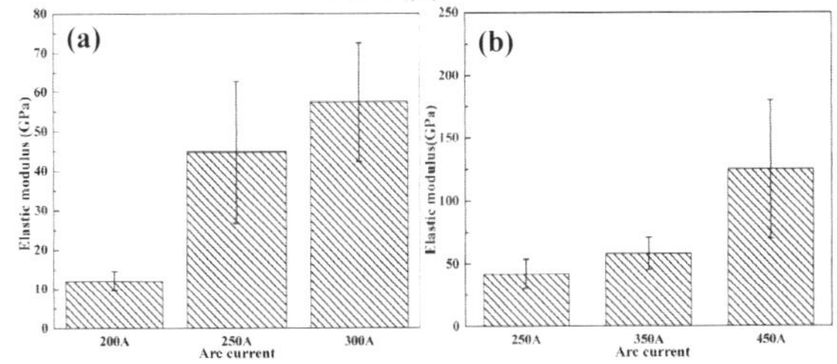

Fig. 8 The elastic modulus of coatings engineered with (a) Gun 1 and (b) Gun 2.

Discussion

The solid spheres of YSZ spray dried powder used in this study are bigger and more porous than those commonly used in APS, and the rate of heat conduction within individual particle would be too slow to ensure a homogeneous heating during their residence in the plasma jet [26], a condition of which is favorable for preserving the original nano character. The heat trajectories of reconstituted powder is a combination effect of plasma jet temperature and feedstock properties. There are comprehensive works on simulation and calculations on thermal and velocity field in the plasma jet space which is helpful in studying the heat trajectories of particles under specified fabrication conditions [27, 28]. Generally the feedstock particles were accelerated by drag force and inertial force. Rajathurai et al. [29] evaluate the resistance behavior of submicron agglomerates stressed by aerodynamic drag forces and Weber number was defined to describe the tensile strength of dry agglomerates

$$W_c = \frac{\rho v^2}{\frac{1-e}{e}\frac{32\pi}{a_g}\frac{h}{d_s}} \qquad \text{Eq.9}$$

Where e is the porosity of agglomerates, ρv^2 as the dynamic pressure of surrounding fluid, d_s is the diameter of average specific surface. It could analogs the disintegration behavior of agglomerates in high velocity shock wave. However, the plasma jet is characterized by severe thermal gradient and cold air entrainment. The conditions are more complicated during the flash flight of agglomerates in the plasma jet. When being fed outwardly, most of the agglomerates were observed to deviate from the jet axis forming a corpuscular cloud dwelling a short distance from the nozzle exit. There was hardly any particles flattened on the substrate at a stand-off distance of 100 mm. Among the forces exerted on an individual particle (the drag force, the Basset force and the thermophoresis force, etc.), the thermophoresis force by temperature gradient would cause the repercussion of agglomerated particles (especially smaller particles) at the edge of plasma plume or bias from the high temperature center. As the plasma jet came out from the nozzle, intense entrainment of cold air caused the formation and movement of eddies at the periphery of plasma jet through laminar shearing effects, as a result there would be vast temperature and velocity decay in a short distance along the jet direction [30-32]. The collisions and severe temperature gradient would accelerate the disintegration of agglomerates, the fine tangles with high velocity and low melting degree could hardly be flattened on the substrate. After assembling a short tube of the same diameter as the anode nozzle, the traveling length of plasma jet inside the gun barrel is comparatively lengthened. The additional tube plays a role of shield between the jet flow and ambient air, which can not only restrain the ambient air diffusing into the working gas, but also weaken the turbulent. The mechanism is to some extent, similar to the gas shell generated to protect plasma generator or arc torch for cutting and welding [33]. The high speed shielding gas worked as a wall to separate the plasma jet and ambient air, and the friction with the boundary layer of jet flow could remarkably improve the jet velocity, as proposed by Cheng et al. [34]. When the porous agglomerates were fed into the tube along radial direction, they would be confined into the jet flow to be heated with higher probabilities. What is more, the particle-to-particle interactions could be enhanced as well as the reflections against the tube wall, in consequence, the particles can obtain higher velocity, which can be evidenced by the sever deformation of the unmelted part along the direction that parallel to the substrate surface. However, the plasma jet properties and particles behavior depend on further detailed measurements. While the nanostructured coating can also be obtained by spraying unsintered agglomerates with internal powder feeding method. Generally, the sintered agglomerated powder tended to be fully melted when using internal powder feeding method. While the nano structure could also be produced by using unsintered agglomerates. According to the theoretical study of Hurevich [7], the porosity of the powder would result in porous coating structure and a higher temperature gradient between its inside and outside, when being subjected to the plasma plume. This is favorable in preserving nanno structure of original powder. Thermophresis caused by the irregular particle shape and low apparent density might also a contribution in preserving the unmelted nanostructure. Though the study of Ercan et al. [35] on spraying plasma spheridized and hollow powders show no difference in thermal conductivity and relaxation of stress comparing to coatings from agglomerated and sintered samples which could be attributed the densification of spherical shell after plasma treatment. In our work, irregular cavities were developed inside the particle and the shell was composed of agglomerated fine precursors. The bimodal structured coating provides certain tolerance to thermal stress, which can give explanations to the differences between the coatings engineered with Gun 1 and 2. However, it needs to point out that though the coating engineered with Gun 2 showed two type of morphology character and presented bimodal distribution of Weibull modulus, the size of the granules embedded into the full molten part was evaluated to be larger than 100 nm when the arc current beyond 350 A.

Conclusion

The unsintered agglomerates were successfully deposited using internal powder feeding method. By using a special gun configuration and porous feedstock powder, the content of nano area were 37%, 35% and 30% respectively. The sever deformation along the direction of substrate surface elucidated a higher deposition efficiency. The average grain size of the as-sprayed coatings were 44 nm, 50 nm and 56 nm respectively. By contrast the unmelted granules fabricated by Gun 2 grew remarkably during deposition, especially when the arc current beyond 350 A. Due to the higher heating efficiency of internal powder feeding method, the coatings had higher level of microhardness, while the elastic modulus are similar to those of the coatings produced by Gun 1 with similar arc currents.

Reference

[1] C. Peter, A. Weber, B. Butz, D. Gerthsen and E. I. Tiffée, Grain-size effects in YSZ thin-film electrolytes, J. Am. Ceram. Soc.73 (2009) 2017-2024.

[2] J. F. Li, H. Liao, X. Y. Wang, B. Normand, V. Ji, C. X. Ding, C. Coddet, Improvement in wear resistance of plasma sprayed yttria stabilized zirconia coating using nanostructured powder, Tribo. Int. 37 (2004) 77-84.

[3] A. K. Keshri1and A. Agarwal, Plasma processing of nanomaterials for functional applications-A review, Nanosci. Nanotechn. Lett. 4 (2012) 228-250.

[4] S.T. Aruna, N. Balaji, J. Shedthi, V. K. W. Grips, Effect of critical plasma spray parameters on the microstructure, microhardness and wear and corrosion resistance of plasma sprayed alumina coatings, Surf. Coat. Technol. 208 (2012) 92-100.

[5] M. Gell, E.H. Jordan, Y. H. Sohn, D. Goberman, L. Shaw, T. D. Xiao, Development and implementation of plasma sprayed nanostructured ceramic coatings, Surf. Coat. Technol. 146-147 (2001) 48-54.

[6] S. Mantry, B. B. Jha, A. Mandal, D. K. Mishrac, B. K. Mishra, M. Chakraborty, Influence of in-flight particle state diagnostics on properties of plasma sprayed YSZ-CeO_2 nanocomposite coatings, Int. J Smart Nano Mater. 5 (2014) 207-216.

[7] C. Lamuta, G. D. Girolamo, L. Pagnott, Microstructural, mechanical and tribological properties of nanostructured YSZ coatings produced with different APS process parameters, Ceram. Int. 41 (2015) 8904-8914.

[8] M. Aliofkhazraei, N. Ali, PVD technology in fabrication of micro- and nanostructured coatings, Compr. Mater. Process. 7(2014) 49-84.

[9] B. Liang, C.X. Ding, H. Liao, C. Christian, Study on structural evolution of nanostructured 3 mol% yttria stabilized zirconia coatings during low temperature ageing, J. Euro. Ceram. Soc. 29 (2009) 2267-2273.

[10] Y. Bai, L. Zhao, Y. M. Qu, Q. Q. Fu, Y. Wang, K. Liu, J. J. Tang, B. Q. Li, Z. H. Han, Particle in-flight behavior and its influence on the microstructure and properties of supersonic atmospheric plasma sprayed nanostructured thermal barrier coatings, J. Alloys Compd. 644 (2015) 873-882.

[11] F. B. Ettouil, O. Mazhorova, B. Pateyron, H. Ageorges, M. E. Ganaoui, P. Fauchais, Predicting dynamic and thermal histories of agglomerated particles injected within a d.c. plasma jet, Surf. Coat. Technol. 202 (2008) 4491- 4495.

[12] L.H. Liang, H. Wei, X. C. Chang, W. Xu, X. N. Li, Y. G. Wei, Enhanced insulation temperature and the reduced thermal conductivity of nanostructured ceramic coating systems, Int. J. Heat Mass Transfer. 65 (2013) 219-224.

[13] H. Jamali, R. Mozafarinia, R. S. Razavi, R. Ahmadi-Pidani, Comparison of thermal shock resistances of plasma-sprayed nanostructured and conventional yttria stabilized zirconia thermal barrier coatings, Ceram. Int. 38 (2012) 6705-6712.

[14] E. P. Song, J. Ahn, S. Lee, N. J. Kim, Microstructure and wear resistance of nanostructured Al_2O_3–8wt.%TiO_2 coatings plasma-sprayed with nanopowders, Surf. Coat. Technol. 201 (2006) 1309-1315.

[15] C. Liu, L. YU, X. Jiang. Particle growth mechanism of nanocrystalline zirconia powder during high temperature heat treatment. Trans. Nonferrous Met. SOC. China 2007, (17):1022-1027

[16] Z. Wang, C. Zhou, H. Xu, S. Gong. Effect of thermal treatment on the grain growth of nanostructured YSZ thermal barrier coating prepared by air plasma spraying. Chinese Journal of Aeronautics, 2004, 17(2): 119-23.

[17] Y. Bai, J.J. Tang, Y.M. Qu, S.Q. Ma, C.H. Ding, J.F. Yang, L. Yu, Z.H. Han. Influence of original powders on the microstructure and properties of thermal barrier coatings deposited by supersonic atmospheric plasma spraying, Part I: Microstructure Ceramics International, 2013,39(5):5113-24

[18] P. Fogarassy, D. Gerday, A. Lodini. Agglomerated nanostructured particles disintegration during the plasma thermal spraying process. Mechanics Research Communications, 2005, (32):221-39

[19] V. Hurevich, I. Smurov, L. Pawlowski. Theoretical study of the powder behavior of porous particles in a flame during plasma spraying. Surf. Coat. Technol.2002, (151-152):370-6

[20] G. Di, Girolamo, F. Marra, C. Blasi, E. Serra, T. Valente. Microstructure, mechanical properties and thermal shock resistance of plasma sprayed nanostructured zirconia coatings, Ceramics International, 2011,37:2711-17

[21] Y. Gao, J. Gao, D. Yang. Equiaxed and Porous Thermal barrier coatings deposited by atmospheric plasma spray using a nanoparticles powder. Advanced engineering materials, 2014, 16(4): 406-12.

[22] H.P. Klug, L.E. Alexande. X-ray diffraction procedures. New York: Wiley; 1974.pp. 618-620.

[23] P. Fogarassy, D. Gerday and A. Lodini, Agglomerated nanostructured particles disintegration during the plasma thermal spraying process, Mech. Res. Commun., 32 [2] 221-39 (2005).

[24] G. D. Girolamo, F. Marra, C. Blasi, E. Serra and T. Valente, Microstructure, mechanical properties and thermal shock resistance of plasma sprayed nanostructured zirconia coatings, Ceram. Int., 37 [27] 11-17 (2011).

[25] R. S. Lima, A. Kucuk and C. C. Berndt, Bimodal distribution of mechanical properties on plasma sprayed nanostructured partially stabilized zirconia, Mater. Sci. Eng., A 327 [2] 224-232 (2002).

[26] Warren BE. X-ray diffraction. New York: Dover; 1990. 251-54.

[27] Gawne, D.T., Zhang, T., Liu, B., Computational analysis of the influence of a substrate, solid shield and gas shroud on the flow field of a plasma jet. Surface and Coatings Technology, 2002,153, 138-47.

[28] A.M. Rajathurai, P. Roth. On the disintegration of submicron agglomerates. Advances in Measurement and Control of Colloidal Processes, 1991, Pages 85-91
[29] B. Ercan, K. J. Bowman, R. W. Trice, H. Wang, W. Porter. Effect of initial powder morphology on thermal and mechanical properties of stand-alone plasma-sprayed 7 wt.% Y_2O_3-ZrO_2 coatings. Materials Science and Engineering A, 2006, (435-436):212-220
[30] E. Pfender, J. Fincke and R. Spores, Entrainment of cold gas into thermal plasma jets, Plasma Chem. Plasma P., 1 [4] 529-543 (1991).
[31] C. H. Chang, E. Pfender, Non-equilibrium modeling of low-pressure argon plasma jets: Part I Turbulent flow, Plasma Chem. Plasma P., 10 [3] 493-500 (1990).
[32] R. Spores, E. Pfender, Flow structure of a turbulent thermal plasma jet, Surf. Coat. Tech., 37 [3] 251-270 (1989).
[33] H. P. Li, Q. Y. Nie, A. Yang, Z. B. Wang and C. Y. Bao, An atmospheric cold plasma jet with a good uniformity, robust stability, and high intensity over a large area, IEEE T. Plasma Sci., 42 [10] 2470-2471 (2014).
[34] K. Cheng, X. Chen, H. X. Wang and W. X Pan, Modeling study of shrouding gas effects on a laminar argon plasma jet impinging upon a flat substrate in air surroundings, Thin Solid Films, 506-507 724-728 (2006).
[35] B. Ercan, K. J. Bowman, R. W. Trice, H. Wang, W. Porter. Effect of initial powder morphology on thermal and mechanical properties of stand-alone plasma-sprayed 7 wt.% Y_2O_3-ZrO_2 coatings. Materials Science and Engineering A, 2006, (435-436):212-220.

Spectroscopic and Electrical diagnostic of the plasma jets characteristic of a new plasma torch for LPPS or VLPPS

C.Q.Sun, Y.Gao and D.M.Yang*
Thermal Spraying Center, Dalian Maritime University, Dalian, P.R. China
**E-mail: gaoyang@dlmu.edu.cn*

Abstract

A novel plasma torch with converging-diverging anode and nozzle used in low pressure plasma spray (LPPS) and very low pressure plasma spray (VLPPS) has been developed, which is expected to improve the quality of the coating. The emission line intensities from Ar I neutral species were carried out using emission spectroscope diagnostic, and electron temperatures were obtained using Boltzmann plots' method. The plasma torch efficiency and electrical characteristic were analyzed. The results indicated that an increase in input power considerably increased the electron temperature, while the increase of detection distance decreased the electron temperature. The torch efficiency increased a little bit with the increase of the current intensity, and the gas composition has a great effect on the current-voltage characteristic.

Introduction

Plasma spraying is a process that the metallic and ceramic powders heated and accelerated in the plasma jet impinges on the substrate to form the coatings. The quality of sprayed coating is dependent on the fundamental plasma characteristics, such as ion and neutral atom temperatures, electron temperature and density. In order to optimize and control the spray processes for desirable coatings, it is necessary to understand the plasma jet properties (Ref 1). Plasma spraying is categorized as either atmosphere plasma spraying (APS) or low pressure plasma spraying (LPPS). In APS, the powder material reacts with oxygen, and the bond strength between the substrate and coating is somewhat low (Ref 2). But LPPS alleviates these drawbacks by reducing the interaction between the plasma jet and the environment. So LPPS has attracted extensively research interest.

Inside the plasma nozzle, the arc root attachment moves continuously along the anode wall, due to the influence to two opposite forces: the drag force and the electromagnetic one. The lengthening of the connecting column favours the arc voltage increase within it. When this voltage becomes higher than a threshold value, a short circuit occurs creating a new arc root and the voltage drops. The arc voltage fluctuations are linked to the plasma gas composition, arc current, plasma torch geometry and the injection gas mode (Ref 3-5). Such phenomena enhance the plasma jet instabilities and the cold-gas entrainment. The coating quality is partly dependent on the arc and jet instabilities of the plasma torch. Different forms of instabilities have been observed with different effects on the coating quality (Ref 6). The large time-dependent effects of the arc root fluctuations on the particle temperature and velocity are experimentally demonstrated. The existence of quiet periods suggests that particles injected at some specific moments in the plasma are not heated sufficiently to be detected (Ref 7). Most of these studies focused on atmospheric pressure plasma arc. Fewer studies were carried out for the voltage fluctuation of LPPS.

On the other hand, the plasma jet characteristic investigations have been performed by means of the optical emission spectroscopy, which is non-intrusive and give the information of the plasma species. The electron temperature can be measured based on the Boltzmann distribution of the excited Ar- I species (Ref 8). Comparison of the temperature values derived from absolute emission coefficients, Stark broadening, with the one from enthalpy probe measurements was carried out. Deviations from LTE were found existing in various regimes of plasma spray jets due to diffusion of high-energy electrons, high velocities in the core of plasma jets and entrainment of cold gas (Ref 9).

Compared to APS, the characteristic of the low pressure plasma jet is very different. In LPPS conditions, the plasma plume was enlarged in length and diameter. Sabsabi et al. (Ref

10) investigated the temperature distributions in a low pressure Ar-H_2 DC plasma jet. They confirm that the local thermal equilibrium (LTE) conditions prevail throughout most of the temperature field. Nityalendra et al. (Ref 11) analyzed the electron temperature and density on a plasma plume under various pressure conditions using emission spectroscopy, and dominant physical effects causing departures from LTE were discussed. Qing-Yu Chen et al. (Ref 12) measured the electron temperature and electron density in the VLPPS jets based on the relative intensity of two spectral lines, H_α and H_β, respectively. They found the electron number density in the plasma jet is higher than the critical value, and the assumption of partial local thermodynamic equilibrium is satisfied.

In this paper, the properties of a new low pressure plasma torch jet were investigated. The first aim of this research work is to discuss the current-voltage characteristic of plasma arc with argon or argon-hydrogen plasma gas, and the effect of the hydrogen volume percentage was analyzed. The emission spectroscopic technique is employed to record and analyze the plasma spectra. The central axial electron temperatures in the plasma jet were determined by using the Boltzmann plot method.

Experimental setup and methods

Very low pressure plasma spray System

A very low pressure plasma spraying equipment was developed by the Thermal Spraying Center of the Dalian Maritime University, which was composed of plasma torch with the nozzle supplied by the working gases, the three dimensional moving device supporting the plasma gun, vacuum chamber and powder feeders and so on (Ref 13). The chamber pressure can be regulated from 30 to 10000Pa by using large capability mechanical booster pumps with a gas flow of 200SLPM. Fig.1 shows the schematic sketh of the torch designed with separated anode and nozzle. Inside diameter of the anode is 6mm and a length of 20mm, and the diameter of the single tugnsten cathode is 9mm, and the anode exit is connected with a trumpet-shaped nozzle and the inside diameter of the nozzle is 8mm. The feed powders are injected inside the nozzle through two little holes with a diameter of ø1.5mm, and the powder portal is located 14mm in front of the cathode, and the distance from the exit nozzle is 35mm, so the powder is heated in high-density plasma inside the nozzle.

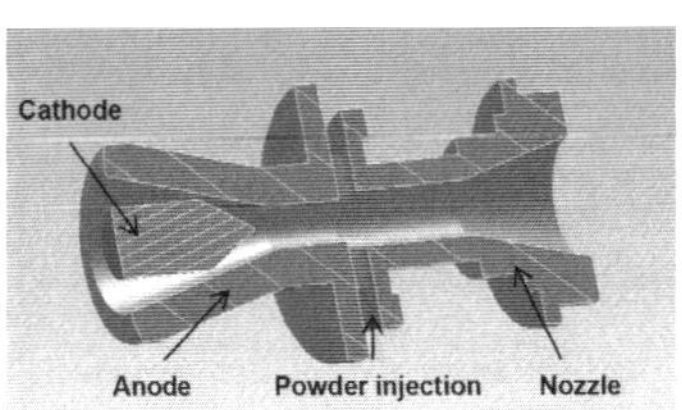

Fig 1: Schematic CAD design of the nozzle

Figure 2 shows the photos of the plasma jets at different current intensity under LPPS (1000Pa, Ar 40L/min, H_2 5L/min, 700A). The reduced pressure compared to atmospheric conditions enlarges the plasma plume length. The structure of the plasma jets exhibit the fetures of an under-expanded jet.

It can be seen from figure 2 that there is a brighter and dense plume exiting at the nozzle exit, which is followed by an expansion wave as evidenced by the increase jet radius, and the expansion waves are then reflected on the jet boundary and the flow is then redirected towards the axis and forms the first compression zone, which can be characterized by a local increase luminosity. The increased luminosity near compression zone reveals that the supersonic flow has transferred a substantial amount of kinetic energy into thermal energy. Subsequent weak expansions/compressions occur until the local jet pressure equilibrates with the chamber pressure. At 700 A, it can be seen that the position z=25 mm corresponds to the expansion zone of the jet in Fig.2b, where a pressure and temperature minimum should occur, with a velocity maximum. Accordingly, the location z=45mm corresponds to the compression zone of the jet in Fig.2b, where a maximum pressure and temperature should occur with a minimum velocity. At 400A, the length of the jet is much smaller, an expansion starts at z=15mm and then the compression zone occurs at z=30mm. The axial position of the expansion / compression zone in the plasma jet is closer to the nozzle. The detection parameters were listed in table 1.

Table 1: Experiment condition

Spraying parameters	
Chamber pressure (Pa)	3000
Current intensity (A)	400, 700
Ar flow rate (L/min)	40
H_2 flow rate (L/min)	8
Detection distance (mm)	10-250

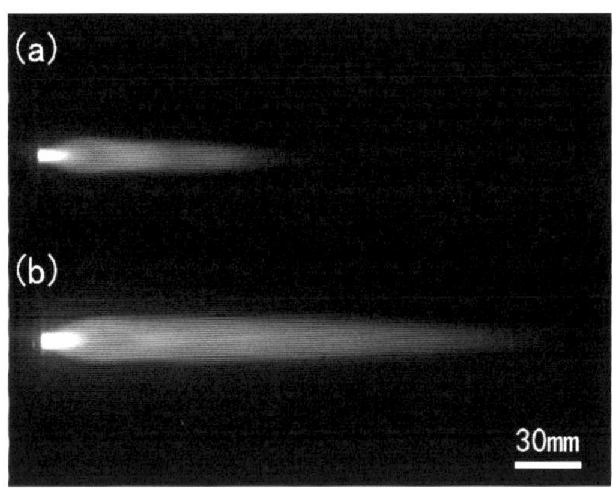

Fig 2: Photos of plasma jet under LPPS: (a) 400A; (b) 700A;

Optical Setup for Emission Spectroscopy

In the experiments, Optical emission spectroscopy (OES, AvaSpec-2048-4-USB2, multichannel spectroscopy, Avantes, The Netherlands) was used to measure spectrum of the plasma jets. The measurement was performed along the flow axis and perpendicular to the axis of the plasma jet. The signals of emission spectra from the plasma were transferred by optical fibers coupled to the spectrometer equipped with a 1200 lines/mm diffraction grating (type Step index Mutimode FC4-UV/VIS -200-2-ME) through a matching fiber adapter. Light was collected and transferred to the entrance slit (10μm) by a multiplexed fiber bundle. The optical fibers (14 fibers 200 μm core, 12 light-fibers, 2 x 1 read fiber, N.A.= 0.22. Standard 2m length, splitting point in the middle) are protected by a silicon inner tube and a flexible stainless steel (BX, O.D. 6.0 mm) or chrome plated brass (ME, 5.0 mm) outer jacket. The jacketing also gives stress relieve. The spectrometer was scanned between 200 and 1100nm by using a 0.2nm step size. A 2048×784 pixel CCD (CCD type SONY2048) collected the signal, and the CCD signals were sent to the computer using the Spectra-AvaSoft for data acquisition. Before performing the experiments, the spectral calibration of the spectrometer system including the CCD chip spectral sensitivity was carried out. The peaks for the Mercury-Argon calibration source (AvaLight-CAL) were by default loaded in AvaSoft. The experimental setup for spectroscopic measurements of plasma temperature is schematically shown in Fig.3. The data of the spectrum was analyzed by using the PlasusSpecline to identify the atomic lines. And the spectral information of the PlasusSpecline comes from the US national standard and technology institute. Figure 4 shows the typical spectral emission at 200A with the pure Ar.

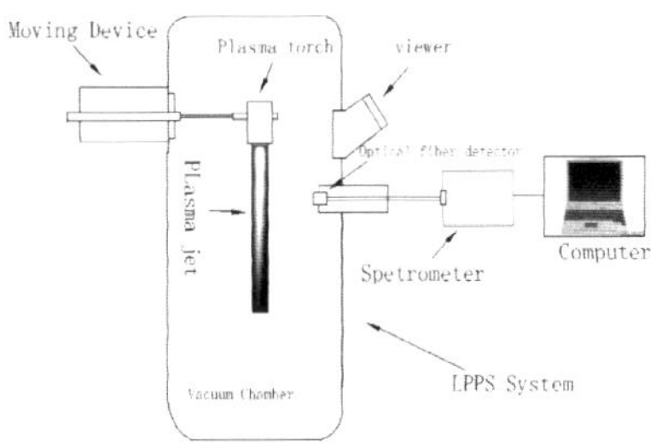

Fig 3 Setup of the emission spectroscopy diagnostics

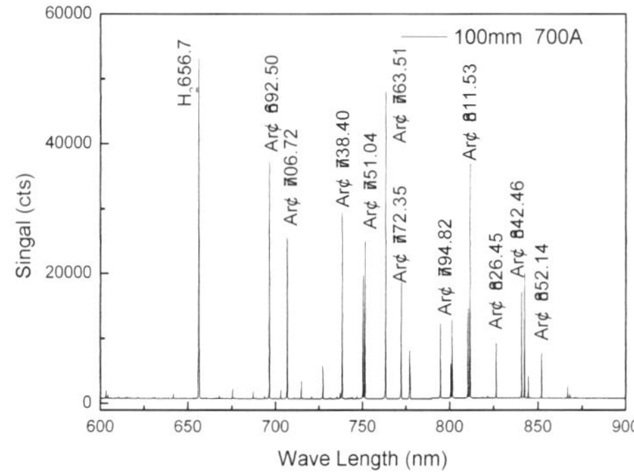

Fig 4 Diagnostic of emission spectra for plasma jet under LPPS

Results and Discussions

Emission spectroscopic measurement

The distribution of various species of ions, atoms, and molecules among all the possible energy states in the plasma is determined in terms of electron temperature and electron density. In general, there are many ways to measure the electron temperature; here the method based on the Boltzmann plot is used.

The intensity of an atomic is proportional to the population of the upper level, to the transition probability and to the energy of the photons,

$$\varepsilon_L = \int \varepsilon_\lambda d\lambda = \frac{1}{4\pi} A_{r,t}^{r,s} n_{r,s} h\nu$$

(Eq1)

Where $A_{r,t}^{r,s}$ is the corresponding transition probability, $n_{r,s}$ is the population density of the atoms, h is the Planck's constant, $h\nu$ is the energy of emitted photons. The integration is performed over the width of the line.

Properly corrected emission intensities can be used to determine the temperatures based on the Boltzmann distribution given by:

$$n_{r,s} = n_r \frac{g_{r,s}}{Z_r} \exp\left(-\frac{E_{r,s}}{k_B T}\right)$$

(Eq2)

Combining Eq1 and Eq2, we can obtain:

$$\varepsilon_L = \frac{1}{4\pi} A_{r,t}^{r,s} n_r \frac{g_{r,s}}{Z_r} \exp\left(-\frac{E_{r,s}}{k_B T}\right) h\nu$$

(Eq3)

Where nr is the population of the chemical species r, Z_r is the partition function of the chemical species r, $g_{r,s}$ is the statistical weight of the upper excited state s of the chemical species r, $E_{r,s}$ is the excitation energy, k_B is the Boltzmann constant, and T is the absolute temperature.

At a fixed axial position from the torch nozzle exit, the optical setup spatially integrates the monochromatic light emission (ε_L), denoted by I(x), so what is observed is the intensity of light integrated along the line of sight. According to the Abel inversion for optically thin plasma, the absolute intensity of the emission line can be written as:

$$I_{t,s} = \frac{l}{4\pi} A_{t,s} n_0 \frac{g_s}{Z} \exp(-\frac{E_s}{k_B T}) hv$$

(Eq4)

Where l is the emission source depth, $I_{t,s}$ is the absolute intensity of the emission line of s to t. Eq4 taking the natural logarithm yields:

$$\ln \frac{I_{t,s}\lambda}{g_s A_{t,s}} = \ln \frac{lhn_0}{4\pi Z} - \frac{E_s}{k_B T}$$

(Eq5)

All other terms in Eq5 are constants except for ($-E_s/k_B T$), and this plot is often referred to as the Boltzmann plot. Eq5 can be simplified as:

$$\ln \frac{I\lambda}{gA} = -\frac{5040}{T} E + C_1$$

(Eq6)

Where C_1 is constant, a plot of the left-hand side versus the excitation energy, E_s, would yield a linear relationship with a slope of (-5040/T) (Ref 14-15).

Due to the strong emission of Ar- Ⅰ species in the visible region ranging from 675 to 820 nm, the spectral emission from the excited Ar- Ⅰ species was selected and the transition probability, statistical weight, excitation energy of the selected Ar- Ⅰ species are listed in table 2. Based on the Boltzmann method, the electron temperature can be obtained (Ref 16). The experimental results on the plasma temperature field based on the full Boltzmann spectral analysis are described and discussed. The measured intensities and the background intensities are used to calculate the absolute intensities as follows:

$$I = I_m - I_b \qquad \text{(Eq7)}$$

Where I is the absolute intensity, I_m is the measured intensity, I_b is the background intensity. Figure 5 shows the Boltzmann plots at a current intensity 700A under detection distances. The resulting Boltzmann plots are of very high quality with the slope of the linear relationship through the data producing an unambiguous plasma temperature. The two examples of Boltzmann plots obtained at different distances indicate that the good capability to measuring temperatures over a wide range.

Table2: Parameters of Ar- Ⅰ spectrum

λ /nm	$A/10^6 s^{-1}$	E_μ/eV	g_μ
675.284	1.93	14.74	5
687.129	2.78	14.71	3
714.704	6.25	13.28	3
727.293	1.83	13.33	3
750.387	4.45	13.48	1
772.35	11.7	13.32	3

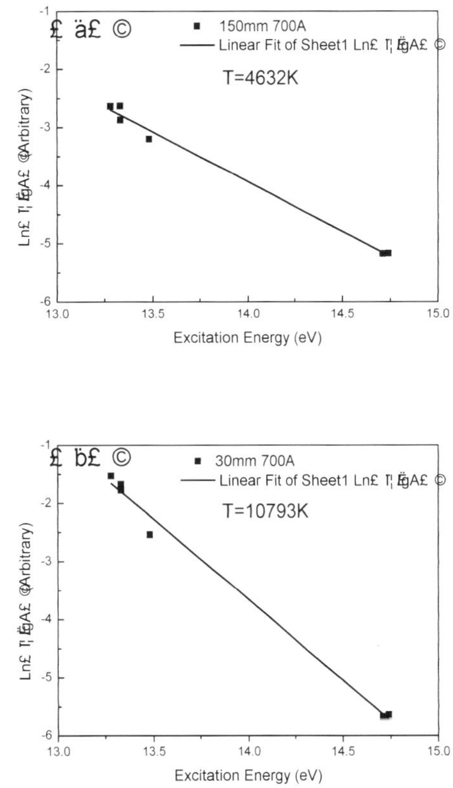

Fig 5: two examples of Boltzmann plots at current intensity of 700A: (a) 150mm; (b) 30mm

Figure 6 shows the axial variation of electron temperature in the centerline of the plasma jet. At current intensity of 700A, the temperature decreases rapidly along the axis from the nozzle exit with increasing downstream distance and the temperature closest to nozzle exit situated near z=5 mm is 12150 K, while, at z=120 mm downstream regions of the plasma jet, the temperature is 7231 K. At the axis position ranged from z=25 mm to z=65 mm, the electron temperatures are almost the same (about 10700K), and then, a gradually decay of temperature with increasing downstream distance was shown in Fig.6. When the current intensity decreased to 400A, the electron temperature is much lower, but the axial variation of electron temperatures is similar to the one at current intensity of 700A. For 400A operating condition, figure 7 also shows the axial temperature slowly decay rate at the axis position from 15 mm to 45 mm due to the change of locations of the expansion and compression zones.

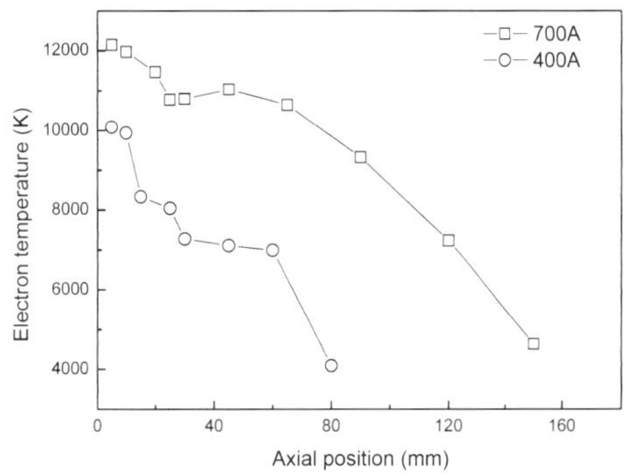

Fig 6: Evolution of electron temperature with detection under different current intensity

The more distant axial positions are from the nozzle exit, the less intense is the collision frequency of the electron, atom, and iron, and the particle population of the excited state decreases, resulting in the decrease of energy density. The engulfment of cold gas eddies into the plasma jet, accelerated by the surrounding plasma jet, is gradually dissolved by convection and molecular diffusion, and consequently, the temperature is substantially reduced. The current intensity has a great influence on the electron temperature axial variation in the plasma jet. The current intensity increase enhances the energy density of the plasma jet, which results in the higher population densities of excited states in the plasma jet due to the more collision among high-energy electrons, atoms and ions. So the higher is the current intensity, the higher the electron temperature in the plasma jet centerline is.

Thermal efficiency and current-voltage characteristics

It is essential that the entire system operate at steady-state conditions prior to, and during, the time when data are taken. Therefore the cooling system must be operating at steady-state conditions prior to arc initiation. Steady-state operation assumes that the coolant temperature, arc current and voltage, and gas and coolant flow rates are steady and not changing with time. Any changes which occur between pre- and post-test values may indicate a buildup of deposits in the cooling system or some other difficulty which could alter the results of the energy balance.

The low pressure DC plasma plume was operated at 3000Pa of chamber pressure and the flow of argon was kept at 40 liters per minute. The total coolant losses were stated as follows:

$$Q = m_w c_p (\Delta T_f - \Delta T_o) \quad \text{(Eq8)}$$

Where m_w is the water mass flow rate, c_p is the water specific heat, ΔT_f is the temperature rise in water with arc operating, ΔT_o is the temperature rise in coolant without arc operating. The thermal efficiency of a plasma jet is defined as the percentage of input power which is contained in the plasma as it leaves the unit,

$$\eta = \frac{IU - m_w c_p (\Delta T_f - \Delta T_o)}{IU} \quad \text{(Eq9)}$$

Where I is the current intensity, U is the arc voltage.

Figure 7 shows effect of the current intensity on efficiency of plasma torch. Some of the factors which can affect efficiency are basic plasma jet design, the size of mixing chambers, anodes and nozzles which are cooled, the amount of secondary gases which are added, the operating conditions.

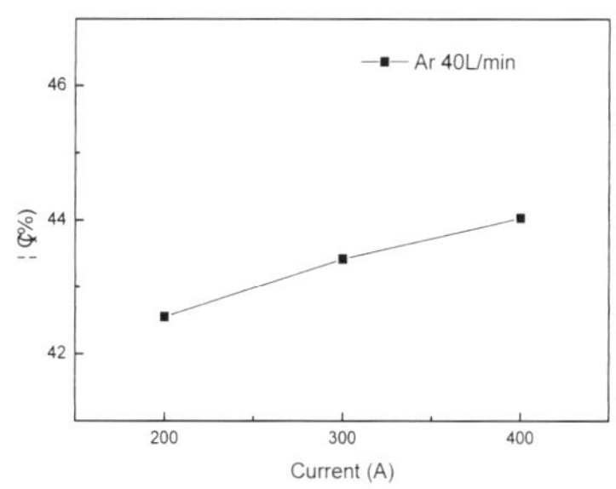

Fig.7 Effect of the current on efficiency of plasma torch

The thermal efficiency of the plasma torch was kept at about 43%, and the most quantities of energy in the plasma jet were transmitted to the cooled walls. With the increase of current intensity, the efficiency increases a little bit. The increase of current improves the energy of the plasma plume, which result in enhancing the arc self-magnetic compression effect, so the energy diffusion of the plasma jet decreases. At the same time, the thermal conductivity of the plasma increases with the increase of the plasma temperature resulted from the increase of current intensity, so the heat exchange with the coolant strengthened. Based on the above reasons, the efficiency was kept about the same.

One of the most important electrical parameters of the plasma arc is its current-voltage characteristics, which depend on the plasma spray parameters, the plasma gas mixture composition, plasma torch configuration and the input power source. The arc voltage was measured using digital multi-meter, and the measurement accuracy is approaching up to 0.1V. During the DC arc plasma spray, the arc root lifetime at the anode wall is limited and new arc roots are continuously created according to restrike, takeover or mixed modes, which results in voltage fluctuations at frequencies in the few thousands hertz range. So the measurement value of arc voltage was actually an average value. Figure 8 shows the effect of the plasma gas composition on the voltage-current characteristic.

The argon arc exhibits a "rising" voltage-current characteristic. For the argon-hydrogen arc, with the increase of the arc current, the arc has a "dropping" voltage-current characteristic in low arc currents but a "rising" voltage-current characteristic in high arc currents. The arc current is about 200A when the arc voltage reaches the minimum. But for argon arc, the inflection point was not observed in fig.8. And at very low arc current, the minimum of the arc voltage could occur.

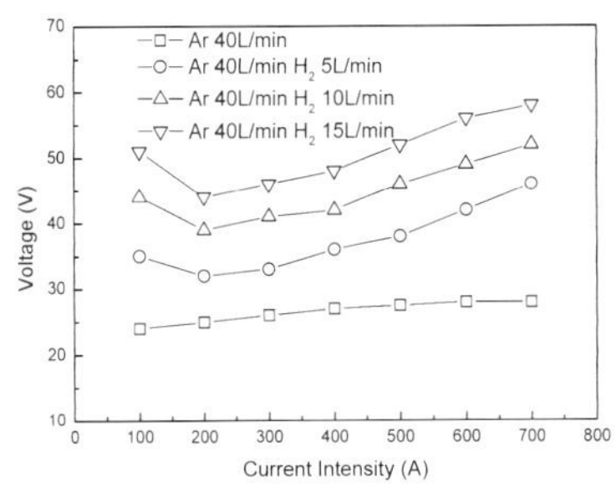

Fig.8 The voltage-current characteristic of the plasma torch

The electric characteristics of the arcs relate to their thermal conductivity. Argon is characterized by relatively low thermal conductivity (0.6W/m • K at 10,000K), while hydrogen has a high thermal conductivity (3.7 W/m • K at 10,000K). So the thermal conductivity of hydrogen is much higher than that of argon. For the argon-hydrogen plasma arc, the recombination effect makes the hydrogen atoms congregate at the fringe of the arc column (Ref 17), which enhances the heat transmission from the arc to the nozzle wall. The hydrogen percentage increase greatly promotes the enhancement of the arc voltage.

The relation between the arc voltage and the current intensity is linked to the mechanisms through which the arc column gains and losses energy. The current-voltage characteristics of a free burning arc generally exhibit a falling tendency, whereas an increasing trend may appear when the arc is crossed by high current intensity or confined by the anode wall. The change trend of current-voltage characteristic can be explained by the differential Ohm's law. The "rising" or "dropping" of the arc voltage depends on the relative speed of the current density and electric conductivity increase. In the low current intensity, the current increase leads to the rapid increase of the electric conductivity, and the increase speed of the electric conductivity is higher than that of the current density. So the arc exhibits a "falling" current-voltage characteristic. With the further increase of the current intensity, the temperature of the arc column does not increase, so the electric conductivity was kept about the same. In this condition, the increase speed of the electric density is higher, which results in a "rising" current-voltage characteristic.

Conclusions

A new plasma torch for LPPS/VLPPS was developed. The characteristic of the plasma jet was investigated by using optical emission spectroscopic method. The central axis electron temperatures were determined using the Boltzmann plot method. The thermal efficiency of the plasma torch was calculated by the energy balance technique. The arc current-voltage characteristic was analyzed. Three important points was stated as follows:

1. The electron temperature was obtained from Ar I neutral species. With the increase of detection distance, the central axis electron temperature drops significantly at the location close to nozzle exit. Successive compression/expansion waves in the supersonic plasma jet have a great influence on the axial variation of the electron temperature. The increase of the current intensity can improve the electron temperature.

2. The thermal efficiency increases slightly with increasing power. The efficiency was maintained at about 42%, and the most energy in the plasma jet was transmitted to the cooling water.

3. Hydrogen can improve the arc voltage. For argon-hydrogen plasma gas, the voltage firstly drops to a minimum and then continuously increases, and the inflection point occurs at the current intensity of 200A. Argon arc exhibits a "rising" voltage-current characteristic in the measuring range.

.

Acknowledgments

This work was supported by the National Natural Science Foundation of China General Project Grant (No.51172033). The authors are grateful for their supports.

Reference

1. E.Pfender, Thermal Plasma Technology: Where Do We Stand and Where Are We Going [J]? *Plasma Chemistry and Plasma Process*, 1999, 19(1), p: 1-31
2. Lech Pawlowski. The Science and Engineering of Thermal Spray Coatigs[M]. France: John Wiley & Sons, 2008: 543-585
3. A. B. Murphy, Demixing in free-burning arcs[J]. *PHYSICAL REVIEW E*, 1997, 55 (6), p: 7473-7494
4. E.Nogues, M.Vardelle, P.Fauchais, P.Granger, Arc voltage fluctuations, comparison between two plasma torch types[J]. *Surface & Coatings Technology*, 2008, 202, p: 4387-4393
5. X Tu, B G Cheron, J H Yan and K F Chen, Electrical and spectroscopic diagnostic of an atmospheric double arc

argon plasma jet[J]. *Plasma Sources Science and Technology*, 2007, 16, p: 803-812

6. Z. Duan and J. Heberlein, Arc Instabilities in a Plasma Spray Torch [J]. *Journal of Thermal Spray Technology*, 2002, 11(1), p: 44-51
7. J.F. Bisson, B. Gauthier, and C. Moreau, Effect of Plasma Fluctuations on In-Flight Particle Parameters [J]. *Journal of Thermal Spray Technology*, 2003, 12(1), p: 38-43
8. S.Semenov, B.Cetegen, Spectroscopic temperature measurements in direct current arc plasma jets used in thermal spray processing of materials[J]. *Journal of Thermal Spray Technology*, 2011, 10(2), p: 326-336
9. W.L.T.Chen, J.Heberlein, and E.Pfender, Diagnostics of a Thermal Plasma Jet by Optical Emission Spectroscopy and Enthalpy Probe Measurements, *Plasma Chemistry and Plasma Processing*, 1993, 14(3), p: 317-331
10. M.Sabsabi, S.Vacquie, D.V.Gravelle, and M.I.Boulos, Emission spectroscopic study of a low pressure supersonic Ar-H2 DC plasma jet[J]. *J.Phys.D:Appl.Phys.*, 1992, 25, p: 425-429
11. Nityalendra Singh, Manitra Razafinimanana and Alain gleizes, The effect of pressure on a plasma plume: temperature and electron density measurements[J]. *J.Phys. D: Appl. Phys*, 1998, 31, p: 2921-2928
12. Qing-Yu Chen, Xiao-Zhuang Peng, Chang-Jiu Li et al., Characterization of plasma jet in plasma spray-physical vapor deposition of YSZ using a <80kW shrouded torch based on Optical Emission Spectroscopy[J]. *Journal of Thermal Spray Technology*, 2015, 24(6), p: 1038-1045
13. Yang Gao, Deming Yang, Chengqi Sun, et al.. Deposition of YSZ Coatings in a Chamber at Pressures below 100 Pa Using Low-Power Plasma Spraying with an Internal Injection Powder Feeding[J]. *Journal of Thermal Spray Technology*, 2013, 22(7), p: 2013—1253
14. Li Tingjun, analysis of emission spectral [M]. atom energy press, 1983, p: 21-45
15. Chen Xinkun, theory of atom emission spectral analysis [M]. Tianjin Science press, Tianjin, 1st, 1991
16. Mauer Georg, Baбen Robert, Stöver Detlev. Plasma and particle temperature measurements in thermal spray: approaches and applications, *Journal of Thermal Spray Technology*, 2010, 20(3), p: 391-405.
17. S.Janisson, A.Vardelle, J.F.Coudert, E.Meillot, Plasma Spraying Using Ar-He-H_2 Gas Mixtures[J]. Journal of Thermal Spray Technology, 1999, 8(4), p: 545-552

Effect of TiO2 and TiB2 Compounds on the Microstructure and Infrared Emissivity of Spinel Ni-Cr Oxide Coatings Prepared by ASP

Jun Zou[a,b], WeiPing Ye[c], ShuRong Dong[a], Zhen Nie[a], XuDong Cheng[a,*]
[a]State Key Laboratory of Advanced Technology for Materials Synthesis and Processing, Wuhan University of Technology, Wuhan 430070, China
[b]College of Mechanical and Power Engineering, China Three Gorges University, Yichang 443002, China
[c]School of Materials Science and Technology, Wuhan University of Technology, Wuhan 430070, China

Abstract: High emissivity coatings used in high temperature condition have attracted a great attention for its significant energy saving properties in the worldwide. In this study, we produced the high emissivity Ni-Cr oxide coatings with TiO_2/TiB2 compounds on stainless steel substrate by spray granulation and atmospheric plasma spraying（APS）method. Compared to the TiB_2-adding composite coatings(NCTB coatings) , X-ray diffraction (XRD) analysis of the TiO_2-adding composite coatings(NCTO coatings) showed remarkable increment in the content of TiO_2. It was revealed that the particle size of the NCTO coatings was larger than the NCTB coatings by scanning electron microscope (SEM). The infrared emissivity of coatings were investigated by fourier transformation infrared radiation (FT-IR) at room temperature. The emissivity of NCTO coatings was higher than NCTB coatings on the band of 2.5-8 μm, similar on the waveband of 8-14 μm, and lower on the wavelength of 14-25 μm. The modified emissivity of NCTO coatings at 800℃ is 0.9201, which is 4.3 percent higher than that of NCTB coatings. The results indicated that NCTO coatings were more suitable for high temperature application than NCTB coatings.

Keywords: APS; TiB_2; TiO_2; infrared emissivity; spinel

1. Intruduction

The emissivity (ε) of a material is the ratio of energy intensity radiated by an actual material to that radiated by a blackbody under the same conditions. High infrared emissivity coatings are widely used in industrial furnaces, which have attracted a great attention recently for the energy saving properties and the furnace lining protective function. Generally, the composition of the coatings high emissivity such as SiC/ZrSiO4-SiO2[1], Fe2O3-MnO2-Co2O3-NiO[2], and ZrO2-Cr2O3-SiC[3] are the mixture of the infrared radiation powder and some binder. Owing to the dilution effect and the ageing of the binder, the emissivity and service life of these coatings are still limited. The problem is hopeful to be solved by using APS method. And next, comparing with magnetron sputtering and multi arc ion plating, APS method is an economical and effective method to produce functional coatings such as thermal wear-resistance coatings, barrier coatings, and abradable sealing coatings. The high emissivity and high bonding strength coating were prepared with pure infrared radiation material by APS method. In addition, the high temperature stability and high emissivity of spinel NiCr2O4 [4]have been investigated by several researchers in many industrial fields[4,5,6,7]. However, the infrared radiation property of spinel NiCr2O4 at high temperature has not been studied in detail. Thanks to the high emissivity of NiO and Cr2O3, spinel Ni-Cr oxide might have high emissivity as well as the spinel compound mentioned above. TiO2 is a common far infrared radiation material, and it also possesses high emissivity on the MIR band.[8] Wu Chen et al.[9] Found that the infrared emissivity of powder increased greatly by adding TiO2 to the spinel Ni-Cr oxide. The optimal content of TiO2 10% was choosed is this study. To compare the influence of TiO2 and TiB2, in the case of titanium content being a constant, we determined the weight propotion of TiB2 was 11.5%.

The purpose of the present work is to fabricate high emissivity coatings with a spinel Ni-Cr oxide on the stainless steel(SS) sheet by APS method, and to study the relationship between the infrared emissivity , surface morphology and crystal structure.

2. Experimental

2.1 Preparation of infrared radiation powders and coatings

In this study, the infrared radiation coatings were prepared by two steps. The first step is the fabrication of the spray powder. Firstly, the feedstock powder, organic binder, and deionized water were mixed in proportion to become a uniform slurry by colloid mill and the slurry was sprayed to aggregated powders in a spray dryer. The aggregated powders were was roasted for 4 h at 1350℃ and chosen with the particle size of 30-75 μm by a vibration sieving machine. The composition of the feedstock powder（named A，and B）is shown in Table 1.

Table 1.The composition of compound powder wt%

	Cr2O3	NiO	Fe2O3	Al2O3	SiO2	TiO2	TiB2
A	53.7	26	0.2	0.2	1.5	10	
B	53.7	26	0.2	0.2	1.5		11.5

And then, a Ni/Al bonding coating with a thickness of 60 μm was deposited on the substrate and CNTO/CNTB coatings were fabricated by the plasma-spraying system (GP-80, Jiujiang Spray Equipment Corporation, China) on the SS substrate. The parameters to prepare the coatings are as follows: the powder feed rate 25±5g/min, spray distance 90-130mm, plasma gas flow 1.4L/min, working voltage 76V and working current 500A. The overall thickness of the coatings is 240μm.

2.2 Characterization

DTA and TG analyses were carried out with the dried powders in a Integrated Thermal Gravimetric Analyzer for Mass Spectrometry（STA449F3) at a heating rate of 10℃/min in air. The phase structures were identified by Rotation Anode High Power X-ray Diffractometer (D/MAX-RB) at the scanning step size 0.02° and the scanning speed of 4°/min. Microstructural observation of the coatings was performed by SEM (JSM-5610LV). The FT-IR spectra were measured by using a Bruker Tensor27 FT-IR spectrometer with a resolution of 1cm^{-1} on the wavelength range of 2.5-25μm at room temperature.

3. Results and discussion

3.1. Thermodynamic analysis

Essential thermodynamic analysis was used to infer the feasibility of reaction and possible products during the experiment. According to the NiO-Cr2O3 phase diagram[10], the sample can transform into liquid phase beyond 2000℃, hence, the reasonable roasting temperature range should be lower than 1700℃. Using relevant thermodynamic data[11], the standard molar Gibbs free energy changes (ΔG=a+bT) of each reaction were calculated as a function of temperature (T), as shown in Fig. 1.

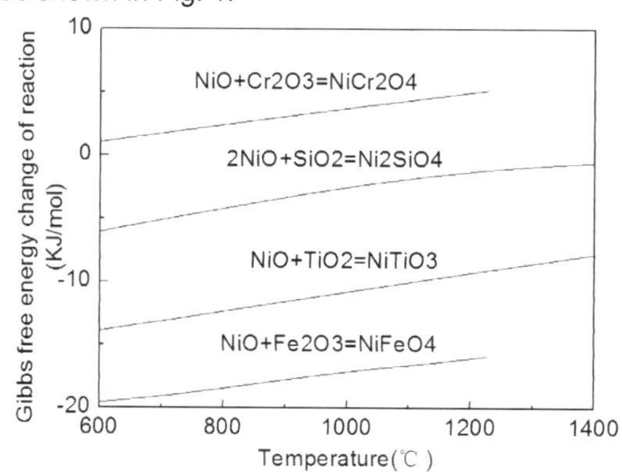

Fig.1. The relationship of Gibbs free energy change with the temperature for different reactions.

But the practical reaction conditions are different from the standard state. In this experiment, the reactant concentration of NiO, Cr_2O_3, and TiO_2/TiB_2 is higher than 1mole per liter. In addition, the synthesis reaction was associated with the decrease of the surface energy and bulk free energy. Considering the reasons mentioned above , the value of ΔG needs to be revised. Generally, we hold that the corresponding ΔG value range of the reversible reaction is between -40KJ and 40KJ. Thermodynamic analysis indicates that the reaction of $NiCr_2O_4$, $NiTiO_3$, and $NiCr_2O_4$ meet the requirement of thermodynamics. Moreover, as the temperature raising，the ΔG value of the reaction

($NiO+Cr_2O3 \rightarrow NiCr2O4$) rise slightly, and the thermal diffusion of reactants rise sharply. This suggested that high temperature is helpful to producing $NiCr_2O_4$. Thus, we choose 1320℃ as the roasting temperature.

3.2. TG-DTA analysis

DTA and TG analyses are carried out from room temperature to 1400℃in air to find the transformation and determine the proper roasting temperature, as shown in Fig. 2., and fig.3. We found that the major part of weight loss occurs below 310°C.

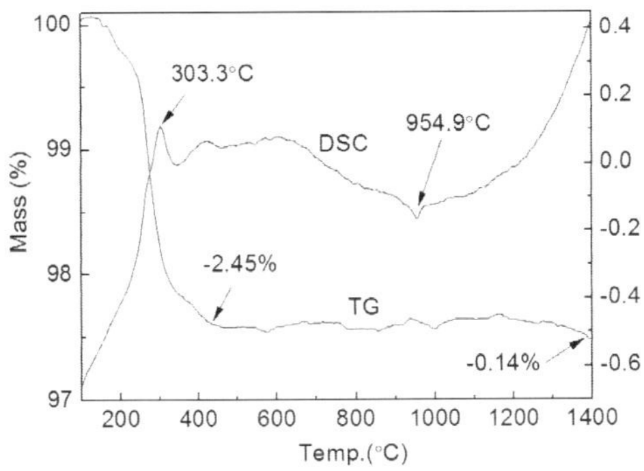

Fig.2. TG and DSC of powder NCTO.

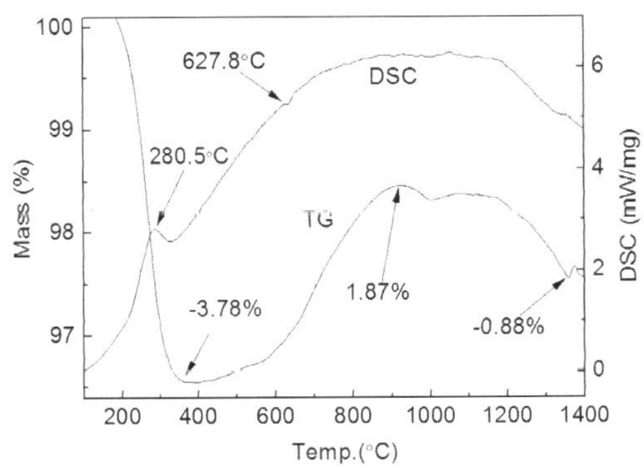

Fig.3. TG and DSC of powder NCTB.

There is a narrow exothermic peak associated with high mass loss ratio at about 300℃ in both two DSC curves, which mainly corresponds to the decomposition of residual organic binder. Because the evaporation and decomposition of trace organic binder are rapid, the exothermic peaks are narrow and weak. A broad endothermic peak in DSC tracing for powder CNTO was observed from 670 to 1400℃, for the initial reaction temperature of $NiCr_2O_4$ is about 660℃ according to the phase diagram of the binary system $NiO\text{-}Cr_2O_3$. A small endothermic peak appeared at 954.9℃, which correspond to the synthesis reaction of $NiO+TiO_2 \rightarrow NiTiO_3$. The result is confirmed by the XRD results that $NiCr_2O_4$ and $NiTiO_3$ is produced after roasted at 1000, and 1320℃ for 4 hours .

The DSC tracing of powder NCTB showed a narrow endothermic peak at 627.8℃, which correspond to the phase transformation of Cr_2O_3. The TG curve showed a rapid weight gain from 600 to 900℃. Combining with XRD analysis, it reveals that TiB_2 is oxidized to TiO_2 and B_2O_3. Evaporation of low melting point boron oxide lead to the weight loss of 0.88% at about 1300℃.

3.3. XRD analysis

The XRD patterns of coating samples using the feedstock powders roasted at different temperatures are shown in Fig.4. and Fig.5.

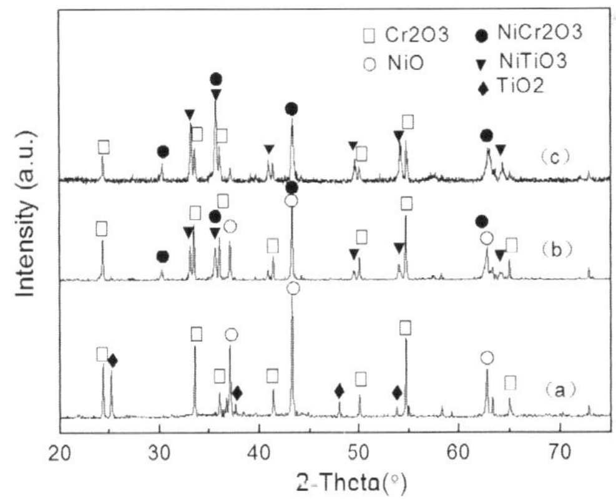

Fig.4. XRD patterns of powder NCTO roasted at 100℃(a),1000℃(b),1320℃(c).

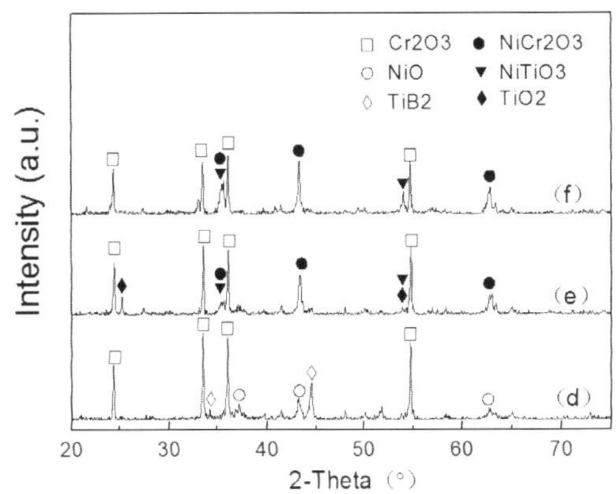

Fig.5. XRD patterns of powder NCTB roasted at 100℃(d),1000℃(e),1320℃(f).

The spinel structure was not found except NiO, TiO_2, TiB_2 and Cr_2O_3 in both samples roasted at 100℃, for the temperature is too low to start the reaction. And after 4-hours roasting at 1000℃ and 1320℃ in air, the spinel $NiCr_2O_4$ is successfully generated in both samples. The diffraction peaks are consistent with the standard JCPDS cards of $NiCr_2O_4$, Cr_2O_3, TiB_2 and NiO, respectively. Moreover, with the increase of temperature, the content of Cr_2O_3 in sample NCTO decreases, while the content of $NiTiO_3$ and $NiCr_2O_4$ increased.

The results indicated that after calcination at 1000℃, there was some TiO_2 existed in NCTB powder, but TiB2 mightbe oxidated into TiO_2 .After calcination at 1320℃ for 4 hours , we found the main components is $TiNiO_3$, $NiCr_2O_4$, and residual Cr_2O_3. It suggested that the adequate reaction between raw materials leads to the decrease of NiO content, which is why there is no diffraction peak of NiO beyond 1000°C.

3.4. Morphology observation

SEM micrographs of the as-prepared coatings are shown in Fig. 6.(a) and(b).

Fig. 6. SEM images of CNTO coating (a), and CNTB coating (b)

Fig. 6.(a) is the surface structure of CNTO coating . It was very rough and porous, which was composed of laminated structure with some smaller spatter grains embedded on the surface. The pore size is about 1-6μm. This morphology features may be explained by the "layer by layer" model proposed by Ng et al.[12] Yu et al. [13] proposed a theoretical model to interpret the effect of porosity on emissivity. The surface roughness of coatings in Fig. 6.(b) is lower than that of Fig. 6.(a). The pore size of CNTB coating was smaller compared with the CNTO coating. The possible reasons of morphology change is that TiB_2 is oxided to B_2O_3.The low melting point B_2O_3 in powder improved the heat transfer between the spray powder and high power plasma jet, and Led to better melting of powder CNTB. Melting powder was sprayed on the SS substrate. In the course of solidification and crystallization, the grain grew outward of the coating surface for the directional cooling of the cold substrate.

3.5. Emissivity performance

Infrared emissivity is primary among the radiative properties in energy-saving applications. The emissivity is determined by many factors, such as chemical composition, crystal structure, and surface morphology. The Planck's law describes the spectral distribution of the energy emitted by a blackbody in thermal equilibrium. And Vien's law shows that The λmax of the wavelength move to the direction of short wavelength. According to the Planck's law, the radiant energy of blackbody on some bands at different temperatures is calculated. The radiant energy proportion of them was obtained. The results are list in Table 2.

Table 2. Radiation energy percentage of subdomain to the whole MIR region at different Temp.

	400℃	600℃	800℃
2.5-8μm	68.80%	79.70%	85.30%
8-14μm	23.20%	15.60%	11.58%
14-25μm	7.95%	4.61%	3.11%

Data in Table 2. shows that the radiation energy distribution shifts to shorter wavelength with the increase of temperature. So, the emissivity on 2.5-8μm band plays a key role in the high temperature IR emissivity of coatings. Fig. 7. shows the infrared emission spectra of coatings on band of 2.5-8 and 8 ~25μm at room temperature. It indicated that two coatings had excellent MIR radiation performance.

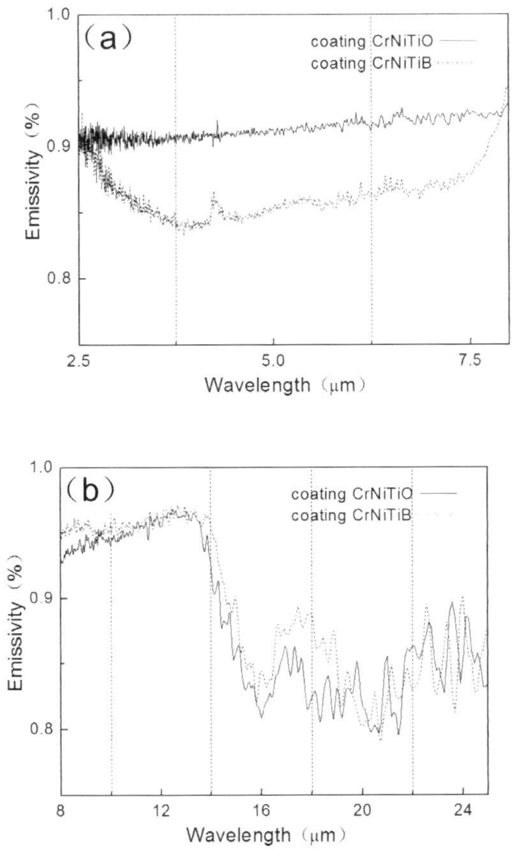

Fig. 7. Infrared spectral emissivity of samples on band of 2.5 ~8μm (a), and 8 ~25μm(b)

The integrated emissivity of CNTO coating on the band of 2.5-8μm ($\varepsilon_{a2.5\text{-}8}$) was 0.919, which was 5.27% higher than that of CNTB coating on the same band($\varepsilon_{b2.5\text{-}8}$). Though we found no significant difference in composition, but the pore size were distinctly different between two Coatings . The pore size of CNTO coating is approximately in the range of 1-6μm,which is the same size as the wavelength of 2.5-8μm. Cavity effect caused by higher roughness and corresponding scale pore size led to the increment of emissivity in this region. The integrated emissivity of two coatings ($\varepsilon_{a8\text{-}14}$,$\varepsilon_{b8\text{-}14}$) on 8-14μm band was similar, which showed reverse trend on the band of 14-25μm. The $\varepsilon_{a14\text{-}25}$ and $\varepsilon_{b14\text{-}25}$ are 0.843 and 0.861, respectively. These phenomena can be explained by the microrough coating surface and compound effect causing by B_2O_3. Considering the difference of contribution rate of each band, we obtained modified emissivity ($\varepsilon_{T2.5\text{-}25}$) at corresponding temperature(T) on the MIR band by using formula 1.

$$\varepsilon_{T2.5\text{-}25} = m_T \times \varepsilon_{2.5\text{-}8} + n_T \times \varepsilon_{8\text{-}14} + k_T \times \varepsilon_{14\text{-}25} \quad (1)$$

Where the coefficients m_T, n_T and k_T are the radiation energy percentage of balckbody at different temperature listing in the table 3. It was calculated that $\varepsilon_{a2.5\text{-}25}$ and $\varepsilon_{b2.5\text{-}25}$ were 0.9196 and 0.8911 at 400 ℃. And εa2.5-25 and εb2.5-25 were 0.9201 and 0.8822 at 800℃. By the modified emissivity we can see that the performance difference between the CNTO coating and the CNTB coating was more significantly as the temperature rising.

4. Conclusion

High emissivity spinel Ni-Cr oxide coatings with TiO2/TiB2 have been fabricated by APS. The main phase composition of both coatings is NiTiO3, spinel Ni-Cr oxide, and residual Cr2O3.The results indicated that TiB2 transformed into TiO2 completely after baking at 1320℃. The emissivity of the TiO2 compound coating on the 2.5~5 waveband is 5.27% higher than that of the TiB2 compound coating. Those may be caused by the surface roughness effects. The rough surface led to the increment of the reflection in the 2.5~8μm scale. The modified emissivity on MIR band of CNTO coating is 4.3% higher than that of CNTB coating, which can be explained by the higher macro-roughness and larger paticle size. The results show that the CNTO coating is more suitable for high temperature IR radiation applications than CNTB coating.

[1] Li, Y. and Z.W. Other: The anti-oxidation behavior and infrared emissivity property of SiC-ZrSiO4-SiO2 coating. Journal of Materials Science 25 (2014), pp. 5433/40.

[2] Zhang, J. and F.X. Other: Ferrites based infrared radiation coatings with high emissivity and high thermal shock resistance and their application on energy-saving kettle. Applied Surface Science 344(2015), pp. 223/29.

[3] Dan, Z. and C.D. Other: Microstructure and properties of high emissivity coatings. Journal of University of Science and Techenology Beijing 15 (2008), Issue 5, pp. 627/32.

[4] Tian, J. and Z.W. Other: Infrared Radiant Emissivities of Ceramics with Spinels Structure. International Journal of

Infrared and Millimeter Waves 14(1993),Issue 9, pp. 1855/63.
[5] Słoczyński, J. and Z.J. Other: Oxidative Dehydrogenation of Propane on NixMg1-xAL2O4 and NiCr2O4 Spinels. Journal of Catalysis 187 (1999), pp. 410/18.
[6] Zhuiykov, S. and N.T. Other: Potentiometric NOx sensor based on stabilized zirconia and NiCr2O4 sensing electrode operating at high temperatures. Electrochemistry Communications 3(2001), pp. 97/101.
[7] Sai, Q. and X.C. Other: Mn, Cr-co-doped MgAl2O4 phosphors for white LEDs. Journal of Luminescence 131 (2011), Issue 11, pp. 2359/64.
[8] Z. Wang;Y. Wang;Y. Liu;J. Xu;L. Guo;Y. Zhou;J. Ouyang and .. Other: Microstructure and infrared emissivity property of coating containing TiO2 formed on titanium alloy by microarc oxidation, Current Applied Physics 6(2011), Issue 6, pp. 1405/09
[9] Chen,W. and Y.W. Other: Preparation, Microstructure and Properties of NiO-Cr2O3-TiO2 Infrared Radiation Coating[J]. Journal of Materials Science & Technology 25(2009), Issue 5, pp. 695/98
[10] Hauck, J. Crystallography and phase relations of MeO M2O3 TiO2 systems (Me = Fe, Mg, Ni; M = Al, Cr, Fe). Journal of Solid State Chemistry 36 (1981), Issue 1, pp. 52/65.
[11] Brain, I. Thermochemiscal Data of Pure Substances. WILEY-VCH Verlag GmbH. 1995.
[12] Ng, H.W. and G.Z. A finite element analysis technique for predicting as-sprayed residual stresses generated by the plasma spray coating process. Finite Elements in Analysis and Design,41(2005), Issue13, pp. 1235.
[13] Yu, H. and X.G. Other: Effects of size, shape and floatage of Cu particles on the low infrared emissivity coatings. Progress in Organic Coatings 66(2009), Issue 2, pp. 173/78.

Preparation, Characterization of $(La_{0.9}Ca_{0.1})(Cr_{0.9}Mg_{0.1})O_3$ High Emissivity Coatings Deposited by APS

Liu Yanbo, Wang Quansheng*, Gao Lihong, Chang Yunpeng, Jia Zijian, Wang Hao

Surface Technology and Engineering Laboratory, School of Materials Science and Engineering, Beijing Institute of Technology, Beijing 100081, China

Abstract: Solid state synthesis was employed to prepare $(La_{0.9}Ca_{0.1})(Cr_{0.9}Mg_{0.1})O_3$ ceramic powders. The $(La_{0.9}Ca_{0.1})(Cr_{0.9}Mg_{0.1})O_3$ coatings were deposited by APS equipment on GH98 Super-alloy substrate. The Mg and Ca atoms entered the lattice of $LaCrO_3$ crystal to form $(La_{0.9}Ca_{0.1})(Cr_{0.9}Mg_{0.1})O_3$ but did not change the orthorhombic perovskite structure. The coatings are composed of splats, pores and micro cracks. The average porosity of $(La_{0.9}Ca_{0.1})(Cr_{0.9}Mg_{0.1})O_3$ coating is 12.4% and the average bond strength is 28.3Mpa.The $(La_{0.9}Ca_{0.1})(Cr_{0.9}Mg_{0.1})O_3$ powders did not occur phase transition during heated process in high temperature plasma plume. No impurity is introduced in the coating during plasma spraying process. Due to the significant distortion of $(La_{0.9}Ca_{0.1})(Cr_{0.9}Mg_{0.1})O_3$ lattice, the normal total emissivity of $(La_{0.9}Ca_{0.1})(Cr_{0.9}Mg_{0.1})O_3$ coatings are 0.88 at 600℃ and 0.89 at 800℃. The crystal structure is stable in the temperature range of thermal shock test from room temperature to 1100℃. And no spalling or fracture occurred within 10 times thermal shock which means the coatings could withstand the impact of thermal stress.

Key words: high emissivity coatings, plasma spraying, the normal total emissivity, thermal-shock resistance

1. Introduction

With the increase of performance requirement, the aircrafts and the engine will reach ultra-high temperature during service process, due to severe drastic aerodynamic heating or high temperature combustion flow heating [1-5]. Providing effective thermal protection for the metal substrate has become an important challenge in research and development process. The current thermal barrier method could not meet the protection requirement of components at ultra-high temperature. Heat transfer by thermal radiation to reducing the temperature of substrate had received great attention and played an important role in high temperature applications. High emissivity materials and coatings depositing process had become the focuses of reach. According previous reports, high emissivity coating has been extensively developed to be used in civil or in military fields, such as electric arc furnaces; annealing furnace; energy conservation equipment for hypersonic vehicles or the Ni-, Fe-, Co-based alloys components in third generation of reused space vehicles which involves high friction heat from acute friction between space vehicle surface and atmosphere [6-10].

Aim at different applications and service temperatures, the high emissivity materials include carbide; violet blue; mullite and transitional metal oxide. Most transitional metal oxide had high melt point and high emissivity and be considered as main candidate materials for high emissivity coatings, such as Cr_2O_3, MnO_2, ZrO_2, NiO, Ni_2O_3 Co_2O_3, Fe_2O_3, TiO_2, ZnO_2 [11-15]. Based on mechanism of infrared radiation of materials, the introduction of defects and impurities will break symmetry of the crystal structure and increase dipole moment, and ultimately improve the emissivity of materials. Therefore, compositing various transitional metal oxides became the main research area to develop high emissivity materials for thermal protected coatings. The current compositing method is heating the mixture of transitional metal oxides at high temperature to initiate doping of reaction between the adjacent components [16-21]. As the important characteristics of raw materials for high emissivity coatings, the components and mixture ratio will have a great impact on the performance of the coatings system.

Another important area for researching and developing high emissivity coatings is depositing process. According the previous studies, the depositing method of the coatings include sol-gel method, EB-PVD, anodic oxidation, high temperature melting process, micro-arc oxidation, thermal spraying process[22-29]. Different depositing process will influence the structure and interface status of coatings, and even impact the thermal performance and mechanical properties of coatings, such as emissivity and thermal shock resistance. As a kind of thermal spraying technologies, Atmosphere Plasma Spraying (APS) had been widely selected to depositing many kinds of coatings, especially be suitable to prepare refractory oxide ceramic coatings because of ultra-high temperature and high velocity plasma plume. The ceramic coatings deposited by APS always had high bond strength, low porosity and long service life. Winnie Tan et al. and Lei Lu et al. had used plasma process to preparing rare-earth modified zirconium diboride high emissivity coatings and ferrite-based amorphous high emissivity coatings [30,31].

Few reports about $(La_{0.9}Ca_{0.1})(Cr_{0.9}Mg_{0.1})O_3$ synthesis and coatings preparation were found in previous studies. In this paper, the possibility of $(La_{0.9}Ca_{0.1})(Cr_{0.9}Mg_{0.1})O_3$ coating deposited by APS as high emissivity coating was explored. The $(La_{0.9}Ca_{0.1})(Cr_{0.9}Mg_{0.1})O_3$ ceramic was attempted to be prepared by solid synthesis method. And sintering process was used to treated the plasma spraying powders. Due to the complexity of phase structure of materials and high temperature of treatment process, the phase stability and structure of as-sprayed powders were characterized and analyzed. In addition, because plasma plume also have ultrahigh temperature then the phase variation and structure characteristics were focused and reached. The normal total emissivity and thermal shock resistance were tested to reflect the properties of the $(La_{0.9}Ca_{0.1})(Cr_{0.9}Mg_{0.1})O_3$ coating. The influence factors of the coatings properties were discussed.

2. Experiment procedure

2.1 Preparation of powders

Solid-phase synthesis was employed to prepare $(La_{0.9}Ca_{0.1})(Cr_{0.9}Mg_{0.1})O_3$ ceramic. The raw materials were all provided by Beijing HWRK Chem Co.,LTD, which included Cr_2O_3 (99% purity), La_2O_3 (99.9% purity), CaO (99% purity), MgO (98% purity). According the molecular formula of $(La_{0.9}Ca_{0.1})(Cr_{0.9}Mg_{0.1})O_3$, the mole ratio of raw materials was 9 (Cr_2O_3): 9 (La_2O_3): 2 (CaO): 2(MgO). After ball-milling and rotary evaporation treatment, the mixture raw particles were calcined at 1300℃ for 1h to react and form $(La_{0.9}Ca_{0.1})(Cr_{0.9}Mg_{0.1})O_3$ ceramic particles. In order to meet the requirement about flowability and size of powders for plasma spraying process, the synthetic particles were agglomerated by spray drying process. PVA was selected as binder for particles agglomerate and the parameters of spray drying device (LGZ-8, Wuxi Dongsheng Spray-granulating and Drying Equipment Plant) were showed in table 1. In order to remove PVA and improve agglomerate strength, the agglomerated powders were sintered at 1100℃ for 1 hours. Before been used to depositing coatings, the agglomerate powders with the size range from 20 μm to 80 μm were extracted by sieving.

Table 1 The parameters of Spray Drying device for preparing agglomerate $(La_{0.9}Ca_{0.1})(Cr_{0.9}Mg_{0.1})O_3$ powders

Inlet temperature(℃)	Outlet temperature(℃)	Feed pump rate(RPM)	Atomizer rotary rate(Hz)
300	125	40	35

2.2 APS process

The coatings were deposited by SG100 Model Plasma Spraying Gun and 5500 Model Computer Controlled Atmosphere Plasma Spraying system which were provided by PRAXAIR-TAFA (USA). The main parameters of APS include current, primary gas and its flow, secondary gas and its flow, carrier gas flow and powder feeder rate. In this paper, the primer gas and carrier gas were Ar, and He was selected as secondary gas. The velocity and trajectory of plasma spraying gun were controlled by 2400 Model Robot (ABB). Based on our experience of high emissivity coatings deposition, the parameters values of APS, in Table 2, for preparing the $(La_{0.9}Ca_{0.1})(Cr_{0.9}Mg_{0.1})O_3$ coatings were selected. The substrate of coatings was GH98 Super-alloy. As the bond coatings of the double coatings system, MCrAlY coatings were deposited by HVOF. For the different properties test, three dimensions of substrate were used including cross-section morphology samples (30×15 ×3mm), bond strength and emissivity test samples (Φ25.4×10mm), thermal shock test samples (Φ23.2× 10mm).

Table 2 The parameters of APS process for preparing $(La_{0.9}Ca_{0.1})(Cr_{0.9}Mg_{0.1})O_3$ coatings

Current (A)	Ar gas flow (SCFH)	He gas flow (SCFH)	Carrier gas flow(SCFH)	Powder feeder rate (RPM)
800	90	35	8	2

2.3 Full radiation emissivity test

As the most important property of high emissivity coatings, the full radiation emissivity was test by IRE-2 Model Infrared Radiation Emissivity Measuring Instrument manufactured by Shanghai institute of Technology Physics of Chinese Academy of Science. In order to get the emissivity at high temperature, the tests were implemented when the samples been heated to 600℃ and 800℃.

2.4 Thermal shock test

The range of temperature for thermal shock test was from room temperature to 1100℃. After been heated in Box Type Resistance Furnace (SX2-5-12 Model, Beijing Electric Furnace Factory) for 10 min at 1100℃, the coating samples were taken out from furnace and dipped room temperature water at once. Till the samples were cooled to room temperature, then been picked out and dried in air. Checked the integrality of coating samples, if no spalling or cracking were observed then put the samples back in the furnace and continue the thermal shock test, and if the area of spalling field exceeded 10% of Original coating area then stopped the test.

2.5 Characterization

The bond strength of coating was tested according ASTM C633 by FM1000 film. The morphology of $(La_{0.9}Ca_{0.1})(Cr_{0.9}Mg_{0.1})O_3$ ceramic powders and coating was observed by scanning electron microscopy with secondary electron image (SEM, S-4800 model, Hitachi, Ltd) and optical microscope(IA32 Image Analysis & Management System, LECO). The porosity of the coatings was also calculated by IA32 Image Analysis & Management System, The elements of $(La_{0.9}Ca_{0.1})(Cr_{0.9}Mg_{0.1})O_3$ ceramic powders were investigated by electron dispersive spectroscopy (EDS) attachment to the SEM equipment. The BrukerD8 Advance X-ray Diffractometer operating under CuK radiation (40 kV, 40 mA) was used to detect phases of coatings and powders.

3. Results and discussion

3.1 Characteristics of $(La_{0.9}Ca_{0.1})(Cr_{0.9}Mg_{0.1})O_3$ powders

3.1.1 Morphology of $(La_{0.9}Ca_{0.1})(Cr_{0.9}Mg_{0.1})O_3$ powders

Fig 1 showed the morphology of $(La_{0.9}Ca_{0.1})(Cr_{0.9}Mg_{0.1})O_3$ powders after been sintering treated. Most of powders were subglobular, no sharp edges or corners were observed (Fig 1 (a)). It means that spray drying process had agglomerated raw $(La_{0.9}Ca_{0.1})(Cr_{0.9}Mg_{0.1})O_3$ ceramic particles successfully. And sintering treatment did not break the shape integrity of agglomerated powders. But the surface of powders was coarse and made up by fine particles. A large number of pores with different size located between adjacent particles and distributed uniformly. First reason for pores formation is that during agglomerate process, the adjacent raw particles did not totally match because of the irregular shapes, so the gaps and pores were formed between the particles. Another reason is that during spray drying process and sintering process, some gases were generated and exhaust overboard though the channels inside of the powders and leave the opening holes to form pores.

At the higher magnification of SEM, the surface particles can be distinguished clearly (Fig 1 (b)). The size range of particles was from 300nm to 1 μ m. The outlines of fine particles with sub-micro size of were relatively smooth and the corners were sleek, but the coarse particles with micro size still had sharp corners and edges. It means that the surface region of fine particles had melted and promoted them to sinter together and form larger clusters. However, the coarse raw particles did not melt or just occurred partially melt at the touch position with adjacent particles. Therefore, the partially sintering between coarse particles and the adherence of fine particles on the coarse particles were arisen at the touched melt position. Another evidence for particles melting is that some micro pores and interface gaps had been sealed by the freezing melt. It can be inferred that during sintering process, the low viscosity melt had been generated at high temperature and filled the defects between the particles by flowing alone the particles surface, after the powders cooling down with the furnace, solidified melt remained in the defects and sealed them.

Fig 1 SEM morphology of $(La_{0.9}Ca_{0.1})(Cr_{0.9}Mg_{0.1})O_3$ powders after been sintered: (a) Shape and surface state of powder; (b) Sintered particles on the surface of powder.

3.1.2 Crystal structure and elements of $(La_{0.9}Ca_{0.1})(Cr_{0.9}Mg_{0.1})O_3$ powders

The EDS tested fields were on the surface of sintered $(La_{0.9}Ca_{0.1})(Cr_{0.9}Mg_{0.1})O_3$ powders and the results were showed in Fig 2. Consistent with the material molecular formula, the diffraction peaks belonged to 5 elements which included La, Cr, Ca, Mg and O. It means that during synthesis process, spray drying process and sintering process, no External impurity was introduced into the products. The atomic percentage of tested powders was showed in table 3. The atomic percentage ration between La, Cr, Ca, Mg was close to the design value. It inferred that most of raw oxide particles were retained as components of final powders, and the impact of raw material loss during synthesis or treatment process on the proportion of components is small. Du to EDS can be applied only to detect the elements and proportion but could not reveal existing state of the elements. Therefore, phases component and crystal structure of final $(La_{0.9}Ca_{0.1})(Cr_{0.9}Mg_{0.1})O_3$ powders should be conformed exactly.

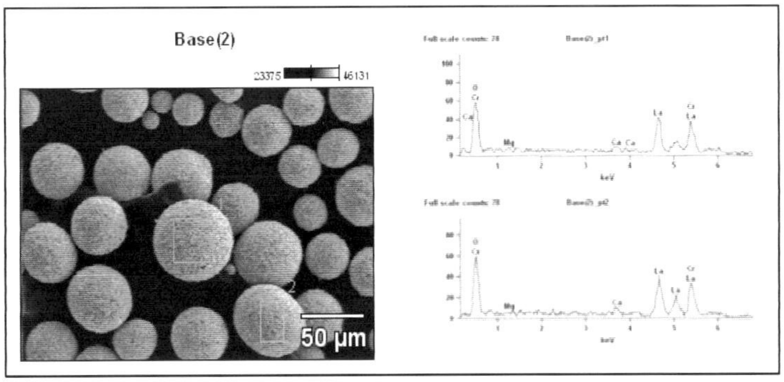

Fig 2 Elements of sintered $(La_{0.9}Ca_{0.1})(Cr_{0.9}Mg_{0.1})O_3$ powders detected by EDS.

Table 3 The atomic percentage of sintered $(La_{0.9}Ca_{0.1})(Cr_{0.9}Mg_{0.1})O_3$ powders detected by EDS

Element	La	Ca	Cr	Mg	O
Field 1	21.18	3.12	24.86	3.23	47.61
Field 2	26.22	2.54	22.02	1.72	47.50
Average	23.70	2.83	23.44	2.48	47.56

Fig 3 (a) shows the XRD pattern of sintered $(La_{0.9}Ca_{0.1})(Cr_{0.9}Mg_{0.1})O_3$ powders, and Fig 3 (b) is the XRD pattern of $LaCrO_3$ powders at different temperature detected by Sapna Gupta et al. Compared two patterns, it can be concluded that the crystal structure of $(La_{0.9}Ca_{0.1})(Cr_{0.9}Mg_{0.1})O_3$ was orthorhombic perovskite structure which is similar with $LaCrO_3$ powders. In addition, the diffraction peaks of $(La_{0.9}Ca_{0.1})(Cr_{0.9}Mg_{0.1})O_3$ powders is clear and no obvious oxide impurity was observed, which inferred that all the raw oxide particles participated in the synthesis reaction and the reaction was sufficient to form high purity final $(La_{0.9}Ca_{0.1})(Cr_{0.9}Mg_{0.1})O_3$ particles. According the similarity of crystal structure (orthorhombic perovskite) between $(La_{0.9}Ca_{0.1})(Cr_{0.9}Mg_{0.1})O_3$ and $LaCrO_3$ as well as no remnant of CaO or MgO components, it can be deduced that the main product and structure were formed by the reaction between the highest molar content of raw materials (Cr_2O_3,La_2O_3). The Mg and Ca atoms entered into the crystal lattice during reaction process and took the place of Cr or La atoms to form Replacement solid solution. So, the Mg and Ca atoms caused distortion of crystal lattice even generated diffraction peaks shifted but did not change the crystal type.

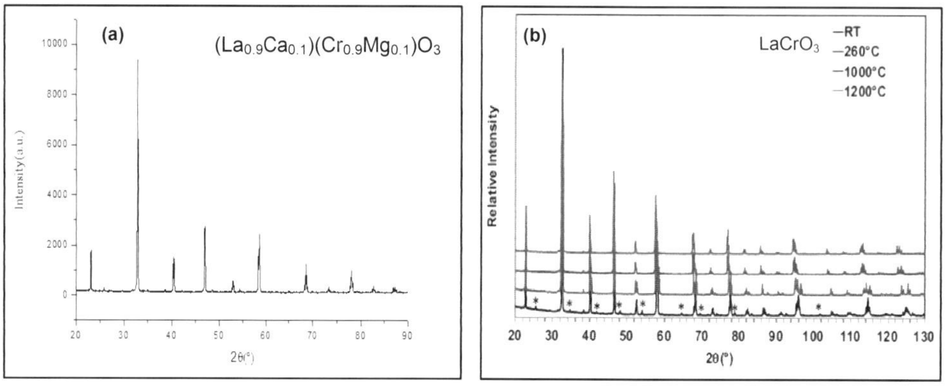

Fig 3 Phase structure of $(La_{0.9}Ca_{0.1})(Cr_{0.9}Mg_{0.1})O_3$ powders: (a) XRD pattern of $(La_{0.9}Ca_{0.1})(Cr_{0.9}Mg_{0.1})O_3$ powders; (b) XRD pattern of $LaCrO_3$ from the related literature [32].

In summary of previous study, after reaction process between raw particles, the element ratio of synthesis product is consisted with experiment design. The Mg and Ca atoms entered the lattice of $LaCrO_3$ crystal to form $(La_{0.9}Ca_{0.1})(Cr_{0.9}Mg_{0.1})O_3$ but did not cause the type change of orthorhombic perovskite structure. The agglomerated powders with good spherical degree which matched the requirement of plasma spraying technology were obtained by spray drying process. The sintering process removed the PVA effectively and improved the strength of powders by cause the sintering between raw particles.

3.2 Characteristics of $(La_{0.9}Ca_{0.1})(Cr_{0.9}Mg_{0.1})O_3$ coating

3.2.1 Morphology and micro structure of $(La_{0.9}Ca_{0.1})(Cr_{0.9}Mg_{0.1})O_3$ coating

Fig 4 showed the appearance of $(La_{0.9}Ca_{0.1})(Cr_{0.9}Mg_{0.1})O_3$ coating on the thermal shock test sample. The surface color of the coating was black which was same with the $(La_{0.9}Ca_{0.1})(Cr_{0.9}Mg_{0.1})O_3$ powders. And the color of the surface was Homogeneous without any mottled area. In addition, the surface of coating was flat like the uncoated sample surface and the roughness was Homogeneous too, no spalling and cracking were observed.

Fig 5 shows the micro structure of plasma sprayed $(La_{0.9}Ca_{0.1})(Cr_{0.9}Mg_{0.1})O_3$ coating surface. From the SEM photograph, it can be found that the surface of coating was rugged and rough because many kinds of different structures mixed together on it such as splats, spherical large particles, fine particles, irregular particles, pores and cracks. According the formation mechanism of plasma sprayed coating, the powders were heated and accelerated in the plasma plume, and before impacted on the surface of substrate they have reached high temperature and velocity. Ideally, the powders have reached semi-molten state and get enough kinetic energy to ensure that they could spread enough and deformed to splats when impact with substrate, then the coating was formed on the surface of substrate by the stacking of deformed splats. But actually, the states of powders were different which can be divided into three kinds included ideal semi-molten state, over molten state and unmolten state. For the unmolten state, the powders could not spread when impacted on the substrate, they just adhered on the surface of the coating with a spherical shape or even broken to form irregular smaller particles, then amount of defects such as pores or cracks were introduced in the coatings. For the over molted state, the powders had form melt drops which would splashed and scattered away because of the impaction, then the splashed droplets frozen and form fine particles sticking on the surface of coating, and the defects were also introduced into the coatings. For the ideal semi-molten state, the powders deformed to form splats with smooth surface which have obviously melting characteristics. Because of the enough spread, the splats always fit well with the substrate. But the pores and cracks still existed at the lap positions or interfaces between adjacent splats.

Fig 4 Appearance of $(La_{0.9}Ca_{0.1})(Cr_{0.9}Mg_{0.1})O_3$ coating.

Fig 5 Surface morphology of $(La_{0.9}Ca_{0.1})(Cr_{0.9}Mg_{0.1})O_3$ coatings.

The morphology of $(La_{0.9}Ca_{0.1})(Cr_{0.9}Mg_{0.1})O_3$ coatings was exhibited in Fig 6. It can be found that the thickness of $(La_{0.9}Ca_{0.1})(Cr_{0.9}Mg_{0.1})O_3$ coating was about 200 μm and bond coating was about 100 μm, the overall thickness of the coatings system was relatively uniform (Fig 6 (a)). All kinds of defects which included pores and cracks distributed uniformly in the cross section of coating and no segregation was observed, which means that the states of powders and spraying system were stable and Homogeneous, then the deposited powders with different states also distributed uniformly. As an important region of coatings system, the interfaces between coatings or coating with substrate were always focused and observed carefully. In the $(La_{0.9}Ca_{0.1})(Cr_{0.9}Mg_{0.1})O_3$ coatings system, the surface coating and bond coating Combined well and no spalling position was found. Though a small amount of pores can be watched at the interface but no crack was observed. The clear interface indicated that most $(La_{0.9}Ca_{0.1})(Cr_{0.9}Mg_{0.1})O_3$ powders were heated and accelerated well and obtained excellent deformation ability. So, the ceramic powders could spread enough alone with the morphology of bond coating surface. As another part of coatings system, the metal bond coating had higher density, few pores or cracks were found in the cross section, the interface between bond coating and substrate was clear and no interface pores or cracks were observed.

Fig 6 (b) shows the micro structured of $(La_{0.9}Ca_{0.1})(Cr_{0.9}Mg_{0.1})O_3$ coating which was observed at higher magnification. The structures of the whole observation area were composed of small density areas, pores and micro cracks. The density areas were formed by the powders with ideal melting state which means semi molten state. After properly heated in the plasma plume, the semi molten powders not only obtain better plasticity and high kinetic energy but also removed the internal original defects. So, after impacted on the substrate, the dense areas could be kept within the splats which came from the semi molten powders. The pores in the coating were irregular and had a wide size range from sub-micrometer to dozens micrometers. The size and directions of micro cracks also changed over a wide range. Some pores had linked by the micro cracks between them. Be same with the surface defects, there were two main reasons for the pores and cracks formation in the coating cross section. First reason is that the melting state of partial powders is not good enough. When impact on the substrate, the over molten powders splashed to cause cracks and fine spherical particles which unable to adhere well with around powders, once the fine particles dropped off then the pores formed at the position. For the unmolten powders, the insufficient spread deformation caused weak bond strength with around powders or even internal fragmentation. The separated particles would create pores and the fragmented powders became the source of crack initiation. Another reason for the formation of defects was also the lap of the deformed powders which create gaps to form pores and interface to form micro cracks.

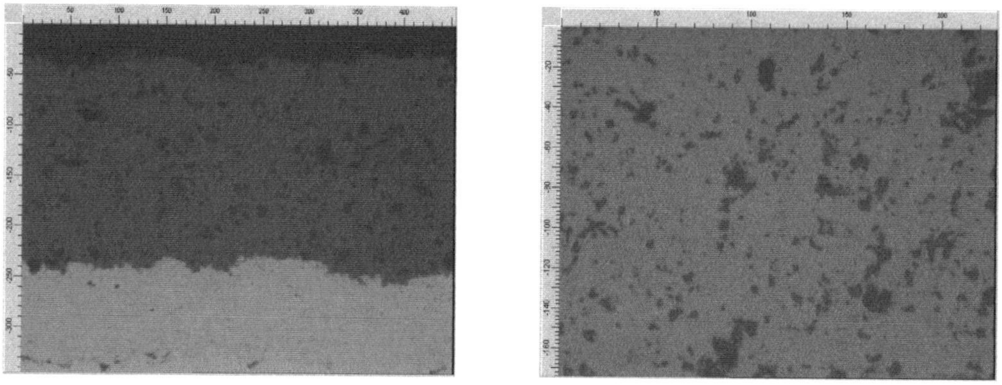

Fig 6 Morphology of cross section of $(La_{0.9}Ca_{0.1})(Cr_{0.9}Mg_{0.1})O_3$ coatings: (a) Overall morphology of cross section of coatings system; (b) Micro structure of cross section of $(La_{0.9}Ca_{0.1})(Cr_{0.9}Mg_{0.1})O_3$ coating.

The porosity and bond strength of the $(La_{0.9}Ca_{0.1})(Cr_{0.9}Mg_{0.1})O_3$ coating were showed in table 4. The average porosity was 12.4% and the average bond strength was 28.3Mpa. High bond strength was relative with the micro structures and interfaces state. Uniform structure, low porosity and compact interface contribute to improve the bond strength of coatings system.

Table 4 the porosity and bond strength of $(La_{0.9}Ca_{0.1})(Cr_{0.9}Mg_{0.1})O_3$ coating

	1	2	3	Average
Porosity (%)	11.7	12.8	12.7	12.4
Bond strength (Mpa)	27.8	26.2	30.8	28.3

3.2.2 Crystal structure of $(La_{0.9}Ca_{0.1})(Cr_{0.9}Mg_{0.1})O_3$ coating

The crystal structure of the $(La_{0.9}Ca_{0.1})(Cr_{0.9}Mg_{0.1})O_3$ coating was detected by XRD, and the result was showed in Fig 7. Compared with the XRD pattern of $(La_{0.9}Ca_{0.1})(Cr_{0.9}Mg_{0.1})O_3$ powders, it can be found that the number and position of diffraction peaks were largely same with each other. It means that the $(La_{0.9}Ca_{0.1})(Cr_{0.9}Mg_{0.1})O_3$ powders did not occur phase change during heated in plasma plume of impact process and still kept of orthorhombic perovskite structure. In addition, no impurity peaks were observed which indicated that no impurity was introduced in the coating during plasma spraying process, and MgO, CaO dopants did not precipitate during rapid Heating and cooling process.

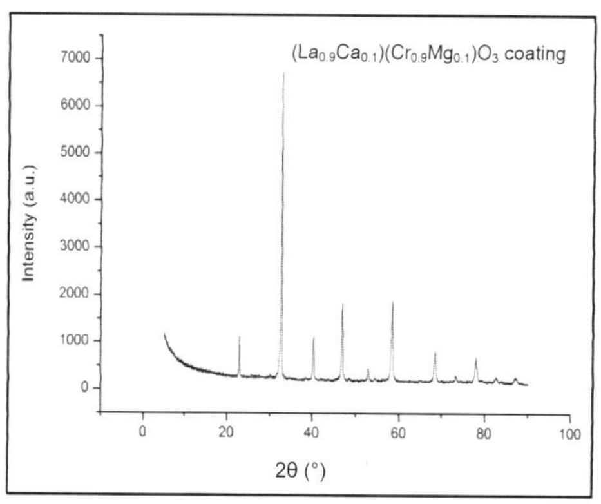

Fig 7 XRD pattern of $(La_{0.9}Ca_{0.1})(Cr_{0.9}Mg_{0.1})O_3$ coating

3.2.3 Emissivity of $(La_{0.9}Ca_{0.1})(Cr_{0.9}Mg_{0.1})O_3$ coating

The normal total emissivity values of $(La_{0.9}Ca_{0.1})(Cr_{0.9}Mg_{0.1})O_3$ coating at 600℃ and 800℃ are exhibited in Table 5. It can be found that the normal total emissivity values at high temperature are close to 0.9. According to Wien's displacement law and Planck's law, the black body radiates most of energy in the wavelength range of 1-5 μm at high temperature, but most materials have quite low emissivity within this range because of the weak intrinsic absorption. And introducing impurities via doping process could effectively enhance the total emissivity especially in the wavelength range of 1-5 μm [7]. For the $(La_{0.9}Ca_{0.1})(Cr_{0.9}Mg_{0.1})O_3$ coating, MgO and CaO could be regarded as the impurities introduced in lattice of $LaCrO_3$. It means that both the La position and Cr position of $LaCrO_3$ lattice occurre iron replacement, i.e., La^{3+} replaced by Ca^{2+} and Cr^{3+} replaced by Mg^{2+}. Due to the significant difference between doping ions and original ions, the distortion of $(La_{0.9}Ca_{0.1})(Cr_{0.9}Mg_{0.1})O_3$ lattice was generated. In addition, the valence state difference between Mg^{2+} and Cr^{3+} caused the Cr^{3+} valence state transition to Cr^{3+} or even caused the generation of oxygen vacancies. So the distortion of $(La_{0.9}Ca_{0.1})(Cr_{0.9}Mg_{0.1})O_3$ lattice is further improved. The serious distorted lattice could increase the infrared active vibration mode, broaden the absorption wavelength range, and eventually increase the normal total emissivity.

Table 5 The normal total emissivity of $(La_{0.9}Ca_{0.1})(Cr_{0.9}Mg_{0.1})O_3$ coating

Temperature (℃)	Emissivity
600	0.88
800	0.89

3.2.4 Thermal shock resistance of $(La_{0.9}Ca_{0.1})(Cr_{0.9}Mg_{0.1})O_3$ coating

For the thermal shock test of $(La_{0.9}Ca_{0.1})(Cr_{0.9}Mg_{0.1})O_3$ coating, a total of 10 cycles were executed. The appearance of tested coating is showed Fig. 8. By comparing with the appearance of original coating, it can be found that the color of coating after thermal shock test is still black, the same as original state and powders. From the XRD results (Fig. 9), it can be concluded that the crystal structure did not change during thermal shock test and no impurity was introduced in coating. The coating shows a good adhesion to the substrate and no spalling field is observed. In addition, no fracture and cracks are detected on the coating surface. It can be concluded that the crystal structure is stable in the temperature range from room temperature to 1100℃, and the coating could withstand the impact of thermal stress within 10 times thermal shock.

Fig. 8 Appearance of $(La_{0.9}Ca_{0.1})(Cr_{0.9}Mg_{0.1})O_3$ coating after 10 times thermal shock.

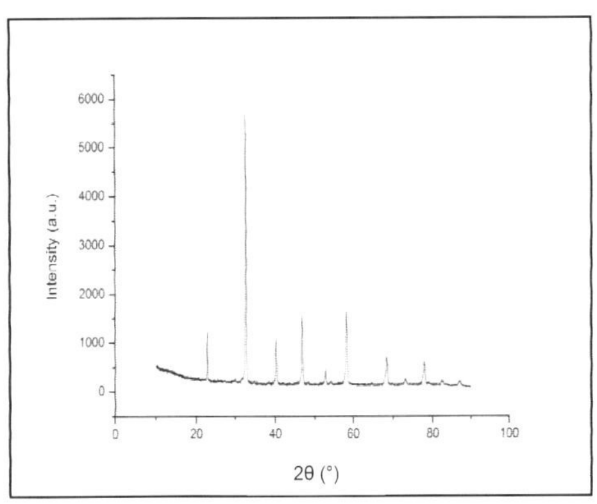

Fig. 9 XRD pattern of $(La_{0.9}Ca_{0.1})(Cr_{0.9}Mg_{0.1})O_3$ coating after 10 times thermal shock

Conclusion

Solid state synthesis was employed to prepare $(La_{0.9}Ca_{0.1})(Cr_{0.9}Mg_{0.1})O_3$ ceramic powders. The raw materials mixture were calcined at 1300℃ for 1h to react and form $(La_{0.9}Ca_{0.1})(Cr_{0.9}Mg_{0.1})O_3$ ceramic powders. After being agglomerated by spray drying process, the powders were sintered at 1100℃ for 1h. The stoichiometric ratio of product is consisted with design. The Mg and Ca atoms entered into the lattice of $LaCrO_3$ crystal to form $(La_{0.9}Ca_{0.1})(Cr_{0.9}Mg_{0.1})O_3$ but did not change the orthorhombic perovskite structure. The agglomerated powders with good spherical degree which meet the requirement of plasma spraying technique were obtained by spray drying process. The sintering process removed the PVA effectively and improved the strength of powders by the sintering among raw particles.

The $(La_{0.9}Ca_{0.1})(Cr_{0.9}Mg_{0.1})O_3$ coatings were deposited by APS equipment on GH98 Super-alloy substrate. The surface of coating is rugged and rough because several kinds of different structures exist such as splats, spherical large particles, fine particles, irregular particles, pores and cracks. The structures of cross section coating are composed of small density areas, pores and micro cracks. The average porosity is 12.4% and the average bond strength is 28.3Mpa.The $(La_{0.9}Ca_{0.1})(Cr_{0.9}Mg_{0.1})O_3$ powders did not occur phase transition during heated process in high temperature plasma plume and still kept the orthorhombic perovskite structure. No impurity was introduced in the coating during plasma spraying process. The normal total emissivity of $(La_{0.9}Ca_{0.1})(Cr_{0.9}Mg_{0.1})O_3$ coating is 0.88 at 600℃ and 0.89 at 800℃. Such high emissivity value is attributed to the lattice distortion caused by the significant difference between doping ions and original ions and the valence state difference between Mg^{2+} and Cr^{3+}. The crystal structure is stable in the thermal shock test executed from room temperature to 1100℃. And no spalling or fracture occurred within 10 times thermal shock which means the coatings could withstand the impact of thermal stress.

Reference

[1] Paul L. Moses, Vincent L. Rausch, Luat T. Nguyen, Jeryl R. Hill. NASA hypersonic Flight demonstrators—overview, status, and future plans. Acta Astronautica, 55 (2004), pp.619–630.

[2] M.Y.M. Ahmed, N.Qin. Recent advances in the aerothermodynamics of spiked hypersonic vehicles. Progress in Aerospace Sciences, 47 (2011), pp. 425–449.

[3] Mehta RC. Numerical heat transfer study over spiked blunt bodies at Mach 6.8. Journal of Spacecraft and Rockets, 2000, 37(5), pp 700–3.

[4] Y. Furuya, K. Kobayashi, M. Hayakawa, M. Sakamoto, Y. Koizumi, H. Harada. High-temperature ultrasonic fatigue testing of single-crystal superalloys. Materials Letters, 69 (2012), pp. 1–3.

[5] R. Savino, G. Festa, A. Cecere, L. Pienti, D. Sciti. Experimental set up for characterization of carbide-based materials in propulsion environment. Journal of the European Ceramic Society, 35 (2015), pp.1715–1723.

[6] Jianping Huanga, Yibin Li, Xiaodong He, Guangping Song, Chenglei Fan, Yue Suna,Weidong Fei, Shanyi Du. Enhanced spectral emissivity of CeO_2 coating with cauliflower-like microstructure. Applied Surface Science, 259 (2012), pp. 301– 305.

[7] Xiaodong He, Yibin Li, Lidong Wang, Yue Sun a, Sam Zhang. High emissivity coatings for high temperature application: Progress and prospect. Thin Solid Films, 517 (2009), pp. 5120–5129.

[8] Lei Lu, Xi'an Fan, Jianyi Zhang, Xiaoming Hu, Guangqiang Li, Zhan Zhang. Evolution of structure and infrared radiation properties forferrite-based amorphous coating. Applied Surface Science, 316 (2014), pp.

82–87.
[9] P.C. Sheil, T.R. Kleeb, High-emissivity coatings for improved performance of electric arc furnaces. Iron Steel Technol, 3 (2006). Pp. 49–53.
[10] R. Scott, S.D. Cherico, High-emissivity coating technology improves annealing furnace efficiency. Iron Steel Technol, 4 (2007), pp. 319–324.
[11] Xia Xin, Jin Yi, Li Wenjun, Tao Dao xian. Study of emissivity improving of infrared coatings at high temperature. infrared technology, 1998, 25(3), pp. 26-30.
[12] Ren Xiaohui, Zhang Xudong, He Wen, Wu Zuogu. Infrared radiating property of cordierite. Norganic Chemicals Industry, 2007, 39(7), pp. 5-7.
[13] JIAO Yongfeng, FU Yida, LU Chanjuan. Research actuality of cordierite base infrared ceramics. Bulletin of the Chinese Ceramic Society, 2008, 27(4), pp. 4-8.
[14] Xu Qing, Chen Wen, Zheng Jinxia, Yuan Runzhang. Development of transitional metal oxides system infrared radiation ceramics. Journal of Ceramics, 2000, 21(1), pp. 18-22.
[15] Xu Qing, Chen Wen, Yuan Runzhang. Microstructure and Infrared Emissivity at Normal Temperature in Transitional Metal Oxides System Ceramics. Journal of Wuhan University of Technology-Materials Science, 2000, 15(2), pp. 15~20.
[16] Lei zhongwei, Zhang ying, Wen dijiang. Preparation and Study of Transitional Metal Oxide Series Infrared Radiate Material. Shandong Ceramics, 2007, 30(2), pp. 3-6.
[17] Takashima H, Matsubara K, Nishimura Y. High efficiency-infrared radiant using transitional element oxide. Yigyo-Kyokai Shi, 1982, 90(7), pp. 373-379.
[18] T. Kleeb, J. Olver. High-emissivity coatings for energy savings in industrial furnaces. Ind.Heating, 2007, 74(6), pp. 57-61.
[19] Hellander J. Ceramic Coatings： Reheat Furance Application. Iron & Steel Engineer, 1987, (6), pp. 40-43.
[20] Krieble, K.Mossbauer. Spectroscopy investigation of Mn-substituted Co-ferrite(CoMn/subx/Fe/sub 2-x/O/sub 4/). Applied Physics, 2005, 97(10), pp. 101-103.
[21] Lee, J.H.Self. Impurity cation diffusion in manganese-zinc-ferrite,Mn1-x-yZnxFe2+yO4. Physics and Chemistry of Solids, 2000, 61(10), pp. 1597-1605.
[22] Liu Chenyu, Xu Yongjun, Jiang Zhaohua, Zhang Zhenwen, Wang Zhijiang. Research Progress of High Emissivity Coatings. Energy Conservation Technology, 2013,31(180), pp. 307-309.
[23] Luo Liechao, Zhao Rong gen, Meng Jia, Chen Jiefeng. White Anodic Coating on Aluminium Surface for Spacecraft. Journal of Inorganic Materials, 2002,17(6); pp. 1269-1276.
[24] Jianping Huang, Yibin Li, Xiaodong He, Guangping Song, Chenglei Fan, Yue Sun, Weidong Fei, Shanyi Du. Enhanced spectral emissivity of CeO_2 coating with cauliflower-like microstructure. Applied Surface Science, 2012 (259), pp. 301-305.
[25] Jianyi Zhang, Xi'an Fan, Lei Lu, Xiaoming Hu, Guangqiang Li. Ferrites based infrared radiation coatings with high emissivity and high thermal shock resistance and their application on energy-saving kettle. Applied Surface Science, 2015 (344), pp. 223-229.
[26] Jianping Huang, Chenglei Fan, Guangping Song, Yibin Li, Xiaodong He, Xinjiang Zhang, Yue Sun, Shanyi Du, Yijie Zhao. Enhanced infrared emissivity of CeO_2 coatings by La doping. Applied Surface Science, 2013 (280), pp. 605-609.
[27] X.Zhao，X.D.He，S Zhang et al. Investigations on B doped SiO_2 thermal protective coatings by hybrid sol-gel method. Thin Solid Films，2011(519). pp. 4849-4854.
[28] Wang Qianping, Zhang Jiasheng,Huang Zhuanhong, et al. Preparation and property of ceramic coatings

for metal substrate using high temperature melting process. NAIHUO CAILIAO, 2005,39(6), pp. 452 -454,459.

[29] Hui Tang, Qiu Sun, Tiezhu Xin et al. Influence of Co(CH3 COO)2 concentration on thermal emissivity of coatings formed on titanium alloy by micro-arc oxidation. Current Applied Physics, 2012(12), pp. 284-290.

[30] Winnie Tan, Christopher A. Petorak, Rodney W. Trice. Rare-earth modified zirconium diboride high emissivity coatings for hypersonic applications. Journal of the European Ceramic Society, 2014(34), pp. 1-11.

[31] Lei Lu, Xi'an Fan, Jianyi Zhang, Xiaoming Hu, Guangqiang Li, Zhan Zhang. Evolution of structure and infrared radiation properties for ferrite-based amorphous coating. Applied Surface Science, 2014(316), pp. 82-87.

[32] Sapna Gupta, Manoj K. Mahapatra, Prabhakar Singh. Phase transformation, thermal expansion and electrical conductivity of lanthanum chromite. Materials Research Bulletin, 2013 (48): pp. 3262–3267.

Vacuum Plasma Spray of Yttria Stabilized Zirconia Coatings

Pengyun Xu, Larry Pershin, Thomas W. Coyle, Javad Mostaghimi, Toronto / CA

Vacuum plasma spray (VPS) has great potential to fabricate dense coatings with stronger inter-splat boundaries and fewer voids and cracks, due to the unique plasma jet properties in a low pressure chamber. In this study, zirconia coatings were fabricated by the vacuum plasma spray process using two modifications of 8 wt% yttria stabilized zirconia (8YSZ) powder: hollow spherical particles (HOSP Metco 204B-NS, Oerlikon Metco) and fused and crushed (40.23.1, GTV Verschleißschutz GmbH). The plasma torch was a F4-VB (Oerlikon Metco) operating with an Ar-H_2 plasma gas mixture. The relationships between the plasma spray process parameters and the in-flight particle velocities and temperatures, under vacuum and atmospheric conditions, were investigated in real time using a DPV 2000. The process parameters yielding dense zirconia coatings were determined. The microstructures of coatings deposited by VPS were correlated with the in-flight particle status and compared with the coatings deposited using the atmospheric plasma spray process.

1 Introduction

Yttria stabilized zirconia (YSZ) is a common material in plasma spray process. By controlling the plasma spray process parameters and powder feedstocks, YSZ coatings with different microstructures and properties can be deposited to satisfy various kinds of application requirements, such as thermal barrier coatings for high-temperature gas turbines and electrolyte for solid oxide fuel cells. Atmospheric plasma spray (APS) is a commonly used process to deposit YSZ coatings. Compared to the APS process, vacuum plasma spray (VPS) is operated in controlled low pressure conditions (usually from 50 mbar to 200 mbar) instead of atmospheric pressure [1]. The reduced pressure causes the plasma jet to enlarge both in length and in diameter and greatly increases the plasma jet velocity due to the weak interaction with the surrounding atmosphere. Particles traveling in the low-pressure plasma jet can achieve a significantly higher velocity. The plasma jet in the VPS process has a larger spray spot, and particles entrained in the plasma exhibit a more homogenous velocity and temperature distribution along the plasma plume [2]. This provides VPS process with great potential to form dense coatings efficiently and effectively. Numerous investigations of APS YSZ coatings have been carried out [3-5]. In contrast, comparatively little work has been reported on VPS YSZ coatings.

In the present work, YSZ coatings were deposited using two kinds of 8wt% YSZ powder in VPS and APS processes. The in-flight particle properties for each plasma spray process were measured in real time. The effects of powder and plasma spray process on coating microstructure, porosity and deposition efficiency were investigated and correlated.

2 Experimental procedure

2.1 Powder

Commercial 8wt% YSZ powder with two different morphologies and structures were used as feedstock. One was agglomerated hollow spherical powder (HOSP powder, product No. Metco 204B-NS, Oerlikon Metco) which contained globular particles 45-75 µm in diameter. The other was a fused and crushed powder (F&C powder, product No. 40.23.1, GTV Verschleißschutz GmbH) which contained dense, angular and blocky particles 20-45 µm in diameter. The typical morphology of each powder is shown in **Fig.1**.

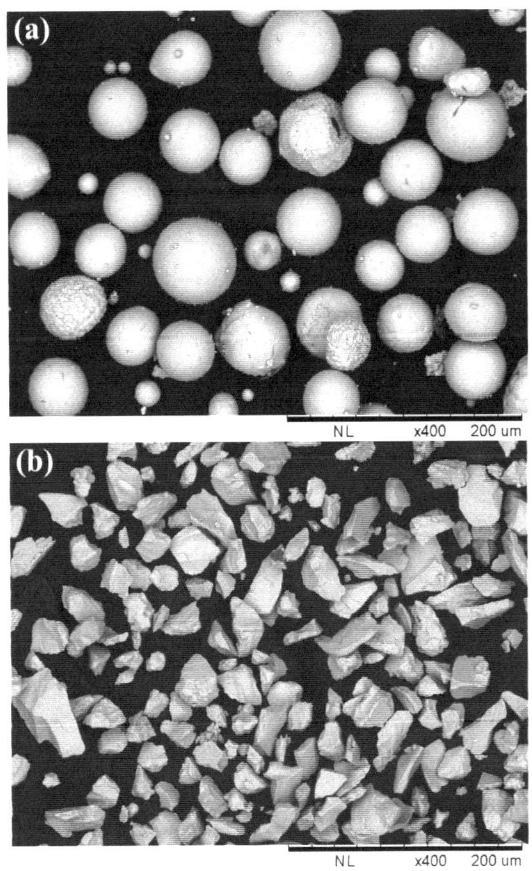

Fig. 1. Morphologies of powder: (a) HOSP, (b) F&C.

2.2 Coating deposition

Using the two powder feedstocks (HOSP and F&C) and two plasma spray processes (VPS and APS), four distinct types of coatings were fabricated and named as HOSP VPS, HOSP APS, F&C VPS and F&C APS. Stainless steel plates 1.5 mm thick were used as substrates and were grit-blasted by abrasive alumina particles before spraying. For all YSZ coatings, a F4-VB torch (Oerlikon Metco) with a 7 mm internal diameter nozzle was used. In all spraying processes, the current input and plasma gas (Ar and H_2) flow rate were kept the same in order to compare the difference between APS and VPS processes using HOSP and F&C powder. In all cases powder was injected radially and upwards into the plasma before the jet exited the nozzle. By watching the powder injection behaviour during the spraying process, the carrier gas (N_2) flow rate was adjusted to make sure the powder was traveling in the plasma core region for all cases. The standoff distance for APS was 100 mm [6-9], which is a typical distance reported in the literature for APS YSZ deposition. For VPS, the standoff distance was 150 mm or longer, as the plasma jet in low pressure is much longer and would overheat the substrate at shorter distances. Longer standoff distances were also used to explore the effect on microstructure variation.

The process parameters were as follows: The plasma forming gas flow rates were 40 lpm Ar and 10 lpm H_2; current input was 600 A; torch voltage was 66 V in the APS process and 70 V in the VPS process. For the same current and plasma gas composition and flow rate, the torch voltage was higher in VPS due to the longer arc formed between the cathode and anode under the lower operating pressure. The carrier gas was N_2 in all cases; in the APS process the powder feed rates were 24 g/min for HOSP and 22 g/min for F&C; in the VPS process powder feed rates were 29 g/min for HOSP and 28 g/min for F&C. The standoff distance in the APS process was 100 mm while in VPS several standoff distances (150, 175, 200, 210, 220, 230 mm) were used. For the VPS process, the chamber pressure was maintained around 100 mbar.

The real time in-flight particle properties (temperature and velocity) measurement was carried out by the DPV-2000 system developed by TECNAR Automation Ltd. (Saint-Bruno, Québec, Canada). In order to capture the particles traveling in the middle of the plasma jet, the sensing head of the DPV-2000 system was aligned before spraying. The spraying process parameters were kept the same as during coating deposition. In order to investigate the particle velocity and temperature evolution with standoff distance, 100 mm and 125 mm were used for the APS process. For the VPS process, the standoff distances ranged from 100 mm to 200 mm, in increments of 25 mm, in order to explore the particle variation under low pressure and compare with the APS process. To study the effect of different powders on the in-flight particle status, HOSP and F&C powder were used in the APS process. To compare the main difference between APS and VPS processes, HOSP powder was used in VPS process.

2.3 Coating characterization

The top surface of as-sprayed coatings were observed under scanning electron microscopy (SEM, Hitachi TM3000, Japan). Then samples were cut by a diamond-coated blade, mounted in epoxy, polished, and then imaged under SEM to analyse the cross-section microstructure. The porosity of each coating was measured by image analysis from SEM images at 500X magnification using ImageJ software. The deposition efficiency of the HOSP VPS and F&C VPS processes was estimated from the coating thickness and the number of torch passes.

3 Results and discussion

3.1 In-flight particle properties

The temperature and velocity variations of the HOSP powder at different standoff distances in APS and VPS are shown in **Fig. 2**. Compared to plasma operated in atmospheric pressure, the plasma jet under low pressure had a much weaker interaction with the ambient atmosphere. This resulted in a large increase of plasma jet velocity. Therefore, particles in the VPS process had a much higher velocity transferred from the high-velocity plasma jet under low pressure. As illustrated in **Fig. 2(a)**, the HOSP powder velocity in VPS reached 397 m/s and 414 m/s at 100 mm and 125 mm standoff distance respectively, while the velocity in APS was only 160 m/s and 148 m/s at the same distances. In addition, the particle velocity in VPS increased slowly from 100 mm and achieved the highest (431 m/s) at 150 mm, then the velocity decreased slowly. In contrast particle velocity in APS process declined gradually from 100 mm. Particle temperature evolution at various standoff distances is shown in **Fig. 2(b)**. Particle temperature in VPS was about 100°C lower than that in APS at the same standoff distances. This was due to the expansion of the plasma plume volume under low pressure, which resulted in a large decrease in plasma mass and energy density [1, 10] and to the shorter residence time of the particle in the plasma jet because of the higher velocity of the plasma jet. It should also be mentioned that the variation in particle temperature with standoff distance was very small. This indicates that little cold gas from the atmosphere surrounding the plasma jet was entrained in the jet, limiting the cooling of the plasma, and therefore of the particles.

The morphology and structure of the powder is closely correlated with the final microstructure of the YSZ coatings, as these powder characteristics influence the interaction with the plasma jet [5]. As shown in **Fig. 2(a)**, the velocity of the F&C powder was a little higher than the HOSP powder in the APS process at the same standoff distances. The velocities of the F&C powder at 100 mm and 125 mm were quite similar, while the HOSP powder velocity decreased gradually with increased standoff distance. This was probably due to the different size and mass of these two powders. The temperature of F&C powder at 100 mm was 3231°C and quickly dropped to 2862°C at 125 mm as shown in **Fig. 2(b)**. The temperature variation was much larger than for the HOSP powder, and requires further investigation and verification. In terms of the comparison of HOSP and F&C powder in the APS process, it was found that the F&C powder attained more heat and momentum transfer from the plasma jet.

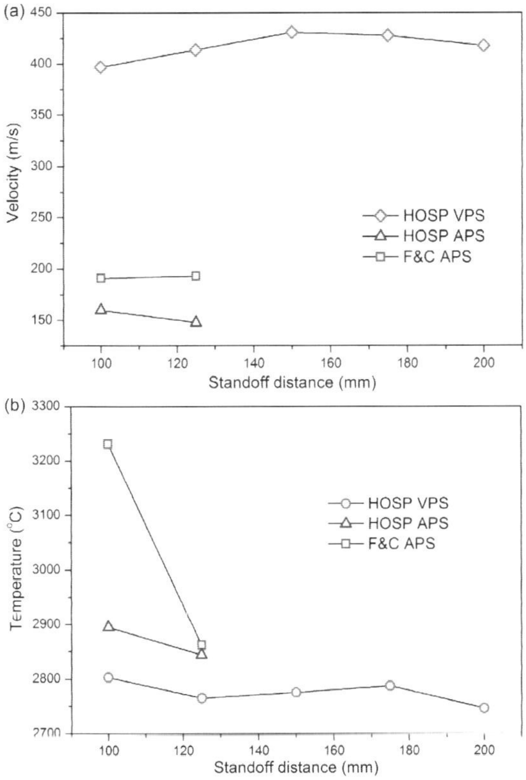

Fig. 2. In-flight particle properties for HOSP and F&C powder in VPS and APS processes: (a) particle velocity, (b) particle temperature

3.2 Coating microstructures

The porosity of coatings deposited under different spray conditions is illustrated in **Fig. 3**. It's demonstrated that the porosity of coatings formed in HOSP APS and F&C APS processes was quite similar, around 4%.

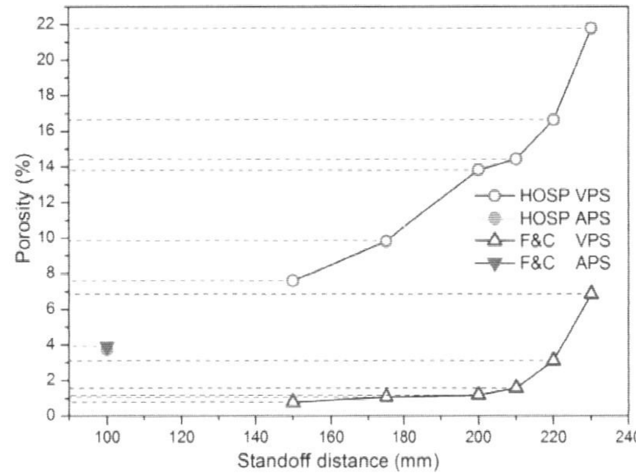

Fig. 3. Coating porosity in different plasma spray processes

The cross-section micrographs of these two APS coatings are shown in **Fig. 4** with the top surface microstructure as insets. Unmelted particles, pores, cracks and splat boundaries were found in both coatings. Compared with the F&C APS coating, the splat boundaries and cracks were more visible in the HOSP APS coating. The lamella structure was more apparent in HOSP APS coating.

For coatings deposited in the VPS process, the HOSP VPS coatings had a higher porosity than APS coatings and F&C VPS coatings. The porosity of HOSP VPS coatings at 150 mm was 7.6%, and climbed quickly with the increase of standoff distance. In contrast to APS coatings and HOSP VPS coatings, the F&C VPS coatings showed a much smaller porosity. The porosity remained at about 1% for standoff distances ranging from 150 mm to 200 mm. When the standoff distance was larger than 200 mm, the porosity increased quickly, which suggested that the melted particles re-solidified at these long standoff distances. It should be noted that although the porosity of the VPS coatings at 150 mm was small, samples were overheated and coatings delaminated from the substrate.

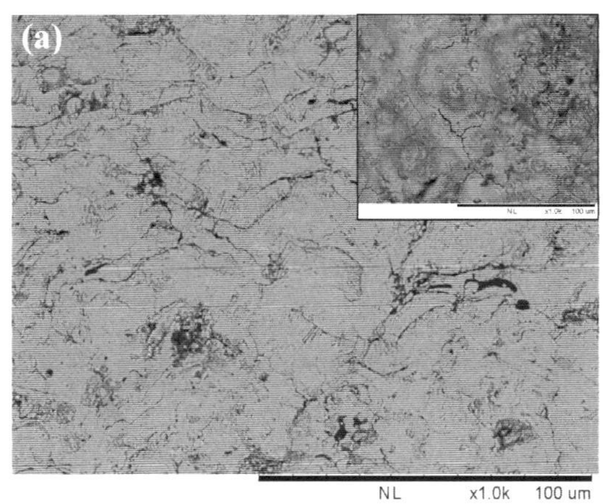

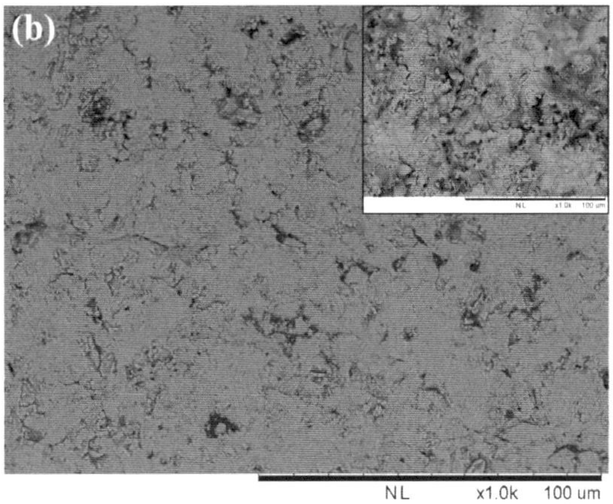

Fig. 4. Cross section micrographs of coatings with the top surface microstructure inset: (a) HOSP APS, (b) F&C APS

The cross sections of the HOSP VPS and F&C VPS coatings at 150 mm, 200 mm and 230 mm were selected as examples and are shown in **Fig. 5**. From the cross-section micrographs of HOSP VPS coatings, it was found coatings became more porous when the standoff distance increased. The amount of pores, cracks and unmelted particle piles grew and the coating thickness declined. For HOSP VPS coating at 150 mm, although the sample was overheated and coating delaminated from the substrate, the microstructure showed a smaller porosity compared with the longer distances. The top surface insets indicated the splat content dropped with the increase of standoff distance and more unmelted angular tiny particles appeared. It was demonstrated from the F&C VPS cross-sections that microstructures at 150 mm and 200 mm were relatively dense and similar. When the standoff distance reached 230 mm, more inter-splats cracks and unmelted particles were observed. This microstructure was likely caused by the resolidification of previously melted particles. From the comparison between HOSP VPS and F&C VPS cross-sections, it was found that F&C VPS coatings were much denser and thicker than HOSP VPS coatings. The cross-sections of F&C coatings exhibited many grains that still had angular shapes, which suggests that those particles were not fully melted and did not form splats. However due to the high velocity in the VPS process the partially-melted particles overlapped tightly and formed a compact and dense microstructure when landing on the surface. The HOSP powder also had a high velocity similar to the F&C powder in the VPS process, as shown in **Fig. 2**. However, the HOSP powder particles are hollow with a fairly dense outer shell of agglomerated fine particles, and the HOSP powder particle size was larger than the F&C power. Even though the DPV real-time measurement showed the HOSP particle surface temperature was above the melting point, it is possible that only the surface was melted while the interior of the shell was not. So when the HOSP powder landed on the surface with a very high velocity, the partially-melted particle ruptured and the small agglomerated particles were scattered on the surface as seen in the top surface microstructure of the HOSP VPS coatings in **Fig. 5**. The fractured agglomerated pieces increased the porosity and void content within the coating. The coating microstructure was, as expected, closely correlated with the powder original morphology and structure.

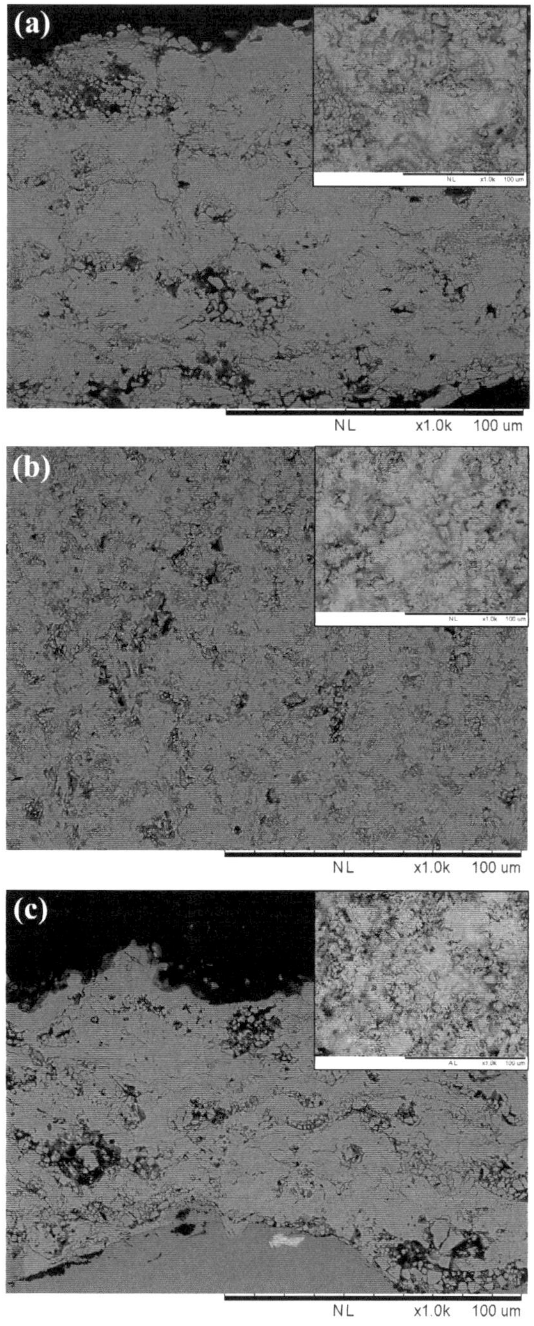

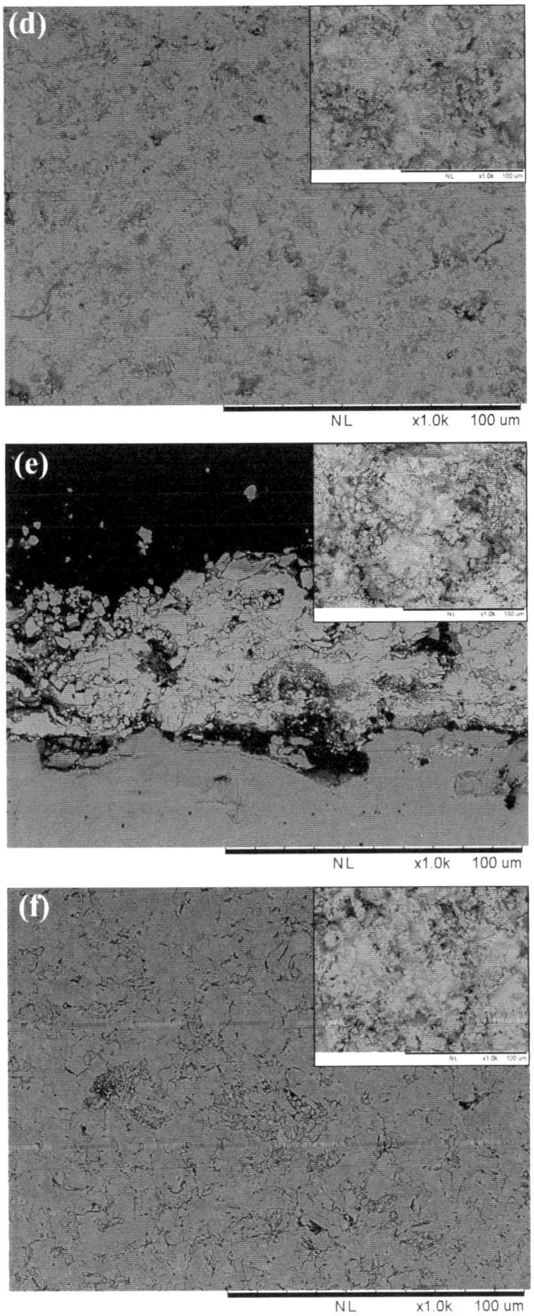

Fig. 5. Cross section micrographs of coatings with the top surface microstructure inset: (a) HOSP VPS at 150 mm, (b) F&C VPS at 150 mm, (c) HOSP VPS at 200 mm, (d) F&C VPS at 200 mm, (e) HOSP VPS at 230 mm, (f) F&C VPS at 230 mm

By using the coating thickness and the number of torch scanning passes, the coating thickness per pass was calculated as shown in **Fig. 6**. This can be used to estimate and compare the deposition efficiency between the two powder feedstocks in the VPS process. It was estimated that the deposition efficiency declined with the increase of standoff distance for both powders in the VPS process. The deposition efficiency was higher with F&C powder. This may be due to the hollow-shell structure of the HOSP powder particles that resulted in partial melting of the large diameter particles as noted above, and the rupture into small agglomerated pieces of the shell which didn't adhere to the surface.

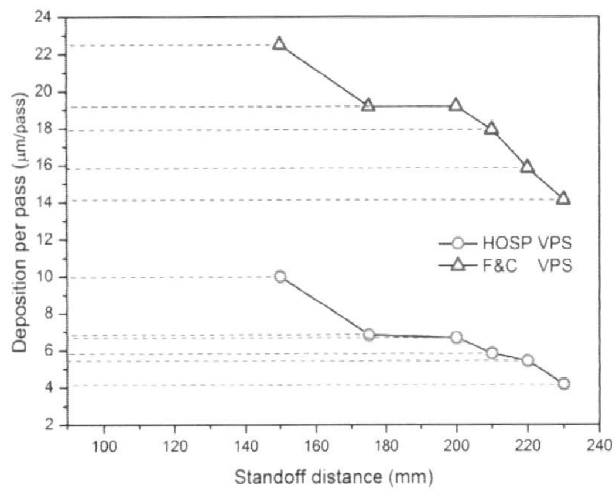

Fig. 6. Coating thickness per pass for HOSP VPS and F&C VPS coatings

4 Conclusions

Yttria stabilized zirconia (YSZ) coatings were deposited using HOSP and F&C powder via atmospheric plasma spray (APS) and vacuum plasma spray (VPS) processes. The in-flight particle properties, coating microstructure, porosity and deposition efficiency under different spraying conditions were investigated and compared. The results claimed that under the same process parameters, VPS endowed particles with a much higher velocity compared with spraying under atmospheric conditions, while particle temperatures were a little lower. YSZ powder morphology and structure played a significant role on the coating microstructure, porosity and deposition efficiency, especially in the vacuum plasma spray process. It was found that the F&C powder resulted in the deposition of denser coatings under the VPS process conditions employed.

5 Literature

[1] Smith MF, Hall AC, Fleetwood JD, Meyer P. Very low pressure plasma spray-a review of an emerging technology in the thermal spray community. Coatings 1 (2011), Issue 2, pp. 117/32.
[2] Von Niessen K, Gindrat M, Refke A. Vapor phase deposition using plasma spray-PVD™. Journal of thermal spray technology 19 (2010), Issue 1-2, pp. 502/9.
[3] Chi W, Sampath S, Wang H. Microstructure-thermal conductivity relationships for plasma-sprayed yttria-stabilized zirconia coatings. Journal of the

American Ceramic Society 91 (2008), Issue 8, pp. 2636/45.
[4] Vaidya A, Srinivasan V, Streibl T, Friis M, Chi W, Sampath S. Process maps for plasma spraying of yttria-stabilized zirconia: An integrated approach to design, optimization and reliability. Materials Science and Engineering: A 497 (2008), Issue 1, pp. 239/53.
[5] Kulkarni A, Vaidya A, Goland A, Sampath S, Herman H. Processing effects on porosity-property correlations in plasma sprayed yttria-stabilized zirconia coatings. Materials Science and Engineering: A 359 (2003), Issue 1, pp. 100/11.
[6] Huang W, Zhao Y, Fan X, Meng X, Wang Y, Cai X, Cao X, Wang Z. Effect of bond coats on thermal shock resistance of thermal barrier coatings deposited onto polymer matrix composites via air plasma spray process. Journal of thermal spray technology 22 (2013), Issue 6, pp. 918/25.
[7] Okajima Y, Nakamura T, Sampath S. Effect of Powder Injection on the Interfacial Fracture Toughness of Plasma-Sprayed Zirconia. Journal of thermal spray technology 22 (2013), Issue 2-3, pp. 166/74.
[8] Keyvani A, Saremi M, Sohi MH. An investigation on oxidation, hot corrosion and mechanical properties of plasma-sprayed conventional and nanostructured YSZ coatings. Surface and Coatings Technology 206 (2011), Issue 2, pp. 208/16.
[9] Pourbafrani M, Razavi RS, Bakhshi S, Loghman-Estarki M, Jamali H. Effect of microstructure and phase of nanostructured YSZ thermal barrier coatings on its thermal shock behaviour. Surface Engineering 31 (2015), Issue 1, pp. 64/73.
[10] Refke A, Barbezat G, Dorier J-L, Gindrat M, Hollenstein C. Characterization of LPPS processes under various spray conditions for potential applications. Proceedings of the 4th ITSC, Orlando, (2003)

Effect of oxidation on the bonding formation of plasma-sprayed stainless steel splats sprayed onto stainless steel substrate

Jun Wang, Chang-Jiu Li
State Key Laboratory for Mechanical Behavior of Materials, School of Materials Science and Engineering, Xi'an Jiaotong University, Xi'an, China / E-mail: licj@mail.xjtu.edu.cn

The properties and performances of plasma spray coatings depend on the coating-substrate interface adhesion and the splat-splat interface cohesion. Oxidation during atmospheric plasma spraying of metal alloy affects significantly the coating cohesion. In this study stainless steel splats were deposited on the preheated stainless steel substrate with the oxide scale of different thicknesses in inert low pressure plasma spay (LPPS) condition to examine the effect of *in-situ* oxidation of previous splat during spraying on the morphology of the later splat and splat bonding formation. The stainless substrate was preheated to 350 °C before splat deposition. The cross sections of splat-substrate interface were prepared by focus-ion-beam (FIB) processing. The splat morphology and splat-substrate interface bonding state were characterized by scanning electron microscopy (SEM). Results show that with the oxide films up to 35 nm thick, the disk-type splats are deposited, which bond to the oxidized substrate well at most splat interface except the periphery region. When oxide scale becomes ~100 nm thick and presents a micrometer rough surface, the splats present fingering shape without any effective bonding at the interface. Therefore, it is clear that the effect of the oxide on substrate surface on splat morphology and splat bonding formation depends on the thickness and surface morphology of oxide.

1 Introduction

Thermal spray processing is an advanced technology for forming protective coatings. Successive impact of ceramic or metal particles heated to fully molten or semi-molten state at accelerated high speeds by various heat sources onto substrate following by spreading and rapid solidification results in the formation of a coating. The properties of the coating are dominated by the lamellar structure [1], which is determined by particle parameters [2-3] including particle size, velocity and phase, and substrate parameters including thermo-physical property, skewness[4] and substrate temperature [5-6].

Based on the fundamental investigations for splat formation, the following important phenomenon was observed. When the substrate surface is preheated above a transition temperature, disk-shaped splats rather than splashed fingering shaped splats are usually observed for the substrate/splat combination in which both materials have a comparable melting point [7], which is known as a transition temperature [8]. With a transition from distorted fingering shape to a disk shape, the bonding strength of coating was found to be improved [5-6]. The adsorption/desorption of adsorbates and condensates [9-11], surface skewness [12], thermal contact resistance [13] and surface wetting ability [14] were investigated to clarify the effect of substrate temperature on the splat formation. Sampath et al. [15] postulated that substrate could be cleaned through desorption/evaporation to eliminate the disturbance during splat spreading. Li et al. [16] also showed that regular disk type splats were formed when preheating temperature exceeds 50 °C over the boiling point of organic substance brushed on substrate surface. Later on, it was argued that after desorption of adsorbates and condensates by preheating, the positive skewness of substrate surface and cooling velocity of splat were increased at least by a factor of 2 and 1.4 to1.7, respectively [17]. Thus, a uniform and fast cooling rate because of a better splat-substrate contact and the improvement of drastic wettability due to positive skewness formation can be confirmed to promote disk splat formation.

With the increase of substrate temperature, the interface temperature between substrate and splats also increases, which improves the bonding quality. Recently, Yang et al. [18] proposed that there was a critical bonding temperature for the formation of effective splat interface bonding in plasma-sprayed ceramic coating. The bonding temperature is ~ 300 °C and ~ 600 °C for Al_2O_3 and YSZ, respectively.

All those investigations suggested that substrate temperature is one of the key factors for suppression of splashing and promotion of the formation of the splat bonding during splatting. Whereas, an oxide layer formation on metallic substrate surface during pre-heating or metal splat surface during deposition at a high deposition temperature prior droplet impact or following splat deposition during the air plasma spraying process is inevitabl, which inhibits the direct contact between fresh metal constituent during metal alloy spraying. Recently many investigations were inducted to clarify the effect of oxide layer on splat shape and interface bonding. Syed et al. [19] observed that entrapped gases in the porous oxide layer on substrate surface cause a porous splat-substrate interface and fingered splats. Trompetter et al. [20] confirmed that thick oxide layer inhibited the bonding formation and effective bonding occurred only when surface oxide was substantially deformed or disrupted to produce mechanically interlocking feature. However, McDonald et al. [21] indicated that the thin oxidation layer on the substrate surface has a little effect on the interface thermal resistance, and disk splats were also formed on oxidized substrate. Along with that, Brossard et al. [22] showed substrate melting occurred for preheated stainless substrate,

but not for aluminium substrate. All these studies show that oxide layer affects the interface bonding formation.
In this paper, using stainless substrate and stainless spray powder particles, the influence of oxide layer thickness on the substrate surface on the splat shape and bonding formation between molten droplet and underlying substrate with identical composition was studied to clarify the effect of *in-situ* oxidation during APS of metal coating on the inter-splat bonding evolution.

2 Experimental procedures

Commercially available 304 stainless steel powder with a size range from 50 to 57 μm was employed to deposit a single layer splat on oxidized AISI304 stainless steel surfaces with the oxide scales of three different thicknesses. To control the oxide thickness, the substrates were first polished with abrasive papers up to a grade No.2000 and finally polished with 1.5 μm diamond. Then they were thermally treated in a pipe type oven at three different conditions as shown in Table 1 as reported by Betz et al. Thus, the stainless substrates with the oxide of three different thicknesses were prepared. The AISI304 stainless steel splats were sprayed onto the prepared substrate surfaces at a deposition temperature of 350 °C by low pressure plasma spraying (LPPS) within the Ar-charged chamber at a pressure less than 50 kPa. The deposition temperature of substrate was accurately controlled by a calibrated heater. Plasma spraying was carried out with a GTV plasma spray system operating at a current of 600 A and a voltage of 67 V. The spray distance between the torch exit and the substrate surface was kept at 100 mm. A plate with small holes in a diameter of 1 mm was placed between substrate and spray torch to allow the deposition of single splat on substrate by avoiding additional heating of plasma flame to the substrate during splat deposition. The plasma jet was generated by using argon and hydrogen at flow rates of 40 and 6 SLPM, respectively. The samples with splats deposited for three oxide thicknesses are referred to as the Sample T1, Sample T2 and Sample T3, respectively. In this study, the splat morphology was examined by using a scanning electron microscope (SEM). The cross-sections of splats were prepared by focus ion beam (FIB) cutting at different splat typical regions for examining the bonding formation at the interface between splat and the substrate by SEM.

Table 1 Substrate thermal heated conditions and estimated oxide thickness [23]

Specimen	Pre-treatment	Oxide thickness
T1	200 °C×1 hour	8 nm
T2	450 °C×1 hour	35 nm
T3	900 °C×0.5 hour	100 nm

3 Results and discussions

3.1 Effect of oxide on splat morphology

The morphology of the splats deposited on the substrates with the oxide of different thicknesses are shown in Fig.1. The disk-type splats are obtained on Specimens T1 and T2, which exhibit a flat central zone taking substantial fraction of splat and periphery rim with a slightly raised height (Fig.1a and Fig.1b). However, with the sample T3, splat with fingered morphology was observed, indicating significant splashing during splatting. Therefore, irregular-shaped splat in Fig.1c indicates that a oxide layer over a certain thickness provokes the splashing during spreading of molten droplet despite higher substrate temperature than the transition temperature.

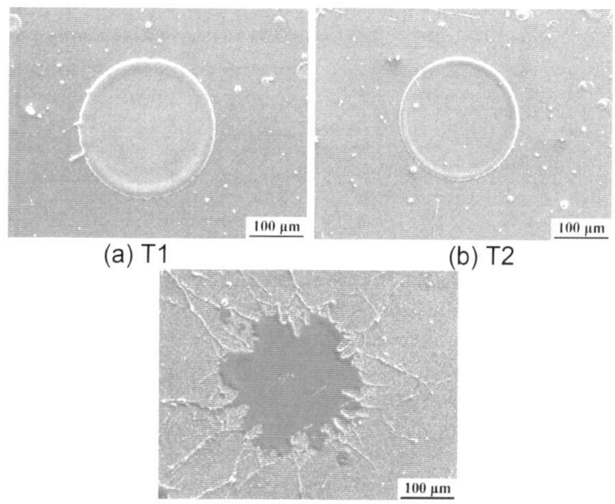

(a) T1 (b) T2

(c) T3

Fig. 1 Morphology of single splat sprayed onto different AISI304 substrates

3.2 Effect of oxide on the formation of splat interface bonding

Three cross sectional samples of splats revealing the interface bonding state were prepared by FIB technique at three typical regions for each sample as shown in Fig.2a, Fig.4a and Fig.5a. The close examination into the interface bonding from the splat cross sections reveals that the bonding is present at the interfaces sampled at the regioins marked A and B, which are located at the inner part of splat specimen T1, while for the interface at the periphery marked C at the rim zone (Fig.2) no effective bonding was observed. An electrolytic etching process was performed to peel off the single splat for further analysing the interface bonding state. The corrosion occurred first in the central zone, where etching pits were concentrated. After the etching was increased to certain time, although partial splat material at the central zone and most of splat at thick periphery were left (Fig.3a), It can be evidently observed that the grain morphologies of substrate microstructure appeared observed through the residual thin splat material in the central zone (Fig.3b). This fact indicates that the perfect bonding formed over these central part of the splat, which is consistent with the results revealed by FIB prepared samples.

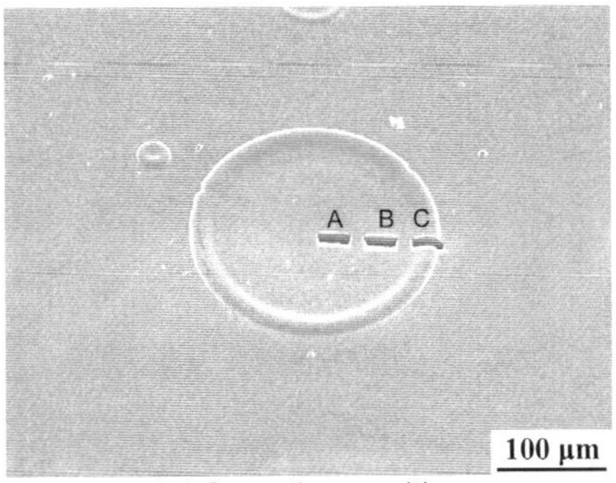

(a) Sampling positions

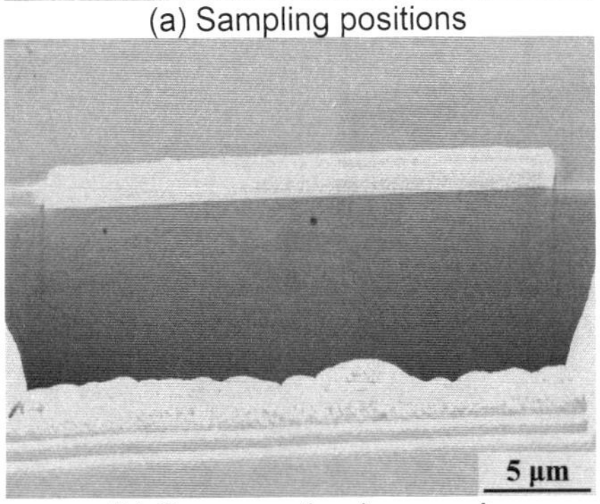

(b) Cross section in zone A

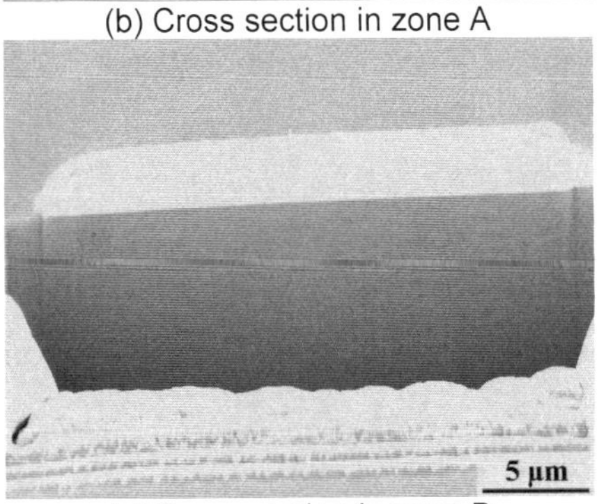

(c) Cross section in zone B

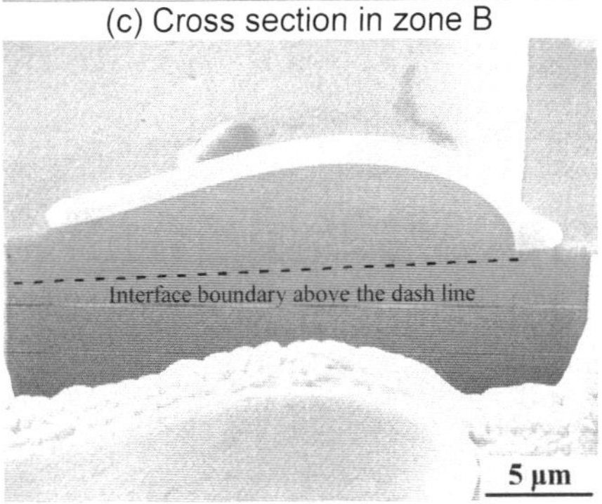

(d) Cross section in zone C

Fig. 2 FIB sampling positions for typical splat interface for the sample T1 (a), cross sections of the sample 1 (b), (c) and (d) showing the bonding state of the interface at typical splat regions marked as A, B and C in (a)

With the oxide thickness increasing to about 35 nm, it was observed that the bonding interface decreased slightly compared with the sample T1. From the cross sections of the splat shown in Fig.4, at the central region of the splat marked A the bonding is present at the whole interface, while at the interface marked C in the periphery of splat no effective bonding was observed. These facts are the same with that observed for the sample T1 with a thin oxide of 8 nm. However, for the interface at the region marked B a small part interface of about 5 µm long at the periphery side presented no ineffective bonding. On the other hand, all cross sections of the splat in the sample T3 despite location either at the central region as marked A, inner region marked B, or periphery region marked as C clearly presented no interface bonding (Fig.5). Therefore, it is clear that the effect of oxide film on the interface bonding formation becomes significant only when its thickness of oxide layer on the substrate is increased over certain value.

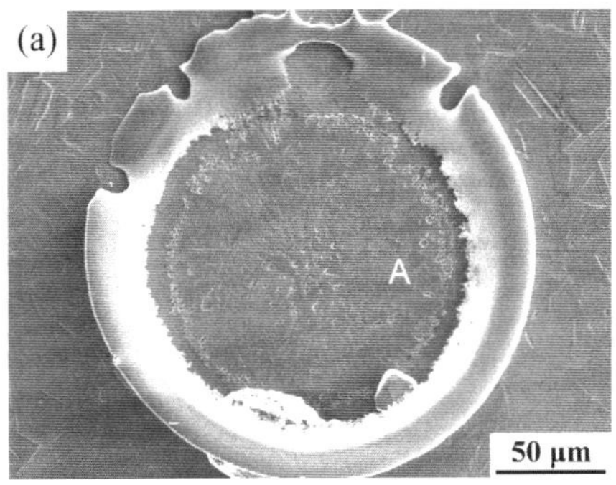

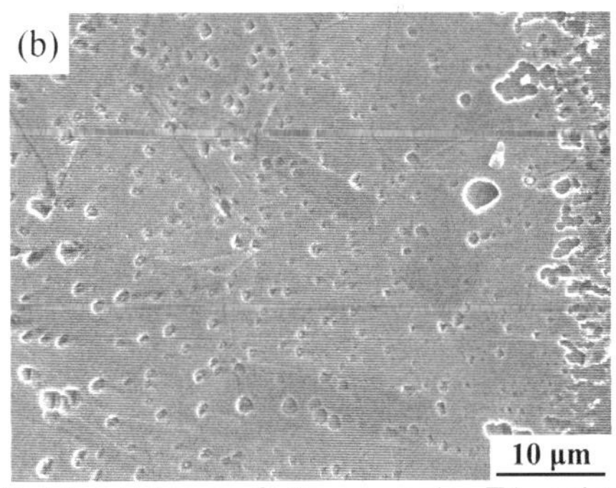

Fig. 3 Morphology of the sample T1 splat after electrolytic etching process (a) entire splat, (b) zone A in (a)

3.3 Discussions

3.3.1 Effect of oxide layer on splat formation

During the splat formation process, the spreading of molten melt is first interfered by the surface roughness [4] and then affected by wetting property [14]. At the same time, splat shape is also determined by the cooling rate of splat which is affected by the interface contact [17]. As a result, the skewness of substrate and splat-substrate interface thermal transfer should be investigated to examine the effect of oxide on the splat formation.

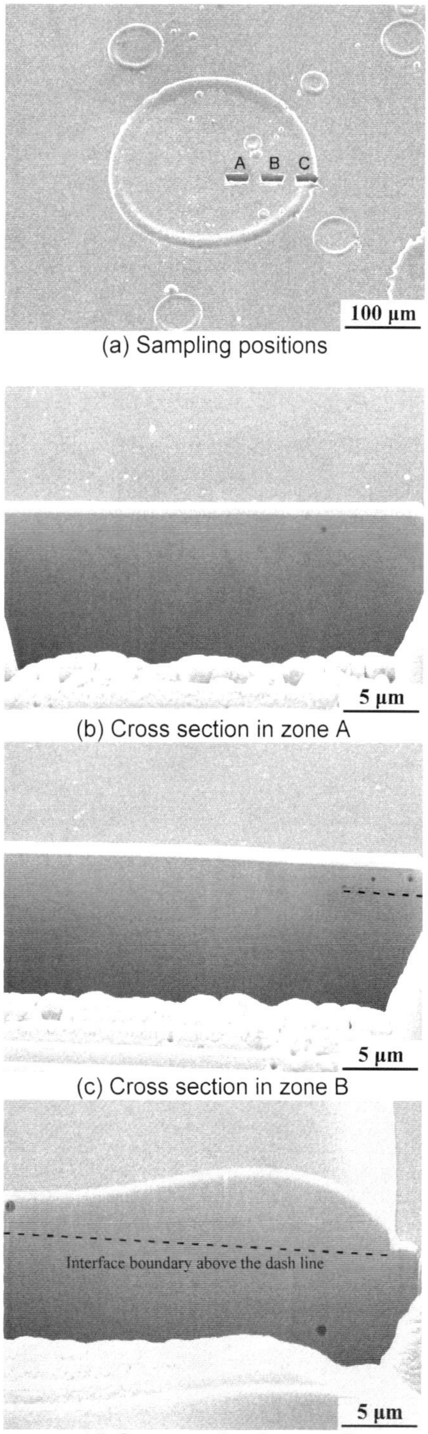

(a) Sampling positions

(b) Cross section in zone A

(c) Cross section in zone B

(d) Cross section in zone C

Fig. 4 FIB sampling positions for typical splat interface for the sample T2 (a), cross sections of the sample 1 (b), (c) and (d) showing the bonding state of the interface at typical splat regions marked as A, B and C in (a)

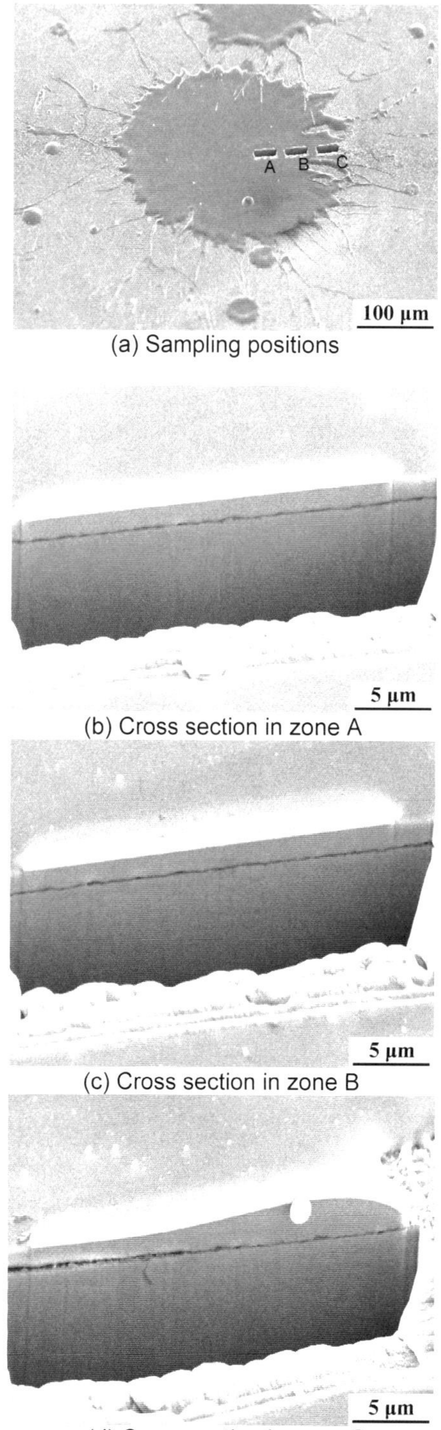

(a) Sampling positions

(b) Cross section in zone A

(c) Cross section in zone B

(d) Cross section in zone C

Fig. 5 FIB sampling positions for typical splat interface for the sample T3 (a), cross sections of the sample 1 (b), (c) and (d) showing the bonding state of the interface at typical splat regions marked as A, B and C in (a)

During the LPPS process, only oxidation in pre-oxidation process is considered except that could be produced by preheating in the argon atmosphere. With the oxide thickness increasing, the substrate surfaces were observed to appear bright, yellow and brown during preheating of substrate for splat deposition for specimen T1, T2 and T3, respectively. No significant surface roughening was observed by SEM examination for the samples T1 and T2 due to nanometer scale roughness. However, the surface tomography of the sample T3 resembles grain structure (Fig.6a), and peaks and valleys in hundreds of nanometers scale were shown at high magnification (Fig.6b).

The disk-type splats are insistently remained on the pre-oxidized surfaces preheated to 350 °C as shown in Fig.1 for the samples T1 and T2, whereas the splats with fragmented morphology were observed on pre-oxidized surfaces for T1 and T2 samples when splat deposition was carried out at room temperature. Those results confirmed that the desorption and reabsorption of adsorbates evidently play a dominant role during splatting, and the effect of oxide layer on the thermal resistance is too less to be significant [13]. Cedelle et al. [12] proposed that positive skewness resulted from pre-treatment accounted for the better wetting of the splat. Based on the interface bonding state revealed by FIB samples shown in Fig.2 and Fig.4 it is clear that during spreading process for the samples T1 and T2 the melt was keeping good contact with solid surface. Therefore, that the disk-shaped splats are formed on the specimen T1 and T2 can be attributed to good melt-substrate contact and thus rapid cooling rates during melt spreading. However, fragment splats are formed on the specimen T3.

Possible explanations for the splashing phenomena include desorption of adsorbates/ condensates through heating the substrates [15-16] and the rapid solidification on-set of splat during flattening [24]. Since the substrate temperature for the specimen T3 exceeds the transition temperature, the former explanation can not be applied to the sample T3. The later theory is related to the thermal contact between spreading melt and solid substrate surface. When the thermal contact resistance decreases slightly on highly oxidized substrate, it results in a higher cooling rate of the splat. The splat periphery may solidify before spreading is complete, which acts as a barrier for the rest fluid spreading, and the flowing over the solidified rim promotes splashing. From the interface bonding revealed by FIB samples shown in Fig.5, it was clearly revealed that over the whole interface no effective bonding was formed. The previous investigations suggested that such poor bonding is associated to a poor contact during molten droplet spreading [25] with a high thermal contact resistance. Under such case, the splashing can not be reasonably explained by the early on-set of local solidification. On the other hand, the significantly roughened surface morphology as observed in Fig.6 makes laterally spreading melt not be able to fill into all asperities, which leads to poor contact and thus high thermal contact resistance. Such effect reduces cooling rate of spreading melt. Thus, the large thermal resistance of the interface provides liquid enough time spreading. The flattening is not restrained because a driving energy for the flattening transfers to viscous energy ineffectively and finally the melt splat disintegrates by the combined effect of inertia and poor wetting. Thus, the poor wetting and poor melt/solid interface contact originated from the roughened surface are responsible for splashing and subsequent formation of irregular splats on the stainless substrate with a thick oxide layer.

Fig. 6 Morphology of substrate surface after peroxidation for specimen T3 in (a) low magnification, (b) high magnification

3.3.2 Effect of oxide layer on the splat bonding formation

Until now, few efforts [19-21] have been made to examine the influence of oxide on the bonding formation. It is usually considered that the oxide layer in the splat-oxide-substrate sandwich-like interface is detrimental to metallic bonding formation. Since the oxidation of metal splat surface during atmospheric plasma spraying is inevitable, the metal-oxide-metal sandwich structure is inherent to thermally spray metal coatings. The first question is how oxide influences the splat bonding formation during splatting. The present investigation was designed to use the stainless steel substrate with the comparable compositions to the spray droplets to simulate the effect of *in-situ* oxidation during spraying on splat bonding formation. The present results clearly showed that no effective bonding was observed at all interface for the sample T3, while the bonding is present at the interface over almost part of splats for sample T1 and T2. Therefore, it is evident that the effect of oxide on the substrate surface on the splat bonding formation depends on the thickness of oxide.

In recent researches, the the bonding formation during splat formation is closely related with the interface temperature. Yao et al. [26] proposed an Intrinsic Bonding Temperature concept for the splat bonding formation of ceramic coatings. The effective bonding forms when the interface temperature exceeds such temperature prior solidification. For three cases in the present investigation, the recognized formation of the effective bonding in the most part of splats for the specimens T1 and T2 reveals that the interface temperature is high enough to form the bonding.

Since a thick oxide on the substrate surface reduces the heat sink effect and thus tends to raise the interface temperature provided the spreading melt is in an intimate contact to the solid substrate. Accordingly, for the sample T3 the splat should be well bonded to the substrate. However, the FIB-prepared cross section samples revealed no bonding formation at the interface for the sample T3. Therefore, there are two possible reasons that result in the unbonding of splat. One is related to the thermal contact resistance at the interface between spreading melt and solid substrate. The rough substrate surface of the specimen T3 includes valleys and peaks, which results in different thermal resistance and overall high thermal resistance. The high thermal resistance reduces heat transfer rate over the interface and thus such micrometre scale roughness causes a decrease of the cooling rate [27]. Another one is related to the poor contact of spreading melt to the micrometer scale rough surface. The valleys on the substrate surface easily entrap the gas to form gas pockets at the interface between the spreading melt and solid surface. Thus, the intimate contact of spreading liquid to the solid surface is hindered, which makes it impossible for spreading melt to form the bonding with the underlying solid substrate surface. Therefore, micrometer scale roughening is responsible for the poor bonding for the sample T3. The future investigation will focus on the understanding the possible critical surface roughness scale below which the bonding formation is not affected and related physical processes in details.

4 Conclusions

Results showed that the effect of the oxide on substrate surface on splat morphology and splat bonding formation depends on the thickness of oxide. Disk splat can be obtained on the preoxided substrate with a thin oxide layer up to ~35 nm. With the increase of oxide thickness ~100 nm, the splats displayed an irregular shape, which is attributed to splashing during spreading of molten droplet. When the thickness of oxide layer on the substrate surface is up to 35 nm, the interface at most part of metal splat was well bonded to the substrate possibly via thin oxide film. The bonding of stainless steel splat is likely perfect to stainless steel substrate through copying of grain morphology of substrate at a deposition temperature of 350°C. It was found that no effective bonding was present at the entire interface of the stainless steel splat deposited on the heavily oxidized stainless substrate of an oxide of ~100 nm thick. It was proposed that the micrometer scale roughened surface with numerous valleys and peaks resulted from oxidation inhinders the effective contact of spreading melt to the underlying solid surface and thus the bonding formation. Moreover, the poor contact during splatting limits the cooling of spreading melt and promotes splashing, resulting in the formation of irregularly shaped of splats.

Acknowledgements

The present project is financially supported by National Natural Science Foundation of China (No. 51474171)

Literature

[1] C.-J.Li, A.Ohmori, Relationship between the structure and properties of thermally sprayed deposits, J. Thermal Spray Technol., 11 (2002), Issue 4, pp. 365-374.

[2] C.-J.Li, G.-J.Y, Relationships between feedstock structure, particle parameter, coating deposition, microstructure and properties for thermally sprayed conventional and nanostructured WC–Co, J. Refractory Metals and Hard Materials, 39 (2013), Issue 1-2, pp. 2-17.

[3] S.-P.Wang, G.-X.Wang, Deposition of a molten layer of high melting point material: substrate melting and resolidification, Mater. Sci. Eng.: A, 262 (1999), Issue 1-2, pp. 25-32.

[4] U.Bardi, L.Carrafiello, R.Groppetti, F.Niccolai, G.Rizzi, A.Scrivani, F.Tedeschi, On the surface preparation of nickel superalloys before CoNiCrAlY deposition by thermal spray, Surf. Coat. Technol., 184 (2004), Issue 2-3, pp. 156-162.

[5] V. Pershin, M. Lufitha, S. Chandra, J. Mostaghimi, Effect of substrate temperature on adhesion strength of plasma-sprayed nickel coatings, J. Thermal Spray Technol., 12 (2003), Issue 3, pp 370-376.

[6] K.Yang, M.Fukumoto, T.Yasui, M.Yamada, Study of Substrate Preheating on Flattening Behavior of Thermal-Sprayed Copper Particles, J. Thermal Spray Technol., 19 (2010), Issue 6, pp. 1195-1205.

[7] C.-J.Li, C.-X.Li, G.-J.Y, Examination of substrate surface melting-induced splashing during splat formation in plasma spraying, J. Thermal Spray Technol., 15 (2006), Issue 4, pp. 717-724.

[8] M. Fukumoto, S. Katoh, I. Okane, Splat behavior of plasma sprayed particles on flat substrate surface, Proceedings of the 14th International thermal Spray Conference, 1995, pp. 353–358.

[9] C.-J.Li, J.-L.Li, W.-B.W, the effect of substrate preheating and surface organic covering on splat formation, Proceedings of 15th International Thermal Spray Conference, 1998, pp. 473-480.

[10] X.-Y.J, Y.-P.W, Herbert Herman, Sanjay Sampath, Role of condensates and adsorbates on substrate surface on fragmentation of impinging molten droplets during thermal spray, Thin Solid Films, 385 (2001), Issue 1–2, pp. 132–141.

[11] C.-J.Li, J.-L.Li, Evaporated-gas-induced splashing model for splat formation during plasma spraying, Surf. Coat. Technol., 184 (2004), Issue 1. pp. 13-23.

[12] J. Cedelle, M. Vardelle, P. Fauchais, Influence of stainless steel substrate preheating on surface topography and on millimeter- and micrometer-sized splat formation, Surf. Coat. Technol., 201 (2006), Issue 3–4, pp. 1373–1382.

[13] A.Mcdonald, C.Moreau, S.Chandra, Effect of substrate oxidation on spreading of plasma-sprayed nickel on stainless steel, Surf. Coat. Technol., 202 (2007), Issue 1, pp. 23-33.
[14] Y.Tanaka, M.Fukumoto, Investigation of dominating factors on flattening behavior of plasma sprayed ceramic particles, Surf. Coat. Technol., 120-121 (1999), pp. 124-130.
[15] S. Sampath, X. Jiang, Splat formation and microstructure development during plasma spraying: deposition temperature effects, Mater. Sci. Eng.: A, 304-306 (2001), Issue 31, pp.144-150.
[16] H.Li, S.Costil, H.-L.Liao, C.-J. Li, M. Planche, C. Coddet, Effects of surface conditions on the flattening behavior of plasma sprayed Cu splats, Surf. Coat. Technol., 200 (2006), Issue 18-19, pp. 5435-5446.
[17] J. Cedelle, M. Vardelle, P. Fauchais, Influence of stainless steel substrate preheating on surface topography and on millimeter- and micrometer-sized splat formation, Surf. Coat. Technol., 201 (2006), Issue 3-4, pp. 1373-1382.
[18] G.-J.Y, C.-X.L, S.H, S.H, Y.-Z.X, E.-J.Y, C.-J.Li, Critical bonding temperature for the splat bonding formation during plasma spraying of ceramic materials, Surf. Coat. Technol., 235 (2013), Issue 25, pp. 841-847.
[19] A.A.Syed, A.Denoirjean, B. Hannoyer, P.Fauchais, P.Denoirjean, A.A.Khan, J.C. Labbe, Influence of substrate surface conditions on the plasma sprayed ceramic and metallic particles flattening, Surf. Coat. Technol., 200 (2005), Issue 7, pp. 2317-2331.
[20] W.Trompetter, M.Hyland, D. McGrouther, P.Munroe, A.Markwitz, The Effect of Substrate Surface Oxides on the Bonding of NiCr Alloy Particles HVAF Thermally Sprayed onto Aluminum Substrates, J. Thermal Spray Technol., 19 (2010), Issue 5. pp 1024-1031.
[21] A.McDonald, C.Moreau, S.Chandra, Effect of substrate oxidation on spreading of plasma-sprayed nickel on stainless steel, Surf. Coat. Technol., 202 (2007), Issue 1, pp. 23-33.
[22] S.Brossard, P.R.Munroe, A.Tran, Hyland.M.M, Study of the Splat-Substrate Interface for a NiCr Coating Plasma Sprayed onto Polished Aluminum and Stainless Steel Substrates, J. Thermal Spray Technol., 19 (2010), Issue 1, pp. 24-30
[23] G.Betz, G.K.Wehner, L.Toth. Composition-ys-depth profiles obtained with Auger electron spectroscopy of air-oxidized stainless-steel surfaces, J. Appl. Phys., 45 (1974), Issue 12, pp. 5312-5216.
[24] S.Goutier, M.Vardelle, P.Fauchais, Understanding of Spray Coating Adhesion Through the Formation of a Single Lamella, J. Thermal Spray Technol., 21 (2012), Issue 3-4, pp. 522-530.
[25] E.-J.Y, X.-T.L, G.-J.Y, C.-X.Li, C.-J.Li, A TEM Study of the Microstructure of Plasma-Sprayed YSZ Near Inter-splat Interfaces, J. Thermal Spray Technol., 24 (2015), Issue 6, pp.907-914.
[26] S.-W.Y, G.-J.Y, C.-X.Li, X.-T.L, C.-J.Li, Understanding the Formation of the Limited Inter-Lamellar Bonding in Thermal Spray Ceramic Coatings Based on the Intrinsic Bonding Temperature Concept, Proceedings of the International Thermal Spray Conference, 2015, pp. 767-773.
[27] D.K.Christoulis, D.I.Pantelis, N.De Dave-Fabregue, Borit, F. Guipont, V.Jeandin, M, Effect of substrate temperature and roughness on the solidification of copper plasma sprayed droplets, Materi.Sci. Eng.: A, 485 (2008), pp. 119-129

Characteristic of NiCoCrAlYTa-10%Al_2O_3 Coating Fabricated by Supersonic Atmospheric Plasma Spraying

Kun Yang[1, 2, 3], Min Liu[1, 2, 3], Changguang Deng[1, 2, 3], Jie Mao[1, 2, 3], Shaopeng Niu[1, 2, 3]

1. Guangdong Institute of New Materials, Guangzhou / China
2. National Engineering Laboratory for Modern Materials Surface Engineering Technology, Guangzhou / China
3. The Key Lab of Guangdong for Modern Surface Engineering Technology, Guangzhou / China

In this study, NiCoCrAlYTa + 10% Al_2O_3 powder was thermally sprayed onto the grit-blasted Cr28Ni48W5 heat resistant steel substrate by supersonic atmospheric plasma spraying (SAPS). The microstructure and properties, like adhesion strength, porosity, oxygen content, micro-hardness and wear resistance of the coating was systematically evaluated. According to the result, the coating structure is dense and adhesive well with the substrate, it also has excellent wear resistance, which could become a potential candidate in the hearth roll coatings.

1 Introduction

In the modern steel plant, hearth roll is the key component in the continuous annealing furnace, which is exposed to a combination of high temperatures, corrosive environments, and wear. In order to improve steel sheet quality and to prolong the roll service life, the surface of hearth roll need to be protected with thick coatings. Thermal spray is such a kind of surface strengthening and surface modification technology technology. The earliest furnace roll coating dates back to the late 1970's in the USA, and wide spread use of coatings began in the early 1980's. Coating use is expanding to the world since the majority of automotive manufacturer in the world demand strip with the highest surface quality. Now, almost all furnace rolls are coated for improving product quality and/or reducing maintenance costs [1]. At temperature up to 850℃, the chromium carbide cermet coating (Cr_3C_2-NiCr) has been extensively used, this coating provides good thermal and wear resistance, moderate build-up resistance, it also has a relative high hardness. At higher temperatures, or after extended periods in service, this coating is prone to spallation as a result of thermal cycling during furnace shut-downs [2], the build-up formation and wear rates are also accelerated at higher temperatures, some other coatings has been recommended because of its superior build-up resistance, but the wear resistance of the coating is unfavorable in most instances [3]. Meanwhile, the oxidation resistance is highly demanded due to the harsh working conditions at high temperature, hence many coatings, such as MCrAlY-ceramic-type cermet, have been developed recently [4, 5].

A supersonic plasma spraying (SAPS) system was developed since 1986 [6, 7]. In this system, hybridizing high voltage extended arc plasma with high gas acceleration in an extended nozzle, which led to gas velocity in a supersonic mode at the nozzle exit, it is beneficial for the adhesion strength, the density and the porosity of the coating. Consequently, the SAPS-coatings exhibits obvious superiority for obtaining coatings with high-quality structure and preferential performance as compared with conventional atmospheric plasma spraying [8, 9].

In this paper, pre-mixed NiCoCrAlYTa + 10% Al_2O_3 powders was thermally sprayed onto the substrate by SAPS, the properties of the coating obtained were evaluated in detail.

2 Experimental procedures

Cr28Ni48W5 heat resistant steel was used as the substrate. The substrate was ultrasonically cleaned and grit-blasted with alumina powder to remove any contamination. The commercially available gas atomized NiCoCrAlYTa alloy powder (AMDRY 997, oerlikon Metco, USA) of particle size lying in the range 5 to 35 μm, and fused Al_2O_3 ceramic powder (AMPERIT ®740, H. C. Starck, Germany) of particle size ranging from 5 to 22 μm was used for coatings. The NiCoCrAlYTa and Al_2O_3 powders were pre-mixed with a ratio of 90 vs. 10%. The metallic component has excellent resistance to both oxidation and hot corrosion, and the ceramic component significantly enhance resistance to wear [10].

The spraying was carried out by SAPS (PLAZER 180 PL–S, Ukraine). The spraying distance between the gun and the substrate was kept at 200 mm. The pre-mixed powders were injected into the plasma jet at a feeding rate of 33 g/min. Air and propane were used as operating gas (referred to as "coating I"). For comparison purposes, some coatings were also fabricated without propane as operating gas (referred to as "coating II"). During the spraying, the substrate was located perpendicular to the spray gun and cooled by compressed air.

The morphologies of the coatings were observed by scanning electron microscope (SEM). The image processing and analysis software: ImageJ (Image Processing and Analysis in Java, National Institutes of Health) was employed to quantify the porosity. The evaluation was based on the SEM images at coating cross section. Vickers hardness of the coating, substrate near the coating-substrate interface, and coating was measured using a load of 300 g for 15 s, the indentations were taken on polished cross-section of the samples, more than 10 locations were measured, and the average value was used to determine the hardness. The oxygen content of the feedstock powder and coatings were measured using TC600 oxygen/nitrogen determinator, the coatings were detached from the substrate for the measurement. Coating performance closely depends on its adhesion to substrate, hence the adhesion strength measurements were carried out using universal testing machine in accordance with the standard American Society for Testing and Materials C633-79. Wear tests were performed with a universal macro-tribometer using the ball-on-disk configuration. The tests were performed at room temperature and 900℃, respectively. The test samples were polished to reach a flat and mirror-like surface finish prior to the measurement. In order to conduct the Fe buildup test, a thin and uniform layer of Fe_3O_4 powders were spread on the coatings, then covered by a steel strip in the center and were fixed by a binder clip. Buildup tests were conducted in a furnace at 1050 ℃ with N_2-5%H_2 for 4 hours. After cooling to room temperature, nonadherent oxide powders were removed and samples were ultrasonically cleansed in alcohol. Cross-sections and element distribution were examined by SEM and electron probe micro analysis (EPMA), respectively.

3 Results and Discussion

3.1 As-sprayed coatings

All coating thicknesses obtained were around 300 μm. The cross section morphologies and Al element distributions of both coating I and coating II were observed as shown in Fig. 1. The microstructure of coating I is lamellar and contained some pores, typical interface composed by thin layer between the lamellar could be found (Fig. 1-a1). According to Fig. 1-a2, continuous thin Al layer between the lamellar could be found, which indicated most of the Al_2O_3 component were fully melted and deposited as splat. On the other hand, some semi-melted and unmelted particles could be found in coating II (Fig. 1-b1), in particular, un-melt Al_2O_3 particles with almost original shape trapped between NiCoCrAlYTa particles could be found in the coating, this can be proved by the Al element distribution (Fig. 1-b2). In the coating II deposition process, there is no propane in the operating gas while other spraying parameters were kept in constant, consequently, the plama temperature should be lower due to the reduced power compared with coating I.

Some more properties of the coatings deposited with/without propane as operating gas were carried out and summarized in Table 1. The porosity of coating I and coating II was 3.0% and 2.2%, respectively. The adsorbed gas condensation on the substrate surface, entrapped environmental gas during the in-flight process and the fragmentation debris from the individual splash splats worked jointly for the pore formation in the coating.

The initial oxygen content of the original powder was as low as 0.18wt. %, while the oxygen content in coating I remarkably increased to 2.5 wt. %, and the value was 1.0 wt.% in coating II. In general, the oxidation is highly dependent on the particle in-flight characteristics, most of the oxidation should happen during the acceleration in the plasma flame. The higher oxygen content in coating I should be caused by the higher plasma flame temperature. The continuing oxidation with the oxygen in the air might also occur even after the coating deposited on the substrate.

The adhesion strength was 46.9 MPa when propane were added in the operating gas (coating I). A significant decrease in the adhesion strength with only air as operating gas was 29.1 MPa (coating II). The images of fracture surfaces after they were pulled apart from substrate are shown in Fig. 2. According to the figure, most of the coatings still remained on the substrate (Fig. 2-a1), which indicated that well adhesion might exist between the coating and substrate in coating I. From the precise observation using SEM, significant deformation could be found and most of the damages occurred inside the individual splat (Fig. 2- a2). The cohesion between the splats is quite favourable. On the other hand, the failure happened at the interface between coating II and substrate (Fig. 2-b1), most of the coatings was removed, and only a small portion of coating remained on the substrate surface. The SEM image shows only few individual splats with pores could be found on the substrate(Fig. 2-b2).

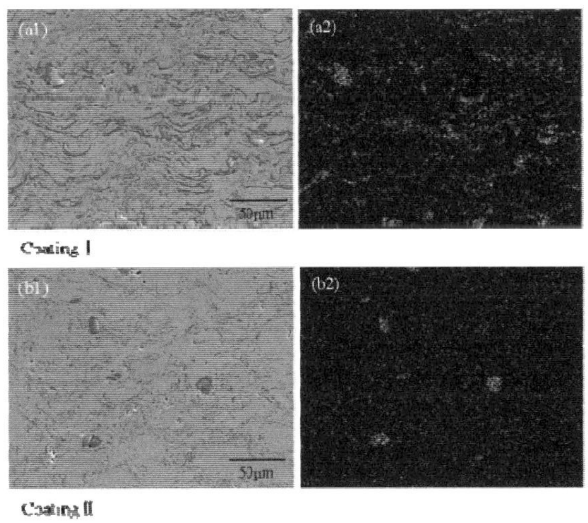

Fig. 1 Cross-section (a1, b1) and Al element map (a2, b2) of as-sprayed coatings

The coating hardness was also measured. The hardness of coating I is considerably higher than coating II. The higher in-flight velocity might promote stronger work hardening, and the uniform distribution of molten Al_2O_3 component could also enhance the hardness. Another phenomenon is the hardness of

Table 1 As-sprayed coating properties

	Porosity (%)	Oxygen content (%)	Adhesion strength (MPa)	Hardness (Hv0.3)		
				Coating	Substrate near interface	Substrate
Coating I	3.0	2.5	46.9	504.7	315.3	187.3
Coating II	2.2	1.0	29.1	336.6	307.3	195.5

the substrate near coating-substrate interface is much higher than the substrate.

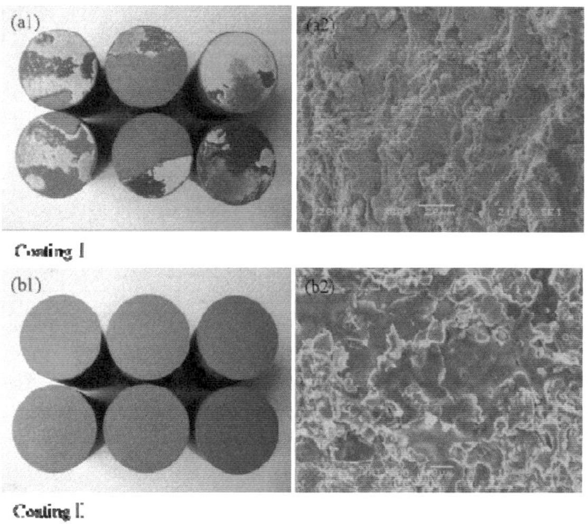

Fig. 2 Fracture microstructures of specimen after tensile test: (a1, b1) overview and (a2, b2) higher magnification

3.2 Wear tests

Firstly, the wear test was conducted under air atmosphere at room temperature, quenched GCr15 steel was used as grinding ball . Under the designated experimental conditions in this study, the mass loss of substrate without coating was 34.9 mg, while the mass loss was remarkably reduced to 0.7 mg when the substrate was coved by coating Ⅰ. Once the coating changed to coating Ⅱ,the mass loss was 3.3mg, which exhibits a larger value than coating Ⅰ, but it is still much smaller than the substrate without coating. The results indicated that the wear resistance at room temperature was significant improved by the obtained coatings.

Figure 3 represents the wear track of substrate without coating, with coating I and coating Ⅱ, respectively. The track width was the widest when the substrate without coating (Fig. 3a), plowing grooves and plastic deformations on the surface could be found from the wear surface micrograph. In the case of coating I (Fig. 3b), the wear trace become narrower and discontinuous, because the ceramic component has favourable wear resistance, and the existence of "island" structure at the coating surface serves to protect the unworn surface of the coating.

When the surface was covered by coating Ⅱ, no typical wear morphology change could be found compared with coating I (Fig. 3c), while the width of the wear track was between the uncoated sample and coating I, which was also proved by the mass loss result.

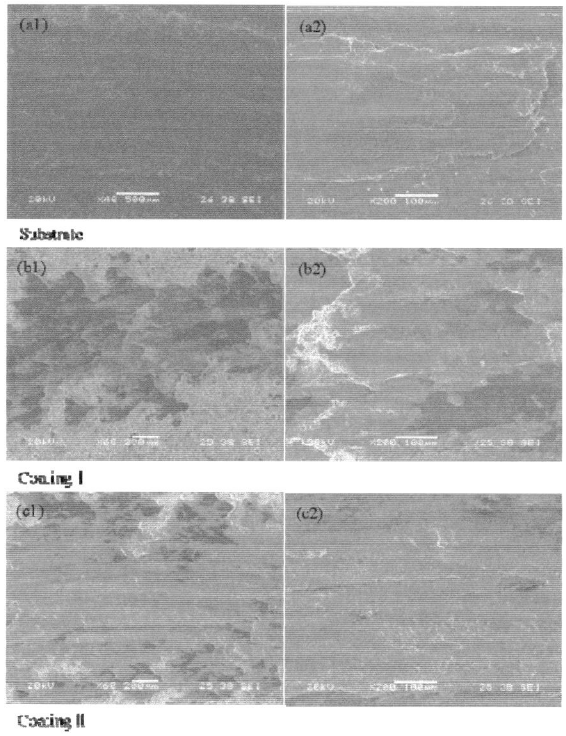

Fig. 3 Wear track at room temperature: (a1, b1, c1) overview and (a2, b2, c2) higher magnification

As the coating was expected to be used in the high temperature environment, the wear test was also conducted at 900℃, Si3N4 was selected as grinding ball. The friction coefficients (COF) were notably smaller than the coatings obtained at room temperature, this difference should be attributed to the softening of the coating material and the reduction of its work hardening ability at high temperatures [11]. Especially, a little noise could be heard for the substrate without coating during the test, while it is very quiet for coating Ⅰ, because most of the Al_2O_3 particles were melted and dispersed uniformly in the coating. To the contrary, the test made a great noise for coating Ⅱ, which might be attributed to the existence of unmelted ceramic particles in the coating. The wear tracks were illustrated in Fig. 4, the tendency of the wear microstructure change was similar with the test carried out at room temperature, However, the wear become more serious, more materials were removed even when the same load, rotational speed and test time were used. In particular, plowing grooves and plastic deformations on the surface also be found from the wear surface of coating Ⅰ (Fig. 4b).

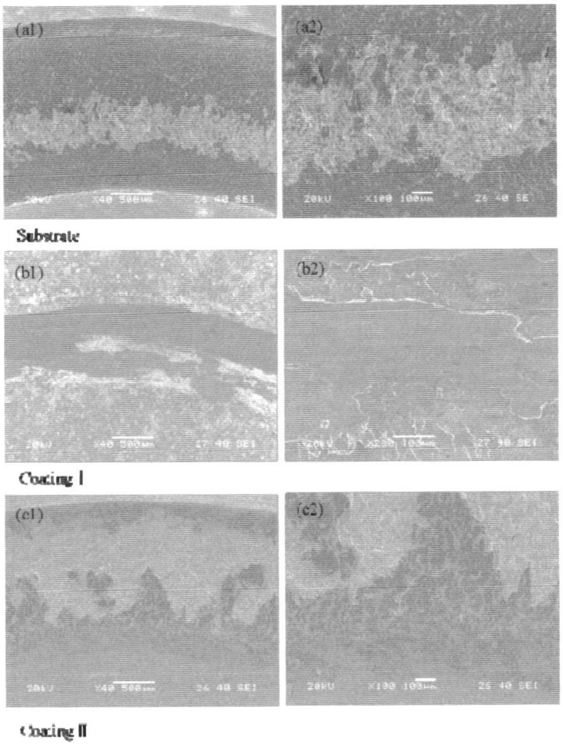

Fig. 4 Wear track at 900℃: (a1, b1, c1) overview and (a2, b2, c2) higher magnification

3.3 Buildup test

Accelerating experiments for the substrate with and without protective coating have been conducted in this study. Cross-sections as well as the element maps after test for 4 hours are shown in Fig. 5. The remained Fe_3O_4 powder and the coating can be easily distinguished from the element maps. As the substrate contain Fe element, O element was detected in the substrate (Fig. 5a). Some micro-cracks could be found at the interface between substrate and Fe_3O_4 powder when the substrate was uncoated. High counts of Fe above the upper part of coating Ⅰ could be found, but has not entered into the coating yet (Fig. 5b). More Fe_3O_4 powder remained on the surface of coating Ⅱ, however,the coating are still intact after the test (Fig. 5c). However, the service time must be much longer than the test time in the actual usage, hence deeper research is necessary in future study.

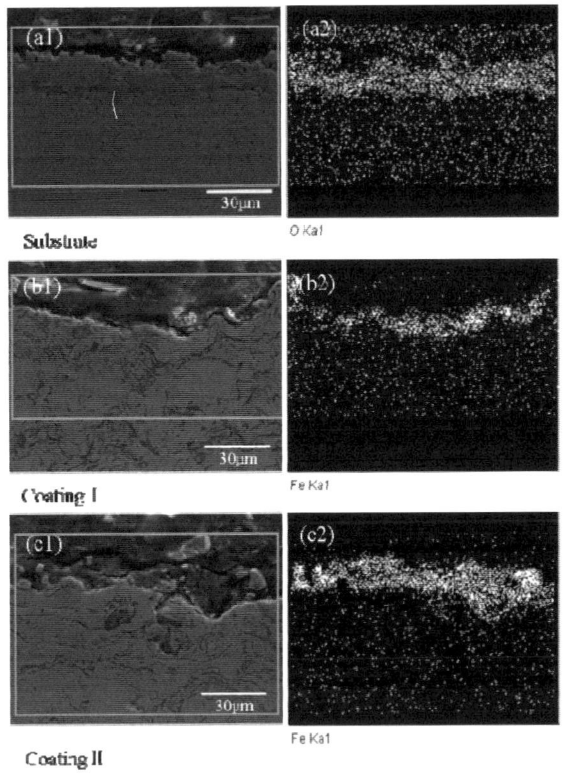

Fig. 5 Cross-section (a1, b1, c1) and element maps (a2, b2, c2) after buildup test

4 Summary

NiCoCrAlYTa alloy powder and Al_2O_3 ceramic powder were pre-mixed and thermally sprayed onto the grit-blasted Cr28Ni48W5 heat resistant steel substrate by supersonic atmospheric plasma spraying. The addition of propane in the operating gas has a significant effect on the coating performance. The favorable property of the coating indicates that the coating could become a potential candicate for high temperature wear applications.

Acknowledgment

This research was supported by Science and Technology Planning Project of Guangdong Province, China (No. 2013B010102023).

Literatures

[1] Harold H. Fukubayashi: Present furnace and pot roll coatings and future development. Proceedings of the 2004 International Thermal Spray Conference (2004)

[2] S. Matthews and B. James: Review of thermal spray coating applications in the steel industry: part 1 -hardware in steel making to the continuous annealing process. Journal of Thermal Spray Technology 19 (2010), pp. 1267-1276.

[3] S. Y. Hwang and B. G. Seong: Characterization of

build-up resistant plasma sprayed coatings for hearth rolls. Proceedings of the 1995 International Thermal Spray Conference (1995), pp. 53-57.
[4] N. Yasuo, T. Suidzu and S. Hashimoto: Development of thermal spray coatings for hearth rolls with superior build-up resistance. The Third Baosteel Biennial Academic Conference 3(2008), pp. J93-J98
[5] R. Hao: Developments andpractices of thermal spraying technology in the iron and steel industry in China. The Third Baosteel Biennial Academic Conference 3(2008), pp.J74-J79
[6] S. Zhu, B.S. Xu and J.K. Yao: High quality ceramic coatings sprayed by high efficiency hypersonic plasma spraying gun. Materials Science Forum 475 - 479 (2005), pp. 3981-3984
[7] Y. Bai, L. Zhao, Y. Wang, D. Chen, B.Q. Li and Z.H. Han: Fragmentation of in-flight particles and its influence on the microstructure and mechanical property of YSZ coating deposited by supersonic atmospheric plasma spraying. Journal of Alloys and Compounds 632 (2015), pp. 794 - 799
[8] Y. Yang, K.Lin, Z. Zhao and H. Li: Ablation resistance of HfC–SiC coating prepared by supersonic atmospheric plasma spraying for SiC-coated C/C composites. Ceramic International 42(2016), pp. 4768–4774.
[9] Y. Bai, Z. H. Han, H. Q. Li, C. Xu, Y. L. Xu, C. H. Ding, and J. F. Yang: Structure-property differences between supersonic and conventional atmospheric plasma sprayed zirconia thermal barrier coatings. Surface and Coatings Technology 205(2011), pp. 3833–3839.
[10] Y. Gao, H. Nitta and R. C. Tucker, Jr., Effect of manganese oxides on the durability of a cobalt based coating on furnace roll in continuous annealing lines, Proceedings of the 1995 International Thermal Spray Conference (1995), pp. 53-57.
[11] Dominique Poirier, Jean-Gabriel Legoux, and Rogerio S. Lima: Engineering HVOF-sprayed Cr_3C_2-NiCr coatings: the effect of particle morphology and spraying parameters on the microstructure, properties, and high temperature wear performance. Journal of Thermal Spray Technology 22 (2013), pp. 280-289

Study on the microstructure and wear behaviour of AlCoCrFeNiTi/NiCrBSi composite coating by atmospheric plasma spraying

Li-Hui Tian*, Wei Xiong, Chuan Liu, Sheng Lu, Zhenjiang/China

The microstructure of the as-sprayed AlCoCrFeNiTi/NiCrBSi composite coating was analyzed and the microhardness and high-temperature wear resistance were investigated. The results show that a dense coating with the adhesion strength of higher than 60 MPa was prepared. The microhardness of the coating was 727 Hv. It was found that with the increase of the wear temperature, the volume loss increased from room temperature to 500°C and decreased from 500°C to 900°C. The volume loss of the as-sprayed coating at room temperature was 21% of that of 316 stainless steel. At the wear temperature of 900°C, the volume loss of the coating was about 5% of that of 316 stainless steel. Abrasion wear was the main failure mechanism of the coatings at room temperature, 500°C, 700°C and 900°C. Adhesion wear occurred at the wear temperature of 900°C. Besides, peeling wear can be found on the worn surface of the as-sprayed coatings.

1 Introduction

The concept of high-entropy alloy was defined by J.W. Yeh in 2004 [1, 2]. High-entropy alloy is a promising high-temperature structural material owing to its excellent properties such as high strength, high hardness, high wear resistance, high corrosion resistance and oxidation resistance. Bulky high-entropy alloys such as $Al_xCoCrCuFeNi$ and $Al_xCo_{1.5}CrFeNi_{1.5}Ti_y$ with high hardness and high wear resistance were prepared by vacuum arc cast [3-5].

Using the high properties high-entropy alloys as coating material can improve the performance of the workpieces, prolong the service life and significantly reduce the manufacturing cost. By now, high-entropy alloy coatings were prepared by electrochemical deposition, magnetron sputtering and laser clad, which were limited by low deposition efficiency, coating thickness and dilution of the substrate. Atmospheric plasma spraying (APS) was widely used due to its high temperature, high deposition efficiency, low cost and so on. However, APS was seldom used to prepare high-entropy alloy coatings [6-8].

Besides, due to its heat resistance, wear resistance and corrosion resistance, NiCrBSi self-fluxing alloy was widely used in petroleum, chemical industry, power and national defence [9-12]. The mechanical properties of AlCoCrFeNiTi/NiCrBSi composite coating such as adhesion strength, microhardness and high-temperature wear resistance must be quite excellent. Therefore, in this study, the as-sprayed AlCoCrFeNiTi/NiCrBSi composite coating was prepared by atmospheric plasma spraying using mechanically blended AlCoCrFeNiTi/NiCrBSi composite powder as the feedstock. The microstructure of the as-sprayed coating was analyzed and the adhesion strength, the microhardness and the high-temperature wear resistance were investigated.

2 Experimental

2.1 Coating preparation

In this study, AlCoCrFeNiTi powder was prepared by ball milling using equimolar Al, Co, Cr, Fe, Ni and Ti element powders as the starting materials. The detailed ball milling parameters are listed in **Table 1**. The ball milled AlCoCrFeNiTi powder in a size range of less than 30 μm was prepared by sieving and then mechanically blended with NiCrBSi powder with an AlCoCrFeNiTi/NiCrBSi weight ratio of 70:30. The morphology of the blended powder is shown in **Fig. 1**. It is seen that AlCoCrFeNiTi powder was angular in shape while NiCrBSi powder was spherical. As expected from the shape, the powder is suitable for spraying due to its good flowability.

Table 1. Ball milling parameters

Parameter	Value
Ball-to-powder weight ratio	10:1
Powder loading (g)	80
Rotation speed (r/min)	800
Milling time (h)	30

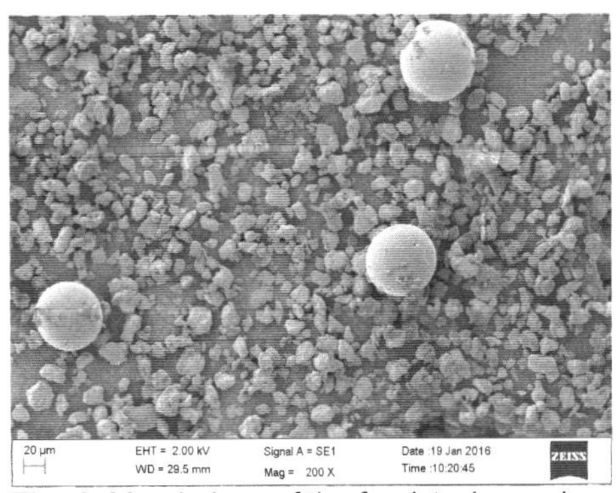

Fig. 1. Morphology of the feedstock powder.

A high-energy plasma spray system (3710, Praxair) was used to deposit AlCoCrFeNiTi/NiCrBSi composite coatings with the feedstock powder. 316 stainless steel plates with dimensions of Φ25×8 mm were used as substrates for the specimens used for microstructure characterization, tensile test, microhardness measurement and high-temperature wear test. The substrates were cleaned with alcohol and sand blasted with white alumina before coating deposition. The thickness of the coating was about 200 μm. The spraying parameters are listed in **Table 2**.

Table 2. Atmospheric plasma spraying parameters

Parameter	Value
Plasma arc power (kW)	45
Spray distance (mm)	100
Gun traverse speed (mm/s)	200
Main gas (Ar) pressure (psi)	60
Secondary gas (N_2) pressure (psi)	65
Powder carrier gas (Ar) pressure (psi)	40
Rotating speed of powder feeder (r/min)	0.8

2.2 Microstructure characterization

Scanning electron microscopy (SEM, ZEISS ΣIGMA) and energy dispersive spectroscopy (EDS) were used to analyze the cross-sectional microstructure of the coatings and morphology of the worn surfaces. The phases present in both the feedstock powder and the as-sprayed coating were examined by an X-ray diffractometer (XRD, XRD-6000) with Cu $K\alpha1$ radiation.

2.3 Tensile test

The adhesion strength of the as-sprayed AlCoCrFeNiTi/NiCrBSi composite coating was tested according to the standard "ISO 14916: 1999, MOD". E-7 high-temperature structural adhesive was used. The tensile speed was 1 mm/s. Three specimens were tested for the coating.

2.4 Microhardness measurement

Microhardness of the as-sprayed AlCoCrFeNiTi/NiCrBSi composite coating was measured by an MH-5D mode Vickers microhardness tester with a load of 200 g and dwelling time of 10 s. For each position on the cross section of the coating, more than 10 points were measured and the results were reported as an average value.

2.5 High-temperature wear test

A high-temperature wear tester (HT-1000) was used to investigate the wear behaviour of the as-sprayed coatings. The upper and lower specimens were fixed on a shaft and a spindle, respectively. The upper specimen was Si_3N_4 balls with a dimension of Φ 5 mm. The coating was deposited on the lower specimen. They can contact with each other with the help of the weights with a load of 5 N. When the load was added, the heating furnace was closed and heated up to the test temperature of room temperature, 500°C, 700°C and 900°C, respectively. The lower specimen rotated at a set velocity of 573 r/min and a friction radius of 5 mm. When 20 min was finished, the lower specimen was taken down and the volume loss which was used to evaluate the wear resistance of the coatings was measured. Two specimens were tested for each temperature.

3 Results and discussion

3.1 Microstructure of AlCoCrFeNiTi/NiCrBSi coating

The XRD patterns of the as-sprayed AlCoCrFeNiTi/NiCrBSi composite coating and the feedstock powder are shown in **Fig. 2**. It can be found that new phases such as CrB, Cr_2B, BCC1, FCC and Cr4Ti4O were formed through atmospheric plasma spraying.

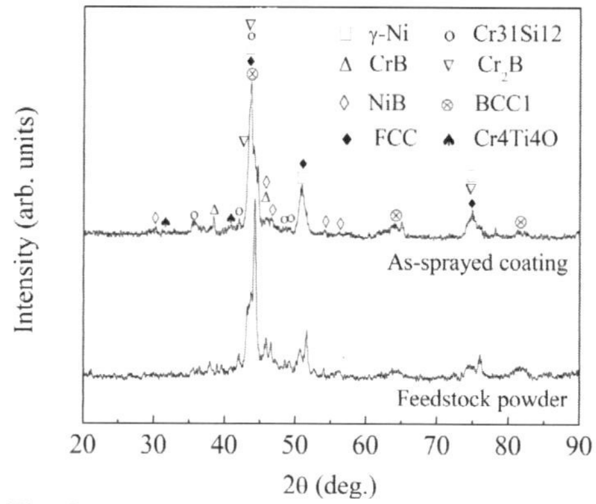

Fig. 2. XRD patterns of the as-sprayed coating and the feedstock powder.

The typical cross-sectional microstructure of the as-sprayed AlCoCrFeNiTi/NiCrBSi composite coating is shown in **Fig. 3** and **Fig. 4**. It can be seen that a dense coating with a thickness of about 200 μm was deposited by atmospheric plasma spraying. The coarse pores in a size less than 20 μm were due to the incomplete wetting and insufficient filling of the fully molten droplets to previously formed rough deposit surfaces [13-16]. It can be seen that a good bonding was obtained between the coating and the substrate, **Fig. 4(a)**. The result of the tensile test shows that the fracture of the specimen occurred in the E-7 adhesive and the tensile strength was 60 MPa, which indicates that the adhesion strength of the coating was higher than 60 MPa. In addition, the as-sprayed AlCoCrFeNiTi/NiCrBSi composite coating exhibited a typical lamellar microstructure. It can be observed that most of the spray particles were present in the coating as well flattened splats which were resulted from fully melted spray particles, **Fig. 4(b)**. NiCrBSi particles were distributed uniformly in the AlCoCrFeNiTi matrix, **Fig. 3**. Moreover, good contact was observed between the melted NiCrBSi and AlCoCrFeNiTi splats, **Fig. 4(b)**.

To determine the chemical composition of typical positions on the cross section of the coating, EDS analysis was conducted and the results are listed in **Table 3**. The regions with a light gray contrast in the coating (+A) was identified to be NiCrBSi splats by EDS. Region B corresponded to phases with Al, Co, Cr, Fe, Ni and Ti elements and it should be BCC1 phase. Region D corresponded to phases without Al element and it should be FCC phase. The regions C, E, F and G with dark gray contrasts, which had higher oxygen contents, should be caused by the oxidation of AlCoCrFeNiTi feedstock powder.

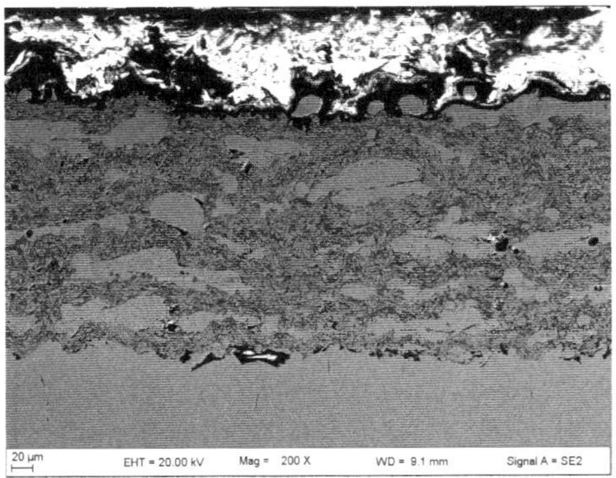

Fig. 3. Image of the as-sprayed coating at a lower magnification.

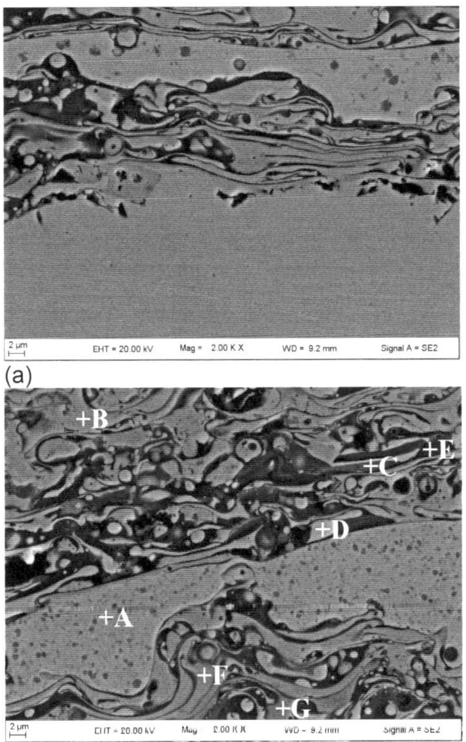

(a)

(b)

Fig. 4. Images of the as-sprayed coating at higher magnifications.

Table 3. EDS results of typical positions in Fig. 4(b) (wt.%)

Position	Al	Co	Cr	Fe	Ni	Ti	Si	O
+A			35.7	5.1	55.8	0.5	2.9	
+B	4.3	24.4	16.4	21.1	24.0	9.7		
+C	5.2	13.1	3.1	31.6	14.0	6.5		26.5
+D		34.8	1.2	5.1	56.5	0.8		1.6
+E	6.9	6.0	8.9	10.2	3.9	27.3		36.8
+F	5.9	20.5	12.3	18.0	20.3	14.1		8.8
+G	4.6	17.1	7.2	11.5	17.3	30.9		11.3

3.2 Microhardness of AlCoCrFeNiTi/NiCrBSi coating

The microhardness of the as-sprayed AlCoCrFeNiTi/NiCrBSi composite coating is shown in **Fig. 5**. It can be seen that the microhardness of the substrate increased with the increase of the distance from the substrate/coating interface. At the distances of -50 and -15 μm, the microhardnesses were 175 Hv and 183 Hv, respectively. However, the microhardness of the substrate at the distance of -5 μm increased to 289 Hv. This may due to the quenching of the substrate during high-temperature plasma spraying [17]. For the coating, with the increase of the distance from the substrate/coating interface, the microhardness increased. At the distance of 150 μm, the microhardness of the coating was 727 Hv. However, at the distance of 200 μm, the microhardness decreased to 639 Hv, which was due to the edge effect of the coating.

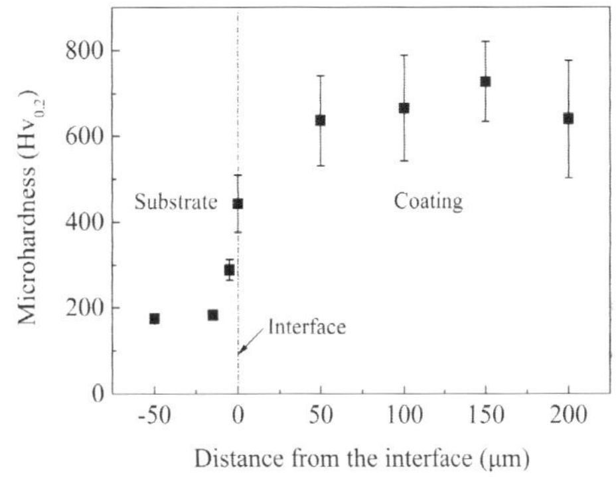

Fig. 5. Microhardness of the as-sprayed coating.

3.3 High-temperature wear resistance of AlCoCrFeNiTi/NiCrBSi coating

The volume loss of the as-sprayed AlCoCrFeNiTi/NiCrBSi composite coating and 316 stainless steel as a function of the wear temperature is shown in **Fig. 6**. It can be found that with the increase of the wear temperature, the volume loss increased from room temperature to 500°C and decreased from 500 to 900°C. In the temperature range from room temperature to 500°C, the increase of the volume loss was due to the decrease of the hardness of AlCoCrFeNiTi/NiCrBSi material. While the decrease of the volume loss in the temperature range from 500°C to 900°C indicates that the wear resistance of AlCoCrFeNiTi/NiCrBSi composite coating was enhanced.

However, the volume loss of 316 stainless steel decreased from room temperature to 500°C and increased from 500°C to 900°C. Comparing the volume loss of the as-sprayed AlCoCrFeNiTi/NiCrBSi composite coating with that of 316 stainless steel, it can be found that the volume loss of the coating was lower than that of 316 stainless steel, which indicates that the wear resistance of the coating was better than that of 316 stainless steel. For example, the volume loss of the as-sprayed coating at room temperature was 21% of that of 316 stainless steel. When the wear temperature was 900°C, the volume loss of the coating was about 5% of that of 316 stainless steel.

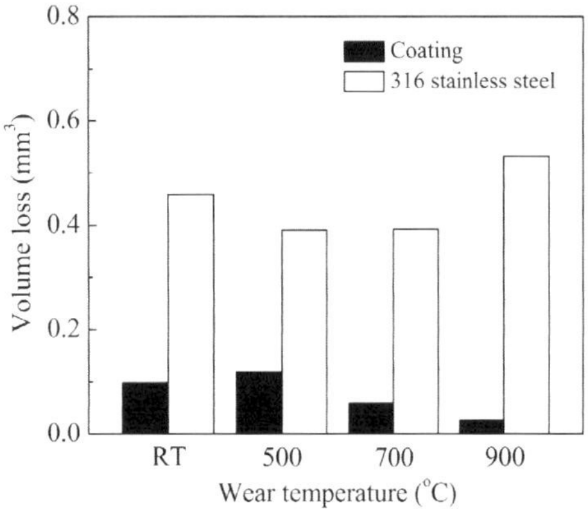

Fig. 6. Effect of wear temperature on volume loss of the as-sprayed coating and 316 stainless steel.

3.4 Wear mechanisms of AlCoCrFeNiTi/NiCrBSi coating

To investigate the wear mechanisms of the as-sprayed AlCoCrFeNiTi/NiCrBSi composite coatings, morphology of the worn surfaces at room temperature, 500°C, 700°C and 900°C was analyzed. It can be seen that the difference of the worn surfaces of the coating was quite obvious, **Figs. 7-10**. It can be found that the wear tracks at room temperature and 500°C were wide and deep, while that at 700°C became narrower and shallower. At 900°C, the wear track became significantly narrow and shallow. These results are consistent with the volume loss data in **Fig. 6**.

In addition, surfaces of the coatings outside the wear tracks were smooth at room temperature and 500°C, while that at 700°C became rougher. At the wear temperature of 900°C, the coating surface outside the wear track was significantly rough. The roughness of the coating surfaces was caused by high-temperature oxidation of the coating material, which was confirmed by EDS.

The morphology of the worn surface of the coating at room temperature is shown in **Fig. 7**. It can be seen that the worn surface was not smooth with large numbers of scraps (point 1) presented on it, which was due to the low plasticity of the AlCoCrFeNiTi/NiCrBSi composite coating, **Fig. 7(b)**. Lots of short, deep and wide furrows can be found on the worn surface. Some cracks (point 2) existed on the worn surface, which resulted from the warping of the flatten splats. Besides, some circular pits (point 3) and pockmarks (point 4) presented on the worn surface, which were caused by the spalling of flatten splats and unmelted particles, respectively. It was reported that the interlamellar gaps had a significant influence on the wear resistance of the thermally-sprayed coatings due to the peeling wear mechanism [18, 19]. Therefore, the main failure mechanisms of the coatings at room temperature were abrasion wear and peeling wear.

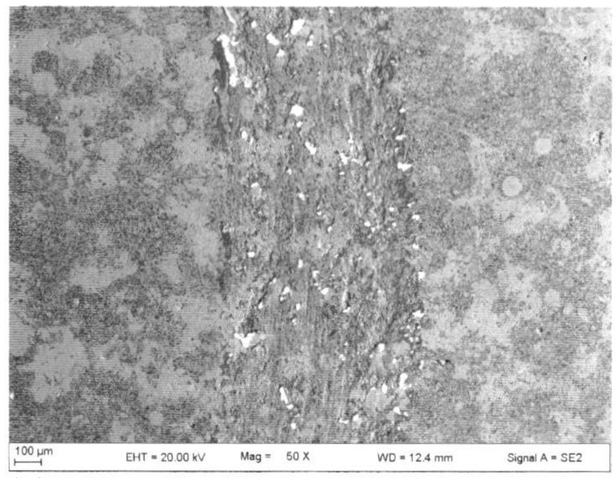

(a)

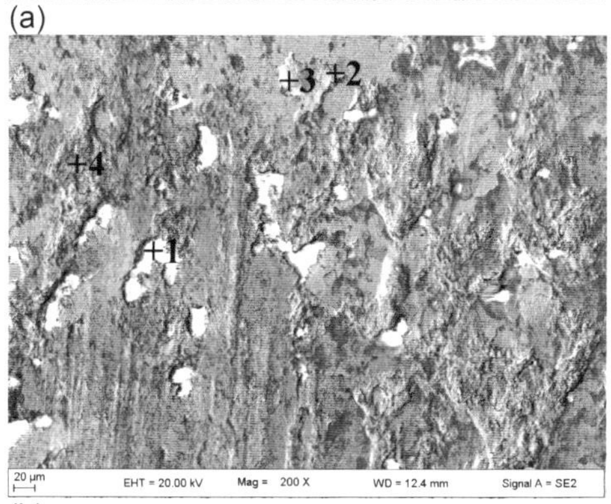

(b)

Fig. 7. Worn surface of the AlCoCrFeNiTi/NiCrBSi coating at room temperature.

The worn surface of the coating at a wear temperature of 500°C is shown in **Fig. 8**. It can be seen that as the wear temperature increased, the worn surface became smoother, which can be caused by the increase of the plasticity of the coating, **Fig. 8(b)**. The furrows became longer, narrower and shallower. Due to the warping and spalling of the splats, cracks (point 5) and pits (point 6) can also be found. NiCrBSi splats (point 7) with light gray contrast which was confirmed by EDS can be observed obviously. Besides, 34.7 wt.% of oxygen element and no silicon element were detected by EDS in the region with a dark contrast (point 8), and this may resulted from the oxidation of the AlCoCrFeNiTi phase. The failure mechanism of the coating at 500°C was mainly abrasion wear.

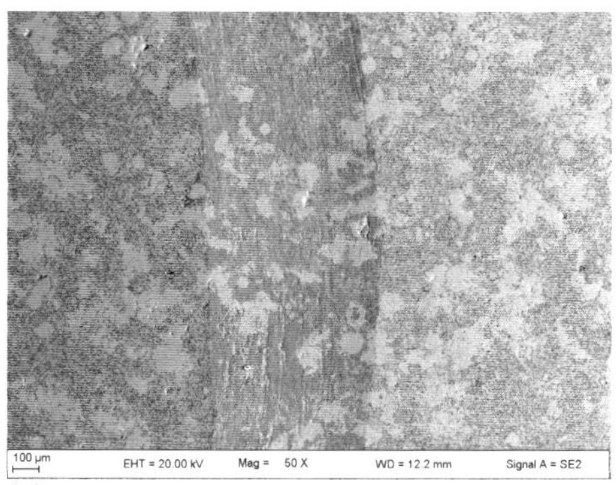

(a)

(b)

Fig. 8. Worn surface of the AlCoCrFeNiTi/NiCrBSi coating at 500°C.

The worn surface of the coating at 700°C is shown in **Fig. 9**. At a wear temperature of 700°C, the worn surface was also smooth and furrows presented on the worn surface were long, narrow and shallow, **Fig. 9(b)**. No cracks were observed on the worn surface. The region (point 9) with 30.8 wt.% of oxygen element and no silicon element should be oxides of AlCoCrFeNiTi, and the region (point 10) with oxygen, Ni, Ti, Co, Fe, Cr, Si and Al elements should be oxides of AlCoCrFeNiTi and NiCrBSi phases. The failure mechanism of the coatings at 700°C was still abrasion wear.

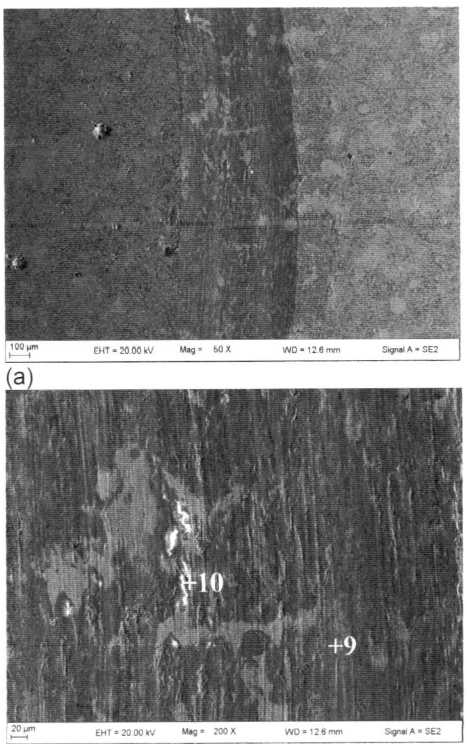

(b)

Fig. 9. Worn surface of the AlCoCrFeNiTi/NiCrBSi coating at 700°C.

Worn surface of the coating at a wear temperature of 900°C is shown in **Fig. 10**. It can be seen that significant oxidation occurred on the coating surface, **Fig. 10(b)**. The content of the oxygen element of point 11 was 34.2 wt.%. The worn surface was quite rough with long, wide and deep furrows. Grindings (point 12) and slices (point 13) which were covered with thin oxide films presented inside and outside the furrows. Spalling of splats occurred and the surface of the underlying splats appeared (point 14). Besides, no cracks were found on the worn surface. The failure mechanisms at 900°C were mainly abrasion wear and adhesion wear.

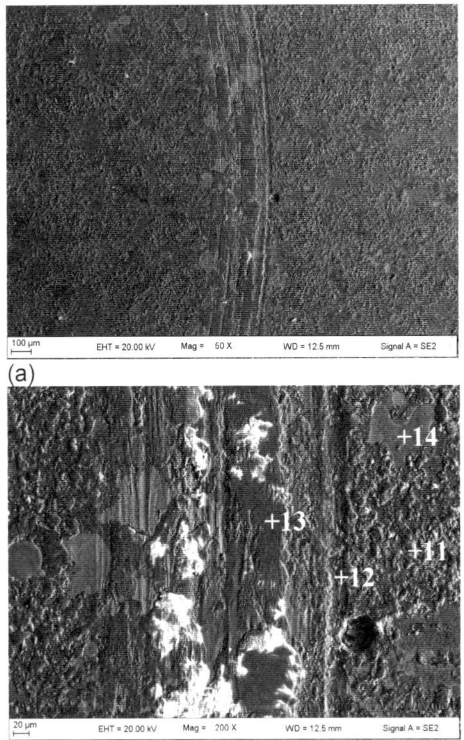

(b)

Fig. 10. Worn surface of the AlCoCrFeNiTi/NiCrBSi coating at 900°C.

4 Conclusions

(1) A dense AlCoCrFeNiTi/NiCrBSi composite coating with the adhesion strength of higher than 60 MPa was prepared by atmospheric plasma spraying.

(2) The microhardness of the as-sprayed coating was 727 Hv.

(3) With the increase of the wear temperature, the volume loss increased from room temperature to 500°C and decreased from 500°C to 900°C. The volume loss of the as-sprayed coating at room temperature was 21% of that of 316 stainless steel. When the wear temperature was 900°C, the volume loss of the coating was about 5% of that of 316 stainless steel.

(4) Abrasion wear was the main failure mechanism of the coatings at room temperature, 500°C, 700°C and 900°C. Adhesion wear occurred at the wear temperature of 900°C. Besides, peeling wear can be found on the worn surface of the as-sprayed coatings.

Acknowledgement

The present work is supported by the National Natural Science Foundation of China (No. 51401091).

References

[1] B. Cantor, I.T.H. Chang, P. Knight and A.J.B. Vincent: Microstructural development in equiatomic multicomponent alloys. Mater. Sci. Eng. A 21 (2004), Issues 375-377, pp. 213/18.
[2] J.W. Yeh, S.K. Chen, S.J. Lin, J.Y. Gan, T.S. Chin, T.T. Shun, C.H. Tsau and S.Y. Chang: Nanostructured high-entropy alloys with multiple principal elements: novel alloy design concepts and outcomes. Adv. Eng. Mater. 6 (2004), Issue 5, pp. 299/303.
[3] J.M. Wu, S.J. Lin, J.W. Yeh, S.K. Chen, Y.S. Huang and H.C. Chen: Adhesive wear behavior of $Al_xCoCrCuFeNi$ high-entropy alloys as a function of aluminum content. Wear 261 (2006), Issue 5, pp. 513/19.
[4] M.H. Chuang, M.H. Tsai, W.R. Wang, S.J. Lin and J.W. Yeh: Microstructure and wear behavior of $Al_xCo_{1.5}CrFeNi_{1.5}Ti_y$ high-entropy alloys. Acta Mater. 59 (2011), Issue 16, pp. 6308/17.
[5] C.Y. Hsu, T.S. Sheu, J.W. Yeh and S.K. Chen: Effect of iron content on adhesive wear behavior of $AlCoCrFe_xMo_{0.5}Ni$ high-entropy alloys. Wear 268 (2010), Issues 5-6, pp. 653/59.
[6] L.M. Wang, C.C. Chen, J.W. Yeh and S.T. Ke: The microstructure and strengthening mechanism of thermal spray coating $Ni_xCo_{0.6}Fe_{0.2}Cr_ySi_zAlTi_{0.2}$ high-entropy alloys. Mater. Chem. Phys. 126 (2011), Issue 3, pp. 880/85.
[7] A.S.M. Ang, C.C. Berndt, M.L. Sesso, A. Anupam, S. Praveen, R.S. Kottada and B.S. Murty: Plasma-sprayed high entropy alloys: microstructure and properties of AlCoCrFeNi and MnCoCrFeNi. Metall. Mater. Trans. A 46 (2015), Issue 2, pp. 791/800.
[8] T.M. Yue, H. Xie, X. Lin, H. Yang and G. Meng: Microstructure of laser re-melted AlCoCrCuFeNi high entropy alloy coatings produced by plasma spraying. Entropy 15 (2013), Issue 7, pp. 2833/45.
[9] C. Navas, R. Vijande, J.M. Cuetos, M.R. Fernández and J. Damborenea: Corrosion behaviour of NiCrBSi plasma-sprayed coatings partially melted with laser. Surf. Coat. Technol. 201 (2006), Issue 3, pp. 776/85.
[10] B. Yin, H.D. Zhou, J.M. Chen and F.Y. Yan: Microstructure and high temperature sliding wear behaviour of laser remelted NiCrBSi/35(WC-12Ni) composite coating. Surf. Eng. 27 (2011), Issue 6, pp. 458/63.
[11] R.G. Rodríguez, M.R. Cadenas, R. Fernández and E. Rodríguez: Wear behaviour of flame sprayed NiCrBSi coating remelted by flame or by laser. Wear 262 (2007), Issue 3, pp. 301/07.
[12] J.M. Miguel, J.M. Guilemany and S. Vizcaino: Tribological study of NiCrBSi coating obtained by different processes. Tribol. Int. 36 (2003), Issue 3, pp. 181/87.
[13] Y.Z. Xing, C.J. Li, C.X. Li and G.J. Yang: Influence of through-lamella grain growth on ionic conductivity of plasma-sprayed yttria stabilized zirconia as an electrolyte in solid oxide fuel cells. J. Power Sources 176 (2008), Issue 1, pp. 31/38.
[14] C.J. Li and A. Ohmori: Relationships between the microstructure and properties of thermally sprayed deposits. J. Therm. Spray Technol. 11 (2002), Issue 3, pp. 365/74.
[15] R. Mcpherson: The relationship between the mechanism of formation, microstructure and properties of plasma-sprayed coatings. Thin Solid Films 83 (1981), Issue 3, pp. 297/310.
[16] C.J. Li, A. Ohmori and R. Mcpherson: The relationship between microstructure and Young's modulus of thermally sprayed ceramic coatings. J. Mater. Sci. 32 (1997), Issue 4, pp. 997/1004.
[17] L. Pawlowski: The science and engineering of thermal spray coatings. China Machine Press, Peking, 2010.
[18] L.H. Tian: Study on the preparation and the properties of FeAl/TiC composite coatings by shrouded plasma spraying. Xi'an Jiaotong University, Xi'an, 2012.
[19] B.S. Xu, Z.X. Zhu, Y. Liu, S.N. Ma and W. Zhang: Fe-Al intermetallics coating produced by high velocity arc spraying. J. Chin. Nonferrous Met. Soc. 14 (2004), Issue 1, pp. 154/58.

Spraying of Aluminum nitride-Alumina-Yttria mixtures in Nitrogen

Mohammed Shahien[1,2], Motohiro Yamada[1], Masahiro Fukumoto[1]
[1] Toyohashi University of Technology, TUT, Toyohashi/JP
[2] Central Metallurgical R & D Institute, CMRDI, Cairo/EG

Aluminum nitride (AlN) high thermal conductivity and chemical stability has attracted wide attention in a diverse range of electrical and electronic applications. However, it was impossible to fabricate the AlN coatings directly in the conventional thermal spray techniques. It is due to the AlN particles decomposition in the high temperature without any stable melting or semi-melting phases, which prevents it deposition. Using the reactivity of the plasma to fabricate the AlN thermal sprayed coatings is a promising solution. This study will investigate the feasibility of reactive spraying aluminum nitride-alumina-yttria mixtures in nitrogen ambient and its effect on the coatings properties. Several AlN based coatings were successfully fabricated. The coatings consist of *h*-AlN, *c*-AlN, Al_5O_6N, γ-Al_2O_3, α-Al_2O_3 and aluminum yttrium oxide. The fabrication process, nitride content and coating microstructure significantly affected with using the yttria addition. However, due to the high oxide content and porosity of the as-sprayed coatings it was difficult to reach high thermal conductivity AlN coatings directly. Increasing the AlN content and assisting the sintering during the heat-treatment were investigated to fabricate AlN coatings with high thermal conductivity (90 W/m.K), high AlN content and low porosity.

1 Introduction

Aluminum nitride (AlN) ceramics is considered as one of the most promising novel ceramics. It possesses a series of excellent properties: high thermal conductivity, appropriate thermal expansion coefficient (close to Si and GaAs), high dielectric resistivity, good dielectric properties, and high resistance to molten metals, wear corrosion and oxidation. In addition, AlN has good physical/chemical stability at fairly high temperature, high hardness, high electrical insulation, high mechanical strength, good resistance to active plasma and attractive piezoelectric properties [1-4].

The combination of these varieties of properties makes AlN a unique material among other ceramic, and it received considerable interests in several applications: as substrates for electronic packages, Heat sink materials for power transistors, Microwave device packages, electronic parts with electrical insulation and heat dissipation, semiconductor processing chamber fixtures and insulators, crucibles for molten metals and vacuum deposition, circuit substrates of semiconductor modules and IC packages, window materials for infrared and radar applications, molten metal handling components, some refractory parts such as protective tubes and a wide range of industrial and structural products [4-6].

Applying AlN and using its promising properties in the field of surface engineering is a big challenge. High quality AlN ceramic coatings are fabricated by vapor deposition techniques including chemical vapor deposition (CVD) and physical vapor deposition (PVD) processes.
However, the relatively low deposition rates of the vapor deposition techniques only enable to fabricate thin AlN layers and it was difficult to fabricate thick AlN layers (more than 100 µm) [7]. The AlN thin layers had received considerable interest in several electrical applications including: microwave filter, surface acoustic wave sensors (SAWs) deposited on silicon wafers, piezoelectric transducers, as thin film bulk acoustic resonator (FBAR) used in mobile phones as well as microelectromechanical systems (MEMs) and as sensors, actuators, ultrasound applications.
On the other hand, applying the AlN thin layers in the structural parts (such as insulating for heat sinks and power devices' applications or as protective layers in the reaction vessels of CVD, PVD or etching), was difficult. Thus, thick AlN layers are required.

Generally thermal spray is one of the most common technologies which had been widely used to produce thick coatings due to its high deposition rates (several µm/min or more). Thermal spray is a generic term for a group of commonly used processes for depositing metallic and non-metallic (ceramic or plastic) or any desired combination of coatings [8]. It is defined as the method of coating application by the deposition of molten or semi-molten (softened) particles propelling onto a substrate.
Furthermore, it is reported that, thick AlN thermal spray coating is a promising candidate as cost-effective solutions in the structural parts and power devices applications [9]. However, due to the deposition mechanism of thermal spray process, fabrication of nitride coatings such as AlN has been considered to be impossible. It is due to the thermal decomposition of the AlN particles during spraying without a stable melting phase.

To fabricate AlN thermal spray coatings, reactive plasma spray (RPS) is a promising technique. It is based on the reaction between the feedstock particles with the surrounding plasma species [10]. In our previous studies, several AlN nitride coatings were successfully fabricated through plasma nitriding of Al powders [11-15], Al-AlN mixtures [16], Al_2O_3 powders [17-19] and Al_2O_3-AlN mixtures [20, 21], in the atmospheric N_2-H_2 plasma. This study will investigate the feasibility of plasma spraying of aluminium nitride-alumina-yttria mixtures in the N_2/H_2 plasma.

2 Experimental Procedures and Materials

2.1 Spray System and Materials

All the spray experiments were carried out by the atmospheric DC plasma spray system (APS: 9MB, Sulzer Metco, Switzerland), as illustrated before [11-21]. All the spray experiments were done using primary gas of N_2 and secondary gas of H_2 at the standard conditions of pressure and flow rate, N_2 (330.9 kPa, 100 SCFH) and H_2 (3344.7 kPa, 5 SCFH), respectively. Table 1.shows the spray parameters.
The effect of post-treatment for the coating was also investigated under N_2 gas flow.

2.2 Feedstock Materials

In this study spray-drying technology were used to prepare the feedstock powder materials. A commercial fine Al_2O_3 powder (1 μm; Kojundo Chemical Lab Co., Ltd., Saitama, Japan) was spray-dried with fine AlN powder (1.14 μm; Tokuyama Co., Tokyo, Japan), using an L-8i spray dryer (Ohkawara Kakohki Co., Ltd., Yokohama, Japan). And the yttrium oxide (Y_2O_3) was added to the initial feedstock during spray-drying.

The chemical composition of the spray-dried mixtures is shown in Fig. 1. The feedstock mixture consists of α-Al_2O_3, AlN and Y_2O_3 without any impurities. Figure 2 shows the SEM morphology of the feedstock mixtures. It is clear that the spray-dried particles have the spherical or donut shape which is common for the spray-drying process.

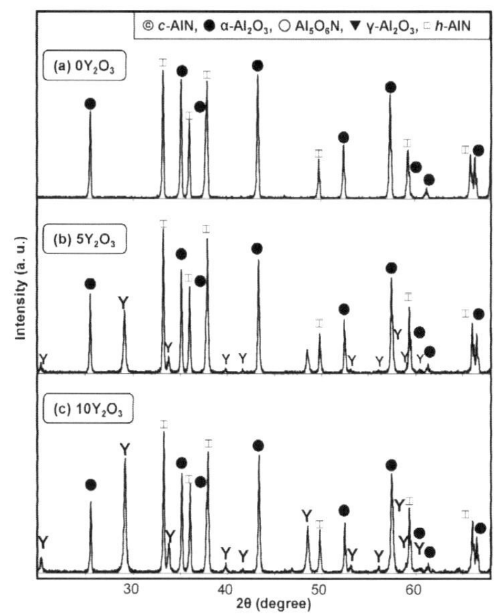

Fig. 1. XRD spectra of the spray-dried feedstock mixtures: 0, 5, 10 Y_2O_3 wt.% respectively.

Table 1 The typical spray conditions

1st gas : Pressure (kPa) : Flow rate (SCFH)	N_2: 330.9 100
2nd gas : Pressure (kPa) : Flow rate (SCFH	H_2: 344.7 5
Spray distance (mm)	100
Carrier gas:	N_2
Substrate materials	SS400/Cu

The SEM images also show that the particle size of the agglomerated mixture is about 55, 65 and 77μm respectively. The feedstock mixture was supplied to the plasma with a carrier gas of N_2 using micro-feeder (FPS-Technoserve, Japan). Grid blasted mild steel plates (50mm × 50 mm × 5mm) were prepared as the substrate material.
For the thermal conductivity measurements free standing coatings were prepared in a disk shape with diameter of 10 mm and thickness about 0.5 mm. A disk-shaped masking was used in front of the pure copper (Cu) substrate to fabricate the free standing disk-shaped coatings. The samples front and back sides were polished in order to provide uniformity of the surface and to remove any Cu contamination from the substrate before measurement.

2.2 Characterization

The phase composition of the fabricated coatings was verified by x-ray diffraction (XRD, RINT-2500; Rigaku, Tokyo, Japan) analysis using Cu Ka radiation.
The microstructure of the feedstock powder and the fabricated coatings were observed by using a scanning electron microscope (SEM, JSM-6390; JEOL, Tokyo, Japan). The coating porosity was estimated by image analysis of SEM cross-section images.

The thermal diffusivity of the free standing coating samples were measured at 25 °C by Laser Flash method using the LFA 447 Nanoflash® instrument (NETZSCH, Inc., Germany). The time that it takes for the heat to travel through the sample and cause temperature increase on the rear face can be used to measure the through-plane diffusivity and calculate the through-plane thermal conductivity. The back side temperature increase was given as follow:

$$\Delta T = \Delta T_m \left\{1 + 2 \sum_{n=1}^{\infty}(-1)^n exp\left(-\frac{n^2 \pi^2 \alpha t}{L^2}\right)\right\}$$

where:

ΔT [T(L, t)] = back side temperature increase (°C)
α= thermal diffusivity (m^2/s)
t = time (s)
L = sample thickness (m)
ΔT_m = back side maximum temperature increase (°C) and it is given by:

$$T_m = \frac{Q}{\rho C_p L}$$

where:

Q = Quantity of input energy on the front face

ρ= density

Cp = specific heat

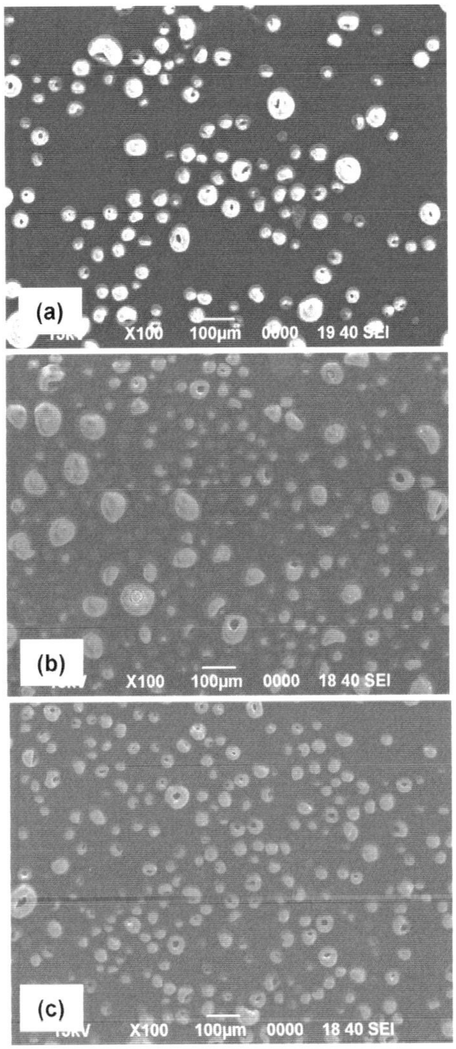

Fig. 2. SEM morphology the spray-dried feedstock mixtures: 0, 5, 10 Y_2O_3 wt.% respectively.

The thermal diffusivity had been deduced using the following equation:

$$\alpha = \frac{0.1388\ L^2}{t_{1/2}} * K$$

where:

$t_{1/2}$ = half maximum temperature increase (is the time required for the back side surface temperature to reach half of the maximum temperature rise: half of ΔTm).

The constant value of approximately 0.139 is known as the Fourier number.

K = the energy loss constant

L = sample thickness (m)

The specific heat capacity of the free standing coating samples was measured at 25 °C by differential scanning calorimeter measurement as follow:

$$C_p = \frac{Q}{m\Delta T}$$

where:

Cp = specific heat (J/kg.K)

Q = Quantity of required energy

m = mass

ΔT = change in the temperature

The bulk density of the samples was calculated from the measured sample volume and mass by using the Shimadzu Co. electronic balance (AuW120D) at 25 °C. The measurement was done as follow:

$$\rho = \frac{m}{S * L}$$

where:

ρ= density (kg/m^3)

m = mass (kg)

S = surface area (m^2)

L = thickness (m)

Finally the thermal conductivity was calculated through the following equation:

$$\lambda = \alpha\ \rho\ Cp$$

where:

λ = thermal conductivity (W/m.K)

α= thermal diffusivity (m^2/s).

ρ= density (kg/m^3)

Cp = specific heat (J/kg.K).

3 Results and Discussion

3.1 Reactive spraying of spray-dried mixtures

Figure 3 show the XRD spectra of the fabricated coatings with using the spray-dried mixtures. It clear that, the coating consists of *h*-AlN, *c*-AlN, Al_5O_6N, γ-Al_2O_3, and a small amount of α-Al_2O_3. The formation of *c*-AlN is related to the nitriding of fine Al_2O_3 particles in the nitrogen plasma as shown before in our previous studies [20, 21]. The *h*-AlN is deposited from the feedstock mixture. Moreover, formation of Al_5O_6N is due to the reaction of the Al_2O_3 with the surrounding nitrogen species. In addition γ-Al_2O_3, phase is due to the transformation from the α-Al_2O_3. It is attributed to the rapid quenching of the molten Al_2O_3 and the rapid solidification rate in the plasma spray process and it occurred easily with using such fine initial particle size under such high temperature and rapid solidification rates of plasma spray process [18, 22].

Furthermore, with using the Y_2O_3 addition in the feedstock mixture, the aluminum yttrium oxide phase appeared in the fabricated coatings and it is common in the Al_2O_3-AlN-Y_2O_3 system at high temperatures. The aluminium yttrium oxide content increased gradually with the Y_2O_3. The amount of *h*-AlN significantly increased in the coatings with using the Y_2O_3 and increasing its content. In addition the Al_5O_6N and γ-Al_2O_3 content reduced significantly with the

Y_2O_3 addition and its content. Moreover, the amount of α-Al_2O_3 gradually increased with the Y_2O_3 addition and increasing its content. It reveals that, using the Y_2O_3 and increasing its content gradually reduced the transformation rate of α-Al_2O_3 to the transient γ-Al_2O_3 phase.

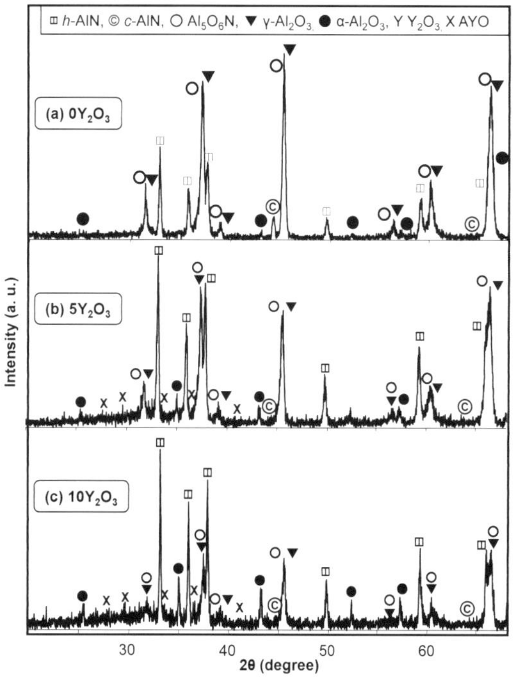

Fig. 3. XRD spectra of the as-sprayed coatings: 0, 5, 10 Y_2O_3 wt.% respectively.

The thermal conductivity of the as-sprayed coatings was 2.43, 1.95 and 3.18 (W/m.K) respectively with spraying 0, 5 and 10 Y_2O_3 wt.% mixtures. The thermal conductivity of the fabricated AlN plasma sprayed coatings was very low compared to the AlN conductivity. It is attributed to the phase composition, oxide content, density, and microstructure. Thus, coatings contain large amounts of aluminium oxides phase (γ-Al_2O_3 and α-Al_2O_3), which are characterized by their low thermal conductivity.

Moreover, the formation of large amount of Al_5O_6N phase in the coatings leads to further decrease in the thermal conductivity due to its low thermal conductivity of oxynitride. Presence of oxygen in the AlN coatings leads to phonon scattering and reducing its thermal conductivity. The oxygen (or Al_2O_3) incorporated in the AlN lattice occupies nitrogen sites and creates aluminium vacancies to balance the electric charge. The Al vacant leads to mass and strain misfits and thereby leads to increasing the scattering cross section of phonons, reducing the phonon mean free path, thereby lowering the thermal conductivity.

In addition, the measured densities of the fabricate coatings were lower than the AlN value. The lower coating density significantly leads to extreme low thermal conductivity. Furthermore, the microstructure of the plasma-sprayed coating strongly affected its conductivity. The coating consists of highly anisotropic layered structure with individual splats oriented parallel to the substrate surface. It leads to lower thermal conductivity, due to the interfacial thermal contact caused by pores or secondary phases. Moreover, the porosity of the plasma sprayed coatings lowered its conductivity due to scattering the phonons. In addition the pores are filled with air (it has poor conductivity), therefore the higher porosity, the lower the conductivity of the coatings.

Figure 4 show the XRD spectra of the coatings after heat-treatment at 1100 °C for 12 h in the N_2 atmosphere, with using the spray-dried mixtures. It is clear that, the coatings consist of *h*-AlN, *c*-AlN, Al_5O_6N, γ-Al_2O_3, and α-Al_2O_3. After heat-treatment the γ-Al_2O_3 phase transformed to the α-Al_2O_3 phase, which is common for the plasma sprayed Al_2O_3 coatings at such conditions. In addition Slight improve in the AlN content. In addition the Al_5O_6N content slightly decreased and it is related to its transformation to AlN phase. Moreover, the aluminium yttrium oxide phase content improved.

On the other hand, the heat-treatment affected the coating microstructure. Figure 5 show the SEM cross-section of the fabricated coatings with using the spray-dried mixtures before and after heat-treatment at 1100 °C for 12 h in the N_2 atmosphere. The as-sprayed coatings are porous coatings and the porosity increases gradually with using the Y_2O_3 and increasing its amount. On the other hand, the porosity decreased and the splats had sintered together to some extent.

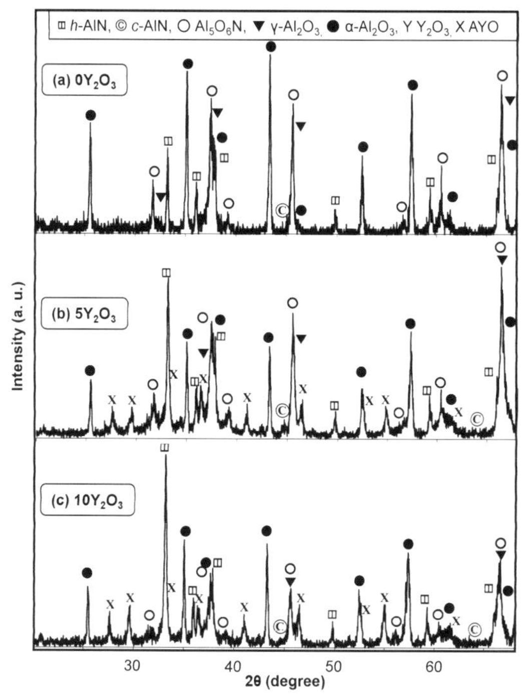

Fig. 4. XRD spectra of the heat-treated coatings at 1100 °C/12 h: 0, 5, 10 Y_2O_3 wt.% respectively.

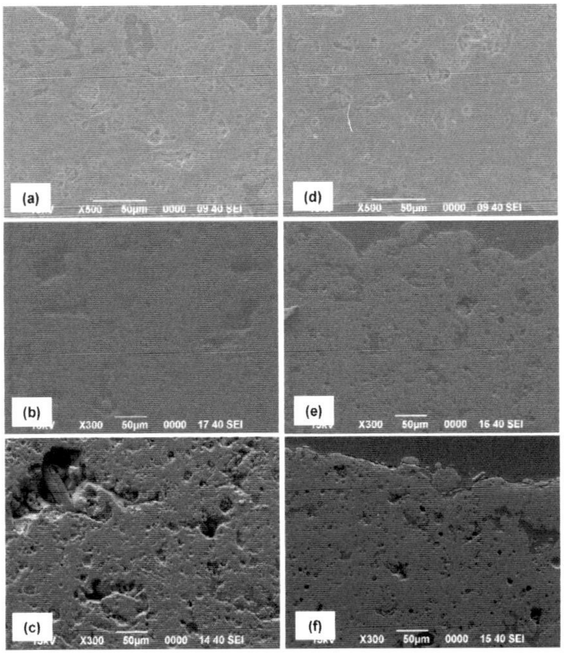

Fig. 5. SEM cross-section of the as-sprayed coatings using: (a) 0 Y_2O_3, (b) 5 Y_2O_3, (c) 10 Y_2O_3 and (d-f) after heat treatment at 1100 °C/12h respectively.

It is well-known in the field of thermal spray that, after heat treatment the coatings porosity decreases and may have an effect on the thermal conductivity. However, the splats are consisting of small and slightly outstretched grains with some extensive distributed internal porosity.

Although, after the heat treatment at 1100 °C for 12 h the nitride content slightly improved and the porosity decreased. The coating still contain high amount of oxide phases and higher porosity, which affected on the thermal conductivity value.

Figure 6 shows the thermal conductivity of the fabricated coatings with using the spray-dried mixtures before and after heat-treatment at 1100 °C for 12 h in the N_2 atmosphere. It is clear that, the heat-treatment showed improvement in the conductivity. Such improvement of the thermal conductivity after heat treatment is well known in thermal spray field. It is mainly attributed to decreasing the coating porosity. Thus, the heat treatment has a tendency to reduce or eliminate the effect of splat-interface thermal resistance on the conductivity.

However, the value itself is still far from the bulk AlN compacts. It attributed to the remaining high oxide content in the coatings (γ-Al_2O_3 and α-Al_2O_3), Al_5O_6N content and the porosity of the coatings. Thus, even after the heat-treatment the splats consist of small and slightly outstretched grains with some extensive distributed internal porosity, which give the low thermal conductivity for the coatings in total.

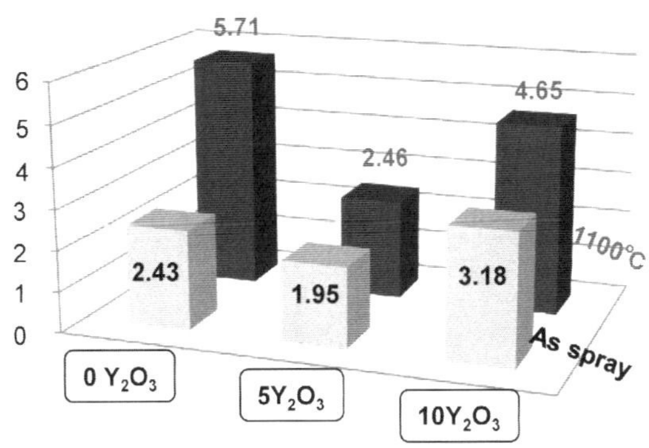

Fig. 6. The thermal conductivity of the coatings before and after heat-treatment at 1100 °C/12h.

3.2 Improved Thermal Conductivity Coating

Several attempts had been investigated to improve the thermal conductivity of the AlN coatings. Increasing the AlN content in the feedstock mixture and increasing the heat-treatment temperature were investigated. Finally AlN coating with high thermal conductivity (more than 90 W/m.K) was successfully fabricated through reactive plasma spraying of AlN-Al_2O_3-Y_2O_3 mixtures in N_2/H_2 plasma.

Figure 7 shows the XRD spectrum, thermal conductivity and the SEM microstructure of the high conductivity AlN coating. It is clear from the XRD spectrum that, the oxide content decreased so much and the coating mainly consists of AlN phase and very small amounts of aluminium yttrium oxide phase and oxynitride. Moreover, the thermal conductivity of as sprayed coating is very low (less than 10 W/m.K) and it gradually improved with increasing the heat treatment temperature.
In addition, the porosity of coating significantly reduced after the high temperature heat-treatment and it is about 2.5% or less. Moreover, the SEM cross-section microstructure of the improved thermal conductivity coating clearly shows the sintering occurred and the nicked-shape sintered particles are significantly shown.

Herein the influence of spraying of aluminum nitride-alumina-yttria mixtures in nitrogen plasma was investigated. It was possible to fabricate AlN coating with high thermal conductivity through controlling the feedstock mixture, coating composition and microstructure.

4 Conclusions

This study investigated the influence of spraying of aluminum nitride-alumina-yttria mixtures in nitrogen plasma. The coating consists of *h*-AlN, *c*-AlN, Al_5O_6N, γ-Al_2O_3, with small amounts of α-Al_2O_3 and aluminum yttrium oxide phases. Although using the Y_2O_3 additives significantly affected the process and the microstructure, it did not show the high thermal conductivity of AlN for the as-sprayed coatings. A high thermal conductivity (more than 90 W/m.K) AlN coating was fabricated through increasing the AlN content and enhancing the sintering during the heat-treatment.

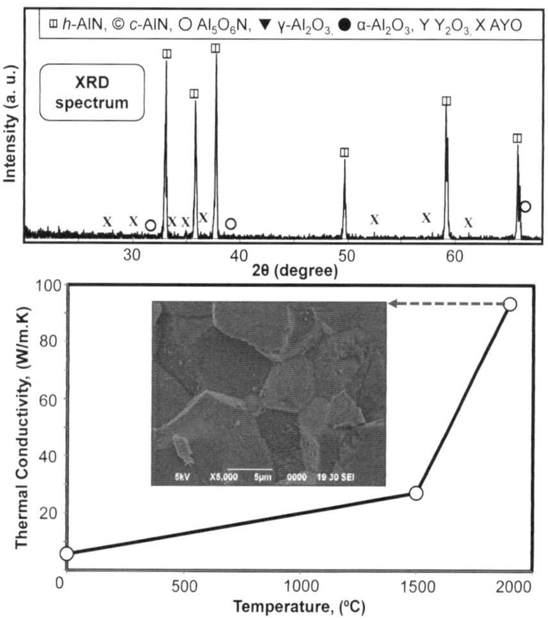

Fig. 7. XRD spectrum, thermal conductivity and SEM cross-section of the high conductivity AlN coating.

5 Literature

[1] Pierson, H.O.: Handbook of Refractory Carbides and Nitrides, Noyes, NJ, 1996, pp. 237-239.

[2] Wemer, A.W.: Carbide, Nitride and Boride Materials Synthesis and Processing, Chapman & Hall, London, 1997, pp. 6-68.

[3] Selvaduray, G. and Sheet, L.: Aluminum nitride: review of synthesis methods, Materials Science and Technology, 9 (1993), pp. 463-473.

[4] Marchant, D.D. and Nemecek, T.E.: Aluminum nitride: preparation, processing, and properties, Advanced Ceramics, 26 (1989), pp. 19-54.

[5] Sheppard, L.M.: Aluminum nitride: a versatile but challenging material, Ceramic Bulletin, 69-11 (1990), pp. 1801-1812.

[6] Mroz, T.J.: Aluminum nitride, Ceramic Bulletin, 71-5 (1992), pp. 782-784.

[7] Sundgren, J.-E., Structure and properties of TiN coatings, Thin Solid Films 128 (1985), pp. 21-44.

[8] Fauchais, P., Vardelle, A. and Dussoubs, B.: Quo vadis thermal spraying?, Journal of Thermal Spray Technology, 10-1 (2001), pp. 44-66.

[9] Krishna, L.R., Sen, D., Rao, Y.S., Rao, G.V.N., and Sundararajan, G.: Thermal Spray Coating of Aluminum Nitride Utilizing the Detonation Spray Technique, J. Mater. Res., 2002, 17(10), pp. 2514-2523.

[10] Yamada, M., Yasui, T., Fukumoto, M. and Takahashi, K.: Nitridation of Aluminum Particles and Formation Process of Aluminum Nitride Coatings by Reactive RF Plasma Spraying, Thin Solid Films, 2007, 515(9), pp. 4166-4171.

[11] Shahien, M., Yamada, M., Yasui, T. and Fukumoto, M.: Reactive Atmospheric Plasma Spraying of AlN Coatings: Influence of Aluminum Feedstock Particle Size, J. Therm. Spray Technol., 2011, 20(3), pp. 580-589.

[12] Shahien, M., Yamada, M., Yasui, T. and Fukumoto, M.: Influence of NH_4Cl Powder Addition for Fabrication of Aluminum Nitride Coating in Reactive Atmospheric Plasma Spray Process, J. Therm. Spray Technol., 2011, 20(1-2), pp. 205-212.

[13] Shahien, M., Yamada, M., Yasui, T. and Fukumoto, M.: Influence of Plasma Gases on Fabrication of AlN Coatings Through Atmospheric Plasma Nitriding Process, Ind. Appl. Plasma Process., 2010, 3, pp. 1-10.

[14] Shahien, M., Yamada, M., Yasui, T. and Fukumoto, M.: Cubic Aluminum Nitride Coating Through Atmospheric Reactive Plasma Nitriding, J. Therm. Spray Technol., 2010, 19, pp. 635-641.

[15] Shahien, M., Yamada, M., Yasui, T. and Fukumoto, M.: Fabrication of AlN Coatings by Reactive Atmospheric Plasma Spray Nitriding of Al Powders, Mater. Trans., 2010, 51, pp. 957-961.

[16] Shahien, M., Yamada, M., Yasui, T. and Fukumoto, M.: In Situ Fabrication of AlN Coating by Reactive Plasma Spraying of Al/AlN Powder, Coatings, 2011, 1-2, pp. 88-107.

[17] Shahien, M., Yamada, M., Yasui, T. and Fukumoto, M.: N_2 and H_2 Plasma Gases Effects in Reactive Plasma Spraying of Al_2O_3 Powder, Surf. Coat. Technol., 2013, 216, pp. 308-317.

[18] Shahien, M., Yamada, M., Yasui, T. and Fukumoto, M.: Synthesis of Cubic Aluminum Nitride Coating from Al_2O_3 Powder in Reactive Plasma Spray process, Mater. Trans., 2013, 54(2), pp. 207-214.

[19] Shahien, M., Yamada, M., Yasui, T. and Fukumoto, M.: Fabrication of AlN/Al_2O_3 Coatings via Atmospheric Reactive Plasma Nitriding of Al_2O_3 Powders, Key Eng. Mater. 2011, 484, pp.166-171.

[20] Shahien, M., Yamada, M., Yasui, T. and Fukumoto, M.: Reactive Plasma Spraying of Fine Al_2O_3/AlN Feedstock Powder, J. Therm. Spray Technol., 2013, 22(8), pp. 1283-1293.

[21] Shahien, M., Yamada, M., Yasui, T., Fukumoto, M., Egota, K. and Okamoto, K.: Reactive Plasma-Sprayed Aluminum Nitride-Based Coating Thermal Conductivity, J. Therm. Spray Technol., 2015, 24(8), pp. 1385-1398.

[22] Zhao, L., Seemann, K., Fischer, A. and Lugscheider, E.: Study on Atmospheric Plasma Spraying of Al_2O_3 Using On-Line Particle Monitoring, Surf. Coat. Technol., 2003, 168, pp. 186-190.

Particle Diagnostics and Coating Properties in Reactive Plasma Spray Process

Mohammed Shahien[1,2], Motohiro Yamada[1], Masahiro Fukumoto[1]
[1] Toyohashi University of Technology, TUT, Toyohashi/JP
[2] Central Metallurgical R & D Institute, CMRDI, Cairo/EG

Reactive plasma spray (RPS) is the promising solution among the thermal spray techniques to fabricate nitride coatings. It is a processing method which combines the *in-situ* synthesis and deposition of reaction products. Generally the RPS process depends on the reaction of the molten or semi-molten particles with the surrounding active species in the plasma jet followed by the deposition of the nitride coatings. Reactive plasma spray process was suitable technology to fabricate the aluminium nitride (AlN) coatings. This paper will discuss the influence of the particle diagnostics during reactive plasma spray process and as well as the properties of the fabricated AlN coatings as function of plasma gases. The N_2 gas flow rate increased the particles' velocity (reduced the particles residence time in the plasma), and thereby affected its melting, nitriding conversion, phase transformation, deposition efficiency and the coating thickness. Using H_2 gas with low flow rate is better for RPS of the fine Al_2O_3/AlN mixture. The H_2 gas flow increased the in-flight temperature and affected the in-flight vaporization, AlN content, phase transformation, deposition efficiency and the thickness of the coatings.

1 Introduction

Thermal spraying is one of the most industrial applicable surface engineering and coating technologies to fabricate thick coatings due to its numerous advantages. Thermal spray had wide variation of techniques (Plasma, Arc, flame, cold spray, warm spray, and aerosol deposition). In addition, it is characterized with wide variation of spray materials (rod, wire, metals, intermetallic compounds, ceramics, plastics and their compounds). Although, these wide variations of techniques and spraying materials, fabrication of thermal sprayed nitride ceramic coatings is big challenge. Thus, the nitride feedstock particles decompose under the high temperature without any stable molten or semi-molten phase. Therefore, it was difficult to deposit most of nitride ceramics in thermal spray process.

Aluminium nitride (AlN), had received much attention among nitride ceramics due to its unique combination of outstanding electrical and mechanical properties [1, 2] The AlN ceramics is well known with its thermal conductivity, thermal expansion coefficient, dielectric resistivity, chemical/physical stability at fairy high applications mechanical strength and high electrical insulation. Aluminium nitride is a promising material for several applications: electronic substrate, heat sink, packaging for electronic components, semiconductor packaging, reaction vessels for etching, crucibles and vessels for handling corrosive chemicals and molten metals, and semiconductor equipment parts. In addition, thick AlN thermal spray coating is a promising candidate as cost-effective solutions in the structural parts and power devices applications [3].

Using the reactivity of the thermal plasma is a promising solution to fabricate AlN coatings. Several AlN based coatings were successfully fabricated through reactive plasma spraying (RPS) of aluminum (Al), and alumina (Al_2O_3) powders in N_2/H_2 plasma [4-13]. The process is based on the reaction of molten particles with the surrounding active nitriding species in the plasma. This study will investigate the influence of the particles in-flight diagnostics during reactive plasma spray as a function of plasma gases. Moreover, the influence of plasma gases, in-flight behaviour on the coating properties will be discussed.

2 Experimental Procedures and Materials

2.1 Spray System and Materials

All the spray experiments were carried out by the atmospheric DC plasma spray system (APS: 9MB, Sulzer Metco, Switzerland), using primary gas of N_2 and secondary gas of H_2. Figure 1 shows schematic diagaram of the APS system and the spray parameters are shown in Table 1.

Figure 2 shows the microstructure of the feedstock powders. The feedstock material is fine Al_2O_3-AlN mixture, Fig. 2 (a). The mixture was prepared through spray-drying of commercial 60 wt.% fine Al_2O_3 powder (1 µm; Kojundo Chemical Lab Co., Ltd., Saitama, Japan) and 40 wt.% fine AlN 40 wt.% powder (1.14 µm; Tokuyama Co., Tokyo, Japan). The mixture had spherical shape with average particle size about 55 µm. The feedstock mixture was supplied to the plasma with a carrier gas of N_2. Grid blasted mild steel plates (50mm × 50 mm × 5mm) were prepared as the substrate material. For the particle diagnostics measurements commercial Al_2O_3 powder (Fujimi Incorporated, Aichi, Japan) was used as the feedstock material. It is fused, crushed powder with polygonal shape and with D_{50} about 26.6 µm Fig. 2 (b).

Table 1 The typical spray conditions

1st gas : Pressure (kPa) : Flow rate (SCFH)	N_2: 330.9 100 – 160
2nd gas : Pressure (kPa) : Flow rate (SCFH	H_2: 344.7 0 – 15
Spray distance (mm)	150
Carrier gas:	N_2
Substrate materials	SS400

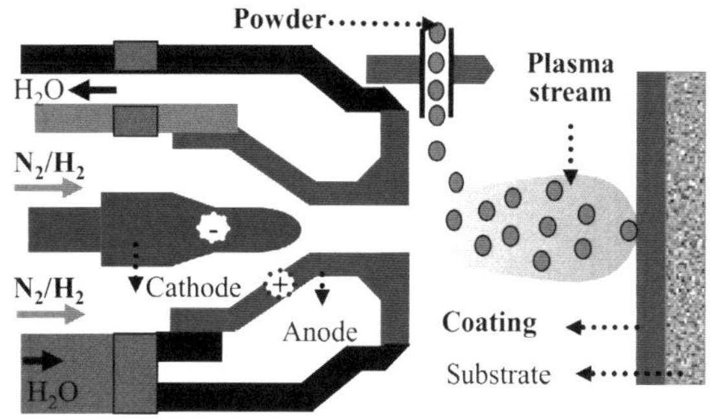

Fig. 1. Schematic diagram of APS system.

2.2 Characterization and properties

The particle temperature and velocity during flight in the plasma were measured by implementing a high-speed two-color pyrometer DPV-2000 system (DPV2000, Tecnar, Canada) at the different conditions. The phase composition of the fabricated coatings was verified by x-ray diffraction (XRD, RINT-2500; Rigaku, Tokyo, Japan) analysis using Cu Ka radiation. The microstructure of the feedstock powder and the fabricated coatings were observed by using a scanning electron microscope (SEM, JSM-6390; JEOL, Tokyo, Japan). The coating Vickers hardness was measured by a micro Vickers hardness tester (Shimadzu, Kyoto, Japan). Hardness tests were carried out using a load of 0.98 N and a holding time of 14 s. The coating porosity was estimated by image analysis of SEM cross-sections. The average Breakdown Strength of the coatings was measured by McKeown method designed on the lab.

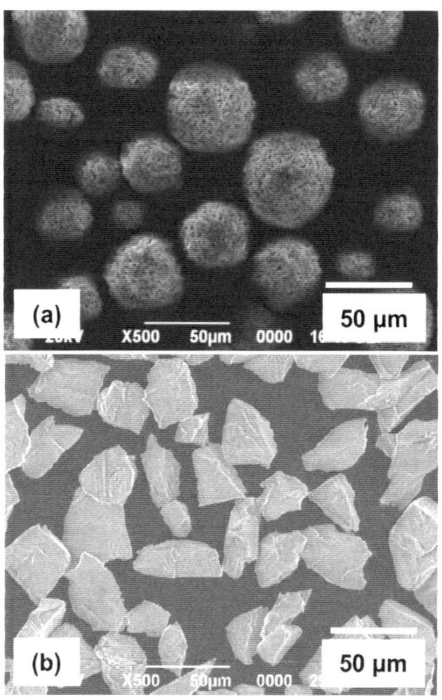

Fig. 2. SEM morphology of the feedstock powders.

The thermal diffusivity and specific heat of the free standing coatings were measured at25 °C by Laser Flash method and differential scanning calorimeter measurement, respectively. The sample bulk density was calculated from the measured sample volume (calculated from the measured sample dimensions) and mass by using the Shimadzu Co. electronic balance (AuW120D) at 25 °C. Finally the thermal conductivity was calculated through the following equation:

$$\lambda=\alpha \rho C_p$$

where:

λ = thermal conductivity (W/m.K)

α = thermal diffusivity (m2/s).

ρ = density (kg/m3)

Cp = specific heat (J/kg.K).

3 Results and Discussion

3.1 RPS and Diagnostics with Nitrogen gas

Figure 3 shows the XRD spectra of the fabricated coatings with spraying the feedstock mixture at the standard condition of N_2 gas flow. It reveals that, the coating consists of *h*-AlN, *c*-AlN, Al_5O_6N, γ-Al_2O_3, and a small amount of α-Al_2O_3. The *c*-AlN is attributed to the plasma nitriding of fine Al_2O_3 particles as explained before in details. The *h*-AlN is attributed to the AlN deposited from the feedstock mixture. The Al_5O_6N formation is due to the reaction of the Al_2O_3 with the surrounding nitrogen species. In addition γ-Al_2O_3, phase is attributed to rapid quenching of the molten Al_2O_3 and the rapid solidification rate in the plasma spray process. The transformation of α-Al_2O_3 to the transient γ-Al_2O_3 occurred easily with using such fine initial particle size under such high temperature and rapid solidification rates of plasma spray process as explained before [11, 14].

Furthermore, the coating phase composition in a good agreement with the the fabricated coatings through plasma nitriding of only Al_2O_3 powder [11, 12]. However, the α-Al_2O_3 content suppressed so much with using the fine particle size and only a very small amount appeared in the coating. It is good indication to the complete melting and transformation of the fine Al_2O_3 particles to the γ-Al_2O_3 phase.

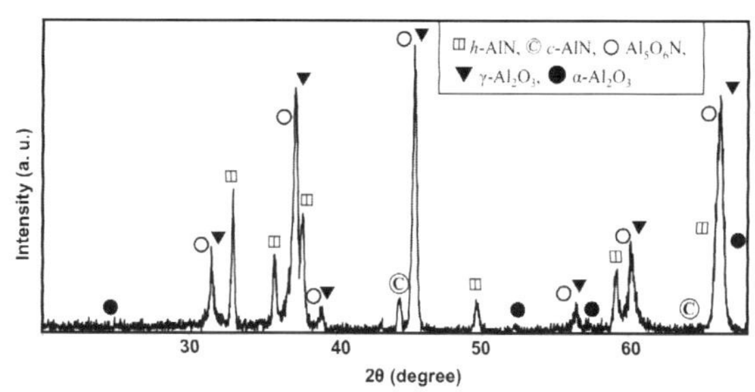

Fig. 3. XRD spectrum of the coating fabricated at the standard N_2/H_2 plasma (N_2: 100 SCFH).

Furthermore, it is clear that, the γ-Al_2O_3 and Al_5O_6N phases are the main phases in the fabricated coating with using the fine particles compared with the large particles. It is a good indication that the feedstock particle size strongly affects the Al_2O_3 particles plasma nitriding and melting behaviour. Thus, during the particle nitriding, the nitriding reaction starts from the particle surface. Therefore, using fine particles and decreasing the particle size of the feedstock powder increased the specific surface area. It led to increasing the possibility of surface nitriding and enhancing the nitriding conversion. In other words, use of the fine particles increased the nitriding area, which enhanced the formation of AlN phase.

Figure 4 shows the influence of the N_2 plasma gas on the fabrication process and phase composition. The relative amounts of the assigned phases in the coatings were estimated by comparing the peak intensities from the XRD data using the following equation [15, 16]:

$$R_X = \frac{I_X}{I_{(113)}^{\alpha-Al_2O_3} + I_{(200)}^{c-AlN} + I_{(400)}^{\gamma-Al_2O_3} + I_{(311)}^{Al_5O_6N}} \times 100\%$$

Where, RX is the concentration ratio, and $I_{(hkl)}$ is the intensity of the peak diffraction for the corresponding plane of *c*-AlN, α-Al_2O_3, Al_5O_6N and γ-Al_2O_3 phases.

Increasing the N_2 gas flow rate enhanced the *h*-AlN content and the *c*-AlN content gradually decreased. It indicate the nitriding conversion during in-flight and after deposition enhanced with the N_2 gas. It is attributed to increasing the nitriding species surrounding the molten particles. Therefore, the formation of Al_5O_6N and its transformation to the *c*-AlN phase were improved, as reported in detail previously [7, 10]. Furthermore, the increase in the nitriding species assisted the complete crystal growth and the formation of *h*-AlN phase after deposition (through plasma irradiation) on the substrate. Furthermore, with using the fine particles the complete crystal growth to the *h*-AlN was clearly exhibited compared to the large-size particles. It is agree with Cao et al. reports [15], regarding improving the nitriding conversion in the plasma spray process with increasing the N_2 gas flow.

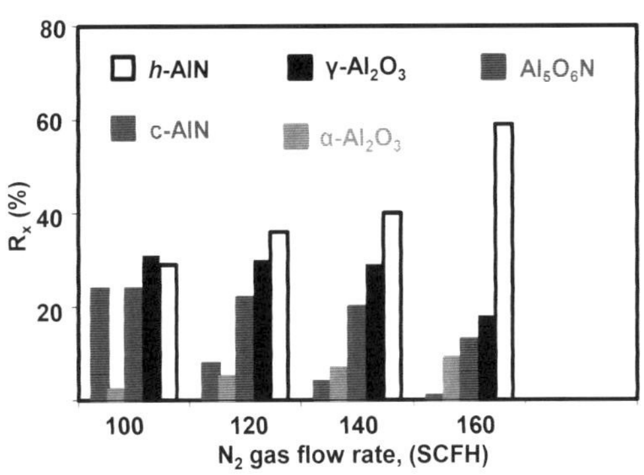

Fig. 4. The phase content (Rx) as a function of N_2 gas.

Figure 5 shows the average particle velocity and temperature with different N_2 gas flow rates. Increasing the N_2 gas flow strongly affects the particle velocity and has no influence on their temperature. Increasing the N_2 gas flow increased the particle in-flight velocity gradually. Increasing the particle velocity with the N_2 gas flow rate is related to the fact that, particles acceleration is linked to the flow momentum. It strongly depends on the mass flow rate of the plasma gas. As the nitrogen ions are much heavier than the hydrogen ions, its mass flow rate was much higher than that of hydrogen. Therefore,N_2 gas is responsible for accelerating the powder particles. On the other hand, the particle velocity increases due to the plasma gas expansion relative to gas flow increase.

The particle in-flight behaviour (velocity and temperature) can explain the effect of N_2 gas during RPS process. Increasing the particle velocity decreased the particles residence time in the plasma which decreased the decomposition of AlN feedstock during spraying. This partially led to increasing the *h*-AlN content in the coating at high N_2 gas flow rate. Besides that, decreasing the particle residence time can be considered as one of the reasons for decreasing the *c*-AlN content and its complete crystal growth at high N_2 gas flow.

On the other hand, the N_2 gas flow and the particle diagnostics significantly affected the coating microstructure and thickness. Thick coatings with uniform and porous structure were fabricated and the thickness increased gradually with the N_2 gas flow as shown in Fig. 6.

Improving the coating thickness with the N_2 gas flow is attributed to increasing the particle velocity and thereby higher acceleration and driving force of the particles to the substrate. Moreover, increasing the velocity led to decreasing the flight time and therefore reducing the evaporation tendency. This reveals that increasing the N_2 gas flow rate strongly affect the particle in-flight diagnostics, enhanced the AlN content and the thickness of the fabricated coatings.

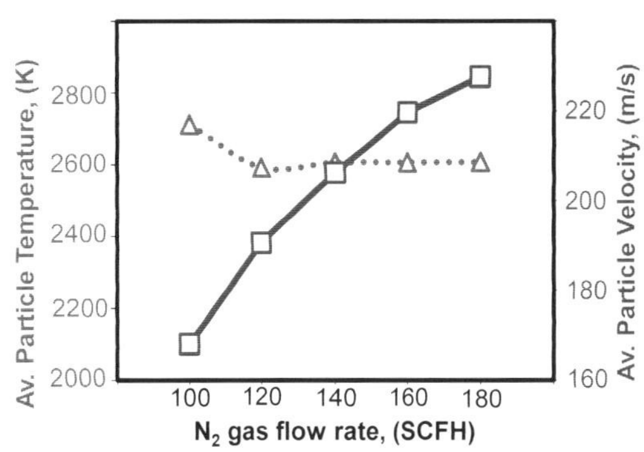

Fig. 5. In-flight diagnostics as function of N_2 gas flow.

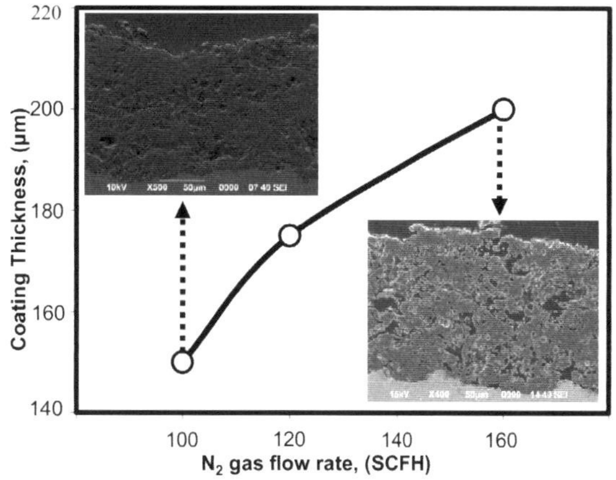

Fig. 6. The coating thickness as a function of N_2 gas.

3.2 RPS and Diagnostics with Hydrogen gas

Figure 7 shows the XRD spectra of the fabricated coatings with spraying the feedstock mixture in only N_2 gas flow (without H_2). The fabricated coating consists of *h*-AlN, Al_5O_6N, γ-Al_2O_3, and a small amount of α-Al_2O_3. It indicates that, the H_2 gas is required for RPS of Al_2O_3 particles.

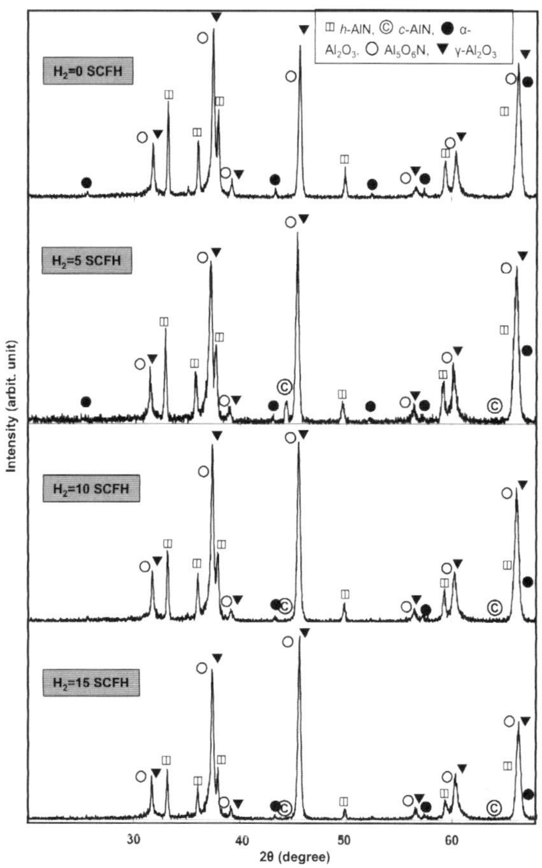

Fig. 7. The XRD spectra of the fabricated coatings as a function of H_2 plasma gas.

On the other hand, *c*-AlN phase started to appear with using the H_2 gas. However, the *c*-AlN as well as the *h*-AlN phases content gradually decreased with the H_2 gas flow rate. On the same time, the α-Al_2O_3 phase transformation to the γ-Al_2O_3 phase was improved with the H_2 gas flow.

In addition using the H_2 gas and increasing its flow rate gradually supressed the coating thickness as shown in Fig. 8. The coating decreased from 300 µm to 110 µm with increasing the H_2 gas flow rate from 0 to 15 SCFH.

To understand the effect of the H_2 gas during the reactive spraying of the fine mixture investigated the particle diagnostics is required. Figure 9 shows the average particle velocity and temperature with different H_2 gas flow rates. The H_2 plasma gas strongly affects the particle in-flight temperature and did not affect its velocity. Thus, using the H_2 gas and increasing its flow rate improved the Particle in-flight temperature. It is due to increasing the enthalpy, thermal conductivity and viscosity of the plasma jet [17]. Increasing the particle temperature is related to increasing the enthalpy through reducing the heavy gas content and increasing the hydrogen percent.

On the same time the particle velocity did not show significant change with the H_2 plasma gas. Thus, the acceleration of the particles is related to the mass flow rate of the plasma gas. The hydrogen ions are lighter than nitrogen ions and its mass flow is lower than the nitrogen. Therefore, the hydrogen gas had no effect on the velocity.

The influence of H_2 gas and its flow rate on the particle in-flight behaviour can be explained its effect on the RPS of fine particles. Decreasing the AlN content in the coating with the H_2 gas flow rate (appeared in the XRD spectra) is attributed to increasing the particle temperature (with the H2 gas). Thus, this enhanced the evaporation of AlN (the feedstock and/or the reacted phase), and thereby led to decreasing the AlN content gradually with increasing the H_2 gas flow especially when using the fine particles.

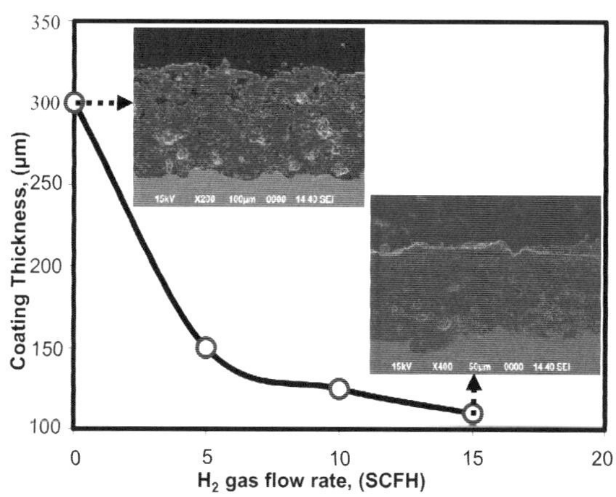

Fig. 8. The coating thickness as a function of H_2 gas.

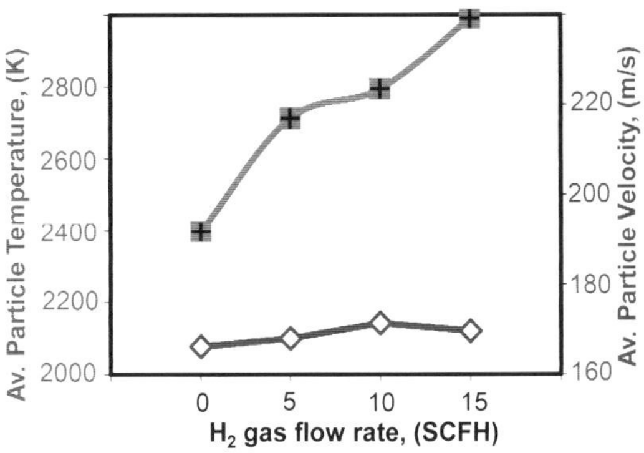

Fig. 9. In-flight diagnostics as function of H_2 gas flow.

Moreover, increasing the particle temperature gradually improved the transformation of the α-Al_2O_3 phase to the γ-Al_2O_3 phase. It is attributed to improving the melting of the particles, especially the fine particles. Besides that, increasing the temperature enhanced the evaporation tendency of such fine particles and thereby suppressed the coating thickness. Therefore, using the H_2 plasma gas with low flow rate is recommended to fabricate high-AlN content coatings.

3.3 Coating Properties

Figure 10 shows the average porosity (estimated by image analysis of SEM cross-section) of fabricated coatings as a function of the N_2 and H_2 plasma gases. The coating porosity gradually increased with increasing N_2 gas. It is due to increasing the particle velocity with the N_2 gas flow rate. This leads to decreasing the particle residence time in the plasma and thereby worse melting behaviour of the particles. Besides that, in the RPS increasing the N_2 gas flow increased the reaction tendency between the particles and the plasma, which lead to increase in the coating porosity.

On the other side, the coating porosity decreased with using the H_2 plasma gas and increasing its flow. It is due to increasing the particle temperature, which improved the melting efficiency of the particles and reduced the porosity of the coating.

Figure 11 shows the average microhardness of fabricated coatings as a function of the N_2 and H_2 plasma gases. The hardness decreased with the N_2 gas flow rate. It is related to increasing the AlN proportion in the coatings (as shown in Fig. 4). It is due to the lower hardness value of AlN compared the Al_2O_3. Moreover, increasing porosity of the coating with the N_2 gas reduced its hardness, due to decreasing the solid area carrying the load. On the other hand, with using the H_2 gas the hardness decreased and it attributed to the particle nitriding and increasing the nitride content. However, with the further increase in the H_2 gas flow, the hardness gradually increased. It is related to the decrease in the AlN content with and the decrease in the porosity of the coating with the H_2 gas flow as presented above.

Figure 12 shows the average thermal conductivity and density of fabricated coatings as a function of the N_2 plasma gas. The low thermal conductivity of the fabricated coatings is strongly related to the high oxide content in the composition (γ-Al_2O_3, Al_5O_6N and α-Al_2O_3), which are characterized by their low conductivity. Furthermore, the plasma spray coating microstructure (high porosity, splats, and interfaces, particularly in RPS process), obviously suppressed the conductivity. The lower density of the coating also lowered its conductivity. In addition, increasing the N2 gas flow leads to further decrease in the thermal conductivity, it is attributed to the further increase in the porosity and further decrease of its density.
Figure 13 shows the average breakdown strength of the fabricated coatings as a function of the N_2 plasma gas. It strongly related to the coating porosity, phase composition (γ-Al_2O_3 and Al_5O_6N) and the OH content on the AlN surface.

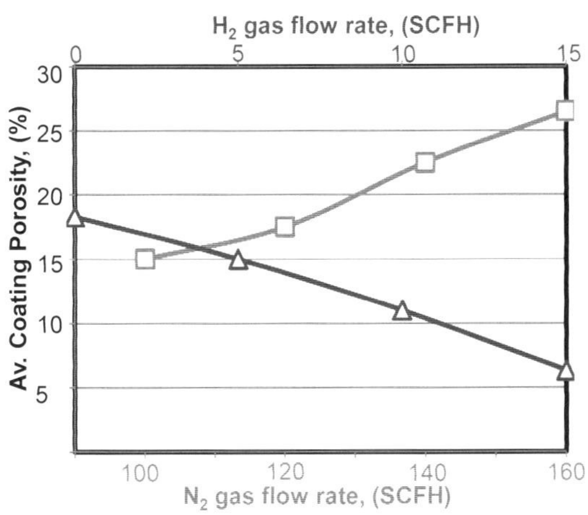

Fig. 10. Average coating porosity with plasma gases.

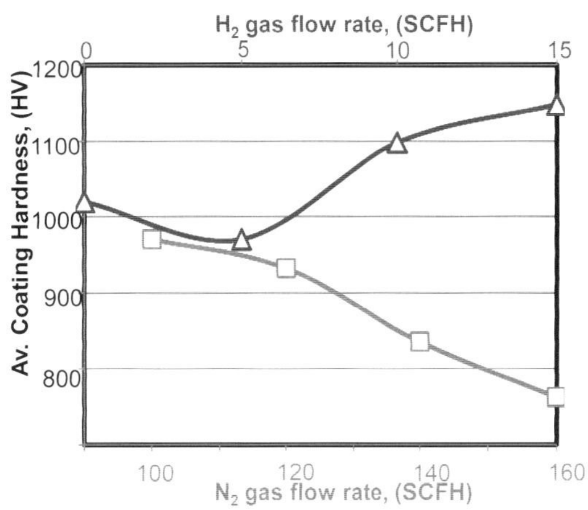

Fig. 11. Average coating hardness with plasma gases.

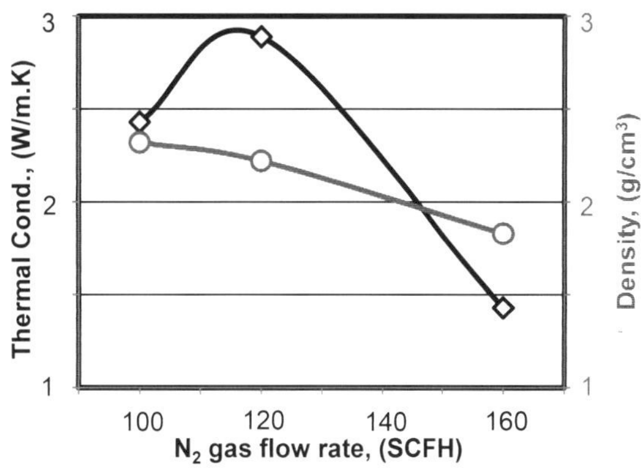

Fig. 12. Coating thermal conductivity and density as function of N_2 gas flow.

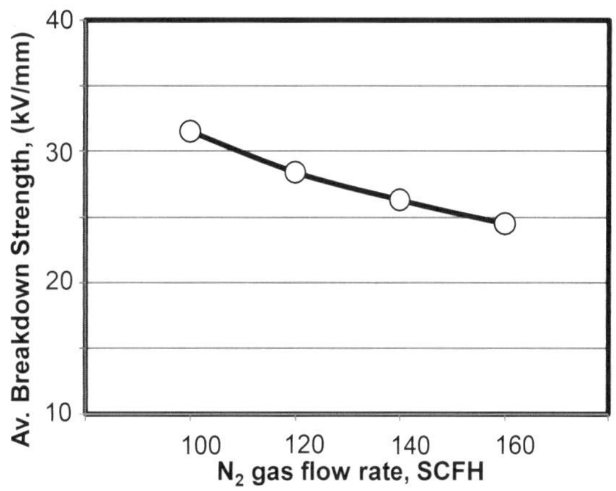

Fig. 13.Average breakdown strength of the coating as function of N_2 gas flow.

4 Conclusions

This study investigated the influence of particle diagnostics during reactive plasma spray. The particle in-flight behaviour (temperature and velocity) strongly affected by the nitriding species (N_2 and H_2) in the plasma. Therefore, the particle behaviour showed a significant effect on the RPS process, nitriding conversion, coating microstructure, thickness, hardness, porosity and thermal conductivity.

5 Literature

[1] Wemer, A.W.: Carbide, Nitride and Boride Materials Synthesis and Processing, Chapman & Hall, London, 1997, pp. 6-68.

[2] Pierson, H.O.: Handbook of Refractory Carbides and Nitrides, Noyes, NJ, 1996, pp. 237-239

[3] Krishna, L.R., Sen, D., Rao, Y.S., Rao, G.V.N., and Sundararajan, G.: Thermal Spray Coating of Aluminum Nitride Utilizing the Detonation Spray Technique, J. Mater. Res., 2002, 17(10), pp. 2514-2523.

[4] Shahien, M., Yamada, M., Yasui, T. and Fukumoto, M.: In Situ Fabrication of AlN Coating by Reactive Plasma Spraying of Al/ AlN Powder, Coatings, 2011, 1-2, pp. 88-107.

[5] Shahien, M., Yamada, M., Yasui, T. and Fukumoto, M.: Reactive Atmospheric Plasma Spraying of AlN Coatings: Influence of Aluminum Feedstock Particle Size, J. Therm. Spray Technol., 2011, 20(3), pp. 580-589.

[6] Shahien, M., Yamada, M., Yasui, T. and Fukumoto, M.: Influence of NH4Cl Powder Addition for Fabrication of Aluminum Nitride Coating in Reactive Atmospheric Plasma Spray Process, J. Therm. Spray Technol., 2011, 20(1-2), pp. 205-212.

[7] Shahien, M., Yamada, M., Yasui, T. and Fukumoto, M.: Influence of Plasma Gases on Fabrication of AlN Coatings Through Atmospheric Plasma Nitriding Process, Ind. Appl. Plasma Process., 2010, 3, pp. 1-10.

[8] Shahien, M., Yamada, M., Yasui, T. and Fukumoto, M.: Cubic Aluminum Nitride Coating Through Atmospheric Reactive Plasma Nitriding, J. Therm. Spray Technol., 2010, 19, pp. 635-641.

[9] Shahien, M., Yamada, M., Yasui, T. and Fukumoto, M.: Fabrication of AlN Coatings by Reactive Atmospheric Plasma Spray Nitriding of Al Powders, Mater. Trans., 2010, 51, pp. 957-961.

[10] Shahien, M., Yamada, M., Yasui, T. and Fukumoto, M.: N_2 and H_2 Plasma Gases Effects in Reactive Plasma Spraying of Al_2O_3 Powder, Surf. Coat. Technol., 2013, 216, pp. 308-317.

[11] Shahien, M., Yamada, M., Yasui, T. and Fukumoto, M.: Synthesis of Cubic Aluminum Nitride Coating from Al_2O_3 Powder in Reactive Plasma Spray process, Mater. Trans., 2013, 54(2), pp. 207-214.

[12] Shahien, M., Yamada, M., Yasui, T., Fukumoto, M.: Fabrication of AlN/Al_2O_3 Coatings via Atmospheric Reactive Plasma Nitriding of Al_2O_3 Powders, Key Eng. Mater. 2011, 484, pp.166-171.

[13] Shahien, M., Yamada, M., Yasui, T. and Fukumoto, M.: Reactive Plasma Spraying of Fine Al_2O_3/AlN Feedstock Powder, J. Therm. Spray Technol., 2013, 22(8), pp. 1283-1293.

[14] Zhao, L., Seemann, K., Fischer, A. and Lugscheider, E.: Study on Atmospheric Plasma Spraying of Al_2O_3 Using On-Line Particle Monitoring, Surf. Coat. Technol., 2003, 168, pp. 186-190

[15] Cao, L.H., Khor, K.A., Fu, L. and Boey, F.: Plasma Spray Processing of Al_2O_3/AlN Composite Powders, J. Mater. Process. Technol., 1999, 89-90, pp. 392-398

[16] Li, Y. and Khor, K.A.: A Study of Processing Parameters in Thermal-Sprayed Alumina and Zircon Mixtures, J. Therm. Spray Technol., 2002, 11, pp.186-194.

[17] Vardelle, M., Vardelle, A. and Fauchais, P.: Spray parameters and particle behavior relationships during plasma spraying, J. Therm. Spray Technol. 2 (1993), pp. 79-91.

Modification in powder feeder for nano-powder for thermal spray thin coatings

Rohit Upadhyaya[1,2], Sharad Shrivastava[1], S.C. Modi[2], Ankur Modi[2]

[1]Birla Institue of Technology and Science,Pilani Rajasthan

[2] Metallizing Equipment Company Pvt Ltd. Rajasthan

Abstract

The feedstock material are an integral part of any coating system. The material properties of coating are generally independent of specific equipment if the powder is optimized with respect to each type of equipment.It is difficult to coat the nano material because of the flowability of powder. Solution Precursor Plasma spray (SPPS) or suspension Plasma spray method are used to coat the nano material.It is difficult to coat the nano material with HVOF system. In this paper we are discussing on the modification of powder feeder to coat the nano material with HVOF , Flame spray and Plasma spray without using liquid. In this we fabricate two different type of powder feeder design to coat the nano carbide material with HVOF process. We also measure micro hardness, roughness, porosity and adhesion strength as per ASTM standard. Scanning electron microscope and XRD used to measure the microstructure of coating.

Introduction

A fundamental requirement of any TS process is to transport the feedstock to the heat source and the direct it towards the substrate. A phenomenon commonly known as the powder-feeding problem is a technological aspect of TS related to the sprayability of the feedstock. Powder feeding often refers only to the intial stage of powder transportation, from powder feeder/hopper to the thermal spray device. Technical difficulties may result in coating of poor quality, regardless of the fact that powder seems to be flowing into the heat source. This aspect of thermal spray technology is the root cause for examining the production and physical aspects of powder preparation.

The powder flows through an orifice or rate controlled restriction and is mixed with a carrier gas. The carrier gas acts as a fluidizing agent and transport the powder through a flexible powder delivery tube with an internal diameter of upto 6 mm. The inside diameter of this tube can have a polymer lining, which decreases powder friction within the confines of the tube.The mass fraction of the carrier gas to the powder mass is approximately 25%. Often the powder feed tube is oriented to take advantage of the natural tendency for powder to flow downward, and powder feeder placed above the gun.

Powder delivery tubes are typically approx 3 m long and end in an approx 25 mm long metal termination called a powder injection port. This port is withering attached directly to the anode or nozzle so that powder flows into its throat, or is directed into the tail flame of the heat source. The powder entered into the gas and injected into the bore of the torch so that it flows coaxially along its longitudinal axis. This method of powder injection claimed to provide consistent "Well mannered" particle injectories and thus enable optimum heat transfer of material. The gun design also plays a very crucial role for spraying the powder. Uneven powder flowing is also a major issue. Increasing the carrier flow rate may smooth out this erratic, irrespective flow behavior. Powder overloading and uneven powder flow will affect the coating quality. It is therefore important to examine the type of powder feed delivery system used and requirement of that constitute good powder feeding.

Feedstock Delivery Systems

Three types of powder-feeder system are most commonly used throughout the thermal spray

a) Gravity based devices,
b) rotating wheel devices,
c) Fluidized bed systems.

In this paper we will discussed the modification on rotation wheel devices. A rotating wheel that delivers precise parcels of powder to the powder feed delivery tube. The wheel can be mounted either vertically or horizontally. When oriented in ther vertical direction the powder is dropped into the carrier gas stream. A gas tube is connected between both sides of the rotation wheel to equalized gas pressure on both sides of the wheel: otherwise the powder would be blown off the pokects that are machined into the outside rim of the wheel.

The horizontally oriented wheel has slots that pass over the gas delivery system . In this fashion, powder is taken out of the slots and enters the carrier gas stream.

In both cases powder delivery rate is controlled by the physical dimension of the slots or pockets that hold the powder and the rotational speed of the wheel. In either geometry , it is easy to overload the carries gas with solid and either block the powder feed tube or cause the saltation effect by using too high powder feed rate or an insufficient carrier gas flow rate.

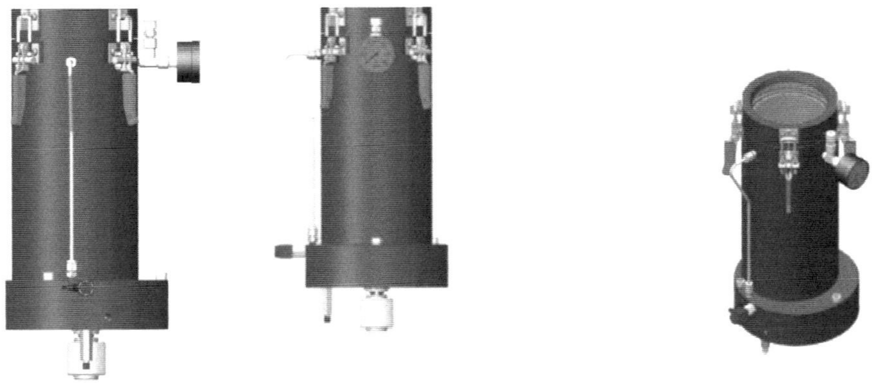

Figure 1:- Powder feeder with side carrier gas injection

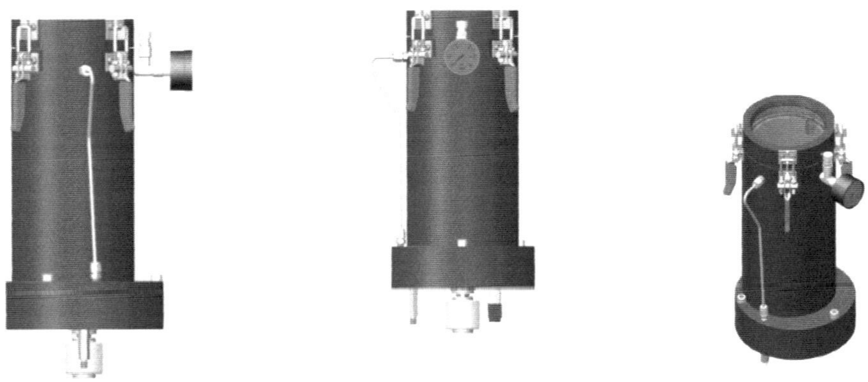

Figure 2 :- Powder feeder with bottom carrier gas injection.

Discussion

In this paper we discuss the design of both the powder feeder and property achieved by this system.

Email:- rohit.upadhyaya@pilani.bits-pilani.ac.in Phone number : +919829094409

AUTHOR INDEX

AUTHOR INDEX

AUTHOR INDEX

AUTHOR INDEX

AUTHOR INDEX

AUTHOR INDEX